# Vegetation and Flora of the Sonoran Desert

*Volume One*

# Vegetation and Flora of the Sonoran Desert

Forrest Shreve and Ira L. Wiggins

*Volume One*

Stanford University Press

*Stanford, California*

Part I, Vegetation of the Sonoran Desert,
*was originally published by the Carnegie Institution of Washington*
*as Carnegie Institution of Washington Publication 591, Washington, D.C., 1951*
*Reprinted here by permission of the Institution*

STANFORD UNIVERSITY PRESS
Stanford, California

Printed in the United States of America
ISBN 0-8047-0163-6
Original edition 1964
Last figure below indicates year of this printing:
01 00 99 98 97 96 95 94 93

# Preface

THE ORIGINAL PLAN, formulated during lengthy discussion with Forrest Shreve, was production of a two-volume work to be called *Vegetation and Flora of the Sonoran Desert.* Shreve's book, *Vegetation of the Sonoran Desert,* was to describe the plant cover from an ecological standpoint in Volume I, and *Flora of the Sonoran Desert* was to appear as Volume II. Early plans called for publication of both volumes by the Carnegie Institution of Washington, with the taxonomic part to follow the survey volume as quickly as possible.

Preliminary estimates of the number of taxa involved were entirely too low, and exigencies of World War II and various other obligations delayed work on *Flora of the Sonoran Desert.* When the manuscript for the latter work was nearing completion, it was apparent that the taxonomic treatment would make a book disproportionately large in comparison with *Vegetation of the Sonoran Desert* and mechanically unwieldy. By that time the first volume was out of print, but a steady flow of inquiries about its availability indicated some demand for reprinting Shreve's book.

Owing to the lapse between publication of Shreve's book and completion of the manuscript for *Flora of the Sonoran Desert,* the Carnegie Institution of Washington was no longer involved directly in studies centering in desert regions and was willing to transfer publishing rights to that book to the Stanford University Press, together wth permission to reprint *Vegetation of the Sonoran Desert.*

Consideration of the bulk involved in the taxonomic treatment led to a decision to combine Shreve's *Vegetation of the Sonoran Desert* with the first part of *Flora of the Sonoran Desert* in Volume I, but to reprint Shreve's book without textual change. Thus, the Foreword by Herman Spoehr appears unaltered, as does the text of Shreve's book. The first 75 families in *Flora of the Sonoran Desert* (Selaginellaceae through Buxaceae), beginning on page 189, complete Volume I, and families 76 through 138 (Anacardiaceae through Compositae) make up Volume II. The index to *Vegetation of the Sonoran Desert* has been moved as a unit toward the end of Volume II rather than being incorporated with the index of the taxonomic part, owing to basic differences in the character of the two works.

When the Stanford University Press acquired publishing rights to the two books, attempts were made to locate the negatives or prints from which the illustrations in *Vegetation of the Sonoran Desert* had been made. These efforts were totally unsuccessful, and it is feared that the 8-by-10-inch glass plates, possibly damaged during the interim between closing of the Desert Laboratory under the control of the Carnegie Institution of Washington and full use of the buildings by personnel from the University of Arizona, were discarded as of no value.

Neither could the halftone plates used in publishing Shreve's book be found, the company that had done the printing in 1950 having gone out of business. It has been necessary to reproduce most of the halftones by rephotographing most of the reproductions printed in *Vegetation of the Sonoran Desert*; others have been replaced by new photographs. Thus the sharpness and quality of some of the original plates have been inescapably reduced, but no other course was practical. I wish to acknowledge with thanks the efforts of Raymond M. Turner of the U.S. Geological Survey and James Rodney Hastings of the University of Arizona in obtaining many of the replacement photographs.

Both Shreve and I fully realized the value of the privilege accorded us by Mexican officials at all levels in allowing us to carry on field work under permits issued by appropriate agencies in Mexico. We appreciated, also, the courtesy and friendliness of the many Mexican citizens whom we met during the field operations. Were he alive, he would join with me in expressing our gratitude to all of them, officials and nonofficial citizens alike. It is unfortunate that our indebtedness can here be acknowledged by me alone.

IRA L. WIGGINS

*Stanford University*

# Contents of Volumes I and II

## *Volume One*

### PART I. Vegetation of the Sonoran Desert



### PART II. Flora of the Sonoran Desert



## *Volume Two*

### PART II. Flora of the Sonoran Desert *(continued)*

# List of Illustrations

## Maps



## Plates

*(Following page 186)*

PART I

# Vegetation of the Sonoran Desert

By Forrest Shreve

# Foreword

Forrest Shreve dedicated virtually his entire scientific career to the study of the botanical features of the arid regions of North America. In this pursuit he exhibited a unique allegiance and fealty to his subject and developed a broad plan of investigation which grew with the years both in intensity and in breadth of vision. During almost forty years he devoted a large portion of his effort to extensive field investigations, carried out under very primitive and rugged conditions. As his knowledge of the southwestern deserts grew, Shreve became more and more convinced of the necessity of living in the region during all seasons in order to gain a full appreciation of the biocoenose of these areas. In reply to a suggestion that it might be more comfortable for him to live in California and visit the deserts on occasional expeditions, he wrote: "I now have at Tucson living material of scores of the outlying parts of our area, from which I am learning much in regard to life histories, ecological habits, and physiological behavior which would remain unknown if they had merely been observed and collected on a field expedition. It is essential that this work be carried on under natural desert conditions. I feel very strongly that it is only by thus learning our plants as intimately as possible that we can hope to make our study of vegetation of the Sonoran Region contribute a fresh and vital view of the play of evolutionary processes under arid conditions." Shreve thus became thoroughly inured to life in the desert, although this was far removed from his native haunts and early environment.

Dr. Shreve was born in Easton, on the eastern shore of Maryland, July 8, 1878. He attended the Johns Hopkins University and was graduated A.B. in 1901. From the same institution he received the Ph.D. in 1905. While holding the Adam T. Bruce Fellowship of the Johns Hopkins University and during two later periods he worked in the Blue Mountain Region at Cinchona, Jamaica, the Tropical Station of the New York Botanical Garden. The results of these investigations were published in *A Montane Rain-Forest* (Carnegie Institution of Washington Publication 199, 1914). Shreve always maintained his interest in the Hopkins, and later collaborated with Professors D. S. Johnson and Burton E. Livingston. Jointly with the latter he published *The Distribution of Vegetation in the United States, As Related to Climatic Conditions* (Carnegie Institution of Washington Publication 284, 1921).

Dr. Shreve served as associate professor of botany in Goucher College 1906 to 1908, and joined the staff of the Desert Laboratory of the Carnegie Institution of Washington, at Tucson, Arizona, in 1909. He retained his connection with the Institution until his retirement in 1945, and continued his work in his chosen field until his death, July 19, 1950. He was editor of *The Plant World* 1911 to 1919, and was president of the Ecological Society in 1921.

It is evident from Shreve's publications on various aspects of the botanical features of the southwestern deserts that his vision of the basic scientific concepts involved underwent profound development. The role of arid conditions in the evolution of the flora of the region and the migration of plants from the humid areas seemed more and more to attract his interest. These investigations entailed the close co-operation of experts on the floristics of the several regions involved and on ecological and plant geographical studies. Shreve accordingly developed a plan to cover one by one the four North American desert areas, the Sonoran, the Chihuahuan, the Mojave, and the Great Basin Deserts, in order to work out the comparative features of distribution, habitat, and behavior of the plants of the four areas. A comprehensive study of the entire North American Desert from the combined viewpoints of vegetation and flora, and a careful examination of all the evidence that present distribution can give as to historical development of desert plants, were to round out the plan.

In a memorandum outlining this plan of investigation Shreve wrote:

> The project . . . provides for a thorough investigation of the taxonomic and ecological features of the plant life of the most arid part of the North American desert. The aims of the work would be the preparation of a flora of the vascular plants, study and description of the principal plant communities and their distribution, comparative investigation of the habitat requirements of characteristic plants in different parts of the area, and determination of some of the most important environmental conditions related to vegetation. The ultimate objective of the work would be a study of the origin and history of the strongly endemic flora of the region and of the highly specialized structural and physiological features which some of its plants possess.
>
> It is only on the basis of an accurate and reasonably complete taxonomic study of the area that any effort can be made to unravel the sources of origin of its plant life, the paths of migration along which various plant races have entered it, and the extent to which these migrations have been accompanied by evolutionary development. There is evidence that the area concerned has been desert at least since the Miocene, and that it has been the arena of a prolonged combat in which a considerable number of plant families have developed species capable of withstanding arid conditions, while other families have contributed members able to persist in favorable spots or at suitable seasons. There appears to be little hope that the paleontological record will give much aid in reading the history of the origin of the present distinctly desert plants of this region. It is believed that a study of the floristic features, interpreted in terms of the ecological behavior of the plants involved, will do as much as can be done toward unraveling the biological history of the area and its relations to the development of similar plant populations in northern central Mexico and in South America. . . .
>
> It is only through a study of the plant communities and the dominant perennials in relation to the conditions of climate and soil that the processes can be evaluated which have given the plant life of the area its distinctive character.
>
> The Chihuahuan and the Sonoran Deserts constitute parallel paths along which plant races have moved or receded under dissimilar sets of conditions. Their histories have undoubtedly been linked by a common participation in the effects of climatic change. The results of investigation of the two regions would reveal many relationships and contrasts of value in the study of each of the regions, and also be of use in deciphering as much as possible of the history of the entire arid region of North America.

In order to carry this program into effect, Shreve enlisted the co-operation of Professor Ira L. Wiggins, of Stanford University, to cover the floristic work on the Sonoran Desert. The results of Professor Wiggins' studies are to appear in volume II of the present publication. The floristic work in the Chihuahuan Desert was undertaken by Professor Ivan Johnston, of Harvard University. Many of the results of the latter investigation have already been published by Professor Johnston in the *Journal of the Arnold Arboretum.*

Dr. Shreve spent considerable time in the Chihuahuan Desert on investigations of the vegetation of this region, and gathered much information in the course of this work. He also did some work in the Mojave and in the Great Basin. He was, however, unable to complete his studies because conditions imposed by World War II made field work very difficult and finally impossible. Some of his collaborators were also called to other lines of duty during this period. Thereafter, Dr. Shreve's impaired health hindered the resumption of his field studies, and his energies went largely into assembling for publication some of the vast amount of material already in hand. It is most fortunate that under these difficult conditions Dr. Shreve was able to complete this, the first, of his contemplated series of publications comprising his ambitious plan for interpreting some of the biological phenomena of the North American deserts on the basis of ecological and floristic investigation.

*January 25, 1951*

H. A. Spoehr
*Chairman Emeritus, Division of Plant Biology, Carnegie Institution of Washington*

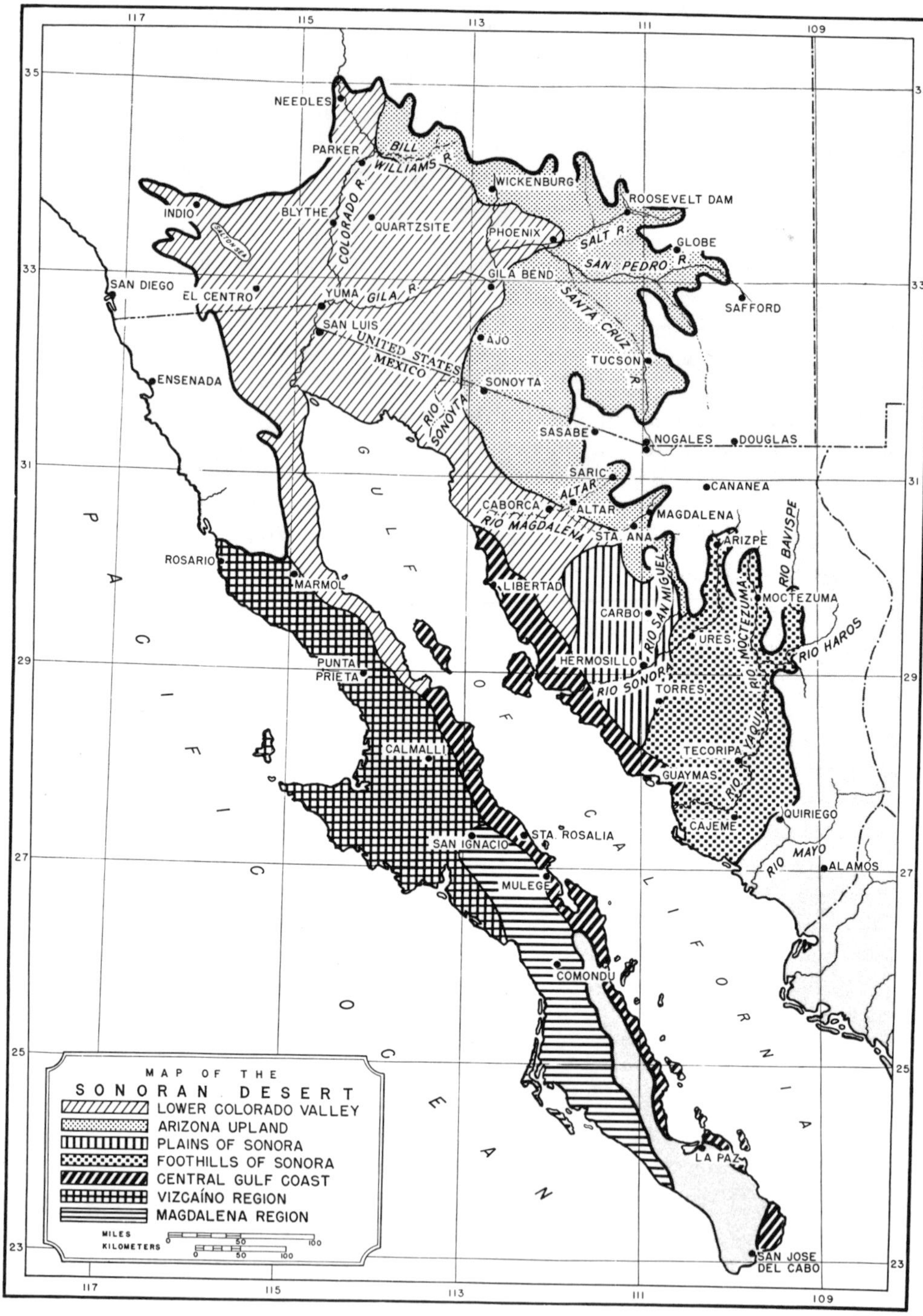


Map 1. The Sonoran Desert and its vegetational subdivisions

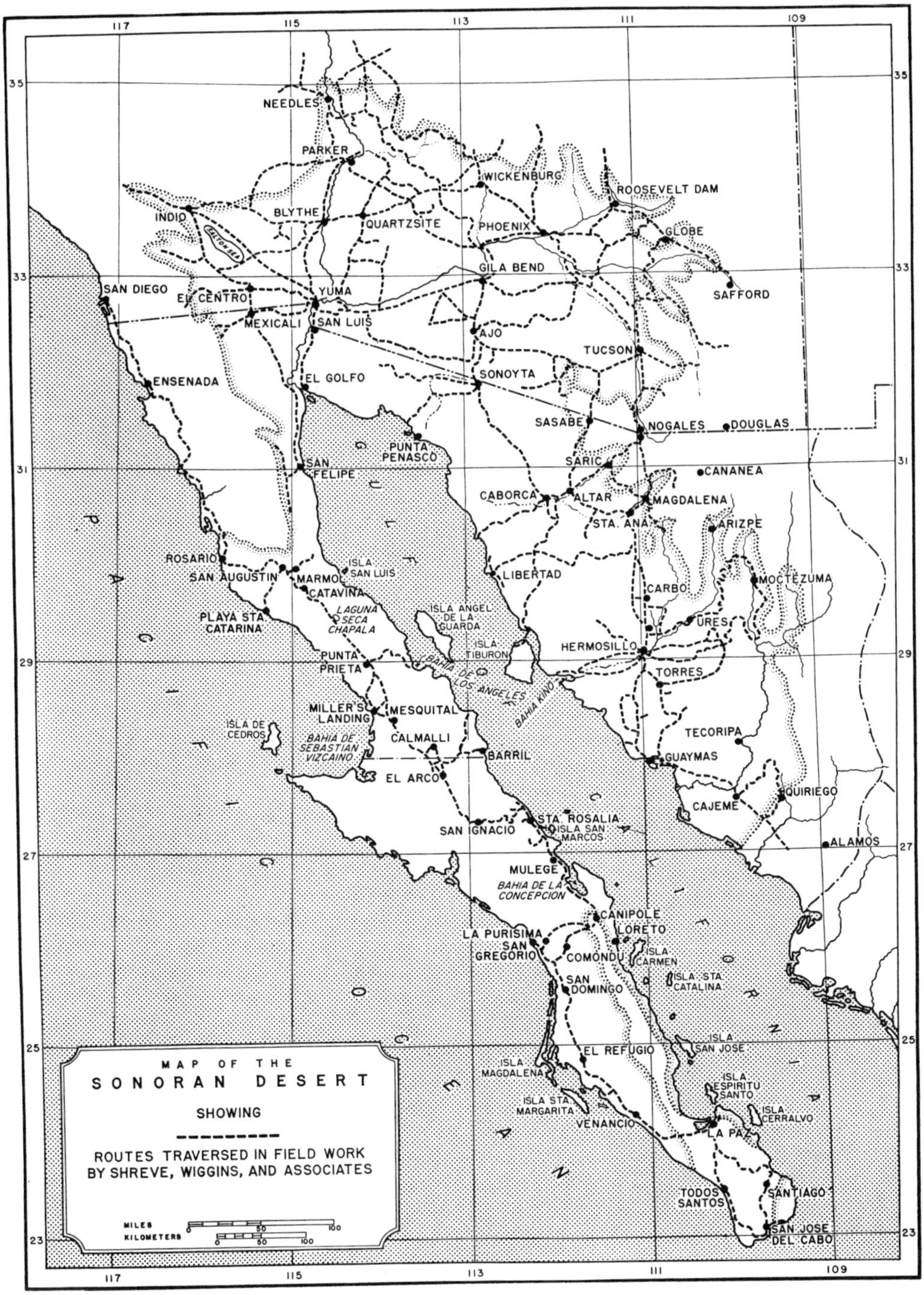


Map 2. Routes traversed in field work by Shreve, Wiggins, and associates

# 1

# Introduction

INVESTIGATION of the botanical features of the Sonoran Desert was undertaken as a logical step in the development of the work of the Desert Laboratory of the Carnegie Institution of Washington, opened in 1903 and located at Tucson, Arizona. Earlier research at the laboratory was chiefly concerned with the physiological and ecological behavior of some of the characteristic plants of the immediate vicinity and with the conditions to which they are subjected. With the exception of work carried out by MacDougal and collaborators in the Salton Basin and by Shreve in the Santa Catalina Mountains, no extended investigations had been made prior to 1932 outside the grounds of the laboratory and their environs.

The Tucson region is typical of a large part of the Sonoran Desert, and its physical conditions and plant life are more or less closely matched throughout the larger area. A comprehensive study of the flora and vegetation of the Sonoran Desert promised a useful addition to knowledge of arid America. Such an investigation seemed to offer opportunities for greatly increasing information on the relation of climatic and soil conditions to the distribution and activities of desert plants. The work also promised an enumeration of the plant species, a much fuller knowledge of their distributional limits, and a basis for the study of their origin, dispersal, and development.

The enlargement of the field of work has done much to elucidate some of the older problems which confronted investigators at the Desert Laboratory. The plant communities of the Tucson area have been followed into drier situations and to the limits of their distribution. It has been possible to study the influences of different degrees of aridity as well as the margins along which desert plants give way to those of different types. Conversely, it was found that the detailed knowledge which had grown from the thirty years of research at the Desert Laboratory was of great help in studying the poorly known parts of the Sonoran Desert. The background with which the work was approached did much to offset the limitations which are inherent in the expeditionary type of research. Early in the work it was realized that very many of the dominant plants in remote parts of the region are known scientifically as to their taxonomic position and in no other respect. An important phase of the investigation has been the effort to learn as much as possible about the life histories of some of these plants. With seeds, living plants, and dry or preserved material collected in the field, it has been possible to study germination, growth, resistance to drought and frost, seasonal habits of foliation and flowering, and anatomical features of stem and leaf.

The project for investigation of the Sonoran Desert was outlined in 1932 and field work was begun in the spring of 1933. The program provided for co-operation between the Desert Laboratory of the Carnegie Institution of Washington and the Dudley Herbarium of Stanford University, the former to be concerned with the study of the vegetation, the latter with that of the flora. The work on vegetation was carried out by Forrest Shreve and Thomas Dwight Mallery, and responsibility for the elaboration of the flora was assumed by Ira Loren Wiggins. Map 2 shows the extent of the field work. The volume now in the hands of the reader embodies the results of the vegetational work. Volume II constitutes the results of the floristic work of Dr. Wiggins and of several specialists whom he invited to contribute the text for certain groups of plants.

The author wishes to acknowledge the very helpful collaboration given throughout the field work by Fred C. Roberts of Tucson, James W. Manson of Nogales, John Davidson of Guaymas, and Enrique Gayazar of Santa Rosalía.

## Boundaries of the Sonoran Desert

The region designated as "Sonoran Desert" was tentatively outlined early in the course of its investigation, and more accurately delimited in later field work. In determining the extent of the region and the location of its boundaries, the only criteria considered were those of its vegetation and flora. The word "desert," which is so commonly used with a definite set of geographic connotations, is here used in a strictly biological sense. The Sonoran Desert is treated as a region of biological unity. The many features of climate, physiography, soils, and hydrography which also distinguish the region from surrounding ones are not a part of its biological character. The word "Sonoran" was used for the area because it has long been employed in physiographic and biological literature in nearly the same sense, because more of the area lies in the Mexican state of Sonora than in any other state, and finally because of its brevity and convenience. There is no relation between the designation "Sonoran Desert" as used in this publication and the "Lower Sonoran Zone" of C. Hart Merriam, although parts of the boundaries of the two are closely coincident.

Harshberger (1911) described the vegetation of North America and published a small-scale map, designating the region here under consideration as "The Sonoran Desert" and giving it approximately the same boundaries that have been adopted in the present publication. Sanders (1921), in discussing the natural regions of Mexico, mapped an area roughly corresponding to the Mexican part of the Sonoran Desert, which he designated as the "Desert of Sonora and Lower California." His desert area extends too far east in Sonora and includes the mountains of Baja California, at the same time that it excludes the coastal fringe of both states. In a study of mammalian distribution and ecology in the southwestern United States, Dice and Blossom (1937) designated as the "Sonoran Province" an area in southern Arizona which corresponds very closely to the Sonoran Desert as delimited in that region. In a publication on the distribution of mammals in Sonora, Burt (1938) outlined, on mammalian evidence, a "Sonoran Biotic Province" which corresponds very closely to the part of the Sonoran Desert that lies in Sonora except that it does not extend so far south in the coastal region. Other publications might be cited which would show that the characteristic plant and animal life of the Sonoran Desert and the position of the boundaries of that area have long been recognized by biological workers.

Each section of the boundary of the Sonoran Desert from 75 to 300 km. in length is the limit of some particular type of vegetation and of a group of species. As the boundary is followed farther, it is found to be the limit for other types of vegetation and for different groups of species. There is naturally no single type of vegetation which ranges unchanged over so large an area. Vegetational types often show a close resemblance in physiognomic features at the same time that they differ greatly in floristic composition.

The boundary of the Sonoran Desert is sharply defined wherever the topography is abrupt, as in Arizona between Safford and Wickenburg, in California along the east base of the San Jacinto and Cuyamaca Mountains, in Baja California along the east foot of the Sierra Juárez and Sierra San Pedro Mártir, and in Sonora near the confluence of the Bavispe and Moctezuma Rivers. In level or rolling regions the boundary is ill defined and there is a gradual transition from the desert to adjacent types of vegetation, as in Sonora between Quiriego and the Gulf of California, in Arizona and California in the vicinity of Needles, and in Baja California to the north of Rosario. From the headwaters of the Salt River, in Arizona, southward to Moctezuma, Sonora, the boundary lies at an elevation of 900 to 950 m. All other sections of the boundary lie at lower levels. The well known influence of slope exposure in determining the distribution of vegetation is responsible for many tongues and islands of desert which give irregularity to the boundary in its rugged sections.

In studying vegetation it is more satisfactory to cover a natural area than a political one, since the boundaries of states and provinces are so often straight and independent of natural features of the sort that limit the range of types of vegetation. In the study of the flora, however, there are some inconveniences in adopting a natural area, and taxonomic literature shows that this has not often been done. Hence, in order to enumerate accurately the plants of the Sonoran Desert, it has been necessary to find the precise locality at which a given plant was collected in any one of the four political divisions, and thus to determine whether it was growing inside or outside the desert boundary.

# 2

# Physical Features

## Geography

As a geographical entity the Sonoran Desert may be most simply described as the region immediately surrounding the head of the Gulf of California. In the broadest parts the inner boundary is 360 km. (224 mi.) from the Gulf coast in Arizona and 280 km. (175 mi.) from it in Sonora. In Baja California the Desert extends from coast to coast for 540 km. (335 mi.), or approximately the central third of the peninsula. Although a considerable part of the Sonoran Desert faces the Pacific Ocean, the Gulf of California is nevertheless its dominant geographical feature as well as the recipient of a very high percentage of its drainage.

In elevation the Sonoran Desert lies below 915 m. (3000 ft.) except for the narrow band along the eastern edge in Arizona and in northern Sonora, where it reaches 1050 m. (3450 ft.). The lowest part of the boundary lies between the place where it crosses the Colorado River, at an elevation of 150 m. (490 ft.), and the place where it most nearly approaches the Salton Sea, at an elevation of 80 m. (260 ft.).

The boundary of the Sonoran Desert is most sharply defined geographically, as well as biologically, near the higher mountain ranges which it approaches in several places. The surface of the desert itself is essentially a series of slightly inclined plains on which are superposed innumerable hills and mountain masses, most of which are less than 915 m. (3000 ft.) in elevation. The percentages of surface occupied by mountain and plain vary in different parts of the area. The Pinal Mountains in Arizona and the Sierra Babiso in Sonora are the only ranges bearing mesic vegetation which are surrounded by desert.

The part of the Sonoran Desert north of the international boundary lies in the southern part of the Basin and Range Province, as defined by Fenneman (1928). It comprises most of the Sonoran Desert Section of this province, all of the Salton Trough Section, and a small part of the Mexican Highland Section. The part of the Desert lying in Sonora is a continuation of the Basin and Range Province, but Baja California is like southern California in its greater physiographic complexity.

The approximate area of the Sonoran Desert is 310,362 sq. km. (119,370 sq. mi.). This is divided among the four states as follows: Sonora, 126,256 sq. km. (48,560 sq. mi.); Arizona, 105,404 sq. km. (40,540 sq. mi.); Baja California, 62,670 sq. km. (24,104 sq. mi.); California, 16,031 sq. km. (6166 sq. mi.). The approximate lengths of shore line are as follows: Gulf coast of Sonora, 950 km. (590 mi.); Gulf coast of Baja California, 1480 km. (920 mi.); Pacific coast of

Baja California, 1370 km. (850 mi.). All parts of Baja California lie within 80 km. (50 mi.) of tidewater. Of the remaining part of the Sonoran Desert, approximately one-half lies within that distance of the Gulf of California.

### Drainage Systems

The Gulf of California is the recipient of all the drainage of the Sonoran Desert except the small fraction that finds its way into the Pacific Ocean. Map 1, at the front of this volume, shows the principal rivers. The largest rivers rise outside the desert in the mountains of Colorado, Arizona, or Sonora, and cross the desert with diminishing volume and constancy. The only large enclosed basin is the Salton Sink, the bottom of which was 88 m. (287 ft.) below sea level before the formation of the present Salton Sea in 1905. There are a number of small undrained basins at low elevations in Arizona and Sonora, and in Baja California there is one in the interior, Laguna Seca Chapala, as well as a large number of shallow or poorly defined ones along the Pacific coast from Vizcaíno Bay to the Magdalena Plain. There are several large semi-bolsons, or nearly enclosed basins, in which a normal dendritic drainage system has developed but rarely carries floods great enough to discharge into its potential outlet. The only river with a perennial surface flow is the Colorado. Before the building of the Roosevelt and Coolidge Dams, the Gila River was a nearly constant stream. The other rivers which rise outside the desert show a seasonal variation in volume and in the distance to which their flood waters are carried after reaching the plains. The largest endogenous desert river is the Sonoyta, in northwestern Sonora; its floods are never discharged to tidewater. The Río Magdalena and Río Sonora rarely discharge their flood waters to the Gulf, but the Río Yaqui does so annually with almost unbroken regularity. The desert tributaries of the largest rivers, as well as the small independent streamways, are in flood for only a few days or even a few hours in the rainy seasons.

Despite the fact that the Colorado River is the largest and most constant stream traversing the Sonoran Desert, the influence of its waters on the vegetation of the desert is very restricted. It enters the desert 25 km. (15 mi.) north of Needles, California, where its mean level is 150 m. (490 ft.) above the sea. It falls at a nearly uniform rate from that place to the head of the Gulf, an airline distance of 358 km. (215 mi.). Parts of its meandering course are bordered by hills or cliffs of old river deposits, and parts by flats subject to inundation and covered with emergent swamp plants or forests of cottonwood. There are few places where the fringe of palustrine vegetation exceeds 3 km. in width.

On its right bank the Colorado receives only small and unimportant tributaries. On the left bank it is fed by the Bill Williams and Gila Rivers, as well as by several small washes. The Bill Williams River carries the drainage of the hilly and broken region between the east side of the Hualpai Mountains and the vicinity of Congress Junction, Arizona. The Gila River rises in the mountains of southwestern New Mexico and is fed on its right by several small rivers which drain the south face of the Colorado Plateau. Its most important tributary on the left is the San Pedro River, which rises in northern Sonora. Whereas the flow of the Colorado is stabilized by the vast area of its distant head, that of the Gila is largely contributed by brief and heavy rainstorms falling near the margin of the desert.

The Río Sonoyta has its head in a large number of small washes which drain valleys in western Pima County, Arizona, and in the northern part of the District of Altar, Sonora. Most of its catchment basin lies between elevations of 500 and 600 m., and none of the small mountains draining into it exceeds 1000 m. It is therefore a river of strictly desert regimen. The upper tributaries converge toward a spot at which the river traverses a pass in the Sierra de Nariz, whence it flows west-northwest for 70 km. In the latter stretch the underflow is augmented by some springs east of the village of Sonoyta. Near the village of Quitovaquito the bed of the Sonoyta broadens and the river veers sharply to the southwest. At this place, known as the "Salada de Sonoyta" (see pl. 7), most of its floods terminate. The remainder of its course of 95 km. is through very arid plains. It has a well defined channel as far as the inner edge of the sand dunes which border the Gulf coast, but the flood waters, so far as is known, never penetrate the dunes to the Gulf. Ives (1936) has published a description of the Río Sonoyta and the behavior of its flood waters.

As one follows the Gulf coast of Sonora southward from the mouth of the Colorado, for the first 265 km. only a few small arroyos discharge to the Gulf. At the distance mentioned is the mouth of one of the important rivers of Sonora, which is designated at its head as the Río Babasac, and through its course in turn as the Río Magdalena, Río Ascensión, and Río de la Concepción. This system will herein be styled the Río Magdalena. It rises in grassy plains west of Cananea at an elevation above 1000 m. All the upper drainage flows through a narrow canyon 15 to 25 km. northeast of Ímuris. In the 55 km. between the canyon and Santa Ana, several small but well watered tributaries occur, and the broadening of the flood plain has favored agricultural development. At Santa Ana the Río Magdalena emerges onto the plains (at 680 m.) and changes in character from a river with nearly constant flow to a floodway of the desert type. Numerous small tributaries reach the river below Santa Ana, the most important being the Río Altar, which joins it just above the village of Pitiquito. With occasional exceptions, the summer floods extend down the Río Magdalena as far as Pitiquito or 16 km. farther to Caborca. In the remaining 120 km. of its course to the Gulf, the river has a fall of 330 m. and only one lowland tributary, the Arroyo Coyote. Only in seasons of very heavy rainfall does the Magdalena debouch into the Gulf.

South of the mouth of the Río Magdalena no important streams enter the Gulf for a distance of 225 km. At that point the Bahía de Kino receives the Río Sonora and several small intermittent streams. The Río Sonora rises within 40 km. of the international boundary in the grassland plains east of Cananea, and flows almost due south for 170 km. It then follows a southwesterly course as far as Hermosillo, emerging from low hills onto the desert plains just below Ures. Closely parallel to the upper course of the Río Sonora is the Río San Miguel, running for a long distance about 50 km. west of the Río Sonora and flowing into it immediately above Hermosillo. Both the parent river and its principal tributary maintain a nearly constant flow to the point of junction. From Hermosillo the Río Sonora follows a west-southwesterly course to the Gulf, and in this section has only a seasonal flow of water. The gradient of the last 75 km. of the Río Sonora is much less than that of the Río Magdalena in its final stretch, and topographic evidence indicates that the former river has built a much larger deltaic plain than the latter.

South of Bahía de Kino the coast is again without important streamways for 190 km., to the point where the mouth of the Río Yaqui is found, now traversing the northern edge of the great delta which it has formed. The Río Yaqui is the most important river of Sonora, with its sources in southern Arizona and northwestern Chihuahua. It drains nearly all the western slope of the Sierra Madre Occidental north of latitude 28° N. Seen at the head of its delta in flood season it rivals the Colorado River in width and volume, but in the dry months its lower course may be without running water for several weeks. The most important tributaries of the Yaqui are the Moctezuma, the Bavispe, and the Haros. The upper course of the Haros is known as the Río Papigochic, and drains the east side of the Sierra Madre in Chihuahua for a distance of 100 km., as well as extensive grassland valleys east of the continental divide.

## Mountain Ranges

The Sonoran Desert is hedged in by mountains which rise gently or abruptly from its borders except along the line of separation from the Mojave Desert between Needles and Indio, California, the line of separation from the Cape Region on the Pacific side of the tip of Baja California, and the line between desert and thorn forest in southern Sonora. The climatic influence of the surrounding mountains is important for the desert, and it has just been shown that the principal streamways of the desert are fed by the mountains. The desert itself is studded with mountains and hills, but, as was mentioned above, only two ranges are high enough to bear mesic vegetation: the Pinal Mountains of Arizona and the Sierra Babiso of Sonora. About 70 per cent of the area of the Sonoran Desert has the basin-and-range type of topography, giving the impression of a great plain on which the mountains rest. Elsewhere the entire surface is hilly or broken, with only narrow bands of level ground along the streams or coast. The latter type of surface extends along the inner edge of the desert from Needles to Quiriego, and covers about 80 per cent of the desert area in Baja California.

A trend from northwest to southeast is strongly marked in the mountains of Arizona and northwestern Sonora. In eastern Sonora the trend is nearly north and south. In Baja California the mountain axis of the peninsula has a northwest-southeast position, but there is great deviation from it in the smaller ranges of volcanic origin.

The writer has shown elsewhere (Shreve, 1922) that the upper limit of desert vegetation on mountains which rise from desert plains is strongly influenced by the summit height of the mountains. The vegetation of the summit of a small mountain of 1500 to 1800 m. is composed of lowland desert plants and a few representatives of desert-grassland or encinal, whereas the same elevations on a large mountain of 2500 to 3000 m. are forested.

Among the mountains which lie wholly within the Sonoran Desert, the vast majority are of such low elevation that their summits are clothed solely by desert plants. There are, however, a number of small ranges which rise 700 to 1200 m. above the surrounding plains and furnish congenial habitats for plants that are common in the larger mountains just outside the desert, but are never or rarely found on the desert floor.

In Baja California the Sierra Juárez and Sierra San Pedro Mártir bear extensive areas of coniferous forest, and their eastern slopes fall abruptly to the very arid deltaic and coastal strip which connects the desert of California, Arizona, and Sonora with the desert part of Baja California. The very irregular series of mountains which forms the backbone of the peninsula rises in several localities to more than 1500 m., but in no case are there large areas above that elevation. The Sierra de Calamajué y San Luis lies between latitudes 29° and 29° 30′ N., and its rocky summits bear a few scattered oaks and junipers. The Sierra San Lino and the peak Las Tres Vírgenes have abrupt volcanic summits on which desert plants are dominant. In the Sierra de las Palmas and the Sierra de Zacatecas, which lie west of Bahía Concepción, the summits are better watered and have restricted areas of open xeric forest resembling that which covers most of the Cape Region.

The mountains in the Californian section of the Sonoran Desert and in the extreme southwestern part of Arizona appear at a short distance to be devoid of vegetation. They do, in fact, support an extremely open stand of highly xeric plants, which are not abundant enough to influence the color given the slopes by the bare rock. These mountains are often narrow and long, maintaining the same width for 100 km. or more, presenting a jagged crest which varies little in height, and meeting the desert bajadas or plains with great abruptness and at a remarkably uniform distance from the crest of the range. Mountains of this type, characteristic of the Basin and Range Province, are not encountered in Baja California (except in the Cucopah and Sierra Pinta ranges west of the Colorado Delta) or in the eastern half of the desert in Arizona and Sonora. Examples of this type of mountain are furnished by the Chocolate Mountains in California, the Mohawk and Sierra Pinta ranges in Arizona, and the Sierra Espuma and Sierra Basura in Sonora.

In the Basin and Range Province in Arizona and Sonora there are numerous mountains which depart widely from the elongated narrow form, often by reason of recent volcanism, but nevertheless have the characteristic highly dissected appearance, with sharp ridges separating the drainages and converging toward the summit. The volcanic mountains assume an infinite variety of forms, with extruded plugs or basaltic mesas contributing to their bizarre outlines. The abrupt walls, bare columns, and deep canyons found in small mountains of this type are well developed in the Ajo, Castle Dome, and Kofa Mountains in Arizona and in the Sierra Picú in Sonora. See plates 9, 10.

Throughout the eastern half of the desert the mountains differ in size, elevation, orientation, and character of rock. The work of erosion has varied with the differences in resistance, and there is nowhere the uniformity of pattern that is so general in the topography of the flanks of the Basin and Range mountains.

## Climate

A relatively uniform type of climate exists throughout the Sonoran Desert, with regional differences due to latitude, elevation, and the geographical configuration of the area. In spite of the nearness of the Pacific Ocean and the intrusion of the Gulf of California, the climate is distinctly of the continental type. This is particularly true of the northernmost part, which lies in the lee of high mountain ranges and is not influenced by the narrow waters of the head of the Gulf of California. Along the Pacific coast of Baja California desert vegetation exists with little modification up to the dunes and strand. The only climatic influences of the sea along the immediate coast are a slight lowering of the diurnal temperature, the intermittent effect of fog, and the almost constant blowing of a strong onshore wind. On the two coasts of the Gulf even these effects are absent except for a slight lowering of the sensible temperature within 100 m. of the shore during the daytime.

The difference of 11° of latitude between the northern and southern ends of the Sonoran Desert underlies a considerable difference between the winter temperatures in the north and south. As regards summer temperatures, the influence of latitude is outweighed by the more continental position of the northern end.

The gradual increase in elevation from the coasts to the interior results in a slight fall in the temperature range and brings the inner half of the desert within reach of occasional severe frosts. The same feature also causes a considerable increase of precipitation in the interior, further augmented by the mountain wall which forms the northern and eastern boundary.

### *Temperature*

Parts of the Sonoran Desert share with Death Valley the highest and most sustained air temperatures in North America. Days with a maximum temperature of 90° or more occur exceptionally in February and may persist as late as December. Periods of 90 consecutive days with a maximum of 100° are not exceptional. There is no question as to the length of the growing season so far as temperature conditions are concerned, for there may be at least some vegetative activity in locally favorable spots throughout the year. The length of the frostless season ranges from 8 to 12 months in different parts of the desert. The duration of the potential growing season is interrupted by the rainless periods, which sharply reduce all plant activity.

In evaluating the effects of high temperature on plant and animal life, consideration must be given to the accompanying dryness of the soil, the aridity of the air, and the long, cloudless days of strong insolation. Wind must also be considered a desiccating agent for plants, though many influences co-operate with it in extracting the last vestiges of moisture from leaves and stems. The days with highest sensible temperature and greatest desiccating effect are in June. The advent of the summer rains lowers the temperature range for a few hours or a few days, according to the duration of rainstorms and cloudiness. With the ending of the rains in September the high temperatures return, and late September and early October are usually as warm and dry as June.

The human reaction to a desert day was aptly expressed by Lumholtz, who stated that he felt as if he were "walking between two great fires."

In table 1 are shown the conventional temperature data for 8 stations in the Sonoran Desert: 3 in Arizona, 3 in Sonora, and 2 in Baja California. These figures show the well marked difference between the three Arizona stations, with their continental position, and the remaining stations situated on or near the coast of the Gulf of California. The Arizona stations show higher maxima in June, July, and August and lower minima during the winter months. The small differences in maxima and minima between Yuma, Phoenix, and Tucson are due to differences in elevation. Although Altar and Hermosillo are both at lower elevations than Tucson, they have lower maximum temperatures and higher minima than does Tucson. The lowest summer temperatures are found at Mulegé and La Paz, except in August and September at Mulegé. In general the monthly minima at Mulegé and La Paz are lower than at Guaymas and considerably higher than at the Arizona stations.

### *Rainfall*

The amount and seasonal distribution of precipitation is the most important physical condition limiting the boundaries of the Sonoran Desert and is likewise the most important in determining the differences to be found in the various parts of that area. In regions of low rainfall the figures showing the monthly or annual totals, or even the averages for 10 or 15 years, have relatively little significance. In its importance for plants there is a wide difference between rain of a given amount in the cool months and in the hot ones. The upbuild of soil moisture effected by a brief torrential downpour is much less than that from a gradual rain of the same amount. Widely different effects are produced by a single rain of 50 mm. and five rains of 10 mm. each on successive days. A rain falling on a moist soil penetrates much more effectively than it would if the soil were dry. These considerations and a number of others make it inadvisable to place much emphasis on small differences in rainfall records for arid regions.

Of particular importance to plants are the number and duration of drought periods, which may be defined as periods without rain or with none of sufficient amount to affect the soil moisture. Such periods with a duration of 30 to 60 days are of almost annual occurrence in the Sonoran Desert. The large nonsucculent perennials are able to withstand them by drawing on the deep-seated moisture.

TABLE 1

AVERAGE MAXIMUM, MINIMUM, AND MEAN TEMPERATURES (IN DEGREES FAHRENHEIT) FOR 8 STATIONS IN THE SONORAN DESERT FOR THE YEARS INDICATED

| MONTH | ARIZONA | | | SONORA | | | BAJA CALIFORNIA | |
|---|---|---|---|---|---|---|---|---|
| | Yuma 1878–1930 | Phoenix 1896–1930 | Tucson 1892–1928 | Altar 1923–1938 | Hermosillo 1923–1938 | Guaymas 1923–1938 | Mulegé 1923–1938 | La Paz 1923–1938 |
| | | | | MAXIMUM | | | | |
| Jan. | 66.7 | 65.0 | 65.2 | 61.9 | 63.3 | 72.3 | 60.3 | 66.0 |
| Feb. | 72.0 | 65.0 | 65.2 | 61.9 | 66.7 | 69.6 | 64.0 | 67.5 |
| Mar. | 78.1 | 74.1 | 73.4 | 65.1 | 73.0 | 74.7 | 69.4 | 72.1 |
| Apr. | 85.4 | 81.8 | 80.7 | 73.6 | 71.6 | 75.6 | 72.1 | 73.8 |
| May | 92.5 | 90.0 | 89.3 | 80.6 | 82.6 | 82.0 | 77.4 | 77.4 |
| June | 101.7 | 100.9 | 98.8 | 87.6 | 89.8 | 87.4 | 83.7 | 82.0 |
| July | 105.5 | 102.7 | 98.6 | 91.0 | 91.8 | 92.1 | 88.7 | 84.9 |
| Aug. | 104.1 | 100.9 | 96.8 | 89.8 | 89.4 | 88.3 | 90.5 | 86.2 |
| Sept. | 99.6 | 96.7 | 94.3 | 87.0 | 89.2 | 87.3 | 87.6 | 85.3 |
| Oct. | 88.0 | 85.9 | 85.1 | 78.6 | .... | 83.5 | 80.4 | 80.8 |
| Nov. | 76.2 | 74.5 | 74.3 | 70.0 | .... | 76.1 | 72.0 | 74.5 |
| Dec. | 67.2 | 65.0 | 64.9 | 64.6 | .... | 69.8 | 61.7 | 69.3 |
| *Year* | *86.4* | *83.9* | *82.4* | *76.1* | .... | *79.5* | *75.6* | *76.6* |
| | | | | MINIMUM | | | | |
| Jan. | 42.2 | 38.7 | 34.3 | 49.5 | 58.3 | 60.1 | 53.8 | 61.9 |
| Feb. | 45.7 | 42.4 | 37.5 | 53.6 | 58.8 | 62.8 | 55.0 | 62.6 |
| Mar. | 49.6 | 46.7 | 41.3 | 56.7 | 62.0 | 65.8 | 60.1 | 64.4 |
| Apr. | 54.1 | 52.3 | 46.1 | 63.0 | 70.5 | 70.5 | 61.3 | 69.4 |
| May | 60.1 | 59.5 | 53.8 | 70.7 | 71.8 | 75.0 | 64.4 | 72.1 |
| June | 67.8 | 68.8 | 63.6 | 76.8 | 85.8 | 82.6 | 76.3 | 75.2 |
| July | 76.6 | 76.5 | 72.2 | 84.2 | 87.8 | 84.6 | 81.7 | 81.7 |
| Aug. | 76.2 | 75.3 | 70.6 | 83.3 | 87.1 | 84.0 | 79.0 | 80.0 |
| Sept. | 69.2 | 68.4 | 64.3 | 79.2 | .... | 83.8 | 78.8 | 78.3 |
| Oct. | 57.5 | 55.6 | 51.6 | 69.1 | .... | 78.8 | 69.8 | 77.2 |
| Nov. | 48.4 | 45.2 | 41.4 | 58.6 | .... | 69.8 | 63.1 | 69.4 |
| Dec. | 42.8 | 38.5 | 35.2 | 53.4 | .... | 63.1 | 53.2 | 64.4 |
| *Year* | *57.5* | *55.7* | *51.0* | *66.6* | .... | *73.4* | *66.4* | *71.4* |
| | | | | MEAN | | | | |
| Jan. | 54.3 | 50.9 | 49.7 | 55.0 | 60.3 | 63.9 | 57.2 | 64.1 |
| Feb. | 58.9 | 55.1 | 49.7 | 59.5 | 63.1 | 66.0 | 60.8 | 64.9 |
| Mar. | 63.9 | 60.5 | 57.3 | 61.7 | 68.0 | 69.3 | 64.0 | 69.1 |
| Apr. | 69.8 | 67.1 | 63.4 | 68.0 | 73.6 | 73.0 | 68.2 | 71.2 |
| May | 76.3 | 74.7 | 71.5 | 75.7 | 79.5 | 77.9 | 73.2 | 74.5 |
| June | 84.7 | 84.4 | 81.2 | 83.7 | 87.8 | 83.8 | 80.9 | 78.1 |
| July | 91.1 | 89.7 | 85.4 | 88.5 | 89.6 | 87.3 | 86.9 | 82.7 |
| Aug. | 90.3 | 88.4 | 83.7 | 86.7 | 88.5 | 86.9 | 86.7 | 84.0 |
| Sept. | 84.4 | 82.0 | 79.3 | 84.0 | 86.4 | 86.4 | 84.2 | 82.9 |
| Oct. | 72.7 | 70.3 | 68.3 | 74.1 | .... | 81.1 | 76.3 | 79.0 |
| Nov. | 62.3 | 59.2 | 57.8 | 64.0 | .... | 72.7 | 67.1 | 72.7 |
| Dec. | 55.2 | 51.9 | 49.9 | 57.5 | .... | 65.5 | 58.8 | 66.4 |
| *Year* | *72.0* | *69.5* | *66.6* | *71.4* | .... | *76.1* | *72.0* | *73.9* |

In soils of fine texture the moisture at a depth of 2 m. is nearly constant throughout the year and is affected from year to year only by very exceptional periods of drought or rain.

The writer has published (Shreve, 1934) a digest of the rainfall, runoff, and soil-moisture records of the Desert Laboratory at Tucson. These data serve to illustrate many features of the rainfall which are characteristic of the entire Sonoran Desert.

An important feature of rainfall in the Sonoran Desert is the change in seasonal distribution which is encountered on going inland from the Pacific coast. West of the drainage divide of Baja California and north of Mulegé, rain is confined to the late winter and early spring. Winter rain forms about 50 per cent of the annual total at Yuma and 40 per cent at Altar. At Tucson the winter and summer rains are nearly equal, and 80 km. farther east, at Benson, Arizona, the winter rain decreases to 25 per cent. The sporadic summer rains of the Gulf coast of Baja California increase to 20 per cent of the annual total at Yuma, to 35 per cent at Gila Bend, and to 50 per cent at Tucson; in Sonora, the percentage at Altar is 66 and at Guaymas 72. There are further increases in the percentage for the summer months in the eastern valleys of Sonora and in southern Sonora and Sinaloa.

The winter period of precipitation extends from December to the end of March or the middle of April. The summer period is sharply limited to July, August, and early September. The winter rains are gentle and prolonged and usually occur over wide areas. The summer rains are commonly brief and torrential and limited to paths or small areas.

The accompanying map (map 3) of rainfall distribution is based on data for 4 stations in California, 17 in Arizona, 16 in Sonora, and 3 in Baja California. The map is purely provisional, as the records vary in length and reliability and considerable interpolation has been necessary. The outstanding feature is the gradual increase from readings of 55 mm. at San Luis, Sonora and 79 mm. at Yuma, Arizona to 459 mm. at Roosevelt Dam, Arizona, 443 at Saric, Sonora, 469 at Moctezuma, and 445 at Tecoripa. It is also noteworthy that most of the stations in northern Sonora have higher rainfall than do Arizona stations located at elevations 300 to 400 m. higher. In northern Sonora the distance from tidewater to the Sierra Madre mountains is great, but in southern Sonora the coastal plain narrows to less than 150 km. This is an important factor in increasing the rainfall of the inner edge of the coastal plain and in determining the southern boundary of the Sonoran Desert. Humphrey has given a detailed account of rainfall in this and other deserts (1933).

## Soils

The origin, mode of formation, and developmental history of soils in a desert region present many features which are common to all soils, at the same time that they exhibit a number of distinctive phenomena and conditions which are to be attributed to the deficiency of rainfall or else to its intermittent and torrential character.

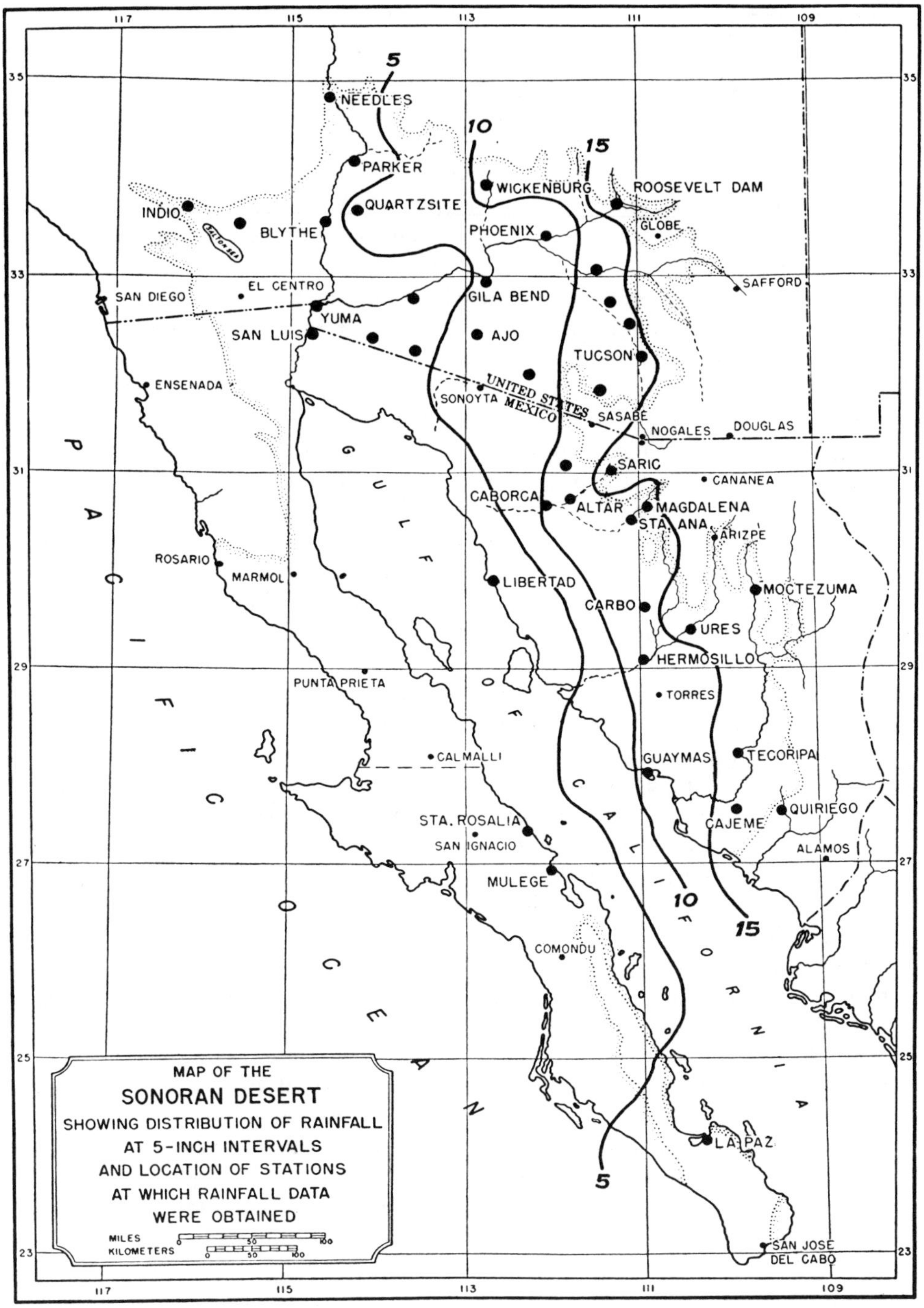


Map 3. Distribution of rainfall at 5-inch intervals. Large dots indicate location of stations at which rainfall data were obtained.

The basic importance of the underlying rock in determining the character of residual soils is as great in the desert as it is elsewhere. The arid climate begins to exert its influence, however, in the very genesis of soil formation, in the weathering of rock in place and the disintegration of rock fragments. In the Sonoran Desert the influence of frost in weathering is negligible. It seems possible that disintegrating action may be exerted on black and very dark-brown basaltic rocks by the sudden falling of rain at a time when the rocks are hot. The first water which falls on them is so greatly heated that its power as a solvent is far greater than that of cold rain water. Rocks with a surface temperature of 150° to 160° F. are often suddenly bathed by rain water with a temperature of 60° to 70°, and the sudden and great temperature changes to which they are thus subjected must do much to aid in the disintegration of the less resistant rocks.

The decomposition of rock fragments and the coarser soil elements, as distinguished from their mechanical disintegration, is accomplished slowly in arid climates as compared with humid ones. This results in the formation of soils which vary in the mineralogical character of the smaller particles, and include many small particles of composite nature.

The disintegration of rock surfaces by wind-borne sand is sometimes locally of greater extent than the weathering due to water. Small hills and isolated rocks lying near the general level of large plains are more subject to wind erosion than are the slopes of larger hills or mountains. Even the very small rock fragments which cover the surface of the ground are worn slowly by shifting sand. Where there is an abundance of fine, loose material, it is often actively moved along the surface of the ground when the wind is not strong enough to raise visible clouds of dust. The cutting action of the sand blast cannot always be gauged by the number of dust storms which occur. Large bodies of shifting sand are found in numerous areas around the head of the Gulf of California. Though there are only a few localities which show conspicuous evidence of wind erosion, the phenomenon is going on to a slight extent throughout the sandy part of the Sonoran Desert.

The general features of the relation of land forms and physiographic processes to soil history are the same in arid and in humid climates. A series of soils ranging from mountain side or talus slope through outwash slope and bajada to playa or flood plain will exhibit a decreasing number of large stones and an increasingly finer texture. Because of torrential rains and the lack of a continuous cover of vegetation, there is a more ready downgrade movement of coarse material in the arid climate, and a consequent coarser texture in the soils of such relatively mature surfaces as the lower bajadas.

McGee (1897) has described the manner in which the runoff takes place over a bajada following a torrential rain. Water flows in a continuous sheet over areas of hundreds of acres, to a depth of 1 cm. or more, rapidly replaced by the precipitation and the drainage from up slope. The cutting of the surface by such drainage has been designated by McGee as "sheetflood erosion." Although there has been some skepticism as to the occurrence of such a phenomenon, the writer can vouch for having witnessed it many times. The sheets of water are of local occurrence and of very brief duration, but there must be very few appropriate areas in the desert that are not visited by them once or twice every year. After the cessation or onward movement of the rain, the sheet of water vanishes from the general surface. The smallest streamways are filled with water for a few minutes longer. The larger streamways, from 1 to 2 m. wide, are in flood for 10 to 20 minutes. Thirty minutes after the cessation of the rain, no free water surfaces are to be found except in the largest streamways.

It is erosion of the type just described that does more than any other one factor to give the bajadas their characteristic uniformity of gradient and their consequent straight-line profiles. This is also the principal factor in the removal of the finest material from the surface and its transportation to playa or flood plain, and it is further the means of removal of leaves, small twigs, fruits, feces, and other materials which serve to build up the organic content of soils in which they find a resting place.

Tolman (1909) contends that such floods constitute "in every case a depositing and not an eroding sheet, save when confined to a valley where it can do some undercutting." A detailed investigation of sheet floods would undoubtedly show that their behavior is in many respects analogous to that of streams. They are merely streams of very intermittent character and slight depth which are not confined to definite banks. The ratio between the erosional and the depositional work which they perform is a function of the character of the surface over which they move, the gradient of the surface, and the load which the sheet is carrying. The movement of such sheets brings the water rapidly into depressions and normal drainageways. The observations of the writer would indicate that the sheet is the stage of most active erosion, and that the rapidly succeeding movement of the flood waters in the streamways or over the flood plains or playas is the stage of greatest deposition.

The removal of the finest soil material leaves a surface covered with rock fragments in a great variety of sizes, depending upon the location. Near localities where there is rock in place the surface fragments are larger and more numerous, and they decrease in size and abundance on passing toward the playas and flood plains. In all but the finest types of desert soils the surface gives the impression that the soil has a much coarser texture than is actually the case. Beneath a surface which is almost completely covered with stones from a few millimeters to several centimeters in diameter will be found a fine-textured soil, often containing a surprisingly low percentage of such coarse material as covers the surface. The distance to which the surface fragments have been transported is often small, and their form is almost invariably angular. Having accumulated on the surface as a result of violent precipitation, these coarse fragments serve as a protection against the impact of rain and the cutting power of the sheet floods. In general, wind plays a minor role in the removal of fine surface material. Some of it is moved from place to place, but it never accumulates after the manner of sand.

The close-set surface covering of small stones has been called "desert pavement," "erosion pavement," or "lag gravel." In some localities its surface has been long undisturbed and slowly worn down to a uniform level by wind erosion. See plate 1; plate 2, figure 1.

A very constant distinguishing feature of desert soils is the low content of organic matter and the consequent small percentages of humus. Primarily this condition is due to the thinness of the plant cover and the small amount of falling and decaying vegetable matter. Sheet floods wash away much of the small plant litter and deposit it in the playas and flood plains. Such organic matter as the soil contains is rapidly oxidized and decomposed, with the consequence that the humus content is low, even in the flood plains, as compared with that of the soils of humid regions. Figures given by Hilgard (1906) show the amounts of humus in soils of the Colorado and Mojave Deserts of California to range from 0.25 to 0.65 per cent, as compared with a range from 3 to 5 per cent in soils of humid regions. To the poverty in organic matter may be attributed in part the light color and hard surface of most desert soils, as well as the absence of any criteria for distinguishing soil and subsoil.

From the reciprocal relation between the water content and the air content of soils, it follows obviously that arid soils are well aerated most of the time. With the prevailingly thin water films that accompany the dry states of the soil, there is nearly the maximum air content for the given type of soil, and good air communication to a depth which probably exceeds that of plant roots. Though such conditions prevail most of the time over most of the desert, however, there are sometimes local conditions which bring about very poor soil aeration. In the flood plains, and more particularly in the poorly drained playas, the flood waters of the rainy seasons may accumulate to such a depth immediately after the rains as to saturate the soil and to keep it in this condition for a number of days. The harmful effect of the saturation may go so far as to kill trees 30 to 50 years old.

The most striking and best-known peculiarity of desert soils is their high content of readily soluble salts. The accumulation of these is very obviously due to the limited extent to which rainfall has been able to bring such salts into solution and to carry them through a constant series of streams to the sea. The existence of many bolsons and semi-bolsons means that flood waters are held and subjected to evaporation near their place of origin, with resulting heavy accumulation of salts in their playas. Some of the largest streams of the Sonoran Desert are unable to carry their heavy loads of alluvium and dissolved salts through to the sea. Also, much of the precipitation which falls on desert soils is destined to penetrate them to a short distance, dissolving the salts present, and then to return to the surface on account of the active evaporation, depositing the dissolved salts again in almost the precise spot which they originally occupied.

The extended comparisons of arid and humid soils made by Hilgard show that the most constant and striking difference lies in the high calcium content of desert soils, with high magnesium content somewhat less important. In the Sonoran Desert the commonest soluble or somewhat soluble salts in the soils are the carbonates, chlorides, and sulfates of calcium, magnesium, and sodium.

Closely associated with the high salt content of desert soils is the occurrence of superficial incrustations or of buried layers of hardpan. In alluvial soils which become very wet at certain seasons, and also in soils which have been irrigated for many years, there may often be found a thin but conspicuous layer of salts uniformly covering the irregularities of the surface. These incrustations are usually white, but are black where there is a high proportion of sodium carbonate in the soil solution, owing to the action of the carbonate on the organic matter of the soil. The layers of hardpan, better known by the widely used Spanish name *caliche*, are of wide occurrence in the North American deserts in certain types of soil and in certain topographic situations. The caliche consists of hard white sheets or layers from a few millimeters to 10 cm. or more in thickness and sometimes as much as 10 to 15 sq. m. in area, or elsewhere serving to cement together the stones and rocks of the subsoil in a discontinuous layer. Caliche may lie near the surface, or may even be exposed in localities where the surface has been actively eroded; on the other hand, the uppermost layers may be as much as 2 m. below the surface, or rarely even deeper. In the localities where caliche reaches its greatest development it occurs in successive layers to a depth of over 30 m., there being from ten to twenty layers to a meter.

Some of the physical properties of caliche have been investigated by Shreve and Mallery (1933), with particular reference to its role in modifying the water relations of the soil and thereby influencing those of the plant. Their work indicates that caliche retards penetration of water and interferes with the development of the root system to an extent which outweighs its influence in retarding soil evaporation. Both experimental and field evidence indicate that its presence hinders the best development of desert vegetation.

# 3

# Perennial Vegetation

## General Characteristics of Deserts

The deserts of the world are as varied in their features as the forests are. Their plants differ in appearance, structure, behavior, and relationships. The feature that is common to all deserts is the limitation that is placed upon plants by an inadequate or uncertain water supply and by all the characteristics of climate, soil, topography, and drainage that are due to a low rainfall or associated with it. The limitation to which desert plants are subjected affects the number of plants per hectare, the size and height of the plants, the number of species per square kilometer, the number of growing days and total growth per year, the size of the seed crop, the opportunities for germination, and the chances for survival of seedlings, as well as the operation of all the processes concerned in evolution.

There is no single criterion by which desert may be recognized and defined in either the geographical or the biological sense. The amount of rainfall coincident with desert varies with temperature and the seasonal distribution of the rain. A true characterization of desert must include evaluation of many variables. It is essentially a region of low and unevenly distributed rainfall, low humidity, high air temperatures with great daily and seasonal ranges, very high surface soil temperatures, strong wind, soil with low organic content and high content of mineral salts, violent erosional work by water and wind, sporadic flow of streams, and poor development of normal dendritic drainage.

From the biological standpoint desert is best defined in terms of the limitations which have just been mentioned. For plants these serve to prevent the full degree of development that would enable them to form a closed covering, attain a considerable size, maintain vegetative activity throughout the year, and meet the environmental conditions without structural features or types of physiological behavior that tend to reduce their maximum performance.

The comparisons of plant life which are an essential feature of the study of its geographical relationships involve three aspects. These concern (*a*) the individual species with reference to its structure and physiological behavior; (*b*) the individual species with reference to its phylogenetic relationship; and (*c*) the communities with reference to their appearance, structure, relationships, environmental requirements, and composition. Examination of these aspects of desert vegetation shows it to have distinctive features in each of them. In greater or less degree, it differs from the vegetation of related regions in its life forms, in its genera and species, and in the character of its communities.

### *Types of Plants*

Certain outstanding types of plants were recognized by the earliest plant geographers as playing an important role in the appearance of the vegetation and contributing to the character of natural landscapes. The occurrence of broad-leaved evergreen trees, palms, century plants, or bamboos was seen to have an important part in determining the general aspect or physiognomy of the vegetation. In later years the study of such types, called life forms or growth forms, has attracted the attention of a large number of plant geographers, and several detailed schemes for their classification have been proposed. A summary of the work in this field, with citation of important papers, has been given by Rübel (1920). Both the earlier and the later classifications of life forms are based upon obvious features of outward appearance, although the ultimate subdivisions begin to be colored by knowledge of the structure and relationship of plants.

Recognition of the outstanding life forms serves a useful purpose in descriptive plant geography. It will be a long time before we have a broad knowledge of the ecological equivalence of unrelated species of the same life form, or the differences in physiological behavior which may distinguish the unlike types. In view of these limitations there is no ground for multiplying the number of life forms beyond the needs of descriptive plant geography or beyond those subdivisions which have an obvious relation to the state of adjustment existing between the plant and its environment.

Consideration of the outstanding life forms is of particular importance in the discussion of deserts, with respect both to the physiognomy of desert vegetation and to the relation of the dominant plants to their environment. The mingling of dissimilar life forms throughout the vegetation is a characteristic of all but the coldest or most arid deserts. It is, in fact, a circumstance which seems to have considerable significance in the constitution of the vegetation of certain deserts. If the great nondesert plant formations of the world are compared, it will be noted that each of them is dominated by plants which are of the same life form, however widely they may differ in phylogenetic relationship. Tropical rain forests are dominated by evergreen broad-leaved trees, temperate rainy regions by deciduous broad-leaved trees, cold regions by needle-leaved evergreen trees, regions with light winter rain by broad-leaved sclerophyllous shrubs, and regions with low but well distributed rainfall by grasses. In each of these regions the conditions for plants are favorable and the competition between individuals, species, and life forms has been keen, resulting in the development of a high degree of uniformity among the dominant plants. In arid regions there has been much less competition. The greatest "struggle" of the plants has not been with one another, but with the environment. Therefore the conditions tending toward the elimination of certain types and the survival and dominance of a relatively uniform one have not been operative. This fact has brought about the diversity of plant types which characterizes many semiarid and arid regions and some extremely arid ones. The opportunity which the desert has offered for the persistence of unstandardized types has undoubtedly resulted in the survival in it of species, and even of genera, for very long periods of geological time.

### *Species of Plants*

Each of the great desert regions of the world has a flora which is distinct from that of adjacent moister regions. Despite the invariable overlapping of ranges along the edge of the desert, and the occurrence of widely distributed aquatic and palustrine plants around oases and along artificial streams, the true desert flora remains distinct and rather strictly confined to its own region. This is, indeed, only a detail of the broader fact that all the great plant formations of the world, distinguished by reason of their unlike vegetation, are also strongly unlike in flora.

The extremely arid deserts, as well as those that are high and cold, have a small flora. As more and more favorable deserts are examined, the flora is found to grow larger and larger. In the richest deserts the size of the flora is to be attributed in part to the relatively favorable conditions of moisture and temperature and in part to historical factors. It is in the deserts with the most favorable conditions and the largest flora that distinctive desert genera are found, and also a very small number of families which are wholly or almost wholly confined to the desert.

The conditions of the desert not only have influenced the vegetative structures of plants, but also have shaped the processes of evolutionary development, the origins, survivals, changes, and movements by which the present flora and its distribution have been determined.

### *Communities of Plants*

The plant communities of the desert have some distinctive features and some which they share with the vegetation of other regions. The outstanding characteristics of most desert communities are the low but unequal stature of the plants, the openness of stand, and the mixture of dissimilar life forms. These features interact in some respects; for example, the inequality of stature is largely due to the mingling of life forms.

From these three primary features of desert communities arise others which are of importance. The low stature is responsible for the scarcity of truly subordinate plants, such as the secondary trees, undershrubs, herbaceous perennials, epiphytes, and lianes which are involved in the layering of the tropical rain forest. The openness gives ample opportunity for the establishment of new individuals of the dominant species or for invasion by other species, but it is an opportunity which is rarely embraced. The lowness and openness combine to cause a relative scarcity of plant litter. The little that is produced is to a great extent eaten by termites and only to a very small extent reduced by fungi and bacteria. By the added agency of winds and sheet floods the plant litter is prevented from permanently enriching the spot where it falls, and ultimately comes to rest on flood plains and along streamways. Consequently this most important of the reactions of plants upon their habitats is reduced to a minimum in the desert. The long existence of a community in a given spot does little if anything to pave the way for the entrance of plants with life requirements different from those of the occupants.

Throughout the North American desert the character of the soil is of primary importance in determining the make-up and distribution of communities. The physical texture of the soil, its depth, and the nature of its surface are equally important. Less important, except in extreme cases, is the salt content. The profound influence of soil upon desert vegetation is to be attributed to its strong control of the amount, availability, and continuity of the water supply. This fundamental requisite of plants is the most effective single factor in the differentiation of desert communities.

The desert is often alluded to as being built on a vaster scale than other types of country. This notion is due to the untrammeled vision which its low vegetation permits. The long vistas of the desert make it easy to observe the uniformity of the plant cover, spread with little variation for long distances over terrain in which the soil is highly uniform. The frequency in the desert of extensive communities which are simple in composition is not due to the poverty of the perennial flora so much as to the severity of the physical conditions. In a given region there are few species able to persist in the least favorable habitats, which are invariably the ones with simplest floristic composition. Close to such habitats, in others of nearly identical climate but different soil, grow numerous other species whose presence is made possible by slightly more favorable conditions. In such simple communities the presence or absence of a particular species, or a change in its relative abundance, becomes important in determining the physiognomy. There are virtually no uncommon plants, except outlying individuals of the dominants of adjacent desert regions. The same few species occur over and over again, and the recurrence of a particular habitat brings a repetition of its characteristic community.

The successional changes which are so important elsewhere may be read into the vegetation of the desert, but the evidence for them would hardly suggest that they are of importance in determining the relations of the communities to one another. The secular changes of surface are accompanied by the shifting of the communities over long periods, by which their locations and areas are changed in the simplest possible manner. If a particular community is destroyed without change in the soil, the earliest stage in the return of vegetation will be the appearance of young plants of the former dominants. Not only do the same species reappear at the outset, but their first individuals ultimately constitute the restored community.

It is not possible to use the term "climax" with reference to desert vegetation. Each habitat in each subdivision of a desert area has its own climax, which must be given an elastic definition and must not be interpreted as having a genetic relation to any other climax. It is merely the particular group of species which, in somewhat definite proportions and with a fairly definite communal arrangement, is able to occupy a particular location under its present environmental conditions.

Analogy with moist regions makes it natural to regard the vegetation of flood plains or river bottoms as the climax of desert regions. This is merely a case in which the plants of highest water requirement are occupying the habitats which have the most liberal water supply. The flood plain is the ultimate feature in the idealized course of physiographic development. At any given period in the history of a desert area the processes of degradation and leveling give a much greater area of upland, with far poorer water supply, and the vegetation of these areas might much more appropriately be regarded as the climax if it seems mandatory that the vegetation be interpreted in terms of physiographic processes. From a purely vegetational standpoint, on the other hand, the true climax that has been achieved by plants in clothing the desert is to be found in the richly diversified communities living in relatively favorable but purely desert environments, in which the plants exemplify the most advanced types of structure and behavior that have been developed to meet desert conditions.

## The North American Desert and Its Parts

A nearly continuous arid region extends through western North America from the valley of the Snake River, in eastern Washington, southward and eastward to the Mexican state of Puebla. This irregular area of over 192,000 sq. km. (69,000 sq. mi.) is mainly continental and separated from the Pacific Ocean or the Gulf of Mexico by mountain ranges. Only in northwestern Mexico, in the state of Sonora and the territory of Baja California, does any part of the region lie at low elevations along the coast. Throughout its extreme length of 3700 km. (2300 mi.) there is to be found the characteristic groundwork of physical and biological features which distinguishes all desert regions. There are great differences, however, in the plant and animal life of the various parts of the North American Desert. With an extent of 27° of latitude, a range of elevation from sea level to 1525 m. (5000 ft.), and the attendant differences in climate, soil, and proximity to the sea, a complex background has been furnished on which plants and animals have followed diverse lines of development and have built up communities of wide variety.

With respect to the character of its vegetation, the North American Desert falls naturally into four areas. Following geographical and biological usage, these are designated Chihuahuan Desert, Great Basin Desert, Mojave Desert, and Sonoran Desert. The first is isolated, but the other three lie in a continuous series. Without regard to their individual features of geologic structure, physiographic development, and climate, the four deserts may be distinguished solely on the basis of their vegetation. The boundaries may be placed with varying degrees of definiteness, depending upon the sharpness with which the controlling conditions for desert and for adjacent types of vegetation meet under the alternation of mountain, plain, and plateau.

### *The Chihuahuan Desert*

The North American Desert is divided, like that of South America, into eastern and western sections separated by the highlands of the continental divide. Between the northern ends of the Chihuahuan and Sonoran Deserts is the lowest part of the continental divide in its whole extent from the Peace River, in Canada, to the Isthmus of Tehuantepec, in southern Mexico. In extreme southeastern Arizona and southwestern New Mexico the separating region is a high plain, studded with small mountains, on which the conditions are arid but not desert and the vegetation is a transition between desert and grassland. Farther south the separation is more completely brought about by the Sierra Madre Occidental and its adjacent foothills and plateaus. Whereas the ranges of many desert plants extend across the lowest part of the divide, with or without interruption, there is a much more effective separation of the Chihuahuan and Sonoran Deserts south of the international boundary.

The Chihuahuan Desert comprises small areas in New Mexico and Texas, the lowlands of the Mexican states of Chihuahua and Coahuila, and parts of Durango, Zacatecas, Nuevo Leon, and San Luis Potosí. The vegetation of the Mexican deserts south of latitude 25° is best known through a publication of Ochoterena (1937), but this area has not been compared with the deserts of northern Mexico in sufficient detail to confirm fully the strong indications of Ochoterena's work that it constitutes a wholly distinct fifth desert region.

The Chihuahuan Desert lies mainly above 1070 m. (3500 ft.). Its precipitation ranges from 70 to 500 mm. (3 to 20 in.), of which 70 to 80 per cent falls between the middle of June and the middle of September. Its winter temperatures are low, with occasional periods of 30 to 72 hours of freezing temperature. The prevailing plants in the most widespread communities are low shrubs (*Larrea tridentata*, *Acacia cymbispina*, *Flourensia cernua*, *Prosopis juliflora* var. *Torreyana* as a shrub) or low leaf succulents (*Agave lechuguilla*, *A. falcata*, *Hechtia* sp.). Trees are found only along the drainageways, and tall cacti are represented only by infrequent cylindropuntias (*Opuntia imbricata*, *O. Kleiniae*). The most abundant low stem succulents are *Euphorbia antisyphilitica* and numerous species of *Echinocactus*, *Echinocereus*, *Coryphantha*, *Mammillaria*, *Escobaria*, *Thelocactus*, and *Neolloydia*. Semisucculents are the commonest tall plants, being represented by several species of *Yucca* (*Y. elata*, *Y. macrocarpa*, *Y. Treculeana*, *Y. carnerosana*, and *Y. australis*). The region is essentially one in which the dominance is shared by microphyllous shrubs and low stem succulents or tall semisucculents.

### *The Great Basin Desert*

This northernmost part of the North American Desert is nearly coincident with the states of Nevada and Utah, with a slender extension into Oregon and Washington. Numerous mountain ranges are scattered over the Great Basin, most of which have either a xeric or a mesic forest cover, while the network of desert occupies the floors of the detrital valleys. The elevation lies above

1200 m. (4000 ft.) except in the extreme north. The precipitation in several localities is higher than in any other part of the North American Desert, but for the Great Basin as a whole it ranges from 100 to 270 mm. (4 to 11 in.), of which about 60 per cent falls in the winter months. The winter temperatures are sometimes very low and the frost periods long.

The vegetation of the Great Basin Desert is dominated by a small number of species of low shrubs, most of which are either wholly or partly deciduous. It is almost devoid of trees and also of shrubs with large evergreen leaves, and is poor in succulents and semisucculents. A distinctive feature of the vegetation is the predominance of communities which are very simple in composition or have as much as 95 per cent of their stand made up of a single species. The commonest dominants of these uniform communities are *Atriplex confertifolia*, *Artemisia tridentata*, *A. nova*, *Chrysothamnus puberulus*, and *Eurotia lanata*. Each of the communities is sharply confined to its particular site and type of soil, and in each of them the stature of the plants is determined by local conditions of water supply. In the simplicity of its vegetation and the small number of life forms it contains, the Great Basin Desert is distinctly of the cold northern type.

### *The Mojave Desert*

The Mojave Desert lies in interior southern California and extreme southern Nevada, and is the smallest of the divisions of the North American Desert. Its elevation falls from 1200 m. (4000 ft.) on the western edge nearly to sea level at the Colorado River. The precipitation decreases from 125 mm. (5 in.) on the west to less than 50 mm. (2 in.) on the east, and occurs only in the late winter. Temperature conditions vary with altitude, so that only the western half is subjected to severe frost.

The Mojave Desert shows its most distinctive development between 600 and 1200 m. elevation (2000 to 4000 ft.). When it is followed thence toward the northeast or southeast, it loses some of its characteristic vegetational features and much of its distinctive flora. The groundwork of the vegetation throughout the Mojave Desert is a very open stand of *Larrea tridentata* and *Franseria dumosa*. On the western edge these plants are joined, and to some extent replaced, by *Artemisia*, *Chrysothamnus*, *Grayia*, *Tetradymia*, *Haplopappus*, and other suffrutescent perennials, and at higher elevations on the north *Coleogyne ramosissima* and *Grayia spinosa* are dominant. The only plants which rise above the low and open shrubbery are *Yucca schidigera* and *Y. brevifolia*, both of which are limited to the higher marginal elevations. Most of the Mojave Desert is very poor in succulents, and only on coarse detrital slopes along its southern and western edges will *Opuntia echinocarpa*, *O. basilaris*, and *Ferocactus acanthodes* be found in great enough abundance to affect the aspect of the vegetation.

### *The Sonoran Desert*

The Sonoran Desert occupies the region in the United States and Mexico which surrounds the upper two-thirds of the Gulf of California. Its elevations range from sea level to 1050 m. (3450 ft.). Its precipitation ranges from almost nothing to 330 mm. (13 in.) on its eastern edge, occurring only in winter on the western half and in winter and summer on the eastern. The winter temperatures are very mild in the south, with short freezing periods in the north.

The Sonoran Desert exceeds the other three deserts in the number and variety of its life forms and in the diversity of its plant communities. On the Pacific coast of Baja California low microphyllous shrubs, the leaf succulents *Agave* and *Dudleya*, several species of stem succulents, and the semisucculent *Yucca valida* are the dominant plants. In the interior, the higher areas support only low, dense stands of microphyllous shrubs, whereas the vegetation of the lower parts exhibits an unrivaled wealth in which nearly all the life forms found in the North American Desert are represented. At the low elevations along the lower Colorado River and the head of the Gulf of California the vegetation is low, open, and simple in composition, some of it being equaled in these respects only by some of the large bolsones of the Chihuahuan Desert. Above 300 m. (1000 ft.) in Arizona and northern Sonora the communities again become richer and more diversified, with many small trees and a strong representation of cacti of many types. In central and southern Sonora trees, shrubs, cacti, and sarcophytes are represented in nearly all the communities. The density is very irregular, but in some communities of shrubs and small trees there is nearly complete coverage.

### *Contrasting Features of the Four Deserts*

The salient differences in the vegetation of the four divisions of the North American Desert are closely correlated with their differences in geographical position, in elevation, and in climate. These groups of physical conditions are strong in their operation even where they are not sharply delimited in area.

The very brief descriptions of the four areas which have just been given are sufficient to support the view that the Sonoran Desert is by far the richest in number of life forms and in variety and development of communities. It has large numbers of both evergreen and deciduous shrubs with a wide range of leaf sizes. It has many species of small trees which mingle with the shrubbery on the open plains and include both highly xeric and somewhat mesic types. It supports a display of succulents of great variety in form and size, as well as leaf succulents, semisucculents, and sarcophytes. Its flora is enriched by two large groups of ephemeral plants which appear separately in the two rainy seasons. The wealth of life forms found in the Sonoran Desert is reduced in number and in physiognomic importance on passing north or east into the other deserts. Lower rainfall, limitation of rain to a single season, and lower winter temperatures are the principal factors involved.

The small trees which form the arboreal deserts of Sonora and Baja California are confined to the Sonoran Desert. The most important of these trees are: *Cercidium microphyllum*, *C. floridum*, *C. sonorae*, *Olneya tesota*, *Jatropha cinerea*, *J. cardiophylla*, *Bursera microphylla*, *B. Hindsiana*, *B. filicifolia*, *Pachycormus discolor, Pithecellobium sonorae,* and *Ipomoea arborescens.* Some of the trees confined to streamways or alluvial plains in the Sonoran Desert are also found in the other deserts. Among these are *Prosopis juliflora* var. *Torreyana, Populus Fremontii, Sapindus Saponaria* var. *Drummondii, Chilopsis linearis,* and *Celtis reticulata.* A large number of leguminous trees which are not of xeric stamp are characteristic of lower elevations in the Sonoran Desert, including *Acacia occidentalis, A. Farnesiana, Caesalpinia pumila, Lysiloma candida, L. Watsoni, L. divaricata, Eysenhardtia orthocarpa, Erythrina flabelliformis,* and *Mimosa laxiflora.* None of these reach the Mojave or the Great Basin, and the only upland trees of this type in the Chihuahuan Desert are *Acacia Farnesiana* and *A. Berlandieri.* In general the arborescent habit is indicative of favorable moisture conditions. In the Sonoran Desert trees are favored as much by the biseasonal occurrence of rain as by its amount. The Mojave and Great Basin are dry during the season of favorable temperatures, and the only arboreal form suited to such conditions is *Yucca brevifolia.*

The number of species of evergreen shrubs in the Sonoran Desert is not large, but they include the ubiquitous *Larrea* as well as *Simmondsia chinensis, Condalia spathulata, Celtis pallida, Cordia sonorae, Sapium biloculare, Atamisquea emarginata,* and *Viscainoa geniculata.* The evergreen habit is indicative of the ability of the plant to carry on its leaf functions not only during the cold months, but during the dry ones as well. The region of mild, rainy winters is the one to which nearly all the desert evergreens are confined. *Larrea* is able to range from this favorable region into parts of the colder Mojave Desert as well as into the Chihuahuan Desert, where the winter rain is very light. It is not able to endure the winter temperatures of the Great Basin Desert. The wide range of this shrub in the drier and somewhat colder deserts may be due to its being a facultative evergreen, sometimes losing four-fifths of its leaf surface in very dry periods. Thus enabled to endure the rainless summers of the Mojave Desert, it finds its most favorable season in the late winter and early spring.

The native American succulents are immune to long seasons of drought, but require favorable periods for the replenishment of the water content of their tissues. In the region of biseasonal rainfall, replenishment is greater in the summer than in the winter, being slight during periods of low soil temperature. All cactus seeds require a soil temperature of 21° to 27° C. (70° to 80° F.) for germination, and most of them germinate more surely and quickly in the range 27° to 33° C. (80° to 90° F.). Mature cacti are subject to damage from frost if the temperature falls below −10° to −5° C. (14° to 23° F.) for a short period or below −5° to −2° C. (23° to 28° F.) for a longer period. Under any given temperature range, turgid plants are much more susceptible to injury than plants with low water content. In spite of immunity to normal drought periods, cacti are abundant only in the regions in which the soil is periodically warm and moist at the same time, favoring both root absorption and germination, and in the regions in which winter minima are neither very low nor very prolonged. Small or procumbent species of cacti endure very cold winters in regions where the frost season is preceded by a dry period, during which the water content of the plants is reduced.

The role played by succulents in the vegetation of the four deserts is sharply determined by the conditions which have just been outlined. Large columnar types (*Pachycereus Pringlei*, *P. pecten-aboriginum*, *Carnegiea gigantea*, *Lophocereus Schottii*, *Lemaireocereus Thurberi*) are all found in the warmer parts of the Sonoran Desert, and none of them beyond it except southward into the thorn forest. The arborescent cylindropuntias (*Opuntia fulgida*, *O. molesta*, *O. spinosior*, *O. cholla*, *O. Thurberi*, *O. echinocarpa*, *O. acanthocarpa*) are sparingly represented in the southern Mojave Desert by the last two, and in the Chihuahuan Desert by *O. imbricata* and *O. Kleiniae*. The platyopuntias are more sensitive to prolonged drought than any other type of cactus, and extensive stretches of the Sonoran Desert are almost devoid of them. The heaviest stands of these plants, *Opuntia discata*, *O. Engelmannii*, *O. phaeacantha*, *O. rhodantha*, *O. fuliginosa*, and others, are found at higher levels along the northern and eastern edge of the Sonoran Desert. They extend beyond the desert with little reduction in abundance into the desert-grassland transition, and are abundant in the Chihuahuan Desert. In the Mojave Desert the platyopuntias are represented almost solely by *O. basilaris*, a species which has little in common with the rest of the genus. In the Great Basin, as well as on the high plains of New Mexico, Colorado, and Nebraska, the platyopuntias are restricted in occurrence and represented by low species. Although they are the most widely distributed type of succulent, the platyopuntias do not range far beyond the Sonoran and Chihuahuan Deserts without great reduction in the number of species and in their prominence in the vegetation, as well as in the size of individual plants. The smaller types of cacti in the genera *Echinocereus*, *Mammillaria*, *Coryphantha*, and *Escobaria*, as well as the larger ones in *Ferocactus* and *Echinocactus*, are particularly important in the Chihuahuan Desert and well represented in the Sonoran Desert, but play a very unimportant part in the vegetation of the Mojave and Great Basin Deserts.

The two groups of herbaceous ephemeral plants found in the Sonoran Desert owe their existence not only to the biseasonal rainfall, but to the mild temperatures which permit growth at the time of the winter rains. In the Great Basin and the northern Chihuahuan Desert the cold winters confine activity of the ephemerals to the summer, whereas in the Mojave Desert the late winter is the only season in which their growth is possible.

## Vegetational Features of the Sonoran Desert

A brief characterization of the Sonoran Desert was given in describing the four major subdivisions of the North American Desert. It is now proposed to treat in more detail the Sonoran Desert and its vegetational subdivisions. On account of the size of this area of nearly 120,000 sq. mi., it has been necessary to confine both the exploration and the descriptive text to the larger features of the plant covering. No precise statistical studies have been made, and no efforts to determine any of the sociological aspects of the smaller units of the vegetation.

Before taking up the regional treatment of the vegetation, it is desirable to discuss some of the aspects of the plant life which are common to the entire area. Certain general features are exhibited by all deserts, but require discussion as exemplified in the Sonoran Desert. Also a description of the life forms is fundamental to a portrayal of the vegetation and its habital and regional differences.

### *General Features*

HEIGHT. The average height of the dominant plants reaches its maximum in the arboreal desert of the foothills of Sonora. The largest individual trees, however, are found in the more open vegetation of the plains of central Sonora. The average height of *Acacia*, *Bursera*, *Lysiloma*, and *Prosopis* in the former region is from 5 to 6 m., but in the latter region the finest examples of *Cercidium*, *Olneya*, *Forchammeria*, and *Ipomoea* reach a height of 10 to 12 m. These dimensions are exceeded only by those of *Populus*, *Prosopis*, and other trees growing along streamways. A height of 11 m. is sometimes reached by *Carnegiea* and 16 to 18 m. by *Pachycereus* and *Idria*. Several yuccas exceed the normal height of the trees, notably *Yucca elata*, *Y. brevifolia*, and *Y. valida*. The plants which exceed a height of 6 m. rarely occur in such abundance that their collective tops seem to form the canopy of the vegetation. It is the trees which form the discontinuous canopy, above which tall cacti and yuccas project with great irregularity of height and spacing. Along the inner edge of the Vizcaíno Desert, in Baja California, *Yucca valida* is more abundant than the trees associated with it. At close range there seems to be great irregularity in its height, but in a longer vista its canopy has as regular a profile as does that of the pines of the southeastern United States, with a height of 7 to 8 m.

More than half of the Sonoran Desert is dominated by plants with a stature of less than 1.5 m. The extensive plains with nearly pure stands of *Larrea*, or with *Larrea* and *Franseria*, *Acacia*, *Lycium*, *Caesalpinia*, and numerous other plants, have a cover varying from 0.5 to 1.5 m. in height. It is only where nearly pure stands prevail or a single species is greatly in the majority that there is uniformity of height.

DENSITY. Openness of stand is almost a universal characteristic of the entire North American Desert. Many valleys in the Great Basin with favorable moisture and soil conditions bear nearly closed stands of *Atriplex confertifolia* or *Artemisia tridentata*. Restricted areas of *Larrea* and of *Acacia* with very little uncovered surface are occasionally seen in the Sonoran Desert. In far more than half of the Sonoran Desert, however, the plants cover less than 30 per cent of the ground area, and in at least one-fifth of it they cover less than 10 per cent of the surface.

These rough estimates for the entire region are based on a number of exact measurements in small areas. They have been made by regarding the plants as covering all of the surface that would be comprised if their tops were projected vertically onto the ground. Where several small shrubs grow under a tree, their areas are included in the coverage of the tree and do not contribute to the total. In their position as subordinates to the tree they do nothing to affect the open or closed appearance of the vegetation.

Greater density is commonly found in places where small shrubs are dominant than in the stands of larger shrubs or small trees. This is particularly true of areas in which *Franseria deltoidea* or *Encelia farinosa* forms a high percentage of the cover. It is also true of many overgrazed areas in southern Arizona which are now dominated by *Haplopappus Hartwegii.* It is not true of the region adjacent to the Colorado River in which *Franseria dumosa* is the dominant small shrub, as this region rarely shows more than a 15 per cent coverage.

Simplicity of composition. The number of species of large perennials composing the vegetation in the most unfavorable habitats is very small, and in the most favorable ones is not large. Simplicity of composition is found not only in the driest habitats, but also in situations where the soil is uniform in level, texture, and surface. This is true of the extensive nearly pure stands of *Larrea* in southwestern Arizona and northwestern Sonora, as well as of many smaller areas of *Atriplex* and *Lycium.*

The simplest composition is found on the low bajadas within 40 to 50 km. of the Colorado River and on the sandy plains of extreme northwestern Sonora, where 2 species, or sometimes 3, form the entire vegetation (*Larrea, Franseria dumosa, Hilaria rigida*). On plains and low bajadas below 300 m. elevation the total perennial flora has rarely been found to exceed 10 species in Arizona and 16 species in western Sonora.

With the prevailing simplicity of composition, it follows that the entrance or exit of a single species, or a great change in the relative abundance of two or more species, may make a profound change in the physiognomy of the vegetation.

Physiographic control of vegetation. So great is the importance to desert plants of the texture, depth, and surface character of the soil that in every area of adequate size there are differences in the vegetation which are closely correlated with these conditions. The same relations of vegetation to soil are presented over and over again in hundreds of valleys and intermont plains. There are hills or mountain slopes with rock in place and pockets of soil, there are pediments, upper bajadas with shallow, coarse soil, lower bajadas or plains with deeper soil of uniform but rather coarse texture, and flood plains or playas with deep, fine soil. The physiographic features of the basin determine the physical character and distribution of the soils, which in turn determine the vegetation.

It is impossible to institute comparisons between different parts of the Sonoran Desert without taking complete drainage basins into consideration, or at least the complete series of physiographic units and soil types found in each basin. In the moister regions on the east and northeast the distribution of vegetation is likewise controlled by the physiographic features, but in the driest parts of the Sonoran Desert and in the eastern Mojave Desert the relations are more weakly manifested. Here a low and open stand of *Larrea* and *Franseria* usually extends from the center of a basin, or the edge of its playa, to the summit of the surrounding hills.

Consistency of local distribution. The trees, shrubs, and smaller perennials of the desert are unfailingly distributed over the areas which are favorable to them, no matter whether the stand is very open or relatively close. There is no suggestion of the mosaic type of distribution. Sporadic groups of the common perennials are rare except near the periphery of their geographic ranges. For a situation of given altitude, physiographic character, and slope exposure, the composition of the vegetation may be predicted with great certainty. It is only around springs and ponds or in relatively moist and shaded canyons that plants are to be found which are not of common occurrence. In this respect the desert with its small flora and simple vegetation forms a marked contrast with the moist temperate and tropical forests, in which scarcely any two hectares have the same flora.

The largest cacti (*Pachycereus*, *Carnegiea*, *Lemaireocereus*) and the solitary and caespitose ones (*Mammillaria*, *Coryphantha*, *Echinocereus*) are like the shrubs in the consistency and uniformity of their occurrence. Certain cylindropuntias which multiply almost solely by fallen joints (*Opuntia fulgida*, *O. cholla*, *O. Bigelovii*, *O. fulgida* var. *mammillata*) are conspicuously colonial in their occurrence. *Opuntia fulgida* and *O. cholla* owe some of their abundance to cattle, and the local dense colonies of these species are the product of favorable soil and the constant loosening of terminal joints by passing cattle. In ungrazed regions they are widely spaced, as are all the other cylindropuntias, in which the joints are not easily detached. *Opuntia Bigelovii* occurs only below 1000 m. and in the warmest situations, either on sandy soil or on the south slopes of hills. It is extremely abundant on certain hills and absent on adjacent ones of identical character. This distribution is apparently due to the increase of a small colony by vegetative multiplication and the inefficiency of dissemination by seed. The only plants showing inconsistent occurrence are these few cacti, which are the only plants of the area that exhibit vegetative multiplication.

Subordinate plants. In localities where there are trees and large shrubs it will be noticed that very many of them are closely surrounded by small perennials. The largest of these subordinate colonies are found under old trees with wide-spreading branches, where there is an accumulation of litter and dead annuals, and some humus in the soil. Close observation shows that a high percentage of the young individuals of the large perennials are in these situations and that many of the small perennials are more abundant there than elsewhere. It is significant that there are no subordinates under young and small perennials, which give little shade and have no accumulation of litter and humus; nor, for the same reason, are there any around *Carnegiea*, *Pachycereus*, *Lemaireocereus*, and *Lophocereus*.

The small perennials or semishrubs (*Encelia*, *Franseria*, *Viguiera*, and others) often shelter the early growth of their own seedlings as well as heavy stands of ephemerals. They do not commonly give shelter to the seedlings of large perennials. However, the period during which *Olneya*, *Bursera*, *Cercidium*, or *Pachy-*

*cormus* would be small enough to derive advantage from a location beneath the branches of *Encelia* or *Franseria* would commonly be much longer than the life of the last-named plants. This makes it difficult to estimate the number of young plants of *Olneya* or *Bursera* which apparently have sprung up in open spaces but really owe their start to the shelter of a semishrub which has long since disappeared.

The shade and proximity of either large or small perennials is a conspicuously favorable habitat for both winter and summer ephemerals (see pl. 30). The condition which is chiefly responsible is the slightly higher moisture content under the bushes and the longer duration of favorable moisture there. Fully as important as the amount of available water is the opportunity for interception and lodgment of wind-blown seeds. These move readily over the relatively smooth surface between bushes, but are caught by the bush itself and by the living or dead ephemerals already clustered about it. At the close of the growing season for ephemerals, the action of wind in disseminating seeds along the surface of the ground may be observed in any open situation, and it can then be seen that most of the crop comes to rest under large or small perennials or on very rough stony surfaces.

In many places in the driest parts of the Sonoran Desert subordinate plants are few under either large or small perennials. This is true of the sandy soils, on which the moisture conditions are much the same in the open and under trees or bushes, and on which the wind moves substratum as well as seeds (see pls. 5, 16, 34). It is also true of flat areas subject to overflow or deposition. Without small irregularities of surface and without the dead herbage of previous years, there is nothing to catch and hold wind-blown seeds on these smooth and level areas.

The colonies of seedling trees, small shrubs, grasses, and vines to be found under and around the large trees or oldest shrubs grow denser toward the central and southern part of Sonora, and include many species not found in the open. As the subordinate colonies increase in density, they also grow in extent and spread beyond the shade of the tree. On the inner half of the Sonoran plains between Hermosillo and the Yaqui River the stand of large perennials becomes denser than it is north or west of there, at the same time that the subordinate colonies spread more widely. As a result it is a common thing for several colonies to coalesce and, with the group of fostering trees, to form a sharply defined motte, or island, of vegetation. Broad expanses of bare or sparsely covered ground separate the mottes. It is obvious that many conditions are concerned in making the establishment of new individuals easier in the mottes or around their edges than in the intervening open areas.

### *Life Forms of the Sonoran Desert*

The vegetation of the Sonoran Desert is distinguished from that of adjacent regions by the wide differences in appearance between its dominant plants. They are unlike in height, bulk, manner and amount of branching, character of stem, size, duration, color, and seasonal behavior of leaves, degree of succulence, and time of flowering. The characteristic appearance of the landscape and the physiognomy of the vegetation are determined by the varied admixture and grouping of plants which exhibit these differences in structure and habit. Certain habitats are occupied by representatives of a large number of these types, particularly the hillsides, mountain slopes, and upper bajadas. The vegetation of plains and lower bajadas in the northwestern part of the Sonoran Desert is often reduced to three or four of the types. It is in the parts of the desert with 150 to 300 mm. (6 to 12 in.) of rain that the greatest number of types is found.

Investigation of representative members of nearly all the life forms of the Sonoran Desert was carried on at the Desert Laboratory for thirty-five years, and has thrown light on the ability of these plants to live in close association without serious disadvantage to any one of the types. In general there is no lack of room for the establishment of new individuals. In the rainy seasons the ground is often thickly covered with herbaceous ephemerals, among which there is competition for space and soil moisture. The seedlings of perennials must compete during their first season, and in some cases for several seasons, with the dense cover of ephemerals. The shoot growth of seedling perennials is outstripped by that of the ephemerals, but the roots of the young nonsucculent perennials penetrate as far as the roots of the ephemerals and in some species farther. The young perennial is therefore not seriously hampered in its first season by its short-lived associates so far as water supply is concerned, and indeed derives considerable advantage from the shade and shelter which they afford while they are alive and for many weeks after they die.

The successful establishment of a young perennial of any type depends upon whether its first few seasons are wetter or drier than normal and whether the particular spot in which it sprang up is sheltered or exposed, provided with deep or shallow soil, and protected or unprotected from a considerable change of surface level. The presence of closely adjacent mature perennials seems never to be an adverse circumstance in the fate of a young plant. On the contrary, the number of young plants that appear and survive close to mature ones greatly exceeds the number that appear and survive on the open or bare spaces.

These facts have an important bearing on the close association of diverse life forms in the vegetation of the Sonoran Desert. Establishment behavior is far from being identical in the various life forms, but no evidence has been found to indicate that any one of them has a more successful course of behavior than the others do; there are a few cases, however, in which individual species have acquired features of behavior which appear to be disadvantageous. In their germination and early growth the several life forms are under more nearly identical conditions than at any subsequent time. The development of their individual features of structure and behavior soon brings each form into a slightly different set of relations to the environment. Differences appear in height and spread of branches, depth and distribution of root system, and season of foliation and defoliation. These serve to separate the activities of some of the life forms, in either time or space, so as to reduce or eliminate the competition between them.

Table 2 (pp. 42–43) indicates the principal life forms of the Sonoran Desert and gives examples of them. Forms have been omitted which are of minor importance in the vegetation, as aquatics, vines, parasites, etc. The subdivision has been carried no farther than is necessary to illustrate the dissimilarity among the common plants of the region. Each of the examples given is a characteristic plant in some association in some part of the Sonoran Desert.

It seems desirable to discuss briefly the 25 life forms enumerated, in order to indicate their differences more clearly than is possible in the abbreviated key. Each of the forms is represented by a few or many clear-cut examples, in addition to which there are plants with intermediate characteristics which are not worthy of recognition as distinct forms in such a classification, but are of great interest as indicating the possible stages by which one life form may have been derived from another.

The EPHEMERALS (life forms nos. 1–3) constitute a large group, merging into the root perennials through a few common species of facultative perennials, the duration of which is controlled by rainfall fluctuations. The winter and summer ephemerals are distinct groups in the northern half of the Sonoran Desert, the time of their appearance being controlled by the prevailing range of surface soil temperature. Although related to one another in duration of life and in having a restricted season of activity, the ephemerals exhibit differences in habit, branching, stem, leaf, and other features which make them very unlike among themselves. For further discussion of these plants see chapter 4.

The ROOT PERENNIALS (4, 5), plants which persist from year to year strictly by underground organs, are all inactive in the dry seasons, and active either in winter or in summer, but in no case in both the moist seasons. There is no discoverable difference between the mechanisms for resting over the cool moist season, the warm moist season, and the dry ones, except that the few bulbous plants of the Sonoran Desert are active only in the cool moist season (*Allium Kunthii*, *A. haematochiton*, *Triteleiopsis Palmeri*, *Hesperocallis undulata*, *Calochortus Kennedyi*). Very large perennial roots characterize most of the Cucurbitaceae (*Cucurbita digitata*, *C. palmata*, *Brandegea Bigelovii*) as well as certain species in other families (*Rumex hymenosepalus*, *Phaseolus* spp.). In the cucurbitaceous genus *Maximowiczia*, part of the stem is enlarged into a turnip-shaped structure which is half buried and half exposed. The perennial organs in most of the members of this group are slender or moderately stout rhizomes; in some, they are the largest roots (*Delphinium*, *Anemone*). The root perennials are closely and perhaps genetically related to the semishrubs (17), in which the root system is perennial and the shoots commonly so but very sensitive to unfavorable moisture or temperature conditions.

Closely related in the manner of perennation are the GRASSES (6), in which the inner basal part of the cluster of shoots and leaves and the upper part of the root system persist under ordinary unfavorable conditions. Unlike the strictly underground perennials, the grasses show vegetative activity in both summer and winter, although flowering is confined to the former season.

Of great importance in the physiognomy of desert vegetation are the CAUDICAL

TABLE 2

LIFE FORMS OF THE SONORAN DESERT

| *Form* | *Examples* |
|---|---|
| Ephemerals: | |
| Strictly seasonal: | |
| Winter ephemerals | 1. Daucus pusillus, Plantago fastigiata |
| Summer ephemerals | 2. Tidestromia lanuginosa, Pectis papposa |
| Facultative perennials | 3. Verbesina encelioides, Baileya multiradiata |
| Perennials: | |
| Underground parts perennial: | |
| Perennial roots | 4. Pentstemon Parryi, Anemone tuberosa |
| Perennial bulbs | 5. Hesperocallis undulata, Brodiaea capitata |
| Shoot base and root crown perennial | 6. Hilaria mutica, Aristida ternipes |
| Shoots perennial: | |
| Shoot reduced (a caudex): | |
| Caudex short, entirely leafy: | |
| Leaves succulent | 7. Agave Palmeri, Dudleya arizonica |
| Leaves nonsucculent | 8. Nolina microcarpa, Dasylirion Wheeleri |
| Caudex long, leafy at top: | |
| Leaves entire, linear, semi-succulent | 9. Yucca baccata, Yucca brevifolia |
| Leaves dissected, palmate, nonsucculent | 10. Washingtonia filifera, Sabal uresana |
| Shoot elongated: | |
| Plant succulent (soft): | |
| Leafless, stem succulent: | |
| Shoot unbranched | 11. Ferocactus Wislizenii, Echinomastus erectocentrus |
| Shoot branched: | |
| Shoot poorly branched: | |
| Plant erect and tall | 12. Carnegiea gigantea, Pachycereus Pringlei |
| Plant erect and low or semi-procumbent and low | 13. Pedilanthus macrocarpus, Mammillaria microsperma |
| Shoot richly branched: | |
| Stem segments cylindrical | 14. Opuntia spinosior, Opuntia arbuscula |
| Stem segments flattened | 15. Opuntia Engelmannii, Opuntia santa-rita |
| Leafy, stem not succulent | 16. Talinum paniculatum, Sedum Wootoni |
| Plant nonsucculent (woody): | |
| Shoots without leaves, stems green | 17. Holacantha Emoryi, Canotia Holacantha |
| Shoots with leaves: | |
| Low bushes, wood soft | 18. Encelia farinosa, Franseria dumosa |
| Shrubs and trees, wood hard: | |
| Leaves perennial | 19. Simmondsia chinensis, Larrea tridentata |
| Leaves deciduous: | |
| Leaves drought-deciduous: | |
| Stems specialized: | |
| Stems indurated on surface | 20. Fouquieria splendens |
| Stems enlarged at base | 21. Idria columnaris, Bursera microphylla |

| *Form* | *Examples* |
|---|---|
| Stems normal: | |
| Stems not green...... | 22. Jatropha cardiophylla, Plumeria acutifolia |
| Stems green......... | 23. Cercidium microphyllum, Parkinsonia aculeata |
| Leaves winter-deciduous: | |
| Leaves large........... | 24. Populus Fremontii, Ipomoea arborescens |
| Leaves small........... | 25. Olneya tesota, Acacia Greggii |

PLANTS (7–10), a group of life forms characterized by a stout or somewhat succulent stem (caudex) in which the internodes are so greatly abbreviated that the leaves are inserted close to one another. Groups 7 to 9 are chiefly made up of members of the monocotyledonous families Bromeliaceae (*Hechtia*), Liliaceae (*Yucca*, *Nolina*, *Dasylirion*), and Amaryllidaceae (*Agave*). Group 7 includes, besides *Agave*, members of the Crassulaceae (*Dudleya*). *Agave* and *Dudleya* are very unlike in size and lignification, yet they are similar in having a short caudex completely covered by leaves, both caudex and leaves being succulent. Although the difference in size gives *Agave* and *Dudleya* distinct roles in the vegetation, nevertheless their similarity of relation to environment is indicated by the coincidence of the areas of maximum speciation and abundance for the two genera in western central Baja California.

The group of caudical plants in which the leaves are not succulent, but long and grasslike (8), is of some importance along the northern and eastern margin of the Sonoran Desert, and much more abundant just outside the desert. Only two genera (*Nolina*, *Dasylirion*) are represented. Far more abundant and characteristic in all parts of the Sonoran Desert is group 9, in which the leaves of mature plants are borne only at the top of the caudex and are either pliant and non-succulent or stiff and succulent. In the Sonoran Desert the group includes one arborescent *Dasylirion* and a number of species of *Yucca*. This essentially monocotyledonous life form is foreshadowed in the smaller mesic liliaceous plants *Xerophyllum* and *Tofieldia*. The characteristics of *Yucca* are intermediate between those of the dicotyledonous tree and those of the columnar cactus. The photosynthetic and transpiring surfaces are confined to the leaves, but are fixed in area so as to admit of no seasonal reduction. The trunk is from 10 to 75 cm. in diameter, and in *Yucca brevifolia* has a definite secondary thickening. The water content of trunk in all species and of leaf in the thick-leaved ones is great enough at the close of a rainy season to suggest the importance of these parts as storage organs, although they are without the mechanism which the Cactaceae have for adjustment to changes of volume.

The palms (10) are represented in the Sonoran Desert by members of only two genera with pinnately compound leaves and three with palmate leaf blades more or less dissected into linear segments toward their margins. The trunks of these trees are unbranched, often attaining a height of 20 m. or more with a diameter that rarely exceeds 1 m. The leaves remain attached to the trunk for a number of years if not removed artificially, or they may drop soon after they have ceased to carry on photosynthesis. In *Washingtonia*, *Erythea*, and *Sabal* the persistent leaves form a dense thatch or "shag" that may hide a considerable portion of the trunk of healthy, unmolested trees; this situation is less pronounced in *Phoenix*.

The palms grow along watercourses, around seeps or potholes in rocky canyons, and on bajadas where the underground water supply is abundant. They rarely occur on dry hillsides or on plains more than one or two kilometers from the mouths of canyons. No special storage organs are present, but the spongy inner tissue of the trunks does hold a considerable amount of water and constitutes a reservoir from which moisture can be drawn for a limited time if the natural water supply is temporarily depleted. The leathery leaves are well adapted to reducing the loss of water by transpiration.

Owing to the rather specialized habitats necessary for the growth of the palms, they constitute a conspicuous element in the flora in only certain parts of the Sonoran Desert. *Erythea* is particularly abundant along the flanks of the southern part of the Sierra San Pedro Mártir in Baja California, and *Sabal* occurs in considerable numbers in the interior of Sonora. *Washingtonia* is less widely distributed and less abundant than either of the other genera just mentioned.

In the deserts of North America the STEM SUCCULENTS (11–15) are nearly confined to the Cactaceae, a family in which a wealth of types has developed. The only important stem succulent in the Sonoran Desert which is not of this family is *Pedilanthus*, if the halophyte *Allenrolfea* is left out of account as not being a strictly desert plant. The Cactaceae are of great importance in giving a characteristic physiognomy to the vegetation in nearly all parts of the Sonoran Desert. The life forms recognizable among the stem succulents may appear to be more distinctive in their mere geometry than in their essential relations to the environment. Study of the geographic and habital distribution of the five life forms has, however, given clear indication that their differences in size, form, and manner of branching are of fundamental importance in their sustained adjustment to different environments.

The barrel type of stem succulent (11) is found only in the genera *Ferocactus* and *Echinomastus* and is recognizable only by the lack of branching in a shoot which may be of any height from 1 dm. to 1.5 m. Injury to the single growing region of this type may stop further height growth or may in rare cases induce abnormal formation of several growing regions near the apex of the shoot. The barrel type is probably a derived rather than a primitive one, since it has few representatives in comparison with the similar branched types, and its simplicity of form is a development away from the prevailing condition in primitive cacti.

The columnar type (12) includes all the tallest and most massive stem succulents of the Sonoran Desert (*Pachycereus*, *Carnegiea*, *Lophocereus*, *Lemaireocereus*). Branching takes place either from mid height of the shoot or from the base. In sheer bulk these plants greatly exceed any others of the region, and there are only three nonriparian trees which reach greater height.

The small columnar type (13) resembles the barrel type, but its members are never more than 2 to 5 dm. high, with the exception of *Pedilanthus macrocarpus*.

This is the only euphorbiaceous plant in the Sonoran Desert which has strongly developed the succulent habit characteristic of so many of the family in the deserts of South Africa. The other members of this group are unbranched in early life and are then analogous to the plants of group 11. In later life they branch, usually from the base, and have from one or two to eight or ten erect or partly procumbent branches. A few forms branch so richly that the mature plant forms a compact hemisphere, superficially resembling alpine polster plants (*Echinocereus maritimus, Mammillaria fasciculata*).

The cylindropuntias (14) constitute a life form which is represented only by members of this section of *Opuntia*. The erect habit and frequent branching give the plants an arborescent form. The stems throughout are circular in cross section and segmented into joints, each of which is the outcome of a single season's growth. The tallest of this group are 2 to 3 m. in height (*O. fulgida, O. acanthocarpa*); others rarely exceed 1.5 m. (*O. Bigelovii, O. arbuscula*).

The platyopuntias (15) are also a life form restricted to the members of a section of *Opuntia*. The plant is either erect or prostrate and built entirely of flat joints of round, oval, or pyriform outline. In the tallest erect forms the oldest joints gradually assume a circular cross section and develop woody tissue of adequate strength for support (*O. chlorotica, O. laevis*). In the radiating branches of the prostrate forms the oldest joints often bear erect shoots of one to several joints (*O. rhodantha, O. phaeacantha*).

The life form in which succulent leaves are borne on a nonsucculent stem (16) is represented by a relatively small number of species which are only locally abundant. The leaves may be terete (*Portulaca lanceolata, Dudleya attenuata*) or broad and entire (*Talinum paniculatum, Nitrophila occidentalis*).

The NONSUCCULENT WOODY PERENNIALS (17–25) embrace the plants of greatest abundance in the make-up of the vegetation as well as the majority of the members of the flora.

The green-stemmed leafless plant (17) is represented in either ultimate, developmental, or incipient form by a large number of trees and shrubs. The ultimate form considered here includes a few trees and shrubs which are totally leafless after the seedling stage (*Canotia, Koeberlinia, Holacantha, Ephedra*). The multiphyletic origin of the leafless green-stemmed habit is obvious; initial stages of it are noted under 23 and may be seen in trees and shrubs which have evergreen leaves under usual conditions, as well as in deciduous shrubs which have well developed chlorenchyma immediately beneath a dark but translucent epidermis.

Among the nonsucculent leafy perennials of the Sonoran Desert are a large number of plants (18) characterized by soft wood, rich branching, indeterminate growth, and great responsiveness to favorable or unfavorable temperature or moisture conditions. The majority are from 5 to 8 dm. in height, but they may exceed 2 m. These will be alluded to as "semishrubs," in distinction from the true shrubs, which have hard wood, restricted branching, determinate growth, and marked ability to withstand cold or dry periods. The commonest semishrubs are *Franseria dumosa, F. deltoidea, F. magdalenae,* and *Encelia farinosa,* four plants which vastly outnumber any others of any life form in the entire region. The semishrubs also include most of the perennial species in the genera *Atriplex*, *Eriogonum*, *Viguiera*, *Salvia*, *Haplopappus*, *Hyptis*, *Acamptopappus*, *Trixis*, *Franseria*, and many others.

The evergreen trees and true shrubs (19) include forms with thick, highly cuticularized leaves which are tenaciously held in long dry periods as well as through the winter (*Simmondsia chinensis*, *Jacquinia pungens*, *Atamisquea emarginata*, *Forchammeria Watsoni*, *Sapium biloculare*), forms with thinner and less xeric leaves which are shed in part during very dry periods (*Larrea tridentata*, *Condalia spathulata*, *Celtis pallida*, *Piscidia mollis*), and also forms with thin compound leaves which are retained throughout the year under favorable conditions, and after severe drought or cold die in place without a definite period of fall (*Olneya tesota*, *Coursetia glandulosa*, *Eysenhardtia orthocarpa*).

The perennials with definitely deciduous leaves (20–25) are divisible into those in which leaf fall is induced by drought, regardless of temperature conditions (20–23), and those in which defoliation occurs at the beginning of winter and new leaves appear in the spring more or less independently of the surface indications of soil-moisture conditions.

The small group of life forms in which the fall and appearance of leaves is most sensitive to fluctuating moisture (20) is also characterized by broad layers of tissue in the inner bark in which all cells are filled with resin. In the species of *Fouquieria*, which are the only examples of this growth form, the leaves are from 2 to 5 sq. cm. in area, thin, and mesic in every particular except the moderately heavy epidermis. When *Fouquieria* is in leaf, its water requirement is very high. When it is not in leaf, the armored stem is effectively protected from even a moderate water loss, and the requirement of the plant is very low. Under exceptional conditions the leaves may be retained for six months. The distribution of rainy periods is sometimes such that six or seven crops of leaves appear in one year.

Stout-stemmed drought-deciduous trees (21), herein designated as "sarcophytes," have few examples in the flora of the Sonoran Desert, but it happens that each of them is very abundant in its particular distributional area. In fact, the members of this group (*Idria columnaris*, *Bursera microphylla*, *B. filicifolia*, *Pachycormus discolor*, *Cyrtocarpa edulis*) and of the preceding form are second only to the stem succulents in giving the vegetation of the Sonoran Desert its characteristic physiognomy. *Idria* and *Pachycormus* are found almost solely in the region of winter rains and are in leaf only in that season. *Bursera* and *Cyrtocarpa* commonly come into leaf in summer as well as in winter. The exaggerated basal diameter of the stem of *Idria* is due to enlargement of the pith, and the surface is heavily indurated. In the other three trees the enlargement is confined to the cortex and periderm, and the smooth surface is not heavily armored. A slight basal enlargement of the trunk, either abrupt or tapering, is manifested by members of other life forms (22, 25), including other species of *Bursera* as well as *Jatropha*, *Ceiba*, and *Erythrina*.

A number of small trees have stems which are normal and large leaves which are drought-deciduous after the summer rainy season and do not appear at all in the winter rainy season. The group is related in behavior to the strictly winter-deciduous trees (24, 25), but is unlike them in producing its leaves at the beginning of the summer rains rather than at the advent of spring, and also unlike them in producing no leaves if the summer rains fail (*Jatropha cinerea*, *J. cuneata*, *J. cordata*, *Plumeria acutifolia*).

The important group of trees and shrubs (23) with green stems and small drought-deciduous leaves has been alluded to under 17 as constituting a stage in the development of the leafless green-stemmed form. In nearly all the members of this group the photosynthetic surface presented by the stems and twigs is greater than that seasonally provided by the leaves. The small-leaved forms produce foliage in summer and in winter (*Cercidium microphyllum*, *C. floridum*, *C. sonorae*, *Dalea spinosa*). Seasonal conditions sometimes enable the summer leaves to persist until the winter rainy season. The large-leaved forms produce foliage only in very favorable seasons and on the wood of that season (*Euphorbia colletioides*, *E. Xanti*, *Beloperone californica*). Many green-stemmed shrubs and semishrubs have small or linear leaves which persist throughout the year but are more numerous in the summer rainy season (*Krameria* spp., *Baccharis sarothroides*, *Porophyllum gracile*, *Bebbia juncea*, *Condalia lycioides*).

The winter-deciduous woody plants (24, 25) are characteristically represented by a group in which the leaves or leaflets are large (over 5 sq. cm.), the autumnal fall is rapid and complete, and the individuals are either confined to moist habitats (*Populus*, *Salix*, *Sapindus*) or else found only on the inner and more elevated edge of the desert (*Juglans*, *Fraxinus*). These are greatly outnumbered in species and individuals by forms of common occurrence in the desert, some of which have a leaf area as great as that of *Salix* or *Sapindus* (*Ipomoea arborescens*, *Ceiba acuminata*, *Tabebuia Palmeri*), but a larger number of which have compound leaves with small leaflets (*Prosopis juliflora* var. *velutina*, *P. juliflora* var. *Torreyana*, *Acacia Farnesiana*, *A. constricta*, *A. Greggii*). There is only a very broad line of separation between woody plants which are promptly and completely winter-deciduous and those that retain more than half their leaves through the winter, albeit in greatly impaired condition for performance of foliar functions. Several species of *Acacia*, *Mimosa*, and *Lysiloma* are more completely deciduous in northern Sonora than in the south. *Prosopis* is wholly deciduous in cold habitats, while at the same time near-by individuals in warmer situations retain many functional leaves. The behavior of all the plants of intermediate character is largely determined by the severity of the winter temperatures.

## Vegetational Subdivisions of the Sonoran Desert

The attempt to subdivide the Sonoran Desert into areas which correspond to the natural botanical features of its several parts may be made from three standpoints: first, that of the geographic and physiographic groundwork which is so important in the diversification of the plant life; second, that of the genera or species of plants which are most abundant in the various parts of the region; third, that of the character and organization of the communities of plants. From these three standpoints it is possible to make respectively geographic, floristic, and vegetational subdivisions. These are distinct in their rational basis even though their outlines happen to coincide to a greater or less extent.

It is obvious that the geographic subdivision relates largely to the environmental conditions, which are not a part of the plant life. Floristic and vegetational subdivisions must be made on the basis of their own intrinsic data. The physical setting is a controlling agent with reference to the vegetation, but is not one of its characteristics. The custom of using the habitat as part of the characterization of a plant community is an illogical procedure of long standing.

The designations for vegetational units or aggregates which are most acceptable in theory are unfortunately too cumbersome for repeated use. The adoption of established local names which owe their origin to the character of the vegetation, as marsh, chaparral, and tundra, has much to recommend it, but such a system would be very incomplete and would complicate the comparison of widely separated areas. In Latin American countries the terms used by the observant natives, as *cedral*, *sabinal*, and *corozal*, are simple and useful; yet these words, although they emphasize the dominance of a particular plant, tell nothing of the character of the community or of the principal associates of the dominant plant.

The subdivisions of the Sonoran Desert used in this publication have been made purely with reference to the vegetation. Names have been given to the subdivisions which refer solely—even if inadequately—to the character of the vegetation. Names have also been given these areas which indicate their preponderant or most distinctive genera. Finally, the areas have also been designated by names derived from their location or some outstanding geographical feature. In this way subdivisions have been given a vegetational basis at the same time that their floristic character has been indicated and brief geographical names have been provided for use in subsequent pages (see map 1).

The three sets of equivalent designations for the seven subdivisions are as follows:

1. Microphyllous desert
   Larrea-Franseria region
   Lower Colorado Valley
2. Crassicaulescent desert
   Cercidium-Opuntia region
   Arizona Upland
3. Arbosuffrutescent desert
   Olneya-Encelia region
   Plains of Sonora
4. Arborescent desert
   Acacia-Prosopis region
   Foothills of Sonora
5. Sarcocaulescent desert
   Bursera-Jatropha region
   Central Gulf Coast

6. Sarcophyllous desert
   Agave-Franseria region
     Vizcaíno Region
7. Arbocrassicaulescent desert
   Lysiloma-Machaerocereus region
     Magdalena Region

*Brief Characterizations of Vegetational Subdivisions*

1. Lower Colorado Valley. This is the largest subdivision of the Sonoran Desert, occupying the lower drainages of the Colorado and Gila Rivers, the Salton Basin, the extreme coast of Baja California as far south as Bahía Los Angeles, and all of Sonora lying below 400 m. elevation as far south as the valley of the Río Magdalena. The area is chiefly drained by the Colorado River and the Salton Sea, but its two southern extensions contribute their flood waters directly to the Gulf. The Lower Colorado Valley is well defined along its western edge and rather sharply marked between Parker and Phoenix as well as between Sonoyta and Santa Ana, but elsewhere is not sharply defined, since its typical vegetation occurs in restricted areas of valley floor in adjacent subdivisions.

In the Lower Colorado Valley the percentage of surface occupied by hills and mountains is less than in the other subdivisions. About 85 per cent of the area outside the delta of the Colorado is comprised in bajadas of low gradient, or in nearly level plains of gravelly outwash or sand. This subdivision is one of the most arid parts of the Sonoran Desert, being low in rainfall and soil moisture, and high in percentage of sunshine and in both summer and winter temperatures. The surface features are more diversified than in any of the other subdivisions, embracing volcanic and granitic mountains, sandy plains and dunes, alkali flats, and soils varying in origin, texture, and depth. The Lower Colorado Valley includes the delta of the Colorado River, where an abundant moisture supply gives rise to a series of vast hydric and mesic communities that have little in common with the surrounding desert vegetation. The history of the delta and its changes in historic time have been ably treated by Sykes (1926, 1937).

The vegetation of the Lower Colorado Valley is distinguished by its simplicity of composition on the gravelly and sandy plains which make up such a high percentage of the area. The upper bajadas, the level fields of recent volcanic eruptives, or malpais, and the hills and mountains support a richer vegetation. The two species which vastly outnumber all others are *Larrea tridentata* and *Franseria dumosa*, from which the designation Larrea-Franseria region is derived. Low, open stands in which these shrubs form from 90 to 95 per cent of the plant population cover thousands of hectares in all the larger intermont plains, and earn for the area its designation as microphyllous desert. On the rocky soil of upper bajadas and pediments are to be found small trees, notably *Prosopis juliflora* var. *Torreyana*, *Cercidium floridum*, *C. microphyllum*, and *Olneya tesota*. Larger individuals of the same trees are also to be found along the drainageways.

The poverty of the vegetation in perennial plants is compensated by a large flora of ephemeral herbaceous species. These have their most certain opportunity and make their greatest display after the late winter rains, particularly on sandy soil west of longitude 114°. Summer ephemerals rarely appear in large numbers in that area, on account of the sporadic character of the summer rains there. On the eastern edge of the Lower Colorado Valley and in its southern extension in Sonora the summer ephemerals are abundant as well as those of the late winter.

The north-and-south extent of the Lower Colorado Valley is 650 km., a distance great enough to restrict some of the important dominants to the northern or the southern part of the area. There are also some notable floristic differences between the eastern and western edges of the area.

2. Arizona Upland. Under this designation is comprised the northeastern part of the Sonoran Desert, mainly in Arizona, partly in northern Sonora, and lying at elevations between 150 and 950 m. On the south and west the area merges into the plains of the Lower Colorado Valley; on the north and east it is flanked by a broken series of mountains, most of which rise above the level of desert conditions and vegetation. The limit of the vegetation of the Sonoran Desert is here clearly marked, although a few of its characteristic plants occur as high as 2000 m. in the mountains.

The principal part of the Arizona Upland drains, actually or potentially, into the Gila River. A small area at the north is drained by the Bill Williams River, a very small part in the southwest by the Sonoyta, and the southernmost extension by the Altar, Magdalena, and San Miguel Rivers. Throughout this area the landscape is studded with small mountains and hills, which cover a much greater percentage of the total surface than in the Lower Colorado Valley. Only along the lower courses of the Santa Cruz and Gila Rivers and southeast of Sonoyta are there extensive valleys, and it is in them that the vegetation of this area most closely approaches that of the Lower Colorado Valley. The rainfall ranges from about 75 to 300 mm. and is biseasonal throughout the area.

The vegetation of the Arizona Upland exceeds that of the Lower Colorado Valley in stature, density, and the number of species which play a dominant or subdominant role. *Larrea* is still an important plant in nearly all situations, but has a greater number of associates. *Cercidium microphyllum*, though absent from the lower bajadas, is elsewhere so abundant as to characterize the area. Other important species are *Prosopis juliflora* var. *velutina*, *Olneya tesota*, and *Fouquieria splendens*. The cacti are numerous and diversified, the most conspicuous ones being *Carnegiea gigantea*, *Ferocactus Wislizenii*, and 12 or 15 species of cylindropuntia and platyopuntia. A considerable number of large and small shrubs occur here and do much to add to the density of the ground cover. Perennial grasses are locally abundant. Both winter and summer ephemerals carpet the ground in favorable seasons. As contrasted with the Lower Colorado Valley, there are fewer species of the former group and more of the latter.

From the floristic viewpoint the Arizona Upland may well be designated the Cercidium-Opuntia region, as these two genera play a more important part in its vegetation than do any others (except *Larrea*) and are more important here than in any other subdivision of the Sonoran Desert. It is from the important role of the succulents, including *Carnegiea*, that the Arizona Upland is called crassicaulescent, or stem-succulent, desert.

3. PLAINS OF SONORA. Lying between the foothills and the coastal strip in central Sonora is an area which has some physical features in common with the Lower Colorado Valley. The general surface is very regular, with an inclination from 750 m. elevation on the inner edge to 100 m. on the outer. The mountains and hills are few and do not exceed 10 per cent of the total area. Only four mountains rise above 1000 m. The surface is coarse or fine gravelly outwash, with only small accumulations of sand marking the terminus of clogged arroyos. Although some of the mountains are volcanic, there are no recent malpais fields. The rainfall is 250 to 375 mm., an amount substantially greater than in the Lower Colorado Valley. Frost is less frequent and less severe than in the areas to the north. Summer temperatures are more moderate than those in the Lower Colorado Valley.

The vegetation of the Plains of Sonora is dominated by trees and shrubs, because of which it is designated arbosuffrutescent desert. The name Olneya-Encelia region is suggested by the dominant tree and shrub. The trees are remarkable for the size which they attain under desert conditions, and also for their abundance on the inner half of the area. *Olneya tesota*, *Cercidium microphyllum*, and *Prosopis juliflora* var. *velutina* are the dominant species. In this area *Larrea* reaches the southern limit of its importance, occurring only locally in the most favorable situations. The dominant shrub is *Encelia farinosa*, which is confined to rocky slopes in the two subdivisions to the north. In this area *Cercidium sonorae* and *Fouquieria Macdougalii* reach their northern limit and are locally abundant. Conspicuous elements of the vegetation are the large columnar cacti *Lemaireocereus Thurberi* and *Lophocereus Schottii*. *Carnegiea* is here greatly reduced in abundance. The only other important representatives of the cacti are four species of *Opuntia* (*O. Thurberi*, *O. arbuscula*, *O. fulgida*, and *O. leptocaulis*) and the tall but slender *Rathbunia alamosensis*. A large number of shrubs and semishrubs are abundant on the Plains of Sonora, growing in the stands of *Encelia* or *Larrea*, in the protection of the larger trees, or on the hills and lower mountain slopes. Many of these also occur in the Arizona Upland, but a high percentage reach their northern limits here or in the Sonoran part of the Central Gulf Coast. Along the banks of the larger and smaller arroyos of the Plains of Sonora the trees and shrubbery are often thickly massed and in strong contrast with the open vegetation between the drainageways. These thickets provide shade and slight betterment of moisture conditions which favor the occurrence of many vines, root perennials, and small shrubs not found in other habitats.

4. FOOTHILLS OF SONORA. The easternmost subdivision of the Sonoran Desert extends from the vicinity of Arizpe, on the Río Sonora, southward to the delta of the Río Yaqui, falling gradually from about 1000 m. to sea level. On the west it

passes with remarkable sharpness into the Plains of Sonora, on the north it follows up the valleys of Río Sonora, Río Moctezuma, and Río Bavispe, on the east it extends to the oak grassland of the lower mountain slopes, and south of 28° N. it merges gradually into thorn forest. Except at its southern end the entire area is rolling or hilly, with only narrow stretches of plain at intervals along the larger rivers. The mineralogical material is chiefly older volcanics or limestone, but there are large lava fields in the Moctezuma Valley and between Río Moctezuma and Río Sonora which appear to be of relatively recent origin. The delta of the Río Yaqui is a featureless alluvial plain over 50 km. in diameter. Part of it is devoted to the cultivation of rice, but much of it is in a natural or nearly natural condition.

The Foothills of Sonora constitute the least desert part of the Sonoran Desert. Incomplete records indicate that the rainfall near the center of the area has, at least in certain years, been slightly more than 500 mm. Whereas summer and winter rains are nearly equal in the Arizona Upland, there is a higher percentage of summer rain in the Foothills of Sonora. The time and duration of the rainfall seasons are not greatly different in the two subdivisions. In spite of the heavy summer rainfall, the Foothills area is subjected to long drought periods, and these are accentuated in their effects on the vegetation by the fact that the warm season is longer in the Foothills and that the drought periods are accompanied by slightly higher temperatures.

The most noticeable features of the vegetation of the Foothills of Sonora are the abundance of small trees (from which it is called arborescent desert), the frequent mottes of dense shrubbery, the amount of grass at higher elevations, the infrequency of cacti, the occurrence of palms, and the appearance among the dominant perennials of many trees and shrubs not found elsewhere in the Sonoran Desert.

Comparable situations in the northern and southern parts of the Foothills differ in their physiognomy and dominant woody plants. Toward the south a number of trees are important which reach their northern limits in the Foothills subdivision. Toward the north the upland vegetation is more open, with a greater amount of grass. On the flood plains and well watered level areas *Prosopis juliflora* var. *velutina* is the dominant tree in the north, but in the south it shares its dominance with *Acacia cymbispina* and *Lysiloma divaricata*, giving rise to the designation Acacia-Prosopis region. Throughout the upper half of the area the secondary trees and the shrubs along the arroyos are chiefly the same that are characteristic of the Arizona Upland. On rocky slopes and coarse outwash at all but the highest elevations of the Foothills, the stand of trees and shrubs is heavier than in any but the most favorable situations in the Arizona Upland. With few exceptions the trees have small leaves or leaflets which are wholly or almost wholly deciduous. The trees of commonest occurrence are *Bursera odorata*, *Jatropha cordata*, *Cercidium sonorae*, *Olneya tesota*, *Fouquieria Macdougalii*, *Bursera laxiflora*, *Guaiacum Coulteri*, and *Ceiba acuminata*. Shrubbery plays an important role in giving a more closed appearance to the vegetation, the principal species being

*Caesalpinia pumila, Dodonaea viscosa* var. *angustifolia, Haematoxylon brasiletto, Celtis pallida, Karwinskia Humboldtiana, Franseria cordifolia,* and *Randia Thurberi.*

The general aspect of the vegetation is more monotonous in the Foothills of Sonora than in the Arizona Upland. Large cacti, yuccas, and agaves are less frequent in the Foothills and not so conspicuous as in more open vegetation. In the southern part, *Pachycereus pecten-aboriginum* is a common associate of acacias and other trees. In many places the palm *Erythea Roezlii* occurs in groups along arroyos or scattered singly over the rolling valleys. The arborescent *Nolina matapensis* is common in scattered colonies.

There is a heavy growth of herbaceous plants in almost all situations during the summer, but the number of species involved in the display is not so great as in the Arizona Upland. There are a few herbaceous plants which are active only in the late winter. The herbaceous root perennials outnumber both groups of ephemerals and are of greater importance in the vegetation of wholly undisturbed situations.

5. Central Gulf Coast. The close similarity of the vegetation on the two sides of the head of the Gulf of California is not maintained in the central part of the Gulf. The vegetation of the opposed shores is, however, very closely related, as well as very distinct in each case from that of the interior. The Central Gulf Coast subdivision embraces the Sonoran coast from the mouth of Río Magdalena nearly to the mouth of Río Yaqui, and in Baja California extends from Isla Angel de la Guarda in a thin and broken strip to San José del Cabo. Throughout the two parts of this area the vegetation is almost identical in physiognomy, and a large number of its component species are the same. There is, however, a floristic difference between the parts such as is only feebly manifested nearer the head of the Gulf.

The rainfall of the Central Gulf Coast is very low and uncertain. The summer temperatures are very high, except when those of the immediate shore are ameliorated by onshore winds. The hills are bare of soil, and the streamways are bordered by aprons of sand and boulders. In the few places where there is an underground seepage of water, the vegetation forms a sharp contrast with that of the surroundings. The activity of the plants is determined by the occurrence of rain, and is sporadic rather than seasonal. Rain may fall either in the latter half of the winter or in midsummer, but is frequently lacking, sometimes for several seasons in succession.

Among the plants which are abundant in the Central Gulf Coast subdivision are the sarcocaulescent trees, with trunks of exaggerated diameter, including *Bursera microphylla, B. Hindsiana, Jatropha cinerea,* and *Idria columnaris.* This has therefore been designated the sarcocaulescent desert, or Bursera-Jatropha region. Although these striking trees are, in their abundance, distinctive of the area, they are actually outnumbered by *Olneya tesota, Cercidium floridum, Fouquieria splendens,* and *Prosopis juliflora* var. *Torreyana,* and by small-leaved shrubs, the identity of which is often determinable only in a favorable season.

The only evergreen shrubs are *Larrea tridentata*, *Viscainoa geniculata*, *Maytenus phyllanthoides*, and *Stegnosperma halimifolium*. The most abundant deciduous shrubs are *Jatropha cuneata*, *Encelia farinosa*, *Lycium Andersonii*, *Euphorbia misera*, and *Solanum Hindsianum*, accompanied by small, shrublike individuals of the trees that have been mentioned. Succulent plants are represented locally but are often lacking over large areas. *Pachycereus Pringlei*, like *Idria columnaris*, is much more local than in the interior of Baja California, and the most common cacti are the cylindropuntias, including *Opuntia Bigelovii*, *O. cholla*, *O. ramosissima*, *O. tesajo*, and *O. clavellina*.

The banks of washes and sandy flood plains have the heaviest vegetation in the Central Gulf Coast subdivision and the largest trees of *Cercidium floridum*, *Prosopis juliflora* var. *Torreyana*, *Bursera Hindsiana*, *Olneya tesota*, and *Dalea spinosa*. All the shrubs mentioned reach greater size and abundance along drainageways. Dense thickets often occur which are made up of very large individuals of *Hymenoclea pentalepis*, *Bebbia juncea* var. *aspera*, *Hyptis Emoryi*, and *Lycium brevipes*, accompanied by a number of other shrubs and a few vines. After heavy rains, ephemeral plants are abundant on sandy soil, but are few on the coarse soils and hillsides.

6. Vizcaíno Region. The central third of the Pacific coast of Baja California presents conditions and plant life very unlike those of the corresponding coast of Sonora. The Vizcaíno Region of the Sonoran Desert extends from the vicinity of Rosario to Punta Pequeña, west of Comondú, and runs inland approximately to the drainage divide of the peninsula, except at the extreme south. This area differs from the desert-chaparral transition region, lying between San Diego and Rosario, in the greater distance to a mountain background. For that reason the maritime influences scarcely extend more than 10 km. inland in any part of the Vizcaíno Region. A light winter rainfall occurs, which is more certain in the north than in the south and is extremely uncertain in much of the interior. Summer storms visit the mountains and hills along the eastern edge of the area on rare occasions. The immediate coast has the poorest vegetation of any part of the area, and no distinctive plants except those of the strand itself. Shrubs, cacti, and stones are heavily covered with lichens, and the epiphytic *Tillandsia recurvata* is very abundant. Although the humidity of the ocean wind is usually between 50 and 60 per cent, it does little to contribute moisture to the soil, and its average velocity is high enough to make it an important factor in transpiration. The strong and almost constant wind is apparently chiefly responsible for the very open and stunted vegetation of the coastal strip.

The Vizcaíno Region presents a hilly or rolling surface with a few small mountain areas and many rough fields of volcanic rock. A coastal plain over 50 km. in length and width has developed at the head of Vizcaíno Bay and extends southeastward toward San Ignacio. It is bounded on the southwest by the rugged Sierra Vizcaíno, and widens into another triangular area around the Laguna de San Ignacio. Outside this area, known as the Vizcaíno Desert, the only level ground with deep soil in this subdivision is to be found in a few very small valleys near the coast. The principal soils are the scanty pockets found on stony slopes, a very fine red volcanic clay, the alkaline clay of the outer Vizcaíno Desert, and the areas of sand along the coast, which appear to have been derived from the beach. The most favorable soils of the area are the coarse outwash of the inner Vizcaíno Desert and the volcanic clay.

The distinctive feature of the vegetation of the Vizcaíno Region is the rich development there of the sarcophyllous or leaf-succulent plants, which play only a minor role in the vegetation of other parts of the Sonoran Desert. The most conspicuous ones are several large species of *Agave*, but the smaller Dudleyas are even more abundant, being numerous in many habitats where *Agave* is uncommon. The most nearly omnipresent of the shrubby perennials is *Franseria chenopodifolia*, which gives a light-gray tone to the landscape in all but the most rocky situations. The sarcophyllous desert may well be designated floristically as the Agave-Franseria region. Almost as important as *Agave* in this area is the giant *Yucca valida*, which grows in great abundance in all areas with deep soil, but is absent from a very considerable part of the interior. The range of *Idria columnaris* coincides closely with the Vizcaíno Region; this species is very abundant on fine and coarse outwash soils, but is absent from the Vizcaíno Desert and the inner hills. The sarcocaulescent *Pachycormus discolor* is also nearly confined to this subdivision, exhibiting its most grotesque forms on rocky slopes near the sea but attaining its greatest height and abundance on gentle slopes in the interior.

The well deserved reputation for uniqueness enjoyed by the vegetation of Baja California is chiefly based on the occurrence of *Idria columnaris*, *Pachycormus discolor*, *Fouquieria peninsularis*, *Pachycereus Pringlei*, *Yucca valida*, *Agave Shawii*, *Viscainoa geniculata*, and *Glaucothea armata* in the optimum habitats of the Vizcaíno Region. It is of interest that all these dominants are confined to this area and the one immediately to the south, except for a few known localities for four of them in Sonora. In spite of the abundance of this galaxy of striking plants in the best situations, in far more than half of the Vizcaíno Region they are all rare or absent. It is distinctly an area of barren slopes and alkaline or extremely arid valleys with a low stand of *Franseria chenopodifolia*, *F. magdalenae*, *Atriplex polycarpa*, *Larrea tridentata*, *Lycium californicum*, *Machaerocereus gummosus*, *Viguiera deltoidea*, and *Lycium Fremontii*. The areas of salitral, or alkaline plain, in the southern part of the Vizcaíno Desert are extremely poor in plants, and some of the plateaus and rolling plains above 500 m. are thinly covered by a simple stand of *Atriplex polycarpa* and *Lycium californicum*. The contrast between the areas of richest and poorest vegetation is greater here than in any other subdivision of the Sonoran Desert.

7. Magdalena Region. This area embraces the desert of the southern third of the peninsula and lies entirely in the Pacific drainage. It extends from the vicinity of San Ignacio nearly to Todos Santos, and throughout its southern half desert interdigitates with the arid type of low forest characteristic of the

Cape Region. North of Comondú the drainage divide of the peninsula and the eastern edge of the Magdalena Region are formed by the Sierra de Santa Lucía, Sierra de las Palmas, and Sierra de Zacatecas. These small mountains have been the centers of vigorous volcanic activity, giving origin to extensive malpais fields which cover the northern part of this subdivision from the mountains nearly to the sea. South of Comondú the eastern boundary of the area lies along the lower slopes of the Sierra de la Giganta, 10 to 20 km. west of the drainage divide. These slopes drain into the Magdalena Plain, which occupies nearly the width of the peninsula between latitude 24° and 26° N. The coast is bordered by lagoons and sand dunes. The interior of the plain has fine or moderately coarse alluvial soil with innumerable playas ranging from several hundred hectares to a few square meters in area. A few of the largest streams draining the Sierra de la Giganta flow through seasonally to the Pacific, but a very large proportion of the drainage is indeterminate or spills from one playa to another. The coastal half of the plain is very poor in small drainageways. At the close of a rainy period the plain abounds in lakes and in meadows which have been formed by the silting of old playas and the invasion of palustrine plants.

The marked difference between the soil conditions of the malpais mesas in the north and the Magdalena Plain in the south results in types of vegetation which differ greatly in physiognomy, stature, and the local allocation of individuals. There is also some difference in the groups of species characteristic of the two soil regions, and a great difference in their relative abundance. With respect to the entire flora there is a very strong floristic difference between the Magdalena and Vizcaíno Regions. In the lists of dominant species which give character to the vegetation the difference is much less. The absence of *Idria columnaris* and the relative scarcity of *Pachycormus discolor*, *Yucca valida*, and *Agave Shawii* in the Magdalena Region does much to distinguish the two areas. The physiognomy of the vegetation in the Magdalena Region is distinctly that of the desert. If the vegetation is viewed only with reference to the habit and form of the plants and the organization of the communities, it strongly resembles that of the hills and plains of central Sonora. As a whole the area exhibits an abundance of large cacti, notably *Pachycereus Pringlei*, *Machaerocereus gummosus*, *Lemaireocereus Thurberi*, and *Opuntia cholla*, together with a frequency of small trees in all situations except malpais, including *Prosopis juliflora* var. *Torreyana*, *Lysiloma candida*, *Cercidium peninsulare*, *Bursera laxiflora*, *Fouquieria peninsularis*, and *Jatropha cinerea*. The most distinctive feature of the Magdalena Region division is the equality of dominance between trees and large succulents, and it is accordingly designated arbocrassicaulescent, or tree and stem-succulent, desert. The examples of these two types of plant which are most abundant in the area are *Prosopis juliflora* var. *Torreyana* and *Pachycereus Pringlei*, but it seems best to derive the floristic designation from two examples which are almost as common and more distinctly characteristic of the area. These are *Lysiloma candida* and *Machaerocereus gummosus*, from which this is styled the Lysiloma-Machaerocereus region.

## *Lower Colorado Valley*

### (Microphyllous Desert. Larrea-Franseria Region)

The strongest impressions given by the vegetation of the Lower Colorado Valley are of uniformity over wide areas, and simplicity of composition. These impressions are strengthened by the fact that they are particularly true of the plains and bajadas which form such a high percentage of the entire surface. Closer examination of the area reveals the importance of the substratum in determining the stature, density, and composition of the vegetation. Though this is true of all parts of the Sonoran Desert, and indeed of almost all vegetational areas, it is less manifest in the two most arid subdivisions (the Central Gulf Coast and the Lower Colorado Valley) than it is in the others. Wide differences exist between the vegetation of rocky slopes and that of gravelly plains, and between that of volcanic and of granitic hills, but there is often a close vegetational similarity between the gravelly and sandy plains. *Larrea* and *Franseria* retain their dominance on a series of soils which differ greatly in depth and texture, and they are also constituents of the vegetation on malpais and volcanic hills. These two plants are either dominant or abundant under differences of substratum which would support dissimilar vegetation under more favorable moisture conditions. This is due to the remarkable physiological constitution of these shrubs, and also to the simplicity of the flora, which has given them little competition. The simplicity of the flora, as a fact in itself, means merely that no other shrubs have developed biological and physiological characteristics so highly suited to survival in this region as are those of *Larrea* and *Franseria*.

Intermont plains and bajadas. These features form a great network throughout the Lower Colorado Valley, with mountains and hills filling the irregular meshes. The plains are commonly separated from the mountains by pediments and bajadas of varying width, but this condition is not invariable. Many hills and some of the small mountains are built of such durable rock that they have yielded almost no outwash material under the prevailing low rainfall, and in such cases the intermont plains extend to the base of the mountain slope. The width and gradient of the zone of pediment and bajada which surrounds most of the mountains is determined by the size of the mountain and the readiness with which its material is eroded. The recent eruptives are very resistant and supply little or no material for the formation of a bajada. Older volcanics and granitic rocks weather more readily and give rise to bajadas which may be short and steep, as in the Californian parts of the area, or long and of low gradient, as is common in Arizona and Sonora. The line separating bajada from intermont plain is never clear-cut, except where the base of a very steep bajada has been accentuated by the development of a drainageway paralleling the base. Under usual conditions the lower part of the bajada merges very gradually into the intermont plain, with only slight change in gradient. In the largest plains there is a fall of 4 to 9 m. to the kilometer.

The nearly pure stands of *Larrea* and *Franseria* which dominate all the intermont plains and lower bajadas in the Lower Colorado Valley are usually very uniform throughout a given plain. The number and spacing of the plants is nearly the same from hectare to hectare, their relative numbers are nearly the same, and their height is uniform. When widely separated plains are compared, considerable difference is often found in the structure of this simple community. Where the rainfall is 100 mm. or less, the stand of plants is very open, as on many of the lower bajadas north of the Chocolate Mountains in California and throughout the extension of the area into Baja California. Along the northeastern edge of the area the stand of *Larrea* is much heavier and the plants are taller, as in McMullen Valley and the Harquahala and Palomas Plains, in Arizona. In general the spacing of individuals is more uniform for *Larrea* than for *Franseria*, and there seems to be no complementary relation between the abundances of the two plants. *Franseria* is most abundant on plains with sandy loam to a depth of at least 1 m. (see pl. 11), and its numbers are much reduced on gravelly bajadas, although the plants found there are large. On the deep clay loam plains of Sonora, south of Río Magdalena, *Franseria* is far less abundant than elsewhere in this area. The plant coverage in the microphyllous desert usually ranges from 3 to 12 per cent.

On the plains and lower bajadas *Larrea* and *Franseria dumosa* form from 90 to 100 per cent of the vegetation. The remaining small percentage is made up of shrubs and cacti, some of which are commoner in other situations. Small trees or shrubs of *Prosopis juliflora* var. *velutina* and *Cercidium floridum* are rare, *Fouquieria splendens* is infrequent and indicates a spot with thin or stony soil, *Atriplex canescens* is infrequent, and *Hilaria rigida* is frequent in parts of Yuma County, Arizona. In the following list of plants characteristic of the plains and lower bajadas, the habitats in which they are more abundant are indicated.

*Plants of intermont plains and lower bajadas in the Lower Colorado Valley*

| | |
|---|---|
| Larrea tridentata | |
| Franseria dumosa | |
| Opuntia ramosissima | |
| Opuntia echinocarpa | Upper bajadas |
| Lycium Andersonii | Margins of washes |
| Condalia lycioides | Margins of washes |
| Franseria deltoidea | Upper bajadas |
| Hilaria rigida | Sandy plains |
| Krameria Grayi | Upper bajadas, hills |
| Prosopis juliflora var. velutina | Margins of washes, flood plains |
| Cercidium floridum | Margins of washes, upper bajadas |
| Echinocereus Engelmannii | Upper bajadas |
| Encelia frutescens | Margins of washes |
| Atriplex canescens | Alkaline flats |
| Hymenoclea monogyra | Margins of washes, sandy arroyos |
| Opuntia Wrightiana | |
| Fouquieria splendens | Upper bajadas, hills |

All evidence as to the physical conditions indicates that the intermont plains and lower bajadas of the Lower Colorado Valley are one of the most unfavorable habitats of the Sonoran Desert from the standpoint of water relations. These conditions are met by the dominance of two perennials and the infrequent occurrence of 15 other species, all but two of which are more abundant in other habitats. The two exceptions, *Opuntia ramosissima* and *O. Wrightiana*, are found elsewhere only on the sandy plains, where they are rare. These two cacti and *Hilaria rigida* are the only plants mentioned above which do not occur in any of the more favorable habitats of the Sonoran Desert, as do even the two dominant shrubs.

The importance of a very slight betterment of the moisture conditions is shown by the vegetation along the margins of very small drainageways. Water flows in these outlets from four or five to twelve or fifteen times each year, to judge from the average number of rains producing runoff. The conditions for penetration of water are better than on the neighboring surfaces, and the sand which commonly forms the bed of the drainageway serves to retard surface evaporation. These circumstances increase the moisture of the soil during the growing periods and also slightly prolong the periods. During much of the year the conditions are not better than elsewhere. Plants are more abundant along the drainageways and reach greater size; also, species occur there which are absent from the level surfaces. The response of the vegetation to the slight and temporary betterment of water supply merely serves to emphasize the adverse conditions to which the plants of the level ground are subjected. In large drainageways the volume of water carried after rainstorms is greater and the bed of sand is thicker. A considerable percentage of the runoff is therefore stored under and along the arroyos, and large trees of *Prosopis*, *Cercidium*, and *Dalea* are found in close proximity to open stands of *Larrea* and *Franseria*.

Not only the size of drainageways, but also the character of their distribution over the land surface is important to the vegetation. Two types of lesser drainage pattern are found in the Lower Colorado Valley. One is the dendritic type, of general occurrence elsewhere, in which the flood waters are gathered into larger and larger streamways and the drainage assumes a definite course. The other is the reticulate type, in which the drainageways anastomose, increasing little in size as they progress down slope, and forming a continuous network of shallow rills which fail to contribute to the formation of a through drainage system. Dendritic drainage develops under heavier rainfall and in regions of greater relief. Reticulate drainage is found under very low rainfall, where the detrital load of the storm water is heavy, and where the goal of the flood water is a playa or a plain rather than a large streamway originating at a distance.

The reticulate type of drainage is common to nearly all of the Californian part of the Lower Colorado Valley and to approximately half of the remainder of the area. Between Yuma and Niland the main line of the Southern Pacific Railroad runs at right angles to the general direction of the reticulate drainage from the Chocolate Mountains. After a torrential rain the flood water is a menace to the railroad embankment, and its disperse distribution has been controlled by placing culverts at regular intervals and connecting them by A-shaped dikes running about 200 m. up the slope.

Where the drainage is dendritic, the slightly heavier vegetation along the drainageways follows them closely and makes a treelike pattern when the plain or bajada is viewed from an elevated spot. Where the drainage is reticulate, the plants of slightly higher water requirement do not form continuous belts along the shallow washes, but nevertheless are often so abundant that they appear to be part of a uniform community covering the entire bajada. The species of plants favored by the two types of drainageway are the same, but their allocation is so different that it affects the physiognomy of the vegetation.

The trees which are most abundant along large drainageways are *Prosopis juliflora* var. *velutina* or *P. juliflora* var. *Torreyana*, *Cercidium floridum*, *Olneya tesota*, and *Dalea spinosa*. The height and size of the trees is roughly proportional to the size of the drainageway. Along the smallest washes they may occur as mere shrubs or may be absent. The commonest plants along the small drainageways are *Hymenoclea monogyra* or *H. pentalepis*, *Baccharis sarothroides*, *Lycium Andersonii*, and *Acacia Greggii*. The following list of plants characteristic of the margins of drainageways covers the plains and lower bajadas of the entire Lower Colorado Valley. Those found only in the northern or southern parts of the area are indicated. The species commonest in sandy washes are also indicated.

*Plants of margins of drainageways in intermont plains and lower bajadas in the Lower Colorado Valley*

Prosopis juliflora var. velutina
Prosopis juliflora var. Torreyana
Cercidium floridum
Olneya tesota
Dalea spinosa (sand)
Baccharis sarothroides
Hymenoclea monogyra (sand)
Hymenoclea pentalepis (southern; sand)
Lycium Andersonii
Acacia Greggii (northern)
Condalia lycioides
Chilopsis linearis
Franseria ambrosioides
Larrea tridentata
Encelia frutescens
Beloperone californica
Anisacanthus Thurberi
Sapium biloculare (southern)
Celtis pallida (northern)
Horsfordia alata (southern)
Franseria cordata (southern)
Phaulothamnus spinescens (southern)
Berginia virgata (southern)
Acalypha californica (southern)
Stegnosperma halimifolium (southern)
Bebbia juncea var. aspera (sand)
Solanum Hindsianum (southern)

In previous discussion of physiographic features (p. 57) mention has been made of the gradual change in the texture, depth, and surface of the soil on nearly all bajadas when they are followed from top to base. Large mountains of great age are surrounded by a zone of pediment, on which the soil is coarse and thin, or in rare cases is confined to pockets in a planed surface of rock in place. There is great variation in the distance to which pediment extends from the base of a mountain and in the depth of the outwash and soil by which it is covered. Throughout the Sonoran Desert, upland vegetation is strongly influenced by the character of the substratum to a depth of 1 m., but it is not influenced by differences in the nature of the soil below 2 m. The plant cover differs as between planed rock surface, shallow soil with a partially weathered rock base, and deeper soil with many included boulders, but no differences can be detected between spots in which the pediment is covered with 2 m. of soil and others in which the valley fill is several hundred meters deep.

On ascending a bajada a gradual change in the vegetation is observable. The stand of plants becomes thicker, the stature greater, the number of trees and large cacti increases, and new species make their appearance. The distance from the base of the mountain at which these changes begin, and the extent of the change, are dependent on the rainfall and the character of the detrital material. Under rainfall of 100 mm. or less, on soil of light texture, the change is least but is still manifested, as may be seen in the Coxcomb and Palen Mountains in California, and in the Sierra Pinta in Baja California. Under rainfall of 150 to 200 mm. on granitic outwash, the change is greater, as in the Harcuvar and Harquahala Mountains in Arizona. In general the amount of change is directly proportional to the size of the mountain from which the bajada descends.

The distinction which is here made between intermont plains, lower bajadas, and upper bajadas is merely the recognition of three phases in a very gradual change of vegetation which accompanies the equally gradual change in soil conditions. The character of the vegetational change is such as to indicate clearly that the water supply for plants grows better as the bajada is ascended. There is an increase in rainfall on going from the center of a valley toward a mountain mass. Added to this initial advantage, the soils of upper bajadas are better suited to the penetration and retention of moisture than are those of lower bajadas and plains. Another feature of the upper bajadas which operates to increase the density and variety of the vegetation is the very coarse character of the soil. Large stones on and near the surface provide the root anchorage which is important for tall, massive cacti. The coarse texture of the soil and good surface drainage obviate the danger of overwetting and poor aeration after periods of heavy rain. The stony surface favors in several ways the establishment of young perennials.

On the ascent of a bajada, the first indications of the change from the lower to the upper type of vegetation are the occurrence of *Olneya tesota* and *Cercidium floridum* or *C. microphyllum* away from streamways, the appearance of *Carnegiea gigantea*, and greater abundance of *Fouquieria splendens*, *Opuntia echinocarpa*, *Ferocactus acanthodes*, and *Acacia Greggii*. *Larrea* continues to be abundant but is no longer the predominant plant. *Franseria dumosa* is still abundant and of greater size, but in many localities is outnumbered or replaced by *Franseria deltoidea*. The number of species and individuals of small perennials gradually increases, and a larger number of plants are found which are mainly confined to the shade of trees. The margins of streamways are lined more heavily by trees and shrubs than on the lower bajadas, but are less conspicuous than they are in the midst of the pure stands of *Larrea* and *Franseria*. When the composition of the vegetation of upper bajadas is compared with that of streamways on the lower bajadas, it is evident that the plants of the latter habitat are mainly a group of representatives of those found in the former. The slight betterment of soil-moisture conditions found in the two habitats is marked by the appearance of many of the same species. For *Olneya*, *Cercidium*, and *Prosopis* the streamsides of the lower bajadas and plains are a more favorable habitat, at least after initial stages of establishment, than the rocky and more shallow soils of upper bajadas, if favorableness may be judged by the maximum size which these trees attain. *Carnegiea*, *Lemaireocereus*, and the large species of *Opuntia* do not descend along the streamways into the lower bajadas, where soil-moisture conditions are not so favorable for shallow-rooted plants. Certain common plants of the streamways of plains and lower bajadas are rarely found on the upper bajadas, as *Dalea spinosa*, *Baccharis sarothroides*, *Chilopsis linearis*, and *Hymenoclea monogyra*. These find their optimum habitat along streamways which have a heavy bed of sand such as rarely develops on the steeper gradients of the upper bajadas.

In the following list, trees and other plants over 2 m. in height are indicated by figures for their usual range of height. The plants are listed in the order of their relative abundance.

*Plants of upper bajadas throughout the Lower Colorado Valley*

Larrea tridentata
Cercidium floridum (4–5 m.)
Olneya tesota (4–5 m.)
Fouquieria splendens (3–4 m.)
Carnegiea gigantea (8–10 m.)
Prosopis juliflora var. velutina (3–6 m.)
Cercidium microphyllum (4–5 m.)
Franseria dumosa
Franseria deltoidea
Acacia Greggii
Opuntia echinocarpa
Encelia farinosa
Ferocactus acanthodes

Sandy plains and dunes. Nearly one-seventh of the Lower Colorado Valley subdivision of the Sonoran Desert is occupied by sandy plains or dunes. The largest areas lie around the head of the Gulf of California and the Salton Basin; smaller areas lie west of the Mohawk Mountains and southeast of Parker, Arizona. The origin of the sand is not definitely known. It seems probable that it has been largely derived from the broad beaches which are exposed by the wide fluctuations of tide in the Gulf, and from the floor of the Salton Basin. Some of it is of local origin, and perhaps much of it is ultimately derived from the granitic mountains lying within or bordering this subdivision. The sand of the area southeast of Parker has probably been derived from local material. See plates 5, 7.

The greatest development of dunes has taken place west of Pinacate Peak, both along the Gulf coast on Adair Bay and in the interior, culminating in great masses of yellow sand piled against the west base of the peak. A belt of dunes known as "the algodones" extends for 90 km. along the eastern edge of the Salton

Basin, and is crossed by the highway between Yuma and El Centro. Innumerable small dunes occur along the west side of the Salton Basin and along the west coast of the Gulf in Baja California. The small ones are often crescentic in shape, and most of them are gradually moving toward the northeast. These dunes and their vegetation have been described by Rempel (1936), who states that their movement is too rapid for them to become stabilized by plants. Along the eastern edge of the sand in Sonora there are a number of long dune areas projecting eastward over the gravelly plains, indicating the extent to which prevailing westerly winds have been responsible for spreading the sand.

At least four-fifths of the sandy area is relatively level, without local dunes, and has a stabilized surface. In these sandy plains there is little evidence of surface drainage and some evidence of gradual leveling from a less stable condition. Stabilization of the dunes is chiefly due to the coarse grass *Hilaria rigida*, which is not usually abundant on the well stabilized sandy plains. Neither it nor the shrubs have been as important in binding the level surfaces as have the blue-green algae and ground lichens, which form a nearly continuous crust on the most stable plains. The commonest lichens involved are species of *Lecidia* and *Acarospora*, whose periods of vegetative activity must be very brief.

In addition to the plains, in which the sand is from 0.5 m. to many meters deep, there is a large adjacent area of gravelly plain on which a small amount of sand is in perpetual motion over the hard surface and accumulates locally in low miniature dune areas. Such deposits are frequently seen in the malpais areas as well as on the gravel plains. They are abundant in the Castle Dome and La Posa Plains in Yuma County, Arizona, and may be detected by their covering of *Hilaria rigida* or *Opuntia Wrightiana*.

Throughout the extreme northwestern corner of Sonora there is often a close juxtaposition of sand and basaltic rock. Accumulations of sand are common at the windward base of hills (see pl. 6) and have encroached on some of the nearly level malpais fields. The volcanic eruptives have a veneer which is very dark brown or almost black and polished by the moving sand. This results in a strong contrast of color in the landscape, as well as a marked dissimilarity of vegetation.

In the areas of active or moderately active movement the dunes are either devoid of plants or dotted very irregularly with perennials which are usually large examples of their species. When sand accumulates around a plant slowly enough for growth to keep pace with it, the conditions of water supply are improved; but when sand is removed, the plant persists only by reason of its deep-seated root system, or frequently perishes. The only perennials commonly found on moderately active dunes are *Larrea tridentata*, *Franseria dumosa*, *Ephedra trifurca*, *Dalea Emoryi*, *Atriplex canescens*, *Coldenia Palmeri*, *Eriogonum deserticola*, and *Petalonyx Thurberi*. Rare examples of other perennials of the surrounding region may be found on unstable sand, including *Cercidium floridum*, *Hymenoclea monogyra*, and *Chilopsis linearis*. The sporadic distribution of perennials indicates clearly that conditions on unstable sand are more difficult for the establishment than for the persistence of a well rooted plant. This results from the instability and prevailing dryness of the surface in contrast with the relatively good moisture supply of lower levels. The absence of cacti from deep sand is significant in this connection.

The occurrences of *Hilaria rigida* indicate that it becomes established only on sandy surfaces that have already passed the stage of most active movement. When this stage is reached, *Hilaria*, *Franseria dumosa*, *Dalea Emoryi*, and *Coldenia Palmeri* increase greatly in abundance and the stabilization of the surface is rapidly accelerated.

Malpais fields and volcanic hills. The Lower Colorado Valley embraces a large number of areas of varying size and topography in which the surface is occupied by volcanic eruptives which either are recent or have undergone very little erosion or degradation. The largest mountain in this group of areas is Pinacate Peak, which is surrounded by smaller cones, volcanic plugs, and an irregular lava field 50 by 75 km. in maximum extent. The largest malpais area including no large hills is located near Sentinel, Arizona, in the southwestern corner of Maricopa County, and is approximately 25 by 40 km. in maximum extent. Other large areas exist, some of which occupy parts of small ranges of hills while others are nearly level malpais fields. There are innumerable smaller areas.

The volcanic surfaces are exposed in every position from vertical cliffs to nearly level areas. A very common form is the gently tilted mesa, extending for several kilometers at the same gradient, and ending in a scarp at the high end and along one or both of the sides. In southern Yuma County and western Pima County, Arizona, many of the basaltic hills consist essentially of older material and are merely covered with a veneer of basalt from 1 to 2 m. in thickness. This is true of Raven Butte and Sierra Pinta, in southern Yuma County, the latter showing vertical bands of black veneer and the nearly white subjacent material. Road excavation on the highway from Gila Bend to Phoenix has cut through the veneer in two localities (pl. 3, fig. 1), exposing a finely divided highly calcareous soil, and showing the fractured condition of the veneer. In the region south of the Gila River, between Sentinel and Wellton, there are a large number of roughly circular volcanic areas which are highest at the center and fall smoothly at a very low gradient. There is little rock in place at the center although the surface is thickly covered, and the amount of surface rock grows less toward the periphery. Surface deposits of volcanic cinder often exceed the amount of lava. These areas are undoubtedly due to explosive eruptions rather than to the wearing down of small hills.

The character of the vegetation on the volcanic areas is controlled by the three sets of soil conditions associated with the types of surface that have been described. On the edges of the flat areas of explosive origin, the conditions of the plain are slightly modified in the direction of conservation of moisture. On the thin but nearly continuous veneer of impervious rock covering a retentive soil, the runoff penetrates the cracks in the lava and is retained by the soil with extremely small loss by evaporation. On the thick beds of basalt and other eruptives, a very fine brown clay is formed by weathering and accumulates in pockets of various sizes. The vegetation betrays the slight betterment of moisture conditions around the edges of the malpais fields and the very much better conditions existing on the thick beds of lava. It is not often possible to determine whether trees growing on a volcanic mesa are drawing water from deep pockets of volcanic soil or from the underlying bed of soil which antedated the eruption of the lava.

On the explosive volcanic areas there is great variation in the amount of rock beneath the surface. On the edges of several such areas in Arizona the soil has been found nearly free of the type of stones which thickly cover the surface, but this is not true of the centers of the areas, where one-half to one-fourth of the volume of the substratum is stone.

On the edges of the explosive volcanic areas the vegetation differs little from that of the gravelly intermont plains, but as the center is approached both *Larrea* and *Franseria* become very much less abundant and much smaller, particularly where there is a heavy deposit of cinder. Other perennials characteristic of the gravelly plains and lower bajadas are absent or very sporadic. Small *Olneya* and *Prosopis* trees occur in slight depressions, and single plants of *Fouquieria* appear from place to place.

The extremely rocky and rough surface of the volcanic hills, tilted mesas, and level malpais areas exhibits a great lack of uniformity in its plant cover. There are about 20 trees to the hectare in some localities and none in others. Shrubs and semishrubs may grow abundantly or may be so few that the black-brown rocks give their color to the landscape. Trees are more abundant on small hills and level malpais than on large hills, and semishrubs are commonly more abundant where the rocks are largest and the surface most broken. The vegetation is usually denser at the base of a hill than on its upper slopes, and invariably taller in the shallow canyons and waterways. On the upper slopes of hills there are frequently extensive areas which are nearly bare, or elsewhere stretches of loose rock, with a depth of from one to several meters, with no soil and no vegetation other than the crustose lichens which cover the rocks.

The sporadic occurrence of plants under these conditions is due to the irregular distribution of bodies of soil, and their dissimilar locations with respect to receiving and retaining moisture under conditions of low rainfall and ready evaporation. The relative abundance of trees on small hills and level malpais appears to be due to the existence of soil beneath the basaltic veneer, and the ability of roots to penetrate the cracks in the veneer. The sporadic occurrence of trees and groups of shrubs on slopes is due to the irregular distribution of deep pockets and veins of soil. At the base of a hill and in a place where the rocks are few and large, the pockets of soil receive a heavy runoff, and the depth to which the water can penetrate is usually limited, since rock in place has no such deep cracks as are found in the veneer.

In spite of the lack of uniformity in the physiognomy of the vegetation in the volcanic areas, the same types of community occur over and over again in similar situations, and the composition of the vegetation is more uniform than its physiognomy. The scarcity of large drainageways and tanks or seepages of water sharply limits the occurrence of trees and other plants of high water requirement.

Plants that are abundant on recent eruptives throughout the Lower Colorado Valley are the following:

| | |
|---|---|
| Cercidium microphyllum | Opuntia acanthocarpa |
| Olneya tesota | Acacia constricta |
| Fouquieria splendens | Bebbia juncea |
| Encelia farinosa | Trixis californica |
| Larrea tridentata | Celtis pallida |
| Heteropogon contortus | Franseria dumosa |
| Hyptis Emoryi | Lycium Andersonii |
| Krameria canescens | Encelia frutescens |
| Echinocereus Engelmannii var. chrysocentrus | Porophyllum gracile |
| | Dyssodia porophylloides |

In the Pinacate lava fields several large perennials are common which are rare or absent north of the Gila River, including *Bursera microphylla*, *Lemaireocereus Thurberi*, *Jatropha cuneata*, and *Solanum Hindsianum*.

Older volcanics. The mountains and hills comprised under this name are those built of rhyolite, andesite, basaltic conglomerate, and all other volcanics except recent basalt. The distinction between "older" and "recent" refers to the stage of physiographic development of their surfaces rather than to their historical sequence. Some of the "older volcanics" are probably contemporaneous with the unmodified areas of basalt.

All the mountains of this class are highly eroded, usually with bare peaks or summits and with very narrow lateral ridges. The surface is either bare rock in place or an accumulation of coarse and unstable angular stones, with a poor development of soil. Since all the mountains and hills of this type in the Lower Colorado Valley lie in the region with less than 150 mm. of rainfall, the conditions for vegetation are extremely unfavorable.

When the older volcanic ranges are viewed from a distance of 2 or 3 km. they appear to be entirely barren of plants, owing to the absence or scarcity of the small trees of the region. On closer approach it is found that there is a very scattered and irregular cover of small perennials and grasses. Even in the canyon bottoms of the ranges near the Colorado River there are few trees, except where a local accumulation of sand or clay has favored *Cercidium floridum* or *Olneya tesota*.

The only plants which are conspicuous on mountain slopes of the most favorable type are *Encelia farinosa* and *Fouquieria splendens*. In many localities the former occurs on the slopes of the mountains and the latter only on the outwash slopes below. In extreme southern Arizona *Lemaireocereus Thurberi* occurs on rocky slopes, usually on a substratum such that it is accompanied by *Olneya* and *Fouquieria*. One of the most ubiquitous plants on the bare slopes is *Hetero-*

*pogon contortus.* The only cacti that are frequent are *Ferocactus acanthodes* and *Echinomastus Johnsonii.*

Other characteristic perennials of the slopes of the older volcanic mountains are:

Franseria dumosa
Fagonia californica
Krameria canescens
Hyptis Emoryi
Trixis californica
Hilaria mutica
Brickellia atractyloides
Peucephyllum Schottii
Agave deserti

Granitic mountains and hills. Approximately half of the mountainous part of the Lower Colorado Valley subdivision is composed of crystalline rock, with granite and gneiss preponderant and with smaller areas of schist. The largest granitic ranges are the Harcuvar and Harquahala Mountains and the Gila and Tinajas Altas Ranges, in Arizona; parts of the Chuckwalla and Chocolate Mountains, in California; and much of the great range which bounds the area on the west in California and Baja California. Both these and the smaller masses of crystalline rock vary in their topographic features and differ in the character of the weathering they have undergone. In general the surfaces have weathered more rapidly than in the older volcanic ranges and far more rapidly than in the recent eruptives. The rate of weathering in the granitic ranges has been determined by the durability of the rocks, which are sometimes reinforced by dikes of very resistant quartz. In general there has been sufficient aerial disintegration to provide a soil, and beneath the soil there is an active region of disintegration sometimes extending to a depth of 1 m. or more. Outcrops of rock and dislocated boulders are abundant on the slopes, but are likewise undergoing slow weathering. The soil of the granitic mountain slopes is a coarse sand at the surface, with many large angular fragments, but weathers to a loam at a depth of a few centimeters.

The granitic mountains differ from those previously described in the physiognomy and structure of their vegetation as well as in their flora. Their distinguishing features seem to be closely correlated with the types of soil to which they give rise. The boulder-strewn slopes and the porous surface tend to lessen the runoff, and the structure of the soil is favorable for the retention of water as well as for its penetration into the highly fractured rock beneath the soil.

When the vegetation of granitic mountains is compared with that of near-by volcanic ranges, in localities with nearly identical rainfall, it will be noted that there is a heavier cover on the former and a greater number of large perennials. The cover is far from uniform, being heavier in the bottoms of canyons, near the bases of slopes, and often along definite lines determined by the structure of the rock. There are invariably many small areas of bare rock, cliffs, and boulders, below or around which shrubs are abundant. On granitic mountains the difference in density due to slope exposure is greater than on volcanics so far as nonsucculent plants are concerned. Both large and small cacti are more abundant on slopes of southern aspect, and the nonsucculents on those of northern.

The most conspicuous large perennials of granitic slopes are those indicated by asterisks in the following list. The vegetation of typical granitic hills in extreme northwestern Sonora is shown in plate 20. The characteristic perennials of granitic mountains are the following, listed in order of approximate abundance:

| | |
|---|---|
| *Cercidium microphyllum | *Bursera microphylla |
| *Olneya tesota | *Encelia farinosa |
| *Simmondsia chinensis | Trixis californica |
| *Fouquieria splendens | Ephedra viridis |
| *Jatropha cuneata | Agave deserti |
| Haplopappus laricifolius | Lemaireocereus Thurberi |
| Acacia Greggii | Nolina Bigelovii |
| Opuntia acanthocarpa | Sapium biloculare |
| Franseria dumosa | Hibiscus denudatus |
| *Carnegiea gigantea | Galium stellatum |

### *Arizona Upland*

(Crassicaulescent Desert. Cercidium-Opuntia Region)

The crassicaulescent desert of the Arizona Upland occupies the inner and more elevated part of the northern half of the Sonoran Desert. Its greatest width is 250 km. in the latitude of Tucson. In its northwestern extension along the foothills of the Colorado Plateau and in its southern extension into Sonora the area narrows to a width of 50 km. or less. The elevation rises gradually from 150 m. at Parker and 335 m. at Phoenix, Arizona, to 950 m. at sections of its boundary near Tucson, Arizona, and Magdalena, Sonora. Approximately 25 per cent of the total area is occupied by hills and mountains. North of the international boundary the drainage of the Arizona Upland is into tributaries of the Colorado; south of the boundary it is carried by Río Magdalena, Río Sonora, and their tributaries. Along the lower courses of the Santa Cruz and Gila Rivers there is a large area of outwash plain in which hills are few and widely spaced.

The rainfall in the Arizona Upland is 125 mm. at Parker and 225 mm. at Ajo, Arizona, and 250 mm. at Altar, Sonora, all these localities being near the lower edge of the area. At Tucson the annual rainfall is 275 mm. and at Magdalena, Sonora, 325 mm., these stations being near the more elevated edge of the subdivision. At the northern end of the area there is a slight preponderance of winter rain over summer, in the center of the area there is approximate equality between them, and at the southern end there is a preponderance of summer rain.

The vegetation of the Arizona Upland is an open stand of microphyllous shrubs and trees with a strong admixture, in certain habitats, of succulents and plants of other types. In the broadest valleys the vegetation is almost as open as on the better-covered gravelly plains of the Lower Colorado Valley. In other situations the cover is heavier, occupying from 20 to 60 per cent of the surface. The broad valleys have a simple vegetation, in which *Larrea* is strongly dominant, with *Franseria dumosa* or *F. deltoidea* occupying a subdominant place. The number of infrequent plants associated with these is greater than in the Lower Colorado

Valley. On coarse outwash plains and slopes *Franseria deltoidea* is usually strongly dominant as to number of individuals, forming the matrix in which *Larrea* is usually likewise abundant, and in which scattered individuals of *Cercidium microphyllum*, *C. floridum*, *Olneya tesota*, and *Prosopis juliflora* var. *velutina* occur, as well as various types of cacti. On the most favorable upper bajadas the number of trees increases, but they never occupy more than 15 per cent of the surface. Under the best conditions for trees the matrix of low shrubbery is usually thinner and broken, and the larger cacti are more abundant.

The Arizona Upland has a maximum length of 650 km. and a north-and-south extent nearly as great as that of the Lower Colorado Valley, yet there is a greater uniformity in the composition of the vegetation in the former area than in the latter. A large number of species are found only in the northern or the southern end of the Arizona Upland; only a few of these play a dominant or subdominant role in the vegetation, except a few species found only west of Wickenburg, Arizona. Owing to the physical features, the vegetation in the Arizona Upland is relatively homogeneous. There are few sharp contrasts in the nature of the surface due to the mineralogical character of the rock, and no large deposits of sand. Throughout the area there are contrasts between the vegetation of volcanic hills and soils and that of granitic ones. With this exception, the differentiation of the vegetation is to be attributed to gradient of slope, depth of soil, physical texture of soil, and surface conditions affecting infiltration. These conditions are closely correlated with the physiographic status of a locality, and result in the consistent repetition throughout the area of closely similar habitats occupied by closely similar vegetation.

The physiographic units with which the vegetation shows close correlation in the Arizona Upland are: plains and lower bajadas; upper bajadas; the slopes of hills and mountains; and streamways. There are some features which distinguish hills and upper bajadas derived from volcanic mountains from those of granitic material. Also the northern and southern ends of the area merit separate treatment.

PLAINS AND LOWER BAJADAS. The plains which border the lower courses of the Gila and Santa Cruz Rivers in this subdivision fall toward the streams at a gradient of 4 to 8 m. per kilometer. Elsewhere, and particularly near the larger mountains, the gradient is steeper, but there are no localities in which lower bajadas are tilted more than 12 m. per kilometer. The soil is commonly a loam or sandy loam, including much coarse gravel, which has a tendency to accumulate on the surface. The structure of the soil is variable, sometimes containing lenses of sand or fine alluvium. Throughout the Arizona Upland the surface is underlain at varying depths by a calcareous hardpan, or caliche, which occurs near the surface as one or more hard, impervious layers, and at a depth of 1 to 3 m. appears as a soft, continuous mass, with included stones, often extending to a depth of several meters. The part played by caliche in the infiltration and evaporation of soil water and in the distribution of roots has been described by Shreve and Mallery (1933). The visible uniformity of the surface and soil of the plains is matched by the continuity of the vegetation, which is broken only by the bands of larger and more close-set plants along the drainageways. Through the dominance of *Larrea* the vegetation is given a uniform height of 1.5 to 2 m.; the coverage varies from 10 to 20 per cent. The commonest associates of *Larrea* which reach its own or greater height are *Acacia constricta*, *A. Greggii*, and *Prosopis juliflora* var. *velutina*. Away from streamways these plants rarely greatly exceed *Larrea* in height. Several species of cylindropuntia are also locally subdominant, *Opuntia versicolor*, *O. arbuscula*, and *O. spinosior* rarely exceeding the general height, but *O. fulgida* and *O. fulgida* var. *mammillata* often greatly exceeding it.

On the largest plains, where the soil contains the smallest amount of coarse material and where the surface is without a continuous cover of gravel, the spaces between bushes are bare or nearly so. Semishrubs, with a height of 2 to 5 dm., occur with great irregularity. The commonest of these are *Franseria dumosa*, *F. deltoidea*, *Krameria canescens*, *Zinnia grandiflora*, and *Haplopappus Hartwegii*. Grasses are scantily represented except in poorly drained areas of heavy soil, in which well closed stands of *Hilaria mutica* are found. Under *Larrea*, and reclining on its limbs, *Muhlenbergia Porteri* is abundant where there has not been recent grazing. Near the flood-plain edge of the gravelly plains bushes of *Prosopis* are often abundant, and on alkaline soil *Atriplex canescens* forms nearly pure stands.

Between the Gila and Salt Rivers, south of Chandler, Arizona, is a plain of very low gradient with deep loam soil. This has largely been brought under cultivation since 1930, but originally bore a heavy cover of *Atriplex polycarpa* and an open forest of *Carnegiea gigantea* (pl. 12). This is the only locality in which *Carnegiea* is known to occur abundantly on a relatively fine, rock-free soil.

West and southwest of Wickenburg there are a number of depressions, all less than 1 km. in diameter, in which *Hilaria mutica* forms a close stand, free of bushes and trees. The only perennial growing with this grass is the erect platyopuntia *O. chlorotica*, which gives the community an appearance like that of the cactus-acacia grassland which is widespread in Zacatecas and adjacent states. In the surrounding areas *O. chlorotica* is infrequent. Its most widespread habitat is the rocky slopes of the mountains which fringe the Sonoran Desert on the north and east, where it is rarely as abundant as in the small areas of *Hilaria*.

Throughout the plains and lower bajadas the monotony of the vegetation is sometimes broken by the occurrence of isolated plants of species which are abundant on the upper bajadas, especially *Fouquieria splendens*, *Ferocactus Wislizenii*, *Carnegiea gigantea*, *Condalia lycioides*, and *Celtis pallida*. Much more infrequent are a few species found in no other habitat in the region, including *Holacantha Emoryi*, *Koeberlinia spinosa*, and *Bumelia occidentalis*.

In the valley of the Bill Williams River the areas of plain and lower bajada are limited in extent. The region north of the river is rugged and rolling and bears the upper bajada type of vegetation. South of the Bill Williams and north of the Harcuvar Mountains is Butler Valley, a plain 30 by 40 km. in extent in which there is one of the heaviest and finest stands of *Larrea* in the Sonoran Desert.

South of Santa Ana, Sonora, the lower bajadas differ from those farther north in the greater number of small trees of *Prosopis* and *Cercidium* growing with the prevailing *Larrea*, and in a stronger representation of grasses.

Upper bajadas. Approximately half the area of intermont plain in the Arizona Upland is here designated as upper bajada. It includes the pediments which often surround the mountains, and a varying percentage of the outwash slope which falls at uniform gradient toward the central drainageway. Where the mountains are widely spaced, the zone of upper bajada may extend only one-fifth of the distance from mountain to drainageway. Where the mountains are closely spaced, the physical and vegetational features of the upper bajada may extend down the entire slope to its base. The boundary between upper and lower bajada, as already stated, is a very broad and indefinite one. Also the vegetation of the upper bajada, especially on pediment, approaches closely that found on the slopes of hills and mountains.

The areas of upper bajada encircle all the isolated mountains in the central part of the Arizona Upland, and include nearly all the intermont areas south and west of the Salt River and north of the Altar River. Whereas the soil of the lower bajadas is often a mixture of material from various sources, that of the upper bajadas is derived from a single mountain range, or part of such mountain, and commonly has distinctive physical features attributable to the character of the rock from which it is derived. The prevailing material from which the soils of the upper bajadas have come consists of basalt, andesite, rhyolite, and volcanic conglomerate, resulting in soils of coarse texture with many angular fragments on and beneath the surface, and very generally cemented by calcareous hardpan. The upper bajadas on which the soil has been derived from rocks of granitic types have a deep, sandy loam, without calcareous hardpan. Much coarse sand is included in the granitic soil and accumulates on the surface, where it clogs the drainage and results in rapid lateral migration of the streamways. Large stones and coarse gravel are relatively infrequent in the soil and on the surface. Such soil has sufficient clay to become very hard when dry, but is friable when moist.

The vegetation of upper bajadas in the Arizona Upland is the highest type found in any part of the Sonoran Desert north of the international boundary and outside the moist flood plains. There are more large perennials per unit area than in any other group of habitats, the stature of the large perennials is greater, the number of associated smaller perennials is greater, and the number of species is larger. The number of species of perennials reaching a height of 1 m. or more is rarely more than 20 in any one locality, and their relative abundance varies from place to place. The most abundant of the 20 (*Larrea*, *Cercidium microphyllum*, *Opuntia versicolor*, *O. Engelmannii*, *Fouquieria splendens*, *Lycium Andersonii*, *Condalia lycioides*, *Celtis pallida*, and *Prosopis juliflora* var. *velutina*) are present on nearly every square kilometer of the area. *Carnegiea* is far less numerous than the above plants, but its height makes it a conspicuous element in the vegetation. *Fouquieria* is particularly abundant on the pediments and very coarse soil. *Larrea* is more abundant on level areas and benches than on the steeper slopes with coarse, shallow soil. As the bajada is ascended, *Cercidium* increases in abundance, and *Prosopis* more frequently reaches the size of a small tree.

With particular reference to the broad part of the Arizona Upland lying between the Salt River on the north and the Altar River on the south, the following perennials more than 1 m. in height are characteristic of the upper bajadas:

Cercidium microphyllum
Larrea tridentata
Acacia constricta
Fouquieria splendens
Prosopis juliflora var. velutina
Opuntia versicolor
Olneya tesota
Opuntia Engelmannii
Acacia Greggii
Opuntia fulgida
Celtis pallida
Lycium Berlandieri
Ferocactus Wislizenii
Simmondsia chinensis
Ephedra trifurca
Opuntia spinosior
Lycium Fremontii
Condalia lycioides
Forestiera neomexicana
Condalia spathulata

The smaller perennials accompanying the above are:

Franseria deltoidea
Encelia farinosa
Calliandra eriophylla
Opuntia leptocaulis
Zinnia grandiflora
Opuntia phaeacantha
Krameria glandulosa
Coldenia canescens
Menodora scabra
Psilostrophe Cooperi†
Brickellia Coulteri
Parthenium incanum
Muhlenbergia Porteri
Sphaeralcea Gooddingii var. grossulariaefolia
Jatropha cardiophylla
Trixis californica
Tragia ramosa
Mammillaria microsperma
Hibiscus denudatus
Ayenia microphylla
Dalea Parryi
Abutilon incanum
Polygala macradenia
Porophyllum gracile
Hibiscus Coulteri
Haplophyton cimicidium

In some of the warmest situations along the western edge of the Upland between Gila Bend, Arizona, and Caborca, Sonora, there are other dominant and subdominant species, most of which are here at the northern limit of their ranges and are not found elsewhere in the United States. These include *Lemaireocereus Thurberi*, *Bursera microphylla*, *Ferocactus Covillei*, *Lophocereus Schottii*, *Jatropha cuneata*, and *Bumelia mexicana*.

†This and several other species grow as annuals if rainfall is abnormally low, but become perennials of short duration if the conditions for growth are more favorable.

The northwestern arm of the Arizona Upland west of Wickenburg is a rugged area of low mountains and pediments with few expanses of plain even along the Bill Williams River. The general features of the vegetation are those of the central part of the subdivision. The stature and density are similar and the composition is somewhat richer. A number of plants playing an important role in the vegetation occur in the Sonoran Desert only in this small section. Several of them are common in the Mojave Desert and are here at their southeastern limit. Others are dominants of the adjacent arid woodland which extend only a few kilometers into the desert in close association with its characteristic plants. Near Wikieup and Signal, Arizona, there are many localities in which *Carnegiea*, *Cercidium*, *Larrea*, and *Encelia* grow with *Yucca brevifolia*, *Juniperus utahensis*, *Quercus turbinella*, and *Berberis haematocarpa*. Exit from the desert on the rolling ground southwest of the south end of the Hualpai Mountains is indicated by a slight increase in stature and density of the plants, by the disappearance of *Cercidium*, *Larrea*, *Carnegiea*, *Encelia*, and *Simmondsia*, and by the dominance of *Yucca brevifolia*, *Juniperus utahensis*, *Yucca baccata*, and *Quercus turbinella* (pl. 13).

The following plants are found in the Sonoran Desert only in the northern arm of the Arizona Upland, where they are commonest on upper bajadas, pediments, and hillsides. They are all either abundant or conspicuous elements in the vegetation.

Yucca brevifolia
Juniperus utahensis
Acamptopappus sphaerocephalus
Thamnosma montanum
Ephedra viridis
Yucca baccata
Opuntia basilaris
Canotia Holacantha
Halliophytum Hallii
Salazaria mexicana
Nolina Bigelovii
Quercus turbinella
Agave deserti
Aster abatus

The southern end of the Arizona Upland, lying south of the Altar River, is similar to the northern arm in its rugged topography and low relief. It lies between the oak-clad mountains which bound the desert and the plains which extend to the Gulf of California. The surface is mainly hill, pediment, and upper bajada, with small areas of lower bajada and flood plain.

In the southern arm of the Upland the stature of the vegetation is similar to that in the main area, but the density is slightly greater, particularly that of the shrubs. The dominance is still held by *Cercidium*, *Opuntia*, and *Carnegiea*, but the composition is different from that in the central area. The dominant perennials are:

Cercidium microphyllum
Prosopis juliflora var. velutina
Acacia constricta
Opuntia Engelmannii
Opuntia spinosior
Larrea tridentata
Carnegiea gigantea
Cercidium floridum
Celtis pallida

(*Continued on following page*)

Encelia farinosa
Olneya tesota
Opuntia fulgida
Jatropha cordata
Acacia Greggii
Dodonaea viscosa
Opuntia versicolor
Mimosa laxiflora
Condalia spathulata
Croton sonorae
Lycium Andersonii
Franseria cordifolia
Coursetia glandulosa
Caesalpinia pumila

South of the Altar River it is noticeable that slight topographic differences affect the vegetation and flora more than is the case farther north. This holds true whether it is moisture or temperature conditions that are concerned.

In a short and very abrupt box canyon on the southwest face of Sierra Babiso, about 35 km. southeast of Magdalena, is a heavy stand of the palm *Sabal uresana.* The soil of the floor of the canyon is kept moist by springs, and the shade of the palms is so dense that the only accompanying plants are small turgescent herbaceous perennials, including mats of *Tradescantia semisomna*; the lower walls of the box are clothed with mesic ferns and *Rhus choriophylla.*

About 15 km. northeast of Ímuris, Sonora, the Magdalena River traverses a canyon about 7 km. in length at an altitude of 1000 m. On the steep but well clothed lower slopes of the canyon a group of plants of southern range is found, few of which are known farther north and several of which are not known for 200 to 300 km. to the south. There is undoubtedly a marked temperature inversion here, as in all situations of the same character (see Shreve, 1912), the walls of the canyon being steep and the air-drainage area being large and of higher elevation. Under such conditions the lowest winter minima on the slopes above the cold-air flow would be higher than on other slopes in the vicinity, as well as higher than the minima above a more shallow flow of cold air.

Among the plants detected in Magdalena Canyon are:

Acacia Farnesiana
Acalypha subviscida
Iresine paniculata
Cocculus diversifolius
Phaulothamnus spinescens
Croton ciliato-glandulosus
Gronovia scandens
Solanum Grayi
Milleria quinqueflora
Tillandsia recurvata
Pseudabutilon sonorae
Tithonia Thurberi
Rothrockia cordifolia
Acalypha polystachya
Elytraria imbricata
Senecio saligna

HILLS AND MOUNTAIN SLOPES. There are many localities in the Arizona Upland where the vegetation of the hills and lower mountain slopes does not differ in any respect from that of the bajadas and pediments which the slopes overlook. This is particularly true along the inner edge of the Upland, where a high percentage of the surface is covered with thin residual soils. The physiognomy of the slope vegetation depends primarily on the kind of rock involved and the manner in which it has weathered, since this determines the amount, depth, and continuity of the soil, as well as its capacity for infiltration and retention of water. In general, the slopes which have the most uniform surface have the most uniform and evenly spaced vegetation. Under these conditions the stand of shrubbery is very open and the trees few or absent. On slopes with a rough surface, broken by large boulders, small escarpments, or extrusions, the plants are irregular in spacing and stature and each species is confined to sites of a definite type. On rough slopes the stature of the largest plants greatly exceeds that on the uniform slopes.

On recent eruptives below 750 m. elevation, *Larrea* is dominant on the lower slopes of many hills and mountains in this area. At higher elevations, and on other soils, particularly granitic loam, *Larrea* is consistently absent from the slopes. The occurrence of *Cercidium microphyllum* is the reverse of this. It reaches its maximum size and density on granitic loam slopes and upper bajadas near the inner edge of the Arizona Upland, and extends up the larger mountains to 1000 to 1300 m. above the edge of the desert.

The lower slopes throughout the central part of the Arizona Upland are dominated by *Cercidium microphyllum*, *Fouquieria splendens*, *Carnegiea gigantea*, *Encelia farinosa*, *Acacia constricta*, and several cylindropuntias and platyopuntias. On upper slopes *Cercidium floridum* is more abundant than *C. microphyllum*, and it is not there confined to the banks of streamways as it is on the floor of the desert; *Prosopis juliflora* var. *velutina* becomes more abundant as a small tree of 2 to 3 m.; and *Carnegiea* and *Fouquieria* are closely restricted to warm sites. In the rugged area drained by the Gila River, *Simmondsia chinensis* is strongly dominant on many north slopes. At desert levels in such situations it is not uncommon to find large perennials which are characteristic and abundant at higher elevations, for example *Dasylirion Wheeleri*, *Haplopappus laricifolius*, *Vauquelinia californica*, *Lippia Wrightii*, *Rhamnus ilicifolia*, and *Quercus turbinella*. Through a vertical range of 400 m. above the limit of the desert there is a gradual disappearance of characteristic desert plants. Isolated individuals or colonies of certain ones may be found as high as 2000 m. The transition from purely desert communities to desert grassland, xeric woodland, or chaparral invariably takes place at lower elevations on granitic slopes than on volcanic eruptives.

A faithful description of the vegetation of a basaltic hill in the Arizona Upland, near Tucson, was published by Spalding (1909), including lists of all the known species found in the several habitats of the area. The vegetation of the desert slopes of the Santa Catalina Mountains, near Tucson, was described by Shreve (1915), who also gave an account of the transition from desert to xeric woodland.

The plants which are most characteristic of hill and mountain slopes and most generally distributed throughout the Arizona Upland are the following:

Cercidium microphyllum
Encelia farinosa
Fouquieria splendens
Acacia constricta
Celtis pallida
Opuntia versicolor
Cercidium floridum
Carnegiea gigantea
Condalia spathulata
Calliandra eriophylla
Jatropha cardiophylla

(*Continued on following page*)

| | |
|---|---|
| Opuntia leptocaulis | Prosopis juliflora var. velutina |
| Menodora scabra | Janusia gracilis |
| Dalea Parryi | Ayenia microphylla |
| Ferocactus Wislizenii | Opuntia Bigelovii |
| Heteropogon contortus | Ephedra trifurca |
| Krameria canescens | Dyssodia porophylloides |
| Muhlenbergia Porteri | Trixis californica |
| Brickellia Coulteri | Lippia Wrightii |
| Parthenium incanum | Hibiscus Coulteri |

In addition to the plants listed, there are over 300 perennials which either occur infrequently and locally or else are very inconspicuous elements of the vegetation. On very rough slopes, particularly those of northern or eastern aspect, *Selaginella rupincola* forms an almost continuous carpet, gray and inconspicuous in the dry months but a vivid deep green during the summer rains. Several grasses which are commonly annuals (notably *Aristida adscensionis*) often find the conditions favorable enough to live through two or more rainy seasons. Under ledges of rock and in the shade of large trees are found small xeric ferns, which produce fresh fruiting fronds twice a year. The most abundant of these are *Cheilanthes Lindheimeri*, *C. Pringlei*, *C. Wrightii*, *Notholaena sinuata*, and *Pellaea mucronata*.

In the northern arm of the Arizona Upland the character of the physiography is such as to obscure the contrast between pediments and slopes, and their vegetation is often identical. On the steepest slopes, however, the vegetation is more open than in the central part of the Upland, and large trees of *Cercidium* and *Olneya* are rare. Near the upper limit of the desert the influence of slope exposure is pronounced. The familiar phenomenon of "inversion of zones" may be noted in many localities, where steep south-facing slopes overlook bajadas with marginal xeric woodland vegetation but are themselves clothed with desert plants.

Nearly all the plants listed as characteristic of slopes in the central part of the Arizona Upland are also found in the northern arm, although several are less frequent, and *Opuntia versicolor*, *Jatropha cardiophylla*, *Ferocactus Wislizenii*, and *Janusia gracilis* appear to be absent. A few species are found abundantly in the northern arm which are not known elsewhere in the Upland:

| | |
|---|---|
| Eriogonum fasciculatum | Rhus choriophylla |
| Ferocactus Lecontei | Galium stellatum |
| Nolina Bigelovii | Ditaxis lanceolata |
| Crossosoma Bigelovii var. glaucum | Physalis crassifolia |
| Echinocereus Engelmannii | Notholaena Parryi |

On going from the central Arizona Upland to its southern end, little change in the slope vegetation is encountered except along the eastern edge of the desert south of Magdalena, Sonora. The principal differences are in certain regions where the shrubbery is very heavy, and others where perennial grasses are abundant. Also there are a few important differences in composition.

There appears to be no definite relation between the density of the open stands of *Cercidium* and *Olneya* and the number of shrubs occurring in the intervals between them. *Larrea* is almost completely absent from slopes in this region, and *Encelia* is here more common on the bajadas and plains than on the hills. The increasing density of the interarboreal shrubbery is not due to greater abundance of the shrubs found farther north so much as it is to the appearance of southern species not found in Arizona or else uncommon there. The principal shrubs in this group are *Dodonaea viscosa*, *Caesalpinia pumila*, *Franseria cordifolia*, *Croton sonorae*, *Jatropha cardiophylla*, *Phaulothamnus spinescens*, and *Tecoma stans*.

STREAMWAYS. In all parts of the Arizona Upland the banks and margins of streamways show vegetational differences from the general surface, therein contrasting with such habitats in the most arid parts of the North American Desert. The differences may be only in the density and size of species common elsewhere, or may consist in the presence of distinctive species. Usually both these features appear along small drainageways.

The converging ribbons of denser vegetation along the small streamways are very conspicuous when a plain or bajada is viewed from an elevation or from the air (see pl. 8). As soon as a drainageway becomes large enough to have interrupted deposits of sand in its bed, the marginal vegetation begins to be heavier. When the sandy bed becomes nearly continuous and reaches a width of 1 to 2 m., the marginal vegetation becomes still more dense; *Larrea* responds notably in height and bulk to the slight improvement in moisture conditions; *Prosopis juliflora* var. *velutina*, *Acacia constricta*, and *A. Greggii* contribute more than any other plants to the increased density. Species then found along the streamways but not away from them are *Cercidium floridum*, *Condalia spathulata*, *Lycium* spp., *Baccharis sarothroides*, *Anisacanthus Thurberi*, and *Franseria ambrosioides*. Several species of *Opuntia*, notably *O. versicolor*, *O. spinosior*, *O. leptocaulis*, and *O. arbuscula*, retain their relative abundance along the streamways.

Broad streamways filled with sand to depths as great as 1 to 3 m. present ideal conditions for a greater soil-moisture supply than that prevailing away from their banks. The runoff into the sandy beds greatly increases the potential water supply at the expense of the upland. The infiltration is rapid and deep. The quick drying of the surface sand retards the loss of water. The presence of large trees of *Cercidium*, *Olneya*, and *Prosopis* along sandy streamways forms a sharp contrast with the low and open stands of *Larrea* which prevail a few meters beyond the edge of the sandy reservoirs of moisture.

The vegetation along larger streamways is determined by the character of their storm floods and by the distance from the headwaters or parent mountain of the stream. There are very few complacent streams in the Arizona Upland and none that flow for more than 6 or 7 months of the year. All of them are subject to violent floods which last only a few hours or a few days and are often very destructive to the marginal vegetation. Flood plains have developed along most of the large streamways, but are rarely more than a few hundred meters in width. The flood bed of the stream commonly lies below the level of the flood plain at a depth ranging from 1 or 2 m. to as much as 12 or 15 m. in exceptional cases. It is believed that the cutting of deep channels by desert streams has taken place largely within the past seventy-five years and is due to disturbance of the surface of the catchment basin by heavy grazing (see Bryan, 1925). Nevertheless, a limited amount of channel cutting has always been a part of the normal history of desert streamways, alternating with periods of lateral cutting, and bringing about constant readjustment of level and equalization of gradient.

Along large streamways the vegetation of the margins and flood plains is determined by the character of the soil, which is roughly of three types. Near the parent mountain the stream bed and the flood-plain soil are full of boulders, and this condition may extend for a few kilometers beyond the base of the mountain. Where the flood plain has been built up over a long period by wide and slowly moving floods, the soil is a fine alluvial clay. This type of flood plain is found along streamways which have a low gradient of fall, unconfined banks, and a heavy seasonal flow from a large drainage basin. The third type of flood plain is the product of flash floods. The surface is sandy, and the subsoil may be sand or a heavier soil laid down under different conditions in the past. Sandy flood plains are found where the channel is shallow or absent and the floods are infrequent, violent, and usually from sources far upstream.

The vegetation of the boulder flood plains is highly mesic and closely related to that of the encinal. It is richly supplied with water at the same time that it is subjected to maximum temperatures as great as those of the floor of the desert. The extent of the bands of mesic vegetation is sharply limited on the sides by contrasting slopes with xeric plants, and gradually wanes downstream as the moisture supply decreases. Between 750 and 950 m. elevation in all the large canyons of the mountains which border the inner edge of the Arizona Upland there is a mingling of encinal and desert trees. The former include *Quercus Emoryi*, *Q. arizonica*, *Q. oblongifolia*, and *Q. hypoleucoides;* the latter include *Populus Fremontii*, *Salix Gooddingii*, *Fraxinus velutina* var. *Toumeyi*, *Juglans major*, *Prosopis juliflora* var. *velutina*, *Platanus Wrightii*, *Sambucus mexicana*, and *Celtis reticulata*. At this elevation the oaks rarely extend beyond the mouth of the canyon, as they do in similar situations at higher elevations.

The shrubs and other perennials which characterize the boulder flood plains are:

Celtis pallida
Acacia Greggii
Franseria ambrosioides
Baccharis sarothroides
Hymenoclea monogyra
Opuntia versicolor
Dodonaea viscosa var. angustifolia
Baccharis viminea
Coursetia glandulosa
Rhamnus californica subsp. ursina
Lycium spp.

After the largest streams of the Arizona Upland have traveled a few kilometers from the base of the mountains in which they originated, there is a great reduction in the number of common trees and shrubs and an approach toward the characteristics of either the alluvial or the sandy type of flood plain. *Prosopis* becomes the dominant tree, and *Populus* and *Salix* are frequent. All the trees occur in belts, in irregular groves, or as isolated trees.

The alluvial flood plains of the Arizona Upland are one of the most favorable desert environments in respect to water supply and depth and texture of soil. The original vegetation was a nearly pure stand of large trees of *Prosopis juliflora* var. *velutina*. These trees long constituted the only source of good fuel in the region, and the soil on which they grew was the most fertile, tillable, and easily irrigated land. For these reasons nearly all the alluvial flood plains have been denuded of trees or now have open stands of small second- or third-growth trees which have sprung from suckers from the original stumps. In some places there is evidence that the changing regimen of the rivers has covered the clay bottoms with sand and converted the flood plains from the alluvial to the sandy type.

The splendid forests of *Prosopis* which once grew in many localities along the Gila, Santa Cruz, San Pedro, and Magdalena Rivers have given place to fields of cotton, alfalfa, or barley. A few small virgin stands of *Prosopis* may still be found on the San Pedro River above Redington, Arizona, on the Altar River near Tubutama, and on the San Miguel River above Horcasitas, Sonora. In these stands the trees are 15 to 18 m. in height with a trunk diameter of 12 to 15 dm. The finest groves that have been examined are nearly pure, with a few *Acacia Greggii* and *Fraxinus velutina* var. *Toumeyi*, and have a clean floor with few shrubs and herbaceous plants. The edges of the *Prosopis* groves and the places disturbed by natural or human agency are dense jungles of young *Prosopis*, *Acacia Greggii*, *Celtis pallida*, *Condalia lycioides*, and *Lycium* spp. In these jungles there are more vines than in any other desert habitat. The commonest is *Clematis Drummondii*, which flourishes in sun and shade. Other climbers found chiefly in this habitat are *Vitis arizonica*, *Funastrum cumanense*, *Echinopepon Wrightii*, and *Cocculus diversifolius*.

The sandy flood plains lie near the level of the stream bed and are subject to overflow at times of heavy rain. They are virtually a seldom used part of the stream bed, submerged only by rapid torrents which overflow the bed itself, whereas the alluvial flood plains were originally built by slowly moving widespread floods in their depositional phase. The substratum of the sandy flood plains is a mixture of sand and coarse, partially worn gravel. The conditions are very unfavorable for the establishment of trees. Here and there are individuals of *Populus*, *Salix*, or *Prosopis*, which usually show distortion and root exposure due to the floods.

The dominant plants of the sandy flood plains throughout the Arizona Upland are *Hymenoclea monogyra* and *Baccharis sarothroides*, both of which are shrubs of rapid growth and of such habit that partial burial by sand does not interfere with their growth. On the Bill Williams River in western Arizona the sandy flood plain is overflowed almost annually and is exceptionally well supplied with moisture. There the commonest plant is *Baccharis viminea*. On the Gila River the natural conditions have been much modified by construction of the Coolidge and Gillespie Dams, and there has been a steady decrease in the extent of sandy flood plain in recent years. In many localities on the Gila River the introduced *Tamarix parviflora* has replaced the native shrubs. The only common woody associate of *Hymenoclea* and *Baccharis* in Arizona is the slender shrub or small tree *Chilopsis linearis*. The semishrub *Franseria ambrosioides* is also abundant, as well as the winter-active, tuberous-rooted *Rumex hymenosepalus*.

### *Plains of Sonora*

(Arbosuffrutescent Desert. Olneya-Encelia Region)

The Plains of Sonora constitute the smallest and least diversified of the vegetational subdivisions of the Sonoran Desert. They occupy an important place in the North American Desert, however, because they are the southern termination of the very arid type of continental desert which extends north to eastern Oregon, and in them are found the first manifestations of botanical features which characterize the west coast of Mexico far to the south.

Over most of the area the impression given by the vegetation is that of a very open forest of small, low-branching trees, with irregularly placed colonies of shrubs which are not tall enough to impair the view, and with large but very widely spaced columnar cacti. The dominants of the Lower Colorado Valley, *Larrea*, *Franseria dumosa*, and *F. deltoidea*, are here of local occurrence. Colonies of *Opuntia* are relatively infrequent. There is a distinct approach toward association of plants which are more uniform in habit than is the case in the subdivisions which have been described. In the Plains of Sonora *Cercidium*, *Olneya*, and *Prosopis* find their optimum conditions and show their maximum size, abundance, and catholicity of habitat. In the central and southern parts of the Plains these trees are joined by a number of subdominants, only one of which, *Bursera microphylla*, extends north of the international boundary. The gradual enrichment of the flora also extends to the shrubs and herbaceous perennials, although not to the ephemerals. In this subdivision low stature, open spacing, and diversity of life forms, which are the cardinal features of the desert, are still prevalent but are beginning to wane.

The northern boundary of the Plains of Sonora is on the long slope which rises from the Río Magdalena southward to the Sierra de la Compañía. The area extends south for about 250 km. to the vicinity of Guaymas. The western edge of the Plains subdivision is a poorly defined transition to the Lower Colorado Valley or to the Central Gulf Coast subdivision at elevations of from 100 to 400 m. and at 30 to 50 km. from the Gulf coast. The eastern edge of the Plains is defined by hilly or mountainous areas lying in the southern extension of the Arizona Upland or in the Foothills of Sonora. The floor of the Plains area reaches its highest elevation of 800 m. at its northeast corner, and the eastern boundary falls to 100 m. at the south.

Only in the proximity of the widely spaced mountains is the gradient of the intermont plains greater than 1 or 2 m. to the kilometer. The only important rivers crossing the area are the Río Sonora and its tributary the Río San Miguel. Many of the short drainageways originating in the Plains of Sonora flow toward the Gulf of California, but their waters are spread and lost in the Gulf coast region.

The soils are prevailingly of mixed material that has been moved far from its place of origin. The mountains are chiefly volcanic extrusives; a few are granite or gneiss; at least two large groups of hills are limestone, and small exposures of limestone are found in many places. The distinction between upper and lower bajadas which has been described for the Arizona Upland is only of local importance here, where there is neither physical nor vegetational basis for its recognition. Sections of the soil exposed by river channels show it to be a loam or clay loam which is deep, relatively uniform, and poor in large stones. Caliche is present along the western edge of the region, but is far less general in occurrence than it is in the Lower Colorado Valley and Arizona Upland.

Both the vegetation and the flora of the Plains of Sonora are influenced by the position of the area at the threshold of the transition from desert to thorn forest. So many species appear which are absent from the northern subdivisions that the contrast between the northern and southern ends of the area is far greater than the contrast in the same distance in the Lower Colorado Valley from Needles to Yuma, or in the Arizona Upland from Wickenburg to Sonoyta. The gradual enrichment of the flora from north to south and its effect on the composition, structure, and physiognomy of the plant communities is an outstanding feature of the area. There is also the universal contrast between the vegetation of hills and that of plains. The fact that many of the newly encountered plants extend much farther north on the hills than on the plains causes an intermingling of the two principal differentiating features.

NORTHERN PLAINS REGION. Throughout the northwestern third of the Plains of Sonora there are numerous areas in which *Larrea* is abundant. Large pure stands are rare, as it is generally accompanied by at least a small representation of *Olneya*, *Cercidium*, *Fouquieria splendens*, and *Encelia farinosa*. Many open areas are encountered in which *Larrea*, trees, and large cacti are absent and shrubs are few. These bare areas commonly have a narrow, irregular form and extend for several kilometers between the well wooded belts along the drainageways.

North of Carbó the vegetation of the Plains of Sonora is little affected by the influx of southern plants. Colonies of *Larrea* alternate with open forests of *Cercidium microphyllum*, *C. floridum*, *Olneya tesota*, *Prosopis juliflora* var. *velutina*, *Fouquieria splendens*, and *Lemaireocereus Thurberi* (pl. 14). The stature of the trees ranges from 4 or 5 m. to 8 or 10 m., and the density is highly variable. The largest trees occur in the heaviest stands, where the coverage is as much as 20 per cent. The smallest trees occur with *Larrea* or on the borders of bare areas, and often number only 5 or 6 to the hectare. The prevailing vegetation is near the average between these extremes. *Encelia farinosa* is the dominant shrub outside the colonies of *Larrea*. Its height is about 1 to 1.5 m., and its density is rarely so great that its branches interlock. *Encelia* is most abundant in the open, and rarely occurs under wide-spreading trees, although the shade of such trees is always light (pl. 15).

There are no subdominants in the northern part of the Plains of Sonora. Large perennials other than those that have been mentioned are either infrequent or else found in scattered colonies. The most common colonial perennials are *Opuntia arbuscula*, *O. fulgida* var. *mammillata*, and *Lophocereus Schottii*. The trees and large shrubs most often found are:

Acacia constricta (shrub)
Bursera microphylla (tree)
Acacia occidentalis (tree)
Caesalpinia pumila (shrub)
Croton sonorae (shrub)
Jatropha cardiophylla (small tree)
Lycium brevipes (shrub)
Krameria parvifolia (shrub)
Celtis pallida (shrub)
Condalia lycioides (shrub)
Opuntia leptocaulis (slender cylindropuntia)
Coursetia glandulosa (shrub or small tree)
Eysenhardtia orthocarpa (shrub)

On volcanic hills the vegetation is usually somewhat heavier than on the plains, but this is not invariably the case. *Cercidium* and *Olneya* are the dominants on hills as well as plains. *Prosopis* and *Acacia occidentalis* are rarely found on hills in this region. The characteristic population of hills includes:

Cercidium microphyllum
Olneya tesota
Fouquieria splendens
Lemaireocereus Thurberi
Jatropha cuneata
Opuntia Engelmannii
Bursera microphylla
Croton sonorae
Opuntia Gosseliniana
Celtis pallida
Ferocactus Covillei
Phaulothamnus spinescens
Carnegiea gigantea
Sapium biloculare
Condalia spathulata
Larrea tridentata
Lycium Berlandieri
Echinocereus Engelmannii
Hyptis Emoryi

About 70 km. southwest of Santa Ana is the largest group of limestone hills that has been found in the Plains of Sonora. The highest is Cerro Santa Rosa, which rises only about 400 m. above the basal elevation. The irregular group of hills surrounding Cerro Santa Rosa is maturely dissected and rounded in outline. The vegetation departs radically from that of the volcanic hills both in physiognomy and in composition. The cover is a dense, uniform stand of large shrubs, without *Cercidium* or *Olneya* and with very few cacti of any type. The dominants are *Acacia constricta*, *Caesalpinia pumila*, *Eysenhardtia orthocarpa*, *Fouquieria splendens*, *Condalia spathulata*, *Bursera laxiflora*, *Croton sonorae*, *Tecoma stans*, and *Jacobinia ovata*.

CENTRAL PLAINS REGION. In the central part of the Plains of Sonora between the latitudes of Carbó and Hermosillo there is an increase in the density of the vegetation, but not in its stature. The areas of open forest become greater, except in the west, where colonies of *Larrea* are still frequent. The shrubbery becomes much more continuous and its stature greater than in the northern end of the Plains. There are still many areas with few trees and shrubs, and occasional ones with no large perennials. The ephemerals and summer-active perennials which occupy the open areas as well as the wooded and shrubby ones are abundant in the late summer and usually completely carpet the soil. Grasses often form as much as 75 per cent of this cover, which gives an aspect of abundant verdure after the period of summer rain. By the middle of November all these plants are dead or inactive, changing the tone of the landscape from green to brown, but still serving to prevent the appearance of bareness that is so common in the areas to the north.

In the central third of the Plains subdivision there is a strong enrichment of composition, due to the entry of many plants of southern range. Several of the new dominants and subdominants are plants of striking habit, which serve to maintain the diversity of life forms. The majority of the less common entrants, however, are more alike and nearer to the generalized types that prevail in dry subtropical regions.

The two levels of vegetation formed by trees and by shrubs throughout the northern part of the Sonoran Desert now become less clearly marked. The stature of *Encelia* increases toward the south and away from the Gulf, and many of its associates exceed it in height. Also several woody perennials occur which in height and form are intermediate between shrubs and the small trees of the desert.

The large perennials entering the vegetation in the central part of the Plains of Sonora occur at first singly and far apart, but usually as large, well developed individuals. The fringe of their ranges is not broad, however, and within 30 or 40 km. of the northernmost individuals, most of the species are abundant or even subdominant. The large perennials which quickly take a prominent place in the vegetation are:

Fouquieria Macdougalii
Bursera laxiflora
Cercidium sonorae
Jatropha cordata
Guaiacum Coulteri
Rathbunia alamosensis
Ipomoea arborescens
Jacquinia pungens
Randia Thurberi
Opuntia Thurberi

The entrance of these plants does little to remove *Cercidium microphyllum*, *Olneya*, and *Prosopis* from their dominant role. The colonies of *Encelia* become less general in occurrence, except toward the Gulf, and the nearly pure stands give way to mixed shrubbery including:

Caesalpinia pumila
Acacia constricta
Coursetia glandulosa
Calliandra eriophylla
Eysenhardtia orthocarpa

(*Continued on following page*)

Phaulothamnus spinescens
Krameria parvifolia
Hyptis Emoryi
Desmanthus Covillei
Lycium exsertum
Mimosa laxiflora
Sapium biloculare
Bursera Hindsiana
Waltheria americana
Lantana horrida
Euphorbia californica
Atamisquea emarginata
Carlowrightia californica
Malpighia ovata
Forestiera neomexicana

With the gradually increasing density of shrubbery and small trees there is an almost complete elimination of the species of cacti which are abundant in the open desert to the north. *Lemaireocereus*, *Lophocereus*, and *Ferocactus Covillei* are much less frequent, colonies of *Opuntia fulgida* var. *mammillata* are still encountered, *Opuntia arbuscula* is rare, *O. Thurberi* is frequent, and *Rathbunia alamosensis* occurs in large but widely scattered thickets. On the western edge of the Plains the cacti are still prominent in the vegetation, particularly the cylindropuntias.

SOUTHERN PLAINS REGION. The southern part of the Plains of Sonora, lying south of the Río Sonora, is a narrow triangle in which there is a rapid transition from very arid plains on the west to a more broken surface and heavier vegetation on the east. In stature and density the vegetation resembles that of the western edge of the Plains north of Río Sonora, but there are a number of further additions to the dominant and subdominant species. The waning colonies of *Larrea* do not extend inland to the eastern edge of the Plains. *Encelia* is less abundant than it is to the north, but frequently reaches a height of over 2 m. *Olneya* decreases rapidly in abundance from west to east. The differentiation of the vegetation along streamways is more conspicuous than it is north of Río Sonora. In some localities the vegetation of hills is heavier than in the northern part of the Plains. On the highest mountain of the area, Sierra Libre, the desert trees and shrubs of the base are still dominant on the highest ridges, 600 m. above. The slopes of this mountain are the southernmost locality in which *Carnegiea* has been observed in abundance.

The dominant large perennials of the Plains of Sonora south of the Río Sonora are:

Olneya tesota
Cercidium floridum
Fouquieria Macdougalii
Bursera laxiflora
Jatropha cordata
Prosopis juliflora var. velutina
Ipomoea arborescens
Lemaireocereus Thurberi
Bursera microphylla
Encelia farinosa
Opuntia Thurberi
Caesalpinia pumila
Croton sonorae
Lantana horrida
Larrea tridentata
Cordia parvifolia
Pithecellobium sonorae
Jacquinia pungens
Coursetia glandulosa
Desmanthus Covillei

Several large perennials first become prominent south of the Río Sonora or do not range north of it. Among these are plants which strongly affect the physiognomy of the vegetation, although, with two exceptions, they are rarely abundant in any one spot. These include:

Forchammeria Watsoni
Ipomoea arborescens
Pachycereus pecten-aboriginum
Acacia Willardiana
Piscidia mollis
Cordia sonorae
Pithecellobium sonorae
Acacia cymbispina
Ceiba acuminata

*Forchammeria*, *Ipomoea*, *Piscidia*, and *Ceiba* are found on the open floor of the plains; *Acacia Willardiana* is found only on rocky slopes; *Pachycereus*, *Cordia*, *Pithecellobium*, and *Acacia cymbispina* are confined to the streamways and their bordering depressions. All these species occur here as scattered individuals, and only the two Acacias are ever sufficiently abundant and closely spaced to become the dominants of their communities. The character of these plants and their role in the vegetation are briefly described in chapter 5.

STREAMWAYS. The streamways of the Plains of Sonora are nearly all shallow. Owing to the low gradient and the abundance of sand and fine gravel in the soil, there has been very little channel cutting. Owing to the relatively heavy cover of vegetation over the general surface, the bands of much heavier vegetation along the streamways are not so conspicuous as in the more arid parts of the Sonoran Desert. Also the influence of the better moisture conditions along the arroyos here extends farther from their banks and ends gradually. Other than the Río San Miguel there are no streamways large enough to have developed extended flood plains, and there are few pure stands of large *Prosopis* such as are found along the rivers which rise outside the desert.

The most common type of vegetation along the streamways is a very irregularly spaced stand of *Prosopis*, *Cercidium floridum*, *Olneya*, *Acacia Greggii*, and *A. occidentalis*, in which the height of the trees is uneven. The stand is closed by shrubs along the larger arroyos but somewhat open along the smaller ones. Nowhere is there any consistent layering, and there are no places with continuous heavy shade and a bare floor. There are more vines in these situations than in any other desert habitat. Ephemerals and small perennials are sporadically abundant in spots with good light.

The commonest shrubs in the almost impenetrable thickets that fringe the streamways are:

Franseria ambrosioides
Acacia constricta
Lycium spp.
Condalia lycioides
Mascagnia macroptera
Anisacanthus Thurberi
Baccharis sarothroides
Aloysia ligustrina
Rathbunia alamosensis
Celtis pallida
Mimosa laxiflora
Lantana horrida
Forestiera neomexicana

Away from the streamways there are few vines in the Plains of Sonora. *Janusia gracilis* is widely distributed as either a vine or a small shrub. Several herbaceous species of *Ipomoea* are common. *Cucurbita digitata* and *C. palmata* run over the ground more often than they climb. The massive half-exposed root of *Maximowiczia sonorae* is frequently found in dry situations under low bushes, on which its stems climb during the summer rainy period.

In the thickets along streamways the vines are numerous in individuals and species. Most of them climb only to the height of the shrubs. A few with woody stems reach the tops of the trees. A nearly complete list of them follows:

Clematis Drummondii
Eupatorium sagittatum
Cardiospermum corindum
Funastrum cumanense
Commicarpus scandens
Nissolia Schottii
Sicyosperma gracile
Merremia Palmeri
Passiflora mexicana
Iresine interrupta
Cocculus diversifolius
Metastelma Watsonianum
Antigonon leptopus
Phaseolus atropurpureus
Calonyction muricatum
Matelea caudata
Marsdenia edulis

### *Foothills of Sonora*

(Arborescent Desert. Acacia-Prosopis Region)

This subdivision of the Sonoran Desert forms the southeastern border of the continental part of that area. Much of its surface is rugged but not mountainous. Two-thirds of the Foothills area lies more than 100 km. inland from the Gulf of California. The rainfall is greater than in any of the other subdivisions, locally and perhaps exceptionally reaching 500 mm., but 70 to 80 per cent of it falls in the three summer months and there is a long period of deficient rainfall usually lasting from January or February to July. Though the vegetation of the area is distinctly desert in character, it is heavier and more continuous than in any of the other subdivisions of the Sonoran Desert.

The Foothills area is 375 km. in length from north to south, and 100 to 150 km. in width. Its northwestern corner is drained by the Río Sonora, and its central and eastern part by the Río Yaqui and its tributaries Río Moctezuma, Río Bavispe, and others. Approximately 25 per cent of the area is occupied by level or gently tilted plains, following the rivers or skirting the coast. There are innumerable hills, lava fields, and small mountains, but only three mountains rising as much as 1000 m. above the surrounding plains. The elevation of the area falls from about 1000 m. near Arizpe to sea level in the delta of Río Yaqui.

The north-and-south extent of the Foothills of Sonora is greater than that of the Plains of Sonora. The differences of 3.5° of latitude and 1000 m. of elevation between the north and south ends of the area are responsible for the conditions which underlie the principal differences of vegetation in the area.

At its northern end the Foothills area extends up the valleys of the Sonora, Moctezuma, and Bavispe Rivers. In these three projections desert vegetation is nearly limited to the valley floor, but many species of desert plants occur above the floor on dry south slopes and in small tributary valleys. The principal valleys are separated by forested mountains. Between the Río Bavispe and the Río Moctezuma is Sierra Oputo (Sierra Moctezuma), between the Río Moctezuma and the Río Sonora is Sierra Santa Margarita, and between the Río Sonora and the Río San Miguel is Sierra Aconchi. Sierra Oputo and Sierra Aconchi have large stands of pine, and Sierra Santa Margarita is heavily forested with oaks. The region north of these mountains and north of the three projections of desert is a rolling plain covered by grassland. This is the southernmost large body of grass, of the low turf-forming type dominated by species of *Bouteloua*, that has been found west of the Sierra Madre Occidental. Each of the mountains just mentioned is flanked by a narrow and sharply dissected bajada, on which the vegetation is cleanly controlled by the topography. The undissected strips of bajada are covered with heavy stands of grass. The north slopes of the deeply cut streamways are either heavily or lightly wooded with evergreen oaks, and the south slopes are dominated by open shrubbery in which many desert plants participate.

At its southern end the Foothills of Sonora merges into thorn forest over a flat terrain which makes the transition very gradual. The southern boundary of the Sonoran Desert has been placed between the lower courses of the Río Yaqui and the Río Mayo, approximately at the place where vegetation with many open spaces occupied by xeric plants gives way to a continuous cover of summer-green trees. There are, however, a few small areas as far south as the Río Fuerte in which the vegetation has the open desert aspect, heightened by an abundance of *Cercidium* and *Opuntia.*

The delta of the Río Yaqui lies mainly south of the present lower and westward course of the river. Much of this extensive area is now in cultivation and produces important crops of wheat, rice, and garbanzo. The original vegetation of the delta is stated to have been like that of the parts which remain uncultivated, a closed or open forest of *Prosopis juliflora* var. *Torreyana* with heavy shrubbery and a very rich representation of cacti (pl. 19). Near the coast there are very few trees, and the areas of dense shrubbery are interrupted by lagoons and small areas of dunes.

The Foothills area occupies an important position in the vegetation of Mexico. On the north it is bounded by a temperate type of grassland, and on the south by arid subtropical thorn forest. On the west it is sharply limited by the level, arid coastal plain, and on the east by the lower slopes of the outlying ranges of the Sierra Madre Occidental, where an increase of rainfall is immediately encountered.

Though the Foothills of Sonora is not a mountainous area, it is sufficiently rugged to present a large variety of habitats. A few of the dominant plants of the Sonoran Desert are more abundant in it than elsewhere, but the majority of them are commoner in adjacent areas and are restricted to the Foothills habitats which most nearly resemble their areas of greatest abundance. The thorn forest has contributed heavily to the flora of the Foothills of Sonora, and many of these plants are conspicuous in the vegetation of the southern half of the area. The great majority of plants which range north from Guerrero and Jalisco as far as latitude 29° to 30° N. reach their northern limit in the Foothills of Sonora and are absent from the same latitude in the Plains of Sonora and the Central Gulf Coast.

NORTHERN FOOTHILLS REGION. The area north of the latitude of Ures is largely comprised in the three nearly parallel valleys of the Sonora, Moctezuma, and Bavispe Rivers. All these rivers carry a heavy summer discharge, which has restricted the formation of flood plains. Their courses frequently run alongside the hills for many kilometers. There are rolling plains surrounding Moctezuma and Ures, and narrow stretches of valley floor at many localities farther north.

Between Cumpas and Moctezuma there is a broad valley in which the vegetation is very open, forming a 20 per cent cover. The dominant plants are widely spaced small trees of *Prosopis juliflora* var. *velutina* and an open stand of *Encelia farinosa*. The principal elements of the vegetation in this valley are:

***Prosopis juliflora var. velutina
***Encelia farinosa
**Cercidium sonorae
**Fouquieria splendens
**Dodonaea viscosa var. angustifolia
**Lemaireocereus Thurberi
**Celtis pallida
*Mimosa laxiflora
*Opuntia fulgida
*Opuntia Thurberi
*Opuntia Engelmannii
Condalia spathulata
Karwinskia Humboldtiana
Forestiera neomexicana
Janusia gracilis

The physiognomy of this area, and similar ones in the adjacent valleys, suggests the Arizona Upland, but the vegetation does not include *Larrea tridentata*, *Cercidium microphyllum*, *C. floridum*, *Carnegiea gigantea*, and *Olneya tesota*. Neither does it include any of the species which are common in the thorn forest.

The elevation of the three valleys is between 500 and 800 m. All of them have extensive drainage areas lying in grassland plains or forested mountain slopes, and all are subject to pronounced inversion of temperature. No adequate temperature data are available for this area, but the topographic situation is identical with that of localities which have been investigated in Arizona (Shreve, 1912; Turnage and Hinckley, 1938). By reason of air drainage from large elevated areas, the inversion of temperature probably results in absolute winter minima in these valleys from 15° to 20° F. lower than the minima on the adjacent bajadas and slopes 50 to 100 m. above. Freezing temperatures occur nearly every winter at Cumpas, Moctezuma, and Ures, but the maximum duration of temperatures below 32° F. is not known.

*Here and below, asterisks indicate relative abundance: three asterisks, greatest abundance, etc.

The behavior of a number of large perennials of southern Sonora which have been under cultivation at Tucson (including *Ceiba acuminata, Bursera microphylla, B. laxiflora, Jatropha cordata, Piscidia mollis, Pithecellobium sonorae*) has shown that they are killed to the ground by the minimum temperature and frost duration of a normal winter. Winter temperature conditions appear, therefore, to be responsible for the absence of some of the characteristic plants of southern Sonora from the valley floors of the northern Foothills of Sonora.

North of Baviacora, Cumpas, and Granados, the south slopes above the influence of temperature inversion afford a safe habitat for plants of southern range which are absent from the valleys. The observations of the author and the collections made by Stephen S. White in the area immediately outside the Sonoran Desert indicate the occurrence of a number of Sonoran Desert plants in warm situations which are well above their distributional limits in the valleys. The following plants have been found at the localities indicated:

| | |
|---|---|
| Ceiba acuminata<br>Bursera laxiflora<br>Lysiloma divaricata<br>Parthenium Stramonium<br>Sapium biloculare<br>Ficus petiolaris | Cañon Higuera, west of Moctezuma |
| Sapium biloculare | Bajada above Baviacora |
| Fouquieria Macdougalii | Several localities above Granados |
| Diphysa Thurberi | Cañon de Huepari, north of Aribabi |
| Acacia cymbispina | Puerto de Huepari, north of Aribabi |
| Ipomoea arborescens<br>Jatropha cordata | Aguaje de Bacatejaca, between Granados and Bacadehuachi |
| Manihot angustiloba | Bajada of Sierra Santa Margarita, west of Cumpas |
| Manihot angustiloba | Cañon Bavispe |
| Erythrina flabelliformis | Río Bonito |

The bajadas and rolling plains lying just above the valley floor of the Río Moctezuma are covered by trees and shrubs with a density which varies from a 20 per cent to a 60 per cent cover. Perennial grasses are here more abundant than in any other part of the Sonoran Desert. As one ascends the bajadas the trees and shrubs become fewer and the patches of grass more continuous. At an elevation of 800 m. the aspect of the vegetation is that of grassland, with an open sod of *Bouteloua radicosa* and *B. eludens* and widely scattered trees of *Prosopis juliflora* var. *velutina* and *Acacia Farnesiana*.

The characteristic perennials of the upper areas of desert in the Sonora, Moctezuma, and Bavispe valleys are listed below. In order to indicate the strong relationship between the northern part of the Foothills of Sonora and the Arizona Upland, the initials of the latter area have been placed after the names of the plants which are common to the two.

Prosopis juliflora var. velutina (AU)
Acacia cymbispina
Fouquieria splendens (AU)
Celtis pallida (AU)
Acacia Greggii (AU)
Cercidium sonorae
Fouquieria Macdougalii
Randia Thurberi
Condalia spathulata (AU)
Jatropha cardiophylla (AU)
Bursera filicifolia
Lemaireocereus Thurberi
Encelia farinosa (AU)
Karwinskia Humboldtiana
Caesalpinia pumila
Eysenhardtia orthocarpa
Opuntia Thurberi
Bursera laxiflora

In the upper levels of the Foothills of Sonora there are many herbaceous perennials and small woody perennials which do much to heighten the aspect of density during the summer months, when they are in leaf and in flower. Many of these plants are confined to the light shade of *Prosopis* or *Acacia* or are found on banks along arroyos. The principal species are:

Turnera diffusa
Franseria ambrosioides
Ruellia nudiflora
Elytraria imbricata
Abutilon sonorae
Brickellia Coulteri
Hintonia latiflora
Phaseolus atropurpureus
Antigonon leptopus
Ayenia pusilla
Carlowrightia californica
Commicarpus scandens
Talinum paniculatum
Desmodium cinerascens
Salvia Seemannii
Rivina humilis
Brongniartia Palmeri

Central Foothills region. The region between Ures and Tecoripa is one of low relief in which the surface is about equally divided between hills or volcanic mesas and plains or slopes of low gradient. It is here that trees become an important feature of the desert vegetation to an extent that is matched in the north only in the most favorable situations of the Arizona Upland. The dominant arborescent forms are small leguminous trees with winter-deciduous compound leaves and leaflets rarely exceeding 20 sq. mm. in area. The crown is flat, and the form of all the trees in a community is remarkably uniform. The height varies from 4 to 12 m. according to the site. Local distribution of the leguminous trees is determined by soil moisture, and there is great variation in the density of the stands. Low situations often support stands in which the crowns meet, whereas in the higher situations the trees stand well apart, often separated by distances of 50 to 75 m. The most abundant trees are *Prosopis juliflora* var. *velutina* and *Acacia cymbispina*. In low situations and along arroyos *Lysiloma divaricata*, *L. Watsoni*, and *Pithecellobium mexicanum* are characteristic. In the western half of the area, *Olneya tesota* is frequent in the driest situations.

On rocky slopes and hillsides the leguminous trees are absent or infrequent. Here the dominant arborescent forms are *Bursera odorata*, *B. laxiflora*, and *Jatropha cordata*, accompanied by a small percentage of other trees and by *Pachycereus pecten-aboriginum* and *Fouquieria Macdougalii*.

In the central Foothills region there are many short valleys or embayments of structural origin into which there is no surface drainage. These occur around the volcanic mesas and also along the major and minor streamways. The bottom of the embayments has an accumulation of soil in which the moisture content is relatively high, and the absence of cold-air drainage from a large or elevated region makes them warm. In such situations along the Río Moctezuma north of its confluence with the Río Bavispe are found the northernmost examples of *Acacia cymbispina*, the palm *Erythea Roezlii*, and the woody composite *Parthenium Stramonium*.

In a few localities in this area above an elevation of 750 m. the upper limit of desert is indicated by the abundance of coarse perennial bunch grasses and the occurrence of open groves of *Quercus chihuahuensis*.

Trees are abundant and ubiquitous in the arborescent desert of the Foothills of Sonora, but shrubs also play an important part in the vegetation. In the most favorable situations, occupied by dense stands of *Acacia* or *Lysiloma*, shrubs are few and the floor is clean (pl. 34). Elsewhere shrubs are abundant, growing in scattered groups in the open, massed under the trees, or forming thickets so dense that they can be penetrated only by following a devious route. In the largest level areas the shrubs are as widely spaced as the trees. Immediately south and southeast of Ures there are extensive plains drained by the Río Sonora. In them there is considerable diversity of substratum, comprising deep granitic soils, shallow volcanic clay and loam, and aggraded limestone areas with no soil other than that which fills the depressions in the rock surface. In this area the distribution of the vegetation is very irregular and its aspect changes radically from place to place.

Several of the dominant plants of the Arizona Upland and the Plains of Sonora are found in the most arid situations along the western edge of the Foothills of Sonora, but all of them are rare or absent along the Yaqui River. These plants include *Cercidium microphyllum*, *C. floridum*, *Carnegiea gigantea*, *Encelia farinosa*, *Acacia Greggii*, and *Lemaireocereus Thurberi*. *Celtis pallida* is infrequent in the central Foothills area, but extends far into the mountains east of the Sonoran Desert. *Olneya* is locally abundant in the western half of the Foothills area, but is nowhere dominant or subdominant. The desert aspect of vegetation and the plants characteristic of the most arid parts of the Sonoran Desert are found chiefly in the part of the Foothills of Sonora which lies north of Tecoripa and west of the Río Yaqui.

The following list of the characteristic large perennials of the plains and bajadas of the central Foothills of Sonora is based on determinations of composition in twelve localities in the leguminous type of arborescent desert:

***Prosopis juliflora var. velutina
***Acacia cymbispina
***Cercidium sonorae
***Haematoxylon brasiletto
***Caesalpinia pumila
***Karwinskia Humboldtiana
**Lysiloma divaricata
**Acacia pennatula
**Opuntia Thurberi
**Cercidium floridum

**Opuntia fulgida
**Croton sonorae
**Lantana Camara
*Olneya tesota
*Lysiloma Watsoni
*Pithecellobium mexicanum
*Jatropha cardiophylla
*Guazuma ulmifolia
*Randia Thurberi
*Sapium biloculare
*Ipomoea arborescens
*Encelia farinosa
*Eysenhardtia orthocarpa
*Pachycereus pecten-aboriginum
*Acacia Farnesiana
Piscidia mollis
Acacia Greggii
Ceiba acuminata
Jacquinia pungens
Guaiacum Coulteri
Schaefferia Shrevei
Caesalpinia pulcherrima
Dodonaea viscosa
Randia laevigata
Celtis pallida
Erythea Roezlii
Nolina matapensis
Bursera odorata

In the central Foothills of Sonora may be noted the almost universal difference in physiognomy and composition between the vegetation of granitic slopes and that of volcanic eruptives. The difference manifests itself in this area by greater irregularity in the emplacement of large plants on granite, frequent changes of dominance, a heavier stand of shrubs, and a greater amount of grass. The eruptives have a simpler total composition, a smaller number of dominants more uniformly distributed, fewer shrubs, and very few grasses.

The granite mountains in the vicinity of Ures have a basal elevation of about 500 m., and their vegetation is desert up to about 1000 m. In the following list are noted the common plants of the lower slopes of these mountains, with indication of their relative abundance. Many other infrequent or scatteringly distributed plants add to the density and variety of the vegetation.

***Cercidium microphyllum
***Fouquieria Macdougalii
**Ipomoea arborescens
**Prosopis juliflora var. velutina
**Cercidium sonorae
**Jatropha cordata
**Franseria cordifolia
**Karwinskia Humboldtiana
**Mimosa laxiflora
**Lemaireocereus Thurberi
**Lantana Camara
*Guaiacum Coulteri
*Ceiba acuminata
*Piscidia mollis
*Lysiloma divaricata
*Cassia biflora
*Encelia farinosa
Erythrina flabelliformis
Opuntia Thurberi
Tecoma stans
Jatropha cardiophylla
Diphysa suberosa
Agave yaquiana
Randia laevigata

The volcanic hills and mesas of the central Foothills of Sonora lie between elevations of 300 and 600 m. On the gentle slopes well provided with soil are found the most uniform stands of *Bursera odorata* and *Jatropha cordata*, growing in stands so open that the crowns of the trees are separated by a distance of one to three times their diameter. The accompanying shrubs are also widely spaced.

The mesas are bordered by sharp but well weathered escarpments. At the base of an escarpment and on its talus the vegetation is relatively dense. The two dominant trees of the gentle slopes are joined by *Ipomoea arborescens*, *Pachycereus pecten-aboriginum*, *Acacia cymbispina*, *Cercidium sonorae*, *Acacia pennatula*, and a number of shrubs, as enumerated below.

The lava fields which form the level surface of the mesas are often more poorly provided with soil than the fields which lie at an accordant level with the surrounding area (as described on pp. 64–65). The vegetation of the mesas is very similar in composition to that of the gentle slopes, but the plants are smaller and more widely and unevenly spaced.

The tops of the volcanic mesas give every indication of providing a poorer water supply than any other habitat in the Foothills region. During the period of very low precipitation from February to June, the moisture supply sinks to a level which prevents all vegetative activity. It is of interest that none of the cacti and few of the nonsucculent plants of the Lower Colorado Valley and the Arizona Upland are found on the arid mesas. None of the strictly evergreen woody plants of neighboring habitats are found, except the partly deciduous *Sapium biloculare* and *Dodonaea viscosa*. Cacti are sparingly represented by a few thorn forest species, including *Pachycereus pecten-aboriginum*, *Opuntia fuliginosa*, and *O. tomentosa*. *Olneya* and *Cercidium microphyllum* are infrequent.

Throughout the Foothills of Sonora the arid aspect of the landscape in winter and spring is relieved only by the appearance of leaves in March on individuals of *Prosopis*, *Lysiloma*, and *Pithecellobium* which are growing where there is a good soil-moisture supply. With the beginning of the summer rains there is a rapid development of foliage by the trees and shrubs and an equally prompt appearance of root perennials and ephemerals. The lush and verdant aspect of the landscape persists until the latter part of September or often to the middle of October. The approach of the dry autumn is first indicated when the leaves of *Jatropha* turn yellow and those of *Bursera* and *Haematoxylon* turn red. At that season the colorful landscape accurately reveals the abundance of these plants and the pattern of their distribution.

The following are the characteristic plants of the volcanic slopes and mesas:

***Bursera odorata
***Jatropha cordata
**Cercidium sonorae
**Fouquieria Macdougalii
**Haematoxylon brasiletto
**Ipomoea arborescens
**Croton sonorae
*Randia Thurberi
*Sapium biloculare
*Acacia pennatula
*Opuntia Thurberi
*Krameria paucifolia
*Caesalpinia pumila
*Opuntia fuliginosa
Abutilon Pringlei
Ceiba acuminata
Dodonaea viscosa
Guazuma ulmifolia

*Lava beds.* At a number of localities in the central Foothills of Sonora there are lava beds several square kilometers in area. The surface of the beds is nearly level and extremely rough, but affords many crevices and pockets of fine clay. One of the largest fields evidently originated by extrusive activity of a small cone 32 km. south-southeast of Moctezuma. From this the lava flowed southwest and spread to a width of about 15 km. along the left bank of the Río Moctezuma, its southern edge being just above Batuc.

The lava beds may be clearly distinguished at a distance by the exposed dark brown basalt and by their heavy cover of vegetation, the color tone of which is a much darker green than that of the surrounding vegetation. There is much bare rock in the lava fields, but plants occupy all the crevices and patches of soil very densely and there are many heavily covered areas of nearly a hectare. As was brought out in connection with the lava beds of the Lower Colorado Valley, the soil-moisture conditions in such a habitat are more favorable than on a deep, homogeneous soil. The character of the surface prevents the formation of drainage systems, and there is almost no runoff. Infiltration along the rock surfaces is good, the fine clay is highly retentive, and evaporation is reduced by partial shading and by the interruption to wind action due to protruding masses of rock. Among the plants characteristic of the lava beds are several which are north of their normal limits. Several considerations indicate that the lava beds are a warmer habitat than the bajadas or plains surrounding them.

Shrubs predominate over trees on the lava beds, but there is no constant relation between the abundances of the two. The profile of the vegetation is very irregular. The following list of characteristic plants is drawn from the examination of several localities in the field near Batuc:

Cercidium sonorae
Ipomoea arborescens
Fouquieria Macdougalii
Opuntia fuliginosa (tall platyopuntia)
Croton sonorae (shrub, 1–2 m.)
Eysenhardtia orthocarpa
Mimosa laxiflora
Esenbeckia Hartmanii (southern)
Randia obcordata (slender tree, 3–5 m.; southern)
Franseria cordifolia (suffrutescent perennial, 1–1.5 m.)
Guaiacum Coulteri (compact tree, 3–4 m.)
Jatropha cardiophylla
Acacia cymbispina (tree, 5–6 m.; southern)

Annuals and herbaceous perennials are very abundant in the shade of trees and rocks in the late summer. Perennial grasses are infrequent, but annual species are common in open situations.

Southern Foothills region. This area comprises the region south of the latitude of Tecoripa and is the southernmost part of the Sonoran Desert on the mainland of Mexico. On the west and north of the Río Yaqui the central feature of the topography is Sierra Bacatete, which rises above the level of desert vegetation and is surrounded by low hills and coastal or fluviatile plains. To the east and south of the Río Yaqui an area of about 1800 sq. km. is occupied by the delta, which has its apex near Cajeme (Ciudad Obregón). South of Cajeme the coastal plain extends inland for 30 to 40 km. North of Quiriego the inner edge of the Foothills of Sonora is a region of low, close-set hills.

In the southern Foothills the prevailing vegetation still exhibits the characteristics of desert, at the same time that 20 to 30 per cent of the area is occupied by littoral or thorn forest vegetation. The littoral communities include the strand, small sand dunes, mangrove swamps, marshes, and saline flats as well as low thickets of shrubbery and open forests of *Prosopis juliflora* var. *Torreyana.* Thorn forest is found along streamways and on flood plains and valley bottoms. The relatively dense thorn forest vegetation almost invariably terminates sharply at a variable distance from a streamway. Beyond these edges are open stands of desert plants.

The hills of the southern Foothills of Sonora are more heavily clothed with trees and tall shrubs than are those of the central Foothills. The coverage is from 75 to 90 per cent. The single canopy of xeric trees has an irregular height of 4 to 10 m. The commonest shrubs range in height from 0.5 to 6 m. Microphyllous trees are dominant, most of them being Leguminosae, but the number of broad-leaved trees is here greater than in any type of vegetation previously described. Three species of large cacti are abundant and others are frequent. *Agave* is uncommon except on rock outcrops.

In this habitat all the criteria of desert, constantly exemplified in the areas to the north, have become so weakly expressed that much of the hill vegetation in the southern Foothills area is more closely allied to thorn forest than it is to desert. The percentage of cover is high, the stature of the plants is consistently greater, and the diversity of life forms is very much less than in the desert. The denser cover results in a controlled runoff and a considerable accumulation of plant litter. Though the surface of the slopes is stony, the soil is several decimeters deep, except in the steepest places, and there is a clear differentiation of soil and subsoil. The margins of streamways are more strongly held in place by the vegetation than is the case in the desert, and the influence of the better moisture conditions of streamways extends farther from the bed of the stream.

The open desert communities which occupy the upland of the southern Foothills of Sonora are dominated by an irregularly spaced stand of trees with scattered shrubs and widely spaced cacti. The coverage may be as low as 40 per cent or as high as 70 per cent. These areas may be readily traversed in any direction, which is not true of the dense growth in the bands of thorn forest. Approximately half the stand of trees is made up of the first 6 species indicated in the following list as being of the first order of abundance. The remainder of the arborescent vegetation is made up of a much larger number of trees, either consistently or sporadically distributed. The most conspicuous cactus is *Lemaireocereus Thurberi,* but the most common one is *Opuntia fulgida* var. *mammillata*, which occurs frequently, sometimes in colonies, rarely exceeds 1 m. in height, and is very sparing in the formation of flowers and fruit.

The characteristic plants of the desert upland of the southern Foothills of Sonora are as follows:

***Prosopis juliflora var. velutina (tree)
***Cercidium floridum (tree)
***Acacia cymbispina (tree)
***Fouquieria Macdougalii (tree)
***Cercidium sonorae (tree)
***Piscidia mollis (tree)

***Eysenhardtia orthocarpa (shrub)
***Karwinskia Humboldtiana (shrub)
***Guazuma ulmifolia (shrub)
**Bursera microphylla (tree)
**Jatropha cinerea
**Opuntia fulgida var. mammillata
**Pithecellobium sonorae
**Bursera laxiflora
**Guaiacum Coulteri
**Lemaireocereus Thurberi
**Janusia gracilis
*Condalia lycioides
*Haematoxylon brasiletto
*Lycium Berlandieri var. peninsulare
*Mimosa laxiflora
Krameria sonorae
Ceiba acuminata
Cassia biflora
Turnera diffusa
Ruellia californica
Fouquieria peninsularis
Desmanthus Covillei
Euphorbia colletioides
Citharexylum flabellifolium
Hyptis Emoryi

Mention has been made of the grouping of herbaceous ephemerals and young perennials in the shade of large perennials, which is common in all but the driest parts of the Sonoran Desert. In the Plains of Sonora and the Foothills of Sonora this feature is constant and clearly marked, and the number of root perennials and young trees and shrubs is greater than the number of ephemerals. These groups of subordinate plants themselves do much to improve the soil-moisture conditions during the summer rains, and the groups very commonly come to extend well beyond the shade of the sheltering trees. Where closely spaced trees are all surrounded by small plants, the groups coalesce into a single area, often 50 m. in diameter. The edge of a group is almost invariably well defined, the coverage dropping from 100 per cent in it to 30 to 40 per cent outside it.

In the subordinate groups are found young plants of the large desert perennials and desert shrubs, many young or small plants characteristic of the bands of thorn forest, and a large number of root perennials and ephemerals which do not occur in the open. Several trees which are abundant in thorn forest and infrequent in desert (*Ipomoea arborescens*, *Ceiba acuminata*, *Cordia sonorae*, *Pithecellobium sonorae*) doubtless owe their desert occurrence to conditions favorable for their germination and early growth in subordinate groups which no longer exist.

A very obvious phase of the transition from desert to thorn forest may be observed in connection with the types of habitats in which the two vegetations occur near their meeting ground. North of Cajeme the outlying bodies of thorn forest occupy the most favorable habitats in a predominantly desert vegetation. South of Cajeme small areas of desert vegetation are found in the driest situations of the prevailing thorn forest.

The bands of thorn forest which follow the flood plains of rivers and the margins of smaller streams are conspicuous by their density and the occurrence of trees more than 8 to 10 m. in height. On account of the irregularity of the canopy and the thin shade of the microphyllous trees, the vegetation shows no layering. There is usually a solid body of foliage from the ground to the treetops. The regular canopy of pure stands of *Acacia cymbispina* here serves merely to form a general level with which a few other trees are accordant and several are not.

Among the latter are the broad-leaved trees *Ipomoea arborescens*, *Sapindus Saponaria*, *Vitex mollis*, and *Cordia sonorae*, all of which occur farther north, and *Cassia emarginata*, *Zizyphus sonorensis*, and *Albizzia sinaloensis*, here at their northern limit. These are the trees which often exceed 10 m. in height (except *Vitex* and *Zizyphus*).

The dense undergrowth of shrubs is commonly dominated by *Franseria cordifolia*, which is from 6 to 10 dm. high and is not accompanied by lower herbaceous perennials. In other situations a low carpet of small perennials covers the ground.

The following list embraces the most generally observed plants found along streamways in the southernmost part of the desert. The broad-leaved trees are indicated by "B."

***Acacia cymbispina
***Franseria cordifolia
**Ipomoea arborescens (B)
**Pithecellobium sonorae
**Cassia emarginata (B)
*Coursetia glandulosa
*Zizyphus sonorensis (B)
*Rathbunia alamosensis
*Cordia sonorae (B)
*Randia echinocarpa (B)
*Sapindus Saponaria (B)
*Celtis pallida
*Pachycereus pecten-aboriginum
*Franseria ambrosioides
*Lantana Camara
*Acacia occidentalis
*Opuntia Thurberi
*Lagascea decipiens
*Lantana velutina
*Brickellia Coulteri
*Plumbago scandens
Vallesia glabra (B)
Alvaradoa amorphoides
Jacquinia pungens
Bumelia occidentalis (B)
Jacobinia ovata (B)
Caesalpinia pulcherrima
Vitex mollis (B)
Opuntia fuliginosa
Cardiospermum corindum
Albizzia sinaloensis (B)
Colubrina glabra
Caesalpinia platyloba (B)
Celosia nitida
Pereskiopsis Porteri
Gouania mexicana
Chiococca petrina
Parthenium Stramonium (B)
Xanthoxylum sonorense
Solanum amazonium (B)
Hintonia latiflora
Plumeria acutifolia (B)

The ecological features of the undisturbed parts of the delta of the Río Yaqui are very complex and have not been examined in detail. Mangrove swamp occupies a few small areas on the coast. *Rhizophora Mangle* is the dominant tree, accompanied by *Avicennia nitida* and *Conocarpus erecta*. *Rhizophora* occurs infrequently around Guaymas and as far north as Tiburón Island, but does not develop its characteristic stands. Small dunes skirt the coast just inside the upper beach. Their total area is very small.

The large level expanse of light gray alluvial soil which covers the delta shows great variety in its local plant communities. Along the coastal edge there is much bare ground and the prevailing plants are halophytic shrubs, including *Atriplex canescens*, *Suaeda fruticosa*, *Lycium carinatum*, *Maytenus phyllanthoides*, *Stegnosperma halimifolium*, *Phaulothamnus spinescens*, and *Lippia Palmeri*. Farther inland the shrubbery becomes denser and taller, and *Prosopis juliflora* var. *Torreyana* more frequent. In its original condition the delta area must have included many heavy stands of *Prosopis*, but there are none now which compare with the groves along the Altar and Magdalena Rivers.

Almost throughout the northern edge of the delta there is a strong representation of cacti in the vegetation (pl. 19). The most abundant are *Lemaireocereus Thurberi* and *Opuntia fulgida* var. *mammillata*. *Rathbunia alamosensis* is very frequent, and *Pachycereus pecten-aboriginum* and *Ferocactus* sp. are often seen. In some localities *Mammillaria Swinglei* is remarkably abundant. The substratum for the cacti is a heavy alluvial clay, which appears to be ill suited in texture and aeration for such rich stands of succulent plants.

### *Central Gulf Coast*

(Sarcocaulescent Desert. Bursera-Jatropha Region)

The part of the Sonoran Desert designated as the Central Gulf Coast lies along the shores of the Gulf of California and is partly in Sonora and partly in Baja California. The Sonoran part of the area extends from the southern end of the dunes and sandy plains of the Lower Colorado Valley to the delta of the Río Yaqui. In Baja California the area extends from Bahía Los Angeles to the vicinity of Cerralvo Island, just south of La Paz. The largest islands in the Gulf, Angel de la Guarda and Tiburón, as well as the numerous smaller islands, are also a part of the area. In Sonora the vegetation of the Central Gulf Coast extends inland for 20 to 30 km. and merges gradually into that of the Lower Colorado Valley or the Plains of Sonora. In Baja California it extends inland for a shorter and variable distance, being bounded by the drainage divide of the peninsula or by the chains of hills lying nearest the coast.

The Central Gulf Coast has been examined at five localities in Sonora and six in Baja California, involving stretches of 30 to 100 km. in each locality. Comparison of these areas shows that there is essential identity in the vegetation on the two sides of the Gulf. Several important dominants of Baja California are very local in their occurrence in Sonora and a few are absent, so far as known. A very small number of large perennials are found in the Sonoran section which are absent from Baja California. In the physiognomy of the vegetation there are no constant differences to distinguish the two coasts. In several cases, to be mentioned later, the same place is taken in the vegetation of the two coasts by different but closely related species. North of the Central Gulf Coast area the flora on the two sides of the Gulf is nearly identical.

The Gulf coast is occupied in part by hills and small mountains rising abruptly from the beach and in part by bajadas extending far inland. The most mountainous section of the Sonoran coast extends for 80 km. north from Guaymas. In Baja California the coast is prevailingly rugged almost throughout its length. There are several broad bajadas between Bahía Los Angeles and the Tres Vírgenes peak, but only narrow wedges of outwash material south of that. South of Bahía

Concepción the Central Gulf Coast area is confined to a narrow strip along the base of Sierra de la Giganta.

The soils which support the vegetation of this subdivision are shallow, coarse, and covered with rock fragments. Only where streamwavs debouch to the Gulf are there areas of boulder-filled sand. Dunes are infrequent and small. Lagoons, marshes, and mangrove swamps are few and small except at several localities in Bahía Concepción. The drainageways of the area are all small and rarely in flood. The Río Sonora is the only large stream which originates inland and cuts across the Central Gulf Coast. At the head of Bahía Concepción there is no evidence that any drainage other than sheet floods ever enters the bay.

The few climatic records available for the Central Gulf Coast, together with the evidence of the vegetation, indicate that it is the driest subdivision of the Sonoran Desert. In Baja California, the annual mean for Santa Rosalía is 138 mm. and for Mulegé 105 mm. In Sonora, the mean for Guaymas is 263 mm. The winter rains of the Pacific coast of Baja California rarely extend inland beyond the mountainous axis of the peninsula, and are exceeded by the winter rains on the coast of Sonora. The rains which sporadically visit the Central Gulf Coast in Baja California result from the northward extension of the storms which commonly visit the mountains of the Cape Region in the summer months.

The vegetation of the Central Gulf Coast exhibits its lowest and most open phase and its most simple composition within 5 km. of the beach at the lower ends of the long bajadas which descend from the interior. Hills and rocky slopes near the strand are also thinly covered with shrubs and small individuals of the large perennials (pl. 20). The margins of streamways and their broad, sandy beds are occupied by scattered stands of *Baccharis*, *Hymenoclea*, *Franseria ambrosioides*, and *Chilopsis*. On both bajadas and hills there is a slight increase in density and a gradual increase in number of species from coast to interior, reaching a stable maximum about 12 to 14 km. inland.

In spite of the similarity of the vegetation on the coasts of Sonora and Baja California, there is nevertheless sufficient difference to warrant separate description of the two.

Gulf coast of Sonora. The transition from the dunes and sandy plains of the Lower Colorado Valley to a hard surface takes place on the Sonoran coast a short distance above Puerto Lobos, which is 45 km. north of Libertad. In this vicinity *Larrea* and *Franseria dumosa* begin to yield their dominance to a group of large perennials in which *Larrea* is uncommon in pure stands and *Franseria fl* only locally abundant. The new group of dominants includes *Cercidium sioridum*, *Olneya tesota*, *Bursera microphylla*, *Jatropha cinerea*, *J. cuneata*, *Opuntia Bigelovii*, and *Fouquieria splendens*. On all the plains and bajadas more than 5 km. from the beach, some or all of these plants form the matrix of the vegetation. In the driest situations there are few shrubs, but their numbers greatly increase along streamways, in depressions, and at the base of hills. Shrubs are found almost exclusively in the open and not closely assembled under the trees. No shrub or root perennial has been noted in this region which occurs only in the shade of large plants, except the vine *Brandegea.*

The belt of vegetation along the immediate coast is low, very open, without trees, and monotonous in its aspect. The coverage rarely exceeds 15 to 20 per cent. The dominant plant is *Frankenia Palmeri,* a microphyllous gray shrub 4 to 8 dm. in height. Within reach of salt spray *Frankenia* and *Atriplex canescens* are commonly the only plants. Farther from the beach other plants appear, and there is a rapid transformation for a distance of 3 to 5 km. *Frankenia* rarely occurs as much as 3 km. inland. The plants which characterize the extreme coastal belt are:

Frankenia Palmeri
Atriplex canescens
Lycium Fremontii
Suaeda ramosissima
Euphorbia misera
Encelia farinosa
Stegnosperma halimifolium
Jatropha cuneata
Franseria dumosa
Errazurizia megacarpa

On the bajadas which extend far inland, *Frankenia, Suaeda, Euphorbia,* and *Stegnosperma* are not found beyond the vicinity of the beach. The vegetation beyond their range is made up of the following species:

Cercidium floridum
Prosopis juliflora var. Torreyana
Fouquieria splendens
Jatropha cuneata
Bursera microphylla
Olneya tesota
Pachycereus Pringlei
Bursera Hindsiana
Encelia farinosa
Larrea tridentata
Solanum Hindsianum
Acacia constricta
Horsfordia alata
Jatropha cinerea
Hyptis Emoryi
Sapium biloculare
Opuntia Gosseliniana
Acacia Greggii
Simmondsia chinensis
Cordia parvifolia
Atamisquea emarginata
Mimosa laxiflora
Ferocactus acanthodes

On the coast 10 km. south of Libertad is the only colony of *Idria columnaris* known on the mainland. The trees occupy the north slope of a low granite ridge which runs east from the beach, and follow the ridge inland for about 3 km. There are none on the south slope of the ridge, and none have been found elsewhere in the vicinity or anywhere else in Sonora. The total number of individuals is probably about 2000. Reproduction is good, and the tallest individuals are 12 to 14 m. high, as compared with the highest known trees, 22 m. *Idria* is very abundant throughout the central third of Baja California, and its occurrence at a single locality in Sonora is probably due to a chance natural introduction rather than to survival from a former wider occurrence.

There is little change in the vegetation southward along the coast. The region opposite Tiburón Island and the island itself closely resemble the vicinity of Libertad. *Cercidium, Olneya, Bursera, Fouquieria,* and *Jatropha* dominate the hills in very open stands of small trees. *Lemaireocereus* and *Opuntia Bigelovii* are the most common cacti. *Pachycereus Pringlei* is abundant, but only on nearly level ground, whether hard or sandy.

The largest streamways are bordered by *Prosopis juliflora* var. *Torreyana*, *Olneya tesota*, *Acacia Greggii*, and *Bursera Hindsiana*. The margins of broad, sandy arroyos are occupied by light or heavy stands of the shrubs *Hymenoclea pentalepis*, *Baccharis Emoryi*, *Franseria ambrosioides*, *Hyptis Emoryi*, *Haplopappus sonoriensis*, *Beloperone californica*, *Bebbia juncea*, and *Lagascea decipiens*.

Tiburón Island and the neighboring coast are of interest as being the southernmost locality in Sonora for *Koeberlinia spinosa*, *Franseria deltoidea*, *Simmondsia chinensis*, *Ferocactus acanthodes*, *Opuntia arbuscula*, and *Atriplex polycarpa*, as well as the northern limit for *Rhizophora Mangle*, *Colubrina glabra*, *Lippia Palmeri*, *Maytenus phyllanthoides*, *Acacia Willardiana*, and *Viscainoa geniculata*. *Viscainoa*, like *Idria*, is abundant in Baja California, but is now known in Sonora only at three stations between Tiburón Island and Guaymas.

For 60 km. south of the mouth of the Río Sonora the Gulf is bordered by the low, featureless Llano de San Juan Bautista, which gives some evidence of deltaic origin. The vegetation of this plain, so far as it has been explored, is a very open stand of *Olneya tesota*, *Cercidium floridum*, and *Bursera microphylla*, with scattered thin stands of *Larrea*. The drainageways fall at a very low gradient and are choked with sand. Sudden changes in the course of the drainage frequently occur, flooding open areas and stopping the water supply of large colonies of *Cercidium* and *Olneya*. Open groves of trees killed in this way are common. The extremely hard heartwood of *Olneya* resists decay and termites for many years. After the death of a tree, the sapwood around the base is soon etched away by the sand which blows along the surface of the ground, but the heartwood is much more resistant and long supports the skeleton of the tree.

From the southern edge of the Llano de San Juan Bautista at the Estero de Tastiota, southward to Guaymas, a distance of 80 km., low mountains rise almost continuously directly from the shore. This section of the coast is difficult of access from the interior and has few suitable places for landing from a boat. The vegetation is poorly known except at the southern end in Bahía San Carlos and in the vicinity of Guaymas. General impressions of this coast have been obtained in skirting it closely by boat along a stretch of 20 km. north of Bahía San Carlos.

The vegetation of this region is open and very xeric except on some of the steep slopes which face immediately on the Gulf. These slopes are well covered by shrubs, which are in leaf from March to November, even in seasons when the plants 2 or 3 km. inland are leafless. Where gentle slopes face the Gulf, the vegetation is low and open, rarely presenting a green appearance. Where the floor of a large valley opens to the Gulf, there is a band of heavier vegetation, which retains a green appearance after the rainy seasons for one or two months longer than the interior plains.

In the vicinity of Guaymas the vegetation of both hills and plains is richer in composition than it is in similar habitats near Tiburón Island. A few level areas with deep soil are occupied by nearly pure stands of *Larrea* or of *Prosopis juliflora* var. *articulata*. The prevailing shallow soils and hillsides are clothed with the following perennials:

***Bursera microphylla
***Prosopis juliflora var. Torreyana
***Acacia Willardiana
**Coursetia glandulosa
**Acacia Farnesiana
**Caesalpinia pumila
**Phaulothamnus spinescens
*Acacia occidentalis
*Jacquinia pungens
*Guaiacum Coulteri
*Randia Thurberi
*Fouquieria peninsularis
*Lemaireocereus Thurberi
*Vallesia glabra
*Desmanthus Covillei
*Ruellia californica
Lippia Palmeri
Echinopterys eglandulosa
Lycium carinatum
Cordia sonorae
Opuntia fulgida
Forchammeria Watsoni
Pachycereus pecten-aboriginum

Among the hills near Guaymas the borders of streamways support heavy stands of shrubs and small trees, particularly in valleys which face the Gulf. In these communities a few species are more abundant than others but there are no true dominants. The composition varies from place to place, and some of the plants are represented by very few widely separated individuals. Among these are several plants not found elsewhere in the latitude of Guaymas or north of there. At Bahía San Carlos and Miramar the following plants have been found in the slightly moist situations:

**Prosopis juliflora var. Torreyana
**Lysiloma divaricata
**Pithecellobium sonorae
**Sapindus Saponaria
*Maytenus phyllanthoides
*Lantana Camara
*Gossypium Davidsonii
*Eupatorium sagittatum
*Mascagnia macroptera
*Franseria ambrosioides
Cordia sonorae
Sapium biloculare
Anisacanthus Thurberi
Vitex mollis
Ruellia californica
Stegnosperma halimifolium
Bumelia occidentalis
Forchammeria Watsoni
Beloperone californica
Vallesia glabra
Viscainoa geniculata
Pisonia capitata
Colubrina glabra
Washingtonia robusta
Zizyphus sonorensis
Acacia Farnesiana
Euphorbia colletioides
Bursera confusa
Abutilon californicum
Benthamantha Wrightii
Cordia brevispicata
Hintonia latiflora

On the thin soil of inland slopes the vegetation is open and very xeric. The slopes rising from small bays and inlets are like those of the interior. The low cover is relatively dense only on slopes that are dominated by xeric semishrubs or *Hechtia*, above which rise widely spaced trees, cacti, or *Fouquieria*. In several localities northwest of Guaymas and within 5 km. of the coast the following species have been found to predominate:

***Bursera microphylla
***Fouquieria peninsularis
***Acacia Willardiana
***Jatropha cinerea
***Jatropha cuneata
***Hechtia montana
**Haplopappus sp.
**Citharexylum flabellifolium
**Bebbia juncea
*Agave Schottii
*Lemaireocereus Thurberi
*Lycium brevipes
*Krameria parvifolia
Ferocactus alamosanus
Carlowrightia californica
Porophyllum pausodynum
Echinocereus scopulorum

Quiet shores of inlets and lagoons are bordered by *Rhizophora Mangle* and *Avicennia nitida*, but there are no extensive areas of these trees. East of Guaymas are several hundred hectares of saline flats very sparsely covered with *Suaeda ramosissima*, *Allenrolfea occidentalis*, *Wislizenia Palmeri*, and *Aster parviflorus*. On the outwash slopes immediately inland from the saline flats is the heaviest stand of *Pachycereus Pringlei* known on the Sonoran coast. It rivals in density the stands at the head of Bahía Concepción, in Baja California, but the individuals are neither so large nor so much branched. *Pachycereus* is here made very conspicuous by a white coating of dung of sea birds which roost on the tops.

Guaymas is cited as the type locality for a large number of plants, a fact which has given rise to the impression that there is a high degree of endemism in the flora of its environs. The circumstance is due largely to the fact that the Guaymas region was far more accessible than any other locality in central Sonora until about 1900, and was the first place visited by active collectors. Very many of the plants first found at Guaymas are now known to have a wide distribution in central and southern Sonora or even in Sinaloa. There remain, however, perhaps 40 species which are not known outside a radius of 50 km. around Guaymas.

Gulf coast of Baja California. North of the Cape Region the Gulf coast of Baja California is the most thinly settled part of the Sonoran Desert, because of the lack of arable soil and the absence of favorable places for the development of water supplies. Almost the entire coast is poorly known and difficult of access by land. A road leads from Mexicali, on the international boundary, to Bahía San Felipe, a distance of 200 km. The coast may be reached from the interior by saddle from Marmol to Miramar (Punta Santa Isabel), by road from Punta Prieta to Bahía Los Angeles, and by road from Pozo Alemán to Barril. The only road paralleling the coast runs from Santa Rosalía to the mouth of Bahía Concepción. All these routes have been traversed in the present work, and the coast has been closely followed by boat from Isla Carmen to La Paz.

The only rainfall data for the coast are those for Santa Rosalía, Mulegé, and La Paz. The northern part of the coast, in the rain shadow of the Sierra Juárez and Sierra San Pedro Mártir, is undoubtedly the driest, and its precipitation is probably below an annual mean of 75 mm. (3 in.). At Santa Rosalía the mean annual rainfall is 138 mm. (5.43 in.), and at Mulegé 104 mm. (4.09 in.).

In spite of the uniformly low rainfall along the entire coast, there is a notable change in the vegetation between Bahía Los Angeles and Barril. This is associated with the change from winter to summer incidence of the principal percentage of the light and uncertain precipitation, and also with the limitation of frost to brief and infrequent periods. The dominant perennials and many of the herbaceous ephemerals of the Lower Colorado Valley persist as far south as Bahía Los Angeles. The large perennials of the sarcophyllous desert (Vizcaíno Region), which is adjacent on the west, either are absent from the Gulf coast north of Bahía Los Angeles or are confined to small favorable habitats. *Pachycereus Pringlei* and *Bursera microphylla* range sparingly as far north as Bahía San Felipe. Near Bahía Los Angeles, *Idria* approaches the coast only in gravelly bottoms of streamways, and *Pachycormus* only on south slopes of boulder-covered hills. *Fouquieria splendens* ranges south to Bahía San Bartolomé; *Larrea* persists on deep, level soils as far south as Bahía Concepción; and *Cercidium microphyllum* extends to Bahía Concepción. With these exceptions the vegetation of the Central Gulf Coast is made up of species which do not occur in abundance north of Bahía Los Angeles. The change in the aspect of the vegetation is from widely spaced slender shrubs to widely spaced shrubs and small sarcocaulescent trees. The dominants are *Jatropha cuneata, Fouquieria peninsularis, Bursera microphylla, Jatropha cinerea*, and *Bursera Hindsiana*.

Immediately north and south of Bahía Los Angeles the coast is bordered by short ranges of hills very lightly clothed with vegetation. To the west of them, and on or near the median line of the peninsula, are Sierra Calamajué (1658 m.), Sierra Yubay (800 m.), and Sierra San Borjas (1200 m.). Between the coastal hills and these mountains are two valleys, 30 and 75 km. long, comprising the Desierto de San Julián, the Desierto de Santa María, and the Llano de San Pedro. These are enclosed basins, the bottoms of which are subject to overflow at very rare intervals. The bottoms are devoid of plants, and the surrounding floors bear only widely scattered individuals of *Atriplex Barclayana*. The slopes of the hills bordering these basins are almost devoid of woody perennials. The landscape here comes nearer than any other in the Sonoran Desert to fitting the European definition of desert as a place with no plants whatever.

On the Gulf coast east of Calmallí, between Cabo San Miguel and Punta San Juan Bautista, long outwash slopes fall from the interior to the coast. The dominant plants on these slopes are *Larrea, Bursera microphylla*, and *Fouquieria splendens. Fouquieria* is less abundant than *Larrea* or *Bursera*, but is conspicuous by reason of its size and branching. The mature individuals are 7 to 9 m. high, with very stout branches which fork frequently at the top. The slender tips of the main arms and branches are gracefully dependent and sinuously curved. This locality is near the southern limit of *Fouquieria splendens* in Baja California, and the specimens here contrast strongly with the small plants along its eastern distributional limit, with their straight, slender, unbranched arms.

The simple vegetation of the bajadas in this locality forms a 15 per cent cover, and its composition is as follows:

***Larrea tridentata
***Bursera microphylla
***Fouquieria splendens
***Jatropha cuneata
***Opuntia cholla
**Olneya tesota
**Bursera Hindsiana
**Opuntia clavellina
Viscainoa geniculata
Fouquieria peninsularis
Ferocactus horridus
Pedilanthus macrocarpus
Pachycereus Pringlei

The influences of slope exposure are responsible for very general differences in vegetation between north and south slopes in the Central Gulf Coast subdivision. Where the same type of rock and soil is involved, the north slopes have a slightly heavier stand of perennials. *Idria* and *Pachycormus* are occasionally found on north slopes although absent from south slopes and level ground. The southern limit of *Idria* is on the north slopes of Cerro Tres Vírgenes, and it is not found on the south slopes. No cases have been observed in which west slopes, facing the Pacific, bear heavier vegetation than adjacent east slopes. Ready access of storms and moist winds from the ocean are of great importance in differentiating the Central Gulf Coast from the Vizcaíno Region, but they do not outweigh the importance of orientation toward or from the midday sun. The abundance of the epiphytic *Tillandsia recurvata* on the desert trees and shrubs in the Vizcaíno Region is an indication of recurrent periods of moist air, but the plant is rare in the interior hills and absent from the Gulf slopes. The influence of the Pacific Ocean on the vegetation of the Gulf coast could scarcely be less if the ocean were 1000 km. away instead of 70 to 100 km.

South of Cerro Tres Vírgenes as far as Bahía Concepción, the drainage divide of the peninsula lies between 500 and 1000 m., with a number of peaks and mesas of higher elevation. Precipices, steep slopes, and strongly tilted bajadas characterize the slopes draining into the Gulf. The landscape is dominated by evidences of recent volcanism. It is stated by the inhabitants that the Tres Vírgenes volcano was in eruption in 1790 (pl. 27). Malpais fields cover large areas, and partially aggraded slopes have a rocky surface and thin soil heavily banded with caliche. The hanging talus which is occasionally seen in Arizona and Sonora (pl. 3, fig. 2) is very common here. In some of the higher valleys the volcanic surface is covered with sand, which is evenly distributed and never accumulated in dunes.

In the area under consideration there is a slightly heavier cover of vegetation at 500 m. elevation than there is on the slopes near the shore of the Gulf. The coverage in the two situations is approximately 20 per cent and 10 per cent respectively. The composition at the higher elevation is richer, but many of the perennials found there are sporadic in occurrence. This is true of common species —*Olneya tesota*, *Encelia farinosa*, *Koeberlinia spinosa*, *Viscainoa geniculata*, *Simmondsia chinensis*, and others—as well as of woody and herbaceous perennials, which are nowhere abundant.

There is a notable scarcity of pure stands or even of communities strongly dominated by two or three species. At all elevations and in nearly all habitats the vegetation is mixed, even if it is very open. The exceptions are small valleys with a level surface and a deep soil overlaid by sand or sandy clay. In these, *Larrea* forms from 80 to 95 per cent of the stand. A single valley about 2 km. wide was seen in which exceptionally large plants of *Encelia farinosa* formed a nearly pure stand.

The following list indicates the composition and relative abundance of the vegetation on the upper slopes of the Gulf coast in the area between Cerro Tres Vírgenes and Arroyo Santa Agueda, which debouches just south of Santa Rosalía.

***Larrea tridentata
***Jatropha cuneata
***Opuntia clavellina
***Franseria magdalenae
***Bursera microphylla
***Fouquieria peninsularis
***Opuntia cholla
**Jatropha cinerea
**Cercidium floridum
**Opuntia calmalliana
**Agave sobria
*Pachycereus Pringlei
*Lycium brevipes
*Machaerocereus gummosus
*Viscainoa geniculata
*Prosopis juliflora var. Torreyana
*Franseria Bryantii
*Opuntia invicta
*Simmondsia chinensis
*Krameria parvifolia
*Haplopappus sonoriensis
Pedilanthus macrocarpus
Solanum Hindsianum
Petalonyx linearis
Atamisquea emarginata
Cercidium sonorae
Bebbia juncea
Acalypha californica
Stillingia linearifolia
Melochia tomentosa
Berginia virgata
Ruellia californica
Cassia confinis

The borders of the shallow streamways of the Tres Vírgenes region support only a slightly heavier stand of large perennials than the general surface does. The habitat is characterized by *Prosopis juliflora* var. *Torreyana*, *P. Palmeri*, *Olneya tesota*, and *Cercidium floridum*, but the principal cover is formed by the shrubs *Hymenoclea pentalepis*, *Bebbia juncea*, *Haplopappus sonoriensis*, *Beloperone californica*, and *Horsfordia Newberryi*.

The middle courses of the streamways which descend across the Gulf slopes are often deeply cut, occupying small barrancas from 10 to 200 m. in width. The floors and banks of these small valleys afford moisture conditions favorable enough to support an open stand of small trees, often accompanied by nearly closed shrubbery. Here the sarcocaulescent trees yield the dominance to microphyllous leguminous trees. The commonest of the latter is *Lysiloma candida*, a slender tree of *Acacia* form with very light-colored bark. Because of the value of its inner bark for tanning, this is the most useful native plant of the peninsula. Cutting has reduced the number of large trees in many localities, but second growth from stumps maintains its original local distribution.

The plants characteristic of the small barrancas are:

***Lysiloma candida
**Prosopis juliflora var. Torreyana
**Vallesia glabra
**Euphorbia californica
**Lantana Camara
*Prosopis Palmeri
*Mascagnia macroptera
Condalia globosa
Hyptis laniflora
Berginia virgata
Ficus Palmeri
Brongniartia peninsularis
Stegnosperma halimifolium
Aeschynomene nivea
Thryallis angustifolia

A very different picture is presented by the lower courses of the large streams which have descended nearly to sea level. High water is rare and brief in these streams, but the floods are extremely violent on account of the steep gradient of the upper tributaries. Among these streamways are the Arroyo Reforma, Arroyo Santa Agueda, Arroyo San José, and Río Magdalena. The Arroyo Reforma has a broad, sandy bed, and the Río Magdalena, which drains the north slopes of Sierra de Zacatecas, crosses the coastal plain as a series of sandy beds meandering among islands of sand and boulders, the whole having a width of nearly 3 km.

The vegetation of the broad stream beds is confined to (1) upland trees and shrubs which are well enough rooted to withstand the floods, (2) short-lived shrubs of species characteristic of such habitats, and (3) small perennials of rapid growth. All these are distributed in extremely irregular fashion. Common in the three groups respectively are:

(1) Lysiloma candida
Prosopis juliflora var. Torreyana
Cercidium floridum
Pithecellobium confine
Sapium biloculare

(2) Hymenoclea pentalepis
Haplopappus sonoriensis
Hyptis laniflora
Bebbia juncea
Aster spinosus

(3) Fagonia californica
Nicotiana trigonophylla
Sympetaleia aurea
Cassia confinis
Sphaeralcea Hainesii
Hofmeisteria fasciculata var. pubescens
Physalis crassifolia

The section of the Gulf coast lying between Mulegé and Canipolé embraces the area around Bahía Concepción and is the broadest part of the coastal belt. North and west of Mulegé are large, nearly level bajadas with sandy surface. On both sides of Bahía Concepción hills closely approach tidewater, occasionally separated by wedges of bajada or canyon floor. In spite of the low rainfall indicated by the record for Mulegé (4.09 in., 104 mm.), the vegetation of the bajadas in its vicinity is slightly taller and denser than it is north of Santa Rosalía, and the vegetation of the hills is approximately twice as dense, as well as richer in small trees. It is difficult to account for this contrast except on the possibility that the Mulegé record is not truly representative of the adjacent region, or else that the mean precipitation on the coast north of Santa Rosalía is lower than the 138 mm. recorded for that town, which may well be the case.

On the sandy bajadas north of Mulegé the vegetation forms a coverage of 30 to 40 per cent and is much more uneven in its height as well as its color tones than the bajada communities which are strongly dominated by *Larrea*. Trees are commonest at the upper edges of the bajadas and along the streamways. *Pachycereus Pringlei* is frequent on the lower bajadas, increases in abundance rapidly as they are ascended, and remains abundant on the lower slopes of the parent hills. Its role in the vegetation is very much the same as that of *Carnegiea* in the Arizona Upland and that of *Idria* in the Vizcaíno Region. Several cylindropuntias are common on the lower bajadas, particularly *Opuntia cholla*, which closely resembles in every respect the equally abundant *Opuntia fulgida* of the northern subdivisions of the Sonoran Desert.

On representative areas of sandy or gravelly bajada away from streamways, the composition of the vegetation in the Mulegé region is as follows:

***Jatropha cuneata
***Larrea tridentata
***Bursera microphylla
***Encelia farinosa
***Opuntia cholla
**Fouquieria peninsularis
**Colubrina glabra
**Jatropha cinerea
**Pachycereus Pringlei
**Cercidium microphyllum
*Opuntia ciribe
*Lycium brevipes
*Pedilanthus macrocarpus
*Simmondsia chinensis
Solanum Hindsianum
Machaerocereus gummosus
Condalia globosa
Opuntia invicta
Bursera Hindsiana
Olneya tesota
Trixis californica

The universally rocky slopes in the vicinity of the mouth of Bahía Concepción differ from the bajadas far more in their physiognomy than in their composition. The plant covering of the hills is open or extremely sparse, varying in density with the gradient of the slopes and the amount of soil which they afford. *Fouquieria peninsularis* is the most conspicuous of the abundant plants, and *Jatropha cuneata* is the commonest. *Bursera microphylla* and *Cercidium microphyllum* are common, but seldom greatly exceed the shrubs in height. *Pachycereus* is infrequent, and other cacti are far less common than on the bajadas. *Larrea* is very infrequent on the slopes and often absent over large areas. Plants which are common on the slopes and infrequent on the bajadas include the following, for which the relative abundance would be difficult to determine:

Abutilon Palmeri
Acalypha saxicola
Atamisquea emarginata
Bebbia juncea
Bourreria sonorae
Cercidium sonorae
Ditaxis Brandegei
Echinocereus Brandegei
Euphorbia tomentulosa
Fagonia californica
Ferocactus rectispinus
Lemaireocereus Thurberi

Opuntia ciribe
Peucephyllum Schottii
Pithecellobium confine
Porophyllum gracile
Randia Thurberi
Ruellia californica
Viguiera deltoidea

Bahía Concepción is 40 km. long and varies from 5 to 10 km. in width. Bluffs, inlets, and islands make the shore-line topography of the north half of the bay intricate and picturesque. In the south half the shores are low and continuous. There are numerous colonies of *Rhizophora Mangle* in quiet corners of the bay. In the northern section the typical vegetation of the slopes and valley bottoms descends nearly to the water. The only woody plant definitely confined to the vicinity of the shore is *Maytenus phyllanthoides*, which in some spots forms 20 per cent of the stand. On the west side the slopes which drain into the bay rise for 15 to 20 km. to the summit of Sierra de Zacatecas, where there is a considerable area over 1000 m. Four large canyons, as well as several small ones, enter the bay from the west. The largest of these is Arroyo Codeje. On the east and at the head of the bay there are no important affluents. There is an area of alluvial bottom land at the mouth of Arroyo Codeje covered by a heavy forest of *Prosopis juliflora* var. *Torreyana* and *Lysiloma candida*. Farther south, the alluvial flats near the bay are covered with thickets of *Maytenus phyllanthoides* or are nearly bare, supporting only scattered plants of *Allenrolfea occidentalis* and *Frankenia Palmeri*. On the inner edge of the alluvial land at the south end of the bay is the heaviest stand of *Pachycereus Pringlei* seen in Baja California. In the center of the area there are magnificent examples of this massive cactus from 12 to 14 m. in height, well branched and growing in an unusually close stand.

South of Punta Pulpito, due east of the head of Bahía Concepción, the coast is closely followed by the escarpment which forms the eastern slope of Sierra de la Giganta. The true desert is here confined to a narrow and interrupted band along the coast, to the low salient ridges of the mountains, and to the islands which hug the coast. The vegetation of the canyon slopes and the steep, rocky beds of the streamways of Sierra de la Giganta belongs to the type of thorn forest characteristic of the Cape Region.

East of La Paz and southward along the Gulf, the vegetation is desert on volcanic soils and thorn forest on the granitic bajadas which descend sharply from the Sierra de la Laguna and Sierra de San Lazaro. In aspect and composition this southernmost tip of the Sonoran Desert closely resembles the region around Bahía Concepción.

On the stony volcanic bajadas and slopes east and southeast of La Paz, most of the vegetation is very open, forming a cover of only 10 to 15 per cent. Few of the trees exceed 4 m. in height, and *Pachycereus* is small and unbranched. The composition here is:

***Jatropha cuneata
***Fouquieria peninsularis
***Opuntia cholla
**Bursera microphylla
**Acacia californica
**Machaerocereus gummosus

| | |
|---|---|
| *Cyrtocarpa edulis | Caesalpinia arenosa |
| *Ruellia peninsularis | Encelia farinosa |
| *Euphorbia californica | Solanum Hindsianum |

### *Vizcaíno Region*

(Sarcophyllous Desert. Agave-Franseria Region)

The lofty, uniform crest of Sierra San Pedro Mártir terminates abruptly just north of the 30th parallel. North of this latitude Baja California has three longitudinal belts of very dissimilar vegetation: chaparral on the coast and westerly slopes, coniferous forest on the mountains, and desert on the Gulf coast. South of this latitude, as far as the northern end of Sierra de la Giganta, the peninsula is desert from coast to coast. South of Sierra San Pedro Mártir the peninsula maintains a fairly uniform width of about 80 km. for a distance of 250 km. It then widens, with a long triangular projection on the Pacific coast, and has its maximum width of 200 km.

The Vizcaíno Region comprises the Pacific drainage of central Baja California from the southern end of Sierra San Pedro Mártir south nearly to latitude 26° N. The designation of this subdivision of the Sonoran Desert is an extension of the use of the name "Vizcaíno Desert," which has long been applied to the plains which occupy the long triangular projection. The Vizcaíno Region includes the rough mountainous interior as well as the Vizcaíno Desert and the additional wedge of coastal plain which extends south to Punta Pequeña.

The great escarpment which is so well defined in northern and in southern Baja California is less clearly marked in the central section, but has served nevertheless to place the drainage divide much nearer the Gulf than the Pacific. The backbone of the peninsula is here formed by Sierra Calamajué, Sierra San Borjas, and Sierra Calmallí. Only a few peaks in these ranges are over 1500 m. in elevation. From Rosario to Calmallí the face of the country is very rugged. The larger mountains run nearly parallel to the axis of the peninsula, the volcanic hills and mesas are very irregular in position, and the level areas with deep soil are few and small even along the coast. A single enclosed basin drains into Laguna Seca Chapala at latitude 29° 25′ N. Approximately 40 per cent of the area north of the Vizcaíno plain lies below 500 m.

On the southern edge of the Vizcaíno cape are two small ranges of grotesquely built mountains, Sierra Pintada and Sierra Santa Clara. If they are left out of account, the rugged axis of Baja California may be pictured as extending with uniform width throughout the Vizcaíno subdivision. Its increased width between latitudes 27° and 28° is chiefly due to the featureless plains of the Vizcaíno Desert and those in the vicinity of Laguna de San Ignacio.

Very little is definitely known about the climatic conditions of the Vizcaíno Region, as there are no official records from it. The summer temperatures of the immediate coast are greatly reduced as compared with those of the interior. The rainfall is almost wholly in the winter and early spring. At long intervals there are summer storms south of the Vizcaíno plain. According to the inhabitants, the winter rainfall is extremely variable from year to year. When the peninsula was visited in 1934 it was very dry, and the residents in several localities stated that it had been seven years since there had been good rains. On the visit of 1935, heavy rains were encountered in February, and the condition of the plants indicated that rain had begun several weeks before. Some of the same informants then alleged that the rains had not been so copious for twelve years. Whatever the conditions may be from January to April, the remaining eight months of the year are almost invariably dry throughout the Vizcaíno Region. During spring and summer there is morning fog along the coast as far south as the Vizcaíno plain, and it extends inland wherever there is a low coastal shelf for a distance of 5 to 6 km. Foggy mornings are said to be most frequent in July and August. A strong wind from the Pacific whips the coast during the day and early part of the night, and is apparently continuous throughout the year.

The vegetation of the Vizcaíno Region comprises relatively heavy stands of large perennials, including some of its most striking plants, in communities of rich composition. It also comprises every phase of the impoverishment of these outstanding communities, some of the impoverished ones being areally the most important. Finally, there are great areas of salitral in the Vizcaíno plain on which there is an extremely scanty cover of halophytic plants.

On purely observational evidence it is possible to determine some of the features which control the conditions underlying the vegetational differences. Six features seem to be particularly important:

Physiography: The greatest contrast in the area is between the mountainous axis of the peninsula and the alluvial deposits of the Vizcaíno plain.

Proximity to the sea: This means little with respect to rainfall, humidity, or soil moisture. The extreme coastal belt is invariably open and barren on account of the incessant high wind.

Character of substratum: This is important in differentiating the vegetation of small areas of deep soil from that of large areas of shallow soil, as well as in making a difference between the vegetation of volcanic soils, that of granitic soils, and that of sand.

Altitude: There is little in which this composite condition outweighs the others. Probably some of the high mesas are visited by winter temperatures which are hostile to some of the common plants. There is nothing to indicate that the precipitation increases appreciably with altitude on the small and low mountains of this region. The principal characteristics of the vegetation at higher levels are low stature, simple composition, and absence of several of the large lowland dominants. These features may be due to high wind and shallow soil as much as to any of the conditions commonly attributable to small increases in elevation.

Water supply: In general the streamways of the Vizcaíno Region make little impress upon the vegetation, but springs or places with a constantly generous supply of water are as striking as they are rare. In every case observed, the limited area of moist soil was heavily clothed with plants, including rare perennials of the region and also extralimital plants at the north or south edge of their ranges.

Orientation of slope: Under comparable conditions of substratum there is an almost universal difference in vegetation between slopes facing north and south. In some places the difference is more conspicuous than in others, depending on the dominants involved, but it can always be detected by careful examination. At the higher elevations the difference in composition on opposed slopes is greater, although sometimes not involving any of the large perennials. On hills lying within 4 or 5 km. of the coast there is a marked difference in vegetation between the slopes facing the sea and those facing inland. The windward slopes are usually without any large perennials and may be scantily covered with low perennials or may be so bare of them that the yellow lichens covering the rocks give the entire slope their color, conspicuous for several kilometers. The lee slopes of coastal hills are closely similar to the inland ones.

Northern Vizcaíno Region. In the northwestern corner of the Vizcaíno Region the vegetation is influenced in physiognomy and composition by the chaparral of the northwestern coast, which is here giving way to desert. The writer has described elsewhere the transition between these formations (Shreve, 1936) and the extension of certain chaparral plants southward into the desert. Some of the plants which are dominant a little farther south are just making their appearance, and some of the widespread desert plants are absent, although well represented in the northeastern corner of the area.

The predominant surfaces in the northern Vizcaíno Region are low, maturely worn hills and gently rolling slopes with 10 to 20 cm. of stony soil and a surface accumulation of coarse fragments. On certain types of basaltic rock the soil is a very fine rusty-red clay of greater depth. The nearly level areas usually have a stony loam soil or less commonly a fine sandy loam. The steep hillsides are either bare or covered with large stones supporting pockets of soil.

On the gentle slopes with loam soil the most abundant plant is *Franseria chenopodifolia*. In many places it outnumbers all other plants combined, and its silvery gray foliage blankets the landscape. The plants have a very uniform height of 5 to 7 dm. and a low spread of branches which covers a little less than a square meter in mature plants. *Franseria chenopodifolia* is very abundant in the desert-chaparral transition, but in the Vizcaíno Region it wanes rapidly on approaching the drainage divide of the peninsula. It is a facultative evergreen and distinctly of the desert rather than the chaparral type. A few desert perennials are lower than *Franseria*, but the majority of its associates protrude well above the canopy which it forms. The well closed appearance which it sometimes gives the vegetation would be greatly altered if all the Franserias were removed, for the associates alone would seldom make more than a 15 per cent cover.

Several of the conspicuous and abundant plants of the Vizcaíno Region reach their northern limits in the southern part of the desert-chaparral transition, including *Machaerocereus gummosus*, *Myrtillocactus cochal*, *Idria columnaris*, *Yucca valida*, and *Pachycereus Pringlei*. Throughout the northernmost part of the Vizcaíno Region all these except *Idria* are infrequent or found only in small colonies. *Idria* at once becomes abundant on gentle south slopes, but does not here reach its maximum size.

On the heavy red clay *Franseria chenopodifolia* is very uncommon and *Franseria camphorata* is extremely abundant. The latter is a small plant of 2 to 3 dm. with gray, fernlike leaves. Its presence is a sure indication of extremely fine volcanic clay. It is known outside Baja California only at a single station on this type of soil near the south end of Sierra Picú, in the District of Altar, Sonora.

The characteristic plants of the prevalent volcanic soils in the northern part of the Vizcaíno Region are:

- ****{Franseria chenopodifolia
- Franseria camphorata}
- ***Agave Shawii
- **Viguiera deltoidea
- **Encelia farinosa
- **Fouquieria splendens
- **Simmondsia chinensis
- **Eriogonum fasciculatum
- **Opuntia prolifera
- **Ephedra californica
- *Idria columnaris
- *Yucca valida
- *Pachycereus Pringlei
- *Machaerocereus gummosus
- *Myrtillocactus cochal
- *Larrea tridentata
- *Encelia californica
- *Atriplex polycarpa
- Ferocactus acanthodes
- Opuntia ciribe
- Euphorbia misera
- Agave deserti
- Yucca Whipplei
- Euphorbia tomentulosa
- Salvia Munzii

The broad-leaved sclerophylls which characterize the chaparral have all but disappeared in the northern Vizcaíno Region, being represented only by *Rhus laurina*, *R. integrifolia*, *Rhamnus crocea*, and *Prunus fasciculata*, all of which are confined to moist situations. On steep north slopes are found a few of the large perennials which are abundant in the desert-chaparral transition, but are not of the chaparral type. These include the deciduous *Aesculus Parryi*, *Galvezia juncea*, *Cneoridium dumosum*, and *Isomeris arborea*, and the green-stemmed leafless *Adolphia californica*.

About halfway across the peninsula from the Pacific coast, and also at elevations above 500 m., the aspect of the vegetation begins to change, chiefly because of the reduction in numbers of *Franseria chenopodifolia* and *Agave Shawii*. New shrubs and cacti begin to prevail, along with some of the more xeric dominants, including *Larrea*, *Cercidium microphyllum*, and *Olneya*. *Yucca valida* and *Pachycereus Pringlei* remain infrequent, but *Idria* is abundant on north slopes. No one plant strongly dominates the vegetation. The density of the shrubs 1 m. or more in height ranges from 40 per cent on deep soil to 15 per cent or less on gentle slopes with shallow soil. Except *Pachycereus* and *Idria*, there are no perennials over 4 or 5 m. in height.

A record follows of the vegetation in localities near latitude 30° N. on the thin loam soil of gentle volcanic slopes and plains above 500 m. elevation and near the median line of the peninsula.

**Larrea tridentata
**Agave deserti
**Franseria dumosa
**Encelia farinosa
**Opuntia prolifera
**Atriplex polycarpa
**Viguiera deltoidea
**Fouquieria splendens
**Eriogonum fasciculatum
*Simmondsia chinensis
*Ephedra aspera
*Opuntia ciribe
*Cercidium microphyllum
*Coldenia canescens
*Condalia lycioides
Franseria chenopodifolia
Krameria Grayi
Yucca valida
Lycium brevipes
Condalia spathulata
Ferocactus acanthodes
Yucca schidigera
Eurotia lanata
Calliandra californica
Acacia Greggii
Echinocereus Engelmannii
Fagonia californica

Hills of granite or gneiss occur at a number of localities in the northern Vizcaíno Region, but the total area covered by them and their derived soils is small. The granite plains, with their rounded boulders and light-colored surface of coarse angular sand, afford a background very unlike the prevailing one. Also the vegetation has several distinctive features. The composition is rich and the species are thoroughly intermingled. A high percentage of the plants are large, and many of them have reached the maximum size of their kind. The cover is variable in density and height. Close stands of shrubs, small colonies of *Agave*, or groups of cacti alternate with meandering bare areas of clean, hard sand. *Larrea* here reverses its behavior in other parts of the Sonoran Desert and is more abundant on granite soils than on volcanic. *Pachycormus discolor* and *Viscainoa geniculata* reach their northern limit for the west side of the peninsula on the granite plains. In spite of the evidence that moisture conditions are slightly better on the granite soils, the latter are not colonized by any of the southward-ranging plants of the desert-chaparral transition except *Franseria chenopodifolia*, which is indeed as much a plant of the desert as of the transition.

The composition of the communities on granitic sandy loam is as follows:

***Encelia frutescens
***Franseria chenopodifolia
***Viguiera deltoidea var. tastensis
**Larrea tridentata
**Opuntia ciribe
**Atriplex polycarpa
**Idria columnaris
**Eriogonum fasciculatum
**Agave deserti
**Lophocereus Schottii
*Prosopis juliflora var. Torreyana
*Lycium californicum
*Simmondsia chinensis
*Eurotia lanata
Franseria dumosa
Encelia californica var. asperifolia
Trixis californica
Viscainoa geniculata
Ephedra californica
Bursera Hindsiana
Solanum Hindsianum
Franseria magdalenae

Beloperone californica
Ferocactus viscainensis
Acalypha californica
Pachycormus discolor
Crossosoma Bigelovii var. glaucum
Isomeris arborea
Abutilon Palmeri

The Vizcaíno Region is the only part of the Sonoran Desert in which sarcophyllous plants are abundant. They are represented by several species of *Agave*, *Dudleya*, and *Mesembryanthemum*, the commonest of these being *Agave Shawii* and *A. deserti*. The former is a large plant with leaves 5 to 8 dm. long and inflorescences 2 to 3 m. high. In many localities near the coast this is one of the dominant plants, particularly on deep, somewhat sandy soils. Toward the center of the peninsula the heavy stands become more local, and near the divide it is seldom seen. The high percentage of plants with flower stalks indicates that *Agave Shawii* is short-lived, probably blooming at the age of 15 or 20 years. In every colony there are many dead plants with leaves and flower stalks in the living position, many leaning and fallen ones, and still other rosettes of old leaves in various stages of decay. Throughout the North American Desert the vegetation as a whole has a neat and well kept appearance, due to the rapid decay of the few defunct plants and fallen parts. The heavy stands of *Agave Shawii*, littered with fallen and decaying plants, are the most untidy of all the desert communities. No evidence was found that the fallen carcasses of *Agave* afford favorable spots for germination and early growth of their associates.

*Agave deserti* is a smaller plant than *A. Shawii* and grows either singly or in groups and rows. It never forms pure stands, but is a common plant on all but the finest soils throughout the Vizcaíno Region. *Mesembryanthemum* is limited to the extreme coast, and the several species of *Dudleya* shade from abundance near the coast to infrequency in the interior.

Twenty-three species of *Agave* have been credited to Baja California, many of them poorly known. The ecological prominence of the genus in Baja California is matched only by the extensive stands of *A. lechuguilla* in Coahuila. All the sarcophylls are absent or very local in other parts of the Sonoran Desert. It is difficult to obtain any evidence indicating why the Pacific coast is a favorable locality for this life form. The one feature in which the coastal strip differs from the rest of the Sonoran Desert is the frequent visitation by fog and the many periods of high relative humidity. The zone affected by these conditions is accurately marked in the Vizcaíno Region by the abundance of *Tillandsia recurvata*, which is not so widely distributed, however, as *Agave deserti*. Not enough is known about the requirements and ecological life history of *Agave* to help in the solution of this problem.

South of Cataviñá, high mesas and rolling plateaus stretch eastward toward Cerro Ugarte (1386 m.) and extend with little change as far south as Laguna Seca Chapala. At 500 to 1000 m. this area is characterized by a low, open cover or by very simple stands of shrubs. In the deeper soil of valley bottoms *Larrea* and *Atriplex polycarpa* form a 20 per cent cover with very few associates. On the bajadas and gentle stony slopes these plants are joined by *Franseria chenopodifolia* and *Lycium californicum*, in a slightly heavier stand which includes occasional individuals of the plants listed below. On the thin soil of somewhat steeper slopes the vegetation is low but denser. There are many areas of 100 sq. m. on which there are no plants more than 5 dm. high. The few individuals of *Idria* and *Pachycereus* are small. The trees common at lower elevations are infrequent here, and *Ferocactus* and *Echinocereus* are more abundant than *Opuntia*.

The plants characteristic here are:

Larrea tridentata
Atriplex polycarpa
Franseria chenopodifolia
Lycium californicum
Agave deserti
Viguiera deltoidea
Haplopappus sonoriensis
Fouquieria splendens
Ferocactus acanthodes
Lycium brevipes
Eriogonum fasciculatum
Yucca Whipplei
Ferocactus viscainensis
Echinocereus maritimus
Simmondsia chinensis
Ephedra aspera
Krameria parvifolia

Southern Vizcaíno Region. On traveling south in Baja California, gradual but continuous changes in the plant life are observable. Among the most impressive are the first appearances of *Idria*, *Pachycereus*, *Yucca valida*, and other plants which mark the entrance into the Vizcaíno Region at the north. Equally noticeable are the changes in the character of the plant communities. Plants which at first are rare become common, plants which are confined to definite habitats in the north become ubiquitous, large and well developed individuals of dominant species become more frequent, and the number of localities increases in which there is a diversified stand with members of many life forms.

There is no particular latitude at which the vegetational change is most obvious. The elevation and the character of the soil constantly influence the change, and in many localities there is a sharp return to poor and simple communities, apparently owing to topographic reduction of rainfall.

The broad boundary between the northern and southern parts of the Vizcaíno Region lies on the descent from Laguna Seca Chapala to Punta Prieta. In this distance of 50 km. there are no important floristic changes, but there is a very noticeable enrichment of the vegetation. In the long granitic valley which descends to Punta Prieta from the north, and on many of the higher volcanic plains and mesas of the neighboring region, are to be found some of the outstanding communities of Baja California. Among them are rich assemblages of many of the striking plants which have given the peninsula its reputation as a botanical wonderland.

Here, as in all other parts of the desert, the landscape is dominated by a few large species, which are usually greatly outnumbered by some of their smaller associates. A strong individuality is given the vegetation of the southern Vizcaíno Region by *Idria columnaris*, *Pachycormus discolor*, *Agave Shawii*, *Pachycereus Pringlei*, and *Yucca valida*. Rarely is their combined total more than 20 per cent of the population. Although these five perennials are closely associated, it is uncommon for them to occur in nearly equal abundance in any one restricted area. *Idria* and *Pachycereus* are the most general in occurrence, a fact which indicates that they are the most catholic members of the group in their habitat requirements. Indeed, it is remarkable that five plants exemplifying such dissimilar life forms are so often able to reach their greatest size and finest development under the same conditions of climate and soil.

*Idria* and *Pachycereus* often attain a height of 14 m., and several exceptional examples have been measured which were over 20 m. *Pachycormus* is commonly 5 to 8 m. high and rarely exceeds the latter height. On the immediate coast and in some of its least favorable inland habitats, *Pachycormus* is a prostrate and grotesquely formed tree with a very thick trunk tapering abruptly into normal slender terminal branches. Trees of this type have commonly been selected by travelers to illustrate *Pachycormus* (Goldman, 1916). Over most of its inland range, however, it is an erect and shapely tree. The basal diameter of the trunk is always greatly exaggerated for a tree of such low stature, and it tapers rapidly among the branches. The smooth, light bark contrasts strongly with the dark foliage to make it a highly ornamental tree. In very rainy seasons *Pachycormus* is heavily infested with *Cuscuta Veatchii*, which, with no vestige of a connection with the ground, covers the foliage with its stout threads so thickly that it is conspicuous for a long distance. The finest stands of *Pachycormus* were found near the divide between Punta Prieta and Bahía Los Angeles at 300 to 400 m. elevation on rusty-red volcanic clay. They here form 75 per cent of the stand of large perennials, being accompanied by *Bursera microphylla*, *Fouquieria splendens*, and *Jatropha cuneata*, with infrequent individuals of *Idria* and *Pachycereus* and no *Agave Shawii*.

The heaviest stands of *Idria* were found on high bajadas near Punta Prieta, on deep, stony volcanic clay (pl. 23). The commonest associates of *Idria* in these situations are *Pachycereus*, *Agave Shawii*, and *Yucca valida*. *Pachycormus* is infrequent in the heaviest stands of *Idria*, but is abundant on adjacent north and west slopes, where *Idria* is more poorly represented. In all the communities visually dominated by *Idria*, *Pachycormus*, *Pachycereus*, *Yucca valida*, and *Agave Shawii*, there is a discontinuous cover of shrubs from 5 to 10 dm. high, with cacti and smaller species of *Agave*, the whole forming a cover of 35 to 60 per cent. In general the cover has a uniform density on clay soils and is scattered and aggregated on granitic sandy loam.

The following list combines the records of composition and frequency for three localities within 25 km. of Punta Prieta on plains and bajadas of clay or loam.

***Idria columnaris
***Pachycereus Pringlei
***Agave Shawii
***Franseria magdalenae
***Viguiera deltoidea
***Encelia frutescens
***Agave deserti
***Opuntia clavellina
**Yucca valida
**Pachycormus discolor
**Franseria chenopodifolia
**Jatropha cinerea

**Opuntia ciribe
**Pedilanthus macrocarpus
**Larrea tridentata
*Lycium californicum
*Lophocereus Schottii
*Fouquieria splendens
*Machaerocereus gummosus
*Prosopis juliflora var. Torreyana
*Atriplex polycarpa
*Franseria camphorata
Lemaireocereus Thurberi
Agave sobria
Ferocactus viscainensis
Simmondsia chinensis
Triteleiopsis Palmeri
Bursera microphylla
Echinocereus Brandegei
Viscainoa geniculata

The Vizcaíno plain is the largest area in Baja California in which a nearly uniform vegetation extends for many kilometers in every direction. Along its inner edge, the plain is entered by a few streamways, all of which spread and disappear in a short distance. The behavior of the local drainage is indicated by bands of heavy vegetation separated by broader bands of nearly bare desert pavement at nearly the same level. At 5 to 8 km. from the edge of the plain there are few traces of the drainage from the interior.

The terraces near the beach in the northern part of the Vizcaíno plain have a low vegetation very unlike that of the interior. There are a few widely scattered wind-stunted plants of *Yucca valida*, *Machaerocereus gummosus*, and *Fouquieria peninsularis*. The commonest shrub is *Frankenia Palmeri*, and *Lycium*, *Atriplex*, *Dudleya*, and *Echinocereus maritimus* are abundant.

A few dunes are found in the northernmost part of the plain near the last of the coastal hills. All the dunes seen were very active, and their only plants were in thickets at the base of their lateral slopes. On the dunes there are none of the characteristic desert plants, but many of the dune and beach plants of northern Baja California. Among them are *Rhus integrifolia*, *Sphaeralcea axillaris*, *Helianthus niveus*, *Oenothera crassifolia*, *Hosackia nivea*, and *Abronia maritima*. Level stretches along the beach have a close sod of *Monanthochloe littoralis* or open stands of *Frankenia*, a prostrate *Lycium*, *Sesuvium portulacastrum*, *Allenrolfea occidentalis*, and *Encelia ventorum*.

The dominant plant of the Vizcaíno plain is *Yucca valida*, which forms from 60 to 80 per cent of the number of large perennials throughout the northern part of the plain. The tallest individuals are from 7 to 10 m. high. The manner of growth is such that there is no suggestion of a canopy of foliage among the trees near the observer. When an expanse of *Yucca* forest is seen from a slight elevation, it gives the impression of having an open canopy not unlike that of the dry pine barrens of the southeastern United States. This resemblance is heightened by the dark green of the leaves, a shade which is rare among desert plants. *Yucca valida* grows chiefly in close groups of 5 to 15 plants. It is erect until it reaches about half its mature height, and then begins to lean. Near the coast the trees are inclined away from the wind, but in the interior they lean and bend grotesquely in every direction. Very rarely do the individuals or groups of *Yucca valida* have the symmetrical and balanced form that is so universal in *Y. brevifolia* in the Mojave Desert.

There appear to be few places on the Vizcaíno plain where the heavy stands of *Yucca valida* extend to within 3 or 4 km. of the coast, at which distance all large plants give way to thin stands of low perennials. *Pachycereus* is relatively infrequent on the Vizcaíno plain and seldom produces branches. *Pachycormus*, *Idria*, and *Agave Shawii* are uncommon there. *Machaerocereus gummosus* is widely distributed in Baja California, but here reaches its maximum abundance. *Fouquieria peninsularis* occurs infrequently in the northern Vizcaíno Region, but is a characteristic plant of the Vizcaíno plain.

The following are the principal perennials of the Vizcaíno plain:

***Yucca valida
***Franseria magdalenae
***Lycium californicum
***Encelia frutescens
**Opuntia calmalliana
**Fouquieria peninsularis
**Viguiera deltoidea
**Jatropha cinerea
**Machaerocereus gummosus
*Atriplex sp.
*Pedilanthus macrocarpus
*Viguiera purisimae
*Euphorbia misera
*Opuntia ciribe
*Prosopis juliflora var. Torreyana
Larrea tridentata
Viscainoa geniculata
Lophocereus Schottii
Lycium Fremontii
Bursera Hindsiana
Lippia fastigiata
Pachycereus Pringlei
Triteleiopsis Palmeri
Solanum Hindsianum

The entire Vizcaíno plain lies below 100 m. elevation, and its inner edge is sharply marked by a sudden rise onto volcanic mesas. From the margin of the mesas there is a gradual rise for 10 or 15 km. to the interior plateau, lying slightly above 500 m. A strong contrast between the vegetation of the plain (pl. 26) and that of the mesas is observable from Pozo Alemán to the vicinity of San Ignacio. On the mesas there are very few large perennials, but there is a continuous stand of shrubs varying greatly in density and showing a tendency to occur in groups. *Yucca valida* is still present, but greatly reduced in size and abundance. *Pachycereus* and *Fouquieria peninsularis* are more abundant than on the plains. *Agave Shawii* and *A. deserti* are absent, but *A. sobria* is infrequently seen. *Idria* and *Pachycormus* are no longer found. Conditions more xeric than those of the plain are indicated by the occurrence of *Larrea*, *Jatropha cuneata*, *Bursera microphylla*, and *Cercidium sonorae*, and by the abundance of *Opuntia ciribe* and *O. cholla*. The coverage varies greatly, being as low as 10 per cent on very shallow soil and locally as high as 60 per cent where the shrubs are closely grouped.

The vegetation of the volcanic mesas east of the Vizcaíno plain may be summarized as follows:

**Jatropha cuneata
**Franseria magdalenae
**Opuntia cholla
**Lycium californicum
**Fouquieria peninsularis
*Encelia farinosa
*Jatropha cinerea
*Opuntia ciribe
*Machaerocereus gummosus
*Larrea tridentata

*Encelia californica
*Franseria Bryantii
Viguiera deltoidea
Atriplex polycarpa
Agave sobria
Bursera microphylla
Yucca valida
Echinocereus Brandegei
Pedilanthus macrocarpus
Opuntia invicta
Bursera Hindsiana
Pithecellobium confine
Mammillaria sp.
Cordia parvifolia
Bebbia juncea
Asclepias subulata
Hibiscus denudatus
Petalonyx linearis
Stillingia linearifolia

### *Magdalena Region*

(Arbocrassicaulescent Desert. Lysiloma-Machaerocereus Region)

The Magdalena Region is the southernmost part of the Sonoran Desert, extending 3° of longitude (350 km.) farther south than the mainland part. The eastern boundary is formed by a more continuous series of higher mountains than is the case in the Vizcaíno Region. The northern half of the region, between San Ignacio and Comondú, is an intricate field of volcanic hills and mesas, with a narrow, sandy coastal plain. Much of the area is occupied by long tongues of volcanic eruptives falling at a uniform gradient from the central series of extinct cones. The volcanic tongues are uniformly nearly level in transverse section, with rocky surface and shallow soil, and are steep on the sides. Between the tongues are narrow barrancas with soil and local seepages of water. Comondú, San Miguel, and Purísima are located in the bottoms of large valleys of this type, in which a constant water supply is available.

South of Comondú the coastal plain widens gradually into the Magdalena Plain, which occupies half the width of the peninsula at latitude 25° N. and then tapers southward to the latitude of La Paz (24° 10′ N.). Offshore at the broadest part of the Magdalena Plain are the islands Magdalena and Santa Margarita, and Cabo San Lazaro, at the end of a long barrier island. These three islands are occupied by low hills, but there are no hills on the Magdalena Plain. It is obvious that the plain has been built by the deposition of outwash material from the Sierra de la Giganta, and the presence of the hilly islands suggests that deposition began in a shallow bay. Innumerable streamways drain the west face of Sierra de la Giganta, and, according to the most detailed maps of Baja California, all gather into four large arroyos which debouch into the coastal lagoons. Examination of the plain after the very rainy winter of 1935 showed clear evidence that part of the mountain drainage reaches playas from which it does not escape. At that time all the large playas and many of the small ones were shallow lakes, and some of them were so thickly covered with plants that they resembled meadows.

Not only are large playas common, but very small ones (*playitas*) are extremely abundant. Away from sandy soil, a single hectare commonly has from twenty to fifty small bare depressions from 5 to 15 m. in diameter (see pl. 29). Their presence explains the lack of small drainageways for local runoff on the floor of the plain. North of Venancio and the Arroyo Flor de Malva there are no small stones on the soil surface in any situations, and even in the largest arroyos stones are few. The prevailing color of the soil is light gray. At the north end of the plain there are several large areas of very light gray volcanic ash. Along the coastal margin of the plain there is much sandy soil, and there are dunes of moderate size at several localities.

In the inland settlements of the Magdalena Region the residents state that there are occasional frosts as far south as Purísima, but only very light and infrequent ones farther south. The mean annual rainfall is 202 mm. at La Paz and is probably of similar amount throughout the lowlands of southern Baja California. On the Pacific coast the rains are almost wholly in the late winter. On the Gulf coast and throughout the interior there are summer thunderstorms as well as gentle winter rains. Recent periods of 3 to 5 years without rain were reported at several localities in 1935. On the Magdalena Plain there is some morning fog in all months, but it is most persistent in April and May. At Comondú and San Miguel there is no fog from the Pacific, but it sometimes spreads inland from the Gulf of California.

NORTHERN MAGDALENA REGION. North of Comondú, volcanic mesas and slopes form the prevailing surface. There are extensive malpais fields with scattered pockets of soil, and also areas of thin but continuous soil strewn with small stones. The surface has undergone little change from erosion, but has many structural rifts and barrancas. The vegetation of the malpais and the thin soil of the mesas is open, irregular, and xeric, closely resembling that of the area east of the Vizcaíno plain. Some of the cliffs and barranca walls apparently receive no seepage of water and are thinly covered or naked. Other steep slopes which have soil, seepage, and northern aspect are heavily clothed with communities in which many of the thorn forest species of the Cape Region make their first appearance and mingle with desert plants. In the largest barrancas the streamways and small shelves of alluvial soil support a heavy mesic vegetation in which there are few desert plants.

South of Purísima and around the north end of the Magdalena Plain the rainfall is evidently very low. The vegetation here is extremely open and almost wholly made up of species characteristic of the very desert areas of Baja California and Sonora. In the most xeric habitats of the northern Magdalena Region there are only a few prominent plants which do not extend north of San Ignacio. The most favorable habitats are not dominated by any one plant, but by an assemblage which is largely made up of plants near their northern limits.

The vegetation of the volcanic mesas varies in its height and density. On the very rough malpais, plants are very irregularly distributed and vary from 1 to 4 m. in height. Rough malpais is very difficult to traverse, as the only places where the footing is good are densely covered with shrubs. On uniform surfaces the spacing is more even, but the stature is similar.

The landscapes of the Magdalena Region are less striking than those of the Vizcaíno Region. This difference is due primarily to the absence of *Idria* and *Pachycormus* and the infrequency and poor development of *Pachycereus* and all the species of *Yucca* and *Agave*. Also there is a greater uniformity in the life forms, with microphyllous trees and shrubs taking the dominant place. On the volcanic mesas cacti are relatively infrequent, with *Lophocereus Schottii*, *Machaerocereus gummosus*, and other tall forms very infrequent. The cylindropuntias are very irregular in occurrence, forming open colonies in one spot and being absent over large areas. The platyopuntias, which are very uncommon in the Vizcaíno Region, are abundant in many parts of the Magdalena Region.

In the hills west of San Miguel, where there is little malpais, the vegetation resembles that of the Gulf coast of Sonora more closely than does that of any other place in Baja California. The coverage is rarely above 5 to 10 per cent. The similarity is due to several dominants which these areas have in common, and to the close resemblance between some of the specifically distinct dominants. In the following list of characteristic species the ones also found in Sonora are marked "S," and after those with close analogues in Sonora the name of the Sonoran plant is given in parentheses.

| | |
|---|---|
| ***Larrea tridentata | S |
| ***Jatropha cuneata | S |
| ***Opuntia cholla | (Opuntia fulgida) |
| ***Atriplex polycarpa | S |
| **Jatropha cinerea | S |
| **Lophocereus Schottii | S |
| **Machaerocereus gummosus | (Rathbunia alamosensis) |
| *Bursera microphylla | S |
| *Fouquieria peninsularis | (Fouquieria splendens) |
| *Lycium brevipes | (Lycium Fremontii) |
| *Pachycereus Pringlei | (Carnegiea gigantea) |
| Lemaireocereus Thurberi | S |
| Acacia Brandegeana | (Acacia occidentalis) |
| Cercidium sonorae | S |
| Ditaxis lanceolata | S |
| Caesalpinia arenosa | (Caesalpinia pumila) |
| Fagonia californica | S |
| Encelia farinosa | S |
| Krameria paucifolia | (Krameria Grayi) |

The barrancas which drain the volcanic mesas have sharp crests, and it is rarely that they are joined by tributary barrancas or receive much lateral drainage. The upper walls of the barrancas are steep, and the only tree suited to gain foothold there is the semiscandent *Ficus Palmeri*. When the descent from the crest of the barranca is more gradual, there is a heavy stand of shrubs with scattered trees and columnar cacti (pl. 28). Such vegetation is very uneven in the distribution of its components. Small colonies may be found of shrubs which are uncommon elsewhere. The Opuntias and smaller cacti are sporadic in occurrence. Trees are very widely spaced except where local conditions are specially favorable. In the rich composition and irregular distribution of their vegetation the barrancas are in sharp contrast with the level mesa above, and bear a strong resemblance to similar situations in the Foothills of Sonora.

In the following list are mentioned only the most common of the many species found in the barrancas, with no attempt to indicate their relative abundance. The first seven are the most conspicuous ones.

| | |
|---|---|
| Pachycereus Pringlei | Franseria ambrosioides |
| Fouquieria peninsularis | Ruellia californica |
| Lemaireocereus Thurberi | Jatropha cinerea |
| Bursera odorata | Opuntia alcahes |
| Lysiloma candida | Brickellia Coulteri |
| Prosopis Palmeri | Schoepfia californica |
| Opuntia comonduensis | Randia megacarpa |
| Viguiera deltoidea | Carlowrightia californica |
| Acacia Brandegeana | Viguiera tomentosa |
| Colubrina glabra | Opuntia sp. (platyopuntia) |
| Berginia virgata | Pithecellobium confine |

The large barrancas, or valleys, differ from the smaller ones in having much more xeric conditions on their slopes and much more favorable moisture conditions along the floor. Dense jungles of trees and large shrubs occupy all the uncultivated areas in the valley of Comondú. The number of small perennials is large, and palustrine and aquatic plants find a few favorable habitats. The dominant trees are *Cercidium peninsulare*, *Prosopis juliflora* var. *Torreyana*, *P. Palmeri*, *Lysiloma candida*, and *Parkinsonia aculeata*, above which rise groups of *Washingtonia robusta*. The shrubbery is chiefly *Tecoma stans*, *Viguiera tomentosa*, *Vallesia glabra*, *Alvordia angusta*, *Euphorbia Xanti*, and *Berginia Palmeri*. Farther north in Baja California and Sonora the well watered habitats of this character are not without a few xeric plants, but here there are none.

The coastal plain west of Purísima and Comondú and north of the Magdalena Plain has a light brownish gray sandy soil, with more open and xeric vegetation than any occurring farther south on the Pacific coast. Just inside the coastal lagoons there is a belt of active dunes from 2 to 3 km. in width. On the sandy plains which extend inland from the dunes the coverage is about 5 to 10 per cent. Sparingly branched plants of *Pachycereus* are the most conspicuous element, and trees are very widely spaced. This simple vegetation is composed as follows:

| | |
|---|---|
| ***Opuntia cholla | *Euphorbia magdalenae |
| ***Franseria magdalenae | *Larrea tridentata |
| ***Encelia farinosa | Machaerocereus gummosus |
| **Pachycereus Pringlei | Opuntia tesajo |
| **Jatropha cinerea | Solanum Hindsianum |
| **Fouquieria peninsularis | Bursera Hindsiana |

Two unusual plants are confined to this immediate vicinity, *Euphorbia magdalenae* and *Machaerocereus eruca*. The former is a very compact shrub with slender green stems bearing a few short-lived leaves on the youngest shoots. It resembles *Euphorbia colletioides* of Sonora in every respect except its dense habit of growth. *Machaerocereus eruca* is in effect a stout columnar cactus which has adopted a prostrate habit of growth. The trunks are rooted throughout their length, and the oldest parts gradually die as the plant slowly advances in the usually straight course determined by the growing tip. The slightly elevated position of the youngest 20 or 30 cm. of the stem has led Goldman to liken the colonies of *Machaerocereus* to groups of huge catepillars (for illustration see Goldman, 1916, pl. 127).

In several small areas on the inner edge of the northern end of the Magdalena Plain the soil is a light gray, impalpably fine volcanic ash. Nothing is known about the behavior of such a surface under light or heavy rain, nor about the infiltration or retention of water. Much of the surface is bare of perennials. When seen in March, after heavy rains, most of the ash was bare, but parts of it were covered with *Plantago*. The only perennials found on it were *Franseria magdalenae*, *Encelia farinosa*, and *Opuntia cholla*.

Southern Magdalena Region. The Magdalena Plain from San Domingo to Venancio, a distance of 180 km., is very uniform in its physical features and monotonous in its vegetation (pl. 29). There are no changes in level until the gradual rise toward Sierra de la Giganta is reached, about halfway across the peninsula. There are very few arroyos and no small streamways. The only interruptions to the uniformity of the vegetation are due to the innumerable playitas and the frequent large playas, some of which are meadows or lakes in wet seasons.

The profile of the vegetation is irregular, being formed by either *Prosopis* or *Lycium* and by the projecting tops of *Pachycereus*. There are no tall plants of this cactus, but many massive ones which have branched just above the ground. Other cacti, *Lemaireocereus*, *Lophocereus*, *Machaerocereus gummosus*, and *Opuntia cholla*, are very abundant, but are outnumbered by the microphyllous trees and shrubs. The absence of perennials from the playitas does much to make the profile of the vegetation irregular. The coverage varies from about 40 to 80 or 90 per cent, but the more open type covers the greatest area. The minimal area is small, as there are few large perennials and these are of infrequent occurrence.

The following conspectus of the vegetation is based on four nearly identical determinations between Médano Blanco and Venancio:

***Prosopis juliflora var. Torreyana
***Lycium brevipes
***Opuntia cholla
**Fouquieria peninsularis
**Machaerocereus gummosus
**Pachycereus Pringlei
**Euphorbia californica
*Lemaireocereus Thurberi
*Franseria Bryantii
*Lycium Andersonii
*Lophocereus Schottii
*Opuntia tesajo
*Prosopis Palmeri
*Franseria magdalenae

Cercidium floridum
Jatropha cinerea
Hyptis laniflora
Maytenus phyllanthoides
Encelia farinosa

Three small arroyos were observed in the Magdalena Plain which were bordered by narrow strips of recently dissected ground. In these situations several xeric perennials were found which are uncommon elsewhere on the Plain, including *Koeberlinia spinosa*, *Olneya tesota*, *Jatropha cuneata*, *Bursera odorata*, and *Simmondsia chinensis*. In reverse of the usual case, the margins of these arroyos appear to be drier than the surrounding surface, doubtless owing to desiccation of the soil from the dissected surfaces.

South of El Refugio there is a large area in which the vegetation is more xeric than elsewhere. *Prosopis* and *Lycium* are infrequent, and the dominant plants are *Larrea tridentata*, *Opuntia cholla*, *Fouquieria peninsularis*, *Franseria magdalenae*, *Jatropha cuneata*, and *Bursera microphylla*, accompanied by other plants in the list above. Small areas of volcanic ash in this region are covered by open, nearly pure stands of *Larrea*.

Between the sandy region around Médano Blanco on the north and the Arroyo Flor de Malva on the south there are many large playas and seasonal meadows. In some cases the floor of the playa is bare and its margin covered with herbaceous perennials and annuals. In other cases the entire playa is heavily covered with grasses, sedges, and herbaceous plants. The composition in the seasonal meadows is rich and varies from playa to playa. The genera represented are those common to moist soils throughout Baja California, Sonora, and Sinaloa, but many of the species are endemic to southern Baja California.

A few of the commonest plants observed in a meadow 9 km. south of Querétaro are:

Leptochloa viscida
Eragrostis cilianensis
Eragrostis diffusa
Cyperus aristatus
Mecardonia exilis
Vicia exigua
Lythrum Bryantii
Verbena bajacalifornica
Eryngium nasturtiifolium
Castilleja Bryantii
Coreocarpus parthenioides
Rumex conglomeratus
Marsilea Fournieri

South of the Arroyo Flor de Malva the coastal plain narrows to a width of a few kilometers, and gentle rocky slopes run down nearly to the sea. The foundation of this area is limestone, as shown in cliffs 30 m. high along the left bank of the arroyo. Volcanic intrusives also occur and contribute coarse angular material, with which are interspersed many waterworn pebbles.

The vegetation of the southernmost tip of the Magdalena Region is dense, low, and poor in large perennials. *Yucca valida* is prominent in some places, particularly in depressions, but is only 2 to 3 m. in height. *Pachycereus* is rare. *Fouquieria peninsularis* is abundant but rarely more than 2.5 m. high. The prevailing shrubs are *Lycium*, *Encelia*, *Simmondsia*, and *Atamisquea*. Near the coast a uniformly gray color is given the landscape by the abundance of the lichen *Rocella*, which covers every branch and twig.

South of Arroyo Flor de Malva there are no coastal lagoons, and a narrow belt of sand dunes lies just inside the beach. Much of the dune surface is very active and bare of plants. Stabilization appears to proceed rapidly, however, when once begun, and to be aided by an unusually large number of species. The mature slopes are covered by dense thickets of *Maytenus phyllanthoides*, *Stegnosperma halimifolium*, *Caesalpinia arenosa*, *Simmondsia chinensis*, and *Lycium brevipes*.

To travel east from the inner edge of the Magdalena Plain toward the slopes of Sierra de la Giganta is to pass rapidly from desert into the Cape thorn forest. The coverage in the lower foothills is 60 to 80 per cent, which is about the same that is found in the sandy lowland parts of the Cape Region. Certain spots are more open and others are very dense. Small trees become rapidly more abundant, and so many of the shrubs are over 2 m. high that it is not possible to see very far through the vegetation. Cacti, yuccas, and *Fouquieria* are still abundant, and the trees are nearly all microphyllous deciduous types. It is manifest that the environment is still an arid one, but one in which the aridity is greatly ameliorated. The average height of the trees is about 5 to 6 m., but the canopy is irregular. There is no layering, and the open branching and sparse foliage of the trees allow strong light to reach the floor of the forest.

The following list of perennials characteristic of the inner edge of the Magdalena Region is based on examination of it at the northern and southern ends. Though many of the same species are found at these extremities, there is some difference in their relative abundance, and a few are found only at the southern end.

Jatropha cinerea
Lysiloma divaricata
Fouquieria peninsularis
Bursera microphylla
Prosopis Palmeri
Randia Thurberi
Lophocereus Schottii
Cyrtocarpa edulis
Bourreria sonorae
Opuntia cholla
Tecoma stans
Bursera odorata
Euphorbia Xanti
Esenbeckia flava
Haematoxylon brasiletto
Yucca valida
Cercidium peninsulare
Forchammeria Watsoni
Melochia pyramidata
Pithecellobium confine
Encelia farinosa
Hyptis laniflora
Larrea tridentata
Lemaireocereus Thurberi
Bursera filicifolia
Ruellia peninsularis
Mimosa laxiflora
Colubrina glabra

# 4

# Ephemeral Herbaceous Vegetation

IN THE SKETCH of the vegetation of the Sonoran Desert in the preceding chapter, attention was confined to perennial plants. The short-lived herbaceous plants, however, form about half the flora and also play a conspicuous role in the vegetation in their appropriate seasons. For several reasons it has seemed better to give separate treatment to the herbs, which are here designated "ephemerals" rather than "annuals" because their life span rarely fills the entire growing season as measured by frostless temperatures.

The desert environment offers particular advantages to the short-lived plant. The restriction of its vegetative activity to brief moist periods almost completely relieves the ephemeral plant from the difficulties of water supply which beset the perennial. By speed of germination, rapidity of growth, and early flowering and maturity during favorable periods the ephemeral escapes the most potent one of the conditions controlling the life and ecological behavior of desert plants. Also the large expanses of bare ground assure ephemerals ample space and little competition with established plants.

The appearance of ephemerals follows periods of moderate or heavy rain, whether this occurs regularly in the cool months or the hot months or at rare and sporadic intervals. There is a minimum of rainfall below which the ephemerals do not appear, but there is no maximum which is deterrent to them. It follows that the geographic distribution of the species of ephemerals is determined by the seasonal incidence of the moist periods. It is scarcely affected by the annual or seasonal totals of rain, by the length and severity of drought periods, by the high maximum temperatures of summer, or by the freezing temperatures of winter. The environment is much more simple for the ephemeral than it is for the perennial. This is shown by the fact that the ephemeral vegetation of the Sonoran Desert falls into a smaller number of regional categories than does the perennial vegetation.

The most favorable local habitats for ephemerals are merely those affording an abundant and briefly sustained moisture supply. The character of the soil is of little importance aside from its penetrability and retentiveness. The depth, structure, and mineralogical origin of soils, all so important to perennials, mean little to the ephemeral. Sand, however, is a particularly favorable soil for the ephemerals of the cool season, because of the deep infiltration of moisture and the rapid warming of the surface early in the season and early in the day. In any part of the Sonoran Desert it would be difficult to make a local differentiation of habitats solely on the basis of the short-lived herbaceous plants. Their local distribution is determined by the moisture of the soil, the character of the soil surface, and the shade of perennials. The character of the surface is important in several respects. On sand the wind-blown seeds are covered, held, and left at a favorable depth for germination. In general the crop of ephemerals is much more evenly distributed over a sandy soil than over any other. On a hard, smooth surface with little vegetation there is rarely a thick crop of ephemerals, as their seeds are either washed away or blown away. A slight depression in such a surface is usually thickly colonized. On undisturbed desert pavement the chances for lodgment of seeds are also poor, and few germinations take place there (see pl. 1). On a surface with small projecting stones there is a light or moderately heavy crop. On slopes covered with large stones the ephemerals are abundant but localized. In the shade of perennial plants the ephemerals are larger and more numerous than in the open. Under low, spreading shrubs the crop is larger than under trees. The larger numbers are due to the arresting of wind-borne seeds by the shrub and its dead undergrowth; the larger size, to conservation of moisture by the shade of the shrub and to the shade effect on the ephemeral (see pl. 30).

In slight depressions or on benches where conditions have been favorable for the accumulation of seeds at the end of the active season, the number of seedling ephemerals may be very great. On a slightly shaded level area of deep, moist clay near Tucson, 1200 seedlings of *Plantago* were counted on a square meter. It is characteristic of such colonies that they are composed of a single species or have one species greatly predominant. This is probably due to the depositing of the seeds by the runoff of a rain which happened to occur just when the seed crop of the predominant species was ripe for distribution. In such heavy stands competition is great, and the vast majority of the seedlings perish. Dense stands of ephemerals are commonest on flood plains and in other situations with level alluvial soil, a fact which points to flood water as the agent in their distribution. Such stands are particularly characteristic of *Monolepis Nuttalliana*, *Pectis papposa*, *Plantago insularis*, and *Amaranthus Palmeri*. On gentle slopes and other surfaces covered with coarse lag gravel the ephemerals are more widely spaced and numerous species more evenly intermingled.

A feature which lends particular interest to the ephemeral plants of the Sonoran Desert is the existence of two distinct groups of species which are respectively confined in their appearance to the warm and the cold rainy periods. The seasonal distribution of rainfall is controlled by broad climatic conditions and appears to have undergone little change in this region, at least since early Pliocene time, whatever fluctuations may have taken place in the amount.

The Pacific coast of Baja California, like that of California, is visited by rain only in the winter and early spring. The winter rains extend eastward across Arizona and Sonora, diminishing in amount. Summer rains are general along the eastern edge of the Sonoran Desert, diminishing toward southern Baja California, northwestern Sonora, and western Arizona, and only infrequently visiting southeastern California. Because of the overlapping of the winter and summer rainfall areas, a wide belt of the Sonoran Desert receives biseasonal precipitation. In the vicinity of Tucson the average amounts of summer and winter rain are nearly equal. South of the international boundary, the winter rain diminishes in amount and in percentage of the annual total. These features of rainfall distribution are critical in controlling the ranges of the two groups of ephemerals.

The fact has long been known, and confirmed by several workers, that the restriction of the two groups of ephemerals to their respective seasons is dependent on the optimum temperature for germination of their seeds. For winter ephemerals the optimum is between 60° and 65° F., and for summer ephemerals between 80° and 90° F. The optimum differs slightly for the different species that have been tested in each group, and for nearly all species in either group germination is fairly prompt and complete within a range of about 5° to 8° above or below the optimum.

The narrow margin of 5° to 10° between the highest possible germination temperature for winter ephemerals and the lowest for summer ephemerals is sufficient for the sharp seasonal separation of the two groups, with certain exceptions to be noted later. In winter the seeds of summer ephemerals remain dormant under favorable conditions of soil moisture, and likewise in summer the seeds of winter ephemerals lie in a moist soil without germinating.

Little is known about the length of the period for which viability is retained by the seeds of desert ephemerals. It has been observed that an exceptionally favorable season will bring forth large crops of species that have been uncommon for ten to fifteen years. Also the relative abundance from year to year of the regularly recurring species does not bear a constant relation to the extent of the population and seed crop of the immediately preceding two or three years.

With their vegetative activity confined to a moist period, the ephemerals are chiefly mesomorphic in foliage. There are great differences among them, however, in habit, in normal limits of height, in manner of branching, and in the size, thickness, surface, and manner of display of the leaves.

Certain ephemerals have a narrow range of variation in height, as *Cryptantha*, *Pectocarya*, *Pectis*, and many species of *Eriogonum* and *Gilia*, whereas others vary from a few centimeters to 2 or 3 m. in height, as *Sphaeralcea Coulteri* and *Amaranthus Palmeri*. Height and bulk are chiefly determined by the amount of soil moisture available during the growing period, and, in the case of winter ephemerals, by the number of warm and sunny days following the rainy periods. Individuals of the same species may come to maturity at a height of 1 cm. with a single flower or may grow to a height of 2 or 3 dm., bearing hundreds of flowers. Such divergence of behavior is due to the amount of precipitation. A recurrence of rain after a short dry period frequently causes a renewal of growth in small plants and may result in a second crop of flowers. The duration of life of the ephemeral is commonly limited to 6 or 8 weeks, but may be prolonged by extraseasonal rains. A few quasi-perennial species germinate in the winter and, with unusually late spring rains, may persist into the summer.

Economy of material in development of stems and exposure of leaf surface has obviously been a factor in the development of the ephemerals. In spite of their lives' being brief and led in the most favorable season, they are nevertheless, desert plants, subjected to strong insolation and dependent on a short-lived and precarious water supply.

Very few species have stems which are thick or provided with a strong development of mechanical tissue. On the other hand, there are numerous prostrate forms, as *Hosackia*, *Pectis*, *Achyronychia*, *Tidestromia*, *Calyptridium*, *Bowlesia*, and many species of *Euphorbia*. In an even larger number of genera there are many semiprocumbent plants, in which the lowest third or half of the stem rests on the ground, as *Nama*, *Gilia*, *Cryptantha*, *Monoptilon*, and *Eriophyllum*. In the erect species the stem is usually slender and in some cases very slender, as in *Nemacladus*, *Eremocarya micrantha*, *Parietaria*, *Greenella*, and many others. The stoutest stems are found in certain species of *Oenothera* and *Mentzelia* and in *Dicoria*, *Lupinus*, *Streptanthus*, *Cirsium*, *Martynia*, and *Helianthus*. A curious type of greatly inflated hollow stem is found in *Eriogonum inflatum;* in the Mojave Desert it occurs in *Caulanthus inflatus* and is weakly developed in *Oenothera decorticans* var. *condensata*. The most ingenious proponents of adaptational suggestions have been unable to ascribe any survival value to the inflated stems, and there is no obvious physiological explanation for their development.

A very common type of ephemeral is one in which a small rosette of prostrate or nearly prostrate leaves is formed, from which one or more shoots arise. The shoots may bear a few leaves, diminishing in size toward the top, may bear small bracts, or may be naked. The formation of basal rosettes is characteristic of many species in *Oenothera*, *Eriogonum*, *Chorizanthe*, *Chaenactis*, *Draba*, and *Malacothrix*, and in *Salvia columbariae* and others. The rosette of leaves usually dies before the seeds mature. The chlorophyll-bearing tissue of the stems and bracts is capable of making some contribution toward the filling of the seeds. In several species of *Eriogonum*, notably *E. deflexum* and *E. polycladon*, the stems are richly branched and remain alive for several weeks after all other ephemerals have died. In these plants the seed crop is much larger than in most of the other ephemerals.

Some of the most aggressive and drought-resistant herbs remain alive and more or less active after the majority of ephemerals have matured and died. The line of demarcation between ephemerals and herbaceous perennials is not sharp, and, indeed, not constant in the same species. Some of the perennating species are very dependent on extraseasonal rains for their survival. Others, like the species of *Eriogonum* mentioned above, lose their rosette of mesic leaves and persist as leafless green-stemmed plants, suggesting miniature trees of *Canotia* or *Holacantha*. Among the summer Compositae there are several species that germinate and grow quickly but bloom less promptly than other ephemerals and often persist to the second year (*Verbesina encelioides*, *Xanthium pennsylvanicum*, *Helianthus petiolaris*).

The disturbed conditions and supplementary water supply on roadsides, canal banks, and fields give an opportunity for perennation to several native and introduced plants that are strictly ephemeral in natural situations.

A few ephemerals may be regarded as recent derivatives from perennial races, in which continued subjection to long drought periods has fixed the ephemeral habit. This suggestion may apply to *Cassia Covesii* (still often truly perennial), *Atriplex elegans*, *A. Wrightii*, and *Haplopappus gracilis*.

Very few ephemerals have leaves larger than 4 sq. cm. (*Amsinckia tessellata*, *Datura*), and in the great majority the leaves are less than 1 sq. cm. in area. Many anatomical features found in the leaves of desert perennials are regarded, with more or less evidence, as concerned in protection from excessive insolation or water loss. The leaves of ephemerals are fully as diversified as those of perennials, including a highly mesic type (*Parietaria floridana*, *Bowlesia incana*), a less mesic type (*Phacelia crenulata*, *Boerhaavia erecta*), lightly cutinized ones (*Hosackia brachycarpa*, *Eremalche rotundifolia*), and still others with heavy epidermis and cuticle (*Perezia nana*, *Chorizanthe rigida*). Entire or slightly cut leaves are rare outside the most mesic types. In nearly all leaves with an area of 1 sq. cm. or more, the blade is divided or deeply dissected (*Phacelia* spp., *Lepidium* spp., *Astragalus* spp.). In the much larger group of ephemerals with leaves less than 1 sq. cm. in area, there are many with narrow entire leaves (*Plantago*, *Platystemon*), or with leaves which are very short as well as narrow (*Pectis* spp., *Pectocarya* spp.).

A long series of ephemerals might be cited in which the leaves have a gray color due to air-filled epidermal cells (*Tidestromia*, *Dicoria*), close pubescence (*Geraea canescens*, *Stylocline micropoides*), or long wool (*Eriophyllum lanosum*, *Trichoptilium incisum*). Succulent leaves are found in a few species (*Portulaca pilosa*), and slightly succulent ones in others (*Calandrinia ciliata*, *Tillaea erecta*). Leaves beset with oil glands are found in a number of ephemerals, especially in the Compositae (*Dyssodia concinna*, *Pectis papposa*).

A striking development of the ephemeral has given rise to a series of plants in which the stem is erect or semierect but very slender, much branched, and either leafless or sparsely covered with very small leaves or bracts. These forms, which may be appropriately called "shadowless plants," expend the minimum amount of material and expose a very small photosynthetic and transpirational surface. Their seeds are very small and their seedlings poorly supplied with parental material for their earliest growth. Common examples of the shadowless plants are *Nemacladus ramosissimus*, *Euphorbia revoluta*, *Mollugo cerviana*, *Linanthus aureus*, *L. dianthiflorus*, and *Houstonia asperuloides*. Further examples are supplied by several species of *Eriogonum* in the mature phase of their development.

## Winter Ephemerals

### *Geographic Distribution of Winter Ephemerals*

In southern California the native herbaceous plants are perforce limited to the winter and early spring, since the summer is rainless. The number of "annuals" credited to this area in Munz's *Flora of southern California* is 965, including 765 species and 200 varieties and forms, and excluding a number of species which sometimes persist beyond their first year. Of this number, 358 species and 111 varieties and forms are restricted to the coast, and 407 species and 89 varieties and forms are found in the Mojave Desert and in the arm of the Sonoran Desert which extends into southeastern California and is often ambiguously called the "Colorado Desert."

The Mojave Desert is particularly rich in ephemerals, probably having 250 species and varieties. Of this number more than one-third are confined to the Mojave Desert, while others extend north into the Great Basin or are found also in the Sonoran Desert. The greatest number of species and the greatest abundance of individuals are found along the elevated southern and western edge of the Mojave Desert. To the north and east of this belt there is a reduction in both species and numbers.

On the Pacific coast of Baja California the ephemerals are also restricted to the winter. In the chaparral of northwestern Baja California the ephemerals correspond closely to those of coastal southern California in both floristic and ecological respects. South of the latitude of Rosario the relation to the northern coastal region becomes much less. The Pacific coast has obviously been the center of development of the ephemerals which respond to a cool, moist period. If the ephemerals of the rich flora of the Mojave Desert are followed eastward, their numbers are found to diminish rapidly through the Colorado Valley and the Arizona Upland. If they are followed south along the Gulf coast, they are found again to diminish on both sides of the Gulf. Both to the east and to the south a few species are added to the group which are not found in the Mojave Desert.

### *Winter Ephemerals of Baja California*

In the desert parts of Baja California there is a rich display of ephemerals on suitable surfaces for several weeks following heavy winter rains. In the very rainy spring of 1935 the most favorable localities were heavily covered with a flourishing growth of ephemerals, from which was obtained most of the information about them in the following paragraphs. In the dry spring of 1934 scarcely 100 individuals of this type of plant were seen between Rosario and Bahía Concepción. The rainfall records indicate that there must be from four to seven such years in every decade.

The relative abundance of the common ephemerals varies greatly. Nearly pure stands of a single species are often seen, but are not usually more than 10 to 15 m. in diameter. Stands in which from 3 to 5 species dominate are more common and larger. Stands with 10 to 15 species are common on the heaviest soils in well watered locations. In 1935 the inner edge of the Vizcaíno plain was covered with a stand of 15 to 20 species so heavily that little bare ground was left exposed (pl. 26). The genera most abundantly represented are *Plantago*, *Cryptantha*, *Perityle*, *Eriogonum*, *Tillaea*, *Lepidium*, *Oenothera*, *Hosackia*, and *Phacelia*. Grasses are conspicuously infrequent and limited to open representation of *Festuca octoflora* on sandy soil and somewhat more abundant occurrence of *Muhlenbergia microsperma* on heavier soils.

The ephemeral flora is like the perennial in changing gradually from north to south, and in showing differences between the northeastern coast and the central and Pacific coast sections, a strong relationship to the Lower Colorado Valley region, and some endemism. The coastal ephemerals of southern California are poorly represented south of the desert-chaparral transition. A very few summer ephemerals of Arizona and Sonora have been found in the winter in central and southern Baja California (*Pectis papposa*, *Bouteloua aristidoides*, *Physalis pubescens*, *Parthenice mollis*). In southern Baja California, on the Gulf coast, the dead remains of unidentifiable species of *Boerhaavia* were found in the spring, indicating the existence of at least a few summer ephemerals where there are occasional summer rains.

The total number of ephemerals native to Baja California is probably between 400 and 500 species, or slightly greater than the number in the desert areas of California. The following list of 114 species includes those that were most abundant or most widely distributed in the spring of 1935. In order to show something of regional differences, the species seen only on the northern Gulf coast are indicated by (G) and those seen only south of San Ignacio by (S). The species which also occur on the coast or desert of California, or the Sonoran Desert in Arizona or Sonora, are so designated. Names followed by "BC" indicate species which are endemic to Baja California so far as known.

| | |
|---|---|
| Festuca octoflora | Cal. coast and desert; Ariz. |
| Bouteloua annua (S) | BC |
| Muhlenbergia microsperma | Cal. coast and desert; Ariz. |
| Bouteloua aristidoides | Cal. desert; Ariz. |
| Cenchrus Palmeri | BC |
| Cyperus aristatus var. inflexus | Cal. coast and desert; Ariz. |
| Parietaria floridana | Cal. coast and desert; Ariz. |
| Eriogonum scalare | BC |
| Eriogonum angulosum | Cal. desert; Ariz. |
| Eriogonum reniforme (G) | Cal. desert; Ariz. |
| Chorizanthe flava | BC |
| Aphanisma blitoides | Cal. coast |
| Amaranthus lepturus | BC |
| Boerhaavia intermedia | Cal. desert; Ariz. |
| Mirabilis oligantha | BC |
| Abronia gracilis | BC |
| Nemacaulis denudata | Cal. coast and desert |
| Atriplex elegans | Cal. desert; Ariz. |
| Pterostegia drymarioides | Cal. coast and desert; Ariz. |
| Drymaria holosteoides (S) | BC |
| Drymaria viscosa | BC |
| Achyronychia Cooperi (G) | Cal. desert; Ariz. |
| Loeflingia squarrosa | Cal. coast |
| Eschscholtzia minutiflora | Cal. desert; Ariz. |
| Dryopetalon Palmeri | BC |
| Lepidium lasiocarpum | Cal. coast and desert; Ariz. |

| | |
|---|---|
| Thelypodium lasiophyllum | Cal. coast and desert |
| Dithyraea californica | Cal. desert; Ariz. |
| Sibara laxa | BC |
| Lyrocarpa Xanti (S) | BC |
| Lyrocarpa Coulteri var. Palmeri | BC |
| Wislizenia refracta | BC |
| Oligomeris linifolia | Cal. coast and desert; Ariz. |
| Tillaea erecta | Cal. coast; Ariz. |
| Hosackia humilis | Cal. desert; Ariz. |
| Hosackia nivea | BC |
| Hosackia tomentella | Cal. desert; Ariz. |
| Lupinus sparsiflorus (G) | Cal. desert; Ariz. |
| Astragalus insularis | Cal. desert |
| Dalea vetula | BC |
| Dalea peninsularis | BC |
| Dalea mollis subsp. pilosa | BC |
| Petalostemon evanescens | BC |
| Cassia confinis | BC |
| Vicia exigua | Cal. coast; Ariz. |
| Polygala desertorum | BC |
| Ditaxis serrata | Cal. desert; Ariz. |
| Stillingia linearifolia | Cal. desert; Ariz. |
| Euphorbia polycarpa | Cal. coast and desert; Ariz. |
| Euphorbia Pondii | BC |
| Euphorbia eriantha | Cal. desert; Ariz. |
| Euphorbia heterocarpa var. eriocarpa | BC |
| Cardiospermum corindum | Son. |
| Sphaeralcea axillaris | BC |
| Sphaeralcea Orcuttii | Cal. desert; Ariz. |
| Sphaeralcea Coulteri | Cal. desert; Ariz. |
| Sphaeralcea Hainesii (G) | BC |
| Mentzelia adhaerens (G) | Son. |
| Mentzelia albicaulis (G) | Cal. desert; Ariz. |
| Lythrum Bryantii (S) | BC |
| Oenothera angelorum | BC |
| Oenothera clavaeformis vars. (G) | Cal. desert; Ariz. |
| Oenothera deltoides (G) | Cal. desert; Ariz. |
| Daucus pusillus | Cal. desert; Ariz. |
| Linanthus Jonesii | BC |
| Nemophila Menziesii var. integrifolia | Cal. coast and desert |
| Nama Coulteri | BC |
| Phacelia distans | Cal. coast and desert; Ariz. |
| Phacelia crenulata (G) | Cal. desert; Ariz. |
| Amsinckia tesselata | Cal. coast and desert; Ariz. |
| Pectocarya recurvata | Cal. desert; Ariz. |
| Pectocarya peninsularis | BC |
| Cryptantha maritima | Cal. coast and desert |
| Cryptantha angelica | BC |

| | |
|---|---|
| Cryptantha fastigiata | BC |
| Cryptantha angustifolia (G) | Cal. desert; Ariz. |
| Cryptantha echinosepala | BC |
| Cryptantha Grayi | BC |
| Verbena bajacalifornica (S) | BC |
| Verbena Shrevei (S) | BC |
| Salvia similis | BC |
| Stachys tenerrima | BC |
| Nicotiana Clevelandii | Cal. coast and desert; Ariz. |
| Nicotiana trigonophylla | Cal.; Ariz. |
| Mohavea confertiflora (G) | Cal. desert; Ariz. |
| Antirrhinum cyathiferum | Ariz. (one locality) |
| Mecardonia exilis | BC |
| Castilleja Bryantii (S) | BC |
| Martynia altheaefolia | Cal. desert; Ariz. |
| Plantago insularis | Cal. coast |
| Plantago fastigiata | Cal. desert; Ariz. |
| Houstonia brevipes (G) | BC |
| Nemacladus glanduliferus var. australis (G) | Cal. desert; Ariz. |
| Eriophyllum lanosum | Cal. desert; Ariz. |
| Ambrosia pumila | Cal. coast |
| Coreocarpus parthenioides | BC |
| Dyssodia anthemidifolia (S) | BC |
| Gnaphalium oxyphyllum (S) | BC |
| Gnaphalium palustre | Cal. coast |
| Malacothrix californica var. glabrata (G) | Cal. desert; Ariz. |
| Malacothrix Xanti (S) | BC |
| Malperia tenuis | BC |
| Nicolletia trifida | BC |
| Palafoxia linearis | Cal. desert; Ariz. |
| Parthenice mollis | Ariz.; Son. |
| Pectis ambigua | BC |
| Pectis papposa | Cal. desert; Ariz. |
| Perityle aurea | BC |
| Perityle Emoryi | Cal. coast and desert; Ariz. |
| Perityle lobata | BC |
| Perityle californica | Son. |
| Senecio mohavensis | Cal. desert; Ariz. |
| Stylocline micropoides | Cal. desert; Ariz. |

The above table serves to suggest the geographic relationships of the ephemerals of Baja California, although it is based only on a fraction of the total number. As listed, the abundant species are about equally divided between endemics and species also occurring in the Sonoran Desert in California and Arizona, chiefly in the Lower Colorado Valley. A large percentage of the latter group extend into the Arizona Upland and northwestern Sonora, but do not range south of Tiburón Island. Most of the endemic ephemerals of Baja California are found in the southern half of the peninsula. Only two species are listed which occur on both sides of the Gulf without being known in California or Arizona.

The distributional features indicate a strong similarity in the winter ephemeral flora throughout the Lower Colorado Valley, the Central Gulf Coast, and the Vizcaíno Region. The Magdalena Region is not so closely related to the areas in the north. The ephemerals are distinctly an element of the desert flora. The ephemerals of coastal California and of the desert-chaparral transition in Baja California have many generic relationships to the desert ephemerals, but the cases of specific identity in the two groups are probably not more than 15 per cent of the number of coastal forms.

### *Winter Ephemerals of California and Arizona*

As an element of the vegetation, the winter ephemerals retain their importance in the Sonoran Desert to its northern and northeastern limit, but the number of species decreases slightly toward the north and considerably toward the east. In favorable seasons they are most conspicuous on the sandy plains of the Lower Colorado Valley, where they often form an open carpet of green, followed by unrivaled displays of flowers. On the active dunes there are few ephemerals, and on the stony plains they are usually widely spaced. Many of the winter ephemerals range only to the limits of the desert or a very short distance beyond it. In the neighboring regions the spring display of verdure and flowers is almost wholly derived from herbaceous perennials.

South of the valley of Río Magdalena and the Plains of Sonora the winter ephemerals become a minor feature of the vegetation. A few species, notably *Perityle californica*, may be locally abundant, but the display is not general and the number of species involved is small.

The most abundant winter ephemerals of the sandy areas around the head of the Gulf of California are the following:

Festuca octoflora
Eriogonum reniforme
Eriogonum Thomasii
Eriogonum trichopes
Chorizanthe rigida
Chorizanthe brevicornu
Chorizanthe Thurberi
Nemacaulis denudata
Abronia villosa
Calyptridium monandrum
Achyronychia Cooperi
Eschscholtzia minutiflora
Dithyraea californica
Hosackia tomentella
Lupinus sparsiflorus var. arizonicus
Lupinus concinnus
Euphorbia melanadenia
Stillingia spinulosa
Stillingia linearifolia
Eremalche rotundifolia
Eremalche exilis
Petalonyx Thurberi
Oenothera decorticans var. condensata
Oenothera brevipes
Oenothera clavaeformis vars.
Nama demissum
Gilia floccosa
Langloisia setosissima
Cryptantha angustifolia
Eremocarya micrantha
Pectocarya linearis var. ferocula
Salvia columbariae
Plantago insularis var. fastigiata
Nemacladus ramosissimus

Monoptilon bellioides
Stylocline gnaphalioides
Dicoria canescens
Geraea canescens
Palafoxia linearis
Eriophyllum Wallacei
Chaenactis stevioides
Gaillardia arizonica
Psathyrotes annua

In the Arizona Upland the abundance of winter ephemerals is, as elsewhere, determined by the amount and frequency of the late winter rains. In very dry seasons the display of ephemerals is limited to small plants of *Plantago insularis* var. *fastigiata*, *Lesquerella Gordoni*, and the introduced *Sisymbrium irio*. In favorable seasons the ephemerals make an almost continuous ground cover on level areas with fine retentive soil, grow abundantly in the light shade of shrubs, and form very open stands on slopes or plains with a gravelly surface. The period of their best development usually extends from early in March to the middle of April. The predominant flower colors are white and yellow. Nowhere in this region are there such patches of vivid color as are made by *Abronia villosa* in the Lower Colorado Valley or by the various species of *Gilia* in the Mojave Desert.

The following list includes the characteristic winter ephemerals of the Arizona Upland. So much interest attaches to the distributional ranges of these plants that the table has been drawn up to indicate those found in the Mojave Desert (MD), those extending south into the inner part of the desert in Sonora (Son.), those occurring in the Great Basin (Utah, Nev.), and those extending east into western Texas (Tex.).

| | | | | |
|---|---|---|---|---|
| Bromus arizonicus | Cal. (coast) | | | |
| Festuca octoflora | MD | | | Tex. |
| Muhlenbergia microsperma | MD | Son. | | |
| Poa Bigelovii | MD | | Nev. | Tex. |
| Parietaria floridana | MD | Son. | | |
| Pterostegia drymarioides | MD | | Utah | |
| Chorizanthe brevicornu | MD | Son. | Nev. | |
| Chorizanthe rigida | MD | Son. | | |
| Eriogonum deflexum | MD | | Utah | |
| Eriogonum trichopes | MD | | Utah | |
| Eriogonum inflatum | MD | | | |
| Eriogonum Abertianum | | | | Tex. |
| Eriogonum polycladon | | | | Tex. |
| Eriogonum angulosum | MD | | Utah | |
| Monolepis Nuttalliana | MD | Son. | Utah | Tex. |
| Calandrinia ciliata | Cal. (var. on coast) | | | |
| Calyptridium monandrum | MD | | | |
| Loeflingia texana | | Son. | | Tex. |
| Myosurus cupulatus | Cal. | Son. | | |
| Platystemon californicus | MD | | | |
| Eschscholtzia mexicana | | | Utah | Tex. |
| Streptanthus arizonicus | | Son. | | |
| Lepidium lasiocarpum | MD | Son. | Utah | Tex. |

| | | | | |
|---|---|---|---|---|
| Dryopetalon runcinatum | | Son. | | Tex. |
| Lesquerella Gordoni | MD | Son. | | Tex. |
| Lesquerella purpurea | | | | Tex. |
| Draba cuneifolia var. integrifolia | MD | Son. | Utah | Tex. |
| Descurania pinnata subspp. | MD | Son. | Nev. | Tex. |
| Thysanocarpus amplectens | | | Utah | |
| Oligomeris linifolia | MD | Son. | | Tex. |
| Tillaea erecta | Cal. (coast) | | | |
| Lupinus sparsiflorus var. arizonicus | MD | Son. | Nev. | |
| Lupinus concinnus | MD | | Utah | |
| Hosackia brachycarpa | MD | | | |
| Hosackia humilis | MD | | | |
| Astragalus austrinus | MD | Son. | Utah (var.) | Tex. |
| Erodium texanum | MD | | | Tex. |
| Euphorbia eriantha | | Son. | | |
| Sphaeralcea Coulteri | | Son. | | |
| Eremalche exilis | MD | | | |
| Mentzelia albicaulis | MD | | Nev. | |
| Mentzelia aspera | | | | |
| Mentzelia nitens | MD | | Utah | |
| Oenothera leptocarpa | MD | | | |
| Oenothera clavaeformis vars. | MD | | | |
| Oenothera chamaenerioides | MD | Son. | | Tex. |
| Oenothera primiveris | MD | Son. | | Tex. |
| Gaura parviflora var. lachnocarpa | | Son. | | |
| Spermolepis echinata | | Son. | | Tex. |
| Daucus pusillus | MD | Son. | | Tex. |
| Bowlesia incana | MD | | | Tex. |
| Linanthus aureus | MD | | | Tex. |
| Linanthus Bigelovii | MD | | Utah | Tex. |
| Gilia filifolia var. diffusa | MD | | | Tex. |
| Gilia stellata | MD | Son. | Nev. | Tex. |
| Gilia sinuata | | | Nev. | |
| Pholistoma auritum var. arizonicum | Cal. (coast, the sp.) | | | |
| Eucrypta chrysanthemifolia var. bipinnatifida | MD | | | |
| Eucrypta micrantha | MD | Son. | | Tex. |
| Phacelia distans | MD | | | |
| Phacelia Popei var. arizonica | | Son. | | Tex. |
| Phacelia crenulata | MD | | Utah | Tex. |
| Nama demissum | MD | | Utah | |
| Nama hispidum vars. | MD (var.) | | | Tex. |
| Harpagonella Palmeri var. arizonica | Cal. (coast, the sp.), Son. | | | |
| Pectocarya heterocarpa | | Son. | Utah | |
| Pectocarya linearis var. ferocula | Cal. (coast, var.) | | | |
| Lappula texana var. coronata | | | | Tex. (the sp.) |
| Lappula Redowskii var. desertorum | MD (the sp.) | | | Tex. (the sp.) |

| | | | | |
|---|---|---|---|---|
| Eremocarya micrantha | MD | | Nev. | Tex. |
| Cryptantha angustifolia | MD | | Utah | Tex. |
| Cryptantha pterocarya | MD | | Nev. | Tex. |
| Cryptantha barbigera | MD | | Utah | |
| Plagiobothrys arizonicus | MD | | Nev. | |
| Plagiobothrys Pringlei | | | | |
| Amsinckia intermedia | MD | | | |
| Amsinckia tessellata | MD | | Utah | |
| Salvia columbariae | MD | | | |
| Petunia parviflora | Cal. (coast) | | Son. | Tex. |
| Orthocarpus purpurascens | MD | | | |
| Plantago Purshii | MD | | | Tex. |
| Plantago insularis var. fastigiata | MD | | Nev. | |
| Galium proliferum | MD | | Utah | Tex. |
| Nemacladus glanduliferus var. orientalis | | | | |
| Greenella arizonica | | Son. | | |
| Monoptilon bellioides | MD | Son. | Utah | |
| Stylocline gnaphalioides | MD | | | Tex. |
| Evax multicaulis | MD | | | Tex. |
| Filago californica | Cal. (coast) | | | |
| Filago arizonica | MD | | | |
| Layia glandulosa | MD | | Nev. | |
| Baeria chrysostoma var. gracilis | MD | | | |
| Eriophyllum lanosum | MD | | Utah | |
| Perityle californica | | Son. | | |
| Chaenactis carphoclinia | MD | | Nev. | |
| Chaenactis stevioides | MD | | Nev. | |
| Gaillardia arizonica | | | | |
| Dyssodia concinna | | Son. | | |
| Pectis Palmeri | | Son. | | |
| Pectis prostrata | | Son. | | |
| Microseris linearifolia | MD | | | |
| Stephanomeria exigua | | | Nev. | Tex. |
| Rafinesquia neomexicana | MD | | Utah | Tex. |
| Malacothrix Coulteri | MD | | Utah | |
| Malacothrix californica var. glabrata | MD | | Utah | |
| Malacothrix sonchoides | MD | Son. | Nev. | |
| Calycoseris Wrightii | MD | Son. | Utah | Tex. |
| Senecio monoensis | MD | Son. | Utah | |

The winter ephemerals just listed show their strongest geographical affinity with those of the Mojave Desert. Out of the total of 109 species in the Arizona Upland, 73 are represented in the Mojave Desert by plants which are specifically identical or of varietal relationship. As there are 496 species and varieties of winter ephemerals in the Mojave Desert, it appears that only one-seventh of them extend eastward into the Arizona Upland, although they form there the dominant element in that group of plants. The Californian element is further strengthened by the occurrence of 8 species in Arizona which are found on the coast but not in the Mojave Desert. Many plants of the Mojave and Sonoran Deserts are found in the Great Basin Desert only in extreme southern Nevada and Utah. Among the 43 ephemerals common to the Great Basin and the Sonoran Desert are a number which are confined to the south, but also several which range north into Oregon and Washington.

There are not more than 12 or 15 winter ephemerals in Arizona which are absent from the Mojave Desert and either are endemic to Arizona or extend east to Texas. There are 42 species which range into western Texas, of which 31 occur in the Mojave Desert or on the California coast. Because of the later arrival of spring, the season of activity for the ephemerals in Texas ranges from the middle of April to the end of May.

Any speculation as to the geographical source of the winter ephemerals must not overlook the fact that the relatively large number of species common to west Texas and the Sonoran Desert is so largely made up of forms extending east from the Mojave Desert.

The number of winter ephemerals in the upland parts of northern Sonora is much smaller than that on the sandy plains of northwestern Sonora. The 38 species that have been recorded are possibly not all the forms that might be found there. Three of them, *Perityle californica*, *Dyssodia concinna*, and *Astragalus austrinus*, are not found in the United States. One of them, *Pectis prostrata*, is a summer ephemeral in Arizona. Several other summer ephemerals of the northern part of the Sonoran Desert have been observed in small colonies south of Guaymas in the late winter, including *Kallstroemia grandiflora*, *Tidestromia lanuginosa*, and *Pectis papposa*.

### Summer Ephemerals

The herbaceous plants which appear after the first heavy rains of summer are as abundant and ubiquitous as the winter ephemerals. After successive copious rains they carpet the plains and gentle slopes, in places where the surface had previously been bare for three or four months. The seedlings begin to appear about three days after the first heavy rain, and their growth is even more rapid than in the case of the winter ephemerals. In the Arizona Upland there are very few years in which conditions are favorable for their start earlier than July 10. Rarely there are heavy showers in May, followed by a short period of moist soil. A few colonies of summer ephemerals have been observed after these pre-seasonal rains, but in no case have the plants persisted until the usual time of the summer rains.

In general, the season for summer ephemerals is longer than that for winter ephemerals, and the moisture of the soil is higher and more continuously maintained. A few summer ephemerals are of low habit, but the majority reach a height of 3 to 4 dm., and a few commonly grow to 2 m. Many of them are more abundant on disturbed ground than in wholly natural situations, and have many of the characteristics of weeds. Grasses are much more richly represented in the summer than in the winter, both in species and in individuals. The number of species of summer ephemerals is smaller than the number of winter species, but their role in the vegetation is just as important.

The foliage of the summer ephemerals is prevailingly mesic, and in only a few cases provided with wool, hair, air-containing epidermis, or heavy cutinization. In a few species the leaves are more than 25 sq. cm. in area (*Helianthus petiolaris*, *Xanthium spinosum*), and in a larger number they are more than 10 sq. cm. (*Datura meteloides*, *Parthenice mollis*, *Amaranthus Palmeri*). Simple leaves are more common than compound, and very small leaves or leaflets are found only in the several species of *Euphorbia* and in a few Compositae.

### *Geographic Distribution of Summer Ephemerals*

In southern California and in Baja California the Sonoran Desert is extremely poor in summer ephemerals. After the summer rains which very infrequently visit the interior of southern California, a few of the commonest herbs of the region farther east may be found. In the Imperial Valley and on the plains west of Blythe, California, the following species have been noted over a period of 10 years:

Pectis papposa
Bouteloua aristidoides
Kallstroemia grandiflora
Amaranthus Palmeri
Boerhaavia coccinea
Euphorbia micromera
Euphorbia pediculifera
Datura discolor

In Baja California summer rains occur exceptionally on the Gulf coast at long intervals, and probably a few of the summer ephemerals are active at such times. No definite observations have been made other than the finding of dead remains of a few recognizable plants, as noted previously.

The following list includes all the widely distributed and frequently noted native summer ephemerals in the Arizona Upland and the Plains of Sonora, and probably comprises 75 per cent of the total number of plants of this group in the Sonoran Desert. Those indicated by the abbreviation (Tex.) range eastward to Texas.

Eragrostis arida (Tex.)
Sporobolus patens
Aegopogon cenchroides
Leptochloa panicea (Tex.)
Leptochloa viscida (Tex.)
Chloris virgata (Tex.)
Bouteloua aristidoides (Tex.)
Bouteloua barbata (Tex.)
Digitaria sanguinalis (Tex.)
Eriochloa gracilis (Tex.)
Panicum arizonicum (Tex.)
Panicum hirticaule (Tex.)
Panicum pampinosum
Setaria Grisebachii (Tex.)
Cenchrus echinatus (Tex.)
Cenchrus pauciflorus (Tex.)
Cenchrus Palmeri
Hackelochloa granularis
Cyperus aristatus var. inflexus
Atriplex elegans (Tex.)
Amaranthus Palmeri (Tex.)
Amaranthus fimbriatus (Tex.)
Amaranthus blitoides (Tex.)
Amaranthus graecizanus (Tex.)

Tidestromia lanuginosa (Tex.)
Allionia incarnata (Tex.)
Boerhaavia erecta (Tex.)
Boerhaavia gracillima (Tex.)
Boerhaavia coccinea (Tex.)
Boerhaavia megaptera
Boerhaavia spicata
Boerhaavia Coulteri
Boerhaavia intermedia (Tex.)
Boerhaavia pterocarpa
Boerhaavia purpurascens (Tex.)
Mollugo verticillata (Tex.)
Mollugo cerviana (Tex.)
Trianthema portulacastrum (Tex.)
Portulaca parvula (Tex.)
Cleome tenuis
Sphinctospermum constrictum
Desmodium procumbens
Crotalaria pumila
Kallstroemia grandiflora (Tex.)
Kallstroemia californica
Euphorbia serpyllifolia (Tex.)
Euphorbia micromera (Tex.)
Euphorbia pediculifera
Euphorbia setiloba
Euphorbia gracillima
Euphorbia florida
Euphorbia polycarpa
Anoda cristata (Tex.)
Quamoclit coccinea (Tex.)
Ipomoea hirsutula (Tex.)
Heliotropium curassavicum (Tex.)
Chamaesaracha coronopus (Tex.)
Quincula lobata (Tex.)
Conobea intermedia
Datura meteloides (Tex.)
Datura discolor
Martynia parviflora (Tex.)
Martynia arenaria (Tex.)
Houstonia prostrata
Diodia teres var. angustata
Mitracarpus hirtus (Tex.)
Haplopappus gracilis (Tex.)
Haplopappus spinulosus subspp. (Tex.)
Psilactis Coulteri (Tex.)
Aster exilis (Tex.)
Parthenice mollis
Iva ambrosiaefolia (Tex.)
Ambrosia aptera (Tex.)
Ambrosia psilostachya (Tex.)
Xanthium spinosum (Tex.)
Xanthium pennsylvanicum (Tex.)
Eclipta alba (Tex.)
Helianthus petiolaris (Tex.)
Psilostrophe Cooperi
Baileya multiradiata (Tex.)
Hymenothrix Wislizeni
Bahia absinthifolia var. dealbata (Tex.)
Pectis linifolia
Pectis papposa (Tex.)

As already noted, the summer ephemerals are poorly represented in both species and individuals west of the Colorado River and the Gulf of California. Nearly all the species listed have been observed or collected in northern Sonora, and several species have been found in Sonora which do not occur in the United States.

The distributional features of the winter and summer ephemerals are exactly opposed. The former are very richly represented in the California deserts and diminish toward the east, whereas the latter are largely found also in the desert areas of western Texas and diminish rapidly in western Arizona. These features are closely correlated with the seasonal distribution of rainfall. It seems highly probable that these two groups of herbaceous plants have retained their distinctive characteristics and their general distributional pattern for a long period of geological time.

# 5

# Ecological Features of Characteristic Species

A COROLLARY to the simplicity of desert vegetation is the extreme abundance of many of its species. Over large areas the same type of situation is characterized by abundant stands of the same small number of species. The uncommon plants occasionally found are usually near the limits of their distributional range, in some part of which they also are very abundant. Nearly every one of the perennials of the Sonoran Desert which is ordinarily infrequent has been found in great abundance in one or more restricted localities, often being the dominant plant. In the desert there is little suggestion of the localized occurrence and limited range which add so much to the diversity of the flora in moist warm regions.

The vegetative dissimilarity of many of the abundant desert plants is accompanied by differences in ecological behavior and physiological response which make almost every species a distinct entity requiring separate investigation. General statements can be made only for the members of a few well represented life forms, such as the deciduous compound-leaved shrubs or the platyopuntias.

The purpose of this chapter is to describe the distribution, habitats, and ecological characteristics of a few of the commonest plants of the Sonoran Desert. The individuality of the desert plant seems to make this a necessary supplement to the description of the vegetation. The uneven treatment of the various species is due to their particular features, as well as to the fact that some have been more accessible for study than others.

The maps showing distribution are based chiefly on notes made by the author in field work in this region. Only a few of the localities are substantiated by herbarium specimens. The clear identity of the species involved made it unnecessary to multiply the number of collections. A few localities have been added on the basis of published records (mostly those of Brandegee and of Goldman in Baja California) or of collections in the herbaria of Stanford University, the University of California, and the University of Arizona.

After the treatments of some of the species are given citations to papers which bear on their ecological features. The copious literature on *Carnegiea* has been listed only in part, since the titles cited contain references to most of the other papers. On many of the plants no specific work has been done, and there is nothing in botanical literature about them except brief taxonomic descriptions.

## *Acacia constricta* (Mimosaceae)

This shrub is abundant throughout the northern part of the Sonoran Desert and sparingly represented farther south. It is abundant along streamways, but grows in all situations in the Arizona Upland. Pure stands are rare, as it is commonly associated with *Larrea* and other shrubs. It is commonly from 2 to 3 m. in height and rarely exceeds 4 m. in the most favorable situations. The bark of the larger stems is rough and gray; that of the twigs and smaller stems is smooth and reddish brown. The form of the crown is extremely variable, and the manner of branching has no unusual features. The young stems are armed with straight white spines from 2 to 5 cm. in length. *Acacia constricta* is deciduous, but is unlike other winter-deciduous shrubs in renewing its foliage in the spring only in case it has an adequate water supply. This means that in a dry spring the favorably situated individuals will come into leaf early in April, whereas the plants growing away from streamways will remain leafless 8 or 10 weeks longer, coming into leaf only after the beginning of the summer rains. The leaves are decompound, the branches bearing from 5 to 8 pairs of leaflets from 1 to 2 mm. in length. Nyctitropic and drought movements are executed by the leaflets. The foliage may begin to fall at any time from late September to late November, according to the autumn rainfall. The flowers of *A. constricta* form small yellow balls borne on pedicels several times longer than their diameter. The fruit is a pod from 5 to 10 cm. long, constricted between the seeds. With ample rainfall the crop of seeds is large and the opening of the pods is immediate. Rodents take their usual heavy toll of the crop, but many seeds fall on the sand of small arroyos and are widely disseminated by succeeding rains. Germination is delayed for at least 12 months by a coating of wax.

### *Acacia cymbispina* (Mimosaceae) — ESPINO

This is the dominant tree of the coastal plain of Sinaloa, and is abundant in the southernmost part of the Sonoran Desert. Its northern limit is a few kilometers northeast of Hermosillo and in the region near the confluence of Río Sonora and Río San Miguel, where it is confined to the margins of streamways but reaches its customary size and features of development.

*Acacia cymbispina* is a widely branched, slender, symmetrical tree, usually with a single trunk and wide-spreading branches. The top of the crown is flat, and the canopy of foliage is so thin and open that only a very light shade is thrown by the largest trees. The stout boat-shaped spines are 2 to 4 cm. in length and dark in color when young, and give the tree a distinctive appearance. The growth of young trees is very rapid and soon closes roads and trails which pass through the heavy stands.

### *Acacia Greggii* (Mimosaceae) — CATCLAW. UÑA DE GATO

Although most commonly seen as a shrub from 1 to 2 m. in height, *Acacia Greggii* is sometimes found as a low-branching tree 5 to 8 m. in height with a trunk 20 to 40 cm. in diameter. It is easily distinguished from *A. constricta* by the more open and spreading habit, by the larger leaflets, and by the gray bark throughout the plant. In all individuals there are many long, unbranched shoots extending out from the crown and armed with short, curved spines. The decompound leaves bear from 3 to 5 pairs of leaflets from 3 to 6 mm. long. Like the other mimosaceous plants, *A. Greggii* shows nyctitropic and drought movements of the leaves. The flowers are borne in long racemes and are pale buff. The pods are 10 to 15 mm. wide and are usually curved in several planes. The seeds are broader and flatter than those of the other acacias, but undergo a similar delay in germination due to a waxy coating.

*Acacia Greggii* is widely distributed throughout the Sonoran Desert, but is most abundant in its eastern half. It is particularly abundant on upper bajadas in the Arizona Upland and along the streamways of that area. It extends far beyond the limits of the Sonoran Desert and often appears as an isolated individual of maximum size. The foliage persists through all but the most severe winters, falling gradually during the growth of new leaves in April or May. A severe frost will kill the leaves, but they will remain on the plant for several weeks or until the advent of spring growth.

*Acacia Willardiana* (Mimosaceae) PALO BLANCO

A slender tree of 6 to 8 m., conspicuous by reason of its white exfoliating bark and long, slender cladophylls (pl. 33). It is found only on rocky slopes and is particularly abundant on crests and ridges which afford little soil. It is rare north of the Río Sonora, but becomes a characteristic plant of the hills in the southern part of the Plains of Sonora. Slopes covered with a very open but evenly spaced stand of trees of uniform height give the first forecast of the extensive acacia thorn forest which lies south of the Sonoran Desert. *Acacia Willardiana* is the only woody plant in the Sonoran Desert with well developed cladophylls. The leaves are reduced to one or three short-lived pinnae, which appear on the seedling and on the first shoots of the seasonal growth. Nearly all the photosynthetic performance of the tree is carried on by the cladophylls.

*Bursera microphylla* (Burseraceae) COPAL

(Map 4)

In its best development this is one of the handsomest and most striking trees of the Sonoran Desert. It is low-branching but reaches a maximum height of 10 to 12 m. The trunk and lower limbs are thickened out of proportion to the height of the tree, but there is no thickening of the endmost branches and twigs. The tree is commonly richly branched, with a smooth gray or bronze-red surface and no tendency to form secondary bark. The leaves are simply pinnate and dark green. The foliage is evergreen but liable to killing in severe winters. If not so killed, the leaves persist from one growing season to the next, and many of them appear to persist for two or even three years.

The thickness of the trunk is due to extensive development of the cortex, which accounts for 60 to 70 per cent of the basal diameter. The wood is soft, and the tissue of the cortex is very soft and impregnated with gum, resin, and latex, which flow copiously from a freshly cut trunk. In the seedling the first indications of the thickening are given by the development of a small, almost spherical enlargement just at the surface of the ground. The excessive development of cortex proceeds gradually up and down from this enlargement.

The seeds of *Bursera microphylla* have a shiny black coat and germinate readily. The number of well established seedlings to be found is greater than in the case of other xeric trees, but there is severe mortality among the young plants from 10 to 40 cm. in height.

The distribution of *Bursera* is almost coincident with the extent of the Sonoran Desert. The tree is absent only from the most northern and eastern parts of the desert, and extends beyond it only in the Cape Region of Baja California. It is uncommon north of Rosario and the Río Magdalena, and shows its best development as to size and numbers in central Baja California and central Sonora. No pure stands have been seen, and there are rarely more than 8 or 10 trees to a hectare.

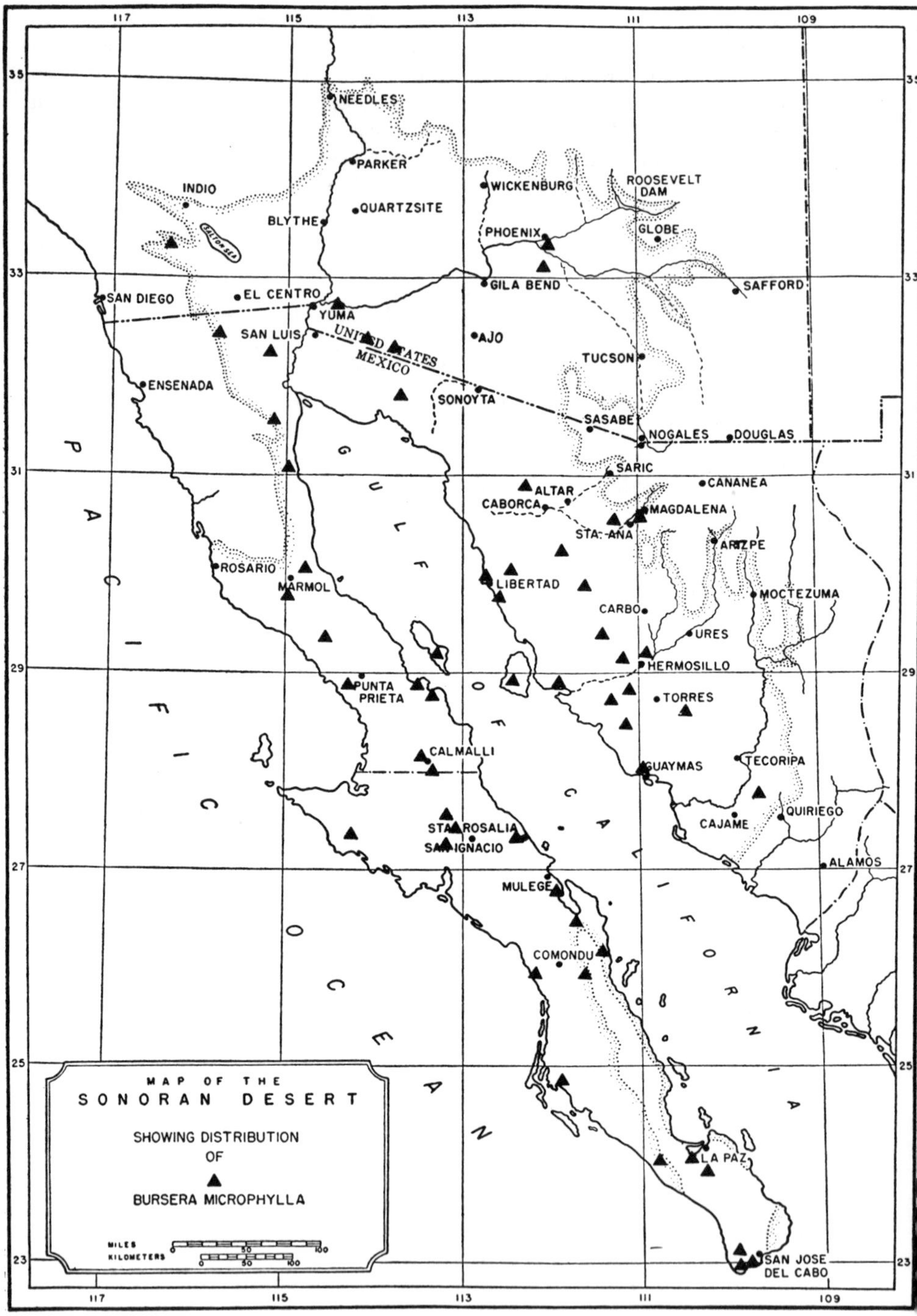


MAP 4. Distribution of *Bursera microphylla*

*Carnegiea gigantea* (Cactaceae) SAHUARO

(Map 5)

*Carnegiea* is the best known of the characteristic plants of the Sonoran Desert. Its size and graceful form make it conspicuous in any locality, and its position as the northernmost of the many massive columnar members of the cactus family has made it the subject of numerous ecological, anatomical, and morphological investigations.

The height of *Carnegiea* at maturity is 10 to 12 m., or slightly more in exceptional cases. The young plant is a grooved cylindrical column, rising from a slender base 15 to 20 cm. in diameter, and reaching a thickness of 30 to 40 cm. at a level well above the middle of the plant. Branches usually appear on plants 5 to 8 m. high, and there are frequently 5 or 6 primary branches, although the number may be much greater in the oldest individuals. Secondary branches are found only on large plants and are never numerous. The water content is from 75 to 95 per cent of the green weight, being greater at the top and less at the base. Old trees with numerous branches have been calculated to weigh as much as 6 or 7 tons.

The root system of *Carnegiea* is little more than 1 m. in depth, but extends laterally near the surface for as much as 30 m. A stout but elastic woody skeleton extends from base to top, and consists of a cylindrical group of rods, anastomosing to middle height and free above. The remainder of the plant is a soft parenchyma of high water content and richly impregnated near the base with crystals of calcium silicate. The chlorophyll-bearing tissue blankets the plant from the top to the rough gray bark which clothes the slender base. Clusters of very stout spines are borne on the ridges of younger plants, but slender, limber ones take their place at higher levels on old trees.

The flowers of the sahuaro are borne near the apex of the trunk and larger branches, or exceptionally at widely separated intervals for a distance of 1 m. or more from the tip. The flowers open during the second or third week of May and are chiefly pollinated by bees. The fruits develop rapidly and are mature by the middle of June. The number of fruits on a large plant may be as great as 200, and there are approximately 1000 seeds in each fruit. The seeds are nearly spherical and about 1.5 mm. in diameter. The abundant seed crop stands in strong contrast to the very small number of new individuals that become established even in the most favorable years. The seeds are borne in a sweet, juicy pulp which is eagerly eaten by birds. In the case of seed-eating birds, this means the destruction of all the seeds eaten. In the case of other birds, the seeds are passed without loss of viability. A heavy toll of the seed crop is taken by ants, which remove the small embryo and store the seeds. Rodents and coyotes eat the seed-bearing pulp and destroy a small percentage of the seeds eaten. The fact that the seed crop matures from 2 to 4 weeks before the first summer rains bring favorable conditions for their germination gives a full opportunity for the various predators to decimate the crop. The seedling is very small and provided with a short, poorly branched root, which results in many seedlings' being covered or washed out during the rains immediately following germination. It is doubtful whether more than half a dozen seedlings persist to the following year from a crop of many thousands of seeds.

The early growth of the seedlings of *Carnegiea* is very slow. After 2 or 3 years they are only a few mm. in height, and after 10 years they are not more than 1.5 to 2 cm. in height. The later growth is so variable that the age of a plant 1 m. high may be from 20 to 50 years. After a height of 2 to 3 m. is reached, the growth rate is about 10 cm. per year in the main trunk and slightly less in the large branches. The age of the

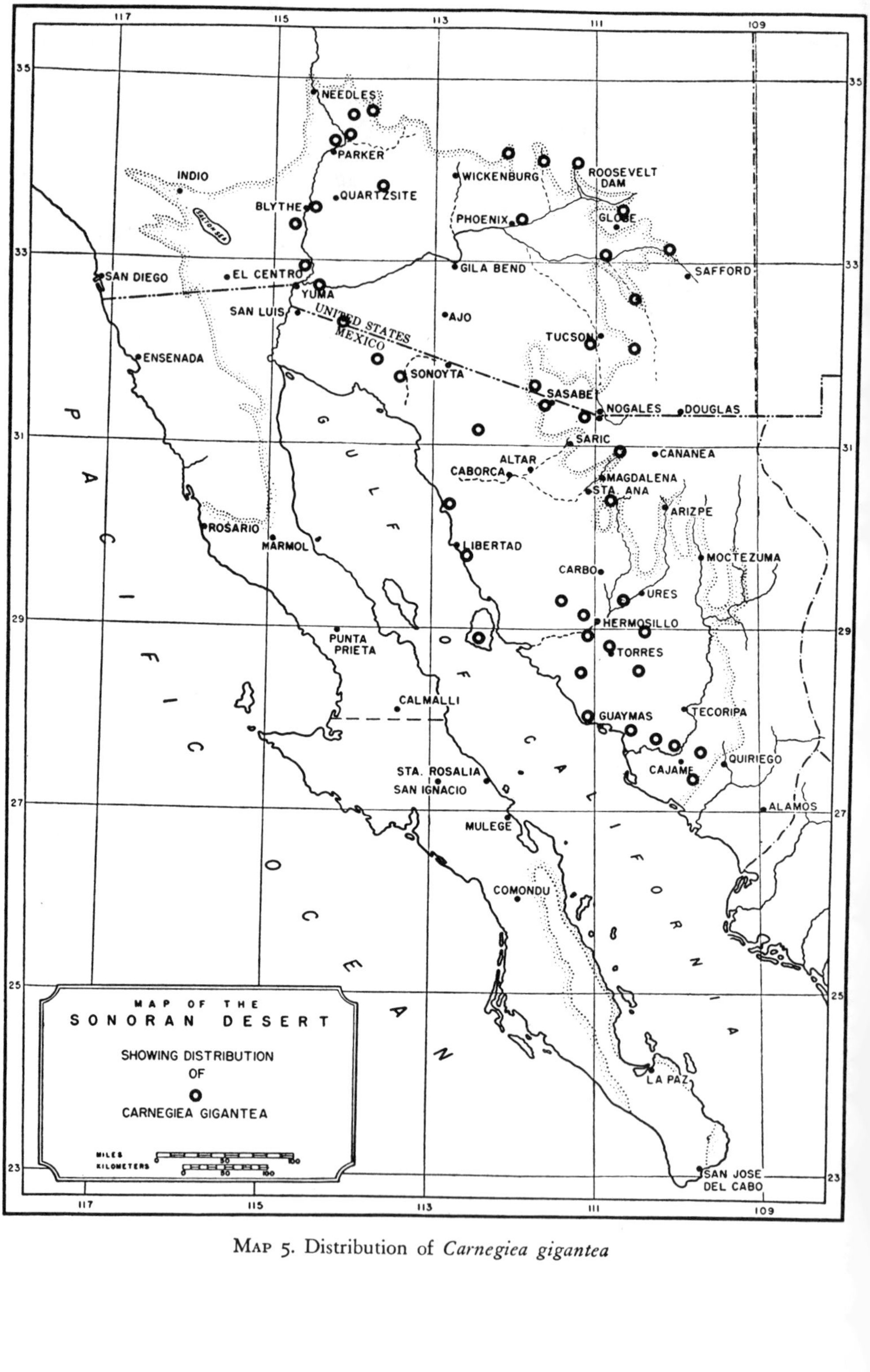


Map 5. Distribution of *Carnegiea gigantea*

largest individuals has been estimated from the known growth rate to be from 150 to 200 years. Wounds and infections are quickly walled off by the formation of callus. Nests made in the sahuaro by woodpeckers are lined with a heavy bag-shaped callus, often to be found in sound condition in the rotting remains of fallen trees. The principal natural cause of injury is the breaking of branches by high winds. The broken surface heals quickly in dry weather, but in a rainy period it usually leads to infection and the rapid death of the entire plant. The principal cause of death is overturning by wind or by the removal of soil and stones from the base of the tree, or because of the softening of the soil by prolonged rain.

The range of *Carnegiea* coincides closely with the boundary of the Sonoran Desert from the Colorado River east and south to the valley of the Río Sonora. Its altitudinal limit in this stretch is about 1200 m., but occasional individuals are found as high as 1500 m. on steep south slopes. In southern California the sahuaro is found in only three localities close to the Colorado River, and in Baja California it has never been observed. In northwestern Sonora it is absent from the sandy plains adjacent to the Gulf, but is abundant on most of the low mountains. Heavy stands are increasingly rare south of Tiburón Island, and it has not been reported near the coast in the area between San Carlos Bay and the mouth of the Río Yaqui. In the interior south of the Río Sonora the stands of sahuaro are infrequent and open. The southernmost individuals that have been noted are on hills about 15 km. south of Cajeme (Ciudad Obregón), Sonora. It has been reported as occurring in the lower valley of the Río Fuerte in northern Sinaloa, but the observation is open to doubt.

The finest stands of sahuaro are on the coarse outwash slopes of mountains in southern Arizona, south of the Gila River and east of Ajo, Arizona. The size and abundance of the plants are greatest on slopes of southern or southwestern aspect, and on the lower slopes of the adjacent hills. In no part of its range is the sahuaro found on flood plains or alluvial flats. Its best habitats are localities in which there is rock in place or a soil well filled with large angular rocks, in either of which situations a firm anchorage is supplied for the root system. A single large colony has been seen on coarse alluvial level soil between Chandler, Arizona, and the Gila River; here the cactus was associated with *Atriplex polycarpa*. This colony has recently been completely destroyed by the clearing of the land for cultivation.

### Literature


Johnson, Duncan S. The influence of insolation on the distribution and on the developmental sequence of the flowers of the giant cactus of Arizona. Ecology **5:** 70–82, 1924.

Shreve, Forrest. The rate of establishment of the giant cactus. Plant World **13:** 235–240, 1910.

——— The influence of low temperatures on the distribution of the giant cactus. Plant World **14:** 136–146, 1911.

Spalding, Effie S. Mechanical adjustment of the sahuaro (*Cereus giganteus*) to varying quantities of stored water. Bull. Torrey Bot. Club **32:** 57–68, 1905.


## *Ceiba acuminata* (Bombacaceae) — Pochote

The pochote, or kapok, is a small tree of 6 to 8 m., sparingly branched and with a very open crown. The trunk is stout at the base, tapers rapidly, and is covered with broad, blunt thorns which arise by secondary thickening from the slender thorns of the young tree. The leaves are palmate, with a spread of 15 cm. and an area of 65 to 100 sq. cm. They are distinctly mesic in structure and are present only in the summer rainy season. *Ceiba* is infrequent in the southern part of the Sonoran Desert, growing on level ground and low hills, but is conspicuous by reason of its light gray bark, tapering trunk, sparse and irregular branching, and large fruits.

### *Cercidium floridum* (Caesalpiniaceae) — BLUE PALO VERDE

There is a strong similarity between young plants of *Cercidium floridum* and those of *C. microphyllum*, but with increasing age a number of obvious differences develop. The former usually has a single trunk to a height of 1 to 2 m. with light gray bark, a rounded crown with many nearly horizontal branches, and stout end twigs bearing scattered thorns. The leaflets are about twice the diameter of those of *C. microphyllum*, and the foliage and branches are blue-green, in strong contrast with the yellow-green of the related tree. The flowers are deep yellow, in contrast with the pale yellow of *C. microphyllum*, and appear 2 or 3 weeks earlier.

The distribution of the two palo verde trees is much the same, but *C. floridum* extends beyond *C. microphyllum* in nearly all parts of their limit, particularly in the south on the coastal plain of Sinaloa. In the mountains *C. floridum* invariably reaches a slightly higher altitude. In the center of the distributional range there is a strong difference in the habitat preferences of the two, *C. microphyllum* being confined to the coarse soil of plains, outwash slopes, and hills, whereas *C. floridum* is most abundant on fine soil, on alluvial plains, and in or near the beds of streams. It is also sparingly found on sandy plains.

The oldest trees of *Cercidium* are 7 to 10 m. high and have a wide-spreading crown. The finest examples of it are found along the streamways of extreme southwestern Arizona and northwestern Sonora. It often occurs in continuous belts along the larger streamways, but large isolated trees are frequently found on small streamways. The seed and its behavior in germination are nearly the same as in *C. microphyllum*, but reproduction is better. In young trees less than 2 m. in height there are numerous horizontal or prostrate branches with very thorny twigs. Only on reaching an age of 30 to 50 years does the single trunk become dominant.

### *Cercidium microphyllum* ((Caesalpiniaceae) — PALO VERDE

(Map 6)

Among the green-stemmed microphyllous and aphyllous perennials of the Sonoran Desert, *Cercidium microphyllum* and *C. floridum* are by far the most abundant. With their almost constant difference of habitat, the common occurrence of these two small trees does much to give perennial greenness to the desert landscape. The presence or absence of the very small compound leaves makes only a slight change in the appearance of the trees except on close examination.

Throughout southern Arizona the distribution of *Cercidium microphyllum* conforms closely to the boundary of the desert, but isolated trees have been found beyond it on south slopes in a number of localities in southeastern Arizona. In southern California the tree is rare outside a few localities near the Colorado River. In Baja California *Cercidium* is nowhere so abundant as it is in Arizona, and it is absent from the Pacific side of the peninsula and the southern end south of Bahía Concepción. It is locally abundant between latitudes 27° and 30° N., but in this region rarely reaches the height

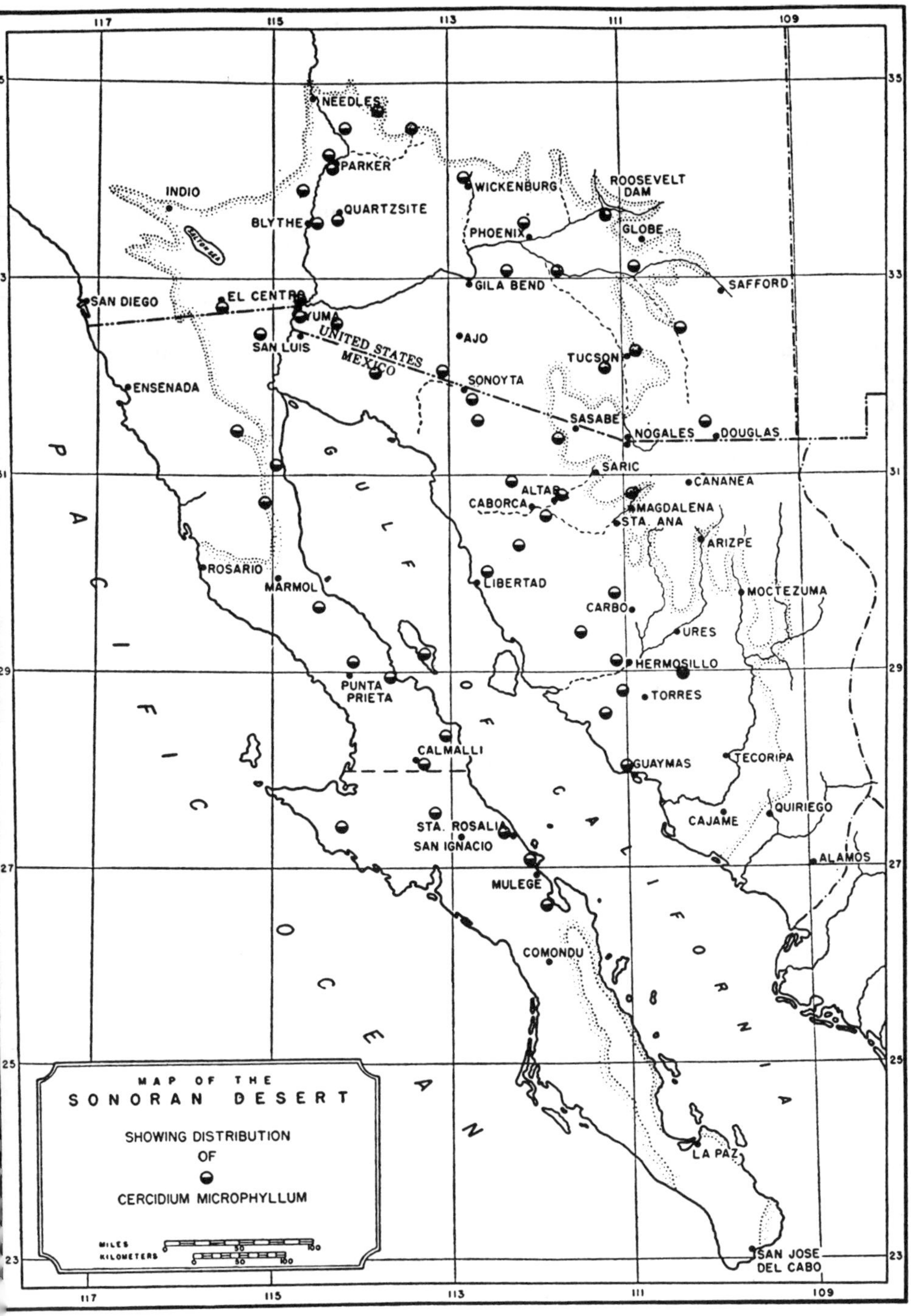


MAP 6. Distribution of *Cercidium microphyllum*

and spread attained on the Plains of Sonora. Except on the sandy plains of extreme northwestern Sonora, *Cercidium* is very abundant in that state as far south as Río Sonora. Toward the interior it becomes very infrequent in the long, narrow desert valleys which penetrate the desert grassland. Between Río Sonora and Río Yaqui *Cercidium microphyllum* is uncommon, and it has not been seen or reported south of the latter river.

*Cercidium microphyllum* commonly reaches a height of 5 to 6 m. and exceptionally 8 to 10 m. The trunk is greatly abbreviated and divides into two or three main branches at 10 to 25 cm. from the ground. The stems are richly branched, but the crown is not dense. The extreme base of the trunk is covered with a smooth, light gray bark which ends above at a very definite horizontal line. Observation of seedlings and young plants shows this line to be at the point of attachment of the cotyledons. Only rarely is this developmental vestige obliterated in the oldest trees. The smooth green surface of the rest of the tree is broken only by small patches of dark bark formed over wounds. The wood is light yellow, uniform in appearance and color from center to surface of the trunk, and very soft. Very close-set rings of seasonal growth are visible in carefully prepared sections.

The leaves of *Cercidium* are compound, with 3 to 5 pairs of nearly round leaflets 1 mm. or less in diameter. The leaves are borne in fascicles on older stems and in pairs on the growth of the season. There are commonly two crops of leaves, following the principal rainy seasons, and failing to appear if the rains do not take place. The leaves rarely persist more than 6 to 10 weeks. On the return of dry conditions the leaflets turn yellow and fall from the rachis, which in turn falls a few weeks later. On a tree in leaf the total area of the leaflets certainly does not exceed the total area of green stem surface, and is probably much less.

Flowers appear on *Cercidium* in April in central Sonora and about the middle of May in the Arizona Upland, provided there has been rainfall of the average amount or more. The trees either do not bloom at all or bloom very profusely, and there is little divergence in their behavior in different situations. The long, constricted pods ripen about 6 weeks after flowering, and without a definite mode of dehiscence soon liberate from 1 to 3 seeds. These are about 6 to 9 mm. in length, very slightly flattened, and with a heavy, dark brown, waxy coat. A large part of the seed crop is consumed by rodents, but the crop is so large that many seeds are left. Fresh seeds will not germinate until the coat has been scarified or the seeds washed in ether. Under natural conditions the wax coating is presumably removed by bacterial action, and the seeds are ready to germinate 12 months after they have been shed—or, in other words, at the first succeeding period of favorable moisture and temperature. The viability of the seeds is high, their germination takes place in about 3 days, and the number of seedlings appearing during the first week of the summer rainy period is very large. Seedlings reach a height of 2 to 3 cm. during July and August and develop a root system from two to three times the length of the shoot. During the dry weeks of autumn the mortality among the seedlings is very high, and it continues relatively high during their first 4 or 5 years.

The root system of *Cercidium* is deep and is also wide-spreading, but not superficially. The distribution of the roots is determined by the nature of the substratum on which a particular tree is growing. On rocky slopes the greatest penetration is through the pockets of soil and narrow fissures in the rocks, where small amounts of water penetrate and are prevented from further infiltration as well as from evaporation. For very young trees, successful establishment is primarily determined by the existence of a substratum which enables the roots to reach the lowest pockets of moist soil.

Exceptionally dry periods result in the death of twigs and branches, and sometimes of entire large limbs. Only very rarely does an entire mature tree die from drought. After a severe loss of branches, a large tree produces strong new shoots from the base. The oldest trees show evidence of having lost large limbs at intervals throughout their lives. The most careful estimate that is possible on large trees indicates that they reach an age of 300 to 400 years.

LITERATURE

SHREVE, EDITH B. The daily march of transpiration in a desert perennial. Carnegie Inst. Wash. Pub. 194, 1914.

SHREVE, FORREST. Establishment behavior of the palo verde. Plant World **14**: 289–296, 1911.

——— The establishment of desert perennials. Jour. Ecol. **5**: 210–216, 1917.

*Cercidium sonorae* (Caesalpiniaceae) BREA

This small tree resembles the other species of *Cercidium* in the complete greenness of its trunk and branches, but has a very different habit of growth (pl. 34). The trunk is erect and usually unbranched to a height of 1 to 2 m., and the branches are nearly horizontal, forming a flat, wide-spreading crown. All the surfaces have a matte deep green color and are covered with a minutely papillose coating of wax. The leaflets are slightly larger than in the other species of *Cercidium*, and the leaves are sparingly distributed throughout the crown. The flowers are borne in short fascicles widely distributed along the branches, in strong contrast with the terminal position of the inflorescences in the other species of *Cercidium*.

The northernmost individuals of the brea are to be found on the Plains of Sonora a few kilometers south of Santa Ana. It is never very abundant, and has not been observed in pure stands. Its greatest numbers and best development are south of Guaymas and in the southern tip of the Sonoran Desert. It has been seen at several localities in central Baja California, but is not abundant in any of them.

*Cordia sonorae* (Boraginaceae) PALO DE ASTA

*Cordia* is a small tree with a very open crown and large, rather mesic, simple leaves about 3 by 6 cm. in size. It is found only along streamways in the southern part of the Plains of Sonora, and is uncommon but very conspicuous on account of its large leaves. The trees are leafless during the dry spring months. Leaves appear during the summer rains and often persist through the winter, finally turning yellow and remaining in place for several weeks.

*Encelia farinosa* (Compositae) WHITE BRITTLEBUSH. INCIENSO

This small shrub has an open, hemispherical crown of foliage borne on sparsely branched stems. The trunk and principal branches are from 5 to 20 mm. thick. The plants rarely exceed 1 m. in height, and smaller plants are much more common. The leaves are light green, oval, petiolate, from 3 to 7 cm. in length. Leaves are borne during both the summer and the winter rains, and the two crops show differences in structure and physiological behavior. In the dry seasons the plant is without functioning leaves and is subject to the death of twigs and branches. Large drops of gum, very commonly used as incense, are exuded from the large stems and harden in place or fall to the ground.

They are particularly abundant in the driest habitats of the plant. The flowers are borne twice a year on long terminal peduncles, extending from 10 to 15 cm. beyond the foliage. Large crops of seeds are produced, from which a relatively large number of germinations take place. The number of young individuals to be found in a given season is closely dependent on the rainfall conditions of the preceding year or two.

*Encelia* occurs throughout the Sonoran Desert and extends into the Mojave Desert and also into the northern edge of the thorn forest. Its greatest abundance and best development are in the northern half of the Sonoran Desert. In Arizona it is characteristic of coarse outwash slopes and bajadas. Farther south it exhibits a complete reversal of habitat and occurs only on level plains with a coarse loam or sandy soil. In the thorn forest it is found only on the sandy soil of alluvial bottoms. It has a wide distribution in Baja California and is frequently seen there in pure stands.

LITERATURE

SHREVE, EDITH B. Factors governing seasonal changes in transpiration of *Encelia farinosa*. Bot. Gaz. **77:** 432–439, 1924.

### *Forchammeria Watsoni* (Capparidaceae) JITO

(Map 7)

This tree reaches a height of 12 to 15 m. and has a compact, rounded crown of dark green foliage (pl. 36). The thickness of the trunk is somewhat greater than the size of the tree would suggest. There are almost no dead limbs, and very few dead branches or twigs. The leaves are thick and coriaceous and are retained throughout the driest months. A general defoliation occurs at flowering time in March and April. *Forchammeria* is not a tree of arroyo margins, but grows with *Olneya* and *Cercidium* on the open desert south of La Palma and Torres, Sonora. The wood is very light in weight and burns so quickly that it is regarded as worthless for fuel. Samples of the wood taken in March showed a moisture content ranging from 47 to 60 per cent. In all the particulars that have been mentioned, except leaf texture, this tree departs so widely from the common characteristics of desert trees that its abundance here is a remarkable phenomenon.

### *Fouquieria splendens* (Fouquieriaceae) OCOTILLO

(Map 8)

The ocotillo is a plant of very distinctive appearance, which has been aptly described as suggesting "a loosely held group of wands." The branches arise from a short crown and grow to a length of 3 or 4 m., usually without forking; they are nearly straight, and the majority of them stand at an angle of more than 45° to the surface of the ground. The successive regions of growth in a long branch may be easily distinguished, and the diameter increases from 1 cm. at the end to as much as 5 to 7 cm. at the base. Shoots of recent growth are dark red, those of the preceding year are light gray with dark longitudinal markings, and the older parts of the stem are dark gray with frequent splits in the bark, revealing the light green layers of tough inner bark mentioned below. Elongation of the stems takes place only during the summer rainy period. The leaves on newly grown shoots have stout petioles about 1.5 cm. long. An elongation of the mechanical tissue of the petiole ends on the lower side of the leaf, near the base, as a small spine. When the leaf falls, it is the blade only which drops, and the petiole remains

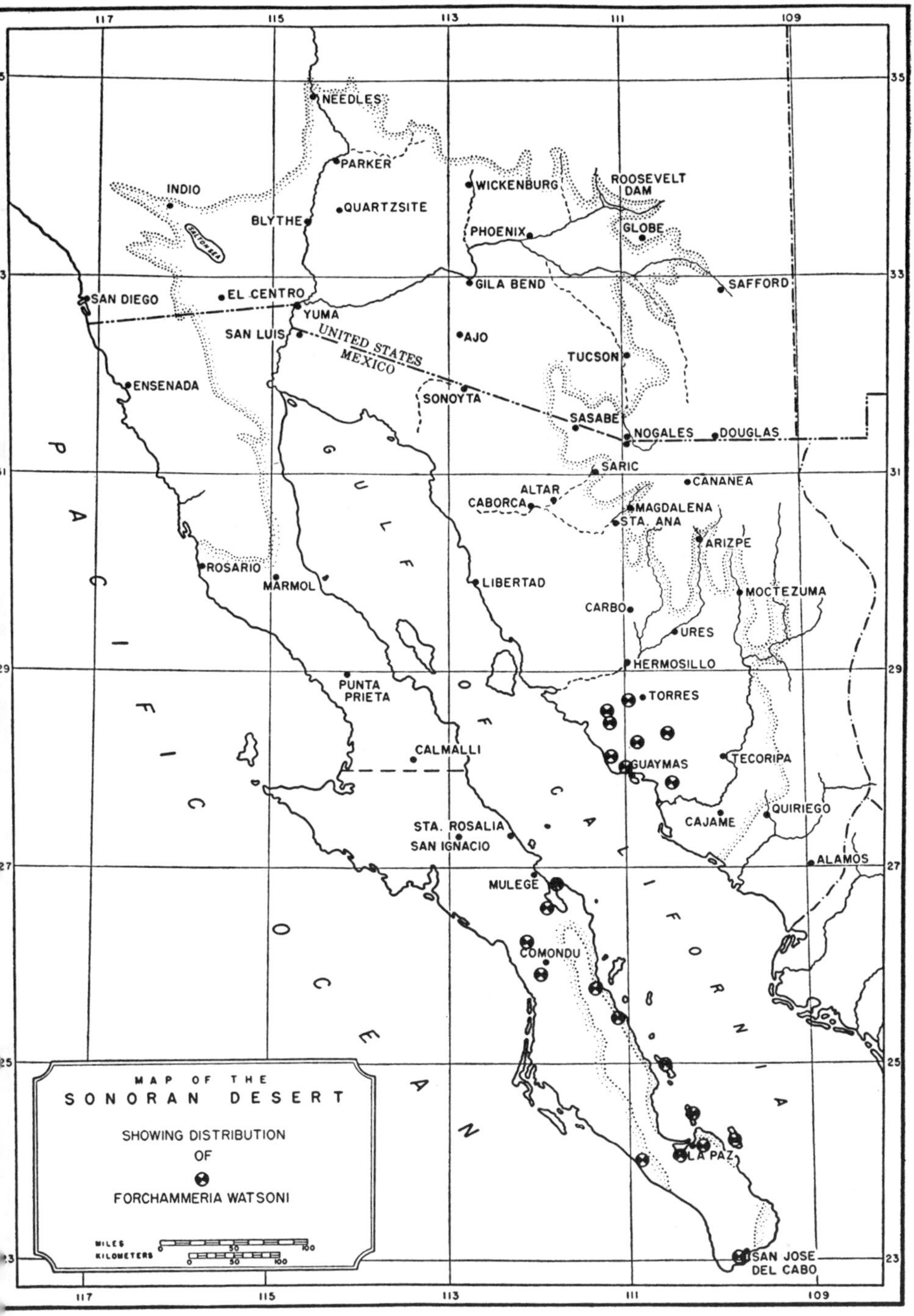


MAP 7. Distribution of *Forchammeria Watsoni*

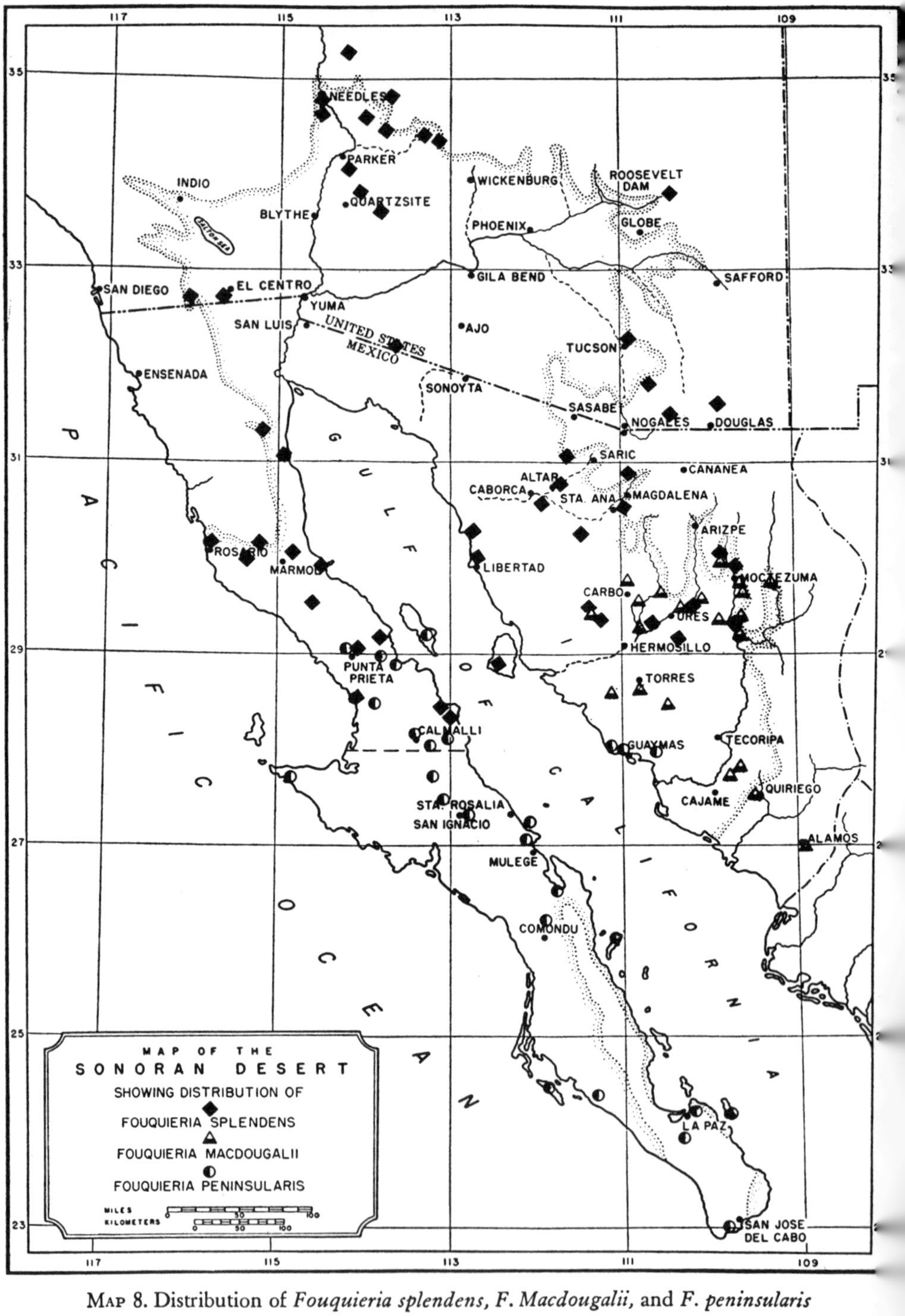


MAP 8. Distribution of *Fouquieria splendens*, *F. Macdougalii*, and *F. peninsularis*

as a slender spine from 1.5 to 2.5 cm. long. The leaves on older parts of the shoot are sessile and are borne in clusters from greatly shortened branches axillary to the spines. Leaves appear only after moderately heavy rains, either in winter or in summer, and the period of their persistence is determined by the duration of the rains. It may happen that there is such a distribution of wet periods as to bring forth five or six crops of leaves during the year. The foliage of *Fouquieria* is highly mesic in appearance, the leaves being thin, smooth, dark green, and with few of the structural features associated with arid conditions, except the depth of the chlorenchyma. The leaves are 1.5 to 4 cm. long and 0.5 to 1.5 cm. wide. *Fouquieria* is the only nonsucculent perennial of the region which can be propagated by cuttings.

The flowers of *Fouquieria* are borne in terminal racemes at the close of the winter rainy season. The light, thin-winged seeds are formed in great abundance and are readily disseminated by the wind. They germinate readily at the opening of the summer rainy season, and immense numbers of seedlings may be found in favorable localities late in July. From many thousands of seedlings, not more than two or three survive until the following summer.

The range of *Fouquieria splendens* slightly exceeds the limit of the Sonoran Desert throughout southern California and southwestern Arizona. Small colonies or single plants occur in many localities on the mountains which overlook the northern edge of the desert. In the Santa Catalina Mountains it occurs at 1800 m., and throughout the desert-grassland and encinal of southeastern Arizona it occurs on south exposures between 1500 and 1800 m. In many localities it is abundant on limestone at higher elevations and absent from neighboring soils founded on other types of rock. In the Swisshelm Mountains, in southeastern Arizona, it reaches the highest recorded altitude for the United States at 2100 m. These higher occurrences carry the continuity of distribution from the Sonoran to the Chihuahuan Desert.

In Baja California *F. splendens* is abundant at desert levels on the Gulf coast as far south as latitude 28°, and on the Pacific coast it extends from the desert boundary near Rosario southward nearly to the Vizcaíno plain. In southern Baja California *F. splendens* is absent and its place in the landscape is taken by *F. peninsularis.* In Sonora *F. splendens* is abundant north of latitude 29°, extending from the shores of the Gulf to the hills bordering the valley of Río Moctezuma. Here again there are many isolated occurrences of the plant far above the desert level. South of Río Sonora this species is infrequent, and it has not been observed south of the vicinity of Torres. In this region *F. Macdougalii* is the dominant member of the genus, but it does not take the place of *F. splendens* in the physiognomy of the vegetation.

Throughout most of its range *Fouquieria splendens* is most abundant on the shallow soil of rocky slopes or coarse outwash slopes. On the broad plains of the Lower Colorado Valley it occurs infrequently in the nearly pure stands of *Larrea.* When colonies of it occur in the *Larrea* plains, they are invariably found to be occupying sites with unusually rocky soil, derived from small volcanic emergences. In all parts of the desert with more than 100 mm. of rain, *Fouquieria* is absent from fine-textured soils and habitats subject to sheet flooding. In the drier parts of the desert it is frequent on finer soils and is occasionally found on sand. The root system of *Fouquieria* is shallow and wide-spreading, much like that of the cacti, and its optimum situation seems to require quick wetting and penetration in rainstorms, combined with good aeration at all times.

In the various parts of its range *F. splendens* differs considerably in height, number of stems, and amount of branching of stems, as well as in density of stand and rate of growth. The largest plants are found in regions where the number of individuals is smallest, namely, in the driest part of the range on both sides of the head of the Gulf of California. A height of 10 m. is attained by many individuals in this region, with great variation in the number of stems, the largest of which have a diameter of 12 to 15 cm. On the coarse outwash slopes leading down to Bahía San Rafael, in Baja California, there is a very open stand of *F. splendens* in which all the individuals are from 6 to 10 m. in height, with very stout stems, commonly branched near the summit, with sinuously curved dependent ends. This is one of the finest stands of the plant in the Sonoran Desert, and the age of the individuals is probably much greater than in somewhat similar stands in areas of greater rainfall.

In the Arizona Upland *Fouquieria* commonly reaches a height of 3 to 6 m. In a representative group of 40 plants examined near Tucson, the average number of stems of all lengths was 25. In this region the stems are nearly straight, and branching is infrequent except on the largest plants. The stoutest stems are conspicuously more slender than in the plants of lower and drier areas, seldom exceeding 6 to 8 cm. in diameter. The amount of elongation in a single season can be easily determined by locating the nodelike places in the stem at which growth has been arrested and an inflorescence formed. Only a few of the stems on a single plant grow in length in a given season, but all of them increase in diameter and undergo surface changes. In the youngest growth the epidermis is mainly intact, the color dark red, and the spines long, slightly curved, dark red, and sharp. After 2 or 3 years the surface and spines become weathered, the epidermis cracks, and the formation of ridges of bark begins. At this time begins the segregation of subepidermal layers of unspecialized cells in which the lumen becomes filled with resin. As the stem advances in age, the layers of resinous tissue become thick and stout and envelop the entire stem in overlapping layers. The cracks in the epidermis and bark reveal the greenish resinous layers, the tough texture of which resembles the chitinous tissues in the joints of large crustaceans. These layers play an important role in protecting the stem from too great or too sudden loss of moisture. At the age of 5 to 6 years the stem has greatly changed from its early appearance, and the spines have become brittle or have weathered to half their original length. All these indications make it easy to distinguish a young, vigorously growing plant from an old, inactive one.

The heaviest stands of *Fouquieria* are found above 1000 m. along the upper edge of the desert and in hills just above the desert. All these stands occupy slopes gently inclined to the south or southwest and with a coarse soil filled and covered with rock fragments. The height of the plants is from 2 to 3 m., the number of stems is from 10 to 15, and the appearance of the stem surface indicates young plants and active growth. In all such stands the plants are so closely spaced that it is impossible to walk freely between them.

### Literature


Humphrey, R. R. Thorn formation in *Fouquieria splendens* and *Idria columnaris.* Bull. Torrey Bot. Club **58:** 263–264, 1931.

——— A study of *Idria columnaris* and *Fouquieria splendens.* Amer. Jour. Bot. **22:** 184–207, 1935.

Shreve, Forrest. The establishment of desert perennials. Jour. Ecol. **5:** 210–216, 1917.

——— Die Fouquieriaceae. Pflanzenareale, ser. 3, no. 1, pp. 3–4, map, 1931.

*Franseria deltoidea* (Compositae)

(Map 9)

In spite of its perennial habit, the stems of *Franseria deltoidea* are slender and brittle, like those of many of the annual composites. Innumerable branches spring from the base and grow in such erect positions as to keep most of the foliage well above the soil surface. The compact crown is made up of many small leaves as well as a smaller number of full-sized ones from 1.5 to 2 cm. in length. Throughout its life the plant carries a large percentage of dead branches and leaves, which are shed only as a result of weathering. During the rainy seasons the plant is covered with large green leaves, which assume a grayish color with age. Very light rains give rise to a large number of small leaves. The fruits are spiny and produced in large numbers after both summer and winter rains. The plant is relatively short-lived. *Franseria deltoidea* occurs only on soil with a content and covering of coarse rock fragments. It is rarely found on sandy soil, and its local distribution is exactly complementary to that of *F. dumosa*.

*Franseria deltoidea* is found throughout the northern part of the Sonoran Desert and is not known beyond its boundaries. It is infrequent in Baja California and is not found in Sonora south of Río Sonora.

*Franseria dumosa* (Compositae) BURRO WEED

(Map 10)

*Franseria dumosa* is the commonest plant in the most arid part of the Sonoran Desert (pl. 37). This small, gray suffrutescent perennial covers the floor of every valley from the eastern Mojave Desert south to the Río Magdalena, but is local in its occurrence above 600 to 700 m. On the plains near the head of the Gulf of California, on either sand or fine-textured soil, it occurs in open stands, either without associates or accompanied by very widely spaced plants of *Larrea*. In the same region it also occurs on extremely barren slopes and on sand dunes, where the individuals are few but large and well developed.

The plants of *Franseria dumosa* are roughly hemispherical, branching at the ground and consisting of an intricate maze of slender stems. The leaves are small and deeply divided and are so dry that it is often difficult to decide whether they are alive or in the first stages of desiccation. In the late spring a large percentage of the leaves die, change to a light fawn color, and either fall or remain attached to the plant for many weeks. The youngest leaves are the most likely to survive, and form a dense cluster at the end of the twig, where growth is quickly renewed after rainfall. The fallen fruits are covered with spiny processes, which improve the chances that the fruit will lodge in a spot favorable for germination. Under favorable conditions the growth of young plants is rapid and early accompanied by the development of a root from 5 to 15 times the length of the stem.

*Franseria dumosa* ranges north to Death Valley, southern Nevada, and southwestern Utah. It extends south along the Gulf in Baja California as far as Bahía Los Angeles and in Sonora as far as Tiburón Island. It has not been observed in southern Arizona above an elevation of 800 m.

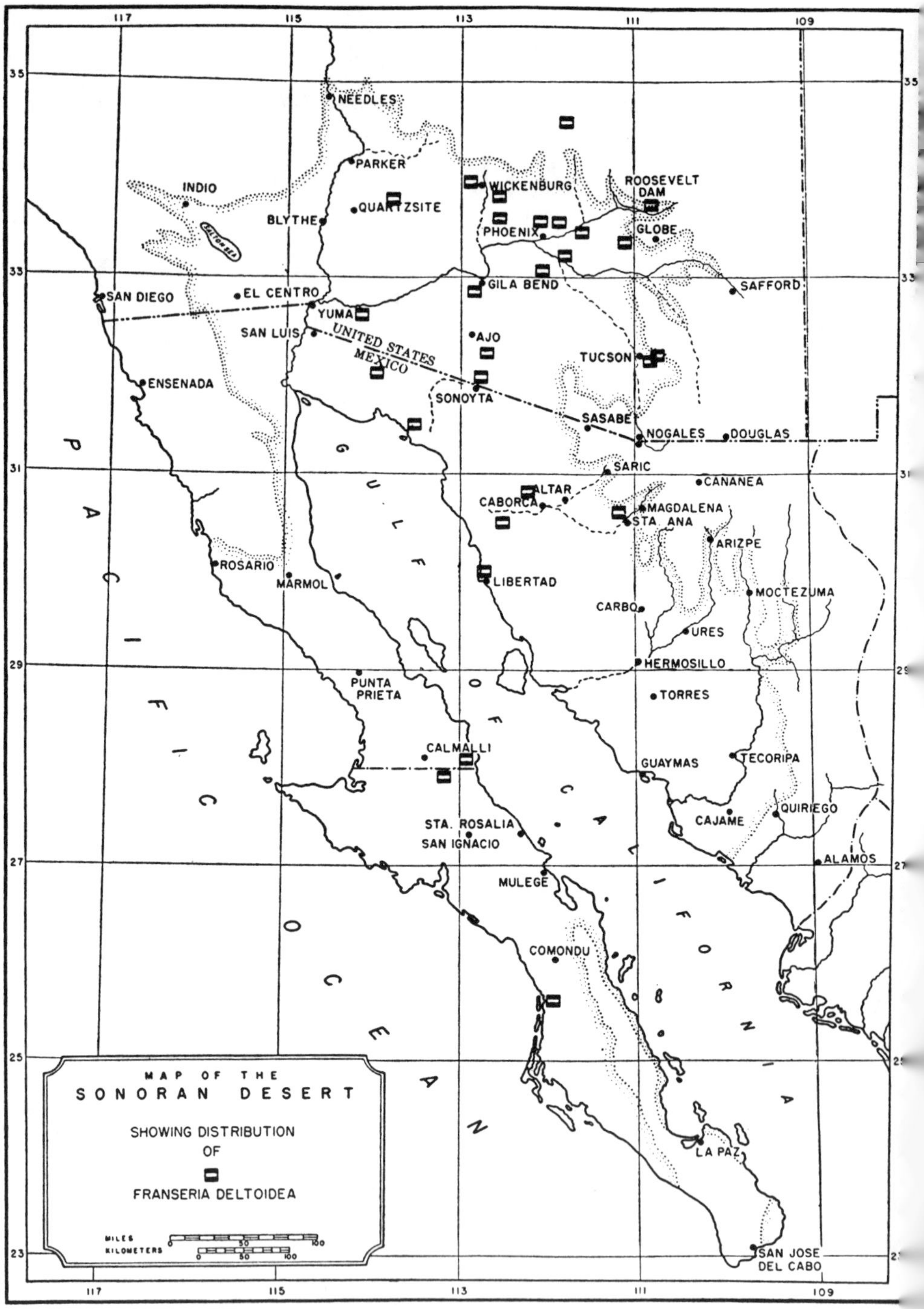


Map 9. Distribution of *Franseria deltoidea*

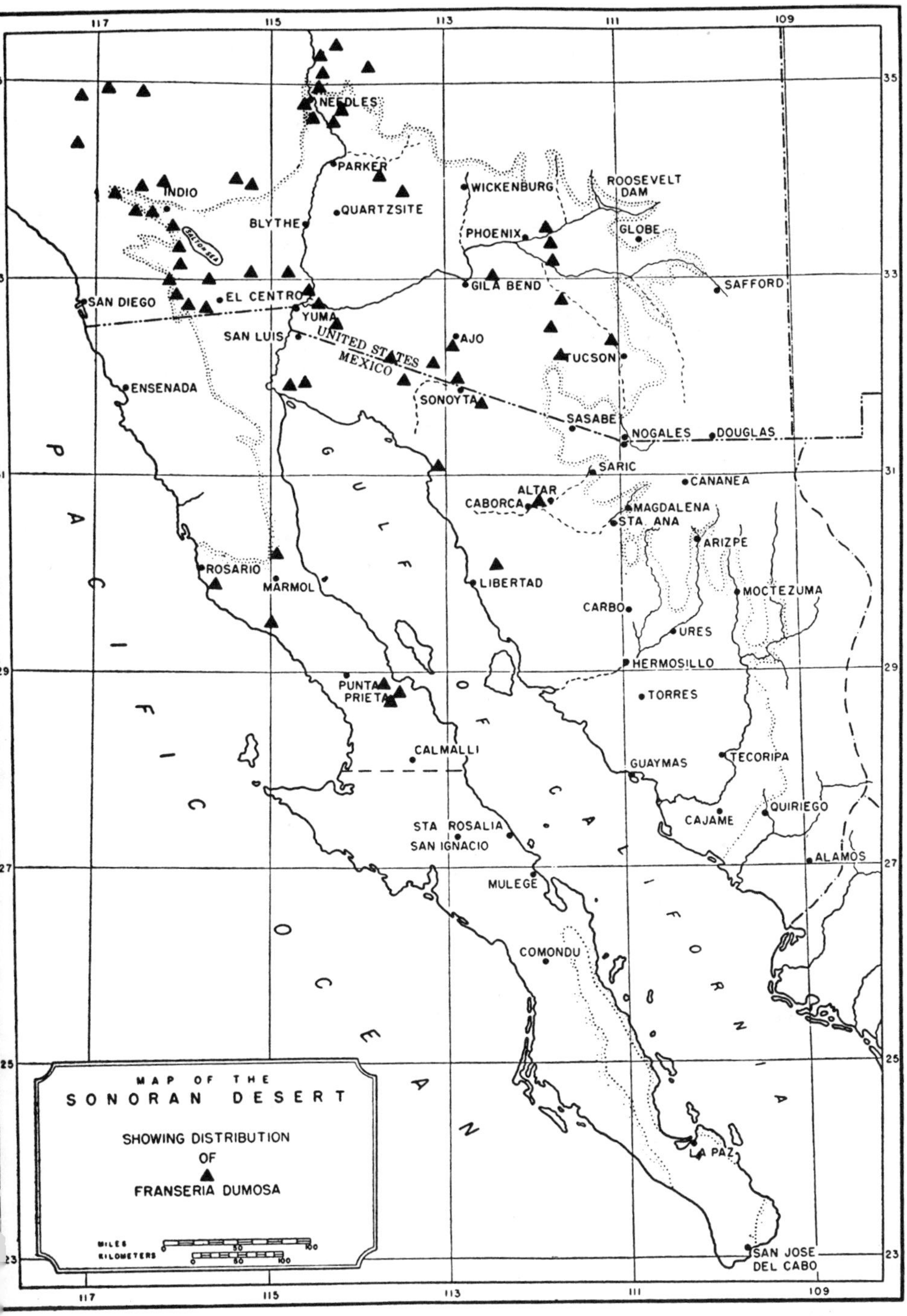


MAP 10. Distribution of *Franseria dumosa*. Also extends north to Death Valley, southern Nevada, and southwestern Utah.

### *Holacantha Emoryi* (Simarubaceae) CRUCIFIXION THORN

(Map 11)

The leafless, green-stemmed *Holacantha* bears some resemblance to *Cercidium microphyllum*, but its habit is somewhat more erect and its ultimate branches are stouter (pl. 35, fig. 1). It is found as an individual tree or as a group probably having a common root system. It never forms pure stands, but is often fairly abundant as an associate of *Larrea*. The scattered colonies never extend more than a few kilometers and are widely separated by areas from which the tree is absent. Its occurrence is limited to outwash plains along the northern edge of the Sonoran Desert. It has been found in a few localities in extreme northern Sonora and is known from several localities in the Mojave Desert. The deeply divided leaves are found on seedlings and very small plants, but never appear on the new growth of mature trees. The fruits are borne in large clusters, and each seed is encased in a very heavy capsule. There is no mechanism for the release of the seeds, and the clusters remain on the tree until they are broken off by wind and the capsules have become gray and weathered. In the natural course of events it is at least 5 to 7 years after the maturing of a seed before it reaches the ground. Germination is further delayed by the thickness of the capsule wall, and can be hastened only by filing through it. When this is done, the seeds germinate readily, with a high percentage of viability. The seedlings at once develop a deep taproot, and grow rapidly if given an adequate water supply.

### *Idria columnaris* (Fouquieriaceae) CIRIO

(Map 12)

*Idria* ranks without rival as the most bizarre plant of the Sonoran Desert (pls. 23, 32). In fact, it is one of the most striking woody plants in the flora of North America. Its vegetative parts as well as its flowers indicate clearly its affinity with *Fouquieria*. In spite of its bulk, the trunk of *Idria* strongly suggests a single stout and erect branch of *Fouquieria splendens*. In mature trees the basal diameter of the trunk is 50 to 60 cm., or exceptionally as much as 75 cm. In many individuals the trunk tapers gradually from the base; in others it increases in diameter to a height of about 1 to 2 m. and then tapers gradually. The extreme top of the tree, from 6 or 8 m. to the tip, maintains a more uniform diameter of 6 to 12 cm. The trunk is brownish gray and covered by a smooth series of epidermal layers which are heavily permeated with mechanical tissue. The trunk is usually single to the top, but is sometimes branched. A wholly distinct type of short, slender branch is also borne on the lower part of the trunk. This type has a horizontal position, is repeatedly branched, and bears leaves but never produces flowers. The branches of the latter type arise from the trunk at regularly spaced intervals on great spirals, as if conforming to a very precise form of phyllotaxy.

LITERATURE

See under *Fouquieria*.

### *Ipomoea arborescens* (Convolvulaceae) PALO SANTO

The palo santo, or morning-glory tree, occurs infrequently north of the Río Sonora along streamways, and becomes increasingly abundant toward the south. It is commonly

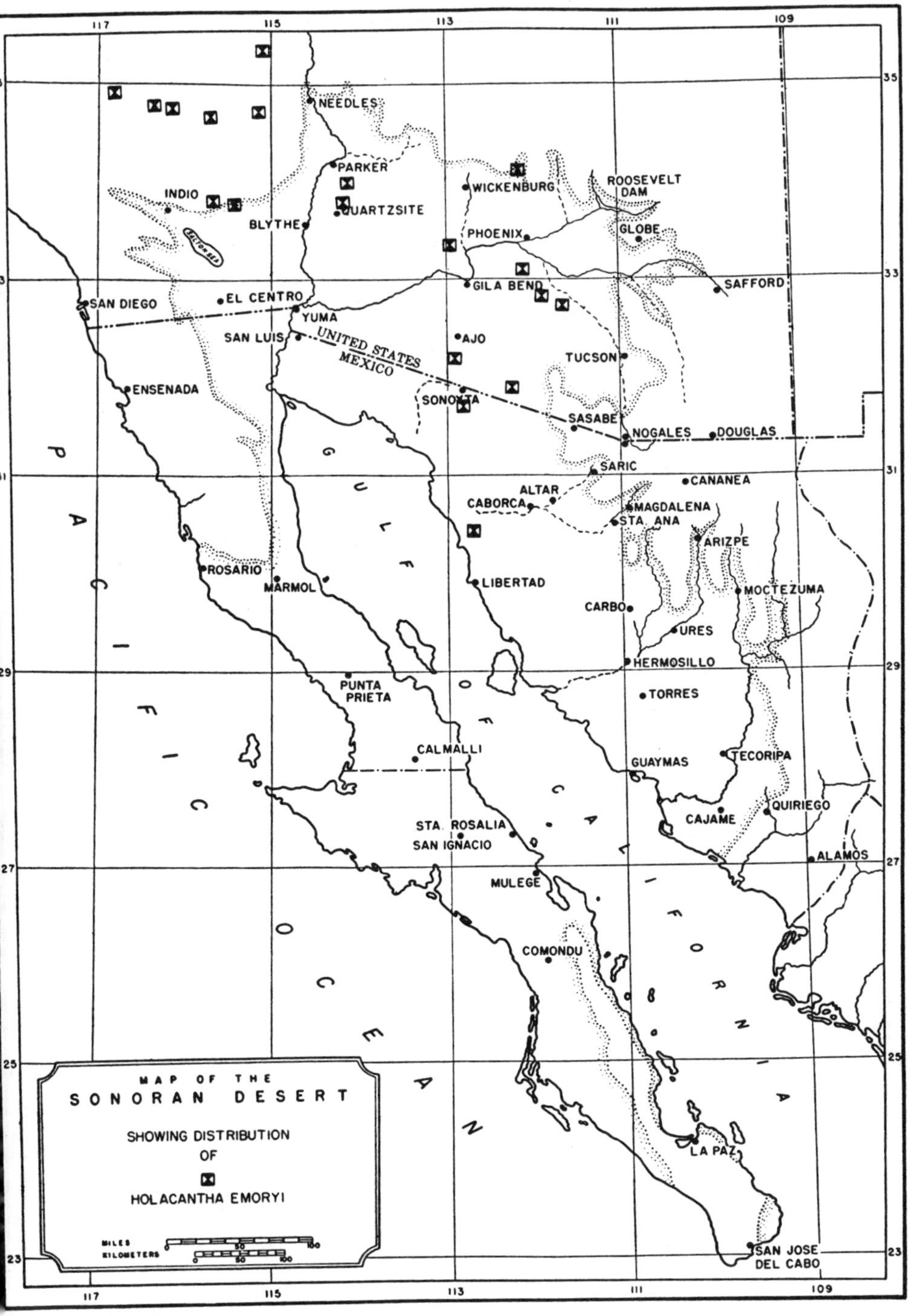


Map 11. Distribution of *Holacantha Emoryi.* Also extends into Chihuahua and Zacatecas.

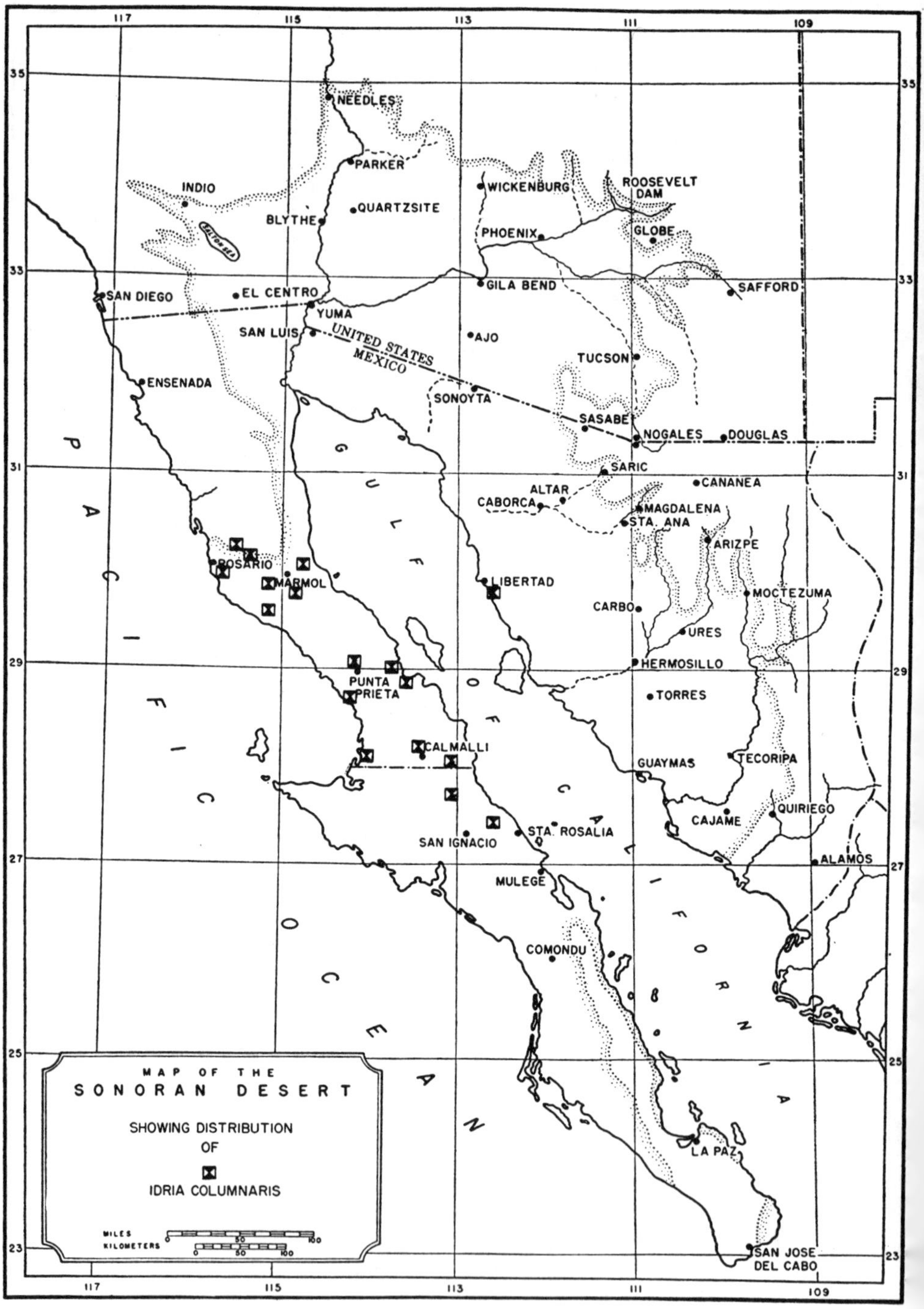


Map 12. Distribution of *Idria columnaris*

10 to 12 m. high and has a very open crown of ascending branches. The trunk is stout, but both trunk and stems taper rapidly. The large, mesic leaves are borne during the summer rainy period, and the conspicuous white flowers appear on the bare trees from December to March. It is a tree of relatively rapid growth.

*Larrea tridentata* (Zygophyllaceae) CREOSOTE BUSH

(Map 13)

*Larrea* is one of the most abundant and widespread shrubs of the Sonoran Desert, and is equally prominent in the vegetation of the Chihuahuan Desert, the Mojave Desert, and a small part of the Great Basin Desert. It ranges from elevations below sea level in Death Valley to 2625 m. in the mountains of Zacatecas. It persists in regions in which 12 months frequently pass without precipitation, and, irrespective of differences in the seasonal distribution of rainfall, occurs in all sections of the desert and invades the edges of grassland and encinal areas in some of which the annual precipitation is about 500 mm. It grows on deep alluvial soil, on the edges of alkaline flats, on sandy plains, and on the rocky slopes of volcanic hills. It is rarely found on granite slopes and is uncommon on granitic soils. Its adaptability to a wide range of conditions is greater than that of any other desert plant. Also its height, density of branching, size of leaves, and rate of growth are very unlike under different conditions, since it responds readily to an increased water supply. Near the edges of its range *Larrea* is restricted to well marked habitats, but throughout the center of the desert it is less restricted, and in the most unfavorable areas it is highly ubiquitous.

The northern limit of distribution of *Larrea* is in southern Nevada and southwestern Utah, where it follows approximately the contour of 1200 m. In New Mexico the northernmost colonies are near Socorro, but isolated individuals are found 18 km. southwest of Santa Fe at an elevation of 1700 m. In Baja California, it is confined to the eastern side of the northern mountains but extends across the peninsula to the ocean near Rosario. It extends south to the vicinity of La Paz and Todos Santos, being absent from the southern mountains. At its eastern limit in Sonora *Larrea* occurs in large or small isolated colonies, as is also the case in the belt which connects the ranges in the Sonoran and Chihuahuan Deserts. The southern limit on the coast of Sonora is about 12 km. southeast of Guaymas.

Mature individuals of *Larrea* consist of a group of slender branches growing vertically or obliquely from the crown of the root. There is no main branch, either in position or in size. Lateral branches are few below but abundant above, and all the branching is dichotomous. On the twigs there is a slightly elevated ring at each node, which is still recognizable after considerable thickening of the limb. Young plants are simple and graceful in form. With advancing age the crown thickens and the number of branches increases. Young branches usually have an erect position, but gradually bend down as additional branches are formed, through their own weight, through the impact of heavy rains, or particularly through the weight of snow. In very old plants the crown is 20 to 30 cm. thick and the number of branches well over 100. In such plants the largest branches are nearly horizontal and mixed with large dead ones, and the younger oblique and vertical branches carry the bulk of the foliage. The age of an individual is indicated by the size of the crown, and is greater than the age of any of the dead or living limbs. It is therefore impossible to estimate the greatest age which *Larrea* may attain, but probably it greatly exceeds 100 years.

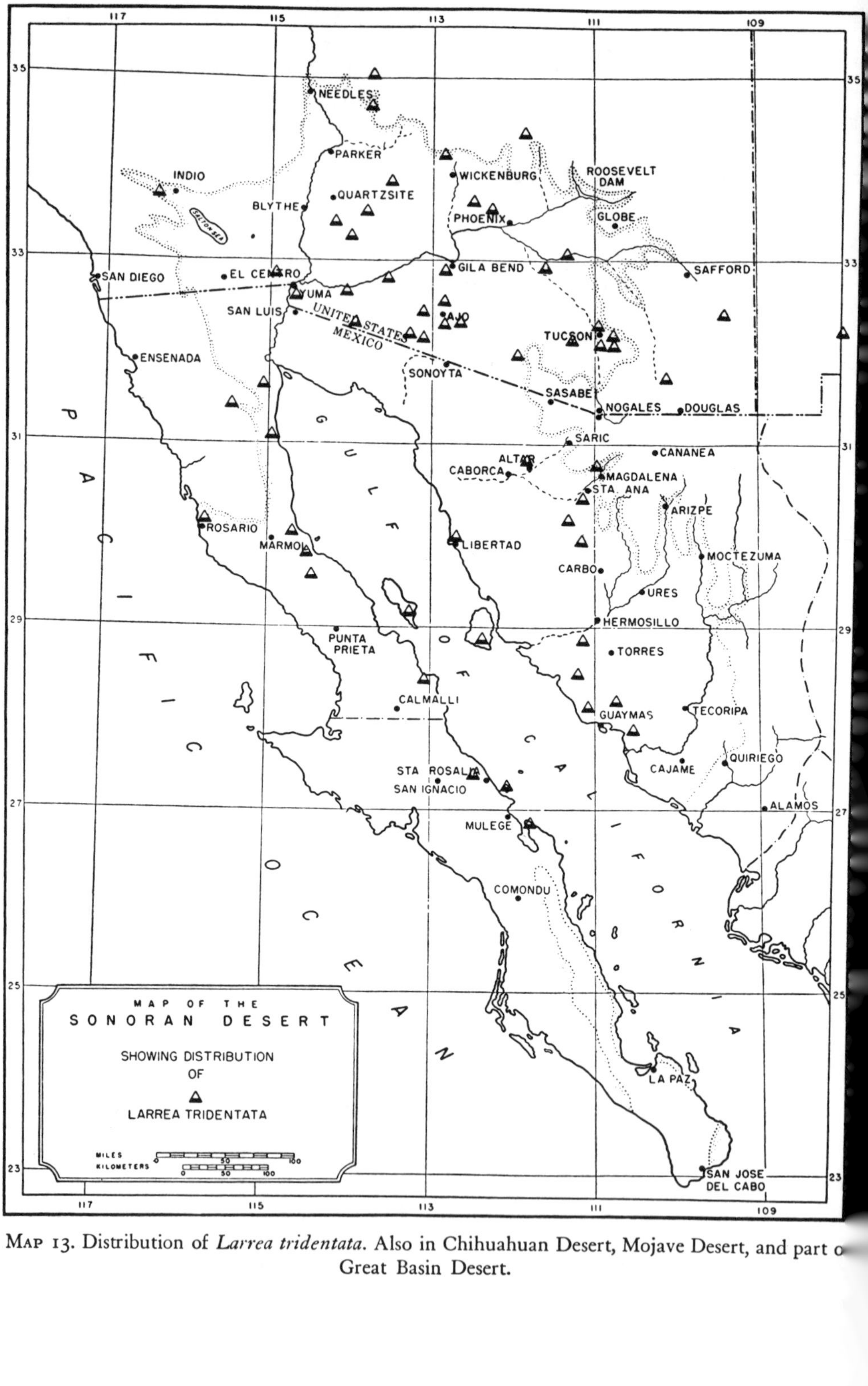


MAP 13. Distribution of *Larrea tridentata.* Also in Chihuahuan Desert, Mojave Desert, and part o[f] Great Basin Desert.

The foliage of *Larrea* is well distributed throughout the crown of the plant, but is heaviest near the ends of the branches. Under the most favorable conditions the foliage is so light that the plant never casts a solid shadow. Under unfavorable conditions many of the leaves are shed, after turning from a deep olive green to a light brownish green. Under very unfavorable conditions tufts of foliage persist here and there on plants which are mainly bare of leaves. A leafless branch is not necessarily dead, but will die if conditions force it to remain leafless for many months.

The leaves of *Larrea* are bifurcated and vary greatly in size according to the moisture conditions at the time of their growth, ranging from 4 to 28 mm. in length. Young leaves which are arrested in growth by drought will usually resume growth on the return of favorable conditions. Under conditions which have not been well determined, the surface of the leaves is covered with a lac or varnish which makes them shiny and viscid. The leaves of *Larrea* are shed copiously during drought periods, but at least 10 to 20 per cent of them are retained during all but the most prolonged rainless periods. They are not winter-deciduous and are rarely injured by low temperatures except at the northern limit.

Flowers are borne sparingly or copiously near the ends of the youngest shoots, and appear after the rains which terminate a drought period, at whatever time of the year they may fall. The principal flowering period follows the late winter rains. The fruits come to maturity in about 6 or 8 weeks and open in a manner which causes some of the seeds to be shed immediately while others remain on the plant. Finally the slender pedicel of the fruit breaks and the entire crop falls to the ground beneath the parent plant. Owing to the lightness of the fruits and their hairy covering, they are blown about to a slight extent and washed away by rain to a much greater extent.

Like those of nearly all other desert perennials, the seeds of *Larrea* germinate in the warm, moist period of July and August. There is a notable scarcity of seedlings in the common stands of mature shrubs. On ground that has been cleared, raked, cultivated, or covered with new soil, the number of seedlings is large. The amount of growth during the first 2 or 3 years varies greatly in different individuals, so that 3-year-old plants may be from 2 or 3 cm. in height to 10 or 15 cm. On disturbed areas a very dense stand of young shrubs may persist until the tallest ones have reached a height of 60 to 80 cm., by which time there is high mortality among the repressed individuals, so that the resulting mature stand is very little denser than is commonly the case.

### Literature


Runyon, E. H. The organization of the creosote bush with respect to drought. Ecology **15:** 128–138, 1934.

Shreve, Forrest. The edge of the desert. Yearbook Assoc. Pacif. Coast Geographers **6:** 6–11, 1940.

——— and T. D. Mallery. The relation of caliche to desert plants. Soil Sci. **35:** 99–112, 1933.

Spalding, V. M. Biological relations of certain desert shrubs. I. The creosote bush (*Covillea tridentata*) in its relation to water supply. Bot. Gaz. **38:** 122–138, 1904.


## *Olneya tesota* (Fabaceae) IRONWOOD. TÉSOTA

(Map 14)

*Olneya tesota* is a leguminous tree belonging to a monotypic genus. It is most abundant in the parts of the Sonoran Desert with less than 200 mm. of rainfall. Even under very

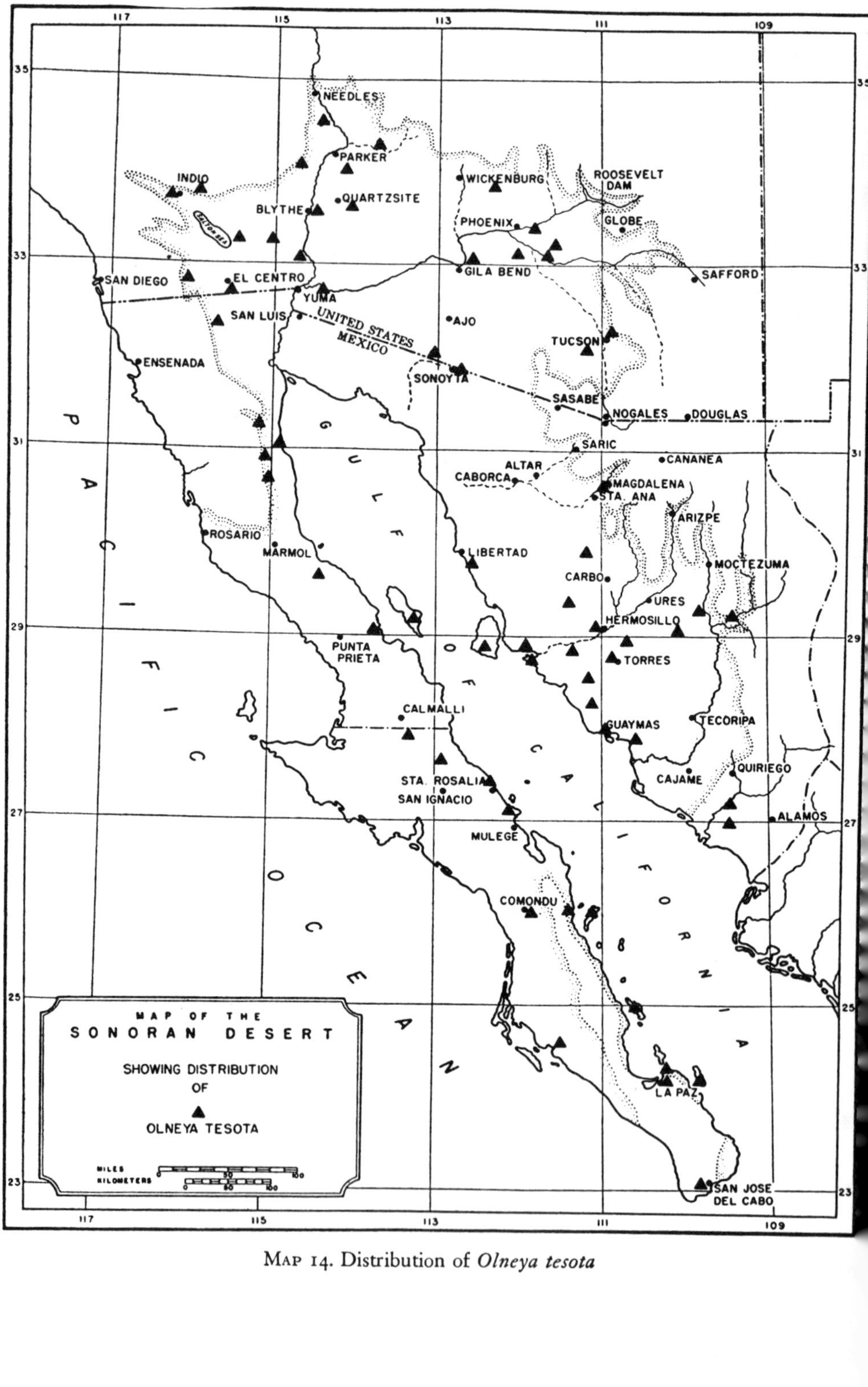


Map 14. Distribution of *Olneya tesota*

adverse moisture conditions it often reaches a height and bulk exceeding those of any other nonsucculent plant in the region.

The distributional limits of *Olneya* correspond closely to the outlines of the Sonoran Desert, but its occurrence is not continuous throughout that area. A few isolated colonies have been noted at short distances beyond the desert at its northern edge, and in the south it occurs in a few localities south of the Mayo River. In Baja California the tree is found throughout the length of the peninsula on the Gulf side but is rare or absent on the Pacific side. In the Arizona Upland *Olneya* occurs in numerous but sharply circumscribed areas, above the lowest plains and on outwash slopes or benches, where temperature inversion does not subject it to the lowest minimum temperatures of the coldest winter nights. In central Sonora it is very abundant near the coast and on the broad plains, but occurs only in scattered colonies along the eastern edge of the desert. The damage suffered by *Olneya* from exceptionally low winter temperatures indicates that they are important in limiting its northern distribution (Turnage and Hinckley, 1938).

The mature trees of *Olneya* have a single trunk or a group of trunks with the lowest limbs either well above the ground or prostrate on it. The main trunk is often surrounded by young upright branches, by suckers from the roots, and by dead trunks of various sizes. Almost every tree shows the loss of small limbs or even of large accessory trunks as a result of drought. Normally the crown is fairly dense, and vigorous trees throw a heavy shade. The leaves are compound, with 12 to 20 leaflets and a total area of about 350 sq. mm. The leaflets are thin and flexible, but the foliage is evergreen and falls in part only as a result of severe drought or heavy frost. In drought the leaflets fall without change of color. After frost the leaves die but do not fall for several weeks. Leafless living trees are seldom seen, and have been noted by the writer only between Ehrenberg and Quartzsite, Arizona, in June 1936.

Flowers appear in March, being very abundant in about 2 years out of 5, and almost wholly absent in the other years. It has not been possible to correlate the appearance of flower crops with the moisture or temperature conditions of the preceding months. The fruit is a short indehiscent pod, usually containing a single seed and very rarely more than two. The pods ripen from about the end of May to the middle of June and fall as soon as mature. The seeds are black, oval, about 8 mm. in length, and provided with a heavy, slightly waxy coat. Like all the other large, nutritious leguminous seeds, those of *Olneya* are eagerly sought by all the desert rodents. At night in season a continuous crackling sound may be heard under every tree as the rodents remove the seeds from the pods. The principal factor favoring the germination of the remaining fraction of the seed crop is the occurrence of unusually early rain. The thin wax coating does not serve to delay the germination of the seeds, which takes place within a few weeks of their fall. A single rain may rupture the seed coat without providing the continued moisture needed for full germination. Large numbers of seeds with ruptured coats have been seen under trees in northwestern Sonora, together with a few in which the radicle and plumule had begun to grow. Although these seeds had been thoroughly dry for a fortnight, they were found to be capable of resuming the process of germination when placed in moist soil. All the seeds with merely ruptured coats that were tested were found viable, but only a few of the slightly more advanced ones with a dried radicle and plumule were capable of growth.

*Olneya* is readily propagated from seeds, but the number of natural seedlings and small plants in all parts of its range is extremely small. With a limited supply of water, greenhouse seedlings will reach a height of 15 to 20 cm. in the first year. Suckers and vigorous young branches on trees in the open make a growth of as much as 20 cm. in a single season with good soil-moisture conditions. Nothing is known of the average growth performance of the tree or of its length of life.

The sapwood of *Olneya* is light yellow, and 2 to 4 cm. thick. The heartwood is dark brown and has the extreme hardness which gives the tree its English name. Rings of growth are discernible in both layers of wood, but are very narrow and greatly interrupted, and serve only as a record of the number of favorable seasons in the life of the tree.

The commonest cause of the death of large trees is a prolonged saturation of the soil, such as frequently occurs during the summer rains as a result of the clogging of former drainage channels and the spreading of the floodwater over areas which offer it no outlet. Small groves of dead trees are frequent along the western edge of the Plains of Sonora. The sapwood of dead trees is quickly consumed by termites, but the heartwood is not attacked. Groups of trees which have been dead for several years consist merely of the abbreviated heartwood skeleton, which withstands weathering for many years. The constant movement of wind-blown sand near the ground level slowly cuts the base of the skeleton until it is small enough to break.

LITERATURE

TURNAGE, W. V., and A. L. HINCKLEY. Freezing weather in relation to plant distribution in the Sonoran Desert. Ecol. Monogr. **8:** 529–550, 1938.

*Pachycereus pecten-aboriginum* (Cactaceae) HECHO

This stout columnar cactus reaches a height of 10 to 12 m. Its branches are numerous and sharply ascending. It is rare in the open desert, and its northernmost representatives are found in thick stands of trees and shrubs. *Pachycereus pecten-aboriginum* occurs sparingly in the southern part of the Plains of Sonora, but becomes abundant in the thorn forest of Sinaloa. The dry fruit is covered with strong, slender spines. The specific name of the plant refers to the use of the fruits as combs, a practice which is still very common.

*Pachycereus Pringlei* (Cactaceae) SAHUASO. CARDÓN

(Map 15)

In size and general form *Pachycereus Pringlei* is similar to *Carnegiea*, but the two can be readily distinguished at a distance. *Pachycereus* is more massive and more compactly branched, its center of gravity is nearer the ground, and the branches are inserted at a sharp angle rather than a sweeping curve.

*Pachycereus Pringlei* occurs almost throughout the desert areas of Baja California and near the coast in central Sonora, only slightly overlapping the range of *Carnegiea* and rarely growing in equal abundance with it. The northernmost occurrence of *Pachycereus* in Baja California is in the foothills west of Bahía San Felipe (latitude 31° N.), where it is very infrequent. On the Pacific coast its northernmost occurrence is about 40 km. north of Rosario. Southward from these localities it rapidly increases in abundance and is found from coast to coast in every area that has been visited below an elevation of about 700 m. In Sonora, *P. Pringlei* is abundant in a coastal belt about 50 km. wide, extending from Sierra Picú, northeast of Puerto Libertad, south to hills east

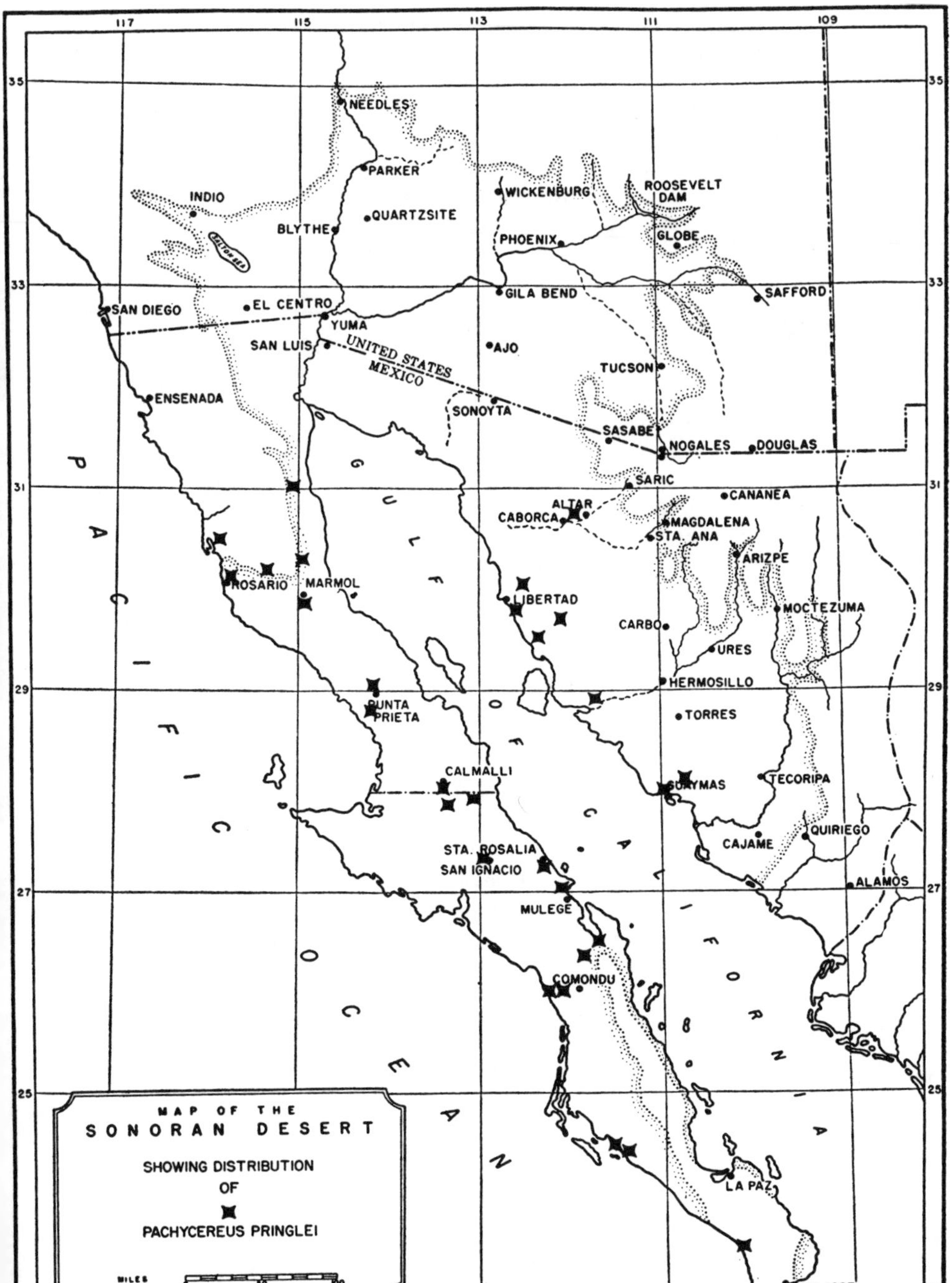


MAP 15. Distribution of *Pachycereus Pringlei*

of Guaymas. A single isolated colony is known about 100 km. from the coast near Pitiquito. Throughout most of the range it occurs as isolated individuals or in very open stands associated with many other plants. Like those of *Carnegiea*, its heaviest stands are on coarse outwash slopes and low hills, but it is much more tolerant of sand and alluvial soil than is *Carnegiea*. The maximum of size and abundance, and the greatest catholicity of habitat, are on the Gulf coast of Baja California in the vicinity of Mulegé and Bahía Concepción. The only heavy stand observed in Sonora is on the outwash slopes overlooking the tidal flats near Empalme.

Young plants of *P. Pringlei* are slender at the base for many years, but increase greatly in thickness after formation of branches begins. The old branched individuals are massive, with the trunk reaching a diameter of 50 to 60 cm. and the branches a diameter of 30 to 40 cm. In favorable situations the plants often reach a height of 10 to 12 m. and produce from 5 or 6 to as many as 30 branches. It is only rarely that the sahuaso retains an upright posture with all its branches in erect position. They usually give the impression of having been vigorously battered by the wind and thrown out of balance by their great weight.

The longitudinal grooves of the stem are at first deep, but become more shallow with increasing age. In young plants the ridges are capped by a close-set series of areoles with white glochids. The glaucous green of the surface is marked by faint dark lines which run from the bottom of the groove to the apex of the ridge. On the trunks of old trees, a smooth gray bark replaces the green surface and there is little or no indication of the ridges, which have been flattened by the secondary thickening. The smooth, gray trunk vividly suggests the leg of an elephant. Higher on the old plants a rough bark is formed, gradually extending down from the tops of the ridges, on which the remaining glochids are now black.

The skeleton of *P. Pringlei* is a cylinder of 12 to 15 ribs which are separate in the upper part of the trunk, but connected below by loose, open woody tissue in which there are longitudinal rows of small holes 1 to 2 cm. apart. In trunks 40 to 50 cm. in diameter the ribs are fused into a solid cylinder 3 to 4 cm. thick, but sometimes still show series of small holes 2 to 4 mm. in diameter, the last vestiges of the areoles. The basal part of the woody cylinder is about 15 cm. thick and occupies most of the cross section of the trunk, with only a small lumen at the center and a thin outer cortical layer.

The flowers of *P. Pringlei* appear in late March or early April, arising from the areoles just below the apex of the stem down to 30 or 40 cm. from the apex. The fruit is relatively dry at maturity and contains about 500 very small seeds, a large number of which are abortive. The features of germination and growth are much like those in *Carnegiea* so far as is known. In comparable stands of the two plants, the number of small individuals is commonly somewhat greater in the case of *Pachycereus*.

### *Pachycormus discolor* (Anacardiaceae) — ELEPHANT TREE. TOROTE

(Map 16)

*Pachycormus* is one of the most striking and beautiful trees of Baja California; also it often assumes very grotesque forms (pl. 35, fig. 2). In all locations it has a trunk and limbs of exaggerated thickness, and near the Pacific coast it is usually prostrate and gnarled. In the interior the trees are upright and freely branched, usually with a clean trunk of 5 to 12 dm. between the ground and the lowest limbs. The trunk and limbs are covered with a smooth, light buff bark which exfoliates repeatedly during the life

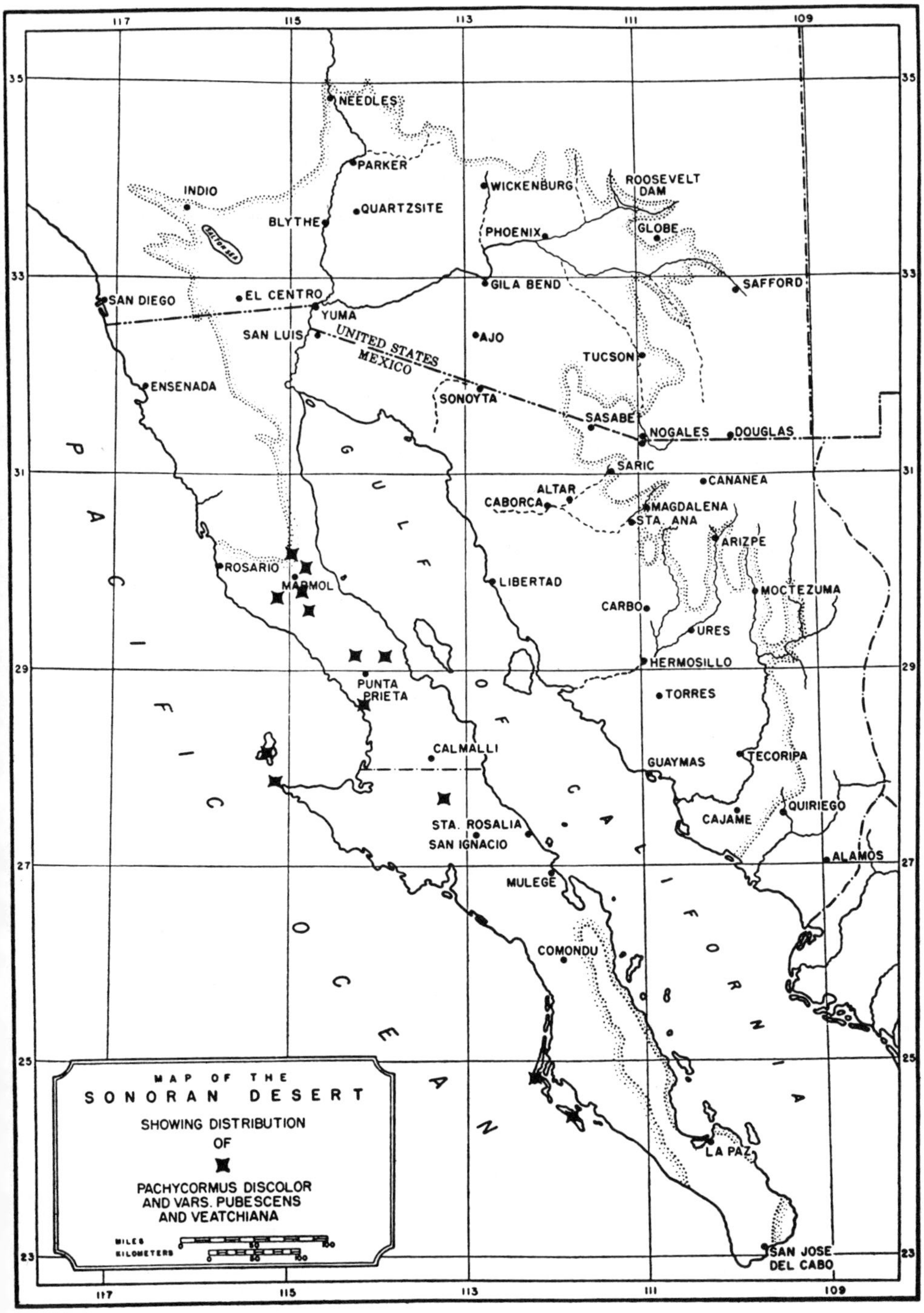


MAP 16. Distribution of *Pachycormus discolor* and vars. *pubescens* and *Veatchiana*

of the tree. The dark green compound leaves are borne during the rainy seasons and shortly thereafter. The flowers are small and inconspicuous. A very stout-stemmed species of dodder infests the trees in rainy years and appears to be responsible for the death of many of the badly infested individuals. *Pachycormus* is found on rocky slopes and hillsides throughout the central third of the peninsula of Baja California. In some localities it is abundant, in others it is absent over large areas. Its best development and greatest abundance are in the hills near the drainage divide of the peninsula, but it extends down to the Gulf coast in several places and is known from Angel de la Guarda Island.

*Piscidia mollis* (Fabaceae)

This is a small tree, rarely exceeding 7 m. in height, with the irregular form and branching of an oak. It has an open crown and large, sericeous, trifoliolate leaves. There is considerable defoliation in the late spring, but the tree is never leafless. *Piscidia* is infrequent but conspicuous in the southernmost part of the Plains of Sonora, where it is confined to the eastern half of the area.

*Pithecellobium sonorae* (Mimosaceae) PALO JOCONO

A slender tree with a compact, narrow crown, often 8 m. in height. Single individuals are less common than close groups of small trees surrounding two or three large ones. The bark is light gray and beset with short, sharp thorns. The leaves are bipinnate, about 4 cm. long, and bear from 4 to 6 pinnae with very small leaflets 3 to 4 sq. mm. in area. This species of *Pithecellobium* is found only in the southern part of the Plains of Sonora, where its tall, narrow crown makes the scattered individuals conspicuous along the streamways.

Maps 17 to 27 show the known distribution of species important in the make-up of the flora of the Sonoran Desert, but not described individually in the text.

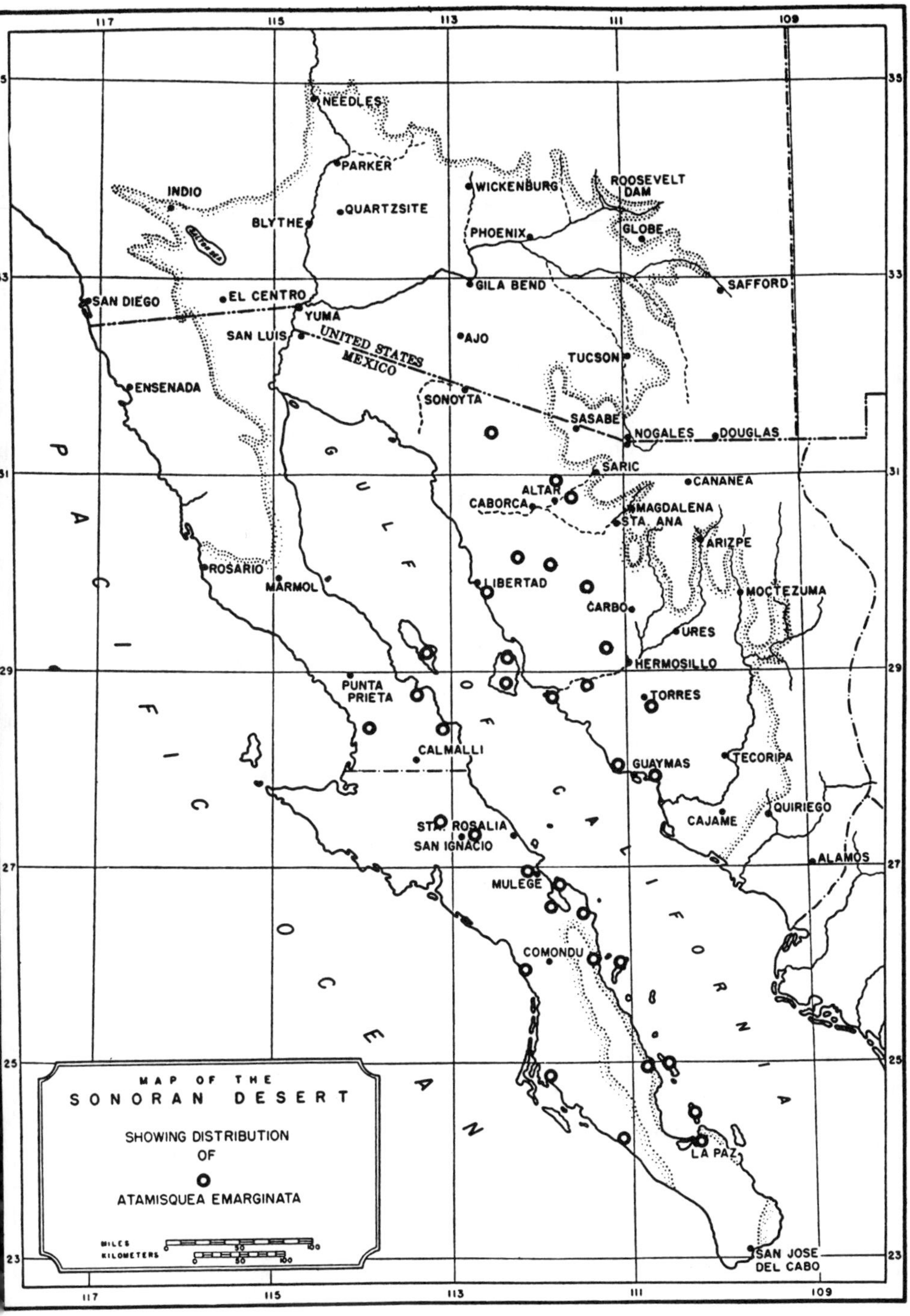


MAP 17. Distribution of *Atamisquea emarginata*

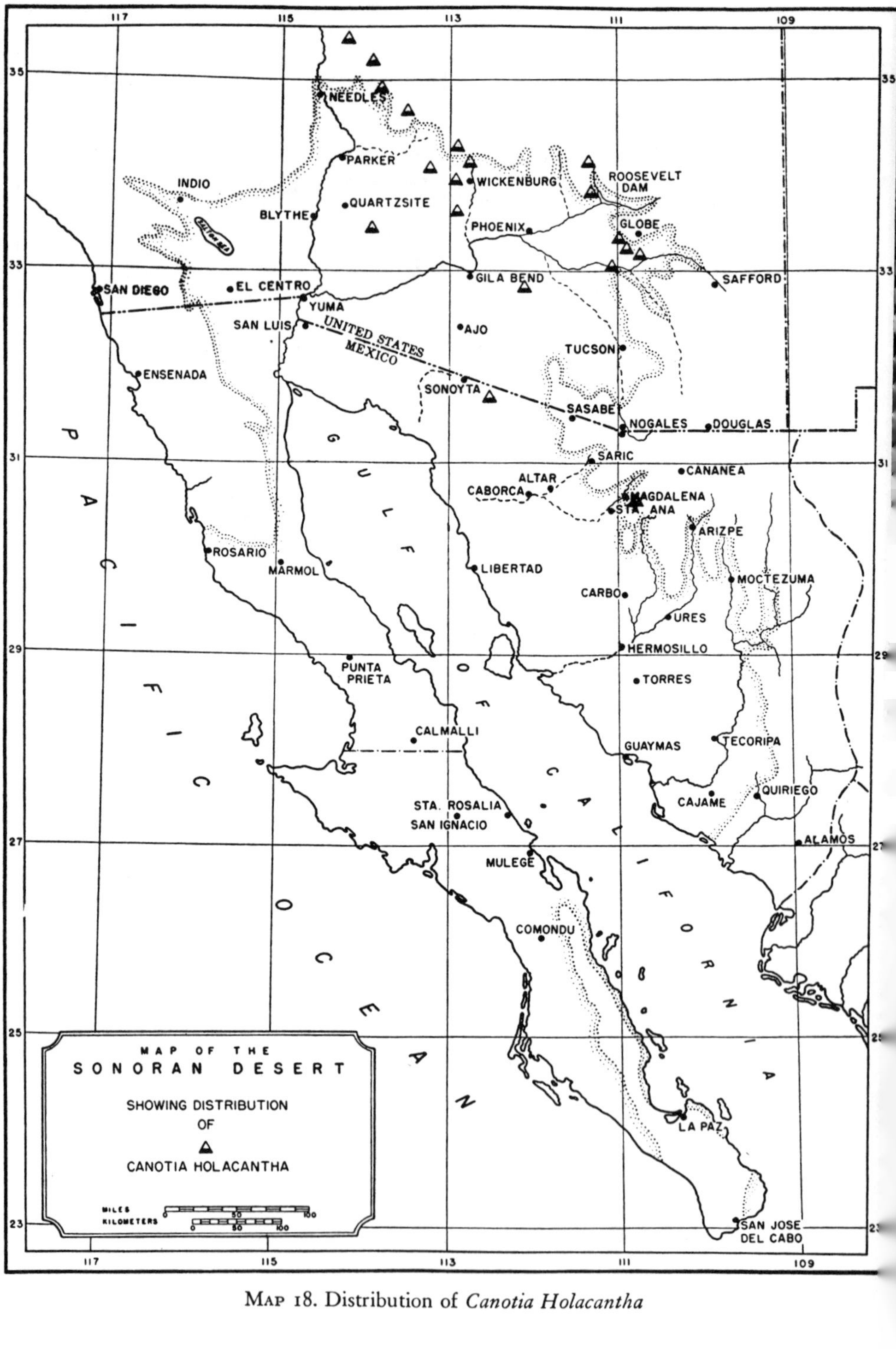


Map 18. Distribution of *Canotia Holacantha*

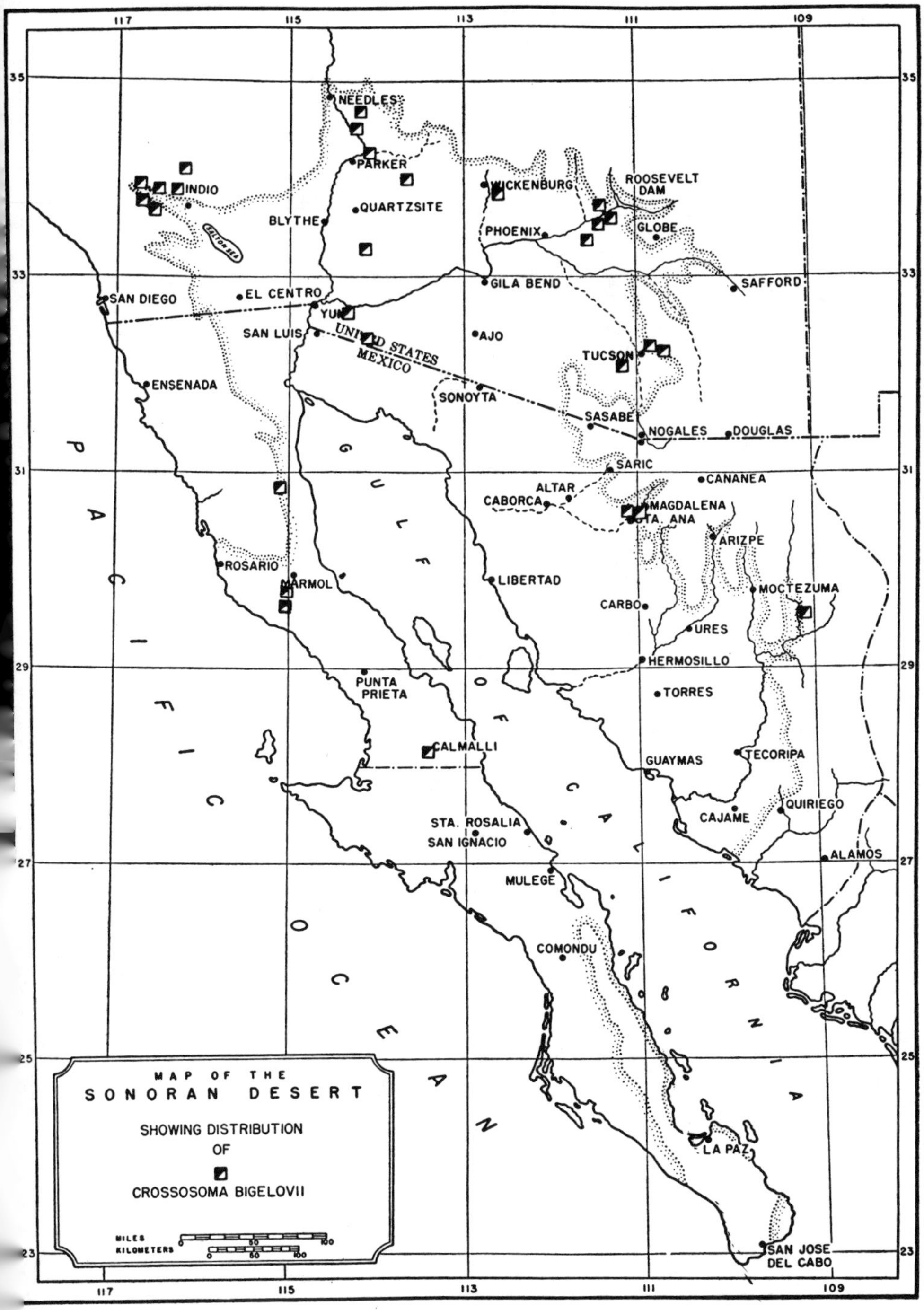


MAP 19. Distribution of *Crossosoma Bigelovii*

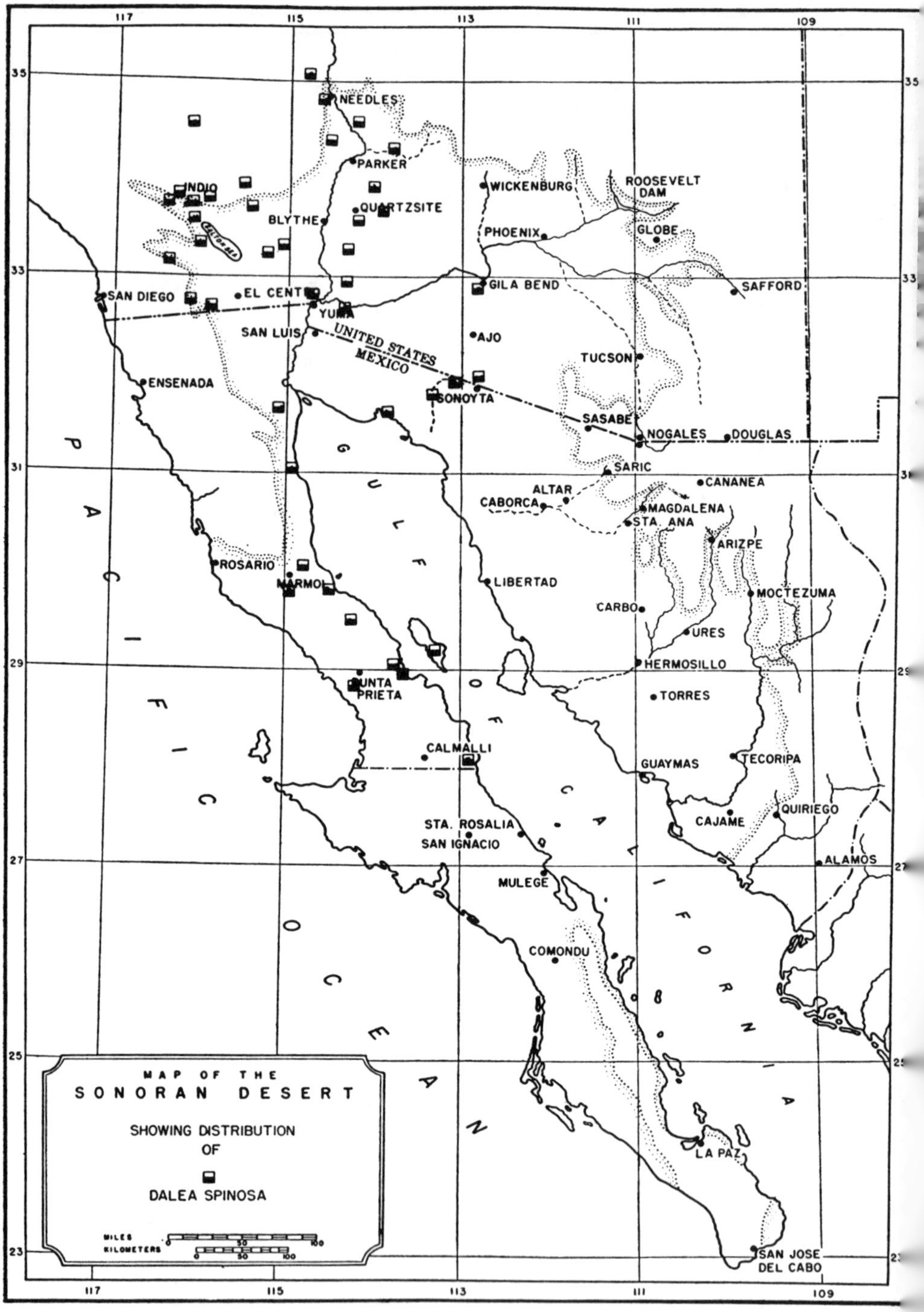


MAP 20. Distribution of *Dalea spinosa*

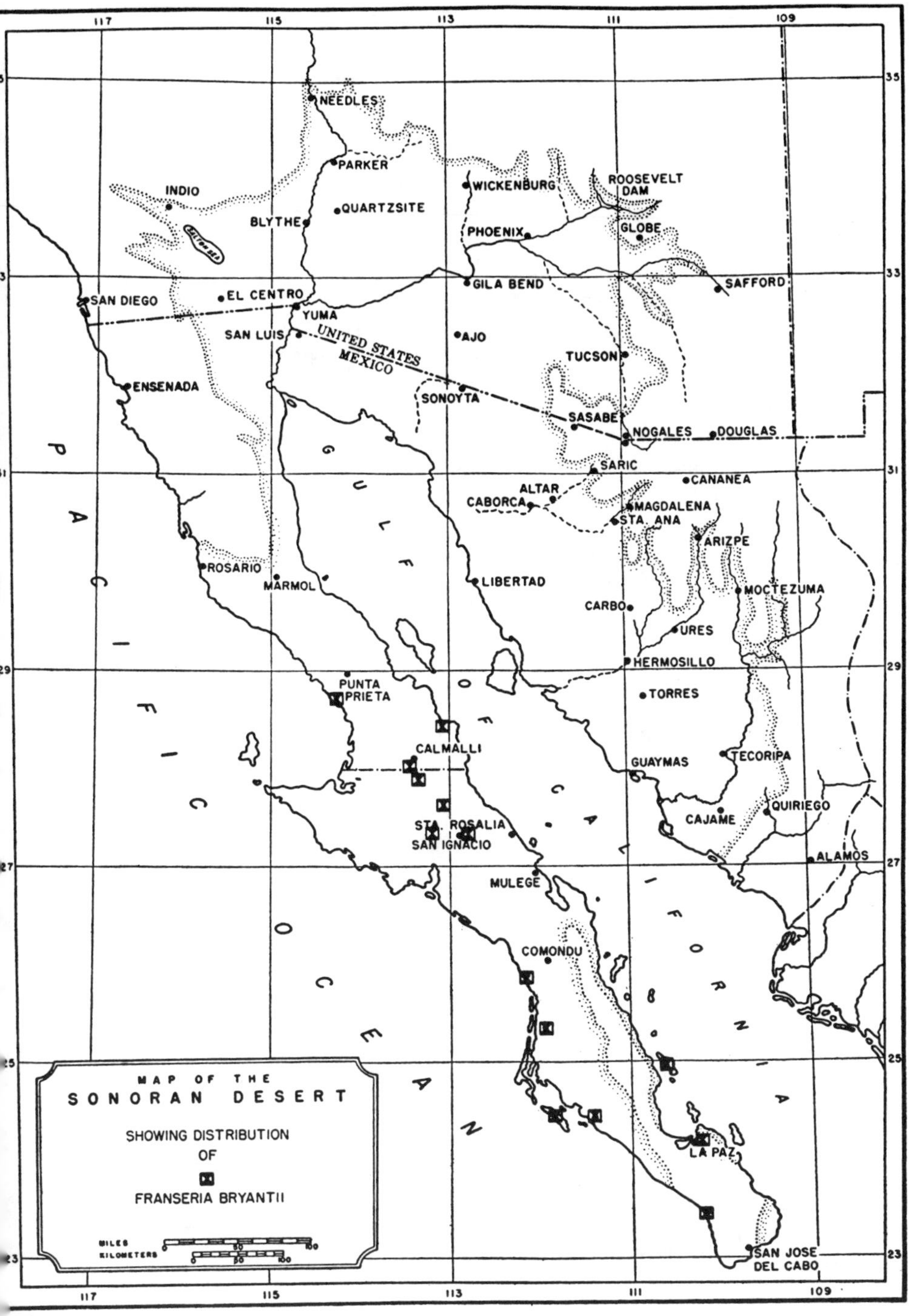


MAP 21. Distribution of *Franseria Bryantii*

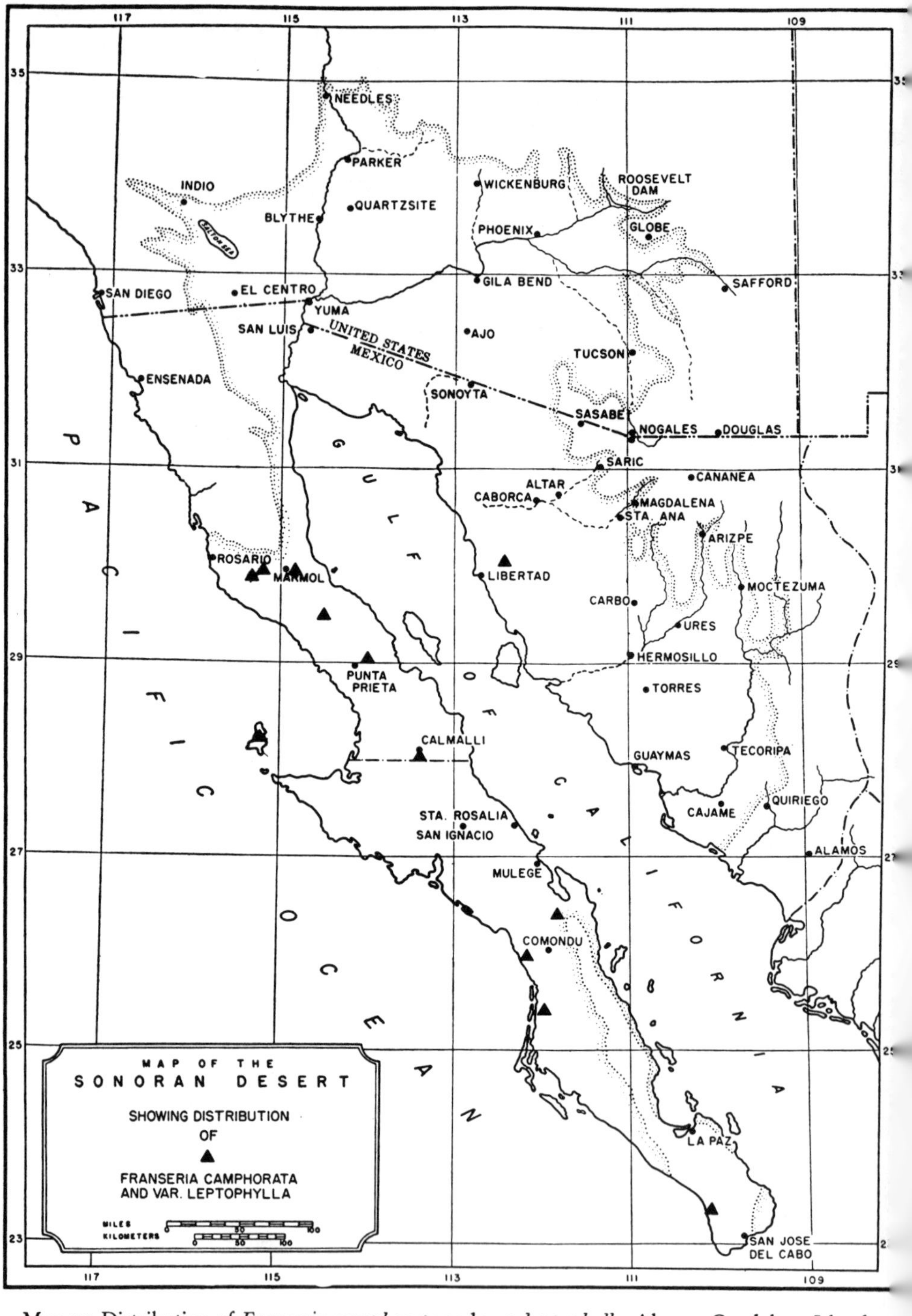


MAP 22. Distribution of *Franseria camphorata* and var. *leptophylla*. Also on Guadalupe Island.

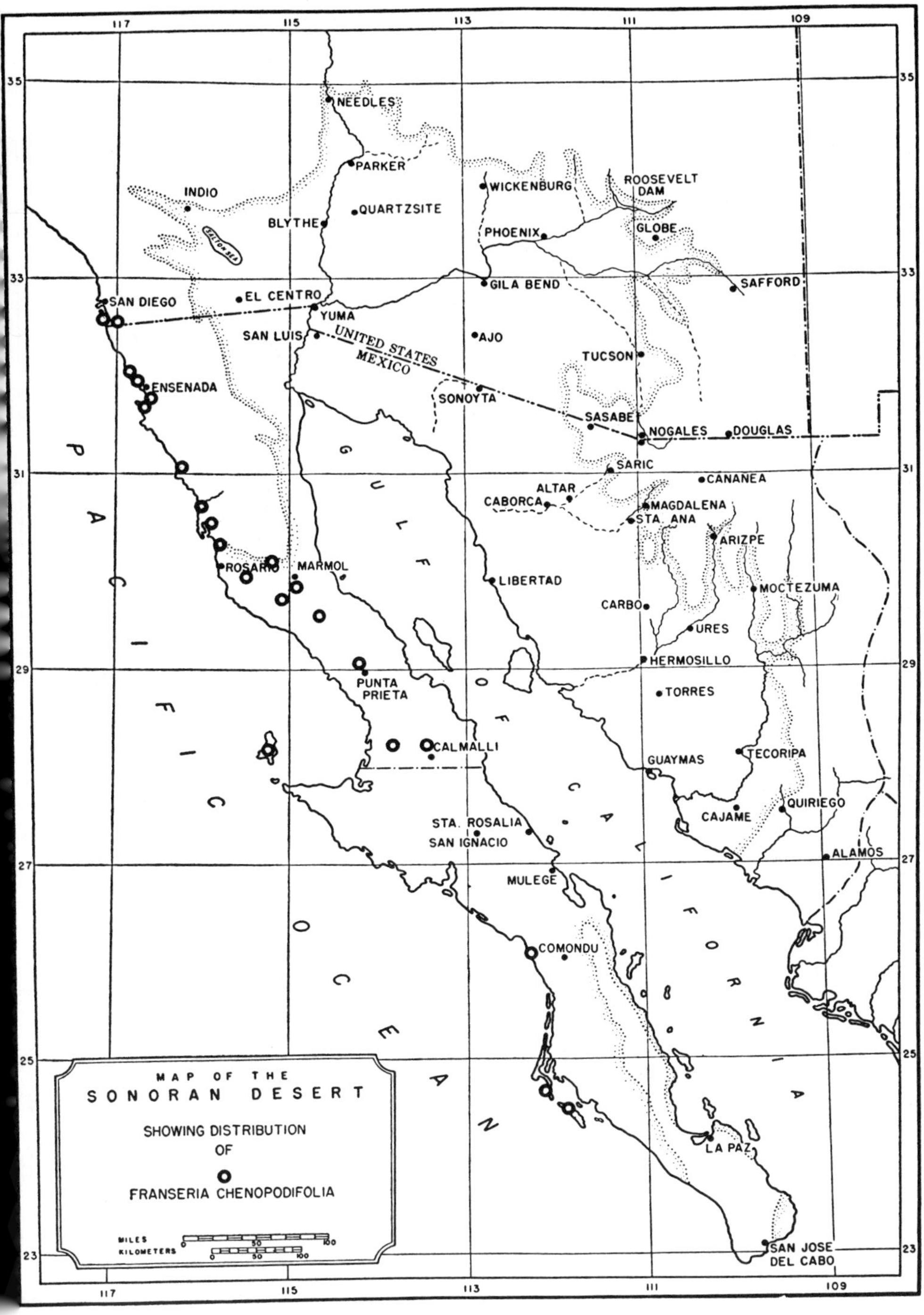


MAP 23. Distribution of *Franseria chenopodifolia*

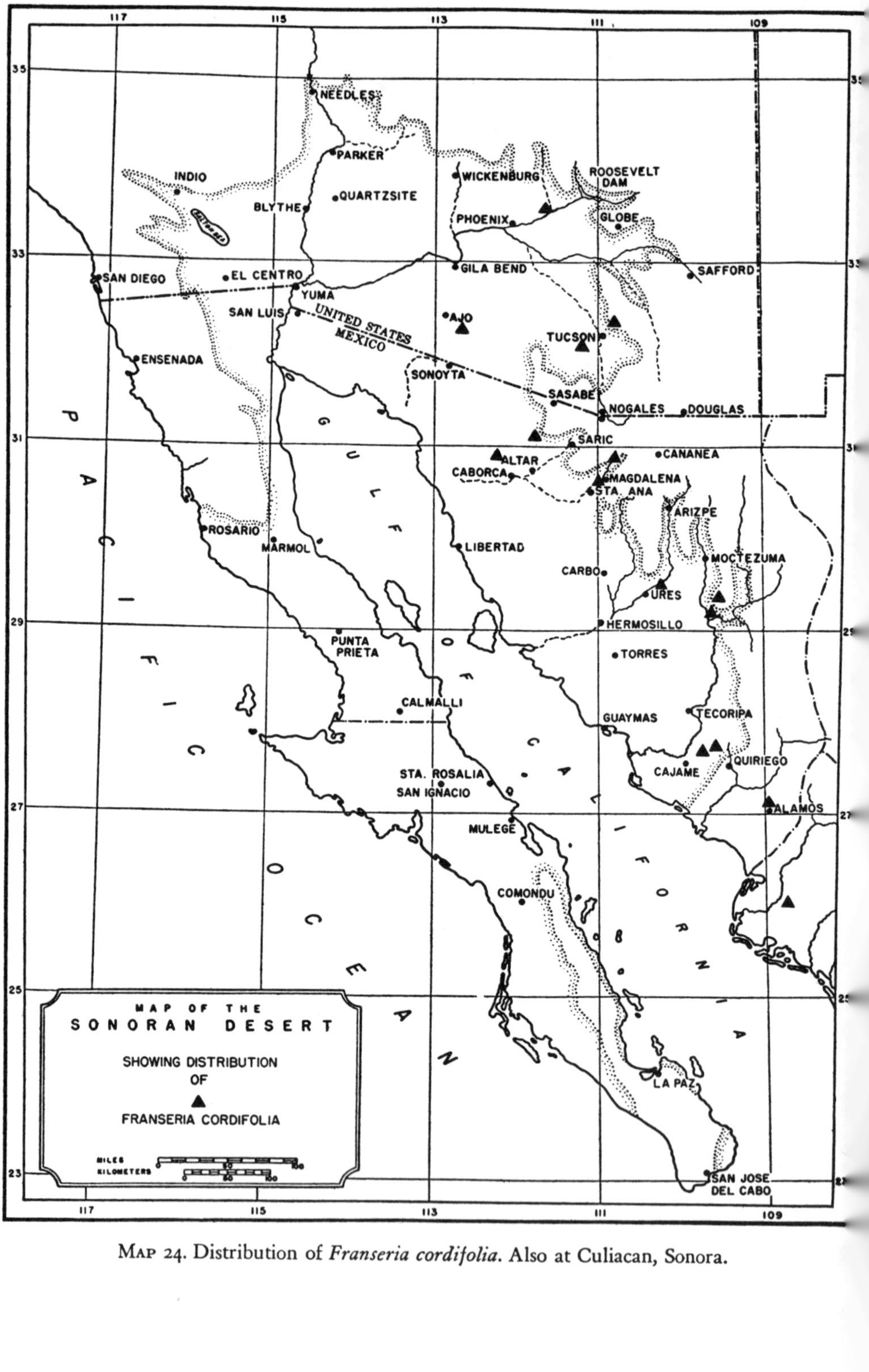


Map 24. Distribution of *Franseria cordifolia*. Also at Culiacan, Sonora.

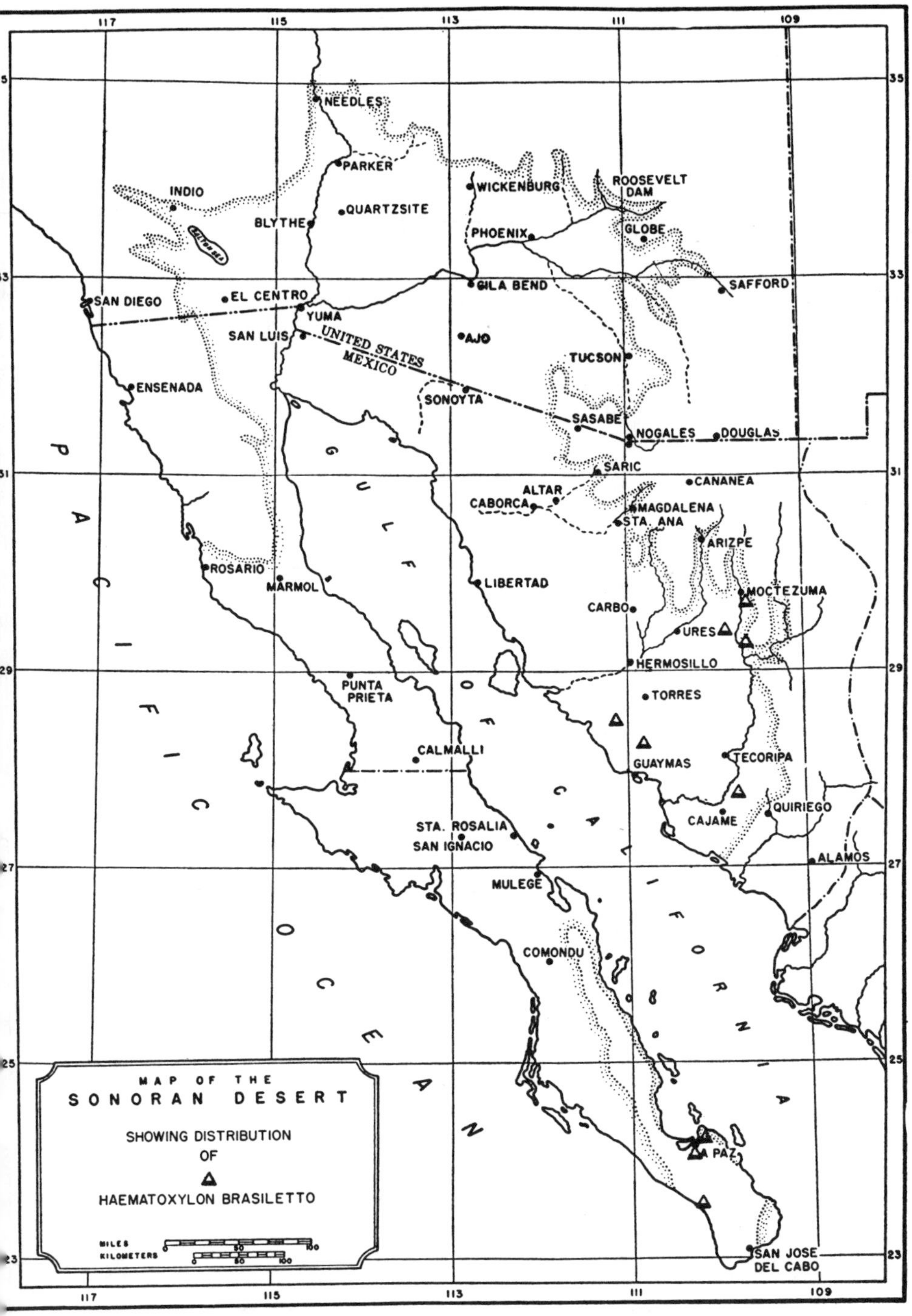


MAP 25. Distribution of *Haematoxylon brasiletto*

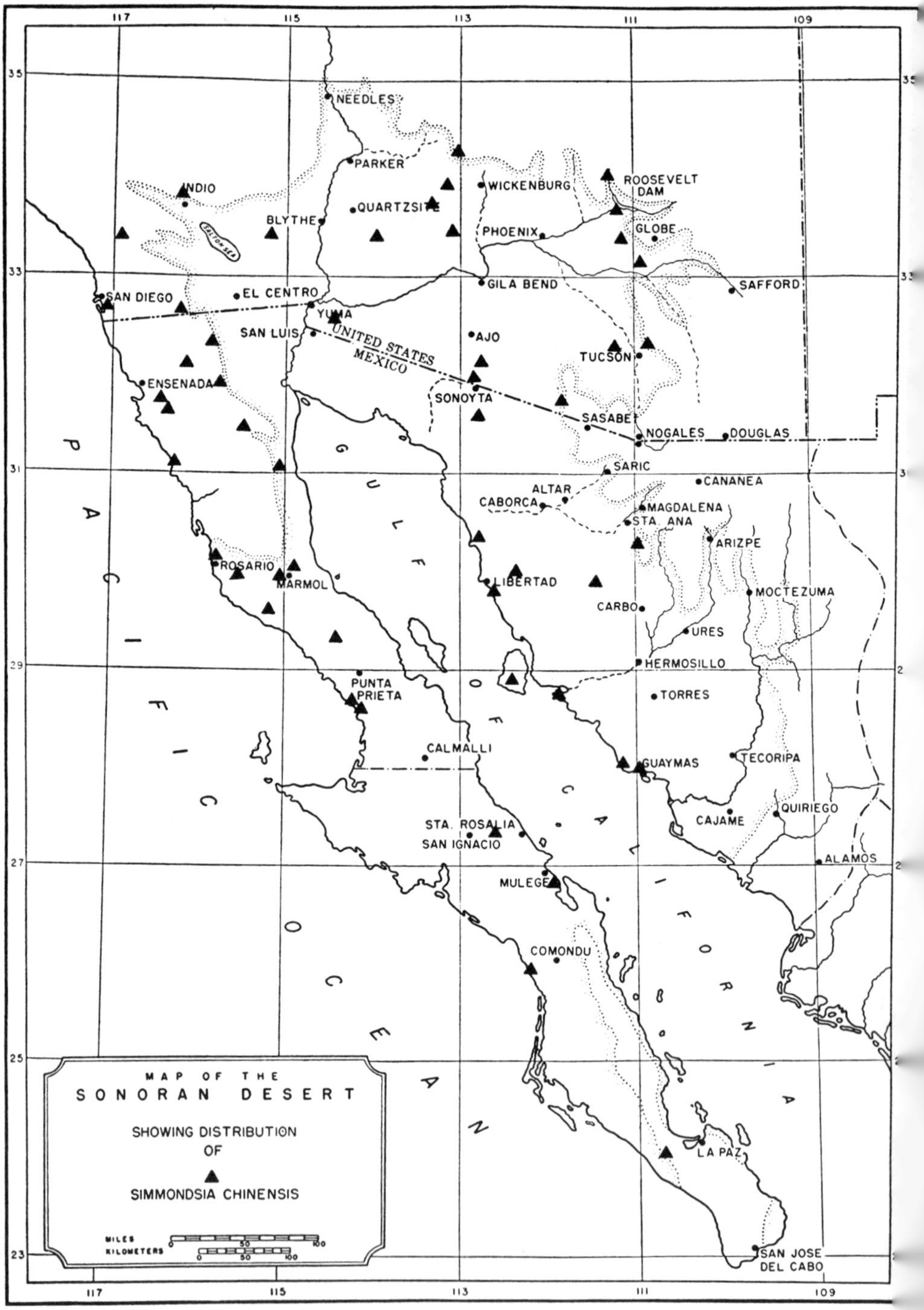


Map 26. Distribution of *Simmondsia chinensis*

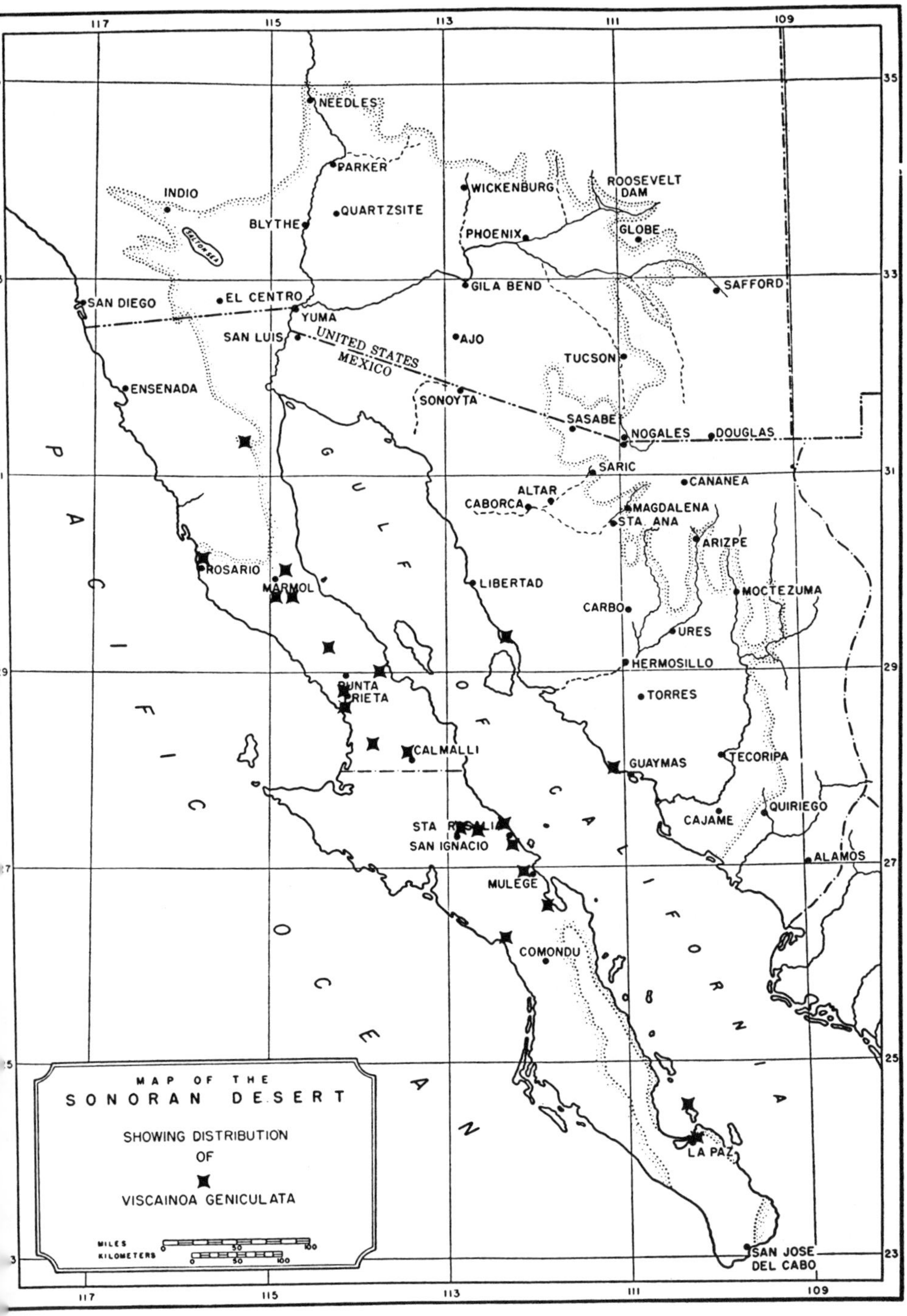


Map 27. Distribution of *Viscainoa geniculata*

# Literature


Bryan, Kirk. 1925. The Papago country, Arizona. U. S. Geol. Surv. Water Supply Paper No. 499.

Burt, William H. 1938. Faunal relationships and geographic distribution of mammals in Sonora, Mexico. Univ. Mich. Mus. Zool. Misc. Pub. No. 39.

Davis, W. M. 1921. Lower California and its natural resources: a review. Geogr. Rev. **11:** 551–562.

Dice, Lee R. 1939. The Sonoran biotic province. Ecology **20:** 118–129.

——— and Philip M. Blossom. 1937. Studies of mammalian ecology in southwestern North America, with special attention to the colors of desert mammals. Carnegie Inst. Wash. Pub. 485.

Fenneman, N. F. 1928. Physiographic divisions of the United States. Ann. Assoc. Amer. Geographers **18:** 261–353, map.

Goldman, E. A. 1916. Plant records of an expedition to Lower California. Contr. U. S. Nat. Herb. **16:** 309–371.

Harshberger, John W. 1911. Phytogeographic survey of North America. *In* Engler and Drude, Vegetation der Erde, vol. 13, map.

Hilgard, E. W. 1906. Soils. 396 pp. New York.

Humphrey, Robert R. 1933. A detailed study of desert rainfall. Ecology **14:** 31–34.

Ives, Ronald L. 1936. Desert floods in the Sonoyta valley. Amer. Jour. Sci., ser. 5, **32:** 349–360.

McGee, W J 1897. Sheetflood erosion. Bull. Geol. Soc. Amer. **8:** 87–112.

Ochoterena, Isaac. 1937. Esquemas biotípicos y sinecias características de las regiones geográfico-botánicas de Mexico. Ann. d. Inst. Biol. Mexico **8:** 463–597.

Rempel, Peter J. 1936. The crescentic dunes of the Salton Sea and their relation to the vegetation. Ecology **17:** 347–358.

Rübel, E. 1920. Über die Entwicklung der Gesellschaftsmorphologie. Jour. Ecol. **8:** 18–40.

Sanders, E. M. 1921. The natural regions of Mexico. Geogr. Rev. **11:** 212–226.

Shreve, Forrest. 1912. Cold air drainage. Plant World **15:** 110–115.

——— 1915. The vegetation of a desert mountain range as conditioned by climatic factors. Carnegie Inst. Wash. Pub. 217.

——— 1922. Conditions indirectly affecting vertical distribution on desert mountains. Ecology **3:** 269–274.

——— 1934. Rainfall, runoff and soil moisture under desert conditions. Ann. Assoc. Amer. Geographers **24:** 131–156.

——— 1936. The transition from desert to chaparral in Baja California. Madroño **3:** 257–264.

——— and T. D. Mallery. 1933. The relation of caliche to desert plants. Soil Sci. **35:** 99–112.

Spalding, Volney M. 1909. Distribution and movements of desert plants. Carnegie Inst. Wash. Pub. 113.

Sykes, Godfrey. 1926. The delta and estuary of the Colorado River. Geogr. Rev. **16:** 232–255.

——— 1937. The Colorado delta. Carnegie Inst. Wash. Pub. 460.

Tolman, C. F. 1909. Erosion and deposition in the southern Arizona bolson region. Jour. Geol. **17:** 136–163.

Turnage, William V., and A. L. Hinckley. 1938. Freezing weather in relation to plant distribution in the Sonoran Desert. Ecol. Monogr. **8:** 529–550.

# Plates

About two-thirds of the plates are reproductions of the printed plates in the original publication. The new photographs were supplied by Ira L. Wiggins, Stanford University (1, 2, 3 [Fig. 2], 5, 26), and by Raymond M. Turner, U. S. Geological Survey, and James Rodney Hastings, University of Arizona (20–24, 27–28). The new pictures were views of similar areas, except for plates 20, 21, 22, and 27, which were taken from the same locations.

Close-set mosaic of coarse gravel covering soil near Papago Wells, Pima County, Arizona

FIG. 1. Bare desert pavement alternating with belts of vegetation near San Felipe, northern Baja California

FIG. 2. Area of fine soil subject to flooding by infrequent rains, Vizcaíno Desert, Baja California. Plants are *Cenchrus Palmeri* and *Phaseolus filiformis*

FIG. 1. Hill of sedimentary material capped by basalt, as shown by road cut. Near Gillespie Dam, Maricopa County, Arizona

FIG. 2. Areas of lag talus on mesas near Rancho Cuarenta, Baja California

PLATE 4

Granitic hills at base of Cuyamaca Mountains, Imperial County, California

Pure open stand of *Larrea tridentata* near Arroyo Calamajué, northern Baja California

PLATE 6

Looking south toward hills 65 km. west of Sonoyta, Sonora. The sandy soil shows strips almost devoid of plants, and the sand

The Salada of the Río Sonoyta, near Quitovaquito, Sonora

Vegetation near Tule Tanks, Yuma County, Arizona. A very open stand of *Franseria dumosa, Jatropha cuneata, Fouquieria splendens,* and

South face of the Kofa Mountains, northern Yuma County, Arizona. The steep slopes, active erosion, and hanging talus result in a very light cover of vegetation.

PLATE 10

A deeply cut canyon on the north face of the Kofa Mountains. This and similar adjacent canyons are the only Arizona localities

Coarse outwash plain near Wickenburg, Maricopa County, Arizona, with vegetation characteristic of large areas in the Arizona Upland desert. *Franseria deltoidea, Cercidium microphyllum, Celtis pallida,* and *Acacia constricta.*

PLATE 12

Vegetation [illegible] north of Aguila, Yavapai County, Arizona, at the southern limit of *Yucca brevifolia*

The upper bajadas near the south end of the Hualpai Mountains, Arizona, are well covered with *Larrea* and *Hilaria mutica,* with infrequent but conspicuous individuals of *Yucca brevifolia, Juniperus utahensis,* and *Yucca schidigera.*

The Plains of Sonora in the vicinity of Carbó are characteristic of a large area in which *Olneya, Cercidium,* and *Encelia* are very abundant and *Larrea* is sharply localized. At right, *Rathbunia alamosensis.*

On the Plains of Sonora 25 km. northwest of Hermosillo. A heavy stand of *Encelia farinosa* with *Prosopis* (leafless), *Guaiacum*, and *Caesalpinia pumila*.

PLATE 16

Low hills 35 km. northeast of Estación Moreno, Sonora. A well mixed stand of *Bursera microphylla, Ipomoea arborescens, Ceiba acuminata,* and *Piscidia mollis.* The only cactus is *Pachycereus pecten-aboriginum.*

In the southernmost arm of the Sonoran Desert 20 km. northeast of Cajeme, Sonora. *Acacia cymbispina, Cercidium sonorae, Jatropha cinerea, Pachycereus pecten-aboriginum, Rathbunia alamosensis, Karwinskia parvifolia* var. *pubescens, Caesalpinia pumila,* and *Opuntia fulgida.*

On the northern edge of the delta of Río Yaqui near Potam, Sonora. With a scattered stand of *Lycium*, *Atriplex*, and *Stegnosperma* is a rich growth of cacti, including *Lemaireocereus*, *Lophocereus*, *Rathbunia*, *Ferocactus*, *Opuntia*, and *Mammillaria*.

Plate 20

Looking north across an arm of the Gulf of California from hills 10 km. south of Puerto Libertad, Sonora. *Idria columnaris* on the slopes.

PLATE 22

Volcanic slopes 25 km. north of Laguna Seca Chapala, Baja California. *Prosopis juliflora* var. *Torreyana*, *Atriplex polycarpa*, *Franseria chenopodifolia*, and *Haplopappus sonoriensis*.

A heavy stand of *Idria columnaris* on coarse volcanic soil 20 km. north of Punta Prieta, Baja California. Among the associated plants are *Pachycereus Pringlei, Yucca valida, Machaerocereus gummosus, Pachycormus discolor, Lophocereus Schotti, Franseria chenopodifolia, Pedilanthus macrocarpus, Euphorbia misera, Franseria magdalenae,* and *Agave.*

Vegetation near the locality shown in plate 23, with characteristic trees of *Pachycormus*

Granite hills and sandy flood plain 25 km. northeast of Pozo Alemán, Baja California. *Yucca valida, Lophocereus Schottii, Opuntia ciribe,* and *Haplopappus sonoriensis.* The sandy surface is well covered with herbaceous ephemerals.

The inner edge of the Vizcaíno Desert 37 km. north of El Arco, Baja California. The perennial vegetation includes *Yucca valida, Fouqueria peninsularis, Machaerocereus gummosus, Franseria magdalenae, Opuntia cholla, O. calmalliana,* and *Larrea.* The exceptionally heavy growth

Edge of the lava at the south base of Tres Vírgenes peak, Baja California. *Bursera, Lycium, Encelia, Solanum Hindsianum.*

Slopes of barranca in volcanic mesa 6 km. north of Comondú, Baja California. *Fouquieria peninsularis, Acacia, Lemaireocereus, Pachycereus,*

Characteristic surface on the Magdalena Plain near El Refugio, Baja California (latitude 24° 50′ N.), showing bare depressions. *Prosopis juliflora* var. *Torreyana, Pachycereus, Lophocereus, Jatropha cinerea, Lemaireocereus, Machaerocereus, Opuntia cholla, Fouquieria peninsularis,* and species of *Lycium* and *Suaeda.*

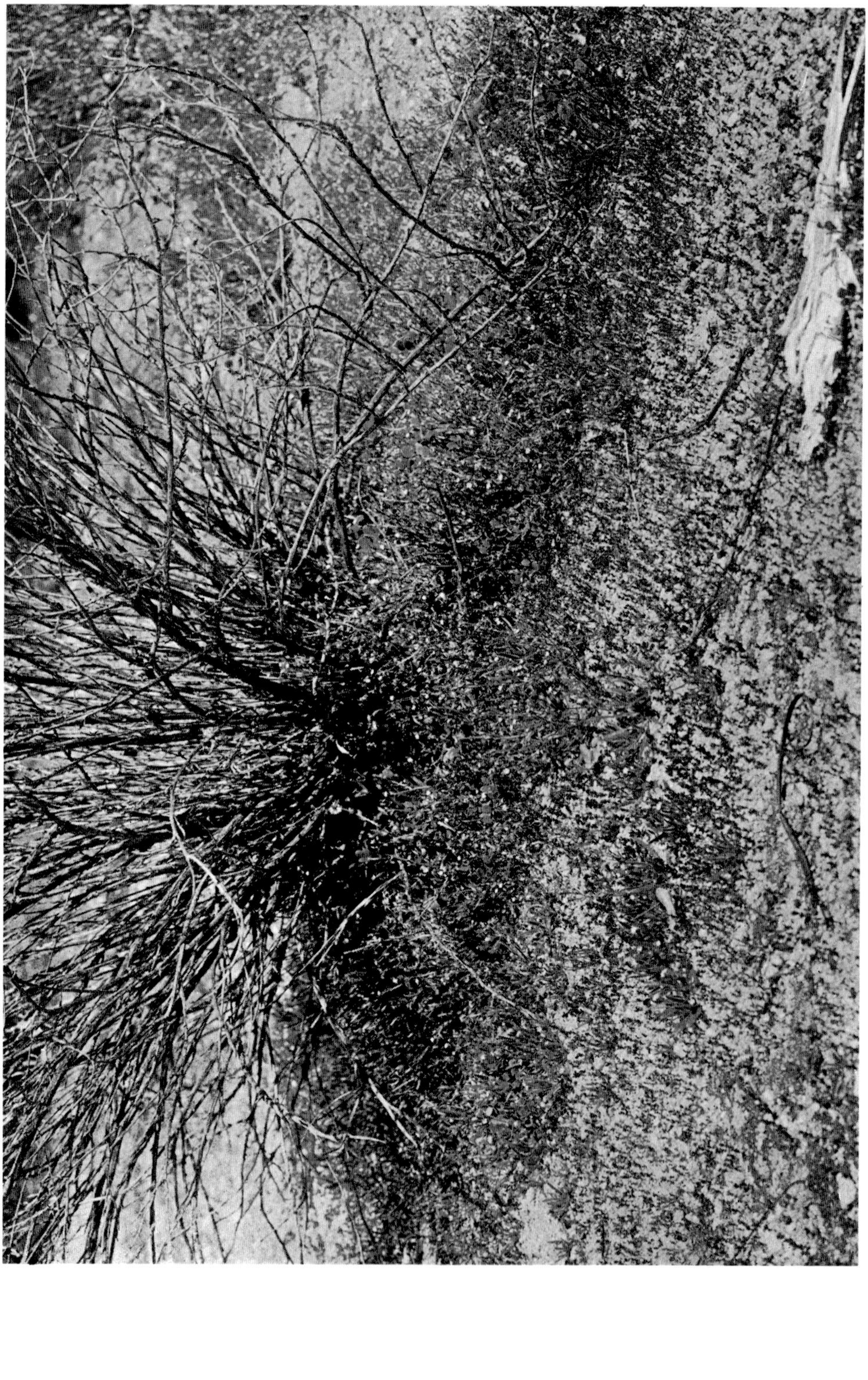

Herbaceous ephemerals around base of *Franseria chenopodifolia* south of San Augustín, Baja California. The tallest plants are *Cryptantha*, the smaller ones *Hosackia*, the smallest *Tillaea*. The outermost dark patches are a blue-green alga.

FIG. 1. Standing skeleton of *Idria columnaris*

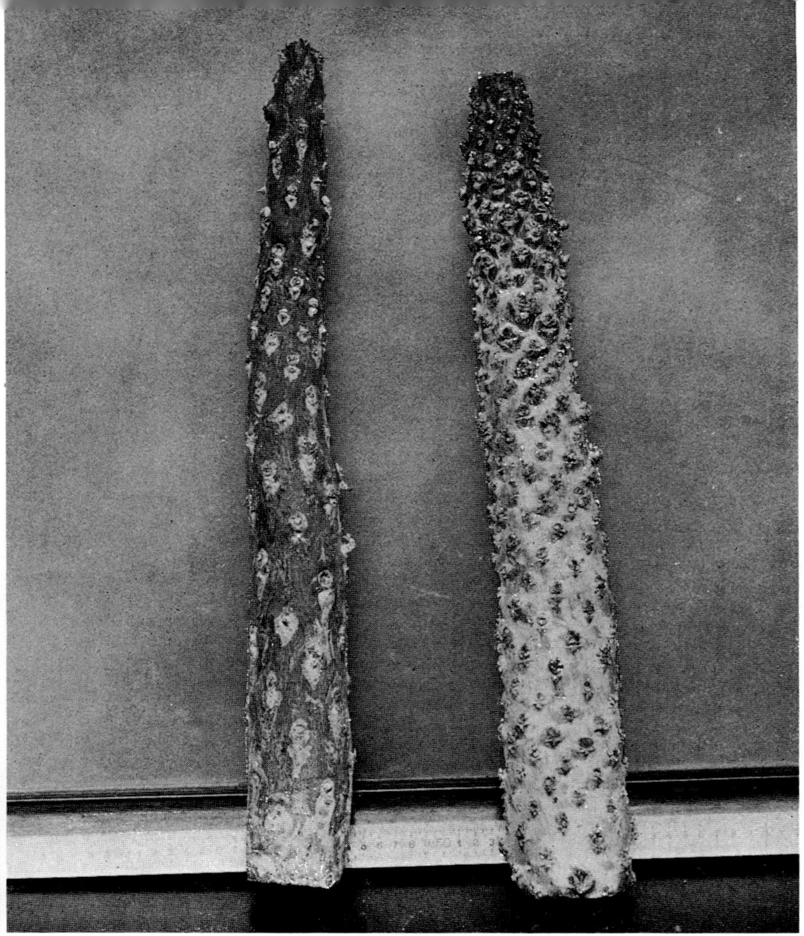

FIG. 2. Growing tips of main stem of *Idria*

FIG. 1. Cross section and surface of trunk of *Idria columnaris*

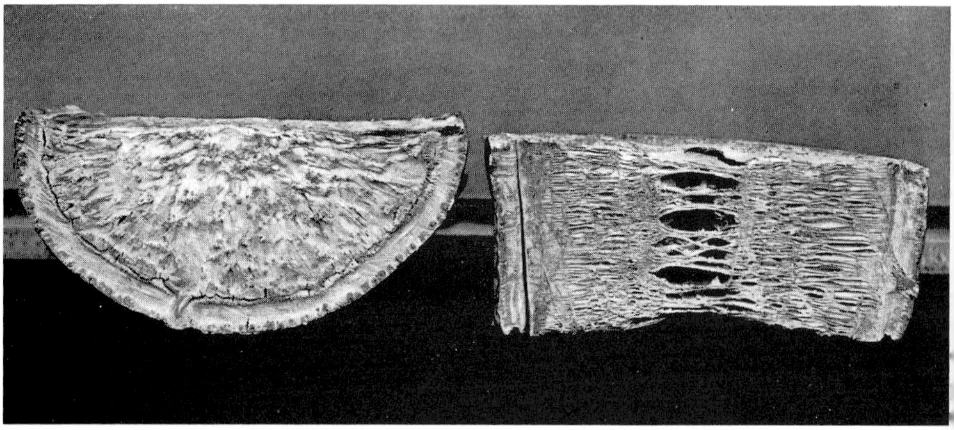

FIG. 2. Cross and vertical sections of trunk of *Idria*, showing large lacunae which accommodate changes in water content of tissue

FIG. 1. Trunk of *Acacia Willardiana*, showing exfoliation of thin layers of white bark

FIG. 2. Mature tree of *Acacia Willardiana*, showing thin crown of cladophylls

Plate 34

*Cercidium sonorae* in leaf and in flower growing with *Prosopis* on the Plains of Sonora

FIG. 1. Old, many-stemmed tree of *Holacantha Emoryi,* showing dark clusters of persisting fruit. Near Casa Grande, Arizona

FIG. 2. Exceptionally heavy stand of *Pachycormus discolor* on drainage divide of Baja California east of Punta Prieta

*Forchammeria Watsoni* on outwash plain near La Palma, Sonora

Well developed single individual of *Franseria dumosa* partially in leaf, near Tinajas Altas, Arizona

PART II

# Flora of the Sonoran Desert

By Ira L. Wiggins

# Introduction

This book has been written in an attempt to provide a systematic account of the vascular plants in a phytogeographic area, the Sonoran Desert, with little attention to political boundaries. The region covered has been described physiographically and as to its boundaries in Part 1 by Forrest Shreve. Inasmuch as no single previous publication covers all vascular plants known to grow naturally in the Sonoran Desert, it is hoped that this book will advance our knowledge of the plants occurring there. Although several works have dealt with the ferns and seed plants in the Colorado Desert of California, and Kearney and Peebles' admirable books* render a signal service to botanists interested in Arizona, it seems advisable to include the desert parts of both southwestern Arizona and southeastern California. In this manner a uniform treatment of the whole phytogeographic province is presented. Such a course, it is believed, will help clarify relationships among plants occurring in the Sonoran Desert.

Collecting in the Sonoran Desert. Botanical collecting in the Sonoran Desert began in 1839 when George Barclay, botanical collector, and Richard Brinsley Hinds, surgeon on H.M.S. *Sulphur,* obtained plant specimens at several points along the west coast of Baja California, notably at Magdalena Bay and at Cabo San Lucas. John Xantus de Vasey lived at Cabo San Lucas as a tidal observer for the United States Government from November 1859 to February 1861, and while there collected plants that were described by Asa Gray. Dr. Edward Palmer collected in Sonora and Baja California between 1875 and 1890. In the 1880's and 1890's several collectors penetrated various parts of our region, among them being such well-known naturalists as T. S. Brandegee, A. W. Anthony, C. G. Pringle, C. R. Orcutt, Marcus E. Jones, C. A. Purpus, S. B. Parish, and W. E. Bryant. Since the turn of the century the collectors working in parts of the Sonoran Desert have been too many to enumerate.

Intensive collecting in the Sonoran Desert was begun in the spring of 1932 by Dr. LeRoy Abrams on a trip with Dr. Forrest Shreve. In October 1932 the present author began field work in the desert parts of Sonora, Arizona, Baja California, and in the Colorado Desert of California. Thirty-three field trips were made, varying in duration from ten days to two and one-half months, the last being concluded in April 1963. Collecting in the Sonoran Desert was carried on by the writer personally during parts of every month in the year. The routes followed crisscrossed the area in order to ensure good coverage. Eight of the field trips were sponsored and financially supported jointly by the Carnegie Institution of Washington and by Stanford University, six were under the sole auspices of the University, three were privately financed, four were underwritten by Dr. Albert M. Vollmer, and fourteen by the Belvedere Scientific Fund. Appreciation is due, and gladly tendered, for the aid that made the field work possible.

---

* Thomas H. Kearney and Robert H. Peebles: *Flowering Plants and Ferns of Arizona,* 1096 pp., Washington: U.S. Government Printing Office, 1942; and *Arizona Flora,* 2d ed., with Supplement by John Thomas Howell, Elizabeth McClintock, and collaborators, viii + 1085 pp., Berkeley: University of California Press, 1960.

Distribution of Specimens to Other Institutions. The first set of most collections made by the writer has been deposited in the Dudley Herbarium at Stanford University. Duplicate sets have been distributed to leading herbaria within the United States. A set has gone to the University of Mexico, and some others to European centers. A nearly complete set has been deposited in the U.S. National Herbarium in Washington, D.C.

Special Support. A grant from the Carnegie Institution of Washington made it possible to study types, historical specimens, and other pertinent material in England. In the United States herbarium studies were carried on at the Gray Herbarium, the Arnold Arboretum, the New York Botanical Garden, Cornell University, the U.S. National Herbarium, Notre Dame University, the Chicago Museum of Natural History, the Missouri Botanical Garden, the California Academy of Sciences, and the University of California, Berkeley. The financial support from the Carnegie Institution of Washington and the courtesies extended by the officers of these herbaria are gratefully acknowledged.

Methods of Treatment. Critical revision of genera has not been attempted. In general, full use has been made of recent monographs and of other taxonomic researches in current literature. Many groups represented in the Sonoran Desert have been examined critically, in whole or in part, by other workers in preparation of manuals and floras, and the advances made through their studies have been of great help in the preparation of this book. Owing to the absence of manuals or other works dealing with the total area covered by the Sonoran Desert, it has seemed desirable to present descriptions of all families, genera, and species included in this volume, and of some infraspecific taxa. This entails repetition of some descriptive matter available in other books, but it seems better to have this overlapping and keep the treatment reasonably uniform than to force workers to consult several publications in order to find descriptions. Keys to all the categories recognized have been provided.

Presentation of synonymy under the species is not always complete and is not wholly uniform in treatment. In those cases in which recent monographs list full synonymy, references to the monographic works are given and only the name-carrying synonym is listed here. In some other instances, the synonyms known to apply and likely to appear in other works partially covering the region are listed; those of uncertain disposition or occurring only in older works not generally available in this country are omitted. Still other species carry a list of synonyms too lengthy to be handled readily in a book of this nature, and in these cases only the important ones are included. It is hoped that this partial presentation of synonymy will not cause serious inconvenience.

The question of including or excluding the names of species whose ranges lie chiefly outside the Sonoran Desert, but from which stragglers may occasionally enter the region, was decided in favor of the broader inclusion. This was done in the belief that it is better to include a few species which may not be encountered in several years' collecting, but which may be picked up along watercourses and in shaded canyons by the intensive collector, than to ignore their sporadic occurrence. In the main, those species known to occur at altitudes below 3,500 feet in and around the immediate boundaries of the Sonoran Desert have been included. Very rarely has a species been inserted unless a herbarium specimen has been available to authenticate its occurrence within the Sonoran Desert.

No one is more keenly aware than the author that species, and even genera, excluded from this book are bound to turn up within the confines of the Sonoran Desert as collecting continues. The large areas involved in, and the variety of habitats represented by, the Sonoran Desert preclude the possibility of a few diligent collectors finding all the species growing there. If this book presents the majority of the species actually growing naturally in the Sonoran Desert and stimulates other workers to search rugged canyons and desert playas in order to make the record complete, it will have served a useful purpose.

Collaborators. The task of completing this publication has been lightened greatly by the generous aid given by botanists who have contributed manuscript for various families or parts of families. Credit for each contribution appears as a footnote following the family heading, but it seems appropriate to acknowledge this help here also. The families contributed, and their authors, are as follows: Gramineae by Jason R. Swallen; Nyctaginaceae and the key to the species of *Cryptantha* by Roxana S. Ferris; Umbelliferae by Mildred E. Mathias and Lincoln Constance; Onagraceae by John Hunter Thomas; Verbenaceae and Avicenniaceae by Harold M. Moldenke; the genus *Gilia* of the Polemoniaceae by Alva Grant; Acanthaceae, with the exception of the genus *Carlowrightia,* by E. C. Leonard; and the Cucurbitaceae by Howard Scott Gentry. To each of these contributors the author admits his indebtedness and extends his deep appreciation.

Acknowledgments. Thanks are due, and willingly given, to numerous other friends and colleagues who have contributed encouragement, advice, and counsel when puzzling taxonomic problems arose. Among these friends are Dr. LeRoy Abrams, who often laid aside his own important work to give me aid; Mrs. Roxana S. Ferris, who has borne much of the load involved in preparing bales of specimens for accession and for distribution to other herbaria, and who has helped in the final checking of the manuscript, in reading proof, and in untangling many taxonomic and bibliographic puzzles; Dr. Reed C. Rollins, who accompanied me on one trying midsummer field trip and who pressed forward his investigation of several groups of Cruciferae strongly represented in the Sonoran Desert in order that his findings might be available for use in this book; Mrs. Sylvia Lowe Vincent and Mrs. Barbara W. Law, who, as secretaries in the Dudley Herbarium during the preparation of the manuscript, have checked bibliographic references and transcribed my semilegible handwriting onto cleanly typed pages. A. C. Smith, Lincoln Constance, Ellsworth P. Killip, T. H. Kearney, P. A. Munz, Herbert L. Mason, W. H. Camp, I. M. Johnston, C. A. Weatherby, J. M. Greenman, C. V. Morton, Jack Whitehead, W. R. Maxon, David D. Keck, M. L. Fernald, E. D. Merrill, W. P. Stockwell, H. A. Gleason, and many others have given me sound advice and helpful encouragement on numerous occasions. To each of these, and to others whose names are not listed, I extend my thanks. Very special appreciation goes to the late Dr. Forrest Shreve, who led the field party on several memorable trips. Special mention should also be made of the constant support given me by the late Dr. Herman A. Spoehr and by Dr. C. Stacy French of the Department of Plant Biology, Carnegie Institution of Washington, when the work on the book lagged or obstacles appeared. T. D. Mallery, who handled the commissary on several of the joint Carnegie-Stanford trips, and Dr. Albert M. Vollmer, whose constant support made possible four extended trips into Baja California and Sonora, also deserve particular mention and my sincere thanks. Mr. K. K. Bechtel, President of the Belvedere Scientific Fund, provided financial support for fourteen field trips and arranged with Stanford University for me to devote full time to the Sonoran Desert since September 1959. This generous aid is most gratefully acknowledged.

The exacting task of checking manuscript and printers' proof for errors, omissions, and inconsistencies has been lightened greatly by generous aid from Roxana S. Ferris, Peter H. Raven, and John Hunter Thomas, who have read critically the entire book. I thank them, collectively and individually.

Especially I thank my wife, Dorothy Bruce Wiggins, for her help and patience. She shared the rigors of most of the field trips after 1958, and handled admirably the commissary details of all field expeditions in which she participated. She gave invaluable assistance throughout the proofreading and in preparing the Index. I am grateful for her constant aid.

To those mentioned above, and to many others, belongs much of whatever credit may accrue upon the publication of this book. Censure for the omission of species, for inconsistencies in treatment, and for other shortcomings should rest on my shoulders, for, although the contributors have been given wide latitude in selecting the species they wished to include and in the general method of treating them, the responsibility for attaining a reasonably uniform treatment is mine.

IRA L. WIGGINS

*Dudley Herbarium*
*Stanford University*

# Flora of the Sonoran Desert: Taxonomy

## Key to the Divisions

Plants mosslike, producing spores of two kinds (microspores and megaspores) in sporangia attached to the adaxial surface of, but not enclosed by, the sporophylls, these in a 4-ranked strobilus; flowers and seeds none ........ Division I. LEPIDOPHYTA, p. 195

Plants not mosslike, producing spores of only one kind, or if heterosporous the sporangia enclosed within sporocarps, or plants producing seeds:

- Seeds never present; sporangia borne on peltate sporangiophores aggregated into a terminal strobilus; each spore bearing 4 slender elaters; stems longitudinally ridged, coated with silica ............................... Division II. CALAMOPHYTA, p. 197
- Seeds none in Class I, present in Classes II and III; sporangia in Class I not aggregated into strobili, borne abaxially on sporophylls or enclosed within sporocarps; spores, when present, without elaters; stems rarely ridged or coated with silica .......... .......................................... Division III. PTEROPHYTA, p. 198

# Division I. Lepidophyta

## Family 1. SELAGINELLACEAE. Selaginella Family

Low, creeping or depressed, freely branching, terrestrial or epiphytic perennial plants of mosslike habit. Leaves numerous, minute, 1-nerved, of 2 kinds in 4 ranks dorsiventrally arranged, or (in ours) all alike and closely imbricated. Sporangia in terminal 4-sided strobili, each sporangium solitary near base of adaxial side of sporophyll. Megasporangia larger and paler than, and usually below the more numerous microsporangia.

### 1. SELAGINELLA Beauv., Prod. Aethog. 101, 1805

Characters of the family.

Stems ascending or erect, rooting at the base only; leaves subglaucous:
- Terminal setae 0.2–0.5 mm. long; megasporophylls 1.7–2.4 mm. long; lateral cilia less than one-fourth as long as width of sterile leaves ................ 1. *S. bigelovii*
- Terminal setae 0.8–1.8 mm. long, megasporophylls 2.4–3 mm. long; lateral cilia equaling half the width of sterile leaves ................................ 2. *S. rupincola*

Stems prostrate, creeping, rooting freely; leaves green, not glaucous:
- Leaves with milk-white terminal setae, these 0.1–0.3 mm. long; leaves on lower side of stem with 16–22 cilia on each side ............................ 3. *S. arizonica*
- Leaves not setigerous; leaves on lower side of stem with 18–25 cilia on each side ...... ...................................................... 4. *S. eremophila*

**1. Selaginella bigelovii** Underw., Bull. Torrey Club **25:** 130, 1898

Cespitose, erect or ascending, repeatedly branching plants 6–25 cm. tall, ultimate divisions of sterile branches terete; leaves oblique, appressed-imbricate, blade 1.6–2.4 mm. long, 0.3–0.6 mm. wide, subulate-attenuate, subglaucous, deeply sulcate on the back, scaberulous; terminal seta 0.2–0.5 mm. long, whitish, rigid; lateral cilia 8–17 on each side of sterile leaves, obliquely spreading, less than 0.1 mm. long; strobili 5–16 mm. long, 4-angled, to twice as thick as sterile branches; sporophylls deltoid-ovate, attenuate apically, 1.7–2.4 mm. long, 0.6–1.2 mm. wide, lateral cilia minute, about 30 on each side; megasporangia orange-yellow, 3-lobed; megaspores reticulate; microsporangia bright orange.

Dry ridges and rocky slopes, Upper Sonoran Zone, San Francisco Bay Region, California, south to the southern Sierra de la Giganta, Baja California; entering the Sonoran Desert along the western margins of the Colorado Desert and the east slopes of the Sierra in Baja California.

**2. Selaginella rupincola** Underw., Bull. Torrey Club **25:** 129, 1898
*Selaginella rupestris* var. *rupincola* Clute, Fern Allies 142, 264, 1905.

Stems ascending or erect, subflexuous, 5–15 cm. tall, rooting only at or near base, considerably branched, ultimate branches 3–15 mm. long, terete; leaves closely appressed-imbricate, glaucous or cinereous-green, linear-lanceolate, sulcate dorsally, tapering gradually to apex; terminal seta white, denticulate or scaberulous, 0.8–1.8 mm. long; lateral cilia 15–18 on each side, equaling half the width of sterile leaves, spreading, conspicuous; strobili 5–12 mm. long; sporophylls slightly broader than sterile leaves; megasporangia dark yellow, often lobed and vertically striate on the outer equatorial band; megaspores 240–270 microns in diameter, reticulate-pitted; microsporangia deep orange.

In crevices on rocky hillsides and canyon walls, mainly Lower Sonoran Zone, south-central Arizona to New Mexico, southwestern Chihuahua, central Sonora, and southern Baja California.

### 3. **Selaginella arizonica** Maxon, Smiths. Misc. Coll. **72**[5]**:** 5, *pl. 3,* 1920

Prostrate, freely branching, matted plant with creeping stems 5–20 cm. long, ultimate branches short, their tips turning upward slightly; leaves arranged dorsiventrally in about 6 ranks, bright green when young, grayish in age; dorsal leaves smaller than ventral ones, closely imbricated, subulate, 1.8–2.2 mm. long, with 4–8 cilia on each side, terminal seta milk-white, 0.1–0.3 mm. long; ventral leaves linear-lanceolate, acuminate or short-setigerous, sulcate on the back, 2.5–2.8 mm. long, bearing 16–22 cilia on each side; strobili ascending, 2–6 mm. long; sporophylls narrowly ovate, 1.5–2 mm. long, subcarinate, short-setigerous, with 18–22 cilia on each side; megasporangia pale yellow; megaspores about 360 microns in diameter, reticulate on outer surface; microspores orange, about 35 microns in diameter.

On rocky hillsides and among rocks of desert ranges, Upper and Lower Sonoran Zones, central Arizona to Texas, south into Chihuahua and Sonora; Sierra de la Giganta near Loreto, Baja California.

### 4. **Selaginella eremophila** Maxon, Smiths. Misc. Coll. **72**[5]**:** 3, *pl. 2,* 1920

Prostrate, strongly dorsiventral, sparsely rooting plants with creeping stems 5–15 cm. long; branches spreading, subequal, 2–3-pinnate, with short, involute ultimate branchlets; leaves crowded, imbricate, ventral ones lanceolate-acute, 0.4–0.6 mm. wide, 1.8–2 mm. long, not setigerous, bearing 18–25 cilia on each side; dorsal leaves nearly erect, deltoid-subulate, 1–1.4 mm. long, bearing 6–12 weak cilia 0.1–0.2 mm. long on each side; strobili numerous, 6–12 mm. long, arcuate-ascending; sporophylls deltoid-ovate, acute, not setigerous, 1.2–1.5 mm. long, with 12–18 cilia on each side; megasporangia light yellow; megaspores about 360–400 microns in diameter, reticulate; microspores light yellow, 35–40 microns in diameter.

Rocky hillsides, mostly in crevices, Lower Sonoran Zone, western borders of the Colorado Desert to the southern Sierra San Pedro Mártir, Baja California.

# Division II. Calamophyta

## Family 2. EQUISETACEAE. Horsetail Family

Rushlike perennial plants with articulated aerial stems from black or brown, wide-creeping, branching rhizomes. Aerial stems annual or perennial, erect, fluted; branches in whorls at nodes or stems unbranched, internodes hollow; harsh deposits of silica on the surface in form of tubercles, transverse ridges, rosettelike aggregations, or uniformly and smoothly distributed; stomata in grooves between the longitudinal ridges. Leaves reduced to small scales coalesced into cylindrical sheaths. Strobili terminal, of shield-shaped, crowded sporophylls. Spores greenish, equipped with 4 hygroscopic elaters. Gametophytes small, subterranean, lobed.

### 1. EQUISETUM L., Sp. Pl. 1061, 1753

Characters of the family.

**1. Equisetum laevigatum** A. Br., Amer. Jour. Sci. **46:** 87, 1944
(For synonymy see Amer. Fern Jour. **52:** 34, 1962.)

Aerial stems slender, often somewhat flexuous, up to 7.5 dm. tall, simple or verticillately branched near base, basal branches mostly sterile; stems annual, 20–30-ridged, ridges with numerous transverse, sharply projecting bands of silica; sheaths narrowly funnelform to campanulate, green or with a dark basal ring, teeth horny, blackish, incurved, especially in age; strobili small, 1–2 cm. long, obtuse or acute but not markedly apiculate at apex.

Damp seeps and bottoms of canyons, mostly Upper Sonoran Zone, sparingly in the Lower Sonoran Zone, Humboldt and Siskiyou Counties, California, south to northern Baja California. A straggler into the edges of the Sonoran Desert.

# Division III. Pterophyta

Herbaceous or woody plants reproducing by spores or seeds. Branch gaps and leaf gaps both present in vascular cylinder of stems. In plants reproducing by spores (Class I) the sporangia on abaxial surface of fronds or sporophylls, or in sporocarps; gametophyte generation distinct, independent of sporophyte generation; fertilization of egg cell by motile antherozoids. In plants reproducing by seeds (Classes II and III) gametophyte generation enclosed by and dependent on parts of sporophyte generation; fertilization of egg cell by growth of a pollen tube; motile antherozoids (in ours) not present.

Plants producing spores, lacking flowers and seeds; gametophytic generation macroscopic, distinct, independent of the sporophytic generation; flagellate motile antherozoids produced by antheridia .................................. Class I. FILICINAE, p. 198

Plants producing seeds; gametophytic generation microscopic, dependent on and enclosed by structures of sporophytic generation; motile antherozoids absent:

Seeds borne on adaxial side of naked scales, the scales forming a cone, or in *Juniperus,* a leathery, berrylike structure; flowers none ...... Class II. GYMNOSPERMAE, p. 221

Seeds enclosed within a carpel or carpels; flowers present, although represented by only one envelope in some ........................ Class III. ANGIOSPERMAE, p. 226

## CLASS I. FILICINAE

Spore-bearing vascular plants with a life cycle of two well-marked phases, the gametophytic and sporophytic. Spores all alike or of two kinds: megaspores which produce female gametophytes, and microspores which produce male gametophytes. Antherozoids motile, coiled, bearing several long flagella.

### Key to the Families of the Filicinae

Plants homosporous, terrestrial; each sporangium provided with a hygroscopic annulus; prothallia developing outside the spore case, macroscopic, up to 1 cm. or more wide:

Fertile and sterile fronds strongly dimorphic, the fertile ones with very narrow, reduced lobes in long-pedunculate spikes; sporangia pyriform, the annulus at one end ....... ............................................ Family 1. *Schizaeaceae,* p. 199

Fertile and sterile fronds (in ours) alike; sporangia borne on abaxial sides of normally shaped blades, lenticular-orbicular, the annulus running part way around the vertical equatorial line ...................................Family 2. *Polypodiaceae,* p. 199

Plants heterosporous, aquatic or in wet places; sporangia borne inside sporocarps, devoid of an annulus; female prothallia developing inside the megaspore case, microscopic:

Leaves petiolate, the blade 4-foliolate, palmate; plants rooting in mud; sporangia 4-many in a bony sporocarp ................................ Family 3. *Marsileaceae,* p. 220

Leaves sessile, 2-lobed; plants minute, floating; sporangia in pairs on basal lobes of leaves, the sporocarps thin ............................... Family 4. *Salviniaceae,* p. 220

## Family 1. SCHIZAEACEAE. Climbing Fern Family

Plants with erect, simple, pinnate or dichotomous, or climbing vinelike leaves, leaf division often palmately lobed, paired or alternate, usually stalked. Sporangia borne in double rows on reduced lobes on special, long-pedunculate, spikelike fronds; sporangial cases ovoid, pyriform, or globose, the annulus apical and dehiscing by a longitudinal slit.

### 1. ANEMIA Swartz, Syn. Fil. 6, 155, 1806

Rhizomes creeping or ascending, densely paleaceous. Fronds cespitose on a horizontal or ascending rhizome or dorsally distichous on a creeping one, erect or ascending, dimorphous with some fronds entirely fertile, or basal pair of pinnae fertile, others of that frond sterile. Fertile pinnae usually surpassing sterile part of blade; sterile part of a blade pinnatifid to pinnately decompound. Sporangia obovoid or obpyriform, borne in 2 rows, one row on each side of the fertile segment, nonindusiate or margins of pinnules forming thin, membranous indusia. Spores triplanate.

**1. Anemia anthriscifolia** Schrad., Gött. Gel. Anz. **1824:** 865, 1824

Rhizomes horizontally creeping, densely covered with castaneous to light cinnamomeous linear-attenuate scales 2–3.5 mm. long, these grading into subulate hairs of equal length at base of stipes; stipes slender, dull stramineous, base slightly darker, deciduously pilose; fertile fronds 15–60 cm. long, sterile blades of the same frond ovate-deltoid, 5–18 cm. long and broad, bipinnate, rachises rusty-pilose and slightly glandular; pinnae 6–11 pairs; lower pinnules petiolate, pinnatifid, upper confluent, pilose-glandular, bright green above, paler beneath; fertile pinnae erect, 5–20 cm. long, branches elongate, sparingly pilose and very glandular; sporangia obpyriform to subglobose.

Widespread in Mexico from central Sonora and southern Chihuahua south; through Central America to Brazil; entering the margins of the Sonoran Desert in Sonora.

## Family 2. POLYPODIACEAE. Fern Family

Leafy ferns of various habit with hairy or paleaceous, creeping to erect rhizomes. Fronds spreading to erect, usually stalked; blades simple to several times pinnate or pinnatifid, the vernation circinate. Sporangia borne on under sides of leaf blades, or on special nonfoliose blades, mostly stalked, provided with an incomplete vertical ring of thickened hygroscopic cells called the annulus, dehiscing transversely. Sori naked or having a membranous indusium. Prothallia green, readily visible without a lens, at least at maturity.

Sori not marginal, separate at least in origin, dorsal on the veins:
- Indusium inferior, attached to leaf beneath the sorus, its segments spreading radially; stipes often jointed near the base .......... 1. *Woodsia*
- Indusium superior, peltately attached to a central stalk, or laterally at the side of sorus; stipes not jointed near the base:
  - Pinnules coriaceous; sori round; indusia peltate .......... 2. *Phanerophlebia*
  - Pinnules filmy-membranous; sori oblong to linear; indusia attached at sides of sori .......... 3. *Asplenium*

Sori marginal or submarginal or in naked lines along the veins:

Sporangia borne nearly full length of veins on fertile segments; no indusia or marginal modifications present:

Blades ceraceous beneath but not hairy; rhizomes nodose-multicipital, not widely creeping; stipes dark brown to purplish, shining .................... 4. *Pityrogramma*

Blades (in ours) densely hairy on both surfaces; rhizomes slender, widely creeping; stipes stramineous to light brown, dull ................................ 5. *Bommeria*

Sporangia submarginal, borne at or near tips of veins; indusia or indusiform marginal modifications often present:

Leaf tissue filmy; plants of moist situations; sporangia borne under reflexed, membranous lobes of leaf ................................................ 6. *Adiantum*

Leaf tissue thicker, often coriaceous; plants of drier situations; sporangia not under reflexed lobes of leaf:

Veins distinctly thickened at the tips; true marginal indusia often present ........... ........................................................ 7. *Cheilanthes*

Veins slightly or not at all thickened at the tips; true marginal indusia wanting:

Blades glabrous or essentially so; margins of segments broadly revolute ... 8. *Pellaea*

Blades ceraceous, hairy or paleaceous, at least beneath; margins narrowly revolute or plane ................................................ 9. *Notholaena*

## 1. WOODSIA R. Br., Prod. **1:** 158, 1810

Small ferns of rocky situations with short-creeping or densely tufted ascending rhizomes. Fronds clustered, erect or spreading-ascending, stipes often jointed near base, more or less scaly; blades narrowly elliptic-lanceolate, pinnate to bipinnate with lobed or pinnatifid membranous segments, veins free. Sori roundish, separate or confluent in age. Indusia inferior, early splitting into several lobes or fine, hairlike divisions, segments mostly concealed by crowded sporangia.

Fronds conspicuously glandular-pubescent; indusia mostly of broad lobes, ciliate or lacerate along margins; segments of fronds not strongly denticulate .......... 1. *W. plummerae*

Fronds glabrous or nearly so; indusia of slender, hairlike segments; ultimate segments of fronds sharply dentate, teeth ending in short, hairlike bristles .......... 2. *W. mexicana*

### 1. Woodsia plummerae Lemmon, Bot. Gaz. **7:** 6, 1882

*Woodsia obtusa* var. *glandulosa* D. C. Eaton & Faxon, Bull. Torrey Club **9:** 50, 1882.

Rhizomes closely clothed with lance-linear scales 3.5–5 mm. long, median part slightly thickened, sclerotic, dark brown to black, margins pale brown, membranous, somewhat sinuate; fronds erect or ascending, 4–15 cm. tall; stipes stramineous to cinnamomeous and sparingly scaly at base, stramineous to pale green or whitish and scaleless above; blades 1.5–4 cm. broad, 3–12 cm. long, moderately to densely glandular-pubescent, pinnate, pinnae 4–12 pairs, subopposite or rarely alternate, pinnatifid, lance-ovate, lobes denticulate with acute to rounded teeth; indusia of broad, concave lobes, their margins ciliate, shallowly lacerate, (or very rarely entire); sporangia crowded and often obscuring indusia in age, bright brown annulus lustrous.

Shaded places under rocks or ledges, mostly Upper Sonoran Zone to higher parts of Transition Zone, Coconino and Apache Counties, Arizona, to western Texas, Chihuahua, and Sonora. Also in the Sierra San Pedro Mártir, Baja California.

### 2. Woodsia mexicana Fée, Mém. Fam. Foug. **7:** 66, *pl. 26, fig. 3,* 1857

Rhizomes clothed with linear-lanceolate scales 3–5 mm. long, basal ones with a median sclerotic, dark brown to black strip, those above concolorous, membranous, pale brown, margins somewhat sinuate; fronds 5–25 cm. tall, ascending-spreading, glabrous or sparsely glandular-puberulent; stipes stramineous or pale green, sparingly scaly below, glabrous above, jointed near base; blades 2.5–5 cm. wide, 6–20 cm. long, pinnate, with deeply pinnatifid lanceolate pinnules; lobes of pinnules distinctly denticulate, each tooth subhyaline-margined and terminating in 1–several stout, whitish bristles 0.2–0.8 mm. long; sori confluent in age, nearly or quite covering segments; indusia of numerous hairlike divisions and a few narrowly lanceolate, attenuate lobes, conspicuous even in mature sori; sporangia pale golden brown, annuli slightly shiny.

Shaded, somewhat moist rocky slopes and ledges, mostly above the Lower Sonoran Zone, rarely dropping to the edge of the desert in eastern Pima and Santa Cruz Counties, Arizona; northern and central Arizona to western Texas and southward into Mexico.

## 2. PHANEROPHLEBIA Presl, Tent. Pterid. 84, 1836

Coarse, stiff ferns of rocky situations with stout, woody, creeping or nearly erect rhizomes sparingly to copiously clothed with paleaceous scales. Fronds rigidly ascending, blades pinnate; pinnules with spinulose margins, coriaceous or herbaceous; veins branched freely. Sori borne in 2–several rows on either side of the heavy midvein. Indusia orbicular, centrally peltate.

### 1. **Phanerophlebia auriculata** Underw., Bull. Torrey Club **26:** 212, *pl. 359, fig. 3–4, pl. 360, fig. 2,* 1899

Rhizomes stout, copiously clothed with paleaceous scales; stipes 2–3 dm. long, light cinnamomeous, dull, distinctly 4–5-angulate, 2–3 mm. in diameter, sparsely clothed at base with lance-ovate, membranous, light reddish brown, subentire to ciliate scales 7–12 mm. long, the bases of scales rounded to subcordate and intermingled with reddish, crinkly hairs; fronds oblong to elliptic, 10–18 cm. wide, 2–3 dm. long, pinnate; pinnules 9–17, sparingly pilose beneath, alternate, rachis ferruginously pilose; lower pinnules 1.5–3 cm. broad, 5–10 cm. long, asymmetrically auriculate at base, acuminate at apex, margins spinulose-serrate; sori orbicular, median, arranged in 1–3 rows on each side of midvein, 1.5–2 mm. in diameter; indusia orbicular, peltate, margins irregularly denticulate or short-ciliate.

In shaded canyons, especially under overhanging rocks, from southern Arizona (Huachuca and Baboquivari Mountains) to central Mexico.

## 3. ASPLENIUM L., Sp. Pl. 1078, 1753

Ferns of rocky ledges with rosulate to erect-spreading fronds and ascending or widely creeping rhizomes, these bearing crowded, rigid, dark scales. Fronds (in ours) delicate and once-pinnate, glabrous, pubescent, or slightly scaly; rachises dark and shining or green and succulent; pinnae entire or incised or lobed. Sori narrowly linear to oblong, borne just below tips of ultimate branches of veins. Indusia lateral, oblong or linear, membranous.

Pinnae distinctly toothed, not auriculate; fronds arcuate and rooting at tips .... 1. *A. palmeri*
Pinnae entire to crenulate-undulate, auriculate; fronds rarely rooting at tips ... 2. *A. resiliens*

### 1. **Asplenium palmeri** Maxon, Contr. U.S. Nat. Herb. **13:** 39, 1909

*Asplenium parvulum* var. *grandidentatum* Goodding, Muhlenbergia **8:** 92, 1912.

Rhizomes ascending, 3–5 mm. in diameter, about as long, bearing closely crowded, lance-linear, blackish scales 1.5–2 mm. long, these dull; fronds oblong to oblong-lanceolate, 5–20 cm. long, spreading-arcuate and tips striking root to form new plants; stipes slender, 1–3 cm. long, dull, purplish black, glabrous; pinnules gradually diminishing in size from middle toward each end, middle ones horizontal, 3–4 mm. wide, 6–10 mm. long, oblong, obliquely subauriculate or more often broadly cuneate at base, regularly serrate along upper margin; sori median, confluent in age; indusia elliptical, glabrous, margins subentire but pushed aside and often obscured by maturing sporangia.

Under overhanging ledges and in shaded ravines, Upper and Lower Sonoran Zones, southern Arizona and northeastern Sonora to central Mexico.

**2. Asplenium resiliens** Kunze, Linnaea **18:** 331, 1844

*Asplenium parvulum* Mart. & Gal., Mém. Acad. Roy. Belg. **15**[5]: 60, *pl. 15, fig. 3,* 1842. Not *A. parvulum* Hook., 1840, nor Wall., 1829.
*Chamaefilix resiliens* Farwell, Amer. Midl. Nat. **12:** 269, 1931.

Rhizomes ascending, closely paleaceous with linear-lanceolate, dull black scales; fronds ascending, oblong, 10–20 cm. long; stipes slender, 1–4 cm. long, black or purplish, sparsely appressed-puberulent with simple, brownish hairs in youth, soon glabrate, dull to slightly shining; pinnules 15–25 pairs, middle ones 6–15 mm. long, basally auriculate on distal side, subentire to crenulate-undulate, reduced in size toward base and apex of frond; sori 2–8 on both distal and proximal sides of pinnae; indusia linear-oblong, essentially entire.

Mostly in crevices of cliffs and canyon walls and under rocks, Arizona to Kansas, Pennsylvania, and Florida; Kofa Mountains, Arizona.

**Asplenium blepharodes** D. C. Eaton, Zoe **1:** 197, 1890. Stipes brownish; indusia distinctly ciliate. Occurs in the Sierra de la Laguna and in the Sierra de la Giganta, Baja California, and may grow on moist canyon walls bordering the Sonoran Desert in southern Baja California.

**Asplenium monanthes** L., Mant. **1:** 130, 1767. Stipes cinnamomeous, lustrous; sori mostly confined to proximal side of fertile pinnules; indusia entire. In the Sierra de la Laguna and may enter the Sonoran Desert in southern Baja California.

## 4. PITYROGRAMMA Link, Handb. Gewächse 3: 19, 1833

Small ferns of shaded banks and rock ledges in arid regions. Rhizomes short-creeping to ascending, clothed with dark, linear-lanceolate, rigid scales. Fronds erect, in close clusters; stipes firm, dark, glossy. Blades 1–3-pinnate, deltoid-pentagonal, glabrous or glandular above, powdery and often glandular beneath. Sori borne along short lateral veins, usually confluent, nonindusiate; spores rounded, deltoid, or trilobate.

**1. Pityrogramma triangularis** var. **maxonii** Weatherby, Rhodora **22:** 119, 1920

Rhizome scales linear-lanceolate with narrow, light brown, thin margins along the black, rigid, central costa, basal scales of stipes broader with less prominent, or lacking the, central thickening; fronds erect, 10–40 cm. high; stipes reddish brown, shining, sparsely scaly at the very base; blades 3–12 cm. long, nearly as wide, nearly tripinnate below, basal pinnules of lower pinnae deeply pinnatifid, other pinnules less deeply lobed to undulate-margined, segments subcoriaceous, sparsely yellow-glandular on upper surface, white-ceraceous and often sparsely glandular beneath; spores usually trilobate.

Upper and Lower Sonoran Zones, western margin of Colorado Desert and in the desert ranges of southwestern Arizona and adjacent Sonora and Baja California.

## 5. BOMMERIA Fourn. ex Baillon, Dict. Bot. 1: 448, 1876

Rhizomes slender, widely creeping, closely paleaceous; fronds erect, distichous. Stipes straight or subflexuous, erect or ascending; blade pentagonal, bipinnately parted at the base, leaf surfaces sparsely to densely (in ours) strigose, often scaly along mid-vein beneath. Venation areolate or free. Sporangia borne on veins, forming marginal bands.

**1. Bommeria hispida** (Mett.) Underw., Bull. Torrey Club **29**: 633, 1902

*Gymnogramme hispida* Mett., in Kuhn, Linnaea **36**: 72, 1869.
*Bommeria schaffneri* Fourn., Bull. Soc. Bot. France **27**: 327, 1880.
*Gymnopteris hispida* Underw., Our Nat. Ferns ed. 6, 84, 1900.

Rhizomes slender, 1–1.5 mm. in diameter, widely creeping, densely paleaceous with minute, lance-attenuate, subpersistent, dark brown scales; stipes slender, erect or subflexuous, 5–15 cm. long, dull to somewhat shining, ferruginous to light brown, strigose in youth, glabrate in age; blade deltoid-pentagonal to deltoid-ovate, 3–8 cm. wide, 4.5–10 cm. long; basal pinnae bipinnatifid, asymmetrical, lower lobes larger and more deeply lobed than upper part of blade, which is pinnate to bipinnatifid, segments obtuse or rounded, light green and moderately strigose above, ferruginous and densely strigose beneath; sporangia in marginal bands 1–3 mm. wide, obscured by the pubescence.

Rocky canyons and hillsides, Lower Sonoran Zone to Arid Transition Zone, southern Arizona to New Mexico, Texas, San Luis Potosí, Valley of Mexico and Durango; Sierra de la Giganta, Baja California.

## 6. ADIANTUM L., Sp. Pl. 1094, 1753

Delicate ferns of moist woods and ravines. Rhizomes slender and creeping, or sometimes stouter, paleaceous. Fronds with slender but rigid stipes and simple, 1–3-pinnate blades, the pinnules thin, stalked. Sori marginal, borne along and between tips of forking veins, the marginal lobes of pinnules sharply reflexed to form narrow, arcuate or oblong, membranous indusia.

Sori lunate or linear-oblong; veins of sterile pinnules terminating in distinct apiculations; frond ovate:
- Pinnules deeply 2–7-lobed one-half to three-fourths of the way to the base; sori mostly lunate; stipes bearing light brown scales on basal 1–5 cm. . . . . . . . . 1. *A. capillus-veneris*
- Pinnules shallowly lobed or merely dentate; sori linear-oblong; stipe bearing brown-black scales at the very base . . . . . . . . . . . . . . . . . . . . . . . . . . . . . . . . . . . . . . . . . . . . . 2. *A. jordanii*

Sori reniform; some veins terminating in the sinuses between the denticulations; frond oblong, several times as long as broad . . . . . . . . . . . . . . . . . . . . . . . . . . . . . . . 3. *A. concinnum*

**1. Adiantum capillus-veneris** L., Sp. Pl. 1096, 1753

*Adiantum modestum* Underw., Bull. Torrey Club **28**: 46, 1901.
*Adiantum rimicola* Slosson, Bull. Torrey Club **41**: 308, *pl. 7, fig. 1,* 1914.

Rhizomes laxly paleaceous, the scales small, thin, light brown, lance-linear; fronds laxly ascending to pendent, 20–70 cm. long; stipes purplish black, equaling or shorter than blades, these ovate to lanceolate-ovate, 12–40 cm. long, 2–3-pinnate at the base;

pinnules obovate, rhombic, or obliquely oblong, 5–30 mm. long, teeth of sterile pinnules acuminate; sori solitary on lobes, oblong-lunate.

In moist places along ravines and shaded banks and in caves, Upper and Lower Sonoran Zones, southern California to Utah, Virginia, Florida, and northern Mexico. Not a member of the truly desert flora but reaching its margins in particularly moist places as far south as the Cape Region, Baja California.

**2. Adiantum jordanii** C. Müll., Bot. Zeit. **1864:** 26, *pl. 1, fig. 1,* 1864

*Adiantum emarginatum* D. C. Eaton, Ferns N. Amer. **1:** 285, *pl. 38, figs. 1–3,* 1879. Not Bory, 1810.

Rhizomes densely paleaceous with large, rigid, dark brown, lance-attenuate scales; fronds erect, 20–50 cm. long; stipes dark brown, nearly equaling blades, these ovate to deltoid-ovate, 12–30 cm. long, 2–3-pinnate at base; pinnules roundish to transversely oval or semicircular, 8–30 mm. broad; sori linear or faintly arcuate.

Mostly Upper Sonoran and Transition Zones, northern Baja California to Oregon, reaching the margins of the Sonoran Desert near Rosario, Baja California, where it grows on shaded banks under *Aesculus parryi*, *Cneoridium dumosum*, and *Idria columnaris*.

**3. Adiantum concinnum** H. & B. ex Willd., Sp. Pl. **5:** 451, 1810

*Adiantum tenerum* Schk., Kr. Gew. **1:** 112, *pl. 121,* 1809. Not *A. tenerum* Swartz, 1788.

Rhizomes creeping, 3–5 mm. in diameter, densely paleaceous with reddish brown, shining, linear-lanceolate, attenuate scales 3–4 mm. long; stipes cinnamomeous to deep brown, shining, naked; blades oblong, 2–6 dm. long, 4–10 cm. wide, 2–3-pinnate at base or whole frond pinnate; pinnae oblong, 2.5–6 cm. long, pinnate or basal ones bipinnate; pinnules rhomboid or obovate, usually broader than long, 5–15 mm. wide, deeply 3–7-lobed; veins of sterile pinnules terminating in sinuses between denticulations; sori solitary, distinct, reniform to orbicular-reniform, borne at apices of lobes; indusia membranous, grayish.

In moist, shaded canyons, mainly Tropical Zone, southern Sonora to Vera Cruz and the Galapagos Islands and the Cape Region, Baja California. Barely straggling into the Sonoran Desert.

## 7. CHEILANTHES Swartz, Syn. Fil. 126, 1806

Small, xerophilous ferns of rocky habitats with massive, nodose-multicipital or creeping rhizomes closely covered with sclerotic or paleaceous scales, and tomentose, glandular-pubescent, paleaceous, ceraceous, or naked stipes and blades. Fronds 1–4-pinnate or simply pinnate and pinnatifid, the segments often small and beadlike. Sori borne at enlarged tips of veins, solitary or confluent. Indusia pouchlike, introrse or composed of thin, revolute margins of lobes or segments.

Rhizome massive, multicipital, divisions decumbent, erect, or short-creeping:
  Segments of blade densely glandular-viscid or glandular-puberulent above:
    Rhizome scales thin, soft, crinkly-sinuate, light cinnamomeous; blades narrowly oblong; glands sessile or nearly so . . . . . . . . . . . . . . . . . . . . . . . . . . . . . . . . . . . . 1. *C. viscida*
    Rhizome scales rigid, straight, black, and shining; blades deltoid or ovate-pentagonal, basal pinnae longest; glands obviously stalked . . . . . . . . . . . . . . . . . . . 2. *C. kaulfussii*

Segments of blade glabrous above or if pubescent not glandular:
Ultimate segments 4–9 mm. broad; stipes very fragile at base with several transverse zones of weakness marked by distinct constrictions ........... 3. *C. brandegeei*
Ultimate segments 1–3 mm. broad; stipes less easily fractured, without weakened constrictions:
Fronds glabrous on the upper surface; ultimate segments ovate to linear-lanceolate, acute to short-subulate at apex; rhizome scales rigid, heavily sclerotic and usually dark-colored to margins:
Pinnules serrate; sori solitary; indusia distinct, roundish-lunate, not linear ........ .................................................. 4. *C. californica*
Pinnules entire; sori contiguous with a common, crenate or shallowly lobed, linear indusium .......................................... 5. *C. pyramidalis*
Fronds hairy on upper surface; ultimate segments broadly ovate to orbicular, obtuse or rounded at apex; rhizome scales sclerotic and dark-colored centrally, margins lighter and thinner:
Rachis scales few or wanting; stipes hairy or sparingly scaly near base and apex:
Segments arachnoid-villous on both surfaces, not densely tomentose; stipes nearly glabrous; rachises of pinnae sparingly scaly .................. 6. *C. parishii*
Segments arachnoid-villous above, densely tomentose beneath; stipes hairy full length; rachises of pinnae scaleless .............................. 7. *C. feei*
Rachis scales numerous, imbricated, tangled; stipes scaly full length:
Scales of the rachises broadly ovate-cordate, 1–2.4 mm. wide, acute to short-acuminate, closely imbricate, obscuring lower surface of segments:
Stipes clothed with very slender, appressed scales; hairs on upper surface of leaf segments few, weak, tortuous, lightly attached and soon deciduous ....... ............................................... 8. *C. myriophylla*
Stipes clothed with oblong, spreading scales in addition to appressed, narrower ones; hairs on upper surface of leaf segments coarse, densely crowded, persistent .................................................. 9. *C. villosa*
Scales of the rachises narrowly lanceolate to linear, rounded to shallowly subcordate at base, 0.2–0.8 mm. wide, attenuate-acuminate, slightly imbricated or tangled, not obscuring segments:
Segments densely tomentose and canescent above; rachis scales light rusty to white, lanceolate ....................................... 10. *C. eatonii*
Segments sparsely castaneous-puberulent and green above; rachis scales cinnamomeous to castaneous .................................. 11. *C. castanea*
Rhizome slender, widely creeping, or if multicipital the divisions distinctly short-creeping:
Blade and upper part of stipe densely glandular-puberulent; rhizome heavy, 3–5 mm. in diameter, its branches short-creeping ............................. 2. *C. kaulfussii*
Blade and other parts not glandular; rhizome slender, 1–2.5 mm. in diameter, widely creeping (short-creeping in *C. peninsularis* and *C. covillei*):
Ultimate segments ovate to elliptic, with comparatively broad bases, not beadlike at maturity; scales or hairs on lower surface of rachises few or lacking, not projecting conspicuously beyond margins of segments:
Rachises clothed with loosely imbricated scales, these linear to ovate-lanceolate, attenuate at apex, rounded to shallowly cordate at base; stipes clothed with linear-to ovate-attenuate scales .................................... 12. *C. pringlei*
Rachises glabrous, sparsely hairy, or bearing a few scattered scales, these narrowly linear-lanceolate to hairlike, rounded at base; stipes glabrous, hairy, or sparingly scaly:
Stipes moderately hairy and sparingly scaly; margins of mature segments faintly revolute, scarcely at all metamorphosed:

Segments glabrous; rachises sparsely scaly; pinnules 0.5–1.5 mm. wide, deeply dentate .......................................... 13. *C. peninsularis*
Segments sparingly hairy with short, coarse hairs; rachises hairy, bearing a few linear scales; pinnules 1–3.5 mm. wide, shallowly crenulate-dentate ........ ............................................... 14. *C. microphylla*
Stipes glabrous, scaleless except at base; margins of fruiting segments strongly revolute, distinctly metamorphosed to form membranous indusia which obscure the sporangia .............................................. 15. *C. wrightii*
Ultimate segments orbicular to broadly obovate, strongly concave-convex, rounded and beadlike from above at maturity, their bases narrow; scales or hairs on lower surface of rachises numerous, projecting conspicuously beyond margins of segments:
Segments canescent, densely canescent-tomentose above, hairs of adjacent segments intertangled ........................................ 16. *C. lindheimeri*
Segments green, glabrous or sparsely villous above, hairs of adjacent segments not intertangled:
Rhizome scales light reddish brown to rusty, scarcely sclerotic; rachis scales firmly attached, rounded or faintly subcordate at base ................ 17. *C. wootoni*
Rhizome scales atropurpureous, shining, heavily sclerotic; rachis scales lightly attached, distinctly cordate at base ............................ 18. *C. covillei*

**1. Cheilanthes viscida** Davenp., Bull. Torrey Club **6:** 191, 1877
*Allosorus viscidus* Farwell, Amer. Midl. Nat. **12:** 286, 1931.

Rhizomes massive, of subglobose multicipital divisions, conspicuously scaly; scales linear, 1 mm. wide or less, 6–8 mm. long, light cinnamomeous, crispate-flexuous; fronds erect, clustered, 10–25 cm. long; stipes 4–10 cm. long, fragile, dark brown, sparsely to densely glandular; blades linear-oblong, acute, 5–15 cm. long, 1.5–3.5 cm. wide, bipinnate; pinnae ovate-deltoid to deltoid, lower 2 or 3 pairs distant, densely covered on upper surface with sessile and short-stalked, whitish, viscid glands, beneath with longer-stalked, yellowish glands; pinnules obliquely pinnatifid, with distant, short, decurrent, shallowly incised segments; sori solitary on the ultimate lobes; indusia roundish to round-reniform.

Shaded crevices in rocks, Panamint Mountains, Inyo County, California, to the western limits of the Colorado Desert.

**2. Cheilanthes kaulfussii** Kunze, Linnaea **13:** 145, 1839
*Cheilanthes viscosa* Link, Hort. Berol. **2:** 43, 1833. Not *C. viscosa* Carmichael, 1818.
*Cheilanthes glandulifera* Liebm., Vid. Selsk. Skr. V. **1:** 258, 1849.

Rhizomes multicipital, short-creeping or ascending, densely clothed with atropurpureous, linear-acicular, slightly sinuate, sublustrous scales 3–4 mm. long, margins not lighter nor membranous; stipes erect, 5–15 cm. long, reddish brown, glandular-puberulent; blades broadly triangular-ovate, 2–8 cm. wide, 1.5–10 cm. long, 3–4-pinnate, densely glandular-puberulent on both surfaces, pinnules pinnatifid, 2–12 mm. long; sori near tips of ultimate segments and enfolded by thin, coarsely crenate indusia formed from modified margins of segments; sporangia light brown, dull, mostly obscured by the membranous indusia.

Shaded spots mostly above the Lower Sonoran Zone, southern Texas and northern Sonora to Venezuela.

**3. Cheilanthes brandegeei** D. C. Eaton, Bull. Torrey Club **17:** 215, *pl. 104,* 1890

Rhizomes massive, nodose-multicipital, the short, stout divisions closely crowded and erect-ascending, densely clothed with light reddish brown, linear-acicular, thin, nonsclerotic scales 2.5–4 mm. long; fronds clustered or solitary, 7–15 cm. high; stipes rather stout, greenish, light castaneous, or reddish brown, dull, glabrous or with a few hairs near apex, slightly flattened, and usually with 1–several distinct constrictions near base, brittle and fragile, breaking readily at these constrictions; blades broadly triangular-deltoid to broadly ovate, 3–7 cm. broad, 4–10 cm. long, bipinnate, ultimate segments broadly ovate to suborbicular, entire or shallowly to deeply 3–5-lobed, 4–9 mm. wide, 5–12 mm. long, bright green, fleshy, glabrous above, sparsely and coarsely pilose beneath; sori nearly continuous around margins of segments; indusia membranous, whitish, reniform-orbicular, attached to margins of segments, sparingly and irregularly ciliate; sporangia yellowish.

Shaded crevices and gulches, mostly in volcanic rocks, Lower Sonoran Zone, vicinity of Cataviñá to Calmallí and on Magdalena and Cedros Islands, Baja California.

### 4. Cheilanthes californica (Hook.) Mett., Abh. Senckenb. Ges. Frankfurt 3: 88, 1859

*Hypolepis californica* Hook., Sp. Fil. **2:** 71, *pl. 88A, figs. 1–2,* 1858.
*Cheilanthes amoena* A. A. Eaton, Fern Bull. **5:** 44, 1897.

Rhizomes thickish, short-creeping; scales rigid, acicular, 2–2.5 mm. long, dark castaneous centrally, margins sometimes lighter; fronds numerous, 6–40 cm. long; stipes 4–25 cm. long, nearly naked, dull castaneous; blades deltoid-ovate to pentagonal, 2.5–15 cm. long, nearly as wide at base, 3–4-pinnate; segments oblique, linear to lance-ovate, decurrent, acute to long-mucronate, ultimate divisions about 1 mm. wide; sori solitary at ends of transversely enlarged vein tips; indusium pale, roundish-lunate, adherent along lower part of margins to a slender, saccate tooth, appearing to be at base of a sinus between ultimate divisions of pinnule; leaf tissue herbaceous, bright green, glabrous.

On cliffs and canyon walls, mainly in the Transition and Upper Sonoran Zones, Humboldt and Butte Counties, California, to northern Baja California. Barely entering margins of the Sonoran Desert.

### 5. Cheilanthes pyramidalis Fée, Mém. Fam. Foug. 7: 38, *pl. 25, fig. 3,* 1857

Rhizomes rather massive, short-creeping, densely clothed with dull black, sclerotic, acicular, often arcuate scales 3–4 mm. long; fronds 2–3.5 dm. tall; stipes erect, atropurpureous, sublustrous, mostly channeled on upper side, 8–20 cm. long, glabrous; blades ovate or narrowly triangular-ovate, 5–12 cm. broad, 10–20 cm. long, tripinnate, lower divisions often 10 cm. long, spreading-ascending; ultimate segments oblong-elliptic to oblong-oblanceolate, 1.5–3 mm. broad, 4–12 mm. long, glabrous above, margins faintly denticulate, bright green above, paler beneath; indusia continuous along margins of segments, sharply incurved, crenate, and long-ciliate, cilia usually meeting from opposite sides along the midvein of segment; sporangia yellowish or light brown.

Mainly south and east of our territory, but one specimen from the Cape Region of Baja California in the U.S. National Herbarium. To be expected along the canyons of the Sierra de la Giganta.

**Cheilanthes pyramidalis** subsp. **arizonica** Maxon, Amer. Fern. Jour. **8:** 116, *pl. 6,* 1918. Indusia less conspicuously lobed and sparsely clothed with short marginal papillae instead of long cilia; lower surfaces of segments dotted with red spots. It occurs in the Huachuca and Chiricahua Mountains of Arizona but has not been found within the limits of the desert.

**6. Cheilanthes parishii** Davenp., Bull. Torrey Club **8:** 61, *pl. 8,* 1881

Rhizomes short-creeping, freely branched; rhizome scales linear-lanceolate, oblique, 3–4 mm. long, attenuate, castaneous along the margins, the central line black; fronds few, erect, 12–15 cm. long; stipes 4–8 cm. long, dark castaneous, sparsely hairy and with a few broader scales near base and apex, these deciduous; blades oblong-lanceolate, 8–10 cm. long, 2–3.5 cm. wide, 3–4-pinnate, coarsely villous on both surfaces with long, flexuous hairs; rachises sparsely paleaceous beneath, scales light castaneous, flexuous; pinnae deltoid to deltoid-oblong; segments pinnately lobed or the larger ones with 1–4 free divisions, 1.5–2.5 mm. wide, terminal ones mostly 3-lobed; leaf tissue herbaceous, gray-green.

In crevices of rocks in the vicinity of Palm Springs on the east side of the San Jacinto Mountains, California, Lower Sonoran Zone. A rare fern, of which there are no specimens in the U.S. National Herbarium or the Dudley Herbarium collected since 1908.

**7. Cheilanthes feei** Moore, Ind. Fil. 38, 1857

*Myriopteris gracilis* Fée, Gen. Fil. 150, *pl. 29, fig. 6,* 1850–1852.
*Cheilanthes gracilis* Riehl, in Mett., Abh. Senckenb. Ges. Frankfurt **3:** 80, 1859. Not *C. gracilis* Kaulf., 1824.
*Allosorus vestitus* Farwell, Amer. Midl. Nat. **12:** 286, 1931, in part.

Rhizomes multicipital with short-creeping or ascending divisions; scales tufted, 5–7 mm. long, linear, cinnamomeous with a sharply marked central black stripe, subflexuous apically; fronds many, erect-spreading, 8–25 cm. long; stipes brown, deciduously pilose; blades linear-oblong to ovate, 5–13 cm. long, 1.5–4 cm. wide, tripinnate; pinnae thinly villous above, hairs whitish, lax, flexuous, under surface densely tomentose and pale fulvous; segments round or oval, simple or the larger lobed or divided, margins narrowly recurved; leaf tissue gray-green, herbaceous.

Under ledges and in rock crevices, mainly Upper Sonoran to Transition Zones, British Columbia to Illinois and Missouri; Providence Mountains, California, to the Santa Catalina, Huachuca, and Chiricahua Mountains of Arizona, along the Mexican boundary to Texas. A rare straggler into the Sonoran Desert along deep, moist canyons.

**8. Cheilanthes myriophylla** Desv., Ges. Naturf. Fr. Berlin Mag. **5:** 328, 1811

*Cheilanthes elegans* Desv., loc. cit.
*Cheilanthes paleacea* Mart. & Gal., Nouv. Mém. Acad. Sci. Brux. **15:** 76, *pl. 21, fig. 2,* 1842.

Rhizomes massive, erect or decumbent, of several to many erect or ascending branches 3–5 mm. in diameter, 5–12 mm. long, densely clothed with linear-subulate, attenuate, nearly black scales 0.3–1 mm. wide, 3–4 mm. long, margins thin, dull brownish, tips often sinuous; fronds erect, closely crowded, to 50 cm. tall; stipes about 1.5 mm. in diameter, cinnamomeous to castaneous, lustrous beneath the scales, these narrowly linear to lance-linear, grayish, to 3 mm. long; blade 3–4-pinnate, 2.5–3.5 cm. wide, 1–2.5 dm. long, densely paleaceous beneath with lanceolate to ovate, slightly attenuate, grayish to ferruginous scales about 1 mm. wide, these cordate at base, firmly attached, subentire to erose-dentate, reduced scales filiform, tortuous; segments beadlike, sessile or short-stalked, 0.4–1 mm. in diameter, bright green and bearing a few tortuous, moniliform, deciduous hairs on upper surface; sporangia obscured by cucullate borders of segments.

Rocky slopes, Lower Sonoran to Tropical Zones, Sierra de la Giganta, Baja California, and Chihuahua south to northern South America.

**9. Cheilanthes villosa** Davenp., Cat. Davenp. Herb. Suppl. 45, 1883

Rhizomes massive, ascending or decumbent, multicipital, densely paleaceous with narrowly subulate, falcate, flexuous scales 5–10 mm. long, margins light-colored and minutely denticulate, central line strongly sclerotic, dark brown; fronds numerous, erect, 10–35 cm. long; stipes 4–15 cm. long, subflexuous, atropurpureous, scaly; blades linear-oblong to narrowly oblong-lanceolate, 8–25 cm. long, 2–5 cm. wide, tripinnate; rachises copiously scaly beneath, scales large, nearly covering pinnae; rachis scales broadly ovate-cordate, erose-denticulate, attached by cordate base, auricles often overlapping; ultimate segments minute, beadlike, densely villous above, glabrous beneath; indusia narrow, formed by slightly modified margin of segments; leaf tissue gray-green, herbaceous.

Under ledges and overhanging rocks, mostly Upper Sonoran Zone, Sierra Estrella, near Phoenix, Arizona, and in the Santa Rita, Huachuca, and Mule Mountains in southern Arizona, eastward to Texas and Chihuahua.

**10. Cheilanthes eatonii** Baker ex Hook. & Baker, Syn. Fil. 140, 1867

*Cheilanthes tomentosa* var. *eatonii* Davenp., Cat. Davenp. Herb. Suppl. 49, 1883.
*Allosorus lindheimeri* var. *eatonii* Farwell, Amer. Midl. Nat. **12:** 285, 1931.

Rhizomes nodose-multicipital, branches decumbent to erect-ascending, closely clothed with flexuous, thin, linear-subulate scales 3–5 mm. long with lightly sclerotic, dark brown central stripes and light brown or rusty, membranous margins; fronds 1–3.5 dm. long; stipes clustered, sparsely to densely invested with linear-subulate, membranous, rusty scales and simple, appressed hairs; blades oblong, 3–4 cm. wide, 5–20 cm. long, tripinnate, pinnae subequal except at apex of fronds, 2–2.5 cm. long; rachises invested with hairs and ovate-attenuate, cordate scales 3–4 mm. long; segments elliptic to ovate, rounded, 1–2.5 mm. long, densely white-tomentose above, ferruginous-tomentose beneath; indusia continuous along membranous margins of segments; sporangia reddish brown, dull, obscured by tomentum and rachis scales.

Rocky slopes and ledges, mostly Upper Sonoran and Transition Zones, south-central Arizona to Colorado, Oklahoma, Texas, and northern Mexico.

**Cheilanthes tomentosa** Link, Hort. Berol. **2:** 42, 1833, differs from *C. eatonii* mainly in having tomentum on upper surfaces of fronds so sparse that fronds are green above. It occurs with *C. eatonii* and also in the Baboquivari Mountains, Arizona, slightly above the Sonoran Desert.

**11. Cheilanthes castanea** Maxon, Proc. Biol. Soc. Wash. **32:** 111, 1919

Rhizomes nodose-multicipital, divisions erect or ascending, closely clothed with linear-lanceolate, attenuate scales 3–4.5 mm. long, with atropurpureous to black, sclerotic median part, and narrow, membranous, light brown margins; fronds 15–35 cm. tall, clustered, erect; stipes light brown, 8–15 cm. long, moderately paleaceous with slender, filiform-subulate, attenuate, flexuous, ferruginous scales 1.5–3 mm. long; blades oblong, 2–3.5 cm. wide, 10–15 cm. long, tripinnate; pinnae 12–24 pairs; rachises paleaceous; segments suborbicular to broadly elliptic, 0.5–1.5 mm. wide, terminal one on each pinnule ovate, 1–3 mm. long, bright green and sparsely castaneously puberulent above, densely castaneous-tomentose beneath; indusia narrow, continuous, entire; sporangia pale castaneous.

Shaded canyon slopes and ledges, mostly Upper Sonoran and Transition Zones, south-central Arizona to Texas and Nuevo León, Mexico. Approaching and probably entering the Sonoran Desert in the Baboquivari Mountains, Arizona.

**12. Cheilanthes pringlei** Davenp., Bull. Torrey Club **10:** 61, *pl. 34,* 1883

*Cheilanthes sonorensis* Goodding, Muhlenbergia **8:** 93, 1912.

Rhizomes slender, creeping, paleaceous with slender, lance-linear, attenuate, light ferruginous to deep red-brown, scarcely sclerotic scales 1.5–3 mm. long; fronds crowded to slightly distant, erect, 5–15 cm. tall; stipes red-brown, sublustrous, sparsely pubescent with crinkled, castaneous hairs and paleaceous with scattered to crowded, linear- to ovate-attenuate scales 2–3 mm. long, these with rounded to cordate bases, margins irregularly sinuous; blades triangular-ovate, 2–3.5 cm. wide, 3–6 cm. long, bipinnate with bipinnatifid pinnules; vestiture on rachises similar to but denser than that on stipes, scales narrower, rounded at base; pinnae 5–9 pairs, next to basal usually longest; segments bright green and glabrous on both surfaces, narrowly oblanceolate and irregularly lobed or incised; sori at tips of segments; indusia pale castaneous, narrow.

Crevices in rocks and under shrubs, Lower Sonoran Zone to lower margins of Transition Zone, south-central Arizona to central Sonora.

**13. Cheilanthes peninsularis** Maxon, Contr. U.S. Nat. Herb. **10:** 496, 1908

Rhizomes short-creeping, closely covered with linear-acicular, dark brown to blackish, attenuate scales 2–2.5 mm. long; fronds 10–18 cm. tall; stipes slender, 5–10 cm. long, purplish brown, sublustrous, sparsely paleaceous with tortuous, appressed, yellowish brown scales; blades narrowly ovate, 4–5 cm. wide, 6–9 cm. long, pinnate; pinnae 4–7 pairs, bipinnatifid, 3–4 cm. long, somewhat distant; segments oblong to ovate, entire or dentate, dark green and glabrous above, pale beneath; rachises sparsely paleaceous or sometimes nearly naked; indusia terminal on teeth and apices of segments, greenish, membranous, margins minutely erose-denticulate; sporangia golden brown.

Crevices in rocks and canyon walls, Lower Sonoran Zone, central Baja California to the Cape Region.

**14. Cheilanthes microphylla** Swartz, Syn. Fil. 127, 1806

*Adiantum microphyllum* Swartz, Prod. Veg. Ind. Occ. 135, 1788.
*Adiantum pubescens* Poir., in Lam., Encycl. Suppl. **1:** 141, 1810. Not *A. pubescens* Schkuhr, 1809.
*Cassebeera microphylla* J. Smith, in Hook., Jour. Bot. **4:** 159, 1841.
*Allosorus microphyllus* Liebm., Vid. Selsk. Vid. V. **1:** 219, 1849.
*Notholaena microphylla* Keyserl., Polyp. Cyath. Herb. Bung. 28, 1873 (as *Nothochlaena*).

Rhizomes slender, widely creeping, clothed with minute, acicular, somewhat tortuous, reddish brown, almost hairlike scales 1–1.5 mm. long; fronds mostly distant, 15–30 cm. tall; stipes slender, dark purple-brown to nearly black, 5–12 cm. long, puberulent with reddish brown hairs and with a few ovate-lanceolate scales near base; blades oblong, 2.5–4 cm. wide, 10–18 cm. long, bipinnate, pinnae distant, 10–15 pairs, upper merely dentate, lower 8–15 mm. wide; rachises puberulent with spreading, crinkly, coarse hairs and often bearing a few linear scales; pinnules ovate to oblong, dentate with rounded teeth, sparsely puberulent with crinkly, coarse hairs on both surfaces or strigillose above, green; sori almost continuous along margin; indusia narrow, dentate-crenulate; sporangia reddish brown, dull.

Mostly in limestone pockets and crevices, Subtropical to Lower Sonoran Zone, Florida to Alabama; Texas to Sonora, Yucatán, Central America, and the Andes.

**15. Cheilanthes wrightii** Hook., Sp. Fil. **2:** 87, *pl. 110A, fig. 1–2,* 1858

*Cheilanthes tenuifolia* subsp. *wrightii* Domin, Bibl. Bot. **20** [Heft 85[1]]: 144, 1915.
*Allosorus wrightii* Farwell, Amer. Midl. Nat. **12:** 284, 1931.

Rhizomes slender, widely creeping, moderately branched, densely clothed with slender, lance-linear, membranous, ferruginous or castaneous, nonsclerotic scales 1.5–2 mm. long; fronds crowded or distant, 5–20 cm. long; stipes slender, 2–9 cm. long, reddish brown, sparsely paleaceous and puberulent or glabrate in age, sublustrous or dull; blades ovate to oblong, 1.5–3 cm. wide, 3–9 cm. long, bipinnate with pinnatifid lobes or lower tripinnate at base, glabrous, dark green; pinnae 5–8 pairs, pinnules pinnatifid or serrate-dentate, oblong-lanceolate, 15 mm. long or less; sori at tips and along sides of pinnule lobes; indusia shallowly dentate or crenate; sporangia pale golden brown, conspicuous.

Rocky slopes, canyon walls, and crevices in ledges, Lower Sonoran Zone to Transition Zone, southwestern Arizona to Texas and central Sonora.

**16. Cheilanthes lindheimeri** (J. Smith) Hook., Sp. Fil. **2:** 101, *pl. 107A, fig. 1–4,* 1858

*Myriopteris lindheimeri* J. Smith ex Seem., Bot. Herald 340, 1854.
*Allosorus lindheimeri* Farwell, Amer. Midl. Nat. **12:** 285, 1931.

Rhizomes slender, widely creeping, branching, densely clothed with closely appressed, broadly ovate, acute, thin, ferruginous scales 0.5–1.5 mm. long, growing tips bearing additional slender, nearly hairlike, attenuate scales; fronds usually crowded, 8–30 cm. tall; stipes 4–20 cm. long, slender, wiry, dark brown to purple-brown, closely invested with crinkly, tangled hairs and narrowly lance-attenuate, whitish or rusty scales or becoming glabrate in age; blades oblong or narrowly triangular-ovate, tripinnate, of 8–20 pairs of lance-ovate pinnae, lower ones often somewhat distant, upper more crowded and contiguous, densely canescent-tomentose above, densely woolly and rachises imbricate-scaly with lance-attenuate, tortuous, membranous, ferruginous scales beneath; ultimate segments orbicular, 0.4–1 mm. wide, crowded; indusia narrow, obscured by tomentum and scales.

Abundant in crevices of rocks and under ledges, Upper and Lower Sonoran Zones, Superstition and Baboquivari Mountains, Arizona, to Texas, Sonora, and San Luis Potosí.

**17. Cheilanthes wootonii** Maxon, Proc. Biol. Soc. Wash. **31:** 146, 1918

*Allosorus myriophyllus* var. *wootoni* Farwell, Amer. Midl. Nat. **12:** 285, 1931.

Rhizomes slender, widely creeping, densely but deciduously scaly with loosely imbricated, oblong-ovate to lance-oblong, distantly denticulate, light reddish brown to rusty, nonsclerotic scales 2–3 mm. long; fronds several, erect, 10–30 cm. long; stipes slender, 5–20 cm. long, paleaceous with subpersistent, linear-attenuate, castaneous scales; blades lance-oblong, 5–18 cm. long, 2–4.5 cm. wide, tripinnate; rachises scaly, scales firmly attached, rounded to faintly subcordate at base, ciliate, closely imbricate; ultimate segments beadlike, glabrous above; sporangia few; leaf tissue herbaceous but rigid, glistening bright green, minutely papillose.

Mountains at moderate altitudes from southern Colorado to Oklahoma, Texas and northern Sonora; a rare straggler along the eastern margins of the Sonoran Desert.

**18. Cheilanthes covillei** Maxon, Proc. Biol. Soc. Wash. **31:** 147, 1918
*Allosorus myriophyllus* var. *covillei* Farwell, Amer. Midl. Nat. **12:** 285, 1931.

Rhizomes short-creeping, branching, appressed-paleaceous with atropurpureous, shining, heavily sclerotic, linear to lance-attenuate scales 1.5–2.5 mm. long; fronds numerous, 10–30 cm. long; stipes slender, 5–18 cm. long, brown to atropurpureous, sparsely paleaceous with small, linear-attenuate, laxly ascending scales; blade oblong to ovate-deltoid, acuminate, 5–15 cm. long, 2–6 cm. wide at base, tripinnate; pinnae 5–12 pairs; rachises and segments imbricate-scaly beneath, scales lightly attached above closed sinus of deeply cordate base, ciliate on basal part only, white to pale castaneous, ovate to ovate-lanceolate; ultimate segments glabrous and naked above; sporangia borne inside recurved, slightly crenulate margins of minute, beadlike segments; leaf tissue spongiose-herbaceous.

On shaded cliffs and under overhanging rocks, Lower and Upper Sonoran Zones, Inyo and Ventura Counties, California, to southern Utah, Arizona, and Bahía de los Angeles, Baja California.

## 8. PELLAEA Link, Fil. Sp. Hort. Berol. 59, 1841

Small, rigid ferns of dry, rocky habitats, with nodose or creeping rhizomes and leathery, essentially glabrous foliage. Blades 1–4-pinnate. Segments linear to ovate, simple or branched; veins free, the immersed tips not thickened. Sori terminal and subterminal, often becoming confluent along, and obscured by revolute, indusiform margin.

Segments decurrent or merely lobed, thin; stipes greenish to tawny ......... 1. *P. seemannii*
Segments distinct, leathery, never decurrent; stipes pinkish, atropurpureous, or castaneous:
  Stipes pinkish; rhizomes slender, creeping:
    Veins prominent beneath; fertile segments emarginate, not apiculate ........................................................................ 2. *P. andromedifolia*
    Veins obscure; fertile segments truncate or roundish at apex, apiculate .. 3. *P. intermedia*
  Stipes atropurpureous or castaneous; rhizomes nodose-multicipital, erect or ascending:
    Segments oblong to lanceolate, 7–35 mm. long; fronds narrowly oblong ............................................................................ 4. *P. ternifolia*
    Segments oval to elliptic, 2–10 mm. long; fronds lance-deltoid to ovate:
      Pinnae once-pinnate; segments broadly ovate, crenulate-sinuate, 5–10 mm. long; mucro conspicuous ........................................ 5. *P. longimucronata*
      Pinnae with some trifoliolate pinnules; segments narrowly ovate, 2–6 mm. long; mucro minute ................................................ 6. *P. mucronata*

**1. Pellaea seemannii** Hook., Sp. Fil. **2:** 141, *pl. 117B, fig. 1–2,* 1858

Rhizomes nodose-multicipital, branches ascending, clothed with linear-subulate, bicolorous, crinkly scales 2–3.5 mm. long, median stripe purplish black, sclerotic, shiny, membranous margins pale brown, entire to faintly erosulate; fronds 1–4.5 dm. tall, ascending or erect; stipes greenish to tawny, slightly glandular, sparsely clothed with filiform-subulate, castaneous scales, 0.2–1 mm. long; blades lance-elliptic, 3–10 cm. wide, 7–20 cm. long, pinnate with coarsely and irregularly 2–15-lobed, decurrent to entire segments, these lanceolate, thin, light green, distinctly veined; sori marginal, nearly continuous; indusia narrow, membranous, inconspicuous; sporangia pale yellowish brown, moderately lustrous.

Shaded cliffs, rocky hillsides, and clay banks, Lower Sonoran and Tropical Zones, central Sonora to southwest Chihuahua and Guerrero.

**2. Pellaea andromedifolia** (Kaulf.) Fée, Gen. Fil. 129, 1850–1852

*Pteris andromedaefolia* Kaulf., Enum. Fil. 188, 1824.
*Allosorus andromedaefolius* Kunze, Linnaea **9:** 56, 1834.
*Platyloma andromedaefolia* J. Smith, in Hook., Jour. Bot. **4:** 160, 1841.
*Notholaena andromedaefolia* Keyserl., Polyp. Cyath. Herb. Bung. 29, 1873 (as *Nothochlaena*).
*Pellaea rafaelensis* Moxley, Amer. Fern Jour. **5:** 107, *pl. 8,* 1915.
*Cassebeera andromedaefolia* Farwell, Amer. Midl. Nat. **12:** 280, 1931.

Rhizomes slender, wide-creeping, bearing attenuate, lance-linear, pale castaneous scales 1.5–3.5 mm. long with atropurpureous, sclerotic, lustrous median stripe; stipes 6–40 cm. long, flesh-colored, slightly glaucous, glabrous or puberulent; blades ovate to deltoid, acuminate, 5–20 cm. broad, 10–40 cm. long, 2–4-pinnate with glabrous or glandular-puberulent rachises; segments coriaceous, dark green above, pale beneath, deltoid-ovate to broadly elliptic, 3–17 mm. long, mostly emarginate at apex, sterile ones flat, fertile ones with strongly revolute margins; sporangia cinnamomeous, dull.

Dry stony ridges, mostly Upper Sonoran Zone, mountains and foothills from southwestern Oregon to central Baja California.

**3. Pellaea intermedia** Mett. ex Kuhn, Linnaea **36:** 84, 1869

*Cassebeera intermedia* Farwell, Amer. Midl. Nat. **12:** 281, 1931.

Rhizomes slender, wide-creeping, closely invested with linear-lanceolate, bicolorous scales 2–3 mm. long with dark brown to black, lustrous, sclerotic median strip and pale cinnamomeous, membranous margins; fronds erect, 2–4 dm. high; stipes 5–18 cm. long, flesh-colored, glabrous except for a few scales at very base, dull or slightly lustrous; blades deltoid-ovate, 4–12 cm. wide, 6–30 cm. long, bipinnate; rachises puberulent; pinnae distant, narrowly lance-ovate to oblong, 1–2.5 cm. wide, 3–7 cm. long; segments 3–6 pairs and a terminal one, coriaceous, broadly elliptic, 4–15 mm. long, rounded to subcordate at base, acute, rounded, or shallowly retuse and short-apiculate at apex; margins revolute, obscuring continuous sori; sporangia pale yellowish brown, dull.

Dry ledges and rocky slopes, mostly Upper Sonoran and Arid Transition Zones, Arizona to Texas and south in Mexico to Nuevo León and Zacatecas.

**4. Pellaea ternifolia** (Cav.) Link, Fil. Sp. Hort. Berol. 59, 1841

*Pteris ternifolia* Cav., Descr. Pl. 266, 1802.
*Pteris subverticillata* Swartz, Syn. Fil. 103, 1806.
*Allosorus subverticillatus* Presl, Tent. Pterid. 153, 1836.
*Platyloma ternifolia* J. Smith, Bot. Mag. **72:** Comp. 21, 1846.
*Allosorus ternifolius* Kunze ex Klinge, Linnaea **20:** 339, 1847.
*Cheilanthes ternifolia* Moore, Ind. Fil. 255, 1861.
*Notholaena ternifolia* Keyserl., Polyp. Cyath. Herb. Bung. 29, 1873 (as *Nothochlaena*).
*Cassebeera ternifolia* Farwell, Amer. Midl. Nat. **12:** 281, 1931.

Rhizomes nodose-multicipital, massive, with short, ascending branches densely matted with subulate-attenuate, bicolorous scales 2–3.5 mm. long, central sclerotic strip atropurpureous, lustrous, margins cinnamomeous, membranous, erosulate-ciliolate; fronds clustered, 1–3.5 dm. tall; stipes atropurpureous to ebony, glabrous or nearly so, shining, one-fourth to two-thirds the length of entire frond; blades narrowly oblong to narrowly deltoid-ovate, 2–7 cm. wide, simple or usually bipinnate, each pinna consisting of 3 linear-oblong segments, or in depauperate specimens or toward apex of frond, of a single segment; segments 3–8 mm. wide, 1–4 cm. long, coriaceous, strongly revolute, mucronate at apex, pale green; sori almost concealed.

Shaded rocky slopes and ledges, Tropical and Lower Sonoran Zones, southeastern Texas to Arizona and south to Guatemala; Cape Region, Baja California.

**5. Pellaea longimucronata** Hook., Sp. Fil. **2:** 143, *pl. 115A, fig. 1–3,* 1858

*Pellaea wrightiana longimucronata* Davenp., Cat. Davenp. Herb. Suppl. 46, 1883.
*Pellaea truncata* Goodding, Muhlenbergia **8:** 94, 1912.
*Cassebeera ternifolia* var. *longimucronata* Farwell, Amer. Midl. Nat. **12:** 282, 1931.

Rhizomes nodose-multicipital, densely clothed with attenuate-subulate, bicolorous scales 2.5–4 mm. long, the sclerotic median strip atropurpureous, the erose-ciliolate margin membranous, cinnamomeous; fronds 2–4.5 dm. tall; stipes dark seal brown to atropurpureous, glabrous, dullish to sublustrous; blades deltoid-ovate, 3–8 cm. wide, 1–3 dm. long, bipinnate; pinnae deltoid-ovate to oblong, of 1–5 pairs and an odd terminal segment; segments moderately coriaceous, pale green, sterile ones broadly elliptic to nearly orbicular, fertile ones elliptic, 5–10 mm. long, acute or rounded at base, mucronate at apex, margins often crinkly or crenate-sinuate before becoming revolute; sori obscured, dull brownish.

Dry rocky ridges and ledges, Upper and Lower Sonoran Zones, Nevada to Colorado, southern Arizona, and central New Mexico; Ivanpah Mountains, San Bernardino County, California, and in the southern Sierra San Pedro Mártir, Baja California.

**6. Pellaea mucronata** D. C. Eaton, in Torr., Bot. Mex. Bound. 233, 1859

*Allosorus mucronatus* D. C. Eaton, Amer. Jour. Sci. II, **22:** 138, 1856.
*Pellaea ornithopus* Hook., Sp. Fil. **2:** 142, *pl. 116A, figs. 1–3,* 1858.
*Cheilanthes mucronata* Moore, Ind. Fil. 45, 248, 1861.
*Allosorus ornithopus* Kuntze, Rev. Gen. Pl. **2:** 806, 1891.
*Cassebeera ternifolia* var. *mucronata* Farwell, Amer. Midl. Nat. **12:** 282, 1931.
*Cassebeera ternifolia* var. *ornithopus* Farwell, loc. cit.

Rhizomes nodose-multicipital, massive, densely clothed with acicular castaneous scales 5–10 mm. long, median strip sclerotic, dark brown, membranous margins light brown, denticulate; fronds crowded, 1–5 dm. tall; stipes purplish brown, 4–20 cm. long, dullish, glabrous; blades ovate-lanceolate, deltoid-oblong, or oblong, 3–15 cm. wide, 8–35 cm. long, 2–3-pinnate; pinnae distant or subcrowded, linear to linear-lanceolate; pinnules 6–15 pairs, larger ones trifoliolate or pinnate, smaller ones simple; segments elliptic to linear-oblong, 2–6 mm. long, apiculate; sporangia mostly concealed by revolute margins, brownish, dull.

Rocky ridges, dry brushy slopes, and dry ledges, mostly Upper Sonoran Zone, northern California to southern Sierra San Pedro Mártir, Baja California. Common along the western margin of the Colorado Desert.

## 9. NOTHOLAENA R. Br., Prod. 1: 145, 1810

Small xerophytic ferns with short and ascending or elongated creeping rhizomes and glandular, ceraceous, tomentose, or scaly, erect or ascending fronds. Blades 1–4-pinnate, variously lobed or pinnatifid, linear, deltoid, or pentagonal. Sori rounded or oblong, borne submarginally at tips of unmodified veins, often confluent. Indusia wanting but margins of pinnules revolute and often partly covering sori; sporangia often concealed or obscured by scaly or hairy covering.

Pinnules noticeably scaly, hairy, or tomentose at least on midveins beneath:
- Stipes and lower surface of pinnules scaly:
  - Scales on stipes and rachises strongly sinuate, entire; pinnae scaly only on midveins, ceraceous elsewhere beneath, sparsely ceraceous above ................ 1. *N. grayi*
  - Scales on stipes and rachises straight to slightly sinuate, ciliolate or nearly entire; pinnules scaly over whole surface beneath, not ceraceous, hairy on upper surface:
    - Fronds distinctly bipinnate; stipes black or nearly so; scales on rhizomes shining black or atropurpureous .................................... 2. *N. aschenborniana*
    - Fronds once-pinnate; stipes light brown or cinnamomeous, dullish; scales on rhizomes cinnamomeous or light brown, dull:
      - Scales on rhizomes pectinate-ciliate along full length; pinnae 1–2.5 cm. long, 4–6-lobed; scales on upper surface of blades mostly reduced to stellate trichomes, usually deciduous .............................. 3. *N. sinuata* var. *sinuata*
      - Scales on rhizomes entire or sparsely ciliolate on distal one-third; pinnae 2.5–8 mm. long, 2–3-lobed; scales on upper surface of blades ovate, usually persistent ............................................ 3*a*. *N. sinuata* var. *cochisensis*
- Stipes and lower surface of pinnae hairy or glabrate, but not scaly:
  - Blades pinnate, linear-lanceolate in outline; pinnae densely rusty-tomentose beneath, the hairs closely felted, loosely villous on upper surface ........... 4. *N. aurea*
  - Blades bipinnate, broadly oblong-lanceolate in outline; pinnules with some spreading hairs loosely arranged, overtopping tomentum in *N. newberryi*:
    - Stipes brownish; rhizome scales blackish, shining; fronds 0.6–3.5 dm. high ...... ...................................................... 5. *N. newberryi*
    - Stipes black, wiry; rhizome scales reddish brown, dull; fronds 0.5–2 dm. high, usually about 1 dm. tall ....................................... 6. *N. parryi*

Pinnules glabrous or ceraceous but not markedly hairy, tomentose, or scaly:
- Blades oblong or oblong-lanceolate .................................. 7. *N. lemmonii*
- Blades ovate, triangular, or pentagonal:
  - Fronds arcuate-spreading, devoid of ceraceous covering; segments few, remote ...... ........................................................ 8. *N. jonesii*
  - Fronds erect or ascending, not strikingly arcuate, always ceraceous beneath and sometimes above; segments numerous, rather crowded:
    - Blades distinctly pentagonal; stipes cinnamomeous or with a reddish brown tinge; mature sporangia black or dark brown:
      - Pinnae divided into discrete ovate to lance-ovate segments; rhizome scales denticulate .................................................... 9. *N. californica*
      - Pinnae bipinnatifid but not divided into separate segments; scales on rhizomes entire to faintly erosulate .............................. 10. *N. standleyi*
    - Blades ovate or triangular; stipes black or atropurpureous; mature sporangia golden brown or light brown:
      - Ultimate segments entire, glaucous above; blades mostly 4-pinnate, at least at the base ................................................ 11. *N. limitanea*
      - Ultimate segments mostly crenate or shallowly lobed, not glaucous; blades mostly bipinnate with pinnatifid lobes ............................ 12. *N. incana*

### 1. **Notholaena grayi** Davenp., Bull. Torrey Club **7**: 50, 1880

*Notholaena hypoleuca* Goodding, Muhlenbergia **8**: 94, 1912. Not *N. hypoleuca* Kunze, 1834.
*Cheilanthes grayi* Domin, Bibl. Bot. **20** [Heft 85[1]]: 133, 1913.

Rhizomes nodose-multicipital or sometimes short-creeping, 2–3.5 mm. in diameter; scales linear-lanceolate, attenuate, somewhat sinuate, 2–3.5 mm. long, subcordate at base, black or atropurpureous, shining, crowded; fronds 10–22 cm. long, erect or ascending; stipes light cinnamomeous, 2–8 cm. long, minutely glandular-puberulent and sparsely to moderately scaly with thin, sinuate, light brown, lanceolate-attenuate scales; blades oblong or lance-oblong, bipinnate, with ultimate divisions dentate to pinnatifid; pinnae 1–2 cm. long, 5–10 mm. wide, sparingly glandular-ceraceous but green above, closely ceraceous beneath and with scales similar to those of rhizome and stipe on midveins and rachises; sori nearly continuous along slightly revolute pinnule margins; sporangia light to dark brown.

On rocky slopes and in sparse chaparral, Upper and Lower Sonoran Zones, south and central Arizona to Texas and northern Sonora.

**2. Notholaena aschenborniana** Klotzsch, Linnaea **20:** 417, 1847

*Notholaena bipinnata* Liebm., Vid. Selsk Skr. V. **1:** 214, 1849.
*Cheilanthes aschenborniana* Mett., Cheil. 21, no. 7, 1859.

Erect plants 1.5–4 cm. tall from nodose-multicipital rhizomes clothed with crowded, atropurpureous or black, shining scales 2.5–3.5 mm. long, these linear-lanceolate, attenuate, with a light-colored, subcordate-rounded base and ciliolate, straight margins; stipes 4–12 cm. long, densely tomentulose with fine-rayed stellate hairs and few to many dark brown, lance-linear, ciliolate scales, eventually glabrate; blades 2.5–4 cm. wide, bipinnate, green but arachnoid-pubescent above with fine-rayed, stellate hairs, densely so and densely scaly with ferruginous, ciliolate scales beneath; sporangia light brown, largely obscured by numerous scales and stellate indument.

Rocky hillsides, Upper and Lower Sonoran Zones, mostly above the desert, Pima County, Arizona, to Texas, Sonora, and Chihuahua.

**3. Notholaena sinuata** (Lag.) Kaulf., Enum. Fil. 135, 1824, var. **sinuata**

*Acrostichum sinuatum* Lag. ex Swartz, Syn. Fil. 14, 1806.
*Gymnogramme sinuata* Presl, Tent. Pterid. 219, 1836.
*Cheilanthes sinuata* Domin, Bibl. Bot. **20** [Heft $85^1$]: 133, 1915.
(For additional synonymy see Lilloa **4:** 258, 1941.)

Erect or arcuately ascending plants 15–50 cm. high from massive, nodose-multicipital rhizomes densely clothed with straight to slightly sinuous, attenuate-linear, thin, light brown or ferruginous, tangled, ciliolate scales 5–10 mm. long and 1–2 cells thick; stipes stout, 1.5–3 mm. in diameter, arachnoid-tomentose and conspicuously scaly, scales similar to but shorter than those on rhizomes; blades pinnate, 1.5–4 cm. wide, pinnae one-third to one-half as wide as long, usually coarsely lobed halfway to midrib, thick, leathery, becoming strongly involute and appressed to rachis with advancing drought, green and sparsely stellate-pubescent above in youth, glabrate toward maturity, heavily clothed on lower side with crowded, ferruginous to whitish, lance-attenuate, ciliolate scales 2–3 mm. long; sori and sporangia mostly or completely hidden by scales.

Dry rocky hillsides and canyon walls, Lower Sonoran Zone, central Arizona to western Oklahoma, Texas, Baja California, the West Indies, and southward at least as far as Chile.

**3*a*. Notholaena sinuata** var. **cochisensis** (Goodding) Weatherby, Jour. Arnold Arb. **24:** 314, 1943

*Notholaena cochisensis* Goodding, Muhlenbergia **8:** 93, 1912.
*Notholaena sinuata* var. *crenata* Lemmon, Ferns Pac. Slope 7, 1882 (nomen nudum).
*Notholaena sinuata* forma *crenata* Clute, Fern Bull. **15:** 46, 1915.
*Notholaena sinuata* var. *integerrima* of authors, not Hook., 1864.

Plants smaller and more slender than in var. *sinuata*; pinnae 2.5–8 mm. long and with 2–3 pairs of broad, short lobes; scales on rhizomes less markedly ciliolate, often entire; scales on upper surface of pinnules narrowly ovate with broad central part, and usually persisting until full maturity of frond.

Rocky hillsides, canyon walls, and under edges of rocks and boulders along desert water courses, Providence Mountains, California, to Arizona, New Mexico, and northern Mexico.

**Notholaena sinuata** var. **integerrima** Hook., Sp. Fil. **5**: 108, 1864, has been reported from the margins of the Sonoran Desert, but no specimens from our area have been seen. It is similar to var. *cochisensis* in habit, but has entire or faintly lobed pinnae; scales on upper surface of pinnules like those on var. *sinuata* and deciduous.

### 4. Notholaena aurea (Poir.) Desv., Mém. Soc. Linn. Paris **6**: 219, 1827

*Pteris aurea* Poir., in Lam., Encycl. **5**: 710, 1804.
*Acrostichum bonariense* Willd., Sp. Pl. **5**: 114, 1810.
*Pellaea ferruginea* Nees, Linnaea **19**: 654, 1847.
*Notholaena bonariensis* C. Chr., Ind. Fil. 459, 1906.

Rhizomes short-creeping, decumbent, freely branching, branches short, covered with closely compacted, lance-linear, bicolorous scales 2–2.5 mm. long with atropurpureous median strip, narrow, light brown, thin margins, filiform-attenuate, conspicuously nodulose apices, and rounded, light-colored to nearly white bases; stipes slender, 3–15 cm. tall, atropurpureous, tomentulose with rusty hairs, finally glabrate; fronds erect or ascending, 10–35 cm. tall, linear-lanceolate, 8–32 mm. wide; pinnae 10–40 pairs, broadly lanceolate to oblong-lanceolate, 3–7 mm. wide, with 5–9 rounded teeth on each side, arachnoid-pubescent above, densely tomentose with tangled ferruginous hairs beneath; margins of lobes sharply but narrowly revolute; sori practically continuous near margins of blunt teeth; sporangia light brown, shining at maturity, obscured by tomentum.

Shaded hillsides and rocky canyons, mostly Upper Sonoran Zone, Arizona to Texas and south to Jamaica and Argentina; in the Sierra de la Giganta, Baja California. Straggling into the Sonoran Desert in deep canyons.

### 5. Notholaena newberryi D. C. Eaton, Bull. Torrey Club **4**: 12, 1873

Ferns of compact habit with fronds 6–35 cm. tall from creeping, profusely branched rhizomes densely clothed with appressed, acicular-lanceolate, attenuate, atropurpureous, entire scales 2–3 mm. long, tips of these sinuous; stipes 4–15 cm. long, wiry, tawny-tomentose in youth but soon glabrate and glossy, purplish brown; blades oblong to oblong-lanceolate, 1.5–5 cm. broad, 3.15–15 cm. long, bipinnate; pinnules deeply pinnatifid, delicately and permanently ashy-tomentose above, densely tawny-tomentose beneath; pinnae 6–10 pairs, narrowly triangular to oblong-ovate; segments minute, rounded-obovate, entire to crenulate; sporangia dark brown, annulus conspicuous and shining, sides lighter brown, dull.

In crevices of rocks and under overhanging ledges, mostly Upper Sonoran Zone, along western slopes of mountains in southern California and Baja California; Guadalupe Island. A straggler into the western edge of the Sonoran Desert.

### 6. Notholaena parryi D. C. Eaton, Amer. Nat. **9**: 351, 1875

*Cheilanthes parryi* Domin, Bibl. Bot. **20** [Heft $85^1$]: 133, 1913.

Rhizomes multicipital, short-creeping, with crowded, appressed, linear-lanceolate scales 3.5–6 mm. long, their central strips atropurpureous, margins thin, cinnamomeous, entire; fronds crowded, arcuate-spreading, 5–20 cm. tall; stipes slender, wiry; blades oblong-lanceolate, 3–8 cm. long, 8–30 mm. wide, bipinnate; pinnules shallowly pinnatifid, 1–4 mm. long, densely woolly, tawny-tomentose beneath, coarsely hirsute-tomentose above with ashy, spreading, simple hairs; pinnae curling upward to form tight, involute balls when drying; sporangia few to numerous, dark brown, dull, partially to almost wholly obscured by dense wool.

In crevices in rocks and under ledges and boulders, Upper and Lower Sonoran Zones, Panamint Mountains, California, to southern Utah, central Arizona, the western edge of the Colorado Desert, and south in Baja California about to Laguna Chapala Seca.

### 7. **Notholaena lemmonii** D. C. Eaton, Bull. Torrey Club **7:** 63, 1880

Rhizomes multicipital, ascending, bearing crowded, appressed, linear-lanceolate scales 2–2.5 mm. long, about 0.5 mm. broad at base, dark purplish brown, nearly concolorous, light brown margins just a faint line, minutely ciliolate on young scales but cilia soon deciduous; fronds graceful, 15–25 cm. tall; stipes arcuate or erect, 3–12 cm. long, atropurpureous, shining, sparsely scaly toward base and with some scattered glands full length, shallowly and broadly sulcate along upper side; blades oblong-lanceolate, 1.5–3.5 cm. wide, 6–18 cm. long; pinnae 5–12 pairs, narrowly triangular to oblong, green and glabrous on upper side, white-ceraceous or golden-ceraceous beneath; segments crenulate; sori continuous along margins of lobes; sporangia light brown, dull.

Rocky slopes and canyons, Lower and Upper Sonoran Zones, in ranges bordering the deserts, southern Arizona to the vicinity of Guaymas, Sonora, and in the Cape Region and Sierra de la Giganta, Baja California.

### 8. **Notholaena jonesii** Maxon, Amer. Fern Jour. **7:** 108, 1917

Rhizomes massive, short, ascending, closely covered with light brown, thin, linear scales 3–4 mm. long with attenuate, flexuous, capillary tips; bases of old stipes persistent; fronds arcuate-spreading to ascending, 3–15 cm. long; stipes reddish brown, dull; blades 2–10 cm. long, triangular to ovate-oblong, bipinnate or less commonly tripinnate; pinnae 3–7 pairs, triangular-oblong, each with 2–5 pairs of entire or crenately lobed, subcordate segments below the larger, broadly triangular, usually 3-lobed terminal segment; all segments fleshy, glabrous, faintly glaucous above but not ceraceous on either surface; sporangia crowded in submarginal bands, dark buff, dull, with darker brown annuli.

Rare and local in crevices of limestone cliffs, Upper and Lower Sonoran Zones, in arid ranges from inner Santa Barbara County, California, to southwestern Utah, western Arizona, and the San Felipe Desert, Baja California.

### 9. **Notholaena californica** D. C. Eaton, Bull. Torrey Club **10:** 27, 1883

*Notholaena sulphurea* var. *californica* Farwell, Amer. Midl. Nat. **12:** 283, 1931.

Delicate plants 5–15 cm. high from nodose-multicipital rhizomes covered with rigidly acicular, dark reddish brown to black, sublustrous, finely denticulate scales 3–4.5 mm. long; fronds usually numerous, crowded, erect or ascending-spreading; stipes 3–12 cm. long, reddish brown to castaneous, dull; blades broadly deltoid-pentagonal, acute, 2–5 cm. long, 2–4 cm. wide, tripinnate in basal part, gradually simpler to pinnate apex; ultimate segments thickly ceraceous, broadly deltoid-ovate to oblong, rounded-obtuse, moderately crowded, sparsely glandular-puberulent above, densely yellow-ceraceous beneath, margins narrowly but sharply revolute, but not concealing the continuous bands of dark brown, dullish sporangia.

In crevices of rocks, Upper and Lower Sonoran Zones, Santa Catalina Island and San Jacinto Mountains, California, to western Arizona, northern Sonora, and central Baja California.

**10. Notholaena standleyi** Maxon, Amer. Fern Jour. **5:** 1, 1915

*Notholaena candida* var. *5-fidopalmata* Hook., Sp. Fil. **5:** 111, 1864.
*Notholaena hookeri* D. C. Eaton, in Wheeler, Rep. **6:** 308, *pl. 30,* 1879. Not *N. hookeri* Lowe, 1856.
*Cheilanthes hookeri* Domin, Bibl. Bot. **20** [Heft 85[1]]: 133, 1913.
*Chrysochosma hookeri* Kumm., Magyar Bot. Lapok **13:** 50, 1914.

Rhizomes nodose-multicipital, densely clothed with acicular, bicolorous scales 3–4.5 mm. long, median ridges of scales black, sublustrous, 0.2–0.4 mm. wide, margins thin, light brown, entire or faintly erosulate, sinuate, half as wide as to equaling the width of the central ridge; fronds 1–3 dm. tall; stipes reddish brown, sublustrous, sparingly ceraceous and bearing a few broadly ovate, subcordate scales 0.4–1.2 mm. wide near their bases; blades broadly pentagonal, 2.5–7 cm. long and wide, divided into 5 distinct pinnae, each of these broadly triangular-lanceolate or triangular-ovate and pinnate with 8–14 pairs of pinnatifid or shallowly pinnately lobed segments, dark green above, glabrous, densely yellow-ceraceous beneath; sori continuous along margins of weakly revolute segments; sporangia dark brown, dull.

Rocky canyons and cliffs, Upper and Lower Sonoran Zones, south-central Arizona to western Texas, northern Mexico, and southern Oklahoma and Colorado; Baja California south of Calmallí.

**11. Notholaena limitanea** Maxon, Amer. Fern Jour. **9:** 70, 1919

Delicate fern 1–2.5 dm. high; rhizomes nodose-multicipital, densely clothed with closely crowded, thin, flexuous, sinuous-margined, linear-attenuate, light brown scales 4–6 mm. long; stipes slender, 5–15 cm. long, atropurpureous and sublustrous or in youth lighter and dullish, glabrous; blades broadly ovate or triangular-ovate, 3–10 cm. wide, 4–15 cm. long, bearing 4–12 pairs of bipinnate pinnae; segments ovate to triangular-oblong, obtusish at apex, rounded at base, 1–3 mm. long, entire, moderately crowded, glaucous above, midrib impressed, densely white-ceraceous beneath; margins of segments subrevolute but not obscuring submarginal sori; sporangia light golden brown.

Dry rocky hillsides and crevices in cliffs and under boulders, Upper and Lower Sonoran Zones, Arizona to southern Utah, New Mexico, and Sonora.

**12. Notholaena incana** Presl, Rel. Haenk. **1:** 19, *pl. 1, fig. 2,* 1825

A delicate fern similar to *N. limitanea,* but with nonglaucous, mostly crenately lobed segments, and blades less intricately lobed or divided; rhizomes multicipital; scales light brown, sinuous, thin, 4–5 mm. long, lance-attenuate, concolorous; stipes atropurpureous or black, sublustrous, 5–12 cm. long; blades lance-triangular, 3–4.5 cm. broad, 6–12 cm. long, bipinnate, with segments further shallowly lobed or crenate, green above, white-ceraceous beneath, 2–4 mm. long; sporangia light brown, dull.

Dry rocky slopes and crevices in cliffs, Upper and Lower Sonoran Zones, Sonora and Chihuahua, southward almost throughout the drier parts of Mexico.

## Family 3. **MARSILEACEAE.** Pepperwort Family

Perennial, herbaceous plants growing in shallow water of sluggish streams or ponds or on mud at their margins, with slender, creeping rhizomes and wiry, fibrous roots. Leaves filiform and lacking blades or 2–4-foliolate, petioles slender, often elongated, and reaching the surface of water 1–5 dm. deep. Sporangia in bony, pedunculate, ovoid or globose sporocarps. Sporocarps arising from the rhizomes at or near the base of vegetative leaves. Megaspores and microspores closely associated in sporangia embedded in mucilaginous rings within the sporocarps.

### 1. **MARSILEA** L., Sp. Pl. 1099, 1753

Leaf blades 4-foliolate. Sporocarps ovoid with 2 small teeth near attachment of peduncle, pubescent, 2-celled vertically, tardily (often after several years) dehiscent into 2 valves and emitting 2 mucilaginous bands containing several transverse groups of sporangia. Megasporangia few, containing solitary megaspores; microsporangia numerous, each containing many minute microspores.

**1. Marsilea fournieri** C. Chr., Ind. Fil. 418, 1906

*Marsilea minuta* Fourn., Bull. Soc. Bot. France **27**: 329, 1880. Not *M. minuta* L., 1771.

Plants small, the leaves in tufts 0.5–3 cm. apart; petioles slender, 1–7 cm. long, spreadingly pubescent at first, soon glabrate except at extreme base and apex; leaflets 4, obovate, 3–12 mm. long, entire, arachnoid-pubescent on both surfaces, gray-green, finally glabrate; fruiting peduncles short, 1–5 mm. long, densely ferruginous-pubescent; sporocarps solitary on each peduncle but clusters of 3–15 at a node common, 3–4 mm. long, 2–3 mm. wide, densely hairy and conspicuously papillate, dorsal tooth 0.4–0.8 mm. long, curved forward, sharp-pointed, ventral tooth blunt, about half as long as dorsal; sori 7–9 in each valve of sporocarp.

Banks of intermittent streams and margins of vernal pools, Lower Sonoran Zone, central Sonora and Baja California south at least to Zacatecas.

## Family 4. **SALVINIACEAE.** Waterfern Family

Small, aquatic, floating plants with dorsiventral, branching stems bearing minute leaves imbricated in 2 ranks along the upper side, and simple, unbranched roots from the lower side. Sporocarps 2 or more on a common stalk situated among lower leaves, thin-walled, each sporocarp bearing 1–several sporangia. Megasporangia containing a solitary megaspore in each, microsporangia containing numerous minute microspores.

### 1. **AZOLLA** Lam., Encycl. **1**: 343, 1783

Characters of the family. Sporocarps of two kinds, those containing megasporangia ovoid, containing 1 megasporangium and several smaller, apparently nonfunctional bodies; those containing microsporangia globose and producing several microsporangia each.

**1. Azolla filiculoides** Lam., Encycl. **1**: 343, 1783

Plants somewhat mosslike, 0.5–2 cm. long and broad, pinnately branched, reproducing vegetatively by breaking into fragments that give rise to new plants, these floating on surface of water or becoming rooted to mud as water level drops; roots simple, solitary at each point of origin; basal leaves and stems often containing blue-green algal cells, upper leaves often deep red; leaves bilobed, dorsal lobe thick, spongy, papillate-hairy above, containing a gas-filled cavity in basal part, ventral lobe submersed, larger and thinner than dorsal, ventral lobes of lowest leaves modified to form sporocarps; megaspores globose, papillate.

Sluggish streams and ponds, Lower Sonoran to Transition Zones, Washington to southern California, Arizona, and south throughout much of Mexico and South America.

## CLASS II. GYMNOSPERMAE

Shrubs or trees with resinous wood made up of tracheids, vessels lacking among ours except in the Ephedraceae. Ovules borne naked on the surface of a sporophyll, or in some apparently at the tip of a branch, the sporophyll then obscure. Microspores dividing into 3 or more cells before the microgametophyte (pollen grain) germinates to produce a pollen tube. Endosperm made up of megagametophytic tissue. Cotyledons 2–20 (in ours 2–6).

KEY TO THE FAMILIES OF THE GYMNOSPERMAE

Branchlets deciduous with their attached leaves; cone scales woody, peltate, each bearing 3–15 seeds; large trees over 20 m. tall at maturity ...... Family 1. *Taxodiaceae*, p. 221

Branchlets not deciduous, the leaves falling from them after 3–7 years; cone scales fleshy or membranous, not peltate, rarely bearing more than 2, and usually only 1 seed each; shrubs or small trees less than 15 m. tall:

- Ovulate cones fleshy, berrylike at maturity; stems not equisetoid; leaves completely covering the twigs; vessels lacking in the xylem ........ Family 2. *Cupressaceae*, p. 222
- Ovulate cones of several pairs or whorls of membranous scales; stems equisetoid; leaves minute at the nodes, covering only a minute fraction of the internodes; vessels as well as tracheids in the xylem .................... Family 3. *Ephedraceae*, p. 224

### Family 1. **TAXODIACEAE.** BALD CYPRESS FAMILY

Monoecious trees with linear or scalelike ovate leaves spirally arranged and decurrent; branchlets (in ours) regularly deciduous at end of growing season or after several years. Staminate cones terminal or on short lateral branchlets; cone scales arranged spirally, each bearing several pollen sacs on abaxial surface. Ovulate cones terminal; scales spirally arranged, more or less peltate, and each bearing several ovules. Mature cones woody, scales peltate, spreading, in ours separating from the axis when mature. Seeds erect, small, irregularly angled, with resin pockets in seed coat; cotyledons 2–several.

#### 1. **TAXODIUM** L. C. Rich., Ann. Mus. Hist. Nat. **16:** 298, 1810

Large trees with deciduous leaves and branchlets; the roots producing erect, conic "knees" when growing in swamps. Leaves often nearly 2-ranked, linear. Staminate aments in pendent, panicled spikes or racemes. Ovulate cones globose to ovoid, scales stalked-peltate, woody, falling away from the axis when mature. Seeds wingless, containing "pitch pockets" in the corky coats.

**1. Taxodium mucronatum** Tenore, Ann. Sci. Nat. III, **19:** 355, 1853
*Taxodium montezumae* Decaisne, Bull. Soc. Bot. France **1:** 71, 1854.
*Taxodium mexicanum* Carr., Trait. Conif. 147, 1855.

Trees 20–30 m. tall, with straight trunk enlarged near the base; bark brownish red, relatively smooth to shallowly furrowed, fibrous, more or less shredded; leaves 6–12 mm. long, linear, spreading in nearly 2-ranked sprays, these and some young branchlets falling with appearance of new growth; staminate cones small, 1.5–2.5 mm. long, ovoid, in slender spikes 5–15 cm. long; ovulate cones subglobose, 1.5–2.5 cm. in diameter; seeds dark reddish brown, 4–8 mm. long, irregularly angular because of crowding.

Chiefly on the Mexican tableland and along the coast of the Gulf of Mexico; reaching the margin of the Sonoran Desert on the banks of the Río Cedros about 25 miles north of Quiriego, Sonora.

## Family 2. **CUPRESSACEAE.** Cypress Family

Monoecious or dioecious shrubs or trees with scaly or shredding bark and opposite or whorled, scalelike (or less commonly subulate-linear) leaves decurrent on and closely investing terete or slightly angled branchlets. Staminate cones small, terminal (in ours), each scale with several pollen sacs borne on under side of the peltate scale. Ovulate cones with several opposite or whorled scales, each bearing 1–many erect ovules at its base. Cones woody or (in ours) fleshy, scales imbricate or peltate, fused into a berrylike fruit in *Juniperus*. Seeds usually angled or winged. Cotyledons 2–several.

### 1. **JUNIPERUS** L., Sp. Pl. 1038, 1753

Evergreen trees or shrubs with thin, shredded bark and close-grained, reddish, soft wood. Leaves persistent for 3–7 years, scalelike or subulate, opposite or in whorls of 3, decurrent. Plants dioecious or monoecious. Staminate cones (in ours) solitary at tips of twigs, with ovate or peltate scales, each scale bearing 3–6 pollen sacs on its abaxial side. Ovulate cones of 2–3 series of fleshy scales, each scale bearing 1–2 ovules, but frequently all save 1 or 2 ovules in each cone abortive; mature cones berrylike by fusion of fleshy scales. Seeds 1–6, wingless; cotyledons 2–6.

Plants monoecious; mature fruits mealy or fibrous, 6–18 mm. in diameter; cotyledons 4–6:
  Leaves with a resinous pit on the back, rounded at the tips, usually arranged in threes; fruit 10–18 mm. long ........................................ 1. *J. californica*
  Leaves without a resinous pit, except on very vigorous branchlets, acute to acuminate, usually arranged in twos; fruit 6–10 mm. long .................. 2. *J. osteosperma*
Plants dioecious; mature fruits succulent, 4–6 mm. in diameter; cotyledons 2 ............ ............................................................ 3. *J. monosperma*

**1. Juniperus californica** Carr., Rev. Hort. IV, **3:** 352, *fig. 21,* 1854
*Sabina californica* Antoine, Cupress. Gat. 52, *pls. 71, 72,* 1857.
*Juniperus cedrosiana* Kell., Hesperian **4:** 2, 1860.
*Juniperus cerrosiana* Kell., Proc. Calif. Acad. **2:** 37, 1860.

Rounded shrub or infrequently a small tree to 12 m. tall with numerous, closely crowded branches forming a rounded or conical crown; leaves scalelike, usually in threes, ovate to broadly oblong, 3–4.2 mm. long, obtuse or subacute with tips strongly incurved and closely appressed, each leaf with a distinct orbicular to ovoid, resinous-glandular pit on the back; staminate cones very numerous, broadly ovoid, 3–5 mm. long; ovulate cones berrylike, globose to globose-oblong, 10–18 mm. long, smooth or with tips of coalesced scales still visible as low umbos, surface bluish with a heavy bloom in youth, later reddish brown, pulp dry, firm, and sweetish; seeds usually 1, sometimes 2, rarely more, ovoid, 5–10 mm. long, usually more or less trigonous, acute at apex, light brown and shining distally, duller and yellowish toward base; cotyledons 4–6.

Hillsides and outwash slopes, mostly Upper Sonoran Zone but a straggler in the Lower Sonoran Zone, desert slopes from northern Coast Ranges and central Sierra Nevada, California, to northern Baja California and Cedros Island, Mexico.

**2. Juniperus osteosperma** (Torr.) Little, Leafl. West. Bot. **5:** 125, 1948

*Juniperus tetragona* var. *osteosperma* Torr., Pacif. R. Rep. **4:** 141, 1857.
*Juniperus californica* var. *utahensis* Engelm., Trans. Acad. St. Louis **3:** 588, 1877.
*Juniperus utahensis* Lemmon, Rep. Calif. State Board Forestry **3:** 183, 1890.
(For additional synonymy see Leafl. West. Bot. **5:** 125, 1948.)

Arborescent shrub or small tree 2–5 m. tall, trunk gnarled and crooked, base to 8 dm. in diameter, crown rounded, often wider than high; bark reddish brown and flaking off in rhomboidal scales; leaves in twos or very rarely in threes, usually 1–1.5 mm. long on young twigs, oldest ones on larger twigs sometimes to 5 mm. long, acute to acuminate, devoid of a dorsal pit or this very inconspicuous, light yellowish green; staminate cones 4–5 mm. long at anthesis of anthers; ovulate cones maturing second year, globose to oblong-globose, 6–10 mm. long, covered with a dense bloom, reddish brown beneath the bloom when mature, flesh thin, sweet; seeds 1–2, ovoid, acute at apex, strongly 1–4-angled, dull hilum extending almost to apex; cotyledons 4–6.

Arid mesas and mountain slopes, mainly Upper Sonoran Zone, Providence and Panamint Mountains, California, to southern Wyoming, Colorado, and northern Arizona, barely touching the fringe of the Sonoran Desert in Mohave County, Arizona.

**3. Juniperus monosperma** (Engelm.) Sarg., Silva N. Amer. **10:** 89, *pl. 522,* 1896

*Juniperus occidentalis* var. β *monosperma* Engelm., Trans. Acad. St. Louis **3:** 590, 1877.
*Juniperus occidentalis* var. *gymnocarpa* Lemmon, West. Amer. Cone-Bearers 80, 1895.

A many-stemmed shrub or occasionally a tree to 15 m. tall with lobed and buttressed trunk to 1 m. in diameter; bark thin, ashy gray, furrowed and shreddy to expose red-brown inner bark; branchlets slender, scaly; leaves in pairs, 2–3.5 mm. long, ovate to ovate-acuminate, thickened and rounded on back, usually eglandular, but on vigorous shoots to 8 mm. long and conspicuously glandular; fruit globose to globose-oblong, 4–6 mm. long, dark blue to coppery, epidermis thick, glaucous, flesh sweetish, thin; seed usually 1 per cone, broadly ovoid, longitudinally 4-angled, slightly grooved between ridges, chestnut-brown, hilum paler, bilobed; cotyledons 2.

Rolling hills and lower mountain slopes, mostly Arid Transition Zone, but straggling through the Upper Sonoran Zone, reaching the upper margins of the Sonoran Desert in Arizona; southern Nevada to Colorado, Kansas, and Texas, southward into Chihuahua, Coahuila, and Nuevo León, Mexico.

## Family 3. **EPHEDRACEAE.** Ephedra Family

Shrubs (in our area) with jointed, opposite or fasciculately whorled branches with chlorenchymatous cortex in twigs and scalelike, opposite or whorled leaves, the latter more or less connate at base and either persistent or tardily deciduous. Flowers unisexual and usually dioecious, borne in axils of decussate, more or less membranous, persistent bracts. Stamens monadelphous, subtended by a 2-lobed, calyxlike, membranous perianth; anthers shedding pollen through terminal pores. Ovulate inflorescence composed of 1–several erect, solitary ovules, each enclosed in an urnlike perianth and subtended by 2–several binately or ternately arranged membranous scales. Seed coat terminating in a slender, stylelike micropyle; embryo axillary; cotyledons 2; endosperm copious.

### 1. **EPHEDRA** L., Sp. Pl. 1040, 1753

Characters of the family.

Leaves and bracts of cones in threes; bracts of at least ovulate cones short-clawed:
  Branches spinose-tipped; leaf scales persistent, erect, acute to acerose at tip, light-colored; seeds 3 to 6 times as long as wide .................................. 1. *E. trifurca*
  Branches blunt-tipped; leaf scales soon splitting to base and recurved, obtuse at apex, dark colored in age; seeds subglobose, nearly as wide as long ............. 2. *E. californica*
Leaves and bracts mostly in twos, rarely in threes; bracts of cones all sessile, often somewhat connate at base:
  Seeds solitary (or, if 2, light in color), circular or faintly trigonous in cross section; leaves 1–3 mm. long:
    Bracts of ovulate cones orbicular; leaf bases brown and persistent; stems often rough; seeds smooth to asperous ....................................... 3. *E. aspera*
    Bracts of ovulate cones elliptical; leaf bases gray and deciduous; stems usually smooth; seeds finely furrowed:
      Seeds 8–13 mm. long ............................. 4. *E. fasciculata* var. *fasciculata*
      Seeds 5–8 mm. long ............................... 4a. *E. fasciculata* var. *clokeyi*
  Seeds in pairs, distinctly trigonous in cross section; leaves 2–4 mm. long:
    Ovulate cones on peduncles to 5 cm. long; young branches pale green, glaucous ...... ........................................................ 5. *E. nevadensis*
    Ovulate cones sessile or essentially so; young branches bright green to yellow-green, not glaucous ....................................................... 6. *E. viridis*

**1. Ephedra trifurca** Torr., in Emory, Notes Mil. Rec. 153, 1848
*Ephedra trifaria* Parl., in A. DC., Prod. **16**²: 359, 1868.

Erect, yellowish green shrubs 5–20 dm. high with nearly smooth, straight branches 1.5–3.5 mm. in diameter; leaf scales persistent, sheathing to about middle or above, 5–13 mm. long, acuminate to acerose, becoming white and shredded; cones sessile, staminate obovate, 6–9 mm. long, with 8–12 whorls of thin, membranous bracts; bracts of ovulate cones in 8–10 whorls, membranous, obovate, slightly clawed, 2–3 mm. wide, 3–4 mm. long; ovulate cones solitary or in whorls, obovate, 10–14 mm. long, sessile or short-pedunculate, mature bracts orbicular, 8–12 mm. long, reddish brown in center and toward base, translucent-yellowish marginally, entire; seeds tetragonal in cross section, 9–14 mm. long, 1.5–3 mm. wide, equaling bracts.

Rocky hillsides, mesas, arroyos, and sandy plains, Lower Sonoran Zone, western Texas and southern Utah to southern California, Chihuahua, Sonora, and Baja California.

**2. Ephedra californica** S. Wats., Proc. Amer. Acad. **14:** 300, 1879

Low, spreading to suberect shrubs 5–12 dm. tall, with smooth or slightly scabrous, pale green, glaucous branches 2–4 mm. in diameter; leaf scales in threes, oblong, acute or obtuse, 2–6 mm. long, sheathing to middle or above, soon splitting to base and parts becoming recurved, persistent, dark brown to black in age; staminate cones sessile, solitary or in whorls, ovate, 6–8 mm. long, with 4–12 whorls of membranous bracts slightly connate at base, orange-yellow except at whitish margins; ovulate cones solitary or several in a whorl, ovate, 7–10 mm. long, on short, scaly peduncles, bracts in 4–6 whorls, orbicular, short-clawed, orange-yellow medianly, yellow-scarious along margins; seeds solitary, slightly exserted, ovoid to subglobose, 7–10 mm. long, nearly as wide.

On gravelly slopes, arroyos, and on rocky mesas and hillsides, Upper and Lower Sonoran Zones, Inyo County and western Fresno County, California, to the vicinity of Cataviñá, Baja California.

**3. Ephedra aspera** Engelm. ex S. Wats., Proc. Amer. Acad. **18:** 157, 1883

*Ephedra peninsularis* I. M. Johnston, Univ. Calif. Pub. Bot. **7:** 437, 1922.
*Ephedra reedii* Cory, Rhodora **40:** 216, 1938.
*Ephedra nevadensis* var. *aspera* L. Benson, Amer. Jour. Bot. **30:** 232, 1943.

Erect shrub to 15 dm. tall, or main branches spreading and the erect shoots 2–3 dm. long; branches stiff, terete, 1.5–3 mm. in diameter, opposite or in whorls at nodes, obtuse at apex; internodes 1–6 cm. long; young stems pale green to dark green, glaucous, strongly asperous to nearly smooth, finely furrowed; bark cinereous and fissured on old stems; leaf scales opposite or rarely in threes, 1–2.5 mm. long, connate from one-half to most of their length, obtuse, sheath splitting but subpersistent; staminate cones usually paired, obovate, 4–7 mm. long, sessile, bracts in 6–10 pairs, obovate, 2 mm. wide, 3 mm. long, membranous, yellow to reddish brown; ovulate cones ovate, 6–10 mm. long, sessile or short-pedunculate, bracts in 5–7 pairs, orbicular, 2–5 mm. in diameter, thickened centrally, scarious toward margins, reddish brown; seeds solitary, light brown, terete or obscurely trigonous in cross section, 2.5–4 mm. wide, 5–8 mm. long, smooth to asperous.

Desert slopes and mesas, Upper and Lower Sonoran Zones, Mojave and Colorado Deserts to Texas, south to southern Baja California, Zacatecas, and Tamaulipas.

**4. Ephedra fasciculata** A. Nels., Amer. Jour. Bot. **21:** 573, 1934, var. **fasciculata**

Prostrate to erect shrubs 2.5–10 dm. tall with flexuous, terete branches 1.5–3.5 mm. in diameter, arranged oppositely or in whorls; young stems pale green, slightly scaberulous to smooth, somewhat glaucous, tips obtusely conical; leaves in pairs, 1–3 mm. long, connate to three-fourths their length, obtusely tipped; sheaths hyaline, white, splitting but subpersistent; staminate cones narrowly elliptic, 4–8 mm. long, sessile, their bracts binate, obovate, about 2 mm. wide, 2–3 mm. long, membranous, slightly connate basally, in 4–8 whorls, light yellow; ovulate cones sessile or short-pedunculate, 6–13 mm. long, bracts elliptic, 2–4 mm. wide, 3–7 mm. long, slightly connate basally, in 4–7 pairs, somewhat thickened and green to light brown on back, hyaline marginally; seeds usually solitary, circular in cross section, longitudinally furrowed, 3–5 mm. in diameter, 8–13 mm. long.

Desert slopes and mesas, Upper and Lower Sonoran Zones, Colorado Desert and eastern Mojave Desert, California, to central Arizona.

**4*a*. Ephedra fasciculata** var. **clokeyi** (Cutler) Clokey, Madroño **8:** 56, 1945

*Ephedra clokeyi* Cutler, Ann. Mo. Bot. Gard. **26:** 402, 1939.

Differing from var. *fasciculata* chiefly in having seeds only 5–8 mm. long.

Desert hillsides, washes, and mesas, Lower Sonoran Zone, Colorado Desert, California, to southern Nevada, southern Utah, and central Arizona.

**5. Ephedra nevadensis** S. Wats., Proc. Amer. Acad. **14:** 298, 1879

*Ephedra antisyphilitica* var. *pedunculata* S. Wats., Bot. King Expl. 330, *pl. 39,* 1871. Not *E. pedunculata* C. A. Meyer, 1846.

*Ephedra nevadensis* subvar. *paucibracteata* Stapf, Denkschr. K. Akad. Wiss. Wien **56**[2]**:** 83, 1889.

Erect, pale green or glaucous shrub 5–12 dm. high with spreading, slightly scabrous, more or less flexuous branches; leaf scales in twos, 2–6 mm. long, sheathing to about middle, obtuse to acute at apex, mostly deciduous in age; staminate cones sessile or short-pedunculate, with 4–9 pairs of obovate bracts 2–3 mm. wide, 3–4 mm. long; ovulate cones subglobose, 5–11 mm. long on peduncles 1–50 mm. long, bracts binate, round-ovate, in 3–5 whorls, firm, scarious on margins, sessile; seeds 1–2 in a cone, 6–9 mm. long, 2–4 mm. wide, exserted, smooth, trigonous, when 2 in a cone, with flattish face and strongly convex or carinate back.

Arid hillsides and plains, Upper and Lower Sonoran Zones, Mojave Desert to southern Utah, south into Sonora and Baja California.

**6. Ephedra viridis** Coville, Contr. U.S. Nat. Herb. **4:** 220, 1893

*Ephedra nevadensis* subvar. *pluribracteata* Stapf, Denkschr. K. Akad. Wiss. Wien **56**[2]**:** 83, 1889.

*Ephedra nevadensis* var. *viridis* M. E. Jones, Proc. Calif. Acad. II, **5:** 726, 1895.

Erect, rigidly branched shrub 5–12 dm. tall with bright green or yellow-green branches 2–3 mm. in diameter, their tips obtuse; leaves in pairs, 1.5–4 mm. long, sharp-tipped, connate from one-third to three-fourths their length, deciduous but leaving persistent brown bases; staminate cones obovate, 5–7 mm. long, sessile, with 6–10 pairs of ovate, yellow or slightly roseate and membranous bracts 2–3 mm. broad, 2–4 mm. long; ovulate cones obovate, 6–10 mm. long, sessile or short-pedunculate, with 4–8 pairs of ovate branches 2–4 mm. wide, 4–7 mm. long; seeds trigonous, smooth or faintly striate, brownish, 5–8 mm. long, 1–2 mm. longer than bracts.

Rocky hillsides, mesas, and washes, Lower Sonoran Zone to Transition Zone, eastern Ventura County, California, to western Colorado, south into Arizona and the Colorado Desert.

## CLASS III. ANGIOSPERMAE

Herbs, shrubs, trees, or vines with vessels, fiber cells, and often tracheids in xylem of vascular tissues. Microspore nucleus usually dividing once before dispersal of microgametophytes (pollen grains), thus the microgametophyte 2-celled or binucleate at time of dispersal. Megasporophylls (carpels) often elongated at apex to form a slender style, this bearing a receptive area (stigma) at or near its summit. Carpels 1–several in each perfect or pistillate flower, if more than 1 either all free and distinct or more or less coalesced and pistil then made up of more than 1 carpel. Styles and stigmas in multicarpellate pistils distinct or coalesced. Ovules borne inside carpels on the placental tissue of ovary, the placenta central, peripheral, or basal; each ovule attached to placenta by a slender stalklike filament, the funiculus. Microgametophytes carried from anthers to stigmas by gravity, wind, water, or insects, or transferred to stigma by mechanical contact owing to growth movements of the style, the stamens, or both, the pollen grains germinating on stigma and producing a pollen tube which grows to ovules in ovary. Microgametes usually producing "double fertilization," one male gamete nucleus uniting with nucleus of egg cell, the other with endosperm nucleus or nuclei to initiate formation of food-supplying endosperm of the seed. Seeds with 1 or 2 coats. Embryo minute to large, containing much stored food, or the latter occurring in the endosperm, or some stored in each tissue. Endosperm copious, scanty, or none. Cotyledons 1 or 2.

KEY TO THE SUBCLASSES OF THE ANGIOSPERMAE

Cotyledon 1; leaves usually parallel-veined; vascular bundles "closed," that is, without a meristem (cambium) between the xylem and phloem in the stem .................................................... Subclass I. MONOCOTYLEDONES, p. 227

Cotyledons 2; leaves usually netted-veined; vascular bundles "open," with a meristem (cambium) between the xylem and phloem in the bundles of the stem .................................................... Subclass II. DICOTYLEDONES, p. 370

## SUBCLASS I. MONOCOTYLEDONES

Stems composed of soft, parenchymatous ground tissue with vascular bundles scattered irregularly throughout the mass. Cambium none, or present in limited areas to form new "closed" vascular bundles in outer part of stems in some arborescent species, no definite differentiation of stem into wood, pitch, and bark. Leaves usually parallel-veined, mostly alternate and entire, or margins dentate, bases often sheathing stem with little or no distinction into blade and petiole. Flowers mostly trimerous, or some whorls suppressed. Embryo with a single cotyledon and first leaves of seedling alternate.

KEY TO THE FAMILIES OF THE MONOCOTYLEDONES

Plants strictly aquatic, floating on surface or submerged in water:
- Plants immersed in water, rooted to bottom:
  - Stigmas sessile or essentially so, capitate or peltate .................................................... Family 3. POTAMOGETONACEAE, p. 230
  - Stigmas slender, subulate:
    - Flowers solitary; leaves spiny-toothed, whorled or opposite, not over 5 cm. long; fruit ellipsoidal, unicarpellate ........................ Family 4. NAIADACEAE, p. 233
    - Flowers borne in rows on a slender spadix; leaves 2-ranked, entire, ribbonlike, to 2 m. long; fruit cordate, bicarpellate ................ Family 5. ZOSTERACEAE, p. 233
- Plants floating on the surface, rarely rooted in mud:
  - Body of plant differentiated into stem, leaves, and roots, the leaves short-petiolate; inflorescence of several staminate and 1 pistillate flower in each spathe .................................................... Family 11. ARACEAE, p. 327
  - Body of plant minute, thalluslike; inflorescence of a solitary stamen and pistil in each minute spathe (plants flowering infrequently) ... Family 12. LEMNACEAE, p. 327

Plants terrestrial, or only partly submerged:

Perianth reduced or lacking, not petaloid, its parts often of bristles or scales and in a single whorl:

Flowers not in the axils of chaffy bracts; stamens 2–7; inflorescences each ensheathed in a spathelike, often fugacious bract:

Flowers in dense terminal spikes; perianth of silky bristles; fruits bearing many buoyant hairs ................................ Family 1. TYPHACEAE, p. 229

Flowers in globose heads, the staminate uppermost; perianth of a few chaffy scales; fruits without buoyant hairs ................ Family 2. SPARGANIACEAE, p. 230

Flowers in the axils of chaffy bracts (glumes); stamens 1–3; inflorescences not ensheathed in spathelike bracts:

Stems mostly with hollow internodes, terete; leaf sheaths split down one side; each flower subtended by 2 chaffy bracts, the lemma and palea (or the latter rarely lacking); leaves mostly 2-ranked ............... Family 8. GRAMINEAE, p. 237

Stems usually solid, often angular; leaf sheaths cylindrical, not split down one side (or tardily ruptured in age); each flower normally subtended by 1 chaffy bract; leaves mostly 3-ranked ........................ Family 9. CYPERACEAE, p. 301

Perianth of 2 distinct whorls, the inner one often petaloid:

Carpels 3–many, distinct:

Plants herbaceous; flowers not in spathes; carpels 6–many:

Petals greenish, similar to the sepals; inflorescences terminal spikes on naked scapes, fruit a capsule ................ Family 6. SCHEUCHZERIACEAE, p. 234

Petals white, larger and thinner than the sepals; inflorescences paniculate or umbellate; fruit an achene ....................... Family 7. ALISMACEAE, p. 235

Plants treelike, with a single, unbranched trunk; flowers in spathes; carpels 3 ...... ......................................... Family 10. PHOENICACEAE, p. 321

Carpels 1–3, united into a single pistil:

Leaves pinnately compound or the blade flabelliform, many-ribbed; plants usually treelike, the trunk unbranched ........... Family 10. PHOENICACEAE, p. 321

Leaves mostly simple, not flabelliform or many-ribbed; plants herbaceous or if treelike the leaves simple, linear:

Herbage bearing water-absorbing scales, at least when young; plants often epiphytic ......................................... Family 13. BROMELIACEAE, p. 329

Herbage not clothed with water-absorbing scales; plants not epiphytic:

Ovary superior:

Perianth segments glumelike, greenish, brownish, or purplish .............. ...................................... Family 16. JUNCACEAE, p. 336

Perianth segments not glumelike, at least the inner whorl petaloid:

Functional stamens 3; inflorescences in spathelike leaves or bracts:

Inner perianth segments irregular; carpels 1–2-ovuled ................. ............................. Family 14. COMMELINACEAE, p. 331

Inner perianth segments regular; carpels many-ovuled ................. ............................. Family 15. PONTEDERIACEAE, p. 335

Functional stamens usually 6; inflorescence not in a spathe:

Inner perianth segments fugacious; carpels 2-ovuled; stamen filaments long-villous below ........... Family 14. COMMELINACEAE, p. 331

Inner perianth segments not fugacious; carpels usually several- to many-ovuled; stamen filaments mostly glabrous:

Styles 3, distinct or nearly so; capsule septicidal .................... ............................. Family 17. MELANTHACEAE, p. 339

Style 1, the stigma often 3-lobed; capsules loculicidal except in *Calochortus,* in which it is septicidal ...... Family 18. LILIACEAE, p. 340

Ovary inferior:

Flowers perfect; plants scapose herbs from bulbs, or perennial plants with rosettes of heavy, fibrous leaves from short-caulescent or acaulescent rootstocks; leaves linear to oblong, not cordate .............................................................. Family 19. AMARYLLIDACEAE, p. 359

Flowers unisexual; plants scandent or climbing vines from fleshy, tuberlike roots; leaves cordate ................ Family 20. DIOSCOREACEAE, p. 369

## Family 1. **TYPHACEAE.** CATTAIL FAMILY

Herbaceous marsh or aquatic plants with creeping rootstocks and glabrous, erect stems. Leaves flat, linear, sheathing at base. Flowers monoecious, borne in dense, cylindrical terminal spikes which are subtended by fugacious spathelike bracts and divided at intervals by smaller caducous bracts, staminate flowers above pistillate. Perianth of several fine hairs or bristles. Stamens 2–7 with connate filaments. Ovary 1–2-celled. Fruit a minute achene bearing silky, buoyant bristles.

### 1. **TYPHA** L., Sp. Pl. 971, 1753

Characters of the family.

Pistillate part of spike dark brown, contiguous to staminate, 4–6 times as long as thick; stigmas ligulate to rhomboid; leaves mostly flat on back; pollen grains in fours .................................................................. 1. *T. latifolia*

Pistillate part of spike buff to light brown, separated from staminate, 6–10 times as long as thick; stigmas linear; leaves convex on back; pollen grains distinct or rarely in pairs .................................................................. 2. *T. domingensis*

#### 1. **Typha latifolia** L., Sp. Pl. 971, 1753

Plants comparatively coarse and stout; stems 1.5–3 m. tall; leaves 6–25 mm. wide, almost flat on back, nearly equaling fruiting spikes; sheaths with scarious margins toward apex, sometimes truncate or faintly auriculate, usually tapering into blade; spikes dark brown, staminate part beginning immediately above, and about one-half as wide as pistillate part, this 20–30 mm. in diameter, 10–20 cm. long; bractlets wanting or hairlike; pistillate spikes compact; staminate spikes loose at anthesis; pedicels of pistillate flowers slender but compound, 2–3 mm. long at maturity; stigmas medium brown to dark brown, rhomboid, lanceolate, or spatulate, fleshy, about half as long as ovary; ovary stipitate, bearing silky hairs attached at base of stipe; seeds bearing numerous, silky, buoyant hairs about equaling stigma.

In fresh and brackish marshes throughout the north temperate regions. Occasional along ditches and about *tinajas* in the Sonoran Desert, as at Tinajas Altas, Arizona, and near El Centro, California. May–July.

#### 2. **Typha domingensis** Pers., Syn. Pl. **2:** 532, 1857

*Typha truxillensis* H. B. K., Nov. Gen. & Sp. **1:** 82, 1816.
*Typha bracteata* Greene, Bull. Calif. Acad. **2:** 413, 1887.
*Typha angustifolia* var. *virginica* Tidestrom, Rhodora **13:** 242, 1911.

Stems slender, 1–2 m. tall, sterile ones with 6–9 leaves and white pith, equaling to slightly surpassing spikes; sheaths tapering to throat, scarious-margined above; pistillate spike light brown, 15–25 mm. in diameter, 15–25 cm. long, separated from staminate spike by 0.5–6 cm. of naked axis; pistillate flowers on stout, compound pedicels, each flower subtended by a linear bract with an obovate, lanceolate, to apiculate, light brown, translucent blade, and bearing slender, silky white hairs at base of stipe; stigma on fertile flowers linear, deciduous; sterile pistillate flowers with obovoid abortive ovaries and bearing hairs in whorls; bracts in staminate spikes linear, cuneate, or laciniate, brown; pollen grains distinct (1-celled) or rarely in pairs.

Marshy areas in the Lower and Upper Sonoran Zones, throughout most of California, east to the Atlantic coast, and south through tropical areas. May–July.

## Family 2. **SPARGANIACEAE.** Bur Reed Family

Herbaceous marsh or pond plants with fibrous roots and creeping rootstocks, erect or floating stems, and alternate, linear, basally sheathing leaves. Flowers monoecious, crowded into distinct globose heads on upper parts of simple or branched stems, sessile or pedunculate, staminate uppermost. Spathes linear to oblong, immediately subtending or somewhat below head. Perianth of chaffy scales. Stamens usually 5; filaments distinct; anthers cuneate or oblong. Ovary 1- or rarely 2-celled, sessile. Fruit nutlike, 1 seed per cell.

### 1. **SPARGANIUM** L., Sp. Pl. 971, 1753

Characters of the family.

**1. Sparganium greenei** Morong, Bull. Torrey Club **15:** 77, *pl. 79, fig. 3,* 1888
*Sparganium eurycarpum* var. *greenei* Graebner, in Engler, Pflanzenreich **4**[10]**:** 13, 1900.

Stem erect, 5–20 dm. tall; leaves 8–12 mm. wide, 3–15 dm. long, keeled, somewhat spongy; bracts similar but smaller; inflorescence branched, lower branches erect, bearing 1–4 (usually 2) pistillate and 6–10 staminate heads, the latter 1.5–2 cm. in diameter; fruiting heads 2–2.5 cm. in diameter; scales clawed, cuneate or rhombic blade about equaling body of achene; achenes sessile, brown, shining, obovoid, 5 mm. in diameter, 6–9 mm. long, obtuse to rounded at summit, the beak 2 mm. long.

Marshes and ditches, Lower Sonoran Zone to Transition Zone, British Columbia to northern Baja California, entering the Sonoran Desert in Imperial Valley, California, and at Valle Trinidad, Baja California. May–Aug.

## Family 3. **POTAMOGETONACEAE.** Pondweed Family

Perennial fresh- and brackish-water plants with jointed, leafy stems and submerged or floating leaves or both, these petiolate or sessile, linear to expanded into a blade. Flowers perfect or monoecious in axillary clusters, or in sessile or pedunculate spikes. Flowers often enclosed in a hyaline sheath; perianth of 4 or 6 valvelike segments or none. Stamens 1–4; anthers extrorse, 1–2-celled. Ovaries 1–6, distinct, 1-celled. Fruit indehiscent, sessile or stipitate.

Flowers perfect; stamens 2 or 4:
  Drupelets sessile; peduncles not coiled; stamens 4 ..................... 1. *Potamogeton*
  Drupelets stipitate; peduncles coiled after anthesis; stamens 2 ............... 2. *Ruppia*
Flowers monoecious; stamen 1 ........................................ 3. *Zannichellia*

### 1. **POTAMOGETON** L., Sp. Pl. 126, 1753

Leaves alternate or uppermost opposite, frequently of two kinds, submerged and floating, the former mostly linear, the floating petiolate, lanceolate to broadly elliptic, coriaceous. Stipules more or less united and sheathing, membranous. Spikes sheathed by stipules in bud, usually reaching the surface on a straight peduncle in flower. Sepals 4, rounded; stamens 4, opposite the sepals. Ovaries 4, sessile, distinct, 1-celled and 1-ovulate; stigma sessile or on a short recurved style. Fruit of 4 drupelets, these somewhat compressed.

Leaves lanceolate or elliptic-oblong, 15–30 mm. wide; peduncles 6–14 cm. long ........... 1. *P. illinoensis*
Leaves linear or capillary, 0.1–2 mm. wide; peduncles 0.3–3 cm. long:
  Stipules free from leaves; leaves 0.5–2 mm. wide, 3–5-nerved; nutlets 3-keeled, middle keel winged and usually somewhat dentate ........................ 2. *P. foliosus*
  Stipules adnate to leaf for half their length; leaves capillary to narrowly linear, rarely 0.75 mm. wide, 1-nerved; nutlets obscurely 2-keeled, rounded and smooth on back ...... 3. *P. pectinatus*

### 1. **Potamogeton illinoensis** Morong, Bot. Gaz. **5:** 50, 1880

*Potamogeton lucens* of American authors. Not *P. lucens* L., 1753.
(For full synonymy see Rhodora **45:** 154–155, 1943.)

Stems thick, profusely branched, often rooting at nodes; leaves all submerged, short-petiolate or sessile, lanceolate to elliptic, 15–30 mm. wide, 6–15 cm. long, acute and cuspidate at apex, rounded to attenuate at base; petioles 4–20 mm. long, often winged above; stipules axillary, free from petioles, obtuse at apex, 2–3 (or –7) cm. long; spikes 3–6 cm. long, cylindrical; peduncles 6–14 cm. long, thicker at apex than at base; nutlets about 3 mm. high, faintly 3-keeled, with a groove running inward from base; style short.

In streams, ponds, and lakes, in several life zones, British Columbia to Quebec southward to Central America and the West Indies; occurring in the Sonoran Desert near Comondú and Loreto, Baja California. July–Oct.

### 2. **Potamogeton foliosus** Raf., Med. Rep. II, **5:** 354, 1808

*Potamogeton pauciflorus* Pursh, Fl. Amer. Sept. 121, 1814.
*Potamogeton pauciflorus* var. *californicus* Morong, Bot. Gaz. **10:** 254, 1885.
*Potamogeton californicus* Piper, Contr. U.S. Nat. Herb. **11:** 98, 1906.
*Potamogeton foliosus* var. *genuinus* Fernald, Mem. Amer. Acad. **17:** 43, 1932.

Stems flattened, leafy, often rooting at the nodes, to 1 m. long; leaves all submerged, narrowly linear, 1–10 cm. long, 0.5–2 mm. wide, acute to inconspicuously cuspidate, 3–5-nerved (rarely 1-nerved); stipules forming tubular blunt sheaths at first, 3–18 mm. long, soon deciduous; peduncles slender, 3–20 mm. long; spike few-flowered, subcapitate to short-cylindric; nutlets green to yellowish, compressed, obovoid to suborbicular, 1.8–2.5 mm. high, 3-keeled, middle keel most prominent, narrowly winged and undulate to dentate; beak 0.2–0.8 mm. long; winter buds developing in late fall or lacking.

In fresh or brackish water, Washington to New Jersey and southward to the West Indies and Central America collected in the Sonoran Desert only along eastern margin of the Cape Region, Baja California.

### 3. **Potamogeton pectinatus** L., Sp. Pl. 127, 1753

*Potamogeton columbianus* Suksdorf, Deutsch. Bot. Monatsschr. **19:** 92, 1901.

Rootstocks and stems very slender, rooting at nodes, dichotomously branching; leaves all submerged, capillary to narrowly linear, attenuate at apex, 0.1–1 mm. wide, 1.5–4 cm. long, 1-nerved, with a few faint transverse veinlets; stipules adnate about half their length, 15–25 mm. long, scarious along margins of their sheaths; peduncles filiform, 5–30 mm. long; flowers in 2–6 whorls; nutlets obliquely obovoid, 3–4 mm. long, smooth, rounded and with 2 obscure keels on back; beak 0.2–0.4 mm. long.

In quiet fresh or brackish water, world wide in distribution. Known in the Sonoran Desert region from Mulegé, Santa Agueda, and Comondú, Baja California. May–Oct.

## 2. RUPPIA L., Sp. Pl. 127, 1753

A submerged aquatic plant with capillary, branching stems and slender, attenuate, nearly capillary, alternate, 1-nerved leaves with membranous sheathing bases. Flowers perfect, at first enclosed within a spathelike leaf base; on capillary, spadixlike peduncles which coil spirally after pollination; anthers 2; pistils sessile, 4 in each flower; stigma sessile, peltate. Fruit small, obliquely ovoid, pedicellate drupes.

**1. Ruppia maritima** L., Sp. Pl. 127, 1753

*Ruppia curvicarpa* A. Nels., Bull. Torrey Club **26:** 122, 1899.

Stems 6–10 dm. long, irregularly branching, internodes 2–8 cm. long; leaves capillary to narrowly linear, 0.1–0.3 mm. wide, 2–10 cm. long, obscurely 1-nerved, acute to attenuate-pungent; stipular sheath 4–10 mm. long, membranous, free part from as long as the tube of the sheath to wanting, rounded or forming 2 acute, minute teeth at apex; peduncle elongating at anthesis to carry flowers to surface of water, later coiling into a loose spiral; anthers 2, sessile, 2-celled, soon deciduous; fruit ovoid-flask-shaped, oblique or equilateral, 2–2.5 mm. long, borne on stipes 1–3 cm. long, fruits of each peduncle forming umbellate clusters; style straight or curved.

In brackish or salt water along the coast and in slightly saline waters inland, widely distributed in the northern hemisphere. In the Sonoran Desert known from Salt Creek, Imperial County, California, from Mulegé and San Gregorio, Baja California, and from Lerdo, Sonora. March–Oct.

## 3. ZANNICHELLIA L., Sp. Pl. 969, 1753

Submerged aquatic with capillary, sparsely branched stems from a creeping rhizome and flat, filiform, 1-nerved leaves. Staminate and pistillate flowers in same axil and enclosed by a hyaline, spathelike sheath in bud; staminate flowers solitary, bearing a single 2-celled anther on a short filament; pistillate flowers 2–5 in each sheath. Ovary flask-oblong, short-stipitate; style short, bearing a cup-shaped stigma with angled or dentate margins. Fruit a flattened, faintly falcate, narrowly ribbed nutlet with sinuate or toothed back.

**1. Zannichellia palustris** L., Sp. Pl. 969, 1753

Stems capillary, slightly branched; leaves 2–10 cm. long, acute at apex; stipules free, scarious, slender; staminate flower of a solitary 2-celled anther on a slender filament, without perianth; pistillate flowers of 2–10 sessile carpels which may become pedicellate after anthesis; nutlet slightly flattened, somewhat falcate, 2–4 mm. long, sinuately ribbed on margins or distinctly but finely toothed on angled, convex edge.

In fresh to brackish water, almost world wide in distribution. In the Sonoran Desert it is known from the vicinity of Tucson and Sacaton, Arizona, from Needles, Palm Springs, Coachella, and several places in Imperial County, California, and from Mulegé and Comondú in Baja California. June–Sept.

## Family 4. NAIADACEAE. NAIAS FAMILY

Slender, submerged aquatic plants with moderately branching stems and opposite or whorled, dentate leaves with broad, sheathing bases. Flowers dioecious or monoecious, axillary and solitary, sessile or short-pedicellate. Staminate flowers with an outer entire or toothed perianth whorl and an inner hyaline one adnate to anthers. Stamen 1; anthers 1–4-celled, 2-lobed or apiculate at summit. Pistillate flowers of a solitary short-styled ovary; stigmas subulate, 2–4. Fruit sessile, ellipsoidal, pericarp crustaceous.

### 1. NAIAS L., Sp. Pl. 1015, 1753

Characters of the family.

Flowers dioecious; internodes and back of leaf spiny; leaves coarsely serrate .... 1. *N. marina*
Flowers monoecious; internodes and back of leaf unarmed; leaves minutely serrulate ...... ........................................................ 2. *N. guadalupensis*

**1. Naias marina** L., Sp. Pl. 1015, 1753
*Naias major* All., Fl. Ped. **2**: 221, 1785.
*Naias marina* var. *californica* Rendle, Trans. Linn. Soc. II, **5**: 398, *pl. 39, fig. 15*, 1899.
*Naias marina* var. *mexicana* Rendle, loc. cit., *fig. 30*.

Stems branched irregularly or dichotomously, spinulose-dentate along the internodes; leaves opposite or rarely alternate, linear, 0.5–3.5 mm. wide, 1–4.5 cm. long, margins coarsely serrate-dentate with triangular, apiculate teeth 0.5–1.2 mm. long, often dorsally toothed along stiff midrib; leaf bases rounded, sheathing, devoid of teeth; flowers dioecious, staminate 3–4 mm. long, anthers 4-locular; pistillate flowers about as long; stigmas 3; fruit 3–4 mm. long, reticulate-tessellated.

In quiet fresh water, widely distributed. In our region known from near Tucson, Arizona, also at Mulegé, near Comondú, and at San Gregorio, Baja California. July–Aug.

**2. Naias guadalupensis** (Spreng.) Morong, Mem. Torrey Club **3**: 60, *pl. 67*, 1893
*Caulinia guadalupensis* Spreng., Syst. Veg. **1**: 20, 1825.
*Naias flexilis* var. *guadalupensis* A. Br., Jour. Bot. Brit. & For. **2**: 276, 1864.
*Naias microdon* var. *guadalupensis* A. Br., loc. cit.

Stems more slender than in the former species; leaves 0.3–1.5 mm. wide, 10–25 mm. long, acute at apex, usually tipped with 1–2 minute apiculations and bearing 20–40 minute marginal serrulations or these wanting; leaf bases oblique to rounded, minutely denticulate; flowers monoecious, 2–3 mm. long, anthers 4-locular; fruit 1.2–1.8 mm. long, ellipsoid, tipped with a stigmatic remnant, dull or the ridges of the minute but distinct, quadrangular reticulations sometimes shining.

In quiet water, from Oregon to Pennsylvania and southward to Central America; in the Sonoran Desert known from ponds between Canipolé and Comondú, at San Evaristo Bay, and in the vicinity of Loreto, Baja California. July–Aug.

## Family 5. ZOSTERACEAE. EELGRASS FAMILY

Submerged perennial marine plants from tough, creeping rootstocks, with flattened, branching stems and ribbonlike, 2-ranked leaves sheathing at base. Flowers monoecious or dioecious, borne in spadices enclosed by spathes. Perianth wanting, the flowers enclosed by a hyaline scale. Staminate flowers in 2 rows on spadix, each comprised of a single sessile anther; pollen filamentous. Pistillate flowers 1-celled with 2 carpels and 2 stigmas.

### 1. **PHYLLOSPADIX** Hook., Fl. Bor. Amer. **2:** 171, 1838

Characters of the family. Plants dioecious.

**1. Phyllospadix torreyi** S. Wats., Proc. Amer. Acad. **14:** 303, 1879

Stem simple or branched, somewhat flattened, 2–6 dm. long, dark brown or nearly black; leaves linear, flat, or complicate to terete in age, 1–2.5 mm. wide or less, up to 2 m. long, with 1–3 primary nerves, or these obscure or lacking in narrower leaves; terminal part of spathe usually loosely twisted spirally; spadices 2 or 3 pairs, borne near apex of stem, 4–5 cm. long, each pair subtended by 2–several fugacious bracts; staminate flowers numerous, each consisting of a single sessile, 2-celled anther; pistillate flowers 3–4 mm. long at anthesis, the overlapping flap obovate to obovate-elliptic, 4–6 mm. long; fruit deeply cordate-sagittate, body 3–4 mm. long, tipped with persistent style and bearing 2 projecting basal wings.

Among rocks in the surf from about low-tide level to a depth of 2–3 fathoms, along nearly all rocky parts of the coast from Bolinas Bay, California, almost to Cabo San Lucas, Baja California. April–Dec.

## Family 6. **SCHEUCHZERIACEAE.** Arrow Grass Family

Small or moderate-sized plants of marshy places with rushlike, terete basal leaves with membranous sheaths. Inflorescences spicate or racemose at tip of a slender scape. Flowers perfect; perianth 3–6-parted, segments in 2 series, deciduous or persistent. Stamens 3–6, filaments elongated or anthers nearly sessile; anthers 2-celled, commonly extrorse. Ovary superior, 1-celled; carpels 3–6, each carpel 1–2-ovuled, closely or imperfectly united. Capsules dehiscent (or indehiscent in some), separating from central axis from base upward and dehiscing by a ventral pore in ours. Embryo straight.

### 1. **TRIGLOCHIN** L., Sp. Pl. 338, 1753

Characters of the family. Perianth segments 6, concave, inner whorl of 3 inserted higher than outer whorl. Ovary in ours 6-carpellate; style short; stigmas 6, plumose. Capsules dry, firm.

**1. Triglochin concinna** Davy, Erythea **3:** 117, 1895

Rootstock slender, stoloniferous or rhizomatous, creeping, 2–4.5 mm. in diameter; caudex short, naked or retaining a few dead leaf bases; leaves spreading-ascending, blade nearly terete, 1–2 mm. in diameter, 5–20 cm. long, sheath 1–4 cm. long, margins hyaline-scarious; ligule 1–2 mm. long, biauriculate; scape slender, 1.5–2 mm. in diameter, 1–3 dm. tall; racemes slender, 5–20 cm. long; pedicels 2–5 mm. long, slightly decurrent; perianth segments about 2 mm. long; anthers sessile; fruit ovoid to oblong-ovoid, 2–3.5 mm. in diameter, 5–6 mm. long, obtuse at base, each carpel acute at apex, trigonous in cross section, separating from central axis.

In saline coastal marshes, Lower Sonoran Zone to Transition Zone, southern British Columbia to central Baja California. Feb.–Aug.

**Triglochin concinna** var. **debilis** (M. E. Jones) J. T. Howell, Leafl. West. Bot. **5:** 18, 1947, with heavier, more densely clothed rootstocks and more robust habit, occurs in saline marshes of the interior from eastern Oregon to North Dakota and southward into northern Arizona, but has not been reported, to date, from the Sonoran Desert.

## Family 7. ALISMACEAE. Water Plantain Family

Marsh or aquatic plants with long-petioled leaves sheathing at the base, stems scapose. Flowers regular, perfect, monoecious or dioecious, borne in racemes or panicles. Perianth segments of 3 outer herbaceous, persistent sepals and 3 deciduous white petals. Stamens 6 or more, included; anthers bilocular. Ovaries numerous, distinct, 1-celled, 1-ovuled, maturing as achenes.

Flowers all perfect; leaves cordate to cuneate at base, not sagittate ......... 1. *Echinodorus*
Flowers dioecious, or monoecious with upper staminate, lower pistillate; some leaves (in ours) sagittate-lobed at base ........................................ 2. *Sagittaria*

### 1. ECHINODORUS Rich. ex Engelm., in A. Gray, Man. 460, 1848

Annual or perennial herbs with elliptic-ovate to lanceolate, often cordate leaf blades, these 3–13-ribbed and mostly punctate with dots or lines. Stems scapose, exceeding leaves, bearing perfect flowers in whorls in panicles or racemes, each whorl subtended by 3 outer herbaceous bracts and several inner bracteoles. Receptacle convex or globose. Stamens 12–30; anthers dehiscing by lateral slits. Ovaries numerous, with obliquely apical, persistent styles. Achenes compressed, ribbed and beaked, in dense, spinose heads.

**1. Echinodorus cordifolius** (L.) Griseb., Abh. K. Ges. Wiss. Gött. **7:** 257, 1857
*Alisma cordifolia* L., Sp. Pl. 343, 1753.
*Echinodorus rostratus* Engelm., in A. Gray, Man. ed. 2, 438, 1856.

Leaves lanceolate to broadly ovate, 1–22 cm. wide, 3–25 cm. long, acute to obtuse at apex, cuneate, truncate, or commonly cordate at base, 3–13-veined, petioles mostly exceeding blades; scapes 1–several, 1–12 dm. tall, simple or paniculately branched; pedicels 7–15 mm. long; bracts lance-linear, 5–25 mm. long; sepals ovate, acute, 4–5 mm. long; petals white, broadly obovate, 6–10 mm. long; fruiting heads burlike, subglobose, 4–8 mm. in diameter; achenes cuneate, flat, 2.5–3 mm. long, prominently ribbed, beak erect, straight or slightly curved, half as long as body.

Margins of ponds and ditches, Upper and Lower Sonoran Zones, central California to Illinois and Missouri to tropical America. Occurring at Comondú, Baja California. April–Aug.

### 2. SAGITTARIA L., Sp. Pl. 993, 1753

Perennial aquatic or marsh plants from nodose or tuber-bearing rootstocks. Scapes erect, decumbent, or floating, ensheathed at base by long, spongy petioles. Flowers pedicellate, borne in tribracteate whorls of threes near summits of scapes, dioecious, or if monoecious, uppermost flowers staminate, lower ones pistillate. Sepals persistent and reflexed or spreading in pistillate flowers. Stamens numerous, inserted on receptacle; anthers bilocular, dehiscing laterally. Ovaries 1-celled, with persistent stigmas, crowded in spherical heads, compressed.

Basal lobes of leaves equaling or shorter than terminal lobe:
- Achenes with a horizontal beak, more broadly winged at top than on margins; fruiting heads 1.5–3.2 cm. in diameter .......... 1. *S. latifolia*
- Achenes with an erect, minute beak, equally winged on both margins; fruiting heads 1–1.5 cm. in diameter .......... 2. *S. cuneata*

Basal lobes of leaves longer than terminal lobe:
- Seed coat smooth; achenes winged laterally; basal lobes of leaves less than twice as long as terminal lobe .......... 3. *S. greggii*
- Seed coat punctate; achenes not winged on lateral faces; basal lobes of leaves two to three times as long as terminal lobe .......... 4. *S. longiloba*

### 1. **Sagittaria latifolia** Willd., Sp. Pl. **4:** 409, 1805

*Sagittaria variabilis* Engelm., in A. Gray, Man. 461, 1848.

Plants in shallow water or muddy places, often wholly emersed; rootstocks bearing winter tubers; leaves 10–40 cm. long, terminal lobe deltoid to ovate or lanceolate, basal lobes ovate-lanceolate, somewhat divergent, from half as long as, to equaling terminal lobe; scape erect, somewhat angled, 1.5–15 dm. tall; bracts ovate to lance-ovate, glabrous, 0.5–1 cm. long, acute to obtuse; whorls of inflorescence 3–15 cm. apart; fruiting heads 1.5–3.2 cm. in diameter; achenes broadly obovate, truncate and broadly winged at apex, 2.5–3.5 mm. high, beak horizontal, 1.5–2 mm. long, acuminate, sides smooth.

In shallow streams and margins of lakes, Lower Sonoran Zone to Boreal Zone, British Columbia to the Atlantic seaboard and southward to Arizona and southern California. Near Tucson, Arizona. June–Sept.

### 2. **Sagittaria cuneata** Sheldon, Bull. Torrey Club **20:** 283, *pl. 159,* 1893

*Sagittaria arifolia* Nutt. ex J. C. Smith, Rep. Mo. Bot. Gard. **6:** 32, *pl. 1,* 1894.
*Sagittaria paniculata* Blankins., Mont. Agr. Coll. Sci. Stud. **1:** 40, *pl. 6,* 1905.

Plant partly emersed, 1.5–4 dm. tall or taller if growing in deep water; leaf blades hastate to sagittate, 6–18 cm. long, abruptly acuminate, basal lobes one-half as long as, to equaling terminal lobe; scapes usually shorter than leaves or equaling them, weak; whorls of flowers 1–several, each subtended by a linear bract about equaling or slightly exceeding pedicel; sepals ovate to ovate-oblong, 6–8 mm. long in fruit; petals white with yellowish or greenish bases, 8–12 mm. long; fruiting heads 1–1.5 cm. in diameter; achenes obovate, 2–2.5 mm. long, wings about equal along each margin, slightly narrower at apex, somewhat callus-thickened; beak minute, erect.

In sluggish streams, ditches, and ponds, Lower Sonoran to Canadian Zones but mainly in the latter, British Columbia to Nova Scotia, south to Connecticut, Michigan, the Rocky Mountains, and central Arizona. July–Aug.

### 3. **Sagittaria greggii** J. G. Smith, Rep. Mo. Bot. Gard. **6:** 43, *pl. 12,* 1894

Plant partly emersed, 2–14 dm. high; leaf blades 20–40 cm. long, terminal lobe lanceolate to ovate, acuminate, basal lobes widely divergent, narrowly lanceolate, exceeding terminal one; submerged leaves without blade or with a narrowly lanceolate one; scape erect, surpassing leaves, paniculately branched, angled below, terete above; bracts lance-acuminate, 15–30 mm. long; pedicels 15–30 mm. long; sepals scaberulous, obovate, 6–8 mm. long; stamens 30 or more; fruiting heads globose, 7–16 mm. in diameter; achenes obovate to orbicular-obovate, 2–3 mm. long, tumid and crested along both margins, sides with a narrowly winged, irregularly tuberculate depression; beak minute, horizontal; seed smooth.

Ditches and sluggish streams, Upper and Lower Sonoran Zones, San Joaquin Valley, Imperial Valley, California, and Arizona to central Mexico. July–Aug.

### 4. **Sagittaria longiloba** Engelm., in Torr., Bot. Mex. Bound. 212, 1859

Plant slender, erect, partly emersed, 3–5 dm. high; leaf blades 10–20 cm. long, terminal lobe linear to ovate-lanceolate, one-half to one-third as long as linear-lanceolate, acuminate, slightly divergent basal ones; scape usually single, slightly surpassing leaves; bracts lance-acuminate, 6–8 mm. long; fertile pedicels 15–35 mm. long; stamens 15–21; fruiting head 10–12 mm. in diameter; achenes about 2 mm. long, narrowly winged on both margins and with a minute horizontal costa from base of minute beak across upper end of body; seed ovoid, about 1 mm. long, minutely punctate.

Margins of ponds and sluggish streams, Lower Sonoran to Transition Zones, Colorado and Kansas to Texas and Sonora. July–Aug.

**Alisma plantago-aquatica** L., Sp. Pl. 342, 1753, has been collected at Victorville, at the southern edge of the Mojave Desert, and may occur in ditches or ponds in Imperial County, California, or in Arizona.

## Family 8. **GRAMINEAE.** Grass Family*

Flowers perfect (rarely unisexual), arranged in spikelets consisting of a shortened axis (rachilla) and 2–many 2–4-ranked bracts, the lowest 2 (glumes, rarely 1 or both obsolete) empty, each succeeding 1 or more (lemmas) bearing in their axils a single flower (1–few of the lower lemmas sometimes barren, and the upper 1 or more often reduced and sterile), and between the flower and rachilla a 2-nerved bract (palea), the lemma, palea, and included flower constituting the floret; stamens 1–many, usually 3, with delicate filaments and 2-celled anthers; pistil 1, with a 1-celled, 1-ovuled ovary, 2 (rarely 1 or 3) styles, and usually plumose stigmas; fruit a caryopsis, the grain rarely free from the pericarp. Herbaceous or sometimes woody plants with round or somewhat flattened, hollow or solid stems (culms), and 2-ranked, usually parallel-veined leaves consisting of a sheath which envelops the culm, a blade, usually flat, and between the two on the inside a membranous or hairy appendage (ligule), this rarely obsolete.

Spikelets in groups of 2–5, falling entire, sometimes enclosed in spiny burs (Zoyseae; see also genus *Cenchrus*):
- Spikelets in groups of 2–5, sessile on a short, zigzag rachis, enclosed in small, spiny burs composed of indurate second glume of 2 lower spikelets ............... 28. *Tragus*
- Spikelets in groups of 3, central one perfect, lateral spikelets staminate, not enclosed in burs:
  - Groups of spikelets sessile, erect on a stiff axis; plants rigid, stiffly erect perennials ........................................................ 29. *Hilaria*
  - Groups of spikelets nodding, pedunculate on a very slender axis; plants delicate, not stiff and rigid ............................................ 30. *Aegopogon*

---

* Contributed by Jason R. Swallen, U.S. National Herbarium, Smithsonian Institution, Washington, D.C.

Spikelets not arranged in groups:
Spikelets dorsally compressed, with 1 terminal fertile floret and a sterile or staminate floret below, the latter usually without a palea; articulation below the spikelets, either in pedicel, in rachis, or at base of a cluster of spikelets (PANICOIDEAE):
Fertile lemma and palea firmer than glumes, usually indurate (PANICEAE):
Spikelets subtended by bristles or enclosed in spiny burs:
Spikelets subtended by bristles (see also *Cenchrus*):
Bristles below only terminal spikelet of each branch .......... 52. *Setariopsis*
Bristles below all spikelets .......................................... 53. *Setaria*
Spikelets enclosed in spiny burs or bristles united only at base ....... 54. *Cenchrus*
Spikelets neither subtended by bristles nor enclosed in spiny burs:
Spikelets in open panicles:
Fruit dark brown; lemma with more or less prominent white margins, these not inrolled .............................................. 46. *Leptoloma*
Fruit pale; margins of indurate lemma inrolled:
Spikelets without woolly hairs; plants herbaceous .............. 49. *Panicum*
Spikelets with a tuft of woolly hairs at tip of second glume and sterile lemma; plants woody, clambering .............................. 50. *Lasiacis*
Spikelets short-pedicellate on one side of panicle branches:
Second glume mucronate, sterile lemma mucronate or awned .. 51. *Echinochloa*
Second glume and sterile lemma awnless:
Spikelets covered with long, silky hairs .................... 44. *Trichachne*
Spikelets glabrous or appressed-pubescent:
Racemes digitate; first glume present; weedy, decumbent, spreading annuals ................................................... 45. *Digitaria*
Racemes paniculate:
First glume and rachilla joint forming a swollen, ringlike callus below spikelet; back of fruit turned away from rachis ........ 47. *Eriochloa*
First glume wanting (occasionally present in *Paspalum distichum*); back of fruit turned toward rachis ........................ 48. *Paspalum*
Fertile lemma and palea thin, hyaline; glumes indurate; sterile lemma like fertile one:
Spikelets unisexual, pistillate ones below, staminate ones above in same inflorescence ........................................................ 61. *Tripsacum*
Spikelets in pairs, one sessile, the other pedicellate, both usually fertile, or sessile spikelet perfect and pedicellate one sterile or staminate, sometimes much reduced (ANDROPOGONEAE):
Spikelets all alike, surrounded by a conspicuous tuft of soft hairs; rachis continuous; fertile lemma awnless ....................................... 55. *Imperata*
Spikelets unlike, sessile one perfect, pedicellate one usually sterile, but sessile spikelet staminate and pedicellate one perfect in genus *Trachypogon*:
Racemes reduced to one or few joints, these pedunculate in a subsimple or compound panicle ............................................ 59. *Sorghum*
Racemes of several to many joints, solitary, digitate, or aggregate in panicles:
Spikelets awnless, perfect spikelet globose; freely branching annual with short racemes partly enclosed in sheaths .................... 56. *Hackelochloa*
Spikelets, or at least some of them, awned:
Awns delicate, glabrous, not more than 2.5 cm. long ....... 58. *Andropogon*
Awns stout, conspicuous, 3–12 cm. long or, if rather slender, then plumose on lower part; racemes solitary:
Awns hirsute; primary spikelet fertile, pedicellate one sterile; lower pairs of spikelets all staminate or neuter ................... 60. *Heteropogon*
Awns plumose; primary spikelet sterile, pedicellate one fertile ........... ............................................ 57. *Trachypogon*

Spikelets more or less laterally compressed; sterile florets, if any, above the fertile ones (except in PHALARIDEAE); articulation above glumes (except in genera *Leersia, Luziola,* and *Spartina*) (FESTUCOIDEAE):
 Spikelets with 2 sterile florets below fertile one (PHALARIDEAE) .......... 41. *Phalaris*
 Spikelets with no sterile florets below the 1 or more fertile florets:
  Glumes wanting; pedicels articulate below the 1-flowered spikelets:
   Lemma (at least the nerves) and palea indurated, strongly flattened and keeled; spikelets perfect (ORYZEAE) .................................... 43. *Leersia*
   Lemma and palea very thin, rounded on back; spikelets unisexual, staminate and pistillate spikelets on separate inflorescences (ZIZANEAE) ......... 42. *Luziola*
  Glumes present:
   Spikelets sessile or subsessile in spikes or spikelike racemes:
    Spikelets on opposite sides of continuous or disarticulating rachis (HORDEAE):
     Spikelets solitary at each node of rachis; first glume wanting except on terminal spikelet; spikelets placed edgewise to rachis ................. 16. *Lolium*
     Spikelets more than 1 at each node of rachis:
      Spikelets 3 at each node of rachis, 1-flowered, lateral ones usually reduced to awns ................................................ 19. *Hordeum*
      Spikelets 2 at each node of rachis:
       Rachis usually continuous; glumes acute to aristate, entire .... 17. *Elymus*
       Rachis readily disarticulating; glumes narrow, extending into long, divergent awns ........................................ 18. *Sitanion*
    Spikelets subsessile on one side of continuous rachis; spikes solitary, digitate, or racemose on a common axis (CHLORIDEAE):
     Plants with unisexual, rarely perfect florets, staminate and pistillate ones in the same spike ........................................ 40. *Cathestecum*
     Plants with perfect flowers:
      Spikelets 1-flowered with no rudimentary florets above perfect one:
       Spikes digitate ........................................ 36. *Cynodon*
       Spikes racemose ........................................ 34. *Spartina*
      Spikelets 2–several-flowered or, if only 1-flowered, then with 1 or more rudimentary florets above fertile one:
       Spikes solitary or racemose:
        Lemmas entire or minutely bifid, awnless or 1-awned; spikelets with 2–several perfect florets:
         Spikes few to several ............................. 31. *Leptochloa*
         Spikes solitary ................................... 35. *Microchloa*
        Lemmas variously lobed or dentate, 3-awned; spikelets with 1 perfect floret and 1 or 2 rudimentary florets above it ......... 39. *Bouteloua*
       Spikes digitate or aggregate on a very short axis:
        Lemmas awnless:
         Rachis prolonged beyond spikelets in a naked point ............... ........................................ 32. *Dactyloctenium*
         Rachis not prolonged beyond spikelets ................ 33. *Eleusine*
        Lemmas awned:
         Lemmas 1-awned; spikes digitate ..................... 37. *Chloris*
         Lemmas 3-awned; spikes aggregate on a very short axis, but scarcely digitate ...................................... 38. *Trichloris*
   Spikelets in open or spikelike panicles:
    Spikelets 1-flowered (AGROSTIDEAE):
     Fruit indurate, terete, awned; callus well developed, oblique, bearded:
      Awn trifid, lateral awns sometimes minute .................. 27. *Aristida*
      Awn simple, with a distinct line of demarcation between lemma and awn ........................................................ 26. *Stipa*

Fruit thin or firm but not indurate; callus not well developed:
Fertile spikelets surrounded by numerous sterile spikelets in the form of bristles or delicate bracts .............................. 23. *Pereilema*
Spikelets all perfect, not surrounded by bristles:
Glumes longer than lemma; lemma and palea much thinner than glumes ....................................................... 22. *Agrostis*
Glumes shorter than lemma, similar in texture:
Lemmas 3-nerved ................................ 24. *Muhlenbergia*
Lemmas 1-nerved ................................. 25. *Sporobolus*
Spikelets 2–several-flowered:
Glumes longer than lowest floret, usually longer than spikelet (Aveneae):
Lemmas awnless; panicles small, dense ..................... 20. *Schismus*
Lemmas awned from back; panicles diffuse ............. 21. *Deschampsia*
Glumes shorter than first floret; lemmas awnless or awned from tip or from a minutely bifid apex (Festuceae):
Plants dioecious; pistillate lemmas with 3 long, twisted, divergent awns, staminate lemmas mucronate ...................... 15. *Scleropogon*
Plants with perfect flowers or, if dioecious, then pistillate spikelets awnless:
Lemmas divided at summit into 5 or more awns or awnlike lobes:
Lemmas divided into 5 awnlike lobes .................... 10. *Orcuttia*
Lemmas divided into 9 or more awns:
Rachilla disarticulating between florets; summit of lemmas with awns intermixed with awned teeth ...................... 11. *Cottea*
Rachilla not disarticulating between florets, all florets falling attached; summit of lemmas with awns only:
Awns nearly equal, plumose .................... 12. *Enneapogon*
Awns unequal, not plumose .................. 13. *Pappophorum*
Lemmas awnless, 1-awned, or, if 3-awned, then lateral awns minute:
Tall, stout reeds with plumelike panicles:
Lemmas hairy; rachilla naked ......................... 7. *Arundo*
Lemmas naked; rachilla hairy ...................... 8. *Phragmites*
Low or rather tall grasses; panicles not plumelike:
Plants dioecious, culms erect from creeping rhizomes; plants of saline soil:
Blades very short, stiff, spreading; panicles small, hidden in axils of leaves ................................ 14. *Monanthochloe*
Blades elongate; panicles exserted:
Staminate and pistillate spikelets similar in appearance .......... ........................................... 5. *Distichlis*
Staminate and pistillate spikelets very unlike, staminate many-flowered, rather distant in a loose spike, pistillate solitary or clustered, nearly hidden in leaves ............... 6. *Jouvea*
Plants with perfect flowers:
Lemmas 3-nerved, nerves usually prominent:
Lemmas glabrous, awnless ...................... 4. *Eragrostis*
Lemmas densely pubescent on nerves, mucronate or awned ...... ............................................ 9. *Tridens*
Lemmas 5–9-nerved:
Lemmas awnless, usually rounded on back; blades with boat-shaped tips .................................. 3. *Poa*
Lemmas mucronate or awned, keeled at least toward apex:
Lemmas awned from between teeth of minutely bifid apex .... ........................................... 1. *Bromus*
Lemmas awnless, or awned from tip .............. 2. *Festuca*

## 1. BROMUS L., Sp. Pl. 76, 1753

Annuals with closed sheaths, linear, flat blades, and open panicles of rather large spikelets; spikelets several-flowered, rachilla disarticulating above the glumes and between florets; glumes acute or acuminate, the first 1- or obscurely 3-nerved, second 3–5-nerved; lemmas keeled, 5–9-nerved, awned from between teeth of bifid apex; palea a little shorter than lemma, thin, ciliate on keels, adherent to caryopsis.

Awn geniculate, twisted below the bend .................................... 1. *B. trinii*
Awn straight:
  Lemmas conspicuously compressed-keeled; awns 7 mm. long or less ...... 2. *B. arizonicus*
  Lemmas rounded on back; awns 3.5–5 cm. long ........................ 3. *B. rigidus*

**1. Bromus trinii** Desv., in Gay, Fl. Chil. **6:** 441, 1853
*Trisetum trinii* Louis-Marie, Rhodora **30:** 243, 1928.

Annual; culms erect or ascending, simple or branching below, 30 cm. to as much as 1 m. tall, or smaller in depauperate plants, usually pubescent at nodes; leaves pilose or nearly glabrous; panicles 5–20 cm. long, erect or somewhat nodding, rather dense, branches ascending or lower ones spreading, stiff or flexuous; spikelets 1.5–2 cm. long, 5–7-flowered; glumes acuminate, first narrow, 8–10 mm. long, 1-nerved, second broader, 12–16 mm. long, 1–3-nerved; lemmas 10–18 mm. long, rather densely pilose to nearly glabrous, with narrow, aristate teeth mostly 2–3 mm. long; awn 6–20 mm. long, geniculate, twisted below the bend.

Waste places, Oregon to Colorado and Baja California; Ecuador to Chile and Argentina.

**2. Bromus arizonicus** (Shear) Stebbins, Proc. Calif. Acad. IV, **25:** 309, 1944
*Bromus carinatus* var. *arizonicus* Shear, U.S. Dept. Agr. Div. Agrost. Bull. **23:** 62, 1900.

Annual; culms erect or ascending, mostly 25–80 cm. tall, or shorter in depauperate plants; sheaths softly hairy or nearly glabrous, mostly longer than internodes; blades as much as 30 cm. long, 2.7 mm. wide, lax, glabrous or sparsely pilose, scabrous on margins; panicles as much as 30 cm. long, branches stiffly ascending or somewhat flexuous; spikelets 2–3 cm. long, 6–8-flowered; glumes acuminate, about half as long as spikelet, the first 3-nerved; lemmas pilose or hirsute toward margins, nearly glabrous on back, rather prominently notched, bearing a scabrous awn mostly 5–7 mm. long.

Stream banks, flats, sand dunes, and sandy land, southern California, Arizona, and Baja California.

**3. Bromus rigidus** Roth, Mag. Bot. Roem. & Ust. **10:** 21, 1790

Annual; culms erect, 20–70 cm. tall; leaves rather densely pilose or papillose-pilose, the blades short, mostly 2–5 mm. wide; panicles open, nodding, the branches bearing 1 or 2 spikelets; glumes attenuate, glabrous, the first 1.5–2 cm. long, second 2.5–3 cm. long; lemmas 2.5–3 cm. long, scabrous or sometimes nearly glabrous, bearing a rather stout, very scabrous awn 3.5–5 cm. long.

Open ground and waste places, British Columbia to Baja California (Santa María Plains, Guadalupe Island) and Arizona; rare eastward; introduced from Europe.

## 2. FESTUCA L., Sp. Pl. 73, 1753

Slender annuals; spikelets in narrow or open panicles, 1–several-flowered; rachilla disarticulating above glumes and between florets; glumes narrow, acute, unequal; lemmas rounded on back, 5-nerved, awned from tip.

Spikelets more than 5-flowered; florets crowded; lemmas 4–5 mm. long, margins inrolled; awns 2–5 mm. long .......................................... 1. *F. octoflora*
Spikelets usually less than 5-flowered or, if more than 5-flowered, lemmas 7–8 mm. long, and awns more than 10 mm. long:
Spikelets glabrous:
Spikelets 3–6-flowered; pedicels appressed .......................... 2. *F. pacifica*
Spikelets 1–2-flowered; pedicels spreading, especially those on upper part of main axis ........................................................ 3. *F. reflexa*
Spikelets hirsute or villous .............................................. 4. *F. grayi*

**1. Festuca octoflora** Walt., Fl. Car. 81, 1788
*Festuca tenella* Willd., Sp. Pl. **1**: 419, 1797.
*Vulpia octoflora* Rydb., Bull. Torrey Club **36**: 538, 1909.

Annual; culms slender, tufted, erect or ascending, mostly 10–30 cm. tall, sometimes dwarfed; leaves glabrous, blades soft and lax, less than 1 mm. wide; panicles 2–8 cm. long, narrow, dense, short branches appressed, often reduced to a single spikelet; spikelets mostly 5–8-flowered, 6–8 mm. long, florets crowded; first glume narrow, subulate, 1-nerved, second a little longer than first, acute or acuminate, 3-nerved, 3–5 mm. long; lemmas commonly 4–6 mm. long, acuminate, glabrous or scabrous, bearing an awn 2–5 mm. long.

Open ground at low altitudes, southern Canada, throughout the United States, and Baja California.

**2. Festuca pacifica** Piper, Contr. U.S. Nat. Herb. **10**: 12, 1906
*Vulpia pacifica* Rydb., Bull. Torrey Club **36**: 538, 1909.

Annual; culms erect or geniculate at lower nodes, commonly 10–30 cm., or as much as 60 cm. tall; blades soft, loosely involute, glabrous, 1–5 cm. long; panicles 3–12 cm. long, branches solitary, stiffly spreading, 1–4 cm. long; spikelets 3–6-flowered, appressed; first glume about 4 mm. long, subulate-lanceolate, second about 5 mm. long, lanceolate-acuminate; lemmas 6–7 mm. long, glabrous or scabrous, bearing an awn 10–15 mm. long.

Open ground and open woods, British Columbia and Montana, south to Baja California, Arizona, and New Mexico.

**3. Festuca reflexa** Buckl., Proc. Acad. Phila. **1862**: 98, 1862
*Vulpia reflexa* Rydb., Bull. Torrey Club **36**: 538, 1909.

Similar to *F. pacifica*; spikelets 1–3-flowered, stiffly divergent; first glume 2–4 mm. long, second 4–5 mm. long; lemmas 5–6 mm. long, glabrous or scaberulous, bearing an awn 5–10 mm. long.

Washington to Utah, south to southern California and central Arizona.

**4. Festuca grayi** (Abrams) Piper, Contr. U.S. Nat. Herb. **10**: 14, *pl. 3*, 1906
*Festuca microstachys grayi* Abrams, Fl. Los Ang. 52, 1904.

Annual; culms erect, 10–50 cm. tall, slender, glabrous, purple toward base; lower sheaths sparsely to densely retrorsely pilose, upper glabrate; blades as much as 10 cm. long, narrow, rather stiff, glabrous beneath, pubescent or pilose; panicles 5–15 cm. long, branches stiffly ascending to spreading, usually spikelet-bearing nearly to base;

spikelets 3–4-flowered, appressed or some of them spreading; glumes villous, first lanceolate-subulate, 1-nerved, 4–5 mm. long, second a little longer, acute or acuminate, 3-nerved; lemmas 5–7 mm. long, acuminate, hirsute or villous, bearing an awn about 1 cm. long.

Dry open ground and rocky slopes, Washington to southern California and Arizona.

## 3. POA L., Sp. Pl. 67, 1753

Annual or perennial grasses with narrow or open panicles and linear blades with boat-shaped tips; spikelets few- to several-flowered, rachilla disarticulating above glumes and between florets; glumes subequal, rounded or keeled, 3-nerved; lemmas awnless, 5-nerved, glabrous, villous on the keel and lateral nerves or pubescent across the back, often with a tuft of long, cottony hairs at base.

Plants annual; lemmas distinctly keeled, villous on keel and marginal nerves, with a tuft of cottony hairs at base .................................................. 1. *P. bigelovii*
Plants perennial; lemmas rounded, not keeled, crisp-pubescent across back toward base or nearly glabrous .................................................. 2. *P. scabrella*

### 1. **Poa bigelovii** Vasey & Scribn., in Vasey, Desc. Cat. Grasses U.S. 81, 1885

Annual; culms rather densely tufted, erect, 5–40 cm. tall; sheaths glabrous; ligule broad, hyaline, 2–4 mm. long; blades 2–12 cm. long, 2–4 mm. wide, glabrous; panicles 3–10 cm. long, narrow, rather dense, branches usually appressed, spikelet-bearing to base; spikelets 4–6 mm. long, 4–5-flowered, short-pedicellate; first glume 3 mm. long, second 4 mm. long; lemmas webbed at base, densely villous on keel and marginal nerves, pubescent on internerves below, lowest one about 3.5 mm. long.

Open ground, southern California and Baja California to western Texas and Chihuahua.

### 2. **Poa scabrella** (Thurb.) Benth. ex Vasey, Grasses U.S. 42, 1883

*Poa tenuifolia* Buckl., Proc. Acad. Phila. **1862:** 96, 1862.
*Atropis scabrella* Thurb., in S. Wats., Bot. Calif. **2:** 310, 1880.
*Poa orcuttiana* Vasey, W. Amer. Sci. **3:** 165, 1887.

Perennial; culms densely tufted, erect, 45–60 cm. or sometimes as much as 1 m. tall, usually scaberulous at least below the panicle; sheaths scabrous, sometimes obscurely so, lower ones usually reddish or purple; ligule 3–5 mm. long, thin; blades mostly basal, lax, flat, more or less scabrous, 1–2 mm. wide; panicles commonly 5–12 cm. long, sometimes longer, narrower, rather dense, branches ascending or somewhat spreading or drooping; glumes 3–4 mm. long, acutish; lemmas 4–5 mm. long, obtuse, rounded, obscurely keeled, puberulent or scabrous on back, crisp-pubescent at the base or sometimes nearly glabrous.

Dry hills, woods, and rocky places at lower altitudes, Washington to northern Baja California (Guadalupe Ranch, Todos Santos Bay; Guadalupe Canyon).

## 4. ERAGROSTIS Host, Ic. Gram. Austr. 4: 14, *pl. 24,* 1809

Annual or perennial grasses with flat blades and open panicles; spikelets several-flowered, rachilla disarticulating above glumes and between florets or continuous, lemma persistent, palea deciduous; glumes acute, 1-nerved, shorter than first floret; lemma 3-nerved, rounded on back; palea about as long as lemma.

Plants perennial; panicles relatively large, diffuse ........................ 7. *E. intermedia*
Plants annual:
  Paleae conspicuously ciliate with long hairs; plants viscid .................. 1. *E. viscosa*
  Paleae not ciliate:
    Lemmas with a few prominent glands on keel; panicles densely flowered; spikelets 2.5–3 mm. wide ............................................ 2. *E. cilianensis*
    Lemmas without glands on keel; panicles open; spikelets much narrower:
      Spikelets about 1 mm. wide; panicles delicate .......................... 3. *E. pilosa*
      Spikelets 1.5 mm. wide or wider; panicles not delicate:
        Panicle branches densely flowered, naked at base, branchlets and spikelets appressed ........................................................ 4. *E. diffusa*
        Panicle branches loosely flowered, branchlets and spikelets spreading:
          Spikelets linear; pedicels stiff, widely divergent ...................... 5. *E. arida*
          Spikelets ovate to ovate-oblong; pedicels spreading, but not stiffly divergent; panicles usually purple ...................................... 6. *E. mexicana*

**1. Eragrostis viscosa** (Retz.) Trin., Mém. Acad. St.-Pétersb. VI, Math. Phys. Nat. **1:** 397, 1830

*Poa viscosa* Retz., Obs. **4:** 20, 1786.

Annual; culms 10–50 cm. high, geniculate-ascending, viscid at least below nodes; sheaths longer or shorter than internodes, glandular toward summit, with a few long hairs in throat; blades 3–12 cm. long, 1–3 mm. wide, rather firm; panicles commonly 5–15 cm. long, 1–2 cm. wide, rather densely flowered, branches stiffly ascending or spreading, viscid; spikelets 3–4 mm. long, pale or purple; lemma and palea obtuse, keels of palea ciliate with long hairs.

Infrequent, Mexico to Panama; introduced from southeastern Asia. In the Sonoran Desert area this species has been found only on sand dunes near Punta de Conejo, Baja California.

**2. Eragrostis cilianensis** (All.) Vign.-Lut., Malpighia **18:** 386, 1904

*Poa cilianensis* All., Fl. Ped. **2:** 246, 1785.
*Eragrostis major* Host, Ic. Gram. Austr. **4:** 14, *pl. 24,* 1809.
*Eragrostis megastachya* Link, Hort. Berol. **1:** 187, 1827.

Annual; culms commonly 15–60 cm. tall, or in dwarf plants as little as 1 cm., erect or geniculate-ascending, branching; sheaths keeled, more or less pilose at mouth, with rather prominent glandular depressions on keel; blades usually 3–15 cm. long, 2–7 mm. wide, acuminate, glabrous, with prominent glandular papillae on margins; panicles 3–15 cm. long, rather dense, branches stiffly ascending or spreading, spikelet-bearing to base; spikelets oblong, 5–15 mm. long, 2.5–3 mm. wide, 10–40-flowered, short-pedicellate, usually divergent; glumes 2 mm. long, acute; lemmas obtuse or subacute, about 2.5 mm. long, keel scabrous and bearing a few prominent glands.

A weed in fields and waste places, southern Canada, throughout the United States except at higher altitudes, southward to Argentina; introduced from Europe.

**3. Eragrostis pilosa** (L.) Beauv., Ess. Agrost. 71, 1812

*Poa pilosa* L., Sp. Pl. 68, 1753.

Annual; culms erect or geniculate-ascending, 15–40 cm. tall; sheaths glabrous, more or less pilose at mouth; blades 6–12 cm. long, 1–2 mm. wide, attenuate, glabrous; panicles usually about half the height of plant, lax, branches ascending or spreading, flexuous; spikelets 2.5–6 mm. long, 1 mm. wide, 3–10-flowered, pale or tinged with purple; first glume 0.3 mm. long, nerveless, second 1–1.2 mm. long, narrow, 1-nerved; lemmas thin, acutish, minutely scabrous on keel, the lowest one 1.5–1.6 mm. long.

Open ground and waste places, California and Arizona; Colorado to Massachusetts, south to Argentina. In the Sonoran Desert this species has been collected only at Superior, Arizona, and Rancho San Carlos west of Hermosillo, Sonora.

### 4. **Eragrostis diffusa** Buckl., Proc. Acad. Phila. **1862:** 97, 1862

*Eragrostis purshii* var. *diffusa* Vasey, Contr. U.S. Nat. Herb. **1:** 59, 1890.

Annual; culms erect, often somewhat geniculate at lower nodes, 20–70 cm. tall, sometimes dwarfed; sheaths glabrous, with a tuft of long hairs at throat; blades mostly 4–20 cm. long, 1–4 mm. wide, lax, glabrous; panicles 10–22 cm. long, 4–14 cm. wide, branches solitary or in pairs, stiffly and widely ascending, naked at base, branchlets appressed; spikelets short-pedicellate, appressed, 4–10 mm. long, 6–17-flowered; glumes subequal, acute, 1–1.4 mm. long; lemmas acutish, finely scabrous on nerves, lowest about 1.6 mm. long.

Open ground and waste places, Idaho to Colorado, south to southern California, Texas, and northern Mexico.

### 5. **Eragrostis arida** Hitchc., Jour. Wash. Acad. **23:** 449, 1933

Annual; culms 20–60 cm. tall, erect or geniculate at lower nodes, simple or sparingly branching; sheaths keeled, glabrous, with a conspicuous tuft of hairs in mouth; blades mostly 4–8 cm. long, 1–3 mm. wide, attenuate, glabrous; panicles as much as 30 cm. long, open, slender branches ascending, usually solitary, axils pilose with long hairs; spikelets 5–10 mm. long, about 1.5 mm. wide, 8–15-flowered, usually spreading, lateral pedicels mostly 2–25 mm. long; glumes acute, first about 1 mm. long, second a little longer and broader; lemmas 1.6–1.8 mm. long, acute; paleae persistent.

Lake shores, stream banks, and waste places, usually in moist soil, Arizona to Louisiana and northern Mexico.

### 6. **Eragrostis mexicana** (Hornem.) Link, Hort. Berol. **1:** 190, 1827

*Poa mexicana* Hornem., Hort. Hafn. **2:** 953, 1815.

Annual; culms erect or geniculate at lower nodes, 20–45 cm. high, glabrous; sheaths glabrous with conspicuous tuft of hairs at throat; blades mostly 5–10 cm. long, 2–3 mm. wide, flat-attenuate; panicles oblong or pyramidal, 5–15 cm. long, loosely flowered, branches ascending or spreading, branchlets and pedicels somewhat divergent; spikelets oblong, usually purple, 4–6 mm. long, 6–10-flowered; lemmas about 2 mm. long, acute or subobtuse.

Open ground, fields, and waste places, Mexico to Costa Rica. In the Sonoran Desert, found only in the Sierra de San Francisquito, Baja California.

### 7. **Eragrostis intermedia** Hitchc., Jour. Wash. Acad. **23:** 450, 1933

Perennial; culms rather densely tufted, erect, commonly 40–80 cm. high; sheaths mostly shorter than internodes, glabrous, conspicuously pilose in throat; blades to 30 cm. long, 1–4 mm. wide, attenuate, pilose on upper surface near base; panicles 15–35 cm. long, open, sometimes diffuse, branches usually slender but rather stiff, somewhat flexuous, ascending or spreading, usually conspicuously pilose in axils, branchlets and pedicels divergent; spikelets 4–10 mm. long, mostly 4–7-flowered; lemmas 1.6–2 mm. long, acute or subobtuse, turgid, nerves obscure.

Sandy prairies, rocky hills, and canyons, Arizona to Missouri and Louisiana; Mexico to Costa Rica. This species usually occurs above the limits of the desert in Arizona, but has been collected within the area in the Sierra de la Laguna, Baja California.

## 5. DISTICHLIS Raf., Jour. Phys. **89:** 104, 1819

Dioecious perennials with creeping, scaly rhizomes, rigid culms, and dense, narrow panicles; glumes broad, acute, keeled, 3–7-nerved, nerves usually faint; lemmas closely imbricate, acute or subacute, faintly 9–11-nerved, pistillate coriaceous, much firmer than staminate; palea shorter to longer than lemma, margins bowed out below, keels minutely ciliate, sometimes winged.

Spikelets small, lemmas 4–5 mm. long; margins of blades scabrous ............ 1. *D. stricta*
Spikelets large, lemmas 1–1.5 cm. long; margins of blades glabrous .......... 2. *D. palmeri*

### 1. Distichlis stricta (Torr.) Rydb., Bull. Torrey Club 32: 602, 1905

*Uniola stricta* Torr., Ann. Lyc. N.Y. **1:** 155, 1824.
*Distichlis spicata* var. *stricta* Scribn., Mem. Torrey Club **5:** 51, 1894.

Perennial; culms usually erect, from strong rhizomes, commonly 10–30 cm., or sometimes to 60 cm., tall; sheaths glabrous or usually with a few long hairs at the mouth; blades 3–9 cm. long, occasionally shorter, 2–4 mm. wide at base, scabrous on margins, scaberulous and sometimes sparsely pilose on upper surface; panicles 2–6 cm. long, pistillate usually congested, staminate much looser; spikelets 8–15 mm. long, rarely longer, 5–12-flowered; glumes subequal, mostly 3–4 mm. long, acute; lemmas 4–5 mm. long, acute, pistillate much firmer than staminate, with broad, papery margins; palea equaling or exceeding lemma, minutely ciliate on keels, keels of pistillate florets often winged and irregularly toothed.

Brackish soil, Washington, Saskatchewan, and Minnesota to central Mexico.

### 2. Distichlis palmeri (Vasey) Fassett ex I. M. Johnston, Proc. Calif. Acad. IV, 12: 984, 1924

*Uniola palmeri* Vasey, Gard. & Forest **2:** 401, *fig. 124,* 1889.

Rhizomatous perennial; culms rather stout, ascending, branched, 25–50 cm. or more high, glabrous; sheaths crowded, glabrous, with a tuft of woolly hairs at the mouth; blades 4–12 cm. long, 3–5 mm. wide at base, firm, stiffly ascending, involute-pointed, glabrous; staminate panicle 6–20 cm. long, exceeding leaves, lower branches to 8 cm. long; pistillate panicle 7–15 cm. long, shorter than upper leaves and partly enclosed in them, branches short, appressed; staminate spikelets 7–9-flowered, about 1.5 cm. long, lemmas acuminate, about 1 cm. long; pistillate spikelets 2.5–3 mm. long, lemmas acuminate, subulate-pointed, very firm, lowest about 1.5 cm. long.

Tidewater marshes, upper Gulf of California, Sonora (Cholla Beach, mouth of Hardy River, Horseshoe Bend).

## 6. JOUVEA Fourn., Bull. Soc. Bot. Belg. **15:** 475, 1876

Tough, rigid, stoloniferous, dioecious perennials; staminate spikelets several- to many-flowered, sessile, somewhat distant on opposite sides of rachis, forming a terminal loose spike, florets closely imbricate; glumes shorter than first floret; lemmas 3-nerved, acute, awnless; pistillate spikelets in axils of leaves, single or in fascicles of 3–4, terete, somewhat curved, acuminate, consisting of 1–3 florets embedded in spongy tissue of thickened rachilla, only summit of floret free; style and base of stigma elongate, 2-lobed.

**1. Jouvea pilosa** (Presl) Scribn., Bull. Torrey Club **23:** 143, 1896

*Brizopyrum pilosum* Presl, Rel. Haenk. **1:** 280, 1830.
*Rhachidospermum mexicanum* Vasey, Bot. Gaz. **15:** 110, *pl. 12,* 1890.

Culms erect or ascending, commonly 15–30 cm. tall, branching from all nodes, branches short, appressed; sheaths of main culm much shorter than internodes, those of branches crowded; blades 4–14 cm. long, 2–4 mm. wide, flat or usually loosely folded, pungent, stiffly spreading, often curved or arcuate, glabrous beneath, scabrous above; staminate spikes terminal and axillary, short-exserted from upper sheaths, few-flowered, spikelets appressed; pistillate spikes partly enclosed in short, congested uppermost sheaths, forming a thorny mass.

Sand dunes and sandy plains, northwestern Mexico to El Salvador. Known within the Sonoran Desert only from the east coast of Baja California south of Mulegé to the Cape Region.

## 7. ARUNDO L., Sp. Pl. 81, 1753

Coarse, tall, perennial reeds with broad, linear blades and plumelike terminal panicles; spikelets several-flowered, disarticulating above the glumes and between florets; rachilla glabrous; glumes membranous, subequal, as long as spikelet, 3-nerved, tapering into a slender point; lemmas thin, 3-nerved, densely long-pilose, tip bifid, bearing a short, straight awn from between lobes; palea much shorter than lemma, densely but minutely ciliate on keels.

**1. Arundo donax** L., Sp. Pl. 81, 1753

Culms usually in large clumps, 2–6 m. tall, from stout, knotty rhizomes; blades of main culm elongate, 5–7 cm. wide, gradually tapering from rounded, subcordate base, glabrous, scabrous on margins, those of branches often much reduced; panicles 30–60 cm. long, dense, erect or branches somewhat drooping; spikelets about 12 mm. long.

Along rivers, streams, and irrigation ditches, and in moist places, southern California to Texas and northern Mexico, occasionally eastward; West Indies; Colombia and British Guiana to Argentina; introduced from the Old World.

## 8. PHRAGMITES Trin., Fund. Agrost. 134, 1820

Coarse, rhizomatous, stoloniferous, reedlike perennials with firm, flat blades and large, silky, plumelike panicles. Spikelets several-flowered, rachilla densely covered with long, fine, silky hairs, disarticulating above glumes and at base of each segment between florets, first floret staminate or neuter, others perfect; glumes unequal, narrow, acute, first 1-nerved, second 3-nerved; lemmas narrow, long-acuminate, each one in a spikelet successively smaller, summits of all of them thus about equal; palea of staminate or neuter floret one-fourth as long as lemma, that of first fertile floret one-half as long as lemma.

**1. Phragmites communis** Trin., Fund. Agrost. 134, 1820

Culms erect, reedlike, 2–5 m. high, usually growing in large colonies; sheaths glabrous, much longer than internodes; blades as much as 50 cm. long, 1–2.5 cm. wide, flat, firm, attenuate-pointed, with very scabrous margins; panicles 25–50 cm. long, plumelike, drooping, branches naked below, densely villous in lower axils; spikelets 12–20 mm. long; glumes unequal, first 4–5 mm. long, second 6–8 mm. long; lemmas 10–13 mm. long, narrowly acuminate; rachilla hairs more than half as long as lemma.

Marshes, irrigation canals, river banks, and wet places in temperate regions around the world.

## 9. TRIDENS Roem. & Sch., Syst. Veg. **2:** 34, 599, 1817

Tufted or stoloniferous perennials with contracted panicles; spikelets few- to several-flowered; glumes subequal, 1-nerved, acute or obtuse; lemmas rounded on the back, obtuse, 3-nerved, densely pilose on nerves, entire, or lobed and midnerve excurrent from between lobes as a short awn, lateral nerves excurrent as minute points.

Plants stoloniferous; panicles very short, usually exceeded by leaves; lemmas lobed .................................................................. 1. *T. pulchellus*
Plants not stoloniferous; panicles elongate; lemmas entire .................. 2. *T. muticus*

**1. Tridens pulchellus** (H. B. K.) Hitchc., in Jepson, Fl. Calif. **1:** 141, 1912
*Triodia pulchella* H. B. K., Nov. Gen. & Sp. **1:** 155, *pl. 47,* 1816.*

Stolons slender, with one elongate internode mostly 5–8 cm. long, sometimes very short, at first erect, producing a new plant at the tip, finally reclining and taking root; culms 1–5 cm. long, mostly produced in new plants at tip of stolons, bearing 1–4 short-pedicellate spikelets, usually exceeded by leaves; sheaths crowded, densely pilose or villous below; blades 0.5–3 cm. long, convolute, scabrous, sharp-pointed, usually arcuate-spreading; spikelets 5–12 mm. long, few- to several-flowered; glumes 6–10 mm. long, acuminate; lemmas 4 mm. long, densely long-pilose on internerves below, cleft to middle, awn a little longer than lobes, finally divergent.

Open ground, mesas, and rocky hills, southern California to Utah, Texas, and central Mexico (Querétaro and Hidalgo).

**2. Tridens muticus** (Torr.) Nash, in Small, Fl. SE. U.S. 143, 1903
*Tricuspis mutica* Torr., Pacif. R. Rep. **4:** 156, 1857.
*Triodia mutica* Scribn., Bull. Torrey Club **10:** 30, 1883.

Culms densely tufted, erect, 10–50 cm. tall; sheaths longer than internodes, glabrous or sparsely pilose, rather densely hairy on collar; blades mostly 8–15 cm. long, flat or loosely rolled, firm, scaberulous and sometimes pilose; panicles 5–15 cm. long, narrow, usually dense, short branches appressed, spikelet-bearing from or near base; spikelets 8–12 mm. long, 6–8-flowered, pale, tinged with purple; glumes obtuse, about as long as lowest floret; lemmas 4.5–6 mm. long, obtuse, entire or with very shallow lobes, awn minute or wanting, lateral nerves not excurrent.

Dry plains and rocky hills, southeastern California to Colorado, Arkansas, and northern Mexico.

---

* References to pages in Nov. Gen. & Sp. as cited in this book, with the exception of those in the Gramineae, are to the folio edition. In the Gramineae they are to the quarto edition. The discrepancies in pagination between the two editions are treated by J. H. Barnhart in "Dates of the 'Nova Genera' of Humboldt, Bonpland and Kunth," Bull. Torrey Club **29:** 585, 1902.

## 10. ORCUTTIA Vasey, Bull. Torrey Club **13:** 219, *pl. 60*, 1886

Tufted annuals with short, flat blades and spikelike racemes; spikelets several-flowered, upper florets reduced; rachilla tardily disarticulating between the florets; glumes subequal, entire, many-nerved; lemmas firm, prominently many-nerved, irregularly toothed and awned at summit; palea about as long as lemma.

### 1. Orcuttia fragilis Swallen, Jour. Wash. Acad. **34:** 308, 1944

Annual; culms 15–40 cm. long, many-noded, erect or usually ascending or decumbent at base, internodes rather short, of nearly equal length, prominently papillose or papillose-pilose, bearing short, appressed flowering branches from upper nodes; leaves 3.5–6.5 cm. long, 6–12 mm. wide, flat, acute, papillose or papillose-pilose, division into sheath and blade not evident except for slight constriction at ligular area; panicles dense, spikelike, all or partly enclosed in upper leaves, exserted part 1–3 cm. long, 5–10 mm. wide, or those on lower branches smaller; spikelets 3–8-flowered, 6–12 mm. long; glumes equal, 7 mm. long, acuminate, margins thin, hyaline; lemmas pubescent, especially toward base, and also sparsely pilose, lowest 6–7 mm. long, acute or subacuminate, others successively smaller, minutely toothed, midnerve excurrent as a short mucro; palea little shorter than lemma, dentate, minutely scabrous on keels, margins broad, thin, hyaline.

Known only from the type locality, Llano Datillare, Magdalena Plain, Baja California.

## 11. COTTEA Kunth, Rév. Gram. **1:** 84, 281, *pl. 52,* 1829

Small tufted perennials with flat blades and narrow or oblong, rather dense panicles; glumes nearly equal, about as long as lowest floret, several-nerved; lemmas rounded on back, villous below, with 9–11 prominent nerves extending into teeth or awns; palea a little longer than body of lemma.

### 1. Cottea pappophoroides Kunth, Rév. Gram. **1:** 84, 1829

Culms erect, 15–60 cm. tall, softly pilose especially below panicle, nodes bearded; sheaths mostly longer than internodes, pilose or nearly glabrous; blades 4–17 cm. long, 2–7 mm. wide, attenuate, pilose on both surfaces; panicles 6–20 cm. long, 2–6 cm. wide, nearly simple branches ascending, more or less flexuous; spikelets 5–10 mm. long, nearly as wide, pedicels divergent.

Dry hills and plains, southern Arizona to western Texas and central Mexico; Ecuador to Argentina.

## 12. ENNEAPOGON Desv. ex Beauv., Ess. Agrost. 81, *pl. 16, fig. 11,* 1812

Slender perennials with narrow blades and dense, spikelike panicles; spikelets 2–3-flowered, rachilla disarticulating above glumes, upper 1 or 2 florets reduced or rudimentary; glumes equal, much longer than body of first lemma, 5–7-nerved; lemmas rounded on back, 9-nerved, nerves extending beyond body as conspicuous, nearly equal, plumose awns; palea as long as lemma, densely short-ciliate on keels.

### 1. Enneapogon desvauxii Beauv. ex Desv., Opusc. 98, 1831

*Pappophorum wrightii* S. Wats., Proc. Amer. Acad. **18:** 178, 1883.

Culms densely tufted, erect or spreading, commonly geniculate at some nodes, softly pilose, nodes bearded; sheaths much shorter than internodes; blades 4–10 cm.

long, not more than 2 mm. wide, usually involute, pilose; panicles 1–8 cm. long, less than 1 cm. wide, dense, lead-colored; glumes 4–5 mm. long, acute, pubescent; body of lemma 1.5 mm. long, densely pilose, awns 3 mm. long.

Rocky hills and plains, Arizona to Texas and central Mexico; Peru to Argentina.

## 13. PAPPOPHORUM Schreb. ex Vahl, Symb. Bot. 3: 10, 1794

Tufted perennials with long, pale, narrow, bristly, spikelike panicles; spikelets 3–5-flowered, rachilla disarticulating above glumes, lower 1 or 2 florets fertile, upper ones reduced; glumes 1-nerved, thin; lemmas very firm, obscurely many-nerved, irregularly divided into many unequal, scabrous awns; palea acuminate, tip divided into two thin, narrow lobes.

Panicle very dense, spikelike; glumes acute, 4–4.5 mm. long; lemmas glabrous on inside .... .......................................................... 1. *P. mucronulatum*

Panicle interrupted; glumes acuminate, 5–7 mm. long; lemmas densely pubescent on inside at base of awns .......................................... 2. *P. subbulbosum*

### 1. Pappophorum mucronulatum Nees, Agrost. Bras. 412, 1829

*Pappophorum vaginatum* Buckl., Prel. Rep. Geol. Agr. Surv. Tex. App. 1, 1866.

Culms erect, commonly 50–70 cm., sometimes as much as 1 m., tall; sheaths glabrous with a small tuft of hairs at throat; blades firm, flat or subinvolute, mostly 10–25 cm. long, 2–5 mm. wide, those of innovations often elongate, scabrous on upper surface and margins; panicles 10–25 cm. long, narrow, dense, pale, tapering at summit; glumes 4–4.5 mm. long, acute; body of first lemma 3 mm. long, densely pilose on midnerve and margins nearly to summit, awns mostly about 5 mm. long, some of them fine and threadlike.

Open ground and low places on plains, Arizona, Texas, and northern Mexico; South America.

### 2. Pappophorum subbulbosum Arech., An. Mus. Nac. Montevideo 1: 405, *pl. 48,* 1896

Culms erect, 30–50 cm. tall, sparingly branching at lower nodes; sheaths glabrous; blades mostly 8–17 cm. long, about 2 mm. wide, flat or loosely inrolled, scabrous on upper surface; panicles 10–18 cm. long, interrupted, somewhat tapering at summit; glumes 5–7 mm. long, acuminate, bearing an awn 1–2 mm. long; body of first lemma about 3 mm. long, densely villous on midnerve margins, densely pubescent on inside at base of awns, awns numerous, at least some of them as much as 1 cm. long, many of them fine and threadlike.

Sandy land, Sonora (Guaymas), Uruguay, and Argentina.

## 14. MONANTHOCHLOE Engelm., Trans. Acad. St. Louis 1: 436, *pl. 13,* 1859

Wiry, creeping perennial with short, crowded, subulate blades; plants dioecious; spikelets 3–5-flowered, uppermost florets rudimentary, rachilla tardily disarticulating in pistillate spikelets; glumes wanting; lemmas rounded, several-nerved, those of pistillate spikelets like blades in texture, uppermost enclosed between keels of palea of floret below, those of staminate spikelets thin; palea as long as lemma, with broad margins, glabrous keels rather close together.

**1. Monanthochloe littoralis** Engelm., Trans. Acad. St. Louis **1:** 437, 1859

Plants with rather strong rhizomes, culms erect or widely creeping with erect branches; blades mostly 5–10 mm. long, crowded, margins incurved, tip cucullate, sharp-pointed; spikelets 1 to few, nearly hidden in leaves at ends of branches, stamens and pistils exserted.

Tidal flats and seashores, southern California, Texas, Florida, and northern Mexico.

## 15. SCLEROPOGON Phil., An. Univ. Chile 36: 205, 1870

Monoecious or dioecious, stoloniferous perennials, with short, flexuous blades and narrow, few-flowered racemes or panicles; staminate spikelets several-flowered, rachilla not disarticulating; glumes equal, acuminate, 1–3-nerved; lemmas similar to glumes, 3–5-nerved, mucronate; pistillate spikelets several-flowered, rachilla disarticulating above glumes, first floret perfect with a sharp, bearded callus, upper ones reduced; glumes acuminate, 3-nerved, first one-half as long as second; lemmas narrow, 3-nerved, nerves extending into long awns.

**1. Scleropogon brevifolius** Phil., An. Univ. Chile **36:** 206, 1870

*Lesourdia karwinskyana* Fourn., Bull. Soc. Bot. France **27:** 102, *pl. 4, fig. 12,* 1880.
*Scleropogon karwinskyanus* Benth. ex S. Wats., Proc. Amer. Acad. **18:** 181, 1883.

Culms mostly 10–30 cm. tall, tufted, erect; stolons wiry, internodes 5–15 cm. long; leaves crowded at base, sheaths glabrous, or densely hairy at base, blades 1–5 cm. long, 1–2 mm. wide, sharp-pointed, usually somewhat curled; staminate panicles with few to sometimes numerous spikelets, these mostly 2–3 cm. long; pistillate racemes with 1–4 spikelets, each subtended by a long, narrow bract; body of spikelets 2.5–3 cm. long, awns 4–10 cm. long, spreading, loosely twisted.

Plains, open slopes, and valleys, Nevada and southern Colorado to Arizona, Texas, and central Mexico. In the Sonoran Desert, known only from Arizona.

## 16. LOLIUM L., Sp. Pl. 83, 1753

Perennials with flat blades and usually long, slender spikes; spikelets several-flowered, placed edgewise, alternate, on opposite sides of the continuous rachis; first glume wanting; lemmas rounded on back, obtuse, awnless.

**1. Lolium perenne** L., Sp. Pl. 83, 1753

Culms erect or geniculate-ascending, mostly 20–50 cm. tall, glabrous; sheaths glabrous, shorter than internodes, auricles small or wanting; blades lax, smooth and shining on lower surface, 6–15 cm. long, 2–4 mm. wide; spike 8–15 cm. long, sometimes longer, erect or somewhat drooping; spikelets 8–14 mm. long, about as long as or somewhat longer than internode of rachis; glume usually about half as long as spikelet, obtuse; first lemma 6–7 mm. long, obtuse or acutish, awnless, or rarely with a short awn.

Meadows and waste places, throughout the United States and temperate regions of the world. Introduced from Europe. A rare waif in the Sonoran Desert.

## 17. ELYMUS L., Sp. Pl. 83, 1753

Rhizomatous perennials with firm, flat blades and narrow, erect spikes; spikelets sessile, in pairs (sometimes solitary), alternate on opposite sides of the continuous rachis, more or less dorsiventral to axis, rachilla disarticulating above glumes and between florets; glumes subequal, firm, somewhat asymmetric; lemmas rounded on back, acuminate or short-awned, 5-nerved.

**1. Elymus triticoides** Buckl., Proc. Acad. Phila. **1862:** 99, 1862
*Elymus condensatus* var. *triticoides* Thurb., in S. Wats., Bot. Calif. **2:** 326, 1880.

Culms 60–120 cm. tall, erect from usually extensively creeping rhizomes; sheaths longer than internodes, glabrous; blades firm, elongate, 2–6 mm. wide, scabrous above and sometimes pilose; spikes mostly 10–15 cm. long, erect, slender to rather dense; spikelets mostly 8–15 mm. long; glumes narrow to subulate, usually nerveless, awn-tipped, 5–10 mm. long; lemmas 6–10 mm. long, glabrous, brownish, awn-tipped.

Washington and Montana south to Baja California, Arizona, and Texas.

## 18. SITANION Raf., Jour. Phys. 89: 103, 1819

Tufted perennials with firm, narrow blades and rather dense, bushy spikes; spikelets 2–few-flowered, in pairs, alternate on opposite sides of readily disarticulating axis; glumes narrow, firm, 1–3-nerved, divided above into 2 long, scabrous, divergent awns, sometimes with a short bristle from margins; lemmas firm, rounded on back, 5-nerved, minutely bifid, midnerve extending into a long, divergent awn.

**1. Sitanion hystrix** (Nutt.) J. G. Smith, U.S. Dept. Agr. Div. Agrost. Bull. **18:** 15, 1899
*Aegilops hystrix* Nutt., Gen. **1:** 86, 1818.

Culms usually erect, rather stiff, 10–50 cm. tall; leaves glabrous to softly pubescent, blades 5–20 cm. long, mostly 1–3 mm. wide, flat or involute, rather stiffly ascending or spreading; spike erect, as much as 10 cm. long; lemmas glabrous to appressed-pubescent, awns of glumes and lemmas 2–10 cm. long, widely spreading.

Open sandy ground, rocky hills, and open pine woods, British Columbia to South Dakota, south to Missouri, Texas, and central Mexico. Probably to be found only near the limits of the Sonoran Desert.

## 19. HORDEUM L., Sp. Pl. 84, 1753

Annual plants with flat blades and dense, bristly spikes; spikelets 1-flowered, 3 at each node of articulate rachis, middle one sessile, perfect, lateral ones pedicellate, usually imperfect; glumes narrow or subulate; lemmas with back turned toward rachis, rounded, obscurely nerved, tapering into an awn.

**1. Hordeum stebbinsii** Covas, Madroño **10:** 17, 1949

Culms erect or geniculate-ascending at base, 10–50 cm., rarely as much as 1 m., tall; sheaths glabrous, lower ones becoming loose and papery; blades 4–25 cm. long, 2–10 mm. wide, sparsely to rather densely pilose; spike 4–10 cm. long, partly included or exserted from the usually inflated uppermost sheath; glumes awned, inner ones of lateral spikelets subulate, scabrous, outer ones and those of central spikelet broader, conspicuously ciliate; lemma of fertile spikelet 7–9 mm. long, glabrous, bearing an awn 1.5–2 cm. long, rachilla produced beyond floret as a slender bristle, more than half as long as floret; lemmas of lateral spikelets larger, strongly scabrous above, bearing an awn about 3 cm. long; awns of upper spikelets much shorter.

Roadsides and waste places, Washington to Utah; southern California to New Mexico and Baja California. Introduced from the Old World.

### 20. SCHISMUS Beauv., Ess. Agrost. 73, *pl. 15, fig. 4,* 1812

Low annuals with short, slender blades and small, rather dense panicles; spikelets several-flowered; glumes acute, subequal, nearly as long as spikelet; lemmas broad, rounded on back, minutely or deeply bidentate, pilose on lower part of margins; paleae broad, hyaline, nerves near margins.

Lemmas emarginate, pilose on margins; glumes usually slightly shorter than spikelet ...... .......................................................... 1. *S. barbatus*

Lemmas lobed; lobes acute, conspicuously pilose across back; glumes usually slightly longer than florets ................................................ 2. *S. arabicus*

**1. Schismus barbatus** (L.) Thell., Bull. Herb. Boiss. II, **7:** 391, 1907

*Festuca barbata* L., Amoen. Acad. **3:** 400, 1756.

Culms slender, in small, dense tufts, erect or usually spreading, 5–20 cm. tall; sheaths shorter than internodes, glabrous, with a few long hairs at mouth; blades as much as 7 cm. long, mostly 1–1.5 mm. wide, sparsely pilose with long hairs or nearly glabrous; panicles 1.5–4 cm. long, branches appressed or ascending; spikelets short-pedicellate, 5–7 mm. long; glumes acute, a little shorter than spikelet; first lemma 2.5 mm. long, emarginate, densely pilose on margins, sparsely on back; palea nearly as long as lemma.

Roadsides and open sandy land, southern California, Arizona, and Utah. Introduced from the Old World.

**2. Schismus arabicus** Nees, Fl. Afr. Austral. **1:** 422, 1841

Very similar to *S. barbatus,* differing in that lemmas are bidentate, lobes acute, and conspicuously pilose across the back.

Southern California and Arizona. Introduced from the Old World.

### 21. DESCHAMPSIA Beauv., Ess. Agrost. 91, *pl. 18, fig. 3,* 1812

Annuals or perennials with narrow blades and narrow or open, shining panicles; spikelets 2-flowered, rachilla prolonged beyond upper floret, sometimes bearing a rudimentary floret; glumes equal, acute or acuminate; lemmas thin, erose, truncate, with a straight or a bent and twisted awn from below middle.

**1. Deschampsia danthonioides** (Trin.) Munro, in Benth., Pl. Hartw. 342, 1857

*Aira danthonioides* Trin., Mém. Acad. St.-Pétersb. VI, Math. Phys. Nat. **1:** 57, 1830.
*Deschampsia gracilis* Vasey, Bot. Gaz. **10:** 224, 1885.

Culms very slender, erect, 10–60 cm. tall; blades few, lax, soft, scaberulous, mostly 2–8 cm. long, less than 1 mm. or rarely as much as 2 mm. wide; panicles 7–25 cm. long, open, slender branches mostly in pairs, rather stiffly ascending, spikelet-bearing above middle; spikelets 4–8 mm. long; glumes acuminate, 3-nerved, much longer than florets, glabrous, or obscurely scabrous on keel; lemmas 2–3 mm. long, smooth, shining, callus and rachilla densely bearded, hairs on the rachilla as long as floret; awns 4–6 mm. long, geniculate, twisted below.

Dry or moist open ground, Alaska to Montana, south to Arizona and Baja California; Texas; Chile.

## 22. AGROSTIS L., Sp. Pl. 61, 1753

Perennials with flat blades and open or dense panicles of small spikelets; spikelets 1-flowered, disarticulating above or below the glumes; glumes equal, longer than floret; lemma thinner than glumes, awned or awnless; palea well developed or obsolete.

Culms usually decumbent; lemma awnless; palea well developed ....... 1. *A. semiverticillata*
Culms erect; lemma awned from the middle, geniculate; palea obsolete ........................................................................ 2. *A. microphylla*

**1. Agrostis semiverticillata** (Forsk.) C. Christ., Dansk Bot. Ark. **4**$^{3}$: 12, 1922
*Phalaris semiverticillata* Forsk., Fl. Aegypt. 17, 1775.
*Agrostis verticillata* Vill., Prosp. Pl. Dauph. 16, 1779.
*Agrostis alba* var. *verticillata* Pers., Syn. Pl. **1**: 76, 1805.

Culms erect to widely decumbent-spreading and rooting at nodes, 15–50 cm. long or sometimes longer; sheaths shorter than internodes, glabrous; blades mostly 4–12 cm. long, 3–6 mm. wide, firm, scabrous; panicles 3–10 cm. long, densely flowered, lobed, branches bearing spikelets to base; spikelets 2 mm. long, usually falling entire; glumes equal, subobtuse, very scabrous; lemma 1 mm. long, truncate, awnless; palea nearly as long as lemma.

Wet ground, especially along streams and ditches, Washington to California, Texas, and Guatemala; Peru to Argentina and Chile; introduced from the Old World.

**2. Agrostis microphylla** Steud., Syn. Pl. Glum. **1**: 164, 1854

Culms tufted, erect, branching at base, mostly 10–25 cm. tall, long and short culms in the same tuft, glabrous; sheaths glabrous, upper ones somewhat inflated; ligules thin, 3 mm. long; blades flat, 1–7 cm. long, rarely longer, 1–3 mm. wide, scabrous; panicles mostly 2–8 cm. long, narrow, dense, lobed; glumes 3–4.4 mm. long, acuminate or awn-pointed, scabrous on keel; lemma 1.7–1.9 mm. long, minutely toothed, awned from middle, awn 3.5–6 mm. long, geniculate; palea obsolete.

Open moist ground, Vancouver Island to northern Baja California (Guadalupe Ranch).

## 23. PEREILEMA Presl, Rel. Haenk. 1: 233, 1830

Annuals with narrow blades and dense, interrupted panicles of small spikelets; spikelets 1-flowered, in dense fascicles along axis of a panicle, surrounded by numerous sterile spikelets in form of bristles; rachilla disarticulating above glumes, not produced beyond floret; glumes equal, 1-nerved, thin, minutely lobed, awned from between lobes, awn slender, straight; lemma terete, pilose below, minutely bifid, midnerve extending into a slender, flexuous, threadlike awn longer than bristles; palea as long as lemma, keels approximate.

**1. Pereilema crinitum** Presl, Rel. Haenk. **1**: 233, *pl. 37, fig. a,* 1830
*Pereilema crinitum* var. *cirratum* Fourn., Mex. Pl. **2**: 93, 1886.

Annual; culms 10–40 cm. high, erect or decumbent at base, branching, especially at lower nodes; sheaths much longer than internodes, scabrous, auriculate, auricles pubescent; ligules very short, truncate; blades 5–20 cm. long, 2–4 mm. wide, gradually tapering to a usually long, petiole-like base, scaberulous; panicles 3–12 cm., rarely 15 cm., long, narrow, dense, interrupted, short branches usually appressed; lemma 1.5 mm. long, scabrous, very fine, flexuous awn about 2 cm. long.

Brushy slopes, moist banks, and waste places, Mexico to Ecuador. Found in the Sonoran Desert only in the Sierra de la Laguna, Baja California.

## 24. MUHLENBERGIA Schreb. ex Gmel., Syst. Nat. ed. 13, **2:** 171, 1791

Annuals or perennials with simple or branching culms and narrow or open panicles; glumes much shorter than the lemma; lemmas firm, 3-nerved, awnless or awned, glabrous, scabrous, or pubescent on midnerve and margins or across the back with a very short, usually pilose callus.

Plants annual:
- Spikelets awnless ........ 1. *M. fragilis*
- Spikelets awned:
  - Panicles narrow, its branches closely appressed; spikelets 4.5–6 mm. long ........ 2. *M. appressa*
  - Panicles open, its branches ascending to spreading; spikelets not more than 4 mm. long:
    - Awn of lemma 10–30 mm. long; spikelets 2–4 mm. long; glumes glabrous ........ 3. *M. microsperma*
    - Awn of lemma not more than 2 mm. long; spikelets 1.6–1.8 mm. long; glumes pilose ........ 4. *M. texana*

Plants perennial:
- Culms stout, woody, profusely branching; panicles small, dense, few-flowered ........ 5. *M. dumosa*
- Culms usually slender, if woody, widely spreading; panicles open, or, if dense, elongate:
  - Culms widely spreading, wiry, woody; awns 5–10 mm. long ........ 6. *M. porteri*
  - Culms erect, densely tufted:
    - Panicles open or at least loose, branches naked at base; sheaths keeled:
      - Blades mostly less than 5 cm. long, with conspicuous white midnerve and margin; plants slender ........ 7. *M. arizonica*
      - Blades elongate; plants coarse ........ 8. *M. emersleyi*
    - Panicles dense, spikelike; sheaths rounded ........ 9. *M. rigens*

### 1. **Muhlenbergia fragilis** Swallen, Contr. U.S. Nat. Herb. **29:** 206, 1947

Annual; culms erect or geniculate-ascending, 10–30 cm. high, freely branching at lower nodes, strigose below nodes, otherwise glabrous; sheaths longer than internodes, scaberulous, with thin, hyaline margins; ligule 1–3 mm. long, thin, hyaline, decurrent; blades to 6 cm. long, 1.5 mm. wide, scaberulous, with prominent, white, thickened midnerve and margins; panicles very diffuse, more than half the length of culm, 4–10 cm. wide, numerous usually solitary slender branches becoming stiffly spreading or even reflexed, with a pulvinus in axils, branchlets divergent; pedicels capillary but usually straight, divergent, mostly 2–5 mm., or sometimes as much as 10 mm., long; spikelets 1–1.1 mm. long; glumes one-half to three-fourths as long as spikelet, obtuse or subacute, glabrous; lemma obtuse, glabrous or densely silky-pubescent on keel and margins; palea as long as lemma, sparsely and minutely pilose on nerves; anthers about 0.4 mm. long; caryopsis elliptic, 0.7 mm. long, somewhat dorsally compressed, reddish brown.

Moist sandy soil and rocky hills, Arizona to western Texas and central Mexico.

**2. Muhlenbergia appressa** C. O. Goodding, Jour. Wash. Acad. **31:** 504, 1941

Annual; culms erect or decumbent, 10–40 cm. tall, branching below, hispidulous below nodes; sheaths shorter than internodes, scabrous; blades 1–4 cm. long, 1–2 mm. wide, scabrous or puberulent; panicles 5–20 cm. long, those of branches shorter, narrow, often included at base, branches closely appressed; spikelets 4.5–6 mm. long; glumes 1–2 mm. long, obtuse, thin, pale; lemma scabrous above, densely pilose on callus and margins at base, tapering into a slender, scabrous awn 1–3 cm. long; palea acuminate, pilose below. Cleistogamous spikelets borne in lower sheaths.

Canyons and slopes, southern Arizona. Usually occurs at higher altitudes but occasionally is found within the desert area (Camp Creek, Maricopa County).

**3. Muhlenbergia microsperma** (DC.) Kunth, Rév. Gram. **1:** 64, 1829

*Trichochloa microsperma* DC., Cat. Hort. Monsp. 151, 1813.
*Muhlenbergia purpurea* Nutt., Jour. Acad. Phila. II, **1:** 186, 1848.
*Muhlenbergia ramosissima* Vasey, Bull. Torrey Club **13:** 231, 1886.

Annual; culms slender, erect or commonly widely decumbent-spreading, 10–60 cm. long, usually freely branching, at least from lower nodes; sheaths glabrous, or scabrous toward summit, shorter than internodes; blades 3–10 cm. long, 1–2 mm. wide, lax, scabrous, more or less pubescent or pilose above; panicles 3–15 cm. long, narrow, rather lax, branches scabrous, stiffly ascending to spreading, spikelet-bearing to base; spikelets short-pedicellate, usually abruptly spreading; glumes obtuse, subequal, 0.5–1 mm. long; lemma narrow, 2–4 mm. long, scabrous, slender awn 1–3 cm. long. Large cleistogamous spikelets borne in lower sheaths.

Rocky slopes and canyons, southern California and Arizona to Guatemala; Colombia and Venezuela to Peru.

**4. Muhlenbergia texana** Buckl., Proc. Acad. Phila. **1862:** 91, 1862

Annual; culms slender, erect, 10–30 cm. high, branching at lower nodes, pubescent; sheaths longer than internodes, pubescent; ligule thin, decurrent, about 1 mm. long; blades 2–5 cm. long, less than 1 mm. wide, scabrous and pubescent; panicles more than half as long as culms, fragile but rather stiff, branches and pedicels spreading; spikelets 1.6–1.8 mm. long; glumes acute, pilose-pubescent, about two-thirds as long as floret; lemma pubescent on nerves, awn minute to 2 mm. long.

Hills and canyons, southern Arizona to western Texas; southern Baja California to Chihuahua, apparently rare.

**5. Muhlenbergia dumosa** Scribn., in Vasey, Contr. U.S. Nat. Herb. **3:** 71, 1892

Perennial; culms solid, woody, erect from a hard, knotty base, profusely branching and rebranching from most nodes, branches decompound, spreading; sheaths of main culm much shorter than internodes, blades early deciduous; blades to 12 cm. long, 2 mm. wide, flat or soon involute, those of branches smaller, usually curved or flexuous; panicles numerous, scarcely exserted from sheaths, 1–3 cm. long; spikelets 3 mm. long; glumes acute or subobtuse, 1 mm. long; lemma subacute, pilose on margins, awn flexuous, 3–5 mm. long; palea pilose between keels.

Rocky canyon slopes and valleys, southern Arizona and Sonora to Mexico.

**6. Muhlenbergia porteri** Scribn. ex Beal, Grasses N. Amer. **2:** 259, 1896

*Podosaemum porteri* Bush, Amer. Midl. Nat. **7:** 36, 1921.

Perennial; culms wiry, woody, widely spreading from a hard, rather knotty base, mostly 35–75 cm. long, often zigzag, freely branching, scabrous; sheaths glabrous, much shorter than subequal internodes; blades 2–8 cm. long, 1–2 mm. wide, lax, attenuate, glabrous, falling early from sheath; panicles 4–12 cm. long, about as wide, rather fragile branches finally widely spreading or even reflexed, few-flowered; spikelets 3–5 mm. long, usually purple; glumes narrow, acuminate, 2–3 mm. long; lemma acuminate, sparsely pilose, with a delicate awn 5–10 mm. long; palea acuminate, as long as or slightly longer than lemma.

Dry rocky hills and plains, southern California to Colorado and western Texas; Sonora to Coahuila and Durango.

**7. Muhlenbergia arizonica** Scribn., Bull. Torrey Club **15:** 8, *pl. 76, fig. A,* 1888

Perennial; culms densely tufted, slender, erect or ascending, 15–40 cm. high, usually branching at lower nodes; sheaths rather sharply keeled, glabrous; ligule hyaline, 1.5–2 mm. long; blades mostly less than 5 cm. long, flat or folded, 1–2 mm. wide, with conspicuous white midnerve and margins; panicles 5–15 cm. long, open, branches slender, almost filiform, ascending or spreading; spikelets long-pedicellate, 3 mm. long; glumes subobtuse, one-third to nearly one-half as long as spikelets; lemma subacute, densely pilose on midnerve and margins below, apex minutely notched, bearing an awn about 1 mm. long; palea a little longer than lemma.

Rocky hills and open woods, Arizona, Sonora, Chihuahua, Sinaloa, and Durango.

**8. Muhlenbergia emersleyi** Vasey, Contr. U.S. Nat. Herb. **3:** 66, 1892

*Epicampes emersleyi* Hitchc., U.S. Dept. Agr. Bull. **772:** 144, 1920.
*Epicampes subpatens* Hitchc., loc. cit.

Densely tufted perennial; culms erect, to 1.5 m. high, glabrous; sheaths rather broad, distinctly flattened and keeled, glabrous, all longer than internodes; ligule thin, lacerate, 1–2 cm. long; blades elongate, conduplicate at base, 2–5 mm. wide, scabrous, especially on margins; panicles to 40 cm. long, dense but rather loose, erect, branches slender, narrowly ascending, 5–10 cm. long, naked below; spikelets 2–2.5 mm. long; glumes equal, as long as or a little shorter than floret, acute, scabrous; lemma scabrous above, villous across back near the base, awn mostly 1–2 cm. long.

Rocky slopes and canyons, southwestern United States to Panama.

**9. Muhlenbergia rigens** (Benth.) Hitchc., Jour. Wash. Acad. **23:** 453, 1933

*Epicampes rigens* Benth., Jour. Linn. Soc. **19:** 88, 1881.

Perennial; culms densely tufted, slender or coarse, 37 cm. to nearly 1 m. high, glabrous; sheaths longer than internodes, rounded, glabrous; ligule about 1 mm. long, decurrent; blades firm, attenuate, strongly nerved, becoming involute, scabrous above, smooth below; panicles erect, dense, spikelike, 12–40 cm. long, branches usually short, appressed; spikelets about 3 mm. long, glumes one-half as long to equaling the floret; lemma scabrous, awnless.

Rocky canyons and hillsides, southern California to Texas and northern Mexico.

The specimens from southern Baja California are not typical. The culms are short and slender, and the glumes are only one-half as long as the floret. They may represent a distinct species.

## 25. SPOROBOLUS R. Br., Prod. 1: 169, 1810

Annuals or perennials with open or spikelike panicles of small, 1-flowered spikelets; glumes equal or usually unequal, the second often as long as the lemma; lemma membranous, 1-nerved, awnless, glabrous; palea as long as lemma; fruit free from lemma and palea.

Plants annual:
- Spikelets 1.5–1.7 mm. long; pedicels short, appressed .................. 1. *S. pulvinatus*
- Spikelets 1.8–2 mm. long; pedicels longer, spreading ...................... 2. *S. patens*

Plants perennial:
- Lower panicle branches in distinct whorls .......................... 3. *S. pyramidatus*
- Lower panicle branches not whorled:
  - Sheaths conspicuously villous on collar:
    - Panicle dense, spikelike, the branches appressed, densely flowered to base ........... ........................................................ 4. *S. contractus*
    - Panicle open, the branches naked at base ........................ 5. *S. cryptandrus*
  - Sheaths not villous on the collar, usually with long hairs at throat:
    - Culms wiry, creeping, with hard, scaly rhizomes; seacoast plants ..... 6. *S. virginicus*
    - Culms cespitose, erect, without rhizomes:
      - Panicle broad, its branches stiffly spreading, relatively few-flowered; blades soon involute ................................................ 7. *S. airoides*
      - Panicle narrow, elongate, its branches ascending, densely flowered nearly to base; blades usually flat .........................................8. *S. wrightii*

**1. Sporobolus pulvinatus** Swallen, Jour. Wash. Acad. **31:** 351, 1941

Annual; culms 5–30 cm. tall in small or rather dense tufts, decumbent-spreading glabrous; sheaths shorter than internodes, glabrous, more or less hispid at throat; ligule ciliate, about 0.5 mm. long; blades mostly 4–7 cm. long, 2–5 mm. wide or smaller in depauperate plants, lanceolate-acuminate, scabrous on both surfaces and on the thick, white margins, uppermost much reduced; panicles 2–5 cm. or rarely as much as 8 cm. long, pyramidal, branches verticillate, appressed, spreading at maturity, naked at base, densely flowered, with scattered but rather prominent glandular areas, and a large pulvinus at base of each; spikelets 1.5–1.7 mm. long, short-pedicellate, appressed; first glume minute; second glume about as long as spikelet, abruptly acute or subobtuse; lemma similar to second glume but somewhat narrower; palea broad, conspicuous, as long as lemma, minutely dentate.

Sandy plains and roadsides, Arizona to Texas and Oaxaca.

**2. Sporobolus patens** Swallen, Jour. Wash. Acad. **31:** 352, 1941

Annual; culms 10–25 cm. tall, slender, erect, in small tufts, glabrous; sheaths shorter than internodes, upper ones elongate, almost bladeless, glabrous, sparsely hispid at throat; ligule ciliate, about 0.5 mm. long; blades 1–2 cm. long, 1–2 mm. wide, lanceolate, flat, scabrous on margins; panicles pyramidal, 2.5–5 cm. long, branches slender, spreading or even reflexed, lower ones subverticillate, upper ones scattered, few-flowered, branchlets abruptly spreading; spikelets 1.8–2 mm. long, pedicels slender, spreading, to 3 mm. long; first glume minute, about 0.3 mm. long; second glume and lemma equal, acute, lemma a little broader than glume; palea somewhat shorter than lemma, broad, truncate, minutely dentate.

Sandy flats, Arizona, and Sonora.

**3. Sporobolus pyramidatus** (Lam.) Hitchc., U.S. Dept. Agr. Misc. Publl. **243:** 84, 1936

*Agrostis pyramidata* Lam., Tab. Encycl. **1:** 161, 1791.
*Vilfa arguta* Nees, Agrost. Bras. 395, 1829.
*Sporobolus argutus* Kunth, Rév. Gram. **1:** Suppl. XVII, 1829.

Perennial; culms densely tufted, erect or somewhat spreading, 10–70 cm. high; sheaths glabrous with a few long hairs in throat; blades 3–15 cm. long, 1–3 mm., rarely as much as 4 mm., wide, glabrous, scabrous on margins, uppermost greatly reduced; panicles 3–12 cm. long, pyramidal, branches ascending or finally spreading, lower ones in whorls, at least some of them naked at base; spikelets commonly 1.7–2 mm. long, sometimes less, short-pedicellate, appressed; first glume 0.3–0.5 mm. long, obtuse; second glume as long as lemma or a little shorter; lemma acute or subacuminate.

Sandy or mucky alkaline soil, Colorado and Kansas to Texas and Louisiana; Baja California to the West Indies and Nicaragua; Colombia to Brazil and Argentina.

**4. Sporobolus contractus** Hitchc., Amer. Jour. Bot. **2**: 303, 1915

*Sporobolus cryptandrus* var. *strictus* Scribn., Bull. Torrey Club **9**: 103, 1882.
*Sporobolus strictus* Merr., U.S. Dept. Agr. Div. Agrost. Circ. **32**: 6, 1901. Not Franch., 1893.

Perennial; culms tufted, erect, 40–140 cm. tall; sheaths conspicuously villous at throat and on collar; blades attenuate, flat, 2–4 mm. wide, scaberulous; panicles dense, spikelike, to 50 cm. long, about 5 mm. wide, partly enclosed in sheath; spikelets 2.5–2.8 mm. long; first glume about 1 mm. long; second glume and lemma equal, acute.

Sand hills, fields, and canyons, southeastern California to Colorado, Texas, and northern Mexico.

**5. Sporobolus cryptandrus** (Torr.) A. Gray, Man. 576, 1848

*Agrostis cryptandra* Torr., Ann. Lyc. N.Y. **1**: 151, 1824.

Plants similar to *S. contractus,* differing in having spreading panicle branches, naked at the base, and smaller spikelets 1.5–2 mm. long.

Dry or moist open, sandy land, Washington to Ontario and Maine, south to North Carolina, Indiana, Louisiana, and northern Mexico.

**6. Sporobolus virginicus** (L.) Kunth, Rév. Gram. **1**: 67, 1829

*Agrostis virginica* L., Sp. Pl. 63, 1753.
*Sporobolus littoralis* Kunth, Rév. Gram. **1**: 68, 1829.

Extensively creeping, tough, wiry perennial with hard, scaly rhizomes; culms slender to rather stout, 50–70 cm., or even to 1 m., long, with short internodes; sheaths glabrous; blades conspicuously distichous, mostly 8–12 cm. long, 3–5 mm. wide at base, very firm, becoming loosely rolled or involute, gradually narrowed to a rather fine but firm point; panicles densely spikelike, 6–10 cm. long, 0.5–1 cm. thick, usually partly enclosed in sheath; spikelets 2.5 mm. long, rarely longer; glumes acute, subequal, second as long as spikelet, first a little shorter.

Salt marshes and sandy seacoasts, Florida and Baja California to the West Indies, Brazil, and Chile.

**7. Sporobolus airoides** (Torr.) Torr., Pacif. R. Rep. **7**: 21, 1856

*Agrostis airoides* Torr., Ann. Lyc. N.Y. **1**: 151, 1824.
*Sporobolus diffusissimus* Buckl., Proc. Acad. Phila. **1862**: 90, 1862.

Perennial; culms usually in large, tough clumps, erect, 30–100 cm. tall; sheaths glabrous, or scabrous on margins, with a few long hairs at throat; blades elongate, less than 4 mm. wide, firm, soon involute, smooth below, scabrous above and on margins, ends slender, flexuous, very scabrous; panicles mostly 20–30 cm. long, to 20 cm. wide, branches and secondary branches stiffly ascending or spreading, branchlets appressed or divergent; spikelets 2–2.5 mm. long; first glume about one-half as long as spikelet, acute, second a little shorter than floret, acute or subobtuse.

Dry hills and meadows, often in saline soil, eastern Washington to South Dakota, south to southern California, Texas, and northern Mexico.

**8. Sporobolus wrightii** Munro ex Scribn., Bull. Torrey Club **9:** 103, 1882

*Sporobolus altissimus* Vasey, in Brandegee, Proc. Calif. Acad. II, **2:** 212, 1889.

Perennial; culms in large, dense clumps, erect, 1–2 m. tall; sheaths longer than internodes, sometimes elongate, sparsely hairy at throat; blades elongate, attenuate, 3–6 mm. wide, usually remaining flat, very scabrous above and on margins; panicles 30–50 cm. long, less than 10 cm. wide, branches numerous, approximate, narrowly ascending, sometimes spreading, densely flowered nearly to base, branchlets appressed; spikelets 2–2.5 mm. long; first glume about one-half as long as spikelet, second one-half to two-thirds as long as floret.

Dry sandy open ground, rocky slopes, and river banks, southern California to Texas and central Mexico.

## 26. STIPA L., Sp. Pl. 78, 1753

Tufted perennials with usually involute blades and narrow or open panicles; rachilla disarticulating above glumes, articulation oblique, leaving a sharp, bearded callus on floret; glumes equal, membranous to papery, thin, narrow, longer than floret; lemma firm to indurated, terete, strongly convolute, terminating in a prominent once- or twice-geniculate awn, twisted below.

Terminal segment of awn plumose, flexuose, to 12 cm. long ............. 1. *S. neomexicana*
Terminal segment of awn not plumose:
  Lower segment of once-geniculate awn conspicuously plumose; hairs 5–8 mm. long ............................................................... 2. *S. speciosa*
  Lower segments of twice-geniculate awn scabrous or pubescent, but not plumose:
    Panicles open or narrow and few-flowered:
      Glumes 1.5 cm. long or more:
        Panicles narrow, few-flowered; terminal segment of awn strongly flexuous ............................................................... 3. *S. comata*
        Panicles open with slender, spreading branches; terminal segment of awn straight ............................................................... 5. *S. pulchra*
      Glumes not more than 10 mm. long; lemma 5–6 mm. long .............. 4. *S. lepida*
    Panicles dense, narrow, many-flowered, the branches appressed ......... 6. *S. diegoensis*

**1. Stipa neomexicana** (Thurb.) Scribn., U.S. Dept. Agr. Div. Agrost. Bull. **17:** 132, *fig. 428,* 1899

*Stipa pennata* var. *neomexicana* Thurb., in Coult. & Nels., Man. Bot. Rocky Mts. 408, 1885.

Culms densely tufted, erect, 20–60 cm. tall; sheaths glabrous or puberulent; ligule about 0.5 mm. long, minutely ciliate; blades elongate, convolute, glabrous, stiff, usually arcuate-spreading; panicles as much as 15 cm. long, narrow, few-flowered, exserted or partly enclosed in sheath; glumes long-attenuate, 4–5 cm. long; lemma mostly 15–18 mm. long, pale, turning brown, pilose with white hairs; culms 4–5 mm. long, sharp-pointed, densely appressed-pilose; hairs brownish; awn to 18 cm. long, obscurely twice-geniculate, lower segment tightly twisted, appressed-villous, terminal segment flexuous, copiously plumose with hairs about 3 mm. long.

Sandy or rocky hills and plains, Arizona to Colorado, Texas, and northern Mexico. This species usually occurs above the Sonoran Desert, but has been collected at Tucson, Arizona.

**2. Stipa speciosa** Trin. & Rupr., Sp. Gram. Stip. 45, 1842
*Stipa speciosa* var. *minor* Vasey, Contr. U.S. Nat. Herb. **3**: 52, 1892.
*Stipa humilis* var. *speciosa* Kuntze, Rev. Gen. Pl. **3**[2]: 371, 1898.

Culms densely tufted, erect, 25–60 cm. tall; basal sheaths pubescent or pilose, densely villous at base, reddish brown; culm sheaths glabrous; ligule 0.5–1.5 mm. long, decurrent, densely ciliate; blades commonly 15–30 cm. long, stiff, incurved, glabrous or scaberulous, minutely pubescent on inner surfaces; panicles 10–15 cm. long, rather dense, branches appressed; glumes 15–20 mm. long, acuminate or attenuate, tips very thin and hyaline; lemma 7–10 mm. long, densely appressed-pilose with white hairs; callus narrow, sharp, glabrous below, awn 3–4 cm. long, with one sharp bend at middle, lower segment tightly twisted, densely long-pilose on lower half or two-thirds, hairs 5–8 mm. long, spreading; upper segment straight, rigid.

Dry rocky hills and canyons, Nevada to Colorado, Arizona, and Baja California; southern South America.

**3. Stipa comata** Trin. & Rupr., Sp. Gram. Stip. 75, 1842

Culms erect, mostly 30–60 cm. tall; sheaths glabrous or minutely pubescent; ligule 3–4 mm. long; blades 10–30 cm. long, 1–2 mm. wide, flat or inrolled, scabrous, usually curved; panicles 10–20 cm. long, partly enclosed by sheath; glumes 1.5–2 cm. long, thin, attenuate; lemma 8–12 mm. long, pale or brownish, pilose to nearly glabrous; callus about 3 mm. long; awn 10–15 cm. long, twice-geniculate, two lower segments twisted, pilose, terminal flexuous, scabrous, very slender.

Dry hills and plains, Yukon Territory to Indiana, south to California, Arizona, and Texas. Usually at higher altitudes within the desert region.

**4. Stipa lepida** Hitchc., Amer. Jour. Bot. **2**: 302, 1915

Culms slender, tufted, erect, 60–100 cm. tall, scaberulous, densely appressed-pubescent below nodes; sheaths glabrous or puberulent; blades 10–30 cm. long, 2–3 mm. wide, flat, becoming involute in drying, scaberulous beneath, densely pubescent above; panicles mostly 15–20 cm. long, open, branches slender, distant; glumes 3-nerved, first 6–10 mm. long, second a little shorter; lemma 5–6 mm. long, sparsely to densely appressed-pilose; callus short, oblique, sharp; awn 2.5–4 cm. long, obscurely twice-geniculate, usually curved or flexuous.

Dry hills and slopes and open woods, central California to central Baja California.

**5. Stipa pulchra** Hitchc., Amer. Jour. Bot. **2**: 301, 1915

Culms densely tufted, erect, 60–100 cm. high, glabrous, or pubescent below nodes; sheaths mostly shorter than internodes, densely pilose on margins and at throat, or sometimes nearly glabrous; blades 8–20 cm. long, 1–2 mm. wide, usually involute, glabrous beneath, pubescent above; panicles 10–20 cm. long, slender branches spreading, flexuous, bearing 2–several spikelets; glumes subequal, about 1.5 cm. long, 3-nerved, acuminate; lemma 7.5–10 mm. long, papillose, sparsely pilose, the constricted summit bearing a row of stiff hairs; awn 5.5–9 cm. long, twice-geniculate, first segment twisted, hispid, terminal slender, straight, scabrous.

Dry hills and deserts, central California to northern Baja California.

**6. Stipa diegoensis** Swallen, Jour. Wash. Acad. **30**: 212, 1940

Culms densely tufted, erect, 70–100 cm. tall, scaberulous, densely pubescent below nodes; lower sheaths longer than internodes, upper ones shorter than internodes, elongate, glabrous or scaberulous, usually densely pilose on collar; ligule 1–2 mm. long, obtuse or truncate, ciliate, pubescent; blades 15–40 cm. long, 2–4 mm. wide, flat or involute, scabrous on lower surface, pubescent on upper; panicles 15–40 cm. long, narrow, dense, interrupted, appressed branches to 10 cm. long; first glume 9–11 mm. long, 1–3-nerved, acuminate; second glume 8–9 mm. long, 3-nerved, acuminate; lemma 6.5–7.5 mm. long, pilose hairs at top of lemma 1–2 mm. long; callus 0.5 mm. long, curved, sharp-pointed, appressed-pilose; awn 2–3.3 cm. long, twice-geniculate, scabrous; palea 4 mm. long, acute, pilose.

Along vernal streams, San Diego County, California, and northern Baja California.

## 27. ARISTIDA L., Sp. Pl. 82, 1753

Tufted annuals or perennials with firm, usually involute blades and narrow or open panicles; glumes equal or unequal, acute, acuminate, or awn-tipped; lemma indurated with a sharp, bearded callus, 3-awned, lateral awns sometimes much reduced, base of awns sometimes undivided, twisted, forming a column.

Lemma articulate with column of awns; culms wiry, freely and divaricately branching from most nodes:
- Plants annual .................... 1. *A. peninsularis*
- Plants perennial:
  - Culms densely pubescent .................... 2. *A. californica*
  - Culms glabrous .................... 3. *A. glabrata*

Lemma not articulate with awns:
- Lateral awns greatly reduced, less than 3 mm. long, usually minute:
  - Central awn erect; column not twisted .................... 4. *A. ternipes*
  - Central awn divergent; column twisted .................... 5. *A. schiediana*
- Lateral awns nearly as long as central awn:
  - Plants annual .................... 6. *A. adscensionis*
  - Plants perennial:
    - Panicles diffuse, branches abruptly spreading, naked below .................... 7. *A. hamulosa*
    - Panicles narrow, often rather dense, branches ascending or appressed:
      - Glumes subequal; panicles dense, purple, branches stiffly ascending, naked at base .................... 12. *A. parishii*
      - Glumes unequal, first one-half as long as second or as much as two-thirds as long in *A. glauca*:
        - Lemma narrowed into a slender beak 5–6 mm. long, equaling or exceeding second glume; awns 1–2.5 cm. long, widely spreading .................... 8. *A. glauca*
        - Lemma not narrowed above, usually shorter than second glume:
          - Panicle branches very slender, more or less flexuous; lemma conspicuously scabrous in lines; awns terete or nearly so at base .................... 9. *A. purpurea*
          - Panicle branches stiffly ascending or appressed; if sometimes lax in *A. longiseta*, lemma nearly glabrous; awns flattened at base:
            - Culms slender to stout, 30–60 cm. tall; panicles 15–30 cm. long, erect, densely flowered; awns 2 cm. long, or central awn to 3 cm. long .................... 11. *A. wrightii*
            - Culms slender, 20–30 cm. tall (to 50 cm. tall in var. *robusta*); panicles open or few-flowered; awns 4–8 cm. long:
              - Awns 6–8 cm. long .................... 10. *A. longiseta* var. *longiseta*
              - Awns 4–5 cm. long .................... 10*a*. *A. longiseta* var. *robusta*

**1. Aristida peninsularis** Hitchc., Contr. U.S. Nat. Herb. **22:** 521, 1924

Plants annual, bushy-branched; culms slender, ascending, branching, pubescent, 10–20 cm. tall; sheaths glabrous, pubescent at throat and slightly so on collar; blades flat or somewhat involute, glabrous beneath, puberulent on upper surface, nerved, 1–3 cm. long; panicles very numerous, mostly 1–3-flowered; glumes unequal, 1-nerved, glabrous or keel of first slightly scabrous toward tip, first about 1 cm. long, acute, second about 2 cm. long, narrowed into a slender tip but not awned; lemma about 8 mm. long, glabrous below, minutely scaberulous toward summit, callus densely pubescent, about 2 mm. long; column about 2 cm. long, twisted, the 3 awns about equal, 4–5 cm. long, diverging but not arcuate-contorted at base.

Sandy beaches, Bahía de los Angeles, Baja California, the type locality. Known only from the type collection.

**2. Aristida californica** Thurb. ex S. Wats., Bot. Calif. **2:** 289, 1880

*Aristida californica* var. *fugitiva* Vasey, Contr. U.S. Nat. Herb. **3:** 49, 1892.

Perennial; culms densely tufted, erect or ascending, 10–30 cm. tall, pubescent, divaricately branching at most nodes; sheaths glabrous or puberulent, usually pubescent at throat and on collar; blades less than 5 cm. long, flat or involute, glabrous or obscurely scabrous beneath, pubescent above; panicles numerous, few-flowered, branches appressed; glumes 1-nerved, first glume about 8 mm. long, second narrower, twice as long; lemma 5–7 mm. long, scabrous above, callus slender, sharp-pointed, 1–1.5 mm. long, densely pilose; column 15–20 mm. long, tightly twisted, free ends of awns 2.5–3.5 cm. long, nearly equal, divergent.

Sand hills and sandy deserts, southern California, Arizona, Baja California, and Sonora.

**3. Aristida glabrata** (Vasey) Hitchc., Contr. U.S. Nat. Herb. **22:** 522, 1924

*Aristida californica* var. *glabrata* Vasey, Proc. Calif. Acad. II, **3:** 178, 1891.

Perennial; culms erect, branching, 20–40 cm. tall, glabrous, first internode elongate; sheaths glabrous, more or less pubescent at throat and on collar; blades 1–5 cm. long, flat or soon involute, scaberulous beneath, puberulent above; panicles 3–6 cm. long, terminal and axillary, sometimes fascicled, narrow, rather dense, branches appressed; glumes 1-nerved, first 5–6 mm. long, second narrower, nearly twice as long; lemma 5–7 mm. long, scabrous in lines toward summit, callus about 1 mm. long, densely short-pilose; column 6–14 mm. long, twisted, scabrous, awns about equal, 1.5–3 cm. long, divergent, sometimes loosely contorted at base.

Dry gravelly slopes and meadows, southern Arizona and Baja California.

**4. Aristida ternipes** Cav., Ic. **5:** 46, 1799

*Streptachne scabra* H. B. K., Nov. Gen. & Sp. **1:** 124, *pl. 40,* 1816.

*Aristida scabra* Kunth, Rév. Gram. **1:** 62, 1829.

Perennial; culms erect, 50–100 cm. tall or sometimes taller; sheaths glabrous; blades elongate, 2–3 mm. wide, scabrous, tapering to a fine point; panicles commonly 30–50 cm. long, rather lax, usually distant, very scabrous branches abruptly spreading, drooping, naked in lower half, spikelets appressed; glumes nearly equal, 8–10 mm. long, scabrous or nearly glabrous; lemma slender, glabrous, narrowed into a scabrous beak, 12–17 mm. long to base of lateral awns, central awn 1–1.5 cm. long, terete, scabrous, lateral ones minute or as much as 3 mm. long, lemma with its awn more or less falcate.

Rocky hills and plains, Arizona and New Mexico to northern South America and the West Indies.

**5. Aristida schiediana** Trin. & Rupr., Sp. Gram. Stip. 120, 1842

Perennial; culms tufted, erect, 30–80 cm. high, several-noded; sheaths much longer than internodes, glabrous, or rather conspicuously villous on the collar; blades firm, elongate, attenuate, as much as 3 mm. wide, lower ones becoming curled with age; panicles open, as much as 30 cm. long, branches solitary or in pairs, distant, spreading, naked at base; glumes subequal, 8–10 mm. or as much as 15 mm. long; lemma narrowed into a slender, twisted column, 10–15 mm. long including column; central awn 5–10 mm. long, divergent, lateral ones minute or obsolete.

Rocky hills and plains, Mexico and Guatemala. Found in the desert region only in the Sierra de la Laguna, Baja California.

**6. Aristida adscensionis** L., Sp. Pl. 82, 1753

*Aristida bromoides* H. B. K., Nov. Gen. & Sp. **1:** 122, 1816.
*Aristida fasciculata* Torr., Ann. Lyc. N.Y. **1:** 154, 1824.

Annual; culms freely branching, erect or geniculate-ascending, glabrous, 10–80 cm. tall; sheaths glabrous; blades to 10 cm. long, 2 mm. wide, flat or often short and involute, scabrous above; panicles mostly 5–15 cm. long, dense or often loose and drooping; glumes 1-nerved, first 5–7 mm. long, obtuse or subacute, second 8–10 mm. long, acute or blunt; lemma 6–9 mm. long, scabrous on keel above, callus subobtuse, densely short-pilose; awns 10–15 mm. long, subequal, equally divergent, flat at base, gradually narrowed to a fine point.

Dry open ground and rocky slopes, California to Missouri and Texas; Mexico and the West Indies to Chile; warmer parts of the Old World.

**7. Aristida hamulosa** Henr., Meded. Rijks Herb. Leiden **54:** 219, 1926

Perennial; culms rather slender, erect, 40–70 cm., or even to 1 m., tall; sheaths glabrous or scaberulous, all longer than internodes; blades to 30 cm. long, less than 3 mm. wide, flat or becoming involute, glabrous beneath, finely scabrous above; panicles commonly 25–40 cm. long, sometimes three-fourths the height of plant, branches stiffly spreading, more or less drooping, naked in lower half, lower ones as much as 20 cm. long; spikelets appressed; glumes 12–15 mm. long, subequal, acuminate; lemma about 12 mm. long, usually strongly scabrous, not twisted, central awn 1.6–2.5 cm. long, lateral ones shorter, somewhat divergent.

Dry slopes and plains, southern California to western Texas and Guatemala.

**8. Aristida glauca** (Nees) Walp., Ann. Bot. [London] **1:** 925, 1849

*Chaetaria glauca* Nees, Linnaea **19:** 688, 1847.
*Aristida reverchoni* Vasey, Bull. Torrey Club **13:** 52, 1886.
*Aristida vaseyi* Woot. & Standl., N. Mex. Coll. Agr. Bull. **81:** 55, 1912.

Perennial; culms slender, very densely tufted, erect, 15–65 cm. tall; leaves crowded toward base, sheaths glabrous with a few hairs at throat, blades firm, involute, curved or flexuous, scabrous, 5–10 cm. long; panicles 8–15 cm. long, narrow, erect, branches appressed; first glume 5–8 mm. long, second twice as long; lemma 10–12 mm. long, narrowed to a slender, scabrous, slightly twisted beak, often extending well beyond glumes; awns 1–2.5 cm. long, equal, flattened at base, scabrous, finally widely spreading.

Dry rocky slopes and plains, southern California to Utah, Texas, and central Mexico.

**9. Aristida purpurea** Nutt., Trans. Amer. Phil. Soc. n.s., **5:** 145, 1837

*Aristida filipendula* Buckl., Proc. Acad. Phila. **1862:** 93, 1862.
*Aristida purpurea* var. *californica* Vasey, Contr. U.S. Nat. Herb. **3:** 47, 1892.

Similar in habit to *A. glauca,* averaging coarser and taller; panicles 10–20 cm. long, lax, nodding, usually purple, branches very slender, usually flexuous; lemma about 1 cm. long, usually exceeded by second glume, strongly scabrous in lines from below middle; awns slender, scabrous, mostly 3–5 cm. long.

Dry rocky hills and plains, southern California to Kansas, Arkansas, and northern Mexico.

**10. Aristida longiseta** Steud., Syn. Pl. Glum. **1:** 420, 1855, var. **longiseta**

*Aristida purpurea* var. *longiseta* Vasey, in Wheeler, Rep. **6:** 286, 1878.

Perennial; culms densely tufted, sometimes in large clumps, 20–30 cm. tall; leaves crowded toward base, blades involute, arcuate-spreading, scabrous, usually less than 10 cm. long; panicles few-flowered, short branches ascending, lower ones more or less curved or flexuous; glumes relatively broad, acute or mucronate, second to 2 cm. long; lemma 12–15 mm. long, scarcely narrowed above, scabrous in lines above middle, sometimes obscurely so; awns 6–8 cm. long, broad and flat at base, usually purple, slender, scabrous.

Dry hills and plains, Arizona to Colorado and Texas.

**10*a*. Aristida longiseta** var. **robusta** Merr., U.S. Dept. Agr. Div. Agrost. Circ. **34:** 5, 1901

Differs from var. *longiseta* in having culms 30–50 cm. tall, leaves less crowded at base, longer and denser panicles, and stiffer awns 4–5 cm. long.

Washington to Minnesota, Arizona, and Texas.

*Aristida longiseta* var. *longiseta* and var. *robusta* occur near the upper limits of the Sonoran Desert, and are found mainly in the area around Tucson at about 3,000 feet and higher.

**11. Aristida wrightii** Nash, in Small, Fl. SE. U.S. 116, 1903

Perennial; culms densely tufted, rather slender to stout, erect, mostly 30–60 cm. tall, glabrous or occasionally puberulent; sheaths glabrous or those of innovations puberulent, villous at throat, with a line of hairs on collar; blades involute, firm, curved or flexuous, 10–15 cm. long, glabrous beneath, finely scabrous above; panicles pale, 15–20 cm. long, usually erect, branches few-flowered, appressed or lower ones sometimes spreading, more or less flexuous; glumes 1-nerved, first 6–7 mm. long, scabrous on keel, second twice as long, nearly glabrous; lemma 10–12 mm. long, scabrous toward summit, callus 1 mm. long, densely pilose; awns about equal, scabrous, 1.5–3 cm. long.

Dry plains and rocky slopes, southern California to Colorado, Texas, and northern Mexico.

**12. Aristida parishii** Hitchc., in Jepson, Fl. Calif. **1:** 101, 1912

Perennial; culms densely tufted, stiffly erect, 30–60 cm. tall; sheaths glabrous, villous at throat, pubescent on collar; blades 15–30 cm. long, firm, soon involute, nearly smooth beneath, finely scabrous above and toward tip; panicles usually dark purple, mostly 15–30 cm. long, rather dense, branches stiffly ascending or appressed; first glume 10–14 mm. long, second 1–2 mm. longer, short-awned; lemma 12–14 mm. long, glabrous below, becoming strongly scabrous toward summit, callus about 1 mm. long, densely pilose; awns very scabrous, about equally divergent, central 2.5–3 cm. long, lateral ones somewhat shorter.

Dry rocky hills, southern California, Nevada, Arizona, to central Baja California.

## 28. TRAGUS Hall., Stirp. Helv. 2: 203, 1768

Low annuals with flat blades and groups of spikelets closely arranged on a slender axis; groups, or spikes, subsessile, falling entire, composed of 2–5 spikelets on a short, zigzag rachis; first glumes small or wanting, second glume of the 2 lower spikelets bearing stout, hooked spines along each side; lemma and palea thin, lemma flat, palea convex.

### 1. **Tragus berteronianus** Schult., Mant. 2: 205, 1824

Culms erect or ascending, often decumbent, 5–40 cm. tall; sheaths much shorter than internodes, glabrous; blades mostly 2–4 cm. long, 2–4 mm. wide, firm, flat, with white, cartilaginous, stiffly hispid-ciliate margins; racemes 4–10 cm. long, 4–5 mm. thick; spikes 2–3 mm. long, subsessile.

Dry open ground, Arizona to Texas, south to Argentina; warmer parts of the Old World.

**Tragus racemosus** (L.) All., Fl. Pedem. 2: 241, 1785, with pedicellate spikes 4–5 mm. long, has been collected at Tucson. Although within the desert region, it can scarcely be considered a part of the flora.

## 29. HILARIA H. B. K., Nov. Gen. & Sp. 1: 116, 1816

Stiff perennials with solid culms, narrow blades, and usually dense, narrow, terminal spikes; spikelets arranged in groups of 3, central spikelet perfect, 1-flowered, lateral ones staminate, 2-flowered, groups falling entire; glumes usually firm, asymmetric, those of lateral spikelets bearing an awn on one side from about middle, those of fertile spikelet divided into awns or lobes; lemma and palea hyaline, nearly equal, lemma 3-nerved, shallowly lobed, awnless or short-awned.

Plants slender, stoloniferous:
  Spikelets pale, not crowded:
    Stolons present; blades 5 cm. long or less ................ 1. *H. belangeri* var. *belangeri*
    Stolons none; blades over 5 cm. long ................ 1a. *H. belangeri* var. *longifolia*
  Spikelets purple or dark brown, crowded ........................... 2. *H. cenchroides*
Plants coarse, not stoloniferous:
  Culms conspicuously lanate; glumes rather thin, faintly nerved, not conspicuously broadened at summit ................................................ 3. *H. rigida*
  Culms glabrous; glumes firm, strongly nerved, conspicuously broadened or flabellate at summit ................................................ 4. *H. mutica*

### 1. **Hilaria belangeri** (Steud.) Nash, N. Amer. Fl. 17: 135, 1912, var. **belangeri**

*Anthephora belangeri* Steud., Syn. Pl. Glum. 1: 111, 1854.
*Hilaria texana* Nash, in Small, Fl. SE. U.S. 68, 1903.

Plants in small tufts, sending out slender, wiry, arching stolons, internodes 5–25 cm. long; culms erect, 10–30 cm. tall, nodes bearded with ascending or spreading hairs;

sheaths glabrous, margins rather broad and hyaline; blades short, mostly less than 5 cm. long, 1–2 mm. wide, scabrous, becoming curled; spikes mostly 2–3 cm. long, bearing 4–8 clusters of spikelets, axis flattened, alternately curved; glumes of lateral spikelets unequal, asymmetrical, first smaller, midnerves of both glumes excurrent as short awns; glumes of fertile spikelet broadened above, very scabrous, deeply lobed, midnerves extending into awns usually longer than staminate spikelets; lemma truncate, minutely erose.

Plains, mesas, and rocky hills, southeastern California, Arizona to Texas and northern Mexico.

**1*a*. Hilaria belangeri** var. **longifolia** (Vasey) Hitchc., Proc. Biol. Soc. Wash. **41:** 162, 1928

*Hilaria cenchroides* var. *longifolia* Vasey ex Beal, Grasses N. Amer. **2:** 69, 1896.

Differs from var. *belangeri* in having elongate blades; stolons wanting.

Rocky hills, Arizona, Texas, and southern Sonora.

**2. Hilaria cenchroides** H. B. K., Nov. Gen. & Sp. **1:** 147, *pl. 37,* 1815

Very similar in habit to *H. belangeri*; internodes of stolons shorter; spikes dense, groups of spikelets crowded, purplish or dark brown; glumes not widened above.

Prairies and rocky hills, Mexico and Guatemala. Apparently rare in the Sonoran Desert, collected only at Comondú, Baja California.

**3. Hilaria rigida** (Thurb.) Benth. ex Scribn., Bull. Torrey Club **9:** 86, 1882

*Pleuraphis rigida* Thurb., in S. Wats., Bot. Calif. **2:** 293, 1880.

Culms erect or ascending from a somewhat woody rhizomatous base, 50–100 cm. tall, rather freely branching, conspicuously lanate; sheaths mostly shorter than internodes, lanate at least on collar; blades 2–12 cm. long, 2–4 mm. wide, flat or rolled, acuminate into a rigid point, scabrous, villous especially on upper surface; spikes 5–10 cm. long, dense or rather loose, groups of spikelets about 8 mm. long, densely bearded at base; glumes of lateral spikelets thin, rather faintly nerved, sparsely long-ciliate, summit divided into 2–4 rounded lobes, some nerves excurrent into slender awns; glumes of fertile spikelet narrow, deeply divided into few to several ciliate lobes and slender awns; lemma slightly longer than glumes, thin, ciliate, 2-lobed, short-awned.

Deserts, plains, and rocky hills, southern California to southern Utah, Arizona, Baja California, and Sonora.

**4. Hilaria mutica** (Buckl.) Benth., Jour. Linn. Soc. **19:** 62, 1881

*Pleuraphis mutica* Buckl., Proc. Acad. Phila. **1862:** 95, 1862.

Culms 30–60 cm. tall, erect or ascending from a tough rhizomatous base, glabrous, nodes pubescent; sheaths much longer than internodes, glabrous or sparsely papillose-pilose; blades firm, flat or rolled, 2–3 mm. wide, scabrous, sometimes pilose above; spikes usually long-exserted, 4–9 cm. long, group of spikelets bearded at base; glumes of lateral spikelets flabellate, ciliate, conspicuous, obscuring fertile spikelet; glumes of fertile spikelet sharply keeled, narrow, divided into few to several ciliate lobes and awns; lemma longer than glumes, awnless or mucronate between 2 rounded lobes.

Dry mesas and hills, Arizona to Texas and northern Mexico. Occurring near the upper limits of the Sonoran Desert.

## 30. AEGOPOGON H. & B. ex Willd., Sp. Pl. 4: 899, 1805

Slender perennials with short, narrow, flat blades and small racemes, groups of spikelets all turned to one side, spikelets in groups of 3, groups pedunculate, spreading, falling entire, arranged alternately on opposite sides of a slender, flattened axis, central spikelet fertile, lateral ones staminate or neuter; spikelets 2-flowered; glumes equal, 1-nerved, notched at apex, awned from between lobes; lemmas 3-nerved, longer than glumes, nerves extending into awns.

### 1. Aegopogon cenchroides H. & B. ex Willd., Sp. Pl. 4: 899, 1805

Slender, spreading perennial; culms 20–50 cm. long, erect, or longer ones spreading, freely branching; sheaths glabrous, keeled; ligule hyaline, 1–2 mm. long, obtuse; blades 2–6 cm. long, 1–2 mm. wide, flat, acuminate, scabrous; racemes slender, 2–5 cm. long, usually purple; spikelets about 3 mm. long, central awn of first floret about 10 mm. long, others half as long or less.

Dry rocky slopes and savannas, Mexico to Bolivia. Found in the Sonoran Desert only in the Sierra de la Giganta, Baja California.

## 31. LEPTOCHLOA Beauv., Ess. Agrost. 71, 166, 1812

Annual or perennial grasses with flat blades (involute in *L. monticola*) and numerous usually slender spikes, scattered along a common axis. Spikelets few- to several-flowered, upper floret reduced to a small awnless rudiment, subsessile or short-pedicellate, crowded or somewhat distant on one side of slender rachis; glumes 1-nerved (or second rarely 3-nerved), second usually broader and longer than first; lemmas 3-nerved, acute to obtuse, often 2-toothed or 2-lobed, awnless, mucronate, or awned, pubescent or pilose on margins and sometimes on internerves.

Plants perennial; lemmas notched at apex, awnless, lobes broad, obtuse ........ 1. *L. dubia*
Plants annual; lemmas mucronate or awned from between teeth of a minutely bifid apex, if awnless then not with broad, obtuse lobes:
- Sheaths sparsely papillose-pilose; spikelets 1.5–2 mm. long, 3- or 4-flowered ........................ 2. *L. filiformis*
- Sheaths glabrous or scabrous; spikelets 3–12 mm. long, 5–12-flowered:
  - Lemmas 5–7 mm. long, mucronate only, scarcely narrowed toward apex, lateral nerves excurrent ........................ 5. *L. uninervia*
  - Lemmas 2–5 mm. long, awned; awns sometimes minute:
    - Lemmas 2 mm. long, acute, viscid on back; panicle usually less than 10 cm. long, tinged with purple ........................ 3. *L. viscida*
    - Lemmas 4–5 mm. long, acuminate, not viscid; panicle more than 10 cm. long, not tinged with purple ........................ 4. *L. fascicularis*

### 1. Leptochloa dubia (H. B. K.) Nees, Syll. Pl. Ratisb. 1: 4, 1824

*Chloris dubia* H. B. K., Nov. Gen. & Sp. 1: 169, 1816.
*Leptochloa pringlei* Beal, Grasses N. Amer. 2: 436, 1896.

Perennial; culms in dense clumps, erect, simple or branching, 40 cm. to more than 1 m. tall, glabrous; sheaths mostly shorter than internodes, glabrous, or sparsely pilose on margins, lower ones keeled, upper ones rounded on back; ligule 1 mm. long, minutely lacerate; blades flat, 10–25 cm. long (or uppermost shorter), 2–6 mm. wide, scabrous, often pilose on upper surface near base; inflorescence 8–20 cm. long, with several ascending to spreading, scattered spikes, lower ones 8–10 cm. long; spikelets

4–10-flowered, florets spreading; glumes acuminate, first 3 mm. long, second a little broader, 3.5–4 mm. long; lemmas 3 mm. long, awnless, minutely pubescent on margins, sparsely pilose on internerves, tip with 2 broad, rounded lobes.

Dry rocky hills and open ground, Arizona to Oklahoma and Texas, southward through Mexico; Argentina.

**2. Leptochloa filiformis** Beauv., Ess. Agrost. 71, 161, 166, 1812

*Festuca filiformis* Lam., Tab. Encycl. **1:** 191, 1791.
*Eleusine mucronata* Michx., Fl. Bor. Amer. **1:** 65, 1803.
*Leptochloa mucronata* Kunth, Rév. Gram. **1:** 91, 1829.

Annual; culms slender to coarse, branching, erect or sometimes geniculate at lower nodes, 10 cm. to more than 1 m. tall; sheaths sparsely papillose-pilose with long hairs, rarely glabrous; ligule membranous, subentire to lacerate-ciliate, 1–2 mm. long; blades flat, lax, 5–30 cm. long, 3–10 mm. wide, glabrous or sparsely papillose-pilose, margins scabrous; inflorescence nearly half the length of culm; spikes several to numerous, racemose on an elongated axis, lax, spreading to reflexed, 2–15 cm. long; spikelets 3–4-flowered, 1.5–2 mm. long, upper floret scarcely exceeding glumes; glumes acuminate, subequal; lemmas 1–1.5 mm. long, obtuse, minutely 2-lobed, awnless, pubescent on nerves and sometimes on internerves toward base.

Fields and waste places, southern California to Kansas and Virginia, southward through Mexico and the West Indies to Argentina.

**3. Leptochloa viscida** (Scribn.) Beal, Grasses N. Amer. **2:** 434, 1896

*Diplachne viscida* Scribn., Bull. Torrey Club **10:** 30, 1883.

Annual; culms in small to rather dense clumps, freely branching, erect or usually decumbent-spreading, 2–40 cm. tall; sheaths scabrous; ligule 1–3 mm. long, hyaline, minutely erose; blades flat, scabrous, 2–15 cm. long, 2–6 mm. wide; inflorescence included at base, very short and dense with reduced spikes, or to 12 cm. long with numerous ascending spikes to 4 cm. long, 4–7-flowered; spikelets 3–5 mm. long; glumes acute, first about 2 mm. long, second broader, 2.5–3 mm. long; lemma 2–3 mm. long, pubescent on nerves, tip minutely lobed, awned from between lobes, awn about 1 mm. long.

Wet places, usually in heavy soil, plains, roadsides, and waste places, California to Texas, Baja California, Sonora, and Chihuahua.

**4. Leptochloa fascicularis** (Lam.) A. Gray, Man. 588, 1848

*Festuca fascicularis* Lam., Tab. Encycl. **1:** 189, 1791.
*Diplachne acuminata* Nash, in Britt., Man. 128, 1901.

Annual; culms erect or decumbent-spreading at base, to more than 1 m. tall, simple or usually freely branching, especially in depauperate plants, glabrous; sheaths rounded on back or keeled only near summit, glabrous or scaberulous; ligule hyaline, lacerate, 3–5 mm. long; blades flat, scabrous, 10–50 cm. long, 1–5 mm. wide; inflorescence 15–30 cm. long (much reduced in depauperate plants); spikes numerous, approximate or rather distant, ascending or spreading, lower ones 8–15 cm. long, becoming shorter upward; spikelets 8–10-flowered, 7–12 mm. long, appressed; glumes acuminate, first narrow, 2.5–3 mm. long, second broader, 3–4 mm. long; lemmas 4–5 mm. long, abruptly acute, minutely notched, mucronate or aristate, densely pubescent on margins and midnerve in lower half.

Wet ground and waste places, throughout the United States, south through tropical America to Argentina.

**5. Leptochloa uninervia** (Presl) Hitchc. & Chase, Contr. U.S. Nat. Herb. **18:** 383, 1917

*Megastachya uninervia* Presl, Rel. Haenk. **1:** 283, 1830.
*Leptochloa imbricata* Thurb., in S. Wats., Bot. Calif. **2:** 293, 1880.

Annual; culms erect, simple or sparingly branched, 30–100 cm. tall, glabrous; sheaths rounded on back, glabrous or scabrous, shorter than internodes; ligule membranous, 2–3 mm. long; blades firm, flat or loosely involute, attenuate-scabrous, 10–45 cm. long, 1–4 mm. wide; panicle 10–30 cm. long; spikes numerous, approximate, stiffly ascending, lower ones 4–9 cm. long, upper ones gradually shorter and closer together; spikelets lead-colored, 6–9-flowered, 5–7 mm. long, appressed; first glume narrow, acute, 1–1.5 mm. long, second much broader, abruptly acute or obtuse, mucronate; lemmas 5–7 mm. long, scarcely narrowed toward tip, abruptly subacute or obtuse, minutely lobed, apiculate, lateral nerves more or less excurrent, margins pilose near base.

Wet ground, ditches, and roadsides, southern California to Colorado, Mississippi, and northern Mexico; Peru to Argentina.

## 32. DACTYLOCTENIUM Willd., Enum. Hort. Berol. 1029, 1809

Annual grasses with comparatively short, broad blades and 2–several short, thick, digitate, ascending or spreading spikes, rachis extending beyond spikelets in a naked point; spikelets compressed, 3–5-flowered, alternate in 2 rows on one side of flattened rachis, horizontally spreading; glumes subequal, 1-nerved, first crescent-shaped, acute, awnless, second obtuse, bearing a short, stout awn; lemmas thin, acute, obovate, 3-nerved, lateral nerves obscure, acuminate; palea about as long as lemma.

**1. Dactyloctenium aegyptium** (L.) Beauv., Ess. Agrost. Expl. Pl. 15, 1812

*Cynosurus aegyptius* L., Sp. Pl. 72, 1753.
*Dactyloctenium mucronatum* Willd., Enum. Hort. Berol. 1029, 1809.

Culms branching, decumbent, spreading at base and rooting at lower nodes, ends ascending, 20–40 cm. long; sheaths overlapping, compressed, glabrous; ligule membranous, minutely erose, about 0.3 mm. long; blades flat, mostly 5–10 cm. long, 2–5 mm. wide, scabrous and papillose on margins, otherwise glabrous or nearly so; spikes 2–5, mostly 2–5 cm. long; spikelets pectinate, about 3 mm. long; first glume 1.5 mm. long, acute, scabrous on keel; second glume a little longer than first with an awn 1–2.5 mm. long from a broad apex; lemmas 2.5–3.5 mm. long, acute or acuminate, scabrous on keel, otherwise glabrous.

Warm regions of both hemispheres.

## 33. ELEUSINE Gaertn., Fruct. 1: 7, 1788

Annual grasses, with compressed culms and 1–several densely flowered, digitate spikes, frequently with an additional spike a short distance below the others; spikelets few- to several-flowered, compressed, sessile in 2 rows on one side of a flattened rachis, rachilla disarticulating above glumes and between florets; glumes unequal, acute, 1-nerved, shorter than first lemma; lemmas broad, 3-nerved, nerves strong, approximate, awnless.

**1. Eleusine indica** (L.) Gaertn., Fruct. **1:** 8, 1788

*Cynosurus indicus* L., Sp. Pl. 72, 1753.

Culms in tough, spreading clumps, somewhat decumbent, 15–70 cm. or rarely to 1 m. tall, compressed, very smooth; sheaths compressed, sharply keeled, sparsely papillose near margins toward summit, otherwise glabrous, margins hyaline; ligule membranous, erose, about 1 mm. long; blades to 25 cm. long, 2–8 mm. wide, flat or conduplicate, tip navicular, glabrous or sparsely pilose on upper surface, margins scabrous; spikes 2–several, 5–10 cm. long, stiffly ascending; spikelets 5 mm. long; lemmas 3 mm. long, broad at base, rather abruptly narrowed to subacute apex, sometimes mucronate.

Introduced; a common weed in the warm regions of both hemispheres.

## 34. SPARTINA Schreb. ex Gmel., Syst. Nat. ed. 13, **2:** 123, 1791

Coarse, usually tall perennial with strong, scaly rhizomes, tough, flat blades, and several to numerous appressed spikes racemose on a common axis, rachis produced beyond spikelets; spikelets 1-flowered, laterally compressed, sessile in 2 rows on one side of a narrow, flattened rachis, disarticulating below glumes; glumes unequal, first shorter, second longer than floret, keeled, 1-nerved; lemma firm but thinner than glumes, 1-nerved, or 3-nerved with lateral nerves obscure, keeled, subobtuse; palea as long as or longer than lemma, broadly acute with thin, very wide margins.

**1. Spartina foliosa** Trin., Mém. Acad. St.-Pétersb. VI, **6**$^{2}$**:** 114, 1840

*Spartina leiantha* Benth., Bot. Sulph. 56, 1844.
*Spartina densiflora* subvar. *brongniarti* f. *acuta* St. Yves, Candollea **5:** 76, 81, 1932.

Culms 70–100 cm. tall, erect from comparatively slender, spreading rhizomes, rooting at lower nodes, leafy to summit; sheaths crowded, much longer than internodes, rounded, usually succulent; blades 15–40 cm. long, 5–10 mm. wide at base, usually stiffly ascending, glabrous on both surfaces, margins scaberulous; panicles spikelike, cylindric, 15–20 cm. long, usually partly enclosed in upper sheath; spikes several to numerous, sessile, crowded, at least at summit of panicle, lower ones 5–7 cm. long, rachis glabrous; spikelets distant, at least in lower spikes, closely appressed; first glume one-half to three-fourths as long as floret, keel hirsute in lower half; second glume acute, equaling or exceeding lemma; lemma 8–10 mm. long, rather thin, subobtuse, obscurely lobed, glabrous except for sparsely ciliate margins; palea narrowed at summit, exceeding lemma.

Salt marshes near the coast, San Francisco Bay, California, to Magdalena Bay, Baja California.

## 35. MICROCHLOA R. Br., Prod. 208, 1810

Densely tufted annuals or perennials with short, narrow or involute blades and solitary, slender, arcuate spikes; spikelets 1-flowered, sessile in 2 rows on one side of a narrow, flattened rachis, rachilla disarticulating above glumes; glumes subequal, longer than floret, acute, 1-nerved; floret with a soft, pointed callus; lemma thin, 3-nerved, flabellate, ciliate on margins; palea narrow, a little shorter than lemma, ciliate on keels.

**1. Microchloa kunthii** Desv., Opusc. 75, 1831

*Paspalum tenuissimum* M. E. Jones, Contr. West. Bot. **18:** 24, 1933.

Perennial; culms very slender, erect in small, dense tufts, 10–30 cm. tall; sheaths, except lowermost, much shorter than internodes, scaberulous; ligule ciliate, 1–1.5 mm. long; blades firm, flat or usually folded, with thick, white, scabrous margins, those of culm 1–2.5 cm. long, those of innovations to 6 cm. long, 1–1.5 mm. wide; spike 6–15 cm. long, arcuate, rachis ciliate; spikelets 2.5–3.5 mm. long; lemma 2–2.5 mm. long, pilose on midnerve, margins densely ciliate with hairs about 1 mm. long.

Sterile ground and rocky hills, Mexico to Argentina. Within the desert area collected only in the Sierra de la Laguna, Baja California.

## 36. CYNODON L. Rich. ex Pers., Syn. Pl. 1: 85, 1805

Stoloniferous perennials with short blades and few to several slender, digitate spikes; spikelets 1-flowered, sessile in 2 rows on one side of the narrow, triangular rachis, rachilla disarticulating above glumes, prolonged beyond spikelet in a naked stipe, sometimes bearing a rudimentary floret; glumes subequal, acuminate, 1-nerved, first lunate, second lanceolate; lemma acute, awnless, 3-nerved, pubescent on nerves; palea narrow, acute, equaling lemma.

**1. Cynodon dactylon** (L.) Pers., Syn. Pl. **1:** 85, 1805

*Panicum dactylon* L., Sp. Pl. 58, 1753.

Rhizomatous, stoloniferous, widely creeping; culms wiry, compressed, 10–40 cm. tall; sheaths usually overlapping, keeled, glabrous or sparsely pilose at throat; ligule membranous, minutely erose, 0.2–0.3 mm. long; blades flat, 2–20 cm. long (usually 5–10 cm.), 2–4 mm. wide, scabrous, especially on margins, sometimes sparsely pilose; spikes 4–7, slender, arcuate, 2–7 cm. long; spikelets 2–3 mm. long.

A common weed on open ground in the warmer regions of the world.

## 37. CHLORIS Swartz, Prod. Veg. Ind. Occ. 25, 1788

Annual or perennial, often stoloniferous grasses, with 2–several digitate, verticillate, or subracemose spikes; spikelets subsessile in 2 rows on one side of rachis, with 1 perfect floret, rachilla articulate above glumes, prolonged beyond fertile floret, and bearing a narrow or club-shaped rudiment composed of 1 or more reduced sterile lemmas; glumes narrow, 1-nerved (rarely 3-nerved), second a little longer than the first; fertile lemma 3-nerved, awnless (*C. submutica*) or awned from the back just below the usually acute tip, callus more or less bearded on sides, keel glabrous or sparsely pilose, margins usually ciliate, hairs on upper part often much longer than the rest.

Plants annual; rudiment broad, truncate, conspicuous; fertile lemma conspicuously ciliate; cilia as much as 4 mm. long ........................................ 1. *C. virgata*

Plants perennial; rudiment narrow, acute, not conspicuously ciliate:

- Awn of fertile lemma 7–15 mm. long; plants with large cleistogamous spikelets borne on slender, branching underground stems ............................ 2. *C. chloridea*
- Awn of fertile lemma less than 1 mm. long; plants without underground spikelets ........ ........................................................ 3. *C. brandegeei*

**1. Chloris virgata** Swartz, Fl. Ind. Occ. **1:** 203, 1797

*Chloris elegans* H. B. K., Nov. Gen. & Sp. **1:** 166, 1816.

*Chloris alba* Presl, Rel. Haenk. **1:** 289, 1830.

Annual; culms erect, or usually decumbent, sometimes rooting at lower nodes, 10 cm. to more than 1 m. tall; sheaths compressed-keeled, glabrous, mostly longer than internodes; ligule very short, minutely ciliate; blades flat, 4–25 cm. long, 3–7 mm. wide, scaberulous or nearly smooth on both surfaces, sometimes more or less papillose-pilose on upper surface near base, margins scabrous; spikes 3–15 (usually about 6), 3–9 cm. (usually 5–6 cm.) long, erect or narrowly ascending, flexuous, crowded in a rather dense, silvery head; spikelets 3–3.5 mm. long; glumes acuminate, 1-nerved, scabrous on keel, first 1.5–2.5 mm. long, second 2.5–3 mm. long, with an awn about 1 mm. long; fertile floret 3–3.5 mm. long, callus bearded, lemma acute, keel bowed out at middle, short-ciliate on margins on lower part, long-ciliate on upper third, hairs to 4 mm. long, awn rather stout, 6–10 mm. long, scabrous; rudiment 2–2.5 mm. long, about 0.7 mm. wide, truncate, awn 5–7 mm. long.

A weed in fields and waste places, southwestern and central United States to Panama and the West Indies; Venezuela to Argentina; in the Old World.

**2. Chloris chloridea** (Presl) Hitchc., Proc. Biol. Soc. Wash. **41**: 162, 1928

*Dineba chloridea* Presl, Rel. Haenk. **1**: 291, 1830.
*Gymnopogon longifolius* Fourn., Mex. Pl. **2**: 144, 1886.
*Chloris clandestina* Scribn. & Merr., U.S. Dept. Agr. Div. Agrost. Bull. **24**: 25, 1901.

Perennial; culms tufted, erect, 50–110 cm. tall, bearing cleistogamous spikelets on slender branches of underground stems; sheaths longer than internodes, glabrous, keeled, keel sometimes obscure; ligule very short, ciliate; blades flat, attenuate-pointed, narrowed toward base, to 30 cm. long, 5–9 mm. wide, glabrous on both surfaces, margins scabrous; spikes 5–13 (usually less than 10), 8–18 cm. (usually about 12 cm.) long, stiffly ascending or spreading, approximate or relatively distant on the axis; spikelets 6–7 mm. long, appressed; glumes acute, 1-nerved, first 1–1.5 mm. long, second broader, 2–3.5 mm. long; fertile floret 6–7 mm. long, lemma acuminate, callus bearded, more or less hispid on margins in upper half, otherwise glabrous, awn 7–15 mm. long, scabrous; rudiment very narrow, about 2 mm. long, awn 4–8 mm. long, scabrous; cleistogamous spikelets solitary, 6–7 mm. long, terminating slender branchlets.

Open ground and waste places, Arizona and Texas; Baja California and Sonora to Tamaulipas and Oaxaca; Honduras.

**3. Chloris brandegeei** (Vasey) Swallen, Amer. Jour. Bot. **22**: 41, 1935

*Diplachne brandegei* Vasey, in Brandegee, Proc. Calif. Acad. II, **2**: 213, 1889.
*Gouinia brandegei* Hitchc., U.S. Dept. Agr. Bur. Pl. Ind. Bull. **33**: 21, 1903.

Perennial; culms tufted, erect, 40–105 cm. tall; sheaths compressed-keeled, lower ones elongate, pubescent or loosely pilose with long hairs, felty-pubescent at base, upper ones usually shorter than internodes, glabrous on back, margins more or less appressed-pilose, with a densely pubescent line on the collar; blades flat, or more or less folded on innovations, to 40 cm. long, 4–7 mm. wide, or narrower on innovations, finely pubescent or pilose on both surfaces with a dense line of long hairs just above ligule; ligule 0.5 mm. long, minutely erose or ciliate; spikes 5–10, 5–12 cm. long, ascending, in small, rather distant fascicles, densely pilose at very base and in axils; spikelets relatively distant, about 6 mm. long, appressed; glumes 1-nerved, first acute, 1–2 mm. long, second broader, acuminate, 2–4 mm. long; fertile florets, 4.5–6 mm. long, callus rather densely bearded on sides, lemma sparsely pilose or nearly glabrous, awn not more than 1 mm. long, scabrous; rudiment very narrow, 2 mm. long, awn 0.5 mm. long, scabrous.

Among rocks, Carmen Island, Purísima, and San Ignacio, south to the Cape Region, Baja California.

## 38. TRICHLORIS Fourn., Mex. Pl. 2: 142, 1886

Leafy, tufted perennial with several narrowly ascending spikes crowded on a short axis; spikelets 2-flowered, uppermost reduced, subsessile on one side of a very narrow rachis; glumes acuminate, 1-nerved, persistent, second longer and a little broader than first; lemma rounded on back, 3-nerved, nerves obscure, 3-awned, lateral awns sometimes short or obsolete; palea comparatively broad, slightly exceeding body of lemma.

**1. Trichloris crinita** (Lag.) Parodi, Rev. Argentina Agron. **14:** 63, 1947

*Chloris crinita* Lag., Var. Cienc. **4:** 143, 1805.
*Chloris mendocina* Phil., An. Univ. Chile **36:** 208, 1870.
*Trichloris mendocina* Kurtz, Mem. Fac. Cienc. Exact. Univ. Cordoba **1896:** 37, 1897.

Culms erect or sometimes decumbent, branching, 40–80 cm. tall; sheaths compressed-keeled, lower ones much longer than internodes, glabrous; ligule ciliate, 1 mm. long; blades 10–25 cm. long, 2–4 mm. wide, conduplicate, attenuate, scabrous, sparsely to densely pilose on upper surface toward base; spikes 10–18, 5–12 cm. long, crowded, closely appressed; spikelets hidden by numerous awns; glumes acute, first about 1 mm. long, short-awned, second 1.5 mm. long, awn 1.5–3 mm. long; lemma of fertile floret 3 mm. long, 3 slender awns 10 mm. long, scabrous; awns of rudimentary floret shorter than those of fertile lemma.

Plains and rocky hills, Arizona to western Texas and northern Mexico.

## 39. BOUTELOUA Lag., Var. Cienc. 4: 134, 1805

Annual or perennial, usually tufted grasses with slender culms and 1–many short one-sided spikes solitary or racemose on a short or often elongated axis; spikelets 1–many in each spike, with 1 fertile floret and 1 or 2 rudimentary florets above it, second floret often staminate and sometimes perfect; glumes 1-nerved (3–5-nerved in *B. nervata*), first narrow, second much broader, twice as long as first, or both broad, similar and subequal; fertile lemma 3-nerved, glabrous or bearded at base, glabrous or pilose on margins and sometimes pilose across back, tip variously lobed or dentate, nerves usually excurrent in short awns; rudiment reduced to 3 (rarely to 1) awns or glumaceous, lobed or dentate with 3 usually conspicuous awns; third floret sometimes present, but very much reduced.

Spikes persistent; spikelets pectinate:
  Plants annual:
    Second glume papillose-hispid on keel; rachis papillose-hispid, ciliate ...... 1. *B. parryi*
    Second glume scabrous on keel; rachis not ciliate ...................... 2. *B. barbata*
  Plants perennial:
    Rachis produced beyond spikelets, pointed; second glume tuberculate-hispid ....... ....................................... 3. *B. hirsuta*
    Rachis terminating in a spikelet, this often rudimentary; second glume glabrous, scabrous, or pubescent (sparsely papillose-pilose in *B. gracilis*):
      Spikes normally 2, rarely 1 or 3 .................................. 4. *B. gracilis*

Spikes 3–8:
  Culms felty-pubescent .......... 5. *B. eriopoda*
  Culms glabrous:
    Plants stoloniferous .......... 6. *B. sonorae*
    Plants not stoloniferous:
      Culms branching; awns of fertile lemma 2–3 mm. long .......... 7. *B. rothrockii*
      Culms simple; awns of fertile lemma about 5 mm. long .......... 8. *B. trifida*

Spikes deciduous; spikelets not pectinate:
 Plants annual:
  Spikes narrow, abruptly spreading or reflexed above; rachis sharp-pointed at base; spikelets appressed .......... 9. *B. aristidoides*
  Spikes broad, usually ascending; rachis not sharp-pointed at base; spikelets spreading .......... 10. *B. annua*
 Plants perennial:
  Spikes usually fewer than 15:
   Second glume hairy:
    Spikes 3–8, rhomboid; margins of rachis densely ciliate; spikelets obscurely pectinate .......... 11. *B. chondrosioides*
    Spikes 10–15, triangular; margins of rachis not conspicuously ciliate; spikelets not pectinate .......... 12. *B. eludens*
   Second glume glabrous or scabrous, sometimes hispid on keel but not hairy over the back:
    Base of plants hard and rhizomatous; sheaths usually broad and conspicuous .......... 13. *B. radicosa*
    Base of plants comparatively soft, not rhizomatous; sheaths inconspicuous .......... 14. *B. filiformis*
  Spikes 15–50, usually more than 20 .......... 15. *B. curtipendula*

## 1. **Bouteloua parryi** (Fourn.) Griffiths, Contr. U.S. Nat. Herb. **14:** 381, 1912

*Chondrosium parryi* Fourn., Mex. Pl. **2:** 150, 1886.

Annual; culms erect, or usually geniculate at lower nodes, 15–60 cm. (usually 30–40 cm.) tall, freely branched; sheaths shorter than internodes, glabrous or sparsely pilose or papillose-pilose; blades flat, 4–15 cm. (usually less than 10 cm.) long, 1–3 mm. wide, usually pilose, margins sparsely papillose-pilose; spikes 4–7, fewer on branches, ascending or spreading, straight or flexuous, 1–3.5 cm. long, peduncle papillose-hispid, margins papillose-hispid-ciliate; first glume acute, hyaline, glabrous, about 2.5 mm. long, second broader, 3 mm. long, papillose-hispid on keel, with an awn 2 mm. long; lemma 3 mm. long, villous over lower half of back, awns about 3 mm. long; rudiment about 2 mm. long, bearded at base, lobes obtuse, awns 3–4 mm. long, usually with a second much reduced club-shaped, awnless floret.

Mesas, washes, and gravelly slopes, Arizona, New Mexico, and northern Mexico.

## 2. **Bouteloua barbata** Lag., Var. Cienc. **4:** 141, 1805

Annual; culms slender, branching, geniculate, decumbent-spreading, mostly 10–30 cm. tall; sheaths glabrous, shorter than internodes; ligule ciliate, nearly 1 mm. long; blades flat, acuminate, 1–6 cm. long, 1–2 mm. wide, scabrous above, glabrous below; spikes 4–6 (sometimes only 3), 1–3 cm. long, ascending to spreading; first glume hyaline, acuminate, 1.5–2 mm. long, second acute or acuminate, sometimes subobtuse and mucronate, 2–3 mm. long, scabrous; lemma 2–4 mm. long, densely villous at least on margins, and often over entire back, awns from shorter than lobes to 3 mm. long;

rudiment more or less bearded at base, deeply cleft, lobes obtuse, awns about as long as those of lemma.

Dry open ground and rocky hills, southeastern California to southern Utah, Texas, and Mexico; Argentina.

**3. Bouteloua hirsuta** Lag., Var. Cienc. **4:** 141, 1805

Perennial; culms tufted, erect, 20–60 cm. tall, scaberulous; sheaths glabrous or scaberulous, more or less papillose-hispid at throat; blades firm, flat, acuminate, 5–15 cm. long, 1–3 mm. wide, sparsely papillose-ciliate; spikes 1–3 (usually 2), mostly 2–4 cm. long, rarely shorter, ascending or spreading, rachis prolonged beyond spikelets in a stiff point; first glume narrow, acuminate, hyaline, about 2.5 mm. long, second broader, acuminate, awn-pointed, pubescent; keel prominently tuberculate-hispid; lemma 5 mm. long, sparsely bearded at base, pilose on margins and lower part of midnerve, 3-lobed, lobes mucronate, middle one somewhat longer than lateral ones; rudiment glabrous or sometimes shortly bearded at base, deeply cleft, awns 2–4 mm. long, with broad scales between them.

Dry prairies, rocky hills, and mesas, South Dakota to Wisconsin, south to Texas, southern California, and central Mexico. Mostly above the Sonoran Desert.

**4. Bouteloua gracilis** (H. B. K.) Lag. ex Steud., Nom. ed. 2, **1:** 219, 1840
*Chondrosium gracile* H. B. K., Nov. Gen. & Sp. **1:** 176, *pl. 58,* 1816.

Perennial; culms densely tufted, erect, 15–60 cm. tall, glabrous; leaves mostly crowded toward base; sheaths glabrous, hispid at throat; blades firm, flat at base, involute-attenuate, curved or flexuous, 5–10 cm. (rarely to 15 cm.) long, 1–2 mm. wide, glabrous or sometimes sparsely pilose, margins scabrous; spikes 1–3, usually 2, ascending or spreading, straight or somewhat curved, 2–4.5 cm. long (mostly 3–4 cm.); first glume narrow, acuminate, 3 mm. long, second broader, acuminate, 4 mm. long, very sparsely papillose-hispid on keel, otherwise glabrous; lemma 4 mm. long, bearded at base, long-pilose on both sides of midnerve, margins short-ciliate, tip shallowly lobed, central awn about 1 mm. long, lateral ones a little longer; rudiment 2 mm. long, densely bearded at base, lobes broad, cucullate, awns about 3 mm. long, an additional awnless rudimentary floret sometimes developed.

Prairies and open ground, Manitoba and Alberta to Wisconsin, to southern California, Texas, and Mexico.

**5. Bouteloua eriopoda** (Torr.) Torr., Pacif. R. Rep. **4:** 155, 1857
*Chondrosium eriopodum* Torr., in Emory, Notes Mil. Rec. 154, 1848.

Perennial; culms wiry, straggling, sparsely stoloniferous, geniculate at nodes, felty-pubescent, base rather hard and knotty; upper sheaths shorter than internodes, glabrous, lowermost crowded, felty-pubescent; blades flat or loosely rolled, attenuate, mostly 5–10 cm. long, 1–2 mm. wide, glabrous; spikes 3–5, mostly 2–5 cm. long, somewhat curved, ascending or sometimes spreading; spikelets rather loosely arranged, nearly appressed to rachis; glumes acuminate, first narrow, 3 mm. long, second broader, 7 mm. long, glabrous; lemma 6–7 mm. long, bearded at base, sparsely pilose on margins and midnerve, acuminate, gradually narrowed into the awn, the awn 3–4 mm. long, lateral awns very short; rudiment short-bearded at top of slender rachilla joint, almost reduced to 3 slender, scabrous awns about 7 mm. long.

Dry gravelly hills, plains, and open ground, Arizona to southern Colorado and Utah, Texas, and northern Mexico.

**6. Bouteloua sonorae** Griffiths, Contr. U.S. Nat. Herb. **14:** 389, 1912

Stoloniferous perennial; culms tufted, branching, geniculate-spreading, or usually solitary, erect from distant nodes of stolons, glabrous; ligule ciliate, 1–1.5 mm. long; blades flat or loosely involute, 2–6 cm. long, 1–2 mm. wide, scabrous on upper surface, glabrous on lower; spikes 4–6, 1–2 cm. long, ascending or spreading, curved; glumes acute, glabrous, or scabrous on keel, first 1 mm. long, second 2 mm. long; lemma 2.5 mm. long, deeply cleft for one-third its length, lobes obtuse or subacute, minutely erose, awns from between the lobes about 1.5 mm. long, scabrous; rudiment cleft to base, intermediate lobes 1 mm. long, obtuse, awns 2–3 mm. long, rachilla with a tuft of spreading hairs at summit.

Gravelly or stony ground near the sea, southern Sonora and Sinaloa.

**7. Bouteloua rothrockii** Vasey, Contr. U.S. Nat. Herb. **1:** 268, 1893

Perennial, or sometimes flowering first year and resembling an annual; culms tufted, erect or rarely decumbent, branching, mostly 20–60 cm. tall (rarely less), glabrous; sheaths shorter than internodes, glabrous; ligule ciliate, 1 mm. long; blades flat, acuminate, 5–15 cm. long, 1–3 mm. wide, scabrous, margins papillose-ciliate toward base; spikes 4–8, appressed to spreading, 1.5–4 cm. long, rachis scabrous; first glume obtuse, hyaline, 1.5 mm. long; second glume broader, 3 mm. long, pubescent, short-awned from between short, irregular teeth; lemma 3 mm. long, densely villous with spreading hairs on lower part of back and margins, awns 2–3 mm. long; rudiment 1.5 mm. long, bearded at base, lobes cucullate, awns about 3 mm. long.

Mesas, meadows, and rocky hills, southern California, Arizona, and northern Mexico.

**8. Bouteloua trifida** Thurb., in S. Wats., Proc. Amer. Acad. **18:** 177, 1883

Perennial; culms densely tufted, erect, 10–35 cm. (mostly about 20 cm.) tall, scaberulous; sheaths glabrous or scaberulous, lower ones longer, upper shorter, than internodes; ligule ciliate, about 0.5 mm. long; blades flat, acute, scabrous, 0.8–5 cm. long, 0.5–1 mm. wide; spikes 3–6, 1–3 cm. long, appressed to spreading; spikelets spreading or nearly appressed; glumes acuminate, glabrous, first 3.5 mm. long, second 4.5 mm. long, scarcely broader than first; body of lemma about 2 mm. long, glabrous, or pubescent toward base, 3 slender lobes gradually narrowed into awns, these about 5 mm. long; rudiment cleft to base, lobes gradually narrowed into awns about as long as those of fertile lemma.

Dry prairies, mesas, and rocky slopes, Arizona to southwestern Utah, Texas, and northwestern Mexico.

**9. Bouteloua aristidoides** (H. B. K.) Griseb., Fl. Brit. W. Ind. 537, 1864
*Dinebra aristidoides* H. B. K., Nov. Gen. & Sp. **1:** 171, 1816.

Annual; culms slender, branching, decumbent-spreading, geniculate, 10–40 cm. (usually 15–25 cm.) tall, glabrous; sheaths much shorter than internodes, glabrous, hispid at mouth; ligule ciliate, about 0.5 mm. long; blades flat or drying involute, attenuate-pointed, 1–6 cm. or rarely to 10 cm. long, 1–2 mm. wide, glabrous below, scabrous and sparsely pilose above; spikes few to several, 1–2 cm. long, abruptly spreading or reflexed, falling entire, sharp-pointed base densely pubescent with stiff, appressed hairs; spikelets 2–6, appressed, distant, lowest one without 3-awned rudiment; first glume narrow or subulate, 1-nerved, 1–3 mm. long, second broader, 3-nerved, 5–6 mm.

long; lemma acuminate, about 5 mm. long, more or less pilose, minutely toothed; awns of rudimentary floret about 5 mm. long, somewhat spreading.

Deserts, dry mesas, and rocky hillsides, southern California to Texas and northern Mexico; Peru and Brazil to Argentina.

**10. Bouteloua annua** Swallen, Jour. Wash. Acad. **25:** 414, 1935

Annual; culms in dense clumps, branching, erect or geniculate-spreading at base, 3–25 cm. tall; sheaths shorter than internodes, glabrous; ligule ciliate, 0.2 mm. long; blades flat, 1–3.5 cm. long, 1.5–2 mm. wide, acute, pubescent on upper surface or nearly glabrous; spikes 2–7, falling entire, rather distant, ascending or spreading, 1.5–2 cm. long, bearing 4–7 usually appressed, green or purple spikelets; glumes acuminate, scabrous on keel, first narrow, 3.5–4 mm. long, second much broader, 6–7 mm. long; lemma 7–9 mm. long, 3-toothed, densely pubescent at base, pilose in lines on both sides of midnerve and margins; rudiment cleft to base, lobes with their awns 5–7 mm. long, sparsely appressed-pilose in lines along nerves, rachilla appressed-pilose on margins.

Rocky hillsides, central Baja California.

**11. Bouteloua chondrosioides** (H. B. K.) Benth. ex S. Wats., Proc. Amer. Acad. **18:** 179, 1883

*Dinebra chondrosioides* H. B. K., Nov. Gen. & Sp. **1:** 173, *pl. 53,* 1816.

Perennial; culms tufted, erect, 15–55 cm. tall, glabrous; leaves clustered toward base, sheaths glabrous, hispid at throat; blades flat, firm, 4–12 cm. (rarely more than 10 cm.) long, 2–3 mm. wide, glabrous or sparsely hispid, margins often papillose-ciliate toward base; ligule ciliate, about 0.5 mm. long; spikes 3–8, falling entire, ascending or spreading, rhomboid, tinged with purple, 1–1.5 cm. (rarely to 2 cm.) long, rachis densely long-ciliate on margin; spikelets crowded, appressed or more or less pectinately arranged; first glume 3–5 mm. long, acuminate, ciliate on keel; second glume 7–8 mm. long, acuminate, hairy; lemma about 7 mm. long, hairy, especially on midnerve and margins, tip 3-toothed, awnless; awns of the rudimentary floret 5–10 mm. long, scabrous.

Rocky hills, southern Arizona, western Texas, and Mexico; Honduras and Costa Rica.

**12. Bouteloua eludens** Griffiths, Contr. U.S. Nat. Herb. **14:** 401, 1912

Perennial; culms densely tufted, erect, 25–70 cm. tall; sheaths glabrous, lower ones crowded, upper ones shorter than internodes; blades firm, flat, attenuate, somewhat flexuous, 5–15 cm. long, 1–2.5 mm. wide, glabrous or pilose on upper surface, margins scabrous; spikes 10–15, about 1 cm. long, roughly triangular, spreading or drooping, falling entire, rachis densely short-pubescent; spikelets 3–6, crowded; glumes acuminate, hirsute, first narrow, about 5 mm. long, second broader, 7–8 mm. long; lemma 6–7 mm. long, 3-toothed, glabrous on lower part of back, hairy on margins and toward summit; second floret often staminate, with 4 slender teeth, awned from between the teeth, awns flattened, stout, hairy, 5–6 mm. long; a rudimentary floret, similar to the second but much reduced, often developed.

Rocky hills, Arizona and Sonora, apparently rare.

**13. Bouteloua radicosa** (Fourn.) Griffiths, Contr. U.S. Nat. Herb. **14:** 411, 1912

*Atheropogon radicosus* Fourn., Mex. Pl. **2:** 140, 1886.

Perennial; culms densely tufted, from a hard, rhizomatous base, simple or with short branches appearing late from lower nodes, 25–80 cm. tall, glabrous; sheaths glabrous, hispid at throat, lower ones crowded, upper ones much shorter than internodes; blades flat, attenuate, flexuous or curled toward tip, 5–15 cm. long, 2–3 mm. wide, margins papillose-ciliate toward base, sometimes papillose only; spikes 5–15, 1–3 cm. long, ascending or spreading, sometimes reflexed, falling entire, rachis ciliate; spikelets bearded at base, ascending; glumes broad, acute, glabrous, first 5–6 mm. long, second 7–8 mm. long; lemma about 8 mm. long, glabrous, 3-toothed, teeth mucronate or awn-tipped; rudiment 5–6 mm. long, usually staminate, teeth awned, 4–8 mm. long, central one longer than lateral.

Rocky hills and canyons, southern California to New Mexico and Mexico, usually at elevations above the Sonoran Desert.

**14. Bouteloua filiformis** (Fourn.) Griffiths, Contr. U.S. Nat. Herb. **14:** 413, 1912
*Atheropogon filiformis* Fourn., Mex. Pl. **2:** 140, 1886.

Perennial; culms in small, dense tufts, simple or branching, erect or decumbent, spreading at base, 15–50 cm. tall, glabrous; sheaths glabrous or sparsely pilose; blades flat, acuminate, mostly 5–10 cm. long, 1–2 mm. wide, glabrous or sparsely pilose, margins scabrous, usually more or less papillose-ciliate; ligule ciliate, less than 1 mm. long; spikes 5–10, ascending or spreading, mostly 1–2 cm. long, falling entire, rachis shortly ciliate, densely pubescent at base; glumes acuminate, subequal or second a little longer, mostly 5–7 mm. long, glabrous, keel scabrous to hispid; lemma about 7 mm. long, glabrous or bearded at base, acuminate, awns about 1 mm. long, scabrous; upper floret usually staminate, sometimes perfect, glabrous or bearded at base, lemma minutely bifid, central awn thickish, 3–8 mm. long, lateral ones more slender, a little shorter than central.

Mesas and rocky foothills, Arizona to Texas and Guatemala; Panama, Colombia, and Venezuela.

**15. Bouteloua curtipendula** (Michx.) Torr., in Emory, Notes Mil. Rec. 154, 1848
*Chloris curtipendula* Michx., Fl. Bor. Amer. **1:** 59, 1803.

Perennial; culms erect from short rhizomes, 30 cm. to more than 1 m. tall; sheaths mostly shorter than internodes, rounded, glabrous or sparsely papillose-pilose, more or less pilose at throat; ligule 0.5 mm. long, ciliate; blades to 25 cm. long, 2–5 mm. wide, flat, attenuate, scabrous, especially on margins, upper surface sometimes pilose; spikes mostly 20–40, 0.5–2 cm. long, spreading or reflexed, falling entire; spikelets crowded, appressed or somewhat spreading; first glume about 4 mm. long, narrow, acuminate, scabrous on keel; second glume 6–8 mm. long, much broader than first, acute, scabrous; lemma 5–6 mm. long, glabrous or sparsely pilose, tip 3-toothed; rudiment from nearly obsolete to 5 mm. long and rather broad, awned from between teeth of bifid apex, awn mostly 3–6 mm. long, lateral nerves excurrent in slender awns, from minute to 4 mm. long, often enclosing a second very small rudiment; spikelets in upper spikes smaller, rudiment inconspicuous or nearly obsolete.

Prairies, rocky hills, and canyons, Montana to Maine, south to southern California, Texas, and South Carolina; Mexico to El Salvador; Colombia to Argentina.

## 40. CATHESTECUM Presl, Rel. Haenk. 1: 294, *pl. 42,* 1830

Annual or perennial grasses, frequently stoloniferous, with relatively short, flat blades and 3–10 V-shaped or rhomboid, spreading spikes evenly arranged on opposite sides of slender, flattened axis; spikes falling entire, consisting of 3 spikelets, lateral ones 2-flowered, staminate or sterile, rarely pistillate, upper floret sometimes much reduced; central spikelet 3-flowered, lower floret usually pistillate, sometimes staminate, very rarely perfect, upper florets staminate or sterile; first glume short, that of central spikelet usually flabellate; second glume about equaling spikelet, acuminate, sometimes minutely lobed and mucronate; lemmas dissimilar, lower ones cleft about one-fourth their length, awns from between lobes equaling or slightly exceeding them, upper ones deeply cleft, awns villous in lower part, extending as much as 3 mm. beyond lobes; in spikes wholly staminate, lemmas all alike; palea nearly equaling lemma, nerves excurrent in short awns.

**1. Cathestecum erectum** Vasey & Hack., Bull. Torrey Club **11:** 37, *pl. 45*, 1884

Stoloniferous perennial, the stolons slender, wiry, conspicuously arching, internodes elongate; culms in small, dense tufts, simple or branching, erect or somewhat geniculate-spreading, 15–30 cm. tall, glabrous; lower sheaths crowded, pilose in throat, lowermost densely villous at base, upper ones more distant, glabrous; blades flat or commonly loosely involute, 3–6 cm. long, 1–1.5 mm. wide, glabrous on lower surface, scabrous and sparsely pilose on upper; spikes usually pale, dimorphous, one form entirely staminate, other with lower floret of central spikelet pistillate, upper floret and those of lateral spikelets staminate or neuter, spikes all of same kind in a single inflorescence, parts of same plant (joined by stolons) sometimes with both kinds of spikes; staminate spikes: first glume 1 mm. long; second glume acuminate, glabrous or nearly so, that of lateral spikelets 3 mm. long, that of central spikelet about 4 mm. long; lemmas similar, 3 mm. long, glabrous, irregularly lobed, awnless or mucronate; pistillate spikes: glumes villous, otherwise like those of staminate spikes; lemmas of lateral spikelets similar, 3 mm. long, glabrous, cleft to middle, awns from between lobes equaling or barely exceeding them, more or less hispid; florets of central spikelet unlike, lower lemma glabrous or sparsely pubescent, cleft one-third of length, awns from between lobes glabrous, subequal or central a little longer, somewhat exceeding lobes; upper florets similar to those of lateral spikelets, awns slightly longer, extending 1–2 mm. beyond lobes; anthers 2 mm. long.

Dry rocky hills and prairies, Arizona, southwestern Texas, northern Mexico, and El Salvador.

## 41. PHALARIS L., Sp. Pl. 54, 1753

Annual with flat blades and erect, dense, spikelike panicles; spikelets laterally compressed, with 1 fertile floret, and 2 much reduced sterile florets below fertile one; fertile lemma coriaceous, shorter than glumes.

**1. Phalaris caroliniana** Walt., Fl. Car. 74, 1788

*Phalaris americana* Ell., Bot. S.C. & Ga. **1:** 101, *pl. 5, fig. 4*, 1816.
*Phalaris occidentalis* Nutt., Trans. Amer. Phil. Soc. n.s. **5:** 144, 1837.

Culms erect, 15–60 cm. tall or sometimes taller, glabrous; blades 3–15 cm. long, occasionally longer, 3–8 mm. wide, acuminate, scabrous; panicles 2–8 cm. long, oblong; glumes 5–6 mm. long, abruptly narrowed to acute apex, narrowly winged from about middle; keel scabrous; fertile lemma 3.5–4 mm. long, appressed-pubescent, sterile lemmas one-third to one-half as long.

Moist places, Oregon and Colorado to Virginia, south to California, Texas, Florida, and northern Mexico.

## 42. **LUZIOLA** Juss. ex Gmel., Syst. Nat. ed. 13, **2:** 637, 1791

Aquatic perennial; monoecious, staminate and pistillate spikelets in separate inflorescences, staminate usually terminal, pistillate from middle and upper nodes; glumes wanting; lemma and palea nearly equal, thin, several- to many-nerved, nerves more prominent in pistillate spikelets; stamens several; stigmas long, plumose; caryopsis free, subspherical, smooth or minutely striate.

**1. Luziola gracillima** Prod., in Mez, Bot. Arch. **1:** 241, 1922

Culms tufted, erect, 30–40 cm. tall, glabrous; sheaths loose, somewhat inflated, longer than internodes, glabrous or papillose, auriculate, auricles united with long ligule; blades 11–22 cm. long, 2–3 mm. wide, acuminate, glabrous, or scabrous on upper surface; panicles 3–7 cm. long; staminate spikelets 5–6.5 mm. long; pistillate spikelets 0.8–1 mm. long, subglobose, scabrous on nerves.

Shallow water and margins of rain pools, Sonora and Jalisco.

## 43. **LEERSIA** Swartz, Prod. Veg. Ind. Occ. 21, 1788

Slender, erect or decumbent perennial with creeping rhizomes and flat, scabrous leaf blades; spikelets 1-flowered, laterally compressed, disarticulating from pedicel; glumes wanting; lemma chartaceous, boat-shaped, hispid, nerves indurated; palea as long as lemma but much narrower, its margins firmly held by margins of lemma.

**1. Leersia oryzoides** (L.) Swartz, Prod. Veg. Ind. Occ. 21, 1788

*Phalaris oryzoides* L., Sp. Pl. 55, 1753.
*Homalocenchrus oryzoides* Poll., Hist. Pl. Palat. **1:** 52, 1776.

Culms weak, 0.5–1.5 m. long, erect or decumbent at base, with slender rhizomes, nodes appressed-pilose; sheaths mostly longer than internodes, strongly retrorsely scabrous; blades acuminate, often elongate, mostly 8–10 mm. wide, very scabrous; panicles terminal and axillary, 10–20 cm. long; flexuous branches slender, scabrous, naked in lower part; spikelets 5 mm. long, abruptly pointed, sparsely hispidulous, hispid-ciliate on keels; axillary panicles usually partly included in sheaths.

Wet ground, marshes, and along rivers and irrigation ditches, Washington to Quebec and Maine, south to southern California, Arizona, and northern Florida; Europe. Introduced in the desert region along irrigation ditches, Fort Lowell, Arizona.

## 44. **TRICHACHNE** Nees, Agrost. Bras. 85, 1829

Perennial, with flat blades and slender, erect or ascending racemes forming a white, silky inflorescence; first glume minute; second glume and sterile lemma equal, covering fruit, conspicuously villous with long hairs; fertile lemma acuminate, brown, with broad, hyaline margins.

**1. Trichachne californica** (Benth.) Chase ex Hitchc., Jour. Wash. Acad. **23:** 455, 1933

*Panicum californicum* Benth., Bot. Sulph. 55, 1844.
*Trichachne saccharata* Nash, in Small, Fl. SE. U.S. 83, 1903.

Culms erect from a swollen, knotty base, 40–100 cm. tall, glabrous; cataphylls densely villous; sheaths longer than internodes, glabrous or sparsely papillose-pilose; blades mostly 6–12 cm. long, 3–5 mm. wide, nearly glabrous to densely minutely pubescent, margins scabrous; panicles mostly 8–15 cm. long, racemes few to several, narrowly ascending or appressed, white-silky, sometimes tinged with violet; spikelets 3–4 mm. long; second glume long-silky across back, sterile lemma long-silky only on margins, hairs often spreading.

Dry open ground, plains, and rocky hills, Arizona to Texas and northern Mexico.

## 45. DIGITARIA Heist. ex Scop., Fl. Carn. ed. 2, **1:** 52, 1772

Decumbent, spreading annuals with flat blades and rather slender, ascending or spreading, digitate racemes; spikelets in pairs on one side of a flat, winged rachis; first glume evident; second glume shorter than nearly equal fruit and sterile lemma; fertile lemma cartilaginous, pale or lead-colored, with hyaline margins.

Second glume two-thirds as long as spikelet; fruit pale .................. 1. *D. adscendens*
Second glume one-half as long as spikelet; fruit lead-colored ............ 2. *D. sanguinalis*

### 1. **Digitaria adscendens** (H. B. K.) Henr., Blumea **1:** 92, 1934

*Panicum adscendens* H. B. K., Nov. Gen. & Sp. **1:** 97, 1815.

Annual; culms erect to widely decumbent-spreading, 10 cm. to as much as 130 cm. long; sheaths sparsely to densely hispid or papillose-hispid, especially toward summit; blades mostly from 4–20 cm. long, 3–8 mm. wide, scabrous and more or less papillose-hispid; racemes 2–9, as much as 15 cm. long, ascending to spreading; spikelets in pairs, 2.8–3.4 mm. long; first glume a small triangular nerveless, glabrous scale, 0.3–0.4 mm. long; second glume about two-thirds as long as spikelet, narrow, 3-nerved, rather densely pilose between nerves; sterile lemma lanceolate, 5-nerved, densely and shortly pilose on margins and between approximate lateral nerves; fruit acuminate, pale.

A common weed in waste and cultivated ground, Massachusetts to Florida and west to California; Mexico and the West Indies to Argentina and Chile; tropics of the Old World.

### 2. **Digitaria sanguinalis** (L.) Scop., Fl. Carn. ed. 2, **1:** 52, 1772

*Panicum sanguinale* L., Sp. Pl. 57, 1753.

Closely resembling *D. adscendens* and often confused with it, but differing in the usually pilose blades, shorter, narrowly oval spikelets 2.8–3 mm. long, and lead-colored fruit.

A common weed, Nova Scotia to Washington south to Texas and Mexico; southern South America; temperate regions around the world.

## 46. LEPTOLOMA Chase, Proc. Biol. Soc. Wash. **19:** 191, 1906

Slender perennial, felty-pubescent at base, with branching, brittle culms, flat leaf blades, and diffuse panicles; first glume minute or obsolete; second glume and sterile lemma nearly equal, appressed-hairy on internerves and margins; fertile lemma elliptic, acute, brown with thin hyaline margins.

### 1. **Leptoloma cognatum** (Schult.) Chase, Proc. Biol. Soc. Wash. **19:** 192, 1906

*Panicum divergens* Muhl. ex Ell., Bot. S.C. & Ga. **1:** 130, 1816. Not *P. divergens* H.B.K., 1815.
*Panicum cognatum* Schult., Mant. **2:** 235, 1824.

Culms erect or usually ascending from a decumbent base, 20–70 cm. tall, branching at lower nodes; sheath mostly longer than rather short internodes, glabrous or softly pilose; blades commonly 2–6 cm. long, 2–4 mm. wide, flat, glabrous or puberulent on upper surface; panicles diffuse, fragile, nearly as broad as long, easily breaking off, flexuous branches stiffly ascending to widely spreading; spikelets solitary on long, capillary pedicels, 2.5–3 mm. long, acute or acuminate.

Rocky slopes and sandy or gravelly plains, Minnesota to New Hampshire, south to Arizona, Texas, Florida, and northern Mexico.

## 47. ERIOCHLOA H. B. K., Nov. Gen. & Sp. 1: 94, *pls. 30–31,* 1816

Erect or decumbent annuals with flat blades and several to numerous erect or ascending racemes, approximate on a common axis; spikelets usually solitary in 2 rows on one side of a narrow rachis; first glume adnate to swollen first joint of rachilla, forming a ringlike callus below second glume; second glume and sterile lemma about equal, acuminate; fruit much shorter than second glume and sterile lemma, minutely rugose, apiculate.

Second glume tapering into a slender awn; spikelets 7–10 mm. long including awn; fruit 3.5 mm. long .................................................. 1. *E. aristata*
Second glume acuminate only or with an awn to 1 mm. long; spikelets 4–5 mm. long; fruit 3 mm. long .................................................. 2. *E. gracilis*

### 1. Eriochloa aristata Vasey, Bull. Torrey Club 13: 229, 1886

*Eriochloa punctata* var. *aristata* M. E. Jones, Contr. West. Bot. **14:** 11, 1912.

Culms erect or ascending at base, branching, 25–80 cm. tall, pubescent below panicle, nodes pubescent; sheaths rounded, glabrous; blades 5–15 cm. long, 5–12 mm. wide, glabrous, or sparsely pilose above, margins scabrous; panicles 8–12 cm. long, composed of several ascending or appressed, overlapping racemes, main axis and rachis of racemes pubescent and pilose; spikelets 7–10 mm. long, appressed-pilose, second glume tapering into an awn, sterile lemma acuminate or short-awned; fruit 3.5 mm. long, apiculate.

Ditches and roadsides, sometimes in partial shade, southeastern California, southern Arizona, and northern Mexico.

### 2. Eriochloa gracilis (Fourn.) Hitchc., Jour. Wash. Acad. 23: 455, 1933

*Helopus gracilis* Fourn., Mex. Pl. **2:** 13, 1886.

Similar to *E. aristata* in habit and appearance; spikelets 4–5 mm. long, appressed-pilose; second glume acuminate, awnless or with an awn to 1 mm. long; sterile lemma a little shorter than glume, acuminate; fruit 3 mm. long, apiculate.

Moist open ground, Arizona to Texas and central Mexico.

## 48. PASPALUM L., Syst. Nat. ed. 10, 2: 855, 1759

Perennials with flat blades and 2–several spikelike racemes, paired or racemose on a common axis; spikelets solitary or paired in 2 rows on one side of rachis; first glume wanting or occasionally present; second glume and sterile lemma equal or subequal, usually covering fruit; fertile lemma indurate, smooth, usually obtuse.

Racemes conjugate or nearly so at summit of culm, usually in pairs; plants markedly stoloniferous:
  Spikelets 2.5–3 mm. long, solitary or occasionally paired; racemes usually 2; sterile lemma glabrous .................................................. 1. *P. distichum*
  Spikelets 3–3.5 mm. long, usually paired; racemes usually 3 or more; sterile lemma usually pubescent .................................................. 2. *P. paucispicatum*

Racemes more than 2, racemosely arranged on axis:
  Spikelets 3–3.2 mm. long:
    Culms decumbent-spreading, rooting at lower nodes ................. 3. *P. pubiflorum*
    Culms usually erect, not rooting at lower nodes ................... 4. *P. hartwegianum*
  Spikelets not more than 2.1 mm. long:
    Spikelets glabrous or nearly so, broadly ovate or obovate ................ 5. *P. virletii*
    Spikelets pubescent, at least on second glume, elliptic or obovate-elliptic:
      Sheaths and blades densely velvety-pubescent, rarely glabrous; spikelets 1.9–2.1 mm. long ................................................ 6. *P. tenellum*
      Sheaths sparsely papillose or papillose-pilose; blades glabrous; spikelets 1.7–1.9 mm. long ................................................ 7. *P. lentiginosum*

**1. Paspalum distichum** L., Syst. Nat. ed. 10, **2:** 855, 1759

*Digitaria paspalodes* Michx., Fl. Bor. Amer. **1:** 46, 1803.
*Paspalum paspaloides* Scribn., Mem. Torrey Club **5:** 29, 1894.

Stolons extensively creeping; rhizomes slender; nodes of stolons bearded; culms erect or ascending, 6–50 cm. tall; sheaths keeled, pilose on margins toward summit; blades 3–12 cm. long, 2–6 mm. wide, flat, acuminate, basally ciliate, sometimes pubescent on upper surface; racemes usually 2, erect or reflexed, 1.5–7 cm. long; rachis 1–1.5 mm. wide, its margins scabrous; spikelets 2.5–3 mm. long, usually solitary, elliptic, abruptly acute; first glume often developed; second glume and sterile lemma equal, glume pubescent, lemma glabrous.

Along ditches and streams, Washington to New Jersey, south to California, Texas, and Florida; Mexico and the West Indies to Argentina and Chile; eastern hemisphere.

**2. Paspalum paucispicatum** Vasey, Contr. U.S. Nat. Herb. **1:** 281, 1893

Similar in habit to *P. distichum*; racemes 2–5, usually 3, spikelets 3–3.5 mm. long, usually in pairs, sterile lemma as well as glume usually pubescent.

Moist sandy ground, southern Sonora to Oaxaca.

**3. Paspalum pubiflorum** Rupr. ex Fourn., Mex. Pl. **2:** 11, 1886

*Paspalum planifolium* Fourn., Mex. Pl. **2:** 10, 1886.

Culms rather robust, 40–100 cm. long, decumbent and often rooting at lower nodes, bearing a few leafy branches, nodes swollen, at least lower ones bearded; sheaths mostly shorter than internodes, rather loose, glabrous or sparsely papillose-pilose; blades 6–25 cm. long, 6–14 mm. wide, acuminate, margins scabrous, sparsely papillose-hispid at base; racemes 2–8, usually 3–5, 2–10 cm. long, rather thick, erect to spreading, with long white hairs in the axils; rachis 1.2–2 mm. wide, scabrous on margins; spikelets usually in pairs, crowded, 3–3.2 mm. long, obovate, obtuse; second glume and sterile lemma subequal, glume softly pubescent with spreading hairs, lemma minutely appressed-pubescent.

Moist open ground, along streams and ditches, and in low woods, Texas and Louisiana; Baja California and Sonora to Oaxaca; Cuba.

**4. Paspalum hartwegianum** Fourn., Mex. Pl. **2:** 12, 1886

*Paspalum buckleyanum* Vasey, Bull. Torrey Club **13:** 167, 1886.

Culms solitary or in hard clumps, 50–150 cm. tall, ascending, decumbent, branching at lower nodes, glabrous; sheaths glabrous, overlapping; blades 10–35 cm. long,

2–6 mm. wide, firm, ascending, margins scabrous; racemes 3–13, commonly 4–7, 2–9 cm. long, ascending, with a few long hairs in axils, rachis 1–1.5 mm. wide, minutely scabrous; spikelets 3 mm. long, in pairs, elliptic, apiculate; second glume and sterile lemma sparsely to rather densely pubescent.

Wet prairies, alkaline meadows, and along irrigation ditches, southern Texas; Sonora to Oaxaca.

**5. Paspalum virletii** Fourn., Mex. Pl. **2:** 12, 1886

Culms tufted, erect, 30–90 cm. tall, glabrous; sheaths compressed, longer or shorter than internodes, pilose or papillose-pilose; blades 6–30 cm. long, 5–14 mm. wide, acuminate, narrowed toward base, flat, lax, softly pilose; racemes 4–8, spreading, rather distant, 3–7 cm. long; rachis slender with a few long hairs at base; spikelets 2 mm. long, in pairs, broadly ovate, strongly plano-convex; second glume thin, a little shorter than fruit, minutely pubescent; sterile lemma as long as fruit, glabrous.

Sandy soil, Arizona, Sonora, and San Luis Potosí.

**6. Paspalum tenellum** Willd., Enum. Hort. Berol. 39, 1809

*Paspalum elegans* Flügge, Monogr. Pasp. 183, 1810.

Culms 30–150 cm. tall, suberect from a knotty base, simple or branching at lower nodes, compressed, glabrous; sheaths mostly shorter than elongate internodes, sparsely to densely velvety-pubescent, rarely glabrous; blades 8–25 cm. long, 8–21 mm. wide, flat, velvety-pubescent to nearly glabrous, margins scarious; panicles long-exserted, with 3–17, commonly 5–9, ascending to spreading racemes, lower 2–8 cm. long; rachis 1 mm. wide, with a few long hairs at base; spikelets 1.9–2.1 mm. long, in pairs on slender pedicels, obovate-elliptic; second glume and sterile lemma equal, glume densely papillose-villous, lemma pubescent with shorter hairs.

Open moist ground, depressions in rocks, and along ditches and roadsides, Sonora to El Salvador; Ecuador to Brazil.

**7. Paspalum lentiginosum** Presl, Rel. Haenk. **1:** 218, 1830

Perennial; culms in small tufts, erect or ascending, 75–140 cm. tall, glabrous, compressed; sheaths loose, keeled, mostly overlapping, glabrous or sparsely papillose or papillose-hispid, more or less hairy on collar; blades 12–25 cm. long, 8–14 mm. wide, flat, with a few long hairs at base, margins scabrous; racemes mostly 5–13, ascending or spreading, lower 5–7 cm. long, rachis slender with a few long hairs at base; spikelets 1.7–1.9 mm. long, in pairs on slender pedicels, subhemispheric or broadly elliptic; second glume and sterile lemma equal, just covering fruit, glume finely papillose-pubescent, lemma nearly glabrous.

Moist open ground, Sonora to Guatemala.

## 49. PANICUM L., Sp. Pl. 55, 1753

Annuals or perennials, very variable in habit, size, and appearance; spikelets paniculate or sometimes short-pedicellate in one-sided racemes; first glume short to nearly as long as the spikelet; second glume and sterile lemma nearly equal, lemma often enclosing a palea; fertile floret indurate, nerves obsolete, margins inrolled, firmly enclosing palea.

Plants annual:
Fruit transversely rugose; spikelets short-pedicellate, obscurely arranged in spikelike racemes:
Spikelets glabrous, strongly reticulate-veined, usually dark brown:
Blades glabrous except on scabrous margins; spikelets usually dark brown, 2.1–2.5 mm. long .............................. 2. *P. fasciculatum* var. *fasciculatum*
Blades pubescent; spikelets pale, 2.6–3 mm. long .................................................................. 2a. *P. fasciculatum* var. *reticulatum*
Spikelets pubescent, reticulate-veined only at apex, green ............ 3. *P. arizonicum*
Fruit smooth and shining; spikelets in open panicles:
First glume short, truncate; plants glabrous or nearly so ......... 4. *P. dichotomiflorum*
First glume one-third to three-fourths as long as spikelet, acute or acuminate; plants, at least the sheaths, usually papillose or papillose-hispid:
First glume subacute or blunt, about one-third as long as spikelet .................................................................. 8. *P. stramineum*
First glume acute or acuminate, more than half the length of spikelet:
Spikelets 4–4.5 mm. long ................................ 6. *P. pampinosum*
Spikelets not more than 3.3 mm. long:
Panicles more or less drooping; spikelets usually brown ........ 7. *P. sonorum*
Panicles erect; spikelets green ................................ 5. *P. hirticaule*
Plants perennial:
Spikelets short-pedicellate along one side of short, appressed racemes:
First glume obtuse or truncate, one-third as long as spikelet; plants not stoloniferous .................................................................. 1. *P. geminatum*
First glume nearly as long as spikelet; plants with long, widely spreading stolons .................................................................. 9. *P. obtusum*
Spikelets in open or contracted panicles:
Spikelets densely villous, 6–7 mm. long; nodes densely bearded ...... 10. *P. urvilleanum*
Spikelets glabrous, 3.5–5 mm. long; nodes glabrous .................. 11. *P. virgatum*

**1. Panicum geminatum** Forsk., Fl. Aegypt. 18, 1775
*Panicum glomeratum* Buckl., Prel. Rep. Geol. Agr. Surv. Tex. App. 3, 1866.

Perennial; culms tufted, to 1 m. tall, erect or decumbent, glabrous, soft but scarcely succulent; sheaths glabrous, overlapping, often shorter than internodes; blades 10–20 cm. long, 3–6 mm. wide, rather stiffly erect or spreading, finely scabrous on upper surface; panicles 10–30 cm. long, short-exserted or partly enclosed in sheath; racemes appressed, lower ones distant, 2.5–3 cm. long, upper ones approximate, gradually shorter; spikelets 2.2–2.5 mm. long, subsessile, turgid, rather abruptly acute, glabrous; first glume clasping, obtuse or truncate, about one-third as long as spikelet; fruit 2.2 mm. long, abruptly pointed, strongly transversely rugose.

Moist or wet ground, mostly near the coast, southern Texas and Florida; Mexico and the West Indies to Peru and Brazil; warmer parts of the Old World.

**2. Panicum fasciculatum** Swartz, Prod. Veg. Ind. Occ. 22, 1788, var. **fasciculatum**
*Panicum fuscum* Swartz, op. cit. 23.
*Panicum fastigiatum* Poir., in Lam., Encycl. Suppl. **4:** 277, 1816.

Annual; culms 15 cm. to more than 1 m. tall, erect or decumbent, branching from lower nodes, scabrous just below panicle, otherwise glabrous or sparsely hispid; blades 4–30 cm. long, 6–20 mm. wide, rather thin, glabrous, margins scabrous; panicles 5–15 cm. long, branches scabrous, stiffly and rather narrowly ascending; spikelets 2.1–2.5 mm. long, rarely 3 mm., pale or usually dark brown, obovate, glabrous; first glume one-third as long as spikelet, clasping, broadly acute; second glume and sterile lemma pointed beyond fruit, strongly transversely veined; fruit obscurely apiculate, rather strongly transversely rugose.

Open sandy ground and waste places, southern Florida and southern Texas, Mexico, and the West Indies to Argentina.

**2*a*. Panicum fasciculatum** var. **reticulatum** (Torr.) Beal, Grasses N. Amer. **2:** 117, 1896

*Panicum reticulatum* Torr., in Marcy, Expl. Red Riv. 299, 1852.

Differs from var. *fasciculatum* in being usually smaller with narrower, pubescent blades, shorter, more compact panicles, and pale spikelets 2.6–3 mm. long.

Prairie fields and waste ground, Arizona to Arkansas, Louisiana, and Mexico.

**3. Panicum arizonicum** Scribn. & Merr., in Merr., U.S. Dept. Agr. Div. Agrost. Circ. **32:** 2, 1901

Annual; culms 20–60 cm. tall, erect or decumbent at base, branching at lower nodes, scabrous and papillose-hispid below panicle; sheaths glabrous to papillose-hispid, rather loose; blades 5–15 cm. long, 5–12 mm. wide, glabrous or scabrous, sometimes sparsely papillose-hispid, white, scabrous margins more or less papillose-hispid-ciliate toward base; panicles as much as 20 cm. long, branches scabrous, sparsely hispid, stiffly ascending; spikelets 3.5–3.8 mm. long, paired, appressed, glabrous or pubescent; first glume about half as long as spikelet, clasping, broadly acute; second glume and sterile lemma pointed beyond fruit, obscurely reticulate above; fruit apiculate, minutely transversely rugose.

Sandy ground and open rocky slopes, southern California to Texas and northern Mexico.

**4. Panicum dichotomiflorum** Michx., Fl. Bor. Amer. **1:** 48, 1803

Annual; culms erect or geniculate-ascending, mostly 50–100 cm. tall, branching, glabrous; sheaths usually longer than internodes, glabrous, or ciliate toward summit; blades 10–50 cm. long, 3–20 mm. wide, glabrous, or sparsely pilose on upper surface; panicles commonly 10–40 cm. long, terminal and axillary, usually included at base, branches rather stiffly ascending; spikelets about 2.5 mm. long, acute, green or purplish, glabrous; first glume about one-fifth as long as spikelet, broadly triangular or truncate; second glume and sterile lemma equal, somewhat pointed beyond fruit.

Moist ground and waste places, Nebraska to Maine, south to Texas and Florida, sparingly introduced westward. In the desert area found only at Tempe, Arizona.

**5. Panicum hirticaule** Presl, Rel. Haenk. **1:** 308, 1830

Annual; culms 15–70 cm. tall, erect or decumbent at base, simple or sparingly branching, nodes bearded with appressed or spreading hairs, otherwise glabrous or nearly so; sheaths mostly much shorter than internodes, papillose-hispid or nearly glabrous; blades 5–15 cm. long, 4–13 mm. wide, subcordate, margins scabrous, usually ciliate toward base; panicles 5–15 cm. long, their finely scabrous branches rather stiffly ascending; spikelets 2.7–3.3 mm. long, acuminate, glabrous; first glume one-half to three-fourths as long as spikelet, acute or subacuminate; second glume and sterile lemma pointed beyond fruit, usually gaping.

Dry open ground and waste places, southern California to Arkansas and Texas; Mexico and the West Indies to Argentina.

**6. Panicum pampinosum** Hitchc. & Chase, Contr. U.S. Nat. Herb. **15:** 66, *fig. 48,* 1910

*Panicum capillare* var. *pampinosum* Gould, Madroño **10:** 94, 1949.

Similar in habit to *P. hirticaule*; culms as much as 40 cm. tall, usually shorter; panicles long-exserted, 6–15 cm. long, their branches solitary, relatively stout, very stiffly ascending; spikelets 4–4.5 mm. long, glabrous; first glume usually more than three-fourths as long as spikelet, acute or acuminate; second glume and sterile lemma acuminate, much exceeding fruit; fruit 2.2 mm. long.

Mesas and rocky slopes, Arizona and New Mexico; Sonora and Chihuahua to Guatemala.

**7. Panicum sonorum** Beal, Grasses N. Amer. **2:** 130, 1896

Annual; culms to 1.2 m. tall, erect, papillose or papillose-hispid; sheaths rather densely papillose-hispid or nearly glabrous; blades to 35 cm. long, mostly 1–3 cm. wide, subcordate at base, glabrous or sparsely papillose-pilose; panicles 15–35 cm. long, branches scabrous, ascending, but drooping at the tips, naked toward base; spikelets 2.8–3.2 mm. long, somewhat crowded toward ends of branchlets, usually brownish; first glume one-half to three-fourths as long as spikelet, broad, clasping, acute; second glume and sterile lemma pointed beyond fruit, usually somewhat gaping.

Cultivated ground and bottom land, Arizona, Mexico, Honduras, and El Salvador.

**8. Panicum stramineum** Hitchc. & Chase, Contr. U.S. Nat. Herb. **15:** 67, *fig. 50,* 1910

Annual; culms 10–50 cm. tall, erect or geniculate at lower nodes, sparingly branching, glabrous, nodes appressed-villous; sheaths glabrous or sparsely papillose or papillose-pilose; blades 10–30 cm. long, 4–15 mm. wide, scabrous on margins and sometimes on upper surface; panicles mostly 10–20 cm. long, branches and branchlets ascending, more or less flexuous; spikelets 3.2–3.7 mm. long, rather long-pedicellate, appressed or somewhat spreading, glabrous; first glume about one-third the length of spikelet, subacute or blunt; second glume and sterile lemma pointed beyond fruit, but not much exceeding it; fruit 2.2 mm. long with rather prominent scars at base.

In fields, along ditches, and in moist places on sandy plains, southern Arizona and northwestern Mexico. In the desert area, collected at Tucson, Guaymas, and on the Río Yaqui.

**9. Panicum obtusum** H. B. K., Nov. Gen. & Sp. **1:** 98, 1816

*Brachiaria obtusa* Nash, in Britt., Man. 77, 1901.

Perennial; culms 20–80 cm. tall, erect or geniculate, loosely tufted from a knotty crown, with long, usually arching stolons conspicuously hairy at nodes, culm nodes glabrous; sheaths glabrous; blades 5–20 cm. long, 2–7 mm. wide, dull green, flat, becoming involute at tip, glabrous or occasionally long-pilose on upper surface toward base; panicles 3–12 cm. long, their few branches narrowly ascending or appressed, densely flowered; spikelets 3–3.8 mm. long, obtuse, appressed, glabrous, first glume nearly as long as spikelet.

Low open ground, mostly along streams and irrigation ditches, Colorado to Missouri, south to Arizona, Texas, and central Mexico.

**10. Panicum urvilleanum** Kunth, Rév. Gram. **2:** 403, *pl. 115,* 1831
*Panicum preslii* Kunth, Rév. Gram. **1:** Suppl. 10, 1830.

Perennial; culms usually solitary, 50–100 cm. tall, erect from a creeping rhizome, nodes densely bearded; sheaths densely retrorsely villous; blades 20–50 cm. long, 4–7 mm. wide, flat, tapering to a long, involute point, glabrous or appressed-pilose; panicles 20–30 cm. long, branches rather stiffly ascending; spikelets 6–7 mm. long, densely villous, first glume from two-thirds to nearly as long as spikelet; fertile lemma ciliate on margins.

Sandy deserts, southern California and Arizona; Argentina and Chile.

**11. Panicum virgatum** L., Sp. Pl. 59, 1753
*Panicum coloratum* Walt., Fl. Car. 73, 1788. Not L., 1767.

Perennial with strong, scaly rhizomes; culms often in large clumps, erect, usually 1–2 m. tall, slender to robust, hard, glabrous; sheaths ciliate, sometimes villous at throat; blades as much as 60 cm. long, 3–15 mm. wide, long-acuminate, flat, usually pilose on upper surface toward base; panicles 15–50 cm. long, open, branches ascending or spreading, naked in lower third or half; spikelets mostly 3.5–5 mm. long, acuminate, strongly nerved; first glume clasping-acuminate, two-thirds to three-fourths as long as spikelet; second glume and sterile lemma acuminate, much exceeding fruit, usually gaping or spreading.

Moist canyons and open, sometimes rocky ground, Montana to Quebec and Maine, south to southern Mexico. Usually at elevations above the desert, but it may enter the upper margins of the Sonoran Desert as a straggler.

## 50. LASIACIS (Griseb.) Hitchc., Contr. U.S. Nat. Herb. **15:** 16, 1910

Coarse, woody, canelike, clambering perennials with fascicled branches and open or contracted panicles terminating the culm and leafy branches; spikelets subglobose, somewhat inflated, placed obliquely on the pedicels; glumes and sterile lemma papery, usually lanate-pubescent at tip; first glume one-third to one-half as long as spikelet, broad and clasping; second glume and sterile lemma nearly equal, a little shorter to a little longer than fertile floret; fruit bony-indurate, blunt, lemma and palea with a tuft of short hairs from a minute depression at tip, margins of lemma firm but not inrolled; palea convex above, the tip often free at maturity.

**1. Lasiacis ruscifolia** (H. B. K.) Hitchc., Proc. Biol. Soc. Wash. **24:** 145, 1911
*Panicum ruscifolium* H. B. K., Nov. Gen. & Sp. **1:** 101, 1815.
*Panicum compactum* Swartz, Adnot. Bot. 14, 1829.
*Lasiacis liebmanniana* Hitchc., Proc. Biol. Soc. Wash. **24:** 145, 1911.

Culms robust, clambering, freely branching; sheaths usually shorter, sometimes a little longer than internodes, glabrous or rarely hispid, margins glabrous or sometimes villous; ligule inconspicuous; blades usually asymmetrical, ovate-lanceolate to lanceolate, acuminate, primary ones 10–15 cm. long, 3–6 cm. wide, those of branches, especially secondary branches, often much smaller, softly pubescent to glabrous on lower surface, upper surface glabrous or scabrous; panicles 5–20 cm. long, narrow, compact, or lower branches somewhat distant and spreading, but densely flowered, whole panicle sometimes open with implicate, loosely flowered branches; spikelets 3–4 mm. long, green, nearly globose, at least at maturity.

Forests, thickets, and brushy slopes, Mexico and the West Indies to Peru and Argentina. This species has been collected near the edge of the desert region at Baviácora, Sonora, and Arroyo Hondo, north side of Cerro de la Giganta, Baja California.

## 51. ECHINOCHLOA Beauv., Ess. Agrost. 53, *pl. 11, fig. 2,* 1812

Slender to coarse annuals with compressed sheaths and few to many densely flowered racemes, distant or approximate on main axis; spikelets subsessile in 2 rows on one side of rachis, sparsely to prominently hispid; first glume broad, triangular, about one-half as long as spikelet; second glume and sterile lemma nearly equal, mucronate, or glume short-awned, sometimes enclosing a palea and staminate flower; fruit minutely crested, margins of lemma not inrolled.

Spikelets 2–2.5 mm. long, hispid-scabrous, the second glume and sterile lemma mucronate only; culms slender, erect to prostrate-spreading ........................ 1. *E. colonum*
Spikelets about 3 mm. long, strongly hispid, the sterile lemma awned or awnless; culms coarse, usually erect:
- Lemmas distinctly tuberculate and hispid on the nerves, more finely hispid between the nerves:
  - Panicles 10–20 cm. long; racemes loosely to moderately flowered; spikelets usually with at least some awned lemmas ........................ 2. *E. crusgalli* var. *crusgalli*
  - Panicles 15–35 cm. long; racemes densely flowered; spikelets devoid of awned florets or lemmas ........................................ 2a. *E. crusgalli* var. *mitis*
- Lemmas hispid but not tuberculate on the nerves, internerves usually essentially glabrous ........................................ 2b. *E. crusgalli* var. *zeylayensis*

### 1. Echinochloa colonum (L.) Link, Hort. Berol. 2: 209, 1833

*Panicum colonum* L., Syst. Nat. ed. 10, **2:** 870, 1759.
*Oplismenus colonum* H. B. K., Nov. Gen. & Sp. **1:** 108, 1816.
*Panicum crusgalli* var. *colonum* Coss. ex Richt., Pl. Eur. **1:** 26, 1890.

Culms slender, prostrate to erect, freely branching at base and lower nodes, 15–70 cm. long; sheaths glabrous; ligule wanting; blades soft and rather thin, lax, 5–10 cm. (rarely 15 cm.) long, 3–6 mm. (rarely 10 mm.) wide, glabrous, margins sparsely scabrous; inflorescences 3–10 cm. long, rarely longer; racemes few to several, ascending, lower ones 1–2 cm. distant, 1–3 cm. long; spikelets 2–2.5 mm. long, scabrous or finely hispid; first glume triangular, acute, about one-third as long as spikelet; second glume and sterile lemma equal, pointed, awnless; fruit 1.8 mm. long, elliptic, acute.

Moist ground, a weed in waste places, Missouri and Virginia, south to Texas and Florida, west to southeastern California; tropical and subtropical regions of both hemispheres; introduced in America.

### 2. Echinochloa crusgalli (L.) Beauv., Ess. Agrost. 53, 161, 1812, var. crusgalli

*Panicum crusgalli* L., Sp. Pl. 56, 1753.
*Panicum muricatum* Michx., Fl. Bor. Amer. **1:** 47, 1803.
*Echinochloa pungens* Rydb., Brittonia **1:** 81, 1931.

Culms 25 cm. to 1.5 m. tall, usually erect, often branching at base, glabrous; sheaths compressed, keeled, glabrous; blades acuminate, often elongate, 5–15 mm. wide, scabrous on margins; panicles 10–20 cm. long, usually erect, racemes ascending or appressed; spikelets crowded, about 3 mm. long excluding awns; first glume a little less than one-half as long as spikelet; second glume and sterile lemma tuberculate-hispid on nerves, hispid between nerves, glume acuminate-pointed, lemma awnless or with an awn 5–10 mm. long, sometimes longer, awns variable in same inflorescence; fruit abruptly pointed, pale, striate.

Moist open ground, fields, and waste places, Washington to New Brunswick, south to Florida, California, and northern Mexico; temperate and subtropical regions of both hemispheres.

**2a. Echinochloa crusgalli** var. **mitis** (Pursh) Peterm., Fl. Lips. 82, 1838

*Panicum crusgalli mite* Pursh, Fl. Amer. Sept. 66, 1814.
*Panicum crusgalli* var. *muticum* Ell., Bot. S.C. & Ga. **1:** 114, 1816.

Racemes mostly longer and more densely flowered than in var. *crusgalli*; spikelets awnless or nearly so.

Moist ground, British Columbia to Massachusetts, south to northern Mexico.

**2b. Echinochloa crusgalli** var. **zeylayensis** (H. B. K.) Hitchc., U.S. Dept. Agr. Bull. **772:** 238, 1920

*Oplismenus zeylayensis* H. B. K., Nov. Gen. & Sp. **1:** 108, 1815.
*Echinochloa zeylayensis* Schult., Mant. **2:** 269, 1824.

Racemes short, appressed, often crowded; spikelets awnless, less strongly hispid than in var. *crusgalli*, not papillose.

Moist, often alkaline soil, Oregon to Oklahoma, south to California and Texas; Mexico to Argentina.

## 52. SETARIOPSIS Scribn., in Millsp., Field Mus. Bot. Ser. 1: 288, *pl. 11*, 1896

Annuals with branching culms, flat blades, and narrow panicles, short branches terminating in a long bristle; spikelets short-pedicellate, inflated; first glume 5–7-nerved, obtuse or broadly acute, about one-fourth as long as spikelet; second glume 11–15-nerved, transversely veined, cordate, very broad, margins spreading, winglike; sterile lemma longer and narrower than second glume, acute, more or less contorted at base; fertile lemma ovate, acute, strongly transversely rugose, margins inrolled.

**1. Setariopsis auriculata** (Fourn.) Scribn., in Millsp., Field Mus. Bot. Ser. **1:** 289, 1896

*Setaria auriculata* Fourn., Mex. Pl. **2:** 43, 1886.
*Setaria pauciseta* Vasey, Bull. Torrey Club **13:** 230, 1886.

Culms branching, erect or ascending, 20–60 cm. tall; sheaths shorter than internodes, keeled, pubescent; blades mostly 10–15 cm. long, 5–10 mm. wide, scabrous or pubescent; panicles 5–15 cm. long, branches ascending, loosely to rather densely flowered, bristles as much as 10 mm. long; spikelets 3–4 mm. long.

Fields, rocky slopes, shady and waste places, Sonora and Yucatán to Honduras; Colombia and Venezuela.

## 53. SETARIA Beauv., Ess. Agrost. 51, *pl. 13, fig. 3*, 1812

Annual or perennial plants with flat leaf blades and spikelike or somewhat open panicles; spikelets subtended by one or more scabrous bristles, the spikelets deciduous, bristles persistent; first glume broad, less than half as long as spikelet; second glume and sterile lemma equal or glume a little shorter; fruit smooth or transversely rugose.

Plants annual; first glume not over one-third as long as spikelet:
  Spikelets 1.5–2.2 mm. long; bristles subtending each spikelet 1–2, usually solitary; panicles loosely flowered, often somewhat interrupted:
    Panicle often nodding, 5–25 cm. long; palea wanting; blades scabrous but not hispid or villous .......... 1. *S. liebmannii*
    Panicle erect, rarely nodding, 3–15 cm. long; palea present; blades scabrous and with some longer hairs:
      Fertile lemma finely rugulose; palea one-third as long as fertile lemma .......... 2. *S. grisebachii*
      Fertile lemma coarsely transversely rugose; palea nearly equaling fertile lemma .......... 3. *S. arizonica*
  Spikelets 3–3.4 mm. long; bristles subtending each spikelet several; panicle densely flowered, not interrupted .......... 4. *S. lutescens*
Plants perennial:
  Spikelets gibbous; fertile lemma strongly undulate-wrinkled; ligules 2–4 mm. long .......... 5. *S. macrostachya*
  Spikelets not gibbous; fertile lemma finely to coarsely transversely rugose; ligules not over 2.5 mm. long, usually shorter:
    First glume about one-third as long as spikelets:
      Axis of panicle scabrous-hispid; bristles subtending each spikelet 5 or more, 5–10 mm. long; spikelet elliptic .......... 6. *S. geniculata*
      Axis of panicle villous; bristles solitary, 15–25 mm. long; spikelets lanceolate-ovate .......... 7. *S. villosissima*
    First glume nearly one-half as long as spikelet:
      Panicles densely flowered; blades 2–5 mm. wide; ligules 1–2.5 mm. long, densely ciliate .......... 8. *S. leucopila*
      Panicles loosely flowered; blades 6–12 mm. wide, rigid; ligules 1 mm. long or less, sparsely ciliate .......... 9. *S. palmeri*

**1. Setaria liebmannii** Fourn., Mex. Pl. **2:** 44, 1886

*Chaetochloa liebmanni* Scribn. & Merr., U.S. Dept. Agr. Div. Agrost. Bull. **21:** 31, 1900.

Annual; culms usually less than 50 cm. tall, sometimes to 1 m. tall, erect or ascending, branching at lower nodes, scabrous below panicle, nodes sometimes appressed-pilose; sheaths glabrous, densely ciliate, collar appressed-pilose or hispid; blades to 20 cm. long, 2 cm. wide, acuminate, narrowed toward base, lax, scabrous; panicles usually 5–25 cm. long, rarely longer, loosely flowered, axis scabrous; bristles 7–15 mm. long, slender, scabrous, 1 or 2 below each spikelet; spikelets 2 mm. long, ovate; first glume one-third as long as spikelet; second glume three-fourths as long as fruit; sterile lemma equaling fruit; fruit pointed, gibbous, strongly rugose transversely.

Open sandy or rocky ground, fields, and waste places, Arizona; Baja California to Chihuahua and Nicaragua.

**2. Setaria grisebachii** Fourn., Mex. Pl. **2:** 45, 1886

*Setaria laevis* Fourn., loc. cit.

*Chaetochloa grisebachii* Scribn., U.S. Dept. Agr. Div. Agrost. Bull. No. **4:** 39, 1897.

Annual; culms erect or spreading, 10 cm. to 1 m. tall, scabrous below nodes and panicle, nodes appressed-pilose; sheaths densely ciliate, pubescent, or pilose on collar; blades to 25 cm. long, 1.5 cm. wide, usually much smaller, rather lax, scabrous and more or less pubescent or pilose; panicles 3–12 cm. long, tapering toward apex, axis scabrous and also villous, branches rather densely flowered, usually 5–10 mm. long or sometimes to 2.5 cm. long; bristles 5–15 mm. long, 1 or 2 below each spikelet, flexuous, scabrous; spikelets 2 mm. long; first glume obtuse or broadly acute, about one-third as long as spikelet; second glume and sterile lemma subequal, or glume a little shorter than spikelet; fruit acute, finely cross-wrinkled.

Open ground, sometimes a weed on cultivated land, Arizona to Texas and Oaxaca.

**3. Setaria arizonica** Rominger, Univ. Ill. Biol. Monogr. No. **29:** 66, *pl.* 2, 1962

Annual; culms erect or geniculate basally, to 50 cm. tall, glabrous except below panicle, where scabrous; nodes villous; leaf sheaths glabrous but margins ciliate above; ligule 1–2 mm. long, densely ciliate; leaf blades flat, 5–8 mm. wide, 7–15 cm. long, upper surface scabrous and papillose-hispid along nerves; panicle loosely flowered, its branches to 1.5 cm. long, scabrous; bristle 1 below each spikelet, 5–15 mm. long, scabrous, flexuous; spikelets ovate, 1.8–2 mm. long; first glume one-third as long as spikelet, 3-nerved; second glume obtuse, about two-thirds as long as lemma, 5-nerved; sterile and fertile lemmas subequal; fertile lemma turgid on convex side, coarsely rugose transversely; palea turgid in upper half, concave below.

In shade on rocky hillsides, vicinity of Tucson west to the Baboquivari Mountains, Arizona, and south into northern Sonora, Mexico.

**4. Setaria lutescens** (Weigel) F. T. Hubbard, Rhodora **18:** 232, 1916

*Panicum lutescens* Weigel, Obs. Bot. 20, 1772.

Annual; culms in large tufts, 30–130 cm. tall, branching at base and sometimes geniculate below; nodes dark, glabrous; leaf sheaths glabrous, strongly compressed, margins glabrous, hyaline; ligule very short, ciliate; leaf blades 4–10 mm. wide, 15–30 cm. long, loosely twisted, scabrous on upper surface, papillose-villous near throat, glabrous beneath; panicle densely flowered, 3–15 cm. long, cylindrical, axis hispid; bristles several below each spikelet, scabrous, 3–8 mm. long, yellow at maturity; spikelets ovate, very turgid on convex side, 3–3.4 mm. long; first glume one-third as long as spikelet, acute, 3-nerved; second glume one-half as long as spikelet, broadly ovate, 5-nerved; sterile and fertile lemmas equal; fertile lemma strongly rugose transversely.

Along ditchs, roadsides, and in cultivated fields; introduced and widely distributed throughout most of the United States, less common but present in Mexico, Central America, and the West Indies.

**5. Setaria macrostachya** H. B. K., Nov. Gen. & Sp. **1:** 110, 1815

*Panicum macrostachyum* Nees, Agrost. Bras. 245, 1829.

*Chaetochloa macrostachya* Scribn. & Merr., U.S. Dept. Agr. Div. Agrost. Bull. No. **21:** 29, 1900.

Perennial; culms 4–12 dm. tall, erect, densely tufted; sheaths keeled, glabrous or scaberulous, sometimes pubescent, margins ciliate; blades 15–40 cm. long, 3–15 mm. wide, scabrous, rarely pubescent; panicles spikelike, 10–25 cm. long, 5–10 mm. in diameter, dense, more or less interrupted, axis pubescent and villous; bristles mostly single below each spikelet, scabrous, 10–15 mm. long; spikelets 2–2.5 mm. long, turgid; first glume about one-half as long as spikelet; second glume two-thirds to three-fourths as long as spikelet; sterile lemma as long as fruit; fertile lemma strongly cross-wrinkled; sterile and fertile paleas about equal.

Dry open ground and dry woods and hillsides, Arizona to Colorado and Texas, south into Baja California and Oaxaca, Mexico.

**6. Setaria geniculata** (Lam.) Beauv., Ess. Agrost. 51, 178, 1812

*Panicum geniculatum* Lam., Encycl. **4:** 727, 1798.

Densely tufted perennial; culms erect or geniculate-spreading, from short, knotty rhizomes, 20 cm. to 1 m. tall, branching at lower nodes, glabrous; sheaths keeled, glabrous or scabrous toward summit; blades mostly 5–15 cm. long, 4–6 mm. wide, flat, acuminate, scabrous, often sparsely villous on upper surface toward base; panicles dense, narrow, spikelike, yellow, purple, tawny, or greenish, 2–6 cm. or sometimes 10 cm. long, axis shortly and densely hispid; bristles below each spikelet 5 or more, 5–10 mm. long; spikelets 2–2.5 mm. long (rarely to 3 mm. long), ovoid; first glume one-third as long as spikelet, 3-nerved; second glume one-half to two-thirds as long as spikelet, 5-nerved; sterile lemma acute, as long as fruit, 5–7-nerved, palea well developed, sometimes with a staminate flower; fruit strongly transversely rugose.

Open ground, cultivated fields, and waste places, Kansas to Massachusetts, south to Texas and Florida; California to New Mexico, south through Mexico and the West Indies to Argentina.

**7. Setaria villosissima** (Scribn. & Merr.) K. Schumann, in Just, Bot. Jahresb. **28:** 417, 1902

*Chaetochloa villosissima* Scribn. & Merr., U.S. Dept. Agr. Div. Agrost. Bull. No. **21:** 34, *fig. 19*, 1900.

Erect perennial; culms 4–10 dm. long, somewhat geniculate at base, compressed, striate, villous-pubescent; nodes glabrous or sparsely appressed-puberulent; sheaths loose, striate, compressed, equaling or exceeding internodes, nearly glabrous below, villous above, long-villous along margins; leaf blades 1.5–3 dm. long, 6–8 mm. wide, long-acuminate at apex, tapering toward base, villous on both surfaces, serrulate-scabrous on cartilaginous margins; panicle 2–3 cm. in diameter at base, tapering toward apex, 6–20 cm. long, axis scabrous and villous; bristles solitary, 15–25 mm. long, flexuous, scabrous; spikelets lanceolate-ovate, 2.5–3 mm. long; first glume one-third as long as spikelet, broadly ovate, 3-nerved; second glume nearly equaling lemma, apiculate, 5–7-nerved; sterile lemma about equaling fertile one, the latter abruptly apiculate and incurved at tip, finely transversely rugose below, striate and punctate toward apex; palea similar to fertile lemma.

On sandy or gravelly soil, usually of granitic origin, southwestern Texas to Arizona and adjacent northern Mexico.

**8. Setaria leucopila** (Scribn. & Merr.) K. Schumann, in Just, Bot. Jahresb. **28:** 417, 1902

*Chaetochloa leucopila* Scribn. & Merr., U.S. Dept. Agr. Div. Agrost. Bull. No. **21:** 26, *fig. 14*, 1900.

Perennial 2–10 dm. tall, usually pale or glaucous; culms erect or geniculate from a cespitose base, compressed, scabrous below panicle and nodes, often pubescent below nodes; sheaths compressed-keeled, glabrous except near scabrous summit and keel, villous along upper margins; ligule 1–2.5 mm. long, densely ciliate; leaf blades 2–5 mm. wide, flat or folded, 8–25 cm. long, scabrous on both surfaces or nearly glabrous beneath; panicles densely flowered, 6–15 cm. long, pale green, columnar, often interrupted below, axis scabrous or villous; bristles mostly solitary below each spikelet, 4–15 mm. long; first glume nearly one-half as long as spikelet, 3-nerved; second glume three-fourths to nearly equaling spikelet, 5-nerved; sterile lemma and fertile one about equal; fertile lemma apiculate, finely rugose transversely; sterile palea one-half to three-fourths as long as fertile palea.

Common on range lands and open ground from southern Colorado to Arizona and Texas, southward into central Mexico.

**9. Setaria palmeri** Henr., Blumea **3:** 415, 1940

*Chaetochloa rigida* Scribn. & Merr., U.S. Dept. Agr. Div. Agrost. Bull. No. **21:** 30, 1900.
*Setaria rigida* K. Schumann, in Just, Bot. Jahresb. **28:** 417, 1902. Not *S. rigida* Stapf, 1899.

Erect, rigid perennial 3–6 dm. tall; culms scabrous below panicles and sometimes just below nodes, otherwise glabrous, striate; sheaths equaling or exceeding internodes, glabrous, margins not ciliate; ligule very short (1 mm. long or less) sparingly ciliate; blades rigid, 10–25 cm. long, acuminate; 6–12 mm. wide, minutely scabrous on both sides or rarely nearly glabrous, serrulate-scabrous on margins; panicles loosely flowered, pale or stramineous, 5–20 cm. long, its axis short-puberulent to villous; bristles solitary below each spikelet, or nearly obsolete, stout, flexuous, 3–10 mm. long; spikelets ovate, acute, 2–2.5 mm. long; first glume about one-half as long as spikelet, acute, 3-nerved, midnerve excurrent; second glume to two-thirds as long as spikelet, 5–7-nerved, apiculate; sterile lemma equaling and slightly enclosing fertile lemma, latter ovate, acute, apiculate, transversely rugose below, smooth and glossy at tip; palea nearly plane, about equaling fertile lemma.

Canyons and rocky hillsides, Cape Region and on Carmen Island, Baja California.

## 54. CENCHRUS L., Sp. Pl. 1049, 1753

Annuals or perennials with terminal racemes of burs; spikelets sessile, enclosed in spiny burs, burs falling entire, body of bur irregularly lobed, lobes rigid, spines barbed, sometimes composed of bristles united only at base; spikelets 1–several in each bur, usually glabrous; first glume narrow, 1-nerved, much shorter than spikelet, sometimes wanting; second glume and sterile lemma nearly equal, 3–5-nerved, lemma enclosing a palea and usually a staminate flower; fruit indurate, lemma acuminate, margins thin, flat.

Bristles united only at base; plants perennial:
  Bristles antrorsely scabrous; spikelets 2–5 in each bur ............... 1. *C. multiflorus*
  Bristles retrorsely scabrous; spikelets solitary in each bur ............ 2. *C. myosuroides*
Bristles united above base, enclosing spikelets; plants annual, or becoming perennial:
  Burs subtended by a ring of slender bristles; spikelets 3–6 in each bur .... 3. *C. echinatus*
  Burs not subtended by a ring of slender bristles:
    Spikelets usually 2 in each bur; burs several in each spike, relatively small .......... ........................................................ 4. *C. incertus*
    Spikelets 4–7 in each bur; burs 1 or 2, sometimes 3 or 4 in each spike, large, conspicuous ........................................................ 5. *C. palmeri*

**1. Cenchrus multiflorus** Presl, Rel. Haenk. **1:** 318, 1830

Culms 0.5–1.5 m. tall, erect, or ascending from a hard, knotty base, simple, or branching at lower nodes, very scabrous below panicle; sheaths keeled, sparsely pilose on margins toward summit; lower ones sometimes papillose-pilose; blades 10–40 cm. long, mostly 5–10 mm. wide, acuminate or attenuate, scabrous, and sometimes papillose-pilose, sometimes smooth beneath; panicles 5–17 cm. long, dense, tawny or purplish, axis strongly angled, pubescent; fascicles subsessile, with ring of slender bristles at base, inner bristles flattened, sparsely ciliate, 5–8 mm. long, innermost sometimes elongate, much longer than others; spikelets 2–5 in each fascicle, 5–6 mm. long; first glume acute, one-third to half as long as spikelet; second glume and sterile lemma acuminate or attenuate, a little shorter than fruit.

Ravines and rocky slopes, Sonora (Mátape and Ures) to Costa Rica.

**2. Cenchrus myosuroides** H. B. K., Nov. Gen. & Sp. **1:** 115, *pl. 35*, 1815

*Panicum cenchroides* Ell., Bot. S.C. & Ga. **1:** 111, 1816.
*Pennisetum myosuroides* Spreng., Syst. Veg. **1:** 303, 1825.
*Cenchrus elliottii* Kunth, Rév. Gram. **1:** 51, 1829.

Perennial; culms solitary or in small clumps, 1–2 m. tall, hard, woody, erect or decumbent at base, usually branching at lower nodes, glabrous, often glaucous; sheaths loose, shorter than internodes, glabrous; blades 15–40 cm. long, 5–12 mm. wide, firm, attenuate, scabrous on upper surface; spikes 10–25 cm. long, 5–9 mm. wide, erect, dense, axis angular, pubescent; burs 5–7 mm. long, 1-flowered, bristles united only at base, retrorsely scabrous, inner ones about as long as spikelet; spikelets 4.5–5.5 mm. long, acuminate; first glume acute, one-third as long as spikelet; second glume and sterile lemma acute or acuminate, strongly nerved, about as long as fruit.

Moist sandy or rocky soil, southern Texas and southern Louisiana, Georgia, and Florida; Mexico and the West Indies to Argentina.

**3. Cenchrus echinatus** L., Sp. Pl. 1050, 1753

*Cenchrus brevisetus* Fourn., Mex. Pl. **2:** 50, 1886.
*Cenchrus echinatus brevisetus* Scribn., in Millsp., Field Mus. Bot. **2:** 26, 1900.

Annual; culms erect to geniculate or decumbent-spreading, often rooting at nodes, branching, as much as 1 m. long, glabrous; sheaths compressed, keeled, glabrous, or ciliate on margins near summit; ligule ciliate, 1 mm. long; blades mostly 6–20 cm. long, 3–8 mm. wide, acuminate, more or less pilose; spikes 3–10 cm. long, axis rather stout, flexuous, scabrous; burs scarcely crowded, 4–7 cm. long, usually broader than long, pubescent, containing 3–6 spikelets, bristles below bur few and stout, lobes or spines erect to spreading, retrorsely barbed at tip; spikelets 4.5–6 mm. long; first glume small, narrow, 1-nerved, second two-thirds to three-fourths as long as nearly equal fruit and sterile lemma; fruit acuminate, nerves of lemma apparent near tip.

Open sandy ground and waste places, Arizona to South Carolina and Florida; Mexico and the West Indies to Argentina.

**4. Cenchrus incertus** M. A. Curtis, Boston Soc. Nat. Hist. Jour. **1:** 135, 1837

*Cenchrus pauciflorus* Benth., Bot. Sulph. 56, 1844.
*Cenchrus albertsonii* Runyon, Amer. Jour. Bot. **26:** 485, *figs. 1–2*, 1939.

Annual, usually becoming perennial; culms 10–30 cm., sometimes as much as 60 cm. tall, erect or ascending, often somewhat spreading at base; blades mostly 10–15 cm. long, 2–4 mm. wide, glabrous, scabrous on margins; spikes usually 3–8 cm. long, short-exserted or partly included in sheath; burs 3–7 mm. wide, pubescent or nearly glabrous, with no ring of slender bristles at base, lobes flat, deltoid, erect or spreading, sometimes inflexed, enclosing 1–3, usually 2, spikelets; spikelets 5–6 mm. long; first glume narrow, about one-third as long as spikelet; second glume and sterile lemma subequal, lemma sometimes as long as acuminate fruit.

Open sandy ground and waste places, Washington to Ontario and Maine, south through Mexico and the West Indies to Argentina.

**5. Cenchrus palmeri** Vasey, in Brandegee, Proc. Calif. Acad. II, **2:** 211, 1889

Annual; culms 6–30 cm. long, erect to prostrate-spreading, simple or branching, pubescent at least below nodes and panicle; sheaths mostly overlapping, loose, retrorsely pubescent or pilose; blades 3–18 cm. long, 3–7 mm. wide, flat, attenuate, scabrous and pubescent or pilose; spikes usually composed of 1 or 2 burs, sometimes 3 or 4; burs 1.5–2.5 cm. high, 2.5–4 cm. wide, subglobose, spines numerous, spreading or reflexed, thickened at base, retrorsely barbed toward tip; spikelets 4–7 in each bur, 7–9 mm. long, 2–3 mm. wide; first glume very narrow, usually wanting; second glume and sterile lemma a little shorter than acuminate fruit.

Sand hills and dry sandy flats, Baja California and Sonora.

## 55. IMPERATA Cyrill., Pl. Rar. Neap. 2: 26, 1792

Slender, deep-rooted perennial from hard, scaly rhizomes, with linear blades narrowed toward base to thickened midrib, and narrow, dense, terminal, silky panicles; spikelets all alike, paired, unequally pedicellate on a continuous rachis, surrounded by long, silky hairs, awnless; glumes about equal, membranous; sterile lemma, fertile lemma, and palea thin and hyaline.

**1. Imperata brevifolia** Vasey, Bull. Torrey Club **13:** 26, 1886

*Imperata hookeri* (Anderss.) Hack., in A. & C. DC., Monogr. Phan. **6:** 97, 1889.

Culms erect, 1–1.5 m. tall, glabrous; leaves mostly crowded toward base, blades 3–12 mm. wide, scabrous on margins, lower ones elongate, those of culms rather short, uppermost sheath elongate with a greatly reduced blade; panicles 15–30 cm. long; spikelets 3 mm. long, glumes sparsely covered with long hairs, basal hairs 2 to 3 times as long as spikelet.

Deserts and rocky canyons, Utah, Nevada, and southern California to western Texas and northern Mexico. Infrequent in the Sonoran Desert (Tucson, Arizona; Hermosillo, Sonora).

## 56. HACKELOCHLOA Kuntze, Rev. Gen. Pl. 2: 776, 1891

Annual grasses with flat blades and rather numerous short, solitary racemes on flowering branches from middle and upper nodes; spikelets very dissimilar, in pairs at nodes of a disarticulating rachis, rachis segments and pedicels grown together, more or less clasped by margins of first glume of sessile spikelet; sessile spikelet globose, first glume indurate, alveolate; pedicellate spikelet conspicuous, staminate or sterile, membranous, flattened, first glume broad, rather strongly nerved, acute, keels winged.

**1. Hackelochloa granularis** (L.) Kuntze, Rev. Gen. Pl. **2:** 776, 1891

*Cenchrus granularis* L., Mant. **2:** 575, 1771.
*Manisuris granularis* Swartz, Prod. Veg. Ind. Occ. 25, 1788.
*Rytilix granularis* Skeels, U.S. Dept. Agr. Bur. Pl. Ind. Bull. **282:** 20, 1913.

Culms erect or ascending, 30–100 cm. high, freely branching, more or less papillose-hispid, especially below nodes; sheaths short, densely and conspicuously papillose-hispid; blades mostly 5–15 cm. long, 5–12 mm. wide, flat, papillose-hirsute on both surfaces, margins ciliate; racemes numerous, short, partly enclosed in subtending spathe; fertile spikelet 1 mm. long, pedicellate spikelet 2 mm. long.

A common weed in open ground and waste places in tropical regions around the world. Found in the Sonoran Desert only in southern Baja California.

## 57. TRACHYPOGON Nees, Agrost. Bras. 341, 1829

Tufted perennials with racemes solitary or few and approximate on a short axis; spikelets in pairs at nodes of a slender, continuous rachis, one nearly sessile, staminate, awnless, the other pedicellate, perfect, long-awned; pedicel of perfect spikelet obliquely disarticulating near base, forming a sharp, bearded callus below spikelet; first glume of perfect spikelet firm, rounded on back, several-nerved, obtuse; second glume firm, obscurely nerved; fertile lemma narrow, extending into a stout, twisted, geniculate or flexuous, often plumose awn; sessile spikelet as large as fertile spikelet, persistent, awnless.

**1. Trachypogon secundus** (Presl) Scribn., in Merr., U.S. Dept. Agr. Div. Agrost. Circ. **32:** 1, 1901

*Heteropogon secundus* Presl, Rel. Haenk. **1:** 335, 1830.
*Andropogon secundus* Kunth, Rév. Gram. **1:** Suppl. 39, 1830.
*Trachypogon plumosus* var. *montufari* subvar. *secundus* Hack. ex Henr., Meded. Rijks Herb. Leiden **40:** 40, 1921.

Culms densely tufted, erect, 5–10 dm. high, nodes densely bearded; sheaths glabrous or papillose-pilose, auriculate; ligule 5–12 mm. long, firm, margins fused with auricles of sheath; blades elongate, 2–5 mm. wide, flat or loosely rolled, attenuate, narrowed to an almost petiole-like base, scabrous or papillose-pilose; racemes 1–2, 10–20 cm. long; fertile spikelets 10–12 mm. long including the sharp, bearded callus; first glume rather densely pubescent to nearly glabrous; awns 5–7 cm. long, the two lower segments conspicuously plumose.

Dry grassy or rocky hillsides, Arizona to Texas, south to Panama. Found in the Sonoran Desert only in southern Baja California.

## 58. ANDROPOGON L., Sp. Pl. 1045, 1753

Perennials with flat, narrow blades and few to numerous racemes, these solitary or several to numerous and approximate on a short or somewhat elongate axis; sessile spikelet perfect, awned, pedicellate spikelet as large as sessile one or narrower and somewhat reduced, awnless; first glume flat or rounded on back, acute or acuminate, several-nerved; sterile lemma shorter than glumes, empty, hyaline; fertile lemma very thin, hyaline, narrow, bearing from the tip a geniculate awn twisted in the lower part.

Racemes few to numerous on a relatively long axis; culms sparingly branched only from base:
  Racemes relatively few on a short axis, panicle short-exserted or partly enclosed in the sheath; nodes densely bearded; sessile spikelets 5–6 mm. long ........ 1. *A. barbinodis*
  Racemes numerous on a relatively long axis, panicle usually long-exserted; nodes glabrous or appressed-hispid; sessile spikelets about 4 mm. long ............ 2. *A. saccharoides*
Racemes solitary or paired; culms branching toward summit:
  Racemes few, solitary on each peduncle:
    Rachis joints and sterile pedicels glabrous, or ciliate only near summit; first glume of sessile spikelet glabrous ........................................ 3. *A. cirratus*
    Rachis joints and sterile pedicels ciliate from base; first glume of sessile spikelet scabrous, hairy ........................................ 4. *A. hirtiflorus* var. *feensis*
  Racemes numerous, paired on each peduncle, forming a dense, feathery inflorescence .... ........................................................ 5. *A. glomeratus*

**1. Andropogon barbinodis** Lag., Gen. & Sp. Pl. 3, 1816

*Andropogon saccharoides* var. *barbinodis* Hack., in A. & C. DC., Monogr. Phan. **6:** 494, 1889.
*Amphilophis barbinodis* Nash, in Small, Fl. SE. U.S. 65, 1903.

Culms tufted, erect, 40–120 cm. tall, simple or sparingly branching below, glabrous, nodes conspicuously bearded; sheaths rounded, glabrous, mostly shorter than internodes; blades elongate, attenuate, 2–7 mm. wide, scabrous; panicles silky-white, short-exserted or partly included in dilated sheath, flabellate, axis usually shorter than long racemes; rachis and sterile pedicel densely long-villous; sessile spikelets 5–6 mm. long, densely bearded at base, first glume acute, scabrous on margins; awns mostly 2–2.5 cm. long; pedicellate spikelet narrow, awnless, about as long as sessile spikelet.

Open sandy or gravelly ground and rocky slopes, California to Oklahoma, Texas, and Mexico.

**2. Andropogon saccharoides** Swartz, Prod. Veg. Ind. Occ. 26, 1788

*Andropogon argenteus* DC., Cat. Hort. Monsp. 77, 1813.
*Andropogon laguroides* DC., ibid. 78.
*Andropogon torreyanus* Steud., Nom. ed. 2, **1:** 93, 1840.

Very similar to *A. barbinodis* in habit and appearance; nodes glabrous or appressed-hispid but not conspicuously bearded; panicle white or tawny, axis long, racemes relatively short; sessile spikelet about 4 mm. long, awns 10–15 mm. long; pedicellate spikelet narrow, somewhat reduced, awnless.

Prairies and rocky slopes, Colorado to Missouri, southern California to Alabama south through Mexico and the West Indies to Argentina. Occurs only toward the upper limits of the desert (Tucson).

**3. Andropogon cirratus** Hack., Flora **68:** 119, 1885

*Schizachyrium cirratum* Woot. & Standl., N. Mex. Coll. Agr. Bull. **81:** 30, 1912.

Culms slender, tufted, erect, 30–70 cm. tall, usually glaucous or purplish, sparingly branching above, glabrous; sheaths obscurely keeled, scaberulous; blades elongate, 1–4 mm. wide, flat, attenuate, scabrous; racemes long-exserted, 3–6 cm. long, straight, rachis joint thickened, glabrous; sessile spikelet 8–9 mm. long, first glume acuminate, scabrous toward summit, awn 5–10 mm. long; pedicellate spikelet scarcely reduced, acuminate, scabrous, awnless, pedicel long-ciliate on one side near summit.

Canyons and rocky slopes, southern California to Texas and northern Mexico. Occurs only near the upper limits of the desert (Tucson).

**4. Andropogon hirtiflorus** Kunth var. **feensis** (Fourn.) Hack., in A. & C. DC., Monogr. Phan. **6:** 372, 1889

*Andropogon feensis* Fourn., Mex. Pl. **2:** 62, 1886.
*Schizachyrium feense* A. Camus, Ann. Soc. Linn. Lyon **70:** 89, 1923.

Culms tufted, slender, erect, commonly 30–70 cm. high; sheaths narrow, compressed but not conspicuously keeled, glabrous; blades 5–12 cm. long, 2–3 mm. wide, acuminate, scabrous, sparsely long-ciliate at base; racemes few, rather distinct, partly enclosed in narrow spathe; sessile spikelets 7–9 mm. long, first glume acuminate, scabrous, sparsely to densely hairy; awn of fertile floret about 1 cm. long, geniculate, twisted below; pedicellate spikelet reduced, bearing a short, straight awn, first glume acuminate, scabrous.

Canyons and rocky slopes, southern Arizona to western Texas and southern Mexico.

**5. Andropogon glomeratus** (Walt.) B. S. P., Prel. Cat. N.Y. 67, 1888

*Cinna glomerata* Walt., Fl. Car. 59, 1788.
*Andropogon macrourus* Michx., Fl. Bor. Amer. **1:** 56, 1803.

Culms in dense clumps, erect, 50–150 cm. high, compressed, freely branching toward summit; lower sheaths broad, strongly flattened and keeled, overlapping, usually pilose at least on margins; blades elongate, 3–5 mm. wide, firm, flat or folded; inflorescence dense, feathery, ultimate branchlets densely hairy below narrow spathes; racemes slender, flexuous, borne on a short peduncle, partly enclosed in spathe; sessile spikelet 3–4 mm. long, awn slender, straight, about 15 mm. long; pedicellate spikelet greatly reduced or wanting, only pedicel present.

Fields and open ground, usually in wet places, eastern and southern United States and the West Indies to Panama. Found in the Sonoran Desert only in southern Baja California.

## 59. SORGHUM Moench, Meth. 207, 1794

Coarse rhizomatous perennial with flat blades and open panicles of short, few-flowered racemes; sessile spikelet ovate with a twisted, geniculate awn, glumes indurate; pedicellate spikelet well developed, lanceolate, awnless, glumes membranous.

### 1. Sorghum halepense (L.) Pers., Syn. Pl. 1: 101, 1805

*Holcus halepensis* L., Sp. Pl. 1047, 1753.

Culms erect from strong, widely spreading rhizomes, 50–150 cm. tall, glabrous; sheaths longer than internodes, glabrous, often densely pubescent or pilose on collar; blades attenuate, 1–2 cm. wide, glabrous; panicles 15–30 cm. long, open or rather dense, branches ascending, naked below; sessile spikelet 4.5–5.5 mm. long, ovate, appressed-pilose, readily deciduous awn 7–15 mm. long; pedicellate spikelet 5–7 mm. long.

A weed in fields, along irrigation ditches, and in waste places, Wyoming to Massachusetts, south to southern California, Texas, and Florida; Mexico and the West Indies to Argentina and Chile.

## 60. HETEROPOGON Pers., Syn. Pl. 2: 533, 1807

Annuals or perennials, with flat blades and solitary, terminal racemes; lower few pairs of spikelets alike, staminate, awnless, remaining sessile spikelets fertile, long-awned, pedicellate spikelets staminate like lower ones; rachis continuous below, bearing fertile spikelets above, disarticulating at base of each joint, the joint forming a sharp, barbed callus below fertile spikelet; glumes of fertile spikelet dark brown, coriaceous, first glume enclosing second one; glumes of staminate spikelet membranous, broad, obscuring fertile spikelets; lemmas hyaline, fertile one with a long, stout, twisted, geniculate awn.

Plant annual; culms usually more than 1 m. high; first glume of staminate spikelet with a row of conspicuous glands on back ............................ 1. *H. melanocarpus*

Plant perennial; culms slender, usually less than 1 m. high; first glume of staminate spikelet glandless, papillose-hispid ............................................ 2. *H. contortus*

### 1. Heteropogon melanocarpus (Ell.) Benth., Jour. Linn. Soc. 19: 71, 1881

*Andropogon melanocarpus* Ell., Bot. S.C. & Ga. 1: 146, 1816.

Culms 1–2 m. tall, erect, freely branching, sometimes with flowering branches even from lower nodes, glabrous; sheaths shorter than internodes, glabrous, keeled above, with a row of glands on keel; blades elongate, attenuate, flat, 5–10 mm. wide; raceme partly enclosed in sheath or spathe; sessile spikelet 9–10 mm. long, dark brown, densely appressed-pilose, awns 10–15 cm. long, lower segments densely hirsute with long hairs; pedicellate spikelet 1.5–2.5 cm. long, acuminate, first glume with a row of prominent glands down middle.

Canyons, fields, and waste places, Arizona; Alabama to Florida; tropical and subtropical regions of both hemispheres. Infrequent in the Sonoran Desert (Ures, Sonora).

**2. Heteropogon contortus** (L.) Beauv., in Roem. & Sch., Syst. Veg. **2:** 836, 1817

*Andropogon contortum* L., Sp. Pl. 1045, 1753.
*Heteropogon firmus* Presl, Rel. Haenk. **1:** 334, 1830.

Culms erect, tufted, 20–80 cm. tall, glabrous, with a few flowering branches at upper nodes; sheaths compressed, keeled, glabrous or scabrous, shorter than internodes; blades 5–15 cm. long, 3–7 mm. wide, flat or usually folded, very scabrous on upper surface and margins, tip boat-shaped; raceme usually long-exserted, 4–7 cm. long, peduncle elongate; sessile spikelets about 7 mm. long, nearly hidden by pedicellate spikelets, first glume obtuse, strongly nerved, hirsute with spreading hairs, awns 5–12 cm. long, hirsute, becoming implicate; pedicellate spikelet about 1 cm. long, first glume sparsely papillose-hispid.

Rocky slopes and canyons, Arizona to Texas; Mexico and the West Indies to Argentina; warmer regions of both hemispheres.

## 61. TRIPSACUM L., Syst. Nat. ed. 10, 2: 1261, 1759

Coarse perennial with flat blades and 1–several racemes in terminal and axillary inflorescences; monoecious, staminate and pistillate spikelets in same inflorescence; staminate spikelets 2-flowered, in pairs at nodes of a continuous rachis, one sessile, the other subsessile or pedicellate; glumes membranous, equal, many-nerved, first 2-keeled, margins rather sharply inflexed; pistillate spikelets solitary on opposite sides on lower part of same rachis, sunken in thickened, hardened disarticulating segments, composed of 1 fertile floret and a sterile lemma; first glume coriaceous, rounded, the margins nearly enclosing spikelet; sterile lemma, fertile lemma, and palea hyaline, each successively smaller.

**1. Tripsacum lanceolatum** Rupr. ex Fourn., Mex. Pl. **2:** 68, 1886

*Tripsacum lemmoni* Vasey, Contr. U.S. Nat. Herb. **3:** 6, 1892.
*Tripsacum dactyloides* var. *lemmoni* Beal, Grasses N. Amer. **2:** 19, 1896.
*Tripsacum dactyloides hispidum* Hitchc., Bot. Gaz. **41:** 295, 1906.

Culms in large, tough, hard clumps 1–2 m. high; blades elongate, attenuate, flat, 0.5–2.5 cm. wide, pubescent or pilose on upper surface, margins scabrous; racemes 1–several; staminate spikelets sessile or subsessile, 5–8 mm. long; pistillate spikelets 5–6 mm. long.

Open ground and brushy slopes, Arizona and Mexico to Panama. Infrequent in the Sonoran Desert (Baviácora, Sonora).

# Family 9. CYPERACEAE. Sedge Family

Annual or perennial, rushlike or grasslike plants, often of moist habitats or growing in water. Culms slender, mostly solid, triquetrous, quadrangular, terete, or flattened, encased at the base by closed leaf sheaths. Leaves narrow or blades reduced or obsolete. Flowers perfect or unisexual, arranged in spikelets, spikelets 1–many-flowered, solitary or arranged in umbels, panicles, or heads, each scale usually subtending a single flower. Scales 2-ranked or spirally imbricated. Perianth hypogynous, composed of interior scales or bristles, rarely calyxlike or wanting. Stamens 1–3, filaments slender. Anthers 2-celled. Ovary superior, 1-celled, 1-ovuled. Style 2–3-cleft or -toothed. Fruit an achene, lenticular, plano-convex, or trigonous. Endosperm mealy, copious. Embryo minute.

Spikelets with perfect flowers; achenes not enclosed in a flasklike perigynium:
- Perianth of a small, hyaline basal scale on adaxial side of the ovary; stamen 1 or rarely 2 .......... 1. *Hemicarpha*
- Perianth of several bristles, or alternating bristles and scales, or none; stamens usually 3, sometimes 2:
  - Scales of spikelets in 2 rows; spikelets often strongly flattened .......... 2. *Cyperus*
  - Scales of spikelets spirally imbricated; spikelets terete or nearly so, not strongly flattened:
    - Perianth bristles present (lacking in *Scirpus cernuus* var. *californicus*):
      - Heads more or less burlike; perianth of 3 stalked scales alternating with as many barbed bristles .......... 3. *Fuirena*
      - Heads or spikelets not burlike; perianth of bristles only, or rarely these lacking:
        - Style not swollen at the base; inflorescence usually of several spikelets, at least 1 involucral leaf present .......... 4. *Scirpus*
        - Style swollen at base, this persistent as a tubercle at summit of achene; inflorescence of a single terminal spikelet; no involucral leaf .......... 5. *Eleocharis*
    - Perianth bristles lacking:
      - Basal empty scales of the spikelet 1–2; leaf margins not strongly serrate:
        - Style's swollen base persistent as a tubercle on the achene .......... 6. *Bulbostylis*
        - Style's swollen base deciduous, achene not tuberculate .......... 7. *Fimbristylis*
      - Basal empty scales of spikelet 3 or more; leaf margins serrate .......... 8. *Cladium*

Spikelets of unisexual flowers, the staminate and pistillate in different spikelets or one type above the other; achenes enclosed in a flasklike perigynium .......... 9. *Carex*

## 1. HEMICARPHA Nees & Arn., Edinb. New Phil. Jour. 17: 263, 1834

Annual sedge with slender, erect or spreading stems and leaves. Spikelets terete, solitary or capitate in clusters of 2–4 spikelets, subtended by 1–3 narrowly linear or linear-subulate, foliaceous bracts. Scales spirally and imbricately arranged, each subtending a perfect flower, deciduous. Perianth of a single minute hyaline scale at base of pistil and between it and rachis of spikelet. Stamen 1 (in ours, to 3 in other species); filaments filiform, glabrous. Bristles none. Style slender, not swollen at base, shallowly to deeply 2-cleft. Achenes oblong, terete or lenticular in cross section.

Tips of scales slightly spreading to faintly recurved; perianth scales shorter than achene, deeply emarginate to bifid, or wanting .......... 1. *H. micrantha* var. *micrantha*

Tips of scales closely appressed; perianth scales about equaling achene, entire or faintly erose .......... 1a. *H. micrantha* var. *drummondii*

### 1. **Hemicarpha micrantha** (Vahl) Pax, in Engler & Prantl, Nat. Pflanzenf. **2**[2]: 105, 1887 var. **micrantha**

*Scirpus micranthus* Vahl, Enum. **2**: 254, 1806.
*Scirpus minimus* Pursh, Fl. Amer. Sept. 55, 1814.
*Scirpus subsquarrosus* Muhl., Descr. Gram. 39, 1817.
*Isolepis micrantha* Roem. & Sch., Syst. Veg. **2**: 110, 1817.
*Isolepis subsquarrosa* var. *minor* Schrad., in Roem. & Sch., Mant. **2**: 64, 1824.
*Hemicarpha subsquarrosa* var. *minor* Nees, in Mart., Fl. Bras. **2**[1]: 62, 1842.
*Hemicarpha micrantha* var. *minor* Friedland, Amer. Jour. Bot. **28**: 860, 1941.

Glabrous annual 3–15 cm. tall with compressed culms grooved along one side; leaves linear-setaceous, often reddish, 1–1.5 cm. long or rarely to 3 cm. long, mostly 3-nerved; spikelets ovoid to cylindric-ovoid, many-flowered, 1.8–5 mm. long (usually about 2 mm. long in flower, nearly twice as long in fruit); involucral leaves 3–15 mm. long; scales brownish, obovate, about 0.5 mm. long, tip short-acuminate, spreading to slightly recurved; stamen 1; style 2-cleft almost to base; achene narrowly obovoid, 0.5–0.7 mm. long, usually reddish brown, minutely but very closely papillate.

Muddy creek banks, margins of temporary pools, or on hillsides following rains, occurring in several life zones, Washington to New Hampshire and south into northern South America, entering the Sonoran Desert near Tucson, Arizona, and in Sonora. July–Aug., usually collected in fruit in Sept.–Nov. in our area.

**1*a*. Hemicarpha micrantha** var. **drummondii** (Nees) Friedland, Amer. Jour. Bot. **28:** 860, 1941

*Hemicarpha drummondii* Nees, in Mart., Fl. Bras. **2**$^{1}$: 62, 1842.

Similar to var. *micrantha* in size and general appearance but perianth scale about equaling achene, entire or faintly erose but not bifid or emarginate.

In wet soil and shallow margins of pools, in several life zones, Ontario to Indiana and southwest to southern Arizona and Texas, entering the margin of the Sonoran Desert at the foot of the Rincon Mountains, Pima County, Arizona.

## 2. CYPERUS L., Sp. Pl. 44, 1753

Annual or perennial sedges with simple, triangular, erect or spreading culms, leafy at base with 1–several involucral leaves at summit subtending a simple or decompound, umbellate or capitate inflorescence. Rays of umbel basally sheathed, unequal, some of spikes or heads often sessile. Spikelets terete or strongly flattened laterally, scales persistent or falling from rachilla as achenes mature, rachilla winged or naked or minutely bracteolate, remaining entire or breaking into 1-seeded segments. Scales concave, conduplicate or keeled, in 2 more or less imbricated ranks, all fertile or lower 1–2 empty. Flowers perfect or rarely dioecious. Perianth lacking. Stamens 1–3. Style 2–3-cleft, deciduous. Achenes lenticular to trigonous, straight or slightly curved.

Style branches 2; achenes lenticular:
- Scales loosely imbricated, their margins white, hyaline-membranous; achenes obovoid-lenticular, widest at summit .............................. 1. *C. albomarginatus*
- Scales closely imbricated, margins not white; achenes ellipsoidal-lenticular, widest about middle:
  - Achenes biconvex, flattened laterally; umbels appearing terminal, subtended by 2–3 involucral leaves .............................. 2. *C. lanceolatus*
  - Achenes plano-convex, flattened dorsally; umbels appearing lateral, inferior involucral leaf a continuation of culm .............................. 3. *C. laevigatus*

Style branches 3; achenes trigonous:
- Flowers dioecious; culms rooting at base of inflorescence if touching soil ...... 4. *C. dioicus*
- Flowers perfect; culms rarely rooting:
  - Spikelets terete, not strongly flattened; achenes more or less enfolded by rachilla wings:
    - Spikelets 10–15-flowered, articulate beneath each flower and breaking into 1-seeded segments .............................. 5. *C. ferax*
    - Spikelets 1–3-flowered, articulate only at base of rachis, spikelet falling entire:
      - Culms and leaf blades asperous on angles and plane surfaces; spikelets closely crowded in cylindrical spikes .............................. 6. *C. asper*

Culms and leaf blades glabrous or slightly asperous on margins only; spikelets loosely arranged in more open spikes:
Rachilla wings almost completely enfolding each achene; spikelets 4–7 mm. long; culms 1.5–3 dm. tall .................................... 7. *C. tenuis*
Rachilla wings enfolding achenes only to middle or below; spikelets 3–5 mm. long; culms 3–10 dm. tall:
Lateral nerves of fruiting glumes 3–5 on each side, conspicuous; leaves of culms 3–8 mm. wide ........................................ 8. *C. pringlei*
Lateral nerves of fruiting glumes obscure; leaves of culms 2–3 mm. wide ...... .................................. 9. *C. subambiguus* var. *pallidicolor*
Spikelets strongly flattened; achenes slightly if at all enfolded by rachilla wings:
Culms 2 dm. tall or less; glumes aristate, excurrent midrib distinctly recurved; stamen 1:
Plants annual; glumes 6–8-nerved; achenes longer than broad ................. ........................................ 10. *C. aristatus* var. *inflexus*
Plants perennial; glumes 5-nerved; achenes as broad as long ........ 11. *C. perennis*
Culms usually 3–10 dm. tall; glumes not aristate, tips appressed or erect; stamens 3:
Spikelets arranged in spikes, not in ovoid to globose heads; plants not glutinous:
Annuals; roots fibrous; tuber-bearing rootstocks lacking:
Rachilla wings well developed but deciduous; achenes ovoid, pointed at both ends; spikelets 0.5–1 mm. wide .................... 12. *C. erythrorhizos*
Rachilla wingless; achenes oblong, obtuse at both ends; spikelets 1.5–2 mm. wide .............................................. 13. *C. parishii*
Perennials; scaly rootstocks tuber-bearing:
Glumes with green midribs, pale stramineous sides; bracteoles subtending spikelets lance-attenuate, 2.5–4 mm. long ................. 14. *C. esculentus*
Glumes with greenish-yellow midribs, pale purplish brown sides; bracteoles subtending spikelets ovate, acute or obtuse but not attenuate, 1.5–2 mm. long ............................................. 15. *C. rotundus*
Spikelets in dense, subglobose heads; lower parts of culms and leaves glutinous-viscid ................................................ 16. *C. elegans*

**1. Cyperus albomarginatus** Mart. & Schrad. ex Nees, in Mart., Fl. Bras. **2**[1]**:** 9, 1842

*Cyperus flavicomus* Vahl, Enum. **2:** 360, 1806. Not Michx., 1803.
*Cyperus expansus* Bojer, Hort. Maurit. 380, 1837.
*Cyperus sabulosus* Britt. & Brown, Ill. Fl. ed. 2, **1:** 300, *fig. 727,* 1913. Not *C. sabulosus* Mart. & Schrad., 1842.

Stout or slender annual 1.5–10 dm. tall; leaves basal, sheaths thin and subscarious, usually reddish, 2.5–4 cm. long, blades flat, 2.5–6 mm. broad, to 3 dm. long, glabrous except for antrorsely scaberulous margins; culms triangular; involucral leaves to 3.5 dm. long, much surpassing inflorescences; umbels usually compound, their several rays to 8 cm. long; spikelets numerous, more or less racemosely arranged on each ultimate ray, 10–18 mm. long, strongly compressed, spreading, dark brown; glumes broadly oblong, obtuse, about 2 mm. long, loosely imbricated and leaving spaces between them, scarious margins white; stamens 3; style branches 2-cleft; achenes obovoid-lenticular, black, broadest at apex, slightly mucronate, nearly equaling scales and often persistent on rachis after scales have fallen, dull.

Moist plains and along ditches and streams, Lower Sonoran Zone to Carolinian and Tropical Zones, southern Arizona to Virginia and Florida and south into South America; widely distributed in tropical and subtropical parts of both the Old and the New Worlds. Aug.–Sept.

**2. Cyperus lanceolatus** Poir., in Lam., Encycl. **7:** 245, 1806
(For extensive synonymy see Pflanzenreich $4^{101}$: 349, 1936.)

Perennial with cespitose culms somewhat decumbent, 1–5 dm. tall from short, slender rhizomes; culms triangular, somewhat canaliculate on one face and clothed in leaves to lower one-third of their length; leaf sheaths brownish, firm, 1–8 cm. long; leaf blades 1–3 mm. wide, lightly but distinctly carinate, glabrous; involucral leaves spreading, 3–12 cm. long; umbels terminal, simple, and spikelets rather densely crowded into heads 1–2.5 cm. in diameter; spikelets sessile to short-pedicellate, oblong to oblong-lanceolate, acute, strongly compressed, 5–9 mm. long, 2–2.5 mm. wide, glumes obtusely ovate, yellowish, but with 3 greenish veins, 1.6–2 mm. long, glabrous and slightly lustrous; stamens 2; style deeply bifid; achenes elliptic- to obovoid-biconvex, about one-half as long as scales, pale gray, dull, broadest at or slightly above middle.

Stream banks, arroyos, and seepy places, Lower Sonoran Zone to Tropical Zone, Louisiana to Georgia and from Baja California into South America; subtropical Africa. Sept.–Oct.

**3. Cyperus laevigatus** L., Mant. **2:** 179, 1771
(For extensive synonymy see Pflanzenreich $4^{101}$: 321, 1936.)

Perennial with ascending to erect, stiffish, trigonous, glabrous culms 5–50 cm. tall from slender, rather short, reddish or brownish rhizomes; leaf sheaths reddish to brown, closely appressed to lax, 5–35 mm. long, truncate at apex; leaf blades glabrous, 1–1.8 mm. wide, 0.3–3.5 dm. long, smooth and glabrous; involucral leaves 2, inferior one erect and forming a continuation of culm, to 8 cm. long, upper one 2–12 mm. long, spreading or somewhat reflexed; spikelets appearing lateral, strongly compressed laterally, 4–10 mm. long, about 2 mm. wide, somewhat turgid, in heads of 2–12 crowded spikelets; glumes broadly ovate, strongly concave, obtuse or rarely mucronulate, about 1.5 mm. long, stramineous with dark brown or reddish brown splotches on sides near apex, closely imbricate; stamens 3, filaments dilated; achenes plano-convex, plane face next rachis, about equaling scales, brownish, minutely and closely puncticulate, shining.

Moist soil, Lower Sonoran Zone to Tropical Zone, central western California southward to Argentina; Mediterranean region and south to south Africa. March–Sept.

**4. Cyperus dioicus** I. M. Johnston, Proc. Calif. Acad. IV, **12:** 988, 1924

Perennial from short, stout rhizomes with spreading, obtusely triangular culms 4–12 dm. long, these sometimes even trailing and then rooting from bases of inflorescences; leaf sheaths firm, chartaceous, or longer ones subscarious, often brownish or reddish, basal ones 1–6 cm. long and tapering to an acute to obtuse, bladeless apex, inner ones greenish or pale stramineous, to 1.5 dm. long; leaf blades flat, stiffish or larger ones lax, 4–8 mm. wide, 10–30 cm. long, margins slightly asperous; involucral leaves 1–3, reflexed, 3–25 mm. wide, 6–15 cm. long; umbels usually of 3–15 compound rays, longest of these to 3 cm. long; spikelets numerous, strongly compressed, 2–3 mm. wide, 8–20 mm. long, 10–60-flowered; glumes closely imbricate, broadly ovate, mucronulate, pale to dark reddish brown on sides, greenish keel faintly serrulate near apex; stamens 3, staminate spikelets slightly thinner than pistillate ones; style 3-cleft at apex, branches shorter than main body; achenes obtusely trigonous-ovoid, 0.6–0.8 mm. long, ashy gray or nearly white, minutely and densely punctulate, mucronulate at apex.

In creek beds and in moist soil along arroyos, Lower Sonoran and Tropical Zones, from vicinity of San Pablo (a few miles south of Campo Alemán) southward to the Cape Region, Baja California. March–May.

**5. Cyperus ferax** L. C. Rich., Act. Soc. Hist. Nat. Paris **1:** 106, 1792
(For extensive synonymy see Pflanzenreich **4**[101]**:** 615, 1936.)

A robust annual; culms triangular, to 12 dm. tall, bulbously swollen basally and canaliculate on the faces; leaf sheaths thin, subscarious, nearly white to bright red, 2–20 cm. long; blades of lower leaves 4–10 mm. wide, 5–50 cm. long, flat, somewhat acuminate, firm and subrigid, glabrous or sparsely and minutely asperous along margins; involucral leaves 6–8, spreading, at least some of them surpassing inflorescences; umbels laxly to rather densely decompound with 5–15 rays, these to 12 cm. long; secondary rays again subtended by smaller involucral leaves, rays to 3 cm. long; spikelets divergently arranged in oblong-cylindric, rather loose spikes 1.5–2 cm. in diameter, 2–4 cm. long; each spikelet linear-terete, about 1 mm. in diameter, 8–16 mm. long, 10–15-flowered; glumes narrowly ovate to ovate-elliptic, obtuse or rarely mucronulate at apex, 2–3.5 mm. long, basal ones shortest, midrib greenish, sides yellowish brown to reddish and 7–9-nerved; rachilla scariously winged, wings enfolding achene nearly to summit; stamens 3; styles 3-cleft; achenes oblong or oblong-ovate, about 2 mm. long, falling inside the wings of disarticulated segment of rachilla.

Along ditches and streams, in meadows, and about margins of lakes and ponds, Lower and Upper Sonoran to Tropical Zones, southern California to Florida; warm and tropical parts of New and Old Worlds. Mainly July–Oct.

**6. Cyperus asper** (Liebm.) O'Neill, in Tidestrom & Kitt., Fl. Ariz. & New Mex. 767, 1941
*Mariscus asper* Liebm., Skrift. Vid. Selsk. V, **2:** 229, 1851.
*Cyperus mutisii* var. *asper* Kükenthal, in Engler, Pflanzenreich **4**[101]**:** 486, 1936.

Moderate-sized perennial from a short, nodose, lignescent rhizome; culms somewhat swollen basally, 3–10 dm. tall, obtusely triangular, antrorsely asperous; leaf sheaths thin, usually reddish or brownish, 3–20 cm. long, truncate at apex opposite blade; leaf blades flat, 2.5–10 mm. wide, 1–3 dm. long, asperous, involucral leaves 4–8, spreading, to 2 dm. long; umbels with 7–10 primary rays to 15 cm. long; spikes usually solitary on a ray, cylindric-oblong, 7–12 mm. in diameter, 1.2–2.5 cm. long; spikelets closely crowded, 1–1.5 mm. in diameter, 3–6 mm. long, mostly 2-flowered; glumes ovate-oblong, obtuse but often mucronulate, 2.5–3.5 mm. long, stramineous or slightly ferruginous, distinctly 7–9-nerved; rachilla scariously winged, the wing enfolding achene full length; achenes obtusely trigonous, oblong, about 2 mm. long, grayish, closely and minutely puncticulate.

Moist ground and hillsides, Lower Sonoran Zone to Tropical Zone, central Sonora to Colima, Mexico. Aug.–Oct.

**7. Cyperus tenuis** Swartz, Prod. Veg. Ind. Occ. 20, 1788
(For synonymy see Pflanzenreich **4**[101]**:** 416, 1936.)

Perennial; culms erect, 1.5–3 dm. tall, slender, triquetrous, slightly swollen basally, glabrous or faintly asperulous on angles; leaf sheaths delicate, reddish or brownish, soon becoming shredded, 1–6 cm. long; leaf blades flat, 1–3 mm. wide, 2–4 dm. long, often markedly surpassing culms; involucral leaves 3–9, spreading, to 3 dm. long;

umbels 3–9-rayed, pedicels 0.2–3.5 cm. long; spikes very loose and open, usually solitary on each ray but sometimes decompound, about 1.5 cm. wide, 2–3.5 cm. long; spikelets terete-linear, about 1 mm. in diameter, 4–7 mm. long, mostly 2-, sometimes 3-flowered; rachilla corky and broadly winged, scarious wings almost completely enfolding achene; glumes oblong-elliptic, obtuse but mucronulate, reddish or brownish, distinctly 7–11-nerved; stamens 3; style 3-cleft about to middle; achenes linear-oblong, trigonous, and somewhat curved, minutely and closely punctulate, apiculate.

Plains and grasslands, borders of rain pools, Lower Sonoran Zone to Tropical Zone, central Sonora to Central America; Australia; tropical West Africa. July–Sept.

### 8. **Cyperus pringlei** Britt., Bull. Torrey Club **12:** 7, 1885

*Cyperus tetragonus* var. *pringlei* Kükenthal, Pflanzenreich $\mathbf{4^{101}}$: 493, 1936.

Erect or ascending perennial with triangular, glabrous culms 3–10 dm. tall with swollen bases from short, nodose rhizomes; leaf sheaths thin, subscarious, yellowish to reddish brown, to 12 cm. long; leaf blades flat, 3–8 mm. wide, glabrous or slightly asperulous along midrib and margins, to 3 dm. long; involucral leaves 5–8, to 2 dm. long, surpassing inflorescences; umbels simply 5–12-rayed, slender rays 1–10 cm. long, spreading; spikes cylindric-oblong, rather loose, 7–10 mm. in diameter, 1.5–3 cm. long; spikelets terete-oblong, 1 mm. in diameter or less, 3–5 mm. long, 2- or less commonly 3-flowered; glumes pale yellow to pale brown, 2.5–3 mm. long, distinctly 7–9-nerved; rachilla scariously winged; stamens 3; style deeply 3-cleft; achenes ellipsoid-oblong, trigonous, basal one-third of margin enfolded by scarious rachilla wings, olive brown, sublustrous, minutely punctulate.

Canyons, moist meadows, and mountain slopes, Lower Sonoran Zone to Transition Zone, southern Arizona to Chihuahua and central Sonora. Aug.–Sept.

### 9. **Cyperus subambiguus** Kükenthal var. **pallidicolor** Kükenthal, in Engler, Pflanzenreich $\mathbf{4^{101}}$: 528, 1936

?*Mariscus schaffnerianus* Schlecht., Linnaea **26:** 379, 1853.

Similar to *C. pringlei* in general appearance and size but leaf blades 2–3 mm. wide; involucral leaves only 4–6; rays of umbels about 1 cm. long; fruiting glumes yellow or stramineous, lateral nerves obscure.

In muddy locations on mountain sides and about meadows, upper margins of the Lower Sonoran Zone to the Transition Zone, southern Arizona to Guatemala. Occurring in the Sonoran Desert near Magdalena, Sonora. Aug.–Oct.

### 10. **Cyperus aristatus** Rottb. var. **inflexus** (Muhl.) Boeck., Linnaea **35:** 500, 1868

*Cyperus inflexus* Muhl., Descr. Gram. 16, 1817.
(For additional synonymy see Pflanzenreich $\mathbf{4^{101}}$: 505, 1936.)

Annual with 1–several subfiliform tufted culms 3–15 cm. long from fibrous roots; leaf sheaths delicate, subhyaline, reddish, soon becoming shredded, 0.5–3 cm. long; leaf blades linear, flat, 0.5–2 mm. wide, about equaling culms, glabrous; umbels sessile or few-rayed, rays to 3.5 cm. long; 2–5 involucral leaves 2–3 mm. wide, to 10 cm. long but usually about 5 cm. long or less; spikes capitate or subcapitate, ovoid to subglobose, dense, 6–12 mm. in diameter; spikelets linear-oblong, strongly flattened, 1–1.8 mm. wide, 4–6 mm. long, 6–10-flowered; glumes light brown with greenish midribs, lanceolate, 1.5 mm. long, midrib tapering into an aristate, firm, recurved awn, sides strongly 6–8-nerved, deciduous from persistent, wingless rachilla at maturity; stamen

1; style 3-cleft; achenes oblong to narrowly obovoid, trigonous, obtuse but minutely mucronulate at apex, ashy to pale brown, minutely punctulate, 0.6–0.8 mm. long.

In moist, sandy soil, Lower Sonoran Zone to Transition and Tropical Zones, British Columbia to New Brunswick southward throughout the continent to the West Indies and temperate South America. Flowering in our area after either the winter or the summer rains.

**11. Cyperus perennis** (M. E. Jones) O'Neill, in Morton, Contr. U.S. Nat. Herb. **29:** 93, 1945

*Cyperus aristatus* var. *perennis* M. E. Jones, Contr. West. Bot. **18:** 25, 1933.

A small perennial closely resembling *C. aristatus,* with slender, triquetrous, tufted culms 1.5–6 cm. tall, their bases bearing many fibers remaining from leaf sheaths of previous year; rhizome lignescent but short; leaf sheaths dull brown, about 1 cm. long; leaf blades setaceous, 1 mm. wide or less, 1.5–8 cm. long, usually surpassing culm; involucral leaves 2–4, unequal, to 9 cm. long; spike a subglobose, dense head 1–1.5 cm. wide, without any long-pedicellate rays; spikelets oblong, 1–1.8 mm. wide, 4–7 mm. long, strongly compressed, 10–15-flowered; rachilla wingless; glumes lance-ovate, body 1.5–1.8 mm. long, tipped with a stiffish, recurved awn 0.5–0.8 mm. long, 5-ribbed, sides membranous, pale brown or stramineous; stamen 1; style deeply 3-cleft; achenes broadly turbinate, about 0.4 mm. broad and high, purplish brown to nearly black, truncate and minutely apiculate at apex, minutely raised-punctulate.

Sandy or clay soil, Lower Sonoran Zone, central and southern Sonora, and southern Baja California. March–Sept.

**12. Cyperus erythrorhizos** Muhl., Descr. Gram. 20, 1817

?*Cyperus tenuiflorus* Ell., Bot. S.C. & Ga. **1:** 70, 1816.
*Cyperus chrysokerkos* Steud., Syn. Pl. Glum. **2:** 316, 1855.
*Chlorocyperus erythrorhizos* Palla, Allg. Bot. Zeit. **9:** 69, 1903.

Robust annual with tufted culms to 9 dm. tall, these triquetrous, glabrous, quite leafy at base; leaf sheaths 2–20 cm. long, firm, usually reddish, delicate and easily ruptured on side opposite blade; leaf blades flat, 3–8 (or rarely to 12) mm. wide, 1–5 dm. long, margins scaberulous; involucral leaves 3–12, unequal, to 4 dm. long; inflorescence decompound, primary rays 3–9, to 1.5 dm. long; secondary rays (if present) to 1 dm. long; spikes oblong-cylindric, 10–16 mm. in diameter, 2–3.5 cm. long; spikelets crowded, linear, strongly compressed, many-flowered, 0.5–1 mm. wide, 6–25 mm. long; glumes chestnut brown, closely appressed, oblong-lanceolate, mucronulate, 1.2–1.4 mm. long, 5-nerved, lateral nerves obscure; rachilla bearing a pair of white, minute wings at base of each achene, deciduous with glumes; stamens 3; style 3-cleft about to middle; achenes oblong to broadly ellipsoidal, sharply trigonous, pale yellow and shining, about 0.8 mm. long, pointed at both ends, faces somewhat concave, puncticulations scarcely visible.

Moist soil, Lower and Upper Sonoran Zones and into the Transition or possibly Canadian Zone, southern Washington to the New England states and south to Florida, Arizona, and southern California. July–Oct.

**Cyperus erythrorhizos** var. **cupreus** (Presl) Kükenthal, Pflanzenreich **4**$^{101}$**:** 59, 1936, with narrower leaves, shorter and sometimes prostrate culms, smaller inflorescences, obovate glumes, and obovoid achenes may also occur within our area. The plants of this species growing along the lower Colorado River seem to represent a particularly robust race, the culms often attaining a height of 10–12 dm. This form almost certainly occurs in northeastern Baja California.

**13. Cyperus parishii** Britt. ex Parish, Bull. S. Calif. Acad. **3:** 52, *pl. 3,* 1904

*Cyperus sphacelatus* Britt., Bull. Torrey Club **13:** 212, 1886. Not *C. sphacelatus* Rottb., 1773.
*Cyperus congestus* var. *parishii* Kükenthal, in Engler, Pflanzenreich $\mathbf{4^{101}}$: 446, 1936.

Annual (or possibly short-lived perennial) from fibrous roots; culms solitary or in small tufts, moderately slender, triquetrous, glabrous, 1–3 dm. tall, with 3–5 basal leaves; leaf sheaths firm, 1–3.5 cm. long, pinkish or brownish; leaf blades flat to involute, 1.5–3 mm. wide, 3–20 cm. long, glabrous except on the asperulous margins; involucral leaves 3–5, to 6 mm. wide, unequal, 2–18 cm. long; umbels 1–6-rayed and usually with 1 or more sessile spikes, pedicels to 5 cm. long; spikes broadly ovoid, about 3 cm. long or less, about as wide; spikelets linear, acute, 1.5–2 mm. wide, 7–20 mm. long, 10–18-flowered; glumes oblong-lanceolate to obovate, obtuse, 2.5 mm. long, 7–9-nerved, midrib region chestnut brown, sides purplish brown; rachilla wingless; stamens 3; stigma 3-cleft; achenes oblong, obtuse at both ends, 1.2–1.4 mm. long, trigonous, dark purplish brown to nearly black, raised-puncticulate.

Creek banks and moist meadows, Lower and Upper Sonoran Zones, San Bernardino County, California, to Washington and east to New Mexico. Aug.–Sept.

**14. Cyperus esculentus** L., Sp. Pl. 45, 1753

(For synonymy see Pflanzenreich $\mathbf{4^{101}}$: 116–117, 1936.)

Stoutish perennial 3–8 dm. tall from scaly, tuber-bearing rootstocks; culms somewhat swollen at base, glabrous; leaf sheaths tawny to light brown, 1–8 cm. long; leaf blades light green, flat, 3–8 mm. wide, 1–4 dm. long, midrib prominent beneath, margins scaberulous; involucral leaves 3–6, unequal, longer ones well surpassing umbels; inflorescence decompound, with several rays to 1.5 dm. long; spikes loose, 1.5–2.5 cm. wide, 2–4 cm. long, spikelets ascendingly spreading, about 2 mm. wide, 1–2.5 cm. long, strongly flattened; glumes stramineous, oblong-ovate, acute, about 2.5 mm. long, 3–5-nerved; rachilla narrowly winged by downward-projected margins of glume of next flower above; stamens 3; style 3-cleft about to the middle; achenes obovoid, obtusely trigonous, 1.2–1.4 mm. long, grayish, impressed-puncticulate.

Moist soil, Lower Sonoran Zone to Canadian Zone, Alaska to New Brunswick southward into temperate South America; widely distributed in the Old World. In our area flowering Aug.–Sept.

**15. Cyperus rotundus** L., Sp. Pl. 45, 1753

(For synonymy see Pflanzenreich $\mathbf{4^{101}}$: 107–108, 1936.)

Erect perennial from a scaly, tuber-bearing rootstock; culms stoutish, 1.5–5 dm. tall, triquetrous, glabrous; leaf sheaths dark brown to stramineous, 2–8 cm. long; leaf blades flat, 3–6 mm. wide, 1–2.5 dm. long, glabrous except for sparsely asperulous margins near tips; involucral leaves 3–5, unequal, 1–10 cm. long; umbels decompound, rays 3–8, these 2–12 cm. long; spikes loose, of 3–15 spikelets, the latter 1.5–2 mm. wide, 8–20 mm. long, strongly compressed, rich purplish brown; glumes ovate, acute, 2.5–3 mm. long, 3–5-nerved, midrib greenish yellow, margins greenish or pale purplish brown; rachilla winged as in *C. esculentus*; stamens 3; style 3-cleft; achenes narrowly ellipsoidal to narrowly obovoid, trigonous, 1.2–1.4 mm. long, pale brown, minutely raised-puncticulate.

In wet soil, often a weed along ditches and in fields and pastures, Tropical Zone to Transition Zone, widely distributed in tropical and warm-temperate regions of the Old and New Worlds; apparently adventive in southern California and Arizona. Flowering July–Oct. in our area.

**16. Cyperus elegans** L., Sp. Pl. ed. 2, 68, 1762

*Cyperus viscosus* Swartz, Prod. Veg. Ind. Occ. 20, 1788.
(For additional synonymy see Pflanzenreich **4**[101]: 164, 1936.)

Annual, or possibly a short-lived perennial, with tufted, triquetrous culms 3–6 dm. tall, these and leaves viscid-glutinous toward base; leaf sheaths leathery, to 8 cm. long, tapering gradually into slender, more or less convolute, transversely septate blades, the latter about equaling culms; involucral leaves 3–5, unequal, spreading, exceeding inflorescences; umbels compound, rays 3–10, to 10 cm. long, each bearing a densely crowded, subglobose head 1–1.5 cm. in diameter; spikelets oblong, 2.5–3 mm. wide, 4–8 mm. long, 8–12-flowered; glumes broadly ovate, acute and slightly mucronate, midrib pale green, sides stramineous, faintly 7–9-nerved; rachilla rigid, wingless; stamens 3; style deeply 3-cleft; achenes broadly obovoid, trigonous, nearly black or with a grayish bloom, minutely raised-puncticulate.

Moist, swampy places and margins of streams, Lower Sonoran and Tropical Zones, Texas to Florida and central Sonora, southward through the West Indies and Central America. A variety occurs in Peru. Aug.–Nov.

## 3. FUIRENA Rottb., Descr. & Ic. Pl. 70, *pl. 19,* 1773

Perennial or annual sedges with closed-sheath leaves, these sometimes reduced to sheaths only. Spikelets terete, many-flowered, often burlike, in close clusters or less commonly solitary. Scales spirally imbricated, awn-tipped. Flowers perfect. Perianth of 3 stalked, more or less petaloid scales, these usually alternating with retrorsely barbed bristles. Stamens 3. Ovary stipitate. Style slender, not swollen basally, deciduous. Stigma 3-cleft. Achenes trigonous, subsessile to distinctly stipitate.

**1. Fuirena simplex** Vahl, Eclog. Amer. **2**: 8, 1798

*Fuirena squarrosa* var. *aristulata* Torr., Ann. Lyc. N.Y. **3**: 291, 1836.
*Fuirena primiera* M. E. Jones, Contr. West. Bot. **18**: 25, 1933.

Perennial from short, frequently tuber-bearing rootstocks; culms slender, 1–6 dm. tall, leafy to inflorescence, glabrous, but almost entirely covered by sheaths; leaf sheaths exceeding next internode above, to 10 cm. long or more, slightly inflated upward, sparsely hispid but eventually nearly glabrate; ligules short, hyaline, rather densely villous-ciliate; leaf blades flat, oblong to lance-linear, 2–15 cm. long, 5–10 mm. wide, glabrous or hispid-ciliate, more or less shining above; spikelets ovoid, acute, 3–4 mm. in diameter, 5–10 mm. long, in 1–several terminal and axillary, pedunculate clusters of 3–10 spikelets; scales broadly ovate to obovate, rusty with greenish or stramineous midrib, body 2.5–3 mm. long, midrib excurrent as a curved or squarrose awn 0.5–1.5 mm. long; perianth scales (sepals) with ovate to oblong blades 0.6–0.8 mm. long, set on a slender claw about as long, emarginate, awned with a smooth or retrorsely barbed bristle 0.2–0.4 mm. long set on the back just below apex, rounded to subcordate at base; bristles between perianth scales retrorsely barbed, equaling or exceeding achene; achene pale yellowish, strongly trigonous, obovoid, about 1 mm. long, short-stipitate, acute at each end.

Moist meadows and margins of streams and ponds, Lower Sonoran Zone to Tropical and Upper Carolinian Zones, Nebraska and Missouri to Texas, Cuba, and Central America. Central Sonora and near Loreto, Baja California. Aug.–Oct.

## 4. SCIRPUS L., Sp. Pl. 47, 1953

Perennials or annuals, stoloniferous or cespitose plants with erect culms sheathed at base by lower part of leaves, blades well developed or reduced to a mucro. Culms terete or triangular, spongy. Inflorescence a terminal or subterminal umbel or panicle, appearing lateral when the main bract is stiff and erect, subtended by 1–several bracts. Spikelets 1–many, sessile or pedunculate, 3–many-flowered. Scales (glumes) imbricated spirally, all perfect. Perianth of bristles or none. Stamens 3. Style 2- or 3-cleft, deciduous. Achenes lenticular, plano-convex, or trigonous, usually apiculate.

Involucre of foliaceous bracts; cauline leaves with blades to 2 dm. long or more .......... 1. *S. paludosus*
Involucre of stiff, nonfoliaceous bracts; cauline leaves consisting of sheaths only or blades greatly reduced:
  Plants annual; culms rarely over 1 dm. tall; bristles none; achenes 0.75–1 mm. long .... 2. *S. cernuus* var. *californicus*
  Plants perennial; culms much over 1 dm., often to 2 m., tall; bristles present; achenes 2–3 mm. long:
    Spikelets few, sessile or very short-pedunculate; culms slender:
      Plants from black, scaly rhizomes; sides of culms concave; scales 3 mm. long, slightly fimbriate; spikelets obtuse .......... 3. *S. chilensis*
      Plants from reddish to brownish, scaleless stolons; sides of culms flat or slightly convex; scales 3–4 mm. long, strongly fimbriate; spikelets acute .......... 4. *S. americanus* var. *polyphyllus*
    Spikelets numerous, usually pedunculate; culms stout:
      Roots fibrous; spikelets ovoid to oblong-conic; achenes nearly as long as subtending scales .......... 5. *S. validus*
      Roots spongy-swollen; spikelets subcylindric; achenes about one-half as long as scales:
        Bristles retrorsely barbed, not plumose; culms terete to apex; spikelets 7–20 mm. long .......... 6. *S. acutus*
        Bristles plumose toward the base; culms trigonous at summit; spikelets 5–10 mm. long .......... 7. *S. californicus*

### 1. **Scirpus paludosus** A. Nels., Bull. Torrey Club **26:** 5, 1899

(For complete synonymy see N. Amer. Fl. **18:** 483, 1947.)

Perennial with erect, smooth, triquetrous culms to 1.5 m. tall from nodose, horizontal rhizomes that frequently bear tubers; roots slightly spongy, swollen; sheaths successively longer so that culms are often sheathed nearly or quite one-half their length, most if not all sheaths bearing blades; leaf blades 5–9 mm. wide, 2–5 dm. long, firm and ascending, glabrous except for margins and midribs, these scabrous near tip; involucral leaves 2–5, unequal, the longest to 3 dm. long, sharply scabrous on margins and midrib toward tip; inflorescence of simple or decompound umbels, primary rays 1–6 cm. long, or some heads sessile; spikelets narrowly ovate to ovate-cylindric, 4–7 mm. in diameter, 8–25 mm. long, acute at apex, obtuse or rounded at base; glumes pale stramineous to light brown, oblong-ovate, 5–6 mm. long, puberulent, midrib excurrent as an erect or faintly reflexed awn 1–2 mm. long, margins usually irregularly erose, at least toward summit; bristles 2–6, about 1.5–2 mm. long, delicate, minutely barbed; stamens 3; style 2-cleft, about 6 mm. long; achenes ovoid-lenticular, light brown to dark brown or nearly black, 3–4 mm. long, about 2 mm. wide, apiculate.

Margins of fresh-water or brackish ponds and lagoons, along streams and irrigation ditches, Lower Sonoran Zone to Transition Zone and in the north into the Canadian Zone, temperate and tropical regions of both North and South America, from southern Canada to Argentina; Australia and the East Indies. Flowering in our area March–July.

**2. Scirpus cernuus** Vahl var. **californicus** (Torr.) Beetle, Rhodora **46:** 145, 1944

*Isolepis leptocaulis* Torr., Pacif. R. Rep. **4:** 153, 1857.
*Isolepis pygmaea* var. *californica* Torr., Bot. Wilkes Exp. **17:** 476, 1874.
*Scirpus subtilis* Boeck., Allg. Bot. Zeit. **2:** 54, 1896.

Annual or short-lived perennial with slender, cespitose culms 3–10 or rarely to 40 cm. tall from fibrous roots or short, spreading rhizomes; culms terete or obscurely triquetrous, smooth and glabrous; sheaths reddish at base, sometimes greenish above, truncate but mucronate-tipped with a short, subulate blade 0.5–2.5 mm. long; involucral bract solitary, erect or nearly so, 0.5–6 mm. long, blunt; spikes solitary or sometimes in pairs, 3–8-flowered, ovoid, acute, 1–2 mm. wide, 2–4 mm. long; scales broadly ovate, pale green to dark purplish brown, midrib greenish or nearly white; bristles none; stamens 3; style 3-cleft virtually to base; achenes broadly obovoid, obtusely trigonous, whitish to dark brown or nearly black, puncticulate-rugulose, minutely apiculate, 0.75–1 mm. long.

Moist soil, chiefly Upper Sonoran Zone to Transition Zone, mostly near the coast, British Columbia to northern Baja California; Florida, but probably adventive there. April–Sept.

**3. Scirpus chilensis** Nees & Meyen, in Nees, Nova Acta Acad. Leop.-Carol. **19** Suppl.: 93, 1843

*Scirpus olneyi* A. Gray, in Engelm. & Gray, Bost. Jour. Nat. Hist. **5:** 238, 1845.
*Schoenoplectus olneyi* Palla, Bot. Jahrb. **10:** 299, 1888.
*Scirpus olneyi* var. *longisetis* Kükenthal, in Urban, Ark. Bot. **22A**[17]**:** 8, 1929.
*Scirpus olneyi* f. *australis* Barros & Osten, An. Mus. Hist. Nat. Montevideo II, **3:** 204, 1931.

Perennial with erect culms to 2 m. tall from scaly, black rhizomes; culms stoutish, sharply triquetrous with faces strongly concave; leaf sheaths brownish or reddish, thin and nearly scarious to firm, terminating as a rounded, deltoid point, hence bladeless or the blade 2–15 cm. long, erect or strongly ascending, glabrous; involucral leaf solitary, 1–5 cm. long; spikelets in capitate clusters of 3–12, ovoid-oblong to oblong, usually obtuse at both ends, 5–8 mm. long, 3–4 mm. wide, pale brown or reddish, scales glabrous, oval to suborbicular, about 3 mm. long, dark brown with a greenish midrib, emarginate or mucronulate; bristles 2–3, retrorsely barbed on upper half; style 2-cleft; achenes obovoid, plano-convex, about 1.5 mm. wide, 2.5 mm. long, smooth, dull to slightly lustrous, apiculate.

In marshy places, usually near the coast but at scattered inland localities, Lower Sonoran Zone to Transition, Carolinian, and Tropical Zones, western America from Oregon to South America, along the Atlantic seaboard from Nova Scotia southward into the West Indies. Occurring at Calamajué, Baja California. March–July.

**4. Scirpus americanus** Pers. var. **polyphyllus** (Boeck.) Beetle, Amer. Jour. Bot. **30:** 399, 1943

(For synonymy see N. Amer. Fl. **18:** 501, 1947.)

Perennial with slender, triquetrous culms 1–10 (usually about 2–5) dm. tall, from tough, reddish to dark brown rhizomes; culms often laxly arching or decumbent, with 1–6 blade-bearing sheaths; leaf blades 2–5 mm. wide, 1–3 dm. long, distinctly keeled on lower side, margins smooth; spikelets few, borne in compact pseudolateral glomerules of 1–7 spikelets, oblong-ovoid, acute, 6–12 mm. long, 2.5–4 mm. wide; involucral leaf slender, 2–10 cm. long, rigidly erect or somewhat flexuous; scales chocolate brown, chestnut brown, or sometimes upper margins slightly lighter but flecked with dark brown, broadly ovate, obtuse to rounded at emarginate summit, 3–4 mm. long, midrib excurrent as an erect awn 0.5–2 mm. long, margins scarious, upper margins often ciliolate-fimbriate; bristles 2–6, usually 4, retrorsely barbed, shorter than or about equaling achene; stamens 3; style 3-cleft; achenes about 3 mm. long, 2 mm. wide, plano-convex, dark brown, smooth, distinctly apiculate.

Sandy or clay soil, Lower Sonoran Zone to Canadian Zone, British Columbia to Ontario, south to Texas, the West Indies, and South America; New Zealand and Tasmania. Occurring in the Sonoran Desert about the margins of Salton Sea and at other localities where saline soil occurs. Feb.–Oct.

**5. Scirpus validus** Vahl, Enum. **2:** 268, 1806

(For synonymy see N. Amer. Fl. **18:** 503, 1947.)

Perennial with stout, terete, smooth, erect culms 1–3 m. tall from slender to stoutish rootstocks or rhizomes closely covered with brownish scales; lower sheaths brown or reddish, usually acute, upper one often with a greenish, broad, hyaline margin and ending in an erect or stiffly ascending blade to 1 dm. long; culms to 2 cm. in diameter near base, spongy, dark green; bract of inflorescence solitary, rigidly erect, 1–7 cm. long, terete or canaliculate on adaxial side, usually much shorter than umbellate inflorescence, this appearing lateral, decompound; slender primary rays 1–7 cm. long; spikelets many, in glomerules of 1–5 on flattish, scabrous-margined pedicels, the spikelets ovoid to oblong-conic, some of them sessile, acute or obtuse, 3–4 mm. in diameter, 5–12 mm. long, pale straw-brown, few to many-flowered; fertile scales about 2 mm. broad, 2.5–3 mm. long, ovate, midrib sometimes excurrent as a weak awn, this straight or slightly recurved, margins strongly fimbriate or entire; bristles 4–6 (usually 6), about equaling achene, retrorsely but finely barbed to base; stamens 3; style 2-cleft about halfway to base, reddish; achenes obovoid, plano-convex, nearly equaling scales, gray to brown, about 2 mm. wide, dullish, apiculate.

In ponds and at margins of marshes and streams, usually in fresh water, Lower Sonoran Zone to Transition Zone, British Columbia to Newfoundland and south to Argentina. April–Aug.

**6. Scirpus acutus** Muhl. ex Bigelow, Fl. Bost. 15, 1814

*Scirpus lacustris* var. *occidentalis* S. Wats., Bot. Calif. **2:** 218, 1880.
*Scirpus occidentalis* Chase, Rhodora **6:** 68, 1904.
*Scirpus occidentalis* var. *congestus* Farwell, Rep. Mich. Acad. **19:** 247, 1917.
*Scirpus acutus* f. *congestus* Fernald, Rhodora **23:** 131, 1921.
*Scirpus malheurensis* Henderson, Rhodora **32:** 20, 1930.

Perennial with erect, terete, green culms 1–3 m. tall, 0.5–2 cm. in diameter, from thick, brown rhizomes with spongy, brown roots; lower sheaths green or pale brown, firm, acute at apex, often splitting away from culm, upper sometimes with green blades to 1.5 cm. wide, 3–10 cm. long; involucral leaf solitary, straight, a continuation of culm, 1.5–10 cm. long; umbels decompound, primary rays 0.5–8 cm. long; spikelets ovate to cylindric-oblong, acute, solitary or in glomerules of 3–5, about 3–5 mm. wide, 7–20 mm. long, many-flowered, pale reddish brown to chestnut; glumes broadly ovate, 2–2.5 mm. wide, 3.5–4.2 mm. long, hyaline, emarginate, viscid at summit, margins fimbriate or entire, midrib puberulent, excurrent as a slender awn 1–1.5 mm. long; bristles 3–6, retrorsely barbed; stamens 3; style 5–6 mm. long, 2-cleft about one-third its length; achenes broadly obovoid, biconvex or slightly flattened on one side, 1.5–1.75 mm. broad, about 2 mm. long, brown to shining black, apiculate.

Margins of ponds and in marshy places, Lower Sonoran Zone to Canadian Zone, southern Alaska to Newfoundland, south to North Carolina, Texas, Arizona, and southern California; barely entering the upper margins of the Sonoran Desert. June–July.

**7. Scirpus californicus** (C. Meyer) Steud., Nom. ed. 2, **2:** 538, 1841
(For synonymy see N. Amer. Fl. **18:** 504, 1947.)

Robust perennials with culms 1–4 m. tall, to 3 cm. in diameter at base, about 3 mm. in diameter at tip, dark green, triangular near summit, usually terete below; roots spongy, reddish; basal sheaths firm, dark brown; inflorescence an open-paniculate, decompound umbel to 2 dm. long, spikelets on smooth primary rays 5–25 mm. long, secondary and tertiary rays 3–5 cm. long; outer involucral bracts 1.5–12 cm. long, smooth, stiffly erect, subspinulose, inflorescence appearing lateral; bractlets lanceolate, 5–15 cm. long; spikelets numerous, narrowly ovate, acute at apex, obtuse to rounded at base, pale chestnut brown, 3–4 mm. in diameter, 6–10 mm. long, frequently drooping; glumes ovate, about 2 mm. wide, 3 mm. long, obtuse to rounded at apex, margins entire, midrib excurrent 0.2–0.6 mm.; bristles 2–4, dark red, plumose toward base, retrorsely barbed toward apex or sometimes plumose throughout; style 2-cleft; achenes broadly ovoid-lenticular, about 1.25 mm. wide, 1.8–2 mm. long, dark brown, smooth, slightly and bluntly apiculate.

In water at the margins of streams and ponds, Upper Sonoran to Tropical Zones, California to South Carolina, south into South America; Hawaiian Islands. Feb.–Nov.

## 5. ELEOCHARIS R. Br., Prod. 224, 1810

Annual or perennial sedges with simple terete, triquetrous, quadrangular, grooved, or flattened culms in tufts or distinctly separated, from rootstocks, rhizomes, or fibrous roots. Leaves usually reduced to sheaths surrounding bases of culms, sometimes lowermost sheaths blade-bearing. Inflorescence of a solitary, erect, terminal spikelet devoid of involucral leaves and producing several to many flowers. Scales spirally imbricated, concave, more or less membranous. Perianth of 1–12 retrorsely barbed bristles or none. Stamens 2–3; anthers usually linear. Style 2-cleft and achene lenticular, or 3-cleft and achene trigonous, the angles then sometimes rounded and achene turgid. Base of style persistent as a terminal tubercle, a constriction often occurring between the achene and the tubercle.

Style branches 2; achene lenticular:
  Spikelets subglobose to broadly ovoid, 3–8 mm. long; achenes shining black, about 1 mm. long; plants annual ........ 1. *E. geniculata*
  Spikelets cylindric-lanceolate, 1–3 cm. long; achenes yellowish to pale brown, 1.2–1.6 mm. long; plants perennial ........ 2. *E. mamillata*
Style branches 3; achene trigonous or plano-convex:
  Tubercle continuous with achene, to one-half as long as achene ........ 3. *E. rostellata*

Tubercle obviously distinct from and one-sixth to one-fourth as long as achene:
Spikelets 2–3 mm. long; achenes longitudinally ribbed and transversely cancellate, nearly white; anthers 0.3–0.4 mm. long .................................... 4. *E. bella*
Spikelets 3.5–15 mm. long; achenes not ribbed or cancellate, grayish, or yellowish to dark brown; anthers 0.8–1 mm. long:
Spikelets linear-lanceolate; bristles white ............................ 5. *E. parishii*
Spikelets ovoid to ellipsoid; bristles brown ..................... 6. *E. montevidensis*

## 1. Eleocharis geniculata (L.) Roem. & Sch., Syst. Veg. 2: 150, 1817

*Scirpus geniculatus* L., Sp. Pl. 48, 1753, in part.
*Scirpus caribaeus* Rottb., Descr. Pl. Rar. Ic. 24, 1772.
*Eleocharis caribaea* S. F. Blake, Rhodora **20:** 24, 1918.
(For complete synonymy see Rhodora **41:** 51–52, 1939.)

Cespitose annual; culms 3–40 cm. tall, firm, 0.4–0.6 mm. in diameter, striate-sulcate; sheaths usually reddish brown at base, becoming stramineous toward the oblique to attenuate apex, 1–2 cm. long; spikelets broadly ovoid to subglobose, 2–3.5 mm. in diameter, 3–8 mm. long; scales ovate-orbicular, mostly stiffish, nearly cartilaginous but sometimes submembranous, yellow to pale brown, sublustrous in age, margins entire or erosulate toward apex; stamens 2–3; anthers linear, about 1 mm. long; style 2-cleft; bristles usually 7, 3 closely crowded near midline on abaxial face, 1 at each lateral angle, and 2 on adaxial face of and equaling or slightly exceeding achene, the latter obovoid, tumid-lenticular with rounded angles to nearly plano-convex, about 1 mm. long, shining black; tubercle whitish, spongy, usually strongly depressed and only slightly constricted at base.

In wet places, Lower Sonoran to Tropical Zones, southern California to Maryland and south to southern Brazil; also in the tropics of the Old World. Occurring in the Sonoran Desert at Palm Springs, Riverside County, California, in Arizona, Sonora, and Baja California. March–Sept.

## 2. Eleocharis mamillata Lindb. f., Acta Soc. Faun. Fl. Fenn. 23$^7$: 7, 1902

*Eleocharis macrostachya* Britt., in Small, Fl. SE. U.S. 184, 1903.

Erect or ascending perennial with soft, flattened culms 1–2 mm. wide, 2–12 dm. tall from creeping, dark brown rootstocks; sheaths reddish at base, pale brown to gray-brown toward nearly truncate apex; spikelets slenderly cylindric to narrowly lanceolate, 2–5 mm. thick at base, 1–3 cm. long, tapering to apex; basal sterile scales 2–3, narrowly ovate to oblong; fertile scales many, membranous, narrowly ovate, subacute to obtuse, 2–4 mm. long, pale brown to purplish; anthers linear, 1.2–2 mm. long; style 2-cleft; bristles 5–6 or rarely to 8 or wanting, often exceeding achene; achene yellowish to light brown, obovoid, 1.2–1.6 mm. long, shining; tubercle depressed-conic, essentially sessile, about as broad as high.

Wet soil along ditches, streams, and lakes, Transition and Canadian Zones to Tropical Zone, British Columbia to Illinois and south to southern California, Texas, and Michoacán; also in Eurasia. Occurring in the Sonoran Desert along the Colorado River, eastern Riverside County, California. March–July.

## 3. Eleocharis rostellata (Torr.) Torr., Fl. N.Y. 2: 347, 1843

*Scirpus rostellata* Torr., Ann. Lyc. N.Y. **3:** 318, 1836.
*Eleocharis rostellata* var. *occidentalis* S. Wats., Bot. Calif. **2:** 222, 1880.
*Eleocharis rostellata* var. *congdonii* Jepson, Fl. Calif. **1$^6$:** 196, 1922.

Perennial with tufted, erect or arching culms 1–2 mm. in diameter and 2–15 dm.

long from short, vertical rootstocks, culms considerably flattened and sulcate, often rooting at tips when these touch ground; sheaths reddish or brownish at base, greenish or stramineous toward truncate or oblique apex, rigid, to 4 cm. long; spikelets fusiform-elliptic, acute, 2–3 mm. in diameter, 6–20 mm. long, 12–20-flowered; scales rather rigid, ovate, light brown, lower ones obtuse or rounded, upper often acute, midrib greenish; stamens 3; styles 3-cleft; bristles light brown, firm, regularly toothed, about equaling achene, this obovoid, obtusely trigonous, or plano-convex, 2–3 mm. long, lightly and finely reticulate, olive, lustrous; tubercle continuous with achene, light green, about one-fourth to one-half as long as body of achene.

Margins of marshy places and along streams and ponds, particularly in alkaline regions, Lower Sonoran Zone to Transition and Tropical Zones, California and Oregon to Nova Scotia, and south into Bermuda, the West Indies, and Mexico. Occurring in the Sonoran Desert at Palm Springs, Riverside County, California. March–July.

**4. Eleocharis bella** (Piper) Svenson, Rhodora **31:** 201, 1929

*Eleocharis acicularis* var. *bella* Piper, in Piper & Beattie, Fl. Palouse Region 35, 1901.
*Eleocharis acicularis* var. *minima* Torr. ex Britt., Jour. N.Y. Micr. Soc. **5:** 104, 1889.
*Eleocharis acicularis* sensu Jepson, Fl. Calif. **1**[6]**:** 195, *fig. 16,* 1922. Not *E. acicularis* Roem. & Sch., 1817.

Densely crowded and tufted annual growing in patches 2–10 cm. in diameter, culms erect or ascending, 1.5–6 cm. tall, about 0.4 mm. in diameter, light green and soft, slightly angled or shallowly sulcate; sheaths thin, loose, slightly inflated at acutish summit, hyaline or nearly so; spikelets narrowly ovoid, 2–3 mm. long, usually acute but sometimes obtuse, several-flowered; scales narrowly ovate-lanceolate, about 1 mm. long, acute to keeled and slightly apiculate at summit, midrib green, broad, margins scarious, whitish or purple-streaked; styles 3-cleft or occasionally 2-cleft; stamens 2–3, anthers 0.3–0.4 mm. long; bristles none or 1–2 present but caducous; achene white to pale yellowish, linear-obovate, 0.7–0.8 mm. long, obtusely trigonous to nearly terete, longitudinally several-ribbed and finely cancellate transversely between ribs; tubercle slenderly ovoid, constricted at base, about one-sixth to one-fifth as long as body of achene.

Along stream banks and in marshy places, Lower Sonoran Zone to Transition Zone, Washington to Montana, south to New Mexico, Arizona, and central Sonora, where it barely enters the upper margin of the Sonoran Desert. March–July.

**5. Eleocharis parishii** Britt., Jour. N.Y. Micr. Soc. **5:** 110, 1889

*Eleocharis disciformis* Parish, Bull. S. Calif. Acad. **3:** 81, *pl. 6,* 1904.

Perennial with slender, striate culms 1.5 mm. in diameter, 1–3 dm. high from slender, creeping, reddish rootstocks but apparently flowering first year of growth; upper sheaths reddish brown below, stramineous to greenish above, brownish again at truncate, minutely toothed apex; spikelets linear-lanceolate, acute, 10–15 mm. long, many-flowered; scales oblong-ovate, stramineous along midrib, dark brown along sides, 2.5–3 mm. long, acute to obtuse, rather rigid; bristles 6–7, white or nearly so, retrorsely barbed, shorter than or exceeding achene; stamens 3; style 3-cleft; achenes obovoid to ellipsoid, 1.2–1.5 mm. long, trigonous with rounded angles or nearly plano-convex, yellow to light brown, faintly reticulate, shining; tubercle conic to subulate, usually constricted at base, paler than achene.

Stream banks, meadows, and marshy places, Lower Sonoran Zone to Transition

Zone, Oregon to Nevada, Arizona, southern California, and northern Chihuahua. March–July.

**6. Eleocharis montevidensis** Kunth, Enum. Pl. **2:** 144, 1837

*Limnochloa montevidensis* Nees, in Mart., Fl. Bras. **2**[1]**:** 99, 1842.
*Eleocharis montana* of authors. Not *E. montana* (H. B. K.) Roem. & Sch., 1817.
*Eleocharis arenicola* Torr., in Engelm. & Gray, Bost. Jour. Nat. Hist. **5:** 237, 1845.
*Eleocharis dombeyana* sensu Boeck., Linnaea **36:** 450, 1869–1870. Not *E. dombeyana* Kunth, 1837.
*Trichophyllum arenicolum* House, Amer. Midl. Nat. **6:** 204, 1920.
*Heleocharis montana* subsp. *montevidensis* Osten, An. Mus. Hist. Nat. Montevideo II, **3:** 183, 1931.

Perennial with erect culms 0.5–4.5 dm. tall from slender, creeping, reddish or reddish brown rhizomes or rootstocks; upper sheath deep brown to bright reddish brown at base, usually stramineous or slightly greenish toward truncate apex, one side bearing an apical, narrowly deltoid tooth 0.5–2 mm. long; culms very slender, about 0.5 mm. in diameter, light green, smooth, terete; spikelets ovoid to oblong, 3.5–13 mm. long, 2–2.8 mm. in diameter, many-flowered, obtuse at apex; scales yellowish to light brown, ovate, obtuse, 1.3–1.6 mm. long, hyaline along entire margins; anthers linear, 0.8–1 mm. long; bristles 4–6, dark reddish brown, equaling or slightly shorter than achenes, minutely toothed or barbed; style 3-cleft; achenes about 1 mm. long, obovoid-trigonous, angles obtuse and rounded, grayish, yellow, or brown, finely reticulate or minutely punctulate and glossy; style base conical, lightly constricted at base or sessile, about one-fifth the length of achene.

In wet soil at margins of streams, ponds, and seepy meadows, Lower Sonoran and Upper Sonoran Zones, Sacramento Valley, California, to South Carolina and south to Guatemala. March–July.

## 6. BULBOSTYLIS Kunth, Enum. Pl. 2: 205, 1837

Annual or perennial sedges with slender, erect culms leafy only at base, sheaths pubescent or ciliate, hyaline-margined; blades filiform to narrowly linear, in ours much shorter than culms. Spikelets umbellate, capitate, or solitary, subtended by 1–several involucral leaves. Scales spirally imbricated, deciduous. Flowers perfect, without a perianth scale. Stamens 2–3. Style deeply 2–3-cleft, glabrous, greatly swollen at base, swollen part persisting as a small tubercle at apex of lenticular or trigonous, turgid achene.

Plants perennial; culms 3–5 dm. tall; bases of stems hardened and bulblike; leaf blades ciliate-hirsutulous; achenes slate-colored ............................ 1. *B. juncoides*
Plants annual; culms 0.3–3 dm. tall; bases of stems not hardened or bulblike; leaf blades glabrous or essentially so; achenes white or pale ashy gray:
  Spikelets 2–5 in each inflorescence, borne only at tips of culms, not sessile in axils of basal leaves; leaf sheaths usually sparsely villous, at least at summit; achenes all alike .... .................................................. 2. *B. capillaris* var. *crebra*
  Spikelets usually solitary at tips of culms, and others sessile in axils of leaves; leaf sheaths glabrous; middle achenes of basal spikelets larger than those of culms ............ ............................................................ 3. *B. funckii*

**1. Bulbostylis juncoides** (Vahl) Kükenthal ex Osten, An. Mus. Hist. Nat. Montevideo II, **3:** 185, 187, 1931

*Schoenus juncoides* Vahl, Enum. **2:** 211, 1806.

Densely tufted perennial with subfiliform, strongly channeled, slightly scaberulous culms 3–5 dm. tall, these swollen and hardened at base into bulblike processes; leaf sheaths reddish, subscarious, 5–15 mm. long, acuminately narrowed to blade, ciliate-villous along margins, back glabrous or puberulent; leaf blades filiform, to 12 cm. long, strongly 3-ribbed beneath, ciliate-hirsutulous on margins and ribs, erect or ascending; spikelets in loose, slightly compound umbels, on slender branches and pedicels to 2.5 cm. long, some spikelets sessile or essentially so just above the 3–5 setaceous involucral bracts, these to 1.5 cm. long; individual spikelets brown, narrowly ovoid, 5–8 mm. long; scales broadly ovate, 1.6–2 mm. long, minutely puberulent, margins erosulate-ciliolate; achenes trigonous, broadly obovoid, 1–1.2 mm. long, minutely rugulose-papillate transversely, pale slate colored, plump.

In seepy ground, Transition Zone to Tropical Zone, Arizona to western Texas and south into South America. Aug.–Sept.

**2. Bulbostylis capillaris** (L.) C. B. Clarke var. **crebra** Fernald, Rhodora **40:** 395, *pl. 510, figs. 4–5,* 1938

Annual plant with subfiliform, erect, glabrous culms 0.3–3 dm. tall from delicate, fibrous roots; leaf sheaths scarious, at least on margins, ferruginous, 5–12 mm. long, glabrous or sparsely villous at or near apex, gradually narrowed to leaf blades, these filiform-setaceous, 1–6 cm. long, 3-nerved, glabrous or essentially so, erect or ascending; spikelets in small umbels of 2–5, or in depauperate plants often solitary at apices of culms, ovoid to oblong-ovoid, 4–7 mm. long, about 2 mm. in diameter, subtended by 1–3 setaceous bracts 1.5–12 mm. long; floral scales broadly ovate, abruptly short-acuminate, dark reddish brown with a greenish midrib, glabrous, deciduous at maturity, margins entire or faintly erosulate at apex; achenes broadly obovoid, turgid, trigonous in cross section, minutely rugulose-papillate in transverse lines, (in ours) white or pale ashy grey, 0.6–0.8 mm. long.

Margins of temporary pools and in seepy places, Lower Sonoran Zone to Carolinian and Transition Zones, Arizona, northern Sonora, and Texas to Georgia and north to Oklahoma, Nebraska, and Maine. Aug.–Sept.

**3. Bulbostylis funckii** (Steud.) C. B. Clarke, in Urban, Symb. Antill. **5:** 290, 1907

*Isolepis funckii* Steud., Syn. Pl. Glum. **2:** 91, 1855.
*Scirpus heterocarpus* S. Wats., Proc. Amer. Acad. **18:** 171, 1883.
*Stenophyllus funckii* Britt., Bull. Torrey Club **21:** 30, 1894.

Differing from *B. capillaris* only in having sessile spikelets produced in axils of leaves, these basal spikelets containing achenes 1–1.2 mm. long; in having leaf sheaths totally glabrous; and in most culms bearing solitary spikelets, occasionally some of them in pairs on erect culms.

Wet soil, chiefly Upper Sonoran Zone to Transition Zone, straggling to the margins of the Sonoran Desert along canyons; Arizona, New Mexico, and Chihuahua south to central South America and in the West Indies. Aug.–Sept.

## 7. FIMBRISTYLIS Vahl, Enum. 2: 285, 1806

Annual or perennial sedges with erect or ascending culms and basal, sheathing leaves. Spikelets terete, ovoid to subcylindric, several- to many-flowered, arranged in umbels or heads, each inflorescence subtended by a 1–many-leaved involucre. Scales spirally imbricated, all fertile, deciduous in fruit. Perianth lacking. Stamens 1–3, filaments glabrous. Bristles none. Style slender, 2–3-cleft, glabrous or pubescent, much enlarged at base but deciduous from apex of achene at maturity. Achenes lenticular or obtusely 3-angled, longitudinally striate or ribbed, cancellate or reticulate between ribs, apex acute to truncate but without a terminal knob.

Culms ascending-spreading, densely tufted, 2–10 cm. long; achenes cancellate but not papillate, 0.6 mm. long; spikes capitately arranged; styles glabrous toward base .......... 1. *F. vahlii*

Culms erect or slightly spreading, only slightly tufted, 10–30 cm. tall; achenes cancellate and papillate on longitudinal ridges, 0.9–1.2 mm. long; styles puberulent to base:

Leaf blades flat, 1–2 mm. wide (or somewhat involute in age); umbels decompound; achenes 0.9–1 mm. long, obscurely tuberculate .................... 2. *F. alamosana*

Leaf blades filiform-setaceous, less than 0.4 mm. wide; umbels simple; achenes 1–1.2 mm. long, distinctly muriculate-papillose on ribs ........................ 3. *F. arenicola*

## 1. **Fimbristylis vahlii** (Lam.) Link, Hort. Berol. **1:** 287, 1827

*Scirpus vahlii* Lam., Tab. Encycl. **1:** 139, 1791.
*Fimbristylis congesta* Torr., Ann. Lyc. N.Y. **3:** 345, 1836.

Densely tufted annual with very slender, compressed, longitudinally striate, nearly glabrous to sparsely scaberulous culms 2–10 cm. long, exceeding or shorter than leaves; leaf sheaths subscarious, margins hyaline, gradually narrowing upward; leaf blades setaceous to subfiliform, 2–6 cm. long, those of involucre nearly erect, to 2.5 cm. long, much exceeding capitate clusters of spikelets; spikelets 4–7 mm. long, 1.5–2.5 mm. wide at anthesis, somewhat longer in fruit, oblong-cylindric, many-flowered; scales acuminate-lanceolate, pale brown with greenish midrib; stamen 1, slightly surpassing style at anthesis; style 2-cleft, glabrous below; achene biconvex, obovoid, cream-colored to nearly white, about 0.6 mm. long, minutely cancellate but not at all papillate.

In moist soil, Lower and Upper Sonoran Zones to Carolinian Zone, Lake and Tulare Counties, California, and at Yuma, Arizona; Texas to Missouri and east to North Carolina and Florida. July–Oct.

## 2. **Fimbristylis alamosana** Fernald, Proc. Amer. Acad. **36:** 491, 1901

Somewhat tufted annual 1–3 dm. tall with densely ciliate to glabrous sheaths and flat, slender leaf blades 0.3–2.5 dm. long, these glabrous or sparsely scaberulous along margins; leaves of involucres unequal, to 1 dm. long; umbels decompound, with 3–9 slender, lax, unequal rays, one spikelet usually sessile at base of umbel, rest on slender pedicels to 3 cm. long; spikelets narrowly ovoid, 3–5 mm. long, scales pale stramineous, 1.5–2 mm. long, broadly ovate, acute to obtuse and slightly mucronate, glabrous and somewhat shining; stamen 1; style compressed, pubescent, at least on margins to base; achenes broadly obovoid, biconvex, narrowed and substipitate at base, 0.9–1 mm. long, white or nearly so, cancellate and longitudinally ribbed, the ribs smooth or sparsely and obscurely tuberculate or muriculate.

Margins of pools and on hillsides, Lower Sonoran Zone, central to southern Sonora. July–Aug.

## 3. **Fimbristylis arenicola** Wiggins, Contr. Dudley Herb. **4:** 15, 1950

Erect, slightly tufted annual with very slender culms 1–2 dm. tall or less; leaf sheaths 1–2 cm. long, reddish, submembranous, villosulous; leaf blades nearly filiform, flat, less than 0.5 mm. wide, 3–7 cm. long, glabrous; spikelets solitary or 2–3 at apex of culms; involucre of a single subfiliform leaf 1.5–4 cm. long; spikelets sessile or on pedicels to 2 cm. long, narrowly ovoid, about 2 mm. in diameter, 3–4 mm. long, stramineous, scales broadly ovate, abruptly acute, about 1.5 mm. long, glabrous; stamen 1; style 2-cleft, scarcely flattened, puberulent to base; achenes broadly obovoid-biconvex, white, cancellate, and longitudinally 7–9-ribbed, ribs sparsely but distinctly muriculate-papillose, base slightly contracted-attenuate.

On gravelly hillsides, Lower Sonoran Zone, between Mátape and Batúc, Sonora. Sept.

## 8. **CLADIUM** P. Br. ex Crantz, Inst. **1:** 362, 1766

Robust, perennial, leafy sedges with fusiform to oblong, few-flowered spikelets in glomerules, panicles, or umbels. Leaves very long. Glumes imbricated in the terete spikelets, lower ones empty, middle ones bearing imperfect or sterile flowers, only the upper subtending fertile flowers. Perianth none. Stamens 2 or less commonly 3. Sterile bristles none. Style 2–3-cleft, branches sometimes again 2–3-parted, deciduous from apex of achene. Achenes ovoid to globose, longitudinally striate or smooth and lustrous, terminal tubercle none.

**1. Cladium californicum** (S. Wats.) O'Neill, in Tidestrom & Kitt., Fl. Ariz. & New Mex. 773, 1941

*Cladium mariscus* var. *californicum* S. Wats., Bot. Calif. **2:** 224, 1880.
*Mariscus californicus* Fernald, Rhodora **25:** 52, 1923.

An erect, densely tufted sedge with stout, obtusely triquetrous culms 1–2 m. tall; leaves linear, 7–14 mm. wide, equaling culms, glabrous on faces, margins antrorsely spinulose-serrulate; umbels decompound, numerous, loosely arranged in panicles 1–4 dm. long, 1–2 dm. wide; spikelets reddish brown, glomerate in loose to compact groups at ends of raylets, 1.5–2 mm. wide, 4–6 mm. long, terete or essentially so, narrowly ovoid or elliptic, acute; glumes lance-ovate, acute, faintly striate, sublustrous, subscarious, lower ones empty or subtending sterile flowers, uppermost bearing a perfect floret; stamens 2; style 2-cleft; achenes ovoid, acute at apex, lustrous, pale brown.

In springy or swampy ground, Lower and Upper Sonoran Zones, Inyo, Los Angeles, and Riverside Counties, California, east into southern Nevada and Arizona. May–July.

## 9. **CAREX** L., Sp. Pl. 972, 1753

Perennial, grasslike, monoecious or dioecious sedges from slender to stout rootstocks. Culms usually triquetrous, leafy or less commonly nearly leafless. Leaves mostly in 3 ranks, blades elongated, or upper ones subtending the spikes short or wanting. Spikes 1–many, either wholly pistillate, wholly staminate, or with both types of flowers androgynously or gynaecandrously arranged in the same spike, sessile or pedunculate. Flowers solitary in axils of scalelike glumes spirally arranged. Staminate flowers of 2–3 stamens with filiform filaments. Pistillate flowers of a single pistil borne inside a flasklike perigynium; styles exserted through a minute terminal pore. Perianth none. Style 2-cleft and achenes lenticular, or 3-cleft and achenes triquetrous or plano-convex, achenes inside perigynium or rarely rupturing it at maturity.

Spikes short, 0.5–1.5 cm. long, crowded and scarcely distinguishable; scales stramineous; leaf sheaths hyaline, not transversely netted ventrally .................... 1. *C. alma*
Spikes elongate, 4–15 cm. long, distinct; scales of pistillate spikes purplish black with lighter midribs; leaf sheaths brownish, breaking into transverse network of fibers ventrally ... ........................................................................ 2. *C. schottii*

**1. Carex alma** L. H. Bailey, Mem. Torrey Club **1:** 50, 1889
*Carex vitrea* Holm, Amer. Jour. Sci. IV, **17:** 302, *figs. 5–7,* 1904.
*Carex arizonensis* C. B. Clarke, Kew Bull. Add. Ser. **8:** 68, 1908, in part.

Cespitose, with triquetrous culms 3–12 dm. tall from black, fibrillose rootstocks; leaves several near base of each culm, blades flat or canaliculate, 3–6 mm. wide, 1–5 dm. long, rough near apex and serrulate-margined; sheaths tight, somewhat septate-ridged dorsally, thin and often purple-dotted ventrally, not splitting into fibrous cords, ligule about as long as wide; heads 2.5–20 cm. long, 1–2 cm. in diameter, compound or decompound, individual spikes closely crowded and scarcely distinguishable, staminate flowers uppermost in each spike, lower ones subtended by bracts 2 cm. long or less, these dilated at base and attenuate at apex; scales stramineous or pale brown, ovate, obtuse to briefly awned, 3–4.2 mm. long, 1.5–1.8 mm. wide; perigynia plano-convex, ovate to ovate-oblong, 3.5–4 mm. long, nearly black at maturity, tapering to a serrulate beak about one-third to one-half the length of body, beak minutely bidentate; style branches 2; achenes oval-lenticular, short stipitate and apiculate, about 1.5 mm. long.

Along streams and in moist canyons, upper margins of Lower Sonoran Zone and Upper Sonoran Zone, central California to northern Baja California, east to southern Nevada and Arizona. Occurring on the western margins of the Colorado Desert. May–July.

**2. Carex schottii** Dewey, in Torr., Bot. Mex. Bound. 231, 1859
*Carex laciniata* Boott, Ill. Carex 175, *pl. 594,* 1867, in part.
*Carex barbarae* Parish, Bull. S. Calif. Acad. **4:** 108, *pl. 14,* 1905. Not *C. barbarae* Dewey, 1858.

Densely cespitose plant from a short-creeping caudex and long stolons; culms stout below, 1–1.5 m. tall, sharply triquetrous, serrulate on angles, clothed on the lower one-third with leaves of current and previous years; leaf blades 6–12 mm. wide, 3–10 dm. long, revolute-margined, puberulent toward base, papillate; sheaths sharply keeled and rough-puberulent dorsally, lower ones breaking and becoming filamentous-netted ventrally, ligules longer than wide; staminate spikes 2–3, upper pedunculate, lower sessile, 4–14 cm. long, about 4 mm. in diameter, scales reddish brown; pistillate spikes usually 3–5, but staminate at apex, 5–6 mm. in diameter, 5–22 cm. long, the scales linear-lanceolate to oblong, purplish black with a lighter median line, narrower than but exceeding perigynium; perigynium about 1.7–1.8 mm. wide, 3–3.5 mm. long, greenish-stramineous, 2-ribbed and puncticulate, base substipitate, minutely beaked, orifice subemarginate; style branches 2; achenes lenticular, obovate, 1.75–2 mm. long, about 1.5 mm. wide, substipitate.

Stream banks and margins of swampy places, Upper Sonoran Zone and upper margins of Lower Sonoran Zone, Santa Barbara County to San Diego County, California, east to the margins of the Mojave and Colorado Deserts. Feb.–June.

## Family 10. **PHOENICACEAE.** Palm Family

Tall or small trees with cylindrical, unbranched trunks and an apical crown of large leaves, these persistent and clothing trunk below the living crown, or deciduous. Leaves fan-shaped or pinnate, short- or long-petiolate; petiolar bases clasping trunk by tough, fibrous sheaths; petioles armed or smooth. Flowers small, unisexual or perfect, in axillary spadices surrounded by large spathes; perianth segments 6, in 2 whorls. Stamens 6–24, filaments dilated basally; anthers introrse. Ovary of 3 distinct or united, 1-ovulate carpels; styles 1–3. Fruit a drupe or berry; endosperm horny, embryo cylindrical, embedded near surface of endosperm.

Fruit 4 cm. long or less, drupaceous or baccate, not hollow; leaves induplicate in bud:
  Seed not grooved, slightly longer than broad or subglobose; leaves flabelliform; fruit with thin, dry flesh:
    Style or stigma basilar in fruit; seed depressed-globose; leaf blade deflexed along extended costa of petiole ........................................ 1. *Sabal*
    Style or stigma terminal in fruit; seed globose to elliptic, not depressed; leaf blade not decurved:
      Perianth segments valvate in bud; leaf sheaths not split at base; fruit 15–24 mm. long at maturity ........................................ 2. *Erythea*
      Perianth segments imbricate in bud; leaf sheaths split at base; fruit 4–10 mm. long at maturity ........................................ 3. *Washingtonia*
  Seed grooved, two to four times as long as broad; fruit fleshy ................ 4. *Phoenix*
Fruit 10–25 cm. in diameter, nutlike, with a bony endocarp, hollow and containing milklike fluid; leaves reduplicate in bud ........................................ 5. *Cocos*

## 1. SABAL Adans. ex Guer., Bull. Sci. Soc. Phil. **87:** 205, *pl.* 25, 1804

Low to tall palms, the trunk erect, or subterranean and sigmoidly curved, with entire-margined petioles and flabelliform leaf blades bearing a long-continued costa from which the segments arise, this strongly arched downward; hastula extended, pointed. Inflorescence horizontal in youth but soon drooping, bearing diffuse, elongated, lateral branches and many flowers. Flowers hermaphroditic; calyx tubular; petals 3, imbricate in bud; stamens 6; pistils of 3 carpels, only 1 of which matures. Fruit thin-fleshed, rather dry, depressed-globose; seed shaped like fruit, raphe short, in a slight basal cavity; micropyle lateral.

**1. Sabal uresana** Trelease, Rep. Mo. Bot. Gard. **12:** 79, *pls.* *35–37,* 1901
*Inodes uresana* Cook, Bull. Torrey Club **28:** 534, 1901.

Tree to 10 m. tall with blue-glaucous leaves and obscure rings on the trunk after the leaves fall; petiole 2–3 m. long, concave on upper surface, 5–6 cm. broad and filiferous at base, this plunging downward along trunk, 2–4 cm. wide at apex; hastula pointed, 10–14 cm. long; costa stout and strongly decurved (a character not evident in most pressed specimens); blade 1–2 m. across, filiferous, central segments 3.5–4 cm. broad at base, lateral ones about half as wide; inflorescence to 2–3 m. long, rachillae terete, glabrous; flowers sessile, 4 mm. long, petals 3 times as long as sepals; fruit 18–20 mm. wide at maturity, 12–14 mm. high, shining light brown, glabrous; seed 12–14 mm. wide, 7–8 mm. high.

In canyons and on tablelands, Lower Sonoran Zone, near Ures, between Batúc and Hermosillo, at San Pedro and San Carlos Bays, near Guaymas, Sonora. Flowering in early spring.

## 2. ERYTHEA S. Wats., Bot. Calif. **2:** 211, 1880

Tree palm with a single erect trunk bearing a close "shag" just below crown but nude below, interfoliar spadices, and flabelliform leaves; blades large, green or glaucous, glabrous to tomentose at base, segments stiff or drooping, filiferous; hastula mostly broad and short; petioles plain or spinose on margins. Spathes tubular but flattened along edge, spadix arched or drooping, persistent after dropping of fruit. Flowers small, sessile, 3 petals valvate in bud; sepals 3, shorter than petals; stamens 6, filaments dilated and connate or distinct below. Ovary of 3 nearly distinct to coherent carpels, of which only 1 matures; style terminal in fruit. Fruit globose or ovoid, drupe-like, slightly asymmetrical or flattened on one face, black or purplish and slightly fleshy at maturity; seed hard, embryo opposite raphe.

Leaves usually conspicuously glaucous on both sides; petioles 3–4 cm. broad at middle; petiolar spines 4–20 mm. long:
- Spadices 4–5 m. long, drooping in flower, pendent in fruit; petiolar spines stout, 4–10 mm. long; fruit flattened on one side .................................. 1. *E. armata*
- Spadices 2–3 m. long, ascending to arched but not pendent in fruit; petiolar spines slender, 8–20 mm. long; fruit not flattened on one side ..................... 2. *E. roezlii*

Leaves moderately glaucous or green above; petioles 1.5–3 cm. broad at middle; petiolar spines 2–7 mm. long:
- Trunk about 1 m. high; spadices much exceeding leaves, conspicuous; fruit pyriform to elliptic .................................................... 3. *E. elegans*
- Trunk 10–12 m. high; spadices equaling or shorter than leaves, inconspicuous; fruit oblong-globose ................................................ 4. *E. brandegeei*

**1. Erythea armata** S. Wats., Bot. Calif. **2:** 212, 1880
*Glaucothea armata* Cook, Jour. Wash. Acad. **5:** 239, 1915.

Large tree with a stout trunk to 15 m. high; leaves heavily blue-glaucous; shag persistent for many years but eventually falling; petiole 1–1.2 m. long, 3–4 cm. wide at middle, irregularly tomentose in brownish patches, armed marginally with curved to hooked, sharp, callous spines 4–10 mm. long, these mostly contiguous; blade 1–2 m. wide, incised about one-half its depth with 40–60 strongly veined, 2-cleft segments 2–3 cm. broad at middle and bearing a few short threads toward apical margins, tips spreading or arching but not pendent; spadices 4–5 m. long, arcuate-drooping in flower, vertically pendent outside the shag in age; rachillae slender, 0.5–1.5 mm. in diameter; flowers subtended by subulate bracts 3–4 mm. long; stamens distinct at base; fruit broadly ovoid to globose, 18–24 mm. long, nearly as wide, splotched and streaked with brown; seed oblong-globose, somewhat flattened on one side, about 15 mm. in diameter, dull brown.

Native in desert and semidesert canyons, Lower Sonoran Zone and dry parts of Upper Sonoran Zone, on the east side of the mountains of Baja California from Cantillas Canyon south almost to San Ignacio. On the west side of the range it occurs less abundantly from Cataviñá southward. Feb.–Mar.

**2. Erythea roezlii** (Linden) Becc., Ann. Roy. Bot. Gard. Calcutta **13:** 320, 1931
*Brahea roezlii* Linden, Ill. Hort. **28:** 38, 1881.
*Erythea armata* var. *microcarpa* Becc., Webbia **2:** 136, 1907.

Tree with trunk to 10–18 m. tall, up to 30 cm. in diameter; shag semipersistent on trunks 5–6 m. high, taller trunks usually naked at base, even when not burned; crown fairly dense and usually bluish glaucous, though sometimes quite green; petiole 1–1.5 m. long, glabrous or nearly so, 3–4 cm. wide at middle, margin bearing slender, mostly straight or slightly curved teeth 8–20 mm. long, these well separated; blade 1–1.75 m. long and wide, segments stiffish, standing in plane of blade or tardily declined, sparsely filiferous; spadices 2–3 m. long, ascending or arcuate-ascending, standing out well beyond leaves but not pendent; rachillae short-divaricate, stout (2–3 mm. in diameter); flowers 2–2.5 mm. long, somewhat crowded on rachillae, bractlets inconspicuous; fruit nearly globose, 18–20 mm. wide, 15–16 mm. high, flesh thin, wrinkled in drying, brownish-striped or -splotched; seed depressed-globose, embryo lateral, endosperm excavated basally and filled with a dark brown "podium."

In canyons and cliffs in desert ranges, Lower Sonoran Zone, in Sonora from near Magdalena to Guaymas, where it comes down almost to salt water.

**3. Erythea elegans** Franceschi ex Becc., Webbia **2:** 138, 1907

*Glaucothea elegans* I. M. Johnston, Proc. Calif. Acad. IV, **12:** 993, 1924.

A dwarf or slow-growing tree; crown large and open; leaves thinly glaucous to green above, more deeply glaucous beneath, sparsely and shortly filiferous; petioles 6–10 dm. long, 1.5–2 cm. broad, glaucous and irregularly tomentose, margins armed with simple or 2-cleft spines 2–5 mm. long; blade 7–12 dm. wide, thin in texture, segments 2–3 cm. wide, 2-cleft at apex, extending about halfway to base of blade; spadices exceeding leaves, 1.5–2.5 m. long, gracefully arched in fruit, heavily flowered; rachillae slender, pubescent-tomentose; flowers 3–4 mm. long, petals narrow, acute; fruit pyriform to narrowly oblong and pubescent when young, globose-oblong and glabrous at maturity, 18–22 mm. long, nearly as wide; seed loose in the crustose epicarp, 8–10 mm. high, 14–15 mm. wide, deeply excavated basally.

Known only from cultivated plants, the seed said to have come originally from Sonora.

**4. Erythea brandegeei** Purpus, Gartenfl. **52:** 11, *figs. 1–2,* 1903

*Glaucothea brandegeei* I. M. Johnston, Proc. Calif. Acad. IV, **12:** 992, 1924.

Slender tree with trunk 10–12 m. tall, obscurely ringed in age; petioles slender, 1–1.5 m. long, 3 cm. or less in width at middle, irregularly tomentose toward base, margin armed with simple and multiple spines 5–7 mm. long; blade dull green above, heavily glaucous beneath, 1–1.3 m. broad, undivided part 35–45 cm. across, segments to 4 cm. wide, slender, sparsely short-filiferous, tips deeply 2-cleft; spadices usually shorter than and obscured by leaves, horizontally arched, pubescent in youth, glabrate in age; rachillae slender; flowers crowded, 3 mm. long, petals broad, blunt; fruit oblong-globose, 15–22 mm. long, 1–4 mm. narrower; pericarp papery, puffy, shining, brown; seed 13–15 mm. wide, 9–10 mm. high, deeply excavated basally.

Growing in canyons and mountains of the Cape Region (Tropical Zone), and Sierra de la Giganta west of Escondido Bay, Baja California. Feb.–March.

## 3. WASHINGTONIA H. Wendl., Bot. Zeit. **37:** 68, 1879

Tall fan palms with unbranched trunks clothed with drooping leaves (unless these are removed by fire or pruning). Leaves on long, spine-margined petioles, segments of blade separating on edges into slender filaments, glabrous or variously tomentose about hastula and on ridges of segments beneath when young. Spadix subtended by 2-edged, flattened, smooth spathe 1–1.5 m. long, 1–3-branched, with flat spathelets at distant nodes; rachillae slender, glabrous. Flowers perfect, numerous, white to pale cream; calyx tubular, nearly equaling strongly reflexed, narrow petals; stamens 6, alternate ones with broader filaments, exserted. Fruit globose to elliptic, small, with thin, dry flesh. Seed brown, shallowly excavated on raphal face, hard endosperm white, plain.

Trunk stout, not flaring at base; leaves and old fruiting branches drooping vertically when dead; petiolar bases not forming a definite pattern on trunk ............ 1. *W. filifera*
Trunk slender, flaring abruptly at base; leaves hanging at an angle of 20–50° from trunk, old fruiting branches often extending horizontally through the shag; petiolar bases forming a crisscross pattern on trunk after leaves fall ................ 2. *W. robusta*

**1. Washingtonia filifera** (L. Linden) H. Wendl., Bot. Zeit. **37:** lxi, 1879

*Washingtonia filifera* var. *robusta* Parish, Bot. Gaz. **44:** 420, *figs. 3–5,* 1907. Not *W. robusta* H. Wendl., 1883.
(For added synonymy see Principes **7:** 170, 1963.)

A stout tree with a trunk to 1 m. in diameter and 15 m. tall with a crown of gray-green, flabelliform leaves; petioles green except for brownish or reddish bases, 1.5–2.2 m. long, 10–15 cm. broad at base, plano-convex in cross section, bearing heavy, simple or double, curved to hooked spines about 1 cm. long on basal 4–5 dm. of margins, then progressively smaller teeth to apex; hastula 6–16 cm. long, margins soon lacerate; blades 1.5–2.3 m. wide, usually slightly longer, segments 50–60, usually extending more than halfway to base, abundantly filiferous, apex mostly frayed; shag dense and persistent, leaves and old fruiting branches hanging vertically; spadix to 3–4 m. long, ultimate flowering branches 30–40 cm. long; rachillae slender, 20 cm. long, glabrous; flowers 8–9 mm. long, 5–6 mm. broad; fruit ellipsoid or ovoid, brown or black, glabrous, 7–10 mm. long, flesh dry and thin, usually wrinkled.

In canyons leading down to the Colorado Desert from Whitewater to Carrizo Creek in southern California, in Cantillas Canyon (and probably others) in northern Baja California, and in a small canyon on the western slopes of the Kofa Mountains, Yuma County, Arizona, all in the Lower Sonoran Zone. Frequently cultivated as an ornamental from Chico, California, and south into northwestern Mexico.

**2. Washingtonia robusta** H. Wendl., Garten-Zeit. **2:** 198, 1883

*Washingtonia sonorae* S. Wats., Proc. Amer. Acad. **24:** 79, 1889.
*Washingtonia gracilis* Parish, Bot. Gaz. **44:** 420, *figs. 8–10,* 1907.
*Washingtonia filifera* var. *sonorae* M. E. Jones, Contr. West. Bot. **15:** 48, 1929.

Slender palm attaining a height of 25 m. with basal part of bole abruptly flaring to 2 or 3 times the diameter of trunk above; petioles 0.8–1.2 m. long, brownish, spines about as in *W. filifera* on young trees, becoming smaller, fewer, or nearly lacking on petioles of mature or large trees; blades circular in outline, 1–1.5 m. wide, bearing a patch of tawny tomentum on lower side about hastula, segments narrow (1–1.5 or 2 cm. wide), abundantly filiferous on young trees, nearly devoid of threads in mature trees; hastula 3–8 cm. long; shag persistent, ragged because of various angles (from 20° to 50° or even 90°) at which dead leaves hang from trunk, dead fruiting branches often extending horizontally through shag; spadix 1.5–2.5 (or –3) m. long, extending horizontally through crown during flowering; flowers numerous, 7–8 mm. long including the exserted stamens; fruit ellipsoid to subglobose, glabrous, dark brown to bluish black, 4–7 mm. long, nearly as wide.

In canyons and along desert streams not far from the coast, Lower Sonoran Zone, in Baja California at Cataviñá and Jaraguay; in similar situations near Guaymas (San Carlos Bay), Sonora.

## 4. PHOENIX L., Sp. Pl. 1188, 1753

Shrubs to robust, tall trees with simple, rough but unarmed trunks and recurved-spreading, pinnate leaves. Petioles short, armed with rigid, spinelike basal leaf divisions, the other leaf divisions numerous, stiff but not spine-tipped. Spadix erect, axillary, ultimately drooping among lower leaves, fastigiately branched, surrounded in bud by a boat-shaped, 2-edged, somewhat flattened spathe. Flowers dioecious, numerous, staminate larger than pistillate; calyx cuplike, with 3 broad, short lobes, these spreading; petals 3, several times as long as sepals, narrowly ovate, spreading-ascending, surpassing stamens. Stamens usually 6, included; anthers narrowly sagittate, on slightly dilated filaments. Carpels 3, distinct, usually only 1 maturing. Fruit terete-oblong, with a yellow, orange, brown, or black, fleshy pericarp. Seed oblong, deeply grooved longitudinally along one side.

### 1. Phoenix dactylifera L., Sp. Pl. 1188, 1753

A tree often 15–20 m. high or more; leaves long-persistent in nature but cut away as they die to facilitate harvesting fruit; reproducing by numerous vegetative offshoots in axils of lower leaves as well as by seed; leaves large, 2–5 m. long, dark green, the segments 2–6 dm. long; spathes 10–18 cm. deep, 1–1.5 m. long at anthesis, it and peduncle of spadix orange-yellow to light brown; spadix 1–2 m. long in fruit, on a peduncle 3–8 cm. wide and often equaling flowering part; inflorescence a large panicle, rachillae 1–4 dm. long, glabrous; fruit 2–4 cm. long, yellow, orange, or light brown; seed 5–8 mm. in diameter, 1.5–3 cm. long.

This palm is cultivated widely in Mexico and grows spontaneously at San Ignacio and Comondú and probably at other places in Baja California. Not seen growing in Sonora except in cultivation.

## 5. COCOS L., Sp. Pl. 1188, 1753

A tall tree with a simple trunk unarmed except for a few needlelike spines among the leaf sheaths, and long, spreading, pinnate leaves forming a graceful, plumose crown. Spadix much shorter than leaves, at first ascending but soon drooping. Perianth segments of staminate flowers valvate, those of pistillate flowers imbricate, becoming greatly enlarged and persisting in fruit. Flowers monoecious, staminate numerous, pistillate few, borne toward bases of rachillae. Stamens mostly 6, spreading, filament slender, scarcely dilated at base, much longer than small, ovoid anther. Carpels 3, coherent. Fruit a trigonous drupe with a fibrous pericarp, a hard, bony endocarp, and an endosperm with a large cavity containing a watery fluid that turns milky at maturity.

### 1. Cocos nucifera L., Sp. Pl. 1188, 1753

A graceful tree to 30 m. tall; trunk swollen, inclined or even decumbent, naked except for plumose crown; leaf blades 3–7 m. long, on a naked petiole 1–2 m. long, segments numerous, somewhat arcuate-drooping, 5–10 dm. long; spathes woody, 1–1.5 m. long, acute, these and the fibrous leaf sheaths persisting until leaves fall; perianth of staminate flowers 9–12 mm. long, of pistillate flowers 2–3 times as large; rachillae paniculately disposed, usually sinuous, slender; drupe ovoid or oval, 2–3 dm. long, green to brownish; endocarp dark brown to black; fleshy endosperm containing copious "milk."

Common along the coasts of Mexico and other tropical shores, Lower Sonoran and Tropical Zones, cultivated as an ornamental in the interior some distance from the sea. Mature trees along the estuary at Mulegé, at Loreto and La Paz, Baja California, and a few near Guaymas, Sonora.

## Family 11. **ARACEAE.** Arum Family

Herbs with mostly basal simple or compound leaves; inflorescence a spadix enclosed within or subtended by a spathe. Spadix few- to many-flowered; staminate flowers above, pistillate below in monoecious plants, or dioecious, or rarely the flowers perfect. Perianth of 4–6 scalelike segments or absent. Stamens 4–10, subsessile or on short, stout filaments; anthers 2-celled, connective usually thick and truncate, thecae opening dorsally by pores or slits. Ovary 1–several-celled, each cell 1-ovulate; styles short or stigma sessile, usually minute. Fruit a berry or an utricle. Endosperm scanty, copious, or none.

### 1. **PISTIA** L., Sp. Pl. 963, 1753

A floating aquatic with a very short stem; leaves in a rosette, the blades entire, floating on the surface, petioles short. Inflorescence axillary, few-flowered, enclosing spathe white. Flowers monoecious; perianth none. Pistil solitary; ovary 1-celled.

**1. Pistia stratiotes** L., Sp. Pl. 963, 1753
(About twenty other names have been applied to this species.)

Leaf blades broadly obovate, 2.5–15 cm. wide, 3.5–25 cm. long, round, truncate or slightly emarginate, 10–15-ribbed, closely and minutely cinereous-puberulent, hairs longer on upper surface than on lower; petioles to 4 cm. long or some leaves essentially sessile; spathes 1–1.5 cm. long, densely villous without, nearly glabrous or merely puberulent within, cylindrical below; expanded limb ovate.

Floating on ponds, streams, and lakes in tropics and subtropics from Texas to Florida, south into South America and in the warmer parts of the Old World; collected in an irrigation ditch at Yuma, Arizona, where it is probably adventive.

## Family 12. **LEMNACEAE.** Duckweed Family

Minute stemless, free-floating plants on the surface of fresh-water ponds and sluggish streams. Plant body a disk-shaped, elliptic, or elongated thallus of loosely arranged chlorophyllous cells; rootless or bearing 1–several simple rootlets. Inflorescence of 1 or more monoecious flowers in a minute, saclike spathe on upper surface or edge of the frond. Flowers of solitary pistil or stamen; anthers 2–4-celled; ovary 1-celled, 1–6-ovulate, flask-shaped. Fruit an utricle.

Rootlets several, fascicled; fronds 5–12 nerved .............................. 1. *Spirodela*
Rootlet 1; fronds 1–5 nerved .............................................. 2. *Lemna*

### 1. **SPIRODELA** Schleid., Linnaea **13:** 391, 1839

Fronds disk-shaped to oblong, 5–12-nerved; rootlets fascicled; reproductive pouches 2, triangular, appearing as clefts in either margin of basal part of the frond. Spadix of 1 pistillate and 2 staminate flowers from the reproductive pouch; anthers 2-celled; ovules 2.

**1. Spirodela polyrhiza** (L.) Schleid., Linnaea **13:** 392, 1839

*Lemna polyrhiza* L., Sp. Pl. 970, 1753.

Fronds solitary or in colonies of 2–5, suborbicular to obovate, 3–6 mm. long, flat and green above, lighter green, purplish, and convex beneath, 5–11-nerved; rootlets 3–12 per frond, several times as long as frond; root cap acute; spathe flasklike, opening by a narrow apical ostiole; seed smooth, slightly compressed.

In sluggish streams, ponds and ditches, widely distributed in Europe, North and South America; within the Sonoran Desert known from near the village of Cedros, Sonora. Flowering time irregular.

## 2. LEMNA L., Sp. Pl. 970, 1753

Fronds disk-shaped or obovate, stipitate or sessile, usually with a central nerve and with or without 2–4-lateral veins; rootlet solitary on each thallus, centrally located, devoid of vascular tissue; root cap blunt or pointed, minute. Spadix containing 1 pistillate and 2 staminate flowers; spathe open or saclike. Anthers 2-celled, transversely dehiscent. Ovules 1–6. Utricle more or less ribbed, ovoid.

Frond 1-nerved or nerveless; spathe open:
- Frond thick, papillate above the midline, often nerveless; root cap straight or faintly curved, cylindric .......... 1. *L. minima*
- Frond thin, without papules; midnerve obscure to evident; root cap curved, tapering to acute point .......... 2. *L. valdiviana*

Frond 3–5-nerved; spathe flasklike or saclike, narrowed toward apex, or open in *L. aequinoctialis*:
- Fronds biconvex, but not conspicuously gibbous beneath, 3-nerved; papillae along midnerve on upper surface:
  - Spathe flasklike or saclike; fruit wingless; seed 12–15-nerved; fronds symmetrical; terminal papilla small, not conspicuous .......... 3. *L. minor*
  - Spathe open; fruit winged; seed 15–18-nerved; fronds asymmetric, laterally curved near base; terminal papilla much larger than others, conspicuous .. 4. *L. aequinoctialis*
- Fronds strongly gibbous beneath, 3–5-nerved; papillae none .......... 5. *L. gibba*

**1. Lemna minima** Phil., Linnaea **33:** 239, 1864

Frond light green, elliptic to oblong, 1.5–2.5 (rarely –4) mm. long, faintly 1-nerved or more commonly nerveless, with a row of minute papules along median line on upper surface, slightly convex beneath; stipe very short and soon rounding out and disappearing; root cap straight or slightly curved, cylindrical, blunt at apex, 0.4–0.7 mm. long; spathe open, its lobes rounded or obtusish; ovary clavate, 1-ovulate; ovule orthotropous or nearly so; seed oblong, minutely 12–16-ribbed.

In pools and slow streams, Lower Sonoran Zone to Transition Zone, California to Wyoming, Kansas, and Florida; South America; in the Sonoran Desert occurring at Carrizo Creek, San Diego County, California, and near Pitiquito, Sonora.

**2. Lemna valdiviana** Phil., Linnaea **33:** 239, 1864

*Lemna cyclostasa* Chev., Fl. Paris **2:** 256, 1827. Nomen nudum.

Fronds elliptic-oblong, 2–3 (or sometimes –4) mm. long, slightly falcate or asymmetrically curved toward the short-stalked base, obscurely 1-nerved, moderately biconvex, smooth on both surfaces; root cap short-tapering to an acute tip, strongly curved; spathe open, saucer-shaped, broadly reniform; fruit ovoid, with persistent style almost a continuation of lateral line; seed minutely 12–19-ribbed, amphitropous.

In pools and slow streams, Lower and Upper Sonoran Zones, widely distributed in North and South America, uncommon north of central California on the Pacific coast. Known from Palm Springs on the western margin of the Colorado Desert.

**3. Lemna minor** L., Sp. Pl. 970, 1753

Fronds broadly oblong, 2–3 (or –4) mm. long, symmetrically rounded at each end, faintly convex above, more strongly so beneath, thickish, green or purplish both above and below, somewhat spongiose beneath, 3-nerved, the laterals obscure, several minute papillae along midnerve; root cap straight or nearly so, acute; spathe saccate or irregularly lobed by rupturing in fruit; fruit wingless; seed solitary, 12–15-ribbed, with numerous minute transverse striations; hilum prominent.

Widely distributed throughout the temperate regions of the world except in South America. Collected at Palm Springs on the margin of the Colorado Desert.

**4. Lemna aequinoctialis** Welw., Ann. Conselho Ultamar. **56:** 545, 1858

Fronds in groups of 3–4 or rarely solitary, thick, asymmetric, 1–2 mm. long, 0.8–1.5 mm. wide, 3-nerved, bearing several papillae along central nerve, the terminal one much enlarged and prominent, each frond short-stipitate and asymmetrically turned to one side near base; root sheath narrowly winged; root about 10 mm. long; root cap 1–1.2 mm. long, tapering to a sharp point; spathe open; stamens 0.7 mm. long; anthers 0.3 mm. long, 0.2 mm. wide; stigma orbicular, 0.1 mm. wide; ovule solitary, orthotropous; fruit asymmetric, narrowly winged, 0.7–0.85 mm. long; seed ovoid, 0.5–0.6 mm. long; testa thick, 15–18-ribbed.

On ponds and in quiet water of streams and ditches, southern Arizona to Argentina; Angola, Africa; Galapagos Islands; Philippine Islands. Oct.–Nov. in our area.

**5. Lemna gibba** L., Sp. Pl. 970, 1753

Fronds obovate to oblong, 2–5 mm. long, symmetrical to strongly asymmetrical, thick, moderately convex and slightly keeled, epapulate and green above, flattish to strongly convex, hemispherical-gibbous, and pale beneath, strongly spongiose-cavernous, especially in the gibbous part, 3–5-nerved; root cap straight to slightly curved, cylindrical; spathe saccate when young, irregularly ruptured in age; fruit with rounded wings on either side of stigma at upper margin, purplish; seeds 1–6, prominently and unevenly ribbed, transversely striate.

In ponds and sluggish streams, Lower Sonoran Zone to Transition and Tropical Zones, California to Nebraska, Texas, and Mexico; also in Europe and Asia. Occurring near the mouth of Sabino Canyon, Arizona; at Palm Springs, California; and in stream 3 miles west of Pitiquito, Sonora.

## Family 13. BROMELIACEAE. Pineapple Family

Herbs or shrubby perennials, mostly epiphytic. Leaves spirally arranged, usually basal, sheathing below, simple, entire to spinose-serrate, bearing peltate, moisture-holding scales, at least when young. Inflorescence of simple or paniculate spikes or racemes, conspicuously bracteate. Flowers perfect or polygamodioecious. Sepals and petals free or connate. Stamens 6, in 2 series. Style 3-parted. Ovary 3-celled, inferior or superior. Placentation axillary. Fruit a capsule or baccate. Seeds naked, winged, or plumose. Embryo small. Endosperm mealy, copious.

Leaves coarsely spinose-serrate; seeds narrowly winged or naked, not plumose; flowers polygamodioecious; terrestrial ........ 1. *Hechtia*
Leaves entire; seeds plumose-appendaged; flowers perfect; epiphytic ........ 2. *Tillandsia*

## 1. HECHTIA Klotzsch, in Otto & Dietr., Allg. Gart. 3: 401, 1835

Stem short; leaves densely rosulate, recurved, coarsely spinose-serrate, pungently tipped, lepidote throughout or on the under side only. Scape erect or ascending; bracts at base of scape foliaceous, upper vaginiform. Inflorescence 2- or 3-paniculate. Flowers subsessile to pedicellate, polygamodioecious, remnants of nonfunctional sex organs usually present in functional flowers. Sepals free, squamiform. Petals naked, free or joined to ring of filaments. Stamens included to exserted. Ovary superior to nearly inferior, glabrous or lepidote-scaled. Capsule ellipsoid to ovoid, septicidal and loculicidal at same time. Seeds oblong, narrowly winged or sometimes nearly naked.

### 1. Hechtia montana Brandegee, Erythea 7: 9, 1899

Plant terrestrial, about 1 m. high; leaves numerous in a dense basal rosette, 15–45 cm. long; sheaths ovate to suborbicular, brown, glabrous or sparingly punctulate, finely ciliolate and denticulate on margins of upper half; blades linear-triangular, 2–3 cm. wide, cinereous-scurfy beneath, lepidote but soon glabrous above, armed with slender, uncinate teeth 2–5 mm. long, 8–30 mm. apart; scape 8–12 mm. in diameter, sparsely lepidote at first, soon glabrous; lower scape bracts 10–15 cm. long, pungent, upper narrower, remote; inflorescence 3–5 dm. high; racemes ascending or spreading, few-flowered; floral bracts linear-lanceolate, acuminate, equaling or exceeding staminate pedicels, half as long as pistillate pedicels; pedicels slender, staminate 2–3 mm. long, pistillate 4–8 mm. long; staminate sepals thin, 2.5 mm. long, pistillate 3 mm. long, thickened at base; staminate petals broadly elliptic, obtuse, 4.5 mm. long; pistillate deltoid, acuminate, 5 mm. long; stamens included; ovary slightly inferior; capsule ovoid, acute, 10–12 mm. long; seeds oblong or falcate, dorsal wing nearly as long as seed.

Rocky cliffs and hillsides, Lower Sonoran Zone, southern Baja California and from central Sonora southward into Sinaloa.

## 2. TILLANDSIA L., Sp. Pl. 286, 1753

Caulescent or acaulescent perennial herbs. Leaves fasciculate, rosulate, or distributed along the stem, entire, ligulate to narrowly triangular. Scape mostly distinct. Inflorescence of distichous spikes or reduced to single polystichous spike or single flower. Flowers perfect. Sepals free or variously joined. Petals free, naked or rarely with vertical folds. Ovary superior, glabrous. Capsule septicidal. Seeds erect, cylindrical to fusiform, bearing a white to tawny straight, plumose, basal appendage.

### 1. Tillandsia recurvata L., Sp. Pl. ed. 2, 410, 1762

*Renealmia recurvata* L., Sp. Pl. 287, 1753.
*Tillandsia uniflora* H. B. K., Nov. Gen. & Sp. 1: 290, 1816.
*Diaphoranthema uniflora* Beer, Bromel. 154, 1857.
*Diaphoranthema recurvata* Beer, ibid. 156.

Densely crowded with simple or few-branched stems 1–10 cm. long; leaves distichously arranged on stem but spreading in all directions and usually recurved, 2–17 cm. long, sheaths closely imbricated over stem, blades terete, linear, 0.5–2 mm. in diameter, densely lepidote-cinereous; scape prominent, terminal, 6–15 cm. long, slender, 1–5-flowered; scape bracts linear-lanceolate, 12–22 mm. long, lepidote-cinereous, 1 or 2 per scape; floral bracts similar to but smaller than scape bracts, from shorter than to slightly exceeding sepals; flowers subsessile, erect; sepals lanceolate, acute, 4–10 mm. long, 3–5-nerved, mostly glabrous; petals white to pale violet, delicate, 12–16 mm. long; stamens included; capsule cylindrical, 2–2.5 mm. in diameter before dehiscence, 1.5–3 cm. long, valves separating into a thin outer membrane and thicker inner part; seeds numerous, 2–2.5 mm. long, the tawny plumose basal appendage 1.5–2 cm. long.

On shrubs and trees and less commonly on rocky cliffs, Lower Sonoran and Tropical Zones, southern Arizona, Texas, and Florida south to northern Argentina and Chile. Abundant in Baja California from vicinity of Miller's Landing south.

Both **Tillandsia exserta** Fernald, Bot. Gaz. **20:** 537, 1895, with leaf blades only slightly wider than those of *T. recurvata,* floral bracts 2–3 cm. long, and long-exserted stamens, and **T. inflata** Mez, in A. & C. DC., Monogr. Phan. **9:** 701, 1896, with flat leaf blades 4 cm. wide, white-margined purple petals 4 cm. long, floral bracts 2–3 cm. long, and exserted stamens, have been reported from Sonora, but the writer has seen no specimens from the desert area.

**Tillandsia ferrisiana** L. B. Smith, Brom. Soc. Bull. **10:** 92, 1960, differs from *T. recurvata* in having very short scapes rarely equaling the leaves, and appressed-lepidote, instead of glabrous, sepals. It has been collected on the western slopes of the Sierra de la Laguna only a few miles outside the Sonoran Desert, Cape Region, Baja California.

## Family 14. **COMMELINACEAE.** Spiderwort Family

Perennial herbs with alternate leaves, their bases sheathing and often scarious. Stems erect, ascending, decumbent, or prostrate, subfleshy. Inflorescence cymose, umbellate, or racemose, often subtended by spathelike or leaflike bracts. Flowers perfect, regular or irregular, outer perianth segments sepal-like, foliaceous, or scarious, hyaline to variously suffused with green, rose, or purple. Inner perianth segments petaloid, showy, fugacious, withering soon after the sun strikes them. Stamens 6, all functional or some sterile. Ovary superior, 3-celled. Capsule 3-celled, loculicidally dehiscent. Seeds few, often roughened or rugose; hilum mostly rather conspicuous.

Flowers irregular; inflorescences subtended by a single ovate, spathelike bract; filaments glabrous; only 3 stamens functional .................................. 1. *Commelina*

Flowers regular; inflorescences subtended by several smaller, foliaceous bracts, these not spathelike; filaments villous below; all 6 stamens functional .......... 2. *Tradescantia*

### 1. **COMMELINA** L., Sp. Pl. 40, 1753

Perennial herbs with weak, procumbent to erect stems, often rooting at the nodes. Roots fibrous or tuberous-thickened and densely covered with subpersistent root hairs. Sheathing bases of leaves usually membranous opposite the blade, summit often ciliate-villous. Spathes strongly compressed, the two halves folded along midvein, sessile to distinctly pedunculate. Flowers irregular, 1–several in each spathe, included or commonly exserted; inner perianth segments delicate, blue or white, the two lateral ones larger than upper one. Stamens 6, 3 of them fertile, others with 4-lobed but empty anthers; filaments naked. Capsule 3-celled, 3-lobed, or globose or 2-lobed if some ovules abortive.

Spathes connate at base, on peduncles about as long as the width of spathe or subsessile; upper petal white:
- Bases of spathes puberulent with short hairs ............ 1. *C. erecta* var. *angustifolia*
- Bases of spathes villous with long, flaccid, white hairs ........... 1a. *C. erecta* var. *crispa*

Spathes not connate at base, borne on peduncles equaling or surpassing length of spathe, not subsessile; all 3 petals blue:
- Stems usually erect; leaves linear, 2–7 mm. wide; spathes attenuate, 2.5–7 cm. long .......................................................... 2. *C. dianthifolia*
- Stems prostrate or procumbent, rooting freely at nodes; leaves ovate to elliptic, 1–2 cm. wide; spathes acute but rarely attenuate, 1–2.6 cm. long ............... 3. *D. diffusa*

## 1. **Commelina erecta** L. var. **angustifolia** (Michx.) Fernald, Rhodora **42:** 439, 1940

*Commelina angustifolia* Michx., Fl. Bor. Amer. **1:** 24, 1803.
*Commelina virginica,* β *angustifolia* C. B. Clarke, in A. & C. DC., Monogr. Phan. **3:** 183, 1881.
*Commelina swingleana* Nash, Bull. Torrey Club **22:** 160, 1895.
*Commelina nashii* Small, Fl. SE. U.S. 242, 1903.

Erect or slightly procumbent, usually much branched herb 2–5 dm. tall, or scandent plants to 8–10 dm. tall, or procumbent stems spreading 2–6 dm.; herbage puberulent to sparsely villous; stems 2–3.5 mm. in diameter; internodes 2–12 cm. long; leaf sheaths 5–18 mm. long, readily ruptured; leaf blades linear-lanceolate to lanceolate, 1–2 cm. wide, 4–7 cm. long (sometimes to 3 cm. wide, 10 cm. long), slightly paler beneath than above; peduncles 1 cm. long or less; spathes ovate-cordate, 1.5–2.5 cm. long, 8–15 mm. high, connate from base almost to top of curved part, lower side nearly straight; flowers short-pedicellate, about equaling spathe; lateral petals blue, 8–10 mm. long, upper one white, slightly shorter; capsules thin, 4–5 mm. wide; seeds subglobose, brownish flecked with purple or darker brown, 2.5–3 mm. in diameter, hilum conspicuous.

Ditch banks, along streams, and around intermittent rain pools, Lower Sonoran, Upper Sonoran, and Carolinian Zones, Arizona to Colorado, the central Atlantic seaboard, and south into Mexico and the Cape Region, Baja California. Flowering following rains, mostly Aug.–Oct.

## 1*a*. **Commelina erecta** var. **crispa** (Woot.) Palmer & Steyermark, Rhodora **40:** 131, 1938

*Commelina crispa* Woot., Bull. Torrey Club **25:** 451, 1898.
*Commelina erecta* var. *angustifolia* forma *crispa* Fernald, Rhodora **42:** 440, 1940.

Differing from var. *angustifolia* in having long, white, flaccid hairs on the base of the spathe and on the adjacent peduncles and leaf sheaths.

Range about the same as, but in our area less common than var. *angustifolia.* Aug.–Oct.

## 2. **Commelina dianthifolia** Delile, in Redoute, Liliac. **7:** *pl. 390,* 1813

Slender, erect herbs 2–5 dm. tall with slender stems and linear leaves; herbage glabrous to sparsely puberulent; internodes to 10 cm. long; leaf sheaths 1.5–3.5 cm. long, often densely speckled with purple, thin-chartaceous but firm, margins not ciliate; leaf blades linear to lance-linear, 2–7 mm. wide, 4–10 cm. long or occasionally to 15 cm. long; flowering branches slender, 3–7.5 cm. long; spathes ovate-attenuate, 8–12 mm. deep, 2.5–5 (or rarely –7) cm. long, lower edge straight or falcately curved, base not at all connate, glabrous and usually shining, margins rarely ciliolate with a single row of short, stoutish hairs; flowers several in each spathe, often exserted 1–2 cm.; petals blue, longer ones obovate, 8–10 mm. long; capsule 5–6 mm. high, acute and slightly apiculate; seeds subreniform but often truncate at one end owing to mutual compression of adjacent seeds, dark brown spots on a lighter background, slightly rugulose, dull.

Commonly in forests and margins of meadows, frequently along streams, chiefly Upper Sonoran Zone, Arizona, New Mexico, and south throughout most of the uplands of Mexico; barely entering the fringe of the Sonoran Desert near Loreto and La Paz, Baja California. Aug.–Oct.

### 3. **Commelina diffusa** Burm. f., Fl. Ind. 18, *pl. 7, fig. 2,* 1768

*Commelina nudiflora* Burm. f., op. cit. 17, and of American authors. Not *C. nudiflora* L., 1753.
*Commelina longicaulis* Jacq., Coll. **3:** 234, 1789.
(For additional synonymy see Jour. Arnold Arb. **18:** 64, 1937.)

Decumbent plant rooted freely at nodes and with numerous erect branches 0.5–3 dm. tall, these weak, 1.5–2.5 mm. in diameter, often tinged with reddish or purplish; herbage glabrous except on sheathing petioles, these scarious, 5–11 mm. long, ciliate-villous along margins; leaf blades elliptic to ovate-elliptic, 7–18 mm. wide, 2–5 cm. long, acute or slightly obtuse at apex, obtuse to rounded at base, lightly 7–12-veined; flowering spathes borne on slender, peduncle-like, axillary branches 0.5–2 cm. long, adaxial side of these branches puberulent; spathes broadly ovate-cordate, 6–8 mm. deep, 10–22 mm. long, rounded base not connate, acute but not attenuate at apex, lower side slightly curved; flowers borne singly or in pairs on slender peduncles 3–11 mm. long; pedicels subfiliform, 2–5 mm. long; outer perianth segments oblong, 1.5–2 mm. long, hyaline; petals blue, smaller 3–4 mm. long, larger 5–8 mm. long; capsules 4.5–5.5 mm. long, oblong, lying within spathe owing to sharp reflexion of peduncles in fruit; seeds brownish, regularly foveolate-reticulate.

Along ditches and creek beds, tropical and warm temperate parts of the world; occurring in the Sonoran Desert in southern Baja California and probably in southern Sonora. In our area flowering nearly throughout the year.

## 2. **TRADESCANTIA** L., Sp. Pl. 288, 1753

Perennial herbs with erect to creeping, subsucculent stems and alternate leaves sessile, with a thin, often hyaline and subscarious, perfoliate, basal sheath. Inflorescence a dichotomous, umbellate cyme, terminal or axillary, each one subtended by 2–3 subequal to very unequal foliaceous or scarious bracts; pedicels subtended by 1–2 hyaline to foliaceous bracteoles. Sepals 3, free, navicular or concave, hyaline, petal-like or foliaceous. Petals 3, regular, ephemeral, becoming twisted and contorted upon withering. Stamens 6, hypogynous, equal and fertile, the filaments pilose; anthers with reniform pollen sacs. Ovary 3-celled, 3–6-ovulate; style filiform; stigma capitate. Capsule dry, loculicidally 3-valved. Seeds rugose, ridged, or pitted.

Leaf blades linear to lance-linear, rarely over 2 cm. wide; flowers well exserted from subtending bracteoles; petals spreading widely, corolla 2–3 cm. in diameter:
- Sepals and pedicels uniformly glandular-pubescent ...... 1. *T. occidentalis* var. *occidentalis*
- Sepals and pedicels wholly glabrous ................ 1a. *T. occidentalis* var. *scopulorum*

Leaf blades lance-ligulate to oblong-spatulate, 2–5 cm. wide; flowers included in swollen leaf sheaths; petals ascending, corolla rarely over 1 cm. in diameter .... 2. *T. semisomna*

**1. Tradescantia occidentalis** Britt., Mem. N.Y. Bot. Gard. **1:** 87, 1900, var. **occidentalis**

(For extensive synonymy see Contr. Arnold Arb. No. **9:** 101, 1935.)

Erect or ascending plant 5–7 dm. tall with rather stout stems and internodes 1–25 cm. long; leaves firm but thin, glabrous and glaucous, blade linear to linear-lanceolate and long-acuminate, 0.2–2 cm. broad, 5–50 cm. long, leaf sheath 5–30 mm. long, 8–35 mm. broad, turgid and somewhat inflated; cymes few- to many-flowered; bracts foliaceous, 4–20 mm. wide, 1.5–20 cm. long, glabrous; pedicels 12–21 mm. long, moderately to densely glandular-puberulent; sepals 4–10 mm. long, glandular-puberulent like the pedicels, elliptic, acute to acuminate, often tinged with rose or purple, usually glaucous; petals bright blue or sometimes rose or magenta, 12–16 mm. long; capsules oblong, obtusely trigonous, 4–7 mm. long; seeds 2–4 mm. long with a linear hilum about as long as seed.

Mostly in sandy soil, Lower Sonoran Zone to Transition Zone, Montana to Wisconsin and south into Texas and southwestern Arizona. April–Aug.

**1*a*. Tradescantia occidentalis** var. **scopulorum** (Rose) Anderson & Woodson, Contr. Arnold Arb. No. **9:** 106, 1935

*Tradescantia scopulorum* Rose, Contr. U.S. Nat. Herb. **5:** 205, 1899.

Stems somewhat more slender than in var. *occidentalis,* 10–35 cm. tall; leaf blades 5–45 cm. long; bracts 4–15 cm. long; pedicels and sepals glabrous; petals 7–10 mm. long, apparently always bright blue.

In moist canyons and along streams, mainly Upper Sonoran Zone into the Transition Zone, known with certainty only from Arizona and barely entering the Sonoran Desert at the foot of Baboquivari Peak. April–Sept.

**2. Tradescantia semisomna** Standley, Field Mus. Bot. Ser. **17:** 228, 1937

Ascending to erect, subfleshy plant with several coarse stems 1–3.5 dm. tall from somewhat tuberous, fleshy roots; internodes 1–4 cm. long, often barely surpassing subtending leaf sheaths, these 1.5–2 cm. deep, 1–3 cm. wide, moderately flaring upward; leaf blades lance-ligulate to oblong-spatulate, 2–5 cm. wide, 5–17 cm. long, acute at apex, finely puberulent on both surfaces, bright green above, paler and somewhat silvery beneath, frequently moderately involute, slightly falcately curved downward; inflorescences, dense axillary cymes; pedicels 1 cm. long or less, thus the flowers crowded and barely equaling rim of leaf sheath; sepals lance-linear to narrowly ovate, 5–9 mm. long, greenish or tinged with rose, puberulent to subglabrous; petals mostly roseate, less commonly blue, ascending or only slightly spreading, 6–10 mm. long; capsules nearly globose, 2.5–3 mm. long; seeds oblong, 1.8–2 mm. long, dull black, transversely rugose, hilum punctate.

Canyon slopes and bottoms, Lower and Upper Sonoran Zones, central Sonora (Baviso Mountains) to southern Sonora and southwestern Chihuahua. Aug.–Sept.

**Tinantia modesta** Brandegee, Proc. Calif. Acad. II, **3:** 175, 1891, with a terminal, racemose inflorescence, occurs in Baja California in the Sierra de la Laguna and on its lower slopes, but has not yet been found in the Sonoran Desert. It might occur in the Sierra de San Francisquito and Sierra de la Giganta.

## Family 15. **PONTEDERIACEAE.** Pickerelweed Family

Perennial aquatic or semiaquatic plants with grasslike or petiolate leaves having broad, thick blades. Flowers perfect, regular or slightly irregular, solitary or spicate, subtended by leaflike spathes. Perianth petaloid, 6-parted, free from ovary, often withering-persistent. Stamens 3 or 6, inserted on tube or base of perianth, usually dissimilar; anthers 2-celled, introrse. Ovary superior, 3-celled with axial placentae or 1-celled with 3 parietal placentae. Style 1; stigma 3-lobed or 6-toothed. Ovules many, anatropous. Fruit a capsule or an utricle. Embryo central, cylindric; endosperm copious.

### 1. **HETERANTHERA** Ruiz & Pav., Fl. Per. Prod. 9, *pl.* 2, 1794

Low herbs of shallow water or mud with creeping, ascending, or floating stems, petiolate, oval to reniform leaves or these linear, sessile, and grasslike. Flowers 1–several in a spathe, small, petaloid, the spathe erupting from base of a sheathing petiole. Perianth lobes equal or nearly so, linear. Tube slender, cylindrical. Stamens 3, inserted in throat, subequal or unequal. Ovary fusiform, 1-celled or incompletely 3-celled by intrusion of placentae; ovules numerous; style filiform; stigma 3-lobed. Capsule many-seeded, enclosed within the withering perianth tube. Seeds ovoid, many-ribbed.

Leaves petiolate, blades ovate to oblong, erect, emersed; stamens distinctly unequal; 2 anthers ovate .......................................................... 1. *H. limosa*
Leaves sessile, grasslike, blades linear, submerged or floating; stamens equal or nearly so; anthers sagittate-linear .................................................. 2. *H. dubia*

#### 1. **Heteranthera limosa** (Sw.) Willd., Ges. Naturf. Fr. Neu. Schr. **3:** 439, 1801

Leaves erect or ascending from a creeping rhizome; stipules sheathing, membranous, 8–10 mm. long; petioles 3–5 cm. long; blades ovate to ovate-lanceolate or oblong, 1.5–4 cm. long, 10–20 mm. wide, rounded to subcordate, obtuse at apex, finely many-veined; spathes 1-flowered, on peduncles 1–2 cm. long; corollas 1.5–2 cm. broad, light blue, thin; posterior anthers ovate, the third sagittate; capsule incompletely 3-celled, fusiform to oblong, 12–15 mm. long; seeds 0.8–1 mm. long, ovoid, brownish, finely ribbed.

Muddy pools or shallow water, Oklahoma to Virginia, to South America. Occurring within the Sonoran Desert in central Sonora. July–Aug.

#### 2. **Heteranthera dubia** (Jacq.) MacMillan, Met. Minn. 138, 1892

*Commelina dubia* Jacq., Obs. Bot. **3:** 9, *pl. 59,* 1768.
*Leptanthus gramineus* Michx., Fl. Bor. Amer. **1:** 25, *pl. 5, fig. 2,* 1803.
*Heteranthera graminea* Vahl, Enum. **2:** 45, 1806.
*Schollera graminea* A. Gray, Man. 511, 1848.
*Zosterella dubia* Small, Fl. Lancaster Co. 68, 319, 1913.

Submerged aquatic grasslike plant with slender, forking stems that often root at nodes; leaves sessile, linear, 1–6 mm. wide, 5–10 cm. long, translucent, often encrusted with lime, finely many-nerved; spathes terminal but becoming lateral by continued growth of stem, 1-flowered; flowers 1–1.5 cm. broad, petals narrowly linear, yellow; tube very slender, cylindrical; stamens subequal; anthers all alike, sagittate-linear, 3.5–4 mm. long; capsule 1-celled, placentae rarely intruding; seeds small, numerous.

In quiet water, Washington to New England and south to Cuba and Mexico. Common in ditches in southern Arizona and Sonora. July–Sept.

## Family 16. JUNCACEAE. Rush Family

Perennial or annual, usually tufted, grasslike herbs of moist habitats. Flowers small, perfect, regular, in heads, panicles, umbels, corymbs, or rarely solitary. Perianth segments 6, glumelike. Stamens 3 or 6, the 2-celled anthers introrse. Pistil superior, tricarpellate, 1- or 3-celled; stigmas 3, filiform. Fruit a loculicidal capsule. Seeds 3–many, subglobose to cylindrical, often with loose seed coat forming a short appendage.

### 1. JUNCUS L., Sp. Pl. 325, 1753

Characters of the family.

Stems leaf-bearing or scapose. Leaf sheaths with free margins, often scarious; leaf blades terete or flattened. Each flower subtended by a bract. Seeds usually ribbed, reticulate, or tailed.

Lowest leaf of inflorescence appearing like a continuation of the terete stem, usually acute, often rigidly spinose; inflorescence appearing lateral:
- Flowers inserted singly on branches of inflorescence, each subtended by 2 bractlets:
  - Stems terete; leaf blade wanting on sheaths of basal leaves ............ 1. *J. balticus*
  - Stems compressed and twisted; leaf blade well developed on uppermost basal sheaths .. ........................................................ 2. *J. mexicanus*
- Flowers in 2–5-flowered paniculate heads, each in axil of a bract but without bractlets:
  - Outer perianth segments obtuse, scarious and often erose at edges; capsule broadly obovoid to subglobose, 1.5–2 times as long as perianth .................. 3. *J. acutus*
  - Outer perianth segments acute, margins firm, entire; capsule ovoid, about equaling perianth ................................................... 4. *J. cooperi*

Lowest leaf of inflorescence pushed aside by inflorescence or, if continuous with stem, deeply grooved on inner side, usually not rigidly spinose; inflorescence appearing terminal:
- Flowers inserted singly, each subtended by a bract and 2 bractlets:
  - Plants annual, rarely over 25 cm. tall; lower leaf of inflorescence not exceeding inflorescence; seeds tailless ........................................ 5. *J. bufonius*
  - Plants perennial, 3–5 dm. tall; lower leaf of inflorescence continuous with stem, deeply grooved on inner face, surpassing short panicles; seeds apiculate at each end .... ............................................ 6. *J. interior* var. *arizonicus*
- Flowers in few- to many-flowered heads, each subtended by a single bract only:
  - Leaf blades gladiate; perianth segments 3–3.5 mm. long, equal .......... 7. *J. xiphioides*
  - Leaf blades not gladiate; perianth segments 4–6 mm. long, inner and outer of unequal length:
    - Leaf blades strongly septate; stems terete; heads many-flowered; outer perianth segments longer than inner ........................................ 8. *J. torreyi*
    - Leaf blades weakly septate; stems compressed; heads 3–8-flowered; outer perianth segments shorter than inner ............................... 9. *J. macrophyllus*

#### 1. Juncus balticus Willd., Berlin Mag. 3: 298, 1809

Stems 2–8 dm. high, 2–3.5 mm. thick, terete, from creeping rootstocks; basal leaf sheaths bladeless or uppermost somewhat mucronate; lowest leaf of inflorescence erect, 3–25 cm. long, spinose; panicle loose, 2.5–7 cm. high; flowers inserted singly, subtended by a bract 2–3 mm. long, and 2 shorter bractlets; perianth segments lance-acuminate, 4–5 mm. long, inner and outer about equal; stamens 2.5–3 mm. long, anthers longer than filaments; capsule narrowly ovoid, mucronate, brownish, 4–5 mm. long; seeds oblong to narrowly obovoid, about 0.8 mm. long, about 40-striate, and with faint transverse lines.

Moist situations, mostly in Upper Sonoran to Boreal Zones, Alaska to Labrador and south to southern California, New Mexico, and Pennsylvania; Europe and Asia. Rare in the Sonoran Desert; Mecca, Imperial County, California.

**2. Juncus mexicanus** Willd., in Roem. & Schult., Syst. Veg. **7:** 178, 1829

*Juncus compressus* H. B. K., Nov. Gen. & Sp. 235, 1816. Not *J. compressus* Jacq., 1762.
*Juncus balticus* var. *mexicanus* Kuntze, Rev. Gen. Pl. **3:** 320, 1893.

Stems 2–6 dm. high, 1–2 or –3 mm. in diameter, distinctly compressed and often somewhat twisted, from creeping rootstock 3–6 mm. thick; basal leaf sheaths stramineous to brown, upper ones producing leaf blades similar to stems and 5–30 cm. long; lowest leaf of inflorescence 3–30 cm. long, spinose-tipped; inflorescence a loose, few-flowered panicle 2–10 cm. high; perianth segments 4–5 mm. long, lance-acuminate, equal, margins scarious; stamens half as long as perianth segments, anthers longer than filaments; capsule ovoid, abruptly but strongly mucronate, 4–5 mm. long, brownish or stramineous, 3-celled; seeds about 0.8 mm. long, irregularly and finely reticulate with narrowly rectangular meshes.

Moist, slightly saline soil in Upper and Lower Sonoran Zones, Monterey County and San Joaquin Valley, California, to Baja California and Sonora. Occasional in Colorado Desert; Mecca, Imperial County, California.

**3. Juncus acutus** L., Sp. Pl. 325, 1753

*Juncus acutus* var. *sphaerocarpus* Engelm., Rep. U.S. Geol. Surv. West 100 Merid. **6:** 376, 1878.
(For added synonymy see Pflanzenreich **4**[36]**:** 149–150, 1906.)

Stout, rigid, spinescent perennial growing in large tufts 5–15 dm. high; stems terete; dark green basal leaf sheaths bearing terete blades similar to and nearly as long as flowering stems; lowest leaf of inflorescence 5–12 (or –15) cm. long, erect, stout, spinescent; panicle of unequal branches 1–20 cm. long; flowers in clusters of 2–4; perianth segments pale brown, 3–4 mm. long, scarious-margined, outer broadly lanceolate, obtuse and erosulate at apex, inner slightly shorter, retuse at apex; stamens about equaling inner perianth segments; capsule broadly obovoid to subglobose, 4–5 mm. long, abruptly apiculate; seeds 0.8–1 mm. long, minutely reticulate, acute to slighty appendaged at each end.

Salt marshes, Upper and Lower Sonoran Zones, San Luis Obispo County and western border of the Colorado Desert, California, south to central Baja California and Cedros Island.

**4. Juncus cooperi** Engelm., Trans. Acad. St. Louis **2:** 590, 1868

Stems erect, 6–10 dm. tall, closely tufted; basal leaf sheaths bearing stiff, terete, spinescent, light green blades about one-half as long as flowering stems; lowest leaf of inflorescence spinescent, 5–11 cm. long; branches of compound panicle 0.5–10 cm. long; flowers subtended by scarious, lanceolate bracts 4–5 mm. long; bractlets lacking; perianth segments oblong-lanceolate, scarious-margined, pale green to stramineous, 5–6 mm. long, outer longer than inner, not erosulate; stamens about equaling inner perianth segments; capsule narrowly oblong, acute, 3-angled, about equaling or slightly exceeding perianth, 3-celled; seeds finely reticulate and with short appendages.

In moist alkaline places, Lower Sonoran Zone, Inyo County, California, to southern Nevada; Imperial County, California, and northern Baja California.

**5. Juncus bufonius** L., Sp. Pl. 328, 1753

*Juncus congdoni* S. Wats., Proc. Amer. Acad. **22:** 480, 1887.

Annual, branching from base, with erect stems 5–25 cm. tall; leaves 1–3 on stem below inflorescence, with flat blades 1 mm. wide or less, 1–5 cm. long, often involute; inflorescence an open panicle forming about one-half the height of plant, with flowers inserted singly and each subtended by a bract and 2 bractlets; perianth segments greenish to pale brown, lance-acuminate, 4–5 mm. long, margins scarious; stamens rarely 3, usually 6, about one-half as long as perianth segments, anthers about equaling or slightly shorter than filaments; capsule oblong, obtuse, 3–4 mm. long, 3-celled, short-mucronate; seeds oblong, 0.3–0.4 mm. long, with minute reticulate meshes in 30–40 longitudinal rows, acute to obtuse or rounded at tips.

Dried-up pools, stream banks, and desert washes, throughout North America except in the extreme north; Europe. Collected in deserts of California, Arizona, Sonora, and Baja California.

**6. Juncus interior** Wieg. var. **arizonicus** (Wieg.) Hermann, Jour. Wash. Acad. **30:** 217, 1940

*Juncus arizonicus* Wieg., Bull. Torrey Club **27:** 517, 1900.

Fibrous-rooted perennial 3–7 dm. high, tufted; leaf sheaths stramineous to light brown, 1.5–5 cm. long, abruptly narrowed to a rounded apex; leaf blades linear, 1–2 mm. wide, 1–3 dm. long, narrowly but distinctly involute; lowest leaf of inflorescence suberect or slightly spreading, 3–7 cm. long; inflorescence a short, moderately compact panicle; bract narrowly lanceolate-acuminate, 4–5 mm. long; bracteoles lance-ovate, short-acuminate, 2.5–3 mm. long; perianth segments about equal, 4–5 mm. long, linear-lanceolate, margins membranous; capsule broadly oblong, about 4 mm. long, rounded at apex, stramineous; seeds oblong, about 0.5 mm. long, reddish brown, longitudinally finely striate, cross reticulations delicate.

Lower and Upper Sonoran Zones, south-central Arizona and northern Sonora. Collected near Tucson.

**7. Juncus xiphioides** Meyer, Syn. Junc. 50, 1822

*Juncus xiphioides* var. *littoralis* Engelm., Trans. Acad. St. Louis **2:** 481, 1868.
*Juncus xiphioides* var. *auratus* Engelm., ibid.

Stems 4–8 dm. high, greatly compressed, acutely 2-edged, from a heavy, widely creeping rhizome and usually forming definite lines or ranks; leaves flattened laterally; sheaths with membranous margins, without auricles; leaf blades 3–10 mm. wide, 1–4 dm. long, often somewhat falcate, moderately transversely septate; inflorescence a many-headed, compound panicle; heads 3–20-flowered; perianth segments 3–3.5 mm. long, acuminate-lanceolate, brownish; stamens 6, about 1.5 mm. long, anthers about equaling filaments; capsule oblong, brown, slightly exceeding perianth, acute and apiculate at apex; seeds ovate, minutely reticulate-areolate.

Wet places, mostly in Upper Sonoran and Transition Zones, northern Sierra Nevada and Coast Ranges to Sierra San Pedro Mártir, but reaching the edge of the Colorado Desert at Palm Springs, Riverside County, and near Mountain Springs, Imperial County, California.

**8. Juncus torreyi** Coville, Bull. Torrey Club **22:** 303, 1895

*Juncus nodosus* var. *megacephalus* Torr., Fl. N.Y. **2:** 326, 1843.
*Juncus megacephalus* Wood, Bot. ed. 2, 724, 1861. Not *J. megacephalus* Curtis, 1835.

Stems 2–7 dm. high, produced singly from tuberlike swellings on slender rootstocks; leaves 1–4 on a stem, terete, 2–5 mm. in diameter, strongly septate, abruptly divergent from stem; auricles hyaline, 3–5 mm. high, obtuse; panicle of 1–20 heads in congested clusters, lowest cluster 1–6 cm. long; heads subspherical, 10–15 mm. in diameter; perianth segments subulate, outer about 5 mm. long, inner 4 mm. long; stamens 6, 2–2.5 mm. long; capsule trigonous-subulate, 1-celled, 4–5 mm. long, beak 1–1.5 mm. long and remaining intact even after dehiscence of body of capsule; seeds oblong-fusiform, longitudinally striate and finely areolate.

Washington to Massachusetts, Alabama, Texas, and southern California. Mostly in the Upper Sonoran and Transition Zones, but occurring in the vicinity of Mecca and Dixieland, Imperial County, California.

**9. Juncus macrophyllus** Coville, Univ. Calif. Pub. Bot. **1:** 65, 1902

*Juncus canaliculatus* Engelm., Bot. Gaz. **7:** 6, 1882. Not *J. canaliculatus* Liebm., 1850.

Stems erect, compressed, 3–10 dm. tall, in small tufts from short rhizomes; basal leaves from half to nearly as long as stems, sheaths scarious-margined; auricles 1–2.5 mm. high, scarious, hyaline or whitish; leaf blades flat, firm, 1.5–4 mm. wide, distinctly but finely striate, attenuate and pungent; stem leaves few, 6–15 cm. long; inflorescence a loose panicle; heads 10–25, 3–8-flowered; perianth segments narrowly lanceolate-ovate, green tinged with brown, inner 6 mm. long, outer about 5 mm. long, hyaline-margined; stamens 6, 2.5–3 mm. long, anthers reddish brown, much longer than filaments; capsules oblong-trigonous, about 4 mm. long, short-beaked; seeds about 0.5 mm. long, with 18–20 fine ribs and lineolate reticulations.

Hillsides and marshes, mainly Upper Sonoran and Arid Transition Zones, southern Sierra Nevada to Sierra San Pedro Mártir, and in suitable situations in Yavapai and Maricopa Counties, Arizona.

## Family 17. MELANTHACEAE. Bunch-Flower Family

Perennial herbs with leafy or scapose stems and elongated or bulblike rootstocks. Leaves alternate or basal, parallel-veined, grasslike or broad. Flowers perfect, polygamous, or dioecious, regular, in terminal spikes, racemes, or panicles, or rarely solitary. Perianth petaloid, of 6 segments, these persistent, distinct or nearly so. Stamens 6; anthers 2-celled and ovate or confluently 1-celled and cordate to reniform. Ovary 3-celled, superior or, in ours, usually partially inferior; ovules few to numerous. Styles 3, distinct or nearly so. Fruit a capsule, septicidally dehiscent. Seeds appendaged.

### 1. ZYGADENUS Michx., Fl. Bor. Amer. 1: 213, 1803

Bulbs membranous-coated. Leaves linear, mainly basal, the cauline ones reduced. Flowers perfect or polygamous. Perianth segments bearing (in ours) a single obovate gland near the base. Stamens distinct from perianth. Cavities of capsule dehiscent to the base.

**1. Zygadenus brevibracteatus** (M. E. Jones) Hall, Univ. Calif. Pub. Bot. **6:** 165, 1915

*Zygadenus fremontii* var. *brevibracteatus* M. E. Jones, Contr. West. Bot. No. **12:** 78, 1908.

Scapes erect, 4–6 dm. tall, glabrous; bulbs ovoid, 1–2.5 cm. thick, 1.5–4.5 cm. long, outer coats brownish; basal leaves 12–25 cm. long, 5–8 mm. wide, faces glabrous but scabrous along margins; cauline leaves 1–6 cm. long, not sheathing basally, linear to subulate; racemes open, simple or compound, 10–25 cm. long; pedicels 1–4 cm. long, rigid, horizontally spreading, subtended by scarious, narrowly ovate, apiculate to attenuate bracts 3–7 mm. long; perianth segments yellowish, 4–7 mm. long; sepals broadly elliptic, short-clawed; petals narrowly elliptic, distinctly clawed; glands green, obovate, with a low ridge along lower margins; stamens two-thirds as long as perianth segments; capsule oblong, 12–15 mm. long; styles persistent.

Rocky hillsides and mesas, Lower Sonoran Zone and desert slopes of Upper Sonoran Zone, Mojave and Colorado Deserts. May.

## Family 18. **LILIACEAE.** Lily Family

Leafy-stemmed or scapose perennial herbs from bulbs, corms, or rootstocks, or in a few genera the caudices woody and often arboreal. Leaves grasslike and ephemeral to large, rigid, and long-persistent. Flowers regular, usually perfect, solitary, racemose, umbellate, or variously clustered. Perianth of 6 nearly distinct segments or these united into a tube, the outer whorl like the inner or quite different from it. Stamens 6 or 3, hypogynous or borne on perianth tube; anthers 2-celled, versatile or basifixed. Ovary superior, 3-celled; ovules few to many in each cell, anatropous or amphitropous. Style 1; stigma capitate or more or less 3-lobed. Fruit a capsule, usually loculicidal. Embryo straight or curved, embedded in rather copious endosperm.

Plants herbaceous, stems from bulbs or corms; leaves not stiff, filiferous, nor dentate-margined:
- Plants from tunicated bulbs:
  - Inflorescences paniculate or racemose; bulbs 4–6 cm. long:
    - Perianth segments distinct, 6–8 mm. long, withering-persistent; inflorescence a widely spreading panicle . . . . . . . . . . 1. *Chlorogalum*
    - Perianth segments united into a tube at base, 3–4 cm. long, not persistent on fruit; inflorescence racemose . . . . . . . . . . 2. *Hesperocallis*
  - Inflorescences umbellate; bulbs rarely over 3.5 cm. long:
    - Bulbs and herbage with an alliaceous odor and taste; ovules 2–several in each cell . . . . . . . . . . 3. *Allium*
    - Bulbs and herbage not alliaceous; ovules several in each cell:
      - Anthers versatile; perianth segments 10–12 mm. long; bracts 12–20 mm. long . . . . . . . . . . 4. *Nothoscordum*
      - Anthers basifixed; perianth segments 2–3 cm. long; bracts 5–7 mm. long . . . . . . . . . . 8. *Milla*
- Plants from solid corms, these not tunicated:
  - Perianth segments all essentially alike, united into a tube, at least at base; conspicuous glands not present on inner perianth segments:
    - Flowers (in ours) blue or purplish; stamens shorter than perianth segments; perianth segments equaling or longer than tube:
      - Stamens opposite inner perianth segments with appendaged filaments; umbels 2–20-flowered; scapes slender, 2–4 mm. in diameter . . . . . . . . . . 5. *Dichelostemma*
      - Stamens not appendaged; umbels 30–100-flowered; scapes stout, 6–15 mm. in diameter . . . . . . . . . . 6. *Triteleiopsis*
    - Flowers red or red-orange; stamens well exserted; perianth segments less than one-fifth as long as cylindrical tube . . . . . . . . . . 7. *Behria*
  - Perianth segments of two kinds, outer much narrower than inner, all distinct to base; a conspicuous gland near base of each inner perianth segment . . . . . . 9. *Calochortus*

Plants with a more or less woody caudex, often arborescent; leaves stiff, swordlike, the margins often filiferous or strongly serrate:
  Ovary 3-celled; leaves with smooth or filiferous margins; flowers perfect or polygamodioecious:
    Ovules many in each cell; flowers all perfect; perianth segments 1.5 cm. long or more, subfleshy, not papery-persistent; seeds strongly flattened, many in each cell ............................................................... 10. *Yucca*
    Ovules 2 in each cell; flowers, or at least many of them, polygamodioecious; perianth segments less than 1 cm. long, papery-persistent; seeds globose, 1 in each cell or only 1–2 ovules in whole flower maturing .......................... 11. *Nolina*
  Ovary 1-celled; leaves with serrate-spiny margins; flowers completely dioecious ......... ........................................................... 12. *Dasylirion*

## 1. **CHLOROGALUM** Kunth, Enum. Pl. **4:** 681, 1843

Bulbs fibrously or (in ours) membranously coated. Leaves mostly basal, long-linear, tufted, cauline ones greatly reduced. Bracts scarious, small. Pedicels articulate apically. Perianth white to purplish, persistent and twisted about the ovary in age, its 6 segments distinct, spreading. Stamens 6, shorter than perianth segments. Style filiform, shallowly 3-cleft. Capsule broadly turbinate, 3-lobed, loculicidally 3-valved. Seeds 1–2 per cell, obovate, testa rugulose.

### 1. **Chlorogalum parviflorum** S. Wats., Proc. Amer. Acad. **14:** 243, 1879

*Laothoe parviflora* Greene, Leaflets Bot. Obs. **1:** 91, 1904.

Bulbs broadly ovoid, 1.5–3 cm. wide, 4–6 cm. long, outer coats dark brown, inner white or nearly so; stem and inflorescence 5–10 dm. high; basal leaves linear, 5–8 mm. wide, 8–20 cm. long, slightly scabrous, margins undulate; panicle 25–50 cm. long, open, branches spreading; bracts narrowly triangular, short-acuminate, 1–4 mm. long, scarious; pedicels slender, 1–4 mm. long; perianth segments oblong-oblanceolate, 6–8 mm. long, recurved, white with deep rose midrib; style exceeding segments; capsule 3–5 mm. wide; seeds black.

Dry hillsides and mesas in heavy soil, Upper and Lower Sonoran Zones, San Diego County into northern Baja California. The small-flowered amole occurs mostly on the seaward side of the crest but occurs sparingly "east of Julian" in San Diego County. May–July.

## 2. **HESPEROCALLIS** A. Gray, Proc. Amer. Acad. **7:** 390, 1867

Erect plants with deep-seated, tunicated bulb, mostly basal, linear, undulate-margined leaves, and large funnelform, white, fragrant flowers in a simple raceme, each flower subtended by a large scarious bract. Pedicels jointed apically. Perianth of 6 obovate-spatulate segments united below the middle to form a tube, withering-persistent. Stamens 6, inserted in throat; anthers linear, versatile. Ovary sessile, containing numerous ovules; style filiform, persistent, equaling perianth; stigma capitate-discoid, shallowly 3-lobed. Capsule subglobose, loculicidally dehiscent. Seeds horizontal, flattened, black.

### 1. **Hesperocallis undulata** A. Gray, Proc. Amer. Acad. **7:** 391, 1867

Bulb ovoid, 4–5 cm. long; outer scales membranous, shining white; stem erect, 3–8 dm. high, stout; basal leaves 8–15 mm. wide, 2–4 dm. long, glabrous, margins strongly undulate; cauline leaves much reduced; raceme 3–30 cm. long, few- to many-flowered, withering flowers often deciduous by means of jointed pedicel; bracts 10–15 mm. long, broadly ovate, short-acuminate; pedicels 6–12 mm. long; perianth segments 3–4 cm. long, white within, green-ribbed without, margins silvery, tube about one-half the length of segments; anthers yellow, 6–7 mm. long; capsule 12–18 mm. long, acute.

Sandy places in the open desert, Lower Sonoran Zone, eastern Mojave Desert to Arizona, south into northern Sonora and central Baja California. March–April.

## 3. ALLIUM L., Sp. Pl. 294, 1753

Slender scapes from a tunicated bulb with linear, basal leaves. Herbage with a strong odor of onions. Flowers in a simple, terminal umbel, subtended by 2–4 membranous bracts. Pedicels slender. Perianth persistent, its segments distinct or sometimes slightly united basally. Stamens inserted on bases of perianth segments, included or rarely exserted; filaments filiform or dilated at base. Capsule obovate to globose, obtusely and shallowly 3-lobed, each lobe often crested, loculicidally dehiscent. Seeds 2–several in each cell, black, wrinkled.

Outer bulb coats membranous, not fibrous; anthers linear to lance-linear, at least twice as long as broad:
- Bulbs oblong, usually clustered, coats deep pink to deep reddish purple, shining ...... ........ 1. *A. haematochiton*
- Bulbs ovoid, mostly solitary, coats white to brownish, dull:
  - Reticulations of inner bulb coats indistinct or none, if present very delicate:
    - Ovary obscurely or not at all crested; peduncles 3–5 dm. high ......... 2. *A. kunthii*
    - Ovary bearing 6 conspicuous oblong crests; peduncles 1–3 dm. high:
      - Stigma with 3 distinct linear lobes; crests of ovary coarsely fimbriate; reticulations on bulbs faint and longitudinal or lacking ................ 3. *A. fimbriatum*
      - Stigma capitate; crests of ovary minutely short-ciliate; reticulations on bulbs transverse ........ 4. *A. palmeri*
  - Reticulations of bulb coats distinct, coarsely quadrate ................ 5. *A. lacunosum*

Outer bulb coats fibrous; anthers broadly elliptic, scarcely one and a half times as long as broad; flowers fragrant ........ 6. *A. macropetalum*

### 1. Allium haematochiton S. Wats., Proc. Amer. Acad. 14: 227, 1879

*Allium californicum* Rose, in Vasey & Rose, Contr. U.S. Nat. Herb. 1: 12, 1890.

Bulb buried 4–12 cm. below surface, narrowly oblong-ovoid, 1.5–3 cm. long, clustered or less commonly solitary, coats membranous, deep pink to deep reddish purple, epidermal cells long and slender, not sinuous; scape (in ours) 5–20 cm. long, faintly 2-edged; leaves 3–7, 1–2 mm. wide, about equaling or shorter than scape; bracts 2–4 (usually 2), connate at base, 6–8 mm. long, obtuse, with a short, acute point, 3–5-nerved; umbel erect; pedicels 6–15 (or rarely 25), slender, 12–18 mm. long; perianth segments 7–9 mm. long, narrowly ovate-lanceolate, acute, inner 3 segments slightly narrower than outer, deep rose to red-purple, midrib darker; stamens and style included, filaments dilated at base; ovary globose-truncate, with short, rounded crests; capsule globose-obcordate, 3.5–4 mm. long.

On dry hillsides, on mesas, and along watercourses, Upper and Lower Sonoran Zones, on western slope of the watersheds, central Baja California to San Luis Obispo County, California. Feb.–May.

**Allium praecox** Brandegee, Zoe 5: 228, 1908, occurs in northern Baja California, but it has been collected only along the coastal slopes, mostly in Upper Sonoran associations. It may occur in particularly favorable spots between Rosario and Punta Prieta.

**2. Allium kunthii** G. Don, Mem. Wern. Soc. **6:** 82, 1827
*Allium scaposum* Benth., Pl. Hartw. 26, 1840.

Bulbs globose to ovoid, 10–15 mm. in diameter, 1.5–2 cm. long, outer coats brownish, reticulations minute, longitudinal, side walls straight or nearly so; leaves 3–several, 3–4 mm. wide, equaling or slightly surpassing scapes, dark green, glabrous; scapes terete, 3–5 dm. high; bracts ovate-attenuate, 1.5–2 cm. long; pedicels slender, 2.5–5 cm. long; umbels loose, 8–20-flowered; perianth segments lanceolate, 4–8 mm. long, acute to acuminate, spreading, white with pinkish midrib, slightly thickened, withering-persistent; stamens shorter than perianth, filaments united in cup below; anthers linear, 2–2.5 mm. long; ovary nearly crestless; seeds faintly pustuliferous.

Heavy soil, mostly Upper Sonoran and Transition Zones, but occurring in the foothills of the Santa Catalina Mountains, Arizona, and eastward to Texas and Chihuahua. Entering the Sonoran Desert occasionally as a straggler. April–May and Aug.–Sept.

**3. Allium fimbriatum** S. Wats., Proc. Amer. Acad. **14:** 232, 1879

Bulb broadly ovoid, 12–15 mm. high, coats reddish brown, without reticulations, rather dull; scapes terete, 5–15 cm. high, shallowly striate; leaf solitary, 1.5–3 mm. wide, usually revolute, equaling or exceeding scape; bracts 2–3, 10–13 mm. long, broadly ovate, abruptly acuminate, 3–7-nerved, often roseate; umbel erect, loose; pedicels 10–20 mm. long, stout; perianth rose-purple, 8–12 mm. long, outer segments ovate-lanceolate, acute to acuminate, inner ones narrower; stamens one-half to two-thirds as long as perianth segments, filament dilated at base; style 4–5 mm. long, stigma with 3 linear lobes 0.3–0.8 mm. long; ovary obovate to globose-truncate, with 6 prominent fimbriate crests 1.5–2 mm. high; seeds ovate, acute at apex, 2–2.5 mm. high.

Usually in rocky, heavy soil, Upper and Lower Sonoran Zones, from Santa Clara County, California, south in the inner Coast Range and along the margins of the Mojave and Colorado Deserts into northern Baja California. April–June.

**4. Allium palmeri** S. Wats., Bot. King Expl. 487, *pl. 37, figs. 10, 11,* 1871

Bulb globose to broadly ovoid; coats thin, white or faintly tinged with red, reticulations minutely transverse, walls slightly if at all sinuous; leaves linear, 1.5–2 mm. wide, 2–3 per scape, 10–15 cm. long, about equaling scape, glabrous; scapes solitary or 2–3 in a cluster, terete; bracts broadly ovate, abruptly acuminate, 12–16 mm. long, white or roseate, 3–5-nerved; umbels few- to many-flowered, loose; pedicels slender, 10–12 mm. long; perianth 6–9 mm. long, segments lance-ovate, acute, white with reddish midnerves; stamens about three-fourths as long as segments, filaments strongly dilated at base; ovary subglobose, bearing 6 oblong, short-ciliolate crests 2–2.5 mm. high; style slender, slightly surpassing stamens, stigma capitate.

Open hillsides and in forest, mainly in the Upper Sonoran and Arid Transition Zones, Arizona to southern Utah, New Mexico, and southern Colorado. May–July.

**5. Allium lacunosum** S. Wats., Proc. Amer. Acad. **14:** 231, 1879

Bulb broadly ovoid, 1.5 cm. high, coats thick, distinctly reticulate with transverse quadrate or oblong pits; scapes 8–20 cm. high, terete; bracts 2, broadly ovate, abruptly acuminate, 10–15 mm. long, white, with 5–9 dark rose to red nerves; umbels 10–20-flowered, open; pedicels 6–15 mm. long; flowers 6–7 mm. long; perianth segments pink with darker red midnerve, oblong-lanceolate, acute to acuminate; stamens nearly equaling perianth segments, filaments narrowly deltoid, dilated below; style about equaling stamens; ovary subglobose, with 3 low, obtuse, thickened ridges near apex of capsule.

Dry interior of California from the western margin of the Colorado Desert north to San Mateo County, California. April–June.

**6. Allium macropetalum** Rydb., Bull. Torrey Club **31:** 401, 1904

*Allium reticulatum* var. *deserticola* M. E. Jones, Contr. West. Bot. **10:** 30, 1902.
*Allium deserticola* Woot. & Standl., Contr. U.S. Nat. Herb. **16:** 114, 1913.

Bulb ovoid, with a rather long neck, deep-seated, as much as 15 cm. below surface; outer bulb coats becoming fibrous by disintegration of thin-walled cells lying within meshes of fibrous reticulum, fibrous covering tawny; leaves 2–5, linear, 2–3 mm. wide, equaling or often surpassing scapes, these terete, 2–4 dm. high; bracts 2–3, ovate-acute to short-acuminate, roseate, 15–18 mm. long, 5–7-nerved; umbels 10–15-flowered; pedicels slender, 8–15 mm. long; perianth 7–9 mm. long, segments lance-ovate, acute, white or pinkish with roseate midnerve; stamens about one-half as long as perianth segments, filaments broadly dilated basally; anthers broadly elliptic, 1.5 mm. long; style about equaling stamens; stigma faintly 3-lobed; ovary depressed-globose, bearing 3 low, rounded, entire, ridgelike crests.

In heavy soils, often among rocks, Lower and Upper Sonoran Zones, western Colorado, eastern Utah, and southward into Arizona, New Mexico, and Texas. Common in the Tucson Mountains, Arizona. March–May.

## 4. NOTHOSCORDUM Kunth, Enum. Pl. 4: 457, 1843

Scapose herbs with a tunicated bulb devoid of an alliaceous odor. Leaves basal, narrowly linear. Inflorescence a 2-bracteate umbel, few-flowered. Perianth segments 6, distinctly united a short distance basally, 1-nerved, withering-persistent. Stamens 6, inserted on bases of perianth segments; filaments linear, slightly dilated at base; anthers versatile, introrsely dehiscent. Ovary sessile, 3-celled; style filiform, jointed near base; stigma capitate. Ovules several in each cell. Capsule shallowly 3-lobed, loculicidal. Seeds flat or angled, black.

**1. Nothoscordum bivalve** (L.) Britt., in Britt. & Brown, Ill. Fl. **1:** 415, 1896

*Ornithogalum bivalve* L., Sp. Pl. 306, 1753.
*Allium striatum* Jacq., Coll. Suppl. 51, 1796.

Bulb globose, 8–20 mm. in diameter, coats membranous, white; leaves narrowly linear, 1–4 mm. wide, acute to obtuse at apex, equaling scape or shorter; scape terete, 1–4 dm. high; bracts lanceolate to lance-ovate, acuminate, membranous, 12–20 mm. long, persistent; umbel 6–12-flowered; pedicels filiform, unequal, 1.5–5 cm. long in fruit; flowers white, 10–12 mm. long, perianth segments oblong-lanceolate, acute; stamens included; capsule obovoid to depressed-globose, 4–6 mm. high, obtusely 3-lobed, equaled or slightly surpassed by style.

In sandy soil from Nebraska to Tennessee, Georgia, and Florida and south to central Mexico and Baja California. April–May.

**Muilla maritima** (Torr.) S. Wats., Proc. Amer. Acad. **14:** 235, 1879, with greenish white, rotate perianth segments and purple anthers, and **M. serotina** Greene, Erythea **1:** 152, 1893, differing from it only in having retrorsely scaberulous leaves, occur in the foothills and near the coast in northern Baja California, almost to Rosario, and may occur farther south within the limits of the Sonoran Desert, but neither has been collected as yet within our region.

**BLOOMERIA** Kell., Proc. Calif. Acad. **2:** 11, 1859. Scapose herbaceous plant from a fibrous-coated corm, with linear, carinate leaves and yellow flowers in a terminal umbel. Perianth of 6 nearly equal, persistent, distinct segments. Filaments slightly appendaged at the base. Capsule subglobose, loculicidal. Seeds angular, black.

This genus, represented in San Diego County and northern Baja California by two species, *B. crocea* (Torr.) Coville, with a solitary leaf and filament appendages weakly bicuspidate, and *B. clevelandii* S. Wats., with several leaves and filament appendages entire and obtuse at the apex, may occur as occasional stragglers along the western margin of the Sonoran Desert in Baja California near Rosario. Neither of these species has been found within our area to date.

## 5. DICHELOSTEMMA Kunth, Enum. Pl. 4: 469, 1843

Scapes erect, tortuous or twining from a fibrous-coated, deep-seated corm. Leaves few to numerous, linear, flattened or subterete. Umbel subtended by 3–several scarious, spathelike bracts. Pedicels jointed at base of flower. Perianth short-tubular at base, tube equaled or surpassed by segments. Stamens 6, inner whorl often consisting of sterile, more or less bidentate staminodia; anthers (in ours) basifixed or versatile; filaments narrow, winged or dilated at base. Ovary 3-celled, sessile or short-stipitate; ovules 3–8 in each cell; style persistent; stigmas short, divergent. Capsule ovate to oblong, loculicidal. Seeds black, angled.

Bracts and perianth segments deep violet; umbels subcapitate; pedicels 2–10 mm. long, slender ........................................ 1. *D. pulchellum* var. *pulchellum*
Bracts and perianth segments pale blue; umbels open; pedicels stout, 6–35 mm. long ...... ........................................ 1a. *D. pulchellum* var. *pauciflorum*

### 1. Dichelostemma pulchellum (Salisb.) Heller, Muhlenbergia 1: 132, 1906, var. pulchellum

*Hookera pulchella* Salisb., Parad. Lond. under *pl. 98*; *pl. 117*, 1808.
*Brodiaea capitata* Benth., Pl. Hartw. 339, 1857.
(For full synonymy see Amer. Midl. Nat. **24:** 468–469, 1940.)

Leaves 2–4, mostly shorter than slender scape, 2–15 mm. wide, margins scaberulous; scapes 2–8 dm. tall; bracts 8–15 mm. long, broadly ovate, abruptly acuminate; pedicels slender, 2–10 mm. long; perianth deep violet-purple, rarely reddish purple or white, 12–18 mm. long, withering-persistent, tube thin, 4–8 mm. long, slightly constricted at throat; perianth segments oblong-ovate, obtuse, 3–4 mm. wide, 6–10 mm. long; stamens 6, outer with short, naked filaments broadly dilated at base, their anthers 2 mm. long, inner anthers 4 mm. long, nearly concealed by erect, oblong-lanceolate, lateral appendages; capsule sessile, ovate, 6–8 mm. long; style slender, persisting as a beak nearly as long as capsule.

Dry open ridges and grassy plains, Lower Sonoran to Transition Zones, southern Oregon to central Baja California. Specimens from the western Colorado Desert often are intermediate between vars. *pulchellum* and *pauciflorum*. Feb.–May.

### 1a. Dichelostemma pulchellum var. pauciflorum (Torr.) Hoover, Amer. Midl. Nat. 24: 471, 1940

*Brodiaea capitata* var. *pauciflora* Torr., Bot. Mex. Bound. 218, 1859.
(For full synonymy see Amer. Midl. Nat. **24:** 471, 1940.)

Umbels open, with 2–12 (mostly 4–7) flowers; pedicels stout, especially in fruit, unequal, 6–35 mm. long; bracts and perianth light blue; capsule 6–10 mm. long.

Rocky hillsides, Lower and Upper Sonoran Zones, western margins of the Colorado Desert to Utah, New Mexico, and Sonora. This variety replaces var. *pulchellum* in the desert hills of Arizona and southeastern California and extends well into the Upper Sonoran Zone in Arizona and New Mexico; occasional in central Sonora. Feb.–May.

### 6. **TRITELEIOPSIS** Hoover, Amer. Midl. Nat. **25:** 98, 1941

Perennial from a stramineous, fibrous-coated corm bearing several small cormlets around base of stem and enclosed within fibrous coat, corms not alliaceous in odor. Stems terete, stoutish. Leaves 3–8 or more, alternate, membranous and hyaline at base, carinate but flat above. Scape pithy, glaucous. Umbel many-flowered, subtended by several scarious, acuminate bracts. Pedicels slender, jointed. Perianth tube narrowly funnelform, about equaling lobes in length. Perianth segments with transverse scale-like appendages free from stamens. Filaments slender, terete, inserted near mouth of perianth tube, adnate parts forming ridges in lower part of perianth tube, unappendaged. Stigma capitate. Ovary slightly stipitate. Capsule 3-celled, loculicidally dehiscent. Seeds flat, black, minutely roughened.

**1. Triteleiopsis palmeri** (S. Wats.) Hoover, Amer. Midl. Nat. **25:** 99, 1941
*Brodiaea palmeri* S. Wats., Proc. Amer. Acad. **24:** 78, 1889.
*Triteleia (?) palmeri* Greene, Pittonia **1:** 292, 1889.

Scape stout, 6–15 mm. in diameter at base, 3–10 dm. high, terete, erect; leaves numerous, broadly V-shaped in cross section, 10–25 mm. wide, 2–5 dm. long, fleshy, glabrous; bracts several, membranous, broadly ovate, abruptly short-acuminate, 12–20 mm. long, pale blue to nearly white; umbels open, 30–100-flowered, 5–10 cm. in diameter; pedicels slender, 1.5–3 cm. long; perianth deep blue, funnelform, 2–2.5 cm. long, tube narrow, not saccate, about equaling segments, not constricted at throat; segments lance-ovate, obtuse; stamens 6, filaments linear, not dilated or winged, inserted in perianth tube, about equaling perianth; anthers linear, 2–2.5 mm. long, basifixed, becoming tightly circinately coiled after dehiscence; ovary lance-ovate, sessile; style equaling perianth; stigma subcapitate, indistinctly 3-lobed.

Sandy desert flats, rarely in rocky or heavy soil, Lower Sonoran Zone, Baja California from the vicinity of Punta Prieta southward to the Magdalena Plain. Feb.–May.

### 7. **BEHRIA** Greene, Bull. Calif. Acad. **2:** 143, 1886

Perennials, scapose plants from a membranous-coated corm. Leaves several, narrowly linear, nearly or quite equaling scape. Inflorescence umbellate, subtended by several scarious bracts, several-flowered. Perianth tubular, bright orange-red with yellow veins, 6-toothed. Stamens 6, somewhat exserted, filaments slender, slightly dilated and coalesced into a short ring at base. Anthers versatile. Ovary 3-celled, short-stipitate; style slender, exserted; stigma small, 3-lobed. Capsule oblong, rupturing persistent perianth. Seeds many, oblong, strongly flattened.

**1. Behria tenuiflora** Greene, Bull. Calif. Acad. **2:** 143, 1886
*Bessera tenuiflora* J. F. Macbride, Contr. Gray Herb. **56:** 11, 1918.

Bulb ovoid, 2–2.5 cm. long, 3–6 cm. below the surface, coats dark reddish brown; leaves 3–10, narrowly linear, 1–2.5 mm. wide, 2–3 dm. long, finely retrorse-scaberulous on margins and midveins; scapes 2–3.5 cm. long, sparsely scaberulous in 1–several lines nearly or quite their full length; flowers 5–15; pedicels slender, 1–4.5 cm. long; bracts lance-subulate, 1–1.5 cm. long, scarious or subscarious; perianth tube 1.5–2 cm. long, 2–2.5 mm. in diameter, glabrous; perianth lobes narrowly ovate-lanceolate, 3–4 mm. long, margins thin, midrib rather thick, slightly recurved; stamens exserted 5–10 mm.; anthers greenish, linear-oblong, 2–2.5 mm. long; style exserted 1–1.5 cm.; capsule about 1.5 cm. long, 5–6 mm. in diameter; seeds 2–2.5 mm. long, about 1 mm. wide, strongly compressed.

Rocky hillsides and sandy plains, Tropical Zone to Upper Sonoran Zone, southern Baja California. Sept.–Oct.

## 8. MILLA Cav., Ic. 2: 76, *pl. 196,* 1793

Bulb tunicate. Leaves basal, linear. Scape slender, naked, bearing 1–5 (usually 2) flowers in an open umbel. Perianth white, salverform, with a narrowly turbinate tube, segments oblong-spatulate, with conspicuous greenish midribs, withering-persistent. Stamens 6, nearly sessile, inserted in throat. Anthers yellow, basifixed. Ovary superior, clavate, trigonous, slender; style exceeding perianth tube; stigma capitate. Capsule oblong to oblong-obovate; seeds ovate, compressed.

### 1. Milla biflora Cav., Ic. 2: 76, *pl. 196,* 1793

Bulb ovoid, 12–18 mm. high, bearing fascicles of fleshy and fibrous roots, its coat brown, membranous to somewhat fibrous; scape solitary, 15–40 cm. long, terete; leaves filiform-linear, 0.6–1.2 mm. wide, equaling to somewhat surpassing scape, retrorsely scabrous on angles of the 6–10 striations; bracts ovate-attenuate, 5–7 mm. long, membranous; pedicels 5–10 cm. long; perianth segments white, with green midribs, 2–3 cm. long, inner linear-oblong, 4–6 mm. wide, acute at apex, outer broadly oblong, 8–12 mm. wide, rounded or obtuse at apex; tube about 1 cm. long; capsule 1.5–2.5 cm. long.

In heavy soil, Upper and Lower Sonoran Zones, southern Arizona and New Mexico, southward to central Mexico. Occurring sparingly in the Sonoran Desert. Near Divisadero, between Ures and Cumpas, Sonora. Aug.–Sept.

## 9. CALOCHORTUS Pursh, Fl. Amer. Sept. 240, 1814

Herbs from perennial tunicated bulbs encased in membranous or fibrous-reticulate coats; herbage glabrous. Stems leafy or scapose, simple or moderately branched, often bulbiferous in axils of leaves above or below surface of soil. Leaves linear (in ours), cauline ones smaller than lower ones. Flowers conspicuous, nodding or (in ours) erect and broadly campanulate, solitary or 2–several on branching stem. Perianth segments 6, outer 3 narrower and less highly colored than inner, inner 3 convolute in bud, obovate, with a slightly to markedly depressed, glandular spot or pit inside near base, this usually more or less covered by hairs. Stamens 6; anthers linear or oblong. Ovary 3-celled, triangular to 3-winged. Stigma 3-branched or 3-toothed. Fruit a 3-celled septicidally dehiscent capsule. Seeds in 2 rows in each locule, flattened or irregularly angled.

Bulb coats membranous; sepals shorter than petals:
  Glands only slightly or not at all depressed, never surrounded by a membrane; flowers white, pink, or lavender:
    Stems usually flexuous-twining; capsules lanceolate; petals each with a transverse yellow band at the gland .......................................... 1. *C. flexuosus*
    Stems erect, not flexuous-twining; capsules linear; petals not transversely yellow-banded at the gland:
      Hairs or processes on glands linear to subclavate, not fungoid-enlarged apically; anthers white, equaling filaments .............................. 2. *C. dunnii*
      Hairs or processes on glands fungoid-expanded apically; anthers blue or purple, shorter than filaments .................................... 3. *C. splendens*
  Glands distinctly depressed, their edges surrounded by a membrane; flowers yellow, orange, or vermilion:
    Petals yellow, with or without a purple band above and/or below gland; hairs on face of petal not thickened apically; anthers oblong .................. 4. *C. concolor*
    Petals orange-red to vermilion with a brownish purple to nearly black spot near base below gland; hairs on face of petal distally thickened; anthers lanceolate ........ ........................................................... 5. *C. kennedyi*
Bulb coats fibrous-reticulate; sepals about equaling petals:
  Petals pink to deep rose, bearded within only below middle; bases of hairs surrounding gland united in fascicles ....................................... 6. *C. plummerae*
  Petals orange-yellow flecked and margined with reddish brown, bearded almost to apex within; bases of hairs surrounding the gland united into a nearly continuous membrane ........................................................... 7. *C. weedii*

### 1. **Calochortus flexuosus** S. Wats., Amer. Nat. **7:** 303, 1873

A flexuously twining or scrambling plant 1.5–6 dm. long, stems usually 2–several-branched; bulb broadly ovoid, 1–2.5 cm. in diameter, 1.5–3 cm. long, 5–20 cm. below surface, membranous coats dark brown; leaves linear to lance-linear, to 7 mm. wide, lower one 5–45 cm. long, upper ones gradually reduced, basally broader in comparison, attenuate; flowers 1–8 (usually 2–4), campanulate, white to pink, petals 3–4.5 cm. long, broadly obovate, with a purple spot on claw and a transverse yellow band through gland, glabrous except near gland; sepals 6–8 mm. wide, 2–2.5 cm. long, acute to obtuse, not attenuate; anthers white, 6–9 mm. long, about equaling or slightly shorter than filaments; gland not depressed, circular to transversely crescentic, densely covered with short, stoutish, hairlike processes; ovary narrowly fusiform to sublinear, triquetrous but not winged; capsule slender, about 10 mm. in diameter, 5–6 cm. long or more; seeds strongly flattened.

Desert mesas and hillsides, Lower Sonoran Zone and more arid parts of Upper Sonoran Zone, central Colorado Desert and eastern Mojave Desert, California, to southern Utah and southwestern Colorado, south into central Arizona and northern Baja California. April–June.

### 2. **Calochortus dunnii** Purdy, Proc. Calif. Acad. III, **2:** 147, 1901

*Calochortus palmeri* var. *dunnii* Jepson & Ames, in Jepson, Fl. Calif. **1:** 294, 1922.

Stems erect, 3–7 dm. tall, glabrous, monochasially 1–5-flowered; bulb ovoid, 1–2 cm. in diameter, 2–3.5 cm. long, coats chestnut brown and submembranous to somewhat chartaceous, 5–12 cm. below surface; lower leaf to 3.5 dm. long, linear, 3–5 mm. wide, upper ones linear-subulate, attenuate, successively reduced; flowers erect, broadly campanulate, white or flushed with pink, a reddish brown spot just above gland on each petal, these 2–2.5 cm. long, sparsely hairy immediately near circular, nondepressed gland; gland densely covered with stoutish, hairlike processes; anthers about equaling filaments; fruit linear, triquetrous, wingless, about 5 cm. long.

On arid, stony ridges, mainly Upper Sonoran Zone but straggling into upper edge of Lower Sonoran Zone, San Diego County near Julian, California, south into northern Baja California. May–June.

**3. Calochortus splendens** Dougl. ex Benth., Trans. Hort. Soc. Lond. II, **1:** 411, *pl. 15, fig. 1,* 1835

(For synonymy see Ann. Mo. Bot. Gard. **27:** 460, 1940.)

Stems erect, 1–5-branched, rarely bulbiferous in lower axils, 3–6 dm. tall; bulb ovoid, 2–3.5 cm. long, membranous coats dark brown; leaves linear, lower often nearly equaling stem, successively reduced upward; inflorescence 1–5-flowered; flowers campanulate, erect; petals pale lavender to deep lavender, 2.5–3.5 cm. long, usually with a deeper purple spot on claw and sometimes with another spot around gland, sparsely hairy below middle with slender, flexuous hairs; gland not depressed, usually densely covered with stoutish hairs fungoid-branched apically; anthers bluish to purplish, usually shorter than filaments; capsule linear, triquetrous, 5–9 cm. long; seeds strongly flattened, nearly white.

Hillsides, usually in grassy areas or in openings in the chaparral, mainly Upper Sonoran Zone but occasionally in the upper parts of the Lower Sonoran Zone along the western margins of the Colorado and San Felipe Deserts; Colusa County, California, southward in the inner Coast Range to the southern end of the Sierra San Pedro Mártir near Rosario, Baja California, and on Santa Catalina and Los Coronados Islands. April–June.

**4. Calochortus concolor** (Baker) Purdy, Proc. Calif. Acad. III, **2:** 135, 1901

*Calochortus luteus* var. *concolor* Baker, Garden **48:** 440, *pl. 1043,* 1895.

Stems stoutish, 3–5 dm. tall, usually simple; bulb broadly ovoid, 2.5–3.5 cm. long, membranous coat dark brown; bulblets rose; basal leaves linear, deeply channeled, nearly equaling stem, upper ones, usually 1–2, much reduced, 3–5 cm. long, recurved; inflorescence subumbellate, 2–4-flowered; sepals 2.5–3 cm. long, yellow within, usually with a purple spot near base; petals broadly cuneate-obovate to flabelliform, 2.5–3.5 cm. long, deep yellow, with or without a purple area above and below gland, clothed with erect, yellow hairs below middle within; gland oblong, depressed; anthers yellow, 8–10 mm. long, on broad filaments 4–6 mm. long; capsule linear, obtusely triquetrous, 6–10 cm. long, attenuate; seeds strongly flattened.

Dry hillsides, Lower and Upper Sonoran Zones, southern San Bernardino County, California, south to northern Baja California. May–June.

**5. Calochortus kennedyi** Porter, Bot. Gaz. **2:** 79, 1877

*Calochortus speciosus* M. E. Jones, Contr. West. Bot. **14:** 27, 1912.
*Calochortus kennedyi* var. *munzii* Jepson, Man. Fl. Pl. Calif. 236, 1923.

Stems simple, stoutish, 1–2.5 dm. tall, distinctly glaucous; basal leaves linear, deeply channeled, 2–5 mm. wide, about equaling stems, glaucous; upper leaves 1–2 or wanting, 3–6 mm. wide at base, 3–5 cm. long, recurved and thickish, attenuate, scarious along margins; umbels 2–4-flowered; sepals broadly ovate to lance-ovate, 2–3 cm. long, 6–10 mm. wide, acute, orange-red within, usually with a brownish spot near base; petals broadly cuneate-obovate, apiculate, 2.5–3.5 cm. long, orange-red with brownish purple claws, bearing a few distally thickened hairs around gland, otherwise naked; gland circular, small, densely covered with short, matted hairs; anthers 6–8 mm. long, about twice as long as filaments; capsule broadly linear-lanceolate, 4–5 cm. long, obtusely triquetrous, slightly attenuate above; seed coats loose.

Gravelly hills, mesas, and outwash slopes, mainly Lower Sonoran Zone, arid sides of Mount Pinos, Ventura County, California, and desert slopes of the Tehachapi Mountains to southern Nevada, Arizona, and northern Sonora. April–June.

**Calochortus ambiguus** (M. E. Jones) Ownbey, Ann. Mo. Bot. Gard. **27:** 505, 1940 [*Calochortus watsonii* var. *ambiguus* M. E. Jones, Contr. West. Bot. **14:** 27, 1912], with pinkish to bluish gray petals, and a depressed, circular to transversely lunate gland surrounded by a fringe of distally branched, stoutish hairs, occurs in the mountains of Arizona just to the east of and a few hundred feet higher in altitude than the Sonoran Desert. It may occur as sporadic waifs along the upper margins of the Sonoran Desert in this area.

### 6. **Calochortus plummerae** Greene, Pittonia **2:** 70, 1890

*Calochortus weedii* var. *purpurascens* S. Wats., Proc. Amer. Acad. **14:** 265, 1879, in part.

Erect, stem simple or sparingly branched, 3–12 dm. tall; bulb ovoid, 2.5–4 cm. long, coats dark brown, fibrous-reticulate; basal leaves 1–several, linear, attenuate above, 3–8 mm. wide, to 5 or 6 dm. long, convolute; upper cauline leaves usually several, to 25 cm. long, narrowly attenuate; inflorescences with 1–5 (usually 2) flowers, often appearing subumbellate; sepals lance-acuminate, 2–4 cm. long, purple-tinged; petals rounded-flabelliform, 3–4 cm. long, usually erose-denticulate above, pink to deep rose, drying purplish, conspicuously bearded with long, yellow hairs to slightly above middle, nearly or quite glabrous at apex; gland orbicular, depressed, densely rimmed with short hairs more or less united into fascicles; anthers lance-linear, acute and short-apiculate, equaling or somewhat exceeding filaments; capsule linear, acuminate above, 4–6 cm. long, 6–8 mm. wide, obtusely triquetrous; seeds flattened.

Arid, rocky slopes, mostly Upper Sonoran Zone, but straggling into upper margins of Lower Sonoran Zone near Beaumont Pass; inner ranges of Los Angeles, San Bernardino, and Riverside Counties, California. May-June.

### 7. **Calochortus weedii** Wood, Proc. Acad. Phila. **1868:** 169, 1868

*Calochortus luteus* var. *weedii* Baker, Jour. Linn. Soc. **14:** 309, 1874.
*Calochortus citrinus* Baker, Bot. Mag. III, **31:** *pl. 6200,* 1875.

Bulb ovoid, 2–2.5 cm. long with thick, fibrous-reticulate coats; stems erect, 3–8 dm. tall, often branched, not bulbiferous; basal leaves broadly linear (to 12 mm. wide), convolute, to 5 dm. long but nearly always withering before flowering occurs, upper leaves reduced, lance-attenuate, to 15 cm. long; inflorescences usually 2–3-flowered; sepals narrowly lanceolate, attenuate, about equaling petals; petals orange-yellow, minutely flecked and marginally suffused with brown, broadly obovate-cuneate, 2.5–3.5 cm. long, conspicuously bearded within almost or quite to dentate or fimbriate apex with long, yellow hairs; gland orbicular, slightly depressed, nearly naked but margined with a dense ring of hairs fused basally into a nearly continuous marginal membrane; anthers oblong, obtuse to acute, 8–10 mm. long, about equaling filaments; capsule linear, acuminate, about 5 cm. long; seeds strongly flattened, minutely reticulate.

Dry rocky hillsides, Upper Sonoran Zone but sporadic as straggler into upper margin of Lower Sonoran Zone in western Riverside County, California, thence south into northern Baja California and west to the coast near San Diego, California. May–June.

## 10. YUCCA L., Sp. Pl. 319, 1753

Acaulescent or arborescent perennials with a simple or branched, woody caudex. Leaves flexuous or usually rigid, pungently tipped, denticulate-serrulate or filiferous on the margins. Inflorescence a large panicle. Flowers open-campanulate to subglobose with incurved perianth segments, these often fleshy. Stamens 6, filaments enlarged above, puberulent with coarse, blunt hairs or glabrous; anthers versatile. Stigma sessile or on a short style, shallowly 6-lobed. Fruit a loculicidally dehiscent capsule or indehiscent, dry and spongy or baccate. Seeds flat and thin, black, dull or lustrous, usually finely wrinkled.

Fruits fully dehiscent:
- Plants acaulescent or the stems not over 1 m. long and procumbent; styles short, less than 3.5 mm. long; fruits 2.5–4 cm. long:
  - Mature leaf blades usually less than 2 cm. wide, about 6 dm. long; plants dying after fruiting .......... 1. *Y. whipplei* subsp. *whipplei*
  - Mature leaf blades usually more than 2.5 cm. wide, 2.5–4.5 dm. long; plants flowering several times .......... 1*a*. *Y. whipplei* subsp. *eremica*
- Plants arborescent when mature; styles 7–13 mm. long; fruits 4–8 cm. long ... 2. *Y. elata*

Fruits indehiscent or very tardily and irregularly dehiscent:
- Fruits fleshy at maturity; perianth segments somewhat spreading and all expanding alike; leaf margins filiferous except in *Y. schottii*:
  - Pistils 4.5–8.2 cm. long at anthesis; filaments of stamens adnate to perianth segments well above base of ovary; fruits 13–23 cm. long:
    - Plants acaulescent, or with procumbent stems; terminal part of filaments bending outward at sharp angle in anthesis:
      - Plants usually simple or in clumps of 2–6 individuals; fibers on leaf margins coarse .......... 3. *Y. baccata* var. *baccata*
      - Plants usually in dense clumps of many individuals; fibers on leaf margins fine and wiry .......... 3*a*. *Y. baccata* var. *vespertina*
    - Plants caulescent with erect-ascending stems to 2 m. tall; terminal part of filaments slightly bent outward or nearly straight at anthesis:
      - Fibers present on margins of young leaves, scanty or none on old; leaves 3–6 dm. long .......... 4. *Y. arizonica*
      - Fibers lacking on margins of young leaves, abundant on mature ones; leaves 3–13 dm. long .......... 5. *Y. thornberi*
  - Pistils 15–33 mm. long at anthesis; filaments adnate to perianth segments barely to base of ovary; fruits 4–12 cm. long:
    - Margins of leaves not filiferous; branches of inflorescence densely short-tomentose .......... 6. *Y. schottii*
    - Margins of leaves more or less coarsely filiferous; branches of inflorescence glabrous or minutely puberulent:
      - Stamens glabrous; leaves 4–8 dm. long .......... 7. *Y. schidigera*
      - Stamens puberulent; leaves 1.5–3.5 dm. long .......... 8. *Y. valida*
- Fruits spongy and dry at maturity; outer perianth segments incurved; corolla nearly globose, inner segments outcurved toward their tips; leaf margins denticulate, not filiferous (plants arborescent) .......... 9. *Y. brevifolia*

### 1. Yucca whipplei Torr., Bot. Mex. Bound. 222, 1859

*Yucca graminifolia* Wood, Proc. Acad. Phila. **1868:** 167, 1868.
*Hesperoyucca whipplei* Baker, Kew Bull. **1892:** 8, 1892.
*Yucca whipplei* var. *typica* Haines, Madroño **6:** 44, 1941.

Acaulescent, simple; leaves in a dense, broad rosette, pointing in all directions, 7–22 mm. wide, 3–7 dm. long, rigid, glaucous, straight or falcate, sharply denticulate, tipped with a needlelike spine; flowering stalk 2–4 m. high, 4–10 cm. in diameter and bearing broadly ovate to lanceolate, leathery to papery, pungently tipped bracts, glabrous; panicle 0.75–1.5 m. long, branches 6–25 cm. long; flowers pendent, fragrant, perianth segments creamy white, often faintly purple-tinged, 3.5–5 cm. long; stamen about equaling pistil, fleshy-enlarged above; capsule 3–4 cm. long, 2.5 cm. in diameter, loculicidally dehiscent; seeds flat, thin, shining black.

Hillsides, mostly in the Upper Sonoran Zone, western slopes of the Sierra San Pedro Mártir, Baja California, northward to Monterey and Tulare Counties, California, and in northwestern Arizona; barely touching the Sonoran Desert in eastern San Diego County, California, and adjacent Baja California. April–June.

**1*a*. Yucca whipplei** subsp. **eremica** Epling & Haines, Brittonia **9**: 172, 1957

*Yucca peninsularis* McKelvey, Yuccas SW. U.S. pt. **2**: 52, 1947.

Acaulescent, usually cespitose plants in clumps of several to many plants that flower and fruit repeatedly before dying; leaves straight or falcate, stiff, 2.5–4 cm. wide at base of blade, tapering gradually upward, 2.5–4.5 dm. long, plano-convex or somewhat keeled beneath, finely striate, usually glaucous over sage green, margins sharply denticulate; terminal spine very sharp, reddish or brownish, 1–1.5 cm. long; scape 1–2.5 m. tall (exclusive of inflorescence), bearing ovate-lanceolate to narrowly lanceolate, somewhat attenuate, spine-tipped bracts to 1.2 dm. long; inflorescence to 1.5 m. long, 2–3.5 dm. wide; perianth segments white or cream, often tinged with purplish, 1–6 (usually 2.5–4.5) cm. long; pistil obovoid, 8–20 mm. long; stamens about equaling pistil, minutely puberulent upward; capsule 2.5–3.5 cm. long, 12–20 mm. wide, loculicidally dehiscent, with conspicuous placental intrusions; seeds black, strongly compressed, 3–5 mm. long.

On arid, usually rocky ridges and mountain slopes, mostly Lower Sonoran Zone, eastern slopes of Sierra Pinal and from vicinity of Rosario and the southern flanks of the Sierra San Pedro Mártir south to near Bahía de los Angeles and Punta Prieta, Baja California. Feb.–March.

**2. Yucca elata** Engelm., Bot. Gaz. **7**: 17, 1882

*Yucca angustifolia radiosa* Engelm., in S. Wats., Bot. King Expl. 496, 1871.
*Yucca radiosa* Trelease, Rep. Mo. Bot. Gard. **3**: 163, 1892.

Caulescent, the trunk simple or branched, to 7 m. high; leaves numerous, forming dense clusters to 1 m. in diameter at tips of branches, old ones persistent for several years as a close, reflexed "shag," individual leaves linear-lanceolate, 3–4 cm. wide at base, abruptly narrowed to 6–8 mm. wide just above base, 8–15 mm. wide above middle, yellow-green, smooth, thin, pungently tipped, margin white, finely and copiously filiferous in age; inflorescence 1–1.5 m. high, scape short and stout, branches spreading, 10–20 cm. long, glabrous; flowers white, campanulate, perianth segments narrowly lanceolate, acute, 4–4.5 cm. long; capsule dry, broadly oblong, 4–8 cm. long, 2.5–3 cm. in diameter, loculicidally dehiscent; seeds obovate, flat, thin, 6–8 mm. long, dull black, finely wrinkled.

Dry plains and stony mesas, Lower and Upper Sonoran Zones, southwestern Arizona to Texas, northern Sonora, and Chihuahua. May–July.

**3. Yucca baccata** Torr., in Emory, Bot. Mex. Bound. 221, 1859, var. **baccata**

Plant usually simple but sometimes in clumps of 2–6 individuals; stems very short or to 1 m. long and procumbent; leaves 5–7.5 dm. long, 2.5–4 cm. wide at about middle, tapering gradually to apex and narrowed slightly toward base, usually somewhat twisted or falcate, rigid except near union with basal part, scabrous on both surfaces, blue-green or yellow-green, moderately glaucous; leaf margins separating into coarse, flattish fibers along upper half of leaf, persistent fibers 1–10 cm. long, recurved; terminal spine 1.5–7 cm. long, stiff; scape 10–15 cm. long, stoutish; inflorescence to 7.5 dm. tall, 2.5–4.5 dm. in diameter, ovoid; pedicels 7–40 mm. long; flowers 5–15 cm. long, campanulate, perianth segments expanding only slightly in anthesis, fleshy, denticulate along margins, inner 4–5 mm. wider than outer ones; stamens 3–4.5 cm. long, clavate tip nearly horizontal at anthesis; pistil 5–8 cm. long; style 5–7 mm. long; fruit fleshy, 12–17 cm. long, 5–6.5 cm. in diameter at base, tapering toward apex.

Desert slopes and mountainsides, Lower and Upper Sonoran Zones, eastern Mojave Desert, California, to southwestern Colorado and south into central Arizona, most of New Mexico, and southwestern Texas. April–July.

**3*a*. Yucca baccata** var. **vespertina** McKelvey, Yuccas SW. U.S. pt. **1:** 45, 1938

Differing from var. *baccata* in having procumbent stems and in growing in large, cespitose clumps.

Arid slopes and mesas, Lower and Upper Sonoran Zones, northern Mohave County, and in Yavapai, Graham, and Apache Counties, Arizona. April–July.

**4. Yucca arizonica** McKelvey, Jour. Arnold Arb. **16:** 270, 1935

*Yucca treleasei* J. F. Macbride, Contr. Gray Herb. **56:** 15, 1918. Not *Y. treleasei* Sprenger, 1906.

Shrub or small tree up to 2.5 m. high or young plants appearing acaulescent; leaves ascending-spreading, or older ones reflexed, 12–30 mm. wide, 2–4 dm. long, yellowish green, smooth, pungently pointed, somewhat falcate, margins gray to brown, finely filiferous in youth, nearly devoid of fibers in age; inflorescence pyramidal, 4–7 dm. high, branches ascending-spreading, glabrous, 10–25 cm. long; pedicels stoutish, 2–3 cm. long; flowers pendent, campanulate, bases narrow, tips of oblanceolate perianth segments flaring; perianth segments 6–8 cm. long, creamy white; filaments 2.5–3 cm. long, densely pubescent; pistil oblong, 4–5 cm. long at anthesis, style about 5 mm. long; fruit baccate, 8–10 cm. long.

Hillsides and mesas, Upper and Lower Sonoran Zones, from the vicinity of Nogales, Arizona, south to the vicinity of Divisadero, Sonora. March–May.

**5. Yucca thornberi** McKelvey, Jour. Arnold Arb. **16:** 268, *pl. 138*, 1935

Plants cespitosely crowded in dense clumps 2–6 m. broad, caulescent, stem to 1.5 m. tall and densely clothed with reflexed to spreading dead leaves; leaf blades lance-linear, 3–13 dm. long, 1.2–4 cm. wide near middle (greatest width), tapering gradually to spine-tipped apex, margins denticulate on young leaves, separating into fine to coarse, loosely curled fibers 5–30 cm. long on older leaves; inflorescence slender, 1–1.3 m. long, on a scape 2.5–5 dm. long; flowers few, large, 7.5–12.5 cm. long, campanulate, expanding very slightly at anthesis, segments united 5–10 mm. at base, their margins irregularly denticulate; filaments 4–5 cm. long, clavate apical part nearly erect; fruit 10–18 cm. long, 3–4.5 cm. in diameter, base fringed by persistent, reflexed perianth segments and filaments.

Arid slopes and mesas, Lower and Upper Sonoran Zones, from flanks of Santa

Catalina and Rincon Mountains near Tucson into extreme southeastern Arizona. April–May.

**6. Yucca schottii** Engelm., Trans. Acad. St. Louis **3:** 46, 1873
*Yucca macrocarpa* Engelm., Bot. Gaz. **6:** 224, 1881.

Caulescent; trunk 2–2.5 m. high, simple or rarely slightly branched above; leaves spreading, at length deflexed and persistent for several years, 2–4 cm. wide, 5–7 dm. long, bluish green, smooth, concave, pungent, margin yellowish to reddish brown, finely filiferous; inflorescence erect, about 1 m. high, its branches 10–20 cm. long, usually finely but closely tomentose at anthesis, later often glabrate; flowers campanulate, white, perianth segments 3–3.5 cm. long, obtuse to acutish; filaments slender, about one-half as long as pistil, puberulent; pistil about 2.5 cm. long at anthesis, oblong; fruit baccate, indehiscent, 10–12 cm. long; seeds obovate, flattened, dull black, somewhat wrinkled.

Hillsides and mountain slopes, mostly in the Upper Sonoran Zone, southern Arizona to northern Sonora and Chihuahua. Possibly not entering the Sonoran Desert, the type locality given as "Santa Cruz River." April–Aug.

**7. Yucca schidigera** Roezl ex Ortgies, Gartenfl. **20:** 110, 1871
*Yucca mohavensis* Sarg., Gard. & Forest **9:** 104, 1896.
(For complete synonymy see McKelvey, Yuccas SW. U.S. pt. **1:** 92, 1938.)

Shrub or small tree 1–5 m. high; trunk simple or branched above, 15–25 cm. in diameter; leaves 30–50 cm. long, 3–4 cm. wide, concave, smooth, stout-pungent at apex, coarsely filiferous along grayish to light brown margins; inflorescence a crowded, erect, sessile or short-scapose panicle 4–6 dm. high; perianth segments creamy white or lightly tinged with purple, 2.5–4.5 cm. long, acute; filaments slender, glabrous except at apex, there somewhat puberulent, equaling pistil; style 0.5–2 mm. long, pistil 1.5–2 cm. long at anthesis; fruit 5–10 cm. long, narrowly ovoid to oblong, often variously constricted, baccate, with white, sweetish flesh; seeds black, lustrous.

Arid hills and along desert washes, Lower and Upper Sonoran Zones, San Diego County, California, to southern Nevada, western Arizona, and northern Baja California. March–April.

**8. Yucca valida** Brandegee, Proc. Calif. Acad. II, **2:** 208, *pl. 11*, 1889

Tree 3–7 m. high with trunk 5–6 dm. in diameter, growing in clumps and branching anywhere from base to apex; leaves spreading, older ones deflexed, persistent for 1–2 m. below living ones, these yellow-green, 15–35 cm. long, 1.5–2.5 cm. wide (rarely to 50 cm. long, 3 cm. wide), lanceolate, smooth, rigid, the tip stout-pungent, the margins gray to brownish, strongly filiferous; inflorescence erect, 30–60 cm. high, the branches ascending-spreading, 10–20 cm. long, puberulent to glabrous; pedicels slender, 1.5–2 cm. long; flowers broadly campanulate, with a dill-like odor; perianth segments elliptic-lanceolate, acute to obtuse, 2.5–3 cm. long, creamy white; filaments densely papillate-puberulent, 10–12 mm. long; pistil equaling stamens, oblong; stigma nearly sessile; fruit oblong, 2.5–4.5 cm. long, moderately fleshy, turning black in drying.

Sandy places and gravelly slopes, Lower Sonoran Zone, from vicinity of Rosario, Baja California, south to the Cape Region. March–April.

**9. Yucca brevifolia** Engelm., in S. Wats., Bot. King Expl. 496, 1871

*Yucca draconis* var. *arborescens* Torr., Pacif. R. Rep. **4:** 147, 1857.
*Yucca arborescens* Trelease, Rep. Mo. Bot. Gard. **3:** 163, 1892.
*Cleistoyucca arborescens* Trelease, Rep. Mo. Bot. Gard. **13:** 41, 1902.
*Cleistoyucca brevifolia* Rydb., Fl. Rocky Mts. 170, 1917.

Tree 5–12 m. high with a roughened trunk 5–9 dm. in diameter and openly branched to form a rounded head in older trees, young trees unbranched and clothed to the ground with persistent, reflexed leaves; leaves rigid, sharp-pointed, radiately spreading at ends of branches, 10–15 mm. wide, 15–30 cm. long, somewhat 3-sided, minutely but sharply denticulate; panicle sessile, dense, 25–40 cm. high, erect, branches often short-hispid; flowers oblong to subglobose, greenish white, perianth segments 25–30 mm. long; fruit ovoid, 5–10 cm. long, 2–4.5 cm. in diameter, indehiscent, dry and spongy; seeds flattened, obovate to nearly orbicular, 7–8 mm. long, black, dull to somewhat lustrous.

Desert slopes and hillsides, Lower Sonoran Zone, mostly in the Mojave Desert but occasionally entering the extreme northern part of the Sonoran Desert near Needles, California. March–May.

## 11. NOLINA Michx., Fl. Bor. Amer. 1: 207, 1803

Acaulescent or strongly caulescent perennial, in the latter the trunk woody and often much dilated at base. Leaves numerous, narrowly linear, stiff to drooping, finely serrulate at least in youth. Inflorescence a compound racemose panicle borne on a nearly naked scape, main branches subtended by deltoid-attenuate bracts. Bractlets minute, scarious, more or less lacerate-ciliate. Pedicels jointed near base or about the middle. Perianth small, persistent, with 6 oblong, 1-nerved segments. Stamens 6, usually abortive in fertile flowers; filaments short, slender, or somewhat dilated at base. Ovary sessile or short-stipitate and abortive in flowers with fertile stamens, deeply 3-lobed; style very short or virtually wanting; ovules 2 in each cell. Fruit a thin-walled capsule, often bursting irregularly. Seed 1 per cell, or usually 1 maturing in a fruit, globose to oblong, light-colored, wrinkled.

Fruit large, 8–15 mm. wide; perianth segments of pistillate flowers 2.5–5 mm. long:
  Plants caulescent, 1–2 m. high; capsules deeply emarginate at both ends, or retuse at base in *N. parryi*; pedicels jointed about the middle:
    Leaf margins becoming filiferous, not serrulate in age; perianth segments of pistillate flowers 2.5–3 mm. long, the outer strongly reflexed and introrsely apiculate at the apex; stigma lobes sessile .......... 1. *N. bigelovii*
    Leaf margins not filiferous, permanently serrulate; perianth segments of pistillate flowers 3.5–5 mm. long, outer spreading or slightly reflexed, not apiculate; stigma lobes on a short style .......... 2. *N. parryi*
  Plants acaulescent; capsules retuse to rounded at base; pedicels jointed near base .......... 3. *N. beldingii* var. *deserticola*
Fruit small, 4–7 mm. wide; perianth segments of pistillate flowers 1.5–3 mm. long:
  Capsules somewhat inflated, the seed not protruding early:
    Plant caulescent, 3–8 m. high; pedicels 1.5–2 times as long as capsules; primary branches of inflorescence 15–25 cm. long; capsules chartaceous, stiff .......... 4. *N. matapensis*
    Plant acaulescent; pedicels about equaling or slightly shorter than capsule; primary branches of inflorescence 15–50 cm. long; capsules papery, delicate .......... 5. *N. microcarpa*

Capsules not inflated, early ruptured by growing seed:
- Pedicels filiform, equal to or surpassing fruit; branches of inflorescence smooth; leaves 8–13 mm. wide:
  - Plants acaulescent .................................. 6. *N. palmeri* var. *palmeri*
  - Plants caulescent, to 5 m. high .................... 6a. *N. palmeri* var. *brandegeei*
- Pedicels stout in fruit, shorter than or barely equaling capsule; inflorescence roughened in lines; leaves 3–4 mm. wide ........................ 7. *N. texana* var. *compacta*

**1. Nolina bigelovii** (Torr.) S. Wats., Proc. Amer. Acad. **14:** 247, 1879
*Dasylirion bigelovii* Torr., Pacif. R. Rep. **4:** 151, 1857.

Caudex stout, 0.6–2 m. high, bearing large tufts of flat, stiff, thick leaves, these 1.5–2.5 cm. wide above broadly dilated base, 8–12 dm. long, margins serrulate at first but soon becoming filiferous, serrulations then disappearing; scapes 5–8 dm. long; bracts deltoid-lanceolate, thin, 4–10 cm. long, 1–2 cm. wide at base, attenuate, soon deciduous; primary branches of the inflorescence slender, 10–20 cm. long, ascending, glabrous, smooth; secondary branchlets 2–8 cm. long; bractlets lance-attenuate, lacerate-margined; perianth segments oblong-linear, 2.5–3 mm. long, outer introrsely short-apiculate, strongly reflexed in pistillate flowers, inner erect or ascending; filaments of abortive stamens dilated below; stigma lobes sessile, capsule deeply emarginate at both ends, 8–12 mm. wide and about as high, its walls thin; seeds ovate to oblong, 2.5–3.5 mm. long, whitish or white-tawny, coat intricately wrinkled.

Gravelly and rocky slopes, Lower Sonoran Zone, western margin of the Colorado Desert, occasional in the mountains of the Colorado Desert, to western Arizona and northern Baja California. May–June.

**2. Nolina parryi** S. Wats., Proc. Amer. Acad. **14:** 247, 1879

Caudex stout, unbranched, 1–2 m. high; leaves rigid, slightly concave on upper side, thick, 8–10 dm. long, 8–15 (or –20) mm. wide above broad base, deeply striate, margins sharply serrate, not filiferous, tips often splitting into brushlike tufts; scapes erect, 5–6 dm. long; bracts subtending lower branches of inflorescence spathaceous, ovate-lanceolate, attenuate, 4–6 cm. wide, 12–20 cm. long; primary branches of inflorescence 20–40 cm. long; bractlets lacerate-ciliate; perianth segments 3.5–5 mm. long, not apiculate, spreading or outer slightly reflexed; style short; capsule emarginate above, mostly retuse at base, 12–15 mm. wide, about as high, its walls thin, light stramineous; seeds subglobose, light brown, 3–3.5 mm. in diameter, finely and intricately wrinkled.

Locally at scattered stations in the Lower and Upper Sonoran Zones, Mohave and Coconino Counties, Arizona, Little San Bernardino and Chuckwalla Mountains, California, western margin of the Colorado Desert, south into northern Baja California; also in the Santa Ana Mountains, Orange County, California. May–June.

**3. Nolina beldingii** Brandegee var. **deserticola** Trelease, Proc. Amer. Phil. Soc. **50:** 424, 1911

Acaulescent; leaves 15–20 mm. wide, 4–5 dm. long, slightly glaucous, sharply serrate; primary branches of inflorescence 8–10 cm. long, ultimate branchlets 1–2 cm. long; fruit depressed, retuse at base, 8–10 mm. high, 12–15 mm. wide; seed 4 mm. wide, 5 mm. high (ex char.).

Collected at Yubay, Baja California, in 1889 by T. S. Brandegee. This may be only a young plant of *N. beldingii* or it may be a distinct species.

**4. Nolina matapensis** Wiggins, Contr. Dudley Herb. **3:** 65, *pl. 14, figs. 1–5,* 1940

Strongly caulescent; trunk 15–45 cm. in diameter, 3–8 m. high, strongly dilated at base, freely branched above and bearing several large tufts of leaves; defoliated trunk checked and fissured in quadrate segments, dark gray to nearly black; leaves narrowly linear, 2.5–4 cm. wide at broadened base, 8–11 mm. wide above base, 6–8 dm. long, light green but not glaucous, moderately stiff, flattened or slightly concave above, with a broad, flat keel beneath, strongly striate, margins finely but sharply serrulate with simple, cartilaginous, forward-pointing teeth; inflorescence 8–12 dm. high on a naked scape 1–3 dm. long; primary branches of fruiting inflorescence 15–25 cm. long, slender, glabrous, mottled with small splotches of red; ultimate branchlets nearly filiform, 1–3 cm. long; pedicels filiform, smooth, 8–10 mm. long, jointed 1–2 mm. above base; perianth segments oblong-linear, 2 mm. long, thin, outer spreading, inner erect between adjacent lobes of capsule; capsule strongly depressed, 4–5 mm. high, 6–7 mm. wide, walls firm, moderately thick, lobes obovate to nearly orbicular, laterally spreading, not ruptured by growing seed, the apical notch broad, 1.5–2 mm. deep; seed subglobose to broadly obovate, 2.5–3 mm. high, dark stramineous, finely and intricately wrinkled; micropyle circular.

Gravelly slopes and along washes near Mátape, between Hermosillo and Suaqui, Sonora.

**5. Nolina microcarpa** S. Wats., Proc. Amer. Acad. **14:** 247, 1879

Acaulescent; leaves 6–12 mm. wide, 6–12 dm. long, margins unevenly serrulate-scabrous, with a broad, flat keel beneath, tips splitting into a brushlike tuft 5–12 cm. long; inflorescence often 1 m. high, primary branches 15–50 cm. long, these again bearing branches often one-half as long; bracts subtending lower branches lance-linear, 5–12 mm. wide, 8–15 cm. long, attenuate; bractlets 1–2.5 mm. long, fimbriate; pedicels slender, about equaling or slightly shorter than perianth segments, these lance-oblong, 1.5–2 mm. long, spreading; fruit subglobose, deeply notched at apex, rounded to shallowly notched at base, 4–5 mm. high, 6–7 mm. wide, papery, cells ruptured only at maturity; seeds obovate, 3–3.5 mm. high, light brown, finely wrinkled.

Exposed mountainsides throughout much of Arizona, mainly if not entirely in the Upper Sonoran Zone and at similar elevations eastward through New Mexico into western Texas and adjacent Chihuahua and northeastern Sonora. Rare along the upper margins of the desert. May–June.

**6. Nolina palmeri** S. Wats., Proc. Amer. Acad. **14:** 248, 1879, var. **palmeri**

Acaulescent cespitose plants with regularly denticulate-serrulate leaves 8–13 mm. wide, 6–12 dm. long, these more or less fibrous-tufted at tips; flowering stalks to 2 m. high, primary branches 12–25 cm. long, branchlets rigidly ascending, 2–6 cm. long; bracts lance-linear, attenuate, 7–18 cm. long, serrulate toward apex; bractlets broadly obovate, shallowly fimbriate; pedicels slender, slightly dilated toward apex, from equal to, to one and one-half times as long as, the fruit; perianth segments broadly oblong, scarcely 2 mm. long, spreading in fruit; capsule early ruptured by seed, 4 mm. high, 4–5 mm. wide, shallowly notched at apex; styles distinct, 0.5–0.75 mm. long; seed globose, 2.5–3 mm. in diameter, light yellow-brown to nearly black, finely wrinkled.

Gravelly slopes and rocky hillsides, Lower Sonoran Zone to Arid Transition Zone, northern Baja California (Cantillas Canyon) southward along the Sierra San Pedro

Mártir and inland foothills. In the Sierra San Pedro Mártir this species ascends to an elevation of nearly 8,000 feet. It is fairly abundant in the pinyon belt of the Sierra Juárez and occurs frequently on the rocky desert slopes of the Sierra Pinal. April–June.

**6*a*. Nolina palmeri** var. **brandegeei** Trelease, Proc. Amer. Phil. Soc. **50:** 420, 1911

Differs from var. *palmeri* in being arborescent and attaining a height of 4–5 m., and in having more glossy leaves only 7–8 mm. wide, and somewhat shorter primary branches of the inflorescence.

Desert ranges between San Augustín and Calmallí; San Julio and Paraíso, Baja California.

**7. Nolina compacta** S. Wats. var. **texana** (Trel.) I. M. Johnston, Jour. Arnold Arb. **24:** 90, 1943

(For synonymy see Jour. Arnold Arb. **24:** 90, 1943.)

Acaulescent; leaves narrowly linear, 3–4 mm. wide, 7–12 dm. long, margins somewhat serrulate-roughened or sometimes nearly smooth; inflorescence slender, with primary branches 10–15 cm. long, branchlets 2–5 cm. long, branches slightly scaberulous in lines; bracts surpassing branches, caudate-attenuate; fruit 4–5 mm. high, 5–7 mm. wide, ruptured early by enlarging seeds; seeds globose, 3 mm. in diameter (ex char.).

Sonoran Zones, southern Arizona and adjacent Sonora to Chihuahua. Mainly a plant of the Upper Sonoran Zone. May–June.

## 12. DASYLIRION Zucc., in Otto & Dietr., Allg. Gart. 6: 258, 1838

Coarse perennial herbs with woody caudices and numerous crowded, spiny-margined, linear leaves. Flowering stems erect, narrowly paniculately branched above, with or without a slender, naked scape below. Flowers dioecious, in crowded panicles. Pedicels jointed at base of flowers. Perianth white, withering-persistent, small; segments distinct. Stamens 6, exserted, represented by staminodia in pistillate flowers. Ovary 1-celled, 3-ovuled. Stigmas 3, on a short style that may equal notch at apex of ovary. Fruit leathery, 3-winged, indehiscent or essentially so. Seed solitary, 2 ovules aborting.

**1. Dasylirion wheeleri** S. Wats. ex Rothr., Wheeler Rep. 272, 1878

Short-caulescent, the caudex 1 m. high or less; leaves linear, 15–20 mm. wide, 6–10 dm. long, ascendingly spreading, crowded, glaucous, minutely roughened, margins bearing forward-curved, rigid, horny, yellowish to brownish prickles 2–4 mm. long and 5–15 mm. apart; inflorescence slender, 3–5 m. high, panicle 1–1.5 m. long, 2–3 dm. wide, borne on a scape 2.5–4 cm. in diameter; pedicels 2–4 mm. long, jointed at apex; staminate flowers densely crowded in catkinlike spikes 6–8 cm. long, subtended by broadly obovate to suborbicular, papery, fimbriolate bractlets; perianth segments 2–3 mm. long, oblong; stamens exserted about 1 mm.; pistillate flowers less crowded, on branches 6–15 cm. long; style stoutish, 0.5–0.8 mm. long; fruit obovate, 5–6 mm. wide, 7–9 mm. high, notch about 1 mm. deep, apices of capsule wings acute to obtuse; seed 3.5–4 mm. long, brown.

On gravelly and rocky hillsides and mesas, upper edges of the Lower Sonoran Zone to the Arid Transition Zone, Baboquivari Mountains and the vicinity of Phoenix, Arizona, to western Texas, Chihuahua, and northern Sonora. May–Aug.

## Family 19. **AMARYLLIDACEAE.** AMARYLLIS FAMILY

Perennial herbs or some tropical representatives arborescent, from bulbs or rootstocks. Flowering stems scapose or leafy. Leaves linear and entire to broadly ample, soft-herbaceous to stiff and very fibrous, then usually quite thick (at least basally) and often spiny-toothed. Flowers perfect, regular or essentially so. Perianth of 6 segments or lobes, outer and inner whorls alike or slightly different, distinct or united into a tube below and adnate to the ovary. Stamens 6, inserted on bases of perianth segments or in tube opposite the lobes. Anthers comparatively large, versatile or basifixed, 2-celled, dehiscent longitudinally. Ovary inferior, 3-celled. Style slender, entire, 3-lobed or cleft into 3 distinct stigmas. Ovules many, numerous (in ours) or less commonly 1–2 in each cavity. Fruit a capsule, usually dry. Seeds (in ours) strongly flattened horizontally, dark gray to black or less commonly greenish; embryo small, straight or nearly so, embedded in fleshy endosperm.

Plants from tunicated bulbs; leaves and scapes ephemeral, appearing only following summer rains; flowering scapes naked except at apex; stamens shorter than perianth segments:
- Perianth segments oblanceolate, 6–12 mm. broad; flowers solitary, pedicellate; filaments free to base, not connate into a tube; seeds black .................... 1. *Zephyranthes*
- Perianth segments narrowly linear, 1.5–2 mm. wide; flowers several, sessile; filaments connate into a fragile tube at base; seeds greenish .................. 2. *Hymenocallis*

Plants from rootstocks; leaves persistent, fibrous, usually very stiff; flowering stalks bearing foliaceous to subscarious bracts full length; stamens surpassing perianth segments .... ................................................................ 3. *Agave*

### 1. **ZEPHYRANTHES** Herbert, App. Bot. Reg. 36, 1821

Perennial herbs from an ovoid, tunicated bulb, outer coats membranous. Leaves few, narrowly linear, slightly fleshy, somewhat channeled. Stems scapose, erect, slender. Flowers solitary, conspicuous, pedicellate or sessile, subtended by a papery, spathelike bract. Perianth funnelform, straight or slightly curved, tube short or slightly elongated, throat ampliate, lobes 6, equal or essentially so. Stamens 6, equal or in two slightly unequal whorls, inserted in the throat; anthers linear. Ovary inferior, 3-celled; style slender; stigma more or less 3-cleft. Ovules many, biseriate in each cell. Capsule subglobose, shallowly 3-lobed, loculicidally dehiscent. Seeds oblong to suborbicular, angled or (in ours) strongly flattened, black. Embryo straight. Endosperm copious.

**1. Zephyranthes longifolia** Hemsl., Diagn. Pl. Mex. 55, 1880

*Zephyranthes arenicola* Brandegee, Proc. Calif. Acad. II, **2:** 205, 1889.
*Atamosco arenicola* Greene, Pittonia **3:** 187, 1897.
*Atamosco longifolia* Cockerell, Canad. Entom. **1901:** 283, 1901.

Bulb ovoid, 1.5–2 cm. in diameter, 2.5–4 cm. long, 6–10 cm. below surface, outer coats dark brown, wrinkled, dull, persisting for long periods and often forming a stout "neck" to 2 cm. wide, extending nearly or completely to surface; inner coats shiny, papery, minutely reticulate in small rectangles parallel with long axis; leaves 3–10, linear to subfiliform, 0.5–2 mm. wide, 5–25 cm. long, equaling or surpassing scapes, these 1–3 from a bulb, to 2.5 dm. tall, 2–3.5 mm. in diameter, glabrous; bracts 1.5–3 cm. long, papery, often purplish or pinkish near apex, usually bifid halfway to base; pedicels 1–2.5 cm. long; perianth yellow or tips whitish or faintly tinged with purple or rose, segments 2–3 cm. long, oblanceolate, outer 6–7 mm. wide, inner 10–12 mm. wide; tube 2–2.5 mm. long, apically fringed with a ring of short, fimbriate processes; stamens inserted just below fimbriate fringe, 10–15 mm. long, anthers orange; capsule glabrous or slightly depressed, about 1.5 cm. in diameter; seeds black, obovoid to semicircular, 5–7 mm. long, strongly flattened, sublustrous, minutely puncticulate.

In sandy or gravelly soil, Lower Sonoran Zone to lower margins of Transition Zone, Pima County, Arizona, to western Texas and central Mexico and from central Baja California to the Magdalena Plain and the adjacent islands. June–July in the southwestern United States; Sept.–Nov. in Baja California.

## 2. **HYMENOCALLIS** Salisb., Trans. Hort. Soc. Lond. **1:** 338, 1812

Perennial herb from a tunicated bulb. Leaves oblong to oblanceolate, somewhat narrowed toward base. Flowering scape terete, slender, about equaling or somewhat exceeding leaves, bearing papery white involucrate bracts at summit subtending umbels. Flowers several, sessile, umbellate. Perianth tube long and slender, strongly or faintly ampliate above; perianth segments 6, linear, spreading or ascending. Stamens 6, equally inserted near apex of perianth tube, free above but connate into a more or less spreading webbed tube at base. Anthers linear, versatile, often curved and frequently twisted in drying. Ovary 3-celled, with 2 ovules in each cell. Capsule ovoid to globose, more or less fleshy. Mature seeds frequently solitary, large, greenish, coat more or less spongy.

### 1. **Hymenocallis sonorensis** Standley, Field Mus. Bot. Ser. **17:** 229, 1937

Bulb ovoid, 5–7 cm. long, outer coats membranous, white or pinkish; leaves to 2.5 dm. long, 8–12 mm. wide a little above middle, narrowly oblanceolate, acute or obtuse at apex, dark green, glabrous, thin, margins smooth; scape 2–3 dm. tall, about 3 mm. in diameter; bracts spathaceous, scarious, 4–5 cm. long; flowers 3–6, sessile; perianth tube very slender, 7–10 cm. long, 1–1.5 mm. in diameter, greenish; perianth segments narrowly linear, 4.5–7.5 cm. long, about 1.5 mm. wide, white; stamineal tube 1.5–2 cm. long, very fragile; free parts of filaments 2–2.5 cm. long, filiform; anthers 1–1.5 cm. long, curved, spirally twisted after dehiscence; ovary ovoid, 1–2 cm. long; style slightly exceeding perianth lobes; mature capsules and seeds unknown.

Along streams and in well watered valleys, Lower Sonoran and Upper Sonoran Zones, northern to southern Sonora. July.

## 3. **AGAVE** L., Sp. Pl. 323, 1753

Perennial plants from a more or less woody caudex or rootstock. Leaves succulent, numerous, imbricated in a closely spiral basal rosette, their margins often armed with rigid, horny teeth or filiferous, tip sharply spinose. Scapes stout and tall, terminating in an elongate raceme or in broad panicles, more or less bracteate. Flowers many, perfect, mostly yellow. Perianth tubular to funnelform, the segments ascending or slightly incurved. Stamens 6, exserted; anthers versatile. Ovary 3-celled, thick-walled, many-ovuled. Capsule loculicidal, dry. Seeds strongly flattened, black or dark brown.

Flowering occurs only after several to many years' growth of the rosette, and there is only moderate correlation between anthesis and the season. In our area, however, the heaviest flowering occurs in March, April, and May, although scattered plants in bloom may be seen during almost any month of the year.

(The treatment of the genus is dishearteningly inadequate. The range of variation within a species is often extremely puzzling, and it is probable that hybridization occurs frequently. These conditions make a satisfactory taxonomic solution exceedingly difficult. A reasonably acceptable classification of the genus *Agave* can be presented only after extensive and painstaking study in the field, prosecuted at all seasons of the year. The keys and recognition of species here are strictly tentative.)

Inflorescence narrowly spicate or racemose, never paniculate, usually much higher than broad:
Margins of leaves unarmed or filiferous, not dentate:
Scapes 1–3 m. tall; leaves 5–15 mm. wide, 5–30 cm. long:
Flowers (including ovary) 10–12 mm. long; perianth tube subglobose, slightly constricted at the apex; capsule 6–8 mm. high, ovoid .............. 1. *A. hartmanii*
Flowers (including ovary) 18–50 mm. long; perianth tube funnelform, not constricted apically; capsules subglobose to broadly oblong, 10–25 mm. long:
Perianth tube 3–4 mm. long; flowers 18–25 mm. long; capsules 9–12 mm. long .... .............................................. 2. *A. toumeyana*
Perianth tube 10–20 mm. long; flowers 30–45 mm. long; capsules 10–15 mm. long .............................................. 3. *A. schottii*
Scapes 2–4 m. tall; leaves 4–6 cm. wide, 5–15 dm. long .............. 4. *A. chrysoglossa*
Margins of leaves armed with teeth or prickles (leaves 2–4 cm. wide; scapes 2–4 m. tall) .............................................. 5. *A. utahensis*

Inflorescence broadly paniculate, often broader than tall:
Leaves with teeth joined by a continuous horny border .................. 6. *A. shawii*
Leaves with the teeth separate or joined by a horny border only on the upper one-fourth to one-third of the leaf:
Tissue of the leaves hard, very fibrous; leaves swordlike or daggerlike, mostly broadest at the base:
Leaves only faintly to moderately glaucous, not falcate:
Terminal spine 10–15 mm. long:
Leaves 5–10 cm. broad; teeth 4–6 mm. long .................... 7. *A. sullivanii*
Leaves 2–3 cm. broad; teeth 20–35 mm. long ....................8. *A. owenii*
Terminal spine 15–30 mm. long:
Leaves numerous, lax .............................................. 9. *A. pacifica*
Leaves few, rigid:
Filaments 30–45 mm. long, exceeding perianth by 20–25 mm.; terminal spine grooved only at the base ...................... 10. *A. datylio* var. *datylio*
Filaments 20–25 mm. long, exceeding perianth by 10–15 mm.; terminal spine grooved to the middle .................... 10*a*. *A. datylio* var. *vexans*
Leaves strongly glaucous, nearly white, lower strongly falcate .... 11. *A. yaquiana*
Tissue of the leaves subfleshy, softer and less fibrous; leaves lanceolate, oblong, or oblanceolate, usually broader about the middle than at the base of the free part:
Leaves large, 6–25 dm. long, 10–25 cm. wide:
Flowering stalks 8–12 m. tall; leaves 10–25 dm. long:
Leaves dark green, 10–15 dm. long; teeth very unequal ........... 12. *A. mescal*
Leaves gray-green, 15–25 dm. long; teeth nearly equal .......... 13. *A. americana*
Flowering stalks 2–6 m. tall; leaves 6–10 dm. long:
Leaves gray-green; flowers golden yellow, chrome yellow, or orange-yellow, not purplish:
Perianth lobes nearly as wide as long .................... 14. *A. promontorii*
Perianth lobes over twice as long as wide:
Flowers golden yellow; perianth lobes over twice as long as the perianth tube .............................................. 15. *A. aurea*
Flowers chrome yellow; perianth lobes only slightly longer than the tube .................................. 16*a*. *A. palmeri* var. *chrysantha*
Leaves blue-green; flowers greenish tinged with purple .................. .................................. 16. *A. palmeri* var. *palmeri*

Leaves small or medium-sized, mostly less than 6 dm. long, under 10 cm. broad:
Teeth firmly attached to margins of leaves; leaves broadly lanceolate to obovate:
Perianth tube shallowly cup-shaped, 2–5 mm. long; leaves 4–6 dm. long, to 10 cm. wide:
Leaf margins nearly straight; perianth tube 4–5 mm. deep ........................................................ 17. *A. sobria* var. *sobria*
Leaf margins deeply repand; perianth tube 2–3 mm. deep ........................................................ 17a. *A. sobria* var. *roseana*
Perianth tube 8–10 mm. deep; leaves 1–2 dm. long, 5–10 cm. wide:
Capsules to 5 cm. long; seeds 3–4 mm. in diameter; leaves twice as long as broad; inflorescence distinctly paniculate .................... 18. *A. margaritae*
Capsules to 6.5 cm. long; seeds 4–5 mm. in diameter; leaves 3 times as long as broad; inflorescence branching only slightly .......... 19. *A. subsimplex*
Teeth rather weakly attached to leaf margins; leaves usually triangular-lanceolate to deltoid in outline:
Teeth minute, about 1 mm. long; plants slightly cespitose .... 20. *A. dentiens*
Teeth medium, 3–8 mm. long; plants densely cespitose ........ 21. *A. deserti*

**1. Agave hartmanii** S. Wats., Proc. Amer. Acad. **26**: 156, 1891

Plants cespitose, rosettes 2–3.5 dm. wide, stiff; leaves linear-oblong, 1–1.5 cm. wide at widest point (about 3 cm. below apex), 8–12 mm. wide just above the dilated base, narrowing abruptly to apex, yellowish green, moderately fleshy, conspicuously marked with gray lines on both surfaces when fully dry but these lines not evident in fresh material, margins bearing 8–15 light gray, recurved threads, terminal 1–2.5 cm. of leaf slightly to strongly involute and tipped with a brownish spine 4–6 mm. long and somewhat grooved on upper side; basal margins minutely but sharply denticulate; flowering scapes solitary, 1–2.5 m. tall, about 1–1.5 cm. in diameter at base, greenish, more or less mottled with reddish or purplish dots; inflorescence proper 3–6.5 dm. long, flowers in short-pedunculate, few-flowered clusters, racemose-virgate; perianth 7–8 mm. long, tube about 3 mm. long and wide; perianth segments erect, outer broadly oblong, 2 mm. broad, obtuse, with a short, blunt apiculation, inner somewhat broader, often shallowly 3-dentate near apex, 2.5–3 mm. wide; stamens 6, about 10 mm. long, inserted at base of tube; stigma faintly 3-lobed; style and stamens curved sharply downward over perianth tube, at least in drying; fruits ovoid, 6–8 mm. long, with 3 grooves indicating septa, loculicidal; seeds ovoid, flattened by mutual compression, about 2 mm. long, black, dorsally reticulate with low, rounded ridges, sides indistinctly ridged and papillate with low, minute mounds, angles narrowly winged.

Gravelly slopes and rocky hillsides, Lower Sonoran Zone, central Sonora. July (?).

**2. Agave toumeyana** Trelease, in Standley, Contr. U.S. Nat. Herb. **23**: 140, 1920

Plants cespitose, rosettes subglobose, 15–40 cm. in diameter; leaves bright green, straight or falcately curved, narrowly lanceolate, 10–22 mm. wide, 8–25 cm. long, margins with fine, white hairs 1–5 cm. long, denticulate only on broadly dilated base; terminal spine 6–15 mm. long; borders of adjacent leaves forming conspicuous white marks on both leaf surfaces in zigzag patterns; flowering scapes 1–3 m. tall, slender; inflorescence spicate-racemose; perianth tube funnelform, 3–4 mm. deep; perianth segments narrowly deltoid, 7–8 mm. long, obtuse and slightly incurved at apex, white tinged with green; stamens inserted at about middle of perianth tube; anthers 7–8 mm. long, greenish or tan; ovary 10–12 mm. long; capsules subglobose to broadly oblong, 7–8 mm. in diameter, 9–12 mm. long; seeds black, 2–5 mm. wide, about 3 mm. long.

Rocky hillsides and mesas, upper margins of Lower Sonoran Zone and lower part of Upper Sonoran Zone, Pinal and Gila Counties, Arizona.

**3. Agave schottii** Engelm., Trans. Acad. St. Louis **3**: 305, 1875

*Agave geminiflora* var. *sonorae* Torr., Bot. Mex. Bound. 214, 1859.
*Agave schottii* var. *serrulata* Mulford, Rep. Mo. Bot. Gard. **7**: 73, 1896.
*Agave mulfordiana* Trelease, in Standley, Contr. U.S. Nat. Herb. **23**: 140, 1920.

Cespitose plants with numerous offshoots; rosettes subglobose, 3–4 dm. in diameter; leaves bright green, usually falcately curved, 6–12 mm. wide at middle, 2–3 dm. long, gradually tapering from dilated, sparsely denticulate base to apex, margins bearing recurved whitish or grayish hairs, surfaces with zigzag white markings made by pressure of adjacent leaves; terminal spine 6–12 mm. long; flowering scapes slender, 15–27 dm. tall, bearing numerous crowded geminate clusters of flowers in a spicate-racemose inflorescence; flowers 3–4.5 cm. long, green ovary 8–10 mm. long; perianth tube broadly funnelform, 10–20 mm. long, somewhat constricted at base; perianth lobes light yellow, 5–6 mm. wide, 12–15 mm. long; stamens inserted near apex of perianth tube; anthers 7–12 mm. long, light yellow; capsule subglobose to subcylindrical, 8–10 mm. in diameter, 10–15 mm. long; seeds 3–4 mm. broad, 4–4.5 mm. long.

Arid, rocky slopes, desert grassland, and oak woodland, upper margins of Lower Sonoran Zone and Upper Sonoran Zone, Pima and Cochise Counties, Arizona, to southwestern New Mexico and southward into northern Sonora.

**4. Agave chrysoglossa** I. M. Johnston, Proc. Calif. Acad. IV, **12**: 998, 1924

Acaulescent; leaves linear-lanceolate, 4–6 cm. wide near the middle and narrower both apically and basally from this point, 5–15 dm. long, pale yellowish green and moderately glaucous, arranged in loose rosettes, margins firm but unarmed, entire; terminal spine subulate, 2.5–3.5 cm. long, brown in youth, later gray, grooved on upper sides to slightly above middle; flowering scape slender, to 4 m. tall, including spicate-racemose inflorescence, this 1–2 m. long, 1–1.5 dm. in diameter; pedicels 4–5 mm. long; bracts subulate to filiform, 2–3 cm. long, broadly dilated at base; flowers geminate, about 3 cm. long, bright yellow; perianth tube cup-shaped, about 3 mm. long; perianth segments linear, obtuse, 3–3.5 mm. wide, 12–15 mm. long; filaments 25–30 mm. long; anthers 8–9 mm. long; capsule 18–20 mm. long, 10–12 mm. in diameter; seeds 2–2.5 mm. wide, dull black.

On rocky hillsides, Lower Sonoran Zone, San Pedro Nolasco Island, Gulf of California, and possibly on the adjacent mainland of Sonora.

**5. Agave utahensis** Engelm., in S. Wats., Bot. King Expl. 497, 1871

?*Agave newberryi* Engelm., Trans. Acad. St. Louis **3**: 309, 1875.
*Agave utahensis* var. *discreta* M. E. Jones, Contr. West. Bot. **17**: 19, 1930, in part.

Mainly solitary, acaulescent plants; rosette subglobose, about 3.5 dm. in diameter; leaves narrowly lanceolate, usually tapering from base to apex and about 2–5 cm. wide near middle, 15–35 cm. long, bright green and scarcely at all glaucous; terminal spine 2–5 cm. long or occasionally to 10 cm. long; lateral teeth few, 1–3 mm. long; flowering scapes 1.5–4 m. tall, slender, bearing clusters of 2–4 flowers at tips of peduncles 2–5 cm. long; flowers many, 22–35 mm. long, perianth tube 4–5 mm. long; perianth segments yellow, 10–15 mm. long; stamens inserted near middle of tube, filaments 15–18 mm. long; anthers 10–12 mm. long; capsules cylindrical to oblong, 10–12 mm. in diameter, 15–30 mm. long; seeds about 3 mm. wide, 4 mm. long.

Arid slopes and ridges, Lower Sonoran Zone to lower part of Transition Zone, eastern Mojave Desert, California, to southwestern Utah and northern Arizona, barely touching the Sonoran Desert in Mohave County. May–July.

**6. Agave shawii** Engelm., Trans. Acad. St. Louis **3:** 314, 1875

*Agave sebastiana* Greene, Bull. Calif. Acad. **1:** 214, 1885.
*Agave orcuttiana* Trelease, Rep. Mo. Bot. Gard. **22:** 47, 1912.
*Agave pachyacantha* Trelease, op. cit. 48.
*Agave goldmaniana* Trelease, op. cit. 49.

Solitary or somewhat cespitose, short-caulescent plants with dense, globular rosettes and stout flowering scapes; leaves ovate, obovate, or oblong, 6–12 cm. wide, 2.5–5 dm. long, glossy green to slightly glaucous; terminal spine stoutish, 3–8 mm. wide, 15–40 mm. long, reddish or eventually dark brown to nearly black, straight or somewhat flexuous; lateral teeth 3–15 mm. long, straight, hooked, or usually recurved, 5–15 mm. apart, connected by a hard, horny, continuous margin of leaf; flowering scapes 1–4 m. tall, broadly paniculate; flowers crowded on spreading branches, greenish yellow or orange, 6–9 cm. long; perianth tube conical, 12–20 mm. deep; perianth segments deltoid-attenuate, about 20 mm. long, about one-half as long as ovary; filaments 5–6 cm. long, inserted about middle of tube; capsule 5–7 cm. long; seeds 7–9 mm. long, 6–7 mm. wide, dull black.

Hillsides and desert plains, Lower Sonoran Zone and lower part of Upper Sonoran Zone, near the coast near Tijuana and south to central Baja California.

**7. Agave sullivanii** Trelease, in Standley, Contr. U.S. Nat. Herb. **23:** 119, 1920

"Leaves green, about 10 cm. wide and 100 cm. long, with a chestnut, very short-conical spine, this abruptly pointed, or less commonly conical, and round-grooved at base, 5 mm. wide and 10–15 mm. long, the teeth triangular, upcurved, 20 mm. apart and 4 to 6 mm. long."

Known only from the vicinity of La Paz, Baja California, where it is cultivated for mescal.

**8. Agave owenii** I. M. Johnston, Proc. Calif. Acad. IV, **12:** 999, 1924

Acaulescent, cespitose plants with numerous offshoots; leaves green and lightly glaucous, 6–8 cm. wide at base, 5–8 dm. long, abruptly contracted immediately above base to a linear, swordlike blade 2–2.5 cm. wide; terminal spine 10–15 mm. long, 3–5 mm. in diameter basally, decurrent about 1 cm., brown to ashy gray, straight; lateral teeth dark brown to nearly black, 2–3.5 cm. long, 1–4 cm. apart, sharply antrorse; flowering scape stout, to 18 dm. tall; panicle open, ovate to oblong; flowers 4 cm. long, pale yellow tinged with green near base; perianth tube saucer-shaped, 3–4 mm. deep, 10–11 mm. wide; perianth lobes linear-lanceolate, 15–17 mm. long, about 4 mm. wide, erect; ovary fusiform, about 2 cm. long; filaments 35 mm. long, inserted in throat; capsule 4–4.5 cm. long, oblong; seeds 5–6 mm. wide, 7–8 mm. long, shining black.

On volcanic rocks, Lower Sonoran Zone, vicinity of Guaymas, Sonora.

**9. Agave pacifica** Trelease, in Standley, Contr. U.S. Nat. Herb. **23:** 118, 1920

Acaulescent; rosette subglobose, to 1.5 m. in diameter; leaves comparatively few, straight, yellow-green, very slightly glaucous and rather irregularly zoned in diagonal bands, 3–5 cm. wide, 5–7.5 dm. long, often abruptly contracted just above dilated base, then gradually widening to near middle and tapering gradually from there to apex; terminal spine conical to broadly subulate, 3–4 mm. broad at somewhat decurrent base, 15–25 mm. long, chestnut to purplish brown; lateral teeth upcurved, triangular or somewhat hooked, 3–5 mm. long, 15–25 mm. apart, leaf margin between them straight or essentially so; flowering stalks 2–3 m. tall, slender; flowers 4.5–5 cm. long, greenish yellow; perianth tube cup-shaped, 8–10 mm. deep, about as wide; perianth lobes linear-oblong, 5–6 mm. wide, 18–22 mm. long, obtuse; filaments inserted at apex of tube, 3–4 cm. long; anthers greenish yellow, 20–25 mm. long; capsules oblong, 2–2.5 cm. in diameter, 4–4.5 cm. long.

Rocky hillsides and desert flats, Lower Sonoran Zone, central Sonora southward to Tepíc. March–June.

**10. Agave datylio** Simon ex Weber, Bull. Mus. Hist. Nat. Paris **8:** 224, 1902, var. **datylio**

Acaulescent, more or less cespitose plant with a subglobose rosette to 1 m. in diameter and nearly or fully as high; leaves lance-linear, straight, 3–4 cm. broad, 3–7.5 dm. long, somewhat fleshy but rigid and swordlike, yellowish green with only a light glaucous covering; terminal spine purplish brown to nearly black, 4–6 mm. in diameter at base, 2–3 cm. long, scarcely at all decurrent, shallowly and broadly grooved near base or nearly terete; lateral teeth deltoid-attenuate, 3–5 mm. long, 2–5 cm. apart, spreading or retrorsely hooked, leaf margin between teeth straight or nearly so; flowering stalks 4–5 m. tall; flowers 4.5–5.5 cm. long, closely crowded on branches of panicles; perianth tube funnelform, 8–12 mm. deep, 6–8 mm. in diameter; perianth segments linear-oblong, 3–4 mm. wide, 15–20 mm. long, greenish yellow; filaments 3–4.5 cm. long, inserted at about middle of perianth tube; anthers yellowish, 2–3 cm. long; capsules oblong to pyriform, 15–20 mm. in diameter, 3.5–4 cm. long, slightly stipitate; seeds about 6 mm. wide, 7 mm. long.

Hillsides and along margins of arroyos, Lower Sonoran and Tropical Zones, southern Baja California. Oct.–Nov.

**10*a*. Agave datylio** var. **vexans** (Trelease) I. M. Johnston, Proc. Calif. Acad. IV, **12:** 1003, 1924.

*Agave vexans* Trelease, Rep. Mo. Bot. Gard. **22:** 62, 1912.
*Agave sleviniana* I. M. Johnston, op. cit. 1000, 1924.

Differing from var. *datylio* mainly in having smaller leaves, mostly 4.5 dm. long or less; terminal spine usually grooved for a greater proportion of its length; filaments only 20–25 mm. long.

Arid slopes and rocky hillsides and mesas, Lower Sonoran Zone, central Baja California southward to the vicinity of Loreto and the adjacent lower mountains.

**11. Agave yaquiana** Trelease, in Standley, Contr. U.S. Nat. Herb. **23:** 120, 1920

Acaulescent, solitary or slightly cespitose plants; leaves spreading-ascending to erect, more or less falcate, 4–5 cm. wide, 6–7.5 dm. long, alternately glaucous and green in irregular transverse bands, rosette rather open; terminal spine brown to dark gray, 3–4 mm. wide at base, 20–25 mm. long, distinctly decurrent; lateral teeth rather heavy, antrorsely curved, triangular, 15–50 mm. apart, 10–36 mm. long, intervening cartilaginous margins slightly hollowed; flowering scape slender, 1.5–3 m. tall; panicle loose, narrow, 1 m. long or less; flowers greenish yellow, 4.5–5 cm. long; perianth segments narrowly lance-ovate, 15–20 mm. long; capsule 4–5 cm. long, obtuse at each end; seeds shining black, 5–6 mm. long.

Hillsides and margins of plains, Lower Sonoran Zone, central Sonora.

**12. Agave mescal** K. Koch, Wochenschr. Gart. & Pfl. **8**: 94, 1865
*Agave hookerii* Jacobi, Hamb. Gart. Blumenzeit. **22**: 168, 1866.

Large, acaulescent plant with a rosette 2–3 m. in diameter or more; leaves oblong to oblong-obovate, 1.5–2.5 dm. wide, 10–15 dm. long, dark green and scarcely ever glaucous, rather thin, flat or distinctly concave on the upper side; terminal spine slender but very stiff, 2–6 cm. long; lateral teeth set on fleshy prominences 1–5 cm. apart and smaller ones on margin between them, triangular to deltoid-attenuate, straight or hooked, to 2.5 cm. long; flowering stalks 8–12 m. tall, 8–15 cm. in diameter near base; panicles to 2 m. broad, branches stout and somewhat flattened; flowers crowded at ends of branches, 5–6 cm. long, greenish yellow, oblong perianth segments longer than cup-shaped tube; capsules 4–6 cm. long, rounded or obtuse at each end; seeds black, 5–6 mm. wide, 7–8 mm. long.

Hillsides and arid plains, often cultivated for mescal, Lower Sonoran Zone to Arid Tropical Zone, southern Sonora to Michoacán.

**13. Agave americana** L., Sp. Pl. 323, 1753
*Agave expansa* Jacobi, Abh. Schles. Ges. Vaterl. Kult. **1868**: 151, 1868.

Large, acaulescent plant often producing offshoots from near base and from spreading rootstocks; leaves oblong, straight or somewhat reflexed or downward-curved at tip, 1.5–2.5 cm. broad, 1.5–2.5 m. long, gray-green and somewhat glaucous; terminal spine conical, 5–6 mm. in diameter at base, 2–3.5 cm. long; lateral teeth triangular or somewhat attenuate, straight-spreading to recurved, 5–8 mm. long, 1.5–6 cm. apart, intervening leaf margin straight or repandly curved; flowering stalks to 12 m. tall, 1–1.8 dm. in diameter at base, broadly paniculate-branched in upper one-fourth to one-third; flowers yellow, tinged green, 5–6 cm. long, very numerous, heavily and disagreeably scented; capsules 2.5–3.5 cm. in diameter, 4.5–6 cm. long, obtuse to rounded.

Planted as an ornamental (and often reproducing naturally in Sonora) in California and Arizona. Originally described from plants grown in Europe, the exact source unknown.

**14. Agave promontorii** Trelease, Rep. Mo. Bot. Gard. **22**: 50, 1912
*Agave brandegeei* Trelease, op. cit. 57.

Solitary, acaulescent plants similar to *A. aurea* in general appearance; leaves more heavily glaucous; flowering stalks and inflorescence 4–6 m. tall, panicle forming one-half to two-thirds of the flowering structure, scape stout; flowers golden yellow or orange-yellow, 5–5.5 cm. long; perianth tube broadly cup-shaped, 10–12 mm. deep; perianth segments 7–8 mm. wide, 10–15 mm. long; ovary 25–30 mm. long, eventually exceeding perianth; filaments 4–4.5 cm. long, about 3 times as long as perianth lobes; capsules pyriform-oblong, 3–3.5 cm. long, 15–20 mm. in diameter, briefly stipitate; seeds 4–5 mm. wide, 6–9 mm. long.

Rocky hillsides and mesas and gravelly slopes, Lower Sonoran and Tropical Zones, southern Baja California.

**15. Agave aurea** Brandegee, Proc. Calif. Acad. II, **2**: 207, 1889

Solitary acaulescent plants with a globose to broadly oblong rosette; leaves grayish green, smooth, lance-oblong, 10–15 cm. wide, 7.5–10 dm. long, acuminate, shallowly concave on upper side; terminal spine conical or broadly subulate, glossy chestnut to gray, 3–5 mm. in diameter at base, 2–3 cm. long, decurrent an equal distance, narrowly grooved below middle; lateral teeth chestnut to gray, triangular or base rounded-broadened, 4–8 mm. long, 1–2 cm. apart, each set on a fleshy prominence; flower stalk, including inflorescence, 2–4 m. tall, stout; flowers golden yellow, 4.5–5 cm. long; perianth tube 7–10 mm. long, cup-shaped; perianth segments 15–20 mm. long, 6–7 mm. wide; filaments inserted below middle of tube, about twice as long as perianth segments; capsules oblong, 4–5.5 cm. long; seeds about 6 mm. wide, 8 mm. long.

Rocky hillsides, mesas, and slopes, Lower Sonoran Zone, central Baja California southward along the western flanks of the Sierra de la Giganta.

**16. Agave palmeri** Engelm., Trans. Acad. St. Louis **3:** 319, 1875, var. **palmeri**

Plants solitary or with a few lateral offshoots; rosette 6–15 dm. in diameter, 4–12 dm. tall; leaves bright blue-green, linear-oblanceolate, 6–10 cm. broad, 2–10 dm. long, stiff and rigid, quite thick, especially at base; terminal spine 2–4 cm. long; lateral teeth 3–6 mm. long, 1–2 cm. apart; flowering stalks 2–3.5 m. tall, often to 12 cm. in diameter at base, branches horizontal; flowers in clusters, 4–5.5 cm. long, perianth tube 10–15 mm. deep; perianth segments about equaling tube, greenish tinged with purple; stamens inserted near middle of tube; filaments 3–5 cm. long, purplish; anthers 12–18 mm. long, greenish yellow; ovary 2–3 cm. long, 6–7 mm. in diameter; capsule cylindrical to narrowly obpyriform, 15–17 mm. in diameter at apex, 5–6 mm. wider at base, 4–6.5 cm. long; seeds about 4 mm. wide, 5–6 mm. long, compressed.

Rocky hillsides and mesas, upper margin of Lower Sonoran Zone and Upper Sonoran Zone, southeastern Arizona into adjacent New Mexico and southward into Sonora. Barely entering the Sonoran Desert as a straggler. June–July.

**16*a*. Agave palmeri** var. **chrysantha** (Peebles) Little, in Benson, Amer. Jour. Bot. **30:** 235, 1943

*Agave chrysantha* Peebles, Proc. Biol. Soc. Wash. **48:** 139, 1935.

Differing from var. *palmeri* chiefly in having fewer marginal teeth, 2–3 cm. apart; somewhat smaller rosettes and broader leaves; ascending branches of the flowering stalks; and chrome yellow perianths lacking purple tinge.

Arid foothills and mountain slopes, Lower and Upper Sonoran Zones, in Maricopa, Pima, Gila, and Pinal Counties, Arizona. June–July.

**17. Agave sobria** Brandegee, Proc. Calif. Acad. II, **2:** 207, 1889, var. **sobria**

*Agave cerulata* Trelease, Rep. Mo. Bot. Gard. **22:** 55, 1912.
*Agave carminis* Trelease, loc. cit.
*Agave affinis* Trelease, op. cit. 56.
*Agave avellanidens* Trelease, op. cit. 60.

Acaulescent plant with rosettes 1–1.8 m. broad, leaves rather openly spreading-ascending; leaves lanceolate to oblong-lanceolate, 6–10 cm. wide, 4–6 dm. long, thick and fleshy at base, tapering from near middle to apex, glaucous; terminal spine 1–1.5 cm. long, scarcely at all decurrent; lateral teeth ashy brown, lanceolate-deltoid, straight or very slightly to strongly hooked, 8–15 mm. long, 0.5–4 cm. apart, set on fleshy prominences and intervening margin straight to curved; flowering stalk 4–5 m. tall, branching panicle 1–1.3 m. long, 7–10 dm. in diameter; flowers 4–5 cm. long, greenish yellow; perianth tube 3–5 mm. deep, broadly cup-shaped; perianth segments oblong to lance-linear, 3–4 mm. wide, 20–25 mm. long; filaments about 3 cm. long; anthers 2–2.5 cm. long; capsules oblong, 1.5–2 cm. wide, 4–5 cm. long.

Mesas and slopes of volcanic hills, Lower Sonoran Zone, from vicinity of Calmallí, Baja California, southward to the Sierra de la Giganta and on the adjacent islands in the Gulf.

## 17*a*. Agave sobria var. roseana (Trelease) I. M. Johnston, Proc. Calif. Acad. IV, **12**: 1002, 1924

*Agave roseana* Trelease, Rep. Mo. Bot. Gard. **22**: 59, 1912.
*Agave connochaetodon* Trelease, op. cit. 58.

Differing from var. *sobria* mainly in having more deeply and conspicuously repand-curved intervals between the teeth, and a very shallow perianth tube only 2–3 mm. long.

Basaltic mesas and hillsides, Lower Sonoran Zone, Sierra de la Giganta and on the adjacent islands in the Gulf of California southward to the vicinity of La Paz, Baja California.

## 18. Agave margaritae Brandegee, Proc. Calif. Acad. II, **2**: 206, 1889

Cespitose, acaulescent plant with subglobose rosettes 30–50 cm. in diameter; leaves dull green, glaucous, and more or less banded transversely, obovate-oblanceolate to nearly orbicular, abruptly acuminate, 6–10 cm. broad, 10–20 cm. long, openly concave above; terminal spine 2–3 cm. long, 3–4 mm. in diameter at base, grooved above toward apex, decurrent about its own length; lateral teeth few, usually 4–10 on each side of leaf, 4–8 mm. long, narrowly triangular, mostly recurved on lower part of leaf and antrorsely curved on upper part; flowering stalk 3–4 m. tall; branches of panicle few-flowered, 15–25 cm. long; flowers light yellow, 4–5 cm. long; perianth tube narrowly funnelform, 10–12 mm. long; perianth segments lance-attenuate, 4–5 mm. broad at base, 12–15 mm. long; filaments 20–25 mm. long, inserted in throat, capsules obpyriform, 8–12 mm. wide near summit, 3–5 cm. long; seeds 3–4 mm. long and wide, black and smooth.

Rocky and sandy hillsides, Lower Sonoran Zone, Magdalena and Santa Margarita Islands, Baja California.

## 19. Agave subsimplex Trelease, Rep. Mo. Bot. Gard. **22**: 60, 1912

"Leaves very glaucous, somewhat transversely banded below, nearly smooth, broadly oblong- or elliptical-lanceolate, slightly acuminate, about 5 × 15 cm.: spine slightly curved, shortly acicular, dull ashen-gray, 3 × 20 mm., round-grooved below the middle, little-decurrent: prickles from blackish purple or through yellow and scarlet becoming similarly colored, 10–20 mm. apart, 5–10 mm. long, narrowly triangular, mostly upcurved above and recurved below, little widened into the tops of rather high fleshy prominences between which the margin is rather straight. Inflorescence slender, narrowly paniculate, the lower branches short-branched, the upper nearly simple: pedicels extremely short. Flowers? Capsules brown, apparently varying greatly and almost inconceivably: short (15 × 30 mm.) and rather obscurely stipitate and beaked; or much larger (20 × 65 mm.) with prominent beak and long stipitate contraction: seeds about 4 × 5 mm."

Arid islands off the coast of Sonora, Lower Sonoran Zone, in vicinity of Tiburón Island.

**20. Agave dentiens** Trelease, Rep. Mo. Bot. Gard. **22:** 51, 1912

*Agave disjuncta* Trelease, loc. cit.

Somewhat cespitose, acaulescent plants; leaves openly spreading, elongated-triangular, 3–6 cm. wide at base, gradually narrowing to an acute or acuminate apex, 3–5 dm. long, very thick and subfleshy, strongly glaucous, gray-green, slightly roughened; terminal spine triquetrously conical, ashen gray to light brown, 3–4 mm. in diameter at base, 2–3 cm. long, narrowly and deeply grooved to a little above middle, long-decurrent; lateral teeth minute, 1 mm. long or less, 3–10 mm. apart, mammiform, weak and readily friable, intervening margin straight or nearly so; flowering stalk and inflorescence 3–4 m. high; flowers unknown; capsules oblong, 18–20 mm. in diameter, 4–5 cm. long, short-stipitate and slightly beaked.

Rocky slopes and along washes, Lower Sonoran Zone, islands of the upper Gulf of California and on San Benito Island, off the west coast of Baja California.

**21. Agave deserti** Engelm., Trans. Acad. St. Louis **3:** 310, 370, 1875

*Agave pringlei* Engelm. ex Trelease, Rep. Mo. Bot. Gard. **22:** 54, 1912.
*Agave consociata* Trelease, op. cit. 53.
*Agave nelsonii* Trelease, op. cit. 61.

Plants usually somewhat cespitose in large colonies but occasionally solitary; rosettes 4–7 dm. in diameter, subglobose; leaves gray-green, plain or transversely banded, triangular-lanceolate, broadest at base or about middle, 5–7 cm. wide, 2–4.5 dm. long, gradually tapering to apex, terminal spine 2–4.5 cm. long; marginal teeth 3–8 mm. long, 1–3 cm. apart; flowering stalks slender, 2–6 m. tall, upper one-fourth to one-third bearing spreading, paniculately arranged branches; flowers 3.5–4 cm. long, yellow; perianth tube funnelform, 5–8 mm. long, segments 15–18 mm. long, about one-half as wide basally; stamens 3–4.5 cm. long, inserted near top of perianth tube, anthers 18–20 mm. long; ovary 15–20 mm. long; capsules oblong, about 1.5 cm. in diameter, 3.5–4.5 cm. long; seeds about 4 mm. broad, 5 mm. long.

Arid hills and rocky or gravelly desert slopes, Lower Sonoran Zone, western margins of the Colorado Desert eastward to western Arizona and southward to central Baja California. May–July.

## Family 20. DIOSCOREACEAE. Yam Family

Scandent or climbing perennial plants, usually from large, fleshy roots, with ample, usually cordate leaves palmately 3–many-ribbed and reticulately veined. Flowers dioecious or rarely monoecious, small, regular, borne in racemes or spikes. Perianth of staminate flowers 6-parted, campanulate, rotate, or tubular. Stamens 6, all perfect or 3 alternate ones reduced to staminodia, filaments distinct or more or less connate into a column, anthers small. Perianth of pistillate flowers 6-lobed; staminodia 3, 6, or none. Ovary inferior, more or less triquetrous, 3-celled, each cell 2-ovuled. Styles 3, short; stigmas 3, each entire or 2-lobed. Capsule loculicidally dehiscent. Seeds globose or compressed, often winged. Embryo small; endosperm cartilaginous.

### 1. DIOSCOREA L., Sp. Pl. 1032, 1753

Twining vine or less commonly scandent. Leaves alternate or opposite, in ours 9–13-nerved. Characters otherwise as in the family.

**1. Dioscorea convolvulacea** Cham. & Schlecht. subsp. **grandifolia** (Schlecht.) Uline, in Kunth, in Engler, Pflanzenreich **4**[43]**:** 99, 1924

*Dioscorea grandifolia* Schlecht., Linnaea **17:** 602, 1843.

Leaves ovate-orbicular, strongly acuminate, 11–13-nerved, at maturity to 3.7 dm. broad, glabrous above, puberulent along nerves beneath; petioles 2–5 cm. long; staminate racemes simple, 1.5–3 dm. long; pedicels 4–8 mm. long; perianths sordid green to reddish, 3–4 mm. in diameter, segments oblong, obtuse, spreading or reflexed; fertile stamens 3, filaments dilated and slightly connate at base; capsule oblong to oblong-elliptic, 8–9 mm. in diameter, 12–15 mm. long; seeds winged.

In deep soil of valleys and lower slopes of adjacent hills, Lower Sonoran Zone to Upper Sonoran Zone and at least approaching the Tropical Zone, southern Sonora and Chihuahua, southward into Jalisco. July–Aug.

## SUBCLASS II. DICOTYLEDONES

Stems of pith, wood, and cortex (bark), a layer of thin-walled, meristematic cells (cambium) lying between the xylem and phloem and adding new cells to both these tissues; the cambium functional throughout the life of the plant. Medullary rays and xylem rays traversing the xylem of the stem. Leaves usually pinnately or palmately veined, the veinlets arranged in a reticulate pattern, rarely parallel. Floral parts mostly in 4's or 5's, rarely in 3's or 6's. Embryo of the seed mostly with 2 cotyledons. First leaves of a young seedling opposite.

### Key to the Series and Families of the Dicotyledones

Petals none; perianth segments all alike in texture and color, or represented by a bract, or none ........................................................ Series 1. *Apetalae*

Petals present; inner perianth whorl (corolla) usually differing markedly in color, or texture, or in both characters, from outer perianth whorl (calyx):

  Petals distinct, free from one another at least at the base and usually the full length .... ........................................................ Series 2. *Polypetalae*

  Petals united into a tube or cup, at least at the base, and usually for much of their length, the corolla lobes either shorter or longer than the tube ........ Series 3. *Gamopetalae*

### Series 1. Apetalae

Spikes subtended by conspicuous white or roseate bracts; minute white bracts scattered among the perfect flowers, these closely crowded on the elongate-conical receptacle ........... ........................................................ Family 21. *Saururaceae,* p. 383

Spikes or racemes not subtended by large white bracts; flowers if subtended by small, scale-like bracts, unisexual, the axis not fleshy-conical:

◆ Flowers (either staminate or pistillate or both sexes) in catkins or catkinlike spikes or racemes:

  Seeds bearing a tuft of long, silky hairs; fruit a small capsule ...................... ........................................................ Family 22. *Salicaceae,* p. 384

  Seeds not bearing silky hairs; fruit not a capsule:

    Pistillate and staminate flowers both in catkins:

      Leaves opposite; fruit simple, bitter or salty:

        ■ Ovary 4-celled; catkins erect; plant of maritime habitats; leaves fleshy ......... ........................................ Family 38. *Batidaceae,* p. 489

■Ovary 1-celled; catkins drooping; shrub of montane habitats; leaves leathery .... .......................................... Family 105. *Garryaceae*, p. 1064

Leaves alternate (in ours); fruit multiple, the flesh sweet .......................... .......................................... Family 26. *Moraceae*, p. 395

Pistillate flowers solitary or in small clusters; staminate flowers in catkins:

Stinging hairs arming the stems and leaves; plants (in ours) herbaceous; fruit an achene .......................................... Family 27. *Urticaceae*, p. 397

Stinging hairs none; plants shrubs or trees; fruit a nut, an acorn, or an achene:

Leaves ample, often lobed or compound, not fleshy; fruit nutlike or a capsule:

Flowers monoecious; leaves alternate; fruit a nut or acorn:

Fruit a hard-shelled nut with a leathery covering, not set in an involucre; cotyledons 2-lobed; leaves pinnately compound ...................... .......................................... Family 23. *Juglandaceae*, p. 389

Fruit a thin-shelled acorn, the base set in a cuplike involucre; cotyledons entire; leaves simple ...................... Family 24. *Fagaceae*, p. 390

Flowers dioecious; leaves opposite; fruit a capsule ........................ .......................................... Family 75. *Buxaceae*, p. 839

Leaves narrow, fleshy, entire; fruit an achene enveloped in a winged calyx; genus *Sarcobatus* .......................... Family 35. *Chenopodiaceae*, p. 437

◆Flowers not in catkins:

Flowers inserted inside a hollow, globose or ovoid, fleshy receptacle; stipules calyptra-like over terminal buds, caducous; genus *Ficus* ....... Family 26. *Moraceae*, p. 395

Flowers variously arranged, but not inside a fleshy receptacle; stipules not calyptra-like:

Ovary inferior, adnate to the calyx if the latter is present:

Plants parasitic on stems or branches of shrubs and trees:

Stems well developed; fruit 1-seeded, berrylike, with sticky, sweetish flesh; plant readily seen throughout the year ........... Family 28. *Loranthaceae*, p. 399

Stems wanting; fruit several to many-seeded, the capsule not fleshy nor sticky; plants visible only during flowering and fruiting period, minute .......... .......................................... Family 33. *Rafflesiaceae*, p. 407

Plants terrestrial, aquatic, or rarely parasitic on roots of other plants:

Ovary 1-celled:

Fruit a capsule; ovary 3-carpelled; styles 3; stamens 8–12 .................. .......................................... Family 96. *Datiscaceae*, p. 955

Fruit a 1-seeded nutlet; ovary 1-carpelled; style 1; stamens 4–5 ............. .......................................... Family 29. *Santalaceae*, p. 403

Ovary 2–several-celled:

Calyx tube extending well beyond the ovary, zygomorphic; stamens in 1–2 whorls around summit of ovary or stylar column ..................... .......................................... Family 32. *Aristolochiaceae*, p. 406

Calyx tube exceeding ovary only slightly or not at all; stamens not surrounding a column:

Leaves fleshy, often with beadlike, water-filled vesicles on epidermis; seeds many; genus *Mesembryanthemum* ....... Family 40. *Aizoaceae*, p. 492

Leaves not fleshy nor bearing vesicles; seeds 2, 1 in each carpel; genus *Galium* .......................................... Family 132. *Rubiaceae*, p. 1397

Ovary superior, free from the calyx or calyx none:

▲Pistils 3–many in each flower, each becoming a laterally compressed or angled achene:

Flowers perfect or dioecious, not in heads; stipules none; herbs or scandent vines; achenes laterally compressed ........... Family 44. *Ranunculaceae*, p. 514

Flowers monoecious, in globose heads racemosely arranged on a slender axis; stipules present, caducous; large trees; achenes 3–4-angled ............. .......................................... Family 54. *Platanaceae*, p. 579

▲Pistil 1 in each flower, of 1-several carpels; fruits various:
- ●Ovary 1-celled:
  - Fruit bearing a long tail; stamens distinctly perigynous ........................................................ Family 56. *Rosaceae,* p. 581
  - Fruit not long-tailed; stamens hypogynous:
    - Shrubs or trees:
      - Fruit a drupe:
        - Stipules none; leaf blades asymmetrical or with 2 glands at juncture with petiole; flowers not in racemes, perfect:
          - Drupes solitary or in small clusters; exocarp fleshy; leaves asymmetrical at base, scabrous; stigmas 2 ..... Family 25. *Ulmaceae,* p. 393
          - Drupes crowded in a head; exocarps leathery; leaves symmetrical, smooth, bearing a pair of glands at juncture of petiole and blade; stigma 1 .................. Family 100. *Combretaceae,* p. 1039
        - Stipules present; leaf blades symmetrical but glandless; flowers in racemes, dioecious .................. Family 30. *Opiliaceae,* p. 404
      - Fruit a pearly, translucent berry or an achene; leaves symmetrical at the base; glandless:
        - Flowers imperfect, monoecious or dioecious; genus ***Atriplex*** ......... .............................. Family 35. *Chenopodiaceae,* p. 437
        - Flowers perfect; genus *Phaulothamnus* ......................... .............................. Family 39. *Phytolaccaceae,* p. 490
    - Herbaceous or slightly suffrutescent plants:
      - Plants submerged aquatics; leaves dissected into filiform segments ...... .............................. Family 43. *Ceratophyllaceae,* p. 513
      - Plants terrestrial, or if aquatic not submerged; leaves entire or lobed but segments not filiform:
        - Fruit circumscissile:
          - Fruit a capsule; seeds several, reniform; perianth not scarious; genus *Trianthema* .................... Family 40. *Aizoaceae,* p. 492
          - Fruit an utricle; seed 1 in each carpel, lenticular-orbicular; perianth segments usually scarious ...... Family 36. *Amaranthaceae,* p. 454
        - Fruit not circumscissile:
          - Fruit indehiscent; 1-seeded:
            - Stinging hairs often present; sepals 4 in staminate flowers ........ ............................. Family 27. *Urticaceae,* p. 397
            - Stinging hairs none; sepals usually 3, 5, or 6 or fused into a tube:
              - Stipules present, frequently sheathing; achenes mostly trigonous; sepals usually 6 in 2 whorls .......................... .......................... Family 34. *Polygonaceae,* p. 408
              - Stipules none; fruit lenticular, ovoid, or globose; sepals commonly 5 or fused into a tube:
                - Calyx not tubular nor brightly colored; leaves rarely opposite; ovary not appearing inferior:
                  - Bracts and sepals fleshy or herbaceous, not scarious; stamens distinct at the base; leaves often scurfy .............. .................. Family 35. *Chenopodiaceae,* p. 437
                  - Bracts and sepals scarious; stamens connate at base; leaves not scurfy .............. Family 36. *Amaranthaceae,* p. 454
                - Calyx tubular and often brightly colored; leaves opposite or basal; ovary often appearing inferior, but not adnate to the calyx tube ......... Family 37. *Nyctaginaceae,* p. 472
          - Fruit dehiscent, several-seeded .. Family 42. *Caryophyllaceae,* p. 503

● Ovary 2–several-celled:
Flowers unisexual, monoecious or dioecious, often enclosed in a calyxlike involucre or borne on axillary bracts:
Staminate flowers of 2–several stamens, inside an involucre or in spikes; ovary 3-celled, stipitate; styles 3; stipules present ................................................... Family 73. *Euphorbiaceae,* p. 773
Staminate flowers consisting of 2 axillary bracts and 1 stamen; ovary 4-celled, sessile; styles 2; stipules none ........................................................ Family 74. *Callitrichaceae,* p. 839
Flowers perfect, not in an involucre nor on axillary bracts:
Fruit capsular, circumscissile ................ Family 40. *Aizoaceae,* p. 492
Fruit various but not circumscissile:
**Plants herbaceous:**
Fruit a capsule, 4–10-valved, not compressed contrary to a partition; stamens 5–10 ............... Family 42. *Caryophyllaceae,* p. 503
Fruit a silicle, 2-valved, compressed contrary to the partition; stamens 2, 4, or 6, tetradynamous when 6; genus *Lepidium* ............................................... Family 49. *Cruciferae,* p. 530
Plants shrubby or arborescent:
Fruit a samara; leaves pinnately compound:
Samara 2–3-winged; stamens 5; genus *Alvoradoa* ........................................... Family 68. *Simarubaceae,* p. 753
Samara terminally 1-winged; stamens usually 2; genus *Fraxinus* .................................... Family 113. *Oleaceae,* p. 1085
Fruit a capsule or a drupe; leaves simple:
Flowers large, 5–8 cm. broad; sepals yellow, petaloid; stamens on a stout column; genus *Fremontodendron* ................................................. Family 84. *Sterculiaceae,* p. 919
Flowers much smaller, rarely over 1 cm. broad; sepals not petaloid; stamens not in a column:
Fruit a winged capsule; embryo spirally coiled; genus *Dodonaea* .............................. Family 78. *Sapindaceae,* p. 851
Fruit a drupe; embryo straight or curved, not spirally coiled:
Stamens 5; drupe globose or depressed-globose; leaves alternate; genus *Rhamnus* ........... Family 79. *Rhamnaceae,* p. 856
Stamens 2–4; drupe ovoid to oblong; leaves opposite or fascicled; genus *Forestiera* ............Family 113. *Oleaceae,* p. 1085

Series 2. POLYPETALAE

◆ Ovary superior, free from the calyx or adnate to it only slightly at the base:
○ Stamens and petals inserted on the receptacle or on a narrow disk at the base of the pistil:
■ Stamens more than 10:
▲ Pistils several and free; fruits achenes or follicles:
Sepals persistent; leaves alternate or basal, or when opposite the plant a woody vine:
Seeds not arillate; plants herbaceous or scandent vines with opposite, compound leaves ............................ Family 44. *Ranunculaceae,* p. 514
Seeds enclosed in a fimbriate aril; plants shrubby to arborescent .................................................. Family 55. *Crossosomataceae,* p. 580
Sepals caducous; leaves opposite or whorled; genus *Platystemon* ................................................... Family 47. *Papaveraceae,* p. 523

▲Pistil 1, of 1–several connate carpels; fruits various:
Ovary 1-celled:
Style 1; sepals 3–5:
Flowers dioecious ........................ Family 50. *Capparidaceae*, p. 564
Flowers perfect:
Leaves simple, entire or dissected:
Petals as many as the sepals; sepals persistent:
Sepals all alike; genus *Corchorus* ......... Family 81. *Tiliaceae*, p. 872
Sepals dissimilar, the outer 2 much smaller than the inner 3 .......... .................................. Family 89. *Cistaceae*, p. 932
Petals half as many as the sepals; sepals caducous (connate into a conical calyptra in *Eschscholzia*) ........... Family 47. *Papaveraceae*, p. 523
Leaves compound:
Leaves bipinnately compound; ovary sessile; sepals and petals usually 5; sepals persistent .................. Family 58. *Mimosaceae*, p. 585
Leaves digitately compound; ovary usually stipitate; sepals and petals 4, sepals deciduous ................. Family 50. *Capparidaceae*, p. 564
Styles 2–8; sepals 2 ......................... Family 41. *Portulacaceae*, p. 496
Ovary 2–several-celled:
Stamens free, not fused into a column:
Sepals in 2 series, the outer 2 smaller than the inner 3 ..................... ........................................ Family 89. *Cistaceae*, p. 932
Sepals alike or essentially so, in a single whorl:
Leaves simple but often lobed; stipules persistent; anthers 2-celled:
Seeds invested in long, silky hairs or by a papery arilloid coat; anthers opening by terminal pores ..... Family 90. *Cochlospermaceae*, p. 934
Seeds naked or short-puberulent; anthers dehiscing longitudinally ....... ......................................Family 81. *Tiliaceae*, p. 872
Leaves digitately compound; stipules caducous; anthers 1-celled ........... .................................... Family 83. *Bombacaceae*, p. 917
Stamens fused into a column, or connate at the base into bundles or clusters:
Ovary 5–30-carpelled; leaves stipulate; style branches simple:
Anthers 1-celled; fruit a 5-carpelled, woody capsule to 15 cm. long; bark on old trunks often bearing conical tubercles .......................... .................................... Family 83. *Bombacaceae*, p. 917
Anthers 2-celled; carpels separating in fruit or capsular, then 5 cm. long or less; bark not tuberculate ............... Family 82. *Malvaceae*, p. 875
Ovary 3-carpelled, 3-celled; leaves estipulate; style branches again 2–3-branched; genus *Ditaxis* ....................... Family 73. *Euphorbiaceae*, p. 773
■Stamens 10 or fewer:
●Corolla irregular:
Leaves compound:
Ovary unicarpelled and 1-celled; fruit a legume, dehiscent or less commonly indehiscent; endosperm usually none:
Sepals not petaloid; upper 3 petals free; legumes mostly dehiscent, not armed with barbed spines; stamens mostly 10:
Flowers only moderately irregular, the lower pair of petals not connate, lower pair of petals outermost in bud .... Family 59. *Caesalpinaceae*, p. 613
Flowers markedly irregular, the lower pair of petals fused along the lower margins near the apex to form a keel; upper pair of petals outermost in bud ................................. Family 60. *Fabaceae*, p. 631
Sepals petaloid; 3 upper petals connate at the base of their claws; legumes indehiscent, globose, armed with subulate or barbed spines; stamens 4 ..... .................................. Family 61. *Krameriaceae*, p. 731

Ovary 2–5-celled; fruit a capsule, follicular, drupaceous or the carpels separating as bony, often spiny nutlets; endosperm usually present:
Flowers perfect; fruit a capsule or of bony nutlets .......................... .................................... Family 66. *Zygophyllaceae,* p. 740
Flowers dioecious; fruit of 1–5 follicles or a 2–5-celled drupe:
Fruit of 1–5 follicles; branches armed with spines or prickles; genus *Zanthoxylum* ............................ Family 67. *Rutaceae,* p. 750
Fruit drupaceous; branches unarmed; genus *Cyrtocarpa* ................ ................................. Family 76. *Anacardiaceae,* p. 841
Leaves simple, although often toothed or lobed:
Fruit a legume; some adnation among petals:
Anthers dehiscing longitudinally; stamens 10; 2 lower petals adnate along the lower margin to form a boat-shaped keel; pod dehiscent except in some species of *Hosackia* ......................... Family 60. *Fabaceae,* p. 631
Anthers dehiscing by terminal pores; stamens 4; 2 lower petals free, upper 3 adnate at base of claws; pod indehiscent, globose ...................... .................................. Family 61. *Krameriaceae,* p. 731
Fruit not a legume; petals free:
Spur or saccate swelling present on either the upper sepal or the lower petal; fruit of 3–5 distinct follicles or a 1-celled, 3-carpelled, elastic capsule:
Fruit of 3–5 distinct follicles; sepals petaloid, the uppermost spurred; petals 4, in 2 pairs; stamens usually many; genus *Delphinium* ............... ................................ Family 44. *Ranunculaceae,* p. 514
Fruit a 1-celled, elastic capsule dehiscing into 3 valves; sepals green, not petaloid nor spurred; petals 5, the lowermost saccate at the base; stamens 5 .................................. Family 92. *Violaceae,* p. 937
Spurs and saccate swellings none; fruits various, but not elastic nor follicular:
Petals 2; fruit a 4-sulcate capsule opening at the apex only; sepals 4 ........ ..................................... Family 51. *Resedaceae,* p. 570
Petals 3 or 5; fruit a samara, nutlet, capsule, or drupe, but if dehiscent splitting nearly full length; sepals 5:
Stamens with filaments more or less united into a tube or sheath; sepals persistent; fruit not stipitate:
Sepals all alike, usually bearing 2–several glands on outer surface; anthers 2-celled; stamens 5 or 10 ........ Family 71. *Malpighiaceae,* p. 762
Sepals of 2 kinds, the outer much smaller than the inner, glandless; anthers 1-celled, opening by terminal pores; stamens 3, 6, or 8, usually 8 ......................... Family 72. *Polygalaceae,* p. 768
Stamens free; sepals deciduous; fruit stipitate; genus *Viscainoa* .......... ............................. Family 66. *Zygophyllaceae,* p. 740
●Corolla regular:
★Flowers dioecious, monoecious, or polygamodioecious:
Pistillate flowers apetalous; leaves stipulate; stamens 4 ...................... .......................................... Family 30. *Opiliaceae,* p. 404
Pistillate flowers bearing 1 or more petals; leaves estipulate; stamens 5–10:
Stamens free, not connate into a tube or cup:
Sepals 6, in 2 series, the inner whorl longer than the outer; seeds reniform; genus *Cocculus* .................. Family 46. *Menispermaceae,* p. 522
Sepals 3–5, essentially alike, in 1 series; seeds not reniform:
Carpels 1-ovuled; seeds not arillate:
Branches spinescent; carpels united only at the base, stellate-divergent in fruit ........................ Family 68. *Simaroubaceae,* p. 753
Branches not spinescent; carpels, if more than 1, united full length, not stellate-divergent in fruit ....... Family 76. *Anacardiaceae,* p. 841

Carpels 2-ovuled; seeds more or less arillate .......................... .................................... Family 69. *Burseraceae,* p. 755

Stamens connate into a tube or their bases fused into a shallow cup:

Pistillate flower consisting of 1 sepal, 1 petal, and 1 pistil; ovary 1-celled; leaves simple; anthers 2 at apex of column; genus *Cissampelos* ........ ................................ Family 46. *Menispermaceae,* p. 522

Pistillate flower with 3–5 sepals and petals, stamens as many as or twice as many as the petals; ovary 2–5-celled; leaves pinnately or bipinnately compound; anthers 5, 8, or 10 at apex of column ........................ ................................... Family 70. *Meliaceae,* p. 760

★Flowers perfect, both stamens and pistil in each flower:

Ovary 1-celled:

Fruit indehiscent:

Anthers opening by terminal or subterminal pores; sepals in 2–3 series, imbricate; leaflets spinose-margined; ovary sessile ...................... .................................. Family 45. *Berberidaceae,* p. 521

Anthers dehiscing longitudinally; sepals 4–5, in a single series; leaflets not spinose-margined; ovary often stipitate:

Plants shrubby or arborescent; stamens free; seeds 1–2 in each fruit; petals 4, in unequal pairs; genus *Atamisquea* .......................... ................................ Family 50. *Capparidaceae,* p. 564

Plants herbaceous or vinelike; stamens connate into an androphore; seeds several to many in each fruit; petals 5, all alike .................... ................................. Family 94. *Passifloraceae,* p. 939

Fruit dehiscent:

Stamens free, their filaments not coalesced:

Sepals 2 ............................ Family 41. *Portulacaceae,* p. 496

Sepals 4–5:

Leaves opposite; petals often with appendages on the claws; herbage not strongly scented:

Plants shrubby; stipules none; stamens inserted on the calyx tube ... ........................... Family 86. *Frankeniaceae,* p. 927

Plants herbaceous; stipules usually present; stamens inserted on receptacle of disk at base of the ovary ......................... ......................... Family 42. *Caryophyllaceae,* p. 503

Leaves alternate; petals without appendages; herbage strongly scented:

Fruit not follicular; flowers perfect; branches not spinescent; stamens 6 .......................... Family 50. *Capparidaceae,* p. 564

Fruit follicular; flowers dioecious; branches spinescent; stamens 3–5; genus *Zanthoxylum* ............... Family 67. *Rutaceae,* p. 750

Stamens connate into a tube or their filaments adnate basally into shallow cup:

Seeds embedded in a fleshy aril; style 2–5-branched; fruit a capsule, 2–5-carpelled; endosperm starchy; leaves estipulate; genus *Stegnosperma* ............................... Family 39. *Phytolaccaceae,* p. 490

Seeds not arillate; style 1, entire; fruit a legume, unicarpelled; endosperm scanty or none; leaves stipulate ... Family 59. *Caesalpiniaceae,* p. 613

Ovary 2–several-celled:

Pistils 3–5, distinct; herbage fleshy; leaves commonly in a dense basal rosette; fruit follicular ......................... Family 52. *Crassulaceae,* p. 571

Pistil 1, of 2–several connate carpels; herbage not markedly fleshy; leaves rarely forming a dense rosette; fruit a capsule, berry, drupe, samara, or achene:

■Fruit indehiscent, or separating into 2–several indehiscent nutlets or achene-like segments:

Stamens distinct, the filaments not connate nor adnate to other structures:

Stipules present; stamens 8–10; interstamineal glands alternating with the filaments of the stamens ....... Family 66. *Zygophyllaceae,* p. 740

Stipules none; stamens 2–4 or 6; interstamineal glands none or inconspicuous:

Fruit a drupe; stamens 2–4; leaves always simple, opposite or in fascicles; genus *Forestiera* ........ Family 113. *Oleaceae,* p. 1085

Fruit 2 nutlets separated by a false partition; leaves 3–5-foliolate or simple by reduction, alternate; genus *Wislizenia* .............. ............................ Family 50. *Capparidaceae,* p. 564

Stamens connate into a tube or coalesced into a shallow cup at the base:

Styles closely adhering to a slender, central column several times as long as the 5 carpels; stamens 5, alternating with 5 staminodia; styles persisting on the achenelike nutlets after the latter fall, hygroscopic ................................ Family 62. *Geraniaceae,* p. 735

Styles not adhering to a column; stamens 5–10, all fertile or 1–2 sterile but not regularly alternating with staminodia; styles not hygroscopic:

Leaves pinnately or bipinnately compound; stipules none; stamens 10–12 ........................... Family 70. *Meliaceae,* p. 760

Leaves simple; stipules present; stamens 5–10:

Petals with a bilobed ligule within; leaves alternate; calyx lobes not glandular without ........ Family 65. *Erythroxylaceae,* p. 740

Petals without ligules within; leaves opposite; calyx lobes often bearing 2–several sessile glands on the outer surface ............ ....................... Family 71. *Malpighiaceae,* p. 762

■**Fruit dehiscent:**

Leaves compound:

Fruit a leathery, drupelike capsule dehiscing tardily; trees with papery-exfoliating bark; juice resinous and aromatic .................... ................................ Family 69. *Burseraceae,* p. 755

Fruit a dry capsule; bark not papery-exfoliating; juice watery, not aromatic:

Stamens free, the bases of their filaments alternating with ovoid glands; leaves opposite; stipules present ............................ .......................... Family 66. *Zygophyllaceae,* p. 740

Stamens connate into a tube or disk near the base; interstamineal glands none; leaves alternate or basal; stipules none:

Leaves digitately 3–4-foliolate; seeds not arillate; plants herbaceous, with acid juice ............... Family 63. *Oxalidaceae,* p. 736

Leaves pinnately compound; seeds embedded in thin, fleshy arils; plants arborescent, juice not acid..Family 70. *Meliaceae,* p. 760

Leaves simple though often more or less lobed or dissected:

Flowers cruciferous, i.e., sepals 4, petals 4, stamens 6 and tetradynamous; fruit a silique or silicle ............ Family 49. *Brassicaceae,* p. 530

Flowers not cruciferous; fruit capsular or drupelike but with a leathery, tardily dehiscent coat:

Stamens connate into a tube or cup:

Leaves opposite; pubescence of 2-rayed hairs attached at or near the middle .................... Family 71. *Malpighiaceae,* p. 841

Leaves alternate; pubescence of simple and stellately several-rayed trichomes ................... Family 84. *Sterculiaceae,* p. 919

Stamens free and distinct:

Fruit a leathery, tardily dehiscent drupelike capsule; seeds arillate on the lower part; bark papery-exfoliating .................. ........................... Family 69. *Burseraceae,* p. 755

♦ Fruit a membranous or woody capsule dehiscing readily; seeds not arillate; bark not papery-exfoliating:

Plants arborescent; capsule woody, echinate or tuberculate; ovary partially embedded in a fleshy disk .................................................. Family 67. *Rutaceae,* p. 750

Plants herbaceous; capsules not woody nor echinate; ovary not embedded in a disk:

Leaves alternate; stipules none or glandular; flowers cymose or racemose .................. Family 64. *Linaceae,* p. 739

Leaves opposite; stipules present; flowers axillary ...................................... Family 85. *Elatinaceae,* p. 926

○Stamens and petals inserted on a pronounced disk lining the calyx cup or on the upper edge of the calyx cup itself, hence noticeably above the base of the ovary and more or less perigynous:

Stamens 15–50:

Fruits achenes or follicles; hypanthium persistent:

Stamens inserted on an evident disk; seeds with a fimbriate aril ................................................... Family 55. *Crossosomataceae,* p. 580

Stamens inserted near the upper edge of the hypanthium, the disk thin, obscure; seeds not arillate ............................ Family 56. *Rosaceae,* p. 581

Fruits drupaceous; hypanthium deciduous ......... Family 57. *Amygdalaceae,* p. 584

Stamens 14 or less:

Stamens as many as the petals and opposite them; petals either hooded or caducous:

Plants shrubby; petals hooded, each more or less enfolding the adjacent stamen; tendrils none ............................ Family 79. *Rhamnaceae,* p. 856

Plants climbing vines; petals not hooded; tendrils present, usually opposite the leaves ...................................... Family 80. *Vitaceae,* p. 869

Stamens as many as the petals and alternate with them, or twice as many as the petals; petals not hooded nor caducous:

Leaves opposite or verticillate:

Plants herbaceous; capsule readily dehiscent; calyx tube often with accessory appendages in the sinuses ..................Family 99. *Lythraceae,* p. 1037

Plants shrubby or arborescent; fruit indehiscent or tardily dehiscent; accessory appendages none:

Ovary 1-celled; seeds not germinating inside the capsule on the tree; calyx 4–5-lobed; fruit 1-seeded ................Family 100. *Combretaceae,* p. 1039

Ovary 1–5-celled; seeds often germinating inside capsule while on tree; calyx 3–14-lobed; fruit 1–10 seeded ...... Family 101. *Rhizophoraceae,* p. 1040

Leaves alternate:

■Ovary 1-celled:

Stipules present; fruit a berry; calyx tube bearing a corona of erect or ascending filaments; fruit (in ours) borne on a stipelike gynophore ................................................. Family 94. *Passifloraceae,* p. 939

Stipules none; fruit a drupe, utricle, or capsule; corona none; ovary sessile, not on a gynophore:

Fruit a drupe or utricle; ovules 1 in each cell; juice markedly resinous or milky .......................... Family 76. *Anacardiaceae,* p. 841

Fruit a capsule; ovules several to many in each cell; juice not markedly resinous or milky:

Leaves minute, scalelike; stigmas broadly capitate, not fringed; seeds villous at the apex or over whole surface ........................................................ Family 87. *Tamaricaceae,* p. 928

Leaves ample, not scalelike; stigmas laciniate-fringed at the apex; seeds naked ........................... Family 93. *Turneraceae,* p. 938

■Ovary 2–several-celled:

Leaves pinnately compound; stamens often excentrically inserted, their filaments usually hairy ........................ Family 78. *Sapindaceae,* p. 851

Leaves simple; stamens uniformly inserted, their filaments glabrous:

Ovary sessile; leaves not scalelike; branches not stiffly spine-tipped; style not subulate .......................... Family 77. *Celastraceae,* p. 849

Ovary short-stipitate; leaves scalelike, soon deciduous; branchlets rigid, spine-tipped; styles subulate, persistent ... Family 91. *Koeberliniaceae,* p. 936

◆ Ovary inferior, adnate to the calyx; stamens borne near the apex of the ovary or above it on the calyx tube:

Stamens numerous, usually over 15:

Ovary 1-celled:

Sepals 2; fruit a circumscissile capsule; seeds with central endosperm; stamens 6–20.. ..................................... Family 41. *Portulacaceae,* p. 496

Sepals 4–many; fruit a berry or a capsule but this never circumscissile; endosperm scanty or none, not wholly central; stamens over 20:

Fruit a capsule; sepals 4–5; petals 4–10; stamens usually in clusters opposite the petals; stems not armed with spines ........... Family 95. *Loasaceae,* p. 943

Fruit a berry, usually armed with spines or glochids; sepals and petals indefinite in number, not sharply differentiated; stamens uniformly distributed in calyx tube and throat; stems usually armed with spines or glochids or both ........... ......................................... Family 98. *Cactaceae,* p. 958

Ovary 2–5-celled:

Petals numerous, linear; flowers perfect; fruits not winged; genus *Mesembryanthemum* ...................................... Family 40. *Aizoaceae,* p. 492

Petals 2–5, elliptic to obcordate; flowers monoecious; fruits with 1 or more membranous to fleshy wings ........................ Family 97. *Begoniaceae,* p. 956

Stamens 15 or less, usually 4–10:

Ovary 1-celled:

Fruit dehiscent; ovules numerous; flowers solitary, cymose, or in few-flowered clusters:

Capsule circumscissile, only partially inferior; leaves somewhat succulent; calyx lobes 2; genus *Portulaca* .................. Family 41. *Portulacaceae,* p. 496

Capsule opening longitudinally or at the apex but not circumscissile; leaves harshly scabrous, not succulent; calyx lobes 4–5 ......... Family 95. *Loasaceae,* p. 943

Fruit indehiscent; ovules few; flowers racemose, spicate, or rarely solitary:

Leaves alternate; stamens 4–5; fruit a berry; style 2-branched ................... ...................................... Family 53. *Grossulariaceae,* p. 578

Leaves opposite; stamens 8–10; fruit a drupe; style 1, entire; genus *Laguncularia*.. ...................................... Family 100. *Combretaceae,* p. 1039

Ovary 2–5-celled:

Inflorescence umbellate; stamens inserted on margin of a distinct epigynous disk:

Ovary 5-celled; styles 5; fruit a berry or drupe; petioles not strongly sheathing .... .........................................Family 104. *Araliaceae,* p. 1063

Ovary 2-celled; styles 2; fruit a schizocarp of 2 one-seeded mericarps; petioles usually sheathing the stem .....................Family 106. *Umbelliferae,* p. 1066

Inflorescence various but not umbellate; stamens inserted on calyx tube or on a less evident disk:

▲Stipules present; ovary 2–3-celled:

Ovary 2-celled; fruit leathery or woody, tardily dehiscent or indehiscent; petals not hooded ........................Family 101. *Rhizophoraceae,* p. 1040

Ovary 3-celled; fruit separating into 3 indehiscent, narrowly winged cocci; petals hooded and partially or wholly enclosing the anthers; genus *Gouania* ...... ........................................ Family 79. *Rhamnaceae,* p. 856

▲Stipules none; ovary 4-celled (2-celled in *Burragea*):

Plants terrestrial; stigma capitate or 4-lobed, not plumose; fruit usually a dehiscent, 4-celled capsule; seeds several to many in each cell (fruit woody and immersed in the stem in *Burragea*) ................ Family 102. *Onagraceae,* p. 1041

Plants aquatic, with immersed leaves dissected; stigmas 4, plumose; fruit separating into 4 indehiscent nutlets; seeds 1 in each cell .................... ...................................... Family 103. *Haloragaceae,* p. 1062

Series 3. Gamopetalae

◆ Ovary superior, not adnate to the calyx tube except at the base:

○Corolla regular, all lobes of each flower essentially of the same size and shape:

Ovary 1-celled, although false partitions partially separate the cells in some species:

Plants strongly shrubby or arborescent; leaves somewhat leathery or rigid:

Fruit a drupe or berrylike; stamens 5, opposite the corolla lobes; petioles not becoming a rigid spine; leaves evergreen:

Ovary partially or wholly embedded in a fleshy disk; staminodia none; leaves not spinose-tipped ............................. Family 31. *Olacaceae,* p. 405

Ovary not embedded in a fleshy disk; staminodia alternating with the stamens; leaves rigid, spinose-tipped ...........Family 108. *Theophrastaceae,* p. 1076

Fruit a capsule; stamens 10–15; abaxial part of each petiole on new growth persisting as a rigid, woody spine; leaves deciduous in drought .................. ...................................... Family 88. *Fouquieriaceae,* p. 930

Plants herbaceous or weakly suffrutescent and scandent; leaves not leathery or rigid:

Stamens opposite the corolla lobes; corolla lobes imbricate in bud; filaments not winged nor corolla lobes fringed:

Seeds several; style 1, entire; plants completely herbaceous; sepals and petals 4–8 (corolla lobes often united only at the base and appearing distinct) ........ ...................................... Family 109. *Primulaceae,* p. 1077

Seed 1, pendulous; styles 5; plants suffrutescent and scandent; sepals and petals 5 (corolla salverform) ................Family 110. *Plumbaginaceae,* p. 1080

Stamens alternate with the corolla lobes; corolla lobes contorted in bud; filaments winged or corolla lobes somewhat fringed.. Family 115. *Gentianaceae,* p. 1092

Ovary 2–15-celled:

Stipules present, often small, inconspicuous; stamens united into a column around slender style; petals distinct at attachment to receptacle but coalesced with stamen column at base of claws .................. Family 82. *Malvaceae,* p. 875

Stipules none (base of petioles sometimes sheathing); stamens free or attached to a corona around the styles; petals united at the base, often forming a tubular, campanulate, or salverform corolla:

■Fruits indehiscent, berrylike or of nutlets:

Plants parasitic on roots of host plants; herbage colorless or at least not green; each cell of ovary breaking into 2 indehiscent nutlets .................... .......................................Family 107. *Lennoaceae,* p. 1075

Plants autotrophic (or if parasitic then growing on leaves and stems of host plant, the plant body orange or yellowish and with a 2-celled ovary maturing as a capsule); each cell of ovary not breaking into 2 nutlets:

Fruit fleshy, a berry or berrylike:

Flowers not arranged in scorpioid spikes or racemes; carpels not on a gynobase; no appendages or swellings in throat of the corolla:

Ovules usually 1–2 in each cell; corolla lobes imbricate in bud; ovary 3–5-celled (in ours):

Flowers perfect; stamens same number as corolla lobes; juice milky .... ................................ Family 111. *Sapotaceae*, p. 1081

Flowers mostly dioecious; stamens usually 2 to 4 times as many as the corolla lobes; juice not milky ...... Family 112. *Ebenaceae*, p. 1083

Ovules usually several to many in each cell (only 1 in *Lycium carinatum*); corolla lobes valvate or contorted in bud; ovary 2-celled or falsely 4-celled ............................ Family 125. *Solanaceae*, p. 1298

Flowers in scorpioid spikes or racemes; carpels borne on a gynobase; corollas often with swellings or appendages partially constricting the throat .... ................................ Family 121. *Boraginaceae*, p. 1202

Fruit of 1–4 nutlets (less than 4 by abortion); (flowers usually in scorpioid spikes or racemes; carpels attached to an elevation or prolongation of the receptacle, the gynobase) ............ Family 121. *Boraginaceae*, p. 1202

■Fruits dehiscent, capsular or follicular:

Carpels 2, distinct or adnate only at the summit in flower, free in fruit; fruits follicular; juice often milky:

Corolla cylindrical-campanulate, tubular-salverform, or funnelform; corona and column none; translators between anthers none ...................... ................................ Family 116. *Apocynaceae*, p. 1095

Corolla rotate or shallowly saucer-shaped; corona of laminate or cucullate appendages usually surrounding the anthers; stamens coherent in a column; translators between adjacent anthers (pollinia) present ................ ................................ Family 117. *Asclepiadaceae*, p. 1104

Carpels 2–5, but united into a single ovary; fruits capsular; juice usually clear (milky in some of the *Convolvulaceae*):

Capsules circumscissile, opening by a lid:

Stamens 2, opposite each other; capsule distinctly didymous, each locule circumscissile or sometimes indehiscent; corolla tube campanulate or urceolate; genus *Menodora* ................. Family 113. *Oleaceae*, p. 1085

Stamens usually 4–5 (sometimes 2 in *Plantago* but corolla in that genus scarious); capsule single, not didymous; corolla funnelform, salverform, or the tube cylindrical and the lobes scarious, spreading:

Corolla funnelform or salverform, the lobes contorted in bud, not scarious; stamens usually 5; flowers not in spikes; plants twining vines or shrubs or trees ...................... Family 118. *Convolvulaceae*, p. 1128

Corolla with cylindrical tube and widely spreading, distinct lobes, these imbricated in bud, scarious; flowers in spikes; herbaceous plants, mostly with basal leaves .............. Family 131. *Plantaginaceae*, p. 1392

Capsules dehiscing longitudinally or irregularly, not circumscissile:

Stamens 4; sepals 4; petals 4; flowers small, in dense cymes or heads arranged in the axils of the leaves ............ Family 114. *Loganiaceae*, p. 1090

Stamens 5; calyx lobes 5; corolla lobes 5; flowers larger, arranged in racemes, panicles, or solitary:

Inflorescence often scorpioid; corolla lobes imbricated or very rarely contorted in bud ............... Family 120. *Hydrophyllaceae*, p. 1180

Inflorescence not scorpioid; corolla lobes contorted in bud:

Ovary 3-celled; capsule 3-valved; styles 3-cleft ...................... ........................... Family 119. *Polemoniaceae*, p. 1160

Ovary mostly 2-celled (sometimes 2–5-celled); capsule 2- or 4-valved; styles entire or 2-lobed:

Styles 2-lobed; base of stamen filaments often dilated; sepals free nearly or wholly to the base, often in 2 whorls; plants often twining vines ....................Family 118. *Convolvulaceae,* p. 1128

Styles entire; stamens rarely dilated at the base; sepals coalesced into a definite cup, tube, or disk, only the lobes free; plants erect or spreading, herbs or shrubs .... Family 125. *Solanaceae,* p. 1298

○Corolla irregular, some lobes of each flower differing in size or shape or in both characters from the other lobes:

Fruit of 2 or 4 nutlets or sometimes drupaceous; stems often quadrangular:

Style entire; ovary not 4-lobed .................. Family 122. *Verbenaceae,* p. 1235

Style 2-cleft or 2-lobed; ovary 4-lobed ................Family 124. *Labiatae,* p. 1264

Fruit capsular; stems usually terete:

Ovary 1-celled (incompletely 2- or 4-celled in *Avicenniaceae*):

Plants shrubby or arborescent, of maritime habitats; capsule densely tomentose, 1-seeded by abortion; placenta central ......Family 123. *Avicenniaceae,* p. 1263

Plants herbaceous, not confined to maritime habitats; capsules puberulent to glabrate but not tomentose, several- to many-seeded; placenta lateral:

Corolla gibbous at the base; plants not parasitic; capsule becoming woody, the slender beak splitting into 2 stiffly recurving, rigid spines; leaves large (up to 3 dm. long) ...................... Family 128. *Martyniaceae,* p. 1369

Corolla not gibbous; plants root parasites, without chlorophyll; capsules not long-beaked; leaves scalelike ............ Family 129. *Orobanchaceae,* p. 1372

Ovary 2-celled:

Flowers not conspicuously bracteate; seeds borne on slender funiculi, not on indurate papillate or hooked processes; stigmas usually distinctly 2-lobed:

Seeds wingless; endosperm present, fleshy; leaves simple but often slightly to deeply dissected; staminodium 1 or absent .............................. ................................Family 126. *Scrophulariaceae,* p. 1332

Seeds (in ours) flattened and winged; endosperm none; leaves usually compound (simple in *Chilopsis*); staminodia often 3..Family 127. *Bignoniaceae,* p. 1366

Flowers often conspicuously bracteate; seeds borne on stout papillae or on hooked projections from the placenta; stigmas entire, cupulate, or inconspicuously 2-lobed; (capsule often stoutly stipitate, much broader near the apex than below) ........................................Family 130. *Acanthaceae,* p. 1375

◆ Ovary inferior, adnate to the calyx tube:

Stamens numerous; petals usually more than 5, appearing distinct and grading into the sepals; plants bearing spines or glochids or both; leaves usually reduced and caducous or lacking (present and fairly persistent in *Pereskiopsis*). Family 98. *Cactaceae,* p. 958

Stamens 5 or fewer; corolla lobes 4–5, obviously connate into a tube or cup at least at the base; plants rarely spiny; leaves mostly moderate to ample in size (infrequently reduced to scales):

Flowers monoecious or dioecious:

Plants usually with tendrils, climbing or scrambling vines; ovary 1–5-celled, of 2–5 carpels, usually 3–many-seeded; flowers not in involucrate heads; fruit usually a pepo ................................ Family 135. *Cucurbitaceae,* p. 1417

Plants without tendrils, rarely vinelike; ovary 1-celled of 1 carpel; flowers crowded in involucrate heads; fruit an achene ........... Family 138. *Compositae,* p. 1445

Flowers perfect, or with some unisexual flowers in a head in some of the *Compositae*:

■Corollas regular:

Stipules present:

Flowers solitary, paniculate, or capitate; corolla lobes 4–10; stipules usually well developed; stigma mostly 2-lobed ..........Family 132. *Rubiaceae,* p. 1397

Flowers cymose; corolla lobes 5; stipules minute or none; stigma usually entire.. ...................................... Family 133. *Caprifoliaceae,* p. 1415

Stipules none:

Flowers not in involucrate heads; calyx well developed; ovary 2–10-celled; fruit capsular or berrylike; anthers free:

Leaves opposite; plants shrubby or scrambling vines; stigma usually entire.... ................................... Family 133. *Caprifoliaceae,* p. 1415

Leaves alternate; plants herbaceous; stigma usually 2–several-lobed .......... ...................................Family 136. *Campanulaceae,* p. 1434

Flowers in involucrate heads; calyx reduced to 1 or more rings of hairy, bristly, or chaffy pappus or none; ovary 1-celled; fruit an achene; anthers connate into a tube around the style .............. Family 138. *Compositae,* p. 1445

■Corollas irregular:

Anthers free, not connate into a tube; stigma entire or essentially so:

Plants mainly shrubby or suffrutescent; stamens 5; corolla tube not spurred or saccate at the base:

Ovules pendulous or horizontal from a central placenta; stigmas not hooded at the apex; leaves opposite .......... Family 133. *Caprifoliaceae,* p. 1415

Ovules erect or ascending; stigmas hooded or indusiate at the apex; leaves alternate .......................... Family 137. *Goodeniaceae,* p. 1444

Plants (in ours) wholly herbaceous; stamens 1–3; corolla tube spurred or saccate at the base .......................... Family 134. *Valerianaceae,* p. 1417

Anthers connate into a tube around the style; stigma usually distinctly 2-lobed:

Fruit a capsule or fleshy, several- to many-seeded; flowers not crowded into an involucrate head; ovary 2–3-celled; calyx herbaceous, with a tube and lobes.. .................................. Family 136. *Campanulaceae,* p. 1434

Fruit an achene, 1-seeded; flowers crowded into involucrate heads; ovary 1-celled; calyx reduced to a pappus or none ........ Family 138. *Compositae,* p. 1445

## Family 21. **SAURURACEAE.** Lizard's-Tail Family

Perennial herbs with broad, petioled, alternate leaves and small, perfect flowers closely spirally arranged in a bracteate spike. Perianth wanting. Stamens 6–8, hypogynous. Ovary 3–4-carpelled, carpels distinct or united, each containing 1–2 orthotropous ovules. Fruit (in ours) a capsule. Seeds globose to ovoid, with copious mealy endosperm; embryo minute, cordate.

### 1. **ANEMOPSIS** Hook. & Arn., Bot. Beechey 390, *pl. 42,* 1841

Stems nodose, scapose, and stoloniferous from creeping, aromatic rootstocks. Leaves chiefly radical, minutely punctate. Flowers in a compact spike subtended by a whorl of 5–8 white or faintly colored involucral bracts, each flower save lowermost also subtended by a smaller, less conspicuous bracteole. Ovary sunk in rachis of spike, 1-celled; stigmas 3–4. Capsule apically dehiscent.

**1. Anemopsis californica** (Nutt.) Hook. & Arn., Bot. Beechey 390, *pl. 42,* 1841

*Anemia californica* Nutt., Ann. Nat. Hist. **1:** 136, 1838.
*Anemoposis bolanderi* C. DC., Linnaea **37:** 333, 1871.
*Houttuynia californica* Benth. & Hook. ex S. Wats., Bot. Calif. **2:** 483, 1880.
*Anemopsis californica* var. *typica* Kelso, Amer. Midl. Nat. **13:** 111, 1932.
*Anemopsis californica* var. *subglabra* Kelso, op. cit. 112.

Radical leaves on petioles 3–30 cm. long, these glabrous or sparsely arachnoid-pubescent, blades elliptic to obovate, obtuse to cordate at base, rounded at apex, entire, 1.5–8 cm. wide, 4–15 cm. long, minutely punctate on both surfaces, glabrous to sparsely puberulent; cauline leaves of flowering scapes sessile, clasping, with 1–3 short-petiolate leaves on a short branch in axil of each cauline leaf; spike 1–5 cm. long; involucral bracts white to pinkish, often deeper red beneath, 1–3.5 cm. long, oblong, rounded at apex; floral bractlets 5–6 mm. long, clawed, white; fruiting spike clothed with deltoid-subulate projections 2–3 mm. long and perforated by apical pores of carpels.

Low alkaline or saline soil, Sacramento and Santa Clara Valleys, California, to Colorado and Texas, south into northern Sonora and to the Cape Region, Baja California. April–Aug.

## Family 22. **SALICACEAE.** Willow Family

Trees or shrubs with smooth or furrowed bark, and simple, alternate, stipulate, deciduous leaves. Flowers dioecious in terminal catkins appearing before or with the leaves and falling as a unit. Each flower subtended by a scale bract, with or without a glandular disk. Stamens 2–many. Ovary 1-celled, many-ovuled, ovoid or globose; stigmas 2–4. Fruit a 2–4-valved capsule containing numerous minute, silky-comose seeds.

Catkin scales fimbriate; disk cup-shaped; buds with several imbricated, resinous scales.......... 1. *Populus*
Catkin scales entire; disk reduced to a gland; buds with a single nonresinous scale .. 2. *Salix*

### 1. **POPULUS** L., Sp. Pl. 1034, 1753

Trees with pale, furrowed bark; soft, light wood; resinous buds; petiolate, broad or narrow leaves; and minute, fugacious stipules. Scales of aments fimbriate, narrowed at base. Disk cup-shaped, lobed or entire. Staminate catkins pendulous; stamens 12–60, the filaments distinct. Pistillate catkins pendulous, erect, or spreading; ovary sessile; style short; stigmas 2–4, entire or 4-lobed. Capsule 2–4-valved; valves recurved at dehiscence.

Leaves dimorphic, those of young shoots linear-lanceolate, those on older growth deltoid; disk 1–2 mm. broad .......... 1. *P. dimorpha*
Leaves not dimorphic, all deltoid or elliptic or ovate; disk 2–6 mm. broad:
- Leaves deltoid, truncate to subcordate at base or occasionally broadly subcuneate; bark of seasonal twigs whitish or yellowish, usually glossy; petioles flattened laterally:
  - Capsules 8–12 mm. long:
    - Disk flattish, 2–3 mm. broad; leaves bluish green; petioles and young shoots densely pilose .......... 2. *P. macdougalii*
    - Disk cup-shaped, 5–6 mm. broad; leaves yellowish green; petioles and young twigs glabrous or puberulent with short hairs of equal length:
      - Twigs and petioles glabrous .......... 3. *P. fremontii* var. *fremontii*
      - Twigs and petioles puberulent .......... 3*a*. *P. fremontii* var. *pubescens*
  - Capsules 4–5 mm. long .......... 4. *P. arizonica*
- Leaves elliptic, cuneate at base; bark of seasonal twigs dark reddish brown, dull; petioles terete:
  - Leaves densely tomentose on both surfaces, not coriaceous; petioles 3–20 mm. long, densely tomentose .......... 5. *P. brandegeei* var. *brandegeei*
  - Leaves glabrous or essentially so, coriaceous; petioles 1.5–5 cm. long, sparsely pubescent in age or subglabrate .......... 5*a*. *P. brandegeei* var. *glabra*

**1. Populus dimorpha** Brandegee, Zoe **5:** 197, 1905

A large tree to 20 m. high with a rough-barked trunk to 2 m. in diameter; young twigs slender, light gray-green to nearly white, minutely puberulent, soon glabrate; leaves of young trees and "suckers" on older trees linear-lanceolate, 2–4 mm. wide, 1–3 cm. long, attenuate into a petiole 1–3 mm. long, acute at apex, minutely denticulate, leaves on older trees deltoid-ovate, 3–7 cm. wide, 5–10 cm. long, broadly cuneate to truncate at base, crenate-serrate, long-acuminate and serrulate at apex; petioles flattened laterally, slender, 2–5 cm. long, finely puberulent; staminate catkins dense, 2–4 cm. long, stamens on each scale numerous, crowded into globose masses, anthers about 1 mm. long; pistillate catkins slender, lax, 5–8 cm. long, pedicels slender, 6–8 mm. long; disk minute, flattish, 1–2 mm. broad, shallowly lobed, caducous; pistils ovoid, 4–5 mm. long, 2–3 mm. wide, glabrous, blunt at apex; stigmas 4-lobed.

Stream banks, Lower Sonoran Zone, southern Sonora into Sinaloa.

**2. Populus macdougalii** Rose, Smiths. Misc. Coll. **61**[12]**:** 1, *pl. 1,* 1913

A large tree to 20 m. high with light gray, deeply fissured bark and slender, tawny, densely pubescent twigs; leaves broadly deltoid, 4–12 cm. wide, 6–10 cm. long, usually slightly longer than broad, truncate or slightly cuneate at base, abruptly short-acuminate at apex, crenate-serrate except at base and near apex, puberulent when young, soon glabrous, bluish green, coriaceous; petioles pubescent, 3–7 cm. long; pistillate catkins 5–12 cm. long, lax; disk of pistillate flowers shallowly cup-shaped, 2–3 mm. broad, shallowly 3–4-lobed or subentire; capsules broadly ovoid, 9–12 mm. long, thick-walled, mostly 3-valved, on pedicels 3–4 mm. long.

Along the Colorado River and about desert springs, Lower Sonoran Zone, southern Nevada to the head of the Gulf of California. Common in Imperial Valley along ditches and planted as a shade tree.

**3. Populus fremontii** S. Wats., Proc. Amer. Acad. **10:** 350, 1875, var. **fremontii**

*Populus fremontii* var. *thornberi* Sarg., Bot. Gaz. **67:** 213, 1919.
*Populus fremontii* var. *toumeyi* Sarg., op. cit. 214.

A large tree up to 35 m. high with gray, fissured bark, a trunk 1–2 m. in diameter, light yellowish green to tawny glabrous twigs, and a broad, open crown; leaves 4–7.5 cm. long, about as wide or slightly wider, deltoid, truncate to subcuneate at the base, abruptly pointed at the apex, coarsely and irregularly crenate-serrate, glabrous or sparingly pubescent when young, bright green, lustrous, coriaceous at maturity; petioles 4–8 cm. long, strongly flattened laterally, glabrous; staminate catkins 4–7 cm. long, lax, rachis glabrous; stamens about 60 in each flower, anthers reddish; pistillate catkins 2.5–10 cm. long, obovate scales deeply fimbriate at apex; disks broadly cup-shaped, membranous, 4–6 mm. wide, shallowly lobed; capsule thick-walled, 8–12 mm. long, the 3 valves spreading at tips, glabrous; pedicels 2–3 mm. long.

Stream banks, bottom lands, and along ditches, Lower Sonoran and Upper Sonoran Zones, upper Sacramento Valley, California, to northern Baja California, southern Nevada, and northern Sonora. Extremely variable in the size and number of serrations of the leaves.

***3a.* Populus fremontii** var. **pubescens** Sarg., Bot. Gaz. **67:** 213, 1919

Differing from var. *fremontii* only in having pubescent twigs, petioles, and basal part of leaves on undersurface, particularly along midrib. With the species in the southern part of its range. Distinguishable from *P. macdougalii* by its more yellow-green leaves and larger disk of pistillate flowers.

**4. Populus arizonica** Sarg., Bot. Gaz. **67:** 210, 1919
*Populus arizonica* var. *jonesii* Sarg., op. cit. 211.
*Populus fremontii* var. *arizonica* Jepson, Man. Fl. Pl. Calif. 268, 1923.

Large tree to 25 m. high with pale gray to nearly white, deeply fissured bark and slender, pale green, pubescent branchlets soon becoming glabrous and light yellowish brown; leaves rhombic to deltoid, acute to short-acuminate, broadly cuneate, truncate, or subcordate at base, 5–10 cm. wide, 4–10 cm. long, crenate-serrate except at base and toward apex, puberulent when young, soon glabrous except ciliate-tomentose margins, coriaceous, yellow-green, lustrous; petioles terete or nearly so, 3–8 cm. long, puberulent at first, soon glabrous; catkins dense, staminate 2.5–3.5 cm. long, pistillate 3–6 cm. long at anthesis, elongating in fruit; pistillate disks cup-shaped, 5–6 mm. in diameter, entire or nearly so; capsule 4–5 mm. long, 2–4-valved, valves spreading; pedicels stout, 2.5–3 mm. long, glabrous.

Along streams in the mountains and foothills, Lower Sonoran Zone to lower parts of Transition Zone, southern Arizona and New Mexico and adjacent northern Mexico; northern Baja California.

**5. Populus brandegeei** Schneider, Handb. Laubholzk. **1:** 23, 1904, var. **brandegeei**
*Populus monticola* Brandegee, Zoe **1:** 273, 1890. Not *P. monticola* Mert. ex Loud., 1844.

A tree to 25 m. high with gray, somewhat roughened bark and rather stout, dark gray-brown branchlets, the latter densely tomentose in youth; leaves elliptic to ovate, 2–3.5 cm. wide, 3–7 cm. long, densely silky-pubescent on both surfaces, cuneate at base, acute at apex, irregularly rounded-crenate; petioles terete, 3–25 mm. long, densely tomentose; catkins short, dense, 2–4 cm. long; scales on staminate catkins orbicular, 1–2 mm. in diameter, ciliolate-pilose on margins, nearly glabrous on back; disk of pistillate flowers flat or nearly so, 2–3 mm. broad; capsules narrowly ovoid, 1–2 mm. wide, 4–5 mm. long, densely white silky-tomentose, 2–3-valved, on pedicels about 1 mm. long, closely crowded, ascending on rachis.

Along streams and arroyos, Lower Sonoran and Tropical Zones, Sierra de San Francisquito, in the Cape Region, Baja California.

**5*a*. Populus brandegeei** var. **glabra** Wiggins, Contr. Dudley Herb. **4:** 16, 1950

Petioles to 5 cm. long, glabrate; leaf blades broadly ovate, rounded to subcordate at base, glabrate, lustrous above, coriaceous, veins more prominent beneath; catkins unknown.

Along streams in the Sierra de la Giganta back of Puerto Escondido and in the Sierra de la Laguna, Cape Region, Baja California.

## 2. SALIX L., Sp. Pl. 1015, 1753

Creeping shrubs to large trees with single bud scales lined with a thin, adherent membrane, mostly narrow and short-petiolate leaves and deciduous or persistent stipules, these minute to large and conspicuous or less commonly completely absent. Flowers dioecious, in sessile to pedunculate aments, erect or spreading, not drooping, appearing before, with, or after the leaves. Flowers consisting of a single pistil or of a cluster of stamens, each subtended by a small bract and 1–2 small glands. Stamens 1–10, commonly 2 or 5, filaments distinct or somewhat united toward base, often villous. Ovary sessile or short-pedicellate, glabrous or pubescent, 2-valved; style elongate or none, entire or bifid; stigmas short or slightly elongated, entire to 2-lobed. Fruit a 2-valved capsule, dehiscing from apex downward. Seeds minute.

Scales blackish, persistent; leaf blades narrowly but distinctly revolute .... 1. *S. lasiolepis*
Scales yellowish, deciduous; leaf blades not revolute:
Leaf blades closely and finely serrulate; stamens 3–6:
Capsules pilose when young, glabrate in age; pedicels slender, 2.5–4 mm. long; leaves same color on both surfaces .................................... 2. *S. gooddingii*
Capsules glabrous; pedicels stout, 1–2 mm. long; leaves more or less glaucous beneath:
Capsules broadly ovate, often nearly as broad as long; leaves glabrous or sparsely pilose in youth; catkins dense, axis hidden by the flowers:
Leaves lanceolate, rounded or obtuse at the base; axis of pistillate catkins nearly glabrous ........................... 3. *S. bonplandiana* var. *bonplandiana*
Leaves linear-lanceolate, cuneate at base; axis of pistillate catkins densely pilose .... .................................... 3a. *S. bonplandiana* var. *toumeyi*
Capsules narrowly ovate, 2 to 4 times as long as broad; leaves densely silky-strigose on undersurface in youth; catkins lax, axis usually visible at anthesis:
Petioles, midrib of leaves, and young branchlets glabrate ...................... ..................................... 4. *S. laevigata* var. *laevigata*
Petioles, base of midrib, and young branchlets pilose .. 4a. *S. laevigata* var. *araquipa*
Leaf blades remotely denticulate or subentire; stamens 2:
Anthers oblong, twice as long as broad; leaves 4–12 cm. long, remotely denticulate; capsules glabrous or sparsely villous; male catkins 2–5 cm. long:
Mature leaves green, nearly glabrous; pistils glabrous; stigmas 0.5–0.8 mm. long .... .................................................... 5. *S. exigua*
Mature leaves silvery, densely tomentose; pistils sparsely villous; stigmas 0.8–1.2 mm. long ................................ 6. *S. hindsiana* var. *leucodendroides*
Anthers globose, as broad as long; leaves 1–3.5 cm. long, remotely denticulate to entire; male catkins 0.5–1.5 cm. long .............................. 7. *S. taxifolia*

**1. Salix lasiolepis** Benth., Pl. Hartw. 335, 1857

Shrub or tree 2–12 m. high with brownish black or sometimes yellowish, usually densely pubescent branchlets and gray, smooth bark on slender trunk; leaves oblanceolate, 1–2.2 cm. wide, 5–10 cm. long, cuneate at base, rounded to acute at apex, margin entire or essentially so, slightly revolute, usually densely tomentulose beneath when mature, glaucous and sparsely puberulent when young; aments precocious, 3–7 cm. long; scales broadly obovate, dark brown to nearly black, densely villous-tomentose; stamens 2, filaments united at base; pistils glabrous, style branches 0.5 mm. long; pedicels 0.5–1.2 mm. long; capsules glabrous, 4–5.5 mm. long.

Along stream banks, Lower Sonoran Zone to Transition Zone, Washington and Idaho to Nevada, New Mexico, and northern Baja California, reaching the margin of the Sonoran Desert near Rosario, Baja California.

**2. Salix gooddingii** Ball, Bot. Gaz. **40**: 376, *pl. 12, figs. 1–2,* 1905

*Salix nigra vallicola* Dudley, in Abrams, Fl. Los Ang. 100, 1904.
*Salix vallicola* Britt., in Britt. & Shafer, N. Amer. Trees 134, *fig. 141,* 1908.

A tree to 12 m. high with fissured bark and yellowish twigs densely pubescent when young; leaves linear-lanceolate, 6–15 mm. wide, 6–12 cm. long, acute to acuminate at apex, acute at base, finely glandular-serrulate, puberulent when young, eventually glabrous and green on both surfaces, prominently veined; petioles 3–10 mm. long, puberulent to glabrous; aments 4–8 cm. long, lax when fully open, on leafy lateral branches 2–4 cm. long; scales yellow, densely villous, staminate bearing 4–5 stamens; pistils sparingly to densely pilose at anthesis, later glabrate; pedicels 2.5–4 mm. long; capsules 5–7 mm. long at dehiscence.

Stream banks, Lower and Upper Sonoran Zones, Tehama County, California, to southern Nevada, Arizona, northern Sonora, and northern Baja California. Abundant along Colorado River from Needles to Yuma.

**3. Salix bonplandiana** H. B. K., Nov. Gen. & Sp. **2:** 20, *pls. 101–102,* 1817, var. **bonplandiana**

*Salix pallida* H. B. K., Nov. Gen. & Sp. **2:** 20, 1817.

A tree to 15 m. high with irregularly furrowed bark, a trunk 20–45 cm. in diameter, and slender, somewhat drooping, reddish branchlets; petioles 6–15 mm. long, glabrous; stipules ovate, minute, glabrous; leaf blades lanceolate, rounded at base, acute at apex, 8–22 mm. wide, 3–12 cm. long, finely serrulate, green above, glaucous beneath, sparsely pilose to glabrous in youth, glabrate at maturity; aments 1.5–4 cm. long, dense, subsessile or on short peduncles 5 mm. long; scales yellow, broadly ovate, acute to obtuse, villous; staminate aments 5–7 mm. in diameter, pistillate 1 cm. in diameter or less at anthesis; stamens 3–5; anthers nearly globose; pistils broadly ovoid, 2–3 mm. long at dehiscence, 1–2.5 mm. wide, glabrous; rachis of pistillate ament glabrous or nearly so.

Stream banks and about desert springs, Lower Sonoran and Tropical Zones, central Baja California (Bahía de los Angeles) and southern Sonora southward throughout Mexico into Guatemala. Feb.–March.

**3*a*. Salix bonplandiana** var. **toumeyi** (Britt.) Schneider, Bot. Gaz. **65:** 20, 1918

*Salix toumeyi* Britt., in Britt. & Shafer, N. Amer. Trees 187, *fig. 145,* 1908.

Leaves linear-lanceolate, 1–1.5 cm. wide, acute at base, acuminate at apex; pistillate aments with puberulent rachises.

Southern Arizona and New Mexico southward into Chihuahua, Sonora, and northern Baja California.

**4. Salix laevigata** Bebb, Amer. Nat. **8:** 202, 1874, var. **laevigata**

A round-crowned tree 5–15 m. high with irregularly fissured bark and yellowish to reddish brown branchlets; stipules rounded, 1 mm. long or less, often broader than long, somewhat glandular on inner surface; petioles stout, 4–10 mm. long, glabrous or with some hairs in groove on upper side; leaf blades oblanceolate to obovate, densely silky-pubescent and cuspidate when young, becoming rounded to subcordate at base, acute to short-acuminate at apex, 1.5–3 cm. wide, 6–15 cm. long, shallowly serrulate or sometimes subentire, coriaceous, glabrous and shining above, densely pubescent to glabrate and glaucous beneath; aments 3–10 cm. long, on leafy peduncles 1–3 cm. long; staminate aments somewhat lax, densely pubescent; scales yellowish, pubescent; stamens 4–6; pistils glabrous, on pedicels 1.5–2 mm. long; capsules ovate, 3.5–5 mm. long.

Along streams, mostly Upper Sonoran and Transition Zones, Siskiyou and Mendocino Counties, California, to southern Nevada, western Arizona, and northern Baja California. Entering the Sonoran Desert in eastern San Diego County, California.

**4*a*. Salix laevigata** var. **araquipa** (Jepson) Ball, Bot. Gaz. **72:** 234, 1921

*Salix laevigata* forma *araquipa* Jepson, Fl. Calif. **1:** 339, 1909.

The young twigs, petioles, and basal parts of the midrib beneath densely pilose.

Abundant in southern California, western Arizona, Sonora and northern Baja California.

**5. Salix exigua** Nutt., N. Amer. Sylva **1:** 75, 1842

*Salix longifolia* var. *exigua* Bebb, in S. Wats., Bot. Calif. **2:** 85, 1880.
*Salix fluviatilis* var. *exigua* Sarg., Silva **9:** 124, 1896. (In part.)

Grayish shrub 2–4 m. high with numerous slender, erect or ascending, reddish brown branches; leaves linear, acute at both ends, 2–10 mm. wide, 5–12 cm. long, remotely denticulate or sometimes entire, silky-canescent with fine silvery hairs when young, usually becoming green and nearly glabrate in maturity; petioles 4–10 mm. long; peduncles 1–7 cm. long, leafy at base; aments dense, staminate 2–4 cm. long, pistillate 3–6 cm. long; scales oblanceolate, acute, white-pilose, often glabrate after anthesis; stamens 2; capsules narrowly lanceolate, subsessile, 4–6 mm. long, sparsely puberulent when young, glabrate at maturity.

Along streams and washes, Sonoran and Transition Zones, from the Fraser River in British Columbia to Alberta, south on east side of the Cascade Mountains and the Sierra Nevada to northern Baja California, Arizona, New Mexico, and northern Chihuahua.

**6. Salix hindsiana** Benth. var. **leucodendroides** (Rowlee) Ball, in Abrams, Ill. Fl. Pacif. States **1:** 491, 1923

*Salix macrostachya* var. *leucodendroides* Rowlee, Bull. Torrey Club **27:** 250, 1900.
*Salix sessilifolia* var. *leucodendroides* Schneider, Bot. Gaz. **65:** 26, 1918.
(For full synonymy and discussion of taxonomic history see Madroño **6:** 232–236, 1942.)

A shrub 2–5 m. high with gray, furrowed bark and tomentose, erect young twigs; leaves linear to linear-lanceolate, 3–8 mm. wide, 4–10 cm. long, acute at both ends, remotely and finely denticulate, villous and somewhat silky on both surfaces or sometimes glabrate above, midrib prominent beneath; petioles 1–6 mm. long, stout; catkins 2–4.5 cm. long, dense, on sparingly leafy peduncles 5–20 mm. long; scales yellowish, densely pilose or rounded tips sparsely pilose to glabrate; stamens 2, filaments hairy; pistils villous-tomentose at anthesis; stigmas about 1 mm. long; capsules 5–6 mm. long, sparsely villous to glabrate at dehiscence.

Sandy washes, ditch banks, and sand bars, Lower and Upper Sonoran Zones, Monterey County, California, south into northern Baja California.

**7. Salix taxifolia** H. B. K., Nov. Gen. & Sp. **2:** 18, 1817

Shrub or tree, reported to attain height of 18 m., with dark reddish branches, seasonal twigs densely and finely silky-tomentulose; leaves linear to linear-lanceolate, 1–3 mm. wide, 1–3.5 cm. long, subsessile, acute at both ends, finely and remotely denticulate to entire, finely cinereous-silky-tomentulose in youth, becoming glabrate and slightly cinereous in age; aments short, 8–15 mm. long, on leafy peduncles to 3 cm. long, dense; scales rounded to abruptly apiculate at apex, villous; stamens 2, anthers globose, about 0.3 mm. long; pistils lance-ovate, cinereous-villous at anthesis, stigma lobes short, 1 mm. long or less; mature capsules 4–5 mm. long, sparingly villous, tips of valves usually recurved at dehiscence.

Mostly well above the Sonoran Desert and possibly not entering it; southeastern Arizona to western Texas and in southern Baja California.

## Family 23. **JUGLANDACEAE.** Walnut Family

Trees or shrubs with alternate, pinnately compound, estipulate leaves, rough bark, and hard, dark-colored wood. Flowers monoecious, bracteolate, staminate in drooping catkins, pistillate solitary or in few-flowered clusters on peduncles of the season's twigs. Staminate flowers bearing 3–many stamens; calyx irregularly lobed and adnate to the bract or wanting. Pistillate flowers bracteate and usually 2-bracteolate, with a 3–5-lobed calyx. Petals present or absent. Ovary inferior, 1-celled; styles 2. Fruit a drupe with a somewhat fleshy exocarp and a hard, rugose or variously sculptured endocarp. Seed 2–4-lobed; cotyledons 2, fleshy and oily, irregularly corrugated.

### 1. **JUGLANS** L., Sp. Pl. 997, 1753

Characters of the family. Petals 4 in pistillate flowers.

**1. Juglans major** (Torr.) Heller, Muhlenbergia **1:** 50, 1904
*Juglans rupestris* var. *major* Torr., in Sitgr., Rep. 171, *pl. 16,* 1853.

A tree 10–20 m. high with a trunk 4–8 dm. in diameter, dark gray to nearly black, deeply furrowed bark, and finely rusty-tomentulose seasonal twigs; leaves 1–3 dm. long; petioles 2–8 cm. long, rusty- or tawny-tomentulose; leaflets 9–15, lanceolate, 1–3 cm. wide, 3–10 cm. long, asymmetrically rounded to subcuneate at base, acute to acuminate at apex, serrate on margins, puberulent with simple and fine stellate hairs when young, often nearly glabrate except on veins and midrib in age; staminate catkins slender, lax, 3–5 cm. long, finely but closely puberulent; anthers 1.5–2 mm. long; peduncles 1–5 cm. long, bearing 1–3 pistillate flowers 1–2 cm. apart; pistillate flowers pubescent, ovoid, yellowish or yellow-brown, with darker, plumose stigmas 5–8 mm. long; fruit globose, 2–2.5 (or 3.5) cm. in diameter; exocarp fibrous, thin-skinned; endocarp longitudinally rugose, ridges rounded.

Along streams and in canyons, mostly in the Upper Sonoran Zone, central Arizona into New Mexico, southward into northern Chihuahua and Sonora.

## Family 24. **FAGACEAE.** Oak Family

Trees and shrubs with deciduous or evergreen, alternate, petiolate leaves and small, deciduous stipules. Flowers monoecious. Petals wanting. Staminate flowers in pendulous or erect aments or in pedunculate heads with 4–7-lobed calyx, stamens 4–20; pistillate flowers with a 4–8-lobed calyx, in small clusters or solitary, subtended by a burlike or cuplike involucre. Ovules 1 or 2 in each cell, pendulous, usually only 1 ovule in each ovary developing. Styles linear. Fruit a 1-seeded nut.

### 1. **QUERCUS** L., Sp. Pl. 994, 1753

Characters of the family.

Staminate aments pendulous. Calyx of staminate flowers usually 6-lobed; stamens 5–12. Fruit an acorn consisting of a 1-seeded coriaceous nut enclosed at the base by a cup consisting of imbricated, more or less united scales or bracts.

Acorns maturing second year; nut tomentose inside:
  Leaves lanceolate, shallowly toothed or entire, margins not undulate, smooth and glossy on both surfaces; scales of cup rounded at apex ........................ 1. *Q. emoryi*
  Leaves broadly elliptic to rounded, pungently toothed, margins undulate, glaucous and dull beneath; scales of cup truncate and abruptly apiculate at apex ......... 2. *Q. palmeri*
Acorns maturing first year; nut glabrous inside:
  Leaves oblong, glaucous above, entire; a tree 10–15 m. high ........... 3. *Q. oblongifolia*

Leaves elliptic, ovate, obovate, or rounded, green or gray-green but not glaucous above, pungently toothed; shrubs 1–6 m. high:
  Acorns broadly ovate, tapering gradually to apex; leaf blades more or less decurrent on petiole; dentation of leaves irregular, teeth acute .................. 4. *Q. dumosa*
  Acorns oblong to narrowly ovate or subcylindric, tapering abruptly to apex; leaf blades not decurrent on petiole, usually subtruncate to subcordate; dentation of leaves regular, teeth spinose-tipped:
    Leaf blades stellate-puberulent, especially beneath, with golden to grayish hairs; acorns 18–25 mm. long, 6–9 mm. in diameter ....................... 5. *Q. turbinella*
    Leaf blades waxy-glaucous beneath, nearly or completely glabrous on both surfaces; acorns 12–15 mm. long, 5–7 mm. in diameter .................. 6. *Q. ajoensis*

## 1. **Quercus emoryi** Torr., in Emory, Notes Mil. Rec. 151, 1848

*Quercus hastata* Liebm., Overs. Danske Vid. Selsk. Forh. **1854:** 171, 1854.

Small deciduous tree with finely puberulent, slender, red twigs that become glabrous or nearly so, and slender, glabrescent, brownish buds 3–8 mm. long; leaves lanceolate, 1–2 cm. wide, 3–8 cm. long, elliptic to lanceolate, truncate to subcordate at base, mucronately acute at apex, scurfy-glabrescent, rigid, glossy, repandly few-toothed to entire; acorn narrowly ellipsoid, 1.5–2 cm. long, one-third included, cup with blunt, thin, closely appressed-puberulent scales.

Dry hillsides and lower mountain slopes, chiefly Upper Sonoran Zone, from the Santa Catalina Mountains, Arizona, east into New Mexico and south into northern Sonora.

## 2. **Quercus palmeri** Engelm., Trans. Acad. St. Louis **3:** 393, 1877

*Quercus dunnii* Kell., Pacif. Rural Press **17:** 371 (June 7), 1879.
*Quercus chrysolepis* var. *palmeri* Engelm., in S. Wats., Bot. Calif. **2:** 97, 1880.

An evergreen shrub to 6 m. high with rigid, divaricate, slender, minutely scurfy twigs and dark reddish gray bark; leaves 1–2 cm. wide, 1.5–3 cm. long, broadly ovate to rounded, minutely golden-scurfy beneath in youth, glabrescent, coriaceous, acute to rounded at apex, subcordate at base, crisped, undulate, or folded, coarsely and pungently dentate, light gray-green; acorns sessile or on stout peduncles 2–10 mm. long, ovoid-oblong, acute, 12–15 mm. in diameter, 2–3 cm. long, set in a shallow, subturbinate, woody, densely tomentulose cup with closely appressed, truncate, abruptly apiculate, more or less coalesced scales.

Eastern slopes of the mountains along the western edge of the Colorado Desert from southern Riverside County, California, into northern Baja California. Common in the Sierra Pinal and Sierra San Pedro Mártir, Baja California.

## 3. **Quercus oblongifolia** Torr., in Sitgr., Rep. 173, 1853

A moderate-sized tree with a rounded crown, slender, glabrescent, somewhat pruinose twigs, and fissured, dark gray bark; leaves oblong, 1–2.5 cm. wide, 3–7 cm. long, glabrous, pruinose, rounded at both ends or subcordate at base, finely but prominently reticulate-veined beneath, margins often slightly revolute, entire or sometimes shallowly few-toothed; petioles stoutish, 3–5 mm. long; acorns ovoid-conic, 10–15 mm. in diameter, 15–20 mm. long, abruptly apiculate at somewhat rounded apex, about one-third included in hemispherical cup; scales of cup ovate, acute, bluntly keeled, closely appressed, finely tomentose with short, golden-brown hairs.

Lower mountain canyons and foothills, mainly Upper Sonoran Zone, Santa Catalina Mountains, Arizona, east into New Mexico and south into northern Chihuahua and Sonora.

**Quercus arizonica** Sarg., Gard. & Forest **8:** 92, 1895, a tree up to 18 m. tall, and **Quercus hypoleucoides** A. Camus, Mus. Hist. Nat. Bull. II, **4:** 124, 1932, with distinctly glaucous leaves, occur in the foothills and margins of mesas in Arizona and extend southward into Mexico but are not normally inhabitants of the Sonoran Desert. They may occasionally occur as sporadic stragglers along waterways in the upper margins of the desert.

### 4. Quercus dumosa Nutt., N. Amer. Sylva 1: 7, 1842

*Quercus acutidens* Torr., Bot. Mex. Bound. 207, *pl. 51,* 1859.

(Several forms and varieties proposed by Trelease are included in this polymorphic species, but they are so inconstant that recognition or citation seems useless.)

A shrub 1–3 m. high with stout, rigid, divaricate branchlets, rusty-tomentose twigs, and light gray-brown, scaly bark; leaves evergreen, elliptic to roundish, 15–25 mm. long, entire, sinuately lobed or spinose-serrate, coriaceous, stellate-puberulent and pale beneath, dark green and glossy above; acorn oval, 1–3 cm. long, 8–15 mm. in diameter, rounded to acute at apex, about one-third enclosed in deep, light brown cup; scales ovate, rusty-tomentulose, united and tuberculate toward base of cup.

Foothills of the Coast Ranges and Sierra Nevada from Lake and Tehama Counties, California, southward in the Upper Sonoran Zone to northern Baja California; barely touching the border of the Sonoran Desert in eastern San Diego County, California, and adjacent Baja California.

### 5. Quercus turbinella Greene, Ill. W. Amer. Oaks 1: 37, 1889

*Quercus dumosa* var. *turbinella* Jepson, Fl. Calif. **1:** 356, 1909.

*Quercus dumosa* var. *turbinella* f. *crispata* Trelease, Mem. Nat. Acad. **20:** 117, *pl. 208,* 1924.

A stiffly branching shrub or small tree 1–5 m. tall with rigid, divaricate branchlets, minutely tomentulose twigs, and light gray-brown, scaly bark; leaves evergreen, coriaceous, ovate, broadly oblong or roundish, 1–2 cm. wide, 1.5–2.5 cm. long, usually dentate, obtuse or roundish at apex, truncate, broadly rounded, or subcordate at base, the blade not decurrent on the petiole, the teeth spinulose-tipped, gray-green above, moderately to densely but finely stellate-puberulent with slightly yellowish or grayish hairs beneath; acorns on peduncles 3–30 mm. long; cup shallowly turbinate and enclosing the nut only at base, or more deeply cup-shaped; scales of cup thin and only slightly if at all tuberculate; nut slender, 6–9 mm. in diameter, 20–25 mm. long, often nearly cylindrical and tapering abruptly to apex.

Arid mountain slopes, Lower Sonoran Zone and into lower edges of Upper Sonoran Zone on dry ridges, eastern slopes of Sierra San Pedro Mártir, Baja California, and in isolated ranges in eastern California, southern Nevada, western Arizona, and eastward to Utah.

### 6. Quercus ajoensis C. H. Muller, Madroño 12: 140, *fig. 1 (in part)*, 1954

Shrub or small tree 3–8 m. tall, branching from near the ground, with thin, gray, scaly or slightly furrowed bark; twigs stellate-puberulent, pubescence persistent or twigs soon glabrate; petioles slender, 2–4 mm. long, stellate-puberulent, often somewhat reddish; leaf blades coriaceous, ovate or rarely lanceolate, 5–20 (rarely to 30) mm. wide, 10–35 (or sometimes to 50) mm. long, cordate to rounded at base, acute to obtuse at apex (excluding the aristate tooth), with 5 or 6 attenuate, spinose-aristate teeth on each side, glaucous-green above, waxy glaucous beneath, glabrous or sparsely stellate-puberulent along midrib beneath, margins flat or slightly crisped between teeth; staminate catkins 2–4 cm. long, their peduncles tomentulose; pistillate peduncles 1.5–7 cm. long, nearly glabrous; fruits maturing in one season, solitary or in pairs; acorn cups thin, 6–8 mm. in diameter, 3–4 mm. high, scales appressed, glabrous at tips, tomentulose at base, not strongly tuberculate; acorns ovoid to oblong, 12–15 mm. long, 5–7 mm. in diameter, abruptly pointed, puberulent at apex around persistent styles, otherwise glabrous.

North-facing canyon slopes and along intermittent water courses, Lower and Upper Sonoran Zones, Ajo Mountains, Arizona.

## Family 25. **ULMACEAE.** Elm Family

Trees or shrubs with watery juice, and terete or slightly angled branchlets arising from lateral, scaly buds. Leaves alternate, simple, serrate or entire, petiolate, the blade asymmetrically oblique at base, deciduous; stipules papery, caducous. Flowers perfect, polygamomonoecious, or monoecious, clustered or pistillate solitary. Calyx 4–9-lobed or -parted. Petals none. Stamens 4–6; style sessile, stigmas 2; ovary 1-celled, 1-ovuled. Fruit a papery samara, a nut, or a small drupe. Seed flat to angular-subglobose with little or no endosperm.

### 1. **CELTIS** L., Sp. Pl. 1043, 1753

Trees or shrubs with thin, smooth or slightly fissured bark and slender, arcuate, spreading or sometimes scandent, armed or unarmed branches. Leaves deciduous or subpersistent, entire or dentate, thin or coriaceous, stipules scarious. Flowers small, polygamomonoecious, appearing with leaves on new wood, pedicellate; staminate cymose or fascicled, pistillate solitary or in axillary few-flowered clusters. Calyx deeply 4–5-lobed, greenish yellow, scarious, deciduous. Ovary ovoid, glabrous or puberulent, with a 2-lobed, subsessile style. Fruit globose or ovoid, with thin flesh and a smooth or rugulose nutlet.

Plants armed with spines; branchlets rigidly divaricate; leaves toothed:
- Spines stout, recurved; fruit sparsely pilose; leaves 3–5 cm. wide, finely serrate or entire ........ 1. *C. iguanaea*
- Spines slender, straight; fruit glabrous; leaves 6–20 mm. wide, coarsely toothed ........ 2. *C. pallida*

Plants unarmed; branchlets not conspicuously divaricate; leaves entire:
- Leaves thick, scabrous above, heavily and conspicuously reticulate-veined beneath; pedicels one and one-half to two and one-half times as long as petioles ....... 3. *C. reticulata*
- Leaves thin, smooth or faintly scaberulous above, lightly reticulate-venulose beneath; pedicels shorter to slightly longer than petioles ............. 4. *C. laevigata* var. *brevipes*

#### 1. **Celtis iguanaea** (Jacq.) Sarg., Silva **7**: 64, 1895

*Rhamnus iguanaea* Jacq., Enum. Pl. Carib. 16, 1760.
*Celtis platycaulis* Greenm., Proc. Amer. Acad. **39**: 78, 1903.
*Momisia iguanaea* Rose & Standl., Contr. U.S. Nat. Herb. **16**: 8, 1912.

Shrub or small tree 2–13 m. high with slender, arcuate, somewhat angled, densely puberulent, often scandent branchlets armed with stout, recurved spines 7–14 mm. long; leaves oval to ovate, 3–5 cm. wide, 3.5–7 cm. long, moderately coriaceous, scabrous above, soft-puberulent and moderately venulose beneath, margins finely serrate to entire, slightly revolute; petioles stout, 4–6 mm. long, densely puberulent; flowers small, yellowish green, the calyx lobes finely puberulent; fruit yellowish, 7–12 mm. long, sparsely short-pilose, on pedicels 5–8 mm. long; seed pale stramineous, irregularly but prominently ridged, subalate along two angles.

Possibly not occurring in the Sonoran Desert, but recorded from near Alamos, Sonora, and to be expected where the desert meets the tropical scrub. Standley gives its range as "Chihuahua to Baja California and southward." March–April.

**2. Celtis pallida** Torr., Bot. Mex. Bound. 203, 1859

*Momisia pallida* Planch., in A. DC., Prod. **17:** 191, 1873.
*Celtis tala* var. *pallida* Planch., loc. cit.
*Celtis spinosa* var. *pallida* M. C. Johnston, Southw. Nat. **2:** 46, 1957.

A divaricately branched, dense, spiny shrub 1–6 m. high with some branches usually long and arcuate; branchlets terete or very slightly angled, finely appressed-canescent-puberulent, armed with straight, slender spines 3–12 mm. long; leaves oval, ovate, or elliptic, rounded to obtuse at base, obtuse to ovate at apex, 0.6–2 cm. wide, 1–3 cm. long, more or less scaberulous on both surfaces, margin coarsely serrate or rarely subentire; petioles 2–3 mm. long, puberulent; flowers in small cymes, greenish yellow; pedicels about equaling petioles; fruit yellow, orange, or sometimes red, 5–8 mm. long, nearly or quite as wide when mature, pulp fairly copious, edible; seed smooth or nearly so.

Along washes and on rocky hillsides and plains, Lower Sonoran Zone and occasionally entering the Upper Sonoran Zone, central Arizona to western Texas, Chihuahua, Oaxaca, and Baja California. March–April.

**3. Celtis reticulata** Torr., Ann. Lyc. N.Y. **2:** 247, 1828

*Celtis mississippiensis* var. *reticulata* Sarg., Silva **7:** 72, 1895. (In part.)

Tree to 15 m. high with short, crooked branchlets, reddish brown bark, and lanceolate-ovate leaves, 1–3 cm. wide, 3–7 cm. long, blades asymmetrically rounded at base, acute to attenuate at apex, thick, leathery, scaberulous on both surfaces, dark green above, lighter beneath, conspicuously and heavily reticulate-veined beneath, margins entire; petioles densely puberulent, 3–8 mm. long; fruiting pedicels slender, 7–12 mm. long, short-pilose; fruit red or orange, globose to obovate, 6–8 mm. in diameter, glabrous, becoming wrinkled in age; seed irregularly ridged and tuberculate.

Colorado, Texas, and Arizona to Coahuila and Baja California, but mostly at elevations above those of the Sonoran Desert. March–April.

**4. Celtis laevigata** Willd. var. **brevipes** (S. Wats.) Sarg., Bot. Gaz. **67:** 226, 1919

*Celtis brevipes* S. Wats., Proc. Amer. Acad. **14:** 297, 1879.

Small tree with slender, reddish brown, glabrous or glabrate branchlets; leaves ovate, 1.5–4 cm. wide, 3–7 cm. long, unequally rounded at base, acute to short-acuminate at apex, dark green and smooth or faintly scabrous above, lighter and glabrous except for tufts of fine hairs in the angles of lateral veins beneath, margins entire, blades stiff but thin and inconspicuously reticulate-venulose beneath; petioles 3–6 mm. long, spreading-puberulent; pedicels slender, pilosulous, equaling to one and one-half times as long as the subtending petioles; fruit globose to oblong, yellow, 5–6 mm. long, glabrous, pulp scanty; seed smooth or faintly ridged.

Canyons and along streams and irrigating ditches, Upper and Lower Sonoran Zones, southern Utah and Arizona into northern Sonora. Specimens with old leaves resemble *C. reticulata,* but the venation is less pronounced than in that species. March–April.

## Family 26. MORACEAE. Mulberry Family

Trees, shrubs, or rarely herbs with milky juice. Leaves alternate, simple, pinnately or palmately nerved; stipules 2, usually caducous and often long-sheathing, leaving a scar. Flowers small, often in heads or in hollow receptacles, monoecious or dioecious. Calyx lobes 4 or absent. Stamens same number as and opposite the sepals. Ovary inferior to superior, bicarpellate or 1 carpel abortive, usually 1-celled. Styles 2, filiform. Fruit a small achene, nut, or drupe. Seed with scanty or no endosperm; embryo curved or straight.

Flowers borne in spikes or aments; fruit a syncarp .......................... 1. *Morus*
Flowers borne inside a hollow receptacle; fruit of numerous achenes inside a fleshy synconium .......................................................... 2. *Ficus*

### 1. MORUS L., Sp. Pl. 986, 1753

Large or small trees with smooth to scaly bark. Leaves thin, deciduous, 3-nerved from the base, dentate or on young branches somewhat lobed. Staminate flowers in cylindrical spikes; sepals 4, distinct; stamens 4, two to three times as long as sepals. Pistillate spikes cylindrical; sepals 4, the 2 lateral ones larger than and somewhat overlapping the other 2, fleshy and very juicy at maturity; stigmas 2, short. Fruit of many drupelets, each surrounded by 4 juice-filled sepals.

**1. Morus microphylla** Buckl., Proc. Acad. Phila. **1862:** 8, 1862
*Morus confinis* Greene, Leaflets Bot. Obs. **2:** 119, 1910.
*Morus microphilyra* Greene, op. cit. 120.

Small tree 4–7.5 m. high with a compact crown, or often an open shrub with several stems from the base; wood light brown, hard, close-grained; bark gray, furrowed, scaly; leaves ovate, rounded to subcordate at base, acute to acuminate at apex, 1–4 cm. wide, 2–3 (or 5.5) cm. long, harshly scabrous on both surfaces, dark green above, somewhat paler beneath, serrate on margins, teeth apiculate, blades sometimes 3-lobed on vigorous shoots; petioles 3–6 mm. long, densely puberulent; stipules lance-linear, 3–5 mm. long, papery, caducous; fruit 1–1.5 cm. long, red to black when mature.

Along streams and canyons, Upper and upper edges of Lower Sonoran Zones, Arizona to western Texas, south into Sonora, Chihuahua, and Durango. April–May.

### 2. FICUS L., Sp. Pl. 1059, 1753

Shrubs, trees, or climbing plants with thick, milky juice and alternate, simple, undivided or 3–5-lobed leaves, blades entire or serrate, glabrous or pubescent. Stipules more or less sheathing, interpetiolar, caducous. Flowers monoecious or rarely dioecious, in hollow receptacles, these subtended by 2- or 3-lobed involucre. Staminate flowers sessile or nearly so, calyx of 2–6 subdistinct sepals or these wanting; stamens 1 or 2, infrequently 3, on short, stout filaments that are somewhat united when more than 1. Pistillate flowers short-pedicellate with fewer and narrower sepals than in the staminate flowers. Ovary 1-celled, sessile; style lateral; stigma clavate, peltate, or 2-lobed. Achenes embedded inside the hollow, fleshy receptacle.

Branchlets densely pilose; receptacle pyriform; leaves pubescent beneath .. 1. *F. palmeri*
Branchlets glabrous or short-puberulent; receptacle globose; leaves glabrous or with tufts of hairs along the costa beneath:

Leaf blades broad as long, orbicular-cordate, with tufts of hairs in axils of main veins; petioles nearly equaling to one and one-half times as long as blade .... 2. *F. petiolaris*
Leaf blades longer than broad, glabrous; petioles one-third as long as blade or less:
Receptacles 8–12 mm. in diameter; ostiole deeply impressed; peduncles 3–8 mm. long; stipules, peduncles, and fruits often minutely puberulent; leaf blades acute to long-acuminate at apex .......................................... 3. *F. padifolia*
Receptacles 12–20 mm. in diameter; ostiole not impressed; peduncles to 2.5 cm. long or more; stipules, peduncles, leaves, and fruits glabrous or nearly so; leaf blades obtuse to rounded at apex .......................................... 4. *F. brandegeei*

## 1. Ficus palmeri S. Wats., Proc. Amer. Acad. **24**: 77, 1889

Small to large tree with yellowish or whitish bark and densely white-pilose branchlets; stipules ovate-triangular, 1.5–2 cm. long, acuminate, thin, brown, glabrous; petioles slender, 1.5–5 cm. long, finely puberulent to short-pilose, longer hairs often intermingled with short; leaf blades broadly ovate-deltoid to oval, rarely subcordate, 4–11 cm. wide, 6–15 cm. long, rounded to acutish at apex, coriaceous, glaucous, densely pubescent on both surfaces in youth, usually glabrate above later; peduncles geminate, 5–25 mm. long, puberulent; involucre bilobate, sparsely puberulent, lobes broadly rounded, 3–4 mm. long; receptacles pyriform to turbinate-subglobose, 12–20 mm. in diameter, densely soft-puberulent, at length glabrate; ostiole prominent, closed by 3–5 broadly triangular, thin brown scales.

Canyon walls and rocky cliffs, Lower Sonoran and Tropical Zones, from Yubay southward to the Cape Region, Baja California, and on Tiburón Island, San Pedro Mártir Island, and Seal Island. Reported from "San Martin Island" by Standley, but this locality is a misprint for "San Pedro Mártir" in the Gulf of California; "San Martin" Island is off the west coast of the peninsula near San Quintín, whence no figs are known.

**Ficus radulina** S. Wats., Proc. Amer. Acad. **26**: 151, 1891, with sericeous-pilose branchlets but with solitary instead of geminate receptacles, trilobate involucres, and elliptic-oblong leaves acute at both ends, occurs at Alamos, Sonora, but is not known from the Sonoran Desert.

## 2. Ficus petiolaris H. B. K., Nov. Gen. & Sp. **2**: 49, 1817

*Urostigma petiolare* Miq., Lond. Jour. Bot. **6**: 527, 1847.
*Ficus jaliscana* S. Wats., Proc. Amer. Acad. **26**: 150, 1891.

Small to large tree with stout, grayish to yellowish, glabrous branches and slender petioles from one-half to one and one-half times as long as blades, these cordate-orbicular, 6–15 cm. broad and long, 5–7-nerved at base, broadly rounded and abruptly short-apiculate at apex, glabrous above, glaucous, tufts of long hairs in axils of larger veins, otherwise glabrous beneath; peduncles geminate, 7–20 mm. long, glabrous; involucre bilobate, lobes short, rounded, sparsely and minutely puberulent near margins; receptacles globose to depressed-globose, 10–15 mm. in diameter, somewhat mottled, densely villous in youth, later glabrate or puberulent; ostiole prominent, closed by several thin, triangular, brownish scales.

Canyon walls and stream banks, Lower and Upper Sonoran Zones to Tropical Zone, northern Sonora to Guerrero and Morelos; in the Sonoran Desert collected near Hermosillo, Ures, Baviácora, Sonora.

**Ficus nitida** Thunb., Diss. Ficus 10, 1786, with narrowly obovate to elliptic, glabrous, shining leaves, is frequently cultivated as an ornamental, and near San Bruno, Baja California, appears to be wild but probably was planted many years ago by the inhabitants of the mission.

### 3. Ficus padifolia H. B. K., Nov. Gen. & Sp. 2: 47, 1817

*Ficus fasciculata* S. Wats., Proc. Amer. Acad. **24:** 78, 1889. Not *F. fasciculata* King, 1888.
*Ficus sonorae* S. Wats., loc. cit.

Small to large tree with pale yellowish bark and slender, fulvous to brownish, glabrous or minutely puberulent branchlets; stipules narrowly lance-acuminate, 5–15 mm. long, brown, glabrous to minutely and sparsely puberulent outside; petioles 5–30 mm. long, glabrous or puberulent, shallowly canaliculate; leaf blades ovate, lance-elliptic to oblong, 1.5–4.5 cm. wide, 4–12 cm. long, rounded to obtuse and 3-nerved at base, acute to long-acuminate at apex, subcoriaceous, bright green, glabrous on both surfaces; peduncles slender, glabrous or sparsely puberulent, 3–8 mm. long; involucre deeply bilobate, lobes 3–4 mm. long, rounded to acute, spreading, mostly glabrous; receptacles globose, 8–12 mm. in diameter, mottled, glabrous to minutely puberulent; ostiole deeply impressed, surrounded by a conspicuous elevated annulus.

Lowlands, canyons, and stream banks, Lower Sonoran Zone to Tropical Zone, central Sonora to Yucatán and Panama. A variable species to which many names have been applied, growing over other trees as young plants and eventually killing the host tree in most instances. The elevated annulus about the ostiole is a striking feature separating this fig from others of the Sonoran Desert and immediately adjacent regions.

**Ficus cotinifolia** H. B. K., Nov. Gen. & Sp. **2:** 49, 1817, with sessile receptacles 6–12 mm. in diameter, and **Ficus goldmanii** Standley, Contr. U.S. Nat. Herb **20:** 32, 1917, with receptacles 2–2.5 cm. in diameter on geminate peduncles 10–13 mm. long, occur at Alamos, Sonora, but neither has been reported from the Sonoran Desert. Both these species have densely sericeous stipules that are short and comparatively broad.

### 4. Ficus brandegeei Standley, Contr. U.S. Nat. Herb. 20: 22, 1917

A broad-crowned tree to 10 m. tall; crown often 6–7 m. wide; bark creamy white on older trunk, this often to 1.5 m. in diameter; branchlets glabrous, scars from stipules of several seasons past distinctly visible; stipules ovate-triangular, 1.5–2.2 cm. long, acuminate, glabrous; petioles mostly 2–4 cm. long, glabrous; leaf blades broadly oval, 5–10 cm. wide, 7–12 cm. long, rounded or broadly obtuse at base, obtuse to rounded at apex, with 6–9 lateral veins on each side of midrib, completely glabrous, sublustrous; peduncles 2–3 cm. long, glabrous or minutely and sparsely puberulent, often bearing scattered, low-conical pustules; involucres of 2 essentially free, broadly ovate bracts 4–5 mm. long, obtuse at apex; receptacle globose, 12–20 mm. in diameter, glabrous, ostiole not impressed, surrounded by 3–5 deltoid bracts 1.5–2 mm. long.

On rocky canyon walls and slopes, Lower Sonoran and Tropical Zones, near Loreto and in the Cape Region, Baja California. Nearly ripe fruit collected in November.

## Family 27. URTICACEAE. Nettle Family

Annual or perennial herbs or some tropical species shrubs or trees with alternate or opposite, mostly stipulate leaves. Flowers small, greenish, monoecious or polygamous, in racemes or fascicles. Calyx 2–5-cleft or of distinct sepals. Petals wanting.

Stamens as many as and opposite the sepals, inflexed. Ovary superior, 1-celled; ovule solitary, erect or ascending. Fruit an achene.

Leaves opposite, stipulate; herbage with stinging hairs ........................ 1. *Urtica*
Leaves alternate, estipulate; herbage without stinging hairs .............. 2. *Parietaria*

## 1. URTICA L., Sp. Pl. 983, 1753

Annual or perennial herbs with stiff, stinging hairs and opposite, petiolate, dentate or serrate, stipulate leaves. Flowers clustered in axillary racemes or heads. Sepals or calyx segments 4, equal in staminate flowers, inner larger and enclosing achenes in Pistillate flowers. Stamens 4. Stigma sessile, tufted. Achenes somewhat flattened laterally, smooth, shining.

### 1. Urtica urens L., Sp. Pl. 984, 1753

Erect annual herb 1–5 dm. high with a simple or basally branched stem and herbage armed with stiff, stinging hairs; leaves broadly ovate to oblong-ovate, 0.5–2 cm. wide, 1–4 cm. long, 3–5-nerved, thin and nearly glabrous except for the stinging hairs, deeply serrate; petioles slender, 5–30 mm. long; stipules deltoid-ovate, acute, 1.5–4 mm. long; flower clusters dense, shorter than or about equalling petioles, both staminate and pistillate flowers in the same cluster, pistillate most numerous; achene 1.5–2 mm. long, equaled by dilated-margined, ovate-rounded sepals.

Occasionally naturalized along ditches and in gardens, Imperial Valley, California, San Quintín and near Rosario, Baja California.

## 2. PARIETARIA L., Sp. Pl. 1052, 1753

Annual or perennial herbs or some shrubs or trees in tropics with alternate, estipulate, entire, 3-nerved, petiolate leaves and without stinging hairs. Flowers perfect and pistillate, in axillary glomerules, subtended by leafy bracts. Perfect flowers with 4-parted calyces, the pistillate with tubular-ventricose calyces with 4 connivent lobes. Stigma spatulate, tufted. Achene ovoid, smooth and shining, surrounded by the dry calyx.

Involucre 2 to 3 times as long as the sepals; stems simple or sparingly branched; leaves ovate-lanceolate to lanceolate ....................................... 1. *P. pennsylvanica*
Involucre only slightly longer than the sepals; stems usually much branched; leaves broadly ovate to suborbicular ............................................ 2. *P. floridana*

### 1. Parietaria pennsylvanica Muhl. ex Willd., Sp. Pl. 4²: 955, 1805

*Parietaria occidentalis* Rydb., Bull. Torrey Club **39:** 306, 1912.

Erect, slender-stemmed annuals 1–4 dm. tall, simple or sparingly branched at base, glabrate or more commonly sparsely to moderately villous with weak, white hairs; leaves lanceolate to lance-oblong, cuneate to slightly rounded at base, acute to acuminate at the apex, puberulent to villosulous, blades 5–12 mm. wide, 2.5–7.5 cm. long, on slender petioles about one-half as long as blade; flowers in glomerules in most axils; bracts linear or lance-linear, 4–6 mm. long, sparsely pubescent; sepals lance-linear, about 2 mm. long, acute; achenes brownish, about 1 mm. long.

Along watercourses and in shade of trees, cliffs, and rocks, Lower Sonoran Zone to Transition and Carolinian Zones, Pima County, Arizona, to British Columbia, northern California, Maine, and northern Mexico. March–June.

**2. Parietaria floridana** Nutt., Gen. **2:** 208, 1818
*Parietaria obtusa* Rydb., in Small, Fl. SE. U.S. 359, 1903.

Annual herb with slender, prostrate to erect, basally branching, somewhat hispid stems 1–4 dm. long; leaves broadly ovate, 5–20 mm. long, rounded to abruptly subcuneate at base, acute to obtuse at apex, finely and sparsely to copiously hispid; petioles slender, frequently equaling blades, weakly hispid; flower clusters dense, axillary; bracts linear to linear-lanceolate, 2–3 mm. long, acute; sepals lanceolate, acute, sparingly ciliolate-margined, 1.5–2.5 mm. long; achene 1 mm. long, ovoid, olivaceous, shining.

On moist or shaded soil and among rocks, Upper and Lower Sonoran Zones, central California to Arizona, Sonora, and central Baja California; southeastern United States. Very common in rocky ravines and under shrubs throughout much of the northern half of the Sonoran Desert. Oct.–March.

## Family 28. LORANTHACEAE. Mistletoe Family

Small shrubs parasitic on the branches of other shrubs or trees, green and containing chlorophyll, or sometimes yellow or brown and devoid of chlorophyll. Leaves mostly opposite, estipulate, thick and leathery or reduced to scales. Flowers small or large, perfect, monoecious or dioecious, clustered or solitary. Petals none. Calyx tube 2–5-lobed, adnate to the ovary. Stamens inserted on the calyx lobes and usually equaling them. Ovary inferior, 1-celled, 1-seeded; style simple or none; stigma 1. Fruit a berry, translucent-white or colored. Seed in a glutinous testa.

Flowers small, less than 1 cm. long, dioecious:
  Calyx lobes 3; flowers not calyculate, embedded in axis of spike; mature berries 3–6 mm. in diameter, pearly white to roseate .......................... 1. *Phoradendron*
  Calyx lobes mostly 6, sometimes 5; flowers calyculate, not embedded in axis; mature berries 8–15 mm. in diameter, orange to red .......................... 2. *Struthanthus*
Flowers large, 3 cm. long or more, perfect:
  Leaves linear, terete; seeds with endosperm .......................... 3. *Phrygilanthus*
  Leaves with flat, elliptic, ovate, or lanceolate blades; seeds without endosperm .................................................................. 4. *Psitticanthus*

### 1. PHORADENDRON Nutt., Jour. Acad. Phila. II, 1: 185, 1848

Woody green plants with brittle, jointed stems and entire, foliaceous or reduced, scalelike, opposite leaves. Flowers embedded in a jointed rachis, 2–30 or more on each joint and usually several in the axil of each bract. Calyx 3–4-lobed, globose, bearing a sessile, transversely 2-celled anther at the base of each lobe in staminate flowers, adnate to the inferior ovary in pistillate flowers. Berry ovoid or globose, sessile, fleshy, 1-seeded.

Leaves reduced to short, thin scales; pistillate flowers 2 on each joint:
  Berries pearly pink:
    Fruiting spikes 5–12 mm. long, whorls of berries crowded .................................................................. 1. *P. californicum* var. *californicum*
    Fruiting spikes 20–30 mm. long, whorls of berries separated .................................................................. 1*a*. *P. californicum* var. *distans*
  Berries white-translucent .......................... 1*b*. *P. californicum* var. *leucocarpum*

Leaves foliaceous, not scalelike; pistillate flowers 6–30 or more on each joint:
Fruiting spikes 3–8 cm. long; leaves large, 4–8.5 cm. long (except in *P. riberense*):
Rachis of spikes densely pubescent:
Leaves 3–8 cm. long; joints of staminate inflorescences 20–30-flowered .............. ........................................................ 2. *P. longispicum*
Leaves 1.5–2.5 cm. long; joints of staminate inflorescences 40–100-flowered ........ ........................................................ 3. *P. riberense*
Rachis of spikes glabrate or sparsely puberulent ................. 4. *P. macrophyllum*
Fruiting spikes mostly 2 cm. long or shorter; leaves small, 1–3 (rarely 4) cm. long:
Berries glabrous except at base of sepals; leaves broadly obovate to suborbicular, densely tomentulose or villous:
Leaves suborbicular, abruptly short-petiolate, tomentulose; pubescence of minute, intricately branched, stellate hairs; calyx lobes of pistillate flowers pubescent base to apex ................................................ 5. *P. coryae*
Leaves obovate, cuneately narrowed to a petiole 5–10 mm. long, short-villous; pubescence of slender, simple hairs, with scattered tufts of same type of hairs closely crowded; calyx lobes pubescent only at base .................... 6. *P. villosum*
Berries sparsely short-villous throughout; leaves elliptic-obovate to broadly spatulate, sparsely and deciduously short-villous, appearing varnished in age ............ ........................................................ 7. *P. diguetianum*

**1. Phoradendron californicum** Nutt., Jour. Acad. Phila. II, **1:** 185, 1848, var. **californicum**

Plants with numerous slender, branching, terete stems forming dense masses of erect, spreading, or drooping branches 1–6 dm. long; internodes 1–3 cm. long, puberulent in youth, eventually glabrate and often reddish brown; leaves reduced to thin, slightly keeled, narrowly deltoid-ovate scales 1–2 mm. long, persistent; spikes axillary, mostly solitary but sometimes 2–3 at a node, 5–12 mm. long with 2–4 joints; berries reddish or pearly pink, glabrous, ovoid, 3–3.5 mm. in diameter, 4–5 mm. long, slightly umbonate owing to persistent, erect calyx lobes.

Parasitic on *Prosopis, Condalea, Olneya, Acacia,* and other angiospermous desert shrubs, from Inyo County, California, and northern Mohave County, Arizona, south in the Lower Sonoran Zone into southern Sonora and central Baja California.

**1*a*. Phoradendron californicum** var. **distans** Trelease, Monogr. Phoradendron 21, *pl. 13,* 1916

Fruiting spikes 20–30 mm. long, whorls of berries distinctly separated.

Occurring over about the same range as var. *californicum* but more abundant than the latter throughout the southern part of the range; Sonora, and Baja California.

**1*b*. Phoradendron californicum** var. **leucocarpum** Jepson, Man. Fl. Pl. Calif. 283, 1923

Berries white-translucent.

On *Prosopis,* Twentynine Palms and Andreas Canyon, California; to be expected elsewhere in the range of the species.

**2. Phoradendron longispicum** Trelease, Monogr. Phoradendron 39, *pls. 38–39,* 1916

Robust plant 3–8 dm. high with stout, yellowish green branches, puberulent, partially glabrate in age, internodes 3–5 cm. long; leaves broadly oblanceolate to elliptic-obovate or nearly orbicular, obtuse, 2–3.5 cm. wide, 3.5–8 cm. long, abruptly cuneately narrowed to a flattish petiole 5–12 mm. long, faintly but distinctly 3–5-veined; spikes solitary or clustered, 15–30 mm. long, elongating to 40–50 mm. in fruit, joints 3–5, densely puberulent, broadly clavate, about 12-flowered in pistillate spikes, 20–30-flowered in staminate; scales rounded, puberulent and ciliolate; berries white-translucent, subglobose, 3.5–4.5 mm. in diameter, glabrous except at base of calyx lobes.

Parasitic on *Populus, Salix, Juglans, Platanus,* and *Fraxinus,* northern Sacramento Valley, California, south into northern Baja California and Arizona.

**3. Phoradendron riberense** Wiggins, Contr. Dudley Herb. **3:** 66, *pl. 15, figs. 1–2,* 1940

Stems slender, terete, without cataphylls, 2–6 dm. long or more, puberulent with short, simple hairs in youth, soon glabrate and appearing somewhat varnished; internodes 2–6 cm. long; leaves obovate, obtuse or rounded at apex, cuneately narrowed to base, 8–15 mm. broad, 1.8–2.5 cm. long, short-villosulous in youth but soon glabrate and shining, petiole broad, 3–5 mm. long; staminate spikes 2–4 in an axil, slender, 4–7.5 cm. long, 6–9-jointed, 2.5–3 mm. in diameter, each joint nearly cylindrical, bearing 40–100 flowers or more; bracts connate, truncate, puberulent; peduncles 2 mm. long or less; rachis densely puberulent; staminate flowers closely crowded, 3–4-merous, puberulent, 0.6–0.8 mm. in diameter at anthesis; pistillate flowers and fruits unknown.

Collected on *Bursera,* plains between Ribera and Santiago, Baja California.

The leaves and pubescence are similar to those of *P. diguetianum,* from which species it differs in the long spikes, more robust habit, and longer internodes.

**4. Phoradendron macrophyllum** Cockerell, Amer. Nat. **34:** 293, 1900

*Phoradendron flavescens* var. *macrophyllum* Engelm., in Rothr., Wheeler Rep. 252, 1878.
*Phoradendron coloradense* Trelease, Monogr. Phoradendron 39, *pl. 37,* 1916.
*Phoradendron macrophyllum* var. *circulare* Trelease, op. cit. 38, *pl. 35.*
*Phoradendron macrophyllum* var. *jonesii* Trelease, loc. cit.
*Phoradendron serotinum* var. *macrophyllum* M. C. Johnston, Southw. Nat. **2:** 46, 1957.

Robust plant with terete, glabrate stems 2–8 dm. long, internodes 2–3 cm. long; leaves broadly elliptic to orbicular, obtuse or rounded, 2–4.5 cm. wide, 3–6 cm. long, on abruptly winged petioles 4–7 mm. long, minutely puberulent in youth, soon glabrate, distinctly 3–7-veined; spikes 1–3 in an axil, 15–30, or sometimes 50 mm. long in fruit, puberulent at first but soon glabrate except for ciliolate margins of scales, joints subglobose, about 6-flowered in pistillate, 18–36-flowered in staminate spikes; berries white-translucent, globose, glabrous, 4–5 mm. in diameter, rather crowded.

Mostly in Upper Sonoran and Transition Zones, on *Salix, Populus, Fraxinus, Celtis, Juglans,* etc., from vicinity of Yuma, Arizona, to New Mexico and northern Sonora.

**5. Phoradendron coryae** Trelease, Monogr. Phoradendron 43, *pls. 5, 10, 44,* 1916

*Phoradendron havardianum* Trelease, op. cit. 44, *pl. 45.*

Stems stout, 2–5 dm. long, densely but finely tomentulose with intricately branched stellate hairs; internodes 2–4.5 cm. long; leaves orbicular to broadly ovate or obovate, 1–2.5 cm. wide, 1–3.5 cm. long, rounded to obtuse at apex, abruptly short-petiolate, persistently tomentulose, obscurely to distinctly 3–5-veined; spikes 1–2 cm. long, 2–4-jointed, staminate with 18–36 mostly 3-merous flowers per joint, pistillate about 12-flowered per joint, rachis puberulent; peduncles 2–3 mm. long; berries subglobose, about 3 mm. in diameter, glabrous except at apex, calyx lobes puberulent from base to apex, erect or nearly so.

Mostly if not exclusively on *Quercus,* Arizona, New Mexico, and western Texas southward at least to southern Sonora. The leaf size, used by Trelease to separate *P. coryae* and *P. havardianum,* is inconstant, leaves from one extreme of his measurements for *havardianum* to the other given for *coryae* occurring on the same plant.

**6. Phoradendron villosum** Nutt., Jour. Acad. Phila. II, **1:** 185, 1848

*Phoradendron flavescens* var. *villosum* Engelm., in Rothr., Wheeler Rep. 252, 1878.

Plants stout, erect, spreading or sometimes partially pendent; stems 3–8 dm. long; internodes 2–3 cm. long, densely short-villous with simple hairs, these crowded into stellate-like scattered tufts on some specimens; leaves obovate to oblanceolate-obovate, 2–4 (rarely 5) cm. long, 1.5–2 cm. wide, cuneately narrowed to broad petioles 3–6 mm. long; spikes solitary to clustered, 10–15 mm. long, short-villous, 2–4-jointed; pistillate joints usually 6-flowered, staminate 12–18-flowered; berries white or pinkish white, 3.5–4 mm. in diameter, in crowded whorls, short-villous at apex and on basal parts of broadly deltoid calyx lobes.

Parasitic on various hosts, Oregon to northern Baja California.

**7. Phoradendron diguetianum** Van Tieghem, Bull. Mus. Hist. Nat. Paris **1:** 31, 1895

*Phoradendron eduardii* Trelease, Monogr. Phoradendron 46, *pl. 47,* 1916.
*Phoradendron globuliferum* Trelease, op. cit. 48, *pl. 51.*
*Phoradendron aureum* Trelease, op. cit. 49, *pl. 52.*
*Phoradendron brachyphyllum* Trelease, op. cit. 49, *pl. 53.*
*Phoradendron tumidum* Trelease, loc. cit.
*Phoradendron peninsulare* Trelease, op. cit. 50, *pl. 55.*
*Phoradendron saccatum* Trelease, loc. cit.

Stems erect or spreading, 2–5 dm. long, sparsely short-villous in youth but soon glabrate and somewhat varnished; internodes 1.5–6 cm. long; leaves narrowly obovate, elliptic, or oblong-spatulate, 5–10 mm. wide, 18–30 mm. long, rounded or obtuse at apex, cuneately narrowed basally to a broad petiole 3–5 mm. long or subsessile, sparsely puberulent with straight, simple hairs when young, soon glabrous and shining; spikes stout, 8–20 mm. long, joints 2–4, globose-clavate, on peduncles 1–3.5 mm. long, 12–15-flowered in pistillate, 18–30-flowered in staminate spikes; berries globose or ovoid, 4–5 mm. in diameter, white, sparsely short-villous.

Parasitic on *Jacquinia, Castela, Jatropha,* and other hosts, central Baja California southward to the Cape Region, on the islands of the Gulf of California, and in the vicinity of Guaymas, Sonora.

## 2. STRUTHANTHUS Mart., Flora 13: 102, 1830

Green, usually glabrous plants parasitic mostly upon dicotyledons. Branches terete, compressed, or narrowly winged-angled. Leaves opposite or sometimes alternate on some plants, well developed. Flowers dioecious, small, in spikes, racemes, or cymes, calyculate. Calyx 6-lobed. Fruit ellipsoid, white, greenish or (in ours) orange, the flesh firm.

**1. Struthanthus haenkeanus** (Presl) Standley, Contr. U.S. Nat. Herb. **20:** 212, 1919

*Spirostylis haenkeana* Presl ex Schult., Syst. Veg. **7:** 163, 1829.
*Loranthus spirostylis* DC., Prod. **4:** 315, 1830.

Branches glabrous, terete, slender, 0.5–2 m. long or more, drooping and sometimes twining; leaves linear, narrowly elliptic to lanceolate, straight or falcate, 6–15 mm. wide, 3.5–10 cm. long, cuneate and subsessile at base, acute at apex, yellowish green, glabrous; flowers in short, pedunculate, axillary racemes, calyculus shallowly cup-shaped, irregularly and shallowly toothed; calyx lobes linear, 2–4 mm. long, cream-colored; style twisted into a short spiral or sigmoidly bent near apex; fruit oblong, 10–13 mm. long, orange or red, drying black.

Parasitic on *Prosopis, Acacia, Taxodium, Celtis, Quercus,* etc., central Sonora southward to Oaxaca.

### 3. **PHRYGILANTHUS** Eichl., in Mart., Fl. Bras. **5**$^2$**:** 45, 1868

Glabrous plant parasitic on dicotyledonous hosts. Branches terete. Leaves opposite or alternate, terete or with flattened blades. Flowers large, showy, solitary or cymose, perfect. Calyculus shallow. Sepals 6. Anthers dorsifixed. Fruit oblong, dark reddish purple or black.

**1. Phrygilanthus sonorae** (S. Wats.) Standley, Contr. U.S. Nat. Herb. **20:** 212, 1919
*Loranthus sonorae* S. Wats., Proc. Amer. Acad. **24:** 73, 1889.

Glaucous plants with stout, stubby branches 1–4 dm. long; leaves terete, alternate or opposite on same plant, 2–3 mm. in diameter, 3.5–8 cm. long, abruptly acute at apex; flowers solitary or in few-flowered axillary cymes; calyculus shallow, 2–2.5 mm. in diameter, thin margin entire; calyx lobes 6 (or rarely 5), 2.8–4 cm. long, dark red, narrowly oblanceolate, acute at apex; stamens about equaling calyx lobes; fruit oblong, smooth, purplish to nearly black, 10–12 mm. long.

Parasitic on *Bursera* and *Cyrtocarpa edulis,* Sonora and Baja California from the vicinity of Bahía de los Angeles southward.

### 4. **PSITTICANTHUS** Mart., Flora **13:** 106, 1830

A green plant with well-developed leaves, parasitic mostly on dicotyledonous trees and shrubs. Flowers large and showy, perfect, in cymes or corymbs; calyx 6-cleft or rarely 4- or 5-cleft; calyculus well developed, entire or toothed. Stamens alternately large and small; filaments filiform; anthers linear. Fruit a globose or elliptic berry. Seed without endosperm.

**1. Psitticanthus calyculatus** (DC.) G. Don, Gen. Hist. Pl. **3:** 415, 1834
*Loranthus calyculatus* DC., Coll. Mém. **6:** *pl. 10,* 1830.

Stems coarse, terete or quadrangular and narrowly winged on the angles, 3–12 dm. long, glabrous; leaves lanceolate or sometimes ovate, mostly somewhat falcate, 1.2–5 cm. wide, 4.5–12 cm. long, cuneately narrowed to base, attenuately narrowed to an acute or less commonly an obtuse apex, glabrous, leathery; petioles 5–10 mm. long, narrowly winged; inflorescences axillary corymbs; pedicels 1–1.5 cm. long; calyculus thin, about 3 mm. in diameter, eñtire and truncate at apex; calyx bright red, 2.5–3.5 cm. long, of 6 distinct sepals, these linear; fruit black, ellipsoid, 12–15 mm. long.

Parasitic on a number of dicotyledonous hosts, southern Sonora (crossing of Río Yaqui near Cajeme, on *Populus dimorpha*) to Chiapas and Yucatán.

## Family 29. **SANTALACEAE.** Sandalwood Family

Herbs or low shrubs with entire, simple, alternate leaves; a root parasite. Flowers small, greenish white, perfect, in small axillary or terminal clusters. Corolla none.

Calyx 4–5-cleft, valvate in bud, lower part of the tube adherent to the ovary. Stamens as many as and opposite the calyx lobes, inserted on the margin of a fleshy disk. Ovary inferior, 1-celled; style 1; stigma capitate. Ovules 2–4, suspended from the apex of a free, central placenta. Fruit 1-seeded, indehiscent, nutlike. Embryo minute, straight; endosperm abundant.

### 1. **COMANDRA** Nutt., Gen. **1:** 157, 1818

Perennial herbs with rootstocks, striate stems, and almost sessile leaves. Flowers in cymes. Calyx campanulate or urceolate, 5-cleft. Disk 5-lobed. Anthers connected to calyx lobes both by a filament and by tufts of threadlike hairs. Fruit crowned by persistent calyx lobes.

**1. Comandra pallida** A. DC., Prod. **14:** 636, 1857

Stems several to many, 1–4 dm. high, glabrous, striate, leafy; leaves sessile or essentially so, lanceolate, 1–3 cm. long, acute to cuspidate, pale to glaucous, glabrous; flowers 4–5 mm. long and nearly as wide; pedicels slender, 1–2 mm. long; calyx lobes narrowly deltoid-lanceolate, 1–1.5 mm. wide, 2–3 mm. long, margins faintly granulose; style slender, about equaling calyx lobes; fruit ovoid to subglobose, 5–6 mm. long, glaucous, ventricose, persistent calyx tube about 0.5 mm. long.

Lower Sonoran Zone to Transition Zone, British Columbia to southern California, Arizona, Texas, and Kansas. Mostly above the Sonoran Desert, but occurring along its margins in the Santa Rita and Pinal Mountains, Arizona.

## Family 30. **OPILIACEAE.** Opilia Family

Trees, shrubs, or woody climbers with alternate, simple, estipulate leaves. Flowers regular, perfect or (in ours) dioecious. Calyx minute, 4–5-lobed. Petals 4–5, narrow, free or somewhat united at base. Stamens as many as petals and opposite them, free or united to base of petals, usually exserted; glands on the disk alternating with stamens. Ovary superior or partially immersed in the fleshy disk, 1-celled; style slender or stigma sessile. Ovule solitary. Fruit drupaceous, fleshy. Endosperm copious; embryo small.

### 1. **AGONANDRA** Miers, in Benth. & Hook., Gen. Pl. **1:** 349, 1862

Ours shrubby or a small tree, glabrous throughout. Branches slender, often pendulous; bark smooth. Leaves entire. Inflorescence axillary racemes. Stamens subtended by 4–5 small scales. Pistillate flowers apetalous, urceolate disk surrounding basal part of ovary.

**1. Agonandra racemosa** (DC.) Standley, Jour. Wash. Acad. **10:** 506, 1920
*Schaefferia racemosa* DC., Prod. **2:** 41, 1825.

Shrub or small tree 4–6 m. tall with wholly glabrous foliage and twigs; petioles 8–12 mm. long; leaf blades entire, lanceolate to broadly ovate or elliptic, 2.5–4 cm. wide, 4–7.5 cm. long, broadly cuneate to rounded at base, acute, acuminate, or obtuse but abruptly short-acuminate at apex, dark green above, pale and slightly glaucous beneath, midrib and lateral veins moderately conspicuous; staminate flowers 2–2.5 mm. long, greenish; fruits subglobose to ellipsoidal, 8–12 mm. long, apex slightly apiculate.

Rocky hillsides and canyons, Lower Sonoran Zone to Tropical Zone, southern Sonora to Guerrero. Fruiting in March.

## Family 31. **OLACACEAE.** Olax Family

Shrubs or small trees with alternate, entire, estipulate leaves, smooth, thin bark, and rigid, unarmed or sometimes spinose branches. Flowers small, perfect, in cymes or racemes; calyx 4–5-dentate; corolla 4–6-lobed, cylindrical or urceolate or lobes separate nearly to base. Stamens inserted on base of perianth tube. Style 1; stigma capitate or 2-lobed. A fleshy disk surrounding base of the style. Fruit a drupe.

Stamens twice as many as the perianth lobes; corolla lobes separate almost to base, densely hairy within ........ 1. *Ximenia*
Stamens as many as perianth lobes; corolla lobed to middle or less, lobes glabrous or sparingly pubescent within ........ 2. *Schoepfia*

### 1. **XIMENIA** L., Sp. Pl. 1193, 1753

Shrubs or small trees with leathery leaves and sometimes with spinose branches. Flowers whitish, solitary or in small, axillary, cymose clusters; calyx 4–5-dentate, glabrous or sometimes pubescent; corolla deeply 4–5-lobed, pubescent within. Fruit ovoid.

**1. Ximena pubescens** Standley, Contr. U.S. Nat. Herb. **20:** 212, 1919

A dense, intricately branched shrub 9–12 dm. high with slender branches armed with stout, straight spines 5–8 mm. long and densely pilose with persistent, short, yellowish hairs; petioles 3–5 mm. long; leaf blades broadly oval to orbicular, 2.2–4 cm. long, 2–3.5 cm. wide, coriaceous, densely puberulent; pedicels and calyx densely puberulent; calyx lobes broadly ovate, obtuse or subacute, minute; petals 4–5 mm. long, acute, puberulent without, densely barbate from base to middle within.

Borders of sand dunes, Lower Sonoran Zone to Tropical Zone, San Nicolas Bay, Baja California; Oaxaca.

### 2. **SCHOEPFIA** Schreb., Gen. Pl. 129, 1789

Glabrous shrubs or small trees with divaricate, brittle, unarmed or sparingly subspinose branches and smooth, gray bark. Leaves coriaceous, often glaucous. Flowers in axillary racemes, solitary or fasciculate in leaf axils; calyx cupulate, small, shallowly denticulate; corolla urceolate, 4–6-lobed, lined inside the base with a thick, fleshy disk. Fruit ovoid to subglobose.

Style equaling the perianth tube; calyx glabrous; disk barely equaling calyx ........ 1. *S. shreveana*
Style shorter than the perianth tube; calyx puberulent; disk twice as long as calyx ........ 2. *S. californica*

**1. Schoepfia shreveana** Wiggins, Contr. Dudley Herb. **3:** 66, *pl. 14, figs. 6–13,* 1940

Rounded shrub 2–6 m. high; branchlets rigid, brittle, slightly zigzag, glabrous; leaves oblanceolate to oblanceolate-elliptic, 2–6 mm. wide, 12–25 mm. long, cuneate-tapering to base, subsessile, acute to rounded at apex, leathery, glaucous, faintly 3–5-veined beneath; flowers solitary or in pairs on slender, glabrous, axillary peduncles

4–9 mm. long; calyx 3–5-dentate, 2 mm. long, 2.5–3 mm. wide, glabrous, equaling or surpassing disk of ovary; perianth cylindric-urceolate, 2.5–3 mm. wide, 5 mm. high, with 5 erect lobes 1.5–2 mm. long; anthers about 1 mm. long, barely equaling perianth tube; ovary inferior, 2–3-celled at the base, 1-celled above, surmounted and surrounded by a fleshy disk through which the style protrudes; style equaling perianth tube; fruit oval, 6–8 mm. long; seed solitary, pendulous.

Alkaline playas in the vicinity of Empalme, Sonora. Feb.–March.

**2. Schoepfia californica** Brandegee, Proc. Calif. Acad. II, **2:** 139, 1889

Rounded, straggly shrub or small tree 2–6 m. high; branches stiff, divaricate, brittle, minutely puberulent in youth but soon glabrate; bark gray to whitish; leaves oblanceolate to obovate-spatulate, 2–5 mm. wide, 10–25 mm. long, cuneate, rounded to subacute at apex, glaucous, minutely puberulent, coriaceous; flowers reddish yellow, solitary or in axillary pairs; peduncles 2–4 mm. long; calyx shallowly 3–5-dentate, teeth low and rounded, minutely puberulent; corolla 5–7 mm. long, lobes rotately spreading; style stout, about half as long as perianth tube; disk of ovary puberulent, about twice as long as calyx; fruit subglobose to ovoid, 6–8 mm. long, tipped by minute remnants of the disk.

Washes and alkaline flats, Lower Sonoran Zone, vicinity of Punta Prieta, Baja California, southward to the Cape Region. Feb.–April.

## Family 32. ARISTOLOCHIACEAE. Birthwort Family

Low caulescent or acaulescent herbs or twining shrubs. Leaves alternate or basal, petiolate, entire, chiefly cordate or reniform. Flowers perfect, regular or irregular, solitary or clustered, axillary or terminal. Calyx 3–6-lobed, the tube mostly adnate to the ovary. Petals none. Stamens 5–12, somewhat united to the style; anthers 2-celled, adnate, extrorse. Ovary partly or completely inferior, 6-celled; placentae parietal; ovules numerous, anatropous. Fruit a 6-celled capsule. Seeds compressed or angled; raphe usually large; embryo minute; endosperm fleshy.

### 1. ARISTOLOCHIA L., Sp. Pl. 960, 1753

Perennial climbing or twining shrubs or herbs. Leaves alternate, cordate to sagittate, entire or somewhat lobed, palmately nerved. Flowers solitary or clustered, irregular, greenish to purple. Calyx tube usually inflated around style and contracted at the throat, curved, deciduous, limb spreading or reflexed, entire or 3–6-lobed. Stamens mostly 6; anthers sessile, adnate to style or stigma. Ovary 6-celled. Capsule septicidally 6-valved. Seeds compressed, with plane or concave sides, dark brown or black.

Leaf blades hastate, tomentulose; capsule narrowly winged along midrib of each valve; calyx limb 1.5–2.5 cm. long, acuminate ................ 1. *A. brevipes* var. *acuminata*

Leaf blades lanceolate, sparsely hirsutulous; capsules not winged; calyx limb 12–15 mm. long, cordate, acute ................................................ 2. *A. batucensis*

**1. Aristolochia brevipes** Benth. var. **acuminata** S. Wats., Proc. Amer. Acad. **18:** 148, 1883

*Aristolochia watsonii* Woot. & Standl., Contr. U.S. Nat. Herb. **16:** 117, 1913.

Stems mostly herbaceous or woody at base, 1–3 dm. long, chiefly scrambling over rocks and earth, slender, finely but closely subtomentose, hairs tawny; petioles 5–15 mm. long, about equaling sinus of blade; blades hastate-lanceolate, 1.5–6 cm. long, 1–4 cm. wide across the broadly ovate, hastately spreading, basal lobes, deeply cordate, acuminate, finely subtomentose on both surfaces; peduncles solitary, axillary, 3–6 mm. long, bearing an ovate bract 2–3 mm. long near base; calyx tube slightly curved, scarcely saccate, 1–1.5 cm. long, limb gradually expanded, 1.5–2.5 cm. long, acuminate, about 8–10 mm. wide at base, finely tomentulose on outside, sparsely hairy inside, greenish with brownish specks inside; capsule 1.5–2 cm. long, narrowly winged along midvein of each valve; seeds strongly compressed, triangular, 4–5 mm. long, edge toward valve curved, black, minutely shining-papillose.

Among rocks and at bases of cliffs, Lower Sonoran Zone to Tropical Zone, extreme southwestern New Mexico, southern Arizona, and adjacent Mexico to the Revillagigedo Islands; Las Animas Bay to the Cape Region, Baja California.

**2. Aristolochia batucensis** Wigg. & Roll., Contr. Dudley Herb. **3:** 269, *pl. 61, fig. 1,* 1943

Scandent vine from a woody rootstock; branches slender, 3–6 dm. long, slightly zigzag, with striate internodes 2–5 cm. long; petioles 7–15 mm. long, sparsely hirsute; leaf blades sagittate, 2–4 cm. wide at base, tapering gradually to acute or abruptly obtuse apex, 4–8 cm. long, dark green above, paler beneath, sparsely hirsutulous on both surfaces; flowers few, solitary, on slender peduncles 8–10 mm. long, each subtended by an ovate bract 2–3 mm. long; calyx limb and anterior half of tube sharply reflexed, brownish purple or greenish toward base of limb, latter narrowly cordate, 12–15 mm. long, inner surface greenish yellow with a few scattered, coarse, purplish hairs borne on heavy tubercle-like bases, tube 12–18 mm. long, greenish or slightly purplish, slightly hirsutulous; ovary sulcate, clavate, 5–7 mm. long at anthesis, hirsutulous; capsule about 15 mm. long, 9–10 mm. in diameter, smooth, margins of the valves forming low, rounded ribs; seeds compressed-triangular, acute at base, rounded at apex, about 4 mm. long, nearly as broad, about 0.8–1 mm. thick, concavo-convex, irregularly crinkled near edges and with a heavy median ridge marking hypocotyl.

On hillsides among shrubs and at margins of openings, upper margin of the Lower Sonoran Zone, Sierra de Batúc, Sonora. Aug.–Sept.

## Family 33. **RAFFLESIACEAE.** Rafflesia Family

Fleshy parasitic plants on the roots, stems, or branches of various trees and shrubs. Leaves scalelike, inconspicuous, only a few of them subtending the flower outside tissue of host plant. Flowers minute (in ours) to very large, solitary or rarely spicate, rarely perfect, usually unisexual by abortion. Calyx somewhat epigynous with 4–10 imbricate or less commonly valvate segments, in ours similar to subtending scales. Anthers sessile, arranged in 1–3 series around a fleshy central column, the latter expanded at apex into a fleshy disk. Pistillate flowers having a ring-shaped stigma around the margin of a fleshy disk surmounting the column. Ovary inferior, 1-celled, with several placentae extending inward almost to center. Ovules numerous, with a single integument. Seeds numerous, minute.

### 1. **PILOSTYLES** Guill., Ann. Sci. Nat. II, **2:** 21, *pl. 1,* 1834

Characters of the family, the flowers minute, with 4–5 perianth segments.

**1. Pilostyles thurberi** A. Gray, Pl. Thurb. 326, 1854

Vegetative part of the plant wholly embedded in the stems of host plant, *Dalea emoryi*; flowers apparently dioecious, brownish, 3–4 mm. long, solitary on fleshy peduncles 0.2–1.5 mm. long; scales subtending flowers 4–7, broadly ovate to suborbicular, 1–1.5 mm. long, fleshy at broad base, gradually thinner toward apex, irregularly erosulate along margins, upper ones hardly distinguishable from less fleshy, broadly obovate perianth segments, these yellowish, 1.5 mm. long; fleshy disk about 1 mm. in diameter.

Parasitic on *Dalea emoryi* and definitely known from Signal Mountain and near Ocotillo on Highway 80 in Imperial County, California, and from near the mouth of the Gila River in Yuma County, Arizona. This inconspicuous little parasite probably is widely distributed in the Sonoran Desert, because the host plant occurs in southwestern Arizona, southeastern California, and south into Sonora and the Cape Region, Baja California. Jan.

## Family 34. POLYGONACEAE. Buckwheat Family

Herbs, shrubs, or less commonly trees or vines. Leaves alternate, opposite, whorled, or fascicled, sometimes all basal, usually entire, simple. Stipules connate and sheathing or absent. Flowers monoecious, dioecious, polygamodioecious, or perfect, on slender, jointed pedicels and variously arranged in clusters, rarely solitary. Calyx 2–6-cleft or -parted, lobes often petaloid, persistent. Petals lacking. Stamens 2–9, inserted near base of calyx or hypanthium. Ovary superior, free from calyx, at least at anthesis, 1-celled, 1-ovulate; ovule orthotropous, pendulous or erect; styles 3, or rarely 2 or 4, usually distinct; stigmas capitate or peltate. Fruit an achene, lenticular, ovoid, or 3–6-angled, usually enclosed in persistent perianth. Embryo straight or curved. Endosperm mealy.

Plants herbs or shrubs; calyx lobes and tubes not markedly accrescent:
- Stipules wanting; leaf blades rarely over 5 cm. long, usually much smaller and coriaceous:
  - Flowers subtended by 1 or more distinct bracts, not in involucres:
    - Bracts solitary, not hooked at tip; stems not conspicuously brittle:
      - Flowers not in heads; bracts glabrous or puberulent but not lanate, bilobed, enlarged and conspicuously reticulate-veined in fruit:
        - Herbs; leaves rounded; bract 2–3 mm. long in fruit ............. 1. *Pterostegia*
        - Shrubs; leaves linear-oblanceolate; bract 1–1.5 cm. long in fruit .. 2. *Harfordia*
      - Flowers in globose heads; bracts lanate, not bilobed nor enlarged in fruit ........ ...................................................... 3. *Nemacaulis*
    - Bracts in whorls, hooked at the tips; stems brittle at the nodes ........ 4. *Lastarriaea*
  - Flowers in cup-shaped to tubular involucres:
    - Involucres with spine- or bristle-tipped teeth, usually 1-flowered; stems very brittle at the nodes when dry ........................................ 5. *Chorizanthe*
    - Involucres without bristles or spines on the teeth, usually several-flowered; stems usually not conspicuously brittle at the nodes ................. 6. *Eriogonum*
- Stipules present, sheathing, usually conspicuous; leaf blades ample, usually well over 5 cm. long, slightly fleshy:
  - Calyx lobes 6, outer 3 spreading, shorter than inner, not petaloid .......... 7. *Rumex*
  - Calyx lobes 5, all erect, often petaloid ................................ 8. *Polygonum*

Plants trees or vines; calyx tube or lobes accrescent, closely investing achene:
Vine; tendrils terminating branches of inflorescence; inflorescence racemose-paniculate; calyx lobes accrescent, red-purple ................................ 9. *Antigonon*
Tree; tendrils wanting; inflorescence spicate; calyx tube accrescent, closely investing achene, lobes not accrescent or brightly colored ......................... 10. *Coccoloba*

## 1. PTEROSTEGIA Fisch. & Mey., Ind. Sem. Hort. Petrop. 2: 48, 1835

Slender scrambling or prostrate annual with diffuse, dichotomously branching stems, opposite leaves, and small, foliaceous bracts. Bract 2-lobed, solitary, sessile, axillary, 2-gibbous on the back, becoming enlarged and scariously reticulate in fruit, loosely investing the triangular, glabrous achene. Perianth mostly 6-parted, slightly surpassing bracts. Stamens 3 or 6.

**1. Pterostegia drymarioides** Fisch. & Mey., Ind. Sem. Hort. Petrop. **2:** 48, 1835
*Pterostegia diphylla* Nutt., Jour. Acad. Phila. II, **1:** 168, 1848.
*Pterostegia diphylla* var. *biloba* Nutt., loc. cit.
*Pterostegia microphylla* Nutt., loc. cit.

Stems slender, usually branching from base, prostrate or scrambling, 1–5 dm. long, sparsely hirsute; lower leaves flabelliform, 3–18 mm. broad, often 2-lobed and minutely crenulate, gradually narrowed to a short petiole; upper leaves orbicular or slightly wider than long, rarely notched, entire or crenulate; bract 2-lobed, 1.5–2 mm. long in flower, 2–3 mm. long in fruit, laciniate-toothed; calyx lobes usually 6, rarely 5, lanceolate, acute, finely and sparsely crispate-hirsutulous; achene trigonous, 1 mm. long, reddish or buff, smooth and shining.

Open or shaded slopes or among rocks, mostly Upper Sonoran Zone, northern California to northern Baja California, to southern Utah and western Arizona. Abundant in canyons along western margins of Colorado Desert; Tempe, Arizona; Superstition Mountains, Arizona. Feb.–May.

## 2. HARFORDIA Greene & Parry, Proc. Davenp. Acad. 5: 27, 1886

Low, rounded shrub with herbaceous or woody branches and small, opposite, spatulate, fleshy leaves. Flowers axillary, short-pedicellate, subtended by a minute, oblong bract with reflexed tip, the bract enlarging and becoming conspicuously bisaccate in fruit, membranaceous, reticulate with conspicuous red veins; calyx lobes 6, lance-elliptic, hirsutulous without. Stamens 6. Achene trigonous, enfolded by central, 2-lobed part of bract, smooth, shining.

**1. Harfordia macroptera** (Benth.) Greene & Parry, Proc. Davenp. Acad. **5:** 27, 1886
*Pterostegia macroptera* Benth., Bot. Sulph. 44, 1844.
*Pterostegia galioides* Greene, Bull. Calif. Acad. **1:** 213, 1885.

A low, rounded shrub, woody at the base and with slender, straggly branches 1–6 dm. long, finely puberulent throughout; leaves somewhat fleshy, oblong-spatulate, 6–15 mm. long, acute to obtuse, narrowing gradually toward base; flowers axillary, 1–4 in an axil, each subtended by a minute, slightly 2-lobed bract, its tip somewhat recurved and bearing 2 saccate evaginations near base on abaxial surface, a small elliptical pore communicating with adaxial surface of bract, sacs enlarging greatly and attaining a length of 1–1.5 cm. in fruit; calyx deeply 6-parted, sparsely hirsutulous without,

1–1.2 mm. long at anthesis, persistent; stamens 6, included; achene trigonous, slender, 4–5 mm. long, smooth, shining, enfolded by inner, 2-lobed part of involucre.

Rocky hillsides and bluffs along the western side of the peninsula, Lower and Upper Sonoran Zones, from San Antonio del Mar to Magdalena Bay, Baja California. Feb.–March.

### 3. NEMACAULIS Nutt., Jour. Acad. Phila. II, 1: 168, 1848

Diffusely branching, slender annual with estipulate, mostly basal leaves and floccose herbage. Flowers perfect, in capitate sessile clusters at nodes, each subtended by a small, free, herbaceous, involucral bract. Calyx 6-parted, delicately colored, enclosing achene. Stamens 3. Styles 3; stigmas capitate. Achenes ovoid, obtusely trigonous.

**1. Nemacaulis denudata** Nutt., Jour. Acad. Phila. II, **1**: 168, 1848

*Nemacaulis foliosa* Nutt., op. cit. 169.
*Nemacaulis nuttallii* Benth., in A. DC., Prod. **14**: 23, 1857.

Stems branching from base, prostrate to ascending, 4–40 cm. long, finely tomentulose in youth, later glabrate; basal leaves spatulate, 1.5–5 cm. long, obtuse to rounded, gradually narrowed to base, densely floccose-tomentose on both sides, cauline leaves mostly reduced to bracts; flower clusters 3–5 mm. in diameter, involucral bractlets obovate-spatulate, 2 mm. long or less, glabrous on outer and tomentose on inner surface, obtuse; calyx lobes yellow, about 1 mm. long, glabrous; stamens included; achenes short-ovoid, obtusely trigonous, or almost circular in cross section, 0.8–1 mm. long, dark brown, shining.

Sandy soil, Lower and Upper Sonoran Zones, mostly near the coast, but also along western edge of Colorado Desert; Los Angeles County, California, to central Baja California. Feb.–May.

### 4. LASTARRIAEA Remy, in Gay, Fl. Chil. 5: 289, *pl. 58,* 1849

Prostrate, slender, brittle-stemmed annual diffusely branched from the base. Leaves mostly basal, linear to narrowly oblanceolate, gradually merging into hook-tipped bracts above. Flowers solitary, sessile, axillary and terminal, subtended by involucre-like whorls of bracts with hooked tips. Calyx tubular, 5–6-cleft, lobes hooked apically. Stamens 3; filaments faintly toothed on each side at base. Achene trigonous, slender, about equaling calyx lobes, often slightly curved.

**1. Lastarriaea chilensis** Remy, in Gay, Fl. Chil. **5**: 290, *pl. 58,* 1849

*Chorizanthe lastarriaea* Parry, Proc. Davenp. Acad. **4**: 63, 1884.
*Lastarriaea chilensis* var. *californica* H. Gross, in Engler, Bot. Jahrb. **49**: 345, 1913.
*Chorizanthe lastarriaea* var. *californica* Goodman, Ann. Mo. Bot. Gard. **21**: 33, 1934.

Decumbent with sparsely hirsutulous stems 5–25 cm. long; leaves 1–2.5 cm. long, basal ones obtuse, upper acute or awn-tipped, hirsutulous; bracts 3–4 mm. long, yellowish green, crowded at ends of short lateral branchlets, becoming reddish in age; calyx of 5 lobes, 3 long and 2 short, these with hooked tips; anthers globose; achene 2–2.5 mm. long, trigonous, smooth, rather dull, angles distinct.

In dry, sandy soil, Lower and Upper Sonoran Zones, central California southward into Baja California. Mostly west of the mountains but occurring near Palm Springs, California, and probably elsewhere along the western margin of the Colorado Desert. March–April.

## 5. CHORIZANTHE R. Br. ex Benth., Trans. Linn. Soc. 17: 416, *pl. 19*, 1836

Annuals or suffruticose perennials with soft-pubescent to tomentose, usually dichotomously branching, erect or procumbent stems. Leaves chiefly basal, petiolate, entire. Inflorescence of spreading, capitate, or uniparous cymes; involucres solitary in axils or crowded at tips of branches, 3- or 6-toothed, and usually 3- or 6-ribbed, each tooth usually tipped with a straight, downward-curved, or uncinate spine, these spreading, ascending or erect. Flowers solitary in each involucre, included or slightly exserted; perianth 6-lobed, lobes entire to laciniate, similar or outer whorl larger. Stamens usually 9, sometimes 3 or 6, inserted at base of perianth or higher on tube; filaments glabrous. Achene 3-angled, glabrous; styles 3; stigmas capitate.

Involucral tube cylindric, transversely corrugated, not strongly angled:
  Involucral teeth 3, all similar, spreading; tube strongly corrugated; blades of basal leaves broadly ovate to orbicular .................................... 1. *C. corrugata*
  Involucral teeth 5, erect or ascending, 1 foliaceous and much larger than the rest; tube obscurely corrugated; basal leaves narrowly oblanceolate-spatulate .... 2. *C. watsonii*
Involucral tube strongly 3- or 6-ribbed:
  Bracts entire; involucres without spurs at base:
    Teeth of the involucres 3 (inner teeth obsolete or essentially so):
      Plants compact, erect; involucral tubes 2–2.5 mm. long ............. 3. *C. rigida*
      Plants openly branched, lax, mostly procumbent; involucral tubes 3–5 mm. long:
        Involucral teeth equal, spreading; perianth lobes similar:
          Perianth lobes entire; outer involucral lobes broadly ovate .......... 4. *C. flava*
          Perianth lobes laciniate; involucral lobes lanceolate .......... 5. *C. pulchella*
        Involucral teeth unequal, 1 much longer than the others, all ascending, rarely spreading; inner perianth lobes narrower and shorter than others ........... 6. *C. turbinata*
    Teeth of the involucres 6, well developed, inner usually shorter than outer:
      Perianth lobes similar or essentially so:
        Involucres slightly arcuately curved; stems very brittle; plant erect ............ 7. *C. brevicornu*
        Involucres straight; stems not markedly brittle; plants procumbent or prostrate:
          Perianth 1.5–2 mm. long; involucral tube about equaling perianth lobes ...... 8. *C. procumbens*
          Perianth 6–7 mm. long; involucral tube 5 mm. long, shorter than perianth ..... 9. *C. mutabilis*
      Perianth lobes dissimilar, inner ones shorter than outer:
        Perianth lobes fimbriate or laciniate:
          Flowers 5.5–7 mm. long; terminal part of lobes linear-oblong, wider than fimbriations ................................ 10. *C. fimbriata* var. *fimbriata*
          Flowers 7–9 mm. long; terminal part of lobes filiform, scarcely wider than fimbriations .............................. 10*a*. *C. fimbriata* var. *laciniata*
        Perianth lobes entire or slightly erose, not strongly fimbriate:
          Outer involucral teeth 1–3 mm. long, uncinate or downward-curved:
            Tubes of involucres sparsely arachnoid-pubescent, about 2 mm. long; ribs not obscured by pubescence ................................ 11. *C. parryi*
            Tubes of involucres densely tomentose, 2.6–3 mm. long, ribs obscured by tomentum .............................. 12. *C. xanti* var. *leucotheca*
          Outer involucral teeth 2.5–4 mm. long, straight ........... 13. *C. inequalis*
  Bracts 3-lobed; involucres with 3 spurs at the base ...................... 14. *C. thurberi*

**1. Chorizanthe corrugata** (Torr.) Torr. & Gray, Proc. Amer. Acad. **8:** 198, 1870
*Acanthogonum corrugatum* Torr., Pacif. R. Rep. **5:** 364, 1856.

Erect annual 5–25 cm. high with silky-pubescent stems branching profusely but branchlets short, forming a compact plant; leaves petiolate, basal and cauline, floccose-tomentose beneath, floccose-villous above, blades broadly ovate to orbicular, 7–20 mm. wide, on villous petioles about as long; bracts opposite, lower spatulate, 6–12 mm. long, acerose and reduced above; lower involucres loosely disposed in axils, densely congested above, involucral tube cylindric, 2.5–4 mm. long, strongly corrugated transversely, floccose-villous in youth, later glabrate, teeth 3, equaling or slightly exceeding tube, those of lower involucres elliptic-foliose, short spines slightly downward-curved; flowers included, 2–2.5 mm. long, segments oblong, obtuse, entire; stamens 6; achenes papillose above, slightly exserted.

Sandy, dry soil, Lower Sonoran Zone, southern Inyo County, California, eastward to New Mexico, southward through the Mojave and Colorado Deserts into northern Sonora. March–April.

**2. Chorizanthe watsonii** Torr. & Gray, Proc. Amer. Acad. **8:** 199, 1870

Erect or ascending, freely branching plants to 1 dm. tall, with canescent, green or reddish stems; leaves basal, 1–3 cm. long, petiolate, blades narrowly oblanceolate-spatulate, obtuse or acute, tomentose on both surfaces; bracts opposite, lower often foliaceous, 7–10 mm. long, upper acerose, 3–6 mm. long; involucres solitary or in small clusters, cymose; involucral tube cylindric, obscurely 5-ribbed, 3–4 mm. long, canescent; involucral teeth 5, erect or ascending, anterior one lanceolate, foliaceous, 3–15 mm. long, the other four 1–3 mm. long, with uncinate tips; flowers included or barely exserted, yellow; perianth tube pubescent, the lobes equal, oblong; stamens 9, inserted near top of perianth tube, filaments 0.5 mm. long.

Arid, sandy soil, Lower Sonoran and arid parts of Upper Sonoran Zone, southeastern Washington and adjacent Idaho, southward east of the Sierra Nevada into the Mojave Desert and northwestern Arizona; near Palm Springs, western Colorado Desert. March–April.

**3. Chorizanthe rigida** (Torr.) Torr. & Gray, Proc. Amer. Acad. **8:** 198, 1870
*Acanthogonum rigidum* Torr., Pacif. R. Rep. **4:** 133, 1856.

An erect, compact annual 1 dm. high or less, the desiccated, spiny plants remaining erect for some time after fruiting; stems robust, arachnoid-pubescent, nearly hidden by involucres, leafy to summit; leaves petiolate, 1–5 cm. long, blades elliptic to broadly ovate, 5–20 mm. wide, 8–30 mm. long, tomentose beneath, sparsely arachnoid-villous to subglabrate above; bracts opposite, foliaceous to acerose, 1–2.5 cm. long, rigidly spine-tipped; involucres compactly crowded, 3-toothed and strongly 3-ribbed, tube about 2–2.5 mm. long, nearly or quite as wide, ventricose-reticulate, lanate in youth, finally glabrate, teeth ascending-spreading, rigid, 3–5 mm. long, straight, unequal; perianth yellow, lobes oblong, barely exserted, pubescent on outside.

Desert regions, Lower Sonoran Zone, Inyo County, California, to southern Nevada and south through the Mojave and Colorado Deserts to northern Sonora and central Baja California. Feb.–May.

**4. Chorizanthe flava** Brandegee, Proc. Calif. Acad. II, **2:** 202, 1889
*Chorizanthe vaseyi* Parry & Rose, Bot. Gaz. **15:** 64, *pl. 10,* 1890.

Decumbent to suberect plants with sparsely puberulent, dichotomously forking stems 0.5–3 dm. long; leaves chiefly basal, long-petiolate, 2–8 cm. long, sparsely lanate below, glabrate above, tardily so beneath, blade broadly ovate, 5–22 mm. wide, 8–28 mm. long, gradually narrowing to slightly winged petiole; bracts opposite, acerose, 4–7 mm. long, reddish; involucres solitary in axils, slightly crowded toward tips of branches, those in basal axils more robust than upper ones; involucral tubes 3-angled, 4–5 mm. long, sparsely arachnoid-pubescent, teeth 3, narrowly to broadly ovate, spreading, 1.5–4 mm. wide, 5–7 mm. long, ventricose-reticulate, rigidly tipped with downward-curved, subuncinate spine; inner teeth obsolete or essentially so; perianth yellow, linear-lanceolate lobes entire, pubescent on back; stamens 9; anthers short-oblong.

Sandy flats and washes, Lower Sonoran Zone, Vizcaíno Desert from vicinity of Calmallí to Lagoon Head and near San Ignacio, Baja California. Feb.–April.

**Chorizanthe jonesiana** Goodman, Ann. Mo. Bot. Gard. **21:** 88, *pl. 2, fig. 2,* 1934, resembles *C. flava* but has 6 teeth, the inner 3 smaller than the outer, and a shorter involucral tube. It has been reported only from the coastal bluffs between Tijuana and Ensenada, but might occur near the sea farther south.

### 5. Chorizanthe pulchella Brandegee, Proc. Calif. Acad. II, 2: 203, 1889

Plants with a single, erect, primary stem 1–6 cm. long, and several spreading-ascending branches 2–15 cm. long from apex or with several ascending-spreading branches from base, these sparsely puberulent to glabrate, reddish; leaves basal, 1–2.5 cm. long, long-petiolate, blades 2–6 mm. broad, 4–8 mm. long, broadly ovate, mostly truncate at base, lanate beneath, slightly arachnoid-pubescent but soon glabrate above; bracts opposite, acerose, 3–4 mm. long; involucres solitary in axils, scarcely crowded at tips of branches, tubes triangular in cross section, 3–3.5 mm. long, lanate, becoming glabrate and reddish; involucral teeth 3, or 6 and the inner 3 minute, spreading, 3–6 mm. long, subequal, spiny tip straight on older, uncinate on young, involucres; perianth white to light pink, lobes deeply and finely laciniate, exserted 1.5–4 mm.; stamens subequal, attached at base of perianth tube; anthers broadly oblong, 0.6 mm. long.

Sandy desert flats, Lower Sonoran Zone, vicinity of Laguna Seca Chapala, Baja California, southward to the northern edge of the Vizcaíno Desert. Feb.–April.

### 6. Chorizanthe turbinata Wiggins, Contr. Dudley Herb. 3: 67, *pl. 15, figs. 3–5,* 1940

A procumbent annual with dichotomously forked branches 1–2 dm. long radiating from an erect primary stem 5–20 mm. long; branches sparsely and finely arachnoid-hirsutulous, reddish, internodes 1–3 cm. long; basal leaves narrowly oblanceolate-spatulate, 2–4 cm. long, 2–6 mm. wide, sparsely puberulent or glabrate; bracts opposite, narrowly subulate to setaceous, rigid, 3–4.5 mm. long; involucres solitary in axils, somewhat crowded near tips of branches, turbinate, strongly 3-angled and alternate ribs much smaller, spineless, curving inward to their rounded to acutish apices, these barely equaling membranous sinuses between outer teeth; involucral tube 3.5–4 mm. long, sparsely puberulent; involucral teeth 3, ascending or spreading-ascending, ventricose-reticulate, each tipped with a straight, subulate spine, 1 tooth about twice as long as and broader than the other 2; perianth white to pink, 4–5 mm. long, sparsely puberulent on tube, borne on a stipe 0.2–0.3 mm. long, lobes entire, outer and inner lobes 1.5 and 1 mm. long, respectively, outer broader than inner and surpassing involucral tube about 1 mm.; stamens 9, unequal, inserted near base of perianth tube, about equaling inner perianth lobes; achene obscurely trigonous, 2.5–3 mm. long, broadest near base, brownish.

Collected in sandy soil on dry slopes along Socorro Wash, between San Quintín and Rosario, Baja California. March–April.

## 7. **Chorizanthe brevicornu** Torr., Bot. Mex. Bound. 177, 1859

Erect annual with several to numerous puberulent, greenish stems branching profusely from base and very brittle at nodes when mature; leaves mostly basal, petiolate, oblanceolate-spatulate, 3–6 cm. long, strigose-puberulent; bracts foliaceous at lower nodes, acerose above, the latter type 3–6 mm. long, reddish; involucres solitary in axils, 3-angled, 6-ribbed toward apex of puberulent tube, this 3.5–6 mm. long, somewhat falcately curved, 6-toothed, teeth short, subequal, 1 mm. long or less, uncinate (lower involucres larger and more strongly curved than upper ones); perianths short-pedicellate, 3–4 mm. long, almost or completely included, greenish yellow, lobes similar, linear-oblong, about 1 mm. long; stamens 3, inserted at base of perianth tube, opposite inner lobes; achenes slightly curved, slightly exserted.

Sandy desert slopes, flats, and washes, Lower Sonoran Zone, Mono County, California, to southern Utah and south to northern Baja California and northwestern Sonora. March–May.

## 8. **Chorizanthe procumbens** Nutt., Jour. Acad. Phila. II, **1:** 167, 1848. Not *C. procumbens* A. Gray, 1861

*Chorizanthe uncinata* Nutt., loc. cit.

Procumbent, considerably branched, sparsely puberulent plant with several stems 0.5–2 dm. long; leaves basal, narrowly oblanceolate, 2.5–7 cm. long, blades 3–9 mm. broad, gradually tapering to narrowly winged petioles, densely pubescent beneath to glabrate; bracts opposite, lower ones often foliaceous, upper ones acerose, 1.5–2 mm. long, hispid-ciliate along lower half of margins; involucres axillary but aggregated into small clusters near tips of branches, 6-ribbed, 6-toothed, sparsely hispidulous in intervals between ribs, tube 1.2–1.8 mm. long; outer involucral teeth subequal, 1.4–2 mm. long, spreading, uncinate, the inner about half as long; perianth yellow, 1.5–2 mm. long, obtuse, oblong lobes slightly exserted, scantily pubescent on the midribs without; achenes included or barely exserted.

Sandy places, Lower and Upper Sonoran Zones, western San Bernardino County, Riverside and Los Angeles Counties, California, south to northern Baja California; mostly west of the Sonoran Desert but occurring at Cabazon, Riverside County, California. March–June.

**Chorizanthe chaetophora** Goodman, Ann. Mo. Bot. Gard. **21:** 89, *pl. 2, fig. 1,* 1934, closely resembling *C. procumbens* but differing from it in having thicker, larger teeth, grayer stems, and perianths with pubescence exceeding perianth lobes, has been reported only from the vicinity of San Quintín, Baja California, but may appear elsewhere to the south within the limits of the Sonoran Desert.

## 9. **Chorizanthe mutabilis** Brandegee, Proc. Calif. Acad. II, **2:** 203, 1889

Procumbent to erect plants with branching stems 1–3 dm. long; stems sparsely and loosely spreading-hirsutulous; leaves basal, narrowly oblanceolate, 1–2 cm. long, petiolate, loosely tomentose beneath, sparsely long-pubescent above; bracts lance-subulate to acicular, 2–5 mm. long; involucres solitary, axillary, mostly remote, appressed-long-pubescent, 6-ribbed, 6-toothed, tube cylindric, 4–5 mm. long; outer teeth spreading, about 2.5 mm. long and margins involute at anthesis, flattening and becoming broadly ovate, 5–8 mm. long at maturity, then ventricose-reticulate, spiny tips straight or slightly downward-curved; inner teeth much smaller, mostly uncinate; perianth 6–7 mm. long with linear-oblong, obtuse, subequal, entire or crenulate, white to rose-colored lobes, pubescent outside; stamens 6–9, inserted near base of and slightly adnate to perianth tube.

Sandy desert, Lower Sonoran Zone, central Baja California. Feb.–March.

**10. Chorizanthe fimbriata** Nutt., Jour. Acad. Phila. II, **1**: 168, 1848, var. **fimbriata**

Erect plants with spreading-ascending branches 1–3 or rarely 5 dm. long, usually trichotomously branching at first node; branches reddish and sparingly crispate-pubescent; leaves basal, broadly oblanceolate or with broadly obovate-spatulate blades 5–12 mm. broad, 1–2 cm. long, tapering into petioles to twice their length, villous-hirsute to sparsely pubescent; bracts opposite, acerose, 3–5 mm. long, mostly spreading; involucres bright red, solitary and remote or clustered near tips of branches, puberulent, 6-ribbed, 6-toothed, tube 3.5–4.5 mm. long; outer teeth spreading, 2.5–4.5 mm. long, tips somewhat downward-curved or uncinate; inner teeth about half as long or less, uncinate; perianth white or pale rose-colored, exserted 1.5–3 mm., oblong lobes fimbriate or sometimes merely crenate toward base, tip of lobe much broader than lateral fimbriations, slightly pubescent along midrib on outer surface; stamens 9, mostly included.

On sandy or clay soil, Lower and Upper Sonoran Zones, western Riverside and southern Los Angeles Counties, California, south into Baja California, mostly toward the coast. April–May.

**10*a*. Chorizanthe fimbriata** var. **laciniata** (Torr.) Jepson, Fl. Calif. **1**: 394, 1914
*Chorizanthe laciniata* Torr., Pacif. R. Rep. 19, 1856.

Differing from var. *fimbriata* mainly in having slightly longer perianth lobes, which are deeply and finely laciniate, terminal lobe no broader than lateral ones.

Sandy or clay soil of interior cismontane valleys, occasionally occurring along the western margin of the Colorado Desert; and south into northern Baja California. April–May.

**11. Chorizanthe parryi** S. Wats., Proc. Amer. Acad. **12**: 271, 1877

Decumbent plants with several branches 1–2 dm. long from base; stems sparsely ascending-pubescent to canescent, greenish or later dull reddish; leaves basal, oblanceolate-spatulate, gradually tapering to a winged petiole, 3–7 mm. wide, 2–7 cm. long, obtuse but mucronulate, pubescent to canescent-tomentulose beneath, sparsely villous to glabrate above; bracts opposite, acerose, 2.5–4 mm. long, spreading, lower ones foliaceous; involucres somewhat crowded toward tips of branches, remote below, urceolate, 6-ribbed, 6-toothed, tube 2–2.5 mm. long, short-villous; teeth spreading, the outer 2–3 mm. long, the spinose tip downward-curved; inner teeth much shorter, uncinate; perianth 2.5–3.2 mm. long, pink, lobes exserted 1 mm. or less, the outer oblong-obovate, obtuse and mostly erose, the inner narrower, slightly shorter, entire; stamens 9, included.

Sandy, arid soil, mainly Upper Sonoran Zone, Los Angeles, western Riverside and San Bernardino Counties, California, south to northern Baja California. Barely reaching the western margin of the Colorado Desert near Banning, California. April–May.

**12. Chorizanthe xanti** S. Wats. var. **leucotheca** Goodman, Ann. Mo. Bot. Gard. **21:** 60, 1934

Ascending, openly branched annuals 1–1.5 dm. high or smaller with reddish, moderately ascending-villous stems; basal leaves petiolate, 1.5–3.5 cm. long, lamina ovate, lanate beneath, pubescent to glabrate above; lower cauline leaves elliptic to linear-lanceolate, 1 cm. long or less, gradually merging into opposite, acerose, spreading bracts with slightly recurved tips; involucres remote below, crowded into loose clusters above, cylindric, 6-ribbed, 6-toothed, tube 2.6–3 mm. long, white-tomentose; outer teeth 1–1.5 mm., inner 0.5–0.8 mm. long, spreading, uncinate; perianth exserted 0.5–1 mm., sparsely puberulent without, lobes oblong-linear, obtuse, entire, inner ones slightly shorter and narrower than outer ones; stamens 9.

Arid sandy or gravelly soil, Lower and Upper Sonoran Zones, western San Bernardino and Riverside Counties, California. April–May.

**13. Chorizanthe inequalis** Stokes, Zoe **5:** 60, 1900

Diffuse annual with spreading-villous stems 0.1–3 dm. long branching from base; leaves basal, petiolate, narrowly oblanceolate or oblanceolate-spatulate, 1.5–3 cm. long, sparsely arachnoid-villous on both surfaces or subglabrate above; bracts opposite, simple, acerose, spreading, 3–8 mm. long, these and older stems bright red; involucres solitary in axils or somewhat crowded at tips of branches, sparsely arachnoid-villous, 6-angled, 6-toothed, urceolate, tube 2–3 mm. long; outer teeth spreading, acerose, straight, 2.5–4 mm. long, inner 0.4–1 mm. long, divergent; perianth white, exserted about 1 mm., sparsely hairy on outer surface, outer lobes obovate or oblong, slightly erose, inner linear-oblong, about two-thirds as long as outer; stamens 9, included, anthers short-elliptic.

Sandy, arid places, mostly near the coast, Lower and Upper Sonoran Zones, vicinity of San Antonio del Mar (Johnson's Ranch) southward at least to Socorro Canyon, Baja California. April–May.

**14. Chorizanthe thurberi** (A. Gray) S. Wats., Proc. Amer. Acad. **12:** 269, 1877

*Centrostegia thurberi* A. Gray ex Benth., in A. DC., Prod. **14:** 27, 1856.
*Chorizanthe thurberi* var. *cryptantha* Curran, Bull. Calif. Acad. **1:** 275, 1885.

Diffusely branched annual with spreading-ascending stems 0.5–2.5 dm. long, glandular-hispid near base, glabrate above; leaves mostly basal, broadly obovate to oblanceolate, 3–15 mm. wide, 5 cm. long or less, glabrous, entire or rarely shallowly and irregularly sinuate-crenate, often slightly glaucous; lower cauline leaves ovate, apiculate; bracts opposite, 3-lobed, lobes ovate to acerose, 3–5 mm. long; involucres solitary in axils, trigonous in cross section, tube 3.5–4.5 mm. long, bearing 3 spreading, spine-tipped, saccate teeth 2–3 mm. long at base of tube, and 5 erect, spinose teeth 0.5–1 mm. long at apex; perianth 3–3.5 mm. long, pubescent externally, included.

Dry, sandy places, Upper and Lower Sonoran Zones, eastern Ventura and Los Angeles Counties, California, to southern Nevada and southwestern Arizona. March–May.

## 6. ERIOGONUM Michx., Fl. Bor. Amer. 1: 246, 1803

Annual or perennial herbs or shrubs with entire, alternate, opposite, whorled, or fasciculate, estipulate leaves. Flowers perfect, borne on slender, articulate pedicels in sessile or pedunculate involucres. Involucres broadly campanulate to cylindrical, 4–8-toothed or -lobed, few- to many-flowered, bearing scarious or setaceous slender bractlets intermingled with the flowers. Calyx 6-parted or deeply 6-cleft, segments more or less scarious, petaloid, equal and similar or the inner differing from the outer. Stamens mostly 9, included or exserted slightly, glabrous or villous near base, inserted at base of calyx. Styles 3-parted to base; stigmas capitate. Achene triangular or lenticular, usually narrowed to a short beak.

Bracts foliaceous; teeth of involucres longer than tube .................. 1. *E. abertianum*

Bracts not foliaceous; teeth of the involucres shorter than or barely equaling tube (except in *E. galioides* and *E. orcuttianum*):

♦ Involucres campanulate, not markedly ribbed nor angulate, teeth broadly rounded:

Plants perennial:

Outer calyx lobes glabrous, saccate and cordate at base; involucres deeply parted .... ........................................................ 2. *E. galioides*

Outer calyx lobes pubescent, not saccate nor cordate; involucres parted halfway to base or less:

Stems not inflated, villous toward base, verticillately branched above; leaves densely pilose on both surfaces, margins finely crispate-undulate ......... 3. *E. scalare*

Stems often inflated toward apices of lower internodes, glabrous throughout, di- or sometimes trichotomously branched; leaves glabrous to sparsely pilose along margins and veins on upper surface, margins coarsely undulate to plane:

Lower internodes robust, inflated toward apex ..... 4. *E. inflatum* var. *inflatum*

Lower internodes slender, not inflated ........... 4*a*. *E. inflatum var. deflatum*

Plants annual:

Leaves present at least at first node above basal rosette:

Plants decumbent; outer calyx lobes broadly hastate; flowers glabrous .......... ........................................................ 5. *E. foliosum*

Plants erect or ascending; outer calyx lobes not hastate; flowers more or less pubescent:

Outer calyx lobes only slightly broader than the inner, densely pubescent, only slightly concave; basal leaves 5–8 cm. long, glabrous ............ 6. *E. ordii*

Outer calyx lobes about twice as broad as the inner, sparsely glandular-puberulent to glabrate, deeply concave; basal leaves 2–4 cm. long, tomentose beneath:

Outer calyx lobes distinctly short-clawed, white, tinged with rose, not conspicuously maculate ....................... 7. *E. angulosum* var. *angulosum*

Outer calyx lobes nearly sessile, yellow, with a deep rose spot in the center .... .................................. 7*a*. *E. angulosum* var. *maculatum*

Leaves all basal, or upper ones reduced to bracts:

Involucres 4-lobed or 4-toothed:

Calyx lobes acute, densely pubescent, the outer about as wide as or wider than the inner; involucres on slender peduncles 5–15 mm. long ...... 8. *E. trichopes*

Calyx lobes obtuse, sparsely pubescent or glabrous, the outer distinctly narrower than the inner; involucres sessile ........................ 9. *E. intricatum*

Involucres 5-lobed or 5-toothed, rarely 4-toothed:

Peduncles filiform to nearly capillary, 5–25 mm. long, erect or ascending; calyces 1–1.5 mm. long:

Calyx lobes glabrous, the outer ones slightly dilated at apex .... 10. *E. capillare*

Calyx lobes puberulent or glandular, at least toward base, not dilated above:

Involucres glandular-puberulent; outer calyx lobes broader than long, abruptly narrowed to a short, distinct claw ............ 11. *E. thurberi*

Involucres glabrous; outer calyx lobes longer than broad, not narrowed to a claw:

Leaf blades reniform; outer calyx lobes not saccate-dilated; nodal bracts usually arachnoid-pubescent ...................... 12. *E. reniforme*

Leaf blades ovate, rarely cordate at base; outer calyx lobes saccate-dilated on each side of cordate base; nodal bracts glabrous ... 13. *E. thomasii*

Peduncles stouter, 1–4 mm. long (1–15 mm. long in *E. deflexum* subsp. *austrinum* and *E. deflexum* var. *brachypodum*), usually strongly reflexed; calyces 2–2.5 mm. long:

Stems and peduncles glabrous, eglandular:

Outer calyx lobes longer than broad; peduncles 1–3 mm. long ........... .................................. 14. *E. deflexum* subsp. *deflexum*

Outer calyx lobes nearly as broad as long, suborbicular; peduncles 2–15 mm. long ............................ 14*a*. *E. deflexum* subsp. *austrinum*

Stems, peduncles, and involucres stipitate-glandular; peduncles 4–15 mm. long ............................. 14*b*. *E. deflexum* subsp. *brachypodum*

◆ Involucres cylindrical or narrowly turbinate-campanulate, usually distinctly ribbed or angulate, teeth acute:

Plants annual; outer calyx lobes flabelliform with sides incurving toward base:

Stems profusely branching, at length incurved, forming dense crown; leaf blades orbicular-cordate, all basal; involucres 1 mm. long ........... 15. *E. nidularium*

Stems branching sparingly, virgate, erect, crown open, slender; leaf blades elliptic to oblong or oblanceolate, cuneate at base, cauline ones present; involucres 1.5–2.5 mm. long:

Leaves elliptic; involucres 2–2.5 mm. long; flowers 2–2.2 mm. long, outer calyx lobes truncate ......................................... 16. *E. polycladon*

Leaves oblong or oblanceolate; involucres 1.5–2 mm. long; flowers 1.5–2 mm. long, outer calyx lobes rounded ................................ 17. *E. gracile*

Plants perennial; outer calyx lobes oblong or obovate, not flabelliform, sides convex:

Stems fastigiate, forming dense mat 0.5–1.5 dm. high, leafy to apex; involucres obscured by closely crowded leaves ......................... 18. *E. fastigiatum*

Stems ascending or erect but not conspicuously fastigiate, usually over 2 dm. high; involucres on naked or nearly leafless scapes, not hidden among leaves:

Involucres crowded into heads on naked scapes; leaves linear to narrowly elliptic, fasciculate:

Leaves glabrous on upper surface; peduncles glabrous or essentially so; involucres and calyces glabrous or nearly so on outer surfaces:

Herbage dark green; plant 1–2 m. high..19. *E. fasciculatum* subsp. *fasciculatum*

Herbage yellow-green; plant usually 3–4 dm. high ......................... ............................. 19*a*. *E. fasciculatum* subsp. *flavoviride*

Leaves pubescent-cinereous on upper surface; peduncles cinereous-pubescent; involucres and calyces strigose-cinereous on outer surfaces ................. .............................. 19*b*. *E. fasciculatum* subsp. *polifolium*

Involucres solitary at nodes of inflorescence, not compacted into heads; leaves not fasciculate, broadly elliptic to orbicular, rarely linear:

■ Calyx lobes densely pubescent on outer surface; leaf blades broadly ovate, obovate, or suborbicular:

Involucres 5-cleft to base; calyx lobes pubescent inside and out; branches of inflorescence zigzag; leaves glabrous above or nearly so, dark green ...... ............................................... 20. *E. orcuttianum*

Involucres shallowly 4–5-toothed; calyx lobes glabrous within except for tuft of hairs near base; branches of inflorescence not zigzag; leaves white-woolly ............................................... 21. *E. deserticola*

■Calyx lobes glabrous on outer surface; leaf blades elliptic to narrowly obovate or oblanceolate:
  Involucres glabrous on outer surface; branches of inflorescence irregularly divaricate .................................. 22. *E. plumatella* var. *jaegeri*
  Involucres tomentose on outer surfaces; branches of inflorescence virgate, ascending:
    Achenes scaberulous; involucres 2–3 mm. long; flowers 2.5–3 mm. long:
      Involucres cylindric-turbinate, on erect, main branches:
        Bases of petioles not persistent as glabrate rings on stems ............ ............................... 23. *E. wrightii* subsp. *wrightii*
        Bases of petioles dilated, persisting as brown, glabrate rings on lower parts of stems ........... 23*a*. *E. wrightii* subsp. *membranaceum*
      Involucres turbinate-campanulate, mostly on spreading, short, lateral branches ...................................... 24. *E. nodosum*
    Achenes glabrous; involucres 1–1.4 mm. long; flowers 1.4–1.8 mm. long... ........................................... 25. *E. pringlei*

## 1. Eriogonum abertianum Torr., in Emory, Notes Mil. Rec. 150, 1848

*Eriogonum abertianum* var. *neomexicanum* Gandoger, Bull. Soc. Bot. Belg. **42:** 185, 1906.
*Eriogonum abertianum* var. *ruberrimum* Gandoger, loc. cit.
*Eriogonum cyclosepalum* Greene, Muhlenbergia **6:** 1, 1910.
*Eriogonum abertianum* subsp. *typicum* Stokes, Gen. Eriog. 36, 1936.
*Eriogonum abertianum* var. *villosum* Fosberg, Madroño **4:** 191, 1938.
*Eriogonum abertianum* var. *gillespiei* Fosberg, op. cit. 193.

Annual, profusely branched from or near base with ascending appressed-hirsute stems 1–4 dm. long; basal leaves petiolate, blade ovate to oblong, 1–2 cm. wide, 1.5–2.5 cm. long, petiole to 6 cm. long, upper leaves usually sessile, obovate-lanceolate to linear, all loosely villous to hoary; peduncles 1–6 cm. long or sometimes wanting; involucre solitary, broadly campanulate, villous-canescent on outside, tube 2–3 mm. long, 5 lobes oblong, 4–6 mm. long, 1–2 mm. wide, often reflexed in maturity; calyx papery, white to pale yellow tinged with rose, midribs often deep rose, outer lobes orbicular-cordate, 3–4 mm. in diameter, sometimes sparingly glandular, inner lobes spatulate, 0.7–1.2 mm. wide, slightly longer than outer; achenes dark brown, smooth and shining or sometimes transversely rugulose with labyrinthine ridges, body subglobose-trigonous, 0.6–0.8 mm. long, narrowing abruptly to a slender, trigonous beak of equal length.

Sandy plains, washes, and granitic hills, Lower and Upper Sonoran Zones, western Pima County, Arizona, eastward to El Paso, Texas, south into northern Sonora and Chihuahua. March–Sept.

**Eriogonum abertianum** subsp. **pinetorum** (Greene) Stokes, Gen. Eriog. 37, 1936, is a form with slightly less hoary leaves and generally more slender stems. It occurs in the Upper Sonoran and Arid Transition Zones from the Santa Catalina Mountains, Arizona, into New Mexico, but is approached by slender specimens from Tumamoc Hill, Tucson, Arizona.

## 2. Eriogonum galioides I. M. Johnston, Proc. Calif. Acad. IV, 12: 1012, 1924

Profusely branching herbaceous perennial from a woody root; stems numerous, mostly trichotomously branching, several from base, spreading to form broadly rounded mass of intricately intertangled stems 1–2 dm. high, 4 dm. wide (or more); basal leaves petiolate, about 2 cm. long, blade broadly ovate, acute or obtuse at apex, subtruncate at base, 5–9 mm. long, 3–6 mm. wide, thick, sparingly pilose; leaves at first two or three nodes reduced, oblong to oblanceolate, 1–2.5 mm. wide, 5–6 mm. long, sessile or short-petiolate, the uppermost reduced to lance-subulate or lance-ovate bracts 3–4 mm. long; peduncles filiform; 2–5 mm. long, minutely and sparsely puberulent-glandular; involucres 4-lobed to base, lobes 1–1.5 mm. long, elliptic-subulate, smooth and glabrous or faintly glandular; pedicels slender, glabrous, about equaling involucral lobes or slightly shorter, recurved; flowers minute, about 1 mm. long, glabrous, outer lobes saccate-distended near base, much narrower above, acute; inner ones narrow, about equaling the outer ones, yellow tinged with rose at base; achenes trigonous, brownish, about 1 mm. long or less.

Known only from the type locality, San Luis Island, near the head of the Gulf of California. March–April.

**3. Eriogonum scalare** S. Wats., Proc. Amer. Acad. **12:** 261, 1877

*Eriogonum irretitum* Brandegee, Proc. Calif. Acad. II, **2:** 202, 1889.
*Eriogonum pilosum* Stokes, Gen. Eriog. 22, 1936.

Herbaceous perennial from a slender, woody caudex; stems several, slender, glaucous, 2–6 dm. long, spreading or ascending, sparsely to densely pilose at base, verticillately branched with 6–30 short, rigidly radiating stems at nodes, tertiary branches often verticillately arranged also; basal leaves oblanceolate to oblong, acute, 5–14 mm. wide, 1–6 cm. long, densely pilose on both surfaces, hairs on veins and margins longest, midvein prominent beneath, margins finely crispate-undulate; cauline leaves reduced, 4–12 mm. long, sessile or nearly so, or in some specimens reduced to bracts; peduncles filiform, 3–15 mm. long, glaucous and glabrous or sometimes sparingly pubescent toward base; involucres campanulate, 1.5 mm. high, glabrous, 4-toothed, teeth broadly triangular, acute, erect; calyx greenish yellow, about 2 mm. long, with a prominent subglabrous midvein, lamina densely pubescent with rather coarse, appressed hairs; achenes ovoid, reddish brown, 1.6–2 mm. long, abruptly narrowed to slender beak.

Sandy and rocky desert slopes and plains, vicinity of El Mármol to eastern and western margins of Vizcaíno Desert, Baja California. March–May.

**4. Eriogonum inflatum** Torr. & Frem., in Frem., Second Rep. 317, 1845, var **inflatum**

*Eriogonum glaucum* Small, Bull. Torrey Club **25:** 51, 1898.
*Eriogonum inflatum* subsp. *typicum* Stokes, Gen. Eriog. 22, 1936.
*Eriogonum trichopes* subsp. *glaucum* Stokes, op. cit. 24.

Erect, sparsely branched perennial 3–8 dm. tall from a woody root; lowest internode 2–5 dm. tall, glabrous, glaucous, conspicuously inflated toward apex, the plant branching trichotomously or more profusely above first joint to form an umbrella-like rounded top often as wide as height of plant; leaves basal, petioles 2–6 cm. long, sparsely villous, blades broadly round-reniform to oblong, 1–2.5 cm. long, hirsute and prominently veined beneath, sparsely villous to glabrate above, undulate-margined; inflorescence a profusely dichotomously and trichotomously branching panicle; peduncles filiform, 5–20 mm. long, straight, becoming rigid; involucres solitary, broadly turbinate, 1.2–1.6 mm. long, 5-lobed about to middle, slightly glaucous and glabrous or scantily stipitate-glandular on lobes; flowers 10–20, on stipitate-glandular pedicels 1.5–2 mm. long; calyx lobes yellow with reddish or greenish midribs, densely short-hirsute, 2–2.5 mm. long, the outer ovate, the inner narrower, acute; achenes triquetrous, about 2 mm. long, the angles acute.

Sandy washes and rocky hillsides, Mojave and Colorado Deserts to southern Nevada, Utah, and central Arizona. March–July.

**4a. Eriogonum inflatum** var. **deflatum** I. M. Johnston, Proc. Calif. Acad. IV, **12:** 1013, 1924

*Eriogonum glaucum* Small, Bull. Torrey Club **25:** 51, 1898.
*Eriogonum lagunense* M. E. Jones, Contr. West. Bot. **18:** 34, 1933.
*Eriogonum inflatum* subsp. *lagunense* Stokes, Gen. Eriog. 22, 1936.

Differs from var. *inflatum* in having more slender, more numerous stems from the woody base, and noninflated internodes. The two varieties intergrade in the Colorado Desert, but var. *inflatum* does not occur along the eastern side of Baja California, where variety *deflatum* is common.

Desert slopes, washes and plains, Lower Sonoran Zone, southern Riverside and Imperial Counties, California, to Mohave County, Arizona, south into northern Sonora and to Loreto, Baja California. Feb.–April.

**5. Eriogonum foliosum** S. Wats., Proc. Amer. Acad. **20:** 371, 1885

*Eriogonum hastatum* Wiggins, Contr. Dudley Herb. **1:** 165, *pl. 12, fig. 2A–B,* 1933.

Small, imperfectly dichotomously branched, decumbent annuals with arachnoid-lanate branches 1–5 dm. long, these finally glabrate; basal leaves narrowly ovate to oblong, 2–8 mm. wide, 1–4 cm. long, sparsely lanate to glabrous above, densely lanate beneath, petioles equaling blade or lacking; cauline leaves in pairs at nodes, ovate, sessile or nearly so; bracts scarious, 1–1.2 mm. long; involucres turbinate, 2–3 mm. long, solitary or in terminal clusters of 1–4, 5-toothed; calyx glabrous, pink, or white with pink veins and base, 1.5–2.2 mm. long; calyx lobes glabrous, dissimilar, the outer broadly hastate at base, obtuse at apex, margins crispate, the inner lobes narrower, broadly oblong, exceeding the outer; filaments scantily pilose at base; achenes lenticular to subtriquetrous, 2 mm. long, purplish black, the beak scaberulent, slightly surpassing the calyx lobes.

Sandy slopes and meadows, mostly above the desert, upper margin of Lower Sonoran Zone to Transition Zone, "Cantillas Mts." near California line southward to southern end of Sierra San Pedro Mártir, Baja California. March–May.

**6. Eriogonum ordii** S. Wats., Proc. Amer. Acad. **21:** 468, 1886

*Eriogonum tenuissimum* Eastw., Proc. Calif. Acad. IV, **20:** 139, 1931.

Slender, erect annual 1–2.5 dm. high with 1–several erect, loosely floccose-tomentose stems branching di- or trichotomously above the basal internodes, the latter 1 dm. long or less; basal leaves petiolate, blades obovate to oblanceolate, 1–1.5 cm. wide, 2–4 cm. long, obtuse to rounded at apex, narrowing gradually to a slender petiole equaling or slightly surpassing blade, glabrous on upper surface, sparingly arachnoid-villous to glabrous beneath, margins plain, entire; leaves of lower 1–3 nodes elliptic to obovate, sessile to short-petiolate, 2–6 mm. wide, 7–12 mm. long, acute to rounded at apex; upper bracts subulate, reduced; peduncles capillary, 5–20 mm. long; involucres solitary, turbinate, 4-toothed, 1–2 mm. high, glabrous, mostly 1–3-flowered; calyx 1.5–2 mm. high, white or pale yellowish tinged with pink, densely short-villous, lobes narrowly ovate to oblong, all similar; achenes trigonous, about 2 mm. long, 0.5 mm. wide, sharply angled to narrowly subulate, stramineous, smooth and shining.

Dry, sandy soil, Lower Sonoran Zone, Colorado Desert to Fort Mohave, Arizona, and to Kern and San Benito Counties, California. April–June.

**7. Eriogonum angulosum** Benth., Trans. Linn. Soc. **17:** 406, *pl. 18, fig. 1,* 1836, var. **angulosum**

A woolly-tomentose annual 0.5–5 dm. high, often twice as wide, with more or less striate-angled stems branching from or near base, branches ascending to spreading; basal leaves oblanceolate to oblong-oblanceolate, 2–4 cm. long, 5–10 mm. wide, short-petiolate, margins crispate-undulate to revolute; cauline leaves sessile, lanceolate to oblong, 1.5 cm. long or less; stipules triangular-subulate, scarious; peduncles capillary, axillary, 1–2 cm. long, glabrous to sparsely arachnoid-villous; involucres solitary, broadly campanulate-turbinate, 2–3 mm. high and broad, with 4 broad, rounded teeth, finely and sparsely puberulent to glabrous; flowers 6–15 in each involucre, rose, tipped with white, 2–2.5 mm. high; outer calyx lobes broadly ovate-elliptic, short-clawed, deeply concavo-convex to cucullate, sparsely and minutely puberulent to glabrous; inner calyx lobes narrowly spatulate, sessile, slightly longer than the outer, especially at maturity; achenes narrowly ovoid-subulate, light brown, smooth and shining, 1.2–1.6 mm. long.

Arid, sandy soils, mostly Lower Sonoran Zone, northern Baja California northward through foothills of Coast Ranges to Alameda County, California. April–Oct.

**7*a*. Eriogonum angulosum** subsp. **maculatum** (Heller) Stokes, Gen. Eriog. 38, 1936

*Eriogonum viridescens* Heller, Muhlenbergia **2:** 25, 1905.
*Eriogonum maculatum* Heller, op. cit. 188, 1906.
*Eriogonum angulosum* var. *maculatum* Jepson, Fl. Calif. **1:** 405, 1914.
*Eriogonum angulosum* subsp. *viridescens* Stokes, Gen. Eriog. 38, 1936.

Similar to typical *E. angulosum* in habit but with more noticeably glandular, nearly sessile, outer calyx lobes, a dark rose spot on each outer calyx lobe, slightly longer inner calyx lobes (these conspicuously exceeding the outer in maturity), and achenes 1–1.3 mm. long.

Desert regions, eastern Oregon, southward east of the Sierra Nevada to western Arizona and northeastern Baja California. May-Oct.

According to Abrams, *E. bidentatum* Jepson, Madroño **1:** 115, 1923, is an autumnal and winter phase occurring both in var. *angulosum* and in var. *maculatum.*

**8. Eriogonum trichopes** Torr., in Emory, Notes Mil. Rec. 151, 1848

*Eriogonum trichopodum* Torr. ex Benth., in A. DC., Prod. **14:** 20, 1856.
*Eriogonum clavatum* Small, Bull. Torrey Club **25:** 50, 1898.
*Eriogonum trichopes* subsp. *clavatum* Stokes, Gen. Eriog. 24, 1936.
*Eriogonum trichopes* subsp. *cordatum* Stokes, loc. cit.

Erect annual 1–4.5 dm. high with simple base and repeatedly branched, slender to nearly filiform branches; first internode 2–15 cm. long, sometimes slightly clavate-inflated above; ultimate branches glabrous to sparsely glandular immediately above nodes, slightly glaucous; basal leaves with orbicular to ovate blades 1–2 cm. wide, 1–4 cm. long, pubescent to pilose beneath, glabrate or sparsely pubescent above, margins crenate to entire, on pilose petioles 1–6 cm. long; upper leaves reduced to triangular-subulate bracts 1–4 mm. long, the larger often pubescent; peduncles capillary, 5–15 mm. long, glabrous; involucres broadly turbinate, 0.75–1 mm. long, shallowly 4-lobed, glabrous; calyx 1 mm. long in flower, yellow to greenish, lobes similar, narrowly ovate, acute, densely hispidulous, becoming 2–2.5 mm. long and midribs more prominent in fruit; achenes 1.8–2 mm. long, included, sharply angled, smooth but dull, light brown.

Sandy slopes and washes, Lower Sonoran Zone, southern Coast Ranges and foothills of the Sierra Nevada, California, to southern Utah, New Mexico, northern Sonora and central Baja California. Jan.–Dec.

**9. Eriogonum intricatum** Benth., Bot. Sulph. 46, *pl. 22,* 1844

Intricately and rather rigidly branched annual from a rather stout taproot, the rounded plant 2–4 dm. tall and about as wide or slightly wider; basal leaves on slender petioles to 1.5 cm. long, blade broadly obovate, crenate-dentate, 3–12 mm. wide, 8–20 mm. long, finely but closely puberulent with straight, spreading hairs; cauline leaves greatly reduced, oblanceolate, acute or obtuse, sessile or short-petiolate, 1–3 mm. wide, rarely over 1 cm. long and usually shorter; internodes of main branches 1–5 cm. long, glabrous, reddish; each node of larger branches with a whorl of smaller, lateral ones; involucres axillary to minute bracts, about 0.5 mm. long, deeply 4-lobed, glabrous; flowers 1.6–1.8 mm. long, lobes rounded-carinate and minutely strigillose on back, rounded at apex, pinkish; pedicels much shorter than involucre; achenes ovoid, terete or subtrigonous in cross section, barely exceeding closely appressed calyx lobes, glabrous.

Sandy and rocky slopes, Lower Sonoran Zone, Cedros Island and on the adjacent mainland along the western margins of the Vizcaíno Desert, Baja California. Aug.–Oct.

**10. Eriogonum capillare** Small, Bull. Torrey Club **25**: 51, 1898

*Eriogonum arizonicum* Stokes ex M. E. Jones, Contr. West. Bot. **11**: 16, 1903.

Slender, acaulescent annual 2–6 dm. tall, branching dichotomously or trichotomously above apex of main stem, this usually 1–1.5 dm. long; leaves basal, blades broadly ovate to orbicular-ovate, 1–2 cm. broad, nearly as long, truncate to subcordate at base, acute or obtuse and abruptly apiculate at apex, sparsely strigose on both surfaces, margins finely undulate-crispate; petioles about equaling blades, tomentose; ultimate divisions of branches 1–4 cm. long, erect or spreading, nearly capillary; involucres solitary at tips of the branches, turbinate, about 1 mm. long, deeply 5-lobed into oblong or ovate-oblong segments; flowers exserted beyond involucre, urn-shaped, 1–1.2 mm. long, pinkish, oblong segments slightly dilated at apex, glabrous, slightly revolute about middle, obtuse or rounded at apex, the outer 3 slightly wider than the inner; achenes ovoid-trigonous, about 1 mm. long, short-beaked, somewhat swollen toward base.

Plains and hillsides, mainly Upper Sonoran Zone, Gila County, Arizona. Sept.–Oct.

**11. Eriogonum thurberi** Torr., Bot. Mex. Bound. 176, 1859

*Eriogonum thurberi* var. *parishii* Gandoger, Bull. Soc. Bot. Belg. **42**: 198, 1906.
*Eriogonum cernuum* subsp. *thurberi* Stokes, Gen. Eriog. 42, 1936.
*Eriogonum cernuum* subsp. *viscosum* Stokes, loc. cit.

Small annual with 1–several stems 1–4 dm. high, diffusely branched above, usually somewhat arachnoid-lanate below, sparingly glandular to glabrate above; leaves basal, oblong-ovate, 0.8–4.5 cm. long, densely white-lanate beneath, sparsely so to glabrate above, slightly rugulose, petioles about as long as blades; bracts deltoid-ovate, 1–1.5 mm. long, connate at base, glandular on outside and lanate within; peduncles 5–25 mm. long, capillary, erect, sparsely glandular-puberulent toward apex; involucres broadly turbinate, 1.8–2 mm. high, minutely glandular-puberulent, 5-lobed, teeth ovate, equaling tube; calyx 1–1.5 mm. long, deep rose with white margins, glandular-puberulent toward base, outer lobes suborbicular to broader than long, usually rotately spreading at apex, abruptly narrowed to a short claw, hairy near the base within; inner calyx lobes narrowly oblanceolate, shorter than outer; achenes broadly ovoid, 0.6–0.8 mm. long, short-beaked, dull gray-brown and minutely flecked with black, smooth and shining.

Sandy washes and slopes, mostly Lower Sonoran Zone, Los Angeles County, California, to Arizona and northern Baja California. March–June.

**12. Eriogonum reniforme** Torr. & Frem., in Frem., Second Rep. 317, 1845

*Eriogonum reniforme* var. *comosum* M. E. Jones, Proc. Calif. Acad. II, **5:** 719, 1895.
*Eriogonum comosum* M. E. Jones, Contr. West. Bot. **11:** 16, 1903.
*Eriogonum reniforme* var. *asarifolium* Gandoger, Bull. Soc. Bot. Belg. **42:** 196, 1906.
*Eriogonum reniforme* subsp. *typicum* Stokes, Gen. Eriog. 35, 1936.

Annual 0.5–4 dm. high with 1–several divergently branched leafless stems somewhat floccose toward base; leaves all basal, reniform, 5–20 mm. broad, irregularly crenate to subentire, densely arachnoid-tomentose beneath, sparsely so to glabrate and green above, on slender, tomentose petioles 0.5–6 cm. long; bracts loosely lanate, at least on inner surface, triangular-subulate, 1–2 mm. long, connate at base; peduncles filiform, 5–15 or rarely 20 mm. long; involucres broadly turbinate, about 2 mm. long, glabrous without, sparsely lanate within, the 5 lobes rounded; calyx yellowish with greenish or reddish midribs, sparsely glandular-puberulent toward base, about 1 mm. long in flower, outer lobes elliptic-ovate, inner ones oblong, becoming 1.5–2 mm. long and roseate in fruit; achenes ovoid, 0.5–0.7 mm. long, dark brown, nearly beakless.

Arid, sandy soil, Lower Sonoran Zone, Inyo County, California, and adjacent western Nevada, the Mojave and Colorado Deserts. March–June.

**13. Eriogonum thomasii** Torr., Pacif. R. Rep. **5:** 364, 1858

*Eriogonum minutiflorum* S. Wats., Proc. Amer. Acad. **26:** 125, 1891.

Annual, 1–3 dm. high with 1–several diffusely branched, delicate, glabrous stems from a basal rosette of broadly oval leaves, stems sometimes sparsely glandular at base; petioles slender, 6–30 mm. long, sparsely pilose to loosely lanate; leaf blades broadly ovate to suborbicular, 5–8 mm. long, rounded at apex, from broadly cuneate to shallowly subcordate at base, arachnoid-lanate on both surfaces or often glabrate and green above, margins usually slightly undulate; bracts broadly deltoid, about 1 mm. long, connate at base, glabrous; peduncles capillary, spreading to faintly recurved, 4–20 mm. long or those at lowest nodes sometimes 2.5–3 cm. long, glabrous; involucres turbinate-campanulate, 0.6–0.8 mm. long, 5-lobed, glabrous; flowers yellow and 0.8–1 mm. long at anthesis, becoming rose or white with red midribs and 1.2–1.8 mm. long in fruit, lobes narrowly ovate to oblong, sparsely glandular-puberulent, similar at first, the outer becoming saccate-lobed on each side of the cordate base in fruit, these saccate lobes usually white or paler rose than the rest of the calyx lobe, the inner lobes surpassing the outer 0.4–0.6 mm. in fruit; achenes turgidly lenticular, 0.8–1 mm. long, dark red-brown or with irregular splotches of gray, smooth, shining, narrowed to a short, blunt beak.

Sandy desert soil, Lower Sonoran Zone from Inyo County, California, south to northern Sonora and central Baja California. March–April.

In early spring, before any achenes have developed, the outer calyx lobes show no evidence of the basal saccate swellings and are very slightly if any shorter than the inner lobes. Such specimens are frequently determined as *E. reniforme* Torr. & Frem. The latter species, however, has arachnoid-lanate bracts and sparse tangles of crinkly hairs inside the involucres. In *E. thomasii,* both bracts and involucres are devoid of these tangled hairs, and a careful examination of the involucres at the basal nodes will often reveal one or two flowers with incipient swellings on each side of the midvein at base of the outer calyx lobes. All Arizona specimens labeled *E. reniforme* seen by the writer thus far, except one collected at Yucca, Arizona, were immature plants of *E. thomasii.*

**14. Eriogonum deflexum** Torr., Ives Rep. 24, 1860, subsp. **deflexum**

*Eriogonum deflexum* subsp. *typicum* Stokes, Gen. Eriog. 44, 1936.

Annual with 1–several ascending, divaricately branching, glabrous stems 1–4.5 dm. high, these forming a spreading, flat-topped crown; leaves basal, blades 1.5–5 cm. broad, broadly ovate or mostly orbicular with truncate to shallowly subcordate bases, densely lanate beneath, less densely so or glabrate above; petioles 1–6 cm. long, often reddish, sparsely to densely woolly; bracts green, deltoid-subulate, 2–4 mm. long, glabrous; peduncles 1–4 mm. long, slender, sharply deflexed or sometimes spreading; involucres solitary, turbinate, shallowly 5-toothed, 1.8–2.2 mm. long, glabrous, several-flowered; calyx pale yellow or white, turning rose in age, glabrous, about 2 mm. high, tube very short; outer calyx lobes elliptic, rounded at apex, shallowly cordate at base, 2–2.5 mm. long in fruit, the inner shorter, narrowly ovate, acutely rounded at apex; achenes turgidly ovoid, 2 mm. high, light brown, smooth but dull, beak stout, triquetrous, about 0.2–0.4 mm. long.

Washes and sandy desert flats, Lower Sonoran Zone, Mojave Desert to western Nevada, Arizona, and northern Baja California. March–July.

**14*a*. Eriogonum deflexum** subsp. **austrinum** Stokes, Gen. Eriog. 44, 1936

Usually a less robust plant than typical *deflexum*; outer calyx lobes suborbicular or sometimes broader than long, more deeply cordate at base; peduncles 3–12 mm. long.

Sandy desert soil, southern Arizona, adjacent Sonora, and probably Baja California.

**14*b*. Eriogonum deflexum** subsp. **brachypodum** (Torr. & Gray) Stokes, Gen. Eriog. 44, 1936

*Eriogonum brachypodum* Torr. & Gray, Proc. Amer. Acad. 8: 180, 1870.
*Eriogonum deflexum* var. *brachypodum* Munz, Man. S. Calif. 119, 1935.

Similar to subsp. *deflexum* but with involucres, peduncles, and branches markedly stipitate-glandular and puberulent.

Sandy flats, arroyos, and rocky hillsides, Lower Sonoran Zone, Inyo County, California, to western Nevada, western Arizona, and northern Baja California. March–June.

**15. Eriogonum nidularium** Coville, Contr. U.S. Nat. Herb. 4: 186, 1893

*Eriogonum densum* Greene, Pittonia 3: 17, 1896.
*Eriogonum nidularium* var. *luciense* M. E. Jones, Contr. West. Bot. 11: 17, 1903.
*Eriogonum vimineum* subsp. *nidularium* Stokes, Gen. Eriog. 49, 1936.
*Eriogonum vimineum* var. *densum* Stokes, loc. cit.

Compact annual 0.5–3 dm. high with numerous repeatedly dichotomously branched stems with short internodes, outer branches often turning inward in age to form dense, intertangled tuft, closely lanate throughout, or involucres at length becoming subglabrate; leaves basal, densely lanate, blades orbicular-cordate, 0.6–2.5 cm. broad, petioles often two to three times as long; bracts deltoid-subulate, 2–3 mm. long, floccose or finally subglabrate without; involucres solitary, sessile, about 1 mm. high, often obscured by bracts, 1–3-flowered; flowers yellow or pink or rarely white, glabrous, about 1.5 mm. long; outer calyx lobes broadly flabelliform-obovate with truncate apex and concave sides, the terminal part often somewhat rotately spreading; inner lobes similar but narrower, about the same length; achenes 1.6–2 mm. long, body ovoid, turgid, beak triquetrous, about equaling the body, scaberulous.

Dry, sandy slopes and washes, Lower Sonoran Zone, western Mojave Desert to Kern and Inyo Counties, California, southwestern Utah, central Arizona, and the southwestern part of the Colorado Desert. April–Oct.

**16. Eriogonum polycladon** Benth., in A. DC., Prod. **14:** 16, 1856

*Eriogonum polycladon* var. *mexicanum* Gandoger, Bull. Soc. Bot. Belg. **42:** 196, 1906.
*Eriogonum polycladon* var. *crispum* Gandoger, loc. cit.
*Eriogonum vimineum* subsp. *polycladon* Stokes, Gen. Eriog. 53, 1936.

Erect annual 3–7 dm. high with 1–several erect, virgate, sparingly branched stems densely clothed with tangled, woolly hairs; leaves, both basal and cauline, elliptic to oblanceolate, 4–20 mm. broad, 1.5–3.5 cm. long, densely white-woolly beneath, less copiously so above, petioles about equaling blades or slightly shorter; upper bracts lance-subulate, 1–2.5 mm. long; internodes of inflorescence 5–15 mm. long; involucres solitary, sessile, 2–2.5 mm. long, strongly ribbed, subglabrous or sparsely arachnoid-pubescent in sinuses and between ribs, several-flowered; flowers 2–2.2 mm. high glabrous, pale rose or with red midribs, outer lobes obovate-flabelliform, truncate at apex, incurved sides narrowing to a clawlike base, tips somewhat spreading, inner lobes similar and nearly as wide; achenes about 2 mm. long, the plump, ovoid body abruptly narrowed to a triquetrous, minutely scaberulous beak of about equal length.

Dry ridges and slopes, mostly in Upper Sonoran Zone, central Arizona to Texas and Chihuahua. June–Nov.

**17. Eriogonum gracile** Benth., Bot. Sulph. 46, 1844

*Eriogonum acetoselloides* Torr. ex Benth., in A. DC., Prod. **14:** 16, 1856.
*Eriogonum leucocladon* Benth., Pl. Hartw. 333, 1857.
*Eriogonum aginum* Greene, Pittonia **2:** 165, 1891.
*Eriogonum vimineum* subsp. *gracile* Stokes, Gen. Eriog. 51, 1936.

Erect, diffusely branched annual 2–5 dm. high with slender, ascending, densely floccose-tomentose stems or these becoming subglabrous with age; basal leaves on slender, floccose petioles equaling or exceeding oblanceolate to oblong blades, the latter 5–10 mm. wide, densely white-lanate beneath, sparsely floccose and green above; cauline leaves mostly sessile, elliptic to oblong; bracts foliaceous at lower nodes, lance-subulate, 2–3.5 mm. long above; involucres solitary, strongly ribbed, 1.5–2 mm. high, glabrous except for tufts of wool in sinuses; flowers 1–5 per involucre, white to pale rose, 1.5–2 mm. long at anthesis, glabrous; outer calyx lobes oblong to obovate, rounded at apex, the inner oblong, narrower; achenes ovoid, 2 mm. long, the triquetrous beak equaling the body.

Roadsides, hillsides, and plains, mainly Upper Sonoran Zone, Sacramento Valley, California, southward into northern Baja California, where it enters the margins of the Sonoran Desert east of Rosario. July–Oct.

**18. Eriogonum fastigiatum** Parry, W. Amer. Sci. **6:** 102, 1889

Matted perennial with numerous closely crowded, fastigiate branches 0.5–1.5 dm. long from prostrate, woody stems; old stems dark red-brown with shreddy bark; branches 0.5–1 mm. in diameter, densely white-tomentulose, the internodes 1–4 mm. long; leaves alternate, linear or linear-elliptic, acute, 3–7 mm. long, revolute, green and sparsely lanate above, white-tomentulose beneath, the base expanded to form a membranous subglabrous ring around stem; involucres solitary or, owing to shortness of internodes, appearing to be in pairs near apex of branches, campanulate, about 2 mm. long, pubescent, strongly 5-ribbed, the intervals membranous, teeth about equaling the shallow tube, acute; flowers several, exserted 1–3 mm. from involucres on slender, glabrous pedicels, white or pale rose with deep rose midribs, 2–2.5 mm. long, glabrous; outer calyx lobes obovate, rounded at apex, inner ones slightly narrower but of equal length; achenes ovoid, about 2 mm. long, minutely papillate, beak triquetrous, slightly shorter than the body.

Sandy clay or rocky soil, Lower and Upper Sonoran Zones, near the west coast of Baja California from near San Antonio del Mar to the mesas 6–8 miles east of Rosario. Barely entering the desert. Feb.–May.

**19. Eriogonum fasciculatum** Benth., Trans. Linn. Soc. **17:** 411, 1837, subsp. **fasciculatum**

*Eriogonum rosmarinifolium* Nutt., Jour. Acad. Phila. II, **1:** 164, 1848.
*Eriogonum fasciculatum* var. *oleifolium* Gandoger, Bull. Soc. Bot. Belg. **42:** 189, 1906.
*Eriogonum fasciculatum* var. *maritimum* Parish, Muhlenbergia **3:** 59, 1907.
*Eriogonum fasciculatum* subsp. *typicum* Stokes, Gen. Eriog. 85, 1936.

Rounded shrub 6–15 dm. high with spreading, leafy, thinly pubescent branches, bark gray, shreddy; leaves arranged in alternating fascicles on short, dwarf branches, linear to linear-oblanceolate, 5–18 mm. long, strongly revolute, leathery, green and glabrous or sparsely villosulous above, white-tomentose beneath; inflorescence cymose on slender, naked peduncles 5–25 cm. long, these glabrous or sparingly puberulent; bracts subtending the cymes usually foliaceous, 3–10 mm. long; involucres crowded together in the cymes, cylindrical-prismatic, 3–4 mm. long, sparsely arachnoid-lanate at first but soon glabrate, shallowly and acutely 5-toothed; flowers numerous, pale pink or white with deeper rose or greenish midribs, 2.5–3 mm. long; outer calyx lobes broadly oblong to obovate, glabrous on outer surface, villous near base within, inner lobes similar or slightly broader; achenes about 3–3.5 mm. long, ovoid, brown, shining, beak triquetrous.

Near the coast, Lower and Upper Sonoran Zones, from the vicinity of Santa Barbara, California, to central Baja California. May–Nov.

**19*a*. Eriogonum fasciculatum** subsp. **flavoviride** (Munz & Jhtn.) Stokes, Gen. Eriog. 86, 1936

*Eriogonum fasciculatum* var. *flavoviride* Munz & Jhtn., Bull. Torrey Club **49:** 350, 1922.

Smaller plants, usually 5 dm. tall or less, with glabrous or subglabrous yellow-green leaves. March.

Northwestern margin of the Colorado Desert and adjacent part of the Mojave Desert. March–May.

**19*b*. Eriogonum fasciculatum** subsp. **polifolium** (Benth.) Stokes, Gen. Eriog. 86, 1936

*Eriogonum polifolium* Benth., in A. DC., Prod. **14:** 12, 1856.

Stems, leaves, and scapes canescent-pubescent; calyces puberulent on outer surface; otherwise as in subsp. *fasciculatum*.

San Joaquin Valley and Inyo County, California, to southern Utah, Arizona, and northern Baja California. April–Nov.

**20. Eriogonum orcuttianum** S. Wats., Proc. Amer. Acad. **20:** 371, 1885

*Eriogonum peninsulare* K. Brandegee, Erythea **5:** 79, 1897.

Low shrub with divaricately branched, zigzag stems closely crowded to form rounded clumps 2–6 dm. high and up to 4–12 dm. wide; older stems with shreddy gray to reddish bark, newer growth cinereous with close, fine, spreading hairs; leaves alternate, blades orbicular to broadly ovate or obovate, 1–2 cm. wide, 1–2.5 cm. long, sparsely villous on both surfaces when young but green and eventually subglabrous, on petioles 3–8 mm. long, these spreadingly cinereous-villous; branches of inflorescence zigzag and divaricate; involucres sessile, mostly solitary in axils and terminal on branchlets, broadly cup-shaped, 5-cleft almost or entirely to base, lobes lance-subulate, pubescent, 2–3 mm. long, spreading, many-flowered; calyces white to pink, rounded-turbinate at base, villous-puberulent within and without, 2.5–3 mm. high, lobes spreading; outer calyx lobes broadly ovate or nearly orbicular, abruptly narrowed to broad base, with greenish midrib, the inner about two-thirds as broad; achenes (immature) ovoid-triquetrous, beak villous.

Desert slopes and washes, Lower Sonoran Zone, eastern flanks of the Sierra Juárez and Sierra San Pedro Mártir to Las Animas Bay and Paraíso, Baja California. March–May.

**21. Eriogonum deserticola** S. Wats., Proc. Amer. Acad. **26:** 125, 1891

A diffusely branched shrub of shifting sand dunes, frequently with a woody stem 2–5 cm. in diameter several meters long covered by the sand and forming arcs 1–2 m. long when the sand flows away; young branches white-tomentose but soon glabrate and light green; leaves hoary-tomentose, blades broadly ovate to suborbicular, rounded at apex, broadly cuneate to truncate at base, 5–15 mm. broad, margins crispate-undulate; petioles equaling or slightly exceeding blades, lanate; leaves soon deciduous, the plant usually leafless at flowering time; bracts broadly deltoid, 0.5–1.5 mm. long, usually with a small tuft of wool persisting in their axils; involucres solitary and subsessile to short-pedunculate in axils of the bracts, turbinate-campanulate, about 2 mm. long, finely lanate, but eventually glabrate, teeth minute, few-flowered; flowers exserted on slender pedicels 3–5 mm. long, yellow with greenish or reddish midribs, 3–3.5 mm. long, loosely lanate on outer surface; outer calyx lobes oblong, rounded at apex, margins membranous, glabrous inside except for tuft of wool toward base; inner calyx lobes similar but slightly narrower; achenes narrowly ovoid-lenticular, 3.5–4 mm. long, reddish brown, body sparsely arachnoid-lanate, beak 1 mm. long, triquetrous, minutely papillate.

Sand dunes and washes (Lower Sonoran Zone) of the Colorado Desert, chiefly along the western margins and in the dune areas between Yuma, Arizona, and Calexico, California. Sept.–Dec.

**22. Eriogonum plumatella** Dur. & Hilg. var. **jaegeri** (Munz & Jhtn.) Stokes ex Munz, Man. S. Calif. 120, 1935

*Eriogonum nodosum* var. *jaegeri* Munz & Jhtn., Bull. Torrey Club **49:** 350, 1922.

Openly branched perennial 3–6 dm. high with several divaricately branched stems, woody at base and densely hoary-tomentose below inflorescence; leaves on lower parts only, narrowly oblong to oblanceolate, 2–4 mm. broad, 6–15 mm. long, hoary-lanate on both surfaces, acute, gradually narrowed to a petiole 1–5 mm. long, margins finely undulate; bracts narrowly deltoid to linear-subulate, 1–7 mm. long, glabrous without, lanate within; the leafless branches of inflorescence glabrous, divaricate-spreading; involucres solitary at proximal nodes and terminal on branchlets, cylindric-campanulate, 2–2.5 mm. long, glabrous, distinctly 5-ribbed, short teeth rounded; flowers barely exserted on filiform pedicels, 2–2.5 mm. high, white or pale yellow, glabrous; outer calyx lobes obovate, erect or slightly spreading, inner ones somewhat narrower, of equal length; achenes ovoid, light brown, 2.5–3 mm. long, the triquetrous beak papillate-scaberulous, about equaling the turgid body.

Known only from the eastern and southern desert slopes of the Little San Bernardino Mountains, western San Bernardino and Riverside Counties, California. Aug.–Nov.

**23. Eriogonum wrightii** Torr. ex Benth., in A. DC., Prod. **14:** 15, 1856, subsp. **wrightii**

*Eriogonum trachygonum* subsp. *wrightii* Stokes, Gen. Eriog. 58, 1936.
*Eriogonum wrightii* subsp. *typicum* Stokes, Leafl. West. Bot. **2:** 47, 1937.

Erect or ascending perennial 3–6 dm. high from woody rootstock and base; older branches deep red-brown with shreddy bark, the younger hoary with fine, appressed tomentum, sparingly and virgately branched, leafy on basal 25–35 cm.; leaves oblanceolate or elliptic, acute, 3–6 mm. broad, 1–2 cm. long, lanate on both surfaces, margins plane or crispate-undulate, gradually narrowed to a petiole 2–6 mm. long or rarely subsessile; inflorescence virgate, branches simple, erect; bracts linear-subulate, 2–4 mm. long; involucres solitary, sessile, cylindric-turbinate, 2–2.5 mm. long, distinctly 5-ribbed and 5-toothed, finely appressed-lanate, teeth minute, erect; flowers several in each involucre, 2.5–3.5 mm. long, white or faint rose with deep rose midribs, exserted from involucres 1–2 mm. on filiform pedicels, glabrous, outer calyx lobes obovate, rounded at apex, 1.8–2.2 mm. wide, the inner slightly narrower; achenes papillate-scaberulous on stoutish, triquetrous beak.

Gravelly slopes and rocky hillsides, Lower and Upper Sonoran Zones, Arizona to western Texas. June–Oct.

**23*a*. Eriogonum wrightii** subsp. **membranaceum** Stokes, Leafl. West. Bot. **2:** 47, 1937

*Eriogonum wrightii* var. *membranaceum* Stokes ex Jepson, Fl. Calif. **1:** 416, 1914.
*Eriogonum trachygonum* subsp. *membranaceum* Stokes, Gen. Eriog. 59, 1936.

Similar to subsp. *wrightii* but with narrower, somewhat more crowded, tightly revolute leaves, more robust stems, and persistent membranous rings formed by expanded base of leaves on lower parts of stems of the final two seasons' growth.

Interior dry slopes, mostly along the desert side, San Bernardino Mountains, California, to the Sierra San Pedro Mártir, Baja California.

**24. Eriogonum nodosum** Small, Bull. Torrey Club **25:** 49, 1898

Low, sparsely leaved shrub 3–10 dm. high with ascending, di- and trichotomously branching, hoary, lanate stems; leaves linear to oblanceolate or elliptic, 1.5–6 mm. wide, 8–25 mm. long, usually revolute, leathery, acute at apex, gradually narrowed to a petiole 2–6 mm. long, closely white-lanate, present whole length of stems except on the flowering branches; inflorescence of slender ascending branches and shorter, more or less spreading ones of 2–10 internodes 3–15 mm. long; bracts lance-deltoid, 1.5–3.5 mm. long, reddish or brown, thinly lanate to subglabrate; involucres turbinate-campanulate, 2–2.5 mm. long, densely lanate between 5 distinct, sparingly lanate ribs; flowers about 3 mm. long, exserted 1–3 mm. on filiform, glabrous pedicels, narrowly turbinate at base, white to pale rose with darker rose midribs, glabrous; outer calyx lobes narrowly obovate, the inner oblong, about one-half to two-thirds as wide as the outer; achenes equaling calyx lobes or slightly exceeding them, triquetrous beak scaberulous.

Dry rocky and gravelly slopes, Lower Sonoran Zone, along the western margin of the Colorado Desert; near Puerto Libertad, Sonora. Aug.–Dec.

**25. Eriogonum pringlei** Coult. & Fish., Bot. Gaz. **17**: 351, 1892

*Eriogonum trachygonum* subsp. *pringlei* Stokes, Gen. Eriog. 59, 1936.
*Eriogonum wrightii* subsp. *pringlei* Stokes, Leafl. West. Bot. **2**: 47, 1937.

Erect or ascending perennial 3–7 dm. high from a woody base, with shreddy, red-brown bark on old wood and dense, appressed white wool on younger growth; leaves few, on basal part of seasonal growth, broadly elliptic, 2–5 mm. wide, 7–15 mm. long, short-petiolate, hoary-lanate, reduced to subulate or deltoid-subulate bracts 2–3 mm. long above; inflorescence of paniculately arranged virgate branches 1–15 cm. long, its internodes 5–12 mm. long; involucres solitary, sessile, cylindric-campanulate, 1–1.4 mm. long, prominently 5-ribbed, arachnoid-lanate or ribs often glabrate and reddish in age, several-flowered, teeth minute, slightly incurved; pedicels filiform, glabrous, equaling or slightly exceeding involucres; calyces glabrous, pale yellow with roseate or deep pink midribs, 1.4–1.8 mm. long at anthesis, lobes oblong-obovate, rounded and somewhat spreading at tips, all similar or the inner slightly narrower than the outer; achenes ovoid with short, triquetrous beaks, glabrous, or beaks minutely and sparingly papillate on angles.

Rocky desert hills, Lower Sonoran Zone, Maricopa, Pinal, Pima, and Yuma Counties, Arizona. March–Oct.

## 7. RUMEX L., Sp. Pl. 333, 1753

Annual or perennial herbs with coarse stems and leaves and small, greenish flowers. Leaves chiefly basal, the petioles of the alternate cauline ones somewhat sheathing at the base. The flowers in several- to many-flowered glomerules and paniculate or paniculate-racemose. Sepals 6, the outer 3 herbaceous, spreading in fruit, often somewhat united at base, the inner 3 longer, somewhat convergent over the achene, venose-reticulate, often bearing a callous grain on the back, usually somewhat colored. Stamens 6, included. Styles 3; stigmas feathery. Achenes trigonous.

Inner fruiting sepals 6–14 mm. long, deeply cordate at base (callous grain absent) ........ .......................................................... 1. *R. hymenosepalus*
Inner fruiting sepals 2–6 mm. long, truncate or rounded at base:
  Vegetative axillary shoots present; stems often decumbent:
    Fruiting calyx lobes 5–6 mm. long; leaves ovate-lanceolate ............. 2. *R. ellipticus*
    Fruiting calyx lobes 2–4 mm. long; leaves lanceolate or linear-lanceolate:
      Valves grainless ............................................ 3. *R. californicus*

Valves, or at least one of them, bearing callous grains:
Inner fruiting sepals 2.4–3 mm. long, usually only 1 callous-grained; grain wider than margin of sepal .................................... 4. *R. salicifolius*
Inner fruiting sepals 4–5 mm. long, 2 or 3 callous-grained; grain narrower than margin of sepal .......................................... 5. *R. mexicanus*
Vegetative axillary shoots rarely present; stems erect:
Valves entire; leaves more or less undulate or crisped along margins:
Leaves strongly undulate-margined; inner fruiting sepals 4–6 mm. long, broadly ovate; callous grains narrower than margins of valves .................. 6. *R. crispus*
Leaves weakly crispate-margined; inner fruiting sepals 2–3 mm. long, narrowly ovate; callous grains broader than margins of valves ............. 7. *R. conglomeratus*
Valves shallowly dentate toward base; leaves plane, margins not crisped..8. *R. violascens*

**1. Rumex hymenosepalus** Torr., Bot. Mex. Bound. 177, 1859
*Rumex saxei* Kell., Pacif. Rural Press **17:** 371 (June 7), 1879.
*Rumex arizonicus* Britt., Trans. N.Y. Acad. **7:** 73, 1889.

Coarse perennial 3–10 dm. high from 1–several tuberous roots; leaves oblong to elliptic-oblong, 1–3 dm. long, 3–10 cm. wide, thick, wavy-margined, on stout petioles shorter than blades; sheathing stipules 1–3 cm. long; panicles compact but profusely branching, 1–4 dm. long, pink to bright red; pedicels slender, 8–12 mm. long, jointed near middle; outer sepals linear or elliptic, 1.5–2.5 mm. long, spreading; inner fruiting sepals cordate-ovate, membranous, reticulate, bright red, 6–14 mm. long, devoid of callous grains; achene reddish brown, shining, 4–5 mm. long.

Arid plains and washes, mostly in sandy or gravelly soil, Upper and Lower Sonoran Zones, southern Kern County, California, to Wyoming, Colorado, Texas, Chihuahua, Sonora, and central Baja California. Jan.–April.

**2. Rumex ellipticus** Greene, Pittonia **4:** 234, 1901

Perennial 2–5 dm. high with flexuous, slender stems usually with vegetative shoots in lower axils and frequently decumbent; stipules pale brown, 1–1.5 cm. long, caducous; leaves lanceolate or ovate-lanceolate, 6–10 cm. long, acuminate at apex, margins plane and entire; petioles about equaling width of blades; inflorescence paniculate, its branches short; glomerules approximate, devoid of leaves; pedicels slender, equaling or shorter than sepals, jointed near base; inner fruiting sepals broadly ovate, 5–6 mm. long, 4–5 mm. wide, reticulate-nerved, devoid of callous grains or 1 valve with a slightly thickened midrib; achene about 3 mm. long.

Fields and ditch banks, Lower and Upper Sonoran Zones, central Arizona to Texas. March–July.

**3. Rumex californicus** Rechinger f., Rep. Spec. Nov. **40:** 297, 1936
*Rumex salicifolius* var. *denticulatus* Torr., Bot. Mex. Bound. 178, 1859.

Perennial with more or less decumbent stems 3–6 dm. long, bearing vegetative shoots in the lower axils; leaves linear-lanceolate, tapering gradually toward each end, plane, 5–15 mm. wide, 3–15 cm. long, on petioles 5–30 mm. long; inflorescence openly paniculate, with reduced leaves subtending lower branches and glomerules; pedicels slender, 1–2.5 mm. long, shorter than or barely equaling fruiting calyces; inner fruiting sepals about 3 mm. long, 2.5 mm. wide, broadly triangular, reddish brown, distinctly reticulate-veined, acute at apex, sometimes minutely and irregularly denticulate at base, devoid of a callous grain; achene dark brown, about 2 mm. long, shining.

Waste places and ditch banks, Lower and Upper Sonoran Zones, southern Oregon to southern California and southwestern Arizona. March–July.

**4. Rumex salicifolius** Weinm., Flora **4:** 28, 1821

Erect or less frequently decumbent perennial 3–9 dm. high with vegetative shoots in lower axils; sheathing stipules 5–10 mm. long, membranous, caducous; leaves plane, narrowly lanceolate, 4–13 cm. long, 7–20 mm. wide, gradually tapering toward each end, on petioles 5–30 mm. long; inflorescence simply paniculate with few lateral branches to 6 cm. long and linear-lanceolate leaves subtending lower branches; pedicels shorter than to slightly exceeding sepals, slender, jointed near base; inner fruiting sepals 2.4–3 mm. long, about 2 mm. wide, triangular, acute at apex, reticulate-veined, usually 1 bearing a large oblong grain wider than the narrow surrounding membranous margin, the other 2 valves usually grainless; achene shining dark brown, 1.5 mm. long and wide.

Fields, roadsides, and waste places, upper margin of Lower Sonoran Zone to Transition Zone, northern California south into northern Baja California and Sonora, Mexico. March–Sept.

**5. Rumex mexicanus** Meisn., in A. DC., Prod. **14:** 45, 1856

Erect or decumbent perennial 3–10 dm. high with vegetative shoots in lower axils; sheathing stipules 5–20 mm. long, membranous, caducous; petioles 1–3 cm. long; leaf blades linear-lanceolate, five to seven times as long as broad, 3–20 cm. long, tapering toward each end, plane or slightly undulate- or crispate-margined, acute to attenuate at apex; inflorescence paniculate, 1–3 dm. long, lateral branches to 10 cm. long, the lower subtended by slightly reduced leaves; pedicels stoutish, about equal to or shorter than fruiting sepals, somewhat thickened and jointed near base; inner fruiting sepals broadly triangular-ovate, acute to obtuse at apex, 4–5 mm. long, strongly reticulate-veined, all 3 callous-grained, the grain narrower than the membranous margins; achene 2–3 mm. long, brownish, shining.

Along ditches, on flats and gentle slopes, Lower and Upper Sonoran Zones into Tropical Zone, New Mexico and south to San Luis Potosí and Michoacán; less abundant in Sonora and Arizona. April–Aug.

**6. Rumex crispus** L., Sp. Pl. 335, 1753

Erect perennial 6–12 dm. high with simple stems usually devoid of axillary vegetative shoots; leaves 1–3 dm. long, oblong-lanceolate, strongly veined, margins strongly undulate; petioles shorter than to equaling the blades; inflorescence strict panicles with ascending branches 1–2 dm. long, the lower subtended by petiolate leaves, glomerules leafless, dense, reddish brown in fruit; pedicels slender, 5–10 mm. long, with swollen joints below the middle; inner fruiting sepals broadly ovate to nearly orbicular, 4–6 mm. long, all with ellipsoid to subglobose, smooth callous grains narrower and shorter than reticulate-veined margins; achenes broadly ovoid-trigonous, 2–3 mm. long, dark brown to nearly black, shining.

Naturalized weed from Europe and found almost throughout the north temperate zone. April–July.

**7. Rumex conglomeratus** Murr., Prod. Fl. Goett. 52, 1770

Erect perennial 8–15 dm. high with open, slenderly branched panicles; sheathing stipules about 1 cm. long, closely investing the stem, caducous; petioles slender, 4–8 cm.

long; leaf blades oblong, obtusely rounded to truncate or cordate at base, acute at apex, margins slightly crispate-undulate; panicles 1–5 dm. long, leafy, lateral branches 1–3 dm. long, ascending; glomerules interrupted; pedicels 2–4 mm. long, jointed near base, geniculate in fruit; inner fruiting sepals narrowly ovate, 2–3 mm. long, rounded at base, acutish at apex, all with prominent, smooth, entire callous grains about equal in width to finely reticulate-veined margins; achene broadly ovoid-trigonous, dark red-brown, 1.2–1.8 mm. long, nearly as wide.

Naturalized from Europe, Lower Sonoran Zone to Transition Zone, British Columbia to northern Baja California and southwestern Arizona. March–July.

### 8. **Rumex violascens** Rechinger f., Rep. Spec. Nov. **39**: 171, 1936

Erect annual or biennial 4–8 dm. high with simple stems; sheathing stipules 5–15 mm. long, caducous; petiole 2–6 cm. long, slender; leaf blades oblanceolate to elliptic-obovate, 1–2.5 cm. wide, 4–10 cm. long, broadly cuneate to truncate at base, acute to obtuse at apex, margin faintly crispate or plane; upper leaves reduced, subsessile; inflorescence paniculate, leafy, glomerules mostly remote, dense; pedicels slender, to twice as long as fruiting valves, jointed very near base; inner fruiting sepals triangular-ovate, 2.5–4 mm. long, acute to obtuse, strongly reticulate-veined, shallowly but distinctly denticulate toward base, unequally callous-grained, the grain smooth dorsally but wrinkled to subdenticulate along sides; achenes 2–2.5 mm. long, light brown or yellowish, smooth, broadest near middle.

Mainly Lower Sonoran Zone, San Joaquin and Imperial Valleys, California, to Texas, Sonora, and central Baja California. April–July.

**Rumex obtusifolius** L., Sp. Pl. 335, 1753, differing from *R. violascens* in having larger lower leaves 10–35 cm. long; less markedly interrupted inflorescences; and valves of fruit hastate, 4–5 instead of 6–9 lateral teeth or bristles, and usually bearing a grain-like callosity on only one of its valves, has been collected recently at Mulegé, Baja California. April–Dec.

## 8. **POLYGONUM** L., Sp. Pl. 359, 1753

Annual or perennial herbs from fibrous roots or heavy rootstocks, with entire, alternate leaves bearing entire, ciliate, or lacerate, scarious sheathing stipules (ocreae). Flowers small, on slender, jointed pedicels, variously disposed. Calyx white, rose, or greenish, 4–6-parted, frequently petal-like, withering or persistent, erect in fruit. Stamens 3–9. Styles 2 and achenes lenticular, or 3 and the achene trigonous, mostly enclosed in the calyx. Embryo curved, filling a groove at an angle of the mealy endosperm.

Leaves small, 0.5–2.5 cm. long; stipules lacerate or 2-lobed; flowers solitary, in pairs, or in few-flowered fascicles in axils of somewhat reduced leaves; plants of roadsides, pathways, and fields:
- Plants prostrate or decumbent; sepals 1.5–2 mm. long; achenes striate, dull . . 1. *P. aviculare*
- Plants erect; sepals 2–3 mm. long; achenes smooth, shining . . . . . . . . . . . 2. *P. argyrocoleon*

Leaves 3.5 cm. long or more; stipules truncate or oblique, not 2-lobed nor regularly lacerate; flowers in terminal or axillary spiciform racemes, these often paniculately arranged; plants growing in moist soil or shallow water:
- Stipule sheaths not ciliate (achenes lenticular; style branches 2) . . . . . . 3. *P. lapathifolium*
- Stipule sheaths bristly-ciliate:
  - Calyx conspicuously punctate-glandular; leaf blades 3.5–8 cm. long, two to four times as long as broad; achenes consistently trigonous . . . . . . . . . . . . . . . . . 4. *P. punctatum*

Calyx not punctate-glandular; leaf blades 6–20 cm. long, six to ten times as long as broad; achenes lenticular or trigonous:
Stipule sheaths erect at summit, closely investing stem; racemes usually several in each inflorescence; plants usually in moist soil:
Leaves obviously strigose over entire lower surface, though more heavily so on midrib and margins; internodes not conspicuously swollen .................... .................................. 5. *P. hydropiperoides* var. *persicarioides*
Leaves glabrous or with few hairs on midrib and margins only; internodes conspicuously swollen, subfusiform ............................ 6. *P. fusiforme*
Stipule sheaths spreading at summit; raceme usually solitary; plants mainly aquatic.. ............................................ 7. *P. natans* f. *hartwrightii*

## 1. **Polygonum aviculare** L., Sp. Pl. 362, 1753

Prostrate or decumbent annual (sometimes biennial) with blue-green, striate, wiry stems 1–6 dm. long; leaves lanceolate, oblong, or oblanceolate, 5–20 mm. long, 1.5–5 mm. wide, acute or rarely obtuse, cuneate at base, glabrous; stipule sheath 3–6 mm. long, more or less lacerate, hyaline to faintly rosaceous; flowers in 1–6-flowered axillary fascicles; pedicels 1–4 mm. long; calyx lobes 1.5–2 mm. long, oblong, greenish with white or pinkish margins, erect, surpassed by tip of the achene; stamens usually 8, included; style branches 3; achene trigonous, 2.2–2.6 mm. long, minutely granular-striate, dark brown, dull.

A weed in waste places, along roadsides and paths, and in dooryards, throughout most of temperate North America; uncommon in the desert. Naturalized from Europe. March–Sept.

## 2. **Polygonum argyrocoleon** Steud. ex Kunze, Linnaea **20**: 17, 1847

Erect annual 1–6 dm. high with simple or moderately branched, finely striate stems; leaves elliptic-lanceolate to oblong, tapering gradually to each end, 1.5–7 mm. wide, 1–3.5 cm. long, glabrous, faintly grayish green, midrib distinct; petioles 1–3 mm. long, narrowly winged; stipule sheaths lacerate, 5–10 mm. long, hyaline, silvery; flowers in axillary fascicles, forming subspiciform terminal inflorescences; pedicels slender, 2–5 mm. long; calyx 2–3 mm. long, the lobes pink to light red with a light green center; stamens 3–6; style branches 3; achenes triquetrous, 2 mm. long or less, dark brown, shining, completely enclosed by the closely investing sepals.

Common weed in alfalfa fields and along ditches in Imperial Valley, sporadic northward into the San Joaquin Valley, California. Naturalized from Persia. April–Sept.

## 3. **Polygonum lapathifolium** L., Sp. Pl. 360, 1753

*Polygonum persicaria* var. *lapathifolium* Meisn., Monogr. 69, 1826.

Stout annual with erect or ascending, simple to moderately branched, glabrous stems 3–15 dm. long; stipule sheaths membranous, truncate, 1–2 cm. long, conspicuously red-veined, friable; leaves lanceolate, broadly cuneate at base, long-acuminate at apex, 6–10 cm. long, 0.5–5 cm. broad, glabrous or with us usually strigose-scabrous below on midrib and along margins, inconspicuously but distinctly punctate on lower surface; peduncles slender, minutely and usually sparsely glandular; spikes slender, 1–8 cm. long, dense, paniculately disposed, erect or nodding; calyx 1.5–2 mm. long, white to pinkish; stamens 6; style branches 2; achene lenticular, more or less biconcave, about equaling sepals, dark brown, moderately shining.

In moist places throughout North America except at high altitudes and latitudes. June–Sept.

**4. Polygonum punctatum** Ell., Bot. S.C. & Ga. **1:** 455, 1817

*Polygonum acre* H. B. K., Nov. Gen. & Sp. **2:** 143, 1817.

Glabrous perennial or often flowering first year with simple to much-branched, slender, erect or ascending stems 3–15 dm. long; stipule sheath cylindrical on young stems, often split on older parts, 1–1.5 cm. long, truncate, glabrous to sparsely strigillose, fringed with bristles to half as long as cylindric part; leaf blades narrowly lanceolate to oblong-lanceolate, 3–15 cm. long, acuminate at both ends, glabrous, margin and midrib subscabrous with short, stout, forward-pointing hairs, short-petiolate; inflorescence paniculate, narrow racemes 3–8 cm. long, erect, pedunculate; ocreolae funnelform, 2–3 mm. long, sparingly bristly-ciliate; pedicels slender, 3–4 mm. long; calyx 2–2.5 mm. long, greenish, conspicuously punctate-glandular, lobes ovate to oblong; stamens 8; style branches 3; achene triquetrous or sometimes lenticular, about 2.5 mm. long, black, smooth, shining.

Margins of ponds, along ditches, and in moist ground, British Columbia to the Atlantic coast and south into South America. Rare in the desert regions. April–Oct.

**5. Polygonum hydropiperoides** Michx. var. **persicarioides** (H. B. K.) Stanford, Rhodora **28:** 27, 1926

*Polygonum persicarioides* H. B. K., Nov. Gen. & Sp. **2:** 143, 1817.
*Persicaria persicarioides* Small, Fl. SE. U.S. 378, 1903.

Decumbent or erect perennial with glabrous to strigillose, simple or branched stems 3–10 dm. long; stipule sheaths cylindrical to narrowly funnelform, 1–2 cm. long, glabrous to strigose, truncate or oblique and bristly-ciliate at apex; petioles 1–2 cm. long, or leaves sometimes subsessile; leaf blades narrowly lanceolate, acuminate at both ends, 5–20 cm. long, finely strigillose, inconspicuously punctate, more densely and coarsely strigose on midrib and margins; inflorescence paniculate; racemes erect, 2–7 cm. long, narrowly oblong to linear, loosely flowered; ocreolae obliquely funnelform, 2.5–3 mm. long, bristle-fringed at summit; pedicels 1–3 mm. long; calyx 2–2.5 mm. long, rosaceous, tinged with green, with oblong, obtuse lobes; stamens 8; achenes 2.5–3 mm. long, lenticular or triquetrous, black, finely granular and shining.

In wet soil or shallow water, from Nebraska and Texas south through Mexico, Chile, and Argentina. Cape Region of Baja California. June–Oct.

**6. Polygonum fusiforme** Greene, Erythea **1:** 259, 1893

Stout, nearly glabrous perennial with decumbent, assurgent stems 8–20 dm. long; internodes conspicuously swollen above nodes, subfusiform; stipule sheaths cylindrical or lower ones funnelform and ruptured by swollen internodes, 1–2 cm. long, long-ciliate at apex or naked owing to rupturing of membranous sheath, strigillose or glabrous; leaves narrowly lanceolate, 1–2 cm. broad, 4–12 cm. long, acuminate at both ends, subsessile or short-petiolate, glabrous or sparsely strigose on midrib and margins; petioles lacking to 1.5 cm. long, strigillose; inflorescence a simple or few-branched panicle, the spicate racemes 1–3 cm. long; ocreolae obliquely funnelform, 2–2.5 mm. long, pinkish, ciliate or sometimes naked; pedicels 1–2 mm. long; calyx 2–2.5 mm. long, 4- or rarely 5-parted, lobes broadly oblong, rose or pink, the outer pair narrower than the inner; stamens 4–7; achene 2–2.2 mm. long, lenticular, dark brown, smooth, shining.

A poorly known plant from along the Colorado River near Needles, California. June–Sept.

**7. Polygonum natans** (Michx.) Eaton f. **hartwrightii** (A. Gray) Stanford, Rhodora **27:** 89, 1925
(For complete synonymy see above reference.)

Perennial, terrestrial form with creeping, ascending, or suberect hispid stems 3–7 dm. long or these glabrous if growing in water; stipule sheaths cylindrical, 1–2 cm. long, hispid, apex expanded-spreading and strongly ciliate; petioles nearly lacking to 3 cm. long, hispid; leaf blades lanceolate, 1–3 cm. wide, 5–18 cm. long, obtuse to acute at base, acute at apex, short-hispid on both surfaces; inflorescence a simple spicate, erect raceme 1–3 cm. long, flowers crowded; ocreolae 1–2 mm. long, obliquely funnelform, ciliate; pedicels 1–2 mm. long; calyx deep rose, 3–4.5 mm. long; stamens 5, slightly exserted; style branches 2; achene biconvex-lenticular, black, smooth and shining or minutely granular and dull.

Margins of ponds and slow-running streams, throughout the United States; "Cantillas Mts.," Baja California. June–Sept.

## 9. ANTIGONON Endl., Gen. 310, 1837

Fruticose at base, a scandent herbaceous vine with branching tendrils in inflorescence. Leaves alternate, deltoid or cordate; ocreae minute. Inflorescence fascicles of flowers in simple or loosely branched racemes, each branch terminating in branched tendrils. Flowers perfect. Calyx lobes 6, erect, the 3 outer cordate, accrescent, often brightly colored, the inner oblong, smaller, one of these often abortive. Stamens 7–8; filaments connate into a ring at base. Ovary 3-angled; styles 3, short; stigmas capitate. Achenes ovoid-trigonous. Seed subglobose, 3–6-lobed. Embryo eccentric, curved, with narrowly oblong cotyledons. Endosperm mealy.

**1. Antigonon leptopus** Hook. & Arn., Bot. Beechey 308, *pl. 69,* 1839–1840
*Antigonon cordatum* Mart. & Gal., Bull. Acad. Brux. **10**[1]**:** 14, 1843.

A scrambling to strongly climbing vine often reaching tops of trees 40 m. high or more; stems minutely and densely puberulent or rarely glabrous; ocreae reduced to mere transverse line with a subulate projection 1–4 mm. high directly above petiole; petioles slender, 1–3.5 cm. long; leaf blades cordate to deltoid, acuminate, 2–12 cm. wide, 4–20 cm. long, not decurrent on petiole, glabrous to sparingly puberulent above, slightly paler and sparsely to densely puberulent beneath, especially along reticulate veins; inflorescence a paniculately arranged raceme, 10–50 cm. long, glabrous to densely puberulent; flowers in fascicles of 2–10, on slender pedicels 3–15 mm. long; calyx lobes broadly elliptic, cordate at base, at first greenish but soon purplish red, showy, 1.5–2.5 cm. long and conspicuously veined at maturity, erect; achenes ovoid, trigonous above, about 1 cm. long, brown, glabrous, shining.

Along washes and canyons, Lower Sonoran Zone, central Sonora to Chihuahua and Oaxaca and from central Baja California to Cape San Lucas. April–Nov.

## 10. COCCOLOBA L., Syst. Nat. ed. 10, 1007, 1759

Evergreen shrubs, trees, or rarely climbing vines with alternate, coriaceous, entire, mostly orbicular leaves and thin bark. Ocreae membranous, funnelform or cylindric, truncate, lobed, or acuminate. Flowers perfect, small and inconspicuous, fasciculate in axils of small bracts arranged on slender axillary or terminal spicate racemes. Sepals 5, green, about equal. Hypanthium fleshy, accrescent. Petals lacking. Stamens 8, with slender filaments inserted on hypanthium. Ovary superior, 3-angled. Styles 3. Achene ovoid or globose, bearing a crustaceous or bony pericarp and invested in the accrescent hypanthium. Ovule erect. Seed 3–6-lobed, testa membranous. Embryo with cordate cotyledons; endosperm mealy.

**1. Coccoloba goldmanii** Standley, Contr. U.S. Nat. Herb. **23:** 245, 1922

Tree to 10 m. high with smooth, light gray bark, trunk up to 1 m. in diameter, and young branches finely puberulent at first but soon glabrate; ocreae funnelform, 4–8 mm. long, acuminate or those at base of rachis truncate, fimbriate-tufted at apex, dark brown, minutely but closely puberulent; petioles stout, 5–12 mm. long; leaf blades orbicular to broadly elliptic, 5–10 cm. long, 4–7 cm. wide, rounded at apex, rounded to emarginate at base, thick-coriaceous, margins entire, somewhat undulate, green and glabrous to sparsely puberulent along veins on upper surface, paler and conspicuously reticulate-veined and short-pilose along main veins beneath; racemes slender, 10–30 cm. long, rachis puberulent to short-pilose with simple, brownish hairs; ocreolae broadly cup-shaped, about 0.5 mm. deep, puberulent; pedicels slender, 0.6–1 mm. long; flowers usually 2 in each ocreola, hypanthium 0.6–1 mm. long at anthesis, stamens exserted, calyx lobes broadly ovate, spreading, 1.2–1.4 mm. long; fruiting hypanthium 7–8 mm. long, closely appressed to ovoid achene, glabrous.

In narrow, shaded canyons, Lower Sonoran and Tropical Zones, at San Pedro Bay near Guaymas, Sonora, southward into Sinaloa. Aug.–Dec.

## Family 35. CHENOPODIACEAE. Goosefoot Family

Herbs, shrubs, or rarely small trees with glabrous or pubescent, often succulent, leaves and frequently articulate stems. Leaves opposite or alternate, estipulate, sessile or petiolate or reduced to scales, entire, dentate or lobed. Flowers perfect, polygamous, monoecious or dioecious, variously disposed; perianth simple or wanting in pistillate flowers, of 2–5 or sometimes only 1 segment. Stamens equaling or fewer than perianth lobes and opposite them, hypogynous or borne on a disk or on base of the perianth; filaments filiform to subulate; anthers dorsifixed. Ovary superior, rarely adnate to base of the perianth, 1-celled. Style short or elongate. Stigmas 2–5, capitate or filiform. Ovule solitary. Fruit a membranous, coriaceous, or fleshy utricle, indehiscent or circumscissile, pericarp free or adherent to seed. Seed lenticular, globose, or reniform, smooth or granulate. Embryo annular or nearly so, or spirally coiled, surrounding fleshy or mealy endosperm, or this nearly lacking.

Sepals imbricate; leaves all opposite; low herb ........ 1. *Nitrophila*
Sepals not noticeably imbricate; leaves alternate or mainly so; herbs or shrubs:
  Stamen 1, or wanting in some flowers:
    Utricle 5-costate, circumscissile; seed horizontal; flowers perfect; calyx 3- or rarely 4–5-cleft ........ 2. *Aphanisma*
    Utricle ecostate, indehiscent; seed erect; flowers polygamous; calyx of 1 segment ........ 3. *Monolepis*
  Stamens 2–5, or, if rarely solitary, the leaves fleshy or reduced to scales:
    Leaves flattened, well developed, not particularly fleshy; stems not articulate:
      Flowers perfect, not enclosed in a pair of bracts ........ 4. *Chenopodium*

Flowers monoecious or dioecious, pistillate enclosed in 2 variously shaped bracts:
Herbage scurfy with inflated hairs; fruiting bracts compressed laterally, not obcompressed ............................................ 5. *Atriplex*
Herbage pubescent with slender, branched hairs or glabrous, not scurfy; fruiting bracts obcompressed:
Fruit glabrous; leaves mostly glabrate; bracts broadly winged dorsally..6. *Grayia*
Fruit hirsute; leaves stellate-pubescent; bracts not winged.............7. *Eurotia*
Leaves fleshy and terete or reduced to minute scales; stems articulate except in *Suaeda*:
Embryo annular; leaves reduced to scales; stems articulate:
Branches alternate; bracts of inflorescence alternate, spirally arranged..8. *Allenrolfea*
Branches opposite; bracts of inflorescence opposite:
Plants shrubby; seeds glabrous; endosperm present .......... 9. *Arthrocnemum*
Plants herbaceous; seeds pubescent; endosperm lacking .......... 10. *Salicornia*
Embryo spiral; leaves fleshy, well developed, terete or only slightly flattened; stems not articulate ............................................ 11. *Suaeda*

## 1. NITROPHILA S. Wats., Bot. King Expl. 297, 1871

Perennial, subsucculent, much branched herbs from horizontal rootstocks, with semiarticulate branches and linear to oblong, entire, opposite, fleshy leaves. Flowers perfect, small, solitary or in threes in axils of the leaves and subtended by 2 leaflike bracteoles of different lengths. Calyx lobes 5, erect, concave, unequal, chartaceous-membranous. Stamens 5, perigynous; filaments distinct, filiform. Style slender; stigmas 2, subulate. Utricle ovoid, membranous, loosely investing the seed and exceeded by calyx. Seed lenticular, vertical; embryo annular, almost surrounding endosperm; radicle inferior.

### 1. Nitrophila occidentalis (Nutt.) S. Wats., Bot. King Expl. 297, 1871

*Banalia occidentalis* Moq., in A. DC., Prod. **13**[2]: 279, 1849.
*Halimicnemis occidentalis* Nutt. ex Moq., in A. DC., loc. cit., as synonym.
*Glaux acutifolia* Heller, Muhlenbergia **2**: 109, 1906.

Plant glabrous and often somewhat glaucous; stem 1–4 dm. long, decumbent or erect; leaves sessile, oblong to linear, 0.5–3 cm. long, semiterete, acute, mucronate, or acuminate, the lower ones clasping; bracts similar to leaves, 3–10 mm. long, the smaller deltoid-subulate; central flower of cluster usually sessile, the lateral often short-pedicellate; calyx lobes whitish or stramineous, ovate to broadly oblong, obtuse or rounded, subequal, 3–4 mm. long, persistent and erect around utricle, this ovoid, about 2 mm. long, dull, loosely investing the dark brown, smooth, shining, ovoid-lenticular seed, the latter 1.2–1.4 mm. long.

Saline soils, Lower and Upper Sonoran Zones, both coastal and inland from eastern Oregon southward into southern California, Nevada, and northern Sonora. March–May.

## 2. APHANISMA Nutt. ex Moq., in A. DC., Prod. 13[2]: 54, 1849

Annual herbs with alternate, sessile or petiolate, slightly succulent, entire leaves. Flowers perfect, sessile, in axillary clusters or less frequently solitary. Calyx lobes 3 or sometimes 4 or 5, green, obovate to oblong, obtuse, 1-nerved, subequal. Stamen 1, hypogynous, with a filiform filament and a globose to ellipsoid anther. Style short; stigmas 3, recurved, short. Utricle globose, faintly 5-costate, exceeding calyx and eventually circumscissile; pericarp green, crustaceous, free from the horizontal, lenticular, minutely rugulose, shining seed; embryo nearly annular, surrounding the endosperm.

**1. Aphanisma blitoides** Nutt. ex Moq., in A. DC., Prod. **13²:** 54, 1849

*Cryptanthus blitoides* Nutt. ex Moq., loc. cit., as synonym.

Semisucculent plant much branched from base with prostrate, decumbent, ascending, or erect stems 1–5 dm. long; lower leaves oblanceolate to spatulate, 1–5 cm. long, obtuse to acute at apex, tapering gradually at base to a short petiole, upper leaves broadly ovate-oblong to cordate-ovate, 5–18 mm. wide, 1–3 cm. long, acute to acuminate, bright green, smooth and glabrous, sessile, base clasping; calyx lobes 0.6–1 mm. long; utricle 1–2 mm. broad, two-thirds as high, pericarp dull, circumscissile halfway between rim and base; seed dark brown to nearly black, 1–1.5 mm. in diameter.

Saline soil, mostly along the coast, Lower and Upper Sonoran Zones, islands of southern California, and Orange County southward into Baja California; Carrizo Creek, Colorado Desert, California, and near Punta Prieta, Baja California. Feb.–May.

## 3. MONOLEPIS Schrad., Ind. Sem. Hort. Gött. **1830:** 4, 1830

Annual herbs with prostrate to erect branches. Leaves alternate, sessile or petiolate, entire or hastately lobed, sparingly scurfy-vesicular. Flowers polygamous, ebracteate, in dense axillary clusters or rarely solitary. Sepal solitary, persistent (rarely absent). Stamen 1 or lacking. Stigmas 2, connate at base. Fruit ovoid-compressed, the pericarp lightly attached to the erect, compressed seed; embryo annular, surrounding endosperm; radicle inferior.

**1. Monolepis nuttalliana** (Schult.) Greene, Fl. Fran. 168, 1891

*Blitum chenopodioides* Nutt., Gen. **1:** 4, 1818. Not *B. chenopodioides* Lam., 1783.
*Blitum nuttallianum* Schult., in Roem. & Sch., Syst. Veg. Mant. **1:** 65, 1822.
*Chenopodium trifidum* Trev., Ind. Sem. Vratisl. 1829.
*Monolepis trifida* Schrad., Ind. Sem. Hort. Gött. **1830:** 4, 1830.
*Blitum nuttallianum* var. *minus* Moq., Chenop. Enum. 45, 1840.
*Monolepis chenopodioides* Moq., in A. DC., Prod. **13²:** 85, 1849.
*Monolepis chenopodioides* var. *trifida* Moq., op. cit. 86.
*Chenopodium arcticum* Moq., loc. cit.
*Monolepis asiatica* Moq., loc. cit.

Stems succulent, decumbent to ascending, 1–3.5 dm. long; herbage moderately scurfy-farinose in youth, soon glabrate; leaves triangular- to lanceolate-hastate, 3–12 mm. wide, 1–6.5 cm. long, on slender petioles 0.5–5.5 cm. long, lobes divergent; upper leaves often entire; flowers in dense, sessile clusters; sepal obovate to spatulate, acute, 1–2 mm. long; utricle 1–1.4 mm. in diameter, about 0.5 mm. thick, pericarp membranous, grayish, minutely faveolate-pitted, rather closely attached to the lenticular-orbicular seed, at least at base, somewhat wrinkled; seed 1 mm. in diameter, dark brown to black, dull with a frostlike bloom, the margin acutish.

In dry or alkaline and often in heavy soil, Lower Sonoran Zone to Transition Zone, Sonora and Arizona to Oregon, Manitoba, Alberta, Missouri, and Texas. Introduced in many parts of the world. Feb.–Oct.

## 4. CHENOPODIUM L., Sp. Pl. 218, 1753

Annual or perennial herbs usually with glandular or farinose and often ill-scented herbage. Leaves alternate, entire, dentate, or pinnatifid, petiolate or sometimes sessile.

Flowers small, perfect or rarely unisexual, ebracteate, usually in axillary glomerules. Calyx lobes 5 or rarely 3 or 4, concave, frequently carinate. Stamens 5 or less; filaments distinct or connate at base. Style short or none; stigmas 2–5. Utricle ovoid and erect or orbicular-lenticular and horizontal; pericarp adherent to or free from the horizontal or vertical seed. Embryo annular or nearly so; endosperm mealy; radicle inferior or centrifugal.

Plants perennial; seeds vertical; flowers in terminal, elongate spikes 5–20 cm. long ........ 1. *C. californicum*

Plants annual; seeds horizontal; flowers in axillary glomerules or, if in spikes, less than 5 cm. long:

- Leaves dark green, shining on upper surface, mealy only when young; margins of seeds acute ........ 2. *C. murale*
- Leaves pale, farinaceous on both surfaces, the mealiness usually persistent; margins of seeds obtuse:
  - Pericarp closely adherent to seed; leaves ovate to ovate-rhombic, not linear:
    - Herbage copiously farinose; seeds 1–1.5 mm. broad; lower leaves usually sinuate-dentate ........ 3. *C. album*
    - Herbage sparsely farinose, soon glabrate; seeds 0.8–1 mm. broad; lower leaves entire or obscurely and bluntly hastately lobed at base ........ 4. *C. arizonicum*
  - Pericarp readily freed from seed; leaves linear to linear-lanceolate .. 5. *C. leptophyllum*

### 1. **Chenopodium californicum** (S. Wats.) S. Wats., Bot. Calif. **2**: 48, 1880

*Chenopodium anthelminticum* var. *hastatum* Moq., in A. DC., Prod. **13**²: 74, 1849.
*Blitum californicum* S. Wats., Proc. Amer. Acad. **9**: 101, 1874.

Several to many decumbent to erect leafy stems from a fleshy perennial root with dark green, sparsely farinose young herbage, this soon glabrate; stems 1–10 dm. long, sparingly branched or simple; petioles of lower leaves as much as 2 dm. long, slender; leaf blades triangular to lance-triangular, 3.5–12 cm. long, 1.5–8 cm. broad, mostly acute at apex, cuneate, truncate, or shallowly cordate at base, coarsely, irregularly, and often deeply sinuate-dentate, deep green, reduced above; flowers in glomerules on dense, nearly leafless, terminal spikes 5–20 cm. long, lower glomerules often pedunculate; flowers sessile or pedicellate; calyx lobes herbaceous, about 1 mm. long, obtuse, irregularly erosulate at apex, equaled or surpassed by the slightly compressed, subglobose, dull fruit; pericarp adhering closely to the compressed, globose, black, puncticulate seed, this 1.5–2 mm. in diameter, vertical.

Mostly in Upper Sonoran Zone from central California to northern Baja California; entering the Colorado Desert in western Riverside County, California. Feb.–July.

### 2. **Chenopodium murale** L., Sp. Pl. 219, 1753

*Atriplex muralis* Crantz, Inst. **1**: 206, 1766.
*Chenopodium murale* var. *albescens* Moq., Chenop. Enum. 32, 1840.

Erect or ascending, bright green, subglabrous annual 1–8 dm. high; leaf blades ovate to ovate-rhombic, 1–5.5 cm. wide, 2.5–9 cm. long, acute or obtuse, cuneate to truncate or subcordate at base, irregularly sinuate-dentate, farinose at least beneath in youth, soon glabrate, especially on shining upper surface; flowers in small axillary and terminal glomerules, farinose; calyx lobes oblong, obtuse, about 1 mm. long, weakly carinate toward apex, incompletely enclosing the fruit; pericarp greenish, closely adherent; seed horizontal, 1.2–1.6 mm. in diameter, dull even under pericarp, finely and shallowly puncticulate, the margin acute.

Introduced from Europe and established as a weed throughout most of North America from southern Canada to central America and the West Indies. Along Gila River, about Phoenix, Arizona; Santa Catarina Landing, and on several islands in Gulf, Baja California. March–Oct.

### 3. **Chenopodium album** L., Sp. Pl. 219, 1753

*Atriplex alba* Crantz, Inst. **1**: 206, 1766.
*Chenopodium album* subsp. *collinsii* Murr, Bull. Herb. Boiss. II, **4**: 990, 1904.
*Botrys alba* Nieuwl., Amer. Midl. Nat. **3**: 276, 1914.
*Botrys alba* var. *pacifica* Lunell, Amer. Midl. Nat. **4**: 305, 1916.

Erect or ascendingly and moderately branched annual 1–20 dm. high with obtusely angled, striate, pale green, and mainly glabrate stems; petioles slender, 0.5–5 cm. long; leaf blades lanceolate, ovate, or more frequently ovate-rhombic, 7–30 mm. wide, 1.5–8 cm. long, obtuse to acute at apex, broadly cuneate and sometimes subhastate at base, entire to sinuate-toothed, pale green, finely farinose beneath; flowers in large, dense glomerules forming paniculate, stout spikes, the inflorescence narrow and compact, nearly leafless; calyx lobes obtuse, densely farinose, strongly carinate toward apex, completely enclosing the fruit; pericarp closely adherent, dull; seeds horizontal, 1–1.5 mm. broad, puncticulate, black, shining after removal of pericarp, margin obtuse.

Introduced from Europe and established as a weed from Alaska and southern Canada to northern Mexico. March–Oct.

### 4. **Chenopodium arizonicum** Standley, N. Amer. Fl. **21**: 19, 1916

Erect, ill-scented annual 3–6 dm. high, simple at base, openly and spreadingly branched above; branches slender, obtusely angled, finely but closely farinose, tardily glabrate; petioles slender, 6–15 mm. long; leaf blades ovate-rhombic to triangular, 4–15 mm. wide, 5–18 mm. long, acute or obtuse and apiculate at apex, broadly cuneate at base, thin, light green, and glabrate on upper surface, finely but densely farinose beneath; flowers in small glomerules arranged in slender, interrupted or dense, paniculate spikes, branches of inflorescence usually flexuous, leafless or nearly so; calyx lobes densely farinose, oblong, obtuse, carinate from lower third about to apex, incompletely covering the fruit; pericarp very thin but tenaciously adherent; seed horizontal, 0.8–1 mm. in diameter, black and shining, minutely puncticulate, margin rounded.

Arid, rocky hills, Lower and Upper Sonoran Zones, southwestern Arizona and adjacent Sonora. July–Oct.

### 5. **Chenopodium leptophyllum** Nutt. ex S. Wats., Proc. Amer. Acad. **9**: 94, 1874

Erect, much-branched annual 2–8 dm. tall with copiously farinose, often eventually glabrate herbage; petioles slender, 2–5 mm. long; leaf blades linear to linear-lanceolate, 1–4 (or rarely –8) mm. broad, 1.5–5 cm. long, acute, apiculate, obtuse, or rounded at apex, attenuate to narrowly cuneate at base, entire or rarely with 1–2 small basal lobes, densely farinose when young, usually glabrate in age, scurf persistent on lower surface; inflorescence of slender or stoutish paniculate spikes of small glomerules; calyx lobes densely farinose, oval to obovate, obtuse, carinate, almost completely enclosing the fruit, margins membranous; pericarp free; seed about 1 mm. in diameter, horizontal, black, moderately shining, puncticulate, margins obtuse.

Arid regions, Manitoba and Alberta to southern California, Sonora, and Chihuahua. May–Sept.

**Chenopodium flabellifolium** Standley, N. Amer. Fl. **21:** 19, 1916, with flabelliform leaf blades mostly broader than long and broadly rounded at the apex, was collected on San Martín Island, Baja California, and may occur on the mainland of the peninsula.

**Chenopodium lanceolatum** Muhl. ex Willd., Enum. Hort. Ber. 291, 1809, has been collected in the Sierra Juárez, in northern Baja California, but has not been found in the near-by desert.

## 5. ATRIPLEX L., Sp. Pl. 1052, 1753

Annual or perennial herbs or shrubs with scurfy-pubescent herbage. Leaves alternate or the lower opposite or rarely all opposite, entire, dentate, or lobed. Flowers monoecious or dioecious, solitary, in axillary glomerules or in terminal spikes or panicles. Perianths of staminate flowers 3–5-parted; pistillate flowers each subtended by 2 accrescent bracts, these distinct or united, entire or variously dentate and appendaged. Stamens 3–5. Styles 2. Seed usually erect, lenticular; embryo annular, surrounding the mealy endosperm.

Plants shrubby, or at least woody at base:
- Bracts orbicular, entire or faintly dentate:
  - Leaves entire or undulate-margined, usually longer than broad; bracts often faintly denticulate toward apex, not stipitate-cordate at base; flowers monoecious or dioecious; twigs spinescent ........ 1. *A. lentiformis*
  - Leaves coarsely dentate, usually as broad as or broader than long; bracts not denticulate, stipitate-cordate at base; flowers dioecious; twigs not spinescent..2. *A. hymenelytra*
- Bracts not truly orbicular, usually distinctly dentate:
  - Leaves 2–4 mm. long, involute-sulcate, sagittate-clasping ........ 3. *A. julacea*
  - Leaves 0.5–5 cm. long, plane or nearly so, not sagittate-clasping:
    - Stems herbaceous above; leaves obovate, 0.5–2.2 cm. broad:
      - Bracts thickened, compressed slightly or not at all, usually longer than wide:
        - Bracts bearing 2–5 small teeth at summit, sides entire ........ 4. *A. barclayana* subsp. *barclayana*
        - Bracts bearing 5–10 slender teeth halfway to base ........ 4*a*. *A. barclayana* subsp. *sonorae*
      - Bracts compressed, usually as wide as long, or wider:
        - Leaves entire, thick, canescent on both surfaces:
          - Bracts longer than broad ........ 4*b*. *A. barclayana* subsp. *palmeri*
          - Bracts broader than long ........ 4*c*. *A. barclayana* subsp. *dilatata*
        - Leaves saliently toothed, thin, greenish above .. 4*d*. *A. barclayana* subsp. *lurida*
    - Stems woody; leaves linear to oblong-spatulate, 2–10 mm. broad:
      - Bracts 2–4 mm. long, not bearing extra wings on lateral faces; leaves 3–18 mm. long ........ 5. *A. polycarpa*
      - Bracts 6–15 mm. long, bearing extra wing on each lateral face; leaves 1.5–5 cm. long:
        - Leaves spatulate to oblong, usually over 3 mm. wide:
          - Bracts 4–10 mm. wide, extra wings conspicuous:
            - Bracts dentate ........ 6. *A. canescens* subsp. *canescens*
            - Bracts laciniate ........ 6*a*. *A. canescens* var. *laciniata*
          - Bracts 1.5–3 mm. wide, extra wings inconspicuous or wanting on some fruiting bracts ........ 6*b*. *A. canescens* var. *macilenta*
        - Leaves linear, 1–2 mm. wide:
          - Wings exceeding terminal free part of bracts; fruits sessile ........ 6*c*. *A. canescens* subsp. *linearis*
          - Wings shorter than terminal free part of bracts; fruits short-stipitate ........ 6*d*. *A. canescens* subsp. *macropoda*

Plants herbaceous, not noticeably woody at base:
- Bracts thickened, red and fleshy, strongly nerved at maturity .......... 7. *A. semibaccata*
- Bracts not thickened nor fleshy, weakly or strongly nerved:
  - Leaves bright green above, sparsely scurfy to green beneath at maturity .. 8. *A. serenana*
  - Leaves green or canescent above but always gray-scurfy beneath:
    - Leaves dentate; stems and branches erect (lower branches sometimes decumbent in *A. elegans*):
      - Bracts orbicular, finely and evenly toothed all around; leaves 2–7 mm. wide, sparsely and finely toothed ............................ 9. *A. elegans* subsp. *elegans*
      - Bracts ovate to obovate, irregularly toothed above middle; leaves 0.5–3 cm. wide, coarsely toothed:
        - Staminate flowers in axillary glomerules and short spikes 1 cm. long or less; leaves ovate to rhombic-ovate, rarely green on upper surface ......... 10. *A. rosea*
        - Staminate flowers in naked terminal panicles 5–30 cm. long; leaves oblanceolate to elliptic-spatulate, glabrate and green on upper surface .... 11. *A. wrightii*
    - Leaves entire; stems or branches prostrate or decumbent:
      - Bracts exactly orbicular, finely toothed all around .. 9*a*. *A. elegans* subsp. *fasciculata*
      - Bracts not exactly orbicular, not toothed on lower part:
        - Staminate flowers intermingled with pistillate flowers in axillary glomerules, at least in part of inflorescence; bracts broadest below middle:
          - Perennial; prostrate; bracts distinct to base; leaves oblong to linear-lanceolate, narrowed to base ................................... 12. *A. californica*
          - Annual; decumbent; bracts united about to middle; leaves broadly ovate, subcordate at base .................................... 13. *A. parishii*
        - Staminate flowers in terminal spikes or on separate plants; bracts broadest at or above middle:
          - Perennial; leaves orbicular to obovate, thick, with persistent dense scurf; bracts elliptic-globose, not compressed, 5–7 mm. long ....... 14. *A. leucophylla*
          - Annual; leaves elliptic to broadly ovate, not noticeably thickened, scurf sparser; bracts compressed, 1–5 mm. long:
            - Flowers monoecious; fruiting bracts 1–2 mm. long; leaves nearly or quite sessile, 2–6 mm. wide ................................ 15. *A. pacifica*
            - Flowers mostly dioecious; fruiting bracts 3–5 mm. long; leaves petiolate, 5–15 mm. wide .............................. 16. *A. magdalenae*

**1. Atriplex lentiformis** (Torr.) S. Wats., Proc. Amer. Acad. **9:** 118, 1874

*Obione lentiformis* Torr., in Sitgr., Rep. 169, 1853.

Erect shrub 1–3 m. high with flexuous or rigid or angular branches forming a dome-shaped outline, the lateral twigs sometimes spine-tipped; leaves deltoid, rhombic, ovate, or oblong, 0.5–2.5 cm. wide, 1–5 cm. long, truncate to broadly cuneate at base, obtuse and often short-mucronate at apex, entire to subhastate, closely and finely gray- or blue-scurfy; flowers dioecious or sometimes monoecious, in crowded clusters along branches of diffuse panicles 1–5 dm. long; bracts orbicular or slightly broader, compressed-lenticular, 2.5–5 mm. broad, margins entire to crenulate, faces smooth, gray-scurfy; seeds 1.2–1.5 mm. long, dark brown.

Alkaline flats and benches, desert outwash slopes and playas, and interior valleys, Lower Sonoran Zone, Salinas and San Joaquin Valleys, California, to southwestern Utah, south through the Mojave and Colorado Deserts to central Baja California, and east to central Arizona and northwestern Sonora. Feb.–April.

**2. Atriplex hymenelytra** (Torr.) S. Wats., Proc. Amer. Acad. **9:** 119, 1874

*Obione hymenelytra* Torr., Pacif. R. Rep. **4:** 129, *pl. 20,* 1857.

Compact, profusely branched shrub 2–12 dm. high with dense, persistent, white scurf on young twigs and leaves; branches somewhat tortuous, stoutish, erect or ascending; bark tan on younger wood, finally gray and more or less checked and fissured; petioles 3–10 mm. long; leaf blades round-ovate to orbicular, 1.5–4.5 cm. wide and long, truncate to subcordate at base, obtuse at apex, saliently and irregularly toothed, silvery with a dense, smooth scurf, turning red to purplish with age; flowers dioecious, in dense, leafy panicles; bracts borne on a turbinate stalk 2–4 mm. long, orbicular, 6–10 mm. long, strongly compressed, distinct except at base, margins entire, faces smooth or reticulately veined on body of the fruit; seeds 1.5–2 mm. long.

Alkaline flats and slopes, Lower Sonoran Zone, southwestern Utah, southern Nevada, and southeastern California from Inyo County south through the Mojave and Colorado Deserts to central Baja California, western Arizona, and northwestern Sonora. Feb.–April.

**3. Atriplex julacea** S. Wats., Proc. Amer. Acad. **20:** 370, 1885

Decumbent, rigidly branched, compact subshrub 1–3 dm. high and 1–5 dm. wide with slender, erect or ascending, terete, gray-scurfy to glabrate twigs; branches brittle, with corky brownish bark on older basal parts; leaves crowded or even imbricate on branches, broadly ovate-triangular, sessile by a cordate-clasping base, 2–4 mm. long, involute-sulcate, thick, entire, obtuse at apex, permanently gray-scurfy, usually more or less arcuately curved upward; flowers dioecious, in few-flowered axillary glomerules or solitary; fruiting bracts sessile, ovate, 3.5–6 mm. long, nearly as broad, united to about middle or slightly above, not compressed, faces ornamented with irregular corky knobs and ridges, the terminal, distinct part usually less corky; seed reddish brown, 1–1.2 mm. long.

Saline or alkaline flats, mostly near the coast, Upper and Lower Sonoran Zones, vicinity of Ensenada southward to the coastal margins of the Vizcaíno Desert, Baja California. Feb.–June.

**4. Atriplex barclayana** (Benth.) Dietr., Syn. Pl. **5:** 537, 1852, subsp. **barclayana**

*Obione barclayana* Benth., Bot. Sulph. 48, 1844.

*Atriplex barclayana* subsp. *typica* Hall & Clements, Carnegie Inst. Wash. Pub. No. **326:** 316, *pl. 49, figs. 8–13,* 1923.

Prostrate, decumbent, or erect shrub or suffrutescent herb 2–15 dm. tall; branches terete or obscurely angled, usually striate and checked after loss of the dense white scurf; leaves short-petiolate, ovate, obovate, or broadly elliptic, abruptly tapering to base, obtuse at apex, 15 mm. wide, 6–20 mm. long, entire, densely white-scurfy on both sides, thickish, indistinctly 1–3-veined; flowers dioecious, staminate in glomerules on terminal branches, pistillate in axillary glomerules on leafy-bracted panicles 0.5–2 dm. long; fruiting bracts globoid, spongiose, 2.5–4 mm. long, nearly as wide, united almost to summit, entire except for 2–5 small apical teeth, sides smooth or sparsely and minutely tuberculate; seeds 1–1.5 mm. long, reddish brown, dull.

Mostly near the coast, occasional inland, Lower Sonoran Zone, from head of the Gulf of California to Sinaloa and the Cape Region, Baja California; Pacific side of Baja California from near Rosario south. Feb.–June.

**4*a*. Atriplex barclayana** subsp. **sonorae** (Standley) Hall & Clements, Carnegie Inst. Wash. Pub. No. **326:** 316, *pl. 49, fig. 16,* 1923

*Atriplex sonorae* Standley, N. Amer. Fl. **21:** 62, 1916.

Differing from subsp. *barclayana* chiefly in having the fruiting bracts with flatter, broader margins that are laciniately toothed from apex about halfway down the sides, and with 2–several toothlike processes on faces.

Coastal regions from near the head of the Gulf of California south to the Cape Region of Baja California and Sinaloa.

**4*b*. Atriplex barclayana** subsp. **palmeri** (S. Wats.) Hall & Clements, Carnegie Inst. Wash. Pub. No. **326:** 315, 1923

*Atriplex palmeri* S. Wats., Proc. Amer. Acad. **11:** 146, 1876.
*Atriplex insularis* Rose, in Vasey & Rose, Contr. U.S. Nat. Herb. **1:** 80, 1890.

Fruiting bracts compressed-lenticular, as long as, or slightly longer than broad, margins dentate from about middle to apex; leaves entire.

Guadalupe Island and from the vicinity of Jaraguay to Magdalena Island, Baja California, on the islands in the Gulf of California, and at least in the vicinity of Puerto Libertad, Sonora.

**4*c*. Atriplex barclayana** subsp. **dilatata** (Greene) Hall & Clements, Carnegie Inst. Wash. Pub. No. **326:** 315, *pl. 49, fig. 17,* 1923

*Atriplex dilatata* Greene, Pittonia **1:** 264, 1889.
*Atriplex rosei* Standley, N. Amer. Fl. **21:** 60, 1916.

Leaves entire; bracts appreciably broader than long, sessile.

Guadalupe and San Benito Islands off the coast of Baja California. A form that is questionably separable as a subspecies.

**4*d*. Atriplex barclayana** subsp. **lurida** (Brandegee) Hall & Clements, Carnegie Inst. Wash. Pub. No. **326:** 315, *pl. 49, figs. 14–15,* 1923

*Atriplex lurida* Brandegee, Proc. Calif. Acad. II, **2:** 200, 1889.

Leaves thinner than those of the other subspecies, acutely dentate with broad, salient teeth or margins of some merely crenate and upper leaves entire; stems usually bright red under scurf; bracts compressed-lenticular, dentate above middle, faces smooth.

Central Baja California from the vicinity of Calmallí to the northern end of the Magdalena Plain. The most sharply marked subspecies in this complex.

**5. Atriplex polycarpa** (Torr.) S. Wats., Proc. Amer. Acad. **9:** 117, 1874

*Obione polycarpa* Torr., Pacif. R. Rep. **4:** 130, 1857.
*Atriplex curvidens* Brandegee, Proc. Calif. Acad. II, **2:** 201, 1889.

Erect, intricately branched shrub 0.8–2 m. high with slender, rigid or flexuous, terete branches, these gray-scurfy in youth but soon smooth and light tan to light brown; leaves crowded on terminal twigs and soon deciduous, mostly sessile, elliptic to spatulate, broadly cuneate to rounded at base, acute or obtuse at apex, 2–5 mm. wide, 3–15 mm. long, or considerably larger on sterile twigs, entire, densely and permanently gray-scurfy; flowers dioecious, staminate in axillary and terminal spicate glomerules, pistillate in profuse terminal panicles; fruiting bracts sessile, only moderately compressed, cuneate-orbicular, 2–4 mm. long, about as broad, margins thin, shallowly to deeply dentate above base, usually united to well above middle, faces usually cristate-appendaged or tuberculate but sometimes smooth; seed 1–1.5 mm. long, pale brown.

Moderately alkaline areas, Lower Sonoran Zone, southern Salinas Valley and southern San Joaquin Valley, California; southern Nevada and Utah south through Sonora and Baja California at least to the northern Magdalena Plain. Feb.–Oct.

**6. Atriplex canescens** (Pursh) Nutt., Gen. Pl. **1:** 197, 1818, subsp. **canescens**

*Calligonum canescens* Pursh, Fl. Amer. Sept. 370, 1814.

*Atriplex canescens* subsp. *typica* Hall & Clements, Carnegie Inst. Wash. Pub. No. **326:** 343, *pl. 58, figs. 1–11,* 1923.

(For full synonymy see Carnegie Inst. Wash. Pub. No. **326:** 344–345, 1923.)

Erect shrub to 2.5 m. high with stout, terete branches, these gray-scurfy in youth, soon glabrate and pale; leaves subsessile or sessile, linear-spatulate to oblong, 2–8 mm. wide, 1.5–5 cm. long, entire, densely gray-scurfy; flowers dioecious, staminate in dense, nearly leafless terminal spikes and panicles, pistillate in dense, leafy-bracted spikes and panicles; fruiting bracts 6–15 (sometimes 20) mm. long, with wings 4–8 mm. wide, shallowly to deeply toothed, similar wings usually along median line of outer face, smooth or slightly appendaged between wings; seeds 1.5–2 mm. long.

Dry alkaline soils, South Dakota to Kansas, western Texas, eastern Washington, and south into Baja California, Sinaloa, and Zacatecas. March–Sept.

**6*a*. Atriplex canescens** var. **laciniata** Parish, in Jepson, Fl. Calif. **1:** 442, 1914

Similar to subsp. *canescens* except that the wings of fruiting bracts are 7–8 mm. wide, deeply laciniate.

Southern Mojave Desert and Salton Basin of the Colorado Desert.

**6*b*. Atriplex canescens** var. **macilenta** Jepson, Fl. Calif. **1:** 442, 1914

Fruiting bracts only 1.5–3 mm. wide, shallowly dentate, the median wings sometimes reduced to a row of tubercles.

Saline soil about southern end of Salton Sea.

**6*c*. Atriplex canescens** subsp. **linearis** (S. Wats.) Hall & Clements, Carnegie Inst. Wash. Pub. No. **326:** 344, *pl. 58, figs. 13–14,* 1923

*Atriplex linearis* S. Wats., Proc. Amer. Acad. **24:** 72, 1889.

*Atriplex canescens* var. *linearis* Munz, Man. S. Calif. 141, 1935.

Leaves linear, rarely over 2 mm. wide; wings on fruit often 1–2 mm. wide or less; otherwise as in subsp. *canescens.*

Colorado Desert, California, into Arizona, southern Sonora, and southern Baja California.

**6*d*. Atriplex canescens** subsp. **macropoda** (Rose & Standl.) Hall & Clements, Carnegie Inst. Wash. Pub. No. **326:** 344, 1923

*Atriplex macropoda* Rose & Standl., N. Amer. Fl. **21:** 72, 1916.

Leaves linear, 1–1.5 mm. wide; wings of bracts 2–4 mm. wide, shorter than terminal, free part of bracts; fruiting bracts borne on slender stalks 4–10 mm. long.

Pinchilinque Island, near La Paz in Gulf of California. A broader-leaved plant with fruits similar to those of subsp. *macropoda* though with shorter stalks occurs near Calmallí, Baja California.

**7. Atriplex semibaccata** R. Br., Prod. 406, 1810

Prostrate or decumbent perennial herb with trailing stems 2–15 dm. long and ascending or erect twigs 1–2 dm. long; leaves numerous, elliptic to elliptic-oblong, 3–12 mm. wide, 1.5–4 cm. long, irregularly and remotely repandly dentate or the smaller ones often entire, deep green on upper surface, finely gray-scurfy beneath, strongly 1-nerved; flowers monoecious, staminate in leafy-bracted glomerules, near tips of branches, pistillate solitary or in few-flowered axillary clusters from base to near tips of branches; fruiting bracts turgidly ovoid and red-fleshy when fresh, becoming nearly flat when dry, 3–6 mm. long, slightly narrower, cuneate at base, the terminal half distinct, entire to denticulate, faces smooth when fresh, strongly 3–5-nerved in drying; seeds either black and 1.5 mm. long or brown and about 2 mm. long; both types of seeds may occur on same plant.

Introduced from Australia. A well-established roadside weed at numerous localities from central California to northern Baja California and east into Arizona. April–Sept.

**8. Atriplex serenana** A. Nels., in Abrams, Fl. Los Ang. 128, 1904

*Obione bracteosa* Dur. & Hilg., Pacif. R. Rep. **5:** 13, 1858.
*Atriplex bracteosa* S. Wats., Proc. Amer. Acad. **9:** 115, 1874. Not *A. bracteosa* Traut., 1870.

Erect or decumbent annual herb 3–10 dm. high, often forming dense mats 1–3 m. in diameter; branches obtusely angled, sparsely scurfy when young, soon glabrate but rarely turning reddish; leaves sessile or nearly so, narrowly ovate, narrowly elliptic, or lanceolate, 4–15 mm. wide, 2–4 cm. long, cuneate at base, acute to acuminate at apex, sparingly dentate to entire, thin, sparsely scurfy and faintly gray when young, soon glabrate and green on both faces; flowers monoecious, staminate in glomerules in terminal panicles or spikes 5–15 cm. long, pistillate in small axillary clusters; fruiting bracts moderately compressed, cuneate-orbicular or flabelliform, 2–3 mm. long, about as wide as long, united to about middle, margins dentate above the middle, faces tuberculate or rarely smooth.

Alkaline soil, Lower and Upper Sonoran Zones, northern Baja California to the Sacramento Valley, California, mostly west of the desert areas but entering the Sonoran Desert near Needles, California. Also in western Nevada. April–Oct.

**9. Atriplex elegans** (Moq.) D. Dietr., Syn. Pl. **5:** 537, 1852, subsp. **elegans**

*Obione elegans* Moq., in A. DC., Prod. **13**[2]**:** 113, 1859.
*Obione radiata* Torr., Bot. Mex. Bound. 183, 1859.
*Atriplex elegans* var. *thornberi* M. E. Jones, Contr. West. Bot. **12:** 76, 1908.
*Atriplex thornberi* Standley, N. Amer. Fl. **21:** 57, 1916.

Erect or decumbent annual 1–6 dm. high with a rounded habit; leaves elliptic-spatulate, oblong, or oblanceolate, cuneate at base, acute to obtuse at apex, 2–7 mm. wide, 5–25 mm. long, entire to remotely dentate, thin, white-scurfy or becoming greenish on upper surface; flowers monoecious, in axillary glomerules, the lower all pistillate, the upper mixed or the terminal ones all staminate; fruiting bracts compressed, short-pedicellate, orbicular, 2–4 mm. in diameter, deeply dentate to laciniate all the way around, teeth to 1 mm. long, faces smooth, tuberculate, or bearing 2 low, laciniate crests; seeds 1–1.4 mm. long, brown.

Desert habitats, Lower Sonoran Zone, eastern Colorado Desert to western Texas, south into northern Chihuahua, Sonora and Durango. March–Aug.

**9*a*. Atriplex elegans** subsp. **fasciculata** (S. Wats.) Hall & Clements, Carnegie Inst. Wash. Pub. No. **326:** 301, *pl. 46, figs. 23–24,* 1923

*Atriplex fasciculata* S. Wats., Proc. Amer. Acad. **17:** 377, 1882.
*Atriplex fasciculata* var. *elegans* M. E. Jones, Contr. West. Bot. **12:** 76, 1908.

Leaves not dentate; fruiting bracts very shallowly dentate or merely crenulate.

Inyo County, California, south through the Mojave and Colorado Deserts to northern Baja California and the western part of Arizona. Abundant on the Colorado Desert.

**10. Atriplex rosea** L., Sp. Pl. ed. 2, 1493, 1763
*Atriplex spatiosa* A. Nels., Bot. Gaz. **34:** 360, 1902.

Erect annual herb to 2 m. high with ascending to divaricate, slightly angled, nearly glabrous branches; leaves petiolate or sessile, ovate, rhombic-ovate, or lanceolate, cuneate to rounded at base, acute or mucronulate at apex, 1–3 cm. wide, 1.5–6 cm. long, remotely sinuate-dentate, usually gray-scurfy on both surfaces, persistent and cartilaginous on drying; flowers monoecious, axillary staminate glomerules and spikes above those of pistillate; fruiting bracts ovate or rhombic from a broad base, 4–6 mm. long, indurated in age, margins greenish, acutely dentate, faces sharply tuberculate; seed 2–2.5 mm. long, dull, dark brown.

Native of Europe, widely established throughout temperate North America as far north as Massachusetts and Wyoming. Known in the Sonoran Desert from Needles, California, and near Ojos Negros, Baja California. July–Sept.

**11. Atriplex wrightii** S. Wats., Proc. Amer. Acad. **9:** 113, 1874
*Atriplex radiata* J. M. Coult., Contr. U.S. Nat. Herb. **2:** 368, 1894.

Erect or ascendingly branched bushy annual 1.5–10 dm. high with deeply grooved, sparsely scurfy branches; leaves oblanceolate to elliptic-spatulate, cuneately narrowed toward base, obtuse and often mucronulate at apex, 5–25 mm. wide, 2–7 cm. long, coarsely sinuate-dentate to repand-dentate or the smaller, acute upper leaves entire, persistently white-scurfy beneath, soon glabrate and green above; flowers monoecious, staminate glomerules borne in terminal, naked panicles 5–35 cm. long, pistillate flowers in axillary glomerules; fruiting bracts sessile or nearly so, compressed, cuneate-obovate to suborbicular, 2–3 mm. long and to 4 mm. wide, margins acutely toothed, green, faces usually smooth, 3-nerved; seed 1–1.5 mm. long, brown, dull.

Roadsides and desert slopes, Lower Sonoran Zone, south central Arizona to southwestern New Mexico and northern Sonora. July–Oct.

**12. Atriplex californica** Moq., in A. DC., Prod. **$13^2$:** 98, 1849

Prostrate perennial herb from a spongiose, fusiform taproot; stems spreading, wiry, 1–2 dm. long, branches erect or ascending; leaves sessile, lanceolate to ovate-lanceolate, cuneate at base, acute at apex, entire, 2–5 mm. wide, 8–20 mm. long, gray-tomentose and scurfy, finally greenish; flowers monoecious, intermingled in axillary glomerules, some staminate flowers in terminal spikes; fruiting blades compressed, sessile, distinct almost to base, ovate, acute at apex, 2.5–4 mm. broad, 3–5 mm. long, margins entire, faces smooth except for a single, inconspicuous nerve; seeds 1.8–2 mm. long, black.

Sandy areas along the coast from vicinity of Bodega Bay, California, to Cedros Island, Baja California. April–Oct.

**13. Atriplex parishii** S. Wats., Proc. Amer. Acad. **17:** 377, 1882
*Atriplex depressa* Jepson, Pittonia **2:** 304, 1892.
*Atriplex minuscula* Standley, N. Amer. Fl. **21:** 51, 1916.

Erect, spreading, or decumbent, annual, profusely branched herb 5–25 cm. high; branches stoutish, brittle, coarsely white-scurfy-pubescent; leaves crowded, opposite, alternate, or both arrangements on the same plant, sessile, ovate to lanceolate, rounded at base, acute at apex, 3–5 mm. wide, 4–10 mm. long, rather rigid, entire, gray to white with a dense, wool-like scurf; flowers monoecious, in axillary glomerules, staminate in upper and middle axils, pistillate in lower and middle axils; fruiting bracts sessile, moderately compressed, united to about middle, ovate to rhomboid, acute at apex, some hastately lobed, 2.5–3 mm. wide, slightly longer, entire or 1–2-toothed on each side, faces smooth or slightly muricate; seed 1–1.2 mm. long, brown to black.

Alkaline and saline soil, Sacramento Valley to San Diego County, California, along the coast and eastward to the western margins of the Mojave and Colorado Deserts. May–Sept.

**14. Atriplex leucophylla** (Moq.) D. Dietr., Syn. Pl. **5**: 536, 1852

*Obione leucophylla* Moq., in A. DC., Prod. **13**²: 109, 1849.

Perennial herb with prostrate to ascending branches 1–3 dm. long, often forming dense mats 2–10 dm. in diameter, densely, persistently, and coarsely white-scurfy throughout, stems often pinkish through scurf; leaves crowded to loosely imbricate, sessile, orbicular-ovate to obovate, 5–15 (or 20) mm. broad, 1–3.5 cm. long, entire, rounded to slightly narrowed toward base, acute to rounded at apex, thick and leathery; flowers monoecious, staminate in globose glomerules in short, terminal spikes, pistillate in few-flowered clusters in axils of upper leaves; fruiting bracts spongiose, ellipitc, 5–7 mm. long, acute and free at apex, united elsewhere, entire or dentate, faces often warty-tuberculate, densely white-scurfy; seeds 2.5–3 mm. long.

Coastal beaches, sand dunes, and occasionally on inland saline flats, Upper and Lower Sonoran Zone, Humboldt County, California to the Vizcaíno Desert, Baja California. April–Oct.

**15. Atriplex pacifica** A. Nels., Proc. Biol. Soc. Wash. **17**: 99, 1904

*Obione microcarpa* Benth., Bot. Sulph. 48, 1844.

*Atriplex microcarpa* D. Dietr., Syn. Pl. **5**: 536, 1852. Not *A. microcarpa* Wald. & Kitaib., 1812, or Benth., 1870.

Prostrate to procumbent annual with much-branched stems 1–5 dm. long, forming dense masses 1–4 dm. high; branches furfuraceous when young, soon glabrate; leaves alternate or some lower ones opposite, sessile to short-petiolate, oblanceolate to spatulate-elliptic, 3–6 mm. wide, 5–20 mm. long, acute to obtuse at apex, green and sparsely scurfy above, gray and densely scurfy beneath; flowers monoecious, staminate flowers in upper axils or in short spikes, pistillate flowers in lower axils; perianth deeply 5-cleft in staminate flowers, lacking in pistillate; fruiting bracts nearly orbicular, 1–2 mm. long, 1–2.2 mm. wide, with a few minute teeth on distal end, the rest of the margins entire, faces smooth or with 1–several minute tubercles, 1-nerved; seed about 0.8 mm. long, light brown.

Sandy soil back of the beaches and along roadsides and trails, Upper and Lower Sonoran Zones, near the coast from Los Angeles County, California, southward to Cedros Island, Baja California. March–Oct.

**16. Atriplex magdalenae** Brandegee, Proc. Calif. Acad. II, **2**: 200, 1889

*Atriplex barclayana* subsp. *magdalenae* Hall & Clements, Carnegie Inst. Wash. Pub. No. **326**: 315, 1923.

Annual decumbent to ascending herb with slender, scantily scurfy stems 1–3 dm. long (possibly a perennial but if so, flowering the first year); leaves short-petiolate, broadly elliptic or broadly ovate, 4–12 mm. broad, 6–25 mm. long, broadly cuneate, narrowed at base, acute or mostly short-mucronate at apex, thin, entire, gray-scurfy on both sides; flowers monoecious, staminate in small glomerules in short, terminal, slightly leafy spikes, pistillate scattered in axils, a few occurring with the lower staminate glomerules; fruiting bracts cuneate-obovate, 3–5 mm. long, about as wide at the dilated, 3-toothed apex, united to middle or slightly above, the faces smooth, 1-nerved from the base.

Sandy plains, mesas, and coastal dunes, Lower Sonoran Zone, western side of Baja California from the vicinity of Miller's Landing to Magdalena Island and the Cape Region. Feb.–May.

### 6. **GRAYIA** Hook. & Arn., Bot. Beechey 387, 1841

Much-branched, erect shrubs with mostly spinose branches and stellate-scurfy herbage. Leaves alternate, entire, sessile, moderately fleshy. Flowers dioecious or infrequently monoecious. Staminate flowers ebracteate, pedicellate in axillary glomerules; calyx lobes 4–5, membranous, obovate; stamens 4–5, usually same number as calyx lobes. Pistillate flowers racemose, each subtended by 2 conduplicate, carinate bracts which become broadly winged dorsally in fruit; stigmas 2. Utricle compressed-orbicular, surpassed by and included in the 2 accrescent, thin bracts. Seed erect, orbicular-lenticular, brown, loosely invested by the pericarp; embryo annular; radicle inferior.

**1. Grayia spinosa** (Hook.) Moq., in A. DC., Prod. **13**[2]**:** 119, 1849

*Chenopodium spinosum* Hook., Fl. Bor. Amer. **2:** 127, 1838.
*Grayia polygaloides* Hook. & Arn., Bot. Beechey 388, 1841.
*Eremosemium spinosum* Greene, Pittonia **4:** 225, 1900.

Intricately branched shrub 3–12 dm. high; branchlets usually spinose, sparsely stellate-scurfy in youth, bark glabrate and dark gray on older stems; leaves oblanceolate to oblong-oblanceolate, 0.5–4 cm. long, rounded or obtuse at apex, gradually narrowed to a short, slender petiole, stellate-pubescent when young, soon glabrate; flowers crowded in terminal spikes or some in axillary clusters; fruiting bracts obovate to orbicular, 3–15 mm. in diameter, lightly veined, lutescent or often tinged with red; seed 1.5–2 mm. in diameter, dull.

Arid plains and rocky hillsides, mostly Upper Sonoran Zone, eastern Washington to western margin of Colorado Desert, eastward to Wyoming, Colorado, and northern Arizona. March–June.

### 7. **EUROTIA** Adans., Fam. Pl. **2:** 260, 1763

Low shrubs with stellate-pubescent herbage. Leaves alternate and fasciculate, entire. Flowers polygamous or dioecious, staminate ebracteate, in dense spikes subtended by leaves, pistillate bibracteate, solitary or in axillary clusters. Calyx lobes of staminate flowers 4, obovate, obtuse, pubescent on outer surface. Stamens 4. Calyx lacking in pistillate flowers, the 2 bracts conduplicate, connate below, pilose. Stigmas 2. Utricle elliptic-compressed, enclosed by the bracts, the latter forming a 2-horned, 4-valvate covering; pericarp free from the seed, membranous. Seed vertical, lenticular; embryo almost annular; radicle inferior.

**1. Eurotia lanata** (Pursh) Moq., Chenop. Enum. 81, 1840

*Diotis lanata* Pursh, Fl. Amer. Sept. 602, 1814.
*Eurotia ceratoides* var. *lanata* Kuntze, Rev. Gen. Pl. **2:** 549, 1891.
*Eurotia subspinosa* Rydb., Bull. Torrey Club **39:** 312, 1912.

Erect shrub 2.5–9 dm. tall with slender, leafy branches densely tawny-stellate-tomentose and with intermingled simple hairs when young, becoming glabrate and with tawny to light gray, thin bark in age; leaves alternate with fascicles of smaller ones in their axils, linear-oblong to linear-lanceolate, 1–4 (or rarely 8) mm. wide, 1–3.5 (or rarely 5) cm. long, densely but finely stellate-lanate, hairs tawny to brownish, margins entire and narrowly revolute, acute at apex, cuneately narrowed at base to petioles 1–5 mm. long or sessile; flowers monoecious, crowded in tight clusters in upper axils; calyx lobes of staminate flowers about 1 mm. long, short-pubescent, margins scarious; fruiting bracts 4–8 mm. long, densely long-villous with rusty, simple hairs 3–6 mm. long; utricle ovoid-lenticular, lanate, seeds free from pericarp.

Arid regions, Lower and Upper Sonoran Zones, eastern Washington to Saskatchewan and southward east of the Sierra Nevada to Texas, Sonora, and central Baja California. April–July.

## 8. **ALLENROLFEA** Kuntze, Rev. Gen. Pl. **2:** 545, 1891

Erect, succulent shrub with much-branched, erect or ascending, glabrous, jointed, green branches and woody older stems with smooth, tawny bark. Leaves scalelike. Flowers perfect, sessile in threes or fives in axils of fleshy scales, forming slender, cylindrical, fleshy spikes. Calyx lobes 4–5, small, angled, concave, somewhat carinate on back. Petals wanting. Stamens 1–2, slightly exserted. Stigmas 2 or sometimes 3. Fruit ovoid, invested with a membranous pericarp.

### 1. **Allenrolfea occidentalis** (S. Wats.) Kuntze, Rev. Gen. Pl. **2:** 546, 1891

*Halostachys occidentalis* S. Wats., Bot. King Expl. 293, 1871.
*Spirostachys occidentalis* S. Wats., Proc. Amer. Acad. **9:** 125, 1874.

Succulent shrub 0.5–1.5 m. high; branches green or somewhat glaucous, jointed and fragile above, woody and brittle in old growth; joints 2–12 mm. long; leaves triangular, fleshy, 0.4–2 mm. long, readily deciduous; terminal and axillary spikes numerous, 5–25 mm. long, 3–4 mm. in diameter; calyces closely crowded, lobes becoming spongy; fruit ovoid-lenticular, 0.8–1.2 mm. long, closely invested by spongy calyx lobes, the stigmas frequently persistent; seeds ovoid-lenticular, about 0.8 mm. long, reddish brown, smooth and shining.

Moist saline soil, mostly Lower Sonoran Zone, San Joaquin Valley, Mojave and Colorado Deserts to eastern Oregon, Utah, western Texas, southern Sonora, and the Cape Region, Baja California. July–Sept.

## 9. **ARTHROCNEMUM** Moq., Chenop. Enum. 111, 1840

Decumbent or erect, succulent, glabrous shrubs with opposite, articulate branches, each joint dilated into a short sheath at apex. Leaves wanting. Flowers perfect, in groups of 3 on opposite sides of the joints of cylindric terminal spikes, rarely only lower joints of stems floriferous. Calyx 3–4-toothed at apex, oval to obpyramidal. Stamens 2, barely exserted. Ovary compressed-oval; stigmas 2. Utricle ovoid-lenticular, included within perianth. Seed erect, oblong-lenticular, glabrous. Embryo curved around mealy endosperm; radicle inferior.

### 1. **Arthrocnemum subterminale** (Parish) Standley, Jour. Wash. Acad. **4:** 399, 1914

?*Arthrocnemum fruticosum* var. *californicum* Moq., in A. DC., Prod. **13**$^{2}$**:** 151, 1849.
*Salicornia subterminalis* Parish, Erythea **6:** 87, 1898.

Densely branched, green, erect or decumbent shrub 2–5 dm. high with joints 2–15 (or rarely 30) mm. long and 1.5–3.5 mm. in diameter, the fleshy younger joints expanded apically into a truncate or shortly bilobate sheath, lobes broadly deltoid, acute or obtuse, about 1 mm. high; flowering spikes terminal or axillary, 0.5–3.5 cm. long, joints about as broad as long; flowers present in axils of lower joints of spike, the upper scales usually sterile; utricle glabrous, thin; seed about 1.2 mm. high, ovoid-lenticular, brown, glabrous.

Saline soil, Lower and Upper Sonoran Zones, San Joaquin Valley, California, coastal marshes from vicinity of Santa Barbara to Baja California and Sinaloa. July–Sept.

## 10. SALICORNIA L., Sp. Pl. 3, 1753

Annual or perennial fleshy herbs with decumbent to erect, opposite, jointed stems, often suffrutescent, glabrous, joints dilated into a short sheath at apex. Leaves none. Flowers perfect or polygamous, embedded in groups of 3 on opposite sides of the joints of cylindrical spikes. Perianth 3–4-dentate, spongy in fruit. Stamens 1–2. Styles 2–3. Utricle ovoid to oblong, included in the perianth. Seed erect, ellipsoid, covered with short hairs; embryo conduplicate, radicle inferior; endosperm lacking.

Plants annual, herbaceous, erect, branching only above; seed 1.5–2 mm. long .. 1. *S. europea*
Plants perennial, suffrutescent, stems decumbent and branching near the base; seeds 1–1.2 mm. long .......................................................... 2. *S. pacifica*

### 1. **Salicornia europea** L., Sp. Pl. 3, 1753

*Salicornia virginica* L., op. cit. 4.
*Salicornia herbacea* L., Sp. Pl. ed. 2, 5, 1762.
*Salicornia annua* Smith, Engl. Bot. *pl. 415*, 1797.
*Salicornia procumbens* Smith, op. cit. *pl. 2475*, 1813.
(For additional synonymy see N. Amer. Fl. **21**: 84, 1916.)

Erect annual 0.3–5 dm. tall, usually simple at base and much branched above, the branches erect or ascending; joints slender, 1–2.5 mm. thick, 4–26 mm. long, lobes of sheath somewhat spreading, broadly deltoid, acute; spikes 1.5–4 mm. in diameter, 2–6.5 cm. long, joints slightly longer than to twice as long as thick; flowers subequal, the central one of each group inserted higher than the other two and reaching at least two-thirds of way to apex of joint; seeds 1.5–2 mm. long, light brown, puberulent with short, curved hairs.

Salt marshes, Canadian Zone to Tropical Zone, Alaska to Baja California; Sinaloa, eastern United States, West Indies, Europe, Africa, and East Indies. In our region, at Punta Peñasca, Sonora. July–Aug.

### 2. **Salicornia pacifica** Standley, N. Amer. Fl. **21**: 83, 1916

Perennial, suffrutescent, glabrous plants 1.5–6 dm. high, usually considerably branched from base, and stems more or less decumbent but not rooting at nodes; joints 2–5 mm. in diameter, 5–25 mm. long, sheaths rounded or with 2 broadly acute lobes; spikes of 6–18 joints, 2.5–3.5 mm. in diameter, 1–6 cm. long, joints nearly as thick as long or sometimes thicker; flowers nearly equal, the central one only slightly higher on the joint than the 2 lateral ones; seeds brownish, 1–1.2 mm. long, densely puberulent with short, curved hairs.

Salt marshes, Transition Zone to Lower Sonoran Zone, British Columbia to Baja California and Sonora and the adjacent islands. July–Sept.

## 11. SUAEDA Forsk., Fl. Aegypt. 69, 1775

Annual or perennial herbs or shrubs with alternate or fasciculate, succulent, often terete leaves. Flowers perfect or polygamous, in few-flowered clusters or solitary in axils of the leaves, bracteate; calyx lobes 5, fleshy, equal (in ours), forming a globose perianth. Stamens 5. Ovary globose or depressed-globose; styles 2 or rarely 2–5, slender, short. Utricle enclosed in the perianth. Seeds horizontal or erect, smooth or roughened; embryo a flat spiral; endosperm scanty or lacking.

Leaves markedly flattened; herbage green; branches of inflorescence very slender:
Seeds 0.8–1 mm. in diameter, smooth; herbage densely but finely pubescent ........... 1. *S. ramosissima*
Seeds 1–1.5 mm. in diameter, minutely tuberculate; herbage glabrous .... 2. *S. torreyana*
Leaves terete or subterete; herbage glaucous, or densely puberulent if green; branches of inflorescence stouter:
Herbage glabrous, at least at maturity; seeds vertical or horizontal; leaves 1–3.5 cm. long:
Seeds 0.6–0.8 mm. broad; flowers usually 2–5 in an axil; leaves not crowded in the inflorescence ........ 3. *S. fruticosa*
Seeds 1.5–2 mm. broad; flowers usually 1–2 in an axil; leaves crowded in the inflorescence ........ 4. *S. californica*
Herbage densely pubescent; seeds mostly horizontal; leaves 3–8 mm. long ........ 5. *S. taxifolia* subsp. *brevifolia*

### 1. Suaeda ramosissima (Standley) I. M. Johnston, Proc. Calif. Acad. IV, 12: 1017, 1924

*Dondia ramosissima* Standley, N. Amer. Fl. **21**: 91, 1916.
*Suaeda torreyana* var. *ramosissima* Munz, Man. S. Calif. 144, 1935.

Green, erect or straggling shrub 3–12 dm. high, sometimes twice as broad, woody and brittle at base, branches slender; herbage finely but densely puberulent throughout; leaves 5–20 mm. long, 0.5–2.5 mm. wide, linear to spatulate, strongly flattened, acute or obtuse, much reduced in inflorescence; flowers in globose glomerules of the elongate, very slender spikes, each flower globose or somewhat depressed, 1–1.5 mm. broad at anthesis; calyx lobes acute or obtuse, rounded on back, densely puberulent, closely incurved; seed vertical or horizontal, about 1 mm. broad, reddish brown to black, smooth.

Alkaline soils, Lower Sonoran Zone, Colorado Desert, adjacent Arizona, and southward at least to the vicinity of Guaymas, Sonora; and to Loreto, Baja California. Aug.–Oct.

### 2. Suaeda torreyana S. Wats., Proc. Amer. Acad. 9: 88, 1874

*Chenopodina linearis* Torr., in Stansbury, Exp. 394, 1853. Not *C. linearis* Moq., 1849.
*Suaeda fruticosa* S. Wats., Bot. King Expl. 294, 1871. Not *S. fruticosa* Forsk., 1775.
*Dondia torreyana* Standley, N. Amer. Fl. **21**: 90, 1916.

A green, openly and slenderly branched, sparsely leafy, glabrous shrub 3–10 dm. high; leaves linear to linear-spatulate, 1–1.5 mm. wide, 1–3 cm. long, distinctly flattened, acute or slightly acuminate, those of the inflorescence 2–5 mm. long and abruptly mucronate or obtuse; flowers globose, usually 1–4 in each axil of the slender, ascending branches of inflorescence, about 1 mm. broad at anthesis; calyx lobes green, closely incurved, rounded on back, and obtuse; utricle obovoid-lenticular, enclosed by calyx lobes; seed vertical or horizontal, 1–1.5 mm. broad, dark brown to black, shining, minutely tuberculate.

Alkaline soil, Upper and Lower Sonoran Zones, eastern Oregon southward east of the Sierra Nevada to Inyo County, California, Arizona, and northern Sonora, east to Wyoming and New Mexico. June–Oct.

**3. Suaeda fruticosa** (L.) Forsk., Fl. Aegypt. 70, 1775

*Chenopodium fruticosum* L., Sp. Pl. 221, 1753.
*Salsola fruticosa* L., Sp. Pl. ed. 2, 324, 1762.
*Suaeda fruticosa* var. *multiflora* Torr., Pacif. R. Rep. **4:** 130, 1857.
*Suaeda intermedia* S. Wats., Proc. Amer. Acad. **14:** 296, 1879.
*Suaeda moquini* Greene, Pittonia **1:** 264, 1889.
*Dondia conferta* Small, Bull. N.Y. Bot. Gard. **1:** 280, 1899.
*Dondia fruticosa* Northrop, Mem. Torrey Club **12:** 35, 1902.

Erect or ascending, glaucous, nearly or quite glabrous shrub 2–10 dm. high with fleshy, densely leafy branches and woody base; leaves 1–1.8 (rarely –3) cm. long, terete or nearly so, acute or obtuse, only slightly reduced in inflorescence; flowers in axillary clusters of 1–5, depressed-globose, 1–1.5 mm. broad; calyx lobes closely incurved, rounded on back, acutish at apex, sparsely and minutely puberulent to glabrous; seeds mostly horizontal, 0.6–0.8 mm. in diameter, smooth, black, and shining.

Alkaline or saline soil, Alberta to central Mexico, southern California, and the West Indies; also in Europe, Africa, and Asia. June–Oct.

**4. Suaeda californica** S. Wats., Proc. Amer. Acad. **9:** 89, 1874

*Dondia californica* Heller, Cat. N. Amer. Pl. 3, 1898.

Glaucous and glabrous, decumbent to ascendingly branched perennial 2–10 dm. tall with stout, densely leafy branches; stems woody and usually leafless toward base, bark thin, tawny or stramineous; leaves terete or nearly so, 1.2–3.5 cm. long, ascending or nearly erect, often falcately curved, acute, closely crowded and only slightly reduced in inflorescence; flowers depressed-globose, usually 1–2 in an axil, 1.5–2.5 mm. broad; calyx lobes rounded on back, obtuse, 1–1.4 mm. long, incurved; anthers 1–1.2 mm. long; seed horizontal or vertical, 1.4–2 mm. broad, smooth, black, and shining.

Coastal salt marshes, Lower and Upper Sonoran Zones, central California to southern Baja California. Feb.–Sept.

**5. Suaeda taxifolia** Standley subsp. **brevifolia** (Standley) Abrams, Ill. Fl. Pacif. States **2:** 96, 1944

*Dondia brevifolia* Standley, N. Amer. Fl. **21:** 92, 1916.

Grayish green, much-branched perennial 2–6 dm. tall with procumbent or ascending stems; herbage densely and minutely puberulent throughout; leaves terete, 3–8 mm. long, obtuse or abruptly acute at apex, cuneately narrowed at base, slightly reduced in leafy inflorescence; flowers globose or slightly depressed, 1.2–2 mm. broad at anthesis, usually 3 in an axil; calyx lobes densely puberulent, rounded on back, 1.8–2 mm. long, obtuse tips incurved; utricle depressed-ovoid; seed usually horizontal, 1.5–2 mm. broad, smooth, black, and shining.

Salt marshes and saline soil along the coast, Lower and Upper Sonoran Zones, Orange County, California, south into Baja California. March–Aug.

## Family 36. AMARANTHACEAE. Amaranth Family

Herbs or shrubs with alternate or opposite, simple, estipulate leaves. Flowers small, usually inconspicuous, often green, perfect or unisexual, in spikes, clusters, or paniculate racemes. Sepals 3–5 or sometimes only 1, more or less scarious, persistent. Petals none. Stamens 5 or fewer. Ovary superior, 1-celled; stigmas 2–3. Fruit an utricle, indehiscent, irregularly dehiscent, or circumscissile. Embryo curved; endosperm copious.

Anthers 4-celled; leaves alternate; ovules erect:
  Ovules 2–several; filaments connate at base, forming shallow cup; flowers perfect; seeds 2–many ........ 1. *Celosia*
  Ovule solitary; filaments distinct; flowers (in ours) monoecious or dioecious; seed one ........ 2. *Amaranthus*
Anthers 1- or 2-celled; leaves opposite (or upper leaves alternate in *Dicraurus*); ovule pendulous from elongated funiculus:
  Stamens perigynous, inserted on perianth tube; plant prostrate, matted .... 3. *Brayulinea*
  Stamens hypogynous; plants erect, ascending, or decumbent but not prostrate-matted:
    Perianth tubular, indurate, bearing spines or cristate wings in age; anthers sessile in sinuses of perianth tube ........ 4. *Froelichia*
    Perianth segments distinct; anthers borne on short or long filaments or on a stamineal tube not attached to perianth segments:
      Leaves subtending flowering glomerules indurate, somewhat involucrate; glomerules few-flowered; stamineal cup very shallow, staminodia wanting .. 5. *Tidestromia*
      Leaves subtending flowering head not indurate nor involucrate; glomerules many-flowered; stamineal tube equaling ovary, or, if shorter, staminodia often present:
        Anthers 1-celled; stamineal tube long, staminodia wanting; lobes of stamineal tube 3-lobed or 3-dentate ........ 6. *Gomphrena*
        Anthers 2-celled; staminodia present except in 2 species of *Iresine*; lobes of stamineal tube entire:
          Flowers perfect; perianth segments 3-nerved and somewhat indurate, stamen tube 4–10-lobed; stigma capitate ........ 7. *Achyranthes*
          Flowers dioecious, polygamodioecious, or rarely perfect; stamen tube usually 5-lobed; stigma 2–3-lobed:
            Leaves all opposite; sepals usually lanate; staminodia entire or wanting ........ 8. *Iresine*
            Leaves mostly alternate, only the lower opposite; sepals glabrous or sparingly villous; staminodia dissected into fine, papillate lobes ........ 9. *Dicraurus*

## 1. CELOSIA L., Sp. Pl. 205, 1753

Annual or perennial, glabrous or pubescent herbs or shrubs. Leaves alternate; flowers perfect, sessile or pedicellate in dense terminal or axillary spikes or in fascicles along flowering stems, white, yellowish, greenish, or red; perianth 5-parted into striate-nerved, scarious segments. Stamens 5; filaments connate with shallow cup at the base; anthers 4-celled. Ovary ovoid to cylindrical; style none to elongate; stigmas 2–3, subulate to capitate; ovules 2–many; utricle included or exserted from the perianth, circumscissile, rupturing irregularly or indehiscent. Seeds 2–many, erect, smooth and shining, lenticular; embryo annular; endosperm mealy.

Sepals obscurely nerved, 1.5–2.5 mm. long; leaf blades often hastately lobed, puberulent beneath ........ 1. *C. floribunda*
Sepals finely but prominently veined, 4.5–5 mm. long; leaf blades not hastately lobed, entire, glabrous beneath ........ 2. *C. nitida*

### 1. Celosia floribunda A. Gray, Proc. Amer. Acad. 5: 167, 1861

Shrub or small tree 1–5 m. high; branches prominently striate, green or glaucous, glabrous below inflorescence; leaves extremely variable, linear, subhastate-oblong, or triangular-ovate, 0.5–11 cm. wide, 3–20 cm. long, attenuate-acuminate to rounded at apex, acute-cuneate to rounded at base, prominently reticulate-veined, scabrous-puberulent beneath, glabrous or nearly so above; petioles 8–40 mm. long, often winged nearly to base; flowers sessile, very numerous in slender, loosely flowered spikes 1.5–5 cm. long that are aggregated in dense panicles 5–30 cm. long; rachises sparsely puberulent; bracts round-ovate, 0.6–1 mm. long; sepals 2 mm. long, ovate to oblong, rounded at apex, papery, white or stramineous, faintly nerved; stamens 5, included; style about equaling or shorter than ovary; stigmas 3, exceeding style; utricle brown, globose to broadly ovoid, about equaling sepals; seeds 2–4, 1.2 mm. in diameter, dark brown to black.

Gravelly plains and washes, central to southern Baja California. March–Oct.

**2. Celosia nitida** Vahl, Symb. Bot. **2:** 44, 1791

*Lestibudesia paniculata* R. Br., Prod. 414, 1810. Not L., 1753.
*Lophoxera paniculata* Raf., Fl. Tell. **3:** 42, 1837.
*Gonufa paniculata* Raf., Sylva Tell. 124, 1838.
*Celosia texana* Scheele, Linnaea **22:** 148, 1849.

Perennial from woody base with slender, erect or ascendingly scrambling stems 3–15 dm. long, these glabrous, green or glaucous, striate or smooth; petioles 0.5–2 cm. long; leaf blades ovate to deltoid, rhombic-ovate or triangular-lanceolate, 1–4.5 cm. wide, 2–7 cm. long, usually asymmetric, obtuse, truncate, or oblique and slightly decurrent at base, obtuse to acuminate at apex, prominently veined beneath; flowers in terminal panicles composed of few, sessile or short-pedunculate spikes; bracts broadly ovate, 1–1.5 mm. long, acute to obtuse, lightly carinate; sepals 4–5 mm. long, oblong, acute to mucronulate, yellow to brown, prominently and finely parallel-veined; style about 1 mm. long, stout; utricle ovoid, about equaling sepals; seeds many, smooth or faintly punctulate, black and lustrous, 0.8–1 mm. in diameter.

In partially shaded canyons, central Sonora to Texas, Yucatan, West Indies, and South America; Florida. June–Oct.

## 2. AMARANTHUS L., Sp. Pl. 989, 1753

Annual, erect or prostrate, glabrous or pubescent herbs. Leaves alternate, petiolate. Flowers small, apetalous, monoecious, dioecious, or polygamous, bracteate and bibracteolate in terminal spikes, axillary glomerules, or both; sepals 5 or rarely 1–3, membranous. Stamens 5 or rarely 1–3, on filiform or subulate distinct filaments; anthers 4-celled. Ovary ovoid; style short; style branches 2–3, papillose or pubescent; ovule 1, erect; utricle entire or 2–3-dentate apically, indehiscent, irregularly dehiscent, or regularly circumscissile, membranous or coriaceous, smooth or rugulose. Seed erect, lenticular-compressed, smooth; embryo annular; endosperm farinaceous; radicle inferior.

Sepals of pistillate flowers spatulate, the claw narrow, the calyx more or less urceolate:
  Utricle indehiscent ........................................ 1. *A. obcordatus*
  Utricle dehiscent:
    Plants dioecious:
      Bracts two to three times as long as pistillate sepals; herbage eglandular; seed oval, 1.3 mm. long ........................................ 2. *A. palmeri*
      Bracts only slightly exceeding sepals; herbage glandular-puberulent; seed orbicular, 0.6 mm. in diameter ........................................ 3. *A. watsonii*

Plants monoecious:
Bracts equaling or exceeding sepals of pistillate flowers:
Stamens 5; sepals of pistillate flowers 2.5 mm. long; seed subglobose, brownish, 1.2 mm. in diameter .................................... 4. *A. lepturus*
Stamens 2–3; sepals of pistillate flowers 1.5–2 mm. long; seeds orbicular-lenticular, black, 1 mm. in diameter .................................... 5. *A. pringlei*
Bracts shorter than sepals of pistillate flowers:
Sepals of pistillate flowers with fimbriate margins; utricle circumscissile near apex ....................................................... 6. *A. fimbriatus*
Sepals of pistillate flowers with crenate margins; utricle circumscissile at middle .. ....................................................... 7. *A. venulosus*
Sepals of pistillate flowers oblong or obovate, not contracted to a narrow claw, the calyx not urceolate; plants all monoecious:
Style branches 2; herbage often somewhat glaucous .................. 8. *A. graecizans*
Style branches 3; herbage not glaucous:
Plants prostrate; stamens 3; leaves often white-margined .............. 9. *A. blitoides*
Plants erect or ascending; stamens 5; leaves not white-margined:
Spines pungent in leaf axils; utricle irregularly dehiscent ........... 10. *A. spinosus*
Spines wanting; utricle regularly circumscissile:
Edges of pistillate sepals overlapping; terminal spike drooping from base; seeds light-colored ................................................ 11. *A. caudatus*
Edges of pistillate sepals barely meeting, not overlapping; terminal spikes erect, or drooping above middle; seeds dark brown to black:
Pistillate sepals much exceeding utricle; terminal spikes drooping from middle ................................................ 12. *A. retroflexus*
Pistillate sepals equaling or shorter than utricle; spikes erect:
Sepals of pistillate flowers oblong, the inner obtuse; bracts not over one and one-half times as long as sepals; utricle conspicuously exceeding sepals ... ................................................ 13. *A. cruentus*
Sepals of pistillate flowers ovate to oblong, acute to acuminate; bracts twice as long as sepals; utricle barely equaling sepals ............ 14. *A. hybridus*

**1. Amaranthus obcordatus** (A. Gray) Standley, N. Amer. Fl. **21:** 107, 1917
*Amblogyna urceolata* var. *obcordata* A. Gray, Proc. Amer. Acad. **5:** 168, 1861.
*Amaranthus urceolatus* var. *obcordatus* Uline & Bray, Bot. Gaz. **19:** 269, 1894.
*Amaranthus urceolatus* var. *jonesii* Uline & Bray, loc. cit.

Stems erect or ascending, simple or branched, 1–6 dm. high, glabrous, whitish or tinged with red; leaves lanceolate, linear-oblong, or narrowly oblong, 1–3.5 cm. long, narrowly cuneate at base, truncately rounded to acute at apex, pale green, glabrous, margins entire; petioles slender, 5–22 mm. long; flowers monoecious, in dense axillary glomerules or in terminal, leafy to nearly naked spikes; bracts broadly ovate, acute-apiculate, one-third to one-half as long as sepals; sepals of pistillate flowers 5, spatulate, 1.5–2 mm. long, united at base, urceolate-spreading at summit, conspicuously veined, margin entire or minutely erosulate, rounded or emarginate at apex; sepals of staminate flowers 1 mm. long, oblong, obtuse or abruptly apiculate-acute, membranous; stamens 3; style branches 3; utricle narrowly oblong, 2 mm. long, rugulose on upper third; seed obovoid, 0.6–0.8 mm. long, dark reddish brown.

Margins of vernal pools and desert playas, Lower Sonoran Zone, western Texas to southern Arizona, Sonora, and Sinaloa. Aug.–Oct.

**2. Amaranthus palmeri** S. Wats., Proc. Amer. Acad. **12:** 274, 1877
*Amaranthus palmeri* var. *glomeratus* Uline & Bray, Bot. Gaz. **19:** 272, 1894.

Stems stout, 1–15 dm. high, simple to profusely branched, branches spreading or ascending, green or frequently tinged with red, glabrous below, often villous in inflorescence; petioles slender, 1–15 cm. long; leaf blades ovate, rhombic-ovate, or lanceolate, 1.5–15 cm. long, acute to abruptly acute-acuminate at apex, rounded to acute-cuneate at base, bright green, glabrous or sparsely puberulent beneath, prominently pinnately veined; flowers diocecious, in slender terminal and axillary spikes 0.5–4 dm. long, erect or drooping; bracts linear to lance-ovate, rigid, spinose at tips, one and one-half to three times as long as flowers; sepals of staminate flowers oblong, 2–3 mm. long, acute, with a single excurrent nerve, glabrous, stramineous; sepals of pistillate flowers spatulate, 2–3.5 mm. long, obtuse, rounded, or emarginate, minutely denticulate, stramineous or reddish, outer ones often acute; stamens 5; style branches 2 or infrequently 3; utricle slightly compressed, 2–2.5 mm. long, circumscissile; seed oval, 1.3 mm. long, dark reddish brown.

In waste ground and cultivated fields and along washes, mostly Lower Sonoran Zone, San Bernardino and Riverside Counties, California, to Kansas, Texas, central Mexico, and the Cape Region, Baja California. April–Oct.

**3. Amaranthus watsonii** Standley, Bull. Torrey Club **41**: 505, 1914

*Amaranthus torreyi* var. *suffruticosus* Uline & Bray, Bot. Gaz. **19**: 272, 1894.

Stout, profusely branched plant with ascending branches 1–10 dm. high, stems obtusely angled, puberulent to short-villous and glandular, densely so above; petioles 0.5–6 cm. long; leaf blades round, ovate, obovate, rhombic-oval, or oblong, 0.6–8 cm. long, obtuse to truncate and emarginate to deeply retuse at apex, broadly cuneate to rounded at base, thick, conspicuously veined beneath, glandular-puberulent on both surfaces or only beneath; flowers dioecious, in axillary glomerules arranged in terminal and axillary, erect or drooping spikes 3–16 cm. long, 5–15 mm. thick; bracts lanceolate, equaling or slightly longer than sepals, subulate-spinose, glandular-puberulent on midrib; sepals of staminate flowers oblong, 3–3.5 mm. long, obtuse to acute, the single nerve short-excurrent; sepals of pistillate flowers broadly spatulate, 2.5 mm. long, broadly rounded and minutely denticulate at apex, emarginate, nerve often excurrent in outer sepals; stamens 5; style branches 2; utricle globose, circumscissile at middle; seeds orbicular-lenticular, 0.6 mm. in diameter, dark reddish brown.

Desert washes and llanos, Lower Sonoran Zone, vicinity of Guaymas, Sonora, and from central Baja California to the Cape Region. Feb.–Sept.

**4. Amaranthus lepturus** S. F. Blake, Jour. Bot. **53**: 104, 1915

Plants slender, about 1 m. high, with ascending to erect branches, glabrous; petioles slender, 1.5–3 cm. long; leaf blades rhombic-oval, 2–2.7 cm. long, about half as broad, obtuse to retuse and cuspidate at apex, cuneate at base, glabrous; flowers monoecious in erect terminal and axillary spikes, these interrupted-leafy below, leafless above, 1–4 dm. long, 0.5–1.5 cm. thick; bracts lance-subulate, about one and one-half times as long as sepals; sepals of staminate flowers ovate-lanceolate, acute, scarious, 2 mm. long; stamens 5; sepals of pistillate flowers 5, oblong, spatulate, 2.5 mm. long, retuse to obtuse or acute, glabrous or sparsely short-villous below, greenish with scarious margins, mucronulate; style branches 3 or sometimes 2; utricle 1.5–2 mm. long, circumscissile near the middle; seed subglobose, 1.2 mm. long, brownish black, shining.

Rocky slopes, Lower Sonoran Zone, vicinity of Calmallí to Bahía Concepción and Magdalena Bay, Baja California. Oct.–March.

**5. Amaranthus pringlei** S. Wats., Proc. Amer. Acad. **22:** 476, 1887

Stout plant with erect, angled, much-branched, glabrous stems 1.5–10 dm. high, often tinged with red; petioles 3–25 mm. long, stout; leaf blades elliptic, oblong, linear-oblong, or upper ones linear, 1–4.5 cm. long, rounded or obtuse at apex, cuneate at base, glabrous; flowers monoecious, in dense axillary glomerules and short terminal spikelike panicles; bracts equaling or slightly exceeding flowers, oblong to oblong-linear, acute, rigidly spinose; sepals 5, those of staminate flowers oblong, acute, scarious; sepals of pistillate flowers broadly spatulate, 1.5–2 mm. long, emarginate or apiculate at apex, conspicuously veined, margins scarious, often tinged with purple; stamens 3; utricle ovoid, circumscissile near middle; seed orbicular-lenticular, 1 mm. in diameter, black and shining.

Mountain slopes, chiefly Upper Sonoran Zone, western Texas to southern Nevada and southeastern Arizona, south to San Luis Potosí. Aug.–March.

**6. Amaranthus fimbriatus** (Torr.) Benth. ex S. Wats., Bot. Calif. **2:** 42, 1880

*Sarratia berlandieri* var. *fimbriata* Torr., Bot. Mex. Bound. 179, 1859.
*Amblogyna fimbriata* A. Gray, Proc. Amer. Acad. **5:** 167, 1861.

Stems erect, simple or branched above, 1–10 dm. high, glabrous or puberulent above; petioles slender, 8–30 mm. long; leaf blades lanceolate, obtuse or acute at apex, narrowly cuneate at base, 2–7 cm. long, apiculate, glabrous; flowers monoecious, in dense axillary glomerules, interrupted below, forming a slender spike above; staminate flowers 1.5 mm. long, sepals oblong, acute, membranous; bracts ovate-acuminate, one-third as long as pistillate sepals; these broadly spatulate, 2.5 mm. long, margin fimbriate, heavily veined, tinged with red; stamens 3; style branches 3; utricle about three-fourths as long as sepals, circumscissile near apex; seeds brown or black, 0.8 mm. in diameter, obovoid-lenticular, shining.

Plains and desert slopes and washes, mostly Lower Sonoran Zone, southern Utah and Nevada, south through southern California and Arizona into Sinaloa and the Cape Region, Baja California. Aug.–Nov.

**7. Amaranthus venulosus** S. Wats., Proc. Amer. Acad. **17:** 376, 1882

*Sarratia berlandieri* var. *denticulata* Torr., Bot. Mex. Bound. 179, 1859.
*Amaranthus fimbriatus* var. *denticulatus* Uline & Bray, Bot. Gaz. **19:** 270, 1894.

Stems slender, erect, striate, 1–6 dm. high, glabrous; petioles slender, 5–25 mm. long; leaf blades lanceolate or linear-oblong, 1–4 cm. long, rounded or abruptly acute at apex, cuneate at base, glabrous; flowers monoecious, in dense axillary glomerules, the upper aggregated into slender, leafy or nearly leafless, spikelike panicle 5–20 cm. long; bracts ovate, acute, about one-fourth to one-half as long as sepals; sepals 5, those of staminate flowers oblong, acute, 1–1.2 mm. long, those of pistillate flowers spatulate, 2 mm. long, slightly united at base, blades rounded and often emarginate and crenulate at apex, spreading in age, conspicuously veined; stamens 3; style branches 3; utricle oblong or ovoid, shorter than sepals, rugulose above, circumscissile about middle; seeds orbicular-lenticular, 0.8 mm. in diameter, dark reddish brown, shining.

Roadsides, stream banks, and washes, Lower Sonoran Zone, vicinity of Tucson, Arizona, south into Sonora and southern Baja California. Aug.–Oct.

**8. Amaranthus graecizans** L., Sp. Pl. 990, 1753

*Amaranthus albus* L., Syst. Nat. ed. 10, 1268, 1759.
*Amaranthus blitum* var. *graecizans* Moq., in A. DC., Prod. **13**[2]**:** 263, 1849.
(For additional synonymy see N. Amer. Fl. **21:** 116, 1917.)

Erect or more commonly decumbent, profusely branched plant with stout branches, 0.5–12 dm. high, glabrous; petioles slender, 0.5–5 cm. long; leaf blades obovate, elliptic, or oblong, rounded or obtuse at apex, cuneate at base, prominently white-veined beneath, pale green to nearly glaucous beneath, mucronate, margins finely crispate-undulate, 5–70 cm. long; flowers monoecious, in dense axillary glomerules; bracts oblong-lanceolate, one and one-half to four times as long as sepals of pistillate flowers, green, rigidly spinose-tipped; sepals and stamens 3; sepals of staminate flowers scarious, oblong, cuspidate, 1 mm. long or less, those of pistillate flowers oblong to linear, acute, unequal, green along nerve, scarious and often tinged with red along margins; style branches 2; utricle subglobose-ovoid, circumscissile just above middle, rugose and thickened above, thin and scarious below line of dehiscence; seeds orbicular-lenticular, black or dark brown, shining.

Waste ground, roadsides, and fields, over most of United States and northern Mexico. Commonly only along streams or irrigating ditches in our area. June–Oct.

**9. Amaranthus blitoides** S. Wats., Proc. Amer. Acad. **12:** 273, 1877
(For synonymy see N. Amer. Fl. **21:** 115, 1917.)

Prostrate annuals with stoutish, much-branched stems 1–6 dm. long, pale green or whitish, glabrous to sparsely pubescent; leaves numerous and usually crowded, particularly near tips of the branches; petioles 2–20 mm. long; leaf blades 0.8–4 cm. long, elliptic, spatulate, obovate, or ovate, attenuate to broadly cuneate at base, acute to rounded at apex, glabrous, pale green, often with a whitish band or line along margins and along prominent veins beneath; flowers monoecious, in compact axillary glomerules, usually shorter than petioles; bracts green, lanceolate to oblong, erect, attenuate into a short, spinose tip, equaling or slightly exceeding the 4–5 sepals; sepals of staminate flowers oblong, acute, scarious, those of pistillate flowers narrowly oblong, acuminate, green with white margins; stamens 3; style branches 3; utricle equaling or slightly exceeding sepals, subglobose, smooth or faintly wrinkled, circumscissile; seeds subglobose, 1.3–1.5 mm. in diameter, dull black.

Roadsides, cultivated fields, and waste ground, Lower Sonoran Zone to Canadian and Carolinian Zones, Washington to Wyoming and south to southern California, Arizona, Texas, and northern Mexico. April–Oct.

**10. Amaranthus spinosus** L., Sp. Pl. 991, 1753
(For synonymy see N. Amer. Fl. **21:** 114, 1917.)

Stems erect or spreading-ascending, much branched, 2–15 dm. high, striate, glabrous below, puberulent in upper parts of branches and inflorescence, succulent; petioles slender, 0.4–10 cm. long, unevenly villous; 2 slender, rigid spines 5–25 mm. long axillary to most leaves; leaf blades ovate to ovate-lanceolate, 1–12 cm. long, rounded or obtuse at narrow apex, acute to cuneate at base, conspicuously veined and pale green beneath, often tinged with red and dark green above, glabrous to sparsely puberulent; flowers monoecious, pistillate in dense, many-flowered, globose, axillary glomerules, staminate in slender, erect or drooping spikes 2–20 cm. long; bracts lanceolate, spinose, one to three times as long as sepals; sepals of staminate flowers oblong, acute, apiculate, vein green, margins scarious, 1.5 mm. long; sepals of pistillate flowers 5, oblong-obovate, obtuse or acutish, usually mucronulate, 1–1.5 mm. long; stamens 5; style branches 3; utricle ovoid, 1–1.2 mm. long, irregularly or incompletely dehiscent; seeds orbicular- to obovoid-lenticular, dark reddish brown to black, 0.6–0.8 mm. in diameter.

Widely distributed from central Mexico to Central America and common in fields in the Cape Region, Baja California. June–Oct.

**11. Amaranthus caudatus** L., Sp. Pl. 990, 1753
(For synonymy see N. Amer. Fl. **21:** 110, 1917.)

Stems stout, erect, 3–20 dm. high, simple or much branched, often tinged with red, glabrous below, puberulent in inflorescence; petioles 2–20 cm. long; leaf blades ovate, lanceolate, or rhombic-ovate, 1–8 cm. wide, 3–20 cm. long, rounded, acute, or acuminate at apex, acute to broadly cuneate at base, prominently veined beneath, glabrous or nearly so, usually tinged with red; flowers monoecious in terminal panicles and axillary glomerules or short spikes, these 2–25 cm. long, drooping from about the base; bracts lance-ovate, one and one-half to two times as long as sepals, crenate, spinose-subulate at apex, finely and unevenly puberulent; sepals of staminate flowers ovate to oblong, carinate, veins excurrent, often tinged with red; sepals of pistillate flowers 2 mm. long, lance-oblong to spatulate, acute to attenuate at apex; stamens 5; style branches 3, elongate, equaling or exceeding ovary, more or less rigid and erect; utricle ovoid, acute at apex and with persistent erect style branches, smooth or slightly rugulose above, circumscissile at middle; seed orbicular-lenticular, yellowish white, dull or sometimes darker-colored, 1–1.2 mm. in diameter.

Occasionally in fields and along ditches as an introduced weed in Arizona; central Mexico to Central America.

**12. Amaranthus retroflexus** L., Sp. Pl. 991, 1753
(For synonymy see N. Amer. Fl. **21:** 113, 1917.)

Stems stout, green or whitish, 2–15 dm. high, erect, single or profusely branched, densely villous about inflorescence, less so or glabrate below, leafy; petioles 1–10 cm. long, usually villous; leaf blades elliptic, rhombic-ovate, or lanceolate, rounded at apex, acute to cuneate at base, glabrous above, 3–12 cm. long, 1–7 cm. wide; flowers monoecious, in dense axillary and terminal paniculate spikes, these 5–20 cm. long and erect or slightly drooping from the middle; bracts ovate-lanceolate, dark green nerve excurrent as a rigid spinose tip, one and one-half to two times as long as sepals, puberulent; sepals of staminate flowers lanceolate to ovate-oblong, acute, 2.5–3 mm. long, scarious, nerve excurrent; sepals of pistillate flowers linear-oblong, 3 cm. long, rounded to truncate and usually emarginate at apex, often mucronate, scarious; stamens 5; style branches 3, shorter than ovary; utricle subglobose-ovoid, shorter than sepals, slightly rugulose or smooth on the upper part, circumscissile at middle; seed lenticular-orbicular, reddish brown to black, 1 mm. in diameter, shining.

A common weed in waste ground and along ditches and roadsides, throughout most of the United States and northern Mexico. Mostly Upper Sonoran and Transition Zones. April–Dec.

**13. Amaranthus cruentus** L., Syst. Nat. ed. 10, 1269, 1759
(For synonymy see N. Amer. Fl. **21:** 111, 1917.)

Stems stout, erect, simple to much branched, sulcate, tinged with red, short-pubescent nearly throughout, 3–20 dm. high; petioles slender, 1–20 cm. long, sparsely pubescent or glabrous; leaf blades ovate, rhombic-ovate, or lance-ovate, 3–30 cm. long, 1–10 cm. wide, rounded or acute at apex, broadly cuneate at base, prominently veined beneath, often tinged with red, farinose-puberulent or glabrous; flowers monoecious in dense, terminal, interrupted panicles comprised of several lateral spikes 2–15 cm. long, terminal spike longer than lateral ones; bracts ovate-acuminate, tapering to a slender, spinose tip, equaling or somewhat surpassing sepals; sepals of staminate flowers ovate to oblong, acute, with excurrent nerve; sepals of pistillate flowers oblong, 1.5 mm. long, inner ones obtuse or rounded, outer ones often acute at apex, scarious, stramineous, greenish, reddish, or purplish; stamens 5; style branches 3, stout, one-fourth as long as ovary, erect; utricle obovoid, about one and one-half times as long as sepals, brownish, scarious, slightly rugulose or striate above, circumscissile about to middle; seed orbicular-lenticular, dark reddish brown to black, 1–1.3 mm. in diameter, shining.

Occasionally escaped from gardens in New Mexico and Arizona but mostly at elevations above the Sonoran Desert.

**14. Amaranthus hybridus** L., Sp. Pl. 990, 1753
(For synonymy see N. Amer. Fl. **21:** 111, 1917.)

Stems stout, erect, simple or more frequently spreadingly branched, 2–2.5 dm. high, striate or sulcate, green or reddish, rough-puberulent below, usually sparsely pubescent above; petioles 1–10 cm. long, pubescent; leaf blades ovate, rhombic-ovate or lance-ovate, 2–18 cm. long, acute to rounded at apex, acute to broadly cuneate at base, bright green above, paler and sparsely puberulent to glabrous and prominently veined beneath; flowers monoecious, in numerous terminal and axillary spikes 2–8 cm. long, erect; bracts lance-subulate, tapering to spinose tip, once to twice as long as sepals, puberulent; sepals of staminate flowers 5, ovate to narrowly oblong, acute, with an excurrent nerve; sepals of pistillate flowers 5 (or 3), linear-oblong, 1.5–2 mm. long, acute, or inner rarely obtuse, dark green nerve excurrent as spinose tip; stamens 5; style branches 3; utricle subglobose-ovoid, about equaling sepals, thin-walled, rugulose or smooth, circumscissile at middle; seeds orbicular-lenticular, 1 mm. in diameter, brownish or black, shining.

Widely introduced as a weed in much of the United States and northern Mexico; roadsides and fields. June–March.

## 3. BRAYULINEA Small, Fl. SE. U.S. 394, 1903

Much-branched, leafy, prostrate, perennial herb with densely lanate stems and undersurfaces of leaves. Leaves opposite, members of each pair unequal, short-petiolate, the petioles connate at the base. Flowers perfect, small, sessile or subsessile in axillary glomerules, bracteate and bracteolate, bracts and bracteoles subequal; perianth campanulate, 5-lobed, membranous, lanate. Stamens 5, perigynous, inserted on tube at bases of lobes; filaments deltoid-subulate; anthers 2-celled. Ovary compressed; style short; stigma inconspicuously 2-lobed; ovule solitary, pendulous. Utricle indehiscent, membranous. Seeds lenticular-orbicular; embryo annular; endosperm mealy; radicle superior.

**1. Brayulinea densa** (Willd.) Small, Fl. SE. U.S. 394, 1903
*Illecebrum densum* Willd. ex Roem. & Sch., Syst. Veg. **5:** 517, 1819.
*Guilleminea densa* Moq., in A. DC., Prod. **13²:** 338, 1849.
*Guilleminea densa* var. *aggregata* Uline & Bray, Bot. Gaz. **20:** 343, 1895.

Stems 3–30 cm. long from lignescent vertical root; basal leaves evanescent, oblanceolate or elliptic, blades 4–9 mm. wide, 1–2.5 cm. long, petioles about equaling blades; petioles of cauline leaves short, winged, 1–4 mm. long, blades elliptic to oval, asymmetric, 3–15 mm. long, 2–10 mm. wide, abruptly narrowed at base, acute or rounded at apex, bright green and glabrous above, densely lanate beneath; glomerules shorter than leaves; bracts ovate-acute, white, scarious, glabrous, 1–1.5 mm. long; calyx 2–2.5 mm. long, lobes lance-ovate, acute, glabrous, white-scarious; utricle glabrous, compressed-ovate, about equaling perianth tube; seeds ovoid-lenticular, 0.5 mm. long, brown, lustrous.

Dry soil, Lower and Upper Sonoran Zones, southern Arizona to western Texas and south to southern Mexico and Argentina. Introduced at Riverside, California. July–Oct.

## 4. FROELICHIA Moench, Meth. 50, 1794

Annual or perennial pubescent herbs with erect or procumbent, simple or branched stems. Leaves opposite, petiolate or sessile. Flowers perfect, sessile, bracteate and bibracteolate, in sessile or pedunculate spikes; perianth tubular, with 5 distinct, scarious, glabrous lobes, tube long-lanate, indurate in age, and bearing longitudinal rows of spines, or entire or cristate wings. Stamens 5; filaments connate to form elongate, 5-lobed tube; anthers 2-celled, sessile in sinuses between lobes of stamineal tube. Ovary ovoid; style elongate or wanting; stigma capitate or penicillate; ovule solitary. Utricle ovoid, indehiscent, membranous, encased by indurate perianth tube. Seed obovoid or lenticular; embryo annular; endosperm mealy; radicle superior.

Appendages on calyx tube 2 rows of distinct spines; plants slender; wool on tube usually white or very light stramineous; basal parts of stems usually glabrate in age .................................................................. 1. *F. gracilis*

Appendages entire or cristate ridges, not of distinct spines; plants stout; wool on perianth white-shiny; stems persistently lanate:

Calyx crests dentate; calyx lobes 2 mm. long; basal spine on 1 or both faces of tube ............................................................ 2. *F. arizonica*

Calyx crests entire or nearly so; calyx lobes 1.5 mm. long; faces of tube spinulose ............................................................ 3. *F. interrupta*

### 1. Froelichia gracilis (Hook.) Moq., in A. DC., Prod. 13²: 420, 1849

*Oplotheca gracilis* Hook., Ic. Pl. under *pl. 256,* 1840.

Annual with erect, ascending, or infrequently procumbent stems, simple to much branched at base and sparsely to densely villous-tomentose, basal parts of stems usually glabrate in age, 1–6 dm. high; leaves short-petiolate or sessile, lance-elliptic, linear-lanceolate, or narrowly oblanceolate, acute to acuminate at apex, gradually narrowing to a cuneate or attenuate base, 3–12 mm. wide, 2–12 cm. long, sparsely to densely lanate-pubescent or glabrate in age on upper surface; spikes terminal and axillary, 1–4 cm. long, 7–16 mm. thick; bracts broadly ovate, acute to short-acuminate, scarious, 1-nerved, fuscous along nerve and on upper half, lighter below, 1–1.5 mm. long; bractlets narrowly ovate, acute, 2–2.5 mm. long, pilose, scarious; calyx lobes linear-oblong, acute, 1–1.5 mm. long; calyx tube with 2 longitudinal rows of distinct spines about 1 mm. long at maturity; sides minutely tuberculate near base, long-lanate with shining white hairs; seeds reddish, subglobose, 1.5 mm. in diameter.

Central Arizona to Colorado, Iowa, Arkansas, and Chihuahua and northern Sonora. Chiefly above the limits of the Sonoran Desert. June–Oct.

### 2. Froelichia arizonica Thornber, in Standley, N. Amer. Fl. 21: 128, 1917

*Froelichia campestris* Small, Fl. SE. U.S. 397, 1903. (In part.)

Perennial from a woody root, with 1–several stout, erect or ascending, sericeous-tomentose stems 2–10 dm. high, branching mostly at or near base; leaves oblanceolate, elliptic, or oblong-lanceolate, 4–25 mm. wide, 2–10 cm. long, acute or obtuse at apex, acute to attenuate-cuneate at base, densely sericeous-lanate beneath, sparsely so to glabrate on upper surface, mostly crowded at or near base of stems; petioles to 1.5 cm. long or blades sessile; peduncles 1–3 dm. long or more; spikes 1–3.5 cm. long, about 1 cm. wide; bracts ovate, acute, apiculate, 1.5–2 mm. long, stramineous or fuscous at apex; bractlets broadly ovate to nearly orbicular, obtuse to rounded at apex, fuscous, scarcely at all stramineous, 2.5–3 mm. long, scarious, glabrous; perianth lobes lance-linear, glabrous, 2 mm. long, stramineous; tube 3 mm. long at maturity, broadly winged, wings lacerate-dentate, about 1 mm. high; one or both sides of tube bearing a short spine or tubercle, densely lanate with yellowish or rusty hairs to 6 mm. long; seed reddish, about 1 mm. in diameter, subglobose.

Dry hillsides and rocky slopes, Lower and Upper Sonoran Zones, Arizona, Texas, and Sonora. Mostly along margins of the desert or in foothills. July–Oct.

**3. Froelichia interrupta** (L.) Moq., in A. DC., Prod. **13**$^{2}$**:** 421, 1849

*Gomphrena interrupta* L., Sp. Pl. 224, 1753.
*Froelichia texana* Coult. & Fish., Bot. Gaz. **17:** 350, 1892.
(For additional synonymy see N. Amer. Fl. **21:** 129, 1917.)

Perennial from a stout, woody root with 1–several erect or ascending, sericeous-tomentose stems 2–10 dm. high, branching at base; leaves opposite, mostly petiolate, blades elliptic, oblanceolate, oval, or nearly orbicular, 0.5–3.5 cm. wide, 2–10 cm. long, closely scaberulous to canescent or later nearly glabrate above, tomentose-sericeous or floccose beneath; petioles short to nearly as long as blades; spikes in pairs or terminal, 1–4 cm. long, 8–12 mm. wide, scattered; bracts broadly ovate, short-acuminate, 1.5–2 mm. long, stramineous; bractlets orbicular to oblong-ovate, rounded or obtuse at apex, 3–4 mm. long, scarious, stramineous or fuscous; calyx lobes lance-oblong, obtuse, 1.5 mm. long, stramineous; calyx tube 2.5–3 mm. long, laterally winged, wings entire or shallowly crenate, faces of tube unappendaged, densely lanate with silky yellowish hairs 5 mm. long; seeds reddish brown, subglobose, 1.2–1.5 mm. in diameter.

Dry, rocky hillsides and grassy slopes, Tropical, Lower Sonoran, and Upper Sonoran Zones, Arizona and Texas to South America and the West Indies. Southern Baja California. March–Oct.

## 5. TIDESTROMIA Standley, Jour. Wash. Acad. 6: 70, 1916

Annual or perennial herbs or suffruticose at base, branched from above the base, pubescent with branched hairs. Leaves opposite. Flowers minute, perfect, in axillary glomerules, bracteate and bibracteolate, bracts and bracteoles hyaline; glomerules included in involucre-like cup formed by the indurate bases of subtending leaves. Perianth segments 5, distinct, 1-nerved, membranous, glabrous or villous. Stamens 5, hypogynous; filaments united at base; staminodia present or wanting; anthers 2-celled. Ovary globose; stigma capitate or 2-lobed; ovule 1, pendulous. Utricle subglobose, glabrous, indehiscent. Seeds globose, brown.

Annual, branches often procumbent; staminodia very short or wanting; pubescence roughish, not velvety-tomentose .............................................. 1. *T. lanuginosa*

Perennial, branches ascending; staminodia one-half to as long as the filaments, deltoid; pubescence dense, velvety-tomentose:

Leaves oblong, 2–15 mm. wide; indurated bases of leaves forming involucres about 1 mm. deep . . . . . . . . . . . . . . . . . . . . . . . . . . . . . . . . . . . . . . . . . . . . . . . . . 2. *T. oblongifolia*

Leaves orbicular, 2–7 mm. wide; indurated bases of leaves forming involucres 3–4 mm. deep . . . . . . . . . . . . . . . . . . . . . . . . . . . . . . . . . . . . 2a. *T. oblongifolia* subsp. *cryptantha*

**1. Tidestromia lanuginosa** (Nutt.) Standley, Jour. Wash. Acad. **6:** 70, 1916

*Achyranthes lanuginosa* Nutt., Trans. Amer. Phil. Soc. II, **5:** 166, 1820.
*Alternanthera lanuginosa* Moq., in A. DC., Prod. **13²:** 359, 1849.
*Cladothrix lanuginosa* Nutt. ex Moq., in A. DC., op. cit. 360, as synonym.

Procumbent, much branched annual with rather slender stems 1–5 dm. long, densely rough-branched-pubescent or glabrate in age or on shaded plants; leaves ovate to ovate-orbicular, 5–50 mm. long, nearly as broad, obtuse or rounded at both apex and base, or broadly cuneate at base of basal leaves, densely pubescent with intricately branched hairs or glabrate on lower surface in age; petioles stout, 2–25 mm. long; glomerules 2- to several-flowered; perianth 1.5–3 mm. long, segments oblong, acute or obtuse, 1-nerved, densely pubescent when young, at least at apex, often glabrate in age, scarious, yellow; bracts ovate, acute, scarious, about 0.5–0.75 mm. long, pubescent; stamens nearly equaling perianth segments; staminodia usually wanting, when present very short, broadly deltoid or a mere apiculation; seed globose-lenticular, dark reddish brown, about 0.6 mm. in diameter.

Dry plains, hillsides, and often on disturbed soil, Tropical, Sonoran, and Carolinian Zones, southern Utah to Kansas and south to Sinaloa and Zacatecas. Aug.–Oct.

**2. Tidestromia oblongifolia** (S. Wats.) Standley, Jour. Wash. Acad. **6:** 70, 1916, subsp. **oblongifolia**

*Cladothrix oblongifolia* S. Wats., Proc. Amer. Acad. **17:** 376, 1882.

Erect, spreadingly branched perennial 1–6 dm. high, often suffruticose at base and densely velvety-pubescent throughout with fine, intricately branched hairs; leaves ovate, ovate-orbicular, or oblong, 3–25 mm. long, 2–15 mm. wide or lower ones sometimes larger, rounded or obtuse at apex, obtuse to broadly cuneate at base, prominently veined beneath, equally pubescent on both surfaces; glomerules few-flowered, subtended by indurate bases of leaves; bracts ovate, acute, about 0.3–0.5 mm. long, lanate; perianth 1–1.5 mm. long, segments oblong, acutish or obtuse, densely pubescent toward apex or nearly glabrate in age, scarious; stamens three-fourths as long as perianth segments; staminodia deltoid-acute, one-half to nearly as long as slender filaments; utricle subglobose, membranous; seed about 0.5 mm. in diameter.

Desert soil, eastern Mojave and Colorado Deserts, and east into Nevada and Arizona. Aug.–Oct.

**2*a*. Tidestromia oblongifolia** subsp. **cryptantha** (S. Wats.) Wiggins, Contr. Dudley Herb. **4:** 16, 1950

*Cladothrix cryptantha* S. Wats., Proc. Amer. Acad. **26:** 125, 1891.

Differing from subsp. *oblongifolia* in having smaller, orbicular to broadly ovate leaves 2–7 mm. wide, 2–10 mm. long; more slender, wiry stems; bases of leaves subtending the glomerules connate into a deeper, indurate involucre often 3–4 mm. deep.

Desert slopes, Lower Sonoran Zone, western Colorado Desert about Salton Sea and adjacent slopes. Aug.–Oct.

## 6. GOMPHRENA L., Sp. Pl. 224, 1753

Annual or perennial herbs with erect, ascending, or decumbent, pubescent stems, nodes usually somewhat swollen. Leaves opposite, petiolate or upper sessile. Flowers perfect, bracteate and bibracteolate, in terminal and axillary heads or short spikes, the terminal ones subtended by 2 or more leaves or naked. Bracts and bractlets concave, keeled, often slightly cristate dorsally; perianth 5-parted, the segments scarious, 1-nerved, equal, lanate at base. Stamens 5 (sometimes 4), filaments united to form a tube that is included in perianth or slightly exserted, 5-lobed at apex, lobes bifid or emarginate, bearing a sessile or short-stipitate anther in sinus of each lobe; anther 1-celled. Ovary turbinate; style slender; stigmas 2, subulate or filiform; ovule 1, pendulous. Utricle ovoid-oblong; seed lenticular-obovoid, smooth; embryo annular; endosperm mealy; radicle superior.

Bractlets strongly cristate along keel, at least twice as long as bracts; style 2 mm. long, twice as long as ovary; annual; stamen-tube lobes very short .................. 1. *G. nitida*

Bractlets not cristate, one to one and one-half times as long as bracts; style 1 mm. long, about equaling ovary; perennial:

- Perianth lobes lanceolate, long-acuminate; stamineal tube and its lobes often exceeding perianth; terminal heads usually subtended by 2 leaves; anthers 1 mm. long; plants not cespitose; nodes swollen ........................................ 2. *G. sonorae*
- Perianth lobes oblong, obtuse; lobes of stamineal tube not equaling perianth; heads naked; anthers 1.5–2 mm. long; plants cespitose; nodes not conspicuously swollen:
  - Leaves cinereous, densely pilose; upper leaves broadly ovate to nearly orbicular, about as broad as long; perianth lobes obtuse or rounded, usually not emarginate; seed 1.5 mm. in diameter ........................................ 3. *G. caespitosa*
  - Leaves bright green, glabrous or very sparsely pilose when young; upper leaves elliptic, twice as long as wide; perianth lobes usually emarginate and serrate; seed 1 mm. in diameter ................................................ 4. *G. viridis*

### 1. Gomphrena nitida Rothr., Wheeler Rep. 233, 1878

*Xeraea nitida* Kuntze, Rev. Gen. Pl. **2:** 545, 1891.
*Gomphrena decumbens* var. *albiflora* Stuchlik, Rep. Spec. Nov. **11:** 158, 1912. (In part.) Not *G. decumbens* f. *albiflora* Chod. & Hassl., 1903.

Annual, erect or decumbent herb 1–8 dm. high with simple or branched stems and swollen nodes; leaves short-petiolate or upper sessile, broadly oblong, oblanceolate, or oval, 4–25 mm. wide, 1–6 cm. long, entire, obtuse to rounded at apex, attenuate-narrowed at base, sparsely to densely appressed-pilose on both surfaces; spikes subglobose to short-cylindric, 1–2 cm. long, 10–15 mm. in diameter, subtended by 2 sessile acute leaves, shorter or slightly longer than heads; bracts ovate-triangular, short-acuminate, white; bractlets two to four times as long as bracts, conspicuously laciniate-cristate on keel, ovate, attenuate, 6–7 mm. long, scarious, white, glabrous; perianth segments linear-lanceolate, long-attenuate, 4–5 mm. long, basally densely long-lanate; stamineal tube united to apex, lobes very short; anthers 1 mm. long or less; style slender, 2 mm. long, longer than ovary; utricle ovate, slightly compressed; seed reddish brown, obovate-lenticular, 1.5 mm. long.

Dry, rocky hillsides and grassy slopes, Lower and Upper Sonoran Zones, Chiricahua and Huachuca Mountains, Arizona, to southwestern New Mexico, south to Durango and Jalisco. July–Nov.

**2. Gomphrena sonorae** Torr., Bot. Mex. Bound. 181, 1859

*Gomphrena decipiens* S. Wats., Proc. Amer. Acad. **21:** 437, 1886.
*Xeraea sonorae* Kuntze, Rev. Gen. Pl. **2:** 545, 1891.
*Gomphrena sonorae* var. *watsonii* Stuchlik, Beih. Bot. Centr. **30:** 405, 1913.

Annual or short-lived perennial with erect, sparingly branched, sparsely villous to glabrate stems with conspicuously swollen nodes; leaves short-petiolate or upper sessile, narrowly elliptic, oblong, or oblanceolate, 3–21 mm. wide, 2–9 cm. long, acute to acuminate at apex, gradually attenuate at base, entire, green, sparsely villous when young, often glabrate in age; heads solitary or in few-headed glomerules at ends of branches and sessile in some axils, subglobose to short-cylindric, about 10–12 mm. in diameter; bracts ovate, short-acuminate, 2.5–4 mm. long, scarious, white or bright stramineous; bractlets ovate-acuminate, 5–6 mm. long, carinate but not cristate, white or tinged with red; perianth lobes lanceolate-acuminate, almost equaling bractlets, densely lanate below; stamineal tube united to apex, lobes 0.4–1 mm. long, linear-subulate; anthers 1 mm. long or less; style 1 mm. long or less, about equaling ovary; utricle ovoid, 1.5–2 mm. long, membranous; seed reddish brown, obovoid-lenticular, 1.5 mm. long.

Stony hills and plains, occasionally on grassy slopes, Lower and Upper Sonoran Zones, foothills of Santa Catalina Mountains, Arizona, to Chihuahua, Nayarit, and the Cape Region, Baja California. Aug.–March.

**3. Gomphrena caespitosa** Torr., Bot. Mex. Bound. 181, 1859

*Xeraea caespitosa* Kuntze, Rev. Gen. Pl. **2:** 545, 1891.

Crowded cespitose perennial from a heavy, woody root and branched caudex; leaves mostly basal; petioles 5–30 mm. long, slender; blades elliptic, oblanceolate, ovate, or suborbicular, 5–20 mm. wide, 1–5 cm. long, densely appressed-pilose on both surfaces, canescent; each flowering branch bearing 2 subsessile, broadly ovate to suborbicular leaves 0.5–3 cm. below the head; heads subglobose to short-cylindric, 1–2.5 cm. long, 1–1.8 cm. wide; bracts suborbicular to ovate-oblong, acute or obtuse, 3–4 mm. long, scarious, white, glabrous; bractlets 6–7 mm. long, oblong, acute or obtusish, white, scarious, glabrous; perianth segments linear-oblong, obtuse or rounded, usually entire, about 5 mm. long, densely fuscous-lanate below; stamineal tube shorter than perianth segments, lobes one-half as long as anthers; anthers 1.5–2 mm. long; style 1 mm. long; utricle flask-ovate; seed reddish brown, shining, 1.5 mm. long.

On grassy and rocky hillsides, mainly in Upper Sonoran Zone, New Mexico, southern Arizona, and northern Sonora. Entering the desert in foothills of Santa Rita Mountains, Arizona. March–June.

**4. Gomphrena viridis** Woot. & Standl., Contr. U.S. Nat. Herb. **16:** 120, 1913

Cespitose perennial from a deep-seated woody root with short, branching caudex and mostly basal, petiolate leaves; leaves oblanceolate, broadly spatulate, oblong, or oval, 3–8 cm. long, 1–3 cm. wide, obtuse at apex, gradually attenuate at base, bright green, glabrous or very sparsely villous when young, on petioles 1–3.5 cm. long; flowering peduncles with one pair of elliptic petiolate leaves 2–3 cm. long, borne below middle; heads terminal, naked, subglobose to short-cylindric, 12–16 mm. wide, 15–20 mm. long; bracts broadly ovate to ovate-orbicular, obtuse, 3.5–5 mm. long; bractlets oblong to ovate-oblong, 6–7 mm. long, obtuse, scarious, white, glabrous; perianth lobes 4.5–5 mm. long, linear-oblong, obtuse, often emarginate and serrulate at apex, densely lanate with yellowish hairs below; stamineal tube about 3.5–4 mm. long, distinctly lobed at apex, lobes one-half as long as the 1.5–2-mm.-long anthers; style 1 mm. long; seed reddish brown, subglobose-lenticular, about 1 mm. in diameter.

Mainly Upper Sonoran Zone, southwestern New Mexico, adjacent Arizona, and south into Chihuahua and Sonora.

## 7. ACHYRANTHES L., Sp. Pl. 204, 1753

Pubescent or glabrous herbs and shrubs of various habits. Leaves opposite. Flowers perfect, sessile or stipitate, bracteate and bibracteolate, capitate or spicate, usually compressed; perianth of 5 unequal, distinct sepals. Filaments united into a tube with 3–5 antheriferous lobes and 3–5 alternating staminodia; anthers 2-celled. Ovary ovoid, obovoid, or globose; stigma capitate; ovule 1, pendulous. Utricle membranous, indehiscent. Seed smooth; embryo annular; endosperm farinaceous; radicle superior.

Stems prostrate, villous; staminodia linear, entire; anthers subglobose, 0.5 mm. long or less, one-half as long as filament; style very short, less than 0.25 mm. long .... 1. *A. repens*

Stems erect or ascending, short-stellate-pubescent; staminodia ligulate, laciniate at apex; anthers oblong, 1–1.5 mm. long, exceeding utricle; style 1 mm. long ... 2. *A. watsonii*

### 1. Achyranthes repens L., Sp. Pl. 205, 1753

(For synonymy see N. Amer. Fl. **21**: 136, 1917.)

Prostrate or decumbent perennial from stout root with spreading-villous stems 0.3–5 dm. long; leaves often crowded; petioles 3–10 mm. long, villous; blades ovate to obovate, acute to cuneate basally, obtuse and mucronulate at apex, bright green, 3–15 mm. wide, 5–25 mm. long, sparsely villous when young, soon glabrate; heads ovoid to short-cylindric, 5–8 mm. in diameter, 5–15 mm. long, solitary or in sessile, axillary glomerules; bracts and bractlets about alike, about three-fourths as long as sepals, ovate, pungently mucronate, glabrous except for the ciliate-villous margins; sepals unequal, outer oval to broadly ovate, 3–5 mm. long, acute, pungently tipped, strongly 3-nerved, villous along the nerves, inner linear-subulate; staminodia linear or subulate, wrinkly, entire, about equaling or shorter than filaments; anthers less than 0.5 mm. long, subglobose; style very short; utricle globose-subtruncate; seed ovate-lenticular, 1.2–1.5 mm. long, reddish brown, shining.

South Carolina and Florida to southern Arizona and south through Mexico and the West Indies to Argentina and Peru. May–Oct.

### 2. Achyranthes watsonii Standley, Jour. Wash. Acad. 5: 73, 1915

*Telanthera stellata* S. Wats., Proc. Amer. Acad. **21**: 436, 1886. Not *Achyranthes stellata* Willd., 1797.
*Alternanthera stellata* Uline & Bray, Bot. Gaz. **20**: 451, 1895.
*Alternanthera stellata* var. *glabrata* Uline & Bray, loc. cit.
*Telanthera stellata* var. *glabrata* Rose ex Uline & Bray, loc. cit., as synonym.

Erect or ascending perennial, often flowering first year, with stellate-pubescent, profusely branched stems 3–10 dm. high; leaves broadly elliptic, ovate, or obovate, 1–3.5 cm. wide, 2–10 cm. long, acute to rounded at apex, usually mucronate, narrowing obtusely to the short, narrowly winged petiole, densely fine-stellate-pubescent with white hairs when young, glabrate in age; heads sessile, axillary, solitary or few-glomerate, subglobose, 5–8 mm. broad and long; bracts and bractlets ovate-acuminate, 1.5–2 mm. long, white, scarious, sparsely villous, carinate, spinose-tipped; sepals unequal, outer about 0.5 mm. long, inner 3.5–4 mm. long, lanceolate, acuminate to a rigid mucronate tip, strongly 3-nerved, nerves glabrous, intervals and margins villous, brownish on lower one-half or two-thirds; filaments linear; staminodia about 3 mm. long, ligulate, equaling or slightly exceeding anthers, lacerate at apex; utricle ovoid, obtuse; style 1 mm. long; seed 1–1.2 mm. in diameter, reddish, shining.

In sandy or rocky soil, Lower Sonoran and Tropical Zones, southern Sonora to southwestern Chihuahua and northern Sinaloa. March–Nov.

## 8. IRESINE P. Br., Nat. Hist. Jamaica 358, 1756

Herbs, shrubs, or small trees of various habit, glabrous or pubescent. Leaves opposite. Flowers perfect, polygamodioecious, or dioecious, in paniculate spikes or heads; perianth terete or 5 distinct sepals, usually pubescent. Stamens 5, united into short tube at base; filaments subulate; staminodia short or wanting; anthers oblong, 2-celled. Ovary somewhat compressed; style short; stigmas 2–3, filiform or in staminate flowers sometimes capitate; ovule solitary, pendulous. Utricle compressed, indehiscent. Seed lenticular, smooth; embryo annular; endosperm farinaceous; radicle superior.

Plants herbaceous; staminodia wanting:
- Sepals obtuse or rounded; lobes of stamineal tube obsolete; leaves thin, all alike .......... 1. *I. celosia*
- Sepals acute to acuminate; lobes of stamineal tube broadly rounded; leaves thick and succulent, lower ones broader and more obtuse than the upper ........ 2. *I. heterophylla*

Plants shrubby or suffrutescent; staminodia present:
- Flowers perfect; plant suffrutescent; staminodia minute, entire; leaves narrowly lanceolate .......... 3. *I. angustifolia*
- Flowers polygamodioecious or dioecious; shrubs 1 m. high or more; staminodia one-half as long as filaments or, if shorter, then denticulate; leaves broadly ovate, elliptic, or ovate-lanceolate:
  - Stigmas short and stout; staminodia nearly equaling filaments; bracts and bractlets one-half to one-third as long as sepals .......... 4. *I. hartmanii*
  - Stigmas slender; staminodia not over one-half as long as filaments; bracts of pistillate flowers about equaling sepals:
    - Shrub scandent or reclining; branches glabrous or sparingly short-villous when young; sepals of pistillate flowers ovate, obtuse; staminodia one-half as long as filaments .......... 5. *I. interrupta*
    - Shrub erect; branches villous-canescent when young; sepals of pistillate flowers lance-attenuate, acute; staminodia very short, dissected or denticulate ...... 6. *I. calea*

### 1. Iresine celosia L., Syst. Nat. ed. 10, 1291, 1759
(For synonymy see N. Amer. Fl. **21**: 166, 1917.)

Annual or short-lived perennial herb with much-branched, erect, ascending, or scrambling stems 4–30 dm. long, usually sparsely villous or, if glabrous, pubescent at nodes; leaves opposite; petioles slender, 0.8–7 cm. long; blades ovate or broadly deltoid-lanceolate, 2–15 cm. long, 1–7 cm. wide, rounded or abruptly acute at base, acute to obtuse or rounded at apex, bright green, thin, glabrous, papillate to soft-pubescent on both surfaces; panicles broadly pyramidal, 1–5 dm. long, 1–3 dm. wide, sparsely to moderately pubescent; spikelets 2–6 mm. long, numerous, alternate, opposite, or verticillate; flowers dioecious, 1–1.5 mm. long; bracts ovate to orbicular-ovate, obtuse or abruptly acute, somewhat pubescent, 0.5–1 mm. long; sepals of pistillate flowers conspicuously 3-nerved, with long, woolly hairs at the base, obtuse or rounded, those of staminate flowers 0.5–0.8 mm. long, glabrous or nearly so, obscurely nerved; stamens nearly or quite equaling sepals; lobes of stamineal cup obsolete; utricle membranous, 0.75 mm. long; seed reddish brown to black, shining, lenticular-orbicular, 0.5 mm. in diameter.

North Carolina to Florida and southern Louisiana near the coast; Sonora to Central and South America. Sept.–March.

**2. Iresine heterophylla** Standley, Contr. U.S. Nat. Herb. **18:** 95, 1916

*Iresine celosioides* var. *obtusifolia* J. M. Coult., Contr. U.S. Nat. Herb. **2:** 364, 1894.
*Iresine paniculata* var. *obtusifolia* J. M. Coult. ex Uline & Bray, Bot. Gaz. **21:** 354, 1896.

Herbaceous perennial from slender, spreading or cespitose rootstocks; stems somewhat striate, erect or ascending, 3–10 dm. long, sparsely villous throughout with larger hairs at nodes; leaves opposite, with smaller ones often fasciculate in axils; petioles often narrowly winged above, 2–25 mm. long; blades extremely variable in outline, broadly rhombic-ovate to broadly ovate-lanceolate, lower ones broader and more rounded than upper, usually acuminate to broadly cuneate and somewhat decurrent at base, acute to rounded at apex, glabrous or nearly so on upper surface, scabrous on margins and veins on lower surface, 0.8–4 cm. wide, 1.5–7 cm. long, somewhat succulent and thick, veins prominent beneath; inflorescence a narrow, dense panicle 1–4 dm. long, 3–10 cm. wide, branches sparsely villous; flowers dioecious, in small alternate or opposite spikes 3–20 mm. long; sepals elliptic-oblong, acute to acuminate, 1–1.3 mm. long, those of pistillate flowers 3-nerved and long-woolly at base; lobes of stamineal cup broadly rounded; utricle shorter than sepals; seed orbicular-lenticular, reddish brown, shining, about 0.6 mm. in diameter.

Along canyons and stream banks, Lower and Upper Sonoran Zones, southern Arizona to Texas and south to central Mexico. Sept.–Feb.

**3. Iresine angustifolia** Euphrasén, Beskr. St. Barthel. 165, 1795

(For synonymy see N. Amer. Fl. **21:** 157, 1917.)

Erect suffrutescent perennial with sparsely short-villous, striate branches 5–10 dm. high, older branches soon glabrate; leaves narrowly lance-ovate to linear-lanceolate, 3–10 cm. long, 0.5–3.5 cm. wide, acute to acuminate-cuneate at base, acuminate to attenuate at apex, green, glabrous above, usually sparsely villous along veins beneath; petioles slender, 5–25 mm. long; inflorescence a pyramidal panicle with flexuous branches bearing numerous spikelets on a lanate rachis; flowers perfect; bracts broadly ovate, abruptly acute, 0.5–0.7 mm. long, bractlets ovate, cuspidate-acuminate, 1.5 mm. long, dark brown, glabrous, rachis of spikes brown-villous; bracts persistent for some time after perianth and utricle have fallen; sepals narrowly ovate, 1.5 mm. long, rounded or obtuse at apex, distinctly 3-nerved, densely long-pilose with white hairs at base; filaments subulate-linear, 0.5 mm. long, one-fourth as long as sepals; staminodia narrowly deltoid, entire, one-third as long as filaments; utricle compressed; seed orbicular-lenticular, reddish brown to nearly black, 0.8–1 mm. in diameter.

In gravelly soil, mostly in canyons, Lower Sonoran Zone, Agua Verde Bay to the Cape Region, Baja California; east coast of Mexico, West Indies, south to Ecuador and Brazil. Jan.–April.

**4. Iresine hartmanii** Uline, Field Mus. Bot. Ser. **1:** 422, *pl. 24,* 1899

An erect or sometimes scandent branched shrub 1.5–3 m. high with slender, stramineous, finely but closely pubescent branchlets, bark on older branches stramineous to light gray, finely striate-fissured by long, narrow lenticels; leaves opposite, ovate-lanceolate, 10–34 mm. wide, 2.5–8 cm. long, rounded at base, acute-attenuate at apex, bright green and sparsely short-villous on upper surface, appressed-pilose-canescent and distinctly pinnately veined beneath; petioles short, 3–8 mm. long, canescent; flowers polygamodioecious, crowded in spikes 5–15 mm. long; infloresence a naked, broadly pyramidal, open panicle 15–30 cm. long, on peduncles 2–6 mm. long; rachises and peduncles canescent-puberulent; bracts and bractlets about equal, ovate to orbicular, rounded to obtuse, 0.75–1 mm. long, sparsely villous to glabrate, scarious, stramineous along the single broad nerve, margins white; sepals 2 mm. long, oblong, obtuse, stramineous or light brown, densely long-villous; filaments linear-subulate, 1–1.2 mm. long, anthers ovoid, one-third as long as filaments, whole stamen nearly equaling sepals; staminodia narrowly deltoid, about one-fourth as long as filaments; style almost wanting, stigmas short and stout; utricle ovoid-compressed; seed unknown.

Rocky canyons, Lower and Upper Sonoran Zones, Sonora to Durango and Sinaloa. Nov.–March.

**5. Iresine interrupta** Benth., Bot. Sulph. 156, 1844
(For synonymy see N. Amer. Fl. **21**: 160, 1917.)

Scandent shrub with profusely branched stems 1–6 m. long; branches terete, finely striate, glabrous or very sparsely short-villous in youth, strongly divergent; leaves petiolate, blades elliptic-lanceolate to ovate-lanceolate, 2–15 cm. long, 0.6–10 cm. wide, rounded and very short-decurrent at base, acute to attenuate at apex, glabrous or very sparsely villous when young, prominently veined beneath; petioles 0.5–5 cm. long; flowers polygamodioecious, in densely-flowered, subcylindrical spikes 4–6 mm. wide and 5–8 mm. long, these arranged in open, broad, pyramidal panicles 3–5 dm. long, rachis glabrate or sparsely puberulent; bracts and bractlets ovate to orbicular, about equal, in staminate flowers about one-third to one-half as long as sepals, nearly equaling sepals in pistillate flowers; sepals oblong, 1.5–2 mm. long in staminate flowers, obtuse, densely short-villous, whitish or stramineous; sepals of pistillate flowers ovate, obtuse, 1.5 mm. long, distinctly 3-nerved, yellowish-greenish, long-villous at base; stamens nearly equaling sepals, filaments linear; staminodia narrowly deltoid to broadly rounded, margins erosulate or denticulate; stigmas slender; utricle orbicular-lenticular, closely investing the shining, dark brown, pendulous seed, this 0.5–0.6 mm. in diameter.

Rocky canyons and hillsides, scrambling through thickets, Lower Sonoran and Tropical Zones, southern Sonora to Guatemala. Nov.–March.

**6. Iresine calea** (Ibáñez) Standley, Contr. U.S. Nat. Herb. **18**: 94, 1916
(For synonymy see N. Amer. Fl. **21**: 163, 1917.)

An erect, profusely branched, finely villous-canescent shrub about 1–3 m. high; leaves broadly ovate or lance-ovate, 1–7 cm. wide, 2–12 cm. long, rounded, obtuse, or rarely subcordate at base, acute-acuminate or rarely rounded at apex, thinly short pubescent and green on upper surface, densely canescent-pubescent beneath, the pubescence velvety; veins prominent beneath; petioles 3–12 mm. long, canescent; inflorescence a broad terminal panicle 2–5 dm. long, openly and divaricately branched, a few small leaves near base or nearly naked, rachises cinereous-lanate; flowers polygamodioecious, in densely flowered ovoid spikes 3–7 mm. long, these sessile or short-pedunculate; bracts and bractlets of staminate flowers ovate-orbicular, rounded or abruptly acute-apiculate, one-fourth to one-third as long as sepals, densely short-villous to nearly glabrous; sepals narrowly oblong, about 2 mm. long, obtuse, 1-nerved, densely pilose-canescent; stamens unequal, one short pair about one-half as long as sepals, one pair three-fourths as long as sepals, and a single one nearly equaling sepals; staminodia less than one-fourth as long as filaments, fimbriate-dissected; bracts and bractlets of pistillate flowers nearly as long as sepals, ovate to oblong; sepals lance-attenuate, 1.5 mm. long, densely pilose at very base or on receptacle, faintly 1-nerved; utricle ovoid, lenticular; seeds closely invested by utricle, orbicular-lenticular, about 0.5 mm. in diameter.

Shaded canyons and rocky slopes, Upper Sonoran to Tropical Zones, east central Sonora and southern Baja California to Coahuila and Costa Rica. Nov.–Feb.

### 9. DICRAURUS Hook. f., in Benth. & Hook., Gen. Pl. 3: 42, 1880

Pubescent, slender-branched shrubs. Leaves mostly alternate, entire. Flowers small, bracteate and bibracteolate, polygamodioecious or dioecious, in small spikes in axillary or terminal panicles; perianth terete or 5 distinct, 1-nerved segments. Stamens 5, united at base into shallow cup, filaments subulate, alternating with minute, papillate-dissected staminodia; anthers 2-celled, oblong. Style short; stigmas 2, subulate; ovule 1, pendulous by a long funiculus. Utricle ovoid, membranous, indehiscent. Seed smooth; embryo annular; endosperm farinaceous; radicle superior.

**1. Dicraurus alternifolius** (S. Wats.) Uline & Bray, Bot. Gaz. **21:** 355, 1896

*Iresine alternifolia* S. Wats., Proc. Amer. Acad. **24:** 72, 1889.
*Iresine pulchella* M. E. Jones, Contr. West. Bot. **18:** 34, 1933.

Shrub 1–2 m. high with terete, densely tomentose branches; bark gray-brown on older branches; leaves near bases of branches usually opposite, the upper alternate, ovate, rhombic-ovate or rounded-deltoid, 1–3.5 cm. wide, 9–25 cm. long, mostly rounded at apex, broadly rounded to truncate at base, bright green and glabrous or sparsely villous above, densely tomentose beneath; petioles 5–10 mm. long; panicles 2–12 cm. long or staminate ones longer, mostly terminal; spikes interrupted, pedunculate or sessile; bracts and bractlets suborbicular, about one-third as long as sepals, scarious, white, glabrous or sparsely villous; sepals elliptic-oblong, about 1.5 mm. long, scarious, short-villous; filaments linear-subulate, almost equaling sepals; staminodia very short, dissected into papillate lobes; sepals of pistillate flowers densely pilose with long, silky hairs; style short; utricle compressed-globose.

Rocky canyons and shaded slopes, Lower Sonoran Zone, western Sonora and southern Baja California. Sept.–March.

## Family 37. NYCTAGINACEAE. Four-o'Clock Family*

Herbs, shrubs, or trees, sometimes scandent, with simple, estipulate, opposite or alternate leaves. Flowers variously disposed, perfect or dioecious with calyxlike involucres, with separate bracts or bracteoles, or sometimes bractless. Perianth usually corolla-like, often brightly colored, funnelform, campanulate, or salverform, lower part persistent, often becoming hardened to form a close covering for the superior, 1-celled, 1-ovuled ovary. Stamens 1–many. Style short or elongate; stigma capitate, peltate, fimbriate, or linear. Fruit an anthocarp, costate, sulcate, or winged. Embryo straight or curved.

---

* Text contributed by Roxana S. Ferris.

Flowers borne on costa of a foliaceous bract, falling with bract when fruiting .......... 3. *Bougainvillea*
Flowers variously arranged, never on costa of a foliaceous bract:
  Perianth not markedly differentiated in fruit into an upper and lower part, the entire perianth of a like texture:
    Flowers unisexual; trees or large scandent shrubs; leaves mostly opposite ... 1. *Pisonia*
    Flowers perfect; herbaceous perennials, often woody at base; leaves alternate ........ 2. *Salpianthus*
  Perianth in fruit markedly differentiated into an upper and a lower part, the upper corolla-like, deciduous or withering-persistent, the lower thickened and adhering to the utricle:
    Flowers merely bracteate or subtended by an involucre of distinct bracts:
      Stigma hemispheric or capitate; inflorescence various, if capitate then short-pedunculate with flowers 3–5 mm. long:
        Perianth slender, long-funnelform, the tube 4–15 cm. long ........ 6. *Acleisanthes*
        Perianth campanulate to subrotate, about 1 cm. long:
          Fruit strongly compressed, the thin margins inflexed; bract half-enclosing the fruit .......... 5. *Allionia*
          Fruit not strongly compressed; bract minute, usually deciduous:
            Fruit scarious-winged, wings at least 2 mm. broad ........... 4. *Ammocodon*
            Fruit wingless (narrow, coriaceous wings present in *Boerhaavia* spp.):
              Plants erect or prostrate, herbs .......... 7. *Boerhaavia*
              Plants subscandent, suffrutescent at base .......... 8. *Commicarpus*
      Stigma linear; inflorescence of long-pedunculate heads; flowers 1.5–3.5 cm. long ... 11. *Abronia*
    Flowers subtended by a gamophyllous calyxlike involucre:
      Involucres membranous, accrescent, and rotate-spreading in age ...... 9. *Oxybaphus*
      Involucres leafy in texture, scarcely accrescent, not rotate-spreading in age ........ 10. *Mirabilis*

## 1. PISONIA L., Sp. Pl. 1026, 1753

Erect or subscandent trees or shrubs, usually more or less spinose with mostly opposite, petiolate, entire leaves. Flowers dioecious, without involucres, maroon to yellowish green in sessile or pedunculate, loose or congested, headlike cymes, each flower subtended by 2–3 bracts. Staminate perianth broadly or narrowly campanulate with a triangular or toothed, expanded limb. Stamens 5–8, exceeding perianth, unequal, filaments slender. Pistillate perianth tubular, 5-toothed, closing over the fruit. Ovary sessile, style little exserted, stigma penicillate. Fruit coriaceous, usually oblong, terete or 5-costate, the costae usually beset with stipitate glands.

Staminate perianth broader than long, lobes 1 mm. long or less; leaves narrowly cuneate at base .......... 1. *P. flavescens*
Staminate perianth longer than broad, lobes about 0.5 mm. long; leaves broadly cuneate to rounded at base .......... 2. *P. capitata*

### 1. **Pisonia flavescens** Standley, Contr. U.S. Nat. Herb. **13:** 389, 1911

Small trees 6–8 m. tall with dark gray branches and divaricate branchlets, spineless or with straight, stout spines 5–6 mm. long; leaves oblanceolate, 3.5–6 cm. long, narrowly acute to acuminate at base, yellowish green, mostly glabrous, slightly hairy beneath, mostly along veins; petioles slender, one-third to one-fourth the length of blades; staminate inflorescence of congested cymules appearing capitate, little expanded at anthesis, about 1 cm. in diameter, peduncle 8–10 mm. long, pedicels very short, closely puberulent; pistillate inflorescence not seen; staminate perianth maroon, basin-shaped, puberulent without, about 3 mm. long and 3.25–3.5 mm. broad, lobes 1 mm. long; pistillate perianth not known; fruit about 10 mm. long, beset along the angles with low, stout, stipitate glands.

Shady canyons, Lower Sonoran Zone, Sierra de la Giganta and San José del Cabo, Baja California. May.

**2. Pisonia capitata** (S. Wats.) Standley, Contr. U.S. Nat. Herb. **13:** 388, 1911

*Cryptocarpus ?capitatus* S. Wats., Proc. Amer. Acad. **24:** 71, 1889.

Dense shrub or small tree with arching, grayish stems and branchlets spreading at right angles, often subscandent, up to 6 m. high, young growth velvety, glabrate in age, stout, recurved spines 7–14 mm. long usually present; leaves broadly obovate, round at apex or sometimes abruptly acute, rounded to broadly cuneate at base, sparsely short-villous above, densely so below, becoming glabrate; staminate inflorescence pedunculate, dense, appearing capitate, cymules somewhat expanded at anthesis; pistillate inflorescence closely puberulent with peduncles 1–1.5 cm. long, congested in anthesis, greatly expanded in fruit, pedicels of mature fruit to 1 cm. long; staminate perianth maroon, puberulent, broadly campanulate, 2.5–3 mm. long, about 2.5–2.75 mm. broad, lobes 0.5 mm. long, subtended by bracts; stamens much exserted, filaments dark red, anthers yellow; pistillate perianth trumpet-shaped, densely puberulent, 2–2.5 mm. long, subtended by bractlets; fruit clavate, 7–10 mm. long, angles with rows of heavy stipitate glands, interspaces puberulent.

Sandy soil in thickets and along watercourses, Lower Sonoran Zone, San Carlos Bay north of Guaymas, Sonora, south to Acaponeta, Tepic. Feb.–April.

## 2. SALPIANTHUS H. & B., Pl. Aequin. 1: 154, 1808

Herbs, often suffrutescent at base, or sometimes shrubs, with alternate, entire, petiolate leaves. Flowers perfect, glomerules in dense racemes or cymes arranged in leafy panicles. Perianth persistent, campanulate or tubular, 4-toothed and usually angulate, often reddish or whitish at tips. Stamens 3 or 4, opposite perianth segments; filaments slender, unequal, exserted. Ovary tapering to a filiform style; stigma acute. Fruit subglobose, dark, usually shining.

**1. Salpianthus macrodontus** Standley, Contr. U.S. Nat. Herb. **13:** 393, 1911

*Boldoa lanceolata* var. *macrodonta* Heimerl, Rep. Spec. Nov. **12:** 222, 1913.

Herbaceous, much-branched plants from suffrutescent base, 1–2 m. high, puberulent throughout and usually somewhat villous on branches of inflorescence; leaves 4–7 cm. long, ovate-lanceolate, attenuate at base; petioles slender, about 2–4 cm. long; flowers in compact cymes at ends of branches; perianth 6–7 mm. long, densely puberulent with some villous hairs, angulate, somewhat expanded above, purplish green, the 4 triangular teeth whitish or purplish; stamens exserted 4–6 mm., unequal; fruit compressed, spherical.

Sandy soil in fields and along stream banks or cliffs, Lower Sonoran and Tropical Zones, Sonora, Sinaloa, and Baja California. Nov.–March.

## 3. BOUGAINVILLEA Commers. ex Juss., Gen. 91, 1789

Scandent shrubs or vines, much branched, often armed with heavy, straight or curved spines; branches villous or glabrate. Leaves alternate. Inflorescence in axillary panicles of brightly colored, leafy, persistent bracts, these in groups of 3, flowers borne singly on inner faces of bracts and pedicels confluent with midveins. Perianth tubular, mostly 5-lobed. Stamens 5–10, included. Style short. Anthocarp fusiform, 5-nerved.

**1. Bougainvillea spectabilis** Willd., Sp. Pl. **2:** 348, 1799

*Bougainvillea bracteata* Pers., Syn. Pl. **1:** 418, 1805.
*Tricycla spectabilis* Poir., in Lam., Encycl. Suppl. **5:** 358, 1817.
*Josepha angusta* Vell., Fl. Flum. 154, 1825.
*Bougainvillea virescens* Choisy, in A. DC., Prod. **13**[2]: 437, 1849.

Woody, much-branched vines usually conspicuously villous on younger parts; leaves broadly ovate, rounded to broadly acute at base, abruptly acute at apex, 5–10 cm. long, more densely villous below than above, petioles slender; bracts purplish red, broadly ovate, veiny, sparsely puberulent to short-villous, smaller than leaves; perianth greenish or yellowish, densely short-villous without, tubular, short lobes usually lighter than tube; fruit grayish green, 11–14 mm. long, oblong-ellipsoid.

Commonly cultivated in towns and villages, and occasionally escaping. From the Sonoran Desert area through tropical and subtropical parts of Mexico and Central America to South America. April–July.

## 4. AMMOCODON Standley, Jour. Wash. Acad. 6: 631, 1916

Pubescent perennial herbs with woody roots and erect or decumbent stems and opposite, petioled leaves. Flowers bracteate, sometimes cleistogamous, arranged in small, clustered umbels on the cymose-branching inflorescence. Perianth induplicate-plicate, campanulate, constricted above ovary; stamens 2 or 3, exserted; style filiform; stigma peltate. Anthocarp with 4–5 hyaline, nerveless wings.

**1. Ammocodon chenopodioides** (A. Gray) Standley, Jour. Wash. Acad. **6:** 631, 1916

*Selinocarpus chenopodioides* A. Gray, Amer. Jour. Sci. II, **15:** 262, 1853.

Plants 1.5–4 dm. high with several widely branching stems from woody root, puberulence under lens appearing scurfy, at least lower part of stems becoming glabrate in age; leaves ovate to ovate-deltoid, petiole 0.5–4 cm. long, blade 1.5–5 cm. long, mostly reduced above, entire or wavy-margined, veins prominent beneath; free part of perianth about 5 mm. long and about as broad, purplish red; anthocarp usually with 5 thin, hyaline wings 5–6 mm. long, about one-half as broad, body of anthocarp sulcate.

In sand, Upper and Lower Sonoran Zones, western Texas to southern Arizona and northern Chihuahua. June–Aug.

## 5. ALLIONIA L., Syst. Nat. ed. 10, 890, 1759

Prostrate annuals or perennials. Flowers perfect, in axillary pedunculate clusters of 3, each subtended by a bract that encloses the fruit. Perianth campanulate-rotate, 4–5-lobed. Stamens 5–7 with unequal filaments, exserted. Anthocarp coriaceous, compressed, the thin margins entire or toothed, usually reflexed; outer surface with 1–2 longitudinal rows of stipitate glands. Embryo curved, the broad cotyledons enclosing the endosperm.

**1. Allionia incarnata** L., Syst. Nat. ed. 10, 890, 1759

*Allionia malacoides* Benth., Bot. Sulph. 44, 1844.
*Allionia incarnata* var. *nudata* Munz, Man. S. Calif. 152, 598, 1935.
*Allionia incarnata* var. *villosa* Munz, loc. cit.
(For additional synonymy see N. Amer. Fl. **21:** 200, 1918, and Rep. Spec. Nov. **31:** 91–98, 1932.)

Winter annuals or perennials with numerous, much-branched, often trailing stems 2–10 dm. long, densely viscid-villous or glandular-puberulent throughout; leaves 2–6 cm. long, petioles 0.2–3.5 cm. long, blades broadly deltoid-orbicular to oval, usually rounded at base, unequal, green above, paler beneath; involucres on slender peduncles 3–5 cm. long, lobes free, ovate-orbicular, 5–9 mm. long; perianth 6–15 mm. long, purplish red, rarely white; fruit 3–4.5 mm. long, inner side 3-nerved, margins with 3–5 broad teeth or sometimes entire, strongly incurved.

Sandy or rocky soil, Upper Sonoran Zone to Tropical Zone, southern Colorado, Utah, and Nevada and southeastern California south through Mexico to Argentina and Chile. April–Sept.

## 6. ACLEISANTHES A. Gray, Amer. Jour. Sci. 15: 259, 1853

Plants perennial, sometimes shrubby, with opposite, petiolate, entire leaves. Flowers white or purple-tinged, subtended by 1–3 small bracts, axillary or terminal, solitary or occasionally in 2–3-flowered cymes. Perianth funnelform, with long tube and shallowly 5-lobed limb, constricted above ovary. Stamens 2–5, unequal, exceeding perianth tube. Ovary ovoid or oblong; style exserted; stigma capitate. Anthocarp narrowly ellipsoid, 5-angled or -ribbed, ribs sometimes ending in conspicuous glands. Embryo curved; cotyledons enclosing endosperm.

**1. Acleisanthes longiflora** A. Gray, Amer. Jour. Sci. II, **15:** 261, 1853

Plants decumbent or ascending from woody root, stems slender, puberulent above, glabrate below; leaves thick, entire, sometimes with crisped margins; petioles 3–8 mm. long; blades deltoid to rhombic-ovate or lanceolate, 1.5–4.5 cm. long; fragrant flowers mostly solitary, sessile; bracts subulate, 2–3 mm. long; perianth white tinged with purple, sparsely puberulent, 10–16 cm. long, limb 1.5–2 cm. broad; anthocarp narrowly oblong, 5–6 mm. long, truncate at both ends, 5-angled, longitudinally striate between ribs.

Desert slopes, Lower Sonoran Zone, from eastern Riverside County, California, to southern Arizona and western Texas and south to Mexico. May–Aug.

## 7. BOERHAAVIA L., Sp. Pl. 3, 1753

Annual or perennial herbs, procumbent to erect, usually branching from base, the stems often with viscous areas on internodes. Leaves entire or sinuate, opposite, petiolate, often brown-punctate. Inflorescence ample, cymosely much branched, the perfect flowers in racemes, umbels, or heads, or solitary at ends of branches. Perianth differentiated into a thickened lower part which becomes the coat of the anthocarp, and an upper part thin, corolla-like, campanulate to rotate. Stamens 1–5, unequal, usually exserted, filaments slender, connate at base. Style filiform, stigma peltate. Anthocarp obovoid or obpyramidal, costate, angled, or winged.

Fruit pubescent; plants perennial:
  Flowers sessile or nearly so in headlike clusters; hairs on fruit dark and blunt, usually gland-tipped .................................................... 1. *B. coccinea*

Flowers solitary on slender pedicels; glandular hairs on fruit whitish, tapering, not gland-tipped ........ 2. *B. gracillima*
Fruit glabrous; plants annual:
Subtending bracts as long as or longer than mature fruit, persistent:
Angles of fruit narrow, sulci broad and smooth; flowers sessile in headlike clusters on ends of branches ........ 3. *B. purpurascens*
Angles of fruit broad, sulci narrow and rugose; flowers in spikes ........ 4. *B. wrightii*
Subtending bracts not more than one-half as long as mature fruit, usually deciduous:
Body of fruit winged for at least part of its length:
Fruit stipelike at base, 4-winged, wings wider above than below; sulci broader than widest part of winged angles ........ 5. *B. pterocarpa*
Fruit without a stipelike base, 5-winged; sulci narrower than winged angles:
Fruit as broad as long ........ 6. *B. alata*
Fruit longer than broad ........ 7. *B. megaptera*
Body of fruit not winged, angles broad or narrow, acute or rounded:
Ultimate branches of the inflorescence umbelliform or the flowers 1–2 on slender pedicels; fruit obpyramidal, truncate:
Flowers solitary, less commonly 2- or 3-clustered on slender pedicels:
Fruit narrowly obpyramidal, 3–4 mm. long, the angles as well as the sulci rugose ........ 8. *B. maculata*
Fruit broadly obpyramidal, 2–2.5 mm. long, the angles smooth; sulci rugose ........ 9. *B. triquetra*
Flowers umbellate or in close cymules:
Fruit 3–4.5 mm. long, usually substipitate at base, dark greenish or brown; branches of inflorescence erect or ascending ........ 10. *B. erecta*
Fruit 2–2.7 mm. long, scarcely substipitate at base, light green to stramineous; inflorescence diffusely branched ........ 11. *B. intermedia*
Ultimate branches of cymose-paniculate inflorescence closely or loosely spiciform; fruit obovoid, rounded at apex (except in *B. rosei*):
Fruit narrowly obpyramidal, truncate ........ 12. *B. rosei*
Fruit clavate or obovoid, rounded at apex:
Angles of fruit acute; sulci broad, rugose:
Flowers 3–5 mm. long; fruit broadly obovoid ........ 13. *B. xanti*
Flowers 1–2 mm. long; fruit narrowly obovoid ........ 14. *B. spicata*
Angles of fruit obtuse; sulci nearly closed; flowers 1–1.5 mm. long ........ 15. *B. coulteri*

## 1. **Boerhaavia coccinea** Mill., Gard. Dict. ed. 8, no. 4, 1768

*Boerhaavia caribaea* Jacq., Obs. Bot. **4:** 5, 1771.
(For additional synonymy see N. Amer. Fl. **21:** 206, 1918.)

Decumbent or prostrate perennial, branching from base with many stout stems 3–14 dm. long, viscid-pubescent and sometimes glandular-hirsute below, more or less glandular above, occasionally glabrate; leaves 2–6 cm. long, ovate-orbicular to oblong, rounded to acute at apex, green above, pale below, with a brown-punctate margin, glabrous to hirsute, often viscid; inflorescence cymose, much branched, branches slender, glandular-pubescent; flowers in heads on slender peduncles; bracts minute, lanceolate; perianth purplish red, 2 mm. long; stamens 1–3, barely exserted; fruit obovoid, 2.5–3.5 mm. long, densely glandular-puberulent with dark, blunt, usually gland-tipped hairs, 5-angled, sulci smooth, broader than angles.

Sandy soil, Lower Sonoran and Tropical Zones, Tulare County, California, east to the southern states and south to the American tropics. Common throughout the Sonoran Desert, often appearing as a weed. June–Sept.

**2. Boerhaavia gracillima** Heimerl, Bot. Jahrb. **11:** 86, 1889

*Boerhaavia anisophylla* var. *paniculata* Coult. & Fish., Bot. Gaz. **17:** 348, 1892.
*Boerhaavia organensis* Standley, Contr. U.S. Nat. Herb. **12:** 385, 1909.
*Boerhaavia gracillima* subsp. *decalvata* Heimerl ex Standley, op. cit. 386.

Perennial with several to many, usually decumbent stems 4–10 dm. long from woody root, much branched, finely puberulent to glabrous, not viscid, leafy below the ample inflorescence; leaves rhombic-elliptic to ovate, paler beneath, leaf margin and petioles ciliate, mostly rounded at base, acute to obtuse at apex, 5–6.5 mm. long; petioles slender, shorter than blade, more reduced in length on upper leaves; inflorescence many-flowered, widely branching, with small bracts at branches of numerous axillary cymes and cymules, branches of inflorescence slender; flowers on filiform pedicels 5–10 mm. long, bract at base of flower minute, soon deciduous; perianth purple-red, about 5 mm. broad; stamens 2–3, exserted; anthocarp 4–5 mm. long, narrowly obovoid, more or less pubescent with short, white, tapering, nonglandular hairs, 5-angulate, angles narrow, rounded, sulci smooth.

Rocky hillsides, Sonoran Zones, Texas and central Mexico west to southern Arizona, Sonora, and Baja California. Sept.–May.

**3. Boerhaavia purpurascens** A. Gray, Amer. Jour. Sci. II, **15:** 321, 1853

Erect, slender annual 1.5–4.5 dm. high, 1–3 stems from base, more or less glandular-hirtellous, branches spreading above into paniculate-cymose inflorescence; leaves brown-punctate and paler beneath, glabrous or nearly so, oblong, acute or rounded at apex, sinuate-margined, 2.5–3 cm. long, upper leaves lanceolate, reduced, and sessile; petioles shorter than blade, 6–11 mm. long; flowers in headlike glomerules at ends of branches of cymose-paniculate inflorescence; perianth rotate-spreading, pink to rose-pink, about 2 mm. long; stamens 3, much exserted; subtending bracts ovate, apiculate, brown-punctate, scarious with short, spreading, villous hairs, as long as or longer than fruit; anthocarp obovoid, 2–2.5 mm. long, 5-angled, angles narrow, acute, and smooth, sulci broad, rugulose in age, and usually darker than angles.

Hill slopes, often in limestone, Upper and Lower Sonoran Zones, western Texas to southern Arizona and in northern Sonora. Aug.–Sept.

**4. Boerhaavia wrightii** A. Gray, Amer. Jour. Sci. II, **15:** 322, 1853

*Boerhaavia bracteosa* S. Wats., Proc. Amer. Acad. **20:** 370, 1885.

Erect annual, 1–3 dm. high, branching from base; stems slender, pubescent below, often with viscid rings; leaves oblong to lanceolate, acute at apex, 2–5 cm. long, blades 1.5–4 cm. long, green above, pale and glandular-punctate beneath; inflorescence glabrous or puberulent, flowers in loose spikes; bracts ovate, two-thirds as long as fruit, persistent, ciliolate; perianth pink, 1–2 mm. long; fruit broadly obovoid, 2 mm. long, at least 1 mm. wide, mostly 4-angled, angles broad but acute, sulci broad, rugose, rugae sometimes extending somewhat into the broad angles.

Gravelly plains and hillsides, Lower Sonoran Zone, Nevada and southeastern California and adjacent Arizona to western Texas and Sonora. Sept.–Oct.

**Boerhaavia torreyana** (S. Wats.) Standley, Contr. U.S. Nat. Herb. **12:** 385, 1909, ranges from northern and central Arizona to western Texas and Coahuila. It has been reported from the Sonoran Desert, but the writer has seen no material of this species collected inside our area.

**5. Boerhaavia pterocarpa** S. Wats., Proc. Amer. Acad. **17**: 376, 1882

Slender, erect or decumbent annual, 1–3 dm. high, branching from base and also above in more robust plants, stems leafy, scabrous-puberulent to glabrate; leaves not or but little reduced above, rhombic-ovate to lanceolate, rounded to obtuse at base, 1–2.5 cm. long; petioles to 1 cm. long; infloresence of few, terminal or axillary, short-peduncled, headlike cymes, flowers subsessile or with pedicels about 1 mm. long; perianth pink, 1–1.5 mm. long; stamens included or nearly so, subtending bracts brown-punctate, lanceolate-attenuate; anthocarp 4-angled, 3.5–4 mm. long, 2.5 mm. wide, truncate or depressed at apex, acute angles smooth, conspicuously wing-margined above, narrowed below to stipelike base, sulci broad, transverse rugae coarse.

Rocky slopes, Lower Sonoran Zone, Chiricahua Mountains, Arizona, west to Pima County, and in Sonora as far south as Altar. Aug.

**6. Boerhaavia alata** S. Wats., Proc. Amer. Acad. **24**: 69, 1889

Slender-stemmed, much-branched annual, mostly glabrous, the internodes with viscid areas; leaves somewhat glandular-punctate, ovate- to lanceolate-oblong, 2–3 cm. long, reduced in size above; inflorescence cymose-paniculate, the flowers solitary on slender pedicels 2–15 mm. long, bracts minute; flowers 3.4–4 mm. long; anthocarp 4 mm. long, sharply and broadly winged, wings thin, but little narrowed below, 1–1.4 mm. wide, sulci narrow, transversely rugose.

Lower Sonoran Zone, known only from a rocky island in Guaymas Harbor, Sonora. Oct.

**7. Boerhaavia megaptera** Standley, Contr. U.S. Nat. Herb. **12**: 379, 1909

Erect annual 3–6 dm. high, branching from base and also above, branches erect to somewhat spreading, scabrous-puberulent to glabrate; leaves ovate-oblong to narrowly elliptic, upper leaves narrowly lanceolate to linear, sessile, obtuse at apex, rounded at base; petioles short, often brown-punctate, paler beneath, 2–2.5 cm. long; inflorescence ample, cymose-paniculate, flowers on ultimate branches in 3–7-flowered umbels, pedicels 1.5–2 mm. long; perianth pink, about 1 mm. long, stamens included, subtending bracts minute, often brown-punctate, lanceolate- to ovate-acuminate; anthocarp broadly obpyramidal, 3.5 mm. long, about 2.5 mm. wide, 5-winged, the wings thin, smooth, narrowed below, somewhat extended above body of fruit above, sulci narrow, smooth.

Lower Sonoran Zone, known only from Pima County, Arizona. Aug.–Sept.

**8. Boerhaavia maculata** Standley, Contr. U.S. Nat. Herb. **12**: 379, 1909

Erect, much-branched annual, the stems glabrous or somewhat puberulent below; leaves lanceolate to broadly linear, 1.5–3.5 cm. long, brown-punctate, short-petiolate or sessile; inflorescence cymose-paniculate, much branched; flowers solitary or in clusters of 2 or 3 on slender pedicels 4–12 mm. long, the subtending bracts minute, ovate-lanceolate; perianth pink, about 1.5 mm. long; stamens included; anthocarp straw-colored, narrowly obpyramidal, 3–4 mm. long, 5-angled, angles sharply acute, somewhat rugulose next to sulci; sulci open, strongly rugose.

Open slopes, Lower Sonoran Zone, vicinity of Guaymas, Sonora. Oct.

**9. Boerhaavia triquetra** S. Wats., Proc. Amer. Acad. **24:** 69, 1889

Slender-stemmed annual, branching from base with ascending branches; leaves oblong to narrowly lanceolate, 1–3 cm. long, paler beneath, sometimes glandular-punctate; inflorescence cymose-paniculate, branches slender, erect-spreading; flowers in open cymules, solitary or sometimes 2 or 3 on pedicels 2–6 mm. long, subtending bracts ovate-lanceolate to lanceolate, scarious, white or pinkish, with or without brown spots; anthocarp broadly obpyramidal, straw-colored to pale green, 2–2.5 mm. long, 1.3–1.8 mm. broad, 3–5-angled, ridges acute, broad, and smooth, sulci open, coarsely and deeply rugose.

Sandy or rocky soil, Lower Sonoran Zone. In the Sonoran Desert it occurs in Riverside County, California, and Yuma County, Arizona, south along the Gulf in Baja California to Bahía Concepción and to Guaymas, Sonora. Oct.–Nov.

**10. Boerhaavia erecta** L., Sp. Pl. 3, 1753

*Boerhaavia elongata* Salisb., Prod. Stirp. 56, 1796.
*Boerhaavia virgata* H. B. K., Nov. Gen. & Sp. **2:** 215, 1817.
*Boerhaavia discolor* H. B. K., loc. cit.
*Boerhaavia atomaria* Raf., Aut. Bot. 40, 1840.
*Valeriana latifolia* Mart. & Gal., Bull. Acad. Brux. **11**$^1$**:** 124, 1844.
*Boerhaavia thornberi* M. E. Jones, Contr. West. Bot. **12:** 72, 1908.
*Boerhaavia erecta* subsp. *thornberi* Standley, Contr. U.S. Nat. Herb. **12:** 381, 1909.

Usually coarse, glaucous annuals 3–10 dm. high, glabrous or hirsutulous below, branching from base and above, branches ascending, basal branches sometimes procumbent; leaves ovate to oblong-ovate, usually acute, rarely rounded at apex, entire or subsinuate, glabrous, lower leaves 2.5–8 (or –10) cm. long, upper ones short, narrowly ovate-lanceolate; inflorescence cymosely paniculate, ample, much branched, branches and branchlets erect or strictly ascending; flowers umbellate, subumbellate, or subracemose in clusters of 3–6 flowers, occasionally in twos or solitary on older inflorescences, fruiting pedicels 1–5 mm. long; perianth greenish or white, sometimes tinged with pink, 1 mm. or more long; stamens 2–3, short-exserted; subtending bracts minute, deciduous; anthocarp greenish or brownish green, 5-angled, 3–4.5 mm. long, narrowly obpyramidal, strongly narrowed below and sometimes appearing substipitate, truncate or sometimes broadly angled above, ridges smooth, obtuse to acutish, sulci narrow, coarsely rugose.

Open places and disturbed areas, Lower Sonoran and Tropical Zones, southeastern United States to southern Arizona, south through Mexico and the West Indies to South America. July–April.

**11. Boerhaavia intermedia** M. E. Jones, Contr. West. Bot. **10:** 41, 1902

*Boerhaavia universitatis* Standley, Contr. U.S. Nat. Herb. **12:** 380, 1909.
*Boerhaavia lateriflora* Standley, op. cit., **13:** 426, 1911.

Erect annual, 2–5 dm. high, branching from base, branches ascending, slender, minutely puberulent throughout; leaves broadly oblong to ovate-lanceolate, obtuse or acute at apex, 4–5 cm. long, 1–5 cm. broad, often brown-punctate, green above, pale beneath; inflorescence cymose-paniculate, diffusely branched, flowers in umbel-like clusters at tips of branches; bracts minute; perianth pink, 1.5–2 mm. long; stamens 2–3, mostly equaling perianth; anthocarp straw-colored or pale green, narrowly obpyramidal, 2–2.7 mm. long, about 1 mm. broad at apex, sharply 5-angled, sulci narrow, rugose.

Sandy desert, Lower Sonoran Zone, Colorado Desert, California, and Baja California, east through Texas and central Mexico. July–Sept.

**12. Boerhaavia rosei** Standley, Contr. U.S. Nat. Herb. **13:** 424, 1911

Slender-stemmed, diffuse, annuals to 5 dm. high, with many-spreading, usually glandular-puberulent branches; leaves 1.5–4 cm. long, but little reduced on upper stem, paler above than below, ovate to deltoid, acute at apex, rounded to truncate at base; petioles slender, one-half to one-third the length of blade; inflorescence cymosely much branched, flowers loosely arranged in slender spikes on ultimate branchlets; perianth scarcely 1 mm. long, stamens included; anthocarp narrowly obpyramidal, truncate above, almost stipelike below, 2.25–2.5 mm. long, pale green or brownish, ridges smooth, acute, sulci open, obscurely rugulose.

Sandy soil, Lower Sonoran Zone, along the Magdalena River and at Ures, Sonora, then extending south into Sinaloa and Nayarit. Sept.–March.

The specimens from the Sonoran Desert are doubtfully assigned to *B. rosei.* They have the obpyramidal truncate fruit and the small flowers and diffuse habit that are attributed to that species in the original description. The specimens collected along the Magdalena River have the flowers in dense, rather than in typical loosely flowered spikes.

**13. Boerhaavia xanti** S. Wats., Proc. Amer. Acad. **24:** 69, 1889

*Boerhaavia fallax* Standley, Contr. U.S. Nat. Herb. **13:** 424, 1911.

Usually decumbent annual branching from base, 2–4 dm. high, branches few to many, ascending, lower stems puberulent and villous, glabrous above; leaves ovate, oval, oblong, and often lanceolate, 2–5 cm. long, margins often strongly sinuate; petioles of lower leaves often as long as the blades; inflorescence cymose-paniculate, flowers on ultimate branchlets in rather dense, short spikes; perianth rotate-spreading, 3–5 mm. long, white or pink; anthers 4–5, long-exserted; bracts subtending flowers small, broadly or narrowly ovate-acuminate, ciliate-margined, scarious and sometimes spotted with brown; anthocarp broadly obovoid, 2.5–3 mm. long, 1.5–2 mm. broad, 5-angled, ridges broad, smooth, often sharply acute, sulci broad, transverse-rugose.

In sand and on stony ridges, Lower Sonoran Zone, Bahía Los Angeles southward to San José del Cabo, Baja California, and the coast of Sonora. Sept.–Jan.

**14. Boerhaavia spicata** Choisy, in A. DC., Prod. **13**$^{2}$**:** 456, 1849

*Boerhaavia palmeri* S. Wats., Proc. Amer. Acad. **24:** 69, 1899. Not *B. palmeri* S. Wats., 1883.

Erect or procumbent annual to 6.5 dm. high, stems densely coarse, glandular-villous except those of upper inflorescence, often reddish; leaves 1.5–5 cm. long, ovate or lanceolate, acute to narrowly acute at apex, the upper reduced and narrowly lanceolate, subsessile, margins entire or subsinuate, puberulent or short-villous, often glandular; petioles usually one-half to one-third as long as blade; inflorescence cymose-paniculate, few branched, the flowers in short, dense, slender spikes at the ends of branchlets; bracts subtending flowers ovate to ovate-attenuate, usually brown-spotted; perianth pink, usually with brown lines, 1–1.5 (or –2) mm. long; stamens 1–3, included; anthocarp narrowly obovoid, 2.5 mm. long, 5-angled, ridges smooth, broad, acute on the margin, sulci open, broad, more or less rugose.

Sandy soil, Lower Sonoran and Tropical Zones, Pinal and Puma Counties, Arizona, south to Sinaloa and Chihuahua. Sept.–Oct.

**15. Boerhaavia coulteri** (Hook. f.) S. Wats., Proc. Amer. Acad. **24:** 70, 1889

*Senkenbergia coulteri* Hook. f., in Benth. & Hook., Gen. Pl. **3:** 6, hyponym, 1880.
*Boerhaavia spicata* var. *palmeri* S. Wats., Proc. Amer. Acad. **24:** 70, 1889.
*Boerhaavia watsoni* Standley, Contr. U.S. Nat. Herb. **12:** 384, 1909.

Slender annual, usually with many branches spreading from base, more or less pubescent and sometimes a little ciliate below, not conspicuously glandular, sometimes with viscid bands; leaves ovate to ovate-deltoid, 1.5–5 cm. long, obtuse, often rounded at apex, mostly truncate at base, margin entire or sinuate, upper leaves smaller, lanceolate to ovate-lanceolate; inflorescence cymose-paniculate, spreadingly much branched, the flowers in loose, slender, interrupted spikes; bracts minute, lanceolate-ovate, usually brown-punctate; perianth 1–1.5 mm. long, white or pale pink; stamens 1–3, included or short-exserted; anthocarp narrowly obovoid, 2.5–3 mm. long, 5-angled, ridges broad, smooth, obtuse, sulci closed or nearly closed, rugulose.

Sandy soil, Lower Sonoran Zone, southwestern Arizona and Sonora and on the Gulf side of Baja California south to the central part. Aug.–Oct.

## 8. COMMICARPUS Standley, Contr. U.S. Nat. Herb. 12: 373, 1909

Weak, reclining or scandent, shrubby perennials with opposite, entire, petiolate leaves. Inflorescence of umbels in terminal bracteate cymes or axillary pedunculate umbels. Flowers perfect, white, pinkish or yellowish green, ephemeral, campanulate-funnelform to rotate, limb shallowly lobed; stamens 2–5; style filiform; stigma peltate; ovary stipitate. Anthocarp fusiform-cylindric, striate, beset with warty glands.

Perianth 3–4 mm. broad, greenish; fruiting pedicels mostly 5–7 mm. long .......... 1. *C. scandens*
Perianth 10–15 mm. broad, white or pinkish-tinged; fruiting pedicels mostly 1 cm. or more long .......... 2. *C. brandegeei*

**1. Commicarpus scandens** (L.) Standley, Contr. U.S. Nat. Herb. **12:** 373, 1909

*Boerhaavia scandens* L., Sp. Pl. 3, 1753.
*Boerhaavia grahamii* A. Gray, Amer. Jour. Sci. II, **15:** 323, 1853.

Suffrutescent or shrubby, much-branched plants with long, weak, slender, pale, reclining branches; leaves scattered, glabrous, 1.5–6 cm. long, ovate to ovate-deltoid, typically cordate to almost truncate at base, attenuate at apex or sometimes merely acute, often apiculate; inflorescence terminal or axillary, flowers arranged in umbels on leafy or bracteate cymose branches; umbels 5–10-rayed, occasionally forked, rays of nearly equal length, 4–8 mm. long, glabrous, subtending bracts 1.5–2 mm. long, lanceolate-attenuate, sparsely hairy, caducous; perianth greenish or yellowish green, rotate-spreading, 3–4 mm. broad; stamens 2, much exserted; style long-exserted; ovary stipitate; anthocarp greenish, narrowly clavate, finely ribbed, glabrous or occasionally finely hirsutulose, 10–12 mm. long, the coarse glands scattered, larger and more abundant at upper end of anthocarp.

Canyons and thickets, Upper and Lower Sonoran and Tropical Zones, western Texas to southern Arizona, south through Mexico and the West Indies to northern South America; in central and southern Baja California. Sept.–April.

**2. Commicarpus brandegeei** Standley, Contr. U.S. Nat. Herb. **12:** 374, 1909

*Boerhaavia elongata* Brandegee, Proc. Calif. Acad. II, **2:** 199, 1889. Not *B. elongata* Salisb., 1796.
*Commicarpus brandegei* subsp. *glabrior* Standley, Contr. U.S. Nat. Herb. **12:** 374, 1909.

Reclining or scandent shrubby plant with slender, weak stems to 1 m. long, glabrous below, somewhat puberulent above; leaves 1.5–4.5 cm. long, ovate to deltoid-ovate, cordate at base, acute and usually apiculate at apex, more or less puberulent on both surfaces, rather thin; petioles 1–10 mm. long, more densely puberulent than blades; inflorescence mostly terminal, usually somewhat puberulent; bracts of branches of inflorescence persistent, firm, much reduced, apiculate; bracts of umbels linear, pubescent, caducous; rays of umbel 3–6, simple or sometimes forked, more or less densely puberulent to glabrate, of unequal length, mostly 1 cm. or more long; perianth white or pink-tinged, 10–15 mm. broad, not so long as broad; anthocarp narrowly clavate, 5–6 mm. long, with fine longitudinal veins, greenish and bearing coarse, sticky glands mostly in horizontal rows on upper end.

Rocky slopes, Upper Sonoran and Transition Zones, Sierra de la Giganta to Cabo San Lucas, southern Baja California. Oct.–March.

## 9. OXYBAPHUS L'Her. ex Willd., Sp. Pl. 1: 185, 1797

Perennial plants, mostly herbaceous from woody root, branches swollen at nodes. Leaves opposite, entire, sessile or petiolate. Flowers perfect, 1–3 in each involucre. Involucres 5-lobed, accrescent, nearly rotate in age and strongly reticulate-veined. Perianth campanulate to short-funnelform, limb 5-lobed. Stamens unequal, 3–5, slightly united at base. Stigma depressed-capitate. Anthocarp narrowly obovoid, constricted at base, conspicuously 5-angled or -ribbed, mucilaginous when wet. Embryo curved; cotyledons enclosing copious endosperm.

### 1. **Oxybaphus coccineus** Torr., Bot. Mex. Bound. 169, 1859

(For complete synonymy see N. Amer. Fl. **21**: 221, 1918.)

Plants 3–6 dm. high with slender, erect or ascending branches, glaucescent, glabrous except on the growing parts; leaves filiform to linear, 3–8 cm. long, 1–6 mm. wide; inflorescence loosely cymose; involucres short-pilose, lobes deeply cut, acute, 3–4 mm. long when young, 5–6 mm. in age; perianth deep red, 12–15 mm. long, tube funnelform; stamens exserted; fruit finely puberulent, 5-ribbed, somewhat rugose.

Desert slopes, Upper Sonoran Zone, southeastern California to southern New Mexico and south to Sonora. Barely within the boundaries of the Sonoran Desert. May–Aug.

**Oxybaphus pumilus** (Standley) Standley, Field Mus. Bot. Ser. **8**: 11, 1930, a pink-flowered, broad-leaved, pilose species with much the same range, has been collected in the Santa Catalina Mountains, Arizona, and is likely to occur as a straggler into the Sonoran Desert.

## 10. MIRABILIS L., Sp. Pl. 177, 1753

Perennial herbs or shrubs with dichotomously branching stems and opposite, entire leaves. Involucres axillary or clustered at ends of branches, 1–10-flowered, somewhat accrescent with age. Perianth surpassing the involucre, long-salverform, funnelform, campanulate, or campanulate-funnelform with a short, broad tube. Stamens 3–5, unequal; filaments capillary. Stigma capitate. Anthocarp ellipsoid, ovoid, or subglobose.

Involucres 3–10-flowered:
- Filaments scarcely if at all exserted; perianth rose-purple, 3.5–5 cm. long .. 1. *M. froebelii*
- Filaments long-exserted; perianth red, 2.5 or less cm. long .............. 2. *M. triflora*

Involucres 1-flowered:
  Perianth funnelform or salverform with a long, slender tube:
    Perianth 3–5.5 cm. long, reddish purple or yellowish .................. 3. *M. jalapa*
    Perianth 8–17 cm. long, white tinged with pink or reddish purple ................ .......................................... 4. *M. longiflora* var. *wrightiana*
  Perianth campanulate, the tube, if at all evident, broad:
    Fruiting involucres somewhat accrescent in age, 8–15 mm. long, lobes about equaling or longer than tube:
      Mature fruit subglobose or oval, smooth, not marked with vertical lines .......... ....................................................... 5. *M. tenuiloba*
      Mature fruit ellipsoid, marked with vertical lines or faintly rugulose:
        Mature fruit 6–8 mm. long; peduncles of involucre 4–15 mm. long ............ ..................................................... 6. *M. oligantha*
        Mature fruit 4–4.5 mm. long; involucre sessile or with peduncles 2–3 mm. long ... ...................................................... 7. *M. laevis*
    Fruiting involucres scarcely accrescent in age, 5–8 mm. long, the lobes usually shorter than the tube:
      Pubescence for the most part of short, conic hairs; lower leaves 2 cm. long or less ... ........................................ 8. *M. californica* var. *cedrosensis*
      Pubescence not of short, conic hairs; lower leaves 3 cm. long or more:
        Upper stems and inflorescence viscid-villous or -pubescent .................... ..................................... 9. *M. bigelovii* var. *bigelovii*
        Upper stems and inflorescence scabrous as well as viscid-pubescent, sometimes glabrate ................................... 9*a*. *M. bigelovii* var. *aspera*

**1. Mirabilis froebelii** (Behr) Greene, Bull. Calif. Acad. **1**: 124, 1885

*Oxybaphus froebelii* Behr, Proc. Calif. Acad. **1**: 69, 1855.
*Mirabilis multiflora* var. *pubescens* S. Wats., Bot. Calif. **2**: 2, 1880.
*Quamoclidion froebelii* Standley, Contr. U.S. Nat. Herb. **12**: 359, 1909.
*Quamoclidion multiflorum* subsp. *obtusum* Standley, loc. cit.

Much-branched perennial 3–8 dm. high, forming large clumps, densely viscid-pubescent throughout; leaves thick, 6–11 cm. long with stout petioles, broadly ovate, sometimes orbicular, mucronate or acute at apex; involucres 5–9-flowered, somewhat accrescent with age, 2–3 cm. long, lobes about one-third as long as involucre; perianth rose-purple, 3.5–5 cm. long, expanded into shallowly 5-lobed limb, usually viscid-pubescent without; anthocarp 8 mm. long, slightly constricted at both ends, brown with darker spots and marked with 10 pale or dark vertical lines.

Dry, sandy soil, Upper Sonoran Zone, southern Nevada and Arizona to Kern County, California, and south to Baja California. April–Aug.

**2. Mirabilis triflora** Benth., Pl. Hartw. 23, 1839

*Quamoclidion nyctagineum* Choisy, in A. DC., Prod. **13**$^{2}$: 429, 1849.
*Quamoclidion triflorum* Standley, Contr. U.S. Nat. Herb. **12**: 358, 1909.

Erect, decumbent, or subscandent shrub 1–2 m. high, viscid-puberulent or short-viscid-villous, becoming glabrate in age, branches slender; leaves ovate, broad, cordate at base, often abruptly attenuate at apex, 3–6 cm. long, including slender petiole; inflorescence densely short-viscid-villous when young, leafy with much-reduced leaves; flowers many, clustered at ends of branches; peduncles mostly slender; involucres 3-flowered, 10–12 mm. long, spreading in age, lobes acute, longer than tube; perianth dark red, viscid-puberulent, tubular, lobes short, obtuse, scarcely spreading; stamens 5, long-exserted; anthocarp rugulose, elliptic, about 5 mm. long.

Hillsides, Upper Sonoran Zone, Sierra de la Laguna at about 2,000 feet elevation, Baja California and in Jalisco. Oct.–April.

**3. Mirabilis jalapa** L., Sp. Pl. 177, 1753

*Mirabilis odorata* L., Cent. Pl. **1:** 7, 1755.
*Jalapa congesta* Moench, Meth. 508, 1794.
(For additional synonymy see N. Amer. Fl. **21:** 238, 1918.)

Much-branched, bushy perennial 6–15 dm. high with erect or ascending branches, pubescence viscid, short-villous or puberulent; leaves ovate, subcordate or rounded at base, acute to short-attenuate at apex, 4–13 cm. long, short-petiolate or subsessile above; inflorescence leafy-bracted, cymose-glomerate at ends of branches; involucre campanulate, lobes longer than tube, to 15 mm. long; perianth 3–5.5 cm. long, reddish purple although whitish and yellowish or variegated forms occur in cultivation; stamens 5, little exserted; anthocarp 7–9 mm. long, obovoid, somewhat rugose and 5-angulate, glabrous or puberulent.

Upper and Lower Sonoran and Tropical Zones, western Texas to southern Arizona, south through Mexico to Central and South America. The common four-o'clock has been collected in southern Arizona and in Sonora and may occur occasionally as a garden escape.

**4. Mirabilis longiflora** L. var. **wrightiana** (A. Gray) Kearney & Peebles, Jour. Wash. Acad. **29:** 475, 1939

*Mirabilis wrightiana* A. Gray ex Britt. & Kearney, Trans. N.Y. Acad. **14:** 28, 1894.

Perennial with erect or ascending, slender branches and long internodes 5–14 dm. high, minutely puberulent especially above, scarcely viscid; leaves ovate, cordate at base, acute or attenuate at apex, 7–12 cm. long, puberulent to nearly glabrous, long-petioled near base of stem, short-petioled above; flowers in dense, leafy clusters at ends of branches, leafy bracts often long-attenuate; involucre densely puberulent, lobes triangular and about as long as tube; perianth 8–17 cm. long, white tinged with pink or purple; stamens 5, long-exserted; anthocarp ellipsoid, tuberculate, about 8 mm. long, puberulent.

Among trees and shrubs, Upper Sonoran Zone, western Texas to southern Arizona and south into Coahuila and Sonora. Aug.–Sept.

**5. Mirabilis tenuiloba** S. Wats., Proc. Amer. Acad. **17:** 375, 1882

*Hesperonia tenuiloba* Standley, Contr. U.S. Nat. Herb. **12:** 363, 1909.

Stout-stemmed perennial 3–6 dm. high, branches few, rather strict, internodes long, shrubby at base, pubescence of inflorescence and new growth densely viscid-villous; leaves thick, 2–5.5 cm. long, reduced above, deltoid to deltoid-ovate, truncate to subtruncate at base, sometimes decurrent on short petioles, these 3–7 mm. long on lower and middle leaves, sessile above, sharply acute to attenuate at apex, short-viscid-villous; flowers many, sessile, densely clustered on upper leafy branches; involucres 9–15 mm. long, acute lobes longer than tube, accrescent in age and mostly erect; perianth 10–18 mm. long, white; stamens scarcely exserted; fruit oval to nearly spherical, 4–6 mm. long, cerebriform-ruminate when moistened, brown.

Sandy and gravelly slopes, Lower Sonoran Zone, western Colorado Desert and the eastern desert slopes of Baja California to Bahía San Francisquito. Feb.–June.

**6. Mirabilis oligantha** (Standley) J. F. Macbride, Contr. Gray Herb. **56:** 23, 1918

*Hesperonia oligantha* Standley, Contr. U.S. Nat. Herb. **12:** 363, 1909.
*Hesperonia polyphylla* Standley, op. cit. 364.
*Mirabilis polyphylla* Standley, Field Mus. Bot. Ser. **8:** 306, 1931.

Much-branched, typically compact, shrubby plant 3–5 dm. high with short internodes except on new growth and more slender stems than preceding species, more or less viscid-puberulent above, glabrous below; leaves 2–5.5 cm. long, broadly ovate or deltoid, rounded or subcordate to broadly cuneate at base, obtuse at apex, petioles one-third to one-fifth as long as blade; flowers many, scattered in upper leaf axils, peduncles 4–15 mm. long; involucre 7–10 mm. long, lobes somewhat accrescent in age, lanceolate, usually as long as or longer than tube; perianth white (or pink?), 12–20 mm. long; anthocarp ellipsoid, 6–8 mm. long, about 3 mm. thick, dark marked with light-colored vertical lines.

Dry slopes, Lower Sonoran Zone, central Baja California, El Mármol south to Calmallí. Feb.–May.

**7. Mirabilis laevis** (Benth.) Curran, Proc. Calif. Acad. II, **1:** 235, 1888

*Oxybaphus laevis* Benth., Bot. Sulph. 44, 1844.
*Quamoclidion laeve* Rydb., Bull. Torrey Club **29:** 687, 1902.
*Hesperonia laevis* Standley, Contr. U.S. Nat. Herb. **12:** 363, 1909.

Weak-stemmed decumbent or trailing plant from woody base, glabrous with a few hairs to short-viscid-villous with some coarser retrorse hairs on upper stems; leaves thin, glabrous below and usually villous above, ovate to deltoid, deeply to shallowly cordate at base, acute or attenuate at apex, 2–6 cm. long, much reduced on inflorescence; flowers clustered at ends of branches or arranged on short, leafy terminal branchlets, sessile or nearly so except those at forks of branches, which have pedicels 2–3 mm. long; involucres accrescent in age, more or less viscid-villous, 8–11 mm. long, lobes lanceolate, normally equaling tube; perianth rose or lavender-pink, 11–13 mm. long; anthocarp dark, faintly rugulose, ellipsoid, narrowed at both ends, 4–4.5 mm. long, 2–2.25 mm. wide.

Banks of arroyos, Lower Sonoran Zone, Tres Vírgines and the vicinity of San Ignacio south and west to Magdalena Bay, Baja California. Nov.–March.

The additional material now available confirms the opinion expressed by J. F. Macbride (Contr. Gray Herb. 56: 20–21, 1918) that this species is related to the *tenuiloba-oligantha* group. This is indicated not only by the longer-lobed involucre, but also by the ellipsoid fruit narrowed at both ends and scarcely half as wide as long. Inland specimens are quite viscid-puberulent to short-villous and often have the coarser and almost retrorse scabrous hairs on the upper stems that occur in forms of *M. californica.*

**8. Mirabilis californica** A. Gray var. **cedrosensis** (Standley) J. F. Macbride, Contr. Gray Herb. **56:** 24, 1918

*Hesperonia cedrosensis* Standley, Contr. U.S. Nat. Herb. **12:** 362, 1909.
*Hesperonia californica* subsp. *microphylla* Standley, op. cit. 365.
*Mirabilis cedrosensis* Jepson, Fl. Calif. **1:** 459, 1914.
*Mirabilis laevis* var. *cedrosensis* Munz, Man. S. Calif. 151, 598, 1935.

Much-branched, shrubby perennial 2–4 dm. high, glabrous below, more or less scabrous above, with short conic or flattened hairs; leaves thick, ovate, 0.5–2 cm. long, subcordate at base, acute at apex, short-petioled, scabrous, reduced leaves of inflorescence sometimes viscid-puberulent; flowers scattered in upper leaf axils, sometimes clustered; peduncles 1–4 mm. long; involucres 4.5–6 mm. long, scabrous and often somewhat viscid-puberulent, lobes obtuse or triangular, typically shorter than tube; perianth rose-purple, 11–14 mm. long; stamens equaling or little exceeding perianth; anthocarp subglobose, 3.5–4.5 mm. long, smooth, dark, with vertical lines usually evident.

Sandy or rocky slopes, Upper and Lower Sonoran Zones, San Clemente Island,

California, south on the western side of Baja California to San Hipólito in the Vizcaíno Desert area. March–April.

**9. Mirabilis bigelovii** A. Gray, Proc. Amer. Acad. **21:** 413, 1886, var. **bigelovii**

*Hesperonia glutinosa* subsp. *gracilis* Standley, Contr. U.S. Nat. Herb. **12:** 365, 1909.
*Hesperonia bigelovii* Standley, N. Amer. Fl. **21:** 235, 1918.

Much-branched, typically stout, erect or spreading perennial, shrubby at base, stems of young growth more or less viscid-villous, often with elongate internodes, suffrutescent at base and in compact forms also suffrutescent above base; leaves, especially petioles, villous, 1.5–5 cm. long, reduced above, usually ovate-deltoid, subtruncate at base, obtuse or broadly acute at apex; inflorescence corymbosely branched; flowers solitary in leaf axils or clustered at ends of branches; involucres 4.5–6 mm. long, viscid-villous, lobes broadly acute to triangular, mostly shorter than tube; perianth 7–11 mm. long, white or pale pink; anthocarp scarcely longer than wide, usually greenish brown and sometimes faintly lineate.

Desert canyons, Lower Sonoran Zone, southwestern Nevada and adjacent California south to Arizona and Sonora; eastern Baja California to the central part of the peninsula. March–May.

**9*a*. Mirabilis bigelovii** var. **aspera** (Greene) Munz, Man. S. Calif. 151, 1935

*Mirabilis aspera* Greene, Erythea **4:** 67, 1896.
*Mirabilis californica* subsp. *aspera* Parish, Muhlenbergia **3:** 125, 1907.
*Hesperonia aspera* Standley, Contr. U.S. Nat. Herb. **12:** 362, 1909.
*Hesperonia aspera* subsp. *villosa* Standley, op. cit. 363.

Much like the preceding, but characterized by scabrous as well as rather sparse glandular-villous pubescence and rose-pink flowers.

Western Mojave and Colorado Deserts, California, and occurring much more commonly than var. *bigelovii* in Baja California as far south as Punta Prieta. Feb.–May.

## 11. ABRONIA Juss. ex Lam., Tab. Encycl. 1: 469, 1791

Annual or perennial branching herbs, erect or prostrate, sometimes appearing acaulescent, usually viscid-pubescent. Leaves opposite, unequal, thick, the margins entire or more or less deeply sinuate. Flowers perfect, in pedunculate heads, subtended by 5–8 scarious bracts. Perianth fragrant, salverform with a slender tube constricted above the ovary, enlarged at the throat, the limb with 4–5 emarginate lobes. Stamens 4–5, unequal, united at base and adnate to perianth tube above, included. Stigma fusiform. Anthocarp turbinate or biturbinate, deeply lobed or winged, when winged showing much variation in an individual head, not mucilaginous when wet; seed elliptic-oblong, dark brown, one cotyledon enclosing the farinaceous endosperm, the other abortive.

Anthocarp wingless or winged (central cavity not extending into the wings); annuals or winter annuals:
  Beak of anthocarp conspicuous, indurated, 1.5–3 mm. long; body of anthocarp more or less deeply rugose-reticulate:
    Flowers 9–18 mm. long; fruit mostly longer than wide, wingless, or, if winged, these not extending above body of fruit ...................... 1. *A. villosa* var. *villosa*
    Flowers 2–3 cm. long; fruit wider than long, broad wings extending above body of fruit ................................................ 1*a*. *A. villosa* var. *aurita*

Beak of the anthocarp scarcely evident, not strongly indurated; body of anthocarp not strongly rugose-reticulate:
Wings of fruit wide, entire, rounded above and but little narrowed below into body of anthocarp; whole fruit broadly cordate in outline:
Slender annuals; leaves sinuately lobed ................ 2. *A. gracilis* subsp. *gracilis*
Coarse annual herbs; leaves mostly merely sinuate .. 2*a*. *A. gracilis* subsp. *platyphylla*
Wings of fruit when present narrow, lobed to body of anthocarp; whole fruit turbinate in outline ................................................ 3. *A. carterae*
Anthocarp lobed (central cavity extending into winglike lobes); coarse perennial ........ ............................................................ 4. *A. maritima*

## 1. Abronia villosa S. Wats., Amer. Nat. 7: 302, 1873, var. villosa

Ascending, decumbent, or procumbent annuals with stout, densely viscid-villous branches 1–4 dm. long; leaves 2–4 cm. long, petioles densely viscid-villous, blades 1.5–3 cm. long, often irregular at base, entire or obscurely undulate-margined, sparingly glandular to glabrate, short-villous on margin; peduncles slender, 2–7 cm. long, usually densely viscid-villous; involucral bracts ovate-lanceolate to lanceolate, 6–8 mm. long; perianth rose-colored, 9–18 mm. long; anthocarp tan or pale brown, a little longer than broad, wings 2–4 when present, body including the evident villous beak 4–8 mm. long, wings narrower below than above, fruit turbinate in outline, wings gradually tapering into beak above; body of fruit coarsely rugose-reticulate, reticulations extending into irregular wings.

Sandy soil, Lower Sonoran Zone, southern Nevada and southwestern Utah west to Inyo County, California, south to Imperial County and southwestern Arizona. March–Oct.

## 1*a*. Abronia villosa var. aurita (Abrams) Jepson, Fl. Calif. 1: 455, 1914

*Abronia aurita* Abrams, Bull. Torrey Club **32:** 537, 1905.

Habit as in var. *villosa,* but bracts lanceolate to linear-lanceolate, to 1 mm. in length; perianth, 2–3 cm. long; anthocarp longer than broad, body normally less rugose-veined than in var. *villosa,* the wings extending above beak, veiny, mostly obtuse, angled apically, usually narrowed below.

Distributed with var. *villosa* from the western Colorado Desert, California, to southwestern Arizona, and northern Baja California.

## 2. Abronia gracilis Benth., Bot. Sulph. 44, 1844, subsp. gracilis

Slender, prostrate-spreading annual with branches 1–2 dm. long, more or less short-viscid-villous especially on upper leaves and stems, sometimes glabrate; leaves mostly oblong or elliptic, obtuse at apex, rounded to cuneate at base, often asymmetrical, 2.5–6 cm. long, petioles equaling to nearly equaling blades; peduncles 3–6.5 cm. long; involucral bracts scarious, villous, ovate to ovate-lanceolate, 4.5–7.5 mm. long; perianth bright pink with a paler or somewhat greenish tube, 12–16 mm. long, limb to 10 mm. broad, lobes deeply notched, tube villous without, sometimes nearly glabrous; anthocarp mostly obcordate in outline, usually wider than long, straw-colored or somewhat darker, about 5–7 mm. long; body of anthocarp slender, beak but little prolonged and not strongly indurated; wings 2–5, rounded above and usually forming a narrow sinus, a little narrowed below or sometimes the width of upper part of wing, thin, obscurely veined.

Coastal plains, Lower Sonoran Zone, northern Vizcaíno Desert to Magdalena Bay, Baja California. Nov.–March.

**2a. Abronia gracilis** subsp. **platyphylla** (Standley) Ferris, Contr. Dudley Herb. **4:** 33, 1950

*Abronia platyphylla* Standley, Contr. U.S. Nat. Herb. **12:** 314, 1909.
*Abronia alba* var. *platyphylla* Jepson, Fl. Calif. **1:** 455, 1914.
*Abronia umbellata* var. *platyphylla* Munz, Man. S. Calif. 149, 598, 1935.

Coarse, spreading annual; leaves entire or merely sinuate, petioles shorter than blades; involucral bracts mostly longer than in subsp. *gracilis*; perianth to 20 mm. long; anthocarp to 8 mm. long, 2–5-winged, wings rounded above and but little narrowed at base of fruit, villous above, thin in texture, veins usually more prominent than in subsp. *gracilis.*

In sand, Lower Sonoran Zone, southern coast of San Diego County, California, to the northern edge of the Vizcaíno Desert and east to Bahía de los Angeles, Baja California. Feb.–May.

**3. Abronia carterae** Ferris, Contr. Dudley Herb. **4:** 32, *pl. 4,* 1950

Coarse, prostrate winter annual (of longer duration?) more or less viscid-villous throughout especially on upper stems and inflorescence, stems 2–4 dm. long; leaves lobed nearly or halfway to midrib with 2–3 pairs of lobes, truncate to broadly angled at base, rounded at apex, 2–5 cm. long, petioles somewhat shorter and more villous than blades; peduncles 4–6 cm. long, surpassing leaves; involucral bracts lanceolate, 7–11.5 mm. long, thin, evidently greenish; perianth 11–15 mm. long, 5–7 mm. broad, tube villous, greenish, limb magenta; anthocarp obpyramidal in outline, sometimes fusiform when wings are not developed above, villous except at extreme base, densely so above, body of anthocarp thick; base of anthocarp strongly ridged, ridges often developing narrow wings which are interrupted and may become mere ridges in central part of fruit, then expanded to wider and often more numerous wings above, giving a lobed appearance to the fruit; wings 5–10, irregular in width and margin, plane or curved, ultimate margin of upper part of wings rather fragile.

Sandy plains, Lower Sonoran Zone, Magdalena Plain from San Jorge south to Almejas Bay and Médano Blanco, Baja California. Nov.–April.

**4. Abronia maritima** Nutt. ex S. Wats., Bot. Calif. **2:** 4, 1880

Prostrate perennial, glandular-puberulent to villous throughout, with fleshy roots and very stout, trailing stems 2–10 dm. long; leaves thick, 4–6 cm. long, petioles 1–2 cm. long, blades thick, 3–6 cm. long, oval or oval-oblong, cuneate or rounded at base, rounded at apex; peduncles 3–8 cm. long; bracts lanceolate to ovate-lanceolate, 6–8 mm. long; perianth 11–14 mm. long, dark crimson, limb 3–5 mm. broad; fruit coriaceous, densely puberulent above and glabrate below, 10–14 mm. long, turbinate, winglike lobes irregular, usually 5, truncate above, scarcely extending above body of fruit, attenuate below; body and base of fruit coarsely reticulate-veined.

Sandy beaches, Upper and Lower Sonoran and Tropical Zones, San Luis Obispo County, California, south to Baja California, Sonora, and Sinaloa. Flowering throughout the year.

## Family 38. BATIDACEAE. Batis Family

Prostrate or low subshrubs of maritime flats and beaches. Leaves opposite, sessile, entire, fleshy. Flowers dioecious, small, crowded in axillary spikes. Staminate strobili 8–12-flowered, with each flower subtended by a persistent scale, these imbricate;

staminate flowers with 2-lobed cup-shaped calyx, 4 stamens and 4 petaloid staminodia. Pistillate spike 4–12-flowered, the scales deciduous; perianth none. Ovary 4-celled, sessile; stigma sessile. Fruit with 1 seed in each cell, bearing a spongy pericarp. Seeds with a membranous testa and no endosperm.

### 1. **BATIS** L., Syst. Nat. ed. 10, 1289, 1380, 1759

Characters of the family.

**1. Batis maritima** L., Syst. Nat. ed. 10, 1380, 1759

Branches prostrate and 1–2 m. long, or ascending and 1–5 dm. long, glabrous, often rooting at nodes; leaves linear, linear-oblanceolate or narrowly spatulate, 8–30 mm. long, acute or obtuse, fleshy; spikes solitary in axils along branches, ovoid, ellipsoid or cylindrical, 3–16 mm. long, staminate spikes mostly sessile and ovoid, pistillate spikes short-pedunculate, mostly elliptical or cylindrical; bracts on staminate spikes suborbicular, rounded or obtuse, exceeding perianth; stamens exserted; fruit 1–2.5 cm. long; seeds ellipsoid, about 3 mm. long.

Coastal mud flats and less commonly on sandy beaches, Upper Sonoran Zone to Carolinian and Tropical Zones, Los Angeles County, California, to Baja California, the coast of Sonora and southward to Central and South America; North Carolina to Florida and Texas and in the Hawaiian Islands. March–Oct.

## Family 39. **PHYTOLACCACEAE.** Pokeberry Family

Herbs, shrubs or small trees with alternate entire leaves, the latter sometimes fleshy. Perfect or dioecious flowers in racemes or panicles. Sepals 4 or 5, distinct. Petals 5 or wanting. Stamens 4-many, hypogynous. Ovary semi-inferior or superior, 1–several-celled. Fruit capsular, baccate, or an achene.

Petals 5; fruit a 1–5-seeded capsule .................................... 1. *Stegnosperma*
Petals wanting; fruit baccate or an achene:
  Pedicels ebracteolate; fruit a translucent white berry; branches spinose..2. *Phaulothamnus*
  Pedicels bracteolate; fruit red or pink berry; branches not spinose:
    Fruit a berry; flowers pedicellate, not appressed to rachis .................. 3. *Rivina*
    Fruit a dry achene; flowers subsessile, appressed to the rachis ............ 4. *Petiveria*

### 1. **STEGNOSPERMA** Benth., Bot Sulph. 17, *pl. 12*, 1844

Shrub 1–5 m. high with alternate, thick, leathery, glaucous, estipulate leaves and slender ascending branches. Flowers perfect, in terminal or axillary racemes or cymes. Sepals 5, ovate, fleshy. Petals 5, short-clawed. Stamens 10; filaments dilated and connate at base to form a narrow ring below ovary. Ovary globose-ovoid, 1-celled, 3–5-ovulate; style branches 3–5, short, recurved. Ovules ascending, amphitropous. Capsule leathery, 5-angled, dehiscing at the angles, 1–5-seeded. Seeds ovoid, smooth, shining, ensheathed in a white fleshy aril that turns red at maturity. Embryo curved, closely applied to seed coat; cotyledons flattened, equal, longer than the radical, curved downward at apex; endosperm starchy.

Inflorescence terminal; petals gradually narrowed toward base ........ 1. *S. halimifolium*
Inflorescence axillary; petals abruptly narrowed at base .................. 2. *S. watsonii*

**1. Stegnosperma halimifolium** Benth., Bot. Sulph. 17, *pl. 12,* 1844

Branches often semiscandent over other trees; leaves oblong, 1–5 cm. long, obtuse at apex, cuneate at base; petioles 1–10 mm. long, red; inflorescence mostly terminal, 6–12 cm. long; pedicels 5–10 mm. long, subtended by 2 subulate bracteoles 1–1.5 mm. long; sepals ovate, 5 mm. long, fleshy medianly, hyaline-membranous along margins; petals about 3–4 mm. long, white; stamens 5 mm. long; fruit capsular, ovoid, 6–8 mm. long, bright red toward apex at maturity.

A shrub of the strand and along both permanent and intermittent watercourses as much as fifty miles from the sea, Lower Sonoran and Tropical Zones, central Baja California and northwestern Sonora to southern Baja California, Mexico. Oct.–May.

**2. Stegnosperma watsonii** D. J. Rogers, Ann. Mo. Bot. Gard. **36:** 475, 1949

Similar to *S. halimifolium* in general habit and character of leaves but differing from that species in the following respects: inflorescence of axillary cymules and small terminal cymes of 1–8 flowers; calyx lobes elliptic or ovate, 3–7 mm. long, 2–4 mm. wide, obtuse or rounded at apex; petals rounded-ovate, abruptly constricted at base; seeds ellipsoidal, 2–3 mm. wide, about 3 mm. long, with a lateral funicular scar.

Along the seacoast and about saline swales and less commonly along washes, Lower Sonoran Zone to Tropical Zone, Sonora and Sinaloa. Oct.–May.

## 2. PHAULOTHAMNUS A. Gray, Proc. Amer. Acad. 20: 293, 1885

Erect, profusely branching shrub with spinose, glabrous, gray branches. Leaves alternate on young growth, in dense fascicles on older stems, glabrous, glaucous. Flowers dioecious, racemose. Sepals 4, distinct. Stamens 12–14, irregularly inserted on base of calyx, ovary none in staminate flowers; ovary conical-oblong, 1-celled, 1- or sometimes 2-seeded, anthers abortive in pistillate flowers; styles 2, filiform, 2–2.5 mm. long. Fruit a pearly white, translucent berry, loosely enfolded at base by ascending calyx lobes. Seed coat irregularly tuberculate and shining. Embryo curved, incompletely annular, the slender radicle about equaling the flattened subequal cotyledons; endosperm starchy.

**1. Phaulothamnus spinescens** A. Gray, Proc. Amer. Acad. **20:** 294, 1885

Shrub 1.5–4 m. high, branches with divaricately spreading, spine-tipped, lateral branchlets; leaves oblong-obovate or oblong-spatulate, chartaceous, glaucous, glabrous, rounded or emarginate at apex, acute to attenuate-cuneate at base, 1–5 cm. long, 5–20 mm. wide; petiole narrowly winged, 1–10 mm. long; inflorescence a few-flowered raceme 0.5–5 cm. long; pedicels 1–2 mm. long, bracts about 1 mm. long; sepals broadly elliptic, 2.5 mm. long, rounded at apex, white-membranous on margins; filaments of staminate flowers 1.5 mm. long; fruit ovoid to globose, 4–5 mm. long, flesh translucent-white, bitter; seed 2.5–3 mm. long, tuberculately roughened, black, shining.

Clay washes and broad flats, Lower Sonoran Zone, vicinity of Magdalena, Sonora, southern half of Baja California, Nayarit, and Tres Marías Islands, Mexico. July–Aug.

## 3. RIVINA L., Sp. Pl. 121, 1753

Herbaceous perennial or suffrutescent plant with robust striate stems and alternate, thin, sparsely pubescent or glabrate leaves. Flowers perfect, in cylindrical terminal and axillary racemes. Sepals 4, distinct. Petals none. Stamens 4, hypogynous, alternating with the sepals. Ovary superior, 1-celled, 1-seeded; style and stigma 1. Fruit a globose, fleshy, red berry. Seed erect, lenticular. Embryo annular, radicle stout; endosperm farinaceous-starchy.

**1. Rivina humilis** L., Sp. Pl. 121, 1753

*Rivina laevis* L., Mant. 41, 1767.
*Rivina portulaccoides* Mult., Trans. Amer. Phil. Soc. **5**: 167, 1837.
*Rivina purpurascens* Schrad., Gen. Pl. 17, *pl. 5,* 1808.

Herbaceous or suffrutescent, 0.3–1 m. high; leaves ovate-acuminate, rounded to subcordate at base, 3–13 cm. long, 1–5 cm. wide, sparsely puberulent to glabrous; petioles slender, 2–4 cm. long; inflorescence erect or somewhat curved, 3–8 cm. long, the rachis green, short-puberulent to sparsely pilose; bracteoles minute; sepals broadly elliptic, 2–3 mm. long, pink or purplish with a greenish midvein, spreading; stamens about equaling or slightly shorter than sepals; fruit 3–4 mm. in diameter; seed black, shining or short-puberulent, 2.5–3 mm. long.

In shade of shrubs or trees in moist places, along watercourses or on flats, where water stands during the rainy season, Lower Sonoran and Tropical Zones, Florida to Arizona and south to northern South America; southern Baja California. Flowers practically throughout the year when moisture is available.

## 4. PETIVERIA L., Sp. Pl. 342, 1753

Erect, slender-branched, suffrutescent plant with alternate, petiolate, single leaves and a distinct alliaceous odor. Flowers perfect, in slender terminal and axillary racemes. Sepals 4, distinct, linear, subequal, erect and closely investing the fruit at maturity. Petals none. Stamens 4–8, irregularly inserted. Ovary superior, oblong, 1-celled, 1-seeded. Stigma sessile. Fruit dry, bilobed pod with 4–6 recurved spicules at the apex. Seed erect, linear. Embryo straight; endosperm farinaceous, scanty.

**1. Petiveria alliacea** L., Sp. Pl. 342, 1753

*Petiveria octandra* L., Sp. Pl. ed. 2, 486, 1762.
*Petiveria ochroleuca* Moq., in A. DC., Prod. **13**$^{2}$: 9, 1849.
*Petiveria hexandria* Sesse & Moc., Fl. Mex. 98, 1894.

Stems angulate-striate, 0.5–1.5 m. high, puberulent when young, glabrate in age; leaves elliptic to ovate, rounded to acuminate at apex, acute to broadly cuneate at base, 3–12 cm. long, 1.5–5 cm. wide, thin, bright green; petiole 3–15 mm. long; stipules triangular-subulate, 1 mm. long; inflorescence erect, to 25 cm. long; pedicels 1 mm. long or less; bracts 1 mm. long and minute bracteoles both present; sepals 4–6 mm. long, white to rose, green in fruit; fruit 6–8 mm. long, appressed to the rachis.

Lower Sonoran and Tropical Zones, southern Baja California and central Mexico to Florida, West Indies, and northern South America. Flowers throughout the year.

## Family 40. AIZOACEAE. Aizoon Family

Mostly prostrate or ascending, often succulent, annual or perennial herbs or shrubs. Leaves opposite, alternate, or falsely whorled, mostly estipulate. Flowers (in ours) perfect, regular, solitary or clustered. Calyx 4- or 5-lobed. Petals none or numerous. Stamens few to many. Ovary superior, inferior or partially inferior, 1–many-celled; styles equaling the number of cells; ovules usually many but sometimes few or solitary in a cell. Capsule loculicidal, circumscissile or dehiscent. Seeds solitary, few or many, obovoid-reniform; embryo curved.

Petals none; calyx free from ovary:
  Capsule dehiscent by vertical valves; herbage not succulent; leaves whorled; calyx lobes not appendaged:
    Seeds not strophiolate; herbage glabrous .............................. 1. *Mollugo*
    Seeds strophiolate; herbage (in ours) puberulent ........................ 2. *Glinus*
  Capsule circumscissile; herbage succulent; leaves opposite or alternate; calyx lobes appendaged on back just below apex:
    Leaves not covered with glistening vesicles; ovary 1–2-celled .......... 3. *Trianthema*
    Leaves covered with minute glistening vesicles; ovary 3–5-celled ........ 4. *Sesuvium*
Petals present, numerous; calyx tube partially or wholly adnate to ovary ..................................................................... 5. *Mesembryanthemum*

## 1. MOLLUGO L., Sp. Pl. 89, 1753

Low, diffusely branched annual (ours) or perennial herbs with glabrous or puberulent herbage. Leaves whorled. Flowers perfect, regular. Sepals 5, scarious-margined. Petals wanting. Stamens 3 or 5, hypogynous. Stigmas 3. Ovary superior, 3-celled, several-seeded. Fruit a loculicidal membranous capsule. Seeds ovate to nearly semicircular, laterally flattened, reddish brown, ridged on back or minutely reticulate on sides.

Seeds 0.25–0.33 mm. long, minutely reticulate laterally; leaves glaucescent .. 1. *M. cerviana*
Seeds 0.5–0.6 mm. long, distinctly 6–7-striate, striae darker than sulci; leaves green ........ ........................................................ 2. *M. verticillata*

### 1. Mollugo cerviana (L.) Ser., in DC., Prod. 1: 392, 1824

*Pharnaceum cerviana* L., Sp. Pl. 272, 1753.

A delicate annual with glabrous, filiform, erect or ascending stems; branches 10–20 cm. long and glaucous, linear; leaves 6–15 mm. long, in whorls of 4–6; flowers on filiform pedicels 5–8 mm. long in 1–3-flowered axillary umbels; peduncles equaling or surpassing pedicels; bractlets minute, hyaline-margined; sepals 5, 1–1.5 mm. long, green on midrib, glabrous, hyaline on margins, erect; stamens 5, alternate with sepals, 1 mm. long; stigma sessile, 3-lobed, each lobe a minutely denticulate, membranous flap 0.2 mm. high and twice as wide; seeds uniformly reddish brown, faintly striate dorsally.

In dry, sandy soil, along arroyos and silted plains, Lower Sonoran Zone, New Mexico and Arizona south into Mexico and the tropics of both Old and New Worlds; infrequent in Texas and southern California. Introduced to the western hemisphere. Aug.–Oct.

### 2. Mollugo verticillata L., Sp. Pl. 89, 1753

Prostrate, forming mats 1–3.5 dm. broad; leaves spatulate, 5–6 in a whorl, 5–40 mm. long, 2–8 mm. wide, basal ones distinctly petiolate; flowers several at a node, pedicels slender, 5–15 mm. long but not pedunculate; sepals oblong, 2–2.5 mm. long, with green midrib and white margins; stamens usually 3, alternating with cells of ovary, about 1.5 mm. long; capsule ovoid, slightly surpassing sepals; seeds numerous, usually visible through distended capsule walls, sulci light reddish brown, striae dark brown.

In sandy and waste soil, Sonoran and Transition Zones, as an occasional naturalized weed from Oregon, Idaho, Illinois, and Virginia to the tropics of the western hemisphere. Native in the tropics of the Old World. Sept.–Oct.

## 2. GLINUS L., Sp. Pl. 463, 1753

Procumbent or ascending herbaceous annuals with much-branched stems and whorled unequal leaves. Flowers in dense axillary glomerules. Sepals 5. Petals none. Stamens 3–5, with short, slender filaments and 2-celled anthers. Ovary 3–5-celled; style short, stigmas 3–5. Capsule loculicidally dehiscent. Seeds numerous, strophiolate, smooth or tuberculate; embryo curved; cotyledons fleshy, oblong.

**1. Glinus radiatus** (Ruiz & Pav.) Rohrb., in Mart., Fl. Bras. **14²:** 238, *pl. 55, fig. 1,* 1872

*Mollugo radiata* Ruiz & Pav., Fl. Peruv. **1:** 48, 1798.
*Mollugo glinoides* Camb., in St. Hil., Fl. Bras. Merid. **2:** 171, *pl. 109,* 1829.
*Glinus cambessidesii* Fenzl, Ann. Wien. Mus. **1:** 358, 1836.
(For additional synonymy see N. Amer. Fl. **21:** 271, 1932.)

Stems 1–3 dm. long; herbage finely stellate-tomentose; leaf blades obovate, elliptic, or sometimes suborbicular-spatulate, 2–15 mm. wide, 5–20 mm. long, acute to rounded at apex, gradually or abruptly narrowing at base to petioles 1–6 mm. long; flowers in axillary clusters of 3–8; sepals elliptic-lanceolate or oblong, 2.5–3 mm. long, acute, stellate-tomentulose, margins hyaline and glabrous; stamens 3–5, about 1.5–2 mm. long; capsule ovoid or ellipsoid, 3–3.5 mm. long, acute, thin-walled; seeds brown, 0.4–0.5 mm. long, smooth and shining, strophiole about half as long, whitish.

Margins of dried pools, at scattered localities in Texas, Mexico, West Indies, Central and South America. Reported from "Baja California" by Brandegee but exact station unknown.

## 3. TRIANTHEMA L., Sp. Pl. 223, 1753

Fleshy prostrate herbs with subopposite leaves. Flowers solitary or in small clusters in axils of the branches, clusters sessile or pedunculate. Calyx 5-lobed, lobes concave and each bearing a short dorsal horn just below apex. Petals none. Stamens 5–10. Ovary 1–2-celled, truncate, few-ovulate, bearing 1 or 3 styles. Capsule thickened above, bearing a short, shallowly lobed appendage at apex, tardily circumscissile. Seeds reniform, rough; embryo annular; endosperm mealy.

**1. Trianthema portulacastrum** L., Sp. Pl. 223, 1753

*Trianthema monogynum* L., Mant. 69, 1767.
*Trianthema procumbens* Mill., Gard. Dict. ed. 8, no. 1, 1768.

Diffusely branched, with succulent branches 2–10 dm. long; leaf blades obovate, suborbicular or elliptic, 1–2 (rarely 5) cm. long, obtuse but sometimes apiculate at apex, cuneate at base, glabrous, on petioles 3–20 mm. long, the latter with bases expanded into thin sheaths; flowers partly concealed by sheaths, purplish within; sepals 4–5 mm. long; capsule 4–5 mm. long, including crest; seeds black, 1.6–2 mm. long, roughened and lightly ridged, dull.

Frequent on waste ground, southern California to Texas, Mexico, Central America, and northern South America. Also in Florida and introduced along the Atlantic seaboard. Common in the Sonoran Desert region about Tucson and Phoenix, Arizona; Imperial Valley, California; and in both Sonora and Baja California. July–Oct.

## 4. SESUVIUM L., Syst. Nat. ed. 10, 1058, 1759

Fleshy annual or perennial herbs or shrubby plants. Leaves opposite, estipulate, but petioles often dilated at base. Flowers sessile or pedicellate, axillary. Sepals 5, usually appendaged near apex. Petals none. Stamens 5–many, perigynous. Ovary 3–5-celled; ovules numerous; styles 3–5. Capsule membranous, circumscissile. Embryo annular.

### 1. Sesuvium verrucosum Raf., New Fl. 4: 16, 1838

*Sesuvium portulacastrum* var. *subsessile* A. Gray, in S. Wats., Bibl. Index 411, 1878.
*Sesuvium sessile* Robinson, in A. Gray, Syn. Fl. N. Amer. 1: 259, 1897. Not *S. sessile* Pers., 1806.

Freely branching, prostrate herb with stems 1–9 dm. long and herbage finely verrucose with minute, crystalline, hemispherical papillae; leaves oblanceolate to obovate-oblong, 0.5–3 cm. long, 2–10 mm. wide, rounded or acutish at apex, cuneately narrowed to slightly clasping base; flowers subsessile to short-pedicellate; calyx lobes ovate-lanceolate to ovate-oblong, 4.5–6 mm. long, shallowly hooded, back prolonged into short beak 1–1.5 mm. long, crystalline-papillate on outside, rose or orange inside; stamens numerous, capsule 3–3.5 mm. in diameter, 4–5 mm. high; seeds reniform-obovoid, 0.8–1 mm. long, black, smooth, sublustrous.

In alkaline or saline soils, Lower and Upper Sonoran Zones, central California to Kansas, Texas, Mexico, and tropical America. March–Nov.

## 5. MESEMBRYANTHEMUM L., Sp. Pl. 480, 1753

Low, succulent, annual or perennial herbs with opposite, estipulate, flat to nearly terete leaves. Flowers axillary and terminal. Calyx 4- or 5-lobed, these often unequal. Petals numerous, linear, united into a tube at the base. Stamens numerous. Ovary and capsule half inferior to nearly superior, 5-celled; stigmas sessile, 5. Capsule 5-valved, variously winged and keeled, or fleshy and indehiscent. Seeds small.

Herbage covered with shining, crystalline or water-filled vesicles; fruit a capsule:
  Leaf blades flat, ovate to broadly spatulate; stems stout, 3–8 mm. in diameter .......... 1. *M. crystallinum*
  Leaf blades linear, almost terete; stems slender, 2–3 mm. in diameter .. 2. *M. nodiflorum*
Herbage smooth, not covered with vesicles; fruit fleshy, indehiscent ...... 3. *M. chilense*

### 1. Mesembryanthemum crystallinum L., Sp. Pl. 480, 1753

*Cryophytum crystallinum* N. E. Brown, in E. P. Phillips, Gen. S. Afr. Pl. 245, 1926.

A profusely branched, procumbent, succulent herb 2–5 dm. broad and nearly as high; herbage covered with small, transparent, glistening vesicles, hence the common name "ice plant"; leaves opposite or sometimes alternate, with flat, undulate, obovate to broadly spatulate blades 0.7–4 cm. wide, 1.5–5 cm. long, tapering to a broad clasping base, lower leaves 5–12 cm. long and short-petiolate; flowers axillary, subsessile, 12–30 mm. broad; calyx tube 6–10 mm. in diameter, campanulate, lobes broadly ovate, subequal, 4–7 mm. long; petals white to reddish, spreading; capsule 8–10 mm. high, nearly as broad; seeds blackish brown, about 1 mm. in diameter, finely tuberculate.

Commonly naturalized in sandy places, Lower and Upper Sonoran Zones, abundant along the coast and sparingly in the deserts of southern California, abundant well inland in central Baja California. Also introduced into South America and the Mediterranean region. Native of South Africa. Feb.–June.

**2. Mesembryanthemum nodiflorum** L., Sp. Pl. 480, 1753

*Cryophytum nodiflorum* L. Bolus, S. Afr. Gard. **17:** 327, 1927.

Decumbent to erect succulent herb with herbage covered with transparent shining vesicles 1–1.5 mm. in diameter; stems slender, rarely over 3 mm. in diameter; leaves linear, nearly terete, 1–2.5 mm. wide, 1–4 cm. long, obtuse at apex, slightly narrowed toward base; flowers subsessile or short-pedicellate and solitary in axils, 1–1.5 cm. broad; calyx 4- or 5-lobed, the broadly campanulate tube 4–6 mm. long, obtusely ribbed; calyx lobes ovate, subequal, 3.5–5 mm. long; petals white, shining, coalescing to form beaklike tip in withering; capsule obovoid, 5–6 mm. high, tardily dehiscent by terminal pores or slits; seeds reniform-ovoid, brownish, 0.6–0.8 mm. long, minutely tuberculate, shining.

Naturalized in sandy soil along coast, Lower Sonoran and Upper Sonoran Zones, Oregon, southern California from Point Mugu southward, on the islands off southern California, and throughout much of Baja California, where it often covers thousands of acres. April–July (rarely to Oct.).

**3. Mesembryanthemum chilense** Molina, Sagg. Chile ed. 2, 133, 1810

*Mesembryanthemum dimidiatum* Torr., Pacif. R. Rep. **4:** 75, 1857.

*Mesembryanthemum aequilaterale* Brewer & S. Wats., Bot. Calif. **1:** 251, 1876. Not *M. aequilaterale* Haw., 1803.

*Carpobrotus chilensis* N. E. Brown, Jour. Bot. Brit. & For. **66:** 324, 1928.

Prostrate, much-branched, succulent, perennial herb with spreading stems up to 3–4 m. long and often forming dense, extensive mats; leaves opposite, connate at base, oblong to linear, 0.5–1.5 cm. wide, 2.5–10 cm. long, triangular in cross section, angles acute or obtusish, straight or arcuate, glabrous and smooth, light green or sometimes faintly glaucous; flowers nearly sessile or pedicellate, terminating short vertical branches 3–15 cm. long, showy, 3–5 cm. in diameter; calyx tube turbinate, 2–2.5 cm. long, 1.5–2 cm. wide, lobes broadly deltoid-ovate, unequal, acute to obtuse, 6–12 mm. long; petals pink to crimson, linear, numerous, spreading; fruit fleshy, indehiscent, 8–16-celled, 3–5 cm. long, turning red at maturity; seeds numerous, dark brown to black.

On coastal sand dunes and slopes, Lower Sonoran and Upper Sonoran Zones, introduced from Chile into southwestern Oregon and along the California coast and southward in Baja California at least to Rosario. April–Sept.

## Family 41. PORTULACACEAE. Purslane Family

Mostly small herbaceous plants, annual or perennial, often succulent. Leaves entire, alternate, opposite or basal. Flowers perfect, regular or slightly asymmetrical; sepals usually 2, caducous or persistent; petals mostly 5 (sometimes 3–16); stamens few to many. Style 2–8-cleft or divided. Ovary superior or partly inferior, 1-celled. Capsule circumscissile or longitudinally dehiscent. Seeds 1–many, suborbicular-reniform or orbicular, slightly compressed; embryo curved.

Capsule 2–3-valved longitudinally; ovary superior:
  Sepals deciduous; roots mostly fleshy and tuberlike ........................ 1. *Talinum*

Sepals persistent; roots not tuberous nor markedly fleshy:
Stem leaves alternate, not connate; seeds 6–35:
Capsule 3-valved, ovate or globose; inflorescence not recurved ....... 2. *Calandrinia*
Capsule 2-valved, oblong or cylindric; inflorescence recurved ...... 3. *Calyptridium*
Stem leaves opposite and connate to form a disk; seeds (in ours) 1–3 ...... 4. *Montia*
Capsule circumscissile; ovary partly inferior ............................... 5. *Portulaca*

## 1. TALINUM Adans., Fam. Pl. 2: 245, 1763

Perennial herbs or shrubby plants, often with tuberlike, fleshy roots and short or elongated stems. Leaves fleshy, alternate or subopposite, with entire, terete, or flat blades. Flowers cymose, paniculate, racemose, or solitary in axils of leaves, frequently showy. Sepals 2, deciduous. Petals 5 (or more), withering early. Stamens 5–many, filaments filiform. Ovary superior; style 3-cleft; ovules and seeds numerous. Capsule 1-celled, ovoid or globose, 3-valved. Seeds ovoid-reniform, somewhat flattened, testa smooth or ridged.

Leaf blades ample, 1–5.5 cm. broad; inflorescence paniculate:
Flowers on slender pedicels; sepals 3–4 mm. long; seeds striolate or minutely tuberculate, uniformly black, shining ..................................... 1. *T. paniculatum*
Flowers sessile or nearly so; sepals 8–9 mm. long; seeds smooth, faintly mottled gray and black, dull ................................................ 2. *T. guadalupense*
Leaf blades linear, not over 6 mm. wide, usually less; inflorescence racemose or cymose:
Seeds concentrically striate and roughened on sides; stamens 15–20 or more; flowers solitary or on 2–3-flowered axillary peduncle:
Stems very slender, rarely over 2 mm. in diameter, glabrous and sublustrous; petals yellow; capsule globose, 4–5 mm. in diameter .............. 3. *T. angustissimum*
Stems stoutish, often 3–5 mm. in diameter, sparsely granulose, dull; petals coppery or orange; capsule ovoid, 5–7 mm. long ........................ 4. *T. aurantiacum*
Seeds smooth; stamens 5–6; flowers in terminal cymes ................ 5. *T. gooddingii*

### 1. Talinum paniculatum (Jacq.) Gaertn., Fruct. 2: 219, 1791

*Portulaca paniculata* Jacq., Enum. Pl. Carib. 22, 1760.
*Talinum reflexum* Cav., Ic. 1: 1, *pl. 1*, 1791.
*Talinum patens* Willd., Sp. Pl. 2: 863, 1800.
*Talinum sarmentosum* Engelm., Bost. Jour. Nat. Hist. 6: 153, 1850.
*Talinum chrysanthum* Rose & Standl., Contr. U.S. Nat. Herb. 13: 288, 1911.
(For complete synonymy see N. Amer. Fl. 21: 282, 1932.)

Root tuberous; stems simple or branched, slender, often woody at base, 2–8 dm. tall, glabrous or slightly granulose, striate, often tinged with pink; leaves elliptic to obovate, 1–5.5 cm. broad, 2.5–12 cm. long, obtuse, rounded or sometimes acute at apex, tapering at base to a broad petiole 0.5–1.5 cm. long; inflorescence a terminal panicle 5–45 cm. long, cymose branches simply or intricately branched; pedicels slender, 1–3 cm. long; sepals ovate to orbicular, 3–4 mm. long; petals elliptic to orbicular, 3.5–5 mm. long, red, pink, or rarely yellowish; stamens 15–20; capsule globose, 3–4.5 mm. in diameter, shining; seeds about 1 mm. in diameter, black, minutely striolate or finely and faintly punctate-pitted or smooth, shining.

Shaded places among shrubs or about rocks, Lower and Upper Sonoran Zones, Arizona to Florida and south to South America and the West Indies. April–Nov.

### 2. Talinum guadalupense Dudley, in D. S. Jordan, Fur Seals N. Pacif. 3: 282, 1899

Stout, fleshy herb 3–5.5 dm. high from large, tuberlike root 3–5 cm. in diameter;

leaves basal, blades obovate to spatulate, 1–1.5 cm. wide, 3.5–6 cm. long, rounded or rarely acute at apex, tapering rather abruptly to short, broad petiole; inflorescence a panicle of short, congested cymes terminating stout branches 4–6.5 cm. long; flowers sessile or essentially so; sepals broadly ovate to suborbicular, rounded at apex, 8–9 mm. long, nearly as broad; petals obovate, 10–15 mm. long, rose-red; stamens numerous, usually over 50; capsule ovoid, 5–7 mm. long, chartaceous; seeds obovoid-reniform, about 1 mm. long, smooth, finely and faintly mottled with dark gray and black, dull.

Known only from Guadalupe Island, mostly at the south end. Jan.–May.

**3. Talinum angustissimum** (A. Gray) Woot. & Standl., Contr. U.S. Nat. Herb. **16:** 120, 1913

*Talinum aurantiacum* var. *angustissimum* A. Gray, Pl. Wright. **1:** 14, 1852.

Erect, slenderly and moderately branched plant 2–5 dm. high from tuberlike or woody root and suffrutescent base, branches rarely over 2 mm. in diameter and often sublustrous; leaves linear, 1–2.5 mm. broad, 1–6 cm. long, acute at apex, slender terete petiole 3 mm. long or less; flowers solitary, axillary; pedicels slender, 0.8–2 cm. long, bibracteate near base, often reflexed in fruit; sepals ovate, 5–6 mm. wide, 6–9 mm. long, acutish at apex; petals yellow, obovate, 10–13 mm. long; stamens 20–30; capsules globose, 4–5 mm. in diameter; seeds black, shining, 1–1.2 mm. long, concentrically ridged on sides, minute striations running transversely on ridges.

Often among bushes, sandy plains and rocky hillsides, Lower and Upper Sonoran Zones, southern Arizona to western Texas and adjacent northern Mexico. Rare in the Sonoran Desert. July–Oct.

**4. Talinum aurantiacum** Engelm., Bost. Jour. Nat. Hist. **6:** 153, 1850

*Claytonia aurantiaca* Kuntze, Rev. Gen. Pl. **1:** 57, 1891.

Erect or ascending, moderately branched herb 1–4 dm. high from tuberlike roots; stems stoutish, often 3–5 mm. in diameter, mostly dull; leaves linear to linear-lanceolate, 1.2–6 mm. wide, 1–6 cm. long, fleshy, narrower ones only slightly flattened; flowers solitary in axils of leaves or sometimes on 2–3-flowered axillary peduncles; pedicels bibracteate near base, usually recurved in fruit; sepals ovate, 3–4.5 mm. broad, 6–9 mm. long, cuspidate at apex; petals orange, reddish, or coppery, obovate, 10–13 mm. long; stamens 20 or more; capsule ovoid, 5–7 mm. long, 4–5 mm. in diameter; seeds black, shining, 1.2–1.4 mm. long, concentrically ridged on sides, ridges transversely wrinkled.

Grassy or shrubby hillsides and plains, mainly Upper Sonoran Zone, but entering upper margins of Sonoran Desert in Sonora; southern Arizona to western Texas, and in adjacent northern Mexico. April–Sept.

**5. Talinum gooddingii** P. Wilson, N. Amer. Fl. **21:** 287, 1932

"An erect herb, 1–2 cm. high, with rather slender stems 2–3 cm. long; leaf blades terete, 1.8–3 cm. long, 1–1.5 mm. thick, acute at the apex; inflorescence a cyme, bracted at the forks; peduncles slender, 9–15 cm. long; pedicels slender; sepals ovate, 3 mm. long, 1.7 mm. broad, with a horn-like appendage on the back at the apex; petals obovate, 4.5–6 mm. long, 2–2.8 mm. broad; stamens 5 or 6, the filaments long and slender, the anthers oblong, 1.2–1.3 mm. long; stigma capitate; capsule ovoid, 5 mm. long, 3 mm. in diameter; seeds black, 1 mm. wide, smooth." (From N. Amer. Fl. **21:** 287, 1932.)

Known only from elevations slightly above the Sonoran Desert in Greenlee and Santa Cruz Counties, Arizona. A possible straggler into the desert.

## 2. CALANDRINIA H. B. K., Nov. Gen. & Sp. **6:** 77, 1823

Annual or perennial herbs with alternate leaves and ephemeral flowers in leafy-bracted or naked panicles or racemes. Sepals 2, persistent. Petals usually red, 3–7 but usually 5. Stamens 3–14, shorter than and rarely the same number as the petals. Style short, divided into 3 short stigmas. Capsule membranous, 3-valved from the apex. Seeds numerous, reniform-ovoid, somewhat flattened, dark brown or black, usually minutely pitted or granulated, strophiolate or naked.

Inflorescence racemose; seeds strophiolate; petals rose-red:
- Herbage green, not glaucous; seeds black and shining, 0.8–1 mm. in diameter, lenticular-subreniform; petals 4–10 mm. long .................... 1. *C. ciliata* var. *menziesii*
- Herbage glaucous; seeds dull, 0.4–0.6 mm. in diameter, ovoid-reniform, not flattened; petals 3.5–5 mm. long ........................................ 2. *C. maritima*

Inflorescence a panicle of umbels; seeds not strophiolate; petals white ...... 3. *C. ambigua*

**1. Calandrinia ciliata** (Ruiz & Pav.) DC. var. **menziesii** (Hook.) J. F. Macbride, Field Mus. Bot. Ser. **11:** 20, 1931

*Talinum menziesii* Hook., Fl. Bor. Amer. **1:** 223, *pl. 70,* 1834.
*Calandrinia elegans* Spach, Hist. Veg. **5:** 232, 1836.
*Calandrinia caulescens* var. *menziesii* A. Gray, Proc. Amer. Acad. **22:** 277, 1887.
*Calandrinia arizonica* Rydb., N. Amer. Fl. **21:** 292, 1932.
*Calandrinia heterophylla* Rydb., op. cit. 293.
*Calandrinia filifolia* Rydb., loc. cit.

Annual decumbent herb, profusely branched basally, with stems 5–25 cm. long; herbage glabrous to coarsely short-ciliate on margins and angles; leaves linear-oblong, spatulate, or oblong-oblanceolate, 2.5–8 cm. long, gradually or abruptly narrowed to narrowly winged petioles; inflorescence racemose, leafy-bracted; pedicels 4–10 mm. long; sepals moderately keeled, 4–6 mm. long, frequently ciliate on keel and margins, acute; petals 5, 4–10 mm. long, rose-red; stamens 7–14, about two-thirds as long as petals; capsule 4–5 mm. high, ovoid; seeds numerous, black, shining, 0.8–1 mm. long, minutely roughened with low, rounded tubercles, distinctly lenticular-compressed.

Open or grassy situations, mostly Upper Sonoran and Transition Zones, British Columbia, to central Baja California and east into Arizona and Sonora. Mostly Feb.–April in our area.

**2. Calandrinia maritima** Nutt., in Torr. & Gray, Fl. N. Amer. **1:** 197, 1838

*Claytonia maritima* Kuntze, Rev. Gen. Pl. **1:** 57, 1891.

Glaucous subsucculent annual, branching basally with scapelike decumbent to ascending stems 0.5–3 dm. long; basal leaves broadly spatulate to broadly oblanceolate, 1–8 cm. long, cauline leaves narrower and reduced or sometimes no more than linear or ovate bracts; inflorescence racemose or cymose, terminating erect or ascending scapes 0.5–2.5 dm. tall; pedicels slender, 3–12 mm. long; sepals ovate, obtusish, 4–5 mm. long; petals 3–5, rose-red, 3.5–5 mm. long; stamens 4–8; capsule ovate, 4.5–6 mm. long; seeds ovoid-reniform, scarcely or not at all compressed, minutely and closely muriculate, grayish, dull.

Near the sea, mostly Upper Sonoran Zone, Santa Barbara County, California, south at least halfway down the west coast of Baja California. Feb.–May.

**3. Calandrinia ambigua** (S. Wats.) Howell, Erythea **1:** 34, 1893

*Claytonia ambigua* S. Wats., Proc. Amer. Acad. **17:** 365, 1882.
*Calandrinia sesuvioides* A. Gray, op. cit. **22:** 278, 1887.

Glaucous, glabrous, succulent annuals with several prostrate or decumbent, leafy stems 2–20 cm. long; leaves linear-spatulate, 3–5 mm. wide, 1–3.5 cm. long; flowers in lateral or terminal, compact, umbel-like clusters, each subtended by a subscarious, ovate or ovate-attenuate, whitish bract 0.5–1.5 mm. long; pedicels slender, 2–7 mm. long; sepals ovate, 3.5–4 mm. long, obtuse or with a minute terminal apiculation; petals white, 3–5, obovate, about equaling or slightly shorter than sepals; stamens 5–8; stigma undivided; capsule ovoid, 5–6 mm. long, many-seeded; seeds black, shining, compressed, 0.8–1 mm. long, 0.4–0.6 mm. broad, without a strophiole, very faintly marked or wholly smooth.

Alkaline soil, Lower Sonoran Zone, Inyo County, California, south through the Mojave and Colorado Deserts into southwestern Arizona and Baja California. Feb.–May.

## 3. CALYPTRIDIUM Nutt., in Torr. & Gray, Fl. N. Amer. 1: 198, 1838

Annual herbs from slender taproots. Basal leaves several to many, forming a rosette; cauline leaves alternate, fleshy. Flowers perfect, small, borne in spikes arranged in cymes or panicles. Sepals 2, unequal, persistent. Petals 2–4, more or less coherent in withering and carried up on apex of enlarging capsule. Stamens 1–3. Ovary 6–24-ovulate; styles short or obsolete; stigmas 2. Capsule flat, oval to linear-oblong, 2-valved. Seeds compressed-orbicular, smooth and shining or muriculate.

**1. Calyptridium monandrum** Nutt., in Torr. & Gray, Fl. N. Amer. **1:** 198, 1838

Branched at base, stems spreading or prostrate, 5–25 cm. long; leaves of rosette spatulate, deciduous, roundish or obtuse at apex, 2–7 mm. wide, 1–6 cm. long, fleshy, petiole decurrent, cauline leaves smaller; inflorescence in terminal panicles 5–15 cm. long, spicate branches 1–3 cm. long, often somewhat scorpioid toward tips; sepals reniform-ovate to reniform-deltoid, 1–2 mm. long; petals usually 3, sometimes 2 or 4, white, 1–1.2 mm. long, soon withering to form a coherent calyptra on top of capsule; stamen 1; capsule linear-oblong, finely striate, usually weakly to strongly falcate, 4–8 mm. long, 5–10-seeded; seeds 0.4–0.6 mm. in diameter, black, smooth and shining except on the blunt, narrow wing which is concentrically lined with minute, dull, shallow pits.

Sandy or clay soil, Lower and Upper Sonoran Zones, Santa Clara County, California, to central Nevada, Arizona, Sonora, and central Baja California. Mostly March–May, occasionally as early as January and as late as July in favored spots.

## 4. MONTIA L., Sp. Pl. 87, 1753

Small, glabrous annual or perennial herbs of various habits, often propagating by stolons or small bulblets. Leaves more or less fleshy, alternate, opposite or basal, stem leaves often in pairs and more or less coalescent. Inflorescence racemose or paniculate; flowers small; pedicels enlarged at base of flower and often recurved in fruit. Sepals 2, subequal, persistent. Petals 2–5, white or pink. Stamens 2–5. Ovules 3 or rarely 4. Styles 3, united below. Capsule ovoid to globose, 3-valved. Seeds 1–3, smooth, tuberculate or foveolate, shining.

**1. Montia perfoliata** (Donn) Howell, Erythea **1:** 38, 1893

*Claytonia perfoliata* Donn, Hort. Cantab. 25, 1796.
*Claytonia cubensis* Bonpl., Ann. Mus. Paris **7:** 82, *pl. 6,* 1806.
*Limnia perfoliata* Haw., Syn. Pl. Succ. 11, 1812.
*Limnia hassei* Rydb., N. Amer. Fl. **21:** 311, 1932.
*Limnia guadalupensis* Rydb., loc. cit.
(Several varieties and subspecies also have been proposed.)

Annual, with erect or spreading, reddish stems 0.5–3 dm. high, and fleshy, ascending leaves; basal leaves lanceolate, spatulate, or rhombic, 1–7 cm. long; cauline leaves 2, united at base to form an orbicular or somewhat angled, peltate disk subtending the racemose inflorescence; flowers white or light pink; sepals ovate, rounded, 2–3.5 mm. long; petals 5, short-clawed, deeply emarginate, 3–4.5 mm. long; capsule globose, about 2 mm. long; seeds 1–3, 1–1.6 mm. in diameter, black and shining, minutely sculptured with low, rounded tubercles.

In shady places, Lower Sonoran Zone to Transition Zone, British Columbia to North Dakota, Arizona, and Baja California. Chiefly March–April in our area.

## 5. PORTULACA L., Sp. Pl. 445, 1753

Prostrate to erect, annual or perennial, fleshy herbs with (in ours) alternate leaves. Stipules scarious, reduced to tufts of hairs, or wanting. Calyx 2-cleft, the tube adnate to ovary below. Petals 4–6, usually 5. Stamens 6–20, inserted on the calyx at the base of the petals. Ovary partly to almost wholly inferior. Capsule 1-celled, membranous, circumscissile. Seeds many, reniform or cochleate, smooth or minutely tuberculate.

Hairs in axils of leaves and around flowers lacking or inconspicuous; seeds gray or black, dull or shining, but not iridescent; leaf blades flattened:
- Capsule without collar, lid conical or hemispherical; ripe seeds black:
  - Style lobes 5–6; stamens 6–10; seeds granular-tuberculate; leaf blades rounded to truncate at apex .......... 1. *P. oleracea*
  - Style lobes 3–4; stamens 7–19; seeds echinate-tuberculate; leaf blades retuse at apex .......... 2. *P. retusa*
- Capsule with narrow collar around the rim; lid saucer-shaped; ripe seeds gray .......... 3. *P. lanceolata*

Hairs in axils of leaves and around flowers conspicuous; seeds iridescent green-black; leaf blades mostly terete:
- Stems somewhat woody at base; leaves 8–30 mm. long; corolla copper color .......... 4. *P. suffrutescens*
- Stems herbaceous; leaves 4–18 mm. long; corolla pink, red, or purplish:
  - Capsule 3–4 mm. broad, circumscissile at about middle; larger leaf blades often somewhat flattened .......... 5. *P. pilosa*
  - Capsule 1.5–2.5 mm. broad, circumscissile slightly below middle; leaf blades all terete .......... 6. *P. parvula*

**1. Portulaca oleracea** L., Sp. Pl. 445, 1753

(For synonymy see N. Amer. Fl. **21:** 332, 1932.)

A glabrous, prostrate to ascending, fleshy annual with stems to 4 dm. long; axillary hairs few and inconspicuous or lacking; leaf blades obovate to spatulate, 2–15 mm. wide, 0.5–3 (rarely –6) cm. long, rounded to almost truncate at apex, broadly cuneate at base; flowers sessile, solitary or in few-flowered clusters, surrounded by a few inconspicuous hairs or these lacking; sepals ovate to orbicular, 2.5–4.5 mm. long, nearly to quite as broad, acute, lightly keeled; petals yellow or coppery, 3–7 mm. long; stamens 6–10; capsule 5–8 mm. high, 2.5–5 mm. in diameter, slightly 2-winged toward apex, circumscissile; seeds black, dull or sublustrous but not metallic-iridescent, granular-tuberculate, 0.7–0.8 (rarely –1.0) mm. wide.

A common weed in the temperate and tropical regions of the world; known in the Sonoran Desert from fields in Arizona, Sonora, and southern half of Baja California. Feb.–Sept.

**2. Portulaca retusa** Engelm., Bost. Jour. Nat. Hist. **6:** 154, 1850

A glabrous, stoutish, fleshy annual similar to *P. oleracea* but with leaf blades 4–25 mm. long, cuneate, emarginate to retuse at the apex; sepals carinate-winged; petals 1.5–3 mm. long, yellow; stamens 7–19; style lobes 3–4; capsule 5–6 mm. high; seeds 0.9–1 mm. wide, echinate-tuberculate.

Known to the writer only through descriptions, and, though reported from Arizona, it may not occur in the Sonoran Desert.

**3. Portulaca lanceolata** Engelm., Bost. Jour. Nat. Hist. **6:** 154, 1850

(For synonymy see N. Amer. Fl. **21:** 330, 1932.)

A glabrous, erect or ascending annual with striate-angled stems; leaves few, mostly borne near tips of branches, blades flat, obovate, oblong or spatulate, 2–12 mm. wide, 1–2.5 cm. long, rounded to acute at apex, gradually cuneate at base; hairs in axils few or none; flowers in small clusters at end of branches; sepals ovate, 1.5–2.5 mm. long, obscurely carinate; petals obovate to spatulate, yellow or orange tinged with red, 3–4 mm. broad; capsule circumscissile slightly above the middle, lid shallowly saucer-shaped, rim crowned by a thin, narrow wing 1–2 mm. wide; seeds dull gray, tuberculate, 0.7–0.9 mm. wide.

Moist, shady places, Lower Sonoran Zone to Tropical Zone, south-central Arizona to Baja California, Sonora, and western Texas; Georgia; Cuba and Jamaica. Occurring along fringe of the Sonoran Desert. July–Sept., in our area.

**4. Portulaca suffrutescens** Engelm., Bot. Gaz. **6:** 236, 1881

(For synonymy see N. Amer. Fl. **21:** 332, 1932.)

Erect or ascending fleshy herb 0.5–3 dm. high from creeping tuberous rootstocks and often somewhat woody at base; hairs in axils of leaves conspicuous or few; leaf blades linear, terete, 1–3 cm. long, 1–2 mm. broad, acute at apex; flowers in few-flowered terminal clusters, surrounded by 6–8 leaves and pale brown hairs 4–6 mm. long; sepals broadly ovate to suborbicular, 5–8 mm. broad, 6–7 mm. long; petals copper- or buff-colored, 7–12 mm. long, emarginate to obcordate; style lobes 5–6; capsule subglobose, 3.5–5 mm. in diameter, 5–6 mm. long, circumscissile slightly below middle; seeds black with a greenish metallic sheen, 0.5–0.8 mm. broad, round-tuberculate.

South-central Arizona to Arkansas, Texas, and northern Mexico. Mostly July–Sept. in our area but flowering in April in the mountains of Arizona.

**5. Portulaca pilosa** L., Sp. Pl. 445, 1753

*Portulaca karwinskii* Fisch. & Mey., in C. A. Meyer, Ind. Sem. Hort. Petrop. **11:** suppl. 60, 1846.
*Portulaca gagatosperma* Millsp., Field Mus. Bot. Ser. **2:** 299, 1909.

Ascending or rarely prostrate fleshy annual with terete stems, 1–3.5 dm. long, and often with conspicuous tufts of tawny hairs in axils of leaves, rootstocks often tuberous-thickened; leaf blades oblong-lanceolate, often slightly flattened, 2–4 mm. broad, 5–18 mm. long, acute; flowers terminal, solitary or in 1–3-flowered clusters, surrounded with white to brownish hairs 3–6 mm. long and 6–10 terete linear leaves; sepals triangular to ovate, 2–3 mm. long; petals purple-pink, obovate, 2.5–4.5 mm. broad, 3–6 mm. long, sometimes obcordate or emarginate; stamens 15–32; style lobes 4–6; capsule 3–4 mm. broad, subglobose, circumscissile about middle, lid turbinate; seeds black, metallic-iridescent, 0.4–0.6 mm. wide, minutely rounded-tuberculate.

Widespread in the tropics and subtropics about the Gulf of Mexico and in northern South America; vicinity of Magdalena, Sonora; southern Baja California. Mostly in sandy or gravelly soil. March–Nov.

**6. Portulaca parvula** A. Gray, Proc. Amer. Acad. **22:** 274, 1887
(For synonymy see N. Amer. Fl. **21:** 334, 1932.)

Annual, somewhat fleshy herb with erect or ascending branches 0.6–2.5 dm. long; leaf blades linear, 0.5–2 mm. in diameter, 4–15 mm. long, terete, obtuse at apex; flowers on terminal and axillary branches, usually in 1–5-flowered clusters surrounded by copious tufts of tawny or white silky hairs and several involucral leaves; sepals triangular to suborbicular, 2–3 mm. broad, 2.5–3 mm. long; petals red, elliptic-obovate or obovate, 1–2 mm. broad, 3–4 mm. long, rounded at apex; stamens 10–22; style lobes 4–5; capsule subglobose, 1.5–2.5 mm. wide, mostly straw-colored, circumscissile slightly below middle; seeds black, metallic-iridescent, 0.4–0.6 mm. wide, with low rounded tubercles.

Sandy soil and grassy slopes, Lower and Upper Sonoran Zones, Arizona to Colorado, Missouri, and Texas south in northern Mexico. July–Sept.

## Family 42. CARYOPHYLLACEAE. Pink Family

Annual or perennial herbs with opposite, entire, often stipulate leaves. Flowers perfect, regular, sometimes apetalous; petals sometimes bearing crests or appendages near apex of claw. Stamens 5 or 10, rarely less. Styles and stigmas 2–5. Ovary superior. Fruit a capsule, mostly many-seeded, dehiscent by valvate teeth at apex; or a 1-seeded utricle or achene.

Fruit a few–many-seeded capsule; petals usually present (sometimes wanting in *Stellaria, Cerastium,* and *Loeflingia*):
- Styles 2–5, usually separate to base; capsule 3–10-valved:
  - Sepals distinct to base or nearly so; petals not appendaged or distinctly crowned:
    - Stipules none; petals white or tinged with purple or rarely wanting:
      - Petals deeply bifid or rarely lacking ............ 1. *Stellaria*
      - Petals entire or shallowly emarginate:
        - Capsule cylindric, splitting into twice as many teeth as styles; styles usually 5 ............ 2. *Cerastium*
        - Capsule broadly ovoid or oblong, splitting into as many teeth as styles; styles usually 3 ............ 3. *Arenaria*
    - Stipules present, conspicuous, scarious; petals pinkish ............ 4. *Spergularia*
  - Sepals united nearly to apex, the calyx synsepalous; petals often appendaged or crowned and usually distinctly clawed:
    - Styles usually 3; calyx with 10 or more conspicuous ribs or nerves; petals with crown-like appendages ............ 5. *Silene*
    - Styles usually 2; calyx inconspicuously 5-nerved, terete or 5-angled; petals (in ours) without appendages ............ 6. *Saponaria*

Style 1, 3-toothed or 3-cleft; capsule 3-valved:
Sepals not recurved or rigid; petals often small but present; ovary usually short-stipitate ........ 7. *Drymaria*
Sepals recurved, rigid and subspinescent in age; ovary sessile ........ 8. *Loeflingia*
Fruit a 1-seeded achene or utricle; petals minute or wanting:
Plants annual; leaves elliptic to spatulate, not subulate:
Sepals united at base into a short tube; stamens 10–15; leaves spatulate; stipules conspicuous ........ 9. *Achyronychia*
Sepals distinct to base; stamens 2–5; leaves elliptic-oblong; stipules minute ........ 10. *Herniaria*
Plants perennial; leaves subulate ........ 11. *Cardionema*

## 1. STELLARIA L., Sp. Pl. 421, 1753

Small, herbaceous, annual or perennial plants with branching or tufted stems and cymose white flowers. Sepals 5. Petals 5 or sometimes lacking, 2-cleft or -parted. Stamens hypogynous, 10 or fewer. Styles usually 3, rarely 4–5. Capsule globose to oblong, dehiscent by twice as many valves as styles. Seeds small, few to several, smooth or roughened.

Stems pubescent with longitudinal lines of hairs, sparsely glandular; leaf blades ovate ........ 1. *S. media*
Stems glabrous except at the base, not glandular; leaf blades lanceolate ........ 2. *S. nitens*

### 1. Stellaria media (L.) Cyrillo, Char. Comm. 36, 1784

*Alsine media* L., Sp. Pl. 272, 1753.

Annual, with weak, decumbent to ascending stems 1–4 dm. long, glabrous except for longitudinal line of pubescence running from axil of each leaf upward to next higher node; leaves broadly ovate to elliptic, blades 3–15 mm. wide, 1–3 cm. long, acute at apex, rounded or broadly cuneate below, upper leaves sessile, lower on petioles often equaling blades, margins of petioles sparsely pubescent; flowers in axillary or terminal leafy cymes, on slender pedicels 5–15 mm. long; sepals lance-oblong, 4–6 mm. long, acute, sparsely pubescent and glandular on outside; petals 2-parted, about half as long as sepals; capsule oblong-ovoid, acute, 5–7 mm. long, shining; seeds orbicular-reniform, about 1 mm. in diameter, dark brown, roughly papillate in concentric lines.

Rare in the desert but in lawns and gardens in Tempe and Phoenix, Arizona; frequently naturalized from southern California to Alaska and eastward to the Atlantic seaboard. Feb.–April.

### 2. Stellaria nitens Nutt., in Torr. & Gray, Fl. N. Amer. 1: 185, 1838

*Alsine nitens* Greene, Man. Bay Reg. 33, 1894.

Annual, with filiform, glabrous (except at base), erect and forking stems 0.5–2.5 dm. high; leaves chiefly basal, lowermost on slender petioles about equaling the elliptic to orbicular, 3–4 mm. long blades, cauline leaves ovate to linear-lanceolate, acute, 6–10 mm. long, sessile; flowers few, mostly solitary in axils on filiform pedicels 5–15 mm. long; sepals lanceolate-attenuate, 3–4 mm. long, setose-tipped, scarious-margined; petals about half as long as sepals or wanting; capsule oblong, acutish, nearly equaling sepals; seeds numerous, reddish brown, ovoid-lenticular, 0.4–0.6 mm. wide, distinctly roughened on dorsal ridge, sides minutely so, dullish.

Grassy situations, mostly Upper Sonoran to Transition Zones, northern Baja California to British Columbia, Arizona, and Utah. Feb.–April in our area.

## 2. CERASTIUM L., Sp. Pl. 437, 1753

Annual or perennial, pubescent or hirsute, often tufted herbs with terminal dichotomous cymes. Leaves opposite, estipulate. Sepals 5 (rarely 4). Petals 5, 2-lobed, 2-cleft, or rarely wanting. Stamens 10 or 5. Styles usually 5, opposite the sepals. Capsule slender, cylindrical, often curved, dehiscent by twice as many valves as styles. Seeds numerous, variously roughened.

### 1. Cerastium texanum Britt., Bull. Torrey Club 15: 97, 1888

Annual, with pubescent and sparsely glandular stems 1–3 dm. tall; leaves oblanceolate to obovate, 1.5–10 mm. wide, 1–5 cm. long, acute to apiculate, glandular-hirsute, the lower narrowing to narrowly winged petioles which constitute to half the length of leaf, the upper sessile or nearly so; pedicels slender, 5–25 mm. long; sepals lanceolate-elliptic, 4–5 mm. long, sparsely glandular-hirsutulous, acute at apex; petals shallowly cleft, 7–8 mm. long; capsule straight, 5–7 mm. long, slender valves strongly recurved at apex; seeds reddish or brown, 0.5–0.8 mm. long, roughened with minute slender papillae.

Sandy soil, mostly in grassy places in Lower and Upper Sonoran Zones, Pima and Maricopa Counties, Arizona, to western Texas. Feb.–April.

## 3. ARENARIA L., Sp. Pl. 423, 1753

Annual or perennial, often tufted or matted herbs with estipulate, subulate to ovate, sessile leaves and cymose, capitate or solitary, axillary flowers. Sepals 5. Petals 5 or wanting, entire or shallowly emarginate. Stamens 10. Styles 3 (rarely 2 or 5). Ovary 1-celled, several-ovuled. Capsule globose to oblong, dehiscent by as many or twice as many valves as styles. Seeds few to many, reniform or globose, roughened or smooth.

Plants annual; seeds smooth; leaves filiform, 5–12 mm. long .............. 1. *A. douglasii*
Plants perennial; seeds tuberculate-roughened; leaves subulate, 1.5–5 cm. long ........... ........................................... 2. *A. macradenia* var. *parishiorum*

### 1. Arenaria douglasii Fenzl, in Torr. & Gray, Fl. N. Amer. 1: 674, 1840

*Alsinopsis douglasii* Heller, Muhlenbergia 8: 20, 1912.
*Minutaria douglasii* Mattf., Bot. Jahrb. 57: Beibl. 126: 27, 1921.

Delicate annuals with slender stems 5–25 cm. tall, sparingly capitate-glandular or glabrate; leaves filiform-subulate, 5–12 mm. long, bases slightly expanded, scarious at margins and connate; flowers loosely cymose; pedicels filiform, 5–45 mm. long; sepals ovate, 2–3 mm. long, margins scarious, midrib often reddish; petals obovate, 3–5 mm. long; capsule subglobose, 4–5 mm. high; seeds reddish brown, about 1 mm. broad, smooth.

Sandy places, mainly Upper Sonoran Zone, Oregon to northern Baja California, reaching the margin of the Colorado Desert at Palm Springs, Riverside County, and Banner, San Diego County, California. March–June.

### 2. Arenaria macradenia S. Wats. var. parishiorum Robinson, Proc. Amer. Acad. 29: 296, 1894

Perennial, from considerably branched, decumbent to prostrate, woody caudex which gives rise to erect, slender, flowering branches; leaves mostly on basal third of branches, linear-subulate, 1.5–5 cm. long, ascending, scarious and connate at bases, minutely ciliolate on margins near base, grooved on upper side, subrigid, and tip armed with a short prickle; pedicels slender, 1–3.5 cm. long; sepals ovate, 5–6 mm. long, 3–5-nerved, apiculate, margins scarious; petals 6–7 mm. long; capsule broadly oblong, 6–7 mm. long, tips of valves slightly spreading; seeds reddish brown, about 1.2 mm. long, finely papillate-roughened.

Desert slopes of San Bernardino and Providence Mountains, mainly Upper Sonoran Zone but entering the Colorado Desert in the mountains between it and the Mojave Desert. April–May.

## 4. SPERGULARIA Pers., Syn. Pl. 1: 504, 1805

Low, annual or perennial herbs, mostly of saline or alkaline situations, with linear, terete or semiterete, scariously stipulate leaves. Flowers in terminal, racemose, leafy or bracteate cymes. Sepals 5. Petals 5 or sometimes absent, white or purplish, entire. Stamens 2–10, mostly 10. Ovary 1-celled; styles 3. Capsule 3-valved to the base. Seeds numerous, reniform-globose, or more or less flattened, often finely sculptured and frequently winged.

Plants perennial; roots fusiform, fleshy; sepals 6–10 mm. long ........... 1. *S. macrotheca*
Plants annual; roots fibrous, not fusiform; sepals 2.5–5 mm. long ........... 2. *S. marina*

### 1. Spergularia macrotheca (Hornem.) Heynh., Nom. 2: 689, 1846

*Arenaria macrotheca* Hornem. ex Cham. & Sch., Linnaea 1: 53, 1826.
*Lepigonum macrothecum* Fisch. & Mey., Ind. Sem. Hort. Petrop. 3: 14, 1837.
*Tissa pallida* Greene, in Britt., Bull. Torrey Club 16: 129, 1889.
*Tissa macrotheca* Britt., loc. cit.
*Tissa talinum* Greene, Erythea 1: 106, 1893.
*Tissa valida* Greene, op. cit. 107.
*Alsine valida* House, Amer. Midl. Nat. 7: 134, 1921.
*Alsine greenei* House, loc. cit.

Perennial, from branched caudex that becomes woody in age; stems prostrate to ascending, 0.3–4.5 dm. long, glandular-pubescent or glabrous below inflorescence; leaves opposite or fasciculate, linear, 0.6–3 mm. wide, 6–50 mm. long, mucronate, glabrous to glandular-pubescent, with conspicuous, triangular-acuminate, scarious stipules 4–11 mm. long; inflorescence glandular-pubescent; sepals broadly lanceolate, 5–10 mm. long, margins scarious; petals pink, rosy, or white, 4–6 mm. long; stamens 10; capsules mostly 6–8 mm. long; seeds reddish brown, dull, smooth or papillose or rounded-tuberculate, usually with a narrow, scarious, white to brownish, entire wing 0.2 mm. wide, or rarely wingless.

Saline or alkaline soil, Lower Sonoran to Transition Zones, mostly near the coast, British Columbia to Guadalupe Island, Baja California. Jan.–June.

### 2. Spergularia marina (L.) Griseb., Spicil. Fl. Rumel. 1: 213, 1843

*Arenaria rubra* β *marina* L., Sp. Pl. 423, 1753.
*Spergularia salina* J. & C. Presl, Fl. Cech. 95, 1819.
*Lepigonum marinum* Wahlb., Fl. Gothob. 47, 1820.
*Spergula marina* Bartl. & Wendl. f., Beitr. 2: 64, 1825.
(For complete synonymy see Rhodora 42: 123–133, 1940.)

Annual, with 1–several erect or ascending stems 0.5–3.5 dm. long and glandular-pubescent or glabrous herbage; leaves linear, 0.6–1.5 mm. wide, 5–40 mm. long, short-mucronate; stipules deltoid, 2–4 mm. long, nearly as wide, acute to short-acuminate; inflorescence cymose, short to long; sepals ovate, acutish or blunt at tip, 2.5–5 mm. long, glandular-pubescent, margins scarious; petals ovate, pink, white, or rosy, 2–4 mm. long; stamens 2–5; capsules 3.6–6.4 mm. long, acutish or obtuse at apex; seeds light or red-brown, dull, smooth or slightly papillose, 0.6–0.8 mm. long, usually wingless, sometimes with narrow erose wing.

Saline and alkaline soils, Lower and Upper Sonoran Zones, widely distributed in both North and South America. Mojave and Colorado Deserts, southwestern Arizona, southward in Baja California at least to Guadalupe Island. March–June.

## 5. SILENE L., Sp. Pl. 416, 1753

Annual or perennial herbs with opposite, estipulate leaves, more or less glandular-viscid herbage, and white, pink, or red flowers in cymes, panicles, or racemes. Calyx cylindric or campanulate, 5-toothed, 10–many-nerved. Petals 5, claws usually long, blades often toothed or cleft and each bearing a scalelike appendage at base. Stamens 10. Styles 3 or sometimes 4 or 5. Ovary 1-celled or incompletely 2- or 4-celled. Capsule dehiscent by 6 (or rarely 3) apical teeth. Seeds small, numerous, usually papillate or finely tuberculate.

Herbage glabrous except for viscid bands on upper internodes; petals barely exceeding calyx; inflorescence paniculate .......................................... 1. *S. antirrhina*

Herbage glandular-pubescent, without viscid bands on internodes; petals exceeding calyx 4–6 mm.; inflorescence racemose .......................................... 2. *S. gallica*

### 1. Silene antirrhina L., Sp. Pl. 419, 1753

*Silene dioica* Cham. & Sch., Linnaea **1:** 38, 1826.

Annual; glabrous except for a brownish viscid band near middle of some or all the upper internodes; stems erect, sparingly branched or simple; leaves linear to oblong-lanceolate, mostly acute at apex (basal leaves broadly oblanceolate-spatulate), often ciliolate along margins near base, 2–5 cm. long, 1.5–3 (rarely 7) mm. wide; pedicels slender, 1–3.5 cm. long; calyx subcylindrical and 6–7 mm. long at anthesis, 10-nerved, broadly ovoid and 7–9 mm. long in fruit, teeth 1–1.5 mm. long; petals exceeding calyx teeth about 1 mm., white, bright pink, or tipped with deep rose, emarginate; capsule 6–8 mm. long, the 6 short teeth somewhat recurved; seeds reniform, laterally flattened, 0.6 mm. long, broadly and shallowly grooved dorsally, papillate in regular rows, light reddish brown.

Open sandy places and hillside burns, Transition and Sonoran Zones, widely distributed across the continent; in Mohave, Pima, and Yuma Counties, Arizona, and south into Sonora. March–May.

### 2. Silene gallica L., Sp. Pl. 417, 1753

*Silene anglica* L., op. cit. 416.

Annual, hirsute and slightly glandular throughout; stems simple or branching, striate, erect; leaves spatulate, 1–4.5 cm. long, 12 mm. or less broad, obtuse and short-mucronate; inflorescence a recurved, terminal raceme; pedicels 3 mm. long or less; calyx cylindric, 7–10 mm. long at anthesis, becoming ovoid and 10–14 mm. long in fruit, teeth subulate, 1–2 mm. long, often spreading or recurved in fruit; petals white or pinkish, blades exserted about 4–6 mm. and twisted through about 90° at anthesis, appendages erect, 1–1.5 mm. high or less, entire or faintly erosulate; capsule broadly ovoid, 7–9 mm. long, teeth slightly spreading at dehiscence; seeds dark reddish brown, laterally flattened, reniform, about 0.8 mm. in diameter, papillate.

Native of Europe but well established as a wayside weed throughout the Pacific Coast area, south into Baja California to the vicinity of Rosario and on Guadalupe Island; in Maricopa and Pima Counties, Arizona. March–May.

**Silene laciniata** Cav., Ic. **6:** 44, *pl. 564,* 1801, occurs in grassy openings in chaparral belt in southern California and northern Baja California, south to the vicinity of San Antonio del Mar Canyon, but has not been collected in the desert proper. The variety *S. laciniata* var. *greggii* (A. Gray) S. Wats., occurs in the mountains of Arizona, but only above the desert.

## 6. SAPONARIA L., Sp. Pl. 408, 1753

Annual or perennial herbs. Leaves opposite, subsessile. Flowers large, in corymbose clusters. Calyx narrowly ovoid to cylindric, 5-toothed, faintly 5-nerved. Petals 5, pink or red, clawed; crested or naked. Stamens 10. Styles 2. Capsule 4-toothed at the apex, 1-celled except at the base where rudimentary partitions occur.

### 1. Saponaria vaccaria L., Sp. Pl. 409, 1753

*Vaccaria vulgaris* Host, Fl. Aust. **1:** 518, 1827.

Glabrous, dichotomously branched annual 3–6 dm. high; leaves lanceolate, 1–3 cm. wide, 3–10 cm. long, sessile, entire, 3–5-veined; inflorescence 1–3 dm. broad; pedicels 1.5–5 cm. long; calyx mostly ovoid, at least soon after flowering, 10–12 mm. long, dark green at angles, lighter between, calyx teeth triangular, 1–1.5 mm. long, reddish or purplish; petals reddish, 1.5–2 cm. long, not crowned, narrow; styles slightly exserted; capsule broadly ovoid, 7–9 mm. in diameter, dilating the calyx tube, opening by 4 valvelike teeth at apex; seeds subglobose, about 2 mm. in diameter, reddish brown, finely granulose-tessellate.

A roadside and grain field weed in most of temperate North America and Europe; reported from Tucson by Kearney and Peebles; rare in the desert. April–May.

## 7. DRYMARIA Willd., in Roem. & Sch., Syst. Veg. 5: 406, 1819

Annual or perennial, prostrate to erect, slightly to profusely branching herbs with linear to cordate leaves and glabrous to densely glandular-pubescent herbage. Stipules usually present, persistent or fugacious. Flowers small, in cymes or racemes; sepals 5, 1–3-nerved, margins scarious-winged; petals (in ours) deeply cleft, often with additional appendages in sinuses; stamens 5, filaments slightly dilated and connate at base; anthers ovate; style slender, shorter than to equaling ovary; stigma 3-lobed or 3-cleft. Capsule ovoid to subglobose, longitudinally dehiscent, mostly 3-valved, 1-celled. Seeds obovoid to obovoid-reniform, variously sculptured.

Leaves about as broad as or broader than long:
- Flowers in short, crowded, umbel-like racemes; plants mostly prostrate or decumbent; leaves leathery or fleshy, often glaucous:
  - Herbage somewhat puberulent, scarcely glaucous; petals 2.8–3.5 mm. long, about equaling sepals; internodes much surpassing leaves; plant not matlike; style 1.4–1.6 mm. long .................................................. 1. *D. holosteoides*
  - Herbage glabrous, conspicuously glaucous; petals 2–2.5 mm. long, shorter than sepals; internodes mostly shorter than leaves; plants matlike; style 0.6–0.8 mm. long .... .................................................. 2. *D. crassifolia*

Flowers in cymes; plants mainly erect or ascending; leaves thin, not fleshy or leathery, rarely glaucous:
  Petals surpassing sepals; sepals not strongly ribbed:
    Sepals oblong-elliptic or narrowly oblanceolate, obtuse at apex; petals 5–6.5 mm. long; pubescence made up of hairs 0.4 mm. long or less .. 3. *D. polystachya* var. *diffusa*
    Sepals broadly ovate, acute at apex; petals 3.8–4.5 mm. long; pubescence of crinkled, spreading hairs to 1 mm. long .................................. 4. *D. debilis*
  Petals shorter than sepals, these strongly 3-ribbed ...................... 5. *D. fendleri*
Leaves linear or narrowly oblong, several times as long as broad:
  Herbage heavily viscid-puberulent throughout; seeds smooth, not tessellate or muriculate, light-colored; petals merely 2-cleft, not appendaged .................. 6. *D. viscosa*
  Herbage viscid-puberulent only on pedicels and sepals; seeds tessellate, muriculate, dark-colored; petals bearing 4 appendages near base of sinus ........ 7. *D. arenarioides*

**1. Drymaria holosteoides** Benth., Bot. Sulph. 16, 1844

*Drymaria veatchii* Curran, Proc. Calif. Acad. II, **1:** 227, 1888.
*Mollugophytum holosteoides* M. E. Jones, Contr. West. Bot. **18:** 35, 1933.

Annual, with slender, spreading, decumbent or prostrate stems 4–40 cm. long; internodes 2–10 cm. long, puberulent and sparsely glandular or glabrate on older stems; stipules represented by narrow scarious wing between bases of leaves; leaves mostly in fascicles of 4–6, blades elliptic-spatulate, 3–7 mm. wide, 4–12 mm. long, fleshy or leathery, faintly 3-nerved, glabrous or sparsely puberulent near base, lightly glaucous; petioles 2–10 mm. long; flowers in dense, short racemes in axils of most leaves, each subtended by a narrow, ovate, hyaline bract about 1 mm. long; pedicels 3–9 mm. long; sepals obovate, obtuse or rounded at apex, 3–3.5 mm. long, middle portion herbaceous, glandular-puberulent at base, margins conspicuously scarious-winged; petals persistent, about equaling sepals, 2.8–3.5 mm. long, white, bifid to middle, bearing 2–4 narrow segments near base of inner margin of each main lobe, claw short, broad, erosulate; stamens about equaling petals, nearly twice as long as ovary; ovary about 1.5 mm. high; style equaling ovary; capsule 1.5–2 mm. high; seeds obovoid-reniform, 0.6–0.8 mm. long, dark brown to black, dull, minutely rounded-tuberculate in longitudinal lines.

Sandy areas, Lower Sonoran Zone, from the vicinity of Calmallí to La Paz; Cedros, Magdalena, and Tiburón Island. Flowering throughout the year when rainfall is favorable.

**2. Drymaria crassifolia** Benth., Bot. Sulph. 16, 1844

*Mollugophytum crassifolium* M. E. Jones, Contr. West. Bot. **18:** 35, 1933.
*Drymaria holosteoides* var. *crassifolia* J. Duke, Ann. Mo. Bot. Gard. **48:** 187, 1961.

Prostrate, matted, glabrous and glaucous perennial; stems profusely branched, 3–20 cm. long, internodes 0.5–2.5 cm. long, mostly obscured by numerous fasciculate leaves; stipules as in *D. holosteoides*; petioles 3–9 mm. long; leaf blades broadly ovate, broadly rhombic-spatulate or elliptic, mostly rounded at apex, broadly cuneate at base, 2.5–8 mm. broad, 4–12 mm. long, fleshy, glaucous, indistinctly 3-nerved beneath; flowers in short racemes; pedicels 5–12 mm. long; sepals slightly dissimilar, broadly elliptic, rounded, acutish, or faintly emarginate, 1.4–1.8 mm. wide, 3–3.4 mm. long, fleshy, glaucous, faintly veined; petals 2–2.5 mm. long, bifid to middle, usually with 2 short appendages at base of inner margin of each main lobe; stamens 1.8–2 mm. long, about equaling ovary; style 0.4–0.6 mm. long, one-fourth as long as ovary; stigma lobes minute; capsule ovoid, equaling sepals; seeds about 0.8 mm. long, obovoid-reniform, minutely reticulate in longitudinal lines, black, dull.

Sandy soil and along washes, Lower Sonoran and Tropical Zones, Cape Region, Baja California. Flowering following rains.

**3. Drymaria polystachya** Brandegee var. **diffusa** (Rose) Wiggins, Proc. Calif. Acad. IV, **25:** 198, *pl. 21, figs. 6–8*, 1944

*Drymaria diffusa* Rose, Contr. U.S. Nat. Herb. **1:** 130, *pl. 12*, 1892.

Annual, with weak spreading-ascending stems 1–3 dm. long, glandular-puberulent throughout; stipules linear-subulate, 1–2 mm. long; petioles very slender, 4–10 mm. long; leaf blades 3–10 mm. wide, 3–8 mm. long, truncate or broadly cuneate at base; flowers in subterminal and terminal cymules; peduncles slender, 1–8 cm. long, a pair of narrowly ovate, scarious bracts 1.6–2.2 mm. long at each node; pedicels subfiliform, 0.6–2 cm. long; sepals narrowly oblanceolate, about 3.5 mm. long; petals bifid about to middle, 4.5–6.5 mm. long, devoid of appendages, claws narrow, about 2 mm. long, naked; stamens about 3 mm. long; ovary nearly equaling stamens; capsule 3.5–4.5 mm. long; seeds obovoid-reniform, slightly compressed laterally, brown, finely tessellate in numerous longitudinal rows.

Sandy areas, Lower Sonoran Zone, Carmen Island and the mainland of Baja California above Punta Escondido in the Sierra Giganta. Flowering after rains.

**4. Drymaria debilis** Brandegee, Proc. Calif. Acad. II, **2:** 131, 1889

Annual; stems subfiliform, weak, spreading-ascending, 1–2.5 dm. long, sparsely pilosulous; internodes 3–4.5 cm. long; stipules linear, tortuous, 1–1.5 mm. long; petioles 2–10 mm. long, pilosulous; leaf blades broadly ovate, 3–10 mm. long and wide, acute or short-acuminate, sparsely pilosulous or glabrate on upper surface; inflorescence loosely cymose; pedicels 4–14 mm. long, filiform; sepals 2.6–3.2 mm. long, pilosulous, ovate-acuminate, wings obtusish, faintly 3-nerved; petals 3.8–4.5 mm. long, bifid to middle, claws narrow and naked; stamens about equaling sepals; capsule ovoid, about equaling sepals; seeds reddish brown, obovoid-reniform, 0.6–0.8 mm. long, minutely tessellate in fine longitudinal lines.

Arroyos and in shade of cliffs, Lower Sonoran Zone, southern Baja California. Nov.–May, or following rains.

**5. Drymaria fendleri** S. Wats., Proc. Amer. Acad. **17:** 328, 1882

*Drymaria fendleri* var. *perennis* M. E. Jones, Contr. West. Bot. **18:** 65, 1933.
*Drymaria glandulosa* var. *fendleri* Fosberg, in Morton, Contr. U.S. Nat. Herb. **29:** 96, 1945.
*Drymaria glandulosa* var. *perennis* Fosberg, loc. cit.

Erect or ascending annual (rarely biennial under favorable conditions) with glandular-puberulent herbage; stems 1–3.5 dm. long; internodes to 12 cm. long; stipules bifid into linear-subulate lobes 1.5–2 mm. long, persistent; leaves thin, broadly ovate or ovate-cordate, 4–15 mm. long, 6–22 mm. wide, acute or rounded but abruptly apiculate at apex, truncate or shallowly cordate at base; petioles 2–7 mm. long; inflorescence cymose; peduncles 1–6 cm. long, glandular-puberulent, bearing a pair of ovate scarious bracts 2–2.5 mm. long at node; pedicels 1–2 mm. long or some flowers sessile; sepals lanceolate, 1–1.4 mm. wide, 4–5 mm. long, strongly 3-ribbed, acute to acuminate, central part green, glandular-puberulent, marginal wings scarious; petals 3.5–4 mm. long, bifid to middle, lobes linear-spatulate, claws narrow and naked; stamens half as long as petals; capsule ovoid, 2–2.5 mm. high, 8–15-seeded; seeds obovoid-reniform, 0.6–0.8 mm. long, reddish brown, minutely tessellate in longitudinal rows.

Foothills and mountains, Lower Sonoran Zone to lower parts of Transition Zone, New Mexico and Arizona to central Mexico, and in the Cape Region of Baja California. Mostly above the limits of the Sonoran Desert. Sept.–April.

**6. Drymaria viscosa** S. Wats., Proc. Amer. Acad. **22:** 469, 1887

*Drymaria viscosa* Orcutt, W. Amer. Sci. **2:** 57, 1886. Nomen nudum.

Annual, with prostrate or spreading branches, dichotomously branched stems 5–20 cm. long and glandular-puberulent leaves, pedicels, and sepals; internodes to 6 cm. long, sparsely glandular-pubescent or glabrate; stipules 1–1.5 mm. long, glandular; leaves often pseudo-verticillate, linear to linear-oblanceolate, 1.5 mm. wide or less, to 15 mm. long, acute or obtuse at apex, gradually narrowing to a slender petiole 1–3 mm. long; inflorescence cymose; peduncles to 3.5 cm. long; pedicels 1–3 mm. long or some flowers subsessile; sepals narrowly to broadly elliptic, 0.6–1.2 mm. wide, about 2.5 mm. long, midrib ending in an excurrent apiculation slightly below apex of scarious, glabrous wings; petals bifid about two-thirds of way to base, 2 mm. long, without appendages, claw naked; stamens about 1 mm. long, equaling ovary; style half as long as ovary; capsule nearly equaling sepals; seeds obovoid-circinate, 0.6 mm. high, pale cream, with light amber angles and dorsal band, smooth or slightly wrinkled but not tessellate or muriculate, dull.

In sandy soil of the desert, mainly Lower Sonoran Zone, from the vicinity of San Quintín south to the Cape Region, Baja California; Magdalena Island. Nov.–June, or following summer rains.

**7. Drymaria arenarioides** Willd., in Roem. & Sch., Syst. Veg. **5:** 406, 1819

*Drymaria frankenioides* H. B. K., Nov. Gen. & Sp. **6:** 21, *pl. 515,* 1823.

Prostrate, glandular-puberulent, canescent, short-lived perennial with profusely branched stems 3–20 cm. long; stipules subulate, 1–1.5 mm. long, persistent; leaves more or less fasciculate, linear-lanceolate to narrowly elliptic, 1–2.5 mm. wide, 5–15 mm. long, granular-puberulent and slightly canescent, acute, tapering gradually to short slender petioles 1–2 mm. long; flowers axillary and in terminal, 1–3 flowered cymules; pedicels 3–10 mm. long, often deflexed in fruit; sepals dissimilar, narrowly lanceolate to ovate-lanceolate, 1–2.2 mm. wide, 5–6 mm. long, acute to short-acuminate, weakly 1-nerved, the margins scarious; petals 5–6.5 mm. long, divided to middle into about 6 lobes, the inner four in 2 pairs slightly shorter than main lobes; stamens about 1.5 mm. long, equaling ovary; style half as long as ovary; capsule ovoid, 4–5 mm. long; seeds orbicular-circinate, slightly compressed laterally with a pronounced, broad, dorsal groove and deep lateral depressions, uniformly and minutely granular-tessellate, dull, brown.

Sandy and gravelly soil on hillsides and flats, Lower and Upper Sonoran Zones, northern Sonora and Chihuahua to central Mexico. Aug.–Nov. and March–April.

## 8. LOEFLINGIA L., Sp. Pl. 35, 1753

Low, diffusely branched, spreading annuals with glandular-pubescent herbage, inconspicuous subulate leaves, and setaceous stipules. Flowers small, sessile, axillary, solitary or in fascicles, greenish. Sepals 5, narrow, attenuate to a setaceous tip, straight or recurved, entire or the outer three with a small slender tooth on each margin. Petals

3–5 and minute, or wanting. Stamens 3–5. Style short or none; stigmas 3; ovary trigonous. Capsule 3-valved. Seeds oblong, several.

Sepals markedly recurved-spreading; capsule oblong, the tip distinctly exserted; seeds oblong or elliptic, 0.5–0.6 mm. long .................................... 1. *L. squarrosa*
Sepals straight or the outer slightly curved, not recurved-spreading; capsule broadly ovoid, not exserted; seeds broadly obovoid, 0.4 mm. long .................... 2. *L. texana*

**1. Loeflingia squarrosa** Nutt., in Torr. & Gray, Fl. N. Amer. **1:** 174, 1838

Glandular-pubescent throughout; stems prostrate or decumbent, 5–15 cm. long; leaves subulate-setaceous, 4–6 mm. long, mostly recurved, moderately rigid; sepals similar to leaves, 4–5 mm. long, recurved, rigid, and bristle-tipped, moderately carinate, whole calyx distinctly squarrose owing to spreading-recurved tips of sepals; petals minute or wanting; stamens 3–5, 0.4–0.7 mm. long and about equaling ovary at anthesis; capsule oblong, triquetrous, 3–3.5 mm. long, slightly surpassing upright portion of sepals; seeds numerous, elliptic, 0.5–0.6 mm. long, pale yellow or light brown, smooth.

Dry, sandy, or clay soil, Upper and Lower Sonoran Zones, central Baja California to the Sacramento Valley, California, but not definitely known from the Colorado Desert; Pima and Pinal Counties, Arizona, according to Kearney and Peebles, but all specimens from that area seen by the writer are *L. texana.* March–April.

**2. Loeflingia texana** Hook., Ic. Pl. **3:** *pl.* 275, 1840

Glandular-pubescent throughout and similar to the last in general appearance, differing in the following characteristics: outer sepals slightly curved outward and less distinctly carinate; capsule broadly ovoid, 2–2.6 mm. long, 1.8–2.2 mm. wide, considerably broader at base than above, distinctly shorter than erect or slightly curved sepals; seeds broadly obovoid, barely 0.4 mm. long, about two-thirds as broad as long.

Dry soil, Lower and Upper Sonoran Zones, central Texas to Nebraska; southwestern Arizona and northern Sonora. March–April.

## 9. ACHYRONYCHIA Torr. & Gray, in A. Gray, Proc. Amer. Acad. 7: 330, 1868

Low or prostrate, glabrous annuals with opposite spatulate leaves, the members of each pair unequal; stipules white-scarious, distinct. Flowers small, in dense axillary cymose clusters. Sepals 5, scarious, united below. Petals wanting. Stamens 10–15, 1–5 fertile, the others with abortive anthers. Style 2-cleft; ovary 1-celled and 1-seeded. Utricle thin, included in the calyx.

**1. Achyronychia cooperi** A. Gray, Proc. Amer. Acad. **7:** 331, 1868

Stems numerous from slender taproot, prostrate, rarely branching, 2–18 cm. long, floriferous nearly full length; leaves spatulate, larger one of pair 8–18 mm. long, 3–5 mm. wide, the smaller about half as large; stipules ovate, about 2 mm. long, erosulate on margins; cymes numerous, borne in axils of all leaves, subsessile or on jointed peduncles 6–15 mm. long; calyx tube green, about 1 mm. long, lobes broadly ovate, 1.6–2 mm. long, on short green claw which forms short green midrib at base, otherwise shining white-scarious; utricle 0.8–1 mm. long, included in calyx tube, which is slightly constricted at summit in fruit; solitary seed obovoid, 0.8 mm. long, shining except for a narrow, dull, yellow-brown line running down one side.

Sandy soil, Lower Sonoran Zone, Mojave and Colorado Deserts to southwestern Arizona, southern Sonora, and southern Baja California. March–April.

## 10. HERNIARIA L., Sp. Pl. 218, 1753

Annual or perennial, profusely branched, prostrate or spreading herbs. Leaves opposite or rarely alternate or fascicled, entire, small. Stipules scarious, entire or (in ours) ciliate. Flowers minute, in small axillary clusters. Calyx 4–5-parted, segments often unequal. Petals none. Stamens 3–5; staminodia 4–6. Style short, bifid. Utricle membranous, included in persistent calyx. Seed erect.

### 1. Herniaria cinerea DC., Fl. Franc. Suppl. 375, 1815

Prostrate spreading annual forming mats 5–25 cm. wide, hispid throughout; stipules ovate, 0.5–0.8 mm. long, ciliate, hyaline; leaves sessile, oblong to oblanceolate, 3–11 mm. long, acutish at apex; flower clusters in all axils; calyx 1.8–2 mm. long, tube campanulate, about 0.4–0.6 mm. long; calyx lobes ovate, somewhat boat-shaped, unequal, 0.8–1.4 mm. long, margins of inner smaller ones narrowly hyaline-winged; utricle broadly ovoid-lenticular, yellowish, granulose-puberulent near apex, about 0.8 mm. high; seeds suborbicular-lenticular, about 0.6 mm. in diameter, dark red-brown, shining, embryo annular, forming a low dorsal ridge.

Sparingly introduced from Europe in the Great Valley of California, southern California, and southern Arizona. March–April.

## 11. CARDIONEMA DC., Prod. 3: 372, 1828

Low tufted or prostrate perennial herbs with more or less woolly stems, the shorter upper internodes mostly covered with persistent, papery-hyaline, bifid stipules. Leaves linear-subulate, pungent. Flowers sessile, in small axillary clusters. Sepals 5, unequal, hooded, the outer three larger than the inner and tipped with a divergent rigid spine, woolly at base. Stamens 3–5, alternate with 5 minute, scalelike petals. Style short, 2-cleft. Utricle enclosed in persistent calyx.

### 1. Cardionema ramosissima (Weinm.) Nels. & Macbr., Bot. Gaz. 56: 473, 1913

*Loeflingia ramosissima* Weinm., Flora 3: 608, 1820.
*Pentacaena ramosissima* Hook. & Arn., Bot. Misc. 3: 338, 1833.

Stems numerous from the woody rootstock, but mostly simple above the perennial part, 5–20 cm. long, prostrate or tips turning upward, forming dense mats, closely covered with silvery-white, papery stipules and green subulate leaves; internodes sparingly to densely arachnoid-tomentulose; leaves 7–15 mm. long, glabrous, 3-ribbed; stipules 4–6 mm. long, erosulate; sepals 3–4 mm. long, woolly below, divergent spines rigid; utricle elliptic, apiculate, 1.2–1.4 mm. long; seed elliptic, shining, red-brown, 1–1.2 mm. long, a darker, dull, shallow groove on one side.

Sandy areas near the coast, Lower Sonoran Zone to Canadian Zone, western Washington to Baja California. March–April.

# Family 43. CERATOPHYLLACEAE. Hornwort Family

Submerged aquatic herbs with whorled, finely dissected, sessile leaves on slender branching stems. Flowers monoecious, minute, sessile and solitary in leaf axils, subtended by 8–12-cleft, calyxlike involucre. Perianth none. Stamens hypogynous, numerous; anthers tipped with 2–3 apiculations. Pistillate flowers a single carpel, superior, 1-celled, 1-seeded. Style filiform. Fruit an indehiscent nutlet, the slender style persistent. Embryo straight, plumule highly developed; endosperm none.

### 1. CERATOPHYLLUM L., Sp. Pl. 992, 1753

Characters of the family.

**1. Ceratophyllum demersum** L., Sp. Pl. 992, 1753

Stems 1 dm. to 1 m. long, according to depth of water; leaves 1–2.5 cm. long, sessile in verticils of 5–12, dichotomously forked into 2–8, filiform or linear, prickly-dentate segments; internodes 0.5–2.5 cm. long; nutlets ovoid or elliptic, about 5 mm. long, smooth or tuberculate, with or without a basal spur on each side, winged or wingless.

In ponds and slow streams, from Lower Sonoran Zone to Boreal Zones on all continents; in streams between Santiago and San Ignacio; Mulegé, Baja California. May–Nov.

## Family 44. RANUNCULACEAE. Crowfoot Family

Annual or perennial herbs or, less commonly, shrubs or woody climbers. Leaves alternate or rarely opposite, simple or compound, estipulate. Flowers regular or irregular, (in ours) hypogynous. Sepals 3–15, more or less petaloid, imbricate or valvate, usually deciduous or caducous. Petals the same number as, or more numerous than, the sepals or none. Stamens 5–many, spirally arranged; anthers basifixed and continuous with the filaments. Carpels distinct, 3–many, or rarely solitary, 1-celled. Ovules 1–many. Fruits achenes, follicles, or berries.

Fruits follicles, few–several-seeded; flowers irregular; upper pair of petals spurred ........ 1. *Delphinium*
Fruits achenes, 1-seeded; flowers regular; petals, when present, not spurred:
  Petals none; stem leaves opposite or whorled:
    Plants herbaceous; leaves whorled; sepals imbricate; styles short, glabrous or pubescent but not plumose ........ 2. *Anemone*
    Plants woody vines; leaves opposite; sepals valvate; styles elongate, plumose ........ 3. *Clematis*
  Petals present (or sometimes lacking in *Myosurus*); stem leaves alternate:
    Sepals spurred basally; fruiting receptacle spikelike ........ 4. *Myosurus*
    Sepals not spurred; fruiting receptacle ovoid ........ 5. *Ranunculus*

### 1. DELPHINIUM L., Sp. Pl. 530, 1753

Annual or perennial herbs with erect, simple or branched stems arising from fleshy-tuberous or woody-fibrous roots. Leaves alternate, or mostly basal, long-petiolate, the blades palmately lobed or divided. Flowers showy, irregular, racemosely arranged. Sepals 5, petaloid, the uppermost prolonged backward into a spur. Petals 2 or (in ours) 4, the upper 2 with slender spurs that extend downward inside hollow spur of sepal, the lower 2 not spurred, the blade about equaling the slender claw, usually notched at apex. Stamens numerous, about equaling petals or slightly exceeding them. Pistils 1–3, sessile, free, many-ovuled, follicular in fruit. Seeds closely crowded, encased in a thin, cellular coat, usually somewhat winged on the angles.

Flowers red or sometimes tinged with yellow .......................... 1. *D. cardinale*
Flowers blue, lavender or purplish, never red:
Follicles puberulent or villous; lower petals shallowly notched, cleft 1–1.5 (rarely 2.5) mm. deep; ultimate divisions of leaves linear or narrowly oblong:
Sepals deep violet-purple; lower petals usually crenulate-margined; pubescence of simple and some stellate trichomes .................................. 2. *D. parryi*
Sepals pale blue, pale lavender or lavender-blue; lower petals not crenulate-margined; pubescence of simple hairs or none:
Follicles finely puberulent, 10–15 mm. long; stems mainly glaucous and glabrous or minutely subappressed-puberulent ........................... 3. *D. parishii*
Follicles sparsely hirsute, 6–10 mm. long; stems hirsute with spreading hairs ....... ................................................. 4. *D. subglobosum*
Follicles glabrous; lower petals deeply notched, sinus 2–3 mm. deep; ultimate divisions of leaves broadly toothed or lobed, not linear ...................... 5. *D. scaposum*

**1. Delphinium cardinale** Hook., Bot. Mag. *pl. 4887,* 1855

*Delphinium coccineum* Torr., Pacif. R. Rep. **4:** 62, 1857.
*Delphinium flammeum* Kell., Proc. Calif. Acad. **2:** 22, 1863.
*Delphinium cardinale* var. *angustifolium* Huth, in Engler, Bot. Jahrb. **20:** 473, 1895.

Tall, slender plant with virgate, fistulous, cinereous, puberulent, to glabrate stems to 1.5–2 m. tall from woody-fibrous rootstocks; petioles 1–2.8 dm. long, or uppermost leaves sessile; leaf blades 1–2 dm. wide, divided into two primary sections and these again divided almost to base into deeply palmatifid segments, ultimate segments conspicuously divergent, acuminate, thinly pilose above along veins; racemes open-interrupted or paniculate in robust plants, with 3–20 flowers, these scarlet or sometimes yellowish-tinged; sepals obovate, 9–13 mm. long, acute to rounded, somewhat incurved, glabrous or very sparsely and finely puberulent, the spur rather stoutish, 12–16 mm. long, abruptly acute; upper petals narrowly oblong, exserted, with a narrowly oblong, slightly retuse upper lobe; lower petals narrowly oblong, entire to faintly emarginate, sinus about 1 mm. deep and closed; follicles broadly oblong, mostly 10–14 mm. long but sometimes only 7 mm. long, erect, distinctly reticulate-veined, with slender, somewhat spreading cusp; seeds quadrangular, 2–3 mm. long, dark gray to nearly black, angles narrowly and inconspicuously winged.

Rocky or gravelly hillsides and broad valley floors, Lower and arid parts of Upper Sonoran Zones, Santa Barbara County, California, south to the southern flanks of the Sierra San Pedro Mártir. May–July.

**2. Delphinium parryi** A. Gray, Bot. Gaz. **12:** 53, 1887

*Delphinium parryi* var. *maritimum* Davidson, Muhlenbergia **4:** 35, 1908.
*Delphinium inflexum* Davidson, Bull. S. Calif. Acad. **26:** 70, 1927.
*Delphinium parryi* f. *maritimum* Ewan, Univ. Col. Studies, ser. D, **2:** 182, 1945.

Simple or slightly branched erect stems 3–8 dm. tall from woody-fibrous root; stems and leaves puberulent, some trichomes stellate; leaf blades 2.5–4 (or in shade to 10) cm. broad, 3–5-parted nearly or completely to base, main divisions deeply cleft once to thrice into narrowly linear lobes; racemes rather densely flowered, rarely over 1.5 dm. long in anthesis, often much longer in fruit; pedicels 1–6 cm. long, ascending; sepals deep violet-purple, broadly elliptic, 10–17 mm. long, puberulent with subappressed hairs on exposed parts of back; spur stoutish, straight or curved slightly upward, about equaling sepals, puberulent; upper petals 9–11 mm. long, white tinged with purple, entire and acute or slightly notched; lower petals violet, blades broadly oval, 7–8 mm. long, shallowly notched, often more or less crenulate on margins, hirsutulous with whitish tangled hairs near base of inner face of blade; follicles 10–15 mm. long, straight and erect or slightly spreading, slenderly cuspidate tips divergent; seeds about 1.5 mm. long, sordid, with a loose, cellular-membranous coat, narrowly winged at the angles.

Arid hillsides and mesas, chiefly Upper Sonoran Zone, Monterey County, California, southward to the western flanks of the Sierra San Pedro Mártir, Baja California. Entering the Sonoran Desert near Palm Springs, Riverside County. April–June.

**3. Delphinium parishii** A. Gray, Bot. Gaz. **12:** 53, 1887

*Delphinium coelestinum* Rydb., Bull. Torrey Club **39:** 320, 1912.
*Delphinium amabile* Tidestrom, Contr. U.S. Nat. Herb. **25:** 206, 207, 1925.
*Delphinium apachense* Eastw., Proc. Calif. Acad. IV, **20:** 142, 1931.
*Delphinium parishii* var. *pallidum* Munz, Bull. S. Calif. Acad. **31:** 61, 1932.
*Delphinium amabile* subsp. *apachense* Ewan, Bull. Torrey Club **63:** 334, 1936.
*Delphinium amabile* subsp. *typicum* Ewan, loc. cit.

Stems 1–several, 3–6 dm. tall, glabrous and glaucous or minutely and sparsely puberulent, arising from stout, woody rootstock, more robust specimens usually with fistulous stems; leaf blades 3–5-parted, 3–5 cm. broad, divisions narrowly to broadly cuneate, toothed or again cleft into narrow often linear lobes, usually glabrous but sometimes sparsely puberulent; racemes open but strict, to 1.5 cm. long, to twice as long in fruit; pedicels 1–2 cm. long, ascending, mostly glabrous; sepals light blue to pale lavender, 6–10 mm. long, finely puberulent without; spur 10–12 mm. long, straight or slightly curved downward, puberulent; upper petals white to pale yellow, entire or faintly emarginate, 5–6 mm. long, glabrous; lower petals blue, slightly longer than the upper, blade broadly obovate, emarginate nearly to middle, hirsutulous with slender, crinkled, appressed hairs without; follicles 10–15 mm. long, erect, sparsely short-pubescent; seeds enclosed in a cellular-membranous coat, narrowly winged at the angles.

Rocky hillsides, mesas, and washes, Lower Sonoran and Upper Sonoran Zones, western margins of the Colorado Desert to Inyo County, California, and east into southern Utah and Arizona. April–June.

**4. Delphinium subglobosum** Wiggins, Contr. Dudley Herb. **1:** 99, *pl.* 7, 1929

*Delphinium parryi* var. *subglobosum* Munz, Bull. S. Calif. Acad. **31:** 61, 1932.
*Delphinium collinum* Ewan, Bull. Torrey Club **63:** 338, 1936.

Stout, erect, simple or branching above from a woody fibrous root; stems glaucous, reddish purple below, greenish blue above; leaves sparsely hirsutulous with spreading hairs, lower leaves 2–6 cm. broad, palmately 3-divided, each division again 2–3-cleft to about middle or a little deeper, lobes linear, divergent, 1–3 mm. broad, acute and often apiculate; upper leaves smaller, palmately 3-divided into linear, entire lobes; petioles 2–8 cm. long, sparsely hirsute on adaxial side, ciliate-margined, slightly dilated at base; raceme loosely 5–15-flowered; pedicels 0.5–3 cm. long or lowermost sometimes markedly elongated, ascending, glabrous at base, sparsely villous-pubescent near summit; bracts linear-lanceolate, 3–4 mm. long, villous on outer surface; flowers bright lavender-blue, about 2 cm. broad; sepals 8–10 mm. long, 4–5 mm. wide, villous on exposed surfaces of outer face, glabrous within and on covered margins; spur appressed-villous, slightly paler than sepals, 10–13 mm. long, blunt, curved slightly downward; upper petals ovate, apex rather acute and shallowly notched, sparsely hispid on outer face, blue shading to bluish white on margins; lower petals with orbicular blades, cleft almost to middle, long-villous on inner face, ciliate-margined along cleft only, nearly glabrous on outer face, bright blue-lavender, blade about 5 mm. long, equaling claw; follicles broadly ovoid, erect, 6–10 mm. high, sparsely villous, thin-walled, forming a subglobose fruit; seeds 2 mm. long, with a thin, clear white, cellular coating forming pronounced wings at the angles.

Rocky slopes and sandy bottoms of washes and valleys, mostly Lower Sonoran Zone, margin of the Colorado Desert in eastern San Diego County, south into northern Baja California. March–June.

### 5. **Delphinium scaposum** Greene, Bot. Gaz. **6:** 156, 1881

*Delphinium decorum* var. *scaposum* Huth, Helios **10:** 33, 1892.

Stoutish to slender, scapose plants with glabrous and lustrous, fistulous stems to 6 dm. tall, lower parts sometimes minutely and softly puberulent; rootstock woody; leaves wholly or mostly basal or subbasal; petioles 3–12 cm. long, glabrous or minutely puberulent, often markedly striate; leaf blades pentagonal, triangular, or semicircular, 2.5–8 cm. broad, with 3–5 primary divisions, then broadly cuneate-ovate to spatulate, entire or more commonly with 1–3 rounded, minutely apiculate lobes or deeply cleft into 1–3 lobes; racemes open, 3–12-flowered, 5–25 cm. long; bracts lance-linear, 5–20 mm. long; flowers 2–2.8 cm. broad; sepals an intense royal blue, 5–6 mm. wide, elliptic-lanceolate to oblong, glabrous within, sparsely puberulent without, 10–13 mm. long, acute to apiculate at apex; spur 10–12 mm. long, purplish to bronzy, often bidentate at tip; upper petals 10–12 mm. long, white with bluish tips, minutely emarginate, glabrous; lower petals blue, on a claw 4–5 mm. long, blade orbicular, 5–6 mm. long, sparsely hirsutulous within, cleft about halfway to base but narrow sinus nearly closed; follicles oblong, 10–20 mm. high, glabrous, slender cusp 3–4 mm. long; seeds dark brown but enclosed in a slightly sordid papery membrane.

Rocky hillsides, gravelly mesas and open deserts, Lower Sonoran Zone and arid slopes in Transition Zone, Pima, Maricopa, and Mohave Counties, Arizona, to southern Nevada, southwestern Colorado, New Mexico, and northern Sonora. March–May.

**Delphinium virescens** var. **wootonii** (Rydb.) Ewan, Univ. Col. Studies, ser. D, **2:** 169, 1945. *Delphinium wootoni* Rydb., Bull. Torrey Club **26:** 587, 1899. This plant, a low perennial rarely over 35 cm. tall, with leaves canescent-puberulent on both sides, pale whitish blue flowers, and corrugate-echinate seeds, occurs in the mountains immediately east of the Sonoran Desert in southern Arizona to elevations of 6,000 feet or more. The writer has seen no specimens from the desert area, but it may occur as a straggler in arroyos and canyons.

## 2. **ANEMONE** L., Sp. Pl. 538, 1753

Erect perennial herbs with basal, compound or divided leaves and 2–3 leaves forming an involucre subtending the flowers. Peduncles 1-flowered, solitary or umbellate. Sepals 4–20, petaloid, often pubescent without, usually showy. Petals none. Stamens numerous, the filaments slender. Pistils numerous. Fruit a dense head of flattened, glabrous or pubescent achenes with short persistent styles.

### 1. **Anemone tuberosa** Rydb., Bull. Torrey Club **29:** 151, 1902

Tuber fusiform, 2–4 cm. long; stems 1–3 dm. tall, glabrous below involucre or sparingly pubescent just below it, striate; basal leaves 1–5 or sometimes more numerous, on petioles 4–7 cm. long, glabrous, ternately parted and each lobe again thrice incised and toothed, blade orbicular-flabelliform in outline, 1.5–3.5 cm. broad; involucral leaves similar but subsessile, with longer segments and teeth, and sparsely pubescent on both surfaces; peduncles 1–2 (rarely 3), elongating to 1.5–2 dm. long in fruit, appressed-pubescent; sepals linear-oblong, 1–2 cm. long, white or tinged with pink or purple, appressed-pubescent on outer surface; fruiting head ellipsoidal, 8–12 mm. in diameter, 1.5–2.5 cm. long; achenes 3.5–4.5 mm. long, densely woolly; filiform styles about 1.5 mm. long.

Rocky hillsides, Lower and Upper Sonoran Zones, Providence and Panamint Mountains, California, to southern Utah, southern New Mexico and western Texas; Sierra Tucson; Coyote Mountains, Arizona. Feb.–April.

## 3. CLEMATIS L., Sp. Pl. 543, 1753

Erect perennial herbs or (in ours) subwoody vines climbing by means of tortile, tendril-like petioles. Leaves opposite, pinnately or bipinnately compound. Flowers perfect, dioecious or polygamodioecious, solitary or cymose on axillary peduncles. Sepals 4–5, petaloid or thick and leathery, valvate, spreading or erect; petals none or rudimentary. Stamens and pistils many. Fruit a globose head of achenes with persistent, elongated, plumose styles.

Plumose tails of achenes 5–10 cm. long; leaflets longer than broad, cinereous, often copiously pubescent; achenes copiously hairy; flowers in panicles ............ 1. *C. drummondii*

Plumose tails of achenes 2–4 cm. long; leaflets mostly as broad as or broader than long, dark green, sparingly pubescent or glabrate; achenes glabrous; flowers 1–3 on a peduncle ... ........................................................ 2. *C. pauciflora*

### 1. Clematis drummondii Torr. & Gray, Fl. N. Amer. **1**: 9, 1838

*Clematis nervata* Benth., Pl. Hartw. 5, 1839.
*Clematis dioica* var. *sericea* subvar. *drummondii* Kuntze, Verh. Bot. Brandenburg **26**: 102, 1885.

Scandent and climbing vine with slender woody stems to 10 m. long or more; bark tawny or light gray, striate and eventually stringy; petioles 3–8 cm. long, sparsely puberulent, coiling like tendrils when in support of vine; leaflets 3–5 (rarely 7), lanceolate to narrowly ovate, 5–15 mm. broad, usually less than 5 cm. long, divergently 1–3-toothed or entire, grayish-pubescent, often copiously so; flowers in cymose panicles, on pedicels 1–2.5 cm. long; sepals obovate to narrowly oblanceolate, 1 cm. long or less, spreading but soon involute and irregularly reflexed; stamens about 7–8 mm. long; achenes narrowly ovoid, about 4 mm. long, pubescent, tails 5–10 cm. long, filiform, shining white but turning slightly rusty in drying.

Climbing over rocks and shrubs, Upper Sonoran Zone, and in favorable places about margin of Lower Sonoran Zone, Maricopa, Pinal, and Glen Counties, Arizona, to Texas, Nuevo León, Tamaulipas, and Baja California, Mexico. March–Sept.

### 2. Clematis pauciflora Nutt., in Torr. & Gray, Fl. N. Amer. **1**: 657, 1840

*Clematis parviflora* Nutt., in Torr. & Gray, Fl. N. Amer. **1**: 9, 1838. Not *Clematis parviflora* DC., 1828.

Climbing or scrambling vine with soft woody stems 1–5 m. long or more; leaflets 3–9 (mostly 5–7), cordate to broadly cuneate-obovate, 1–2 cm. long and often fully as wide, mostly 3-toothed or 3-lobed, glabrous to sparsely silky-pubescent; flowers solitary or on 1–3-flowered panicles, dioecious; pedicels slender, 2–10 cm. long, silky-puberulent; sepals oblong-oblanceolate, 8–12 (or 17) mm. long, thin, white, densely silky-tomentose on outer surface, spreading; stamens 10–12 mm. long; achenes ovoid-lenticular, body 4–5 mm. long, 2.5–3 mm. broad, reddish brown with a lighter marginal line, glabrous, narrowing gradually to plumose tail, which is 2–4 cm. long.

On hillsides and in canyons, chiefly Upper Sonoran Zone, reaching the western edge of the Sonoran Desert in southern California and northern Baja California, Mexico. Feb.–May.

## 4. MYOSURUS L., Sp. Pl. 284, 1753

Dwarf annual, acaulescent herbs. Leaves basal, linear to linear-spatulate, entire. Flowers solitary on very short to elongated, slender scapes, inconspicuous. Sepals 5 or rarely 6–7, basally spurred, caducous. Petals same number as sepals or lacking, greenish yellow, bearing a minute nectar pit at apex of the claw. Stamens 5–25, shorter than or equaling the sepals. Pistils numerous, on a slender cylindrical axis, which becomes spikelike in fruit. Achenes apiculate to aristate, often ridged or winged.

Mature achenes with a cuplike border surrounding base of the beak, this usually lighter in color than rest of achene .......................................... 1. *M. cupulatus*
Mature achenes without cup at base of beak, achene concolorous .......... 2. *M. minimus*

### 1. **Myosurus cupulatus** S. Wats., Proc. Amer. Acad. **17:** 362, 1882

Leaves narrowly linear to narrowly linear-spatulate, 2–8 cm. long, 1–2 mm. wide, faintly 3–5-nerved; scapes slender, 1.5–9 cm. long; sepals submembranous, about 2.5 mm. long (exclusive of spur), slender spur 1.5 mm. long; petals about equaling sepals, narrowly linear; spikes 1.5–5 cm. long, 2–2.5 mm. in diameter at maturity; achenes roundish, 1–1.3 mm. long, base of beak surrounded by a thickened, light yellow or buff, cuplike rim, beak about 0.6–0.8 mm. long.

Margins of vernal pools and in moist sand along streams, Lower Sonoran Zone to Transition Zone, Providence and Little San Bernardino Mountains in southern California; central and southern Arizona to New Mexico and northern Sonora. Feb.–March.

### 2. **Myosurus minimus** L., Sp. Pl. 284, 1753

*Myosurus shortii* Raf., Amer. Jour. Sci. **1:** 379, 1819.
*Myosurus major* Greene, Pittonia **3:** 257, 1898.

Leaves filiform to narrowly linear, 3–10 cm. long, blunt or rounded at apex, indistinctly 1–3-nerved; scapes slender to stoutish, 3–15 cm. long; sepals oblong or broadly ovate, 2.5–3 mm. long (exclusive of the spurs); spurs slender, 1–2 mm. long; petals narrowly spatulate or lacking, 2.5–3 mm. long when present; stamens about equaling petals; spikes 2–3 mm. in diameter, 1–5 cm. long in fruit; achenes flat or slightly depressed dorsally, not winged marginally, quadrate or rhomboidal, about 1.4–1.8 mm. long, favose-reticulate on sides, beak minute, less than 0.5 mm. long.

Margins of vernal pools and moist soil, mostly Upper Sonoran Zone to Transition Zone, British Columbia to Baja California, eastward across the continent. Nearly cosmopolitan. Reported from Sacaton, Arizona, by Kearney & Peebles. Feb.–April.

## 5. RANUNCULUS L., Sp. Pl. 548, 1753

Glabrous or pubescent annual or perennial herbs from fibrous roots and erect or procumbent stems 0.5–12 dm. long. Basal leaves entire, 3-lobed, 3-parted, or 3-divided or pinnately compound; cauline leaves mostly alternate, similar to or simpler than the basal. Flowers terminal. Sepals 5, usually deciduous. Petals 1–18, white to yellow, usually short-clawed, with a nectariferous gland usually covered by a scale near the base. Stamens 10–many. Pistils 5–many, 1-ovuled, the ovule subbasally attached. Achenes capitate or spicate, flattened or turgid, variously sculptured, ridged, or tuberculate or smooth, glabrous or puberulent, beak mostly persistent, coat usually firm, rarely loose or utricular.

Petals 8–18, 1–1.5 cm. long; achenes 3–4 mm. long, with slender, persistent beak about as long; fruiting heads globose .................................. 1. *R. macranthus*
Petals 5, 3–5 mm. long; achenes about 0.8–1.5 mm. long, beaks minute, capitate, subsessile; fruiting heads cylindric, two to four times as long as broad:
 Leaf blades crenate but not divided; achenes strongly striate laterally; plants perennial, markedly stoloniferous, glabrous .............. 2. *R. cymbalaria* var. *saximontanus*
 Leaf blades deeply 3-parted, the divisions again lobed and toothed; achenes smooth, not striate; plants annual, not stoloniferous, slightly pubescent ..................... .............................................. 3. *R. sceleratus* var. *multifidus*

**1. Ranunculus macranthus** Scheele, Linnaea **21:** 585, 1848

*Ranunculus repens* var. *macranthus* A. Gray, Bost. Jour. Nat. Hist. **6:** 141, 1845.

Perennial 3–9 dm. high, moderately hirsute throughout or flowering scapes appressed-pubescent; petioles 0.5–2.5 dm. long; leaf blades ovate in outline, 5–10 cm. wide, ternately compound or divided, main divisions broadly cuneate at base, trifid and coarsely toothed at apices, dark green above, paler beneath, upper cauline leaves merely lobed or uppermost linear, entire, 2–3.5 cm. long; sepals broadly lance-ovate, 6–8 mm. long, hirsute externally, scarious-margined; petals 8–18, bright yellow, glossy, 10–15 mm. long, rounded or obtuse at apex, the scales of nectary free laterally most of their length; fruiting head globose or nearly so, about 1 cm. in diameter; achenes ovoid to discoid, body 3–4 mm. long, smooth, tipped with slender persistent style nearly as long, this straight or only slightly recurved.

Grassy hillsides, meadows, and along streams, mostly in Upper Sonoran and Transition Zones, eastern Arizona to southern Texas and adjacent Mexico. Entering the Sonoran Desert along Río de los Alisos, Sonora. June–Sept.

**2. Ranunculus cymbalaria** Pursh var. **saximontanus** Fernald, Rhodora **16:** 162, 1914

*Ranunculus tridentatus* var. *major* H.B.K., in DC., Syst. **1:** 253, 1818. Not *R. cymbalaria* var. *major* Hook. f. & Thompson, 1855.

Glabrous perennial with scapes 0.5–3 dm. high and stolons to 5–6 dm. long; petioles 1–15 cm. long, leaf blades broadly ovate to cordate-orbicular, 12–20 (or 25) mm. broad, 15–25 mm. long, shallowly crenately lobed, rounded at apex; sepals 5, thickish, 3–8 mm. long, yellow, often fading nearly white in age; stamens about 20; achenes 100–300 in a cylindrical head 3–5 mm. in diameter and 5–12 mm. long, body oblong, about 1.5 mm. long, 0.8 mm. broad, compressed, narrower at base than at apex, longitudinally about 4-striate on each face; beak deltoid, about 0.2 mm. long.

In mud, wet sand, or margins of pools and streams and ocean inlets, Lower Sonoran Zone to Transition Zone, drier parts of western North America, Alaska to northern Baja California and eastward to Nebraska, Kansas, and central Mexico; northern Baja California. April–Aug.

**3. Ranunculus sceleratus** var. **multifidus** Nutt., in Torr. & Gray, Fl. N. Amer. **1:** 19, 1838

*Ranunculus eremogenes* Greene, Erythea **4:** 121, 1896.

Annual with erect stems 1–10 dm. long and more or less fistulous, herbage sparingly pubescent or glabrate; leaf blades reniform to broadly ovate in outline, 1–3 cm. long, 1.5–5 cm. wide, deeply 3-parted or 3-divided, primary divisions again deeply parted or divided; sepals 5, spreading, three-fourths as long as to exceeding petals; petals 5, light yellow, 2–3 or sometimes 5 mm. long; fruiting head 2–7 mm. in diameter, 3–10 mm. long; achenes obovate, 0.8–1 mm. long, turgid, pericarp smooth or with a few minute depressions on each face.

Moist places, particularly in alkaline soil, mostly Upper Sonoran and Transition Zones, eastern Washington and Oregon to northern California, to Great Basin and Great Plains; also collected near Sacaton, Arizona. May–Aug.

## Family 45. BERBERIDACEAE. Barberry Family

Shrubs (or rarely herbs) with basal or alternate, simple or compound, mostly estipulate leaves. Flowers perfect, usually terminal, racemose or solitary. Sepals and petals in imbricated series, hypogynous, usually in threes, often subtended by 2–3 bractlets. Petals with glands near their bases. Stamens hypogynous, as many as and opposite the petals; anthers extrorse, opening by 2 terminal or subterminal valves. Ovary 1-celled; styles short or stigma sessile, peltate; ovules 2–many. Fruit a berry or capsule.

### 1. BERBERIS L., Sp. Pl. 330, 1753

Spineless shrubs with yellow wood, bitter inner bark, and pinnately compound, persistent leaves, the leaflets coriaceous, spinose-dentate, and strongly reticulate-veined. Flowers yellow, racemose. Filaments dilated, in ours with a pair of toothlike appendages near the summit, sensitive and closing around the style when touched on inner basal surface. Fruit a berry, red, white, or blue-black, often with a bloom. Seeds 1–few.

Berries at maturity red, without bloom; leaflets mostly 5, two to five times as long as wide, strongly glaucous .......................................... 1. *B. haematocarpa*

Berries at maturity blue-black, with pronounced bloom; leaflets usually 3, rarely over three times as long as broad, not glaucous .......................... 2. *B. harrisoniana*

**1. Berberis haematocarpa** Wooton, Bull. Torrey Club **25:** 304, 1898

*Mahonia haematocarpa* Fedde, Bot. Jahrb. **31:** 100, 1901.

*Odostemon haematocarpus* Heller, Muhlenbergia **7:** 139, 1912.

Erect or rounded shrub 1–2 m. high with rigid, often straggly branches; leaflets 3–7, usually 5, narrowly ovate to lanceolate, 1–1.5 cm. wide, 2–5 cm. long, lateral ones with 1–3 teeth on each margin, terminal leaflet longer and with 3–6 teeth on each margin, bright green, distinctly glaucous, teeth 2–5 mm. long, pungent; racemes on slender peduncles 2–3 cm. long, terminating numerous dwarf branches, 4–8-flowered; bracts subulate, 3–3.5 mm. long; pedicels slender, 1–2 cm. long, glabrous; petals 6, 5–6 mm. long, bright yellow, obtuse or rounded; berries bright red, translucent, 6–8 mm. in diameter, juicy, devoid of bloom, 2–several-seeded.

Rocky hillsides and occasionally along washes, Upper and sometimes Lower Sonoran Zones, northern Arizona into New Mexico and south into northern Sonora. Mostly in the oak belt, above the Sonoran Desert. Feb.–April.

**Berberis fremontii** Torr., Bot. Mex. Bound. 30, 1859, with comparatively broader leaflets and blue-black berries that become dry and somewhat inflated when mature, occurs in the mountains bordering the Sonoran Desert from southern California to Utah, Arizona, and New Mexico, and may occur infrequently in canyons debouching upon the desert. It usually occurs at elevations of 4,000–7,000 feet.

**2. Berberis harrisoniana** Kearney & Peebles, Jour. Wash. Acad. **29:** 477, 1939

A low, rounded, straggly, or subprostrate shrub 2–7 dm. tall with light gray, lightly fissured bark; leaves trifoliate, leaflets sessile, rhombic or triangular-obovate, 2.5–3.5 cm. broad, 3–5 cm. long, cuneate at base, 1–2 spinose teeth on each margin, deep green, not at all glaucous but slightly paler beneath than above, petiole equaling or slightly longer than terminal leaflet; inflorescence corymbose-racemose, terminal on dwarf branches; peduncle short, 5–10 mm. long; bracts ovate, about 1 mm. long; outer whorl of sepals 1.5–2 mm. long, inner about twice as long; petals 5–6 mm. long, rounded, bright yellow; berries globose, blue-black, 5–6 mm. in diameter, bearing a distinct bloom.

On cliffs and talus slopes, Lower Sonoran Zone, Palm Canyon, Kofa Mountains, and Pitahaya Canyon in the Ajo Mountains, Arizona. Feb.–March.

## Family 46. MENISPERMACEAE. Moonseed Family

Twining vines or sometimes erect shrubs or small trees. Roots bitter. Leaves alternate, estipulate, simple, or rarely trifoliolate. Inflorescence cymose, paniculate, fasciculate, or rarely flowers solitary; flowers inconspicuous, dioecious, regular or slightly irregular. Male flowers with imbricate sepals in 2–4 series, petals smaller than the sepals or absent; stamens 3 or 6, opposite the petals when few. Female flowers with or without staminodia, carpels usually 3 or 6, free, sessile or stipitate, stigma entire or lobed; ovules 2, one soon aborting. Fruit drupaceous, the seed often curved, endosperm ruminate, smooth, or lacking.

Carpel 1; leaves often peltate; bracts of inflorescence leaflike, often 1 cm. wide or more; stamens connate . . . . . . . . . . . . . . . . 1. *Cissampelos*
Carpels usually 3; leaves not peltate; bracts of inflorescence minute, usually less than 2 mm. in diameter; stamens free . . . . . . . . . . . . . . . . 2. *Cocculus*

### 1. CISSAMPELOS L., Sp. Pl. 1031, 1753

Slender vines with broad, mostly cordate and more or less peltate, palmately veined leaves. Staminate flowers cymose-paniculate, pistillate flowers in racemose clusters, with conspicuous leaflike bracts. Staminate flowers with 4 sepals, petals united below into a shallow cup, 2–4 anthers sessile at peltate apex of stamineal column. Pistillate flowers consisting of a rudimentary scalelike perianth of 1 sepal and 1 petal and a single carpel with a tridentate style. Fruit globose; seed compressed, concave on both sides, tuberculate on the back.

**1. Cissampelos pareira** L., Sp. Pl. 1031, 1753

*Cissampelos caapeba* L., Sp. Pl. 1032, 1753.
*Cissampelos tomentosa* DC., Syst. **1:** 535, 1818.
*Cissampelos acuminata* Benth., Pl. Hartw. 58, 1839. Not DC., 1818.

*Cissampelos benthamiana* Miers, Ann. Mag. Nat. Hist. III, **17:** 134, 1866. Nomen nudum; Contr. Bot. **3:** 144, 1871.

Climbing vine with slender stems to 5 m. long or more, young stems, petioles, inflorescences, and undersurfaces of leaves densely tomentose; leaves broadly cordate, orbicular, or subreniform, slightly peltate or not peltate; petioles 1–8 cm. long; racemes of pistillate flowers 5–8 cm. long, pedicels 2–5 mm. long; bracts orbicular to broadly cordate, 5–15 mm. broad, sepals about 1 mm. long; staminate flowers in panicles to 8 cm. long, flowers numerous, on filiform pedicels 1–3 mm. long; fruits in crowded racemes, 4–5 mm. in diameter, hirsute, fleshy; seed rugulose or transversely ridged around the margin.

In moist places, mostly in shade and scrambling over bushes. Reported from Sonora by Standley, but no specimens from the Sonoran Desert seen.

## 2. COCCULUS DC., Syst. 1: 515, 1818

Slender scandent shrubs with thin, entire, or shallowly lobed leaves. Inflorescence axillary cymes or panicles, rarely elongated, sometimes reduced to a solitary flower on pistillate plants. Sepals 6, biseriate, the inner the longer. Petals 6, sometimes smaller than sepals, entire or bifid, concave or sometimes involute around stamens. Staminate flowers with 6 free stamens. Pistillate flowers with 6 staminodes or none, carpels 3, styles straight or curved. Fruit obovate or globose, compressed and excavated laterally. Seed horseshoe-shaped, flattened.

### 1. Cocculus diversifolius DC., Syst. 1: 523, 1818

*Cocculus oblongifolius* DC., op. cit. 529.
*Cebatha diversifolia* Kuntze, Rev. Gen. Pl. **1:** 9, 1891.

Characters of genus; leaves extremely variable, ranging from narrowly linear to broadly ovate, 0.3–3 cm. wide, 2–7 cm. long, rounded and mucronulate to acute at apex, obtuse or rounded at base, entire, glabrous or sparsely pubescent, especially along margins and veins beneath; pedicels 1–3 mm. long, pubescent; sepals ovate, outer about 1 mm. long, inner about 1.5 mm. long; petals 2–2.5 mm. long, rounded at apex, cucullate-concave; fruit globose, 5–6 mm. in diameter, purplish black, fleshy, and staining deeply; seeds 3–4 mm. in diameter, about 2.5 mm. thick, excavated laterally and strongly rugose-ridged transversely.

Mostly in shade of trees and larger shrubs, Lower and Upper Sonoran Zones, Baboquivari Mountains, Pima County, Arizona, to Texas, south to Oaxaca and Tamaulipas. May–Aug.

## Family 47. PAPAVERACEAE. Poppy Family

Herbs or shrubs with watery, yellow, or milky sap and alternate or rarely opposite, estipulate, entire or dissected, decompound leaves. Flowers perfect, regular. Sepals 2 or rarely 3–4, distinct or united into a caplike calyptra. Petals 4 or 6 or rarely wanting, deciduous. Stamens commonly numerous, distinct, hypogynous; anthers dehiscing by longitudinal slits. Pistil 2–several carpels united into a central 1–several-celled ovary or sometimes becoming distinct in fruit; stigma entire or divided, on a short style or sessile; ovules mostly numerous on parietal placentae. Fruit a capsule, usually opening by valves or by terminal or subterminal pores.

Carpels separating into distinct, torulose follicles in fruit; leaves entire, opposite or whorled ........ 1. *Platystemon*

Carpels united into single capsule; leaves lobed or dissected, alternate:
- Herbage smooth, not prickly; petals not undulate and crepelike along margins; capsules dehiscing from base or by means of small subterminal openings:
  - Receptacle hypanthium-like; sepals united into a calyptra; capsules dehiscing full length from base upward .......... 2. *Eschscholzia*
  - Receptacle not hypanthium-like; sepals distinct; capsules dehiscing by small subterminal pores .......... 3. *Stylomecon*
- Herbage strongly prickly; petals undulate and crepelike along margins; capsules dehiscing from apex downward by 4–6 valves .......... 4. *Argemone*

## 1. PLATYSTEMON Benth., Trans. Hort. Soc. Lond. II, 1: 405, 1835

Low villous annuals with opposite, entire, narrow leaves and cream-colored flowers on slender erect peduncles. Sepals 3, distinct. Petals 6. Stamens numerous with flattened, slightly dilated filaments. Carpels 6–25, connivent or coherent at first, separating in fruit; stigmas free, linear. Fruit consisting of several torulose indehiscent pods that break transversely between seeds.

### 1. **Platystemon californicus** Benth., Trans. Hort. Soc. Lond. II, **1:** 405, 1835, var. **californicus**

*Platystemon arizonicus* Greene, Pittonia **5:** 190, 1903.
*Platystemon confinis* Greene, op. cit. 176.
*Platystemon mohavensis* Greene, loc. cit.

More or less decumbent and branching from base, or in dry situations simple and unbranched, villous stems 5–40 cm. long; leaves broadly linear to oblong-spatulate, 3–8 cm. long, sessile or clasping, villous; peduncles erect, 3–20 cm. long; sepals broadly obovate, villous, 5–7 mm. long, forming globose or ovoid buds, deciduous at flowering; petals cream-colored and frequently darker yellow at base, 6–20 mm. long, broadly obovate; carpels forming an oblong to cylindrical head 1–1.8 cm. long, filiform persistent styles forming a tuft of free beaks.

Sandy soil of hillsides and valleys, Lower and Upper Sonoran Zones, southern Oregon to southern end of the Sierra San Pedro Mártir, Baja California; east to Utah and Arizona. Rare in the desert but common along its western margin and along the foothills in Arizona and Nevada. March–May.

**Platystemon californicus** var. **crinitus** Greene, Fl. Fran. 282, 1892, differs chiefly in being more densely hirsute with longer hairs on the peduncles, and in having deeper-yellow petals. It occurs on slopes in the Upper Sonoran Zone bordering the Mojave and Colorado Deserts.

## 2. ESCHSCHOLZIA Cham., in Nees, Hor. Phys. Ber. 73, 1820

Annual (in ours) or perennial herbs with colorless sap and mainly glabrous, glaucous, ternately dissected leaves. Receptacle dilated and forming a funnelform torus around the base of the pistil. Sepals 2, but coherent into a pointed cap and pushed off by the unfolding petals. Petals usually 4, sometimes 6 or 8, bright yellow or orange, broadly obovate. Stamens many; the filaments short; anthers linear. Ovary 1-celled, cylindrical, biplacentate. Styles short; stigma of 4–6 linear lobes. Capsule 10-nerved, narrowly cylindric or linear, dehiscent from base; seeds globose or ovoid, reticulate or tuberculate or pitted.

Torus with an outer spreading rim:
  Stems branching more or less, not entirely scapose; cotyledons 2-cleft or notched ........................................................ 1. *E. californica* var. *peninsularis*
  Stems scapose, the plant mostly acaulescent except for flowering scapes; cotyledons entire ........................................................ 2. *E. mexicana*
Torus without an exterior spreading rim:
  Seeds smooth or reticulate, not deeply pitted, blackish; stems usually leafy:
    Capsules mostly 5–7 cm. long; herbage conspicuously glaucous; petals with orange base ........................................................ 3. *E. elegans*
    Capsules mostly less than 5 cm. long; herbage slightly glaucous; petals yellow:
      Petals 3–10 mm. long; stamens 4–15 .......................... 4. *E. minutiflora*
      Petals 8–22 mm. long; stamens 16–37 .......................... 5. *E. parishii*
  Seeds deeply pitted with angular depressions, gray; stems scapose .... 6. *E. glyptosperma*

**1. Eschscholzia californica** Cham. var. **peninsularis** (Greene) Munz, Man. S. Calif. 181, 1935

*Eschscholtzia peninsularis* Greene, Bull. Calif. Acad. **1:** 68, 1885.

Annual from a slender taproot with cespitose leafy stems 2–4 dm. high; leaves ternately compound, 3–25 cm. long with linear or oblong segments 2–10 mm. long, slightly glaucous, glabrous, or sparingly and finely puberulent, the dissected blade equaled or surpassed by slender, slightly succulent petiole; peduncles 3–10 (rarely 15) cm. long; calyptra broadly ovoid-turbinate, 0.8–2.5 cm. long; petals flabelliform, 1–3.5 cm. long, deep orange to yellow, or rarely white, glossy; outer rim of torus about 1 mm. wide; seeds nearly black, dull, reticulate; cotyledons 2-cleft or notched.

Sandy soil and gravelly washes, Lower and Upper Sonoran Zones, central San Joaquin Valley, California, south through the Coast Ranges to Playa Santa Catarina, Baja California. Feb.–May.

**2. Eschscholzia mexicana** Greene, Bull. Calif. Acad. **1:** 69, 1885

*Eschscholtzia aliena* Greene, Pittonia **5:** 260, 1905.
*Eschscholtzia jonesii* Greene, op. cit. 261.
*Eschscholtzia arizonica* Greene, loc. cit.

Annual with leaves forming rosette, stems mainly scapose; leaves 3–15 cm. long, segments oblong, mostly about 1 mm. wide, faintly glaucous, glabrous, flabelliform dissected blade usually one-third as long as petiole or less; calyptra broadly ovoid-turbinate, 12–18 mm. long, apiculate; exterior rim of torus narrow, often cuspate or undulate, mostly 1 mm. wide or less; petals flabelliform-obovate, 1.5–3.5 cm. long, yellow to deep orange or rarely white tinged with pink; capsule 4–6 cm. long, longitudinally ridged, the grooves glaucous; seeds 0.8–1 mm. in diameter, subglobose, black, reticulate; cotyledons entire.

Sandy and gravelly soil, Sonoran Zones, Arizona to Utah, Texas, and Sonora. Feb.–May.

**3. Eschscholzia elegans** Greene, Bull. Calif. Acad. **1:** 182, 1885

*Eschscholtzia elegans* var. *ramosa* Greene, loc. cit.
*Eschscholtzia ramosa* Greene, Bull. Torrey Club **13:** 217, 1886.
*Eschscholtzia pseudopraecox* Fedde, Notizblatt **4:** 153, 1904.
*Eschscholtzia quadrangularis* Greene, Pittonia **5:** 264, 1905.
*Eschscholtzia exilis* Greene, op. cit. 265.
*Eschscholtzia humilis* Greene, loc. cit.
*Eschscholtzia crassula* Greene, op. cit. 266.
*Eschscholtzia wrigleyana* Millsp. & Nutt., Field Mus. Bot. Ser. **5:** 109, *pl. 1,* 1923.

Annual with stems 2–4 dm. high branching above the base, glabrous, scarcely or not at all glaucous, somewhat fleshy; leaves flabelliform-obovate, finely dissected, glaucous and glabrous or slightly scabrous, blade 1–5 cm. wide, slightly longer, on petioles one to five times as long; torus broadly turbinate, upper margin scarious, erect; calyptra ovoid-turbinate, 1 cm. long or less, with apiculation 1–6 mm. long at apex; petals 1–2 cm. long, yellow with orange spot at base; seeds ovoid-subglobose, about 1 mm. long, dull black, reticulate.

Grassy slopes and openings in chaparral, mainly Upper Sonoran Zone, Santa Rosa Island, California, to Guadalupe and Cedros Islands, Baja California. March–June.

**4. Eschscholzia minutiflora** S. Wats., Proc. Amer. Acad. **11:** 122, 1876

*Eschscholtzia modesta* Greene, Pittonia **1:** 169, 1888.
*Eschscholtzia ludens* Greene, op. cit. **5:** 272, 1905.
*Eschscholtzia micrantha* Greene, op. cit. 277.

Glabrous or sparsely puberulent annual with moderately glaucous, branching, leafy stems 1–3.5 dm. high; leaves finely to coarsely dissected, rather stiffly ascending, rarely forming a basal rosette, blade flabelliform in outline, 1–2.5 (–4) cm. broad, as wide as or broader than long; petioles of all except basal leaves about equaling or slightly surpassing blades; flowers on slender peduncles 1–8 cm. long scattered along leafy stems; torus short-turbinate, without a spreading rim; calyptra broadly ovoid-turbinate, short-apiculate, 4–7 mm. long; petals yellow, 3–10 mm. long; stamens 4–15, 2–4 mm. long; capsules slender, narrowly fusiform, 2–3.5 (rarely –5) cm. long; seeds globose, greenish black, dull, reticulate.

Dry plains and desert slopes and washes, Lower and Upper Sonoran Zones, Inyo County, California, to Washoe County, Nevada, southern Utah, Arizona, Sonora, and central Baja California. Common on the Colorado Desert. March–May.

**5. Eschscholzia parishii** Greene, Bull. Calif. Acad. **1:** 183, 1885

*Eschscholtzia minutiflora* var. *parishii* Jepson, Fl. Calif. **1:** 573, 1922.

Differs from *E. minutiflora* only in having petals 8–22 mm. long, stamens 16–37, 4–7.5 mm. long, correspondingly larger calyptras, and capsules 3.5–5 cm. long.

Desert slopes and flats, eastern Kern County; southern San Bernardino County, California, to central Baja California. March–May.

**6. Eschscholzia glyptosperma** Greene, Bull. Calif. Acad. **1:** 70, 1885

*Eschscholtzia paupercula* Greene, Pittonia **5:** 262, 1905.

Annual with numerous erect, slender, naked scapes from a dense basal rosette of leaves with broadly flabelliform, finely dissected blades 1–3 cm. broad; petioles slender, 1–5 cm. long, herbage glaucous and glabrous, frequently tinged with red; scapes 1–2.5 dm. high; calyptra ovoid-turbinate, 8–12 mm. long, gradually narrowed to a stout apiculation 2–3 mm. long; torus turbinate, about 2 mm. long, without a spreading rim; petals broadly obovate, yellow, 1–2 cm. long; capsules linear-fusiform, 3–7 cm. long; seeds ovoid-subglobose to globose, 1.2–1.5 mm. in diameter, grayish, coarsely and irregularly favose-pitted, not reticulate.

Desert slopes and plains, Lower Sonoran Zone, Inyo County, California, to northeastern Colorado Desert and east to southern Utah. March–May.

## 3. STYLOMECON G. Taylor, Jour. Bot. Brit. & For. **68:** 140, 1930

Annual herbs with simple or branching stems and yellow juice. Leaves pinnately parted or divided. Flowers nodding in bud, erect after flowering. Sepals 2. Petals 4, usually showy. Stamens many. Style persistent, short, tipped with a globose mass of stigmas. Fruit a capsule, 1-celled, with 4–11 placentae, dehiscing apically by as many short teeth or valves as placentae. Seeds numerous, reniform, reticulate-rugose.

**1. Stylomecon heterophylla** (Benth.) G. Taylor, Jour. Bot. Brit. & For. **68:** 140, 1930

*Meconopsis heterophylla* Benth., Trans. Hort. Soc. Lond. II, **1:** 408, 1835.
*Meconopsis crassifolia* Benth., loc. cit.
*Papaver heterophyllum* Greene, Man. Bay Reg. 9, 1894.
*Papaver heterophyllum* var. *crassifolium* Jepson, Fl. W. Mid. Calif. 209, 1901.

Annual with simple or branching erect stems 2–6 dm. tall and glabrous or sparsely pilose herbage; leaves pinnately divided or parted, 2–5 cm. wide, to 10 cm. long or more, mostly petiolate and lobes often again shallowly lobed or sinuately toothed, somewhat succulent; peduncles slender, often 2–4 dm. long; petals broadly obovate, 1–2 cm. long, brick red, with green claw and purple spot just above it; capsule obovate or turbinate, 1–1.5 cm. long, placentae conspicuously indicated by light longitudinal lines, valves rounded, operculae situated immediately beneath placental arches; seeds about 0.4 mm. in diameter, black, dull.

On shaded slopes and in canyons, mainly Upper Sonoran Zone, Lake and Fresno Counties, California, to south end of Sierra San Pedro Mártir, Baja California. April–May.

## 4. ARGEMONE L., Sp. Pl. 508, 1753

Annual or biennial prickly herbs, or one Mexican species a shrub, with acrid, white or yellow juice, and alternate, sinuate to pinnatifid, spinose leaves. Flowers erect in bud, showy. Sepals 2–3, each bearing a spine-tipped appendage at or near the summit. Petals 4 or 6, white, yellow, or orange. Stamens many. Stigmas sessile or nearly so, more or less radiate; ovary with 4–6 slender placentae. Capsules oblong or fusiform, opening apically by 4–6 valves. Seeds numerous, globose, favose, reticulate.

Petals pale yellow to orange .......... 1. *A. mexicana*
Petals white, rarely tipped with rose:
  Horns of sepals slender, devoid of lateral prickles; spines on capsules sparse, tending to spread horizontally early in development of capsule:
    Leaves sinuate-pinnatifid; flowers few, on elongated branches, not corymbose .......... 2. *A. intermedia* var. *intermedia*
    Leaves repand-dentate or repand-sinuate, obovate to ovate; flowers numerous, in corymbose clusters .......... 2*a*. *A. intermedia* var. *corymbosa*
  Horns of sepals deltoid-pyramidal, bearing several small lateral bristles near base; spines on capsule densely crowded, tending to be ascending until capsule is mature:
    Herbage devoid of hirsute covering between heavier bristles .......... 3. *A. platyceras* var. *platyceras*
    Herbage densely hirsute among heavier bristles .......... 3*a*. *A. platyceras* var. *hispida*

**1. Argemone mexicana** L., Sp. Pl. 508, 1753

*Echtrus trivialis* Lour., Fl. Cochinch. **1:** 344, 1790.
*Argemone spinosa* Moench, Meth. 227, 1794.

*Argemone versicolor* Salisb., Prod. Stirp. 376, 1796.
*Argemone sexvalvis* Stokes, Bot. Mat. Med. **3:** 195, 1812.
*Argemone vulgaris* Spach, Hist. Veg. **7:** 26, 1839.
*Argemone mexicana* α *lutea* Kuntze, Rev. Gen. Pl. **1:** 12, 1891.

Erect herb 2–10 dm. tall with glabrous, glaucous, but bristly or spinose stems, leaves, peduncles, and calyces; leaves sinuate-pinnatifid, sessile and more or less clasping, to 8 cm. wide, 6–20 cm. long, sparingly spinose on margins and whitish veins, spines 2–8 mm. long; buds ovoid, 10–18 mm. long, 6–8 mm. in diameter, sepals sparsely bristly, horns slightly divergent, ascending, 5–8 mm. long, spine-tipped but devoid of lateral bristles; petals yellow to orange, broadly obovate, 2–3 cm. long; style about 2 mm. long, stigma lobes suberect, 1–1.5 mm. long; capsule ovoid-fusiform, 2–4 cm. long, sparingly bristly, bristles coarse, 5–9 mm. long, early spreading; seeds dark brown, 1.6–1.8 mm. in diameter.

Sandy arid flats and slopes, mainly Lower Sonoran Zone, vicinity of Tucson, Arizona, where it may be introduced, and Sonora south through Mexico and Central America to Chile and Brazil. Widely introduced in tropics and subtropics of both hemispheres. April–May.

**2. Argemone intermedia** Sweet, Hort. Brit. ed. 2, 585, 1830, var. **intermedia**

*Argemone intermedia* var. *typica* Prain, Jour. Bot. Brit. & For. **33:** 363, 1895.
*Argemone arida* Rose, Contr. U.S. Nat. Herb. **8:** 26, 1903.
*Argemone delicatula* Small, Fl. SE. U.S. 462, 1903.

Annual or biennial herb to 1 m. high with glabrous, glaucous, spinose-bristly stems, peduncles, calyces, and herbage; leaves sinuate-pinnatifid, 2–5 cm. wide, 6–20 cm. long, with ovoid to oblong segments and spines 2–4 mm. long tipping marginal teeth and lobes, midvein whitish, sparsely but coarsely bristly on lower surface; buds globose, 10–15 mm. in diameter, coarsely but sparsely bristly; sepals with naked, ascending or moderately spreading, spine-tipped horns 4–6 mm. long; petals white, suborbicular to broadly obovate, 3–4 cm. long; stigmas sessile or on stout styles 0.5–1.5 mm. long; capsules ovoid, 2–6 cm. long, 8–20 mm. in diameter, moderately bristly with spreading bristles 5–8 mm. long; seeds 1.5–2 mm. in diameter.

Mostly in sandy or rocky soil, Lower Sonoran Zone to Transition Zone, Wyoming to South Dakota, south to Arizona, Texas, and northern Sonora; and in central Baja California. Flowering during much of the year.

**2*a*. Argemone intermedia** var. **corymbosa** (Greene) Eastw., Erythea **4:** 96, 1896

Differs chiefly in bearing the flowers in many-flowered corymbose clusters, and in having shorter, ovate or obovate leaves less deeply pinnatifid. Mojave Desert and Hayfields, northern part of Colorado Desert, California.

**3. Argemone platyceras** Link & Otto, Ic. Pl. Rar. **1:** 85, *pl. 43,* 1828, var. **platyceras**

*Argemone munita* Dur. & Hilg., Jour. Acad. Phila. II, **3:** 37, 1855.
*Argemone platyceras* var. *typica* Prain, Jour. Bot. Brit. & For. **33:** 367, 1895.
*Argemone sanguinea* Greene, Pittonia **4:** 68, 1899.

Stout, strict, or more frequently branched annual or biennial 5–10 dm. high, prickly throughout with yellow spines 3–8 mm. long; leaves oblong to oblong-obovate, sinuate-pinnatifid, 5–25 cm. long, 2–6 cm. wide, glaucous beneath; inflorescence leafy, paniculate; buds broadly ovoid to globose, 12–20 mm. in diameter; sepals bristly, horns 5–9 mm. long, narrowly deltoid-pyramidal, spine-tipped, and with 1–several weaker lateral bristles below the apical spine; petals white, broadly obovate, 3.5–5 cm. long (sometimes smaller on depauperate plants); stigmas nearly globose, roughened, nearly sessile; capsules ovoid-elliptic, 2–5 cm. long, rather densely bristly or spinose, bristles ascending on young capsules, becoming spreading only when capsules approach maturity; seeds 1.5–2 (rarely 2.5) mm. in diameter.

Arid sandy soil, Sonoran Zones, Sacramento Valley, California, to central Baja California, Mexico, and eastward to Wyoming, Nebraska, and Texas. Flowering practically throughout the year.

**3*a*. Argemone platyceras** var. **hispida** (A. Gray) Prain, Jour. Bot. Brit. & For. **33:** 367, 1895

Differs only in having a dense hirsute mat of slender yellowish hairs intermingled with coarser bristles. Range similar to the species but less common. The Sonoran Desert representatives are rarely as hirsute as some from Nevada and eastern California.

## Family 48. FUMARIACEAE. Fumewort Family

Herbaceous plants with watery juice, alternate or basal, much-dissected, decompound, estipulate leaves. Flowers irregular; sepals 2, small and scalelike; petals 4, erect-connivent, in 2 dissimilar pairs, the outer with spreading tips and one of them (sometimes both) spurred or strongly gibbous at base, inner pair narrower, the callous-crested tips lightly coherent over the enclosed stigma. Stamens 6, in 2 sets of three, each set opposite one of the outer petals. Pistil bicarpellate, biplacentate; stigma bilobed. Ovary 1-celled, with 1–several ovules. Fruit 1–2-valved, several-seeded capsule or a 1-seeded indehiscent nut. Embryo minute, embedded in a fleshy endosperm.

### 1. CORYDALIS DC., Syst. 2: 113, 1821

Herbaceous biennial or short-lived perennial with glabrous glaucous herbage. Leaves bipinnately or ternately dissected into numerous small segments. Flowers in spikelike clusters terminating slender to stoutish scapes, irregular; sepals 2; petals 4, erect-connivent, one of the outer ones strongly gibbous at the base, the other not gibbous, the inner pair carinate or winged dorsally. Stamens 6, included within the fold of the petals. Ovary with 2 narrow placentae, 1–many-ovulate; style filiform; stigmas entire or lobed. Capsule ovate, oblong, or linear, sometimes inflated, 2-valved. Seeds black, shining.

**1. Corydalis aurea** Willd. var. **occidentalis** Engelm., in A. Gray, Man. ed. 5, 62, 1867

*Corydalis montana* Engelm., in A. Gray, Pl. Fendl. 6, 1849.
*Capnoides montanum* Britt., Mem. Torrey Club **5:** 166, 1894.

Biennial (often flowering first season), 0.6–3 dm. tall, stems usually erect or ascending; faintly glaucous; leaves 3–6 cm. long, segments elliptic to oblong-spatulate, 1–2.5 mm. broad; racemes 2–5 cm. long; bracts lance-elliptic, acute, 4–8 mm. long; flowers somewhat crowded, 10–12 mm. long, spur 4–6 mm. long; capsule about 2–2.5 cm.

long, slightly curved, ascending, on an ascending pedicel; seeds black and shining, about 2 mm. in diameter.

Mainly in Upper Sonoran and Transition Zones, occasional on north slopes of desert hills between Tucson and Sells, Arizona. Feb.–June.

## Family 49. BRASSICACEAE. Mustard Family

Annual or biennial herbs or rarely suffrutescent perennials with acrid colorless juice and alternate leaves. Flowers minute to large and showy, in racemes or corymbs. Sepals 4, deciduous or persistent, all slender and alike, or the inner concave or saccate at base. Petals 4, or less commonly 2 or lacking, hypogynous, equal or essentially so, cruciate, usually distinctly clawed. Stamens 6, adaxial and abaxial ones paired, lateral stamens single, or by reduction 4 or 2, hypogynous. Nectaries commonly situated near bases of some stamens or petals or both. Pistil 1, of 2 united carpels; placentae parietal, but united by a dissepiment that separates fruit into 2 cells, or in a few genera fruit 1-celled. Style present or stigma sessile; stigma discoid or slightly bilobed. Fruit a silique or silicle, mostly 2-celled, 2-valved, rarely 1-celled, rarely indehiscent, 1–many-seeded. Endosperm none. Cotyledons incumbent, accumbent, or conduplicate.

Stigma lobes situated over valves:
  Fruits siliques, several to many times as long as broad; anthers sagittate at base, often twisted or spirally coiled:
    Siliques long-stipitate; stamens strongly exserted, often twice as long as sepals; anthers spirally coiled .......... 1. *Stanleya*
    Siliques short-stipitate or nonstipitate; stamens only slightly exserted, only slightly longer than sepals; anthers straight or twisted but not spirally coiled:
      Petals crisped along margins; anthers often curved or twisted; siliques not strongly beaked:
        Siliques rarely flattened; tips of sepals curved inward, calyx more or less urn-shaped; seeds rarely flattened .......... 2. *Caulanthus*
        Siliques flattened parallel to the partition; sepals erect, calyx not urn-shaped; seeds usually flattened .......... 3. *Streptanthus*
      Petals not crisped; anthers straight, not curved or twisted; siliques strongly beaked, the beak stylelike .......... 4. *Streptanthella*
  Fruits silicles, as broad as long or not over two to three times as long as broad; anthers not sagittate, not twisted or coiled:
    Silicles flattened contrary to partition, emarginate at apex or didymous; stigmas sessile or essentially so:
      Petals linear to ligulate, often crisped or contorted, sordid yellow tinged with brown or purple; silicle lyre-shaped, emarginate, each cell several-seeded.. 13. *Lyrocarpa*
      Petals broadly spatulate, not crisped or contorted, white or faintly tinged with purple or pink; silicles didymous, each cell 1-seeded .......... 14. *Dithyrea*
    Silicles inflated, globose or nearly so, not compressed, rounded at apex; stigmas on slender styles one-half as long as pod to equaling it .......... 15. *Lesquerella*
Stigma lobes situated above placentae:
  Fruits siliques, more than twice as long as wide:
    Siliques moderately to conspicuously beaked; cotyledons conduplicate:
      Beak strongly flattened, beak commonly containing a seed; valves 3-nerved..7. *Sinapis*
      Beak not markedly flattened, devoid of any seed; valves usually 1-nerved:
        Seeds in 1 row in each cell; pods terete .......... 8. *Brassica*
        Seeds in 2 rows in each cell; pods slightly flattened parallel with partition .......... 9. *Diplotaxis*

Siliques not beaked; cotyledons incumbent or accumbent:
Plants glabrous or bearing only simple hairs:
Plants semiaquatic; stems often decumbent at base, rooting freely at lower nodes .. ........................................................ 11. *Rorippa*
Plants of dry or moist habitats but not aquatic, not rooting at nodes:
Sepals spreading, not markedly scarious- or hyaline-margined; seeds often in 2 rows in each cell:
Cotyledons incumbent; plants not confined to marshy habitats; siliques not ovoid ........................................... 6. *Sisymbrium*
Cotyledons accumbent; plants of marshy areas; siliques usually ovoid ....... ........................................................ 11. *Rorippa*
Sepals erect, distinctly scarious- or hyaline-margined; seeds in single row:
Petals entire or faintly emarginate, not lobed; leaf segments linear, not over 2 mm. wide, the terminal not wider than the laterals ......... 12. *Sibara*
Petals 3–9-lobed; leaf segments broader, the terminal much larger than the laterals ........................................... 10. *Dryopetalon*
Plants more or less pubescent with forked to many-rayed, dendritically branched hairs:
Leaves finely dissected, or if more coarsely lobed, then bipinnatifid; glandular hairs intermingled with forked ones; flowers small, rarely over 5 mm. long ....... ................................................... 21. *Descurainia*
Leaves coarsely lobed or singly pinnatifid to entire; glandular hairs absent; flowers usually larger, often 8–12 mm. long or more:
Petals white, tinged with pink or purple; seeds faintly to broadly winged all around; cauline leaves often auriculate-clasping ................ 22. *Arabis*
Petals yellow, orange, or brownish; seeds wingless or winged only at apex; cauline leaves not auriculate-clasping ........................... 23. *Erysimum*
Fruits silicles, rarely more than twice as long as broad:
Silicles dehiscent; ovary 2-celled; several seeds in each cell except in *Lepidium*:
Seeds 1 in each cell; pubescence of simple hairs only when present, often lacking .... ........................................................ 5. *Lepidium*
Seeds 2–many in each cell; pubescence of forked or branched hairs when present:
Silicles strongly compressed contrary to partition:
Pods elliptic; valves strongly 1-nerved ....................... 16. *Hutchinsia*
Pods cuneate-obcordate or triangular-obcordate; valves not prominently nerved ........................................................ 17. *Capsella*
Silicles compressed parallel to partition or turgid and not compressed:
Silicles turgid, ovoid, not compressed; pubescence of simple hairs; leaves pinnatified ................................................ 11. *Rorippa*
Silicles strongly compressed; pubescence of forked and branched hairs as well as simple ones; leaves entire or dentate .......................... 18. *Draba*
Silicles indehiscent; ovary 1-celled, 1-seeded:
Pod not winged or margined, body beset with uncinate hairs ......... 19. *Athysanus*
Pod winged or thin-margined, wing often notched or perforated; body glabrous or puberulent with fine, tangled hairs, not uncinate-hispid ..... 20. *Thysanocarpus*

## 1. STANLEYA Nutt., Gen. 2: 71, 1818

Annual or perennial, herbaceous to suffrutescent plants with simple or paniculately branched stems and glabrous to pubescent herbage. Leaves forming a basal rosette or only cauline, petiolate, sessile, or auriculate, entire to deeply dissected. Inflorescences racemose, the flowers usually rather closely crowded. Sepals linear-oblong, spreading to reflexed at anthesis. Petals white or yellow, usually conspicuous. Stamens subequal, single ones usually surrounded basally by glandular tissue, this obsolete or only on underside of paired stamens. Gynophore present. Silique linear, flattened parallel to the septum or nearly terete. Stigma sessile or nearly so. Seeds oblong, not margined. Cotyledons varying from incumbent to accumbent, sometimes both types in seeds of single species.

**1. Stanleya pinnata** (Pursh) Britt., Trans. N.Y. Acad. **8:** 62, 1889, var. **pinnata**

*Cleome pinnata* Pursh, Fl. Amer. Sept. **2:** 739, 1824.
(For extensive synonymy see Lloydia **2:** 116–117, 1939.)

Suffrutescent perennials with several stems 4–15 dm. tall from woody base, glabrous or sparsely pilose, glaucous; no rosette leaves present; lower cauline leaves pinnatifid, sometimes bipinnatifid or rarely nearly entire, 2–8 cm. wide, 5–20 cm. long, glabrous or sparsely pilose with short, flattish hairs; upper cauline leaves ovate to linear, mostly petiolate, entire or occasionally pinnatifid, 3–20 mm. wide, 3–12 cm. long; inflorescence a dense raceme 5–6 cm. in diameter, 1–3 dm. long; pedicels divergent, 5–12 mm. long, glabrous to sparsely pilose; sepals yellowish, 1–1.5 cm. long, about 2 mm. wide, spreading-reflexed in anthesis; petals 10–18 mm. long, blade yellow, 1.5–3 mm. wide, 3–9 mm. long, claw brownish, villous on inner face, slightly dilated at base; stamens to 2.5 cm. long, their filaments pilose at base; siliques 3–8 cm. long, spreading and more or less arcuate, on a slender gynophore 1–2 cm. long; seeds about 1 mm. wide, 1.8–2 mm. long, brownish, wingless.

Arid hillsides and along washes, Lower Sonoran Zone, northwestern margin of the Colorado Desert to Ventura County, California, and to southern Idaho, western North Dakota, south to central Arizona and the Big Bend region, Texas. March–June.

## 2. CAULANTHUS S. Wats., Bot. King Expl. 27, 1871

Annual (in ours) or less commonly perennial herbs, pubescent with simple hairs or glabrous and frequently glaucous. Basal leaves usually petiolate and often forming a rosette in youth, the cauline ones gradually reduced, often more or less auriculate-clasping. Flowers in racemes, white, yellow, or purple. Calyx urceolate, tips of sepals pinched in; sepals equal or unequal, often yellow or purplish. Petals frequently narrow, with a crisped blade. Single stamens often surrounded by subtending gland. Siliques erect, divaricate or reflexed, terete or nearly so, often torulose, usually sessile or short-pedicellate. Style short. Stigma entire or 2-lobed with the lobes extended over tips of valves. Cotyledons usually obliquely incumbent.

Leaves, at least upper ones, distinctly auriculate; petals yellow or purplish, 8–14 mm. long:
  Pedicels 4–18 mm. long; pods 5–12 cm. long; leaf blades more or less pinnatifid, hispid; sepals yellowish or purple:
    Pods pendent; sepals purple, 8–9 mm. long; petals purplish-veined, 12–14 mm. long ... .......... 1. *C. heterophyllus*
    Pods divaricate; sepals yellowish, about 6 mm. long; petals yellowish, 8–9 mm. long ... .......... 2. *C. hallii*
  Pedicels 1–3 mm. long; pods 2–4.5 cm. long; leaves entire to sinuately lobed, glabrous; sepals greenish .......... 3. *C. cooperi*
Leaves petiolate, not auriculate; petals white, about 6 mm. long:
  Pods reflexed, nearly straight; leaves more or less hispid, blades often thickish .......... 4. *C. lasiophyllus* var. *lasiophyllus*
  Pods reflexed but tips curved outward; leaves glabrous, thin .......... 4*a*. *C. lasiophyllus* var. *utahensis*

**1. Caulanthus heterophyllus** (Nutt.) Payson, Ann. Mo. Bot. Gard. **9:** 298, 1923

*Streptanthus heterophyllus* Nutt., in Torr. & Gray, Fl. N. Amer. **1:** 77, 1838.

Erect, moderately hirsute annual with simple or sparingly branched stems 3–10 dm. tall; leaves lance-linear, 5–15 mm. wide, 3–12 cm. long, subentire, sinuate-dentate or pinnatifid with strongly divaricate lobes 5–9 mm. long, lowermost petiolate and not amplexicaul, the rest sessile and distinctly auriculate-clasping, sparsely to moderately hirsute or at length nearly glabrate; racemes lax; sepals purplish, linear-lanceolate, 8–9 mm. long, subequal, sparsely hirsute, outer ones not noticeably saccate; petals pale cream or nearly white with purplish veins, 12–14 mm. long, blade somewhat recurved or strongly spreading; filaments distinct, glabrous; pedicels 4–10 mm. long, hirsute in youth, subglabrate and reflexed in fruit; siliques linear, pendent, slightly compressed, 1.5–2 mm. wide, 5–8 cm. long.

Grassy hillsides and in openings in brush, Upper Sonoran Zone and margins of Lower Sonoran Zone, Los Angeles County to northern Baja California. March–May.

**2. Caulanthus hallii** Payson, Ann. Mo. Bot. Gard. **9:** 290, 1923

*Streptanthus hallii* Jepson, Fl. Calif. **2:** 23, 1936.

Erect annual 3–15 dm. tall, branching from near base upward and with herbage glabrous or nearly so; leaves mostly basal or subbasal, oblanceolate, 5–18 mm. wide, 4–16 cm. long, irregularly laciniate-pinnatifid or upper ones dentate with a few coarse lobes, sparsely hirsutulous or hispidulous, upper ones sessile, linear to lanceolate, subentire or entire; racemes loose, few-many-flowered; pedicels 6–18 mm. long, widely divergent; sepals yellowish, 5–6 mm. long, sparsely hispidulous, not saccate; petals pale yellowish to white, 8–9 mm. long, narrowly spatulate; siliques terete, divaricately spreading-ascending, often curved or flexuous, 7–12 cm. long, glabrous; style 1.5–2 mm. long; stigma cupulate.

Arid hillsides and along washes, Lower Sonoran and Upper Sonoran Zones, western margin of the Colorado Desert, California. April–May.

**3. Caulanthus cooperi** (S. Wats.) Payson, Ann. Mo. Bot. Gard. **9:** 293, 1923

*Thelypodium cooperi* S. Wats., Proc. Amer. Acad. **12:** 246, 1877.

*Guillenia cooperi* Greene, Leaflets Bot. Obs. **1:** 228, 1906.

Slender annual with a flexuous, simple or sparingly branched stem 2–6 dm. long, glabrous or sparingly pubescent, slightly glaucous; lower leaves oblong to oblanceolate, 2–6 cm. long, entire to sinuately and shallowly lobed, often reddish, obtuse at apex, gradually narrowed to a broad petiole below; upper stem leaves sagittate, entire or nearly so, acute at apex, auriculate-clasping at base, mostly deciduous before fruits are mature; racemes lax; pedicels 1–3 mm. long, recurved; sepals greenish, 6–7 mm. long, very faintly if at all saccate; petals pale yellow, linear-spatulate, 8–10 mm. long; siliques terete, 2–4.5 cm. long, glabrous or short-puberulent, deflexed, subsessile; style 1–2 mm. long; cotyledons oblique.

Arid hillsides and desert plains, Lower and Upper Sonoran Zones, Mojave and Colorado Deserts and adjacent southern Nevada and western Arizona. March–April.

**4. Caulanthus lasiophyllus** (Hook. & Arn.) Payson, Ann. Mo. Bot. Gard. **9:** 303, 1923, var. **lasiophyllus**

*Turritis ?lasiophylla* Hook. & Arn., Bot. Beechey 321, 1841.

*Thelypodium lasiophyllum* Greene, Bull. Torrey Club **13:** 142, 1886.

(For additional synonymy see Ann. Mo. Bot. Gard. **9:** 303–304, 1923.)

Erect annual with simple or ascendingly branched stems 3–20 dm. tall, these more or less hirsute with simple or forked, spreading hairs; leaves petiolate, lower oblong to oblanceolate, 3–5 cm. long, irregularly pinnatifid with obtuse to acute, divaricate segments, these often again toothed or slightly lobed; upper leaves entire or shallowly sinuate-dentate, not auriculate; racemes more or less corymbose; pedicels 2–4 mm. long, ascending in flower but strongly reflexed in fruit; sepals oblong, obtuse, 3–4 mm. long, greenish; petals narrowly spatulate, white or pale yellow, 5–7 mm. long; siliques linear, terete, straight or curved, 3–6 cm. long, subsessile, deflexed; style about 1 mm. long; seeds oblong, often more or less apiculate; cotyledons oblique.

Grassy hillsides and opening in chaparral, Lower and Upper Sonoran Zones, western Washington to northern Baja California. Jan.–June.

**4a. Caulanthus lasiophyllus** var. **utahensis** (Rydb.) Payson, Ann. Mo. Bot. Gard. **9:** 307, 1923

*Thelypodium utahense* Rydb., Bull. Torrey Club **29:** 233, 1902.

Plants glabrous or essentially so; leaves very thin, nearly membranous, lobes mostly rounded and obtuse; pods reflexed but with tips usually curved outward; otherwise as in var. *lasiophyllus.*

Along washes, under trees and shrubs and about rocks, mainly Lower Sonoran Zone, Colorado and Mojave Deserts, California, east to southern Utah, western Arizona, Sonora and northeastern Baja California. Jan.–April.

## 3. STREPTANTHUS Nutt., Jour. Acad. Phila. 5: 134, 1825

Annual or perennial herbs with glaucous or simply pubescent stems and herbage. Cauline leaves often clasping stems, basal ones sometimes doing so. Inflorescence racemose or rarely corymbose, usually elongating in fruit. Sepals ovate to oblong, sometimes outer and inner pairs dissimilar, often colored. Petals narrow, blade often undulate-crisped. Stamens often in pairs in three different lengths. Stigma capitate, scarcely lobed. Pods linear to narrowly oblong, flattened parallel to the septum, erect, spreading, or reflexed, sessile or on a short, stoutish stipe. Seeds flattened, usually winged.

Pods divergent-ascending, 1.5–2 mm. wide; sepals usually purple; seeds 2.5 mm. long or less . . . . . . . . . . . . . . . . . . . . . . . . 1. *S. campestris*
Pods erect, 4.5–6 mm. wide; sepals yellowish or ochroleucous; seeds 4–5 mm. long:
  Style less than 1 mm. long; sepals ochroleucous . . . . . . . . . 2. *S. arizonicus* var. *arizonicus*
  Style 1–2 mm. long; sepals bright yellow when fresh . . . . . . . . 2a. *S. arizonicus* var. *luteus*

**1. Streptanthus campestris** S. Wats., Proc. Amer. Acad. **25:** 125, 1890

Annual or biennial, nearly glabrous and glaucous plant 5–18 dm. tall; basal leaves oblanceolate, 1.5–6 cm. wide, 5–22 cm. long, rounded at apex, sinuately dentate to subentire, gradually or abruptly narrowed to broad petiole, margins more or less setose-ciliate with stout, rather remote hairs; cauline leaves much smaller, lanceolate, cordate-clasping at base, acute at apex, entire or sinuate-dentate; pedicels 5–15 mm. long in flower, spreading-ascending, to 2.5 cm. long in fruit; sepals 9–11 mm. long, usually dark purple, bearing 1–several stoutish setae at apex; petals narrow, purplish, 10–12 mm. long, tips somewhat recurved; styles 1–2 mm. long; pods slightly flattened, 1.5–2 mm. wide, 7–15 cm. long, curved, spreading-ascending; seeds 1.5–2.5 mm. long, narrowly winged.

Rocky hillsides and openings in brush or forest, mainly Upper Sonoran and Arid Transition Zones, Santa Barbara County, California, to southern flanks of the Sierra San Pedro Mártir, Baja California; barely entering the margins of the Sonoran Desert in eastern San Diego County and adjacent Baja California. April–June.

**2. Streptanthus arizonicus** S. Wats., Proc. Amer. Acad. **25:** 125, 1890, var. **arizonicus**

*Disaccanthus arizonicus* Greene, Leaflets Bot. Obs. **1:** 225, 1906.

Annual with erect and simple or ascending and moderately branching stems 2–5 dm. tall, herbage glabrous and glaucous; basal leaves broadly oblanceolate to obovate, 1–5 cm. wide, 3–25 cm. long, subentire, remotely and weakly toothed or lyrately lobed, petiolate; upper cauline leaves lanceolate-ovate to sagittate, cordately clasping at base, entire or nearly so, acute to attenuate at apex; racemes loose, few-flowered, 1 dm. long or less, elongating in fruit; pedicels spreading-ascending, 3–7 mm. long at anthesis, only slightly longer but considerably stouter in fruit; sepals whitish to ochroleucous, 8–10 mm. long, saccate at base; petals yellow, strongly veined with purple, exceeding sepals 3–6 mm., recurved, blades crisped; pods erect or nearly so, 4.5–6 mm. wide, 5–7 cm. long, strongly flattened, midnerve prominent; style about 1 mm. long or less; seeds nearly orbicular, 4.5–5.5 mm. in diameter, the thin wing often 1 mm. wide.

Rocky hillsides and mesas, Lower Sonoran and Upper Sonoran Zones, Mohave County to Greenlee and Pima Counties, Arizona. March–April.

**2*a*. Streptanthus arizonicus** var. **luteus** Kearney & Peebles, Jour. Wash. Acad. **29:** 478, 1939

Differing from var. *arizonicus* in having brighter yellow sepals (that fade in drying until they are indistinguishable from those of the typical plant); and a style 1–2 mm. long.

Known only from Canyon Diablo, the type locality, in the Ajo Mountains, Pima County, Arizona.

## 4. STREPTANTHELLA Rydb., Fl. Rocky Mts. 364, 1917

Annual herbs with entire to shallowly dentate leaves, stems branched above, and glabrous herbage. Inflorescence a lax, rather short raceme. Flowers small. Lateral pair of sepals saccate at base. Petals linear-spatulate, slightly exserted. Pods slender, pendent or recurved, strongly compressed, apex elongated into a slender beak, dehiscent only below the beak, valves remaining attached to septum at apex. Seeds flattened and narrowly winged; cotyledons oblique.

Plants distinctly glaucous; lower leaves merely dentate or repand .................. 1. *S. longirostris* var. *longirostris*

Plants not glaucous; lower leaves pinnatifid with a few spreading lobes .................. 1*a*. *S. longirostris* var. *derelicta*

**1. Streptanthella longirostris** (S. Wats.) Rydb., Fl. Rocky Mts. 364, 1917, var. **longirostris**

*Arabis longirostris* S. Wats., Bot. King Expl. 17, *pl. 2,* 1871.
*Streptanthus longirostris* S. Wats., Proc. Amer. Acad. **25:** 127, 1890.
*Guillenia rostrata* Greene, Leaflets Bot. Obs. **1:** 228, 1906.
*Thelypodium longirostris* Jepson, Man. Fl. Pl. Calif. 413, 1925.

Erect, much-branched above, 2–6 dm. tall, lower part of stem and leaves distinctly glaucous; lower leaves narrowly oblanceolate, 1–9 mm. wide, 1.5–7 cm. long, sinuate-dentate or repand, acute to obtuse at apex; upper leaves linear to narrowly oblong, entire or essentially so; sepals 3–6 mm. long, greenish, sometimes purplish near tips; petals pale yellow veined with brown or purplish, exceeding sepals 1–2 mm., blades narrow and distinctly crisped; pedicels 2–5 mm. long and recurved in fruit; pods 1–2 mm. wide, 3–7 cm. long, beak 3–6 mm. long; seeds 1.5–2 mm. long, brownish, wing about 0.2–0.3 mm. wide, hyaline-membranous.

Open hillsides, mesas, desert slopes and plains, Lower and Upper Sonoran Zones, eastern Washington to northern Baja California east of the Sierra Nevada, and east to Wyoming, Colorado, Arizona, and northern Sonora. March–June.

**1*a*. Streptanthella longirostris** var. **derelicta** J. T. Howell, Leafl. West. Bot. **2**: 57, 1937

Differing from var. *longirostris* in having bright green stems and leaves without a glaucous coating and lower leaves remotely pinnatifid with narrow, divaricate lobes.

Desert areas, mostly in sandy soil, Lower Sonoran Zone, Colorado Desert in Riverside, San Diego, and Imperial Counties, California, and into northeastern Baja California, western Arizona, and northwestern Sonora.

## 5. LEPIDIUM L., Sp. Pl. 643, 1753

Erect, decumbent, or diffuse annuals, biennials, or perennials with pinnatifid or lobed lower leaves and reduced, often entire, upper leaves. Flowers small, white, yellowish, or greenish, arranged in terminal, compact to loose racemes, these elongating considerably in fruit. Sepals obtuse or rounded at apex, often thin and subscarious at margins, caducous. Petals small, sometimes wanting. Stamens 6, 4, or 2. Styles slender, one-half as long as ovary or shorter, or stigma sessile. Fruit a silicle, this orbicular, oblong, or obovate, flattened contrary to partition, more or less emarginate and often somewhat winged at apex, the valves separating from septum at dehiscence. Seeds 1 in each cell, somewhat flattened; cotyledons incumbent or less commonly oblique or accumbent.

Plants perennial; petals 3–4 mm. long:
- Stems wholly herbaceous, from a deep-seated rootstock; silicles cordate in outline, somewhat inflated; leaves 1–5 cm. wide, coarsely dentate .................. 1. *L. draba*
- Stems woody at the base; silicles obcordate, not inflated; leaves narrowly linear and entire, 1–2 mm. wide .......................................................... 2. *L. fremontii*

Plants annual or rarely biennial; petals 1–3 mm. long or lacking:
- Herbage yellow-green; flowers yellow; styles one-third to one-half as long as silicle:
  - Silicles 2–3 mm. long; styles one-half as long as fruiting body .. 3. *L. flavum* var. *flavum*
  - Silicles 3–3.5 mm. long; styles one-third as long as fruiting body .................. ............................................... 3*a*. *L. flavum* var. *felipense*
- Herbage dark green or gray-green, not yellowish; flowers white or greenish; styles usually not over one-fifth as long as silicle or stigmas sessile:
  - Style 0.4–0.6 mm. long; plants hirsute to pilose, longer hairs flattened .. 4. *L. thurberi*
  - Style less than 0.2 mm. long or stigma sessile; plants glabrous or pubescent but hairs not flattened; silicles mostly lenticular, somewhat convex on both surfaces:
    - Pedicels terete or only slightly flattened:
      - Sepals persistent until fruit is nearly fully grown; cauline leaves usually pinnatifid; petals shorter than sepals or lacking ......... 5. *L. oblongum* var. *oblongum*

Sepals caducous, falling soon after anthesis; cauline leaves merely dentate or entire; petals exceeding sepals, 2–3 mm. long:
Stems and inflorescences glabrous, at least above base ........................................................... 6. *L. medium* var. *medium*
Stems and inflorescences finely but densely puberulent throughout ............................................... 6a. *L. medium* var. *pubescens*
Pedicels strongly flattened, two to five times as wide as thick:
Silicles glabrous, smooth and shining, 3.5–6 mm. long; lower leaves deeply pinnatifid to bipinnatifid; seeds 2–3 mm. long .................. 7. *L. nitidum*
Silicles puberulent to hispidulous on both sides, about 2.5–4 mm. long; lower leaves entire, dentate, incised or rarely pinnatifid; seeds 1.8–2.2 mm. long:
Hairs not strongly pustulate basally; glands between stamens 0.1–0.2 mm. long:
Pedicels not over three times as broad as thick:
Hairs on stems often 0.5–1 mm. long; pedicels puberulent on both surfaces ................................ 8. *L. lasiocarpum* var. *lasiocarpum*
Hairs on stems usually less than 0.5 mm. long; pedicels usually glabrous beneath:
Plants erect or branches decumbent-ascending; racemes to 20 cm. long; silicles 3–4 mm. long; divisions of leaves often over 2 mm. wide:
Cauline leaves 5–20 mm. broad, lobed or dentate; racemes loosely flowered, 5–20 cm. long ...... 8*a*. *L. lasiocarpum* var. *latifolium*
Cauline leaves usually less than 5 mm. wide, usually pinnatisect; racemes compactly flowered, mostly less than 10 cm. long ....................................... 8*b*. *L. lasiocarpum* var. *georginum*
Plants prostrate; racemes 1–5 cm. long; silicles less than 3 mm. long; divisions of leaves usually less than 2 mm. broad .................................................. 8*c*. *L. lasiocarpum* var. *rosulatum*
Pedicels three to five times as wide as thick ... 8*d*. *L. lasiocarpum* var. *palmeri*
Hairs markedly pustulate basally; glands between stamens 0.2–0.5 mm. long .......................................... 8*e*. *L. lasiocarpum* var. *wrightii*

**1. Lepidium draba** L., Sp. Pl. 645, 1753
*Cardaria draba* Desv., Jour. Bot. **3:** 163, 1814.

Perennial from a deep-seated rootstock with numerous erect or ascending stems 2–5 dm. tall; herbage sparsely and finely puberulent to densely puberulent and cinereous; leaves oblong to oblong-obovate or upper sometimes lanceolate, 1–5 cm. wide, 2.5–10 cm. long, denticulate to coarsely dentate, lower with terete petioles to 1 cm. long, upper sessile with clasping bases; racemes corymbosely arranged, many-flowered; pedicels slender, terete or essentially so, 7–10 mm. long; sepals whitish, obovate, 1–1.5 mm. long, glabrous; petals white, short-clawed, 3–4 mm. long; silicles inflated, triangular-ovate to rounded-ovate, 4–6 mm. wide, 3–5 mm. long, cordate at base, without a terminal notch or faintly retuse; styles slender, 1–1.3 mm. long; seeds 1–2 in each cell, dark brown, about 2 mm. long; cotyledons incumbent.

Waste places and in cultivated fields, Lower Sonoran Zone to Transition Zone, Washington and Oregon south to southern California; east into Idaho, Utah, Arizona, and New Mexico; introduced from Europe. April–July.

**2. Lepidium fremontii** S. Wats., Bot. King Expl. 30, *pl. 4, figs. 3–4,* 1871
*Nasturtium fremontii* Kuntze, Rev. Gen. Pl. **2:** 937, 1891.

Suffrutescent, much-branched and rounded plant 2–8 dm. tall and often fully as wide, herbage glabrous and more or less glaucous; leaves narrowly linear and entire,

1–2 mm. wide, 2.5–6 cm. long, or with 1–2 pairs of linear lobes or teeth to 2 cm. long, acute at apex of leaves and lobes; racemes numerous, borne terminally and in axils of upper leaves, compact in flower, more open and elongated to 1 dm. or more in length in fruit; pedicels slender, terete or essentially so, spreading, 4–7 mm. long in fruit; sepals whitish, submembranous along margins, 1.6–2 mm. long, rounded at apex; petals white, 3–4 mm. long, rounded-obovate; silicles shallowly obcordate, 5–8 mm. broad, thin, light stramineous; style about 0.5 mm. long; stigma capitate, 0.3–0.5 mm. in diameter; seeds brownish, 2.5–3 mm. long, narrowly winged.

Rocky or gravelly slopes, Lower Sonoran Zone, Mojave and Colorado Deserts to southern Utah and Arizona. Feb.–May.

**3. Lepidium flavum** Torr., Pacif. R. Rep. **4:** 67, 1857, var. **flavum**

*Sprengeria flava* Greene, Leaflets Bot. Obs. **1:** 198, 1906.
*Sprengeria watsoniana* Greene, op. cit. 199.
*Sprengeria minuscula* Greene, loc. cit.
*Lepidium flavum* var. *apterum* Henrard & Thell., Medelell. Rijks Herb. Leid. **34:** 1, 1918.
*Lepidium flavum* var. *typicum* C. L. Hitchcock, Madroño **3:** 298, 1936.

Prostrate to decumbent annual with yellow-green, glabrous herbage and stems 0.3–3 dm. long; lower leaves spatulate, oblong-lanceolate or narrowly oblanceolate, 1.5–5 cm. long, irregularly lobed, toothed or pinnatifid; upper leaves smaller and less deeply lobed or entire; racemes loose to crowded, sometimes subumbellate; pedicels slender, terete or only faintly flattened, 1.5–3 mm. long, spreading or slightly ascending; sepals about 1.5 mm. long, obtuse or rounded, yellow; petals sulfur-yellow, about 2 mm. long; stamens 6; silicles oval, 2–3 mm. long, 1.5–2.2 mm. wide, apical wings thin, usually slightly divergent, 0.2–0.5 mm. long, faces distinctly reticulate-nerved, glabrous; styles slender, 1–1.5 mm. long, usually nearly or quite one-half as long as silicle; cotyledons incumbent.

Sandy or somewhat saline soils, Lower Sonoran Zone, Humboldt and Elko Counties, Nevada, southwest to southern California and northern Baja California. Commoner in the Mojave Desert than in the Colorado Desert. March–June.

**3*a*. Lepidium flavum** var. **felipense** C. L. Hitchcock, Madroño **3:** 299, 1936

Differing from var. *flavum* in having silicles 3–4 mm. long, 3–3.5 mm. wide, with very small apical wings; styles 1–1.5 mm. long, hence less than one-half as long as body of fruit.

Arid slopes and saline soil, Lower Sonoran Zone, western margin of the Colorado Desert near San Felipe and Borrego Springs and on adjacent mountain slopes.

**4. Lepidium thurberi** Wooton, Bull. Torrey Club **25:** 259, 1898

Rank annuals or apparently biennial under favorable conditions with erect or ascending, freely branched stems 1–6 dm. tall, stems hirsute-canescent to pilose throughout, longer hairs obviously flattened, shorter ones usually papilliferous or clavate; lower leaves oblanceolate in outline, 3–7.5 cm. long, 1–1.5 cm. wide, pinnatifid into 3–8 pairs of more or less lobed or dissected segments, ultimate divisions linear to obovate, acute, or apiculate; cauline leaves somewhat reduced but similar to lower, or uppermost entire; racemes many-flowered, often elongating to 1.5 or 2 dm. in fruit; pedicels slender, spreading, and 6–9 mm. long in fruit, slightly flattened, pilosulous along margins; sepals broadly ovate, white-margined, 1–1.5 mm. long, caducous; petals white, 2–3 mm. long; stamens 6, filaments glabrous; silicles ovate to suborbicular, 2–2.5 mm. wide, 2–3 mm. long, very shallowly and narrowly notched, narrowly winged near apex, glabrous, slender style 0.4–0.6 mm. long; seeds light brown, 1–1.5 mm. long; cotyledons incumbent.

Waste places, roadsides, along washes, and on desert grassland, Lower and Upper Sonoran Zones, Pima County to Yavapai County, Arizona, and east into southwestern New Mexico. Feb.–Sept.

**5. Lepidium oblongum** Small, Fl. SE. U.S. 468, 1903, var. **oblongum**

*Lepidium oblongum* var. *typicum* C. L. Hitchcock, Madroño **8**: 125, 1945.

Prostrate to ascending annuals 1–3 dm. tall with hirtellous to villose-hirsute stems usually much branched from base; leaves pinnatifid or bipinnatifid, 4–9 mm. wide, 8–20 mm. long, narrowly elliptic to linear segments 0.6–3 mm. wide, acute; racemes 2–10 cm. long, rather compact; pedicels 2–3.5 mm. long, ascendingly spreading, moderately flattened, pubescent-strigillose on upper surface and narrowly winged margins, glabrous to faintly strigillose beneath; sepals green, broadly ovate to suborbicular, 0.8–1 mm. long, sparsely puberulent to glabrate; petals minute, shorter than sepals or lacking; stamens 2; silicles broadly elliptic to obovate-elliptic, 2.5–3.2 mm. long, shallowly notched, faintly veined, usually with a few marginal hairs but sometimes glabrous; stigma sessile; cotyledons incumbent.

In waste ground and on grassy hillsides, Lower Sonoran Zone to Carolinian Zone, southern California to Kansas, Arkansas, Texas, and Guatemala. Feb.–May.

**6. Lepidium medium** Greene, Erythea **3**: 36, 1895, var. **medium**

*Lepidium intermedium* A. Gray, Pl. Wright. **2**: 15, 1853. Not *L. intermedium* A. Rich., 1847.
*Lepidium idahoense* Heller, Bull. Torrey Club **26**: 312, 1899.
*Lepidium glaucum* Greene, Pittonia **4**: 312, 1901.
*Lepidium virginicum* var. *medium* C. L. Hitchcock, Madroño **3**: 285, 1936.

Erect to decumbent, strict to much-branched annuals with stems 1–3.5 dm. long; at least upper parts of stems and pedicels glabrous or essentially so; lower leaves spatulate to oblanceolate in outline, 5–14 mm. wide, 2–8 cm. long, pinnatifid into remote or crowded dentate segments; upper leaves reduced, merely dentate or slightly lobed or uppermost entire; racemes 1.5–8 cm. long; pedicels terete or nearly so, slender, 3–6 mm. long, spreading or slightly ascending at tips; sepals obovate to suborbicular, white-margined, about 1 mm. long; petals white, 2–3 mm. long; stamens usually 2, rarely 6; silicles suborbicular, 2.5–3.2 mm. in diameter, glabrous, obscurely reticulate-veined; apical notch about 0.2 mm. deep; stigma sessile; seeds reddish brown, about 1.2 mm. long; cotyledons incumbent or slightly oblique.

Sandy gentle slopes and plains, Lower and Upper Sonoran Zones, British Columbia to Montana and south to southern California, Arizona, and Texas. Feb.–Aug.

**6a. Lepidium medium** var. **pubescens** (Greene) Robinson, in A. Gray, Syn. Fl. N. Amer. **1**[1]: 127, 1895

*Lepidium intermedium* var. *pubescens* Greene, Bot. Gaz. **6**: 157, 1881.
*Lepidium virginicum* subsp. *texanum* var. *pubescens* Thell., Monog. Lepid. 224, 1906.
*Lepidium bernardinum* Abrams, Bull. Torrey Club **37**: 149, 1910.
*Lepidium hirsutum* Rydb., Bull. Torrey Club **39**: 322, 1912.
*Lepidium virginicum* var. *pubescens* C. L. Hitchcock, Madroño **3**: 283, 1936.

Differing from var. *medium* in having the entire stem and pedicels finely but densely puberulent and often cinereous.

Open grassy hillsides and plains, Lower and Upper Sonoran Zones, Washington to Wyoming and south to northern Baja California, Arizona, and Texas.

**7. Lepidium nitidum** Nutt., in Torr. & Gray, Fl. N. Amer. **1:** 116, 1838

*Lepidium leiocarpum* Hook. & Arn., Bot. Beechey 324, 1840.
*Lepidium nitidum* var. *insigne* Greene, Fl. Fran. 274, 1891.
*Lepidium nitidum* var. *typicum* C. L. Hitchcock, Madroño **3:** 292, 1936.

Simple and erect to much-branched from base and moderately procumbent annual 1–3.5 dm. tall, stems puberulent in inflorescences, glabrate to puberulent below; lower leaves 3–10 cm. long, deeply pinnatifid or bipinnatifid into linear segments, these 0.2–1.5 mm. wide, 3–15 mm. long, finely and sparsely puberulent to glabrous, soon turning reddish and deciduous; upper leaves smaller, entire to pinnatifid, their central rachis 1–4 mm. broad; pedicels strongly flattened, 3–4 mm. long, spreading, puberulent on edges; sepals thin, ovate, 0.8–1 mm. long, white to purplish, glabrous to puberulent on back; petals white, spatulate, 1–1.5 mm. long or sometimes minute and inconspicuous; stamens 6 or sometimes 4; silicles broadly ovate to nearly orbicular, 3.5–6 mm. long, shallowly and narrowly but distinctly notched apically, planoconvex to concave-convex, smooth and shining, reddish, with fine but distinct reticulate veining; stigma almost sessile in notch; seeds brownish, 2–3 mm. long; cotyledons incumbent.

Open grassy hillsides or plains, Lower and Upper Sonoran Zones, Washington to the southern Sierra San Pedro Mártir, Baja California; rare in the desert regions. Feb.–April.

**8. Lepidium lasiocarpum** Nutt., in Torr. & Gray, Fl. N. Amer. **1:** 115, 1838, var. **lasiocarpum**

*Lepidium lasiocarpum* subsp. *wrightii* var. *pubescens* Thell., Monog. Lepid. 267, 1906.
*Lepidium lasiocarpum* var. *typicum* C. L. Hitchcock, Madroño **3:** 289, 1936.

Annual 0.5–2.5 dm. tall with 1–many erect to decumbent stems from base, herbage hispidulous throughout; lower leaves oblanceolate to obovate, 0.3–2.2 cm. wide, 2–15 cm. long, subentire, dentate, incised or sometimes pinnatifid, narrowing cuneately to a slender petiole below, lobes or teeth usually rounded or obtuse and rarely over 5 mm. wide, lower leaves usually deciduous rather early; upper leaves smaller and merely toothed or less commonly pinnatifid; racemes erect, numerous, to 1.5 dm. long; pedicels strongly flattened, divaricately spreading, 1–3 mm. long; sepals broadly oblong or narrowly oblanceolate, about 1 mm. long, purplish with membranous white margins; petals minute or lacking; pods suborbicular or broadly ellipsoid-orbicular, about 3 mm. long, shallowly and narrowly notched apically, often purplish or reddish, puberulent to hispidulous on both surfaces, thin at apex; stigma sessile; seeds brownish, 1.8–2.2 mm. long, very narrowly winged.

Mainly sandy soils, Lower and Upper Sonoran Zones, coastal southern California to Colorado and south into Coahuila, Sinaloa, and central Baja California. Feb.–May.

**8*a*. Lepidium lasiocarpum** var. **latifolium** C. L. Hitchcock, Madroño **8:** 136, 1945

Plants sparsely puberulent with short hairs, usually much less than 0.5 mm. long; cauline leaves 5–20 mm. broad, thinnish, sinuately and shallowly lobed; racemes loosely flowered, 10–20 cm. long; pedicels two to three times as broad as thick, puberulent to glabrous beneath; silicles about 3.5 mm. long, pubescent on all surfaces or on margins only.

Sandy arid soils, Lower Sonoran Zone, southwestern Arizona, Baja California, and Sonora.

**8*b*. Lepidium lasiocarpum** var. **georginum** (Rydb.) C. L. Hitchcock, Madroño **3:** 291, 1936

*Lepidium georginum* Rydb. Bull. Torrey Club **30:** 253, 1903.
*Lepidium lasiocarpum* subsp. *georginum* Thell., Monog. Lepid. 267, 1906.

Plants softly puberulent or hispidulous or occasionally subglabrous; cauline leaves 1–5 mm. wide, entire to slightly toothed; racemes compactly flowered; pedicels glabrous on lower or both surfaces; silicles 3–3.5 mm. long, puberulent on upper surface or only on margins, or fruits less commonly wholly glabrous.

Arid slopes and flats, Lower Sonoran Zone, Inyo County, California, to southern Utah, western Arizona, and northwestern Sonora.

**8*c*. Lepidium lasiocarpum** var. **rosulatum** C. L. Hitchcock, Madroño **8:** 137, 1945

Plant prostrate, forming a rosette 0.5–2 dm. in diameter, herbage strigillose to hirsutulous; leaves 1–2 cm. long, pinnatifid into segments 1–2 mm. wide; racemes very densely flowered, 1–5 cm. long; pedicels only slightly flattened, about twice as wide as thick, usually puberulent on upper surface only; silicles oblong-obovate, 2–3 mm. long, pubescent, especially on margins or sometimes only margins puberulent or entire pod glabrous.

Hillsides and plains, Lower Sonoran Zone, southwestern Arizona and adjacent Sonora.

**8*d*. Lepidium lasiocarpum** var. **palmeri** (S. Wats.) C. L. Hitchcock, Madroño **8:** 140, 1945

*Lepidium palmeri* S. Wats., Proc. Amer. Acad. **24:** 39, 1889.
*Lepidium lasiocarpum* subsp. *palmeri* Thell., Monog. Lepid. 267, 1906.

Branches distinctly decumbent, hirsutulous with hairs to 0.5 mm. long or slightly more; leaves pinnatifid or bipinnatifid with segments 1–3 mm. broad; racemes compact, densely flowered, 1–7 cm. long; pedicels strongly flattened, three to five times as wide as thick; usually puberulent on both surfaces; pods oblong-ovate, about 4 mm. long, densely puberulent to glabrous.

Arid hillsides and sandy flats, Lower Sonoran Zone, vicinity of Bahía de los Angeles, Baja California.

**8*e*. Lepidium lasiocarpum** var. **wrightii** (A. Gray) C. L. Hitchcock, Madroño **3:** 289, 1936

Plant densely hispidulous with pustulous-based hairs 0.5–1.2 mm. long; cauline leaves usually pinnatifid into segments 1–5 mm. wide; pedicels four to six times as wide as thick; glands at bases of stamens 0.2–0.5 mm. long; silicles 3.5–4 mm. long, densely hispidulous.

Sandy and rocky soil, mainly Lower Sonoran Zone, south-central Arizona to southwestern Texas and adjacent Coahuila.

**Lepidium montanum** Nutt. var. **canescens** (Thell.) C. L. Hitchcock, Madroño **3:** 304, 1936. This peppergrass occurs commonly in northern Arizona, but the writer has seen no specimens of it from the Sonoran Desert.

## 6. SISYMBRIUM L., Sp. Pl. 657, 1753

Annual or biennial, simple to branched herbs with entire, dentate, or pinnatifid leaves, lower ones petiolate, cauline leaves often auriculate and sessile. Inflorescences racemose, elongating greatly in fruit. Sepals spreading at anthesis. Petals yellow or purplish. Stamens 6. Siliques terete, elongate, appressed to rachises or spreading, dehiscent. Style short or 2-lobed stigma sessile. Seeds oblong, in 1 or 2 rows in each cell, wingless. Cotyledons incumbent.

Siliques erect, closely appressed to rachis, subulate-linear, acuminate, 10–15 mm. long .... ........................................................ 1. *S. officinale*
Siliques ascending or spreading, narrowly linear, 2–10 cm. long:
  Petals 3–4 mm. long; upper leaves coarsely pinnatifid; siliques 0.5–0.8 mm. in diameter, 2–5 cm. long ........................................................ 2. *S. irio*
  Petals 6–8 mm. long; upper leaves pinnatifid into linear segments; siliques 1–1.2 mm. in diameter, 5–10 cm. long .................................... 3. *S. altissimum*

### 1. Sisymbrium officinale (L.) Scop., Fl. Carn. ed. 2, 2: 26, 1772

*Erysimum officinale* L., Sp. Pl. 660, 1753.
*Sisymbrium leiocarpum* Jordan, Diag. **1**: 139, 1864.

An erect, slenderly simple or much-branched herb 3–15 dm. tall with rather rigid stems and sparsely rough-hispid herbage; leaves lyrate-pinnatifid to pinnately parted, 7–15 cm. long, larger to 7 cm. broad at widest part, segments coarsely dentate, lower leaves on petioles equaling blades; upper leaves smaller, lanceolate to hastate; pedicels stout, about 2 mm. long in fruit; sepals about 2 mm. long, minutely hispid; petals yellow, about 3 mm. long, blades only slightly spreading; siliques strictly erect, appressed to rachis, 10–15 mm. long, about 1.5 mm. wide at base, gradually narrowed to subulate tip, seeds in 1 row in each cell.

A common weed along roadsides, ditches, and in fields, Lower Sonoran Zone to Transition Zone, Washington to northern Baja California, and at scattered points along the Atlantic seaboard; introduced from Europe. Known from the Sonoran Desert only along its western margins in southern California and adjacent Baja California. March–July.

### 2. Sisymbrium irio L., Sp. Pl. 659, 1753

*Norta irio* Britt., in Britt. & Brown, Ill. Fl. ed. 2, **2**: 174, 1913.

Erect annual, strict or branching from above base, 2–10 dm. tall, glabrous throughout; leaves ovate to lanceolate in outline, to 20 cm. long, petiolate, runcinately pinnatifid into 1–3 pairs of entire to slightly dentate lobes below large terminal lobe which is usually several times as large as lateral ones, petioles 1.5–3.5 cm. long; inflorescence dense at first, later elongating; sepals elliptic-lanceolate, 2–2.5 mm. long; petals narrowly oblanceolate, 3–4 mm. long, pale cream, and fading almost white before falling; stamens exserted 1–2 mm. beyond petals; pedicels slender, about 0.5 mm. in diameter, spreading-ascending, 0.5–2 cm. long in fruit; siliques slenderly linear, 0.5–0.8 or rarely 1 mm. in diameter, 2–5 cm. long, usually arcuately curved upward, each valve finely but distinctly 3-nerved; seeds in a slightly zigzag row, about 0.8 mm. long.

Wayside weed, Lower and Upper Sonoran Zones, a native of Europe, naturalized from Ventura County, California, to southern Utah and southeastern Texas south into southern Sonora. Fairly common in the Colorado Desert and in adjacent Arizona. Feb.–May.

**3. Sisymbrium altissimum** L., Sp. Pl. 659, 1753

*Norta altissima* Britt., in Britt. & Brown, Ill. Fl. ed. 2, **2:** 174, 1913.

Erect, widely branching annual 4–15 dm. tall, the stems glabrous or nearly so; leaves pinnately parted, lower ones into lobes 5–25 mm. broad, upper into linear lobes 1–4 mm. wide, large lower leaves to 15 cm. long, often sparsely hispid beneath; sepals yellowish, 3–4.5 mm. long; petals yellow, 6–8 mm. long; stamens about equaling sepals; pedicels stoutish, ascending-spreading, becoming rigid in fruit, 6–8 mm. long; siliques linear, 5–10 cm. long, valves 1-nerved.

Introduced European weed, mainly Upper Sonoran and Transition Zones, Washington to southern California and less commonly into adjacent western states. Occurring within the Sonoran Desert only rarely and then as a straggler. April–July.

## 7. SINAPIS L., Sp. Pl. 668, 1753

Annual or biennial, erect, moderately to much-branched, more or less hispid herbs with lobed to pinnatifid leaves and showy yellow flowers in terminal racemes. Sepals ascending to slightly spreading. Petals narrowly clawed. Stamens 6. Siliques terete or nearly so, on short pedicels, usually constricted between the seeds, tipped with an angled or flattened beak, valves usually 3-nerved. Stigma faintly 2-lobed. Seeds in 1 row in each cell, globose. Cotyledons conduplicate.

**1. Sinapis incana** L., Cent. Pl. **1:** 19, 1755

*Hirschfeldia adpressa* Moench, Meth. 264, 1794.
*Sinapis geniculata* Desf., Fl. Atlantica **2:** 98, 1798.
*Brassica adpressa* Boiss., Voy. Espag. **2:** 38, 1839.
*Brassica incana* Meigen, Deutsch. Fl. **3:** 270, 1842. Not *B. incana* Tenore, 1811–1813.
*Hirschfeldia incana* Lagreze-Fossat, Fl. Tarn et Garonne 19, 1847.

Erect, much-branched biennial 3–8 dm. tall with sparsely hispidulous stems and densely hispidulous leaves; lower leaves pinnatifid, 4–15 cm. long, with large, ovate to suborbicular, terminal lobe and several pairs of smaller, ovate to oblong, lateral lobes; upper leaves much reduced, denticulate, short-petioled; pedicels appressed closely to rachis, 1–4 mm. long; sepals 2.5–3 mm. long; petals 5–6 mm. long; siliques torulose, closely appressed, 6–10 mm. long, the slenderly conical beak 1–2 mm. long; valves faintly 3-nerved.

Adventive weed from Europe, Lower and Upper Sonoran Zones, central and southern California; rare in the desert regions. May–Aug.

## 8. BRASSICA L., Sp. Pl. 666, 1753

Annual, biennial, or sometimes perennial herbs with erect or ascendingly branched stems and alternate leaves, the basal pinnatifid or pinnately divided, upper only toothed or entire and frequently sessile. Inflorescences racemose, more or less elongated, especially in fruit. Sepals slightly dimorphic, the lateral ones somewhat gibbous basally. Petals yellow, large, the long claw abruptly expanding into a broad blade. Nectary glands 4, papilla-like, green, episepalous. Silique terete or slightly angled, terminating in a stoutish, subulate to conical beak; valves 1–3-nerved. Stigmas truncate or 2-lobed. Seeds in 1 row in each cell, globose. Cotyledons conduplicate, incumbent.

Siliques closely appressed to the rachis; beaks 3–4 mm. long; stems not glaucous .. 1. *B. nigra*
Siliques ascending-spreading, not appressed to rachis; beaks 8–12 mm. long; stems slightly to pronouncedly glaucous:

Cauline leaves auriculate-clasping; seeds 1 mm. in diameter ............ 2. *B. campestris*
Cauline leaves petiolate or sessile but not clasping; seeds 1.4–1.8 mm. in diameter ...... ........................................................ 3. *B. juncea*

**1. Brassica nigra** (L.) Koch, in Roehl., Deutsch. Fl. ed. 3, **4:** 713, 1833
*Sinapis nigra* L., Sp. Pl. 668, 1753.

Annual, much branched from base unless closely crowded, then often nearly strict, 3–25 dm. tall, sparsely hispid, especially on lower sides of larger leaves, to glabrate; lower leaves pinnatifid, with large, dentate, terminal lobe and several smaller, denticulate, lateral ones; blade 2–15 cm. wide, 5–45 cm. long, petioles shorter than blades; upper leaves smaller, short-petioled, pinnatifid or dentate, uppermost lanceolate, often entire; sepals yellowish, 3.5–4 mm. long; petals bright yellow, oblong-spatulate, 6–8 mm. long; siliques slender, about 1 mm. in diameter, 8–20 mm. long, subulate-conic beak about 3 mm. long, ascending and closely appressed to rachis; seeds dark brown, about 1 mm. long.

A weed in fields and wasteland, mainly Upper Sonoran Zone, adventive from Europe and widely distributed in North America south of the boreal areas; reaching only the margins of the Sonoran Desert. April–Sept.

**2. Brassica campestris** L., Sp. Pl. 666, 1753

Erect, sparingly branched or strict herb 3–20 dm. tall with rather succulent, glabrous and glaucous herbage or lowest leaves scantily hispid; lower leaves lyrate-pinnatifid, to 5.5 cm. wide, 10–25 cm. long, lobes irregularly dentate or serrate, petiolate; cauline leaves lance-oblong, sessile and clasping by an auricled base; inflorescence dense, elongating markedly after flowering; sepals yellow, narrowly oblong; petals bright yellow, 6–7 mm. long, spatulate; siliques terete, 5–7 cm. long, about 3 mm. in diameter, on spreading pedicels 10–25 mm. long; beak of silique 8–12 mm. long; stigma truncate.

Roadsides and fields, chiefly Upper Sonoran Zone to Transition Zone, adventive from Europe and widely established in the Pacific States, less commonly eastward; occasional in irrigated desert regions of Arizona and Mexico. Flowering almost throughout the year after sufficient moisture.

**3. Brassica juncea** (L.) Cosson, Bull. Soc. Bot. Fr. **6:** 609, 1859
*Sinapis juncea* L., Sp. Pl. 668, 1753.

Erect, moderately to much-branched annual with stoutish, glabrous or sparingly hirsute, slightly glaucous stems 1.5–12 dm. tall; lower leaves lyrate-pinnatifid and crenate, long-petiolate, 2–10 cm. wide, 5–30 cm. long, usually sparsely but coarsely hispid, particularly on midrib and main veins, rounded at apex; upper leaves reduced, usually simple and remotely toothed to entire, linear-lanceolate to narrowly oblanceolate, sessile but not auriculate-clasping; racemes 5–20 cm. long in fruit; flowers pale yellow, petals 5–8 mm. long; pods 3–5 cm. long, 2–3 mm. in diameter, nearly terete, each valve with a conspicuous median vein, glabrous, beak 8–12 mm. long, empty or containing solitary seed in basal part; seeds reddish brown, 1.4–1.8 mm. in diameter, finely granular-papillose.

Native of Asia. Introduced as a weed and more or less naturalized from Washington to southern California and sparingly into Arizona, Lower Sonoran Zone to Transition Zone; also from Saskatchewan to the New England states. Flowering Feb.–July in our area.

## 9. DIPLOTAXIS DC., Syst. 2: 628, 1821

Annual, biennial, or perennial herbs with branching stems and glabrous to hispid herbage. Leaves basal and cauline, dentate, lobed, or pinnatifid. Inflorescences terminal racemes, showy, but flowers comparatively few. Petals yellow. Stamens 6. Nectar glands 4. Pods linear, somewhat flattened parallel to the partition, short-beaked or beakless, dehiscent, valves 1-nerved. Seeds wingless, in 2 rows in each cell. Cotyledons conduplicate.

### 1. Diplotaxis muralis (L.) DC., Syst. 2: 634, 1821

*Sisymbrium murale* L., Sp. Pl. 658, 1753.

Annual or less commonly biennial, branched from base, the curving-ascending branches 1.5–6 dm. tall, glabrous to sparsely hispid; leaves all at or near base, oblanceolate in outline, 1–2.5 cm. wide, 3.5–10 cm. long, shallowly pinnately lobed to pinnatifid, on slender petioles 1–3 cm. long; racemes loose, few-flowered, elongating in fruit; sepals purplish, 3–4 mm. long, very sparingly hispid; petals purplish, 6–8 mm. long; pedicels ascending-spreading, slender, 10–22 mm. long in fruit; pods linear, about 2 mm. wide in middle, tapering slightly toward each end, 2–3.5 cm. long; seeds pale reddish brown, about 0.8–1 mm. long.

An adventive from Europe and established as a weed from Oregon to southern California and east to the middle western states, south into northern Mexico. Fairly common in irrigated regions in Arizona. March–May.

## 10. DRYOPETALON A. Gray, Pl. Wright. 2: 11, 1853*

Annual or biennial herbs with strict or moderately branched terete stems and entire to dissected leaves, lower ones petiolate, upper cauline ones short-petiolate to sessile. Flowers in dense to loose racemes. Sepals oblong, obtuse at the apex, with wide hyaline margins. Petals white to purple, obovate to spatulate in outline, shallowly to deeply 3–9-lobed, glabrous or sometimes papillose near base. Stamens glabrous or pilose basally. Siliques linear, erect or somewhat arcuate, terete to slightly flattened parallel to septum, short-stipitate to sessile. Style short or stigma sessile. Seeds in a single row in each locule, oblong. Cotyledons accumbent or incumbent.

Cotyledons accumbent; siliques sessile; basal leaves subpersistent, 5–15 cm. long:
- Inflorescence dense; stems hirsute below; sepals usually sparsely pubescent .............................................................. 1. *D. runcinatum* var. *runcinatum*
- Inflorescence lax, few-flowered; stems glabrous or very sparsely pubescent below; sepals glabrous .................................... 1a. *D. runcinatum* var. *laxiflorum*

Cotyledons incumbent; siliques stipitate; basal leaves caducous or none:
- Stamens and petals pilose at base; siliques 1–1.5 cm. long; fruiting pedicels and siliques strongly ascending ............................ 2. *D. crenatum* var. *racemosum*
- Stamens and petals glabrous; siliques 1.5–5 cm. long; fruiting pedicels and siliques divaricately ascending:
  - Petals white to slightly purplish, blade about one-half as long as claw; siliques less than 1.5 mm. wide; lower fruiting pedicels about 1 cm. long; leaves repand to shallowly lobed ..................................................... 3. *D. palmeri*
  - Petals purple, blade about equaling claw; siliques 2 mm. wide; lower fruiting pedicels 1.5–2 cm. long; leaves deeply lobed to pinnately dissected ...... 4. *D. purpureum*

---

* Treatment adapted from Reed G. Rollins, "The Cruciferous Genus *Dryopetalon*," Contr. Dudley Herb. 3: 199–207, *pls. 48–49,* 1941.

**1. Dryopetalon runcinatum** A. Gray, Pl. Wright. **2:** 12, *pl. 11,* 1853, var. **runcinatum**

*Sisymbrium umbrosum* Robinson, Bot. Gaz. **30:** 60, 1900.
*Coelophragmus umbrosus* O. E. Schulz, Pflanzenreich **4**[105]**:** 158, 1924.

Annual or biennial with several stems from simple caudex; stems hirsute with white simple trichomes below, glabrous above; basal leaves runcinate to pinnately or bipinnately divided, petiolate, densely to sparsely hirsute, oblong to obovate, 3–8 cm. wide, 4–20 cm. long, short-petiolate; cauline leaves sessile or very short-petiolate, 1–5 cm. wide, 3–15 cm. long, usually more sharply divided than basal leaves, hirsute or uppermost glabrous; inflorescence crowded, flowers numerous; sepals oblong to obovate, obtuse, with 1–several bristlelike trichomes near apex or rarely glabrous, 3.5–4.5 mm. long, 1.5–2.5 mm. wide; petals white to purplish, spatulate to narrowly obovate, 2.5–3.5 mm. wide, 6–9 (12) mm. long, with 5–12 lobes, upper lobe larger than others; stamens minutely papillose to sparsely pilose at base; nectaries small, subtending single stamens; fruiting pedicels obliquely spreading, 5–20 mm. long, glabrous to sparsely hirsute; siliques sessile or essentially so, about 1 mm. wide, 2–6 cm. long, straight to arcuate, divaricate to slightly ascending; stigma sessile or style to 1 mm. long; seeds about 1 mm. long; cotyledons accumbent.

Rocky slopes and sides of cliffs, upper margins of Lower Sonoran Zone to Transition Zone, Pima County, Arizona, east to western Texas and south into Chihuahua and northern Sonora. Barely entering the upper margins of the Sonoran Desert. March–April.

**1*a*. Dryopetalon runcinatum** var. **laxiflorum** Rollins, Contr. Dudley Herb. **3:** 201, *pl. 48, fig. 2,* 1941

Differing from var. *runcinatum* in having lax, few-flowered instead of densely crowded, many-flowered racemes; pedicels and sepals glabrous, and pedicels more slender than in var. *runcinatum.*

Rocky hillsides and cliffs, Lower Sonoran Zone, southern Sonora into Sinaloa, Mexico. Feb.–March.

**2. Dryopetalon crenatum** (Brandegee) Rollins var. **racemosum** Rollins, Contr. Dudley Herb. **3:** 203, 1941

An annual herb with purplish, rigidly branching stems and crowded, rigidly ascending, flowering and fruiting pedicels; only flowers on the type specimen small and aberrant; siliques 1–1.5 cm. long, strongly ascending.

Known only from Isla Espíritu Santo, Baja California. Aug.–Sept.

**3. Dryopetalon palmeri** (S. Wats.) O. E. Schulz, Notizblatt **10:** 561, 1929

*Cardamine palmeri* S. Wats., Proc. Amer. Acad. **24:** 38, 1889.
*Sibara palmeri* Greene, Pittonia **3:** 12, 1896.

Annual with erect, somewhat branched, glabrous to sparsely pilose, slightly glaucous stems 2–5 dm. tall; leaves petiolate, shallowly cordate, ovate, 1.5–3 cm. wide, 3–8 cm. long, glaucous, sinuate-dentate to deeply lobed; inflorescence lax; sepals oblong, obtuse, 1.5–2 mm. wide, 3.5–4 mm. long, outer pair only faintly saccate, glabrous; petals white to purplish, spatulate, 5–7-lobed, 2–3 mm. wide, 6–8 mm. long, claw nearly twice as long as blade; stamens glabrous; glands surrounding single stamens, subtending paired ones; fruiting pedicels divaricately ascending, 8–12 mm. long, glabrous or sparsely pilose; siliques divaricate to erect, glabrous, 2.5–5 cm. long, 1–1.5 mm. wide, on a stipe about 1 mm. long; style 1–2 mm. long; seeds brown, about 1.5 mm. long, nearly 1 mm. wide; cotyledons incumbent.

Shaded canyon slopes and rocky cliffs, Lower Sonoran Zone, central Baja California, and western Sonora. Feb.–March.

**4. Dryopetalon purpureum** Rollins, Contr. Dudley Herb. **3:** 204, *pl. 48, fig. 3,* 1941

Annual with a single stem 2–4 dm. tall, branching above, herbage glabrous or nearly so; leaves petiolate, ovate to lanceolate, 1–2 cm. wide, 2–5 cm. long, deeply lobed to pinnately parted, lobes usually dentate, sometimes with a few scattered simple hairs along petiole and rachis; inflorescence laxly racemose; flowering pedicels 5–15 mm. long, divaricate, glabrous to pilose; sepals purplish, glabrous, 1.5–2 mm. wide, about 4 mm. long, obtuse; petals obovate to broadly spatulate, 3–6-lobed, 2.5–3.5 mm. wide, 6–8 mm. long, short claw about equaling or slightly shorter than blade; stamens glabrous; glandular tissue surrounding single stamens, subtending paired ones; fruiting pedicels 1–2 cm. long; siliques linear, 1.5–2.5 cm. long, about 2 mm. wide, on a stipe about 0.5 mm. long; seeds oblong, about 1.5 mm. long, nearly 1 mm. wide, inconspicuously reticulate on surface; cotyledons incumbent or slightly obliquely so.

Rocky slopes and canyons, Lower Sonoran Zone, Magdalena and Santa Margarita Islands, Baja California. Jan.–April.

## 11. RORIPPA Scop., Fl. Carn. 520, 1760

Annual or perennial herbs of moist or wet habitats, usually with branched stems. Leaves simple or lobed, pinnate, or dissected, rarely entire. Racemes short at first, markedly elongating with age. Sepals spreading. Petals white or yellow. Stamens 6 or often less. Styles short; stigmas subentire to shallowly but distinctly 2-lobed. Pods globose to cylindric, terete or essentially so. Seeds small, turgid, usually in 2 rows. Cotyledons accumbent.

Plants semiaquatic; petals white; leaves pinnately compound ... 1. *R. nasturtium-aquaticum*
Plants terrestrial, in moist habitats; petals yellow or yellowish; leaves pinnatifid:
 Pods subglobose, ovoid, or short-ellipsoidal, not over twice as long as wide; stems hispid-pubescent, branching only above ...................... 2. *R. islandica* var. *hispida*
 Pods cylindric, usually more than twice as long as broad; stems glabrous or essentially so, branching from base ............................................ 3. *R. obtusa*

**1. Rorippa nasturtium-aquaticum** (L.) Schinz & Thell., Fl. Schweiz. ed. 3, 240, 1909

*Sisymbrium nasturtium-aquaticum* L., Sp. Pl. 657, 1753.
*Nasturtium officinale* R. Br., in Ait., Hort. Kew. ed. 2, **4:** 110, 1812.
*Radicula nasturtium-aquaticum* Britt. & Rendle, List Brit. Seed-Plants 3, 1907.

Stems floating, tips ascending or mostly decumbent with erect tips if water level drops, glabrous stems rooting at nodes; leaves pinnate, 5–15 cm. long, of 3–9 leaflets, these elliptic to ovate or suborbicular, 0.5–3 cm. long, terminal one larger and usually broader than others, margin entire or more commonly sinuate; sepals elliptic, 2.5–4 mm. long, thin; petals clear white or veined with rose or purple, claws equaling sepals, blades 1–2.5 mm. wide, 3–5 mm. long; pedicels slender, 1–2 cm. long, ascending or spreading; pods cylindric, 2–2.5 mm. in diameter, 1–3 cm. long, terete, straight, or more commonly slightly curved upward; seeds reddish brown, 0.6–0.8 mm. long.

In quiet water along margins of streams, ditches, and springs, Lower Sonoran Zone to Canadian Zone; introduced from Eurasia and naturalized throughout most of temperate North America. Flowering much of the year.

**2. Rorippa islandica** (Oeder) Borbas var. **hispida** (Desv.) Butters & Abbe, Rhodora **42:** 26, 1940

*Brachylobus hispidus* Desv., Desv. Jour. Bot. **3:** 183, 1814.
*Sisymbrium hispidum* Poir., in Lam., Encycl. Suppl. **5:** 161, 1817.
*Nasturtium hispidum* DC., Syst. **2:** 201, 1821.
*Nasturtium terrestre* var. *hispidum* Fisch. & Mey., Ind. Sem. Hort. Petrop. **3:** 41, 1837.
*Nasturtium palustre* var. *hispidum* A. Gray, Man. ed. 2, 30, 1856.
*Rorippa hispida* Britt., Mem. Torrey Club **5:** 169, 1894.
*Radicula hispida* Britt., Torreya **6:** 30, 1906.
*Radicula palustris* var. *hispida* Robinson, Rhodora **10:** 32, 1908.

Annual or biennial with 1–several erect stems 3–12 dm. tall, branching above; stems, branches, petioles, and veins on lower surfaces of leaf blades hispid with scattered spreading hairs; lower leaves lyrately pinnatifid, 4–15 cm. long, 1–3 cm. wide, terminal part of blade irregularly dentate or sinuate instead of divided; petioles slightly winged, 1–4 cm. long; sepals broadly obovate, cuculate, about 1 mm. long; petals narrowly obovate, only slightly surpassing sepals, pale yellow to nearly white; racemes greatly elongated in fruit; pedicels spreading-ascending, 5–8 mm. long; pods ovoid, not compressed, 4–6 mm. long, glabrous; seeds about 0.8 mm. long.

Along stream beds in our area, in marshy places northward, Lower Sonoran Zone (as waif) to Hudsonian Zones, Alaska to New Brunswick, southward to California, Arizona, New Mexico, and Florida. Occurring in the Sonoran Desert only sparingly. May–Aug.

**3. Rorippa obtusa** (Nutt.) Britt., Mem. Torrey Club **5:** 169, 1894

*Nasturtium obtusum* Nutt., in Torr. & Gray, Fl. N. Amer. **1:** 74, 1838.
*Radicula obtusa* Greene, Leaflets Bot. Obs. **1:** 113, 1905.

Glabrous annual branching diffusely from base, branches 1–3 dm. long; leaves pinnatifid, oblong to oblanceolate in outline, 0.5–2.5 cm. wide, blades to 6 cm. long, petioles sometimes nearly equaling blades, wingless, lobes oblong to elliptic; upper leaves merely lobed or toothed or sinuate; sepals and petals about 1 mm. long, petals pale yellow to nearly white; pedicels arcuately spreading-ascending, 2–4 mm. long; pods ovoid to short-cylindric, 5–8 mm. long, 1–2 mm. in diameter, straight or slightly curved upward, turgid; persistent style about 1 mm. long, stigma no wider than style.

Stream beds and meadows, Lower Sonoran Zone (as a waif) to Canadian Zone, Washington to Michigan and south to California, Arizona, Texas, and Missouri. Occurring in the Sonoran Zone. May–Sept.

**Rorippa curvisiliqua** (Hook.) Bessey, in Britt., Mem. Torrey Club **5:** 169, 1894, also occurs in the mountains above the limits of the Sonoran Desert and has appeared along stream beds only slightly outside the limits of our area.

## 12. SIBARA Greene, Pittonia 3: 10, 1896

Annual or biennial herbs with 1–several stems branching divaricately from the base and glabrous to pubescent, glaucous herbage. Lower leaves runcinate-pinnatifid to (in ours) pectinately divided into narrow segments; cauline leaves smaller but very rarely entire. Inflorescences lax racemes, elongating moderately in fruit. Flowers small; sepals ovate to oblong, outer pair slightly saccate in some species. Petals spatulate to oblong, white or tinged with pink or purple, entire, denticulate at the base or constricted near the middle. Nectar glands absent near paired stamens, subtending or surrounding base of single stamens. Siliques slender, terete or slightly flattened parallel to partition, valves nerveless or nerved only toward base. Style stout or slender. Stigma entire, flat. Seeds in a single row, oblong to suborbicular, in ours wingless. Cotyledons accumbent or incumbent.

Mature siliques 2–5 cm. long; cotyledons accumbent; styles slender:
- Petals not constricted at middle; siliques usually arcuate; leaf segments to 2 mm. wide ... ........ 1. *S. angelorum*
- Petals markedly constricted near middle; siliques usually straight; leaf segments rarely over 1 mm. wide ........ 2. *S. pectinata*

Mature siliques usually less than 2 cm. long; cotyledons incumbent; styles stout:
- Siliques somewhat flattened; pedicels 5–10 mm. long, usually sparsely hispidulous in youth; styles narrowing from base to apex; seeds 1–1.5 mm. long ........ 3. *S. laxa*
- Siliques terete; pedicels 1.5–4 mm. long, glabrous; styles clavate, slightly expanded toward apex; seeds 0.8–1 mm. long ........ 4. *S. brandegeana*

### 1. **Sibara angelorum** (S. Wats.) Greene, Pittonia **3:** 11, 1896

*Cardamine angelorum* S. Wats., Proc. Amer. Acad. **24:** 39, 1889.

Erect annual 3–7 dm. tall with glabrous and glaucous herbage, stem branching above only; basal leaves caducous, cauline petiolate, 4–10 cm. long, pinnately divided into linear segments 0.75–2 mm. wide, 1–2.5 cm. long, these entire or rarely again remotely lobed; sepals narrowly ovate, 2.5–3.5 mm. long, scarious-margined, nonsaccate; petals broadly spatulate, 4.5–5.5 mm. long, pink to purplish; pedicels divaricately spreading or slightly arched downward, 10–15 mm. long; siliques moderately flattened, about 1.5 mm. wide, 2.5–4.5 cm. long, slightly arcuately curved, valves nerved only below; style 1–1.5 mm. long, slender; seeds oblong, about 1.5 mm. long, wingless; cotyledons accumbent.

Growing among shrubs or under trees in sandy or rocky areas, Lower Sonoran Zone, central Baja California from about Bahía de los Angeles to San Ignacio and the Vizcaíno Desert. Feb.–April.

### 2. **Sibara pectinata** (Greene) Greene, Pittonia **3:** 11, 1896

*Arabis pectinata* Greene, Pittonia **1:** 287, 1889.

Glabrous and glaucous annual 1–4 dm. tall, single stem branching from near base upward; basal leaves caducous, 3–10 cm. long, pinnately divided into narrowly linear, entire segments to 1.5 cm. long, petiolate; cauline leaves similar, gradually reduced in size toward apex of stem; sepals narrowly ovate to oblong, 2–3 mm. long, scarious-margined, nonsaccate; petals pinkish to purplish, 5–6 mm. long, shallowly emarginate at apex, slightly constricted at summit of claw; stamens about equaling sepals; pedicels 8–15 mm. long, divaricately spreading; siliques spreading, 1.5–2 mm. wide, 2.5–4.5 cm. long, moderately flattened, straight or slightly curved, valves nerveless above, faintly nerved near their bases; style 1.5–2 mm. long; seeds wingless, about 1 mm. wide, 1.5 mm. long; cotyledons accumbent.

Sandy and rocky places, Lower Sonoran Zone, west central Baja California from near Punta Prieta to Bahía San Bartolomé, and on Cedros Island. Feb.–April.

### 3. **Sibara laxa** (S. Wats.) Greene, Pittonia **3:** 11, 1896

*Nasturtium (?) laxum* S. Wats., Proc. Amer. Acad. **24:** 39, 1889.

Erect, single-stemmed annuals 3–7 dm. tall with few to many lax weak branches and glabrous and glaucous herbage; basal leaves caducous; cauline leaves 3–5 cm. long, petiolate, pinnate with 3–10 pairs of linear, spreading, entire segments 0.5–1.2 mm. wide and 1–2 cm. long; inflorescence very lax, often 10–12 cm. long in anthesis; sepals oblong, narrowly ovate, 2.5–4 mm. long, scarious-margined, glabrous; petals purplish, 4–6 mm. long, oblong to broadly spatulate, sometimes faintly narrowed near middle; pedicels 5–10 mm. long, slender, sparsely pubescent with a few scattered simple hairs at anthesis, soon glabrate, or less commonly remaining puberulent, reflexed in fruit; siliques about 1.5 mm. wide, 1–1.5 cm. long, moderately flattened, valves nerveless; style thick, abruptly narrowing; seeds oblong, about 1–1.5 mm. long; cotyledons incumbent.

Sandy or rocky soil, usually in shade, Lower Sonoran Zone, vicinity of Punta Prieta and Bahía de los Angeles south at least to Santa Rosalia and west to the coast. Feb.–April.

### 4. **Sibara brandegeana** (Rose) Greene, Pittonia **3:** 11, 1896

*Sisymbrium brandegeanum* Rose, Contr. U.S. Nat. Herb. **1:** 10, 1890.
*Microsisymbrium lasiophyllum* f. *brandegeanum* O. E. Schulz, Pflanzenreich $4^{105}$: 163, 1924.

A lax annual with 1–several spreading to erect stems 1–4 dm. long, stems simple or branched above; stems and herbage glabrous, often reddish or purplish, somewhat glaucous; basal leaves caducous, pinnate, with 2–5 pairs of filiform segments 1–4 mm. long; cauline leaves pinnate, 1–3 cm. long, with narrowly linear, obtuse segments 5–10 mm. long; stems beginning to produce flowers while only 1–4 cm. long, inflorescence greatly elongated with age; sepals oblong, 1.5–2 mm. long, scarious-margined, often purplish or reddish; petals purplish, 3–4 mm. long, oblong, minutely denticulate near base; pedicels stoutish, 1.5–4 mm. long, glabrous, geniculately reflexed in fruit; siliques nearly terete, 1–1.5 cm. long, reflexed or spreading, usually arcuately curved upward but sometimes straight; style 1.5–2 mm. long, thick, only slightly narrower than silique; seeds wingless, about 1 mm. long; cotyledons incumbent.

Sandy soil, Lower Sonoran Zone, vicinity of San Quintín, Baja California, south at least to the Vizcaíno Desert. Feb.–April.

## 13. **LYROCARPA** Hook. & Harv. ex Harv., Lond. Jour. Bot. **4:** 76, 1845*

Herbaceous annuals or slightly suffrutescent perennials densely pubescent throughout with branched hairs. Leaves petiolate, repand to pinnately divided. Flowers fragrant, in loose elongated racemes. Calyx cylindrical, outer pair of acute sepals subsaccate basally and exceeding the inner pair, connivent. Petals much exceeding sepals and stamens, blades attenuate to obtuse at apex and often twisted. Stamens tetradynamous; 2 discoid glands at base of each single stamen. Ovary canescent; style short or obsolete; stigma bifid, densely papillate. Fruits obcordate to panduriform, flattened contrary to the septum, pubescent. Ovules 3–10 in each cell.

Plants annual; petals spatulate, abruptly acute to obtuse, bright purple; leaves shallowly lobed or repand; fruiting pedicels often reflexed .......................... 1. *L. xanti*

---

* Treatment adapted from Reed C. Rollins, "A Revision of Lyrocarpa," Contr. Dudley Herb. **3:** 169–173, *pl. 46, figs. 4, 5a, 5b,* 1941.

Plants perennial; petals linear, usually acuminate, brownish to dull purple; leaves deeply lobed; fruiting pedicels spreading to ascending:
Leaf segments not apiculate; petals usually 2.5–3 mm. wide; leaves thickened, the lower always divided:
Fruits panduriform, longer than broad; lower leaves 2–4 cm. wide, segments nearly ovate; ovules 4–8 in each cell ........................ 2. *L. coulteri* var. *coulteri*
Fruits broadly obcordate, nearly as broad as long; lower leaves usually less than 1 cm. broad, segments linear; ovules 3–5 in each cell ....... 2*a*. *L. coulteri* var. *palmeri*
Leaf segments apiculate; petals less than 1.5 mm. wide; leaves thin, the lowermost nearly entire ............................................ 2*b*. *L. coulteri* var. *apiculata*

**1. Lyrocarpa xanti** Brandegee, Proc. Calif. Acad. II, **2:** 127, 1889

Annual with simple or branched, erect or decumbent stems 2–5 dm. tall, pubescent to glabrous above; leaves mostly basal, ovate-cordate, obtuse, repand or rarely slightly lobed, 3–6 cm. wide, 3–10 cm. long, sparsely to densely but evenly pubescent; sepals 1–1.5 mm. wide, 6–8 mm. long, connivent above, inner pair flat; petals spatulate to obovate-lanceolate, 2.5–4 mm. wide, 14–18 mm. long, bright purple; fruiting pedicels spreading to reflexed; silicles panduriform but only slightly constricted below the lobes, 4–8 mm. wide, 15–25 mm. long, densely pubescent; ovules 6–10 in each cell; seeds brown, orbicular, about 2 mm. in diameter, the seed coat mucilaginous when moistened; cotyledons accumbent.

Sandy plains and along washes, Lower Sonoran and Tropical Zones, western side of the peninsula near San Gregorio south to the Cape Region, Baja California. Feb.–March.

**2. Lyrocarpa coulteri** Hook. & Harv. ex Harv., Lond. Jour. Bot. **4:** *76, pl. 4, figs. 1–6,* 1845, var. **coulteri**

*Lyrocarpa coulteri* var. *typica* Rollins, Contr. Dudley Herb. **3:** 171, *pl. 46, fig. 5b,* 1941.

Perennial, herbaceous or somewhat suffrutescent plants, irregularly branched stems often scrambling through shrubs and to 1 m. long; herbage densely pubescent with dendritic trichomes; leaves petiolate, linear to nearly ovate, 0.5–6 cm. wide, 1–15 cm. long, deeply parted to moderately lobed, segments nearly ovate, acute but not apiculate, densely pubescent but not canescent; pedicels 3–7 mm. long, densely pubescent, widely spreading to ascending in fruit; sepals 8–11 mm. long, about 1.5 mm. wide, acute; petals linear to oblanceolate, acute to acuminate, limb usually twisted, 1–3 mm. wide, 1.5–2.5 cm. long, brown to dull purple; silicles panduriform, 1–2 cm. long, 5–15 mm. wide; ovules 4–8 in each cell; style obsolete or to 1 mm. long; seed coat not mucilaginous when moistened; cotyledons accumbent.

Sandy plains, rocky hillsides, and along washes, Lower Sonoran Zone, Pinal and Pima Counties, Arizona, south to west-central Sonora and in east-central Baja California. March–April and Aug.

**2*a*. Lyrocarpa coulteri** var. **palmeri** (S. Wats.) Rollins, Contr. Dudley Herb. **3:** 171, *pl. 46, fig. 5A,* 1941

*Lyrocarpa palmeri* S. Wats., Proc. Amer. Acad. **11:** 123, 1876.

Differing from var. *coulteri* in being densely canescent; lower leaves usually less than 1 cm. broad, with linear instead of ovate segments; silicles obcordate, nearly as broad as long; and 3–5 ovules in each cell.

In sandy or gravelly soil and among boulders, Lower Sonoran Zone, Colorado

Desert in Riverside, Imperial, and eastern San Diego Counties south to central Baja California. Jan.–April.

**2*b*. Lyrocarpa coulteri** var. **apiculata** Rollins, Contr. Dudley Herb. **3:** 172, 1941

Differing from var. *coulteri* in having larger lower leaves 3–6 cm. broad, the segments apiculate, thin and greenish, uniformly and sparsely pubescent; and petals linear, 1–1.5 mm. wide.

Sandy or rocky soil, Lower Sonoran Zone, vicinity of Santa Rosalia to San Gregorio, Baja California. Feb.–April.

## 14. DITHYREA Harv., Lond. Jour. Bot. 4: 77, *pl.* 5, 1845

Annual or perennial herbs with branching, erect or ascending stems and stellate-pubescent herbage. Flowers perfect, fragrant, in rather dense short racemes that elongate in fruit. Sepals erect to connivent apically, ovate to oblong. Petals white or faintly purplish, narrowly clawed, blades broadly spatulate. Stamens 6; anthers linear-sagittate. Ovary didymous, strongly compressed contrary to the very narrow partition, cells suborbicular and prominently corded along the margin; stigma helmetlike, sessile. Seeds solitary in each cell. Cotyledons accumbent.

Fruit prominently notched both above and below; petals 8–12 mm. long; pedicels 2–3 mm. long in fruit; margins of carpels much more densely puberulent than faces:
  Carpels 4–6 mm. in diameter; styles 0.2–0.4 mm. long ... 1. *D. californica* var. *californica*
  Carpels 3–4 mm. in diameter; styles 0.6–1 mm. long ...... 1*a*. *D. californica* var. *clinata*
Fruit truncate, slightly apiculate or very shallowly notched above; petals 5–8 mm. long; pedicels 10–15 mm. long in fruit; margins of carpels no more densely puberulent than the faces ........................................................ 2. *D. wislizenii*

**1. Dithyrea californica** Harv., Lond. Jour. Bot. **4:** 77, 1845, var. **californica**
*Biscutella californica* Brewer & Wats., Bot. Calif. **1:** 48, 1876.

Annual with 1–several ascending stems 0.5–4 dm. tall branching from base, sparsely stellate-puberulent with fine-rayed trichomes, green or less commonly cinereous; basal leaves petiolate, 2–13 cm. long, 0.5–7 cm. broad, oblong-obovate to ovate, coarsely and shallowly few-toothed, usually obtuse or rounded at apex, cuneately narrowing to petioles nearly equaling blades, minutely puberulent with slenderly rayed, stellate trichomes; cauline leaves ovate, 1–3 (6) cm. long, mostly sessile and entire or sinuate-margined; racemes dense, elongating in fruit; pedicels spreading, 2–3 mm. long, puberulent; sepals lanceolate, 5–8 mm. long, stellate-puberulent; petals white or diffused with purple, slender claws equaling sepals, blades rotately spreading, 4–6 mm. long; fruits about equally notched above and below, carpels 4–6 mm. in diameter, heavily corded submarginally, margins densely tomentose with stout, simple, whitish hairs about 0.3 mm. long, faces sparsely puberulent, lightly reticulate-veined.

Sandy or silty soils, Lower Sonoran Zone, Mojave and Colorado Deserts in California east into Nevada and Arizona and south into Baja California, where it grades into the following. Feb.–June.

**1*a*. Dithyrea californica** var. **clinata** (Macbr. & Pays.) Wiggins, comb. nov.
*Dithyrea clinata* Macbr. & Pays., Ann. Mo. Bot. Gard. **5:** 150, 1918.

Differing from var. *californica* in usually being smaller, fewer-stemmed; in having smaller pods, each valve being only 3–4 mm. in diameter at maturity; and in having a slender style nearly or quite 1 mm. long.

Sandy soils, Lower Sonoran Zone, desert areas of central Baja California from about Punta Prieta south to the southern end of the Magdalena Plain. Feb.–May.

**2. Dithyrea wislizenii** Engelm., in Wisliz., Mem. Tour North. Mexico 95, 1848

*Dithyrea griffithsii* Woot. & Standl., Contr. U.S. Nat. Herb. **16:** 124, 1913.
*Dithyrea wislizenii* var. *griffithsii* Payson, Ann. Mo. Bot. Gard. **5:** 148, 1918.

Erect or ascending, simple or branched, annual or possibly biennial herb to 6 dm. tall, minutely stellate-puberulent to subtomentose throughout or fruits sometimes glabrous; lower leaves lanceolate to ovate-lanceolate, 1–2.5 cm. wide, to 20 cm. long, margins shallowly sinuate to distantly pinnatifid; petioles 0.5–3 cm. long; racemes rather dense, elongating to 1.5–3 dm. in fruit; pedicels spreading to slightly deflexed, turned upward abruptly at apex, flower and fruit thus erect; sepals ovate, 2–3 mm. long, obtuse at apex, ascending to somewhat spreading in anthesis; petals white or slightly purplish or roseate, 5–8 mm. long, blade usually exceeding claw; fruiting pedicels 10–15 mm. long; fruit truncate or faintly notched at apex, cordate at base, valves each 5–8 mm. broad, moderately corded submarginally, sparsely puberulent or occasionally glabrous, thin margin not more densely pubescent than faces; style about 0.5 mm. long or less.

Sandy soil, Lower and Upper Sonoran Zones, Mohave and Yuma Counties, Arizona, to southern Utah, Oklahoma, Colorado, and western Texas; south into northern Sonora. Feb.–Aug.

## 15. LESQUERELLA S. Wats., Proc. Amer. Acad. 23: 249, 1888

Small annual or perennial herbs with simple leaves, stellate pubescence, and chiefly yellow flowers in compact racemes that elongate markedly in fruit. Sepals elliptic or oblong, mostly rounded or obtuse at apex. Petals obovate to spatulate, exceeding sepals. Stamens 6, filaments slender; anthers sagittate. Style slender; stigma capitate, entire or essentially so. Pod subglobose to oblong, usually distinctly inflated, with nerveless valves; septum thin, orbicular to elliptic, nerved from apex nearly or quite to middle, often with tortuous areolae. Seeds several to many in each cell, moderately to strongly flattened, marginless (in ours) to narrowly winged.

Plants perennial; petals white, pinkish or purplish; style 1.5–2 mm. long ... 1. *L. purpurea*
Plants annual; petals yellow; style 2–3 mm. long:
  Pods on short stipes 0.5–0.8 mm. long, glabrous; funiculus attached to the septum about one-third its length ........................................... 2. *L. gordonii*
  Pods sessile, stellate-puberulent; funiculus attached to septum one-half its length ....... ........................................................ 3. *L. palmeri*

**1. Lesquerella purpurea** (A. Gray) S. Wats., Proc. Amer. Acad. **23:** 253, 1888

*Vesicaria purpurea* A. Gray, Pl. Wright. **2:** 14, 1853.
*Vesicaria purpurea* var. *albiflora* A. Gray, Ive's Rep. 6, 1860.
*Alyssum purpureum* Kuntze, Rev. Gen. Pl. **2:** 931, 1891.

Decumbent to partially erect perennial with silvery-stellate herbage and stems 1–4 dm. long; basal leaves oval to oblanceolate, 2–12 cm. long, repand to lyrately pinnatifid, gradually narrowed toward base to a slender petiole one-half as long as blade or less, obtuse to rounded at apex; cauline leaves oblanceolate, 7–25 mm. long, remote, acute to obtuse at apex, entire; inflorescence elongated in fruit, erect or ascending; petals 7–9 mm. long, distinctly clawed, blade white, pink, or purplish; fruiting pedicels slender, horizontal to recurved, 5–15 mm. long; styles 1.5–2 mm. long; pods globose, 4–6 mm. in diameter, horizontal to pendent, glabrous; septum strongly nerved, areolae tortuous or diagonal; ovules 2–5 in each cell; seeds flat, wingless.

Mountainsides and outwash slopes and mesas, Lower and Upper Sonoran Zones, Pima County, Arizona, to southwestern Texas and adjacent northern Mexico. April–June and Sept.–Oct.

**Lesquerella arizonica** S. Wats., Proc. Amer. Acad. **23:** 254, 1888. Cespitose perennial densely stellate-pubescent throughout, with oblanceolate leaves 5–25 mm. long; yellow petals 6–7 mm. long; straight or slightly curved pedicels 4–8 mm. long; ovate pods 4–6 mm. long, acute at apex; ovules 4 in each cell.

Occurring mainly on rocky mesas and slopes above 3,500 feet; southern Utah to central Arizona, and may occur as a straggler along the margin of the Sonoran Desert in Mohave County, Arizona.

## 2. **Lesquerella gordonii** (A. Gray) S. Wats., Proc. Amer. Acad. **23:** 253, 1888

*Vesicaria gordoni* A. Gray, Bost. Jour. Nat. Hist. **6:** 149, 1850.
*Alyssum gordonii* Kuntze, Rev. Gen. Pl. **2:** 931, 1891.

Densely stellate-canescent annual with several to many decumbent stems 1–3 dm. long; basal leaves narrowly oblanceolate to spatulate, entire to slightly repand (rarely somewhat lyrate), 1.5–3.5 cm. long, acute at apex, gradually narrowed to a slender petiole sometimes nearly equaling blade; cauline leaves numerous, 1–3 cm. long, linear to narrowly oblanceolate, entire or faintly repand; racemes rather compact in flower, later elongated; pedicels stoutish, sigmoidly curved and horizontally spreading to slightly recurved, 7–10 mm. long; petals yellow, narrowly obovate, claw slightly dilated basally; pod globose and glabrous, 3.5–4 mm. in diameter on a stipe 0.5–0.8 mm. long; style about 3 mm. long; septum with slightly tortuous areolae; ovules 4–10 in each cell, funiculi attached to septum about one-third their lengths; seeds flat, wingless.

Sandy plains and mountain slopes, Lower and Upper Sonoran Zones, Pima County, Arizona, to western Oklahoma and Texas, northern Chihuahua, and Sonora. Feb.–June.

## 3. **Lesquerella palmeri** S. Wats., Proc. Amer. Acad. **23:** 255, 1888

*Lesquerella gordoni* var. *sessilis* S. Wats., op. cit. 253.
*Lesquerella tenella* A. Nels., Bot. Gaz. **47:** 426, 1909.

Annual, sparsely to densely invested with many-rayed, fine, stellate hairs, and finely glandular; stems slender, 1–4 dm. long, decumbent to erect, often branched in larger plants; basal leaves oblanceolate, entire to lyrately lobed, 1–5 cm. long, slenderly petiolate; cauline leaves linear to oblanceolate, entire to obscurely repand, 1–5 cm. long, acute to obtuse at apex; racemes dense in early flower but soon elongating; pedicels ascending, horizontal, or recurved, more or less sigmoid, about 1 cm. long in fruit; petals yellow, broadly spatulate; pods globose to slightly ellipsoidal, 3–5 mm. in diameter, erect, horizontal, or very rarely somewhat pendent, sparsely stellate-puberulent; style 2–3 mm. long; areolae of septum somewhat tortuous; ovules 4–6 in each cell, funiculi attached to septum about one-half their lengths; seeds flattened but wingless.

Desert slopes and plains, mainly Lower Sonoran Zone, Colorado Desert, California, to southern Utah, southern Nevada, western Arizona, and northern Baja California. Feb.–May.

## 16. HUTCHINSIA R. Br., in Ait., Hort. Kew. ed. 2, 4: 82, 1812

Low, diffusely branched, annual (in ours) herbs with glabrous to stellate-puberulent leaves and stems. Leaves entire to pinnately lobed. Flowers minute, racemose, white. Sepals reddish or purplish, submembranous. Stamens 6. Silicles ellipsoidal, only slightly compressed contrary to the septum, rounded at the apex. Stigma sessile or essentially so. Seeds several in each cell. Cotyledons incumbent or accumbent.

### 1. Hutchinsia procumbens (L.) DC., Desv. Jour. Bot. 3: 168, 1814

*Lepidium procumbens* L., Sp. Pl. 643, 1753.
*Capsella elliptica* C. A. Mey., in Ledeb., Fl. Alt. 3: 199, 1831.
*Capsella procumbens* Fries, Novit. Fl. Suec. Mant. 1: 14, 1832.
*Hymenolobus divaricatus* Nutt., in Torr. & Gray, Fl. N. Amer. 1: 117, 1838.
*Hymenolobus erectus* Nutt., in Torr. & Gray, loc. cit.
*Capsella divaricata* Walp., Rep. 1: 175, 1842.
*Capsella erecta* Walp., loc. cit.

Stems 1–several, slender, often somewhat zigzag, branching from base, 3–15 cm. long, often reddish, glabrous or sparsely and minutely puberulent; basal leaves lance-elliptic to narrowly oblanceolate, 1–3.5 mm. wide, 3–15 mm. long, dentate, pinnatifid, or rarely entire, petiole usually about one-third to one-half as long as blade; cauline leaves reduced and usually entire or lacking, sessile or nearly so; racemes elongating as long as flowering continues; pedicels slender, spreading to ascending, 3–12 mm. long in fruit; sepals about 0.5 mm. long, pinkish; petals white, 0.6–1 mm. long; pods ellipsoidal or ovoid, 2–3.5 mm. long, rounded or rarely slightly emarginate at apex, glabrous, finely reticulate-veined.

In our area usually in shade of rocks or on playas, often in moist places farther north, Lower Sonoran Zone to Canadian Zone, British Columbia to Labrador, south to central Baja California, northern Arizona, and Colorado; also in Asia, Europe, and Australia. Feb.–May.

## 17. CAPSELLA Medic., Pflanzeng. 1: 85, 1792

Erect or ascending, annual herbs with scattered forked hairs and mainly basal, lanceolate to oblanceolate, lyrately pinnatifid to dentate leaves on petioles shorter than blades. Inflorescence racemose, elongating in fruit. Flowers white, small. Pods strongly flattened contrary to narrow septum, cuneate-cordate, many-seeded. Style 0.5 mm. long or less.

### 1. Capsella bursa-pastoris (L.) Medic., Pflanzeng. 1: 85, 1792

*Thlaspi bursa-pastoris* L., Sp. Pl. 647, 1753.
(About 200 names have been applied to this polymorphic species.)

Stems erect or ascending, strict to branching, 5–40 cm. tall, pubescent near base, glabrous or essentially so above; basal leaves in a rosette, mostly oblanceolate, dentate to lyrate-pinnatifid, 1–2.5 cm. wide, 2–12 cm. long; cauline leaves lanceolate, auriculate, dentate to entire, smaller than basal ones and sessile; pedicels slender, 6–15 mm. long, ascending to spreading; flowers about 2–2.5 mm. broad; sepals obovate, rounded at apex, pinkish; petals white, oblong or oblanceolate, about twice as long as sepals; pods triangular-cordate or cuneate-cordate, shallowly to deeply emarginate at apex, 4–6 mm. long, nearly as wide.

A cosmopolitan weed, Lower Sonoran Zone to Hudsonian Zone, naturalized widely in North and South America from Europe, in our area occurring in Imperial Valley near Westmoreland. Feb.–May.

## 18. DRABA L., Sp. Pl. 642, 1753

Low tufted annual or perennial herbs, mostly with leafy or scapose stems, stellate-puberulent herbage and simple leaves. Inflorescences racemose, elongating considerably in fruit. Flowers rather small; petals entire, emarginate or bifid, white or yellow. Style short or slender; stigma entire or faintly bilobed. Silicles oval, oblong or linear-oblong, flattened parallel to the septum, the latter membranous. Valves nerveless. Seeds in 2 rows in each cell, wingless (in ours). Cotyledons accumbent or very rarely incumbent.

Pods stellate-puberulent or glabrous; plants often branched from base; pedicels about one-half as long as pods .............................. 1. *D. cuneifolia* var. *integrifolia*
Pods hispidulous with simple hairs; plants rarely branched from base; pedicels nearly equaling pods ........................................ 1a. *D. cuneifolia* var. *platycarpa*

**1. Draba cuneifolia** Nutt. ex Torr. & Gray var. **integrifolia** S. Wats., Proc. Amer. Acad. **23:** 256, 1888

*Draba sonorae* Greene, Bull. Calif. Acad. **2:** 59, 1886.
*Draba integrifolia* Greene, Pittonia **4:** 201, 1900.
*Draba cuneifolia* var. *sonorae* Parish, Bull. S. Calif. Acad. **2:** 81, 1903.
*Draba sonorae* var. *integrifolia* O. E. Schulz, Pflanzenreich **4**$^{105}$**:** 336, 1927.
*Draba cuneifolia* var. *brevifolia* S. Wats. ex Munz, Man. S. Calif. 200, 1935.

Small annual 2–25 cm. tall with a simple stem erect or branching from base, stem leaves and calyces hirsute with branched or forked trichomes, somewhat cinereous; stems leafy on lower one-third; leaves ovate, obovate, or oblanceolate, 3–20 mm. wide, 6–50 mm. long, dentate or denticulate at least on upper one-half or smaller leaves sometimes entire; inflorescence to as much as one-half total height of plant, 5–70-flowered, stellate-puberulent throughout; pedicels 2–5 mm. long, spreading to ascending; sepals 1.5–2.5 mm. long; petals white, 3.5–5 mm. long, spatulate and emarginate or reduced to linear entire rudiments 1–2 mm. long or completely wanting; pods linear, elliptic or oblong-obovate, 1.7–5 mm. wide, 4–15 mm. long, glabrous or more commonly stellate-puberulent; styles evident as small apiculations; seeds about 0.7 mm. long, 25–200 in a fruit.

Arid plains, slopes, and mesas, Lower and Upper Sonoran and Tropical Zones, southern California to southwestern Utah, south into Baja California, Sonora, and Nuevo León, Mexico. Feb.–May.

**1*a*. Draba cuneifolia** var. **platycarpa** (Torr. & Gray) S. Wats., Proc. Amer. Acad. **23:** 256, 1888

*Draba platycarpa* Torr. & Gray, Fl. N. Amer. **1:** 108, 1838.
*Draba viperensis* St. John ex Payson & St. John, Proc. Biol. Soc. Wash. **43:** 104, 1930.

Stems leafy nearly to inflorescence; plants usually simple from base; pedicels usually equaling pods.

This variety occurs infrequently in Arizona at elevations somewhat above the Sonoran Desert, and the writer has seen no specimens from the Desert referable to it.

## 19. ATHYSANUS Greene, Bull. Calif. Acad. 1: 72, 1885

Slender annual diffusely branched from base, hirsute with forked hairs. Leaves simple, few-toothed or entire, confined to lower parts of branches. Inflorescences greatly elongated, racemes unilateral. Flowers minute. Pedicels slender, recurved in a uniform, moderate to sharp curve. Sepals ovate, minute, spreading at anthesis, green with white margins. Petals linear, or lacking. Stamens 6. Ovary 1-celled, 2-ovuled. Style stoutish, minute, persistent. Silicle broadly elliptic to orbicular, compressed, wingless, closely beset with stiff, simple, sharply hooked hairs, indehiscent, 1-seeded.

### 1. Athysanus pusillus (Hook.) Greene, Bull. Calif. Acad. 1: 72, 1885

*Thysanocarpus pusillus* Hook., Ic. Pl. 1: *pl. 42*, 1837.
*Thysanocarpus oblongifolius* Nutt., in Torr. & Gray, Fl. N. Amer. 1: 118, 1838.
*Athysanus pusillus* var. *glabrior* S. Wats. ex Robinson, in A. Gray, Syn. Fl. N. Amer. $1^1$: 113, 1895.

Stems branching from base, weak, 5–25 cm. long, unilaterally racemose from near base, puberulent with forked to few-rayed hairs below, usually glabrous above; lowermost leaves oblanceolate to narrowly obovate, 5–20 mm. long, sparingly toothed to entire, petiolate, hispidulous; cauline leaves few, sessile, few-toothed or entire; pedicels 1–3 mm. long, mostly uniformly recurved throughout their length, sparsely and minutely puberulent to glabrous; sepals about 0.5 mm. long; petals white, about twice as long as sepals, or often lacking; silicles 1.5–2 mm. long, nearly or fully as wide, uncinate hairs about 0.2–0.3 mm. long.

Mostly in sandy or gravelly soils, hillsides, and washes, often entangled in grasses and other herbs, chiefly Upper Sonoran and Transition Zones, British Columbia and Idaho to northern Baja California and Arizona, reaching the margins of the Sonoran Desert in eastern San Diego County, California, and in central Arizona. Feb.–March in our area.

## 20. THYSANOCARPUS Hook., Fl. Bor. Amer. 1: 69, *pl. 18, fig. A*, 1830

Slender, erect or ascending, strict to moderately branched annuals with petiolate lower leaves and sessile, often auriculate-clasping, upper ones. Inflorescences racemose, elongating rapidly immediately after anthesis. Flowers minute. Sepals ovate, spreading or spreading-ascending. Petals white or rose, spatulate, about equaling or slightly surpassing the sepals. Stamens usually 6, sometimes only 4. Ovary 1-celled, 1-seeded. Silicle orbicular or broadly elliptic, strongly compressed, the margins extended into a membranous wing, this entire, crenately notched or perforated, with or without distinct radiating nerves. Style often set in the bottom of a terminal notch in the wing, shorter than the depth of the notch, equaling it or surpassing it.

Stigma sessile or essentially so, surpassed by full width of wing; pedicels ascending, curved only at apex, often nearly erect; upper cauline leaves denticulate ......... 1. *T. erectus*

Stigma borne on short but evident style, surpassed by only portion of wing or equaling or exceeding the latter; pedicels spreading to gradually recurved; upper cauline leaves entire:

- Lower leaves not rosulate, laciniate into remote, nearly linear segments; fruits slightly longer than broad, broadly elliptic .................. 2. *T. laciniatus* var. *crenatus*
- Lower leaves usually rosulate, oblanceolate, lanceolate, or oblong-spatulate, remotely dentate or remotely and shallowly lobed; fruits orbicular in outline:

Wings of fruits with distinct, often conspicuous, rays, the intervening membrane perforated or regularly crenate .......................... 3. *T. curvipes* var. *elegans*
Wings of fruits rayless, the membrane entire or essentially so, not perforated or regularly crenate .................................. 3a. *T. curvipes* var. *eradiatus*

**1. Thysanocarpus erectus** S. Wats., Proc. Amer. Acad. **11**: 124, 1876

Strictly erect plant or more robust ones ascendingly branched from base or in upper parts, 1–3.5 dm. tall, with glabrous, slightly glaucous, or wholly green herbage; lower leaves oblanceolate to elliptic in outline, 4–25 mm. wide, 2–11 cm. long, saliently and coarsely toothed to subpinnatifid, teeth or lobes narrowly triangular, often again faintly denticulate with 1–2 teeth on one or both sides; cauline leaves sessile, 2–10 mm. wide, 1.5–5 cm. long, auriculate-clasping, shallowly but distinctly denticulate with several spreading teeth or the uppermost entire, lanceolate in outline; inflorescence often branched, branches 2–12 cm. long; sepals rose-purple, about 1 mm. long, obovate-elliptic, ascending-spreading; petals about equaling sepals or slightly exceeding them; fruiting pedicels slenderly filiform, 2.5–4 mm. long, erect at first but at maturity somewhat recurved-spreading and fruits then nearly pendent; silicles orbicular or slightly narrowed, 3.5–4.5 mm. wide, body about 2 mm. in diameter, tumidly convex on both sides, finely to densely puberulent or glabrous, reticulately veined; wings with 5–9 rays on each side, very narrowly notched apically, otherwise entire or essentially so; stigma sessile at bottom of the apical notch in wing.

Hillsides and rocky washes, Lower Sonoran Zone, Cedros and Guadalupe Islands, and in the peninsula of Baja California from the vicinity of Rosario south at least to Jaraguay. Feb.–March.

**2. Thysanocarpus laciniatus** Nutt. var. **crenatus** (Nutt.) Brewer, in Brewer & Wats., Bot. Calif. **1**: 49, 1876

*Thysanocarpus crenatus* Nutt., in Torr. & Gray, Fl. N. Amer. **1**: 118, 1838.
*Thysanocarpus ramosus* Greene, Bull. Calif. Acad. **2**: 390, 1887.
*Thysanocarpus desertorum* Heller, Muhlenbergia **2**: 47, 1905.
*Thysanocarpus laciniatus* subsp. *desertorum* Abrams, Ill. Fl. Pacif. States **2**: 299, 1944.

Erect slender plant 1–3 dm. tall with glabrous and glaucous herbage; leaves linear to narrowly oblanceolate, 1–6 mm. broad, 1–4 cm. long, lower sparingly short-toothed with slender divaricate teeth, upper leaves entire, faintly auriculate-clasping; inflorescences slender, to 10 cm. long; sepals about 1 mm. long, narrowly obovate-elliptic, cupped, spreading in anthesis; petals white, about 1.5 mm. long; pedicels slender, 3–8 mm. long, spreading or slightly arcuate to distinctly curved either at tips or less commonly their whole length; silicles broadly elliptic, 2.5–4 mm. wide, 3–5 mm. long, body usually moderately short-puberulent or sometimes glabrous, wing rayed and shallowly but regularly crenate, the outer margins white to pinkish; style 0.75–1 mm. long, slightly exceeding terminal notch in wing.

Grassy slopes and washes, Lower and Upper Sonoran Zones, Monterey County to southern California, where it reaches the western margins of the Colorado Desert. March–April.

**3. Thysanocarpus curvipes** Hook. var. **elegans** (Fisch. & Mey.) Robinson, in A. Gray, Syn. Fl. N. Amer. **1**[1]: 114, 1895

*Thysanocarpus elegans* Fisch. & Mey., Ind. Sem. Hort. Petrop. **2**: 51, 1835.
*Thysanocarpus amplectens* Greene, Pittonia **3**: 87, 1896.

Erect to ascendingly branched herbs 1–5 dm. tall with sparsely hirsute to glabrous stems and lower leaves; basal leaves more or less rosulate, narrowly oblong to narrowly oblanceolate in outline, 3–9 mm. wide, 2–6 cm. long, remotely but divergently dentate to shallowly pinnatifid, mostly glabrous in our area but often sparsely hirsute on plants growing in higher hills and in coastal California; cauline leaves narrowly lance-sagittate, 1–3 cm. long, slightly to distinctly auriculate-clasping, mostly glabrous; branches of inflorescences 5–15 cm. long; sepals about 1.5 mm. long, white-margined, mid section pinkish or pale green; petals white, slightly exceeding sepals; pedicels filiform, 3–6 mm. long, strongly recurved throughout their full length; silicles orbicular or essentially so, 4–6.5 mm. in diameter, body reticulate-veined, puberulent or glabrous, wings 5–8 rayed on each side, faintly crenate and usually perforated with elliptic holes between rays, outermost edge of wings pinkish or purplish, membranous; styles about 0.5 mm. long, set in an apical notch just about as deep.

Grassy slopes, canyons, and among rocks, Lower and Upper Sonoran Zones, central to southern California, east into Arizona and New Mexico, and south into northwestern Sonora. Feb.–May.

Most of the Arizona material has glabrous lower leaves, whereas the plants in the California Coast Ranges nearly always have at least a few scattered hairs on the lower leaves. The silicles are either glabrous or puberulent throughout its range.

**3*a*. Thysanocarpus curvipes** var. **eradiatus** Jepson, Man. Fl. Pl. Calif. 447, 1925
*Thysanocarpus laciniatus* var. *eremicola* Jepson, Fl. Calif. **2:** 100, 1936.

Differing from var. *elegans* in having less obviously toothed, often broader leaves, upper ones being entire and only faintly auricled or not auricled; and in having silicles with very thin, membranous, entire wings devoid of rays.

Desert slopes and canyons, Lower Sonoran Zone, Inyo County, California, south to the western margin of the Colorado Desert. March–April.

## 21. DESCURAINIA Webb & Berthel., Phytogr. Canar. 1: 72, 1836

Annual or biennial herbs with pinnate to tripinnatifid leaves, subglabrate to moderately dense-pubescent herbage, and racemose inflorescences elongating considerably in fruit. Flowers numerous, small, tetradynamous. Sepals acute-ovate, green, yellowish, or tinged with rose. Petals white to cream, narrowly clawed, with obovate-obtuse blades. Stamens included to slightly exserted. Siliques narrowly linear to clavate, terete or nearly so, valves more or less 1-nerved and separating from partition from bottom upward. Style short or the truncate to capitate stigma sessile. Seeds in 1 or 2 rows in each locule, lenticular-elliptic, yellowish to brownish.

Fruiting pedicels 12–30 mm. long; stigmas broadly capitate; seeds 24–32 in each locule, closely crowded .................................. 1. *D. obtusa* subsp. *adenophora*
Fruiting pedicels 3–12 mm. long; stigmas truncate or obscurely capitate; seeds 6–20 in each locule, not markedly crowded:
  Petals 2–3 mm. long, bright yellow; sepals 2–2.5 mm. long, yellowish ................. ................................................ 2. *D. pinnata* subsp. *menziesii*
  Petals 1–2 mm. long, whitish or pale yellow; sepals 0.75–2 mm. long, often rosy:
    Stems, leaves, and inflorescences moderately to densely canescent; seeds 0.75–1 mm. long; petals whitish:
      Pubescence dense, nonglandular; siliques mostly 9–13 mm. long; fruiting pedicels averaging about 6 mm. long ................ 2*a*. *D. pinnata* subsp. *ochroleuca*

Pubescence moderate to dense, often more or less glandular; siliques mostly 5–10 mm. long; fruiting pedicels averaging about 10 mm. long ........................................................................ *2b. D. pinnata* subsp. *halictorum*

Stems and leaves lightly to moderately canescent; inflorescences glabrous or nearly so; seeds 0.5 mm. long; petals pale yellow ............... *2c. D. pinnata* subsp. *glabra*

**1. Descurainia obtusa** (Greene) O. E. Schulz subsp. **adenophora** (Woot. & Standl.) Detling, Amer. Midl. Nat. **22:** 498, 1939

*Sophia adenophora* Woot. & Standl., Contr. U.S. Nat. Herb. **16:** 127, 1913.
*Descurainia adenophora* O. E. Schulz, Pflanzenreich $4^{105}$: 321, 1924.
*Descurainia obtusa* var. *adenophora* Kearney & Peebles, Jour. Wash. Acad. **29:** 478, 1939.

Strict or moderately branched, erect plants 3–8 dm. tall with sparsely canescent stems and moderately canescent leaves, often sparsely to moderately densely stipitate-glandular, especially in inflorescence; leaves 3–7 cm. long, obovate to oblanceolate in outline, pinnatifid into 3–5 pairs of pinnae and these entire to again pinnatifid into acute to obtuse segments 1–6 mm. long; sepals oblong, 2–2.5 mm. long, somewhat yellowish, sparsely glandular-puberulent; petals about as long as sepals or very slightly longer, oblanceolate, pale yellow or whitish; style about 0.5 mm. long; stigma flat-capitate, about twice as broad as diameter of the short style; fruiting pedicels 12–30 mm. long, spreading-ascending at an angle of about 45° to 65°, straight; siliques straight to slightly arcuately curved upward, 15–20 mm. long, each locule containing 24–32 seeds 0.75–1 mm. long, these often closely crowded and appearing uniseriate.

Hillsides and meadows, mainly Upper Sonoran and Transition Zones, Los Angeles, Riverside, and San Bernardino Counties, California, to western New Mexico and south into Chihuahua and northern Baja California. May–Sept.

**2. Descurainia pinnata** (Walt.) Britt. subsp. **menziesii** (DC.) Detling, Amer. Midl. Nat. **22:** 508, 1939

*Cardamine ?menziesii* DC., Syst. **2:** 267, 1821.
(For additional synonymy see Amer. Midl. Nat. **22:** 508, 1939.)

Erect, simple, or somewhat branched annuals 1–8 dm. tall with moderately pubescent to nearly glabrous stems and densely pubescent, pinnate to tripinnatifid leaves; leaf blades obovate to oblanceolate in outline, 1–10 cm. long, ultimate segments linear to oblanceolate; inflorescence subglabrous to moderately pubescent, with a few glandular hairs intermingled with dendritic ones or nonglandular; sepals bright yellow or only faintly greenish, 2–2.5 mm. long (rarely only 1.5 mm. long); petals bright yellow, broadly oblanceolate, slightly exceeding sepals, rather conspicuous; styles obsolete or nearly so; stigmas truncate, little if any broader than faintly developed style; fruiting pedicels 5–10 mm. long, spreading at 45°–90° angle; siliques 5–12 or rarely to 15 mm. long, usually slightly curved upward; seeds 0.75–1 mm. long, about 6–10 in each locule, not markedly crowded.

Hillsides, meadows, and along streams, Lower Sonoran Zone to Transition Zone, central California south to the Mojave and Colorado Deserts and southern flanks of the Sierra San Pedro Mártir, Baja California. Feb.–June.

**2*a*. Descurainia pinnata** subsp. **ochroleuca** (Wooton) Detling, Amer. Midl. Nat. **22:** 504, 1939

*Sophia ochroleuca* Wooton, Bull. Torrey Club **25:** 455, 1898.
*Sisymbrium ochroleucum* K. Schum., in Just, Bot. Jahresb. $26^{1}$: 348, 1900.
*Descurainia menziesii* var. *ochroleuca* O. E. Schulz, Pflanzenreich $4^{105}$: 329, 1924.
*Descurainia pinnata* var. *ochroleuca* Shinners, Field & Lab. **17:** 145, 1949.

Differing from subsp. *menziesii* chiefly in having smaller, pale yellow flowers, whole plant densely canescent and nonglandular; fruiting pedicels averaging about 6 mm. long and spreading at an angle of 40°–80°; seeds 7–12 in each locule of the clavate, slightly arcuate siliques, these 9–13 mm. long.

Plains and low hills, Lower Sonoran and Upper Sonoran Zones, southern Arizona to Texas and Chihuahua, barely entering the Sonoran Desert. March–May.

**2*b*. Descurainia pinnata** subsp. **halictorum** (Cockerell) Detling, Amer. Midl. Nat. **22:** 505, 1939

*Sophia halictorum* Cockerell, Bull. Torrey Club **25:** 460, 1898.
*Descurainia pinnata* var. *halictorum* M. E. Peck, Man. Pl. Oregon 340, 1941.
(For additional synonymy see Amer. Midl. Nat. **22:** 505, 1939.)

Annual 1.5–5 dm. tall with moderately to densely pubescent, often somewhat glandular stems and inflorescences; leaves moderately canescent, lanceolate to oblanceolate in outline, pinnate to bipinnatifid, the segments linear to narrowly ovate; calyx 1–2 mm. long, often rose-tinged; petals slightly exceeding the sepals, whitish or pale yellowish; fruiting pedicels 8–12 mm. long, spreading at 65°–90° angle; siliques 5–10 mm. long; seeds 0.75–1 mm. long, 8–20 in each locule.

On hillsides and plains, and along washes, Lower Sonoran Zone and Upper Sonoran Zone, eastern Oregon to Arkansas and Texas and south along the eastern flank of the Sierra Nevada to central Baja California, southern Sonora, and central Chihuahua. Jan.–June.

**2*c*. Descurainia pinnata** subsp. **glabra** (Woot. & Standl.) Detling, Amer. Midl. Nat. **22:** 507, 1939

*Sophia glabra* Woot. & Standl., Contr. U.S. Nat. Herb. **16:** 127, 1913.
*Descurainia pinnata* var. *glabra* Shinners, Field & Lab. **17:** 145, 1949.

Annual 1–5 dm. tall with stems usually branching from near base; leaves and lower part of stems lightly canescent with fine nonglandular hairs, inflorescence and upper parts of stems glabrous; sepals 0.75–1.5 mm. long, yellowish or rose-colored; petals pale yellow, barely exceeding or only equaling sepals; fruiting pedicels 4–12 mm. (usually 6–10 mm.) long, spreading 60°–90° from the rachis; siliques 5–8 mm. long; seeds about 0.5 mm. long, 8–12 in each locule.

Desert plains, washes, and hillsides, Lower and Upper Sonoran Zones, Kern and Inyo Counties, California, to New Mexico, south into Chihuahua, Sinaloa, and central Baja California. Feb.–May.

## 22. ARABIS L., Sp. Pl. 664, 1753

Biennial or perennial herbs or plants with lignescent bases, herbage glabrous to pubescent with simple, forked, branched, or stellate hairs. Stems simple or branched, terete. Basal leaves petiolate, entire to dissected. Cauline leaves usually sessile, entire to dentate. Inflorescences racemose, ebracteate. Flowers erect or reflexed at anthesis, tetradynamous. Sepals oblong to subovate, erect, the margins often white. Petals spatulate, oblong or obovate, white to purple or less commonly stramineous. Stamens straight or single ones curved. Siliques linear, straight or curved, erect to reflexed, flattened parallel to partition or rarely terete, 2-valved. Style prominent or none. Stigma entire or faintly 2-lobed. Seeds pendulous, numerous, lenticular-orbicular to oblong, winged or wingless. Cotyledons accumbent.

Siliques 1.2–3.5 mm. wide; seeds narrowly winged, to 2.5 mm. in diameter; valves of siliques not glaucous:
Seeds uniseriate; sepals not saccate:
Pedicels slender, glabrous; petals 6–9 mm. long, 1.5–2.5 mm. wide ... 1. *A. perennans*
Pedicels stout, appressed-pubescent; petals 8–12 mm. long, 2–4 mm. wide ........... ........................................ 2. *A. sparsiflora* var. *californica*
Seeds biseriate; outer sepals more or less saccate:
Siliques puberulent, strictly appressed, on geniculately reflexed pedicels at maturity; herbage densely puberulent with appressed hairs ...... 3. *A. pulchra* var. *pulchra*
Siliques glabrous, pendulous on arched pedicels, not strongly appressed; herbage glabrous or nearly so toward the inflorescence, lower part with sparse, spreading hairs ............................................. 3*a*. *A. pulchra* var. *gracilis*
Siliques 5–8 mm. wide; seeds broadly winged, 5–6 mm. in diameter; valves of siliques glaucous ................................................ 4. *A. glaucovalvula*

**1. Arabis perennans** S. Wats., Proc. Amer. Acad. **22:** 467, 1887

*Arabis arcuata* var. *perennans* M. E. Jones, Proc. Calif. Acad. II, **5:** 621, 1895.
*Arabis gracilenta* Greene, Pittonia **4:** 194, 1900.
*Arabis eremophila* Greene, loc. cit.
*Arabis recondita* Greene, op. cit. 195.
*Arabis angulata* Greene, in Woot. & Standl., Contr. U.S. Nat. Herb. **16:** 123, 1913.

Perennial herb with several to many ascending or erect branches 1.5–6 dm. tall; stems pubescent below with several-times branched coarse hairs, glabrous or glabrate above; basal leaves several to numerous, petiolate, oblanceolate to obovate, 4–10 mm. wide, 2–6 cm. long, usually dentate, densely pubescent with coarse, several-branched hairs; cauline leaves lanceolate-auriculate and sometimes sagittate, 2–8 mm. wide, 1–3.2 cm. long, entire to sparsely dentate, lower ones pubescent, upper ones glabrous; sepals pubescent, 1–1.5 mm. wide, 3.5–4.5 mm. long; petals purple or pinkish, 1.5–2.5 mm. wide, 6–9 mm. long, with a narrow claw; pedicels spreadingly arched downward, 1–2 cm. long, glabrous; siliques spreading to pendulous, 1.2–2 mm. wide, 4–6 cm. long, usually nerveless, glabrous; seeds orbicular-lenticular, 1–1.5 mm. wide, uniseriate in each locule, winged.

Arid slopes and flats, often growing through shrubs, Lower Sonoran Zone to Transition Zone, Inyo County, California, to southern Utah, Colorado, northern New Mexico, and south to southern Arizona and northern Baja California. April–May.

**2. Arabis sparsiflora** Nutt. var. **californica** Rollins, Rhodora **43:** 402, 1941

Perennial herbs 2.5–9 dm. tall with more or less branched caudex, branches densely pubescent throughout with closely appressed, dendritically branched hairs; basal leaves numerous, 3–10 mm. wide, 3–10 cm. long, coarsely dentate, densely pubescent with moderately fine dendritic hairs; cauline leaves linear-lanceolate to broadly lanceolate, sagittate-auriculate, 3–10 mm. wide, 2–8 cm. long, obtuse, finely but densely appressed-pubescent; pedicels 5–15 mm. long, appressed-pubescent; sepals oblong, sparsely to moderately pubescent, 1.5–2 mm. wide, 4–6 mm. long; petals pink to purple, 2–4 mm. wide, 6–15 mm. long; nectaries subtending all stamens; siliques usually widely pendulous, 6–12 cm. long; seeds orbicular, narrowly winged, in a single row.

Mountainsides and washes, upper edge of Lower Sonoran Zone and throughout Upper Sonoran Zone, Los Angeles and San Bernardino Counties, California, south along the inner margins of the coastal ranges into northern Baja California. April–May.

**3. Arabis pulchra** M. E. Jones, in S. Wats., Proc. Amer. Acad. **22:** 468, 1887, var. **pulchra**

*Arabis pulchra* var. *typica* Rollins, Rhodora **43:** 459, 1941.

Perennial with a lignescent base, simple or considerably branched, stems erect, simple or branched, 2–8 dm. tall, densely pubescent with minute, dendritic, appressed hairs; basal leaves linear, obtuse, entire to obscurely dentate, 3–6 mm. wide, 4–8 cm. long, densely pubescent with minute, branched hairs; cauline leaves linear, 3–4 mm. wide, 2–6 cm. long, not auriculate, densely pubescent; pedicels erect to divergent at anthesis, geniculately reflexed in fruit, densely pubescent, 8–20 mm. long; sepals more or less purplish, 1.5–2.5 mm. wide, 5–8 mm. long, outer pair slightly saccate; petals white, reddish, or purplish, 3–5 mm. wide, 8–20 mm. long; nectaries continuous beneath stamens; siliques reflexed, straight, 2.5–3.5 mm. wide, 4–7 cm. long, densely puberulent, valves nerved to middle or above; stigma sessile or essentially so; seeds suborbicular, 1.5–2 mm. broad, distinctly winged, in 2 rows.

Rocky hillsides and along arroyos, mainly upper parts of Lower Sonoran Zone and adjacent Upper Sonoran Zone, west-central and southern Nevada, and from Inyo County to western San Diego County, California, south into northern Baja California. March–May.

**3*a*. Arabis pulchra** var. **gracilis** M. E. Jones, Contr. West. Bot. **8:** 41, 1898

*Arabis trichopoda* Greene, in Fedde, Rep. Nov. Spec. **5:** 242, 1908.
*Arabis pulchra* var. *glabrescens* Wiggins, Contr. Dudley Herb. **1:** 100, 1929.
*Arabis pulchra* var. *viridis* Jepson, Fl. Calif. **2:** 70, 1936.

Differing from var. *pulchra* in having coarser and sparser pubescence on lower leaves and stems, upper parts of stems glabrous; pedicels arched downward but never geniculately reflexed; siliques glabrous or nearly so.

Meadows, hillsides, and washes, Lower and Upper Sonoran Zones, southern Nevada to eastern San Diego County, California. March–May.

**4. Arabis glaucovalvula** M. E. Jones, Contr. West. Bot. **8:** 40, 1898

Perennial, fairly robust plant from a ligneous branching caudex with 1–several stems 1.5–4 dm. tall, hoary-pubescent throughout; basal leaves linear to narrowly oblong, entire, 2–5 mm. wide, 2–5 cm. long, densely hoary-pubescent with dendritic hairs; cauline leaves linear-lanceolate to lanceolate, 2–5 mm. wide, 1–4 cm. long; pedicels stoutish, strongly recurved; sepals 2 mm. wide, 4–5 mm. long, nonsaccate, reddish, pubescent; petals pink to whitish, 6–8 mm. long, about 2 mm. wide, blade narrowed to a slender claw; nectaries beneath single stamens, nearly or quite obsolete under paired stamens; siliques reflexed, oblong, 5–8 mm. wide, 2–4.5 cm. long, obtuse at each end, glabrous and distinctly glaucous, 1-nerved to middle or full length of valves; style nearly 1 mm. long; seeds orbicular, 5–6 mm. broad including the conspicuous wings, in 2 rows.

Hillsides and mesas, mainly Lower Sonoran Zone, Inyo County to northern Riverside County, California; once reported from southern Nevada. April–May.

## 23. ERYSIMUM L., Sp. Pl. 660, 1753

Annual, biennial, or short-lived perennial herbs or rarely suffruticose plants with more or less pubescent stems and foliage. Leaves simple, entire to deeply dentate, or lobed, often forming a basal rosette, petiolate or the upper ones subsessile. Racemes dense in anthesis, elongating markedly in fruit. Sepals erect, narrow, lateral ones often slightly saccate at base. Petals yellow to orange, claws slender, expanding into obovate, horizontally spreading limbs. Stamens 6, filaments about equaling claws of petals. Style slightly exceeding the stamens; stigma shallowly bilobed. Silique linear, 4-angled to terete, each valve with a prominent midrib, valves deciduous. Seeds somewhat compressed, oblong, often narrowly winged or wingless. Cotyledons accumbent.

**1. Erysimum capitatum** (Dougl.) Greene, Fl. Fran. 269, 1891

*Cheiranthus capitatus* Dougl., in Hook., Fl. Bor. Amer. **1:** 38, 1829.
*Erysimum elatum* Nutt., in Torr. & Gray, Fl. N. Amer. **1:** 95, 1838.
(About a score of additional names and combinations have been applied to the various subspecific taxa constituting this polymorphic species, but they are omitted here in the interest of economy of space and because most of them do not apply to var. *capitatum,* which represents the species within our area.)

Biennial with a short robust caudex and 1-several erect branches 3–15 dm. tall; lower leaves linear-oblanceolate to lanceolate, 3–25 mm. wide, 4–15 (rarely to 23) cm. long, dentate or sometimes entire, cinereous-puberulent with 1–several-rayed trichomes; racemes to 8 dm. long but usually 1–4 dm. long; pedicels stoutish, spreading-ascending, 4–25 mm. long; sepals 8–10 mm. long, finely puberulent, lateral two to three times as broad as outer and inner ones and distinctly saccate at base, obtuse at apex; petals about 5–10 mm. wide at widest part of blade, 18–25 mm. long, yellow to orange; siliques ascending to divergent, 4–12 cm. long, about 2 mm. wide, puberulent; seeds slightly compressed, lightly winged at apex and faintly keeled along margins, 2–3 mm. long, about 1–1.2 mm. wide.

Hillsides and among rocks in washes, mainly Upper Sonoran and Transition Zones, Washington to Saskatchewan and south to northern Baja California, Arizona, New Mexico, and Missouri. Entering the Sonoran Desert only along its margins as an occasional straggler. Feb.–Sept.

## Family 50. **CAPPARIDACEAE.** CAPER FAMILY

Herbaceous or woody plants, usually with ill-scented leaves, bark, or wood. Leaves alternate, simple or palmately compound. Flowers perfect, regular or essentially so, usually in bracteate racemes. Sepals and petals 4. Stamens 6–many, mostly well exserted. Ovary 1–2-celled, stipitate. Fruit a 2-valved capsule, berrylike, drupelike, or of 2 nutlets.

Plant a small tree or large woody shrub; leaves mostly simple; fruit drupelike or berrylike:
  Flowers dioecious; leaves sometimes 3-foliolate, margins strongly revolute; bark light gray or chalky gray ........................................ 1. *Forchammeria*
  Flowers perfect; leaves always simple, margins not revolute, sometimes involutely folded; bark dark gray or blackish ................................ 2. *Atamisquea*
Plants herbaceous or a low rounded shrub with trifoliolate leaves; leaves mostly trifoliolate; fruit capsular:
  Shrubby plant; fruit an inflated, tardily dehiscent capsule .................. 3. *Isomeris*
  Herbaceous plants; fruit not inflated:
    Fruit didymous, the nutlets indehiscent, each falling with the solitary seed enclosed ... ........................................................ 4. *Wislizenia*
    Fruit capsular, not didymous; capsules dehiscent, several-seeded:
      Capsules broader than long, quadrangular or triangular; petals always yellow ...... ........................................................ 5. *Cleomella*

Capsules linear to oblong, much longer than broad; petals white, pink, or in one species pale yellow:
  Stamens 6; herbage not clammy nor glandular-villous; petals pink or yellowish ... ........................................................ 6. *Cleome*
  Stamens 8–32; herbage glandular-villous and clammy; petals white or nearly so ... ........................................................ 7. *Polanisia*

## 1. FORCHAMMERIA Liebm., Nat. For. Kjøbenhavn Vid. Medd. 1853: 93, 1854

Glabrous or pubescent trees with simple or compound petiolate leaves. Flowers small, in racemes or panicles, dioecious or occasionally a pistillate flower among staminate or abortive stamens around bases of a few pistillate flowers. Calyx shallowly cuplike, faintly 4–8-toothed. Staminate flowers with 8–24 stamens inserted on a fleshy disk; pistillate flowers with 4–10 or more subsessile, apparently sterile stamens inserted on a fleshy, slightly lobed disk. Ovary 2-celled. Fruit ovoid to subglobose, subdrupaceous, 1-celled by abortion; seed solitary.

### 1. Forchammeria watsonii Rose, Contr. U.S. Nat. Herb. 1: 302, *pls. 24–25,* 1895

A tree 3–8 m. high with smooth, light gray bark and a dense, usually spreading crown; leaves coriaceous, linear or narrowly linear-oblong, 4–12 mm. wide, 5–12 cm. long, dark green and sparsely puberulent above, paler, strongly reticulate, and more densely puberulent beneath, often subcordate at base, the margins revolute; flowers in racemes 1.5–6 cm. long; stamens in staminate flowers 16–24, filaments distinct, 2–4 mm. long; pistillate flowers bearing 5–16 sessile, abortive stamens with acute anthers 1.5–2 mm. long; fruit pyriform, 8–10 mm. long, orange to purplish.

Desert plains and hillsides, Lower Sonoran Zone, central Sonora to northern Sinaloa; near San Ignacio to the Cape Region, Baja California. March–April.

## 2. ATAMISQUEA Miers, Trav. Chile 2: 529, 1826

A shrub or small tree with divaricate, rigid, brittle branches and ill-scented wood and herbage. Leaves simple, alternate, coriaceous, dark green and glabrate above, silvery-lepidote and pustulate-glandular beneath, narrowly oblong. Flowers solitary or in few-flowered fascicles. Sepals 4, in two pairs, the outer valvate and larger than the spatulate inner ones. Petals 4, small, in two unequal pairs. Stamens 6, with 1–6 alternating staminodia. Ovary stipitate, unilocular, with 4 parietal placentae. Ovules several; stigma sessile. Fruit drupelike, ovate to globose, 1–2-seeded, short-stipitate. Seeds partially surrounded by a pulpy aril.

### 1. Atamisquea emarginata Miers, Trav. Chile 2: 529, 1826

Densely branched shrub or rarely a tree 1–8 m. high; branches rigid, lateral branchlets forming nearly right angles with larger branches, tawny- or silvery-lepidote and sprinkled with small sessile glands; leaves linear to linear-oblong, 0.8–1.5 cm. long, emarginate at apex, often involutely folded; outer sepals about 6–7 mm. long, lepidote on outside, pilose within, cuculate-valvate, inner pair spatulate, 3 mm. long, lepidote without, pilose within; petals 7–8 mm. long, linear-lanceolate, lepidote without, glabrous within, densely pilose-ciliate on margins; filaments of stamens lepidote at base, glabrous above; staminodia 1–3, pilose at base; stipe 4–8 mm. long; fruit drupelike, ovoid, about 6–8 mm. long.

Sandy or gravelly bajadas and plains, Lower Sonoran Zone, Baja California from the vicinity of San Ignacio southward and in southwestern Pima County, Arizona, to central Sonora; Tiburón Island. Also in Argentina. March–April.

### 3. **ISOMERIS** Nutt., in Torr. & Gray, Fl. N. Amer. **1**: 124, 1838

Brittle rounded shrubs with puberulent branchlets and ill-scented herbage and wood. Leaves digitately 3-foliolate. Flowers large, yellow, axillary or in dense to sparse, bracteate racemes. Sepals 4, erect, persistent. Petals 4, subequal, but usually the upper pair connivent and scarcely spreading, the lower pair divergent. Stamens 6, inserted on receptacle, which is dilated into a hemispheric torus. Ovary stipitate, 1-celled, many-ovuled; style short; stigma capitate. Capsule inflated, dehiscing tardily. Seeds subglobose to obovoid, smooth.

Capsules strongly inflated, tapering abruptly at the apex ......... 1. *I. arborea* var. *arborea*
Capsules weakly inflated, tapering gradually at each end ...... 1a. *I. arborea* var. *angustata*

**1. Isomeris arborea** Nutt., in Torr. & Gray, Fl. N. Amer. **1**: 124, 1838, var. **arborea**
*Cleome isomeris* Greene, Pittonia **1**: 200, 1888.
*Isomeris arborea* var. *insularis* Jepson, Fl. Calif. **2**: 11, 1936.

Divergently branching, loose to compact shrub 0.5–3 m. high; branchlets finely but closely puberulent; leaflets oblong to elliptic, 4–8 mm. wide, 1–2.5 cm. long, entire, mucronulate, dark or grayish green, glabrous or nearly so, about equaling petioles; racemes stout to elongate, to 10 or 15 cm. long, upper bracts unifoliolate, sessile or petiolate, equaling leaflets or reduced to 4–5 mm. in length; calyx lobes deltoid-ovate, attenuate, 6–8 mm. long; petals oblong, 10–16 mm. long; capsule leathery, strongly inflated, 3.5–6 cm. long, abruptly tapering at apex, long-attenuate at base, stipe 1.5–3 cm. long; seeds obovoid to subglobose, 5–9 mm. long, yellowish or tawny.

Dry hillsides and along washes, Lower and Upper Sonoran Zones, Monterey County south in the Coast Ranges and the western edge of the Mojave and Colorado Deserts to central Baja California; Cedros Island. Flowering almost throughout the year.

**1*a*. Isomeris arborea** var. **angustata** Parish, Muhlenbergia **3**: 128, 1907

Differs from var. *arborea* only in having weakly inflated pods usually 10 mm. broad or less, and gradually tapering toward both ends.

Mainly Lower Sonoran Zone, rare in the western Mojave Desert, commoner in the western Colorado Desert and south into Baja California.

### 4. **WISLIZENIA** Engelm., in Wisliz., Mem. Tour North. Mexico 99, 1848

Erect, slender or profusely branching, ill-scented, annual herbs with petiolate, 1–3-foliolate leaves. Stipules bristlelike. Flowers in terminal racemes, yellow. Sepals 4, distinct except at base. Petals 4, short-clawed. Stamens 6, well exserted. Pods didymous, the valves divergently to horizontally spreading and each valve closely contracted upon the solitary seed and falling with it. Valves cylindric to obovoid, smooth, reticulate, tuberculate, or horned terminally, several tubercles or horns present around the distal rim. Stipe elongated and usually strongly refracted in fruit. Style slender, bristlelike, persistent.

Valves of capsules reticulate, smooth, or faintly tuberculate distally; mature nutlets subglobose to obovoid, rarely over 1.5–2 mm. long; leaflets mostly broadly obovate ...... ............................................ 1. *W. refracta* var. *refracta*
Valves of capsules strongly tuberculate or horned; mature nutlets cylindric to cylindric-obovoid, mostly 2.5–5 mm. long; leaflets linear, lanceolate, or narrowly oblanceolate:
  Leaves mostly 2–3-foliolate; nutlets bearing several fingerlike, slender horns or projections around the distal rim, these horns usually considerably longer than broad .......... ............................................ 1a. *W. refracta* var. *mamillata*
  Leaves mostly simple, occasionally bifoliolate; nutlets mostly strongly tuberculate, sometimes stoutly horned ............................ 1b. *W. refracta* var. *palmeri*

**1. Wislizenia refracta** Engelm., in Wisliz., Mem. Tour North. Mexico 99, 1848, var. **refracta**

*Wislizenia scabrida* Eastw., Bull. Torrey Club **30:** 490, 1903.
*Wislizenia californica* Greene, Proc. Biol. Soc. Wash. **19:** 130, 1906.
*Wislizenia divaricata* Greene, loc. cit.
*Wislizenia pacalis* Greene, op. cit. 131.
*Wislizenia costellata* Rose, in Greene, op. cit. 132.

Stout, much-branched, erect annual 5–20 dm. high with sparsely leafy branches and 3-foliolate leaves, leaflets oblanceolate to obovate, 1–3 cm. long, obtuse or rounded or rarely acute at apex, gradually narrowed to a cuneate base, entire, glabrous or scaberulous along midvein; petioles about equaling leaflets; inflorescences 1–4 cm. long, elongating in fruit; sepals ovate, about 1 mm. long, acute; petals yellow, 5–6 mm. long; stamens 6–10 mm. long; styles slender, 4–5 mm. long, persistent; capsules stipitate and refracted in fruit; nutlets spreading, 1.5–2 mm. long, obovoid or subglobose, tuberculate, reticulate, or smooth.

Sandy places, roadsides, washes, and waste places, mainly Lower Sonoran Zone, San Joaquin Valley, southern California, and southern Nevada to western Texas and south into Mexico. May–Sept.

Fruit and leaf characters variable.

**1*a*. Wislizenia refracta** var. **mamillata** (Rose) Wiggins, Contr. Dudley Herb. **4:** 17, 1950

*Wislizenia mamillata* Rose ex Greene, Proc. Biol. Soc. Wash. **19:** 132, 1906.

Leaflets 1–3, mostly 2 or 3, linear-oblanceolate or narrowly elliptic; nutlets 2.5–4 mm. long, surmounted by slender, fingerlike projections 0.5–1.5 mm. long, projections 3–8, erect or somewhat spreading.

Sandy soil, often in saline situations, Lower Sonoran Zone, vicinity of Guaymas, Sonora; grading into the next variety northward.

**1*b*. Wislizenia refracta** var. **palmeri** (A. Gray) I. M. Johnston, Proc. Calif. Acad. IV, **12:** 1028, 1924

*Wislizenia palmeri* A. Gray, Proc. Amer. Acad. **8:** 622, 1873.
*Wislizenia fruticosa* Greene, Proc. Biol. Soc. Wash. **19:** 131, 1906.

Leaves mostly simple, sometimes a few bifoliolate and rarely trifoliolate leaves on the same plant with predominantly simple leaves; leaflets (or blade of simple leaf) linear or narrowly lanceolate, 1–3 cm. long, 1.5–4 mm. wide; nutlets 3–5 mm. long, tuberculate or bearing hornlike projections distally, these stout, usually nearly or fully as wide as long.

Sandy or saline places, Lower Sonoran Zone, from near the mouth of the Colorado River in southeastern California, south almost to Guaymas, Sonora.

## 5. CLEOMELLA DC., Prod. 1: 237, 1824

Erect or ascending, branched annuals with digitately 3-foliolate, alternate leaves. Flowers yellow, nearly regular, in terminal or subterminal racemes or in axillary clusters. Sepals 4, distinct and deciduous. Petals 4. Stamens 6. Capsule rhomboidal, broader than long, pendent or spreading on slender pedicels and stipes, few-seeded, the valves often hoodlike or slightly horned.

**1. Cleomella obtusifolia** Torr. & Frem., in Frem., Second Rep. 311, 1845

*Cleomella taurocranos* A. Nels., Proc. Biol. Soc. Wash. **18**: 172, 1905.
*Cleomella obtusifolia* var. *pubescens* A. Nels., loc. cit.
*Cleomella obtusifolia* var. *jonesii* Crum, in Jepson, Fl. Calif. **2**: 13, 1936.
*Cleomella obtusifolia* var. *florifera* Crum, in Jepson, loc. cit.

Freely branching from near the base and forming compact rounded plant 1–3.5 dm. high; petioles 5–15 mm. long, glabrous or sparsely puberulent; leaflets obovate to oblong, 3–6 mm. wide, 5–15 mm. long, dark green, obtuse, often with a slender mucro at apex, scaberulous to coarsely pubescent; stipules fimbriate, 2–3 mm. long and often white-hairy; flowers in terminal racemes or sometimes in axillary fascicles; petals linear or oblong, 4–5 mm. long, slightly mealy-puberulent without; style about 2 mm. long, persistent; capsule 5–10 mm. broad, valves inflated and each bearing a lateral hornlike process; fruiting pedicels 6–8 mm. long, recurved slender stipe usually about two-thirds as long.

In alkaline soil, Lower Sonoran Zone, Sacramento and San Joaquin Valleys; Inyo County, California, and adjacent Nevada to the southwestern part of the Colorado Desert. March–Sept.

## 6. CLEOME L., Sp. Pl. 671, 1753

Annual or perennial herbs, or shrubs, or trees, with simple or digitately 3–5-foliolate, compound leaves. Sepals 4, distinct or basally united, deciduous or persistent. Petals 4, clawed, entire. Stamens 6 or rarely 4, often distinctly exserted. Ovary stipitate, with a basal gland. Fruit usually a long-stipitate, linear to oblong, many-seeded capsule; seeds obovate-reniform, dull.

Petals 3–5 mm. long; leaflets linear to linear-oblong; capsule not stipitate; sepals distinct or nearly so at base, deciduous ........ 1. *C. tenuis*
Petals 10–12 mm. long; leaflets lanceolate or oblong; capsule long-stipitate; sepals united at base and persistent:
  Petals purple, pink, or white, 3–4 mm. wide; leaflets 3; stipes about equaling or shorter than pedicels ........ 2. *C. serrulata*
  Petals pale yellow, 1.5–2 mm. wide; leaflets 5; stipes one and one-half times to twice as long as pedicels ........ 3. *C. jonesii*

**1. Cleome tenuis** S. Wats., Proc. Amer. Acad. **24**: 39, 1889

Slender annual 1–10 dm. high (usually 1–5) with slender, erect or ascending branches and glabrous herbage except for sparsely pubescent petioles; petioles mostly short, 2–10 mm. long, slender; leaflets 3–5, linear to linear-oblanceolate, 4–25 mm. long, obtuse, often curling into a close ball in dry weather; racemes few-flowered, lax, elongating only slightly in fruit; pedicels slender, 3–9 mm. long; sepals 1–1.5 mm. long, distinct, ovate, apiculate, deciduous; petals inconspicuous, pale yellow, 3–5 mm. long; capsules on slender spreading pedicels, 1.5–3.5 cm. long, slightly striate, not torulose, sessile on pedicel or stipe, 1–1.5 mm. long; style slender, 3–4 mm. long; seeds subglobose, transversely rugose dorsally, dark brown, dull.

In heavy clay soil or sometimes on sandy slopes, Lower Sonoran Zone, central Sonora, and southern Baja California, Mexico. May–Aug.

**Cleome sonorae** A. Gray, Pl. Wright. **2:** 16, 1853, with small pinkish or whitish flowers and short-stipitate pods, occurs in southeastern Arizona and adjacent Sonora but has not been reported from the Sonoran Desert.

**Cleome ephemera** Brandegee, Proc. Calif. Acad. II, **3:** 112, 1891, has small pinkish flowers similar to those of *C. sonorae,* but the leaves are simple, linear, 1–3 cm. long, and the capsules sessile instead of stipitate. It is known only from the vicinity of San José del Cabo, but it may occur on sandy soil north of the Cape Region.

## 2. Cleome serrulata Pursh, Fl. Amer. Sept. 441, 1814

*Peritoma serrulatum* DC., Prod. **1:** 237, 1824.
*Cleome integrifolia* Torr. & Gray, Fl. N. Amer. **1:** 122, 1838.

Annual glabrous herb 5–12 dm. high, openly branched above; leaflets 3, glabrous, lanceolate to oblong-lanceolate, up to 2 cm. wide, 7.5 cm. long, entire or sometimes serrulate, lower leaves on petioles equaling or surpassing leaflets, upper nearly or quite sessile; racemes dense in flower, elongated in fruit; pedicels slender, 1–2 cm. long in flower, only slightly elongated in fruit; sepals pinkish, united at bases, ovate-lanceolate, acuminate, about 3–4 mm. long, persistent as a shallow cup at base of stipe; petals pink or purplish, rarely white, oblong or obovate, 3–4 mm. wide, 10–12 mm. long, obtuse at apex, short-clawed; pods linear, 2.5–5 cm. long, 4–5 mm. in diameter; stipes equaling or shorter than pedicels; seeds dark brown, 2.5 mm. wide, slightly wrinkled-vesicular dorsally.

Plains, Upper and Lower Sonoran Zones, eastern Washington to Saskatchewan, northeastern California, Kansas, New Mexico, and northern Arizona. Introduced in the Mojave Desert and to be expected in the northwestern part of the Sonoran Desert. June–Sept.

## 3. Cleome jonesii (J. F. Macbride) Tidestrom, Proc. Biol. Soc. Wash. 48: 39, 1935

*Cleome lutea* var. *jonesii* J. F. Macbride, Contr. Gray Herb. **65:** 39, 1922.

Annual, erect, openly branching herb 5–12 dm. high, glabrous or stipes sparsely and minutely puberulent; petioles 5–35 mm. long, leaflets mostly 5, oblanceolate or elliptic-oblong, obtuse and apiculate, or acute, 5–15 mm. wide, up to 6 cm. long, slightly lighter beneath than above, on petiolules 1–4 mm. long; inflorescence crowded, greatly elongate in fruit, about 3–4 cm. broad; bracts elliptic to narrowly obovate, apiculate, 3–5 mm. long; pedicels stoutish, about 1 cm. long in fruit; sepals ovate-acuminate, 2–3 mm. long, persistent; petals yellow, about 2 mm. broad, 10–12 mm. long; stamens 2 cm. long; capsules linear, 4–6.5 cm. long, on stipes 1.5–2.5 cm. long; seeds dark brown with lighter brown vesicular tubercles and ridges on the back.

Sandy soil, mostly along streams, Lower and Upper Sonoran Zones, Santa Cruz and Pima Counties to Mohave and Yavapai Counties, Arizona, and in northern Baja California. May–Sept.

### 7. POLANISIA Raf., Jour. Phys. 89: 98, 1819

Openly branching annual herbs with viscid-pubescent, ill-scented herbage. Leaves palmately 3–5-foliolate or rarely simple. Flowers in terminal, bracteate, often dense racemes. Sepals 4, deciduous. Petals 4, slender, or broader and distinctly clawed. Stamens 8–32, unequal, often surpassing the petals and conspicuous. Ovary stipitate or sessile, a gland borne on the receptacle at base of upper side of ovary. Pods elongated, cylindrical or compressed, dehiscing by 2 valves from summit. Seeds annular-reniform, smooth, reticulate, or rugose, numerous.

**1. Polanisia trachysperma** Torr. & Gray, Fl. N. Amer. **1:** 669, 1840
*Jacksonia trachysperma* Greene, Pittonia **2:** 175, 1891.

Usually considerably branched, 2–8 dm. high; petioles 1–5 cm. long; leaflets 3, oblanceolate to narrowly ovate or rarely broadly elliptic, acute or obtuse, 1–3 cm. long, to 1.5 cm. wide, petiolulate; sepals lanceolate, 4–5 mm. long, purplish; petals oblanceolate, 8–12 mm. long, emarginate at apex, tapering to a slender claw, yellowish white; stamens 10–18 mm. long, exserted, purplish; pods 3–6 cm. long, narrowed toward base but scarcely stipitate; seeds plump, 1.5–2 mm. in diameter, dark brown, usually reticulate, verrucose or rugose transversely.

Sandy or gravelly soils, often along washes, mostly Upper Sonoran and Transition Zones, eastern Washington and Oregon to northeastern California, Minnesota, Texas, Arizona, and northern Mexico. Occurring irregularly along the margins of the Sonoran Desert in Pima County, Arizona, and near Cucurpe, Sonora. June–Sept.

## Family 51. RESEDACEAE. Mignonette Family

Annual or perennial herbs or subwoody plants with fasciculate or alternate leaves and glandlike stipules. Flowers small, greenish, in bracteate spikes or racemes, somewhat dimorphic. Calyx 4–7-parted. Petals usually as many as calyx lobes. Receptacle often with a one-sided hypogynous disk. Stamens 3–many, unequal. Carpels 3–6; styles and stigmas 3, or stigmas sessile; ovules numerous. Fruit capsular. Seeds reniform, without endosperm.

### 1. OLIGOMERIS Camb., in Jacq., Voy. Ind. Bot. 4: 23, *pl. 25,* 1844

Small glaucous annual with linear entire leaves. Sepals 4. Petals 2, entire or 2-lobed; disk none; stamens 3–10. Ovary 4-angled and 4-beaked. Capsule opening at apex, 4-sulcate. Seeds many.

**1. Oligomeris linifolia** (Vahl) J. F. Macbride, Contr. Gray Herb. **53:** 13, 1918
*Reseda subulata* Delile, Fl. Aegypt. Ill. 15, 1813. Nomen nudum.
*Reseda linifolia* Vahl, in Hornem., Hort. Hafn. 501, 1815.
*Ellimia ruderalis* Nutt., in Torr. & Gray, Fl. N. Amer. **1:** 125, 1838.
*Oligomeris glaucescens* Camb., in Jacq., Voy. Ind. Bot. **4:** 24, *pl. 25,* 1844.
*Oligomeris subulata* Webb, Fragm. Aethiop. 26, 1854.
*Dipetalia subulata* Kuntze, Rev. Gen. Pl. **1:** 39, 1891.

Erect and strict or profusely branched from base, 0.5–4 dm. tall; leaves linear, 1–3.5 cm. long, often in fascicles, slightly fleshy, somewhat glaucous or green; spikes densely flowered, bracted, 1–10 cm. long, terminating branches; flowers greenish, 1–1.5 mm. long; petals white or greenish white, about 1 mm. long, oblanceolate or oblong, nearly entire or faintly lobed; stamens 3, slightly exserted; capsule 4-lobed,

4-beaked, 1.5–2 mm. wide, broader than high; seeds obovoid-reniform, about 0.4 mm. long, black, smooth, and shining.

Common on desert flats and along margins of washes, particularly abundant in saline situations, Lower and Upper Sonoran Zones, southern California to Nevada, Texas, and south throughout Sonora and Baja California. In dry years often flowering and fruiting when only 3–5 cm. high. March–June.

**Reseda lutea** L. has been collected at Tucson, but there is nothing on the label to indicate whether the plant was an escape or was collected in a garden. If established as an escape, it apparently is uncommon in Arizona.

## Family 52. CRASSULACEAE. Stonecrop Family

Chiefly fleshy or succulent herbs or slightly shrubby plants with thick, estipulate, (in ours) sessile leaves. Inflorescences cymose or less commonly racemose or solitary. Flowers regular, perfect. Calyx 4–5-lobed or 4–5-parted, free from the ovary. Petals 4–5, distinct or united into a ring or tube below, rarely lacking. Stamens same number as the petals or twice as many, mostly shorter than petals. Carpels as many as the calyx lobes, distinct or slightly united at base, usually each one subtended by a small scale; styles as many as carpels, filiform or subulate; ovules usually many, arranged along ventral placenta, rarely few or solitary. Fruit follicular. Seeds minute. Endosperm fleshy; embryo straight.

Stamens 3–5; plants minute annuals; seeds 1–2 in each carpel; leaves not in basal rosettes ... .................................................................... 1. *Tillaea*
Stamens 8–10; plants perennial; seeds many in each carpel; leaves forming a basal rosette:
  Petals erect or flaring only near apex, not transversely marked; stamens erect or nearly so .................................................................. 2. *Dudleya*
  Petals widely spreading, with transverse markings; stamens strongly reflexed after dehiscence of anthers ............................................ 3. *Graptopetalum*

### 1. TILLAEA L., Sp. Pl. 128, 1753

Minute annual plants with erect or strongly ascending, simple or branching stems and slightly fleshy, minute, glabrous, opposite leaves with their bases connate into a minute shallow cup. Flowers minute, in small clusters in axils of all except lowermost leaves. Calyx 3–5-parted, sepals fleshy, erect or nearly so. Petals 3–5, distinct or slightly connate at the very base, erect. Stamens as many as petals and alternate with them, free. Carpels 3–5, distinct, styles short-subulate. Fruits follicular, each carpel 1–2-seeded. Seeds ellipsoidal, minutely grooved longitudinally.

Pedicels mostly 1 mm. long or less ............................ 1. *T. erecta* subsp. *erecta*
Pedicels to 3.5 mm long, very slender ......................... 1*a*. *T. erecta* var. *eremica*

**1. Tillaea erecta** Hook. & Arn., Bot. Beechey 24, 1830, subsp. **erecta**

*Tillaea minima* Miers ex Hook. & Arn., Bot. Misc. **3**: 338, 1833.
*Tillaea leptopetala* Benth., Pl. Hartw. 310, 1849.
*Tillaea minima* var. *subsimplex* S. Wats., Bot. Calif. **1**: 208, 1876.
*Crassula minima* Reiche, Fl. Chile **2**: 369, 1898.

Plants small, 1–5 cm. tall, smaller ones often simple and unbranched, larger branching profusely from base and densely tufted; plants at first green, later becoming reddish; leaves broadly ovate, 1–1.5 mm. long, connate bases subhyaline, main blade fleshy, strongly convex on abaxial side, slightly concave on upper side, acute or obtuse but minutely apiculate; flowers clustered in all except lowermost axils, 3–10 flowers in a cluster; pedicels slender, mostly 1 mm. long or less; sepals commonly 4, ovate, acute and abruptly apiculate, 1–1.2 mm. long, erect or apices slightly incurved; petals as many as sepals, 0.6–0.9 mm. long, lance-linear, white but soon becoming red in a median spot or throughout, erect; stamens about equaling petals; pistils 3–5, commonly 4, nearly equaling or slightly exceeding sepals in fruit, splitting down inner face, back rounded, glabrous or minutely papillate on lower part of back, bright red in fruit; seeds 0.4–0.5 mm. long, reddish brown, obtuse at each end, longitudinal striations fine but obvious under a good lens.

On barren soil, among herbaceous plants or in openings in brushy areas, frequently in sandy soil, Lower and Upper Sonoran Zones, southern Oregon to northern Baja California and east into southern Arizona. Also in Chile. Feb.–May.

**1*a*. Tillaea erecta** subsp. **eremica** (Jepson) Wiggins, status nov.
*Tillaea erecta* var. *eremica* Jepson, Man. Fl. Pl. Calif. 450, 1925.

Differing from subsp. *erecta* only in averaging slightly more slender and in having pedicels as much as 3.5 mm. long.

Sandy soil, Lower Sonoran Zone, along the western margins of the Colorado Desert in California, and south into adjacent Baja California. Feb.–April.

## 2. DUDLEYA Britt. & Rose, Bull. N.Y. Bot. Gard. 3: 12, 1903*

Glabrous succulent perennials. Caudex simple or branched dichotomously, each branch bearing a usually evergreen rosette of ovate to linear succulent leaves and covered below with old persistent leaves or leaf bases. Each rosette leaf attached by the whole area of its broad base, veins several, parallel toward base or somewhat diverging, with as many separate bundle scars. Floral stems axillary, strictly annual, bearing sessile leaves smaller than rosette leaves and diminishing upward. Inflorescence a cyme whose ultimate branches are cincinni. Flowers 5-merous. Calyx fleshy, divided nearly to base, lobes erect. Petals convolute in bud, united below or to above the middle, erect and appressed to form a tube or variously spreading above. Stamens 10, inserted on base of corolla. Nectar scales truncate, two or three times wider than high, upper surface transversely oblong. Carpels erect or suberect in flower, slightly separated to widely spreading in fruit, styles slender. Seeds many, brown or reddish, narrowly ovoid, longitudinally striate.

Petals ascending to spreading from near middle, mostly 5–8 mm. long; predehiscent follicles ascending-spreading; leaves terete, acute [Subgenus STYLOPHYLLUM] .......... .......... 1. *D. attenuata*
Petals erect, appressed to form a tube, only tips sometimes somewhat spreading; predehiscent follicles erect; leaves strongly flattened or, if rarely subterete above, then commonly acuminate [Subgenus DUDLEYA]:
  Corolla lobes shorter than tube, red or sometimes red and yellow, rarely yellow; pedicels often bending upward in fruit; caudex simple or few-branched:
    Floral stems 7–15 cm. tall, with about 4–6 leaves in the upper half; pedicels 4–6 mm. long; sepals united about 1 mm. .......... 9. *D. rubens*
    Floral stems 15–100 cm. tall, with 15–60 leaves, some of which are well below the middle; sepals essentially free:

* Contributed by Dr. Reid V. Moran.

Caudex 4–8.5 cm. thick; rosette leaves 30–80, 8–27 cm. long, 3–10 cm. wide; floral stems 3–10 dm. tall; flowers mostly spreading to pendent at anthesis, becoming erect in fruit by a conspicuous bending of pedicel; corolla red, 11–19 mm. long .. .................................... 10. *D. pulverulenta* subsp. *pulverulenta*

Caudex 1–4 cm. thick; rosette leaves 10–35, 3–17 cm. long, 1–5 cm. wide; floral stems 1.5–6 dm. tall; flowers usually erect in anthesis and fruit, though pedicels often somewhat bent in fruit; corolla apricot yellow to (mostly) red, 9–15 mm. long .. .................................... 10*a*. *D. pulverulenta* subsp. *arizonica*

Corolla lobes exceeding the tube or if rarely about equaling tube then white or pink; corolla never uniformly deep red; flowers erect from beginning or becoming erect by unrolling of cincinnus; pedicels straight:

Petals united one-fifth their length or less, bright yellow to greenish yellow:

Rosette leaves 20–40, 2.5–6 cm. long; floral stems 6–17 cm. tall, with 5–8 leaves mostly in upper third; pedicels 2–6 mm. long ......................... 2. *D. linearis*

Rosette leaves 10–25, 4–15 cm. long; floral stems 10–40 cm. tall, with 5–15 leaves, the lowest usually below middle; pedicels 5–20 mm. long:

Corolla 8–12 ( –15) mm. long, but if more than 12 mm., then greenish yellow .... ............................................. 3. *D. saxosa* subsp. *aloides*

Corolla 12–18 mm. long, bright yellow ............ 3*a*. *D. saxosa* subsp. *collomiae*

Petals united for more than one-fifth and usually for more than one-third their length:

Corolla bright to pale yellow, often red-lineolate and often drying reddish; rosette leaves mostly 10–20; caudex 1–3 cm. thick:

Rosette leaves oblong-lanceolate or tapering from base, mostly 6–12 mm. wide; petals pale yellow, thin and often erose, broadly acute ......................... ...................................... 4. *D. abramsii* subsp. *abramsii*

Rosette leaves ovate to triangular-ovate, 15–30 mm. wide; petals bright yellow, entire, slender-acute ........................................... 5. *D. acuminata*

Corolla white to pink; rosette leaves 10–60; caudex 1–6 cm. thick:

Rosette leaves turgid, subterete toward acuminate or slender-acute apex, 5–15 mm. wide; plants often branching to form large clusters; corolla pure white ...... ...................................................... 6. *D. albiflora*

Rosette leaves strongly flattened, not subterete toward apex, 10–35 mm. wide; plants mostly solitary or few-branched; corolla white or pink:

Leaves of rosettes linear-oblong, 1–2 cm. wide; lower pedicels 5–20 mm. long ... ...................................................... 7. *D. ingens*

Leaves of rosettes suborbicular or triangular-ovate to broadly oblong, 1.5–6 cm. wide; lower pedicels 4–12 mm. long ....................... 8. *D. gatesii*

**1. Dudleya attenuata** (S. Wats.) Moran, Desert Pl. Life **14:** 191, 1943

*Cotyledon attenuata* S. Wats., Proc. Amer. Acad. **22:** 472, 1887. Not *Cotyledon attenuata* Lindberg f., 1932.
*Stylophyllum attenuatum* Britt. & Rose, Bull. N.Y. Bot. Gard. **3:** 36, 1903.
*Stylophyllum orcuttii* Rose, in Britt. & Rose, loc. cit.
*Cotyledon orcuttii* Fedde, in Just, Bot. Jahresb. **31**[1]**:** 829, 1904.
*Cotyledon edulis* var. *attenuata* Jepson, Man. Fl. Pl. Calif. 453, 1925.
*Echeveria attenuata* Berger, in Engler & Prantl, Nat. Pflanzenf. ed. 2, **18a:** 481, 1930.
*Echeveria orcuttii* Berger, loc. cit.
*Echeveria edulis* var. *attenuata* Jepson, Fl. Calif. **2:** 116, 1936.
*Dudleya attenuata* subsp. *orcuttii* Moran, Desert Pl. Life **14:** 191, 1943.
*Dudleya attenuata* subsp. *typica* Moran, op. cit. **15:** 41, 1943.

Caudex 5–30 cm. long, 3–20 mm. thick, branched to form dense clumps 1–4 dm. wide; rosette leaves 5–20, erect, glaucous at least when young, linear to linear-oblanceolate, acute, terete, 2–10 cm. long, 2–5 mm. wide, base lenticular to orbicular, 5–15 mm. wide, 2–8 mm. high; floral stems 1–2.5 dm. tall, with turgid ascending leaves mostly less than 2 cm. long; cincinni 1–4, ascending, 2–11 cm. long, with 3–15 flowers; pedicels erect, 0.5–3 mm. long; sepals triangular-ovate, acute, 2–4 mm. long, 1–2 mm. wide; petals white or yellow, marked with red, elliptic, acute, 5–10 mm. long, spreading from near middle; ovaries 3–6 mm. long; styles 1–2.5 mm. long; follicles ascending-spreading.

Mesas and hillsides, Lower and Upper Sonoran Zones, western slopes of the Sierra San Pedro Mártir foothills from near San Antonio del Mar south to the southern end of the range east of Rosario, Baja California. May–June.

**2. Dudleya linearis** (Greene) Britt. & Rose, Bull. N.Y. Bot. Gard. **3:** 16, 1903

*Cotyledon linearis* Greene, Pittonia **1:** 285, 1889.
*Echeveria linearis* Berger, in Engler & Prantl, Nat. Pflanzenf. ed. 2, **18a:** 478, 1930.

Caudex 1–2 cm. thick, cespitosely branched to bear 10–20 rosettes; rosette leaves 20–40, erect or somewhat spreading, green, oblong to oblong-oblanceolate, acute to acuminate, 2.5–6 cm. long, 5–10 mm. wide, 2–3 mm. thick, margins rounded, base 5–8 mm. wide; floral stems 6–17 cm. tall, with 5–8 ascending lanceolate leaves mostly in upper third; inflorescence glaucous; cincinni mostly 2–3, ascending, 2–5 cm. long, with 2–10 flowers; lower pedicels 2–6 mm. long; sepals triangular-lanceolate, acute, 2.5–4 mm. long, 1.5–2 mm. wide; petals yellow, 7–12 mm. long, 2.5 mm. wide, connate 1.5–2.5 mm.; ovaries 6 mm. high; styles 3 mm. long.

Known only from the San Benito Islands, Baja California. March–May.

**3. Dudleya saxosa** Britt. & Rose subsp. **aloides** (Rose) Moran, Madroño **14:** 108, 1957

*Dudleya aloides* Rose, in Britt. & Rose, Bull. N.Y. Bot. Gard. **3:** 15, 1903.
*Dudleya grandiflora* Rose, in Britt. & Rose, op. cit. 16.
*Dudleya delicata* Rose, in Britt. & Rose, op. cit. 24.
*Cotyledon aloides* Fedde, in Just, Bot. Jahresb. **31**[1]**:** 826, 1904.
*Cotyledon delicata* Fedde, loc. cit.
*Cotyledon grandiflora* Fedde, loc. cit.
*Echeveria aloides* Berger, in Engler & Prantl, Nat. Pflanzenf. ed. 2, **18a:** 477, 1930.
*Echeveria grandiflora* Berger, op. cit. 478.
*Echeveria delicata* Berger, op. cit. 480.
*Echeveria lanceolata* var. *aloides* Munz, Man. S. Calif. 598, 1935.
*Echeveria lanceolata* var. *composta* Jepson, Fl. Calif. **2:** 115, 1936.

Caudex simple or cespitosely branched, short, 1–3 cm. thick; rosette leaves 10–25, erect or ascending, glaucous, oblong-lanceolate, acute to subacuminate, 4–15 cm. long, 6–20 mm. wide, 2–5 mm. thick, base 1–2.5 cm. wide; floral stems 1–3.5 dm. tall, 4–9 mm. thick, often reddish, with triangular-lanceolate leaves; cincinni usually 3–5, somewhat sinuate when old, 1–12 cm. long, with 2–20 flowers; lower pedicels 1–2 cm. long; sepals triangular-ovate, acute, 3–5 mm. long, 2.5–4 mm. wide; petals bright yellow or greenish, elliptic, acute, 8–15 mm. long, 2.5–4 mm. wide, connate 1.5–3 mm.; ovaries 4–8.5 mm. long; styles 1–1.5 mm. long.

Rocky hillsides and canyon walls, Lower Sonoran Zone, western margins of the Colorado Desert and probably in northern Baja California. May–June.

**3*a*. Dudleya saxosa** subsp. **collomiae** (Rose) Moran, Madroño **14:** 108, 1957

*Dudleya collomae* Rose, in Morton, Desert Pl. Life **6:** 68, 1934.
*Echeveria collomae* Kearney & Peebles, Jour. Wash. Acad. **29:** 479, 1939.

Caudex short, 1.5–3 cm. thick, cespitosely branched to bear 5–10 rosettes; rosette leaves 12–20, erect or spreading, glaucous when young, oblong-lanceolate, acute, 6–15 cm. long, 1–2 cm. wide, 3–6 mm. thick, base 1–2.5 cm. wide; floral stems 1.5–4 dm. tall, 3–6 mm. thick, with ascending, triangular-lanceolate leaves; inflorescence reddish, obpyramidal; cincinni mostly 4–8, 3–12 cm. long, with 4–12 flowers; lower pedicels 5–15 mm. long; sepals triangular-ovate, acute, 4–7 mm. long, 2.5–4 mm. wide; petals bright yellow, oblong-lanceolate, sharply acute, 12–18 mm. long, 3–4 mm. wide, connate 1–4 mm.; ovaries 5–7.5 mm. long; styles 1–2 mm. long.

Rocks and hillsides, Lower and Upper Sonoran Zones, central Arizona, from 2,000 to 6,000 feet elevation. March–May.

**4. Dudleya abramsii** Rose, in Britt. & Rose, Bull. N.Y. Bot. Gard. **3:** 14, 1903, subsp. **abramsii**

*Dudleya tenuis* Rose, in Britt. & Rose, loc. cit.
*Cotyledon abramsii* Fedde, in Just, Bot. Jahresb. **31**[1]**:** 826, 1904.
*Cotyledon tenuis* Fedde, loc. cit.
*Echeveria abramsii* Berger, in Engler & Prantl, Nat. Pflanzenf. ed. 2, **18a:** 477, 1930.
*Echeveria tenuis* Berger, in Engler & Prantl, loc. cit. Not *Dudleya tenuis* Rose, 1903.

Caudex short, 1–1.5 cm. thick, mostly cespitosely branching; rosette leaves 10–20 (–30), glaucous, oblong-lanceolate or tapering from the base, acute to subacuminate, 1.5–6 (–11) cm. long, 6–12 (–19) mm. wide, 2–4 mm. thick, the base 5–15 (–20) mm. wide; floral stems 2–15 cm. tall, 1–3 mm. thick, with 5–18 triangular-ovate to triangular-lanceolate leaves; cincinni 2–6, ascending, 2–10 cm. long, with 2–11 flowers; lower pedicels 0.5–5 mm. long; sepals triangular-ovate to triangular-lanceolate, acute, 2–5 mm. long, 1.5–3 mm. wide; petals pale yellow, red-lineolate on keel, elliptic, acute, thin, often erose, 8–12.5 mm. long, 2–2.5 mm. wide, connate 2–4.5 mm.; ovaries 3–5.5 mm. long; styles 0.5–1.5 mm. long.

In rock crevices and around boulders, especially in granitic soil, mostly Upper Sonoran Zone to Arid Transition Zone, San Bernardino, San Jacinto, and Laguna Mountains, California, to Sierra Juarez, Baja California, occasionally down to 3,000 feet elevation on desert slopes. April–June.

**5. Dudleya acuminata** Rose, in Britt. & Rose, Bull. N.Y. Bot. Gard. **3:** 26, 1903

?*Dudleya brandegei* Rose, in Britt. & Rose, op. cit. 21.
?*Dudleya brevipes* Rose, in Britt. & Rose, op. cit. 24.
*Cotyledon acuminata* Fedde, in Just, Bot. Jahresb. **31**[1]**:** 826, 1904.
?*Cotyledon brandegeei* Fedde, loc. cit.
?*Cotyledon breviceps* Fedde, loc. cit.
?*Echeveria brandegeei* Berger, in Engler & Prantl, Nat. Pflanzenf. ed. 2, **18a:** 478, 1930.
*Echeveria acuminata* Berger, in Engler & Prantl, op. cit. 479.
?*Echeveria brevipes* Berger, in Engler & Prantl, op. cit. 478.

Caudex short, 1–2.5 cm. thick, simple or cespitosely branched to bear 2–6 rosettes; rosette leaves 10–25, pale green or slightly glaucous, ovate to triangular-lanceolate, usually widest at or just above base, acute to acuminate, 2–10 cm. long, 1.5–3 cm. wide, 2–5 mm. thick; floral stems 5–40 cm. tall, 2–5 mm. thick, with 8–25 triangular-ovate leaves; cincinni 2–6, ascending, 1–12 cm. long, with 5–13 flowers; lower pedicels 2–10 mm. long; sepals triangular-ovate, acute, 3–5 mm. long, 2–3 mm. wide; petals yellow, sometimes red-lineolate, drying reddish, oblanceolate, sharply acute, 10–14 mm. long, 2–2.5 mm. wide, connate 4–7 mm.; ovaries 4.5–7 mm. long; styles 1–2 mm. long.

Plains and hillsides mostly below 2,000 feet altitude, Lower Sonoran Zone, central Baja California, and on Cedros Island. April–May.

**6. Dudleya albiflora** Rose, in Britt. & Rose, Bull. N.Y. Bot. Gard. **3:** 13, 1903

*Cotyledon albiflora* Fedde, in Just, Bot. Jahresb. **31**[1]**:** 826, 1904. Not *C. albiflora* Hemsl., 1878.
*Echeveria albiflora* Berger, in Engler & Prantl, Nat. Pflanzenf. ed. 2, **18a:** 477, 1930.
?*Dudleya moranii* Johansen, Cactus & Succ. Jour. **4:** 244, *figs. 4–6,* 1932.

Caudex 1–2 cm. thick, to 3 dm. long, usually branched cespitosely, clumps sometimes 7 dm. wide and of 200 or more rosettes; rosette leaves 10–35, green or glaucous, lanceolate to triangular-linear, acuminate or slender-acute, 2–6 cm. long, 5–15 mm. wide, 2–6 mm. thick, base 0.5–2 cm. wide, apex terete; floral stems 0.5–4.5 dm. tall, 2–4 mm. thick, with 10–20 triangular-ovate to -lanceolate leaves; cincinni mostly 2–4, ascending, 3–20 cm. long, with 3–25 flowers; lower pedicels 2–6 mm. long; sepals triangular, acute, 3–6.5 mm. long, 2–3 mm. wide; petals white, oblong-oblanceolate, acute, 11–15 mm. long, 1.5–3 mm. wide, connate 3–4.5 mm.; ovaries 5.5–8 mm. long; styles 1.5–3 mm. long.

Rocky slopes and hilltops, Lower Sonoran Zone, central and southern Baja California, and on Cedros, Natividad, and Espíritu Santo Islands. April–June.

**7. Dudleya ingens** Rose, in Britt. & Rose, Bull. N.Y. Bot. Gard. **3:** 18, 1903

Caudex short, simple or few-branched, 2–4 cm. thick; rosette leaves 20–40, green or slightly glaucous, linear-oblong, oblong, or tapering from the base, sharply acute, 7–15 cm. long, 1–2 cm. wide, 2–7 mm. thick, the base 1–3.5 cm. wide; floral stems 2.5–6 dm. tall, 3–7 mm. thick; cincinni mostly 3–8, ascending, 4–15 cm. long, with 6–15 flowers; lower pedicels 5–20 mm. long; sepals triangular, acute, 4–7 mm. long, 2–4 mm. wide; petals pink, sometimes slightly yellowish, oblong, acute, 10–15 mm. long, 2–3 mm. wide, connate 4–6 mm.; ovaries 5–9 mm. long; styles 1–1.5 mm. long.

Rolling hills and mesas, Lower and Upper Sonoran Zones, San Vicente to San Fernando, northern Baja California. April–June.

**8. Dudleya gatesii** Johansen, Cactus & Succ. Jour. **4:** 286, 1932

Caudex simple or 1–2-times branched, 2–6 cm. thick, 1–5 dm. tall; rosette leaves 25–40 or more, pale green, glaucous, triangular to oblong, acuminate, 6–15 cm. long, 1.5–3.5 cm. wide, 5–12 mm. thick, drying thick and leathery, base 3–5 cm. wide; floral stems 2–7 dm. tall, 5–10 mm. thick, with 20–40 triangular acute leaves; cincinni 6–12, spreading to ascending, 4–14 cm. long, with 7–17 flowers; lower pedicels 6–12 mm. long; sepals triangular, acute, 4–5 mm. long, 2–4 mm. wide; petals white or pink, narrowly oblong, acute, 10–14 mm. long, about 2 mm. wide, connate 5–6 mm.; ovaries 5–7.5 mm. long; styles 1–2 mm. long.

Dry granitic hillsides, Lower Sonoran Zone, between about 28° and 29° N. lat., Baja California. April–May.

**9. Dudleya rubens** (Brandegee) Britt. & Rose, Bull. N.Y. Bot. Gard. **3:** 23, 1903

*Cotyledon rubens* Brandegee, Proc. Calif. Acad. II, **2:** 155, 1889.
*Echeveria rubens* Berger, in Engler & Prantl, Nat. Pflanzenf. ed. 2, **18a:** 479, 1930.

Caudex short, 1.5–2 cm. thick, branching; rosette leaves about 20, glaucous, oblong-lanceolate, acute, 3.5–7.5 cm. long, 10–12 mm. wide; floral stems 7–15 cm. tall, 2–3 mm. thick, with 4–6 triangular-lanceolate, acuminate leaves in upper half; cincinni 2–3, ascending, 3–6 cm. long, with 7–12 flowers; pedicels 4–6 mm. long; sepals triangular-ovate, acute, 4–5 mm. long, 2.5 mm. wide, apparently connate 1 mm.; corolla dull red, contracted at mouth, tube 5–7 mm. long, lobes oblong, acute, about 4 mm. long.

Known only from the type collection from cliffs near San Esteban, Lower Sonoran Zone, central Baja California. April–May.

**10. Dudleya pulverulenta** (Nutt.) Britt. & Rose, Bull. N.Y. Bot. Gard. **3:** 13, 1903, subsp. **pulverulenta**

*Echeveria pulverulenta* Nutt., in Torr. & Gray, Fl. N. Amer. **1:** 560, 1840.
*Cotyledon pulverulenta* Baker, in Saunders, Ref. Bot. **1:** *pl. 66,* 1869.
*Dudleya pulverulenta* subsp. *typica* Moran, Desert Pl. Life **15:** 70, 1943.

Caudex mostly unbranched, 4–8.5 cm. thick, to 4 dm. long; rosette leaves 30–80, chalky-pulverulent, oblong, acuminate to cuspidate, 8–27 cm. long, 3–10 cm. wide, 3–10 mm. thick, base 3–8 cm. wide; floral stems 3–10 dm. tall, 5–15 mm. thick, with about 30–60 ovate to suborbicular, acuminate leaves; cincinni 3–6 or several, spreading or ascending, 1–4 dm. long, with 10–30 flowers; lower pedicels 1–3.5 cm. long; flowers mostly pendent in bud, declined or pendent at anthesis, becoming erect by a bending of pedicel near or above middle; sepals triangular-ovate, acute, 4–8 mm. long, 3–4 mm. wide; corolla red, 11–19 mm. long, tube 6–10 mm. long, lobes oblong, acute to obtuse, 5–9 mm. long, 2–4 mm. wide; ovaries 6–12 mm. long; styles 1–2 mm. long.

Arid hillsides, mainly Upper Sonoran Zone, San Luis Obispo County, California, to central Baja California; east in San Diego County to the western margin of the Colorado Desert, where apparently it intergrades with the subsp. *arizonica.* May–Aug.

**10*a*. Dudleya pulverulenta** subsp. **arizonica** (Rose) Moran, Desert Pl. Life **15:** 72, 1943

*Dudleya arizonica* Rose, Addisonia **8:** 35, *pl. 274,* 1923.
*Echeveria lagunensis* Munz, Bull. S. Calif. Acad. **31:** 64, 1932.
*Dudleya lagunensis* E. Walther, Leafl. West. Bot. **1:** 29, 1932.
*Echeveria arizonica* Kearney & Peebles, Jour. Wash. Acad. **29:** 479, 1939. Not *Echeveria arizonica* Hort. ex Berger, 1930.
*Echeveria pulverulenta* subsp. *arizonica* Clokey, Madroño **8:** 56, 1945.

Caudex mostly unbranched, 1–4 cm. thick, to 15 cm. long; rosette leaves 10–35, glaucous to farinose, oblong to oblong-obovate, mostly acuminate, 3–17 cm. long, 1–5 cm. wide, 2–4 mm. thick, base 1–3.5 cm. wide; floral stems 1.5–6 dm. tall, 2–6 mm. thick, with about 15–25 ovate, acute to acuminate leaves; cincinni usually 3–6, ascending, 4–27 cm. long, with 4–20 flowers; lower pedicels 5–15 or rarely 20 mm. long; flowers mostly erect or ascending in bud, anthesis, and fruit, rarely spreading or pendent in flower and erect in fruit, but fruiting pedicels often somewhat bent near or above middle; sepals triangular-ovate, acute, 3–6 mm. long, 2–4 mm. wide; corolla yellow to mostly red, 9–15 mm. long, tube 4–8 mm. long, segments oblong, obtuse to sharply acute, 1.5–2 mm. wide; ovaries 6–9 mm. long; styles 1–2 mm. long.

Rocky slopes of desert ranges, mainly Lower Sonoran Zone, southern Nevada and western Arizona to the western margin of the Colorado Desert and south into northern Baja California and coastal Sonora. April–Aug.

**Dudleya rigidiflora** Rose, in Britt. & Rose, Bull. N.Y. Bot. Gard. **3:** 18, 1903, is known only from floral stems without rosette leaves, collected at Playa María, Baja California, by Anthony and by Pond. The scanty material does not permit an evaluation of the species.

## 3. GRAPTOPETALUM Rose, Contr. U.S. Nat. Herb. 13: 296, 1911

Acaulescent to short-caulescent plant with thick, acute to acuminate leaves forming a compact rosette. Inflorescence a few-flowered, slightly branched, open cyme. Sepals 5, erect, nearly distinct. Corolla with a short tube and 5 rotately spreading to slightly reflexed lanceolate lobes, these more or less banded or spotted with red or nearly all red. Stamens 10, inserted high on corolla tube, at first erect or slightly spreading, strongly recurved after anthers dehisce. Carpels 5, erect, not spreading at maturity.

**1. Graptopetalum rusbyi** (Greene) Rose, Addisonia **9:** 31, *pl. 304,* 1924
*Cotyledon rusbyi* Greene, Bull. Torrey Club **10:** 125, 1883.
*Dudleya rusbyi* Britt. & Rose, Bull. N.Y. Bot. Gard. **3:** 13, 1903.
*Echeveria rusbyi* Nels. & Macbr., Bot. Gaz. **56:** 476, 1913.
*Graptopetalum orpetii* E. Walther, Cactus & Succ. Jour. **1:** 185, *figs. 1–5,* 1930.

Cespitose with several acaulescent or short-caulescent rosettes in a compact cluster; rosette leaves oblong-oblanceolate to ovate, 3–10 mm. wide, 5–35 mm. long, acute and acuminate or attenuate-acuminate, green, glabrous but minutely and closely papillate; stem leaves narrowly deltoid-ovate, 2–4 mm. long; flowering stem erect, 1–2 dm. tall, 1–3-branched into short secund racemes; pedicels slender, 2–12 mm. long; calyx lobes 5–6, erect, narrowly ovate, turgid; corolla 8–10 mm. long, strongly 5–6-angled when in bud, narrowly lanceolate lobes keeled without and rotately spreading in flower, yellow at base, spotted or with crossbars of red about middle, solid red at apex; corolla tube yellow, slightly shorter than sepals to about equaling them; stamens 5–7 mm. long, strongly recurved after dehiscence; carpels slenderly subulate, 6–7 mm. long, erect, tinged with red or brownish.

Rocky canyons, Lower and Upper Sonoran Zones, Greenlee, Graham, and Maricopa Counties, Arizona, south into northern Sonora. April–May.

## Family 53. GROSSULARIACEAE. Gooseberry Family

Erect, ascending, or trailing shrubs with or without nodal spines. Leaves alternate, simple, petiolate, rarely entire; stipules adnate to base of petioles or absent. Inflorescence terminal or short axillary spurs, racemose; flowers small, regular, bracteate, usually perfect; hypanthium elongated, short, or obsolete; sepals usually 5, sometimes 4; petals small, as many as and alternate with the sepals; stamens 5 (or 4), opposite the sepals. Ovary 1-celled, inferior; style 2-lobed to 2-divided, rarely 2-cleft. Placentae parietal. Fruit a berry, smooth, glandular, or spinose.

Pedicels jointed, bearing pair of bractlets immediately below joint, or bractlets wanting; plants without nodal spines, or if nodal spines present, hypanthium obsolete ... 1. *Ribes*
Pedicels not jointed, bractlets at base of pedicel or absent; nodal spines present, hypanthium always well developed ........ 2. *Grossularia*

### 1. RIBES L., Sp. Pl. 200, 1753

Shrubs or rarely trailing vines, usually unarmed, sometimes bristly. Leaves usually lobed, mostly deciduous. Racemes few- to many-flowered. Pedicels jointed beneath ovary and usually bearing a pair of bractlets just below joint. Hypanthium tubular, cupuliform, or in the bristly species obsolete. Ovary glandular or smooth, not bristly. Fruit pulpy, disarticulating from the pedicel.

**1. Ribes tortuosum** Benth., Bot. Sulph. 17, 1844
*Ribes palmeri* Vasey & Rose, Proc. U.S. Nat. Mus. **11:** 529, 1889.

An intricately branched, unarmed shrub to 2 m. high, with tortuous branches; young shoots whitish-puberulent; leaves nearly orbicular, 1–3.5 cm. broad, shallowly

3–5-lobed, irregularly crenate, thin, finely puberulent on both sides; racemes 1–2.5 cm. long on short, stiff, usually erect or spreading peduncles, 10–20-flowered; pedicels 1.5–2.5 mm. long, stiff, glandular-puberulent; hypanthium cylindric, 2–4 mm. long, glabrous or sparsely glandular; sepals oblong or obovate, 1–1.5 mm. long, reflexed; petals about 1 mm. long, erect or slightly spreading, pink to deep rose, equaling stamens; ovary glandular-puberulent but eventually glabrate; berry smooth, 4–8 mm. in diameter, translucent, red.

Rocky or sandy soil at low elevations, lower Upper Sonoran and Lower Sonoran Zones, from Ensenada south to the vicinity of Punta Prieta and Rancho Mesquital. Jan.–April.

**Ribes viburnifolium** A. Gray, Proc. Amer. Acad. **17:** 202, 1882, a straggling, unarmed vine with resinous-glandular twigs and thick leaves that are dark green and shining above and paler and resinous-dotted beneath, extends from Santa Catalina Island and the mainland of southern California southward in Baja California at least to San Vicente and on Cedros Island. It may occur near the coast south of San Quintín but has not yet been discovered in the desert.

## 2. GROSSULARIA Mill., Gard. Dict. ed. 7, 1759

Shrubs with ascending, erect, or trailing branches armed with simple or 3-forked nodal spines, or rarely spineless. Racemes few-flowered, these bracteate; pedicels not jointed; bractlets minute and situated at base of pedicel or absent. Hypanthium evident. Ovary spiny or smooth. Fruit not disarticulating from the pedicel.

### 1. Grossularia quercetorum (Greene) Cov. & Britt., N. Amer. Fl. 22: 220, 1908

*Ribes quercetorum* Greene, Bull. Calif. Acad. **1:** 83, 1885.
*Ribes congdoni* Heller, Muhlenbergia **1:** 101, 1904.
*Ribes leptanthum* var. *quercetorum* Jancz., Mém. Soc. Genève **35:** 380, 1907.
*Ribes velutinum* var. *congdoni* Jancz., op. cit. 381.

Spreading to ascendingly branched shrub to 1.5 m. high with puberulent to sparingly bristly young shoots; nodal spines mostly solitary, 3–12 mm. long, curved or straight, yellowish; leaf blades suborbicular, thin, 1–2 cm. wide, 3–5-lobed, dentate, puberulent to glabrate and eglandular or sparsely glandular on both surfaces; petioles pubescent, shorter than blades; peduncles 2–3-flowered, bracts shorter than pedicels; hypanthium 2.5–3 mm. long, yellow, pubescent; sepals equaling or slightly exceeding hypanthium, spreading or ascending; petals white, shorter than sepals, surpassing stamens; berry smooth, 6–8 mm. in diameter, black and shining.

Mostly Upper Sonoran Zone, Monterey and Tuolumne Counties to Kern County and south from near the coast to the western edge of the Colorado Desert to southern part of Sierra San Pedro Mártir, Baja California. March–April.

## Family 54. PLATANACEAE. Plane Tree Family

Large trees with thin, light-colored, exfoliating bark and alternate stipulate, long-petioled, deciduous, palmately lobed leaves. Buds concealed within hollow bases of petioles. Flowers minute, monoecious, in dense, spherical, pedunculate heads. Sepals minute, 3–8, scalelike. Petals minute, greenish, 3–6, rounded in pistillate flowers. Carpels same number as sepals, 1-celled, subtended by coma of persistent hairs and a few staminodia. Style terminal, stigmatic on ventral suture. Ovule orthotropous, rarely 2. Fruit a globose head of achenes intermingled with numerous tawny hairs and staminodia. Seeds pendulous; endosperm fleshy; embryo straight.

### 1. PLATANUS L., Sp. Pl. 999, 1753

Characters of the family.

**1. Platanus wrightii** S. Wats., Proc. Amer. Acad. **10:** 349, 1875

Tree 10–25 m. high; leaves 3–7-lobed, 10–20 cm. long and wide, lobes narrow deltoid-lanceolate to lanceolate, usually three to five times as long as broad, entire or distantly serrate, tomentose with intricately branching hairs when young, later glabrate at least above; fruiting heads 2–2.5 cm. in diameter, or some of them on lateral branches of the peduncle 1–5 cm. long.

Along streams from the margin of the Sonoran Desert in Arizona and Sonora to Texas and Chihuahua. April–May.

## Family 55. CROSSOSOMATACEAE. Crossosoma Family

Glabrous, stragglingly branched shrubs with alternate estipulate leaves and gray, finely striate bark. Flowers solitary, borne terminally on short lateral branchlets, regular. Hypanthium turbinate. Sepals 5, persistent. Petals 5, white. Stamens 15–numerous in 2–3 irregular series on a thin disk lining the hypanthium; filaments slightly dilated below. Pistils 1–6, distinct, short-stipitate. Stigmas capitate. Fruit follicular. Seeds several, each with a fimbriate basal aril. Embryo curved.

### 1. CROSSOSOMA Nutt., Jour. Acad. Phila. II, 1: 150, 1848

Only one genus in the family.

Follicles oblong-cylindrical, faintly glaucous, 9–12 mm. long .. 1. *C. bigelovii* var. *bigelovii*
Follicles ovoid, heavily glaucous, 7–9 mm. long ............ 1a. *C. bigelovii* var. *glaucum*

**1. Crossosoma bigelovii** S. Wats., Proc. Amer. Acad. **11:** 122, 1876, var. **bigelovii**

Erect, openly and irregularly branched shrub 1–2 m. high, with long, straggling, main branches and short, rigid, lateral branchlets 1–10 cm. long; leaves solitary on young and fascicled on older growth, subsessile, glaucous, obovate to oblanceolate or elliptic, obtuse to rounded, apiculate, 5–20 mm. long; peduncles 5–15 mm. long; flowers white to rose or faintly purplish, fragrant; sepals deltoid-ovate, rounded, 4–5 mm. long and broad; petals oblanceolate, short-clawed, 10–12 mm. long; stamens 15–30; follicles 1–6, usually 2–3, 9–14 mm. long; seeds reniform-lenticular, 2 mm. long, 1 mm. thick, black, minutely papillate-tessellate, papillae in regular rows; segments of aril about equaling seed, puberulent, yellow.

Rocky canyons, washes, and hillsides, Lower Sonoran Zone, Ord and Sheephole Mountains, Mojave Desert; Whitewater and Warren's Well, Colorado Desert; to the margins of the Sonoran Desert in Arizona and south to Magdalena, Sonora, and Calmallí, Baja California. Feb.–May.

**1*a*. Crossosoma bigelovii** var. **glaucum** (Rydb.) Kearney & Peebles, Jour. Wash. Acad. **29:** 480, 1939

Similar to var. *bigelovii* and differing from it in having slightly less spinescent branches, somewhat smaller and more frequently obtuse leaf blades, and shorter, broader, more heavily glaucous follicles.

Rocky slopes and on cliffs, Lower and Upper Sonoran Zones, eastern margin of the Sonoran Desert and in the adjacent foothills and mountains of eastern Arizona. Feb.–May.

## Family 56. **ROSACEAE.** Rose Family

Herbs, shrubs, or trees, usually with alternate leaves and evident stipules. Flowers regular, mostly perfect. Hypanthium well developed, saucer-shaped, urceolate, or tubular. Sepals 5, or rarely 4, alternating with bractlets when these are present. Petals 5 or lacking. Stamens chiefly 10 to many, inserted in 1 or more whorls on hypanthium or on edge of a well-defined disk within hypanthium. Pistils 1 to many, distinct or coalescent, distinct from or adnate to hypanthium. Ovary superior or inferior. Fruit achenes, follicles, or drupelets which are often united into an aggregate fruit, or pomes or drupes.

Fruit follicular, woody at maturity; flowers in flat panicles ................ 1. *Vauquelinia*
Fruit indehiscent, consisting of achenes or drupes, not woody; flowers solitary, in racemose clusters, or in corymbs:
- Carpels not enclosed in fleshy hypanthium; leaves entire to 3-lobed:
  - Petals wanting; styles more or less villous:
    - Hypanthium well developed, the limb 4–6 mm. broad in flower but soon deciduous; sepals not petaloid; stamens inserted on limb of hypanthium; style terminal, silky-plumose full length, 5–12 cm. long; branches and leaves alternate .. 2. *Cercocarpus*
    - Hypanthium poorly developed, about 1 mm. deep; sepals petaloid; stamens inserted on base of a slender tube surrounding pistil; style lateral, villous only at base, 1 cm. long or less; branches and leaves opposite ................. 3. *Coleogyne*
  - Petals present; styles glabrous or slightly puberulent ...................... 4. *Purshia*
- Carpels enclosed in fleshy hypanthium that becomes reddish at maturity; leaves pinnately compound, with 3–7 leaflets ............................................. 5. *Rosa*

### 1. **VAUQUELINIA** Correa ex Humb. & Bonpl., Pl. Aequin. **1:** 140, *pl. 40,* 1808

Large shrub or small tree with hard close-grained wood and evergreen, simple, serrate-margined leaves. Flowers in flat-topped cymose panicles, numerous; hypanthium turbinate. Sepals ovate, obtuse or acutish, erect. Petals 5, white, spreading to reflexed, rounded. Stamens 15–30. Pistils 5, basally coherent, follicular and somewhat woody in fruit.

#### 1. **Vauquelinia californica** (Torr.) Sarg., Gard. & Forest **2:** 400, 1889

*Spiraea californica* Torr., in Emory, Notes Mil. Rec. 140, 1848.
*Vauquelinia torreyi* S. Wats., Proc. Amer. Acad. **11:** 147, 1876.

Shrub or small tree to 6 m. tall with dark gray, nearly smooth bark; leaves lanceolate, 0.8–2 cm. wide, to 10 cm. long, bright green and glabrate above, finely cinereous-tomentulose beneath, margins shallowly serrate, petioles 0.6–2 cm. long; flowers in terminal cymose panicles; hypanthium about 2 mm. deep, slightly wider, cinereous-tomentose; calyx lobes broadly ovate, 1–1.5 mm. long; petals white, oblong to obovate, 3–4 mm. long, spreading to reflexed; capsule subwoody, of 5 follicle-like erect locules about 4 mm. high, rusty-puberulent.

Margins of the desert, but mostly in the Upper Sonoran Zone, Baboquivari Mountains, central Arizona along the Gila Valley, and south into Baja California and Sonora. May–June.

## 2. CERCOCARPUS H. B. K., Nov. Gen. & Sp. 6: 232, 1823

Shrubs or small trees with hard dark-colored wood and smooth thin bark. Leaves alternate, simple, entire to dentate, evergreen, with short petioles and deciduous stipules. Flowers small, apetalous, solitary or fascicled in axils of leaves or at tips of branchlets. Hypanthium salverform, with deciduous campanulate limb and persistent tubular base; sepals 5, triangular to subulate. Stamens 10–45, inserted in 2–3 rows on limb of hypanthium, filaments subulate to filiform. Pistil 1; style terminal, elongate, silky-plumose, persistent; ovule solitary, ascending. Fruit a terete achene, villous, included within elongating hypanthium; seed linear.

**1. Cercocarpus betuloides** Nutt. ex Torr. & Gray, Fl. N. Amer. **1:** 427, (June) 1840

*Cercocarpus betulaefolius* Nutt. ex Hook., Ic. Pl., *pl. 322,* (Oct.) 1840.
*Cercocarpus parvifolius* var. *glaber* S. Wats., Bot. Calif. **1:** 175, 1876.
*Cercocarpus montanus* var. *glaber* F. L. Martin, Brittonia **7:** 101, 1950.
(For additional synonyms see Brittonia **7:** 101–102, 1950, under the last-named above combination.)

Shrubs or small trees 2.5–8 m. tall with smooth, pale gray to reddish gray bark, older branches and trunk more or less flecked with scales, younger twigs glabrous to sparsely puberulent; leaves obovate to ovate, 1–5 cm. long, cuneate at base, rounded apically and serrate above middle, sparsely appressed-villous in youth, soon glabrate and dark green above, paler and microscopically puberulent in islets of veinlets beneath; hypanthium finely silky-tomentose, limb 4–6 mm. in diameter, soon deciduous, tube 8–16 mm. long in fruit; sepals broadly triangular, about 1 mm. long, spreading or reflexed; fruiting styles 5–12 cm. long.

Hillsides and along washes, mainly Upper Sonoran Zone, but entering the upper margins of the Sonoran Desert in eastern San Diego and Riverside Counties, California, and in Maricopa County, Arizona; ranging from Arizona to northern Baja California. March–July.

## 3. COLEOGYNE Torr., Pl. Frem. 8, 1854

Intricately branched, spreading shrub with spinescent branches and opposite, entire, coriaceous, small leaves. Flowers solitary and terminal on short spur branchlets, each subtended by 1–2 pairs of 3-lobed bracts, central lobe greatly exceeding lateral pair. Hypanthium 1 mm. long or less, cup-shaped. Sepals 4, petaloid, yellowish and glabrous within, cinereous-puberulent without, persistent, spreading. Petals none. Stamens 20–40, basally inserted on outside of tubular sheath enclosing ovary. Pistil 1, with a lateral, twisted, exserted, persistent style dorsally pubescent at base. Fruit a glabrous achene.

**1. Coleogyne ramosissima** Torr., Pl. Frem. 8, *pl. 4,* 1854

Rounded or straggly shrub 0.3–2 m. high; bark gray, finely striate-fissured; leaves appearing fasciculate but opposite and closely crowded on lateral spurs, linear-oblanceolate, 4–11 mm. long, obtusish and often mucronulate at apex, plane or nearly so above, 4-grooved beneath, canescent with centrally attached, 2-branched hairs lying horizontal with surface; stipules subtriangular, 1 mm. long or less, persistent after leaf blade falls; calyx lobes 6–8 mm. long, outer pair ovate to lanceolate, acute, inner pair ovate to obovate, obtusish or acute, abruptly mucronulate; sheath membranous, about equaling calyx lobe, gradually tapering toward 5-toothed apex, silky-villous within, glabrous without; fruit an ovate, slightly curved, glabrous achene 6–8 mm. long.

Sandy or gravelly soil on desert slopes and adjacent low mountains, Lower and desert edges of the Upper Sonoran Zones, western margin of the Colorado Desert, east across southern Nevada, Utah, and northern Arizona to southern Colorado. Barely entering the Sonoran Desert in eastern San Diego County, California. March–May.

## 4. **PURSHIA** DC., Trans. Linn. Soc. **12:** 157, 1817

Small to large shrub with mainly fascicled leaves with cuneate, tridentate, or trilobate leaf blades. Flowers solitary and terminal on short lateral spurs. Hypanthium turbinate to funnelform. Sepals 5, about half as long as tube. Petals 5, clawed, slightly exceeding sepals, yellow or creamy white. Stamens 18–25, inserted near apex of hypanthium. Pistil 1 or rarely 2. Fruit an oblong pubescent achene with slender, persistent, nearly straight style.

### 1. **Purshia glandulosa** Curran, Bull. Calif. Acad. **1:** 153, 1885

*Kunzia glandulosa* Greene, Pittonia **2:** 299, 1892.
*Purshia tridentata* var. *glandulosa* M. E. Jones, Proc. Calif. Acad. II, **5:** 680, 1895.

Dark green shrub 1–3 m. high with dark red bark copiously supplied with conspicuous stipitate glands when young, later becoming gray; leaf blades 3-lobed or rarely 5-lobed, 3–10 mm. long, green and nearly glabrous above, loosely tomentose and with conspicuous veins beneath, the revolute margins stipitate-glandular; hypanthium tube 6–8 mm. long, puberulent and glandular without, lobes broadly ovate, about 3 mm. long; petals obovate, short-clawed, sparingly pubescent near apices on outer surface, 5–6 mm. long; stamens about equaling sepals; achenes obovoid, body about 6 mm. long, puberulent, persistent style equaling or slightly exceeding body, curved slightly to one side.

Mainly in the Upper Sonoran Zone or in higher parts of the Mojave Desert, but entering the Colorado Desert at Piñon Wells; to Inyo and Mono Counties, California, and southern Nevada. March–June.

## 5. **ROSA** L., Sp. Pl. 491, 1753

Prickly shrubs with odd-pinnate leaves, stipules adnate to petioles, and solitary or corymbosely arranged flowers. Hypanthium urceolate to globose. Bractlets none. Sepals 5. Petals 5, rounded, spreading, usually caducous. Stamens many on a silky disk lining hypanthium. Pistils several to many, free and distinct but included in hypanthium; styles subterminal; ovules pendulous, solitary. Achenes bony, enclosed by the enlarged, fleshy, berrylike or bristly, often colored hypanthium.

### 1. **Rosa minutifolia** Engelm., in Parry, Bull. Torrey Club **9:** 97, 1882

Small, dense shrub 2–6 (rarely to 15) dm. tall, with dark gray, puberulent, slender shoots densely covered with straight to slightly curved, reddish to pale gray, acicular spines 1–8 mm. long; stipules 2–4 mm. long, 0.5–0.8 mm. wide at tip, rounded, marginally stipitate-glandular, puberulent; leaves solitary at nodes on young growth, in fascicles on short spurs on older growth, 1–2.5 cm. long, with 3–7, but mostly 5, leaflets; rachises puberulent and with scattered yellowish, stipitate glands; leaflets obovate, 2–5 mm. long, 5–7-denticulate apically, tips of teeth markedly glandular-viscid, blade green above, tomentose and gray beneath; pedicels 5–10 mm. long, densely puberulent; hypanthium subglobose, 4–5 mm. long, densely puberulent and closely armed with yellow to reddish bristles 2–3 mm. long; sepals 6–10 mm. long, acuminate, outer ones pinnatifid with lobes 0.5–2.5 mm. long; petals rose-pink, 8–15 mm. long, broadly obovate to suborbicular, caducous; fruit persistently bristly, 6–8 mm. in diameter.

Rocky to heavy clay soil on hillsides and valley floors, Lower and Upper Sonoran Zones, near Ensenada south to red clay hills near Aguajitos, Baja California. Jan.–June.

## Family 57. **AMYGDALACEAE.** Peach Family

Shrubs (in ours) or trees with alternate, simple, petiolate, entire to serrate leaves and small, early-deciduous stipules; bark exuding a gum, and the bark, leaves, and seeds bitter, containing prussic acid. Flowers solitary or in axillary or terminal racemes or corymbs, usually perfect. Hypanthium turbinate to campanulate, free, deciduous. Sepals 5, deltoid to ovate. Petals 5. Stamens 10–25, inserted in 1–3 rows on the hypanthium interior to the petals. Pistil (in ours) 1; ovary 1-celled, 2-ovuled. Fruit a fleshy or leathery drupe with a bony stone; seed usually 1, rarely 2.

### 1. **PRUNUS** L., Sp. Pl. 473, 1753

Shrubs or small trees with alternate, simple, entire, or variously toothed leaves and small deciduous stipules. Flowers solitary or in clusters, regular and usually perfect. Hypanthium cup-shaped, with narrow disk at or near mouth bearing 15–30 stamens. Sepals 5, soon deciduous. Petals 5, imbricate in bud, white. Pistil 1; ovules 2, one usually abortive, pendulous. Fruit a drupe; stone smooth, globose or somewhat compressed.

Leaves broadly obovate to orbicular or rarely lanceolate, distinctly serrulate, glabrous; flowers pedicellate, 10–14 mm. broad; bark on young twigs reddish ........... 1. *P. fremontii*

Leaves spatulate, entire or nearly so, puberulent; flowers sessile, 3–6 mm. broad; bark on young twigs gray ................................................ 2. *P. fasciculata*

**1. Prunus fremontii** S. Wats., Bot. Calif. **2:** 442, 1880

*Amygdalus fremontii* Abrams, Bull. N.Y. Bot. Gard. **6:** 385, 1910.
*Prunus eriogyna* Mason, Jour. Agr. Research **1:** 168, *fig. 5*, 1913.
*Prunus fremontii* var. *pilulata* Jepson, Man. Fl. Pl. Calif. 507, 1925.

Deciduous shrub or small tree 1.5–5 m. high with stiff, intricately forking, zigzag, spinose branches; bark smooth and reddish in youth but soon grayish and lightly fissured; leaves mostly ovoid but sometimes broadly lanceolate or suborbicular, 1–3 cm. long, cuneate to subcordate at base, serrulate with short, rounded teeth; petioles one-third to one-half as long as blade; flowers in axillary clusters on pedicels 4–10 mm. long; calyx 4–5 mm. long, glabrous, broadly ovate sepals about two-thirds as long as hypanthium and puberulent within; petals obovate to suborbicular, 3.5–5 mm. long, white; stamens 3–7 mm. long; fruit elliptic-ovoid, slightly compressed, or sometimes nearly globose, 8–14 mm. long, densely and finely puberulent at first, at length sparsely so, exocarp thin and leathery.

Canyons and outwash fans, Lower Sonoran Zone, Colorado Desert from Palm Springs to Ensenada and near Cataviñá, Baja California. Jan.–March.

**2. Prunus fasciculata** (Torr.) A. Gray, Proc. Amer. Acad. **10:** 70, 1874
*Emplectocladus fasciculatus* Torr., Pl. Frem. 10, *pl. 5,* 1854.
*Amygdalus fasciculata* Greene, Fl. Fran. 49, 1891.
*Lycium spencerae* J. F. Macbride, Contr. Gray Herb. **53:** 18, 1918.

Rounded deciduous shrub 1–3 m. high with divaricate, thorny, stiff, straight branches and light gray bark; leaves fasciculate on warty, short spur branches, oblanceolate-spatulate, 5–20 mm. long, 1.5–3 mm. wide, entire or with 1–2 minute teeth on one or both sides near acute to obtuse apex, minutely pubescent on both sides; flowers solitary or in few-flowered fascicles on short spurs, sessile or subsessile, 3–6 mm. broad, more or less dioecious; hypanthium 1–1.5 mm. high, finely puberulent; petals about 2 mm. long, often puberulent, caducous; stamens 1.5–3 mm. long; fruit ovoid, slightly compressed laterally, 6–10 mm. long, densely tomentulose with tangled, brownish or grayish hairs; exocarp thin and leathery.

Desert slopes and arid foothills, Lower and Upper Sonoran Zones, northern and northwestern borders of the Colorado Desert to Inyo County, California, southern Utah, and adjacent northern Arizona. March–May.

## Family 58. MIMOSACEAE. Mimosa Family

Trees, shrubs, or herbs, often armed with stipular spines. Leaves usually (always in ours) bipinnate. Flowers small to large, in dense heads, spikes, or racemes, often showy. Calyx 4–5-lobed or -toothed. Petals 4 or 5, free or somewhat connate toward base. Stamens long-exserted, distinct or united into tube at base; anthers oval, small, with minute apical gland or eglandular. Pistil stipitate or sessile; stigma capitate. Fruit a legume, dehiscent, tardily dehiscent, or rarely indehiscent.

A large family well represented in the tropics and warm temperate parts of both hemispheres.

Stamens numerous, more than 10, usually more than 30, mostly over 1 cm. long:
  Filaments united into tube about equaling or slightly surpassing corolla:
    Margins of the pod separating from valves .............................. 1. *Lysiloma*
    Margins of the pod not separating from valves:
      Pods not elastically dehiscent from apex, mostly 1.5–3 cm. wide:
        Pods circinate, indehiscent .................................. 2. *Enterolobium*
        Pods not circinate, dehiscent, although sometimes tardily so:
          Plants usually spiny; pod woody or coriaceous, often thickened .............. ................................................ 3. *Pithecellobium*
          Plants unarmed; pod thin, not thickened ......................... 4. *Albizzia*
      Pods elastically dehiscent from apex, mostly less than 1.5 cm. wide .... 5. *Calliandra*
  Filaments free, or united into very short tube less than 2 mm. long ........... 6. *Acacia*
Stamens 10 or fewer, usually as many as corolla lobes or twice as many:
  Anthers tipped with minute gland between thecae; flowers usually spicate:
    Plants armed with spines; fruits septate between seeds, curved or tightly coiled ...... ................................................................ 7. *Prosopis*
    Plants unarmed; fruit not septate, straight or nearly so .................. 8. *Neptunia*
  Anthers not gland-tipped; flowers capitate or short-spicate:
    Valves of pod separating from persistent, cordlike margin; plants usually spiny ...... ................................................................ 9. *Mimosa*
    Valves of pod not separating from margins, this not cordlike; plants unarmed:
      Trees or large shrubs; seeds arranged transversely in pods ............ 10. *Leucaena*
      Herbs or small shrubs; seeds arranged longitudinally or obliquely in pods ........ ...................................................... 11. *Desmanthus*

## 1. LYSILOMA Benth., Lond. Jour. Bot. 3: 82, 1844

Unarmed shrubs or trees with bipinnate leaves, small flowers in heads or short spikes, and flat pods in which thin valves separate from cordlike margins at maturity. Calyx 5-lobed or 5-toothed, campanulate, mostly pubescent. Corolla lobes 5, exceeding calyx lobes, frequently puberulent. Stamens mostly 15–30, exserted, united into short tube below. Ovary sessile or short-stipitate; ovules several. Pods oblong or linear. Seeds flattened, hard, smooth.

Bark on older branches and trunks white; leaflets 10–22 mm. long, elliptic to obovate; pinnae 1–2 pairs; stipules rounded at apex, often orbicular .................... 1. *L. candida*
Bark gray or brownish; leaflets 3–10 mm. long, linear or oblong; pinnae mostly 3–10 pairs; stipules setaceous to lance-ovate, acute:
  Leaflets puberulent on both surfaces; petiolar gland 1–2.5 mm. high, conical; glands on rachis 1–5, conspicuous; stipules narrowly lance-subulate, 5–6 mm. long .......... ............................................................ 2. *L. watsonii*
  Leaflets glabrous or only ciliate or sparsely puberulent along veins beneath; petiolar glands usually smaller, flattened; glands on rachis lacking or inconspicuous; stipules ovate or ovate-lanceolate, 9–12 mm. long:
    Rachises and peduncles hirsutulous, later glabrate; leaflets glaucous, chartaceous, glabrous; pods glaucous ......................................... 3. *L. thornberi*
    Rachises and peduncles appressed-puberulent or glabrous; leaflets green, not chartaceous, usually ciliate and sometimes puberulent along veins beneath when young; pods reddish brown, not glaucous ................................. 4. *L. divaricata*

### 1. Lysiloma candida Brandegee, Proc. Calif. Acad. II, 2: 153, 1889

A tree 3–10 m. high with smooth white bark on all except the youngest branches and twigs, the latter glabrous to finely and transiently puberulent; stipules 6–15 mm. long, orbicular to oblong, rounded at apex; petioles 0.5–1.5 cm. long, bearing an orbicular, slightly elevated gland dorsally at summit; pinnae usually 1 pair or occasionally 2 pairs; leaflets 2–8 pairs, broadly elliptic, obovate, or rarely broadly oblong, 4–12 mm. wide, 10–22 mm. long, rounded or shallowly emarginate at apex, chiefly broadly cuneate below, chartaceous, glabrous, lightly but distinctly pinnately veined; flowers in globose heads, these racemosely disposed on peduncles 1–3 cm. long, and appearing before new vegetative growth; calyx about 2 mm. long, papery, puberulent but all except the lobes eventually glabrate; stamens white, about 1 cm. long; pods coppery red and spirally twisted when young, straightening as they mature, 2–3 cm. wide, 8–20 cm. long, on stipes 1.5–3 cm. long; seeds oval, flattened, 6–10 mm. long.

Along arroyos and on rocky hillsides, Lower Sonoran Zone, from about 25 miles north of San Ignacio to the Cape Region, Baja California, and on the adjacent islands in the Gulf. March–May.

The young pods, while they are still spirally twisted and the ovules are only 1–2 mm. long, are eaten avidly by cattle. The lower branches are frequently broken from the trees or stripped of pods. The leaves are eaten less frequently and with no great relish.

### 2. Lysiloma watsonii Rose, Contr. U.S. Nat. Herb. 1: 99, 1891

A tree 5–15 m. high with dark gray rough bark and a wide-spreading, often irregular crown; twigs, peduncles and leaf rachises densely pilose; stipules 5–6 mm. long, narrowly lance-subulate, pilosulous; petioles 1–2 cm. long, bearing conical gland 1–2.5

mm. high just below lower pair of pinnae, 1–3 or rarely –5 similar glands borne on rachis between upper of the 4–8 pairs of pinnae; leaflets 15–45 pairs, oblong, 4–10 mm. long, 2–4.5 mm. wide, puberulent on both sides, with spreading to subappressed whitish hairs, obtuse but sometimes minutely apiculate at apex; peduncles stoutish, 2.5–4.5 cm. long; flowers pale cream, calyces about 2 mm. long, puberulent near tips of the broadly ovate-triangular teeth; corolla lobes 3–4 mm. long, lance-oblong, acute, puberulent externally; filaments 8–10 mm. long; pods linear-oblong, 1.5–2.5 cm. wide, 12–22 cm. long, dark reddish brown, glabrous and slightly glaucous, acute to attenuate at apex; stipe 0.5–1.5 cm. long; seeds oval, 8–10 mm. long.

On rocky hillsides, gravelly slopes, and along canyons, Lower Sonoran Zone, northern Sonora from the vicinity of Magdalena and Bavispe south to northern Sinaloa. The tree apparently blooms chiefly following the summer rains; less profusely March–April.

### 3. **Lysiloma thornberi** Britt. & Rose, N. Amer. Fl. **23:** 83, 1928

A shrub or small tree 2–5 m. high with hard brittle wood, pilose twigs, and glaucous leaves; stipules ovate, 10–12 mm. long, acute; petioles 1–1.5 cm. long, bearing a small conical gland 1 mm. high or less 3–5 mm. below lowest of the 4–8 pairs of pinnae; leaflets 20–35 pairs, linear-oblong, 4.5–6 mm. long, rounded or obtuse, glaucous above, brownish beneath, glabrous; peduncles 2–3 cm. long, pilose at first but eventually glabrate; flowers white, capitate; calyx 1.5–2 mm. long, puberulent; corolla about 3 mm. long; pods oblong, 1–2 cm. wide, 10–20 cm. long, glaucous, acute to apiculate at apex, gradually narrowing to a stipe 1–1.5 cm. long; seeds dark brown, oval, 10–13 mm. long, 8–9 mm. wide, smooth, lustrous over much of the surface.

Known only from the foothills of the Rincon Mountains, Pima County, Arizona, at the upper edge of the Lower Sonoran Zone and extending into the Upper Sonoran Zone. July–Sept.

### 4. **Lysiloma divaricata** (Jacq.) J. F. Macbride, Contr. Gray Herb. **59:** 6, 1919

*Mimosa divaricata* Jacq., Pl. Hort. Schönbr. **3:** 76, 1798.
*Acacia divaricata* Willd., Sp. Pl. **4:** 1070, 1806.
*Lysiloma ortegae* Britt. & Rose, N. Amer. Fl. **23:** 82, 1928.
*Lysiloma cayucensis* M. E. Jones, Contr. West. Bot. **18:** 39, 1933.

A large shrub or tree to 15 m. high with grayish bark and glabrous to sparsely puberulent twigs, peduncles, petioles, and rachises; stipules ovate, acute to short-attenuate at apex, asymmetrically rounded at base, 4–6 mm. wide, 9–12 mm. long, paler beneath than above, glabrous; petioles slender, 0.5–2 cm. long, bearing flattened gland about 0.5 mm. high at or near attachment of lowest pair of pinnae, similar but smaller glands often between bases of uppermost pair of 4–9 (12) pairs of pinnae; leaflets 20–30 pairs, linear-oblong, 1–1.5 mm. wide, 3–8 mm. long, ciliate on margins and sparsely puberulent along veins beneath, or glabrous; peduncles 1–3 cm. long; often bearing an ovate bract 4–6 mm. long above middle; calyx about 2 mm. long, puberulent, the lobes about 0.5 mm. long; corolla 3–3.5 mm. long, puberulent; stamens 7–8 mm. long; pods linear-oblong, 1–2 cm. wide, 8–15 cm. long, glabrous, acute at apex, broadly cuneate at base, stipe 5–12 mm. long; seeds 4–5 mm. wide, 8–9 mm. long.

Central Sonora and southern Baja California to Vera Cruz and Oaxaca; also in Costa Rica and Nicaragua. July–Sept.

## 2. **ENTEROLOBIUM** Mart., Flora 20²: Beibl. 117, 1837

Unarmed trees with bipinnate leaves, petiole bearing 1 or more crateriform glands, and perfect flowers in capitate peduncled clusters. Calyx 5-toothed or -lobed. Corolla 5-lobed. Stamens numerous, coalesced into short tube at base; anthers small. Ovary sessile. Pod indehiscent, somewhat fleshy, flattened and spirally curved into a nearly complete circle. Seeds transversely arranged, flattened, smooth.

**1. Enterolobium cyclocarpum** (Jacq.) Griseb., Fl. Brit. W. Ind. 226, 1860
*Mimosa cyclocarpa* Jacq., Fragm. 30, 1809.
*Inga cyclocarpa* Willd., Sp. Pl. **4:** 1026, 1806.
*Pithecolobium cyclocarpum* Mart., Flora **20²:** Beibl. 115, 1837.
*Mimosa parota* Sessé & Moc., Pl. Nov. Hisp. 177, 1890.
*Feuilleea cyclocarpa* Kuntze, Rev. Gen. Pl. **1:** 184, 1891.

A round-crowned tree to 25 m. tall with gray fissured bark and glabrous or puberulent twigs and foliage; stipules minute, caducous; petioles 2–6 cm. long, usually bearing a sessile crateriform gland 1–1.5 mm. in diameter about halfway between base and lowest of the 4–15 pairs of pinnae; leaflets 15–30 pairs, linear-oblong, asymmetrically truncate at base, acute apically, 2.5–5 mm. wide, 8–20 mm. long, dark green and shining above, pale beneath, margins usually ciliate; heads many-flowered, 1.5–2.5 cm. in diameter; peduncles 1.5–5 cm. long; calyx tubular-campanulate, 2.2–2.5 mm. long, strigose; corolla 5–6 mm. long, strigose, pale cream; stamens about 1 cm. long, tube shorter than corolla; pods 3–4.5 cm. wide, curved into an almost complete circle 7–12 cm. in diameter, dark red-brown, glabrous; seeds 1.5–2 cm. long.

Planted as a shade tree and as a source of food for both cattle and humans, probably not reproducing spontaneously. Sinaloa to South America. Southern Baja California; Guaymas, Sonora. March–April.

## 3. **PITHECELLOBIUM** Mart., Hort. Monac. 188, 1829

Shrubs or trees with bipinnate leaves, mostly spinescent stipules, and small perfect flowers in pedunculate heads or spikes. Calyx 5–6-toothed. Corolla 5–6-lobed. Stamens numerous, exserted, united into a short basal tube. Ovary sessile or stipitate, several- to many-ovuled. Pod turgid or flattened, often woody and curved or contorted, the valves twisting upon dehiscence. Seeds hard, more or less arillate.

Leaflets 1 pair, 2–5 cm. long; rachis of inflorescence 10–30 cm. long or more; fruit coiled or contorted; tree 5–20 m. tall ........ 1. *P. dulce*
Leaflets 2–several pairs, 2–18 mm. long; rachis of inflorescence (when racemosely arranged heads) rarely over 10 cm. long; fruit curved, undulate, or straight, but not strongly coiled; shrubs or small trees:
  Pinnae 1 pair; valves of pod hard and woody; leaflets obovate to suborbicular, broadly cuneate at base; branches rigidly divaricate ........ 2. *P. confine*
  Pinnae 2–several pairs; pods not woody; leaflets linear or oblong, rounded at base; branches not rigidly divaricate:
    Flowers glabrous except on margins of lobes; fruit 6–10 mm. wide, arcuately curved; leaflets 5–18 mm. long ........ 3. *P. undulatum*
    Flowers puberulent; fruit 1–2.5 cm. wide, straight; leaflets 2–8 mm. long:
      Flowers sessile; corolla lobes erect; leaflets 2–3.5 mm. long; heads solitary on axillary stoutish peduncles ........ 4. *P. sonorae*
      Flowers short-pedicellate; corolla lobes spreading; leaflets 3–8 mm. long; heads in naked racemes on slender peduncles ........ 5. *P. mexicanum*

**1. Pithecellobium dulce** (Roxb.) Benth., Lond. Jour. Bot. **3:** 199, 1844

*Mimosa dulcis* Roxb., Pl. Corom. **1:** 67, 1795.
*Inga dulcis* Willd., Sp. Pl. **4:** 1005, 1806.
*Inga pungens* H. & B. ex Willd., op. cit. 1004.
*Mimosa pungens* Poir., in Lam., Encycl. Supp. **1:** 36, 1810.
*Inga javana* DC., Prod. **2:** 436, 1825.
*Acacia obliquifolia* Mart. & Gal., Bull. Acad. Brux. **10**$^{2}$**:** 317, 1843.
*Feuilleea dulcis* Kuntze, Rev. Gen. Pl. **1:** 184, 1891.

Large tree to 15 m. or more with glabrous to puberulent twigs and young leaflets, and densely puberulent inflorescences; stipular spines absent or to 1.5 cm. long and ascending; petioles 1.5–3 cm. long; pinnae 1 pair; leaflets 1 pair, ovate, oblong, or obovate, mainly 2–5 cm. long, often rounded, glabrous to puberulent; inflorescence a raceme to 4 dm. long of numerous short-pedunculate heads, rachis and peduncles densely canescent; calyx broadly bowl-shaped, 1–2 mm. deep; corolla 2.5–4 mm. long, white or pale cream; stamens 8–10 mm. long, tube exceeded by corolla; pods curved, coiled, or contorted, 10–18 cm. long, about 1 cm. wide, glabrous.

Mainly, if not entirely, a planted shade and food tree, Lower Sonoran Zone to Tropical Zone, from vicinity of Guaymas, Sonora, and Comondú, Baja California, to northern South America. Nov.–June.

Introduced from the East Indies. The young pods are cooked and eaten. Both foliage and fruits are used for cattle feed.

**2. Pithecellobium confine** Standley, Contr. U.S. Nat. Herb. **20:** 191, 1919

*Ebenopsis confinis* Britt. & Rose, N. Amer. Fl. **23:** 33, 1928.

An intricately branched, rounded shrub to 4 m. tall with rigid, divaricate, zigzag, densely puberulent but soon glabrate twigs; stipular spines slender, subulate, straight, 6–15 mm. long; leaves puberulent or the leaflets glabrate; petioles 3–15 mm. long; pinnae 1 or rarely 2 pairs; leaflets 2–5 pairs, cuneate-obovate to nearly orbicular, coriaceous, 2.5–10 mm. long, rounded to truncate at apex, strongly pinnately veined and pale beneath; peduncles stout, 5–10 mm. long, densely puberulent; heads about 1.5 cm. in diameter; flowers glabrous or nearly so, pale purplish or lavender to rose; calyx 1–1.2 mm. long; corolla 2.5–3 mm. long; pods woody, turgid, 2–3.5 cm. in diameter, 5–12 cm. long, dark brown, puberulent but eventually glabrate and scaly; seeds hard, 1–1.5 cm. long.

Desert arroyos, rocky hillsides, and gravelly slopes, Lower Sonoran Zone, vicinity of Calmallí south to Cape Region, Baja California. Feb.–April.

The seeds are roasted, ground, and used as an adulterant in coffee and chocolate. The pods, soaked in water, provide a brownish black ink.

**3. Pithecellobium undulatum** (Britt. & Rose) Gentry, Carnegie Inst. Wash. Pub. **527:** 121, 1942

*Chloroleucon undulatum* Britt. & Rose, N. Amer. Fl. **23:** 37, 1928.

A rounded shrub or small tree to 5–6 m. with puberulent to glabrous twigs, smooth brownish gray bark, and sparsely puberulent to glabrous foliage; stipular spines mostly lacking; petioles 1.5–2.5 cm. long, bearing a small sessile gland near middle; pinnae 3–5 pairs, 3–10 cm. long; leaflets 5–12 pairs, oblong, 5–18 mm. long, rounded or obtuse at each end, strongly pinnate-veined; peduncles 1.3 cm. long; corollas glabrous except on margins of lobes, 5–7 mm. long; pods linear, 6–10 mm. wide, 1–2.5 dm. long, arcuately curved and margins undulate, transversely reticulate-veined; seeds broadly ovoid, 6–7 mm. long, pale pea green, smooth.

Along arroyos or wooded slopes and rocky valleys, Lower Sonoran Zone to Subtropical Zone, southern Sonora and Baja California to Colima; Tres Marías Islands, Nayarit. Aug.–Sept.

**4. Pithecellobium sonorae** S. Wats., Proc. Amer. Acad. **24:** 49, 1889

*Havardia sonorae* Britt. & Rose, N. Amer. Fl. **23:** 42, 1928.

A rounded shrub to a tree 12 m. tall with canescent twigs and foliage, gray bark, and conical or slightly curved stipular spines 1–3 mm. long; petioles 2–8 mm. long, bearing 1 or 2 crateriform glands; pinnae 1–4 pairs, 1–2 cm. long; leaflets 10–20 pairs, oblong, obtuse, 2–3.5 mm. long, canescent with spreading hairs on both sides; peduncles stoutish, 5–15 mm. long; flowers sessile; calyx 2–2.5 mm. long, densely puberulent; corolla 5–6 mm. long, canescent; stamens about 1 cm. long, short tube puberulent; pods broadly oblong, 1–2 cm. wide, 6–18 cm. long, short-stipitate and apiculate, puberulent but eventually glabrate; seeds suborbicular to somewhat quadrate, 6–9 mm. in diameter, flattened, dark reddish brown.

Along arroyos, on low hills and flats, Lower Sonoran and Subtropical Zones, from vicinity of Guaymas, Sonora, south into Sinaloa. July–Sept.

**5. Pithecellobium mexicanum** Rose, Contr. U.S. Nat. Herb. **1:** 100, 1891

*Havardia mexicana* Britt. & Rose, N. Amer. Fl. **23:** 41, 1928.

A tree 5–10 m. tall with hirsutulous but soon glabrous twigs and slightly recurved stipular spines 2–4 mm. long or these often lacking; herbage pilosulous; petioles 1–2 cm. long, bearing a small sessile gland at or near lower of 2–4 pairs of pinnae, these 1.5–2.5 cm. long; leaflets 3–10 pairs, obovate to oblong, obtuse to rounded, pilosulous on both sides, 3–8 mm. long; heads in racemes or panicles, 1.5–2 cm. in diameter; peduncles slender, 5–25 mm. long; flowers yellowish; pedicels 1.5–2 mm. long; calyx broadly cup-shaped, 1–1.5 mm. deep, densely puberulent; corolla about 3 mm. long, lobes spreading, densely strigulose, at least near tips; stamens about 1 cm. long; pod oblong, 2–2.5 cm. wide, 6–10 cm. long, glabrous, few-seeded.

Along arroyos and on desert slopes and valley floors, Lower Sonoran Zone, from vicinity of Nória, Sonora, south into Sinaloa, and in southern Baja California. The material from Baja California is less pubescent and the pedicels slightly longer than in specimens from Sonora. March–April.

## 4. ALBIZZIA Durazz., Mag. Tosc. 3: 11, 1772

Unarmed trees or shrubs with smooth bark, bipinnate leaves, and small, mostly perfect flowers in solitary or paniculate heads, umbels, or spikes. Calyx 5-toothed, usually campanulate. Corolla 5-lobed, funnelform. Stamens many, long-exserted, united into short tube below; anthers small, ovoid. Legume flat, thin, broadly linear to oblong, indehiscent or tardily dehiscent. Seeds flattened, suborbicular to oblong.

Leaflets obovate, 1.5–4 cm. long, in 2–6 pairs on each pinna; twigs, petioles, and peduncles glabrous or glabrate; pods 3–4 cm. wide, 15–30 cm. long ........... 1. *A. occidentalis*

Leaflets linear-oblong, 1–1.5 cm. long, in 12–16 pairs on each pinna; twigs, petioles, and peduncles tomentulose; pods 2–2.5 cm. wide, 10–15 cm. long ........ 2. *A. sinaloensis*

**1. Albizzia occidentalis** Brandegee, Proc. Calif. Acad. II, **3:** 222, 1892

An open-crowned tree 5–20 m. high with smooth gray bark and glabrous or glabrate twigs, peduncles, and petioles, the last 2–5 cm. long and bearing an oval depressed gland 1–1.5 cm. above base; pinnae 2–4 pairs; leaflets 2–6 pairs on each pinna, oval to obliquely obovate, 1.5–5 cm. long, terminal pair usually largest, dark green above, slightly lighter beneath, acute to apiculate, glabrous or slightly pubescent, or densely puberulent beneath when young, reticulate-veined, midvein eccentric; flowers capitate on axillary peduncles, yellowish; calyx narrowly campanulate, about 3 mm. long, with erect, broadly triangular, acute teeth; corolla 8–9 mm. long, with lanceolate lobes 3–6 mm. long; stamen tube nearly equaling corolla, filaments exserted 1.5–2 cm.; pods oblong-linear, 3–4 cm. wide, 15–30 cm. long, transversely reticulate-veined, on a stipe 1–3 cm. long, dehiscent, margins cordlike; seeds compressed, broadly oblong, 8–10 mm. long, dark brown along margins, a lighter gray-brown elliptic spot 3–4 mm. long on each side.

Mountain slopes and along washes, Lower Sonoran Zone, southern Baja California, Sinaloa, and the Tres Marías Islands, Nayarit. Flowering dates uncertain, probably June–July.

**2. Albizzia sinaloensis** Britt. & Rose, N. Amer. Fl. **23:** 45, 1928

Small to large tree to 20 m. tall with smooth yellowish bark and densely tomentulose twigs, petioles, leaf rachises, and peduncles; petioles 2.5–5 cm. long with a flattish oblong gland 1.5–4 mm. long situated 3–12 mm. from base; pinnae 4–9 pairs, rachillae pubescent; leaflets 12–16 pairs, acute to obtuse, 2–5 mm. wide, 10–15 mm. long, light green, glabrous or glabrate above, finely puberulent below, leathery, distinctly veined beneath, midvein nearly central; inflorescence capitate on axillary peduncles 2–3 cm. long; flowers unknown; legumes broadly linear, 1.8–2.5 cm. wide, 8.5–15 cm. long, finely puberulent, tardily dehiscent, transversely veined, veins distinct only near strongly corded margins, tapering to acute base and apiculate apex, apiculation 2–6 mm. long; seeds oval, flattened, 6–8 mm. long, pale to dark brown, smooth, funiculus folded abruptly against base of the seed.

Along streams and in fertile valleys, Lower Sonoran Zone, southern Sonora into Sinaloa. Flowering dates uncertain.

## 5. CALLIANDRA Benth., in Hook., Jour. Bot. 2: 138, 1840

Perennial herbs, shrubs, or trees with bipinnate eglandular leaves and perfect flowers in capitate axillary clusters or in terminal racemes or heads. Calyx 5-toothed. Corolla usually funnelform, 5-lobed to middle or less. Stamens numerous, long-exserted, filaments coalescent into a tube below; anthers small, glabrous, glandular, or hirsute. Ovules several to many. Pod linear to oblong, flat, straight or nearly so, the stiff valves with cordlike margins and elastically dehiscent from apex. Seeds orbicular-lenticular to ovate-lenticular, smooth, hard.

Leaflets thin, not leathery; stipules ovate to lanceolate; calyx two-fifths to one-half as long as corolla:
  Plants perennial herbs from woody rootstock; leaflets prominently veined beneath; calyces puberulent .......... 1. *C. humilis*

Plants shrubby; leaflets faintly veined beneath; calyces and corollas glabrous:
Leaflets glabrous, 6–13 mm. wide, 0.5–2 cm. long; lateral veins visible ....... 2. *C. rosei*
Leaflets puberulent on both surfaces or at least ciliate, 3–5 mm. wide, 5–10 mm. long ... .......................................................... 3. *C. schottii*
Leaflets leathery; stipules setaceous to subulate; calyces one-eighth to one-fifth as long as corolla:
Flowers glabrous or essentially so; young twigs, peduncles, and rachises densely glandular-puberulent with brown simple hairs and with gland-tipped hairs, viscidulous ....... ........................................................ 4. *C. brandegeei*
Flowers strigulose to hirsutulous; twigs, peduncles, and rachises puberulent to glabrous but not glandular:
Leaflets 18–25 pairs, three to five times as long as broad; stipules 6–8 mm. long ....... ..................................................... 5. *C. peninsularis*
Leaflets 5–15 pairs, one and one-half to three times as long as broad; stipules 2–4 mm. long:
Stamens pink to pale rose; leaflets spreadingly hispidulous with hairs 0.6–1 mm. long, glabrous above when young ............................... 6. *C. eriophylla*
Stamens deep rose to purplish red; leaflets appressed-strigose with hairs 0.2–0.4 mm. long, puberulent above when young ......................... 7. *C. californica*

## 1. **Calliandra humilis** (Schlecht.) Benth., Lond. Jour. Bot. **5:** 103, 1846

*Acacia humilis* Schlecht., Linnaea **12:** 567, 1838.
*Calliandra ?herbacea* Engelm. ex A. Gray, Pl. Fendl. 39, 1849.
*Anneslia herbacea* Britt. & Rose, N. Amer. Fl. **23:** 57, 1928.

Herbaceous or slightly woody stems 1–3 dm. long from a stout woody rootstock, with sparsely pilose or glabrate herbage; stipules ovate-lanceolate, 2–4 mm. long; petioles 1–3 cm. long; pinnae 3–9 pairs, 1–3 cm. long; leaflets 10–18 pairs, oblong, 1.5–5 mm. long, 3-veined and somewhat reticulate-veined beneath; peduncles 1–3 cm. long; heads about 1 cm. in diameter, exclusive of filaments; calyx about 2 mm. long, puberulent mostly near tips of lobes; corolla 4–6 mm. long, sparsely lanate near tips; stamens 12–15 mm. long; pod 5–6 mm. wide, 4–7 cm. long, puberulent in youth, later glabrate.

In dry soil, mostly Upper Sonoran Zone and into the Arid Transition Zone, Cochise, Santa Cruz, and Pima Counties, Arizona, to western Texas and adjacent northern Mexico. April–July.

## 2. **Calliandra rosei** Wiggins, Contr. Dudley Herb. **4:** 17, 1950

*Acacia gracilis* Mart. & Gal., Bull. Acad. Brux. **10$^2$:** 311, 1843.
*Anneslia gracilis* Britt. & Rose, N. Amer. Fl. **23:** 68, 1928.
*Calliandra gracilis* Standley, Field Mus. Bot. Ser. **4:** 309, 1929. Not *C. gracilis* Griseb., 1861; nor Klotzsch, 1871.

Slender, erect-branching shrub 1–2 m. high, glabrous throughout; stipules 4–5 mm. long, narrowly ovate, striately nerved; petioles slender, 1.5–3 cm. long; pinnae 1–2 pairs, 3–4 cm. long; leaflets 5–7 pairs, obovate to oblong, thin, obtuse or rounded and sometimes apiculate at apex, 5–15 (rarely –20) mm. long, truncate and often asymmetrical at base; peduncles axillary, slender, 2–5 cm. long; calyx about 1 mm. long, glabrous; corolla 2–2.5 mm. long; filaments 1–2 cm. long; pod straight, 5–10 mm. wide, 5–10 cm. long, short-attenuate at base, glabrous.

Rocky hillsides, arid slopes, and along arroyos, Lower Sonoran Zone and into the subtropics, Sonora to central Mexico. July–Aug.

**3. Calliandra schottii** Torr. ex S. Wats., Proc. Amer. Acad. **20:** 364, 1885
*Anneslia schottii* Britt. & Rose, N. Amer. Fl. **23:** 67, 1928.

A small slender-branched shrub 1.5 m. high or less with puberulent or rarely glabrate twigs and leaves; stipules ovate to lanceolate, 3–4 mm. long, striate; petioles slender, 1.5–2.5 cm. long; pinnae 1–3 pairs, 1.5–3.5 cm. long, lower ones shorter than upper; leaflets 4–7 pairs, narrowly obovate to oblong, obtuse or rounded at apex, 3–5 mm. wide, 5–10 mm. long, faintly glaucous beneath; peduncles slender, 2–7 cm. long; calyx 1–1.5 mm. long, glabrous; corolla 2–3 mm. long, yellowish; stamens white or faintly tinged with rose, 12–15 mm. long; pod 4–6 cm. long, 6–8 mm. wide, sparsely puberulent in youth, soon glabrate.

Mountain slopes and canyons, mainly Upper Sonoran Zone but reaching the upper parts of the desert; southern Arizona to Chihuahua and Sonora. July–Aug.

**4. Calliandra brandegeei** (Britt. & Rose) Gentry, Allan Hancock Pacif. Exped. **13:** 130, 1949
*Anneslia brandegeei* Britt. & Rose, N. Amer. Fl. **23:** 62, 1928.

Small shrub with slender branches and glandular-puberulent twigs, petioles, peduncles, and rachises, the hairs and glands brown; stipules linear-lanceolate, 2–5 mm. long, pilosulous; petioles 1–3.5 cm. long; pinnae 4–6 pairs, 1.5–4 cm. long; leaflets 10–22 pairs, oblong, 1.5–2 mm. wide, 5–9 mm. long, acute to obtuse, often mucronulate, shining above, paler and dull beneath, ciliate or soon glabrate, reticulate-veined, midvein eccentric; peduncles 1–3.5 cm. long; calyx 1.5–2 mm. long, glabrous or ciliate along margins of minute triangular lobes; corolla purplish red, 6–7 mm. long, glabrous or bearing a few scattered appressed hairs or cilia near tips of lobes; filaments about 2 cm. long, purplish red, tube equaling corolla; pod straight, 1–1.8 cm. wide, 7–10 cm. long, gradually attenuate at base, rounded and apiculate at apex, short-pubescent to glabrate.

Mountain slopes and canyons, Sonoran Zones, Baja California from the Sierra de la Giganta southward. Aug.–Oct.

**5. Calliandra peninsularis** Rose, Contr. U.S. Nat. Herb. **5:** 135, 1897
*Anneslia peninsularis* Britt. & Rose, N. Amer. Fl. **23:** 58, 1928.

Openly and straggly branched shrub with slender stiffish twigs and puberulent young twigs, petioles, peduncles, and leaflets; petioles stoutish, about 1 cm. long; pinnae 5–6 pairs; leaflets 18–25 pairs, linear-oblong, 4–6 mm. long, acute or obtuse and somewhat apiculate, glabrous above, appressed-pubescent beneath, midvein eccentric; calyx 1–1.5 mm. long, puberulent; corolla 6–8 mm. long, bright red or slightly purplish, loosely strigose-pubescent; stamens 1.5–2 cm. long, tube shorter than corolla; pod straight or nearly so, 8 mm. wide, 6–9 cm. long, densely appressed-puberulent, long-tapering at base, obtuse and apiculate at apex.

Along dry watercourses and on rocky hillsides, Lower Sonoran Zone, vicinity of La Paz, Baja California. Nov.–March.

**6. Calliandra eriophylla** Benth., Lond. Jour. Bot. **3:** 105, 1844
*Calliandra chamaedrys* Engelm. ex A. Gray, Pl. Fendl. 39, 1849.
*Calliandra conferta* Benth., in A. Gray, Pl. Wright. **1:** 63, 1852.
*Feuilleea texana* Kuntze, Rev. Gen. Pl. **1:** 187, 1891.
*Feuilleea eriophylla* Kuntze, loc. cit.
*Anneslia eriophylla* Britt. & Rose, Trans. N.Y. Acad. **14:** 32, 1894.

An intricately branched dense shrub 2–10 dm. high with slender, stiff, appressed-pubescent twigs and pubescent herbage; stipules subulate, 3–4 mm. long; petioles short, 1 cm. long or less; pinnae 1–7 pairs, 1–2.5 cm. long; leaflets 5–15 pairs, linear-oblong, 1–2 mm. wide, 3–5 mm. long, acute to obtuse, obscurely to clearly pinnately veined, sparsely strigose with fine white hairs to 1.2 mm. long or often glabrate or glabrous above; flowers distinctly strigose, deep rose; calyx 1–1.5 mm. long; corolla 4–6 mm. long, more densely strigose near tips than below; stamens 1.5–2.2 cm. long, tube shorter than corolla; pod 5–7 mm. wide, 3–7 cm. long, densely short-pubescent with white hairs, cordlike margins often glabrate.

Along arroyos, on hillsides, and in desert canyons, Lower Sonoran Zone, Colorado Desert to New Mexico, Jalisco, and Puebla. Feb.–May, occasionally during summer months.

**7. Calliandra californica** Benth., Bot. Sulph. 14, *pl. 11,* 1844

*Feuilleea californica* Kuntze, Rev. Gen. Pl. **1:** 187, 1891.
*Anneslia californica* Britt. & Rose, N. Amer. Fl. **23:** 59, 1928.

Straggly, divaricately branched shrub 1 m. tall with appressed-pubescent twigs, leaves, and inflorescences; stipules subulate, stiff, 2–3 mm. long, 1-nerved; petioles 0.4–1 cm. long; pinnae 1–4 pairs, 0.6–2.5 cm. long; leaflets 5–15 pairs, oblong, 0.6–2.5 mm. wide, 3–6 mm. long, obtuse or rounded at apex, finely appressed-puberulent on both surfaces but soon glabrate and shining on upper surface, dull and paler beneath, subcoriaceous, obscurely veined beneath; calyx strigose, 1.5–2 mm. long; corolla strigose, purplish red, 6–7 mm. long; filaments purplish red, 18–20 mm. long, tube shorter than corolla; pod 5–7 mm. wide, 4–6 cm. long, densely and persistently puberulent.

Desert washes, hillsides, and gravelly flats, Lower Sonoran Zone, southern Sierra San Pedro Mártir to the Cape Region, Baja California. Flowering practically throughout the year.

## 6. ACACIA Mill., Gard. Dict. abr. ed. 4, 1754

Trees, shrubs, or some only suffrutescent perennials with bipinnate leaves, usually small and numerous leaflets, or in some modified into simple flat phyllodia. Stipular spines frequently present. Flowers small, in heads or spikes or spiciform racemes, usually white or yellow. Calyx 4–5-toothed, or rarely the sepals distinct. Petals 4–5, distinct or more or less united at the base or slightly above, rarely lacking. Stamens numerous, exserted; filaments filiform, distinct; anthers oval, minute, often with a glandular appendage between the thecae. Pod linear, oval, or oblong, flat or turgid, frequently more or less constricted between the seeds.

Inflorescence spicate or racemose, two to ten times as long as broad:
- Peduncles naked on basal 1–4.5 cm., devoid of bract at the base of inflorescence; floral bracteoles caducous:
  - Petioles flattened, phyllode-like, 1.5–4 mm. wide, 6–30 cm. long; pinnae 1–4 pairs, caducous; branchlets drooping; bark of trunk white and smooth .... 1. *A. willardiana*
  - Petioles not flattened, not over 5 cm. long; pinnae (1)3–9 pairs, persistent; branchlets not drooping; bark gray or brown, fissured and flaking:
    - Branches unarmed or the stipular spines straight, ascending; small tree to 8 m. tall; leaflets 10–40 pairs:
      - Spikes 6–10 cm. long; stipules minute, setaceous, caducous ......... 2. *A. coulteri*
      - Spikes 2.5–5 cm. long; stipules acicular, persistent ............... 3. *A. millefolia*

Branches armed with curved persistent spines; shrubs 1–2 m. tall; leaflets 2–8 pairs:
Leaflets 4–6 mm. long; calyx lobes incurved at apex; ovary pubescent at anthesis; floral bractlets fleshy, spoon-shaped .......................... 4. *A. greggii*
Leaflets 6–10 mm. long; calyx lobes erect; ovary glabrous at anthesis; bractlets lance-elliptic, attenuate-acuminate ............................... 5. *A. wrightii*
Peduncles floriferous to base, bearing a cupulate bract at its base; floral bracteoles subpersistent:
Cupule at base of spike sparsely puberulent to glabrous; calyx glabrous, 1–1.2 mm. long; leaflets obovate, 8–25 mm. long .............................. 6. *A. californica*
Cupule densely strigose-puberulent; calyx puberulent, 0.5–0.8 mm. long; leaflets oblong, 3.5–7 mm. long ........................................ 7. *A. brandegeana*
Inflorescence capitate or subcapitate, seldom longer than broad:
Flowers sessile or nearly so, borne on a globose, elliptic or short-oblong, pitted receptacle; bractlets forming a circle at base of receptacle, reflexed, often coalesced into a disk:
Receptacle ovoid to oblong, 3–6 mm. long; valves of legumes thin, 1.5–2 cm. wide, 4–10 cm. long; stipular spines unguiculate:
Leaflets linear to oblong, 1–2 mm. wide; pinnae 2–4 pairs; pedicels 1 mm. long or less; petioles densely hirsutulous ........................... 8. *A. occidentalis*
Leaflets obovate or cuneate-spatulate, much narrower at base than at apex; pinnae 1–2 pairs; pedicels 1–2 mm. long; petioles glabrous or sparsely hirsutulous ......... .................................................... 9. *A. peninsularis*
Receptacle globose or subglobose, 2–3.5 mm. long, nearly or fully as wide; legumes mostly less than 2 cm. wide, their valves thick and leathery; stipular spines straight or boat-shaped:
Pod linear, 3–6 mm. wide, constricted between seeds, readily dehiscent; calyx glabrous except for sparse marginal ciliolation; corolla glabrous ........ 10. *A. constricta*
Pod oblong, 9–25 mm. wide, not constricted, tardily dehiscent; calyx and corolla lobes puberulent, at least toward tips:
Pinnae 2–6 pairs; pod terete; corolla puberulent full length; stipular spines slender, 0.5–5 cm. long, 0.5–1.75 mm. in diameter at base .......... 11. *A. farnesiana*
Pinnae 6–60 pairs; pod flattened; corolla puberulent only on backs of lobes; stipular spines stout or concave, 3–10 mm. wide at base:
Stipular spines often concave, boat-shaped, 0.5–6 cm. long, 3–10 mm. wide; leaves and twigs glabrous or nearly so; pod glabrous; calyx as broad as long ..... ................................................ 12. *A. cymbispina*
Stipular spines not boat-shaped, 0.5–4 cm. long, 3–5 mm. wide at base; leaves and twigs densely puberulent with yellowish green hairs; pod 1.5–2.5 cm. wide, puberulent; calyx one and one-half to two times as long as broad .......... ................................................ 13. *A. pennatula*
Flowers pedicellate, receptacle oblong, ridged and often retaining persistent pedicels; basal bracteoles scattered or wanting, not forming disk or ring:
Leaflets broadly oblong-obovate, 9–20 mm. long, distinctly and strongly pinnately veined, the laterals prominent ........................................ 14. *A. crinita*
Leaflets linear to linear-oblong, rarely over 8 mm. long, with a single, eccentric, median vein, the laterals obscure:
Inflorescence mainly of heads axillary to leaves; leaflets 15–30 pairs:
Branchlets, calyx, petioles, and rachises glabrous or essentially so; pedicels slender ................................................ 15. *A. goldmanii*
Branchlets, calyx, petioles, and rachises pilose or puberulent; pedicels stout ....... ................................................ 16. *A. mcmurphyi*
Inflorescence of numerous globose heads arranged in terminal racemes; leaflets 30–60 pairs:

Shrubs 2–5 m. high; branchlets not markedly striate-ridged:
Leaflets 1–1.2 mm. wide, 1-veined; pods 8–12 mm. wide; pinnae 9–25 pairs .... ................................. 17. *A. angustissima* subsp. *angustissima*
Leaflets 1.5–2.5 mm. wide, mostly delicately pinnately veined; pods 6–9 mm. wide; pinnae 5–12 pairs ............. 17*a*. *A. angustissima* subsp. *lemmonii*
Suffrutescent plants mostly less than 1 m. high; branchlets distinctly striate-ridged ................................. 17*b*. *A. angustissima* subsp. *suffrutescens*

**1. Acacia willardiana** Rose, in Vasey & Rose, Contr. U.S. Nat. Herb. **1:** 88, 1890
*Prosopis ?heterophylla* Benth., Lond. Jour. Bot. **5:** 82, 1846. Not *Acacia heterophylla* Willd., 1805.
*Senegalia heterophylla* Britt. & Rose, N. Amer. Fl. **23:** 114, 1928.

A slender unarmed tree 3–9 m. tall with drooping glabrous branches and smooth, white to pale cream, papery bark; petioles phyllodelike, 1.5–4 mm. wide, 6–30 cm. long; pinnae 1–4 pairs but soon deciduous; leaflets 4–15 pairs, linear or linear-obovate, 2–5 mm. long, acute, glabrous, soon deciduous, leaving only slender, phyllodelike petioles on branchlets; inflorescence a panicle of slender spikes, these 3–6 cm. long; flowers pale yellow; calyx glabrous, about 2 mm. long; petals 2.8–3.2 mm. long, glabrous, acute, slightly united at or near base; stamens often over 100, 6–8 mm. long; pods broadly linear, 1–1.8 cm. broad, 10–20 cm. long, glabrous, acute to obtuse at apex, usually cuneate and short-stipitate at base; seeds 5–15, orbicular-lenticular, 9–10 mm. in diameter, dark brown, smooth.

Rocky hillsides, Lower Sonoran Zone, vicinity of Hermosillo south about to Guaymas; Tiburón Island, Sonora. Feb.–April.

**2. Acacia coulteri** Benth., in A. Gray, Pl. Wright. **1:** 66, 1852
*Senegalia coulteri* Britt. & Rose, N. Amer. Fl. **23:** 112, 1928.

An unarmed tree to 8 m. tall with gray, faintly fissured bark and glabrous or sparsely ciliolate foliage; stipules ovate-setaceous, about 1 mm. long, caducous; petioles slender, 2.5–4.5 cm. long, glabrous or essentially so, bearing small, flattened-crateriform gland about middle or near lowest of the 3–9 pairs of pinnae, these 3–6.5 cm. long; leaflets 10–31 pairs, linear-oblong, 4–6 mm. long, acute, glabrous above, sparsely puberulent to glabrate beneath and marginally, midvein eccentric; spikes slender, 6–10 cm. long; calyx about 1 mm. long, finely puberulent; corolla 2–3 mm. long, lobes distinct above basal 1–1.5 mm., minutely puberulent; pod oblong, 15–25 mm. broad, 7–18 cm. long, glabrous, cuneate at base, rounded and apiculate at apex, the valves thin, transversely veined.

Rocky hillsides and along canyons, mainly in the Tropical Zone but occasionally in the Lower Sonoran Zone, vicinity of Ures, Sonora, south and east to Hidalgo, San Luis Potosí, Nuevo León, and Tamaulipas. March–May.

**3. Acacia millefolia** S. Wats., Proc. Amer. Acad. **21:** 427, 1886
*Senegalia millifolia* Britt. & Rose, N. Amer. Fl. **23:** 111, 1928.

Shrub or small tree to 4–5 m. tall with slender, sparsely puberulent but soon glabrate branchlets and acicular, straight, stipular spines 2–3 mm. long, ascending and persistent; leaves 5–15 cm. long; petioles and rachises sulcate, bearing 1–5 crateriform glands between the pinnae or eglandular; pinnae 5–10 pairs, 2–5 cm. long; leaflets 20–40 pairs, linear, acute to apiculate, 2–3 mm. long, puberulent but glabrate in age; peduncles 5–10 mm. long; spikes 2.5–5 cm. long, about 12 mm. broad; flowers pale cream, sessile; calyx broadly cup-shaped, 0.6–1 mm. deep, glabrous or margins ciliolate; petals about 2 mm. long, glabrous; pod flat, 1–1.8 cm. broad, 7–15 cm. long, slightly constricted between seeds or margins straight, strongly transversely veined, acute to obtuse apically, cuneate at base, glabrous, stipe 5–7 mm. long.

Rocky canyons and hillsides, Upper and Lower Sonoran Zones, southeastern Arizona, Sonora, and Chihuahua. July–Aug.

**4. Acacia greggii** A. Gray, Pl. Wright. **1:** 65, 1852

*Acacia durandiana* Buckl., Proc. Acad. Phila. **1861:** 453, 1862.
*Senegalia greggii* Britt. & Rose, N. Amer. Fl. **23:** 110, 1928.

A rounded shrub or rarely a small tree to 4 m. tall (usually 2 m. or less); branches grayish brown, armed with somewhat flattened recurved spines about 3 mm. long; petioles slender, about 1 cm. long, spreading-puberulent; pinnae 1–3 pairs, 1–1.5 cm. long; leaflets 3–8 pairs, 4–6 mm. long, obovate to oblong, obtuse and sometimes minutely apiculate, ashy, minutely pilosulous, with a prominent eccentric midvein and a secondary one running diagonally from base; peduncles pubescent, 1–2 cm. long; spikes 1–4 cm. long; flowers pale yellow; calyx about 1 mm. long, short-triangular lobes incurved, puberulent; corolla about 2 mm. long, puberulent; pods flat, 12–15 mm. wide, 3–15 cm. long, glabrous, faintly glaucous, somewhat constricted between seeds and curved or tortulose, cuneate and stipitate at base.

Mainly along washes but also on rocky hillsides, Lower Sonoran Zone, western Colorado Desert, California, to Texas, Chihuahua, Sonora, and Baja California. April–Oct.

**5. Acacia wrightii** Benth., in A. Gray, Pl. Wright. **1:** 64, 1852

*Senegalia wrightii* Britt. & Rose, N. Amer. Fl. **23:** 110, 1928.

A shrub or sometimes a small tree to 3–4 m. tall, glabrous throughout, with gray bark and slightly recurved spines 2–4 mm. long or occasionally branchlets unarmed; leaves usually fascicled on older branches; petioles 3–12 mm. long, bearing a small crateriform gland near top; pinnae 1–2 pairs, 1–2 cm. long; leaflets 2–6 pairs, obliquely obovate, obtuse, truncate or retuse, often minutely apiculate, 3–6 mm. wide, 6–10 mm. long, midvein eccentric and with 1–2 additional veins nearly as prominent running diagonally upward from base, all veins prominulous; spikes 2–3 cm. long, on peduncles nearly as long, 1–4 in an axil; calyx about 2 mm. long, lobes flaring or erect, lanate-ciliate marginally; petals about 3 mm. long, lanate-ciliolate; pod flat, 1.5–2 cm. wide, 5–8 cm. long, more or less rounded at both ends, somewhat curved to tortulose, short-stipitate.

Rocky hillsides, desert canyons, and along washes, Lower Sonoran Zone, Texas to Nuevo León and Tamaulipas; central Baja California. March–June.

**6. Acacia californica** Brandegee, Proc. Calif. Acad. II, **3:** 221, 1892

*Acacia sonorensis* Rose, Contr. U.S. Nat. Herb. **8:** 31, 1903.
*Acaciopsis sonorensis* Britt. & Rose, N. Amer. Fl. **23:** 95, 1928.
*Acaciopsis californica* Britt. & Rose, op. cit. 96.

Tree to 12 m. tall with checked dark gray bark, puberulent foliage, and stiff, straight, stipular spines 1–2 cm. long; petioles 1–1.5 cm. long, bearing a flattened-crateriform gland just above base; pinnae mostly 1 pair; leaflets 2 pairs, broadly oval to short-oblong, 10–18 mm. wide, 1–2.5 cm. long, thin, rounded at apex, obtuse to obliquely and broadly cuneate at base, pinnately veined, the midvein slightly eccentric; spikes 4–10 cm. long, sessile, bearing several ovate bracts in a basal ring; flowers numerous, sessile, white or cream; calyx about 1.2 mm. long, glabrous; petals 2.5–3 mm. long, glabrous, tips spreading; stamens about 4 mm. long; pods linear, 4–5 mm. wide, 10–20 cm. long, glabrous, acute at both ends, short-stipitate, dark red-brown, longitudinally striate-veined; seeds elliptic-obovate, flattened, 8–10 mm. long.

Sandy or gravelly valley soils, Lower Sonoran Zone, southern Baja California and vicinity of Guaymas, Sonora. Feb.–March.

**7. Acacia brandegeana** I. M. Johnston, Contr. Gray Herb. **75:** 27, 1925

*Acaciopsis brandegeana* Britt. & Rose, N. Amer. Fl. **23:** 94, 1928.
*Pithecollobium acuminatum* M. E. Jones, Contr. West. Bot. **18:** 38, 1933.

A shrub or small tree to 4 or 5 m. tall, with gray, somewhat fissured bark, straight stipular spines to 1.5 cm. long and puberulent twigs and foliage; petioles 2–5 mm. long, bearing a flattened-crateriform gland about middle or below; pinnae 1 pair, 12–20 mm. long; leaflets 4–7 pairs, oblong, obtuse to rounded, minutely mucronulate, 1.5–3 mm. broad, 3.5–7 mm. long, prominently pinnate-veined, midvein only slightly eccentric; spikes sessile, 2–2.5 cm. long; flowers light yellow, sessile, crowded, glabrous; calyx about 0.8 mm. high; corolla 1.5–1.8 mm. long; stamens 4–5 mm. long; bractlets spoon-shaped, subfleshy, pods linear-falcate, 4–6 mm. broad, 4–15 cm. long, puberulent or eventually glabrate, acute; stipes 3–6 mm. long.

Sandy plains and arroyos, Lower Sonoran Zone, central Baja California south to the Cape Region. Feb.–March.

**8. Acacia occidentalis** Rose, Contr. U.S. Nat. Herb. **8:** 32, 1903

*Senegalia occidentalis* Britt. & Rose, N. Amer. Fl. **23:** 117, 1928.

Dense shrub or tree to 12 m. tall with gray reticulately fissured bark and densely puberulent, unguiculate-spined branches, spines 3–4 mm. long, dark red-brown; petioles 4–6 mm. long, densely puberulent or pilosulous; pinnae 2–4 pairs, 0.5–1.5 cm. long; leaflets 5–15 pairs, linear, 0.6–1 mm. broad, 3–5 mm. long, finely puberulent but evenually glabrate or with a few scattered marginal hairs, faintly veined, midvein eccentric; heads 1–5 in axillary clusters; peduncles 1–2 cm. long, pubescent; heads subglobose, about 1.5 cm. in diameter, dense; flowers glabrous, white to pale cream; calyx tubular-campanulate, about 1.5 mm. long; petals about 2.5 mm. long; pods flat, thin, about 2 cm. broad, 4–8 cm. long, curved to tortulose, irregularly constricted between some of the seeds, dark red-brown, faintly glaucous; seeds lenticular-suborbicular, 8–9 mm. wide.

Sandy plains, Lower Sonoran Zone, vicinity of Magdalena, Sonora, south into Sinaloa. Abundant near Hermosillo and southwest toward the coast. March–April.

**9. Acacia peninsularis** (Britt. & Rose) Standley, Field Mus. Bot. Ser. **11:** 158, 1936

*Senegalia peninsularis* Britt. & Rose, N. Amer. Fl. **23:** 116, 1928.
*Senegalia confusa* Britt. & Rose, loc. cit.

A shrub or small tree to 10 m. tall with a dense rounded crown, dark brown slightly fissured bark, and glabrous to puberulent foliage; stipular spines slightly recurved, 2–3 mm. long, dark brown; petioles 5–10 mm. long, bearing a small gland near middle; pinnae 1–2 pairs, 1.5–2.5 cm. long; leaflets 3–8 pairs, spatulate-obovate, 2.5–6 mm. broad, 5–10 mm. long, cuneate at base, rounded to truncate and minutely mucronulate apically, strongly 2-veined from base; flowers on pedicels 0.2–1 mm. long, greenish white, fragrant, closely crowded in globose heads on peduncles 1–3 cm. long, 1–5 heads in an axil; calyx 1.2–1.5 mm. long; petals 2.5–3 mm. long; stamens

5–6 mm. long; pods flat, oblong, 5–10 cm. long, obtuse to rounded at apex, glabrous, short-stipitate.

Along arroyos and on sandy plains, Lower Sonoran Zone, vicinity of Santa Rosalia, Baja California, to the Cape Region. March–April.

**10. Acacia constricta** Benth., in A. Gray, Pl. Wright. **1:** 66, 1852

*Acacia constricta paucispina* Woot. & Standl., Bull. Torrey Club **36:** 105, 1909.
*Acaciopsis constricta* Britt. & Rose, N. Amer. Fl. **23:** 96, 1928.
*Acaciopsis constricta* var. *paucispina* Moldenke, Rev. Sudam. Bot. **4:** 15, 1937.

A straggly open shrub 0.5–2 or occasionally 4 m. tall with puberulent and usually somewhat glandular twigs, pubescence soon disappearing from older twigs; spines linear-subulate, straight, 0.5–2 cm. long, grayish or reddish; leaves fascicled on older growth, 2.5–4 cm. long; petioles 6–10 mm. long, these and rachises pilosulous; pinnae 3–9 pairs; leaflets 4–16 pairs, linear-oblong, 1.5–3.5 mm. long, about 1 mm. broad, sparsely ciliolate marginally, otherwise glabrous, veins obscure; heads globose; peduncles about 1 cm. long, bearing a ring of minute bractlets about middle, puberulent; flowers yellow; calyx 1–1.5 mm. long; corolla about 1 mm. longer, both glabrous; pods linear, flattened, arcuate, 3–6 mm. wide, 5–12 cm. long, constricted between seeds, reddish brown, sparsely puberulent but soon glabrate, dotted with minute, transparent, reddish glands; seeds ovoid-lenticular, about 5 mm. long.

Along washes, on plains, and on rocky hillsides, Lower and Upper Sonoran Zones, western Pima County, Arizona, to Texas, Sonora, San Luis Potosí, and Puebla. May–Nov.

**11. Acacia farnesiana** (L.) Willd., Sp. Pl. **4:** 1083, 1806

*Mimosa farnesiana* L., Sp. Pl. 521, 1753.
*Acacia acicularis* Willd., Enum. 1056, 1809.
*Pithecollobium minutum* M. E. Jones, Contr. West. Bot. **18:** 38, 1933.

Shrub or small tree to 9 m. tall with glabrous to puberulent twigs and gray to whitish, straight, stipular spines 0.5–5 cm. long; leaves 2–10 (usually about 6) cm. long; petioles 5–15 mm. long, these and rachis pilosulous; pinnae 2–6 pairs, 1.5–3.5 cm. long; leaflets 8–25 pairs, linear to oblong, 2–6 mm. long, obtuse to rounded, pinnately veined, thickish, puberulent but usually glabrate except for marginal cilia; peduncles 1–4 cm. long, pubescent; receptacle globose; flowers yellow, moderately fragrant; calyx 1.5–2 mm. long, glabrous to puberulent, especially on lobes; petals 2.8–3.2 mm. long, puberulent on lobes; pods terete, turgid, 4–7 cm. long, 1–1.5 cm. in diameter, acute to apiculate longitudinally and finely reticulate-veined, tardily dehiscent, glabrous.

Along arroyos and on plains and hillsides, Sonoran Zones to Tropical Zone, Pima County, Arizona, to Florida and south to Argentina; Baja California; San Diego County, California (where it is probably introduced); Old World tropics. April–Nov.

**12. Acacia cymbispina** Sprague & Riley, in Riley, Kew Bull. **1923:** 394, 1923

*Acacia cochliacantha* S. Wats., Proc. Amer. Acad. **21:** 427, 1886. Not *A. cochliacantha* H.B.K., 1806.
*Acacia cymbiacantha* Zucc. ex Benth., Trans. Linn. Soc. **30:** 501, 1875. (Hyponym.)
*Poponax cymbispina* Britt. & Rose, N. Amer. Fl. **23:** 90, 1928.

A tree to 8 m. tall, often with several branches from base to form a flaring, vaselike habit; bark deeply fissured, light gray to almost black, lenticels numerous; spines straight, slightly flattened to cochleate-concave, 0.5–6 cm. long, smooth, light gray in age; twigs puberulent at first, then glabrous; leaves 4–12 cm. long; petioles about 1 cm. long, finely puberulent, later glabrate; pinnae 5–25 pairs, 1.5–2.5 cm. long; leaflets 15–35 pairs, closely crowded, oblong-linear, about 0.5 mm. broad, 1.5–2 mm. long, acute at apex, obliquely rounded at base, glabrous, vein median; peduncles 2–2.5 cm. long, bearing a minute suborbicular bract about 4–5 mm. below the head, puberulent; head globose, 6–7 mm. in diameter; calyx 0.4–0.5 mm. high; corolla 1.5–1.8 mm. long, both sparsely puberulent on lobes; pods linear-oblong, flattened, 8–12 mm. wide, 4–14 cm. long, acute or obtuse, cuneately narrowed to a stipe 1–1.5 cm. long, smooth and shining, dark red-brown, tardily dehiscent; seeds ovoid-lenticular, 5–6 mm. long, embedded in fleshy pulp.

Rocky hillsides and sandy or adobe plains, Lower Sonoran Zone, Sonora and Chihuahua to Vera Cruz, Oaxaca, and Guerrero. March–May, less commonly July–Nov.

**13. Acacia pennatula** (Cham. & Sch.) Benth., Lond. Jour. Bot. **1:** 390, 1842

*Inga pennatula* Cham. & Sch., Linnaea **5:** 593, 1830.
*Poponax pennatula* Britt. & Rose, N. Amer. Fl. **23:** 88, 1928.
*Pithecollobium minutissimum* M. E. Jones, Contr. West. Bot. **18:** 38, 1933.

Usually a tree 3–8 m. tall with a dense rounded crown and dark gray fissured bark; stipular spines stout, 0.5–4 cm. long; young twigs lanate with tawny fine hairs; leaves narrow, 3–4 cm. broad, 12–18 cm. long, rachises densely lanate; petioles stoutish, 1–1.5 cm. long, bearing flattened cup-shaped gland near base; pinnae to 50 or 60 pairs, 1.5–3 cm. long; leaflets 40–70 pairs, minute, linear, 1–2 mm. long, closely crowded and imbricate, puberulent, or in age only lanate-ciliate on margins; peduncles stoutish, often in axillary clusters, 1–2.5 cm. long, lanate; heads globose; flowers yellow, fragrant, densely crowded; calyx about 1.5 mm. long; corolla 2 mm. long; pods oblong, flattened, 1.5–2 (–2.5) cm. wide, 6–15 cm. long, brown-tomentulose at first, eventually glabrous or nearly so, black and smooth, valves woody, tardily dehiscent, fleshy within; seeds 6–7 mm. long.

Stony hillsides, margin of Subtropical Zone and Lower Sonoran Zone, foothills near Colorado, east of Hermosillo, Sonora, south to northern South America, east to Tamaulipas and Vera Cruz. Feb.–July.

**14. Acacia crinita** Brandegee, Zoe **5:** 198, 1905

*Acaciella crinita* Britt. & Rose, N. Amer. Fl. **23:** 105, 1928.

An unarmed shrub to 4 or 5 m. tall with yellow-pilose twigs, peduncles, and sometimes petioles, and glabrous leaflets; stipules lance-linear, 4–7 mm. long, hispid, deciduous; petioles 1.5–4.5 cm. long, usually hispid at base, often glabrous above, eglandular; pinnae 2, or less commonly 3, pairs 4–10 cm. long; leaflets 4–8 pairs, broadly elliptic, ovate, or obovate, 6–13 mm. wide, 9–20 mm. long, obtuse, rounded or retuse at apex, obliquely rounded at base, glabrous, dark green and shining above, pale, dull green and conspicuously pinnately reticulate-veined beneath; inflorescences often paniculate; heads globose, about 1.5 cm. in diameter; peduncles 2–3 cm. long; receptacle oblong; flowers pale yellow, glabrous; pedicels 1–1.5 mm. long, persistent; calyx broadly bowl-shaped, 0.4–0.6 mm. deep, 1.2–1.5 mm. broad; corolla 2.5–3 mm. long, petals basally fused into a tube 0.5–1 mm. long; stamens many, about 6 mm. long; pods oblong-linear, flattened, 7–10 mm. broad, 5–7 cm. long, acute and apiculate at apex, broadly cuneate to a slender stipe 5–10 mm. long, transversely reticulate-veined.

Rocky and usually shaded slopes, Lower Sonoran Zone, vicinity of Ures, Sonora, into Sinaloa and Durango. Sept.–March.

**15. Acacia goldmanii** (Britt. & Rose) Wiggins, Contr. Dudley Herb. **3**: 68, 1940

*Acaciella goldmanii* Britt. & Rose, N. Amer. Fl. **23**: 99, 1928.

A slenderly branched, essentially glabrous shrub 2–4 m. tall with reddish brown to gray bark and unarmed, lightly striate twigs; leaves 5–15 cm. long; petioles 1–2.5 cm. long, sulcate and sparsely strigose to glabrous; pinnae 5–8 pairs, 3.5–6 cm. long; leaflets 15–30 pairs, linear-oblong, 1–1.5 mm. broad, 4–7 mm. long, obliquely rounded at base, acute at apex, bright green above, paler beneath, faintly pinnately veined; inflorescences globose, 1.5–2 cm. in diameter, usually 2–4 in an axil; flowers cream; peduncles slender, 1–1.5 cm. long, glabrous or sparsely strigulose at base; pedicels about 2 mm. long, slender; calyx broadly cup-shaped, 0.8–1 mm. high, glabrous; petals 2.6–3 mm. long, glabrous; stamens 6 mm. long; pods oblong, 10–12 mm. broad, 3–5 cm. long, obtuse and apiculate at apex, broadly cuneate at base, transversely veined; stipe 5–8 mm. long; seeds ovoid-lenticular, greenish brown, about 4 mm. long.

Mostly in shade on rocky slopes and canyons, Lower Sonoran Zone, northern Sierra Giganta south to Cape Region, Baja California. March–May; Sept.–Oct.

**16. Acacia mcmurphyi** Wiggins, Contr. Dudley Herb. **3**: 68, *pl. 15, figs. 6–11,* 1940

An unarmed, ascendingly branched shrub to 5 m. tall with pilose branchlets and densely pilosulous to glabrate foliage; stipules linear-acuminate, 3–3.5 mm. long, deciduous; petioles 1.5–3 cm. long, eglandular or with a small flattened gland near apex; pinnae 3–9 pairs, 2.5–4 cm. long; leaflets 15–23 pairs, oblong, 1–2 mm. wide, 4–8 mm. long, obliquely rounded basally, acute to obtuse at apex, distinctly veined beneath, surfaces glabrate but margin ciliate; inflorescence globose-capitate, 1–1.5 cm. in diameter; peduncles slender, 5–15 cm. long; pedicels stout, about 0.5 mm. long, bearing numerous, minute, reddish brown, vesicular glands; calyx broadly turbinate, 1.5–1.7 mm. wide, 2–2.2 mm. long, glabrous except for ciliate margins of lobes; petals 5 mm. long, granular-puberulent at tips; stamens 8–10 mm. long; pods narrowly oblong, acute at both ends, 6–7 mm. wide, 4–6 cm. long; seeds elliptic-lenticular, about 3 mm. long.

Rocky canyons and hillsides, Lower Sonoran Zone, central Baja California. April–Oct.

**17. Acacia angustissima** (Mill.) Kuntze, Rev. Gen. Pl. **3**: 47, 1898, subsp. **angustissima**

*Mimosa angustissima* Mill., Gard. Dict. ed. 8, no. 19, 1768.
*Acacia glabrata* Schlecht., Linnaea **12**: 569, 1838.
*Acaciella breviracemosa* Britt. & Rose, N. Amer. Fl. **23**: 99, 1928.
*Acaciella angustissima* Britt. & Rose, op. cit. 100.
*Acaciella salvadorensis* Britt. & Rose, op. cit. 101.
*Acacia angustissima* subsp. *typica* Wiggins, Contr. Dudley Herb. **3**: 229, 1942.

Openly branched shrub or small tree 2–4 m. tall with striate-angulate, glabrous to hirsute branches; stipules linear to subulate, 4–8 mm. long, fugaceous; leaves 10–25 cm. long, the rachis glabrous to moderately hirsute; pinnae 9–25 pairs, 2.5–4.5 cm. long; leaflets 30–60 pairs or more, linear, 4–6.5 mm. long, glabrous on both sides, appressed-ciliate on margins, slightly eccentrically 1-veined, acute; heads globose, in terminal and axillary racemes 1–3 dm. long; peduncles 1–2.5 cm. long, striate, glabrous to hirsutulous; pedicels slender, about 1 mm. long; calyx broadly turbinate, 0.8–1 mm. high, glabrous or with a few appressed hairs near base; petals 1.8–2.4 mm. long, acute, glabrous, imperfectly separated near base; stamens white or pale cream, rarely faintly pinkish, 5–6 mm. long; pod 8–12 mm. broad, 4–7 cm. long, broadly cuneate and short-stipitate at base, acute and often slightly apiculate at apex, reddish brown, glabrous or bearing a few scattered simple hairs; seeds ovoid, 3.5–4 mm. long, two-thirds as broad, moderately flattened, brownish mottled with buff or greenish flecks, sublustrous.

Hillsides and gravelly slopes, Lower Sonoran Zone, central Sonora to Texas and south to Guatemala and Costa Rica. April–Oct.

**17*a*. Acacia angustissima** subsp. **lemmonii** (Rose) Wiggins, Contr. Dudley Herb. **3:** 230, 1942

*Acacia lemmonii* Rose, Contr. U.S. Nat. Herb. **12:** 409, 1909.
*Acaciella lemmonii* Britt. & Rose, N. Amer. Fl. **23:** 103, 1928.
*Acaciella shrevei* Britt. & Rose, op. cit. 105.
*Acacia hirta* var. *shrevei* Kearney & Peebles, Jour. Wash. Acad. **29:** 482, 1939.

This subspecies differs from subsp. *angustissima* in having slightly larger leaflets, broader pods, fewer pinnae per leaf, and delicately but distinctly pinnate-veined leaflets.

Mountains of southern Arizona into northern Sonora, mostly in the Upper Sonoran Zone, but well into the desert in north-central Sonora.

**17*b*. Acacia angustissima** subsp. **suffrutescens** (Rose) Wiggins, Contr. Dudley Herb. **3:** 232, 1942

*Acacia suffrutescens* Rose, Contr. U.S. Nat. Herb. **12:** 409, 1909.
*Acaciella suffrutescens* Britt. & Rose, N. Amer. Fl. **23:** 103, 1928.
*Acacia hirta* var. *suffrutescens* Kearney & Peebles, Jour. Wash. Acad. **29:** 482, 1939.

The plants of subsp. *suffrutescens* are rarely woody except at the very base, plants usually not over 1 m. tall and dying back each year nearly or quite to the ground; twigs, petioles, rachises, and peduncles sparsely pilose; pinnae 5–9 pairs, 1.5–3 cm. long; leaflets linear, 3.5–5 mm. long, 0.8–1 mm. wide, 1-veined or faintly pinnately veined, margins sparsely appressed-ciliate; pod 3–6 cm. long, 6–8 mm. wide, sparsely stigulose to nearly glabrous.

Mainly along watercourses, Lower Sonoran Zone, Pima County, Arizona, to New Mexico and northern Sonora.

## 7. PROSOPIS L., Mant. 1: 10, 1767

Spiny shrubs or trees with bipinnate, deciduous leaves bearing small crateriform glands between the paired leaflets. Flowers perfect, small, numerous, in capitate clusters or spikelike racemes, or sessile and in dense spikes. Calyx 5-toothed. Corollas regular, 5 petals distinct or sympetalous. Stamens 10, exserted; anthers terminally glandular. Ovary pubescent; style straight or curved; stigma concave. Fruit indehiscent or tardily dehiscent, usually with fleshy or gummy dissepiments between the seeds, curved, undulate, or tightly coiled spirally. Seeds few to many, compressed and often angled, hard, smooth.

Corollas sympetalous, at least to apex of calyces; spines truly stipular, attached laterally to bases of petioles; petals densely puberulent without, glabrous within; flowers sessile:
- Pods spirally coiled; leaflets green or gray-green but not markedly glaucous; stamens 4–6 mm. long . . . . . . . . . . . . . . . . . . . . 1. *P. pubescens*
- Pods straight or curved; leaflets strongly glaucous; stamens 8–10 mm. long . . 2. *P. palmeri*

Corollas choripetalous, or some weakly coalescent at base; spines axillary to petioles; petals glabrous without, villous to lanate within; flowers pedicellate:

Pods glabrous or puberulent but not glandular-scurfy; leaflets acute to rounded at apex, upper ones not markedly broader than lower ones on each pinnule:

Leaflets linear to narrowly oblong, four to nine times as long as broad; twigs, pods, and young herbage glabrous or sparsely ciliolate:

Calyx teeth, leaflets, and young twigs ciliolate; leaflets mostly 3–6 mm. wide, four to six times as long as broad .................................... 3. *P. juliflora*

Calyx teeth, leaflets, and young twigs usually glabrous; leaflets 2–3 mm. wide, six to nine times as long as broad .................... 3*a*. *P. juliflora* var. *torreyana*

Leaflets three to four times as long as broad; twigs, pods, and young herbage densely puberulent:

Pods slightly compressed but turgid, not at all or only slightly and irregularly constricted between the seeds ...................... 3*b*. *P. juliflora* var. *velutina*

Pods strongly compressed, regularly and deeply constricted between seeds .................................................. 3*c*. *P. juliflora* var. *articulata*

Pods glandular-scurfy, undulate; leaflets rounded to emarginate, upper ones on each pinnule often nearly orbicular, broader and shorter than lower ones ..... 4. *P. pazensis*

**1. Prosopis pubescens** Benth., Lond. Jour. Bot. **5:** 82, 1846

*Prosopis odorata* Torr. & Frem., in Frem., Second Rep. 313, 1845. (Nomen confusum.)
*Strombocarpa pubescens* A. Gray, Pl. Wright. **1:** 60, 1852.
*Strombocarpa odorata* A. Gray, Bot. Wilkes Exp. 475, 1854.

Openly branching tree to 10 m. tall with thin flaky bark, slender twigs, and subulate stipular spines 0.8–2 cm. long and puberulent herbage; petioles slender, 0.5–2 cm. long; pinnae 1 or 2 pairs, 2–3 cm. long; leaflets 4–9 pairs, oblong, obtuse, 2–5 mm. wide, 6–10 mm. long, finely puberulent to glabrate, obscurely veined; peduncles 1–3 cm. long; spikes dense, 2–5 cm. long, rachises puberulent; flowers sessile, finely sericeous-puberulent without; calyx broadly bowl-shaped, about 1 mm. long, teeth minute; petals 2.5–3 mm. long, glabrous within; stamens 4–6 mm. long; pods tightly coiled spirally, sessile, 5–7 mm. in diameter, 3–6 cm. long, puberulent; seeds 1.5–2 mm. long.

Stream banks, bottomlands, and about water holes, Lower Sonoran Zone, Mojave and Colorado Deserts east to southern Utah and south into Texas, Chihuahua, Sonora, and Baja California. Also in western Fresno County, California. April–June.

**2. Prosopis palmeri** S. Wats., Proc. Amer. Acad. **24:** 48, 1889

*Sopropis palmeri* Britt. & Rose, N. Amer. Fl. **23:** 183, 1928.

Open shrub or broad-crowned tree to 6–8 m. tall with dark gray fissured bark and puberulent glaucous foliage; spines lateral to or just below leaf scars, subulate, 0.5–2.5 cm. long, white or light gray; petioles 3–4 mm. long, bearing a terminal subulate appendage; pinnae 1 pair, 1.5–3 cm. long; leaflets 3–10 pairs, broadly oblong to elliptic, 1.5–3 mm. wide, 3–6 mm. long, rounded, minutely mucronulate, obscurely veined; peduncles 0.5–2 cm. long; spikes dense, 2–6 cm. long; flowers sessile, fragrant, yellow; calyx 1.5–2 mm. long, puberulent; corolla 4–4.5 mm. long, sympetalous, puberulent without, glabrous within; stamens 8–10 mm. long, filaments stout; pod about 1 cm. in diameter, 5–12 cm. long, curved, moderately compressed but turgid, sometimes nearly terete in cross section, minutely puberulent; seeds 8–30, brown, smooth, shining, flattened or irregularly angled, 5–7 mm. long, more or less separated by a thick, gummy, reddish brown substance.

Along arroyos and canyons and on desert playas, Lower Sonoran Zone, Baja California south of Mulegé and Purísima. Sept.–March.

**3. Prosopis juliflora** (Sw.) DC., Prod. **2:** 447, 1825, var. **juliflora**

*Mimosa juliflora* Sw., Prod. Veg. Ind. Occ. 85, 1788.
*Acacia juliflora* Willd., Sp. Pl. **4:** 1076, 1806.
*Mimosa salinarum* Vahl, Eclog. Amer. **3:** 35, 1807.
*Prosopis domingensis* DC., Prod. **2:** 447, 1825.
*Neltuma juliflora* Raf., Sylva Tell. 119, 1838.
*Algarobia juliflora* Benth. ex Heynh., Nom. **2:** 18, 1840.
*Neltuma occidentalis* Britt. & Rose, N. Amer. Fl. **23:** 185, 1928.
*Prosopis chilensis* in part, of North American authors. Not *P. chilensis* (Molina) Stuntz, based on *Certonia chilensis* Molina.

Tree to 12 m. with a broad rounded crown, smooth or lightly fissured brown bark, and nearly glabrous twigs; spines axillary, 1–4 cm. long; petioles 1–4 cm. long, glabrous or ciliolate in youth; pinnae 1 or 2 pairs, 3–8 cm. long; leaflets 8–20 pairs, linear-oblong, 8–16 mm. long, obtuse to mucronulate at apex, strongly veined, glabrous or sparsely ciliolate along margins and on midvein; flowers greenish yellow, in dense spiciform racemes 5–12 cm. long; calyx 0.8–1 mm. high, ciliolate on minute teeth; petals 2.5–3 mm. long, villous within, glabrous without; stamens 4–5 mm. long; pod compressed, 1–2.5 dm. long, 1–1.5 cm. wide, somewhat falcate, glabrous; seeds flattened, brown, smooth.

Flats, lower mountain slopes, and along streams and arroyos, Lower Sonoran Zone, throughout most of Mexico and the West Indies, entering the Sonoran Desert in southern Baja California and possibly in southern Sonora. April–Aug.

***3a.* Prosopis juliflora** var. **torreyana** L. Benson, Amer. Jour. Bot. **28:** 751, *fig. 4,* 1941

*Prosopis juliflora* var. *glandulosa* of authors, in part.

Similar to var. *juliflora* but with comparatively narrower, longer, glabrous leaflets and glabrous instead of ciliolate calyx teeth.

Valleys and desert slopes and arroyos, Lower Sonoran Zone, Colorado and Mojave Deserts to southern Nevada, southern Texas, and northern Mexico; south into Baja California at least to Comondú. Introduced along the western side of the San Joaquin Valley, California.

***3b.* Prosopis juliflora** var. **velutina** (Wooton) Sarg., Silva **13:** 15, *pl. 628,* 1902

*Prosopis velutina* Wooton, Bull. Torrey Club **25:** 456, 1898.
*Prosopis chilensis velutina* Standley, Contr. U.S. Nat. Herb. **23:** 1658, 1926.
*Neltuma velutina* Britt. & Rose, N. Amer. Fl. **23:** 186, 1928.

Leaflets 6–13 mm. long, not over four to six times as long as broad; young twigs, petioles, leaflets, rachises, and calyces finely but closely puberulent.

Desert washes and plains, Lower Sonoran Zone, Yavapai County, Arizona, east and south to western Texas and northern Mexico. Common in Sonora at least as far south as Guaymas.

***3c.* Prosopis juliflora** var. **articulata** (S. Wats.) Wiggins, Contr. Dudley Herb. **4:** 17, 1950

*Prosopis articulata* S. Wats., Proc. Amer. Acad. **24:** 48, 1889.
*Neltuma articulata* Britt. & Rose, N. Amer. Fl. **23:** 187, 1928.

Similar to var. *velutina* in foliage but pods regularly and deeply constricted between seeds and more strongly flattened than in the other varieties, 7–9 mm. wide.

Desert flats, Lower Sonoran Zone, west-central Sonora, and, in southern Baja California and the adjacent islands in the Gulf of California.

**4. Prosopis pazensis** (Britt. & Rose) Wiggins, Contr. Dudley Herb. **4:** 18, 1950

*Neltuma pazensis* Britt. & Rose, N. Amer. Fl. **23:** 187, 1928.

Shrub or small tree to 5 m. high with slender, greenish, glabrous twigs and yellow-green, glabrous, or sparsely puberulent but eventually glabrate herbage; spines straight, 4–8 mm. long, greenish or white; petioles 1–2.5 cm. long; pinnae 1 or 2 pairs, 2.5–8 cm. long, rachises angled, glabrous or glabrate; leaflets 8–20 pairs, lower ones oblong, 1–1.5 cm. long, upper obovate to suborbicular, often not over 2–4 mm. long, deeply emarginate, strongly veined; peduncles to 3 cm. long; racemes dense, 3–8 cm. long; pedicels 0.4–1 mm. long; calyx 0.8–1 mm. long, about 1.5 mm. broad, teeth triangular, ciliolate or glabrous; petals lance-oblong, 1.8–2 mm. long, sparsely villous below and densely lanate-villous at tip within; stamens 5–6 mm. long; pod 20–25 cm. long, 7–9 mm. wide, sinuately curved, strongly compressed, slightly constricted between seeds, cuneately narrowed at base to a stipe 5–8 mm. long, minutely glandular-scurfy.

Sandy plains, Lower Sonoran Zone, from San Evaristo Bay, Baja California, southward to the tip of the Cape. Apparently flowering at any season after good rains.

Marshall C. Johnson, in Brittonia **14:** 72–90, 1962, has presented a preliminary realignment of the mesquites that departs considerably from the above. Further supporting evidence is awaited before his conclusions can be evaluated.

## 8. NEPTUNIA Lour., Fl. Cochinch. 653, 1790

Floating or prostrate to ascending, perennial, unarmed herbs or shrubs with bipinnate leaves. Flowers perfect and imperfect in dense pedunculate heads, the staminate imperfect flowers below the perfect; peduncles often bearing 1–2 ovate bracteoles near base. Calyx campanulate, with 5 short, broadly triangular teeth. Petals 5, distinct or united near base. Stamens usually 10, distinct, exserted; anthers bearing a globose stalked gland terminally between thecae. Ovary short-stipitate, several- to many-ovuled. Style filiform; stigma capitate. Pod broadly linear to oblong, flat, impressed between seeds, frequently curved, 2-valved. Seeds lenticular, hard, smooth.

**1. Neptunia plena** (L.) Benth., in Hook., Jour. Bot. **4:** 355, 1842

*Mimosa plena* L., Sp. Pl. 519, 1753.
*Desmanthus plenus* Willd., Sp. Pl. **4:** 1045, 1806.
*Desmanthus punctatus* Willd., op. cit. 1047.
*Desmanthus comosus* A. Rich., Ess. Fl. Cub. 475, 1845. (According to Griseb.)

Suffruticose, prostrate to ascending perennial with slender branches 1–6 dm. long, glabrous throughout; stipules broadly ovate, 5–7 mm. wide, 8–12 mm. long, thin, crispate-margined, oblique at base, short-attenuate at apex; leaves 1 dm. long or less, petiole 1–2.5 cm. long; pinnae 2–4 pairs, 2–3.5 cm. long, 1–1.5 cm. apart on rachis; leaflets 10–20 pairs, oblong-linear, 1–2 mm. wide, 4–10 mm. long, obtuse, slightly glaucescent above, dull beneath, obscurely 1-veined; peduncles slender, 5–12 cm. long, usually with 1–2 cordate bracts similar to stipules at or below middle; flower heads ovoid, 1–2 cm. long; flowers yellow, lower imperfect ones with stamens or staminodia 5–8 mm. long, well exserted; pods 7–8 mm. wide, 2–5 cm. long, slightly falcate, acute or short-apiculate; seeds ovoid-lenticular, about 4 mm. long, dark brown.

In moist places, chiefly in Tropical Zone, Cape Region, Baja California, to northern South America and the West Indies. Sept.–Oct.

## 9. MIMOSA L., Sp. Pl. 516, 1753

Herbs, vines, shrubs, or trees with bipinnate, sometimes sensitive, leaves and small, regular, 4–5-merous, perfect flowers in pedunculate heads or spiciform racemes, each flower subtended by a caducous or deciduous bracteole. Calyx small, with minute teeth; corolla lobes valvate. Stamens 4, 5, 8, or 10, distinct, well-exserted; anthers small, glandless. Ovary sessile, 1–many-ovulate. Pods linear to oblong, flat, wingless, transversely jointed, the joints usually separating from the continuous marginal cords; seeds lenticular, smooth.

Inflorescence of globose or subglobose heads; bracteoles persisting until anthesis or later:
- Pinnae 1 pair; flowers mostly 4-merous; corolla four to six times as long as calyx; leaflets 6–11 mm. long:
  - Petioles, leaflets, and peduncles pubescent and prickly-setose; bracteoles fimbriate-ciliate, exceeding the corollas .......... 1. *M. xanti*
  - Petioles, leaflets, and peduncles setose-prickly but without finer hairs; bracteoles not fimbriate, shorter than corollas .......... 2. *M. margaritae*
- Pinnae 3–10 pairs; corolla lobes and calyx lobes usually 5; stamens 10; corolla one and one-half to two times as long as calyx; leaflets 1.5–6 mm. long:
  - Leaflets distinctly pinnately and reticulately veined; pods 7–10 mm. wide, not constricted between seeds; leaflets 4–6 mm. long:
    - Pods glabrous or nearly so, sparingly prickly along margins or spineless; calyx lobes glabrous, nonciliate .......... 3. *M. grahamii* var. *grahamii*
    - Pods, twigs, and leaflets densely pubescent; pods copiously spinescent marginally; calyx lobes puberulent, ciliate .......... 3*a*. *M. grahamii* var. *lemmonii*
  - Leaflets 1-veined, lateral veins obscure; pods 3–5 mm. wide, lightly to deeply constricted between seeds; leaflets 1.5–4 mm. long .......... 4. *M. biuncifera*

Inflorescence spicate; bracteoles deciduous early, seldom persisting until anthesis:
- Flowers on pedicels 0.5–1.5 mm. long, glabrous; calyces less than one-fifth as long as corolla; legumes stipitate in fruit:
  - Pods glabrous, unarmed; branches few-spined .......... 5. *M. laxiflora*
  - Pods puberulent, setose-prickly on margins and sides; branches many-spined .......... 6. *M. brandegeei*
- Flowers sessile or pedicels less than 0.5 mm. long, sparsely to densely puberulent, at least on calyx lobes and corolla lobes; calyx about one-fourth to one-half as long as corolla; pods sessile:
  - Leaflets obovate to suborbicular, rarely twice as long as broad, sparsely puberulent to glabrous; pods sparsely puberulent to glabrate .......... 7. *M. purpurascens*
  - Leaflets linear, two to four times as long as broad, densely pubescent, at least beneath; pods densely pubescent, at least in youth:
    - Pods usually with a few marginal spines; leaflets densely pubescent on both surfaces; corollas 3.5–4 mm. long; stem spines yellowish .. 8. *M. dysocarpa* var. *dysocarpa*
    - Pods unarmed; leaflets glabrous to glabrate above; corollas 2.5–3 mm. long; stem spines brown or black .......... 8*a*. *M. dysocarpa* var. *wrightii*

### 1. Mimosa xanti A. Gray, Proc. Amer. Acad. **5:** 157, 1861

*Mimosa lagunensis* M. E. Jones, Contr. West. Bot. **18:** 39, 1933.

A rounded, stiffly branched shrub 1–2 m. high with densely hispid-strigose twigs sparingly armed with straight, yellowish or greenish spines 4–8 mm. long; stipules setaceous, 4–6 mm. long; petioles 1–2.5 cm. long, hispid; pinnae 1 pair, 3–6 cm. long; leaflets 4–9 pairs, oblong, acute to obtuse and apiculate, 3–8 mm. wide, 8–15 mm. long, appressed-pubescent on both surfaces, dark green above, paler and dull beneath, reticulate-veined, margins with appressed setose bristles; peduncles 1–3 cm. long, pilose; flower heads 1.5–2 cm. in diameter; calyx about 0.2–0.5 mm. long; corolla 2–2.5 mm. long, 4-parted, puberulent on lobes; stamens 4, about 6 mm. long, pink; pod oblong, 2–2.5 cm. long, 2–4-jointed, apiculate, finely pubescent and setose-bristly on margins and sides.

On sandy plains and along washes, Lower Sonoran Zone, southern Baja California. March–May; Sept.–Oct.

**2. Mimosa margaritae** Rose, in Britt. & Rose, N. Amer. Fl. **23:** 171, 1928

Slenderly branched shrub with reddish brown bark which soon becomes finely fissured, glabrous; spines slightly recurved, 2–8 mm. long, flattened at base, in pairs or borne singly; stipules subulate, 2–4.5 mm. long, setose-hirsute and somewhat glandular; petioles 0.5–1.5 cm. long, short-hispid; pinnae 1 pair, 3–4 cm. long; leaflets 3–12 pairs, lanceolate-oblong to linear-oblong, oblique and unequal at base, acute to obtuse and apiculate at apex, 1.5–4 mm. wide, to 9 mm. long, glabrous, shining above, dull beneath, margins and sometimes lower surface armed with stout, rigid, yellowish, appressed prickles 0.2–0.8 mm. long; peduncles slender, 1.5–3 cm. long, glabrous; heads globose, about 1 cm. in diameter; calyx minute, about 0.5 mm. long, glabrous; corolla 2.4–2.8 mm. long, lobes 4, glabrous or minutely puberulent at apex; stamens 4, about 6–8 mm. long, pale lavender; pods linear-oblong, 2–3 cm. long, about 5 mm. wide, glabrous but armed marginally and sparsely so on sides with stout, subconical, stiff prickles; beak 3–5 mm. long.

Sandy arroyos, Lower Sonoran Zone, Santa Margarita Island and the Cape Region, Baja California. April–May.

**3. Mimosa grahamii** A. Gray, Pl. Wright. **2:** 52, 1853, var. **grahamii**

*Mimosopsis grahami* Britt. & Rose, N. Amer. Fl. **23:** 178, 1928.

A low spreading shrub with prickly, glabrous, slightly angled twigs and glabrous foliage; petioles 8–10 mm. long; rachis bearing stiff short prickles; pinnae 6–8 pairs; leaflets 8–15 pairs, oblong, acute or rounded at apex, 4–6 mm. long, glabrous or sparsely and minutely puberulent, distinctly reticulate-veined; peduncles slender, 2–3.5 cm. long; heads about 1.5 cm. in diameter; calyx 0.8–1 mm. long, glabrous or sparsely ciliolate; corolla about 3 mm. long, glabrous or minutely and sparsely puberulent at apex; stamens 7–9 mm. long; pods oblong, rounded at apex, 7–9 mm. broad, 3–4.5 cm. long, glabrous or nearly so, sparsely prickly along margins or unarmed.

Rocky hillsides and washes, chiefly Upper Sonoran Zone, Cochise, Pima, and Santa Cruz Counties, Arizona, to southeastern New Mexico and adjacent Mexico. Rarely entering the upper limits of the desert in arroyos. April–Aug.

***3a.* Mimosa grahamii** var. **lemmonii** (A. Gray) Kearney & Peebles, Jour. Wash. Acad. **29:** 482, 1939

*Mimosa lemmoni* A. Gray, Proc. Amer. Acad. **19:** 76, 1883.
*Mimosopsis lemmonii* Britt. & Rose, N. Amer. Fl. **23:** 176, 1928.

Differs from var. *grahamii* only in being densely pubescent on the twigs, rachises, peduncles, leaflets, and inflorescences, and in having more copiously spinescent pods. The range is similar to that of var. *grahamii*.

**4. Mimosa biuncifera** Benth., Pl. Hartw. 12, 1839

*Mimosa biuncifera* var. *glabrescens* A. Gray, Pl. Wright. **2:** 51, 1853.
*Mimosopsis biuncifera* Britt. & Rose, N. Amer. Fl. **23:** 176, 1928.

A straggly shrub about 1 m. tall with angled pubescent twigs and pubescent or glabrate foliage; spines paired or solitary, somewhat recurved, 5–8 mm. long, broad at base, pubescent in youth but soon glabrate, yellowish or pale brown; petioles 1–5 mm. long; pinnae 2–10 pairs, 5–15 mm. long; leaflets 5–12 pairs, crowded, pubescent, strongly reticulate-veined beneath, linear-oblong, 1.5–4 mm. long; peduncles often equaling or slightly exceeding subtending leaf; calyx and corolla puberulent, corolla about 2 mm. long; pod linear-oblong, usually falcate, 3–5 mm. wide, 3–4 cm. long, prickly along lower margin, puberulent to glabrate.

Desert arroyos and dry rocky hillsides, Sonoran Zones, Apache and Pima Counties, Arizona, to western Texas and south into Sonora and Chihuahua. May–Aug.

**5. Mimosa laxiflora** Benth., Lond. Jour. Bot. **5:** 93, 1846

An openly branched shrub to 3 m. tall with slender glabrous twigs sparsely armed with small, yellowish brown, curved spines 2–5 mm. long; bark pale yellow, later gray; stipules setaceous, yellow, 2–3 mm. long; petioles slender, 2–4 cm. long; pinnae 2–4 pairs, 1.5–3 cm. long; leaflets distant, 2–5 pairs, obovate to broadly elliptic, obtuse to rounded, 6–12 mm. long, pale green above, lighter and subglaucescent beneath, glabrous or sparsely puberulent when young, more or less 2–3-veined from base; racemes spikelike, paired or solitary, 1.5–5 cm. long; peduncles 5–10 mm. long; pedicels 0.5–1.5 mm. long; calyx 0.2–0.4 mm. long; corolla 1–1.5 mm. long, glabrous; stamens pale lavender, 6–8 mm. long; pod linear-oblong, somewhat tortulose, 3–5 cm. long, 5–6 mm. broad, glabrous, breaking transversely into 1-seeded segments, stipe 3–4 mm. long.

Along arroyos and on alluvial flats, Lower Sonoran Zone, Pima County, Arizona, to Sinaloa and Chihuahua. Aug.–Sept.

**6. Mimosa brandegeei** Robinson, Proc. Amer. Acad. **33:** 316, 1898

A slenderly branched, open shrub 2–3 m. tall with scattered yellowish spines 3–6 mm. long and finely puberulent to glabrate twigs and herbage; stipules subulate, 2–3.5 mm. long; petioles slender, 2–2.5 cm. long, finely puberulent to glabrate, sometimes sparsely prickly; pinnae 2–4 pairs, 1.5–3 cm. long; leaflets 2–4 pairs, broadly obovate, cuneate at base, rounded at apex, minutely mucronulate, 3–8 mm. wide, 5–12 mm. long, minutely appressed-puberulent on both surfaces, the upper at length glabrate, pinnae and leaflets distant, whole leaf often 10 cm. long; inflorescence spiciform, 2.5–7 cm. long; peduncles slender, 1–2 cm. long; pedicels about 1 mm. long; calyx about 0.6 mm. long, puberulent; corolla 1.8–2 mm. long, minutely appressed-puberulent, pinkish; stamens 7–8 mm. long; pods narrowly oblong, acute at apex, 2.5–4 cm. long, closely puberulent and somewhat setose-prickly.

Brushy hillsides and dry canyons, Lower Sonoran and Subtropical Zones, southern Baja California and Sinaloa. Sept. and probably other months.

**7. Mimosa purpurascens** Robinson, Proc. Amer. Acad. **33:** 317, 1898

A straggly, openly branched shrub 2–3 m. tall with slender, tawny, finely puberulent twigs sparingly armed with recurved, broad-based spines 3–6 mm. long; stipules linear-subulate, 3–5 mm. long; petioles 1–2.5 cm. long; pinnae 2–5 pairs, 1–2.5 cm.

long, lower shortest; leaflets 2–6 pairs, broadly elliptic, oblong, or obovate, acute to rounded, often apiculate, 2–7 mm. broad, 4–12 mm. long, appressed-puberulent on both surfaces, faintly glaucous beneath, 1–3-veined from base; inflorescences paired or solitary, 3.5–5 cm. long; peduncles 1.5–2 cm. long; flowers subsessile, densely strigose; calyx 0.8–1 mm. long, broadly cup-shaped; corolla 2–2.4 mm. long; stamens 8, pale lavender, fading to white, 6–7 mm. long; pods linear-oblong, often somewhat falcate, 5–6 mm. broad, 4–5 cm. long, acute, finely but densely puberulent, sessile, 6–10-seeded, breaking transversely into as many squarish segments; seeds dark brown, 2.5–3 mm. in diameter.

Rocky canyons and hillsides, Lower Sonoran Zone, vicinity of Bahía de la Concepción, Baja California, south to the Cape Region. Oct.–May.

**8. Mimosa dysocarpa** Benth., in A. Gray, Pl. Wright. **1:** 62, 1852, var. **dysocarpa**

A straggly shrub 0.5–2 m. tall with strongly angled, heavily spined, brown-villous branches and puberulent foliage; petioles 2–8 mm. long; pinnae 5–12 pairs, 1–2 cm. long; leaflets 5–18 pairs, linear-oblong, acute, 1–1.5 mm. wide, 3–4 mm. long, appressed, pubescent on both surfaces, dark green above, paler beneath; inflorescences 3–10 cm. long; flowers subsessile, purplish pink, strigose; calyx 1.6–2 mm. long; corolla 3.5–4 mm. long; stamens 10–12 mm. long, pinkish; pods linear-oblong, often somewhat falcate, densely pubescent, 5–6 mm. wide, 2.5–5 cm. long, somewhat constricted between seeds, sessile, prickly along margins.

Canyons and hillsides, mainly Upper Sonoran Zone, south-central Arizona to Texas and south to Durango. Barely enters the limits of the Sonoran Desert. July–Sept.

**8*a*. Mimosa dysocarpa** var. **wrightii** (A. Gray) Kearney & Peebles, Jour. Wash. Acad. **29:** 482, 1939

*Mimosa wrightii* A. Gray, Pl. Wright. **2:** 52, 1853.

Differs from var. *dysocarpa* in having unarmed pods, leaflets glabrous or glabrate above, darker spines on twigs, and slightly smaller corollas. Southern Arizona to Sonora and Chihuahua.

## 10. LEUCAENA Benth., in Hook., Jour. Bot. 4: 416, 1842

Trees or shrubs with bipinnate leaves, unarmed branches, glanduliferous petioles, and small white flowers in dense pedunculate heads. Calyx 5-toothed, narrowly campanulate. Petals 5, distinct. Stamens 10, distinct, well exserted, often pubescent; anthers without glands. Ovary short-stipitate. Pod linear to oblong, flat, 2-valved, thin. Seeds ovate to obovate, flat, smooth.

Leaflets distinctly puberulent on both sides; legumes finely puberulent, 12–20 mm. wide:
- Leaflets two and one-half to three times as long as broad; apex of legume acute, the tip not eccentric; rachilla tipped with conic gland about 1 mm. long ... 1. *L. lanceolata*
- Leaflets not over twice as long as broad; pod rounded at apex, tip eccentric; rachilla tipped with subulate gland 2–3 mm. long ............................ 2. *L. brandegeei*

Leaflets glabrous or sparsely puberulent along veins beneath and on margins; legumes glabrous, 10–15 mm. wide:
- Pinnae 4–5 pairs; leaflets 3–6 pairs, oblong-lanceolate, acute; pod 12–22 cm. long ........ ........................................................ 3. *L. palmeri*
- Pinnae 2–3 pairs; leaflets 1–3(–6) pairs, elliptic, obovate, or suborbicular, obtuse or rounded at the apex; pod 6–13 cm. long .............................. 4. *L. microcarpa*

**1. Leucaena lanceolata** S. Wats., Proc. Amer. Acad. **21:** 427, 1886

*Leucaena sonorensis* Britt. & Rose, N. Amer. Fl. **23:** 122, 1928.

Large shrub or small tree to 15 m. high with grayish bark on main trunk, brownish on branches; twigs and leaves finely pubescent; pinnae 2–5 pairs; petioles 1.5–5 cm. long, bearing an elliptic, flattened or conical gland about halfway between base and lower pair of pinnae; leaflets 3–7 pairs, broadly lanceolate or lower ones obovate to ovate, acute to short-acuminate or apiculate, 0.8–1.5 cm. wide, 1.5–4 cm. long, minutely puberulent on both surfaces, bright green above, paler beneath, distinctly veined; rachillae tipped with a slender conical gland about 1 mm. long; peduncles 1–5 cm. long, stoutish in fruit; involucre appressed to base of head; bracteoles orbicular; calyx 2–2.5 mm. long, glabrous; petals about twice as long as calyx; stamens 8–9 mm. long; anthers pubescent; legumes broadly linear, 1.5–2 cm. wide, 15–20 cm. long, reticulate-veined, somewhat shining, acute at apex, narrowed to a stipe 1–2 cm. long at base, reddish brown, finely puberulent in youth, but eventually glabrate; seeds ellipitc, 7–8 mm. long, reddish brown, shining.

Along canyons and in valleys, southern Sonora, Chihuahua, and northern Sinaloa. Absent in drier parts of the Sonoran Desert. Aug.–Sept.

**2. Leucaena brandegeei** Britt. & Rose, N. Amer. Fl. **23:** 122, 1928

Small tree with smooth bark and densely pilose twigs, rachises, petioles, and peduncles; petioles 1–3 cm. long, bearing an elliptic flattened gland just below lower of the 3–6 pairs of pinnae; leaflets 3–6 pairs, ovate, obovate, or broadly elliptic, 1.6–4 cm. long, usually about half as wide, acute or abruptly apiculate, broadly cuneate at base, dark green above, paler beneath, distinctly but finely veined, appressed-puberulent on both sides; peduncles 2–2.5 cm. long; flowers unknown; pod 12–20 cm. long, 1.5–1.8 cm. wide, velutinous, rounded and apiculate at apex, gradually narrowed to a slender stipe 1–2 cm. long at base, transversely veined, seeds outlined by elliptic depressions in pod.

Known only from the foothills in the Cape Region, Baja California, but to be expected in similar situations in the Sierra de la Giganta north of La Paz. Flowering after summer rains.

**3. Leucaena palmeri** Britt. & Rose, N. Amer. Fl. **23:** 123, 1928

Shrub or small tree 3–8 m. high with smooth bark and glabrous herbage; petioles 2–3.5 cm. long, bearing an elliptic compressed gland 0.5–1.5 cm. above base; pinnae (2–) 4–5 pairs; leaflets 3–6 pairs, elliptic to oblong-lanceolate or oblong-oblanceolate, mostly acute at apex, abruptly short-apiculate, delicately veined, 2–3.5 cm. long, dark green above, paler and sometimes sparingly puberulent along veins beneath; rachilla bearing a terminal subulate gland 2–3 mm. long; heads about 1.5 cm. in diameter, on slender axillary peduncles 3–4 cm. long, these elongating in fruit; flowers white; anthers pubescent; pods 12–22 cm. long, 1–1.5 cm. wide, obtuse and apiculate or rarely acute at apex, gradually narrowed to a slender stipe 1.5–2 cm. long, glabrous, reddish brown.

Southern Sonora and Sinaloa.

**4. Leucaena microcarpa** Rose, Contr. U.S. Nat. Herb. **5:** 141, 1897

A tree to 6–8 m. high with smooth bark and glabrous herbage and pods; pinnae 2–3 pairs; petioles 2.5–5 cm. long, bearing an elliptic flattened gland just below lower pair of pinnae; leaflets 2–6 pairs, elliptic, obovate, or suborbicular, 1.5–4.5 cm. long, 1.2–2 cm. wide, obtuse or acute at apex, glabrous or sparsely puberulent along veins beneath; peduncles 2–4 cm. long in flower, elongating somewhat in fruit; heads about 2 cm. in diameter; corollas white, about 6 mm. long, puberulent on lobes; pods broadly linear, 1–1.4 cm. wide, 6–13 cm. long, glabrous, acute at apex, gradually narrowing to a slender stipe 1–2 cm. long.

Cape Region, Baja California, and in Sinaloa. Sept.–Oct.

## 11. DESMANTHUS Willd., Sp. Pl. **4:** 1044, 1806

Perennial herbs or shrubs with bipinnate leaves, chiefly setaceous or filiform stipules, and small, greenish or white, regular flowers in peduncled heads. Petioles or rachises usually bearing 1 or more orbicular to lenticular glands. Flowers sessile, perfect, or the lowest apetalous or staminate or neuter. Calyx short-toothed, campanulate. Petals 5, distinct or slightly joined below. Stamens 5 or 10, distinct, exserted. Ovary sessile or nearly so, few- to many-ovuled. Pod linear or oblong, flat, straight or sometimes curved, the valves membranous or coriaceous.

Flowers minute; corolla 1.5–2 mm. long; stamens 5; pods 1–3-seeded .. 1. *D. oligospermus*
Flowers 3–4 mm. long; stamens 9–10; pods several-seeded:
  Petiolar gland below 2 accessory stipules, nearly basal; leaflets 25–50 pairs .................................................................. 2. *D. bicornutus*
  Petiolar gland usually between lower pair of pinnae, without accessory stipules; leaflets 5–20 pairs:
    Pinnae 3–9 pairs; leaflets 5–10 mm. long:
      Legumes acute or obtuse at apex, somewhat arcuate, 5–6 mm. wide; leaflets 7–10 mm. long; staminodia 12–15 mm. long, flattened .................. 3. *D. fruticosus*
      Legumes subulate-attenuate at apex, straight, 3–4 mm. wide; leaflets 4–6 mm. long; staminodia 6–8 mm. long .................................... 4. *D. subulatus*
    Pinnae 1–3 pairs; leaflets 3–5 mm. long:
      Pods 2.5–5 cm. long; staminodia 12–18 mm long .................... 5. *D. palmeri*
      Pods 8–10 cm. long; staminodia 6–8 mm. long:
        Petiolar gland short-stipitate, orbicular, about 1 mm. in diameter .............................................................. 6. *D. covillei* var. *covillei*
        Petiolar gland sessile, elliptic, 1–1.5 mm. long ...... 6a. *D. covillei* var. *arizonicus*

### 1. **Desmanthus oligospermus** Brandegee, Proc. Calif. Acad. II, **3:** 132, 1891

*Acuan oligospermum* Britt. & Rose, N. Amer. Fl. **23:** 135, 1928.

Small shrub with decumbent slender branches 2–5 dm. long, these puberulent in youth but soon glabrate; bark reddish brown; stipules setiform, 1–3 mm. long; leaves 2–4 cm. long, rachis and petiole puberulent; minute, circular, crateriform gland between basal of 2–4 pairs of pinnae; leaflets 5–12 pairs, linear-oblong to oblong-obovate, 3–4 mm. long, sparsely puberulent or ciliolate, obtuse or rounded, midvein obscure; peduncles filiform, 2–3 cm. long, glabrous or nearly so; heads few-flowered, corollas 1.5–2 mm. long, stamens 5; pods 5–8 mm. long, about 2 mm. wide, 1–3-seeded, glabrous.

Sandy or gravelly places, Lower Sonoran Zone, Cape Region from vicinity of La Paz southward. Flowering after rains.

### 2. **Desmanthus bicornutus** S. Wats., Proc. Amer. Acad. **21:** 426, 1886

*Mimosa canonis* M. E. Jones, Contr. West. Bot. **12:** 6, 1908.
*Acuan bicornutum* Britt. & Rose, N. Amer. Fl. **23:** 136, 1928.

Erect perennial herb 1–1.5 m. tall with glabrous herbage or minutely puberulent on youngest parts, then soon glabrate; stipules filiform, 4–5 mm. long, a second pair 1–2 mm. long borne just above a crateriform gland 4–6 mm. above base of petiole; leaves to 15 cm. long; pinnae 5–15 pairs, 2–3 cm. long; leaflets 25–50 pairs, linear, 2–4 mm. long, glabrous, acute, midvein obscure; peduncles slender, 2–4 cm. long; heads many-flowered, axillary and in terminal racemes; calyx 2.5 mm. long, narrowly obconic; corolla about 3 mm. long, glabrous; stamens 9–10, 7–8 mm. long; pods linear, 3–7 cm. long, 3–3.5 mm. wide, glabrous, 6–15-seeded, narrowly acute at apex.

In shade along streams and at margins of ponds, Lower Sonoran and Subtropical Zones, Sonora and Chihuahua. Aug.–Oct.

**3. Desmanthus fruticosus** Rose, Contr. U.S. Nat. Herb. **1**: 131, 1892

*Acuan fruticosum* Standley, Contr. U.S. Nat. Herb. **23**: 366, 1922.

A slender shrub 1–4 m. tall with glabrous branches and foliage or leaflets sparsely ciliolate in youth; bark reddish on twigs, later gray; stipules setaceous-subulate, 2.5–4 mm. long; petioles 1–2.5 cm. long, bearing an orbicular-crateriform gland 0.8–1 mm. in diameter between lowermost of 3–4 pairs of pinnae, these 2–3.5 cm. long; leaves 6–10 cm. long; leaflets 10–20 pairs, linear-oblong, 7–10 mm. long or lowest pair only 5 mm. long, obtuse, sparsely and minutely ciliolate to glabrous, 1-veined; peduncles 2–5 cm. long; heads many-flowered, about 3 cm. in diameter; imperfect flowers with slightly flattened stamens and staminodia 12–15 mm. long; calyx obconic, 2–2.2 mm. long, corolla 3.5 mm. long, both glabrous; pods linear, 5–6 mm. broad, 5–9 cm. long, acute but not attenuate or subulate at apex, finely reticulate-veined, glabrous; seeds obovoid, 3.5–4 mm. long, dark brown, smooth and shining.

Open hillsides and bluffs, Lower Sonoran Zone, San Francisquito Bay south to La Paz and on Carmen Island, Baja California. Aug.–Nov.

**4. Desmanthus subulatus** (Britt. & Rose) Wiggins ex Turner, Field & Lab. **18**: 128, 1950

*Acuan subulatum* Britt. & Rose, N. Amer. Fl. **23**: 136, 1928.

Suffrutescent, with slender glabrous branches 0.5–2.5 m. tall; stipules filiform-setaceous, 2–4 mm. long, reddish; petioles 5–10 mm. long, nearly filiform, bearing an orbicular-crateriform gland 0.5–0.8 mm. in diameter between lowermost of the 4–9 pairs of pinnae, rachis sometimes sparsely puberulent; leaves 4–7 cm. long; leaflets 8–18 pairs, linear-oblong, 4–6 mm. long, acute or obtuse, midvein delicate, nearly central; heads several-flowered, 12–18 mm. in diameter; peduncles 2–4 cm. long; flowers white, glabrous; calyx about 2 mm. long; corolla 3 mm. long; stamens 6–8 mm. long, those of lower flowers not conspicuously longer than upper ones; pods several, linear, 3–3.5 mm. broad, 4–10 cm. long, finely reticulate-veined, dark brown, subulate beak 5–9 mm. long; seeds 3–3.2 mm. long, ovoid-lenticular, dark brown, bearing several lines of minute, low-rounded papillae on each side.

On hillsides and in shaded arroyos, Lower Sonoran Zone, central Sonora into Sinaloa and Nayarit. Aug.–Oct.

**5. Desmanthus palmeri** (Britt. & Rose) Wiggins ex Turner, Field & Lab. **18**: 122, 1950

*Acuan palmeri* Britt. & Rose, N. Amer. Fl. **23**: 136, 1928.

Ascendingly branched shrub 1–1.5 m. tall with slender glabrous branches and brownish to gray bark; stipules acicular, 2–3 mm. long; petioles 1–1.5 cm. long, these and rachis sparsely puberulent; pinnae 1–2 pairs, 1–2 cm. long, a suborbicular gland between lower pair and sometimes a smaller one between upper pair; leaflets 5–10 pairs, oblong-obovate to linear-oblong, 3–5 mm. long, obtuse or rounded, sparsely ciliolate but some glabrate, midvein slightly eccentric, faint; peduncles 1–2 cm. long, glabrous; flowers glabrous, white or cream; calyx 2–2.4 mm. long; corolla 3.2–3.5 mm. long; stamens or staminodia of lower flowers 12–18 mm. long; pods straight, linear, 2.5–5 cm. long, about 3 mm. wide, short-acuminate, glabrous; seeds ovoid-quadrate, 3.5–4 mm. long, dark reddish brown, smooth.

Known only from the type collection, from Guaymas, Sonora. Aug.–Oct.

**6. Desmanthus covillei** (Britt. & Rose) Wiggins ex Turner, Field & Lab. **18:** 128, 1950, var. **covillei**

*Acuan covillei* Britt. & Rose, N. Amer. Fl. **23:** 135, 1928.

Slender, openly branched shrub similar to *D. palmeri* in all respects except that stamens and staminodia of lower flowers are only 6–8 mm. long and pods are more numerous and 7–12 cm. long. Occasionally a single pod in a head may be no more than 3–4 cm. long.

Hillsides among boulders, Lower Sonoran Zone, central Sonora south into Sinaloa and at northern end of Sierra de la Giganta, Baja California. Aug.–Oct.

**6*a*. Desmanthus covillei** var. **arizonicus** Turner, Field & Lab. **18:** 129, 1950

Differing from var. *corvillei* only in having a sessile, narrowly elliptic, petiolar gland instead of an orbicular, short-stipitate one and in consistently having 2–3 pairs of pinnae instead of sometimes having only 1 pair of pinnae on a leaf.

Known only from the type locality, Fresnal Canyon, Baboquivari Mountains, Pima County, Arizona.

**Desmanthus cooleyi** (Eaton) Trelease, Ark. Geol. Surv. **4:** 178, 1891 (*Acacia cooleyi* Eaton, Man. ed. 5, 89, 1829; *Desmanthus jamesii* Torr. & Gray, Fl. N. Amer. **1:** 402, 1840; *Acuan cooleyi* Britt. & Rose, N. Amer. Fl. **23:** 134, 1928), differs from *D. covillei* chiefly in being decumbent instead of erect or ascending and in having small stipules only 1–2 mm. long or these lacking. It occurs at elevations generally above the Sonoran Desert from Pima County to Nebraska and northern Mexico and may occur sporadically as a straggler in canyons leading into our area.

## Family 59. **CAESALPINIACEAE.** Senna Family

Herbs, shrubs, or trees with simple or compound stipulate leaves. Flowers irregular or nearly regular, perfect or rarely monoecious, dioecious, or polygamodioecious, small to large, solitary or in racemes or clusters. Calyx tube shallow, often oblique; calyx lobes 5 or appearing fewer owing to irregularity in opening. Petals 5, upper overlapped by lateral ones in bud, sometimes only 1. Stamens 5–10, or in a few genera numerous; filaments distinct or united at base; anthers dehiscent longitudinally or opening by terminal pores. Ovary 1–many-ovuled. Fruit a legume, dehiscent or indehiscent.

Leaves simple, entire or 2-lobed; calyx limb spathaceous, nearly closed in bud .. 1. *Bauhinia*
Leaves pinnate or bipinnate, not simple; calyx not spathaceous, lobes evident in bud:
  Anthers basifixed, opening by terminal pores or slits; leaves once-pinnate:
    Pods not elastically dehiscent; valves not strongly coiling at dehiscence; stipules obscurely veined .......................................................... 2. *Cassia*

Pods elastically dehiscent; valves strongly coiling at dehiscence; stipules prominently veined .......... 3. *Chamaecrista*

Anthers versatile, dehiscing longitudinally; leaves bipinnate except in *Tamarindus*:

Pods turgid; seeds separated by fleshy acidulous pulp; leaves once-pinnate .......... 4. *Tamarindus*

Pods compressed, or if terete in cross section around seeds then strongly constricted between the seeds; pulp lacking; leaves bipinnate:

Heartwood yielding a reddish or purplish dye; pods splitting longitudinally through sides of valves, sutures indehiscent .......... 5. *Haematoxylon*

Heartwood not yielding a dye; pods dehiscing along sutures or indehiscent:

Plants herbs or low suffrutescent subshrubs, not spiny; pods elastically valved; pedicels not jointed .......... 6. *Hoffmanseggia*

Plants shrubs or trees, frequently spiny; pods not elastically valved, sometimes indehiscent; pedicels jointed:

Calyx lobes strongly imbricate, lower sometimes lacerate or prickly; seeds without endosperm; pods often glandular .......... 7. *Caesalpinia*

Calyx lobes valvate or only slightly imbricate, not lacerate or prickly; seeds with endosperm; pods not glandular:

Leaf rachises usually over 5 mm. long, not spine-tipped; leaflets persistent; rachises of pinnae not flattened, 1 mm. or less in width ..... 8. *Cercidium*

Leaf rachises very short, distinctly spine-tipped; leaflets caducous, rachis of each pinna flattened, 3–4 mm. wide .......... 9. *Parkinsonia*

## 1. BAUHINIA L., Sp. Pl. 374, 1753

Trees, shrubs, or climbing vines, with deeply bilobed or rarely bifoliolate leaves, caducous stipules, and mostly showy large flowers in racemes or corymbs, or rarely solitary. Calyx tubular, oblong, or campanulate, limb more or less spathaceous. Petals 5 or sometimes only 2, subequal. Antheriferous stamens 1, 3, 5, or 10; filaments distinct or somewhat united at base, sterile filaments usually shorter than antheriferous ones; anthers versatile. Ovary usually stipitate. Legumes linear, flat, mostly elastically dehiscent, valves thin to coriaceous. Seeds ovoid to orbicular-lenticular.

**1. Bauhinia peninsularis** Brandegee, Univ. Calif. Pub. Bot. **10:** 183, 1922
*Casparea peninsularis* Britt. & Rose, N. Amer. Fl. **23:** 211, 1930.

Openly branching shrub 2–3 m. tall with yellow-tomentose branchlets; petiole 1–1.4 cm. long; leaves 3–6 cm. wide, 3–5 cm. long, cordate basally, apically notched about one-third of length; 7–9-nerved, tomentose beneath, glabrous above, lobes rounded, petals about 2 cm. long, glabrous, white; pod oblong, 1–1.8 cm. wide, 5–8 cm. long, acute at both ends, glabrous; stipe 2–2.5 cm. long.

Coastal plains, Lower Sonoran Zone, southern Baja California, common in the Cape Region. Aug.–Sept.

## 2. CASSIA L., Sp. Pl. 376, 1753

Annual or perennial herbs or shrubs or trees with pinnate leaves and flowers in racemes, panicles, or solitary in the axils. Stipules fugacious or deciduous. Sepals 5, subequal. Petals 5, slightly irregular, clawed, imbricated in bud, yellow. Stamens 5 or 10, usually unequal, lower 3 often longer and bearing larger anthers than others, some often sterile. Pod flat or turgid, thin, coriaceous or woody, dehiscent or indehiscent, mostly pendulous. Seeds transversely arranged in the pods.

Shrubs or trees, 1 m. high or more:
Branchlets and tips of midribs of leaves spinescent; leaflets fleshy, distantly arranged on fleshy eglandular rachis ........ 1. *C. armata*
Branchlets and midribs not spinescent; leaflets not fleshy, approximate; petioles and rachises glandular or eglandular:
Pods conspicuously margined with a wing 2–3 mm. wide, breaking into several 1-seeded loments, wing remaining intact; leaflets narrowly oblong, four to five times as long as broad ........ 2. *C. goldmanii*
Pods flat or turgid, wingless, not separating into distinct loments; leaflets obovate to oblanceolate, rarely over twice as long as broad:
Leaflets densely tomentose beneath, 5–15 cm. long; fruit woody, indehiscent, 1–1.5 cm. wide, 15–40 cm. long ........ 3. *C. emarginata*
Leaflets glabrous or sparsely puberulent, 0.4–2 cm. long; fruit thin or turgid, not woody, 5–12 mm. wide, 3–15 cm. long:
Pods more or less turgid, though compressed, bursting irregularly or subdehiscent, 3–7 cm. long; leaflets distinctly glaucous ........ 4. *C. purpusii*
Pods thin and flat, regularly dehiscent along both sutures, 6–15 cm. long; leaflets not glaucous:
Leaves bearing a cylindrical or clavate gland between basal pair of leaflets; ovary densely appressed-pubescent at anthesis ........ 5. *C. biflora*
Leaves glandless; ovary glabrous ........ 6. *C. wislizenii*
Annual or perennial herbs or, if suffrutescent, less than 1 m. high:
Pods linear, 10–25 cm. long; perfect anthers of 2 sizes; leaflets 3–8 pairs ........ 7. *C. leptocarpa* var. *glaberrima*
Pods oblong or linear-oblong, mainly 2–5 cm. long; perfect anthers essentially alike; leaflets 1–4 pairs:
Legume deeply impressed between seeds, finally breaking transversely; seeds quadrate, separated by complete dissepiments ........ 8. *C. uniflora*
Legume not impressed between seeds, dehiscent along both sutures; seeds rounded or angular, not quadrate; dissepiments lacking or evanescent:
Leaflets 1 pair; seeds finely marked with labyrinthine ridges:
Pods appressed-strigose; leaflets lanceolate ........ 9. *C. roemeriana*
Pods pubescent with spreading-ascending hairs; leaflets elliptic to obovate-oblong ........ 10. *C. bauhinioides*
Leaflets 2–4 pairs; seeds deeply ridged with convoluted folds:
Pods 4–6 mm. in diameter; pods densely puberulent with appressed hairs; leaflets 1–3 cm. long; surfaces of main folds on seeds smooth ........ 11. *C. covesii*
Pods 6–7 mm. in diameter, sparsely puberulent with spreading hairs, and intermingled, appressed, shorter hairs; leaflets 2–5 cm. long; surfaces of main fold on seeds again finely ridged ........ 12. *C. confinis*

**1. Cassia armata** S. Wats., Proc. Amer. Acad. **11**: 136, 1876
*Xerocassia armata* Britt. & Rose, N. Amer. Fl. **23**: 246, 1930.

A much-branched shrub to 1 m. tall with striate, pale green, glabrous or sparsely puberulent but soon glabrate branches; leaves 1.5–8 cm. long; stipules none; leaflets 2–3 distant pairs, ovate, elliptic, or suborbicular, acute or obtuse, often mucronulate, glabrous or puberulent, 1.5–6 mm. long; flowers in a terminal raceme 5–15 cm. long; pedicels 1–2 cm. long, slender, glabrous or with a few scattered, somewhat appressed hairs; sepals yellowish, obtuse, 5–6 mm. long, sparsely pilosulous; petals bright yellow, 8–12 mm. long, often with brownish veins; pods linear-oblong, turgid, 4–6 mm. in diameter, 2–4 cm. long, subulate-tipped, sparsely appressed-puberulent to glabrate.

Desert washes and plains, Lower Sonoran Zone, Mojave and Colorado Deserts into western Nevada, Arizona, and the deserts east of the Sierra San Pedro Mártir, Baja California. March–June.

**2. Cassia goldmanii** Rose, Contr. U.S. Nat. Herb. **10:** 98, 1906

*Pterocassia goldmanii* Britt. & Rose, N. Amer. Fl. **23:** 244, 1930.

A slender shrub or small tree 3–6 m. tall with gray or brownish or mottled bark, glabrous foliage, and linear-subulate stipules 1.5–3 mm. long; petioles 1–1.5 cm. long, eglandular or with a few minute reddish glands near base; leaves 6–7 cm. long; leaflets 5–12 pairs, narrowly obovate-oblong, 3–5 mm. wide, 10–16 mm. long, broadly cuneate at base, rounded and apiculate at apex, lightly pinnate-veined; petiolules 0.5–1 mm. long; flowers unknown; fruiting pedicels 1–2 cm. long; pods about 2 cm. wide, 8–12 cm. long, bearing marginal wings 2–5 mm. wide, the broader one along ventral suture, valves breaking transversely into segments 5–8 mm. wide, slightly glaucous, glabrous, transversely veined; seeds much compressed, 6–7 mm. in diameter, orbicular, black.

In shaded canyons and on mountain slopes, Lower Sonoran Zone, upper slopes of Sierra de la Giganta, Baja California. June–July.

**3. Cassia emarginata** L., Sp. Pl. 376, 1753

*Cassia atomaria* L., Mant. 68, 1767.
*Cassia arborescens* Mill., Gard. Dict. ed. 8, no. 15, 1768.
*Cassia elliptica* H. B. K., Nov. Gen. & Sp. **6:** 356, 1824.
*Isandrina arborescens* Raf., Sylva Tell. 126, 1838.
*Cassia emarginata* var. *subunijuga* Robins. & Bartlett, Proc. Amer. Acad. **43:** 53, 1907.
*Isandrina emarginata* Britt. & Rose, in Britt. & Wilson, Sci. Surv. Porto Rico & Virgin Isl. **5:** 374, 1924.

A tree to 20 m. tall with gray bark, densely pubescent twigs, and rounded crown; stipules setaceous, 3–4 mm. long, deciduous; petioles to 10 cm. long or more, puberulent; leaflets 2–5 pairs, oval, oblong-elliptic, or ovate-orbicular, 3–7 cm. wide, 5–15 cm. long, acute, obtuse, rounded or retuse at apex, obtuse or rounded at base, softly appressed-puberulent beneath, puberulent to glabrate above; petioles 3–5 mm. long; peduncles 5–12 cm. long, few- to several-flowered; sepals obovate to oblong, rounded, 6–9 mm. long, densely puberulent; petals yellowish green with brown veins, 12–20 mm. long; larger anthers 4–5 mm. long, shorter ones 1–1.5 mm. long; ovary glabrous; pods linear, 10–15 mm. wide, to 4 dm. long, indehiscent, breaking irregularly transversely, valves hard, subwoody.

Plains, hillsides, and along washes, Tropical Zone to Lower Sonoran Zone, Cape Region, Baja California, southern Sonora, and Tamaulipas to northern South America and the West Indies. Flowering throughout the year.

**4. Cassia purpusii** Brandegee, Zoe **5:** 234, 1906

*Adipera purpusii* Britt. & Rose, N. Amer. Fl. **23:** 240, 1930.

A straggly-branched shrub 1–2 m. tall with gray-black, slightly scaly bark, sparsely minute-puberulent twigs, and glaucous blue-green foliage; stipules linear-subulate, 3–4 mm. long, caducous; leaves 2–4 cm. long, petioles to 1 cm. long, it and rachis bearing 1–3 conical or stalked glands 1–3 mm. long between 1 or more pairs of leaflets; leaflets 2–4 pairs, broadly oval, 6–12 mm. wide, 9–20 mm. long, rounded or slightly retuse at apex, sparsely appressed-puberulent at first but some glabrate, glaucous, thickish, and leathery; inflorescences terminal panicles, few-flowered; peduncles 1–3 cm. long; buds nodding; sepals 5–7 mm. long, rounded, sparsely appressed-pubescent, reddish; petals bright yellow with purplish brown veins, 8–12 mm. long; 3 lower anthers equaling style, 4 upper erect, about half as long, 3 sterile ones reduced to small knobs about 1 mm. long; pods flat but turgid, 10–12 mm. wide, 3–7 cm. long, glabrous, 2–6-seeded, reddish, transversely veined, acute to obtuse at each end, borne on a stipe 5–7 mm. long and tipped with a slender beak 4–5 mm. long.

Arroyos and coastal mesas, Lower Sonoran Zone, vicinity of Punta Prieta and Calmallí, Baja California. Feb.–April; Sept.–Oct.

### 5. **Cassia biflora** L., Sp. Pl. 378, 1753

*Cassia tenuissima* L., Sp. Pl. 378, 1753.
*Cassia pallida* Vahl, Eclog. Amer. **3:** 12, 1807.
*Cassia semperflorens* DC., Cat. Hort. Monsp. 90, 1813.
*Cassia geminiflora* Moc. & Sessé, in Collad., Hist. Cass. 103, *pl. 3,* 1816.
*Cassia nemorosa* H. B. K., Nov. Gen. & Sp. **6:** 353, 1824.
*Cassia acapulcensis* H. B. K., loc. cit.
*Cassia biflora* var. *semperflorens* DC., Prod. **2:** 496, 1825.
*Panisia biflora* Raf., Sylva Tell. 129, 1838.
*Cassia biflora* var. *angustisiliqua* Griseb., Fl. Brit. W. Ind. 208, 1860.
*Peiranisia biflora* Pittier, Trab. Mus. Com. Venez. **3:** 158, 1928.

Shrub or small tree to 5 m. with glabrous or puberulent twigs, inflorescences, and foliage; stipules linear-subulate, 2–3 mm. long, caducous; petioles 3–8 mm. long, bearing a clavate or cylindrical gland about 1 mm. long between lowest of 4–10 pairs of leaflets; leaflets obovate, oblanceolate, or sometimes oblong, thin, 6–9 mm. wide, 1–2 cm. long, sparsely and minutely pubescent or glabrous, obtuse, rounded, or retuse at apex, often broadly cuneate at base, pale beneath; peduncles slender, 5–10 mm. long, 2–3- (usually 2-) flowered; pedicels about twice as long as peduncle; sepals thin, green, 4–9 mm. long, obtuse or rounded; petals bright yellow, to 2 cm. long, brown-veined; perfect anthers usually of 2 kinds, thecae of larger 2 or 3 to 5 mm. long, bearing an apical pore tube 2–2.5 mm. long, other perfect anthers 3–4 mm. long, their pore tubes stouter, less than 1 mm. long; ovary densely pubescent; pods linear, 5–6 mm. wide, 6–15 cm. long, compressed, dark brown, glabrous; seeds suborbicular-lenticular, 3–3.5 mm. in diameter.

Hillsides, plains, and along arroyos, Tropical Zone to Lower Sonoran Zone, northwestern Sonora (vicinity of Nória, Ures), southern Baja California, and Chihuahua to northern South America and the West Indies. Oct.–April.

### 6. **Cassia wislizenii** A. Gray, Pl. Wright. **1:** 60, 1852

*Palmerocassia wislizenii* Britt., in Britt. & Rose, N. Amer. Fl. **23:** 254, 1930.

A rounded shrub to 3 m. tall with short stiff branches, grayish bark, appressed-puberulent twigs, and sparsely puberulent foliage; stipules setose, 2–5 mm. long, persistent; leaves 1.5–3 cm. long; petioles 4–7 mm. long, glandless; leaflets 2–3 pairs, 3–6 mm. wide, 4–10 mm. long, obovate, broadly cuneate at base, rounded or shallowly retuse at apex, pinnately veined; peduncles axillary or terminal, stout, 1–3 cm. long, few-flowered; pedicels slender, about 1 cm. long; sepals thin, oblong to ovoid, 6–10 mm. long, yellowish, rounded at apex, glabrous or nearly so; petals 2–2.5 cm. long; lower anthers on filaments 7–9 mm. long; ovary glabrous; pods linear, 5–6 mm. wide, to 1.5 dm. long, dark brown to nearly black, flattened, slightly constricted between some of the seeds; seeds more or less quadrate, about 3 mm. wide, 4–5 mm. long, dark brown, minutely punctate-pitted in several lines on each face.

Hillsides and canyons, mainly Upper Sonoran and upper parts of Lower Sonoran Zones, southeastern Arizona to Texas and south into Sonora (Trincheras) and Chihuahua. June–Sept.

**7. Cassia leptocarpa** Benth. var. **glaberrima** M. E. Jones, Contr. West. Bot. **12:** 7, 1908

*Ditremexa glaberrima* Britt. & Rose, N. Amer. Fl. **23:** 256, 1930.

A perennial herb or slightly woody at base, simple or branched above, about 1 m. tall, glabrous or sparsely puberulent; stems ridged; stipules linear or lance-subulate, 5–10 mm. long, caducous; leaves to 4 dm. long; petioles and rachis ridged, petiole bearing a stout, conical to oblong gland 1–2.5 mm. long just above base; leaflets 3–8 pairs, lanceolate, 7–25 mm. wide, 4–10 cm. long, glabrous except on puberulent margins, bright green; peduncles axillary and occasionally terminal, few-flowered; buds nodding; pedicels 5–25 mm. long; sepals thin, oblong to obovate, rounded, 7–10 mm. long; petals to 1.5 cm. long, usually 2 lower anthers on filaments equaling entire stamens of upper smaller group, equaling or surpassing petals; ovary appressed-puberulent; pods linear, 4.5–6 mm. wide, to 2.5 dm. long, glabrous or essentially so; seeds arranged transversely, ends mutually compressed, surface minutely reticulate-lined, dark gray, dull.

Along washes and streams, Sonoran and Subtropical Zones, Cochise, Pinal, and Pima Counties, Arizona, to New Mexico, San Luis Potosí, and Sinaloa. July–Sept.

**8. Cassia uniflora** Mill., Gard. Dict. ed. 8, no. 5, 1768

*Cassia ornithopoides* Lam., Encycl. **1:** 644, 1785.
*Cassia sericea* Swartz, Prod. Veg. Ind. Occ. 724, 1788.
*Cassia sensitiva* Jacq., Coll. **2:** 362, 1788.
*Cassia ciliata* Hoffmanns., in DC., Prod. **2:** 493, 1825.
*Cassia monantha* DC., op. cit. 506.
*Sericeocassia uniflora* Britt., N. Amer. Fl. **23:** 246, 1930.

Annual herb 3–10 dm. tall with appressed-villous, brownish or rusty stems and foliage; stipules linear-lanceolate, 5–20 mm. long, often somewhat stipitate-glandular; leaves 6–15 cm. long, bearing 1–several stipitate glands between leaflets; leaflets 2–4 pairs, oblong to obovate, 2–5 cm. long, subsessile, acute to rounded and mucronate apically, narrowed toward base, appressed-villous beneath, pubescent above at first, soon glabrate; sepals 5–6 mm. long, rounded, thinnish; petals 10–12 mm. long; pods linear, compressed, 2.5–5 cm. long, 4–5 mm. wide, deeply impressed between seeds, eventually breaking transversely; seeds quadrate, completely separated by dissepiments.

Sandy ground and waste places, Lower Sonoran Zone to Tropical Zone, southern Sonora to northern South America and the West Indies. July–Oct.

**9. Cassia roemeriana** Scheele, Linnaea **21:** 457, 1848

*Earleocassia roemeriana* Britt., in Britt. & Rose, N. Amer. Fl. **23:** 247, 1930.

Herbaceous perennial from woody root with erect, puberulent, sulcate-angled stems 3–5 dm. high and finely puberulent-ashy foliage; stipules linear-subulate, 5–8 mm. long, subpersistent; petioles 2.5 cm. long or shorter, surpassed by leaflets, bearing a small clavate, apical gland; leaflets 1 pair, lanceolate, 6–12 mm. wide, 2–6 cm. long, asymmetrically rounded at base, acute and mucronulate at apex, appressed-puberulent on both surfaces, midrib slightly eccentric at base; inflorescence about twice as long as subtending leaves, 2–5-flowered; sepals 5–7 mm. long, rounded or obtuse at apex, appressed-pubescent below but gradually becoming glabrous and membranous toward tips; petals 10–12 mm. long, strongly veined, these becoming brown in drying; anthers nearly alike; ovary densely appressed-pubescent; pod oblong, slightly compressed or nearly terete in cross section, 6–7 mm. wide, 2–3.5 cm. long, appressed-strigose; seeds

3–3.5 mm. long, covered with low, rounded, labyrinthine ridges; hilum on a short boss.

Flats and valley floors, mainly Upper Sonoran Zone, Texas to Coahuila, Nuevo León, and northern Sonora. March–April; Sept.–Oct.

**10. Cassia bauhinioides** A. Gray, Bost. Jour. Nat. Hist. **6:** 180, 1850

*Earleocassia bauhinioides* Britt., in Britt. & Rose, N. Amer. Fl. **23:** 248, 1930.

Herbaceous perennial from slender woody rootstock, with several ascending-spreading branches 1–3 dm. tall and soft-pubescent foliage; stipules lance-linear, 3–6 mm. long, persistent; petioles slender, 1–3 cm. long, cylindric gland just below apex; leaflets 1 pair, oblong, 5–12 mm. broad, 1–2.8 cm. long, obtuse or rounded at both ends, silky-pubescent and slightly cinereous on both surfaces; peduncles 2.5–4 cm. long, usually 2-flowered; sepals 6–8 mm. long, oblong, thinnish, green with yellowish margins, pubescent outside, glabrous within; petals 12–18 mm. long, distinctly brown-veined; anthers essentially alike; ovary densely appressed-hirsute; pods oblong, 5–6 mm. in diameter, 2–3.5 cm. long, scarcely compressed, dark brown, spreadingly pubescent; seeds about as in *C. roemeriana.*

Gravelly flats and plains and along washes, Sonoran Zones, Arizona to Texas, Sonora (near Magdalena), Hidalgo, and Zacatecas. April–Aug.

**11. Cassia covesii** A. Gray, Proc. Amer. Acad. **7:** 399, 1868

*Earleocassia covesii* Britt., in Britt. & Rose, N. Amer. Fl. **23:** 249, 1930.

Herbaceous perennial from slightly woody base and woody rootstock, with spreading and ascending branches 3–6 dm. long and grayish to tawny-velvety pubescence throughout; stipules linear, 5–12 mm. long, caducous; petioles 2–5 cm. long, bearing a small apiculate gland on a stalk to 3 mm. long between lowest of 2–3 pairs of leaflets; leaflets elliptic to oblong or oblong-obovate, to 1.5 cm. wide, 1–3 cm. long, rounded and mucronulate at apex; peduncles 3–7 cm. long, normally surpassing leaves, 3–7-flowered; pedicels 1–1.5 cm. long; sepals 6–8 mm. long, densely hirsutulous, rounded at apex, thin; petals 10–16 mm. long, dark-veined; anthers alike; ovary silky-strigose; pod oblong, 5–6 mm. wide, 2–3.5 cm. long, moderately appressed-pubescent, bearing a subulate tip 2–4 mm. long; seeds 3–4 mm. wide, deeply folded and convoluted, surface of convolutions smooth, ridges much coarser than in *C. bauhinioides* or *C. roemeriana.*

Desert plains and washes, Lower Sonoran Zone, western margins of the Colorado Desert to southern Nevada, Arizona, Sinaloa, and the central part of the state of Baja California. April–Oct.

**12. Cassia confinis** Greene, Pittonia **3:** 225, 1897

*Earleocassia confinis* Britt., in Britt. & Rose, N. Amer. Fl. **23:** 249, 1930.

Herbaceous or suffrutescent perennial with canescent-villous but finally pale greenish stems to 1 m. tall; foliage canescent-tomentulose; stipules linear, 6–10 mm. long, caducous; petioles 2–4 cm. long, bearing a cylindrical or clavate gland 1–2 mm. long between lower pair of leaflets; leaflets 2 pairs or rarely 1 or 3 pairs, broadly elliptic, elliptic-obovate, or suborbicular, 2–5 cm. long, to 2.5 cm. wide, asymmetrically rounded at base, rounded and mucronulate at apex; peduncles 5–12 cm. long, 3–10-flowered; petioles slender, 1–1.5 cm. long; sepals oblong, 6–7 mm. long, densely villous; petals 10–15 mm. long, dark-veined; anthers nearly alike; ovary densely pubescent; pods oblong, turgid, 6–7 mm. in diameter, 2.5–3 cm. long, bearing a subulate tip 3–4 mm. long, spreadingly hirsute with scattered hairs 1–1.5 mm. long with numerous short appressed hairs among larger ones; seeds in 2 rows, coarsely convolute-ridged, puberulent but later glabrate, surfaces of convolutions again finely ridged.

Desert washes and plains, Lower Sonoran Zone, Bahía de los Angeles to Cabo San Lucas, Baja California. Jan.–March; Oct.

## 3. CHAMAECRISTA Moench, Meth. 272, 1794

Herbs, low shrubs, or sometimes small trees with pinnate, usually glandular, and sometimes sensitive leaves, and strongly nerved persistent stipules. Flowers yellow, solitary or in axillary clusters. Sepals 5, usually acuminate. Corolla irregular; petals 5, one larger than other four. Stamens 5 or 10; anthers opening by terminal pores. Pods linear to oblong, compressed, usually elastically valved, valves twisted after dehiscence. Seeds flattened, often more or less quadrate.

Plants perennial; pedicels 3–4 cm. long; petals 10–12 mm. long; stipules glabrous or sometimes slightly short-ciliate ........................................ 1. *C. wrightii*
Plants annual; pedicels 3–5 mm. long; petals 3–6 mm. long; stipules strongly hispid-ciliate ........................................................ 2. *C. leptadenia*

### 1. Chamaecrista wrightii (A. Gray) Woot. & Standl., Contr. U.S. Nat. Herb. **19:** 335, 1915

*Cassia wrightii* A. Gray, Pl. Wright. **2:** 50, 1853.

An erect or ascending perennial herb or suffrutescent, 2–6 dm. tall with slender branches longitudinally obscurely striate, finely puberulent, but soon glabrate; stipules lanceolate-acuminate, 3–5 mm. long, glabrous or sparsely short-ciliate; petioles 3–7 mm. long, sparsely puberulent at base, bearing a stalked gland 0.5–1 mm. tall, about 0.2 mm. in diameter near middle; leaves 2–5 cm. long; leaflets 6–9 pairs, oblong, oblique at base, obtuse and apiculate at apex, 1.5–3 mm. wide, 6–13 mm. long, subchartaceous, pinnately veined, midvein eccentric, glabrous or essentially so; pedicels 3–4 cm. long, shallowly grooved, minutely puberulent but soon glabrate; sepals broadly ovate, thin, yellowish, 7–10 mm. long; petals 10–12 mm. long; pod linear-oblong, 4–5.5 mm. wide, 2.5–4.5 cm. long, acute, short-beaked, finely and sparsely appressed-puberulent; seeds dark brown, 2 mm. wide, 2.5–3 mm. long, quadrate, drawn out to a short tip beside hilum, obscurely punctate-glandular.

Mainly in granitic gravelly soil, slopes and hillsides, Sonoran Zones, southern Arizona (Santa Cruz County) south at least to central Sonora. Aug.–Sept.

### 2. Chamaecrista leptadenia (Greenm.) Cockerell, Muhlenbergia **4:** 68, 1908

*Cassia leptadenia* Greenm., Proc. Amer. Acad. **41:** 238, 1905.
*Cassia leptadenia* var. *mensalis* Greenm., loc. cit.

An erect, strict or sparingly branched annual 1–7 dm. tall with sparsely puberulent, often glabrate herbage; stipules narrowly lanceolate-acuminate, 5–7 mm. long, strongly hispid-ciliate; petioles 5–6 mm. long, bearing a stalked gland 0.2–0.3 mm. broad from about middle to just below lowest of the 8–16 pairs of leaflets; leaves 3.5–7 mm. long, spreading or ascending; leaflets 7–12 mm. long, 1–3 mm. wide, rounded at base, acute and mucronate at apex, bright green above, paler beneath, faintly pinnately veined, the midvein only slightly eccentric; pedicels 3–5 mm. long; sepals lanceolate, short-acuminate, 4–5 mm. long, pubescent; petals 3–6 mm. long; stamens 5–8; pods linear, 3–3.5 mm. wide, 3–5 cm. long, appressed-puberulent, acute and short-mucronate; seeds quadrate, about 2 by 2.5 mm., dotted with small punctate glands on both faces.

Sandy or gravelly soil, Sonoran Zones, southern and central Arizona to western Texas, and Sonora; Cape Region, Baja California. Aug.–Oct.

## 4. TAMARINDUS L., Sp. Pl. 34, 1753

A large unarmed tree with evenly pinnate leaves, minute caducous stipules, and irregular showy flowers borne in racemes. Calyx tube turbinate; sepals 4, membranous and colored. Petals 5, upper three subsequal, yellow veined with red, lower pair reduced to bristles at base of stamineal tube. Fertile stamens 3, coalesced into a sheath; staminodia borne at tip of sheath. Ovary stipitate, several-ovuled. Stigma subcapitate. Fruit oblong, indehiscent, slightly compressed with a crustaceous epicarp, pulpy mesocarp, and a coriaceous endocarp forming septae between the hard compressed seeds.

**1. Tamarindus indicus** L., Sp. Pl. 34, 1753

*Tamarindus occidentalis* Gaertn., Fruct. **2:** 310, *pl. 146*, 1791.
*Tamarindus officinalis* Hook., Bot. Mag. **77:** *pl. 4563*, 1851.
*Lonchocarpus capensis* M. E. Jones, Contr. West. Bot. **18:** 46, 1933.

Tree 10–15 m. tall or more, the crown rounded; leaves 5–10 cm. long; leaflets 8–20 pairs, oblong, 1–2 cm. long, rounded or obtuse at apex, apiculate, smooth above, distinctly reticulate-veined and dull beneath, glabrous; flowers yellow, veined with red; pods 1.5–3 cm. broad, 3–10 cm. long, brown and somewhat scurfy, 1–7-seeded; seeds obovate or somewhat quadrate, about 1.5 cm. long, flattened, dark brown, embedded in juicy, reddish or red-brown, acidulous pulp.

A native of Africa and southern Asia but commonly planted as a shade tree throughout the tropics and subtropics. Sparingly naturalized along roadsides and ditches, La Paz, Guaymas, and elsewhere in the southern part of the Sonoran Desert. Flowering after rains.

**Ceratonia siliqua** L., Sp. Pl. 1026, 1753, the St. John's bread or carob tree, is cultivated at Guaymas, Mulegé, Comondú, and elsewhere in northwestern Mexico. The pods contain a reddish brown edible pulp.

## 5. HAEMATOXYLON L., Sp. Pl. 384, 1753

Glabrous trees or shrubs, with thorny or unarmed branches, evenly pinnate leaves, yellow or dark red wood, and fissured, flaky gray bark. Stipules small, persistent or deciduous. Flowers yellow, showy, in axillary racemes. Calyx tube shallowly bowl-shaped, somewhat oblique, lobes 5, subequal, imbricate in bud. Petals 5, oblong-obovate, slightly unequal, short-clawed. Stamens 10, distinct, pilose at base of filaments. Ovary 2–3-ovuled, short-stipitate. Pod flat, tough-papery, oblong, indehiscent along edges of valves but splitting longitudinally about one-third of way up from ventral suture. Seeds oblong-lenticular.

**1. Haematoxylon brasiletto** Karst., Fl. Columb. **2:** 27, *pl. 114*, 1862–69

*Haematoxylon boreale* S. Wats., Proc. Amer. Acad. **21:** 426, 1886.

Tree or shrub 3–7 m. tall with often zigzag branches bearing rigid spines 0.5–2 cm. long; bark reddish brown on young branches, gray on older stems; stipules linear-acicular, 4–7 mm. long, slightly curved, ascending, deciduous; petioles 3–20 mm. long, slender; leaflets 1–3 pairs, obovate-cuneate, 5–25 mm. long, about three-fourths as wide, emarginate to obcordate, thick and leathery, finely pinnately veined; stipels about 1 mm. long, broadly ovate and fimbriate; racemes 3–6 cm. long, mostly few-flowered; pedicels slender, 1.5–2.5 cm. long; calyx 6–7 mm. long at anthesis; lobes thin, yellowish, minutely ciliolate on margins, caducous; petals 6–10 mm. long, pale yellow tinged with red near base; pods broadly elliptic to oblong, 8–15 mm. wide, 2–6 cm. long, diagonally veined from sutures toward line of dehiscence and forward, obtuse, rounded, or acute, deep red-brown; seeds broadly obcordate to obreniform, greatly compressed, 7–12 mm. wide, 4–6 mm. long.

Gravelly and rocky ridges and slopes, Lower Sonoran Zone to Tropical Zone, central Sonora to Chihuahua, Baja California, and Colombia. Flowers after rains.

## 6. HOFFMANSEGGIA Cav., Ic. 4: 63, *pl. 392; pl. 393, fig. 1*, 1798

Herbs or small shrubs with bipinnate glandular-punctate leaves and small caducous or deciduous stipules. Flowers in simple, axillary, and terminal racemes. Calyx 5-parted, lobes subequal. Petals 5, subequal, yellow veined with red. Stamens 10, distinct; filaments often glandular near base. Ovary subsessile. Pod flat, linear to ovate, straight or lunate, several-seeded, dehiscently 2-valved.

Pinnae 3; pods lunate:
- Terminal pinna two to three times as long as lateral pinnae; stipules and bractlets acuminate; branchlets densely appressed-puberulent ................ 1. *H. microphylla*
- Terminal pinna equal to or only slightly longer than lateral pair; stipules and bractlets acute; branchlets glabrous or sparsely spreading-puberulent:
  - Branchlets and inflorescence glabrous; flowers distant; plant shrubby; leaflets glandless ........................................................ 2. *H. intricata*
  - Branchlets and inflorescence spreading-puberulent; flowers crowded; plant herbaceous; leaflets glandular-margined ................................ 3. *H. peninsularis*

Pinnae 5–11; pods falcate:
- Petals 10–12 mm. long, claws almost equaling blades and glandular; pods acute or obtuse and apiculate; inflorescence glandular and pubescent .............. 4. *H. densiflora*
- Petals 5–8 mm. long, claws very short, eglandular; pods rounded at apex; inflorescence eglandular ................................................ 5. *H. drepanocarpa*

### 1. Hoffmanseggia microphylla Torr., Bot. Mex. Bound. 58, 1859

*Caesalpinia virgata* E. M. Fisher, Bot. Gaz. **18:** 123, 1893.
*Larrea microphylla* Britt., in Britt. & Rose, N. Amer. Fl. **23:** 310, 1930.

Virgate shrub 0.5–2 m. tall with slender, densely puberulent, broomlike branches; leaves sparse, eglandular, 3-pinnate; petioles 10–18 mm. long; lateral pinnae 5–10 mm. long, with 4–9 pairs of leaflets, terminal one 15–40 mm. long, with 7–15 pairs of leaflets; leaflets oblong, 2–5 mm. long, acute, sparsely puberulent to glabrate, racemes 1–6 dm. long, many-flowered; bractlets ovate-lanceolate, attenuate, 2–3 mm. long, caducous, puberulent, margins stipitate-glandular; pedicels 3–8 mm. long; calyx puberulent, lobes lance-oblong, 4–5 mm. long, subimbricate, spreading to reflexed in age, glandular-margined; petals 5–7 mm. long, yellow, glandular-dotted on backs; filaments glandular and pilose near base; pod narrowly lunate, 15–25 mm. long, closely set with sessile or short-stipitate, reddish or brownish glands.

Sandy or gravelly soils, Lower Sonoran Zone, Colorado Desert to southwestern Arizona and adjacent northern Sonora, and Baja California to Guardian Angel Island. March–Oct.

**2. Hoffmanseggia intricata** Brandegee, Proc. Calif. Acad. II, **2:** 151, 1889

*Hoffmanseggia microphylla* var. *glabra* S. Wats., Proc. Amer. Acad. **24:** 47, 1889.
*Hoffmanseggia glabra* E. M. Fisher, Contr. U.S. Nat. Herb. **1:** 147, 1892.
*Hoffmanseggia glabra* var. *intricata* E. M. Fisher, loc. cit.
*Caesalpinia intricata* E. M. Fisher, Bot. Gaz. **18:** 123, 1893.
*Caesalpinia intricata* var. *glabra* E. M. Fisher, loc. cit.
*Larrea intricata* Britt., in Britt. & Rose, N. Amer. Fl. **23:** 311, 1930.

An intricately branched, glabrous shrub 4–6 dm. tall, the tips of the branches somewhat spinescent; pinnae 3, about equal, 5–15 mm. long; leaflets 4–10 pairs, broadly oblong, 2–3.5 mm. long, glabrous; racemes 5–20 cm. long; bracts ovate, 1.5–2.5 mm. long, glabrous but ciliate-glandular along the margins; pedicels 2–5 mm. long, slender, glabrous but occasionally with a few stipitate glands; calyx lobes 4–5 mm. long, linear-oblong, glabrous, reddish, and sparsely ciliate-glandular on margins; petals about 5 mm. long, pale yellow, sparsely dotted on backs with hyaline or pale yellow glands; stamens pilose and glandular; pods narrowly but strongly lunate, 3.5–5 mm. wide, 10–20 mm. long, obtuse at apex; cuneate at base, reticulately veined, gland-dotted.

Sandy and rocky soils, Lower Sonoran Zone, central Baja California, San Marcos and Tiburón Islands. March–Oct.

**3. Hoffmanseggia peninsularis** (Britt.) Wiggins, Contr. Dudley Herb. **4:** 18, 1950

*Larrea peninsularis* Britt., in Britt. & Rose, N. Amer. Fl. **23:** 311, 1930.

Herbaceous perennial or somewhat suffrutescent, 2–3 dm. tall, with crowded, intertangled, tortuous, sparsely pilose branches; stipules 1–1.5 mm. long, broadly ovate; pinnae 3, about equal; leaflets 3–6 pairs, oblong-linear, 3–4 mm. long, sparsely pubescent to glabrate, closely crowded, gland-dotted near apex; racemes 4–5 cm. long, few-flowered, sparsely pilose and with a few scattered stipitate glands; calyx lobes 2–3 mm. long, ovate, ciliate; petals 4–5 mm. long, narrowing gradually to short claws 1–2 mm. long, ciliate-glandular on margins below; filaments pubescent; pod narrowly lunate, 4–5 mm. wide, 1–1.5 cm. long, apiculate, puberulent and with a few scattered, yellowish, subsessile glands.

Known only from the type locality, Lower Sonoran Zone, Cocopah Mountains, northeastern Baja California.

**4. Hoffmanseggia densiflora** Benth., in A. Gray, Pl. Wright. **1:** 55, 1852

(For lengthy synonymy see N. Amer. Fl. **23:** 311, 1930, under *Larrea densiflora* Britt.)

Herbaceous perennial 1–4 dm. high with several stems from a heavy woody rootstock, puberulent with minute, strongly incurved hairs; leaves 5–13 cm. long, with 5–11 pinnae 1–2.5 cm. long; leaflets 4–12 pairs, oblong, glabrous or minutely puberulent and eventually glabrate, obtuse, 3–8 mm. long; inflorescence terminal, 1–2 dm. long, pubescent and densely stipitate-glandular; pedicels 2–7 mm. long; calyx lobes linear-oblong, acute, 6–7 mm. long, pubescent and glandular; petals orange-yellow, often veined with red, 10–12 mm. long, claws about equaling blades and densely glandular; stamens slightly shorter than petals, densely stipitate-glandular almost to anther; pod falcate, 5–8 mm. wide, 2–4 cm. long, reticulate-veined, obtuse and apiculate at apex, glandular-dotted.

Sandy and alkaline soils, Lower Sonoran Zone, upper San Joaquin Valley, Inyo County, and Colorado Desert, California, to Texas and northern Mexico as far south as central Baja California, Sonora, and San Luis Potosí. April–Sept.

**5. Hoffmanseggia drepanocarpa** A. Gray, Pl. Wright. **1:** 58, 1852

*Caesalpinia drepanocarpa* E. M. Fisher, Bot. Gaz. **18:** 122, 1893.

*Larrea drepanocarpa* Britt., in Britt. & Rose, N. Amer. Fl. **23:** 312, 1930.

Low herb 1–3 dm. high from woody perennial root, stems and leaves finely puberulent but wholly eglandular; stipules ovate-triangular, acute; petioles 2–6 cm. long; pinnae 5–11, 5–20 mm. long; leaflets 4–10 pairs, broadly oblong, rounded, 2–5 mm. long, minutely and sparsely puberulent to glabrate, veins obscure; racemes erect, surpassing leaves, few-flowered; pedicels 2–5 mm. long; calyx lobes linear-oblong, acute, 5–8 mm. long, pubescent; petals 5–8 mm. long, very short-clawed, eglandular or nearly so; stamens shorter than petals; pods falcate, often forming a semicircle, 6–7 mm. wide, 2–3.5 cm. long, rounded and minutely apiculate, transversely reticulate-veined, puberulent, with a few scattered sessile glands.

Rocky soil, mainly Upper Sonoran Zone, Arizona to Colorado, western Texas and Chihuahua. Barely reaching the upper margins of the Sonoran Desert in Mohave and Gila Counties. May–Sept.

**Hoffmanseggia jamesii** Torr. & Gray, Fl. N. Amer. **1:** 393, 1840, occurs in Arizona but above the Sonoran Desert. Its leaflets are conspicuously dotted with black glands, as are also the flowers, stems, and pods.

## 7. CAESALPINA L., Sp. Pl. 380, 1753

Chiefly unarmed shrubs and trees with bipinnate leaves and showy flowers in racemose inflorescences. Pedicels jointed, subtended by caducous or deciduous bracts. Calyx tube short, oblique, lobes imbricated, lower often slightly to conspicuously longer than other 4 and somewhat cucullate, partially enfolding upper lobes in bud. Petals 5, broad, imbricated, unequal, often papillate near claw. Stamens 10, distinct, long-exserted or sometimes only equaling petals; filaments villous near base; anthers dehiscent longitudinally. Ovary sessile or stipitate. Legume coriaceous or chartaceous, more or less compressed or flattened, elastically dehiscent or in some indehiscent or tardily dehiscent. Seeds few to several, flattened, smooth.

Pod indehiscent or essentially so, usually falling unopened; lower calyx lobe conspicuously lacerate:
  Leaflets orbicular or suborbicular, 0.5–2 cm. long; pod broadly elliptic, 2–3.5 cm. long; seeds 1–2 .......... 1. *C. pumila*
  Leaflets elliptic to oblong, 2–6 cm. long; pod oblong, 5–10 cm. long; seeds 4–8 .......... 2. *C. platyloba*
Pod dehiscent; lower calyx lobe entire:
  Petals 1.5–3 cm. long; leaves 1–3.5 dm. long; stamens long-exserted:
    Petals clear yellow, 2.5–3 cm. long, short-clawed; pods and inflorescence heavily glandular-dotted; plants not prickly; calyx lobes essentially alike .......... 3. *C. gilliesii*
    Petals orange-yellow, variegated, 1.5–2.5 cm. long, with narrow claws 1.5–1.8 mm. long; pods and inflorescence glabrous, eglandular; lower calyx lobe cucullate, longer than others .......... 4. *C. pulcherrima*
  Petals 6–12 mm. long; leaves 1 dm. long or less; stamens shorter than to barely exceeding petals:
    Leaflets obscurely veined, at least lateral veins not prominent; pod lunate, oblong-lunate or oblong-falcate, 2.5–5 cm. long:
      Pods puberulent, but eglandular; calyces and pedicels glabrous, eglandular; leaflets (6–) 10–18 mm. long .......... 5. *C. californica*

Pods glandular, with or without some additional eglandular hairs; leaflets mostly less than 12 mm. long; calyces glandular or puberulent, rarely glabrous:
Calyces and pedicels stipitate-glandular, not puberulent:
Leaflets entire; pedicels jointed below middle; stamens one and one-half times as long as petals .......................................... 6. *C. pannosa*
Leaflets crenate; pedicels jointed above middle; stamens about equaling or slightly shorter than petals .................................. 7. *C. placida*
Calyces and pedicels puberulent to pilosulous, eglandular:
Leaflets chartaceous, glabrous or glabrate, not glandular; midvein not impressed; petals gland-dotted over lower three-fourths of outer surface .. 8. *C. palmeri*
Leaflets subfleshy, densely puberulent, sparsely punctate-glandular beneath, midvein impressed beneath, obscure above; petals gland-dotted over whole outer surface ................................................ 9. *C. arenosa*
Leaflets strongly reticulate- and pinnate-veined; pod straight, 4.5–6.5 cm. long, not lunate .................................................. 10. *C. caladenia*

## 1. Caesalpinia pumila (Britt. & Rose) Hermann, Jour. Wash. Acad. 38: 237, 1948

*Caesalpinia gracilis* Benth., in Hemsl., Diagn. Pl. Mex. 9, 1878. Not *C. gracilis* Miq., 1855.
*Guaymasia pumila* Britt. & Rose, N. Amer. Fl. **23**: 322, 1930.

Open, slenderly branched shrub 2–3 m. tall with sparsely strigulose but soon glabrate twigs and leaves; bark red-brown to dark gray with small, circular, white lenticels; petioles slender, 1–4.5 cm. long; pinnae 2–3 pairs, 5–10 cm. long; leaflets 2–4 pairs, broadly obovate to nearly orbicular, 0.5–2 cm. long, broadly cuneate at base, sparsely strigulose when young, soon glabrate and eventually chartaceous, yellow-green, distinctly pinnately veined; racemes few-flowered; pedicels 1–2 cm. long; calyx pubescent, upper lobes oblong, subequal, 6–7 mm. long, lower narrower, 10–12 mm. long; petals yellow, 8–9 mm. long; pod broadly elliptic to suborbicular, 1.5–2 cm. wide, 2–3.5 cm. long, apiculate, glabrous, reddish and faintly glaucous, distinctly veined transversely, 1–2-seeded; seeds suborbicular, about 7–10 mm. wide, dark red-brown.

Hillsides, gravelly plains, and margins of arroyos, Lower Sonoran Zone, mainly in granitic soils, central Sonora. July–Oct.

## 2. Caesalpinia platyloba S. Wats., Proc. Amer. Acad. 21: 425, 1886

*Brasilettia platyloba* Britt. & Rose, N. Amer. Fl. **23**: 321, 1930.

Small tree 4–8 m. tall with brownish gray or reddish twigs with scattered lenticels, puberulent when young but later glabrate; petioles 3–4 cm. long, densely puberulent; pinnae 2–4 pairs, 1–3 dm. long; rachis loosely pubescent to glabrate; leaflets 4–7 pairs, ovate, elliptic, or oblong, often obliquely unequal at base, 1–3 cm. wide, 2–6 cm. long, pubescent when young, later glabrate, shining-green above, duller and clearly pinnately veined beneath; racemes 5–15 cm. long, few-flowered; pedicels about 1 cm. long, densely puberulent, jointed near apex; calyx lobes oblong, lower 6–7 mm. long, conspicuously lacerate and glandular-margined, upper four 5–6 mm. long, eglandular, all densely puberulent; petals 7–8 mm. long, dotted with minute sessile glands on outer surfaces; pod straight, oblong, 2–3 cm. broad, 5–10 cm. long, on a stipe 0.5–1 cm. long, dark reddish brown, finely but closely puberulent; seeds suborbicular-lenticular, 7–8 mm. in diameter, greenish brown, smooth.

Gravelly slopes and hillsides, Sonoran Zones, Chihuahua and southern Sonora (Tesopaco) to Colima and possibly to Yucatan. July–Sept.

**3. Caesalpinia gilliesii** Wall., in Hook., Bot. Misc. **1**: 129, *pl. 34,* 1830

*Poinciana gilliesii* Hook., loc. cit.
*Caesalpinia macrantha* Delile, Ind. Sem. Hort. Monsp. **1838**: 3, 1838.
*Erythrostemon gilliesii* Klotzsch, in Link, Klotzsch & Otto, Ic. Pl. **2**: 97–100, *pl. 39,* 1844.

Large shrub or tree to 10 m. tall with densely glandular young twigs and inflorescences; stipules ovate to reniform or orbicular, 3–5 mm. long; leaves 1–3 dm. long; pinnae 7–25, 2–3 cm. long; leaflets 7–15, oblong, 5–8 mm. long, obtuse or rounded at apex, glabrous, paler and with 1–several black sunken glands near margins beneath; inflorescence 1–3.5 dm. long; pedicels 1.5–3 cm. long; calyx lobes nearly equal, 12–15 mm. long, fimbriate to lacerate toward tips, densely glandular-pubescent; petals clear yellow, 2.5–3 cm. long; filaments 8–12 cm. long, bright red; pod oblong, somewhat falcate, 6–12 cm. long, about 1.5 cm. wide, glandular-dotted.

Escaped from cultivation in Mohave, Gila, Pima, and Cochise Counties, Arizona, and in New Mexico, Texas, and Sonora. April–Sept.

**4. Caesalpinia pulcherrima** (L.) Swartz, Obs. Bot. 166, 1791

*Poinciana pulcherrima* L., Sp. Pl. 380, 1753.

Rounded shrub or small tree to 5 m. tall with glabrous or sometimes hirsute-prickly young stems and glabrous herbage; leaves 2–3 dm. long; petioles stoutish, 2–5 cm. long; rachises slender; pinnae 3–9 pairs, 3–6 cm. long, an acicular bract 1–2 mm. long subtending each pinna; leaflets 5–12 pairs, oblong to oblong-obovate, 5–10 mm. wide, 7–20 mm. long, obtuse to broadly and abruptly cuneate at base, rounded to subtruncate and mucronulate at apex, bright green above, paler and dull beneath, reticulately veined; racemes terminal and axillary, 1–4 dm. long, often bearing nearly mature fruit and unopened buds at same time; pedicels 2–5 cm. long at anthesis, to 10 cm. long in fruit; lower calyx lobe 12–15 mm. long, cucullate, others about 1 cm. long; petals red-orange to yellow, 1.5–2.5 cm. long; stamens 5–6 cm. long, bright red, hairy at base; legumes oblong, somewhat oblique, 15–18 mm. wide, 7–13 cm. long, coriaceous, often somewhat glaucous, glabrous; seeds obovate to suborbicular, 10–12 mm. long, often minutely pitted near hilum.

Desert washes and slopes, Lower Sonoran Zone, central Sonora southward into the tropics. An escape, but thoroughly naturalized in Sonora and in nearly all tropical areas. July–Oct.

**5. Caesalpinia californica** (A. Gray) Standley, Contr. U.S. Nat. Herb. **23**: 426, 1922

*Caesalpinia mexicana* var. *californica* A. Gray, Proc. Amer. Acad. **5**: 157, 1862.
*Poinciana californica* Rose, Contr. U.S. Nat. Herb. **13**: 303, 1911.
*Poincianella californica* Britt. & Rose, N. Amer. Fl. **23**: 331, 1930.

Small, openly branched shrub 2 m. high or less with glabrous twigs, leaves, and inflorescences; petiole 1–2.5 cm. long, rarely with a few scattered, stipitate glands near base; pinnae 3–7, lower pair often shorter than upper; leaflets 3–4 pairs, oblong to broadly oblong-obovate, 2–8 mm. wide, 6–18 mm. long, obtuse to retuse at apex, dull beneath, lateral venation inconspicuous or none; inflorescence 1–2.5 dm. long or more; pedicels slender, 1–2 cm. long, jointed at or slightly below middle; calyx glabrous or lobes slightly ciliolate, 6–7 mm. long, subequal; petals 7–9 mm. long, bearing numerous dark red, sessile glands; stamens 10–12 mm. long, curved upward at tips; pods oblong-lunate, 1–1.5 cm. wide, 4–5 cm. long, apiculate, puberulent or glabrate, eglandular; seeds 1–3, subcordate, 6–8 mm. in diameter, dark reddish brown, smooth and shining.

Dry hillsides and along washes, Lower Sonoran Zone, vicinity of La Paz and southward in Baja California. Feb.–March, and occasionally in the fall after rains.

**6. Caesalpinia pannosa** Brandegee, Proc. Calif. Acad. II, **2:** 150, 1889

*Poinciana pannosa* Rose, Contr. U.S. Nat. Herb. **13:** 303, 1911.
*Poincianella pannosa* Britt. & Rose, N. Amer. Fl. **23:** 331, 1930.

An unarmed shrub to 1.5 m. tall with light gray or nearly white thin bark and glabrous leaflets; petioles 1–3 cm. long; pinnae 3–7; leaflets 2–5 pairs, oblong to obovate, retuse or rounded at apex, 3–15 mm. long, midrib distinct but lateral veins obscure; inflorescence 2–35 cm. long; pedicels 6–15 mm. long, jointed below middle, sparsely stipitate-glandular or sometimes eglandular; calyx tube stipitate-glandular, lobes oblong, subequal, 5–6 mm. long, ciliate or glabrous, eglandular or slightly glandular at base; petals 6–7 mm. long, bearing numerous sessile, reddish glands; stamens 9–10 mm. long; legume sublunate-oblong, 1–1.5 cm. wide, 2.5–3.5 cm. long, dotted with reddish brown to black sessile glands, puberulent to glabrous; seeds obovate, 6–8 mm. long, dark reddish brown, shining.

Arroyos and dry hillsides, Lower Sonoran Zone, Cape Region, Baja California, but entering the Sonoran Desert west of La Paz. Dec.–Feb.

**7. Caesalpinia placida** Brandegee, Proc. Calif. Acad. II, **3:** 131, 1891

*Poinciana placida* Rose, Contr. U.S. Nat. Herb. **13:** 303, 1911.
*Poincianella placida* Britt. & Rose, N. Amer. Fl. **23:** 331, 1930.

Openly branched shrub 1–2 m. high with dark brown, glabrous twigs and stipitate-glandular pedicels, petioles, and rachises; petioles 1–1.5 cm. long; pinnae 2–3, about 1.5 cm. long; leaflets 3–7 pairs, oblong, 1.5–2.5 mm. broad, 4–7 mm. long, obtuse or rounded at both ends, crenulate at least toward apex, subpapillose on both surfaces; raceme 4–15 cm. long, few-flowered; pedicels 1–1.5 cm. long, jointed above middle; calyx lobes ovate, 5–6 mm. long, obtuse, dark red with yellow margins, stipitate-glandular and somewhat puberulent, soon deciduous; petals 7–9 mm. long, densely glandular-dotted below; stamens about equaling petals; ovary densely stipitate-glandular and puberulent; pod oblong-falcate, 9–10 mm. broad, 3.5–4 cm. long, closely stipitate-glandular, glands dark red.

Rocky canyons and hillsides, Lower Sonoran Zone, Cape Region, from the vicinity of La Paz, Espíritu and Pichilinque Islands, and south in Baja California. Feb.–March.

**8. Caesalpinia palmeri** S. Wats., Proc. Amer. Acad. **24:** 47, 1889

*Poinciana palmeri* Rose, Contr. U.S. Nat. Herb. **13:** 303, 1911.
*Poincianella palmeri* Britt. & Rose, N. Amer. Fl. **23:** 332, 1930.
*Poincianella arida* Britt. & Rose, loc. cit.

Low, openly branched shrub 2–3 m. tall with glabrous or slightly puberulent but soon glabrate twigs, petioles, and leaflets; bark on branchlets dark gray with obvious, elliptic to orbicular, light gray lenticels; petioles slender, 1–2 cm. long; pinnae 3–7, distant in age; leaflets 2–5 pairs, elliptic to oblong-elliptic, or obovate, obtuse to rounded, to 6 mm. wide, 6–10 mm. long, glabrous or sparsely puberulent in youth, soon glabrate, chartaceous, eglandular; inflorescence 2–6 cm. long, few-flowered, densely puberulent but mainly eglandular; pedicels 6–10 mm. long, jointed near tip; calyx lobes oblong, 6–8 mm. long, densely puberulent, margins often glandular-ciliolate; petals 8–9 mm. long, orange-yellow, densely dotted with sessile glands on lower three-fourths of outer surface; stamens 10–12 mm. long; legume oblong-lunate, 12–14 mm. broad at widest part, 2.5–4 cm. long, densely puberulent and stipitate-glandular when young, sparsely puberulent to glabrate and dotted with reddish glands in age.

Rocky hillsides and sandy slopes, Lower Sonoran Zone, vicinity of Carbo, Sonora, south into Sinaloa. Feb.–April; July–Sept.

**9. Caesalpinia arenosa** Wiggins, Contr. Dudley Herb **3:** 68, *pl. 16, figs. 1–3,* 1940

Openly branched shrub 1.5–2 m. tall with puberulent to pilose twigs, leaves, and inflorescences; twigs tortulose; petioles 1–3 cm. long; pinnae 1–3 pairs; leaflets 2–3 pairs, subfleshy, elliptic, 3–7 mm. wide, 5–12 mm. long, rounded at apex, irregularly and sparsely punctate-glandular, puberulent on both surfaces, midvein impressed above, lateral veins obscure; inflorescence 5–12 cm. long, densely pubescent; calyx densely puberulent, tube 1.5–2 mm. deep, lobes oblong, 4–5 mm. long, about equal; petals narrowly obovate, 6–8 mm. long, densely gland-dotted on outer surface, glands reddish purple or brownish red; stamens 9–10 mm. long; legume obovate-lunate, 2.5–3.5 cm. long, 8–11 mm. wide, apiculate, stipitate-glandular; seeds lenticular, 5–6 mm. long, yellowish brown.

Along sandy washes and dunes, Lower Sonoran Zone, from the vicinity of Comondú and Loreto, south into the Cape Region, Baja California. Jan.–March, and occasionally in summer and fall after rains.

**10. Caesalpinia caladenia** Standley, Contr. U.S. Nat. Herb. **20:** 214, 1919

*Poincianella caladenia* Britt. & Rose, N. Amer. Fl. **23:** 329, 1930.

Shrub or small tree to 5–6 m. tall, twigs and leaves glabrous or essentially so, occasionally sparsely stipitate-glandular; petioles 1–3 cm. long, pubescent to glabrate; pinnae 3–9; leaflets 2–4 pairs, oval, elliptic-oblong to oblong-obovate, 5–17 mm. wide, 1–3 cm. long, often oblique at base, rounded at apex, glabrous, chartaceous, veins evident; inflorescence 6–20 cm. long; pedicels 5–16 mm. long, jointed near apex, short-pilose and stipitate-glandular; calyx tube 5–7 mm. broad, lobes oblong, 7–9 mm. long, pilose, marginally glandular; petals 10–12 mm. long, glandular only on lower part of outer surface; stamens 12–15 mm. long; pod straight, 12–16 mm. wide, 4.5–6.5 cm. long, velvety-pubescent and stipitate-glandular.

Rocky mountain sides and canyons, Lower Sonoran Zone, Sonora to Colima. Feb.–March.

## 8. CERCIDIUM Tulasne, Arch. Mus. Paris 4: 133, 1844

Trees or large shrubs with smooth greenish bark and twigs armed with short, nearly straight, stipular spines or branchlets spinose-tipped. Leaves bipinnate with 1–2 pairs of pinnae and small fugacious leaflets. Flowers medium-sized, yellow, more or less flecked with red, in short axillary racemes. Calyx tube broadly campanulate; lobes 5, valvate. Petals 5, slightly unequal. Stamens 10, distinct; filaments pilose near base; anthers versatile. Ovary few-ovuled. Pods flattened, oblong to linear-oblong, coriaceous. Seeds 1 to few, somewhat compressed.

Leaflets minute, 1–2 mm. long; twigs bright yellow-green, spinose terminally but stipular spines lacking; pods swollen, moniliform ........................ 1. *C. microphyllum*

Leaflets larger, 2–10 mm. long; twigs bluish green, grayish, or whitish, armed with nodal spines or completely unarmed; pods flattened, sometimes narrowed between the seeds but not moniliform:

Branches slender, drooping, densely strigose-lanate, unarmed ............... 2. *C. molle*

Branches stouter, erect or ascending, not drooping, puberulent or glabrous, not lanate, armed with nodal spines:

Calyx uniformly puberulent; pinnae 1–2 pairs; leaflets 4–8 pairs, oblong or elliptic:
Pods elliptic to oblong, not constricted or narrowed between seeds; ovary glabrous at anthesis; leaflets oblong, 5–10 mm. long, tardily glabrate ......... 3. *C. praecox*
Pods oblong, narrowed or somewhat constricted between seeds; ovary puberulent at anthesis; leaflets elliptic, 2–4 mm. long, cinereous-puberulent ...... 4. *C. sonorae*
Calyx glabrous or sparsely puberulent near base of tube and tips of lobes; pinnae 1 pair; leaflets 2–4 pairs, obovate:
Branchlets glabrous or sparsely puberulent and soon glabrate; pedicels slender, 5–16 mm. long, glabrous or sparsely appressed-puberulent ........... 5. *C. floridum*
Branchlets persistently spreadingly pilosulous; pedicels stoutish, 3–6 (or rarely 10) mm. long, spreadingly pilosulous ......................... 6. *C. peninsulare*

**1. Cercidium microphyllum** (Torr.) Rose & Johnston, in I. M. Johnston, Contr. Gray Herb. **70:** 66, 1924

*Parkinsonia microphylla* Torr., Pacif. R. Rep. **4:** 82, 1857.
*Cercidiopsis microphylla* Britt. & Rose, N. Amer. Fl. **23:** 306, 1930.

Shrub or tree to 8 m. tall with slender, bright yellow-green, erect or strongly ascending branchlets, often presenting a somewhat "broomlike" habit; twigs spinose apically but without nodal spines, puberulent when young but soon glabrate; pinnae 1 pair, 1.5–2.5 cm. long, virtually sessile so no petiole on common rachis evident; leaflets 4–6 pairs, broadly elliptic to broadly oblong, 1–2 mm. long, puberulent when young; racemes few-flowered, axillary, rarely over 2.5 cm. long; pedicels slender, 5–10 mm. long, sparsely puberulent; calyx 5–6 mm. long, nearly glabrous, tube 1–1.5 mm. long; petals yellow, upper paler than lower, obovate-flabelliform, 5–8 mm. long; claws pilosulous marginally; stamens 8–10 mm. long; ovary usually puberulent at anthesis; pods swollen, moniliform, 4–8 cm. long, closely investing the brown, ovoid-elliptic seeds, these 8–10 mm. long.

Sandy plains, washes, and sometimes hillsides, Lower Sonoran Zone, Colorado Desert, California, western Arizona, and south throughout arid parts of Sonora and Baja California. March–May.

**2. Cercidium molle** I. M. Johnston, Proc. Calif. Acad. IV, **12:** 1038, 1924

An unarmed tree to 6 m. tall with slender, densely strigose-lanate, drooping branchlets and finely strigose foliage; petioles 5–8 mm. long; pinnae 1 pair, 3–4 cm. long; leaflets 4–6 pairs, oblong, 6–10 mm. long, 2.5–4.5 mm. wide, truncate to faintly retuse at apex, obliquely narrowed or rounded at base, eventually glabrate except along margins; inflorescence few-flowered, 2–3.5 cm. long; pedicels about 1 cm. long, geniculately jointed just above middle; sepals oblong, 7–8 mm. long, puberulent outside and very sparsely so inside; petals lemon-yellow, lower one 13 mm. long, upper four 10 mm. long; filaments about 1 cm. long; ovary strigose at anthesis; pods compressed, 6–8 mm. wide, 4–9 cm. long, 1–4-seeded, narrowed between seeds, constricted places about 3 mm. wide.

Known only from the type locality, Agua Verde Bay, Baja California, Lower Sonoran Zone. March–May.

**3. Cercidium praecox** (Ruiz & Pav.) Harms, Bot. Jahrb. **42:** 91, 1908

*Caesalpinia praecox* Ruiz & Pav., in Hook. & Arn., in Hook., Bot. Misc. **3:** 208, 1833.
*Cercidium spinosum* Tulasne, Arch. Mus. Paris **4:** 134, 1844.
*Rhetinophloeum viride* Karst., Fl. Columb. **2:** 25, *pl. 113,* 1862.
*Cercidium viride* Karst., Bot. Jahrb. **8:** 346, 1887.
*Cercidium plurifoliolatum* M. Micheli, Mém. Soc. Genève **34:** 269, *pl. 18,* 1903.
*Cercidium unijuga* Rose, Contr. U.S. Nat. Herb. **8:** 301, 1905.
*Cercidium goldmanii* Rose, loc. cit.

A round-crowned tree to 9 m. tall with smooth bluish-green bark and essentially glabrous twigs armed with straight nodal spines 5–20 mm. long; petioles 5–12 mm. long, somewhat pilosulous or eventually glabrate; pinnae 1–2 pairs; leaflets 5–8 pairs, oblong, 5–10 mm. long, obtuse to retuse at apex, spreadingly pilosulous on both surfaces, tardily glabrate at least above; pedicels 6–10 mm. long, pilosulous; calyx 6–7 mm. long, uniformly fine-puberulent; petals 8–15 mm. long, golden-yellow; ovary glabrous at anthesis; pods oblong to linear-oblong, 8–10 mm. wide, 5–8 cm. long, narrowed toward each end, not constricted between seeds.

Arid plains of deserts, Sonora and Baja California to Venezuela and Peru. Chiefly March–April.

**4. Cercidium sonorae** Rose & Johnston, in I. M. Johnston, Contr. Gray Herb. **70:** 66, 1924

A smooth-barked tree 6–8 m. tall with ascending or spreading, divaricate twigs puberulent and armed with nearly straight spines 5–15 mm. long; petioles 1–5 mm. long, pilosulous; pinnae 1–2 pairs, 12–20 mm. long, rachis hairy; leaflets 4–6 pairs, rounded-oblong, 1.5–2 mm. wide, 2–4 mm. long, puberulent with subappressed hairs on both surfaces; racemes few-flowered, 3 cm. long or less; pedicels 6–10 mm. long, densely pubescent, jointed near tip; calyx uniformly but sparsely puberulent with spreading hairs, lobes 5–6 mm. long; petals 8–12 mm. long, upper 1 white or pale cream and dotted with red-brown flecks, lower 4 pale yellow; stamens 10–12 mm. long; ovary densely strigose at anthesis; pods compressed, 7–8 mm. wide, 4–7 cm. long, somewhat narrowed between seeds, attenuately narrowed at each end; seeds 1–4, oblong-lenticular.

Arid plains and gentle slopes, Lower Sonoran Zone, from the vicinity of Magdalena, Sonora, south to the plains south of Guaymas. March–May.

**5. Cercidium floridum** Benth. ex A. Gray, Pl. Wright. **1:** 58, 1852

*Parkinsonia florida* S. Wats., Proc. Amer. Acad. **11:** 135, 1876.
*Parkinsonia torreyana* S. Wats., loc. cit.
*Cercidium torreyanum* Sarg., Gard. & Forest **2:** 388, 1889.

A rounded-crowned tree to 12 m. tall with smooth, bluish-green, slightly glaucous bark and glabrous or sparsely puberulent but soon glabrate twigs armed with straight nodal spines 4–10 mm. long; leaves sparingly puberulent when very young, soon glabrous or sparsely ciliolate on margins of leaflets and on rachises; petiole 5–10 mm. long; pinnae 1 pair, 1.5–3 cm. long; leaflets 2–4 pairs, obovate, 3–5 mm. wide, 4–8 mm. long, rounded or faintly retuse; racemes glabrous or nearly so, 4–12 cm. long; pedicels slender, 5–16 mm. long, glabrous; calyx glabrous or sparsely and minutely puberulent near base and at tips of lobes, these about 6 mm. long; petals 9–12 mm. long, yellow; ovary glabrous at anthesis; pods linear-oblong or oblong, 1–5-seeded, 8–15 mm. wide, 4–10 cm. long, only slightly if at all narrowed between seeds, narrowed gradually to each end.

Sandy plains and washes, occasionally on rocky hillsides, Lower Sonoran Zone, western edges of Colorado Desert into desert areas of central Arizona, and south into Sonora and Baja California. March–May.

**6. Cercidium peninsulare** Rose, Contr. U.S. Nat. Herb. **8:** 301, 1905

Small, openly branched tree to 6 m. tall with densely short-puberulent twigs armed with straight, stoutish, nodal spines 3–7 mm. long, pubescence spreading; bark greenish on young branches, gray and lightly fissured on old trunks; petioles 2–5 mm. long;

pinnae 1 pair, rachis 4–10 mm. long, spreadingly puberulent; leaflets 2–3 pairs, obovate, rounded or shallowly retuse, 3–5 mm. wide, 6–12 mm. long, spreadingly pubescent on both surfaces at first, glabrate or nearly so above in age; racemes few-flowered, about 2 cm. long; pedicels stoutish, 3–6 (or sometimes 10) mm. long, pilosulous; calyx glabrous or sparingly puberulent at tips of lobes, these 5–6 mm. long; petals 8–11 mm. long, upper 1 pale cream or nearly white, lower 4 yellow; ovary glabrous; pods oblong, 8–12 mm. wide, 2.5–5 cm. long, narrowly cuneate at each end, 1–2-seeded, slightly narrowed between seeds.

Sandy valley floors, arroyos, and hillsides, Lower Sonoran Zone, vicinity of Comondú, Baja California, south into the Cape Region. March–April.

## 9. PARKINSONIA L., Sp. Pl. 375, 1753

A large shrub or tree to 12 m. tall with smooth, yellowish green bark becoming brown, and broken into small flakes in age, and bipinnate, nearly sessile leaves. Branches spiniferous. Common rachis of leaves very short; leaflets many, fugacious, the rachis flattened, persistent, to 6 dm. long. Flowers yellow, in showy racemes. Calyx tube slightly oblique, shallow, lobes 5, subequal, rounded, glabrous without, pubescent within. Petals 5, subequal, spreading. Stamens 10, distinct; filaments pilose at base; anthers versatile. Ovary short-stipitate, pubescent in youth. Pod linear, torulose, longitudinally striately veined, 2-valved, coriaceous. Seeds oblong, longitudinally arranged, smooth, hard, and shiny.

### 1. Parkinsonia aculeata L., Sp. Pl. 375, 1753

Branches slender, the twigs drooping, minutely puberulent when young, armed with spines 0.4–3 cm. long; bark green, smooth; petiole 1–3 mm. long, indistinguishable from short spine-tipped rachis, or rarely 2 pairs of pinnae and rachis slightly longer; leaflets 10–40 pairs or possibly more, short-petiolulate, elliptic to narrowly obovate, 2–8 mm. long, caducous, persistent rachises functioning as chief photosynthetic tissues; racemes several-flowered, 2–4 dm. long; pedicels slender, to 2 cm. long; calyx tube about 2 mm. deep, twice as broad, lobes 4–6 mm. long, rounded at apex, reflexed; petals flabelliform, 10–14 mm. long, erosulate, claws about half as long as limbs; pods linear-cylindrical, 3–18 cm. long, deeply constricted between seeds, these ovoid, 4–5 mm. in diameter, 8–10 mm. long, greenish, mottled with brown or purplish flecks.

Along arroyos and on sandy bottomland, Lower Sonoran and Tropical Zones, Colorado Desert, California, to Florida and south through Mexico, the West Indies, and Bahamas to South America. March–Aug., chiefly in April–May.

## Family 60. FABACEAE. Pea Family

Herbaceous, suffrutescent, or woody plants with alternate, stipulate, usually compound leaves; leaflets often subtended by stipels, petiolulate or sessile. Flowers mostly in axillary and terminal racemes, spikes, pseudoumbels, or solitary, perfect or rarely polygamodioecious. Hypanthium obsolete or reduced and obscure. Calyx 4–5-toothed, 2-lipped or teeth subequal and tube uniformly cleft between lobes. Corolla irregular, usually distinctly papilionaceous; petals usually 5, inserted on rim of short hypanthium, on receptacle, or on stamineal tube; banner or standard (uppermost petal) enclosing 2 lateral wing petals in bud, these in turn enclosing 2 lower keel petals, the latter usually adnate along lower margin for varying distances but claws free. Stamens usually 10, sometimes 5 or 9, their filaments all united into a stamineal tube (stamens then monodelphous) or uppermost (vexillar) filament wholly or partly free from tube (then considered diadelphous); anthers all alike or alternate ones larger than others. Pistil 1, simple; ovary superior, 1-celled or sometimes 2-celled by intrusion of sutures, 1–many-ovuled. Fruit a legume or loment, 1–many-seeded, dehiscent by 2 valves or indehiscent. Seeds with large cotyledons, mostly without endosperm, estrophiolate or less commonly strophiolate.

Stamens monodelphous or pseudomonodelphous (vexillar stamen connate with tube near middle but free both above and below point of connation):
  Calyx distinctly to strongly 2-lipped, one or both lips usually lobed or toothed:
    Leaves palmately or digitately compound, leaflets attached at about same level near or at tip of petiole:
      Pods inflated; corollas mostly yellowish .......................... 1. *Crotalaria*
      Pods compressed, not inflated; corollas mainly blue or lavender ......... 2. *Lupinus*
    Leaves odd-pinnate, pseudodigitately 3-foliolate with leaflets attached fairly closely together or more remotely near tip of petiole, or simple:
      Vexillar stamen connate full length of stamineal tube; pod 2–6-jointed and joints separating from one another:
        Upper lip of calyx consisting of connate upper 4 calyx lobes, tube very slender, only lower lobe free; ovary 2–3-ovulate; terminal joint of pod often abortive ...... .................................................... 27. *Stylosanthes*
        Upper lip of calyx consisting of connate upper pair of lobes, laterals and lowermost free; ovary 5–many-ovulate; terminal joint of pod usually containing a seed ... ...................................................... 28. *Zornia*
      Vexillar filament free at base, attached to stamineal tube near middle, again free above attachment; pods not jointed, often partly septate between seeds .... 34. *Galactia*
  Calyx not 2-lipped, lobes equal or subequal:
    Herbage and calyces more or less gland-dotted; stamens wholly monodelphous; pods neither dehiscent nor samaroid:
      Petals hypogynous, not adnate to stamineal tube; ovary 1-ovulate ........ 8. *Psoralea*
      Petals, or some of them, with claws adnate to base of stamineal tube; ovary mostly 2–3-ovulate:
        Wing petals and keel petals inserted on hypanthium or near base of stamineal tube:
          Flowers strongly papilionaceous; petals thin, clawed; pods usually included in calyx; banner inserted on hypanthium ......................... 10. *Dalea*
          Flowers indistinctly or not at all papilionaceous; petals fleshy, clawless or nearly so; pods usually exserted from calyx; banner inserted on base of stamineal tube .................................................... 11. *Errazurizia*
        Wing petals and keel petals all inserted on stamineal tube:
          Stamens 5; wing and keel petals inserted at mouth of stamineal tube, alternating with free portions of stamen filaments; banner inserted on hypanthium, long-clawed ............................................ 12. *Petalostemon*
          Stamens 9–10; wing and keel petals not inserted alternately with stamens; banner inserted at base of stamineal tube, short-clawed ........... 13. *Thornbera*
    Herbage and calyces glandless or at least not conspicuously so; pods dehiscent, or in *Nissolia* indehiscent but samaroid:
      Pods dehiscent; plants herbs, shrubs, or trees, not vines:
        Leaves stipulate; plants herbaceous to shrubby but not arborescent; ovary usually glabrous ................................................ 15. *Tephrosia*
        Leaves estipulate; plants arborescent; ovary usually strigose ....... 17. *Willardia*
      Pods indehiscent, with a terminal samaralike wing; plants prostrate to scandent woody vines ...................................................... 32. *Nissolia*

Stamens clearly diadelphous, vexillar stamen wholly free from stamineal tube:
  Pods indehiscent or tardily dehiscent:
    Calyx not distinctly 2-lipped, lobes equal or subsequal; pods not breaking into loments or joints at maturity:
      Margins of the leaflets serrulate, at least toward rounded apex; leaflets 3, laterals inserted only slightly below terminal leaflet:
        Pod spirally coiled or strongly falcate-arcuate ...................... 3. *Medicago*
        Pod straight or nearly so, not strongly coiled:
          Petals all free from stamineal tube; pod leathery, not membranous; petals and calyx not marcescent ...................................... 4. *Melilotus*
          Petals, or at least lower ones, adnate to base of stamineal tube; pod more or less membranous; petals and calyx marcescent, including pod ...... 5. *Trifolium*
      Margins of leaflets entire or essentially so, not denticulate-serrulate; leaflets usually more than 3, pinnately arranged, or leaves simple:
        Leaves pinnately compound; pods not moniliform; plants spinose:
          Pods terete to flattened, not inflated:
            Inflorescence umbel-like or of a single flower; stipules (in ours) reduced to small dark glands ...................................... 6. *Hosackia*
            Inflorescence racemose or spicate-racemose; stipules not glandular:
              Leaves not glandular-punctate; pubescence of malpighiaceous hairs (attached in middle); stipules somewhat adnate ................. 7. *Indigofera*
              Leaves and stems, and usually calyces, glandular-punctate; pubescence of basally attached hairs; stipules free from petioles:
                Calyx lobes subequal; ovary 1-ovulate; leaflets 5 or fewer ...... 8. *Psoralea*
                Calyx cleft more deeply between posterior lobes than between anterior ones; ovary 2–4-ovulate; leaflets 9–41 .............. 9. *Eysenhardtia*
          Pods inflated or turgid, papery or leathery:
            Plants shrubby; banner shorter than keel; styles hairy just below stigma ...... ............................................ 23. *Sutherlandia*
            Plants herbaceous or bases slightly woody; banner equaling or surpassing keel; style glabrous ...................................... 24. *Astragalus*
        Leaves simple; pods moniliform, but joints not separating at maturity; branchlets spinescent ............................................... 25. *Alhagi*
    Calyx deeply to shallowly (but distinctly) 2-lipped; pods breaking transversely into joints (loments) at maturity:
      Pods wingless; plants herbaceous or suffrutescent, not arborescent; wings free from keel or adnate to it only at base:
        Leaflets numerous (21–51), linear; loments pubescent with straight or curved but hookless hairs ...................................... 26. *Aeschynomene*
        Leaflets 1–3, linear-lanceolate to ovate; loments glabrous or puberulent with minute hooked hairs ........................................ 29. *Desmodium*
      Pods winged along sutures; plants arborescent; wings adnate to keel near middle ... ........................................................ 30. *Piscidia*
  Pods readily dehiscent at maturity, often elastically:
    Stipules represented by small, conical to rounded glands ................. 6. *Hosackia*
    Stipules scalelike, foliaceous, or lacking, but not glandular:
      Seeds strophiolate:
        Leaves with several to many linear to lanceolate or oblong leaflets; stipels present; banner not resinous-glandular; herbage not gland-dotted .... 14. *Brongniartia*
        Leaves 3-foliolate, ovate to rhombic-ovate; stipels none; banner often resinous-glandular on back; leaves gland-dotted at least beneath ...... 35. *Rhynchosia*
      Seeds estrophiolate:
        Leaves simple; seeds quadrate, notched opposite hilum and grooved between notches ................................................ 16. *Sphinctospermum*

Leaves compound, of 3–many leaflets; seeds not constricted centrally:
Branchlets or petioles or both more or less spinose:
Leaflets 4–12 pairs, about 2 cm. long or less; seeds brown to black, 8–9 mm. long; banner equaling or only slightly exceeding wings ...... 18. *Olneya*
Leaflets 3, in ours to 7.5 cm. long; seeds dark red, 12–18 mm. long; banner much larger than other petals ........................... 33. *Erythrina*
Branchlets and petioles unarmed:
Leaves odd-pinnately compound, with 7–35 leaflets or more:
Plants shrubby or arborescent:
Pods strongly inflated, bladderlike; stipels none; 2 bracteoles beneath each flower; style glabrous ............................... 19. *Diphysa*
Pods not inflated; stipels present; bracteoles none; style bearded ........ ............................................... 20. *Coursetia*
Plants herbaceous or suffrutescent:
Stipules inconspicuous, caducous; leaves without tendrils; seeds quadrate to oblong-cylindric; pods septate between seeds:
Style bearded along upper side near apex; lower calyx lobes equaling or longer than tube; plants perennial, suffrutescent..21. *Benthamantha*
Style glabrous; lower calyx lobes shorter than tube; plants (in ours) annual herbs ..................................... 22. *Sesbania*
Stipules conspicuous, persistent, semisagittate; leaves with terminal tendrils; seeds globose or subglobose; pods not septate between seeds...31. *Vicia*
Leaves 3-foliolate:
Shrubs or trees; stipels represented by conical glands; calyx tube subtruncate; banner much longer than other petals .................. 33. *Erythrina*
Herbs, vines, or suffrutescent perennials; stipels bractlike, setaceous, or lacking; calyx tube usually distinctly toothed; banner usually not greatly exceeding other petals:
Style glabrous, straight or curved but not spirally twisted or coiled; keel curved but not twisted or coiled:
Leaflets not gland-dotted; ovary several- to many-ovuled; pod often somewhat septate between seeds ........................ 34. *Galactia*
Leaflets gland-dotted at least beneath; ovary 1–2-ovuled, not septate .... ........................................... 35. *Rhynchosia*
Style bearded, twisted, or spirally coiled; keel twisted or coiled like style ............................................. 36. *Phaseolus*

## 1. CROTALARIA L., Sp. Pl. 714, 1753

Annual or perennial herbs or low erect shrubs with digitately trifoliolate or rarely simple leaves. Stipules present. Flowers in few- to many-flowered lateral and terminal racemes. Calyx somewhat 2-lipped. Petals yellow or tinged with red; keel curved and often exserted beyond wings. Stamens diadelphous, alternate stamens long with basifixed anthers, others short with versatile anthers. Pods inflated, subglobose to oblong, several- to many-seeded.

Leaves simple; plant prostrate or decumbent; stipules decurrent on stem ..... 1. *C. angulata*
Leaves trifoliolate; plants mostly erect or ascendingly branched; stipules linear, not decurrent on stems:
Pod pubescent with spreading hairs; calyx 6–9 mm. long, tube one-fourth to one-fifth as long as lobes; leaflets broadly elliptic to obovate ...................... 2. *C. incana*
Pod appressed-puberulent; calyx 3–4 mm. long, tube equaling lobes; leaflets narrowly elliptic to linear-oblanceolate ........................................ 3. *C. pumila*

**1. Crotalaria angulata** Mill., Gard. Dict. ed. 8, no. 9, 1768
(For extensive synonymy see Rhodora **41:** 341, 1939.)

Decumbent or prostrate, perennial herb from a slender taproot with spreadingly hirsute stems 1–4 dm. long; stipules of upper leaves narrowly triangular, 3–8 mm. long and decurrent on stem 1–2.5 cm., those of lower leaves smaller and not decurrent or lacking; petioles 1–2 mm. long or leaves subsessile; leaves simple, broadly ovate at base of stem, gradually narrower toward tip of stem to narrowly lanceolate, 1–4.5 cm. long, villous with tawny or brownish hairs; peduncles slender, 1–3 cm. long at anthesis, 1–5-flowered, elongating in fruit; calyx lobes linear-lanceolate, 5–6 mm. long, tube 1.5–2.5 mm. long; corolla cream-yellow, only slightly exceeding calyx lobes; pods broadly oblong, 7–9 mm. in diameter, 1.5–2.5 cm. long, rounded at both ends, pubescent in youth but soon glabrate, reticulately veined.

In sandy to clay soil, Mississippi to North Carolina, and in Mexico from central Sonora southward to Guatemala. Collected in sandy soil in the Sonoran Desert near Mátape and near Cedros. July–Oct.

**2. Crotalaria incana** L., Sp. Pl. 716, 1753
(For extensive synonymy see Rhodora **41:** 350, 1939.)

Erect or ascending annual 2–6 dm. tall with puberulent to glabrate stems and glabrous or glabrate leaflets; stipules linear or filiform, 4–6 mm. long, caducous; petioles to 4 or 5 cm. long, puberulent to glabrate; leaflets 3, obovate to suborbicular, 1–4 cm. long, glabrous, or somewhat hairy along midvein beneath; peduncles to 2 or 3 dm. long, terminal and lateral, few- to many-flowered; calyx lobes narrowly lance-triangular, 6–9 mm. long, tube 1–2.5 mm. long; corolla 10–13 mm. long, yellow; pods oblong, about 1 cm. in diameter, 2.5–4 cm. long, tomentulose with spreading brownish hairs; seeds greenish brown, about 3 mm. long.

A widely spread weed in the southern United States from Texas to Florida and through Chihuahua, southern Sonora and southern Baja California, to northern South America and the West Indies. It is also in Hawaii, the Philippines, the East Indies, and Africa. July–Oct. in our area.

**3. Crotalaria pumila** C. G. Ortega, Hort. Mat. Dec. **2:** 23, 1797
(For extensive synonymy see Rhodora **41:** 358, 1939.)

Erect, ascending, or rarely decumbent annual which may become a short-lived perennial under favorable conditions; stems, petioles, peduncles, and calyces appressed-puberulent; stipules filiform, 2–3 mm. long, tardily deciduous; petioles 2.5 cm. long or less; leaflets 3, narrowly elliptic to linear-oblanceolate, 3–15 mm. wide, to 6 cm. long, bright green and glabrous above, slightly glaucous and appressed-puberulent beneath; racemes to 3 dm. long, many-flowered; calyx 3–4 mm. long, lobes triangular, equaling tube; petals about 10 mm. long, bright yellow, keel and sometimes banner faintly washed with red and drying reddish to purple; acumen of keel slender, 7–8 mm. long, conspicuously exserted beyond wings; pods ovoid or broadly rounded-oblong, 6–9 mm. in diameter, 10–15 mm. long, appressed-strigose; style persistent, forming a slender beak 1.5–2.5 mm. long; seeds yellow to red-brown, reniform, about 3 mm. long, smooth.

Arizona and Southern Utah to Florida and south through Mexico and the West Indies into South America. This weedy species is the commonest *Crotalaria* in the Sonoran Desert. It inhabits sandy, gravelly, or clay soils on flats and along the margins of washes, but is absent from the drier areas of the desert. In Arizona and Sonora it extends into the Upper Sonoran Zone. Aug.–Oct.

## 2. LUPINUS L., Sp. Pl. 721, 1753*

Annual, biennial, or perennial herbs or shrubs to over 2 m. tall. Leaves alternate, palmately compound or rarely unifoliolate, with short to elongated petioles, these often more or less dilated at base, stipulate; leaflets 3–17, usually 5 or over. Flowers 0.5–2 cm. long, racemose, sometimes verticillately interrupted, on pedicels 1–12 mm. long. Calyx bilabiate, lips entire, toothed, or cleft, interstitial bracteoles often present. Banner commonly with a ventral median sulcus, more or less umbonate near center on each side of sulcus, sides frequently reflexed; wings usually glabrous, equaling or shorter than keel; keel arcuate to nearly straight along the upper margin, often ciliate along one or both margins, ciliation on lower margin usually confined to area near claws. Stamens 10, monodelphous; anthers usually dimorphic. Carpel 1, becoming a flattened legume; ovules 2–12. Seeds hard, with a sunken hilum often surrounded by a thickened ring.

Keel nonciliate:
  Herbage densely villous with spreading hairs 1–2.5 mm. long; banner rounded or emarginate at apex; flowers deep blue or purplish:
    Flowers 7–9 mm. long; banner 4–5 mm. wide, 7–8 mm. long .......... 1. *L. concinnus*
    Flowers 6–7 mm. long; banner 3 mm. wide, 6 mm. long.... 1*a*. *L. concinnus* var. *orcuttii*
  Herbage appressed-pubescent; banner angled at apex; flowers white or pale bluish or pale yellowish:
    Leaflets spatulate, 5–6; petals white or pale bluish ........ 1*b*. *L. concinnus* var. *pallidus*
    Leaflets oblanceolate, 6–9; petals white or pale yellowish .......................... ........................................ 1*c*. *L. concinnus* var. *desertorum*
Keel ciliate, at least along one margin:
  Upper calyx lip shallowly bidentate or entire:
    Plant glabrate or sparsely appressed-pubescent; keel ciliate only near claw ........... ........................................ 2. *L. succulentus* var. *succulentus*
    Plant densely pubescent with spreading hairs; keel ciliate only along upper margin from apex to about middle, and on free lower margin near claws .................. ........................................ 3. *L. grayi* var. *medius*
  Upper calyx lip cleft or bifid:
    Keel ciliate along upper margin only between middle and apex:
      Flowers 4–5 mm. long; banner longer than wide; keel nearly straight; lower calyx lip tridentate, rarely entire; pod 10–15 mm. long ..... 4. *L. bicolor* var. *microphyllus*
      Flowers 10–18 mm. long; banner as wide as long; lower calyx lip bidentate; pods 30–50 mm. long:
        Shrubby; flowers 10–14 mm. long .................. 5. *L. excubitus* var. *excubitus*
        Herbaceous perennial, stems dying back to root crown each year; flowers 14–18 mm. long ............................... 5*a*. *L. excubitus* var. *austromontanus*
    Keel ciliate between middle and claws, on either lower or upper margin or both:
      Lower calyx lip entire; ciliation on keel constantly present on both upper and lower margins:
        Flowers 12–17 mm. long; keel arcuate .......... 2. *L. succulentus* var. *succulentus*
        Flowers 9–10 mm. long; keel nearly straight along upper margin ............... ........................................ 2*a*. *L. succulentus* var. *brandegeei*

---

* The key is adapted from that of C. P. Smith, Species Lupinorum no. 2.

Lower calyx lip tridentate; ciliation variable:
Ovules only 2; flowers suberect in anthesis .......................... 6. *L. ruber*
Ovules 3–12; flowers spreading in anthesis:
Plants hispid with stinging hairs 3–5 mm. long; leaflets to 2 cm. wide .......... ................................................ 7. *L. hirsutissimus*
Plants without stinging hairs, the pubescence made up of hairs 2 mm. long or less; leaflets not over 12 mm. wide:
Flowers spreading or drooping after anthesis ................ 8. *L. truncatus*
Flowers ascending after anthesis:
Leaflets acute or obtusely angled at apex, not truncate or rounded:
Largest leaflets 2–4 mm. wide; stem usually slender:
Flowers 11–13 mm. long ........... 9. *L. sparsiflorus* var. *sparsiflorus*
Flowers 8–10 mm. long ............ 9*a*. *L. sparsiflorus* var. *arizonicus*
Largest leaflets 6–12 mm. wide; stem stoutish, succulent or fistulous:
Leaflets mostly 6–9 .................9*b*. *L. sparsiflorus* var. *barbatulus*
Leaflets mostly 4–6, usually 5 ......... 9*c*. *L. sparsiflorus* var. *insignitus*
Leaflets truncate or rounded at apex:
Plants branched near base; stems slightly decumbent .................. ...................................9*d*. *L. sparsiflorus* var. *pondii*
Plants branched near middle; stems erect or ascending ................ ..............................9*e*. *L. sparsiflorus* var. *inopinatus*

**1. Lupinus concinnus** Agardh, Syn. Gen. Lup. 6, 1835, var. **concinnus**

Erect annual 5–30 cm. tall with few to many branches from base and at higher levels; herbage densely villous throughout with spreading hairs; petioles exceeding leaflets, slender, lower ones 4–8 cm. long; stipules lance-linear, 5–9 mm. long; leaflets 5–8, oblanceolate, obtuse to rounded at apex, 3–6 mm. wide, 1–2 cm. long, deep green through pubescence; racemes erect, 3–10 cm. long or sometimes of only 2–3 flowers and surpassed by leaves; bracts linear, 2–3 mm. long, persistent; pedicels slender, 1–2.5 mm. long; calyx densely villous, cup shallow, 1–2 mm. deep; upper calyx lip cleft nearly or fully halfway to base, 4–5 mm. long, lower one tridentate, 5–6 mm. long; petals 7–9 mm. long, lilac or bluish, edged with deep purple; banner obovate, rounded or emarginate at apex; keel nonciliate, nearly straight; pods 10–18 mm. long, straight, densely villous, 2–4-seeded; seeds 2–3 mm. long, squarish, strongly flattened, pale buff, mottled with irregular darker spots and splotches.

Dry soils, mainly Upper Sonoran Zone but occasionally in Lower Sonoran Zone, Monterey County, California, through Arizona into New Mexico, northern Sonora, and Baja California at least to the northern tip of the Sierra San Pedro Mártir. Less common in the desert than some of the following varieties. March–May.

**1*a*. Lupinus concinnus** var. **orcuttii** (S. Wats.) C. P. Smith, Bull. Torrey Club **48:** 225, *fig. 69,* 1921

*Lupinus orcuttii* S. Wats., Proc. Amer. Acad. **20:** 359, 1885.
*Lupinus micensis* M. E. Jones, Proc. Calif. Acad. II, **5:** 630, 1895.

Plant usually somewhat more compact than var. *concinnus*; flowers 6–7 mm. long; banner about 3 mm. wide.

Dry hillsides and flats, Sonoran Zones, Monterey County, California, to southern Nevada and Utah, New Mexico, northern Sonora, and northern Baja California. This variety is more common in the desert areas and has slightly larger flowers than var. *concinnus*.

**1*b*. Lupinus concinnus** var. **pallidus** (Brandegee) C. P. Smith, Bull. Torrey Club **48:** 229, *fig.* 72, 1921

*Lupinus pallidus* Brandegee, Zoe **4:** 203, 1893.

Pubescence mainly appressed but with a few spreading hairs; leaflets 5–6, spatulate, sometimes subglabrate in age; petals white or pale bluish.

Sandy soil, often along arroyos and stream banks, Lower Sonoran and lower edge of Upper Sonoran Zones, southern San Diego and southwestern Imperial Counties, California, south into Baja California at least to Socorro, between San Quintín and Rosario; along the Magdalena River in northern Sonora.

**1*c*. Lupinus concinnus** var. **desertorum** (Heller) C. P. Smith, Bull. Torrey Club **48:** 230, *fig.* 73, 1921

*Lupinus desertorum* Heller, Muhlenbergia **2:** 72, 1905.

Pubescence similar to that of var. *pallidus*; leaflets 6–9, oblanceolate; petals white or tinged with pale yellow.

Desert slopes and hillsides, Lower Sonoran Zone, chiefly in Kern and San Bernardino Counties, but one collection that seems to belong here came from Borrego Valley, eastern San Diego County.

**Lupinus concinnus** var. **agardhianus** (Heller) C. P. Smith, Bull. Torrey Club **48:** 228, *fig.* 71, 1921, has been collected near San Quintín, Baja California, and one specimen on a sheet of Orcutt's supposedly from "southwestern Colorado Desert" has been identified as this species. In general, however, this variety is distributed mainly in the chaparral regions above the desert. It differs from var. *concinnus* chiefly in having shorter, sparser spreading hairs and bright blue petals or these edged with purple, and in having a narrow banner nearly twice as long as wide. It probably occurs sparingly along the western margins of the Colorado Desert.

**2. Lupinus succulentus** Dougl. ex K. Koch, Wochenschr. Gärt. & Pfl. **4:** 277, 1861, var. **succulentus**

*Lupinus affinis* of American authors. Not *L. affinis*, Agardh, 1835.

Stout succulent or fistulous-stemmed annual 2–6 dm. tall, considerably branched and sparsely appressed-pubescent to subglabrous throughout; petioles 6–12 cm. long; stipules linear-subulate, 10–12 mm. long; leaflets 7–9, cuneate-obovate, 6–16 mm. wide, 2–7 cm. long, rounded, truncate, or emarginate and slightly apiculate at apex; peduncles 2–8 cm. long; racemes to 30 cm. long; flowers in whorls, spreading in anthesis; bracts linear, 6–10 mm. long, deciduous; pedicels 4–6 mm. long, pubescent with short spreading hairs; calyx appressed-pubescent, upper lip bifid, 4–5 mm. long, lower lip entire or tridentate, 7–8 mm. long; banner suborbicular, 13–14 mm. long, blue with yellow center or sometimes bluish white; wings and keel 12–14 mm. long, blue or rarely whitish, keel ciliate near claw on both margins; pods 4–5 cm. long, 9–10 mm. wide, spreadingly villous with brownish hairs, 8–10-seeded; seeds 3.5–5 mm. long, marbled with dark brown.

Hillsides, canyons, and open fields, Sonoran and Arid Transition Zones, from Butte County to northern Baja California. Introduced in isolated spots in Arizona. Feb.–April.

**2*a*. Lupinus succulentus** var. **brandegeei** C. P. Smith, Bull. Torrey Club **49:** 203, *fig.* 77, 1922

Similar to var. *succulentus* but flowers smaller, only 9–10 mm. long; upper margin of keel almost straight instead of arcuately curved and less copiously ciliate above; wings rounded instead of truncate at apex, and rounded instead of acutely angled on upper margin just above claw.

Sandy washes or among lava boulders, Lower Sonoran Zone, central Baja California in the vicinity of Comondú and San Júlio.

**3. Lupinus grayi** (S. Wats.) S. Wats. var. **medius** (Jepson) C. P. Smith, Spec. Lupinorum 28, 1938

*Lupinus albifrons* var. *medius* Jepson, Fl. Calif. **2:** 252, 1936.

Erect or ascendingly branched, to 5 dm. tall, finely and densely pubescent with short but spreading hairs throughout; petioles slender, lower 1–2 dm. long, somewhat shorter above; leaflets 5–9, oblanceolate, obtuse or rounded and minutely apiculate at apex, 4–8 mm. wide, 2.5–3.5 cm. long; peduncles robust, 1–1.5 dm. long, topped by racemes to 2.5 dm. long; bracts deciduous; pedicels 2–4 mm. long; flowers 12–15 mm. long, subverticellate; calyx densely pubescent, upper lip bidentate (in ours) about 4 mm. long, lower lip entire (rarely tridentate), 5–6 mm. long; petals blue, banner with a yellow center, suborbicular and often somewhat pubescent on back; keel ciliate along upper margin from apex nearly to middle and near claw on lower margin; pods 6–8 mm. wide, 2.5–3 cm. long, yellowish, densely pubescent; seeds 4–6, about 3 mm. wide, 4 mm. long, mottled and lined.

Rocky slopes, upper edge of Lower Sonoran Zone, eastern San Diego County and southward along the eastern escarpment into Baja California. Known from Mountain Springs Grade, California, and from La Rumarosa, Baja California. Feb.–April.

**4. Lupinus bicolor** Lindl. var. **microphyllus** (S. Wats.) C. P. Smith, Bull. Torrey Club **50:** 382, *fig. 89,* 1923

*Lupinus micranthus* var. *microphyllus* S. Wats., Proc. Amer. Acad. **8:** 535, 1873.
*Lupinus rostratus* Eastw., Proc. Calif. Acad. II, **6:** 424, *pl. 56,* 1896.

Branches decumbent, ascending, or erect, 1–4 dm. tall, hirsutulous throughout with fine, moderately spaced, spreading hairs 1 mm. long or less; petioles slender, 2–5 cm. long; leaflets 5–7, linear-oblanceolate, narrowly oblanceolate or sometimes cuneate, 1.5–3 mm. wide, 6–25 mm. long; peduncles slender, 2–5 cm. long, racemes short, 1–2 cm. long, flowers verticillate or subverticillate, 4–5 mm. long; calyx pubescent, upper lip 1.5–2 mm. long, bifid or deeply bidentate, lower one 2.5–3 mm. long, trifid; banner elliptic to orbicular, 2–4 mm. wide, 5–7 mm. long, acute or rounded at apex, white with blue margins and dark spots in white field; keel straight to arcuate, ciliate along upper margin from middle nearly to slender acumen; pods 10–15 mm. long, pubescent; seeds 3–4, about 2 mm. long, pale buff or sometimes mottled with darker spots.

Hillsides and slopes in grassy places, Sonoran Zones, central California to central Baja California and sparingly in Sonora. March–April.

**5. Lupinus excubitus** M. E. Jones, Contr. West. Bot. **8:** 26, 1898, var. **excubitus**

Erect or ascendingly branched plant 3–15 dm. tall with silky-appressed pubescence throughout; stems woody at base and persistent; petioles 5–25 cm. long, upper shorter; leaflets 8–11, oblanceolate, 8–25 mm. wide, 6–15 cm. long, obtuse or rounded and often apiculate at apex; peduncles stout, 5–10 cm. long; racemes 1–2.5 dm. long, flowers weakly to strongly verticillate; bracts lanceolate, 3–5 mm. long, deciduous; pedicels densely pubescent, 4–6 mm. long; upper calyx lip deeply notched, 5–6 mm. long, lower entire, as long as upper or slightly longer; petals blue or with intermixture of violet and a yellow center in banner, 10–14 mm. long; banner more or less pubescent on back, suborbicular; wings broad; keel slightly arcuate, ciliate on most of upper margin and sometimes near claw on lower free edges, acumen blunt; pods 3–5 cm. long, yellowish, pubescent, reticulately veined; seeds 3–8, 3–4 mm. long, flattened, pale with light brown lateral lines.

Rocky slopes and along washes, mainly Upper Sonoran Zone, desert ranges from Inyo and Kern Counties to northern Baja California. Entering our area near Mountain Springs, San Diego County. Feb.–April.

### 5*a*. **Lupinus excubitus** var. **austromontanus** (Heller) C. P. Smith, in Jepson, Man. Fl. Pl. Calif. 532, 1925

*Lupinus austromontanus* Heller, Muhlenbergia **2:** 69, 1905.

Similar to var. *excubitus* but stems mainly herbaceous and dying back to base each year, crowded with leaves near base; flowers larger, 14–18 mm. long; keel nearly straight instead of arcuate along upper margin.

Rocky and grassy slopes, mainly Upper Sonoran Zone, but descending into the Lower Sonoran Zone along arroyos in Baja California; Kern County at least to the vicinity of Cataviñá, Baja California.

### 6. **Lupinus ruber** Heller, Muhlenbergia **2:** 73, 1905

*Lupinus microcarpus* var. *ruber* C. P. Smith, Bull. Torrey Club **45:** 10, *fig. 4,* 1918.

Annual, usually with several branches from base, 1–3.5 dm. tall, spreadingly hispid on stems, petioles, peduncles, and calyces; petioles to 1.5 dm. long, slender; stipules papery at base, lance-linear tips free, 5–10 mm. long; leaflets 5–10, oblanceolate to spatulate, 3–8 mm. wide, 1–3 cm. long, rounded at apex, glabrous above, villous beneath and along margins; racemes 3–6 cm. long, in 2–8 verticils, shorter than leaves or usually surpassing them and on peduncles to 15 cm. long; flowers 8–10 mm. long, ascending, subtended by persistent bracts 6–10 mm. long; pedicels 1 mm. long or less; upper calyx lip 1.5–2 mm. long; lower lip bidentate, 6–8 mm. long; petals pink to dull red; banner ovate-lanceolate, 2–4 mm. wide; wings narrow; keel straight, ciliate near claws on upper edge; pods 10–18 mm. long, 6–9 mm. wide, villous, biovulate, 1–2-seeded; seeds 3 mm. long, rugose.

Dry soils, San Benito County, southward in the inner Coast Ranges to the margins of the San Felipe Desert, Baja California. Feb.–March.

The central raceme is often short-pedunculate and surpassed by the lateral floriferous branches. During an unusually dry season the central raceme commonly is the only one to mature fruit.

### 7. **Lupinus hirsutissimus** Benth., Trans. Hort. Soc. Lond. II, **1:** 141, 1835

Erect, stout, often fistulous plant to 6 dm. tall, copiously hirsute with stiff, nettle-like hairs 2–5 mm. long; petioles to 12 cm. long; leaflets 5–8, broadly obovate, 15–20 mm. wide, 2–5 cm. long, cuneate at base; peduncles 3–25 cm. long; racemes 10–35 cm. long; bracts linear, 4–8 mm. long, caducous; flowers 12–15 mm. long, scattered, spreading, on hirsute pedicels 3–5 mm. long; upper calyx lip 3–4 mm. long, cleft, lower lip 6–8 mm. long, entire or tridentate; petals lilac to violet; banner suborbicular, nearly erect; keel stout, nearly straight along upper margins, ciliate along lower edges near claws; pods 2.5–3.5 cm. long, spreadingly hirsute, 3–7-seeded; seeds rhomboidal, about 3 mm. long, dark brown, obscurely to strongly mottled with olive green or darker brown, dull.

Foothills and valleys, mainly Upper Sonoran Zone, but entering the margins of the Colorado Desert; San Mateo County to southern flanks of the Sierra San Pedro Mártir, Baja California. Feb.–April.

**8. Lupinus truncatus** Nutt. ex Hook. & Arn., Bot. Beechey 336, 1838

Annual 2–6 dm. tall with sparsely appressed-pubescent or commonly glabrate stems, petioles, and leaves; petioles ascending, 2–8 cm. long, usually distinctly flattened; leaflets 5–7, linear to linear-spatulate, 1–3 mm. wide, 2–4 cm. long, truncate, emarginate or dentate at apex; peduncles 3–10 cm. long, bearing lax, few-flowered racemes nearly twice as long; flowers 9–12 mm. long, spreading to slightly reflexed after anthesis, on pedicels 2–4 mm. long; bracts linear, about equaling pedicels, subpersistent; calyx somewhat more densely appressed-pubescent than rachis, upper lip cleft, 1–2 mm. long, lower lip entire or tridentate, 3–4 mm. long; petals violet to pale purple, banner suborbicular, slightly paler near center; keel blunt and stout, ciliate along upper edges and near claws on lower edges; pods spreading-ascending, 2.5–3 cm. long, appressed-villous, 5–7-seeded; seeds rhomboidal, moderately flattened or oftener plump, 3–3.2 mm. long, light buff, faintly to strongly mottled.

Grassy slopes and openings in chaparral, mainly Upper Sonoran Zone, Monterey County to northern Baja California. Entering the Sonoran Desert near Whitewater and Cabazon, Riverside County, California. Feb.–March.

**9. Lupinus sparsiflorus** Benth., Pl. Hartw. 303, 1848, var. **sparsiflorus**

*Lupinus subhirsutus* Davidson, Bull. S. Calif. Acad. **18:** 80, 1919.

Annual 2–4 dm. tall with slender, erect or ascending stems clothed with appressed and some spreading hairs; petioles 3–8 cm. long; leaflets 5–9, linear to linear-oblanceolate, acute, or, in some varieties, rounded at apex; peduncles 2–10 cm. long; racemes 5–20 cm. long, open; bracts 5–7 mm. long, lance-linear, deciduous; flowers 11–13 mm. long, spreading at anthesis but soon ascending; pedicels 2–3 mm. long; petals blue to purple, a yellow central spot on suborbicular banner; keel strongly curved, ciliate near claws along lower edges and often on upper edges also; pods strongly ascending, 1–2 cm. long, appressed-hirsute, 3–6-seeded; seeds 2–3 mm. long, rounded or angled, dark-mottled on a paler background.

Grassy hillsides, Sonoran Zones, Ventura County, California, to southern Nevada, Arizona, and northern Baja California and Sonora. Jan.–May.

**9*a*. Lupinus sparsiflorus** var. **arizonicus** (S. Wats.) C. P. Smith, Bull. Torrey Club **47:** 495, *fig. 55,* 1920

*Lupinus concinnus* var. *arizonicus* S. Wats., Proc. Amer. Acad. **8:** 537, 1873.
*Lupinus arizonicus* S. Wats., op. cit. **12:** 250, 1877.
*Lupinus sparsiflorus* var. *setosissimus* C. P. Smith, Bull. Torrey Club **47:** 496, *fig. 56,* 1920.

Plants usually smaller, 1–2 dm. tall; flowers 8–10 mm. long; plants otherwise as in var. *sparsiflorus.*

Sandy desert plains and slopes, Lower Sonoran Zone, southern Nevada to northwestern Sonora, the Colorado Desert, and throughout the dry ridges in Baja California.

**9*b*. Lupinus sparsiflorus** var. **barbatulus** Thornber, in C. P. Smith, Bull. Torrey Club **47:** 497, *fig. 57,* 1920

*Lupinus arizonicus* var. *barbatulus* I. M. Johnston, Proc. Calif. Acad. IV, **12:** 1045, 1924.

Stems stoutish and fistulous; leaflets 6–9, larger ones 6–12 mm. wide; petals pale lilac, pinkish, purplish, or sometimes white.

Desert plains, dry washes, and on rocky slopes, Lower Sonoran Zone, southwestern Arizona and adjacent California southward into Sonora and throughout the deserts of Baja California to Cabo San Lucas.

**9*c*. Lupinus sparsiflorus** var. **insignitus** C. P. Smith, Bull. Torrey Club **47:** 499, *fig.* 58, 1920

*Lupinus lagunensis* M. E. Jones, Contr. West. Bot. **15:** 141, 1929.
*Lupinus bartolomei* M. E. Jones, op. cit. 142.

Stems stout and usually fistulous, subsucculent, sparsely appressed-pubescent below, more copiously pubescent and some hairs spreading near tips of branches; leaflets 4–6 (usually 5), broadest near middle, largest 6–12 mm. wide; pods about 4 mm. wide, 1.5 cm. long, 5–6-seeded; seeds marbled with dark brown or olive on a light tan background.

Desert plains and along washes, Lower Sonoran Zone, Santa Margarita Island and the adjacent Magdalena Plain to San José del Cabo, Baja California.

**9*d*. Lupinus sparsiflorus** var. **pondii** (Greene) C. P. Smith, Bull. Torrey Club **47:** 501, *fig.* 60, 1920

*Lupinus pondii* Greene, Pittonia **1:** 288, 1889.

Plants branched at or near base; stems more or less decumbent, 1–3 dm. tall, sparsely pubescent with stiff spreading hairs; leaflets notched or truncate at apex, 2–2.5 mm. wide, 10–12 mm. long; racemes densely flowered, nearly sessile; flowers 10–12 mm. long; seeds about 2 mm. long.

Grassy hillsides and desert plains, Lower and Upper Sonoran Zones, margins of the Colorado Desert, San Diego County, southward to central Baja California.

**9*e*. Lupinus sparsiflorus** var. **inopinatus** C. P. Smith, Bull. Torrey Club **47:** 499, *fig.* 59, 1920

Similar to var. *pondii* except that stems branch above instead of at base and are ascending instead of decumbent; banner slightly narrower and with a faintly more obvious claw than in var. *pondii*.

Hillsides and washes, Sonoran Zones, Cajon Pass, San Bernardino County, California, to vicinity of Calmallí and Lagoon Head, Baja California.

## 3. MEDICAGO L., Sp. Pl. 778, 1753

Annual or perennial herbs with pinnately trifoliolate leaves, dentate leaflets, and adnate, often dentate, stipules. Flowers small, yellow or bluish, in axillary heads or racemes or in few-flowered umbels. Calyx teeth 5, subequal. Banner obovate to oblong; wing petals oblong; keel obtuse. Stamens diadelphous. Styles subulate. Pods curved or coiled spirally, indehiscent, 1–several-seeded, spinose on the margins or reticulate.

Plants perennial; flowers blue or violet; pods unarmed; stipules entire ......... 1. *M. sativa*
Plants annual; flowers yellow; pods armed with hooked prickles; stipules dentate......... ........................................................... 2. *M. hispida*

**1. Medicago sativa** L., Sp. Pl. 778, 1753

*Medicago media* Pers., Syn. Pl. **2:** 356, 1807.
*Medica media* Fourr., Ann. Soc. Linn. Lyon II, **16:** 359, 1868.

Perennial to 1 m. tall from a deep taproot, much branched from base and decumbent to ascending, glabrous or sparsely pubescent on young growth; stipules linear-lanceolate, 1–2 cm. long, entire; petioles stoutish, flattened, to 2.5 cm. long; leaflets obovate to oblanceolate, 1–3 cm. long, dentate on upper one-half or one-third, cuneate at base, obtuse to emarginate at apex; racemes dense, 1–5 cm. long, on a peduncle about as long; calyx 7–9 mm. long, subulate teeth about equaling tube; petals blue or purplish, 8–10 mm. long; pod coiled into 2–3 spirals, puberulent, unarmed, 3–6 mm. in diameter.

Widely cultivated as a forage crop (alfalfa) and escaped along irrigation ditches and roadsides; established on ditch banks near El Centro, California. April.

**2. Medicago hispida** Gaertn., Fruct. **2:** 349, 1791
*Medicago denticulata* Willd., Sp. Pl. **3:** 1414, 1800.

Decumbent annual with numerous spreading branches to 8 dm. long and glabrous stems and foliage or these puberulent in youth but soon glabrate; stipules whitish, asymmetrically ovate-lanceolate, to 1 cm. long, bearing several slender teeth 2–3 mm. long; petioles 1–2 cm. long; leaflets obovate to obcordate or suborbicular, 5–13 mm. wide, 10–15 mm. long, broadly cuneate to obtuse at base, dentate almost to base; peduncles 2–5-flowered, 2 cm. long or less; calyx about 5 mm. long; petals yellow, only slightly exceeding calyx; pods to 1 cm. in diameter, tightly coiled into 2–3 spirals, reticulate on sides, margins keeled and keel armed on each side by a row of curved or hooked prickles 2–3 mm. long.

A weed introduced from Europe and widely established, mostly in the Upper Sonoran Zone, Pacific Coast states to Texas, sparingly established in the Sonoran Desert in Sonora and Baja California; also at Tempe and Tucson, Arizona. March–April.

## 4. MELILOTUS Juss., Gen. 356, 1789

Annual or biennial herbs with pinnately trifoliolate leaves, adnate stipules, and fragrant, glabrous or sparsely pubescent foliage. Flowers white or yellow, small, in slender racemes. Calyx teeth subequal. Petals obovate or oblong; keel obtuse. Stamens diadelphous. Ovary straight, ovoid, 1–few-ovuled; style filiform, nearly straight. Pod ovoid, indehiscent or tardily dehiscent.

Flowers yellow, about 2.5 mm. long .................................. 1. *M. indica*
Flowers white, 4–6 mm. long .................................. 2. *M. alba*

**1. Melilotus indica** (L.) All., Fl. Ped. **1:** 308, 1785
*Trifolium melilotus indica* L., Sp. Pl. 765, 1753.

Erect annual to 1 m. tall with glabrous herbage or leaves and inflorescences sparsely appressed-pubescent when young; stipules subulate or narrowly lanceolate, 3–8 mm. long; petioles slender, to 5 cm. long; leaflets cuneate-oblong to obovate, 3–12 mm. wide, 1–2.5 cm. long, obtuse, rounded or truncate, denticulate; peduncles surpassing subtending leaves; racemes numerous, 2–10 cm. long, about 5 mm. in diameter; flowers 2.5 mm. long, calyx half as long, its teeth triangular, sparsely ciliolate; petals yellow; pods ovoid, 2–2.5 mm. long, reticulate, glabrous, usually 1-seeded.

A ruderal introduced from Europe and common in the Sonoran Zones from Oregon into northern Baja California; less frequent in the desert but occurring sparingly along ditch banks, in cultivated fields, and along streams to central Baja California (near San Ignacio), Arizona, and through Mexico into South America. April–Sept.

**2. Melilotus alba** Desv. ex Lam., Encycl. **4:** 63, 1796

*Melilotus vulgaris* Willd., Enum. Hort. Berol. 790, 1809.

Erect annual with stems to 3 m. tall; foliage glabrous or finely puberulent when very young; stipules linear-subulate to filiform, 3–6 mm. long; petioles usually shorter than leaflets, slender; leaflets obovate to oblong-oblanceolate, 3–8 mm. wide, to 3 cm. long, rounded to truncate at apex, narrowly to broadly cuneate at base, serrulate; racemes numerous, 3–10 cm. long in flower, elongating somewhat in fruit; pedicels slender, about 2 mm. long; flowers drooping-reflexed; calyx 1.5–2 mm. long, triangular teeth about equaling tube; petals white, 4–6 mm. long; pods ovoid, 2–3 mm. long, apiculate, reticulately ridged on sides, glabrous.

Introduced from Eurasia and widely distributed in the Pacific States from eastern Washington south into northern Baja California, east to Texas, and occasionally to the Atlantic seaboard; southward into Coahuila. It is rare in the Sonoran Desert and there confined to irrigated fields, ditch banks, or rarely along streams; collected at Tucson and near Phoenix. April–Sept.

## 5. TRIFOLIUM L., Sp. Pl. 764, 1753

Annual, biennial, or perennial herbs with palmately trifoliolate leaves, adnate stipules, and usually denticulate leaflets. Flowers in globose heads or short spikes, white, yellow, or purplish. Calyx 5-cleft or 5-toothed; teeth equal or subequal. Petals mostly persistent; claws usually adnate at base to stamineal tube, or claw of banner free. Stamens diadelphous. Pods small, thin, usually terete, indehiscent or tardily dehiscent, 1–few-seeded.

Heads distinctly involucrate:
- Involucre cup-shaped; calyces and involucres hirsute .............. 1. *T. microcephalum*
- Involucre rotate; calyces and involucre glabrous:
  - Leaflets serrate but not deeply so, not spinulose-cuspidate:
    - Calyx teeth dilated at base; corollas 10–14 mm. long ........................................ 2. *T. tridentatum* var. *aciculare*
    - Calyx teeth subulate-setaceous, not dilated at base; corollas 6–9 mm. long .................................................. 3. *T. variegatum*
  - Leaflets deeply serrate-spinulose, often strongly cuspidate ............. 4. *T. lacerum*

Heads without evident involucres:
- Plants annual; stems erect or sometimes procumbent-ascending, not rooting at nodes; corolla rose, pink, or purplish:
  - Leaflets obovate, retuse at apex, serrations minute, not setaceous ...... 5. *T. gracilentum*
  - Leaflets oblong, narrowly elliptic-oblong, or narrowly lanceolate, acute and aristate at apex, serrations distinctly setaceous .............................. 6. *T. palmeri*
- Plants perennial; stems usually procumbent, rooting at nodes; corollas white or faintly tinged with pink ................................................ 7. *T. repens*

**1. Trifolium microcephalum** Pursh, Fl. Amer. Sept. **2:** 478, 1814

*Trifolium microcephalum* f. *velutinum* McDermott, N. Amer. Trifol. 118, *pl. 43,* 1910.

Ascending or erect, sparsely villous annual 1–6 dm. tall, the more robust plants branching from base; stipules lanceolate or ovate and acuminate, to 1 cm. long, entire; leaflets obcordate or cuneate-obovate, 5–15 mm. long, emarginate, serrate; heads about 1 cm. in diameter or slightly less; involucre 7–10-lobed, these lanceolate-acuminate, with whitish scarious margins, mostly 3-veined; calyx 4–5 mm. long, villous, teeth subulate, about equaling tube, margins scarious and erosulate-serrulate; corolla white to pale rose, barely exceeding calyx teeth; pod papery, 1–2-seeded.

Grassy plains and open hillsides, chiefly Upper Sonoran and Transition Zones, British Columbia and Montana to Nevada, southern California, and northern Baja California. It has been collected in San Andreas Canyon along the western margin of the Colorado Desert in California; near Williams, Arizona; and at several points along the western slopes of the Sierra San Pedro Mártir in Baja California. March–June.

**2. Trifolium tridentatum** var. **aciculare** (Nutt.) McDermott, N. Amer. Trifol. 26, *pl.* 7, 1910

*Trifolium aciculare* Nutt., in Torr. & Gray, Fl. N. Amer. **1:** 319, 1838.

Glabrous annual with erect or sometimes decumbent stems 1–5 dm. long; stipules lanceolate-acuminate, 1–1.5 cm. long below, ovate and 6–9 mm. long above, upper ones more copiously spinulose-toothed than nearly entire lower ones; petioles 1–2 cm. long; leaflets linear-lanceolate, serrulate, to 3 cm. long; peduncles 5–10 cm. long; heads about 1–3 cm. in diameter; involucres irregularly laciniate-toothed but not lobed; calyx 6–7 mm. long, 15–25-nerved at least below, teeth 2–2.5 mm. long, dilated at base, entire; corolla 10–14 m. long, reddish purple with darker tips on wings; banner with pale to nearly white tips.

Mainly in open areas and on grassy slopes, Upper Sonoran and Transition Zones, northern Baja California to British Columbia, but an occasional waif along the western margins of the Colorado Desert. March–May.

**3. Trifolium variegatum** Nutt., in Torr. & Gray, Fl. N. Amer. **1:** 317, 1838

Glabrous annual with several decumbent to ascending stems 1–6 dm. long from base; stipules broadly ovate, laciniately toothed, 3–6 mm. long; petioles slender, to 4 cm. long or much shorter, usually exceeding leaflets; leaflets obovate to obcordate or sometimes oblong-oblanceolate, 5–15 mm. long, serrulate, narrowly to broadly cuneate at base; peduncles slender, equaling or exceeding subtending leaves; heads few-flowered, 6–12 mm. in diameter; involucres 5–9-lobed and laciniately toothed, about 5 mm. in diameter; calyx 4–5 mm. long, teeth subulate-setaceous, equaling or slightly longer than tube; corolla purplish with whitish tips, 6–9 mm. long; pods ovate, 1–2-seeded.

In moist heavy soil, mainly Upper Sonoran and Transition Zones, northern Baja California to British Columbia, Arizona, and Montana; margins of the Sonoran Desert east of Rosario, Baja California, and near Jacumba, San Diego County, California. March–May.

**4. Trifolium lacerum** Greene, Erythea **2:** 182, 1894

Glabrous annual with decumbent or prostrate slender branches 1–3 dm. long; internodes 1–2 cm. long or less; stipules 5–7 mm. long, acuminate-ovate; petioles 1–1.5 cm. long; leaflets linear-lanceolate, 2–2.5 cm. long, acute and aristate at apex, fimbriate-lacerate along margins; peduncles 2–3 cm. long; heads 10–15 mm. broad; involucre narrower than head, lacerately toothed or cleft nearly to base; calyx 6–7 mm. long, tube about 2 mm. long, teeth subulate-aristate; corolla 7–9 mm. long, reddish purple.

Moist meadows and stream banks, mainly Upper Sonoran and Transition Zones,

Arizona and New Mexico, but descending to the upper margins of the Sonoran Desert in Pima County, Arizona. March–Aug.

**5. Trifolium gracilentum** Torr. & Gray, Fl. N. Amer. **1:** 316, 1838

*Trifolium denudatum* Nutt., Jour. Acad. Phila. II, **1:** 152, *pl. 24,* 1848.
*Trifolium exile* Greene, Pittonia **1:** 6, 1887.
*Trifolium gracilentum* var. *exile* Kennedy, Muhlenbergia **5:** 10, 1909.
*Trifolium gracilentum* var. *reductum* Parish, Bot. Gaz. **65:** 338, 1918.

Glabrous annual with erect or ascendingly spreading stems 1–5 dm. long; stipules ovate-lanceolate, thin and papery, entire, 5–15 mm. long, margins slightly whitish; petioles slender, 1–3 cm. long; leaflets obovate, 6–15 mm. long, rounded, truncate, or slightly emarginate at apex, minutely serrulate; peduncles three to four times as long as subtending leaves; heads few- to many-flowered, ovoid to globose, devoid of an involucre, flowers strongly reflexed in age; calyx 5–6 mm. long, teeth subulate-setaceous, 3–3.5 mm. long; corolla pink to reddish purple, 6–8 mm. long; pod obovoid, somewhat flattened, about 5 mm. long, 1–2-seeded; seeds ovoid, 1.5–2 mm. long.

Grassy hillsides, mostly Upper Sonoran and Transition Zones, foothills and mountains of northern Baja California to British Columbia and Arizona, barely entering the Sonoran Desert near Banner, San Diego County, California. March–May.

**6. Trifolium palmeri** S. Wats., Proc. Amer. Acad. **11:** 132, 1876

*Trifolium gracilentum* var. *palmeri* McDermott, N. Amer. Trifol. 300, *pl. 126,* 1910.

A glabrous annual similar to *T. gracilentum*; stipules lanceolate, to 2 cm. long, subhyaline, membranous, acuminate to attenuate; petioles 1.5–6 cm. long; leaflets linear-lanceolate, narrowly elliptic-oblong or oblong, 3–6 mm. wide, 1.5–3.5 cm. long, acute and short-aristate at apex, setaceous-serrate along margins; flowering peduncles twice to three times as long as subtending leaves; heads globose or subglobose, 15–18 mm. wide; pedicels slender, 2–3.5 mm. long; flowers reflexed in age; calyx 5–6 mm. long, teeth subulate, subequal, longer ones twice to three times as long as tube; corolla purplish, 6–7 mm. long; pods mostly 2-seeded.

Grassy slopes, Sonoran Zones, Guadalupe Island, Baja California, and San Clemente Island off the coast of Orange County, California. March–May.

**7. Trifolium repens** L., Sp. Pl. 767, 1753

A glabrous or sparsely villous perennial with prostrate creeping stems rooting at nodes and erect or ascending leaves and peduncles; stipules lance-ovate, acute or acuminate, entire or essentially so, 3–10 mm. long, membranous; petioles 3–15 cm. long or more; leaflets broadly obovate, emarginate to obcordate, 6–20 mm. long, broadly cuneate at base, finely denticulate; peduncles to 2 dm. tall or more; heads subglobose, 2–3 cm. in diameter, about 1.5–2 cm. tall, flowers reflexed in age; pedicels 2–4 mm. long; calyx tube narrowly campanulate, 2.5–3 mm. long, teeth lance-deltoid, 1–1.5 mm. long, deep green; corollas 7–12 mm. long, white or faintly tinged with pink or rose; pods papery, 1–2-seeded.

Introduced from Europe and naturalized in meadows in many parts of North and South America, rarely established in arid regions but occasionally thriving along stream banks and irrigation ditches in and about Tucson, Arizona, and near Caborca, Sonora. March–Sept.

## 6. HOSACKIA Dougl. ex Benth., Bot. Reg. 15: *pl. 1257*, 1829

Annual or perennial herbs with odd-pinnate leaves (or these sometimes reduced to a single leaflet); stipules scarious, herbaceous, or (in ours) reduced to minute dark glands. Flowers solitary or in few- to several-flowered axillary umbels on short, naked or bracteate peduncles, or sessile. Calyx campanulate to tubular, teeth 5, subequal. Corolla yellow or rosy, or yellow streaked with red; banner erect, or somewhat reflexed, suborbicular; wings usually nearly equaling curved keel. Stamens diadelphous. Ovary sessile, 1–several-ovuled. Pod linear to ovoid, subterete or slightly compressed, dehiscent or indehiscent.

Pods dehiscent, straight or slightly curved; beak short, nearly straight:
 Plants perennial; flowers mostly over 1 cm. long:
  Pubescence appressed, of white hairs; stems erect, ascending or procumbent:
   Calyx teeth about equaling tube; pubescence dense throughout:
    Pods 3–4 cm. long; stems not zigzag; corollas 15–20 (–25) mm. long:
     Pods shining, glabrous, 3–5 mm. broad; seeds granular or papillose; corolla yellow or cream .......... 1. *H. rigida*
     Pods puberulent, 2.5–3.5 mm. wide; seeds smooth; corolla rose or white tinged with pink .......... 2. *H. nivea*
    Pods 1.5–2.5 cm. long; stems zigzag; corollas 10–12 mm. long ...... 3. *H. flexuosa*
   Calyx teeth about one-half as long as tube; pubescence dense only at tips of branches, sparse below .......... 4. *H. plebeia*
  Pubescence spreading, of tawny or brownish hairs; stems mainly prostrate .......... 5. *H. greenei*
 Plants annual; flowers less than 1 cm. long, usually less than 8 mm. long:
  Seeds granular or rugose; style slightly hairy at base of stigma:
   Seeds quadrate, granulose or tuberculate and often rugose:
    Pubescence closely appressed .......... 6. *H. strigosa* var. *strigosa*
    Pubescence spreading .......... 6*a*. *H. strigosa* var. *hirtella*
   Seeds subglobose, granular but rarely rugose .......... 7. *H. tomentella*
  Seeds smooth; style glabrous:
   Pods 1.5–2 mm. in diameter, 1–2.5 cm. long, glabrous or glabrate; herbage appressed-strigose, green .......... 8. *H. humilis*
   Pods 2–3 mm. broad, mostly less than 12 mm. long, spreadingly villous; herbage villous with spreading hairs .......... 9. *H. brachycarpa*
Pods indehiscent, often arcuate and reflexed; beak half as long as, to equaling body, often strongly curved or coiled:
 Annuals (or short-lived perennials); pods constricted between seeds:
  Pods bent upward sharply just above base; beak elongated, hooked or coiled at tip; umbels sessile .......... 10. *H. micrantha*
  Pods curved upward into a semicircle; beak shorter than body, not hooked or coiled; umbels on slender peduncles .......... 11. *H. prostrata*
 Perennials; pods not constricted:
  Keel distinctly longer than banner and wings; plants not cinereous:
   Umbels 1–6-flowered; flowering branches to 4 dm. long; stems broomlike .......... 12. *H. glabra* subsp. *brevialata*
   Umbels 1–2-flowered; flowering branches not conspicuously elongated, 3–6 cm. long; branches curved, tangled .......... 13. *H. nudata*
  Keel equaling or shorter than wings and banner; plants more or less cinereous, at least on young growth:
   Pubescence spreading; pods bearing brownish glands in transversely arranged V's; herbage at length glabrate .......... 14. *H. tomentosa* subsp. *glabriuscula*

Pubescence appressed, often silvery-silky; pods without glands; herbage persistently pubescent:
 Umbels mostly bractless; leaflets 2–3 mm. long; flowers usually solitary or in pairs; pubescence sparse .......................................... 15. *H. haydonii*
 Umbels usually bracteate; leaflets to 12 mm. long; flowers usually 3-many in an umbel; pubescence dense:
  Pods linear, 2–3-seeded, body two to three times as long as calyx; stems erect and recurving; lateral branches distichous ..................... 16. *H. disticha*
  Pods oblong-arcuate, 1-seeded, body about equaling calyx; stems prostrate or decumbent, lateral branches not distichous ................ 17. *H. argophylla*

## 1. **Hosackia rigida** Benth., Pl. Hartw. 305, 1848

*Lotus rigidus* Greene, Pittonia **2:** 142, 1890.
*Lotus argensis* Coville, Contr. U.S. Nat. Herb. **4:** 83, 1893.
*Anisolotus rigidus* Rydb., Bull. Torrey Club **33:** 144, 1906.
*Anisolotus argensis* Heller, Muhlenbergia **9:** 67, 1913.

Perennial; stems several to many, branching from base, ascending or spreading-ascending, 3–7 dm. long, pale green, glabrous or soon glabrate, internodes much longer than leaves; petioles 1–5 mm. long; leaflets 3–5, oblong-linear to oblong-oblanceolate or sometimes obovate, 1.2–5 mm. wide, 5–15 mm. long, finely strigose or eventually glabrate, green; peduncles 3–12 cm. long, 1–3-flowered, bearing a small leafy bract 3–6 mm. long just below flowers, or bractless; pedicels appressed-pubescent, 1–3 mm. long; calyx strigose, 7–9 mm. long; corolla 15–20 or rarely to 25 mm. long, yellow or cream, slightly suffused with red; pod 3–5 mm. broad, 2.5–4.5 cm. long, stramineous, shining and glabrous; seeds subglobose, finely granular or papillose.

Along washes and on rocky hillsides, Lower Sonoran Zone or sometimes in Upper Sonoran Zone, Mojave and Colorado Deserts to southern Utah, Arizona, and central Baja California. It occurs sparingly on dry ridges of the Transition Zone in the Sierra San Pedro Mártir, Baja California. Feb.–May.

Alice M. Ottley has reported natural hybrids between this and several other species (Brittonia **5:** 98, 1944). The hybrids are often extremely difficult or impossible to run through a key to the proper species.

## 2. **Hosackia nivea** S. Wats., Proc. Amer. Acad. **22:** 470, 1887

*Hosackia bryanti* Brandegee, Proc. Calif. Acad. II, **2:** 144, 1889.
*Lotus bryantii* Ottley, Brittonia **5:** 98, 1944.

Decumbent to ascending perennial with several stems 1–5 dm. long branching from root crown, finely strigose-cinereous throughout, often flowering first year and simulating an annual habit; petioles to 6 mm. long; rachises shorter than or equaling petiole; leaflets 3–5, narrowly obovate to oblong-linear, 2–6 mm. wide, 5–25 mm. long; peduncles 1–8 cm. long, 1–several-flowered, bearing a simple, oblong to obovate bract immediately below flowers or this rarely reduced to a small spur; calyx 78 mm. long, lobes subulate, equaling or slightly shorter than tube; corolla 12–16 mm. long, rose, or white tinged with pink or red on back, or sometimes yellowish; banner erect or nearly so; pods straight or slightly curved, 2.5–3.5 mm. in diameter, 2–3.5 cm. long, strigose; seeds subglobose, smooth or slightly wrinkled but not granular.

Near the coast on dunes or gravelly and rocky benches, mainly Lower Sonoran Zone, vicinity of San Quintín south to Magdalena Island, Baja California. Jan.–April.

**3. Hosackia flexuosa** Greene, Bull. Calif. Acad. **1:** 82, 1885

*Lotus cedrosensis* Greene, Pittonia **2:** 144, 1890.

Stems spreading-ascending to decumbent, 1–3 dm. long, from a perennial taproot, zigzag, finely and densely strigose-canescent, internodes 1–2 cm. long; leaflets 3–5, obovate, obtuse, rounded or truncate at apex, 2–6 mm. wide, to 12 mm. long, densely strigose, silvery or cinereous; umbels mostly 1-flowered, without a foliose bract; peduncles 1–2 mm. long; calyx narrowly turbinate, puberulent, tube slightly longer than oblong-linear teeth; corolla white or yellow, tinted with rose on outside of petals and near base of blades; pods 1.5–2.5 cm. long, slender, usually about 2 mm. wide, appressed-pubescent, silvery or canescent.

Known only from the arid parts of Cedros Island, Baja California.

**4. Hosackia plebeia** Brandegee, Proc. Calif. Acad. II, **2:** 144, 1889

*Lotus oroboides* var. *plebeius* Ottley, Brittonia **5:** 107, 1944.

Much branched from the base with stems 2–3 dm. long, minutely appressed-strigose throughout; leaves nearly sessile; rachises flattened, 7–10 mm. long, one leaflet virtually at base on lower side, its mate lacking; leaflets 4–6, usually 6, broadly obovate, 3–5 mm. wide, 5–9 mm. long, obtuse to rounded and apiculate at apex, broadly cuneate at base, appressed-strigose with minute white hairs but scarcely cinereous; peduncles 1–2.5 cm. long in flower, to 6 cm. long in fruit; bracts minute or lacking; flowers solitary or in pairs, 8–9 mm. long; calyx 5–6 mm. long, appressed-strigose; teeth linear-lanceolate, nearly equaling tube; corolla 7–8 mm. long, banner curled upward; pods 18–25 mm. long, slightly upwardly curved near tip, brownish, sparsely strigose but finally subglabrate, 6–12-seeded; seeds subglobose, slightly compressed, smooth, dark brown, 1.5–2 mm. in diameter, dull.

Known only from the type locality, in sandy soil, El Rancho Viejo, Baja California.

**5. Hosackia greenei** (Woot. & Standl.) Wiggins, Contr. Dudley Herb. **4:** 19, 1950

*Hosackia mollis* Greene, Bull. Calif. Acad. **1:** 185, 1885. Not *H. mollis* Nutt., 1838.
*Lotus mollis* Greene, Pittonia **2:** 143, 1890. Not *L. mollis* Balf. f., 1882.
*Lotus neo-mexicanus* Greene, Pittonia **2:** 144, 1890.
*Anisolotus mollis* Heller, Muhlenbergia **8:** 60, 1912.
*Anisolotus neo-mexicanus* Heller, loc. cit.
*Anisolotus greenei* Woot. & Standl., Contr. U.S. Nat. Herb. **16:** 135, 1913.
*Lotus greenei* Ottley ex Kearney & Peebles, Jour. Wash. Acad. **29:** 483, 1939.

Mainly prostrate with spreading stems to 2 dm. long, finely pubescent throughout with spreading, usually tawny or slightly brownish hairs, those on lower parts of stems and calyces particularly brownish; leaflets 3–7, commonly 3–4, elliptic to narrowly obovate, 1.5–3.5 mm. wide, 3.5–8 or rarely to 11 mm. long, acute at apex, on short petioles and rachises 4 mm. long or less; peduncles 1–2.5 cm. long, 1–4-flowered, oftenest 2-flowered; calyx tube subcampanulate, about 3 mm. long, lobes lance-linear, equaling or slightly exceeding tube; corolla 9–12 mm. long, yellow on face, suffused with salmon-pink on back; pod linear, 1.5–2 mm. broad, 1.5–2 cm. long, nearly straight, sparsely puberulent, brown; seeds greenish brown, smooth.

On gravelly and rocky slopes, mainly Upper Sonoran Zone, but entering upper margins of the desert near Nogales, Arizona; New Mexico and eastern Arizona and adjacent Chihuahua and Sonora. March–May; Aug.

Hybrids between this species and *H. puberula* (*Lotus oroboides*) and with *H. rigida* have been reported from Arizona by Ottley. Some of the hybrids may occur within the limits of the desert.

**6. Hosackia strigosa** Nutt., in Torr. & Gray, Fl. N. Amer. **1:** 326, 1838, var. **strigosa**

*Hosackia nudiflora* Nutt., loc. cit.
*Lotus nudiflorus* Greene, Pittonia **2:** 141, 1890.
*Lotus strigosus* Greene, loc. cit.
*Lotus strigosus* var. *nudiflorus* Jepson, Fl. W. Mid. Calif. 302, 1901.
*Anisolotus strigosus* Heller, Muhlenbergia **3:** 101, 1907.
*Anisolotus nudiflorus* Heller, op. cit. **7:** 139, 1912.

Annual with several to many slender, decumbent or ascending stems 1–2 dm. long from a slender taproot; herbage finely appressed-strigose; stipules glandular; leaf rachises flattened, 1 mm. wide or less, 5–15 mm. long; leaflets mostly 5–7, sometimes 3–9, oblong-oblanceolate, acute to cuneate at base, rounded to retuse at apex, 1–3 mm. wide, 3–12 mm. long, terminal usually considerably larger than others; peduncles 1.5 cm. long or less, 1–3-flowered, lower ones usually 1-flowered; bracts 1–3-foliolate, sometimes lacking; calyx tube campanulate, about 2 mm. long, teeth triangular-subulate to ovate, acute, about equaling tube; corolla 5–10 mm. long, yellow or tinged with red, often drying purplish; pods 2–2.5 mm. wide (sometimes to 3 mm. wide), 2–3 cm. long, sparsely strigose, tip usually curved upward; seeds 1–1.2 mm. wide, quadrate, granular and rugose or tuberculate, notched at hilum, pale greenish, or buff.

Dry, sandy or gravelly soil, Upper and Lower Sonoran Zones, from vicinity of San Francisco Bay and Tuolumne County, California, to northern Baja California, entering the Sonoran Desert in eastern San Diego County and adjacent Baja California. Feb.–June.

**6*a*. Hosackia strigosa** var. **hirtella** (Greene) Hall, Univ. Calif. Pub. Bot. **4:** 199, 1912

*Lotus hirtellus* Greene, Pittonia **2:** 142, 1890.
*Lotus strigosus* var. *hirtellus* Ottley, Univ. Calif. Pub. Bot. **10:** 214, *pl. 71, fig. 12,* 1923.

Similar to var. *strigosa* except pubescence is spreading instead of appressed, and seeds are often less strongly rugose, although quadrate and granular.

Dry, gravelly soils, Upper and Lower Sonoran Zones, central California to San Diego County. Occurring in the Sonoran Desert at Palm Springs.

**7. Hosackia tomentella** (Greene) Abrams, Ill. Fl. Pacif. States **2:** 543, *fig. 2725,* 1944

*Lotus tomentellus* Greene, Pittonia **2:** 140, 1890.

Prostrate or decumbent annuals with several spreading branches 0.5–3.5 dm. long, sparsely strigose but only youngest parts cinereous; stems essentially glabrate or nearly so; rachises of leaves 3–8 (–15) mm. long; leaflets 5–7 or rarely only 3, broadly obovate, cuneate at base, retuse or truncate-rounded at apex, 1–5 mm. wide, 3–10 mm. long; peduncles slender, equaling or slightly exceeding leaves, 1–2-flowered; bracts unifoliolate or reduced to a gland; calyx tube narrowly campanulate, 1.5–2.5 mm. long, teeth triangular-subulate to lanceolate, 1–2 mm. long; corolla 4–5 mm. long, yellow, tinged with rose or drying rose; pods 2–2.5 mm. wide, 1–2 cm. long, nearly straight or gently curved upward toward apex, strigose; seeds ovoid to subglobose, finely granular, pale buff or light green or occasionally finely mottled with purplish brown.

Sandy or gravelly soil, Lower Sonoran Zone, southern Mojave Desert eastward into central Arizona and south into central Baja California and the coastal plain of northwestern Sonora. Feb.–May.

**8. Hosackia humilis** (Greene) Abrams, Ill. Fl. Pacif. States **2:** 545, *fig.* 2727, 1944

*Lotus humilis* Greene, Pittonia **2:** 140, 1890.
*Lotus salsuginosus* var. *brevivexillus* Ottley, Univ. Calif. Pub. Bot. **10:** 217, *pl.* 73, *figs.* 2–9, 1923.

Prostrate annual with several stems 0.5–2.5 dm. long, sparsely strigose to glabrate throughout; leaflets thick and slightly succulent; petioles nearly terete, 3–6 mm. long; rachis of leaves scarcely flattened, rarely to 1 cm. long; leaflets 3–5, cuneate-obovate to oblong-obovate, 2–7 mm. wide, 3–10 mm. long, acute to rounded at apex, bright green, sometimes minutely flecked with red; peduncles slender, 1–2.5 cm. long, 1–4-flowered; bract simple, ovate to suborbicular or sometimes 3-foliolate; calyx tube campanulate, about 1 mm. long, strigose, teeth subulate, equaling or slightly shorter than tube; corolla 3.5–4 mm. long, yellow with a faint flush of rose on banner; keel 0.5–1 mm. longer than wings and banner; pods nearly straight, 1–2.5 cm. long, 1.5–2 mm. wide, brownish and glabrous, somewhat constricted between seeds; beak minute, turned abruptly downward; seeds ovoid or globose, 0.7–1 mm. in diameter, smooth, dark olive green or brownish.

Dry desert soils and rocky hillsides, Lower Sonoran Zone, eastern Mojave Desert and western margins of Colorado Desert into the desert regions of Arizona, Sonora, and Baja California at least to Calmallí and on San Benito Island. Feb.–May.

**9. Hosackia brachycarpa** Benth., Pl. Hartw. 306, 1848

*Lotus humistratus* Greene, Pittonia **2:** 139, 1890.
*Lotus trispermus* Greene, Erythea **1:** 258, 1893.
*Anisolotus brachycarpus* Rydb., Bull. Torrey Club **33:** 144, 1906.
*Anisolotus trispermus* Woot. & Standl., Contr. U.S. Nat. Herb. **16:** 135, 1913.

Annual with erect, ascending, or decumbent stems 0.5–2 dm. long, whole plant densely villous with white to slightly tawny, soft hairs; petioles short, rarely over 5 mm. long; rachises of leaves flattened, 5–8 mm. long; leaflets 3–5, broadly elliptic to obovate, 2–6 mm. wide, 4–15 mm. long, cuneate at base, acute to rounded at apex; flowers subsessile, solitary or in pairs in axils of leaves; calyx tube 2–2.5 mm. long; teeth lance-linear, about equaling or slightly exceeding tube; corolla 3–4 mm. long, yellow tinged with red or rose; pods 2–3 mm. wide, 5–10 mm. long, densely villous; seeds lenticular-elliptic, 1–1.6 mm. long, smooth, dark brown to nearly black, semi-lustrous.

Dry ridges and gravelly slopes, mostly Sonoran Zones, Humboldt County and the northern Sacramento Valley, California, to northern Baja California, Arizona, New Mexico, and northern Sonora along the Magdalena River. March–July.

**10. Hosackia micrantha** Nutt., in Torr. & Gray, Fl. N. Amer. **1:** 324, 1838

*Syrmatium micranthum* Greene, Bull. Calif. Acad. **2:** 147, 1886.
*Lotus hamatus* Greene, Pittonia **2:** 150, 1890.

Annual with several slender branches ascending, decumbent, or prostrate, 1–6 dm. long; herbage sparsely villous when young but not cinereous; internodes longer than leaves; leaflets 3–6, obovate to cuneate-oblong, 2–5 mm. wide, 4–10 mm. long; rachises slightly flattened; umbels sessile or nearly so, ebracteate, 3–several-flowered; calyx slender, 1.5–2 mm. long, sparsely and finely villous, teeth subulate, less than half as long as tube; corolla 3–3.5 mm. long, yellow or tinged with pink or rose, banner slightly shorter than keel; pod sharply bent upward near base, linear, finely strigose, 2–3-seeded, constricted between seeds; beak elongated, slender, hooked or coiled at tip; seeds 2.5–3 mm. long, terete in cross section.

Dry gravelly plains and hillsides, mainly Upper Sonoran Zone, Los Angeles County, California, into northern Baja California; western margin of the Colorado Desert near Palm Canyon and near Mission San Fernando, Baja California. March–June.

**11. Hosackia prostrata** Nutt., in Torr. & Gray, Fl. N. Amer. **1:** 325, 1838
*Syrmatium prostratum* Greene, Bull. Calif. Acad. **2:** 147, 1886.
*Lotus nuttallianus* Greene, Pittonia **2:** 150, 1890.

Annual or short-lived perennial with numerous, slender, prostrate or decumbent branches 1–8 dm. long; stems and herbage sparsely strigose when young, glabrate or nearly so in maturity; leaflets 3–5, broadly elliptic to oblong-obovate, 2–4 mm. wide, 3–10 mm. long; petiole short; slightly flattened rachis 2–6 mm. long; peduncles slender, 5–30 mm. long, bearing a simple obovate bract 1–2.5 mm. long or sometimes bractless; umbels 1–10-flowered; calyx about 2 mm. long, glabrous or strigose near base and on angles, teeth triangular, 0.3–0.5 mm. long; corolla 4–5 mm. long, yellow with reddish tips on wings and banner; pod slenderly linear, arcuately curved, 5–10 mm. long, usually 2-seeded and constricted between seeds; beak slender, usually considerably shorter than body.

Sandy and gravelly soils near the coast, Sonoran Zones, San Diego County south to the vicinity of Rosario, Baja California. March–Sept.

**12. Hosackia glabra** Torr. subsp. **brevialata** (Ottley) Abrams, Ill. Fl. Pacif. States **2:** 552, 1944
*Lotus scoparius* var. *brevialatus* Ottley, Univ. Calif. Pub. Bot. **10:** 229, *pl. 77, figs. 10–16,* 1923.

Suffrutescent, with numerous erect or ascending, broomlike branches to 1 m. high; herbage strigose in youth but some glabrous or essentially so; stipular glands dark brown to black, usually conspicuous; pedicels 2–5 mm. long; leaflets usually 3, linear to oblong-linear or elliptic, acute or obtuse, sparsely strigose to glabrate, 3–8 mm. long; umbels sessile, 1–6-flowered; calyx strigose to tardily glabrate, 3–4 mm. long, subulate teeth one-fourth to one-half as long as tube; corolla yellow, often slightly tinged with red, the keel 6–7 mm. long, wings and banner only 3–4 mm. long, conspicuously shorter than keel; pods arcuately curved, glabrous, two to four times as long as calyx, 2-seeded, gradually tapering to slender subulate beak.

Desert slopes among rocks and along washes, Lower Sonoran Zone, eastern side of San Bernardino Mountains southward to the Sierra Laguna, southern Baja California. March–Sept.

**13. Hosackia nudata** (Greene) Vasey & Rose, Contr. U.S. Nat. Herb. **1:** 14, 1890
*Syrmatium nudatum* Greene, Pittonia **1:** 173, 1888.
*Lotus nudatus* Greene, op. cit. **2:** 148, 1890.

An intricately branched perennial with slender, flexuously curved, and somewhat tangled branches 2–5 dm. long, stems minutely and sparsely strigose and finely striate; stipular glands black, conspicuous; leaves minute; leaflets 1–3, broadly ovate to suborbicular, 1.5–2.5 mm. long, minutely strigulose; flowering branches 5–6 cm. long; umbels sessile, of 1–2 flowers separated by internodes about 1 cm. long; calyx 1.5–2 mm. long, finely strigose, teeth lance-subulate, about one-fourth to one-half as long as tube, erect; corolla yellow, 5–6 mm. long; keel about 1 mm. longer than wings and banner; pods arcuately curved, 6–9 mm. long, strigulose, gradually tapering to slender, upwardly curved beak, this 2–3 mm. long.

Dry rocky hillsides, Lower Sonoran Zone, chiefly on Cedros Island but also reported from the adjacent mainland.

**14. Hosackia tomentosa** Hook. & Arn. subsp. **glabriuscula** (Hook. & Arn.) Abrams, Ill. Fl. Pacif. States **2:** 546, 1944

*Hosackia decumbens* var. *glabriuscula* Hook. & Arn., Bot. Beechey 137, 1832.
*Hosackia heermannii* Dur. & Hilg., Jour. Acad. Phila. II, **3:** 39, 1854.
*Syrmatium heermanni* Greene, Bull. Calif. Acad. **2:** 148, 1886.
*Lotus heermanni* Greene, Pittonia **2:** 150, 1890.
*Lotus eriophorus* var. *heermannii* Ottley, Univ. Calif. Pub. Bot. **10:** 232, *pl. 79, figs. 10–17,* 1923.

Perennial with numerous slender, prostrate, villous-tomentose stems forming a mat 2–8 dm. in diameter, stems usually glabrous or nearly so relatively early; leaflets 3–5, obovate to suborbicular, 3–10 mm. wide, 5–12 mm. long, villous-tomentose, acute, obtuse, or rounded but apiculate at apex; umbels on peduncles 2–8 mm. long, 5–several-flowered; calyx villous, about 3 mm. long, teeth linear-subulate, about half as long as tube, somewhat spreading; corolla yellow with reddish tips on keel and banner, 4–5 mm. long; pods ovoid, slightly arcuate, 3–4 mm. long, usually 1-seeded, sparsely villous and bearing transverse V-shaped lines of glands; seed narrowly ellipsoid, 2 mm. long, olive green or brownish green.

Sandy dry soil, Sonoran Zones, San Luis Obispo County, California, southward into northern Baja California, entering our area near Palm Springs, Whitewater, and Coyote Wells, California. March–Oct.

**15. Hosackia haydonii** Orcutt, W. Amer. Sci. **6:** 63, 1889

*Lotus haydoni* Greene, Pittonia **2:** 149, 1890.
*Syrmatium haydoni* Heller, Muhlenbergia **9:** 67, 1913.
*Lotus spencerae* J. F. Macbride, Contr. Gray Herb. **53:** 13, 1918.

A much-branched perennial 1–4 dm. tall with erect, ascending or spreading-ascending, slender branches, at first minutely strigulose but branches soon glabrate; leaves minute, distant; leaflets 3, broadly elliptic, 2–3 mm. long, strigulose in youth, soon glabrate but remaining grayish; flowers solitary or in pairs, ebracteate; calyx cylindric, 3–3.5 mm. long, strigulose, teeth narrowly triangular-subulate, about one-fourth as long as tube, erect; corolla 4–5 mm. long, yellow, faintly tinged with red; pods slightly curved, 4–5 mm. long, tapering toward slender short beak, strigulose but soon glabrate and finely veined.

Dry rocky slopes, Sonoran Zones, mountains bordering the western edge of the Colorado Desert and southward into Baja California. Rarely collected. March–May.

**16. Hosackia disticha** Greene, Bull. Calif. Acad. **1:** 186, 1885

*Syrmatium distichum* Greene, op. cit. **2:** 148, 1886.
*Lotus distichus* Greene, Pittonia **2:** 149, 1890.

Suffrutescent with erect or ascending branches to 1 m. long, with distichous lateral branches, curving downward in upper part, densely silky-pubescent with short appressed hairs throughout; internodes much exceeding leaves; leaflets 2–5, elliptic to obovate, 2–3 mm. wide, 3–6 mm. long, acute or rounded and apiculate at apex; peduncles 1–3 mm. long or the umbels sessile, 2–3-flowered; calyx cylindric-campanulate, about 4 mm. long, silky-strigose; teeth triangular-subulate, 2 mm. long or less, erect; corollas 7–8 mm. long, reddish or yellow deeply tinged with red, banner usually closely enfolded over keel and wings; pods strigose, weakly arcuate, about 1 cm. long, with a slender beak nearly as long as body.

Sandy soil near the coast, Lower Sonoran Zone, western side of middle Baja California. Jan.–May.

**17. Hosackia argophylla** A. Gray, Pl. Thurb. 316, 1854

*Hosackia argentea* Kell., Proc. Calif. Acad. **3:** 38, *fig. 8,* 1863.
*Syrmatium argophyllum* Greene, Bull. Calif. Acad. **2:** 147, 1886.
*Lotus argophyllus* Greene, Pittonia **2:** 149, 1890.

Suffrutescent with numerous ascending or decumbent branches 2–6 dm. long, silvery with dense, silky-appressed pubescence throughout; leaves equaling or shorter than internodes; leaflets 3–5, broadly obovate to oblanceolate, 3–8 mm. wide, 4–12 mm. long, rounded or obtuse at apex; umbels sessile or essentially so, 3–10-flowered, bearing a single obovate bract 3–5 mm. long; calyx tomentose, 3.5–4.5 mm. long, teeth subulate, half as long as tube or shorter; corolla yellow, 6–8 mm. long, banner oblong, ascending; pod ovoid, slightly arcuate, barely exserted from calyx, densely pubescent, 1-seeded, slender beak equaling body, slightly curled at tip.

Rocky hillsides or on gravelly outwash fans, mainly Upper Sonoran Zone, southern Sierra Nevada and Monterey County south into northern Baja California; entering the Colorado Desert near Julian and Jacumba, San Diego County. April–July.

**Hosackia ornithopus** Greene, Bull. Calif. Acad. **1:** 185, 1885, was described from Guadalupe Island and occurs on the Channel Islands off the coast of southern California. It has slightly larger corollas, with the claw of the corolla exceeding the blade. It may occur within our area but has not been reported from the desert with certainty.

## 7. INDIGOFERA L., Sp. Pl. 751, 1753

Herbs or shrubs with pinnately compound leaves, slightly adnate stipules, spicate or racemose inflorescences, and caducous or deciduous floral bracts. Calyx small, with 5 subequal teeth. Corolla pink, red, or purple, more or less strigose with 2-branched appressed hairs. Banner broad, short-clawed. Wings oblong to oblanceolate, short-clawed, slightly adnate to keel near base, basal auricle often thickened. Keel petals with a long claw, acute to rounded at apex. Stamens diadelphous, stamineal sheath constituting three-fourths their length. Ovary sessile, 1–many-ovuled; stigma capitate. Pod globose to linear, tardily dehiscent. Seeds globose, cylindric, or prismatic with truncated ends.

Pods 1–2 cm. long, strongly curved; racemes dense, many-flowered, peduncle shorter than floriferous part .......... 1. *I. suffruticosa*
Pods 2–5 cm. long, straight or only slightly curved; racemes mostly openly few-flowered, peduncle usually longer than floriferous part:
  Leaflets 5, glabrous above; calyx about 3 mm. long, teeth about twice as long as tube; pods about 2 mm. wide .......... 2. *I. laevis*
  Leaflets 7–17, strigose on both surfaces; calyx 1.5–2 mm. long, teeth about equaling tube or shorter; pods 2.5–5 mm. in diameter:
    Pods elliptic in cross section, 2.5–3 mm. in diameter; racemes surpassing leaves .......... 3. *I. fruticosa*
    Pods subrhombic in cross section, 4–5 mm. in diameter; racemes shorter than leaves .......... 4. *I. nelsonii*

**1. Indigofera suffruticosa** Mill., Gard. Dict. ed. 8, no. 2, 1768

*Indigofera anil* L., Mant. 272, 1771.
*Indigofera divaricata* Jacq., Pl. Hort. Schönbr. **3:** 61, *pl. 365,* 1798.
*Indigofera truxillensis* H. B. K., Nov. Gen. & Sp. **6:** 456, 1824.
*Anila tinctoria* var. *media* Kuntze, Rev. Gen. Pl. **1:** 160, 1891.
*Anila tinctoria* var. *vera* Kuntze, loc. cit.

A shrub to 2 m. tall with erect, angled, strigose but eventually glabrate branches;

stipules subulate-setaceous, 3–4 mm. long; leaves to 15 cm. long; petioles and rachises densely strigose, stipels 0.5–1 mm. long; leaflets 7–15, elliptic, oblanceolate or narrowly obovate, 5–15 mm. broad, 1–3.5 cm. long, dark green and glabrous or sparsely strigose above and densely strigose beneath, acute at base, obtuse or rounded and mucronate at apex; racemes densely many-flowered, 2–5 cm. long; calyx about 1.5 mm. long, densely strigose, lobes about equaling or slightly shorter than tube; corolla 5–6 mm. long, orange; pods 1–2 cm. long, reflexed but strongly arcuate and tips spreading-ascending, dark brown, sparsely strigose.

In subtropical and tropical regions from South Carolina to Texas and southward through the West Indies and Mexico into South America, entering the Sonoran Desert in southern Baja California. March–Nov.

**2. Indigofera laevis** Rydb., N. Amer. Fl. **24:** 144, 1923

Suffrutescent or shrubby plant 0.5–1 m. tall with slender, faintly striate, appressed-strigose stems and more or less strigose foliage; stipules filiform-subulate, 2–3 mm. long; petioles 1–2 cm. long, strigose; leaflets usually 5, broadly elliptic to oval, 4–15 mm. wide, 1–3 cm. long, rounded at both ends, minutely mucronulate at apex, light green and glabrous above, paler and appressed-strigose beneath, on petioles 0.5–1.5 mm. long; peduncles slender, 1–6 cm. long, bearing slender, few-flowered racemes elongating to 1–2 dm. in fruit; calyx densely strigose, tube 1 mm. long, lobes 2–2.5 mm. long, subulate-lanceolate; corolla 4–6 mm. long, banner and wings vermilion, keel white with 2 small lateral appendages; pods linear, 1.5–2 mm. wide, 2.5–5 cm. long, sparsely strigose, reflexed, straight or slightly curved, sutures distinctly carinate; seeds yellowish or reddish, 2–2.5 mm. long, ends truncate, separated by membranous septa.

In shade of larger shrubs and trees on rocky hillsides, Lower Sonoran Zone, from vicinity of Ures southward at least to Alamos, Sonora. Aug.–Sept.

**3. Indigofera fruticosa** Rose, Contr. U.S. Nat. Herb. **5:** 140, 1897

A much-branched shrub 1–1.5 m. tall with dark brownish gray, faintly checked bark on older branches, younger growth rusty brown and appressed-strigose; stipules lance-subulate, 3–4 mm. long, persistent; leaves 4–10 cm. long, slender petioles usually 1.5 cm. long or less, these and rachis densely strigose and sparsely stipitate-glandular; leaflets 7–11, oval, elliptic, or the terminal one often narrowly obovate, 5–10 mm. wide, 1–2 cm. long, rounded and mucronulate at apex, obtuse to broadly cuneate at base, finely but densely strigose on both sides, paler beneath than above; racemes to 8 cm. long, usually exceeding subtending leaves; calyx about 2 mm. long, teeth subulate, about equaling tube, strigose; corolla 6–7 mm. long, salmon-pink, banner densely strigose on back, appendages on keel erosulate; pods linear, elliptic in cross section, 2.5–3 mm. in diameter, 2–4 cm. long, sparingly strigose, keel short, stout; seeds 2–2.5 mm. long, ends truncate.

Among other shrubs, usually in shade, rocky hillsides and canyons, Sonoran Zones, southern Baja California. Aug.–Dec. and probably in spring.

**4. Indigofera nelsonii** Rydb., N. Amer. Fl. **24:** 150, 1923

*Indigofera argentata* I. M. Johnston, Proc. Calif. Acad. IV, **12:** 1043, 1924. Not *I. argentata* Rydb., 1923.

An erect shrub to 2.5 m. tall with strigose-cinereous branchlets and foliage and brownish bark on older stems; stipules lance-subulate, 2.5–3.5 mm. long; leaves 3–10 cm. long; leaflets 9–17, oblong, elliptic or oval, 4–10 mm. wide, 1–3 cm. long, rounded and apiculate at apex, strigose-canescent on both sides, on petiolules about 1 mm. long; racemes 3–15 cm. long; calyx densely strigose-canescent, about 2 mm. long; corolla 8–10 mm. long; banner and wings rose or pink, keel white; pods 4–5 mm. wide, 3–4 cm. long, brown, sparsely strigose, straight or slightly curved, pendent, sutures somewhat rounded-carinate.

Rocky hillsides and arroyos, Sonoran Zones, southern Baja California. Aug.–Dec.

**Indigofera sphaerocarpa** A. Gray, occurs in Arizona, but all specimens examined were from areas above the Sonoran Desert.

## 8. PSORALEA L., Sp. Pl. 762, 1753

Perennial herbs or low shrubs with dark or pellucid glandular dots on stems and foliage. Leaves alternate, simple to 5-foliolate, with large stipules. Flowers white, pink, blue, or purple, in pedunculate spikes or racemes. Calyx lobes subequal or upper ones slightly smaller and in some species united. Petals not attached to stamineal tube; banner ovate to orbicular, clawed; wings oblong, straight or falcate; keel obtuse, incurved. Stamens monodelphous or diadelphous. Ovary sessile or nearly so, 1-ovulate. Pod ovoid, indehiscent.

Leaflets obovate, rounded or retuse at apex; calyx 15 mm. long; herbage silky-canescent .... .......................................................... 1. *P. castorea*

Leaflets rhombic-ovate, all except those of lowermost leaves acute; calyx 5–7 mm. long; herbage appressed-strigose but green, not canescent .................. 2. *P. rhombifolia*

### 1. Psoralea castorea S. Wats., Proc. Amer. Acad. **14:** 291, 1879

*Lotodes castoreum* Kuntze, Rev. Gen. Pl. **1:** 194, 1891.
*Pediomelum castoreum* Rydb., N. Amer. Fl. **24:** 22, 1919.

A perennial herb from a deep-seated fusiform taproot which normally produces a single shoot during a season; underground part of stem 1–2 dm. long, glabrous, slender, bearing 1–3 scarious, oblong, stipular bracts near surface of soil; aerial stem 1–6 cm. tall, erect, closely crowded with leaves, racemes, and occasionally with 1–several spreading decumbent branches; herbage densely silky-canescent; leaves 4–5-foliolate; petioles stoutish, often fistulous, 5–15 cm. long; stipules narrowly triangular, somewhat attenuate, 6–10 mm. long; leaflets cuneate-obovate, 12–25 mm. wide, 1.5–4 cm. long, rounded and emarginate to apiculate at apex; peduncles 2–5 cm. long; racemes dense, 2–6 cm. long; calyx about 15 mm. long, lower lobe oblanceolate or oblong-spatulate, others subulate, tube 3–5 mm. deep; petals barely exceeding calyx, banner narrowly obovate; pods 6–7 mm. long, papery, densely strigose-canescent.

Desert plains and gravelly hillsides, Lower Sonoran Zone, from the vicinity of Barstow in the Mojave Desert to southern Utah, entering the Sonoran Desert only in its northernmost tip in northern Mohave County, Arizona. April–May.

### 2. Psoralea rhombifolia Torr. & Gray, Fl. N. Amer. **1:** 303, 1838

*Lotodes rhombifolium* Kuntze, Rev. Gen. Pl. **1:** 194, 1891.
*Pediomelum rhombifolium* Rydb., N. Amer. Fl. **24:** 23, 1919.

Perennial herb from a deep-seated fusiform root; stem branching at surface of soil, branches 1–5 dm. long, decumbent, finely striate and moderately strigose; herbage strigose; stipules lance-triangular to subulate, 3–5 mm. long, persistent; leaves trifoliolate, petioles 0.5–5 cm. long; leaflets rhombic-ovate, or those of lowermost leaves sometimes suborbicular, rhombic ones 5–15 mm. wide, 1–3 cm. long, acute at apex, strigose but green; peduncles slender, to 10 cm. long, bearing subcapitate spikes 1–2 cm. long; bracts lanceolate, 3–5 mm. long; calyx tube 2–3 mm. long, lobes lanceolate, lower 4 mm. long, upper four about 3 mm. long; corolla reddish to dark blue, 6–7 mm. long, banner obovate, retuse or rounded; pods ovoid, strigose, body 4–5 mm. long, tipped with a straight beak 5–8 mm. long; seed brown and shining.

Sandy and gravelly soil, Sonoran Zones, Texas to Louisiana and south into Tamaulipas and Oaxaca, and in southern Baja California. This species is not a regular inhabitant of the desert and may be confined to the higher mountains of the Sierra de la Giganta in Baja California. It is included because the desert slopes of the mountains between Comondú and La Paz are poorly known botanically, and it may well occur there.

## 9. EYSENHARDTIA H. B. K., Nov. Gen. & Sp. 6: 489, 1824

Ascendingly branched shrubs or small trees with odd-pinnate leaves, oblong to elliptic or oval, glandular-punctate leaflets, and terminal, spiciform, racemose inflorescences. Calyx unequally lobed, tube often more deeply split between posterior than anterior lobes. Petals nearly equal, banner slightly to distinctly the largest, truncate or emarginate. Stamens 10, diadelphous, stamineal tube equaling free part of filaments. Ovary sessile; style bearing a gland near apex or glandless; stigma capitate. Ovules 2–4. Pods indehiscent, 1-seeded.

Leaflets 21–41, mostly 8–20 mm. long; calyx shallowly split between posterior lobes:
- Leaflets equally fine-puberulent on both surfaces, faintly reticulate on upper surface, obscurely glandular .......... 1. *E. orthocarpa*
- Leaflets glabrous or glabrate, at least above, strongly reticulate on both surfaces, conspicuously glandular .......... 2. *E. reticulata*

Leaflets 9–11, rarely over 5 mm. long; calyx split nearly to base between posterior lobes .......... 3. *E. peninsularis*

### 1. Eysenhardtia orthocarpa (A. Gray) S. Wats., Proc. Amer. Acad. 17: 339, 1882

*Eysenhardtia amorphoides* var. *orthocarpa* A. Gray, Pl. Wright. 2: 37, 1853.
*Viborquia orthocarpa* Cockerell, Bull. Amer. Mus. Nat. Hist. 24: 97, 1908.

A profusely branched small tree or shrub 2.5–6 m. tall with erect or strongly ascending slender branches and puberulent to canescent-pubescent foliage and young growth; leaves 5–15 cm. long; stipules subulate, 2–3 mm. long; leaflets 21–41, oblong, 8–20 mm. long, puberulent on both surfaces, more heavily so and punctate with minute dark glands beneath, faintly reticulate above; racemes 5–12 cm. long, bracts small, caducous before anthesis; calyx 2.8–3.8 mm. long, tube 2–3 mm. long, glandular-dotted, ridged, uniformly puberulent, rarely sparsely so; petals white, 5–7 mm. long; style upcurved and glandular near tip; pods straight or barely upcurved at tip, 10–15 mm. long, thin, puberulent and punctate-glandular, reflexed or reflexed-spreading at maturity; seed about 5 mm. long.

On rocky and gravelly ridges and along foothill watercourses, upper part of Lower Sonoran and through the Upper Sonoran Zones, Pinal and Cochise Counties, Arizona, south through Sonora and into Chihuahua. May–Aug.

### 2. Eysenhardtia reticulata Pennell, N. Amer. Fl. 24: 36, 1919

Similar in general appearance to *E. orthocarpa* but with more lightly pubescent to glabrate young growth and foliage; leaflets 25–37, narrowly oblong to oblong,

8–17 mm. long, loosely pubescent when young but soon nearly or quite glabrous and strongly reticulate on both surfaces; calyx 5–7.5 mm. long, tube narrower than in *E. orthocarpa,* sparsely puberulent or nearly glabrous, lobes 3–5 mm. long; bracts of inflorescences reddish brown, subulate, 1–3 mm. long, caducous during and after anthesis; petals, style, and pods similar to those of preceding species.

Rocky hillsides, near upper margin of Lower Sonoran Zone and into Tropical Thorn Forest, near Ciudad Obregon, Sonora, to Zacatecas and Jalisco. May–Aug.

This species is less common than *E. orthocarpa* and may be only a variant of that or of *E. polystachya,* but it is readily distinguishable both in the field and in the herbarium.

### 3. **Eysenhardtia peninsularis** Brandegee, Univ. Calif. Pub. Bot. **4:** 180, 1911

An intricately branched shrub 2 m. high or more with gray bark, cinereous-strigose young growth, and small leaves 1–1.5 cm. long; leaflets 9–11, oblong, 4–5 mm. long, finely puberulent to glabrate on both surfaces, slightly paler and punctate with small brownish glands beneath, thickish, no reticulations evident above; racemes to 5 cm. long; bracts lanceolate-attenuate, 1–2 mm. long, persisting after anthesis; calyx 3–3.5 mm. long, tube split nearly to base on posterior side, lobes about 1 mm. long, subequal, puberulent; petals 4–5 mm. long, exserted 3–3.5 mm. beyond calyx; style upcurved at apex, gland small and inconspicuous; immature pods prominently veined, mature ones unknown.

Known only from Paraiso, the type locality, in the Lower Sonoran Zone, central Baja California.

## 10. **DALEA** Juss., Gen. 355, 1789

Annual or perennial herbs or shrubs with more or less gland-clothed branches, leaves, calyces, and pods. Leaves pinnately compound, alternate, stipulate and stipellate. Inflorescence racemose or spicate. Calyx campanulate to subcylindrical, 10-ribbed, 5-lobed. Corolla papilionaceous, at least in youth. Banner inserted on hypanthium, its blade usually cordate. Wings and keel petals inserted on hypanthium or more frequently on stamineal tube, blades mostly obliquely obovate to oblong. Stamens 9 or 10, monadelphous, tube split on upper side. Ovary 1–2-ovulate. Stigma capitate. Pod indehiscent, obliquely obovate, usually about equaling calyx and hairy.

◆ Petals inserted on hypanthium at base of stamineal tube:
  Flowers short-pedicellate; inflorescence short-racemose; twigs spinescent:
    Glands arranged in circle near apex of calyx tube, one in each interval between ridges; branchlets stiff, ascending-crowded, strongly spinescent; blade of banner 4–5 mm. long .......... 1. *D. spinosa*
    Glands irregularly arranged on calyx tube; branchlets less rigid, not conspicuously crowded, slightly spinescent; blade of banner 5–7 mm. long:
      Leaves simple; calyx sparsely puberulent to glabrate, not sericeous:
        Branches bright green, glabrate; leaves sparsely puberulent .......... 2. *D. schottii* var. *schottii*
        Branches and leaves densely canescent-strigose ....... 2*a*. *D. schottii* var. *puberula*
      Leaves pinnate, leaflets 1–11; calyx and young branches densely canescent-strigose, at least while young .......... 3. *D. californica*
  Flowers sessile; inflorescence a subcapitate spike, rarely over 1 cm. long; twigs not spinescent:

Calyx lobes narrowly subulate, equaling or exceeding the tube ........... 4. *D. emoryi*
Calyx lobes lance-triangular or lance-subulate, shorter than tube:
Calyx lobes about equal, closely short-canescent, not pilosulous; leaves and stems puberulent; stems stout:
Stems and leaves distinctly canescent; lateral leaflets rarely more than twice as long as wide ..................................... 5. *D. tinctoria* var. *tinctoria*
Stems and leaves sparsely puberulent but green; lateral leaflets usually three to six times as long as wide ......................... 5*a*. *D. tinctoria* var. *arenaria*
Calyx lobes unequal, lowest one and one-half times as long as others, pilosulous; leaves and stems glabrous; stems slender, wiry .................... 6. *D. juncea*

◆ Petals, except the banner, inserted on stamineal tube:
■ Flowers racemose, distinctly pedicellate, reflexed, at least in age:
● Calyx lobes ovate to deltoid, shorter than or about as long as tube:
Calyces hairy on part of outer surface:
Stems 1 mm. in diameter or less, wiry:
Stems, leaves, and calyces white-hirsutulous with spreading hairs ............. ............................................... 7. *D. peninsularis*
Stems glabrous to strigose-puberulent; leaves and calyces sparsely strigose ...... ............................................... 8. *D. chrysorhiza*
Stems stouter, mostly 2 mm. in diameter or more, not wiry:
Leaves glabrous or essentially so; calyx tube nearly glabrous, lobes pubescent .... ................................................. 9. *D. palmeri*
Leaves pubescent, at least along margins and midvein; both calyx tube and lobes pubescent:
Pod puberulent between rows of glands; calyx uniformly canescent; blade of wings 3 mm. long .................................. 10. *D. parryi*
Pod glabrous; calyx more densely pubescent on lobes than at base; blade of wings 4 mm. long ................................ 11. *D. maritima*
Calyces glabrous on outer surface (lobes sometimes pubescent on margins):
Leaves more or less pubescent; calyx lobes short-acuminate ..................... ....................................12*a*. *D. divaricata* subsp. *anthonyi*
Leaves glabrous; calyx lobes acute to obtuse, not acuminate:
Rachis of leaves not winged; leaflets 2–5 mm. long:
Calyx lobes thick, heavily nerved; leaflets entire ......................... ................................. 12. *D. divaricata* subsp. *divaricata*
Calyx lobes thin, slightly curved; leaflets crenulate .................. 13. *D. crenulata*
Rachis of leaves narrowly winged; leaflets 5–10 mm. long ........ 14. *D. nutans*
● Calyx lobes oblong to lance-subulate, distinctly longer than tube:
Plants annual; leaflets orbicular to obcordate, 1–2 mm. long; petals shorter than calyx lobes, subequal in length .................................... 15. *D. vetula*
Plants perennial; leaflets obovate to oblong, 2–15 mm. long; petals exceeding calyx lobes, keel petals longer than wings and banner:
▼ Calyx lobes lanceolate to oblong, not attenuate-acuminate; pubescence of inflorescence short, tangled, not plumose-silky; flowers not subtended by conical glands:
Calyx lobes broadly lanceolate, obscured by dense pubescence, lowest lobe one and one-half times as long as others; leaflets 2–3.5 mm. long ................ ............................................... 16. *D. alamosana*
Calyx lobes narrowly lanceolate, not obscured by pubescence; leaflets 2–5 mm. long:
Stipules lance-acuminate, 2–4 mm. long; flowers 7–10 mm. long ............ ................................................ 17. *D. calycosa*
Stipules subulate, 1 mm. long or less; flowers 5–6 mm. long ..... 18. *D. orcuttii*

▼Calyx lobes deltoid-ovate, drawn out into an attenuate-acuminate tip several times as long as basal part; pubescence of inflorescence plumose-silky; each flower subtended by 1–3 small conical glands:

Flowers 4–5 mm. long; calyx 3–4 mm. long; pubescence of whitish hairs, rarely turning rusty upon drying .................. 19. *D. mollis* subsp. *mollis*

Flowers 6–8 mm. long; calyx lobes about equaling corolla; pubescence of inflorescence often turning rusty-brown in drying:

Bracts lanceolate; stems prostrate or decumbent, glandular but not conspicuously pustulate:

Wing petals oblong, rounded or retuse and nearly as broad at summit as at base; keel straight on upper margin:

Bracts narrowly linear-lanceolate, acuminate; wings retuse with a small gland at base of notch ............... 19*a*. *D. mollis* subsp. *pilosa*

Bracts lanceolate, acute; wings rounded, not retuse nor glandular at tip .. ......................... 20*a*. *D. neo-mexicana* subsp. *mollissima*

Wing petals ovate, nearly acute at apex, broadest at base; keel petals undulate-crispate along upper margin .................................. ......................... 20. *D. neo-mexicana* subsp. *neo-mexicana*

Bracts ovate; stems erect, glandular-pustulate .......... 21. *D. lachnostachys*

■Flowers spicate or capitate, sessile, not reflexed:

Plants annual:

Corollas purple, rose, or tinged with purple or rose; leaves and stems glabrous:

Leaflets 15–35; heads cylindrical, several times as long as broad:

Bracts with scarious white margins, subpersistent, glabrous above; stamineal tube dark blue on upper part ............................... 22. *D. leporina*

Bracts not scarious-margined, caducous, puberulent throughout; stamineal tube stramineous ..................................... 23. *D. alopecuroides*

Leaflets 3–11; heads capitate, nearly as wide as long; bracts not scarious-margined:

Calyx lobes 4–5 mm. long, nearly twice as long as tube; bracts narrowly lanceolate, villous-ciliolate along margins, deciduous ........... 24. *D. lemmonii*

Calyx lobes 2.5–3 mm. long, only slightly longer than tube; bracts broadly ovate, glabrous, subpersistent ............................... 25. *D. brachystachys*

Corollas ochroleucous; leaves and stems short-villous ................. 26. *D. flava*

Plants perennial, often shrubby or suffruticose:

Calyx lobes ovate to deltoid-lanceolate, acute to short-acuminate, shorter than tube:

Bracts cucullate, covering flowers, persistent; leaflets linear; plants slightly suffrutescent, 1–3 dm. high ................................. 27. *D. lumholtzii*

Bracts plane or nearly so, not covering flowers, deciduous; leaflets obovate to cuneate-oblong; plants shrubby, 5–15 dm. high:

Leaves and stems densely short-pubescent .... 28. *D. seemannii* subsp. *seemannii*

Leaves and stems glabrous or nearly so ..... 28*a*. *D. seemannii* subsp. *trochilina*

Calyx lobes subulate-attenuate, longer than tube:

Leaves and stems pubescent:

▲Plants distinctly shrubby; leaflets rounded to retuse at apex, not acute:

Leaflets 2–4 mm. long; younger stems mostly stramineous:

Stems finely but roughly glandular-tuberculate; leaflets 3–7 .. 29. *D. greggii*

Stems glandular-dotted but not tuberculate or roughened; leaflets 11–23 .... ....................................... 30. *D. wislizenii* var. *sessilis*

Leaflets 4–10 mm. long; younger stems purplish:

Petals each bearing a terminal gland; lowest calyx lobe conspicuously longer than upper four ............................... 31. *D. occidentalis*

Petals without a terminal gland; calyx lobes subequal, lower only slightly longer than others:

Bracts 3–5 mm. long, purplish and sparsely pubescent above middle; herbage sparingly strigose:
Branches slender, virgate; calyx lobes 2–2.5 mm. long, lance-subulate .............................. 28. *D. seemannii* subsp. *seemannii*
Branches stouter, intricately branched; calyx lobes 3.2–4 mm. long, linear ................................................ 32. *D. purpusii*
Bracts 5–7 mm. long, stramineous, densely silky-villous over entire back; herbage densely sericeous .................. 33. *D. megalostachys*
▲Plants herbaceous above woody base; leaflets often acute at apex:
Leaflets glabrous or sparsely puberulent above; stipules 2–3 mm. long; bracts 4–5 mm. long ........................... 34. *D. nana* var. *carnescens*
Leaflets densely sericeous on both sides; stipules 4–5 mm. long; bracts 6–9 mm. long ............................................ 35. *D. wrightii*
Leaves and stems glabrous:
Stems woody; bracts silky-villous, narrowly ovate; spike densely crowded in fruit .................................................. 36. *D. brandegeei*
Stems herbaceous above base; bracts broadly ovate, cucullate, glabrous; spike usually rather lax in fruit .......................... 37. *D. pogonathera*

**1. Dalea spinosa** A. Gray, Pl. Thurb. 315, 1854

*Dalea spinescens* Hemsl., Biol. Centr. Amer. Bot. **1:** 247, 1880.
*Parosela spinosa* Heller, Cat. N. Amer. Pl. ed. 2, 7, 1900.
*Psorodendron spinosum* Rydb., N. Amer. Fl. **24:** 45, 1919.

Intricately branched tree 4–9 m. high with slender spinescent branchlets that are minutely canescent early in season's growth, later glabrate, conspicuously glandular-pustulate; leaves simple, 1.5–15 cm. long, obovate to oblanceolate, white-strigose, irregularly dentate on young plants, early deciduous; racemes spinescent, 2.5–5 cm. long; bracts ovate, about 1 mm. long, gland-tipped, canescent, caducous; pedicels 1–1.2 mm. long; calyx densely white-strigose, 10-ribbed, with a single gland in each interval near apex of tube; calyx tube turbinate, 3–3.5 mm. long, lobes ovate, 1.5–2 mm. long, acute to obtuse; corolla dark blue; blade of banner obcordate, 4–5 mm. long; blades of wing and keel petals obliquely oval to obovate, with prominent basal lobes, 4–5 and 5.5–6.5 mm. long respectively, claws about 2 mm. long; pod obovate, 6–8 mm. long, canescent on apical 1–2 mm., glabrous below, bearing 2 rows of conspicuous glands around terminal third.

Along arroyos and on gravelly hillsides, Lower Sonoran Zone, southern Mojave Desert near Daggett, throughout Colorado Desert, southwestern Arizona, northern Sonora, and to central Baja California. April–May.

**2. Dalea schottii** Torr., Bot. Mex. Bound. 53, 1859, var. **schottii**

*Parosela schottii* Heller, Cat. N. Amer. Pl. ed. 2, 6, 1900.
*Psorodendron schottii* Rydb., N. Amer. Fl. **24:** 44, 1919.

Intricately branched spinescent shrub 1–4 m. high with slender, bright green and glabrate or minutely strigose branches; leaves simple, linear, 1–2 mm. wide, 1–2.5 cm. long, prominently glandular-pustulate along margins, canescent when young, often revolute, in fascicles on old dwarf branchlets; racemes lax, 3–8 cm. long; bracts subulate, 1–1.5 mm. long, usually with a pair of toothlike, lateral, conical glands, caducous; pedicels 1–2 mm. long, canescent; calyx sparingly strigose near apex when young, soon glabrate, tube turbinate, 10-ribbed, irregularly glandular-dotted, 3–3.5 mm. long; calyx lobes broadly ovate, 1–2 mm. long, acute to obtusish, ciliolate-margined; corolla blue; blade of banner obcordate, 5–7 mm. long, with minute basal lobes; blades of wing and keel petals obliquely obovate, about 6 and 7 mm. long respectively, their basal lobes rounded, claws 2–2.5 mm. long; pod obliquely ovate, 8–10 mm. long, sparingly strigose at apex or soon glabrate, conspicuously glandular-dotted over all; beak straight, 2–3 mm. long.

Arroyos, canyons, and mesas, Lower Sonoran Zone; western Colorado Desert from vicinity of Palm Springs eastward into Arizona and south into Baja California to the vicinity of El Mármol. April-May.

### 2*a*. **Dalea schottii** var. **puberula** (Parish) Munz, Man. S. Calif. 263, 1935

Differs from var. *schottii* in having the young branches, leaves, and calyces persistently canescent-puberulent.

Western edge of the Colorado Desert.

### 3. **Dalea californica** S. Wats., Proc. Amer. Acad. **11:** 132, 1876

*Parosela californica* Vail, Bull. Torrey Club **24:** 17, 1897.
*Parosela californica* var. *simplifolia* Parish, Bot. Gaz. **55:** 309, 1913.
*Psorodendron californicum* Rydb., N. Amer. Fl. **24:** 43, 1919.

An intricately branched, somewhat spinescent shrub 0.6–2 m. high with ashy, sparingly glandular-dotted branches densely strigose-canescent in youth, later glabrate; leaves pinnate, 1–5 cm. long, densely strigose; leaflets 1–11, oblanceolate, oblong, or linear, 5–15 mm. long, usually decurrent and upper often confluent; racemes lax, 2–6 cm. long; bracts lanceolate, 1.5–2 mm. long, caducous; calyx densely canescent, irregularly glandular-dotted, lightly 10-ribbed, tube turbinate, about 3 mm. long; calyx lobes subequal, upper two ovate, lower three narrower, somewhat longer, 2–2.5 mm. long; corolla dark blue, about 8 mm. long, petals subequal; banner obcordate; blades of wing and keel petals obliquely obovate, with rounded apex and basal lobes; pods obliquely ovoid, body 6–8 mm. long, sparsely strigose, conspicuously glandular-dotted; beak 4–6 mm. long, straight or slightly curved.

Washes and canyons, Lower Sonoran Zone, western Colorado Desert from Palm Springs to Shaver's Well and Cottonwood Mountains. Mainly March–June but flowering sporadically throughout the year.

### 4. **Dalea emoryi** A. Gray, Pl. Thurb. 315, 1854

*Parosela emoryi* Heller, Cat. N. Amer. Pl. ed. 2, 6, 1900.
*Psorothamnus emoryi* Rydb., N. Amer. Fl. **24:** 47, 1919.

An intricately branched shrub 3–15 dm. high with white-tomentose, glandular-dotted branches, glands orange; leaves 1–10 cm. long, white-tomentose, finely glandular-dotted; leaflets 1–13 (mostly 3–9), linear-oblong to obovate, entire to irregularly sinuate-dentate, lateral ones 3–15 mm. long, terminal one often 20–25 mm. long; spikes globose or nearly so, dense, 10–18 mm. broad, on peduncles 1–6 cm. long; bracts lanceolate, 3–4 mm. long, densely white-villous and sparsely glandular-dotted; calyx densely villous, 10-ribbed, with 6–12 deep-orange conical glands in each interval, tube turbinate, 2.5–3.5 mm. long; calyx lobes linear-subulate, equaling or slightly surpassing tube, short-plumose; corolla blue-purple, 5–6.5 mm. long; banner oblong-cordate, faintly retuse, 4–4.5 mm. long, claw 1.5–2 mm. long; blades of wing and keel petals obovate, 3.5–4 and 4.5–5 mm. long respectively, basal lobes conspicuous, rounded-acute, blades of keel petals confluent for short distance along lower edges, soon distinct and blades widely spreading; pods obliquely obovate, densely villous, glandular-dotted, 3–4 mm. long.

Along washes and on gravelly plains, Lower Sonoran Zone, throughout the Colorado Desert into southwestern Arizona, northwestern Sonora, and to the Cape Region of Baja California. March–May; Aug.–Sept.

### 5. **Dalea tinctoria** Brandegee, Proc. Calif. Acad. II, **2:** 147, 1889, var. **tinctoria**

*Psorothamnus dentatus* Rydb., N. Amer. Fl. **24:** 47, 1919.
*Psorothamnus tinctorius* Rydb., loc. cit.

Intricately but openly branched shrub 0.5–1.5 m. high, often fully as wide, rounded, with gray-velutinous, gland-dotted branches and leaves; leaves 1.5–4 cm. long; leaflets 3–15, entire to dentate, glandular along margins, lateral ones obovate to elliptic, 3–10 mm. long, terminal one oblanceolate to oblong, 12–30 mm. long; peduncles 2.5–15 cm. long; spikes ovoid to subglobose, 1–1.5 cm. long; bracts ovate-lanceolate, 2.5–3 mm. long, caducous; calyx densely silky-canescent, 10-ribbed, glandular-dotted in intervals, tube turbinate, 2–2.5 mm. long, lobes lance-triangular, 1–1.4 mm. long, acute, gland-tipped; corolla dark blue or keel petals lighter; blade of banner 3–4 mm. long, orbicular-cordate, slightly retuse, with a brownish or reddish gland at base of apical notch, claw 2–2.5 mm. long; wing and keel petals 4 and 5 mm. long respectively, strigose on outer surface, former oblong, latter obovate, twice as broad as wing petals; pods obovate, glandular and hairy at tip, glabrous below.

Rocky hillsides and sandy washes, Lower Sonoran Zone, from between Comondú and San Gregorio to Loreto and southward to the Cape Region, Baja California. Nov.–April.

*Psorothamnus dentatus* is a vigorous plant in which the leaflets are more numerous and more conspicuously dentate than in typical *D. tinctoria*. However, the range of variation is continuous from one to the other.

### 5*a*. **Dalea tinctoria** var. **arenaria** Brandegee, Proc. Calif. Acad. II, **2:** 147, 1889

Differs in having narrower, longer, glabrate leaflets and glabrate green stems.
About the same range as var. *tinctoria*.

### 6. **Dalea juncea** (Rydb.) Wiggins, Contr. Dudley Herb. **3:** 49, 1940

*Psorothamnus junceus* Rydb., N. Amer. Fl. **24:** 48, 1919.

Open straggling shrub with glabrous, densely glandular-dotted, green, slender stems 1 m. long or less; leaves 1–2.5 cm. long, early deciduous; leaflets 1–7, oblong to oblanceolate, 10–16 mm. long, entire or slightly sinuate-dentate; spikes subglobose, 1–1.5 cm. long, on peduncles 4–12 cm. long; bracts narrowly ovate-lanceolate, 2.5–3.5 mm. long, caducous; calyx turbinate, pilosulous, 3–5 red or dark orange glands in each interval, tube 3 mm. long, lowest lobe narrowly lanceolate, 2–2.4 mm. long, others triangular, acute to short-acuminate, 1.2–1.5 mm. long; corolla dark blue; blade of banner suborbicular-cordate, 4 mm. long, with a solitary gland at apex, claw 2 mm. long; blades of wing and keel petals obovate, 5 and 6 mm. long respectively, former with an apical notch and gland, latter glandless and broader than wing petals; pods about 3–3.5 mm. long, glabrous at base, glandular and hairy at apex.

Sandy washes and rocky slopes, eastern side of the Sierra San Pedro Mártir from the vicinity of San Matías Pass southward. Nov.–April.

**7. Dalea peninsularis** (Rose) Bullock, Kew Bull. **1939:** 197, 1939

*Dalea canescens* Benth., Bot. Sulph. 12, 1844. Not *D. canescens* Mart. & Gal., 1843.
*Parosela peninsularis* Rose, Contr. U.S. Nat. Herb. **8:** 304, 1905.

Prostrate, diffusely branched annual with many filiform pilosulous stems 1–6 dm. long and strigose-canescent leaves 1–2 cm. long; stipules filiform-subulate, 1.5–2 mm. long; rachises pilosulous; stipels subulate, minute; petioles 2–4 mm. long; leaflets 9–15, obovate or oblong, obtuse to retuse, 2–4 mm. long, hirsutulous, sparingly glandular-punctate beneath; peduncles slender, 1–4 cm. long; racemes dense, 1–2 cm. long; bracts lanceolate-acuminate, hirsutulous and somewhat glandular, caducous; calyx hirsutulous, sparingly glandular between ribs, tube turbinate, 10-ribbed, 1–1.5 mm. long, lobes lance-ovate, acute, equaling or slightly surpassing tube, subequal; corolla 4 mm. long, purple, tinged with yellow on lower side of keel; banner reniform, 2 mm. high, 3 mm. wide; blades of keel and wing petals 3 mm. long, on claws 1 mm. long; pod obliquely and broadly obovate, puberulent, glandular-punctate, about equaling tube.

Sandy places, Lower Sonoran Zone, vicinity of Mulegé to the Cape Region, Baja California, and west-central Sonora. Nov.–April.

**8. Dalea chrysorhiza** A. Gray, Proc. Amer. Acad. **5:** 156, 1861

*Parosela chrysorhiza* Rose, Contr. U.S. Nat. Herb. **8:** 304, 1905.

Prostrate winter annual with numerous, diffusely branched, filiform, sparsely pilose to glabrous, light-orange stems; leaves 1–2 cm. long, delicate, spreading; stipules lanceolate, 1–3 mm. long; rachis sparsely pilosulous to glabrous; leaflets 11–25, narrowly oblong-cuneate, minutely emarginate, revolute, 1.5–3 mm. long, glandular-punctate and sparingly pilosulous beneath; peduncles slender, 1–3 cm. long; racemes few-flowered; bracts ovate, 1 mm. long or less, glabrous, caducous; pedicels bearing 1–4 minute, orange, conical glands; calyx canescent, tube turbinate, 10-ribbed, 1–1.5 mm. long, sparingly punctate between ribs; calyx lobes oblong-lanceolate, equaling or slightly surpassing tube, glandular-apiculate; corolla purple; banner reniform, 1.5 mm. long, claw 1.5 mm. long; blades of wing and keel petals 3 mm. long, claws 1.5 mm. long; pod obliquely obovate, hirsutulous and glandular-punctate; beak incurved, short.

Open sandy places, Lower Sonoran Zone, vicinity of La Paz and south through the Cape Region. Oct.–May.

**9. Dalea palmeri** (Rose) Riley, Kew Bull. **1923:** 337, 1923

*Parosela palmeri* Rose, Contr. U.S. Nat. Herb. **8:** 304, 1905.

Suffruticose perennial 3–6 dm. high with slender, glabrous, glandular-punctate, ascending stems; stipules subulate, minute; leaves 2–4 cm. long with glandular-punctate rachises and 11–21 elliptic, obovate or suborbicular, glabrous leaflets 1.5–2 mm. long, these rounded or slightly retuse, sparingly punctate-glandular beneath; racemes 4–6 cm. long, lax, glabrous; bracts lance-acuminate, glabrous, sparingly glandular-serrulate, caducous; calyx crispy-hairy on 10 ribs and lobes, tube turbinate, 1.5 mm. long, glandular-punctate between ribs above middle; lobes ovate, obtuse, 0.6–1 mm. long or lowest one slightly narrower and longer; corolla about 4 mm. long, ochroleucous, tipped and margined with purple, glandular-punctate on lighter areas; blade of banner reniform, 1 mm. long, 2 mm. wide; blades of wings 2.5 mm. long, those of keel 3–5 mm. long, claws of all 1–1.2 mm. long; pod broadly oblique-obovate, glandular-punctate, glabrous, about equaling calyx.

Dry hills, Lower Sonoran Zone, Alamos, Sonora, and south into Sinaloa. Probably flowering Aug.–Oct.

**10. Dalea parryi** (A. Gray) Torr. & Gray, in A. Gray, Proc. Amer. Acad. **7:** 397, 1868

*Dalea divaricata* var. *cinerea* A. Gray, Proc. Amer. Acad. **7:** 335, 1868.
*Parosela parryi* Heller, Cat. N. Amer. Pl. ed. 2, 6, 1900.
*Dalea angulata* M. E. Jones, Contr. West. Bot. **16:** 25, 1930.

Suffruticose perennial 2–8 dm. high with ashy-strigose, glandular-punctate branches; stipules subulate, 1–2 mm. long; leaves 3.5–6 cm. long with minute stipels and glandular-punctate rachises; leaflets 15–35, suborbicular to obovate, emarginate, strigose and glandular-punctate on both surfaces, 2–6 mm. long; racemes 2–10 cm. long, many-flowered; bracts lance-acuminate, strigose, glandular-denticulate; calyx strigose-canescent, tube turbinate, 10-ribbed, 1.6–2.2 mm. long, glandular-punctate between ribs; calyx lobes deltoid-ovate, upper four lobes half as long as tube, acute to obtuse, lower one slightly longer, lance-ovate, acute; corolla violet to purple; banner reniform, 1.5 mm. long, 2–2.5 mm. wide, claw 1 mm. long; blades of wing petals 3 mm. long, paler to nearly white along lower margin; keel petals 4 mm. long, obliquely obovate, slightly puberulent, glandular-punctate; beak short.

Rocky slopes and washes, Lower Sonoran Zone, from southeastern Mojave Desert and vicinity of Desert Center, Colorado Desert, to southern Nevada, southern Arizona, southern Sonora, and central Baja California. March–June; Oct.

**11. Dalea maritima** Brandegee, Proc. Calif. Acad. II, **3:** 125, 1891

*Parosela maritima* Rose, Contr. U.S. Nat. Herb. **8:** 304, 1905.

Suffruticose, openly branched perennial with nearly prostrate, densely white-lanate, glandular-punctate branches 1.5–4 dm. long; leaves sparse, 2–3 cm. long; stipules filiform-subulate, 1.5–2.5 mm. long; petioles 2–4 mm. long; rachises densely white-canescent and punctate-glandular; leaflets 15–25, broadly obovate or oblong-obovate, 2.5–4 mm. long, rounded to shallowly retuse at apex, sparingly glandular-punctate beneath, densely canescent on both surfaces, readily falling from subpersistent rachises; peduncles stout, 5–30 mm. long; racemes dense, 1–3 cm. long; bracts lanceolate-acuminate, canescent, about equaling calyces; flowers 5–6 mm. long, purple; calyx tube broadly turbinate, 2 mm. long, strongly 10-ribbed, glabrous and scarcely glandular, lobes deltoid-ovate, acute, hirsutulous along their margins, upper four lobes one-half as long as tube or less, lower lanceolate, 1.5 mm. long; banner reniform, 2 mm. high, 3 mm. wide; blades of wing petals 4 mm. long, of keel 5 mm. long, claws 2 mm. long; pod obliquely and broadly obovate, glabrous, with two rows of prominent glands on each side, about equaling calyx.

Sandy areas, Lower Sonoran Zone, Cape Region from La Paz southward. Nov.–March.

**12. Dalea divaricata** Benth., Bot. Sulph. 12, 1844, subsp. **divaricata**

*Parosela divaricata* Rose, Contr. U.S. Nat. Herb. **8:** 305, 1905.
*Parosela variegata* Rydb., N. Amer. Fl. **24:** 55, 1919.
*Parosela oculata* Rydb., op. cit. 60.

Suffruticose perennial diffusely branched from base with glabrous, glandular-punctate, ascending stems 2.5–6 dm. long; stipules subulate, 1–2.5 mm. long; petioles 2–5 mm. long; leaves 1–4 cm. long, rachises glabrous or essentially so, sparingly glandular; leaflets 9–21, oblong to obovate, 2–4 mm. long, glabrous, obtuse to emarginate at apex, sparingly glandular and slightly paler beneath than above; peduncles 0.5–2.5 cm. long; racemes dense, 1.5–3 cm. long; bracts lance-acuminate, 4–5 mm. long, sparsely ciliolate-denticulate on lower half, surpassing calyces; calyx tube turbinate, 10-ribbed, 1.5–1.8 mm. long, glabrous, bearing 1–3 prominent glands between each pair of ribs toward apex; upper four lobes ovate, acute, half as long as tube, lower one narrower, acuminate, equaling tube; corolla blue-purple, with whitish margins on lower sides of wings and keel; banner reniform, 1.5 mm. high, 3 mm. broad; blades of wing petals 3–3.5 mm. long, of keel about 4.5 mm. long, claws 1–1.5 mm. long; pod obliquely obovate, glabrous, bearing prominent glands in 2 rows and a few scattered ones on each side.

Sandy places, Lower Sonoran Zone, Magdalena Island and adjacent mainland southward to San José del Cabo. Nov.–May.

**12*a*. Dalea divaricata** subsp. **anthonyi** (Brandegee) Wiggins, Contr. Dudley Herb. **3:** 50, 1940

*Dalea anthonyi* Brandegee, Erythea **7:** 2, 1899.
*Parosela anthonyi* Rose, Contr. U.S. Nat. Herb. **12:** 272, 1909.

Differing only in having sparsely puberulent leaflets and slightly longer calyx lobes. Southern Cape Region.

**13. Dalea crenulata** Hook. & Arn., Bot. Beechey 285, 1836

*Parosela crenulata* Rose, Contr. U.S. Nat. Herb. **8:** 306, 1905.

An erect shrub 4–10 dm. tall with slender, glabrous, virgate branches; stipules subulate, about 1 mm. long; leaves 1–2 cm. long, glabrous; leaflets 7–19, elliptic, 2–5 mm. long, crenulate, emarginate at apex, glandular-punctate beneath, midvein impressed on upper surface; peduncles 2–10 mm. long, on short lateral branches; racemes dense, 1–2.5 cm. long; bracts ovate, 1–1.5 mm. long, glabrous, ciliolate-dentate and glandular-punctate; calyx broadly turbinate, glabrous, tube 2.5–3 mm. long, strongly 10-ribbed with a row of small glands between each pair of ribs; calyx lobes ovate-lanceolate, thin, minutely serrulate, nearly equaling tube; corolla purple, 7–8 mm. long; banner reniform, 2 mm. high, 3 mm. wide; blades of wing petals 3.5–4 mm. long, rounded, those of keel petals 4.5–5 mm. long, 3–3.5 mm. wide, all on claws 1.5 mm. long; pods broadly and obliquely obovate, glabrous, with 3 rows of glands on each side, about equaling calyx tube.

Hillsides in vicinity of Alamos, Sonora, and south to Guerrero. The thin, crenulate calyx lobes and crenulate leaflets are distinctive. Feb.–April.

**14. Dalea nutans** (Cav.) Willd., Sp. Pl. **3:** 1339, 1803

*Psoralea nutans* Cav., Ic. **3:** 1, 1795.
*Dalea melilotoides* Sessé & Moc. ex DC., Prod. **2:** 246, 1825, as synonym.
*Parosela nutans* Rose, Contr. U.S. Nat. Herb. **8:** 306, 1905.

Erect, openly branched shrub 1–2 m. high with glabrous purplish branches irregularly dotted with small pustulate glands; leaves 2.5–6 cm. long, rachis very narrowly winged; stipules mere apiculations 1 mm. long or less; leaflets 11–41, oblong to cuneate-oblong, 5–10 mm. long, glabrous, dark green and eglandular above, glaucous and sparsely glandular-dotted beneath, margins gland-dotted and more or less crenulate; peduncles slender, 2 cm. long or less, terminating lateral branches; racemes lax, few-flowered; bracts ovate, membranous, glandular-dentate, glabrous; flowers 7–8 mm.

long; pedicels filiform, 2 mm. long; calyx tube broadly turbinate, strongly ribbed, 2–2.5 mm. long, closely crowded glands forming a row between each pair of ribs; lobes deltoid, thin, acute, usually serrulate, about 0.5 mm. long, lower one slightly narrower and longer than upper; corolla rose-purple; banner cordate, 2 mm. long; blades of wing petals oblong, rounded at apex, 3.5–4.5 mm. long, blades of keel petals 4–5 mm. long, blunt or rounded at apex, claws 0.5–0.8 mm. long; pods broadly obovate, glabrous, nearly black, densely glandular-dotted, 3 mm. long.

Arroyos, canyons, and hillsides, margins of the Lower Sonoran Zone and into the Tropical Zone, southern Sonora to Yucatán and Costa Rica. Nov.–March.

Similar to *D. crenulata* but with larger leaflets, laxer racemes, shorter calyx lobes, and more densely glandular pods.

**15. Dalea vetula** Brandegee, Proc. Calif. Acad. II, **2:** 146, 1889

*Parosela vetula* Rose, Contr. U.S. Nat. Herb. **10:** 105, 1906.

Erect, slenderly branched, glabrous annual 2–4 dm. tall with sparsely leafy, sparingly glandular-punctate stems; leaves 2–4 cm. long; stipules filiform-subulate, 1.5–2 mm. long; leaflets 19–25, orbicular to obcordate, 1–2 mm. long; peduncles terminal, 1–7 cm. long; racemes dense, 2–9 cm. long (in age); bracts ovate-lanceolate, 2–3.5 mm. long, subglabrous to sparsely pilosulous on back, ciliolate-pilose on margins; calyx tube broadly campanulate, 1–1.5 mm. long, silky-pilose, lobes lance-attenuate, 3–3.5 mm. long, pilose, with a row of small glands on each side of midrib; corolla pale purple, concealed by long, pilose calyx lobes; banner cordate-suborbicular, 1.5 mm. wide, claw 1.5 mm. long; blades of wing and keel petals obliquely obovate, 2 mm. long on claws less than 1 mm. long; pods broadly and obliquely obovate, 2.5 mm. long, sparsely pubescent, tawny, with 3 rows of conspicuous orange glands on each side.

Sandy soil, Lower Sonoran Zone, Calmallí to Comondú and San Gregorio, central Baja California. Nov.–March.

**16. Dalea alamosana** (Rose) Bullock, Kew Bull. **1939:** 194, 1939

*Parosela alamosana* Rose, N. Amer. Fl. **24:** 63, 1919.

Prostrate or decumbent perennial herb with numerous slender, sparsely strigose, and glandular-pustulate branches 1–3 dm. long; leaves 1.5–3 cm. long; stipules subulate, 1–1.5 mm. long; leaflets 9–21, oblong, 2–3.5 mm. long, striate and glabrate above, closely strigose-canescent and sparingly punctate-glandular beneath, rounded at each end; peduncles terminal or subopposite the leaves, 0.5–2 cm. long; racemes dense, 1–2 cm. long; bracts ovate to lanceolate, acute to acuminate, 1.5–2 mm. long, pilose, caducous; calyx tube turbinate, 1.5 mm. long, 10-ribbed, pilose, punctate-glandular between ribs; calyx lobes lanceolate, acute, upper four about 2 mm. long, lower one 2.5–3 mm. long and narrower; corolla 5–6 mm. long, white tinged with light purple; banner reniform, 1.5 mm. high, 2.5 mm. wide, claw 2 mm. long; blades of wings 3 mm. long, those of keel 4 mm. long, claws 1–1.5 mm. long; pod obliquely obovate, pilosulous, gray-green, with a row of glands along mid-lateral line and another along straight (upper) margin, short-beaked.

Hillsides and gravelly valley floors, Lower and Upper Sonoran Zones, southern Sonora into Sinaloa. Oct.–April.

**17. Dalea calycosa** A. Gray, Pl. Wright. **2:** 40, 1853

*Parosela calycosa* Heller, Cat. N. Amer. Pl. ed. 2, 5, 1900.

Decumbent to ascending perennial herb with slender, somewhat strigose stems

1–3 dm. long; stipules lance-acuminate, 2–4 mm. long, sparsely strigose on outer surface; leaves 1–3 cm. long, petioles 4–7 mm. long; leaflets 15–29, oblong to obovate, 2–5 mm. long, rounded, obtuse, or retuse at apex, glabrous on upper surface, strigose and sparsely and minutely glandular-punctate beneath; peduncles 1–4 cm. long; racemes dense, 1.5–4 cm. long; bracts lanceolate-attenuate, 3.5–4.5 mm. long, sparsely pubescent, caducous; flowers 7–10 mm. long; calyx tube turbinate, 2–2.5 mm. long, strongly 10-ribbed, pilose, glandular between ribs; calyx lobes linear-lanceolate, 4–5 mm. long, densely pilose without, glabrous within; corolla purple and white; banner reniform, about 3 mm. high, 4 mm. wide; blades of wings and keels obliquely ovate, 5 and 6 mm. long respectively, rounded at apex, on claws 1–2 mm. long; pod obovate, 3–3.5 mm. long and wide, sparsely pilosulous, with 2 rows of glands on each side; beak straight, 1–1.5 mm. long.

Dry slopes and washes, Lower Sonoran Zone, southern Arizona to New Mexico, Sonora, and Nuevo León. April–Sept.

This and *D. alamosana* look nearly alike superficially, but the longer calyx lobes and longer pod of *D. calycosa* sharply separate them.

**18. Dalea orcuttii** S. Wats., Proc. Amer. Acad. **20**: 359, 1885

*Parosela orcuttii* Parish, Bot. Gaz. **55**: 305, 1913.

Slightly suffruticose perennial with numerous, sparsely strigose-canescent, decumbent to ascending stems 1–2 dm. long; leaves 1–2 cm. long; stipules ovate-acuminate, 1 mm. long or less; petioles 2–5 mm. long; leaflets 9–17, oblong to oblong-obovate, 2.5–4 mm. long, rounded at apex, glabrous above, short-strigose-canescent and with 1 prominent gland at apex of midrib beneath; peduncles 0.5–1.5 cm. long, strigose; racemes short, dense, 1–2 cm. long; bracts linear-lanceolate, 3–3.5 mm. long, strigose, tips curved upward; flowers 5–6 mm. long; calyx tube 1.5 mm. long, short-turbinate, 10-ribbed, strigose, sparingly glandular between ribs; calyx lobes lance-acuminate, 2–2.5 mm. long; corolla purple and white; banner purple, 1 mm. long, 1.5 mm. broad, claw 2 mm. long; wings and keel petals white with purple along upper margin, blades of wing petals oblong, 2.5–3 mm. long, apex rounded; blades of keel petals 3.5–4 mm. long, 2–2.5 mm. wide, claws 1–1.5 mm. long; mature pods unknown.

Lower Sonoran Zone, northern Baja California. Probably Palmer's collection from "Big Canon, San Diego County, California," 1875, was actually collected in Baja California. Feb.–May.

**19. Dalea mollis** Benth., Pl. Hartw. 306, 1848, subsp. **mollis**

*Parosela mollis* Heller, Cat. N. Amer. Pl. ed. 2, 6, 1900.

Prostrate or decumbent perennial herb with numerous, slender, densely pilose, conspicuously glandular-pustulate stems 1–2.5 dm. long; herbage silky-pilose to canescent, glandular-pustulate; leaves 1–3.5 cm. long; stipules lance-subulate, 1.5 mm. long or less; petioles 3–6 mm. long; leaflets 11–19, broadly obovate, truncate to emarginate at apex, densely silky-pilose and glandular-punctate on both surfaces; peduncles 0.5–3 cm. long; racemes dense, conical, 1–2 cm. long; bracts narrowly lanceolate, 4–5 mm. long, pilose, sparingly glandular; flowers 4–5 mm. long; calyx tube campanulate, 2 mm. long, 10-ribbed, densely pilose, 1–3 glands between each pair of ribs; calyx lobes deltoid at base, attenuate, 1.5–2 mm. long; corolla white, tinged with rose-purple along tips and margins of petals; banner cordate, 2–2.2 mm. long, glandular-punctate near base, claw 1.5 mm. long; wing petals ovate, 2.2–2.6 mm. long, minutely retuse on lower side near apex and bearing a single gland at base of this notch, sparingly glandular at base, claw 1 mm. long; keel petals 2.5–3 mm. long, 1.6–1.8 mm. wide, claw 1.2–1.5 mm. long; pod obliquely obovate, 2 mm. long, sparsely pilosulous, sparingly glandular; beak straight.

Sandy soil, Lower Sonoran Zone, Colorado Desert, extreme southwestern Arizona, northwestern Sonora, and southward in Baja California almost to the Cape Region. Dec.–May.

**19*a*. Dalea mollis** subsp. **pilosa** (Rydb.) Wiggins, Contr. Dudley Herb. **3:** 51, *pl. 10, figs. 8–16,* 1940

*Parosela pilosa* Rydb., N. Amer. Fl. **24:** 64, 1919.

Differing from subsp. *mollis* mainly in having larger flowers 6–8 mm. long; calyx lobes nearly equaling corollas; and pubescence of stems and inflorescences usually more or less rusty hued in age or when dried.

In sandy soil along washes and on desert slopes, Lower Sonoran Zone, eastern part of Baja California from near the International Boundary to Calmallí; near Guaymas, Sonora. Dec.–April.

**20. Dalea neo-mexicana** (A. Gray) Cory, Rhodora **38:** 406, 1936, subsp. **neo-mexicana**

*Dalea mollis ?neo-mexicana* A. Gray, Pl. Wright. **1:** 47, 1852.
*Parosela neo-mexicana* Heller, Cat. N. Amer. Pl. ed. 2, 6, 1900.

Prostrate or decumbent, perennial herb with slender, sparsely to densely pilose, glandular-pustulate stems 1–3.5 dm. long; stipules subulate, 1.5 mm. long or less; leaves 1.5–3 cm. long, leaflets 7–13, obovate, retuse, 3–8 mm. long, villous and glandular-dotted on both surfaces; racemes dense, 1–2.5 cm. long, rachis conspicuously glandular-pustulate; bracts lanceolate, 6–8 mm. long, glandular-dotted and villous; flowers 7–8 mm. long, corolla obscured by the villous-plumose calyx lobes; calyx tube villous, 2.5–3 mm. long, bearing 8–15 glands between each pair of ribs; lobes 4.5–5 mm. long; corolla yellowish tinged with light purple; banner cordate, 3–3.5 mm. long, glandular near base; blades of wing petals 3–3.5 mm. long, obcuneate-ovate, with a conspicuous, rounded, basal lobe, upper margin slightly undulate, 1–2 glands at base, rounded at apex, claw 6–8 mm. long; blades of keel petals obovate, blunt at apex, 3–3.5 mm. long, upper margin distinctly undulate, claw 1.5–2 mm. long; pod 2.5–3 mm. long, villous, beak curved downward.

Sandy soil, Lower Sonoran Zone, southern Arizona, southern New Mexico, western Texas, and adjacent Sonora, Chihuahua, Coahuila, and Tamaulipas. Jan.–May.

**20*a*. Dalea neo-mexicana** subsp. **mollissima** (Rydb.) Wiggins, Contr. Dudley Herb. **3:** 52, *pl. 11, figs. 7–14,* 1940

*Parosela mollissima* Rydb., N. Amer. Fl. **24:** 64, 1919.
*Dalea mollis* var. *mollissima* Munz, Man. S. Calif. 263, 1935.

Prostrate or decumbent, perennial herb with numerous, densely silky-pilose and glandular-dotted stems 1–4 dm. long, these branching at base; stipules linear-subulate, 1.5–2.5 mm. long, villous; leaves 2–4 cm. long; leaflets 11–15, cuneate-obovate, truncate to retuse at apex, 4–8 mm. long, sparsely pilose or glabrous above, densely villous and punctate-glandular beneath, often involute; peduncles 0.3–1.5 cm. long; racemes dense, 2–3 cm. long, elongating to 6–7 cm. in fruit, rachis conspicuously glandular-pustulate; flowers 7–8 mm. long, corolla usually shorter than calyx lobes; calyx tube turbinate, 3–3.5 mm. long, densely villous, with 3–6 glands between each pair of ribs, lobes broadly deltoid at base, filiform-attenuate, 3.5–4.5 mm. long, plumose-villous; corolla white to rose; banner orbicular-cordate, 2–2.2 mm. long, glandular at base, claw 2 mm. long; blades of wing petals oblong, rounded at apex, 2–2.2 mm. long, sparingly glandular near base, claw 1.2 mm. long; blade of keel petals obovate, blunt at apex, 2.8–3 mm. long, basal lobe obtusely angled, claw 1.5 mm. long; pod 3 mm. long, villous.

Sandy desert slopes, Lower Sonoran Zone, southern Nevada, Inyo County, California, eastern Mojave Desert, Colorado Desert from Indio southward into northern Baja California, and extreme southwestern Arizona. Jan.–May.

**21. Dalea lachnostachys** A. Gray, Pl. Wright. **1:** 46, 1852

*Parosela lachnostachys* Heller, Cat. N. Amer. Pl. ed. 2, 6, 1900.

Herbaceous perennial with decumbent to ascending, villous, conspicuously glandular-pustulate, stramineous stems 1–3 dm. long; stipules subulate, 3–5 mm. long; leaves 4–8 cm. long, petiole and rachis conspicuously glandular-pustulate; leaflets 7–11, oval to obovate, rounded at apex, 6–15 mm. long, densely silky-villous on both surfaces, glandular-dotted along margins beneath; peduncles stout, 2–6 cm. long; racemes dense, cylindrical, 3–8 cm. long, 1.5–2 cm. thick; bracts ovate-lanceolate, attenuate, 10–12 mm. long, somewhat concave, scarious-margined, sparsely villous and glandular-dotted, surpassing buds; flowers 8–10 mm. long; calyx densely silky-villous with brownish or rusty hairs, tube turbinate, 2.5–3 mm. long, sparingly glandular between ribs, lobes deltoid at base, linear above, 4.5–6 mm. long, plumose-villous; corolla purple, banner ovate-cordate, about 4 mm. long, claw 2.5–3 mm. long; blades of wing petals oblong, rounded at apex, 3.5–4 mm. long, those of keel petals about 4–5 mm. long, blunt, claws about 3 mm. long; pod densely villous, 3.5 mm. long, beak straight.

Dry hills, Sonoran Zones, southern Arizona to western Texas and Chihuahua. Aug.–Sept.

Somewhat resembles *D. neo-mexicana*, but the larger leaflets, racemes, flowers, and especially the larger conspicuous bracts which are deciduous rather than caducous readily distinguish it.

**22. Dalea leporina** (Ait.) Bullock, Kew Bull. **1939:** 196, 1939

*Psoralea leporina* Ait., Hort. Kew. **3:** 81, 1789.

*Dalea alopecuroides* Hemsl., Biol. Centr. Amer. Bot. **1:** 237, 1880. Not *D. alopecuroides* Willd., 1803.

*Dalea annua* var. *alopecurodes* Kuntze, Rev. Gen. Pl. **1:** 178, 1891, in part.

*Dalea annua* var. *ebracteata* Kuntze, loc. cit.

Erect, much-branched annual 3–15 dm. high with glabrous, sparsely gland-dotted, striate stems and branches; leaves ascending, 3–10 cm. long; stipules 1.5–3 mm. long, subulate; leaflets 19–35, linear-oblong or narrowly obovate, 5–8 (or 10) mm. long, rounded or faintly retuse at apex, glabrous, dark green above, pale and glandular-punctate beneath; peduncles terminal on branches, 3–15 cm. long; spikes dense, cylindrical, 2–6 cm. long, 8–12 mm. in diameter; flowers 5–6 mm. long; bracts ovate, acuminate, equaling flowers, purplish with white scarious margins, sparsely short-villous, deciduous; calyx tube turbinate, 2–2.5 mm. long, silky-villous; lobes deltoid at base, attenuate, about equaling tube, dark purplish; corolla blue or purplish with white on basal parts of petals; stamineal column conspicuously exserted after fall of petals; banner subcordate, about 2 mm. long, claw 2 mm. long; blades of wing petals oblong, rounded at apex, 2 mm. long; those of keel petals 2.5 mm. long, blunt; pod obovate, 2.5–3 mm. long, densely villous and gland-dotted at apex; beak curved sharply downward.

Southern Arizona and New Mexico southward to Guatemala. Chiefly at elevations well above the Sonoran Desert. June–Oct.

**23. Dalea alopecuroides** Willd., Sp. Pl. **3:** 1336, 1803

*Psoralea alopecuroides* Poir., in Lam., Encycl. **5:** 695, 1804.
*Petalostemon alopecuroides* Pers., Syn. Pl. **2:** 268, 1806.
*Dalea pedunculata* Pursh, Fl. Amer. Sept. 474, 1814.
*Parosela dalea* Britt., Mem. Torrey Club **5:** 196, 1894, in part.
*Parosela alopecuroides* Rydb., N. Amer. Fl. **24:** 78, 1920.

Erect annual with stramineous, striate, glandular-dotted, glabrous stems 3–10 dm. high; leaves ascending, 3.5–8 cm. long; stipules subulate, 1.5–2.5 mm. long; leaflets 15–23, oblong, 6–8 mm. long, rounded at apex, bright green and glabrous above, faintly glaucous and sparsely glandular-punctate beneath; spikes dense, 2–8 cm. long, terminal on branches, peduncles 3–10 cm. long; bracts lanceolate, acuminate, sparsely puberulent on back, ciliolate, caducous; flowers 5–7 mm. long; calyx turbinate, silky-villous, tube 2.5–3 mm. long; calyx lobes lance-attenuate from an ovate base, about equaling tube; corolla rose or white tinged with rose; banner 2.5–3 mm. long, hastately obovate with base truncate, basal lobes acute; blades of wings oblong, 2 mm. long, those of keel petals obovate, blunt at apex; pods obovate, villous at apex.

Mostly outside the Sonoran Desert, southern Arizona and northern Sonora to Alabama, Illinois, and South Dakota. June–Sept.

**24. Dalea lemmonii** Parry ex A. Gray, Proc. Amer. Acad. **17:** 200, 1882

*Parosela lemmoni* Heller, Cat. N. Amer. Pl. ed. 2, 6, 1900.

Slender annual 1–3 dm. high with slender, glabrous, greenish, striate stems; stipules dark brown, ovate-subulate, spreading, 1 mm. long; leaves 1–3 cm. long, glabrous; leaflets 3–11, linear or cuneate-oblong, 4–8 mm. long, often involute, dark green above, pale and sparingly punctate-glandular along margins beneath; peduncles slender, 1–6 cm. long; spikes dense, globose to ovoid, 1–1.5 cm. long; bracts narrowly lance-attenuate, 5–6 mm. long, sparingly villous, ciliate on margins, deciduous; flowers about 6 mm. long; calyx tube turbinate, 2 mm. long, 10-ribbed, silky-villous, lobes filiform-setaceous, 4–5 mm. long, slightly surpassing corolla; corolla rose-purple; banner cordate, 2 mm. long, claw slightly longer; blades of wing petals elliptic, rounded at apex, 3 mm. long, those of keel petals 3.5–4 mm. long, blunt, claws 2 mm. long; pod obovate, villous, 2.5 mm. long.

Mostly at elevations above the Sonoran Desert, southern Arizona to Chihuahua and Puebla. Aug.–Sept.

**25. Dalea brachystachys** A. Gray, Pl. Wright. **2:** 39, 1853

*Parosela brachystachys* Heller ex Rose, Contr. U.S. Nat. Herb. **10:** 105, 1906.

Ascendingly branched annual with glabrous, sparingly glandular-punctate, striate stems 1–4 dm. long; leaves ascending, 1–3 cm. long; stipules dark, subulate, 1.5–2 mm. long; leaflets 5–11, cuneate-oblong to linear, 5–10 mm. long or terminal one to 15 mm., glabrous, dark green above, paler and copiously glandular-dotted beneath, rounded to shallowly retuse at apex; peduncles terminal and opposite the leaves, 0.5–6 cm. long; spikes globose to ovoid, dense; bracts ovate, short-acuminate, about 3 mm. long, purplish, conspicuously glandular-dotted, subpersistent; flowers 5–6 mm. long, corolla surpassing calyx lobes; calyx tube turbinate, about 2–2.5 mm. long, silky-villous and glandular-dotted, glands obscured by hairs, lobes deltoid at base, attenuate, 2.5–3 mm. long; corolla rose or yellowish tinged with rose; banner subcordate, about 2 mm. long and wide; blades of wing petals 2–2.5 mm. long, obovate-oblong, basal lobe rounded-acutish; blades of keel petals broader, 3–3.5 mm. long, basal lobe smaller and blunter; pods 2.5 mm. long, obovate, densely villous, beak straight.

Plains and mountain slopes, mostly above the Sonoran Desert, southern Arizona and New Mexico to northern Sonora, southeastward to San Luis Potosí. Aug.–Sept.

**26. Dalea flava** Mart. & Gal., Bull. Acad. Brux. **10**$^{2}$**:** 40, 1843

*Parosela pauciflora* Rose, Contr. U.S. Nat. Herb. **10:** 106, 1906.
*Parosela flava* Rydb., N. Amer. Fl. **24:** 79, 1920.

Ascending, openly branched annual (or perennial) 3–15 dm. high with striate, sparsely villous branches; leaves spreading, 1.5–4 cm. long; stipules subulate, 2–4 mm. long; leaflets 11–25, oblong or elliptic, rounded and often apiculate at apex, 4–8 mm. long, short-villous on both surfaces, glandular-dotted beneath; peduncles terminal, 1–2 cm. long; spikes dense, cylindrical, 1.5–4 cm. long, 10–12 mm. in diameter; bracts lance-attenuate, 5–6 mm. long, short-villous, subpersistent; flowers about 6–6.5 mm. long; calyx villous, tube 2–2.5 mm. long, heavily glandular between ribs; calyx lobes ovate at base, attenuate, about equaling tube, plumose-villous; corolla ochroleucous; banner nearly orbicular, subcordate, acute at apex, 1.5 mm. long, claw 2–2.5 mm. long; blades of wing petals oblong, 2 mm. long, nearly equaling keel petals, these blunt at apex; pods obovate, membranous, villous and glandular-dotted at apex, 2–2.5 mm. long.

Along streams and on moist slopes, southern Sonora to Vera Cruz and Puebla. Nov.–March.

The glands on the calyx are more conspicuous than in any of the preceding three species.

**27. Dalea lumholtzii** Robins. & Fern., Proc. Amer. Acad. **30:** 115, 1894

*Parosela arizonica* Vail, Bull. Torrey Club **24:** 14, 1897.
*Parosela lumholtzii* Vail, op. cit. **26:** 117, 1899.
*Dalea arizonica* K. Schumann, in Just, Bot. Jahresb. **27**$^{1}$**:** 493, 1901.

Erect perennial 1.5–3 dm. high from a woody base; stems slender, stramineous, sparsely puberulent and glandular-dotted, branching sparingly above; stipules filiform-setaceous, 1.5–3 mm. long; leaves 2–4.5 cm. long; leaflets 15–27, linear, obtuse at apex, 3–6.5 mm. long, puberulent when young but soon glabrate, glandular-punctate; peduncles terminal, 2–7 cm. long; spikes ovoid to oblong, dense, 0.6–1 cm. long; bracts broadly obovate, short-acuminate, somewhat cucullate, enclosing flowers, purplish and conspicuously glandular and glabrate on back, silky-villous at base and along margins, persistent; calyx sparingly villous, tube 1.5–2 mm. long, lobes ovate, acute, 1 mm. long or less; banner obovate-subcordate, 2.2–2.5 mm. long; blades of wing and keel petals 2.5 and 3 mm. long respectively; pods pubescent at apex, about 2 mm. long.

Lower Sonoran Zone, southern Arizona in the vicinity of Tucson and south into northern Sonora. An inconspicuous species and apparently collected only rarely. Aug.–Oct.

**28. Dalea seemannii** S. Wats., Proc. Amer. Acad. **22:** 470, 1887, subsp. **seemannii**
?*Dalea ehrenbergii* Schlecht., Linnaea **12:** 290, 1838.
*Parosela seemanni* Rose, Contr. U.S. Nat. Herb. **10:** 106, 1906.

Slenderly branched shrub 5–15 dm. high with short-villous, heavily glandular-tuberculate stems; stipules subulate, 2 mm. long; leaves 1–2.2 cm. long; leaflets 7–13, obovate to cuneate-oblong, 3–5 mm. long, obtuse to rounded at apex, strigose-silky on both surfaces, glandular-dotted and pale beneath; peduncles 1–3 cm. long, spikes dense, oblong, 2–5 cm. long; bracts broadly ovate, acuminate, 3–4 mm. long, densely short-pubescent, glandular-dotted, caducous; flowers 7–8.5 mm. long; calyx campanulate, villous, tube 3 mm. long, lobes lance-deltoid, short-acuminate, 2–2.5 mm. long; banner cordate, 3 mm. long, yellowish, tinged with rose; wing and keel petals rose-purple, blades 4 and 5 mm. long respectively, obtuse at apex, claws about 2 mm. long; pods 3 mm. long, villous at summit, shining below.

Dry slopes and washes, Lower Sonoran Zone, northern Baja California to Zacatecas and Hidalgo. Dec.–May.

**28*a*. Dalea seemannii** subsp. **trochilina** (Brandegee) Wiggins, Contr. Dudley Herb. **3:** 53, 1940
*Dalea trochilina* Brandegee, Proc. Calif. Acad. II, **3:** 220, 1892.

Differs from subsp. *seemannii* only in having glabrous instead of puberulent stems and leaves. A specimen from Arroyo Hondo, Sierra Giganta, *Gentry 4162*, is intermediate between typical *seemannii* and Brandegee's *trochilina* in that it has sparser puberulent stems and leaves than *seemannii*. This tends to emphasize the tuberculate glands so conspicuous in subsp. *trochilina*, and the relationship readily becomes apparent.

**29. Dalea greggii** A. Gray, Pl. Thurb. 314, 1854
*Parosela greggii* Heller, Cat. N. Amer. Pl. ed. 2, 6, 1900.

Spreadingly branched shrub 6–12 dm. high with silky-strigose, slender, glandular-tuberculate branches; stipules broadly subulate, 1–2 mm. long; leaves 1.5 cm. long or less; leaflets 3–7, obovate, often involute, 2–4 mm. long, tomentulose-sericeous, glandular-dotted beneath; spikes dense, globose to oblong, short-pedunculate to sessile, mostly about 1 cm. long; bracts lance-ovate, short-acuminate or acute, conspicuous in bud, 4.5–5 mm. long, sericeous on back, deciduous; flowers 7–9 mm. long at anthesis; calyx villous, tube turbinate, about 2 mm. long, sparingly glandular between ribs; calyx lobes subulate, silky-pilose, 2–3 mm. long; corolla rose-purple; banner rhombic-ovate, 3–3.5 mm. high, slightly wider, claw 2 mm. long; blades of wing and keel petals oblong to obliquely obovate, 4 and 5 mm. long respectively, rounded to obtuse at apex, claws 1.5 mm. long; pods obovate, gland-dotted and pubescent at apex, 2.5–3 mm. long.

Chiefly an undershrub in the Upper Sonoran and Lower Transition Zones, but approaching or entering the Lower Sonoran Zone on the west slope of the Catalina Mountains near Tucson, Arizona, and on the lower slopes of the Baboquivari Mountains; Arizona and Sonora to Nuevo León and San Luis Potosí. Feb.–May.

**30. Dalea wislizenii** A. Gray var. **sessilis** A. Gray, Proc. Amer. Acad. **16:** 105, 1880
*Parosela wislizeni sessilis* Vail, Bull. Torrey Club **24:** 15, 1897.
*Parosela sanctae-crucis* Rydb., N. Amer. Fl. **24:** 103, 1920.
*Parosela sessilis* Rydb., op. cit. 104.

Openly branched shrub 5–12 dm. high with slender, stramineous, finely puberulent, minutely glandular-dotted branches; stipules broadly subulate, 1.5–2 mm. long; leaves 1–3 cm. long; leaflets 11–23, oblong or oblong-cuneate, 2–4 mm. long, sparingly puberulent above, more densely so and glandular-dotted beneath; spikes dense, 1–2.5 cm. long, short-pedunculate at ends of main branches, usually sessile on short lateral branches; bracts lanceolate-attenuate, 5–6 mm. long, short-villous and sparingly glandular-dotted on back; calyx tube turbinate, 2.5–3 mm. long, strongly 10-ribbed, with 1–4 dark, comparatively large glands between each pair of ribs, lobes deltoid below, filiform, usually bearing 2 subulate, glandlike lobes near base, silky-villous, 3–4 mm. long; banner yellowish, blade round-cordate, 3.5–4 mm. long; blades of wing and keel petals rose-purple, 4–5 and 5–6 mm. long respectively, each with a gland near apex, claws 2.5–3 mm. long; pods sparsely glandular and densely villous at summit.

Mostly in the mountains above the limits of the Sonoran Desert, occasionally straggling downward along streams and canyons, Arizona and New Mexico to Sonora and Durango.

Rydberg's *P. sanctae-crucis* appears to be a vigorous plant with some short-pedunculate spikes on the lateral branches, but the two intergrade completely, both types of inflorescence often being found on the same plant. The differences in length of petals given in his descriptions are not constant. April–Oct.

**31. Dalea occidentalis** (Rydb.) Wiggins, Contr. Dudley Herb. **3:** 53, 1940

*Dalea domingensis* Rose, Contr. U.S. Nat. Herb. **1:** 96, 1891, in part. Not *D. domingensis* DC., 1825.
*Parosela occidentalis* Rydb., N. Amer. Fl. **24:** 115, 1920.
*Dalea blasiana* M. E. Jones, Contr. West. Bot. **15:** 138, 1929.

Erect, openly branched shrub 6–12 dm. high with slender, short-pilose and minutely glandular-tuberculate stems bearing numerous lateral branchlets 5–20 cm. long above; stipules narrowly subulate, 1.5–2 mm. long; leaves ascending, 2.5–6 cm. long; leaflets 9–15, broadly elliptic to obovate, obtuse to slightly retuse at apex, 6–10 mm. long, short-villous on both sides, copiously glandular-dotted beneath; peduncles short, mostly under 1 cm.; spikes dense, ovoid, 1–2 cm. long; bracts lanceolate-acuminate, 5–6 mm. long, short-villous and sparingly glandular-dotted; calyx tube campanulate, 2.6–3 mm. long, strongly ribbed, villous, 2–5 conspicuous glands between each pair of ribs; lobes lance-attenuate, upper four about 3 mm. long, lowest 4–4.5 mm. long, plumose-villous; corolla yellow tinged with rose-purple; blade of banner cordate, about 2 mm. long, equaling claw; wing and keel petals 2.5 and 3.5 mm. long respectively, all petals bearing a prominent gland near tip; pods obliquely obovate, villous and glandular at tip.

Dry hillsides, vicinity of Alamos, Sonora, and southward to southern Sinaloa. Dec.–March.

The long, slender, lower calyx lobe and the comparatively large leaflets are a combination that, among the species of our area, signally separates this species from the others.

**32. Dalea purpusii** Brandegee, Erythea **7:** 2, 1899

*Parosela purpusi* Rose, Contr. U.S. Nat. Herb. **10:** 106, 1906.

Intricately to slenderly branched shrub 3–8 dm. high with strigose, finely glandular-tuberculate stems; leaves 1–3 cm. long; leaflets 3–13, obovate, obtuse, rounded, or faintly retuse, 4–8 mm. long, minutely strigose on both sides, finely but distinctly glandular-dotted beneath; peduncles 0.5–5 cm. long; spikes ovoid to cylindrical, 1.5–4.5 cm. long; bracts broadly ovate, acuminate, 4.5–6 mm. long, sparingly silky and prominently glandular above middle, long-villous below, especially along margins, deciduous; calyx tube campanulate, 2.5–3 mm. long, long-villous, glandular between ribs, but glands obscured by pubescence, lobes linear from a deltoid base, 3.2–4 mm. long, long-villous; corolla rose to rose-purple; banner cordate, about 3 mm. long; wing and keel petals obovate, 4 and 5 mm. long respectively, obtuse to rounded at apex, basal lobe of former acutish; pods densely villous and with numerous small glands on upper two-thirds.

Along arroyos and rocky hillsides, Lower Sonoran Zone, from near Rosario, Baja California, south to Calmallí and San Esteban. Dec.–May.

This resembles *D. seemannii* somewhat but is readily separated from that species by the longer leaflets, much longer calyx lobes, and the more copiously glandular and more generally villous pods.

**33. Dalea megalostachys** (Rose) Wiggins, Contr. Dudley Herb. **3:** 54, 1940

*Parosela megalostachys* Rose, in Rydb., N. Amer. Fl. **24:** 102, 1920.

A robust, ascendingly branched shrub 1 m. high or more with densely villous, conspicuously glandular-tuberculate branches; leaves 2.5–5 cm. long; leaflets 7–13, obovate or broadly cuneate-obovate, densely sericeous on both sides, copiously glandular-dotted beneath, 6–10 mm. long; peduncles stout, 0.5–3 cm. long; spikes dense, ovoid to cylindrical, 1.5–5 cm. long; bracts lanceolate-attenuate, densely sericeous, 5–7 mm. long, glandular-punctate on back, but glands obscured by pubescence; calyx tube campanulate, strongly ribbed, densely villous, 3–3.5 mm. long, glandular between ribs; lobes deltoid at base, filiform above, 3.5–4.5 mm. long, villous-plumose; corolla rose, or yellowish at base of blades; blade of banner cordate, 4 mm. long, equaling claw; blades of wing and keel petals 5 and 6 mm. long respectively, former broadly oblong, latter obovate, claws 2–2.5 mm. long; pods densely villous except at very base, sparingly glandular near summit.

Desert plains, Lower Sonoran Zone, central Baja California, in the vicinity of San Esteban. Dec.–April.

**34. Dalea nana** Torr. ex A. Gray var. **carnescens** (Rydb.) Kearney & Peebles, Jour. Wash. Acad. **29:** 483, 1939

*Parosela rubescens* A. Nels., in Coult. & Nels., Man. Bot. Rocky Mts. 298, 1909. Not *P. rubescens* Vail, 1894.

*Parosela carnescens* Rydb., Fl. Rocky Mts. 483, 1917.

*Dalea carnescens* Bullock, Kew Bull. **1939:** 195, 1939.

Herbaceous perennial from a woody rootstock, branched at base; stems decumbent, slender, stramineous, strigose-canescent, 1–3.5 dm. long; stipules subulate, spreading, 2–3 mm. long; leaves spreading, 2–3 cm. long; leaflets occasionally 3, usually 5, oblanceolate, 8–15 mm. long, rounded to acute and often mucronulate at apex, bright green and glabrate above, sericeous and sparsely glandular-dotted beneath; peduncles stout, 1–2 cm. long; spikes dense, cylindrical, 10–12 mm. in diameter, 2–5 cm. long; bracts ovate-lanceolate, acuminate, 4–5 mm. long, villous; calyx densely silky-villous, tube turbinate, 2.5–3 mm. long, glands obscured by hairs; calyx lobes filiform-setaceous from a broader base, plumose, 3–3.5 mm. long; corolla yellowish with a tinge of rose; blade of banner reniform, 1.5 mm. high, 2 mm. broad, claw 3 mm. long; wing and keel petals inserted slightly above middle of stamineal column, blades 2.5 and 3 mm. long respectively, claws 1–1.5 mm. long; pods 3 mm. long, densely villous on terminal one-third.

Chiefly, and possibly entirely, above the Sonoran Desert, Arizona to Colorado, Chihuahua, and Coahuila. May–June.

Peculiar in that the leaflets turn bright pea-green in drying, and in the high insertion of the petals on the stamineal tube.

**35. Dalea wrightii** A. Gray, Pl. Wright. **1:** 49, 1852

*Parosela wrightii* Vail, Bull. Torrey Club **24:** 16, 1897.
*Dalea parrasana* Brandegee, Univ. Calif. Pub. Bot. **4:** 179, 1911.
*Dalea sabulicola* Brandegee, loc. cit.

Perennial herb from a short woody caudex; stems several, densely silky-canescent and leafy, striate, 5–12 cm. long, erect; stipules spreading, subulate, 4–5 mm. long; leaves 2–4 cm. long; leaflets 3–7, elliptic to oblanceolate, 5–20 mm. long, usually acute at both ends, densely silky-canescent on both sides, not perceptibly glandular; spikes dense, 16–20 mm. in diameter, 2–6 cm. long; bracts narrowly lance-acuminate, 6–9 mm. long, sparingly villous, green, sparingly glandular-dotted near margin; calyx densely silky-villous, tube turbinate, 3–3.6 mm. long, lobes deltoid at base, filiform above, 5–7 mm. long, densely plumose; corolla yellowish or tinged with rose, or purple; blade of banner cordate, 3 mm. long, claw about twice as long; wings and keel inserted slightly above middle of stamineal column, blades 3.5 and 4 mm. long respectively, claws 1–1.5 mm. long; pods villous above.

Dry hillsides and plains, mostly above the Sonoran Desert, Arizona to Texas, Coahuila, and Zacatecas. May–Oct.

**36. Dalea brandegeei** (Rose) Bullock, Kew Bull. **1939:** 195, 1939

*Dalea ramosissima* Benth., Bot. Sulph. 11, *pl. 10,* 1844. Not *D. ramosissima* Mart. & Gal., 1843.
*Parosela ramosissima* Heller, Cat. N. Amer. Pl. ed. 2, 6, 1900.
*Parosela brandegei* Rose, Contr. U.S. Nat. Herb. **10:** 106, 1906.

Spreadingly branched shrub 6 dm. high or less with glabrous, reddish, conspicuously glandular-tuberculate, decumbent to ascending, slender branches, these becoming tawny or grayish in age; leaves 1.5–3.5 cm. long, conspicuously glandular-tuberculate on petioles and rachises; leaflets 11–17, narrowly obovate, retuse to rounded at apex, 2.5–6 mm. long, glabrous, prominently glandular-dotted beneath; peduncles 1–5 cm. long, glandular-tuberculate; spikes dense, cylindrical, 10–12 mm. in diameter, 1.5–5 cm. long; bracts ovate-lanceolate, acuminate, stramineous at base, dark purplish and glandular-dotted on upper half, sparingly villous, 4–5 mm. long; calyx silky-villous, tube turbinate, 2.5–3 mm. long, intercostate glands inconspicuous; lobes lanceolate-subulate, 3–4 mm. long, plumose; corolla rose-colored to magenta; blade of banner reniform, 3 mm. high, 4 mm. wide, the claw 2 mm. long; blades of wings and keel petals 4 and 5 mm. long respectively, basal lobe of wings narrow but prominent; pods villous and prominently glandular-dotted on upper half, 2.5–3 mm. long.

Sandy soil, Lower Sonoran Zone, on Magdalena Island and the adjacent mainland of Baja California. Jan.–May.

**37. Dalea pogonathera** A. Gray, Pl. Fendl. 31, 1849

*Parosela pogonathera* Vail, Trans. N.Y. Acad. **14:** 34, 1894.

Erect perennial herb with many branches from a woody caudex; stems 1–3 dm. long, glabrous, greenish, sparingly glandular-dotted; leaves 1–2 cm. long, glabrous;

leaflets 5–7, linear to oblong-cuneate, 5–10 mm. long, rounded to retuse at apex, prominently glandular-dotted beneath; peduncles terminating branches, 2–5 cm. long; spikes dense in bud and flower, rachis elongating somewhat in fruit, 2–6 cm. long at anthesis; bracts broadly ovate, abruptly acuminate, shallowly cucullate, glabrous, scarious-margined, greenish and glandular-dotted on back, 4.5–6 mm. long, persistent and conspicuous; calyx silky-villous, tube broadly turbinate to campanulate, 3–3.2 mm. long, intercostal glands obscured by pubescence but readily visible from side, lobes deltoid at base, attenuate-filiform, 4–5.5 mm. long, villous-plumose; corolla purple; blade of banner reniform, 2.2 mm. high, 3 mm. broad, on a claw 2.5–3 mm. long; wings inserted at about middle of stamineal tube, keel petals a little higher, blades 2.5–3 and 4–4.5 mm. long respectively; pods 3.5–4 mm. long, silky-villous along upper margin, sparingly glandular.

Dry slopes and plains, on subdesert and desert areas, Lower and Upper Sonoran Zones, southern Arizona to Texas, Zacatecas, and San Luis Potosí. March–Sept.

## 11. **ERRAZURIZIA** Phillipi, Anal. Univ. Chile **1872:** 688, 1872

Low spreading shrubs with spinescent stipules and divaricate, brittle, white-tomentose branches sparsely to heavily pustulate with golden glands. Leaves pinnate, leaflets orbicular. Flowers spicate, dense in flower, lax in fruit; bracts linear, gland-tipped. Calyx campanulate, 5-lobed. Petals yellow, distinct, fleshy, clawless, symmetrical, inserted at base of stamineal tube, banner broadest. Stamens 9, monodelphous, tube very short, filaments stoutish and graduated or somewhat alternately longer and shorter. Stigma capitate. Pod ovoid to ellipsoid, exserted, biovulate, 1- or 2-seeded, glandular, the beak straight, slender.

Stipules 3–9 mm. long; terminal leaflet retuse, usually longer than lateral ones; wing and keel petals spatulate, retuse .................................... 1. *E. benthamii*

Stipules 1–2.5 mm. long; all leaflets retuse, about equal in size; wing and keel petals rounded or obtuse at apex, not retuse .................................... 2. *E. megacarpa*

### 1. **Errazurizia benthamii** (Brandegee) I. M. Johnston, Proc. Calif. Acad. IV, **12:** 1043, 1924

*Dalea megacarpa* Greene, Pittonia **1:** 202, 1888. Not *D. megacarpa* S. Wats., 1885.
*Dalea benthami* Brandegee, Proc. Calif. Acad. II, **2:** 148, 1889.
*Dalea megacarpa* var. *biuncifera* Greene, in Vasey & Rose, Contr. U.S. Nat. Herb. **1:** 15, 1890.
*Psorobatus benthami* Rydb., N. Amer. Fl. **24:** 40, 1919.

Low, intricately branched shrub 3–5 dm. high with rigid, white-tomentose, glandular-pustulate branches; leaves 2–5 cm. long, tomentulose; stipules spinescent, curved, 3–9 mm. long; leaflets 7–11, orbicular to broadly obovate, 4–10 mm. long, terminal one usually longer than others and often shallowly retuse, white-tomentulose on both surfaces though slightly greenish and glandular-dotted beneath; spikes terminal on stoutish branches, 2–6 cm. long, 1.5 cm. in diameter, somewhat lax; bracts subulate-acuminate, 4–5 mm. long, subspinescent, gland-tipped and glandular-pustulate near base; calyx white-tomentose, glandular-dotted, tube campanulate, 4–5 mm. long, often slightly constricted at apex, indistinctly 10-ribbed, irregularly glandular-pustulate, lobes oblong, 2.5–3 mm. long, acute to obtuse; corolla light yellow; petals 7–8 mm. long, retuse at apex; banner obovate; keel and wing petals spatulate; pod ovoid, about 10 mm. long, pubescent and prominently glandular.

Sandy soil, Cedros, Santa Margarita, and Magdalena Islands; San Bartolomé Bay, Baja California. Feb.–May.

**2. Errazurizia megacarpa** (S. Wats.) I. M. Johnston, Proc. Calif. Acad. IV, **12:** 1042, 1924

*Dalea megacarpa* S. Wats., Proc. Amer. Acad. **20:** 359, 1885.
*Psorobatus megacarpus* Rydb., N. Amer. Fl. **24:** 41, 1919.

Low, spreading, divaricately branched shrub 2–6 dm. high, usually two to three times as wide as high, with white-tomentose, glandular-pustulate branches; stipules spinescent, 1–2.5 mm. long; leaves 3–8 cm. long, white-tomentose; leaflets 9–13, orbicular to ovate-orbicular, retuse at apex, 4–10 mm. long, glandular-dotted beneath; spikes short-pedunculate to sessile, 3–6 cm. long in flower, elongating to 15 cm. and becoming lax in fruit; bracts lance-linear, 4–6 mm. long, curved upward, gland-tipped and sparingly glandular-dotted near base; calyx canescent to white-tomentose, broadly campanulate, moderately 10-ribbed, tube 5–6 mm. long, lobes lanceolate or lance-oblong, 2–3.5 mm. long at anthesis, acute, one of them gland-tipped; corolla yellow, petals distinct, 6–7 mm. long, clawless, eglandular, banner obovate, 4–5 mm. wide, retuse at apex, other petals narrowly oblanceolate, obtuse at apex; pod ovoid, 8–11 mm. long, canescent and prominently glandular, beak 1–2 mm. long.

Desert washes, hillsides, and plains, Lower Sonoran Zone, vicinity of El Mármol and southern San Felipe Desert to San Ignacio, and San Marcos Island, Baja California. Nov.–March.

## 12. PETALOSTEMON Michx., Fl. Bor. Amer. 2: 48, 1803

Perennial or annual herbs similar to *Dalea.* Flowers in dense terminal spikes or racemes; bracts deciduous. Calyx turbinate or campanulate, 10-nerved, lobes deltoid to lanceolate, lowest longer than others, rarely exceeding tube. Corolla white, pink, yellow, or purple, indistinctly papilionaceous, keel petals adnate along lower margin at anthesis but soon distinct; banner long-clawed, inserted on hyanthium, other petals short-clawed, inserted at apex of stamineal tube and alternating with free parts of filaments. Stamens 5, monodelphous but with the upper one-fifth to one-third distinct. Ovary 2-ovuled; style slender; stigma capitate, minute. Pod 1–2-seeded, membranous, usually hairy at apex.

Plants annual, prostrate or decumbent; leaflets 2–6 mm. long; calyx tube strigose .......... 1. *P. evanescens*
Plants perennial, erect or ascending; leaflets 1–3 cm. long; calyx tube glabrous .......... 2. *P. candidus*

**1. Petalostemon evanescens** (Brandegee) Rose, in Rydb., N. Amer. Fl. **24:** 129, 1920

*Dalea evanescens* Brandegee, Proc. Calif. Acad. II, **2:** 146, 1889.

Low, nearly prostrate annual with minutely strigose stems 6–35 cm. long; stipules linear-subulate, 1–1.5 mm. long; leaves 1–2.5 cm. long; leaflets 7–15, cuneate-obovate, retuse, 2–6 mm. long, dark green, minutely and sparsely strigose, sparingly gland-dotted beneath; peduncles 0.5–2.5 cm. long; inflorescence cylindrical, dense, 1–1.5 cm. long; bracts ovate, acute to short-acuminate, 1–1.5 mm. long; calyx strigose, tube about 1.2 mm. long, lowest lobe lanceolate, 0.8–1 mm. long, others shorter, deltoid;

corolla rose-purple; blade of banner broadly reniform, 1 mm. long or less, about 1.5 mm. wide, gland-dotted; blades of wing and keel petals obliquely obovate, 1–1.5 mm. long, gland-dotted, claws about 0.5 mm. long or less; pod slightly compressed, bearing 3–5 conspicuous glands on each side, pubescent at apex.

Sandy soil, Lower Sonoran Zone, central Baja California to the Cape Region. Feb.–March; Aug.–Oct.

*P. sonorae* Rydb., *P. exile* A. Gray, *P. rothrockii* Rydb., and *P. oligophyllus* (Torr.) Rydb. occur in Arizona or Sonora, but none of these seems to enter the desert regions.

**2. Petalostemon candidus** (Willd.) Michx., Fl. Bor. Amer. **2:** 49, *pl. 37, fig. 1,* 1803

*Dalea candida* Willd., Sp. Pl. **3:** 1337, 1803.
*Psoralea candida* Poir., in Lam., Encycl. **5:** 694, 1804.
?*Petalostemon virgatus* Nees & Schw. ex Maximilian, Reise N. Amer. **2:** 432, 1841.
*Kuhnistera candida* Kuntze, Rev. Gen. Pl. **1:** 192, 1891.

Erect or ascending perennial with several branches 3–10 dm. tall from a woody base and rootstock, stems strongly striate to slightly angulate, glabrous; leaves scattered; stipules narrowly subulate, 1–5 mm. long; petioles 5–8 mm. long, minutely gland-dotted; leaflets 5–9, linear-oblanceolate to oblong, 2–6 mm. wide, 1–3 cm. long, acute to mucronate, liberally dotted with minute brownish glands beneath, nearly glandless above, glabrous; peduncles 3–15 cm. long, striate; bracts lanceolate-subulate, 4–6 mm. long; calyx 2.5–3 mm. long, 10-ribbed and 10-angled, glabrous except on margins and tips of lance-deltoid, villous-ciliate teeth; corolla white; banner reniform-cordate, blade about 2 mm. long, 3 mm. wide, claw 2–2.5 mm. long; wings and keel petals elliptic to oblanceolate, blades about 2.5 mm. long; pods obovoid, about 3 mm. long.

On mesas and openings in forests and brush, mainly Upper Sonoran and Transition Zones, occasionally straggling to upper edge of Lower Sonoran Zone, southern Arizona to Chihuahua, Texas, Indiana, and Saskatchewan. May–Sept.

## 13. THORNBERA Rydb., Jour. N.Y. Bot. Gard. **20:** 66, 1919

Erect or ascending annual or perennial herbs with glandular-dotted stems and leaves. Leaves pinnate, stipulate, with glandlike stipels. Inflorescence densely spicate. Calyx narrowly campanulate, 10-ribbed, usually glandular-dotted and hairy. Corolla white, yellow, rose, purple, or blue; banner inserted in bottom of calyx tube, its blade ovate or cordate; wings and keel petals oblong, oblanceolate, or obovate, on short claws, inserted at mouth of stamineal tube, distinct, not papilionaceous. Stamens 9–10, monodelphous, filaments distinct above. Ovary 2-ovuled; style filiform, curving downward; stigma minute. Pod thin, obliquely obovate, pubescent and often glandular above, glabrous toward base, included in the calyx.

Bracts narrowly lanceolate, attenuate-acuminate, broader part inconspicuous in bud; leaflets puberulent or villous on both sides; stems woody:
  Leaflets 21–31, oblong, 5–8 mm. long, sparsely villous; blade of banner 3 mm. long .......... 1. *T. albiflora* subsp. *albiflora*
  Leaflets 13–19, oval to obovate, 3–5 mm. long, densely villous; blade of banner 2 mm. long .......... 1a. *T. albiflora* subsp. *villosa*
Bracts ovate, short-acuminate, conspicuous in bud; leaflets glabrous or glabrate; stems herbaceous:
  Leaflets obtuse to emarginate at apex; bracts 2–2.5 mm. wide; petals blue .. 2. *T. pringlei*

Leaflets acute to mucronulate at apex; bracts 1–1.5 mm. wide; petals white or creamy:
Leaflets plane or involute, 2–4 mm. long; calyx lobes 1.6–2 mm. long, about equaling tube .......................................................... 3. *T. grayi*
Leaflets revolute, 4–10 mm. long; calyx lobes 1–1.2 mm. long, shorter than tube ...... ............................................................ 4. *T. revoluta*

**1. Thornbera albiflora** (A. Gray) Rydb., Jour. N.Y. Bot. Gard. **20:** 66, 1919, subsp. **albiflora**

*Dalea albiflora* A. Gray, Pl. Wright. **2:** 38, 1853.
*Parosela albiflora* Vail, Trans. N.Y. Acad. **14:** 34, 1894.

Low perennial shrub 2–6 dm. high with moderately puberulent, erect or ascending branches; stipules linear-subulate, 1.5–3 mm. long; leaves 2–5 cm. long; leaflets 21–33, oblong, obtuse or slightly emarginate, 5–8 mm. long, sparsely short-villous on both sides, gland-dotted beneath; spikes dense, subglobose to cylindric, 0.8–7 cm. long; bracts lanceolate-attenuate, equaling or slightly exceeding calyx, deciduous; calyx tube 2–2.5 mm. long, gland-dotted on ternate lobes, these equaling tube, villous; corolla white; blade of banner subcordate-orbicular, 3 mm. long, gland-dotted near base, claw 3 mm. long; wings and keel petals obliquely oblong or obovate, 2–2.5 mm. long, claws 1 mm. long; pod obliquely obovate, glabrous below, gland-dotted or hirsute at apex.

Mostly in the Upper Sonoran Zone, but fringing the desert in the Santa Catalina and Santa Rita Mountains, Arizona; eastward to New Mexico and Chihuahua, southward into northeastern Sonora. April–Oct.

**1*a*. Thornbera albiflora** subsp. **villosa** (Rydb.) Wiggins, Contr. Dudley Herb. **3:** 54, 1940

*Thornbera villosa* Rydb., N. Amer. Fl. **24:** 118, 1920.

Differing from subsp. *albiflora* mainly in being more densely villous and having smaller, somewhat less numerous leaflets and slightly smaller corollas.

Chiefly above the desert in the Santa Catalina and Santa Rita Mountains, Arizona.

**2. Thornbera pringlei** (A. Gray) Rydb., N. Amer. Fl. **24:** 119, 1920

*Dalea pringlei* A. Gray, Proc. Amer. Acad. **17:** 201, 1882.
*Parosela pringlei* Heller, Cat. N. Amer. Pl. ed. 2, 6, 1900.

Upright perennial, slightly suffrutescent, with green, herbaceous, glabrous, and glaucous stems 1.5–6 dm. high; stipules linear-subulate, 2 mm. long; leaves 1.5–6 cm. long; leaflets 15–47, oblong to elliptic, acute to obtuse at both ends, glabrous, glandular-punctate beneath; spikes dense, 2–10 cm. long, 7–10 mm. thick; bracts ovate, abruptly short-acuminate, 4–5 mm. long, densely silky-villous on margins and lower part, gland-dotted on back, dark-colored; calyx campanulate, densely villous, tube 2–2.2 mm. long, with rows of small glands between prominent ribs, calyx lobes lance-subulate, pilose, 2.5–3 mm. long; corolla rose or purplish, blade of banner ovate, about 1.5 mm. long, claw 2.5 mm. long; blades of wings and keel petals obliquely elliptic to ovate, 1.8–2 mm. long, claw about 0.5 mm. long; pod obovate, densely pilose, gland-dotted.

Upper Sonoran and upper fringe of Lower Sonoran Zones, Santa Catalina and Baboquivari Mountains, Arizona, into Chihuahua and Sonora to Guaymas. April–Oct.

**3. Thornbera grayi** (Vail) Rydb., N. Amer. Fl. **24:** 119, 1920

*Dalea laevigata* A. Gray, Pl. Wright. **2:** 38, 1853. Not *D. laevigata* Moc. & Sessé, 1852.
*Parosela grayi* Vail, Bull. Torrey Club **24:** 14, 1897.

Perennial herb 1.5–5 dm. high, somewhat suffrutescent, with pubescent inflorescence, otherwise glabrous; stipules linear-subulate, 1–2.5 mm. long; leaves 1.5–4 cm. long; leaflets 31–41, oval to oblong, mucronulate, 2–4 mm. long, glandular-punctate beneath; spikes ovoid to cylindrical, 1–3.5 cm. long; bracts ovate, short-acuminate, sparsely glandular-dotted and villous, subpersistent; calyx 10-ribbed, faintly glandular-punctate in intervals, tube 1.8–2 mm. long; calyx lobes ovate, acuminate, about equaling tube; corolla white to cream; blade of banner ovate, about 2 mm. long on a claw 2.5–3 mm. long; blades of wings and keel petals distinct, oblong, about 2.5 mm. long; pod glabrous at base, pubescent and minutely glandular-punctate above, about equaling calyx.

Mountain sides and rocky places, mostly Upper Sonoran Zone, on the fringe of the desert in the foothills of the Santa Rita Mountains, Arizona, eastward into New Mexico, southward into Chihuahua and Sonora. May–Sept.

**4. Thornbera revoluta** (S. Wats.) Rydb., N. Amer. Fl. **24:** 119, 1920

*Dalea revoluta* S. Wats., Proc. Amer. Acad. **22:** 404, 1887.
*Parosela revoluta* Rose, Contr. U.S. Nat. Herb. **10:** 105, 1906.

Erect perennial herb 2.5–6 dm. high from a cespitose, woody caudex, with pubescent inflorescence and glabrous stem and foliage; stipules filiform-subulate, 2.5–3 mm. long; leaves 1–3 cm. long; leaflets 15–23, linear, 4–10 mm. long, revolute, copiously glandular-dotted; spikes cylindrical, 1.5–4 cm. long, 6–8 mm. in width; bracts ovate, short-acuminate, puberulent and densely glandular-punctate on back, villous-ciliate at base, persistent; calyx densely short-pilose, with a row of glands in each interval, tube 2 mm. long, lobes ovate-lanceolate, 1–1.2 mm. long; corolla white except for rose-tinged banner; blade of banner 2 mm. long, ovate, truncate at base, claw about 2.5 mm. long; blades of wings and keel petals oblong, about equaling banner but on shorter claws; pod obliquely obovate, densely pubescent and glandular above.

Mainly in Upper Sonoran Zone, southern and eastern Sonora, southward into Jalisco. Sept.–Nov.

## 14. BRONGNIARTIA H. B. K., Nov. Gen. & Sp. 6: 453, 1824

Shrubs or trees with odd-pinnate leaves and caducous or deciduous stipules and often petiolulate and stipellate leaflets. Flowers solitary or in few-flowered axillary umbels or falsely racemose. Bractlets 2, immediately beneath calyx, or represented by tufts of hairs. Calyx somewhat 2-lipped, upper pair of lobes united larger part of their length, lower three usually free to base. Corolla papilionaceous, with a basal, short-clawed banner, more or less falcate wings, and keel petals united from about their middle to tip. Stamens diadelphous. Ovary short-stipitate, several-ovuled; style filiform. Pod flat, elongate, upper suture often wing-margined. Seeds ovoid, strophiolate, on a short funiculus.

Leaflets 5 or more; pods not wing-margined:
Flowers in pseudoracemose inflorescences, floral leaves reduced to large semicordate stipules; leaflets oblong ............................................ 1. *B. palmeri*

Flowers all axillary to foliage leaves; leaflets lanceolate or linear:
Branches and foliage densely pubescent or strigose; calyx lobes 8–10 mm. long; leaflets lanceolate or elliptic, 5–8 mm. broad, not revolute ............. 2. *B. peninsularis*
Branches and foliage glabrous; calyx lobes 4–6 mm. long; leaflets linear-filiform, revolute ........................................................ 3. *B. shrevei*
Leaflets 3; pod narrowly wing-margined along upper suture ............. 4. *B. trifoliata*

**1. Brongniartia palmeri** Rose, Contr. U.S. Nat. Herb. **1:** 97, 1891

Shrub 1 m. tall or less with sparingly pilose young branchlets and leaves 5–10 cm. long; stipules conspicuous, ovate to ovate-cordate, 10–16 mm. long, about two-thirds as broad, sparsely pilose-ciliate near tips and on veins beneath; petioles 1.5–3.5 cm. long, pilose; leaflets 7–17, oblong to lanceolate, apiculate to attenuate, 1–2 cm. long, glabrous above, pilose on margins and midrib beneath, thin, finely and faintly reticulate-veined; inflorescence pseudoracemose, each umbel of 2–5 flowers in axil of ovate-cordate stipules; pedicels slender, glabrous, 1–1.8 cm. long; calyx glabrous, broad tube about 3 mm. deep, lobes 6–9 mm. long, broadly lanceolate, acute; corolla 1.5–1.8 cm. long, rose-purple; banner obovate-orbicular, 18 mm. long; wings obliquely oblanceolate, auricle deltoid; keel petals oblanceolate; pod broadly oblong, 3–6 cm. long, stipe about equaling calyx tube; seeds 4–5 mm. wide, 6–7 mm. long.

Gravelly or rocky soil and hillsides, Lower Sonoran Zone vicinity of Hermosillo to Alamos, Sonora. April–May.

**2. Brongniartia peninsularis** Rose, Contr. U.S. Nat. Herb. **12:** 268, 1909

Open shrub 2–3 m. tall with densely silky-strigulose branchlets and light brown, smooth bark; leaves 10–13 cm. long; stipules broadly lanceolate or nearly elliptic, 6–10 mm. long; petioles 1–1.5 cm. long; leaflets 5–13, elliptic-lanceolate, 5–8 mm. wide, 10–25 mm. long, acute or minutely apiculate, minutely strigulose on both surfaces, distinctly pinnately reticulate-veined, midvein prominent beneath; flowers axillary on pedicels 2–4 cm. long, strigose; bractlets minute, deciduous; calyx glabrous outside, puberulent within, tube 4–5 mm. deep, lower lobe narrowly lanceolate, 8–10 mm. long, upper lobes obtuse; corolla about 1.5 cm. long; pod glabrous and shining, 12–16 mm. wide, 4–5 cm. long, stipe 5–7 mm. long; seeds suborbicular-lenticular, 6–7 mm. long, pale brown, smooth and shining.

Along arroyos and on rocky hillsides, Lower Sonoran Zone, mountains between Mulegé and Comondú, Baja California, southward to the Cape Region. Sept.–Nov.

**3. Brongniartia shrevei** Wiggins, Contr. Dudley Herb. **4:** 18, *pl. 1, figs. 1–4,* 1950

Erect shrub 1–1.2 m. tall with several slender branches from base and fine glabrous foliage; branchlets greenish, glabrous; stipules narrowly ovate, short-attenuate, 2–2.5 mm. long, deciduous; leaves 6–10 cm. long; petioles 3–9 mm. long; leaflets 35–55, linear-filiform, revolute, 0.4–0.6 mm. wide, 3–10 mm. long, acute at both ends, glabrous; flowers axillary, solitary; pedicels 8–10 mm. long in flower, 1.5–2 cm. long in fruit; bractlets minute, ovate, caducous; calyx tube about 3 mm. long, 3–4 mm. wide, oblique, glabrous without, puberulent within, lobes narrowly ovate-lanceolate, 4–6 mm. long, attenuate at apex; petals yellow, 10–12 mm. long; banner erect, suborbicular, claw short; wings obovate, blades about 8 mm. long, with a round auricle at base, claw 1.5–2 mm. long; keel petals blunt at apex, acumen slightly notched on lower, strongly curved side, straight along upper side; stamens slightly curved upward, barely exserted; pod broadly elliptic, 12–15 mm. wide, 2.5–3 cm. long, glabrous, wingless, short-apiculate at apex, stipe 7–9 mm. long; seeds 2–5, obovoid, about 6 mm. wide, 7–8 mm. long, buff along edges, light brown on faces, smooth and shining.

Arid rocky hillsides, Lower Sonoran Zone, central Sonora. Sept.–Dec.

**4. Brongniartia trifoliata** Brandegee, Zoe **5:** 105, 1901

Spreading shrub to 4 m. tall with pubescent branchlets, rachises, and petioles, the last 1–1.5 cm. long; leaves 6–10 cm. long, trifoliolate; leaflets coriaceous, glabrous, cuneate to cordate at base, obtuse, rounded or retuse at apex, lateral ones 2–4 cm. long, 1.5–3 cm. wide, terminal ones 3–5 cm. wide and about as long; petiolules densely pubescent, 2–3 mm. long; flowers not known; fruiting pedicels 3–3.5 cm. long; pod 1.5–2 cm. wide, 3–3.5 cm. long, glabrous, apiculate at apex, a narrow wing along upper suture; stipe about 3 mm. long; seeds 1 or 2.

Known only from the type locality, Sierra de San Francisquito, Baja California.

## 15. TEPHROSIA Pers., Syn. Pl. 2: 328, 1807

Herbs, suffruticose subshrubs or shrubs with odd-pinnate leaves, setaceous to ovate stipules, and leaflets with conspicuous diagonal lateral veins. Flowers in axillary and terminal racemes. Bractlets none. Calyx campanulate, lobes subequal, upper pair somewhat united. Banner suborbicular, pubescent on outer surface. Wing petals obliquely obovate to oblong, slightly adnate to keel, basally auriculate on upper side. Stamens pseudomonodelphous, upper filament attached to stamineal sheath at base, then free, attached near middle, and free again at apex. Ovary sessile; style incurved, usually glabrous. Pod sessile, flat, 2-valved, several- to many-seeded.

Style hairy along upper side; flowers 11–19 mm. long; stems woody or suffrutescent:
  Pods straight or nearly so; seeds brown, mottled with black:
    Leaflets 9–19, glabrous on upper surface; pods glabrous on sides; stipules 4–12 mm. long ........ 1. *T. leiocarpa*
    Leaflets 7–15, densely strigose-pubescent on both surfaces; pods strigose-pubescent on sides; stipules 5 mm. long or less ........ 2. *T. palmeri*
  Pods curved upward in distal one-half; seeds buff, mottled with olive green or black ........ 3. *T. cana*
Style glabrous; flowers 7–8 mm. long; stems herbaceous ........ 4. *T. purpurea*

**1. Tephrosia leiocarpa** A. Gray, Pl. Wright. **2:** 36, 1853

*Tephrosia affinis* S. Wats., Proc. Amer. Acad. **21:** 424, 1886.
*Cracca leiocarpa* Kuntze, Rev. Gen. Pl. **1:** 175, 1891.
*Tephrosia viridis* M. E. Jones, Contr. West. Bot. **12:** 7, 1908.
*Cracca affinis* Rose, Contr. U.S. Nat. Herb. **12:** 269, 1909.
*Cracca calva* Rydb., N. Amer. Fl. **24:** 161, 1923.

A perennial, often suffrutescent herb 3–10 dm. tall with angled strigose stems and filiform-setaceous stipules 4–12 mm. long; leaflets 9–19, oblong to narrowly oblanceolate, 4–10 mm. wide, 1.5–3.5 cm. long, obtuse to emarginate and apiculate at apex, green, glabrous above, minutely and sparsely puberulent beneath; petiolules 1.5–2.5 mm. long; racemes short-peduncled, 1–2.5 dm. long; bracts setaceous, 2.5–4 mm. long, caducous; pedicels slender, 3–5 mm. long; calyx tube 2–3 mm. long, lobes about same, purplish, strigulose; corolla 15–18 mm. long, pink to deep red or purplish, banner densely strigulose without, suborbicular, petals short-clawed, keel semicircular; pods 5–6 mm. wide, 2.5–5 cm. long, straight or slightly curved, glabrous; seeds lenticular-orbicular, about 3 mm. in diameter, smooth.

Among shrubs and under trees, Lower and Upper Sonoran Zones, southern Arizona, where it occurs in the Upper Sonoran Zone, to southern Sonora, Chihuahua, and Jalisco. June–Sept.

**2. Tephrosia palmeri** S. Wats., Proc. Amer. Acad. **24:** 46, 1889

*Tephrosia purisimae* Brandegee, Proc. Calif. Acad. II, **2:** 149, 1889.
*Cracca palmeri* Rose, Contr. U.S. Nat. Herb. **12:** 270, 1909.
*Cracca hamata* Rydb., N. Amer. Fl. **24:** 177, 1923.

Openly branched shrub to 1.5 m. tall with slender, strigose-silvery branches; leaves 8–15 cm. long; stipules lance-subulate, stiffish, 4–5 mm. long, strongly 3-ribbed, persistent, pubescent; leaflets 7–15, linear, linear-oblong, or linear-oblanceolate, 3–8 mm. wide, 1.5–5 cm. long, densely silky-strigose on both surfaces, acute to slightly retuse, terminal mucro straight or somewhat reflexed; racemes on slender peduncles 1–4 dm. long, few-flowered, flowers 15–18 mm. long, 1 or 2 at each node; pedicels 3–5 mm. long; calyx canescent, tube and lobes each about 3 mm. long; banner orbicular, densely puberulent without, light yellow suffused with purple; wing petals oblong-oblanceolate, basal auricle prominent; pods 4–5 mm. wide, 4–6 cm. long, straight or nearly so, minutely but closely strigulose; seeds ovoid-lenticular, about 4 mm. long.

Desert mountain slopes and canyons and outwash fans, Lower Sonoran Zone, vicinity of Guaymas, Sonora, and from central Baja California southward to the Cape Region. Aug.–Oct.; Jan.–May.

**3. Tephrosia cana** Brandegee, Proc. Calif. Acad. II, **3:** 126, 1891

*Cracca californica* Rydb., N. Amer. Fl. **24:** 176, 1923.
*Cracca cana* Rydb., loc. cit.

A suffrutescent perennial to about 1 m. tall with densely strigose-canescent stems, leaves, and peduncles; stipules subulate, 5–9 mm. long, strigose beneath, glabrous above, strongly 3-ribbed, usually deflexed; leaves 6–15 cm. long; leaflets 9–21, oblong-elliptic, to 4 cm. long, acute at each end, greenish and sparingly strigose above, densely so and canescent beneath; racemes 1–1.5 dm. long, few-flowered; bracts lance-subulate, 5–7 mm. long, persistent or subpersistent; pedicels 3–5 mm. long; calyx 6–7 mm. long, sericeous, lobes slightly longer than tube; corolla 12–15 mm. long, yellowish tinged with purple; banner obovate-orbicular; wings obliquely and broadly oblanceolate; pods 4–5 mm. wide, 6–8 cm. long, straight or slightly curved upward; seeds obovoid-lenticular, 4–5 mm. long.

Mountainsides and canyons, Sonoran Zones, southern Baja California, and collected at Los Barriles on the Gulf between Bahía de los Angeles and Santa Rosalia. March–April; Aug.–Oct.

**4. Tephrosia purpurea** (L.) Pers., Syn. Pl. **2:** 329, 1807

*Cracca purpurea* L., Sp. Pl. 752, 1753.
*Galega purpurea* L., Syst. Nat. ed. 10, 1172, 1759.
*Tephrosia ascendens* Macf., Fl. Jam. 257, 1837.
*Tephrosia tenella* A. Gray, Pl. Wright. **2:** 36, 1853.
*Cracca villosa* var. *purpurea* Kuntze, Rev. Gen. Pl. **1:** 174, 1891.
*Cracca tenella* Rose, Contr. U.S. Nat. Herb. **12:** 271, 1909.

Perennial herb from a woody rootstock with numerous, angled, flexuous stems 3–7 dm. tall, sparingly strigulose to glabrate; stipules linear-subulate, 3–5 mm. long; leaves 2–10 cm. long; leaflets 7–21, linear to narrowly oblanceolate, 2–8 mm. wide, 1.5–3.5 cm. long, abruptly cuneate or acute at base, acute to retuse at apex, green, glabrous or glabrate above, appressed-strigulose beneath; petiolules 0.5–2.5 mm. long; racemes to 2 dm. long, mostly opposite leaves, few-flowered; bracts setaceous-subulate, 2–3 mm. long, subpersistent; flowers solitary or in pairs at nodes, 7–8 mm. long; pedicels 2–3 mm. long; calyx about 4 mm. long, tube broadly squarrose, about equaling lobes; corolla purplish; pods 3–4 mm. wide, 3–4.5 cm. long, straight or slightly curved upward, minutely strigulose.

Mountain slopes, canyons, and mesas, Lower and Upper Sonoran Zones, southern Arizona to Texas and southward into northern South America; naturalized from India. July–Sept. in our area.

The plants in the northern part of its range (*T. tenella* A. Gray) seem to have narrower, more acute leaflets than those farther south, but the two forms intergrade completely.

## 16. SPHINCTOSPERMUM Rose, Contr. U.S. Nat. Herb. **10:** 107, 1906

Annual herbs with simple linear leaves. Flowers solitary or sometimes in pairs, axillary, without bractlets. Calyx turbinate, lobes 5, subequal. Corolla pink. Banner obcordate, short-clawed. Wing petals obliquely oblong, auricle acute. Keel petals lunate, united above, with a prominent basal lobe. Stamens diadelphous. Ovary sessile; style slender, pilose near apex. Pod flat, linear, transversely septate between seeds, these quadrate-prismatic, constricted at middle, granular-papillate.

**1. Sphinctospermum constrictum** (S. Wats.) Rose, Contr. U.S. Nat. Herb. **10:** 107, *pl. 34, figs. a–h,* 1906

*Tephrosia constricta* S. Wats., Proc. Amer. Acad. **24:** 46, 1889.

Stems simple or sparingly ascendingly branched, 2–6 dm. tall, slender, sparingly strigose to glabrate; leaves simple, sessile, linear, 1–3 mm. broad, 2–8 cm. long, glabrous or with a few scattered hairs, pale green; pedicels 1–5 mm. long, jointed near base; calyx sparsely strigulose to glabrous, tube 1–1.5 mm. long, lobes lance-subulate, slightly spreading, equaling or slightly exceeding tube; corolla about 5 mm. long; pod linear, 2.5–3 mm. wide, 3–4 cm. long, straight, glabrous, pendent at an angle of about 40 degrees; seeds 10–16, about 2 mm. long, dark green to greenish brown.

Grassy hillsides and plains, mostly in sandy soil, Lower Sonoran Zone, or occasionally into the Upper Sonoran Zone, Pima County, Arizona, southward through Sonora, and in southern Baja California. July–Sept.

## 17. WILLARDIA Rose, Contr. U.S. Nat. Herb. **1:** 97, 1891

A small tree with odd-pinnate, estipulate leaves, nearly smooth gray bark, and lilac or bluish flowers in axillary racemes. Calyx broadly turbinate-truncate, teeth small and about equal. Banner orbicular, erect or nearly so; wings oblong-falcate; keel weakly incurved. Stamens diadelphous at base but axillary one adnate to tube from near middle upward; anthers similar. Ovary subsessile, several-ovuled; style incurved; stigma capitate, minute. Pod linear-oblong, strongly compressed, dehiscent. Seeds reniform, flattened, smooth and sublustrous or dull.

**1. Willardia mexicana** (S. Wats.) Rose, Contr. U.S. Nat. Herb. **1:** 98, 1891

*Coursetia ?mexicana* S. Wats., Proc. Amer. Acad. **21:** 424, 1886.

A small tree to 9 m. tall with smooth gray bark, and minutely strigose leaves and young twigs; leaves about 1 dm. long, of 5–15 oblong-elliptic leaflets, these obtuse or rounded at each end, often subemarginate at apex, 1–1.5 cm. wide, 2–4 cm. long, margins often loosely revolute; racemes 6–12 cm. long, several-flowered; pedicels 4–7 mm. long, puberulent; calyx tube 3–4 mm. long, nearly as broad, sparsely strigose, teeth minute; corolla 10–15 mm. long, lilac to bluish; pods about 1 cm. wide, 5–12 cm. long, minutely and sparsely strigose, margins often sinuous; seeds 6–7 mm. wide, 10–12 mm. long, salmon-buff.

Slopes of stony foothills, margin of Lower Sonoran Zone and into the subtropical thorn forest, southern Sonora and Chihuahua into Sinaloa.

## 18. OLNEYA A. Gray, Pl. Thurb. 328, 1854

Spinescent canescent trees with pinnate or odd-pinnate leaves; obsolete stipules. Bark gray, fissured on older growth. Calyx campanulate, subequally 5-lobed. Banner short-clawed, blade emarginate, reflexed, with 2 basal callosities. Wings obliquely obovate, falcate, with small auricles, short-clawed. Keel petals short-clawed, blades lunate to nearly semicircular, basal auricle broad. Stamens diadelphous, borne on a curved sheath; anthers uniform. Ovary short-stipitate, several-ovuled; style pubescent above; stigma capitate. Pod 2-valved, firm, somewhat torulose, glandular-hispid, with shorter nonglandular hairs intermingled with longer ones. Seeds ellipsoid-lenticular, estrophiolate.

### 1. **Olneya tesota** A. Gray, Pl. Thurb. 328, 1854

A tree 5–10 m. high with densely canescent, striate, spiny branches; spines 5–10 mm. long, paired or solitary below leaves, straight or usually slightly curved, tips brown or black; leaves 3–10 cm. long; leaflets 4–12 pairs, oblong or cuneate-obovate, 5–20 mm. long, densely canescent; racemes axillary, 2–6 cm. long; bracts 0.5 mm. long, deciduous; calyx narrowly turbinate, densely canescent, tube 3–4 mm. long, lobes triangular-ovate to oblong-ovate, 2–3 mm. long, nearly as wide, abruptly acute; corolla 9–10 mm. long, yellow, tinged with purple or maroon; pod 8–9 mm. thick, 3–6 cm. long, torulose, canescent and glandular-pubescent; stipe and beak broad; seeds brownish to black, shining, 5–6 mm. broad, 8–9 mm. long.

Desert washes and low hills in gravelly or silty soil, Lower Sonoran Zone, western and southern Arizona and the Colorado Desert into southern Sonora and southern Baja California. May–June.

## 19. DIPHYSA Jacq., Enum. Pl. Carib. 7, 1760

Trees or shrubs with alternate, odd-pinnate, short-petioluled leaves and small caducous stipules. Stipels none. Inflorescence racemose on a jointed peduncle; bracts caducous; bractlets 2 beneath each flower, caducous. Calyx tube turbinate; calyx lobes 5, unequal, lowest one longest. Petals 5, unequal, banner longest and keel petals shortest, short-clawed; blade of banner suborbicular, more or less reflexed; wings asymmetrically obovate, somewhat falcate; keel petals lunate, strongly falcate, acute to rostrate, free at the apex. Stamens diadelphous, upper filament distinct. Ovary stipitate, several-ovuled. Style subulate, strongly curved, glabrous. Pods stipitate, oblong, somewhat inflated, suture winged or naked, valves partially coherent transversely between seeds. Seeds oblong, arranged transversely, estrophiolate.

Inflorescence and branchlets glabrous; leaflets glabrous and eciliate; pods with short stipes barely equaling or shorter than calyces:

Floral bractlets broadly ovate-acuminate, 5–6 mm. long; leaflets prominently veined beneath; pods 8–10 mm. wide .................................... 1. *D. suberosa*

Floral bractlets ovate, 1–2 mm. long; leaflets faintly veined; pods 15–22 mm. broad ...... ...................................................... 2. *D. occidentalis*

Inflorescence and branchlets viscid-glandular, or pubescent with heavy, conical, spinelike hairs; leaflets ciliate or puberulent throughout; stipes of pods one and one-half to two times as long as calyx:

Leaflets ciliate with very heavy, conical hairs, glabrous on upper surface; inflorescence, petioles, rachis, and branchlets densely pubescent, not glandular; pods gradually narrowing to base, leathery, slightly constricted between seeds ....... 3. *D. thurberi*

Leaflets puberulent-glandular throughout; inflorescence, petioles, rachis, and branchlets densely glandular-puberulent; pods truncate-rounded at base and apex ........... ...................................................... 4. *D. racemosa*

## 1. **Diphysa suberosa** S. Wats., Proc. Amer. Acad. **22:** 405, 1887

Tree or large shrub with corky, fissured, yellowish bark; stipules lance-setaceous, 2–4 mm. long; leaves once-pinnate, petioles slender, glabrous, 5–12 mm. long; leaflets 4–8 pairs and 1 terminal, oblong-elliptic or terminal one obovate, 4–8 mm. wide, 6–20 mm. long, rounded at apex, rounded to broadly cuneate at base, glabrous, dark green above, pinnately veined, lighter beneath; petiolules 1–1.5 mm. long; inflorescence lax, few-flowered, of axillary racemes 1–2 cm. long; pedicels very slender, glabrous, 18–30 mm. long, bearing a pair of membranous, ovate-acuminate bractlets 2.5–3 mm. wide, 5–6 mm. long just beneath calyx; calyx vasiform, 8–10 mm. long, lobes subequal, membranous, veined, lower acute-acuminate, others rounded and broader, glabrous; corolla 12–15 mm. long; banner suborbicular, folded inward over wings in bud, erect, abruptly angled with claw in anthesis, thus standing about 8 mm. above calyx; keel turned upward and backward at apex; pod 8–10 mm. wide, flattened contrary to partition, membranous-inflated, sessile, 4–7 cm. long, glabrous, somewhat undulate along margin, stramineous.

Hillsides and valley floors, Lower Sonoran Zone to Tropical Zone, southern Sonora to Jalisco. April–July.

## 2. **Diphysa occidentalis** Rose, Contr. U.S. Nat. Herb. **12:** 271, 1909

A shrub or small tree similar to *D. suberosa* but leaflets thinner and less distinctly veined; rachis slightly hairy beneath; bractlets smaller, 1–2 mm. long; pods 15–22 mm. broad, more papery and less angular.

Rocky hillsides and gravelly outwash slopes, Lower Sonoran Zone to Tropical Zone, from the vicinity of Ures, Sonora, south into Guerrero. April–July.

## 3. **Diphysa thurberi** (A. Gray) Rydb., in Standley, Contr. U.S. Nat. Herb. **23:** 479, 1922

*Daubentonia ?thurberi* A. Gray, Pl. Thurb. 313, 1854.

Shrub 2–6 dm. high; branchlets, petioles, and inflorescence densely glandular-hirsutulous; stipules lance-deltoid to lanceolate, 5–7 mm. long, ciliate but otherwise glabrous, deciduous; petioles 8–15 mm. long, sparingly glandular; leaflets 7–17 pairs, oblong-ovate or oblong, broadly cuneate to rounded at base, rounded at apex, dark green and glabrous above, lighter, reticulate-veined, and glabrous beneath except for midrib, 4–7 mm. wide, 9–15 mm. long, margins ciliate with heavy, short, spinelike hairs; flowers on axillary racemes 5–10 cm. long; floral bracts ovate-lanceolate, 2–3 mm. long, membranous, ciliolate; pedicels slender, 8–12 mm. long, articulate at apex; calyx stipitate at base, 8–10 mm. long, tube 3–4 mm. long, 2.5–3 mm. wide, lobes ovate, rounded, subequal, ciliolate at margins, otherwise glabrous; flowers about 1 cm. long, banner erect or ascending, its blade 7–8 mm. long; wings 0.5–1 mm. shorter than banner; keel narrow, 7 mm. long, acumen moderately acute; pod papery, 7–8 mm. wide, 6–8 cm. long, turgid, somewhat hairy with stout hairs, slightly constricted between seeds.

Mountain slopes, mainly Upper Sonoran Zone, southeastern Arizona southward into Sonora and Chihuahua. Dec.–Feb.

**4. Diphysa racemosa** Rose, Contr. U.S. Nat. Herb. **1:** 97, *pl. 3,* 1891

Shrub 1.5–3 m. high; branchlets densely glandular-viscid; leaflets 9–17, broadly obovate to oblong, 3–7 mm. long, obtuse at apex and base, midrib prominent only beneath, viscid-pubescent; racemes 10–20 cm. long; peduncle and pedicels densely glandular-viscid; bracts oblong to obovate, 4–5 mm. long, densely glandular; bractlets ovate to obovate, 4–5 mm. long, densely glandular; calyx 7–8 mm. long, glandular without, lobes subequal, upper two ovate, acute, 3.5–4 mm. long, lateral ones lanceolate, acute, about same length, lowest lanceolate, acuminate, 4.5–5 mm. long; corolla 10–12 mm. long, 12–15 mm. broad at anthesis; pod 3–4 cm. long, 15–18 mm. wide, on a stipe 10–12 mm. long, truncate to subcordate at base and apex, glandular-pubescent and black-dotted; seeds creamy, 6–7 mm. long, about 3 mm. wide.

Valley floors and hill slopes, Lower Sonoran Zone, southern Sonora and adjacent Sinaloa. March–May.

## 20. COURSETIA DC., Ann. Sci. Nat. 4: 92, 1825

Shrubs or small trees. Leaves pinnate or odd-pinnate, stipulate, with many petiolulate leaflets; stipels minute. Flowers in axillary racemes, solitary in axils of small caducous bracts. Calyx tube campanulate to turbinate, as broad as high; lobes subequal, upper two often united higher than adjacent lobes. Corolla showy; petals subequal, or wings shorter. Banner suborbicular, short-clawed. Wings obliquely oblanceolate to oblong, short-clawed, with a basal auricle. Keel petals somewhat longer-clawed, blades broader, lunate to semiorbicular, obtuse to acuminate. Stamens diadelphous. Style hairy above on upper side. Pod linear, compressed, often constricted between seeds and torulose, sessile or short-stipitate. Seeds suborbicular-lenticular, estrophiolate.

**1. Coursetia glandulosa** A. Gray, Proc. Amer. Acad. **5:** 156, 1861
*Coursetia microphylla* A. Gray, op. cit. **17:** 201, 1882.

Open shrub 2–5 m. high; branchlets sericeous and somewhat glandular when young, soon glabrate, brownish to purplish, faintly striate; stipules linear-subulate, 3–5 mm. long, ascending; leaves 2–5 cm. long; rachis strigose, sulcate on upper side; leaflets 8–18, oblong to elliptic, rounded to acute, often apiculate at apex, 5–15 mm. long, finely but sparsely strigulose on both surfaces, lateral veins inconspicuous, glabrate in age; racemes short, 2–4 cm. long, often 2–5 in an axil, 3–6-flowered; peduncle and its branches densely glandular-puberulent; bracts lance-subulate, 1–2 mm. long, deciduous; calyx densely glandular-pubescent, turbinate, tube 2.5–3.5 mm. long and wide, lobes deltoid to deltoid-lanceolate, 2–4 mm. long, acute to short-acuminate; corolla 10–12 mm. long, banner suborbicular, shallowly retuse, maroon shading to yellow on margins; wings oblong, with a large basal auricle, yellow tinged with red, claws about 2 mm. long; keel petals white tinged with yellow and red at tip, rounded at apex, auricle broad, claw 3–5 mm. long; pod 4–7 cm. long, torulose, glandular-pubescent, 5–7 mm. wide, 2–12-seeded; seeds compressed, 4–5 mm. in diameter, brownish.

Hillsides, arroyos, and canyons, Lower and Upper Sonoran Zones, from the Santa Catalina and Tortolita Mountains, Arizona, through Sonora and Baja California into Sinaloa. March–April.

## 21. BENTHAMANTHA Alef., Bonplandia **10:** 264, 1862

Suffrutescent perennials with odd-pinnate leaves bearing stipules. Flowers in axillary racemes, inconspicuous; bracts setaceous or subulate, caducous. Corolla yellowish, white, or one or more petals purplish. Banner with reflexed margins. Wings oblong, free. Keel petals obliquely and broadly obovate, united and acute or acuminate at apex. Stamens diadelphous. Ovary sessile, many-ovuled; style inflexed, bearded along upper side above; stigma capitate. Pod linear, flat, 2-valved, constricted externally and partitioned internally between the subquadrate nonstrophiolate seeds.

Pods 6–10 cm. long; lower calyx lobe nearly twice as long as the tube .... 1. *B. brandegeei*
Pods 4–6 cm. long; lower calyx lobe about equaling the tube:
  Banner purplish; plant from woody rootstock ........... 2. *B. edwardsii* var. *edwardsii*
  Banner yellow with purple veins; plant from cormlike caudex ........................................................ 2a. *B. edwardsii* var. *glabella*

### 1. Benthamantha brandegeei Rydb., N. Amer. Fl. **24:** 246, 1924

Low suffrutescent plant 3–7 dm. high with slender, densely appressed-pubescent stems soon glabrate and straw-colored; foliage finely appressed-pubescent; stipules filiform-setaceous, 5–15 mm. long, stiffly ascending; leaves 5–10 cm. long, rachises slender, pilosulous with spreading or slightly ascending hairs; stipels subulate, 0.6–1 mm. long; leaflets 9–17, elliptic to oblong, 4–10 mm. broad, 1–3 cm. long, acute to rounded at base, mostly obtuse but with a mucro 1–2 mm. long at apex, dark green above, paler beneath, closely strigose on both surfaces; racemes 3–15-flowered, 3–10 cm. long; bracts filiform-setaceous, 3–7 mm. long; calyx pilose, tube 2–2.5 mm. long, lobes linear-attenuate, 5–7 mm. long; corolla yellowish or banner usually purple; petals 6–7 mm. long; pod 3–3.5 mm. broad, 6–10 cm. long, 15–30-seeded, finely puberulent; seeds about 1.5 mm. wide, 2–2.5 mm. long, ochroleucous mottled with purple, reniform-quadrate.

Chiefly in shade on rocky slopes, Lower Sonoran Zone, vicinity of Loreto to the Cape Region and southern Sonora to Guerrero and Morelos. July–Sept.

### 2. Benthamantha edwardsii (A. Gray) Rose, Contr. U.S. Nat. Herb. **10:** 99, 1906, var. edwardsii

*Cracca edwardsii* A. Gray, Pl. Wright. **2:** 35, 1853.
*Benthamantha wrightii* Rydb., N. Amer. Fl. **24:** 246, 1924.

Suffruticose perennial 3–8 dm., or in shrubs and shade to 2 m., tall; branches pilosulous with spreading or ascending hairs; stipules subulate-setaceous, 4–8 mm.

long; leaves 6–10 cm. long, rachis pilose; stipels filiform, 0.4–1 mm. long; leaflets 7–13, elliptic, 8–18 mm. broad, 1–3.5 cm. long, obtuse at both ends, minutely apiculate or nonapiculate, minutely and sparsely strigulose and dark green above, densely silky-strigulose and paler beneath; racemes 3–6 cm. long, few-flowered; bracts subulate, 2–3.5 mm. long; calyx strigose, tube about 2 mm. deep, lobes lanceolate-acuminate, 2–3 mm. long; corolla 6–7 mm. long, wings and keel yellowish, banner purple; pod 5–6 cm. long, about 3 mm. wide, minutely strigulose; seeds 15–20, reniform-quadrate, about 2 mm. square, dark green or deep buff mottled with purple.

Bushy slopes, Lower Sonoran and more commonly Upper Sonoran Zones, southern Arizona southward through Sonora. Usually growing in shade. July–Sept.

**2*a*. Benthamantha edwardsii** var. **glabella** (A. Gray) Wiggins, comb. nov.

*Cracca edwardsii* var. *glabella* A. Gray, Proc. Amer. Acad. **17**: 201, 1882.
*Benthamantha grayi* var. *glabella* Britt. & Baker, Jour. Bot. Brit. & For. **38**: 19, 1900.
*Benthamantha glabella* Rydb., N. Amer. Fl. **24**: 247, 1924.

Suffruticose perennial 2–5 dm. tall from a fusiform, cormlike caudex; stems sparsely pilose; stipules subulate-filiform, 1.5–2 mm. long; leaves 4–7 cm. long; stipels minute; leaflets 13–21, oval to elliptic, 5–15 mm. long, rounded or obtuse at both ends, glabrate or sparsely pilosulous along veins above, strigulose and often purplish along veins beneath, mucronulate; racemes 4–7 cm. long; bracts filiform, 2–2.5 mm. long; calyx silky-strigose, tube 1.8–2 mm. long, lobes lanceolate-acuminate, about 3 mm. long; corolla 6–7 mm. long, yellowish, banner purple-veined only; pod 3 mm. wide, 4–5 cm. long, puberulent.

Shaded rocky slopes, Upper Sonoran and occasionally upper margin of Lower Sonoran Zones, southern Arizona to Sonora, Chihuahua, and Durango. July–Sept.

## 22. SESBANIA Scop., Introd. 308, 1777

Erect herbs or shrubs (ours annual herbs) with pinnate leaves, numerous leaflets, and small caducous stipules. Flowers in axillary, few-flowered racemes, pedicels bearing a caducous bract and 2 bractlets. Calyx turbinate-campanulate, tube about as broad as long, deltoid lobes shorter than tube. Corolla yellow or with spots and streaks of purple or black on banner. Stamens diadelphous. Ovary many-ovuled, often stipitate; style incurved, glabrous. Pods slender, terete or lightly compressed, many-seeded and septate between seeds; dehiscent. Seeds numerous, oblong-cylindric, smooth, estrophiolate.

**1. Sesbania exaltata** (Raf.) Cory, Rhodora **38**: 406, 1936

*Darwinia exaltata* Raf., Fl. Ludov. 106, 1817.
*Sesbania macrocarpa* var. *picta* S. Wats., Proc. Amer. Acad. **24**: 46, 1889.
*Sesban exaltatus* Rydb., N. Amer. Fl. **24**: 204, 1924.
*Sesban sonorae* Rydb., op. cit. 205.

Erect annual to 4 m. tall with glabrous, terete or slightly angled stem and lanceolate-subulate stipules 4–10 mm. long; leaves 5–30 cm. long; petioles 5–10 mm. long; leaflets 8–35 pairs, linear-oblong, rounded and mucronulate at apex, broadly cuneate to rounded at base, 3–6 mm. wide, 1–2.5 cm. long, dark green above, paler beneath, glabrous; racemes 2–10 cm. long on peduncles about as long or shorter; bracts and bractlets linear-subulate, 2–7 mm. long; pedicels subfiliform, 5–9 mm. long; calyx 5–6 mm. long, lobes 1.5–2 mm. long; corolla 12–15 mm. long, bright yellow with purplish or blackish streaks and spots on banner; pods linear, 3–4 mm. wide, 1–2 dm. long, glabrous; seeds 1.5–2 mm. wide, 3–4 mm. long, dark brown to dull black.

Bottomland along the Colorado River and occasionally along irrigation ditches and in cultivated fields in the Lower Sonoran and Subtropical Zones, to Florida and to central Mexico; southern Baja California. Aug.–Oct.

## 23. SUTHERLANDIA R. Br., in Ait., Hort. Kew. ed. 2, 4: 327, 1812

A canescent shrub with odd-pinnate leaves, minute stipules, and showy red flowers in axillary racemes. Calyx campanulate, 5-toothed. Banner oblong, sides reflexed, shorter than keel; wings barely equaling or slightly exceeding calyx teeth; keel boat-shaped, claws equaling wings. Stamens diadelphous. Ovary stipitate. Style hairy along upper side and just below minute capitate stigma on lower side. Pod papery, inflated, oblong, indehiscent or tardily and irregularly dehiscent. Seeds numerous, reniform.

### 1. Sutherlandia frutescens R. Br., in Ait., Hort. Kew. ed. 2, 4: 327, 1812

Openly branched shrub to 2 m. tall or more with slender stems and strigose to canescent foliage and young twigs; stipules linear-subulate, 2–3 mm. long, brownish; leaves 5–8 cm. long; petioles 6–10 mm. long; leaflets 19–25 (in our specimens), elliptic to oblong, 3–5 mm. wide, lateral ones 8–10 mm. long, terminal leaflet one-half again as long, rounded to faintly emarginate above, obtuse to acute at base; peduncles 1–2 cm. long, 5–20-flowered; pedicels 5–8 mm. long; calyx tube 7–8 mm. long, 5 mm. in diameter, sparingly strigose with a mixture of black and white hairs, teeth deltoid, 2–3 mm. long; petals scarlet, banner 2.5 cm. long, keel about 3 cm. long; pods oblong, 2.5–3 cm. in diameter, about twice as long, glabrous, papery and shining, pale yellowish green, stipe about 1 cm. long; seeds dark brown, smooth, dull, 3–4 mm. long.

Escaped from cultivation and growing spontaneously in a few places in arid parts of Mexico. Between San Augustín and Cataviñá, Baja California. March–April in Baja California.

## 24. ASTRAGALUS L., Sp. Pl. 755, 1753*

Annual or perennial herbs from a slender to stout taproot, sometimes with a short woody caudex. Stems 1 to many, erect, ascending, or decument, glabrous to densely pubescent. Leaves odd-pinnate, with several to many leaflets. Stipules persistent or rarely deciduous. Inflorescence racemose or spicate-racemose, rarely umbellate or of a single flower. Calyx tube campanulate to cylindrical, teeth 5, deltoid to linear-subulate. Petals white, yellowish, pinkish, purple, or rarely crimson; banner obovate, slightly arched to strongly reflexed, usually equaling or surpassing keel and wings; wing petals oblong to lunate, with a slender claw and obvious basal auricle; keel petals obtuse to acute, usually more or less arcuate, rarely the tip prolonged into a porrect beak. Stamens 10, diadelphous. Style slender; stigma terminal, minute. Ovary sessile or stipitate, several-ovuled. Pod thin and strongly inflated, lightly compressed laterally, or coriaceous and turgid, 1-celled, or becoming partially or wholly 2-celled by intrusion of lower suture or both sutures, readily dehiscent, tardily dehiscent, or sometimes indehiscent. Seeds few to many, small, irregularly or symmetrically reniform.

---

* Manuscript reviewed and key to species provided by R. C. Barneby.

◆ Pods bladdery-inflated:
Pods sessile or elevated out of calyx on a gynophore and disjointing from it when ripe:
Pods fully bilocular:
Valves of pods membranous-papery:
Herbage silvery-canescent or white-tomentose; leaflets pubescent on both surfaces:
Pods white-tomentose; calyx tube 4–5 mm. long; calyx teeth 1.5–3 mm. long ... ........................................ 1. *A. lentiginosus* var. *coachellae*
Pods strigose; calyx tube 3–3.5 mm. long; calyx teeth 0.75–1.5 mm. long ...... .................................... 1*a*. *A. lentiginosus* var. *borreganus*
Herbage strigose or glabrous, green, not sericeous or white-tomentose; leaflets glabrous at least on upper surface:
Keel 10–12 mm. long; corolla white, ochroleucous or faintly diffused with mauve ...................................... 1*b*. *A. lentiginosus* var. *yuccanus*
Keel 12–15 mm. long; corolla purple .......... 1*c*. *A. lentiginosus* var. *australis*
Valves of pods firm and coriaceous ............. 1*d*. *A. lentiginosus* var. *maricopae*
Pods unilocular:
Flowers 2.1–2.8 cm. long ..................................... 2. *A. crotalariae*
Flowers much smaller:
Stipules all, or at least those near base of stems, fully amplexicaul and connate into a sheath, opposite the petiole; San Quintín Bay south to vicinity of Rosario:
Pods 2.7–4 cm. long, tomentulose like herbage; ovules 30–40 .. 3. *A. anemophilus*
Pods about 1.4–2 cm. long, loosely strigulose (leaves tomentulose); ovules 24–27 ...................................................... 4. *A. harbisonii*
Stipules all free opposite petiole, mostly semiamplexicaul or less:
■ Racemes all (or all but some subterminal and obviously depauperate ones) over 10-flowered:
Pods elevated out of calyx tube on a pubescent gynophore 3–9 mm. long, disjointing from it when ripe ............................. 5. *A. fastidius*
Pods sessile (either disjointing or persistent):
Plant silvery- or satiny-pubescent throughout; coastal and desert, but if away from desert, terminal leaflet confluent with rachis:
Leaflets 15–23, terminal one not decurrent on rachis, not markedly longer than adjacent pair; pods 5–12 mm. in diameter; seeds 1.5–3.3 mm. long; rachis terete; peduncles equaling or slightly surpassing subtending leaves:
Seeds 1.5–2 mm. long; petals purple marginally, paler in center and toward base; banner 9–11 mm. long; stems spreading-diffuse ...... ............................ 6. *A. magdalenae* var. *magdalenae*
Seeds 2.5–3.3 mm. long; petals pale lilac; banner 12–12.5 mm. long; stems clumped, erect ............... 6*a*. *A. magdalenae* var. *niveus*
Leaflets 9–13, terminal one decurrent at tip of rachis, longer than distal pair of leaflets; pods 13–21 mm. in diameter; seeds 4.5–5.5 mm. long; peduncles usually shorter than leaves ............................ ............................... 6*b*. *A. magdalenae* var. *peirsonii*
Plant variably pubescent, green, cinereous or silvery, but if silvery-canescent not coastal, and if of the interior then terminal leaflet jointed to rachis:
Flowers and pods crowded into oblong or subglobose heads; plants of alkaline flats and seeps; coarse annuals of long duration, becoming tumbleweeds in fall ....................... 7. *A. hornii* var. *minutiflorus*
Flowers and pods loosely racemose; not halophytic, and if annual then obviously so and of short duration:
● Herbage and pods villosulous or hirsutulous with spreading or loosely incumbent hairs ............................... 8. *A. prorifer*

● Herbage and pods strigulose with appressed or subappressed hairs:
Pod very small, subglobose, scarcely or very shortly beaked, 6–13 mm. long, 6–10 mm. in diameter, 8–11-ovulate; drainage of Gila River ............................................ 9. *A. thurberi*
Pod larger, mostly 1.5–4 cm. long, or if in the same length range strongly beaked, beak one-third to one-half as long as body; body obliquely ovoid; if pod as small as above, range otherwise:
Pods subsymmetrically ovoid, ovoid-ellipsoid, or subglobose; sutures subequally convex as seen in profile; beak short and obscure, not well differentiated from body ................ 10. *A. wootonii*
Pods oblique in profile, half-ovoid or -ellipsoid; ventral suture straight, concave, or less strongly convex than dorsal one; beak developed and well differentiated from body:
Petals ochroleucous or greenish white:
Ovary and pod strigulose (pod rarely glabrate), 27–70-ovulate; racemes mostly 15–40-flowered; from about 28° 50′ N. Lat. northward:
Pod 2.5–6 cm. long, 42–70-ovulate; calyx teeth deltoid or broadly triangular:
Stems diffuse and weakly assurgent, mostly 2–6 dm. long; peduncles incurved-ascending; fruits commonly humistrate; pods 2.5–4(5) cm. long; widely distributed from northern Baja California to San Bernardino Mountains ................. 11. *A. douglasii* var. *parishii*
Stems stiffly erect-ascending in clumps, 4–10 dm. tall; peduncles erect even in fruit; pod (3.5)4–6 cm. long; local in southeastern San Diego County near Jacumba .................. 11*a*. *A. douglasii* var. *perstrictus*
Pod 1.6–2.8 cm. long, 27–40-ovulate; calyx teeth subulate; mountains around head of Coachella Valley, California ....................... 12*a*. *A. palmeri* var. *johnstonii*
Ovary and pod glabrous, 22–36-ovulate; racemes mostly 8–15-flowered; Sierra San Pedro Mártir and south on desert to about 29° 30′ N. Lat. ... 11*b*. *A. douglasii* var. *glaberrimus*
Petals pink-purple, drying bluish:
Pods mostly 0.9–2.3 cm. long, if longer the pedicels deciduous; southern California and Baja California:
Ovules 12–31; pods papery but valves firm, not markedly transparent; along western edge of Colorado Desert south inland, and along west coast of Baja California at about 28° N. Lat.:
to northern Sierra San Pedro Mártir, at about 30° N. Lat.
Pods mostly 5–15 mm. long; fertile pedicels persisting; leaflets 11–21; mostly above 1,000 feet elevation from Sierra San Pedro Mártir northward ............... ...................... 12. *A. palmeri* var. *palmeri*
Pods mostly 2–3 cm. long; fertile pedicels disjointing with fruit; leaflets 17–25; Pacific coast of Baja California at Lagoon Head ..................... 13. *A. piscinus*
Ovules 8–12; pods extremely thin-walled, valves subdiaphanous when ripe .................... 14. *A. idrietorum*
Pods mostly (2)2.5–4 cm. long; fertile pedicels persistent after fall of pods; Arizona to western Texas ... 15. *A. allochrous*

■Racemens 2–10-flowered:

Pods ascending from raceme, not greatly inflated, 1–1.7 cm. long, 4.5–7 mm. wide, silky-canescent like foliage; flowers very small, whitish or pink-tinged; keel 3.6–5 mm. long; ovules 3–7; obligate annuals of low elevations in Colorado Desert, southeastern California and adjoining Arizona and Mexico . . . . . . . . . . . . . . . . . . . . . . . . . . . . . . . . . . . . . . . . 16. *A. aridus*

Pods spreading or declined, glabrous, strigulose, or villous, not canescent; ovules (6)7–28; if in range of *A. aridus* either flowers a little larger or pod villosulous:

Pods villosulous or hirsute, hairs at least 0.7 mm. long; valves of pods firmly papery, not translucent . . . . . . . . . . . . . . . . . . . . . . . . . . 17. *A. sabulonum*

Pods strigulose, hairs shorter and appressed; valves of pod thinly papery, translucent:

Leaflets mostly 7–13; banner 8–11 mm. long; ovules 19–24; entering northern Colorado Desert from north and rare in our area . . . . . . . . . . . . . . . . . . . . . . . . . . . . . . . . . . . . . . . . . . . . . . . . . . . . . . . . . . . . 18. *A. nutans*

Leaflets mostly 9–21; banner 5.5–7.5 mm. long; ovules 7–14; Colorado Desert and Baja California:

Pods obliquely ovoid-acuminate, contracted distally into a well-defined deltoid or triangular-acuminate beak 2.5–7 mm. long; ventral suture of pod commonly salient, much less strongly convex than dorsal one; widespread but on Pacific coast of Baja California unknown north of 29° N. Lat.:

Plants thinly strigulose; herbage deep green; leaflets acute-mucronulate, glabrous above and only thinly ciliate along midrib and toward margins beneath; Cedros Island and adjacent peninsula, Baja California . . . . . . . . . . . . . . . . . 19. *A. insularis* var. *insularis*

Plants strigulose throughout or nearly so; herbage cinereous; leaflets obtuse or emarginate, equally pubescent on both sides or only medially glabrescent above; Colorado Desert, southeastern California and adjoining Arizona and Sonora, south to San Lorenzo Island at 28° 40′ N. Lat. . . . . . . . . . 19*a*. *A. insularis* var. *harwoodii*

Pods subsymmetrically ovoid or subglobose, obscurely beaked; beak broadly deltoid, at most 2.5 mm. long; ventral suture depressed, about equally convex with dorsal one; local about San Quintín Bay, south to edge of desert in Lat. 30° to 30° 30′ N. . . . . . . . . . . . . . . . . . . . . . . . . . . . . . . . . . . . . . . . . . . . . . . 19*b*. *A. insularis* var. *quentinus*

Pods elevated out of calyx on a genuine stipe continuous with body of fruit, persistent or falling only as a unit with pedicel and calyx . . . . . . . . . . . . . . . . . . . . 20. *A. leucopsis*

◆Pods not bladdery-inflated; body of pod laterally compressed or lower suture inflexed, then a groove between two valves:

Pod clearly 1-celled, lower suture never inflexed to form medial partition:

Corollas ochroleucous, 10–12 mm. long; pods strongly compressed laterally, dehiscent first at base; racemes lax, 7–25-flowered; calyx teeth deltoid, 0.5–1 mm. long . . . . . . . . . . . . . . . . . . . . . . . . . . . . . . . . . . . . . . . . . . . . . . . . . . . . . . . . . . . . . . . . 21. *A. filipes*

Corollas crimson, 2–2.5 cm. long; pods not compressed laterally, dehiscent first from apex or indehiscent; racemes subcapitate, 2–8-flowered; calyx teeth subulate, 4–9 mm. long . . . . . . . . . . . . . . . . . . . . . . . . . . . . . . . . . . . . . . . . . . . . 22. *A. coccineus*

Pod completely or almost completely 2-celled by inflexing of one or both sutures:

▲Pod borne on a gynophore, although gynophore often shorter than, and obscured by, calyx tube:

Gynophore 4–8 mm. long, obvious; corolla clear lemon yellow; pod 15–25 mm. long; leaflets obtuse but rarely retuse . . . . . . . . . . . . . . . . . . 23. *A. pachypus* var. *jaegeri*

Gynophore 1–2.5 mm. long, often hidden within calyx; corolla white to pale cream; pods 25–42 mm. long; leaflets usually retuse ............... 24. *A. tricarinatus*

▲Pods sessile, gynophore not developed or so short that it is not evident:

Pods not transversely wrinkled; valves 1.5 cm. long or more, linear to oblong in profile, straight or arcuately curved:

Keel not beaked; calyx white-strigose, black hairs few or none:

Plants perennial, sometimes flowering first year; leaflets strigulose and green, silvery-strigose, or silky-canescent on both surfaces:

Pods 1.5–2 cm. long, deeply sulcate beneath; corolla blue or deep purple:

Petals purple; leaves 2–10 cm. long; leaflets 3–11, 3–18 mm. long; pod densely strigulose, brownish ............................. 25. *A. mohavensis*

Petals pale blue; leaves 10–18 cm. long; leaflets 15–25, 2–8 mm. long; pods glabrate, stramineous .......................... 26. *A. orcuttianus*

Pods 1.5–3 cm. long, flat or faintly sulcate beneath; corolla white, flushed with green, mauve, or pale purple ....................... 27. *A. arizonicus*

Plants annual; leaflets glabrous or glabrate on upper surface, glabrate or sparsely strigose beneath:

Calyx pubescent with relatively long hairs to 0.7–1.2 mm. long; calyx teeth mostly 1.8–2.8 mm. long, hispid-ciliate; south-central Arizona to southern Oklahoma and western Texas, south to Puebla ....................... .................................. 28. *A. nuttallianus* var. *austrinus*

Calyx strigulose or more closely appressed-pubescent with shorter hairs 0.35–0.7 mm. long; calyx teeth 1–1.7 mm. long, but hispid-ciliate:

Leaflets subhomomorphic, all elliptic, those of lower leaves sometimes longer than upper but none retuse or emarginate; calyx tube deeply campanulate, 1.9–2.8 mm. long and wide; western and central Arizona to the Mojave Desert and Utah ........ 28*a*. *A. nuttallianus* var. *imperfectus*

Leaflets dimorphic, those of lowest leaves small and retuse, of upper leaves elliptic and acute; calyx tube shallowly campanulate, 1.4–1.7 mm. long and broad; Colorado Desert and into Baja California and northern Sonora ......................... 28*b*. *A. nuttallianus* var. *cedrosensis*

Keel prolonged into an acute beak; calyx tube with more black hairs than white:

Calyx tube 4–5 mm. long, twice as long as teeth; corolla 8–10 mm. long ........ ............................................... 29. *A. nothoxys*

Calyx tube 1.5–2 mm. long, about equaling teeth; corolla 4–5 mm. long ....... ............................................... 30. *A. acutirostris*

Pods transversely wrinkled, 2.5–4 mm. long, ovoid to subglobose:

Leaflets 11–17, linear, glabrous above; pods subglobose ......................... ................................ 31. *A. didymocarpus* var. *didymocarpus*

Leaflets 5–13, cuneate-obovate to oblong, white-puberulent on both surfaces; pods ovoid, longer than wide ............... 31*a*. *A. didymocarpus* var. *dispermus*

## 1. **Astragalus lentiginosus** Dougl. var. **coachellae** Barneby*

*Astragalus lentiginosus* var. *coulteri* sensu Barneby, Leafl. West. Bot. **4:** 129, 1945. Not *A. coulteri* Benth., 1848.

A short-lived perennial, often flowering first season; stems stout but flexuous, finely

---

* **Astragalus lentiginous** var. **coachellae** R. C. Barneby, var. nov., ab arcte affinis var. *variabilis* Barneby formis holosericeis praesertim dentibus calycinis elongatis (± 1.5–3 mm. longis), a var. *borregano* Jones (= *A. coulteri* Benth. vero) legumine inflato late ovoideo-acuminato absimilis.—CALIFORNIA: Palm Springs, Riverside County, April 30, 1913, *Alice Eastwood 3116*. Holotypus, NY; isotypi CAS, GH.

*Astragalus lentiginosus* var. *coulteri* sensu Barneby, Leafl. West. Bot. **4:** 129, *pl. IV, figs. 14–17*, 1945 (non *A. coulteri* Benth., 1848).—R. C. Barneby.

and densely white-sericeous to tomentulose, 2–5.5 dm. long, ascending or decumbent, striate; stipules deltoid-lanceolate, acuminate, 4–5.5 mm. long, silky-strigose with white hairs intermingled with black ones; leaves spreading, 5–10 cm. long, petiole and rachis densely sericeous; leaflets 7–10 pairs, broadly oval to obovate, obtuse, rounded or retuse at apex, 4–12 mm. wide, 5–20 mm. long, finely white-silky on both sides; peduncles erect or ascending, flexuous, 3–10 cm. long; racemes nearly or fully equaling peduncles, 13–25-flowered, mostly 6–9 cm. long in fruit; bracts lanceolate to subulate, 2–3 mm. long, sericeous; calyx tube narrowly campanulate, 4–5 mm. long, densely silky-villous, few to many hairs black; teeth lance-subulate, 1.5–3 mm. long; corolla reddish purple, 12–15 mm. long; banner obovate, 14–15 mm. long, slightly arched; wings nearly straight, 12–13 mm. long, blade obtuse, oblong-obovate, with a large auricle; keel rounded at apex, 10–11.5 mm. long; pods obliquely ovoid, papery, 1.5–2.2 cm. long, about 1 cm. wide, firm, often mottled with purple, silky-white-tomentose, slightly sulcate along both sutures; beak short-conical.

Sandy washes, plains and hillsides, Lower Sonoran Zone, Colorado Desert around head of Coachella Valley and adjacent mountain slopes. March–June.

**1*a*. Astragalus lentiginosus** var. **borreganus** M. E. Jones, Contr. West. Bot. **8:** 3, 1898

*Astragalus coulteri* Benth., Pl. Hartw. 307, 1848.
*Astragalus arthu-schottii* A. Gray, Proc. Amer. Acad. **6:** 209, 1864, in part.
*Tragacantha coulteri* Kuntze, Rev. Gen. Pl. **2:** 944, 1891.
*Cystium coulteri* Rydb., Bull. Torrey Club **40:** 50, 1913.
*Astragalus aginus* Jepson, Man. Fl. Pl. Calif. 577, 1925.
*Cystium agninum* Rydb., N. Amer. Fl. **24:** 408, 1929.
*Cystium arthu-schottii* Rydb., op. cit. 409.

Habit similar to that of var. *coachellae*; annual or biennial; stems 1.5–3 dm. long; leaflets 15–19, 6–15 mm. long; racemes 15–35-flowered, 6–12(25) cm. long in fruit; calyx tube 3–3.5 mm. long, white-silky, often with some black hairs among white; calyx teeth 1.5 mm. long or less; banner about 13 mm. long; wings and keel slightly shorter; pods narrowly ovoid to lance-acuminate, falcate, 1.5–2 cm. long; septum often incomplete.

Arid desert sands, Lower Sonoran Zone, Colorado and Mojave Deserts, California, into southwestern Arizona, northern Baja California and Sonora. Feb.–May.

**1*b*. Astragalus lentiginosus** var. **yuccanus** M. E. Jones, Contr. West. Bot. **8:** 3, 1898

*Cystium yuccanum* Rydb., N. Amer. Fl. **24:** 407, 1929.
*Astragalus yuccanus* Tidestrom, Proc. Biol. Soc. Wash. **48:** 40, 1935.
*Astragalus fremontii* var. *yuccanus* Tidestrom, in Tidestrom & Kitt., Fl. Ariz. & New Mex. 216, 1941.

Winter annual 6–10 dm. tall with sparsely strigose to glabrate stems; leaves 8–15 cm. long; leaflets 13–25, oval to obovate, 8–12 mm. wide, 10–20 mm. long, obtuse, rounded or retuse at apex, sparsely strigose along midrib beneath, less often on lamina, glabrous above and glabrate beneath in age; peduncles 10–12 cm. long; racemes lax, 8–10 cm. long, to 12 cm. long in fruit; calyx tube 3.5–4.5 mm. long, sparsely pubescent; teeth 1–1.5 mm. long; corolla white or ochroleucous, sometimes faintly tinged with mauve; banner 11–13.5 mm. long; wings 11–12.5 mm. long, obliquely oblanceolate; keel to 10–12 mm. long; pods broadly ovoid to subglobose, papery, 1.5–2.5

cm. long, about 1.5 cm. thick, short-acuminate and slightly turned upward at apex, glabrous or sparsely and minutely strigose, lustrous.

Sandy desert washes and hillsides, Lower Sonoran Zone, Mohave and Maricopa counties and Western Pima County, Arizona. Feb.–June.

**1c. Astragalus lentiginosus** var. **australis** Barneby, Leafl. West. Bot. **4:** 117, *pl. 3, fig. 15–19,* 1945

Perennial with stout stems 3–6 dm. long; leaves 8–12 cm. long; leaflets 19–27, obovate to oblong, (7)10–21 mm. long, glabrous above, finely strigose beneath; peduncles stout, 8–12 cm. long; racemes 6–12 cm. long, many-flowered; calyx tube 4–5 mm. long, strigose; teeth 1.5–2 mm. long; corolla pale purple, at least at tips of petals; banner 14–16 mm. long; wings 13–15 mm. long; keel 12–15 mm. long; pods membranous to thinly chartaceous, obliquely and narrowly ovoid, strongly curved, slightly to strongly inflated, 13–20 mm. long, glabrous or rarely puberulent.

Sandy desert plains and washes, Sonoran Zones, southern Arizona to southwestern New Mexico, south into Sonora and Chihuahua. Feb.–June.

**1*d*. Astragalus lentiginosus** var. **maricopae** Barneby, Leafl. West. Bot. **4:** 140, 1945

Glabrous perennial with stout, erect stems 3–4 dm. tall; leaves short-petiolate, 7–15 cm. long; leaflets 15–23, obovate to elliptic, obtuse, truncate or apiculate, 7–22 mm. long, glaucescent, glabrous or sparsely strigose beneath; peduncles 5–9 cm. long, shorter than leaves; racemes lax, 10–20 cm. long, 15–20-flowered; calyx tube 5–6 mm. long, strigose with white or black hairs or with both; teeth 1.5–2 mm. long, subulate or deltoid; corollas ochroleucous; banner 13–15 mm. long; wings 12–13 mm. long; keel 11–11.5 mm. long; pods strongly ascending, lanceolate and gradually acuminate, coriaceous, smooth and glabrous, castaneous, about 2 cm. long.

Sandy plains, washes, and valley floors, Lower Sonoran Zone, lower parts of Salt River Valley, near Tempe, Arizona. Feb.–May; Sept.–Oct.

**2. Astragalus crotalariae** (Benth.) A. Gray, Proc. Amer. Acad. **6:** 216, 1864

*Phaca crotalariae* Benth., Pl. Hartw. 307, 1848.
*Tragacantha crotalariae* Kuntze, Rev. Gen. Pl. **2:** 944, 1891.
*Astragalus limatus* Sheldon, Minn. Bot. Studies **1:** 126, 1894.
*Astragalus preussii* var. *limatus* Jepson, Man. Fl. Pl. Calif. 566, *fig. 555,* 1925.

Perennial with a short caudex; stems branching from base, decumbent to ascending, 1–6 dm. long, strongly sulcate, ascendingly pubescent in youth, at length subglabrate; stipules ovate-deltoid, 5–6 mm. long, more or less reflexed; leaves 10–15 cm. long, ascending or spreading-ascending; leaflets 9–19, broadly obovate, rounded to retuse at apex, mostly obtuse at base, 5–16 mm. wide, 1–3.5 cm. long, glabrate or glabrous above, finely strigose beneath; peduncles erect, 10–15 cm. long; racemes 5–10 cm. long; bracts ovate-lanceolate, 2–3.5 mm. long, strigose without; calyx tube cylindrical, purplish, 6–7 mm. long, finely and densely strigulose; teeth lanceolate to narrowly deltoid, 2–2.5 mm. long; corolla purple, 18–20 mm. long; banner obovate, retuse, only slightly arched; wings about 3–4 mm. shorter than banner; keel weakly arcuate; pod on an obscure gynophore about 2 mm. long, body ovoid to broadly ellipsoid, 2.5–3 cm. long, 1.5–1.8 cm. wide, firm and leathery, sparsely strigose, dehiscing apically.

Arid plains and washes, Lower Sonoran Zone, central and western part of the Colorado Desert, south into adjacent northern Baja California. Feb.–March.

**3. Astragalus anemophilus** Greene, Bull. Calif. Acad. **1:** 186, 1885

*Phaca vestita* Benth., Bot. Sulph. 13, 1844. Not *Astragalus vestitus* Boiss., 1842.
*Astragalus vestitus* S. Wats., Bibl. Index 202, 1878. Not *A. vestitus* Boiss., 1842.
*Tragacantha vestita* Kuntze, Rev. Gen. Pl. **2:** 949, 1891.
*Astragalus crotalariae* var. *anemophilus* M. E. Jones, Contr. West. Bot. **10:** 59, 1902.

Perennial; stems decumbent, 3–5 dm. long, strigose-canescent to tomentulose; stipules amplexicaul, connate into a sheath opposite petioles, at least on lower leaves, deltoid, 3–5 mm. long, tomentose; leaves 5–12 cm. long; leaflets 19–31, obovate to broadly oblong, 2.5–5 mm. wide, 6–15 mm. long, rounded or obtuse at apex, obtuse to broadly cuneate at base, densely white-villous with fine, tangled, appressed or subappressed hairs; peduncles 5–10 cm. long, erect; racemes dense, 3–5 cm. long; bracts lance-subulate, 2–3 mm. long, densely villous without, glabrous within; calyx tube 4–5 mm. long, densely white-villous, teeth deltoid, 1–1.5 mm. long; corolla 10–12 mm. long, ochroleucous or greenish yellow; banner obovate, gradually narrowing toward base, only slightly exceeding wings, these with obtuse, lunate blades and reflexed basal auricles; keel strongly arcuate along lower edge, blunt at tip; pods 1-celled, ovoid, 2.7–4 cm. long, 1.5–2.5 cm. wide, densely tomentulose in youth, often sparsely pubescent to subglabrate at maturity, upper suture nearly straight; ovules 30–40; seeds reniform, 2.5–3.2 mm. long, brownish.

Sandy soil near the coast, Lower and Upper Sonoran Zones, vicinity of San Quintín Bay, Baja California. Feb.–May; Sept.–Oct.

Bentham reported this species (as *Phaca vestita*) from Magdalena Bay, and San Quintín, but according to R. C. Barneby (personal communication) there is only one specimen of Bentham's *Phaca vestita* in the herbarium at Kew, that being from San Quintín. Not since the voyage of the Sulphur has it been found at Magdalena Bay. Presumably a mixing of labels occurred and the type of *Phaca vestita* came from San Quintín.

**4. Astragalus harbinsonii** Barneby, Leafl. West. Bot. **7:** 34, 1953

Perennial; densely villosulous-tomentulose with short curly and some longer ascending hairs; stems diffuse, ascending, 1.5–3 dm. long; stipules 3–5 mm. long, connate opposite petiole bases; leaves 5–9 cm. long; leaflets (13)17–23, broadly to narrowly elliptic, acutish, 3–11 mm. long; peduncles mostly longer than subtending leaves; racemes 13–23-flowered; flowers ascending, axis of raceme becoming 1.5–5.5 cm. long in fruit; calyx tube 3.5–4 mm. long; teeth subulate, 1.7–2.5 mm. long, villous-tomentulose with mixed black and white hairs; petals whitish, more or less tinged with lavender, banner recurved through 45°, about 12 mm. long; pod ascending, sessile, obliquely ellipsoid, bladdery, 1.4–2 cm. long, 8–11 mm. in diameter, contracted distally into a short-triangular beak; ventral suture less convex than dorsal one or almost straight; valves papery-membranous, semi-transparent; ovules 24–29.

Sandy coastal areas, Lower Sonoran Zone, vicinity of Punta Baja southwest of Rosario, Baja California. Type collected in flower in April.

**5. Astragalus fastidius** (Kell.) M. E. Jones, Contr. West. Bot. **8:** 7, 1898

*Phaca fastidia* Kell., Hesperian **4:** 145, 1860.
*Astragalus fastidiosus* Greene, Bull. Calif. Acad. **1:** 186, 1885.
*Astragalus leucopsis* var. *fastidius* M. E. Jones, Contr. West. Bot. **10:** 62, 1902.

A short-lived perennial, often flowering first season; stems erect or ascending, 2–5 dm. tall, moderately branched from near base or strict, cinereous-puberulent with closely appressed, more or less crinkly silvery hairs; stipules deltoid-acuminate, 3–5 mm. long, 2.5–4 mm. wide, puberulent without, glabrous within, often reflexed; leaves 4.5–15 cm. long, petioles and rachises cinereous-puberulent; leaflets 19–27 on middle and upper leaves, fewer on lowermost leaves, ovate, broadly elliptic, or obovate to nearly orbicular, 2.5–9 mm. wide, 5–17 mm. long, rounded to retuse at apex, puberulent but eventually subglabrate and bright green above on many plants; peduncles 4–10 cm. long, elongating only slightly in fruit; racemes erect or ascending, 3–8 cm. long, 11–19(–23)-flowered; calyx tube 2.5–3.2 mm. wide, 3.5–4.6 mm. long, puberulent with white hairs intermingled with black ones, teeth subulate to narrowly lanceolate, 2.2–2.8 mm. long; corolla yellow on plants from Cedros Island, purple on peninsula; banner 13–15 mm. long, blade only slightly reflexed; wings 12–13 mm. long; keel nearly equaling wings; gynophore slender, 3–9 mm. long; pod asymmetrically ovoid-acuminate, 9–14 mm. in diameter, 2.2–3.5 cm. long, upper suture nearly straight, beak 5–6 mm. long, tapering gradually; ovules 11–27.

Sandy and loamy soil, Lower Sonoran Zone, margins of dunes, inside barrier beaches, along arroyos, and on benches and mesas, usually within a few miles of the coast, Cedros Island, and along western coast of the Vizcaíno Desert, Baja California. Feb.–April; Sept.–Oct.

### 6. **Astragalus magdalenae** Greene, Pittonia **1:** 162, 1888, var. **magdalenae**

*Phaca candidissima* Benth., Bot. Sulph. 13, 1844.
*Astragalus candidissimus* S. Wats., Bibl. Index 191, 1878. Not *A. candidissimus* Ledeb., 1829.
*Tragacantha californica* Kuntze, Rev. Gen. Pl. **2:** 940, 1891. Not *Astragalus californicus* Greene, 1885.
*Astragalus crotalariae* var. *magdalenae* M. E. Jones, Contr. West. Bot. **10:** 59, 1902.

Perennial but often flowering first season; base of stem ultimately woody; herbage white-strigose with silky hairs throughout; stems decumbent, 5–8 dm. long, lightly striate; stipules broadly deltoid, 1–5 mm. long, sericeous, soon deciduous; leaves 4–15 cm. long; leaflets 15–23, obovate to oblong, 3–8 mm. wide, 5–18 mm. long, rounded to retuse at apex, cuneate to obtuse at base, strigose-silvery on both sides; peduncles 5–15 cm. long, erect; racemes lax, 3–7 cm. long; bracts lance-subulate, 1–3 mm. long; calyx tube 3–4 mm. long, strigose with silvery hairs only or with mixture of black hairs among white ones; teeth subulate, 1–1.5 mm. long; corolla purplish, 10–12 mm. long, banner narrowly obovate, tapering gradually below, slightly longer than lunate wings; keel curved, blunt at apex; pods sessile, ovoid-elliptic to broadly ovoid, 2–2.5 cm. long, 12–15 mm. wide, canescent-strigose; beak conical, about 3 mm. long.

Arid sandy soil or among rocks in sandy places near the coast, Lower Sonoran Zone, Magdalena Plain and along Gulf coast from vicinity of Barril around head of the Gulf to Puerto Libertad, Sonora. Nov.–May.

### 6*a*. **Astragalus magdalenae** var. **niveus** (Rydb.) Barneby, Aliso **4:** 135, 1958

*Phaca nivea* Rydb., N. Amer. Fl. **24:** 348, 1929.
*Astragalus niveus* Barneby, Leafl. West. Bot. **4:** 54, 1944.

Differing from var. *magdalenae* in having pale lilac instead of deep purple petals, slightly longer banner, larger seeds, and stiffly erect instead of spreading-diffuse stems; from var. *peirsonii* in its more numerous leaflets, more slender pods, smaller seeds, upper leaflets nearly alike, and terete instead of flattened leaf rachises.

Sandy desert flats and arroyos, Lower Sonoran Zone, around head of the Gulf of California in northeastern Baja California and northwestern Sonora. Jan.–May.

**6b. Astragalus magdalenae** var. **peirsonii** (Munz & McBurney) Barneby, Aliso **4:** 135, 1958

*Astragalus crotalariae* var. *piscinus* Jepson, Fl. Calif. **2:** 350, 1936. Not *A. douglasii* var. *piscinus* M. E. Jones, 1902.

Differing from var. *magdalenae* and var. *niveus* in having leaves with distinctly flattened rachises, leaflets 9–13, terminal leaflet decurrent at tip of rachis; peduncles shorter than subtending leaves; larger pods (13–21 mm. in diameter); and distinctly larger seeds (4.5–5.5 mm. long).

Sand dune areas, Lower Sonoran Zone, eastern San Diego County, California, to vicinity of Yuma, Arizona, and south into northeastern Baja California. Jan.–May.

**7. Astragalus hornii** A. Gray var. **minutiflorus** M. E. Jones, Proc. Calif. Acad. II, **5:** 677, 1895

*Astragalus bajaensis* Sheldon, Minn. Bot. Studies **1:** 169, 1894.
*Astragalus miserandus* Greene, Erythea **3:** 76, 1895.
*Astragalus hornii* var. *bahaeensis* M. E. Jones, Contr. West. Bot. **10:** 62, 1902.

Decumbent perennial with striate, glabrate or sparsely pubescent stems 2–4 dm. long; stipules deltoid, about 3 mm. long, often reflexed; leaves ascending, 7–10 cm. long; leaflets 21–35, obovate or oblong, 3–6 mm. wide, 8–15 mm. long, retuse, sparsely appressed-strigose in youth, soon glabrate; peduncles 5–7 cm. long; racemes 1–2 cm. long, dense, nearly capitate in fruit; bracts ovate, 1.5–2 mm. long; calyx appressed-pubescent with white hairs, tube about 3 mm. long, teeth 1–1.5(–1.7) mm. long, lower three lance-subulate, upper two deltoid; corolla 6–7 mm. long, white; banner obovate; wing petals with large auricles; pods ovoid, 9–17 mm. long, 7–8 mm. in diameter, sparsely short-strigose; beak 2–5 mm. long.

Alkaline flats and around seeps, Lower and Upper Sonoran Zones, central Baja California. Sept.–Oct.

**8. Astragalus prorifer** M. E. Jones, Zoe **4:** 275, 1893 (as *A. proriferus*)

*Astragalus julianus* M. E. Jones, Proc. Calif. Acad. II, **5:** 667, 1895.
*Phaca juliana* Rydb., N. Amer. Fl. **24:** 355, 1929.
*Phaca prorifera* Rydb., loc. cit.

Annual or short-lived perennial branching at base; stems hirsute, 1–3 dm. long; leaves 5–10 cm. long, ascending, hirsute; stipules lance-deltoid, 2–3 mm. long; leaflets 13–19, lance-elliptic to lance-linear, 1–3(4) mm. wide, 8–12 mm. long, acute and minutely apiculate at apex; peduncles 3–8 cm. long; racemes lax, 3–8 cm. long, surpassing leaves; bracts subulate, 1.5–2 mm. long; calyx hirsute with fine white hairs, tube 2–2.5 mm. long; teeth subulate, greenish, equaling tube; corolla white or tips of petals suffused with pale purple, darkening in drying, 6–7 mm. long; banner surpassing keel; tip of keel rounded; pod obliquely ovoid, 10–15 mm. long, 7–8 mm. wide, short-hirsute, upper suture nearly straight, tip not flattened.

Sandy benches and rocky slopes, Lower Sonoran Zone, mountain slopes along western margin of San Felipe Desert from north end of Sierra San Pedro Mártir south to San Júlio, Baja California. Feb.–May.

**9. Astragalus thurberi** A. Gray, Pl. Thurb. 312, 1854

*Tragacantha thurberi* Kuntze, Rev. Gen. Pl. **2:** 948, 1891.
*Phaca thurberi* Kearney, Trans. N.Y. Acad. **14:** 34, 1894.

Perennial with grayish-strigose to glabrate, striate, decumbent to ascending stems 1–3 dm. long; leaves ascending, 3.5–10 cm. long; stipules deltoid, 3–5 mm. long, about as wide; leaflets 11–17, oblong to linear-oblong, obtuse to slightly retuse, 3–6 mm. wide, 8–18 mm. long, green, sparsely strigose to glabrate above; peduncles erect or ascending, 2–12 cm. long, striate; racemes compact, 2–8 cm. long, 10–30-flowered; calyx tube 2–2.2 mm. long, finely silvery-strigose, lobes subulate, 1–1.5 mm. long; corolla pale yellowish, more or less tinged with purple, 6–7 mm. long; banner nearly erect; pods sessile, subglobose, 6–13 mm. long, abruptly short-apiculate, sparsely appressed-strigose; seeds 8–11, obliquely reniform, 1.8–2 mm. long, dull dark brown.

Dry slopes and mesas, Lower Sonoran and Upper Sonoran Zones, Pima, Graham, and Cochise Counties, Arizona, into New Mexico and northern Sonora. Mainly at elevations above the Sonoran Desert. March–May.

**10. Astragalus wootonii** Sheldon, Minn. Bot. Studies **1:** 138, 1894

*Phaca wootoni* Rydb., N. Amer. Fl. **24:** 350, 1929.

Annual or sometimes a short-lived perennial with several stems 1–3 dm. long, erect or decumbent, finely striate, very sparingly strigose on striations; stipules deltoid, short-acuminate or acute, 3–5 mm. long, papery, sparsely strigose; leaves 5–12 cm. long, ascending; leaflets 9–19, oblong to linear-oblong, 1.5–3 mm. wide, 1–2 cm. long, acute to cuneate at base, chiefly obtuse at apex, glabrous above, strigose beneath; peduncles slender, 1.5–5 cm. long; racemes 5–10-flowered, moderately lax; calyx tube campanulate, 2–3 mm. long, moderately strigose with a mixture of black and white hairs or hairs all white, teeth linear-subulate, 2–2.5 mm. long; corolla white, pink, or purplish, 6–8 mm. long, banner narrowly obovate, retuse, moderately to strongly reflexed; wing and keel petals only slightly shorter than banner; pods broadly ovoid or somewhat obovoid, 2–2.5(–3) cm. long, 1–1.5 cm. wide, sparsely strigulose; beak conical, minute to nearly lacking.

Arid, sandy or rocky plains, mainly Lower Sonoran Zone, south-central Arizona to western Texas and south into Chihuahua and Sonora, at least as far as the vicinity of Magdalena. March–May.

**11. Astragalus douglasii** (Torr. & Gray) A. Gray var. **parishii** (A. Gray) M. E. Jones, Contr. West. Bot. **8:** 6, 1898

*Astragalus parishii* A. Gray, Proc. Amer. Acad. **19:** 75, 1883.
*Phaca pseudoocarpa* Rydb., N. Amer. Fl. **24:** 343, 1929.
*Phaca topensis* Rydb., loc. cit.
*Phaca parishii* Rydb., op. cit. 344.

Perennial with several procumbent stems forming a clump or loose mat 4–10 dm. wide; herbage sparsely strigose and green to densely silvery-strigose, often glabrate in age; stipules free, lower papery and more or less amplexicaul, upper herbaceous, often reflexed, 4–5 mm. long; leaves 5–15 cm. long, ascending; leaflets 11–25, ovate to oblanceolate, 6–25 mm. long, 2–8 mm. wide, obtuse to retuse; peduncles 2.5–10 cm. long, incurved-ascending; racemes lax, 10–35-flowered, 2.5–12(–17) cm. long in fruit; flowers spreading; calyx tube campanulate, 3–4 mm. long, white-silky, teeth deltoid to subulate, 1.2–2 mm. long, from broadly triangular bases, sinuses outlined with shorter, silvery hairs; petals ochroleucous, banner 8–12 mm. long, suborbicular; pod sessile, obliquely ovoid-ellipsoidal, 2.5–4.5(–5) cm. long, valves papery, strigulose, often glabrate in age, 27–70-ovuled.

Brushy hillsides and flats, mostly Upper Sonoran Zone but straggling to the edge of the desert, San Bernardino Mountains south into northern Baja California. March–Aug.

**11*a*. Astragalus douglasii** var. **perstrictus** (Rydb.) Munz & McBurney, in Munz, Bull. S. Calif. Acad. **31:** 65, 1932

*Phaca perstricta* Rydb., N. Amer. Fl. **24:** 344, 1929.
*Astragalus parishii* var. *perstrictus* Abrams, Ill. Fl. Pacif. States **2:** 583, 1944.

Differing from var. *parishii* mainly in having a stricter, more upright, tufted habit with stems 4–10 cm. tall, sparsely strigose in youth, often finely speckled with purple, subglabrate in age; leaflets 11–19; peduncles erect even in fruit; pods to 6 cm. long.

Open hillsides, fields and stony outwash slopes, Lower and Upper Sonoran Zones, southeastern San Diego County, near Jacumba and Mountain Springs. April–Aug.

**11*b*. Astragalus douglasii** var. **glaberrimus** M. E. Jones, Proc. Calif. Acad. II, **5:** 645, 1895

*Phaca glabberrima* Rydb., N. Amer. Fl. **24:** 345, 1929.

Differing from var. *parishii* and var. *perstrictus* in having glabrous ovary and pods, fewer flowers in a raceme, and only 22–36 ovules in a pod.

Desert washes, sandy plains and flanks of mountains, mainly Lower Sonoran Zone, but ascending into Upper Sonoran Zone at scattered localities, west-central Sierra San Pedro Mártir, south to the vicinity of Cataviñá, Baja California. Feb.–June.

**12. Astragalus palmeri** A. Gray, Proc. Amer. Acad. **7:** 398, 1868, var. **palmeri**

*Astragalus vaseyi* S. Wats., Proc. Amer. Acad. **17:** 370, 1882.
*Tragacantha palmeri* Kuntze, Rev. Gen. Pl. **2:** 947, 1891.
*Astragalus metanus* M. E. Jones, Proc. Calif. Acad. II, **5:** 666, 1895.
*Phaca vaseyi* Rydb., N. Amer. Fl. **24:** 354, 1929.
*Phaca metana* Rydb., loc. cit.
*Phaca palmeri* Rydb., op. cit. 355.
*Astragalus vaseyi* var. *metanus* Munz & McBurney, in Munz, Bull. S. Calif. Acad. **31:** 66, 1932.

Short-lived perennial, often flowering first season, silvery-strigose throughout or leaflets glabrescent beneath; stems much-branched, 2–6 dm. long; leaves 3–15 cm. long; stipules triangular, 2.5–3 mm. long; leaflets 11–21, oblong to narrowly elliptic, 2–6 mm. wide, 5–25 mm. long, acute to obtuse at both ends; peduncles 4–13 cm. long; racemes 5–10 cm. long, (10–)15–40-flowered; calyx silvery- to black-strigose, tube campanulate, 2.5–3.5 mm. long, teeth subulate, 1–2.5 mm. long; corolla 8–10 mm. long, magenta-purple; banner obovate, retuse, slightly exceeding wings; keel petals rounded at apex; pods sessile, ascending or spreading, ovoid-ellipsoidal, 5–15 mm. long, 5–8 mm. in diameter, finely strigose, flattened into a thin, triangular beak on terminal fourth or third; upper suture nearly straight.

Arid rocky slopes, Lower Sonoran Zone, western margin of the Colorado Desert to northwestern Sonora and south into Baja California along the western margin of the San Felipe Desert. Feb.–May.

**12*a*. Astragalus palmeri** var. **johnstonii** (Munz & McBurney) Barneby, comb. nov.*

*Astragalus vaseyi* var. *johnstonii* Munz & McBurney, in Munz, Bull. S. Calif. Acad. **31:** 66, 1932.

Stems and leaves often nearly glabrous; calyx almost wholly black-strigose, teeth one-fourth to one-third as long as tube; petals often ochroleucous tinged with lavender or light purple; pods 16–28 mm. long, 27–40-ovuled, scarcely flattened at tip.

Arid slopes and outwash bajadas, Upper and Lower Sonoran Zones, around head

* Combination and synonymy provided by R. C. Barneby.

of Coachella Valley and south to the Sierra San Pedro Mártir, Baja California, gradually merging with var. *palmeri* in the southern part of its range. Feb.–May.

**13. Astragalus piscinus** (M. E. Jones) Barneby, comb. nov.*

*Astragalus piscinus* M. E. Jones, Proc. Calif. Acad. II, **5:** 61, 1902, nom. provis.
*Phaca piscina* Rydb., N. Amer. Fl. **24:** 345, 1929.
*Astragalus crotalariae* var. *piscinus* Jepson, Fl. Calif. **2:** 350, 1936, as to name only.

Apparently perennial, thinly strigulose with straight subappressed hairs; leaves green, leaflets glabrous above; stems about 3 dm. long, ascending; stipules 3–5 mm. long, semi-amplexicaul, free; leaves 6–13 cm. long; leaflets 17–25, rather distant, linear-oblanceolate or linear-elliptic, obtuse and mucronulate, 7–30 mm. long; racemes borne on peduncles about 12–15 cm. long, about 20-flowered, loose; calyx tube 3.5 mm. long, loosely black- and white-strigulose, teeth subulate, 1.5 mm. long; petals purple; banner recurved through 45° angle, 10–12.8 mm. long; pod obliquely ellipsoid, bladdery-inflated, 2–3 cm. long, 10–12 mm. in diameter, contracted distally into a triangular-acuminate beak 5.5–7 mm. long; ventral suture nearly straight, dorsal one low-convex; valves papery-membranous, thinly strigulose; ovules 18–25.

Known only from the type locality, Lagoon Head, Baja California. Type collected in March.

**14. Astragalus idrietorum** Barneby, sp. nov.†

Short-lived perennial flowering the first season; stems erect or ascending, branching freely from base, 1.5–5 dm. long; herbage green, sparsely strigulose throughout; stipules deltoid, 2–4.5 mm. long, papery marginally, green in center; leaves 5–13 cm. long; leaflets 9–19, linear-lanceolate to narrowly lance-ovate, 5–23 mm. long, 1.5–4 mm. wide, obtuse to broadly acute at base, acute to short-apiculate at apex; peduncles 3–8 cm. long; racemes (8–)10–20-flowered, about equaling peduncles; calyx campanulate, tube 2.4–3.2 mm. long, strigose with white hairs only, or more commonly with some purplish or blackish hairs at base of teeth and bottom of tube, teeth subulate, 1.3–2.2 mm. long, purplish; corolla purple, banner 7–9 mm. long, blade reflexed about 45° to claw; pods ascending, sessile, papery, ovoid, 6.5–13 mm. in diameter, 1.2–2.2 cm. long, rounded at base, sparsely strigose with white hairs, faintly veined with pink or pale lavender, upper suture nearly straight, beak conical, 3–5 mm. long, obvious.

Along arroyos and on rocky hillsides, Lower Sonoran Zone, between Lat. 29° and 30° N., Baja California. March–April.

**15. Astragalus allochrous** A. Gray, Proc. Amer. Acad. **13:** 366, 1878

*Phaca allochroa* Rydb., Fl. Rocky Mts. 487, 1917.

Perennial from a short woody caudex; stems erect or ascending, 3–5 dm. tall, sparsely strigose, usually more or less purplish; stipules broadly deltoid, acute to acuminate, 2–5 mm. long, scarious, sparsely pubescent to subglabrate; leaves 10–12 cm.

---

* Synonymy and description contributed by R. C. Barneby.

† Description and Latin diagnosis prepared by R. C. Barneby.

**Astragalus idrietorum** sp. nov., *A. palmeri* S. Wats. proxime affinis, legumine tenui membranaceo subdiaphano ovulisque paucis (8–12, nec 12–31), vexillos minus recurvo absimilis. Ab *A. insularis* Kell. varietatibus sympatricis racemo plurifloro, ab *A. piscino* (Jones) Barneby pedicellis persistentibus, floribus paullo minoribus, ovulisque paucies facilius separata.—Estado BAJA CALIFORNIA (Norte): 10 miles s. of El Mármol, March 6, 1930, *Ira L. Wiggins 4377*. Holotypus, CAS; isotypi, DS, GH, NY, POM.—R. C. Barneby.

long, ascending, rachis strongly grooved above; leaflets 11–19, obovate to oblong, 2–4 mm. wide, 8–25 mm. long, rounded to retuse at apex, acute at base, sparsely strigose but green on both surfaces; peduncles 3–8 cm. long; racemes 3–10 cm. long, 8–15-flowered; bracts lance-subulate, strigose, 2–3 mm. long; calyx tube campanulate, 2–3 mm. long, purplish, strigose with white and black hairs; teeth narrowly subulate, 1.5–2 mm. long; corollas 6–8 mm. long, purplish; banner obovate, slightly exceeding wings; blade of wings slightly longer than claw; keel often ochroleucous, strongly curved, obliquely obovate, rounded at apex; pods sessile, ovoid, 2.5–4 cm. long, 1.5–2 cm. wide, strigulose, often slightly tinged with purple.

Desert plains and mesas, Lower Sonoran Zone, western and southern Arizona, into New Mexico and northern Sonora. March–May.

**16. Astragalus aridus** A. Gray, Proc. Amer. Acad. **6:** 223, 1864

*Tragacantha arida* Kuntze, Rev. Gen. Pl. **2:** 943, 1891.
*Astragalus albatus* Sheldon, Minn. Bot. Studies **1:** 128, 1894.
*Phaca arida* Rydb., N. Amer. Fl. **24:** 354, 1929.

Erect annual 1–3 dm. tall, densely white-strigose, stems well branched; leaves 1.5–7 cm. long, strigose; stipules deltoid-acuminate, 2–3 mm. long, erect; leaflets 9–15, oblong, acute to broadly cuneate at base, obtuse to retuse at apex, 1.5–4 mm. wide, 4–15 mm. long, white-strigose on both sides or slightly greenish above; peduncles 2–5 cm. long; racemes 4–9-flowered, lax; calyx tube campanulate, 2 mm. long, white-silky; teeth subulate, 1–1.5 mm. long; corolla ochroleucous, 5–6 mm. long; keel strongly arcuate, 3.6–5 mm. long, equaling banner; pods 10–17 mm. long, 4.5–7 mm. wide, papery, densely strigose-canescent, obliquely ovoid, upper suture straight, lower strongly curved; seeds 3–7, 2 mm. long, brownish.

Sandy desert plains and hillsides, Lower Sonoran Zone, Colorado Desert and western Arizona, northwestern Sonora and northern Baja California. March–May.

**17. Astragalus sabulonum** A. Gray, Proc. Amer. Acad. **13:** 368, 1878

*Astragalus virgineus* Sheldon ex Coville, Contr. U.S. Nat. Herb. **4:** 88, 1893.
*Phaca sabulonum* Rydb., Bull. Torrey Club **40:** 47, 1913.
*Phaca arenicola* Rydb., N. Amer. Fl. **24:** 356, 1929.
*Phaca lerdoensis* Rydb., loc. cit.

Annual or short-lived perennial with several decumbent to ascending branches 1–3.5 dm. long; stems, leaves, calyces, and pods densely cinereous-silky with short spreading hairs; stipules lance-deltoid, 1–3.5 mm. long, sparsely black-strigose on outer surface, glabrate or essentially so within; leaves 3–7 cm. long; petioles 0.5–2.5 cm. long; leaflets 9–15, oblong to oblong-lanceolate, 2–7 mm. wide, 8–14 mm. long, broadly cuneate at base, obtuse to rounded at apex, canescent with fine hairs on both surfaces; bracts subulate, 2–3 mm. long; peduncles 1–4 cm. long; racemes short, 3–4 (–6)-flowered; calyx tube 2–2.5 mm. long, narrowly campanulate, silky-strigose, often faintly reddish or purplish; teeth linear-subulate, 2–3 mm. long; corolla white, ochroleucous or diffused with lavender, or banner streaked with purple, 6–8 mm. long; keel obtuse, strongly arcuate, slightly shorter than banner; pods sessile, spreading or moderately deflexed, obliquely and narrowly ovoid, 7–8 mm. wide, 10–12 mm. long, densely cinereous-hirsute, abruptly narrowed to a short, conical beak 1–2.5 mm. long; seeds 6–28.

In light or sandy soil, Lower Sonoran Zone, Coachella and Imperial valleys, California, into extreme northwestern Sonora and adjoining Baja California; disjunctly from northwestern Arizona and southern Nevada to east-central Utah and northwestern New Mexico. Jan.–June.

**18. Astragalus nutans** M. E. Jones, Rev. N. Amer. Astrag. 108, *pl. 12, fig. 53,* 1923

*Astragalus deserticola* Jepson, Man. Fl. Pl. Calif. 565, 1925.
*Phaca nutans* Rydb., N. Amer. Fl. **24:** 342, 1929.
*Astragalus chuckwallae* Abrams, Ill. Fl. Pacif. States **2:** 586, *fig. 2834,* 1944.

Short-lived perennial flowering first season, with stems branching from base, 1–2 dm. long; herbage finely silvery-strigose throughout; stipules lance-subulate to narrowly deltoid, 3–4 mm. long, free, sparsely strigose with silvery hairs only or these intermingled with black hairs extending downward on stems about 5 mm. below base of petioles; leaves 3–8 cm. long; leaflets 7–13, oblong to oblanceolate-elliptic, 2.5–5.5 mm. wide, 5–15 mm. long, broadly cuneate at base, obtuse to rounded and minutely apiculate at apex; peduncles 2–5 cm. long; racemes loosely 6–10-flowered; calyx tube campanulate, 1.5–3 mm. long at anthesis, silvery- to black-strigose; teeth subulate, 1.5–2 mm. long; corolla purplish; banner 8–11 mm. long, its tip slightly reflexed; keel arcuate; pods spreading, sessile, ovoid, membranous, inflated, 10–12 mm. in diameter, 18–20 mm. long, sparsely and finely strigose, faintly tinged with purple, abruptly narrowed to a short conical beak 3–4 mm. long.

Sandy washes, Lower and Upper Sonoran Zones, Chuckwalla Mountains, Riverside County, California, north into eastern Mojave Desert. March–June; Oct.

**19. Astragalus insularis** Kell., Bull. Calif. Acad. **1:** 6, 1884, var. **insularis**

*Astragalus triflorus* var. *insularis* M. E. Jones, Proc. Calif. Acad. II, **5:** 637, 1895.
*Phaca insularis* Rydb., N. Amer. Fl. **24:** 353, 1929.

Annual with erect, ascending or spreading, branching stems 1–2 dm. long, sparsely strigose throughout, but green, not canescent; stipules deltoid-acuminate, 2–3 mm. long; leaves 4–7 cm. long, surpassing racemes; leaflets 9–17, lanceolate-oblong to narrowly elliptic, 1–3 mm. wide, 6–10 mm. long, glabrate and green above, sparsely strigose beneath, acute to cuneate at each end; peduncles 1–4 cm. long; racemes lax, 3–9-flowered; calyx strigose with black hairs, or a mixture of black and white hairs, tube campanulate, 1.5–2 mm. long, lobes linear-lanceolate or linear-subulate, equaling tube; corolla purple, 5–7 mm. long; banner obovate, equaling arcuate keel; wings obliquely oblanceolate, slightly shorter than banner, with acute auricles; pods sessile, sparsely strigose, 10–15 mm. long, 7–9 mm. wide, broadly and obliquely ovoid, tinged with red along upper suture or finely mottled with reddish purple; upper suture nearly straight; ovules 7–14.

Arid sandy plains and gentle slopes, mostly Lower Sonoran Zone, Cedros Island and peninsula of Baja California from San Quintín to Vizcaíno Desert. Feb.–May.

The hairs on the calyx vary from a few black ones scattered along the upper side to nearly all of the pubescence on the entire calyx.

**19*a*. Astragalus insularis** var. **harwoodii** Munz & McBurney, in Munz, Bull. S. Calif. Acad. **31:** 66, 1932

*Astragalus harwoodii* Abrams, Ill. Fl. Pacif. States **2:** 536, *fig. 2836,* 1944.

With habit of var. *insularis*; stems strigose-canescent, often purplish beneath pubescence; leaves 4–8 cm. long; leaflets often folded, 1.5–4 mm. wide, 8–18 mm. long, obtuse to emarginate at apex, strigulose on both surfaces; calyx tube 1.5–2.5 mm. long, white-strigose; banner exceeding keel about 1 mm., pod 15–20 mm. long.

Arid washes, flats, and gently sloping hillsides, Lower Sonoran Zone, central Colorado Desert, Imperial County, California, south into adjacent northern Baja California and to the vicinity of Altar and Punta Peñasca, Sonora. Feb.–April.

**19*b*. Astragalus insularis** var. **quentinus** M. E. Jones, Contr. West. Bot. **8:** 6, 1898

*Astragalus pondii* Greene, Pittonia **1:** 288, 1889.
*Astragalus insularis* var. *pondii* M. E. Jones, Rev. N. Amer. Astrag. 104, 1923.
*Phaca pondii* Rydb., N. Amer. Fl. **24:** 351, 1929.

Differing from var. *insularis* in having slightly longer, more conspicuously striate stems 1–5 dm. long; leaflets 9–21, to 20 mm. long, and subsymmetrically ovoid to subglobose pods with obscure beaks not over 2.5 mm. long, both sutures about equally convex; from var. *harwoodii* in being sparingly strigose, having acute leaflets, and smaller pods with equally convex sutures.

Sandy areas, margin of Upper and into Lower Sonoran Zones, vicinity of San Quintín, Baja California, south into the desert areas to about Lat. 30° N. Feb.–May.

**20. Astragalus leucopsis** Torr., Bot. Mex. Bound. 56, 1859

*Phaca canescens* Nutt., in Torr. & Gray, Fl. N. Amer. **1:** 344, 1838. Not *Astragalus canescens* DC., 1802.
*Tragacantha leucopsis* Kuntze, Rev. Gen. Pl. **2:** 946, 1891.
*Astragalus leucopsis* var. *lonchus* M. E. Jones, Rev. N. Amer. Astrag. 119, 1923.
*Phaca encenadae* Rydb., N. Amer. Fl. **24:** 336, 1929.

Perennial to 5 dm. tall with tomentose-canescent stems and herbage, stems particularly hoary with appressed, closely matted tomentum; stipules deltoid-subulate, 5–7 mm. long, often deflexed; leaves to 15 cm. long, petioles 8–15 mm. long; leaflets 10–20 pairs, oval to oblong, obtuse or subretuse at apex, 5–25 mm. long, 2–6 mm. wide, white-tomentose on both sides when young, at length green above, canescent beneath; racemes 4–10 cm. long, dense; peduncles often much longer than racemes; bracts lance-subulate, 2 mm. long; calyx tube 4–5 mm. long, canescent with white or sordid hairs; lobes narrowly subulate, 1.5–2.5 mm. long, acuminate; corolla ochroleucous, 11–12 mm. long; banner closely enfolding other petals, equaling or slightly surpassing them; stipes slender, 7–12 mm. long; body of pod ovoid-ellipsoidal, 2–3.5 cm. long, 1.2–2 cm. thick, sparsely strigose or sometimes glabrous, chartaceous-papery, very tardily dehiscent.

Plains and open hillsides, Lower and Upper Sonoran Zones, Santa Barbara County, California, south into Baja California, common to El Rosario, less frequent to about 25 miles southwest of Punta Prieta. March–June.

**21. Astragalus filipes** Torr., Proc. Amer. Acad. **6:** 226, 1864

*Astragalus stenophyllus* of authors, not T. & G., 1838.
*Tragacantha filipes* Kuntze, Rev. Gen. Pl. **2:** 944, 1891.
*Homalobus macgregorii* Rydb., Bull. Torrey Club **50:** 270, 1923.
*Astragalus filipes* var. *residuus* Jepson, Man. Fl. Pl. Calif. 571, 1925.
*Astragalus stenophyllus filipes* Tidestrom, Proc. Biol. Soc. Wash. **50:** 20, 1937.
*Astragalus macregorii* Tidestrom, loc. cit.
*Astragalus stenophyllus* var. *residuus* Barneby, El Aliso **2:** 203, 1950.

Perennial with several erect or ascending, slender stems 3–6 dm. long from a short woody base; herbage green, strigulose to glabrate, or less commonly cinereous-strigose, stems finely striate, sometimes purplish at base; lower stipules connate into a papery sheath, upper ones free, erect, triangular-subulate, thinnish, sparsely strigose to glabrate, about 3 mm. long; leaves 3–10 cm. long, ascending; leaflets 9–19, linear to narrowly oblong, 5–25 mm. long, rounded to retuse at apex, gray-green, finely strigose below, usually glabrate above; peduncles erect, 8–25 cm. long; racemes lax, 7–25-flowered, 3–15 cm. long, flowers spreading to nodding; pedicels slender, curved, 2–5 mm. long; calyx tube campanulate, 3–4 mm. long, thin, minutely strigose with very fine, white and blackish hairs; teeth deltoid, 0.5–1 mm. long; corolla ochroleucous, 10–12 mm. long; banner exceeding keel 1–3 mm., slightly arched backward; pods pendulous, compressed laterally, linear-oblong, 4–5 mm. wide, 1.5–2.5 cm. long, straight, abruptly acute at each end, valves papery, glabrous to puberulent, lustrous; stipe filiform, 8–15 mm. long.

Arid slopes, mainly Upper Sonoran Zone and lower part of Arid Transition Zone, British Columbia to Idaho and northern California, disjunctly in interior mountains of California from Mount Pinos, Ventura County, to Santa Rosa Mountains in eastern San Diego County, where it enters the Sonoran Desert, and to northern Baja California. April–May.

**22. Astragalus coccineus** (Parry) Brandegee, Zoe **2:** 72, 1891

*Astragalus purshii* var. *?coccineus* Parry, W. Amer. Sci. **7:** 10, 1890.
*Xylophacos coccineus* Heller, Muhlenbergia **2:** 217, 1906.

Caespitose perennial; rootstock heavy; stems stout, short, erect, or spreading-decumbent; herbage white-woolly; leaves densely crowded in tufts, 5–10 cm. long, ascending; petioles usually equaling rachises; stipules lanceolate, 4–6 mm. long; leaflets 9–17, elliptic to broadly obovate, acute, obtuse, or rounded at apex, 5–9 mm. wide, 6–12 mm. long; peduncles equaling or shorter than leaves; racemes subcapitate, 2–8-flowered; bracts linear-lanceolate, 6–10 mm. long; calyx tube cylindrical, 3.5–4.5 mm. wide, 12–15 mm. long, villous; teeth subulate, 4–9 mm. long, usually with some black hairs near tips; corolla crimson, exserted 2–2.5 cm. beyond calyx; banner arched 1–1.5 cm. above nearly straight wings and keel; pods oblong, 2.5–3.5 cm. long, slightly curved upward, densely long-villous with yellowish or whitish hairs 2–3 mm. long.

Rocky or sandy soil, Upper Sonoran and upper fringes of Lower Sonoran Zone, Inyo County south through the desert ranges into northern Baja California, and Kofa Mountains, western Arizona, April–May.

**Astragalus newberryi** A. Gray, Proc. Amer. Acad. **12:** 55, 1876, with leaflets 5–7 or rarely to 9; corolla tinged with purple; pods 1.5 cm. wide, 2 cm. long, slightly sulcate on both sutures, occurs in Arizona north of the Sonoran Desert and may enter our area as a straggler in Gila County.

**23. Astragalus pachypus** Greene var. **jaegeri** Munz & McBurney, in Munz, Bull. S. Calif. Acad. **31:** 67, 1932

Perennial with erect or ascending stoutish stems 2–6 dm. long, base woody; branches strigose and often branching above as well as at base; stipules ovate-deltoid, about 1.5 mm. long, ascending, sparsely strigose to glabrate; leaves 10–15 cm. long; leaflets 17–27, narrowly linear, 1.5–2.5 mm. wide, 8–25 mm. long, rounded to retuse at apex, strigose-cinereous on both surfaces; peduncles 1–2 dm. long; racemes lax, 5–10 cm. long; bracts lance-subulate, 1–1.5 mm. long; pedicels stout, about 2 mm. long; calyx tube broadly cylindro-campanulate, about 4 mm. long, densely strigose with white and some black hairs; teeth subulate, 1.5–2 mm. long; corolla yellow, 15–17 mm. long; keel 4–5 mm. shorter than banner, moderately recurved; pods thick-coriaceous, 5–7 mm. wide, 5–25 mm. long, strongly compressed, completely 2-celled, glabrous, transversely wrinkled, acute at both ends, sutures prominent, cordlike; stipe 4–8 mm. long, slender.

Sandy or gravelly valleys and hillsides, mainly Upper Sonoran Zone, but entering the upper edge of the Lower Sonoran Zone in western Riverside County, near Beaumont and Banning, California. March–May.

**24. Astragalus tricarinatus** A. Gray, Proc. Amer. Acad. **12:** 56, 1876

*Hamosa tricarinata* Rydb., Bull. Torrey Club **54:** 20, 1927.

Low, bushy perennial; stems flexuous, 1–2.5 dm. long, shorter than inflorescences, sparsely strigulose, green; stipules free, reflexed, deltoid-apiculate, 1.5–2 mm. long; leaves 7–20 cm. long, ascending; leaflets 17–27, distant, elliptic to narrowly obovate, often retuse, 3–12 mm. long, green and sparsely strigose beneath, closely silvery-strigose above, deciduous from stiff, persistent petioles and rachises; peduncles stout, erect, 8–20 cm. long; racemes loosely 6–17-flowered, 5–18 cm. long in fruit; calyx tube broadly campanulate, 4–5 mm. long, 3–3.5 mm. in diameter; strigose with coarse black and silvery hairs; teeth slender, 2–3 mm. long; petals white or pale cream, drying ochroleucous; banner 13–16 mm. long; wings 11–12 mm. long, keel 9–10 mm. long; pods moderately arcuate, erect or ascending, borne on, and easily detached from, a short gynophore 1–2.5 mm. long, body 4–5.5 mm. wide, 25–42 mm. long, obtusely triquetrous, fully 2-celled by intrusion of sutures, stramineous, glabrous and sublustrous at maturity.

Gravelly arroyos, canyons, and rocky hillsides, Lower Sonoran Zone, northwestern Colorado Desert, California. Feb.–May.

**25. Astragalus mohavensis** S. Wats., Proc. Amer. Acad. **20:** 361, 1885

*Brachyphragma mohavense* Rydb., N. Amer. Fl. **24:** 400, 1929.

Annual or short-lived perennial with densely silvery-strigulose herbage; stems diffuse, mostly 0.5–3.5 dm. long, sometimes flowering when only 1–2 cm. tall; stipules triangular-lanceolate, 1–3.5 mm. long, semi-amplexicaul; leaves 2–10 cm. long; leaflets 3–11, rhombic-elliptic to suborbicular, obtuse or emarginate, 3–18 mm. long; peduncles 1.5–7 cm. long; racemes loosely 3–16-flowered; calyx tube white- or both black- and white-strigulose, 2.5–4.5 mm. long, teeth subulate, 1.8–2.8 mm. long; petals purple, banner recurved through about 45°, 9–12.5 mm. long; pod broadly and turgidly oblong to oblong-ellipsoid in profile, sometimes narrowly clavate, straight or a trifle incurved, abruptly cuspidate at apex, subterete and fleshy when first formed, becoming either bicarinate by sutures or obscurely trigonous and then carinate by ventral suture and depressed dorsally, densely stigulose but eventually brownish, leathery, bilocular.

Rocky slopes and flats, in boulder-strewn washes, and on rock ledges of canyon walls, Lower and Upper Sonoran Zones, mostly from 2,500 to 7,500 feet elevation, northern Colorado Desert to Death Valley and southern Nevada. March–May.

**26. Astragalus orcuttianus** S. Wats., Proc. Amer. Acad. **20:** 361, 1885

*Astragalus orcuttianus* var. *gregorianus* M. E. Jones, Contr. West. Bot. **10:** 63, *pl. 12,* 1902.
*Tium orcuttianum* Rydb., N. Amer. Fl. **24:** 397, 1929.

Compact perennial from a slender taproot with many ascending stems 2–3.5 dm. tall; herbage green, sparsely strigose with short, stoutish, white hairs; stipules broadly ovate to deltoid, 1.5–2.5 mm. long; leaves numerous, 10–18 cm. long; leaflets 15–25, obovate to nearly orbicular, acute to obtuse at base, rounded or faintly retuse at apex,

2–8 mm. long; peduncles ascending, equaling leaves; racemes 3–10 cm. long, laxly 8–13-flowered, elongating in fruit; pedicels 1.5–2.5 mm. long, strigose with white and black hairs; calyx tube campanulate, 3–3.5 mm. long, strigose with mixture of black and white hairs; teeth deltoid-lanceolate, 1.6–2.2 mm. long; corolla pale blue, 10–12 mm. long; banner reflexed slightly, 2–2.5 mm. longer than wing; keel strongly curved upward, 1–1.5 mm. shorter than wing petals; pods sessile, lunately curved, 1.5–2 cm. long, 2.5–3 mm. in diameter, deeply sulcate beneath; placentae deeply intruded, forming a nearly complete partition; valves very firm, dehiscing first at apex, stramineous, glabrate.

Desert washes and rocky slopes, Lower Sonoran Zone, eastern side of Sierra Juarez and Sierra San Pedro Mártir, Baja California. March–May.

**27. Astragalus arizonicus** A. Gray, Proc. Amer. Acad. **7:** 398, 1868

*Tragacantha arizonica* Kuntze, Rev. Gen. Pl. **2:** 943, 1891.
*Hamosa arizonica* Rydb., Bull. Torrey Club **54:** 22, 1927.

Prostrate to decumbent perennial from a tough taproot; stems many, more or less flexuous, 1–5 dm. long, finely white-strigose; stipules broadly deltoid, 1.5–3 mm. long, usually as wide; leaves 5–10 cm. long, ascending; leaflets 9–17, linear, oblong, or lance-oblong, 2–3 mm. wide, 8–15 (30) mm. long, mostly acute at apex, finely silky-strigose on both surfaces; peduncles 4–15 cm. long, canescent; racemes 3–8 cm. long, lax, several- to many-flowered; bracts lanceolate, 2.5–3 mm. long, villous without; calyx tube cylindro-campanulate, 3–4 mm. long, white-strigose; teeth deltoid, subulate-acuminate, about 2 mm. long; corolla white, more or less suffused with greenish yellow, purplish, or maroon, 9–11 mm. long; banner obovate, arched but not reflexed; wings slightly shorter than banner and keel, falcate, obtuse at apex; keel broadly lunate, apex blunt, strongly arcuate; pods linear, ascending, 1.5–3 cm. long, about 3 mm. wide, slightly arcuate, flat to slightly sulcate along lower suture, finely white-strigose.

Grassy hillsides and adobe flats, Sonoran Zones, northern Arizona into central Sonora, east into New Mexico. March–May.

**28. Astragalus nuttallianus** DC. var. **austrinus** (Small) Barneby, comb. nov.*

*Hamosa austrina* Small, Fl. SE. U.S. 618, 1332, 1903.
*Astragalus nuttallianus* var. *canescens* Torr. & Gray ex S. Wats., Bibl. Ind. 198, 1878, nom. confus.

Slender annual with strigose stems 1–5 dm. long; stipules deltoid-lanceolate, 2–5 mm. long; leaves 2–5 cm. long; leaflets 5–17, mostly oval to elliptic or oblong, sometimes linear, 3–12 mm. long, strigose on both surfaces or sometimes glabrous above, acute to obtuse at apex; peduncles 3–10 cm. long, slender, mostly surpassing leaves; racemes subcapitate, compact, 3–8-flowered; bracts subulate to lanceolate, 1–2 mm. long; calyx tube broadly campanulate, 1.5–2.5 mm. long, strigose; teeth subulate, 1.8–2.8 mm. long; corolla white or tips tinged with purple, 4–7 mm. long; banner obovate, retuse, slightly exceeding wings; keel petals 4–5.5 mm. long, broadly lunate, acute; pods horizontally spreading to ascending, moderately arcuate, 1.5–2 cm. long, 2–2.5 mm. wide, about 3 mm. deep, strigose to glabrous; septum complete.

Arid plains and hillsides, along arroyos, Lower and Upper Sonoran Zones, northeastern Arizona to Texas, south into northern Sonora. Feb.–May; Sept.–Oct.

**28a. Astragalus nuttallianus** var. **imperfectus** (Rydb.) Barneby, Leafl. West. Bot. **3:** 109, 1942

*Hamosa imperfecta* Rydb., Bull. Torrey Club **54:** 329, 1927.

* Combination and synonymy furnished by R. C. Barneby.

Differing from var. *austrinus* in having shorter, more closely appressed to slightly ascending hairs 0.35–0.7 (rarely 0.8) mm. long on calyx, teeth short (only 1–1.7 mm. long); pods often imperfectly bilocular owing to partial inflexing of sutures; from var. *cedrosensis* in having all leaflets elliptic, nearly of same size on upper and lower leaves; calyx tube longer (1.9–2.8 mm. long) in contrast to 1.4–1.7 mm. long in var. *cedrosensis*.

Gravelly and stony hillsides, Lower Sonoran Zone, western and central Arizona to the Mojave Desert, southern Utah, and southern Nevada. March–June.

**28*b*. Astragalus nuttallianus** var. **cedrosensis** M. E. Jones, Rev. Gen. Astrag. 270, *pl. 68,* 1923

*Astragalus pertenuis* Greene, Leaflets Bot. Obs. **2:** 42, 1910.
*Hamosa pertenuis* Rydb., Bull. Torrey Club **54:** 329, 1927.

Slender annual with habit of var. *austrinus;* stems nearly filiform; leaflets of lower leaves 2–7, cuneate-obcordate, retuse, 3–5 mm. long, those of upper leaves often acute, 5–10 mm. long; calyx tube shallowly campanulate, 1.9–2.8 mm. long, about as wide; calyx teeth 1–1.6 mm. long; pods 1.5–2 mm. wide, 10–12 mm. long, strigose to glabrate; septum present, usually complete.

Desert flats, rocky arroyos, and open hillsides, Lower Sonoran Zone, western edge of the Colorado Desert, south into northern Sonora and to the Cape Region, Baja California. Jan.–May; Sept.–Oct.

**29. Astragalus nothoxys** A. Gray, Proc. Amer. Acad. **6:** 232, 1864

*Tragacantha nothoxys* Kuntze, Rev. Gen. Pl. **2:** 946, 1891.
*Oxytropis nothoxys* M. E. Jones, Proc. Calif. Acad. II, **5:** 677, 1895.
*Spiesia nothoxys* M. E. Jones, loc. cit.
*Aragallus nothoxys* Heller, Cat. N. Amer. Pl. 4, 1898.
*Hamosa gooddingii* Rydb., Bull. Torrey Club **54:** 20, 1927.
*Hamosa nothoxys* Rydb., op. cit. 330.
*Astragalus gooddingii* Tidestrom, Proc. Biol. Soc. Wash. **48:** 40, 1935.

Annual or short-lived perennial; stems several, decumbent or prostrate, 1–4 dm. long; herbage green and sparingly strigose, or youngest leaves cinereous; stipules lanceolate, 2–4 mm. long, ciliate-margined; leaves 3–8 cm. long; leaflets 13–19, oval to obovate, 2–5 mm. wide, 4–10 mm. long, glabrous and bright green above, paler and strigulose beneath, rounded to retuse at apex; peduncles 5–15 cm. long; racemes 2–8 cm. long, compact in flower, elongating in fruit; bracts lance-subulate, 2 mm. long; calyx tube campanulate, 4–5 mm. long, strigose with white and often a few black hairs; teeth subulate, 1.5–2 mm. long; corolla white or petals purple-tipped, 8–10 mm. long; banner obovate, retuse, abruptly curved upward, slightly exceeding wings; keel petals semiorbicular, abruptly narrowed to an acuminate, erect beak about 0.75 mm. long, much shorter than wings and banner; pods ascending to erect, 3–3.5 mm. wide, 15–20 mm. long, slightly arcuate, deeply sulcate along lower side, finely strigose, acuminate; beak 1.5–2.5 mm. long, straight.

Lower mountain slopes and arroyos, mainly Upper Sonoran Zone but reaching the upper edges of the Lower Sonoran Zone, Pima County, Arizona, into northern Sonora and Chihuahua. March–May.

**30. Astragalus acutirostris** S. Wats., Proc. Amer. Acad. **20:** 360, 1885

*Oxytropis acutirostris* M. E. Jones, Proc. Calif. Acad. II, **5:** 677, 1895.
*Aragallus acutirostris* Heller, Cat. N. Amer. Pl. 4, 1898.
*Hamosa acutirostris* Rydb., Bull. Torrey Club **54:** 331, 1927.
*Astragalus nuttallianus* var. *acutirostris* Jepson, Fl. Calif. **2:** 379, 1935.

Decumbent to prostrate annual with several–many, slender, striate stems 1–3 dm. long; herbage cinereous with closely appressed hairs; stipules deltoid, membranous, 1.5–2 mm. long, often somewhat broader than long, sparsely strigose or glabrate; leaves 1.5–2.5 cm. long, petioles 5–8 mm. long; leaflets 9–13, oblong-oblanceolate, 2–3 mm. wide, 4–10 mm. long, acute to cuneate at base, usually retuse at apex, cinereous with stiffish, ascending, subappressed hairs; peduncles slender, 3–7 cm. long, spreading or ascending; racemes lax, 2–8-flowered; calyx tube shallowly campanulate, 1.5–2 mm. long, pubescent with black hairs along angles and on teeth, white hairs between angles; teeth lance-deltoid to subulate, about equaling tube; corolla white or purple-tipped, about 5 mm. long; banner slightly arched upward, barely exceeding keel; keel short, acute, tip curved upward; pods spreading to ascending, linear, slightly arcuate, 2.5–3 mm. deep, 12–18 mm. long, deeply sulcate along lower suture, finely and sparsely strigose.

Arid desert slopes and washes, Lower Sonoran Zone, southern Nevada and Inyo County, California, south through the Mojave Desert, infrequent along the western margin of the Colorado Desert into northern Baja California. April–May.

**31. Astragalus didymocarpus** Hook. & Arn., Bot. Beechey, 334, 1828, var. **didymocarpus**

*Astragalus catalinensis* Nutt., Jour. Acad. Phila. II, **1:** 152, 1848.
*Tragacantha didymocarpa* Kuntze, Rev. Gen. Pl. **2:** 944, 1891.
*Hesperastragalus didymocarpus* Heller, Muhlenbergia **2:** 87, 1905.
*Hesperastragalus compactus* Heller, op. cit. **2:** 218, 1906.
*Hesperastragalus catalinensis* Rydb., Bull. Torrey Club **53:** 169, 1926.

Slender annual branching from base; stems 0.5–3 dm. long, sparsely strigose with ascending hairs; stipules lance-deltoid to ovate-deltoid, 1–2 mm. long, ciliate; leaves 1.8–6 cm. long, rachis and petiole slender; leaflets 11–17, narrowly oblong to linear, 3–10 mm. long, 1–2 mm. wide, emarginate at apex, finely strigose on both surfaces or glabrate above; peduncles slender, 1–4 cm. long; racemes compact, spiciform, 1–2.5 cm. long, many-flowered; bracts lanceolate, about 1 mm. long, black-strigose on back; calyx tube broadly campanulate, 1–2.5 mm. long, strigose with many black and a few white hairs scattered among black; teeth deltoid, about equaling tube; corolla pale blue-purple, 3–6 mm. long; pods ovoid to subglobose, 2.5–4 mm. long, 2–2.5 mm. in diameter, deeply didymous, gray or greenish gray, strongly wrinkled transversely or diagonally, puberulent with spreading to subappressed hairs; each cell 1-seeded.

In both heavy and light soils, coastal benches, inland arroyos, mesas and hillsides, Lower and Upper Sonoran Zones, Contra Costa County, California, to eastern Mojave Desert and western Nevada, south to northern Sonora and at least to vicinity of Rosario, Baja California. Feb.–May.

**31*a*. Astragalus didymocarpus** var. **dispermus** (A. Gray) Jepson, Fl. Calif. **2:** 376, 1936

*Astragalus dispermus* A. Gray, Proc. Amer. Acad. **13:** 365, 1878.
*Hesperastragalus dispermus* Heller, Muhlenbergia **1:** 137, 1906.
*Astragalus dispermus* var. *albus* L. E. James, Contr. Dudley Herb. **4:** 68, 1951.

Habit of var. *didymocarpus;* herbage silvery-strigose; stipules 1–1.5 mm. long; leaves 2–4 cm. long; leaflets 5–13, cuneate-obovate to oblong-obovate, 1–3.5 mm. wide, 3–8 mm. long, retuse; bracts subulate, 1.5–2 mm. long, greenish, nearly glabrous; calyx tube 1–1.5 mm. long, densely villous with white hairs; teeth subulate, 1.5–2.5 mm. long; corolla white, bluish, or purple, often obscured by villous calyx; pods ovoid, about equaling or slightly exceeding calyx, slightly longer than broad, deeply sulcate, strongly wrinkled transversely or diagonally, short-pubescent or rarely glabrate.

Arid sandy soils, Upper and Lower Sonoran Zones, mostly in desert regions but crossing the coastal ranges to approach the coast in southern California; Colorado and Mojave Deserts, east to Pima County, Arizona, and south in Baja California to desert hills south of Sierra San Pedro Mártir. April–May.

## 25. ALHAGI Adans., Fam. Pl. 2: 328, 1763

Rigid, much-branched shrubs with stiff axillary spines and simple, entire, small leaves. Stipules minute. Flowers red, in axillary racemes, rachises spinose at tip, rigid. Bracts minute. Calyx turbinate, with subequal, broadly deltoid, short teeth. Banner obovate, short-clawed, apex somewhat recurved; wings oblong, falcate; keel incurved, obtuse. Stamens diadelphous, anthers similar. Ovary subsessile, few- to many-ovuled. Style filiform; stigma terminal, minute. Pod linear-moniliform, terete to lightly compressed, indehiscent. Seeds reniform.

### 1. **Alhagi camelorum** Fisch., Hort. Gorenki ed. 2, 72, 1812

Low shrub to 1 m. tall from extensive, deep-seated roots; branches pale green, glabrous, finely striate; spines slender, 1–2.5 cm. long; stipules reddish, 1 mm. long or less; petioles 1–2.5 mm. long, glabrous or sparsely puberulent when young; leaves linear, elliptic, or elliptic-oblanceolate, 8–20 mm. long, acute to rounded and tipped with a short recurved apiculation at apex, finely and sparsely puberulent in youth, soon glabrate above; flowering racemes 1–4 cm. long, usually 4–6-flowered, spine-tipped, spreading; pedicels 2–3 mm. long; calyx about 2 mm. long, teeth about 0.2 mm. long; corolla 7–9 mm. long, red or red-purple; pod stipitate, 1–3 cm. long, 3–4 mm. in diameter over the few seeds, strongly constricted between them.

Along banks of canals and streams and on stony or sandy soil at scattered localities, Lower Sonoran Zone, Imperial and Riverside Counties, California, and in Navajo, Coconino, and Mariposa Counties in Arizona. May–July.

Not as yet extensively established. Viewed with apprehension because it might become a serious pest if once firmly entrenched.

## 26. AESCHYNOMENE L., Sp. Pl. 713, 1753

Perennial or annual herbs or small, suffrutescent, nonclimbing plants with odd-pinnate leaves and setaceous to lanceolate stipules. Flowers in axillary racemes or rarely solitary, yellow, often with purplish stripes. Calyx somewhat 2-lipped but lobes (in ours) subequal. Banner orbicular, short-clawed; wings obovate, broad, nearly equaling banner; keel obovate and slightly curved or narrow and strongly curved or subrostrate. Stamens diadelphous. Ovary stipitate, 2–many-ovuled. Style curved upward, glabrous; stigma small, terminal. Pod flattened, undulate-articulated along lower side and breaking into 2–several, 1-seeded, hairy loments.

### 1. **Aeschynomene nivea** Brandegee, Proc. Calif. Acad. II, 2: 150, 1889

A silvery-sericeous shrub to 1 m. tall, with erect or ascendingly spreading, slender branches; stipules lance-subulate, 3–5 mm. long, appressed; leaves 3–4.5 cm. long; petioles 2–4 mm. long; leaflets about 21–51, linear-lanceolate, 1–1.5 mm. wide, 3.5–10 mm. long, midvein eccentric, rounded at base, obliquely acute and mucronulate at apex; flowers in 2- to few-flowered axillary clusters, yellow, striped or suffused with purple, 7–10 mm. long; pedicels slender, 1–1.5 cm. long; calyx campanulate, tube about 3 mm. long, teeth deltoid, subequal, 1.5–2 mm. long; banner erect, 6–7 mm. high, puberulent on back, purple-striped; pod 1–3-jointed, 4–5 mm. wide, 1.5–2.5 cm. long, finely and densely puberulent.

Rocky places, Lower Sonoran Zone, central Baja California to the Cape Region. Nov.–April.

**Aeschynomene vigil** Brandegee, Proc. Calif. Acad. II, **3:** 128, 1891, differing from the above in having 7–11 leaflets with central instead of eccentric midveins and smaller purple flowers, was described from San José del Cabo. It may enter the southern fringes of the desert in Baja California, but no specimens from such areas have been seen.

**Aeschynomene americana** L., Sp. Pl. 713, 1753, an annual herb with slender, erect stems and sparsely pubescent green herbage, occurs at altitudes of about 3,500 feet in Santa Cruz County, Arizona, and may get into the Sonoran Desert at times as a waif.

## 27. STYLOSANTHES Swartz, Prod. Veg. Ind. Occ. 108, 1788

Perennial herbs or subshrubs with digitately trifoliolate leaves and adnate striate stipules. Flowers small in dense terminal or axillary spikes, each subtended by a foliaceous bract, or solitary. Calyx tube slender, bilabiate, upper lip 4-parted, lower lip entire and elongate. Corolla yellow, banner suborbicular; wings oblong; keel incurved, shorter than wings. Stamens monodelphous, tube split on upper side. Ovary sessile. Pod a loment, usually with 2 segments, one often reduced and sterile or sometimes absent.

### 1. Stylosanthes viscosa Swartz, Prod. Veg. Ind. Occ. 108, 1788

*Stylosanthes prostrata* M. E. Jones, Contr. West. Bot. **15:** 135, 1929.

A prostrate to ascending, much-branched, suffrutescent perennial with branches 1–4 dm. long, scaberulous-pubescent and viscid-glandular throughout; stipules ovate-acuminate, 5–9 mm. long; petioles about equaling or slightly surpassing stipules; leaflets elliptic or narrowly elliptic-ovate or narrowly obovate, 3–5 mm. wide, 10–15 mm. long, acute to mucronate; flowers in terminal and axillary spikes 1–1.5 cm. long; bracts ovate-acuminate, 7–9 mm. long; pedicels slender, 3–4 mm. long, glabrous; calyx 2.5–3 mm. long, lower lip about three times as long as teeth of upper lip; corolla 7–8 mm. long, yellow; pod usually reduced to a single, obovate, somewhat quadrate, strongly veined loment 3–4 mm. long, style persistent, strongly curved to tightly coiled.

Mainly in sandy soil, Lower Sonoran to Tropical Zones, southern Baja California and Sinaloa; widely distributed in the American tropics. Feb.–Sept.

**Stylosanthes biflora** (L.) B.S.P. was distributed by *W. F. Parish, No. 314,* from "Arizona or New Mexico," but it is doubtful that this plant enters the Sonoran Desert. It has not been found in Arizona by recent collectors.

## 28. ZORNIA J. F. Gmel., Syst. Nat. 2: 1076, 1791

Perennial herbs or subshrubs with digitately compound leaves with 2–4 pellucid-dotted leaflets and broad, rigid, peltate stipules. Flowers small in terminal or lateral spicate racemes, each usually subtended by a pair of stiff, broad, connivent bracts. Calyx bilabiate, the upper lip emarginate, the lower 3-toothed. Corolla small, orange or yellow; banner suborbicular, the sides reflexed; wings oblong; keel curved upward.

Stamens monodelphous, with alternate anthers smaller than the others. Pods sessile, strongly compressed, 2–6-jointed, often hispid.

**1. Zornia reticulata** Sm., in Rees, Cycl. **39:** 2, 1819

*Zornia diphyllum* of American authors, not Pers., 1807.

(For extensive synonymy see Mohlenbrock, Webbia **16:** 90–91, 1961.)

Herbaceous, erect or ascending stems 1–5 dm. tall from woody root crowns, stems and leaves essentially glabrous or sparsely hispid; stipules of lower leaves narrowly elliptic, peltate, 1–2 mm. wide, 6–8 mm. long, attached a little below middle, those of inflorescence larger (3–8 mm. wide; 10–18 mm. long), connivent and nearly enclosing flowers, petioles and leaflets in inflorescence usually reduced so stipules are bractlike; petioles of lower leaves slender, 1–2 cm. long; leaflets 2, narrowly ovate, lanceolate, or linear, 1.5–8 mm. wide, 1.5–6 cm. long, glabrous, or ciliate marginally, or sparsely hispidulous on one or both surfaces, pellucid-dotted; flowers mainly solitary, 8–10 mm. long; calyx about 4 mm. long, strongly veined and ciliate-hispid; corolla yellow; pods 4–6-jointed, 1–1.5 cm. long, 2–2.4 mm. wide, bearing numerous reddish, retrorsely barbed spines about 0.5 mm. long.

Gravelly slopes and rocky hillsides and mesas, Lower Sonoran to Tropical and Transition Zones, southern Arizona to South America. Aug.–Oct.

## 29. DESMODIUM Desv., Jour. Bot. 1: 122, *pl. 5,* 1813

Annual or perennial herbs, suffrutescent plants or less commonly herbs with herbage more or less pubescent with straight or uncinate hairs or the vestiture consisting of both types of trichomes. Leaves usually trifoliolate but sometimes unifoliolate, stipulate. Flowers small to medium in size, often showy, paniculately or racemosely arranged; calyx short, more or less 2-lipped and each lip 2- or 3-dentate; banner oblong, ovate, or orbicular; wings oblong, oblique; keel nearly straight, obtuse at apex. Stamens 10, monodelphous or diadelphous. Ovary 2- to many-ovulate. Fruit a flat loment of 2 to several joints, these usually coriaceous and commonly more or less puberulent with uncinate hairs, plane or revolute-involute along margins, usually indehiscent but readily separating from one another, each joint usually 1-seeded.

Leaves unifoliolate:
- Stems moderately puberulent to glabrescent; fruits sessile; leaflets narrowly linear, grasslike, 2–5 mm. wide . . . . . . . . . . 1. *D. angustifolium* var. *gramineum*
- Stems finely but densely uncinulate-puberulent; fruits on stipes about equaling lowest joint; leaflets ovate to ovate-lanceolate, 10–28 mm. wide . . . . . . . . 2. *D. psilophyllum*

Leaves trifoliolate, or predominantly so:
- Bracts ovate-acuminate, densely imbricate and conspicuous before anthesis; stems trigonous . . . . . . . . . . 3. *D. intortum*
- Bracts mostly lanceolate to subulate, not markedly conspicuous before anthesis; stems terete or several-angled:
  - Fruits less deeply notched above than below, not moniliform; banner 6–12 mm. long; flowers relatively conspicuous:
    - Banner 10–12 mm. long; leaflets orbicular or nearly so, 2.5–6 cm. broad; stipules 10–14 mm. long . . . . . . . . . . 4. *D. prostratum*
    - Banner 6–8 mm. long; leaflets ovate to lanceolate, 5–28 mm. wide; stipules 2–5 mm. long:
      - Stipules 2–3 mm. long; stems sparsely puberulent; joints of fruit glabrate or sparsely puberulent . . . . . . . . . . 5. *D. batocaulon*

Stipules 4–5 mm. long; stems densely uncinate-pubescent; joints of fruits densely puberulent .......... 6. *D. prehensile*

Fruits equally notched along both margins; banner 2–5 mm. long; flowers inconspicuous:

Segments of fruit not revolute or involute; loment not appearing twisted:

Leaflets ovate, 1–6 cm. wide; stipules of lower leaves ovate-lanceolate, attenuate, 8–15 mm. long, margins somewhat dentate; joints of fruits 6–10 mm. long .......... 7. *D. psilocarpum*

Leaflets linear to linear-lanceolate, 1.5–6 mm. wide; stipules subulate, 1–3.5 mm. long, entire; joints of fruits 3–4 mm. long .......... 8. *D. rosei*

Segments of fruit revolute-involute; loment appearing more or less twisted:

Plants perennial; stipules broadly ovate but markedly attenuate, 6–12 mm. long, strongly veined, margins pilose .......... 9. *D. tortuosum*

Plants annual; stipules linear-subulate to narrowly ovate-attenuate, 6 mm. long or less, weakly veined, rarely pilose:

Terminal joint of fruit triangular, 7–8.5 mm. long, much larger than others, plane; leaves nearly glabrous .......... 10. *D. scopulorum*

Terminal joint of fruit rhombic, elliptic or orbicular, 3–4 mm. long, not longer than others, usually contorted; leaves sparsely to densely puberulent:

Basal leaves mostly trifoliolate; terminal leaflet not wider than long .......... 11. *D. procumbens* var. *procumbens*

Basal leaves often unifoliolate; if 3-foliolate terminal leaflet transversely elliptic or rhombic, broader than long:

Fruit appearing spirally twisted; upper leaflets lance-linear to narrowly oblong; corolla 3–4 mm. long .......... 11*a*. *D. procumbens* var. *exiguum*

Fruit not markedly spirally twisted; upper leaflets lanceolate to ovate; corolla 2.5–3 mm. long .......... 12. *D. neo-mexicanum*

**1. Desmodium angustifolium** DC. var. **gramineum** (A. Gray) Schubert, Contr. Gray Herb. **129:** 28, 1940

*Desmodium gramineum* A. Gray, Pl. Wright. **2:** 46, 1853.
*Meibomia graminea* Kuntze, Rev. Gen. Pl. **1:** 198, 1891.

Slender, sparsely branched, erect or ascending annual with few leaves and sparsely to moderately uncinate-puberulent herbage; leaves unifoliolate; leaflet linear and grass-like, 2–5 mm. wide, 2–12 cm. long, obtuse at base and apex, subcoriaceous, minutely uncinate-hispidulous on both surfaces, sparsely strigose on veins beneath; petiole stout, 6–8 mm. long; racemes lax, rachis sparsely hirtellous; flowers purple, 2–3 mm. long, inconspicuous, on slender pedicels 10–15 mm. long; loment sessile or essentially so, 5–7-jointed, uncinate-hispid and sparsely strigose, shallowly crenate-lobed on both margins, not twisted, joints symmetrical, about 2.5 mm. wide.

Grassy hillsides and openly wooded slopes, mainly Upper Sonoran Zone, but entering the Sonoran Desert sparingly along the upper margins of the Lower Sonoran Zone, Pima, Cochise, and Santa Cruz Counties, Arizona, southward to northern South America. Aug.–Sept.

**2. Desmodium psilophyllum** Schlecht., Linnaea **12:** 310, 1838

*Desmodium wrightii* A. Gray, Bost. Jour. Nat. Hist. **6:** 177, 1850.
*Meibomia psilophylla* Kuntze, Rev. Gen. Pl. **1:** 198, 1891.
*Meibomia wrightii* Kuntze, loc. cit.

Erect, slenderly branched shrubs 3–8 dm. tall, stems terete, minutely uncinate-pubescent in youth, glabrate with pale brownish, faintly reticulate bark when older;

leaves unifoliolate, lance-oblong to nearly linear-lanceolate, 1–2.8 cm. wide, 2–7 cm. long, apex obtuse or less commonly acute, base broadly rounded to subtruncate, dark green and nearly glabrous above, pale green to subglaucous and sparsely pubescent along distinct veins beneath; petioles slender, 4–8 mm. long, puberulent; stipules lance-subulate, 1–2 mm. long, strongly striate; inflorescence racemose-paniculate, 5–10 cm. long, puberulent; pedicels slender, 4–7 mm. long at anthesis, puberulent; calyx about 1.5 mm. long, puberulent; corolla purplish, petals 4–5 mm. long; loment 2–5-jointed, on a stipe 2–4 mm. long, notches along lower side of loment much deeper than those on upper margin, joints distinctly asymmetrical, 2.5–4 mm. wide, 4–6 mm. long, delicately but distinctly reticulate transversely, finely and closely uncinate-puberulent.

Hillsides, in thickets or forest, upper margins of Lower Sonoran Zone into lower parts of Transition Zone, Santa Cruz and Pima Counties, Arizona, to Texas and southward into Guatemala. Aug.–Sept.

**3. Desmodium intortum** (Mill.) Urban, Symb. Antill. **8:** 292, 1920

*Hedysarum intortum* Mill., Gard. Dict. ed. 8, no. 11, 1768.
*Desmodium sonorae* A. Gray, Pl. Wright. **2:** 47, 1853.
*Meibomia uncinata* Kuntze, Rev. Gen. Pl. **1:** 197, 1891.
*Meibomia hjalmarsonii* Schindl., Rep. Spec. Nov. **20:** 143, 1924.
*Desmodium hjalmarsonii* Standley, Field Mus. Bot. Ser. **18:** 534, 1937.

Herbaceous perennials with scandent to procumbent, trigonous, densely uncinate-pubescent stems to 2 m. long, lowest nodes often pilose; petioles 2–5 cm. long, pubescent; stipules 5–7 mm. long, brown and dry, usually deciduous; leaflets 3, ovate to elliptic, acute or less commonly obtuse at apex, 4–7 cm. long, uncinate-pubescent and often glandular-puberulent beneath, green and appressed-pilose above, paler beneath; pedicels slender, 6–8 mm. long, puberulent, bracts ovate-acuminate, 8–10 mm. long, early deciduous; calyx 3–5 mm. long, puberulent, with longer hairs on teeth; corolla rose-purple, 8–9 mm. long; loment 5–9-jointed, short-stipitate, deeply crenate-lobed on both margins, joints semiorbicular to semirhomboid, 2–3 mm. wide, 3.5–4 mm. long, closely uncinate-pubescent, not twisted at maturity.

Rocky hillsides, fence rows, stream banks and in thickets, upper margins of Lower Sonoran Zone into the Arid Transition and Tropical Zones, southern Arizona to northern South America. Flowering almost throughout the year.

**4. Desmodium prostratum** Brandegee, Proc. Calif. Acad. II, **3:** 128, 1891

Herbaceous perennial with slender prostrate stems 0.5–1.5 m. long, closely covered with slender, uncinate, slightly tawny hairs; stipules persistent, broadly ovate, overlapping at base, abruptly attenuate, 4–7 mm. wide, 10–14 mm. long, narrow part often exceeding the broad portion, finely pubescent and with longer slender hairs sparsely scattered among finer hairs, particularly on lower surface; petioles 3.5–8 cm. long, uncinate-puberulent and sparsely pilose; leaflets 3, lateral pair broadly elliptic to suborbicular, 1.5–3.5 cm. wide, 2.5–6 cm. long, subtruncate at base, rounded and minutely mucronate at apex, bright green and moderately appressed-strigose above, glaucous and more densely strigose beneath, veins distinct but not conspicuous; terminal leaflet larger, mostly suborbicular, 3.5–6.5 cm. broad and long, usually broadly cuneate at base; petiolules 8–12 mm. long; inflorescence racemose-paniculate, 1–3 dm. long; bracts similar to stipules, deciduous; pedicels slender, 1–2.5 cm. long; calyx about 4 mm. long, puberulent, upper teeth deltoid-lanceolate; lowermost tooth slenderly linear-lanceolate, 1–2 mm. longer than others; corolla pink to rose-purple, 10–12 mm. long; loment 2–2.5 cm. long, 5–6-jointed, deeply lobed on lower side, shallowly so on upper side, densely pubescent with short, yellowish, uncinate hairs.

Canyons and stream banks, Lower Sonoran and Tropical Zones, mountains in southern Baja California. Sept.–Nov.

**5. Desmodium batocaulon** A. Gray, Pl. Wright. **2:** 47, 1853

*Meibomia batocaulis* Kuntze, Rev. Gen. Pl. **1:** 197, 1891.

Perennial herbaceous plants from a short woody caudex, stems slender, scandent or prostrate, 3–15 dm. long, densely uncinate-pilose when young, subglabrate in age, tawny to dark reddish brown; stipules deltoid-ovate, attenuate, 2–3 mm. long, striate on upper surface, persistent; petioles slender, 1–3.5 cm. long, sparingly pilose or densely pilose near base; leaflets 3, lance-ovate to lance-oblong, 5–25 mm. wide, 1.5–6 cm. long, acute to obtuse and minutely apiculate at apex, rounded at base, dark green and sparsely strigose with slender hairs seated on pustulate bases above, glaucous and more densely strigose beneath, veins obvious but scarcely conspicuous; inflorescences racemose-paniculate, 0.5–2.5 dm. long, rachis pilose; pedicels subfiliform, 5–10 mm. long, sparsely pubescent to glabrate; calyx nearly glabrous, 4 mm. long, teeth deltoid-lanceolate, lowest 2 mm. long, others 1 mm. long; corolla purplish, banner 6–8 mm. long; loments 4–7-jointed, deeply notched beneath, shallowly so on upper side, joints ovoid-suborbicular to rhomboidal, 3–3.5 mm. wide, 3–5 mm. long, distinctly reticulate-veined, sparsely uncinate-puberulent to subglabrate.

Rocky slopes and often in pine forests, mainly Upper Sonoran and Transition Zones, but entering upper margin of Lower Sonoran Zone along canyons, central and southern Arizona into New Mexico and south into northern Mexico. June–Sept.

**6. Desmodium prehensile** Schlecht., Linnaea **12:** 315, 1838

Perennial herbs with slender, uncinate-pilose, procumbent to scandent stems 5–15 dm. long; stipules lanceolate, 4–5 mm. long, striate and pilose on outer surface, nearly glabrous within; petioles slender, 1.5–6 cm. long, moderately to densely pilose with tawny uncinate hairs; leaflets 3, linear-oblong to elliptic or lanceolate, acute to obtuse and minutely apiculate at apex, rounded at base, 8–28 mm. wide, 2.5–7 cm. long, bright green and sparsely appressed-strigose above, sparsely to densely strigose and gray-green beneath, lateral leaflets only slightly smaller than terminal one; inflorescence racemose-paniculate, 1–3 dm. long, rachis pilose; pedicels subfiliform, 4–8 mm. long, sparsely uncinate-pilose; calyx sparsely puberulent to subglabrate, cup about 1 mm. deep, upper teeth deltoid, 1–1.5 mm. long, lowermost teeth narrower, nearly twice as long; corolla rose-purple, banner 6–8 mm. long; loments 6–8-jointed, deeply crenate-lobate on lower margin, faintly lobed above, joints rhomboidal, 2.5–3.5 mm. long, 1.8–2.5 mm. wide, densely uncinate-puberulent.

Margins of meadows, streambanks, and in shaded canyons, Lower and Upper Sonoran Zones into the Arid Transition Zone, vicinity of Magdalena, Sonora, south into Guatemala. July–Sept.

**7. Desmodium psilocarpum** A. Gray, Pl. Wright. **2:** 48, 1853

*Meibomia psilocarpa* Kuntze, Rev. Gen. Pl. **1:** 198, 1891.

Perennial herb from a woody base with 1–several erect or ascending, much branching, uncinate-puberulent and sparsely pilose stems to 1 m. tall; stipules narrowly to broadly ovate-lanceolate, often asymmetrical at base, attenuate toward slender apex, 3–7 mm. wide at base, 5–15 mm. long, strongly striate and glabrous on both surfaces, spreadingly pilose on margins, bases of adjacent stipules often overlapping; petioles 2.5–5 cm. long, sparsely uncinate-puberulent and pilose; petiolules of lateral leaflets 2–4 mm. long, densely puberulent, those of terminal leaflets 1–3 cm. long; stipels linear-subulate, 5–10 mm. long, 1 mm. broad or less; leaflets ovate, elliptic, or broadly ovate-lanceolate, 1–6 cm. wide, 1.5–10 cm. long, rounded to subcuneate at base, obtuse to rounded at apex, often minutely apiculate, pale green and subglabrate on upper surface, slightly paler and sparsely pubescent beneath; inflorescence racemose-paniculate, 1–4 dm. long; bracts lanceolate, strongly striate, 0.5–0.8 mm. wide, 4–5 mm. long, nearly glabrous on back, margins pilose; pedicels filiform, 1–2 cm. long; calyx 2–2.5 mm. long, minutely puberulent; corolla rose-purple, drying greenish, banner 3.5–5 mm. long; loment 3–6-jointed, 2.5–5 cm. long, joints elliptic-rhomboidal, 3–5 mm. wide, 6–10 mm. long, reticulate and glabrate to sparsely puberulent, notches between joints nearly as deep along upper as along lower margins of loment.

Hillsides and along streams and canyons, usually in shade, Lower and Upper Sonoran Zones, Pima and Santa Cruz Counties, Arizona, south into Sonora and Chihuahua. July–Sept.

**8. Desmodium rosei** Schubert, Contr. Gray Herb. **129:** 22, 1940

*Desmodium neo-mexicanum* var. *glabellum* Vail, Bull. Torrey Club **19:** 116, 1892, in synonymy.
*Desmodium neo-mexicanum* of authors. Not of A. Gray, 1852.

Erect herbaceous plant 1–5 dm. tall with slender, faintly striate, sparsely uncinate-puberulent, slightly branching stems; stipules linear-lanceolate, 1.5–3 mm. long, strongly striate, minutely uncinate-puberulent on both surfaces and ciliate-hispid along margins; petioles slender, 1–3 cm. long, striate to costate, uncinate-pubescent; petiolules 0.5–2.2 mm. long, uncinate-pubescent; leaflets 3, linear to linear-lanceolate, 1.5–6 mm. wide, 1.5–7.5 cm. long, acute to obtuse and minutely apiculate at apex, broadly cuneate to rounded at base, bright green and sparsely pilose to subglabrate above, paler and slightly more pubescent beneath, lateral veins looping upward and inward to join next lateral above, which forms distinct islets; inflorescence racemose-paniculate, to 1.5 or 2 dm. long; pedicels subfiliform, 5–25 mm. long, finely uncinate-puberulent; calyx minute, about 1.5 mm. high, minutely and closely puberulent, lower lobes nearly twice as long as upper ones; corolla about 3–3.5 mm. long; loments 2–4-jointed, equally notched along both margins, joints orbicular or nearly so, 3–4 mm. in diameter, plane or essentially so, reticulate-veined and virtually glabrous.

Hillsides and canyons, mainly Upper Sonoran Zone, Pima and Gila Counties, Arizona, eastward into New Mexico and southward into Sonora, Chihuahua, and Zacatecas. Aug.–Sept.

**9. Desmodium tortuosum** (Sw.) DC., Prod. **2:** 332, 1825

*Hedysarum tortuosum* Sw., Prod. Veg. Ind. Occ. 107, 1788.
*Hedysarum purpureum* Mill., Gard. Dict. ed. 8, no. 6, 1768.
*Meibomia tortuosa* Kuntze, Rev. Gen. Pl. **1:** 198, 1891.
*Meibomia purpurea* Vail, in Small, Fl. SE. U.S. 639, 1903.
*Desmodium purpureum* Fawc. & Rendle, Fl. Jam. $4^2$: 36, 1920. Not *D. purpureum* Hook. & Arn., 1832.

Herbaceous perennials with erect or ascending branches 0.5–1.5 m. tall; stems and branches terete, moderately uncinate-pubescent to subglabrate; stipules narrowly lanceolate-attenuate, 6–12 mm. long, striate, subglabrous on both surfaces, ciliate-hispid along margins; petioles 1–5 cm. long; petiolules 2–4 mm. long, rachis between lateral leaflets and terminal one 1–2 cm. long; stipels linear-subulate, 3–6 mm. long;

leaflets 3, elliptic to rhombic-ovate or sometimes lanceolate, 1–6 cm. broad, 1.5–12 cm. long, obtuse and apiculate at apex, broadly cuneate to rounded at base, bright green, sparsely pubescent and distinctly reticulate-veined above, paler and somewhat more densely pubescent beneath; inflorescence often paniculately much branched, 1–5 dm. long; bracts linear-lanceolate, 3–5 mm. long, striate, ciliate-margined, deciduous; pedicels subfiliform, 8–18 mm. long; calyx about 2.5 mm. long, puberulent and hirsute; corolla purplish, banner 3.5–4 mm. long; loments about equally notched along both margins, 3–6-jointed, joints rhomboidal to suborbicular, 2.5–3 mm. broad, 3–5 mm. long, densely to sparsely uncinate-puberulent, reticulate-veined, margins more or less twisted or folded.

Meadows, stream banks, hillsides, canyons, and margins of fields, Lower Sonoran and Tropical Zones, central Sonora and Florida southward through Mexico and the West Indies into South America. Aug.–Oct. in our area.

Two or three specimens from central Sonora are provisionally placed here even though the plants appear to be annual. It is probable that under favorable growing conditions this species flowers the first season as well as subsequently. If so, some "annual" plants may be properly identified with this species.

**10. Desmodium scopulorum** S. Wats., Proc. Amer. Acad. **24:** 47, 1889

*Meibomia scopulorum* Rose & Standl., Contr. U.S. Nat. Herb. **16:** 212, *pl. 5, fig. b,* 1913.
*Desmodium wigginsii* Schubert, Contr. Gray Herb. **129:** 25, 1940.

Erect or ascending, moderately branched annual 1–6 dm. tall with striate to angulate, sparsely to moderately subuncinate-pilose stems and branches; stipules linear-lanceolate, 1.5–2.5 mm. long, base ovate, abruptly narrowed to attenuate, slightly ciliate apex, often reflexed; petioles slender, striate, 1–8 cm. long, moderately uncinate-hispidulous with short and long hairs; petiolules 1.5–2 mm. long, pubescent like petioles; leaflets 3, linear- to ovate-lanceolate, acute to slightly acuminate at apex, rounded at base, 6–25 mm. wide, 1.5–8 cm. long, dark green and sparsely uncinate-hispidulous to glabrate above, only slightly paler and scarcely more densely pubescent beneath, veins distinct but not conspicuous; inflorescence racemose-paniculate, to 3 dm. long; bracts 1–2.5 mm. long, lanceolate-attenuate, striate and sparsely pilose, deciduous; pedicels slender, 5–15 mm. long; calyx 1.5–2 mm. long, minutely puberulent; corollas greenish yellow to roseate, banner about 3 mm. long; loment 2–4-jointed, about equally notched above and below, joints rhomboidal, all except terminal one 4–7 mm. long and more or less contorted, terminal one triangular, plane, 7–8.5 mm. long, 4–5 mm. wide, minutely and sparsely uncinate-puberulent on the reticulate faces, more densely so along margins.

Rocky slopes, canyon bottoms, and grassy places, Lower and Upper Sonoran Zones, southern Arizona south at least to the vicinity of Guaymas, Sonora. Sept.–Oct.

**11. Desmodium procumbens** (Mill.) A. S. Hitchcock, Rep. Mo. Bot. Gard. **4:** 76, 1893, var. **procumbens**

*Hedysarum procumbens* Mill., Gard. Dict. ed. 8, no. 10, 1768.
*Hedysarum spirale* Sw., Prod. Veg. Ind. Occ. 107, 1788.
*Desmodium spirale* DC., Prod. **2:** 332, 1825.
*Desmodium sylvaticum* Benth., Pl. Hartw. 116, 1843.
*Meibomia spiralis* Kuntze, Rev. Gen. Pl. **1:** 197, 1891.
*Meibomia procumbens* Schindl., Rep. Spec. Nov. **20:** 151, 1924.
*Meibomia procumbens* var. *sylvatica* Schindl., loc. cit.
*Desmodium procumbens* var. *typicum* Schubert, Contr. Gray Herb. **129:** 5, 1940.

Erect or ascending annual with subangulate, sparsely to densely uncinate-hispidulous stems and branches to 1 m. tall or more; petioles slender, 1–5 cm. long, puberulent to subglabrate; petiolules short, rarely over 2 mm. long; stipules narrowly ovate-acuminate or attenuate, 1–4.5 mm. long, ciliate-hispidulous along margins; leaflets 3, thin, elliptic, ovate, or rhombic, acute or obtuse at apex, broadly cuneate to rounded at base, 0.5–3.5 cm. wide, 1–7 cm. long, glabrous to moderately strigose or uncinate-puberulent on both surfaces, margins slightly ciliate; inflorescence racemose-paniculate, branches to 2 dm. long; pedicels subfiliform, 3–12 mm. long; calyx 1.5–2 mm. long, minute teeth short-ciliate; corolla yellowish or greenish, 3–4 mm. long; loment 3–6-jointed, substipitate, moniliform, joints 1.5–3 mm. wide, 2.5–4.2 mm. long, margins alternately involute and revolute, loment appearing spirally twisted, uncinate-puberulent.

Hillsides, canyons, and mountain slopes, Lower Sonoran to Tropical Zones, central Sonora to northern South America, the West Indies, and in Africa and the Philippine Islands. It is probably introduced in the latter two regions. Aug.–Sept. in our area.

**11*a*. Desmodium procumbens** var. **exiguum** (A. Gray) Schubert, Contr. Gray Herb. **129:** 12, 1940

*Desmodium exiguum* A. Gray, Pl. Wright. **2:** 46, 1853.
*Meibomia exigua* Kuntze, Rev. Gen. Pl. **1:** 198, 1891.
*Desmodium spirale* var. *exiguum* Robins. & Greenm., Proc. Amer. Acad. **29:** 385, 1894.

Annual herb 1–4 dm. tall; basal leaves unifoliolate, leaflet broadly rhombic to transversely elliptic, entire or faintly 3-lobate, 5–15 mm. wide; upper leaves mostly 3-foliolate, their leaflets linear to linear-lanceolate, 3–12 mm. wide, 1–6 cm. long, glabrous, sparsely pilose or uncinate on both surfaces, gray-green; flowers and fruits essentially as in var. *procumbens*.

Grassy plains or hillsides, Lower Sonoran and Tropical Zones, southern Arizona and Baja California south through Mexico and Central America to northern South America. Sept.–Oct. in our area.

**12. Desmodium neo-mexicanum** A. Gray, Pl. Wright. **1:** 53, 1852

*Desmodium bigelovii* A. Gray, op. cit. **2:** 47, 1853.
*Desmodium neo-mexicanum* var. *bigelovii* S. Wats., Bibl. Index 217, 1878.
*Meibomia bigelowii* Kuntze, Rev. Gen. Pl. **1:** 197, 1891.
*Meibomia neomexicana* Kuntze, op. cit. 198.
*Desmodium spirale* var. *bigelovii* Robins. & Greenm., Proc. Amer. Acad. **29:** 385, 1894.
(For further synonymy see Contr. Gray Herb. **129:** 15, 1940.)

Erect or ascending, moderately to much branched, annual herbs 1–4.5 dm. tall, stems and branches slender but deeply grooved, terete or subangulate, sparsely to densely uncinate-puberulent; stipules slenderly linear-lanceolate, attenuate, 1.5–6 mm. long, minutely striate, glabrous above and below but ciliate-hispid along margins; petioles slender, grooved, 0.5–5 cm. long, uncinate-puberulent; leaflets 3, linear-lanceolate to ovate or rhombic-lanceolate, obtuse and mucronulate at apex, 2–21 mm. wide, 1.5–6 cm. long, terminal ones similar in shape but somewhat larger than lateral, light green and sparsely strigose to glabrate above, slightly paler and sparsely strigose beneath; inflorescence to 1 dm. long; pedicels filiform, 7–12 mm. long; calyx small, about 2–2.5 mm. long; corolla about 2.5–3 mm. long; loment 2–5-jointed, sessile or faintly stipitate, joints rhomboidal, reticulate and uncinate-hispidulous, 2.5–3 mm. wide, 3–4 mm. long, margins slightly to strongly folded or revolute, terminal joint slightly larger than others.

Mountainsides, mesas, and canyons, mainly Upper Sonoran and Transition Zones, southeastern Arizona to Texas and southward into South America; occurring in the Laguna Mountains, southern Baja California. Occasionally entering the upper margins of the Sonoran Desert in canyons and along streams. Aug.–Sept.

## 30. PISCIDIA L., Syst. Nat. ed. 10, 1155, 1759

Trees or shrubs with odd-pinnate leaves and flowers appearing before leaves in lateral panicles. Calyx campanulate, obscurely 2-lipped, upper lip with 2, and lower with 3, deltoid teeth. Corolla white, reddish, or sometimes yellowish; banner oval-obovate to suborbicular, emarginate, short-clawed; wings falcate-oblong, long-clawed, adnate to keel near middle; keel obtuse, auriculate-sagittate near base on upper side. Stamens diadelphous, vexillar one free at least at base. Ovary sessile, several-ovuled; style slender, glabrous; stigma terminal, small. Pod indehiscent, narrowly to broadly 2-winged on each suture, breaking transversely in age.

### 1. **Piscidia mollis** Rose, Contr. U.S. Nat. Herb. **1:** 98, 1891

*Ichthyomethia mollis* S. F. Blake, Jour. Wash. Acad. **9:** 246, 1919.

A tree 3–9 m. tall with a rounded, moderately dense crown and fissured whitish bark; branchlets cinereous-tomentulose, at length glabrate; leaves 8–25 cm. long, densely cinereous-tomentulose; leaflets 7–13, oval to ovate, 2–4 cm. wide, 3–8 cm. long, acute to rounded at apex, rounded to subtruncate at base, prominently pinnately and reticulately veined beneath, greenish to cinereous above; calyx on fruiting specimens 6–7 mm. long; flowers unknown; pods 2–10 cm. (mostly 5–6 cm.) long, 3.5–5 cm. wide, 1–6-seeded, densely cinereous-tomentulose, wings several times as broad as body; stipe about 1 cm. long; seeds buff, about 1 cm. long, 4–5 mm. wide, smooth, sublustrous.

Dry hillsides, bajadas, and near streamways, Lower Sonoran Zone, from near Ures southward to Alamos in the inner foothills.

## 31. VICIA L., Sp. Pl. 734, 1753

Annual or perennial, herbaceous vines with pinnate leaves bearing terminal tendrils and conspicuous stipules. Flowers solitary, in pairs, or racemose in axils, white, yellowish, or purplish. Calyx with a slender, slightly oblique tube and subequal to markedly unequal, slender teeth. Petals short-clawed; banner obovate, emarginate; wings oblong, somewhat adherent to slightly shorter, oblong keel. Stamens diadelphous; anthers alike. Ovary sessile to stipitate, several- to many-ovuled. Style slender, bearing a tuft or ring of hairs at apex. Pod compressed, dehiscent. Seeds subglobose, smooth, estrophiolate.

Flowers 5–7 mm. long, solitary or in pairs; plants annual .................. 1. *V. exigua*
Flowers 15–18 mm. long, in 2–8-flowered racemes; plants perennial ....................
.................................................. 2. *V. americana* var. *linearis*

### 1. **Vicia exigua** Nutt., in Torr. & Gray, Fl. N. Amer. **1:** 272, 1838

Annual with slender, erect or ascending stems 3–8 dm. high, glabrous to sparingly strigulose throughout; stipules semisagittate, each lobe linear-subulate, 3–7 mm. long; leaves about equaling or shorter than internodes, rachis extending 1–3 cm. beyond upper of 2–5 pairs of linear leaflets, these 1–2.5 cm. long, rounded, obtuse, or rarely acute or emarginate at apex, minutely and sparsely strigose to glabrate beneath; peduncles slender, 2–4 cm. long, 1-flowered or infrequently 2-flowered; calyx tube 1–1.5 mm. long, strigulose or glabrate, teeth equaling or slightly exceeding tube, usually sparsely strigulose; corolla 5–7 mm. long, white suffused with purple; pods 5–7 mm. wide, 2–3 cm. long, obliquely acute at each end, 3–8-seeded, glabrous.

Among shrubs, mainly Upper Sonoran and Transition Zones but barely entering the Sonoran Desert in a few localities in southern Baja California; Oregon to Baja California. March–May.

A form approaching var. *hassei* S. Wats., with linear-oblong emarginate leaflets, was collected near Queretaro, Baja California (*Whitehead 793*).

**2. Vicia americana** Muhl. var. **linearis** S. Wats., Proc. Amer. Acad. **11**: 134, 1876

*Vicia linearis* Greene, Fl. Fran. 3, 1891.

Perennial, with scrambling or climbing stems to 1 m. or more, sparsely pubescent to glabrate throughout; stipules semisagittate and often incisely toothed near base, to 1.5 cm. long; leaves about 1 dm. long, downwardly curved; leaflets 8–16, narrowly linear, 1.5–5 cm. long, acute or some of them truncate and emarginate at apex, all apically cuspidate, sparsely strigose beneath, distinctly ascendingly veined; peduncles 3–6 cm. long; racemes 2–8-flowered; calyx tube about 4 mm. long, strongly veined, sparsely pubescent or glabrate, teeth lance-linear to deltoid, unequal, 1–4 mm. long; corolla purplish, 1.5–1.8 cm. long; pod 4–6 mm. wide, 3–4 cm. long, glabrous.

Among shrubs and on grassy slopes, chiefly Upper Sonoran and Transition Zones, approaching if not entering the desert proper near its northern margin on the coastal side of Baja California. May–Sept.

## 32. NISSOLIA Jacq., Enum. Pl. Carib. 7, 1760

Suffrutescent, herbaceous, scrambling or prostrate vines with odd-pinnate leaves and setaceous stipules. Leaflets mostly 5. Flowers yellow, in racemes or verticils in leaf-axils. Bracts subulate. Calyx campanulate, truncate with subequal, deltoid to setaceous teeth. Banner orbicular or nearly so, erect or reflexed; wings oblong-falcate; keel obtuse. Stamens pseudomonodelphous, vexillar stamen connate with middle of tube. Ovary subsessile, few-ovuled; style filiform. Pod indehiscent, linear to oblong, with a samara-like, terminal, heavily veined wing.

Calyx teeth acute but not setose-tipped; fruits glabrous or sparsely puberulent, not setose-hispid .......... 1. *N. schottii*

Calyx teeth tipped with yellowish, conspicuous setae; fruits setose-hispid with yellow, bristle-like hairs .......... 2. *N. setosa*

**1. Nissolia schottii** (Torr.) A. Gray, Jour. Linn. Soc. **5**: 26, 1861

*Chaetocalyx schottii* Torr., Bot. Mex. Bound. 56, *pl. 18*, 1859.

A scrambling and twining vine with slender woody stems climbing to 5–6 m.; stipules 2–4 mm. long, caducous; leaves 4–12 cm. long; petioles 2–5 cm. long, slender, minutely and sparsely puberulent to glabrate; leaflets mainly 5, ovate, broadly elliptic to obovate, 7–18 mm. wide, 1–3 cm. long, rounded at base, rounded to obtuse and apiculate apically, glabrous, slightly paler beneath than above; flowers in axillary verticils; pedicels filiform, to 1 cm. long; calyx tube 3–3.5 mm. long, filiform teeth about 2 mm. long, calyx and teeth glabrous except for a ring of wool just within lips of tube, or setose with a few, bristlelike, yellow hairs; banner and wings yellow; keel white or nearly so; pods oblong-falcate, body 4–7 mm. wide, heavily veined longitudinally, 1–3-seeded, wing to 12 mm. wide, 1.5–2.5 cm. long, reticulately veined.

Scrambling through and over other shrubs, mainly Lower Sonoran Zone, southern Arizona into Sonora and Baja California. July–Sept.

**2. Nissolia setosa** Brandegee, Proc. Calif. Acad. II, **3:** 127, 1891

A much-branched, slender-stemmed, scrambling and twining vine 3–4 m. high with glabrous leaves and bright yellow flowers; stipules filiform-setose, 3–5 mm. long; leaves 3–8 cm. long or more; leaflets 5, broadly elliptic to orbicular, 1–2 cm. long, rounded or retuse and mucronulate at apex; pedicels filiform, to 1.5 cm. long; calyx tube 2.5–3 mm. deep, nearly as wide, glabrous; teeth filiform-subulate, 3–3.5 mm. long, spreading to reflexed, each tipped with a yellow setose hair about 1 mm. long; corolla rich yellow, 8–10 mm. long; banner puberulent on back; pod puberulent and sparingly yellow-setose, body 1–2-jointed; wing curved upward slightly, about 1 cm. long.

Growing among and over other shrubs, Lower Sonoran and Subtropical Zones, southern Baja California. July–Sept. This may be confined to the Cape Region and possibly is not specifically distinct from *N. schottii.*

## 33. ERYTHRINA L., Sp. Pl. 706, 1753

Shrubs or trees; branches and petioles usually prickly or armed with heavy conical spines. Leaves pinnately trifoliolate; leaflets large; stipules large, stipels of conical glands only. Flowers large and showy, in axillary or terminal racemes. Calyx tube nearly truncate, slightly 2-lipped. Banner much larger than other petals, wings sometimes lacking. Stamens diadelphous, tube usually longer than free filaments. Ovary short-stipitate or sessile, several-ovuled. Pods large, torulose, dehiscent. Seeds large, often bright red or orange, estrophiolate.

**1. Erythrina flabelliformis** Kearney, Trans. N.Y. Acad. **14:** 32, 1894
*Erythrina purpusi* Brandegee, Zoe **5:** 158, 1903.

Shrub or small tree to 8 m. tall armed with short spines 2–10 mm. long on branches and petioles; petioles 5–24 cm. long, tomentulose with fine hairs when young, eventually glabrate; leaflets broadly ovate to suborbicular-deltoid, 3.5–11 cm. broad, 2.5–7.5 cm. long, obtuse, rounded or shallowly retuse at apex, truncate to broadly cuneate at base, pubescent when young, finally glabrous or nearly so, pinnately veined; racemes to 3.5 dm. long; calyx tube about 1 cm. long, reddish, sparsely puberulent to subglabrate; banner red, 4–7 cm. long, to 18 mm. broad; wings 9–16 mm. long; keel petals slightly shorter than wings; ovary densely strigose-pubescent; pods subwoody, 1–3.5 dm. long, 1.5–2.5 cm. broad, constricted between seeds, 1–several-seeded; seeds 12–18 mm. long, two-thirds as broad, dark red, with a black line near hilum.

Hillsides, along arroyos, and rocky canyons, Lower and Upper Sonoran Zones, Cochise and Pinal Counties, Arizona, to Chihuahua, Durango, Zacatecas, and in southern Baja California. March–May; Sept.

## 34. GALACTIA P. Br., Nat. Hist. Jamaica 298, 1756

Herbaceous to suffrutescent or twining perennials with slender stems and 3-foliolate, odd-pinnate leaves. Stipules small and deciduous; leaflets with small stipels. Inflorescences axillary in pedunculate racemes; bracts small, setaceous. Calyx tube campanulate, lobes unequal, upper two connate into a single, entire or minutely bidentate, short lobe, lateral lobes slightly longer and lowermost one surpassing all others. Corolla pale yellow, greenish and purple; banner ovate to orbicular, free; wing petals linear to obovate, adnate to keel near its base; keel equaling or slightly surpassing wings. Stamens diadelphous or vexillar stamen connate with stamineal tube near middle; anthers uniform. Ovary subsessile, several- to many-ovulate. Style filiform, glabrous; stigma terminal, small. Pod linear to oblong-linear, straight or incurved, moderately compressed, dehiscent, often partially to fully septate between seeds. Seeds estrophiolate.

Leaflets grayish, closely sericeous on both surfaces; calyx 7–8 mm. long ..... 1. *G. wrightii*
Leaflets green, not sericeous on either surface; calyx 4.5–5 mm. long ......... 2. *G. striata*

**1. Galactia wrightii** A Gray, Pl. Wright. **1:** 44, 1852

*Galactia tephrodes* A. Gray, op. cit. **2:** 34, 1853.

Trailing, scandent, or ascendingly twining, suffrutescent plant with slender, minutely and closely puberulent-cinereous stems to 1.5 m. long from a woody rootstock; herbage cinereous and finely puberulent throughout; stipules narrowly lance-deltoid, 1–1.5 mm. long; petioles 1–2.5 cm. long, rachis about one-third to one-half as long; leaflets 3, oblong to elliptic-lanceolate, 4–18 mm. wide, 2–6 cm. long, distinctly veined beneath, obtuse to acute at each end; racemes slender, 0.5–2.5 dm. long; pedicels 1–4 mm. long; calyx 7–8 mm. long, tube 2–2.5 mm. deep, lower and lateral lobes narrowly lanceolate, attenuate; corollas 10–12 mm. long, banner surpassing, and nearly paralleling, wings, only slightly recurved; pods falcately curved upward near tip, 5–6 mm. wide, 4–5 cm. long, minutely but closely puberulent, gradually short-attenuate at apex, elastically 2-valved; seeds reniform, dark brown mottled with small black flecks, 2.5–3 mm. wide, 4–5 mm. long; hilum black.

Arid slopes and among rocks, mainly Upper Sonoran Zone but extending upward into Arid Transition Zone and occasionally into upper margins of Lower Sonoran Zone, southern Arizona to western Texas and adjacent northern Mexico. July–Sept.

**2. Galactia striata** (Jacq.) Urban, Symb. Antill. **2:** 320, 1900

*Glycine striata* Jacq., Hort. Vindob. **1:** 32, *pl. 76*, 1770.
*Galactia brevistyla* Schlecht., Linnaea **12:** 288, 1838.
*Adonia retusa* Rose, Contr. U.S. Nat. Herb. **10:** 102, 1906.

Slender, much-branched, scandent-twining, suffrutescent plant with stems to 1.5 m. long or more; herbage finely puberulent but green except on youngest leaves and branchlets; stipules lance-deltoid, about 1–1.5 mm. long, caducous; petioles 1.5–3 cm. long, rachis about one-third as long; leaflets 3, terminal one often larger than laterals, these 6–17 mm. wide, 1.5–3.5 cm. long, terminal ones to 20 mm. wide, 4.5 cm. long, distinctly pinnately veined beneath with about 7 pairs of veins, slightly more densely puberulent beneath than above; racemes slender, 1–2 dm. long, on peduncles 3–8 cm. long; pedicels 1–2 mm. long or the flowers subsessile; calyx 4–5 mm. long, densely puberulent; petals 8–10 mm. long, lavender, banner suffused with green near base; pods 4–6 cm. long, 4–5 mm. wide, minutely and moderately puberulent, elastically 2-valved; seeds reniform, about 2 mm. wide, 3–3.5 mm. long, dull black.

Growing over shrubs on grassy slopes and valley floors, Upper and Lower Sonoran Zones to Tropical Zone, southern Sonora to Central America and the West Indies. Aug.–Oct.

## 35. RHYNCHOSIA Lour., Fl. Cochinch. 460, 1790

Perennial, twining or suffrutescent, erect herbs. Leaves pinnately trifoliolate, leaflets bearing small resinous glands at least beneath, stipels lacking. Stipules ovate to lanceolate. Flowers mainly yellow in axillary, usually geminate racemes; bracts caducous. Calyx campanulate, 4–5-toothed, upper pair more or less connate. Banner obovate or orbicular, erect or reflexed, resinous-glandular and pubescent externally in some, occasionally purplish; wings narrow; keel falcately curved upward, obtuse. Stamens diadelphous, anthers alike. Ovary subsessile, 2-ovuled; style filiform, glabrous. Pod oblong, often asymmetrical, compressed, 2- or rarely 1-seeded, dehiscent. Seeds compressed-globose to subreniform, strongly strophiolate or estrophiolate.

Seeds black with bright red basal portion; pods distinctly constricted between seeds, 8–15 mm. wide .................................................. 1. *R. pyramidalis*
Seeds black or dark brown, concolorous; pods not constricted between seeds, 3–8 mm. wide:
  Flowers 1.5–2.5 cm. long; leaflets conspicuously rugulose-reticulately veined beneath; bracts 5–6 mm. long ......................................... 2. *R. macrocarpa*
  Flowers 5–6 mm. long; leaflets lightly reticulately veined but not conspicuously rugulose; bracts 2–3 mm. long:
    Leaflets broadly rhombic-ovate, usually about as broad as long, 1–3 cm. wide; calyces puberulent with short, mostly appressed hairs .................... 3. *R. minima*
    Leaflets narrowly ovate-lanceolate, usually two to four times as long as wide, 1 cm. wide or less; calyces short-villous with spreading hairs .................. 4. *R. texana*

### 1. Rhynchosia pyramidalis (Lam.) Urb., Rep. Spec. Nov. **15:** 318, 1918

*Dolichos pyramidalis* Lam., Encycl. **2:** 296, 1786.
*Glycine phaseoloides* Swartz, Prod. Veg. Ind. Occ. 105, 1788.
*Glycine precatoria* Humb. & Bonpl., in Willd., Enum. Hort. Ber. 755, 1809.
*Rhynchosia phaseoloides* DC., Prod. **2:** 385, 1825.
*Rhynchosia precatoria* DC., loc. cit.
*Rhynchosia erythrinoides* Cham. & Sch., Linnaea **5:** 587, 1830.
*Dolicholus vailiae* Rose, Contr. U.S. Nat. Herb. **10:** 101, 1906.

A climbing vine with stems to 5 m. or more from a woody base, velvety-pubescent throughout; petioles stoutish, 1–5 cm. long; leaflets broadly ovate, somewhat attenuate, 3–7 cm. wide or sometimes only 1.5–2 cm. wide, 5–10 cm. long, velvety-puberulent but green on both surfaces; racemes 1–3 dm. long or less; bracts ovate-lanceolate, 2–3 mm. long; pedicels 2–3 mm. long; calyx puberulent, tube 1.5–2 mm. long, lower teeth linear-subulate, 4–5 mm. long, upper ones about half as long; corolla yellow striped with brown, 7–9 mm. long; banner puberulent and resinous-dotted on back; keel green; pods broadly oblong, constricted between seeds, 8–15 mm. broad, 2–2.5 cm. long, puberulent; seeds subglobose, 5–6 mm. in diameter, bright red around hilum, shining-black on upper part, proportions of black and red areas variable.

Climbing over shrubs, Lower Sonoran to Tropical Zones, central Sonora to San Luis Potosí and tropical South America. Occurring in the Sonoran Desert near Hermosillo, Batúc, and southward, and in southern Baja California. April–Sept., and occasionally at other seasons.

### 2. Rhynchosia macrocarpa Benth., Pl. Hartw. 11, 1839

*Dolicholus macrocarpus* Rose, Contr. U.S. Nat. Herb. **10:** 101, 1906.

A climbing vine scrambling to 4–5 m. or more with densely puberulent-velvety stems and herbage or stems eventually glabrate; petioles mostly less than 5 cm. long, petiolule of terminal leaflet 1–2 cm. long; leaflets ovate-attenuate, 2–4 cm. wide, to 6 cm. long, lateral leaflets asymmetrically rounded at bases, green above, somewhat tawny or grayish and conspicuously rugulose-reticulate beneath; racemes to 3 dm. long; bracts ovate-lanceolate, 5–6 mm. long, green; calyx densely puberulent, tube 4–5 mm. long, lower tooth 5–6 mm. long, upper ones about equaling tube; corolla yellow or banner faintly striped with brown, or with a brown spot at base of blade, puberulent and resinous-dotted without, 1.5–2.5 cm. long; pods slightly falcate, asymmetrically oblong, 6–8 mm. wide, 3.5–4 cm. long, finely puberulent and resinous-dotted; seeds dark brown or black.

Among shrubs, Lower Sonoran and Subtropical Zones, central Sonora (near Ures) to Chihuahua, Durango, Hidalgo, and Jalisco. August–March.

**3. Rhynchosia minima** (L.) DC., Prod. **2:** 385, 1825

*Dolichos minimus* L., Sp. Pl. 726, 1753.
*Glycine reflexa* Nutt., Gen. **2:** 115, 1818.
*Glycine lamarckii* H. B. K., Nov. Gen. & Sp. **6:** 424, 1823.
*Glycine punctata* DC., Mém. Leg. 365, 1823.
*Rhynchosia caribaea* DC., Prod. **2:** 384, 1825.
*Glycine littoralis* Vahl, in DC., op. cit. 385.
*Rhynchosia ervoidea* DC., op. cit. 386.
*Phaseolus caribaeus* Eaton & Wright, N. Amer. Bot. ed. 8, 353, 1840.

Prostrate to twining or scrambling, perennial vine, or occasionally flowering first year, finely puberulent throughout; stipules lance-subulate, 3–4 mm. long; petioles 1–5 cm. long; leaflets rhombic-ovate, 1–3 cm. wide, often no longer than broad, acute to obtuse or rounded at apex, lateral ones asymmetrical at base; racemes 1–2 dm. long, on peduncles to 1 dm. long; bracts ovate, 2–2.5 mm. long, brownish; calyx 3–4 mm. long, including teeth, these about equaling tube or lower one slightly exceeding it; corolla 5–6 mm. long, yellow or tinged with brown or red; pods oblong, slightly falcate, 4–5 mm. wide, 1.5–2 cm. long, dark brown, finely and densely puberulent; seeds dull black, 3–3.5 mm. long.

On grassy slopes and among shrubs, tropical and subtropical regions, South Carolina to Florida and New Mexico and southward to South America; southern Baja California. Flowering nearly throughout the year after rains.

**4. Rhynchosia texana** Torr. & Gray, Fl. N. Amer. **1:** 687, 1840

*Dolicholus texensis* Vail, Bull. Torrey Club **26:** 108, 1899.

A slender, twining or scrambling vine with stems 1–20 dm. long from a woody root, minutely puberulent throughout; stipules lance-setaceous, 2–2.5 mm. long; petioles slender, 5–12 mm. long, petiolule of terminal leaflet from half as long as to equaling petiole; leaflets narrowly ovate-lanceolate to lanceolate, 3–10 mm. wide, 1–5 cm. long, rounded at base, acute and mucronulate at apex; flowers solitary to few-flowered clusters in axils; pedicels 4–5 mm. long; calyx tube 1.5 mm. long, finely villous with short spreading hairs, teeth linear-subulate, 1.5–2.5 mm. long; corolla yellow, 5–6 mm. long; pods lunate, 5–7 mm. wide, about 2 cm. long, finely and closely puberulent, minute beak turned abruptly downward; seeds suborbicular-lenticular, 3–3.5 mm. wide, black with mottling of minute gray or brown flecks.

Margins of open grassy plains and among shrubs, Upper and Lower Sonoran Zones, Pima and Santa Cruz Counties, Arizona, to Texas, Chihuahua, San Luis Potosí, and Sonora. Mainly Upper Sonoran Zone but entering the desert in Sonora between Magdalena and Cucurpe. April–Sept.

## 36. PHASEOLUS L., Sp. Pl. 723, 1753

Annual or perennial, scandent, twining or sometimes prostrate herbs from fibrous, woody, or tuberous roots. Leaves stipulate, pinnately 3-foliolate. Inflorescences pedunculate, fasciculate-racemose along upper parts of peduncle, nodes more or less swollen. Bracts subtending flowers minute or stipuliform, caducous or persistent. Calyx tubular to broadly campanulate or bowl-shaped, two upper teeth connate or distinct, lowest tooth often longer than others. Banner oblong to orbicular, usually recurved-spreading, margins often inflexed basally; wings equaling or surpassing banner; keel linear to obovate, obtuse beak usually spirally twisted or coiled. Stamens 10, diadelphous. Ovary few- to many-ovuled. Style twisted or coiled with keel, longitudinally bearded toward apex; stigma oblique. Legume linear to falcate, compressed or terete, 2-valved, valves twisting elastically at dehiscence. Seeds compressed to turgid, often squarish at ends, hilum circular to linear, not strophiolate.

Pods 6–12 cm. long; corollas large and showy, 1.5–4 cm. long at anthesis:
- Leaflets 2–3-lobed; calyx tubular or narrowly campanulate, longer than broad; corollas dark purple, 1.5–2 cm. long at anthesis:
  - Leaflets thin; herbage only slightly sericeous ... 1. *P. atropurpureus* var. *atropurpureus*
  - Leaflets thick; herbage densely tomentose-sericeous ............ 1a. *P. atropurpureus* var. *sericeus*
- Leaflets entire; calyx broadly campanulate at least as broad as long; corollas lavender or bluish, 3–4 cm. long at anthesis ............ 2. *P. speciosus*

Pods 1–6 cm. long; corollas smaller, less conspicuous, not over 1 cm. long at anthesis:
- Plants annual:
  - Seeds dark brown, smooth and shining or coarsely and indistinctly ridged, not reticulate-rugose; pods about same width near base and apex; leaflets entire or essentially so:
    - Leaflets linear to linear-lanceolate, 3–12 mm. wide, the lateral ones sometimes asymmetrically sublobate on one side; petioles slender, 1–3 cm. long ............ 3. *P. acutifolius* var. *tenuifolius*
    - Leaflets broadly ovate to ovate-rhombic, 4–6 cm. broad, only slightly longer than wide; petioles stouter, 4–5 cm. long ............ 3a. *P. acutifolius* var. *latifolius*
  - Seeds pale buff or flesh-colored, sparingly mottled with few darker spots, dull, markedly reticulate-rugose; pods tapering gradually toward the base, hence wider near apex than basally; leaflets nearly all distinctly trilobate ............ 4. *P. filiformis*
- Plants perennial from a woody root or from a fleshy tuber:
  - Leaflets, or at least the terminal one, trilobate:
    - Seeds reticulate-rugose; pods soon glabrescent; peduncles rarely over 6–8 cm. long; longest leaflets rarely over 3 cm. long ............ 4. *P. filiformis*
    - Seeds smooth; pods subpersistently pubescent, at least along margins; peduncles often 10–20 cm. long; leaflets to 5 cm. long, often over 3 cm. long .... 5. *P. grayanus*
  - Leaflets all entire or occasionally some of them barely subhastate:
    - Roots woody; stems scandent or climbing, not rooting regularly at nodes; leaflets ovate to rhombic, not over twice as long as wide, equaling or nearly equaling petioles:
      - Stipules ovate, 3–5 mm. wide, 4–6 mm. long; pods villous with ascending tawny hairs; calyx teeth acute, villous ............ 6. *P. anisotrichos*
      - Stipules subulate, about 1.5 mm. long or less, not over 1.5 mm. wide at base; pods glabrous; calyx teeth obtuse to rounded, glabrous or sparsely puberulent ............ 7. *P. ritensis*
    - Root a fusiform, semifleshy tuber; stems strictly prostrate, regularly rooting at nodes; leaflets linear to linear-lanceolate, or lance-elliptic, three to twelve times as long as broad, rarely as long as petioles ............ 8. *P. supinus*

**1. Phaseolus atropurpureus** Sessé & Moc. ex DC., Prod. **2:** 395, 1825, var. **atropurpureus**

*Phaseolus dysophyllus* Benth., Pl. Hartw. 287, 1848.
*Phaseolus semierectus* γ *atropurpureus* Gomez, Anal. Hist. Nat. Madrid **23:** 295, 1894.
*Macroptilium atropurpureum* Urban, Symb. Antill. **9:** 457, 1928.

Herbaceous vine from a woody root with rather stout, densely pilose stems to 5 m. long or more; indument of whitish, chiefly reflexed, slender hairs, this becoming sparser with age; stipules subulate to lance-deltoid, 0.5–1.5 mm. wide, 3–5 mm. long, strongly striate, glabrous within, densely pilose without; petioles rather stoutish, mostly 1–3 cm. long; leaflets suborbicular to rhombic-ovate or oblong, usually shallowly lobed on outer sides of lateral and on both sides of terminal leaflets, 2–8 cm. long, usually about twice as long as broad, broadly cuneate to subtruncate at base, obtuse at apex, thin to moderately thick and sparsely to moderately soft-pilose on both sides; peduncles 1.5–3 dm. long; racemes few-flowered; calyx campanulate-tubular, 2.5–3 mm. wide, 4–5.5 mm. long, densely strigose-pubescent with white or fulvous hairs, two upper teeth deltoid, acute, three lower ones acute, narrowly triangular to subulate, about equaling tube; corolla black-purple, banner long-clawed, 1.5–2 cm. long; pod linear, 3–4 mm. wide, 6–9 cm. long, finely strigose, slenderly attenuate toward apex; seeds several–many, 2.5–3 mm. wide, 4–4.5 mm. long, turgid, brownish and finely speckled with darker flecks, shining, somewhat squarish at ends; hilum narrowly elliptic, about 0.5 mm. long.

Moist to dry thickets, usually twining over and through shrubs, Lower Sonoran and Tropical Zones, southwestern Texas, central Sonora, and central Baja California, south through Mexico and Central America into South America. March–Oct. in our area, depending on the rains.

**1*a*. Phaseolus atropurpureus** var. **sericeus** A. Gray, Proc. Amer. Acad. **5:** 156, 1861

Differing from var. *atropurpureus* in possessing thicker, more leathery leaves which, with stems, petioles, and peduncles, are densely covered with a fine sericeous tomentum.

Habitats similar to those occupied by var. *atropurpureus,* Lower Sonoran and Tropical Zones, central Baja California, Sinaloa, and the Revillagigedo Islands. Flowering after rains.

**2. Phaseolus speciosus** H. B. K., Nov. Gen. & Sp. **6:** 452, 1824

A climbing or scandent vine with herbaceous stems to 5 m. long or more from a woody root; stems pilose with spreading hairs or subglabrate in age; stipules lanceolate-subulate to lance-oblong, 1–2 mm. wide at base, 3–5 mm. long, striate, acute or obtuse, sparsely pilose without and on margins or subglabrate; petioles comparatively short, 1–5 cm. long, densely pilose with tangled or somewhat spreading tawny hairs; leaflets ovate to rhombic-ovate, 1–5 cm. wide, 2.5–11 cm. long, rounded, broadly cuneate or nearly truncate at base, acute to slightly attenuate at apex, dark green and sparsely pilose on upper surface, moderately to densely pubescent with appressed to spreading hairs beneath; peduncles 1–2.5 dm. long, bearing 1–6 flowers clustered at or near apex; flowers short-pedicellate; calyx broadly bowl-shaped, 5–7 mm. deep, 5–10 mm. wide, densely pubescent to glabrate, three lower teeth 2–4 mm. long, obtuse, upper lip slightly shorter, often nearly twice as broad as long; corolla lavender to pale purple, 3–4 cm. long; keel strongly sigmoid-coiled, tubular; pods 4–5 mm. wide, 9–11 cm. long, compressed, attenuate at both ends, few- to many-seeded, strigose to glabrate;

seeds purplish brown with minute tawny flecks, 2–2.8 mm. wide, 3–4 mm. long, smooth and somewhat shining, squarish at ends; hilum elliptic, 0.7–1 mm. long, whitish.

Mostly scrambling over dry to moist thickets, Lower Sonoran and Tropical Zones, southern Baja California and Tamaulipas south through Mexico and Central America into northern South America. Flowering after rains.

**3. Phaseolus acutifolius** A. Gray var. **tenuifolius** A. Gray, Pl. Wright. **2:** 33, 1853

*Phaseolus montanus* Brandegee, Proc. Calif. Acad. II, **3:** 130, 1891.
*Phaseolus tenuifolius* Woot. & Standl., Contr. U.S. Nat. Herb. **16:** 140, 1913.

A slender annual with twining, sparsely puberulent to glabrate stems 2–12 dm. long; if puberulent, hairs slender and mostly retrorse or recurved; stipules subulate, 0.4–1.5 mm. wide, 2–5 mm. long; leaflets thin, linear-lanceolate to narrowly lanceolate, 3–12 mm. wide at base, 3–7 cm. long, lateral ones sometimes slightly subhastate on one side, sparsely scaberulous to nearly glabrous, attenuate at apex, venation prominent; petioles shorter than leaflets, 1–3 cm. long; peduncles slender, 2–6 cm. long; pedicels puberulent, 1–5 mm. long; calyx broadly campanulate, 2.5–3.5 mm. long, sparsely puberulent to glabrate, deltoid or narrowly ovate teeth shorter than tube; corollas white, yellowish, greenish yellow, or pale purple; banner 5–8 mm. long; keel strongly twisted or coiled; pod slightly falcate, 3–5 mm. wide, 4–6 cm. long, strongly compressed, sparsely to moderately but finely puberulent, attenuate at apex, distinctly reticulate-venose; seeds dark reddish brown, faintly speckled or solid in color, 2.5–3 mm. wide, 3.5–5 mm. long, squarish, flattened, slightly shining; hilum white, elliptic, about 0.5 mm. long.

Rocky hillsides, in low brush, and dry mesas, Lower Sonoran, Tropical, and arid parts of Upper Sonoran Zones, southern Arizona to Texas and south at least to Guatemala. Aug.–Oct.

**3*a*. Phaseolus acutifolius** var. **latifolius** Freeman, Bull. Ariz. Agric. Exp. Sta. **68:** 589, *pls. 8–9,* 1912

Differing from var. *tenuifolius* in having broadly ovate to ovate-rhombic leaflets 5–8 cm. long, 4–6 cm. wide, subtruncate to rounded or broadly subcuneate at base, slightly attenuate or acuminate at apex, thin, bright green, sparsely scaberulous to glabrescent; petioles 4–5 cm. long; other characters about as in var. *tenuifolius.*

Habitat and range similar to those of the var. *tenuifolius.* Cultivated as the "tepary bean"; drought resistant.

**4. Phaseolus filiformis** Benth., Bot. Sulph. 13, 1844

Annual or apparently short-lived perennial from a slender taproot under favorable conditions; stems slender, twining or prostrate, 1–15 dm. long, sparsely and finely puberulent to glabrate, 1 to many from a single root; stipules narrowly ovate to deltoid, 1–2 mm. long, weakly striate-nerved, sparsely puberulent to glabrate except along puberulent margins; petioles slender, 0.5–3.5 cm. long; leaflets variable, from ovate and entire to rhombic and shallowly to deeply trilobate, 1–2.5 cm. wide, to 3.5 cm. long, lobes broad and rounded to narrow and acute, bright green, sparsely scaberulous to subglabrate, lamina thin to thickish, rather inconspicuously veined; peduncles much surpassing leaves, to 10 cm. long, mostly few-flowered; pedicels slender, 2–5 mm. long; calyx campanulate, about 2–2.5 mm. long, finely puberulent, teeth broadly deltoid and much shorter than tube; corolla pink to pale salmon, sometimes faintly tinged with lavender, 5–8 mm. long; pods falcate to lunate-falcate, 3.5–5 mm. wide, 1.5–5 cm. long, finely and sparsely puberulent to glabrate, about 2–6-seeded, compressed; seeds pale buff, flesh-colored, or light brown, sparsely flecked with darker spots, about 2.5 mm. wide, 3–3.5 mm. long, flattened, reticulate-rugose along outer margin, dull.

Sandy washes and rocky hillsides, Lower Sonoran Zone, southwestern Arizona southward at least to southern Sonora and the Cape Region, Baja California. Flowering during any month if the moisture supply is adequate.

*P. wrightii* A. Gray seems to be closely related to, and may not be distinct from, this species.

**5. Phaseolus grayanus** Woot. & Standl., Contr. U.S. Nat. Herb. **16:** 139, 1913

*Phaseolus wrightii* var. *grayanus* Kearney & Peebles, Jour. Wash. Acad. **29:** 485, 1939.

Slender, twining or prostrate vine with minutely puberulent, striate stems 2–15 dm. long, arising from a gnarled woody root; stipules lanceolate to oblong, 1–2 mm. wide, 3–5 mm. long, minutely puberulent without, glabrate within; petioles 1–5 cm. long; leaflets broadly ovate to rhomboidal in outline, deeply 3-lobed, 1.5–4 cm. wide, 2–5 cm. long, lobes broad, lateral ones often squarish or rounded, central one usually broadly lanceolate, acute, dark green and sparsely scaberulous above, slightly paler, more densely puberulent, and prominently veined beneath; peduncles greatly surpassing leaves, often 1.5–2 dm. long, apically few-flowered; pedicels finely puberulent, 2–9 mm. long, subfiliform; calyx campanulate, 1.5–2.5 mm. long, minutely puberulent, teeth much shorter than tube; corolla purplish or lavender, 8–10 mm. long, keel pale yellow or cream, strongly coiled, nearly equaling wings; pods 3.5–4.5 cm. long, slightly falcate, puberulent, strongly flattened; seeds 2–2.5 mm. wide, 3–4 mm. long, flattened, dark brown, speckled.

Rocky hillsides and at margins of thickets, Lower Sonoran Zone to Arid Transition Zone, Arizona to Texas and southward into adjacent Mexico. Flowering almost throughout the year when moisture supply is adequate.

**6. Phaseolus anisotrichos** Schlecht., Linnaea **12:** 326, 1838

*Phaseolus leptostachyus* Benth., Ann. Wien. Mus. **2:** 136, 1840.

A slender, much-branched, puberulent to pilose, twining or scandent vine with stems to 1.5 m. long; stipules broadly ovate, obovate, or nearly deltoid, 3–5 mm. wide, 4–6 mm. long, margins pilose-ciliate, back and face strongly striate-veined, glabrous or nearly so; petioles 2.5–6 cm. long, markedly short-pilose with tawny or rufous hairs; leaflets broadly ovate to deltoid-ovate, 1.5–6 cm. wide, 2–9 cm. long, broadly rounded to subtruncate at base, acute or obtuse at apex, hirsute-pilose on both faces, at least when young, veins moderately conspicuous on paler lower surfaces; peduncles surpassing leaves, to 1.5 or 2 dm. to lowermost flowers; racemes many-flowered, frequently over 1.5 dm. long; pedicels very short (1–2.5 mm. long) and ascending at anthesis, reflexed in fruit; calyx broadly campanulate to bowl-shaped, 2–3 mm. long, nearly as wide, almost glabrous except on teeth, these pilose; corollas pale lavender to purple, 8–10 mm. long; keel strongly coiled; pod 3.5–4 mm. wide, 2–2.5 cm. long, strongly compressed, falcate, broadest near apex, minutely puberulent to densely hirsute-pilose.

At margins of wet or dry thickets, margins of meadows, and along streams, Lower Sonoran to Arid Transition Zones, northern Sonora to Honduras and Costa Rica. Flowering after rains.

**7. Phaseolus ritensis** M. E. Jones, Contr. West. Bot. **12:** 14, 1908

Prostrate, scandent or twining perennial with slender, striate, sparsely puberulent stems 2–10 dm. long, older ones often strongly zigzag; root woody, to 1.5 cm. in diameter; stipules oblong-lanceolate to narrowly lanceolate, 0.8–1.5 mm. wide, 3–5 mm. long, strongly striate, finely and sparsely puberulent to subglabrate; petioles 1–3 cm. long, sparsely puberulent; leaflets ovate to broadly rhombic-ovate, symmetrical or nearly so, entire, 1.5–3.5 cm. wide, 2–5.5 cm. long, rather leathery, light green, subglabrate and finely reticulate-veined above, slightly paler, sparsely puberulent and conspicuously veined beneath, acute to obtuse and slightly emarginate at apex, rounded to broadly cuneate at base; peduncles usually surpassing leaves, to 3 dm. long, flowering racemes equaling or exceeding basal portion; pedicels slender, 3–5 mm. long and ascending at anthesis, to 1.5 cm. long and reflexed in fruit; calyx broadly campanulate, 2–3 mm. high, fully as wide, subglabrate, teeth broader than long, shorter than tube; corolla pale lavender, 6–9 mm. long; pods 4–7 mm. wide, 3–4 cm. long, more or less falcate, very abruptly turned upward at base, abruptly rostrate at apex, strongly flattened, glabrous; seeds chestnut brown, squarish and strongly flattened, 4–5 mm. wide, 5–6.5 mm. long, notched at elliptic hilum, faintly polished.

Valley floors and rocky hillsides, mostly Upper Sonoran and Transition Zones in the pine and oak belts, but occasionally straggling to the upper margins of the desert, southern Arizona south through the mountains and higher foothills to northern Sinaloa. Aug.–Sept.

**8. Phaseolus supinus** Wigg. & Roll., Contr. Dudley Herb. **3:** 270, *pl. 60,* 1943

Perennial with slender prostrate stems 2–12 dm. long from an elliptic or oblong, brownish tuber 4–7 cm. long, the stems hirsutulous, rooting at nodes; stipules lance-subulate, 5–6 mm. long, hirsutulous-ciliolate; petioles erect, 4–8 cm. long; leaflets linear-lanceolate, 4–7 mm. wide, 3–8 cm. long, distinctly veined, sparsely scaberulous above, densely so along veins beneath, narrowly revolute, acute to mucronate at apex; flowers of two kinds, ordinary open ones and small, prostrate or shallowly subterranean, cleistogamous ones, latter 5–6 mm. long, yellowish with a slight salmon tinge, its calyx broadly campanulate, about 2.5 mm. long; chasmogamous flowers on erect slender peduncles 1.5–2.5 dm. tall, these 2–8-flowered; calyx broadly campanulate, 2.5–3 mm. long, sparsely hirsutulous, teeth subequal, subulate, 0.8–1 mm. long; banner deep salmon, erect, 5 mm. high; wings 9–10 mm. long, yellowish, blades 4–5 mm. broad, 5–6 mm. long; keel coiled, 4–5 mm. long, greenish yellow; pods oblong, about 4 mm. broad, 12–18 mm. long, 2–4-seeded, hirsutulous; seeds 2–2.5 mm. broad, 3–4 mm. long, pale buff, irregularly mottled with gray-purple to purple flecks; hilum white, surrounded by a purple-brown spot.

Known only from the type specimen, collected on a gentle gravelly slope in the Lower Sonoran Zone between Colorado and Mazatán, Sonora. In flower and fruit in Sept.

## Family 61. **KRAMERIACEAE.** Krameria Family

Shrubs or perennial herbs with alternate, simple, entire or rarely trifoliolate, pubescent leaves. Flowers large, irregular, in terminal racemes or axillary. Peduncles usually bearing a pair of small foliaceous bracts. Sepals 4 or 5, unequal. Petals 5, 3 upper distinct or united at base, long-clawed, lower 2 smaller, broader, thicker, sessile, and often covered with sessile, rounded, glandular papillae. Stamens 4 (in ours), free, or borne on united claws of upper petals; anthers 2-celled, opening by terminal pores. Ovary 1-celled, biovulate. Fruit globose, indehiscent, spiny, 1-seeded. Cotyledons thick. Endosperm lacking.

## 1. **KRAMERIA** Loefl., in L., Sp. Pl. ed. 2, 177, 1762

Characters of the family.

Stems herbaceous, prostrate to ascending, but not divaricate-spinescent .... 1. *K. lanceolata*
Stems woody, rigid, mostly divaricately branching and spinescent:
  Spines of fruit subulate, barbless or nearly so; body of fruit glabrous or sparingly strigose ........................................................ 2. *K. erecta*
  Spines of fruit acicular, barbed, at least at tip; body of fruit strigose to tomentose:
    Upper petals distinct at base; spines with barbs at apex only:
      Herbage and twigs densely canescent when young; sepals all canescent; leaves often slightly crowded ........................................ 3. *K. grayi*
      Herbage and twigs strigose but greenish; lower sepals sparsely strigose, upper more densely so; leaves remote:
        Leaves lance-elliptic, 6–15 mm. long; twigs and leaves not oily-glandular ....... ................................................ 4. *K. sonorae*
        Leaves ovate-lanceolate, 2–8 mm. long; twigs and leaves oily-glandular ......... ................................................ 5. *K. paucifolia*
    Upper petals united below middle; spines with barbs on upper one-half to one-third:
      Leaves linear, mostly 1 mm. wide or less; herbage strigose but usually greenish ..... ........................................ 6. *K. parvifolia* var. *parvifolia*
      Leaves lanceolate, elliptic, or oblanceolate, usually over 1 mm. wide; herbage heavily strigose to canescent:
        Peduncles and sepals eglandular ................ 6*a*. *K. parvifolia* var. *imparata*
        Peduncles and sepals stipitate-glandular ........ 6*b*. *K. parvifolia* var. *glandulosa*

### 1. **Krameria lanceolata** Torr., Ann. Lyc. N.Y. **2:** 168, 1827

*Dimenops lanceolata* Raf., Atl. Jour. 144, 1832.
*Krameria beyrichii* Sporleder, in O. Berg, Bot. Zeit. **14:** 752, 1856.
*Krameria secundiflora* O. Berg, loc. cit. Not *K. secundiflora* DC., 1824.
*Krameria secundiflora* var. *angustifolia* Chodat, Arch. Sci. Phys. III, **24:** 498, 1890.
*Krameria secundiflora* var. *intermedia* Chodat, loc. cit.
*Krameria secundiflora* var. *lanceolata* Chodat, loc. cit.

Perennial herbaceous-stemmed plant from woody root; stems slender, prostrate to slightly ascending, 2–5 dm. long, strigose to sparsely canescent; leaves linear to oblong, 1–3 mm. wide, 6–20 mm. long, strigose, acute, sessile; peduncles axillary, equaling or slightly exceeding leaves, slender; bracts foliaceous, borne near upper fourth of peduncle; sepals obtuse to acute, 8–13 mm. long, strigose, purplish on exposed parts; lower petals reniform-orbicular, 2.5–3 mm. broad; upper petals united to middle or slightly above, limbs shorter than claws, reniform-orbicular; bases of stamens adnate to claws of upper petals; fruit ovoid to globose, body 5–8 mm. in diameter, loosely to densely woolly, spines heavy, subulate, 2–4 mm. long, barbless or merely roughened near apex.

Hillsides and mesas, Lower and Upper Sonoran Zones, Pima County, Arizona, to Texas, Kansas, Arkansas, and northern Chihuahua. May–Aug.

**2. Krameria erecta** Willd., in Roem. & Sch., Syst. Veg. Mant. **3:** 303, 1827
*Krameria rosmarinifolia* Pav., in Chodat, Arch. Sci. Phys. III, **24:** 499, 1890.
*Krameria palmeri* Rose, Contr. U.S. Nat. Herb. **1:** 304, *pl. 27,* 1895.

Intricately branched, rounded shrub 5–10 dm. high with slender, spinescent, strigose twigs; leaves linear, mostly less than 1 mm. wide, 5–15 mm. long, acute, densely strigose but nearly glabrate in age; peduncles somewhat exceeding leaves, bearing 1 pair of bracts 2–3.5 mm. long; sepals oblong, 6–7 mm. long, obtusish or abruptly acute, thin, sparsely strigulose or nearly glabrous; lower petals about 2 mm. long, broadly obovate to suborbicular, 1.5–2 mm. broad; upper petals united below middle, 4–5 mm. long, blades 1–2 mm. wide; stamens inserted on, and about equaling upper petals; fruit broadly ovoid to globose, glabrous or sparingly strigose when young, 6–9 mm. in diameter, slightly compressed; spines subulate, 2.5–4 mm. long, naked or rarely with an occasional weak barb near tip.

On rocky hillsides and desert slopes, Lower Sonoran Zone, central Sonora and south into Sinaloa. March–Oct.

**3. Krameria grayi** Rose & Painter, in Rose, Contr. U.S. Nat. Herb. **10:** 108, 1906.
*Krameria canescens* Gray, Pl. Wright. **1:** 42, 1852. Not *K. canescens* Willd., 1825.

Intricately branched shrub 2–6 dm. high with densely canescent twigs, peduncles, and leaves; leaves linear to oblong, sessile, acute to obtuse, often apiculate, 1–3 mm. wide, 5–10 mm. long; peduncles 15–25 mm. long, sericeous, bracts foliaceous, borne at about middle of peduncle; sepals lanceolate, acute, purplish, 8–15 mm. long, canescent on exposed parts; lower petals 2.5–3 mm. long, suborbicular, often bearing numerous small tubercles; upper petals spatulate, 4–5 mm. long, claws slender, distinct; stamens distinct, shorter pair equaling and longer pair surpassing upper petals; fruit broadly ovoid to globose, body densely woolly, spines acicular, 3–4 mm. long, hairy below, glabrous toward apex, bearing 2–5 retrorse, rather stout barbs in a terminal whorl.

Dry desert slopes, washes, and hillsides, Lower Sonoran Zone, southern California to Nevada, Texas, Chihuahua, Sonora, and central Baja California. April–Sept.

**4. Krameria sonorae** Britt., N. Amer. Fl. **23:** 197, 1930

Straggly green shrub 0.5–1.5 m. high and usually one and one-half to two and one-half times as wide; young stems strigose but green, older stems yellowish green, glabrous and shining; leaves lance-elliptic, 1–2 mm. wide, 6–15 mm. long, acute, moderately strigose but green and often appearing slightly varnished, mostly about equaling or somewhat shorter than internodes; peduncles 5–9 mm. long; bracts similar to leaves, 4–7 mm. long; sepals 5, ovate-lanceolate, lower one 3.5–4 mm. wide, 9–10 mm. long, abruptly acute, pinkish, strigulose, upper 4 about 3 mm. wide, 7–8 mm. long, slightly arcuate, strigose; upper petals distinct to base, elliptic-oblanceolate, acute, middle one 2.5–3.5 mm. long, lateral ones 3.5–4 mm. long, membranous and glabrous, purplish or pinkish; lower petals suborbicular, 2–2.5 mm. broad, fleshy, tuberculate with whitish vesicular papillae on outside, smooth and with a small tuft of short white hairs at base inside, deep purple; stamens free, 4–6.5 mm. long, curved, distinctly exceeding upper petals; fruit broadly ovoid to globose, 8–10 mm. in diameter, densely tomentulose, apex abruptly pointed; spines bright red to reddish brown, acicular, 3.5 mm. long, each with 2–4 (mostly 3) recurved, stoutish barbs at tip only.

Desert llanos and foothill slopes and canyons, Lower Sonoran Zone, central to southern Sonora. April–Nov.

This species often has been confused with *K. grayi* but differs from that species in having green shiny branches; less heavily strigose leaves and branchlets; narrower, greener, and less crowded leaves; and stamens that, in comparison with the length of the upper petals, are slightly longer than those of *K. grayi.*

**5. Krameria paucifolia** Rose, Contr. U.S. Nat. Herb. **10:** 108, 1906

*Krameria canescens* var. *paucifolia* Rose, in Vasey & Rose, op. cit. **1:** 66, 1890.

Openly but intricately branched, rounded shrub 0.3–1.5 m. high, as much as 3 m. wide, with slender, strigose-sericeous, and oily-glandular branches that eventually are glabrate; leaves ovate-lanceolate to ovate, distant, acute to apiculate, strigose when young and oily-glandular, 2–8 mm. long; peduncles much surpassing leaves, 10–25 mm. long, strigose, oily-glandular; bracts 2–5 mm. long; sepals ovate-lanceolate, 6–12 mm. long, outer ones sparsely strigose, purplish or reddish, inner densely sericeous; lower petals flabellate-orbicular, tuberculate, slightly hairy within, 2.5–3 mm. long; upper petals distinct, oblanceolate-spatulate, acute, 4–5 mm. long; longer stamens exceeding upper petals; fruit ovoid to subglobose, body densely velutinous-sericeous, 7–10 mm. in diameter, spines acicular, sparsely hairy at base, bearing 1–5 rather stout retrorse barbs at apex.

Gravelly slopes and rocky hillsides, Lower Sonoran Zone, southern Baja California. Feb.–Aug.

**6. Krameria parvifolia** Benth., Bot. Sulph. 6, *pl. 1,* 1844, var. **parvifolia**

Intricately branched shrub 3–6 dm. high, somewhat depressed, with stout, gray-barked branches and slender, short-strigose twigs; leaves linear, sessile, acute, 1 mm. wide or less, 8–12 mm. long, short-strigose, green; peduncles 1–2 cm. long; bracts linear, 4–6 mm. long; sepals narrowly oblong, obtusish, 7–10 mm. long, reddish at base, sparingly strigose; lower petals nearly rectangular, 3 mm. long, smooth, truncate; upper petals united only at base, 4–5 mm. long, blade of middle one narrowly oblong, those of lateral ones expanded; stamens borne on short claws of upper petals, about equaling petals; fruit somewhat compressed, subcordate, 7–8 mm. broad, 5–7 mm. thick, apex acute, sparsely strigose; spines acicular, 2.5–3.5 mm. long, glabrous, bearing 2–4 retrorse barbs at two or three levels near apex.

Dry washes and rocky hillsides, southern Baja California from the vicinity of Magdalena Bay southward. March–Sept.

**6*a*. Krameria parvifolia** var. **imparata** J. F. Macbride, Contr. Gray Herb. No. 56: 52, 1918.

*Krameria imparata* Britt., N. Amer. Fl. **23:** 199, 1930.

Differing from the var. *parvifolia* in having leaves narrowly oblanceolate and usually 1.5–2.5 mm. wide; herbage more heavily canescent; and fruits more densely tomentulose. In some specimens the barbs on the spines of the fruit are more numerous and occur nearer the base than they do in var. *parvifolia.*

Desert slopes and washes, Lower Sonoran Zone, Colorado Desert and southern Mojave Desert and the desert parts of Arizona to central Sonora and Baja California. April–Oct.

**6b. Krameria parvifolia** var. **glandulosa** J. F. Macbride, Contr. Gray Herb. No. 56: 52, 1918.

*Krameria parvifolia* A. Gray, Pl. Wright. **1:** 41, 1852. Not *K. parvifolia* Benth., 1844.
*Krameria glandulosa* Rose & Painter, in Rose, Contr. U.S. Nat. Herb. **10:** 108, 1906.

Differing from var. *imparata* in having distinctly glandular peduncles and sepals, glands dark red to reddish purple, and heavily stipitate.

Desert slopes, arroyos, and hillsides, Lower Sonoran Zone, western margin of the Colorado Desert to southern Nevada, southern Utah, Texas, Sonora, and central Baja California. April–June.

## Family 62. GERANIACEAE. Geranium Family

Annual, biennial, or perennial herbs or subshrubs. Leaves usually opposite, stipulate; blades single or compound, toothed, lobed, or divided. Flowers cymose, regular or nearly so. Sepals 5, imbricated, persistent. Petals 5. Receptacle bearing 5 small glands alternate with petals or eglandular. Stamens 5 or 10, filaments united at base, often somewhat winged. Ovary superior, of 5 carpels with their slender styles adnate to a long slender column but separating from it at maturity, each carpel 2-ovuled but 1-seeded at maturity. Endosperm scanty. Cotyledons incumbent.

### 1. ERODIUM L'Her., in Ait., Hort. Kew. 2: 414, 1789

Annual herbs with opposite or alternate leaves, usually scarious stipules, and axillary, essentially regular flowers. Leaves simple or pinnate. Sepals 5. Petals 5, hypogynous, two upper slightly smaller than other three. Stamens 5, alternating with 5 sterile filaments. Glands 5. Ovary 5-lobed, 5-celled, styles united into a slender beak several times as long as carpels, each cell 2-ovuled. Carpels each 1-seeded at maturity, sharp-pointed at base, tipped by long persistent style, this bristly-hairy on inner side and spirally coiled as it separates from axial column.

Carpels 7–8 mm. long, truncate at apex, densely silky-hairy; sepals 10–12 mm. long in fruit; leaf blades cordate-reniform .................................. 1. *E. macrophyllum*
Carpels 4–6 mm. long, acute or rounded at apex, sparsely hairy; sepals 5–7 mm. long in fruit; leaf blades ovate or oblong in outline:
  Sepals acute but not bristle-tipped; herbage appressed-pubescent, canescent; leaf blades lobed but not pinnately divided; petals 8–12 mm. long .............. 2. *E. texanum*
  Sepals bristle-tipped; herbage sparsely glandular-villous; leaf blades pinnately divided; petals 4–6 mm. long .......................................... 3. *E. cicutarium*

**1. Erodium macrophyllum** Hook. & Arn., Bot. Beechey 327, 1838

Leaves basal; petioles 3–20 cm. long; stipules scarious, hyaline, ovate-lanceolate, 4–8 mm. long; leaf blades cordate-reniform, crenately 3–7-lobed, 2.5–6 cm. wide, about one-half to two-thirds as long, lobes crenulate with short rounded teeth, puberulent on both surfaces with simple and glandular hairs; flowering stems equaling to one and one-half times as long as petioles, often bearing 1–5 reduced leaves near apex, 2–8-flowered; pedicels densely glandular-pubescent, 1–2 cm. long and deflexed in age, fruits erect; sepals oblong, 10–12 mm. long, 5–6 mm. wide, acute or apiculate at apex but not bristle-tipped, densely pubescent with simple and glandular hairs; petals 10–15 mm. long, white; filaments orbicularly winged at base; achenes 7–8 mm. long, 1.5–2 mm. broad, densely hairy, truncate at apex, style departing from inner margin; style column 3.5–6 cm. long in fruit.

Hillsides and fields, mostly in open grassy places, Upper and Lower Sonoran Zones, along west side of Sacramento and San Joaquin Valleys, coastal southern California, and south near the coast to the margins of the desert in Baja California. Not common in any part of its range. Feb.–April.

### 2. Erodium texanum A. Gray, Gen. Ill. 2: 130, *pl. 151,* 1849

Stems several from base, leafy, 5–30 cm. long, erect or procumbent; herbage minutely appressed-pubescent, canescent; stipules narrowly triangular-ovate, 2–3 mm. long; petioles slender, 1–5 cm. long; leaf blades broadly ovate or oblong-ovate, truncate to subcordate at base, to 2.5 cm. long, shallowly to deeply lobed or sometimes divided, lobes shallowly and irregularly toothed; flowers 2–3 on short axillary peduncles; sepals oblong, 5–7 mm. long, apiculate or acute, silvery-pubescent; petals purple, 8–12 mm. long; carpels slender, about 6 mm. long, sparsely hairy, fusiform; style branch terminal.

Sandy and gravelly soil, mainly Lower Sonoran Zone, Colorado and Mojave Deserts to southern Nevada and east to Texas, south into central Baja California and northern Sonora. Feb.–March.

### 3. Erodium cicutarium (L.) L'Her., in Ait., Hort. Kew. 2: 414, 1789

*Geranium cicutarium* L., Sp. Pl. 680, 1753.

Leaves all forming a compact rosette at first, later some on erect to ascending, reddish stems 1–3 dm. tall; herbage moderately pubescent with simple and glandular hairs; leaves pinnately compound, 2–15 cm. long; leaflets 3–10 pairs, distinct except at apex of leaf, narrowly elliptic-oblong, 6–15 mm. long, deeply incised-pinnatifid, acutely toothed or bipinnatifid; pedicels slender, to 1.5 cm. long, deflexed in fruit; sepals 5–6 mm. long, each tipped with 1 or 2 acicular, hyaline setae; petals rose or purplish, 4–6 mm. long; glands dark; filaments scarcely dilated at base; carpels slenderly fusiform, acute at apex, 4–5 mm. long, sparsely hairy; style attached terminally, 3–3.5 cm. long.

Barren hillsides and plains, often very abundant in heavily grazed pastures, Lower Sonoran to the Arid Transition Zones, introduced from Europe and widely established in the United States and Mexico. Oregon to Baja California; common throughout Arizona and south into Sonora at elevations almost to 7,000 feet. Feb.–March.

Several species of *Geranium* occur in the mountains adjacent to the Sonoran Desert but no record is at hand of a member of this genus growing in the desert. *G. flaccidum* Small was collected at La Chuparosa in the Cape Region of Baja California a short distance outside the Sonoran Desert.

## Family 63. OXALIDACEAE. Woodsorrel Family

Mostly perennial, herbaceous, caulescent or acaulescent plants with acid juice, from bulbs or creeping rootstocks. Leaves digitately compound, with 3 (or rarely 4) cuneate to obreniform leaflets. Flowers perfect, regular. Sepals 5, basally united into a tube. Petals 5, yellowish, pink, or purplish. Stamens 10, fused at base, alternate filaments longer than others. Ovary superior. Styles distinct or united at base. Fruit in ours a columnar capsule, 5-celled, longitudinally dehiscent. Seeds strongly sculptured transversely.

## 1. **OXALIS** L., Sp. Pl. 433, 1753

Characters of the family.

Sepals without apical tubercles; plants caulescent; corollas yellow; longer filaments glabrous, not appendaged:
- Pedicels and stems appressed-pubescent or glabrate; root not a taproot:
  - Shorter filaments glabrous; sepals not purplish nor reddish; roots fibrous:
    - Leaflets more or less pubescent on both surfaces; sepals 3–4 mm. long; stems usually creeping ........ 1. *O. repens*
    - Leaflets glabrous or nearly so; sepals 4–5 mm. long; stems erect or at length decumbent ........ 2. *O. stricta*
  - Shorter filaments puberulent; sepals purplish or reddish at tips; roots fusiform-tuberous ........ 3. *O. albicans*
- Pedicels and stems pilose with spreading or slightly recurved hairs; root a slender fusiform taproot ........ 4. *O. pilosa*

Sepals bearing 2 lance-ovate tubercles at tip of each; plants acaulescent, from a fleshy bulb; corollas pink or lavender; filaments all pubescent, longer ones appendaged on back from near base to about middle ........ 5. *O. amplifolia*

### 1. **Oxalis repens** Thunb., Diss. Oxalis 16, 1781

Stems creeping and prostrate or rarely only decumbent, branched at base, 1–6 dm. long; herbage loosely pubescent with scattered subappressed hairs or subglabrate; pedicels slender, 1–5 cm. long; leaflets 3, deep green, blades cuneate-obcordate, 3–12 mm. broad, ciliate and often with a few scattered hairs on both surfaces; peduncles equaling or surpassing petioles; pedicels strigose, 5–15 mm. long; sepals narrowly ovate, 3–4 mm. long; petals yellow, 6–10 mm. long; filaments glabrous; ovary pilosulous; capsules abruptly narrowed at apex, pubescent, 1–2 cm. long.

A native of the Old World and sparingly established in lawns and waste places near dwellings in Arizona but more extensively introduced in other parts of the United States. March–Sept.

A closely related plant, *O. corniculata* L., has erect or slightly decumbent stems and is also established in the United States and other parts of the world. *O. repens* has been reduced to a synonym under *O. corniculata* by some authors and is recognized as a variety of that species by others.

### 2. **Oxalis stricta** L., Sp. Pl. 435, 1753

?*Oxalis lyoni* Pursh, Fl. Amer. Sept. 322, 1814.
?*Oxalis furcata* Ell., Bot. S.C. & Ga. **1:** 527, 1821.
*Oxalis corniculata* var. *dillenii* Trelease, in A. Gray, Syn. Fl. N. Amer. **1**$^1$**:** 365, 1897.
*Xanthoxalis stricta* Small, Fl. SE. U.S. 667, 1903.

Stems erect, 1–2 dm. long, or when longer (to 5 dm.) becoming decumbent, rather densely crowded with leaves, usually densely strigillose petioles 1–4 cm. long, erect or ascending; leaf blades 15–20 mm. wide, broadly obcordate, glabrous or sparsely pubescent and ciliate, peduncles mostly surpassing petioles; pedicels 5–18 mm. long, strigose; sepals narrowly lance-ovate, 4–5 mm. long, pubescent; petals yellow, 6–10 mm. long; filaments slender, all glabrous; capsules 3.5–4 mm. in diameter, 1.5–3 cm. long, abruptly pointed at apex, closely puberulent.

Mainly in sandy or light loam soil, chiefly Upper Sonoran Zone to Carolinian Zone, Pima and Santa Cruz Counties, Arizona, to Florida and Nova Scotia. Possibly not entering the Sonoran Desert, but occurring at altitudes of 3,000 feet near Tucson. April–June.

**3. Oxalis albicans** H. B. K., Nov. Gen. & Sp. **5:** 244, 1822

*Oxalis wrightii* A. Gray, Pl. Wright. **1:** 27, 1852.
*Xanthoxalis wrightii* Abrams, Bull. Torrey Club **34:** 264, 1907.
*Xanthoxalis albicans* Small, N. Amer. Fl. **25:** 54, 1907.

Stems erect or somewhat decumbent, 1–4 dm. long, from fusiform-tuberous roots; herbage and stems usually densely appressed-pubescent; petioles slender, 3–10 cm. long; blades of leaflets obcordate, 5–15 mm. wide, about as long, pale or dark green, sparsely pubescent with subappressed hairs on both surfaces; peduncles equaling or slightly exceeding petioles; pedicels strigose, to 1.5 cm. long, strongly deflexed in fruit; sepals 4–5 mm. long, lanceolate, reddish or purplish on upper third or at tips; petals yellow or faintly suffused with purple, 6–10 mm. long; longer filaments glabrous, shorter ones puberulent; capsule stoutish, 1.5–2 cm. long, minutely puberulent, gradually narrowed at apex.

Usually in seepy places and often in shade, Lower and Upper Sonoran Zones, Pima County to Apache County, Arizona, to Texas and northern Mexico. March–Sept.

**4. Oxalis pilosa** Nutt., in Torr. & Gray, Fl. N. Amer. **1:** 212, 1838

*Xanthoxalis pilosa* Small, N. Amer. Fl. **25:** 54, 1907.

Stems 1–4 dm. long, erect or decumbent, densely pilose with spreading or somewhat retrorse hairs, growing in tufts from a slender, woody, fusiform, or tapered taproot; petioles 1–5 cm. long; leaflets bright green, 5–18 mm. broad, about one-half to two-thirds as long, pubescent on both surfaces; peduncles from slightly shorter to somewhat longer than petioles; pedicels hirsute, strongly deflexed in fruit; sepals linear-lanceolate, 4.5–6 mm. long, at times faintly tinged with red near tips; petals yellow, 8–12 mm. long; longer filaments glabrous, slender, shorter ones puberulent; capsules stoutish, 12–18 mm. long, hirsutulous, abruptly narrowed at apex.

Mostly in light soil and often in partial shade, chiefly Upper Sonoran and Transition Zones, near the coast in central California to northern Baja California, and sparingly to Arizona and northern Sonora. Not definitely known from the Sonoran Desert proper but occurring in canyons along its upper margin in Arizona. May.

**Oxalis californica** (Abrams) Knuth differs chiefly in having closely appressed pubescence. It occurs on the western slopes of the mountains in southern California and northern Baja California but has not been found within the desert. It may occur as a straggler in canyons leading down to the Colorado Desert.

**5. Oxalis amplifolia** (Trelease) Knuth, Notizblatt No. 67: 314, 1919

*Oxalis latifolia* Trelease, Mem. Bost. Soc. Nat. Hist. **4:** 91, 1888. Not *O. latifolia* H. B. K., 1822.
*Oxalis divergens* var. *amplifolia* Trelease, in A. Gray, Syn. Fl. N. Amer. **1**[1]**:** 368, 1897.
*Ionoxalis amplifolia* Rose, Contr. U.S. Nat. Herb. **10:** 110, 1906.

Acaulescent, presumably from a scaly bulb; petioles erect, 1.5–2 dm. tall, glabrous; leaflets 3, obreniform, 1.8–4 cm. broad, 1–2.2 cm. long, dark green, glabrous, or with a few scattered, closely appressed hairs along veins beneath, notch broad and shallow; peduncles about equaling or slightly shorter than petioles, glabrous, 6–20-flowered; pedicels slender, 1–2 cm. long, glabrous; sepals linear-lanceolate, 4–5.5 mm. long, glabrous, each bearing 2 brownish tubercles 0.4–0.6 mm. long at apex, tubercles entire or sometimes bilobed at base; petals pinkish lavender, 10–12 mm. long; stamens minutely puberulent above middle with brownish hairs, longer filaments each bearing an oblong appendage from near base to about middle, shorter ones unappendaged; styles densely puberulent, ovary sparsely so; capsule stout, 3–3.5 mm. in diameter, 7–11 mm. long, sparingly and minutely puberulent.

In shady places in rich soil, mainly Upper Sonoran Zone, southern Arizona to western Texas. Aug.–Sept.

Specimens from among lava rocks on the mesa 5 miles north of Comondú, Baja California, resembling this species have shorter, stouter filaments with less prominent appendages and more numerous flowers on a peduncle. They may represent a distinct species but the material is incomplete and a definite disposition of them must await more ample specimens.

## Family 64. **LINACEAE.** Flax Family

Annual or perennial herbs, mostly with alternate entire leaves. Stipules lacking or represented by a gland. Flowers cymose or racemose, perfect, regular. Sepals 5. Petals 5, deciduous early. Styles 2–5, distinct. Ovary superior, 5-celled or 10-celled by a complete or incomplete false partition from back of each cell. Fruit a capsule, dehiscing through the false partition and usually septicidally also. Seeds (in ours) flattened, oily.

### 1. **LINUM** L., Sp. Pl. 277, 1753

Characters of the family. Petals blue, or rarely white in ours.

Sepals, at least inner ones, ciliate; plants annual .................... 1. *L. usitatissimum*
Sepals entire, not ciliate; plants perennial (sometimes flowering first season) .. 2. *L. lewisii*

**1. Linum usitatissimum** L., Sp. Pl. 277, 1753

Annuals to 8 dm. tall with striate, lightly angled stems and erect or ascending, linear-lanceolate, sessile leaves 1–3 cm. long; bracts similar to leaves but 1 cm. long or less; pedicels subfiliform, to 3 cm. long; outer sepals elliptic to lanceolate, eciliate, inner sepals ovate, 5–6 mm. long in flower, 7–9 mm. in fruit, papery-hyaline near margins; petals blue, 1–1.5 cm. long; styles distinct or essentially so; capsule broadly ovoid, 8–11 mm. high; seeds light brown, smooth and shining, 4–5 mm. long.

Widely cultivated in the temperate zones, sparingly escaped near dwellings and cultivated fields at a few localities in the Sonoran Desert (near Indio). March–April.

**2. Linum lewisii** Pursh, Fl. Amer. Sept. 210, 1814

*Linum decurrens* Kell., Proc. Calif. Acad. **3:** 44, *fig. 11,* 1863.
*Linum lyallanum* Alef., Bot. Zeit. **25:** 251, 1867.

Perennial to 8 dm. tall, usually considerably branched at base, stems faintly striate; leaves linear or narrowly lanceolate-linear, 1–2 cm. long, acute, sessile, ascending or erect and usually longer than internodes, glabrous; bracts 5–8 mm. long; pedicels slender, 0.5–3 cm. long; sepals ovate, eciliate, 5–7 mm. long, inner slightly broader than outer, mucronulate; petals blue or sometimes white, 1–1.5 (or 2) cm. long; styles distinct; capsules 5–7 mm. long, broadly ovoid, acute at apex, fruiting pedicels spreading horizontally or curved outward; seeds dark brown, 2–2.2 mm. wide, 4–4.5 mm. long, smooth and shining.

Hillsides and mountain slopes, chiefly Transition Zone to Hudsonian Zone, but rarely straggling into margins of the Sonoran Desert (Cunningham Pass, Yuma County, Arizona), southern Alaska to Texas and northern Mexico. March–Sept.

**Linum puberulum** (Engelm.) Heller, Plant World **1:** 22, 1897, a low annual 1–3 dm. tall with grayish, minutely puberulent herbage, linear-lanceolate sepals 7–9 mm. long, bearing marginal glandular cilia, and with yellow petals, was collected not far from Tucson in 1884 by W. F. Parish and labeled "Tucson, Ariz." It has not been found at low elevations in Arizona since, and Parish's plants probably were collected some distance from that city.

## Family 65. ERYTHROXYLACEAE. Coca Family

Glabrous or subglabrous trees or shrubs. Leaves alternate, simple, entire, petioled, with small, often imbricated stipules. Flowers small, solitary or in few-flowered axillary fascicles. Calyx 5-lobed or 5-cleft. Petals 5, each bearing a bilobed ligule within. Stamens 10, coalesced into a tube at base. Ovary 3-celled or rarely 4-celled; ovules 1–2 in each cell; styles 3 or sometimes 4, distinct or somewhat united. Fruit a drupe, usually 1-seeded through abortion of most of ovules. Endosperm mealy or scanty. Embryo straight.

### 1. ERYTHROXYLON L., Syst. ed. 10, 1035, 1759

Characters of the family.

**1. Erythroxylon mexicanum** H. B. K., Nov. Gen. & Sp. **5:** 178, 1822

Large shrub or small tree with stiff stoutish twigs and dark or light gray bark; leaves obovate to oblong-obovate, 1.5–3 cm. wide, 2.5–7 cm. long, broadly cuneate at base, rounded to shallowly retuse at apex, subcoriaceous, dark green above, paler and reticulately veined beneath, glabrous; stipules ovate, acute, 2–3 mm. long; petioles about equaling stipules; flowers in few-flowered axillary fascicles; pedicels slender, 2–3.5 mm. long; calyx lobes narrowly triangular-lanceolate, about 1 mm. long, equaling calyx cup; petals ovate, about 3 mm. long; stamens 5–6 mm. long, filaments filiform; drupe scarlet, ellipsoid, 9–11 mm. long.

Open slopes and valley floors, mainly subtropical and tropical areas, southern Sonora to Tamaulipas, Vera Cruz, and Guerrero, and on the Tres Marías Islands. July–Oct.

## Family 66. ZYGOPHYLLACEAE. Caltrops Family

Annual or perennial herbs, shrubs, or trees, often strong-scented and yielding a bitter gum or resin. Leaves opposite or alternate, simple, digitately or pinnately compound, leaves or leaflets usually entire. Flowers axillary, perfect, regular or essentially so. Sepals 4, 5, or 6. Petals 5, or rarely 4 or 6, hypogynous. Stamens twice or sometimes three times the number of the petals; filaments simple or appendaged near base. Ovary of 2–5 united carpels, 2–5-celled, superior. Styles united; stigma capitate, clavate, or slightly lobed. Fruit capsular or carpels falling entire.

Leaves alternate, simple or odd-pinnate; flowers mostly in few-flowered clusters, rarely solitary .......................................................... 1. *Viscainoa*
Leaves opposite, or rarely appearing alternate by abortion of one of a pair, even-pinnate, or digitately 3–5-foliolate; flowers solitary in the axils:
  Stipules acicular-spinose; leaves digitately 3–5-foliolate .................... 2. *Fagonia*

Stipules not spine-tipped; leaves even-pinnate:
Corollas blue or bluish; stipules deciduous; plant is a tree .............. 3. *Guaiacum*
Corollas yellowish, orange, or rarely whitish; stipules usually persistent; plants slender-stemmed shrubs or herbs:
Fruit pilose-woolly; stamens with scale-like appendages nearly equaling filaments; leaflets 1 pair, glutinous; a shrub to 4 m. high ....................... 4. *Larrea*
Fruit puberulent but not pilose; stamens without appendages; leaflets 3–9 pairs, not glutinous; herbs usually less than 0.5 m. tall:
Stems prostrate; fruit spiny, dividing into as many nutlets as carpels, each 3–5-seeded ....................................................... 5. *Tribulus*
Stems decumbent or ascending; fruit ridged or tuberculate, not spiny, dividing into twice as many 1-seeded nutlets as carpels ................... 6. *Kallstroemia*

## 1. VISCAINOA Greene, Pittonia 1: 163, 1888

Rounded compact shrubs with minutely canescent-pubescent leaves and brittle ascending branches. Leaves alternate, simple or odd-pinnate, with 3–5 leaflets. Peduncles 1- to few-flowered. Sepals 5, deciduous. Petals 5, short-clawed. Stamens 10. Ovary stipitate, 3–5-celled. Styles 3–5, persistent. Fruit a 3–5-lobed, moderately inflated capsule. Seeds mostly 2 in each cell.

Leaves simple, cinereous; flowers pale yellow; filaments brown-glandular near base, filiform ............................................................ 1. *V. geniculata*
Leaves odd-pinnate, green; flowers bright yellow; filaments not glandular, dilated at base .. ................................................ 1a. *V. geniculata* var. *pinnata*

### 1. **Viscainoa geniculata** (Kell.) Greene, Pittonia **1:** 163, 1888, var. **geniculata**

*Staphylea geniculata* Kell., Proc. Calif. Acad. **2:** 22, 1859.
*Chitonia simplicifolia* S. Wats., in Orcutt, W. Amer. Sci. **2:** 58, 1886.

Cinereous-pubescent, 2–6 m. tall, branches stoutish, rigid, and crooked; leaves simple, elliptic, ovate, or obovate, 8–30 mm. wide, 1.5–5 cm. long, entire or sometimes slightly and irregularly lobed, coriaceous; peduncles 2–5 cm. long; pedicels 1–1.5 cm. long, slender; sepals usually 4, broadly oblong, 7–9 mm. long, deciduous; petals obovate, 10–15 mm. long, pale yellow or cream, strongly veined; stamens 5–7 mm. long; capsules usually 4-celled and 4-angled, 1–1.5 cm. wide, 1.5–2 cm. long, styles persistent, forming a beak 3–5 mm. long; seeds obovoid, 4–5 mm. long, black, with an oily testa.

Desert washes and slopes, Lower Sonoran Zone, from vicinity of Rosario to La Paz, Baja California, on many of the islands in the Gulf of California, and near Guaymas, Sonora. Jan.–May.

### 1*a*. **Viscainoa geniculata** var. **pinnata** I. M. Johnston, Univ. Calif. Pub. Bot. **7:** 439, 1922

*Viscainoa pinnata* Gentry, Madroño **5:** 161, 1940.

A shrub 2–3 m. tall with stems slender, green, sparsely puberulent but eventually glabrate except at nodes; stipules subulate, 4–6 mm. long; petioles 1–1.5 cm. long; leaflets 3 or 5 or some leaves simple, leaflets elliptic to ovate, 2–5 cm. long, 0.8–2.5 cm. wide, green, entire, mucronulate at tip, pinnately veined; peduncles 1–3 cm. long; pedicels slender, 8–12 mm. long, pubescent; sepals oblong, 7–9 mm. long, tomentulose, obtuse or rounded at apex; petals bright yellow, obovate, 10–12 mm. long; stamens

7–8 mm. long, filaments slightly dilated at base, not glandular; capsule obovoid, 12–15 mm. broad, 2–3 cm. long, green, sparsely puberulent.

Known only from the western side of the peninsula of Baja California near San Juanico.

## 2. FAGONIA L., Sp. Pl. 386, 1753

Diffusely branching, prostrate to ascending subshrubs from a more or less woody base, with angled stems. Herbage glabrous, pubescent, or glandular-glutinous. Leaves opposite, pinnately 3–7-foliolate, or in one species simple, callous or spinose at tips of leaflets. Stipules linear-subulate, spinose. Flowers solitary in their axils, rose-colored, purplish, or rarely yellowish or nearly white. Sepals 5, caducous. Petals 5, caducous. Stamens 10, borne on an inconspicuous disk. Stigma simple. Ovary 5-celled, each cell 2-ovuled. Fruit a pyramidal-ovoid capsule, 5-angled. Seeds ellipsoid.

Leaves 5–7-foliolate . . . . . . . . . . . . . . . . 1. *F. palmeri*
Leaves 3-foliolate:
  Stipules acerose, ascending, as long as or exceeding petioles . . . . . . . . 2. *F. densa*
  Stipules subulate to linear-subulate, reflexed to spreading, not equaling petioles:
    Stipules subulate, rarely spreading, usually 1–3 (never over 6) mm. long; plants upright to prostrate; flowers about 1 cm. in diameter:
      Branches glabrous to stipitate-glandular, sometimes with some nonglandular hairs:
        Leaflets glandular; branches rarely scabrous; fruits and ovaries glandular and puberulent:
          Pedicels 1–6 mm. long at anthesis, 2–10 mm. long in fruit; lateral leaflets narrower than terminal one, to 9 mm. wide . . . . . . . . 3. *F. californica*
          Pedicels 6–15 mm. long at anthesis, 8–20 mm. long in fruit; lateral leaflets scarcely narrower than terminal one, 1–3.5 mm. wide . . . . . . 4. *F. longipes*
        Leaflets glabrous; branches soon scabrous; fruits and ovaries puberulent but never glandular . . . . . . . . . . . . . . . . 5. *F. laevis*
      Branches villous, never glandular . . . . . . . . . . . . 6. *F. villosa*
    Stipules linear-subulate, rarely reflexed, to 8 mm. long or more; plants prostrate-spreading; flowers about 1.5 cm. in diameter:
      Branches bearing subsessile to short-stipitate glands to 1 mm. in diameter toward ends, glabrous at base . . . . . . . . . . . . 7. *F. pachyacantha*
      Branches bearing many stipitate glands to 0.5 mm. in diameter, and nonglandular hairs, glands most numerous toward tips, hairs toward bases of branches . . . . . . . . . . . . . . . . . . . . 8. *F. barclayana*

**1. Fagonia palmeri** Vasey & Rose, Contr. U.S. Nat. Herb. **1:** 82, 1890

*Fagonia chilensis* var. *palmeri* I. M. Johnston, Proc. Calif. Acad. IV, **12:** 1052, 1924.

Flat-topped to rounded prostrate to upright perennial to 4.5 dm. tall, 1 m. wide; young herbage densely stipitate-glandular, older branches glabrate; stipules acerose, ascending, 2–9 mm. long; leaflets 5–7 (1–2 sometimes aborting), subfleshy, flattened, linear to linear-lanceolate, about 1 mm. in diameter, 2–14 mm. long; petioles 4–12 mm. long; flowers about 1.5 cm. in diameter; pedicels 2–7 mm. long, glandular-stipitate; sepals oblong-lanceolate, glandular-stipitate, 1.5–2.5 mm. wide, 4–5 mm. long; petals rose, 2–4 mm. wide, 5–8 mm. long; filaments 4 mm. long; ovary pubescent, style 2–3.5 mm. long; fruiting pedicels 3–8 mm. long; fruit puberulent and glandular, 4–6 mm. in diameter, 4–5 mm. high; beak 3–4 mm. long.

Washes and hillsides or gypsum cliffs, Lower Sonoran Zone, vicinity of Santa Rosalia and on Isla San Marcos, Baja California. March–May.

**2. Fagonia densa** I. M. Johnston, Proc. Calif. Acad. IV, **12:** 1052, 1924

Dense, compact, upright, globose subshrub 2.5–8 dm. tall, to 1 m. in spread; herbage densely short-stipitate-glandular and bearing many subsessile glands, glabrate in age; younger herbage highly glutinous; stipules acerose to linear-lanceolate, ascending, 3–21 mm. long, together with whorled, acerose leaves crowded at nodes, internodes relatively short; leaflets 3 or rarely 4, acerose to linear-lanceolate, to 1 mm. wide, 1–11 mm. long, deciduous; petioles 3–18 mm. long; pedicels 2–3 mm. long; sepals oblong-lanceolate to oblong, green to yellow, 1.5 mm. wide, 4–5 mm. long; corollas about 1.5 cm. in diameter, petals pink to purple, 2–4 mm. wide, 6–9 mm. long; filaments 4–5 mm. long; ovary pubescent and with small subsessile glands moderately numerous; style 3–4 mm. long; fruiting pedicels 2–7 mm. long; fruit about 6 mm. in diameter, 4–5 mm. high, puberulent and glandular; beak 3–4.5 mm. long; seeds usually 2 in each locule.

Granitic hillsides and in gypsum soils, Lower Sonoran Zone, Isla de la Guarda and vicinity of Bahía de los Angeles to Isla San Lorenzo and San Francisquito, Baja California. March–May.

**3. Fagonia californica** Benth., Bot. Sulph. 10, 1844

*Fagonia californica* var. *hindsiana* Benth., loc. cit.
*Fagonia rosei* Standley, Proc. Biol. Soc. Wash. **24:** 247, 1911.
*Fagonia cretica* var. *californica* Engler, in Engler & Drude, Veg. der Erde **9**[3]**:** 731, 1915.
*Fagonia chilensis* var. *rosei* I. M. Johnston, Proc. Calif. Acad. IV, **12:** 1051, 1924.
*Fagonia californica* subsp. *rosei* Wiggins, Contr. Dudley Herb. **4:** 19, 1950.

Prostrate-spreading perennial herb with little or no woody tissue; herbage densely stipitate-glandular and with scattered simple, glandless hairs; leaves and older branches at length glabrate, branches more or less scabrous; stipules subulate, reflexed to spreading, 1–5 mm. long; leaflets 3, elliptic to oblong-lanceolate or oblong, glandular in youth, soon glabrate, apical leaflet to 12 mm. wide, 18 mm. long, lateral leaflets to 9 mm. wide, 15 mm. long, commonly caducous; petioles 1–17 mm. long; flowers about 1 cm. in diameter; pedicels 1–6 mm. long at anthesis, 2–10 mm. long in fruit; sepals oblong-lanceolate, glandular, green to purple, 1–1.5 mm. wide, 2–3.5 mm. long, at length glabrate; petals pink to dark red-purple, 2–4 mm. wide, 5–7 mm. long; filaments about 4 mm. long; ovary pubescent and glandular, style 2 mm. long; fruit puberulent and glandular, 3–7 mm. wide, 3–5 mm. high, beak 1.5–2.5 mm. long; seed 1 per locule.

Sandy and rocky soils, hillsides and along arroyos, Lower Sonoran Zone, northern Magdalena Plain to the Cape Region, Baja California. Sept.–April.

**4. Fagonia longipes** Standley, Proc. Biol. Soc. Wash. **24:** 250, 1911

Prostrate to spreading-ascending suffrutescent plant 1–4 dm. tall, often 6–10 dm. wide, woody at base, but stems slender; herbage moderately to sparsely stipitate-glandular, glands small, often subsessile; petioles 9–20 mm. long, minutely scaberulous on angles; stipules linear-subulate, spreading, 1–3.2 mm. long, usually slightly curved downward, stramineous; leaflets 3, essentially alike, linear to linear-elliptic, 1–3.5 mm. wide, 5–16 mm. long, acute to apiculate at apex; pedicels slender, 6–15 mm. long and erect at anthesis, 8–20 mm. long and often deflexed in fruit, distinctly glandular; sepals elliptic-ovate, acuminate, roseate, 0.8–1.2 mm. wide, 3–3.6 mm. long, stipitate-glandular on backs, margins subscarious; petals magenta-rose, 7–8 mm. long, 4–5 mm. wide; fruit 5–6 mm. wide, 4–5 mm. high, finely stipitate-glandular, often with scattered, non-glandular hairs intermixed with glands, especially along upper angles; beak 2–2.4 mm. long; seed 1 in each locule, 1.8–2 mm. long, brown, shining.

Rocky hillsides, gravelly plains and along arroyos, Lower Sonoran Zone, near mouth of Bill Williams River, Arizona, and northeastern Colorado Desert, south into northern Sonora. Feb.–April.

**5. Fagonia laevis** Standley, Proc. Biol. Soc. Wash. **24:** 249, 1911

*Fagonia chilensis* var. *aspera* I. M. Johnston, Proc. Calif. Acad. IV, **12:** 1051, 1924. In part.
*Fagonia chilensis* var. *laevis* I. M. Johnston, loc. cit.
*Fagonia californica* subsp. *laevis* Wiggins, Contr. Dudley Herb. **4:** 19, 1950.

Upright to spreading plant to 6 dm. tall, 1 m. wide; herbage essentially glabrous except pedicels, sepals, and occasionally stipules sometimes with small stipitate glands, some specimens entirely glabrous; older branches scabrous, often becoming stoloniferous, prostrate main branches bearing numerous erect branches; stipules subulate, reflexed to spreading, 1–6 mm. long; leaflets 3, linear-elliptic, apical leaflet to 5 mm. wide, 18 mm. long, lateral leaflets to 3 mm. wide, 15 mm. long, one or both laterals commonly caducous; petioles 2–15 mm. long; flowers about 1 cm. in diameter; pedicels 1.5–11 mm. long; sepals elliptic to lanceolate, green to purple, about 1 mm. wide, 2–3 mm. long; petals pink to dark red-purple, 1.5–3 mm. wide, 4–7 mm. long; filaments 3–4.5 mm. long; ovary glabrous to pubescent, style 1–2 mm. long; fruiting pedicels glabrous to glandular, 1.5–11 mm. long; fruit glabrous to puberulent, 3–6 mm. wide, 4–5 mm. high; beak 1–5.2 mm. long; locules 1-seeded.

Desert hillsides, arroyos and outwash slopes, Lower Sonoran Zone, southern Nevada, Mojave and Colorado Deserts and northwestern Arizona to vicinity of Guaymas, Sonora, and central Baja California. Oct.–July.

**6. Fagonia villosa** Porter, Contr. Gray Herb. **192:** 120, 1963

Prostrate-spreading plant to 4 dm. tall, 10–12 dm. wide, with densely to sparsely villous herbage, glandular hairs none; older branches at length glabrate or sometimes scabrous; stipules spreading to reflexed, subulate, 1–6 mm. long; petioles 1–10 (15) mm. long; leaflets 3, ovate to obovate or elliptic, terminal one 1–8.5 mm. wide, 2–23 mm. long, lateral ones to 4 mm. wide, 3–16 mm. long; pedicels 1–7 (rarely to 9) mm. long; sepals linear-oblong, villous, 1–1.5 mm. wide, 2–4 mm. long, green to purple; petals pink to magenta-purple, 1–3 mm. wide, 4–6 mm. long; filaments 3–4 mm. long; ovary pubescent, style 1–2 mm. long; fruiting pedicels 1–6 (rarely to 10) mm. long; fruit puberulent, 3.5–6 mm. wide, 3.5–4.5 mm. high; beak 1–2 mm. long; seeds 1–2 in each locule.

Rocky hillsides, desert plains and mesas, Lower Sonoran Zone, southern flanks of Sierra San Pedro Mártir to the Cape Region, Baja California. Feb.–May.

**7. Fagonia pachyacantha** Rydb., in Vail & Rydb., N. Amer. Fl. **25:** 105, 1910

*Fagonia californica* var. *glutinosa* Pringle ex Vail, Bull. Torrey Club **22:** 229, 1895. Not *F. glutinosa* Delile, 1813.
*Fagonia viscosa* Rydb., in Vail & Rydb., N. Amer. Fl. **25:** 104, 1910. Not *F. viscosa* Hochst. ex Boiss., 1867.
*Fagonia chilensis* var. *glutinosa* I. M. Johnston, Proc. Calif. Acad. IV, **12:** 1051, 1924.
*Fagonia chilensis* var. *pachyacantha* I. M. Johnston, loc. cit.
*Fagonia californica* subsp. *pachyacantha* Wiggins, Contr. Dudley Herb. **4:** 19, 1950.

Prostrate-spreading plant to 1.5 m. wide; younger parts of branches closely beset with large short-stipitate to subsessile, globular golden glands that become cup-shaped when dry, older parts at length glabrate; younger growth often appearing yellow owing to numerous glands; stipules stout, linear-subulate, spreading to slightly reflexed, glandular to glabrate, 3–16 mm. long; petioles glandular to glabrate, 2–16 mm. long; leaflets 3, ovate to elliptic, slightly obovate or linear and terete, glandular but at length glabrate, apical leaflet to 10 mm. wide, 1–26 mm. long, lateral ones to 7 mm. wide, 1–20 mm. long, one or both laterals often caducous; pedicels glandular, 1–7 mm. long, not elongating significantly in fruit; flowers about 1.5 cm. in diameter; sepals ovate, 1–1.5 mm. wide, 2–2.5 mm. long, green to purple, glandular to glabrate; petals light to dark red-purple, 2.5–5.5 mm. wide, 5–8 mm. long; filaments 3.5–5 mm. long; ovary pubescent and glandular, 4–5 mm. wide, 3.5–4 mm. high; beak 1.5–3.5 mm. long; seed solitary in each locule.

Sandy flats and along arroyos, Lower Sonoran Zone, Colorado Desert and southwestern Arizona into northern Sonora and to central Baja California. Feb.–May.

**8. Fagonia barclayana** (Benth.) Rydb., in Vail & Rydb., N. Amer. Fl. **25:** 104, 1910

*Fagonia californica* var. *barclayana* Benth., Bot. Sulph. 10, 1844.
*Fagonia insularis* Standley, Proc. Biol. Soc. Wash. **24:** 247, 1911.
*Fagonia chilensis* var. *barclayana* I. M. Johnston, Proc. Calif. Acad. IV, **12:** 1051, 1924.
*Fagonia chilensis* var. *insularis* I. M. Johnston, loc. cit.
*Fagonia californica* subsp. *laevis* Wiggins, Contr. Dudley Herb. **4:** 19, 1950.

Prostrate-spreading plant to 1.5 m. in diameter; herbage densely stipitate-glandular with scattered simple hairs among glands on young growth, more markedly pubescent with nonglandular hairs and fewer glands in age, at length more or less glabrate; stipules stout, linear-spatulate, spreading, 3–9 mm. long; petioles 3–15 mm. long; leaflets 3, linear to broadly ovate, glandular, pubescent, apical leaflet to 19 mm. wide, 28 mm. long, lateral leaflets to 10 mm. wide, 20 mm. long; flowers about 1.5 cm. in diameter; pedicels glandular and pubescent, 3–14 mm. long; sepals ovate, 1.5–2 mm. wide, 2.5–4 mm. long, green to purplish, glandular on backs; petals rose-magenta to purple, 2.5–5.5 mm. wide, 5–9 mm. long; filaments 4–6 mm. long; ovary pubescent and glandular, style 2.5–4 mm. long; fruits densely glandular and sparsely pubescent, 4–6 mm. wide, 3–5 mm. high; beak 2–4 mm. long; seed solitary in each locule.

Rocky arroyos, canyons, hillsides and mesas, Lower Sonoran Zone, vicinity of Mulegé south, mainly along eastern side of the peninsula, to the Cape Region; also on several islands in the Gulf and in the vicinity of Guaymas, Sonora. Nov.–June.

## 3. GUAIACUM L., Sp. Pl. 381, 1753

Trees or shrubs with stout branches, hard resinous wood, and closely crowded, deep green, opposite, pinnate leaves. Leaflets 2–6 pairs, entire. Flowers pedunculate, regular. Sepals 4–5. Petals 4–5, blue or purple. Stamens twice as many as petals, inserted on an inconspicuous disk; filaments filiform or slenderly subulate, naked. Ovary 2–5-celled, obovate to clavate, 2–5-lobed, short-stipitate. Stigma entire or 2–5-toothed. Ovules 8–10 in each cell. Fruit coriaceous, smooth, septicidal, glabrous or pubescent, angles wing-like. Seeds solitary.

Leaflets 6–10 per leaf, glabrous; large shrub or small tree to 8 m. tall:
  Ovary glabrous .............................................. 1. *G. coulteri* var. *coulteri*
  Ovary puberulent .......................................... 1a. *G. coulteri* var. *palmeri*
Leaflets only 2 per leaf, puberulent; low shrub rarely over 2 m. tall ....... 2. *G. unijugum*

**1. Guaiacum coulteri** A. Gray, Pl. Thurb. 312, 1855, var. **coulteri**
*Guaiacum afrum* Sessé & Moc., Fl. Mex. 118, 1894. Not *G. afrum* L., 1753.

A shrub or small tree to 8 m. tall with light gray bark; leaves 3–6 cm. long, crowded; leaflets 6–10, linear-oblong to elliptic-oblong, 3–6 mm. wide, 1–2.5 cm. long, glabrous, mucronulate, petioles and rachises sparsely to densely pubescent; flowers 3–12 on axillary peduncles; pedicels 1–1.5 cm. long, usually pubescent; sepals 5, broadly ovate to oblong, 4–7 mm. long, rounded at apex, puberulent within and without; petals 5, blue, broadly obovate, 8–15 mm. long; ovary usually 5-celled, obovate, somewhat compressed, glabrous or essentially so; fruit green or tinged with purple, obcordate, 1–1.5 cm. long, usually slightly broader than long, narrowly winged; seeds 10–12 mm. long, ellipsoidal, covered with a pale orange or yellow aril.

Gravelly bajadas and plains, Lower Sonoran Zone, from near Carbo and Ures, Sonora, to Guerrero, Tepíc, and Oaxaca. March–July.

**1*a*. Guaiacum coulteri** var. **palmeri** (Vail) I. M. Johnston, Proc. Calif. Acad. IV, **12:** 1053, 1924

Differing from var. *coulteri* in having slightly broader, suborbicular petals, and pubescent ovaries and fruits.

Gravelly plains and gentle slopes, Lower Sonoran Zone, in the vicinity of Guaymas (San Pedro and San Carlos Bay) Sonora, and less commonly north to Hermosillo and south into Sinaloa. March–July.

**2. Guaiacum unijugum** Brandegee, Univ. Calif. Pub. Bot. **6:** 183, 1915

A rigidly branched divaricate shrub 1–2 m. tall, with brownish gray bark; youngest twigs finely puberulent, soon glabrate; leaves mostly on short spurs 1.5–4 mm. long; stipules ovate-acuminate, 2–2.5 mm. long, puberulent, persistent on spurs, light tan; petioles 3–5 mm. long; leaflets 2, asymmetrically rhomboid-ovate, 4–6 mm. wide, 8–11 mm. long, rounded at base, acute and mucronulate at apex, finely puberulent, lightly veined; flowers on short spurs, usually solitary on a spur; pedicels slender, 7–12 mm. long, puberulent; calyx lobes broadly ovate, cucullate, 4–6 mm. long, thin, moderately puberulent, yellowish; petals blue, broadly oblanceolate to obovate, 9–12 mm. long; stamens 5–6 mm. long, glabrous; ovary obovoid, 5-lobed, puberulent, on a gynobase 1–1.5 mm. long; style slenderly subulate, 4–5 mm. long, persistent; fruit 15–20 mm. long, slightly wider, puberulent; seeds rarely more than one in each fruit.

Sand dunes and adjacent arid hills near the coast, Lower Sonoran Zone, eastern side of the Cape Region, Baja California. Aug.–Sept.

## 4. LARREA Cav., Anal. Hist. Nat. (Madrid) 2: 119, *pls. 18–19,* 1800

Strong-scented evergreen shrubs with several to many brittle, strongly nodose, slender branches from a branching root crown. Leaves opposite, of 2 leaflets connate at base, sessile or short-petiolate. Stipules small, persistent. Flowers solitary, axillary. Sepals 5, imbricate, woolly outside, caducous. Petals 5, yellow, obovate to spatulate, hypogynous. Stamens 10, each adnate basally to outside of a membranous, 2-toothed or 2-cleft, sometimes laciniate scale nearly as long as filament. Ovary short-stipitate, 5-celled, globose to ovoid, pilose, each cell 5–6-ovuled. Stigmas 5. Fruit globose to obovoid, 5-angled, separating into 5, indehiscent, 1-seeded carpels.

**1. Larrea tridentata** (Sessé & Moc. ex DC.) Coville, Contr. U.S. Nat. Herb. **4:** 75, 1893

*Zygophyllum tridentatum* Sessé & Moc. ex DC., Prod. **1:** 706, 1824.
(For added synonymy see Contr. Gray Herb. **192:** 110–111, 1963.)

Diffusely branched, usually rounded, 0.5–4 m. tall, old bark grayish, reddish brown on younger stems; ultimate branchlets and stems to 3 or 4 years old, angled and densely silky-pubescent, hairs white, somewhat flattened, closely appressed and pointing toward tips of branchlets; stipules ovate, 1.5–2.5 mm. long, acute to short-acuminate; petioles 1–1.5 mm. long or leaflets nearly sessile; leaflets obliquely lanceolate or lunate, often slightly falcate, 2–4 mm. wide, 4–10 mm. long, coriaceous, resinous-glutinous, silky-pubescent, hairs more crowded and larger along the 3–4 veins; pedicels 8–15 mm. long, slender; sepals 3–6 mm. long, of 2 sizes, broadly ovate; petals 8–12 mm. long, twisted at claw to stand parallel with axis; fruit 6–8 mm. long, mostly obovoid, silky-pilose.

Plains, outwash bajadas, and rolling hills, Lower and Upper Sonoran Zones, Utah to Mojave and Colorado Deserts, California, south to Texas, Tamaulipas, Querétaro, and Baja California. Flowering throughout year, but mostly Feb.–April.

Comparison of North American material with specimens from Argentina reveals that in addition to the characters pointed out by B. L. Robinson (in A. Gray, Syn. Fl. N. Amer. **1**$^{1}$: 356, 1897)—namely, that the leaflets are narrower, longer, and less falcate in the South American *L. divaricata* Cav. than in our material—the herbage of *L. tridentata* is more glutinous, less copiously pubescent, has less conspicuous veins, and the stipules are clearly different. In *L. tridentata* the stipules are 1.5–2.5 mm. long, narrowly ovate, abruptly narrowed to an acute or short-acuminate tip which is ascending and quite free from the stem or the minute petiole. Those of *L. divaricata,* on the other hand, are broadly ovate to nearly orbicular; the apex is rounded or broadly obtuse and is closely infolded against the stem.

The two seem readily distinguishable and worthy of specific segregation.

## 5. TRIBULUS L., Sp. Pl. 386, 1753

Annual or perennial herbs with prostrate, much-branching stems and pubescent herbage. Leaves opposite. Stipules membranous, lanceolate or subulate. Flowers solitary in axils, regular. Sepals 5, caducous. Petals 5, yellow or orange (rarely white), deciduous. Stamens 10, hypogynous, five inner opposite the sepals, each subtended by a small gland. Ovary 5-celled, set in a 10-lobed disk. Styles adnate into a stout column. Stigmas 5. Ovules 5–10 in each cell. Fruit 5-angled and spinose, breaking into 5 hard carpels. Endosperm lacking.

Petals 4–5 mm. long; plants annual; peduncles shorter than subtending leaves .......... 1. *T. terrestris*
Petals 1.5–2 cm. long; plants perennial; peduncles surpassing subtending leaves .......... 2. *T. cistoides*

**1. Tribulus terrestris** L., Sp. Pl. 387, 1753

A prostrate annual herb with diffusely branching stems 1–8 dm. long; herbage sparsely silky-strigose throughout or upper surfaces of leaflets nearly glabrous; stipules subulate, 2–3 mm. long; leaves 2–5 cm. long, with 3–9 pairs of elliptic or oblong leaflets 3–13 mm. long, these oblique, acute to obtuse at apex; peduncles usually axillary to the shorter of the pair of leaves and exceeded by subtending leaf; sepals narrowly lance-ovate, 3–3.5 mm. long, caducous; petals pale yellow, 4–5 mm. long; fruit 7–12 mm. broad, exclusive of spines; carpels mostly 3-celled by abortion, each bearing 2 divergent spines 3–6 mm. long and several sharp tubercles or smaller spines, sparsely pubescent or hispid.

Introduced from southern Europe and widely distributed from Oregon southward to Argentina. Abundant in cultivated areas and along roads throughout most of our area. July–Oct.

**2. Tribulus cistoides** L., Sp. Pl. 387, 1753

*Kallstroemia cistoides* Endl., Ann. Nat. Mus. Wien **1**: 184, 1836.
*Tribulus terrestris* var. *cistoides* Oliver, Fl. Trop. Afr. **1**: 284, 1868.

Perennial from a woody rootstock, stems decumbent, to 8 dm. long or more, appressed-pubescent or some hairs spreading; stipules 5–8 mm. long; leaves 1–6 cm. long; leaflets 5–9 pairs, obliquely oblong or elliptic, 5–15 mm. long, 2–9 mm. wide, acute, obtuse or rounded and often mucronate at apex, densely silky-pubescent beneath, sparsely so to glabrate above; peduncles axillary to shorter leaves, 3–5 cm. long; sepals lanceolate, 8–9 mm. long, strigose-silky; petals 1.5–2 cm. long, light yellow; fruit 1–1.5 cm. broad, 6–10 mm. high, hirsute; each carpel 3–5-seeded, bearing 2 stout spines 3–5 mm. long and several smaller ones or latter reduced to tubercles.

Gravelly slopes and plains, Lower Sonoran to Tropical Zones, introduced into the southern United States and appearing occasionally southward into South America. Collected near Llano and Nória, Sonora.

## 6. KALLSTROEMIA Scop., Introd. 212, 1777

Annual herbs with decumbent hirsute branches. Leaves evenly pinnate, opposite, one of each pair smaller than other or sometimes lacking. Stipules subulate to narrowly lance-ovate, persistent. Flowers solitary in axils, orange, yellow, or nearly white. Sepals 5–6, linear to lanceolate, mostly persistent and somewhat marcescent. Petals 4–6, obovate to obcordate, caducous. Stamens 10–12, those opposite sepals subtended by a small gland and smaller than others. Ovary superior, 8–12-celled. Styles united into a column or conic axis. Stigma capitate. Fruit with twice as many vertical angles as carpels, roughened or tuberculate, breaking into twice as many bony, indehiscent, 1–2-seeded nutlets as carpels; central axis persistent.

Flowers large, deep orange; petals 1.5–3 cm. long; stems copiously short-pilose; leaflets 8–25 mm. long .................................................... 1. *K. grandiflora*
Flowers small, pale orange-yellow to nearly white; petals 5–12 mm. long; stems usually strigose to glabrate; leaflets 3–15 mm. long:
  Petals 7–12 mm. long; beaks of mature fruit longer than body of fruit, only slightly broadened at base ........................................ 2. *K. parviflora*
  Petals 4–6 mm. long; beaks of mature fruit shorter than or about equaling body, broadly conical at base:
    Leaflets 4–7 pairs; nutlets with sharp tubercles and faintly reticulate sides ........... ........................................ 3. *K. californica* var. *californica*
    Leaflets 3–4 pairs; nutlets with rounded tubercles and strongly reticulate sides ....... ........................................ 3a. *K. californica* var. *brachystylis*

**1. Kallstroemia grandiflora** Torr., in A. Gray, Pl. Wright. **1**: 28, 1852

*Kallstroemia grandiflora* var. *detonsa* A. Gray, loc. cit.
*Tribulus grandiflorus* Benth. & Hook. ex Brewer & S. Wats., Bot. Calif. **1**: 91, 1876.
*Tribulus fisheri* Kell., Proc. Calif. Acad. **7**: 162, 1877.
*Kallstroemia grandiflora* var. *arizonica* Cockerell, Bull. Torrey Club **27**: 87, 1900.

A diffuse annual with hirsute, spreading to ascending stems 1–4 dm. long; stipules linear-subulate, 4–6 mm. long, hirsute; leaves 1.5–7 cm. long, petioles mostly less than 1 cm. long; leaflets 5–9 pairs, obliquely oblong, 2–5 mm. wide, 8–25 mm. long, obtuse or acute, asymmetrical at base, usually glabrous above, pubescent with both coarse and fine hairs or glabrate beneath; pedicels to 6 cm. long, thickened upward, hirsute; sepals lance-linear, 2–2.5 mm. wide, 10–12 mm. long, hirsute, persistent and shriveling in age; petals obovate, 1.5–3 cm. long, deep orange, or pale with a deep orange spot at base; fruit 8–12 mm. wide, nutlets 4–5 mm. high, finely appressed-puberulent, rounded-tuberculate; beak 8–12 mm. long, columnar.

Desert flats and gravelly slopes, Lower and Upper Sonoran Zones, northern Arizona to Texas, Colima, and in the southern part of Baja California. Feb.–Sept.

**2. Kallstroemia parviflora** Norton, Rep. Mo. Bot. Gard. **9:** 153, 1898

*Kallstroemia laetevirens* Thornber, in Woot. & Standl., Contr. U.S. Nat. Herb. **16:** 143, 1913.

Branches spreading to ascending, to 1 m. long, striate, angled, and sparingly hirsute with coarse hairs and usually somewhat strigose with smaller appressed hairs; stipules 6–7 mm. long, linear-lanceolate, persistent; leaves 1–3 cm. long; leaflets 3–5 pairs, 1–3.5 mm. wide, 7–10 mm. long, acute at both ends, hirsute beneath, glabrous above; pedicels 1–2 cm. long at anthesis, to 4 cm. long in fruit; sepals narrowly linear-lanceolate, about 5 mm. long, hirsute, persistent; petals pale yellow, narrowly obovate to spatulate, 7–12 mm. long, 2–4 mm. wide; fruit 5–6 mm. wide, 4–4.5 mm. high, minutely appressed-pubescent, carpels bluntly tuberculate on backs; beak 5–8 mm. long, puberulent, columnar.

Mainly on gentle slopes or plains, Upper and Lower Sonoran Zones, Mohave and Pima Counties, Arizona, to Mississippi and south to Durango, Mexico. Aug.–Oct.

**3. Kallstroemia californica** (S. Wats.) Vail, Bull. Torrey Club **22:** 230, 1895, var. **californica**

*Tribulus californicus* S. Wats., Proc. Amer. Acad. **11:** 125, 1876.

Stems decumbent or rarely ascending, 1–6 dm. long, appressed canescent-pubescent or rarely somewhat hirsute or glabrate in age; stipules linear-subulate to ovate, 1.5–3 mm. long, usually caducous; leaves 1.5–6 cm. long; leaflets 4–7 pairs, broadly elliptic, 1.5–5 mm. broad, 4–10 mm. long, obtuse or rounded at apex, obliquely obtuse at base, strigose-canescent or tardily glabrate above; pedicels slender, 1.5 cm. long or less; sepals narrowly lance-ovate, 3–4 mm. long in flower, usually deciduous in fruit; petals 4–6 mm. long, yellow; fruit 3–4.5 mm. in diameter, 3–3.5 mm. high, puberulent, carpels with sharp tubercles, often to 1.5 mm. long, on backs, faintly reticulate on sides; beak 3–4 mm. long, conic at base, glabrous in fruit.

Sandy or gravelly flats and gentle slopes, Lower Sonoran Zone, southern Mojave Desert south into northeastern Baja California and southwestern Arizona. May–Oct.

**3*a*. Kallstroemia californica** var. **brachystylis** (Vail) Kearney & Peebles, Journ. Wash. Acad. **29:** 485, 1939

*Kallstroemia brachystylis* Vail, Bull. Torrey Club **24:** 206, 1897.
*Tribulus brachystylis* Robinson, in A. Gray, Syn. Fl. N. Amer. **1:** 354, 1897.

Leaflets 3–4 pairs, 6–15 mm. long; carpels bearing low, rounded, dorsal tubercles or these sometimes sharper, sides of carpels strongly reticulate, otherwise as in var. *californica.*

Sandy or gravelly desert plains and rocky hillsides, Lower and occasionally Upper Sonoran Zones, central Arizona to southern Colorado, western Texas, Sonora, and southern Baja California.

## Family 67. **RUTACEAE.** Rue Family

Shrubs, small trees, or some only perennial herbs, from woody rootstocks, strongly scented. Young twigs, leaves, and often petals glandular-punctate. Leaves simple or compound, rachises frequently winged. Flowers perfect, dioecious or polygamodioecious, regular. Sepals 3–5, small. Petals 3–5, distinct or in a few genera sympetalous. Stamens hypogynous or inserted in corolla tube, same number or twice as many as petals. Style filiform to clavate; stigma capitate, entire or lobed. Ovary superior, sometimes immersed in disk, 1–5-celled. Fruit a capsule, follicle, samara, drupe, or berry. Seeds 1–several in each cell.

Fruit follicular; stems (in ours) usually sparingly armed with spines or prickles; flowers dioecious ........................................................ 1. *Zanthoxylum*
Fruit capsular; stems unarmed; flowers perfect:
  Petals spreading to reflexed in age; capsules mostly 5-celled, echinate or tuberculate, woody; plants shrubs or trees 2–8 m. tall; ovary immersed in disk ............ 2. *Esenbeckia*
  Petals erect; capsules 2-celled, smooth or at least not echinate or strongly tuberculate, leathery; low shrubs or perennial herbs less than 1 m. tall; ovary not immersed in disk ........................................................ 3. *Thamnosma*

### 1. ZANTHOXYLUM L., Sp. Pl. 270, 1753

Shrubs or trees with aromatic bark, twigs often armed with sharp curved spines and trunk unarmed or armed with corky, pyramidal-rounded, or warty projections. Leaves alternate, unifoliolate, odd-pinnate or evenly pinnate, usually with stipular spines; leaflets with pellucid glands, entire to crenulate; petioles and rachis often winged. Inflorescences of various types. Flowers dioecious or polygamodioecious. Calyx hypogynous or wanting; sepals 3–5. Petals 3–10. Stamens hypogynous, 3–5, reduced or lacking in pistillate flowers. Pistils 1–5. Ovaries 1-celled, each 2-ovuled. Styles somewhat united near summit. Fruits follicular; follicles sessile or stipitate. Seeds ovoid, globose or oblong, often hanging from carpel on a slender funiculus at maturity; seed coat black, brownish, or reddish, shining.

Leaflets 3–7.5 cm. long, velvety-tomentulose beneath; follicles subsessile; sepals usually 5 ........................................................ 1. *Z. arborescens*
Leaflets 0.5–2 cm. long, hirtellous to glabrate beneath; follicles stipitate; sepals usually 4 ........................................................ 2. *Z. sonorense*

**1. Zanthoxylum arborescens** Rose, Contr. U.S. Nat. Herb. **5**: 112, 1897
*Zanthoxylum peninsulare* Brandegee, Zoe **5**: 249, 1908.

Shrub or tree to 6 m. tall with unarmed branches or these armed with scattered, slightly curved, stout spines, often with heavier conical spines on trunk; twigs, petioles, rachises of leaves, and branches of paniculate inflorescence hispidulous; leaves odd-pinnate, 1–2 dm. long; leaflets 3–7 (usually 5), oblong-elliptic to obovate, 1.5–4 cm. wide, 3–7.5 cm. long, cuneate at base, acute to obtusely short-acuminate at apex, entire to crenulate, tomentulose beneath, hispidulous to glabrate above; sepals 5 or sometimes 4, triangular-ovate, 1.5–2 mm. long; petals 2.5–3 mm. long, greenish yellow; stamens

5; follicles ellipsoid to subglobose, 4–6 mm. long; seeds ellipsoid, 3.5–4.5 mm. long, black and shining.

Rocky canyons, dry arroyos, and valley floors, Lower Sonoran and Subtropical Zones, southern Baja California and in Sinaloa. Sept.–Oct.

**2. Zanthoxylum sonorense** Lundell, Phytologia **1:** 369, 1940

A closely branched shrub 2–3 m. tall with slender hirtellous twigs armed with slender, nearly straight spines 3–8 mm. long; leaves 2.5–5 cm. long, hirtellous, with winged rachises and petioles, these 5–15 mm. long; leaflets 5–9, oblanceolate, obovate, or sometimes broadly elliptic, 2.5–9 mm. wide, 5–20 mm. long, a callosity at cuneate base, obtuse to rounded or sometimes acute at apex, crenulate and glandular along margins, sparsely hirtellous above, densely so beneath, eventually glabrate or pubescence persisting along midrib beneath; inflorescence of axillary short racemes 1–2 cm. long; sepals 4, about 0.5 mm. long or less, rounded, puberulent; petals ovoid, about 2 mm. long, sparsely and minutely puberulent near apex exteriorly; follicles 1 or 2, on a slender stipe 3–3.5 mm. long, follicles broadly ellipsoid, 3–3.5 mm. in diameter, 4–5 mm. long, with scattered tubercle-like glands; seeds ellipsoid, 2–2.5 mm. in diameter, 3–3.5 mm. long, black and shining.

Rocky hillsides and ridges, Lower Sonoran Zone, foothills in central Sonora near Ures, and in southern Baja California. Feb.–March.

The specimens from Baja California have broader, more leathery, and less densely puberulent leaflets than the Sonoran plants.

**Choisya arizonica** Standley, Proc. Biol. Soc. Wash. **27:** 222, 1914, and **C. mollis** Standley, op. cit. 223, both occur in dry, chaparral-covered slopes contiguous to the upper margins of the Sonoran Desert in Pima and Santa Cruz Counties, Arizona, and might occur along watercourses extending into our area. No such stragglers have been reported to date, however.

Two or three specimens of **Ptelea** with winged, nearly orbicular-flattened samaras also occur in the Upper Sonoran Zone along the margins of the Sonoran Desert and may at times occur within the limits as stragglers but have not been collected in our regions as yet.

## 2. ESENBECKIA H. B. K., Nov. Gen. & Sp. 7: 246, *pl. 655,* 1825

Trees or shrubs with alternate or sometimes opposite, simple (in ours), or 1–5-foliolate leaves and pellucid-punctate leaflets or leaf blades. Flowers perfect, paniculate. Calyx 4–5-lobed or 4–5-parted, deciduous. Petals 4 or 5. Disk cup-shaped or flat, entire or with twice as many lobes as petals. Stamens 4–5, subulate filaments lying in grooves in disk, alternate with petals; anthers cordate, mucronulate. Ovary subglobose, somewhat depressed, sessile or immersed in disk, with as many cells and lobes as there are petals, often warty. Style basal; stigma capitate, entire or 4–5-lobed. Capsule echinate, carpels separating septicidally and dehiscing loculicidally, each carpel usually 1-seeded. Seeds enclosed in a chartaceous endocarp which ejects seeds when dry.

Mature capsules 1.5–2.5 cm. broad when open; petioles 1–12 mm. long ... 1. *E. hartmanii*
Mature capsules 3–5 cm. broad when open; petioles 1–2.5 cm. long ............ 2. *E. flava*

**1. Esenbeckia hartmanii** Robins. & Fernald, Proc. Amer. Acad. **30:** 115, 1894

Shrub with numerous branches from the base, 2–5 m. tall with grayish, fairly smooth bark and bearing leaves only at tips of thickish branchlets; petioles 1–12 mm.

long, short-pubescent with spreading gray hairs; leaves simple, obovate, elliptic, or oblong-elliptic, entire, 1.5–6 (or rarely 8) cm. long, mostly 1–3 cm. broad, retuse or rounded at apex, thickish, smooth and deep green above, pubescent and slightly paler beneath, finely reticulate; inflorescence 2–5 cm. long, terminal; pedicels slender, 5–20 mm. long; petals (on old flowers) 4–5 mm. long, densely hirsutulous outside, dotted with small dark glands within; stamens nearly equaling petals, accumbent in very shallow grooves in the disk, in age reflexed; young capsule densely echinate, some of the larger papillae laterally tuberculate; style obclavate; mature capsule 1.5–2.5 cm. in diameter, coarsely warty-roughened; seeds dull black, subglobose, 7–10 mm. in diameter, hilum white, linear, running nearly halfway around seed.

Stony hillsides and along rocky dry arroyos, Lower Sonoran Zone and subtropical thorn forest, central Sonora (near Divisadero) south into Sinoloa. Aug.–Sept.

The ovary is usually 5-celled, but one or more cells frequently are abortive. When only one or two cells produce seeds, the fruit lies within the smaller measurements; when three or more carpels are fertile the fruit is often fully 2.5 cm. broad.

### 2. Esenbeckia flava Brandegee, Zoe **1:** 378, *pl. 12,* 1891

Shrub or small tree to 8 m. tall with smooth, dark gray bark and tomentulose twigs, petioles, inflorescences, and lower surfaces of leaves; petioles 1–2.5 cm. long; leaves simple, elliptic, oblong, or obovate, entire or somewhat sinuously margined, 5–10 cm. long, 2.5–5 cm. wide, rounded or obtuse and usually retuse at apex, acute or obtuse and often asymmetrical at base, finely reticulate, minutely puberulent to glabrate on upper surface, somewhat ashy beneath; panicles terminal and axillary, terminal ones 10–20 cm. long, lateral usually much shorter; pedicels thickish, usually 1 cm. long or less; flowers 13–15 mm. broad, petals oblong-ovate, 4–6 mm. long, pubescent outside, smooth inside; filaments 6–7 mm. long, accumbent in shallow grooves in disk; ovary densely echinate, immersed in disk; capsule 3–5 cm. broad, verrucose-echinate with slender stiff processes; seeds ovoid-conic, 10–13 mm. in diameter, dark brown, dull, hilum brown.

Dry canyons, hillsides, and rocky mesas, Lower Sonoran and Subtropical Zones, southern Baja California. Aug.–Sept.

The echinate processes on the capsules are larger and more slender in this species than in *E. hartmanii.*

## 3. THAMNOSMA Torr. & Frem., in Frem., Rep. 313, 1845

Glandular-dotted shrubs or subshrubs with a strong disagreeable odor. Leaves simple, alternate, linear, entire (often early deciduous or reduced to scales), or trifoliate. Flowers perfect, racemose. Sepals 4. Petals 4, erect or only slightly spreading in anthesis. Stamens 8, with subulate or filiform filaments and inserted on an entire or crenate, cup-shaped disk. Ovary 2-celled (rarely 3-celled), bilobed, sessile or stipitate; style slender; stigma capitate. Ovules 5–6 in each cell. Capsule opening apically. Seeds reniform, smooth or rough.

Branches stoutish, mostly over 2 mm. in diameter; capsules on stipes 4–10 mm. long; seeds smooth or wrinkled but not tuberculate, 4–6 mm. long . . . . . . . . . . . . . . 1. *T. montana*

Branches slender, 0.8–1 mm. in diameter; capsules sessile or short-stipitate; seeds tuberculate, 1.2–2.5 mm. long:

  Stems erect and ascending; leaves simple . . . . . . . . . . . . . . . . . . . . . . . . . . . . . . 2. *T. texana*

  Stems prostrate; leaves trifoliolate . . . . . . . . . . . . . . . . . . . . . . . . . . . . . . . . . 3. *T. trifoliata*

**1. Thamnosma montana** Torr. & Frem., in Frem., Rep. 313, 1845

An erect rigid shrub to 8 dm. tall with stoutish, yellow-green, often spine-tipped branches; leaves distant, soon deciduous, linear to narrowly oblong, 5–15 mm. long, sessile, subfleshy and conspicuously pellucid-dotted; sepals deltoid to broadly ovate, 2.5–3.5 mm. long, rounded or infrequently acute at apex; petals almost erect, purplish, oval to oblong, 8–14 mm. long, 4–5 mm. wide, glandular-pellucid; stamens about equaling petals; anthers apiculate; style filiform, exserted 3–6 mm.; ovary stipitate; capsule 10–13 mm. wide along greatest diameter, 6–9 mm. high, on a stout stipe to 10 mm. long; seeds 1–3 in each cell, cochleate-reniform, 4–6 mm. long, smooth or finely wrinkled, brown to black, dull.

Desert slopes and rocky mesas, mainly Lower Sonoran Zone, western margins of the Colorado Desert, Mojave Desert, and to southern Utah, southwestern Arizona, and northern Baja California. Feb.–April.

**2. Thamnosma texana** (A. Gray) Torr., Bot. Mex. Bound. 42, 1859
*Rutosma texana* A. Gray, Gen. Ill. **2:** 144, *pl. 155,* 1849.

A low subshrub with many slender, erect or ascending branches 1–5 dm. tall and usually about 1 mm. or less in diameter; stems light green, pellucid-punctate; leaves filiform to linear-oblong, sessile, entire, 5–15 mm. long; racemes terminal, somewhat interrupted, mainly less than 5 cm. long; pedicels slender, 2–8 mm. long; sepals broadly ovate or rounded-deltoid, 1–2 mm. long, sometimes retuse; petals ovate to broadly elliptic, 2–3 mm. wide, 3–5 mm. long, greenish yellow or white, slightly suffused with purple, pellucid-punctate; stamens slightly shorter than petals; anthers broadly ovate, apiculate; ovary sessile; style about two-thirds as long as petals; capsule coriaceous, deeply obcordate, 3–7 mm. wide and about as high, sessile or on a stipe about 1 mm. long; seeds 4–6 in each cell, reniform, flattened and somewhat concavo-convex laterally, tuberculate, 1.2–2 mm. long, buff to pale brown.

Dry rocky slopes and mesas, Lower and lower parts of Upper Sonoran Zones, Pima County, Arizona, east to Texas. March–June.

**3. Thamnosma trifoliata** I. M. Johnston, Proc. Calif. Acad. IV, **12:** 1055, 1924

A prostrate perennial herb or subshrub with wiry stems 3–6 dm. long; leaves trifoliolate, remote; petioles slender, 1–5 mm. long; leaflets elliptic to oblong, 3–8 mm. wide, 5–14 mm. long, lateral ones usually somewhat oblique, rounded at apex, finely crenate along margins, light green, paler beneath than above; flowers unknown; young fruits scattered on slender pedicels 1–4 mm. long; sepals broadly ovate to rounded-deltoid, 1–1.5 mm. long; capsule obcordate, deeply notched, 4–5 mm. high, 4–6 mm. wide, sessile or very short-stipitate; ovules 4–5 in each cell but usually only 2 seeds in each locule maturing, 2–2.5 mm. long, about 1 mm. broad, somewhat arched laterally, unevenly roughened with coarse tuberculations, pale brown or deep buff, dull.

Known only from the type locality, Agua Verde Bay (Lower Sonoran Zone), Baja California.

## Family 68. **SIMAROUBACEAE.** Simarouba Family

Shrubs or trees with bitter bark and sometimes bitter wood. Leaves alternate or rarely opposite, simple or pinnately compound. Flowers perfect, monoecious, dioecious, or polygamodioecious, small, variously arranged. Sepals 3–7, free or united at base. Petals 3–8 or wanting. Stamens as many or twice as many as sepals or petals. Carpels 2–5, distinct or united. Styles distinct or united, or wanting. Stigmas terminal or introrse. Ovules 1–2, or rarely several. Fruit a berry, drupe, samara, or capsule. Seed usually solitary.

Carpels 1-ovuled; leaves simple; fruit drupaceous:
  Leaves with well-developed blades; not reduced to scales in older plants; petals 4, red; carpels not stellately divergent .......................................... 1. *Castela*
  Leaves reduced to scales on older plants; petals 7–8, yellowish green; carpels stellately divergent .................................................... 2. *Holacantha*
Carpels 2-ovuled; leaves odd-pinnate; fruit a 2–3-winged samara ............ 3. *Alvaradoa*

## 1. CASTELA Turpin, Ann. Mus. Paris **7:** 78, *pl. 5,* 1806

Shrubs or small trees with numerous thorny branchlets and alternate, simple, rather thick, small leaves. Flowers dioecious, solitary to several in the axils. Sepals 4, distinct or essentially so. Petals 4. Stamens 8; filaments subulate or subfleshy; anthers ovoid, sometimes apiculate at tip. Ovary 4-celled and 4-lobed, 1-ovuled carpels nearly or quite distinct, sessile. Styles very short or introrse stigmas sessile. Drupes solitary or 2, 3, or 4 somewhat united at base, flattened.

### 1. **Castela peninsularis** Rose, Contr. U.S. Nat. Herb. **12:** 278, 1909

*Castelaria peninsularis* Small, N. Amer. Fl. **25:** 231, 1911.

A much-branched, very thorny shrub 2–3 m. tall, branches and twigs rigidly divaricate, velutinous-tomentulose with yellowish hairs except near tips of thorns which are almost glabrous; leaves elliptic to obovate, 5–10 mm. broad, 9–20 mm. long, entire or coarsely dentate, acute to cuneate at base, rounded to emarginate and often apiculate at apex, leathery, finely puberulent to glabrate above, tomentulose and conspicuously veined beneath, on petioles 1.5–3 (or 5) mm. long; pedicels slender, glabrous or nearly so, 2–3 mm. long; sepals triangular, about 0.6 mm. long, sparsely ciliolate; petals red, oblong, 3–5 mm. long, glabrous; stamens about equaling petals, yellow, filaments pubescent; drupes subglobose, somewhat flattened laterally, bright red, bitter, 7–10 mm. long; seeds lenticular-orbicular, 6–8 mm. in diameter, margin sharply angular.

Hillsides and gentle slopes, Subtropical and Lower Sonoran Zones, southern Baja California and the islands along the Gulf side at least as far north as Isla San José. Staminate flowers, Aug.–Nov., pistillate mostly Nov.; also March–April.

## 2. HOLACANTHA A. Gray, Pl. Thurb. 310, 1854

Nearly leafless shrubs with stiff, thornlike branchlets and hard bitter wood. Leaves alternate, early deciduous or in mature plants reduced to mere scales. Flowers dioecious, solitary to glomerulate. Sepals 7–8, ovate. Petals 7–8, surpassing sepals. Stamens 12–16 in male flowers; filaments subulate; anthers oblong. Disk crenulate. Carpels 6–10, imperfectly united and separating more or less toward maturity. Styles divergent; stigmas subulate, introrsely curved. Drupes dry, stellately divergent at maturity, each 1-seeded.

**1. Holacantha emoryi** A. Gray, Pl. Thurb. 310, 1854

A rigid, much-branched shrub to 5 m. tall, densely and finely canescent-pubescent on young branches; branchlets terete, 5–15 cm. long, spinose; leaves linear to lanceolate, 1–1.5 cm. long in seedlings, entire to repand or basally lobed, only subulate or ovate scales in older plants; flowers subsessile; staminate flowers 6–7 mm. in diameter; sepals triangular-ovate, 2–2.5 mm. long, canescent; petals oblong to ovate-oblong, 2.5–3 mm. long, densely pubescent without; filaments pubescent; pistillate flowers 3–4.5 mm. in diameter, sepals comparatively broader and shorter than in staminate flowers, 2–2.5 mm. long; drupes ovoid but somewhat compressed laterally, 6–8.5 mm. long, 5–6 mm. wide, sessile, buff or light brown; clusters of fruits often 10–25 cm. long, about one-half to two-thirds as wide, fruits crowded.

Desert plains and along washes, Lower Sonoran Zone, Hayfields and near Amboy, California, in southwestern Arizona, and northwestern Sonora, but nowhere abundant. April–July.

## 3. ALVARADOA Liebm., Kjoeb. Vidensk. Meddel. 1853: 100, 1854

Slenderly branched shrubs or small trees with alternate, odd-pinnate leaves. Leaflets numerous, usually larger near apex of rachis than below, short-petiolulate. Flowers dioecious, crowded in slender racemes. Sepals 5, broadly ovate, slightly united at base. Petals none. Stamens 5, pubescent filaments exserted; anthers ovate. Ovary superior, 2–3-celled. Styles as many as cells in ovary, somewhat recurved. Ovules 2 in each cell. Fruit a 2–3-winged samara, pubescent.

**1. Alvaradoa amorphoides** Liebm., Kjoeb. Vidensk. Meddel. **1853:** 101, 1854

A shrub or small tree mostly less than 5 m. tall (1–2 m. in specimen from our area) with dark gray, slightly shriveled bark and finely but densely puberulent young branchlets; leaves 1–2 dm. long, rachis yellowish orange and slender; leaflets 17 to about 51, suborbicular to obovate below, oblong to obovate, cuneate, or oblong-ovate toward tip of rachis, 0.5–2 cm. long, deep green and glabrous or minutely and sparsely puberulent on upper surface, finely appressed-puberulent and pale beneath; racemes to 1.5 dm. long in flower, elongating somewhat in fruit; staminate flowers sessile, stamens 4–5 mm. long; sepals 1.5–2 mm. long, puberulent; samaras oblong-lanceolate, 1–1.5 cm. long, velvety-puberulent, on slender pedicels about as long or slightly longer.

Rocky hillsides and along washes and canyons, Lower Sonoran Zone to Tropical Zone, southern Chihuahua, and central Sonora to Costa Rica, and in Florida and the Bahama Islands. Collected near Divisadero, Sonora, at an altitude of 2,000 feet. Sept.–Dec.

# Family 69. BURSERACEAE. Torchwood Family

Shrubs or trees with aromatic, smooth, often exfoliating bark. Leaves alternate, deciduous, usually pinnate but sometimes unifoliolate. Flowers perfect or polygamodioecious, inconspicuous; inflorescence more or less paniculate. Sepals 3–5, united at base into a short tube or free. Petals 3–5, alternating with sepals. Disk annular. Stamens 8–10; filaments free, naked. Ovary superior, 4–5-celled or fewer-celled by abortion. Style short; stigma broad. Ovules usually 2 in each cell, pendent. Fruit drupaceous, containing 1–5 stones, part or all of each one encased in a fleshy aril. Endosperm none.

## 1. BURSERA Jacq., in L., Sp. Pl. ed. 2, 471, 1762

Characters of the family. Ovary 3-celled. Drupe more or less 3-angled; pericarp splitting into 2 or 3 leathery valves.

Leaves or part of them unifoliolate or simple:
  Leaves all simple, glabrous or nearly so, shining; lateral veins broadly spreading ........ ........................................................ 1. *B. cerasifolia*
  Leaves simple to 5-foliolate, pubescent, usually dull; lateral veins 5–12 pairs, ascending:
    Inflorescence on stout peduncles, usually much shorter than petioles; petals broadly oblanceolate, markedly broader above than at base, connivent; sepals broadly lance-deltoid; fruits 9–12 mm. long ................................ 2. *B. hindsiana*
    Inflorescence on slender peduncles, surpassing petioles and usually longer than leaves; petals oblong, only slightly broader above than at base, spreading; sepals lance-linear; fruits 5–9 mm. long .................................... 3. *B. epinnata*
Leaves always pinnately compound:
  Drupes 2-valved; flowers tetramerous:
    Leaflets sparsely puberulent or glabrate beneath; bark on young twigs dark red ....... ........................................................ 4. *B. laxiflora*
    Leaflets finely but closely puberulent to tomentose beneath; young twigs pale, yellowish or pale buff ............................................ 5. *B. filicifolia*
  Drupes 3-valved; flowers pentamerous:
    Leaflets 7.5–10 cm. long, about half as wide; drupes pubescent ..... 6. *B. grandifolia*
    Leaflets rarely over 6 cm. long, usually less; drupes glabrous:
      Bark of one-year twigs dark red; older bark exfoliating in small scales; leaflets oblong to linear, rarely over 1 cm. long, usually less ................ 7. *B. microphylla*
      Bark of one-year twigs pale orange, buff, or grayish, not dark red; mature bark exfoliating in large papery sheets; leaflets lanceolate or ovate, often 2.5–6 cm. long:
        Leaflets ovate, apices acute or rounded; rachises wingless .......... 8. *B. odorata*
        Leaflets ovate-lanceolate, acute to attenuate at apices; rachises usually slightly winged:
          Leaflets entire, 1.2–4 cm. long .............................. 9. *B. confusa*
          Leaflets serrulate, 2.5–6 cm. long ............................ 10. *B. fragilis*

### 1. Bursera cerasifolia Brandegee, Proc. Calif. Acad. II, 3: 121, 1891

*Terebinthus cerasifolia* Rose, Contr. U.S. Nat. Herb. **10:** 119, 1906.
*Elaphrium cerasifolium* Rose, N. Amer. Fl. **25:** 244, 1911.

Shrub or small tree to 8 m. tall; young twigs glabrous, reddish brown, stoutish; leaves simple, mainly clustered at tips of branches, ovate to ovate-lanceolate, 1–2 cm. wide, 4–6 cm. long, glabrous and shining, acute at apex, margins shallowly crenate, lateral veins spreading almost at right angles, numerous; peduncles slender, equaling to surpassing leaves, few- to many-flowered; sepals 4, subequal, about 1 mm. long; petals 2.5–3 mm. long; drupe obovate, 5–6 mm. long; seeds black, basal one-third to one-half embedded in an orange-colored aril.

Along arroyos and on gentle slopes, Lower Sonoran Zone, southern Baja California northward on the islands off La Paz at least to Espíritu Santo Island. Dec.–March.

**2. Bursera hindsiana** (Benth.) Engler, in A. & C. DC., Monogr. Phan. **4:** 58, 1883

*Elaphrium hindsianum* Benth., Bot. Sulph. 11, *pl. 8,* 1844.
*Elaphrium rhoifolium* Benth., op. cit. 10, *pl.* 7.
*Bursera hindsiana* β *rhoifolia* Engler, in A. & C. DC., Monogr. Phan. **4:** 59, 1883.
*Terebinthus rhoifolia* Rose, Contr. U.S. Nat. Herb. **10:** 121, 1906.
*Terebinthus macdougali* Rose, Torreya **6:** 170, 1906.
*Elaphrium macdougalii* Rose, N. Amer. Fl. **25:** 255, 1911.
*Elaphrium goldmani* Rose, op. cit. 256.
*Bursera rhoifolia* I. M. Johnston, Proc. Calif. Acad. IV, **12:** 1058, 1924.

Shrub or small tree to 3.5 m. tall with stoutish, divaricately branching twigs and dark red to reddish gray bark, twigs puberulent in youth, soon glabrous; leaves mostly at tips of short, lateral, spur branches, simple to trifoliolate or rarely with 5–7 leaflets, blades broadly oval to oblong, 1–2.5 cm. wide, to 4.5 cm. long, terminal leaflet much larger than lateral ones in compound leaves, pubescent on both surfaces or glabrate on upper surface in some trees, margins strongly crenate; lateral veins 5–10 pairs, ascending; peduncles to 2 cm. long, usually shorter, stoutish, puberulent, comparatively few-flowered; sepals 4, lance-deltoid, about 1 mm. long, spreadingly hirsutulous; petals 2–2.5 mm. long, erect to connivent, hirsutulous on outer surface; drupes ovoid to subglobose, 9–12 mm. long, glabrous, leathery exocarp 2-valved; seeds ovoid, slightly flattened and 2-angled, pale above, embedded one-half to two-thirds their length in an orange-colored aril.

Along washes and on desert hillsides, Lower Sonoran Zone, Los Angeles Bay and the Vizcaíno Desert region southward to Magdalena Bay and the Cape Region in Baja California, and from Puerto Libertad to Tiburón Island in Sonora. Sept.–Dec.

**3. Bursera epinnata** (Rose) Engler, in Engler & Prantl, Nat. Pflanzenf. II, **19a:** 425, 1931

*Elaphrium epinnatum* Rose, N. Amer. Fl. **25:** 243, 1911.
*Bursera nesopola* I. M. Johnston, Proc. Calif. Acad. IV, **20:** 66, 1931.

A gnarled spreading shrub or small tree 1–3.5 m. tall with reddish brown, smooth bark on young branches, grayish red on older stems, herbage and bark strongly aromatic; young twigs sparingly puberulent but soon wholly glabrate; leaves simple or rarely a few of them trifoliolate, petioles 5–20 mm. long, finely hirsutulous, blades ovate, 5–23 mm. wide, 1.5–3.5 cm. long, acute at apex, rounded to subcordate at base, crenulate, dark green and finely hirsutulous to subglabrate above, paler and more densely puberulent beneath, lateral veins 5–12 pairs; inflorescences borne on short spur branches, slender, ascending, 1–4 cm. long in flower, elongating and often to 10 cm. long in fruit, peduncles 0.5–0.8 mm. in diameter, sparsely hirsutulous; flowers 5–21 on a peduncle, cream to pale greenish white; sepals lance-linear, 1.2–1.5 mm. long, hirsutulous; petals oblong, 1.5 mm. wide, 3 mm. long, sparsely puberulent without, spreading, tips often slightly recurved, base nearly as broad as rounded, abruptly acute apex; fruits obovate, 5–9 mm. long, 3–6 mm. wide, exocarp glabrous, 2-valved; seed black, 4–6 mm. long, 2-angled, lower one-third to one-half embedded in orange-colored aril.

Along washes, on alkaline flats and rocky hillsides, Lower Sonoran Zone, Bahía de la Concepción to tip of the Cape Region in Baja California, and apparently on Socorro Island in the Revillagigedo Islands. Sept.–Dec.

**4. Bursera laxiflora** S. Wats., Proc. Amer. Acad. **24:** 44, 1889

*Terebinthus laxiflora* Rose, Contr. U.S. Nat. Herb. **12:** 279, 1909.
*Elaphrium laxiflorum* Rose, N. Amer. Fl. **25:** 253, 1911.
*Bursera concinna* Sandw., Kew Bull. **1926:** 434, 1926.

A round-crowned tree to 6 m. tall; trunk often 3–3.5 dm. in diameter, with dark gray, slightly scaling bark; young branches slender, dark reddish gray, finely pubescent; leaves pinnate (or often bipinnate on young, vigorously growing shoots), 5–7 or rarely to 10 cm. long; rachis narrowly winged; leaflets 5–11, ovate, obovate, or oblong, 1–2 cm. long, few-toothed or less commonly some of smaller ones entire, narrowly to broadly cuneate at base, minutely pubescent on both surfaces; peduncles slender, drooping, glabrous, 1- to few-flowered; pedicels filiform, 6–12 mm. long; sepals 4, lance-linear, 1.5–2 mm. long, glabrous, free to base; petals linear-oblanceolate, 3–4 mm. long; drupes ovoid, 6–8 mm. long, 2-valved, glabrous; seeds 2-angled, slightly flattened, black above, lower one-half covered with a yellowish orange aril.

On gravelly and rocky hillsides, Lower Sonoran Zone, central Sonora southward into Sinaloa. Sept.–Dec.

**5. Bursera filicifolia** Brandegee, Zoe **5:** 248, 1908

*Elaphrium filicifolium* Rose, N. Amer. Fl. **25:** 254, 1911.

A small to large tree to 10 m. tall, with light gray (less commonly dark gray) bark on mature trunk, and pale yellow-gray, minutely puberulent, slender twigs; leaves pinnate, 1–10 cm. long; rachis very narrowly winged, puberulent; leaflets 5–13, obovate (or terminal sometimes flabelliform), 3–20 mm. long, 1.5–12 mm. wide, terminal leaflet sometimes twice to thrice as wide as lateral ones, shallowly few-toothed above middle, rounded to broadly cuneate at base, acute to obtuse at apex, finely but often densely puberulent on both surfaces; peduncles slender, 1–3-flowered, 2–5 cm. long; sepals 1.5–2 mm. long, puberulent; petals about 4 mm. long; drupes ovoid to obovoid, to 1 cm. long; exocarp reddish, 2-valved; seeds black or dark brown, an orange-red aril enveloping lower two-thirds.

Hillsides and outwash slopes, Lower Sonoran and Subtropical Zones, Sierra de la Giganta to the Cape Region, Baja California. Aug.–Sept.

The leaf form is similar to that of *B. laxiflora* but the light color of the young twigs, and often the densely puberulent leaflets, readily separate it from *B. laxiflora.*

**6. Bursera grandifolia** (Schlecht.) Engler, Bot. Jahrb. **1:** 43, 1881

*Elaphrium grandifolium* Schlecht., Linnaea **17:** 249, 1843.
*Bursera cinerea* Engler, in A. & C. DC., Monogr. Phan. **4:** 43, 1883.
*Terebinthus cinerea* Rose, Contr. U.S. Nat. Herb. **10:** 119, 1906.
*Terebinthus grandifolia* Rose, loc. cit.
*Elaphrium cinereum* Rose, N. Amer. Fl. **25:** 245, 1911.
*Elaphrium occidentale* Rose, op. cit. 246.
*Bursera occidentalis* Riley, Kew Bull. **1923:** 168, 1923.

Tree to 8 m. or more tall with purplish or reddish gray bark; young branches stout, densely puberulent; leaves pinnate, to 4 dm. long or more; rachis terete, not winged; leaflets 3–7, oblong-ovate, broadly elliptic, to obovate, to 6 cm. wide and 10 cm. long, abruptly attenuate from a rounded or acute apex, pubescent on both surfaces, margins entire; inflorescences racemose-paniculate, 10 cm. long, many-flowered; pedicels 5–8 mm. long; hirsutulous; sepals 5, deltoid, 1–1.2 mm. long, puberulent; petals 5, spatulate, 3–3.5 mm. long, glabrous, reflexed at anthesis; drupes ovoid, 8–9 mm. long, minutely puberulent; exocarp 3-valved; seeds 3-angled, the hilum narrowly cordate in outline.

Slopes and canyons, mainly Subtropical and Tropical Zones, southern Sonora to Vera Cruz, Guerrero, and Colima. July–Sept.

**7. Bursera microphylla** A. Gray, Proc. Amer. Acad. **5:** 155, 1861

*Bursera morelense* Ramirez, Anal. Inst. Med. Nac. Mex. **2:** 17, *pl. 3,* 1896.
*Terebinthus microphylla* Rose, Contr. U.S. Nat. Herb. **10:** 120, 1906.
*Terebinthus multifolia* Rose, loc. cit.
*Elaphrium microphyllum* Rose, N. Amer. Fl. **25:** 250, 1911.
*Elaphrium morelense* Rose, loc. cit.
*Elaphrium multifolium* Rose, loc. cit.

A round-topped tree to 6 or 8 m. tall with glabrous, cherry-red branches, the bark exfoliating from the older branches and trunks in thin extensive sheets; leaves pinnate, 3–8 cm. long or more; rachis narrowly winged; leaflets 7–35, oblong to linear, 1–2.5 mm. wide, 6–12 mm. long, obtuse, sessile, glabrous, dark green and shining; flowers appearing before leaves; peduncles 1–3 cm. long, 1–4-flowered; pedicels slender, 3–5 mm. long; sepals 5, ovate, 1–1.2 mm. long, glabrous; petals 5, 3–4 mm. long; drupes 3-angled, glabrous; exocarp 3-valved; seeds yellowish, 5–6 mm. long, hilum a triangular depression.

Hillsides, outwash slopes, and flats, Lower Sonoran Zone, southern Arizona and western margin of the Colorado Desert southward throughout Baja California and Sonora to Zacatecas. July.

**8. Bursera odorata** Brandegee, Proc. Calif. Acad. II, **2:** 138, 1889

*Terebinthus odorata* Rose, Contr. U.S. Nat. Herb. **10:** 121, 1906.
*Elaphrium covillei* Rose, N. Amer. Fl. **25:** 250, 1911.
*Elaphrium odoratum* Rose, loc. cit.

Shrub or tree to 4 m. tall with pale, orange-red bark that peels off in papery sheets from older branches and trunks; leaves pinnate, 3–6 cm. long, rachis faintly or not at all winged; leaflets 3–9, usually 7, elliptic to ovate, 5–10 mm. wide, 1.2–2 cm. long, acute at apex, acute to rounded at base, entire, glabrous, veins often reddish; inflorescence 1- to several-flowered; sepals 5, narrowly triangular, about 1 mm. long, united below to form a tube equaling free portion, glabrous; petals linear-spatulate, 3.5–4 mm. long, spreading; drupes ovoid to globose, 5–6 mm. long, glabrous, slightly glaucous, exocarp 3-valved; seed lightly 3-angled.

Rocky hillsides and slopes, Lower Sonoran Zone, southwestern Arizona to the Cape Region, Baja California, and possibly in Sonora and Sinaloa. July–Aug.

**9. Bursera confusa** (Rose) Engler, in Engler & Prantl, Nat. Pflanzenf. II, **19a:** 426, 1931

*Bursera tenuifolia* Rose, Contr. U.S. Nat. Herb. **3:** 314, 1895. Not *B. tenuifolia* Engler, 1891.
*Terebinthus tenuifolia* Rose, op. cit. **10:** 122, 1906.
*Elaphrium confusum* Rose, N. Amer. Fl. **25:** 251, 1911.
*Elaphrium tenuifolium* Rose, op. cit. 252.
*Bursera lonchophylla* Sprague & Riley, Kew Bull. **1923:** 168, 1923.

A tree to 6 m. tall with slender, pale orange or yellowish, glabrous twigs and darker —frequently very dark—gray bark on older twigs, smooth, yellowish and papery bark on trunk; herbage glabrous throughout; leaves pinnate, 6–15 cm. long (usually 5–8 cm. long); rachises narrowly but distinctly winged; leaflets 5–11, lanceolate and often somewhat attenuate, 4–9 mm. wide, 1.2–4 cm. long, entire, acute to cuneate at base; flowers in clusters, mostly on short lateral spur branches, appearing with young leaves; sepals ovate-triangular, about 1 mm. long; petals 3.5–4 mm. long; drupes ovoid, acute at apex, 7–8 mm. long, exocarp 3-valved; seeds pale yellowish, 3-angled, acute at apex.

Rocky hillsides, Lower Sonoran and Tropical Zones, central Sonora southward to Guerrero. June–July.

**10. Bursera fragilis** S. Wats., Proc. Amer. Acad. **21:** 422, 1886.

*Terebinthus fragilis* Rose, Contr. U.S. Nat. Herb. **10:** 119, 1906.
*Elaphrium fragile* Rose, N. Amer. Fl. **25:** 252, 1911.

Small tree to 8 m. tall with slender, pale orange or orange-gray, glabrous, young twigs and dark, reddish gray bark on older twigs; bark on trunk reddish, exfoliating in papery sheets; leaves pinnate, to 12 cm. long, rachises very narrowly winged between upper leaflets; petioles slender, usually equaling rachises; leaflets 5–11, lanceolate, 6–15 mm. broad, 2.5–6 cm. long, acute at base, slightly attenuate at apex, serrate from base to apex, glabrous; peduncles to 6 cm. long, few- to many-flowered, paniculate; sepals triangular-ovate, about 3 mm. long, glabrous; petals to 1 cm. long; drupes ovoid, acute at apex, 6–10 mm. long, glabrous exocarp 3-valved; seeds broadly ovoid, pale yellowish, aril reduced to a vestigial basal ring, or absent.

Dry ridges and rocky hillsides, Lower Sonoran Zone, central Sonora to southern Chihuahua. June–July.

## Family 70. MELIACEAE. Chinaberry Family

Trees or shrubs with smooth or fissured bark. Leaves alternate, estipulate, usually pinnate. Flowers perfect or less commonly polygamodioecious. Calyx 4–5-lobed or 4–5-parted, sepals imbricate or rarely valvate. Petals 4–6, free or adnate to disk. Stamens 5, 8, 10, or rarely 16–20, usually united into a tube at least at base, but free in *Cedrella.* Disk annular to tubular. Ovary superior, free, 2–5-celled; ovules mostly 2 but sometimes as many as 12 per cell. Style simple. Stigma discoid or pyramidal. Fruit a berry, a capsule, or rarely a drupe.

Leaves bipinnate; fruit a globose drupe .......................................... 1. *Melia*
Leaves once-pinnate; fruit a capsule .......................................... 2. *Trichilia*

### 1. MELIA L., Sp. Pl. 384, 1753

A tree with weak brittle wood and bipinnate leaves. Calyx 5–6-parted, the lobes imbricate. Petals 5–6, linear-spatulate, contorted. Stamen tube subcylindric. Anthers 10–12. Ovary subglobose, 3–6-celled; ovules 2 in each cell. Stigma 3–6-lobed. Drupe subfleshy, maturing locules 1–6, usually 1, and the locule 1-seeded.

Represented in North America by a single species, having escaped sparingly from cultivation in the southwestern United States and southward.

## 1. **Melia azedarach** L., Sp. Pl. 384, 1753

A round-crowned tree to 15 m. tall but usually smaller, the bark smooth on twigs and young branches, fissured and grayish on mature trunks; leaves 1.5–5 dm. long, petiole about one-fifth leaves' total length; leaflets ovate-lanceolate, more or less alternate, 1–2.5 cm. wide, 3–6 cm. long, attenuate at apex, rounded to broadly cuneate at base, serrulate to serrate, glabrous or minutely puberulent along margins, faintly glaucous beneath; peduncles 3–10 cm. long, paniculately branching, many-flowered; sepals ovate, 2–2.5 mm. long, puberulent; petals about 1 cm. long, white tinged with purple; stamineal column 7–8 mm. tall, anthers borne inside lobes; drupes 12–15 mm. in diameter, yellowish, flesh mealy.

Planted extensively in parks and as a street-side shade tree in many places within our range. Occasionally escaping along streams and ditches. Jan.–June.

## 2. **TRICHILIA** L., Syst. Nat. ed. 10, 1020, 1759

Shrub or small tree with even-pinnate or rarely odd-pinnate leaves. Flowers in axillary or rarely terminal panicles. Calyx 4–5-toothed or 4–5-lobed, sometimes with 5 distinct sepals. Petals 4 or 5 or rarely 3. Filaments united into a tube or coalesced only at base. Anthers borne at apex of lobes of tube or of filaments. Ovary 2–4-celled, ovules 1, or 2 with one superimposed upon the other or both collateral. Capsule usually 3-valved, 3- or rarely 4-seeded. Seed encased in a thin fleshy aril.

Leaflets 2.5–8 cm. long; drupes essentially glabrous at maturity; anthers pubescent ............ 1. *T. hirta*
Leaflets 5–12 cm. long; drupes hirsutulous at maturity; anthers glabrous ... 2. *T. colimana*

### 1. **Trichilia hirta** L., Syst. Nat. ed. 10, 1020, 1759

*Trichilia spondioides* Jacq., Enum. Pl. Carib. 20, 1760.
*Cupania trachycarpa* Griseb., Mem. Amer. Acad. II, **8:** 168, 1860.
*Trichilia schiedeana* var. *purpusii* Brandegee, Univ. Calif. Pub. Bot. **4:** 274, 1912.

A tree to 8 m. tall or sometimes taller; young twigs puberulent-strigose; leaves 2.5–4 dm. long, petioles and rachises hirsutulous to nearly glabrous; leaflets 9–21, ovate, lanceolate, or ovate-lanceolate, acute to acuminate or short-attenuate at apex, 1.5–3.5 cm. wide, 2.5–8 cm. long, hirsutulous beneath, especially along midrib and veins; panicles hirsutulous to glabrous, shorter than leaves; calyx lobes triangular-ovate, acute, essentially glabrous, 1.5–2.5 mm. long; petals to 5–6 mm. long with conspicuous veins; stamens united almost to middle; capsule subglobose, 10–14 mm. broad, nearly glabrate at maturity; seeds 2 in each cell, superposed, about 6 mm. in diameter.

Foothills or rocky slopes along eastern border of the Sonoran Desert in southern Sonora to South America and Jamaica. June–Nov.

### 2. **Trichilia colimana** C. DC., Bot. Gaz. **19:** 40, 1894

*Trichilia langlassei* C. DC., Ann. Conserv. & Jard. Bot. Genève **10:** 160, 1907.

Tree with smooth gray bark; leaves to 4 dm. long or more; petioles and rachises hirsutulous; leaflets 9–15, elliptic, ovate-lanceolate, or rarely somewhat oblanceolate, to 5 cm. broad, 5–12 cm. long, acuminate, acute, or slightly attenuate at apex, cuneate to obliquely and unequally rounded at base, hirsutulous to nearly glabrous and dark green above, hirsute or hirsutulous and paler beneath, veins more heavily invested than rest of blade; inflorescence 3–15 cm. long, hirsute, but nearly glabrous in age; calyx lobes pubescent; anthers glabrous; capsule subglobose, 12–22 mm. in diameter, 2–4-valved, wrinkled and finely but closely hirsutulous; seeds 2 in each cell, superposed, dark yellow, 5–7 mm. in diameter.

Mostly in shaded canyons, foothills of the Sierra Madre along eastern margin of the Sonoran Desert in southern Sonora. Gentry gives its altitudinal range as from 1,000 to 3,000 feet. Aug.–Oct.

## Family 71. **MALPIGHIACEAE.** Malpighia Family

Herbs, vine, shrubs, or trees with opposite, simple, stipulate leaves. Flowers perfect, solitary, or in cymes or racemes. Cleistogamous flowers accompanying other flowers or absent. Sepals 5, some or all of them bearing 1–2 sessile or stipitate glands or rarely glandless. Petals 5, distinctly clawed, blade often toothed, fimbriate, or lobed. Stamens 5 or 10, or fewer by abortion; filaments united below middle or toward base or rarely distinct. Ovary superior, sessile, frequently lobed, 2–5-carpellate but usually with 3 winged or unappendaged carpels. Fruit a capsule, drupe, nut, or samara, latter variously winged and often crested.

Fruit strongly winged samaras; calyx bearing 8–10 glands (on cup or sepals or divided between them), rarely lacking in *Heteropteris*:
- Samaras with lateral wings, these in ours united to form an orbicular wing:
  - Fertile stamens 10; styles distinct; lateral samaras 4–5 cm. wide, rounded at base ............ 1. *Mascagnia*
  - Fertile stamens (in ours) 3 or 2; styles united; lateral samaras 8–10 mm. wide, mucronate at base ............ 2. *Gaudichaudia*
- Samaras with dorsal wings only:
  - Wings of samaras distinctly thickened along lower (or outer) edge; fertile stamens 10; styles 3, distinct ............ 3. *Heteropteris*
  - Wings of samaras not noticeably thickened along either edge; fertile stamens 2–7; styles united ............ 4. *Janusia*

Fruit of nutlets, a capsule, or a drupe; calyx glandless or essentially so in *Echinopterys* and *Thryallis,* the others with 6–10 glands on calyx-cup or sepals:
- Calyx glandless; fruit a capsule or of bristly nutlets:
  - Style 1; fruit of 3 bristly nutlets; panicles slender, elongate ............ 5. *Echinopterys*
  - Styles 3; distinct; fruit a capsule; panicle broadly pyramidal ............ 6. *Thryallis*
- Calyx bearing 6–10 glands; fruit a drupe:
  - Styles 3, distinct; stigmas minute, inconspicuous; flowers in short axillary cymes ............ 7. *Malpighia*
  - Styles united into a single column; stigma distinctly 3-lobed; flowers in panicles ............ 8. *Bunchosia*

### 1. **MASCAGNIA** Bertero, in Colla, Hort. Ripul. 86, 1824

Climbing or trailing vine or upright shrubs. Leaves opposite, simple, leathery or thin, entire. Sepals 5, persistent; calyx bearing 8–10 glands. Petals yellow, blue, violet, or purple, blades broad, margins various. Stamens 10; filaments lanceolate or subulate. Ovary superior, 3-lobed. Styles 3, distinct. Fruit of 3 samaras, with distinct or united lateral wings, dorsal wing smaller than laterals or absent. Cotyledons nearly equal, foliaceous.

**1. Mascagnia macroptera** (Sessé & Moc.) Niedenzu, Gen. Masc. 27, 1908

*Hiraea macroptera* Sessé & Moc. ex DC., Prod. **1:** 586, 1824.
*Hiraea septentrionalis* Juss., Ann. Sci. Nat. II, **13:** 259, 1840.
*Hiraea greggii* S. Wats., Proc. Amer. Acad. **17:** 333, 1882.
*Hiraea mexicana* Rose, Contr. U.S. Nat. Herb. **1:** 312, 1895.
*Mascagnia septentrionalis* Niedenzu, in Engler & Prantl, Nat. Pflanzenf. **3**[4]**:** 55, 1896.

A trailing or scrambling vine or an erect shrub to 2 m. tall with slender stems and pale smooth bark; petioles 3–5 mm. long; leaf blades ovate, oblong, or oblong-lanceolate, 5–22 mm. wide, 2.5–7 cm. long, acute at apex, acute to rounded at base, entire, sparsely strigose to glabrate beneath, several small glands borne at various points on petiole, at base of blade, or sometimes lacking; sepals broadly ovate, 2.5–3.5 mm. long, tomentulose on outer surface; glands oval, 0.6–1.5 mm. long, dark reddish purple to nearly black; petals yellow, 6–12 mm. long, markedly different in size, blade ovate, orbicular, to obovate, erose-denticulate, claw short; samaras 4–5.5 cm. wide, slightly longer, flabellate lateral wings undulate and entire to coarsely erose, sparsely appressed-pubescent.

Along rocky washes, on hillsides and sandy plains, Lower Sonoran and Tropical Zones, central Baja California and central Sonora, almost throughout Mexico. Flowering following rains.

## 2. GAUDICHAUDIA H. B. K., Nov. Gen. & Sp. 5: 156, 1822

Trailing or climbing vines or shrubs, stems much branched. Leaves opposite; blades entire, cordate, auricled at base, or merely narrowed into petiole, mostly pubescent. Calyx with 8 or 10 glands. Sepals 5, persistent, each bearing 2 glands near base, or one or two sepals lacking one or both glands. Petals 5, yellow, toothed or lacerated. Stamens 5, two or three sometimes sterile. Ovary 3-lobed. Samaras usually 3, lateral wings united or free. Abnormal flowers with some parts reduced or absent often occur in same inflorescence as normal ones.

**1. Gaudichaudia mucronata** (Sessé & Moc.) Juss., Ann. Sci. Nat. II, **13:** 253, 1840

*Hiraea mucronata* Sessé & Moc. ex DC., Prod. **1:** 586, 1824.
*Gaudichaudia filipendula* Juss., Ann. Sci. Nat. II, **13:** 252, 1840.

A twining vine with slender, appressed-strigose branches; petioles slender, 3–6 mm. long, densely strigose with yellowish hairs; leaf blade oblong-ovate, 12–22 mm. wide, 3–8 cm. long, weakly auricled or rounded at base, acute and mucronate at apex, dark green and sparsely strigose above, cinereous-tomentose beneath, a few stiff, yellowish, appressed hairs along midrib and main laterals; sepals ovate, 3–4 mm. long, strigose, glabrous glands about 1 mm. long; petals yellow, to 6–8 mm. long, blades ovate to suborbicular, erose-denticulate; samaras suborbicular, 8–12 mm. high, rounded at apex, narrowly keeled at base, coarsely strigose in youth but eventually nearly glabrous, purplish.

Hillsides, canyons, and rocky slopes, mostly Upper Sonoran Zone, southeastern Sonora eastward into Chihuahua and south and east to Vera Cruz and Oaxaca. Feb.–April.

**Gaudichaudia schiediana** Juss. was reported from Sonora by Standley but I have seen no specimens from that area.

## 3. HETEROPTERIS H. B. K., Nov. Gen. & Sp. 5: 163, 1822

Vines, shrubs, or trees. Leaves opposite, leathery, entire or mostly so. Inflorescence paniculate or cymose. Calyx with 8 glands or these lacking. Sepals 5, persistent. Corolla yellow, orange, or purple. Petals 5. Stamens 10, all functional; filaments subulate. Ovary 3-lobed. Styles 3, distinct, subequal. Samaras 2–3 together or solitary, wing thickened along outer edge.

### 1. Heteropteris palmeri Rose, Contr. U.S. Nat. Herb. 1: 311, 1895.

*Heteropteris portillana* Rose, op. cit. 95, 1891. Not *H. portillana* S. Wats., 1887.
*Banisteria palmeri* C. B. Robinson, in Small, N. Amer. Fl. 25: 135, 1910.

Leaf blades ovate to lanceolate, bearing 2 (rarely 1) stipitate glands on margins near base, 8–15 mm. wide, 2–7 cm. long, mucronulate at apex, bright green above, pale and finely veined below, glabrous or glabrate, margin often faintly revolute; sepals ovate, 2.5–3 mm. long, puberulent, glands at least half as long; larger petals 5–6 mm. long, pink; samaras 1.5–2.5 cm. long, reddish, minutely pubescent, slightly auricled at base on the inner side; body of the fruit with a low denticulate crown on each side and parallel to wing, 1–1.5 mm. high.

Canyons and valleys, upper margin of Lower Sonoran and into Upper Sonoran Zone, southeastern Sonora and southward into Sinaloa. Following rains.

## 4. JANUSIA Juss., Ann. Sci. Nat. II, 13: 250, 1840

Somewhat woody vines. Leaves opposite, entire, short-petioled. Inflorescences of axillary clusters, or flowers solitary. Calyx with 8 (or 10) glands. Sepals 5, narrow, persistent. Petals 5, yellow, unequal, long-clawed, blades broad, undulate-margined. Glandless cleistogamous flowers with reduced petals often occurring with petaliferous flowers. Stamens 5–7 or reduced to 3. Ovary usually 3-lobed. Styles united. Stigma entire. Samaras usually 2 or 3, sometimes only one developing, each with a dorsal wing.

Leaves linear or lanceolate; larger petals 3–5 mm. long:
- Leaves narrowly lanceolate, plane, bearing 1–5 short-stipitate or sessile, marginal glands near base; samaras 9–12 (–16) mm. long .......................... 1. *J. gracilis*
- Leaves linear, strongly revolute, no glands visible; samaras 7–10 mm. long .. 2. *J. linearis*

Leaves broadly elliptic, ovate, or oval; larger petals 5–7 mm. long ......... 3. *J. californica*

### 1. Janusia gracilis A. Gray, Pl. Wright. 1: 37, 1852

A slender twining vine to 3 m. tall with strigose-cinereous branches; leaf blades linear-lanceolate, 3–7 mm. wide, 1.5–4 cm. long, or upper ones smaller, plane or margins faintly revolute, acute to acuminate, sparsely appressed-pubescent above, somewhat more densely so and slightly paler beneath, 1 to several small marginal glands near base; pedicels slender, 2–5 mm. long in flower, 1–1.5 cm. long in fruit; sepals ovate-oblong, about 2.5 mm. long, glands 0.8–1 mm. long; petals yellow, occasionally suffused with red or turning red in age, larger ones 4–5 mm. long, blades rhombic or broadly ovate; bodies of samaras coarsely veined, wings 9–12 mm. long or rarely to 16 mm. long.

Rocky hillsides and gravelly slopes and along arroyos, Lower and Upper Sonoran Zones, central Mohave and Cochise Counties, Arizona, to western Texas, Chihuahua, Sonora, and central Baja California. March–Oct.

**2. Janusia linearis** Wiggins, Contr. Dudley Herb. **3:** 69, *pl. 16, figs. 10–14,* 1940

A scandent vine with slender, densely silvery-strigose branches 1–2 m. long, underlying bark reddish; leaves narrowly linear, 1–3.5 mm. wide, 3–6 cm. long, acute at apex, acute or rounded at base, dark green and sparsely strigose or later glabrate above, densely silvery-strigose beneath, margins strongly revolute; petioles 1–2 mm. long; pedicels 3–8 mm. long, bibracteate about middle; sepals broadly ovate, 0.6 mm. wide, 1.5–2 mm. long, acute, densely puberulent, basal glands about 0.6–0.8 mm. in diameter; petals 3–4.5 mm. long, blade ovate to suborbicular, somewhat cucullate; fertile stamens 1–3; samaras reddish, strigose but eventually nearly glabrate, wings 7–10 mm. long.

Rocky hillsides in interior foothills, Lower Sonoran Zone, from the vicinity of Carbó and Ures to hills near Navajoa, Sonora. March–Oct.

**3. Janusia californica** Benth., Bot. Sulph. 8, *pl. 4,* 1844

A twining or scandent vine with slender, pale-pubescent branches to 3 m. long; petioles slender, strigose, 1–5 mm. long; leaf blades ovate, elliptic, or oval, 5–20 mm. wide, 1–4 cm. long, thinly strigose on both surfaces, rounded, truncate, or shallowly cordate at base, acute or rounded and apiculate at apex; sepals oblong-ovate, 2.5–3 mm. long, glands about 1 mm. long; petals yellow, turning reddish or brownish, larger ones 5–7 mm. long, blades suborbicular, crispate-margined but not markedly cucullate; samaras 9–11 mm. long, often reddish, pubescent but eventually glabrate or nearly so.

Gravelly slopes, arroyo banks, and rocky hillsides, Lower Sonoran Zone, western Sonora south of Isla Tiburón and central to southern Baja California. A shade form with exceptionally large leaves (2.5 cm wide, 4.5 cm. long) has been collected near Loreto, Baja California. Jan–Sept.

## 5. ECHINOPTERYS Juss., Arch. Mus. Paris 3: 342, 1843

Shrubs with slenderly branched, pliable stems. Leaves alternate or sometimes subopposite, blades thickish, entire. Inflorescences terminal, elongate panicles. Calyx glandless. Sepals 5. Petals 5, yellow, unequal, blades dentate. Stamens 10, all functional; filaments broad, pubescent. Ovary 3-lobed. Styles united. Stigma 3-lobed. Fruit 3-lobed, hirsutulous and bristly with terminally hooked, laterally puberulent, weak spines, separating into 3 nutlets at maturity.

**1. Echinopterys eglandulosa** (Juss.) Small, N. Amer. Fl. **25:** 148, 1910

*Bunchosia eglandulosa* Juss., Ann. Sci. Nat. II, **13:** 325, 1840.
*Echinopterys lappula* Juss., Arch. Mus. Paris **3:** 342, 1843.
*Coelostylis eglandulosa* Kuntze, Rev. Gen. Pl. **1:** 88, 1891.

Shrub to 2 m. tall, young branches strigose-cinereous; leaf blades ovate, ovate-lanceolate, or oblong, 1–3 cm. long, abruptly acute to mucronate, sparsely strigose in youth; panicles 1–4 dm. long; sepals ovate to oblong, 3–4 mm. long, appressed-pubescent, eglandular; petals 4–10 mm. long, longer ones with cordate blades; carpels 4–5 mm. long, 2.5–3 mm. wide, pubescent and bristly.

Sandy washes and plains and rocky hillsides, Lower Sonoran Zone to subtropics, vicinity of Altar, Sonora, to Morelos, Zacatecas, and Oaxaca. Sept.–May.

## 6. THRYALLIS L., Sp. Pl. ed. 2, 554, 1762

Shrubs or small trees. Leaves opposite, blades entire. Inflorescences racemelike panicles with deciduous bracts and bractlets. Calyx glandless or with small glands. Sepals 5, persistent. Petals 5, yellow or turning red, blades abruptly narrowing to claws. Stamens 10, all functional; filaments united basally, glabrous. Ovary 3-lobed. Styles 3, distinct. Fruit a 3-lobed capsule.

Leaves ovate, oblong, or oval; stems, leaves, and ovary glabrous ............. 1. *T. glauca*
Leaves linear, linear-lanceolate, or lowermost sometimes ovate; stems and ovary pubescent (leaves glabrous in *T. angustifolia*, pubescent in *T. vestita*):
  Petals 4–6 mm. long, not apiculate; leaves glabrous; sepals reddish, essentially glabrous .. ........................................................ 2. *T. angustifolia*
  Petals 6–8 mm. long, minutely apiculate; leaves densely strigose; sepals green, pubescent ............................................................ 3. *T. vestita*

### 1. Thryallis glauca (Cav.) Kuntze, Rev. Gen. Pl. 1: 89, 1891

*Galphimia glauca* Cav., Ic. **5**: 61, *pl. 489*, 1799.
*Malpighia glauca* Pers., Syn. Pl. **1**: 506, 1805.
*Galphimia gracilis* Bartl., Linnaea **13**: 552, 1839.
*Galphimia latifolia* Bartl., op. cit. 553.
*Galphimia grandiflora* Bartl., op. cit. 554.
*Galphimia humboldtiana* Bartl., op. cit. 555.
*Galphimia paniculata* Bartl., op. cit. 556.

Shrub 1–3 m. tall with glabrous stems, leaves, and petioles; leaf blades ovate, oblong, or oval, 1–3 cm. wide, 1–7 cm. long, obtuse to acute, cuneate to rounded at base, conspicuously glaucous beneath, 1–2 sessile glands borne near base of margin or toward upper end of slender petiole; panicle 1–3 dm. long, often considerably branched, pyramidal; pedicels slender, 1.5–2.5 cm. long, these and branches of panicle rusty-pubescent; sepals lanceolate to oblong, 3–5 mm. long, glabrous; petals yellow, ovate, 10–13 mm. long, claw 1–2 mm. long; ovary glabrous; capsule subglobose, shallowly 3-lobed, 5–6 mm. in diameter, slightly exceeding sepals.

Canyons and open rocky slopes, mainly Upper Sonoran Zone, southeastern Sonora to Central America and the West Indies. Not an element of the flora of the desert proper. Flowering almost throughout year.

### 2. Thryallis angustifolia (Benth.) Kuntze, Rev. Gen. Pl. 1: 89, 1891

*Galphimia angustifolia* Benth., Bot. Sulph. 9, *pl. 5*, 1844.
*Galphimia linifolia* A. Gray, Gen. Ill. **2**: 196, 1849.
*Galphimia linifolia* β *oblongifolia* A. Gray, Pl. Wright. **1**: 36, 1852.
*Galphimia angustifolia* var. *oblongifolia* S. Wats., Proc. Amer. Acad. **24**: 42, 1889.
*Thryallis linifolia* Kuntze, Rev. Gen. Pl. **1**: 89, 1891.
*Thryallis angustifolia oblongifolia* Vail, Bull. Torrey Club **22**: 228, 1895.

A tufted shrub to 1 m. tall with closely strigose, slender stems, young branchlets usually reddish; petioles 4–7 mm. long; leaf blades linear to lanceolate or lower ones ovate, 4–10 (rarely –15) mm. wide, 1–4 cm. long, acute to obtuse at apex, glabrous, paler beneath than above, bearing 2 sessile, marginal glands near base; panicle few-flowered, strigose; sepals lanceolate to oblong, 3–4 mm. long, glabrous, reddish; petals yellow, turning reddish, to 6 mm. long, tips obtuse or rounded, margins minutely erose; capsules broadly obovoid, 3–3.5 mm. wide, 4–5.5 mm. high, glabrous or nearly so; seeds yellowish, shining.

Rocky or gravelly slopes and arroyos, mainly Lower Sonoran Zone, vicinity of

Santa Rosalia to Cabo San Lucas, Baja California, and in Sonora and adjacent northern Mexico to Texas. Sept.–March.

**3. Thryallis vestita** (S. Wats.) Rose, Contr. U.S. Nat. Herb. **12:** 281, 1909
*Galphimia vestita* S. Wats., Proc. Amer. Acad. **21:** 421, 1886.

Tufted suffrutescent plant 2–4 dm. tall from a heavy woody taproot, densely strigose-cinereous throughout; petioles 1–4 mm. long or leaves subsessile; leaf blades linear-lanceolate, 3–7 mm. wide, 1–5 cm. long, attenuate toward apex, acute to cuneate at base, heavy vestiture obscuring 2 sessile marginal glands near base; inflorescence 1–1.5 dm. long; bracts linear, 5–10 mm. long; pedicels slender, 5–6 mm. long at anthesis, nearly twice as long in fruit; sepals linear-lanceolate, 4–5 mm. long, strigose, green; petals (in ours) 6–8 mm. long, yellow but turning red or pink, blade ovate, minutely emarginate-apiculate, claws 1–1.5 mm. long; capsules 3–5.5 mm. high, about 4 mm. in diameter, sparsely puberulent; seeds dark brown to nearly black, shining.

Sandy arroyos and arid slopes, Sonoran Zones, east-central and southeastern Sonora (near Divisadero and in Río Mayo region) and southwestern Chihuahua. Until recently known only from the type locality in Chihuahua. Flowering in September in Sonora.

## 7. MALPIGHIA L., Sp. Pl. 425, 1753

Shrubs or small trees with opposite, glabrous, or variously pubescent, entire or toothed, simple leaves. Inflorescences cymose, axillary. Sepals 5, persistent, not appreciably accrescent, bearing 6–10 sessile glands. Petals 5, white, pink, or red, blades undulate, erose-dentate or fimbriate, sometimes concave or keeled. Stamens 10; filaments united almost or quite to middle, glabrous. Ovary 3-lobed, glabrous. Styles distinct. Stigmas small. Fruit a depressed or slightly elongate, red or orange drupe.

Petals 9–10 mm. long; anthers globose; sepals tomentose, not bearded .... 1. *M. diversifolia*
Petals 5–6 mm. long; anthers cordate; sepals bearded ........................ 2. *M. ovata*

**1. Malpighia diversifolia** Brandegee, Zoe **5:** 104, 1901

A widely branching shrub to 3 m. tall with slender twigs densely but finely tomentulose in youth, later glabrate and brown; petioles 1–3 mm. long, tomentulose; leaf blades broadly obovate to orbicular, 2–4.5 cm. long, 1.5–3.5 cm. wide, tomentose on both surfaces when young, green and nearly or quite glabrous above when older, margins entire, glandless; peduncles slender, 1–2.5 cm. long, tomentulose; pedicels 8–11 mm. long, bibracteate about at middle, densely pubescent above bracts, sparsely so to glabrate above them; sepals lance-ovate, 3–4 mm. long, pubescent, glands 1–1.5 mm. long; petals pink, larger ones to 1 cm. long, claws 3 mm. long, blade suborbicular, fimbriate, base truncate or nearly so; drupes bright red, strongly depressed, about 1 cm. in diameter, glabrous.

Along arroyos, in canyons, and on hillsides, Lower Sonoran Zone, southern Baja California. Aug.–Oct.

**2. Malpighia ovata** Rose, Contr. U.S. Nat. Herb. **1:** 310, 1895
*Bunchosia parvifolia* S. Wats., Proc. Amer. Acad. **24:** 42, 1889. Not *Malpighia parvifolia* Juss., 1843.
*Malpighia watsoni* Rose, Contr. U.S. Nat. Herb. **1:** 310, 1895.

An open shrub 1–2 m. tall or less with strigose branchlets; leaves ovate, oblong-ovate, or orbicular, quite variable as to size and shape, 1–6 cm. long, acute to obtuse or rounded at apex, bright green and sparsely pubescent above, paler and slightly more densely pubescent beneath; peduncles shorter than leaves, densely pubescent; sepals oblong-ovate, 2.5–3 mm. long, bearded, each sepal bearing 2 glands 1–1.5 mm. long; petals pink, larger ones 5–6 mm. long, blades ovate; drupes bright red, depressed-ovoid, 8–9 mm. high, to 1.5 cm. in diameter.

Hillsides and rocky arroyos, Lower Sonoran Zone, western Sonora south to Oaxaca. Aug.–Oct.

**Malpighia umbellata** Rose, Contr. U.S. Nat. Herb. **1:** 310, 1895, occurs in southern Sonora in the vicinity of Alamos and southward into Sinaloa, but it appears to occur outside the limits of the Sonoran Desert.

### 8. BUNCHOSIA Rich. ex Juss., Ann. Mus. Paris **18:** 481, 1811

Shrubs or trees. Leaves opposite, usually bearing glands on petioles or near base of entire blades. Inflorescences paniculate. Calyx with 8–10 glands, slightly accrescent. Petals 5, white or yellow, concave. Stamens 10. Ovary 2–3-lobed. Styles united. Stigma 3-lobed, capitate or discoid. Fruit an ovoid, globose, or reniform drupe.

**1. Bunchosia sonorensis** Rose, Contr. U.S. Nat. Herb. **1:** 94, 1891

A shrub to 3–4 m. tall with minutely strigulose young twigs; leaf blades ovate, oblong, or lanceolate, 1–3 cm. wide, 2–7 cm. long, acute or obtuse at apex, acute to rounded at base, closely pubescent on both sides when young, later glabrate; pedicels stoutish, 1–4 mm. long, densely puberulent; sepals deltoid-ovate to oblong-ovate, 4–6 mm. long, tomentulose, glands 2–3 mm. long; petals yellow, larger ones 9–10 mm. long, inner petal broadly spatulate, others ovate to suborbicular; ovary and style pubescent; drupe 2-lobed, 12–15 mm. in diameter, dark red.

Rocky slopes and mesas, Lower Sonoran Zone, southern Sonora. The flowers are reported to be very fragrant. Sept.–Jan.

## Family 72. POLYGALACEAE. Milkwort Family

Herbs, shrubs, trees, or scandent vines, often with pellucid glands in leaves and sometimes in flowers or fruits also. Leaves alternate, opposite, or whorled, simple and entire. Flowers perfect, irregular, each subtended by a bract or 2 bractlets. Sepals 5, two lower often united, two interior usually larger and petaloid. Petals 3–5, hypogynous, anterior one (called the keel) boat-shaped, often with a terminal crest, upper two lingulate or ovate, usually attached to stamineal sheath at base. Stamens 3–8, usually 8, filaments coalesced most of their length to form a sheath; anthers 1-celled, opening by a subterminal pore. Disk present, reduced to a gland or wanting. Ovary of 1–5 carpels; stigma 2-lobed. Ovules usually solitary. Fruit a capsule, samara, or drupe, dehiscent or indehiscent. Seeds usually solitary in each cell, usually arillate and pubescent; endosperm present but often scanty. Embryo axial, straight.

Ovary 2-celled or sometimes 1-celled by abortion; fruit a dehiscent capsule ..... 1. *Polygala*
Ovary 1-celled; fruit a winged indehiscent samara .......................... 2. *Monnina*

### 1. POLYGALA L., Sp. Pl. 701, 1753

Herbs, shrubs, or trees with alternate or whorled, entire leaves. Stipules wanting or rarely represented by glands. Inflorescence usually racemose. Sepals 5. Petals normally 3, united at base, keel clawed, unappendaged or with an apical beak or crest; two upper petals attached to stamineal tube or to keel or to both; lateral petals very small when present. Stamens mostly 8, filaments coalesced into a sheath nearly to anthers, sheath split on upper side; anthers opening by an apical or subapical introrse pore. Ovary 2-celled; ovules solitary, pendulous; stigma 2-lobed, often tufted. Capsule compressed contrary to partition, winged, or margined, usually dehiscent. Seeds usually pubescent, arillate; endosperm scanty or none.

Keel blunt, without a beak or crest:
  Calyx with sepals all free; sepals and petals deciduous; wings 6 mm. long or less:
    Leaves and capsules not glandular:
      Leaves oval, upper and lower ones nearly alike; keel pilose .......... 1. *P. xanti*
      Leaves somewhat dimorphous, upper ones distinctly narrower and longer than lower ones; keel glabrous or slightly puberulent, not pilose:
        Aril veil-like, margins scarious, covering about half of seed ......... 2. *P. obscura*
        Aril not veil-like, covering less than one-third of seed:
          Outer sepals acuminate; capsule 7–7.5 mm. long ............. 3. *P. russelliana*
          Outer sepals acute, not acuminate; capsule 7–9.5 mm. long:
            Capsule puberulent on sides at maturity; aril minute, 0.7–0.9 mm. high; sepals 2–2.2 mm. long; keel 3.5–4 mm. long ............... 4. *P. magdalenae*
            Capsule only ciliate at maturity, sides glabrous; aril 1–1.2 mm. long; sepals 2.5–3.5 mm. long; keel 4–5.8 mm. long .................. 5. *P. longa*
    Leaves and capsules conspicuously glandular ...................... 6. *P. macradenia*
  Calyx with 2 lower sepals connate; sepals and petals persistent; wings 12–15 mm. long in fruit ............................................... 7. *P. apopetala*
Keel with a conic or cylindric beak or with a fimbriate crest:
  Flowers 4–12 mm. long; keel beaked but not crested; plants perennial, not glandular:
    Sepals and wings deciduous; aril 3–4 mm. long, umbo pilose; seed 4.5–5 mm. long ... ............................................................ 8. *P. fishiae*
    Sepals and wings persistent; aril 1.2–1.3 mm. long, umbo corneous; seed 3.2–3.5 mm. long ............................................... 9. *P. desertorum*
  Flowers 2.5–3.5 mm. long; keel with a fimbriate crest; plants annual, stipitate-glandular:
    Leaves 3–13 mm. long; flowers rosy purple, rosy, rarely whitish; crest of 3–4 lobes on each side; capsule about 1.5 mm. long ...................... 10. *P. glochidiata*
    Leaves 5–20 mm. long; flowers white or pinkish; crest of 2 (rarely 3) lobes on each side; capsule 3–3.4 mm. long .............................. 11. *P. berlandieri*

## 1. **Polygala xanti** A. Gray, Proc. Amer. Acad. **5:** 153, 1861

Suffrutescent, with procumbent to suberect branches 1–2.5 dm. long, densely pubescent throughout with fine incurved hairs; leaves oval, 6–18 mm. long, 4–12 mm. wide, rounded or obtuse at both ends, sometimes minutely apiculate at apex; racemes loose, 1–4.5 cm. long; flowers white or bluish, keel often darker blue or purple and tinged with yellow; herbaceous sepals 2.5–3.2 mm. long, lance-elliptic, puberulent; wings ovate to rhombic-oval, 3–3.5 mm. wide, 5.3–6 mm. long, rounded or obtuse at apex, broadly cuneate at base, puberulent on back, margins ciliate; keel 5–5.5 mm. long, dark blue or purple on tip, yellowish at base, pilose along angle near base; lateral petals minute, 0.4–0.5 mm. long; capsule ovate, 7–8 mm. long, 4–5 mm. wide, notched at apex, rounded at base, one cell slightly larger than other, narrowly winged, densely puberulent with incurved hairs; seeds 4–4.3 mm. long, silky-pilose with closely appressed, stiff hairs; aril 1–1.2 mm. long, white, margin scarious, faintly repand, umbo about 0.5 mm. long, minutely pilose.

Along washes, usually under larger shrubs, Lower Sonoran Zone, southern Baja California, mainly south of La Paz. Following rains, mostly summer and fall.

**2. Polygala obscura** Benth., Pl. Hartw. 58, 1840

*Polygala puberula* A. Gray, Pl. Wright. **1**: 40, 1852.
*Polygala neomexicana* Woot. & Standl., Contr. U.S. Nat. Herb. **16**: 144, 1913.
*Polygala obscura* var. *genuina* S. F. Blake, Contr. Gray Herb. **47**: 30, 1916.
*Polygala vagans* Brandegee, Univ. Calif. Pub. Bot. **6**: 365, 1917.
*Polygala laeta* Brandegee, op. cit. 366.
?*Polygala obscura* var. *macrophylla* Chod., in Loesner, Verh. Bot. Brandenburg **58**: 148, 1917.

Suffrutescent perennial with numerous, erect or ascending stems 1–6 dm. tall, grayish-puberulent with incurved or some spreading hairs; lower leaves oval to oblong, 8–28 mm. long, 4–7 mm. wide, acute or rounded at apex, broadly cuneate at base, middle and upper leaves elliptic-lanceolate to linear-lanceolate, to 4.5 cm. long, 1.5–6 (or rarely to 12) mm. wide, obtuse to acuminate, puberulent; racemes loose, 3–20 cm. long; flowers purplish; sepals narrowly lanceolate, acute, 3–3.5 mm. long, spreadingly puberulent; wings oval-oblong to elliptic, 2–3.5 mm. wide, 4.5–6 mm. long, rounded at apex, sparsely puberulent along middle and toward apex or nearly glabrous; keel 4.8–5 mm. long, glabrous or essentially so; capsule broadly oval, 8–10 mm. long, 6–8 mm. wide, densely puberulent when young, merely ciliate at maturity; seeds 4–4.8 mm. long, pilose; aril 1.5–3 mm. long, veil-like, margin finely crenate or irregularly lobulate, scarious; umbo pubescent.

Rocky hillsides and gravelly slopes, Sonoran Zones, southern Arizona to Texas and southward to Oaxaca. Collected near Mátape, Sonora. May–Sept.

**3. Polygala russelliana** S. F. Blake, N. Amer. Fl. **25**: 326, 1924

Suffruticose, with several erect stems 1–3 dm. tall, densely puberulent throughout with incurved fine hairs; leaves elliptic to lanceolate, 1.5–4 cm. long, 5–10 mm. wide, acute to acuminate at both ends; racemes loose, 1.5–6 cm. long; flowers yellowish, at least when dried; sepals ovate-lanceolate, acuminate, 2.8–3 mm. long, puberulent; wings 5–6 mm. long, 2.5–3.2 mm. wide, cuneate at base, rounded to faintly apiculate at apex, sparsely puberulent and ciliate-margined; keel 5–5.6 mm. long, glabrous; capsule ovate, narrowly winged, 7–7.5 mm. long, puberulent; seeds 4.2–4.5 mm. long, pilose; aril about 1.2 mm. long, scarious margin 3-lobed, lobes deltoid to oblong, obtuse, obscurely crenate, dorsal lobe slightly longer than lateral; umbo pilose.

Among shrubs and on rocky hillsides, Lower Sonoran and Subtropical Zones, southern Sonora to Morelos. Not definitely known from the Sonoran Desert proper but to be expected toward its southern limits. July–Sept.

**4. Polygala magdalenae** Brandegee, Univ. Calif. Pub. Bot. **4**: 182, 1911

Perennial, with prostrate to erect stems 1–3 dm. long, cinereous-puberulent throughout; lower leaves ovate-oblong, 5–7 mm. wide, middle and upper ones oblong-ovate to linear, 2–8 mm. wide, 8–20 mm. long, rounded, retuse, or acute at apex, densely puberulent, veins inconspicuous and not evidently reticulate; racemes loose, 4–12 cm. long; flowers yellowish; outer sepals lanceolate, 2–2.2 mm. long, puberulent; wings oval-obovate, about 2.8 mm. wide, 4.5 mm. long, obtuse at apex, puberulent near tip; keel 3.5–4 mm. long; capsule oval, 7.5–9.5 mm. long, densely appressed-puberulent; seeds about 4 mm. long, appressed-strigose; aril minute, 0.7–0.9 mm. high, notched at base on side next to placenta, narrow margin faintly crenulate, umbo pubescent.

Plains, often under shrubs, Lower Sonoran Zone, Magdalena and Santa Margarita Islands and in the Cape Region, Baja California. March–May.

**5. Polygala longa** S. F. Blake, Contr. Gray Herb. **47:** 29, 1916
*Polygala tenuiloba* S. F. Blake, op. cit. 36.

Suffruticose below, with several erect stems 1–4.5 dm. tall from a woody rootstock; herbage densely cinereous-puberulent with incurved hairs; lower leaves broadly elliptic to oblong-oval, 8–18 mm. long, 5–7 mm. wide, middle and upper ones progressively narrower and longer, oblong-lanceolate to linear-lanceolate, 1.5–4 mm. wide, 1.5–3.5 cm. long, acute to acuminate, or rarely obtuse; racemes loose, 2–10 cm. long; flowers purplish; sepals lanceolate, 2.2–3.5 mm. long, minutely puberulent; wings suborbicular to broadly oval, 2.4–4 mm. wide, 3.5–5.5 mm. long, rounded at apex, puberulent over most of outer surface and near apex within; keel 4–5.7 mm. long, glabrous or essentially so; capsule oval, 7–8 mm. long, ciliate on margins, sides glabrous and finely reticulate-veined; seeds 4–4.5 mm. long; aril about 1–1.2 mm. high, dorsal margin about 1.5 mm. long; umbo about 0.4 mm. high, rounded, pilose.

Banks of arroyos and rocky hillsides, Sonoran Zones, southern Arizona to Texas, south into Chihuahua and Sonora. Entering the Sonoran Desert 8–10 miles east of Magdalena, Sonora. March–May; Sept.–Oct.

**6. Polygala macradenia** A. Gray, Pl. Wright. **1:** 39, 1852
*Polygala macradenia* var. *genuina* S. F. Blake, Contr. Gray Herb. **47:** 36, 1916.

Suffrutescent, with numerous erect or ascending, stiffish, crowded stems 3–25 cm. long from a woody root; stems densely leafy, flexuous, densely canescent-puberulent; leaves essentially alike, oblong-lanceolate to ovate-lanceolate, 0.5–1.5 mm. wide, 2–7 mm. long, acute to obtuse at apex, rounded or obtuse at base, thick, densely canescent-puberulent and gland-dotted; flowers 1–2 in upper axils, purplish; sepals ovate-oblong, obtuse, about 2 mm. long, puberulent and gland-dotted like leaves; wings obovate, about 2.5 mm. wide, 5–5.5 mm. long, cuneate at base, rounded at apex, sparsely puberulent on back, ciliate, purple; keel about 4.5 mm. long; capsule oblong-ovate, canescently puberulent and gland-dotted, 5–6 mm. long; seeds 4 mm. long, densely silky-pilose, hairs forming a dense tuft at apex; aril 1.2–1.6 mm. high, umbo corneous, vaulted, sparsely puberulent, 0.9–1.1 mm. high, free portion of 3 short, rounded, subappressed lobes, one dorsal and two lateral.

Rocky hillsides, Sonoran Zones, Arizona to Texas and south to Coahuila and Sonora. Collected at several stations in the desert area of northern Sonora. March–July.

**7. Polygala apopetala** Brandegee, Proc. Calif. Acad. II, **2:** 130, *pl. 3*, 1889

A sturdy shrub or small tree to 3 m. tall; branchlets minutely strigillose-puberulent; leaves ovate to ovate-lanceolate, 7–22 mm. wide, 2–5 cm. long, broadly cuneate at base, acute, truncate, or retuse at apex, sparsely strigillose to glabrate and shining, reticulate-veined; petioles 2–3 mm. long; racemes loose, 5–20 cm. long; pedicels slender, 8–13 mm. long; flowers pinkish purple; sepals oblong-ovate, 4.5 mm. long, densely ciliate but otherwise glabrous, pinkish; wings suborbicular, rounded or retuse-rounded at apex, 12–15 mm. long, nearly as wide, sparsely and minutely ciliate-glandular along margins, markedly reticulate in fruit, persistent; keel 8–9 mm. long, tip blunt, strongly turned upward; capsule quadrate-orbicular, 12–19 mm. long, deeply emarginate at apex, glabrous; seed ellipsoid, 8–9 mm. long, about 6 mm. wide, pilosulous; aril minute, corneous, about 0.5 mm. long.

Rocky hillsides and canyon walls, Lower Sonoran Zone, vicinity of Comondú and south to the Cape Region, Baja California. Said to be cultivated as an ornamental in La Paz. Feb.–May and probably after summer rains.

**8. Polygala fishiae** Parry, Proc. Davenp. Acad. **4:** 39, 1884

*Polygala cornuta* var. *fishiae* Jepson, Man. Fl. Pl. Calif. 594, 1925.

Much-branched, frutescent plant to 1.6 m. tall with sparsely strigillose to glabrous leaves and branches; leaves narrowly oblong, or oblong-lanceolate, 3–14 mm. wide, 1.5–5 cm. long, rounded to broadly cuneate at base, acute, rounded, or retuse at apex; racemes loose to moderately dense, 2.5–20 cm. long; pedicels 5–12 mm. long in fruit, glabrous or sparsely strigillose; flowers purplish, whitish and yellowish; sepals elliptic, obtuse, 2.5–3.5 mm. long, densely ciliate, deciduous; wings purplish, obovate, 4–4.2 mm. wide, about 8.5 mm. long, rounded at apex, cuneate at base, finely ciliolate, sides glabrous; keel yellow, 7.5–8 mm. long, with a short curved beak; capsule suborbicular, 7–7.5 mm. wide, 7.5–8.5 mm. long, slightly narrower at apex than at base, shallowly emarginate, glabrous; seeds silky-pilose, 4.5–5 mm. long; aril 3–4 mm. long, umbo corneous, about 3 mm. long, pilose, scarious border irregularly dentate-lobulate, about 1 mm. high.

Oak woodland and clearings in chaparral, mainly Upper Sonoran Zone, Ventura and Los Angeles Counties southward into northern Baja California. June–Sept.

**9. Polygala desertorum** Brandegee, Proc. Calif. Acad. II, **2:** 130, 1889

Many-stemmed, frutescent plant 2–6 dm. tall with sparingly leafy, bright green, sparsely strigillose, slender branches; leaves linear to linear-lanceolate, 1–3 mm. wide, 8–20 mm. long, often slightly arcuately curved, sessile, acute or lower obtuse, appressed-puberulent; racemes loose, 5–10 cm. long, few-flowered; flowers purple, yellow, and white; sepals oblong to elliptic, 5–7.5 mm. long, glabrous, purple-veined; wings purple, oblong-obovate, 3.2–3.4 mm. wide, 7–8 mm. long, rounded at apex but sides inrolled, glabrous, strongly veined; keel yellow with white base, 8–8.5 mm. long, beak blunt, 2–2.2 mm. long; capsules ovate-oblong, 3.5–4 mm. wide, 5–5.5 mm. long, glabrous; seeds puberulent, 3.2–3.5 mm. long; aril 1.2–1.3 mm. long, glabrous, corneous, lateral lobes oblong.

Along arroyos, rocky hillsides and canyons, Lower Sonoran Zone, from near Rosario south to central Baja California at least to Punta Prieta and Bahía de los Angeles. Oct.–May.

**10. Polygala glochidiata** H. B. K., Nov. Gen. & Sp. **5:** 400, 1821

*Polygala glochidiata* f. *typica* S. F. Blake, Contr. Gray Herb. **47:** 91, 1916.
*Polygala glochidiata* f. *leucantha* S. F. Blake, loc. cit.

Slender annual to 3 dm. tall, simple or freely branched, sparsely and minutely stipitate-glandular on stems, leaves, and calyces; leaves from about middle of plant and below in whorls of 4 or 5, lowest lanceolate to oblanceolate, about 1 mm. wide, 3–4.5 mm. long, middle and upper ones linear to linear-lanceolate, 0.5–1.5 mm. wide, 5–13 mm. long, slightly revolute, cuspidate; racemes narrow, loose, 1–8 cm. long, 4.5–6 mm. wide; pedicels 1 mm. long or less; flowers rose, rose-purple, or rarely almost white; sepals about 1 mm. long; wings obovate, 1.3 mm. wide, 2.2–2.5 mm. long, rounded; keel about 2 mm. long, crest with 3–4 lobes on each side; capsule broadly elliptic, 1.3–1.5 mm. long, narrowly margined, glabrous; seeds about 1 mm. long, black or dark brown, strigillose with white hairs; aril none.

On plains and under trees, mainly in Upper Sonoran Zone, southern Sonora and Baja California to Guatemala and Cuba, and in South America. Entering the margins of the Sonoran Desert between Cumpas and Baviácora, Sonora. Aug.–Nov.

**11. Polygala berlandieri** S. Wats., Proc. Amer. Acad. **21:** 416, 1886

Freely branched, slender annual to 2 dm. tall with densely but minutely stipitate-glandular herbage; leaves in whorls of 4 or 5 near base, linear, 0.5–1.5 mm. wide, 5–20 mm. long, acute or acuminate, cuspidate, plane or slightly revolute; racemes loose, cylindrical, 2–5 cm. long; pedicels 0.8–1 mm. long; flowers white or faintly pinkish; sepals broadly ovate to oblong-ovate, 1–1.2 mm. long, acute to obtuse; wings narrowly obovate, about 1 mm. wide, 2–2.3 mm. long, rounded at apex; keel about equaling wings, crest of 2 or rarely 3 lobes on each side; capsule narrowly elliptic, 1.2–1.3 mm. wide, 3–3.4 mm. long, glabrous; seeds black, about 2.5 mm. long, appressed-silky-pubescent; aril about 0.5 mm. long, narrowly 2-lobed.

Plains and hillsides, Lower Sonoran and Subtropical Zones, southern Baja California and Tamaulipas to Salvador. Sept.–Nov.

Several additional species of *Polygala* approach the upper margins of the Sonoran Desert but they are inhabitants of the Upper Sonoran and Arid Transition Zones and are excluded since no bona fide records of their entrance into the Sonoran Desert have been found.

## 2. MONNINA Ruiz & Pav., Syst. Veg. Fl. Peruv. & Chil. 169, 1798

Herbs or shrubs with alternate entire leaves with stipules lacking or reduced to glands. Flowers in racemes. Sepals 5, deciduous, outer three herbaceous and free or lower two united, inner two petaloid, larger than outer. Petals 3, lower one keel-like, sessile, upper two united to stamineal sheath below. Stamens 6 or 8, united into a sheath split along lower side; anthers 1-celled, opening by an apical pore. Ovary (in ours) 1-celled; style lobes 2, dissimilar; ovule 1, pendulous. Disk represented by a gland. Fruit samaroid or drupaceous. Seeds exarillate, glabrous; endosperm scant or none.

**1. Monnina wrightii** A. Gray, Pl. Wright. **2:** 31, 1853

Erect, simple or slightly branched annual to 5 dm. high, sparsely and minutely appressed-puberulent with curved hairs; lowest leaves obovate, obtuse, those above gradually becoming linear-lanceolate, 1.5–10 mm. wide, 1.5–6 cm. long, acuminate to attenuate; petioles 1–2.5 mm. long; peduncles 1–6 cm. long; racemes slender, 3–8 mm. thick, 5–30 cm. long; bracts ovate-subulate, 2.5–3 mm. long, caducous; flowers subsessile, cream-colored or greenish, drying purplish or bluish; lower sepals elliptic-ovate, acute to acuminate, 1–1.8 mm. long; wings obovate, broadly rounded, 2 mm. wide, 3 mm. long, ciliate below or wholly glabrous; keel 3–3.5 mm. long; stamineal sheath pilose-ciliate, united with upper petals, auriculate at base of first filament, pilose at apex; fruit ovate, asymmetrical, compressed, faintly cordate at base, winged along both margins, 2.5–4 mm. wide, 3–5 mm. long, puberulent to glabrous, reticulate-veined.

Gravelly and rocky slopes, mainly Upper Sonoran Zone, southern Arizona to New Mexico and south into Chihuahua and Sonora. Collected in the Sonoran Desert near La Mina Verde, 31 km. west of Cumpas, Sonora. Sept.–Nov.

# Family 73. EUPHORBIACEAE. Spurge Family

Herbs, shrubs, vines, or trees with acrid or milky juice, often containing poisonous substances in some or all parts. Leaves simple or rarely compound, alternate, opposite, or whorled, entire, toothed, or lobed. Stipules present and caducous to persistent or wholly lacking. Flowers monoecious or dioecious, either the staminate or pistillate or both bearing small inconspicuous petals or more commonly apetalous; calyx lacking in some genera or species, then the flowers of both sexes often inserted in green or colored involucres. Stamens 1 to many, filaments free to more or less united, if free often jointed and somewhat fleshy, sometimes repeatedly branched; anthers 2-celled, basifixed or dorsifixed, cells closely adnate to connective or separate and attached only by tips. Ovary usually 3–4-celled, less commonly 1–2-celled or many-celled, each cell with 1 or 2 ovules; styles equaling number of cells, simple to many times divided. Fruit usually 3-lobed, each carpel 2-valved and separating from a persistent axis. Seeds carunculate or ecarunculate, endosperm fleshy or oily. Embryo straight or curved.

Ovules 2 in each cell; stamens (or at least those of outer whorl) opposite sepals:
  Leaves alternate; seeds estrophiolate:
    Flowers petaliferous; plants (in ours) herbaceous, stipitate-glandular, ovary 3-celled . . . . . . . . . 1. *Andrachne*
    Flowers apetalous; plants (in ours) shrubby or arborescent, glabrous or puberulent but not stipitate-glandular; ovary 3- to many-celled . . . . . . . . . 2. *Phyllanthus*
  Leaves usually opposite, often in fascicles; seeds strophiolate . . . . . . . . . 3. *Tetracoccus*
Ovules solitary in each cell; stamens (or at least outer whorl) alternate with sepals, or perianth lacking:
  Flowers not surrounded by involucre:
    Twigs, flowers, and herbage bearing forked, malpighiaceous, stellate, or scale-like hairs (some simple hairs often present also):
      Stamens strongly inflexed in bud; tips of anthers pointing downward and often folded against base of filament:
        Capsule 3-lobed, 3-seeded; seeds carunculate; staminate flowers usually petaliferous . . . . . . . . . 4. *Croton*
        Capsule ellipsoidal, 1-seeded; seeds ecarunculate; flowers apetalous . . . . . . . . . 5. *Eremocarpus*
      Stamens erect or only slightly inflexed in bud; tips of anthers pointing upward:
        Pubescence containing some malpighiaceous hairs; both staminate and pistillate flowers usually petaliferous; stamens united in a column; carpels not carinate . . . . . . . . . 6. *Ditaxis*
        Pubescence of simple and stellate hairs, malpighiaceous ones lacking; flowers apetalous; stamens distinct; carpels carinate . . . . . . . . . 7. *Bernardia*
    Twigs, flowers, and herbage glabrous or bearing only unbranched hairs, some of these often gland-tipped:
      Flowers dioecious; leaves often pellucid-dotted . . . . . . . . . 8. *Adelia*
      Flowers monoecious or rarely dioecious; leaves not noticeably pellucid-dotted:
        Calyx lobes valvate in bud:
          Filaments of stamens not branched; leaves entire, serrate, or 3-foliolate, not peltate nor palmately lobed; capsules not spiny:
            Styles distinct; stinging hairs lacking; anther cells separated, attached only at tip . . . . . . . . . 9. *Acalypha*
            Styles united; stinging hairs usually present; anther cells adnate to connective nearly their full length:
              Stamens (in ours) 3; pistillate bracts small, inconspicuous; staminate calyx not markedly vasiform . . . . . . . . . 10. *Tragia*
              Stamens 10–90; pistillate bracts often accrescent and conspicuous; staminate calyx globose-vasiform . . . . . . . . . 11. *Dalechampia*
          Filaments of stamens repeatedly branched; leaves peltate and palmately lobed, ample; capsules usually spiny . . . . . . . . . 12. *Ricinus*

Calyx lobes imbricate in bud:
Stamens 8–10, staminodia often present:
Flowers petaliferous; pistillate calyx usually persistent .......... 13. *Jatropha*
Flowers apetalous; pistillate calyx caducous (persistent in *Cnidoscolus palmeri*):
Stinging hairs present; stipules lacerate; inflorescences cymose; stamens united into an androphore with a ring of hairs at base ...... 14. *Cnidoscolus*
Stinging hairs lacking; stipules entire to ciliolate; inflorescences racemose; stamens inserted between lobes of disk ................ 15. *Manihot*
Stamens 1–4, staminodia lacking:
Seeds carunculate; calyces deeply 3–6- (usually 3-) parted .... 16. *Sebastiania*
Seeds (in ours) ecarunculate; calyces lobed or dentate:
Stamens 2; staminate calyces 2-lobed; seeds smooth:
Plants herbaceous or suffrutesent; flowers solitary in axil of each bract; carpels usually 3, rarely 2 ......................... 17. *Stillingia*
Plants shrubby or arborescent; staminate flowers often several in axil of a bract; carpels usually 2, rarely 3 ....................... 18. *Sapium*
Stamen 1; calyx consisting of a single cochleate sepal; seeds with crustose testas ......................................... 19. *Dalembertia*
Flowers surrounded by involucre, both staminate and pistillate in same involucre:
Involucre cuplike, regular, often bearing petal-like appendages on outer margins of glands ................................................ 20. *Euphorbia*
Involucre oblique, irregular, without appendages .................. 21. *Pedilanthus*

## 1. ANDRACHNE L., Sp. Pl. 1014, 1753

Plants herbaceous or suffruticose. Leaves alternate, small, thin. Flowers monoecious, petaliferous, male flowers in axillary clusters, pistillate solitary. Staminate flowers with 5–6-lobed calyx, petals shorter than or equaling sepals; stamens 5–6, alternate with petals; anthers dehiscing longitudinally. Pistillate flowers with minute petals or these lacking. Ovary 3-celled, each cell biovulate; style branches 3, short, 2-lobed. Capsule dry, each cell bivalvate. Seeds curved, rugulose, ecarunculate. Endosperm fleshy.

**1. Andrachne ciliato-glandulosa** (Millsp.) Croizat, Jour. Wash. Acad. **33:** 11, 1943
*Phyllanthus ciliato-glandulosus* Millsp., Proc. Calif. Acad. II, **2:** 219, 1889.

A low, spreading to decumbent herb 1–3 dm. tall, often somewhat wider than tall; stems, petioles, and pedicels stipitate-glandular; petioles slender, 2–10 mm. long; leaf blades orbicular to broadly ovate, 3–16 mm. wide, 6–18 mm. long, thin, light green, slightly paler and puberulent along veins beneath; flowers small; pedicels subfiliform, 3–7 mm. long; sepals about 1 mm. long, lanceolate; petals slightly shorter than sepals; fruit subglobose, slightly depressed, 3–3.5 mm. wide; seeds 1.5–2 mm. high, angled, rounded dorsally, smooth, reddish brown.

In shade of rocks and at bases of cliffs, Lower Sonoran Zone, from Calmallí to the Cape Region, Baja California. Oct.–April.

## 2. PHYLLANTHUS L., Sp. Pl. 981, 1753

Usually glabrous shrubs or small trees but a few herbaceous species in the New World. Leaves alternate, small to large, entire, stipulate, commonly distichously arranged. Flowers small, apetalous, monoecious or rarely dioecious. Staminate flowers with 4–6 sepals, these distinct or rarely connate at base, more or less biseriate; stamens

2–5 or rarely numerous, filaments free or connate into a tube. Pistillate flowers with sepals similar to those of staminate. Ovary 3- to many-celled; locules bivalvate; style distinct or connate; stigma terminal or along margins of style branches. Capsule usually breaking into 3 segments or loculicidally dehiscent or rarely indehiscent. Seeds ecarunculate; endosperm fleshy.

Leaves oblanceolate to obovate, broadest near apex, nearly sessile; anthers dehiscing longitudinally .................................................. 1. *P. galeottianus*
Leaves ovate, acute or acuminate, broadest at or below middle, on petioles 1–4 mm. long; anthers dehiscing by oblique slits .............................. 2. *P. acuminatus*

**1. Phyllanthus galeottianus** Baillon, Adansonia **1:** 32, 1860
*Phyllanthus peninsularis* Brandegee, Erythea **7:** 8, 1899.

A small glabrous shrub (or sometimes nearly herbaceous) 2–5 dm. tall with several to many slender, flexuous, ascending branches from a woody taproot; stipules lance-subulate, 2–2.5 mm. long, reddish, white-margined near base; petioles 1 mm. long or less; leaves oblanceolate to obovate, 3–15 mm. wide, 1–3 cm. long, acute to obtuse and minutely apiculate at apex, dark green above, pale beneath, margins faintly revolute; flowers clustered in axils, on slender pedicels 2–10 mm. long; staminate flowers several in a cluster; sepals oblong, 1–2 mm. long; anthers minute, dehiscing by longitudinal slits; usually only 1 pistillate flower among several staminate, surpassing the staminate; sepals 6, obovate, 2.5–4 mm. long, narrowly white-margined; capsule depressed-globose, 3–4 mm. broad, 1.5–2.5 mm. high, faintly and finely reticulate; seeds about 1.5 mm. high, 3-angled, rounded dorsally, dark brown to dull black, minutely papillose.

Rocky slopes and along arroyos, Lower Sonoran Zone, southern Baja California and vicinity of Magdalena, Sonora, to Michoacán and Vera Cruz. Aug.–April.

**2. Phyllanthus acuminatus** Vahl, Symb. Bot. **2:** 95, 1791
*Phyllanthus sessei* Briq., Ann. Conserv. & Jard. Bot. Genève **4:** 224, 1900.

Shrub or small tree to 6 m. tall with slender, green, distinctly angled stems; herbage glabrous or sparsely and minutely puberulent with a few short scattered hairs; stipules ovate-subulate, 1–1.5 mm. long; petioles 1–4 mm. long; leaf blades broadly ovate-acuminate, 1–3 cm. broad, 2–5 cm. long, dark green above, only slightly paler beneath; flowers in axillary clusters, usually only 1 pistillate among several male flowers; staminate flowers minute on filiform pedicels 1–3 mm. long, sepals 0.4 mm. long or less; anthers dehiscing by oblique slits; pistillate flowers on slender pedicels 5–9 mm. long, sepals about 1 mm. long, deltoid-triangular; fruit subglobose, 3.5–4 mm. in diameter, nearly as high; seeds 2–2.5 mm. high, reddish brown, minutely reticulate but not papillose.

Wooded or brushy slopes, mainly Tropical and Subtropical Zones, southern Baja California (in shaded canyons) to the West Indies and northern South America. In our area this species probably occurs only in well-watered canyons extending a short way into the desert and is not a typical constituent of the desert flora. Oct.–April.

## 3. TETRACOCCUS Engelm. ex Parry, W. Amer. Sci. 1: 13, 1885

Small shrubs with alternate, opposite or fascicled, simple leaves. Flowers dioecious, apetalous, staminate ones in compact to loose clusters; sepals 4–10. Stamens 4–9, inserted just exterior to and below lobed disk lining calyx. Pistillate flowers solitary; calyx 6–10-parted or -toothed, lined with a disk. Ovary 3–4-celled, each cell 2-ovuled;

styles 3–4, entire, linear or somewhat dilated distally. Capsule lobed, separating from column at maturity. Seeds 1–2 in each cell, shining, carunculate.

Leaves linear, 1.5–3 cm. long, glabrous; twigs reddish, not spinescent; ovary 4-celled...... ........................................................................ 1. *T. dioicus*
Leaves elliptic to oblanceolate, 1 cm. long or less, puberulent beneath, at least in youth; twigs spinescent; ovary 3-celled ........................................ 2. *T. hallii*

### 1. **Tetracoccus dioicus** Parry, W. Amer. Sci. **1:** 13, 1885

*Tetracoccus engelmanni* S. Wats., Proc. Amer. Acad. **20:** 373, 1885.

Shrub to 1.5 m. tall with glabrous, slender, and grayish branches or youngest growth reddish; leaves linear-spatulate, 2–3.5 mm. wide, 1.5–3 cm. long, entire; segments of staminate calyx narrowly ovate-elliptical, about 1 mm. long; filaments 2–3 mm. long; segments of pistillate calyx linear, 3–5 mm. long; capsule 8–10 mm. high; seeds smooth.

On brushy hillsides, mainly Upper Sonoran Zone, western San Diego County, California, into northern Baja California west of the Sierra San Pedro Mártir; barely reaching the edge of the Sonoran Desert along the southwestern flanks of the range. March–April.

### 2. **Tetracoccus hallii** Brandegee, Zoe **5:** 229, 1906

*Securinega hallii* I. M. Johnston, Univ. Calif. Pub. Bot. **7:** 442, 1922.
*Halliophytum hallii* I. M. Johnston, Contr. Gray Herb. **68:** 88, 1923.
*Securingea fasciculata* var. *hallii* Jepson, Man. Fl. Pl. Calif. 595, 1925.

Divaricately branched, spinescent shrub 0.5–2 m. tall with grayish bark; leaves oblanceolate, 3–10 mm. long, 1–3 mm. wide, in crowded fascicles on stubby dwarf branches closely invested with short white tomentum, blades glabrate, on slender minute petioles 1 mm. long or less; staminate flowers several in axils of leaves, pedicels slender, 1–6 mm. long; stamens 4–6, filaments filiform, 1–2 mm. long, sepals 6, deltoid-ovate, 0.3–0.4 mm. long, reddish; pistillate flowers solitary, sessile or on stout pedicels 2–3 mm. long; ovary densely white-tomentose when young, subglobose to oblong; styles linear-spatulate, reddish, 1–2 mm. long; capsules 6–9 mm. long and nearly glabrate at maturity; seeds dark brown, 5–6 mm. long, obovoid.

Desert slopes and washes, Lower Sonoran Zone, Colorado Desert in Riverside and Imperial Counties, California, and in adjacent southwestern Arizona; south sparingly to eastern flank of Sierra San Pedro Mártir. April–May.

## 4. **CROTON** L., Sp. Pl. 1004, 1753

Annual or perennial herbs, shrubs, or sometimes trees with strong-scented, stellate-pubescent herbage and monoecious or dioecious flowers. Leaves alternate or rarely opposite, entire, toothed or lobed. Flowers in axillary or terminal clusters, staminate flowers uppermost in monoecious species. Calyx 5-parted or 5-lobed; petals usually present but small or rudimentary, usually 5 and alternating with sepals. Stamens 5 or more, moderately to strongly inflexed. Pistillate flowers with 5–10-parted calyx; petals usually lacking. Ovary 3-celled, 3-ovuled. Styles cleft once to several times. Capsule 3-lobed, subglobose, dehiscent; seeds smooth to minutely pitted, with a small caruncle.

Plants dioecious, annual or perennial herbs or suffrutescent:
  Staminate flowers apetalous:
    Annual herbs; seeds 3–4 mm. long, gray to brown, dull .............. 1. *C. texensis*
    Perennial subshrub; seeds 6–9 mm. long, mottled, lustrous ............ 3. *C. wigginsii*
  Staminate flowers with minute but distinct petals:
    Leaves cinereous to silver or pale olive-green above; petals trifle shorter than to barely exceeding staminate sepals:
      Stems usually erect or nearly so; sepals 2–3 mm. long .......................................................... 2. *C. californicus* var. *californicus*
      Stems usually decumbent; sepals 1.5–2 mm. long:
        Leaf blades cinereous above, narrowly oblong to oblong-lanceolate, to 5 cm. long ...................................... 2a. *C. californicus* var. *tenuis*
        Leaf blades olive-green above, broadly oblong to oblong-elliptic, rarely over 2 cm. long ................................ 2b. *C. californicus* var. *mohavensis*
    Leaves dark green above; petals about twice as long as sepals of staminate flowers........................................................... 4. *C. corymbulosus*
Plants monoecious or mainly so, commonly lignescent shrubs:
  Stems mainly suffrutescent from woody base or rootstock; petals of staminate flowers twice as long as sepals .......................................... 4. *C. corymbulosus*
  Stems woody or only terminal twigs herbaceous during growing season; petals only slightly longer than sepals (twice as long in *C. ciliato-glanduliferum*):
    Stipules fimbriate-dissected; margins of leaves conspicuously ciliate-glandular ............................................................ 5. *C. ciliato-glanduliferum*
    Stipules not fimbriate, often inconspicuous or obsolete; leaf margins not ciliate-glandular:
      Style branches thrice-dichotomous into 24 ultimate divisions; stipules orbicular, persistent ................................................ 6. *C. alamosanum*
      Style branches once-dichotomous, with 6 ultimate divisions; stipules deciduous or obsolete:
        Capsules keeled on each locule; petals copiously villous-ciliate on margins, nearly glabrous on face and back:
          Seeds 6–8 mm. long; leaves to 7 cm. wide ....... 7. *C. flavescens* var. *flavescens*
          Seeds 4–5 mm. long; leaves to 10 cm. wide .. 7a. *C. flavescens* var. *brandegeanus*
        Capsules not keeled, locules rounded; petals glabrous, sparsely ciliate, or villous on outer surface:
          Petals glabrous; capsule papillose at maturity .................. 8. *C. fragilis*
          Petals villous or ciliate-margined; capsules smooth or verrucose at maturity, moderately to densely pubescent:
            Leaves 2–9 cm. wide, moderately to densely pubescent above; seeds 7–9 mm. long:
              Petals villous on outer surfaces; capsules 10–15 mm. in diameter, hispid with hairs 1–2 mm. long; style branches narrowly subulate, 4–5 mm. long ................................................. 9. *C. magdalenae*
              Petals ciliate-villous on margins; capsules 7–8 mm. in diameter, hispid-tomentose with hairs mostly less than 1 mm. long; style branches filiform, about 3 mm. long ...................................... 10. *C. caboensis*
            Leaves 0.6–2 cm. wide, subglabrate above; seeds 4–5 mm. long .. 11. *C. sonorae*

**1. Croton texensis** (Klotzsch) Muell. Arg., in A. DC., Prod. **15**$^2$**:** 692, 1866

*Croton muricatum* Nutt., Trans. Amer. Phil. Soc. II, **5:** 173, 1837. Not *C. muricatus* Vahl, 1807.
*Hendecaudra texensis* Klotzsch, in Wiegm., Archiv Naturgesch. **7:** 252, 1841.
*Oxydectes texensis* Kuntze, Rev. Gen. Pl. **2:** 613, 1891.

An erect or ascendingly branched annual 0.5–1.5 m. tall with slender-rayed, stellate pubescence scattered over stems and leaves; stems slender, yellowish green; plants mainly dioecious; stipules lacking; petioles slender, 5–20 mm. long, rather densely stellate-pubescent; leaf blades linear, lanceolate, or ovate-lanceolate, 0.5–5 cm. wide, 2–12 cm. long, obtuse to acuminate at apex, obtuse to rounded at base, entire, upper surface green, with scattered stellate trichomes that rarely overlap, lower surface more closely stellate-puberulent and more or less cinereous, especially on young growth; staminate flowers in short racemes, subtended by minute setaceous bracts; sepals about 1 mm. long, thin, densely pubescent without, glabrous within; petals none; stamens 8–15, about 1–1.5 mm. long, filaments sparsely villous with crispate hairs; pistillate flowers 1–5 per inflorescence, sepals narrowly deltoid, about 2 mm. long, tomentulose without; styles cleft nearly to base, style branches 12–24 or more, about 2 mm. long; capsule 3-celled, ovoid-globose, 4–6 mm. tall, stellate-scurfy; seeds orbicular-sublenticular, rounded on back, obtusely angled on inner face, gray to brown, dull, 3–4 mm. long and broad, caruncle light-colored, rather prominent, about 1 mm. wide.

Roadsides, fields, ditch banks, and along arroyos, Lower and Upper Sonoran Zones, central and southwestern Arizona to Wyoming, Illinois, Arkansas, Texas, and northern Mexico. May–Nov.

**2. Croton californicus** Muell. Arg., in A. DC., Prod. **15²:** 691, 1866, var. **californicus**

*Croton californicus* var. *major* S. Wats., Bot. Calif. **2:** 69, 1880.

Erect, ascendingly branched or spreading, perennial herb or suffrutescent plant 2–12 dm. tall with stellate, scurfy pubescence throughout; petioles slender, 0.5–3 cm. long; leaf blades oblong to broadly ovate, 1.5–5 cm. long, entire, densely stellate-pubescent and hoary or silvery beneath, less densely pubescent and yellowish green above; staminate flowers many in an elongating raceme, lower flowers maturing and falling before terminal ones open, their sepals 2–3 mm. long, petals about equaling or shorter than sepals; stamens 12–15, slightly exceeding sepals; pistillate flowers few in each inflorescence, sepals broadly ovate, 2–3 mm. long, about two-thirds as wide as long; petals none; pedicels 2–5 mm. long in both pistillate and staminate flowers; capsules subglobose, 5–6 mm. long, densely stellate-pubescent, scurfy; seeds broadly ellipsoidal, 3–4 mm. long, rounded on back, slightly flattened on inner face, smooth, pale-stramineous to light gray, lightly mottled with purplish or brownish, irregular streaks and spots; caruncle about 0.5 mm. long or less.

Sandy or gravelly hillsides, slopes, and arroyos, Lower and Upper Sonoran Zones, Contra Costa County, California, south to southern Baja California and east into Arizona and Sonora. Feb.–Oct.

**2*a*. Croton californicus** var. **tenuis** (S. Wats.) Ferg., Rep. Mo. Bot. Gard. **12:** 64, *pl. 27, fig. 1,* 1901

*Croton tenuis* S. Wats., Proc. Amer. Acad. **14:** 297, 1879.

Stems slender, more or less decumbent; leaves narrowly oblong or oblong-lanceolate; flowers somewhat smaller than those in var. *californicus*; capsules 4–5 mm. long.

Sandy soil, Upper and Lower Sonoran Zones, Santa Barbara, California, south to southern Baja California and east to western Arizona and adjacent Sonora.

**2*b*. Croton californicus** var. **mohavensis** Ferg., Rep. Mo. Bot. Gard. **12:** 65, 1901

Stems slender, about as in var. *tenuis;* leaves pale olive-green, the blades broadly oblong to broadly elliptic, 0.5–2 cm. long.

Kern County, California, east and south throughout the Mojave Desert and into the Colorado Desert and adjacent Arizona. This is the commoner form in Arizona.

**3. Croton wigginsii** L. C. Wheeler, Contr. Gray Herb. **124:** 37, 1939

*Croton arenicola* Rose & Standl., Contr. U.S. Nat. Herb. **16:** 12, 1912. Not *C. arenicola* Small, 1905.

Cespitose subshrub 1 m. tall or less, often slightly wider than high, usually densely stellate-lepidote with whitish to silvery hairs throughout; petioles 5–15 mm. long; leaf blades linear-lanceolate to oblong-lanceolate, 2–12 mm. wide, 1.5–6 cm. long, acute to obtuse at apex, cuneate to broadly rounded at base, entire, midrib prominent beneath, densely stellate-lepidote on both surfaces, or in some inland specimens sparsely so, and bright green above; flowers dioecious, both sexes apetalous, staminate in slender racemes 1–3 cm. long; sepals deltoid, about 2–3 mm. long, densely stellate-lepidote, pistillate 4–5 mm. long in fruit; stamens 18–30, slightly exceeding sepals; capsules subglobose, 8–11 mm. high, densely stellate-tomentulose and with a few lepidote trichomes; seeds oblong, 6–9 mm. long, mottled with light brown and gray, smooth and somewhat lustrous, rounded on back, scarcely flattened on inner face; caruncle about 1 mm. long, slightly stipitate.

Sand dunes and along sandy arroyos, Lower Sonoran Zone, southeastern Imperial County, California, and in northwestern Sonora as far south as San Carlos Bay, near Guaymas. March–May.

Closely related to *C. californicus* and its varieties but consistently differing in having larger seeds, lanceolate leaves below the terminal twigs, and apparently wholly apetalous flowers.

**4. Croton corymbulosus** Engelm., in Rothrock, Wheeler Rep. 242, 1878

*Croton eremophilus* Woot. & Standl., Contr. U.S. Nat. Herb. **16:** 144, 1913. Not *C. eremophilus* Muell. Arg., 1874.

Erect suffrutescent plant 1–5 dm. tall with few to many stems from a woody base and rootstock; growing stems, leaves, and outer floral parts closely covered with fine, stellate, slightly lepidote, grayish to silvery pubescence; stipules subulate to ovate-deltoid, 1–1.5 mm. long, deciduous; petioles slender, 0.5–2 cm. long, tomentulose; leaf blades ovate to oblong, 5–18 mm. wide, 1–4.5 cm. long, acute to mucronulate at apex, rounded at base, entire, greenish and lepidote-pubescent above, gray to silvery and densely stellate-tomentulose beneath, pinnately 5–9-veined; flowers monoecious or sometimes dioecious; staminate flowers on pedicels to 6 mm. long, calyx densely tomentulose without, lobes narrowly deltoid, 1.5–2 mm. long; petals spatulate, nearly twice as long as sepals, bearded with slender crisped hairs; stamens 6–18, filaments villous near base; pistillate flowers at base of inflorescence if both sexes present, apetalous, lobes lance-ovate, 2–2.5 mm. long; style branches usually 6, about 2.5–3 mm. long, cinereous-puberulent; capsules globose to ovoid, 4–7 mm. long; seeds oblong, 4–5 mm. long, rounded on back, grayish brown, lustrous, minutely papillose-reticulate, caruncle short-stipitate, yellowish, about 1 mm. wide.

Arid rocky slopes, Lower and Upper Sonoran Zones, southern Arizona to Texas and northern Mexico. Occurring in our area mainly along the upper margins of the Sonoran Desert. March–Oct.

**5. Croton ciliato-glanduliferum** C. G. Ortega, Hort. Mat. 51, 1797

Shrub to 1 m. tall with pale gray to buff bark on older stems and stellate tomentum on young twigs and undersides of leaves and calyces; foliage with a mint-like odor and flowers delicately fragrant; stipules laciniate into filiform, gland-tipped divisions 3–10 mm. long; petioles slender, 1–5 cm. long, sparsely to densely stellate-pubescent; leaf blades ovate-lanceolate to broadly ovate, 1–6 cm. wide, 2–8 cm. long, broadly cuneate to truncate-rounded at base, acute to acuminate at apex, evenly ciliate-glandular along margins, bearing a tuft of filiform-glandular segments 3–6 mm. long at juncture of petiole and blade, upper surface of blades green and sparsely beset with fine minute stellae, under side moderately to densely stellate-pubescent or tomentose, gray; veins pinnate, evident but scarcely prominent; flowers monoecious; racemes 1–5 cm. long, erect; staminate flowers on pedicels to 1 cm. long, densely stellate-tomentose without, sepals deltoid, 2–3 mm. long; petals obovate to oblong, 4–6 mm. long, glabrous or nearly so; stamens 20–30, exceeding sepals 2–3 mm.; pistillate flowers 1–6 at base of raceme, on short stout pedicels; sepals oblong-spatulate, acute, 1.5–2 mm. wide, 5–7 mm. long, stellate-tomentose; style branches 12–24, filiform, 5–10 mm. long; capsules globose, 6–8 mm. in diameter, finely stellate-tomentulose; seeds rounded on back, obtusely and roundly angled on inner face, dark gray, sublustrous, minutely roughened with low blunt papillae; caruncle sessile, about 2 mm. wide, thin.

Rocky hillsides and canyons, Lower and Upper Sonoran Zones to Tropical Zone, southern Arizona and central Baja California to Central America. March–Nov.

**6. Croton alamosanum** Rose, Contr. U.S. Nat. Herb. **1:** 111, 1891

*Croton blasianus* M. E. Jones, Contr. West. Bot. **18:** 49, 1933.

Slender shrub 1–2 m. tall with many strict, ascending branches, younger twigs reddish, villous-pubescent; leaves ovate-lanceolate, oblong-lanceolate or elliptical, 1.5–5 cm. wide, 6–10 cm. long, acute to attenuate at apex, densely stellate-pubescent beneath, dark green and puberulent above, minutely serrate, slightly glandular toward base, with 8–15 pairs of veins conspicuous beneath; petioles 5–8 mm. long, densely stellate-tomentose or some leaves subsessile; stipules orbicular, inconspicuously dentate-glandular; flowers monoecious; staminate inflorescences compact, spikelike in youth, 3–10 cm. long and racemose when fully expanded; staminate flowers with 5 white petals, glabrous without, pubescent within; pistillate flowers solitary or in clusters of 2–5, mainly at base of staminate peduncles, but sometimes throughout staminate raceme; calyx deeply 5-parted, each sepal 4–6 mm. long; petals none; styles 3, each branch thrice-dichotomous, thus 24 ultimate stigmatic lobes; capsules depressed-globose, 7–8 mm. high, densely tomentose; seeds about 2 mm. wide, 5 mm. long, rounded dorsally, smooth, pale brown, lustrous.

Valley benches and adjacent slopes, Lower Sonoran and Upper Sonoran Zones, southern Sonora and northern Sinaloa. Aug.–Oct.

**7. Croton flavesens** Greenm., Proc. Amer. Acad. **39:** 81, 1903, var. **flavescens**

Erect shrub 1–4 m. tall with older bark mottled light and dark gray, slightly striate; current twigs pale brown, densely stellate-pubescent in youth, stellae contiguous but rarely overlapping by end of growing season; petioles slender, 1–3 cm. long, stellate-pubescent and somewhat scurfy when young; leaf blades ovate-lanceolate, orbicular-ovate or ovate-attenuate, 2.5–7 cm. wide, 5–12 cm. long, rounded to subcordate at base, rarely obtuse, commonly acute to attenuate at apex, entire, dark green and sparsely stellate-puberulent above, densely or moderately stellate-puberulent and paler beneath, glands at base lacking; flowers monoecious; racemes 2–5 cm. long, staminate sepals deltoid to broadly ovate, densely stellate-puberulent, 1.5–2 mm. long, petals spatulate, slightly exceeding sepals, copiously villous-ciliate; stamens 12–18, equaling or slightly exceeding petals; filaments villous at base; sepals of pistillate flowers ovate-lanceolate, 3.5–5 mm. long; capsules ovoid-globose, 6–9 mm. high, sharply keeled on each valve, densely stellate-pubescent, seeds 6–8 mm. long, mottled brown and gray; caruncle about 1 mm. wide.

Valley floors and adjacent hillsides, Lower Sonoran and Upper Sonoran Zones to Tropical Zone, southwestern Chihuahua and adjacent Sonora to Guerrero. Aug.–Oct.

### *7a.* **Croton flavescens** var. **brandegeanus** Croizat, Jour. Arnold Arb. **26:** 187, 1945

Differing from var. *flavescens* in having smaller seeds only 4–5 mm. long and generally somewhat broader leaves.

Foothills and adjacent flats, Lower Sonoran Zone, central Sonora (between Mátape and Mazatán).

### 8. **Croton fragilis** H. B. K., Nov. Gen. & Sp. **2:** 75, 1817

*Croton sericeus* Schlecht., Linnaea **5:** 85, 1830. Not *C. sericeus* Lam., 1786.

Slenderly branched shrub 1–2 m. tall with brownish gray bark on older twigs, current ones closely stellate-canescent, stellae slightly overlapping; petioles 5–20 mm. long, stellate-canescent; leaf blades broadly ovate-acuminate to ovate-lanceolate, rounded to subcordate at base, 2–6 cm. wide, 3–10 cm. long, dark green and moderately stellate-puberulent above, densely pubescent and cinereous to canescent beneath; flowers monoecious; inflorescences slender, 2–10 cm. long; staminate sepals broadly deltoid, about 1 mm. long, canescent-stellate without; petals slightly exceeding sepals, spatulate, glabrous; stamens 15–18, filaments slightly villous at base; pistillate flowers about same size as staminate, apetalous; style branches 6, slender, glabrous, 2–3 mm. long; capsules subglobose, 6–7 mm. in diameter, 7–9 mm. high, stellate-pubescent, papillose after hairs have fallen; seeds finely mottled with dark gray and deep brown, 2.5–3 mm. wide, 4–5 mm. long, lustrous; caruncle light buff, subsessile, about 1 mm. broad.

Rocky hillsides and canyons, Lower Sonoran Zone to Tropical Zone, southern Sonora to Vera Cruz and Oaxaca. March–Oct.

### 9. **Croton magdalenae** Millsp., Proc. Calif. Acad. II, **2:** 220, 1889

Ascendingly branched shrub 1–1.5 m. tall with gray, finely striate bark on older branches, younger ones pale yellow to stramineous and densely stellate-tomentose; petioles rather stout to slender, 1–3.5 cm. long, densely stellate-tomentose; leaf blades cordate, acute to apiculate at apex, 1–9 cm. wide, 2–12 cm. long, densely mottled-tomentose with 5-rayed stellate hairs when young, in age upper surface becoming green as stellate hairs are separated by growth of blade, underside cinereous and densely tomentose even at maturity, veins prominent beneath; flowers monoecious; racemes 3–10 cm. long; staminate flowers numerous on pedicels 1–5 mm. long, calyx lobes deltoid, 1–1.5 mm. long, densely stellate-tomentose; petals 5, linear-spatulate, about 2 mm. long, villous on outer surface; stamens 15–18, slightly exceeding calyx lobes, filaments sparsely villous; pistillate flowers 2–6 near base of each rachis, sepals 2–3 mm. long, densely stellate-tomentose, petals none; capsule ovoid-globose, 10–15 mm. in diameter, densely stellate-hispid, with hairs to 1–2 mm. long; style branches 6, narrowly subulate, 4–5 mm. long, style divided almost to base, branches stellately spreading; seeds 7–9 mm. long, 4–6 mm. wide, dark buff, mottled with brown, shining on back, slightly angular on inner face; caruncle thin, horny, 1.5–2 mm. wide, subsessile.

Rocky canyons and shaded slopes, Lower Sonoran Zone, Vizcaíno Desert and San Felipe Desert, Baja California, southward to the Cape Region and the adjacent islands. Jan.–June.

**Croton boregensis** M. E. Jones, Contr. West. Bot. **18:** 48, 1933, with longer petioles to 5 cm. long; floccose tomentum on young twigs, leaves, and inflorescences, with each cluster of hairs made up of a single spreading, central hair many times as long as the much smaller basal ones; and leaves longer and more truncate at base than those of *C. magdalenae,* occurs on the western flanks of the Sierra de la Victoria in the Cape Region, and may be expected along the southern borders of the Sonoran Desert in Baja California. It has not been found within our area yet.

## 10. **Croton caboensis** Croizat, Jour. Arnold Arb. **26:** 181, 1945

An erect shrub 1–2 m. tall with grayish bark and current branches densely velutinous-hispid with pale stellate hairs; petioles slender, 5–25 mm. long, stellate-tomentose; leaf blades lanceolate, ovate or ovate-oblong, 2–5 cm. wide, 5–8 cm. long, thin, slightly cuneate to rounded at base, short-acuminate at apex, entire, moderately pubescent above with more or less appressed, simple hairs, stellate-pubescent and grayish beneath; veins about 7–9 pairs; flowers monoecious; staminate flowers on slender pedicels about 3 mm. long, sepals broadly deltoid, 1.5–2 mm. long, petals about equaling sepals; stamens 12–15; pistillate flowers with sepals about 3 mm. long and wide, apetalous; styles 3, branches about 3 mm. long, dichotomous nearly to base; capsule subglobose, 7–8 mm. high, verrucose and hispid-tomentose; seeds 3–3.5 mm. wide, about 7 mm. long, pale gray, obscurely and minutely rugulose, lustrous; caruncle about 1 mm. wide.

Hillsides, moderate slopes, and along washes, Lower Sonoran and Tropical Zones, southern Baja California. Aug.–Oct.

This species occurs mainly in the Cape Region somewhat south of the Sonoran Desert.

## 11. **Croton sonorae** Torr., Bot. Mex. Bound. 194, 1859

*Croton pringlei* S. Wats., Proc. Amer. Acad. **20:** 373, 1885.
*Croton gonzalezii* Greenm., op. cit. **39:** 81, 1903.
*Croton attenuatus* M. E. Jones, Contr. West. Bot. **18:** 47, 1933.

Slenderly to intricately branched shrub 1–2 m. tall with dark gray bark on older branches, seasonal twigs finely and closely stellate-pubescent with fine ashy hairs; stipules obsolete; petioles 3–8 mm. long, stellate-canescent; leaf blades thin, ovate to oblong-ovate, rounded at base, acute to acuminate at apex, 6–20 mm. wide, 1–5 cm. long, green and nearly glabrate above, sparsely to moderately scurfy-stellate and paler beneath, margins entire to slightly and irregularly crenate-sinuous; flowers monoecious; racemes slender, 1–8 cm. long; staminate flowers with broadly deltoid sepals about 1 mm. long, these subglabrous, some few-rayed stellae near bases and along margins; petals about equaling sepals; stamens 12–18, filaments pubescent at base; pistillate sepals 1–1.5 mm. long, nearly glabrous; style branches 6, about 1 mm. long, glabrous; capsule subglobose, 5–6 mm. wide, 6–7 mm. high, stellate in youth, subglabrate at maturity; seeds dark brown, sublustrous, about 3 mm. wide, 4–5 mm. long; caruncle about 1 mm. wide, set in a broad shallow depression.

Rocky hillsides, Lower Sonoran Zone, southern Arizona to Sinaloa. March–Oct.

## 5. EREMOCARPUS Benth., Bot. Sulph. 53, *pl. 26,* 1844

Pungently scented annual herbs with slender but firm, stellate and simple, glandular hairs. Leaves alternate, petiolate, entire, estipulate. Flowers monoecious, apetalous, produced in axillary cymes or some pistillate, earlier ones solitary. Calyx 5–6-lobed and slightly imbricate in staminate flowers, absent in pistillate. Stamens 6–7, inserted centrally on pubescent receptacle, exserted. Ovary 1-celled, 1-ovuled, bearing 4–5 small basal glands; style simple, slender, apically stigmatic. Capsules oblong-obovoid, bivalvate. Seed ellipsoidal, smooth and shining, with a fleshy endosperm.

### 1. Eremocarpus setigerus (Hook.) Benth., Bot. Sulph. 53, *pl. 26,* 1844

*Croton ?setigerum* Hook., Fl. Bor. Amer. **2:** 141, 1838.
*Piscaria setigera* Piper, Contr. U.S. Nat. Herb. **11:** 382, 1906.

Plants cinereous-lanate with intermingled stellate and simple hairs; stems simple to dichotomously or trichotomously much branched, 0.5–3.5 dm. tall; petioles hirsute, 0.5–5 cm. long; leaf blades ovate to rhombic-ovate, 1–4 cm. wide, 1.5–6 cm. long, strongly 3-veined; staminate flowers numerous in terminal cymes, each flower 1.2–2 mm. long on pedicels about 3 mm. long; pistillate flowers 1–3 in axils, sessile or essentially so; capsule 3–4.5 mm. long, densely lanate and hirsute; seed 3–4 mm. long, trigonous with obtuse angles in cross section, mottled with gray and brown or black, shining.

Arid slopes and valley floors, mainly Upper Sonoran Zone, interior Washington to northern Baja California, where it barely enters the Lower Sonoran Zone as an occasional waif. May–Sept.

## 6. DITAXIS Vahl ex Juss., Euph. Tent. 27, *pl. 7,* 1824

Monoecious or rarely dioecious, annual or perennial, herbaceous or shrubby plants. Leaves alternate, petiolate or subsessile, entire to serrate-dentate, usually 3-nerved from base. Stipules small. Herbage usually pubescent, hairs predominantly appressed-malpighiaceous. Inflorescence racemose and axillary, pistillate flowers usually basal, staminate above, rachis often short, few-flowered. Sepals and petals of staminate flowers 5, petals usually equaling or surpassing sepals and androphore; glands of disk opposite sepals. Sepals and petals of pistillate flowers 5, petals usually shorter than sepals. Stamens (in ours) 10, in two whorls of 5, united into a column (androphore), a third whorl sometimes represented by sterile filaments. Ovary 3-celled, 3-lobed; styles 3, once- or twice-cleft, branches linear or slightly dilated toward apex. Capsule 3-lobed, each locule 1-seeded. Seeds ovoid, globose, or ovoid-pyramidal, smooth to reticulate-faveolate, sometimes slightly scurfy, ecarunculate.

Gland-tipped fimbriae on margins of leaves, bracts, and sepals of pistillate flowers; younger branches moderately pubescent with slender, curved, nonappressed hairs .......... 1. *D. adenophora*
Gland-tipped fimbriae lacking; younger branches invested with closely appressed, setose hairs or glabrous:
- Plants shrubby or suffrutescent:
  - Leaves and stems glabrous or very sparsely pubescent with closely appressed, coarse, malpighiaceous hairs; herbage purplish and slightly glaucous; capsule 8–10 mm. wide; seeds trigonous-pyramidal, truncate:
    - Leaves and capsules glabrous .......... 2. *D. brandegeei* var. *brandegeei*
    - Leaves sparsely beset with coarse, appressed, malpighiaceous hairs; capsules densely setose-hispid .......... 2a. *D. brandegeei* var. *intonsa*

Leaves and stems closely appressed-pubescent except near base; herbage green or grayish, not glaucous; capsules 3–5 mm. wide; seeds not trigonous, rounded, not truncate:
Leaves narrowly lanceolate to elliptic, 1- or 3-veined from base, margins entire:
Glands between bases of petals attenuate and free at apex; leaves linear-lanceolate to narrowly elliptic, densely sericeous-strigose, only midrib obvious .......... .................................................. 3. *D. lanceolata*
Glands between bases of petals acute but not attenuate, tip adnate to androphore; leaves broader, distinctly 3-veined from base .................. 4. *D. palmeri*
Leaves broadly ovate, pinnately 5–9-veined, margins bearing fine, gland-tipped serrations ............................................ 5. *D. manzanilloana*
Plants herbaceous, annual or perennial:
Pistillate calyx with a conspicuous, glandlike callosity just below each intersepalary sinus; capsule glabrous or nearly so; pistillate petals glabrous .......... 6. *D. californica*
Pistillate calyx without intersepalary callosities; capsules hirsute, usually densely so; pistillate petals more or less hirsute on back:
Leaves persistently hirsute at least beneath, not over 4 cm. long; sepals of pistillate flowers ovate-lanceolate, 2.5–6 mm. long:
Seeds lightly reticulate, nearly smooth; pubescence of mixed setae and finer, more or less crinkled hairs; sepals of pistillate flowers 2.5–3 mm. long .. 7. *D. serrata*
Seeds strongly reticulate-faveolate, the depressions marked by fine radiating lines; pubescence of coarse setose hairs only; sepals of pistillate flowers 5–6 mm. long in fruit .................................... 8. *D. neomexicana*
Leaves soon subglabrate, 5–6 cm. long; sepals of pistillate flowers linear-lanceolate, 6–8 mm. long ............................................. 9. *D. gracilis*

## 1. **Ditaxis adenophora** (A. Gray) Pax & K. Hoffm., in Pax, Pflanzenreich $4^{147.vi}$: 65, 1912

*Argyrothamnia adenophora* A. Gray, Proc. Amer. Acad. **8**: 294, 1870.
*Argythamnia clariana* Jepson, Fl. Calif. **2**: 419, 1936.

Erect or ascending, much-branched, perennial herbs 3–6 dm. tall, finely puberulent with simple hairs in youth; stems glabrate, somewhat purplish; leaves elliptic to oblanceolate, 3–20 mm. wide, 1–5 cm. long, finely puberulent with short simple hairs and a few scattered setose ones or these lacking, margins finely serrate, each tooth tipped with a tack-shaped gland; racemes 0.5–2 cm. long, usually congested, bracts 1–4 mm. long, marginally stipitate-glandular; staminate flowers 4–6 mm. long, pilose, without stipitate glands; petals roseate to purplish, slightly exceeding sepals; disk-glands adnate to androphore; stamens 10, verticels crowded; pistillate sepals ovate-lanceolate, 5–7 mm. long, stipitate-glandular on margins, backs pilose, glabrous within; petals of pistillate flowers lanceolate, equaling sepals, minutely puberulent without, glabrous within; capsules 5–8 mm. broad, hispid at first but often eventually glabrate; seeds ovoid-trigonous, truncate at base, about 3 mm. long, faveolate-rugose and papillose.

Gravelly slopes and rocky hillsides, Lower Sonoran Zone, western Colorado Desert to western Arizona and south to southern Sonora. Feb.–Sept.

## 2. **Ditaxis brandegeei** (Millsp.) Rose & Standl., Contr. U.S. Nat. Herb. **16**: 13, 1912, var. **brandegeei**

*Argythamnia brandegei* Millsp., Proc. Calif. Acad. II, **2**: 220, 1889.

An openly branched, glabrous shrub to 2.5 m. tall with a few spreading to drooping branches and more or less purplish, glaucous, young growth; leaves restricted to 1–2 dm. nearest tips of branches; petioles 7–15 mm. long; leaf blades lanceolate, 0.5–2 cm. wide, 2–8 cm. long, acute at apex, cuneate at base, with 3 prominent veins from base and several smaller, pinnately arranged veinlets from each main vein, margins finely and evenly serrulate; racemes at first dense, later elongating to 1.5–3 cm. long; bracts linear-lanceolate, 3–5 mm. long; staminate flowers 7–8 mm. long, petals nearly equaling linear-lanceolate sepals; glands tridentate at apex, adnate to androphore; pistillate flowers with serrate acuminate sepals 7–8 mm. long, petals lanceolate, purple-striate, one-third to one-half as long as sepals, glands auriculate, entire; stigmas slightly clavate-dilated toward apex; capsule glabrous, 8–10 mm. broad; seeds trigonous-pyramidal, truncate at base, acute at apex, 2.5–3 mm. wide, 4–5 mm. long, lightly reticulate toward base.

Alkaline flats, sandy washes, and rocky hillsides and mesas, Lower Sonoran Zone, central Baja California and adjacent islands in the Gulf of California. March–Oct.

**2*a*. Ditaxis brandegeei** var. **intonsa** I. M. Johnston, Proc. Calif. Acad. IV, **12:** 1062, 1924

Differing from var. *brandegeei* only in possessing a few scattered, very coarse, appressed, malpighiaceous hairs on the leaves and young stems, and in the densely setose-hispid capsules.

On rocky hillsides, Lower Sonoran Zone, islands and coastal headlands from Carmen Island to Guadalupe Point, Baja California, and in southern Yuma County, Arizona. It probably occurs in intermediate localities but if so it has not been collected there.

**3. Ditaxis lanceolata** (Benth.) Pax & K. Hoffm., in Pax, Pflanzenreich **4**$^{147.vi}$**:** 71, 1912

*Serophyton lanceolatum* Benth., Bot. Sulph. 52, 1844.
*Aphora lanceolata* Engelm. & Gray, Pl. Lindheim. 25, 1845.
*Argyrothamnia lanceolata* Muell. Arg., Linnaea **34:** 148, 1865.
*Argythamnia sericophylla* A. Gray, in S. Wats., Bot. Calif. **2:** 70, 1880.
*Argythamnia sericophylla* var. *verrucosemina* Millsp., Proc. Calif. Acad. II, **2:** 221, 1889.
*Ditaxis sericophylla* Heller, Cat. N. Amer. Pl. 5, 1898.

Densely strigose-cinereous shrubs with few to many ascending slender branches 1–5 dm. long; plants apparently sometimes dioecious; petioles 1–3 mm. long; leaf blades lanceolate-acuminate to broadly lanceolate, mostly 2–6 mm. wide, 1–2.5 cm. long but in vigorous plants to 1.5 cm. wide, 2–3.5 cm. long, densely pubescent with coarse, shining, closely appressed, setose hairs, 3-veined from the base but usually only midrib readily visible; margins entire, inflorescence sessile in axils, bearing 3–4 staminate flowers and a single pistillate flower or pistillate flower often lacking; staminate flowers 4–5 mm. long, petals short-clawed, ovate-lanceolate, strigose without and surpassing sepals; glands ovate-attenuate, rounded or truncate at base, free above; pistillate flowers with lanceolate sepals 3–4 mm. long, sepals sometimes very narrowly white-margined; pistillate petals ovate-lanceolate, clawed, about equaling sepals, adnate to the broadly triangular glands at base of strigose-hispid ovary; style branches broadly dilated; capsules depressed-globose, 3–5 mm. wide, strigose; seeds trigonous-ovoid, truncate at base, acute at apex, 1.5–2 mm. in diameter, 2–2.5 mm. high, brownish, faveolate-reticulate, the depressions with fine radiating lines; margins sometimes scurfy or minutely crenulate.

Rocky slopes and along washes, Lower Sonoran Zone, Colorado Desert in California eastward into western Arizona and south to southern Baja California and northwestern Sonora. Feb.–Oct.

**4. Ditaxis palmeri** (S. Wats.) Pax & K. Hoffm., in Pax, Pflanzenreich $4^{147.vi}$: 64, 1912

*Argythamnia palmeri* S. Wats., Proc. Amer. Acad. **24:** 77, 1889.

An openly branching, low shrub 5–10 dm. tall with ascending branches and closely strigose young herbage, becoming subglabrate and pale green before maturity of leaves and branchlets; petioles short, 2–3 mm. long; leaf blades lanceolate to oblong-lanceolate, acute and more or less attenuate at apex, mostly broad-cuneate at base, 3-veined, 1.5–4 cm. long, margins entire; flowers apparently dioecious, staminate 4–4.5 mm. long, with strigose sepals about equaling petals, these oblanceolate, slightly erose, and apiculate, finely strigose without, base of claws adnate to androphore; interpetalary glands ovate, acute but not attenuate at apex, adnate to disk or base of androphore, tip not free; pistillate flowers solitary, on pedicels 3–4 mm. long, sepals about 3 mm. long, nearly twice as long in fruit, petals pilose without, ovate-lanceolate, slightly acuminate, shorter than or about equaling sepals; style branches dilated at apex; seeds ovoid to subglobose, about 2 mm. long, reticulate-faveolate.

Rocky slopes, Lower Sonoran Zone, vicinity of Guaymas, Sonora.

Poorly known and possibly only a variant of *D. lanceolata,* which exhibits a tendency toward a dioecious habit and sometimes has broader, greener, less densely pubescent leaves when growing in favorable sites.

**5. Ditaxis manzanilloana** Pax & K. Hoffm., in Pax, Pflanzenreich $4^{147.vi}$: 59, 1912

A slender shrub 1–1.5 m. tall with angular-striate branches sparsely pubescent with appressed malpighiaceous hairs, angles pale-stramineous to nearly white, tardily glabrescent; leaves ovate, 1–3.2 cm. wide, 3–7 cm. long, dark green and sparsely appressed-pilose above, paler and more densely strigose with a mixture of coarse and fine, light-colored, crispate hairs beneath, pinnately 5–9-veined, margins minutely and subremotely serrulate, each tooth tipped with a small, conical, whitish gland; petioles 1–4 mm. long; racemes short and congested, subglomerate; bracts about 2 mm. long, pilose, broadly ovate to suborbicular; staminate flowers 5–6 mm. long; sepals lanceolate, green, finely appressed-pilose; petals equaling or slightly exceeding sepals, densely appressed-pilose on back, adnate to the column along claws; glands broadly ovate, acute and free at apex; pistillate flowers with linear-lanceolate, briefly acuminate petals 3–3.5 mm. long; disk deltoid, free; capsule 5–6 mm. in diameter, sericeous-pilose, nearly equaled by accrescent sepals; seeds ovoid-globose, about 2–2.5 mm. long, dark brown, strongly faveolate-reticulate, with fine but distinct, radiating lines in each depression, ridges rather prominent.

Rocky hillsides, Lower Sonoran Zone, southern Sonora to Colima. Specimens examined flowering in Nov.

**6. Ditaxis californica** (Brandegee) Pax & K. Hoffm., in Pax, Pflanzenreich $4^{147.vi}$: 70, 1912

*Argythamnia californica* Brandegee, Zoe **5:** 230, 1906.

Erect annual 1–3 dm. tall with several to many divergent branches, tips of these and young leaves purplish, older parts soon pale green, glabrous or sparsely appressed-strigose with rather slender, malpighiaceous hairs; leaves oblanceolate to broadly elliptic, 6–12 mm. wide, 1–4.5 cm. long, margins serrulate; petioles 2–5 mm. long; racemes short and subglomerate; staminate flowers 2–2.5 mm. long, petals glabrous, equaling or slightly exceeding sepals; glands linear, erect, about 0.6 mm. long, obtuse at apex, free; pistillate flowers with linear-attenuate sepals 3.5–4.5 mm. long, narrowly white-margined and margin bearing a few short-stalked glands, basal fold of whitish margins just below sinus forming a conspicuous callosity; pistillate petals oblanceolate to broadly deltoid, clawed, apiculate at apex, nearly equaling sepals; style branches linear, not dilated; capsule 3–4 mm. wide, glabrous or nearly so; seeds brown, faveolate-reticulate, depressions faintly marked with fine radiating lines or ridges.

Desert slopes and washes, Lower Sonoran Zone, Colorado Desert, California. March–May.

**7. Ditaxis serrata** (Torr.) Heller, Cat. N. Amer. Pl. 5, 1898

*Aphora serrata* Torr., Bot. Mex. Bound. 197, 1859.
*Argyrothamnia serrata* Muell. Arg., Linnaea **34:** 147, 1865.
*Argythamnia serrata* var. *magdalenae* Millsp., Proc. Calif. Acad. II, **2:** 221, 1889.
*Ditaxis odontophylla* Rose & Standl., Contr. U.S. Nat. Herb. **16:** 12, 1912.

Annual or occasionally short-lived perennial herbs with densely pubescent, decumbent to ascending stems 1–2.5 dm. long; investiture a mixture of coarse, appressed, malpighiaceous hairs and finer, simple, more or less crispate, intertangled ones; leaves rather crowded, usually longer than internodes, oblong to obovate, obtuse to rounded at apex, gradually narrowing to a cuneate base, 5–8 or rarely to 15 mm. wide, 1–3.5 cm. long, usually moderately serrate over rounded apex but sides mostly entire, rather prominently pinnate-veined, sericeous; racemes glomerate-congested in leaf axils; staminate flowers 3–3.5 mm. long, petals sparsely to densely pilose on back, narrowly oblanceolate, equaling or slightly exceeding sepals; glands broadly oblong, obtuse and free at apex; pistillate petals about three-fourths as long as sepals, densely pilose without; glands thin, obdeltoid, about 0.5 mm. long; stigmas linear or faintly dilated; capsules 3–5 mm. wide, pilose; seeds ovoid-globose, about 2–2.5 mm. long, nearly smooth to faintly reticulate-faveolate with low ridges and corrugations, sometimes minutely papillose.

Arid slopes and sandy flats and washes, Lower Sonoran Zone, Colorado Desert east into Arizona and south into northern Sonora and to the Cape Region of Baja California. Feb.–Nov.

**8. Ditaxis neomexicana** (Muell. Arg.) Heller, Cat. N. Amer. Pl. 5, 1898

*Argyrothamnia neo-mexicana* Muell. Arg., Linnaea **34:** 147, 1865.

Annuals or rarely short-lived perennials with spreading to decumbent branches 1–3 dm. long or simple and slightly branched, erect, to 3.5 dm. tall, sparsely to moderately strigose with appressed malpighiaceous hairs; leaves broadly elliptic to narrowly or broadly oblanceolate, acute to obtuse at apex, cuneate at base, entire or less commonly serrulate, 5–10 mm. wide, 1–3.5 cm. long; petioles 2–5 mm. long; racemes glomerate-congested in leaf axils; staminate flowers 2–2.5 mm. long with lanceolate-acuminate, pilose sepals, and glabrous petals barely exceeded by sepals; gland ovate-lanceolate, acute and free at apex; pistillate petals rhomboid-oblanceolate, acute, about one-half to three-fourths as long as sepals, pilose on back, glabrous within; pistillate sepals elongating to 5–6 mm. long in fruit; style branches linear, not dilated; capsule depressed-globose, 3–4 mm. wide, densely long-pilose; seeds globose-ovoid, truncate or rounded at base, acute at apex, brownish, faveolate-reticulate with minute radiating ridges in depressions.

Sandy and rocky desert slopes and washes, Lower Sonoran Zone, Mojave and Colorado Deserts eastward to western Texas and south into central Sonora and Baja California. Flowering nearly throughout the year.

### 9. **Ditaxis gracilis** Rose & Standl., Contr. U.S. Nat. Herb. **16:** 12, 1912

"Low, sparingly branched, slender annual, 30 to 40 cm. high; stems sparingly pilose, pale green; leaves lanceolate or elliptic-lanceolate, 5 to 6 cm. long and 20 to 25 mm. wide, acute, somewhat attenuate at the base, thin, bright green, all except the youngest glabrous, all on slender petioles 10 to 18 mm. long; racemes few-flowered and axillary; staminate flowers with linear, acute, sepals, their pale green petals broadly lanceolate and one-half longer; pistillate flowers with linear-lanceolate, attenuate sepals 6 to 8 mm. long, the petals spatulate, very short and inconspicuous; style tips not enlarged; capsule hirsute, the mature ones and seeds not seen."

Known only from the type collection taken at Guaymas, Sonora, by Dr. Edward Palmer in 1887. (Description quoted from Rose and Standley's original publication.)

## 7. **BERNARDIA** Houst. ex P. Br., Nat. Hist. Jamaica 361, 1756

Shrubs, suffrutescent perennials, or rarely annual herbs, with alternate stipulate leaves and monoecious or dioecious inflorescences. Stipules small. Staminate flowers in short to elongated, axillary spikes or spicate racemes, mostly bracteate; calyx flask-like to globose, 3–4-toothed. Stamens distinct, inserted on receptacle. Pistillate flowers few, axillary or clustered at ends of branches, 4–6-toothed. Ovary 3-celled, subtended by annular hypogynous, glandular disk or distinct glandular lobes, or these rarely lacking. Style short, entire to laciniately parted. Ovules solitary in each cell. Capsule dehiscent into 3 cocci. Seeds trigonous-prismatic, ecarunculate.

Capsule 7–9 mm. in diameter; leaves 0.5–2.5 cm. long or less ................ 1. *B. incana*
Capsules 12–20 mm. in diameter; leaves 3–12 cm. long:
  Pubescence of stellate hairs; pistillate flowers in spikes; capsules not markedly rugose:
    Capsules 3-celled, 3-lobed; seeds acutely carinate dorsally ........... 2. *B. mexicana*
    Capsules 2-celled, 2-lobed; seeds rounded or obtusely carinate on back .. 3. *B. lagunensis*
  Pubescences of simple and few-forked hairs; pistillate flowers solitary, axillary; capsules distinctly rugose ............................................ 4. *B. cinerea*

### 1. **Bernardia incana** C. V. Morton, Jour. Wash. Acad. **29:** 376, 1939

Dioecious (or sometimes monoecious), divaricately much-branched shrub 1–3 m. tall with tomentulose branchlets and glabrate, tawny to dark gray branches; leaves ovate to obovate, obtuse, crenate, 3–18 mm. wide, 0.5–2.5 cm. long and subremote on vigorous young shoots, closely crowded on short spur branches and then 3–8 mm. wide, 6–15 mm. long, densely short stellate-pubescent and cinereous beneath, darker green and less densely puberulent above; stipules narrowly ovate-lanceolate, blackish or purplish, 1 mm. long or less; petioles 1–3 mm. long; staminate racemes slender, 1.5 cm. long or less, flowers 0.6–1 mm. long, their sepals 3; stamens 5–8; pistillate flowers sessile in axils or terminal, 1–2 mm. long, minutely stellate-puberulent, sepals 5; ovary 3-lobed; stigmas short, laciniate; capsules 7–9 mm. wide, about as long, finely stellate-puberulent; seeds pale brown, 4–5 mm. long, somewhat carinate.

Arid rocky hillsides, Lower Sonoran Zone and in very dry situations in Upper

Sonoran Zone bordering the deserts, southern San Bernardino County, California, east into central Arizona, and south along the eastern side of the coastal ranges to the southern end of the Sierra San Pedro Mártir, Baja California. May–June.

**2. Bernardia mexicana** (Hook. & Arn.) Muell. Arg., Linnaea **34:** 171, 1865
*Hermesia ?mexicana* Hook. & Arn., Bot. Beechey 309, 1838.
*Bernardia viridis* Millsp., Proc. Calif. Acad. II, **2:** 223, 1889.
*Bernardia brandegei* Millsp., in Brandegee, op. cit. **3:** 172, 1891.
*Bernardia mexicana* α *genuina* Muell. Arg., Linnaea **34:** 171, 1865.
*Croton crenulatus* M. E. Jones, Contr. West. Bot. **18:** 48, 1933.

An erect, much-branched shrub 1–3 m. tall with young twigs tomentose with whitish or yellowish hairs but sooner or later glabrate; stipules lanceolate-acuminate, about 2 mm. long; petioles stoutish, 4–9 mm. long, tomentose; leaf blades ovate, elliptic, oblong, or ovate-orbicular, acute to obtuse at apex, acute to subcordate at base, 1–6.5 cm. wide, 3–12 cm. long, irregularly cuneate-denticulate and glandular along margins, slightly to markedly scabrous on upper surface, cinereous, velvety-tomentose and prominently veined beneath; staminate spikes axillary, 1–4 cm. long, tomentose; bracts crowded, horizontally spreading, about 3 mm. wide; pistillate flowers sessile in short spikes, subtended by 2–4 bracts, these broadly ovate, 3–3.5 mm. long; staminate sepals 3, ovate, acute, subfleshy, densely tomentose without; pistillate sepals 5; ovary tomentose; capsule 7–9 mm. long, 12–14 mm. wide, shallowly 3-lobed, tomentose; seeds obovoid, 4–4.5 mm. wide, 5–6 mm. long, acutely carinate dorsally.

Plains, hillsides, and margins of washes, Lower Sonoran to Tropical Zones, southern Baja California from Loreto to the Cape Region; San Luis Potosí, Mexico, to northern South America. Flowering after rains.

**3. Bernardia lagunensis** (M. E. Jones) L. C. Wheeler, in Morton, Contr. U.S. Nat. Herb. **29:** 106, 1945
*Croton lagunensis* M. E. Jones, Contr. West. Bot. **18:** 55, 1933.

An erect, openly branched shrub 1–2 m. tall, with gray, smooth bark; branchlets short, leafy mostly at tips only; petioles stout, 8–10 mm. long, finely tomentose; leaves lance-ovate, elliptic, or broadly ovate, 4–7 cm. long, 1.5–3 cm. wide, broadly acute to rounded at base, acute at apex, shallowly serrate, lower surface densely stellate-canescent, upper surface darker, green, but bearing short-rayed, stellate, yellowish pubescence, each serration usually terminated by a tuft of short hairs; veins prominent beneath; staminate flowers unknown; fruiting pedicels stout, 10–12 mm. long; fruits didymous, 2-celled (in known specimens), 12–15 mm. wide, 10–20 mm. high, densely white-stellate, each locule dorsally ridged; seeds suborbicular to broadly ovoid, nearly as long as locules, mottled light gray and dark gray, smooth, rounded or obtusely and faintly carinate on back.

Rocky hillsides, Lower Sonoran and Tropical Zones, Cape Region, Baja California. Fruiting in Sept.

**4. Bernardia cinerea** Wigg. & Roll., Contr. Dudley Herb. **3:** 273, *pl. 61, figs. 2–3,* 1943

Simply branched shrub 1–2 m. tall; branches glabrous except near tips, leaves ovate, dentate, prominently veined and 3–5-costate from base, 2–4 cm. wide, 3–7 cm. long, densely and velvety-pubescent with simple hairs on both surfaces, green above, cinereous beneath; petioles 4–8 mm. long, densely pubescent; stipules subulate, deciduous, about 2 mm. long; staminate flowers unknown; pistillate flowers axillary, solitary; sepals 5, equal, broadly lanceolate, acute, 3–4 mm. long, densely pubescent; styles 3, each bifid and laciniate at apex; capsules strongly 3-lobed, densely tomentose and distinctly rugose, about 1 cm. long, 1.5 cm. wide; seeds dark brown to nearly black, 6–8 mm. long, about 6 mm. wide, 5 mm. from back to interior angle, ecarunculate.

Rocky hillsides, Lower Sonoran Zone, central Sonora. Flowers in anthesis, unknown; young fruit, Aug.–Sept.

## 8. ADELIA L., Syst. Nat. ed. 10, 1298, 1759

Shrubs or small trees with glabrate branches and often spinescent twigs and alternate, short-petiolate, often pellucid-dotted leaves. Flowers dioecious, small, apetalous. Staminate flowers 4–5-toothed, ovate-pyriform in bud; stamens 8–17, free or connate into a column; anthers versatile, dehiscing longitudinally; gland annular or less commonly represented by 5 distinct glands, adnate to calyx tube. Pistillate flowers pedicellate, with 5–7 narrow sepals; gland annular, puberulent. Ovary 2–3-celled; styles distinct or nearly so, laciniate. Ovules solitary in each locule. Capsule trilobate, pubescent; cocci separating from a central column, this usually persistent. Seeds subglobose to elliptic, ecarunculate; cotyledons broad, plane; endosperm fleshy.

Cocci 6 mm. high, about 5 mm. wide; leaves densely pubescent; branchlets usually not spinescent . . . . . . . . . . 1. *A. obovata*
Cocci 8–9 mm. high, about 7 mm. wide; leaves glabrous or very sparsely pubescent; branchlets usually spinescent . . . . . . . . . . 2. *A. virgata*

### 1. Adelia obovata Wigg. & Roll., Contr. Dudley Herb. 3: 271, *pl. 62, fig. 2,* 1943

Dioecious straggly shrub 1–2 m. tall with gray bark and rigid branches and deciduous, entire, subfasciculate leaves; branchlets pubescent when young but soon glabrate, not markedly spinescent; leaves obovate, rounded or rarely emarginate at summit, cuneate at base; blades 5–12 mm. wide, 8–24 mm. long, grayish green but densely pilose on both sides with simple hairs, 3-veined from base but with additional pinnately arranged, lateral veins branching from the three primary ones; petioles slender, 1–2 mm. long; staminate flowers unknown; pistillate flowers apparently solitary in axils on woolly spur branches, on densely pubescent peduncles 4–8 mm. long; sepals 5, ovate-lanceolate, acute, 1–2 mm. long, densely pubescent; capsule deeply trilobate, 3-celled, densely puberulent with simple short hairs; cocci about 5 mm. broad, 6 mm. high, equal or one or two smaller than others; styles 3-parted, each branch bilobed and ultimately laciniately divided, basal portions densely pubescent up to primary branches; seeds solitary in each locule, dark brown, globose, ecarunculate, 4–5 mm. in diameter, very minutely and narrowly ridged down ventral side.

Rocky hillsides, Lower Sonoran Zone, central Sonora. Aug.–Sept.

### 2. Adelia virgata Brandegee, Zoe 4: 406, 1894

Ascendingly branched shrub 2–3 m. tall with long simple branches rather closely studded with very short, stout, more or less tomentose spurs bearing crowded leaves, only youngest twigs tomentulose and these soon glabrate; leaves oblong to oblanceolate, 8–12 mm. wide, 1.5–3 cm. long, cuneate at base, rounded and usually shallowly emarginate at apex, glabrous or very sparsely appressed-pubescent; petioles slender, 3–5 mm. long; flowers borne on spur branches, usually in clusters of 4–6; pedicels

5–10 mm. long; sepals 2–3 mm. long, densely to moderately villous on both surfaces; staminate flowers with a glandular ring exterior to base of filaments; ovary 2–3-celled, usually 2-celled, sparsely pilose with subappressed whitish hairs, each locule about 7 mm. wide, 8–9 mm. high; seeds subglobose, 6–7 mm. in diameter, dark brown, dull, with a light brown stripe running halfway around circumference from hilum.

Arid rocky hillsides and along washes, Lower Sonoran Zone, southern Baja California. Aug.–Oct.

## 9. ACALYPHA L., Sp. Pl. 1003, 1753

Monoecious or rarely dioecious herbs or shrubs with alternate, petiolate, stipulate leaves. Inflorescences spicate or spikelike racemes. Staminate flowers bracteate, apetalous, small, in glomerules on slender pedunculate spikes. Sepals 4, valvate. Stamens 8 to many, with slender filaments and distinct anther cells. Pistillate flowers bracteate, apetalous, in slender spikes or racemes, or 1 to several below male flowers on staminate spikes. Sepals 3–5, slender. Ovary 3-celled, 3-ovuled. Styles 3, free or more or less joined at base, laciniately divided. Capsules 3-celled, usually trilobate, dehiscent. Seeds ovoid to subglobose, with a small caruncle or essentially ecarunculate.

Plants annual, rarely over 8 dm. tall:
  Pistillate bracts rhomboid-ovate, central tooth elongate, markedly longer than others .... ........................................................ 1. *A. neomexicana*
  Pistillate bracts rounded, reniform or flabelliform, teeth equal or essentially so:
    Stems usually subfleshy and fistulous; divisions of pistillate bracts bearing conspicuous glands at tips of subfiliform processes; pistillate bracts mostly 1-flowered ........ ........................................................ 2. *A. polystachya*
    Stems not appreciably fleshy nor fistulous; divisions of pistillate bracts granulose-glandular or only ciliate with shorter stipitate glands; pistillate bracts usually 2–3-flowered:
      Terminal spikes bearing 1–3 indehiscent fruits, those below dehiscent; bracts shallowly 5–9-dentate, puberulent-ciliate along margins, otherwise glabrous; seeds 1.2–1.5 mm. long ........................................ 3. *A. aliena*
      Terminal spikes bearing only dehiscent fruits; bracts divided one-half to two-thirds to base into 11–15 subulate-linear divisions, these closely set with short-stipate, nearly granular glands; seeds 1.8–2 mm. long .............. 4. *A. ostryaefolia*
Plants shrubby or fruticose:
  Herbage and bracts more or less glandular to viscid:
    Pistillate bracts solitary, inserted on peduncle of staminate spikes; vestiture of leaves and young twigs wholly or predominantly of short, tack-shaped glands, surfaces conspicuously viscid ............................................ 5. *A. saxicola*
    Pistillate bracts in separate spikes, or only a few inserted at base of staminate spikes; vestiture of leaves and twigs composed of both glandular and eglandular hairs, less markedly viscid:
      Leaves thickish, acute or obtuse at apex, rarely over 3.5–5.5 cm. long; pistillate bracts closely crowded, 1.5–3.5 mm. long:
        Twigs bearing some spreading, eglandular hairs twice as long as those forming bulk of pubescence; leaves often cordate at base; foliage not markedly viscid ........................................................ 6. *A. pringlei*
        Twigs bearing eglandular hairs of uniform length; leaves truncate or rounded at base, rarely cordate; foliage often distinctly viscid in youth .. 7. *A. californica*
      Leaves thin to submembranous, acuminate to attenuate at apex, to 8–9 cm. long; pistillate bracts lax or less strongly crowded, 3–6 mm. long:
        Stipules 2–3 mm. long; capsules 1–1.2 mm. high, sparsely hirsute, slightly muricate; seeds 0.6–0.8 mm. long ................................ 8. *A. subviscida*

Stipules 5–10 mm. long; capsules 2–2.5 mm. high, minutely puberulent, smooth; seeds 1.4–1.8 mm. long .............................. 9. *A. comonduana*

Herbage and bracts pubescent but eglandular, not viscid:

Capsules distinctly papillose; seeds 1.5–1.7 mm. long; pistillate bracts toothed subequally one-third to one-half of way to base, central tooth not exceeding others ... ........................................................ 10. *A. papillosa*

Capsules hirsute but not papillose; seeds about 1 mm. long; pistillate bracts shallowly toothed, central tooth often twice as long as others ................ 11. *A. vagans*

## 1. Acalypha neomexicana Muell. Arg., Linnaea **34:** 19, 1865

*Ricinocarpus neomexicanus* Kuntze, Rev. Gen. Pl. **2:** 618, 1891.

Erect or ascendingly branched annual 1–4 dm. tall with thin, bright green leaves, slender petioles and stems, and finely puberulent to subglabrate foliage; petioles 1–3 cm. long, moderately puberulent with fine, spreading, slightly curved hairs; leaf blades ovate, 8–30 mm. wide, 1–4.5 cm. long, rounded at base, acute to faintly acuminate at apex, thin, serrate, often tinged with red when growing in full sun, becoming subglabrate and slightly papillose on lower surface, veins puberulent beneath; staminate spikes slender, 5–12 mm. long, axillary but inserted at base of pistillate spikes; latter both terminal and axillary, 1–5 cm. long, with rhomboid-ovate bracts 5–11 mm. long, 7–9-dentate, central tooth prolonged and to five times as long as lateral teeth, distinctly veined but glabrous or essentially so on back, short-ciliolate along margins; capsules depressed-globose, 3–4 mm. wide, 2–2.5 mm. high, distinctly 3-lobed, sparsely hirsute and faintly muricate dorsally; seeds ovoid, 1.2–1.4 mm. long, brownish or gray, often finely mottled on back, smooth, dull, microreticulate.

In moist situations, Lower Sonoran Zone to Transition Zone, Arizona and New Mexico and south into Chihuahua and to central Sonora. Aug.–Nov.

**Acalypha obscura** Muell. Arg., Linnaea **34:** 163, 1865, resembles *A. neomexicana* but has slightly smaller, somewhat more pubescent leaves and pistillate bracts which are only 4–5 mm. long, nearly orbicular in outline, and coarsely 3–7-toothed, median tooth not produced beyond others. It may reach the edges of the Sonoran Desert along its eastern margin in Sonora, for it has been collected a few miles east of our area and at elevations only slightly above 3,500 feet.

## 2. Acalypha polystachya Jacq., Hort. Schönbrunn. **2:** 64, *pl. 246,* 1797. Not Sessé & Moc., 1894

*Acalypha filifera* S. Wats., Proc. Amer. Acad. **22:** 451, 1887.
*Ricinocarpus polystachyus* Kuntze, Rev. Gen. Pl. **2:** 618, 1891.

Erect or ascendingly branched annual 3–6 dm. tall with subfleshy fistulous stems, youngest stems and branches somewhat puberulent but soon glabrate; petioles slender, 4–12 cm. long; leaf blades ovate-acuminate, 3–9 cm. wide, 4–10 cm. long, subtruncate to contracted-acute at base, margin entire near base and at apex but serrate-dentate between, lamina thin, punctate, sparsely pilose, especially along the veins beneath, and ciliate along margin; staminate spikes axillary, 2–4 cm. long, 1.5–2.5 mm. in diameter, slenderly pedunculate, numerous flowers closely crowded; pistillate spikes terminal and often some in upper axils, 7–15 cm. long, and about 1 cm. wide in fruit; pistillate bracts 8–12 mm. long in fruit, parted almost to base into 9–11 flexuous, subsetaceous divisions, these bearing scattered, spreading, tack-shaped glands, each bract subtending a single flower; capsules subglobose, shallowly 3-lobed, 4–5 mm. in diameter, glabrous and smooth; seeds tan, buff, or light brown, 2.5–3 mm. long, ovoid, distinctly scrobiculate-roughened or rugulose.

Along streams and in shaded places, Lower Sonoran Zone to Tropical Zone, southern Sonora to Central America. June–Oct.

**Acalypha gentryi** Standley, Field Mus. Bot. Ser. **22:** 34, 1940, seems to be closely related to *A. polystachya* but the inflorescences on the type collection are too immature for a positive determination. It has been reported from southwestern Chihuahua, and adjacent Sonora in the Río Mayo region.

### 3. Acalypha aliena Brandegee, Proc. Calif. Acad. II, 3: 172, 1891

Slender, erect annual 2–4 dm. tall, minutely pubescent to subglabrate throughout; petioles slender, 1–3.5 cm. long, puberulent with fine, arcuately incurved, subappressed, simple hairs; leaf blades ovate to ovate-lanceolate, 1.5–4 cm. wide, 2–7 cm. long, truncate to rounded at base, acute at apex, very shallowly serrate, thin, dark green above, paler beneath, minutely puncticulate and nearly glabrous except along veins beneath, 3–5-nerved at base; staminate spikes axillary, very slender, mostly 1 cm. long or less, often inserted in same axil with a pistillate spike; pistillate spikes terminal and axillary, terminal ones to 7 cm. long, axillary rarely over 2–3 cm. long; bracts more or less spreading, broadly ovate-flabelliform, 5–7 mm. long and wide, 5–9-dentate, acute to bidentate at apex, distinctly 5–9-nerved, glabrous except along puberulent-ciliate margin; styles dissected into several filiform segments; capsules depressed-globose, 1.5–2 mm. high, 3–4 mm. in diameter, strongly trilobate, sparsely hispid on upper two-thirds; capsules in uppermost 1–3 bracts of terminal spikes each bearing 3 indehiscent muricate cocci separated from each other nearly to base; seeds ovoid, dull, 1.2–1.5 mm. long, brownish or gray or mottled gray on brown.

In shaded areas along arroyos, ditches, and margins of meadows, Lower Sonoran Zone, central Sonora and Southern Baja California. Aug.–Oct.

### 4. Acalypha ostryifolia Ridd., Syn. Fl. West. States 33, 1835

*Acalypha caroliniana* Ell., Bot. S.C. & Ga. **2:** 645, 1824. Not *A. caroliniana* Walt., 1788.
*Ricinocarpus carolinianus* Kuntze, Rev. Gen. Pl. **2:** 617, 1891.

Erect, simple or ascendingly branched annual 1–8 dm. tall with dark green, minutely puberulent to glabrate stems and foliage; petioles slender, 1–7 cm. long; leaf blades ovate, 2–6 cm. wide, 2.5–12 cm. long, obtuse, rounded, or sometimes subcordate at base, acute to short-acuminate at apex, serrate, thin, becoming sparsely puberulent to nearly glabrous and puncticulate in age; staminate spikes axillary, slender, 1.5–2.5 mm. in diameter, 1–3 cm. long, flowers separated below, densely crowded above middle; pistillate spikes terminal on branches, 2–7 cm. long, bracts 5–9 mm. in diameter, lobed one-half to two-thirds of way to base into 11–15 narrowly subulate-linear lobes, these closely beset with short-stiped, nearly granular glands; capsules depressed-globose, 3.5–4.5 mm. wide, 2–3 mm. high, strongly 3-lobed, papillose and muricate on upper surface, wrinkled on outer shoulders; seeds ovoid, brownish, 1.8–2 mm. long, irregularly rugulose.

Along washes, in shade of shrubs and in wet cienagas, and along fence rows, Lower Sonoran and Carolinian Zones southward to Tropical Zone, Oklahoma to New Jersey and southward to Florida, Arizona, and Mexico. Along streams, ditches, and other wet areas in the Sonoran Desert. June–Oct.

### 5. Acalypha saxicola Wiggins, Contr. Dudley Herb. 3: 70, *pl. 16, figs. 4–5,* 1940

Small spreading shrubs 3–6 dm. tall with slender, divaricate, tawny to purplish-brown, sparsely puberulent but soon glabrate branches; all foliage and young twigs glandular-papillose and strikingly viscid; stipules narrowly subulate setae 1–1.5 mm.

long, caducous; petioles slender, sparsely short-villous, 2–6 mm. long, viscid; leaf blades broadly ovate, elliptical or suborbicular, 5–18 mm. wide, 9–24 mm. long, rounded to broadly cuneate at base, obtuse to rounded at apex, crenate-dentate with 9–15 rounded teeth, 2–5 pairs of pinnately arranged veins, sparsely puberulent to glabrate and papillose-glandular on both surfaces; inflorescences axillary, each consisting of a solitary, 1-flowered, pistillate bract at or near base of a short peduncle 1–2 mm. long, and a densely flowered staminate spike 2.5–5 mm. long on a slender continuation of peduncle 3–5 mm. long; pistillate bracts flabelliform, 3–5 mm. long, glabrous but papillose-glandular and viscid; ovary sparsely puberulent and glandular-papillose; mature fruits and seed unknown.

Arid rocky hillsides, Lower Sonoran Zone, near Coyote Bay, Bahía de la Concepción, Baja California. March–June.

**6. Acalypha pringlei** S. Wats., Proc. Amer. Acad. **20:** 373, 1885

Erect or ascendingly branched shrub or fruticose perennial 4–10 dm. tall with reddish, striate-wrinkled bark on older branches, younger twigs moderately but closely hirsute with slender, spreading, nonglandular hairs 0.5–0.8 mm. long and about equal numbers of shorter nonglandular hairs and stouter, gland-tipped ones, both types of shorter hairs 0.2–0.3 mm. long; stipules subulate-linear, 4–5 mm. long, hispid and glandular-pubescent on both surfaces, reddish; petioles slender, 1–2.2 cm. long, 0.3–0.6 mm. in diameter, hispid and glandular like young branches; leaf blades broadly ovate to ovate-cordate, 1.5–3 cm. wide, 1.5–3.5 cm. long, rounded, truncate or shallowly subcordate at base, mostly obtuse at apex, regularly crenate-dentate with 11–17 teeth on each margin, moderately hispid and glandular-pubescent on both surfaces, long slender hairs conspicuous along veins beneath; terminal spikes and some axillary ones pistillate, 1.5–2 cm. long; staminate and androgynous ones with 1–3 pistillate flowers near base, all axillary, 1–4 cm. long, on peduncles about one-half as long as floriferous portion; pistillate bracts 1.5 mm. long, 2–3 mm. wide, 9–11-toothed; mature fruit and seeds unknown.

Rocky hillsides, Lower Sonoran Zone, Pima County, Arizona, to central Sonora. March–Dec.

**7. Acalypha californica** Benth., Bot. Sulph. 51, 1844

*Ricinocarpus californicus* Kuntze, Rev. Gen. Pl. **2:** 617, 1891.
*Acalypha stokesii* Pax & Hoffm., Pflanzenreich **4**$^{147.\text{xvi}}$: 138, 1924.

A low, divaricately branched shrub 0.5–10 dm. tall; young twigs slender, tawny, closely puberulent to subtomentose with an admixture of simple, spreading, nonglandular hairs and shorter, gland-tipped ones; older twigs and branches glabrate, light gray to dark gray with small, oval, tawny lenticels; leaves alternate and remote or smaller and crowded on short, lateral spur branches; petioles slender, puberulent, 1–12 mm. long; leaf blades ovate to deltoid-ovate, rounded, truncate, or cordate at base, acute at apex, 0.5–3 cm. wide, 0.8–5.5 cm. long, crenulate to dentate on margins, puberulent and glandular on both surfaces, dark green above, paler beneath, at length often subglabrate; staminate spikes slender, 1–2.5 cm. long; pistillate flowers solitary at base of staminate spikes or in short spikes 1 cm. long or less, bracts minutely 15–21-denticulate, glandular-pubescent, coarsely glandular-ciliate along margins, 2–3.5 mm. long; styles 3, each much branched into numerous filiform segments, yellow to bright red; capsules depressed-globose, 2–3 mm. in diameter, pubescent; seeds dark brown to nearly black, smooth, 1–1.5 mm. long.

Dry hillsides and along arroyos, Lower Sonoran Zone, and on arid brushy hillsides in lower parts of Upper Sonoran Zone, San Diego County, California, southward in Baja California to the Cape Region, and central Sonora. Feb.–Nov.

**8. Acalypha subviscida** S. Wats., Proc. Amer. Acad. **21:** 440, 1886

*Acalypha adenostachya* of authors. Not *A. adenostachya* Muell. Arg.

Small shrub or fruticose perennial 2–6 dm. tall; slender branches tawny, densely pubescent with intermixed simple, arcuately curved hairs and gland-tipped, straight, spreading ones; stipules lance-subulate, 2–3 mm. long, sparsely hispid, obvious on young branches but soon deciduous; petioles slender, 1.5–4 cm. long, glandular-puberulent; leaf blades ovate, 1–4 cm. wide, 3–8.5 cm. long, rounded to shallowly cordate at base, attenuate at apex, crenate-serrate with obtuse teeth pointing forward slightly, lamina thin, green, sparsely hispid and with scattered sessile glands on both surfaces; terminal spikes androgynous or pistillate, 4–15 cm. long; axillary spikes staminate or pistillate, pistillate spikes with spreading separated bracts, bracts flabelliform-cordate, 3–5 mm. long, slightly wider, 17–21-dentate, distinctly nerved or ridged, moderately clothed on outer surface with stipitate-glandular and nonglandular hairs, margins ciliate with tack-shaped glands; capsules depressed-globose, 1–1.2 mm. high, 3-lobed, sparsely but coarsely hirsute-glandular, slightly muricate; seeds ovoid, 0.6–0.8 mm. long, grayish brown, microscopically puncticulate-reticulate, caruncle pale, flat.

Stream banks and hillsides, Lower Sonoran and Subtropical Zones, southern Sonora and adjacent Chihuahua southward into central Mexico. Aug.–Dec.

All of the specimens from the Sonoran Desert identified as *A. adenostachya* that I have examined are *A. subviscida* S. Wats.

**9. Acalypha comonduana** Millsp., Proc. Calif. Acad. II, **2:** 222, 1889

A low leafy shrub 5–12 dm. tall with slender, minutely pubescent branches with or without gland-tipped hairs intermingled with simple, spreading, nonglandular ones; stipules lance-linear, 5–10 mm. long, deciduous, pubescent; petioles slender, 2–6 cm. long; leaf blades ovate, ovate-lanceolate, or rhombic-ovate, 3–5.5 cm. wide, 4–9 cm. long, abruptly and broadly cuneate to rounded at base, acute to acuminate at apex, crenate-dentate with rounded, abruptly apiculate teeth, 5-nerved at base, thin, sparsely puberulent and with scattered stipitate glands to subglabrate except on veins beneath; staminate spikes axillary, 1–4 cm. long, on slender peduncles 1–3 cm. long and usually bearing 1–3 pistillate bracts at base of floriferous portion; pistillate spikes terminal and axillary, much more numerous than staminate ones on some plants, 2–10 cm. long in fruit, bracts crowded in youth, well separated in fruit, 2-flowered, orbicular-flabelliform to reniform-flabelliform, 4–6 mm. long, 5–8 mm. wide, shallowly 15–21-toothed, distinctly ribbed, puberulent, often with stipitate glands intermingled with nonglandular hairs, margins sometimes distinctly ciliate with stipitate glands; capsules subglobose, 2–2.5 mm. high, minutely and moderately puberulent; seeds ovoid to obovoid, 1.4–1.8 mm. long, dark tan to reddish brown or grayish, smooth, dull.

Canyons and rocky hillsides, Lower Sonoran Zone, central Baja California to the Cape Region. March–Oct.

**10. Acalypha papillosa** Rose, Contr. U.S. Nat. Herb. **1:** 358, 1895

Well-branched shrub 1–2 m. tall with slender, gray-brown, glabrate, older twigs and tawny, finely appressed-pubescent, younger ones; stipules filiform-linear, 3–5 mm.

long, bearing a few scattered marginal hairs or subglabrate, deciduous; petioles 0.8–1.5 cm. long, subappressed-puberulent; leaf blades ovate-acuminate, 1.5–2.5 cm. wide, 2.5–6 cm. long, obtuse to rounded and 3–5-nerved at base, membranous, serrate, dark green above, paler beneath, sparsely pilose along veins but glabrous between veins on both surfaces; staminate spikes axillary, 1 cm. long or less, short-pedunculate, flowers densely crowded; pistillate spikes terminal, floriferous part 1–1.5 cm. long, bracts broadly flabelliform, about 1.5–2 mm. long, 2–3 mm. wide, coarsely and irregularly 7–11-toothed, sinuses extending one-third to one-half to base, sparsely hispid, eglandular; capsules depressed-globose, 3-lobed, 2–2.5 mm. high, hirsute and distinctly papillose-muricate on upper portions; seeds ovoid, 1.5–1.7 mm. long, orange to pale orange-brown, dull, smooth, caruncle a pale, inconspicuous, V-shaped, horny structure.

Along streams and canyons, Tropical Zone and Lower and Upper Sonoran Zones, vicinity of Ímuris and Bavispe, Sonora, southward into Sinaloa. June–Sept.

**11. Acalypha vagans** Cav., Ic. **6:** 47, *pl. 569, fig. 1,* 1801

*Ricinocarpus vagans* Kuntze, Rev. Gen. Pl. **2:** 618, 1891.
*Acalypha vagans* α *genuina* Muell. Arg., Linnaea **34:** 161, 1865.

An erect, ascendingly branched shrub with spreadingly pubescent, eglandular stems, petioles, and leaf blades; petioles slender, 1–3 cm. long, hairs about 0.6 mm. long and uniform in length; leaf blades ovate to ovate-lanceolate, 1.5–5 cm. wide, 3–10 cm. long, shallowly cordate and 5-nerved at base, acuminate to acute at apex, crenate-serrate, membranous, often appressed-pubescent beneath, especially along veins; staminate spikes axillary, usually with 1–5 pistillate bracts below male flowers or if only staminate flowers present peduncle nearly as long as floriferous part; pistillate spikes terminal and some axillary, 5–6 cm. long, rather laxly flowered, bracts 4–6 mm. wide and nearly as long, 2–3-flowered, 17–21-toothed, central tooth often somewhat longer than other, equal ones, moderately hirsute with simple eglandular hairs; capsules subglobose, 2–3 mm. in diameter, sparsely pubescent; seeds ovoid, about 1 mm. long.

Hillsides and canyons, Lower Sonoran Zone and Tropical Zone, vicinity of Guaymas, Sonora, and southward to southern Mexico. Aug.–Dec.

## 10. TRAGIA L., Sp. Pl. 980, 1753

Monoecious or less commonly dioecious, perennial herbs, shrubs, or vines with alternate stipulate leaves, usually copiously invested with stiffly spreading, stinging hairs. Leaf blades simple or compound. Inflorescence racemose, bracteolate, each bract bearing 1, or less commonly 2–3, pistillate flowers below, several to many staminate, apetalous flowers with jointed pedicels above. Staminate calyx 3–6-parted; lobes incurved to recurved; stamens 1 to many. Pistillate calyx 3–6(8)-parted; segments entire, erect. Ovary 3-celled, 3-ovuled; style branches 3. Capsule 3-lobed; carpels 3, bivalved, separating from the central column. Seeds globose, smooth.

Leaves simple; blades dentate-serrate:
- Rachis of inflorescence and pedicels bearing tack-shaped glands; pistillate pedicels 2–3 times as long as ovary; branches scrambling or twining ......... 1. *T. amblyodonta*
- Rachis of inflorescence and pedicels hirsute but not glandular; pistillate pedicels about equaling ovary; branches erect or ascending, not twining ........................ ............................................ 2. *T. nepetifolia* var. *nepetifolia*

Leaves palmately 3-foliolate; leaflets laciniately incised-dentate ............. 3. *T. laciniata*

**1. Tragia amblyodonta** (Muell. Arg.) Pax & Hoffm., in Engler, Pflanzenreich **4**$^{147.ix}$: 51, 1919

*Tragia nepetaefolia* γ, *amblyodonta* Muell. Arg., in A. DC., Prod. **15**$^{2}$: 934, 1866.

Slender, scrambling or twining shrub 1–2 dm. tall or when in the open branches 2–5 dm. long; young branches pubescent, sometimes sparsely hispid; stipules lance-linear or subulate, 1–3 mm. long, hispidulous; petioles slender, 8–16 mm. long, spreadingly hirsutulous; leaf blades narrowly ovate to flabelliform-suborbicular, truncate to subcordate at base, 0.7–3 cm. wide, 1–5 cm. long, coarsely dentate, thin, distinctly but sparsely hispid; racemes slender, somewhat flexuous, hispid and stipitate-glandular, 3–6 cm. long, each bearing a solitary pistillate flower near base and 10–25 staminate flowers, or rarely pistillate flower abortive; staminate bracts 1–2 mm. long, lanceolate, hirsute with slender white hairs; staminate flowers 1.2–1.5 mm. in diameter, on filiform pedicels to 3 mm. long, these jointed near middle and sparsely beset with tack-shaped glands; sepals 3, stamens 3, filaments carnose, spreading, anthers inflexed, white, not quite equaling sepals; pistillate bracts subulate, 2–3 mm. long; pistillate flowers 2.5–3 mm. in diameter, on slender pedicels 5–10 mm. long; sepals 6, linear-elliptic, about 2 mm. long; ovary densely hispid with erect white hairs; styles somewhat fleshy, 1–1.5 mm. long, branches recurved; capsule 4–5 mm. in diameter, depressed-globose, distinctly 3-lobed, hispid to glabrate; seeds globose, about 2 mm. in diameter, smooth but dull, mottled reddish brown and silvery gray, punctate with reddish resinous specks.

Sandy flats, arroyos, and under shrubs, often in shade, Lower and Upper Sonoran Zones, central Sonora to New Mexico, Texas, and Nuevo León. Aug.–Nov.

**2. Tragia nepetifolia** Cav., Ic. **6**: 37, *pl. 557, fig. 1,* 1801, var. **nepetifolia**

*Tragia angustifolia* Nutt., Trans. Amer. Phil. Soc. II, **5**: 172, 1837.
*Tragia urticifolia* Benth., Pl. Hartw. 14, 1839. Not *T. urticifolia* Michx., 1803.
*Tragia brevispica* Engelm. & Gray, Bost. Jour. Nat. Hist. **5**: 262, 1845.
*Tragia nepetaefolia* ζ, *genuina* Muell. Arg., in A. DC., Prod. **15**$^{2}$: 934, 1866.
*Tragia nepetaefolia* β, *latifolia* Muell. Arg., loc. cit.
*Tragia nepetaefolia* ε, *ramosa* Muell. Arg., loc. cit.
*Tragia nepetaefolia* var. *setosa* S. Wats., Proc. Amer. Acad. **22**: 451, 1887.

Erect or ascendingly branched, suffrutescent plant 1–4 dm. tall with slender, often somewhat flexuous stems bearing stiff, spreading, stinging hairs; rootstocks and basal parts of stems woody, most branches subherbaceous; stipules lance-triangular, 1.5–2.2 mm. long, hispid; petioles 3–20 mm. long, lower much longer than upper, puberulent and with scattered spreading hairs; leaf blades lanceolate-ovate to triangular-lanceolate or lowermost sometimes broadly ovate, 4–20 mm. wide, 1.5–6 cm. long, coarsely serrate-dentate or uppermost sometimes incised-serrate, mostly acute at apex, rounded, truncate or shallowly cordate at base, sparsely hispid along veins, otherwise mostly glabrate, 3-nerved basally; racemes opposite upper leaves, 0.5–4.5 cm. long, sparsely to moderately hispid, bearing a single, short-pedunculate, pistillate flower at base of floriferous part; staminate bracts lanceolate, slightly cupped, 1–1.5 mm. long; pistillate bracts broadly ovate, 1–2 mm. long; staminate flowers 1.5–2 mm. wide, sepals 3, obovate, acute, tip slightly incurved, glabrous or essentially so; stamens 3, filaments fleshy, extrorsely spreading, about one-half as long as sepals; anthers ovoid, 0.5–0.7 mm. long, sharply introrse from tips of filaments; pistillate sepals 6, ovate-lanceolate, slightly unequal, 1.5–2 mm. long, ascending and closely appressed to hirsute ovary; styles 3-lobed above basal one-third, 1–1.5 mm. long; capsules depressed-globose, 5–7.5 mm. wide, deeply 3-lobed, hispid; seeds globose, 2–3 mm. in diameter, mottled gray and brownish.

Canyons, hillsides, and valley floors, Lower and Upper Sonoran Zones, eastern Yuma County, Arizona, to Texas and south into central Mexico. March–Nov.

**Tragia nepetifolia** var. **dissecta** Muell. Arg., in A. DC., Prod. **15**$^{2}$: 933, 1866, differs only in having the leaves somewhat more deeply incised-dentate.

Apparently more or less throughout the range of var. *nepetifolia* and not a well-marked taxon.

### 3. Tragia laciniata (Torr.) Muell. Arg., Linnaea 34: 182, 1865

*Tragia urticifolia* var. *laciniata* Torr., Bot. Mex. Bound. 200, 1859.

Slenderly graceful herb 1–3.5 dm. tall from a slender woody rootstock that often appears rhizomatous; stems moderately puberulent with minute, simple, crisped or incurved hairs and a few scattered, coarser, simple ones 0.5–1 mm. long; stipules lance-ovate, 1–1.3 mm. long, sparsely ciliate with coarse white hairs; petioles 2–15 mm. long, investiture similar to that of twigs; leaf blades palmately 3-foliolate, lateral leaflets 4–25 mm. long, each bearing a distinct exterior lobe at base, laciniately toothed, central leaflet to 8 mm. wide, 1–4 cm. long, deeply laciniately toothed, all leaflets dark gray-green and sparsely pubescent above, paler and with coarse white hairs along veins beneath; racemes short, bearing 3–6 minute staminate flowers and a solitary pistillate one below staminate portion; staminate flowers with 3 stamens, filaments fleshy; sepals of pistillate flowers 1–1.5 mm. long, lanceolate, acute; capsules 5–6 mm. in diameter, 3–4 mm. high, sparsely to densely hirsute; seeds globose, 2–2.5 mm. in diameter, smooth, closely puberulent with reddish brown, short hairs.

Along streams and on shaded hillsides, upper parts of Lower Sonoran Zone and Upper Sonoran Zone, southern Arizona into New Mexico, south into Sonora and Chihuahua. April–May; Aug.–Sept.

**Tragia stylaris** Muell. Arg., Linnaea **34**: 180, 1865, is similar in general habit and appearance to *T. nepetifolia* but has staminate flowers nearly twice as large, and 4–7 erect instead of extrorse stamens in each flower; style divided to the base, 2–3 mm. long.

On arid slopes, Upper Sonoran Zone, from the eastern Mojave Desert, California, to Texas but consistently above the limits of the Sonoran Desert. It might occur as a waif along the upper margins of our area in central Arizona.

## 11. DALECHAMPIA L., Sp. Pl. 1054, 1753

Shrubs, vines, or rarely perennial herbs with glabrous to hispid twigs and herbage, sometimes bearing stinging hairs. Leaves alternate, stipulate, petiolate, simple, and entire or 3–5-lobed or compound with 3–5 digitately arranged leaflets. Flowers monoecious, apetalous. Inflorescence of terminal or axillary racemes or sometimes fasciculate, with various arrangements of flowers and bracts; flowers short-pedicellate, pedicels often jointed. Staminate calyx globose-vasiform, 4–6-parted; stamens 10–90, inserted on a rounded or column-like receptacle; disk none. Pistillate flowers with 5–12 narrow, valvately arranged, usually pinnatifid sepals, these often accrescent and indurated. Ovary 3- or rarely 4-celled, each cell 1-ovuled; style often excavated at apex. Capsule 3-lobed, locules dehiscing into 2 valves. Seeds globose, ecarunculate.

**1. Dalechampia scandens** L. var. **fimbriata** (H. B. K.) Muell. Arg., in A. DC., Prod. **15**$^{2}$: 1244, 1866

*Dalechampia fimbriata* H. B. K., Nov. Gen. & Sp. **2:** 100, 1817.

*Dalechampia brevipes* Briq., Ann. Conserv. & Jard. Bot. Genève **4:** 232, 1900. Not *D. brevipes* Muell. Arg., 1865.

*Dalechampia passiflora* Chod. & Hassl., Bull. Herb. Boiss. II, **5:** 609, 1905.

Rarely perennial scandent herb, more commonly scandent shrub to 2 m. or more in height with slender, finely puberulent stems, usually bearing at least a few spreading, coarser, stinging hairs; bark becoming striate and glabrate; stipules ovate, acute to short-acuminate, 2–6 mm. long, 1.5–3 mm. wide, ciliate-hispid; petioles slender to coarse, 2–10 cm. long; leaf blades palmately trilobate, lobes subequal, oblong-ovate to obovate, 1.5–4.5 cm. wide, to 10 cm. long, minutely serrate-denticulate, short-acuminate at apex, blade cordate at base, dark green, rugulose and sparsely pubescent above, paler and more markedly hispid along larger veins beneath, lateral lobes asymmetrical at base; pistillate flowers subtended by broadly ovate to trilobate, whitish bracts 1–2 cm. long in flower, accrescent in fruit and then often 3.5–4.5 cm. long and wide; sepals of pistillate flowers linear, 6–12 mm. long, generously beset with stalked glands 0.2–0.6 mm. long along margins, hirsute on outside, glabrous within; capsules depressed-globose, 12–15 mm. in diameter, 8–10 mm. high, 3-lobed, dark brown, glabrous or glabrate; seeds globose, 3–3.5 mm. in diameter, mottled brown and gray or whitish, ecarunculate but hilum conspicuous.

Rocky hillsides and shady slopes, Lower Sonoran Zone to Tropical Zone, Sonora to tropical South America and Cuba. Aug.–Dec.

## 12. RICINUS L., Sp. Pl. 1007, 1753

A monoecious herb or under favorable conditions becoming a shrub or small tree. Leaves alternate, peltate, palmately lobed, toothed, margins and petioles bearing large, stalked to subsessile glands. Flowers many, small, apetalous, arranged in terminal racemes with staminate flowers below pistillate. Calyx of staminate flowers 3–5-parted; stamens numerous, crowded, filaments branched. Calyx of pistillate flowers caducous. Ovary 3-celled, each cell 1-ovuled; styles 3, bright red, 2-cleft. Capsule often prickly, globose or ovoid, each carpel bivalvate. Seeds oblong, somewhat compressed, mottled, carunculate.

**1. Ricinus communis** L., Sp. Pl. 1007, 1753

Usually 2–3 m. tall but sometimes to 10 m. tall and with trunk to 2.5–3 dm. in diameter, glabrous and red-tinged throughout; leaves to 3.5 dm. broad; racemes often erect; capsules 1–1.5 cm. high, smooth to strongly prickly; seeds 8–10 mm. long, mottled with gray and dark brown or black, with a conspicuous caruncle 1.5–2.5 mm. long, 1.5–3 mm. wide.

Escaped from cultivation and occasionally established along watercourses and near dwellings throughout the warm temperate and subtropical areas of both hemispheres. Flowering throughout the year.

## 13. JATROPHA L., Sp. Pl. 1006, 1753

Herbaceous, suffrutescent, shrubby, or arborescent plants with alternate, petiolate, or sessile, stipulate leaves and dichotomously racemose inflorescences. Flowers monoecious or rarely dioecious. Sepals free to connate at base. Petals 5, distinct or connate basally into a tube. Glandular disk entire or of 5 distinct glands. Stamens 8–10, more or less united below into a column; staminodia present or absent. Ovary 2–5-celled, usually 3-celled, each cell 1-seeded. Styles 1–3, entire or shallowly 2-lobed. Fruit capsular, locules often angled dorsally. Seeds carunculate.

Leaves shallowly to deeply lobed or dissected:
  Plants herbaceous, from a tuberous root; leaves palmately and deeply lobed or parted .... .......... 1. *J. macrorhiza* var. *septemfida*
  Plants shrubby, from woody fibrous roots; leaves shallowly 3–5-lobed:
    Stipules dissected into fimbriae, these gland-tipped; herbage glabrous; leaves irregularly dentate .......... 2. *J. purpurea*
    Stipules not dissected nor gland-tipped; twigs and leaves velvety-tomentose; leaves not dentate .......... 3. *J. platanifolia*
Leaves entire or dentate, but not markedly lobed:
  Leaf blades 3–6 mm. wide, cuneate-obovate; petiole 1 mm. long or less, leaves often sessile .......... 4. *J. cuneata*
  Leaf blades mostly 2 cm. wide or more, ovate, cordate or suborbicular; petioles mostly 1 cm. long or more:
    Pistillate sepals and leaf margins edged with glands:
      Plant a low, openly branched shrub less than 2 m. tall; spur branches conspicuous, much roughened by old leaf bases; glands on leaf margins sessile or essentially so .......... 5. *J. cardiophylla*
      Plant more or less arborescent, 2–6 m. tall; spur branches lacking or less conspicuous, moderately roughened when present; glands on leaf margins stalked or sessile:
        Staminate sepals ciliate; leaf blades usually acuminate .......... 6. *J. cordata*
        Staminate sepals eciliate; leaf blades acute or rounded .......... 7. *J. vernicosa*
    Pistillate sepals and leaf margins eglandular .......... 8. *J. cinerea*

**1. Jatropha macrorhiza** Benth. var. **septemfida** Engelm., Wheeler Rep. 243, 1878
*Jatropha arizonica* I. M. Johnston, Contr. Gray Herb. **68:** 89, 1923.

An erect, nearly glabrous herb 2–5 dm. tall from a tuberlike root, this often 4–6 cm. in diameter, 1–2.5 dm. long; stipules dissected into several finely branched fimbriae to 1 cm. long; petioles 3–12 cm. long, glabrous; leaf blades suborbicular in outline, to 1.5 dm. broad, palmately 7–9-parted or -lobed from one-half to five-sixths to base, lobes lanceolate to obovate-oblong, saliently and irregularly toothed, each tooth usually tipped with a slender flexible bristle; cymes densely flowered, on peduncles 1–6 cm. long, branches of inflorescence sparsely puberulent; sepals of staminate flowers lanceolate, entire, 4–6 mm. long, about half as long as petals; stamens united near base into a column, inner ones longer than outer whorl; sepals of pistillate flowers usually larger and spinulose-dentate, to 1 cm. long; capsules 10–13 mm. long, nearly or fully as wide, obtusely 3-lobed, glabrous; seeds 5–6 mm. wide, 7–9 mm. long, smooth, buff with brownish or reddish flecks and spots, a distinct line along ventral face, caruncle flabelliform-cupulate, thin, margin papery, laciniate, crinkled.

Rocky hillsides, sandy canyon beds, and tops of mesas, upper margin of Lower Sonoran Zone and Upper Sonoran Zone, southern Arizona, New Mexico, and adjacent Mexico. Barely touching the edge of the Sonoran Desert. May–Oct.

**2. Jatropha purpurea** Rose & Pax, in Pax, Pflanzenreich $4^{147}$: 42, *fig. 15,* 1910

An openly branching shrub 1–2.5 m. tall with papery brown bark, glabrous leaves and twigs; stipules dissected into subfiliform, gland-tipped segments 5–6 mm. long;

petioles slender, 2–3 cm. long; leaf blades 3–4 cm. wide, about as long or slightly longer, more or less pentagonal, shallowly but usually distinctly 3-lobed, ciliate-glandular along lower truncate to subcordate margin, median lobe narrowly triangular, acuminate, it and shorter lateral lobes irregularly dentate but less conspicuously glandular than basal margin; cymes short-pedunculate, subtended by lance-linear, glandular-ciliate bracts 5–6 mm. long; staminate sepals ovate, obtuse, 4–5 mm. long; pistillate sepals 7–9 mm. long, lance-linear, glandular-ciliate; petals obovate, about 7–8 mm. long, those of staminate flowers usually slightly longer than pistillate ones; glands of disk distinct in staminate flowers, connate in pistillate; ovary glabrous, 3-lobed; mature fruit unknown.

Dry gravelly or rocky hillsides, Lower Sonoran Zone to Tropical Zone, southern Sonora and southern Baja California south into Sinaloa. July–Aug.

**3. Jatropha platanifolia** Standley, Field Mus. Bot. Ser. **22:** 38, 1940

An openly, regularly branched shrub with rather fleshy branchlets and purplish bark; young branchlets densely tomentose with very short, grayish hairs; leaves confined to tips of branches and a few stubby spurs near top of plant; petioles stout, 5–20 mm. long, densely tomentose-velutinous; leaf blades 3–7 cm. wide, about as long, truncate to shallowly cordate at base, trilobed about to middle or sometimes two additional very short lobes at base; lobes narrowly to broadly ovate, acute to acuminate; margins entire or slightly undulate, 5-veined from base, densely tomentose-velutinous on both surfaces; cymes short-pedunculate and few-flowered; bracts triangular-lanceolate; sepals 3–4 mm. long, oblong to oblong-oblanceolate, sparsely hirsute without; corolla 5–6 mm. long, petals rounded at apex, slightly connate at base, glabrous outside, pilose within; mature fruit unknown.

On rocky, brushy hillsides, Lower Sonoran and Upper Sonoran Zones, southern Sonora. Probably flowering Aug.–Sept.

**4. Jatropha cuneata** Wigg. & Roll., Contr. Dudley Herb. **3:** *272, pl. 62, fig. 1,* 1943

Intricately branched shrub 1–2 m. tall with each branch or twig bearing numerous dwarf spurs and several canes from crown; leaves fascicled, sessile or with a petiole to 1 mm. long, cuneate, emarginate or rarely entire at apex, glabrous, 1–1.5 cm. long, 3–6 mm. wide, mostly borne near tips of spur branches; flowers pinkish, borne on tips of dwarf branches, short-pedicellate to sessile, staminate fasciculate, pistillate solitary or in pairs; sepals of staminate flowers about 3 mm. long, ovate-lanceolate, acute, silvery-puberulent without, glands of disk distinct; stamens 10, connate into a column, petals connate to middle, spatulate-obtuse; sepals of pistillate flowers lanceolate, acuminate, 5–7 mm. long, petals about as in staminate flowers; capsules 1–3-locular, about 2.5 cm. broad, about one-half as long, usually 1-seeded; seeds gray and shining, globose, bearing a 2-lobed basal caruncle.

Arid slopes and mesas, Lower Sonoran Zone, southwestern Arizona south into Sonora and Baja California. July–Aug.

**5. Jatropha cardiophylla** (Torr.) Muell. Arg., in A. DC., Prod. **15[2]:** 1079, 1866
*Mozinna cardiophylla* Torr., Bot. Mex. Bound. 198, 1859.

An openly branched, low shrub with several to many, erect or ascending, slightly zigzag, glabrous, slender stems with reddish bark arising from root crown; dwarf spur branches conspicuous, 5–25 mm. long, much roughened by persistent bases of the petioles, these separating from rest of slender petiole 1–1.5 mm. above base; petioles mostly 5–10 mm. long, glabrous; leaf blades cordate-ovate, suborbicular, or slightly reniform, 1.5–2.5 cm. wide, usually about as long, acute, acuminate, or rarely obtuse at apex, cordate at base, margins shallowly crenate and moderately beset with small, globose, sessile glands especially near base, bright green on both surfaces but usually lighter beneath than above; cymes 1–4 cm. long, usually dichotomously branched; sepals oblong, rounded, about 2 mm. long, distinct; petals oblong-spatulate, connate at base, 4–5 mm. long, pinkish; capsule 12–15 mm. high, often 1-loculed, then about as broad as long, slightly obovoid, sometimes attenuate at one or both ends, glabrous, on pedicels about 1 cm. long.

Arid plains, foothills, and mesas, Lower Sonoran Zone, Maricopa and Pima Counties, Arizona, southward to central Sonora. June–Aug.

### 6. Jatropha cordata (C. G. Ortega) Muell. Arg., in A. DC., Prod. **15**[2]: 1078, 1866

*Mozinna cordata* C. G. Ortega, Hort. Mat. 107, 1798.
*Loureira glandulosa* Cav., Ic. **5**: 18, *pl. 430,* 1799.

Shrub or small tree to 8 m. tall with yellowish juice, glabrous foliage, and shining, red, glabrous twigs; bark on older stems and branches smooth, pale red to yellowish, exfoliating as papery sheets; stipules reduced to a slender stipitate gland 1–2 mm. long; petioles slender, 2–4.5 cm. long, dark red; leaf blades broadly ovate to ovate-cordate, 1.5–4.5 cm. wide, 2.5–6 cm. long, truncate to shallowly cordate at base, acute and usually somewhat acuminate at apex, shallowly serrate with a short-stipitate gland at tip of each serration, glabrous on both surfaces, bright green above, paler beneath, palmately 5-veined from base; cymes short-pedunculate, laxly branching, subtended by oblong to narrowly oblanceolate bracts to 1.5 cm. long, these conspicuously glandular-stipitate along margins; staminate sepals 2–3 mm. long, narrowly ovate to spatulate, eglandular or less commonly sparsely glandular-ciliate, petals 4–5 mm. long, pinkish; pistillate sepals glandular-ciliate, accrescent and to 2 cm. long in fruit; fruits 2–3-lobed or rarely reduced to a single functional carpel, 10–12 mm. high, 15–20 mm. wide, grayish, glabrous; seeds obovoid, 8–9 mm. wide, 10–12 mm. long, dull black, caruncle small, light-colored.

Rocky hillsides and mesas, mainly Lower Sonoran Zone, central Sonora and southern Chihuahua to Jalisco. Aug.–Sept.

### 7. Jatropha vernicosa Brandegee, Zoe **5**: 206, 1905

A glabrous dioecious shrub 2–4 m. tall with smooth papery bark and pale yellowish juice; stipules reduced to sessile glands; petioles slender, 3–7 cm. long; leaf blades ovate-cordate, acute or rarely somewhat acuminate at apex, 3–5 cm. wide, 5–7 cm. long, dark green and shining above, paler beneath, margin subremotely and shallowly serrulate, tip of each serrulation tipped with a sessile gland; venation pinnate; staminate flowers in loose cymes, sepals elliptic, about 5 mm. long, eciliate, connate at base; pistillate flowers usually solitary at tips of branches on pedicels 2–3 cm. long, calyx lobes distinct to base, ovate, 5–6 mm. long, margins bearing sessile glands; petals spatulate, distinct to base, white or faintly purplish, equaling or slightly surpassing sepals; capsule subglobose, about 2 cm. in diameter, glabrous.

Mountain ridges, Lower Sonoran Zone and parts of Tropical Zone, southern Baja California. July–Aug.

**8. Jatropha cinerea** (C. G. Ortega) Muell. Arg., in A. DC., Prod. **15**$^2$: 1078, 1866

*Mozinna cinerea* C. G. Ortega, Hort. Mat. 108, 1798.
*Mozinna canescens* Benth., Bot. Sulph. 52, *pl. 25*, 1844.
*Jatropha canescens* Muell. Arg., in A. DC., Prod. **15**$^2$: 1079, 1866.

Widely branching shrub to small tree 1–7 m. tall with clear acrid juice and smooth, brownish or yellowish bark; stipules reduced to inconspicuous sessile glands; petioles slender, 2–6 cm. long, minutely puberulent but soon glabrate, reddish; leaf blades ovate-pentagonal, suborbicular, or subreniform, distinctly but shallowly 3-lobed to entire, 3–6.5 cm. wide, nearly or fully as long, truncate to shallowly but broadly cordate at base, margin entire and eglandular, minutely puberulent-cinereous to glabrous or glabrate, bright green to cinereous above, usually paler beneath; staminate sepals lanceolate, 2–3 mm. long, cinereous-puberulent, tips often slightly recurved; petals 6–8 mm. long, connate into a tubular corolla, only terminal 1–3 mm. free; pistillate sepals similar but 3–4 mm. long, slightly accrescent to 5–6 mm. long in fruit; capsules 1–3-locular, globose to 3-lobed, about 10 mm. high, to 2.8 cm. wide, slightly glaucous but glabrous; seeds globose to broadly obovoid, 8–10 mm. in diameter, light gray, often speckled with red or purple, with a distinct brownish vertical line along inner face; caruncle deltoid, whitish, about 2 mm. long, margin sinuate-erosulate.

Plains, hillsides, and mesas, Lower Sonoran Zone, northern Sonora and central Baja California to northern Sinaloa. Aug.–Nov.; April.

## 14. CNIDOSCOLUS Pohl, Pl. Brasil. Icon. & Descr. **1**: 56, 1827

Herbaceous, suffrutescent, shrubby, or arborescent plants armed with long stinging hairs. Leaves long-petiolate, palmately divided, lobed, or dissected. Stipules lacerate, thin. Inflorescences cymose at tips of branches. Flowers apetalous, monoecious or dioecious. Staminate calyx petaloid, 5-lobed, white. Stamens 10; filaments united into a column with a ring of hairs at base; anthers in 2 whorls of 5; staminodia 3, filiform; glands of staminate flowers adnate to androphore just below ring of hairs. Pistillate calyx petaloid, white; 5 sepals distinct, often caducous. Ovary 3-celled, 3-ovuled. Styles 3, united below, each one twice-bifid above. Seeds large, carunculate.

Leaves lobed nearly halfway to middle, 8–15 cm. wide; pistillate calyx caducous .......... 1. *C. angustidens*
Leaves merely repand-dentate, not lobed nor dissected, 2.5–6.5 cm. wide; pistillate calyx with a persistent ring of tissue at base, lobes caducous .................. 2. *C. palmeri*

**1. Cnidoscolus angustidens** Torr., Bot. Mex. Bound. 198, 1859

*Jatropha angustidens* Muell. Arg., in A. DC., Prod. **15**$^2$: 1102, 1866.

A robust herb or sometimes slightly suffrutescent plant 1.5–12 dm. tall from a fleshy rhizome with stems, leaves, and inflorescences sparsely to densely armed with stiff stinging hairs 4–8 mm. long, each hair surmounting a conspicuous, rounded, whitish pustule; stipules papery, whitish, 5–6 mm. long, irregularly lacerate; petioles 2–10 cm. long; leaf blades orbicular to subreniform in outline, 8–15 cm. wide, 3–5-lobed nearly to middle, broadly and shallowly subcordate at base, lobes coarsely incised-dentate, teeth often 1–2.5 cm. long, blade glabrous between bases of coarse stinging hairs; peduncles stoutish, 1–5 cm. long; flowers monoecious, pistillate solitary at bases of branches of the inflorescence, staminate in broad cymes to 8 cm. wide; staminate calyx white, papery, 9–14 mm. long, funnelform, armed with spreading stinging hairs 2–5 mm. long; pistillate calyx of 5 distinct, ligulate to oblanceolate segments 10–12 mm.

long, sparsely hispid with stinging hairs without, caducous; capsules 10–12 mm. long, hispid; seeds 6–7 mm. wide, 9–11 mm. long, about 4–4.5 mm. thick, rounded on back, slightly carinate along inner face, dull brown, slightly mottled, caruncle white or yellowish, fleshy, about 3 mm. wide, 1.5–2 mm. long.

Plains and hillsides, mainly Lower Sonoran Zone, Pima County to Cochise County, Arizona, and southward into Sonora, Baja California, and Sinaloa. May–July.

**2. Cnidoscolus palmeri** (S. Wats.) Rose, Contr. U.S. Nat. Herb. **12:** 282, 1909
*Jatropha palmeri* S. Wats., Proc. Amer. Acad. **24:** 76, 1889.

A weak, openly branched shrub 5–15 dm. tall, sparsely to moderately clothed with stiff stinging hairs and with velvety puberulence of minute soft hairs on twigs and both surfaces of leaves; stipules lance-deltoid, 0.5–1.5 mm. long; petioles stoutish, 0.5–2 cm. long, velvety-puberulent between bases of stinging hairs, at length subglabrate; leaf blades suborbicular to reniform, 1.5–6.5 cm. wide, nearly or sometimes fully as long or one and one-half times as wide as long, base rounded, subtruncate, or shallowly and broadly cordate, margins irregularly and shallowly dentate with spreading deltoid teeth 2–5 mm. long, not lobed, bright yellowish green above, paler and more closely hispid beneath; staminate cymes 1–3 cm. broad, on stout branching peduncles; staminate calyx 10–12 mm. long, hispid and minutely puberulent, lobes broadly ovate, about 2 mm. long and wide; pistillate flowers subsessile to short-pedicellate, at base of branches of staminate cymes, calyx lobes greenish white, about 10 mm. long, caducous except for a narrow ring at base of ovary; capsules 3-lobed, 10–13 mm. high, puberulent.

In rocky crevices on canyon walls, Lower Sonoran Zone, islands along the east coast of Baja California, on the adjacent peninsula, and on rocky slopes near Guaymas, Sonora. May–June.

## 15. MANIHOT Adans., Fam. Pl. 2: 356, 1763

Herbs, shrubs, or trees with glabrous to pubescent twigs and foliage and alternate, petiolate, or rarely sessile leaves, blades in ours deeply lobed. Stipules mostly small. Inflorescences racemose, monoecious, staminate above, pistillate near base. Flowers apetalous, staminate ones bearing a campanulate, more or less 5-lobed, petaloid, perianth cup. Stamens 10, alternately long and short and inserted between lobes or glands of disk; anthers dorsifixed, longitudinally dehiscent. Pistillate calyx caducous, disk entire or lobed; ovary 3-celled, 3-ovuled. Capsules bivalvate. Seeds large, carunculate, with fleshy and oily endosperm; cotyledons flat and comparatively wide.

Primary lobes of leaves laterally lobed to sinuate; caruncle not extending beyond end of seed, hence invisible from outer side:
  Primary lobes of leaves lanceolate, tapering to the apex, 3–12 mm. wide; margins of seeds rounded, not noticeably angled .............................. 1. *M. angustiloba*
  Primary lobes of leaves obovate to oblanceolate, often more or less panduriform, 1.5–6 cm. wide; margins of seeds often moderately angled ................ 2. *M. chlorosticta*
Primary lobes of leaves entire, rarely faintly sinuate; caruncle usually extending beyond tip of seed, conspicuous from outer side ............................... 3. *M. isoloba*

**1. Manihot angustiloba** (Torr.) Muell. Arg., in A. DC., Prod. **15²:** 1073, 1866
*Janipha manihot* var. *angustiloba* Torr., Bot. Mex. Bound. 199, 1859.

Suffrutescent perennial or shrub 0.3–2 m. tall with glabrous, lightly pruinose, young branches, glabrous foliage, and woody rootstocks; stipules 1–2 mm. long, caducous; petioles slender, 2.5–12 cm. long; leaf blades divided nearly to base into 5–7 lanceolate to linear-lanceolate lobes 3–12 mm. wide, 4–15 cm. long, all or most of primary lobes sinuate to again laterally lobed, secondary lobes to 7 mm. wide, 5–25 mm. long, acute, divergent-ascending, lamina dark green above, slightly paler and prominently nerved beneath; inflorescences few-flowered, staminate calyx pale greenish to greenish yellow, 9–11 mm. long, glabrous within and without, on slender pedicels 6–8 mm. long; capsules glabrous but somewhat rugose to low-tuberculate, 12–15 mm. high, to 2.5 cm. wide, 3-lobed or through abortion of one or two ovules bilobed or subspherical, cells loculicidally 2-valved; seeds scaraboid (sometimes nearly orbicular in outline), 8–10 mm. wide, 10–12 mm. long, strongly rounded on back and nearly as much so on inner face, lateral margins obtusely and bluntly angled, grayish to pale buff with dark brown to olive green mottling and streaks; caruncle orbicular to oblong, about 1.5 mm. in diameter, invisible from back of seed.

Arid flats, arroyos, hillsides, and outwash slopes, Lower Sonoran Zone and lower margins of Upper Sonoran Zone, southern Arizona south to southern Sonora and east into Chihuahua. June–Oct.

### 2. **Manihot chlorosticta** Standl. & Goldm., Contr. U.S. Nat. Herb. **13:** 375, 1911

Shrub or small tree or sometimes scrambling and vinelike, 1–3 m. tall; foliage glabrous; older branches reddish brown with small, circular, pale buff lenticels; young twigs herbaceous, green, glabrous and often faintly glaucescent; stipules acicular-subulate, 1.5–2.5 mm. long, caducous; petioles 5–15 cm. long; primary lobes of blades 5–7, oblong, oblong-lanceolate, or oblanceolate, 1.5–6 cm. wide, 6–18 cm. long, dark green above, paler beneath, acute, attenuate, or aristate at apex, strongly sinuate, shallowly lobate, or panduriform; racemes 5–15 cm. long, glabrous; staminate calyces greenish, 8–10 mm. long, glabrous within and without, pedicels 5–20 mm. long, whole flower falling soon after anthesis; pistillate flowers 1–5 at and near bases of racemes; capsules 18–22 mm. high, 2–2.5 cm. wide, orbicular, bilobed, or shallowly trilobed, slightly rugose to tuberculate, glabrous; seeds scaraboid, ovoid to ellipsoid, 9–10 mm. wide, 11–13 mm. long, back rounded, inner face less strongly so and longitudinally ridged along midline, mottled and streaked with brown or gray and lighter brown on back, less so on inner face, margins faintly winged and moderately angled; caruncle flabelliform, 2–2.2 mm. long, to 4 mm. wide, reddish brown, invisible from outer surface of seed.

Hillsides and along rocky washes, Lower Sonoran Zone, southern Baja California and southern Sonora into Sinaloa. June–Oct.

**Manihot davisiae** Croizat, Jour. Arnold Arb. **23:** 224, 1942, from the Santa Catalina Mountains in Arizona, is very close to if not identical with *M. chlorosticta.* The seed characters used by Croizat to separate *M. davisiae* from *M. angustiloba* do not seem to hold, for nearly orbicular seeds occur in *M. angustiloba* also. The lack of adequate flowering and fruiting material from intermediate localities and altitudes prevents a decision on the relationship between the two populations at this time.

### 3. **Manihot isoloba** Standley, Field Mus. Bot. Ser. **17:** 197, 1937

A slender, many-stemmed shrub 1–2.5 m. tall, glabrous throughout; younger branches more or less fleshy, rusty-red and often more or less glaucous; stipules filiform-acicular, 2–3 mm. long, caducous; petioles slender, 5–14 cm. long; leaf blades membranous, light green above, glaucous beneath, 5–12 cm. long and wide, 5–9-parted to base into oblong-linear lobes with nearly parallel sides, these primary lobes 1–2 cm. wide, entire, obtuse to acute but tipped with a filiform projection 4–7 mm. long; staminate racemes 2–3 cm. long, few-flowered, pedicels slender, 5–13 mm. long, glaucous; calyx narrowly campanulate, 10–12 mm. long, glabrous; pistillate flowers axillary or in short racemes; capsules subglobose, shallowly trilobate, 15–18 mm. high, 20–27 mm. wide, glaucous, finely rugose-wrinkled when dry; seeds grayish to pale brown, mottled with dark brown, oblong-ellipsoidal, 6–8 mm. wide, 10–12 mm. long, well rounded on back, sides nearly parallel, inner side flatter than back but convexly curved, inner midline conspicuous full length of testa; caruncle reniform, 2.5–3 mm. wide, projecting inward almost at right angles to long axis of seed 1–1.5 mm. and usually extending beyond tip of seed, pale buff, horny.

Grassy slopes and along canyons, mostly Upper Sonoran Zone but reaching the upper margins of the desert along streams; known from a few collections from vicinity of lower Río de Bavispe to southern Sonora and southwestern Chihuahua. Aug.–Sept.

## 16. SEBASTIANIA Spreng., Neue Entd. 2: 118, *pl. 3,* 1821

Herbs, shrubs, or trees, sometimes with spinose branches. Leaves alternate (or rarely opposite), small, short-petiolate, pinnately veined, serrulate or entire. Inflorescence monoecious, spicate, usually terminal, sometimes opposite leaves, usually slender; bracts biglandular at base. Flowers apetalous, without a disk; staminate small, their calyces usually 3-lobed or -parted, lobes subequal; stamens 2–4, usually 3, filaments connate at base or free; pistillate flowers with 3 sepals, larger than staminate. Ovary 2–3-celled, each cell 1-seeded; style branches spreading or recurved, entire. Capsule subglobose to deeply 3-lobed, smooth or spinose, carpels 2-valved along juncture with columella. Seeds subglobose to oblong or cylindrical, more or less carunculate.

### 1. Sebastiania pavoniana Muell. Arg., in A. DC., Prod. $15^2$: 1189, 1866

*Gymnanthes pavoniana* Muell. Arg., Linnaea **32**: 106, 1863.
*Sebastiania palmeri* Rose, Contr. U.S. Nat. Herb. **1**: 112, 1891.
*Sebastiania pringlei* S. Wats., Proc. Amer. Acad. **26**: 149, 1891.
*Sebastiania ramirezii* Maury, Naturaleza II, **2**: 406, 1894.
*Gymnanthes pringlei* S. Wats., in Pax, Pflanzenreich $4^{147.v}$: 147, 1912, in synonymy.

Erect or ascending, glabrous shrub 0.5–2.5 m. tall; stipules linear-lanceolate, about 1 mm. long, scarious and sublacerate along margins; petioles slender, 5–10 mm. long; leaf blades ovate-lanceolate, thin, 1.2–3.5 cm. wide, 3–8 cm. long, obtuse at base, acuminate or short-cuspidate at apex, 6–8-nerved, remotely and minutely serrulate, minutely subulate-glandular near base and on adjacent petiole; spikes terminal, 3–5.5 cm. long, flowers crowded, only basal pistillate; bracts ovate, lacerate-denticulate, basal ones often plicate-rugose, those in staminate part of spike each subtending 2–3 flowers; staminate flowers short-pedicellate, with triangular to setaceous-acuminate sepals, or sepals sometimes obsolete on lateral flowers; pistillate flowers with apiculate-ovate, denticulate sepals; capsules 10–13 mm. long, shallowly trilobate and 3-ridged on back, 12–15 mm. in diameter; cocci rounded and slightly keeled on outer surface, with two flat faces.

Slopes and along streams, Lower Sonoran Zone to Tropical Zone, southern Sonora to San Luis Potosí and south into Tepíc. Aug.–Nov.

## 17. STILLINGIA L., Mant. 19, 1767

Herbs or shrubs with alternate or sometimes opposite, entire or toothed leaves and glabrous stems and foliage, leaves often bearing 2 glands at base. Flowers monoecious, apetalous, in terminal or axillary spikes, subtended by biglandular bractlets. Staminate flowers solitary in axils of bractlets; calyx 2-lobed; stamens 2, exserted. Pistillate flowers with a 3-parted calyx or calyx none, solitary in axils of bracts. Ovary 3-celled, 3-ovuled; styles 3, filiform, entire. Capsule 2–3-lobed, separating into as many bivalved carpels, separating from 3-lobed gynophore. Seeds subglobose to ovoid, carunculate or ecarunculate.

Leaves ovate, strongly 3-nerved, spinulose-serrate; spikes axillary, shorter than leaves; usually annual ........................................................ 1. *S. spinulosa*
Leaves linear, 1-nerved, entire or with few teeth near base; spikes axillary, terminal, equaling or surpassing leaves; perennial:
Spikes dense; leaves densely crowded, 2–3-toothed; capsule 3–4.5 mm. high, not glaucous ........................................................ 2. *S. paucidentata*
Spikes lax; leaves not crowded, entire; capsule 2.5–3 mm. high, glaucous ........................................................ 3. *S. linearifolia*

### 1. Stillingia spinulosa Torr., in Emory, Notes Mil. Rec. 152, 1848

*Sapium ?annuum* Torr., Bot. Mex. Bound. 201, 1859.
*Stillingia ?annua* Muell. Arg., in A. DC., Prod. **15**[2]: 1160, 1866.

Tufted, leafy, winter annual or short-lived perennial 5–40 cm. tall, bearing several to many ascending branches from base; leaves ovate to elliptic, alternate at apex, decurrent on petiole at base, 5–15 mm. wide, 1.5–5 cm. long, conspicuously 3-veined, spinulosely toothed along margin, teeth 0.5–1 mm. long, 8–15 on each side; spikes bracteate, to 2 cm. long, usually markedly shorter than leaves; flowers mostly solitary in axils of bracts; calyx of male flowers 1.1–1.3 mm. long, shallowly and irregularly lobed, subtending bract broadly ovate, acuminate, about 1 mm. long, 2 stalked glands borne at base, one on each side, half as long as, to equaling bract; stamens 2, 2–2.5 mm. long; pistillate calyx none; capsule 3.5–5 mm. high, nearly or fully as broad, smooth, semilustrous; seeds 3–3.5 mm. high, about 2 mm. in diameter, grayish or somewhat marbled with brown, minutely lined with very fine papillae set in longitudinal, slightly wavy rows.

In sandy desert soils, Lower Sonoran Zone, southern Nevada and Death Valley region to central Imperial County and in southwestern Arizona. Feb.–Oct.

### 2. Stillingia paucidentata S. Wats., Proc. Amer. Acad. 14: 298, 1879

*Stillingia linearifolia* var. *paucidentata* Jepson, Man. Fl. Pl. Calif. 598, 1925.

Erect or ascendingly much-branched plant 2–4 dm. tall from a stout perennial root; herbage green but not glaucous; leaves numerous, usually crowded toward tips of branches, linear-attenuate, 1.5–5 mm. wide, 3–8 cm. long, with a few scattered, spine-tipped teeth 0.5–2 mm. long, mostly below middle; spikes many, equaling or surpassing leaves and 4–7 mm. in diameter when in full flower; staminate flowers crowded, their calyces 1.4–1.6 mm. long, bilobed, subtended by an ovate-acuminate bract about 1 mm. long; glands subsessile, 1–1.5 mm. in diameter; stamens 2, 3–3.4 mm. long; pistillate flowers 3–9, crowded near base of spike, without calyces, bracts, and glands as in staminate flowers; capsules 3–4.5 mm. high, slightly broader, green or purplish; column of the gynophore persistent; seeds obovoid, 2.5–3 mm. long, 1.5–2 mm. in diameter, grayish or tan, dull, ecarunculate.

Sandy and gravelly desert slopes, Lower Sonoran Zone, southern Inyo County to Riverside and Los Angeles Counties, California, and barely entering Arizona near the mouth of the Bill Williams River. March–June.

**3. Stillingia linearifolia** S. Wats., Proc. Amer. Acad. **14:** 297, 1879

*Stillingia gymnogyna* Pax & K. Hoffm., in Pax, Pflanzenreich $4^{147.\mathrm{v}}$: 196, 1912.

Generously branched subshrub 1–5 dm. tall; branches several to many, ascending or spreading-ascending, slender; leaves linear, entire, 1–4 cm. long, subsessile, usually with 2 minute stipitate glands on basal margins and sometimes with a few similar glands on margins above base; spikes 2.5–7 cm. long, lax, 2–2.5 mm. in diameter in staminate portion; staminate flowers numerous, only one in axil of each minute triangular bract; stamens 2, divergent, 2–3 mm. long; glands on stoutish stalks, about equaling bracts; calyx 2-lobed, about 0.5 mm. long; pistillate flowers 2–7 at base of each spike, subtending bracts and flanking glands about 1 mm. long; capsule slightly depressed-subglobose, 2.5–3 mm. high, 3.5–4 mm. in diameter, faintly reticulate-veined, glaucous; column of gynophore persistent; seeds about 2 mm. high, obovoid, gray to brownish, smooth but dull, mainly ecarunculate.

Washes and sandy flats and gentle slopes, Lower Sonoran Zone, San Bernardino County, California, into western Arizona and south into northern Sonora and central Baja California. March–May.

Some specimens approach *S. paucidentata* in possessing a few minute glandular teeth along the margins of the leaves but that species has thicker denser spikes and nonglaucous herbage.

## 18. SAPIUM Jacq., Stirp. Amer. 249, 1763

Shrubs or small trees with milky juice and alternate, petiolate, coriaceous, usually glandular-dentate leaves. Foliage and stems glabrous. Stipules triangular, oblique, small. Flowers monoecious, small, apetalous, staminate borne (in ours) in terminal spikes, often several flowers in axil of each bract, pistillate flowers near base of spikes. Staminate calyces 2-lobed; stamens 2, filaments briefly united at base. Ovary 2–3-celled, each cell 1-ovuled. Capsule globose or pyriform, each locule 2-valved or irregularly dehiscent. Seeds ecarunculate.

**1. Sapium biloculare** (S. Wats.) Pax, Pflanzenreich $4^{147.\mathrm{v}}$: 221, 1912

*Sebastiania ?bilocularis* S. Wats., Proc. Amer. Acad. **20:** 374, 1885.
*Sapium biloculare* var. *amplum* I. M. Johnston, Proc. Calif. Acad. IV, **12:** 1077, 1924.

A shrub or small tree to 6 m. tall with glabrous leaves and smooth, tan or grayish bark; stipules about 1 mm. long; petioles 1–6 mm. long; leaf blades lance-linear or narrowly lanceolate, 5–15 mm. wide, 3–7 cm. long, shallowly glandular-serrulate, acute to obtuse at apex, obtuse to cuneate at base; spikes 1–4 cm. long; bracts broadly ovate, staminate ones subtending 3–7 flowers; stamens 2–3; ovary 2-celled; capsule about 1 cm. long, broadly ovate, acute at apex, smooth; seeds ecarunculate, 6–7 mm. long and nearly as wide, dark brown.

Along washes and on stony hillsides or occasionally on level plains, Lower Sonoran Zone, from vicinity of Gila Bend, Maricopa County, Arizona, south through the desert areas of Sonora and Baja California. March–Nov.

The shrubs or small trees are usually widely scattered, individuals often separated by a number of miles.

**Sapium biloculare** var. **amplum** I. M. Johnston was based on a specimen from Loreto having slightly broader, larger leaves and somewhat longer spikes than many of the plants in Sonora. When a large suite of specimens from both sides of the Gulf is examined, segregation of the Baja California material as a distinct variety appears unwarranted.

## 19. DALEMBERTIA Baillon, Étud. Gén. Euph. 545, *pl. 5, figs. 11–15*, 1858

Shrubs with glabrous or pilose twigs and herbage, hairs simple. Leaves ample, alternate, entire, dentate or deeply lobed, 3–7-nerved, caducously stipulate. Inflorescence spicate, pedunculate, bisexual; flowers apetalous, staminate numerous, bracteate, short-pedicellate; pistillate flowers few, at base of staminate portion of spike or less commonly solitary, each subtended by a solitary bract. Staminate flower consisting of a single stamen, partially enwrapped by a cochleate solitary sepal. Pistillate flowers with 3 imbricate, basally biglandular sepals. Ovary 3-celled; each locule 1-ovuled; style branches 3, each branch entire, recurved; capsule 3-lobed, cocci vertically dehiscent along inner face. Seeds ecarunculate, globose, with a crustose testa.

### 1. Dalembertia hahniana Baillon, Adansonia **11**: 125, 1873

Stems glabrous, slender, older bark gray; stipules lance-subulate to oblong, 1.5–2 mm. long, caducous; petioles 2–4.5 cm. long, glabrous or nearly so; leaf blades broadly ovate, obscurely to distinctly 3-lobed, 2.5–10 cm. wide and long, lateral lobes short and near base, acute or obtuse, central lobe much larger, acute and usually short-acuminate, margins of all lobes entire or in some leaves sinuately, shallowly, again lobed, bright green to reddish, glabrous and slightly impressed-veined above, paler and sparsely hispid along veins beneath; peduncles 1.5–4 cm. long, erect or ascending, slightly clavate-swollen toward apex; capsules depressed-globose, 12–15 mm. in diameter, glabrous but slightly rugulose; seeds globose, 4–4.5 mm. in diameter, mottled dark brown and black, punctate, with golden brown, densely crowded, short hairs in bottoms of pits.

Identity of the single specimen from Sonora is uncertain, for no authentic material of *D. hahniana* Baillon from the type locality has been available. If the tentative determination is correct, it extends the range from Xochilcaco, Morelos, into southern Sonora. Quiracoba, Sonora, *H. S. Gentry 2956.*

## 20. EUPHORBIA L., Gen. Pl. ed. 5, 208, 1754

Herbs or shrubs (or trees in other areas) with milky juice. Leaves alternate, opposite, whorled, or absent, simple and entire to variously toothed or lobed, petiolate or sessile. Stipules present or absent. Flowers monoecious in ours, several to numerous staminate and one centrally located pistillate flower surrounded by an involucre to form a cyathium. Involucre gamophyllous, usually bearing 5 small lobes alternating with 5 naked or appendaged glands, or one or more glands lacking. Staminate flowers consisting of a solitary stamen borne on a pedicel subtended by a bracteole or bracteole rarely wanting. Pistillate flower borne on a central pedicel, soon exserted from involucre, often reflexed, sepals and petals wanting. Ovary 3-celled. Style entire, 2-branched or more commonly 3-branched. Capsule 3-celled, often angled, dehiscent (in ours). Seeds carunculate, or in ours, usually without a caruncle, smooth or variously ridged, pitted, or wrinkled.

Seeds carunculate; glands of involucres without petaloid appendages:
  Stipules present, glandlike; glands deeply cupped or concealed by inflexed lacerate lobes of margin; styles entire .................................. Subgenus I. POINSETTIA
  Stipules none; glands convex to flat but never concealed by overarching flaps of margin; styles bifid .................................................. Subgenus II. ESULA
Seeds ecarunculate; glands of involucres with petaloid or fimbriate appendages (if lacking, leaves opposite and stipules present):
  Leaves symmetrical at base; glands 5, all alike; stipules glandlike or none ............. ................................................. Subgenus III. AGALOMA
  Leaves asymmetrical at base; glands 4, fifth represented by a minute lobe, a notch in side of involucre, or wholly missing; stipules present, usually subulate but not glandular ................................................ Subgenus IV. CHAMAESYCE

I. POINSETTIA

Capsules higher than broad, obscurely lobed, nearly terete in cross section; caruncle obvious, short-stipitate; seeds flattened from front to back, wider than thick; involucral glands concealed by overarching lobes .................................... 1. *E. eriantha*
Capsules broader than high, obviously to strongly 3-lobed; caruncle minute and sessile or lacking; seeds not flattened from front to back, about as thick as broad; involucral glands cupped, not concealed by lobes:
  Leaves all opposite or essentially so; seeds 2.4–2.8 mm. long; involucral gland or glands stipitate, equaling or surpassing lobes; floral leaves green or only slightly whitish or reddish at base:
    Leaves ovate to oblanceolate, distinctly dentate; seeds not strongly angled, usually ovoid and rounded ...................................... 2. *E. dentata* var. *dentata*
    Leaves lance-linear to narrowly linear, less deeply toothed or entire; seeds obtusely to strongly 4-angled ............................. 2*a*. *E. dentata* var. *cuphosperma*
  Leaves alternate between the basal pair and the floral whorls; seeds 3–3.5 mm. long; involucral gland sessile or essentially so; floral leaves usually pink, red, or purplish near base:
    Leaves ovate, obovate, to broadly oblanceolate, some often more or less lyrately lobed:
      Capsules glabrous or nearly so; tubercles on seeds not white-farinose on tips ........ .................................... 3. *E. heterophylla* var. *heterophylla*
      Capsules densely tomentose in youth, sparsely so later; tubercles on seeds white-farinose on tips ............................. 3*b*. *E. heterophylla* var. *eriocarpa*
    Leaves narrowly lanceolate to linear, none of them lobed ......................... .................................... 3*a*. *E. heterophylla* var. *graminifolia*

## 1. **Euphorbia eriantha** Benth., Bot. Sulph. 51, 1844

*Euphorbia exclusa* S. Wats., Proc. Amer. Acad. **18:** 150, 1883.
*Poinsettia eriantha* Rose & Standl., Contr. U.S. Nat. Herb. **16:** 13, 1912.

Erect annual or biennial 1–6 dm. tall with a solitary stem at base but ascendingly branched from second or third node upward, branches glabrous, faintly striate; lower leaves alternate, linear, entire, 0.5–2.5 mm. wide, 2–7.5 cm. long, sparsely pubescent along midrib beneath, eventually glabrate, margins often narrowly revolute; petioles 1–4 mm. long; uppermost leaves forming a whorl subtending inflorescences; involucres in glomerules of 1–5, turbinate, 1.5–2 mm. in diameter, about as long, densely cinereous-strigose; glands submarginal, 3–5, circular to radially oval, 1–1.4 mm. wide, concave, margin of involucre bearing 5–7 incurved, densely strigose flaps that overarch and obscure glands; reduced glands none; staminate flowers 23–36 per involucre; capsules oblong, 4–5.5 mm. long, 3–4 mm. in diameter, shallowly and roundly 3-lobed, densely strigose with very fine, ascending, white hairs; seeds 3.5–4.2 mm. long, 2–2.4 mm.

wide, about 1 mm. thick, broadly rounded on back, truncate at apex, densely puberulent with minute, grayish or whitish hairs; caruncle circular, excavate, buff or tawny, about 1 mm. in diameter, notched on inner angle of seed.

Gravelly washes and rocky hillsides, Lower Sonoran Zone, Colorado Desert, California, to Maricopa and Graham Counties, Arizona, southward throughout Baja California and at least to Bahía Kino, Sonora. Feb.–Nov.

**2. Euphorbia dentata** Michx., Fl. Bor. Amer. **2:** 211, 1803, var. **dentata**

*Euphorbia dentata* β *rigida* Engelm., in Torr., Bot. Mex. Bound. 190, 1859.
*Poinsettia dentata* Kl. & Garcke, in Kl., Monatsb. Preuss. Akad. Berlin **1859:** 253, 1859. [Preprint.]
*Poinsettia schiedeana* Kl. & Garcke, in Kl., loc. cit.
*Euphorbia dentata* var. *gracillima* Millsp., Pittonia **2:** 90, 1890.

Erect to ascendingly branched, dull green annual 1–6 dm. tall with finely striate stems and moderately pubescent herbage; leaves mostly opposite; petioles slender, 0.5–2 cm. long; leaf blades ovate, obovate, to oblanceolate or lanceolate, 0.5–4.5 cm. wide, 1.5–10 cm. long, dull green above, paler and somewhat glaucous beneath, scaberulous, margins irregularly dentate, teeth acute to rounded; floral leaves whorled, similar to cauline or some of them subsessile and elliptic, sometimes white or pale pink near base; involucres congested at tips of branchlets, campanulate, about 2 mm. long, green, glabrous, lobes 0.4–0.8 mm. long, laciniately 3–6-toothed; involucral glands 1–4, short-stipitate, circular or elliptic, about equaling lobes, 0.2–0.5 mm. in diameter; capsules depressed-globose, 3–4 mm. high, 4–5.5 mm. in diameter, distinctly 3-lobed, glabrous; seeds ovoid, 2.4–2.8 mm. long, scarcely or only obtusely angled, from minutely to moderately tuberculate, black or dark brown, caruncle sessile and inconspicuous or none, if present not in a shallow excavation in most of our specimens.

Arid slopes and washes, grassy plateaus and prairies, Lower Sonoran, Upper Sonoran, and Canadian Zones, Utah to the Atlantic seaboard and south into Mexico. Aug.–Oct.

**2*a*. Euphorbia dentata** var. **cuphosperma** Engelm., in Torr., Bot. Mex. Bound. 190, 1859

*Euphorbia cuphosperma* Boiss., in A. DC., Prod. **15**$^{2}$**:** 73, 1862.
*Poinsettia cuphosperma* Small, Fl. SE. U.S. 721, 1903.
*Euphorbia dentata* forma *cuphosperma* Fernald, Rhodora **50:** 148, 1948.

Leaves narrowly lanceolate to linear, faintly dentate to entire; capsules frequently sparsely pubescent; seeds mostly obtusely 4-angled, pale brown, tuberculate.

Range and flowering habit approximately the same as in *E. dentata* var. *dentata.*

**3. Euphorbia heterophylla** L., Sp. Pl. 453, 1753, var. **heterophylla**

*Poinsettia heterophylla* Kl. & Garcke, in Kl., Monatsb. Preuss. Akad. Berlin **1859:** 253, 1859. [Preprint.]

Erect annual 3–7 dm. tall, beginning to branch ascendingly near base; stems light green, faintly striate, glabrous or essentially so; leaves opposite below, alternate from above first to third nodes to floral whorls; petioles slender, 0.5–3 cm. long, often gradually widening to cuneate base of leaf blade, latter ovate, obovate, elliptic, or lanceolate in outline, 1–6 cm. wide, to 10 cm. long, bright green above, pale to slightly glaucous beneath, margin entire or shallowly dentate and often lyrately lobed, lobes acutely to obtusely deltoid; floral leaves similar to cauline ones but usually some or all of them white, pink, or scarlet near base; involucres clustered at tips of branches, 1.5–2 mm. high, narrowly campanulate, green and glabrous, bearing a single unappendaged, discoid, sessile gland and deeply fimbriate-dentate lobes which slightly surpass gland; capsules depressed-globose, strongly 3-lobed, 3.5–4.5 mm. high, 5–6 mm. in diameter, glabrous; seeds obovoid-quadrate, 3–3.5 mm. long, dark brown, mottled with lighter brown or buff, truncate and slightly concave-excavate at distal end, tuberculate, caruncle sessile and flattened or lacking, set in a shallow depression at narrow end of the seed.

Margins of streams, washes, and hillsides, Lower Sonoran Zone to Transition, Carolinian, and Tropical Zones, Nebraska to the Atlantic seaboard south through Mexico and Central America into South America; occurring in Arizona, Sonora, and Baja California. Aug.–Oct.

**3*a*. Euphorbia heterophylla** var. **graminifolia** (Michx.) Engelm., in Torr., Bot. Mex. Bound. 190, 1859

*Euphorbia graminifolia* Michx., Fl. Bor. Amer. **2:** 210, 1803.

Leaves linear to narrowly lanceolate, entire, usually bearing curved subappressed hairs along the margins, otherwise as in var. *heterophylla.*

Range and flowering period about as in var. *heterophylla.*

**3*b*. Euphorbia heterophylla** var. **eriocarpa** Millsp., Proc. Calif. Acad. II, **2:** 230, 1889

Stem almost or totally devoid of leaves between basal pair and floral whorl; leaf blades elliptical to obovate, entire or slightly lobed, sparsely pubescent with subappressed scattered hairs; peduncles, involucres, and capsules densely pubescent in youth, sparsely so at maturity; seeds tuberculate; tubercles white-farinose on their tips.

Rocky hillsides, canyon bottoms, and arid plateaus, Lower Sonoran Zone, vicinity of Comondú to the Cape Region, Baja California. Sept.–Dec.

II. ESULA

Plants annual; seeds 1–2 mm. long; stamens 5–15 in each involucre:
  Capsules narrowly 2-winged along each angle, smooth above; glands each bearing 2 curved horns; faces of seed 2–4-pitted .......... 4. *E. peplus*
  Capsules wingless along angles, verrucose or tuberculate above; glands hornless; seeds faintly rugose-reticulate .......... 5. *E. spathulata*
Plants perennial; seeds 2.8–3 mm. long; stamens 16–23 in each involucre:
  Rays of inflorescence usually 3; involucres 2.5–3.5 mm. high; cauline leaves usually broadest at or above middle:
    Herbage and involucres glabrous .......... 6. *E. incisa* var. *incisa*
    Herbage and involucres puberulent .......... 6*a*. *E. incisa* var. *mollis*
  Rays of inflorescence usually 5; involucres 2–2.2 mm. high; cauline leaves usually broadest below middle .......... 7. *E. robusta*

**4. Euphorbia peplus** L., Sp. Pl. 456, 1753

*Tithymalus rotundifolius* Lam., Fl. Franc. **3:** 100, 1778.
*Tithymalus peplus* Hill, Hort. Kew. 172, 1768.
*Esula peplus* Haw., Syn. Pl. Succ. 158, 1812.

Erect or ascendingly branched, glabrous annual 1–4.5 dm. tall; stems frequently reddish; stipules none; cauline leaves alternate, short-petiolate to subsessile, orbicular to broadly ovate, 0.5–3.5 cm. long, entire, those of umbels similar but usually smaller;

floral leaves ovate-cordate, 0.4–2 cm. long, sessile, entire; involucres campanulate, about 1 mm. long, yellowish to pale green, glabrous; 4 glands crescent-shaped, with body 0.3–0.5 mm. wide and 2 spreading, slender, slightly curved horns, glabrous, the 5th gland a short, deltoid, hairy lobe without appendages; staminate flowers 10–15 in each involucre; capsules depressed-globose, 1.8–2 mm. long, 3-angled and each angle bearing 2 parallel, thin, narrow wings from base to apex; seeds oblong, obscurely 6-angled, 1–1.3 mm. long, lateral and dorsal faces 2–4-pitted, ventral faces shallowly grooved longitudinally, coats whitish.

Gardens and margins of fields, usually in shade of shrubs, Lower and Upper Sonoran Zones, Washington to southern California and at many sites eastward to the Atlantic seaboard; introduced from Europe. Established in Tucson, Arizona. Feb.–Oct.

**5. Euphorbia spathulata** Lam., Encycl. **2:** 428, 1788

*Euphorbia dictyosperma* Fisch. & Mey., Ind. Sem. Hort. Petrop. **2:** 37, 1836.
*Euphorbia arkansana* Engelm. & Gray, Bost. Jour. Nat. Hist. **5:** 261, 1845.
*Tithymalus dictyospermus* Heller, Muhlenbergia **1:** 56, 1904.

Erect glabrous annual 1–3.5 dm. tall, usually simple below and several-branched into an umbellate habit above but in robust plants sometimes many-branched from near base, ultimate rays mostly trichotomously arranged; cauline leaves obovate-spatulate, 0.8–3 cm. long, serrulate on distal one-half, subsessile to short-petiolate, cuneate toward base; upper cauline leaves sessile, proportionately broader than lower ones, to 2.5 cm. long; leaves of inflorescence ovate-elliptic, sessile, serrulate nearly to base or margins sometimes entire; involucres broadly campanulate, 0.8–1 mm. in diameter; glands transversely oblong, 0.5–0.7 mm. long, entire; sinus broad, ciliate; staminate flowers 5–8; capsule depressed-globose, slightly 3-lobed, glabrous, smooth below, verrucose or finely tuberculate toward apex, 2.5–2.8 mm. in diameter; seeds lenticular-ovoid to subglobose, about 1.5–2 mm. long, brown, faintly reticulate with rather coarse meshes.

Grassy hillsides, stream banks, and often in disturbed or cultivated areas, Lower Sonoran and (mainly) Upper Sonoran Zones, Washington to southern California and east to Iowa and Alabama. March–Aug.

**6. Euphorbia incisa** Engelm., Ives Rep. **4:** 27, 1860, var. **incisa**

*Euphorbia schizoloba* Engelm., in A. Gray, Proc. Amer. Acad. **5:** 173, 1861.
*Tithymalus schizolobus* Norton, in Tidestrom, Contr. U.S. Nat. Herb. **25:** 343, 1925.

Glabrous, erect to ascending perennial with several to many branches 1–3.5 dm. long from near base or apex of root crown, stems and lower leaves often purplish, slightly glaucous; cauline leaves alternate, 8–15 mm. long, ovate-elliptical to broadly oblanceolate, sessile to short-petiolate, acute to apiculate at apex, entire, usually rounded at base; floral leaves deltoid-ovate to cordate, sessile, 5–10 mm. long, broadly rounded and minutely apiculate at apex, sometimes puberulent at base; main rays of inflorescence 3–5, usually 3; involucres broadly campanulate, 2.5–3.5 mm. in diameter, 3–3.2 mm. long, glabrous; lobes narrowly deltoid, 1.5–2 mm. long, ciliolate; glands transversely oblong-crescentic, 2–2.5 mm. wide, brownish, hornless or with 2 lateral short horns, distal margin shallowly and irregularly toothed across full width; stamens 16–23 per involucre; capsules oblong-ovoid, roundly 3-lobed, glabrous, 3.5–4 mm. long; seeds ashy to white, 2.8–3 mm. long, irregularly and faintly ridged, coat microreticulate.

Arid hillsides, as undercover among shrubs, and rarely along washes, Lower and Upper Sonoran Zones, Inyo and San Bernardino Counties, California, to Nevada and Arizona; barely reaching the northern margins of the Colorado Desert in Riverside County. Feb.–Aug.

**6a. Euphorbia incisa** var. **mollis** (Norton) L. C. Wheeler, Jour. Wash. Acad. **30:** 473, 1940

*Euphorbia schizoloba* var. *mollis* Norton, Ann. Rep. Mo. Bot. Gard. **11:** 43, *pl. 43*, 1899. [Preprint]
*Euphorbia yaquiana* Tidestrom, Proc. Biol. Soc. Wash. **48:** 41, 1935.

Differs from var. *incisa* in being slightly more robust and in having finely puberulent herbage and involucres.

Mainly rocky hillsides, chiefly Upper Sonoran Zone, Yavapai, Graham, Gila, Cochise, and Pima Counties, Arizona, occasionally descending to about 3,000 feet under bushes. April–Sept.

**7. Euphorbia robusta** (Engelm.) Small, in Britt. & Brown, Ill. Fl. **2:** 381, 1897

*Euphorbia montana* β *robusta* Engelm., in Torr., Bot. Mex. Bound. 192, 1859.
*Tithymalus robustus* Small, in Rydb., Fl. Colo. 224, 1906.

Erect or ascending perennial with several to many stems from a heavy rootstock, the herbage glabrous and glaucous or sometimes finely soft-puberulent; branches 1–3 dm. tall, striate to slightly angular, scaly below, each surmounted by a 3–5-rayed, umbellate inflorescence; cauline leaves alternate, broadly ovate-cordate, 5–15 mm. long, nearly as broad, evident near base, sessile, obtuse to apiculate at apex; floral leaves orbicular-cordate, in whorls of 3 or opposite, 5–15 mm. long, usually acute or apiculate at apex, sessile; involucres campanulate, 2–2.2 mm. high, glabrous or sometimes puberulent; lobes broadly ovate to oblong, dentate to shallowly lacerate, pubescent within; glands 4, crescent-shaped, brownish, 1–1.5 mm. wide, with short lateral horns, distal margin occasionally irregularly crenate; capsules globose-ovoid, 3–3.5 mm. high, roundly 3-lobed; seeds obovoid-oblong, 2.8–3 mm. long, grayish to brown, minutely pitted.

Sandy slopes and plains, Lower and Upper Sonoran Zones, Montana to South Dakota and southward to Arkansas and Arizona; reported from Organ Pipe National Monument (Pima County), Arizona, by Kearney and Peebles. March–July.

**Euphorbia lurida** Engelm., Ives Rep. 26, 1860, similar to *E. robusta* but with cauline leaves oblanceolate to spatulate, occurs in the mountains from southern Utah to New Mexico, and Arizona and extends downward to an elevation of approximately 3,500 feet on the slopes of the Catalina Mountains in Pima County, Arizona. It does not form a significant element in the flora of the Sonoran Desert.

III. AGALOMA

Annual herbs; leaves thin, not at all leathery; seeds 1–2.2 mm. long; pits on seeds centrally punctate:
- Seeds 2–2.2 mm. long; appendages white, petaloid, entire, 1–1.4 mm. long; capsules 3–3.5 mm. high .......... 8. *E. colimae*
- Seeds 1–1.5 mm. long; appendages 3–7 slender setae 0.5–0.8 mm. long; capsules 1.5–2.5 mm. high:
  - Leaf blades lance-ovate to oblong; involucres greenish; appendages 5–7, greenish .......... 9. *E. sonorae*
  - Leaf blades reniform to suborbicular; involucres green suffused with purple; appendages 3–5, purplish .......... 10. *E. subreniforme*

Perennial shrubs; leaves thicker, more or less leathery; seeds 2.5–3.5 mm. long; seeds reticulate or pitted but pits not centrally punctate:
- Branches of shrubs coarse, not wandlike, tortuous; leaves alternate, mostly on short lateral spurs:
  - Herbage, involucres, and ovary puberulent; base of leaf blade attached to side of upper part of petiole, not continuous over adaxial side of petiole; petiole slightly canaliculate .......... 11. *E. misera*
  - Herbage, involucres, and ovary glabrous; base of leaf blade continuous over adaxial side of petiole, this not canaliculate adaxially:
    - Petioles very slender, 0.4 mm. in diameter or less, usually equaling or exceeding blade; leaf blade thin, dark green .......... 12. *E. californica* var. *californica*
    - Petioles stouter, usually 0.5–1 mm. in diameter, shorter than blade; leaf blade thick, strikingly leathery, pale yellowish green ...... 12*a*. *E. californica* var. *hindsiana*
- Branches slender, erect or ascending, wandlike; leaves opposite or verticellate, on main stems and branches; lateral spurs absent:
  - Glands distinctly plicate tangential to involucral rim, inner lobe shorter than outer; leaves usually obtuse to emarginate, broadest at distal end, midrib very prominent beneath .......... 13. *E. plicata*
  - Glands not plicate, usually cupped, outer and inner parts alike; leaves acute or rarely obtuse distally, midrib less conspicuous beneath:
    - Appendages 1–1.5 mm. long, shorter than tube of involucre; leaves 2.5–8 cm. long; seeds regularly rugose-pitted .......... 14. *E. colletioides*
    - Appendages 3–3.5 mm. long, equaling or exceeding involucral tube; leaves 1.5–3.5 cm. long; seeds rugose-reticulate .......... 15. *E. xanti*

## 8. **Euphorbia colimae** Rose, Contr. U.S. Nat. Herb. **1**: 356, 1895

*Eumecanthus colimae* Millsp., Field Mus. Bot. Ser. **2**: 414, 1916.

Erect or ascending, weak annual 1–6 dm. tall; stems green, glabrous or very sparsely puberulent with internodes to 10 cm. long; leaves alternate below upper branches, these opposite; stipules inconspicuous, setaceous, deciduous, 0.5 mm. long; petioles slender, 0.5–3 cm. long; leaf blades lanceolate to ovate or obovate, 0.5–3.5 cm. wide, 1.5–6 cm. long, entire, acute to rounded at base and apex, bright green above, paler beneath, moderately and minutely puberulent to slightly villous or glabrate; involucres solitary to few, broadly campanulate, about 1 mm. long, green, sparsely pubescent; glands 5, tangentially elliptical, slightly cupped, greenish; appendages ovate, white, 1–1.4 mm. long, petaloid, entire; involucral lobes about 0.5 mm. long, obtuse, pectinate- or fimbriate-margined, whitish; peduncles slender, 2–8 mm. long in fruit; capsules 3-lobed, subglobose, about 3 mm. high, 3–3.5 mm. wide, greenish, smooth and glabrous; seeds ashy to stramineous, ovoid-truncate, about 1.4 mm. wide, 2–2.2 mm. long, regularly reticulate with each broadly crateriform pit punctate in center, intervening ridges regularly set with small circular tubercles.

In shaded places, Lower Sonoran Zone to Subtropical Zone, central Sonora southward into Colima. July–Jan.

## 9. **Euphorbia sonorae** Rose, Contr. U.S. Nat. Herb. **1**: 356, 1895

Erect, slender, rather weak annual with striate, light green, nearly glabrous stems 1–8 dm. tall and ascending branchlets 0.5–3 dm. long; stems and branches glabrous or essentially so; basal 1–3 pairs of leaves opposite, middle and upper ones alternate; stipules setaceous, 1–1.4 mm. long, distinct; petioles slender, 1–4 cm. long, glabrous; leaf blades lance-ovate to oblong, entire, rounded to broadly acute at base, acute and tipped with a slender seta 1–2 mm. long at apex, sparsely villosulous to glabrate on one or both surfaces, ciliate-villous on margins when young, soon glabrate, bright green above, paler beneath; involucres solitary at tips of very slender branchlets, narrowly campanulate, 0.6–0.8 mm. long, green; glands 5, tangentially elliptical, 0.2–0.3 mm. wide, cupped, greenish; appendages of 5–7 slender setae 0.5–0.7 mm. long; capsules depressed-globose, 2–2.5 mm. high, 3–3.5 mm. in diameter, distinctly 3-lobed, sparsely villous; seeds obovoid, black, dull, 0.7–0.8 mm. in diameter, about 1.5 mm. long, regularly pitted-reticulate, bottom of each pit punctate and ridges set with several light gray, translucent tubercles.

Usually in shade of shrubs along river banks and washes, Lower Sonoran Zone, vicinity of Ímuris southward in Sonora, and presumably into Sinaloa. Sept.–Oct.

**10. Euphorbia subreniforme** S. Wats., Proc. Amer. Acad. **21:** 439, 1886

*Eumecanthus subreniforme* Millsp., Field Mus. Bot. Ser. **2:** 415, 1916.

Very slender, erect annual 1–3 dm. tall with striate, slightly angled, glabrous stems and nearly filiform branchlets, latter sparingly villous in youth; basal 1–3 pairs of leaves opposite, rest of cauline leaves alternate except those near tips of flowering branches, where they are again opposite; stipules minutely setaceous, fugaceous, about 0.3 mm. long; petioles filiform, 3–5 mm. long, sparsely villous but soon glabrate; leaf blades suborbicular to broadly reniform, 3–9 mm. long, 4–12 mm. wide, very thin and delicate, entire, sparingly villous with very delicate, white, crispate hairs; involucres solitary at tips of branches, campanulate, about 1 mm. long, sparsely villous, green suffused with red or purple; glands 4, short-stipitate, purplish, elliptical, long diameter tangential to involucral rim, about 0.3 mm. wide; appendages of 3–5 purple, spreading setae 0.6–0.8 mm. long; capsules depressed-globose, villous, 1.5 mm. long, 1.6–1.8 mm. wide, shallowly 3-lobed; seeds obovoid, light gray to nearly black, dull, about 0.6–0.8 mm. in diameter, 1–1.2 mm. long, regularly pitted-reticulate, each shallow pit punctate in the center and rimmed with pale tubercles.

Moist places in deep shade of trees or shrubs, Lower Sonoran and Upper Sonoran Zones, northern Sonora to southern Chihuahua and Guadalajara. Sept.–Nov.

**11. Euphorbia misera** Benth., Bot. Sulph. 51, 1844

*Trichosterigma misera* Kl. & Garcke, in Kl., Abh. Akad. Berlin Phys. **1859:** 42, 1860.
*Euphorbia benedicta* Greene, Pittonia **1:** 263, 1889.

A compact to straggly shrub about 0.5–1.5 m. tall with thick, usually tortuous branches and pale gray bark; young twigs reddish, finely but closely puberulent with short, simple, spreading hairs, becoming more or less scurfy and markedly nodose; leaves mostly on short spur branches, short-puberulent, petioles 2–5 mm. long, faintly to distinctly canaliculate along upper side; leaf blades orbicular to oblong or cuneate-obcordate, 5–15 mm. long, rounded to retuse at apex, cordate to cuneate at base, margin often forming a narrow short wing attached to petiole alongside the dorsal groove, never continuous across petiole; involucres usually solitary, open-campanulate, puberulent, 3–7 mm. wide or rarely to 12 mm. wide (including appendages); glands transversely oblong, maroon or rarely green, 0.8–2 mm. wide, bearing conspicuous white appendages to 2.5 mm. long, these crenulate to bluntly dentate; inner lobes dentate, laterally ciliolate; staminate flowers 30–50; capsules eventually glabrate, depressed-globose, 3-lobed, 3.5–5 mm. long; seeds white, 2.5–3 mm. long, irregularly faveolate, ecarunculate.

Hillsides and along washes, Lower Sonoran Zone and arid parts of Upper Sonoran Zone, Orange County, California, to central Baja California and on islands in the Gulf, in northwestern Sonora at least to Puerto Libertad; Guadalupe Island. Flowering more or less throughout the year.

**12. Euphorbia californica** Benth., Bot. Sulph. 49, *pl. 23, figs. B, B1–5,* 1844, **var. californica**

*Trichosterigma californicum* Kl. & Garcke, in Kl., Abh. Akad. Berlin Phys. **1859:** 42, 1860.
*Euphorbia comonduana* Millsp., Proc. Calif. Acad. II, **2:** 229, 1889.

Openly, often somewhat tortuous, branched shrub 0.5–2 m. tall with smooth to faintly scurfy, pale to reddish bark and wholly glabrous twigs, herbage, and involucres; branches usually slender; petioles very slender, mostly 0.3 mm. or less in diameter, 8–15 mm. long, terete in cross section; blades orbicular-obcordate to oblong-cuneate, 7–19 mm. long, usually fully as wide, rounded to cordate at apex, lower margin continuous across upper side of petiole, sometimes forming a narrow but distinct flap or ridge over petiole; involucres 1–5 on a spur branch, open-campanulate, 3–5 mm. wide, including appendages; glands 5, transversely oblong, 1–1.5 mm. wide, yellowish to maroon, appendages white to greenish, 1 mm. long or less, bluntly crenulate; inner involucral lobes dentate, not ciliolate marginally; staminate flowers 25–35; capsules 3-lobed, depressed-globose, 4–5 mm. wide, 3–4 mm. long, reddish, glabrous; seeds about 3 mm. long, white or pale buff, faveolate.

Sandy or gravelly soil along washes, hillsides, or desert flats, Lower Sonoran Zone, central Baja California and vicinity of Hermosillo, Sonora, south into Sinaloa, the tip of the Cape Region, Baja California, and on Clarion Island. Nov.–March.

**12*a*. Euphorbia californica** var. **hindsiana** (Benth.) Wiggins, Contr. Dudley Herb. **4:** 346, 1955

*Euphorbia hindsiana* Benth., Bot. Sulph. 51, *pl. 24,* 1844.
*Trichosterigma hindsianum* Kl. & Garcke, in Kl., Abh. Akad. Berlin Phys. **1859:** 42, 1860.

Similar in habit to var. *californica* but with distinctly thicker, pale yellow-green leaves, stouter and comparatively shorter petioles; involucral appendages in this variety are often longer than broad and the glands bright yellow.

On granitic, rocky ridges near Cabo San Lucas, Baja California.

**13. Euphorbia plicata** S. Wats., Proc. Amer. Acad. **21:** 438, 1886

*Aklema plicata* Millsp., Field Mus. Bot. Ser. **2:** 417, 1916.

Erect shrub to 2.5 m. tall with slender, slightly striate, glabrous branches naked except at and near their tips, internodes to 3 dm. long; leaves in threes; petioles 5–12 mm. long, finely and sparsely puberulent; leaf blades oblong-lanceolate to linear-oblong, 4–10 mm. wide, 2–8 cm. long, entire, cuneate at base, obtuse to emarginate or less commonly acute at apex, dark green and glabrous above, pale and sparsely hirsutulous and midrib prominent beneath; involucres in clusters at tips of branches, densely puberulent, campanulate, 3–4 mm. long and wide; involucral lobes obovate, about 0.5 mm. long, lacerate at tips; glands 5, elliptical and plicate tangential to involucral rim, 0.8–1 mm. wide, 1.2–1.6 mm. long, inner fold one-half as long as outer or less; appendages nearly orbicular in outline, 1.2–1.5 mm. wide, white, margins minutely crenulate; capsules subglobose, slightly wider at base than above, 5–6 mm. in diameter, glabrous; gynophores straight, elongating to 1 cm. long in fruit; seeds irregularly tuberculate.

Along washes, Lower Sonoran Zone, central Sonora, southwestern Chihuahua, and south into Jalisco. Feb.–March (in Sonora).

**14. Euphorbia colletioides** Benth., Bot. Sulph. 163, 1844

*Euphorbia padifolia* Brandegee, Univ. Calif. Pub. Bot. **6**: 54, 1914.
*Aklema colletioides* Millsp., Field Mus. Bot. Ser. **2**: 416, 1916.

Erect, ascendingly branched shrub to 2.5 m. tall with striate glabrous stems and opposite to verticellate, elliptic, sparsely pubescent leaves, nodes often puberulent in youth; stipules reduced to elliptical glands 1–1.5 mm. broad; petioles slender, 0.5–1.5 cm. long, sparsely pubescent to glabrate except along lateral margins, these densely puberulent-tomentulose; leaf blades elliptic, 0.8–3 cm. wide, 2.5–8 cm. long, obtuse to acute at each or either end, usually glabrate above, sparingly puberulent to glabrate and pale beneath; involucres few to several in axillary and terminal cymes, campanulate, 1.5–2 mm. wide, 2–3 mm. long, finely and closely puberulent; glands 5, tangentially elliptic, deeply cupped; appendages white to roseate, 1–1.5 mm. long, 2–2.2 mm. wide, entire to faintly erose-denticulate; capsules depressed-globose, deeply 3-lobed, about 4 mm. high, 5–6 mm. wide, glabrous, style branches barely 1 mm. long; seeds ovoid to ellipsoid-truncate, about 2 mm. in diameter, 2.5–3 mm. long, distinctly and regularly rugose-pitted, whitish to pale gray, ecarunculate.

Rocky hillsides, canyons, and desert washes, Lower Sonoran Zone, northern Sonora to Oaxaca. Aug.–Feb.

**15. Euphorbia xanti** Engelm. ex Boiss., in A. DC., Prod. **15²**: 62, 1862

*Euphorbia gymnoclada* Engelm., in A. Gray, Proc. Amer. Acad. **5**: 171, 1861. Not *E. gymnoclada* Boiss., 1860.
*Aklema xanti* Millsp., Field Mus. Bot. Ser. **2**: 417, 1916.
*Euphorbia corallifera* M. E. Jones, Contr. West. Bot. **18**: 50, 1933.

Erect, sparingly to intricately branched shrub 0.5–2 m. tall; branchlets smooth and glabrous, internodes 2–8 cm. long; leaves opposite or in whorls of 3–6; stipules reduced to inconspicuous glands; petioles slender, 4–9 mm. long, glabrate; leaf blades linear to elliptic, ovate, or obovate, acute at base, acute to emarginate at apex, 1–24 mm. wide, 1.5–3.5 cm. long, sparsely and minutely puberulent with subappressed hairs to glabrate; involucres few, in terminal cymes, broadly campanulate, 3–3.5 mm. wide (exclusive of appendages), about 3 mm. long, minutely and sparsely puberulent but soon glabrous; glands 5, tangentially elliptical, deeply cupped, about 1 mm. wide; appendages orbicular to rounded-obovate, 3–3.5 mm. long, slightly wider, white or rose, margin slightly erose-denticulate; capsules depressed-globose, deeply 3-lobed, 3–4 mm. long, 4–5.5 mm. wide, smooth and glabrous, stramineous to greenish; style 1.5–2 mm. long; seeds ovoid, about 2 mm. wide, 3 mm. long, acute at micropylar end, rounded at other, regularly reticulate-rugose, ashy.

Washes, mesas, and rocky hillsides, Lower Sonoran Zone, from vicinity of Playa Santa Catarina, Baja California, south to the Cape Region. Dec.–April.

IV. CHAMAESYCE

◆Leaves, or at least the larger ones, serrulate or crenulate, smaller ones often entire:
■Perennials (*E. thymifolia* flowers first year but is short-lived perennial); involucres often rather crowded on short branches:
Shrubs or subshrubs; capsules glabrous:
Stipules united, a woolly cushion in the axils of each; capsules about 2 mm. high; leaf blades 2.5–7 mm. wide, oblong-obovate ................ 16. *E. magdalenae*

Stipules distinct, glandular marginally; capsules 2–2.8 mm. high; leaf blades 5–13 mm. wide, broadly oval to suborbicular .................. 17. *E. tomentulosa*

Herbs; capsules more or less puberulent to villous:

Plants ascending or erect; leaf blades ovate to linear-lanceolate, 4–25 mm. long; stipules lacerate into several segments ............................ 18. *E. capitellata*

Plants prostrate or decumbent; leaf blades ovate to elliptical, oblong, or suborbicular, 3–16 mm. long; stipules subulate, entire or essentially so:

Stems somewhat compressed dorsiventrally; herbage not glandular and coated with sand grains; involucres not over 1 mm. long; leaf blades elliptical to oblong, not closely crowded .................................... 19. *E. thymifolia*

Stems terete; herbage glandular-glutinous and covered with sand grains; involucres 1.2–1.6 mm. long; leaf blades broadly ovate to suborbicular, closely crowded ....................................................... 20. *E. leucophylla*

■Annuals; involucres usually not crowded:

Herbage and involucres glabrous (pilose within in some); capsules glabrous:

Leaves linear-oblong to lance-oblong, often falcate; seeds 1–1.4 mm. long, faces faintly rugulose or punctate:

Plants prostrate; leaf blades 3–14 mm. long, often falcate, serrulate only toward apex ................................................ 21. *E. serpyllifolia*

Plants erect; leaf blades 5–35 mm. long, straight or falcate, serrulate along both margins ............................................ 22. *E. hyssopifolia*

Leaves linear to lanceolate-ovate, not falcate; seeds 1.6–2.3 mm. long, faces transversely ridged or grooved:

Capsules 2.2–2.5 mm. high; appendages wider than glands; leaf blades revolute in age, remotely serrulate on both margins; seeds 1.6–2 mm. long .. 23. *E. florida*

Capsules 3–3.4 mm. high; appendages as wide as or narrower than glands; leaf blades not revolute, closely serrulate along one margin, other entire; seeds 2–2.3 mm. long .................................... 24. *E. trachysperma*

Herbage and involucres puberulent to villous, at least in youth; mature capsules pubescent except in *E. abramsiana* and *E. serrula*:

Capsules glabrous at maturity; involucres often glabrous at anthesis:

Stems and leaves minutely puberulent with short hairs, often glabrate in age; appendages as wide as or slightly wider than glands; seeds 1–1.4 mm. long, transversely rugose on faces; stipules 0.5 mm. long ............ 25. *E. abramsiana*

Stems and leaves villous with longer spreading hairs; appendages as wide to three times as wide as glands; seeds 1.5–2 mm. long, faces smooth; stipules 1.5–2 mm. long ............................................. 26. *E. serrula*

Capsules puberulent to villous at maturity; involucres pubescent at anthesis and usually at maturity:

Leaf blades ovate to orbicular; stipules 0.2–0.6 mm. long; involucres broadly campanulate; plants of southern Baja California:

Pubescence of tangled arachnoid tomentum; leaf blades 2–6 mm. long; stems subfiliform; stipules obscured by tomentum ............ 27. *E. brandegeei*

Pubescence of straight or curved but not tangled hairs; leaf blades 5–14 mm. long; stems slender but not subfiliform; stipules not obscured by pubescence ....................................................... 28. *E. dentosa*

Leaf blades elliptical, oblong, lanceolate, or oblanceolate, only smallest ones ovate; involucres obconical or turbinate; plants continental:

▲Glands stipitate and cyathia in dense heads, or appendages of dissimilar pairs, and leaves often falcate:

Capsules 1–1.2 mm. long; leaves 8–30 mm. long; glands stipitate; appendages all alike; cyathia in dense heads; seed 0.7–0.9 mm. long .... 29. *E. hirta*

Capsules 1.4–1.6 mm. high; leaves 4–8 mm. long; glands sessile; appendages of dissimilar pairs, proximal twice as long as the distal ones; cyathia not densely crowded; seeds 1–1.2 mm. long ................ 30. *E. indivisa*

▲Glands sessile; cyathia solitary or few at a node; appendages all alike; leaves not markedly falcate:

Stipules on ventral side of stem often united; leaves not spotted with red; leaves elliptic, elliptic-oblong, or obovate:

Seeds 0.8–1 mm. long; faces of capsular lobes glabrate; seeds white, pinkish, or sordid .................................... 31. *E. chamaesyce*

Seeds 1.2–1.4 mm. long; faces of capsular lobes strigose; seeds mottled white and brown .................................... 32. *E. stictospora*

Stipules all distinct; leaves often spotted with red on upper surface, elliptic-ovate ............................................ 33. *E. maculata*

◆Leaves all entire:

Appendages none on involucral glands:

Stems glabrous but glutinous, with adhering sand grains; capsules 3–4 mm. high:

Annual; seeds 2.4–3 mm. long; peduncles to 5 mm. long; glands 0.8–1 mm. wide, slightly elliptical ...................................... 34. *E. platysperma*

Perennial; seeds 1.6–2.2 mm. long; peduncles 1–2.5 mm. long; glands 0.5 mm. in diameter, discoid ........................................... 35. *E. incerta*

Stems glabrous or puberulent but not glutinous; capsules 1–1.8 mm. high:

Leaves narrowly ovate to oblong or linear and revolute; plants annual:

Involucres crispate-hirsutulous in youth; leaves ovate to oblong, 1.5–7 mm. long; stems prostrate .................................... 36. *E. micromera*

Involucres glabrous; leaves linear and revolute, to 2.5 cm. long; stems erect or ascending ............................................ 37. *E. revoluta*

Leaves broadly ovate; plants perennial:

Seeds 1.3–1.5 mm. long; leaves 2–4 mm. long; involucres glabrous without, 1–1.2 mm. long; stipules 1 mm. long, not obscured by pubescence on stems ........ ................................................ 38. *E. parishii*

Seeds 0.7–0.9 mm. long; juvenile leaves 6–9 mm. long; involucres puberulent without, 0.4–0.6 mm. long; stipules about 0.2–0.3 mm. long, obscured by pubescence on stems and petioles .............................. 39. *E. petrina*

Appendages present on at least some of involucral glands on each cyathium:

●Leaves, stems, involucres, and capsules glabrous:

Leaves linear, revolute; stems erect or ascending, not prostrate:

Appendages shorter than diameter of gland, entire or faintly crenulate; leaves to 2.5 cm. long; glands usually discoid ....................... 37. *E. revoluta*

Appendages equaling or slightly exceeding diameter of gland; at least some of appendages bifid; leaves 5–12 mm. long; glands oval .......... 40. *E. gracillima*

Leaves broader; stems prostrate (except in var. of *E. fendleri*):

Stipules conspicuous, united into white, glabrous, membranous scale:

Capsules 1.7–2.3 mm. long; involucres campanulate, 1.5–2 mm. long; appendages 1–1.6 mm. wide; perennial .................. 41. *E. albomarginata*

Capsules 1–1.2 mm. long; involucres turbinate, 0.8–1 mm. long; appendages less than 0.8 mm. wide; annual ....................... 42. *E. serpens*

Stipules inconspicuous, distinct, or if connate, lacerate into narrow divisions:

Capsules and seeds less than 2 mm. long; stipules on ventral side of stems united:

Annual; glands discoid, less than 0.5 mm. in diameter; appendages minute, shorter than diameter of gland ...................... 36. *E. micromera*

Perennial; glands oblong, 0.5–0.7 mm. long; appendages equaling or slightly exceeding length of gland ................... 43. *E. polycarpa* and vars.

Capsules and seeds 2 mm. long or over; stipules distinct:

Glands elliptic; leaves ovate to ovate-orbicular .. 44. *E. fendleri* var. *fendleri*
Glands deltoid; leaves lanceolate .......... 44*a*. *E. fendleri* var. *chaetocalyx*
● Leaves, stems, involucres, and usually capsules pubescent:
Involucres urceolate to long-turbinate; strongly constricted above:
Appendages entire to crenate; pubescence of clavate hairs; perennial ........................................................ 45. *E. arizonica*
Appendages divided to base into several subulate segments; hairs not clavate; annual ................................................ 46. *E. setiloba*
Involucres campanulate, obconical, or turbinate, not constricted at apex:
Capsules glabrous at maturity:
Annual herb; prostrate; stipules subulate, longer than broad, lacerate ........................................................ 25. *E. abramsiana*
Perennial subshrub; stems ascending to decumbent; stipules connate into a narrow subscarious scale, wider than long, erosulate ...... 16. *E. magdalenae*
Capsules puberulent to strigose at maturity:
Seeds obviously to strongly ridged transversely, distinct grooves between ridges:
Pubescence of spreading hairs or arachnoid-tangled; seeds quadrangular in cross section, grooves broad and shallow; leaves 2–6 mm. long:
Lateral branches not congested; pubescence arachnoid, tangled hairs ....................................................... 27. *E. brandegeei*
Lateral branches short and congested; pubescence sparse, of simple spreading hairs ...................................... 47. *E. bartolomaei*
Pubescence of appressed fine hairs; seeds nearly terete in cross section, grooves deep and very narrow between transverse ridges:
Leaves ovate to narrowly oblong, to 20 mm. long ...................................................... 48. *E. pediculifera* var. *pediculifera*
Leaves linear, 15–37 mm. long ........ 48*a*. *E. pediculifera* var. *linearifolia*
Seeds weakly wrinkled or faces smooth, without distinct transverse grooves:
Herbage and capsules closely tomentose; appendages about twice as wide as glands (see also *E. polycarpa* var. *intermixta*) ...... 49. *E. melanadenia*
Herbage and capsules puberulent to glabrous; appendages usually equaling or only slightly wider than glands .............. 43. *E. polycarpa* and vars.

## 16. Euphorbia magdalenae Benth., Bot. Sulph. 50, 1844

*Euphorbia blepharostipula* Millsp., in Vasey & Rose, Contr. U.S. Nat. Herb. **1:** 77, 1890.
*Euphorbia watsonii* Millsp., Zoe **1:** 347, 1891.
*Chamaesyce magdalenae* Millsp., Field Mus. Bot. Ser. **2:** 410, 1916.
*Chamaesyce watsoni* Millsp., op. cit. 412.
*Euphorbia espirituensis* M. E. Jones, Contr. West. Bot. **18:** 50, 1933.

Low, intricately branched subshrub 2–10 dm. tall or subprostrate with slender spreading branches 1–6 dm. long; young growth and leaves sparsely and minutely puberulent to glabrate, young stems reddish; stipules united into a narrow subscarious scale 0.3–0.5 mm. high, to 1.5 mm. wide, entire or margin ciliate-erosulate or obscurely and remotely toothed, deciduous, a cushion of fine white hairs usually in their axils; petioles about 1 mm. long or less; leaf blades oblong-obovate, 2.5–7 mm. wide, 5–13 mm. long, rounded to faintly emarginate at apex, slightly inequilateral at base, margin entire or faintly and remotely serrulate, teeth inconspicuously gland-tipped; involucres scattered on short lateral branchlets or in axils of upper leaves, narrowly obconic, 0.6–0.9 mm. long, glabrous or essentially so; lobes narrowly deltoid, about equaling glands, these 0.2–0.4 mm. wide, orbicular or transversely oblong, brownish to nearly black, appendages white, from narrow, to often nearly twice as wide as glands, sometimes nearly obsolete; capsules about 2 mm. high, glabrous.

Gravelly washes and rocky hillsides, Lower Sonoran Zone, from the vicinity of Punta Prieta, Baja California, southward to the Cape Region. Feb.–March.

**Euphorbia anthonyi** Brandegee, Erythea **7:** 7, 1899 (*Euphorbia clarionensis* Brandegee, loc. cit.), is listed "Baja California; type from San Benito Island" in Standley's *Trees and Shrubs of Mexico* (Contr. U.S. Nat. Herb. **23:** 602, 1923). This is misleading, for the "San Benito" is a typographical error for "San Benedicto." San Benito Islands, off the west coast of Baja California, consist of three small rocky islands west of Sebastian Vizcaíno Bay, the most easterly one lying 14.5 miles west of the northern end of Cedros (Cerros) Island. The largest, West San Benito, is at Lat. 28° 18′ N., Long. 115° 35′ W. San Benedicto Island, however, is in the Revillagigedo Islands and located at Lat. 19° 20′ N., Long. 110° 48′ W. The species has not, to my knowledge, been reported from any point outside the Revillagigedo Islands of San Benedicto and Isla Clarion.

### 17. **Euphorbia tomentulosa** S. Wats., Proc. Amer. Acad. **22:** 476, 1887

*Chamaesyce tomentulosa* Millsp., Field Mus. Bot. Ser. **2:** 412, 1916.

A low, rounded to flat-topped shrub 3–7 dm. tall, often broader than tall, with repeatedly dichotomously branched stems; bark on young twigs reddish but more or less cinereous with minute, crowded, simple but crispate hairs, older ones pale gray to nearly chalky; stipules lance-subulate, 1.2–2 mm. long, distinct, entire to ciliate on margins, tomentulose, slightly glandular marginally; petioles 1–3 mm. long, cinereous-puberulent; leaf blades oval to suborbicular, asymmetrical at base, acute to rounded at apex, serrulate, 5–13 mm. wide, 5–17 mm. long, minutely and sparsely to densely cinereous-puberulent; inflorescences crowded at tips of branches; involucres obconical, 1.7–2 mm. long, puberulent, subulate lobes 0.4–0.6 mm. long; glands transversely oval, 0.3–0.5 mm. wide, reddish brown to sordidly greenish brown; appendages reniform to suborbicular, white to pinkish or purplish, 1–2.2 mm. wide, entire or rarely notched nearly or fully to base; capsules depressed-globose, 3-lobed, 2–2.8 mm. high, 2–3.2 mm. wide, glabrous; seeds pale stramineous, finely reticulate-rugulose on faces, 0.3–0.5 mm. wide, 0.9–1.2 mm. long.

Sandy washes, mesas, and rocky hillsides, Lower Sonoran Zone, San Felipe Desert and Rosario, Baja California, southward to the Cape Region, and from the Altar district southward in Sonora. Sept.–March.

### 18. **Euphorbia capitellata** Engelm., in Torr., Bot. Mex. Bound. 188, 1859

(For synonymy see Rhodora **43:** 174, 1941.)

Perennial with ascending to erect, slender stems 3–40 cm. long, 0.5–1.5 mm. in diameter, from a slender woody taproot; herbage puberulent to glabrous; leaves ovate to linear-lanceolate, 4–25 mm. long, puberulent to glabrous, gray-green, markedly asymmetrical basally, acute to obtuse at apex, usually serrate along lower margin, entire on upper, less commonly alike on both margins; stipules subulate-attenuate, 1.5–2 mm. long, divided into a few linear segments; cyathia rarely solitary, usually in cymose glomerules; peduncles 0.5–3 mm. long; involucres campanulate to obconic, 1.3–1.7 mm. in diameter, hairy on inside of narrowly triangular lobes, these more or less exceeding glands; glands orbicular to transversely oval, 0.2–0.5 mm. wide, stipitate; appendages entire, white or pinkish, glabrous; staminate flowers 28–41 per cyathium, sparsely pubescent androphores 1.6–1.9 mm. long; ovary shallowly 3-lobed, costae obtuse, usually puberulent; capsules 1.3–1.9 mm. long; seeds quadrangular in cross section, ovate vertically, 1.2–1.4 mm. long, 0.6–0.8 mm. wide, obtuse to truncate basally, back somewhat wrinkled or with small, irregular, transverse depressions, reddish brown to grayish.

Hillsides, washes, and arid plains, Lower Sonoran and lower parts of Upper Sonoran Zones, Arizona to Texas and southward to Coahuila, Sinaloa, and Baja California. March–Oct.

**19. Euphorbia thymifolia** L., Sp. Pl. 454, 1753

*Anisophyllum thymifolium* Haw., Syn. Pl. Succ. 160, 1812.
*Chamaesyce thymifolia* Millsp., Field Mus. Bot. Ser. **2:** 412, 1916.

Annual or probably short-lived herbaceous perennial with moderately compressed, pilose branches 1.3 dm. long, prostrate; stipules distinct, 1–1.5 mm. long, linear-subulate, sparsely to densely pilose; petioles pilose, 0.5–1 mm. long; leaf blades elliptical to oblong, 3–10 mm. long, asymmetrical at base, acute to rounded at apex, denticulate-serrate, sparsely villous to subglabrate beneath, nearly glabrous above; involucres 2 to several at nodes of lateral branches, obconical, 0.8–1 mm. long, densely puberulent, sinus deep, opened by growing ovary so all 4 glands appear on one side of capsule; glands reddish, subcircular to transversely oval, 0.2–0.3 mm. wide; appendages on glands nearest sinus asymmetrically elongated toward sinus, about twice as long as glands, glabrous, white to roseate, entire or slightly crenulate; those on distal glands symmetrical, from one and one-half times as wide as gland to rudimentary or lacking entirely; staminate flowers 3–5 per involucre; capsule broadly ovoid, 3-angled, 1.1–1.5 mm. long, nearly as broad at base, densely strigose; seeds about 0.5 mm. wide, 0.8–1 mm. long, ovoid-triangular, 4-angled, transversely 3–5-sulcate with shallow rounded sulci; seed coat pinkish, smooth or nearly so, dull.

Along streams, arroyos, hillsides, and in fields, Lower Sonoran Zone to Tropical Zone, southern Sonora to South America and tropics of Old World. Aug.–April.

**20. Euphorbia leucophylla** Benth., Bot. Sulph. 50, 1844

*Euphorbia velutina* Greene, Bull. Calif. Acad. **2**(5): 57, 1886.
*Euphorbia biserrata* Millsp., Zoe **1:** 347, 1891.
*Chamaesyce leucophylla* Millsp., Field Mus. Bot. Ser. **2:** 410, 1916.

Procumbent to prostrate perennial herb 2–6 dm. broad with stems sometimes 2 dm. tall but more commonly partially buried in drifted sand and ascending 1–5 cm. at tips; herbage minutely but usually densely puberulent, occasionally glabrous or glabrate; dorsal stipules lance-subulate, 0.5–0.8 mm. long, puberulent, ventral ones smaller and often difficult to distinguish; petioles 0.5–1.5 mm. long, densely puberulent to glabrate; leaf blades broadly ovate to broadly elliptic or suborbicular, 3–14 mm. wide, 3–16 mm. long, closely crowded and exceeding internodes, or remote and one-half to one-fourth as long as internodes, margins finely serrulate; involucres solitary in axils of upper leaves, broadly obconic, 1.2–1.6 mm. long, densely puberulent without; glands transversely elliptic, 0.5 mm. radially, 0.8–1.1 mm. tangentially, purplish or reddish; appendages white, about 1 mm. long, 2 mm. wide, usually erosulate-dentate; capsules puberulent, 2.5–3.2 mm. wide, 2–2.5 mm. high; seeds 1.5 mm. long, pale orange with varying amounts of white, testa sometimes obscuring darker underlying color.

Sandy inner beaches, Lower Sonoran Zone, vicinity of Magdalena Bay southward to Cabo San Lucas, Baja California, and Tiburón and other islands in the Gulf of California, and at Bahía Kino, Sonora. Probably throughout the year, and at least Aug.–May.

**21. Euphorbia serpyllifolia** Pers., Syn. Pl. 2: 14, 1806
(For extensive synonymy see Rhodora **43:** 229–231, 1941.)

Annual with prostrate to ascending branches 0.5–3.5 dm. long, herbage glabrous throughout; stipules distinct, linear, entire or somewhat lacerate; petioles subfiliform, 0.5–0.8 mm. long; leaf-blades linear, oblong, or narrowly obovate, cuneate at base, rounded and usually serrulate toward apex, 3–14 mm. long, larger ones often slightly falcate; involucres solitary at nodes, campanulate to obconical, 0.8–1.2 mm. long and wide, glabrous; glands transversely oblong tangential to involucral rim, 0.2–0.5 mm. long; appendages white, about 0.2 mm. long, entire to subdentate; stamens 5–18 per involucre; capsules slightly depressed-ovoid, 1.5–2 mm. long, broadest below, glabrous, distinctly 3-angled; seeds more or less quadrangular in cross section, 1–1.4 mm. long, ovate, with smooth to punctate or rugulose faces, coat whitish to brown.

Sandy flats and canyon floors, Lower and Upper Sonoran and Arid Transition Zones, British Columbia to Michigan and southward to Texas, Arizona, and northern Baja California. Mostly in the mountains above the Sonoran Desert but occasionally occurring as a straggler along its northern margins. March–Oct.

**22. Euphorbia hyssopifolia** L., Syst. Nat. ed. 10, **2:** 1048, 1759
(For synonymy see Rhodora **43:** 141, 1941.)

Erect to strongly ascending annual 5–60 cm. tall with mostly glabrous but sometimes sparsely pilose herbage, stems simple to moderately branched at or near base; leaves lanceolate to oblong, straight or falcate, 5–35 mm. long, 1.5–13 mm. wide, strongly asymmetrical to symmetrical at base, acute at apex, usually serrulate along margins, usually paler beneath than above; stipules triangular, about 1 mm. long, margin ciliate to slightly lacerate, cyathia solitary or in few-flowered cymes; peduncles 0.5–2 mm. long; involucres obconic-campanulate, 0.7–0.9 mm. in diameter, glabrous, except at base of lobes inside and on stipe; lobes deltoid-attenuate; glands circular to transversely elliptical, 0.1–0.3 mm. long; appendages white to reddish; staminate flowers 4–15 per cyathium; androphores 0.9–1.4 mm. long, glabrous; capsule strongly 3-lobed, 1.6–2.1 mm. long, with obtuse to rounded angles, glabrous; seeds ovoid-quadrangular in cross section, 1–1.4 mm. long, 0.7–1.1 mm. wide, back with shallow transverse depressions, grayish white to chocolate brown.

Valley floors, grassy slopes, washes, and rocky hillsides, Lower and Upper Sonoran and Carolinian Zones to Tropical Zone, Arizona to South Carolina and south to central South America. Flowering almost throughout the year under favorable conditions of temperature and moisture.

**23. Euphorbia florida** Engelm., in Torr., Bot. Mex. Bound. 189, 1859
*Chamaesyce florida* Millsp., Field Mus. Bot. Ser. **2:** 409, 1916.

Erect annual 1–6.5 dm. tall, glabrous, stem usually branched from near base, slender branches green or faintly tinged with red, internodes 1.5–5 or occasionally to 7 cm. long; stipules linear-subulate, 1–2 mm. long, 1–2 lateral basal lobes; leaves linear, 0.5–3 mm. wide, 0.5–6 cm. long, remotely serrulate, revolute in age, on petioles 1.5–2.5 mm. long; cyathia 1 to several at nodes, on peduncles 1–3 (or –6) mm. long; involucre campanulate, 1.5–2 mm. in diameter, about 2 mm. high at anthesis, glabrous without, densely pubescent with white straight hairs around inner rim; lobes of cyathia triangular-acuminate, entire or toothed, about 0.3 mm. high; 4 of glands 0.5 mm. in diameter, discoid, 5th filiform, nearly equaling lobes; appendages white or pink, obovate to elliptic, 1–2.8 mm. long; staminate flowers 9–12 in each fascicle, 30–60 per cyathium; androphores glabrous, included; gynophore glabrous, exserted, reflexed; capsule rounded-triangular in cross section, oblate-globose, 2.5 mm. in diameter, glabrous; seeds subtetragonal, 1.6–2 mm. long, slightly winged on lateral and dorsal angles, facets smooth but with 2–3 low transverse ridges, white to sordid, coat minutely reticulate.

On arid flats, washes, and hillsides, Lower Sonoran Zone, southern Arizona southward through Sonora into Sinaloa. Aug.–Nov.

**Euphorbia parryi** Engelm., Amer. Nat. **9:** 350, 1875, is similar in general appearance to *E. florida* but differs from that species in having entire instead of serrate leaves only 5–28 mm. long, slightly smaller involucres, white to yellowish instead of reddish appendages, and seeds with smooth instead of transversely wrinkled faces.

Mojave Desert to Colorado, Utah, western Texas, and northern Chihuahua; and from northern Mohave County to Cochise and Graham Counties in Arizona. It has not been reported from the Sonoran Desert but might be expected along the upper margins of our area flanking the mountains of Arizona and northern Sonora.

**24. Euphorbia trachysperma** Engelm., in Torr., Bot. Mex. Bound. 189, 1859

*Chamaesyce trachysperma* Millsp., Field Mus. Bot. Ser. **2:** 412, 1916.

Erect glabrous annual 1–4 dm. tall with stems simple or branching from base; leaf blades linear to ovate-lanceolate, obtuse at apex, 1–4 cm. long, serrulate along one margin, entire on other; stipules 1–1.4 mm. long, distinct, 2–3-cleft above; cyathia solitary at nodes; peduncles 1–5 mm. long; involucres 1.6–2 mm. wide, 3–3.5 mm. long, glabrous without, puberulent on upper portion within, lobes deltoid, acuminate, exceeding glands; glands suborbicular to transversely oval, 0.5–0.7 mm. long; appendages white, entire, as wide as or narrower than glands; sinus broadly V-shaped, hairy at its base; staminate flowers 50–65 per cyathium, in fascicles of 10–13; androphores 2–2.8 mm. long, glabrous; gynophore usually glabrous but sometimes sparsely hirsute; capsule glabrous, 3–3.4 mm. high and about as wide; triangular in cross section, angles acute to rounded; seeds quadrate in cross section, oblong to ovate-oblong vertically, 2–2.3 mm. high, 1.3–1.6 mm. wide, faces with a few minute pits, angles irregularly notched or beaded, sometimes with 1–3 faint transverse grooves.

Sandy or gravelly areas, Lower Sonoran Zone, southern Arizona southward into southern Sonora. March–Oct.

**25. Euphorbia abramsiana** L. C. Wheeler, Bull. S. Calif. Acad. **33:** 109, 1934

*Chamaesyce saltonensis* Millsp., in Parish, Cat. Pl. Salton Sink 6, 1913. [Preprint of Carnegie Inst. Wash. Pub. No. 193, 1914.]

*Euphorbia pediculifera* var. *abramsiana* Ewan, in Jepson, Fl. Calif. **2:** 427, 1936.

Prostrate annual forming open to moderately dense mat 0.5–4.5 dm. in diameter or tips of stems spreading-ascending; herbage and stems finely pubescent to glabrate; stipules whitish, about 0.5 mm. long, 2- to several-parted, margins sparsely ciliate or smooth; petioles 0.5–1 mm. long, sparsely puberulent to glabrate; leaf blades elliptical to oblong-lanceolate, 2–12 mm. long, obtuse to rounded at apex, obliquely unequal and rounded at base, revolute, minutely serrulate to entire, paler beneath than on upper surface; peduncles about 1 mm. long, glabrous; cyathia mostly on congested lateral branchlets but also solitary at nodes; involucres obconical, 0.6–0.8 mm. long, glabrous except on inner basal surfaces of lobes, these exceeding glands, 2–4-parted; glands orbicular to transversely elliptical, 0.1–0.2 mm. long; appendages white to pinkish, entire or bilobed, 0.3–0.5 mm. wide; staminate flowers 3–5 in each cyathium; capsules ellipsoid-oblong, 3-angled, 1.3–1.7 mm. long, glabrous at maturity; seeds quadrangular, 1–1.4 mm. long, about one-half as wide as long, with 4–6 transverse rugae on faces; coat microreticulate, white or buff to orange.

Desert slopes, washes, and flats, Lower Sonoran Zone, Colorado Desert, California, and southwestern Arizona into desert parts of Baja California, Sonora, and Sinaloa. July–Oct.

**26. Euphorbia serrula** Engelm., in Torr., Bot. Mex. Bound. 188, 1859

*Chamaesyce serrula* Woot. & Standl., Contr. U.S. Nat. Herb. **16:** 144, 1913.

Prostrate to decumbent annual with pilose stems to 2.5 dm. long; leaf blades broadly elliptic to obovate-oblong, 2–8 mm. wide, 3–11 mm. long, strongly unequal at base, sharply serrate along both margins in larger leaves, often entire or with 2–5 small serrations near apex on smaller ones; stipules distinct, deltoid-attenuate, trilobate, 1.5–2 mm. long; peduncles about 1 mm. long, glabrous to sparsely pilose; cyathia solitary in nodes, crowded near tips of branches; involucres obconical-campanulate, about 1 mm. in diameter, glabrous to sparsely pilose; lobes deltoid, 2–3-parted and exceeding glands; glands transversely elliptic to oblong, slightly concave; appendages white, as wide as to thrice as wide as glands, entire to crenulate; staminate flowers 7–13 per cyathium; androphore glabrous, about 1 mm. long; gynophore glabrous to sparsely pilose, long-exserted and reflexed; capsules obtusely triangular in cross section, 2–2.6 mm. long, usually slightly wider than long, glabrous; seeds 1.5–2 mm. long, 1–1.5 mm. wide, chalky to slightly sordid, angles sharp, faces smooth.

Gravelly soil, Lower and Upper Sonoran Zones, Arizona to western Texas and south to Durango and Zacatecas. March–Sept.

**27. Euphorbia brandegeei** Millsp., Proc. Calif. Acad. II, **2:** 226, 1889

*Euphorbia pediculifera* var. *minor* Millsp., op. cit. 227.
*Chamaesyce brandegeei* Millsp., Field Mus. Bot. Ser. **2:** 408, 1916.

Prostrate annual with slender, nearly filiform, spreading, dichotomously branched stems 2–20 cm. long; herbage densely to sparsely arachnoid-tomentulose but eventually subglabrate; stipules rudimentary subscarious scales about 0.4 mm. long or less and obscured by tomentum; petioles filiform, 0.5–0.8 mm. long; leaf blades ovate to orbicular, 1.5–4 mm. wide, 2–6 mm. long, very thin, entire or remotely and minutely few-toothed, dark green and sparsely pubescent above, densely tomentulose and cinereous beneath; peduncles slender, 0.5–1.2 mm. long, sparsely arachnoid-pubescent with spreading crisped hairs; involucres solitary in axils of upper leaves, open-campanulate, 0.7–1 mm. in diameter, about 0.8 mm. long, moderately tomentulose; lobes deltoid, about equaling reddish, transversely oblong glands, these 0.3–0.5 mm. tangentially, about 0.2 mm. radially; appendages white, 1–1.2 mm. tangentially, 0.6–0.8 mm. radially, irregularly 2–8-toothed, teeth blunt; capsules 3-angled, about 1.2 mm. long, sparsely hairy; seeds quadrangular in cross section, 0.8–1 mm. long, one-half as wide, faces and angles transversely wrinkled into low ridges, outer coat white microreticulate.

Sandy flats and hillsides, Lower Sonoran Zone, Magdalena and Santa Margarita Islands and presumably the adjacent mainland, Baja California. Aug.–March.

**28. Euphorbia dentosa** I. M. Johnston, Univ. Calif. Pub. Bot. **7:** 439, 1922

*Euphorbia setiloba* β, *dentata* Engelm. ex Boiss., in A. DC., Prod. **15**²: 44, 1862.

Prostrate, slender-stemmed annual with sparsely to densely villous branches 0.5–3 dm. long; stipules slenderly subulate, distinct, 0.2–0.6 mm. long, minutely villous; petioles slender, 0.5–2 mm. long, villous; leaf blades obliquely ovate to suborbicular, 4–12 mm. wide, 5–14 mm. long, finely serrate, glabrous to sparsely villous and dark green above, sparsely to densely villous and glaucous beneath; involucres solitary or in small clusters at nodes, campanulate, 0.8–1 mm. long, villous, on slender peduncles 0.5–2 mm. long; glands brownish, elliptic tangentially to involucral rim, about 0.2 mm. wide, 0.4–0.6 mm. long, appendages white or roseate, slightly wider than glands, 0.3–0.5 mm. long, lacerate; capsules slightly depressed, subglobose, 1–1.2 mm. long, about as wide, sparsely villous to subglabrate; seeds ovoid-prismatic, 0.7–0.9 mm. long, ashy, irregularly transversely rugulose, angles rather marked.

Sandy or gravelly areas, Lower Sonoran Zone, Cape Region, Baja California. Aug.–March.

**29. Euphorbia hirta** L., Sp. Pl. 454, 1753
(For synonymy see Rhodora **43:** 170–171, 1941.)

Few-stemmed, erect to decumbent, sparingly branched annual 2–25 cm. tall; stems strigose to pilose with yellowish hairs; petioles stoutish, 1–2 cm. long, pilose; leaf blades narrowly lanceolate, rhombic-lanceolate, or ovate, 3–12 mm. wide, 8–30 mm. long, acute at apex, asymmetrical at base, sharply to inconspicuously serrulate, sparsely hispid but soon glabrate above, appressed to spreadingly hirsute beneath, often reddish or with reddish splotches on lower surface; stipules triangular-attenuate, about 1 mm. long or less; cyathia in dense pedunculate heads; peduncles sparsely strigose to glabrate; involucres obconic-campanulate, 0.6–0.9 mm. in diameter, strigose with ascending hairs without, usually glabrous within; lobes triangular, about equaling glands, ciliate to lacerate into filiform segments marginally; glands stipitate, orbicular to transversely oval; appendages white or sometimes absent; staminate flowers 2–8 per cyathium; androphores about 1 mm. long; gynophores glabrous, short-exserted, reflexed; ovary slightly strigose toward apex; capsules 1–1.2 mm. long, truncate at base; seeds sharply quadrangular, 0.5–0.6 mm. wide, 0.7–0.9 mm. high, with low, smooth wrinkles, white; coat microreticulate, thin.

On sandy or light soil, Upper and Lower Sonoran, Carolinian, and Tropical Zones, South Carolina to Arizona and south to Argentina; introduced in Michigan, New York, and in Europe. Mainly above the Sonoran Desert in our area but entering it along its upper eastern margins. March–Sept.

**30. Euphorbia indivisa** (Engelm.) Tidestrom, Proc. Biol. Soc. Wash. **48:** 40, 1935
*Euphorbia dioica* var. *?indivisa* Engelm., in Torr., Bot. Mex. Bound. 187, 1859.
*Euphorbia adenoptera* β, *indivisa* Boiss., in A. DC., Prod. **15**[2]**:** 49, 1862.
*Chamaesyce indivisa* Millsp., Field Mus. Bot. Ser. **2:** 387, 1914.

Prostrate or decumbent annual with pilose stems 3–25 cm. long, internodes near tips greatly shortened and leaves and cyathia crowded; stipules distinct, linear-subulate, entire, pubescent, 1–1.4 mm. long; petioles 0.5–1 mm. long; leaf blades asymmetrically ovate-deltoid, oblong, or linear-oblong, often falcate, 4–8 mm. long, 1.5–4 mm. wide, serrulate, sparsely hirsutulous to glabrous above, sparsely hirsutulous to glabrate beneath; involucres cylindrical-campanulate, about 1 mm. long, pubescent with crispate hairs; lobes subulate, slightly shorter than to barely exceeding glands on each involucre; glands transversely oblong to oval, 0.2–0.6 mm. long, proximal glands about twice as long as distal ones, appendages white to reddish, wavy-margined, proximal oblong to spatulate, 1–1.6 mm. long, distal ones flabelliform, 0.2–0.3 mm. long, 5th gland linear, half as long as involucral lobes; staminate flowers 5–15 per involucre; capsule subpyriform, 3-lobed, 1.4–1.6 mm. long, strigose; seeds 1–1.2 mm. long, about 0.4–0.6 mm. wide, transversely 3–5-ridged on faces and through angles; seed coat microreticulate, white, thin.

Hillsides and arid flats, mainly Upper Sonoran Zone but extending into upper parts of Lower Sonoran Zone, southeastern Arizona to western Texas and south to Coahuila and San Luis Potosí. Aug.–Nov.

**31. Euphorbia chamaesyce** L., Sp. Pl. 455, 1753

(For extensive synonymy see Rhodora **43:** 265–266, 1941.)

Prostrate to decumbent annual with crispate-villous to glabrate stems to 4 dm. long; stipules narrowly triangular-subulate and distinct on dorsal side of stem, ventral ones often united, 0.5–1 mm. long, short-pubescent and sometimes lacerate; petioles slender, 1 mm. long or less; leaf blades elliptical to elliptic-oblong, or less commonly ovate or obovate-spatulate, 3–11 mm. long, asymmetrical at base, rounded or obtuse and serrulate at apex, glabrate above, usually short-villous beneath; involucres mostly on short lateral branches, solitary at nodes, obconical, 0.8–1 mm. long, 0.6–0.9 mm. wide, sparsely crispate-villous to glabrate without, lobes hairy, rounded-triangular, about equaling glands, these transversely oblong, 0.2–0.3 mm. wide; appendages white, equaling to twice as wide as gland, very short (0.2–0.4 mm.), entire or faintly denticulate; staminate flowers 4 per involucre; capsule ovoid, slightly depressed, widest below, 3-angled, 1–1.4 mm. long, persistently but sparsely villous along angles, faces glabrate; seeds about 1 mm. long, 0.6–0.7 mm. wide, transversely marked with low irregular ridges, coat white, pink, or grayish, microreticulate.

Various habitats, Lower Sonoran Zone to Transition, Carolinian, and Tropical Zones, Arizona to South Carolina, Florida, northern Mexico, and Jamaica; introduced in Virginia, Massachusetts, and the Old World. Occurring in the Sonoran Desert at Tucson and near Florence, Arizona. March–Oct.

**32. Euphorbia stictospora** Engelm., in Torr., Bot. Mex. Bound. 187, 1859

*Chamaesyce stictospora* Small, Fl. SE. U.S. 714, 1903.
*Anisophyllum senile* Kl. & Garcke, in Kl., Abh. Akad. Berlin Phys. **1859:** 28, 1860.

Prostrate to ascending annuals with stems 5–25 cm. long and clothed with white spreading hairs; leaves ovate to oblong or obovate, 2–5.5 mm. wide, 3–10 mm. long, sparsely crispate-villous to glabrate above, more densely pubescent beneath, oblique at base, rounded at apex, sharply serrate; stipules distinct or united, deltoid-attenuate, about 1 mm. long; cyathia mostly on short, leafy, lateral branches but sometimes solitary in axils; peduncles pubescent, 1–2 mm. long; involucres obconical, 0.8–1 mm. long, pubescent without, glabrous within; glands transversely elliptical, oblong, or orbicular, 0.1–0.3 mm. long; appendages white, narrow, glabrous, entire to sharply toothed; staminate flowers 3–7 per cyathium; androphores glabrous, about 1 mm. long; capsule triangular in cross section, 1.4–1.9 mm. long, faces strigose, basal angles with spreading hairs; seeds 1.2–1.4 mm. long, 0.6–0.8 mm. wide, acutely quadrangular in cross section, acute at apex, shallowly pitted to transversely rugose on faces, mottled white and brown.

Rocky hillsides and washes, mainly Upper Sonoran Zone, Wyoming and South Dakota to Arizona and south to Zacatecas and Durango; barely approaching the upper fringes of the Sonoran Desert. May–Oct.

**33. Euphorbia maculata** L., Sp. Pl. 455, 1753

*Euphorbia supina* Raf., Amer. Monthly Mag. **2:** 119, 1817.
*Euphorbia depressa* Torr., Cat. Pl. N.Y. 45, 1819.
(For additional synonymy see Rhodora **43:** 254–255, 1941; Webbia **17:** 187–205, 1962.)

Prostrate to ascendingly branched annual with pubescent stems to 4.5 dm. long; stipules distinct, linear-subulate, sparsely hirsutulous-ciliate, about 1 mm. long, often 2–3-divided; petioles 0.5–2 mm. long, pubescent; leaf blades elliptic-ovate, 4–17 mm. long, serrulate to entire, larger often bearing 1–2 red spots on upper surface; involucres obconical, 0.7–0.9 mm. in diameter, sparsely villous; glands transversely elliptical, about 0.2 mm. long; appendages narrow, about equaling or slightly surpassing width of glands, white, crenulate along distal margins; staminate flowers usually 4–5 per cyathium; capsules sharply 3-angled, 1.2–1.4 mm. long, pubescent to subglabrous; seeds quadrangular in cross section, about 1 mm. long, faces bearing low transverse ridges; seed coat white, grayish, or brownish, microreticulate.

An introduced weed, native from North Dakota to Texas and east to the Atlantic seaboard; occurring sparingly in Pima County and elsewhere in Arizona and west into California, Oregon, and Washington. June-Oct.

**34. Euphorbia platysperma** Engelm., in S. Wats., Bot. Calif. **2:** 482, 1880

*Euphorbia eremica* Jepson, Man. Fl. Pl. Calif. 600, 1925.

Prostrate annual forming mats 1–3 dm. in diameter; stems lightly striate and somewhat glutinous; stipules lance-deltoid, 1.5–2 mm. long, 2–3-parted; petioles 1.5–2.5 mm. long; leaf blades oblong to obovate, 5–10 mm. long, entire, rounded to mucronulate at apex, glabrous; peduncles 1–5 mm. long, glabrous; involucres solitary in axils of leaves, 1.5–2 mm. in diameter, glabrous without, lobes deltoid-truncate, entire or bifid; glands equaling lobes, 1 mm. wide, outer margin lighter brown than remainder and slightly emarginate but scarcely appendaged; staminate flowers about 50 per involucre; capsule globose-ovoid, 3.5–4 mm. long, shallowly 3-lobed, glabrous; seeds 2.4–3 mm. long, 1–1.7 mm. wide and thick, smooth on back and side, whitish, microreticulate.

Known only from two collections, one near Coachella and the other from southern Arizona near the mouth of the Colorado River.

**35. Euphorbia incerta** Brandegee, Proc. Calif. Acad. II, **3:** 171, 1891

*Chamaesyce incerta* Millsp., Field Mus. Bot. Ser. **2:** 409, 1916.

Erect or ascending, much branched perennial from a deep perennial taproot, lower part (up to one-half their length) of branches buried in sand, forming tufts 1–2 dm. high, or less commonly procumbent branches forming mats 4–6 dm. in diameter; stems glabrous but glutinous and closely covered with adhering sand grains; ventral stipules united, 1–1.5 mm. long, broadly linear and bifid, dorsal ones distinct, 1.5 mm. long, 0.6 mm. wide, lacerate into 4–5 slender lobes; petioles 1.5–2.5 mm. long, glabrous; leaf blades oblong-elliptic, 3–5 mm. wide, 5–14 mm. long, entire and glabrous, midrib impressed above, prominent beneath; involucres solitary in axils, turbinate, 1.5 mm. in diameter, 1.5–2 mm. long, glabrous without, on peduncles 1–2.5 mm. long, glands discoid or nearly so, about 0.5 mm. in diameter, very short-stipitate, without appendages; capsules depressed-globose, obviously 3-lobed, about 3 mm. high, 3–4.5 mm. in diameter, glabrous; seeds ovoid-spherical, 1.6–2.2 mm. long, about two-thirds as wide as long, rounded distally and acute to apiculate at proximal end, whitish, smooth, microreticulate.

Sandy beaches, and dunes, Lower Sonoran Zone, southern Baja California, and adjacent islands, near Bahía Kino, Sonora, and in Sinaloa and the Revillagigedo Islands. March–Nov.

**36. Euphorbia micromera** Boiss., in Engelm., in A. Gray, Proc. Amer. Acad. **5:** 171, 1861

(For full synonymy see Rhodora **43:** 194, 1941.)

Prostrate, glabrous to puberulent annual with stems 2–25 cm. long and internodes extremely variable in length; stipules triangular to lanceolate, distinct to partially united, 0.5–0.8 mm. long, sparsely ciliate; petioles about 0.5 mm. long, slender; leaf blades 1.5–7 mm. long, ovate to oblong, base oblique in larger leaves, rounded in smaller ones, glabrous or sparsely puberulent, margins entire; peduncles to 1.2 mm. long, pubescent to glabrate; involucres campanulate, slightly constricted above, about 1 mm. long, 0.9 mm. in diameter, crispate-hirsutulous without, or less commonly glabrous, lobes minute, deltoid, about equaling glands; glands discoid to transversely oblong, pink to red, 0.1–0.2 mm. in diameter; appendages absent or sometimes represented by a minute white margin; staminate flowers 2–5 per involucre; gynophores usually reflexed at maturity; capsule globular, 3-angled, 1.2–1.4 mm. long, puberulent to glabrous; seeds quadrangular-oblong, 1–1.3 mm. long, 0.4–0.5 mm. wide, angles rather sharp, facets smooth or faintly wrinkled, dull white but brown of inner tissues often showing through.

Desert flats, washes, outwash bajadas and hillsides, chiefly Lower Sonoran Zone, Inyo County, California, to Utah, western Texas, and south into Mexico; occurring also in Peru, according to L. C. Wheeler. Flowering more or less throughout most of the year.

**37. Euphorbia revoluta** Engelm., in Torr., Bot. Mex. Bound. 186, 1859

*Chamaesyce revoluta* Small, Fl. SE. U.S. 711, 1903.

Spreadingly erect, glabrous annual 0.5–2.5 dm. tall with very slender branches, forking repeatedly, internodes up to 3 cm. long, and narrowly linear, tightly revolute leaves to 2.5 cm. long; petioles about 1 mm. long, subfiliform; stipules linear-subulate, attenuate, 0.4–0.8 mm. long, entire; leaf blades glaucous beneath, pale green above; peduncles 1.5 mm. or less in length; involucres turbinate to obconical, about 1 mm. in diameter, glabrous inside and out or with minute pubescence around inner side of rim; involucral lobes triangular-acuminate; glands orbicular, 0.1–0.3 mm. in diameter, faintly cupped; appendages a small white scale about as long as gland or almost none; staminate flowers 5–10 per involucre; capsules ovoid, truncate at base, 1.2–1.4 mm. long, 3-angled, glabrous; seeds 1–1.3 mm. long, 0.6–0.9 mm. wide, faces smooth or faintly 1–2-ridged transversely; seed coat microreticulate, white, but with brown of deeper layers showing through.

Hillsides and arid slopes, Lower Sonoran Zone and dry habitats in Upper Sonoran Zone, southern Arizona to Colorado, New Mexico, and Chihuahua and probably eastern Sonora. Aug.–Oct.

**38. Euphorbia parishii** Greene, Bull. Calif. Acad. **2:** 56, 1886

*Chamaesyce parishii* Millsp., in Parish, Cat. Pl. Salton Sink 6, 1913. [Preprint of Carnegie Inst. Wash. Pub. No. 193, 1914.]

*Euphorbia patellifera* J. T. Howell, Leafl. West. Bot. **1:** 53, 1933.

Prostrate, spreading, or low, rounded perennial with stems 1–2.5 dm. long, glabrous or sparsely puberulent on leaves and ciliolate on larger stipules; internodes slender, 5–20 mm. long; stipules about 1 mm. long, sparsely ciliolate, upper distinct, lower more or less united; petioles glabrous, 0.5–1 mm. long; leaf blades ovate, entire, 2–4 mm. long, glabrous or sparingly puberulent beneath; involucres solitary, turbinate, 1–1.2 mm. long, glabrous without, puberulent just inside rim; glands discoid, pale yellow to faintly reddish, 0.5 mm. in diameter; appendages none; stamens 40–50; capsule pronouncedly 3-angled, 1.5–1.8 mm. long, glabrous; seeds quadrangular, 0.6–0.7 mm. wide and deep, 1.3–1.5 mm. long, sharply angled on back, faintly wrinkled on faces, coat white, microreticulate.

Desert slopes and along washes, Lower Sonoran Zone, Inyo County, California, south to the western margin of the Colorado Desert. March–June.

### 39. **Euphorbia petrina** S. Wats., Proc. Amer. Acad. **24:** 75, 1889

*Chamaesyce petrina* Millsp., Field Mus. Bot. Ser. **2:** 411, 1916.
*Euphorbia polycarpa* var. *petrina* I. M. Johnston, Proc. Calif. Acad. IV, **12:** 1072, 1924.

Prostrate perennial herb forming dense mats 1–4.5 dm. in diameter (often flowering first season and having the appearance of an annual); herbage moderately pubescent with fine, spreading, whitish hairs; stipules lance-linear, minute, about 0.2–0.3 mm. long and obscured by pubescence of stems and petioles; leaf blades broadly ovate, asymmetrical basally, juvenile ones 3–4 mm. wide, 6–9 mm. long, those of mature stems 2–5 mm. long, entire; involucre solitary in axils, mostly near tips of branches, puberulent, 0.4–0.6 mm. long, on pubescent peduncles 1 mm. or less in length; stamens about 5 per involucre; glands transversely oblong, about 0.2–0.3 mm. wide, reddish, devoid of appendages, or rarely with a very narrow white rim along outer edge; capsules subglobose, about 1 mm. long, 1.2 mm. wide, faintly 3-lobed, puberulent; seeds quadrangular, 0.7–0.9 mm. long, slightly more than one-half as wide, whitish, faintly ridged transversely.

Sandy flats and gentle slopes, Lower Sonoran Zone, vicinity of Bahía Kino, Sonora, and the adjacent islands, south into Sinaloa. Nov.–April.

### 40. **Euphorbia gracillima** S. Wats., Proc. Amer. Acad. **21:** 438, 1886

*Chamaesyce gracillima* Millsp., Field Mus. Bot. Ser. **2:** 409, 1916.

An erect, repeatedly forking, glabrous annual 5–20 cm. tall with very slender, graceful branches; upper internodes about 0.1 mm. in diameter, reddish; leaves linear, 5–12 mm. long, 0.3–0.6 mm. wide, entire, revolute; stipules linear-subulate, entire, distinct, about 0.5 mm. long; cyathia solitary in forks of branches, on peduncles 0.3–1 mm. long; involucre turbinate, 0.5–0.7 mm. in diameter, 0.8–0.9 mm. long, glabrous without, sparsely puberulent within near summit; lobes deltoid-acuminate, equaling or slightly exceeding transversely oval glands, these 0.1–0.3 mm. long; appendages obtuse to bifid, about as long as glands, white or roseate; staminate flowers 5–12 per cyathium; androphores about 0.6 mm. long, glabrous; ovary 3-angled, glabrous; capsules 1.2–1.4 mm. long, truncate at base, rounded above, 3-angled; seeds about 1 mm. long, 0.5–0.6 mm. wide and deep, angle acute, base truncate, coat thin and whitish, smooth on all surfaces but minutely linearly reticulate.

On arid gravelly slopes, planes, and stream banks, Lower Sonoran Zone, Pima County, Arizona, south through the desert area of Sonora into Sinaloa. Aug.–Oct.

**41. Euphorbia albomarginata** Torr. & Gray, Pacif. R. Rep. **2:** 174, 1855
(For synonymy see Rhodora **43:** 202, 1941.)

Prostrate perennial with glabrous herbage and a slender woody taproot; stems to 4 dm. long, often rooting at nodes; leaves orbicular to oblong, 3–6 mm. long, entire, base symmetrical to obliquely unequal, apex rounded; stipules united into a deltoid, membranous, white scale, its margins usually lacerate-ciliate; cyathia solitary at nodes with 15–30 staminate flowers in fascicles of 3–6 flowers; peduncles 2–4 mm. long; involucres 1.5–2.2 mm. in diameter, campanulate-obconic, glabrous except on somewhat puberulent, deltoid-attenuate lobes; glands transversely oblong, 0.5–1 mm. long, usually maroon, sometimes yellowish; appendages white, 1–1.6 mm. wide, entire to slightly cuneate; androphores glabrous, 1–1.3 mm. long; gynophore exserted, usually reflexed; capsule ovoid, 1.7–2.3 mm. long, acutely 3-angled, glabrous; seeds 1.2–1.7 mm. long, 0.7–0.9 mm. wide, narrowly oblong to oblong-ovate, smooth but microreticulate on faces, coat thin, whitish.

Sandy or rocky hillsides, washes, and valley floors, Lower and Upper Sonoran Zones and in arid sites in Transition Zone, Kern and Inyo Counties, California, to Utah, Oklahoma, and Texas south into Baja California and Tamaulipas. March–Nov.

**42. Euphorbia serpens** H. B. K., Nov. Gen. & Sp. **2:** 52 (quarto), 41 (folio), 1817
(For extensive synonymy see Rhodora **43:** 199, 1941.)

Prostrate, moderately to densely leafy, glabrous annual with slender stems to 5 dm. long, often rooting at nodes; stipules united into a deltoid white membrane 0.5–1 mm. long, margin erosulate to shallowly lacerate; petioles slender, 0.8–1.2 mm. long; leaf blades ovate-orbicular to oblong, usually obliquely asymmetrical at base, 2–9 mm. long, 1–7 mm. wide, entire, often bearing an irregular maroon spot above and slightly glaucous beneath; peduncles slender, to 5 mm. long; involucres solitary at nodes, turbinate, about 1 mm. long and wide; lobes acute, deltoid, equaling to slightly exceeding glands; glands transversely oblong, about 0.2 mm. wide, with white, glabrous, crenately margined appendages slightly wider than glands; staminate flowers 5–10 per involucre; capsule depressed-ovoid, bluntly 3-angled, glabrous, 1–1.2 mm. long; seeds narrowly ovoid, 0.8–1 mm. long, angles usually rounded, smooth; seed coat whitish, microreticulate.

On various soils throughout several life zones, Ontario and Montana to South America and introduced in the New England states and in Europe. May–Sept.

**43. Euphorbia polycarpa** Benth., Bot. Sulph. 50, 1844, var. **polycarpa**
*Chamaesyce polycarpa* Millsp. ex Parish, Cat. Pl. Salton Sink 6, 1913. [Preprint of Carnegie Inst. Wash. Pub. No. 193, 1914.]
*Euphorbia polycarpa* var. *typica* L. C. Wheeler, Bull. Torrey Club **63:** 408, 1936.

Prostrate or less commonly erect perennial herb from a slender woody taproot, forming a flat mat or a low rounded bush to 2.5 dm. tall or wide, stems sometimes markedly zigzag; herbage glabrous to slightly puberulent; leaf blades orbicular to lance-oblong, oblique at base, 1–10 mm. long, on petioles 1–2 mm. long; stipules deltoid, 0.3–0.5 mm. long, ventral ones united, usually ciliate-margined; involucres campanulate, 1–1.5 mm. wide, lobes narrowly to attenuately deltoid, about equaling narrow, transversely oblong glands, the latter 0.5–0.7 mm. long; appendages usually present, white to reddish, equaling or slightly exceeding glands, entire to crenate; bracteoles opposite each gland; staminate flowers 15–32 in each cyathium; gynophores exserted and reflexed at maturity, often puberulent above; capsules 3-angled, globose, 1–1.5 mm. in diameter, glabrous; seeds about 0.5 mm. in diameter, 0.9–1.3 mm. long, acute at apex, angular, truncate to obtuse at base, white, their faces smooth or slightly wrinkled, microreticulate.

Desert slopes and washes, Lower Sonoran Zone, Mojave and Colorado Deserts, California, into southern Nevada and southward into Sonora and Baja California. Flowering following rains.

KEY TO VARIETIES OF E. POLYCARPA

Leaves not markedly of dissimilar sizes on main and lateral branches; lateral branches not markedly congested:
- Plants herbaceous except bases of stems and rootstocks; appendages glabrous:
  - Larger leaves to 10 mm. long; appendages about as long as width of glands:
    - Herbage and capsules glabrous or very sparsely puberulent when very young ...... ........................................ 43. *E. polycarpa* var. *polycarpa*
    - Herbage and capsules densely but finely puberulent .............. 43*a*. var. *hirtella*
  - Leaves 2–5 mm. long; appendages reduced to mere line, rarely as long as width of gland ................................................ 43*b*. var. *carmenensis*
- Plants rounded shrubs to 3.5 dm. tall; appendages puberulent on upper surface ......... ....................................................... 43*c*. var. *johnstonii*

Leaves of dissimilar sizes, those on main stems about twice as large as those on lateral congested branches; lateral branches with very short internodes; leaves congested:
- Leaves obtuse or rounded at apex; involucres and stipes not glandular; appendages two to three times as wide as glands .............................. 43*d*. var. *intermixta*
- Leaves acute at apex; involucres and stipes glandular-puberulent, often encrusted with sand grains; appendages one to one and one-half times as wide as glands ........... ....................................................... 43*e*. var. *mejamia*

### 43*a*. Euphorbia polycarpa var. hirtella Boiss., in A. DC., Prod. **15**[2]: 44, 1862

*Chamaesyce polycarpa* var. *hirtella* Millsp. ex Parish, Cat. Pl. Salton Sink 6, 1913. [Preprint of Carnegie Inst. Wash. Pub. No. 193, 1914.]

Differing from var. *polycarpa* in having somewhat narrower appendages and finely puberulent herbage and capsules. Range about as that of var. *polycarpa*. Flowering following rains.

### 43*b*. Euphorbia polycarpa var. carmenensis (Rose) L. C. Wheeler, Bull. Torrey Club **63**: 410, 1936

*Euphorbia carmenensis* Rose, Contr. U.S. Nat. Herb. **1**: 133, 1892.
*Chamaesyce carmenensis* Millsp., Field Mus. Bot. Ser. **2**: 408, 1916.

A prostrate-spreading, many-branched and matted plant to 3–4 dm. in diameter from a stout perennial taproot; stems slender; leaves oblong-lanceolate, 0.5–1.2 mm. wide, 2–4 mm. long; stipules ciliate; involucres 0.6–1 mm. long; appendages a narrow line of white along outer edge of small maroon glands to 0.5 mm. long, crenulate, entire, or occasionally divided or lacking.

On arid slopes and benches, Lower Sonoran Zone, islands along the southeastern coast of Baja California. Flowering following rains.

### 43*c*. Euphorbia polycarpa var. johnstonii L. C. Wheeler, Bull. Torrey Club **63**: 410, 1936

Low, rounded, suffrutescent bush 1.5–3.5 dm. tall from a woody taproot 5–7 mm. in diameter; branches 2–3 mm. in diameter at their bases; herbage glabrous or nearly so; stipules puberulent, 0.3 mm. long; appendages densely puberulent on upper surfaces, linear.

Known from Carmen and Monserrate Islands off the southeast coast of Baja California. Flowering following rains.

**43*d*. Euphorbia polycarpa** var. **intermixta** (S. Wats.) L. C. Wheeler, Bull. Torrey Club **63:** 412, 1936

*Euphorbia intermixta* S. Wats., Proc. Amer. Acad. **24:** 74, 1889.
*Chamaesyce intermixta* Millsp., Field Mus. Bot. Ser. **2:** 409, 1916.

Differing from var. *polycarpa* mainly in having markedly dimorphic, ovate to oblong, obtuse or rounded leaves, large ones 5–10 mm. long on main stems where internodes are 2–3 cm. long, and small ones less than 5 mm. long on short lateral branches with internodes 3–5 mm. long; plant glabrous; involucres often congested on lateral branches; appendages two to three times as wide as glands.

Rocky and gravelly slopes, hillsides, and mesas, Lower Sonoran Zone, from vicinity of La Paz to San José del Cabo and in the vicinity of Guaymas, Sonora. Flowering after rains.

**43*e*. Euphorbia polycarpa** var. **mejamia** L. C. Wheeler, Bull. Torrey Club **63:** 411, 1936

Differing from var. *intermixta* in being sparsely to moderately glandular-puberulent, particularly on involucres and their stipes, less often on stems and leaves; and in having ovate-acuminate to ovate-oblanceolate leaves, majority of both small and large leaves at least acute at apex.

Sandy washes and hillsides near the coast, from the vicinity of La Paz to Cabo San Lucas. This variety is included because, although not yet definitely reported from our area, it approaches it so closely near La Paz that it can be reasonably expected slightly farther north and west. Flowering following rains.

**44. Euphorbia fendleri** Torr. & Gray, Pacif. R. Rep. **2:** 175, 1855, var. **fendleri**

*Euphorbia rupicola* Scheele, Linnaea **22:** 153, 1849. Not *E. rupicola* Boiss., 1838.
*Anisophyllum fendleri* Kl. & Garcke, in Kl., Abh. Akad. Berlin Phys. **1859:** 26, 1860.
*Chamaesyce fendleri* Small, Fl. SE. U.S. 710, 1903.
*Euphorbia fendleri* var. *dissimilis* Payson, Bot. Gaz. **60:** 379, 1915.
*Chamaesyce gooddingii* Millsp., Field Mus. Bot. Ser. **2:** 405, 1916.
*Euphorbia fendleri* var. *typica* L. C. Wheeler, Bull. Torrey Club **63:** 444, 1936.

Decumbent to erect, glabrous perennial with several to many slender stems 5–15 cm. long from a woody taproot; stipules distinct, linear, 0.5–1 mm. long, glabrous to sparingly hirsutulous; petioles about 1 mm. long, glabrous; leaf blades ovate-orbicular to ovate-lanceolate, entire, obliquely asymmetrical at base, 3–11 mm. long; involucres solitary at nodes, campanulate, 1.2–1.7 mm. long, lobes deltoid, about equaling transversely elliptical, reddish glands, these 0.8–1 mm. wide; appendages white, about as wide as glands, 0.3–0.5 mm. long, entire to sinuate-margined; staminate flowers 25–35 per involucre; capsules depressed-ovoid, 3-angled, 2.2–2.5 mm. long, glabrous; seeds quadrangular-ovoid, 1–1.2 mm. wide and deep, 2–2.2 mm. long, prominently angled, outer faces mostly smooth, inner ones slightly wrinkled; seed coat microreticulate, whitish.

Sandy and gravelly desert soil, Lower Sonoran Zone and drier parts of Upper Sonoran Zone, San Bernardino and Inyo Counties, California, to Oklahoma, western Nebraska, Texas, and northern Mexico. April–Sept.

**44a. Euphorbia fendleri** var. **chaetocalyx** Boiss., in A. DC., Prod. **15**[2]: 39, 1862

*Chamaesyce chaetocalyx* Woot. & Standl., Contr. U.S. Nat. Herb. **16:** 144, 1913.
*Euphorbia chaetocalyx* Tidestrom, Proc. Biol. Soc. Wash. **48:** 40, 1935.

Differing from var. *fendleri* only in being erect and in having lanceolate instead of ovate to ovate-orbicular leaf blades and deltoid subentire glands two to three times as long as they are wide.

Virtually throughout the range of the species within Arizona and east to western Texas. Mainly north and east of the Sonoran Desert portion of Arizona and barely, if at all, reaching our area. May-Sept.

**45. Euphorbia arizonica** Engelm., in Torr., Bot. Mex. Bound. 186, 1859

(For synonymy see Rhodora **43:** 243, 1941.)

Perennial from a slender woody taproot but often flowering first season; stems prostrate to nearly erect, 0.5–3 dm. long, pilose with slender, often clavate hairs; leaf blades ovate, ovate-oblong, or deltoid-ovate, acute to rounded at apex, somewhat oblique at base, 1–10 mm. long, margins entire and often faintly revolute; stipules minute; peduncles 1–4 mm. long, cyathia solitary or nearly so; involucres long-turbinate to urceolate, slightly constricted above, 0.6–0.8 mm. in diameter, 1.2–1.5 mm. long, sparsely pilosulous without, glabrous within, subulate lobes equaling glands, ciliate-margined; glands red, transversely oblong; appendages oval, 0.5–1 mm. long, white with a roseate tinge, sometimes cordate; staminate flowers 5–12 per cyathium; androphores glabrous; gynophore glabrous, exserted, reflexed; capsule subspherical, obtusely triangular, about 1.5 mm. in diameter, somewhat flattened basally, sparsely to moderately hirsutulous; seeds quadrangular in cross section, 1–1.2 mm. long, 0.5–0.6 mm. wide, acute at apex, rounded to truncate at base, with low rounded ridges on faces; coat white, very thin.

Gravelly slopes and rocky hillsides, Lower Sonoran Zone and lower parts of Upper Sonoran Zone, Colorado Desert in southern California to Texas and south into Chihuahua, northern Durango, and central Baja California. March–Nov.

**46. Euphorbia setiloba** Engelm., in Torr., Pacif. R. Rep. **5:** 364, 1858

*Chamaesyce setiloba* Millsp., in Parish, Cat. Pl. Salton Sink 6, 1913. [Preprint of Carnegie Inst. Wash. Pub. No. 193, 1914.]
*Euphorbia floccosiuscula* M. E. Jones, Contr. West. Bot. **15:** 145, 1929.

Prostrate to decumbent annual, puberulent throughout with very fine, white, spreading hairs; herbage often reddish toward end of growing season; internodes to 2 cm. long but often greatly shortened near tips of stems to form headlike tufts of leaves and involucres; stipules none; petioles about 1 mm. long, slightly swollen at base; leaf blades oblong to oblong-ovate, 1.5–4 mm. wide, 2–7 mm. long, entire, rounded at apex, slightly oblique at base; cyathia solitary at nodes, on peduncles 1–1.5 mm. long; involucres urceolate, 0.8–1 mm. in diameter, 1–1.3 mm. long, constricted just beneath rim, hirsutulous without, glabrous within; lobes of involucres narrowly deltoid, entire but ciliate, about equaling red, transversely oblong to discoid glands, these 0.1–0.2 mm. wide; appendages white, 0.6–1 mm. long, divided nearly to base into 3–5 attenuate segments, glabrous; staminate flowers 3–7 in each cyathium; gynophore glabrous or nearly so; capsules nearly globose but sharply 3-angled, about 1–1.2 mm. long; seeds 0.9–1 mm. long, 0.5–0.6 mm. wide, sharply angled, irregularly wrinkled on whitish, microreticulate faces, truncate basally.

Desert slopes, flats, and hillsides, Lower Sonoran Zone, Inyo County, California, to New Mexico and Texas south into Sinaloa and central Baja California. March–May; Sept.–Dec.

**47. Euphorbia bartolomaei** Greene, Pittonia **1:** 290, 1889

*Chamaesyce bartolomaei* Millsp., Field Mus. Bot. Ser. **2:** 408, 1916.

Prostrate to slightly procumbent annual with spreading dichotomous branches 3–30 cm. long; herbage puberulent with sparse, simple, spreading hairs; stipules on lower side of stems united, linear, 0.3–0.5 mm. long, puberulent, those on upper side of branches distinct, linear, 0.4–0.6 mm. long, puberulent; petioles stoutish, 0.6–1 mm. long, somewhat more densely puberulent than stems and leaf blades, latter suborbicular to ovate, asymmetrical at base, acute to rounded at apex, 2–5 mm. long, entire; peduncles 0.4–1 mm. long, sparsely hispidulous; involucres solitary in axils of leaves on short, congested, lateral branches, broadly campanulate, 1–1.5 mm. wide, about as long as or slightly shorter than wide, sparsely hispidulous without, glabrous within; lobes obtuse, about equaling glands, ciliate; glands reddish, transversely oblong, 0.3–0.5 mm. tangentially, 0.2–0.3 mm. radially; appendages glabrous, white, or faintly tinged with rose, 0.4–1 mm. long, 0.6–2 mm. wide, shallowly 2–5-toothed, teeth broad and blunt; sinus narrow and shallow; bracteoles about two-thirds as long as androphores; capsules sharply 3-angled, 1.2–1.4 mm long and wide, hirsutulous, exserted and reflexed at maturity; seeds 0.8 mm. wide, about 1.3 mm. long, quadrangular in cross section, shallowly grooved transversely on faces and angles, chalky white, microreticulate.

Sandy flats and gentle slopes, Lower Sonoran Zone, Natividad Island and the Vizcaíno Desert area southward to Magdalena Island, Baja California. Flowering much of the year.

**48. Euphorbia pediculifera** Engelm., in Torr., Bot. Mex. Bound. 186, 1859, var. **pediculifera**

*Euphorbia pediculifera* var. *typica* L. C. Wheeler, Bull. Torrey Club **63:** 442, 1936.
(For complete synonymy see Rhodora **43:** 183, 1941.)

Prostrate perennial forming a mat 1–4 dm. in diameter or less commonly stems ascending to erect; herbage closely appressed-puberulent; stems glabrate in age; leaf blades 2–20 mm. long, 1–10 mm. wide, ovate, oblong, or rarely nearly linear, base oblique to symmetrical, entire; petioles 1–2 mm. long; stipules 0.2–0.5 mm. long, upper distinct, ventral ones united; cyathia solitary at nodes, on peduncles to 1.5 mm. long; involucres campanulate, 1.5–2 mm. long, appressed-puberulent to glabrate, lobes equaling glands, deltoid; glands about 0.5 mm. wide, 0.75–1.25 mm. long, reddish purple, bearing entire or slightly lobed appendages to 2 mm. wide, 3 mm. long, or appendages lacking; staminate flowers 20–25 in a cyathium; gynophore exserted and reflexed at maturity, puberulent to glabrate; capsule about 2 mm. in diameter, strongly and obtusely trilobate; seeds 1–1.3 mm. long, about 0.6 mm. in diameter, nearly terete in cross section, transversely 4–5-ridged, grooves between ridges compressed to mere line; seed coat white or ashy.

Arid washes, plains, and hillsides, Lower Sonoran Zone and occasionally dry parts of Upper Sonoran Zone, Colorado Desert to Arizona and south throughout desert parts of Baja California and Sonora into Sinaloa. Feb.–Sept.

**48*a*. Euphorbia pediculifera** var. **linearifolia** S. Wats., Proc. Amer. Acad. **24:** 76, 1889

Differing from var. *pediculifera* in having only linear leaves 15–37 mm. long and bracteoles less conspicuous.

Arid hillsides among rocks, Lower Sonoran Zone, in the vicinity of Guaymas, Sonora.

### 49. Euphorbia melanadenia Torr., Pacif. R. Rep. 4: 135, 1857

Decumbent to ascending perennial with closely and minutely tomentose stems to 2 dm. long, lower parts soon glabrate; leaves ovate to ovate-lanceolate, 2–9 mm long, oblique basally, entire, closely and finely tomentose on both surfaces; petioles 1–2 mm. long; stipules to 1 mm. long, ventral ones usually connate, dorsal ones distinct; peduncles very short; cyathia solitary; involucres campanulate, 1.2–1.5 mm. in diameter, appressed-strigulose without and just below glands within, lobes narrowly deltoid, equaling glands, hairy; glands reddish, oblong; appendages usually twice as wide as glands, white, entire to crenate-margined, or appendages sometimes lacking; sinus U-shaped, densely hairy; staminate flowers 3–4 in each fascicle, 15–20 in a cyathium; gynaphore exserted and reflexed at maturity; ovary tomentose, 3-lobed; styles 0.5–0.8 mm. long, divided almost to base, hairy below; capsule 1.5–1.7 mm. long, short-tomentose; seeds 1.2–1.5 mm. long, about 0.6 mm. in diameter, quadrangular, faces smooth to faintly wrinkled; coat microreticulate, whitish.

Mountain slopes, hillsides, and along washes, Lower and Upper Sonoran Zones, Los Angeles County, California, to central Arizona and south into northern Sonora and central Baja California. Flowering more or less throughout the year.

## 21. PEDILANTHUS Poit., Ann. Mus. Hist. Nat. Paris 19: 388, 1812

Shrubs with fleshy turgid branches and milky juice. Leaves alternate, deciduous. Plants monoecious; flowers small, borne in oblique to nearly regular, often vase-like, bright-colored or green involucres, the margins of these dentate to shallowly 5-lobed or -parted; glands 2–6 or none. Male flowers in each involucre numerous, pedicellate, each consisting of 1 stamen; anthers subglobose, locules distinct. Pistillate flower solitary and central in each involucre, its perianth of 3 minute scales or none. Ovary 3-celled, 3-ovuled, borne on a stoutish gynophore; styles connate to form a long column, its stigmatic branches 3, each branch entire or bilobed. Capsule 3-lobed, often 3-keeled, each coccus 2-valved along inner angle at separation from persistent axis. Seeds smooth, ecarunculate. Cotyledons broad. Endosperm abundant.

### 1. Pedilanthus macrocarpus Benth., Bot. Sulph. 49, *pl. 23, fig. A,* 1844

An erect shrub to 1 m. tall with several to many fleshy jointed stems 6–15 mm. in diameter from a woody root crown; stems encrusted with a white waxy cuticle; leaves ovate to oblong, 1–2.5 cm. long, sessile or nearly so, soon deciduous; involucres 2–2.5 cm. wide, about 1 cm. high, bright red, base conical; capsule 1.5–2 cm. broad, bright red, 1–3-seeded; seeds ovoid, 7–10 mm. long.

Mainly on desert flats and llanos, Lower Sonoran Zone, central Baja California and Sonora to Colima. Feb.–May; Aug.–Oct.

The turgid stems contain a large quantity of thick milky juice which is reputed to contain rubber.

## Family 74. **CALLITRICHACEAE.** Water-Starwort Family

Herbs of aquatic or muddy habitats with very slender stems. Leaves opposite, entire, estipulate, often dimorphic, submerged ones linear to spatulate, emersed terminal ones broader and shorter. Flowers small and axillary, perfect or monoecious, subtended by 2 saccate bracts or bractless, apetalous. Stamen 1; filament elongate-filiform; anther cordate, 2-celled, dehiscing longitudinally. Pistil 1, sessile or pedunculate; styles 2. Ovary 4-celled, each cell 1-ovuled. Fruit 4-lobed, separating into 4 flattened nutlets. Seeds pendulous, anatropous. Endosperm fleshy; embryo straight or nearly so.

### 1. **CALLITRICHE** L., Sp. Pl. 969, 1753

Characters of the family.

**1. Callitriche palustris** L., Sp. Pl. 969, 1753

*Callitriche verna* L., Fl. Suec. ed. 2, 2, 1755.
*Callitriche palustris* var. *verna* Fenley, in Jepson, Fl. Calif. **2:** 435, 1936.

Slender perennial herb growing in water or on mud of vernal pools or receding lake shores and stream banks; submerged leaves linear, 1–2 cm. long, 1-nerved, acute to retuse at apex, floating leaves obovate, and sessile or narrowed to a narrowly winged petiole, shorter than to equaling blade, latter 2–4 mm. wide, 3–10 mm. long, dotted with minute stellate scales, pinnately veined; flowers bibracteate; fruit sessile, oblong-obovoid, 1–1.5 mm. long, bearing a thin, narrow, subscarious keel on back and usually narrowly winged at apex of each nutlet; bract about 0.5 mm. long, white, scarious; styles erect, 0.5–0.8 mm. long.

In quiet water or stranded on mud, Lower Sonoran Zone to Boreal Zone, mostly in Transition Zone to Boreal Zone, southern California and the vicinity of Tucson, Arizona, to Alaska and the Atlantic seaboard; also in Eurasia. July–Sept.

## Family 75. **BUXACEAE.** Box Family

Monoecious or dioecious herbs, shrubs, or trees with alternate or opposite, simple, mostly evergreen leaves. Flowers regular, solitary or in clusters, subtended by bracts. Calyx present or lacking. Petals none. Staminate flowers with 4–12 stamens, with or without a rudimentary pistil. Ovary 2–4-celled (usually 3-celled), with 1–2 ovules in each cell; styles as many as carpels. Fruit a capsule or drupe; endosperm (in ours) fleshy or scanty; embryo straight.

### 1. **SIMMONDSIA** Nutt., Lond. Jour. Bot. **3:** 400, *pl. 16,* 1844

Evergreen, dioecious shrub with opposite leathery leaves. Flowers short-pedunculate in axils; staminate flowers in clusters, pistillate often solitary but sometimes in clusters of 2–5. Sepals 4–6, usually 5. Stamens 10–12. Ovary 2–3-celled. Fruit a leathery-walled capsule, basal one-third to one-half enclosed by persisting enlarged sepals.

**1. Simmondsia chinensis** (Link) Schneider, Handb. Laubholzk. **2:** 141, 1907

*Buxus chinensis* Link, Enum. Hort. Ber. **2:** 386, 1822.
*Simmondsia californica* Nutt., Lond. Jour. Bot. **3:** 401, *pl. 16,* 1844.

Shrub 1–5 m. tall with rigid knotty branches and finely puberulent branchlets and peduncles; leaves ovate to oblong-elliptic, 1–2 cm. broad, 2–5 cm. long, acute to obtuse at both ends, dull green, leathery, somewhat glaucous or canescent, sparsely puberulent or glabrate in age; sepals of staminate flowers oblanceolate or oblong, 3–4 mm. long, densely puberulent without; stamens 3–4 mm. long, filament about 1 mm. long or less; sepals of pistillate flowers coalescent at base to form a cup 1.5–3 mm. deep, broadly lanceolate, 8–10 mm. long at anthesis, to 1.5 cm. long in fruit; capsule oblong-ovoid, 15–20 mm. long, nutlike.

Dry hillsides, outwash slopes, and along arroyos, Lower Sonoran Zone and occasionally in the Upper Sonoran Zone, southwestern San Diego County to central Arizona and southward into central Baja California and Sonora. Dec.–July.

This shrub is often described as being monoecious but I have never seen a plant producing both staminate and functional pistillate flowers. A rudimentary pistil sometimes is present in a few staminate flowers but they do not develop.

The seeds are eaten by the Indians of the southwest and have a pleasant rich flavor. The endosperm is waxy white or yellowish, and contains a clear white, liquid wax.

FOR FAMILIES 76–138
SEE VOLUME II

# Vegetation and Flora of the Sonoran Desert

*Volume Two*

# Vegetation and Flora of the Sonoran Desert

Forrest Shreve and Ira L. Wiggins

*Volume Two*

Stanford University Press

*Stanford, California*

Stanford University Press
Stanford, California

Printed in the United States of America
ISBN 0-8047-0163-6
Original edition 1964
Last figure below indicates year of this printing:
01 00 99 98 97 96 95 94 93

# Contents of Volumes I and II

## *Volume One*

### PART I. Vegetation of the Sonoran Desert



### PART II. Flora of the Sonoran Desert



## *Volume Two*

### PART II. Flora of the Sonoran Desert *(continued)*

## PART II

# Flora of the Sonoran Desert

*(continued)*

## Family 76. ANACARDIACEAE. Sumac Family

Trees, shrubs, or vines with resinous or milky juice. Leaves alternate or rarely opposite, simple, trifoliolate or pinnately compound. Stipules none. Flowers regular, perfect or polygamodioecious, small. Sepals 3–7. Petals same number as sepals, or rarely lacking, imbricate in bud. Stamens as many or twice as many as petals, infrequently fewer or more, inserted at base of lobed or simple disk. Ovary 1–5-celled; styles 1–3; carpels 1-ovuled. Fruit a small drupe or rarely utricular. Seed coat bony or crustaceous.

Ovary 2–5-celled; fruit to 2 cm. long, its flesh copious .................. 1. *Cyrtocarpa*
Ovary 1-celled; fruit usually 1 cm. long or less (to 1.5 cm. in *Rhus lentii*), its flesh scanty.
  Fruit drupaceous; petals not carinate nor scarious; shrubs, vines, or if trees trunk and limbs not tortuous and swollen ........................................ 2. *Rhus*
  Fruit utricular; petals carinate and scarious; trees with swollen tortuous trunk and limbs .................................................... 3. *Pachycormus*

### 1. CYRTOCARPA H. B. K., Nov. Gen. & Sp. 7: 19, 1825

Trees with pinnately compound, thin, densely pubescent, alternate leaves and entire leaflets. Flowers polygamodioecious, small, in axillary panicles. Sepals 5, ovate, imbricate in bud. Petals 5, oval, spreading. Disk 5-lobed. Stamens 10. Ovary 2–5-celled. Fruit an oblong drupe, pubescent. Embryo straight.

**1. Cyrtocarpa edulis** (Brandegee) Standley, Contr. U.S. Nat. Herb. **23:** 659, 1923
*Tapirira edulis* Brandegee, Zoe **5:** 78, 1900.

A rounded, openly branched tree to 10 m. tall with a trunk to 30 cm. in diameter and smooth, light gray bark; leaves 5–10 cm. long, densely short-pubescent; leaflets 7–11, broadly elliptic to oval, 1–3 cm. long, rounded to obtuse at apex, subsessile on a terete rachis; flowers white or slightly greenish; sepals about 1 mm. long, densely pilosulous, margins thin and nearly glabrous, ciliate; petals 3–3.5 mm. long; fruit oblong-ovoid, 1.5–2 cm. long, clear yellow, velvety-puberulent, flesh with a pleasantly acid taste.

Along arroyos and gentle slopes in sandy soil, Lower Sonoran Zone and margins of Tropical Zone, southern Baja California. May.

Collected in the Sonoran Desert between Arroyo Seco (on west coast) and La Paz.

**Mangifera indica** L., Sp. Pl. 200, 1753, is a tropical tree to 15 m. tall with simple, entire, deep green, shining leaves forming a dense crown, the leaf blades 1–2 dm. long; whitish green flowers in large terminal panicles; drupe to 10 or 15 cm. long and pale yellow; seed fibrous-coated, oblong-lenticular.

The mango is planted as a shade and fruit tree in both Sonora and Baja California as well as generally throughout the tropics. In southern parts of our area it thrives where water is available but does not reproduce spontaneously.

## 2. RHUS L., Sp. Pl. 265, 1753

Vines, shrubs, or small trees with alternate, simple or compound, petiolate leaves. Flowers regular, perfect or polygamodioecious, small, borne in thyrsoid or paniculate, terminal or axillary inflorescences. Calyx 3–7-lobed or -parted. Petals 3–7, imbricate in bud. Stamens usually as many or twice as many as petals, inserted just below a small annular disk. Ovary 1-celled (in ours); styles 1–3; carpels 1-ovuled. Fruit a small drupe, glabrous or puberulent. Seeds bony, often flattened, enclosed in scanty mesocarp.

Most of the species in our area occur most abundantly on the mountains and foothills just above the desert and descend into the latter chiefly along washes or in protected watered canyons.

Leaves simple, entire or toothed but not compound, rarely deeply lobed:
- Drupes glabrous, shining, not flattened laterally; leaves ovate-lanceolate, to 10 cm. long ........ 1. *R. laurina*
- Drupes glandular-pubescent, flattened laterally; leaves ovate to suborbicular, rarely over 6 cm. long:
  - Inflorescence an open thyrse; leaves glaucous beneath; fruit 1.5 cm. long .. 2. *R. lentii*
  - Inflorescence of simple or compound spikes; leaves not glaucous beneath; fruit 5–8 mm. long:
    - Sepals and bracts ciliate with stipitate-glandular hairs; upper surface of leaves usually scurfy-crustose:
      - Inflorescence compact; leaves obtuse .......... 3. *R. integrifolia* var. *integrifolia*
      - Inflorescence diffuse; leaves subacute .......... 3*a*. *R. integrifolia* var. *cedrosensis*
    - Sepals and bracts ciliate with nonglandular hairs; upper surfaces of leaves not scurfy-crustose:
      - Leaves leathery, thick, entire or regularly dentate; sepals 2–2.5 mm. long, nearly or fully as broad as long:
        - Petioles 1.5–2 cm. long; bracts glabrous within; filaments equaling or slightly longer than sepals ........ 4. *R. ovata*
        - Petioles 3–9 mm. long; bracts puberulent on both surfaces; filaments half as long as sepals ........ 5. *R. kearneyi*
      - Leaves thinner, not markedly leathery, crenulate-margined to distinctly lobed or even trifoliolate; sepals 1–1.2 mm. long, about one-third as wide (mesocarp not waxy) ........ 9*b*. *R. trilobata* var. *simplicifolia*

Leaves trifoliolate to pinnately compound, leaflets sometimes deeply lobed:
- Inflorescence thyrsoid; fruit bearing glandular hairs; mesocarp not waxy; leaves pinnately 5–11-foliolate except in varieties of *R. trilobata* and *R. choriophylla*:
  - Leaves evergreen; flowers appearing with young leaves:
    - Lenticels not conspicuous; bracts acuminate, 3 mm. long; filaments shorter than sepals ........ 6. *R. choriophylla*
    - Lenticels conspicuous; bracts acute, 1–1.4 mm. long; filaments longer than sepals ........ 7. *R. tepetate*

Leaves deciduous; flowers usually appearing before leaves:
Branchlets spinescent; leaves 5–9-foliolate; rachis winged ........ 8. *R. microphylla*
Branchlets not spinescent; leaves 3-foliolate or sometimes simple but lobed; rachis not winged:
Leaves predominantly trifoliolate:
Terminal leaflet not deeply lobed; leaflets eventually glabrate.................. .......................................... 9. *R. trilobata* var. *anisophylla*
Terminal leaflet deeply lobed; leaflets persistently puberulent ............... .......................................... 9*a*. *R. trilobata* var. *quinata*
Leaves usually unifoliolate but more or less deeply lobed, sometimes a few leaves on a plant trifoliolate ..................... 9*b*. *R. trilobata* var. *simplicifolia*
Inflorescence paniculate; fruit smooth, glabrous; mesocarp more or less waxy; leaves trifoliolate:
Panicles loose; leaflets obtuse or rounded at apex; petiolule of terminal leaflets 5–15 mm. long ................................................ 10. *R. diversiloba*
Panicles dense; leaflets acute to acuminate at apex; petiolule of terminal leaflets 1–4.5 cm. long:
Seeds not constricted at sides ........................ 11. *R. radicans* var. *radicans*
Seeds constricted at sides ......................... 11*a*. *R. radicans* var. *divaricata*

**1. Rhus laurina** Nutt., in Torr. & Gray, Fl. N. Amer. **1:** 219, 1838
*Lithraea laurina* Walp., Rep. **1:** 551, 1842.
*Toxicodendron laurinum* Kuntze, Rev. Gen. Pl. **1:** 154, 1891.
*Malosma laurina* Nutt. ex Abrams, Fl. Los Ang. ed. 2, 220, 1917.

Evergreen shrubs to 6 m. tall with slender glabrous branches that are usually red near apex; petioles 1–4 cm. long; leaves elliptic-ovate to lanceolate or oblong, 2–4 cm. broad, 4–10 cm. long, somewhat arcuately curved dorsiventrally, leathery, smooth, entire, obtuse to truncate at base, acute to emarginate at apex; flowers polygamodioecious, in a terminal thyrse to 15 cm. long; bracts spatulate; sepals 5, ovate, glabrous; petals white, about 1 mm. long; ovary 1-celled; fruit ovoid, about 1.5 mm. in diameter, glabrous, terete in cross section.

On brushy hillsides, in canyons, and along rocky washes, mainly Upper Sonoran Zone but entering the margins of the Lower Sonoran Zone, Santa Barbara County, California, southward in the coastal foothills to Cedros Island and the Sierra San Francisquito, Baja California. Toward the southern limits of its range it is virtually confined to the intermittent watercourses. March–Aug.

**2. Rhus lentii** Kellogg, Proc. Calif. Acad. **2:** 16, 1860
*Toxicodendron lentii* Kuntze, Rev. Gen. Pl. **1:** 154, 1891.

A low rounded shrub to 3 m. tall with puberulent but soon glabrate, maroon branchlets and simple, entire, broadly ovate to orbicular leaves; petioles stout, 2–10 (or rarely 40) mm. long; leaf blades 1.5–5 cm. wide, 1.5–6 cm. long, obtuse to subcordate at base, abruptly acute to rounded at apex, densely and minutely tomentulose and glaucous beneath; thyrse terminal, 2–4 cm. long; bracts ovate, about 2 mm. long and broad, abruptly acute at apex, puberulent, deciduous; sepals suborbicular, about 3 mm. long, pubescent on margins and outer surface; petals greenish or yellowish white, about 3 mm. wide, 4 mm. long, sparsely pubescent within, glabrous externally; fruits reddish, glandular-pubescent and pilose, 10–15 mm. long and broad, strongly flattened, about 5–6 mm. thick.

Lower Sonoran Zone, mainly on Cedros Island but also at San Pablo Point and San Bartólome Bay on the peninsula, Baja California. Feb.–March.

**3. Rhus integrifolia** (Nutt.) Benth. & Hook. f. ex Rothrock, Wheeler Rep. 84, 1878, var. **integrifolia**

*Styphonia integrifolia* Nutt., in Torr. & Gray, Fl. N. Amer. **1:** 220, 1838.
*Styphonia serrata* Nutt., in Torr. & Gray, loc. cit.
*Rhus hindsiana* Engler, in A. & C. DC., Monogr. Phan. **4:** 388, 1883.
*Toxicodendron hindsianum* Kuntze, Rev. Gen. Pl. **1:** 154, 1891.
*Toxicodendron integrifolium* Kuntze, loc. cit.
*Neostyphonia integrifolia* Shafer, in Britt. & Shafer, N. Amer. Trees 612, 1908.

Evergreen aromatic shrub 1–3 m. tall with stout stiff branchlets at first densely gray-pubescent, later glabrate; petioles 2–10 mm. long, pubescent, stoutish; leaves simple or rarely trifoliolate, ovate to elliptic-lanceolate or obovate, subcuneate to obtuse at base, rounded to acute at apex, 1–3.5 cm. broad, 1.2–6.5 cm. long, subrevolute margin entire or repand-serrate, upper surface silvery, scurfy-crustose, pubescent when young but later glabrate; inflorescence a compound terminal spike 2–3 cm. long; bracts triangular, 2–3.5 mm. long, 1–2 mm. wide, pubescent and with a few interspersed glandular hairs on outer surface, ciliate with glandular and nonglandular hairs; sepals suborbicular, 3 mm. long, pilose near base and ciliate like bracts; petals pink or reddish, about 4 mm. long; fruit ovoid-lenticular, glandular and puberulent, heavily coated with a colorless to white, viscid, acidulous exudate, 7–10 mm. long, flattened.

Coastal bluffs, hillsides, and along washes, Sonoran Zones, Santa Barbara County, California, to Cedros Island, Baja California. It is rare or absent between San Quintín and Cedros Island. Oct.–May.

**3*a*. Rhus integrifolia** var. **cedrosensis** Barkley, Ann. Mo. Bot. Gard. **24:** 363, *pl. 20, fig. 1*, 1937

A small tree with nearly or quite entire, narrower leaves, a more open inflorescence, and ovate, more quickly deciduous bracts.

Known only from Cedros Island, Baja California. Feb.–April.

**4. Rhus ovata** S. Wats., Proc. Amer. Acad. **20:** 358, 1885

*Neostyphonia ovata* Abrams, Bull. N.Y. Bot. Gard. **6:** 403, 1910.

A dense leafy shrub or small tree to 5 m. tall with brownish branches, these at first puberulent, later glabrate; petioles purplish, stoutish, 1.5–2 cm. long, glabrous or essentially so; leaf blades ovate-elliptic to broadly ovate, 3–5.5 cm. wide, 5–9 cm. long, leathery, glabrous, acute to acuminate at apex, cuneate to obtuse or rarely subcordate at base, entire or sometimes repand-serrate, dark green above, slightly paler beneath; inflorescence 3–5 cm. long, glabrous or glabrate; bracts broadly ovate to suborbicular, 1.5–2 mm. long, nearly or fully as broad, sparsely pubescent without, glabrous within, ciliation on margin nonglandular; sepals 2.5 mm. long and broad, ciliate like bracts, otherwise nearly glabrous; petals broadly ovate, 4–5 cm. long, glabrous without, sparsely pubescent basally within; filaments equaling sepals; fruit orbicular- or ovoid-lenticular, 6–8 mm. long and broad, glandular-viscid and puberulent, pulp and exudate sweet.

Hillsides and along washes, chiefly Upper Sonoran Zone and barely entering the edges of the Sonoran Desert, Santa Barbara County, California, to southern part of the Sierra San Pedro Mártir, Baja California; eastward in the mountains to central Arizona, where it ranges from 3,000 to 5,000 feet above sea level. Feb.–May.

**5. Rhus kearneyi** Barkley, Ann. Mo. Bot. Gard. **24:** 363, *fig. 15; pl. 19, fig. 2,* 1937

Small tree with minutely puberulent but later glabrate, gray or tan branches; petioles stoutish, 3–9 mm. long; leaf blades ovate to elliptic-oblong, 1.5–3.5 cm. wide, 2.5–5 cm. long, leathery, entire, acute to rounded at apex, subcordate at base, glabrous or scantily beset with a few glandular hairs, margin revolute; bracts ovate-deltoid, densely pilose on both surfaces, ciliate with nonglandular hairs, persistent; sepals deltoid, 2 mm. long, about 2.5 mm. wide at base, pubescent without, glabrous within, ciliate like bracts; petals 2 mm. broad, 3.5 mm. long, sparingly pilose basally on outer surface; filaments about half as long as sepals; fruit oblong-lenticular, about 10 mm. long, pubescent with glandular and nonglandular hairs.

Dry cliffs, Lower Sonoran Zone, at Tinajas Altas, Yuma County, Arizona, the only known locality. March–April.

**6. Rhus choriophylla** Woot. & Standl., Contr. U.S. Nat. Herb. **16:** 146, 1913

Low shrub to 2 m. tall with brownish, gray-puberulent branchlets; petioles to 3 cm. long, minutely puberulent; leaflets 3–5, ovate, acute to slightly acuminate, subcuneate to obtuse or rounded at base, entire, glabrous except for a few scattered, reddish, subsessile, minute glands; terminal leaflet 2.5–5 cm. long, lateral ones usually less than 4.5 cm. long, on petiolules 2–4 mm. long; bracts ovate-acuminate, about 3 mm. long, pubescent without, glabrous within, marginal ciliation of simple and glandular hairs; sepals triangular-ovate, rounded at apex, about 2 mm. long, glabrous except along ciliate margins; petals white, rhombic-ovate, 2.8–3 mm. long, glabrous except near base within, eciliate; filaments 1–1.5 mm. long; fruit ovoid-lenticular, 6–7 mm. long, pubescent with glandular and simple hairs closely intermingled.

Mostly in brushy areas or beneath large trees, chiefly Upper Sonoran Zone, New Mexico, Arizona, Sonora, and southern Baja California. July–Sept.

Collected within the limits of the Sonoran Desert at the southern end of Sierra Babiso, eleven miles west of Cucurpe, Sonora.

This is very close to *R. virens* and perhaps is no more than a variety of that species.

**7. Rhus tepetate** Standl. & Barkley, in Barkley, Ann. Mo. Bot. Gard. **24:** 382, *fig. 18; pl. 22, fig. 2,* 1937

Shrub to 2 m. tall with stoutish branchlets bearing prominent lenticels; twigs maroon, with a sparse minute puberulence but soon glabrate; petioles stout, 1–2 cm. long; leaflets 7–11, oblong to oblong-ovate, obtuse to acute at apex (rarely submucronate), subcuneate to slightly cordate at base, leathery, entire, sparsely and minutely puberulent on both surfaces, especially on veins beneath, veins markedly whitish above, terminal leaflet 3.5–6.5 cm. long, to 4 cm. broad, lateral leaflets to 4 cm. broad and 5.5 cm. long; inflorescences lateral, 4–5.5 cm. long; bracts deltoid-ovate, about 1–1.4 mm. long, glabrous within, pubescent with glandular and nonglandular hairs without, margins ciliate; sepals suborbicular, about 1.4 mm. long, glabrous except for marginal ciliation; petals about 3.5 mm. long, eciliate; filaments longer than sepals; fruit still unknown.

Open ridges and hillsides, Lower and Upper Sonoran Zones, southern Baja California and southern Sonora. July–Sept.

This also is closely related to *R. virens* and may not be specifically distinct from it.

**8. Rhus microphylla** Engelm., in A. Gray, Pl. Wright. **1:** 31, 1852
(For synonymy see Ann. Mo. Bot. Gard. **24:** 387, 1937.)

Shrub or small tree to 5 m. tall with more or less spinescent branchlets, these with prominent lenticels and fine puberulence but eventually glabrate, brownish; petioles terete, 2–6 mm. long; leaflets 5–9, elliptic-oblong to lance-ovate, obtuse and mucronate to acute, cuneate to asymmetrically rounded at base, thin, sessile, entire, appressed-pilose on both surfaces, margins subrevolute; terminal leaflet 3–4 mm. broad, 4–12 mm. long, lateral ones 1.5–4 mm. broad, 4–10 mm. long; segments of rachis narrowly but distinctly winged; inflorescence lateral, dense, 4–5 cm. long; bracts deltoid-ovate, about 1 mm. long, pilose and glandular without, glabrous within, each flower subtended by a bract and 2 bractlets; sepals rounded-deltoid, 1 mm. long, glabrous except for marginal cilia; petals 2–2.3 mm. long, pilose within and ciliate; filaments exceeding sepals; fruit orbicular-lenticular, 5 mm. long, reddish, glandular-puberulent and pilose.

Along rocky washes and hillsides, Lower and Upper Sonoran Zones and lower parts of Arid Transition Zone, Texas to Arizona, south to Nuevo León and Zacatecas, and on Cedros Island and at San Bartólome Bay, Baja California. Feb.–May.

**9. Rhus trilobata** Nutt. ex Torr. & Gray var. **anisophylla** (Greene) Jepson, Man. Fl. Pl. Calif. 608, 1925
*Schmaltzia anisophylla* Greene, Leaflets Bot. Obs. **1:** 136, 1905.
(For synonymy see Ann. Mo. Bot. Gard. **24:** 411, 1937.)

Openly branched shrub to 1.5 m. tall with slender, gray, puberulent branches; leaflets 3, broadly ovate, cuneate-dentate, rounded to acute at apex, abruptly cuneate at base, sparsely puberulent in youth, later glabrate, terminal leaflet 0.8–4.5 cm. broad, 1–6 cm. long, lateral ones 5–30 mm. wide, to 3 cm. long; inflorescence dense; bracts triangular, 1.5 mm. long, pubescent without and ciliate; pedicels 2–5 mm. long; sepals 1.2 mm. long, glabrous except along ciliate margins, petals yellowish or greenish, obovate, 2.6–2.8 mm. long, pilose within, eciliate; filaments about equaling sepals; fruit suborbicular-lenticular, about 6 mm. long, 7 mm. wide, sparsely pilose and densely glandular-puberulent, viscid, exudate slightly acidulous.

Rocky washes and hillsides, Lower and Upper Sonoran Zones, southeastern Oregon, California, mainly east of the Sierra Nevada and in interior southern California, to Utah, New Mexico, Baja California, Durango, and Jalisco. Feb.–Aug.

The fruits are gathered, dried, and ground and used as seasoning by the Indians, at least in parts of the range of this and the varieties listed below.

**9*a*. Rhus trilobata** var. **quinata** Jepson, Erythea **1:** 141, 1893
*Rhus canadensis* var. *quinata* A. Gray, Syn. Fl. N. Amer. **1:** 386, 1897.
*Schmaltzia quinata* Greene, Leaflets Bot. Obs. **1:** 139, 1905.

Similar to the above but terminal leaflet usually deeply 3-lobed, 2–5 cm. long and leaflets more persistently puberulent.

Rocky hillsides and washes, Lower and Upper Sonoran Zones, south central Oregon to northern Baja California and eastward to Nevada and New Mexico. Fairly common along the western margin of the Colorado Desert.

**9*b*. Rhus trilobata** var. **simplicifolia** (Greene) Barkley, Ann. Mo. Bot. Gard. **24:** 410, 1937
*Rhus canadensis* var. *simplicifolia* Greene, Bull. Torrey Club **17:** 13, 1890.
(For synonymy see Ann. Mo. Bot. Gard. **24:** 410, 1937.)

Leaves mainly unifoliolate, although occasionally some trifoliolate leaves are present among the simple ones; blades usually persistently puberulent; otherwise as in var. *anisophylla.*

Arid rocky hillsides and washes, Lower and Upper Sonoran Zones, southern Utah, Oklahoma, and Colorado to southeastern California and eastern slopes and escarpment of Sierra San Pedro Mártir, Baja California.

**10. Rhus diversiloba** Torr. & Gray, Fl. N. Amer. **1:** 218, 1838

*Toxicodendron diversilobum* Greene, Leaflets Bot. Obs. **1:** 119, 1905.
(For synonymy see Ann. Mo. Bot. Gard. **24:** 422–423, 1937.)

Rounded shrub or a vine climbing by adhesive tendrils; branchlets slender and puberulent; leaflets 3, or rarely 5, ovate, crenate, dentate, bluntly lobed or sometimes entire, soon glabrate on both surfaces, rounded to broadly cuneate at base, terminal leaflet 1.5–4 cm. broad, 2–6.5 cm. long, lateral leaflets 1–5 cm. wide, 2.5–6 cm. long, sessile or short-petiolulate; flowers pale yellowish green, in loose lateral panicles; bracts oblanceolate, about 1 mm. long, glabrate but margin ciliate; sepals ovate, about 0.5 mm. wide, 1 mm. long, glabrate; petals 3 mm. long; fruit whitish or yellowish, glabrous or nearly so, nearly globose, 5–7 mm. in diameter, mesocarp somewhat fibrous; seeds suborbicular-lenticular, about 5 mm. wide, 3.5–4 mm. high.

In chaparral on hillsides and along draws, practically confined to the Upper Sonoran Zone, Washington to Baja California, in southern Sonora and Michoacán. Barely entering the western margins of the Sonoran Desert in southern California and Baja California. Feb.–May.

**11. Rhus radicans** L., Sp. Pl. 266, 1753, var. **radicans**

*Toxicodendron radicans* Kuntze, Rev. Gen. Pl. **1:** 153, 1891.
(For extensive synonymy see Ann. Mo. Bot. Gard. **24:** 425–427, 1937.)

Shrubs or commonly vines with densely puberulent to glabrate branches; leaflets 3, or sometimes 4 or 5, ovate, entire to irregularly serrate, acute to acuminate at apex, subcuneate to rounded at base, glabrate above, pubescent to glabrate beneath, terminal leaflet 1–13 cm. wide, 3–20 cm. long; lateral leaflets asymmetrical at base, 1–10 cm. wide, 3–15 cm. long; panicles lateral, loose; bracts triangular, about 1 mm. long, ciliate along margin, otherwise glabrate; sepals ovate, about 1 mm. long, glabrate; petals oblanceolate, 1 mm. broad, 3 mm. long, glabrous; fruit glabrous, whitish; seeds about 3 mm. long, 4 mm. broad.

Hillsides, stream sides, and level ground, Lower Sonoran Zone to Canadian Zone, British Columbia to Nova Scotia south through Washington, Oregon, Utah, and Arizona into southern Mexico and the West Indies. Barely entering the Sonoran Desert in Pima County, Arizona, where it occurs infrequently down to the 3,000 foot level on the slopes of the Catalina Mountains. March–Sept.

**11*a*. Rhus radicans** var. **divaricata** (Greene) Fernald, Rhodora **43:** 596, 1941

*Toxicodendron divaricatum* Greene, Leaflets Bot. Obs. **1:** 122, 1905.
*Rhus divaricata* McNair, Field Mus. Bot. Ser. **4:** 69, 1925. Not *Rhus divaricata* Eckl. & Zeyh., 1834.
*Rhus greenei* McNair, loc. cit.
*Toxicodendron radicans* var. *divaricata* Barkley, Ann. Mo. Bot. Gard. **24:** 433, 1937.

Similar to the species; leaflets ovate to lanceolate, acute to acuminate, sparsely puberulent; seeds constricted on one or both sides.

Along stream courses and desert arroyos and canyons, Lower Sonoran Zone, central to southern Baja California and vicinity of Magdalena, Sonora; possibly also in Chihuahua.

### 3. PACHYCORMUS Coville, Cent. Dict. rev. ed. 6708, 1911

Tree with short thick trunks and numerous large, thick, and usually very crooked branches. Leaves unevenly pinnately compound, deciduous. Flowers small, dioecious, in loose or dense panicles. Sepals 5, ovate, subvalvate in bud. Petals 5, ovate-oblong, scarious, persistent, somewhat accrescent. Stamens 10. Ovary 1-celled; styles 3; stigmas capitate. Fruit utriclelike.

Leaves 5–8 cm. long, 2–3 cm. broad .......................... 1. *P. discolor* var. *discolor*
Leaves 1.5–4.5 (usually less than 3) cm. long, 7–12 (–18) mm. broad:
  Flowers in broadly pyramidal, loose panicles; corollas 4–4.5 mm. long; sepals about 2 mm. long, pale rose or white ............................ 1*a*. *P. discolor* var. *pubescens*
  Flowers in compact subcylindrical panicles; corollas 5.5–6 mm. long; sepals 3–3.5 mm. long, deep rose .................................. 1*b*. *P. discolor* var. *veatchiana*

**1. Pachycormus discolor** (Benth.) Coville, in Goldm., Contr. U.S. Nat. Herb. **16:** 344, *pl. 118, figs. A–B,* 1916, var. **discolor**

*Schinus ?discolor* Benth., Bot. Sulph. 11, *pl. 9,* 1844.
*Veatchia discolor* Brandegee, Proc. Calif. Acad. II, **2:** 140, 1889.

Tree 3–5 or occasionally to 10 m. tall; trunk often to 6–8 dm. in diameter; bark white or yellowish white, thick and fleshy, 1–2.5 cm. thick, outer layer shed as paper-like thin sheets, pulpy inner tissue copiously ramified with latex ducts containing yellowish, reddish, brown, or milky juice that oozes out for an hour or more when bark is wounded; branches usually tortuous, swollen; leaves odd-pinnate, to 8 cm. long, 2–3 cm. wide, terminal leaflet obovate to cuneate-flabellate, toothed or lobed above middle; lateral leaflets 2–10, broadly elliptic to oblong, 1–1.5 cm. long, bright green and sparsely appressed-puberulent above, pale to ashy and densely appressed-puberulent beneath; sepals triangular-ovate, acute, about 2 mm. long, petals 4–4.5 mm. long, rosy or whitish, puberulent, especially along acute median keel; utricle about 3–3.5 mm. long, pubescent.

Sandy and rocky slopes and flats, Lower Sonoran Zone, Magdalena and Santa Margarita Islands and Sierra de la Giganta. May–Sept.

**1*a*. Pachycormus discolor** var. **pubescens** (S. Wats.) Gentry, Allan Hancock Pacif. Exped. **13:** 141, 1949

*Bursera pubescens* S. Wats., Proc. Amer. Acad. **24:** 44, 1889.
*Veatchia discolor* var. *pubescens* I. M. Johnston, Proc. Calif. Acad. IV, **12:** 1079, 1924.

Differing from var. *discolor* in having smaller leaves, these 7–12 (sometimes –18) mm. wide, 1.5–4.5 cm. long, and terminal leaflets less deeply lobed or toothed.

Occurring along washes but mostly on rocky slopes and hillsides, Lower Sonoran Zone, vicinity of El Mármol and San Augustín southward to foothills of Sierra de la Giganta, Baja California.

**1*b*. Pachycormus discolor** var. **veatchiana** (Kell.) Gentry, Allan Hancock Pacif. Exped. **13:** 55, 1949

*Rhus veatchiana* Kell., Proc. Calif. Acad. **2:** 24, 1860.
*Veatchia cedrosensis* A. Gray, Bull. Calif. Acad. **1:** 4, 1884.
*Veatchia discolor* var. *veatchiana* I. M. Johnston, Proc. Calif. Acad. IV, **12:** 1081, 1924.

Leaves as in var. *pubescens*; flowers in dense cylindrical panicles 2–7 cm. long, 2–3 cm. in diameter, lateral branches very short; pedicels hirsute, 6–8 mm. long; sepals 3–3.5 mm. long, deep rose to red, pilosulous; petals 5.5–6 mm. long, light rose or pink.

Cedros Island and the western margin of Vizcaíno Desert (Turtle Bay), Baja California.

## Family 77. CELASTRACEAE. Staff-Tree Family

Trees, shrubs, or woody vines with alternate or opposite, simple, deciduous or evergreen leaves. Stipules small, caducous, or lacking. Flowers regular, mostly perfect; pedicels commonly jointed. Calyx lobes 4–5, persistent, imbricate in bud. Petals 4–5. Disk broad, flat, and lobed. Stamens same number as petals on a hypogynous disk. Ovary sessile, adherent to disk at base or free, 2–5-celled; style solitary, short; stigma entire or 3–5-lobed. Carpels 2- to several-ovuled. Fruit a berry, capsule, samara, or drupe.

Fruit a 3-winged samara ............................................ 1. *Wimmeria*
Fruit not samaroid:
  Fruit a dehiscent capsule; leaves subfleshy, leathery ..................... 2. *Maytenus*
  Fruit not a dehiscent capsule, either a berry or indehiscent; leaves not fleshy, less markedly leathery:
    Flowers 4-parted; fruit scantily fleshy, berrylike, usually 2-seeded; calyx not tubular .. ........................................................ 3. *Schaefferia*
    Flowers 5-parted; fruit dry, indehiscent, 1-seeded; calyx tubular-obconic .. 4. *Mortonia*

### 1. WIMMERIA Schlecht. & Cham., Linnaea 6: 427, 1831

Shrubs or small trees with alternate, serrate to cuneate, estipulate leaves. Flowers in axillary cymes, small. Calyx lobes 5. Petals 5, inserted at base of annular disk. Ovary 3-celled, each cell 6–8-ovuled. Fruit 1-celled by abortion, 1–2-seeded, bearing 3 broad longitudinal wings.

**1. Wimmeria confusa** Hemsl., Diagn. Pl. Mex. 6, 1878
*Wimmeria pallida* Radlk., Sitzungsb. Akad. Wiss. München 8: 379, 1878.

A shrub or tree to 8 m. tall with light gray bark, slender, dark gray or gray-brown branchlets and moderately conspicuous, light tan lenticels; petioles slender, 6–8 mm. long, glabrous; leaf blades lance-elliptic, lanceolate, oblanceolate, or obovate, 1–2 cm. wide, 1–5 cm. long, glabrous, finely reticulately veined, narrow cuneate base gradually tapering to petiole, crenulate or subentire, apex acute, obtuse, or rounded; sepals 1–1.5 mm. long, rounded, minutely ciliolate; petals about 3 mm. long; fruit 1–1.5 cm. broad, 8–10 mm. high, wings 3–6 mm. wide, glabrous, greenish white; style solitary, 2 mm. high; stigma capitate.

Shrubby slopes and margins of grassy openings, Lower Sonoran and Tropical Zones, southern Sonora to Chihuahua, Jalisco, and Aguascalientes. Barely reaches the fringe of the desert in southern Sonora. March–April.

### 2. MAYTENUS Molina, Sagg. Chile 177, 1782

Trees or shrubs with alternate, entire or dentate, evergreen leaves and minute deciduous stipules. Flowers polygamodioecious, solitary or in axillary panicles. Calyx lobes 5. Petals 5, spreading. Stamens 5, inserted below annular disk; margin of disk undulate. Ovary 2–4-celled, its base embedded in and adherent to disk; cells 1–2-ovuled; style short or stigma sessile, 2–4-lobed. Fruit a coriaceous 1–3-locular capsule. Seeds erect, embedded in a fleshy aril.

**1. Maytenus phyllanthoides** Benth., Bot. Sulph. 54, 1844
*Tricerma crassifolium* Liebm., Nat. For. Kjøbnh. Vid. Medd. **1853:** 98, 1854.

Shrub or rarely a small tree, usually 2–4 m. tall in our area but to 8 m. tall in favorable localities; branches often straggly, crown rounded; bark light gray, smooth; leaves fleshy-leathery, obovate, 2–4 cm. long, 1–2 cm. wide, obtuse, rounded, or shallowly emarginate at apex, broadly cuneate basally, entire, glabrous and glaucous, veins inconspicuous; flowers in few-flowered axillary clusters, short-pedicellate; sepals broadly triangular-ovate, 0.5 mm. long or less; petals about 1.5 mm. long; capsules triangular in cross section, ovoid, 7–10 mm. long, reddish brown, coriaceous, glabrous; seeds enclosed in a bright red, fleshy aril.

Chiefly along the coast but also in saline soil inland, Lower Sonoran and Tropical Zones, from the vicinity of Magdalena Bay and Isla de la Guarda, Baja California, and Tiburón Island, Sonora, southward to Yucatán and Puebla, the West Indies, and southern Florida. A native near Santa Rosalia reported that the arils are a source of ink; the leaves used as a medicine for several ailments. April–Nov.

## 3. SCHAEFFERIA Jacq., Enum. Pl. Carib. 10, 1760

Shrubs with alternate or fasciculate, coriaceous, entire, estipulate leaves. Flowers solitary or in few-flowered axillary clusters, dioecious. Calyx lobes 4. Petals 4, oblong. Stamens 4, inserted below or at margin of small disk. Ovary sessile, free, 2-celled; style short; stigma 2-lobed, ovules solitary in each cell. Fruit a 2-seeded, scantily fleshy drupe.

Foliage, branchlets, and pedicels glabrous; fruit sessile or subsessile ........ 1. *S. cuneifolia*
Foliage, branchlets, and pedicels spreadingly puberulent; fruits on pedicels 4–7 mm. long ........ 2. *S. shrevei*

**1. Schaefferia cuneifolia** A. Gray, Pl. Wright. **1:** 35, 1852

Shrub 1–2 (rarely –3) m. tall with intricately branched, rigid, subspinescent twigs; bark light gray; spur branches mere knobs 1–2 mm. high; leaves cuneate-obovate, sessile, obtuse to shallowly emarginate at apex, 2–8 mm. wide, 5–15 mm. long, glabrous, coriaceous, pale green or somewhat glaucous; flowers sessile or on pedicels to 2 mm. long; sepals about 0.5 mm. long; petals about 3 mm. long; fruit subglobose, slightly obcompressed, shallowly bisulcate longitudinally between 2 seeds, 3–5 mm. long and wide, 2.5–4 mm. thick, flesh scanty, orange to bright red, translucent, shining.

Rocky hillsides and mesas, Lower Sonoran Zone, Texas and Coahuila to Tamaulipas and in central Baja California. April–July.

**2. Schaefferia shrevei** Lundell, Contr. Univ. Mich. Herb. No. 4: 16, 1940

Shrub to 4 m. tall with slender semirigid branchlets and dark gray or brownish bark; spur branches bearing fascicles of leaves less pronounced than in last; leaves cuneate-obovate to cuneate-oblanceolate, 5–12 mm. wide, 1–2.2 cm. long, on pedicels 1.5–3 mm. long, persistently puberulent with spreading hairs, pale green on both surfaces, rounded or faintly emarginate at apex; pedicels 4–7 mm. long in fruit, puberulent; sepals 4, ovate, about 0.5 mm. long, persistent; petals unknown; fruit ovoid to subglobose, 5–7 mm. long, sparsely puberulent with spreading hairs, usually 2-seeded, shallowly bisulcate longitudinally between seeds; flesh scanty, red or orange.

Dry hillsides, mesas, and rocky canyons, Lower Sonoran Zone, central Sonora and central Baja California. (Near Horcasitas, Divisadero, and Mátape, Sonora, and above Loreto in Sierra de la Giganta, Baja California.) May–July?

### 4. **MORTONIA** A. Gray, Pl. Wright. **1:** 34, *pl. 4,* 1852

Intricately branched, low shrubs with crowded, evergreen, coriaceous, alternate leaves, margins revolute. Stipules gland-like, minute, caducous. Flowers white, small, in terminal thyrse-like cymes. Calyx tube 10-ribbed; lobes 5. Petals 5. Stamens 5. Ovary 5-celled; style short; stigma 5-lobed; ovules basal, erect, 2 in each cell. Fruit dry, indehiscent, 1-celled and 1-seeded by abortion. Seed oblong, without an aril.

**1. Mortonia scabrella** A. Gray, Pl. Wright. **2:** 28, 1853

Yellowish green shrub to 2 m. tall with erect, stiff, rather brittle branches, yellowish gray bark and moderately crowded, subsessile leaves, scaberulous throughout; leaves broadly elliptic, 2.5–5 mm. wide, 3–8 mm. long, spreading or slightly ascending, curving upward and margins revolute; inflorescences 1–2 cm. wide, 1.5–3 cm. long; calyx about 2 mm long, lobes broadly rounded-deltoid, about 0.5 mm. high, margins whitish and subscarious; petals obovate-oblong, 2–2.5 mm. long; fruit ovoid, longitudinally 6–10-ribbed, 2–2.5 mm. wide, 3.5–4.5 mm. long, greenish, crustaceous, style short, persistent.

Dry plains, hillsides, and mesas, Lower and Upper Sonoran Zones, in Gila, Cochise, and Pima Counties, Arizona, to western Texas and adjacent northern Mexico. Barely entering the Sonoran Desert in southern Pima County. March–Sept.

**Glossopetalon nevadense** A. Gray, Proc. Amer. Acad. **11:** 73, 1876, a shrub with rigid branchlets, small oblanceolate leaves, 4–5-merous axillary flowers, and fruit an ovoid follicle about 5 mm. long, occurs in the Mojave Desert and in northern Arizona, but present records do not reveal it within the limits of the Sonoran Desert.

## Family 78. **SAPINDACEAE.** Soapberry Family

Small trees, shrubs, vines, or sometimes herbs, often with tendrils. Leaves usually alternate, usually compound but sometimes simple, usually persistent. Flowers small, polygamodioecious or less commonly perfect. Sepals 4 or 5, distinct or somewhat connate. Petals 3–5 or wanting. Stamens mostly 8. Ovary superior, 1–4-celled. Fruit a capsule, drupe or berry, or samara.

Flowers irregular; stamens 8, excentric; leaves odd-pinnate:
  Fruit samaroid, indehiscent .................................................. 1. *Serjania*
  Fruit capsular, dehiscent:
    Leaves without stipules; capsule inflated, loculicidal; seeds not arillate ...............
    ........................................................ 2. *Cardiospermum*
    Leaves stipulate; capsule not inflated, septicidal; seeds slightly arillate ..... 3. *Paullinia*
Flowers regular; stamens 5–10, attached centrally; leaves simple or even-pinnate, terminal leaflet absent:
  Petals present; fruit fleshy, not winged; leaves pinnate ................... 4. *Sapindus*
  Petals lacking; fruit a septicidal capsule, winged; leaves simple ........... 5. *Dodonaea*

### 1. **SERJANIA** Schumach., Skrivt. Naturh.-Selsk. (Copenhagen) **3**[2]**:** 125, 1794

Scandent to erect shrubs or sometimes vine-like, with alternate, estipulate, trifoliolate, biternate, or unevenly pinnate leaves; leaflets usually pellucid-dotted. Flowers yellowish, irregular, polygamodioecious. Sepals 5, exterior two smaller than inner three. Petals 4. Scales more or less petaloid. Disk undulate. Stamens 8, excentric; filaments connate at base. Ovary sessile or short-stipitate, 3-celled, each cell 1-ovuled. Fruit of 3 samaras; seed borne at apex, wings at base of samaras.

Leaves and twigs puberulent to hispidulous; sepals puberulent:
  Twigs 8–10-sulcate or 8–10-striate; leaves 5-foliolate ...................... 1. *S. palmeri*
  Twigs 5–6-sulcate; leaves trifoliolate ...................................... 2. *S. albida*
Leaves, twigs, and sepals glabrous or glabrate .......................... 3. *S. californica*

**1. Serjania palmeri** S. Wats., Proc. Amer. Acad. **24:** 45, 1889

A stout, scrambling or climbing plant, pubescent throughout with soft spreading hairs; leaves bipinnate, lowest pinnae usually 5-foliolate, middle one trifoliolate, and terminal one a single leaflet; leaflets entire or occasionally toothed, elliptic to oblong-lanceolate, acute, the terminal one 1–1.5 cm. long, 3–4.5 mm. wide, lateral leaflets about half as large as terminal, cuneate at base; peduncles equaling or exceeding leaves, shorter than narrow panicles; sepals short-puberulent with crisped hairs, outer two one-half to three-fourths as long as inner, latter 3 mm. long; petals 3.5–4 mm. long, densely glandular within, glabrous and eglandular without; stamens about equaling petals, pubescent; fruit 18–25 mm. long, cells terminal, subglobose, narrowly crested on back, wings broadly expanded to a rounded base 6–9 mm. wide; seeds globose, filling cavity in samara.

On rocky hillsides, Lower Sonoran Zone, vicinity of Guaymas and southward into southern Sonora and northern Sinaloa. Nov.–April.

The flowers are very fragrant.

**2. Serjania albida** Radlk., Contr. U.S. Nat. Herb. **1:** 367, 1895

Scandent suffruticose plant 1 m. tall or less with slender, 5–6-angled stems, ridges of angles glabrous, grooves between them finely puberulent; bark whitish, papery; leaves trifoliolate, 3–5 cm. long, terminal leaflet 1.5–2 cm. long on a petiolule 1–1.5 cm. long, lateral leaflets smaller, ovate, 3-toothed or shallowly 3-lobed, glabrous or sparsely appressed-puberulent, punctate-pellucid; thyrses about equaling leaves; sepals puberulent; additional details of flowers and fruits unknown, hence the true position of this plant is uncertain.

Known only from the type locality, Santa Agueda, Baja California.

**3. Serjania californica** Radlk., Sitzungsb. Akad. Wiss. München **8:** 222, 1878

Scandent or shrubby, with glabrous branches and foliage; branchlets slender and lightly 6-sulcate; leaves pinnately 5-foliolate, about 3.5 cm. long; leaflets about 1 cm. long, ovate, obtuse, entire or shallowly lobed or dentate, terminal gradually narrowing to petiolule, lateral ones subsessile, all of them pale green, thin, glabrous; inflorescence about 2 cm. long, on a short peduncle tipped with 2–3 tendrils; sepals small, minutely puberulent to glabrous within, glabrous on outside; petals narrowly obovate, glanduliferous within; scales of upper petals obcordate, those of lower petals cristate; stamens villous on lower parts of filaments; fruits unknown.

Southern Baja California. Dec.–Feb.

Owing to the absence of fruit on all known specimens, the generic position of this plant is uncertain.

## 2. CARDIOSPERMUM L., Sp. Pl. 366, 1753

Herbaceous or fruticose vines with estipulate, 3-foliolate or decompound leaves. Leaflets deeply crenate or serrate, often pellucid-punctate. Inflorescence a raceme or corymb, axillary. Flowers irregular, plants polygamodioecious. Sepals 4–5, two outer smaller than inner 2 or 3, or rarely nearly equal. Petals 4. Disk unequally developed, lobed or undulate. Stamens 8, eccentric. Ovary 3-celled; style short, 3-parted. Ovules solitary in each cell. Capsule 3-angled, inflated, loculicidally dehiscent. Seeds globose.

Plants shrubby, not climbing; sepals 5:
- Tendrils well developed, coiled ........................................ 1. *C. tortuosum*
- Tendrils reduced to spines, straight or curved ........................ 2. *C. spinosum*

Plants vinelike, climbing over shrubs; sepals 4 ......................... 3. *C. corindum*

### 1. Cardiospermum tortuosum Benth., Bot. Sulph. 9, *pl. 6,* 1844

*Paullinia tortuosa* Brandegee, Zoe **2:** 74, 1891.

A rigidly branching shrub with tortuous branches and subflexuous, slender, terminal twigs densely tomentulose; leaves pinnately compound or bipinnatifid, about 5 cm. long, on petioles 1–2 cm. long; leaflets 3–7, terminal one 1–1.5 cm. long, 8–10 mm. wide, 3-lobed to 3-divided, glabrous to minutely puberulent; lateral leaflets 6–8 mm. long, 2–3 mm. wide, entire or rarely faintly lobed near apex; inflorescences 3–4 cm. long; flowers 4.5–5 mm. long; sepals 5, two outer shorter than inner three, inner ones oblong, 3–3.5 mm. long; petals 4, obovate to spatulate, 4–4.5 mm. long, claws 1–1.5 mm. long; scales about half as long as petals, margins minutely ciliolate; stamens minutely hairy; capsule papery, angled, 1–1.3 cm. high, about 1.5–2 cm. broad, short-stipitate, minutely puberulent; seeds about 7 mm. in diameter, hilum semilunate, 4–4.5 mm. wide.

Desert plains and washes, Lower Sonoran Zone, Magdalena Bay, San José del Cabo, and Isla Cerralvo, Baja California. Apparently flowering in any month if sufficient moisture available.

### 2. Cardiospermum spinosum Radlk., Contr. U.S. Nat. Herb. 1: 368, 1895

*Paullinia spinosa* I. M. Johnston, Proc. Calif. Acad. IV, **12:** 1083, 1924.

A rounded, low, compact shrub to 1 m. tall with rather slender branches with puberulent to glabrate twigs and leaves; leaves 2.5–3 cm. long; leaflets 3–5, crenate, terminal one shallowly 3-lobed, 8–15 mm. long, lateral leaflets crenate, 1 cm. long or less; inflorescence to 1.5 cm. long, few-flowered, on a peduncle 1–2.5 cm. long, 2 short, curved or weakly coiled spines 4–8 mm. long replacing tendrils just below flower cluster; sepals 5, inner three about 4 mm. long, outer about two-thirds to three-fourths as long, all of them minutely puberulent; petals 4, obovate, 3.5–4 mm. long, minutely glandular within; scales about half as long as petals, densely hairy on lower portion; stamens puberulent; capsule 3-angled, closely pubescent.

Rocky slopes and hillsides, Lower Sonoran Zone, vicinity of La Paz and near southern end of Bahía de la Concepción, Baja California. Any month.

### 3. Cardiospermum corindum L., Sp. Pl. ed. 2, 526, 1762

*Cardiospermum palmeri* Vasey & Rose, Proc. U.S. Nat. Mus. **13:** 147, 1890.

A suffruticose vine or scrambling subshrub with slender, glabrous to densely tomentulose branches and 2 strongly coiled tendrils just below flower clusters near tip of each peduncle; leaves extremely variable as to shape, size, and pubescence, 3–10 cm.

long, pinnately or bipinnately compound with crenate to lobed leaflets, primary pinnae 3–9, divisions broadly ovate to lanceolate, 5–20 mm. long, sparsely puberulent to densely tomentulose or hispidulous; tendrils strongly coiled; sepals 4, outer pair 1–1.5 mm. long, ovate, inner pair 3–4 mm. long, each bearing 2 callous glands at base; petals white or pinkish, 4–5 mm. long; scales about two-thirds as long as petals, ovate, petaloid or club-shaped and slightly flattened; fruit papery, 2–3 cm. long, to 3.5 cm. wide, glabrate and strongly veined; seeds globose, 4–5 mm. in diameter, hilum suborbicular but shallowly emarginate.

Along arroyos, on hillsides, and among shrubs on arid plains, Lower Sonoran Zone, Sonora and Baja California and southward into tropical America and the West Indies. Throughout the year.

This is an extremely variable species and numerous varietal, as well as specific, names have been applied to the variants. The writer prefers to assign the Sonoran Desert material to the species only. Radlkofer (Pflanzenreich **4**[165.1]**:** 397–407, 1932) recognized fourteen "formas." For the extensive synonymy see Radlkofer's paper cited above.

Much of the material from Mexico that belongs in this species has been referred to *C. halicacabum*. That species is an annual herbaceous vine or scrambler. All of the material from our area that has come to my notice is definitely woody, at least at the base, and perennial.

## 3. PAULLINIA L., Sp. Pl. 365, 1753

Scandent shrubs or vinelike with alternate, stipulate, pinnately or ternately compound leaves, leaflets dentate or entire, often punctate-glandular. Flowers slightly irregular, polygamodioecious, in axillary racemes. Sepals 5, concave and imbricate, two upper ones more or less connate. Petals 4, two larger bearing scales longer than those of smaller pair. Stamens 8, excentric. Ovary sessile or stipitate, 3-celled, each cell 1-ovuled. Fruit 3-angled, scarcely inflated, septicidally dehiscent. Seeds globose, slightly arillate at base.

### 1. **Paullinia sonorensis** S. Wats., Proc. Amer. Acad. **24:** 45, 1889

A subscandent shrub 1 m. tall or less; leaves pinnate, to 8 cm. long, glabrous or glabrate, basal pinnae sometimes pinnate; leaflets ovate, 8–14 mm. wide, 1–2.5 cm. long, crenate or sublobed, narrowly to broadly cuneate at base, dark green above, slightly paler beneath, rachis often narrowly winged; peduncles 3–5 cm. long, bearing 2 coiled tendrils just below flowers, sparsely puberulent; sepals ovate and orbicular, shorter about 2 mm. long, inner three nearly 3 mm. long, minutely puberulent; petals about equaling longer sepals; capsules somewhat angular, depressed-globose, 8–10 mm. long, tomentulose.

Rocky slopes, bases of cliffs, and dry canyons, Lower Sonoran Zone, Sonora and central and southern Baja California. Aug.–Nov.

## 4. SAPINDUS L., Sp. Pl. 367, 1753

Trees or shrubs or rarely subscandent. Leaves alternate, estipulate, simple or (in ours) pinnately compound. Flowers regular, the plants polygamodioecious. Sepals 4–5. Petals 4–5. Disk annular or shallowly cup-shaped. Stamens mostly 8–10, centrally arranged; filaments distinct, often hairy. Ovary entire or 2–4-lobed, 2–4-celled;

stigma 2–4-lobed; ovules 1 in each cell. Fruit a berry with yellowish or amber, translucent flesh. Seeds globose.

Leaflets obtuse to acute at apex, not acuminate; petals obovate to orbicular-obovate, about one-third longer than sepals .......................... 1. *S. saponaria* var. *saponaria*
Leaflets acuminate at apex; petals rhombic-lanceolate, about twice as long as sepals ........ ............................................. 1a. *S. saponaria* var. *drummondii*

**1. Sapindus saponaria** L., Sp. Pl. 367, 1753, var. **saponaria**

*Sapindus marginatus* Willd., Enum. Hort. Ber. 432, 1809.
*Sapindus inaequalis* DC., Prod. **1:** 608, 1824.
*Sapindus amolli* Sessé & Moc., Pl. Nov. Hisp. 60, 1888.

A tree as much as 16 m. tall with a broad dense crown and dark gray, fissured or flaky bark; leaves pinnate, 12.5 dm. long; leaflets 5–11, elliptic-lanceolate, 1.5–4.5 cm. wide, 5–15 cm. long, acute to obtuse at apex, asymmetrically acute to rounded at base, entire, somewhat coriaceous, dark green and sparsely puberulent along midrib above, yellowish green and minutely puberulent beneath, rachis faintly to distinctly winged between leaflets; inflorescence of terminal panicles 1–2 dm. long, rachis puberulent; sepals minute, about 1.5 mm. long; petals obovate to orbicular-obovate, strongly cupped, whitish, about 2 mm. long, margins ciliolate; berries 10–15 mm. in diameter, yellow, often darkening in drying.

At low altitudes in valleys and along watercourses, less commonly on hillsides, Lower Sonoran and Tropical Zones, central Sonora and southern Baja California south and east throughout Mexico, the West Indies, and the southern tip of Florida. Occurring in the Sonoran Desert only along watercourses. May–Aug.

**1*a*. Sapindus saponaria** var. **drummondii** (Hook. & Arn.) L. Benson, Amer. Jour. Bot. **30:** 239, 1943

*Sapindus drummondi* Hook. & Arn., Bot. Beechey 281, 1836.

Leaves 1–2 dm. long; leaflets 9–17, lanceolate-acuminate, 4–16 mm. wide, 3–12 cm. long, gradually tapering to acuminate apex, unequally and abruptly cuneate at base, glabrate above, sparsely puberulent beneath; petals broadly rhombic-lanceolate, copiously white-ciliate near base, about 2 mm. long, twice as long as orbicular sepals; otherwise as in species.

Along streams and canyons, Lower and Upper Sonoran Zones, Arizona to Oklahoma, Texas, Tamaulipas, and northern Sonora. May–Aug.

This narrow-leaved form grades into the blunter larger leaflet of the typical plants.

## 5. DODONAEA Jacq., Enum. Pl. Carib. 19, 1760

Erect shrubs or small trees with simple (in ours) or pinnate, resinous-glutinous leaves. Stipules lacking. Inflorescence of small lateral corymbs. Flowers inconspicuous, greenish white. Sepals 2–5; imbricate or valvate. Petals wanting. Stamens 5–8, centrally attached; anthers linear-oblong; disk lacking in staminate flowers. Ovary sessile, 3–6-angled and 3–6-celled; style 3–6-branched; ovules 2 in each cell; disk present but small in pistillate flowers. Capsule membranous or leathery, angles narrowly winged, septicidally dehiscent. Seeds lenticular to globose, without an aril. Embryo spirally coiled.

Leaves broadly oblong-spatulate, to 1.5 or 2 cm. broad, widest above the middle .......... .................................................... 1. *D. viscosa* var. *viscosa*
Leaves linear or narrowly oblong, mostly 5 mm. or less in width, widest about at the middle .............................................. 1a. *D. viscosa* var. *angustifolia*

**1. Dodonaea viscosa** Jacq., Enum. Pl. Carib. 19, 1760 var. **viscosa**

*Dodonaea schiedeana* Schlecht., Linnaea **18:** 33, 1844.
*Dodonaea lagunensis* M. E. Jones, Contr. West. Bot. **18:** 57, 1933.

An erect shrub 1–5 m. tall; leaves oblong or oblong-spatulate, 1–1.5 (or –2) cm. broad, 3–7 cm. long, gradually tapering to cuneate base, abruptly acute to short-acuminate at apex, resinous-glutinous and shiny above, slightly duller beneath, pellucid-punctate, margins faintly subrevolute; pedicels slender, 3–6 mm. long, gland-dotted; sepals oblong, about 2–2.5 mm. long, greenish white, anthers about equaling sepals, subsessile; fruits 8–12 mm. high, 10–18 mm. wide, including wings, style short, set in bottom of a broad notch 3–5 mm. deep, wings 3, papery, veined, 2–6 mm. wide beyond body of capsule; seeds black, lenticular, 2.5–3.5 mm. wide, dull.

Gravelly plains, along arroyos, and on rocky hillsides, Lower Sonoran and Tropical Zones, Baja California (southern Sierra San Pedro Mártir), Hidalgo, and Tamaulipas south to South America; widely distributed in the tropics. Feb.–Oct.

An extremely variable species, at least as to leaf form, and numerous varieties and forms have been described.

**1*a*. Dodonaea viscosa** var. **angustifolia** (L. f.) Benth., Fl. Austral. **1:** 476, 1863

*Dodonaea angustifolia* L. f., Suppl. Syst. Veg. 218, 1781.
*Dodonaea arizonica* A. Nels., Amer. Jour. Bot. **21:** 576, 1934.
*Dodonaea viscosa* var. *linearis* f. *angustifolia* Sherff, op. cit. **32:** 214, 1945.
*Dodonaea viscosa* var. *linearis* f. *arizonica* Sherff, loc. cit.

Differing from var. *viscosa* only in having linear to linear-oblong leaves rarely over 5 mm. wide; flowers and fruit as in var. *viscosa.*

Gravelly plains, arroyos, and rocky slopes, Lower Sonoran Zone, Arizona to northern Sinaloa.

**Thouinia villosa** Sessé & Moc. ex DC., Prod. **1:** 612, 1824, has been reported from Sonora, but the writer has seen no specimens collected within the limits of the Sonoran Desert.

## Family 79. **RHAMNACEAE.** Buckthorn Family

Shrubs or small trees with simple, usually alternate but sometimes opposite leaves and small, deciduous or caducous stipules. Flowers small, regular, perfect, or polygamodioecious, in small umbels, these often arranged in racemes or panicles. Calyx lobes 4–5, persistent or deciduous; calyx tube lined with a disk, turbinate to obconic. Petals 4–5 or wanting, often cucullate and usually clawed. Stamens 4–5, inserted on edge of disk, opposite petals and often enfolded by them. Ovary superior but often somewhat embedded in disk, sometimes inferior, 2–4-celled; ovules 1–2 in each cell. Fruit a drupe or capsule. Seeds with a cartilaginous or bony testa.

Shrubs or trees; tendrils none; ovary superior; fruit not winged:
  Fruit with a single 1–4-celled stone, drupaceous:
    Leaves 3–5-nerved from base; some stipular spines present .............. 1. *Zizyphus*
    Leaves pinnately veined; stipular spines none:
      Branches spine-tipped; each cell of ovary 1-ovuled; leaves small, usually fasciculate:
        Petals none; fruit 3–6 mm. long ................................. 2. *Condalia*

Petals present; fruit 6–15 mm. long .............................. 3. *Condaliopsis*
Branches not spine-tipped; each cell of ovary 2-ovuled, one often abortive; leaves ample, alternate or subopposite, not fasciculate ................ 4. *Karwinskia*
Fruit capsular or drupaceous but containing 2–4 distinct or only slightly coherent stones:
Leaves ample (mostly over 8 mm. long), persistent; calyx tube not 10-ribbed:
Flowers sessile, paniculate; branches pungently spine-tipped ........... 5. *Sageretia*
Flowers pedicellate, solitary or in small umbels; branchlets unarmed or bluntly spine-tipped (rigidly so in *Rhamnus crocea*):
Fruit drupelike; ovary 2–4-celled; leaves more or less translucent; ovary wholly superior, free from calyx ...................................... 6. *Rhamnus*
Fruit capsular, dry at maturity; ovary 3-celled, splitting into 3 cocci; leaves not translucent; ovary embedded in disk and calyx on lower one-third to one-half:
Calyx lobes petaloid; petals white, bluish, or lavender, often spreading away from stamens ................................................ 7. *Ceanothus*
Calyx lobes not petaloid; petals yellowish or greenish, enfolding or standing close to stamens ................................................ 8. *Colubrina*
Leaves small (mostly less than 8 mm. long), soon deciduous; calyx tube distinctly 10-ribbed ........................................................ 9. *Adolphia*
Vines, bearing tendrils; ovary inferior; fruit 3-winged ...................... 10. *Gouania*

## 1. ZIZYPHUS Mill., Gard. Dict. Abridg. Ed. 4, 1754

Shrubs or trees with alternate or opposite, coriaceous, entire or crenate, 3–5-nerved, deciduous leaves; stipules often developing into spines. Flowers in few-flowered axillary cymes. Calyx tube obconic; lobes 5, deltoid. Petals 5, cucullate, deflexed. Stamens 5. Ovary 2-celled or rarely 3–4-celled, adnate to disk at base. Drupe globose or oblong; nut woody or bony, 1–3-celled and 1–3-seeded. Seeds plano-convex.

### 1. Zizyphus sonorensis S. Wats., Proc. Amer. Acad. **24:** 44, 1889

*Zizyphus seleri* Loesener, Verh. Bot. Brandenburg **51:** 29, 1909.
*Zizyphus endlichii* Loesener, Rep. Spec. Nov. **8:** 296, 1910.

Shrub or small tree to 12 m. tall; young branches green, stoutish, puberulent; stipular spines 2–8 mm. long or wanting; leaves ovate to orbicular-ovate, 3–6 cm. long, 2–3.5 cm. wide, entire or remotely crenate, rounded, obtuse or emarginate at apex, obtuse to rounded at base, densely pubescent to glabrate, veins prominent beneath; peduncles equaling or surpassing leaves; calyx lobes 1–1.5 mm. long, puberulent, soon deciduous; fruit 8–10 mm. long, globose or subglobose, reddish.

Bottom lands, desert flats, and along arroyos, Lower Sonoran Zone, southern Baja California and Sonora to Oaxaca. Aug.–Sept.

## 2. CONDALIA Cav., Anal. Hist. Nat. (Madrid) **1:** 39, 1799

A rigidly branched shrub or small tree; twigs all spine-tipped. Leaves alternate, subsessile, coriaceous, entire, deciduous. Stipules minute, deciduous. Flowers solitary or fascicled in axils, small, short-pedicellate. Calyx tube broadly obconic, shallow; lobes 5, ovate to deltoid. Petals none. Disk flat, fleshy, 5-angled on free margin. Stamens 4–5; filaments subulate. Ovary 1–2-celled, 2-ovuled, free from disk; style short, 2–3-lobed or stigmas 3 and distinct. Fruit ovoid to globose, drupaceous, usually 1-locular. Seeds compressed or subglobose.

Fruiting pedicels 0.5–8 mm. long; branches slender, often flexuous; leaves with broad, conspicuous lateral veins:
  Sepals persistent in fruit; berries ovoid to ellipsoid, edible, not bitter; shrub to 3 m. tall:
    Fruits ellipsoid, 3–6 mm. long, flesh scanty; pedicels 0.5–3 mm. long ................ 1. *C. warnockii* var. *kearneyana*
    Fruits obovoid, 6–9 mm. long; flesh copious; pedicels 2.5–8 mm. long ................ 2. *C. brandegeei*
  Sepals deciduous in fruit; berries globose or subglobose, very bitter; plant usually a small tree to 4 m. tall:
    Leaves, twigs, pedicels, and calyces glabrous or very early glabrate ................ 3. *C. globosa* var. *globosa*
    Leaves, twigs, pedicels, and calyces persistently puberulent ................ 3a. *C. globosa* var. *pubescens*
Fruiting pedicels 0.5 mm. long or less, fruit usually nearly sessile; branches stout, rigid; lateral veins inconspicuous ................ 4. *C. correllii*

**1. Condalia warnockii** M. C. Johnst. var. **kearneyana** M. C. Johnston, Brittonia **14:** 354, 1962

*Condalia spathulata* of authors, in part, not A. Gray, 1852.

An intricately branched, compact shrub 1–3 m. tall with slender, divaricately spreading, somewhat flexuous, spine-tipped branches and twigs, these and foliage puberulent or sometimes glabrous; stipules about 1 mm. long; petioles 0.5–0.8 mm. long, puberulent; leaf blades narrowly cuneate to spatulate, 1–2.5 mm. wide, usually 3–8 mm. long, acute, obtuse, or rounded and minutely mucronulate at apex, somewhat wrinkled beneath; lateral veins broad, conspicuous; flowers in small clusters or solitary on short spurs; pedicels slender, 0.5–2 mm. long (sometimes to 3 mm. long), puberulent; calyx lobes deltoid, 0.8–1.5 mm. long, suberect, puberulent without, mostly persistent in fruit; petals none; stamens slightly shorter than to nearly equaling calyx lobes, inflexed; disk purplish, thin; drupe ellipsoid or narrowly obovoid, 3–6 mm. long, beakless or short style subpersistent; pericarp purplish or blackish, juicy and edible, but scanty.

Sandy slopes and mesas of desert grasslands, Upper and Lower Sonoran Zones, Cochise, Pima, Pinal and Santa Cruz Counties, Arizona, south into central Sonora. July–Sept.

It is sometimes difficult to distinguish between this and *C. globosa* var. *pubescens* on the basis of herbarium sheets, but the intensely bitter drupes of *C. globosa* set the two apart in the field. The sepals of *C. warnockii* var. *kearneyana* usually persist in fruit, while they are generally deciduous in *C. globosa* var. *pubescens.*

**2. Condalia brandegeei** I. M. Johnston, Proc. Calif. Acad. IV, **12:** 1088, 1924

Intricately branched shrub 1–3 m. tall; branches slightly zigzag, with dark gray, slightly fissured, thin bark and red-brown, small lenticels; spines 5–18 mm. long, smooth and shining, reddish brown, not striate; young branchlets very slender, yellowish green, about 1 mm. in diameter; stipules linear-subulate, 1–1.5 mm. long, ciliate-hirsute; petioles slender, 1.5–3(5) mm. long, slightly puberulent on upper surface; leaf blades leathery, obovate to suborbicular, broadly cuneate at base, rounded to faintly emarginate and minutely apiculate at apex (apiculation 0.2–0.6 mm. long), 5–14 mm. wide, 8–20 mm. long, conspicuously and finely reticulate-veined and pale beneath, dark green, shining above; flowers solitary or in 2–3-flowered fascicles; cup 2.5–2.8 mm. in diameter, glabrous, greenish; sepals deltoid, 1–1.5 mm. long, nearly as wide at base, brownish, acute, glabrous; stamens about half as long as sepals; fruiting pedicels 2.5–8 mm. long; fruits elliptic, 4–5.5 mm. in diameter, 6–9 mm. long, acute at each end; style persistent, forming an apiculation 0.6–0.8 mm. long; flesh moderately copious, purplish, not glaucous.

Canyons and outwash slopes, Lower Sonoran Zone, central Baja California, Mexico. Feb.–April.

### 3. **Condalia globosa** I. M. Johnston, Proc. Calif. Acad. IV, **12**: 1086, 1924, var. **globosa**

An intricately branched, compact, rounded shrub or small tree to 4 m. tall, with slender, spine-tipped twigs, reddish and slightly pruinose, glabrous young branches; stipules deltoid, 0.6–0.8 mm. long; petioles slender, 0.5–2.5 mm. long; leaves in dense fascicles on short spurs (alternate on new twigs), 1–5 mm. broad, 4–10(14) mm. long, entire, acute to minutely emarginate at apex, acuminate-cuneate at base, lateral veins 3–4 pairs, broad and conspicuous beneath; flowers in small clusters or solitary on spurs; pedicels very slender, 4–7 mm. long; sepals narrowly deltoid, 0.6–1 mm. long, acute, glabrous, spreading or ascending at anthesis, deciduous in fruit; petals none; drupe globose, rarely slightly elongated, 3.5–5.5 mm. in diameter, blackish, devoid of bloom; style often persistent, about 1 mm. long, filiform; pericarp juicy, thin, very bitter.

Gravelly slopes and flats, Lower Sonoran Zone, central Baja California to the Cape Region, and in central Sonora. Feb.–April; Oct.–Nov.

### *3a.* **Condalia globosa** var. **pubescens** I. M. Johnston, Proc. Calif. Acad. IV, **12**: 1087, 1924

Differing from var. *globosa* only in having densely short-puberulent leaves, twigs, pedicels and calyces.

Colorado Desert to Pima County, Arizona, and south into Sonora and central Baja California. Flowering following rains.

### 4. **Condalia correllii** M. C. Johnston, Brittonia **14**: 357, 1962

*Condalia mexicana* of authors, in part, not Schlecht., 1841.

A rigidly branching shrub 1–3.5 m. tall; twigs very stout and rigid, 3–3.5 mm. in diameter, puberulent in youth and often persistently so; petioles 1–3 mm. long, puberulent; leaves alternate, less commonly in fascicles; blades coriaceous, obovate to oblanceolate, 3–6 mm. wide, 7–15 mm. long, obtuse, rounded or emarginate at apex, cuneate at base, entire, puberulent or at length glabrate; veins 4–5 pairs, narrow and inconspicuous; flowers nearly or quite sessile, in axils of short spurs; calyx lobes deltoid, 1–1.5 mm. long, purplish green, puberulent without; petals none; stamens shorter than sepals, introrse; drupes sessile or subsessile, obovoid to broadly elliptic or nearly globose, bluish black, 4–8 mm. long; style slender, persistent or tardily deciduous, forming a short beak or fruit beakless.

Rocky slopes and canyons, mainly Upper Sonoran Zone, but entering the upper edges of the Sonoran Desert in the Baboquivari Mountains, Arizona, to eastern Arizona and south into Sonora, Chihuahua, and Coahuila, Mexico. July–Oct.

**Condalia ericoides** (A. Gray) M. C. Johnston, Brittonia **14**: 364, 1962 (*Microrhamnus ericoides* A. Gray, Pl. Wright. **1**: 34, 1852), in general habitat is similar to *Condalia correllii* but its leaves are linear, needlelike, 3–7 mm. long, tightly revolute and forming 2 narrow furrows between the midrib and the margins, these furrows densely tomentose and sometimes obscured when the margins are closely inrolled; drupes ellipsoidal, about 1 cm. long.

This plant occurs in the Santa Catalina, Santa Rita, and Baboquivari Mountains in Arizona, east to Texas and south into Chihuahua. It occurs commonly in the Upper Sonoran Zone, but may grow as a straggler along watercourses at the upper margins of the Sonoran Desert.

## 3. CONDALIOPSIS Suesseng., in Engler & Prantl, Nat. Pflanzenf. ed. 2, 20d: 134, 1953

Shrubs with stiff to rigid branches; some or all twigs spine-tipped. Leaves alternate, small to moderate in size, entire, pinnately veined; stipules never developing into spines. Flowers solitary or fascicled in axils on short spur branchlets; pedicels short. Sepals 5. Petals 5. Disk lining hypanthium, entire or nearly so. Stamens 5, ascending or somewhat introrse. Ovary free, 2-celled by intrusion and overlapping of two opposite placentae. Fruit ellipsoid or obovoid, with dry-fleshy or juicy pericarp, short-beaked by persistent style or soon beakless; kernel bony or woody, 2-celled by fusion and lignification of placentae. Seeds two, or sometimes one by abortion of one ovule.

Fruit 1–2 cm. long, ellipsoid; pericarp dry-fleshy, not juicy, greenish; beak 2–4 mm. long, usually persistent; pedicels 8–20 mm. long in fruit, stoutish .............. 1. *C. parryi*
Fruit 6–8 mm. long, obovoid; pericarp juicy, black; beak 1 mm. long, or early deciduous; pedicels 2–6 mm. long, slender:
  Leaves 15–25 mm. long; stems and leaves glabrous even in youth; fruits without a bloom; young branchlets green, not white-crustose .......................... 2. *C. rigida*
  Leaves 8–19 mm. long; stems and leaves usually white-puberulent in youth; fruits with a bluish bloom; young branchlets canescent or white-crustose:
    Branchlets and leaves soon glabrate; leaves thin, 5–7 mm. wide .......................................................................... 3. *C. lycioides* var. *lycioides*
    Branchlets and leaves persistently canescent-tomentulose; leaves thickish, 5–9 mm. wide ............................................................. 3a. *C. lycioides* var. *canescens*

### 1. Condaliopsis parryi (Torr.) Suesseng., in Engler & Prantl, Nat. Pflanzenf. ed. 2, 20d: 135, 1953

*Zizyphus parryi* Torr., Bot. Mex. Bound. 46, 1859.

A glabrous shrub or small tree to 6 m. tall; ultimate twigs stiffly divaricate and strongly spine-tipped; leaves mainly fascicled on short spurs, elliptic to obovate, 8–20 mm. long, 5–12 mm. wide, entire, glabrous, bright green above and below, lateral veins slender, inconspicuous; petioles slender, 2–7 mm. long; flowers in small fascicles on short spurs; pedicels slender, 2–6 mm. long at anthesis, stoutish and 8–20 mm. long in fruit; calyx lobes narrowly deltoid, 1–1.5 mm. long, deciduous in fruit; petals narrowly oblanceolate, cucullate, slightly exceeding calyx lobes; drupe broadly ellipsoid, 1–2 cm. long, tipped with a broadly subulate beak 2–4 mm. long or this sometimes absent; pericarp thin and dry, greenish or tawny.

Desert hillsides and bajadas, Lower Sonoran Zone, western edge of the Colorado Desert southward into Baja California. Feb.–April.

**Condalia parryi** var. **microphylla** I. M. Johnston, Univ. Calif. Pub. Bot. **7:** 439, 1922 (*Ziziphus parryi* var. *microphylla* M. C. Johnston, Brittonia **14:** 368, 1962), was based on plants from Las Huevitas and from Cedros Island, Baja California and differs from var. *parryi* in having smaller leaves and slightly smaller fruits. Insufficient material is available to permit critical evaluation of the form.

**2. Condaliopsis rigida** (Wiggins) Wiggins, comb. nov.

*Condalia rigida* Wiggins, Contr. Dudley Herb. **4:** 20, *pl. 1, figs. 5–8,* 1950.
*Ziziphus obtusifolia* var. *canescens* M. C. Johnston, Brittonia **14:** 367, 1962, in part.

A rigidly branched, glabrous shrub to 2.5 m. tall with bright green branches and spine-tipped twigs 1–3 cm. long or more, spreading at nearly right angles from larger branches; leaves narrowly to broadly elliptic or sometimes obovate, 5–15 mm. broad, 15–25 mm. long, entire or obovate leaves slightly emarginate, bright green on both surfaces, veins narrowly and intricately reticulate beneath, obscure above; petioles 2–6 mm. long; flowers in dense fascicles or paniculate cymes; peduncles 3–5 mm. long, bifurcate near apex; pedicels slender, glabrous, 1–5 mm. long, slightly longer and stoutish in fruit; calyx lobes broadly deltoid, about 1.5 mm. long and wide, glabrous, deciduous in fruit; petals suborbicular, 1.5–1.8 mm. long, strongly cucullate, claw slender; drupes obovoid, 4.5–5.5 mm. in diameter, 6–8 mm. long, black or blue-black, without bloom; pericarp juicy, moderately copious.

Sandy washes and plains, Lower Sonoran Zone, southern Baja California, Mexico. In flower and fruit in April.

**3. Condaliopsis lycioides** (A. Gray) Suesseng., in Engler & Prantl, Nat. Pflanzenf. ed. 2, **20d:** 135, 1953, var. **lycioides**

*Zizyphus lycioides* A. Gray, Bost. Jour. Nat. Hist. **6:** 168, 1850.

An intricately branched shrub 1–3 m. tall with stout rigid spines at ends of twigs and spreading at right angles from larger branches; branches and twigs glabrous or sparsely puberulent with short spreading hairs, but soon glabrate; bark white-crustose; leaves narrowly elliptic to narrowly oblong, 5–7 mm. wide, 8–15 mm. long, entire or rarely slightly denticulate, thin, lateral veins broad, prominent and sometimes evidently reticulate; flowers fascicled; pedicels 2–4(6) mm. long; sepals deltoid, 1.8–2 mm. long, deciduous in fruit, glabrous or sparsely puberulent and soon glabrate; petals about 1.5 mm. long, white; drupes broadly ellipsoid or obovoid, 6–8 mm. long, black, with a heavy bluish bloom at maturity; pericarp juicy, fairly copious.

Desert slopes and mesas and into chaparral-covered hills and meadows, Lower and Upper Sonoran Zones, southeastern Arizona to western Texas and south into Chihuahua and Sonora. March–Sept.

Some material from the Sonoran Desert in central Sonora placed here may belong under the following variety, for the degree of persistence of puberulence is difficult to determine in some herbarium specimens. However, several specimens from near Magdalena, Ures, and Horcasitas, Sonora, have glabrous or glabrate twigs and leaves and appear to be typical *C. lycioides.*

**3*a*. Condaliopsis lycioides** var. **canescens** (A. Gray) Suesseng., in Engler & Prantl, Nat. Pflanzenf. ed. 2, **20d:** 135, 1953

*Zizyphus lycioides* var. *canescens* A. Gray, in Rothrock, Wheeler Rep. **6:** 82, 1878.
*Condalia lycioides* var. *canescens* Trel., in A. Gray, Syn. Fl. N. Amer. **1**[1]**:** 403, 1897.
*Condalia divaricata* A. Nels., Bot. Gaz. **47:** 427, 1909.

Branches, twigs, leaves, inflorescences, and sepals densely and usually persistently short-puberulent with spreading, white hairs; leaves 5–9 mm. wide, 8–19 mm. long, slightly thicker than in var. *lycioides.*

Mostly on desert flats, Lower Sonoran Zone, northwestern Colorado Desert to southern Nevada, central Arizona, and south into Sonora and the Cape Region, Baja California. April–June.

The flesh of the fruit of both varieties, although insipid, is edible.

## 4. KARWINSKIA Zucc., Abh. Akad. München. 2: 349, 1832

Shrubs or small trees with alternate to subopposite, short-petiolate or sessile, entire, pinnately veined leaves having small pellucid dots scattered on blade. Stipules minute, deciduous. Flowers axillary in cymes, umbels, or solitary. Calyx obconic; tube turbinate or hemispheric; lobes 5, acute, deltoid, spreading or ascending. Petals 5, cucullate. Stamens 5, slightly surpassing petals; filaments subulate. Ovary subglobose, immersed in disk at base but free and superior, 2–3-celled, 2 ovules in each cell. Drupe ovoid or subglobose, usually 1–2-locular, each locule 1-seeded.

Leaves mostly 4–8 cm. long, acute to acuminate at apex, veins usually flecked with dark spots beneath; sepals persistent in fruit; fruit 7–10 mm. long ........... 1. *K. humboldtiana*
Leaves 1–4.5 cm. long, mostly obtuse, rounded, or broadly acute at apex, dark dots almost confined to margins of blade; sepals deciduous in fruit; fruit 4–6 mm. long:
Leaves, twigs, and flowers glabrous ..................... 2. *K. parvifolia* var. *parvifolia*
Leaves, twigs, and flowers pubescent .................. 2a. *K. parvifolia* var. *pubescens*

### 1. Karwinskia humboldtiana (Roem. & Sch.) Zucc., Abh. Akad. München. 2: 351, 1832

*Rhamnus humboldtiana* Roem. & Sch., Syst. Veg. 5: 295, 1819.
?*Rhamnus biniflorus* DC., Prod. 2: 26, 1825.
*Karwinskia glandulosa* Zucc., Flora 7²: Beibl. 71, 1832.
*Karwinskia affinis* Schlecht., Linnaea 15: 460, 1841.
?*Karwinskia biniflora* Schlecht., loc. cit.
?*Karwinskia subcordata* Schlecht., op. cit. 462.
*Rhamnus maculata* Sessé & Moc., La Naturaleza II, 1: app. 38, 1888.

An open shrub or small tree 1–8 m. tall with a trunk to 2 dm. in diameter; leaves glabrous, elliptic, ovate, or oblong-elliptic, 4–8 cm. long, acute to acuminate at apex, rounded or obtuse at base, bright green above, paler beneath, lateral veins narrow but fairly conspicuous beneath, straight, often bearing 1 to several dark reddish to purplish black, linear spots along some or all lateral veins; flowers in short-pedunculate or sessile umbels, glabrous; pedicels 1–3 mm. long in flower, to 6 mm. long in fruit; calyx lobes broadly deltoid, about 2 mm. long, persistent in fruit; petals about equaling or slightly shorter than sepals; drupes globose or broadly obovoid, 7–10 mm. long, smooth and shining, epicarp thin; seed lenticular, 5–8 mm. long.

Along arroyos and on rocky slopes, Lower Sonoran Zone, central Baja California and southern Sonora to Tamaulipas, Yucatán, and Oaxaca. March–Oct.

### 2. Karwinskia parvifolia Rose, Contr. U.S. Nat. Herb. 1: 315, *pl. 31*, 1895, var. parvifolia

An openly branched shrub to 5 m. tall with slender, reddish brown or grayish twigs and glabrous leaves, flowers, and branchlets; leaves broadly elliptic to lance-elliptic, 7–12 mm. wide, 10–25 mm. long, acute, obtuse or rounded at apex, obtuse or rounded at base, yellow-green above, slightly paler beneath, minute black dots irregularly dispersed around margin, veins narrow and obvious but not conspicuous beneath; petioles slender, 2–4 mm. long; flowers in small, sessile or subsessile umbels; calyx lobes narrowly deltoid, about 1 mm. long; petals cucullate, about equaling sepals; drupes globose, 4–6 mm. in diameter, purplish, sometimes with a bluish bloom; pericarp copious and quite juicy.

Dry hillsides and gravelly slopes, Lower Sonoran Zone, central Sonora southward into northern Sinaloa. April–July.

**2a. Karwinskia parvifolia** var. **pubescens** (Standley) Wiggins, Contr. Dudley Herb. **4:** 20, 1950

*Karwinskia pubescens* Standley, Contr. U.S. Nat. Herb. **23:** 716, 1923.

Similar to var. *parvifolia* in general appearance, shape of leaves, and character of flowers and fruit except leaves are slightly larger (to 4.5 cm. long) and leaves, twigs, and flowers are minutely but densely pubescent, leaves glabrous or glabrate above.

Dry hillsides, Lower Sonoran Zone, Baja California and central Sonora to Coahuila, Nuevo León, and Oaxaca.

## 5. SAGERETIA Brongn., Ann. Sci. Nat. 10: 359, 1827

Shrubs or small trees with slender, spinose or unarmed branchlets and subopposite, short-petiolate leaves with pinnate veins and entire or serrate margins. Flowers small, in terminal panicles or axillary glomerules. Calyx tube hemispheric or urceolate; lobes 5, ovate. Petals 5, clawed and cucullate. Stamens 5, about equaling petals. Ovary 3-celled, each cell 1-ovuled. Drupes globose, containing 3 stones, these coriaceous and indehiscent. Seeds oblong.

### 1. Sageretia wrightii S. Wats., Proc. Amer. Acad. 20: 358, 1885

Densely branched shrub to 5 m. tall with spinose branchlets, spines to 1.5 cm. long; leaves elliptic to obovate, 4–10 mm. wide, 7–15 mm. long, thin, serrulate to entire, sparsely short-villous when young but soon glabrate, lustrous; inflorescence about equaling or slightly exceeding leaves; calyx lobes narrowly deltoid, about 1 mm. long, persistent in fruit; petals white; fruit obovoid, 5–6 mm. long, smooth and glabrous.

Grassy slopes, hillsides and canyons, mostly Upper Sonoran Zone, but entering margin of the Sonoran Desert in the Baboquivari and Superstition Mountains, Arizona, to western Texas and south through Sonora to Jalisco. Feb.–Sept.

## 6. RHAMNUS L., Sp. Pl. 193, 1753

Shrubs or small trees with flexuous or rigid branches, unarmed or ultimate twigs weakly spine-tipped. Leaves simple, alternate or fasciculate, finely to coarsely toothed or entire. Flowers solitary or fascicled in sessile or short-pedunculate umbels. Calyx tube turbinate or obconic; lobes 4–5, deltoid, deciduous in fruit. Petals 4–5 or absent. Stamens 4–5. Ovary superior but coherent with disk at base, 2–4-celled, 2–4-ovuled. Drupe fleshy and juicy (barely so in *R. crocea*), with 2–4 cartilaginous separate nutlets or these slightly fused, often plano-convex.

Leaves evergreen, spinose-dentate, not over 2.5 cm. long; fruits red at maturity; flowers mostly 4-merous:
- Branchlets divaricate, ending in stiff spines; teeth of leaf blades tipped with a dark-colored gland .......... 1. *R. crocea* subsp. *crocea*
- Branchlets not divaricate, not spinose-tipped; teeth of leaf blades not dark-colored .......... 1a. *R. crocea* subsp. ***insula***

Leaves deciduous, entire or serrulate, 3–7 cm. long; fruits black at maturity; flowers mostly 5-merous .......... 2. *R. californica* subsp. ***ursina***

### 1. Rhamnus crocea Nutt., in Torr. & Gray, Fl. N. Amer. 1: 261, 1838, subsp. crocea

*Rhamnus crocea* subsp. *typica* C. B. Wolf, Mon. Rancho Santa Ana Bot. Gard. Bot. Ser. **1:** 34, 1938.

Low, intricately branched shrub 1–2 m. tall, often broader than tall, branchlets tipped with short stiff spines; leaves persistent, coriaceous, elliptic, ovate, or obcordate to suborbicular, spinosely serrate or rarely entire, glabrous to sparsely pilose, 10–18 mm. long, 3–15 mm. wide, each tooth tipped with a dark-colored gland; petioles 1–10 mm. long, usually pubescent; pedicels 1–6 mm. long, slender; flowers polygamodioecious, staminate larger and brighter yellow than pistillate ones; calyx lobes usually 4, rarely 5, 1.5–2 mm. long; petals none; ovary 2-celled; fruits obovoid to globose, 5–8 mm. long, glabrous, translucent-red, usually with 2, rarely 3, nutlets, cocci splitting longitudinally along inner face, 4–6 mm. long.

Mainly in chaparral on rolling hillsides, Upper Sonoran Zone, but barely touching the margins of the Sonoran Desert in Baja California. Feb.–May.

According to Wolf the subspecies *crocea* extends southward only to Rosarita Beach, about twenty miles south of Tijuana on the west coast of the peninsula. A single specimen collected by C. Epling and Wm. Robison at "Sacatan" southeast of Rosario seems to be this plant. It may, however, belong under the following subspecies.

**1*a*. Rhamnus crocea** subsp. **insula** (Kell.) C. B. Wolf, Mon. Rancho Santa Ana Bot. Gard. Bot. Ser. **1:** 36, 1938

*Rhamnus insulus* Kell., Proc. Calif. Acad. **2:** 20, 1860.

Shrub to 2.5 m. tall; branchlets gray, pubescent, not spine-tipped; leaves 10–25 mm. long, conspicuously repand-denticulate without glands at tips of teeth; pedicels 4–6 mm. long; fruits 4–6 mm. long; seeds 4 mm. long and about half as wide.

Hillsides and along arroyos, Lower Sonoran Zone, at isolated stations at low altitudes from the vicinity of Santo Tomás, Baja California, to Cedros Island.

**2. Rhamnus californica** Esch. subsp. **ursina** (Greene) C. B. Wolf, Mon. Rancho Santa Ana Bot. Gard. Bot. Ser. **1:** 74, 1938

*Rhamnus ursina* Greene, Leaflets Bot. Obs. **1:** 63, 1904.
*Rhamnus californica* var. *ursina* McMinn, Ill. Man. Calif. Shrubs 329, 1939.

Rounded shrubs 2–5 m. tall with several to many branches from base and grayish bark; young twigs puberulent or glabrous; leaves ovate, oval, or elliptic, 1–3 cm. wide, 3–7 cm. long, usually acute at each end, entire, serrate or faintly dentate, dull to bright green and glabrous or minutely puberulent on upper surface, white-tomentose beneath with an intermixture of long and short hairs, margins plane to slightly revolute; flowers usually 5-merous, borne in peduncled umbels of 6–50 on pedicels to 10 mm. long; calyx 1–2 mm. deep, the lobes 2 mm. long; petals about 2 mm. broad when spread out, shallowly notched at apex; ovary 2–3-celled, glabrous; fruit globose, 10–14 mm. in diameter, black when mature; seeds 2–3, smooth, 5–7 mm. wide, 7–9 mm. long, flattened on back or obtusely angled if 3 seeds occur in a fruit.

Creek bottoms and ravines, mainly Upper Sonoran Zone, eastern Mojave Desert, southern Nevada, and southeasterly to Pima County, Arizona, and southwestern New Mexico. Mainly above the Sonoran Desert but sporadic along its northeastern margin. May–July.

## 7. CEANOTHUS L., Sp. Pl. 195, 1753

Shrubs or rarely small trees with alternate or opposite, deciduous or evergreen leaves and unarmed or spinescent branches. Stipules thin and deciduous or in opposite-leaved section of genus corky and persistent. Flowers white, blue, or purple, in panicles or cymes. Calyx 5-lobed, lobes usually somewhat colored, petal-like, and deciduous. Petals 5, cucullately hooded, long-clawed. Stamens 5, about equaling petals. Disk adnate to calyx at base. Ovary partially sunken in disk and adnate to it at base, 3-lobed. Style branches 3. Capsule more or less 3-lobed, dorsally crested or horned or less commonly smooth, separating into 3 one-seeded nutlets at maturity. Seeds sometimes with a small aril; cotyledons flat.

Leaves plane or the margins turned down slightly, mostly entire (sometimes a few faintly toothed near base) .................................... 1. *C. greggii* var. *greggii*
Leaves, or some of them, concave on upper surface, if toothed the teeth near apex or along most of margins:
  Margins of leaves entire or toothed lightly all around, mostly oblong-obovate, gray-green ....................................................... 1*a*. *C. greggii* var. *vestitus*
  Margins of leaves more deeply and more heavily toothed, particularly near apex, broadly obovate to suborbicular, yellow-green .............. 1*b*. *C. greggii* var. *perplexans*

### 1. **Ceanothus greggii** A. Gray, Pl. Wright. **2:** 28, 1853, var. **greggii**

*Ceanothus verrucosus* var. *greggii* K. Brandegee, Proc. Calif. Acad. II, **4:** 208, 1894.
*Ceanothus australis* Rose, Contr. U.S. Nat. Herb. **12:** 283, 1909.

Erect, rigidly and intricately branched shrub with a rounded outline, 0.5–2 m. tall; branches stiff but not spinescent, bark gray, finely marked with reticulate-striate pattern; twigs minutely and densely puberulent but eventually glabrate; stipules corky, ovate, 1–1.5 mm. long, dark reddish brown, persistent; leaves opposite; petioles stoutish, finely appressed-puberulent, 1–2.5 mm. long; leaf blades narrowly ovate, elliptic, or slightly obovate, 3–8 (or sometimes –10) mm. wide, 5–15 mm. long, plane or faintly revolute, entire or with 1–3 minute teeth near base, dark gray-green above, pale gray-green and distinctly pinnate-veined beneath, finely and moderately puberulent above, densely matted-lanate in islets between veins beneath; flowers white, in small, umbel-like clusters; pedicels slender, 3–9 mm. long, glabrous; calyx lobes deltoid, 1.5–2 mm. long, whitish; petals slightly longer than calyx lobes, claws very slender; capsule globose, faintly 3-lobed, 3–5 mm. in diameter, hornless or with 3 short, erect to slightly divergent horns to 1 mm. long about halfway between calyx rim and apex.

Hillsides and ridges, chiefly Upper Sonoran Zone but occasionally along the upper fringes of the Lower Sonoran Zone, Mohave, Cochise, Graham, and Pima Counties, Arizona, to Texas and southward to Oaxaca, Mexico. Feb.–April.

### 1*a*. **Ceanothus greggii** var. **vestitus** (Greene) McMinn, Ill. Man. Calif. Shrubs 312, 1939

*Ceanothus vestitus* Greene, Pittonia **2:** 101, 1890.
*Ceanothus greggii* var. *orbicularis* Kelso, Rhodora **39:** 151, 1937.

Differing from var. *greggii* mainly in having somewhat broader, upwardly cupped leaves, these often with 3–11 low but sharp teeth from base to apex of margins (although entire leaves are common in this variety), blades sometimes to 3 cm. long; both twigs and leaves somewhat more densely puberulent and on young growth grayer than in var. *greggii*.

Arid mountain slopes, mainly Upper Sonoran Zone, southern San Luis Obispo County and in Mono County south into Los Angeles and Kern Counties, California, east into Nevada, Utah, and northern Arizona. Feb.–April.

### 1*b*. **Ceanothus greggii** var. **perplexans** (Trel.) Jepson, Man. Fl. Pl. Calif. 623, 1925

*Ceanothus perplexans* Trel., in A. Gray, Syn. Fl. N. Amer. $1^1$: 417, 1897.

Leaves broadly obovate to suborbicular, 1–2 cm. long, concavely cupped, soon glabrous and yellow-green on upper surface, canescent and more or less tomentose beneath; margins rarely entire, more commonly rather coarsely dentate along whole length.

Mountain slopes, especially along rocky washes, Upper Sonoran Zone and upper fringe of Lower Sonoran Zone, San Bernardino County and central and eastern San Diego County, California, south along the desert slopes of the coastal ranges into northern Baja California, where it extends to at least the central parts of the Sierra San Pedro Mártir; and in western and central Arizona.

**Ceanothus integerrimus** Hook. & Arn. var. **californicus** (Kell.) G. Benson, Contr. Dudley Herb. **2:** 120, 1930, with somewhat spinose branchlets, alternate glabrous leaves 1.5–4 cm. long and 3-nerved from base, and white flowers in panicles to 10 cm. long, occurs in Arizona in Coconino, Cochise, and Pima Counties but at an elevation of about 3,500 to 7,000 feet. It may rarely occur along the upper fringes of the Sonoran Desert, but I have seen no specimens of it from within our area.

## 8. COLUBRINA L. Rich., in Brongn., Ann. Sci. Nat 10: 368, 1827

Unarmed shrubs or trees with alternate, petiolate, entire to serrate leaves. Stipules small and soon deciduous. Flowers in axillary fascicles or cymes, greenish or yellowish, perfect. Calyx obconic to hemispheric; calyx lobes 5, ovate to deltoid. Petals 5, inserted below disk, narrowly clawed, cucullate. Disk 5- or 10-lobed. Stamens 5, included in cucullate petals. Ovary 3-celled, basal part immersed in and adnate to disk. Fruit subglobose, shallowly 3-lobed; cocci separating from each other and from central axis, crustaceous.

Leaves entire, usually less than 3 cm. long, pinnately nerved; plant a shrub; branches often rigidly divaricate:
- Leaves, twigs, and calyces glabrous or sparsely puberulent and soon glabrate ........... 1. *C. glabra*
- Leaves, twigs, and calyces densely and persistently puberulent .......... 2. *C. californica*

Leaves serrate, 4–12 cm. long, strongly 3-nerved from base; plant a small tree; branches not stiff and divaricate .............. 3. *C. glomerata*

### 1. Colubrina glabra S. Wats., Proc. Amer. Acad. 24: 44, 1889

*Phyllanthus viridis* M. E. Jones, Contr. West. Bot. **18:** 47, 1933.

A stiffly branching shrub to 2 m. tall, twigs more or less divaricate and stiff, glabrous or if puberulent when young soon glabrate, bark gray; stipules narrowly deltoid, 1–1.5 mm. long, reddish, deciduous; leaves obovate to suborbicular or rarely elliptic, 5–16 mm. wide, 5–30 mm. long, thin, bright green, glabrous or nearly so, on slender petioles 2–10 mm. long; flowers axillary, in few-flowered glomerules or solitary; pedicels slender, 3–12 mm. long, glabrous; flowers 5–6 mm. broad; calyx lobes deltoid, 2 mm. long and broad, spreading; petals about equaling calyx lobes, yellowish white; capsules globose, 4–6 mm. in diameter, slightly depressed or excavated at apex, glabrous.

Bajadas, plains, hillsides, and along arroyos, Lower Sonoran Zone, central Sonora and central Baja California south to the Cape Region and southern Sonora. Feb.–May; Sept.–Oct.

### 2. Colubrina californica I. M. Johnston, Proc. Calif. Acad. IV, 12: 1085, 1924

*Colubrina texensis* var. *californica* L. Benson, Amer. Jour. Bot. **30:** 630, 1943.

Moderately dense shrubs to 2 m. tall with intricate and rigidly divaricate, gray-tomentulose branches; leaves broadly elliptic to obovate, 8–12 mm. wide, 10–20 mm. long, entire, thickish, finely but densely puberulent, cuneate to rounded at base, obtuse to rounded and mucronulate at apex, obscurely pinnately veined, dull; petioles 1–3 mm. long, densely puberulent; flowers crowded in axillary clusters; pedicels 1–2 mm. long, at anthesis, puberulent; calyx tomentulose without, lobes broadly deltoid, spreading, 1–1.5 mm. long; petals short-clawed, about 1 mm. long, erect, yellowish; capsule depressed-obovoid, 6–10 mm. high, 4–6 mm. wide, obscurely 3-lobed, glabrous.

Canyon floors and hillsides, Lower Sonoran Zone, southeastern Arizona, Colorado Desert, in southeastern California, and south into Baja California. Not yet reported from Sonora but to be expected in the northwestern part of that area. May–Aug.

**3. Colubrina glomerata** (Benth.) Hemsl., Biol. Centr. Amer. Bot. **1:** 200, 1880

*Rhamnus glomerata* Benth., Pl. Hartw. 9, 1839.
*Zizyphus acuminata* Benth., Bot. Sulph. 78, 1844.
*Barcena guanojuatensis* Dugès, Rev. Cienc. Mex. **1:** 8, 1879.
*Colubrina arborea* Brandegee, Zoe **4:** 401, 1894.
*Colubrina mexicana* Rose, Contr. U.S. Nat. Herb. **3:** 315, 1895.

Large shrub or small tree to 8 m. tall with slender, nondivaricate, glabrous branches; leaves lanceolate to broadly ovate, 1.5–6 cm. wide, 4–12 cm. long, acute to long-acuminate at apex, obtuse or rounded at base, coarsely serrate, prominently 3-veined from base, glabrous or with a few scattered hairs along veins below, bright green; petioles to 4 cm. long; flowers crowded in axillary glomerules; pedicels 1–3 mm. long at anthesis, sparsely pubescent with rufous hairs; calyx sparsely short-villous without, cup about 2 mm. in diameter; calyx lobes triangular-deltoid, 0.8–1 mm. long, nearly as wide; petals about equaling sepals; capsule 6–8 mm. in diameter, slightly depressed, subglobose.

Canyons and hillsides, Lower Sonoran Zone to Tropical Zone, southern Sonora to Chihuahua, Puebla, and Oaxaca, and in southern Baja California. Sept.–Dec.

## 9. ADOLPHIA Meisn., Pl. Vasc. Gen. Comm. 70, 1837

Profusely branched, rigid, spinescent shrubs with small deciduous leaves and green branchlets. Leaves opposite, oblong to ovate, stipulate, early deciduous. Flowers in small axillary clusters or solitary, small, greenish or yellowish. Calyx bowl-shaped, 10-ribbed, with 5 small triangular lobes. Petals 5, spatulate and deeply cucullate. Stamens about equaling petals, inserted at margin of disk. Ovary subglobose, 3-celled, adnate to disk only at base; style slender; stigma 3-lobed. Fruit a coriaceous drupe, adnate to calyx at base, 3-lobed. Seeds plano-convex.

Ultimate twigs stoutish, 1–1.5 mm. in diameter; calyx bowl broadly obconic, sides nearly straight; calyx lobes triangular-deltoid, as broad as long, margins straight ........... 1. *A. californica*

Ultimate twigs slender, mostly 0.5–1 mm. in diameter; calyx bowl hemispheric, sides convex; calyx lobes ovate, nearly twice as long as wide, margins curved ...... 2. *A. infesta*

**1. Adolphia californica** S. Wats., Proc. Amer. Acad. **11:** 126, 1876

A dense shrub forming rounded clumps to 1 m. tall and 2 m. broad, or more straggly; branches bright green, minutely puberulent to glabrate, ultimate twigs rigidly divaricate and spine-tipped, spine brownish or yellowish; stipules red-brown, minute, 0.5–0.8 mm. long; leaves elliptic-oblong to obovate, 3–9 mm. long, obtuse to acutish at apex, cuneately narrowed at base to slender 1–2.5 mm. long petiole, entire; flowers fasciculate in axils; pedicels slender, 3–6 mm. long, slightly dilated toward apex; calyx bowl broadly obconic, 2.5–3 mm. wide, 1.5–2 mm. deep, sides nearly straight; calyx lobes triangular-deltoid, 1.5 mm. long and broad, margins straight or nearly so; petals 2–2.5 mm. long, blade deeply cucullate, completely enfolding anthers at anthesis; fruits depressed-subglobose, 4–6 mm. broad, 3–4 mm. high; style jointed at base, therefore capsule not apiculate.

Along arroyos and on hillsides near springs and seeps, Lower and Upper Sonoran Zones, chaparral areas of southwestern San Diego County, California, south on the western slopes of the foothills to the vicinity of Jaraguay, Baja California. Feb.–May; Nov.

Abundant in several canyons in the transition from Upper to Lower Sonoran Zones, southwestern slopes of the Sierra San Pedro Mártir.

**2. Adolphia infesta** (H. B. K.) Meisn., Pl. Vasc. Gen. Comm. 70, 1837

*Ceanothus infestus* H. B. K., Nov. Gen. & Sp. **7:** 61, *pl. 614,* 1825.
*Colletia ?multiflora* DC., Prod. **2:** 29, 1825.

Differing from the last mainly in having more slender branches and twigs, ultimate ones usually less than 1 mm. in diameter, more densely and more persistently puberulent branchlets, and smaller, differently shaped flowers; pedicels mostly 1–2.5 mm. long; calyx bowl hemispheric-campanulate, 1.5 mm. wide, about as deep, sides convexly curved; calyx lobes triangular-ovate, 1–1.5 mm. long, about half as wide, margins convexly curved; petals 1.2–1.5 mm. long, shallowly cucullate, slightly exceeded by stamens at anthesis; fruit slightly depressed-subglobose, 4–5 mm. wide, 3–4 mm. high, apex tipped by persistent base of style, hence short-apiculate.

Arroyos and hillsides, Lower and Upper Sonoran Zones, central Sonora (southern flank of the Sierra Babiso) to Chihuahua, western Texas, Zacatecas, Hidalgo, and Oaxaca. March–Sept.

This species is a trifle deeper green in color and noticeably more slender and more densely puberulent than *A. californica.*

## 10. GOUANIA Jacq., Stirp. Amer. 263, 1763

Usually scandant shrubs with tendrils borne on peduncles of inflorescences. Leaves alternate, entire or dentate. Stipules deciduous (in ours) or rarely persistent. Inflorescences terminal or axillary racemes or spikes. Flowers polygamodioecious. Calyx tube short-obconic, adnate to ovary; lobes 5. Petals 5, cucullate. Stamens 5, more or less enfolded by petals. Ovary inferior, 3-celled; style 3-lobed or -parted. Fruit coriaceous, narrowly to strongly 3-winged, 3 cocci indehiscent but separating from central axis.

**1. Gouania rosei** Wiggins, Contr. Dudley Herb. **4:** 20, 1950

*Gouania mexicana* Rose, Contr. U.S. Nat. Herb. **3:** 314, 1895. Not *G. mexicana* Sessé & Moc., 1894.

A stout woody vine climbing to 10 m. over shrubs and trees; young twigs densely tawny-tomentose with short tangled hairs, older ones glabrate, with rich, reddish brown bark and numerous light tan to nearly white lenticels in longitudinally arranged lines; stipules lance-subulate, 2–3 mm. long, deciduous; leaves ovate to obovate, 2–5 cm. wide, 4–8.5 cm. long, remotely and shallowly dentate, obtuse to subcordate at base, acute, acuminate or rarely rounded at apex, prominently veined and tomentose beneath, bright green and thinly pubescent above; petioles 6–15 mm. long, tomentose; inflorescences 5–15 cm. long, rachises and calyces densely pubescent; calyx lobes triangular-deltoid, about 1 mm. long and wide, glabrous within; petals equaling calyx lobes, enclosing anthers; fruits 4–5 mm. high, 5–6.5 mm. wide, tomentose, angles bearing wings about 1 mm. wide.

Hillsides and along washes, scrambling over shrubs and trees, Lower Sonoran Zone, central Sonora (near Ures) southward into Sinaloa. July–Sept.

A young flowering specimen from San José del Cabo, Baja California, collected by Brandegee and determined by him as "probably" *G. tomentosa* Jacq., is tentatively placed here also. Fruiting material may prove it some other species, for the pubescence on the slightly thinner leaves is sparser than that on the Sonoran plants.

## Family 80. **VITACEAE.** Grape Family

Climbing, scrambling, or erect vines or shrubs with conspicuously swollen nodes, alternate petioled leaves, and small flowers borne in cymes, racemes, or panicles. Tendrils usually present opposite leaves. Calyx entire or 4–5-lobed or -toothed. Petals 4–5, distinct or coherent near apex, caducous. Stamens 4–5, opposite petals; filaments inserted at base of or between lobes of disk. Disk present or rarely wanting. Ovary usually embedded in disk but eventually growing above it, 2–6-celled; each cell 1–2-ovuled, ovules ascending. Fruit a juicy berry or dry at maturity. Seed coat bony.

Petals coherent at apex into a calyptra; bark shreddy; lenticels inconspicuous or none; berries edible; inflorescence paniculate ........................................ 1. *Vitis*

Petals free or essentially so; bark smooth, with many lenticels; fruits inedible; inflorescence umbellate-cymose:

  Fruit fleshy at maturity; leaves 3–7-foliolate; disks usually present on some tendrils; flowers 5-merous ........................................ 2. *Parthenocissus*

  Fruit dry at maturity; leaves simple to 3-foliolate; disks none on tendrils; flowers 4-merous ........................................ 3. *Cissus*

### 1. **VITIS** L., Sp. Pl. 202, 1753

Climbing or scrambling woody vines, usually with tendrils opposite some leaves and small dioecious, polygamodioecious, or rarely perfect flowers. Leaves simple, mostly dentate and often palmately lobed. Stipules small (in ours), caducous. Flowers in panicles or racemes. Calyx minute; limb entire. Petals 4–5, hypogynous or perigynous, deciduous in a coherent cup without expanding. Stamens 4–5, opposite petals. Ovary 2- or rarely 3–4-celled, each cell 2-ovulate. Berry ovoid to globose, pulpy and juicy.

Leaves persistently lanate-floccose beneath, 10–18 cm. broad; plants climbing strongly; tendrils well developed; berries black, not glaucous ........................ 1. *V. girdiana*

Leaves having tomentum only along ribs or completely glabrate in age, 3–8 cm. broad; plants climbing only slightly; tendrils poorly developed; fruits lightly glaucous ........................................ 2. *V. arizonica*

#### 1. **Vitis girdiana** Munson, Proc. Soc. Prom. Agr. Sci. 59, 1887

Woody vine with slender stems climbing to 15 m. or more; young growth densely white-tomentose, this becoming thinner and somewhat patchy in age; leaves round-cordate or obscurely pentagonal, sometimes deeply 3–5-lobed, 10–18 cm. broad, about as long as broad or slightly longer, basal sinus rounded but usually narrowing outward from petiole; leaf blade firm in texture, green and glabrous above, floccose-tomentose beneath, teeth usually 5–10 mm. broad at base, apiculate or acute; petioles 3–8 cm. long, tomentose; tendrils 2–3-branched, 5–10 cm. long; panicles decompound, 10–15 cm. long, rachis and branches floccose; berries globose, black and rarely faintly glaucous, 6–8 mm. in diameter; seeds pyriform, 4–5 mm. long, hard.

Mostly along streambanks, Lower and Upper Sonoran Zones, Santa Barbara and Inyo Counties, California, and southward on both the coastal and desert slopes of the mountains but not in the main desert areas into northern Baja California. Entering the Sonoran Desert near Palm Springs and probably elsewhere along the eastern slopes of the mountains near the mouths of canyons supporting semipermanent streams. April–July.

**2. Vitis arizonica** Engelm., Amer. Nat. **2:** 321, 1868

*Vitis arizonica* var. *glabra* Munson, U.S. Dept. Agr. Div. Pom. Bull. **3:** 10, 1890.

Scrambling, bushy, or sometimes weakly climbing plants with poorly developed tendrils; stems much branched and floccose-tomentose in youth; bark reddish brown, becoming grayish and shreddy on older stems; leaves broadly cordate, 3–6 cm. long (rarely –8 cm. long and wide), acute to short-acuminate, coarsely dentate and rarely shallowly 3-lobed, teeth broad and low, apiculate, densely floccose-tomentose beneath and less heavily so above when young, later glabrate or more frequently with irregular patches of tomentum on blades and along veins beneath, glabrate above, basal sinuses broadly U-shaped, widest away from petiole; petioles 1–2.5 or rarely –4 cm. long; tendrils 1–2-branched, often not functional; panicles 2–4 cm. long in flower, branches and rachises floccose-tomentose at first, soon glabrate; berries globose, 8–10 mm. in diameter, black and lightly glaucous; seeds 5–6 mm. long.

Mainly in canyons and along streams, chiefly Upper Sonoran Zone but descending into Lower Sonoran Zone along watercourses in Pima County, Arizona; to Texas, southern Utah, Tamaulipas, and in Sonora. April–July.

Reported from Baja California, but the grape found in the Sierra de la Laguna of the Cape Region has been named *V. peninsularis* M. E. Jones, Contr. West. Bot. **18:** 57, 1933. Material of the type collection has leaves with very broad, almost truncate basal sinuses; persistent wool on lower sides of leaves; on a fruiting stem darker, more fibrous bark; tendrils are well developed, appearing to be thoroughly functional, and small fruits (5–6 mm. in diameter) show no evidence of a bloom. It may prove to be conspecific with *V. arizonica,* but the material at hand appears to be distinct.

## 2. PARTHENOCISSUS Planch., in A. & C. DC., Monogr. Phan. $5^2$: 447, 1887

Scrambling, climbing, or trailing vines or erect shrubs with tendrils opposite some leaves, some or all tendrils forming adhesive disks. Leaves digitately compound; leaflets deciduous. Flowers perfect, in umbellate cymes. Calyx cup-shaped or saucer-shaped, shallowly 5-lobed. Petals 5, free (in ours) or sometimes slightly coherent at tip, spreading in anthesis and soon deciduous. Stamens 5, inserted beneath a hypogynous disk lining base of calyx, or disk obsolete. Ovary tipped with a short fleshy style, 2-celled; locules 2-ovulate. Fruit a thin-fleshed berry, 1–4-seeded through partial abortion of ovules. Seeds plano-convex to subglobose, obtusely keeled along inner face.

**1. Parthenocissus inserta** (Kerner) Fritsch var. **laciniata** (Planch.) Rehder, Jour. Arnold Arb. **20**: 419, 1939

*Parthenocissus quinquefolia* δ *laciniata* Planch., in A. & C. DC., Monogr. Phan. **5**[2]: 449, 1887. (For synonymy see Jour. Arnold Arb. **20**: 418–419, 1939.)

Perennial plants with scrambling, climbing, or trailing stems 1–5 m. long (occasionally much longer); bark on older branches pale buff to dark brown, striately wrinkled in drying and with rather conspicuous, elliptical, scattered, dark reddish brown lenticels; twigs and herbage glabrous; tendrils slender, branched one to three times, with few or no clinging disks; petioles 4–15 cm. long; leaflets 3–7 (usually 5), lanceolate-ovate to broadly obovate, 1–6 cm. wide, 3–10 cm. long, coarsely and laciniately toothed, each tooth abruptly short-apiculate, terminal one lance-attenuate, 1–2.5 cm. long; petiolules 2–15 mm. long; inflorescence dichotomously branched, branches of each order about equal, on a peduncle 2–20 cm. long; pedicels 1–5 mm. long in flower, 5–15 mm. long in fruit; calyx saucer-shaped, 1–1.5 mm. broad; petals obovate, about 1.5–2 mm. long, deciduous; berries globose, 6–10 mm. in diameter, blue-black, bearing a faint glaucous bloom, flesh thin over the 1–4 seeds, these obovoid, plano-convex with a low keel along flattened side, 5–7 mm. long.

Canyons and along stream banks, mainly Upper Sonoran Zone and Arid Transition Zone, but occasionally as low as 3,000 feet along streams in the Lower Sonoran Zone, Wyoming and Colorado to Pima and Cochise Counties, Arizona, and in adjacent New Mexico. May–Sept.

## 3. CISSUS L., Sp. Pl. 117, 1753

Vines with herbaceous or woody, subfleshy stems usually from large fleshy tubers and with coiled tendrils. Leaves simple or 3-foliolate, thin or succulent. Flowers usually perfect, 4-merous. Calyx short, entire or 4-toothed. Petals 4, spreading, free at anthesis. Ovary 2-celled, base somewhat immersed in disk. Berries 1–4-seeded; flesh dry, inedible.

Leaves simple, ovate, not lobed nor parted .............................. 1. *C. sicyoides*
Leaves palmately trifoliolate or deeply 3-lobed:
  Leaflets 3, lance-ovate, coarsely dentate to base or some leaves deeply 3-lobed; branches of fruiting panicles reddish; stems not fleshy ........................ 2. *C. trifoliata*
  Leaflets 5, oblong to linear-oblong, entire; branches of panicles greenish; stems subfleshy ........................................................ 3. *C. mayoensis*

**1. Cissus sicyoides** L., Syst. Nat. ed. 10, **2**: 897, 1759

*Cissus elliptica* Schlecht. & Cham., Linnaea **5**: 221, 1830.

A climbing or scrambling vine growing over rocks and trees, sometimes to 20 m. or more; leaves ovate, 4–15 cm. long, 3–10 cm. wide, acute to acuminate at apex, obtuse to subcordate at base, sharply and finely but remotely serrate along margins, rather conspicuously pinnately veined beneath, glabrous to densely puberulent on lower or both surfaces, deep green; inflorescences opposite leaves, decompound, in subglobose open cymes; pedicels glabrous, 1–4 mm. long at anthesis; calyx disk slightly undulate; petals about 2 mm. long, caducous; fruits ovoid to globose, 5–8 mm. long, black, 1-seeded.

Along streams and in shaded canyons, margins of Lower Sonoran and Tropical Zones, southern Sonora, where it barely approaches the Sonoran Desert along streams, and southward to tropical America where it is widely distributed. Oct.–Nov.

**2. Cissus trifoliata** (L.) L., Syst. Nat. ed. 10, **2:** 897, 1759
*Sicyos trifoliata* L., Sp. Pl. 1013, 1753.
*Cissus acida* L., op. cit. ed. 2, 170, 1762.

Trailing over rocks or clambering into shrubs and trees; bark smooth, tawny or greenish, with rather conspicuous lenticels; tendrils simple or 2-branched; leaves trifoliolate or rarely only deeply 3-parted; petioles slender, to 10 cm. long or more; leaflets broadly lance-ovate, 1–5 cm. wide, 3–10 cm. long, broadly cuneate to obtuse and often oblique at base, acute to acuminate at apex, glabrous or sometimes sparsely pubescent, serrations coarse, 5–15 mm. apart; inflorescences equaling or surpassing leaves, their branches turning bright red in fruit; pedicels stoutish, 5–8 mm. long in fruit, glabrous; berries globose, black, 6–9 mm. in diameter.

Canyons and along streams, Lower and Upper Sonoran Zones to Tropical Zone, southern Arizona (Pima and Santa Cruz Counties) to Baja California, Oaxaca, and Yucatán. May–Aug.

Both this and the previous species cause dermatitis similar to that produced by poison ivy. The tubers of both are poisonous if eaten.

**3. Cissus mayoensis** Gentry, Carnegie Inst. Wash. Pub. No. **527:** 175, 1942

A scrambling vine with stems to 5–8 m. long, woody at base, glabrous throughout; bark gray, with prominent lenticels; branches striate, subfleshy, internodes 2–10 cm. long; petioles slender, 2–10 cm. long; leaflets 5, oblong to oblong-linear, entire, pinnately veined, glabrous and slightly glaucous, obtuse and sometimes apiculate at apex, cuneate at base, 6–25 mm. wide, 3–10 cm. long; inflorescence cymose, on peduncles 1–2 cm. long; calyx nearly obsolete, faintly lobed; petals greenish white, broadly oblong to ovate, 1.5–2 mm. long, cupped, caducous; disk 4-lobed; pedicels 4–8 mm. long in fruit; fruits obovoid, purplish, 4–6 mm. long, 1-seeded; seeds obovoid, wrinkled, 3.5–5 mm. long.

Along streams and scrambling over rocky hillsides, Lower Sonoran Zone, northern to southern Sonora. It was not abundant at any of the stations where collected by me. July–Sept.

## Family 81. **TILIACEAE.** Linden Family

Trees, shrubs, or sometimes herbs with simple, alternate, entire or lobed, stipulate leaves that usually are deciduous. Pubescence mostly of stellate or branched hairs. Flowers perfect or less commonly polygamodioecious. Sepals 5 or sometimes 3 or 4, free or coherent, usually valvate. Petals as many as sepals and alternating with them or wanting. Stamens 10 to many. Ovary superior, 2–10-celled, or 1-celled in fruit by abortion. Fruit drupaceous or (in ours) dry, dehiscent or indehiscent, often flattened.

Fruit a silique-like capsule, unarmed; plants herbaceous (or rarely suffrutescent) in ours . . . . . . . . . . . . . . . . . . 1. *Corchorus*
Fruit ovoid or orbicular, armed with bristles or spines; plants shrubby or trees:
  Fruit not compressed, bristly on all sides . . . . . . . . . . 2. *Triumfetta*
  Fruit strongly compressed, plumose-hairy on the margins . . . . . . . . 3. *Heliocarpus*

### 1. CORCHORUS L., Sp. Pl. 529, 1753

Herbs or shrubs with simple, serrate or crenate, alternate leaves. Flowers small, yellow, on short, 1- to few-flowered peduncles opposite leaves or axillary. Sepals 5 or rarely 4. Petals same number as sepals. Stamens numerous or rarely 10. Ovary 2–5-celled, each cell many-ovuled. Capsule long and silique-like or in some subglobose and echinate, cells dehiscing loculicidally, transverse septa occurring between seeds in a few species. Seeds many, pendulous or horizontal. Embryo curved.

### 1. **Corchorus orinocensis** H. B. K., Nov. Gen. & Sp. **5:** 337, 1821

Erect, slightly branched herb 2–6 dm. tall; stem with lines of dense pubescence running from one node to next higher one, rest of stem nearly glabrous; leaves elliptic-ovate to lanceolate, 1–2 cm. wide, 2.5–10 cm. long, obtuse to rounded at base, acute to acuminate at apex, crenate-serrate, dark green and nearly glabrous above, pale and sparsely appressed-puberulent along veins or glabrous beneath; petioles 6–12 mm. long, densely hirsutulous; stipules linear-subulate, 3–5 mm. long; flowers yellowish, opposite leaves, solitary; sepals 3.5–4 mm. long, short-acuminate, sparsely pubescent; petals about equaling sepals; stamens numerous; capsules 2-celled, straight or slightly curved, ascending, 4–5.5 cm. long, sparsely appressed-pubescent, beak 5–8 mm. long, single; seeds terete with truncate ends, 0.8–1.2 mm. long, 0.5–0.7 mm. in diameter, black, minutely tuberculate-roughened.

Rocky hillsides, on clay flats and along arroyos, Lower Sonoran Zone to Tropical Zone, southern Arizona to Florida, the West Indies and South America. Aug.–Sept.

The plant collected in Arizona by Pringle has been determined as *C. hirtus* var. *glabellus* A. Gray. This may be its proper disposition, but the taxonomy of this weedy genus is in need of revision, and pending a revisionary study I am identifying all of the Sonoran Desert material of the genus as *C. orinocensis.*

## 2. **TRIUMFETTA** L., Sp. Pl. 444, 1753

Shrubs or less commonly herbs with stellate-pubescent branchlets, inflorescences, and leaves; leaves simple, often 2–5-lobed. Flowers axillary or opposite leaves, solitary or cymose. Sepals 5, distinct, appendaged near apex. Petals 5 or rarely lacking. Stamens 10 to many. Ovary 2–5-celled, superior, each cell biovulate; style filiform; stigma 2–5-toothed. Fruit a small, subglobose, setose or echinate capsule, indehiscent or breaking into cocci. Seeds solitary or 2 in each cell, if latter seeds separated by a false partition. Embryo straight.

Flowers in fascicles opposite leaves; larger leaves trilobate; sepals 5–7 mm. long; spines of fruit retrorsely hooked at tip .................................... 1. *T. semitriloba*
Flowers in terminal panicles; leaves rarely trilobate; sepals 15–17 mm. long; spines of fruit bearing spreading or ascending, stout hairs, not retrorsely hooked .... 2. *T. goldmanii*

### 1. **Triumfetta semitriloba** Jacq., Stirp. Amer. 147, 1763

(For synonymy see Ann. Mo. Bot. Gard. **37:** 373–374, 1950.)

A slender shrub to 3 m. tall; leaves ovate to lance-ovate, larger slightly 3-lobed, lateral lobes short and ascending, middle lobe or whole leaf acuminate, rounded or broadly obtuse at base, 1–5 cm. wide, 3–10 cm. long, irregularly serrate, 3-veined from base, minutely stellate-pubescent on both surfaces, dark green and sparsely pubescent above, paler and more densely pubescent beneath; petioles to 5 cm. long; inflorescences of few-flowered fascicles opposite leaves; pedicels 3–6 mm. long, peduncles to 1 cm. long; sepals minutely stellate-puberulent, 5–7 mm. long, each bearing a filiform subapical appendage 1.5–2 mm. long; petals yellow, nearly equaling sepals; stamens about 20, about 6–8 mm. long; styles 2–3 mm. longer than stamens; fruit 6–8 mm. in diameter, body glabrate at maturity, spines slender and retrorsely hooked at tip.

Hillsides and along streams, mainly Tropical Zone, southern Baja California and Sinaloa to Yucatán and Chiapas, West Indies, and South America. Aug.–April.

**Triumfetta discolor** Rose, Contr. U.S. Nat. Herb. **12:** 285, 1909, with ovate unlobed leaves, gray-tomentose beneath; glabrate fruits 1 cm. in diameter; and with straight, unbarbed hairs on fruit, occurs in southern Sonora but has not been reported from within the limits of the Sonoran Desert.

### 2. Triumfetta goldmanii Rose, Contr. U.S. Nat. Herb. 12: 285, 1909

*Triumfetta sanctae-luciae* Sprague, in Riley, Kew Bull. **1923:** 114, 1923.
*Triumfetta quercetorum* Bullock, in Bullock & Sprague, op. cit. **1937:** 295, 1937.

A slender, ascendingly branched shrub 1–1.5 m. tall with reddish brown, sparsely stellate to subglabrate twigs; stipules lanceolate to subulate, 1–2 mm. long, caducous; petioles 0.2–2 cm. long, closely stellate-puberulent; leaf blades ovate-lanceolate, rounded at base, gradually acuminate at apex, cuneately dentate, 0.5–2 cm. wide, 1.5–7 cm. long, dark green and sparsely puberulent with minute stellate trichomes above, paler and more densely stellate-pubescent with interspersed simple hairs along veins beneath, pinnately veined but basal pair of lateral veins conspicuous and strongly ascending one-third to one-half way toward tip; inflorescence paniculate with umbel-like clusters of 2–7 on ultimate branches; pedicels 2–10 mm. long, slender, stellate-puberulent; calyx 15–17 mm. long, sparsely and minutely stellate-puberulent, erect subterminal appendages 0.5–1 mm. long; petals yellow, linear-oblanceolate, about equaling sepals; stamens and style about 2 cm. long; fruits (immature) apparently indehiscent, ovoid to spheroid, 4–5 mm. long, rather closely invested with stout prickles 0.5–0.8 mm. long, the body of the fruit sparsely pubescent with simple spreading hairs, the bristles slightly curved to faintly hooked and lighter colored at the tips, bearing a few stout, spreading or ascending, simple hairs from base to near tip.

Arroyos and shaded slopes, Lower and Upper Sonoran Zones, southern Sonora and adjacent southern Chihuahua southward into central Mexico. Sept.–Dec.

## 3. HELIOCARPUS L., Sp. Pl. 448, 1753

Trees or shrubs with simple, entire or lobed, petioled, palmately veined leaves and more or less stellate-pubescent branchlets, foliage, and inflorescences. Flowers perfect or usually polygamodioecious, borne in small to moderately sized panicles. Sepals 4–5, valvate, flat or hooded and often with a small appendage near apex. Petals 4–5, linear-oblong, more or less glandular and pubescent near base. Stamens 14–40, inserted on receptacle below ovary; filaments often present but devoid of anthers in pistillate flowers. Ovary superior, compressed laterally, 2-celled, each cell biovulate; edges of compressed ovary bearing a fringe of plumose hairs. Fruit indehiscent, fringe of plumose hairs conspicuous.

Body of fruit orbicular, stellate-puberulent but not tomentose on faces; fringe equaling or shorter than diameter of body; leaves nearly or quite glabrous ...... 1. *H. polyandrus*
Body of fruit oval, tomentose on faces; fringe exceeding diameter of body; leaves sparsely to densely stellate-pubescent or stellate-tomentose .................... 2. *H. attenuatus*

### 1. Heliocarpus polyandrus S. Wats., Proc. Amer. Acad. 21: 420, 1886

*Heliocarpus glaber* Brandegee, Zoe **5:** 209, 1905.
*Heliocarpus laevis* Rose, Contr. U.S. Nat. Herb. **8:** 317, 1905.

Shrub or small tree with finely stellate-puberulent branchlets and petioles; stipules lance-subulate, 3–8 mm. long; petioles 2–6 cm. long; leaf blades ovate to ovate-cordate, acuminate, 5–10 cm. wide, 8–15 cm. long, serrate, very sparsely stellate-puberulent to glabrous on both surfaces, thin, dark green above, paler beneath; panicles open, to 15 cm. long; sepals narrowly obovate, 3.5–4 mm. long, stellate-puberulent, appendages spreading, about 0.5 mm. long; petals linear-oblong, about equaling sepals; stamens about 40; fruit orbicular, body 3–3.5 mm. in diameter, faces sparsely stellate-puberulent and slightly ridged, fringing hairs 2–3 mm. long.

On shady slopes in forests, mainly Upper Sonoran Zone, southern Sonora, southwestern Chihuahua, and northern Sinaloa. Sept.–Oct.

In Sonora known definitely only from the vicinity of Alamos, outside the limits of the Sonoran Desert, but to be expected in shaded canyons near Cedros, Queriego, and other foothill areas in the southern part of the state.

**2. Heliocarpus attenuatus** S. Wats., Proc. Amer. Acad. **21:** 420, 1886

Slenderly branched shrub or small tree with softly stellate-tomentose branches and petioles, pubescence more or less tawny; stipules narrowly linear, 2–3 mm. long; petioles 8–18 mm. long; leaf blades ovate-lanceolate, gradually long-acuminate, subcordate to rounded at base, 1.5–4 cm. wide, 3.5–10 cm. long, densely stellate-pubescent on both surfaces, more densely so beneath, margins finely serrate; flowers greenish yellow; sepals linear-lanceolate, about 5 mm. long, hooded, appendage erect, 0.4–0.6 mm. long; petals about 2 mm. long, ciliate on basal one-fourth; stamens 15–20; style about twice as long as ovary, simply bifid about one-fourth its length; fruit oval, body about 2 mm. wide, 3 mm. long, faces densely hirsute-tomentose; hairs of marginal fringe 2.5–3.5 mm. long, tawny.

Canyon slopes, Lower and Upper Sonoran Zones, mostly if not entirely above the limits of the desert, southwestern Chihuahua, southern Sonora, and adjacent Sinaloa. Known with certainty in Sonora only from the vicinity of Alamos, although a sterile shrub near Queriego may belong here. Sept.–Oct.

**Heliocarpus virida** E. E. Wats., Bull. Torrey Club **50:** 120, 1923, has shallowly 3-lobed, stellate-puberulent leaves and ovoid fruits about the same size as those of *H. attenuatus* but hairs of the fringe are about 4 mm. long. It was described from "Sonora," with no more definite locality stated, and may occur within our area, but probably came from the vicinity of Alamos.

## Family 82. MALVACEAE. Mallow Family

Herbs or shrubs with mucilaginous juice, stellate pubescence, and alternate, palmately veined, commonly lobed or divided leaves. Stipules small, deciduous or caducous. Flowers regular, perfect or polygamodioecious. Calyx 5-lobed, valvate in bud, often involucellate-bracteate at base. Petals 5, hypogynous, convolute in bud, fused with stamineal tube at or near base. Stamens numerous, hypogynous, forming a monodelphous tube about pistil. Carpels several to many, commonly same number as style branches; ovules 1 to several in each carpel. Ovary superior. Fruit a loculicidal capsule or (more frequently) carpels falling separately. Seeds reniform. Embryo curved; cotyledons plicate or conduplicate; endosperm scanty.

Carpels arranged around an obvious central axis, separating from each other and from axis at maturity; seeds not woolly:
- Carpels 2–9-ovuled, 1- to several-seeded:
  - Bracteoles none; carpels smooth on lateral walls or reticulate in *Horsfordia*:
    - Carpels not falsely 2-celled:
      - Ovules 2–9; carpels wingless, smooth on sides:
        - Carpels chartaceous or leathery, not inflated; fruit hemispherical-truncate, apically umbilicate .......... 1. *Abutilon*
        - Carpels membranous, strongly inflated; fruit subglobose, not umbilicate apically .......... 2. *Bogenhardia*
      - Ovules 3; carpels dorsally winged, reticulate on sides .......... 3. *Horsfordia*
    - Carpels falsely 2-celled by intruding lateral constrictions or by a flap attached to dorsal wall:
      - Partition between false cells formed by lateral constrictions; carpels usually 3 in species of Sonoran Desert .......... 4. *Wissadula*
      - Partition formed by a dorsally attached flap, this free along sides; carpels 5–11 .......... 5. *Pseudabutilon*
  - Bracteoles 3; carpels reticulate laterally below but not winged dorsally .......... 6. *Sphaeralcea*
- Carpels 1-ovuled, 1-seeded:
  - Ovules ascending:
    - Style branches filiform, stigmatic along inner surface .......... 7. *Malva*
    - Style branches stouter, stigmatic at truncate or capitate tips only:
      - Carpels smooth on sides and angles; shrubs .......... 8. *Malvastrum*
      - Carpels reticulate or radiately grooved on sides or angles; annual herbs .......... 9. *Eremalche*
  - Ovules pendulous:
    - Dorsal wall firmly attached to lateral walls of carpels; seeds not enclosed in a loose coat derived from inner wall of carpel .......... 10. *Sida*
    - Dorsal wall separating from lateral walls of carpels; seeds often enclosed in a loose coat formed by inner wall of carpel .......... 11. *Anoda*

Carpels forming a capsule; central axis obscured or inconspicuous; carpels usually remaining attached to receptacle at maturity; seeds woolly or hirsute in *Hibiscus* and *Gossypium*:
- Carpels 1-ovuled; seeds glabrous or minutely puberulent:
  - Bracteoles none; seeds pendulous, filling carpel only partially .......... 12. *Bastardia*
  - Bracteoles 7–10 (sometimes reduced to small stubs); seeds ascending, nearly or completely filling carpel .......... 13. *Kosteletzkya*
- Carpels 3- to many-ovuled; seeds more or less woolly or hirsute:
  - Stigma lobes globose-capitate; bractlets of involucel several, linear or lanceolate; hairs on seeds 2–6 mm. long .......... 14. *Hibiscus*
  - Stigma lobes clavate; bracteoles 3, cordate-foliaceous; hairs on seeds to 2 cm. long or more .......... 15. *Gossypium*

## 1. ABUTILON Adans., Fam. Pl. 2: 398, 1763

Herbs or shrubs (rarely small trees) usually with dense, felt-like, stellate pubescence. Leaves alternate, often cordate, angular, or lobed, rarely linear-lanceolate. Stipules small, deciduous. Flowers axillary or in terminal panicles. Bracteoles none. Calyx lobes 5, equaling or exceeding tube. Petals 5. Stamens numerous, coalescent into a tube, this adnate to petals near their bases. Ovary 5–15-celled, each locule 2- to several-ovuled. Carpels rounded to rostrate-apiculate or muticous at apex, 2-valved, often puberulent dorsally. Seeds subreniform, upper usually ascending, lower ones horizontal or pendulous. Embryo curved.

Calyx small, much shorter than mature carpels:
Carpels 7–11; larger leaves usually over 10 cm. long except in *A. parishii*:
Larger leaves usually 10–25 cm. long; carpels 8–12 mm. high, their beaks 1 mm. long or less; petals yellow, cream, or outside slightly washed with orange:
Calyx lobes acute to acuminate, 3–6 mm. long; stems and petioles persistently pubescent to lanate:
Petals 6–10 mm. long, yellow within, suffused with orange without; leaf blades usually 3-lobed; calyx lobes 3–4 mm. long ........................ 1. *A. sonorae*
Petals 7–8 mm. long, cream, not suffused with orange; leaf blades only rarely 3-lobed, usually crenate; calyx lobes 4–6 mm. long .............. 2. *A. carterae*
Calyx lobes obtuse or at most subacute, not acuminate, 2.5–3.5 mm. long; stems and petioles often glabrate in age ................................ 3. *A. reventum*
Larger leaves rarely over 7 cm. long; carpels 6–7 mm. high, their beaks 1–3 mm. long; petals orange within and without ................................ 4. *A. parishii*
Carpels 4–5 (or rarely 6); leaves less than 6 cm. broad and 15 cm. long:
Stems and branches trigonous; leaves long-acuminate ............. 5. *A. trisulcatum*
Stems and branches terete; leaves acute to short-acuminate:
Leaves ashy on both sides, ovate-lanceolate, twice as long as wide; stellate pubescence minute, velvety; petals yellow to orange, 4–10 mm. long:
Petals deep orange, a deep maroon spot at base, 4–6 mm. long; calyx lobes round-deltoid, 1–1.2 mm. long at anthesis .......................... 6. *A. pringlei*
Petals orange-yellow, slightly darker at base, 6–10 mm. long; calyx lobes ovate, 1.5–3 mm. long at anthesis .................................... 7. *A. incanum*
Leaves ashy beneath, green above, broadly ovate; pubescence scaberulous; petals pink to brick red, 3–5 mm. long ................................ 8. *A. parvulum*
Calyx nearly equaling or surpassing mature carpels:
Carpels 4–5; leaves bright green, sparsely pubescent with simple, forked, and few-rayed hairs 1–3 mm. long; carpels with aristate beaks 2–3 mm. long ........ 9. *A. thurberi*
Carpels 7–15; leaves densely pubescent with fine, many-rayed hairs, often canescent; carpels usually cuspidate, the beaks 1–2 mm. long or less:
Sepals cordate-ovate; calyces strongly plicate-angled; branches and calyces stellate-pubescent or former glabrate, not hirsute:
Leaves nearly concolorus, somewhat tawny; pubescence somewhat scurfy, harsh; sepals barely equaling or slightly shorter than carpels ............ 10. *A. californicum*
Leaves dark green above, nearly white beneath; pubescence very fine, soft-velvety; sepals exceeding carpels .................................. 11. *A. lignosum*
Sepals ovate, angles of calyces not plicate; young branches or calyces or both somewhat hirsute with spreading hairs:
Flowers axillary; leaves not trilobate, 2–5 cm. long ................. 12. *A. wrightii*
Flowers in umbels, panicles, or racemes; leaves often shallowly trilobate, 3–20 cm. long:
Petals pale yellow ................................................ 13. *A. xanti*
Petals deep yellow:
Carpels long-hirsute, their beaks 1.5–2 mm. high; stems green, often slightly viscid; petals 10–12 mm. long ............................ 14. *A. palmeri*
Carpels short-hirsute, beaks 1 mm. high; stems reddish, not viscid; petals 15–25 mm. long ........................................... 15. *A. aurantiacum*

## 1. **Abutilon sonorae** A. Gray, Pl. Wright. **2:** 23, 1853

Herbaceous or suffrutescent perennial to at least 2 m. tall; stems, branches, and petioles hirsute with simple spreading hairs 2–3.5 mm. long; stipules lance-linear, 3–5 mm. long, caducous; petioles 5–15 cm. long; leaf blades broadly ovate-cordate to orbicular-cordate, 5–20 cm. long and wide, shallowly trilobate or margins merely irregularly rounded-dentate, sinus 1.5–3.5 cm. deep, finely and densely stellate-pubescent on both surfaces, dark green above, canescent and velvety beneath; flowers in axillary and terminal panicles; peduncles 1–3 cm. long at anthesis, 1-flowered, pedicels 3–6 mm. long; calyx finely stellate-puberulent, lobes ovate, 3–4 mm. long; petals deep yellow with orange tinge outside, 6–10 mm. long; carpels 7–11, stellate-pubescent, 8–10 mm. high, beak spreading-ascending at about a 45° angle, 0.5–1 mm. long; seeds usually 3 in each locule, about 2 mm. long, dark brown, dull, netted with a reticulum of minute, conical, somewhat curved hairs surmounting low narrow ridges, hairs light brown.

Often under trees or growing up through shrubs, on desert flats, hillsides, stream banks, and canyons, Lower Sonoran Zone, southern Arizona south through Sonora and Chihuahua to Guerrero. Aug.–Oct.

### 2. **Abutilon carterae** Kearney, Madroño **12:** 114, 1953

Herbaceous annual or short-lived perennial 0.5–1 m. tall; stem 4–5 mm. in diameter at base, densely white-lanate with short, forked, and simple hairs, these all approximately same length; stipules filiform, caducous; petioles stoutish, about equaling blades; leaf blades suborbicular, deeply cordate, and abruptly acuminate, 8–17 cm. wide, to 20 cm. long, crenulate-denticulate or sometimes slightly 3-lobed, dark green and tomentulose above, white-tomentose beneath, simple and forked hairs all same length; inflorescence an elongated open panicle; calyx broadly campanulate, 6–8 mm. high, lobes deltoid-acuminate, 4–6 mm. long, loosely scurfy-tomentose; petals cream, 7–8 mm. long; fruit short-cylindric, truncate at apex, with 8–9 carpels, these 9–10 mm. high, each bearing a slender awn 0.5–1 mm. high, villous with long, mainly simple hairs; seeds mostly 2 in each carpel, about 2.5 mm. long, irregularly stellate-puberulent with very fine, short, dark brown hairs.

Rocky hillsides and arroyos, Lower Sonoran Zone, along flanks of Sierra de la Giganta, Baja California. Flowering dates not positively known.

### 3. **Abutilon reventum** S. Wats., Proc. Amer. Acad. **21:** 418, 1886

Similar in general appearance, size, and shape of leaves to *A. sonorae* but pubescence of fine stellate hairs only; petioles and stems often glabrate; leaf blades cordate to reniform-cordate, 6–25 cm. long, abruptly acuminate at apex, shallowly trilobate or lobes none, margin irregularly crenate-dentate to subentire, greenish above, velvety beneath with a white or canescent tomentum; calyx lobes broadly ovate, 2.5–3.5 mm. long, apex subacute to obtuse, not acuminate; petals 10–12 mm. long, bright yellow, without orange flush on outside; carpels 7–8, stellate-pilosulous dorsally, 10–12 mm. long; each carpel 3–4-seeded, and slenderly muticous at apex.

Hillsides and canyons, Lower Sonoran Zone, usually near its upper limits, southern Arizona to Chihuahua, Oaxaca, Querétaro, and Zacatecas. April–Sept.

This species is much less common within the Sonoran Desert than *A. sonorae,* occurring mostly in the scrub formations just above the true desert.

### 4. **Abutilon parishii** S. Wats., Proc. Amer. Acad. **20:** 357, 1885

A low suffrutescent plant 1–6 dm. tall from a woody rootstock; branches and petioles densely stellate-tomentose and somewhat hirsute with slightly reflexed, simple hairs 1–2 mm. longer than tomentum; petioles slender, to 7 cm. long; leaf blades ovate to suborbicular, cordate at base, obtuse, acute or obscurely acuminate at apex, 2–5 cm.

wide, 3–7 cm. long, coarsely and irregularly dentate, densely velvety-tomentose on both surfaces, green above, white to tawny beneath; peduncles 1–3 mm. long; pedicels to 6 mm. long; calyx finely stellate-puberulent, at length subglabrate, lobes ovate-deltoid, acute or apiculate, 3–4 mm. wide, 4–5 mm. long; petals orange, 8–12 mm. long; fruit 7–8 mm. high, 10–11 mm. wide, thinly pubescent with simple spreading hairs along dorsal sutures of oblong carpels, these 6–7 mm. high, beak divergent, 1–3 mm. long; seeds 3 in each carpel, light brown, irregularly puberulent.

Along canyons among rocks, Lower Sonoran Zone, foothills of Santa Catalina Mountains, Arizona, and south into Sonora. March–May.

This species has been relegated to the synonymy of *A. palmeri* but is more closely related to *A. wrightii* than to the former. It is distinguishable from *A. wrightii* most readily by the shorter calyx lobes which are less than half as long as the carpels in fruit.

### 5. **Abutilon trisulcatum** (Jacq.) Urban, Rep. Spec. Nov. **16**: 32, 1919

*Sida trisulcata* Jacq., Enum. Pl. Carib. 26, 1760.
*Sida triquetra* L., Sp. Pl. ed. 2, 962, 1763.
*Abutilon triquetrum* Sweet, Hort. Brit. 53, 1827.
?*Abutilon floribundum* Schlecht., Linnaea **11**: 366, 1837.
*Bastardia triquetra* S. A. Morales, in G. Maza, Anal. Soc. Esp. Hist. Nat. **19**: 218, 1890.

Suffrutescent or shrubby plants 0.3–4 m. tall, stems, branches, and foliage, velvety canescent with almost microscopic stellate hairs; stems and branches trigonous; stipules lance-subulate, 1–2 mm. long, caducous; petioles slender, 1–4 cm. long; leaf blades broadly ovate, long-acuminate (or basal ones sometimes nearly reniform), 1–6 cm. wide, 4–15 cm. long, shallowly cordate at base, crenate, dentate, or subentire, pale green above, canescent beneath; flowers axillary or in vigorous plants openly paniculate; peduncles and pedicels slender, 1–2 cm. long in flower, to 3 cm. long in fruit; calyx lobes ovate-acuminate, 1.5–2 mm. long; petals orange-yellow, slightly darker at base, 5–6 mm. long; carpels mostly 5, erect, 6–7 mm. high, minutely stellate-puberulent, about 3-seeded, apical mucronation about 0.5 mm. long; seeds 1.2–1.5 mm. long, ovoid-reniform, nearly black, dull, irregularly puberulent.

Among shrubs and trees on hillsides, flats, and along arroyos, Lower Sonoran Zone to Tropical Zone, central Sonora to Central America and the West Indies. Commoner farther south than in the Sonoran Desert. Sept.–Jan.

### 6. **Abutilon pringlei** Hochr., Ann. Conserv. & Jard. Bot. Genève **6**: 14, 1902

Erect, ascendingly branched shrub with very slender branchlets, to 2 m. tall; herbage and young branches minutely tomentulose-velvety; stipules linear, 2–3 mm. long; petioles slender, 5–25 mm. long; leaf blades narrowly ovate to lanceolate, 8–25 mm. wide, to 5 cm. long, cordate at base, acute to gradually short-acuminate at apex, crenate-serrate, pale yellowish green above, canescent beneath; flowers axillary on slender lateral branches, sometimes simulating small-leafed panicles; peduncles and pedicels slender, 1–1.5 cm. long; calyx lobes broadly ovate, short-muticous, 1–1.2 mm. long at anthesis; petals 4–6 mm. long, orange, striped with salmon, a maroon spot at base; fruit slightly narrower at apex than at base; carpels 5–6 mm. high, minutely but closely stellate-puberulent, rounded or obtuse at apex; seeds subglobose-reniform, 1.6–2 mm. long, dark brown, dull, puberulent with minute, simple, grayish hairs.

Along arroyos, on hillsides and open desert, Lower Sonoran and into lower part of Upper Sonoran Zones, central and southern Arizona into Sonora at least to vicinity of Torres, south of Hermosillo. March–May.

**7. Abutilon incanum** (Link) Sweet, Hort. Brit. 53, 1827

*Sida incana* Link, Enum. Hort. Berol. **2:** 204, 1822.
*Abutilon ramosissimum* Presl, Rel. Haenk. **2:** 116, 1835.
*Abutilon texensis* Torr. & Gray, Fl. N. Amer. **1:** 231, 1838.

Slenderly branched shrub 0.5–2 m. tall with minutely stellate-tomentose, velvety pubescence on stems, leaves, and inflorescences; leaves ovate to lance-ovate, cordate at base, acute to short-acuminate or sometimes obtuse at apex, 0.5–3 cm. wide, 1.5–6 cm. long, crenate or dentate, minutely grayish-tomentose on both surfaces, slightly paler beneath than above; flowers axillary or subpaniculate near tips of branches; peduncles and pedicels together 1–2.5 cm. long, slender; calyx lobes ovate, abruptly mucronate, 2.5–3 mm. long, reflexed in fruit; petals orange-yellow, slightly darker at base but not deeply spotted, 6–10 mm. long; fruit 7–8 mm. high, 8–9 mm. wide, truncate at apex; carpels with apiculation 0.2–0.5 mm. long at exterior apical angle, densely stellate-puberulent; seeds dark brown, 1.5–2 mm. long, dull, sparsely arachnoid-puberulent with light gray, slender hairs.

Dry rocky slopes, Sonoran Zones, western Arizona to Texas and southward into Baja California, Nuevo León, San Luis Potosí, and Sinaloa. April–Oct.

**8. Abutilon parvulum** A. Gray, Pl. Wright. **1:** 21, 1852

Herbaceous or suffrutescent perennial from a woody rootstock, with slender stems 1–6 dm. tall, finely stellate-puberulent throughout; stems and lower surfaces of leaf blades often canescent; stipules 1–2 mm. long; petioles 5–20 mm. long; leaf blades ovate-cordate to triangular-cordate, 1–4 cm. broad, only slightly longer, obscurely 3-lobed or not lobed, irregularly but distinctly serrate, lower surfaces paler than upper, distinctly green, tissue of leaf visible through pubescence; flowers mainly or entirely axillary, solitary on pedicels 2–3 mm. long; peduncles 5–20 mm. long; calyx 4–5 mm. high, lobes lance-ovate, 2–2.5 mm. long, spreading or reflexed in fruit; petals 3–5 mm. long, pink or brick-red; fruit 6–8 mm. high, densely stellate-puberulent; carpels usually 5, with short, acute, erect beaks 1–2 mm. long; seeds dark brown, minutely puberulent.

Rocky slopes, Lower Sonoran and lower parts of Upper Sonoran Zones, from the Providence Mountains, California, to southern Colorado, western Texas, and northern Sonora. April–Oct.

This species can be distinguished from *A. incanum* more readily when fresh than after drying. The green of the leaf tissue showing through the sparser, longer-rayed pubescence on the lower sides of the leaf blade readily sets it apart in the field.

**9. Abutilon thurberi** A. Gray, Pl. Thurb. 307, 1854

Erect or procumbent, suffrutescent plant 3–6 dm. tall from creeping woody rootstocks; stems and foliage bright green, hirsute throughout with simple hairs; stipules linear, 10–12 mm. long, horizontally spreading, hirsute, and persistent; petioles slender, 1–4 cm. long; leaf blades thin, ovate-cordate, 2–5 cm. wide, 2.5–8 cm. long, gradually acuminate, serrate-dentate, concolorous; flowers mainly axillary, solitary; peduncles and pedicels together 1–2.5 cm. long, very slender; calyx lobes ovate-acuminate, 4–6 mm. long at anthesis, hirsute; petals orange-yellow, 4–6 mm. long; carpels usually 4–5, sparsely hirsute, oblong, body 5–6 mm. long, tipped with a slender, awnlike beak about half as long as body; seeds dark brown, 1.75–2 mm. long, scantily stellate-puberulent.

Canyons, often in partial shade, Lower Sonoran Zone, Baboquivari Mountains in southern Arizona south at least to central Sonora. Aug.–Oct.

The subpersistent, green, horizontally spreading stipules and the thin, hirsute, dark green leaves are distinctive.

### 10. Abutilon californicum Benth., Bot. Sulph. 8, 1844

*Abutilon lemmonii* S. Wats., Proc. Amer. Acad. **20**: 357, 1885.

Erect, openly branched shrub 1–2 m. tall with densely stellate-tomentose, slightly scurfy leaves, young stems, and inflorescences; stipules linear, to 15 mm. long, caducous, densely tomentulose; petioles 1–2.5 cm. long; leaf blades broadly ovate-cordate, 1.5–5 cm. long (Arizona plants with narrower, larger leaves to 10 cm. long), rounded, acute or subacuminate at apex, indistinctly dentate-crenate, only slightly paler beneath than above, thickish; peduncles and pedicels together 1–3 cm. long; buds strongly plicate-angled; calyx lobes cordate-ovate, 6–8 mm. wide, 4–7 mm. long, acuminate, nearly 10 mm. long in fruit; petals yellow-orange, 1–1.5 cm. long; fruits 10–12 mm. wide, 9–10 mm. high; carpels oblong or oblong-ovate, body about 7–8 mm. long, stellate-puberulent, surmounted by a slender, erect beak 1.8–2.2 mm. long.

Rocky hillsides, valleys and canyons, Sonoran Zones, southern Arizona to Sinaloa and the Cape Region, Baja California. March–Sept.

### 11. Abutilon lignosum (Cav.) G. Don, Gen. Hist. Pl. 1: 501, 1831

*Sida lignosa* Cav., Diss. **1**: 28, *pl. 6, fig. 2,* 1785.
*Abutilon jacquinii* G. Don, Gen. Hist. Pl. **1**: 503, 1831.
*Abutilon berlandieri* A. Gray, in S. Wats., Proc. Amer. Acad. **20**: 358, 1885.
*Abutilon scabrum* S. Wats., op. cit. **24**: 41, 1889.

Suffrutescent plants to 2 m. tall with slender, at first densely or sparsely stellate-puberulent and at length glabrate stems; herbage densely but finely stellate-tomentulose; stipules linear, 7–10 mm. long, tawny-tomentulose; petioles 2–10 cm. long; leaf blades broadly ovate, acuminate, cordate with a very narrow sinus at base, rarely faintly trilobate, 4–10 cm. broad, 5–15 cm. long, shallowly crenate or dentate, dark green above, densely but finely tomentose and canescent beneath; pedicels and peduncles to 5 cm. long; sepals cordate-ovate, acuminate, 7–10 mm. long, 6–8 mm. wide, canescent; petals 1–1.5 cm. long; fruit 10–12 mm. high, nearly as broad, truncate; carpels acute but scarcely cuspidate at apex, densely stellate-pubescent.

Along streams and in shaded canyons, Lower Sonoran Zone to Tropical Zone, Sonora to Chihuahua and southern Mexico; southern Texas, southern Florida, and in the West Indies. Aug.–Dec. in our area.

### 12. Abutilon wrightii A. Gray, Bost. Jour. Nat. Hist. 6: 162, 1850

A small suffrutescent plant 2–8 dm. tall with hirsute stems and petioles, often with fine, short-rayed, stellate hairs intermingled with the spreading simple ones; stipules filiform to linear, 3–8 mm. long; petioles slender, 1–4 cm. long; leaf blades broadly ovate-cordate, 2–5 cm. long, nearly or fully as wide, acute, obtuse, or rounded at apex, crenate or dentate, sinus narrow, densely stellate-pubescent but dark green above, stellate-tomentose and whitish beneath; flowers axillary; peduncles and pedicels slender, 1–2.5 cm. long; sepals ovate-acuminate, 3–4 mm. wide, 5–8 mm. long in anthesis, to 1.5 cm. long in fruit, tomentose and hirsute with some spreading hairs; petals 9–15 mm. long, bright yellow or pale yellow; capsules 10–12 mm. long, usually as wide; carpels 7–12, densely stellate-pubescent; beak erect or nearly so, 2–3.5 mm. long; seeds nearly black, about 2 mm. long, back arachnoid-pubescent with fine white hairs about 0.5 mm. long.

Open ground, hillsides, and shady slopes, Sonoran Zones, mainly upper parts of Lower Sonoran Zone to Tropical Zone, western Texas and adjacent Mexico to Sonora, where it has been collected near Costa Rica, east of Hermosillo. March–June.

**13. Abutilon xanti** A. Gray, Proc. Amer. Acad. **22:** 301, 1887

An openly branched, sometimes straggly shrub 1–3.5 m. tall with rusty, stellate-tomentose stems or these glabrate in age; foliage densely stellate-tomentose with very fine hairs; stipules lance-subulate, 2–2.5 mm. wide, 7–9 mm. long; petioles 5–10 cm. long, minutely stellate-tomentose; leaf blades cordate, broadly ovate to suborbicular in outline, sometimes faintly 3-lobed, acuminate, 2–12 cm. wide, 3–20 cm. long, finely crenate-dentate to subentire, green above, densely stellate-tomentose and canescent beneath; flowers often crowded into clusters near tips of branches, those below eventually racemosely arranged; peduncles and pedicels 1–3 or rarely –5 cm. long; sepals ovate-lanceolate, about 3 mm. wide, 4–5 mm. long at anthesis, 8–10 mm. long in fruit, short-acuminate, barely equaling or slightly shorter than carpels; petals 10–15 mm. long, pale yellow; fruit 9–12 mm. wide, 10–12 mm. high; carpels oblong, gradually pointed and cuspidate above, sparsely stellate-hirsute.

Hillsides, arroyos and plains, Lower Sonoran Zone, southern Baja California. Sept.–Jan.

**14. Abutilon palmeri** A. Gray, Proc. Amer. Acad. **8:** 289, 1870

*Abutilon macdougalii* Rose & Standl., Contr. U.S. Nat. Herb. **16:** 13, *pl. 4,* 1912.

Erect shrubs 1–2.5 m. tall with sparsely stellate-puberulent and hirsute stems, petioles, and peduncles or the stellate pubescence sometimes lacking, often somewhat viscid; stipules lanceolate, 2–2.5 mm. wide, 8–14 mm. long, caducous; petioles slender, 3–15 cm. long; leaf blades cordate-ovate to cordate-orbicular, acuminate, 3–15 cm. wide, 4–20 cm. long, coarsely dentate, often shallowly trilobate, densely stellate-tomentose, green above, velvety-canescent and prominently veined beneath; flowers in pyramidal panicles or a few solitary in axils below; peduncles 1–3 cm. long; pedicels 0.5–2 cm. long; calyx densely stellate-hispid, lobes lance-ovate, 3.5–5 mm. wide, 5–9 mm. long, acuminate; petals 10–12 mm. long, orange-yellow; fruit 8–10 mm. wide and high; carpels oblong, 8–9 mm. high, about 4 mm. wide, bearing an erect terminal cusp 1.5–2 mm. long, densely stellate-pilose with soft hairs; seeds dark brown, about 2 mm. high, finely stellate-puberulent in a reticulate pattern.

Desert washes and rocky hillsides, Lower Sonoran Zone, southwestern Arizona, Sonora, northeastern Baja California, and desert slopes in eastern San Diego County, California. March–May.

The material named *A. macdougalii* Rose & Standl. has fewer simple hairs and a more uniform covering of stellate puberulence on the stems and petioles than does the type specimen but the distinction is too slight to warrant taxonomic recognition.

**15. Abutilon aurantiacum** S. Wats., Proc. Amer. Acad. **20:** 357, 1885

Erect, openly branched shrub 1–1.5 m. tall with stellate-pubescent younger stems and leaves, some spreading, simple, or few-rayed hairs extending beyond tomentum; older stems becoming glabrate and reddish-tinged; stipules lance-linear, 5–8 mm. long, velvety-puberulent, caducous; petioles velvety and hirsute with longer hairs,

2–10 cm. long, eventually subglabrate; leaf blades broadly ovate to orbicular, acute or abruptly short-acuminate, sometimes obscurely trilobate, 2–6 cm. broad, to 8 cm. long, finely dentate or crenate-dentate, densely and velvety-stellate-tomentose on both surfaces, slightly paler beneath than above, tawny; peduncles to 5 cm. long; pedicels 8–15 mm. long; calyces densely stellate-tomentose, hairs near base spreading about 1 mm., lobes ovate-deltoid, abruptly short-acuminate, 5–7 mm. wide, and long in anthesis, to 1 cm. long in fruit; petals golden-yellow 1.5–2.5 cm. long; fruit hemispheric-truncate, 10–14 mm. wide, 10–11 mm. high; carpels oblong, 9–10 mm. high, bearing an erect or ascendingly spreading beak about 1 mm. long, stellate-hirsute along dorsal surface; seeds 2.5 mm. long, dark brown with irregular patches of light brown, stellate puberulence.

Hillsides, usually near a drainage course, Lower Sonoran and lower margin of Upper Sonoran Zones, Baja California from the vicinity of Tijuana southward to Cataviñá. March–May.

**Abutilon theophrasti** Medic., Malvenfam. 28, 1787, a robust annual 1–2 m. tall with leaves to 2 dm. wide, yellow flowers, and carpels each bearing spreading beaks 3–5 mm. long, is an introduction from India that occurs commonly as a roadside and fence-row weed from the Atlantic seaboard to Texas and at several points on the Pacific Coast. Kearney collected it in a cottonfield near Glendale, Maricopa County, but it has not been reported from other areas within the Sonoran Desert. It may, however, become fully naturalized in irrigated districts within the Sonoran Desert.

## 2. BOGENHARDIA Reichb., Rep. Herb. Nom. Gen. Pl. pt. 1, 200; pt. 2, 48, 1841

Perennial subshrub or herb from woody root with stellate and intermingled simple pubescence. Leaves alternate, cordate, petiolate. Stipules linear, deciduous. Flowers axillary on slender jointed peduncles. Bracteoles none. Calyx lobes 5, equaling or exceeding tube. Petals 5. Stamens many, coalesced into a tube; tube adnate to petals near their bases. Ovary 8–15-celled, each locule 2- to several-seeded. Carpels bladdery-inflated, rounded at apex, dehiscent dorsally, side walls membranous, smooth. Seeds subreniform. Embryo curved.

### 1. Bogenhardia crispa (L.) Kearney, Leafl. West. Bot. **7:** 120, 1954

*Sida crispa* L., Sp. Pl. 685, 1753.
*Abutilon crispum* Medic., Malvenfam. 29, 1787, (as *cryspum*).
*Beloere crispa* Shuttlew. ex A. Gray, Pl. Wright. **1:** 21, 1852.
*Gayoides crispum* Small, Fl. SE. U.S. 764, 1903.
*Pseudobastardia crispa* Hassler, Bull. Soc. Genève II, **1:** 211, 1909.

Diffuse, slenderly branched perennial from a woody rootstock; stems to 1 m. long, spreading to ascending or procumbent, slender, minutely and closely stellate-puberulent, with additional spreading simple hairs to 2 mm. long or these sometimes lacking (the plant then often given varietal rank); stipules linear to subfiliform, 3–6 mm. long, deciduous; petioles to 4 cm. long or upper leaves sessile; leaf blades ovate-cordate, usually acuminate, 1.5–4.5 cm. wide, 2–7 cm. long, crenate-dentate, sinus deep and narrow, pubescent on both surfaces with both simple and few-rayed, slender, stellate hairs, thin, slightly paler beneath than above; peduncles axillary, solitary, slender to subfiliform, 1–20 mm. long in flower; pedicels 2–9 mm. long, more densely hirsute than peduncles, often sharply geniculate at joint with peduncle; calyx hirsute, lobes ovate, 2.5–3 mm. long at anthesis, spreading or reflexed in fruit; petals pale yellow,

5–8 mm. long; fruit subglobose, 1–1.5 cm. in diameter, hirsute; carpels 8–15, bladdery, side walls membranous, dehiscent dorsally.

Hillsides and sandy flats, often in shade of shrubs, Lower Sonoran Zone to Tropical Zone, southern Arizona to Texas and south into tropical America; southern Florida. Flowering practically throughout the year.

## 3. HORSFORDIA A. Gray, Proc. Amer. Acad. 22: 296, 1887

Herbs or shrubs densely stellate-tomentose or stellate-scabrous throughout. Flowers moderately showy, solitary or in terminal and axillary panicles. Calyces obconic; lobes 5, valvate. Bractlets none. Petals 5, obovate, entire, or emarginate. Stamens numerous, at apex of a stoutish stamineal column. Carpels 8–12, lower part 1–3-seeded and coalescent, upper portion accrescent and developing into 2 small scarious wings. Seeds reniform or ovate-reniform. Embryo curved.

Leaves thin, cordate-orbicular, 3.5 cm. or less in length; wings of carpels erect, lanceolate, somewhat longer than broad .................................... 1. *H. rotundifolia*
Leaves thick, ovate to lanceolate, 2–10 cm. long; wings of carpels spreading or ascending, as broad as long:
  Tomentum rusty-yellow, velvety; leaves mostly lanceolate; petals yellow or orange, 8–10 mm. long ........................................................ 2. *H. newberryi*
  Tomentum grayish, harsh; leaves ovate to lance-ovate; petals pink to reddish lavender, 12–18 mm. long ................................................... 3. *H. alata*

### 1. Horsfordia rotundifolia S. Wats., Proc. Amer. Acad. 24: 40, [Jan.] 1889

*Horsfordia purisimae* Brandegee, Proc. Calif. Acad. II, 2: 135, [Nov.] 1889.

Erect or ascending, slenderly branched shrub 1–1.5 m. tall with very fine-rayed stellate puberulence generally covering branches, foliage, and all of inflorescence except petals; stipules lance-linear, 3–4 mm. long; petioles slender, 1–3 cm. long; leaf blades broadly ovate to suborbicular, obtuse, rounded, or rarely acute at apex, shallowly cordate at base, 1–3 cm. wide, 1.5–3.5 cm. long, finely dentate, smoothly stellate-puberulent on both surfaces, yellowish green, slightly paler beneath than above; some flowers solitary in axils but mostly in open, terminal, naked and narrow panicles; peduncles about 1 cm. long, pedicels 3–5 mm. long, articulation with peduncle often obscure; calyx about 5 mm. high at anthesis, lobes deltoid-ovate, 2–3 mm. long, to 4 mm. long in fruit; petals 6–10 mm. long, light orange or yellow-orange but drying bluish; carpels 7–8 mm. high, exposed parts sparsely and minutely stellate-puberulent, basal part fenestrate-reticulate and membranous on sides, wings erect, 3–4 mm. high, to 2 mm. wide; seeds brownish, 2–2.5 mm. long, minutely pustulate with very small, conical, simple hairs.

Mesas, rocky hillsides, and along arroyos, Lower Sonoran Zone, vicinity of Guaymas, Sonora, and from central Baja California to the Cape Region. Aug.–March.

### 2. Horsfordia newberryi (S. Wats.) A. Gray, Proc. Amer. Acad. 22: 297, 1887

*Abutilon newberryi* S. Wats., Proc. Amer. Acad. 11: 125, 1876.

Erect and often strict, sparingly branched perennial 1–3 m. tall with a dense covering of yellowish or slightly rusty-yellow, stellate tomentum, rays fine and closely intertangled; stipules lance-linear, 3–5 mm. long, deciduous; petioles stout, 5–15 mm. long; leaf blades lance-ovate to narrowly lanceolate, shallowly cordate at base, acute or obtusish at apex, 1–4 cm. wide, 3–10 cm. long (rarely to 15 cm. long), very thick, yellowish above, slightly paler beneath, margins faintly crenate to subentire, veins prominent beneath; peduncles and pedicels stout, 5–10 mm. long in flower, to 2 cm. long in fruit; calyx turbinate to broadly hemispheric, 4–6 mm. high, deltoid-ovate lobes 2–3 mm. long, densely pubescent; petals 8–10 mm. long, yellow or pale orange; carpels 9–10 mm. tall, reticulate below but sides not transparent, wings spreading-ascending, 3–4 mm. wide and high; seeds puberulent with fine whitish hairs.

Desert washes and hillsides, Lower Sonoran Zone, Colorado Desert in southeastern California, southwestern Arizona, and south to central Sonora and Baja California. March–Oct.

The yellowish stellate tomentum and the very thick leaves are strikingly characteristic. The natives in Baja California claim that the hairs on the leaves and stems cause sore eyes if one remains in the vicinity of the plants more than a few minutes.

**3. Horsfordia alata** (S. Wats.) A. Gray, Proc. Amer. Acad. **22:** 297, 1887

*Sida alata* S. Wats., Proc. Amer. Acad. **20:** 356, 1885.
*Horsfordia palmeri* S. Wats., op. cit. **24:** 40, 1889.

Erect shrub or suffrutescent perennial 1–4 m. tall, sparingly branched, densely and harshly stellate-tomentose; stems tawny; leaves green or slightly yellowish green and slightly viscid; petioles 1–2.5 cm. long, stout; leaf blades ovate to lance-ovate, 1–3 cm. wide, 3–5 cm. long, acute to acuminate at apex, subcordate at base, margins serrulate or denticulate, midvein impressed above, veins prominent beneath; peduncles 6–10 mm. long; pedicels 2–4 mm. long in anthesis, to 8 mm. long in fruit; calyx lobes broadly ovate, acuminate, 2.5–3 mm. wide, 3–5 mm. long; petals pink to reddish lavender, 12–18 mm. long; carpels 7–8 mm. long, basal part containing a single seed, laterally reticulate-fenestrate, ridges coarse, upper part usually empty, wings about 3 mm. high and practically as broad.

Desert washes and stony slopes, Lower Sonoran Zone, southern Yuma County, Arizona, southeastern California and south into central Sonora and at scattered localities throughout the desert regions of Baja California. March–Oct.

## 4. WISSADULA Medic., Malvenfam. 24, 1787

Suffrutescent to shrubby plants, commonly stellate-pubescent on younger stems, leaves, and calyces. Stipules mostly caducous. Leaves mostly ample, petiolate, truncate to cordate, entire or dentate. Flowers axillary or in terminal panicles. Calyx commonly 5-lobed to middle, densely stellate-pubescent and sometimes with some spreading simple hairs. Corollas 1–3 cm. in diameter. Petals 5, yellow, pink, or orange, pilose on margins of claws. Stamineal column short; stamens numerous. Ovary mostly 5-celled, rarely 3- or 6-celled; ovules usually 3 in each carpel, pendulous. Style branches same number as carpels; stigmas capitate. Fruit turbinate to subglobose; carpels 3–5 or rarely 6, outer apical tip angulate, apiculate, or rostrate, laterally constricted horizontally to form 2 nearly distinct cells in each carpel, upper one dehiscent, lower indehiscent. Seeds usually 3, one in lower portion, two in upper part of carpel, glabrous to puberulent. Embryo curved. Endosperm scanty.

**1. Wissadula cincta** (Brandegee) Rose, Contr. U.S. Nat. Herb. **5:** 178, 1899

*Abutilon cinctum* Brandegee, Zoe **3:** 348, 1893.
*Wissadula incana* Brandegee, op. cit. **5:** 210, 1905.

Open, slenderly branched shrub to 1.5 m. tall with dense velvety tomentum of minute, stellate, tawny to incanous hairs; stipules linear, 2–3 mm. long, caducous; petioles slender, 1–3 cm. long; leaf blades ovate to orbicular-cordate, acute to short-acuminate at apex, cordate at base, 8–25 mm. wide, 1–3 cm. long, minutely stellate-puberulent on both surfaces, finely serrate-dentate, slightly lighter and veins moderately prominent beneath; flowers axillary or in few-flowered, narrow, terminal racemose panicles; pedicels 1.5–5 mm. long; calyx turbinate, strongly 5-angled in bud, 3–7 mm. long, stellate-tomentulose and with simple spreading hairs on tube, lobes triangular-ovate, acute, 3–5 mm. long, ascending in fruit; petals broadly obovate, 6–9 mm. long, pale lavender; fruit mostly 3-carpellate, 3–5 mm. in diameter, obconic; carpels obovate, 3–4 mm. high, rounded at apex, beakless, densely stellate-tomentulose; seeds obovate-pyriform, 1.5–2 mm. high, dark gray, minutely and sparsely puberulent to glabrous.

Rocky hillsides and at the margins of openings in the brush. Lower Sonoran Zone, central Sonora to Sinaloa and on the Tres Marías Islands. Aug.–March.

## 5. PSEUDABUTILON R. E. Fries, Kgl. Sv. Vet.-Akad. Handl. **43:** no. 4, 96, 1908

Annual or perennial, suffrutescent herbs or low shrubs with stellate-tomentulose or simple pubescent foliage. Leaves petiolate, cordate, dentate, membranous, and sparingly pubescent to thick and densely tomentose. Flowers perfect, without bracteoles, axillary or in terminal spikes or panicles. Calyx turbinate to campanulate; tube about equaling 5 ovate to deltoid lobes. Corolla small to medium in size, pink, lavender, or yellow; petals rounded at apex, claw hirsute. Stamineal column usually pilose below; stamens numerous. Ovary depressed-globose. Carpels 5–11, a transverse partial partition attached only to dorsal wall and imperfectly separating each carpel into a basal, 1-seeded portion and a terminal, 2-seeded chamber. Styles same number as carpels; stigmas capitate. Mature carpels stellate-pubescent, apiculate or rostrate. Seeds 3, ovoid-reniform. Embryo curved. Endosperm scanty.

Carpels with slender beaks 1.5–2 mm. long; stems, pedicels, and calyces pilose with spreading simple hairs; plants annual .................................... 1. *P. sonorae*
Carpels muticous, beak not over 0.5 mm. long; pedicels and calyces stellate-tomentulose; plants perennial, shrubby .......................................... 2. *P. scabrum*

### 1. **Pseudabutilon sonorae** Wiggins, Contr. Dudley Herb. **3:** 70, *pl. 16, figs. 6–9,* 1940

Erect annual herb to 1 m. tall with terete, sparsely hispid branches; stipules linear, 5–7 mm. long, sparsely hispid, fugacious; petioles slender, 2–8 cm. long, sparsely hispid; leaf blades broadly ovate, 2.5–6 cm. wide, 3–10 cm. long, rounded to subcordate at base, acuminate at apex, sparsely hispid with 2–4-rayed stellate hairs, these 1–2 mm. long and spreading on nerves, margins dentate with rounded teeth, distinctly 7-nerved beneath; inflorescence terminal, narrowly paniculate, 1–3 dm. long; pedicels slender, 1–2 cm. long, jointed near apex, spreadingly hispid; calyces ovate, 4–5 mm. high, hispid, lobes ovate, 2–3.5 mm. wide, 3–5 mm. long, abruptly attenuate-acuminate; corolla 10–12 mm. wide; petals obovate, yellowish, 4 mm. wide, ciliate-villous on claws; stamineal column 1.5–2.5 mm. high, glabrous; fruit subglobose, truncate, 6–8 mm. wide, about 5 mm. high, shallowly umbilicate, 5-carpellate; carpels oblong-obovate, abruptly rostrate with a beak 1.5–2 mm. long, sparsely hispid and with a few smaller stellate hairs among bases of simple ones; seeds 1.5–2 mm. high, dark brown, glabrous, irregularly and finely papillate.

Along watercourses and margins of fields, Lower Sonoran Zone, Sonora. Collected in flower and fruit in Sept.

**2. Pseudabutilon scabrum** (Presl) R. E. Fries, Kgl. Sv. Vet.-Akad. Handl. **43:** no. 4, 103, 1908

*Wissadula scabra* Presl, Rel. Haenk. **2:** 117, *pl. 69,* 1835.

Shrub 1–3 m. tall with slender, buff- or gray-barked branches, younger ones finely stellate-puberulent; stipules filiform, 3–6 mm. long; petioles 1.5–5 cm. long, stellate-pubescent, at length glabrate; leaf blades ovate-cordate, 2.5–9 cm. wide, 3.5–12 cm. long, attenuate at apex, serrate-dentate, dark green and sparsely stellate-scabrous above, cinereous and densely but finely stellate-scabrous beneath, prominently 5–7-veined; inflorescence oblong-ovoid, terminally paniculate, sometimes crowded; pedicels 6–10 mm. long, stellate-tomentulose, jointed near apex; calyx broadly campanulate, densely stellate-puberulent, 7–9 mm. high, lobes broadly triangular-ovate, acuminate, 5–7 mm. long, velvety-stellate within; petals yellow, 2.5–3.5 cm. long; stamineal column 6–8 mm. high, densely stellate-pubescent; fruit depressed-subglobose, 6–8 mm. high, 9–10 mm. wide; carpels little wider at top than at bottom, 3–4 mm. wide, stellate-puberulent, obtuse to rounded at apex, barely muticous; seeds tumidly ovate-reniform, 2 mm. high, dark brown, minutely and sparsely scurfy.

Among shrubs and rocks on hillsides and along arroyos, Lower Sonoran and Tropical Zones, southern Sonora to Guatemala. Sept.–March.

## 6. SPHAERALCEA St. Hil., Fl. Bras. Merid. 1: 209, 1827*

Annual, biennial, or perennial herbs or suffrutescent plants with stellate-pubescent herbage. Leaves alternate, angulate or lobed. Flowers short-pedicellate, axillary or in terminal racemes, panicles, or thyrses. Bracteoles 3, distinct. Calyx 5-lobed. Petals obovate, rounded, or emarginate. Stamineal column usually glabrous. Ovary several-loculed, each carpel 2–3-ovulate. Style branches slender, as many as carpels; stigmas clavate or narrowly subcapitate. Carpels ovate, rounded, or deltoid at apex, oblong-reniform, terminal part dehiscent, with smooth sidewalls, basal part reticulate laterally and indehiscent. Seeds 1–3, reniform. Embryo curved; endosperm scanty.

◆Plants annual or biennial; terminal, dehiscent part of carpel minute, forming less than one-third of carpel:
- Carpels 1.2–2.5 mm. high; leaf blades thin and flexible; pubescence grayish:
  - Leaf blades nearly as wide as, to wider than, long, cordate to truncate at base, usually obtuse at apex, 3-lobed .......... 1. *S. coulteri* var. *coulteri*
  - Leaf blades usually longer than wide, subcuneate to truncate at base, usually acute at apex, shallowly lobed or merely angled:
    - Leaf blades finely crenate; inflorescence many-flowered .......... 1*a*. *S. coulteri* var. *californica*
    - Leaf blades coarsely crenate-dentate; inflorescence few-flowered .......... 1*b*. *S. coulteri* var. *margaritae*
- Carpels 2.5–3 mm. high; leaf blades thick and firm; pubescence yellow, dense and scurfy .......... 2. *S. orcuttii*

* The key to the species adapted from that in Thomas H. Kearney's monograph, *The North American species of Sphaeralcea subgenus Eusphaeralcea.* Univ. Calif. Pub. Bot. **19:** 1–128, *pls. 1–12, fig. 1,* 1935.

♦Plants perennial; terminal, dehiscent part of carpel forming from one-third to nine-tenths of carpel:
Carpels with very thin, nearly scarious walls, areoles of the basal reticulate part transparent or nearly so:
Inflorescence an open, comparatively few-flowered panicle:
Stems and leaves usually whitish-canescent or whitish-tomentose; calyx about as pubescent as stems and leaves .......................................... 3. *S. laxa*
Stems and leaves usually green and sparsely pubescent; calyx markedly more densely pubescent than stems and leaves ...................... 4. *S. rusbyi* var. *gilensis*
Inflorescence a narrow, interrupted, many-flowered thyrse:
Indehiscent part of carpel about as wide as dehiscent part:
Petals white or pink, drying yellowish, lavender, or violet:
Carpels 5–7 mm. high, about 2 mm. wide, narrowed to an obtuse apex, mucronate or cuspidate .................................. 5. *S. axillaris* var. *axillaris*
Carpels 2–3.2 mm. high, two-thirds as wide or wider, broad and rounded at apex, usually muticous ............................ 5*a*. *S. axillaris* var. *violacea*
Petals grenadine, not changing color appreciably in drying:
Leaf blades not less than three-fourths as wide as long ... 7*b*. *S. emoryi* var. *arida*
Leaf blades not more than one-half as wide as long .......................... ............................................ 7*c*. *S. emoryi* var. *nevadensis*
Indehiscent part of carpel usually narrower than dehiscent part, finely reticulate:
Larger leaf blades 6–10 cm. long, usually considerably longer than wide; seeds copiously puberulent; carpels often connate basally at maturity ...... 6. *S. hainesii*
Larger leaf blades 3–5 cm. long, as wide as long or nearly so; seeds glabrous or essentially so; carpels not connate basally ................. 7*b*. *S. emoryi* var. *arida*
Carpels with chartaceous thicker walls, areoles usually opaque or nearly so:
■Indehiscent part of carpel usually rugose or muricate dorsally, reticulations prominent and usually coarse:
Fruit truncate-conical; carpels usually cuspidate, not conspicuously galeate:
Pedicels usually persistent; inflorescence a many-flowered thyrse; stems seldom less than 6 dm. high:
Leaf blades merely angulate or toothed near base, thickish .................... ............................................ 7. *S. emoryi* var. *emoryi*
Leaf blades distinctly lobed or parted, often thin .... 7*a*. *S. emoryi* var. *variabilis*
Pedicels deciduous promptly at maturity of fruit; inflorescence racemose or subthyrsoid; stems not over 5 dm. tall ........... 8. *S. subhastata* var. *thyrsoides*
Fruit hemispherical or nearly so; carpels usually muticous or mucronulate, conspicuously galeate:
Leaf blades longer than broad, usually subcuneate and lobed near base; inflorescence a narrow, interrupted, many-flowered thyrse .............. 9. *S. fulva*
Leaf blades about as wide as long, subcordate to deeply cordate, lobed near middle; inflorescence an open panicle, or if thyrsoid then few-flowered:
Pubescence of leaves concolorous:
Stems usually suffrutescent and more than 5 dm. tall; inflorescence an open, long-branched panicle; pubescence whitish:
Petals grenadine to grenadine-pink ........... 10. *S. ambigua* var. *ambigua*
Petals pink or lavender, often drying violet .... 10*a*. *S. ambigua* var. *rosacea*
Stems usually herbaceous and less than 5 dm. long; inflorescence racemose or narrowly thyrsoid; pubescence of stems yellowish ..................... ........................................ 10*b*. *S. ambigua* var. *rugosa*
Pubescence of leaves conspicuously varicolored, yellowish above, white beneath, orange-yellow on margins ................ 10*c*. *S. ambigua* var. *versicolor*

Indehiscent part of carpel smooth or nearly so dorsally, reticulations usually fine, distinct but rarely prominent:
Carpels about one-half as wide as high; inflorescence a few-flowered, open, long-branched panicle or narrower and subthyrsoid:
Stems and leaves usually whitish-canescent or whitish-tomentose; calyx about as pubescent as stems and leaves .................................... 3. *S. laxa*
Stems and leaves usually green and sparsely pubescent; calyx markedly more densely pubescent than stems and leaves .................. 4. *S. rusbyi* var. *gilensis*
Carpels much more than half as wide as high if the inflorescence is few-flowered and racemose; inflorescence usually many-flowered, thyrsoid-glomerate:
Leaf blades much longer than wide, not pedately parted; fruit truncate-conical:
Leaf blades usually at least one-half as wide as long; carpels bearing cusps 1–2 mm. long:
Pubescence not velutinous; plants greenish or grayish; carpels 4–5 mm. high; cusp 1–1.2 mm. long ..................... 11. *S. fendleri* var. *fendleri*
Pubescence velutinous; plant whitish; carpels 5–6 mm. high; cusp 1–2 mm. long .................................. 11*a*. *S. fendleri* var. *albescens*
Leaf blades usually much less than one-half as wide as long; carpels rounded or muticous; cusp less than 1 mm. long:
Petals pink, lavender, or grenadine, 10–20 mm. long; carpels usually obtuse at apex, minutely muticous ............. 12. *S. angustifolia* var. *angustifolia*
Petals always grenadine, 7–12 mm. long; carpels usually acute at apex, short-cuspidate .......................... 12*a*. *S. angustifolia* var. *cuspidata*
Leaf blades about as long as wide, pedately parted; fruit hemispherical ........... .......................................... 13. *S. grossulariifolia* var. *pedata*

**1. Sphaeralcea coulteri** (S. Wats.) A. Gray, Proc. Amer. Acad. **22:** 291, 1887, var. **coulteri**

*Malvastrum coulteri* S. Wats., Proc. Amer. Acad. **11:** 125, 1876.

Annual 1–15 (but usually 1–3) dm. tall, with green or slightly grayish foliage; pubescence stellate, sparse to moderately interlaced; stipules subfiliform, 3–5 mm. long, deciduous; petioles slender, shorter than blades; leaf blades ovate to suborbicular, 1.5–3 cm. long, 1–3.4 cm. wide, shallowly to deeply 3–5-lobed, base cordate to truncate, obtuse to barely acutish at apex, sparsely to moderately stellate-puberulent; inflorescence thyrsoid, to 10 cm. long; pedicels shorter than to slightly exceeding calyx; calyx turbinate, 5–7 mm. long, lobes lance-ovate, attenuate, 3–5 mm. long; petals orange to salmon-orange, 8–15 mm. long; stamineal column conspicuously pubescent; fruit hemispherical, 3–3.5 mm. high; carpels 12–22, 1.8–2.5 mm. high, almost as wide, reniform, shallowly notched between indehiscent and dehiscent parts, latter part irregularly quadrangular-truncate or rounded apically, forming one-fourth to one-third of carpel, much narrower than indehiscent part; sides of basal indehiscent portion prominently reticulate-fenestrate; seed 1, sparsely puberulent.

Desert flats and slopes, Lower Sonoran Zone, southeastern California, southern Arizona, and south into Baja California and through Sonora to Sinaloa. Jan.–June.

**1*a*. Sphaeralcea coulteri** var. **californica** (Rose) Kearney, Jour. Wash. Acad. **29:** 486, 1939

*Sphaeralcea californica* Rose, Contr. U.S. Nat. Herb. **1:** 66, 1890.
*Sphaeralcea coulteri* subsp. *californica* Kearney, Univ. Calif. Pub. Bot. **19:** 32, 1935.

Stems to 3.5 m. tall; leaf blades 3–7 cm. long and more angulate-lobed than in var.

*coulteri,* acute or acutish at apex, finely crenate; pedicels very short (1–3 mm.); carpels 1.8–2 mm. high.

Sandy soil, Lower Sonoran Zone, vicinity of Santa Rosalia, Baja California, south to the Cape Region. Jan.–April.

**1*b*. Sphaeralcea coulteri** var. **margaritae** (Brandegee) Kearney, Jour. Wash. Acad. **29:** 486, 1939

*Sphaeralcea margaritae* Brandegee, Zoe **5:** 156, 1903.
*Sphaeralcea coulteri* subsp. *margaritae* Kearney, Univ. Calif. Pub. Bot. **19:** 33, 1935.

Differs from var. *coulteri* in having short hairs only 0.3 mm. long instead of 0.4–0.8 mm. long; narrower leaf blades only two-thirds to three-fourths as wide as long; a more open, fewer-flowered inflorescence, with peduncles surpassing subtending leaves; and small carpels only 1.2–2 mm. high.

The Magdalena Plain and on Santa Margarita Island, Baja California. Jan.–April.

**2. Sphaeralcea orcuttii** Rose, Contr. U.S. Nat. Herb. **1:** 289, 1893

Erect and often strict annual or biennial to 12 dm. tall, densely stellate-tomentulose with yellowish-canescent, 12–20-rayed hairs; stipules lance-subulate, 5–7 mm. long, caducous; petioles stoutish, 1–3 cm. long; leaf blades deltoid-ovate, 2–4 cm. wide, 3–6 cm. long, shallowly 3-lobed near base with rounded lobes, subcordate to truncate, irregularly crenulate on margins, rugose or nearly plane, moderately to densely stellate-tomentulose on both surfaces, canescent; inflorescence a narrow, many-flowered, glomerate thyrse or elongate lower branches often racemose; pedicels to 5 mm. long or some flowers subsessile; calyx densely stellate-puberulent, 4.5–6.5 mm. long at anthesis, lobes ovate, acuminate, 3–4 mm. long; petals orange to flame-scarlet, 8–12 mm. long; fruit hemispherical, usually about 3 mm. high, 4–6 mm. in diameter; carpels 2.5–3 mm. high, 2–2.8 mm. wide, reniform, deeply notched, dehiscent part ascending, quadrate to subdeltoid, sometimes muticous, indehiscent portion forming three-fourths to four-fifths of carpel, much wider than dehiscent portion, prominently reticulate-fenestrate; seeds 1–2, usually sparsely puberulent.

Sandy desert flats and rocky slopes, Lower Sonoran Zone, Colorado Desert, California, southwestern Arizona, northwestern Sonora, and Baja California east of the main mountain ranges at least to the vicinity of Bahía de los Angeles. March–May.

**3. Sphaeralcea laxa** Woot. & Standl., Bull. Torrey Club **36:** 108, 1909

*Sphaeralcea incana* var. ?β *dissecta* A. Gray, Pl. Wright. **1:** 21, 1852.
*Sphaeralcea ribifolia* Woot. & Standl., Bull. Torrey Club **36:** 109, 1909.

Perennial from a woody crown and stout taproot; stems erect, to about 1 m. tall, densely stellate-tomentose or tomentulose with short-rayed, white hairs; stipules filiform, 3–5 mm. long, caducous; petioles slender, 0.5–3 cm. long; leaf blades ovate to deltoid in outline, truncate to deeply cordate at base, acute to obtuse at apex, shallowly 3-lobed to almost 3-parted near base, lateral lobes rounded, central lobe about twice as long as laterals, margins irregularly crenate to coarsely dentate, varying considerably in size, 8–50 mm. long, nearly as wide, green above with stellate hairs not quite interlacing, paler beneath and hairs interlaced; inflorescense a lax, few-flowered panicle; pedicels usually 2–5 mm. long in flower, to 1.5 cm. in fruit, densely tomentulose; calyx lobes lance-ovate to oblong-lanceolate, acuminate, 5–8 mm. long, stellate-tomentulose; bracteoles filiform, 3–5 mm. long, often dark red; petals grenadine, 10–18 mm. long; fruit truncate-ovoid, 5–6 mm. high, 5–9 mm. wide; carpels 4–6 mm. high, 2–3 mm.

wide, thin-walled, broadly and shallowly notched, empty dehiscent part erect, ovate to ovate-deltoid, acute or obtuse, with cusp to 1 mm. long or sometimes barely mucronulate, basal indehiscent part forming one-fifth to two-fifths of carpel, finely but obviously reticulate, areolae transparent or nearly so; seeds 1–3, usually copiously puberulent.

Plains and foothills, usually on calcareous soils, Sonoran Zones, central Arizona to Texas and south into Sonora. March–Nov.

**4. Sphaeralcea rusbyi** A. Gray var. **gilensis** (Kearney) Kearney, Jour. Wash. Acad. **29:** 486, 1939

*Sphaeralcea rusbyi* subsp. *gilensis* Kearney, Univ. Calif. Pub. Bot. **19:** 54, 1935.

Perennial from a thick crown, green and sparsely to moderately stellate-pubescent or at length glabrate; stems 1 to several, erect or ascending, to 1 m. tall, to 5 mm. in diameter at base; stipules filiform to linear, 2–4 mm. long, caducous; petioles 1–2 cm. long; leaf blades thin, broadly ovate, deltoid, or suborbicular in outline, subcordate to deeply cordate with a broad sinus, acute to truncate at apex, pedately 3–5-parted, divisions cuneate-obovate to oblanceolate and few-cleft or coarsely toothed, middle lobe 5–7 mm. wide at base or sometimes basal leaves scarcely lobed, whole leaf blade 1.5–3.5 cm. long; inflorescence subthyrsoid to paniculate, few-flowered; pedicels 1–8 mm. long; bracteoles subfiliform, 4–6 mm. long, often conspicuously red; calyx lobes deltoid to lance-deltoid, 4–10 mm. long in flower, acuminate; petals grenadine, 14–20 mm. long; fruit ovoid-truncate, 5–7 mm. in diameter, 5–7 mm. high; carpels 10–15, very shallowly notched, 4–5 mm. high, dehiscent part erect, ovate to deltoid-ovate, frequently cuspidate, finely but prominently reticulate on basal one-fifth to two-fifths, indehiscent part; seeds usually copiously puberulent.

Hillsides and along watercourses, mainly Upper Sonoran Zone but entering the Lower Sonoran Zone along the Gila River and near Tucson; to date known only from south-central Arizona. April–Sept.

**5. Sphaeralcea axillaris** S. Wats., Proc. Amer. Acad. **24:** 41, 1889, var. **axillaris**

Erect or ascending perennial to 2.5 m. tall, distinctly lignescent at base, densely stellate-canescent with short, 12–20-rayed hairs; stipules subfiliform, 3–4 mm. long; petioles slender, 1–2.5 cm. long; leaf blades thin to moderately thick, deltoid-ovate to subrhombic in outline, broadly cuneate to cordate at base, angulate or shallowly 3-lobed below middle, shallowly to deeply crenate and often undulate-crispate, longer blades 3–9 cm. long, 1–7 cm. broad, prominently 3–5-veined beneath; inflorescence a many-flowered thyrse, lower branches racemose; pedicels short, rarely equaling fruiting calyx; calyx densely stellate-pubescent with tawny, somewhat scurfy hairs, lobes ovate-lanceolate, acuminate, about 2 mm. wide, 5–9 mm. long; bracteoles subfiliform, 5–7 mm. long; petals white or pink, often drying violet or yellowish, 12–20 mm. long; stamineal column stellate-hispid; fruit to 8 mm. high and to 10 mm. broad; carpels to 7 mm. high, oblong, usually deeply and narrowly notched, dehiscent part ovate to deltoid, erect, obtuse and muticous, mucronate or cuspidate with cusp to 1 mm. long at apex or cusps reduced to mere points; indehiscent part finely reticulate-fenestrate forming one-third to one-half of carpel; seeds 1–2, somewhat puberulent.

Desert slopes, rocky hillsides, and along sandy watercourses, Lower Sonoran Zone, central Baja California, from the vicinity of Punta Prieta to Mulegé and possibly beyond. Feb.–April.

**5a. Sphaeralcea axillaris** var. **violacea** (Rose) Wiggins, comb. nov.

*Sphaeralcea violacea* Rose, Contr. U.S. Nat. Herb. **1:** 81, 1890.
*Sphaeralcea albiflora* Rose, loc. cit.
*Sphaeralcea axillaris* subsp. *violacea* Kearney, Univ. Calif. Pub. Bot. **19:** 36, 1935.
*Sphaeralcea axillaris* var. *rosacea* Kearney, Jour. Wash. Acad. **29:** 486, 1939. [Varietal name applied in error.]

Similar to var. *axillaris* except fruit is shorter in comparison to width, carpels being only 2–3.2 mm. high and from two-thirds to fully as wide, dehiscent part often ascending rather than erect, very obtuse at apex and barely muticous; indehiscent part of carpel often forming as much as two-thirds of carpel and sometimes slightly wider than dehiscent part.

Chiefly in light sandy soil, Lower Sonoran Zone, vicinity of Santa Rosalia south to the Cape Region, Baja California. Jan.–April.

**6. Sphaeralcea hainesii** Brandegee, Proc. Calif. Acad. II, **2:** 136, 1889

Erect or ascending, robust perennial (or sometimes flowering first season) to 2 m. or nearly so, densely stellate-tomentulose with whitish or yellowish hairs; stipules linear-subulate, 4–7 mm. long, caducous; petioles stout, 1–6 cm. long or lower ones longer; leaf blades ovate-oblong to broadly ovate or lower suborbicular, truncate or cordate at base, acute to obtuse or rounded at apex, 5–7-veined, shallowly 3–5-lobed, crenulate to coarsely crenate on margin, stellate-puberulent and green on upper surface, rays of hairs interlacing only slightly, more densely puberulent and rays closely interlaced and imparting a grayish or yellowish cast to lower side, larger blades 6–10 cm. long, 3–10 cm. wide; inflorescence glomerate-thyrsoid, usually narrow; pedicels stout, 1–5 mm. long; calyx 8–13 mm. high at anthesis, densely stellate-pubescent and slightly scurfy, lobes lance-ovate, usually more or less acuminate, 5–9 mm. long; bracteoles subfiliform, 3–5 mm. long; petals deep grenadine to grenadine-pink, 12–20 mm. long; fruit truncate-conical, 6–8 mm. high; carpels 10–16, with thin walls, often connate at base at maturity, 4–6 mm. high, about one-half as wide, dehiscent part ovate, obtuse at apex, minutely cuspidate or muticous, indehiscent part forming one-fifth to two-fifths of carpel, finely but distinctly reticulate; seeds 2–3, copiously puberulent.

Sandy soil on flats, gentle slopes, and along arroyos, often as a weed in fields, Lower Sonoran Zone, general vicinity of Bahía de los Angeles, Baja California, south almost to the latitude of Magdalena Bay. Feb.–April.

A slightly aberrant form with white flowers was collected near the ocean 30 miles south of Punta Prieta.

**7. Sphaeralcea emoryi** Torr., in A. Gray, Pl. Fendl. 23, 1849, var. **emoryi**

Densely grayish-tomentose or canescent perennial to 12 dm. tall or more; stipules narrowly subulate, 3–6 mm. long; petioles stout, 1–2.5 (sometimes –9) cm. long; leaf blades thickish, flat or somewhat rugose, broadly ovate to ovate-oblong, usually more or less cordate at base, acute to obtuse at apex, angulate to shallowly lobed near base with lateral lobes broad and rounded, crenulate to coarsely dentate, larger leaves 3–9 cm. long; inflorescence a many-flowered, interrupted thyrse, rarely subpaniculate; pedicels stout, usually shorter than calyx; calyx densely stellate-tomentulose, lobes lance-ovate or lanceolate, acuminate, 4–6 mm. long at anthesis, only slightly longer in fruit; petals grenadine, pink, or lavender, 1–2 cm. long; fruit truncate-conical; carpels 11–16, 3.5–6 mm. high, two-fifths to three-fifths as wide, deeply notched, dehiscent part ovate, acute or obtuse at apex, cuspidate or mucronate, or rarely only acute, forming one-half to two-thirds of carpel, indehiscent part prominently and usually coarsely reticulate, inner angle just below notch acute, beaklike, walls chartaceous, seeds usually 2, copiously puberulent.

In sandy or loam soil, Lower Sonoran Zone, southern Nevada, eastern Riverside and Imperial Counties, California, western Arizona, and south into eastern Baja California to the southern foothills of the Sierra San Pedro Mártir. Feb.–May.

***7a.* Sphaeralcea emoryi** var. **variabilis** (Cockerell) Kearney, Jour. Wash. Acad. **29:** 486, 1939

*Sphaeralcea variabilis* Cockerell, Amer. Nat. **34:** 291, 1900.
*Sphaeralcea fendleri* var. *californica* Parish, Zoe **5:** 71, 1900. Not *S. californica* Rose, 1890.
*Sphaeralcea fendleri* subsp. *variabilis* Cockerell, Bull. S. Calif. Acad. **1:** 108, 1902.
*Sphaeralcea emoryi* subsp. *variabilis* Kearney, Univ. Calif. Pub. Bot. **19:** 39, 1935.

Stems and leaves less densely pubescent so greener than in var. *emoryi;* leaf blades thinner, 3-parted or -cleft, with lateral lobes sometimes deeply cleft and acute, midlobe three to eight times as long as lateral ones; margins of leaves more deeply crenate or dentate; carpels frequently long-cuspidate.

In sandy or light loam soil, Lower Sonoran Zone, southern part of the Mojave and the Colorado Deserts, California, and western and southern Arizona. Not yet reported from Sonora nor Baja California but to be expected in both areas.

***7b.* Sphaeralcea emoryi** var. **arida** (Rose) Kearney, Jour. Wash. Acad. **29:** 486, 1939

*Sphaeralcea arida* Rose, Contr. U.S. Nat. Herb. **5:** 177, 1899.
*Sphaeralcea emoryi* subsp. *arida* Kearney, Univ. Calif. Pub. Bot. **19:** 41, 1935.

Leaf blades not over 5 cm. long and usually almost or fully as wide, cleft well above base; carpels thin-walled, only 8–12 in a fruit, finely reticulate-fenestrate on basal part, this often narrower than dehiscent part, sometimes nearly smooth dorsally; seeds usually glabrous.

Plains and gentle slopes in light soil, Lower Sonoran Zone, southern Nevada into extreme southeastern California through western Arizona and Sonora into Sinaloa.

***7c.* Sphaeralcea emoryi** var. **nevadensis** (Kearney) Kearney, Jour. Wash. Acad. **29:** 486, 1939

*Sphaeralcea emoryi* subsp. *nevadensis* Kearney, Univ. Calif. Pub. Bot. **19:** 40, 1935.

Leaf blades only one-third to one-half as wide as long; apex of leaf blades very obtuse or retuse; carpels thin-walled, very finely reticulate.

Along arroyos and on sandy plains, Lower Sonoran Zone, southern Nevada to northern Arizona, and the Colorado Desert in southeastern California. It probably extends south into Baja California but has not been reported from that area.

**8. Sphaeralcea subhastata** J. M. Coulter var. **thyrsoidea** (Kearney) Kearney, Jour. Wash. Acad. **29:** 486, 1939

*Sphaeralcea subhastata* subsp. *thyrsoidea* Kearney, Univ. Calif. Pub. Bot. **19:** 77, 1935.

Perennial to 4.5 dm. tall, stems about 3 mm. in diameter, stellate-pubescent with 18–25-rayed hairs, stems decumbent to erect; leaf blades thickish, flat or rugose, oblong-ovate to broadly ovate, 5-nerved from base, deeply 3-incised or -parted, lateral divisions often deeply cleft, usually more than one-half as wide as long, or if not deeply incised margins dentate to irregularly crenate, often somewhat crispate-undulate, larger blades

3–4 cm. long, sparsely stellate-pubescent above, somewhat more densely so beneath but green on both surfaces; inflorescences thyrsoid-racemose, 20–30-flowered, often 2–3 flowers at a node; pedicels slender, 3–12 mm. long; calyx 5–9 mm. high at anthesis, lobes lance-ovate or deltoid-ovate, 4–6.5 mm. long, acuminate; petals grenadine, 10–18 mm. long; fruits truncate-conical, 5–6 mm. high, 6–7 mm. wide at base; carpels 10–17, with thickish chartaceous walls, 4–6 mm. high, half as wide, deeply and narrowly notched, often faintly constricted on back opposite notch, dehiscent part bearing cusps 1–2 mm. long, indehiscent part forming one-third to one-half of carpel, coarsely reticulate laterally, often nigrescent; seeds usually 2, glabrous to puberulent.

In fine-grained, clayey soils, Lower Sonoran Zone, southern Arizona. March–Oct.

**9. Sphaeralcea fulva** Greene, Pittonia **1:** 201, 1888

Erect or ascending perennial to 13 dm. tall, densely stellate-tomentulose throughout with yellowish, many-rayed hairs; petioles slender to stout, 6–25 mm. long; leaf blades usually very rugose, thick, veins prominent beneath, oblong-ovate to deltoid-ovate, usually broadly subcuneate but sometimes truncate or even subcordate at base, obtuse or acute at apex, shallowly 3-lobed near base, lobes angulate or rounded, margin subentire to coarsely and irregularly crenate-dentate, larger blades 3–5 cm. long, 1.5–4 cm. wide, scurfy stellate-pubescent, more heavily so beneath than above; inflorescence usually narrowly glomerate-thyrsoid; pedicels stout, 1–5 mm. long; calyx 7–12 mm. high, lobes from deltoid-ovate and acute to lanceolate-acuminate, 3–9 mm. long; petals grenadine or less often pink or white, 1–2 cm. long; column densely and coarsely stellate-pubescent; fruit hemispherical, 5–6 mm. in diameter, 4–5 mm. high; carpels 12–15, with chartaceous firm walls, 4–5 mm. high, helmet-shaped, dehiscent part deltoid-ovate, obtuse and minutely muticous or merely rounded at apex, indehiscent part forming one-fourth to one-half of carpel, prominently and coarsely reticulate laterally; seeds 1–2, somewhat puberulent.

Silty, sandy, or clay soil, usually near the sea, Lower Sonoran Zone, west coast of Baja California from Colnett Wash to northern edge of the Vizcaíno Desert and on Cedros Island. March–April.

A plant that may be an aberrant form of this species was collected 5 miles inland from the Gulf of California near Barríl.

**10. Sphaeralcea ambigua** A. Gray, Proc. Amer. Acad. **22:** 292, 1887, var. **ambigua**

*Sphaeralcea emoryi* A. Gray, Ives Rep. 8, 1860. Not *S. emoryi* Torr., 1849.
*Sphaeralcea macdougalii* Rose & Standl., Contr. U.S. Nat. Herb. **16:** 13, *pl. 5,* 1912.
*Sphaeralcea ambigua* var. *keckii* Munz, Bull. S. Calif. Acad. **31:** 68, 1932.

Erect or ascending perennial with several to many suffrutescent or suffruticose stems to 1.5 m. tall from a woody crown, whitish or yellowish, stellate-tomentulose throughout; pedicels stout, rarely over 4 cm. long and usually considerably shorter; leaf blades thickish, more or less rugose, broadly ovate, deltoid, or suborbicular in outline, subcordate to deeply cordate at base, sinus open, obtuse, truncate or slightly retuse at apex, crenulate to coarsely, irregularly crenate, shallowly 3-lobed near middle, lobes rounded, larger blades to 6 cm. long, nearly to fully as wide or sometimes wider than long; inflorescence few- to many-flowered panicles; pedicels stout and shorter than calyces; calyx densely stellate-pubescent, 6–20 mm. high in flower, lobes lanceolate to lance-oblong, attenuate, usually two to four times as long as tube; petals grenadine to grenadine-pink, 1.5–3.5 cm. long; fruit nearly hemispherical, about 8–12 mm. wide, 5–7 mm. high; carpels 12–16, with chartaceous walls, 3.5–6 mm. high, galeate, narrowly and deeply notched, dehiscent part obtuse, muticous or mucronulate near apex, indehiscent part forming about one-third of carpel, coarsely reticulate laterally; seeds usually 2, more or less puberulent.

Rocky ridges, dry slopes, and along washes, Lower Sonoran Zone, southern Nevada and southwestern Utah through Arizona and southeastern California to Sonora and Baja California at least to Bahía de los Angeles. Flowers practically throughout the year.

**10*a*. Sphaeralcea ambigua** var. **rosacea** (Munz & Jhtn.) Kearney, Jour. Wash. Acad. **29**: 486, 1939

*Sphaeralcea rosacea* Munz & Jhtn., Bull. Torrey Club **49**: 353, 1923.
*Sphaeralcea purpurea* Parish, in Jepson, Man. Fl. Pl. Calif. 635, 1925.
*Sphaeralcea ambigua* subsp. *rosacea* Kearney, Univ. Calif. Pub. Bot. **19**: 46, 1935.

Differs from var. *ambigua* mainly in having somewhat larger average size, pale purplish pink petals, and dark purple anthers; stems usually more densely white-tomentulose; sepals more conspicuously attenuate in many specimens.

Similar habitats to those of typical *S. ambigua,* around the Colorado Desert in southeastern California, southwestern Arizona, and northeastern Baja California at least to the northeastern escarpment of the Sierra San Pedro Mártir.

**10*b*. Sphaeralcea ambigua** var. **rugosa** (Kearney) Kearney, Jour. Wash. Acad. **29**: 486, 1939

*Sphaeralcea ambigua* subsp. *rugosa* Kearney, Univ. Calif. Pub. Bot. **19**: 49, 1935.

Differing from var. *ambigua* in having a narrowly thyrsoid inflorescence; usually more numerous but smaller flowers; and usually much more finely reticulate lateral walls of basal part of carpel.

Sandy or gravelly soil, Lower Sonoran Zone to lower edge of Arid Transition Zone, San Jacinto Mountains, California, southward into Baja California at least as far as the Sierra San Pedro Mártir.

**10*c*. Sphaeralcea ambigua** var. **versicolor** (Kearney) Kearney, Jour. Wash. Acad. **29**: 486, 1939

*Sphaeralcea ambigua* subsp. *versicolor* Kearney, Univ. Calif. Pub. Bot. **19**: 50, 1935.

Differing from var. *ambigua* in having more markedly woody stems; smaller leaf blades (mostly 1–2 cm. long) with a bright orange-yellow marginal fringe of stellate pubescence, yellowish pubescence on upper and white pubescence on lower surface; petals usually less than 15 mm. long and calyx only 7–8 mm. high at anthesis; carpels 3.5–4 mm. high, notch in carpel cuts about halfway through body.

Silty flats, sandy washes, and rocky slopes, Lower Sonoran Zone, near the northwest coast of Sonora and on San Estéban and Angel de la Guarda Islands in the Gulf of California.

**11. Sphaeralcea fendleri** A. Gray, Pl. Wright. **1**: 21, 1852, var. **fendleri**

*Sphaeralcea miniata* A. Gray, Pl. Fendl. 19, 1849. Not *S. miniata* (Cav.) Spach., 1834.
*Sphaeralcea leiocarpa* Woot. & Standl., Bull. Torrey Club **36**: 107, 1909.

Erect or ascending perennial to 14 dm. tall, densely grayish to whitish stellate-pubescent on stems, less densely so on leaves; petioles slender, usually less than half as long as blades; leaf blades oblong-ovate to broadly ovate, usually cuneate at base, acute to obtuse at apex, 3- or rarely 5-veined from base, shallowly to deeply 3-lobed below middle with narrow, triangular to rectangular, acute, slightly ascending lateral lobes about one-fourth as long as middle one, latter few-toothed or cleft or merely coarsely to finely crenate, larger blades 3–6 cm. long; inflorescence a many-flowered, interrupted thyrse; pedicels usually equaling or slightly exceeding length of calyx; bracteoles usually green, 2–4 mm. long; calyx 4–6 mm. high at anthesis, lobes deltoid, acute, or rarely lance-ovate and short-acuminate, 1–2.5 mm. long; petals grenadine or sometimes pinkish, 8–13 mm. long; fruit truncate-conical, 5–6.5 mm. high; carpels 11–15, 4–5 mm. high, often somewhat connate at maturity, shallowly notched, dehiscent part erect, ovate, obtuse, and cuspidate with cusps to 1.2 mm. long at apex, indehiscent part forming one-fifth to one-third of carpel, finely and faintly reticulate laterally; seeds 2, puberulent.

Open places in forests and brush, mostly Upper Sonoran Zone to lower parts of Arid Transition Zone, but stragglers enter the upper margins of the Sonoran Desert. It has been collected along the Río Magdalena at Ímuris, Sonora. March–Oct.

**11*a*. Sphaeralcea fendleri** var. **albescens** (Kearney) Kearney, Jour. Wash. Acad. **29:** 486, 1939

*Sphaeralcea fendleri* subsp. *albescens* Kearney, Univ. Calif. Pub. Bot. **19:** 62, 1935.

Differs from var. *fendleri* in having velutinous whitish pubescent leaves; usually thin leaf blades; carpels 5–6 mm. high, bearing cusps 1–2 mm. high.

Mostly along streams in the Upper Sonoran Zone, but extending into the Lower Sonoran Zone along the Río Magdalena at least as far as Magdalena, Sonora.

**12. Sphaeralcea angustifolia** (Cav.) G. Don, Gen. Hist. Pl. **1:** 465, 1831, var. **angustifolia**

*Malva angustifolia* Cav., Diss. **2:** 64, *pl. 20, fig. 3(e–g)*, 1786.
*Sphaeroma angustifolia* Schlecht., Linnaea **11:** 353, 1837.
*Sphaeralcea angustifolia* var. *violacea* Davy, Erythea **3:** 118, 1895.

Erect perennial almost to 2 m. tall, densely stellate-pubescent throughout with grayish or slightly yellowish, 8–20-rayed hairs; petioles slender to moderately stout, 0.5–3 cm. long; leaf blades lanceolate to lance-oblong, cuneate at base, gradually tapering to an acute or acuminate apex, angulate or subhastately toothed near base, margin usually finely and regularly crenate to crenate-dentate; inflorescence a long, narrow interrupted, many-flowered thyrse, with leaves nearly to apex; pedicels stout and usually shorter than calyx; calyx 5–9 mm. long at anthesis, lobes lanceolate-acuminate, 3–7 mm. long; petals usually lavender but occasionally white or grenadine, 1–2 cm. long; fruit truncate-conical, 4–7 mm. high; carpels 10–16, with chartaceous walls, usually connate at maturity, 3.5–6.5 mm. high, shallowly notched, dehiscent part ovate, obtuse or rarely acute, muticous at apex, indehiscent part forming one-tenth to two-fifths of carpel, finely and often faintly reticulate laterally; seeds 1–3, usually glabrous.

Chiefly a weed of roadsides and fields, mostly Upper Sonoran Zone to Arid Transition Zone, southwestern Texas to Vera Cruz, Puebla and Michoacán but entering the Sonoran Desert in southern Sonora at Ciudad Obregón, possibly as a waif. May–Oct.

**12*a*. Sphaeralcea angustifolia** var. **cuspidata** A. Gray, Proc. Amer. Acad. **22:** 293, 1887

*Sida stellata* Torr., Ann. Lyc. N.Y. **2:** 171, 1827. Not *S. stellata* Cav., 1785.
*Sphaeralcea stellata* Torr. & Gray, Fl. N. Amer. **1:** 228, 1838.
*Malva stellata* D. Dietr., Syn. Pl. **4:** 816, 1847.
*Sphaeralcea cuspidata* Britt., in Britt. & Brown, Ill. Fl. **3:** 519, 1898.

Differs from var. *angustifolia* in being more densely canescent-pubescent; leaf blades thicker, with veins more prominent beneath, usually narrower and margins often revolute; petals 7–12 mm. long; calyx lobes deltoid and merely acute; carpels less connate at maturity, more distinctly reticulate laterally.

In habitats similar to those of the typical form, mainly Upper Sonoran Zone to Arid Transition Zone, Colorado and western Kansas to eastern Arizona, western Texas, and southward into San Luis Potosí and Zacatecas, but entering the Sonoran Desert along the Santa Cruz River near Tucson, Arizona, and at Indio, Riverside County, California. It may have been a transient waif at Indio.

**13. Sphaeralcea grossulariifolia** (Hook. & Arn.) Rydb. var. **pedata** (Torr.) Kearney, Jour. Wash. Acad. **29:** 486, 1939

*Sphaeralcea pedata* Torr., in A. Gray, Pl. Fendl. 23, 1849.
*Sphaeralcea grossulariaefolia* subsp. *pedata* Kearney, Univ. Calif. Pub. Bot. **19:** 88, 1935.

Perennial from a woody taproot and crown, with few erect or ascending stems 5–11 dm. tall, stellate-tomentulose throughout with whitish hairs; leaves deltoid to broadly ovate in outline, usually cordate at base, 2–4.5 cm. long, pedately deeply cleft or parted, divisions usually coarsely and irregularly toothed; inflorescence thyrsoid-glomerate; calyx 5–10 mm. high, lobes ovate-lanceolate, acuminate; petals grenadine, 8–20 mm. long; fruit hemispherical, 5–8 mm. wide; carpels broadly ovate in outline, 2.5–3.5 mm. high, three-fourths to three-fifths as wide as high, acute at apex, often mucronate or even cuspidate; seeds usually 2, glabrous to puberulent.

Mesas, mountain sides, and rarely plains, chiefly Upper Sonoran Zone to Arid Transition Zone, eastern Oregon to Utah, western New Mexico, and Arizona. Occurring in the Sonoran Desert near Sacaton and in the Buckskin Mountains, Yuma County, Arizona. April–Oct.

## 7. MALVA L., Sp. Pl. 687, 1753

Annual, biennial, or perennial herbs with procumbent to erect stems. Leaves alternate, with pubescent, lobed or dissected, suborbicular blades; stipules membranous. Flowers perfect, solitary or in axillary clusters or in terminal spikes, subtended by 2–3 distinct bractlets. Calyx 5-cleft; lobes broad. Petals 5, emarginate, white to purple. Fruit discoid. Carpels several, 1-celled, indehiscent, reniform, beakless, 1-seeded. Ovule ascending.

Calyx lobes spreading at maturity; petals scarcely equaling calyx lobes, claws glabrous; carpels strongly reticulate dorsally and radiately ridged laterally ...... 1. *M. parviflora*
Calyx lobes incurved at maturity; petals about twice as long as sepals, claws bearded; carpels smooth or only faintly reticulate dorsally, smooth laterally ............ 2. *M. neglecta*

**1. Malva parviflora** L., Amoen. Acad. **3:** 416, 1756

*Malva obtusa* Torr. & Gray, Fl. N. Amer. **1:** 225, 1838.

An erect, spreadingly branched annual or biennial 2 dm. to 2 m. tall with sparsely stellate-pubescent or glabrate foliage and young stems; stipules triangular- to ovate-lanceolate, 5–8 mm. long; petioles 3–15 cm. long; leaf blades suborbicular, shallowly 5–7-lobed, cordate at base, dentate to crenate, 1.5–10 cm. broad, or sometimes to 20 cm. broad; flowers solitary or usually in axillary clusters; pedicels 2–12 mm. long; bractlets linear-lanceolate, 3–5 mm. long, distinct; calyx 4–6 mm. long at anthesis, lobes deltoid-ovate, spreading to form a rotate scarious disk 12–16 mm. broad in fruit; petals obovate, emarginate, 4–6 mm. long, white tinged with rose or purple along veins and at tips, claw glabrous; carpels 8–12, reniform-orbicular, transversely reticulate dorsally, radiately ridged laterally, angles dentate, puberulent on back; seeds minutely papillate-puberulent.

Wayside weed, Lower and Upper Sonoran Zones, Humboldt County California, to Baja California, Arizona, and Sonora; adventive at numerous localities along the Atlantic seaboard and in the Mississippi Valley. April–Nov.

**2. Malva neglecta** Wallr., Syll. Ratisb. **1:** 140, 1824

*Malva vulgaris* Fries, Novit. Fl. Suec. 219, 1814–1823.
*Malva rotundifolia* of authors, not *M. rotundifolia* L.

Much branched, procumbent annual with sparsely stellate-pubescent to glabrate herbage; stipules ovate, 3–5 mm. long, ciliate; petioles slender, 5–10 cm. long, erect; leaf blades round-reniform, 3–6 cm. broad, very shallowly 5–7-lobed, lobes rounded, crenate; flowers solitary or in few-flowered axillary clusters; pedicels 5–15 mm. long; bracts linear-lanceolate, 3–4 mm. long; calyx lobes ovate, acute, 4–5 mm. long, slightly accrescent in fruit; petals obovate, emarginate, 8–10 mm. long, pale lilac or whitish, claws villous-ciliate; fruit discoid, central axis broad and conspicuous; carpels 12–15, rounded, puberulent to glabrate, smooth or inconspicuously reticulate on back, angles rounded, lateral faces smooth.

Roadsides, fields, and waste places, mainly Upper Sonoran and lower Transition Zones, an adventive weed from Washington to Massachusetts, south to central California, Arizona, and North Carolina; occasional as a waif along irrigation ditches and about dwellings, Imperial Valley, California, and northern Mexico. July–Sept.

## 8. MALVASTRUM A. Gray, Pl. Fendl. 21, 1949

Herbs and shrubs with simple, orbicular, angled or lobed, alternate leaves, more or less stellate-pubescent throughout. Inflorescence capitate, interrupted-spicate, or paniculate, few- to many-flowered. Involucellate bractlets 1–3, or rarely wanting. Flowers perfect, white, pink, or rose-purple. Stamineal column antheriferous at summit. Style branches filiform; stigmas capitate. Fruit depressed-globose or discoid, pubescent at summit, at least in youth. Carpels several, 1-ovuled, dehiscent; seeds ascending, irregularly and minutely puberulent, filling carpel; embryo curved.

Stems strigose with 4-rayed hairs; rays of hairs pointing in pairs toward apex and base of stem; carpels tricuspidate, 2 cusps on outer angles, 1 erect at inner dorsal margin .................... 1. *M. coromandelianum*
Stems stellate-pubescent with several-rayed hairs; rays of hairs radiating and those of adjacent hairs more or less interlaced; carpels bicuspidate, cusps radiate-spreading from outer marginal angles .................... 2. *M. bicuspidatum*

**1. Malvastrum coromandelianum** (L.) Garcke, Bonplandia **5:** 295, 1857

*Malva coromandeliana* L., Sp. Pl. 687, 1753.
*Malva tricuspidata* Ait., Hort. Kew. ed. 2, **4:** 210, 1812.
*Malvastrum tricuspidatum* A. Gray, Pl. Wright. **1:** 16, 1852.

Short-lived, fruticose perennial but often flowering first season; stems decumbent to erect, 2–10 dm. long, these and foliage rather harshly strigose with 4-rayed hairs, rays in pairs, one pair pointing toward apex, the other toward base of stem, leaf or pedicel; stipules lance-subulate, 3–5 mm. long, subscarious, deciduous; petioles slender, about one-half to three-fourths as long as blades; leaf blades broadly rhombic-ovate to lance-ovate, rounded or obtuse at base, obtuse, rounded, or sometimes acute at apex, regularly crenulate-dentate, green, thinly strigose on both surfaces, slightly more densely so beneath than above, densest along veins, 2–5 cm. long, 1–4 cm. wide; flowers mostly solitary in axils; pedicels 2–8 mm. long; calyces sparsely strigose-hispid with coarse stiff hairs, lobes deltoid-ovate, abruptly acuminate, 2–4 mm. wide, 4–6 mm. long; petals yellow, 6–8 mm. long; fruit discoid, 5–7 mm. in diameter; carpels 9–15, brown, ovate-reniform, notch at narrow end of ovoid body, a short cusp (to 1 mm. long) radiating horizontally from each of outer marginal angles, a third slightly longer one curving slightly toward outer margins of fruit from a point about 0.5 mm. back from inner dorsal angle, upper surface of carpel hispid, sides radiately ridged.

About rocks, among shrubs, and along streams and fence rows, Lower Sonoran Zone to Tropical Zone, central Sonora to South America and widely distributed in tropical and subtropical regions of the world. July–Nov.

**2. Malvastrum bicuspidatum** (S. Wats.) Rose, Contr. U.S. Nat. Herb. **12:** 286, 1909

*Malvastrum tricuspidatum* var. *bicuspidatum* S. Wats., Proc. Amer. Acad. **21:** 417, 1886.

Fruticose or shrubby plants 1–2 m. tall with erect, ascending or arcuate-spreading branches, stellate-pubescent throughout with several-rayed hairs that overlap and interlace moderately; stipules lance-subulate, 5–9 mm. long, brownish, sparsely hispid, deciduous; petioles 0.5–4 cm. long, sparsely to densely stellate-tomentulose; leaf blades ovate, often broadly so, thin to thick, rounded to subcuneate at base, acute to short-acuminate at apex, regularly crenate-dentate, sparsely (or rarely densely) stellate-pubescent on both surfaces, 0.8–4.5 cm. wide, 1–8 cm. long; flowers solitary in axils or subspicate at tips of branches; pedicels 1–2 mm. long or flowers subsessile; bracteoles lance-elliptic, 4–5 mm. long; calyx lobes deltoid-ovate, acute to abruptly acuminate, 4–6 mm. long, sparsely to densely stellate-pubescent without, minutely and closely stellate-tomentulose with white hairs on inner surface; petals yellow, 5–7 mm. long; fruit discoidal, 6–8 mm. in diameter; carpels 9–15, ovoid-reniform with notch at end of ellipse, about 2 mm. high, 3 mm. long, bicuspidate, one cusp on each outer dorsal margin, pubescent on upper surface with stellate or simple or usually both types of hairs, radiately ridged laterally, brownish.

Stony hillsides and along washes and canyons, Lower Sonoran Zone to Subtropical Zone, Sonora and Chihuahua to Morelos and Oaxaca. March–Oct.

In habit this resembles *M. coromandelianum*, with which it is often confused.

## 9. EREMALCHE Greene, Leaflets Bot. Obs. 1: 208, 1906

Low annual herbs with alternate, orbicular, or palmately parted leaves, stellate-pubescent throughout. Flowers solitary or in pairs in axils of upper leaves. Involucellate bractlets 3, distinct, persistent. Sepals 5. Petals 5, white to rose-purple, claws ciliate. Stamineal column simple, glabrous. Style branches filiform, as many as carpels, to twice as long as stamineal column; stigmas capitate. Fruit discoid. Carpels 10–40, 1-ovulate, indehiscent, reticulate or transversely ridged on back and angles, glabrous. Seed completely filling cavity of carpel. Embryo forming an incomplete circle; endosperm oily, scanty.

Leaves reniform-orbicular, crenate; petals 10–25 mm. long, rose-purple with a crimson basal spot, carpels strongly flattened laterally, angles acute ............... 1. *E. rotundifolia*

Leaves palmately 3–5-lobed; petals 2–5 mm. long, white or faintly tinged with pink; carpels slightly flattened laterally, angles rounded ............................... 2. *E. exilis*

**1. Eremalche rotundifolia** (A. Gray) Greene, Leaflets Bot. Obs. **1:** 208, 1906

*Malvastrum rotundifolium* A. Gray, Proc. Amer. Acad. **7:** 333, 1868.
*Sphaeralcea rotundifolia* Jepson, Man. Fl. Pl. Calif. 633, 1925.

Erect or sparingly branched annual 1–4 dm. tall with sparsely hispid stems and sparsely stellate-hispid leaves and petioles; stipules triangular-lanceolate, 3–4 mm. long; petioles 1–8 cm. long; leaf blades orbicular-cordate, crenate, 1–4 cm. broad; flowers axillary, solitary or in pairs; pedicels 2–6 cm. long; involucellate bractlets linear-lanceolate, 5–10 mm. long; calyx densely and finely stellate-pubescent, lobes broadly ovate-lanceolate, 4–6 mm. broad, 8–12 mm. long, acuminate; petals obovate, 1–2.5 cm. long, rose-purple with a crimson subbasal spot; fruit 10–15 mm. broad; carpels 20–40, orbicular, strongly flattened laterally, 3–3.5 mm. in diameter, dark brown to black, angles acute.

Mainly on sandy soil, Lower Sonoran Zone, Inyo County, California, through the Mojave and Colorado Deserts to southwestern Arizona and adjacent Baja California and Sonora. March-May.

**2. Eremalche exilis** (A. Gray) Greene, Leaflets Bot. Obs. **1:** 208, 1906

*Malvastrum exile* A. Gray, Ives Rep. 8, 1860.
*Sphaeralcea exilis* Jepson, Man. Fl. Pl. Calif. 633, 1925.

Erect or decumbent annual with stems 1–4 dm. long, finely stellate-puberulent but green throughout; stipules papery, subulate, 2–3 mm. long; petioles 1–4 cm. long; leaf blades palmately 3–5-lobed or lower ones irregularly crenate or dentate, broadly cuneate to truncate at base, 1–2.5 cm. broad; flowers solitary in upper axils; pedicels 5–30 mm. long; bractlets linear-lanceolate, 3–7 mm. long; calyx 4–6 mm. long, 2.5–3 mm. wide in flower, lobes ovate-lanceolate, acuminate, stellate-puberulent inside and out, erect in fruit; petals obovate, 2–5 mm. long, white or faintly tinged with purplish or pink; fruit 5–7 mm. broad; carpels 10–15, orbicular-reniform, transversely ridged on back and rounded on angles, only lightly compressed laterally, usually dark brown.

Sandy or fine-grained soil, Lower Sonoran Zone, Inyo and Los Angeles Counties, California, to Arizona, northern Baja California, and Sonora. Feb.–May.

## 10. SIDA L., Sp. Pl. 683, 1753

Perennial herbs or shrubs with stellate-scurfy herbage. Leaves alternate, lobed, toothed, or entire. Involucellate bractlets 1–3, distinct, linear. Flowers perfect, solitary or clustered in axils; pedicels articulate to peduncles similar to pedicels. Calyx turbinate or campanulate-hemispheric, 5-lobed. Petals 5. Stamineal column simple, antheriferous at summit. Styles filiform; stigmas capitate. Carpels 5–12, 1-celled, 1-seeded, indehiscent or incompletely 2-valved, terminal portion often prolonged inward over central axis to form a short blunt beak. Seeds pendulous, 3-angled. Embryo curved; endosperm scanty.

Petals densely stellate-pubescent on portions of backs exposed in bud; at least some lepidote scales present; leaves very asymmetrical at base:
- Leaves as broad as or broader than long, rounded at apex; peltate hairs obscured by stellate ones; bractlets 1–2, inconspicuous .............................. 1. *S. hederacea*
- Leaves rarely as broad as long, obliquely ovate-deltoid to sagittate, acute at apex; peltate hairs obvious; bractlets none:
  - Leaves obliquely ovate to ovate-deltoid; pedicels often equaling or exceeding subtending leaves ........................................... 2. *S. lepidota* var. *lepidota*

Leaves sagittate-lanceolate, irregularly lobed at base, usually hastate; pedicels rarely equaling subtending leaves . . . . . . . . . . . . . . . . . . . . . . 2a. *S. lepidota* var. *sagittifolia*

Petals glabrous except on claws; pubescence not lepidote; leaves symmetrical or nearly so at base:

Sepals cordate-ovate, 1–1.5 cm. wide and marcescent in fruit, thin and nearly membranous; root fusiform-fleshy; carpels black . . . . . . . . . . . . . . . . . . . . . . . . . . . . . . . 3. *S. physocalyx*

Sepals ovate to ovate-lanceolate or deltoid, not cordate nor marcescent, 6 mm. wide or less; roots not fleshy; carpels stramineous, gray, or brown:

Leaves, or most of them, distinctly cordate at base, larger ones 3–10 cm. wide; shrubs or subshrubs 1–2.5 m. tall:

Flowers in dense glomerules on paniculate branches; calyces densely stellate-pubescent; sepals thick; carpels 7–12:

Stems and calyces densely stellate-pubescent but not hirsute; pedicels jointed near base . . . . . . . . . . . . . . . . . . . . . . . . . . . . . . . . . . . . . . . . . . . . . . . . 4. *S. cordifolia*

Stems and calyces hirsute as well as stellate-tomentulose; pedicels jointed near apex . . . . . . . . . . . . . . . . . . . . . . . . . . . . . . . . . . . . . . . . . . . . . . . . . . . . . 5. *S. setifera*

Flowers axillary or spreadingly paniculate, not glomerate; calyces sparsely puberulent to subglabrate; sepals thin; carpels 5–6:

Stems stellate-pubescent; sepals subglabrous; pedicels and upper stems not glandular . . . . . . . . . . . . . . . . . . . . . . . . . . . . . . . . . . . . . . . . . . . . . . . . . . . . 6. *S. glabra*

Stems hirsute; sepals sparsely hirsute and glandular-puberulent; pedicels and upper parts of stems viscid-glandular . . . . . . . . . . . . . . . . . . . . . . . . . . . 7. *S. glutinosa*

Leaves truncate, rounded, or acute at base, rarely subcordate, seldom 3 cm. wide; plants mostly perennial herbs or suffrutescent, mostly 1 m. tall or less:

Sepals ovate-lanceolate in fruit; calyces 8–15 mm. long at anthesis; flowers 3.5–5 cm. broad; plants woody . . . . . . . . . . . . . . . . . . . . . . . . . . . . . . . . . . . . . . . . . 8. *S. xanti*

Sepals deltoid-ovate; calyces 3–8 mm. long at anthesis; flowers 8–25 mm. broad; plants herbaceous or woody only at base:

Pedicels jointed near base; leaves distichous; stipules linear to lanceolate; leaves subglabrous . . . . . . . . . . . . . . . . . . . . . . . . . . . . . . . . . . . . . . . . . . . . . . . 9. *S. acuta*

Pedicels jointed above middle or not jointed; leaves not distichous, puberulent to tomentulose, at least beneath:

Carpels 5–6; fruit 3–4 mm. in diameter:

Leaves obscurely serrate, appearing entire; stipules stiffly linear-subulate; inflorescence paniculate, nearly leafless . . . . . . . . . . . . . . . . . . . 10. *S. lodiegensis*

Leaves obviously serrate or crenate; stipules flexuous, or at least not stiff; inflorescence of solitary or few-flowered axillary corymbs:

Pedicels several times as long as calyces; stems sparsely hispid; leaves ovate-lanceolate to oblong-lanceolate; stems procumbent . . 11. *S. procumbens*

Pedicels not over twice as long as and usually shorter than calyces; stems puberulent, not hispid; leaves linear to narrowly ovate-lanceolate; stems erect or ascending . . . . . . . . . . . . . . . . . . . . . . . . . . . . . . 12. *S. angustifolia*

Carpels 7–14; fruit 5–9 mm. in diameter:

Leaves entire; inflorescence a terminal corymb; plants annual or biennial . . . . . . . . . . . . . . . . . . . . . . . . . . . . . . . . . . . . . . . . . . . . . . . . . . 13. *S. linifolia*

Leaves finely to coarsely serrate or dentate; inflorescence paniculate or flowers solitary in axils, rarely corymbose; plants perennial:

Leaves rounded to subcordate at base, coarsely crenate-serrate to base; calyx 5-angled in bud; stems and petioles hirsute with spreading simple hairs extending beyond stellate puberulence . . . . . . . . . . . . . . . 14. *S. tragiifolia*

Leaves cuneate at base, dentate-crenate above middle; calyx 10-angled in bud; stems and petioles appressed stellate-puberulent, not hirsute . . . . . . . . . . . . . . . . . . . . . . . . . . . . . . . . . . . . . . . . . . . . . . . . . . . . . . . . . . . 15. *S. rhombifolia*

**1. Sida hederacea** (Dougl. ex Hook.) Torr. ex A. Gray, Pl. Fendl. 23, 1849

*Malva hederacea* Dougl. ex Hook., Fl. Bor. Amer. **1:** 107, 1831.
*Sida ?obliqua* Nutt., in Torr. & Gray, Fl. N. Amer. **1:** 233, 1838.
*Disella hederacea* Greene, Leaflets Bot. Obs. **1:** 209, 1906.

Perennial herb from a deep-seated taproot; stems 1–5 dm. high; herbage densely stellate-scurfy and canescent throughout; stipules linear-subulate, 3–4 mm. long, marcescent; petioles rather stout, 1–2 cm. long; leaf blades 1–5 cm. broad, ovate-subcordate to obliquely subreniform, crenate, prominently veined beneath; flowers solitary or in few-flowered axillary clusters; bractlets linear, 3–5 mm. long; calyx campanulate to turbinate, lobes ovate to ovate-lanceolate, 2–3 mm. broad, 6–10 mm. long, short-acuminate; petals 1–2 cm. long, pale cream, dotted or suffused with lavender or rose, narrowly obovate, stellate-puberulent outside on part exposed in bud; stamineal column short, glabrous; fruit conical-truncate, 5–8 mm. wide, 4–5 mm. high; carpels 6–10, dark brown, triangular, sparsely stellate-puberulent on backs, falling from central axis tardily.

Mainly in heavy alkaline soils, Sonoran Zones, Washington to Baja California, Utah, Texas, and northern Mexico. Feb.–Oct.

The stellate puberulence on the backs of the petals is a striking character.

**2. Sida lepidota** A. Gray, Pl. Wright. **1:** 18, 1852, var. **lepidota**

*Disella lepidota* Greene, Leaflets Bot. Obs. **1:** 209, 1906.

Decumbent or ascending perennial herb from a slender rootstock, stellate-lepidote with silvery scales throughout; stems 1–5 dm. long; stipules linear-subulate, 2–3 mm. long; petioles slender, 8–15 mm. long; leaf blades ovate-lanceolate to narrowly ovate, obliquely asymmetrical at base, 8–22 mm. wide, 1–3.5 cm. long, irregularly serrate along whole margin or upper portion entire, acute at apex, sometimes faintly subhastate on one side at base; flowers mostly solitary, on scurfy pedicels 5–15 mm. long; calyx hemispherical, 4–5 mm. wide, lobes ovate-acuminate, 4–6 mm. long at anthesis, densely silvery-lepidote; petals pale yellow, broadly obovate, 10–15 mm. long, sparsely and finely stellate-puberulent along one margin on back; fruits 6–8 mm. broad, carpels rounded and finely reticulate on back, narrowed to a triangular, smooth tip, these connivent, lateral faces fenestrate-reticulate, membranous.

Alkaline, heavy soils, Sonoran Zones, Arizona to Texas and adjacent northern Mexico. March–Oct.

The broad-leaved typical form is commoner at higher elevations than in the desert areas, but one specimen labeled "Tucson" is surely this. It might have been collected in the mountains some distance from Tucson and above the limits of the Sonoran Desert. The following variety is common in our area.

**2a. Sida lepidota** var. **sagittifolia** A. Gray, Pl. Wright. **1:** 18, 1852

*Disella sagittifolia* Greene, Leaflets Bot. Obs. **1:** 209, 1906.
*Sida sagittaefolia* Cory, Rhodora **38:** 407, 1936.

Differing from var. *lepidota* in having leaves with narrowly lanceolate, often sagittately lobed blades 4–8 mm. wide, 1.5–3 cm. long, main portion of blade entire, basal lobes often few-toothed; otherwise as in var. *lepidota.*

Heavy or fine alkaline soil, Sonoran Zones, Arizona to southern Colorado, western Texas, Chihuahua, and into southern Sonora.

**3. Sida physocalyx** A. Gray, Bost. Jour. Nat. Hist. **6:** 163, 1850
*Sida hastata* St. Hil., Fl. Bras. Merid. **1:** 190, 1827. Not *S. hastata* Willd., 1800.

Herbaceous perennial from a tuber-like, often fusiform root; stems slender, 1–8 dm. long, erect or procumbent, sparsely hispid with simple and few-rayed, stellate hairs 1–2.5 mm. long; stipules lance-linear, 4–7 mm. long, persistent; petioles 1–5 cm. long, hispid; leaf blades orbicular to oblong or lance-oblong, 1.5–3 cm. broad, 2–8 cm. long, acute to rounded at apex, subcordate to divaricate at base, crenate-dentate, 3–5-veined from base, bright green, sparsely pubescent with few-rayed, stellate hairs; flowers solitary in axils; pedicels slender, 1–2 cm. long at anthesis, to 3 cm. long in fruit; calyx lobes cordate, 1–1.5 cm. long and wide in fruit, papery and veiny, sparsely stellate-hispid; petals yellow, about equaling or slightly surpassing calyx; carpels black, about 3 mm. high, subreniform but upper portion drawn into a stout blunt beak, 1.5–2.2 mm. long, beaks connivent around summit of central axis, sides of body finely reticulate radially, angles and back slightly grooved transversely.

In rich soil on flats and canyons, Lower and Upper Sonoran Zones to Tropical Zone, Arizona to Texas and south into tropical America. March–Oct.

**4. Sida cordifolia** L., Sp. Pl. 684, 1753
*Sida conferta* Link, Enum. Hort. Berol. **2:** 207, 1822.

Perennial herb or subshrub 5–13 dm. tall, densely stellate-pubescent throughout, stems eventually glabrate; stipules filiform to linear, 3–7 mm. long, caducous; petioles stoutish, usually about half as long as blade; leaf blades ovate-cordate to broadly cordate, acute or short-acuminate, serrate, 2–8 cm. long, 1.5–7 cm. broad, minutely stellate-tomentulose, velvety on both surfaces; inflorescence a terminal panicle of glomerules and with solitary flowers below, or sometimes reduced to latter only; peduncles and pedicel 3–15 mm. long, joint near base (peduncle shorter than the pedicel); calyx 6–7 mm. tall in flower, only slightly longer in fruit, tube distinctly 10-angled, densely tomentose, lobes deltoid, 3–3.5 mm. long and broad; petals yellow, 8–10 mm. long; carpels 7–12, 3–4 mm. long, hairy on backs, beaks erect, spine-like, retrorsely hispid, about as long as body of carpel or half as long, reticulate on sides and back of body.

Commonly along streams and a weedy invader of fields, Lower Sonoran and Tropical Zones, southern Sonora to tropical America, the West Indies; in southern Florida and in tropical Africa and Asia. Sept.–Jan.

**5. Sida setifera** Presl, Rel. Haenk. **2:** 105, 1835

Slender shrub; stems and petioles densely stellate-tomentose with very fine hairs and with simple spreading hairs 1.5–2.5 mm. long extending far beyond velvety tomentum; stipules filiform or linear-setaceous, 4–7 mm. long, deciduous; petioles 1.5–4 cm. long; leaf blades broadly ovate-cordate, 2–6 cm. broad, 1.5–7.5 cm. long, acute at apex, distinctly cordate at base, finely crenate or dentate, stellate-tomentose on both surfaces, slightly tawny, prominently veined beneath; flowers mostly in axillary glomerules; pedicels and peduncles combined rarely over 1 cm. long and often much shorter; calyx 5-angled in bud, hispid and tomentose, lobes deltoid, 2.5–3 mm. long; petals yellow, 7–8 mm. long; carpels 7–12, 2.5–3 mm. tall, barely acute at apex, beakless, back finely striate-reticulate longitudinally, sides thin, reticulate-veined, glabrous except for a few minute hairs on apical portion.

Valleys, flats, and low rolling hills, mostly in good soil, Lower Sonoran Zone to Tropical Zone, southern Sonora southward near the coast into Tepic. Late summer and autumn.

**6. Sida glabra** Mill., Gard. Dict. ed. 8, *Sida* no. 14, 1768

*Sida ulmifolia* Cav., Diss. **1:** 15, *pl. 2, fig. 4,* 1785.
*Sida arguta* Swartz, Prod. Veg. Ind. Occ. 101, 1788.
*Sida alamosana* S. Wats., Proc. Amer. Acad. **26:** 133, 1891.

Erect or ascendingly branched shrub or suffrutescent perennial herb 4–12 dm. tall; branches slender and moderately pubescent with fine, few-rayed, stellate hairs; stipules filiform-setaceous, 1–3 mm. long, usually spreading, subpersistent; petioles slender, 8–30 mm. long; leaf blades ovate-lanceolate, acuminate, 1–3 cm. broad, 2–7 cm. long, cordate and sometimes asymmetrical at base, serrate-crenate, very sparsely stellate-pubescent on lamina, veins often somewhat more densely pubescent; flowers solitary in axils; pedicels and peduncles slender, 1–2.5 cm. long, jointed 2–5 mm. below calyx; calyx 5–6 mm. long, lobes ovate-acuminate, thin, nearly glabrous; petals yellow to orange, 5–6 mm. long; carpels 5–6, about 2.5 mm. high, glabrous when mature, longitudinally striate-reticulate dorsally, beaks shorter than body, minutely puberulous.

A wayside weed, Lower Sonoran Zone to Tropical Zone, southern Sonora and Vera Cruz to tropical America and the West Indies. Flowering following rains.

Not common in our area, barely entering it from the south.

**7. Sida glutinosa** Commers ex Cav., Diss. **1:** 16, *pl. 2, fig. 8,* 1785

*Sida nervosa* DC., Prod. **1:** 465, 1824.
*Sida endlicheriana* Presl, Rel. Haenk. **2:** 111, 1835.

Herbaceous or suffrutescent perennial with slender stem to 12 dm. tall, younger parts pubescent with short glandular hairs, sparsely arranged stellate hairs, and long, simple, spreading ones; stipules filiform, 1–2.5 mm. long, subpersistent; petioles slender, 1.5–3.5 cm. long, hispid; leaf blades ovate-acuminate, shallowly cordate at base, irregularly serrate, 1–3 cm. broad, 2–6 cm. long, thin, stellate-puberulent on both surfaces but denser and slightly paler beneath; flowers solitary in axils or ultimately paniculate; pedicels slender, 1–2.5 cm. long, puberulent and often glandular, jointed above middle; calyx 4–5 mm. high, puberulent with few-rayed stellate hairs, sparingly glandular, and hispid along margins of triangular lobes, these subacuminate, 2.5–3 mm. long at anthesis; petals pale yellow, 4–5 mm. long; carpels 5, body glabrous at maturity, 2–2.5 mm. high, minutely puberulent beaks aristate, about 1 mm. long, erect.

Among larger shrubs and invading clearings, Lower Sonoran Zone to Tropical Zone, southern Sonora to Central America, the West Indies, and the Bahamas. Flowering almost throughout the year.

**8. Sida xanti** A. Gray, Proc. Amer. Acad. **22:** 296, 1887

Erect shrubby plant about 1 m. tall; branches sparsely to densely stellate-puberulent, eventually glabrate and sometimes viscid; stipules filiform, 5–6 mm. long, deciduous; petioles slender, 5–25 mm. long, scurfy-stellate-puberulent; leaf blades lance-ovate to lanceolate, 5–20 mm. wide, 3–10 cm. long, obtuse to acuminate (usually acute), serrate-dentate, rounded to truncate at base, green, sparsely stellate-pubescent beneath, subglabrous above; flowers solitary in axils below, in panicles or cymose clusters at tips of branches; peduncles 1–3 cm. long, pedicels 3–6 mm. long at anthesis, both elongating somewhat in fruit; calyx turbinate in bud, cup hemispheric in flower, lobes ovate, acuminate or attenuate, 5–10 mm. long at anthesis, sides straight or slightly curved, stellate-pubescent; petals orange-yellow, 1.5–2 cm. long; carpels 10–12, 3–3.5 mm. high, about 1 mm. of this constituting an erect, blunt, dehiscent beak, angles rugose-knobby, back grooved and transversely ridged, glabrous, sides reticulate.

Along arroyos and on rocky hillsides, Lower Sonoran Zone, vicinity of Loreto to the Cape Region, Baja California. Flowering throughout year.

There is wide variation in size of the sepals but the intergradation among the various forms is complete.

**9. Sida acuta** Burm. f., Fl. Ind. 147, 1768

*Sida ulmifolia* Mill., Gard. Dict. ed. 8, *Sida* no. 1, 1768.
*Sida carpinifolia* L. f., Suppl. Syst. Veg. 307, 1781.
*Sida stipulata* Cav., Diss. **1**: 22, *pl. 3, fig. 10*, 1785.
*Sida balbisiana* DC., Prod. **1**: 460, 1824.
*Sida trivialis* Macf., Fl. Jam. **1**: 78, 1837.
*Sida acuta* var. *carpinifolia* K. Schum., Fl. Bras. **12**[3]: 326, 1891.

Perennial herb or subshrub, sometimes flowering first season, with prostrate to erect, leafy stems to 1 m. long, minutely stellate-puberulent; stipules lanceolate to linear, 7–15 mm. long, ascending, usually 3-nerved from base; petioles 4–5 mm. long; leaf blades lanceolate, ovate, or oblong-lanceolate, rounded to cuneate at base, 1.5–9 cm. long, 5–30 mm. wide, finely and irregularly serrate, green on both surfaces, virtually glabrous at maturity; flower solitary or later simulating an umbellate inflorescence; pedicels jointed near base, 2–5 mm. long; calyx 6–8 mm. long, 10-nerved, lobes broadly ovate, acuminate, 2.5–3 mm. long at anthesis, often sparsely hispid below sinuses and along margins; petals white to orange-yellow, 10–12 mm. long; carpels 7–12, 3–4 mm. long, margins of back low-crested, short beaks connivent, minutely puberulent.

Hillsides and plains, often among dense shrubby or forest cover, Lower Sonoran Zone to Tropical Zone, central Sonora to tropical America and the West Indies; Texas to Florida. Sept.–March.

**10. Sida lodiegensis** Baker f., in Rose, Contr. U.S. Nat. Herb. **3**: 311, 1895

A slender-stemmed, erect or ascendingly branched perennial to 1.5 m. tall, woody only at base, branches and petioles sparsely stellate-puberulent; stipules subulate-setaceous, stiff, ascending or ascendingly spreading, 3–10 mm. long, persistent; petioles 2–5 mm. long; leaf blades lanceolate to linear-lanceolate, 5–15 mm. wide, 5–8 cm. long, minutely and obscurely serrulate, cuneate at base, acute to attenuate at apex, dark green and glabrous above, paler and sparsely stellate-puberulent beneath, 3-nerved from base; inflorescence racemose-paniculate, leaves greatly reduced, branches slender, stellate-puberulent; pedicels slender, to 1 cm. long or uppermost flowers nearly sessile, not obviously jointed; calyx turbinate, terete, 2.5–4 mm. long at anthesis, densely stellate-puberulent, lobes broadly ovate to deltoid, 1.5–2 mm. long; petals salmon color, about 4 mm. long; carpels 5, tumidly rounded on back, beak scarcely distinguishable from body, broadly triangular, turned inward, rounded back regularly stellate-puberulent with rays not quite interlacing.

Mainly in openings on brushy slopes or canyons, Lower Sonoran Zone and adjacent scrub-thorn formations, southern Sonora into Sinaloa. Aug.–Feb.

The tumid carpels with their regularly spaced, stellate hairs, the subentire leaves, and the bristlelike stipules are sharply diagnostic.

**11. Sida procumbens** Swartz, Prod. Veg. Ind. Occ. 101, 1788

*Sida pilosa* Cav., Diss. **1:** 9, *pl. 1, fig. 8,* 1785. Not *S. pilosa* Mill., 1768.
*Sida ovata* Cav., Diss. **6:** 350, *pl. 196, fig. 2,* 1788.
*Sida supina* L'Hér., Stirp. pt. **5:** 109 *bis, pl. 52 bis,* 1789.
*Sida diffusa* H. B. K., Nov. Gen. & Sp. **5:** 257, 1822.
*Sida filiformis* Moric., Pl. Amer. Rar. 10, *pl. 8,* 1830.

Perennial herb from a woody rootstock or some plants flowering first season; stems slender, prostrate to procumbent, 1–6 dm. long, minutely stellate-puberulent and hispid with simple spreading hairs; stipules filiform, 1–2 mm. long; petioles 3–15 mm. long; leaf blades lanceolate-ovate to suborbicular, obtuse, rounded, or shallowly cordate at base, acute to obtuse at apex, 3–20 mm. wide, 6–30 mm. long, minutely stellate-puberulent, slightly paler beneath than above, crenate-serrate; flowers usually solitary in axils; pedicels slender, to 5–10 mm. long in flower, jointed just beneath calyx; calyx 5–6 mm. high, lobes ovate-acuminate, 3–4 mm. long, hispid; petals pale yellow to nearly white, 5–7 mm. long; carpels 5, 2–3 mm. long including short triangular beak, rugose-reticulate on back and sides.

Open grassland, slopes, hillsides, and along arroyos, Lower Sonoran Zone to Tropical Zone, from Arizona, Texas, and southern Florida southward through the West Indies and Mexico to South America. March–Oct.

An inconspicuous weed easily overlooked.

**12. Sida angustifolia** Lam., Encycl. **1:** 4, 1783

*Sida linearis* Cav., Ic. **4:** 6, *pl. 312, fig. 1,* 1797.
*Sida hyssopifolia* Presl, Rel. Haenk. **2:** 109, 1835.

Erect herb to 1 m. tall from a perennial root; stems and leaves minutely stellate-tomentulose; stipules setaceous, 2–4 mm. long, erect or ascending; petioles 4–7 mm. long; leaf blades linear to linear-oblong, acute to obtuse at apex, rounded or obtuse at base, crenate or serrate, 3–8 mm. wide, 2–7 cm. long, minutely stellate-puberulent on both sides or upper surface at length glabrate; flowers solitary in axils or on older plants in small, axillary, corymblike, few-flowered clusters; pedicels 2–8 mm. long, joint about at middle; calyces 4–5 mm. high, puberulent, lobes triangular, acute, 1.5–2 mm. long; petals pale yellow, 4–6 mm. long; carpels 5–6, about 2.5 mm. high, including apiculate erect beak, back and sides reticulate-rugulose, glabrous, beaks sparsely and minutely puberulent.

Along streams, on grassy flats, and about clearings, Lower Sonoran Zone to Tropical Zone, Arizona to Texas to South America, and in the tropics of the Old World. May–Jan.

**13. Sida linifolia** Juss. ex Cav., Diss. **1:** 14, *pl. 2, fig. 1,* 1785

Erect, simple or ascendingly branched annual or biennial from a slender taproot; younger parts sparsely pubescent with simple, forked, and few-rayed, rather coarse, stellate hairs; stipules linear to lance-linear, 4–7 mm. long, ascendingly spreading; petioles 5–10 mm. long, hispid with ascending yellowish hairs; leaf blades linear to narrowly lanceolate, 3–20 mm. wide, 2–14 cm. long, entire, obtuse to rounded at base, gradually narrowed to acute apex, sparsely pubescent with simple or forked, subappressed hairs on both sides, somewhat denser beneath; flowers in terminal corymbs or 2–3 solitary in upper axils; pedicels to 2 cm. long in fruit, shorter in flower, jointed a little above middle, hispidulous; calyx about 5 mm. high at anthesis, pubescent with short coarse hairs, lobes ovate, acute, often apiculate; petals white with a dark purple base, 6–8 mm. long; carpels 7–9, glabrous, dark brown, about 3 mm. high, beak incurved, blunt, 0.5–0.8 mm. long.

Along streams, on shrubby hillsides and in fields, Lower Sonoran and Tropical Zones, southern Sonora to tropical South America, the West Indies, and tropical Africa. Flowering following rains.

The entire leaves and the annual or biennial habit set this species apart from all others entering our area. It, like several others, barely reaches the southern boundaries of the Sonoran Desert.

### 14. **Sida tragiifolia** A. Gray, Bost. Jour. Nat. Hist. **6:** 164, 1850

Suffrutescent or herbaceous plants from a perennial taproot; stems erect or ascending, 2–6 dm. tall, minutely but closely stellate-puberulent, basal parts eventually glabrate or nearly so; stipules linear, 3–5 mm. long, caducous; petioles 5–10 mm. long; leaf blades oblong to oblong-lanceolate, 1.5–5 cm. long, acute at apex, rounded or rarely faintly subcordate at base, crenate-serrate, minutely stellate-puberulent on both surfaces, more densely so beneath than above; flowers solitary in upper axils; peduncles about 1 cm. long at anthesis, to 2.5 cm. long in fruit, pedicels 3–5 mm. long, stoutish; calyx strongly 5-angled, 5–7 mm. long in bud, stellate-tomentulose, lobes ovate-acuminate, 3–5 mm. long, nearly as wide; petals yellow-orange, 10–12 mm. long; carpels 10–12, about 4 mm. high, about half of this constituting dehiscent blunt beak, which is tipped with a minute apiculation at outer angle, body sharply reticulate both dorsally and laterally, outer angles of beak minutely puberulent.

Canyons and mountain slopes, Sonoran Zones, southern Arizona (near Tucson), western Texas, and northeastern Mexico. Aug.–Nov.

### 15. **Sida rhombifolia** L., Sp. Pl. 684, 1753

*Sida ruderata* Macf., Fl. Jam. **1:** 80, 1837.

Perennial herb or subshrub, 3–10 dm. tall, usually profusely branched, minutely stellate-puberulent or the stems eventually glabrate; stipules setaceous, 3–4 mm. long, usually ascending, deciduous; petioles 3–5 mm. long; leaf blades lanceolate, oblong, or rhomboid-oblong, cuneate at base, 2–8 cm. long, 5–10 mm. wide, acute or obtuse at apex, irregularly dentate-crenate above middle, pale beneath, dark green above; flowers solitary in axils or terminal flowers sometimes corymbose; peduncles 2–3.5 cm. long, jointed above middle; calyx 10-angled in bud, 6–7 mm. long, lobes broadly ovate, abruptly apiculate, 2–3 mm. high and wide, minutely puberulent, becoming thin and somewhat papery; petals yellow or orange, 4–5 mm. long; carpels 10–14, 3–4 mm. high including slender, spinelike, nearly glabrous beak, smooth or nearly so on sides and back or latter longitudinally reticulate.

Common wayside weed and in waste places, on hillsides and grassy slopes, Lower Sonoran Zone to Tropical Zone, central Sonora south to the tropics and the West Indies; from Texas to Florida. Widely distributed in the warm regions of the world. Flowering following rains.

## 11. **ANODA** Cav., Diss. **1:** 38, *pl. 10, fig. 3,* 1785

Herbaceous or rarely suffrutescent plants with stellate or simple hairs or both or glabrous. Leaves ovate, hastate, or lobed, entire or variously toothed or crenate. Flowers violaceous, yellow, or white, pedunculate, axillary and solitary or in terminal panicles or racemes. Bracteoles none. Calyx 5-lobed. Petals 5. Stamineal column often hirsute; anthers many. Ovary 5- to many-celled, each cell 1-ovuled. Style branches as many as carpels. Stigmas capitate. Carpels rounded or umbonate to spurred dorsally, side walls usually disappearing by maturity, an inner layer of carpel wall forming a partial or complete envelope around seed in some species, not apparent in others. Seeds pendulous, reniform-obovoid.

Seeds enclosed in a black, fenestrate, loose envelope composed of inner layer of carpel:
  Carpels sharply angled to short-spurred on back; petals yellow to orange; upper leaves hastate . . . . . . . . . . 1. *A. crenatiflora*
  Carpels rounded, not angled nor spurred on back; petals purple; upper leaves narrowly 3–5-lobed . . . . . . . . . . 2. *A. reticulata*
Seeds incompletely enclosed or closely invested by a smooth, aril-like coating, which is not reticulate-fenestrate (except in *A. acerifolia*):
  Fruit flattened, disk-like; calyx strongly spreading in fruit; petals purple, lavender, or bluish:
    Carpels bearing a radiating spur one-half as long as to equaling body; carpel-coat not splitting into two portions . . . . . . . . . . 3. *A. cristata*
    Carpels merely umbonate or with a blunt spur less than one-half length of body; carpel coat bilamellar, inner one partially surrounding back of seed, whitish to yellowish, fenestrate . . . . . . . . . . 4. *A. acerifolia*
  Fruit hemispherical, not disk-like; calyx lobes ascending in fruit; petals orange-yellow:
    Leaves ovate-acuminate, crenate-dentate, deeply cordate at base; stems densely but minutely stellate-tomentulose; sides of carpels persistent, fragile . . . . 5. *A. abutiloides*
    Leaves narrowly hastate or hastately 3-lobed, entirely to faintly crenate, truncate to cuneate at base; stems sparsely stellate-pubescent or glabrate; sides of carpels disappearing before maturity . . . . . . . . . . 6. *A. pentaschista*

### 1. **Anoda crenatiflora** C. G. Ortega, Hort. Mat. Dec. Fascl. **8**: 96, 1798

*Anoda parviflora* Cav., Ic. **5**: 19, *pl. 431*, 1799.
*Sida crenatiflora* Pers., Syn. Pl. **2**: 247, 1806.
*Sida parviflora* Willd., Enum. Hort. Berol. 726, 1809.
*Anoda ortegae* Spreng., Nachtr. Bot. Gart. Halle 11, 1801.

Annual herb with erect, minutely stellate-puberulent stems 2–4 dm. tall or more; stipules filiform, 2–4 mm. long, caducous; petioles 2–6 cm. long, minutely puberulent; leaf blades broadly ovate to hastate, lower ones rounded or angulate, shallowly cordate to truncate, 1–5 cm. wide, 1.5–8 cm. long, crenate to subentire, upper hastate, central lobe triangular, to 10 cm. long, remotely crenate-dentate to entire, thin, sparsely pubescent with simple or few-rayed, stellate hairs to glabrate, slightly paler beneath than above; peduncles solitary in axils, 5–15 cm. long at anthesis, 3–10 cm. long in fruit; calyx finely but densely stellate-tomentulose and often with a few longer spreading hairs near base, lobes lance-triangular, 3.5–4.5 mm. long in flower, to 10 mm. long and spreading in fruit; petals cream-yellow, 5–7 mm. long; fruit 6–9 mm. in diameter exclusive of spurs, discoid, densely stellate-puberulent and with some fine spreading hairs as well; carpels 8–10, back bearing a spur 1–1.5 mm. long; seed enclosed in black, reticulate-fenestrate, inner lamella of carpel.

Hillsides under trees and about rocks, along arroyos and in canyons, Lower Sonoran Zone, southern Arizona and northern Mexico. Collected near Ímuris and in the Sierra Batuc, east of Hermosillo, in Sonora. Aug.–Sept.

### 2. **Anoda reticulata** S. Wats., Proc. Amer. Acad. **17**: 368, 1882

Annual or short-lived perennial with slender, erect, stellate-puberulent stems; stipules subulate, 1–1.5 mm. long; petioles 1.5–4 cm. long, stellate-puberulent; leaf blades at base of plant broadly ovate, 1.5–3 cm. wide, 2.5–3.5 cm. long, shallowly cordate, those at middle hastate, central lobe lanceolate, to 4 cm. long, upper leaves narrowly 3–5-lobed, lobes linear, 1–4 mm. wide, 1–3.5 cm. long, finely stellate-puberulent on both surfaces, only lower leaves with crenate-dentate margins; inflorescence paniculate; peduncles greatly exceeding subtending leaves or bracts, 1–2.5 cm. long in flower, elongating to 7 cm. in fruit, densely stellate-puberulent, jointed near apex; calyx lobes ovate, 1.5–2 mm. wide, 2.5–3 mm. long in flower, only slightly longer in fruit, acute or short-apiculate, stellate-tomentulose; petals 5–7 mm. long, purple or blue-purple; carpels rounded on back, without an umbo or spur, closely and finely stellate-puberulent, bilamellar, inner coat thicker, black, fenestrate, completely enclosing dark brown seed.

Granitic hillsides and canyons, Lower Sonoran Zone but near its upper limits, Pima and Santa Cruz Counties, Arizona, and between Ures and Baviácora, Sonora. Imperfectly known; collected in flower in Sonora in Sept.

### 3. **Anoda cristata** (L.) Schlecht., Linnaea **11:** 210, 1837

*Sida cristata* L., Sp. Pl. 685, 1753.
*Anoda lavateroides* Medic., Malvenfam. 19, 1787.
*Anoda arizonica* A. Gray, Proc. Amer. Acad. **22:** 298, 1887.
(For synonymy see Hochr., Ann. Conserv. & Jard. Bot. Genève **20:** 40–45, 1916.)

Annual to 1 m. tall; stems sparsely setose-hispid but eventually glabrate; stipules subulate, 3–5 mm. long, deciduous; petioles 1–6 cm. long, hispid to glabrate; leaf blades ovate to hastate, to 10 cm. long and sometimes nearly as wide, truncate to broadly cuneate at base, margins entire to coarsely dentate-serrate, sparsely villous to glabrate, hairs mostly simple; flowers axillary; peduncles 1–3 cm. long in flower, to 6 or 8 cm. long in fruit; calyx lobes lance-triangular, acuminate to attenuate, 5–8 mm. long in flower, 8–12 mm. long and spreading in fruit, sparsely hispid at first, eventually subglabrate; petals purple or lavender, 8–20 mm. long; fruit about 1–1.5 cm. in diameter exclusive of spurs, sparsely hispid; carpels 9–20, each bearing a horizontal spur 2–4 mm. long, sides of carpels disappearing before maturity, dorsal walls not bilamellar; seeds brownish, naked, finely papillate.

A common wayside weed and on gentle slopes and flats, Sonoran Zone to Tropical Zone, Arizona to Texas and southward through Mexico and the West Indies to South America. Aug.–Oct.

The leaf form and density of pubescence are extremely variable. Hochreutiner recognized four varieties within the species.

### 4. **Anoda acerifolia** (Zucc.) DC., Prod. **1:** 459, 1824

*Sida acerifolia* Zucc., in Roem., Collect. 148, 1807–1810.
*Sida zuccagnii* Spreng., Syst. **3:** 121, 1826.
*Anoda hastata* Schlecht., Linnaea **11:** 214, 1837. Not *A. hastata* Cav., 1785.

Annual herb to 6 dm. high or decumbent with branches 1–4 dm. long; stems and branches sparsely hispid, mostly with simple, slightly reflexed hairs or eventually glabrate; stipules slender-subulate, 3–5 mm. long; petioles hispid, 1.5–5 cm. long; leaf blades variable, deltoid, to pentagonal, often more or less palmately to narrowly hastately lobed, to 10 cm. long, truncate or slightly cordate at base, margin entire to serrate, dark green above, pale beneath, sparsely subappressed-hispid along veins at least beneath; peduncles slender, to 3 cm. long in flower, longer in fruit; calyx sparsely hispid to nearly glabrous, lobes about 5 mm. long in flower, to 15 mm. long and spreading in fruit; petals lavender, purple, or bluish, 9–12 mm. long; fruit discoid,

10–14 mm. in diameter, hispid above; carpels 7–15, dorsally gibbous or short-mucronate, walls bilamellar, inner portion forming a white to yellowish, fenestrate hood over dorsal part of seed; seed dropping free at maturity.

Openings in the scrub or on grassy hillsides, Lower Sonoran Zone to Tropical Zone, central Baja California and Sonora through Mexico and the West Indies to South America. Aug.–Nov.

**5. Anoda abutiloides** A. Gray, Proc. Amer. Acad. **22:** 300, 1887

Erect, suffrutescent plant to 1 m. tall with densely velvety stellate-tomentose stems, branches, and petioles with a few simple hairs also; stipules caducous very early; petioles 4–8 cm. long, spreading, fine, simple hairs more numerous than on stems; leaf blades broadly ovate, acuminate at apex, deeply cordate at base, 3–10 cm. wide, to 15 cm. long, or upper ones lance-ovate and attenuate, margins serrate-dentate, minutely stellate-velvety on both surfaces, 7–9-veined; flowers in terminal panicles, 3–4 dm. long; peduncles 1–1.5 cm. long at anthesis; calyx stellate-puberulent, hemispherical, lobes deltoid-ovate, 2–2.5 mm. long, scarcely at all enlarged and erect in fruit; petals orange or yellow but usually drying pink or purple, 8–10 mm. long; fruit 5–7 mm. broad at maturity, minutely puberulent; carpels 5, about 3–3.5 mm. high, slightly umbonate dorsally, sides persistent but membranous and usually torn at maturity.

In rich soil in canyons, Sonoran Zones, Baboquivari and Santa Catalina Mountains, Arizona, and Sierra Madre, Chihuahua. May–Oct.

**6. Anoda pentaschista** A. Gray, Pl. Wright. **2:** 22, 1853

*Sidanoda pentaschista* Woot. & Standl., Contr. U.S. Nat. Herb. **19:** 427, 1915.

Annual herb to 1 m. tall or more; young branches, tips of stems, and petioles minutely stellate-puberulent to glabrate; stipules lance-linear, 2–3 mm. long, fugacious; petioles 5–20 mm. long; leaf blades at base of plant ovate to suborbicular, 0.5–3 cm. long and wide, faintly crenate to entire, upper ones lanceolate to linear-lanceolate, usually hastately lobed at base, central lobe 3–10 mm. wide, 2–6 cm. long, entire or essentially so, stellate-tomentulose on both surfaces, ashy beneath; flowers solitary, but inflorescence becoming paniculate by development of short, lateral, flowering branches in upper axils of main stem and branches; peduncles slender, 1–3 cm. long in flower, sparsely stellate-puberulent; pedicels to 1 cm. long, densely stellate-puberulent and eventually markedly viscid-glandular; calyx hemispheric, 2.5–3 mm. wide, lobes deltoid to ovate, 2–2.5 mm. long, nearly as wide at anthesis, densely stellate-puberulent, enlarging very slightly and erect in fruit; petals yellow-orange, 4–6 mm. long; fruit hemispheric, 4–5 mm. in diameter, minutely and sparsely puberulent with white crisped hairs; carpels 5–7, bluntly umbonate dorsally, 1-nerved on back, side-walls disappearing before maturity; seeds brown, minutely but closely papillate.

Commonly a weedy invader of roadsides and fields, mainly Lower Sonoran Zone, Arizona to Texas south to Michoacán. June–Oct.

## 12. BASTARDIA H. B. K., Nov. Gen. & Sp. 5: 254, 1822

Herbs or shrubs with cordate, entire, or dentate-crenate leaves and yellow or violet flowers on 1-flowered peduncles. Bracteoles none. Calyx lobes 5. Petals 5, obovate. Ovary 5-celled, each cell 1-ovulate. Style branches 5; stigmas capitate. Capsule depressed-globose, acutely to obtusely 5-angled, loculicidally 5-valved. Carpels beakless or slenderly beaked. Seeds pendulous, only partially filling cell.

**1. Bastardia bivalvis** (Cav.) H. B. K., Nov. Gen. & Sp. **5:** 255, 1822

*Sida bivalvis* Cav., Diss. **1:** 13, *pl. 11, fig. 3,* 1785.
*Bastardia berlandieri* A. Gray, Proc. Amer. Acad. **22:** 295, 1887.

Slenderly branched shrub to 1.5 m. tall with finely stellate-puberulent stems and foliage, rays of hairs interlacing only slightly; stipules linear-subulate, 4–6 mm. long, subpersistent; petioles to 10 cm. long; leaf blades broadly ovate- to orbicular-cordate, acute to short-acuminate, crenate-dentate, thin, dark green and sparsely stellate-puberulent (chiefly along veins) above, paler and with slightly coarser stellate hairs beneath; flowers solitary in axils of main branches and of small lateral branches arising in same axils as primary flowers; peduncles 3–6 mm. long, densely stellate-tomentulose, no joint evident; calyx stellate-tomentulose, lobes ovate-attenuate, about 2 mm. wide, 6–7 mm. long at anthesis, 1–3 mm. longer in fruit; petals yellow, 10–12 mm. long; capsules 8–10 mm. wide, stellate-puberulent; carpels each bearing an aristate erect beak about 2 mm. long, beak splitting at dehiscence of carpel; seed dark brown, minutely and irregularly puberulent.

Canyons and shaded hillsides, Lower Sonoran to Tropical Zones, vicinity of Divisadero, Sonora, and from Michoacán and Oaxaca to Vera Cruz, the Greater Antilles, and South America. Aug.–Feb.

**Bastardia violacea** Standley, Field Mus. Bot. Ser. **22:** 41, 1940, has small violet flowers with petals 1 cm. long, ovate leaves 1.5–2.5 cm. long, and beakless rounded carpels. It was described from specimens collected at Chinobampo, along the Río Mayo in southern Sonora, and, although still known only from the type locality, may occur in the southern part of our area.

## 13. KOSTELETZKYA Presl, Rel. Haenk. 2: 130, *pl. 70,* 1835

Perennial herbs or suffrutescent, with alternate, sagittate, ovate, or angularly lobed or parted leaves. Inflorescence paniculate or flowers solitary in axils. Bractlets 7–10 or rarely reduced to mere stubs. Calyx 5-lobed. Petals erect or spreading, rose-purple or yellowish. Stamineal column entire or 5-toothed just below numerous filaments. Ovary 5-celled, each cell 1-ovuled. Style branches 5; stigmas capitate. Capsule somewhat depressed, strongly 5-angled, loculicidally dehiscent. Seeds reniform, ascending.

Petals erect, forming a loose tube, 1.8–3 cm. long:
- Stems with spreading and stellate hairs; leaves (except uppermost) 5-lobed, lobes lance-triangular; sepals 5–7 mm. long at anthesis ........ 1. *K. paniculata*
- Stems stellate-puberulent, without spreading hairs; leaves ovate and merely dentate or shallowly 3-lobed, lobes usually broader than long; sepals 3–5 mm. long at anthesis:
  - Carpels glabrate on back, hispid on angles; leaves mostly 3-lobed ....... 2. *K. thurberi*
  - Carpels stellate-puberulent on backs as well as hispid on angles; leaves broadly ovate, not lobed, or basal ones shallowly 3-lobed ........ 3. *K. malvaviscana*

Petals spreading, not convolute, 1.5 cm. long or less:
- Leaves only hastately lobed at base; seeds minutely puberulent ........ 4. *K. hastata*
- Leaves mostly deeply 3–5-lobed or -parted; seeds glabrous:
  - Petals yellowish or white faintly tinged with rose, 12–15 mm. long; stems hispid with spreading simple hairs ........ 5. *K. coulteri*
  - Petals purplish with a yellow basal spot, 8–10 mm. long; stems puberulent with small stellate hairs ........ 6. *K. digitata*

**1. Kosteletzkya paniculata** Benth., Pl. Hartw. 285, 1848

An open, slenderly branched shrub to 2 m. tall with moderately scattered, harsh, stellate hairs and two lines of finer, densely crowded, stellate hairs running up stems; stipules linear or subfiliform, 3–5 mm. long, hispidulous; petioles harshly stellate-pubescent, 2–4 cm. long; leaf blades pentagonal, 6–12 cm. long, 5-lobed about halfway to center or slightly more, lobes lance-triangular to ovate-lanceolate, dentate with teeth 2–3 mm. long and wide, harshly stellate-pubescent on both surfaces; inflorescence a naked panicle; pedicels 10–14 mm. long at anthesis, jointed to a peduncle of about equal length; bractlets 8–10, subfiliform-setaceous, 2–3 mm. long; calyx densely but minutely stellate-puberulent, lobes lance-ovate, 2.5–3 mm. wide, 5–7 mm. long; petals convolute, 2–2.5 cm. long, pink or rose-purple; fruit about 1 cm. wide, minutely puberulent and setose-hispid on angles and dorsal surfaces; seeds dark brown, dull, 3 mm. high, glabrous except for curving longitudinal lines of microscopic, golden-brown hairs.

Chiefly in the thorn-scrub of the foothills, southern Sonora southward through Sinaloa to Jalisco. Late summer and autumn.

### 2. **Kosteletzkya thurberi** A. Gray, Proc. Amer. Acad. **22:** 314, 1887

Perennial herb to 3 m. tall with finely stellate-puberulent stems and foliage, a denser line of minute stellate hairs running from base of one leaf to that above it; stipules acicular, 4–5 mm. long, rigid; petioles 1–3.5 cm. long, stellate-puberulent; leaf blades rounded-cordate, angulate, 3-lobed or upper ones merely ovate, 2.5–7 cm. wide, 3–7 cm. long, dentate, teeth rounded and abruptly apiculate, stellate-puberulent on both surfaces, coarser stellate hairs intermingled with fine ones beneath; peduncles 4–12 mm. long; pedicels 2–4 mm. long; bractlets 7–8, filiform, 2.5–3.5 mm. long; calyx stellate-puberulent, lobes 3.5–5 mm. long; petals erect, 2–3 cm. long, purplish when dry, pink when fresh, sparsely stellate-puberulent on outer surfaces; stamineal column exceeding petals; carpels 5–6 mm. high, finely stellate-puberulent but glabrate between angles, these coarsely hispid-ciliate; seeds 2.5–3 mm. long, dark brown, with grayish striae curving longitudinally over back and sides.

Mostly, or perhaps entirely, above the Sonoran Desert in canyons and on hillsides, from northern to southern Sonora and in adjacent Chihuahua. Late summer and autumn.

### 3. **Kosteletzkya malvaviscana** Rose, Contr. U.S. Nat. Herb. **8:** 319, *pl. 67,* 1905

Erect shrub or suffrutescent perennial to 2 m. tall, stems sparsely stellate-puberulent, with a denser line of crinkled white hairs from leaf base to leaf base; stipules 2–6 mm. long, acicular; petioles densely and harshly stellate-pubescent, 0.5–15 cm. long; leaf blades broadly ovate or ovate-lanceolate, 2–8 cm. wide, 4–13 cm. long, or basal ones shallowly 3-lobed near apex and nearly orbicular, to 18 cm. long and wide, subcordate to broadly cuneate at base, crenate-dentate, stellate-pubescent but green on both surfaces, 7-veined; panicles 2–4 dm. long; peduncles to 1 cm. long; pedicels 3–8 mm. long; bractlets 6–9, subulate, 3–4 mm. long; calyx stellate-puberulent, lobes lance-ovate, acute or short-acuminate, 4–4.5 mm. long; petals erect, 18–25 mm. long, purplish, stellate-puberulent without; fruit 8–11 mm. broad; carpels densely stellate-puberulent on backs, hispid-ciliate along margins; seeds about 2 mm. long, glabrous.

Moist canyons, chiefly Upper Sonoran Zone, entering margins of Sonoran Desert along canyons leading into its eastern margin from central to southern Sonora. Late summer and autumn.

**4. Kosteletzkya hastata** Presl, Rel. Haenk. **2:** 130, 1835

*Kosteletzkya hispida* Presl, Rel. Haenk. **2:** 132, 1835.
*Kosteletzkya cordata* Presl, loc. cit.
*Hibiscus tampicensis* Moric., Mém. Soc. Genève **7:** 260, *pl. 14*, 1836.
*Kosteletzkya tampicensis* Rose, Contr. U.S. Nat. Herb. **8:** 319, 1905.
*Kosteletzkya violacea* Rose, loc. cit.

Perennial herbaceous plant 1–2.5 m. tall with slender stems sparsely hispid with coarse simple hairs and dense lines of smaller whitish hairs from leaf base to leaf base; stipules filiform-setaceous, 3–6 mm. long; petioles 1–3 cm. long, hispid; leaf blades lance-oblong to cordate, 3–7 cm. long, lower usually shallowly hastate-lobed at base, sharply serrate-dentate, sparsely pubescent on both surfaces with few-rayed, coarse, stellate hairs; peduncles and pedicels together 5–25 mm. long; bractlets 8–12, acicular, 2–3 mm. long; calyx sparsely stellate-pubescent, lobes about 3 mm. long; petals more or less spreading, 10–12 mm. long, yellowish, sparsely stellate-pubescent without; fruit 6–8 mm. broad, carpels transversely lightly ridged and minutely puberulent on back, angles coarsely ciliate with hooked hairs; seeds minutely puberulent.

Margins of fields, along streams, and in canyons, Sonoran Zones to Tropical Zone, southern Sonora to Tamaulipas and southward to Nicaragua. Sept.–Nov.

**5. Kosteletzkya coulteri** A. Gray, Pl. Wright. **1:** 23, 1852

A straggly subshrub 1–2 m. tall with sparsely hispid stems, petioles, and peduncles, stems with an additional narrow line of closely crowded, finer hairs from leaf base to leaf base; stipules lance-linear, 3–8 mm. long; petioles 3–8 cm. long or upper ones shorter; leaf blades deeply 3–5-lobed or -parted, lobes linear to narrowly lanceolate, jaggedly serrate, 5–12 mm. broad, 1.5–6 cm. long, sparsely and harshly stellate-pubescent with yellowish hairs; peduncles and pedicels stout, short, 3–10 mm. long; bractlets 5–9, 1–1.5 mm. long; calyx stellate-puberulent with a few coarser stellate hairs surpassing finer ones, lobes lance-ovate, 3–4 mm. long; petals white or yellowish, sometimes tinged with pale rose, 12–15 mm. long; fruits 7–8 mm. broad, about 4 mm. high, faintly ridged transversely, puberulent between angles, hispid-ciliate along angles with slightly hooked, coarse hairs; seeds glabrous.

Along watercourses and in moist canyons, Sonoran Zones, mainly above the Desert but extending from the vicinity of Hermosillo, Sonora, into Sinaloa. Flowering almost throughout the year.

**6. Kosteletzkya digitata** A. Gray, Proc. Amer. Acad. **8:** 289, 1870

Herbaceous perennial of uncertain size but probably 1–2 m. tall, minutely stellate-puberulent with mostly appressed hairs throughout but with a few spreading ones on some stems and petioles; petioles short, mostly less than 2 cm. long; leaf blades 3–5-parted, lobes linear-lanceolate, 3–6 mm. broad, 3–8 cm. long, sharply serrate, mostly 1-nerved from base; peduncles and pedicels 5–15 mm. long, ascendingly hispidulous; bractlets 7–9, acicular, 1–2 mm. long; calyx sparsely stellate-puberulent, lobes lance-ovate, 2.5–3 mm. long; petals spreading, 8–10 mm. long, pale pink with yellow bases; capsules 7–8 mm. broad, minutely puberulent and faintly lined transversely, angles hispid-ciliate with coarse, yellowish, hooked hairs; seeds dark brown, 2.5 mm. long, glabrous.

In moist soil, Sonoran Zones to Tropical Zone, southern Sonora to Sinaloa. Aug.–Nov. and probably throughout most of the year.

## 14. HIBISCUS L., Sp. Pl. 693, 1753

Herbs, shrubs, or trees with entire, lobed, or parted, alternate leaves, usually more or less pubescent, rarely glabrous. Bracts usually numerous, rarely 3–5, entire, distinct or slightly coalesced at base. Calyx 5-toothed or -parted. Petals 5, variously colored, often spotted near base. Stamineal column bearing many stamens. Ovary 5-celled; locule 3- to many-seeded. Style branches 5, spreading to suberect; stigmas globose-capitate. Seeds reniform to subglobose, glabrous to tomentose.

Bractlets 1–3 mm. long; leaves all ovate to obovate, shallowly serrate, densely stellate-tomentose; corollas light lavender .................................... 1. *H. denudatus*
Bractlets 1–3.5 cm. long; upper leaves 3–5-lobed to -parted, stellate-hirsute, not tomentose; corollas white to yellow:
  Stems uniformly pubescent with appressed, 4–6-rayed hairs, not hirsute; lobes of upper leaves narrow, mostly oblanceolate .............................. 2. *H. coulteri*
  Stems hirsute with simple and 2–3-rayed, erect hairs and with 2 lines of fine stellate pubescence; lobes of upper leaves usually broader, ovate .................. 3. *H. biseptus*

### 1. Hibiscus denudatus Benth., Bot. Sulph. 7, *pl. 3*, 1844

*Hibiscus denudatus* β *involucellatus* A. Gray, Pl. Wright. **1:** 22, 1852.

Suffrutescent, 3–10 dm. tall, with erect or ascending, densely stellate-tomentose, tawny branches; petioles stout, 4–8 mm. long; leaf blades broadly ovate to oblong-ovate, truncate to subcordate at base, obtuse or acute at apex, 0.8–2.5 cm. broad, 1.5–3.5 cm. long, serrate-dentate, finely and densely stellate-tomentose, yellowish; flowers solitary; peduncles stout, 3–10 mm. long; bractlets 6–10, 1–3 mm. long, narrowly triangular; calyx deeply 5-parted, lobes lanceolate, 5–10 mm. long, stellate-pubescent; petals 1.8–2.5 cm. long, pale lavender; capsule ovoid, 8–10 mm. long, glabrous or nearly so; seeds reniform, about 3 mm. long, bearing moderate covering of fulvous to grayish hairs 2–4 mm. long.

Desert washes, hillsides, and mesas, vicinity of Palm Canyon, Riverside County, California, through the Colorado Desert to western Texas, Chihuahua, Coahuila, Durango, and southern Baja California. Jan.–Oct.

### 2. Hibiscus coulteri Harv. ex A. Gray, Pl. Wright. 1: 23, 1852

*Hibiscus coulteri* var. *brevipedunculata* M. E. Jones, Contr. West. Bot. **12:** 4, 1908.

Straggly shrub 0.3–10 dm. high with erect or ascending, dark gray, stellate-pubescent branches and coarsely stellate-hirsute herbage; leaves dimorphic, lower ovate and dentate or shallowly 3-lobed, 1–3.5 cm. long, upper usually deeply 3-parted, lobes obovate to oblanceolate, dentate, 1–3.5 cm. long; peduncles 1–4 cm. long; bractlets 10–15, linear-lanceolate, 1–2 mm. wide, 1–2.5 cm. long, sparsely hirsute; calyx 1.5–2 cm. long, lobes lanceolate, attenuate, sparingly hirsute on margins and veins; petals white to pale yellow, usually with maroon or purplish spot at base of each, 2.5–3.5 cm. long; capsules broadly ovoid, about 1 cm. high, apiculate, glabrous or nearly so; seeds reniform, 3–3.5 mm. long, brownish, lanate with spreading hairs 3–5 mm. long.

Desert slopes and hillsides, southwestern Arizona (Tucson) to western Texas, Chihuahua, Hidalgo, and Sonora. Flowering throughout the year.

### 3. Hibiscus biseptus S. Wats., Proc. Amer. Acad. 21: 418, 1886

Suffrutescent or stems herbaceous, 1 m. tall or less, with scattered, spreading, simple hairs and 2 lines of fine stellate pubescence running from one node to next above; herbage sparsely stellate-hirsute; stipules lance-linear, 5–6 mm. long; petioles 1.5–2.5 cm. long, stellate-hirsute; upper leaf blades deeply 5-lobed, lobes oblong-lanceolate, serrate,dentate, 3–10 mm. wide, 1–3.5 cm. long; lower leaf blades merely dentate or shallowly lobed, 3–12 cm. long; peduncles 2–10 cm. long; pedicels 0.5–1 cm. long, rusty-hirsute; involucels of 10–15 linear, sparsely hirsute bractlets 1.5–3.5 cm. long; sepals narrowly lanceolate, attenuate, 2–3.5 cm. long; petals 3–4 cm. long, cream-yellow with purple spot at base; capsule 12–15 mm. high, glabrous; seeds reniform, finely hirsute with hairs 2–4 mm. long.

Arroyos and gravelly slopes, mostly Lower Sonoran Zone, southern Arizona to Chihuahua, Sinaloa, and Jalisco, and in Baja California south of Loreto. April–Oct.

**Hibiscus ribifolius** A. Gray, Proc. Amer. Acad. **5**: 154, 1861, with nearly glabrous or sparsely hirsute, ovate, 3-lobed leaves and stiff subulate stipules 7–9 mm. long, occurs in the Cape Region of Baja California and may be expected on the margins of the desert somewhat to the northward. It has been collected near Agua Caliente, Todos Santos, Cabo San Lucas, and San José del Cabo.

## 15. GOSSYPIUM L., Sp. Pl. 693, 1753

Herbs, shrubs, or small trees with ample, palmately 3–9-lobed or rarely entire leaves. Involucel of 3 entire or variously lobed or laciniate, united or separate, subcordate bracts. Calyx cuplike, truncate or shallowly 5-lobed. Petals 5, white, yellow, or rosaceous, often puberulent on outer surface and with a dark spot near base. Stamineal column short, antheriferous on apical two-thirds. Style 3–5-parted, clavate apically. Capsule ovoid to globose, loculicidally 3–5-celled. Seeds numerous, subglobose to angulate, finely puberulent to densely lanate.

Bracts laciniate, cordate, 1.5–2.5 cm. wide; leaf blades densely stellate-puberulent; capsules 1.6–2.5 cm. long .................................................. 1. *G. davidsonii*

Bracts entire or nearly so, truncate or rounded at base, 2–9 mm. wide; leaf blades glabrous or nearly so; capsules 1–2 cm. long:

  Leaves entire or some shallowly 3-lobed; capsules devoid of long hairs along inner margins of sutures:

    Bracts persistent until anthesis or later, 10–25 mm. long; twigs and petioles stellate-puberulent; some leaves shallowly 3-lobed .................... 2. *G. harknessii*

    Bracts fugacious before anthesis, 2–5 mm. long; twigs and petioles glabrous; leaves entire ................................................ 3. *G. armourianum*

  Leaves deeply 3–5-lobed; capsule bearing long white hairs along inner margins of sutures ............................................................ 4. *G. thurberi*

### 1. Gossypium davidsonii Kellogg, Proc. Calif. Acad. 5: 82, 1873

Rounded shrub 1–2.5 m. high with cinnamon-brown bark and minutely but closely stellate-puberulent twigs; petioles 1.5–3 cm. long, puberulent; leaves broadly ovate-cordate, 2–5 cm. wide, 2.5–7 cm. long, acute to acuminate, minutely but closely stellate-puberulent, 5–7-nerved; bracts broadly ovate- to orbicular-cordate, 1.5–2.5 cm. wide, 2.5–3.5 cm. long, laciniate, stellate-puberulent, distinct; calyx 5–6 mm. long, about as wide, truncate or undulate at apex, finely and densely puberulent, with numerous small, purple-black glands in irregular vertical lines; corolla sulphur-yellow, 3.5–5 cm. wide; petals obovate, 3–3.5 cm. long, puberulent on external portions exposed in bud, each with a deep maroon spot at base; capsules broadly ovoid to nearly spherical, acute to short-acuminate at apex, 1.6–2.5 cm. high, heavily glandular-punctate, 3–5-celled, inner margins of sutures hirsute-woolly; seeds obovate, 5–8 mm. long, light brown, finely short-puberulent.

Canyons and hillsides, vicinity of Comondú and San Gregorio to the Cape Region, Baja California; vicinity of Guaymas, Sonora. Sept.–May.

**Gossypium barbadense** L., Sp. Pl. 693, 1753, differs from *G. davidsonii* in having more deeply 3-lobed, nearly glabrous, longer leaves; nearly glabrous bracts; and seeds bearing long woolly hairs but devoid of more closely attached short hairs. It is cultivated in Mexico and has escaped to a limited extent near Comondú, Purísima, Loreto, on the Magdalena Plain, and in the Cape Region of Baja California and at many places in Sinaloa, Jalisco, and Vera Cruz.

**Gossypium herbaceum** L., Sp. Pl. 693, 1753, an herbaceous plant with 5-lobed leaves, larger capsules, and with the long hairs firmly attached to the seeds, has also escaped from cultivation in the Cape Region and in southern Sonora.

### 2. **Gossypium harknessii** Brandegee, Proc. Calif. Acad. II, **2:** 136, 1889

*Ingenhouzia harknessii* Rose, U.S. Dept. Agr. Bur. Pl. Ind. Bull. No. **131:** 54, 1908.

Loosely but intricately branched shrub 0.5–1 m. high with reddish brown to dark gray bark; twigs minutely puberulent; petioles 1–1.5 cm. long, finely stellate-puberulent and sparsely punctate- or punctulate-glandular; leaf blades broadly cordate, entire or faintly to shallowly 3-lobed, 1.5–4 cm. wide, acute to short-acuminate, glabrous or with some hairs along main veins beneath, sparsely punctate-glandular; bracts ovate-lanceolate, 7–9 mm. wide, 10–25 mm. long, purplish, punctate-glandular, deciduous soon after anthesis; calyces 5–7 mm. high, truncate to denticulate, sparsely and minutely stellate-puberulent, gland-dotted; corolla 4–5 cm. broad; petals yellow with dark red or maroon spot at base, 2.5–3 cm. long, stellate-puberulent and gland-dotted externally; capsule broadly ovoid to subglobose, 1–1.5 cm. long, punctate-glandular, apiculate, inner margins of sutures lanose; seeds obovate, 6–7 mm. long, golden brown, densely puberulent with tawny silky hairs.

Mostly near the sea in sandy or rocky soil, Lower Sonoran Zone, Loreto southward to the Cape Region, on several islands off the gulf of California, and on Santa Margarita Island, Baja California. Oct.–May.

### 3. **Gossypium armourianum** Kearney, Jour. Wash. Acad. **23:** 558, 1933

A rounded shrub 0.5–1.2 m. high and usually somewhat broader with leathery shining leaves and reddish, somewhat glaucous, glabrous twigs; petioles glabrous, equaling or slightly exceeding thick, glabrous, broadly subcordate leaf blades, these 1.8–3.5 cm. long, obtuse to abruptly acute or mucronate at apex; peduncles 2–4 cm. long; bracts caducous, 2–5 mm. long, distinct; calyces 4–5 mm. high, glabrous, glandular-punctate, shallowly 5-toothed, teeth 1–2 mm. long, triangular or subulate; petals 2.5–4.5 cm. long, sulphur-yellow with a deep red or maroon spot near base or rarely unspotted; capsule 10–15 mm. long, ovoid, abruptly acuminate, glandular-punctate, 3–4-celled, inner margins of sutures glabrous; seeds 7–9 mm. long, obovate, densely covered with yellowish, crisped, appressed hairs.

Known only from San Marcos Island, Baja California. March–Sept.

### 4. **Gossypium thurberi** Tod., Prod. Gossyp. 7, 1878

*Thurberia thespesioides* A. Gray, Pl. Thurb. 308, 1854.

Erect or ascending, openly branched shrub 1–3.5 m. high; petioles slender, 2–8 cm. long, gland-dotted, glabrous; leaf blades deeply 3–5-lobed, lobes 1–3 cm. wide. 3.5–10 cm. long, lanceolate, entire, long-acuminate, gland-dotted, bright green above, usually slightly glaucous beneath, glabrous or sparingly puberulent along main veins beneath; flowers solitary or subcorymbose; peduncles 1–3.5 cm. long; bracts 3, ovate-lanceolate, entire or 3-toothed, 6–15 mm. long, minutely puberulent along margins; calyx broadly cup-shaped, 4–5 mm. high, truncate to sinuately 5-toothed, gland-dotted; petals 2–3 cm. long, white or cream with faint purple tinge at base, turning purplish in age, profusely purple- or black-dotted; capsule broadly ovoid to subglobose, 1–2 cm. long, abruptly apiculate, closely gland-dotted, inner margins of sutures bearing tufts of long white hairs; seeds 4–5 mm. long, dark brown to nearly black, turbinate, angled, finely pubescent with appressed, short, golden-brown hairs and bearing a few longer white hairs.

Along arroyos, canyons, and rocky hillsides, Lower and Upper Sonoran Zones, Superstition and Rincon Mountains, Arizona, to Chihuahua and Jalisco. Aug.–Nov.

## Family 83. **BOMBACACEAE.** Cotton-Tree Family

Trees with simple to digitately compound, alternate, deciduous leaves. Pubescence chiefly of stellate hairs. Flowers perfect, small or large. Calyx 5-lobed, truncate, or opening irregularly. Petals 5, obovate to linear-oblanceolate. Stamens 5 to many; filaments distinct or united into a tube. Ovary superior, 2–5-celled. Fruit dry or fleshy, dehiscent or indehiscent. Seeds 2 to many in each cell, obovoid, subglobose or subspirally twisted.

Stamen tube very short, giving rise to numerous distinct slender filaments; trunk unarmed; leaflets entire .................................................... 1. *Bombax*

Stamen tube longer, dividing into 5 parts, each with several sessile anthers at its summit; trunk and branches often armed with coarse conical spines; leaflets often serrate ..... .................................................................... 2. *Ceiba*

### 1. **BOMBAX** L., Sp. Pl. 511, 1753

Large unarmed trees with spreading crowns and sturdy straight trunks; bark smooth or flaky. Leaves alternate, digitately compound with 3–7 (usually 5) leaflets. Flowers solitary or in fascicles. Calyx truncate. Petals 5, oblong-linear to obovate, often more or less pubescent. Stamen tube very short, giving rise to many simple filaments. Ovary 5-celled, each cell many-seeded. Fruit a 5-celled, 5-valved, woody or leathery capsule, densely silky-lanate within. Seeds 4–6 mm. long, obovoid, embedded in "cotton."

**1. Bombax palmeri** S. Wats., Proc. Amer. Acad. **22:** 399, 1887

A tree 8–10 m. tall or possibly larger; trunk smooth and unarmed, to 7 dm. in diameter; bark green or slightly glaucous-green; leaflets 3–5, subsessile, obovate to suborbicular, 10–25 cm. long, rounded or acute and usually apiculate at apex, tomentose on both surfaces in youth, glabrate above in age, prominently veined beneath, reticulate network conspicuous, margins entire; calyx 1–1.5 cm. long, densely silky-lanate without in bud but soon glabrate; petals 10–17 cm. long, 1–1.5 cm. wide, finely but densely whitish-lanate or ferruginous on both surfaces, pubescence denser without than within; stamens white, pink, or purplish, 10–15 cm. long; fruit ellipsoid-oblong, 10–12 cm. long, silky fibers brownish.

Along streams and on hillsides, mainly subtropical areas, southern Sonora into Jalisco. March.

Not a normal constituent of the Sonoran Desert, barely entering it at the southern moister border.

## 2. CEIBA Medic., Malvenfam. 15, 1787

Small to very large trees with trunk and branches often armed with heavy conical spines, often extensively buttressed with thin woody plates at base of trunk. Leaves alternate, digitately compound, with 3–7 serrate or sometimes entire leaflets. Flowers small to large, solitary. Calyx truncate or 5-lobed. Petals 5, oblong to linear, velvety-tomentulose on outside, deciduous. Stamineal tube short, dividing into 5 branches, each of which bears a few crowded anthers at apex. Ovary superior, 5-celled, densely silky-hairy with long hairs within. Capsule woody, splitting into 5 valves. Seeds obovoid to subglobose, small, glabrous or minutely puberulent.

Leaflets glabrous, even in youth; flowers 3–3.5 cm. long, white or pink .. 1. *C. pentandra*
Leaflets finely but distinctly puberulent, at least in youth; flowers 10–15 cm. long, yellowish or ferruginous .............................................. 2. *C. acuminata*

### 1. Ceiba pentandra (L.) Gaertn., Fruct. **2**: 244, *pl. 133,* 1791

*Bombax pentandrum* L., Sp. Pl. 511, 1753.
*Ceiba caesaria* Medic., Malvenfam. 16, 1787.
*Eriodendron anfractuosum* DC., Prod. **1**: 479, 1824.
*Eriodendron occidentale* G. Don, Gen. Hist. Pl. **1**: 513, 1831.

Large trees to 40 m. tall with massive trunks often supported by extensive basal buttresses; bark greenish or gray, smooth but more or less covered with large conical spines 1–3 cm. wide at base, nearly or fully as high, these often somewhat laminated; leaflets 5–7, oblanceolate to oblong-obovate, rather abruptly acuminate, 8–20 cm. long, glabrous, on petioles about equaling leaflets; calyx 1–1.5 cm. long, lobes broad, very low, 1–2.5 mm. high; petals 1.5–2 cm. long, obovate, broadly and shortly clawed, white or pinkish and glabrous inside, densely woolly outside with white or slightly yellowish, silky hairs; fruit ellipsoid-oblong, 10–12 cm. long; seeds brown, 3–4 mm. long, embedded in silky hairs.

Widely distributed in tropical and subtropical America, Asia, and Africa, cultivated as a shade tree in Sonora at least as far north as Hermosillo. Feb.–March.

### 2. Ceiba acuminata (S. Wats.) Rose, Contr. U.S. Nat. Herb. **8**: 320, 1905

*Eriodendron acuminatum* S. Wats., Proc. Amer. Acad. **21**: 418, 1886.
*Eriodendron tomentosum* Robinson, Proc. Amer. Acad. **29**: 314, 1894.
*Ceiba tomentosa* Britt. & Baker, Jour. Bot. Brit. & For. **34**: 175, 1896.

Medium-sized to large tree to 20 m. or perhaps more; trunk only slightly if at all buttressed at base, more or less covered with conical spines to 2 cm. long; bark smooth, greenish; leaflets 5–8, usually 7, oblanceolate to lance-elliptic, cuneate at base, rounded but abruptly short-acuminate at apex, densely and finely stellate-velutinous below, sparsely so along veins on upper surface, 1–6 cm. wide, 3–15 cm. long, entire to serrate-dentate; calyx 1–1.5 cm. long, lobes low and broad; petals 10–15 cm. long, densely silky with ferruginous hairs on outside, nearly glabrous within; fruit 12–18 cm. long, the silky hairs white and glistening to light brown; seeds dark brown, sublustrous, obovoid, 2.5–5 mm. long.

On wooded or bushy hillsides and in the adjacent canyons, Lower Sonoran Zone to Subtropical Zone, central Sonora and southern Baja California to Chihuahua, Tamaulipas, Oaxaca, and Chiapas. March.

## Family 84. **STERCULIACEAE.** Sterculia Family

Herbs, shrubs, or trees usually more or less stellate-pubescent and frequently with simple hairs intermingled with stellate ones. Leaves alternate, mostly simple, usually serrate or dentate. Stipules usually present. Flowers perfect or unisexual, chiefly axillary. Calyx 3–5-lobed or -parted. Petals 5, hypogynous, or rarely wanting. Stamens 5, alternating with or opposite petals, usually united into a column, at least below. Ovary 3–5-celled (or rarely to 12-celled), superior; ovules 1 to many in each cell. Fruit a capsule or in some a nut or berry. Seeds with a bony or membranous coat; endosperm present or lacking.

Flowers very showy, 3–7 cm. in diameter; petals none; sepals brightly colored; leaves usually more or less 3-lobed ........................................ 1. *Fremontodendron*
Flowers smaller, not over 2.5 cm. in diameter, often much less; petals present; leaves usually not lobed:
  Petals plane or slightly curved, not cucullate nor bearing apical or dorsal appendages; anthers 5, 2-celled:
    Capsule (in ours) bristly along angles, each bristle bearing 2–5 curved, barblike hairs apically; seeds numerous in each cell .......................... 2. *Hermannia*
    Capsule not armed on margins; seeds 1–2 in each cell:
      Capsule 5-celled, angulate ........................................ 3. *Melochia*
      Capsule 1-celled, 1-seeded, not angulate, scarious below .............. 4. *Waltheria*
  Petals cucullate, often bearing apical or dorsal appendages; anthers 15, or if 5 then each usually 3-celled:
    Capsule chartaceous, subglobose, 1 cm. in diameter or less; petals adnate to stamineal column at their apices; anthers 5, each usually 3-celled; ovary more or less stipitate ........................................................ 5. *Ayenia*
    Capsule woody, oblong, 2–4 cm. long, half as wide; petals not adnate to stamineal column; anthers 15, in 5 groups of 3, each 2-celled; ovary sessile ..... 6. *Guazuma*

### 1. **FREMONTODENDRON** Coville, Contr. U.S. Nat. Herb. **4:** 74, 1893

Evergreen shrubs or small trees with thick leaf blades more or less 3–5-lobed and densely stellate-tomentulose to tomentose on young twigs and foliage. Flowers perfect, solitary, showy. Sepals bright yellow or orange-yellow, tomentulose without, each with a large gland near its base. Petals none. Stamineal column short, conical, dividing into 5 stamens with short filaments. Ovary 4–5-celled. Fruit a capsule, loculicidally dehiscent, each cell several-seeded. Seeds ovoid, bony. Endosperm oily.

Glands at base of sepals hairy; flowering period short; sepals lemon-yellow, not tinged with orange ...................................................... 1. *F. californica*
Glands at base of sepals glabrous; flowering period several weeks long; sepals deep yellow tinged with orange ............................................. 2. *F. mexicana*

**1. Fremontodendron californicum** (Torr.) Coville, Contr. U.S. Nat. Herb. **4:** 74, 1893

*Fremontia californica* Torr., Pl. Frém. 6, *pl.* 2, 1854.
*Chiranthodendron californicum* Baillon, Hist. Pl. **4:** 70, 1872.

Large shrub or small tree to 8 m. tall with roughish, dark gray bark and densely stellate-pubescent twigs; leaf blades ovate or broadly so, shallowly lobed or less commonly subentire, subcordate at base, usually acute but sometimes rounded at apex, margins often slightly undulate-crisped, larger ones 1–3 cm. long, green, with scattered stellate hairs above, densely stellate-pubescent and canescent beneath; petioles usually one-half to as long as leaf blades; bractlets lance-subulate, 4–7 mm. long; flowers 3–5 cm. broad, yellow; sepals broadly ovate, mucronate, densely stellate-pubescent without, glabrous except for bristly or densely pubescent basal area around gland within; capsule ovoid, 2–2.5 cm. long, densely hispid-bristly; seeds short-pubescent.

Mainly well-drained, rocky hillsides, Upper Sonoran Zone to Transition Zone, but entering edge of the Sonoran Desert in various localities in Arizona and along the western margin of the Colorado Desert in Imperial County, California, and northern Baja California. The general distribution extends from Tehama County, California, south through the Sierra Nevada and from San Luis Obispo County south to the Sierra San Pedro Mártir by discontinuous local colonies. April–June.

**2. Fremontodendron mexicanum** Davidson, Bull. S. Calif. Acad. **16:** 50, 1917

*Fremontia mexicana* J. F. Macbride, Contr. Gray Herb. **53:** 14, 1918.
*Fremontia californica* var. *mexicana* Jepson, Man. Fl. Pl. Calif. 637, 1925.

Shrub or small tree to 7 m. tall; branches densely stellate-tomentose; hairs long-rayed and yellowish or ferruginous but often darkened in age; leaf blades ovate to orbicular in outline, usually 5-lobed, lobes rounded, deeply cordate at base, 2.5–3.5 cm. long, thick, dark green above, with scattered stellate hairs, paler and densely stellate-tomentose beneath, all sinuses narrow; petioles equaling or shorter than blades, stoutish, densely stellate-tomentose; calyx 6–7 cm. in diameter, orange or deep yellow, tinged with deep orange or reddish toward base, stellate-pubescent without, glabrous within, pit-like glands glabrous; capsule ovoid-acuminate, 3–4 cm. long, densely hispid-bristly; seeds black, shining.

Dry chaparral-covered ridges, chiefly Upper Sonoran Zone, southern San Diego County, California, south at least to San Antonio del Mar Canyon, Baja California. March-Aug.

## 2. HERMANNIA L., Sp. Pl. 673, 1753

Suffruticose herbs or subshrubs with foliaceous or rarely small stipules (or these lacking) and stellate pubescence. Leaves dentate or incised, petiolate. Flowers yellow or reddish, axillary or in racemes or thyrses. Calyx lobes 5. Petals 5, obovate to oblong. Stamens 5, connate at base. Ovary sessile or short-stipitate, superior, 5-celled; cells several- to many-ovuled. Capsule loculicidally 5-valved, apex naked or 5-crowned, often bristly. Seeds reniform, numerous in each cell. Embryo curved; endosperm present.

Marginal spines of fruits 2.5–4 mm. long; anthers subsessile, 4–5.5 mm. long; calyx lobes 4–5 mm. long . . . . . . . . . . . . . . . . . . . . . . . . . . . . . . . . . . . . 1. *H. palmeri*
Marginal spines of fruits less than 2 mm. long; anthers about 2 mm. long, on filaments 4–5 mm. long; calyx lobes 2–2.5 mm. long . . . . . . . . . . . . . . . . . . . . 2. *H. pauciflora*

**1. Hermannia palmeri** Rose, in Vasey & Rose, Contr. U.S. Nat. Herb. **1:** 67, 1890

Low suffrutescent plants from woody roots with slender stems, 1–4 dm. long, spreading or ascending over other shrubs; stems brownish, moderately pubescent with slender-rayed, stellate hairs; stipules lanceolate or lance-ovate, 2.5–3 mm. long; petioles slender, 5–9 mm. long, pubescent; leaf blades cordate-ovate to rhombic-ovate or suborbicular-cordate, 10–18 mm. wide, 12–25 mm. long, densely and roughly stellate-pubescent on both surfaces, dark green above, paler beneath, regularly crenate-dentate, 3–5-veined, veins prominent beneath; flowers solitary or in pairs on an axillary peduncle 2–2.5 cm. long; bractlets linear or subfiliform, 1–1.5 mm. long; calyx densely stellate-pubescent, tube campanulate, 1–1.5 mm. long, lobes lanceolate-attenuate, 4–5 mm. long, hairy within; petals 5–8 mm. long; anthers 4–5.5 mm. long, subsessile, connivent; capsule obovoid, 10–14 mm. long, 8–10 mm. in diameter, on a stipe 1.5–2.5 mm. long, stellate-puberulent, angles bearing spines 2.5–4 mm. long, each one with a ring of 3–4 reflexed, stoutish, curved hairs.

Rocky hillsides and along arroyos, Lower Sonoran Zone, Baja California from the vicinity of La Paz southward; also between Calmallí and Barríl. Jan.–April.

**2. Hermannia pauciflora** S. Wats., Proc. Amer. Acad. **17:** 368, 1882

Small shrub to 3 dm. tall from a woody root; stems reddish brown, stellate-puberulent; stipules lance-linear, 3–4 mm. long; petioles slender, 4–10 (or sometimes –15) mm. long; leaf blades broadly ovate, 5–12 mm. wide, 8–16 mm. long, prominently veined and stellate-canescent beneath, sparsely stellate-puberulent and green above, acute at apex, rounded truncate or subcordate at base, margins finely dentate; flowers solitary in axils, pedicels slender, 2–3 mm. long; peduncles 5–8 mm. long; bractlets ovate, acute, 1–2 mm. long; calyx narrowly campanulate, 3.5–4 mm. long, lobes lanceolate-attenuate, 2–2.5 mm. long, puberulent; petals yellow, 6–8 mm. long; anthers linear, 2 mm. long; filaments 4–5 mm. long; fruit obovoid, 12–15 mm. long, about 10 mm. wide, wings thin, zigzag-toothed, each spine 1–2 mm. long and bearing 3–5 stellately arranged, horizontal or slightly recurved, stout hairs.

Dry rocky hillsides, Sonoran Zones, southern Arizona and south into Sonora. Feb.–April.

## 3. MELOCHIA L., Sp. Pl. 674, 1753

Herbs, shrubs, or rarely trees with narrowly to broadly ovate, serrate leaves and more or less stellate-pubescent twigs and foliage. Flowers small, in axillary glomerules, in spikelike thyrses, or in lax cymose panicles. Calyx campanulate, 5-lobed, sometimes inflated. Petals 5, spatulate to oblong, persistent. Stamens 5, opposite petals, connate at base or to middle. Ovary sessile or short-stipitate, 5-celled, each cell 2-ovuled; styles 5. Capsule loculicidally 5-valved, each cell 1-seeded. Seeds obovoid. Embryo straight. Endosperm scanty.

Flowers mostly 1–3 in an axil; leaves thin, glabrous or sparsely pubescent with rays of stellate hairs rarely interlacing, lateral edges of capsule narrowly triangular-attenuate, surface sparsely stellate-puberulent or glabrate ..................... 1. *M. pyramidata*

Flowers in 1–12-flowered axillary glomerules; leaves thickish, stellate-puberulent to tomentulose with rays of stellate hairs nearly always interlaced; lateral edges of carpels broadly deltoid to rounded, not attenuated into sharp spurs, surface closely stellate-puberulent:

Lateral margins of carpels deltoid; leaves densely tomentulose and canescent, at least beneath; veins conspicuously impressed above; petals 8–10 mm. long .. 2. *M. tomentosa*

Lateral margins of carpels rounded or angled; leaves closely stellate-puberulent but only slightly grayish beneath; veins only slightly impressed; petals 12–18 mm. long ...... ............................................................ 3. *M. speciosa*

**1. Melochia pyramidata** L., Sp. Pl. 674, 1753

Openly branched, slender shrub to 2 m. tall, often with several stems from woody rootstock, sometimes herbaceous; entire plant glabrous or very sparsely pubescent; stipules setaceous, 2–3 mm. long; petioles slender, 5–20 mm. long; leaf blades elliptic-ovate to broadly ovate, rounded, obtuse or subtruncate at base, acute at apex, serrate,

0.5–3 cm. wide, 1–6 cm. long, green, thin, glabrous or sparsely pubescent; flowers solitary or in few-flowered axillary corymbs; pedicels 3–5 mm. long, usually slightly longer than peduncle; calyx narrowly campanulate, tube and lance-acuminate lobes each about 2 mm. long; petals violet or rose-purple, 6–7 mm. long; fruit 5–7 mm. long, 6–9 mm. wide, lateral angles drawn out into triangular to attenuate, spreading spurs, abruptly acuminate at apex, sparsely and minutely stellate to glabrate.

Usually in shaded spots in canyons, along streams, and about fields, Lower Sonoran and Tropical Zones, Sonora southward; widely distributed in the tropics and subtropics of both hemispheres. Flowering practically throughout the year.

**2. Melochia tomentosa** L., Syst. Nat. ed. 10, 1140, 1759
*Melochia plicata* Presl, Rel. Haenk. **2:** 145, 1835.

Shrub to 2.5 m. tall with dark gray bark and stellate-tomentulose twigs and foliage, stipules lance-subulate, 3–4 mm. long; petioles stoutish or sometimes slender, 5–20 mm. long; leaf blades broadly ovate-lanceolate, ovate, or suborbicular, 1.5–4 cm. wide, 2.5–6 cm. long (or much reduced on some plants), subcordate to rounded at base, obtuse, rounded, or acute at apex, regularly dentate or crenate-dentate, densely stellate-tomentose on both surfaces, canescent and often velvety beneath, green but with ashy cast and deeply impressed veins above; flowers in dense glomerules, often 10–15 borne on one peduncle, pedicels to 6–8 mm. long; calyx densely stellate-tomentulose, 5–6 mm. high, lobes lance-triangular, equaling tube; petals pink to rose-purple, 8–10 mm. long; capsule densely stellate-puberulent, pyramidal, body 5–7 mm. high, 7–10 mm. wide, angles drawn out into a broadly deltoid wing 1–1.5 mm. long, beak 2.5 mm. long; seeds black, about 2 mm. high, glabrous.

Rocky hillsides, mesas, and in fields, Lower Sonoran and Tropical Zones, southern Baja California, Tamaulipas, and Nayarit south throughout Mexico and widely distributed in the tropics and subtropics of the western hemisphere. Flowering practically throughout the year.

**3. Melochia speciosa** S. Wats., Proc. Amer. Acad. **24:** 42, 1889
*Melochia arida* Rose, Contr. U.S. Nat. Herb. **8:** 321, 1905.

A graceful shrub with slender branches 1–1.5 m. tall or more, with reddish brown bark and minutely stellate-puberulent twigs and young leaves; stipules lance-subulate, 3–5 mm. long; petioles 5–15 mm. long; leaf blades ovate, 1–3 cm. broad, 2–5.5 cm. long, rounded or obtuse at base, usually acute at apex, regularly but often coarsely crenate-dentate, green but minutely stellate-puberulent above, veins not impressed, more densely puberulent and slightly canescent beneath; flowers in dense axillary glomerules; calyx moderately stellate-puberulent, 6–7 mm. long, tube about one-half as long as lance-attenuate lobes; petals 12–18 mm. long, pink to deep violet; capsule closely stellate-puberulent, pyramidal, 7–8 mm. wide, about as high, angles rounded or merely angulate near base, not drawn out into lateral spurs or deltoid wings; beak abruptly attenuate, 2–4 mm. long; seeds about 2 mm. long, minutely striate.

Rocky hillsides, thickets, and valley bottoms, Lower Sonoran Zone, vicinity of Guaymas, Sonora, south into Sinaloa. Flowering practically throughout the year.

## 4. WALTHERIA L., Sp. Pl. 673, 1753

Herbs, shrubs, or small trees with stellate pubescence often mixed with simple hairs. Stipules narrow. Leaves petiolate, serrate. Flowers small, in axillary glomerules, on peduncles or sessile. Calyx 5-lobed. Petals 5, spatulate, persistent. Stamens 5, opposite petals; filaments connate at least at base. Ovary sessile, 1-celled, 2-ovuled but 1-seeded. Style clavate or fimbriate at apex. Capsule 1-seeded, dehiscing into 2 valves.

**1. Waltheria americana** L., Sp. Pl. 673, 1753

*Waltheria indica* L., loc. cit.
*Waltheria detonsa* A. Gray, Pl. Wright. **2:** 24, 1853.

An open, gracefully branched shrub to 2.5 m. tall; branches sublanate, erect, ascending or arcuately decumbent; stipules lance-linear, 4–6 mm. long, pubescent; petioles stoutish, to 12 mm. long; leaf blades oblong to broadly ovate or elliptic, obtuse to subcordate at base, crenate-dentate, 1–4 cm. wide, 2.5–8 cm. long but usually somewhat smaller, densely stellate-pubescent and often tomentose, at least on lower side; flowers in dense axillary clusters, sessile or on peduncles 2–30 mm. long; bracts narrowly lance-ovate, 2.5–3 mm. long; calyx spreadingly hirsute with simple and few-rayed, stellate hairs, 4–5 mm. high; lobes ovate-attenuate, slightly longer than tube; petals yellow, 5–6 mm. long; capsule obovoid, 2.5–3 mm. high, apically puberulent, acute, scarious below; seed dark brown, glabrous, dull, about 1.5 mm. long.

Dry hillsides, rocky mesas, along washes, and on arid flats, Lower Sonoran Zone to Tropical Zone, southern Arizona, southern Baja California, Texas, and southern Florida southward into tropical America; widely distributed in warmer regions of both hemispheres. Flowering practically throughout the year.

**Waltheria alamosana** Standley, Contr. U.S. Nat. Herb. **23:** 800, 1923, has been collected twice in the vicinity of Alamos, Sonora, but it has not been detected within the limits of the Sonoran Desert.

## 5. AYENIA Loefl., Iter. Hisp. 199, 1758

Annual or mostly perennial herbs or low shrubs with stellate-pubescent, hirsute, or glabrate stems and foliage. Leaves alternate, serrate or crenate. Flowers small, solitary or on few-flowered axillary peduncles. Sepals 5. Petals 5, slenderly long-clawed; limb more or less cucullate, apex inflexed over anthers and adnate to margin of stamineal column, often bearing a dorsal appendage. Stamineal column short, more or less funnelform, bearing 5 fertile stamens and 5 alternating staminodia. Anthers 3-celled. Ovary stipitate, 5-celled, each cell 2-ovuled. Stigma capitate or flattened and faintly 5-lobed. Capsule subglobose, muricate or short-papillate; carpels 5, separating septicidally but also splitting loculicidally. Seeds usually 1 in each carpel, strongly rugose transversely. Endosperm none.

Petals cordate or deeply notched at the base of blade; stipe not over 0.8–1 mm. long at anthesis; dorsal appendages on petals minute or lacking; staminodia fleshy; sepals spreading or reflexed, 2.8–3.8 mm. long:

- Leaves large, 4–8 cm. long, acuminate, sparsely stellate-puberulent to glabrate; staminodia rhombic, forming broad platform about style; dorsal appendages lacking on petals ........ 1. *A. glabra*
- Leaves small, 0.5–1.5 cm. long, rounded or acute, closely stellate-tomentulose, usually canescent; staminodia reniform, not forming a distinct platform; dorsal appendages on petals deltoid, recurved ........ 2. *A. microphylla*

Petals oblanceolate, cuneate at base of blades; stipe 1.8–2.2 mm. long at anthesis; dorsal appendages on petals claviform, 0.4–0.6 mm. long; staminodia minute, toothlike, not fleshy; sepals ascending or erect, 2.2–3 mm. long:

Fruits muricate with papillae 1–1.4 mm. long or more; leaves lance-ovate or linear, 1–5 cm. long:
Leaf blades usually densely stellate-puberulent, at least beneath, stiffish, not reflexed at tips of petioles; petals brownish purple ........................... 3. *A. pusilla*
Leaf blades sparsely puberulent, mainly along veins, thin, sharply reflexed at tips of petioles; petals cream-yellow .................................. 4. *A. reflexa*
Fruits merely short-tuberculate; papillae mostly about 0.5 mm. long or less; leaves broadly ovate, 0.5–1.5 cm. long ........................................ 5. *A. compacta*

## 1. Ayenia glabra S. Wats., Proc. Amer. Acad. 22: 399, 1887

*Ayenia truncata* Rose, Contr. U.S. Nat. Herb. **1**: 94, 1891.

A graceful shrubby plant 1–4.5 m. tall; branches and leaves sparsely stellate-puberulent when young, at length glabrate or nearly so; petioles 1–6 cm. long; leaf blades broadly ovate to lance-ovate, 2.5–8 cm. wide, 4–10 cm. long, subcordate at base, acute or usually acuminate at apex, crenate-dentate, green on both surfaces but paler beneath; flowers 1–4 on slender peduncles 5–12 mm. long, several peduncles often in a single axil; pedicels subfiliform, 5–9 mm. long; sepals brown or purplish, rarely greenish, narrowly lanceolate, 1–1.2 mm. wide, 3.4–3.8 mm. long, spreading to reflexed, subscarious, sparsely stellate-puberulent without; stipe 0.8–1.2 mm. long, white or pale green; stamineal column funnelform, about 1 mm. high, 2.2–3 mm. wide at summit; claws of petals brownish, about 2 mm. long, blades deeply cordate-reniform, suborbicular in outline, each lobe about 0.6 mm. wide, 1 mm. long, horizontally spreading, dorsal appendages none, apex closely crowded between fleshy rhombic staminodia, which form a conspicuous platform about style; stigma shallowly 5-lobed; capsule 5–6 mm. in diameter, short-muricate and minutely puberulent; seeds about 4 mm. high.

Hillsides, rocky slopes, canyon bottoms, and among shrubs on grassy plains, Lower Sonoran Zone to Tropical Zone, central Sonora south to Guerrero. Aug.–Nov.

**Ayenia paniculata** Rose, loc. cit., is similar in general appearance but with densely pubescent leaves that are whitish beneath, sepals 5 mm. long, and capsules 8–10 mm. in diameter.

This species occurs in the vicinity of Alamos, Sonora, but has not been found within the Sonoran Desert proper.

## 2. Ayenia microphylla A. Gray, Pl. Wright. 1: 24, 1852

A small, dense, intricately branched shrub 2–5 dm. tall with distinctly woody branches; leaves broadly ovate, rounded or obscurely subcordate at base, acute to rounded at apex, 4–7 mm. wide, 6–12 mm. long, minutely but densely stellate-puberulent on both surfaces, veins prominent beneath, margins finely dentate; flowers few, on short axillary peduncles, brownish or purplish; pedicels and calyx densely but minutely stellate-puberulent; sepals spreading or slightly reflexed, ovate, slightly attenuate, 1.4–1.6 mm. wide, 2.8–3.2 mm. long; blades of petals reniform-deltoid, distinctly cordate at base, 1.4–1.6 mm. broad, 1–1.2 mm. long, the dorsal appendage a minute deltoid apiculation about 0.3 mm. long and broad, sharply recurved; stamineal column funnelform, 1.8–2.2 mm. long; staminodia fleshy, reniform, 0.5–0.6 mm. wide; stipe only 0.2–0.6 mm. long at anthesis; capsule 4–5 mm. tall, sparsely short-tuberculate.

Dry mesas and rocky slopes, Lower Sonoran Zone, southern Arizona (near Tucson), western Texas, and southward into Chihuahua, Coahuila, and Sonora. Feb.–April; Aug.–Sept.

Often confused with and distributed as *A. pusilla,* but the character of the dorsal appendages, the fleshy staminodia, the reniform petals, and the short stipes set this apart from *A. pusilla.*

**3. Ayenia pusilla** L., Syst. Nat. ed. 10, 1247, 1759

*Ayenia filiformis* S. Wats., Proc. Amer. Acad. **24:** 42, 1889.
*Ayenia dentata* Brandegee, Univ. Calif. Pub. Bot. **6:** 56, 1914.
*Ayenia cuneata* Brandegee, op. cit. 369, 1917.

Slender annual or perennial plants to 4 dm. tall with slender, stellate-puberulent branches and densely stellate-puberulent foliage; leaves lanceolate to suborbicular, 1–5 cm. long, obtuse to cordate at base, acute, obtuse, or rounded at apex, finely dentate or serrate, thickish and firm, often tinged faintly with purple; flowers 1–5 on axillary peduncles 1–5 mm. long; pedicels subfiliform, 2–5 mm. long; sepals subscarious, brownish, erect or ascending, lance-ovate, about 1 mm. wide, 2.4–3 mm. long, sparsely and minutely stellate-puberulent without; claws of petals filiform, 5–7 mm. long, wavy, blades rhombic-obovate, 0.8–1 mm. wide, 1.6–1.8 mm. long, cuneate at base, slightly notched at apex, dorsal appendage attached about 0.4 mm. distal to apical notch, narrowly clavate, erect or recurved, about 0.6 mm. long; stipe 1.4–1.8 mm. long at anthesis; stamineal column funnelform, 0.8–1 mm. long; staminodia minute apiculations or sometimes apparently lacking; fruits 5–6 mm. in diameter, finely stellate-puberulent and muricate with greenish or purplish processes to 1.4 mm. long.

Widely distributed on rocky or fine-grained soil, from Sonora and Chihuahua southward in the warmer parts of the western hemisphere. Feb.–Nov.

**4. Ayenia reflexa** Brandegee, Univ. Calif. Pub. Bot. **6:** 369, 1917

A slender, ascendingly branched annual or short-lived perennial to 5 dm. tall with sparsely and finely stellate-puberulent to glabrate stems and leaves; stipules lance-linear to subfiliform, 3–4 mm. long; petioles slender, about 1 cm. long, spreading-ascending; leaf blades lanceolate to broadly ovate, mostly sharply reflexed from tips of petioles, rounded to obtuse at base, acute at apex, serrate-dentate, bright green, thin, glabrate or nearly so; flowers on 1–3-flowered, slender, axillary peduncles; flower structures and measurements essentially as in *A. pusilla,* petals cream-yellow instead of purplish brown; fruit essentially as in *A. pusilla.*

Bottom lands, canyons, and hills of volcanic origin, Lower Sonoran Zone, central Sonora (near Hermosillo) southward into Sinaloa. July–Nov.

This may not be specifically distinct from *A. pusilla,* but it has a very different aspect in the field. The color of the flowers is strikingly different from the brownish purple of those of *A. pusilla.*

**5. Ayenia compacta** Rose, Contr. U.S. Nat. Herb. **8:** 321, 1905

A low, slenderly and profusely branched shrub 1–4.5 dm. tall; branches finely tomentulose with gray-green, curved, simple, forked, and few-rayed hairs; stipules 1 mm. long or less; petioles slender, 2–7 mm. long; leaf blades ovate to elliptic-ovate, 5–15 mm. long, serrate, minutely stellate-puberulent and often slightly canescent on both surfaces, slightly more densely so beneath; flowers mainly solitary in axils; pedicels subfiliform, 2–6 mm. long; flower structure and size of parts essentially as in *A. pusilla;* capsule subglobose, about 4 mm. in diameter, minutely stellate-puberulent and sparsely papillate with dark and purplish or blackish, blunt glands; seeds nearly black, dull, 2–2.5 mm. long, coarsely rugose transversely.

Rocky hillsides and canyons, Lower Sonoran Zone, western margin of the Colorado Desert, California, into western Arizona (Kofa Mountains; Harquahala Mountains) and southward into northwestern Sonora and in Baja California almost, if not entirely, to the Cape Region. Sept.–April.

**Ayenia peninsularis** Brandegee, Univ. Calif. Pub. Bot. **6:** 368, 1917, is described as having only 2 anther cells and pubescence made up of simple hairs only. It may be not more than a variant of the above, but if these characters hold there may be others also that will set it aside as a distinct species. It was collected by Brandegee in Canyon San Bernardo and at San José del Cabo, just beyond the limits of the Sonoran Desert.

### 6. GUAZUMA Adans., Fam. Pl. 2: 382, 1763

A moderate-sized tree with a sturdy trunk, checked and fissured bark, a dense, ovoid or spreading crown, and stellate-pubescent to glabrate foliage. Leaves oblique at base, unequally dentate. Flowers small, on short axillary peduncles in few- to many-flowered cymes. Calyx 2–3-parted. Petals 5, cucullate-concave and inflexed at apex, short-clawed. Stamineal tube urceolate; stamens 5, partially enclosed by petals. Ovary sessile, 5-celled, 5-lobed, many-ovuled. Capsule woody, subglobose to oblong, more or less echinate or muricate, imperfectly 5-valved at apex. Seeds small; embryo slightly curved.

**1. Guazuma ulmifolia** Lam., Encycl. **3:** 52, 1789

*Theobroma guazuma* L., Sp. Pl. 782, 1753.
*Guazuma polybotrya* Cav., Ic. **3:** 51, *pl. 299,* 1794.
*Guazuma tomentosa* H. B. K., Nov. Gen. & Sp. **5:** 320, 1823.
*Guazuma guazuma* Cockerell, Bull. Torrey Club **19:** 95, 1892.

A tree to 20 m. high or sometimes only a shrub; bark light gray; petioles stout, 5–15 mm. long, densely stellate-tomentose with ferruginous hairs; leaf blades broadly ovate to oblong, 2–6 cm. wide, 4–16 cm. long, rounded to deeply cordate and usually strongly oblique at base, acute to long-acuminate at apex, serrulate, stellate-tomentose at least beneath and often on both surfaces or frequently becoming dark green and glabrate above; flowers yellowish green or whitish; calyx stellate-puberulent, irregularly 2–3-parted, lobes about 3 mm. long; main blade of petals oblong, about 3 mm. long, puberulent, apex sharply inflexed and giving rise to a dorsal appendage about one and one-half times as long as blade that divides midway to the apex into 2 ligulate branches that form a "V" 2.5–3 mm. high, base of the appendage and the tip of blade forming a saddle-like notch fitting closely over a short filament of anther; staminodia narrowly deltoid, 1–1.3 mm. high; stamineal tube 1–1.4 mm. long; each "stamen" bearing 3 subsessile anthers, each with 2 divergent subpeltate cells; fruit oblong or oblong-elliptic, 2–4 cm. long, closely and sharply tuberculate, black or dark brown; seeds brown, about 3 mm. long, tuberculate or somewhat muricate, dull, obovoid.

In our area growing along the banks of streams or arroyos or in the bottoms of canyons, nearly throughout Mexico with the exception of Baja California, and widely distributed in the tropical areas of South America. Collected in Sonora near Moreno, and about 25 miles south of Divisadero. April–Nov.

## Family 85. ELATINACEAE. Waterwort Family

Herbs, or in tropics some woody species, with verticillate or opposite, entire to serrate leaves and small inconspicuous stipules. Flowers small, perfect, regular, solitary or in axillary fascicles. Sepals 2–5, imbricate. Petals 2–5, hypogynous. Stamens of same number as petals or twice as many. Ovary superior, 2–5-celled; styles of same number as cells, each cell many-ovuled. Capsule ovoid or globose; placentae central; dehiscence septicidal. Seeds ribbed or reticulately rugose.

### 1. **BERGIA** L., Mant. 152, 241, 1771

Herbs or slightly woody with much-branched, ascending stems. Foliage more or less pubescent. Leaves opposite, serrate or entire. Flowers 5-merous, or rarely 3–4-merous, in axillary clusters. Capsule ovoid, 5-valved. Seeds numerous, reticulately ridged into small quadrate sections.

**1. Bergia texana** (Hook.) Seub. ex Walp., Rep. **1:** 285, 1842
*Merimea texana* Hook., Ic. Pl. **3:** *pl. 278,* 1840.

Herbaceous plant 1–3 dm. tall; stems scabrous with stiff, spreading, glandular hairs; leaves obovate to ovate, acute at apex, cuneately narrowed below to a petiole 3–8 mm. long, blade 4–12 mm. wide, 5–25 mm. long, finely but sharply serrate; flowers 1–3 in axils; pedicels 1–2.5 mm. long; sepals lance-acuminate, 2.5–3 mm. long, midrib scabrous, margins white-scarious; petals about 2 mm. long; capsules subglobose-ovoid, about equaling sepals; seeds light reddish brown, shining, about 0.5 mm. long, microscopically reticulate in longitudinal rows.

Moist ground, often about ephemeral pools, Lower and Upper Sonoran Zones, Sacramento and San Joaquin Valleys to southern California and east to Nevada, Illinois, and Texas; collected near Comondú, Baja California, the only locality known within the Sonoran Desert. June–Nov.

## Family 86. **FRANKENIACEAE.** Frankenia Family

Perennial herbs or small shrubs of low habit. Leaves opposite, sessile, bases membranous and usually united; blades narrow, entire, frequently revolute. Stipules none. Flowers small, perfect, solitary and sessile or fasciculate in axils. Calyx 4–5-lobed, with a prismatic or tubular, furrowed tube; lobes short. Petals 4–5, inserted on edge of calyx tube; claw appendaged on its inner surface. Stamens 4–7, inserted on calyx tube. Ovary superior, of 2–4 carpels but 1-celled, with 2–4 parietal placentae. Style branches 2–4, filiform. Capsule surrounded by persistent calyx. Seeds few to many; seed coat crustaceous. Endosperm mealy.

### 1. **FRANKENIA** L., Sp. Pl. 331, 1753

Characters of the family.

Leaves oblanceolate, the larger 5–8 mm. wide; styles 3-cleft; seeds several, ovules numerous; stems mainly herbaceous above base:
- Petals exserted 2–4 mm. beyond calyx tube, deep rose; leaves only slightly revolute, 6–18 mm. long .................................... 1. *F. grandifolia* var. *grandifolia*
- Petals exserted 1–2 mm. beyond calyx tube, white to pale rose; leaves more strongly revolute, 4–8 mm. long ........................... 1a. *F. grandifolia* var. *campestris*

Leaves linear or oblong, not over 2 mm. wide, nearly terete; styles 2-cleft; seeds 2–3; stems distinctly lignescent ............................................... 2. *F. palmeri*

**1. Frankenia grandifolia** Cham. & Schlecht., Linnaea **1:** 35, 1826 var. **grandifolia**
*Velezia latifolia* Eschsch., Mém. Acad. St. Pétersb. **10:** 286, 1826.
*Frankenia latifolia* Presl ex Schultes f., Syst. Veg. **7:** 1620, 1830.

Profusely branched from a heavy woody root and crown; stems procumbent to erect, slender, 1–4.5 dm. long, finely puberulent to glabrous, often reddish; leaves oblanceolate, plane or somewhat revolute, 2–8 mm. wide, 6–18 mm. long, dull green beneath, brighter green above, often encrusted with crystals of salt in dry weather, connecting membrane between bases of leaves of a pair sparsely ciliate; calyx tube 5–6.5 mm. long, strongly 4(5)-furrowed, lobes narrowly deltoid, about 0.5 mm. long; petals spatulate, exserted 2–4 mm., pale to deep rose, appendages on claws hispid; stamens 4–7, about equaling petals; capsule linear, several-seeded.

Salt marshes near the coast but always above mean high water, Lower Sonoran Zone to Transition Zone, western coast of central Baja California to Marin County, California. June–Oct.

### 1*a*. **Frankenia grandifolia** var. **campestris** A. Gray, Syn. Fl. N. Amer. **1**[1]: 208, 1895

Usually more compactly tufted and slightly taller than typical form; leaves 4–8 mm. long, strongly revolute; petals white to pale rose, exserted 1–2 mm. beyond calyx lobes; otherwise as in above-described variety.

In moist alkaline soils of the interior, Sacramento Valley to Inyo and Riverside Counties, California, in southern Nevada and in northern Baja California.

### 2. **Frankenia palmeri** S. Wats., Proc. Amer. Acad. **11**: 124, 1876

Low, divaricately branched shrub 1–3 dm. tall with rather brittle, stiff, creeping, spreading or ascending branches; young shoots finely and moderately densely puberulent with spreading, grayish, simple hairs; leaves in crowded fascicles, 2–4 (or under some conditions to 9) mm. long, strongly revolute and nearly terete, canescent with a short fine puberulence; calyx tube 3–3.5 mm. long, puberulent, lobes narrowly deltoid, about 0.5 mm. long, acute; petals linear, 1.5–2 mm. long, white or suffused with pink; stamens 4, about equaling petals; capsule narrowly ovoid, about 2 mm. long, 1–3-seeded.

Coastal salt marshes and subsaline or alkaline soil away from the coast, Lower Sonoran Zone, San Diego County, California, southward to the Cape Region, Baja California, on islands in the Gulf of California, and along the coast of Sonora to Kino Bay. Nov.–May.

One luxuriant specimen, collected at Punta Peñasca, has longer leaves that suggest those of *F. jamesii* Torr. & Gray. Unfortunately it is sterile and its identity is in doubt. Another specimen from Punta Peñasca has two or three pairs of leaves longer than average for *F. palmeri*, whereas most of them are typically short and strongly revolute. So the longer leaves may appear under unusually favorable conditions of growth, such as occur during a season of abundant rain or in a site with rich soil.

## Family 87. **TAMARICACEAE.** Tamarisk Family

Shrubs or suffrutescent plants or rarely herbs or small trees. Leaves small and scalelike, entire, alternate. Stipules none. Flowers perfect, in racemose spikes or solitary, terminal or axillary. Sepals 4–5, connate at base or free. Petals 4–5, distinct, inserted under margin of disk. Stamens 4 to many, on a hypogynous disk lining calyx cup. Ovary superior, 1-celled; placentae 3–5, parietal. Styles 3–5. Ovules few to many. Fruit a coriaceous capsule, dehiscent by valves. Seeds often beaked and bearing plumose hairs.

## 1. TAMARIX L., Sp. Pl. 270, 1753

Shrubs or small trees; leaves scalelike, minute, deciduous or persistent. Flowers white to rose, in dense spikes or racemose on lateral branchlets. Stamens 5 or 10, rarely 4 or 12, distinct or connate at base. Capsule attenuate at apex. Seeds many, bearing plumose hairs.

Flowers 4-merous; petals broadly spreading to slightly reflexed; tips of larger leaves brownish, scarious .......... 1. *T. parviflora*

Flowers 5-merous; petals ascending-spreading, not reflexed; tips of leaves not brownish nor scarious:

Leaves ovate to lanceolate, to 1.5 mm. long; branchlets pale green, not articulate; flowers pedicellate .......... 2. *T. pentandra*

Leaves reduced to minute cusps; branchlets gray-green, distinctly articulate; flowers sessile in the axil of a semisheathing, broadly ovate bract .......... 3. *T. aphylla*

### 1. **Tamarix parviflora** DC., Prod. **3:** 97, 1828

A shrub or small tree to 6 m. tall with slender, purplish, usually arching branches; leaves scalelike, ovate-acuminate, 1.5–2 mm. long, green except for brownish subscarious tips of mature ones, deciduous; racemes densely flowered, 2–3.5 cm. long, lateral on twigs of previous season; flowers appearing before leaves, in axils of linear-oblong bracts about equaling sepals; sepals 4, obtuse, about 1 mm. long; petals 4, light pink, 1.6–2 mm. long, spreading; stamens 4, inserted on edge of disk, filaments continuing from 5 lobes, anthers distinctly apiculate; style branches 3, about half as long as ovary or slightly less; capsule 3–4 mm. long, somewhat angular.

Cultivated in California and established as an escape from Lake County to Imperial County. March–May.

A native of southeastern Europe and central Asia. Nonflowering plants of what appeared to be this species were seen between Mexicali and Rumarosa (formerly called Alaska) in northern Baja California.

### 2. **Tamarix pentandra** Pall., Fl. Ross. **1**$^2$**:** 72, *pl.* 79, 1788

*Tamarix pallasii* Desv., Ann. Sci. Nat. **4:** 349, 1825.

A shrub or small tree 5–7 m. tall with glabrous, more or less purplish branches; leaves ovate-lanceolate to lanceolate, pale glaucous-green, 1–1.5 mm. long, acute or slightly acuminate; flowers rosy pink, in dense slender racemes 3–8 cm. long, these forming large terminal panicles several dm. long; bracts ovate-lanceolate, acute to acuminate, slightly exceeding pedicels; sepals 5, ovate, about 1 mm. long; petals 5, elliptic, connivent, 1.5–2 mm. long; disk with 5 shallowly bilobed lobes; stamens inserted below disk with filaments issuing between lobes; styles with connivent, linear-spatulate branches, these about half as long as ovary; capsules slenderly ovoid, about 4 mm. long.

Frequently cultivated in California and Arizona and extensively escaped along irrigation ditches and streams in southern Nevada, western Arizona, and southeastern California; very common in Imperial Valley south into Baja California and Sonora. March–Aug.

### 3. **Tamarix aphylla** (L.) Karst., Deutsch. Fl. 641, 1882

*Thuya aphylla* L., Cent. Pl. **1:** 32, 1755.
*Tamarix orientalis* Forsk., Fl. Aegypt. 206, 1775.
*Tamarix articulata* Vahl, Symb. Bot. **2:** 48, *pl.* 32, 1791.
*Tamarix furas* Buch.-Ham. ex Royle, Illus. Bot. Himal. 213, 1839.
*Tamarix usneoides* E. Mey., in Drège, Zwei Pfl. Docum. 225, 1843.

Tree to 20 m. or more; bark reddish or reddish brown on branches; branchlets very slender, those of the season about 1 mm. in diameter, gray-green, slightly constricted and ringed by the minutely cuspidate sheathing leaf at each node, leaf blade reduced to a cusp less than 0.5 mm. long, scarious; flowers sessile on lateral branchlets 2–6 cm. long, these forming a terminal panicle; bracts shallowly cup-shaped at base, partially sheathing branchlet, projecting 0.4–0.6 mm., tip lance-triangular; flowers 5-merous, sepals suborbicular, about 1 mm. long; petals white or pale pink; stamens inserted on margin of a 5-lobed disk, lobes extending into filaments; capsule 2–3 mm. long.

Commonly planted as a shade tree and windbreak in southern Arizona but reproducing spontaneously much less commonly than the preceding two species. Less common in southeastern California; Borrego Valley. March–Aug.

A native of northeastern Africa and western Asia.

## Family 88. **FOUQUIERIACEAE.** Ocotillo Family

Shrubs or trees with alternate, entire, petiolate, glabrous leaves, petioles developing into stiff spreading spines that become apparent after leaf blade falls; leaf blades deciduous when water supply wanes. Flowers perfect, in spikes, racemes, or panicles, these terminal. Sepals 5, imbricate in bud. Corolla sympetalous, tubular; lobes 5, imbricate. Stamens 10–15, hypognous, unequal, somewhat exserted beyond corolla tube. Ovary superior, 3-celled, each cell 4–6-ovuled. Capsule 3-valved, ovoid to cylindric-ovoid. Seeds compressed, winged, wing broad, eventually splitting into radiating filaments.

Plant a tree with a thick columnar trunk to 4.5 dm. in diameter; pithy medulla of trunk very large; flowers creamy yellow; bark of trunk white or whitish ............ 1. *Idria*

Plant a shrub or a tree with a very short trunk and many ascending branches; pithy medula small; flowers pink or red; bark green, gray, or yellowish green ........ 2. *Fouquieria*

### 1. **IDRIA** Kell., Hesperian **4:** 101, 1860

A tree with a slenderly tapering, columnar trunk with thick whitish bark, a few twiglike, lateral branches along simple or sparingly branched trunk and usually a tuft of slightly longer ones at apex. Woody cylinder thin, about as thick as bark, greatly exceeded by large central pith. Leaves alternate, obovate or oblanceolate, entire. Flowers yellow, borne in large panicles at summit of trunk, subsessile. Sepals 5, rounded. Corolla narrowly campanulate. Stamens 10. Capsule oblong, 3-valved. Seeds oblong, greatly compressed, membranously winged.

**1. Idria columnaris** Kell., Hesperian **4:** 101, 1860

*Fouquieria columnaris* Kell. ex Curran, Bull. Calif. Acad. **1:** 133, 1885.
*Fouquieria gigantia* Orcutt, W. Amer. Sci. **2:** 48, 1886.

Characters of the genus.

Trunk to 20 m. tall; lateral branches usually 1 m. or less in length; leaves 1–2 cm. long, rounded to acute at apex; panicles at tips of peduncles 2–6 dm. long; sepals 3–4 mm. long; corolla 6–8 mm. long; stamens exserted about 5 mm.; capsule 12–16 mm. long; seeds 6–8 mm. long, 1.5–2 mm. wide, wing 3–3.5 mm. wide, 9–12 mm. long, apparently remaining intact.

On rocky hillsides and alluvial plains, Lower Sonoran Zone, southern end of the Sierra San Pedro Mártir foothills to near north slopes of Volcan de Tres Vírgenes, on

Isla de la Guarda, Baja California, and on a small range of granitic hills about 8–9 miles south of Puerto Libertad, Sonora. Flowering after rains.

This fascinating tree reaches its finest development on the granitic soil a few miles north and south of Cataviñá, and again in an interior valley a few miles north of Punta Prieta where it is associated with *Yucca valida, Fouquieria splendens* and *F. diguetii.*

## 2. FOUQUIERIA H. B. K., Nov. Gen. & Sp. 6: 81, *pl.* 527, 1823

Spiny shrubs or trees with a short but distinct trunk and numerous ascending, comparatively slender branches from near or at base, these leafless for much of year, putting out new foliage only after rain. Leaves in fascicles in axils of spines after first year's growth. Flowers (in ours) red or pink, in racemes or thyrsoid panicles. Sepals 5. Petals 5, imbricate, spreading at apex of corolla tube. Stamens 10–15; filaments slightly fleshy and often pilose near base, attached beneath disk and slightly coalesced at base. Ovary obtuse; style branches 3, free or more or less coalescent. Capsule 3-celled, 3-valved, valves persistent. Seeds broadly and fimbriately winged.

Corolla tubular, bright red, the tube 1.5–2 cm. long:
  Plants usually without a trunk; panicles to 2.5 dm. long; filaments of stamens each bearing a small hairy scale at apex of pilose portion; stamens usually 12–15 .......................... 1. *F. splendens*
  Plants normally with a short but distinct trunk; panicles usually 1 dm. long or less; filaments hairy at base but devoid of a scale; stamens usually 10:
    Branches of panicle short and stout, subfleshy; pedicels 2–6 mm. long; panicle conical .......................... 2. *F. diguetii*
    Branches of panicle long and slender, not fleshy; pedicels 5–30 mm. long; panicle corymbiform .......................... 3. *F. macdougalii*
Corolla salverform, pale salmon-pink, tube 7–8 mm. long .................. 4. *F. burragei*

### 1. Fouquieria splendens Engelm., in Wisliz., Mem. Tour North. Mexico 98, 1848

Shrub to 6 m. tall with several slender simple branches arising from a woody root crown, but without a common aerial trunk; stems ridged by decurrent bases of stiff porrect spines, forming strong ridges and grooves; bark light gray to nearly black, smooth; primary leaves obovate to spatulate, 2–3 cm. long, entire, blades early-deciduous from persistent petioles which form spines 1–2.5 cm. long; secondary leaves 1–1.8 cm. long, in axillary fascicles; panicles 5–25 cm. long, 4–8 cm. wide; pedicels 4–8 mm. long; sepals broadly ovate to suborbicular, 5–8 mm. long, margins subscarious; corolla bright red, tube 1.5–2 cm. long, lobes obtuse or rounded, spreading or recurved, 5–7 mm. long; filaments hairy at base, each bearing a small, obovate, hairy scale 0.5–1 mm. long at apex of hairy part; capsule 10–15 mm. long; seeds flattened, 4–6 mm. long, wings membranous, about 1 cm. long, at length sometimes splitting into filaments.

Desert slopes and plains, Lower Sonoran Zone, southern California to western Texas, Coahuila, Zacatecas, and southern Baja California. Flowering any month following rains.

### 2. Fouquieria diguetii (Van Tieghem) I. M. Johnston, Contr. Gray Herb. 75: 30, 1925

*Bronnia digueti* Van Tieghem, Jour. Bot. **13**: 297, 1899.
*Bronnia thiebauti* Van Tieghem, loc. cit.
*Fouquieria peninsularis* Nash, Bull. Torrey Club **30**: 455, 1903.

Shrub or small tree 2–8 m. tall arising from a short trunk often not over 2 dm. tall and fully as large in diameter; similar to *F. splendens* except that branches are thicker and usually more tortuous, leaves thicker and more leathery and often elliptic to ovate and acute at apex, and in character of inflorescence and flowers; panicles conical, short, usually 5–15 cm. long, flowers crowded and several panicles often arising from tip of a single branch; branches of panicles and pedicels stoutish, subfleshy; pedicels 2–6 mm. long; sepals 4–5 mm. long, rounded and subscarious at tips; corolla 1.5–2 cm. long, bright red, lobes about one-third as long as tube; filaments hairy at base but without scales; stamens 10; capsule and seeds as in *F. splendens.*

In clay or granitic soil on alluvial plains or low hills, Lower Sonoran Zone, from about 20 miles north of Punta Prieta and the vicinity of Bahía de los Angeles, southward in Baja California, and from central Sonora between Hermosillo and Guaymas south into Sinaloa. Flowers some throughout the year, but mainly Feb.–May.

### 3. **Fouquieria macdougalii** Nash, Bull. Torrey Club **30:** 454, 1903

*Fouquieria jaboncillo* Loesener, Rep. Spec. Nov. **9:** 356, 1911.

Small tree to 8 m. tall with a short stocky trunk to 3 dm. thick and 1 m. tall; bark smooth, yellowish green on trunk and bases of larger branches, brownish or reddish on younger ones; spines 1–2 cm. long, light gray; leaves lanceolate to broadly obovate, 2–4 cm. long, rounded or acute at apex, broadly cuneate at base and gradually tapering to slender, 1–2.5 cm. long petiole; panicles corymbose, 1–1.5 dm. long and usually fully as broad, branches slender; pedicels very slender, 5–30 mm. long; sepals 4.5–6 mm. long, obtuse to acute, subscarious along margins and at tips; corolla bright red, 2–2.5 cm. long, lobes one-fourth to one-third as long as tube; stamens 10, basal 2–3 mm. hairy but not appendaged; carpels 2–2.5 cm. long; seeds about 6 mm. long, membranous wing 12–15 mm. long, becoming fimbriate or laciniate at ends.

Mostly on desert flats and gentle slopes, Lower Sonoran Zone, from vicinity of Carbó, Sonora, southward into Sinaloa. Feb.–Sept. following rains.

### 4. **Fouquieria burragei** Rose, Jour. N.Y. Bot. Gard. **12:** 267, 1911

A small tree 3–4 m. tall with a short stocky trunk usually less than 6 dm. tall and many crooked, spreading-ascending branches; general appearance and character of leaves very similar to those of *F. diguetii*; spines 1.5–2.5 cm. long; panicles 12–20 cm. long, 2–3 cm. wide, its side branches strictly ascending, 5–15 mm. long, sparsely flowered; sepals ovate to suborbicular, 4–5 mm. long, white below, tinged with rose at apices; corolla salverform, 10–12 mm. long, pale salmon-pink, tube about 4 mm. in diameter, 7–8 mm. long, lobes 2–4 mm. long, spreading; filaments 8–16 mm. long, white, basal portion pilose; style divided halfway to base or more; capsule 1.5–2 cm. long.

On low hills east of La Paz, Baja California, and on Pichilinque, Ballena, and Partida Islands. Mainly June–Sept. but throughout the year following rains.

## Family 89. **CISTACEAE.** Rockrose Family

Small shrubs or perennial herbs from a woody rootstock. Leaves alternate or opposite, simple, entire. Flowers regular, perfect, in racemes, panicles, fascicles, or solitary. Sepals 3 or 5, outer two smaller when five are present. Petals 3, 5, or none. Stamens many, hypogynous. Ovary superior, 1- to several-celled, sessile; stigma entire or 3-lobed. Fruit a capsule, dehiscent into valves. Seeds several to many, with a slender embryo and mealy endosperm.

Petals 5, yellow, longer than sepals; seeds numerous; stems erect or ascending ........................................................................ 1. *Helianthemum*
Petals 3, red, shorter than sepals; seeds 4–6; stems procumbent ................ 2. *Lechea*

## 1. HELIANTHEMUM Mill., Gard. Dict. abr. ed. 4, 1754

Leaves alternate. Flowers with showy yellow petals or apetalous, cleistogamous ones in addition to showy ones. Sepals 5. Petals 5, fugacious. Stamens numerous. Carpels 3. Ovary with 3 parietal placentae forming false partitions.

Stems sparsely leafy, mostly herbaceous, strictly erect; flowers in open, nearly leafless, corymbose panicles; leaves narrowly linear, 1–3 (or sometimes –6) cm. long .. 1. *H. aldersonii*
Stems densely leafy, distinctly woody, spreading-ascending; flowers axillary to upper leaves; leaves linear-oblanceolate, 2–8 mm. long ............................ 2. *H. nutans*

### 1. Helianthemum aldersonii Greene, Erythea 1: 259, 1893

*Halimium aldersonii* Standley, Contr. U.S. Nat. Herb. **23**: 832, 1923.
*Crocanthemum aldersonii* Janchen, in Engler & Prantl, Nat. Pflanzenf. ed. 2, **21**: 305, 1925.
*Helianthemum scoparium* var. *aldersonii* Munz, Man. S. Calif. 316, 1935.

Erect, broomlike, herbaceous branches 5–8 dm. high from a perennial woody base; leaves linear or lowermost at times narrowly oblanceolate, 0.5–2.5 mm. wide, 1–3 (or –6) cm. long, sparsely stellate-puberulent beneath or glabrate, pale green, deciduous; panicle often corymbose, to 2 dm. long; outer sepals about 2 mm. long, linear; inner 3 sepals ovate, acute or short-acuminate, 5–6 mm. long, finely puberulent; petals bright yellow, 8–12 mm. long; stamens slightly shorter than sepals; capsules ovoid, acute, 4–5 mm. high, smooth and shining; seeds brown, dull, minutely puberulent.

Rocky or sandy slopes and ridges, mainly Upper Sonoran Zone, Cajon Pass south through the foothills into northern Baja California, mainly west of the divide but entering the edge of the Colorado Desert below San Felipe and Jacumba, eastern San Diego County, California, and in adjacent Baja California. Feb.–July.

### 2. Helianthemum nutans Brandegee, Proc. Calif. Acad. II, 2: 129, 1889

*Halimium nutans* Standley, Contr. U.S. Nat. Herb. **23**: 834, 1923.

Densely leafy, woody shrub 1.5–3 dm. tall with somewhat crooked, ascending-spreading branches with light gray bark; leaves narrowly oblanceolate or smaller narrowly obovate, 2–6 (or primary ones –14) mm. long, minutely but closely stellate-puberulent on both surfaces, pale green and slightly cinereous; flowers in upper axils; pedicels slender, 5–10 mm. long, recurved; outer sepals linear, 2.5–3 mm. long, puberulent, inner 3 ovate-acuminate, 5–6.5 mm. long, minutely and densely stellate-puberulent; petals yellow, 7–8 mm. long; seeds covered with white, strap-shaped papillae.

Rocky slopes and llanos, Lower Sonoran Zone, central Baja California. Feb.–May.

**Helianthemum glomeratum** Lag. ex DC., Prod. **1**: 269, 1824, a shrub or suffrutescent plant to 5 dm. tall with broadly oblanceolate leaves to 6 mm. wide and glomerules of cleistogamous flowers in addition to or replacing the petaliferous ones, has been collected several times in the Sierra de la Laguna in the Cape Region but always at an altitude well above the Sonoran Desert.

## 2. LECHEA L., Sp. Pl. 90, 1753

Perennial, herbaceous to suffruticose plants with procumbent to erect, numerous, slender branches. Leaves alternate to pseudoverticillate. Flowers small. Sepals 5 in two series, two outer sepals narrower than three inner ones. Petals 3, dark red, usually shorter than sepals, marcescent. Stamens 3–25. Ovary superior, short-stipitate. Style short or lacking. Stigmas 3, fimbriate-plumose. Carpels 3, each 2-ovuled. Capsule ovoid, obovoid, or depressed-globose, 3-valved. Embryo straight to curved, slender. Endosperm hard.

**1. Lechea tripetala** (Sessé & Moc.) Britt., Bull. Torrey Club **21**: 252, 1894

*Helianthemum tripetalum* Sessé & Moc. ex Dunal, in DC., Prod. **1**: 284, 1824.
*Lechea skinneri* Benth., Bot. Sulph. 66, 1844.
*Lechea depressa* M. E. Jones, Contr. West. Bot. **18**: 59, 1933.

Usually a decumbent herbaceous perennial from a woody root and base; branches slender, densely strigose or slightly spreading-pilose, 1–2 dm. long; flowering stems ascending, developing from decumbent older stems, densely leafy; leaves linear, 6–12 mm. long, pilose; flowers in short, somewhat leafy panicles; pedicels 2–3 mm. long, spreading to erect, substrigose; calyx 2–2.2 mm. long, subappressed-pilose; outer sepals 2–3.5 mm. long, linear; inner sepals ovate, 1.8–2 mm. long, rounded to acute, keeled; petals about equaling or slightly shorter than inner sepals; capsule depressed-globose, 2–2.3 mm. wide, about 2 mm. high; seeds usually 6, sometimes only 4, brown, smooth, about 1 mm. long.

Almost entirely above the Sonoran Desert, north-central Mexico, southern and central Baja California (near Comondú) to Guatemala, and at one locality in Oklahoma. July–Nov.

## Family 90. **COCHLOSPERMACEAE.** Cochlospermum Family

Trees, shrubs, subshrubs, or perennial herbs with colored juice, alternate palmatilobed leaves, and subpersistent stipules. Flowers perfect, showy, paniculate, racemose, or rarely solitary in upper axils. Sepals 5, imbricate, deciduous. Petals 5. Stamens numerous, filaments distinct; anthers 2-celled, linear, opening by terminal, pore-like slits. Ovary superior, 1-celled, with parietal placentae projecting into cell, or completely 3-celled; carpels 3, each carpel producing several to numerous ovules. Style simple, with a minutely toothed stigma. Fruit a 3–5-valved capsule. Seeds glabrous or bearing woolly hairs, globose, ovoid, or cochleate-reniform; often outer coat forming a thin arillode. Embryo straight to strongly curved; endosperm copious.

Plant a tree; leaves 1–2 dm. in diameter; fruit globose-obovoid, depressed at apex; seeds covered with long, shining, white hairs ........................ 1. *Cochlospermum*
Plant a small subshrub or herb; leaves usually less than 1 dm. broad; fruit elliptic or ovoid, acute at apex; seeds surrounded by a papery arillode, this short-puberulent or pilose ........................................................ 2. *Amoreuxia*

### 1. **COCHLOSPERMUM** Kunth, Malvac. 6, 1822

Trees (in ours) or shrubs. Leaves palmately deeply lobed. Flowers yellow, in terminal panicles or in upper axils. Sepals 5, deciduous. Petals 5, contorted-imbricate. Stamens inserted on an eglandular disk. Capsule incompletely 3–5-celled, inner layer of endocarp membranous. Seeds cochleate-reniform or spirally coiled or curved, long-pilose with long, shining, white or tawny hairs; seed coat corneous. Embryo curved.

**1. Cochlospermum vitifolium** (Willd.) Spreng., Syst. **2**: 596, 1825

*Bombax vitifolium* Willd., Enum. Hort. Berol. 720, 1809.
*Cochlospermum hibiscoides* Kunth, Syn. Pl. Aequin. **3**: 214, 1824.
*Cochlospermum serratifolium* DC., Prod. **1**: 527, 1824.
*Maximiliana vitifolia* Krug & Urban ex Urban, Bot. Jahrb. **15**: 293, 1892.

A tree to 20 m. tall with gray fissured bark on trunk and reddish brown, smooth branches; leaves with petioles 1–2.5 dm. long, blades deeply palmately 5-lobed, 1–3 dm. broad, lobes 2–6 cm. wide, acuminate, crenate-serrate, sparsely puberulent when young but soon glabrate; flowers in terminal clusters, 7–15 cm. in diameter; sepals broadly ovate, 1.5–3 cm. long, acute or obtuse or rounded, minutely but densely appressed-puberulent with silky fine hairs; petals broadly obovate, emarginate, bright yellow, 3.5–6 cm. long; anthers about 6 mm. long, slightly curved; capsule globose-obovoid, 6–9 cm. long, 4–6 cm. in diameter, depressed at apex, densely but finely velvety-pubescent, striate longitudinally; seed 4–5 mm. long, 2.5–3 mm. broad across back, swathed in silky hairs 1–1.5 cm. long.

Canyons, rocky slopes, wooded hillsides, and along intermittent streams and in rich bottom land, Lower Sonoran Zone to Tropical Zone, southern Sonora to Yucatán and northern South America. Cultivated in the tropics and subtropics. Nov.–April.

## 2. AMOREUXIA Sessé & Moc. ex DC., Prod. 2: 638, 1825

Low subshrub or perennial herbs from a stout, tuberous or woody rootstock. Leaves alternate, long-petioled, palmatifid into 3–9 serrate or crenate lobes, lower surface often dotted and streaked with resin cells. Stipules conspicuous. Flowers large and showy, in terminal racemes and sometimes in upper axils also. Sepals 5. Petals 5, obovate. Stamens many, in two sets, one with short filaments, other with longer filaments; anthers linear, opening (in ours) by two terminal slits or pores. Ovary superior, 3-celled, each cell several-ovuled; stigma punctiform. Capsule pendulous, ovoid to elliptical, loculicidally dehiscent; wall composed of a thin, papery, inner layer and a leathery outer one. Seeds globose to reniform, enclosed in a thin loose coat (arillode). Embryo curved; endosperm oily-fleshy.

Seeds globose; arillode loosely fitting and often shed before dehiscence of capsule; leaves 5–7-lobed; capsule ellipsoid, 4.5–8 cm. long ........................ 1. *A. gonzalezii*

Seeds strongly reniform; arillode closely fitting and persistent; leaves 7–9-lobed; capsule broadly ovoid, 3–4 cm. long .................................. 2. *A. palmatifida*

### 1. Amoreuxia gonzalezii Sprague & Riley, Kew Bull. 1922: 102, 1922

Perennial herb from a fusiform tuberous rootstock; stems 1 to several, erect, 2.5–3.5 dm. tall; stipules linear-subulate, 4–7 mm. long; petioles slender, to 10 cm. long, sparsely and minutely puberulent; leaf blades 3–6 cm. wide, deeply 5–7-parted, lobes broadly oblanceolate, 1.5–4 mm. wide at base, 5–15 mm. wide above, obtuse or acute, irregularly serrate from about middle outward, glabrous or a few crisped hairs along midrib beneath, dark green above, paler and with scattered, dark brown dots and lines beneath; pedicels 1–2 cm. long, sepals narrowly lance-attenuate, 12–15 mm. long; petals oblanceolate, about 3 cm. long, bright orange-yellow with 1 or 2 brownish carmine spots near base; shorter stamens 10–12 mm. long, longer ones 15–18 mm. long, opening by 2 terminal pores; fruit pendent, ellipsoidal, 4.5–8 cm. long, longitudinally striate, brownish, finely and moderately puberulent; seeds globose, arillode hairy but readily removed, often lacking when seeds are shed, these 3–4 mm. in diameter, black, dull or slightly shining.

Rocky hillsides and on talus slopes, Lower Sonoran Zone, southern Arizona and central Sonora (Babiso Mountains) southward into Sinaloa. July–Sept.

**2. Amoreuxia palmatifida** Sessé & Moc. ex DC., Prod. **2:** 638, 1825

Similar in size and general appearance to the last; leaf blades 7–9-parted, lobes to 2 cm. wide, coarsely serrate, rounded or obtuse at apex; sepals narrowly lanceolate, acute or somewhat attenuate, 15–20 mm. long; petals orange-yellow with 1–2 maroon spots at base of each, 2.5–3 cm. long; longer stamens to 2.5 cm. long; capsule ovoid, 3–4 cm. long, pendant, weakly striate, finely and sparsely puberulent and with scattered, sessile, reddish glands intermingled with fine crisped hairs; seeds strongly reniform, 4–5 mm. long, light brown, puberulent arillode fitting closely to seed and persistent.

Hillsides and plains, in rocky and clayey soil, Lower and Upper Sonoran Zones, southern Arizona southward through Sonora. July–Sept.

A single collection (*Hammerly 168*) from near Comondú, Baja California, is placed here provisionally. It has the leaf characters of *A. palmatifida,* but no fruits were collected so its identity is uncertain.

## Family 91. **KOEBERLINIACEAE.** Junco Family

Almost leafless trees or shrubs with pale green, spine-tipped, stiff branchlets. Leaves alternate, minute, scalelike, and soon deciduous. Flowers small, in small racemes axillary to leaves or leaf scars. Sepals 5, minute. Petals 5, shortly clawed, on a hypogynous disk lining calyx, deciduous. Stamens same number as petals, hypogynous. Ovary of 2–5 united carpels, superior, borne on a short gynophore; ovules 2–8 in each cell. Style subulate, persistent. Fruit a berry or capsule; cells 1–2-seeded. Seeds hard; endosperm scanty.

Fruit a 2-celled berry; seeds wingless; sepals and petals 4; stamens twice as many .......... .......... 1. *Koeberlinia*
Fruit a 5-celled capsule; seeds winged; sepals, petals, and stamens 5 .......... 2. *Canotia*

### 1. **KOEBERLINIA** Zucc., Flora **15²:** Beibl. 73, 1832

Shrubs or small trees with rigid, often interlocking, spinose branches. Racemes several- to many-flowered; pedicels slender. Sepals 4, deciduous. Petals 4, narrowly obovate, greenish white. Stamens 8. Ovary 2-celled; ovules numerous in each cell. Berry globose or subglobose; cells 1- or 2-seeded by abortion. Seeds cochleate-lenticular.

**1. Koeberlinia spinosa** Zucc., Flora **15²:** Beibl. 74, 1832, var. **spinosa**

Mostly a low shrub but at times a tree to 10 m. tall with thin, smooth, green or yellowish bark on young branches, thicker, grayish, flaky bark on older trunks; young twigs sparsely and minutely puberulent with short spreading hairs; leaves linear-subulate, 1.5–2.2 mm. long, caducous; racemes 1–6 cm. long, rachis spine-tipped; pedicels slender, 4–6 mm. long at anthesis, to 1.5 cm. long in fruit; sepals broadly ovate, 1–1.5 mm. long, puberulent, caducous; petals 4–4.5 mm. long, rounded at apex; stamens equaling petals; berry globose, 4–5 mm. in diameter, green or suffused with red.

Desert washes, plains, and hillsides, Lower and Upper Sonoran Zones, Chocolate Mountains, Imperial County, California to Texas, Hidalgo, and Baja California. May–July.

**1*a*. Koeberlinia spinosa** var. **tenuispina** Kearney & Peebles, Jour. Wash. Acad. **29:** 486, 1939

Similar to var. *spinosa* but with longer, more slender branches, bluish green instead of yellowish green bark, slightly narrower petals, and longer filaments.

Confined to the Lower Sonoran Zone in southwestern Arizona, northwestern Sonora, and in the Chocolate Mountains, Imperial County, California. The two forms seem to maintain their identity, but none of the characters separating them seems to be constant. All of the specimens seen from Baja California are var. *spinosa*.

## 2. CANOTIA Torr., Pacif. R. Rep. 4: 68, 1857

Shrubs or small trees with slender, rigid, spine-tipped branches. Leaves soon deciduous, leaving triangular, minutely papillose elevations as scars. Flowers in small racemes. Pedicels jointed and bracteate just above base, gradually enlarged toward fruit. Sepals 5, persistent. Petals 5, narrowly oblong. Stamens 5. Ovary 5-celled, on a short thick gynophore; ovules 4–6 in each cell. Fruit an oblong capsule, fleshy externally at first, later drying over woody inner coat, each valve splitting apically to form 2 subulate beaks; each cell 1–2-seeded. Seeds oval to oblong, flattened, membranously winged below; embryo straight; endosperm scanty.

**1. Canotia holacantha** Torr., Pacif. R. Rep. **4:** 68, 1857

Mostly shrubby, 2–5 m. tall but infrequently to 10 m. tall; branchlets pale green; leaves triangular-ovate, obtuse at apex, about 1–1.5 mm. long; racemes 1–3 cm. long; pedicels 2–6 mm. long; calyx lobes deltoid, 1 mm. long, subfleshy, margins subscarious and finely erose; petals 3.5–4 mm. long, oblong to narrowly oblanceolate, white; stamens equaling petals; capsules ovoid-elliptic, dark brown, faintly glaucous, body 6–8 mm. in diameter, 9–13 mm. long, beaks subulate, 5–10 mm. long; seeds ovoid, body 3–3.5 mm. long, dark brown, minutely faveolate, wing thin, as wide as body and about as long, pale.

Desert slopes and hillsides, Lower Sonoran Zone, northern Mohave County to Cochise County, Arizona, and southward at intervals into northern Sonora. Reported from California by early explorers but not found recently within that state. May–Aug.

# Family 92. VIOLACEAE. Violet Family

Herbs or shrubs mostly with alternate stipulate leaves. Flowers regular or irregular, perfect, sometimes cleistogamous. Sepals 5. Petals 5, lowermost often larger than others and somewhat spurred. Stamens 5, mostly hypogynous; abaxial stamen often spurred at the base; anthers erect, more or less connivent in a ring around ovary, introrse, splitting lengthwise. Ovary sessile, free, 1-celled, 3–5-carpellate, and with 3–5 parietal placentae. Ovules 1–2 or several on each placenta. Capsule elastic or berrylike. Seeds sometimes winged or tomentose. Endosperm fleshy; embryo straight.

## 1. HYBANTHUS Jacq., Enum. Pl. Carib. 2, 1760

Shrubs or herbs, when shrubby with whitish bark; branches often spinescent. Leaves often fascicled, entire or toothed. Flowers slightly irregular; lower petal largest and slightly saccate at base. Sepals subequal. Anthers with connectives dilated and connate below. Capsule 3-valved, 3-seeded.

**1. Hybanthus fruticulosus** (Benth.) I. M. Johnston, Proc. Calif. Acad. IV, **12:** 1097, 1924, var. **fruticulosus**

*Ionidium fruticulosum* Benth., Bot. Sulph. **7,** *pl. 2,* 1844.
*Ionidium fruticulosum* var. *dentatum* A. Gray, Proc. Amer. Acad. **5:** 154, 1861.
*Hybanthus tenuifolius* Standley, Jour. Wash. Acad. **17:** 168, 1927.
*Hybanthus peninsularis* M. E. Jones, Contr. West. Bot. **18:** 59, 1933.
(Additional synonyms are listed by C. V. Morton in Contr. U.S. Nat. Herb. **29:** 78, 1944.)

A sparingly to copiously branched, perennial herb from a woody rootstock or sometimes flowering first season; stems 1–5 dm. tall, rather closely pubescent with fine, curved, simple hairs; stipules oblong, subscarious, 1.5–2 mm. long; leaf blades variable, under favorable conditions at beginning of growing season lance-elliptic, 5–15 mm. wide, 2–5 cm. long, coarsely serrate, later these fall off and subsequent ones are lance-linear, 2–6 mm. wide, 2–4 cm. long, finely serrate to entire, petioles slender, 2–6 mm. long; flowers solitary in axils of upper leaves, uppermost of these reduced almost to bracts; pedicels slender, 5 mm. long or less, flowers often abruptly nodding; sepals ovate, about 1.5 mm. long, puberulent; petals white, 2.5–3 mm. long; capsule 3–3.5 mm. high, finely and inconspicuously reticulate; seeds yellow, 1.5–2 mm. long, minutely irregular-reticulate.

Along arroyos and among rocks on steep slopes, often in shaded places, Lower Sonoran Zone, central to southern Baja California. July–Nov.

**1*a*. Hybanthus fruticulosus** var. **flavescens** (Dowell) I. M. Johnston, Proc. Calif. Acad. IV, **12:** 1097, 1924

Differs from the peninsular form in having yellowish green instead of deep green leaves, and in being glabrous instead of puberulent.

Near the coast in the vicinity of Guaymas, Sonora, on hillsides and along arroyos.

## Family 93. **TURNERACEAE.** Turnera Family

Herbs or occasionally shrubs with entire to lobed, alternate, estipulate leaves. Flowers perfect, regular, yellowish. Calyx tubular or campanulate, 5-toothed. Petals 5, inserted on calyx tube, distinct, contorted in bud. Stamens 5, inserted at base of corolla tube; anthers 2-celled, splitting lengthwise. Ovary superior, 1-celled, with 3 parietal placentae. Style branches 3; stigmas fringed apically. Capsule loculicidally 3-valved, a placenta in middle of each valve. Seeds pitted, arillate. Embryo straight. Endosperm horny or fleshy.

### 1. **TURNERA** L., Sp. Pl. 271, 1753

Characters of the family. Leaves serrate, often bearing 2 glands at base. Flowers axillary and usually solitary. Capsule thin-walled. Aril submembranous.

**1. Turnera diffusa** Willd., in Schult., Syst. Veg. **6:** 679, 1820

*Turnera microphylla* Desv., in Hamilt., Prod. Pl. Ind. Occ. 33, 1825.
*Bohadschia humifusa* Presl, Rel. Haenk. **2:** 98, *pl. 68,* 1835.
*Turnera humifusa* Endl., in Walp., Rep. **2:** 230, 1843.
*Turnera aphrodisiaca* L. F. Ward, Virginia Med. Monthly 49, 1876.
*Turnera diffusa* var. *aphrodisiaca* Urban, Jahrb. Bot. Gard. Berlin **2:** 127, 1883.
*Turnera pringlei* Rose, Contr. U.S. Nat. Herb. **5:** 166, 1899.

A shrub 0.3–2 m. tall with slender brownish branches; leaves short-petiolate, oblong to rhombic-ovate, 1–2 cm. long, cuneate at base, acute to obtuse at apex, pilose to tomentose beneath, dark green and often glabrous or essentially so above, crenate-serrate to dentate; calyx 4–5 mm. long, yellowish, the narrowly lance-triangular teeth nearly as long as tube; petals bright yellow or orange-yellow, 8–12 mm. long; capsule 2.5–5 mm. long, 3 valves widely spreading at dehiscence.

Dry ridges and hillsides, Lower Sonoran Zone to Tropical Zone, southern Baja California, central Sonora, and Texas south through Mexico, the West Indies, and Central America to South America. Jan.–July.

## Family 94. **PASSIFLORACEAE.** Passion Flower Family

Herbaceous or woody vines climbing by tendrils or subscandent shrubs with a terete or 3–5-angled stem. Leaves alternate, entire, lobed or rarely compound; petioles often beset with glands in various positions, shapes and sizes; stipules setaceous to broadly ovate, their margins entire, toothed or deeply cleft. Flowers solitary or in pairs in axils; peduncle bearing a bract in varying positions below flower, sometimes highly colored and forming an involucre-like structure below flower. Calyx tube small and cup-shaped to cylindrical, green or variously colored. Sepals 5, usually same color as calyx tube. Petals 5, inserted at margin of calyx tube or lacking, usually smaller than sepals. Corona formed of 1-several concentric rings of filaments or processes inserted just interior to petals on calyx tube. A glandular nectar-ring lining floor of tube, with an "operculum" just above it, a "limen" just interior to it, these structures consisting of annular flaps of entire to serrate, often plicate membranes. Stamens 5, borne near apex of columnar gynophore, and just below ovary, their filaments coalesced in their lower parts and closely surrounding gynophore. Styles 5, projecting at an angle from top of ovary. Ovary superior, 1-celled. Fruit a berry, with parchmentlike, leathery, or brittle rind. Seeds with transverse ridges, or pitted, embedded in a thin, fleshy aril.

### 1. **PASSIFLORA** L., Sp. Pl. 955, 1753

Characters of the family.

Fruits glabrous; operculum plicate; elliptical glands near apex of petiole or on blade between petiole and notch in leaf blade:
- Seeds reticulate-pitted; herbage hispidulous; fruit ovoid to ellipsoidal, about 2.5 cm. wide, 3.5 cm. long; petioles biglandular near apex; blade bilobed ....... 1. *P. bryonioides*
- Seeds coarsely transversely ridged and the ridges again transversely more finely ridged; herbage glabrous; fruit globose, 6–12 mm. in diameter; petioles eglandular, but 2–7 reddish, ellipsoidal glands between midnerve and adjacent nerves on underside of bilobed blade .......... 2. *P. mexicana*

Fruits distinctly pilosulous to tomentose; operculum denticulate but not plicate; petioles and blades without sessile, conspicuous glands:
- Plants bushy with an erect or ascending caudex and looping, short branches; middle lobe of leaf blade only slightly if at all elongate:
  - Flowers 5–7 cm. wide; sepals 3–3.5 cm. long; petals linear; gynophore 2–2.5 cm. long .......... 3. *P. palmeri*
  - Flowers 2.5–3 cm. in diameter; sepals 1–1.5 cm. long; petals oblong or oblong-lanceolate; gynophore 1–1.5 cm. long .......... 4. *P. fruticosa*
- Plants slender, climbing vines; middle lobe of leaf blade often distinctly elongate:
  - Bracts densely pilose but not glandular; leaf blades densely lanate (longer, spreading hairs extending beyond wool in var. *cerralbensis*):

Leaf blades 3-lobed:
Indument of closely packed woolly hairs, these more or less appressed .................................................. 5. *P. arida* var. *arida*
Indument of erect, nearly straight, soft hairs, not lanate ............................................. 5*a*. *P. arida* var. *cerralbensis*
Leaf blades 5-cleft to base, divisions again sinuately lobed ............................................. 5*b*. *P. arida* var. *pentaschista*
Bracts sparsely to densely pilose and bearing stalked, elliptical glands in addition; leaf blades hirsutulous but scarcely lanate:
Basal leaf lobes suborbicular, not again lobed; middle lobe not conspicuously narrowed toward base:
Peduncles 2–5 cm. long; plant usually very viscous .. 6. *P. foetida* var. *gossypifolia*
Peduncles 5–10 cm. long; plant rarely viscous .. 6*a*. *P. foetida* var. *longipedunculata*
Basal leaf lobes bilobed; middle lobe conspicuously narrowed toward base, lobes again irregularly serrate or lobate ...................... 6*b*. *P. foetida* var. *arizonica*

**1. Passiflora bryonioides** H. B. K., Nov. Gen. & Sp. **2:** 140, 1817
(For synonymy see Field. Mus. Bot. Ser. **19:** 109, 1938.)

Stems angular to subterete, hispidulous; stipules ovate-cuspidate, 2.5 mm. wide, 5 mm. long, somewhat ciliate; petioles 2.5–5 cm. long, hispidulous to hirsute, with 2 clavate glands about 0.8 mm. long and 0.9–1 mm. in diameter within 1 cm. of the apex; leaf blades 4–7 cm. long, 5–9 cm. wide, deeply trilobed, cordate at base, 3–5-nerved, entire to sparingly denticulate, hispidulous on both surfaces, middle lobe usually narrowed toward its base, lateral ones often again bilobed; peduncles 2–3 cm. long, solitary; bracts setaceous, 3–4 mm. long, deciduous; flowers 2–3 cm. wide; sepals lance-ovate, 8–13 mm. long, 3–5 mm. wide, acute or obtuse, greenish yellow; petals linear to lance-linear, 1 mm. wide, about 4 mm. long, white; corona a single series of filiform filaments 6–7 mm. long and purplish at base; operculum arising at base of corona, strongly incurved and plicate; nectar-ring annular; limen attached to floor of calyx, membranous, margin free and entire; ovary glabrous, narrowly ovoid; fruit about 2–5 cm. in diameter, 3–3.5 cm. long; seeds 2.3 mm. wide, 4 mm. long, ovate, closely reticulate.

Growing among shrubs and over rocks along watercourses, mainly Upper Sonoran Zone, Pima County, Arizona, south to southern Mexico. Aug.–Oct.

**2. Passiflora mexicana** Juss., Ann. Mus. Hist. Nat. **6:** 108, *pl. 38, fig. 2,* 1805
*Passiflora contrayerva* J. E. Sm., in Rees, Cycl. **26:** *Passiflora* No. 23, 1819.
*Monactineirma mexicana* Bory, Ann. Gén. Phys. **2:** 138, *pl. 22, fig. 6,* 1819.
*Cieca mexicana* M. Roem., Fam. Nat. Syn. **2:** 146, 1846.

Slender, climbing, glabrous vine with slightly 5-angled stem grooved on faces; stipules setaceous to narrowly linear, 1.5–2 mm. long; petioles 1–2 cm. long, eglandular; leaf blades bilobed one-half their length or more, 3–8 cm. wide, rounded or truncate at base, lobes oblong to linear, 0.5–4 cm. wide, nearly parallel or slightly divergent, sometimes slightly dilated toward apex and emarginate, dark green above, paler beneath, bearing 1–3 pairs of ocellae beneath; peduncles solitary or in pairs, 2–3 cm. long; bracts setaceous, minute, deciduous; flowers 2.5–4 cm. wide, calyx shallow, about 1 cm. wide, deep reddish purple within; sepals narrowly lanceolate, 1–1.5 cm. long, 4–5 mm. wide, white inside, green to reddish without; petals recurved, shorter than sepals; filaments of corona in 2 series, filiform, outer whorl about 1.5 cm. long, pink to crimson, subreflexed, inner whorl erect, about 2 mm. high; operculum strongly incurved, plicate, white with pink tinge toward margin and minutely fimbriate; limen 1.5 mm. high or less, white; ovary globose, glabrous; fruit globose, 6–12 mm. in diameter; seeds about 2 mm. wide, 3 mm. long, transversely ridged with 7–8 irregular rugae.

Climbing over shrubs and trees, often along watercourses in our area, Lower Sonoran Zone to Upper Sonoran and Tropical Zones, southern Arizona to central Mexico. Aug.–Nov.

### 3. **Passiflora palmeri** Rose, Contr. U.S. Nat. Herb. **1:** 131, *pl. 14,* 1892

A flat-topped or rounded, malodorous shrub 4–5 dm. tall and as much as 12 dm. broad with short, viscous branches spreading from main stem and lopping over older parts or adjacent rocks and bushes; branches densely pilose with white to yellow, unequal hairs; stipules running part way around stem, deeply cleft into filiform, gland-tipped segments; petioles 1–1.5 cm. long, pilose with numerous coarse, gland-tipped hairs, devoid of sessile, petiolar glands; leaves broadly cordate-ovate, 3-lobed, 1–3 cm. long, 2–4 cm. wide, 3–5-nerved, serrulate and densely glandular-ciliate along margins, hirsute-tomentose with brownish hairs, glutinous-oily on both surfaces; peduncles 4–5 cm. long; bracts 8–12 mm. wide, 2–4 cm. long, deeply bipinnatisect into capillary, gland-tipped segments 2–5 mm. long; flowers 5–7 cm. wide; sepals linear, about 4 mm. wide, 3–3.5 cm. long, bearing a slender awn dorsally just below apex, glabrous and white within, greenish and pilose-glandular without; petals linear, 2.5–3 mm. wide, 2–3 cm. long, white; corona in several series of filaments, outer two capillary, about 2.5 mm. long, erect, light blue to purple, inner series of minute filaments 0.5 mm. long or less; operculum not plicate, minutely denticulate; limen erect and membranous, 2.5–3 mm. high, margin recurved and minutely crenulate; gynophore about 2 cm. long, violet below, white above, glabrous; ovary subglobose; fruit globose, 2.5–3 cm. in diameter, densely pilose; seeds 2 mm. wide, 6 mm. long, 3-toothed at apex, finely reticulate on each face.

Along arroyos and on rocky hillsides, Lower Sonoran Zone, mainly eastern coast of Baja California from Isla Angel de la Guarda south to Cape Region. March-Oct.

### 4. **Passiflora fruticosa** Killip, Jour. Wash. Acad. **12:** 256, 1922

Shrubby, sprawling plant 2–5 dm. tall, spreading to 1–2 m. from side to side; branches and tendrils densely white-tomentose, often without tendrils; stipules partially encircling stem, deeply dissected into filiform, gland-tipped segments; petioles 5–15 mm. long, densely lanate, devoid of petiolar glands but with numerous glandular-capitate hairs extending beyond wool; leaf blades orbicular in outline, 0.5–2 or rarely to 3.5 cm. wide, subequally trilobed, 3–5-nerved, lanate with white to brownish, soft wool and densely glandular-ciliate, markedly glutinous; peduncles 1–2.5 cm. long; bracts to 4 cm. long, about 2.5 cm. wide, bipinnatisect, hirsute, ultimate segments bearing stalked, elliptical glands; flowers 2.5–3 cm. wide; sepals ovate-lanceolate, 5–6 mm. wide at base, 1–1.5 cm. long, velvety-pubescent without, glabrous within; petals 5–7 mm. long, 3.5–4 mm. wide, sordid or greenish white, obovate, glabrous; corona in several series of filaments, outer series about 1 cm. long, inner 1.5–2 mm. long; operculum membranous, minutely denticulate at apex; limen 1.5 mm. high, entire, with a recurved margin; ovary subglobose, sericeous; fruit subglobose, about 2.5 cm. in diameter, densely sericeous; seeds oblong, 2.5 mm. wide, about 5 mm. long, minutely 3-toothed at apex, truncate basally, reticulate with about 25 meshes on each face.

On sandy and rocky soil, Lower Sonoran Zone, central Baja California south to the vicinity of Magdalena and Santa Margarita Islands and on islands along the eastern coast of the southern one-third of the peninsula. March–Sept.

**5. Passiflora arida** (Mast. & Rose) Killip, Jour. Wash. Acad. **12:** 256, 1922, var. **arida**

*Passiflora foetida* var. *arida* Mast. & Rose, in Rose, Contr. U.S. Nat. Herb. **5:** 182, 1899.

A rounded shrub or spreading, scrambling vine profusely branched at base; herbage densely lanate throughout with short, white or slightly ferruginous wool, ill-smelling; stipules running about halfway around stem, deeply laciniate into minute segments 0.5–2.5 mm. long; petioles 5–35 mm. long, without petiolar glands; leaf blades 3–4 cm. long, 4–5 cm. wide, 3-lobed with oblong to suborbicular lobes, these nearly equal or central one about one-half again as large as lateral ones, cordate at base, densely tomentose on both surfaces with a matted grayish, brownish or reddish indument; peduncles 1–3 cm. long; bracts lanate with whitish or brownish hairs, bipinnatifid into narrow segments 1–5 mm. long, glandless; flowers 2.5–3.5 cm. in diameter, sepals ovate-lanceolate, about 7 mm. wide, 1.5 cm. long, obtuse, short-aristate below apex on back, white and glabrous within, greenish and hirsute on outer surface; petals oblong, 4–5 mm. wide, about 1 cm. long, white, thin and spreading; corona of several series, outer two rows of filaments about 8 mm. long, violet-blue, those of inner 3–5 series capillary, 1 mm. long or less; operculum about 1 mm. high, finely denticulate; limen cup-shaped, 1.5 mm. high, its margin entire, recurved; gynophore 10–12 mm. long, with purple or magenta spots on a white background; fruit subglobose, 2.5–3 cm. in diameter, pubescent; seeds 2–3 mm. wide, 3–5 mm. long, 3-toothed at apex, truncate at base, reticulate with meshes about 0.5 mm. in diameter.

Along arroyos and on rocky hillsides and cliffs, Lower Sonoran Zone, from the vicinity of Hermosillo and Santa Rosalia southward into Sinaloa. Feb.–Oct.

**5*a*. Passiflora arida** var. **cerralbensis** Killip, Field Mus. Bot. Ser. **19:** 470, 1938

Stems, petioles and peduncles densely covered with soft spreading hairs but not lanate; petioles without petiolar glands; leaves hirsute-lanuginous with golden brown lustrous hairs; bracts with filiform segments and pilose; otherwise as in var. *arida.*

Hillsides and arroyos, Lower Sonoran and Tropical Zones, southern Baja California, between La Paz and San Pedro, and on Cerralvo Island in the Gulf of California.

**5*b*. Passiflora arida** var. **pentaschista** Killip, Field Mus. Bot. Ser. **19:** 470, 1938

Leaves 5-cleft to base, division again sinuately lobed; otherwise as in var. *arida.*

Hillsides and along arroyos, Lower Sonoran and Tropical Zones, southern Baja California.

**6. Passiflora foetida** L. var. **gossypifolia** (Desv.) Masters, Trans. Linn. Soc. **27:** 631, 1871

*Passiflora gossypifolia* Desv., in Hamilt., Prod. Pl. Ind. Occ. 48, 1825.
*Dysosmia gossypifolia* M. Roem., Fam. Nat. Syn. **2:** 149, 1846.

Plant villous or hirsutulous throughout with grayish, brownish or yellow-brown hairs about 1.5 mm. long or less, often viscous from gland-tipped hairs intermingled with nonglandular ones, with a strong, disagreeable odor; petiolar glands none; leaf blades broadly ovate or more or less pentagonal, about 5 cm. long and wide (at widest part), basal lobes usually semicircular in outline, rounded to obtuse at apex, all lobes entire to crenate-serrulate along margins; bracts 2–3 cm. long, bi- or tripinnatisect into filiform, gland-tipped segments that are free and not intertwined; flowers 2–5 cm. wide, white, pink, or purplish; sepals ovate-oblong to ovate-lanceolate; corona filaments in several series, outer 2 of filiform segments about 1 cm. long, others capillary, 1–2 mm. long; operculum denticulate, erect; ovary sparsely to densely pilosulous; fruit subglobose, 2–2.5 cm. in diameter, yellow to greenish yellow spotted with red, pilosulous; seeds ovate-cuneiform, coarsely reticulate on each face, faintly 3-toothed at apex.

On rocky or sandy soil and along watercourses, Lower Sonoran Zone to Tropical Zone, Texas to central and southern Sonora and south to northern South America and the West Indies. March–Oct.

***6a.* Passiflora foetida** var. **longipedunculata** Killip, Field Mus. Bot. Ser. **19:** 487, 1938

Stems densely and softly pilosulous with whitish hairs averaging less than 1 mm. in length; leaves hastate, with length and greatest width about equal, margins entire but slightly undulate, surface densely pilosulous; peduncles 6–7 cm. long, very slender; bracts bipinnatisect; ovary white-tomentose.

Known from only two areas, the vicinity of Mulegé, Baja California, and from Cerro del Chino, Tamaulipas, Mexico.

***6b.* Passiflora foetida** var. **arizonica** Killip, Field Mus. Bot. Ser. **19:** 490, 1938

Plant densely gray-villous throughout; leaves palmately 5-lobed, lobes sinuate or again slightly sublobate, denticulate, markedly ciliate-margined, middle lobe narrowed at base; bracts pinnatisect, segments relatively few, divergent at nearly right angles to main rachis; ovary pilose; fruit about 3 cm. in diameter, yellowish and pilose.

Hillsides and canyons, Lower Sonoran Zone, southern Arizona and southward at least to central Sonora. Aug.–Sept.

## Family 95. **LOASACEAE.** Loasa Family

Erect or scandent herbs or shrubs more or less clothed with barbed, or stinging, or glandular hairs or combinations of these. Leaves alternate or opposite. Stipules none. Flowers regular, perfect, in racemes or cymes. Sepals 4–5. Petals 4–5 or sometimes 10, yellow or reddish. Stamens 5 to many, usually in clusters opposite petals; outer filaments often dilated and more or less petaloid; anthers introrse, dehiscent longitudinally. Ovary 1-celled (in all of ours), or rarely 2–3-celled; placenta 2–5, parietal, or a single ovule pendulously attached to apex of locule. Styles slender, entire to 3-lobed. Fruit a capsule, 1- to many-seeded, dehiscent apically, longitudinally, or irregularly. Seeds small, with little or no endosperm.

Ovary 1-ovuled; stamens 4–5; ovule pendent from apex of locule:

Plant a scandent vine; leaves petiolate; blades to 15 cm. long; herbage with stinging hairs ........ 1. *Gronovia*

Plants erect or ascending herbs or shrubs; leaves sessile, 1–4 cm. long; herbage without stinging hairs ........................................ 2. *Petalonyx*

Ovary several- to many-ovuled; stamens many; ovules attached to parietal placentae:

Placentae 3; style branches 3; seeds angulate to pyramidal, not elliptic nor longitudinally striate; capsule urceolate to clavate ........................ 3. *Mentzelia*

Placentae 5; style branches 5; seeds narrowly elliptic, longitudinally and slightly striate spirally; capsule subglobose to obovoid:

Stinging hairs present; petals united only at very base; stamens inserted in clusters on bases of petals ........................................ 4. *Eucnide*

Stinging hairs absent; corolla sympetalous into a distinct tube about equaling or surpassing lobes; stamens inserted in 2 whorls in tube and throat of corolla ........................................ 5. *Sympetaleia*

## 1. GRONOVIA L., Sp. Pl. 202, 1753

Herbaceous perennials with trailing stems, beset with stiff stinging hairs. Leaves alternate, petiolate, broadly cordate and more or less 5-lobed. Flowers small, in subcorymbose clusters on peduncles opposite the leaves; pedicels bearing linear bracteoles. Calyx tube short, subangulate, 5-lobed. Petals 5, inserted in throat of calyx. Stamens 5; staminodia none. Ovary inferior, 1-celled, 1-ovuled. Style filiform; stigma capitate. Fruit subglobose, small, indehiscent. Seed filling locule, its coat membranous. Endosperm none.

### 1. Gronovia scandens L., Sp. Pl. 202, 1753

*Gronovia humboldtiana* Roem. & Sch., Syst. Veg. **5:** 492, 1819.
*Gronovia jacquiniana* M. Roem., Fam. Nat. Syn. **2:** 108, 1846.

Stems prostrate and vinelike, scrambling over rocks and among shrubs, to 5 m. long or more, pilose with slender, spreading, stinging hairs; leaf blades broadly cordate-ovate in outline, 4–10 cm. wide, 3–15 cm. long, shallowly to deeply 3–7-lobed, lobes acute or acuminate, often coarsely and irregularly dentate, pilose like stems but somewhat less densely so; petioles 1.5–5 cm. long; inflorescences on stoutish peduncles opposite leaves and about equaling petioles, 5–12-flowered; calyx tube 2.5–3.5 mm. long at anthesis, densely appressed-pubescent, lobes lance-triangular, 3–5 mm. long; petals yellow, linear-oblong, 1.5–2 mm. long, slightly shorter than stamens; fruit ovoid or obovoid, 4–5 mm. high, slightly angled, angles pilose and often narrowly winged, intervening spaces sparsely pubescent to subglabrous.

Scrambling over rocks and shrubs, Lower Sonoran Zone to Tropical Zone, northern Sonora and Chihuahua to northern South America. June–Nov.

## 2. PETALONYX A. Gray, Pl. Thurb. 319, 1854

Perennial herbs or low shrubs with cinereous-scabrous or pilose branches and small, alternate, sessile leaves. Flowers small, in spikes or heads, white, bibracteolate. Calyx tube short-cylindrical, 4–5-lobed, these about equaling tube and deciduous. Petals 4–5, epigynous, long-clawed. Stamens 4–5. Staminodia none. Ovary 1-celled, 1-ovuled. Fruit oblong; coat fragile and irregularly dehiscent. Seed conforming closely to shape of locule. Cotyledons fleshy; endosperm none.

Leaves sessile or subsessile, not shining; petals 4–6 mm. long:

Leaves broadest at base; floral bracts triangular-ovate, dentate near base; herbage gray with retrorse, stiff hairs ........................................ 1. *P. thurberi*

Leaves linear to elliptic, broadest about middle; floral bracts ovate-cordate, entire; herbage green, scabrous with nearly straight, spreading hairs ................ 2. *P. linearis*
Leaves distinctly petioled, coarsely dentate, shining; petals 7–10 mm. long ..... 3. *P. nitidus*

**1. Petalonyx thurberi** A. Gray, Pl. Thurb. 319, 1854

Perennial herb to 1 m. tall from a heavy rootstock; stems and leaves harshly scabrous with recurved, stiff hairs; leaves ovate to lance-ovate, broadest at base, 3–10 mm. broad, 6–40 mm. long, thick and stiff, spreading or ascending, rounded at sessile base, acute to attenuate above, upper ones entire, lower ones irregularly dentate; racemes 1–4 cm. long; bracteoles ovate-lanceolate, 1–4-toothed near base, 4–6 mm. long, acute above; calyx densely scabrous, tube about 4 mm. long; petals white or cream, 4–6 mm. long; capsule hidden by bract and calyx, about 2 mm. long.

Dry hillsides and sandy areas, chiefly Lower Sonoran Zone but also in dry areas in Upper Sonoran Zone, southern Nevada and Mojave Desert, California, through the desert areas of Arizona and California into northern Sonora and Baja California. May–June.

**2. Petalonyx linearis** Greene, Bull. Calif. Acad. **1:** 188, 1885

Low, very scabrous shrub 1–4 dm. tall; hairs short and spreading, not noticeably recurved; leaves elliptic to lance-elliptic, 1–3 cm. long, 2.5–7 mm. wide, entire, thick and stiff, mostly obtuse at both ends, subsessile, green; racemes 1–6 cm. long, several- to many-flowered; bracts ovate-cordate, becoming somewhat papery, obtuse or acute at apex, 7–12 mm. long, scaberulous; calyx lobes lanceolate, 1.5–2 mm. long; petals oblanceolate or obovate, distinctly and narrowly clawed, 4–5 mm. long, surpassing calyx lobes; fruit narrowly ovoid, 2.5–3 mm. long, densely puberulent.

Rocky and sandy places, Lower Sonoran Zone, Mojave Desert and south into central Baja California and east into western Arizona and northwestern Sonora (Punta Peñasca). Flowering some the year around but mainly Jan.–April.

**3. Petalonyx nitidus** S. Wats., Amer. Nat. **7:** 300, 1873

*Petalonyx thurberi* var. *nitidus* M. E. Jones, Contr. West. Bot. **12:** 16, 1908.

Low, profusely branched shrub 2–4.5 dm. tall and often fully as broad; young branches and leaves short-pubescent with spreading stiff hairs, hence more or less scabrous; leaves ovate to lanceolate in outline, 1.5–3 cm. long, 8–18 mm. wide, crenate-serrate, shining green, abruptly narrowed at base to a petiole 3–7 mm. long; racemes 2–6 cm. long; bracts lance-ovate, 4–6 mm. long, acute to acuminate at apex, entire or slightly toothed near base, becoming somewhat papery and to 1 cm. long in fruit; sepals 4–5 mm. long, subulate; petals pale greenish yellow, 7–10 mm. long, sparsely puberulent externally; stamens exserted, 1–3 mm. long; fruit narrowly ovoid, 2–2.5 mm. long, puberulent.

Rocky and sandy desert soil, Lower Sonoran Zone, Inyo and San Bernardino Counties, California, into southern Nevada and northwestern Arizona. May–June.

Barely entering the margin of the Sonoran Desert in Mohave County, Arizona.

## 3. MENTZELIA L., Sp. Pl. 516, 1753

Annual or perennial herbs or some species in tropics becoming shrubs or trees. Stems often white and shining, these and herbage pubescent with various types of barbed or curved, nonstinging hairs. Leaves alternate or rarely opposite, sessile or petiolate, at least lower usually toothed or lobed. Flowers chiefly yellow, solitary in axils of branches or in terminal cymes or both, each subtended by a bract. Hypanthium ovoid, turbinate, or cylindric. Sepals 5, persistent or deciduous. Petals 5–10, distinct or slightly connate at base, imbricated. Stamens numerous, distinct or united into clusters opposite petals, filaments filiform or outer more or less petaloid, sometimes somewhat dilated at base; anthers introrse. Style branches 3. Ovary inferior, 1-celled, few- to many-ovuled. Capsule 3–5-valved. Seeds smooth or variously sculptured. Endosperm scanty to copious.

Filaments linear or dilated but if dilated lobes rounded, not cuspidate; bracts small, not obscuring calyx tube except in *M. micrantha*; petals 2–25 mm. long:
 Plants annual or some biennial; seeds 1–40, wingless, pendulous in capsule; stamens all antheriferous:
  Seeds few, 1–15; placentae broad and plicate:
   Outer filaments dilated and petaloid; capsule sessile; petals 4–8 mm. long .......... 1. *M. aspera*
   Outer filaments linear, not dilated; capsule pedicellate; petals 10–15 mm. long ...... 2. *M. adhaerens*
  Seeds few to many, 10–40; placentae narrow, not plicate:
   Outer filaments slightly dilated, distinctly broader than inner; bracts leaflike, obscuring calyces; inflorescence congested, many-flowered .......... 3. *M. micrantha*
   Outer filaments linear or filiform, not broadened; bracts smaller than flowers, linear or lanceolate; inflorescence open, not congested:
    Petals 2–4 mm. long; stems nearly glabrous above; sepals 2–2.5 mm. long ........ 4. *M. albicaulis*
    Petals 5–25 mm. long; stems sparsely to densely scabrous-puberulent above; sepals 2–10 mm. long:
     Seeds irregularly to sharply angled, grooved on one or more angles; petals 4–8 mm. long:
      Seeds grooved on 3 angles; sepals 5–6 mm. long ............... 5. *M. affinis*
      Seeds grooved on 1 angle only; sepals 2–3 mm. long ........ 6. *M. veatchiana*
     Seeds not grooved on angles; petals 8–25 mm. long:
      Petals coppery red at base; sepals 5–6 mm. long ............. 7. *M. pectinata*
      Petals golden yellow throughout; sepals 6–10 mm. long:
       Petals 15–20 mm. long .......................... 8. *M. nitens* var. *nitens*
       Petals 8–12 mm. long:
        Stems stout, white .......................... 8*a*. *M. nitens* var. *jonesii*
        Stems slender, greenish .................. 8*b*. *M. nitens* var. *leptocaulis*
 Plants perennial; seeds 50–80, narrowly but distinctly to broadly winged, horizontal in capsule; outer stamens often sterile:
  Petals 15–20 mm. long; capsules acute at base ...................... 9. *M. multiflora*
  Petals 8–15 mm. long; capsules rounded at base:
   Sepals attenuate; petals acute:
    Capsules 15–20 mm. long; leaves 3–10 cm. long ........ 10. *M. pumila* var. *pumila*
    Capsules 10–12 mm. long; leaves 3–5 cm. long ....... 10*a*. *M. pumila* var. *procera*
   Sepals acute or short-acuminate; petals obtuse .................... 11. *M. puberula*
Filaments, at least outer, strongly broadened or dilated and tricuspidate at apex, only central one bearing an anther; bracts large, usually obscuring calyx tube; petals 2–4.5 cm. long:
 Bracts green, narrowly ovate or lanceolate:
  Flowers pedicellate; petals 2–2.5 cm. long; stamens 80–100 ........... 12. *M. tricuspis*
  Flowers sessile; petals 3–3.5 cm. long; stamens 100–300:

Middle lobe of stamens equaling or shorter than lateral ones ..................... .................................... 13. *M. hirsutissima* var. *hirsutissima*

Middle lobe of stamens twice as long as lateral ones .......................... .................................... 13*a*. *M. hirsutissima* var. *stenophylla*

Bracts whitish, membranous-scarious, broadly ovate:

Petals 2–2.5 cm. long .......................... 14. *M. involucrata* var. *involucrata*

Petals 3.5–4.5 cm. long ...................... 14*a*. *M. involucrata* var. *megalantha*

## 1. Mentzelia aspera L., Sp. Pl. 516, 1753

*Mentzelia stipitata* Presl, Rel. Haenk. **2:** 40, 1831.
*Mentzelia pedicellata* Presl, Epimel. Bot. 246, 1849.
*Acrolasia squalida* Hook. f., Trans. Linn. Soc. **20:** 222, 1851.

Annual herb with spreading, procumbent to ascending, longitudinally striate stems 1–4 dm. long, these and foliage puberulent with straight or slightly curved, spinelike hairs and retrorsely barbed, shorter ones; leaves lanceolate to ovate-lanceolate, or broadly ovate, 10–15 cm. long, to 8 cm. broad, acute to acuminate at apex, more or less cuneate to subcordate at base, serrate, serrate-dentate, or deeply incised or sometimes shallowly 3-lobed, dark green on upper surface, paler and more densely scabrous beneath; calyx tube 6–11 mm. long, attenuate at base; calyx lobes lanceolate, 3–4 mm. long, 1–1.5 mm. wide, acuminate; petals 5, obovate to obovate-orbicular, 4–8 mm. long; filaments 20–30 in 1–2 series, exterior dilated and petaloid but antheriferous; capsule subcylindrical, 1.5–2.5 cm. long, 3–5 mm. in diameter, terete, obtuse at base, papery, sessile; seeds 6–9 on 3 placentae, 2.5–3.5 mm. long, half as broad, minutely undulate-striate, wingless.

Sandy or rocky soil, Lower Sonoran and Tropical Zones in dry places, southern Arizona to South America and the West Indies. Present in southern Baja California. July–Oct.

**Mentzelia asperula** Woot. & Standl., Contr. U.S. Nat. Herb. **16:** 148, 1913, is similar to *M. aspera* in general appearance, but the outer filaments are not dilated, the leaves are 3–5 cm. long, and the capsules are not sessile. Reported from the Baboquivari Mountains, Arizona, but at elevations somewhat above the Soronan Desert.

## 2. Mentzelia adhaerens Benth., Bot. Sulph. 15, 1844

Biennial or sometimes annual, 1–7 dm. tall, with scabrous-hispidulous stems and foliage; stems slender, usually trichotomously branched near base, light-colored but not white-shining; leaves ovate to lance-ovate, 1–5 cm. broad, 2–8 cm. long, irregularly and shallowly 3–5-lobed or merely coarsely dentate or sinuate-lobate, dark green, acute at apex, broadly cuneate at base, upper ones sessile, lower on petioles 1–6 mm. long; flowers solitary in axils of dichotomous branches, subtended by small ovate bracts; calyx tube obconic to turbinate, 5–7 mm. long; calyx lobes lanceolate, 7–9 mm. long, scabrous; petals 5, yellow, obovate to oblong-obovate, 1–1.5 cm. long, 5–10 mm. wide, spreading, short-apiculate and slightly pilose; stamens 35–40 in 2 series, 7–8 mm. long, all antheriferous and none dilated; capsule obconic to turbinate, pedicellate, 10–13 mm. long, acute at base; seeds 10–15, pendulous, ovate to round-ovate, irregularly verrucose or striate, wingless.

Rocky hillsides, canyon walls, and bottoms of washes, Lower Sonoran Zone, from Cedros Island and Bahía de los Angeles, Baja California, south to the Cape Region. Oct.–June.

**3. Mentzelia micrantha** (Hook. & Arn.) Torr. & Gray, Fl. N. Amer. **1:** 535, 1840

*Bartonia micrantha* Hook. & Arn., Bot. Beechey 343, *pl. 85,* 1840.
*Acrolasia micrantha* Rydb., Bull. Torrey Club **30:** 278, 1903.
*Acrolasia micrantha stricta* Davidson, Bull. S. Calif. Acad. **5:** 15, 1906.
*Mentzelia micrantha* var. *stricta* Davidson, in Davids. & Moxley, Fl. S. Calif. 239, 1923.
*Acrolasia catalinensis* Millsp., Field Mus. Bot. Ser. **5:** 177, 1923.

An herb 3–6 dm. tall, branching slightly below, more profusely and compactly above; branches pale yellow or whitish, moderately puberulent below, hispidulous and puberulent with barbed hairs above; leaves ovate to broadly lanceolate, 1–5 (or rarely –9) cm. long, acute to acuminate at apex, sinuate-dentate to entire, lower ones petiolate, middle and upper ones sessile, scaberulous; flowers sessile in compact, cymose or rarely corymbose clusters; bracts ovate to suborbicular, 3–14 mm. long; calyx tube 3–4 mm. long, densely scabrous; calyx lobes lance-subulate, about 2 mm. long, acute, erect in fruit; petals 5, obovate, 3–3.5 mm. long; stamens 10–25, the 5 stamens opposite petals 2.5 mm. long, 1.5 mm. wide; capsule cylindrical, quadrangular, 6–12 mm. long; seeds 5–8, minutely punctate, about 1.5 mm. long, wingless, sides faintly muriculate.

Arid gravelly and sandy soil, mainly Upper Sonoran Zone, northern California and south through the Coast Ranges to Baja California; Guadalupe Island. May–July.

**4. Mentzelia albicaulis** Dougl. ex Hook., Fl. Bor. Amer. **1:** 222, 1833

*Bartonia albicaulis* Dougl. ex Hook., loc. cit.
(For complete synonymy see Ann. Mo. Bot. Gard. **21:** 182, 1934.)

Annual with shining white stems branching from base or sometimes only above, 4–6 dm. tall, stems and branches sparsely pubescent to glabrous; leaves sessile, scabrous-puberulent, the lower linear-lanceolate, 3–5 cm. long, entire or dentate or sometimes sinuate-pinnatifid, the middle pinnatifid or with several pairs of linear or oblong lobes, the upper linear to lanceolate, usually less deeply lobed than middle ones or entire; flowers sessile, axillary, the terminal ones in corymbose clusters of 2–3; calyx tube 10–15 mm. long, spreadingly hispid-scabrous; calyx lobes lance-subulate, 2–2.5 mm. long; petals 5, yellow, 2–4 mm. long, usually prominently veined and emarginate; filaments filiform; capsule linear-cylindric, 1–1.5 cm. long, slender, sessile; seeds irregularly angled, granular-tuberculate, processes low and rounded.

Arid soil, mainly Upper Sonoran Zone, eastern British Columbia to Arizona, New Mexico, and central Baja California, entering the Sonoran Desert in southeastern California, Arizona, and Baja California. June–Aug.

**5. Mentzelia affinis** Greene, Pittonia **2:** 103, 1890

*Acrolasia affinis* Rydb., Bull. Torrey Club **30:** 278, 1903.

Annual, 1–5 dm. tall, with white, shining stems usually glabrous below, sparsely scabrous-puberulent above; leaves lanceolate or sometimes lance-ovate, 2–9 cm. long, rarely over 1.5 cm. wide, usually deeply sinuate-pinnatifid, sessile, scabrous-puberulent on both surfaces; flowers terminating branches and solitary in upper forks, sessile; calyx tube linear, 15–25 mm. long, scabrous; calyx lobes lance-subulate, 5–6 mm. long, spreading to reflexed at maturity; petals pale yellow, sometimes nearly white, obovate, 6–8 mm. long; stamens 25–40, filaments linear; capsule 1.5–2.5 cm. long, often somewhat curved; seeds prismatic, strongly angled, grooved on 3 angles, more or less granulate-muriculate on the sides, ends truncate.

Arid sandy or stony ridges, hillsides, and plains, Lower and Upper Sonoran Zones, central California south to central Baja California and east in Arizona to the vicinity of Tucson. March–June.

**6. Mentzelia veatchiana** Kell., Proc. Calif. Acad. **2:** 99, 1861

*Mentzelia albicaulis* var. *veatchiana* Urban & Gilg, Nov. Act. Nat. Cur. **76:** 28, 1900.
*Acrolasia veatchiana* Rydb., Bull. Torrey Club **30:** 278, 1903.
*Mentzelia qracilenta* var. *veatchiana* Jepson, Man. Fl. Pl. Calif. 652, 1925.

Slender, erect or spreadingly ascending annual branching from base or only above, 1.5–6 dm. tall; stems yellowish white or greenish, glabrous to sparsely scaberulous-puberulent; lower leaves narrowly oblanceolate to elliptic-lanceolate, narrowed to a narrowly winged petiole, sinuately dentate to pinnatifid, 5–15 cm. long, to 3 cm. wide, middle and upper leaves usually pinnatifid, somewhat smaller, uppermost commonly entire or nearly so, 1.5–3 cm. long; flowers in small clusters terminating branches; bracts ovate to lanceolate, 4–10 mm. long, entire or slightly toothed; sepals lance-subulate, 2–3 mm. long; petals yellow, 4–6 mm. long, strongly veined; capsule slenderly clavate-cylindrical, 2–2.5 cm. long, nearly terete, scabrous; seeds obscurely prismatic, distinctly grooved along one angle, tuberculate on sides with rounded, distinctly separated, low bosses.

In light, arid soils, Lower and Upper Sonoran Zones, southeastern Oregon to Utah, Arizona, and through the eastern slopes of the Sierra Nevada into the desert regions of southern California and the southern flanks of the Sierra San Pedro Mártir, Baja California. Also in the cross ranges at the southern end of the San Joaquin Valley and eastern San Luis Obispo County, California. April–Aug.

**7. Mentzelia pectinata** Kell., Proc. Calif. Acad. **3:** 40, 1863

*Mentzelia albicaulis* var. *pectinata* Urban & Gilg, Nov. Act. Nat. Cur. **76:** 29, 1900.
*Acrolasia pectinata* Rydb., Bull. Torrey Club **30:** 278, 1903.
*Mentzelia gracilenta* var. *pectinata* Jepson, Man. Fl. Pl. Calif. 652, 1925.

Annual with rather stout, more or less spreading branches 1–4 dm. long, greenish or yellowish white, glabrous to puberulent at base, usually closely puberulent-scabridulous above; leaves sessile, lanceolate in outline, 3–12 cm. long, to 2 cm. wide at base, pectinately lobed, lobes narrowly triangular or pinnatifid, acuminate at apex, sparingly scabridulous along margins and on midrib; flowers in small terminal clusters or solitary; bracts lanceolate, few-toothed, leaflike, 1–2 cm. long; sepals lance-ovate, acuminate, 5–6 (–10) mm. long; petals broadly obovate, rounded at apex, 10–15 mm. long, yellow with a copper-red spot near base; filaments to 1 cm. long; capsule clavate-cylindrical, 2–2.5 cm. long; seeds irregularly angled, minutely tuberculate on sides, angles rarely grooved, often narrowly winged.

Dry, sandy, gravelly, or rocky slopes, Lower and Upper Sonoran Zones, southern San Joaquin Valley and adjacent foothills to the western Colorado Desert, California. Often growing in canyons opening onto the Mojave and Colorado Deserts. March–June.

The bright reddish spot at the base of the petals is distinctive.

**8. Mentzelia nitens** Greene, Fl. Fran. 234, 1891, var. **nitens**

*Acrolasia nitens* Rydb., Bull. Torrey Club **30:** 278, 1903.
*Mentzelia gracilenta* var. *nitens* Jepson, Fl. Calif. **2:** 533, 1936.

Annual or sometimes perennial, with smooth, shining, white stems 3–5 dm. long, somewhat decumbent; leaves sessile, lanceolate, to about 10 cm. long, pinnatifid with linear or oblong lobes, acuminate at apex, uppermost often entire, short-pubescent and scabrous; flowers terminal and in upper axils; sepals lanceolate, acuminate, 6–10 mm. long, spreading to ascending in fruit; petals golden yellow, 15–20 mm. long, obovate, rounded to emarginate at apex; stamens slightly dilated at base, to one-half as long as petals; capsules clavate-cylindrical, 1.5–2.5 cm. long, hispid-scabrous; seeds irregularly angled, faces tuberculate.

Sandy hillsides, plains, and ridges, Lower Sonoran Zone, Mono County, California, into adjacent Nevada and south into Arizona and the Colorado Desert, California, and northeastern Baja California. April–May.

**8*a*. Mentzelia nitens** var. **jonesii** (Urban & Gilg) Darlington, Ann. Mo. Bot. Gard. **21**: 198, 1934

*Mentzelia albicaulis* var. *jonesii* Urban & Gilg, Nov. Act. Nat. Cur. **76**: 29, 1900.
*Mentzelia albicaulis* var. *spectabilis* M. E. Jones, Contr. West. Bot. **12**: 16, 1908.

Differing from var. *nitens* in having petals only 8–12 mm. long, stouter stems, and more of the upper leaves entire or nearly so.

Arid hillsides and canyons, Lower Sonoran Zone, Owens Valley, California, southern Nevada and southern Utah, and south in western Arizona to the vicinity of Wickenburg.

**8*b*. Mentzelia nitens** var. **leptocaulis** Darlington, Ann. Mo. Bot. Gard. **21**: 199, 1934

A rather poorly defined variety with flowers like those of var. *jonesii* and slender flexuous stems having a greenish instead of a yellow-white tinge.

Arid sandy soils, Lower Sonoran Zone, Bill Williams Fork of the Colorado River. Known from one old collection only and probably scarcely worthy of nomenclatorial rank.

**9. Mentzelia multiflora** (Nutt.) A. Gray, Pl. Fendl. 48, 1849

*Bartonia multiflora* Nutt., Jour. Acad. Phila. II, **1**: 180, 1848.
*Mentzelia pumila* var. *multiflora* Urban & Gilg, Nov. Act. Nat. Cur. **76**: 93, 1900.
*Mentzelia aurea* Osterh., Bull. Torrey Club **28**: 644, 1901. Not *M. aurea* Nutt., 1818.
*Mentzelia speciosa* Osterh., op. cit. 689.
*Mentzelia longiloba* Darlington, Ann. Mo. Bot. Gard. **21**: 176, 1934.
(For additional synonymy see Ann. Mo. Bot. Gard. **21**: 164–165, 1934.)

Perennial 4–10 dm. tall, with lower stems sparsely scabrous to smooth, upper parts and foliage densely scabrous-pubescent, yellowish white; leaves lanceolate, sinuate to deeply pinnatifid, 5–7.5 cm. long, 1–2 cm. broad, sessile; flowers in terminal corymbs, 3–4 in a cluster; calyx tube urceolate, 15–16 mm. long, scabrous; sepals lance-subulate, somewhat reflexed, 9–12 mm. long, persistent; petals 10, 15–20 mm. long, obtuse at apex, bright yellow, inner ones smaller than the outer; filaments all antheriferous, outer ones slightly broader than the inner; capsule urceolate, 6–8 mm. wide, 15–20 mm. long, 3–4-valved, scabrous; seeds 50–80 in 2 series, margins winged; placentae fleshy.

Sandy open ground, Lower and Upper Sonoran Zones, Inyo County, California, to southern Wyoming, Colorado, western Texas, and south into Chihuahua, Coahuila, Sonora, and Baja California. March–Sept.

**10. Mentzelia pumila** Nutt., in Torr. & Gray, Fl. N. Amer. **1**: 535, 1840, var. **pumila**

*Mentzelia wrightii* Porter & Coulter, Fl. Colo. 47, 1874. Not *M. wrightii* A. Gray, 1848.
*Mentzelia pumila* var. *genuina* Urban & Gilg, Nov. Act. Nat. Cur. **76**: 93, 1900.
*Hesperaster pumilus* Cockerell, Torreya **1**: 143, 1901.
*Touterea pumila* Rydb., Bull. Torrey Club **30**: 277, 1903.
*Nuttallia pumila* Greene, Leaflets Bot. Obs. **1**: 210, 1906.

Biennial with more or less tortuous, grayish to yellowish white, striate, stout stems finely scaberulous with retrorsely barbed hairs, epidermis splitting and peeling away on older parts; leaves linear-lanceolate to oblanceolate, 1.5–4 or rarely –10 cm. long, irregularly sinuate-dentate to pinnatifid, obtuse or acute at apex, attenuate toward base, scaberulous; flowers in terminal, few-flowered cymes, each subtended by 1–2 linear bracts at base of calyx tube, this clavate to cylindrical, 5–15 mm. long, scabrous; calyx lobes ovate-oblong, attenuate, 8–10 mm. long; petals 10, yellow, lanceolate to obovate, acute, 9–15 mm. long; stamens 60–120, in 3–4 series, outer series slightly broadened and longer than the inner, linear, three series; capsule subcylindrical, clavate to oblong, attenuate below, 15–20 mm. long, scabrous and striate; seeds mottled brown, minutely punctate-granulate, broadly winged on angles.

Arid slopes and sandy plains, Lower and Upper Sonoran Zones, Wyoming to Arizona, New Mexico, and Texas. Feb.–Oct.

**10*a*. Mentzelia pumila** var. **procera** (Woot. & Standl.) Darlington, Ann. Mo. Bot. Gard. **21:** 169, 1934

*Nuttallia procera* Woot. & Standl., Contr. U.S. Nat. Herb. **16:** 150, 1913.

Differing from var. *pumila* in being stricter in its branching, its white papery epidermis which is puberulent only on young growth, its somewhat smaller leaves, and an oblong-cylindric capsule 10–12 mm. long.

Arid, sandy, and rocky slopes, hillsides, and washes, Lower and Upper Sonoran Zones, Nevada to Utah and south to western Texas, Nuevo León, Coahuila, New Mexico, and Arizona. Fairly common near Yuma, Arizona, but not reported from the California side of the Colorado River.

**11. Mentzelia puberula** Darlington, Ann. Mo. Bot. Gard. **21:** 177, 1934

Perennial 1–4 dm. tall with dull white, stout stems and branches closely scaberulous with antrorsely barbed hairs; leaves ovate-lanceolate to elliptic-oblong, 1–3.5 cm. wide, 2–8 cm. long, coarsely sinuate-dentate, acute to obtuse at apex, broadly cuneate to rounded at base, the upper sessile, those below on petioles to 1.5 cm. long, midrib stout and prominent, white and scaberulous; flowers in few-flowered clusters at tips of branches, sessile or pedicellate; calyx tube obconic, 5–6 mm. long, scaberulous; calyx lobes lanceolate or lance-ovate, attenuate, 6–8 mm. long; petals bright yellow, oblanceolate, 8–10 mm. long, obtuse at apex, short-clawed; outer stamens petaloid, inner filiform; capsule turbinate, 9–11 mm. long, scabrous, obscurely veined; seeds broadly winged, about 3 mm. long, minutely tessellate with low, rounded, minute tubercles.

Gravelly and rocky areas, Lower Sonoran Zone, Mojave and Colorado Deserts east into western Arizona, and south into northeastern Baja California. March–April.

**12. Mentzelia tricuspis** A. Gray, Amer. Nat. **9:** 271, 1875

*Bicuspidaria tricuspis* Rydb., Bull. Torrey Club **30:** 275, 1903.
*Acrolasia tricuspis* Davidson, Bull. S. Calif. Acad. **9:** 71, 1910.
*Acrolasia tridentata* Davidson, loc. cit.
*Mentzelia tricuspis* var. *brevicornuta* I. M. Johnston, Univ. Calif. Pub. Bot. **7:** 444, 1922.

Annual 5–25 cm. tall with spreading or erect stems that are whitish or yellowish and closely hispid with simple and retrorsely barbed hairs, more or less striate; leaves linear-lanceolate to lance-ovate, or oblanceolate, short-petiolate or some sessile, 2–7 cm. long, 1–2 cm. wide, nearly entire to coarsely and saliently toothed, acute at apex; flowers terminal on short branches; bracts lance-linear, sinuate-dentate, 1–2 cm. long,

2–4 mm. wide; sepals lance-attenuate, 7–13 mm. long, persistent; calyx tube 10–15 mm. long, densely scabrous-hirsute; petals obovate, 2–2.5 cm. long, apiculate; stamens 90–100, in 4 series, outer 8–9 mm. long, slightly dilated toward tricuspidate apex, lateral cusps exceeding or barely equaling central, anther-bearing one; capsule 10–15 mm. long, 5–8 mm. in diameter, hirsute, usually reflexed at maturity; seeds irregularly angled, rugulose, not winged, 2–2.5 mm. long, grayish.

Dry hillsides and mesas, Lower Sonoran Zone, Colorado and Mojave Deserts, California, to southern Utah and Mohave County, Arizona. April–May.

*Acrolasia tridentata* Davidson and *M. tricuspis* var. *brevicornuta* I. M. Johnston were applied to an inconstant form in the Mojave Desert having reduced rounded lobes on the outer stamens.

### 13. **Mentzelia hirsutissima** S. Wats., Proc. Amer. Acad. **12:** 252, 1877, var. **hirsutissima**

*Bicuspidaria hirsutissima* Rydb., Bull. Torrey Club **30:** 275, 1903.

Annual or biennial herb with stout, erect or ascending branches to 4 dm. tall; stems and branches whitish and subglabrous, hirsute and scabrous-hirsute when young; leaves sessile, lanceolate-linear, acuminate at apex, irregularly and saliently pinnatifid, 3–8 cm. long, 1–1.5 cm. wide, hirsute above and along midvein beneath with spreading spinelike hairs; floral bracts acuminate, narrowly lobed or toothed; flowers sessile, axillary and terminal; calyx tube 10–14 mm. long; calyx lobes 15–20 mm. long, lance-attenuate, somewhat hirsute; petals yellow, 3–3.5 cm. long, acute; stamens 100–300, in 4–5 series, filaments orange near apex, 10–12 mm. long, shortly tricuspidate; capsule oblong, about 1.5 cm. long, hirsute.

Known only from the type locality, Isla Angel de la Guarda in the Gulf of California. Nov.–April.

### 13*a*. **Mentzelia hirsutissima** var. **stenophylla** (Urban & Gilg) I. M. Johnston, Univ. Calif. Pub. Bot. **7:** 443, 1922

*Mentzelia stenophylla* Urban & Gilg, Nov. Act. Nat. Cur. **76:** 80, 1900.

Differing from var. *hirsutissima* only in having the middle, anther-bearing cusp of stamen elongated to about twice the length of the lateral ones.

Rocky hillsides, canyons, and desert mesas, Lower Sonoran Zone, southeastern San Diego County (Mountain Springs) to San Quintín and Isla de la Guarda, Baja California. Common and more widely distributed than var. *hirsutissima.*

**Mentzelia hirsutissima** var. **nesiotes** I. M. Johnston, Univ. Calif. Pub. Bot. **7:** 443, 1922, differs from var. *hirsutissima* only in having thinner, coarsely toothed leaves and more deeply orange flowers. The constancy of these characters is open to question. Known only from Cedros, San Benito, and La Natividad Islands.

### 14. **Mentzelia involucrata** S. Wats., Prod. Amer. Acad. **20:** 367, 1885, var. **involucrata**

*Bicuspidaria involucrata* Rydb., Bull. Torrey Club **30:** 275, 1903.
*Nuttallia involucrata* Davids. & Moxley, Fl. S. Calif. 240, 1923.

Stout annual with stems 15–35 cm. high branching from base and usually white and hispid; leaves lanceolate, acute at apex, the basal petiolate, the upper sessile, 3–8 cm. long, irregularly sinuate-dentate, densely short-hispid and scabridulous; flowers terminating ascending branches; bracts broadly ovate or ovate-lanceolate, 1.5–2.5

cm. long, pectinately toothed, white-scarious in center, green along margins; sepals lance-attenuate, 10–18 mm. long; petals yellow, 2–2.5 cm. long, narrowly obovate to oblanceolate, apiculate at apex; outer filaments tricuspidate at slightly dilated apex, central cusp anther-bearing, lateral ones linear, equaling or exceeding anther; capsule subcylindrical, 5–8 mm. in diameter, 12–15 mm. long; seeds horizontally flattened to ovoid, irregularly angled, 3–3.5 mm. long, closely tuberculate.

Arid hillsides and desert plains, Lower Sonoran Zone, Mojave Desert, California, to western Arizona, and south into Baja California and Sonora. Feb.–May.

**14*a*. Mentzelia involucrata** var. **megalantha** I. M. Johnston, Univ. Calif. Pub. Bot. **7:** 443, 1922

Differing only in having larger, more showy flowers with petals 3.5–4.5 cm. long.

Desert flats and hillsides, Lower Sonoran Zone, Colorado Desert (mainly in the western portion), California, and in adjacent Baja California.

## 4. EUCNIDE Zucc., Linnaea 18: 508, 1844

Herbs or small shrubs with stiff barbed hairs and some straight stinging ones. Leaves alternate, petiolate, lobed or toothed or both. Flowers small or large, in bracted terminal cymes. Calyx tube adnate to ovary, pubescent. Sepals 5, persistent. Petals 5, coalesced at base. Stamens many, inserted on base of petals; filaments filiform. Ovary 1-celled but with 5 placentae, many-ovuled. Style 5-cleft. Capsule obovoid, 5-valved at apex. Seeds many and minute, striate lengthwise.

Leaves ovate to suborbicular, not cordate; petals 3–4 cm. long; stamens half as long as petals ........................................ 1. *E. urens*
Leaves cordate-orbicular; petals 1.5–2 cm. long; stamens longer than petals ... 2. *E. cordata*

**1. Eucnide urens** Parry, Amer. Nat. **9:** 144, 1875
*Mentzelia urens* Parry ex A. Gray, Proc. Amer. Acad. **10:** 71, 1874.

A shrub to 5 dm. tall with rather stout, decumbent, spreading or ascending branches generously clothed with straight, sharp-pointed, white stinging hairs to 4 mm. long, and slightly shorter ones bearing 1–5 whorls of minute, retrorsely curved barbs, and finer, shorter, simple hairs about the base of the other two kinds; leaf blades broadly ovate to suborbicular, 2.5–5 cm. long, nearly or quite as wide, irregularly and coarsely toothed, pubescent in same manner as stems, the upper usually sessile and more or less clasping, the lower on petioles 2–5 cm. long; pedicels stout, 1–2.5 cm. long; sepals lance-oblong, 2–2.5 cm. long, attenuate; petals pale yellow, tinged with green toward base and on back, obovate, 3–4 cm. long, to 2 cm. wide, mucronate and hispid at tips; stamens about half as long as petals; capsule obovoid, 1.5–2 cm. long; seeds brown, fusiform, 0.6–0.8 mm. long.

Rocky hillsides and sandy plains and washes, Lower Sonoran Zone, Inyo and San Bernardino Counties, California, to southern Utah and south into northern Yuma County, Arizona. April–Sept.

**2. Eucnide cordata** Kell. ex Curran, Bull. Calif. Acad. **1:** 137, 1885
*Mentzelia cordata* Kell., Proc. Calif. Acad. **2:** 33, 1860.

A coarse perennial herb 2–10 dm. tall with stout ascending branches, these and leaves closely clothed with a few simple hairs and many of about the same length bearing only a single terminal whorl of recurved, hooklike barbs; petioles 1–3 cm. long; leaf blades 2–10 cm. wide, as long or slightly longer, orbicular to ovate in outline, cordate at base, shallowly 5–9-lobed, lobes rounded and irregularly dentate; flowers in terminal cymes or racemes; pedicels 5–12 mm. long; sepals oblanceolate, acuminate, 1–1.5 cm. long; petals white or pale cream, 1.5–2 cm. long, minutely apiculate and slightly puberulent on back at tip; stamens exserted about 1 cm.; capsule obpyriform, 6–8 mm. high; seeds pale yellow, 0.4–0.6 mm. long, with minute, longitudinal, brownish striations and a pale brown apiculation at one or both ends.

Rocky hillsides, cliffs, or gravelly desert slopes, Lower Sonoran Zone, southern flanks of the Sierra San Pedro Mártir to La Paz, Baja California, Cedros Island, and near the coast in the vicinity of Guaymas, Sonora. Flowers practically the year around.

## 5. **SYMPETALEIA** A. Gray, Proc. Amer. Acad. **12:** 161, 1877

Annual herbs with petiolate, hirsute, and more or less viscid, alternate leaves. Flowers axillary in axils or in terminal cymes. Calyx tube obconic to globose; lobes 5. Corolla sympetalous; lobes 5, rounded, imbricate in bud. Stamens many, inserted in 2 whorls in throat and tube of corolla; filaments short, subfiliform; anthers subreniform. Ovary inferior, 1-celled, many-ovuled. Style filiform; stigma branches 5, connivent. Placentae 5. Capsule subglobose to obovoid, dehiscing apically. Seeds many, oblong, finely striate spirally. Embryo straight. Endosperm scanty.

Corolla tube 8–15 mm. long; stamens included; corolla lobes greenish or bright rouge red:
  Corolla lobes greenish, erect or only slightly flaring; calyx lobes 6–9 mm. long .......... ............................................................ 1. *S. rupestris*
  Corolla lobes bright orange-red, spreading; calyx lobes 2–3 mm. long ........ 2. *S. aurea*
Corolla tube 1–1.5 mm. long; stamens exserted; corolla lobes yellow to ochroleucous ...... ............................................................ 3. *S. tenella*

### 1. **Sympetaleia rupestris** (Baill.) A. Gray ex S. Wats., Proc. Amer. Acad. **24:** 50, 1889

*Loasella rupestris* Baill., Bull. Men. Soc. Linn. Paris **1:** 650, 1887.

Rounded, bright green and glistening herb 1–4 dm. high, often fully as broad; stems stout, hispid with spreading stiff hairs and with intermingled, shorter, columnar ones tipped with a whorl of retrorsely curved barbs; petioles 1–5 cm. long; leaf blades orbicular-cordate, 2–8 cm. wide, shallowly lobed and irregularly toothed, sinus broad, pubescent in the same manner as stems; flowers solitary in upper axils and forming cymose clusters; pedicels 1–5 mm. long at anthesis; calyx tube obconic, 4–6 mm. long, hispid; lobes oblong-spatulate, 6–9 mm. long, erect or nearly so; corolla tube 8–10 mm. long, 2–2.5 mm. in diameter, pubescent, yellowish green, lobes 2–3 mm. long, bright green, erect; stamens included; capsule about 1 cm. long; seeds spirally striate.

Sides of cliffs and among rocks, Lower Sonoran Zone, near Coyote Wells, Colorado Desert, south to San Pablo, Baja California, and in the Pinacate Mountains, northwestern Sonora. Nov.–May.

### 2. **Sympetaleia aurea** A. Gray, Proc. Amer. Acad. **12:** 161, 1877

A low, rounded, nearly or quite prostrate herb with stems 1–3 dm. long, these and leaves beset with sharp, stiff, pricklelike hairs and shorter, retrorsely barbed ones;

petioles 1–3.5 cm. long; leaf blades broadly ovate to suborbicular or somewhat reniform, 2–6 cm. wide, to 5 cm. long, shallowly 5–9-lobed, truncate to cordate at base, lobes rounded and coarsely rounded-dentate; flowers solitary in axils, pedicels slender, 5–25 mm. long in flower, to 8 cm. long in fruit; sepals linear-oblong, 2–3 mm. long; corolla bright orange-red or vermilion, tube to 1.5 cm. long, 1.5–2 mm. in diameter, corolla lobes spreading, broadly ovate, 4–6 mm. long, nearly as broad, hispid-ciliate; stamens included; capsule broadly obovoid or hemispherical, 4–5 mm. high, hispid, about 10-ribbed; seeds broadly ellipsoid, about 0.6 mm. long, pale yellowish.

Rocky talus slopes and in crevices on the face of cliffs, Lower Sonoran Zone, mainly the eastern side of Baja California between Santa Rosalia and Agua Verde Bay and adjacent islands, but occasional in rocky canyons near Pacific. Nov.–May.

**3. Sympetaleia tenella** I. M. Johnston, Proc. Calif. Acad. IV, **12:** 1106, 1924

Prostrate, somewhat viscid annual with bright green leaves and stems 1–2.5 dm. long, branching from base; petioles 1–2.5 cm. long; leaf blades suborbicular to reniform in outline, 1.5–2.5 cm. long, about as wide, shallowly 5-lobed, lobes unequal and coarsely crenate or dentate, villous-hispid; flowers solitary in axils; pedicels 4–5 mm. long at anthesis, to 7 cm. long in fruit; calyx tube depressed-globose, about 1.5 mm. high; sepals oblong, 1.5–2 mm. long; corolla yellow to ochroleucous, tube 1–1.5 mm. long, lobes spreading, oblong, 3.5–4 mm. long, 2–2.2 mm. wide, hispid without; stamens 15–25, attached at and below middle of tube, filaments 4–6 mm. long, exserted most of their length; capsule 2–3 mm. long, hispid; seeds oblong-ellipsoid, about 0.3 mm. long, spirally striate and apiculate at one or both ends.

Rocky canyons, Lower Sonoran Zone, along eastern slopes of Sierra de la Giganta from vicinity of Agua Verde Bay south at least to San Evaristo, Baja California. Nov.–May.

## Family 96. DATISCACEAE. Datisca Family

Herbs (in ours) or trees with alternate, simple or pinnately compound leaves. Plants dioecious, or flowers rarely perfect. Hypanthium short in staminate flowers, adnate to ovary in pistillate flowers. Sepals 3–9, slightly unequal. Petals none or 8. Stamens few to many, usually fewer in perfect flowers than in staminate flowers of dioecious species. Styles 3–8. Ovary inferior, 1-celled, with 3 parietal placentae alternating with the sepals. Fruit a capsule, dehiscing apically between styles. Seeds numerous, small, punctate-reticulate and often striate, anatropous, with a distinct strophiole. Embryo straight; endosperm present.

### 1. DATISCA L., Sp. Pl. 1037, 1753

Robust perennial herbs with unequally pinnatifid leaves. Flowers completely or incompletely dioecious, a few perfect flowers often occurring on predominantly pistillate plants, perfect flowers with 1–5 anthers borne at apex of hypanthium; inflorescences of axillary glomerules or racemes. Staminate flowers with very short hypanthium; petals none; sepals 4–9; stamens 10–25, filaments much shorter than their anthers. Pistillate flowers with hypanthium adnate to ovoid, obscurely 3-angled ovary; petals none; sepals 3; styles 3–4, each style bifid, its branches slenderly subulate. Capsule oblong, coriaceous, 3–5-ribbed.

**1. Datisca glomerata** (Presl) Baillon, Hist. Pl. **3:** 407, 1872

*Tricerastes glomerata* Presl, Rel. Haenk. **2:** 88, *pl. 64,* 1835.

Glabrous perennial herb 0.6–2 m. tall, erect, strict or sparingly branched; leaves ovate to narrowly lanceolate in outline, 6–20 cm. long, 1–15 cm. wide, unequally and laciniately pinnatifid, segments again irregularly toothed to shallowly pinnatifid, acute to somewhat attenuate at apex; floral leaves greatly reduced, often only 5–10 mm. wide, 3–6 cm. long, serrate and with a single pair of narrow basal lobes; flowers forming elongated axillary racemes; plants incompletely dioecious, staminate plants bearing only stamens, pistillate often with some flowers bearing 1–5 normal stamens at apex of hypanthium; staminate flowers with slender sepals 0.6–1.2 mm. long, spreading; anthers 3–5 mm. long, yellow, borne on filaments 0.2–1 mm. long; pistillate flowers with deltoid-subulate sepals 1.5–2 mm. long; styles slender, equaling to twice as long as ovary at anthesis; capsules ovoid-oblong, 6–11 mm. long, 3–4.5 mm. in diameter, obtusely 3–5-angled, 6–10-nerved, truncate and dehiscing between styles at apex; seeds ellipsoidal, 0.6–0.8 mm. long, finely reticulate-punctate in longitudinal rows, pale to dark brown.

Stream banks and margins of moist meadows, upper margins of Lower Sonoran Zone into the Transition Zone, northern California south through the Coast Ranges and Sierra Nevada into northern Baja California and east into western Nevada; reaching the margins of the Sonoran Desert along the western edge of the Colorado and San Felipe Deserts. May–Oct.

## Family 97. BEGONIACEAE. Begonia Family

Chiefly creeping to acaulescent, subsucculent, monoecious herbs or low shrubs with conspicuously jointed stems and alternate, more or less 2-ranked leaves. Leaves petiolate; blades simple, entire, toothed, or lobed, palmately nerved, usually markedly asymmetrical at base. Stipules present, mostly caducous or early deciduous. Flowers unisexual, actinomorphic or zygomorphic. Perianth of staminate flowers of 2 valvate, petaloid sepals and 2 smaller valvate petals. Perianth of pistillate flowers of 2 to many imbricated, petaloid undifferentiated segments. Stamens many; filaments distinct or slightly connate in tufts at base, arranged in several whorls; anthers 2-celled, basifixed, often with an exserted connective. Pistil one; ovary inferior or partially inferior, often 1–3-winged, usually trilocular; styles 2–5, free or connate at base. Placentation generally axial. Ovules many, anatropous. Fruit a loculicidal capsule or less commonly a berry. Seeds numerous, small. Embryo straight, oily. Endosperm none.

### 1. BEGONIA L., Gen. Pl. ed. 2, 516, 1742

Characters of the family. Flowers zygomorphic. Ovary wholly inferior. Fruit a capsule, dehiscing longitudinally below apex, or less commonly berrylike and irregularly dehiscent.

Leaves ovate-lanceolate, 1–5 cm. wide, irregularly toothed but scarcely lobed; wings of fruits deltoid, acute at apex, decurrent on pedicel, 6–13 mm. wide ........................................ 1. *B. gracilis* var. *nervipilosa*

Leaves broadly reniform, 6–15 cm. wide, shallowly but distinctly 5–7-lobed; wings of fruits lunate-rounded, 3–6 mm. wide, cordate at base, not decurrent ........ 2. *B. californica*

**1. Begonia gracilis** H. B. K. var. **nervipilosa** A. DC., Prod. **15**[1]: 309, 1864

A subsucculent perennial herb, from 1–several depressed-globose, pinkish, cormlike tubers 6–10 mm. in diameter; stems erect, slender, 1–3 dm. tall, sparsely hirsute with coarse, somewhat tangled hairs; stipules asymmetrical, 2–4 mm. long, minutely and irregularly denticulate; petioles slender, 3–25 mm. long, sparsely pilose; leaf blades asymmetrically ovate-lanceolate, 1–5 cm. wide, 2–8 cm. long, irregularly dentate, subattenuate at apex, glabrous except along veins, these sparsely pilose beneath, nearly glabrous above; inflorescence racemose, 5–15 cm. long, erect, each pedicel bearing a pair of broadly ovate, acute, basal bracts 1–2.5 mm. long; pedicels slender, 3–8 mm. long at anthesis, to 2.5 cm. long in fruit, spreading-horizontal but abruptly reflexed at apex; flowers white or sometimes tinged with pink, tepals translucent-waxy; staminate flowers with outer pair of segments suborbicular, 11–12 mm. in diameter, or to 15 mm. long in some, the inner pair narrowly ovate, oblong, or narrowly obovate, 4–5 mm. wide, 9–12 mm. long; pistillate flowers with 5 tepals similar to those of staminate flowers, but innermost somewhat shorter; anthers bright yellow, styles 2–3 mm. long, nearly peltate and papillose at apex; capsules chartaceous-stramineous, 10–13 mm. long, 6–8 mm. in diameter, bearing 3 unequal, broadly ovate-deltoid wings, these slightly decurrent along pedicel, nearly truncate across top of fruit, 6–13 mm. wide, acute, distinctly reticulate-veined, firm-papery.

In rich soil in moist places in shaded canyons and seepy areas, Lower Sonoran and Tropical Zones, vicinity of Cumpas, Sonora, south into central Mexico. July–Aug. (Nearly mature fruit collected in late September.)

**2. Begonia californica** Brandegee, Proc. Calif. Acad. II, **3**: 140, 1891

Erect perennial herb 2–5 dm. tall, from a cormlike, pinkish tuber to 15 mm. in diameter; stems subfleshy, erect or slightly decumbent at base, glabrous, pale green or irregularly tinged with purple; leaves cordate-reniform, the lower ones 5–8 cm. long, 10–15 cm. wide, shallowly but distinctly 5–7-lobed, moderately to strongly asymmetrical, faintly dentate-serrulate and ciliate-setiform along margins, sparsely pilose on veins; petioles slender, lower ones 5–12 cm. long, sparingly pilose toward apex; upper leaves smaller, more strongly asymmetrical, and lobes narrower, with petioles 1–3 cm. long; stipules broadly lanceolate, 2–3 mm. wide, 6–7 mm. long, ciliate-margined, tardily deciduous to subpersistent; inflorescence a few-flowered cyme; peduncles slender, ascending, 1–3 cm. long; bracts 1.5–3 mm. long, broadly ovate-lanceolate, acute to obtuse, ciliate; pedicels 2.5–4 cm. long, slender, bibracteolate at or slightly below middle; flowers pink or with a purplish tinge; staminate perianths of 4 tepals, lower pair broadly ovate, 7–9 mm. long, 5–7 mm. wide, minutely denticulate, upper pair narrower (3–4 mm. wide); pistillate perianths of 5 tepals, outer ones 5–6 mm. long, inner smaller; styles 2–4 mm. long; capsules 12–20 mm. long, chartaceous, 2–3-winged; wings roundish-lunate, largest 5–7 (12) mm. wide, smallest 2–3 mm. wide, shallowly cordate at base of fruit, rounded at summit of fruit, not decurrent along pedicel.

Shaded canyons and cliffs, Lower Sonoran and Tropical Zones, Sierra San Francisquito and Sierra de la Laguna, Baja California. Sept.–Oct.

**Begonia portillana** S. Wats., Proc. Amer. Acad. **22**: 414, 1887, is similar in appearance to *B. californica,* but the fruit bears a single wing which is truncate at the apex. Reported from Río Mayo area by Gentry, and to be expected along the eastern margins of the Sonoran Desert in southern and central Sonora.

## Family 98. **CACTACEAE.** Cactus Family

Succulent perennials with columnar, globose, flattened, angled, or terete stems, these simple or branching, erect, ascending, scandent, or prostrate. Leaves present only in *Pereskiopsis*, *Opuntia*, and *Marenopuntia*; ample and eventually deciduous in *Pereskiopsis*, subulate and caducous in *Opuntia* and *Marenopuntia*. Spines, leaves (when present), and flowers produced by specialized structures called areoles, these often arranged on ribs or tubercles. Flowers perfect, solitary, or in a few genera several from the margins of an areole. Sepals and petals several to many, commonly intergrading and more or less imbricated, inserted on a cup or tube adherent to the inferior ovary. Stamens many; filaments inserted on throat of perianth tube, usually included or barely equaling perianth segments, sometimes exserted, frequently sensitive to mechanical stimulus. Style 1; stigmas 2 to several or many. Ovary inferior, 1-celled; ovules numerous, parietally arranged. Fruit dry, or a juicy or pulpy berry, indehiscent, or irregularly dehiscent laterally, or by a basal pore, or circumscissile. Seeds numerous; testa thin and variously sculptured, or bony and smooth. Embryo curved. Endosperm present, often copious.

Foliage leaves present, at least on new growth; glochids present; seeds flattened, covered with a bony aril:
- Flowers usually clustered at tips of flowering branches or around margins of "pads," not solitary; ovary free, i.e., not buried in tissue at tip of stem; indehiscent:
  - Leaves ovate, obovate, or oblong, 5–15 mm. wide or more, moderately persistent but eventually deciduous . . . . . . . . 1. *Pereskiopsis*
  - Leaves subulate, rarely over 2 mm. wide, caducous . . . . . . . . 2. *Opuntia*
- Flowers solitary at tips of flowering branches; ovary partially buried in tissue at tips of branches . . . . . . . . 3. *Marenopuntia*

Foliage leaves none; glochids none; seeds reniform to ovoid, not strongly flattened, testa thin, not bony-arillate:
- ♦ Ovaries spiny, at least in youth; fruit either persistently or caducously spiny:
  - Fruit dry, abundantly spiny or bristly when mature:
    - Plants treelike, 5–15 m. tall or more, often with several to many branches; spines grayish, angular in cross section; perianth segments white to purplish or brownish; flowers mostly nocturnal . . . . . . . . 4. *Pachycereus*
    - Plants smaller, stems rarely over 4 m. tall; spines golden yellow in youth, terete; perianth segments yellow to greenish; flowers mostly diurnal:
      - Stems 2.5–4 cm. in diameter, rarely over 2 m. tall; flowers 4–5.5 cm. long, 3–4.5 cm. wide; stamens 8–12 mm. long . . . . . . . . 6. *Bergerocactus*
      - Stems 7–10 cm. in diameter, to 4 m. tall; flowers 6–7 cm. long, 5–6 cm. wide; stamens 4–6 mm. long . . . . . . . . 5. × *Pachgerocereus*
  - Fruit juicy or pulpy, often sparsely spiny to nearly naked at maturity:
    - Ribs 12–25; plants usually moderately to much branched; stems and branches columnar, erect, candelabrum-like, frequently to 6 m. tall or more:
      - Flowers 6–8 cm. long, purplish; seeds 1.5–2 mm. long; ribs rounded, 1–1.5 cm. wide; branches 1–2 dm. thick . . . . . . . . 7. *Lemaireocereus*
      - Flowers 10–12 cm. long, greenish to nearly white; seeds 0.6–0.8 mm. long; ribs obtuse, 2–4 cm. wide; branches 3–6 dm. thick . . . . . . . . 11. *Carnegiea*
    - Ribs usually less than 12; plants less branched, often prostrate, scrambling, or cespitose, if erect and candelabrum-like less than 3 m. tall:
      - Roots fleshy and tuberlike; spines either weak and appressed or bulbous at base; stems 5–30 mm. in diameter, usually growing among supporting shrubs:

Stamens and styles distinctly exserted; stems terete or nearly so, 6–9-ribbed; roots fusiform-tuberous; spines weak, closely appressed, mostly directed parallel to long axis of stem .......................................... 8. *Wilcoxia*

Stamens and styles included or slightly exserted at full anthesis; stems 3–5-angled; root beetlike, 5–10 cm. in diameter, 15–30 cm. long or more; at least one spine per areole porrect .......................................... 9. *Peniocereus*

Roots fibrous or woody, at least not conspicuously fleshy and tuberlike; spines robust, not markedly bulbous at base; stems stouter, often to 10 cm. in diameter, not scandent on other shrubs:

Stems prostrate or archingly scrambling but neither candelabrum-like nor cespitose; perianth tube slender, equaling or exceeding throat:

Limb of corolla spreading, not oblique; hilum not markedly enlarged ....... .......................................... 10. *Machaerocereus*

Limb of corolla slightly to strongly oblique; hilum large and conspicuous .... .......................................... 12. *Rathbunia*

Stems erect or ascending, moderately branched to densely cespitose but not scrambling or prostrate; perianth tube stout, shorter than to equaling throat:

Stems simple to few-branched, 1–7 m. tall; older stems bearing closely crowded, slender bristles on upper parts; fruit globose, 1.5–3 cm. in diameter, weakly spiny .......................................... 13. *Lophocereus*

Stems branching at or near base, often densely cespitose, 1–4 dm. tall; bristles absent, spines all moderately to strongly rigid; fruits ovoid, more strongly spinose .......................................... 15. *Echinocereus*

◆ Ovaries naked or scaly but not bearing spines; fruit smooth or scaly, spineless at maturity:

Flowering areoles bearing 1 to several flowers each; plants columnar, 1–7 m. tall:

Older stems bearing crowded bristles 2–20 cm. long on upper parts; seeds 2.5–3 mm. long .......................................... 13. *Lophocereus*

Older stems devoid of slender bristles; seeds 1.2–1.5 mm. long ... 14. *Myrtillocactus*

Flowering areoles each bearing a solitary flower; plants globose to low-columnar, mostly less than 1 m. tall but 2–3 m. tall in some species of *Ferocactus*:

Plant body distinctly ribbed, the ribs undulate, nearly straight, or tuberculate at the areoles; flowers 4–6 cm. long, often fully as wide:

Nectariferous gland spines present in young areoles, persisting as dwarf spines; hypocotyl comparatively large .......................................... 16. *Ferocactus*

Nectariferous gland spines lacking or ephemeral; hypocotyl not markedly enlarged:

Fruit scaly; seeds muricate or tuberculate:

Central spine often curved but not hooked; flowers borne at base of a woolly groove; axils of scales on ovary naked ............... 17. *Echinomastus*

Central spine strongly hooked; flowers not in a woolly groove; axils of scales on ovary woolly .......................................... 19. *Sclerocactus*

Fruit densely white-woolly; seeds smooth and shining or papillose ........... .......................................... 18. *Echinocactus*

Plant body not ribbed, tubercles arranged spirally; flowers mostly 1.5–3.5 cm. long and wide:

■ Stems cylindrical, three to five or more times as long as thick:

Tubercles not grooved; flowers in axils of upper mature tubercles; fruits usually bright red to orange:

Fruits globose to obovoid, with large terminal scars; stamens and style exserted .......................................... 20. *Cochemiea*

Fruits mostly clavate, with very small terminal scars; stamens and style included .......................................... 23. *Mammillaria*

Tubercles grooved from base to apex when mature; flowers in axils of young tubercles; fruits greenish to yellow .......................... 21. *Coryphantha*

■Stems globose to short-oblong, usually less than three times as long as thick:
  Tubercles grooved; fruit greenish to yellow; flowers in axils of young tubercles ............................................... 21. *Coryphantha*
  Tubercles not grooved; fruit usually bright red to orange; flowers in axils of upper mature tubercles:
    Fruit circumscissile, basal part remaining embedded among tubercles for long periods ............................................ 22. *Bartschella*
    Fruit indehiscent or opening by basal pore, falling from plant entire ......... ............................................... 23. *Mammillaria*

## 1. PERESKIOPSIS Britt. & Rose, Smiths. Misc. Coll. 50: 331, 1907

Shrubs and small trees with well-defined trunk (at least at base) and weakly to strongly ascending, jointed branches, the old trunk forming a woody cylinder and smooth to moderately scaly bark. Areoles circular, bearing hairs, wool, glochids, and spines or in some areoles the spines lacking, less commonly glochids absent from older areoles. Flowers campanulate, with scaly ovary and sepals intergrading into petals; perianth tube short; ovary bearing leaves and areoles similar to those on stem, or leaves reduced to scales. Stamens many, inserted on throat of perianth tube, sensitive. Ovary sessile or rarely pedunculate; style 1; stigma lobes several, short and stout. Fruit red to yellow-orange, juicy, indehiscent. Seeds few, bony, covered with minute matted hairs.

### 1. Pereskiopsis porteri (Weber) Britt. & Rose, Smiths. Misc. Coll. 50: 332, 1907

*Opuntia rotundifolia* Brandegee, Zoe **2:** 21, 1891. Not *Pereskia rotundifolia* DC., 1828.
*Opuntia porteri* Brandegee ex Weber, in Bois, Dict. Hort. **2:** 899, 1898.
*Opuntia brandegeei* Schumann, Gesamtb. Kakteen 653, 1898.
*Pereskiopsis brandegeei* Britt. & Rose, Smiths. Misc. Coll. **50:** 331, 1907.

A straggly to scrambling plant or less commonly with a sturdy main stem 3–7 cm. in diameter; branches 0.3–4 meters long, with smooth, greenish to reddish brown bark, this splitting vertically above and below larger areoles on older stems; joints 1–25 cm. long, easily disarticulated when young; areoles broadly elliptical, 1.5–2 mm. long on young stems, to 8 mm. long when older, bearing numerous brownish glochids and 1–12 straight, stiff, dark brown spines 1–4 cm. long; leaves oblanceolate to obovate, 5–15 mm. wide, 2–5 cm. long, entire and glabrous, acute, attenuate, or rounded at apex; flowers 5–8 cm. in diameter, petals obovate, 8–15 mm. wide, 2–3.5 cm. long, pale yellow, rounded to slightly apiculate at apex, margins entire; stamens numerous, 4–6 mm. long; ovary obovoid to subclavate, 2–5 cm. long at anthesis, bearing woolly, glochid-filled areoles 6–10 mm. apart; fruit ellipsoidal to obovoid, 3–5 cm. long, yellow-orange to red-orange, bearing both gray wool and reddish brown glochids in the areoles, somewhat fleshy, sometimes proliferating; seeds moderately flattened, 3–4 mm. in diameter, brownish, horny.

Along arroyos and among shrubs on desert flats, Lower Sonoran Zone, southern Baja California and from vicinity of Queriego, Sonora, south into Sinaloa. Sept.–Oct.

## 2. OPUNTIA Mill., Gard. Dict. abr. ed. 4, 1754

Fleshy cacti with more or less woody skeletons and jointed cylindrical, clavate, or flattened stems and branches. Roots fibrous or fleshy-tuberous. Areoles axillary, bearing short, readily detached, barbed bristles or glochids and usually 1 to several stout spines. Spines naked or ensheathed in dry papery coverings. Leaves usually small and terete, early deciduous. Flowers borne in areoles of year-old growth; perianth tubes cup-shaped, short. Ovary areolate, 1-celled, many-ovulate. Sepals green, grading into colored petals. Stamens numerous, shorter than petals, sensitive; stigma lobes short. Fruit fleshy or dry, spiny or spineless. Seeds covered with a bony aril, light-colored, flattened.

◆Joints terete, not flattened; tubercles evident, often prominent (*Cylindropuntia*):
 Main spines usually strongly flattened, marginally serrulate; sheaths quickly caducous; fruits generally persistently spiny or bristly:
  Joints 1–1.5 cm. in diameter at greatest thickness; fruits 2.5–4.5 cm. long:
   Central spines to 3.5 cm. long; tubercles 10–15 mm. long; spines dark gray to nearly black .................... 1. *O. kunzei*
   Central spines 6–15 mm. long; tubercles 6–10 mm. long; spines light gray to nearly white .................... 2. *O. reflexispina*
  Joints 2–6 cm. in diameter; fruits 4–6 cm. long:
   Joints very closely crowded, giving plant appearance of an *Echinocereus*; spines of adjacent areoles interlocking only slightly:
    Tubercles not confluent to form uninterrupted ribs, only 2–5 cm. long; joints 4–6 cm. in diameter, 8–15 cm. long; spines strongly flattened, 1–5 cm. long; petals clear yellow .................... 3. *O. invicta*
    Tubercles confluent to form continuous riblike ridges, each tubercle 10–15 mm. long; joints 3–4 cm. in diameter, 10–30 cm. long; spines slightly flattened, 1–2.5 cm. long; petals yellow with roseate base .................... 4. *O. santamaria*
   Joints less crowded, not simulating *Echinocereus*; spines of adjacent areoles strongly interlocking and obscuring surface of joints:
    Tubercles 1.5–3 cm. long; fruits heavily armed with persistent rigid spines 1–2.5 cm. long; seeds 4.5–6 mm. long .................... 5. *O. wrightiana*
    Tubercles 0.5–1 cm. long; fruits armed only with slender glochids 3–9 mm. long; seeds 3.5–5 mm. long .................... 6. *O. parishii*
 Main spines terete or essentially so, finely banded uniformly around outer part; sheaths subpersistent but always apparent on younger spines; fruits less densely spiniferous, not burlike:
 ■Joints 4–15 mm. in diameter; spines few, 1–7 per areole or areoles spineless, not obscuring joints; plants mostly less than 1 m. tall:
   Tubercles very low, inconspicuous; plants usually with a definite, although often short, main trunk; only few areoles spiniferous:
    Fruit dry, spiny, and burlike; tubercles covering joints with diamond-shaped plates; areoles compressed into narrow slit during age .................... 7. *O. ramosissima*
    Fruit more or less fleshy, not densely spiny; tubercles less sharply delimited; areoles orbicular or oval:
     Surface of joints glabrous, yellow-green to slightly blue-green; mature joints readily disarticulated; fruits somewhat proliferous:
      Fruit purplish, 2–3 cm. long; joints 6–9 mm. in diameter, 5–10 cm. long; spines light gray to brown below the tips .................... 8. *O. arbuscula*
      Fruit reddish, 1–1.8 cm. long; joints 4–6 mm. in diameter, 10–35 cm. long; spines chalky white below the tips .................... 9. *O. leptocaulis*
     Surface of joints puberulent or scurfy, gray-green; mature joints firmly attached, disarticulating with difficulty; fruits rarely proliferous:

Spines 1–6 cm. long, cinereous to pale purplish, yellowish only at tips; flowers 2 cm. in diameter; sheaths stramineous .................... 10. *O. tesajo*

Spines 1.5–3 cm. long, pale yellow, pruinose at base only; flowers 3–3.5 cm. in diameter; sheaths white .............................. 11. *O. cineracea*

Tubercles more prominent, 2–3.5 mm. high; plants without definite main trunk; nearly all areoles spiniferous:

Inner petals greenish yellow, tinged with purple on inner surface; seeds discoidal, distinctly flattened; spines 1–6 per areole, 0.7–4 cm. long .. 12. *O. tetracantha*

Inner petals brownish yellow; seeds subglobose, scarcely flattened; spines 3–7 per areole, 5–12 mm. long .................................. 13. *O. thurberi*

■Joints, except youngest ones, over 1.5 cm. in diameter; spines several to many in each areole (mostly 8–25); plants mostly 1–3 m. tall:

●Fruits fleshy at maturity, often proliferous:

Seeds numerous in mature fruits; proliferation of fruits rare or lacking:

Fruits bearing spines at least in upper areoles; seeds 4–6 mm. long, markedly flattened:

Flowers yellow, tinged with red; petals obtuse; spines brownish; tubercles 15–25 mm. long ........................................ 14. *O. parryi*

Flowers purplish bronze; petals rounded but apiculate; spines reddish with yellowish tips; tubercles 1.5–4 cm. long ................. 15. *O. molesta*

Fruits spineless or essentially so; seeds 3–4.5 mm. long, angular, tumid, slightly flattened to subglobose:

Sheaths ferruginous; fruits subglobose, 1–2 cm. long; plant rarely over 1 m. tall; joints blue-green and glaucous .................. 16. *O. brevispina*

Sheaths whitish or stramineous; fruits obovoid, 2.5–3.5 cm. long; plants 1–4 m. tall; joints yellow-green tinged with purplish:

Flowers 4–5 cm. in diameter at full anthesis; spines interlocking and somewhat obscuring underlying surfaces; joints lightly purplish ........... ................................................ 17. *O. spinosior*

Flowers 2.5–4 cm. in diameter; spines less closely interlocking, the underlying surfaces only slightly obscured; joints strongly purplish ......... ................................................ 18. *O. versicolor*

Seeds few or lacking in most fruits; fruits usually forming chains by proliferation:

Plants 0.5–2 m. tall; spines divergent-interlocked on younger joints:

Joints 3–5 cm. in diameter, dark green under spines; fruits globose, 2–3 cm. in diameter; sheaths sordid-yellow to rusty; seeds 5–6 mm. long .......... ................................................ 19. *O. prolifera*

Joints 1.5–3.5 cm. in diameter, light green to blue-green, glaucous; fruits obovoid, 2–4 cm. long; sheaths glistening-white or pale-stramineous; seeds 3.5–4 mm. long ........................................ 20. *O. ciribe*

Plants 1–5 m. tall; spines less crowded and scarcely interlocking:

Tubercles more or less mammillate, broad and rounded; petals 5–12, 8–10 mm. long:

Petals pale pink or white streaked with purple; fruits nearly flat at apex, not deeply excavate:

Spines to 20 in each areole; fruit nearly smooth, scarcely tuberculate ..... ..................................... 21. *O. fulgida* var. *fulgida*

Spines 2–10 in each areole; fruit distinctly tuberculate ................. .............................. 21*a*. *O. fulgida* var. *mammillata*

Petals peach-pink to pale purplish; fruits distinctly excavate at apex ........ ................................................ 22. *O. cholla*

Tubercles narrow, not mammillate; petals more numerous, 1.5–2 cm. long ... ................................................ 18. *O. versicolor*

- ● Fruits dry at maturity, rarely proliferous:
  - Spines closely interlocking; ultimate joints readily detached, turgid:
    - Fruits spineless or essentially so at maturity; seeds (if present) 3–4 mm. long, margins rounded or ridged:
      - Joints stout, 3–5 cm. in diameter; fruit not proliferous; plants erect or ascending:
        - Green stems not showing through spines; trunks usually single; terminal joints only moderately crowded ...................... 23. *O. bigelovii*
        - Green stems showing through less crowded spines; main trunk often branching at or near base; terminal joints closely crowded:
          - Shrub 2–3 m. tall; tubercles only slightly longer than wide ............................................................ 24. × *O. fosbergii*
          - Shrub to 4 m. tall; tubercles twice as long as wide ..... 25. × *O. munzii*
      - Joints slender, 1–2 cm. in diameter; fruits sometimes proliferous; plants usually spreading to sprawling:
        - Flowers red-brown to purple-brown, 2–4 cm. broad; seeds 3–4 mm. broad; longer spines about 2 cm. long; plant sprawling .... 26. *O. burrageana*
        - Flowers greenish yellow, about 2 cm. broad; longer spines only 1.5 cm. long; plant rather compact ................... 27*b*. *O. echinocarpa* var. *nuda*
    - Fruits very spiny on upper two-thirds; seeds 5–6 mm. long, the margins grooved:
      - Flowers 2–3.5 cm. long; fruit 1.5–2.5 cm. long; plant intricately branched ..................................... 27. *O. echinocarpa* var. *echinocarpa*
      - Flowers 4–5 cm. long; fruit 3–3.5 cm. long; plant less branched ........................................... 27*a*. *O. echinocarpa* var. *parkeri*
  - Spines less crowded, and if interlocking not obscuring underlying surface of joints; joints less easily detached, less turgid:
    - Tubercles not confluent into undulate, riblike ridges; flowers 5–7 cm. in diameter; spines 6–28 per areole:
      - Joints twice to eight times as long as thick, strongly armed; spines mostly over 2.5 cm. long, with conspicuous sheaths; fruits spiny from middle upward:
        - Crown open; younger branches ascending-divaricate, forming acute angles with one another; joints mostly over 2.5 cm. in diameter; spines 8–25 per areole:
          - Surface of joints dull green; spines 8–18 per areole ............................................ 28. *O. acanthocarpa* subsp. *acanthocarpa*
          - Surface of joints bright green; spines 20–25 per areole ............................................. 28*a*. *O. acanthocarpa* subsp. *ganderi*
        - Crown densely branched; younger branches spreading at various angles; joints mostly less than 2.5 cm. in diameter; spines 6–12 per areole ...................................... 28*b*. *O. acanthocarpa* var. *ramosa*
      - Joints four to twenty times as long as thick, weakly armed; spines less than 2.5 cm. long, sheaths inconspicuous; fruit spiny only near apex ............................................................ 29. *O. thornberi*
    - Tubercles confluent to form undulate, riblike ridges; flowers 3–3.5 cm. in diameter; spines 4–7 per areole .............................. 30. *O. rosarica*

- ◆ Joints flat, relatively thin pads; tubercles rarely apparent, areoles elevated only slightly above surface of joint (*Platyopuntia*):
  - ▲ Fruits dry at maturity:
    - Pads short-puberulent, spineless; areoles more or less sunken; flowers deep rose to rose-purple, 10–15 cm. in diameter:
      - Joints orbicular to broadly obovate; stems not branching above base ........................................................ 31. *O. basilaris* var. *basilaris*

Joints narrowly obovate to oblong; stems branching above, 3–6 joints high ......... ............................................31*a*. *O. basilaris* var. *ramosa*

Pads glabrous between areoles, copiously spiniferous; areoles slightly elevated, not sunken; flowers yellow suffused with red, 6–7 cm. in diameter ...... 32. *O. ursina*

▲Fruits fleshy:

Plants prostrate-creeping, or if some branches erect, these only 2–4 joints tall, less than 1 m. high; some joints usually lying on edge on ground in chainlike arrangement:

Spines 7–12 per areole, present from base to apex of vigorous joints; joints minutely puberulent ........................................ 33. *O. pycnantha*

Spines 1–4 per areole, present only on upper two-thirds or less of vigorous joints; joints glabrous:

Fruits 5–7 cm. long; flowers clear yellow; spines light gray or slightly darker at base ................................................ 34. *O. megacarpa*

Fruits 3.5–4 cm. long; flowers yellow suffused with red; spines reddish brown .... ................................................ 35. *O. phaeacantha*

Plants erect or ascendingly branched, rarely prostrate, joints not forming chains on ground; plants mostly to or well over 1 m. tall at maturity:

Plants shrubby or treelike, with a definite main trunk or stem at base:

Joints spiniferous over whole face or essentially so:

Main stem bearing 15–40 bright yellow spines per areole, these closely crowded; young joints glabrous, bearing 3–7 spines per areole; spines bright yellow ... ................................................ 36. *O. chlorotica*

Main stem spineless or bearing same number of spines per areole as in younger joints; young joints minutely puberulent; spines 1–4 per areole, white or pale gray ................................................ 37. *O. tapona*

Joints with spines confined to areoles on upper three-fourths or less of face; areoles near base of pads bearing glochids only:

Young joints puberulent .................................... 37. *O. tapona*

Young joints glabrous:

Fruits 5–11 cm. long, very juicy; plants 3–6 m. tall; pads to 60 cm. long, not tinged with purple:

Spines white; fruit reddish purple, 5–9 cm. long; joints 15–30 cm. long ..... ........................................... 38. *O. ficus-indica*

Spines brownish; fruit yellow or tinged with orange-red, 7–11 cm. long; joints 30–60 cm. long ........................... 39. *O. megacantha*

Fruits 2.5–6 cm. long, fleshy but not markedly juicy; plants 1–2 m. tall at maturity; pads 10–35 cm. long, tinged with purple:

Spines terete, reddish, brownish, or yellowish in youth, darker in age; joints 10–15 mm. thick:

Fruits 4–6 cm. long; joints 12–30 cm. long, usually spiniferous on upper half of pads only ........................... 40. *O. gosseliniana*

Fruits 2–2.5 cm. long; joints 10–20 cm. long, spiniferous only along upper margins and in a few areoles near apex of faces .... 41. *O. santa-rita*

Spines pale gray in age, often pale yellowish in youth; joints only 5–6 mm. thick ........................................ 42. *O. bravoana*

Plants more widely spreading at the ground level, not bushy or treelike, often with numerous rooted stems forming a colony, without a definite trunk or basal stem:

Inner flesh of fruits yellowish; glochids reddish or reddish brown, to 10 mm. long:

Joints spiniferous at most areoles from top to base of pad; longer spines 5–6 cm. long; surface of pads deep green ..... 43. *O. engelmannii* var. *engelmannii*

Joints spiniferous only in areoles on upper third of pad; longer spines not over 2.5 cm. long; surface of pads yellow-green ............................ ................................... 43*a*. *O. engelmannii* var. *flavescens*

Inner flesh of fruit reddish or purplish, not markedly yellowish; glochids mostly less than 5 mm. long, occasionally to 7 mm. long:
  Major spines white or pale gray; surface of joints light yellow-green:
    Fruit 5–9 cm. long, turbinate-cylindrical, with a stipelike base; pads often nearly spineless, narrowly ovate to oblong, usually asymmetrical ... 44. *O. laevis*
    Fruit obovoid-turbinate, 5–6 cm. long, without stipelike base; pads bearing spines on upper two-thirds, pads oval to orbicular, symmetrical .................................................. 45. *O. gilvescens*
  Major spines yellowish, brownish, or nearly black; surface of joints dark green or blue-green tinged with purple:
    Spines 1–4 per areole, 3.5–10 cm. long; joints 8–10 mm. thick, tinged with purple near the margins .......................... 46. *O. macrocentra*
    Spines 2–9 per areole, 1–3 cm. long; joints 2–5 cm. thick, dull green .................................................. 47. *O. littoralis*

**1. Opuntia kunzei** Rose, in Britt. & Rose, Smiths. Misc. Coll. **50:** 505, 1908
*Opuntia stanlyi* var. *kunzei* L. Benson, Proc. Calif. Acad. IV, **25:** 248, 1944.

A low-growing cholla with few to many clavate joints 3–12 cm. long branching candelabra-like to form a plant 1–3 dm. tall; joints firmly attached and difficult to remove, 5–8 mm. in diameter at base, gradually enlarging toward apex to a diameter of about 1.5 cm.; tubercles rather narrow, 10–15 mm. long, very pronounced; areoles circular, younger ones containing short, closely packed wool and 10–20 spines, older ones with pale brownish glochids 5–8 mm. long; larger spines strongly flattened, to 3.5 cm. long, dark gray to nearly black, margins minutely serrulate, 1–3 of lower spines in each areole strongly reflexed, others spreading in various directions, those of adjacent joints interlocking in an impenetrable armament; flowers 2.5–3.5 cm. in diameter, petals yellow; fruit cylindrical to stoutly clavate, bearing numerous areoles 3–8 mm. apart filled with white glochids and a few brownish bristles, mature fruit 12–15 mm. in diameter, 2.5–4.5 cm. long; seeds numerous, discoid, about 4 mm. in diameter, 1.2–1.5 mm. thick, smooth, yellowish to brownish, margins rounded, not angled.

In fine, often heavy soil, Lower Sonoran Zone, Maricopa County to Pima and Pinal Counties, Arizona, and south into adjacent Sonora. June–July.

**2. Opuntia reflexispina** Wigg. & Roll., Contr. Dudley Herb. **3:** 275, *pl. 59, figs. 1–2,* 1943

Low cespitose plant 2–4 dm. tall, forming clumps 3–10 dm. in diameter, moderately branched with erect or ascending, firmly attached joints, somewhat candelabra-like; joints cylindrical to narrowly clavate, 5–10 cm. long, 8–15 mm. in diameter, glabrous, pale, yellowish green; tubercles narrow, 6–10 mm. long, about 3 mm. wide and about as high; areoles ovate to elliptic, 3–5 mm. long, arranged 5–15 mm. apart in spirals of high pitch, white-woolly with crinkled tangled hairs, these at length deciduous; central spines 2–5, narrowly subulate, flattened, scaberulous with minute papillae, bulbous at base, upper spine porrect, grayish with a yellowish to brownish tip, 6–10 mm. long, lower spine (sometimes several of them) reflexed, 8–15 mm. long, gray to nearly white, often tinged with yellow or pink at apex; radial spines about 10, acicular, terete or slightly flattened, 4–6 mm. long; glochids numerous in old areoles, dull yellow, 1–7 mm. long; flowers numerous, terminal and erect on ultimate joints; petals linear to narrowly oblanceolate, 10–15 mm. long, 1–4 mm. wide, outer ones greenish, inner ones yellow; ovary narrowly clavate, about 1 cm. in diameter at apex, 3–4.5 cm. long, closely armed with spine-filled areoles similar to those on stem and apparently persistent; fruit dry, greenish; seeds dark brown to nearly black, dull, slightly granulose, 3 mm. wide, 4 mm. long, about 1.5 mm. thick, edges rounded.

Sandy plains, Lower Sonoran Zone, between main Hermosillo-Guaymas highway and the Gulf of California in the vicinity of Arrieros, Sonora. Aug.–Sept.

**3. Opuntia invicta** Brandegee, Proc. Calif. Acad. II, **2:** 163, 1889

*Corynopuntia invicta* Knuth, in Backeb. & Knuth, Kaktus-ABC 114, 1935.

Very succulent, cespitose plant 2–4.5 dm. tall in dense clumps to 2 m. in diameter, often with a growth habit suggestive of *Echinocereus*; branches mostly of 2–4 joints, ascending or spreading; joints broadly obovoid to broadly clavate, dark green, 4–6 cm. in diameter, 8–15 cm. long, strongly tuberculate; tubercles rounded or somewhat flattened dorsally, 1–2 cm. wide, about 1 cm. high, 2–5 cm. long; areoles nearly orbicular, 8–15 mm. in diameter, closely crowded with fine grayish wool between bases of spines, the wool eventually nearly disappearing; spines 10–25, very rigid, divaricately spreading to deflexed, strongly flattened dorsiventrally but 4-angled, to 3 mm. wide at base, 1–5 cm. long, pale gray or with tips slightly reddish or brownish, sheaths present only on very young red spines and soon deciduous; glochids practically none; flowers 4–6 cm. in diameter, petals canary yellow, obovate, 8–12 mm. wide, apiculate to shallowly emarginate; filaments maroon; anthers pale yellow; stigma lobes light yellow tinged with rose without; fruit moderately fleshy, broadly ovoid, 2.5–3.2 cm. broad, 4–5 cm. long, densely spiny with interlocking, rufous to grayish, acicular spines to 2.5 cm. long, not excavate at the top; seeds numerous, tumidly discoidal, about 4 mm. long and wide, 1.5–2 mm. thick, tawny, smooth, with narrow but pronounced hilar notch.

In sandy or silty soil, Lower Sonoran Zone, the Magdalena Plain near the western coast of Baja California. April–May.

**4. Opuntia santamaria** (E. M. Baxter) Wiggins, comb. nov.

*Opuntia santa-maria* E. M. Baxter, Cactus & Succ. Jour. **4:** 283, 1932. Nomen nudum.
*Grusonia santamaria* E. M. Baxter, op. cit. **6:** 60, 1934.

Low, moderately branched cactus 3–6 dm. tall with cylindrical joints 3–4 cm. in diameter and 1–3 dm. long; tubercles arranged spirally but nearly confluent and forming 8–9 ribs with areoles 10–15 mm. apart; areoles at upper end of each tubercle, orbicular to broadly elliptical, containing wool, caducous glochids, and 12–20 spines, spines subulate, 10–15 mm. long, plum-colored, thinly sheathed in youth but soon naked, spines spreading, rigid, very sharp, margins serrulate-barbed; flowers near apex of previous year's joints; petals greenish yellow to sulphur-yellow, innermost with a roseate blush along midvein, about 2.5 cm. long, 2 cm. wide, obovate, truncate-mucronate, to retuse with an apiculation in apical notch; outer segments more deeply suffused with rose; ovary strongly tuberculate, glochidiate and bristly, about 2.5 cm. tall; mature fruit unknown. Flowers reported to turn red and to be nearly closed by end of third day.

Mesa along southern shore of Santa María Bay, Magdalena Island, Baja California. Flowered in southern California in June.

**5. Opuntia wrightiana** (E. M. Baxter) Peebles, Desert Pl. Life **9:** 43, 1937

*Grusonia wrightiana* E. M. Baxter, Calif. Cactus 58, 1935.
*Opuntia stanlyi* var. *wrightiana* L. Benson, Proc. Calif. Acad. IV, **25:** 248, 1944.
*Opuntia kunzei* var. *wrightiana* Peebles, Leafl. West. Bot. **5:** 192, 1949.

Closely cespitose plant 2–4 (rarely –6) dm. tall in clumps to 10 m. in diameter; joints cylindric to ellipsoidal-oblong, 2.5–3.2 cm. in diameter, 8–15 cm. long, erect, ascending, or lowermost somewhat procumbent, very rigidly joined together and removable only with considerable difficulty; tubercles very pronounced, 3–7 mm. wide, 4–9 mm. high, 1.5–3 cm. long; areoles broadly elliptical to suborbicular, 3–8 mm. long, filled with spines, some very short, closely packed, sordid hairs and rarely with glochids 3–12 mm. long; spines variable in number, length, and color, usually 1 central 2.5–5 cm. long, 3–6 subcentrals nearly as long, all these reddish or brownish, rigidly spreading and strongly flattened-subulate, their thin margins roughened; radial spines 8–20, flattened or smaller ones nearly terte, 1–3.5 cm. long, usually spreading nearly at right angles to central spine, often becoming sordid-white in age, very firmly attached; flowers solitary or 2–3 borne terminally or on uppermost areoles, petals yellow, 1–1.5 cm. long, mostly acute to acuminate; fruit oblong-obovate, yellowish, 5–6 cm. long, about 2 cm. in diameter, rather strongly tuberculate, heavily armed with 8–20 spines 1–2.5 cm. long and a few yellow-brown glochids in an areole, these 5–12 mm. apart; seeds obovoid-discoid, 3.5–4.5 mm. wide, 4.5–6 mm. long, about 2 mm. thick, slightly notched at base, smooth, yellowish brown.

Desert flats in fine silty soil and on sandy and rocky slopes, Lower Sonoran Zone, western Pima County and central Yuma County, Arizona, south into northern Sonora. May–June.

**6. Opuntia parishii** Orcutt, W. Amer. Sci. **10:** 1, 1896

*Opuntia parryi* Engelm., Pacif. R. Rep. **4:** 48, *pl. 22, figs. 4–7, pl. 24, fig. 7,* 1856. Not *O. parryi* Engelm., 1852.
*Corynopuntia parishii* Knuth, in Backeb. & Knuth, Kaktus-ABC 115, 1935.
*Opuntia stanlyi* var. *parishii* L. Benson, Cacti of Ariz. ed. 2, 39, 1950.

A low, creeping, matted plant rooting along under surface, forming mats up to 1.5 m. across, with erect clavate branches 5–15 cm. high; joints 2–3 cm. in diameter, nearly hidden by dense interlocking spines; tubercles 5–10 mm. long, narrow, 3–6 mm. high, prominent; areoles filled with white wool and yellowish glochids; spines sheathless, reddish fading to ashy, centrals 3–4, divergent, strongly flattened, subulate, 2–4 cm. long, lowest broadest and longest; radials 6–12, acicular, terete, spreading; sheaths early caducous; glochids numerous, 2–8 mm. long, light yellowish brown; flowers yellow to reddish; fruit dry, 4–6 cm. long, narrowly ovate, narrowly tuberculate, areoles 3–4 mm. in diameter, filled with white wool and radiating, acicular, yellowish brown glochids 3–9 mm. long that conceal surface; seeds flattened-ovoid, smooth, turgid, 3.5–5 mm. long, yellowish white.

Infrequent in loose sandy or gravelly soil, upper part of Lower and lower part of Upper Sonoran Zones, southern Nevada to the eastern Mojave and western Colorado Deserts, and the Little San Bernardino Mountains, California. May.

**7. Opuntia ramosissima** Engelm., Amer. Jour. Sci. II, **14:** 339, 1852

*Opuntia tessellata* Engelm., Proc. Amer. Acad. **3:** 309, 1856.
*Cylindropuntia ramosissima* Knuth, in Backeb. & Knuth, Kaktus-ABC 122, 1935.

Low bush 0.4–2 m. high with gray-green, widely spreading branches 0.5–1 cm. in diameter; joints 2.5–10 cm. long; tubercles low, crowded, 5–8 mm. long, nearly as wide, 4- or 6-angled, covering surface with diamond-shaped plates; leaves ovoid, 1–3 mm. long, acute; areoles circular when young, compressed into a narrow slit in age, with white to tawny wool and pale yellow glochids; spines 1–4 at an areole (1 large,

0–3 smaller), often wanting, acicular, porrect to spreading, 1–6 cm. long, reddish when young, nearly white in age, covered by a loose, yellow, papery sheath; flowers 3–4 cm. long including ovary, sepals subulate, the petals obovate, 9–12 mm. long, aristate, greenish yellow tinged with red; stamen filaments greenish yellow, anthers orange; style and stigma cream-colored; ovary narrowly obconic, covered with low emarginate tubercles, areoles filled with wool, glochids, and 10–15 unsheathed spines; fruit dry, 1–2.5 cm. long, spines making it appear burlike; seeds few, lenticular, 3–5 mm. wide, stramineous.

Low hills and desert flats, Lower Sonoran Zone, from the vicinity of Victorville, San Bernardino County, California, and western Colorado Desert to southern Nevada, Arizona, and northwestern Sonora. May–Sept.

**8. Opuntia arbuscula** Engelm., Proc. Amer. Acad. **3:** 309, 1856

*Opuntia neoarbuscula* Griff., Rep. Mo. Bot. Gard. **19:** 260, *pl. 23,* 1908.
*Opuntia vivipara* Rose, Smiths. Misc. Coll. **52:** 153, *pl. 12,* 1908.
*Cylindropuntia arbuscula* Knuth, in Backeb. & Knuth, Kaktus-ABC 123, 1935.

A cholla of treelike form 5–25 dm. tall with many intricately branched stems from a sturdy main trunk to 1.5 dm. in diameter; ultimate joints cylindrical, 6–9 mm. in diameter, 5–10 cm. long, surface smooth, tubercles so low and rounded that in turgid plants they are scarcely visible; areoles broadly elliptic to orbicular, 1–2 mm. in diameter, 4–15 mm. apart, containing many stramineous to brownish glochids 0.5–2.5 mm. long, usually only a few areoles spiniferous; spines usually solitary but sometimes 3–4 in an areole, porrect, 0.5–4 cm. long, with whitish or stramineous, rather inconspicuous sheath, spines light to dark gray or brown near base, yellowish toward apex; flowers about 2.5–3 cm. in diameter, petals moderately spreading, broadly obovate, yellow, greenish yellow, or yellow tinged with red; fruit obovoid, truncate and deeply excavated at summit, beset with glochid-bearing areoles 3–10 mm. apart, green tinged with red or nearly purple at maturity, 2–3 cm. long and two-thirds as wide, spineless, sometimes proliferous; seeds pale buff, about 3 mm. long and wide, irregularly angular but not markedly compressed, smooth.

Valleys, plains, and mesas in heavy soil and along sandy washes, Lower Sonoran Zone, central Arizona south into northern Sonora. May–June.

**9. Opuntia leptocaulis** DC., Mém. Mus. Hist. Nat. Paris **17:** 118, 1828

(For synonymy see Britt. & Rose, Cactaceae **1:** 47, 1919.)

Small bushy cholla usually less than 1 m. tall but occasionally scrambling over shrubs to a height of 2 m., from a definite woody trunk or spreading out to form thickets several meters in diameter, often straggly; branches cylindrical, 4–6 mm. in diameter, main ones 1–3.5 dm. long, lateral ones 2–5 cm. long, spreading nearly at right angles to main axis, smooth, commonly nearly spineless, easily detached; leaves subulate, acute, 5–12 mm. long, soon deciduous; areoles elliptical, 2.5–3 mm. long, 3–15 mm. apart, bearing a marginal ring of a few brownish glochids 1–2.5 mm. long, a central pad of grayish wool, and those on main branches with a solitary spine; spine sheaths delicately papery, whitish at base, stramineous toward apex, deciduous; spines 2–5 cm. long, chalky white or pale yellow toward tips; flowers crowded along stems, 12–20 mm. in diameter and about as long, petals greenish yellow; outer perianth segments ovate, acute to cuspidate; fruit fleshy, red, obovoid, 8–10 mm. in diameter, 10–18 mm. long, spineless but bearing some glochids, persistent until second season and sometimes bearing new flowers and fruits around upper margins; seeds 3–4 mm. long, strongly compressed, margins acute.

Desert flats and mesas, usually in heavy soil, Lower Sonoran Zone, Yuma and Mohave Counties, Arizona, to western Texas and south to Puebla, Mexico. April–June.

**10. Opuntia tesajo** Engelm. ex J. M. Coulter, Contr. U.S. Nat. Herb. **3**: 448, 1896

*Cylindropuntia tesajo* Knuth, in Backeb. & Knuth, Kaktus-ABC 122, 1935.

Much-branched, brushy plant 3–6 dm. tall and nearly as wide, or sometimes scrambling through other shrubs to 1.5 m., branches diverging from one another at angles of 30–45°, slender, gray-green, inconspicuously tuberculate; ultimate joints 2–5 or sometimes –10 cm. long, 7–10 mm. in diameter, minutely scurfy, rather firmly attached and difficult to disarticulate; tubercles very low and rounded, about 1.5 mm. wide, 1 mm. high or less, 4–10 mm. long; areoles circular to subelliptic, 1–1.5 mm. in diameter, densely crowded with numerous light brown glochids 1–2.5 mm. long; spines 1–4, these 5–15 mm. long, apparently deciduous and replaced by 1–2 slender, acicular but rigid spines to 6 cm. long; spine sheaths dark-stramineous, soon ruptured at base by rapidly elongating spines, deciduous; mature spines sheathless, cinereous or pale purplish with yellowish tips; flowers yellow, about 2 cm. in diameter; fruits bright red, fleshy, 2–2.5 cm. long, broadly ovoid, bearing both glochids and solitary spines in areoles, these 3–7 mm. apart; seeds very few (many fruits sterile), tumidly discoid, 4–4.5 mm. in diameter, 1.5–2 mm. thick, margin rounded.

Desert washes and hillsides, Lower Sonoran Zone, vicinity of El Mármol south almost to the Cape Region, usually rather near the western coast. May–June.

**11. Opuntia cineracea** Wiggins, Contr. Dudley Herb. **3**: 285, *pl. 65, figs. 1–5*, 1943

*Cylindropuntia tesajo* var. *cineracea* Backeb., Cact. **1**: 173, 1958.

Low, rounded or flat-topped, shrubby plant with woody base and fibrous roots, rarely over 5 dm. tall; joints slender, 8–15 mm. in diameter, 4–8 cm. long, minutely but closely cinereous-tomentulose with short, stout, simple, white hairs; leaves terete, 3–5 mm. long, caducous; tubercles low and narrow, 7–8 mm. long, 1–2 mm. wide and high; areoles narrowly elliptical, 1.5–2.5 mm. long, 0.8–1.2 mm. wide, filled with tufts of brownish straight hairs when young, a central cluster of whitish glochids gradually crowding out most of the hairs; spines few, confined to uppermost 4–8 areoles of older joints, solitary in any one areole, spreading, 1.5–3 cm. long, pale yellow or slightly pruinose near base, terete-acicular or slightly flattened, covered with thin, white, deciduous sheath; flowers terminal on joints of previous season, 1–6 on a single joint; outer perianth segments greenish tinged, with rose toward tips, 5–8 mm. long, inner ones 12–18 mm. long, oblong to oblanceolate-spatulate, abruptly acuminate, bright yellow, glossy, often faintly tinged with green on midrib and at base; stamens 5–8 mm. long; style greenish white tinged with purple-brown; ovary turbinate, 1–1.5 cm. long, 8–10 mm. in diameter, crowded with orbicular areoles 1–1.5 mm. in diameter, areoles filled with brown wool and stramineous glochids 1–6 mm. long; mature fruit and seeds unknown.

Gravelly slopes and mouths of arroyos, Lower Sonoran Zone, eastern foot of the Sierra San Pedro Mártir, San Felipe Desert, Baja California. April–May.

**12. Opuntia tetracantha** Toumey, Gard. & Forest **9**: 432, 1896

*Cylindropuntia tetracantha* Knuth, in Backeb. & Knuth, Kaktus-ABC 124, 1935.

Brushy plant 1 m. tall or less with numerous branches but without definite trunk; stems branching at angle of 30–45°, cylindrical, joints 8–12 mm. in diameter, 1–2.5 dm. long, easily detached; tubercles prominent but low and rounded, 3–4 mm. wide, 10–12 mm. long; areoles orbicular or nearly so, 2–5 mm. in diameter, closely packed with sordid-gray to light brown wool, light brown glochids 1–2 mm. long, and bearing 1–6 (often 4) porrect to slightly deflexed spines 0.7–4 cm. long, these encased in yellow-brown, papery sheaths when young, soon naked and then gray except at stramineous tip, often minutely striate and slightly angled; flowers 1.5–2 cm. in diameter, petals greenish yellow below, tinged with purple on upper surface, obovate, obtuse but usually abruptly short-apiculate; fruit elliptic-obovoid, 2–3 cm. long, reddish at maturity, spineless but bearing minute glochids in orbicular areoles; seeds subdiscoid, about 4 mm. in diameter, 1.5–2 mm. thick, yellowish, smooth, a rounded, marginal ridge set in a broad shallow groove.

Infrequent in sandy soil of desert plains and foothills, Lower Sonoran Zone, Cochise, Pima, and Pinal Counties, Arizona. May–June.

**13. Opuntia thurberi** Engelm., Proc. Amer. Acad. **3:** 308, 1856

*Cylindropuntia thurberi* Knuth, in Backeb. & Knuth, Kaktus-ABC 124, 1935.

Straggly, bushy plant 1.5–4 m. tall with slender, elongate, cylindrical joints 1–2.5 dm. long, 10–15 mm. in diameter, pale green or tinged with red or purplish, often somewhat glaucous; tubercles narrow, about 3 mm. wide, 2–3.5 mm. high, 1.5–2 cm. long, distinctly flattened laterally; areoles broadly elliptic, borne so high on tubercles that they face upward at angle of 15–35° from central axis, 3–3.5 mm. in diameter, bearing closely packed, yellowish wool (this later turning dark gray) and few or no glochids; spines 3–7, spreading, 5–12 mm. long, terete and slender, encased in thin, papery, brown sheaths only in youth, soon naked and pale yellow to dark brown; flowers about 3 cm. high, 3–3.5 cm. in diameter, petals brownish yellow, obovate, rounded but apiculate at apex; fruits 2–3 cm. long, obovoid, deeply excavate terminally, spineless but tawny wool and minute glochids in rather crowded areoles; seeds subglobose, 3.5–4 mm. in diameter, only slightly flattened.

Arid plains and hillsides, Lower Sonoran Zone, western Sonora and Sinaloa. Resembling *O. versicolor* but more straggling, more slenderly branched, and with slightly longer spines. March–April.

**14. Opuntia parryi** Engelm., Amer. Jour. Sci. II, **14:** 339, 1852

*Opuntia bernardina* Engelm., in Parish, Bull. Torrey Club **19:** 92, 1892.
*Cylindropuntia parryi* Knuth, in Backeb. & Knuth, Kaktus-ABC 124, 1935.

An erect or ascending, openly, sparingly to profusely branched shrub 0.5–1.5 m. high; joints slender, 7–30 cm. long, 1.5–2 cm. in diameter; tubercles 1.5–2.5 cm. long, narrow, with 1–5(8) unequal, slender, brownish spines 0.5–3 cm. long at apex of each, longest spine usually porrect or deflexed; glochids brown, fading to yellow or ash-colored; flowers in clusters of 3–8 at ends of older stems, 2–3 cm. long, sepals green to reddish, petals obovate, obtuse, yellow tinged with red; fertile fruit 2–3 cm. long, ovoid, deeply and broadly umbilicate, more or less tuberculate above middle, areoles bearing yellowish glochids, the upper ones 1–7 acicular spines 5–12 mm. long; sterile fruit subglobose to obovoid, fleshy, less spiny; seeds few, whitish, 4–6 mm. broad, margin shallowly grooved.

Dry gravelly fans and washes and interior valleys, Lower and Upper Sonoran Zones, Cuyama Valley, Santa Barbara County, Los Angeles County in the vicinity of

San Fernando, to the San Bernardino Valley and eastern San Diego and western Imperial Counties, California. April–June.

**15. Opuntia molesta** Brandegee, Proc. Calif. Acad. II, **2:** 164, 1889

*Opuntia clavellina* Engelm. ex J. M. Coulter, Contr. U.S. Nat. Herb. **3:** 444, 1896.
*Opuntia calmalliana* J. M. Coulter, op. cit. 453.

Openly branched, ascending, shrubby plant 1–2 m. tall with several branches from base; joints cylindrical to narrowly clavate, 1.5–2.5 cm. in diameter, 5–40 cm. long, green but distinctly glaucous, firmly attached and removed only with difficulty; tubercles long and narrow, 3–5 mm. wide, 2.5–5 mm. high, 1.5–4 cm. long; areoles broadly elliptic, 2–4 mm. wide, 3–6 mm. long, bearing tawny wool, rufous glochids 1–2.5 mm. long, and several spines; main spines 1–4, reddish with yellowish tips or in age purplish to nearly black, 2–5 cm. long, central one longest and often flattened dorsiventrally; secondary spines 2–7, similar to main ones but more slender and 5–12 mm. long, all with tawny papery sheaths when young; flowers 3–5 cm. in diameter, petals purplish bronze, rounded but apiculate at apex; fruits broadly obovoid, fleshy, 2.5–3.5 cm. long, yellow, strongly tuberculate, strongly excavate at apex, upper areoles bearing 1–5 spines 5–25 mm. long, these usually deciduous by maturity; seeds elliptic, flattened, 4–6 mm. long, strongly beaked at base.

Along desert washes and on plains, hillsides, and mesas, Lower Sonoran Zone, central Baja California. April–May.

**16. Opuntia brevispina** H. E. Gates, Cactus & Succ. Jour. **10:** 98, *2 figs.*, 1938

Erect shrubby plant to 1 m. tall with spreading, irregularly spaced branches forming a crown to 1 m. wide from an erect central trunk, this blue-green on younger growth, brownish near base; joints at first clavate, later cylindrical, 1.5–4 cm. in diameter, 3–25 cm. long, with rather prominent tubercles 5–10 mm. wide, 1–2 cm. long, blue-green and slightly glaucous; areoles at upper summit of tubercles, oval, 3–4 mm. long, 2–3 mm. wide, bearing dense pads of rusty to ashy wool and deciduous reddish glochids 1–1.5 mm. long when young, these soon disappearing; radial spines 5–7, ferruginous or soon ashy, 1–5 mm. long, very slender, central spine usually 1, sometimes 2, 1–8 mm. long, sheaths pale ferruginous, delicate, soon deciduous; flowers diurnal, about 3 cm. long, 2.5–3 cm. wide at anthesis, ovary tuberculate, each areole bearing wool, glochids, and several slender acicular bristles 2–10 mm. long; outer perianth segments about 10, broadly obovate, rounded to truncate, 8–10 mm. long and about as wide, reddish purple with a tinge of bronze, margins entire; inner segments about 5, 1.5–2 cm. long, 8–10 mm. wide, truncate to emarginate, bronzy red-purple; filaments light green, anthers yellow; style light green at base, reddish above; stigma lobes greenish white; fruit 1–2 cm. long, subglobose, yellow, areolate and bearing slender deciduous bristles, fleshy; seeds few, subglobose, irregular by mutual crowding, stramineous.

Rocky hillsides, cliffs, and islets, Lower Sonoran Zone, on Isla Espírito Santo, Isla Ballena, and Isla Partida in the southern part of the Gulf of California off the coast of Baja California, and possibly on other islands about halfway up the east coast of the peninsula. March–June.

**17. Opuntia spinosior** (Engelm.) Toumey, Bot. Gaz. **25:** 119, 1898

*Opuntia whipplei* var. *spinosior* Engelm., Proc. Amer. Acad. **3:** 307, 1856.
*Opuntia spinosior* var. *neomexicana* Toumey, Bot. Gaz. **25:** 119, 1898.
*Cylindropuntia spinosior* Knuth, in Backeb. & Knuth, Kaktus-ABC 126, 1935.

Shrubby or arborescent plant 1–3.5 m. tall, usually with a definitely woody, nearly spineless trunk 2–8 dm. tall and rather open, spreading or arching branches, main stems bearing whorls of lateral branches or joints; joints cylindrical, 2–3 cm. in diameter, 0.5–3 dm. long, strongly tuberculate and often purplish tinged beneath sheathing of spines; tubercles crowded, flattened laterally, 2.5–4 mm. wide, 5–6 mm. high, 6–18 mm. long; areoles broadly elliptic, about 2.5 mm. wide, 3–4.5 mm. long, grayish; spines 6–20, divergently spreading, 5–10 mm. long, very slender but stiff, gray, tinged with red or purplish toward tips, sheaths very thin, whitish, deciduous first year; flowers 4–5 cm. in diameter, petals yellow to various shades of red, maroon, burgundy, copper, or rarely nearly white; fruit strongly tuberculate but spineless, obovoid, 2.5–3.5 cm. long, moderately fleshy, deeply excavate at apex, deep yellow or faintly tinged with green or orange, often persisting through winter; seeds tumidly discoid, 3–4 mm. long, nearly as wide, 2–2.5 mm. thick, buff to dark brown, margin rounded.

Desert plains, hillsides, and mountain sides, Lower Sonoran Zone to Transition Zone, Yavapai County, Arizona, south into southern Sonora and east into New Mexico. May–June.

**18. Opuntia versicolor** Engelm. ex J. M. Coulter, Contr. U.S. Nat. Herb. **3:** 452, 1896

*Cylindropuntia versicolor* Knuth, in Backeb. & Knuth, Kaktus-ABC 125, 1935.

Shrubby to arborescent plant 1–4 m. tall with intricately and openly branched stems from a well defined, woody trunk to 1 dm. in diameter or more and 0.4–1.5 m. high; joints cylindrical, slender, green or tinged with purplish, 1.5–2.5 cm. in diameter, 0.5–6 dm. long; tubercles prominent, flattened laterally, 3–5 mm. wide, 2–5 mm. high, 12–20 mm. long, rather closely crowded laterally in dry seasons, broader and more rounded following rains; areoles broadly elliptic to suborbicular, 3–4 mm. wide, bearing grayish wool, rufous or tawny glochids to 1.5 mm. long, and 6–12 acicular gray to purplish spines 4–15 mm. long, sheaths stramineous, whitish or grayish, soon deciduous; flowers 2.5–4 cm. in diameter, petals obovate, rounded, apiculate or shallowly retuse, yellow, orange, greenish, red, brown, or bronze, varying strikingly within a single colony but uniform on each plant; fruit fleshy, broadly obovoid, subtruncate at apex but with an abrupt excavation 3–6 mm. wide and fully as deep in center, tuberculate and bearing wool and a few glochids on areoles but rarely spiny, 2.5–3.5 cm. long, 2–2.5 cm. in diameter, usually persisting though next winter, sometimes proliferating; seeds numerous, tumid, and somewhat angular by compression, 3–4.5 mm. long, tawny, embedded in yellowish or greenish pulp.

Along washes, arroyos, and mouths of canyons, Lower Sonoran Zone, Pinal and Pima Counties, Arizona, south through much of Sonora. April–May.

**19. Opuntia prolifera** Engelm., Amer. Jour. Sci. II, **14:** 338, 1852

*Opuntia alcahes* Weber, Bull. Mus. Hist. Nat. Paris **1:** 321, 1895.
*Cylindropuntia prolifera* Knuth, in Backeb. & Knuth, Kaktus-ABC 126, 1935.

Bushy or with a well defined trunk 1–2.5 m. high, the trunk and older branches woody; joints 3–15 cm. long, 3–5 cm. in diameter, fleshy, turgid, easily detached, dark green; tubercles short, usually low and inconspicuous; areoles bearing yellowish brown felt, light yellow glochids 1–2.5 mm. long, and 5–12 rusty to nearly black spines 8–25 mm. long, these more or less divergently interlocked on young joints, sparser and fewer to lacking on older joints because of weathering; sheaths from sordid-yellow to rusty-brown; flowers 1 to several at ends of branches, 2–3 cm. long (including ovary); sepals green tinged with rose; petals few, obovate, obtuse, 5–8 mm. long, rose to rose-purple; filaments greenish; ovary tuberculate, areoles rather crowded, filled with light brownish glochids, upper containing also 1–5 ascending, acicular, brown spines 5–18 mm. long; fruit globose, 2–3 cm. long, turgid, usually spineless, proliferous, usually seedless; seeds when present 5–6 mm. long, ovate, flattened.

Arid hills and mesas near the coast, Upper Sonoran Zone, Santa Rosa, Santa Catalina, San Clemente, and Anacapa Islands and vicinity of Ventura, California, south into coastal Baja California at least to Rosario; occasionally inland to limits of ancient beach, and on Cedros and Guadalupe Islands. April–July.

**20. Opuntia ciribe** Engelm. ex J. M. Coulter, Contr. U.S. Nat. Herb. **3:** 445, 1896

Shrubby plant 0.5–2 m. tall with numerous stout, densely spiniferous branches from a stout, fairly well defined trunk; older joints firmly attached, younger ones readily detached; terminal and subterminal joints broadly ovoid to cylindric, 1.5–3.5 cm. in diameter, 3–12 cm. long, light green to blue-green, more or less glaucous, nearly hidden by close investiture of spines; tubercles numerous, closely crowded, 3–5 mm. wide, about as high, 6–12 mm. long, somewhat mammillate; areoles circular to broadly elliptic, 2–3 mm. in diameter, closely packed with rufous wool and red-brown glochids 1–2 mm. long, all except a few of lowermost on a joint heavily armed with 3–6 straight acicular spines 1–3 cm. long and 1–10 shorter, more slender marginal ones, each with a fairly persistent, stramineous to glistening-white, papery sheath, these smelling faintly like a skunk when brushed or bruised; flower 2.5–3 cm. in diameter, inner petals greenish with purplish tips, outer yellow to purplish; filaments green; anthers yellow; fruit broadly obovoid, fleshy, 2–4 cm. long, strongly tuberculate, areoles bearing glochids but spineless, apex truncate, and deeply and broadly excavate; seeds few, often only 1 or entirely lacking in a fruit, tumidly discoid, about 4 mm. long, 1.5–2 mm. thick, pale yellow, smooth (fruits and detached joints giving rise to new plants by vegetative propagation).

Rocky hillsides and desert plains, Lower Sonoran Zone, central Baja California from vicinity of Bahía de los Angeles south to the Cape Region. April–May.

**21. Opuntia fulgida** Engelm., Proc. Amer. Acad. **3:** 306, 1856, var. **fulgida**

*Cylindropuntia fulgida* Knuth, in Backeb. & Knuth, Kaktus-ABC 126, 1935.

Arborescent cholla to 3.5 m. tall with a sturdy main trunk often to 2 dm. in diameter at base and 0.5–1 m. to the 3–8 branches, its bark blackish and spineless, main branches sturdy, producing a rounded, often irregular crown; joints cylindrical, 3–5 cm. in diameter (exclusive of spines), terminal ones rather easily detached; tubercles mammillate, 4–5 mm. wide, 4–6 mm. high, 9–18 mm. long, crowded; areoles broadly ellipsoidal, to 5 mm. long, each bearing 2–20 conspicuous, straw-colored, spreading spines 2–3 cm. long, these interlocked to form a dense formidable armament, sheaths pale stramineous to nearly white, usually persisting for two or more years and very conspicuous; flowers about 2 cm. in diameter, with only 5–8 pinkish or whitish petals streaked with lavender; fruits green or slightly yellowish, broadly obovoid, 2.5–4 cm. long, smooth and spineless, usually persistent and bearing new flowers and fruits from their margins, forming conspicuous, drooping and branching chains of fruits; seeds few, many fruits apparently seedless.

Mostly sandy or gravelly soils of desert plains and moderate slopes, Lower and Upper Sonoran Zones, from Bill Williams River in Arizona south to Sinaloa. June–Aug.

**21*a*. Opuntia fulgida** var. **mammillata** (Schott ex Engelm.) J. M. Coulter, Contr. U.S. Nat. Herb. **3:** 449, 1896

*Opuntia mammillata* Schott ex Engelm., Proc. Amer. Acad. **3:** 308, 1856.

Differing from var. *fulgida* only in being more succulent, having longer tubercles to 1.5 cm. high and less closely crowded; areoles bearing fewer (rarely over 10) spines, and in having somewhat larger, more tuberculate fruit.

Desert flats and gently sloping hillsides, Lower and Upper Sonoran Zones, southern Arizona south through much of Sonora. June–Aug.

**22. Opuntia cholla** Weber, Bull. Mus. Hist. Nat. Paris **1:** 320, 1895

Arborescent, 1–5 m. tall from a sturdy central trunk to 1.5 dm. in diameter and 1.5 m. tall; trunk at first very spiny but eventually bark becoming smooth, brownish yellow and spineless; main branches ascending or spreading, often bearing whorls of joints 0.5–3 dm. long, 2–3 cm. in diameter; tubercles broad and rounded, to 1 cm. broad across base, about one-half as high, 1.5–3 cm. long, not crowded, more or less glaucous; areoles orbicular to elliptical, 3–4 mm. wide, containing wool and numerous yellowish glochids; spines 5–13 per areole (mostly 7–10), yellowish when young, later dark gray, 3–25 mm. long, porrect or spreading at an angle; sheaths golden yellow, conspicuous on young spines, later deciduous; flowers 2.5–3 cm. broad, petals few, peach-pink to purplish, with spines from upper areoles of ovary often projecting between spreading petals; fruits broadly obovoid, fleshy but scarcely juicy, usually greenish, often persistent and proliferous, growing readily when falling to the ground, 2–3 cm. in diameter, 3–5 cm. long, moderately excavate at apex, usually spineless when mature; seeds 2–3 mm. long, about 1 mm. thick, stramineous, often abortive.

Desert flats and hillsides, Lower Sonoran Zone, central Baja California to the Cape Region. April–May.

**23. Opuntia bigelovii** Engelm., Proc. Amer. Acad. **3:** 307, 1856

*Cylindropuntia bigelovii* Knuth, in Backeb. & Knuth, Kaktus-ABC 125, 1935.

Sturdy erect plant, usually with a central spiny trunk 7–10 cm. in diameter, 1–2.5 m. high, and numerous short lateral branches, these soon deciduous below, and lower trunk turning black; joints 5–20 cm. long, 3.5–5 cm. in diameter, turgid, with crowded tubercles and dense, closely interlocking armament, terminal joints very easily detached, primary ones persistent, turning sooty-black; tubercles more or less 4-sided, 8–10 mm. long, about as broad, low; spines and persistent sheaths shining, pale yellow or roseate on young growth, 1.5–2.5 cm. long, 7–12 per areole, divergent; glochids and felt yellowish; flowers in clusters at ends of joints, 2.5–4 cm. long, 2.5–3.5 cm. in diameter, greenish yellow, pale green, or white streaked with pale lavender, petals rather few in number; ovary bearing areoles filled with brown wool, glochids, and 1 to several small acicular spines 5–15 mm. long; fruit deeply umbilicate, dry, greenish, nearly or quite spineless; seeds flattened, angulate, greenish white, seldom fertile.

Forming extensive stands on dry hillsides, outwash slopes, mesas, and stabilized sand dunes, Lower Sonoran Zone, southern Nevada to the western Colorado Desert,

California, in northern Sonora, and east of the mountains in northern Baja California; San Estéban Island. Feb.–May.

**24. X Opuntia fosbergii** C. B. Wolf, Occ. Papers Rancho Santa Ana Bot. Gard. **1:** 79, *fig. 22,* 1938. (*O. bigelovii* X *O. echinocarpa*)

*Opuntia bigelovii* var. *hoffmannii* Fosberg, Bull. S. Calif. Acad. **32:** 121, 1933.

Trunk vigorous, 2–3 m. high, often 2 to several branches from base; shorter terminal joints more crowded than in *O. bigelovii*; tubercles essentially as in *O. echinocarpa*, twice as long as broad; green stem showing through armament; fruit as in *O. bigelovii*.

Lower Sonoran Zone, Mason Valley to Vallecitos, San Diego County, California. March–April.

**25. X Opuntia munzii** C. B. Wolf, Occ. Papers Rancho Santa Ana Bot. Gard. **1:** 79, *fig. 23,* 1938. (*O. bigelovii* X *O. acanthocarpa*)

Plant to 4 m. tall, with general habit of X *O. fosbergii*; tubercles narrower, two to three times as long as broad; spines more slender and less crowded.

Lower Sonoran Zone, Chocolate Mountains, Colorado Desert, California. March–April.

**26. Opuntia burrageana** Britt. & Rose, Cactaceae **1:** 70, *pl. 14, fig. 1,* 1919

Freely branching, irregular shrub, often with some branches arcuate-sprawling, others erect, 0.5–1 m. tall, 1–2 m. in diameter, rather open and irregular in outline, older plants with dark gray to nearly black trunk 5–8 cm. in diameter, 1–3 dm. tall; joints cylindrical to cylindroclavate, 1–2 cm. in diameter, 2–15(25) cm. long, densely spiny, pale green or tinged with red or purple during extended dry seasons; tubercles low, 2.5–3.5 mm. wide, 5–10 mm. long, virtually disappearing on old, practically terete joints; areoles suborbicular to oval, closely set, 2–3.5 mm. long, nearly as wide, containing rufescent wool, a tuft of fine, yellowish to grayish glochids at upper edge, and 5–9(11) slender, divaricately spreading, yellow-sheathed spines 0.5–2.2 cm. long, sheaths on some plants somewhat rufous, spines horny to golden yellowish when freshly unsheathed, at length dark gray or ashy; flowers usually few, 2–4 cm. broad in anthesis, outer perianth segments purplish or reddish, greenish at base, inner ones red-brown to purple-brown above, somewhat greenish at base, 8–10 mm. long, lance-ovate, acute to short-acuminate; filaments green; stigmas nearly white; ovary obovate-truncate, 1.5–2 cm. long at anthesis, prominently low-tuberculate and bearing many glochids, often with loosely attached, very slender spines 0.5–1.5 cm. long when young, these mostly or wholly deciduous in fruit; fruit nearly dry, 1.5–2 cm. long, rarely somewhat proliferous; seeds few, 3–4 mm. in diameter, stramineous.

Valleys and hillsides, Lower Sonoran and Arid Tropical Zones, Cape Region and on such islands in the Gulf as Ildefonso, Carmen, Cerralvo, and Pichilinque. Possibly on Isla Partida del Norte (near Isla Angel de la Guarda), but this plant is imperfectly known and its taxonomic position uncertain. April–May.

**27. Opuntia echinocarpa** Engelm. & Bigelow, in Engelm., Proc. Amer. Acad. **3:** 305, 1856, var. **echinocarpa**

*Opuntia echinocarpa* var. *major* Engelm. & Bigelow, in Engelm., loc. cit.
*Opuntia echinocarpa* var. *robustior* J. M. Coulter, Contr. U.S. Nat. Herb. **3:** 446, 1896.
*Opuntia deserta* Griff., Monats. Kakteenk. **23:** 132, 1913.
*Cylindropuntia echinocarpa* Knuth, in Backeb. & Knuth, Kaktus-ABC 124, 1935.

Erect, ascending or spreading, branched plant 0.5–1.5 m. high; joints turgid, 8–25 cm. long, 1.5–2.5 cm. in diameter; tubercles prominent, broadly ovate, about 1 cm. long; areoles 2–6 mm. long, bearing stout spines 2–3 cm. long and 6–12 shorter, more slender ones, fine yellowish wool, and glochids; spines interlocking; flowers clustered at end of older branches, 2–3.5 cm. long, yellow tinged with red; ovary short-turbinate, tuberculate, and densely clothed on upper two-thirds with acicular spines 8–20 mm. long; fruit 1.5–2.5 cm. long, very spiny over upper two-thirds, 6–12 spines in each areole; seeds numerous, 5–6 mm. long, margins grooved.

Desert areas and dry interior foothills, Lower and dry ridges in lower Upper Sonoran Zones, Mojave and Colorado Deserts and adjacent western foothills to Utah, Sonora, and northern Baja California. March–April.

**27*a*. Opuntia echinocarpa** var. **parkeri** (Engelm.) J. M. Coulter, Contr. U.S. Nat. Herb. **3:** 446, 1896

Plant more robust, less intricately branched; joints 10–25 cm. long; flowers and fruits larger.

Borrego Valley, foot of Mountain Springs Grade, Imperial County, California. March–April.

**27*b*. Opuntia echinocarpa** var. **nuda** J. M. Coulter, Contr. U.S. Nat. Herb. **3:** 446, 1896

Rather compact shrub about 1 m. tall with the aspect of var. *echinocarpa*, but spines 1–1.5 cm. long, reddish brown with lighter tips; spine sheaths brownish yellow; flowers 3 cm. tall, 2 cm. in diameter at anthesis, greenish yellow; fruit obovate, 2 cm. long, shallowly umbilicate, nearly or entirely spineless at maturity; seeds 3.5–5 mm. long, flattened.

Mesas and hillsides, Lower Sonoran Zone, northern part of the Magdalena Plain. Feb.–April.

**28. Opuntia acanthocarpa** Engelm. & Bigelow, in Engelm., Proc. Amer. Acad. **3:** 308, 1856, subsp. **acanthocarpa**

*Cylindropuntia acanthocarpa* Knuth, in Backeb. & Knuth, Kaktus-ABC 124, 1935.

Erect, slender, terminally branched shrub 1–2 m. high; terminal joints 4–25 cm. long, 2–3.5 cm. in diameter; tubercles prominent, 2–2.5 cm. long, and laterally flattened; spines 8–18, unequal, 1.5–3.5 cm. long, yellowish to dark brown, each covered with a light yellow or whitish sheath; glochids numerous, yellow; flowers reddish to brownish yellow, 5 cm. long and wide when expanded; perianth segments broadly obovate, obtuse; ovary short-turbinate, with a few prominent tubercles bearing 8–12 rigid acicular spines from middle to apex; fruit dry, 2.5–3.5 cm. long, naked below, tuberculate and spiny above middle; seeds crowded, irregularly angled, 4–6 mm. long, light yellowish.

Desert mesas and slopes, Lower Sonoran Zone, eastern Mojave and Colorado Deserts, California, to southern Utah, central Arizona, Sonora, and northern Baja California. April–May.

**28*a*. Opuntia acanthocarpa** subsp. **ganderi** C. B. Wolf, Occ. Papers Rancho Santa Ana Bot. Gard. **1:** 75, *fig. 20*, 1938

Plant of vigorous growth; joints brighter green; spines 20–25, more slender, lower ones deflexed; flowers smaller.

Lower Sonoran Zone, San Felipe Valley, San Diego County, and southward along the east base of the Laguna Mountains and desert slopes of the San Jacinto Mountains, California. April–May.

**28*b*. Opuntia acanthocarpa** var. **ramosa** Peebles, Cactus & Succ. Jour. **9**: 37, *1 fig.*, 1937

Plant branching close to ground, to 1 m. tall; joints 5–30 cm. long, 17–20 mm. in diameter, gray-green; tubercles 15–25 mm. long, about 5 mm. high and wide; spines 6–12 per areole, loosely and inconspicuously sheathed, pale gray to pale reddish, 5–33 mm. long but mostly less than 25 mm. long; flowers about 6 cm. in diameter, red or variegated red and yellow; fruit obovate, 2–3 cm. long, spiny on upper half or two-thirds; seeds 3.5–4 mm. long, margins angled.

Desert flats and valleys, Pima and Pinal Counties, Arizona. April–May.

**29. Opuntia thornberi** Thornber & Bonker, Fantastic Clan 133, 148, *fig.*, 1932

*Opuntia acanthocarpa* var. *thornberi* L. Benson, Proc. Calif. Acad. IV, **25**: 247, 1944.

Openly branched, rather straggly shrub with several erect or ascending, glaucous, green stems to 1.5 m. tall without a main trunk; main joints slenderly cylindrical, about 1.5–2.5 cm. in diameter, 1–4.5 dm. long, with few short, divergent lateral branches 5–15 cm. long, not easily detached; tubercles laterally compressed, 4–5 mm. wide and high, 2–5 cm. long; areoles broadly oval to orbicular, 3–4 mm. in diameter, crowded with tawny, eventually gray wool; glochids few and insignificant; spines 6–10, spreading to deflexed, 1–2.5 cm. long, tawny or grayish with red-brown tips, youngest ones with thin, yellowish or sordid-whitish sheaths, glochids few, yellow-brown; flowers 5–7 cm. in diameter, with petals varying from light yellow to deep red or purplish; fruits dry, broadly obovoid, truncate and not excavate at apex, 2–2.2 cm. wide, 2.5–3.5 cm. long, bearing a few spines in upper areoles or sometimes spineless when fully ripe; seeds pale tawny, turgidly discoid, about 5 mm. in diameter, about one-half as thick, smooth, marginal ridge low and rounded, slightly sinuous.

Outwash slopes and gravelly washes, Lower and arid lower parts of Upper Sonoran Zones, Pima, Yavapai, and Graham Counties, Arizona. April–May.

**30. Opuntia rosarica** G. Lindsay, Cactus & Succ. Jour. **14**: 56, *fig. 32*, 1942

Low bushy plant to 1 m. tall and about as wide or less; branches spreading to ascending; joints cylindrical, 2–5 cm. in diameter, 1–2 dm. long, blue-green to olive green, or brownish in age; tubercles low and long, often longitudinally confluent to resemble undulate ribs, each tubercle 5–8 mm. wide, 2–3 cm. long; areoles deltoid-ovate, 3.5–4 mm. wide, 4.5–5 mm. long, felted, each bearing numerous golden yellow glochids 2–4 mm. long around top of areole, 4–7 unequal sheathed spines 1–3.5 cm. long, and several slender bristles; central spine porrect, to 3.5 cm. long, radials 1–2.5 cm. long, all acicular, brown-colored; flowers numerous at tips of branches, about 4.5 cm. long, 3–3.5 cm. broad; outer perianth segments 1–2 cm. long, 1–1.5 cm. wide, rounded at apex, yellow with a reddish blush in center, inner segments to 2.5 cm. long, clear yellow with a roseate blush near tips; fruit dry, 1–1.5 cm. long, tuberculate, armed with spines, bristles, and glochids in rather crowded areoles; seeds 4–5 mm. in diameter, irregularly angled, about 3 mm. thick, tawny.

On mesas and valley floors, mostly in tension zone between chaparral and desert, Lower and Upper Sonoran Zones, western foothills of Sierra San Pedro Mártir from about San Telmo south to the hills east of Rosario. June–Aug.

**31. Opuntia basilaris** Engelm. & Bigelow, in Engelm., Proc. Amer. Acad. **3:** 298, 1856, var. **basilaris**

Low, prostrate to erect plant, usually growing in clumps, branching mostly from base, seldom over 2–5 dm. high; joints flattened, orbicular to broadly obovate, 7–25 cm. long, 1–1.5 cm. thick, short-puberulent, pruinose to glabrate in age, occasionally tinged with red or purplish; areoles numerous, circular to elliptic, 2–3 mm. in diameter, spineless, filled with dingy wool when young, this soon replaced by erect, yellowish brown glochids 1–3 mm. long; flowers borne on upper margins of joints, somewhat clustered, 10–15 cm. broad, deep rose to rose-purple, a conspicuous velvety sheen on petals; fruit dry at maturity, 5–6 cm. long, globose to obovoid, spineless, areoles and glochids as on joints; seeds 6–10 mm. broad, more or less angled.

Arid mountain slopes and desert washes, Lower Sonoran Zone to Arid Transition Zone, Mojave and Colorado Deserts and adjacent mountain slopes to southern Utah, Arizona, and northern Sonora. March–June.

**31*a*. Opuntia basilaris** var. **ramosa** Parish, Bull. Torrey Club **19:** 92, 1892

Joints oblong to narrowly obovate, 5–8 cm. wide, up to 30 cm. long; branching freely above; stems frequently 3–6 joints high.

Lower Sonoran Zone to Arid Transition Zone, northwestern Los Angeles County east to the western edge of the Mojave Desert south to western Riverside and northeastern San Diego Counties; Caliente and Democrat Hot Springs, Kern County, California. March–April.

**32. Opuntia ursina** Weber, in Bois, Dict. Hort. **2:** 896, 1898

*Opuntia erinacea* var. *ursina* Parish, in Jepson, Fl. Calif. **2:** 542, 1936.

A low, decumbent, branching plant 2–5 dm. tall, forming dense mats 1–6 dm. wide; joints obovate to oblong, 5–8 cm. wide, 8–15 cm. long, 8–12 mm. thick, pale green, glabrous; areoles orbicular to ovoid, about 3 mm. wide, 3–5 mm. long, 6–10 mm. apart, woolly in youth, filled with numerous yellow to brownish glochids 1–5 mm. long; spines 8–15, ashy gray to white, 2–20 cm. long, bristlelike, reflexed to ascending, flexuous, usually nearly obscuring surface of joint; flowers 6–7 cm. long and wide, petals yellow suffused with red, 10–18 mm. long, apiculate to short-acuminate, margins often minutely erosulate; fruit dry, obovoid-truncate, 3–4 cm. long, its areoles filled with glochids and acicular spines 8–20 mm. long.

Rocky hillsides and outwash slopes, upper edges of Lower Sonoran Zone into Arid Transition Zone, Mohave County, Arizona, into southern Nevada and adjacent southeastern California. April–May.

**Opuntia erinacea** Engelm. & Bigelow, in Engelm., Proc. Amer. Acad. **3:** 301, 1856, differs from *O. ursina* mainly in having shorter, slightly stouter, nonflexuous spines less crowded in the joints, and occurring at higher elevations in the same general vicinity of northwestern Arizona, but to date has not been discovered within the limits of the Sonoran Desert.

**33. Opuntia pycnantha** Engelm., in J. M. Coulter, Contr. U.S. Nat. Herb. **3:** 423, 1896

*Opuntia pycnantha* var. *margaritana* J. M. Coulter, op. cit. 424.
*Opuntia margaritana* E. M. Baxter, Cactus & Succ. Jour. **2:** 388, 1931.

Low creeping plant with several to many pads standing on edge in a continuous chain along the ground with an occasional 1–3 pads standing erect from the upper side of the one on the ground, lateral "chains" often 1–3 m. long, terminal 1–3 pads held a few cm. above the soil, eventually settling to the ground under weight of additional growth; joints orbicular to short-oblong, 8–13 cm. wide, 10–18 cm. long, 1.5–2.5 cm. thick, minutely puberulent with straight, erect, simple hairs between areoles, latter orbicular, 3–7 mm. in diameter, cushionlike, 3–5 mm. high, closely crowded, 4–8 mm. apart, each areole containing a small amount of grayish wool, numerous erect, yellowish to reddish glochids 2–5 mm. long, and 7–12 yellow to pale reddish brown, terete, straight, strongly deflexed spines; spines unequal, 0.5–3 cm. long, central one usually longest, often slightly flattened and less sharply deflexed than others; flowers 4–6 cm. wide, inner petals obovate-acuminate, 2–2.5 cm. long, yellow or suffused with red (reddish flowers on plants with rufescent spines); ovary almost completely covered with closely crowded areoles bearing glochids and bristlelike spines to 2 cm. long; mature fruit and seeds unknown.

Sandy and gravelly flats and gentle slopes, Magdalena and Santa Margarita Islands, Baja California. May–June.

**34. Opuntia megacarpa** Griff., Rep. Mo. Bot. Gard. **20:** 91, *pls. 2, 13,* 1909

*Opuntia engelmannii* var. *megacarpa* Fosberg, Bull. S. Calif. Acad. **33:** 100, 1934.

Decumbent spreading shrub 5–8 dm. high, forming open clumps up to 2.5 m. wide, some spreading branches with joints lying on edge in chainlike arrangement on the ground; joints broadly oval to suborbicular, 9–12 cm. wide, 15–17 cm. long, thin, rarely over 2 cm. thick, bluish green; areoles broadly elliptic, 3–4 mm. long, bearing grayish wool and comparatively few yellow-brown glochids 1–4 mm. long; spines 1–2 (occasionally 4), terete or somewhat flattened, dull, light gray, or sometimes darker at base, 1–4 cm. long, mainly on upper two-thirds of joint; flowers yellow, 5–6 cm. broad; fruits fleshy, subglobose to obovoid, shallowly umbilicate, 5–7 cm. long, bright reddish purple; seeds 4–5 mm. long.

Western edge of Mojave and Colorado Deserts, upper Lower Sonoran Zone to Arid Transition Zone, Los Angeles County to San Diego County, California. May–July.

**35. Opuntia phaeacantha** Engelm., in A. Gray, Pl. Fendl. 51, 1849

(For synonymy, see Britt. & Rose, Cactaceae **1:** 144–145, 1919, and L. Benson, Cacti of Ariz. ed. 2, 56–57, 1950.)

Prostrate or sprawling, bushy plant 2–6 joints up to 1 m. high and as much as 3 m. in diameter; branches often of 10 or more joints lying on edge on the ground, with or without lateral chains arising from some of the older pads; joints broadly obovate or some suborbicular, 1–1.8 dm. wide, 1.2–3 dm. long, about 1.5–2 cm. thick, smooth, pale green; areoles comparatively remote, 2–3 cm. apart, suborbicular to ovoid, 2.5–3 mm. wide, 4–6 mm. long, woolly and bearing numerous yellow to rusty glochids 1–7 mm. long, often only those on upper part of pad bearing 1–4 reddish brown spines, these 1.5–6 cm. long, spreading at various angles, tapering, slightly flattened, straight or a few slighty curved; flowers yellow with a flush of red toward base of inner segments, 6–9 cm. in diameter, inner petals obovate-spatulate, slightly short-acuminate, 3–3.5 cm. long; stigmas green; fruit fleshy, wine-colored, subcylindrical to obovoid, 2–2.5 cm. in diameter, 3.5–4 cm. long, areoles bearing glochids but no spines; seeds reniform-obovoid, flattened, 4–5 mm. long, margin distinctly ridged.

Desert hillsides and mesas, Lower Sonoran Zone and arid ridges in lower parts of Upper Sonoran Zone, southern Nevada to Texas, Sonora, and Chihuahua. May–June.

**36. Opuntia chlorotica** Engelm. & Bigelow, in Engelm., Proc. Amer. Acad. **3:** 291, 1856

*Opuntia tidballii* Bigelow, Pacif. R. Rep. **4:** 11, 1856.
*Opuntia curvospina* Griff., Bull. Torrey Club **43:** 88, 1916.

Erect plant with ascending branches 1–2.5 m. high, usually with a definite woody trunk covered with persistent, scaly, gray or brownish bark and densely armed with rigid deflexed spines; joints orbicular to ovate, 12–20 cm. long, occasionally broader than long, light green, faintly glaucous; leaves subulate, 4–6 mm. long; areoles 1–3 cm. apart, conspicuous, extending 2–4 mm. above surface, 5–8 mm. in diameter on young joints, becoming larger in age, containing a conspicuous tuft of sordid-yellow wool, numerous yellow glochids, and 3–7 unequal spines; spines on joints 1.5–4 cm. long, bright yellow, terete, mostly reflexed; those on main stem 15–40 in an areole, yellow, flattened, 2–5 cm. long, stellately radiating and covering whole surface; flowers yellow, 6–7 cm. broad; petals oblong, 2.5–3 cm. long, 12–15 mm. wide; fruit fleshy, reddish purple, with greenish flesh, 4–5 cm. long, bearing rather crowded areoles containing wool, glochids, and occasionally a few short spines; seeds small.

Occasional in canyons and on rocky slopes, Lower and lower parts of Upper Sonoran Zones, northeastern San Bernardino County to western edge of the Colorado Desert, California, to southern Nevada, New Mexico, Sonora, and northern Baja California. April–June.

**37. Opuntia tapona** Engelm., in J. M. Coulter, Contr. U.S. Nat. Herb. **3:** 423, 1896

*Opuntia angustata* var. *comonduensis* J. M. Coulter, Contr. U.S. Nat. Herb. **3:** 425, 1896.
*Opuntia comonduensis* Britt. & Rose, Smiths. Misc. Coll. **50:** 519, 1908.

Low bushy plant 0.5–1 m. tall, often with a heavy short trunk 1–2 dm. tall and nearly as thick, closely covered with a dense armament of somewhat reflexed spines 3–5 cm. long, branches spreading-ascending or sprawling, of 2–5 joints, plants often forming dense thickets of considerable extent; joints orbicular to broadly obovate, 6–18 cm. wide, 15–30 cm. long, pale green to somewhat ashy and glaucous, finely puberulent with short, straight, erect hairs when young, often glabrate in age; areoles 1.5–3 cm. apart, broadly oval to orbicular, 2–3.5 mm. wide, to 6 mm. long, rufiginous to deep red-brown or in age the wool grayish, some of those on lower part of pad often containing only many reddish glochids 1–4 mm. long, or nearly all bearing 1–4 stiff but slender, usually deflexed, pale gray to reddish spines 1–5 cm. long, when more than one spine borne in an areole subcentrally located one usually longer than others; flowers borne near apex of pads, large, 4–6 cm. in diameter, inner petals broadly obovate, truncate and abruptly apiculate at apex, spreading nearly at right angles to axis of ovary in anthesis, about 2–2.5 cm. long; ovary spineless but bearing numerous glochid-filled areoles 3–12 mm. apart, broadly ovate-truncate to oblong; fruit fleshy, red, 3.5–5.5 cm. long, 2–4 cm. in diameter, truncate and shallowly umbilicate at apex; seeds yellowish, smooth, 3.5–4 mm. long, nearly as wide, marginal ridge paler than rest, low and rounded.

Mesas, hillsides, among rocks, and on gravelly slopes and rocky washes, Lower Sonoran Zone, central Baja California from the vicinity of Comondú and Loreto south to the Cape Region and on most of the adjacent small islands along the Gulf coast. April–May.

The presence or absence of puberulence on the joints, used by Britton and Rose to separate *O. tapona* and *O. comonduensis,* seems to be associated with no other character or with geographical distribution. The range of variation involved among the numerous colonies of this complex observed in the field seems well within the normal intraspecific range.

**38. Opuntia ficus-indica** (L.) Mill., Gard. Dict. ed. 8, no. 2, 1768.

*Cactus ficusindica* L., Sp. Pl. 468, 1753.
*Cactus opuntia* Guss., Prod. Sic. 559, 1827–8. Not *C. opuntia* L., 1753.
*Opuntia vulgaris* Tenore, Syll. Fl. Neap 239, 1831. Not *O. vulgaris* Mill., 1768.
*Opuntia ficus-barbarica* Berger, Monats. Kakteenk. **22:** 181, 1912.

Large, spreadingly branched shrub or small tree up to 5 m. high with a definite, spineless, woody trunk; joints obovate to oblong or spatulate-oblong, 15–50 cm. long, slightly glaucous; areoles circular or broadly elliptic, containing a tuft of short brownish wool, numerous yellow-brown deciduous glochids, and 1–3 spines or sometimes spineless; spines white, rigid, subulate, slightly flattened, unequal, 1.5–4 cm. long, subspreading; flowers 6–10 cm. broad, bright yellow, sepals and outer petals sometimes faintly tinged with red; fruit obovate, 5–9 cm. long, often faintly pruinose, red-purple throughout, flesh juicy, umbilicus deeply depressed; seeds 4–5 mm. broad.

Widely cultivated in tropical and subtropical countries, and occurring as an occasional escape about old gardens and ranches in coastal southern California and Baja California. March–June.

**39. Opuntia megacantha** Salm-Dyck, Hort. Dyck. 363, 1834

*Opuntia robusta* var. *megacantha* Schelle, Handb. Kakteenkultur 57, 1907.
*Opuntia castillae* Griff., Rep. Mo. Bot. Gard. **19:** 261, *pl. 24,* 1908.
*Opuntia incarnadilla* Griff., op. cit. **22:** 27, *pls. 4–5,* 1912.

Arborescent, 3–6 m. tall, with a definite, woody, spineless trunk; joints obovate to oblong, 3–6 dm. long, often asymmetrical and becoming concave-convex on lateral branches, pale green, somewhat glaucous; areoles 2.5–4 mm. in diameter, 5–8 cm. apart on large joints, bearing brown wool, a few light brown glochids, and 1–5 spines or spineless; spines 1–5, brown, slightly spreading to subdeflexed, 2–3 cm. long, confined to upper and marginal areoles in most plants; glochids usually deciduous though sometimes reappearing on older joints; flowers 6–9 cm. broad, sepals greenish yellow, tinged with orange on midribs, petals yellow to deep orange; fruit obovate, 7–11 cm. long, clear yellow or tinged with orange-red, fleshy, edible; seeds yellowish white, 4–6 mm. broad.

Cultivated and occasionally escaped, mainly Upper Sonoran Zone, in the coastal area from Santa Barbara, California, south into Baja California at least as far as Rosario. March–July.

**40. Opuntia gosseliniana** Weber, Bull. Soc. Acclim. France **49:** 83, 1902

Shrubby, ascendingly branched plant 1–1.5 m. tall from a short but definite main trunk bearing numerous acicular spines to 10 cm. long, with 20 or more spines in an areole; joints broadly ovate to obovate, 1–2 dm. wide, 1.2–3 dm. long, 1–1.5 cm. thick, tinged from light to deep purple and often slightly glaucous over the purple, usually bearing spines at most areoles above lower one-fourth of pad; areoles 1–3 cm.

apart, orbicular to broadly elliptical, 2–4 mm. wide, cushionlike, filled with short gray wool and numerous yellowish to reddish brown glochids 1–6 mm. long; spines terete-acicular, 2–8(10) cm. long, pale horn-colored to brown in youth, usually dark brown or purplish brown in age, usually 1, sometimes 2, and rarely 3 in an areole, mostly porrect or somewhat ascending, larger ones often slightly sinuous; flowers yellow, 5–7.5 cm. in diameter, subpersistent upon withering, inner petals rhombic-obovate, 1.5–2 cm. wide, 2–3 cm. long, acute to apiculate at apex; fruit fleshy, ovoid-subcylindrical, 4–6 cm. long, about 2–2.5 cm. in diameter, bearing numerous glochids but no spines in areoles, purplish.

Gravelly slopes, rocky hillsides, and outwash bajadas, Lower Sonoran Zone, western Sonora from near Magdalena south nearly to Guaymas. April–May.

**41. Opuntia santa-rita** (Griff. & Hare) Rose, Smiths. Misc. Coll. **52:** 195, *pl. 15,* 1908

*Opuntia chlorotica* var. *santa-rita* Griff. & Hare, N. Mex. Agr. Exp. Sta. Bull. **60:** 64, 1906.
*Opuntia shreveana* C. Z. Nels., Galesburg Register, July 20, 1915.
*Opuntia gosseliniana* var. *santa-rita* L. Benson, Cacti of Ariz. ed. 2, 65, 1950.

Erect bushy plant 1–1.5 m. tall from a short, stout, basal trunk and with crowded ascending branches; joints orbicular to broadly obovate, 1–1.8 dm. wide, 1–2 dm. long, 10–12 mm. thick, smooth, bluish green but tinged with purple around areoles and along margins or over whole pad during dry or cold seasons; areoles orbicular, 2–3 mm. in diameter, 1.5–2 cm. apart, or those along upper margin closely crowded and to 6 mm. in diameter, filled with grayish wool and numerous yellowish to rufescent glochids 1–7 mm. long, longest ones in upper marginal areoles; spines none in most areoles, usually 1 but occasionally to 3 in upper marginal areoles, 1–5(10) cm. long, terete, slender, tapering only at apex, reddish at base or throughout, on younger pads sometimes yellowish toward tips; flowers 6–7.5 cm. in diameter, petals clear yellow, withering-persistent; fruit fleshy, lilac to purple, about 1.5 cm. in diameter, 2–2.5 cm. long, often curved, spineless but bearing fine glochids in small areoles.

Foothills and outwash plains, Lower and arid parts of Upper Sonoran Zones, Pima and Cochise Counties, Arizona, south into northern Sonora. April–May.

**42. Opuntia bravoana** E. M. Baxter, Cactus & Succ. Jour. **5:** 436, *1 fig.,* 1933

Upright, freely branching plant 1–2 m. tall, to 2 m. wide; joints thin, oblong to obovate, 10–15 cm. wide, 15–35 cm. long, 5–6 mm. thick, glabrous, light green with a purplish spot below each areole, marginal portions often tinged with purplish or whole pad purplish when exposed to full sun or excessively dessicated; areoles mostly oval, 2–2.5 mm. wide, 3–5 mm. long, rather distantly arranged 2.5–4 cm. apart, often somewhat larger and more closely arranged (1–2 cm. apart) around margins, bearing short gray wool and numerous yellow to rufescent glochids 1–4 mm. long; most areoles spineless, the upper, particularly the marginal, areoles bearing 1–5 slightly flattened, slender, yellowish but in age grayish, porrect to deflexed spines 0.5–6 cm. long, largest heaviest spines usually in areoles along margins of pads; flowers (including ovary) 5–6 cm. high, borne on sides of pads near upper end as well as on margins, petals bright yellow, widely spreading, broadly obovate, rounded to subtruncate and undulate-erose but not acuminate or apiculate, forming a corolla 5–8 cm. wide when fully open; filaments reddish at base, yellow above; style whitish, stigma-lobes 5–8, pale greenish yellow; fruit fleshy, obovoid, 3.5–4 cm. long, the woolly, glochid-bearing areoles prominent, 5–10 mm. apart; seeds dark brown, about 3 mm. in diameter, margins rounded.

Chiefly in partial shade of trees and large shrubs, Lower Sonoran and Arid Tropical Zones, on both Pacific and Gulf sides of Cape Region, Baja California, and on Cerralvo and San Pedro Nolasco Islands. May–July.

**43. Opuntia engelmannii** Salm-Dyck, in Engelm., Bost. Jour. Nat. Hist. **6:** 207, 1850, var. **engelmannii**

(For synonymy see Britt. & Rose, Cactaceae **1:** 147–148, 1919.)

Large shrubby plant 1.5–2 m. tall, 1.5–5 m. in diameter, with ascending branches but no definite central trunk; joints obovate to suborbicular, 1.5–2 dm. wide, 1.5–3 dm. long, 2–2.5 cm. thick, pale green, smooth, and glabrous; areoles orbicular, 3–9 mm. in diameter, in age cushionlike, 3–5 mm. high, 1.5–3 cm. apart, filled with gray wool in center and with numerous rufescent glochids 1–10 mm. long around margins; spines 1–5, usually present in all but a few basal areoles, white, pale gray, or light horn-colored tinged with white when young, often darker when older or at base and tip, terete or moderately flattened, 1.5–6 cm. long, spreading at various angles or longer ones more or less deflexed; flowers 6–8 cm. in diameter, clear yellow, inner petals broadly obovate, rounded at apex, slightly erose; stigmas bright green; filaments and anthers yellow; fruit obovoid, 2.5–4 cm. in diameter, 4–6 cm. long, externally red or purple but with yellow flesh; seeds reniform-obovoid, 3–4 mm. wide, strongly flattened, margin ridged.

Arroyos, washes, hillsides, and desert grasslands, Lower Sonoran and lower arid parts of Upper Sonoran Zones, southern Nevada and Utah south into Texas, Durango, and Sonora. April–June.

**43*a*. Opuntia engelmannii** var. **flavescens** (Peebles) L. Benson, Cacti of Ariz. ed. 2, 58, 1950

*Opuntia flavescens* Peebles, Cactus & Succ. Jour. **9:** 67, 1937.

Differing from var. *engelmannii* in having slightly smaller, more markedly yellow-green joints; only apical areoles bearing spines 2.5 cm. long or less; fruit red without, flesh yellow (indicating relationship to *O. engelmannii* var. *engelmannii*).

Hillsides and valleys, Lower Sonoran Zone, western Pima County, Arizona. March–April.

**44. Opuntia laevis** J. M. Coulter, Contr. U.S. Nat. Herb. **3:** 419, 1896

Loosely few-branched, prickly pear 0.6–4 m. tall or forming a low dense bush in full sunshine; branches several joints long, erect to widely spreading to form clumps several meters in diameter; joints narrowly obovate to oblong, often somewhat asymmetrical, 10–14 cm. wide, 12–20 cm. long (or rarely to 30 cm. long), 10–18 mm. thick, light green, spineless or with a few upper areoles bearing 1–3 spines; areoles orbicular to oval, 3–4 mm. wide, 2–3.5 cm. apart, mostly woolly and bearing numerous slender glochids 1–6 mm. long; spines when present pale gray to white, suffused with red near base, subulate-acicular, 1–1.5 cm. long, spreading to reflexed, slightly flattened; flowers about 6–7.5 cm. in diameter at anthesis, petals obovate, obtuse to retuse at apex, pale yellow, 1.5–2.5 cm. wide, 3–3.5 cm. long; stigma lobes green; fruit fleshy, reddish to purple, turbinate-cylindrical, with a stipelike base, 5–9 cm. long, 2–3.5 cm.

in diameter, areoles bearing long glochids but spineless; seeds 4–5 mm. long, margins ridged.

On rocky cliffs and steep slopes of canyons in protected places, Lower and Upper Sonoran Zones, Pima to Pinal and Cochise Counties, Arizona, and in the Grand Canyon area of northern Arizona. April–May.

**45. Opuntia gilvescens** Griff., Rep. Mo. Bot. Gard. **20:** 87, *pls. 2, 5, 7, 13,* 1909

Branching, more or less bushy, prickly pear to 1.5 m. tall; joints orbicular to obovate, symmetrical, yellowish green, 10–13 cm. wide, 15–23 cm. long, 10–13 mm. thick; areoles orbicular to oval, 2–3.5 cm. apart, bearing wool and numerous yellowish to yellow-brown glochids 1–5 mm. long, those on upper two-thirds of joint each containing 1–3 spines, these 2–3 cm. long, pale gray, subulate, spreading to somewhat reflexed; flowers 6–7 cm. in diameter at anthesis, petals yellow; fruit fleshy, reddish or purplish, obovoid to turbinate, not contracted to a stipe-like base, 5–6 cm. long, spineless but areoles containing glochids.

On mesas and open grassland, Lower and Upper Sonoran Zones, Pima and Santa Cruz Counties, Arizona. April–May.

**46. Opuntia macrocentra** Engelm., Proc. Amer. Acad. **3:** 292, 1856

Shrubby, considerably branched, prickly pear 6–12 dm. tall, to 3 m. wide, with orbicular to oblong or obovate joints 10–20 cm. long, plant usually 4–10 joints high; joints thin, 6–10 mm. thick, smooth, blue-green, usually tinged with purple especially near margins; areoles orbicular to oval, 1.5–4 mm. wide, 1.5–2.5 cm. apart, short-woolly, with numerous yellow-brown glochids 1–7 mm. long, only upper ones bearing spines; spines 1–4 per areole, dark brown, dark reddish brown or nearly black, often paler near tips, terete or essentially so, acicular and tapering only near tips, 3.5–10 cm. long, mostly spreading but a few sometimes reflexed; flowers 3–3.5 cm. high, to 5 cm. wide at anthesis, petals obovate, yellow suffused with red toward base, abruptly apiculate at apex, about 1.5 cm. long; ovary bearing few areoles, these 1–1.5 mm. in diameter, containing short brown glochids but spineless; fruit fleshy, red to purplish at maturity, ovoid, about 2 cm. in diameter, 2.5–4 cm. long, falling during autumn; seeds asymmetrically reniform, 4–4.5 mm. long, 3–3.5 mm. wide, margin strongly ridged.

Foothill slopes and along arroyos, Lower and Upper Sonoran Zones, Pima to Greenlee Counties, Arizona, east to western Texas and south into adjacent Mexico. April–May.

**47. Opuntia littoralis** (Engelm.) Cockerell, Bull. S. Calif. Acad. **4:** 15, 1905

*Opuntia engelmannii* var. *littoralis* Engelm., in Brewer & Wats., Bot. Calif. **1:** 248, 1876.
*Opuntia lindheimeri* var. *littoralis* J. M. Coulter, Contr. U.S. Nat. Herb. **3:** 422, 1896.

Bushy spreading plant 0.6–1.5 m. tall, often forming impenetrable thickets over several acres; branches ascending, of 3–10 joints; main stem sometimes woody and becoming subterete and to 2 dm. in diameter but often not apparent; older basal joints or those coalesced to form short (2–3 dm.) stem dark gray to nearly black, spineless or nearly so; joints orbicular, obovate, or oblong, 8–20 cm. wide, 15–30 cm. long, 2–5 cm. thick, dull green, glabrous but scarcely glossy; areoles orbicular, 3–8 mm. in diameter, 1–3 cm. apart, filled with tawny to grayish wool and numerous yellowish glochids 1–4 mm. long, and bearing 2–9 (usually 7 or less) unequal, slightly flattened or terete, clear yellow spines 1–3 cm. long; spines of marginal areoles spreading, those on faces of pads mostly deflexed, straight or curved, sometimes slightly twisted; flowers 5–8 cm. broad, petals bright yellow, obovate, 2–4 cm. long, 1–1.5 cm. wide, obtuse, rounded, or minutely apiculate, outer ones smaller and often suffused with red toward base, lower midrib faintly greenish; fruit fleshy, subglobose to broadly obovoid, 4–5.5 cm. long, shallowly umbilicate to nearly truncate, bright red, bearing fine glochids in areoles 1–1.2 cm. apart; seeds strongly ridged marginally, 3–4 mm. wide, stramineous.

Hillsides, draws, and coastal bluffs, mostly Upper Sonoran Zone, coastal strip from Santa Barbara County into northern Baja California and on the adjacent islands at least to Aguajito, about twenty miles south of Rosario. April–July.

## 3. MARENOPUNTIA Backeb., Desert Pl. Life **22:** 27, 1950

Low shrubby plants from fusiform, tuberous, dahlia-like roots with ascending, moderately branched stems; branches terete-cylindrical and bearing both spines and short fine glochids. Leaves subulate, caducous. Flowers solitary at apex of each flowering branch, ovary buried in apical tissue of stem and not apparent at anthesis. Stamens numerous, short. Style slightly exceeding stamens, stigmas several-lobed. Fruit embedded in tip of stem, dry, bursting irregularly along the side. Seeds reniform.

### 1. Marenopuntia marenae (S. H. Parsons) Backeb., Desert Pl. Life **22:** 27, *2 figs.,* 1950

*Opuntia marenae* S. H. Parsons, Desert Pl. Life **8:** 10, *figs. 1–3,* 1936.

Plant 1.5–6 dm. tall, from 1 to several tuberous roots 0.5–2 cm. in diameter, 6–18 cm. long; stems 8–15 mm. in diameter; joints (from one bifurcation to next above or tip of branch) 5–20 cm. long, slightly tuberculate in youth but soon terete, tubercles 2–3 mm. wide, about 1–1.5 cm. long while still apparent; leaves 5–10 mm. long, purplish to green; areoles orbicular, 2–3 mm. in diameter; radial spines 7–12, white, appressed, slenderly acicular, 3–10 mm. long; central spines 2–6, bulbous at base, longest one porrect at first, later deflexed, white, 1–2 cm. long, second longest spine from upper edge of areole, dark brown to black, first porrect, then upwardly appressed and caducous; corollas white, 6–8 cm. wide, diurnal; inner petals obovate, rounded at apex, slightly crenulate; outer perianth segments successively smaller, purplish-tinged, apiculate to subulate; filaments pale green, to 2 cm. long; anthers yellow; stigma lobes cream; fruit dry, spiny; seeds reniform, about 2 mm. wide, 3 mm. long, white to yellowish gray.

Known only from the type locality, near the mouth of the Bacuachic River, Kino Bay, Sonora. Collected in flower in May.

## 4. PACHYCEREUS Britt. & Rose, Contr. U.S. Nat. Herb. **12:** 420, 1909

Very large, massive, treelike plants to 15 m. tall with definite trunks and usually several to many heavy, erect-ascending, columnar branches. Ribs coarse, each one situated exterior to a slender wand of woody tissue that does not anastomose with the adjacent ones except at very base of branch or trunk. Areoles distant, felted, bearing straight stout spines. Flowers borne on upper parts of all ribs, diurnal, tubes short, perianth segments moderately spreading, spatulate. Stamens numerous, included, inserted along corolla throat. Style 3–6-branched, included. Tube of flower and ovary bearing scales with felt and bristles in their axils. Fruit burlike, dry, eventually breaking up into numerous clusters of spines, bristles, and felt. Seeds black, comparatively large.

Ribs 11–17; spines 20–30 in young areoles; bristles of fruit rarely over 1 cm. long, usually shorter ........................................................ 1. *P. pringlei*
Ribs 10–12; spines 8–12 in an areole; bristles of fruit 2–6 cm. long .. 2. *P. pecten-aboriginum*

**1. Pachycereus pringlei** (S. Wats.) Britt. & Rose, Contr. U.S. Nat. Herb, **12:** 422, 1909

*Cereus pringlei* S. Wats., Proc. Amer. Acad. **20:** 368, 1885.
*Cereus calvus* Engelm., in J. M. Coulter, Contr. U.S. Nat. Herb. **3:** 409, 1896.
*Cereus titan* Engelm., in J. M. Coulter, loc. cit.
*Pilocereus pringlei* Weber, in Bois, Dict. Hort. **2:** 966, 1898.
*Pachycereus calvus* Britt. & Rose, Contr. U.S. Nat. Herb. **12:** 420, 1909.
*Pachycereus titan* Britt. & Rose, op. cit. 422.

A massive, many-branched, columnar plant to 15 m. tall or rarely to 20 m., with pale green older branches, grayish trunk, and usually glaucous young growth; ribs 11–17, obtuse and broad when water is abundant, becoming narrow and acutely angled in drought, 2–4 cm. high; areoles distant, or deciduous on older growth, brown-felted when young, felt later becoming gray; spines 20–30 in an areole in young plants, fewer, stouter, and shorter in older specimens, 1–3 cm. long, or to 12 cm. long in young plants, subulate, stiff, dark gray, brownish, or nearly black, usually lighter near base; flowers borne from near tip downward 0.5–5 m. of flowering branches, areoles in this region nearly or quite confluent and nearly spineless; flowers 6–8 cm. long; tube and ovary with small acute scales 3–5 mm. long, these virtually hidden by axillary tufts of copious, light brown hairs; perianth segments white, oblong, spreading; stamens white or faintly tinged with rose; anthers golden; fruits globose, 5–7 cm. in diameter, bristles crowded, mostly only 5–6 mm. long, embedded in light yellow-brown felt, these tufts breaking apart at maturity; seeds black, testa thin, seeds edible.

On rocky hillsides, desert bajadas, and level plains, Lower Sonoran Zone, Río Santo Domingo and San Felipe Desert south in Baja California, and from vicinity of Puerto Libertad south in Sonora. April–June.

The freshly opened flowers are visited by large numbers of small beetles that work among the numerous filaments and become liberally dusted with pollen. The flowers have a delicate fragrance similar to that of a freshly cut, ripe watermelon.

**2. Pachycereus pecten-aboriginum** (Engelm.) Britt. & Rose, Contr. U.S. Nat. Herb. **12:** 422, 1909

*Cereus pecten-aboriginum* Engelm., in S. Wats., Proc. Amer. Acad. **21:** 429, 1886.

A treelike, columnar cactus 5–10 m. tall with numerous erect or ascending branches 1.5–2.5 dm. in diameter; trunk 3–5 dm. in diameter and 1.2–5 m. tall; ribs 10–12, usually 1.5–2.5 cm. high, obtuse; areoles distant below, closely crowded to nearly confluent near tips of branches, 6–10 mm. in diameter, felted with reddish brown wool that becomes gray with age, a narrow groove below each areole extending downward almost to one below; spines 8–12, one to three of central porrect or spreading, remaining radials horizontal to surfaces of stem, all comparatively short and stout, 0.5–1 cm. long or in young or very spiny plants to 3 cm. long, grayish with black tips; flowers on distal 1 m. or less of branches, 5–7.5 cm. long; outer perianth segments purplish, inner white, fleshy; stamens many; stigma lobes about 10; ovary and tube densely woolly, sparsely bristly in flower; fruit depressed-globose, 6–7.5 cm. in diameter (exclusive of bristles), covered with slender, yellowish or light brownish bristles 1–6 cm. long, fruit resembling a huge chestnut; seeds dull black.

Mostly on gentle slopes or desert llanos where the soil is of fine texture but occasionally ascending the sides of rocky hills, Lower Sonoran Zone, southern Chihuahua into central and southern Sonora, the southern part of Baja California, and south at least into Colima. March–April.

The huge bristly fruits and the more slender branches set this species sharply apart from *P. pringlei,* with which it is associated in Sonora and Baja California. It does not range as far north as *P. pringlei.*

## 5. X PACHGEROCEREUS Moran, Cactus & Succ. Jour. 34: 93, 1962

Shrubby plant with a stout, stocky trunk bearing few to many erect, unbranched stems. Ribs 14–18, spinose, rounded, somewhat undulate-serrate from areole to areole; intervals V-shaped. Areoles distinct to nearly confluent, bearing numerous acicular spines and dense mats of wool; spines golden yellow in youth, gray to black in age, bulbous at base, terete. Flowers borne on upper parts of stem, diurnal, funnelform, yellow. Fruit globose, spiny. Seeds small, black and shiny. An intergeneric hybrid between *Pachycereus* and *Bergerocactus.*

**1. X Pachgerocereus orcuttii** (K. Brandegee) Moran, Cactus & Succ. Jour. **34:** 93, 1962

*Cereus orcuttii* K. Brandegee, Zoe **5:** 3, 1900.
*Pachycereus orcuttii* Britt. & Rose, Contr. U.S. Nat. Herb. **12:** 422, 1909.

Shrubby plant to 4 m. tall, 2 m. wide; trunk to 2 dm. in diameter; branches erect, up to 30 from top of trunk, unbranched above, constricted at end of season's growth, 7–10 cm. in diameter, yellowish green; ribs 8–12 mm. high, 10–14 mm. wide; areoles orbicular to broadly elliptic, 4–6 mm. wide, 4–8 mm. long, 2–9 mm. apart or nearly confluent near tip of stem, each bearing 25–100 spines, often a spineless area at upper edge; spines poorly differentiated into radials and centrals, 5–10 centrals longer than radials, to 7 cm. long; flowers on joints of current and previous season, funnelform, 6–7 cm. long, 5–6 cm. wide, light yellow or with tinge of green on outer segments; tube 2.5–3 cm. long, obconic, white within, pale green outside; scales on tube 2–5 mm. long, acuminate, white-lanate in axils; stamens many, 4–6 mm. long (including anthers); style about 3 cm. long; stigma lobes 7–10, yellow, 4–5 mm. long; fruit densely felted and spiny, 4–5 cm. in diameter, splitting irregularly into 3–4 segments; seeds 2.8–3 mm. long, 2–2.3 mm. wide, 1.5–2 mm. thick, shallowly reticulate-pitted, many viable.

Rolling hills and canyons, Lower Sonoran Zone, southeast of Rosario, Baja California. April–June.

## 6. BERGEROCACTUS Britt. & Rose, Contr. U.S. Nat. Herb. 12: 435, 1909

Low colonial cactus with branching, cylindrical, erect, ascending and procumbent stems. Ribs many, low; spines many and closely interlaced, bright yellow, acicular. Flower small, with widely spreading, pale yellow petals; scales on short tube and ovary small, bearing wool and slender spines in their axils; perianth segments obtuse. **Fruit** globular, dry, closely covered with straight slender spines. Seeds obovate, black, pitted; embryo curved.

**1. Bergerocactus emoryi** (Engelm.) Britt. & Rose, Contr. U.S. Nat. Herb. **12:** 435, 1909

*Cereus emoryi* Engelm., Amer. Jour. Sci. II, **14:** 338, 1852.
*Echinocereus emoryi* Rümpler, in Förster, Handb. Cact. ed. 2, 804, 1885.

Branches 0.2–1.5 m. long, stout, erect, ascending or procumbent, spreading then rooting on lower side, 2–4 cm. in diameter, entirely clothed with a dense mat of bright yellow spines, turning gray to nearly black in age; ribs 18–25, 2–3 mm. high, somewhat tuberculate; spines 10–30 at an areole, slender-acicular, 1–6 of central spines sometimes 6 cm. long; flowers 3–4.5 cm. broad when expanded; outer perianth segments obtuse, obovate, 5–8 mm. long, lemon yellow, tinged with green, inner segments narrower, oblong, 8–10 mm. long, almost or quite devoid of greenish tinge; fruit globular, 2–3.5 cm. in diameter, covered with spines 0.5–1.5 cm. long, dry; seeds oblong, black, shining, pitted, about 2.5–3 mm. long.

On coastal bluffs and hillsides within a few miles of the sea, Lower Sonoran and lower Upper Sonoran Zones, from the vicinity of San Diego and Santa Catalina Island, California, south a few miles beyond Rosario, Baja California. April–May.

## 7. LEMAIREOCEREUS Britt. & Rose, Contr. U.S. Nat. Herb. 12: 424, 1909

Tall columnar plants with many erect branches or in some species scrambling and low in stature. Ribs small, narrow, and many, or coarser and fewer in number. Areoles felted, bearing several to many stout spines. Flowers diurnal or nocturnal, solitary at a flowering areole, tubular-funnelform to campanulate. Stamens numerous, inserted in throat. Ovary tubercled, bearing scales with felt in their axils and soon developing spines. Fruit ovoid to globose, pulpy and juicy, often edible, cracking irregularly when ripe, spiny at first but spines weakly attached and usually deciduous at or soon after maturity. Seeds numerous, small, black, with a thin testa.

Stems 15–20 cm. in diameter, often branching well above the base; fruit 4–7.5 cm. in diameter ........................................... 1. *L. thurberi* var. *thurberi*
Stems 5–7 cm. in diameter, branching only at the base; fruit 2.5–3 cm. in diameter ....... ...................................................... 1*a*. *L. thurberi* var. *littoralis*

**1. Lemaireocereus thurberi** (Engelm.) Britt. & Rose, Contr. U.S. Nat. Herb. **12:** 426, 1909, var. **thurberi**

*Cereus thurberi* Engelm., Amer. Jour. Sci. II, **17:** 234, 1854.
*Pilocereus thurberi* Rümpler, in Förster, Handb. Cact. ed. 2, 689, 1885.

A many-branched, erect plant with no definite trunk or a very short one, bearing 5–35 or more erect or ascending branches 15–20 cm. in diameter, up to 3–8 m. tall from base; ribs 12–19, usually about 1 cm. high but sometimes twice that high, rounded and separated by very narrow, groovelike intervals; areoles 1–3 cm. apart, to 1 cm. in diameter, brown-felted and usually exuding minute viscid droplets that coalesce to form a waxy mass; spines numerous, acicular to subulate, 1–5 cm. long, brownish to black or becoming gray in age; flowers from apex to 3 dm. below tip of flowering branches, 6–8 cm. long; outer perianth segments broadly oblong, reddish purple, gradually merging into scales of tube, inner perianth segments light purple with pale margins; filaments 2–2.5 cm. long; ovary scaly with minute spines and hairs in axils;

fruit globose, 4–7.5 cm. in diameter, spiny but in age rupturing irregularly and becoming naked, pulp red or purplish red within, tinged with olive green on outside; seeds black and shining, 1.8–2 mm. long.

Hillsides and alluvial plains, Lower Sonoran Zone, southwestern Arizona south through western Sonora and from the vicinity of Cataviñá, Baja California, south to the Cape. April–July.

**1*a*. Lemaireocereus thurberi** var. **littoralis** (K. Brandegee) G. Lindsay, Desert Pl. Life **12:** 186, 1940

*Cereus thurberi* var. *littoralis* K. Brandegee, Zoe **5:** 191, 1904.

Stems 5–7 cm. in diameter, branching at very base, growing outward nearly horizontally and then curving sharply upward, or sometimes sprawling and straggling; entire plant smaller and more slender than var. *thurberi,* often only 1–2.5 m. tall; flowers deeper red than in var. *thurberi*; fruit globose, 2.5–3 cm. in diameter.

Known only from coastal bluffs between San José del Cabo and Cabo San Lucas, Baja California. April–July.

## 8. WILCOXIA Britt. & Rose, Contr. U.S. Nat. Herb. 12: 434, 1909

Slenderly branched plants from tuberous, dahlia-like roots. Ribs few and low, flattened, separated by very narrow grooves. Areoles circular or elliptic, rather distant, bearing small, weak, bristlelike, mostly appressed spines. Flowers diurnal or nocturnal, very large in comparison to slender stems, funnelform-campanulate, red-purple or white tinged with rose or pink, solitary at a flowering areole; tube and ovary bearing areoles containing wool and weak spines. Stamens many, inserted in throat, exserted. Style exserted. Fruit pyriform, bristly, scarlet; spines eventually deciduous; fruit indehiscent, juicy. Seeds small, black, rugose-tuberculate, hilum basal, large.

Flowers diurnal; perianth segments purple or rose-purple .................. 1. *W. striata*
Flowers nocturnal; perianth segments white or outer ones tinged with pink or rose ............................................................ 2. *W. diguetii*

**1. Wilcoxia striata** (Brandegee) Britt. & Rose, Contr. U.S. Nat. Herb. **12:** 434, 1909

*Cereus striatus* Brandegee, Zoe **2:** 19, 1891.

Roots tuberous, ovoid to fusiform, brownish, 3–10 cm. long, 0.5–2 cm. in diameter; stems slender, erect, ascending or scrambling and vinelike, 0.5–2.5 m. tall, branches 5–15 mm. in diameter, grayish, with 8–9 low flattened ribs separated by very narrow grooves; areoles 5–8 mm. apart, circular, felted when young, bearing 6–10 slender acicular spines 1.5–3 mm. long, these usually closely appressed, light brown or gray with slightly darker tips, eventually deciduous; flowers narrowly funnelform, 10–12 cm. long, purplish, diurnal, limb 5–7 cm. in diameter; tube and ovary bearing areoles armed with fine, bristlelike spines similar to those on stems and with tufts of long wool in axils of scales; fruit pyriform, scarlet, 3–4.5 cm. long, bristly when young but nearly naked at maturity; seeds minutely pitted, shining-black, about 2 mm. long.

Sandy soil or alluvial plains, Lower Sonoran Zone, central Sonora and Baja California from San Francisquito Bay and the Vizcaíno Desert south to the Cape Region. July.

**2. Wilcoxia diguetii** (Weber) Dig. & Guillaumin, Arch. Hist. Nat. Soc. d'Acclim. **4:** 222, 1928

*Cereus digueti* Weber, Bull. Mus. Hist. Nat. Paris **1:** 319, 1895.
*Neoevansia diguetii* Marshall, in Marshall & Bock, Cactaceae 84, 1941.

Plants 2–7 dm. tall with 1 to several branches above base, often scrambling through shrubs; tubers 10–50, to 7 cm. in diameter, 12–15 cm. long; cortex greenish gray to brownish; stems 5–8 mm. in diameter; ribs 6–9, 1–2 mm. wide, separated by narrow grooves 2 mm. wide and 0.5 mm. deep; areoles 5–15 mm. apart, sparsely lanate when young, each bearing 8–13 spines, these mainly appressed, 3–4 mm. long, or with 1–2 porrect, mainly white or pale gray, tipped with black, all eventually deciduous on older parts of stems; flowers fragrant, 8–9.5 cm. long, limb 5–6 cm. in diameter; perianth segments 5–8 mm. wide, to 4 cm. long, spreading; filaments white, to 7 cm. long; fruit pyriform, 3–5 cm. long, 1.5–2.5 cm. in diameter, bright scarlet, shining, covered with areoles bearing about 15 bright yellow or black-tipped, bristly spines 3–6 mm. long; seeds 2–2.5 mm. long, rugose and tuberculate, hilum basal, about 0.6–0.8 mm. long.

In sandy or alluvial soil, Lower Sonoran Zone, southwestern Arizona in Yuma and Pima Counties southward in Sonora at least to Kino Bay and possibly in central Baja California. July.

**Wilcoxia mariana** Gentry, Río Mayo Plants 191, *pl. 25, fig. 2,* 1942. There is considerable uncertainty concerning the generic position of this plant because the characters of the roots and of the fruits and seeds are unknown. The flowers are similar to those of *Peniocereus johnstonii,* but the stems and spines are somewhat coarser than those of *P. johnstonii.* The type specimen was collected near Aquibiquichi, Sonora, in flower, on February 12, 1937, and presumably has not been re-collected since. Taxonomic assignment can be made with confidence only after additional field observations and study.

## 9. PENIOCEREUS Britt. & Rose, Contr. U.S. Nat. Herb. 12: 428, 1909

Small, slender, sparingly branched plants from a huge, fleshy, turniplike root, the stems 3–5-angled (usually 4–5-angled), the ribs somewhat undulate. Areoles small, with comparatively few spines. Flowers large, funnelform, white, with a tinge of red on outer perianth segments, nocturnal, very fragrant; tube long and slender, bearing scattered scales with spines in their axils below and on ovary, long hairs in axils of upper ones. Stamens just about equaling throat or slightly exserted. Fruit ovoid, bright scarlet, fleshy, and edible, bearing clusters of small weak spines that are easily removed at maturity. Seeds black, finely rugose; hilum large and oblique.

Seeds dull; inner perianth segments about 4 cm. long; young stems pubescent .. 1. *P. greggii*
Seeds shining; inner perianth segments about 3 cm. long; young stems glabrous .......... ........................................................ 2. *P. johnstonii*

### 1. **Peniocereus greggii** (Engelm.) Britt. & Rose, Contr. U.S. Nat. Herb. **12:** 428, *pls. 74–75,* 1909

*Cereus greggii* Engelm., in Wisliz., Mem. Tour North. Mex. 102, 1848.
*Cereus pottsii* Salm-Dyck, Cact. Hort. Dyckensi Cultae **1849:** 208, 1850.
*Cereus greggii* var. *cismontanus* Engelm., Proc. Amer. Acad. **3:** 287, 1856.
*Cereus greggii* var. *transmontanus* Engelm., loc. cit.
*Cereus greggii* var. *roseiflorus* Kunze, Monats. Kakteenk. **20:** 172, 1910.

Root fleshy, sometimes to 6 dm. in diameter, usually 15–20 cm. long, 5–8 cm. in diameter, 1 to several stems from a root, these 1–5 dm. or in brush to 3 m. tall, 2–2.5 cm. in diameter, minutely pubescent on younger growth; areoles small, 3–8 mm. apart, sparsely arachnoid-lanate in youth, bearing 6–11 blackish, bulbously based, subulate radical spines 2–6 mm. long and 1 porrect or slightly recurved central spine 3–10 mm. long, or sometimes a second slightly shorter central spine present; flowers solitary at an areole, 15–20 cm. long, tube slender, abruptly flaring to a short funnelform throat about 2 cm. long; perianth segments lanceolate, acute at apex, 3.5–4 cm. long, outer ones often reflexed, inner ones spreading; filaments erect, slightly exserted when flower is fully open; stigma lobes 8–10 mm. long; fruit 8–10 cm. long, beak nearly as long as ellipsoidal body, the body bearing scattered areoles containing wool and a few small spines, spine clusters easily brushed off at maturity; seeds about 2 mm. long, dull.

Under trees and among shrubs, Lower Sonoran Zone, southern Arizona to Chihuahua, Zacatecas, and Sonora. June–July.

**2. Peniocereus johnstonii** Britt. & Rose, Jour. Wash. Acad. **12:** 329, 1922

Clambering or scandent, perennial plant from a turniplike root 5–20 cm. in diameter, 10–40 cm. long, or probably larger, stems 1 to several from each root, glabrous, 3–5-ribbed, nearly terete, 5–15 mm. in diameter, 0.5–3 m. long, ribs rather low, undulate-concave between areoles, these 6–15 mm. apart on each rib; areoles broadly elliptical, 1.5–3 mm. long, bearing short, fine, gray wool; spines 7–12, reddish brown to black or some radials sordid-yellow to gray, slightly bulbous at base, straight, radials 1.5–9 mm. long, appressed or subappressed; central spines 1–3, subulate, porrect to slightly reflexed, 3–8 mm. long; flowers nocturnal, 12–15 cm. long, tube slender, bearing few elevated, globose, woolly areoles 1–1.5 mm. in diameter, throat broadly funnelform, 2–2.5 cm. long, 1.5–2 cm. wide at mouth, woolly, with reddish brown bristles 1.2–2 cm. long from its areoles; perianth segments white tinged faintly with rose on outer surface, lower ones with short straight bristles on outside, longer crinkled bristles from axils, segments lanceolate, attenuate, margins entire or faintly erose; stamens and anthers yellow, the latter 1.5–2 mm. long; ovary with crowded areoles bearing short wool and several slender, dark reddish brown spines 5–10 mm. long, spine clusters easily removed; fruit ovoid to oblong, 5–6 cm. long, deep red; seeds oblong, 3–3.5 mm. long, shining.

In sandy soil among shrubs, Lower Sonoran Zone, near the coast from Bahía de la Concepción south along the Gulf of California and on the adjacent islands to San Josef Island. Feb.–March.

## 10. MACHAEROCEREUS Britt. & Rose, Cactaceae 2: 114, 1920

Prostrate or low and bushy plants with arching branches, heavily armed with stiff interlacing spines throughout. Ribs low and rounded or obtuse. Areoles large, felted, and spiny. Spines several to many; centrals flattened and daggerlike; radials subulate. Flowers diurnal, slenderly funnelform; limb spreading and symmetrical, perianth withering and persisting on fruit. Stamens many, borne in narrow throat. Ovary and lower portion of tube bearing numerous small scales with tufts of fine hairs in their axils in youth, felt later replaced by small spines. Fruit globose, bearing clusters of small spines when young, eventually naked or nearly so, flesh pulpy, edible. Seeds dull black, punctate-reticulate.

Stems prostrate, 10–18 cm. in diameter including spines; perianth segments rose-pink with white bases ........................................................ 1. *M. eruca*
Stems ascending or arched, tips then resting on the ground or on other branches, 5–8 cm. in diameter; perianth segments purple ................................ 2. *M. gummosus*

**1. Machaerocereus eruca** (Brandegee) Britt. & Rose, Cactaceae **2:** 115, *figs. 171–172,* 1920

*Cereus eruca* Brandegee, Proc. Calif. Acad. II, **2:** 163, *pl. 7,* 1889.
*Lemaireocereus eruca* Britt. & Rose, Contr. U.S. Nat. Herb. **12:** 425, 1909.

Prostrate but terminal 2–4 dm. slightly ascending, this portion gradually becoming prostrate as tip grows forward, owing to weight of fleshy stem; branches 1–5 m. long, 4–12 cm. in diameter exclusive of spines, rooting from lower surface, dying at older end as growing tip advances; ribs 10–12; areoles about 2 cm. apart, 8–12 mm. in diameter; spines unequal, central spines flattened-subulate, 3–6 cm. long, one of them strongly reflexed, radials to about 20, terete-subulate to acicular, some of them somewhat flattened, 1–2.5 cm. long, all pale gray to nearly white or yellowish; flowers nocturnal, 10–12 cm. long, limb 8–10 cm. in diameter, tube slender; perianth segments white at base, gradually shading to rose-pink and often with a touch of lavender at tips; stamens and style pale yellow; ovary globose, spiny; fruit 3–4 cm. in diameter, scarlet, bearing clusters of short spines; seeds dull black, pitted.

Alluvial soil, Lower Sonoran Zone, the Magdalena Plain and Magdalena Island, Baja California. Flowering dates uncertain but ripe fruit collected in November.

**2. Machaerocereus gummosus** (Engelm.) Britt. & Rose, Cactaceae **2:** 116, *figs. 173–175,* 1920

*Cereus gummosus* Engelm. ex Brandegee, Proc. Calif. Acad. II, **2:** 162, 1889.
*Cereus cumengei* Weber, Bull. Mus. Hist. Nat. Paris **1:** 317, 1895.
*Cereus flexuosus* Engelm., in J. M. Coulter, Contr. U.S. Nat. Herb. **3:** 411, 1896.
*Lemaireocereus cumengei* Britt. & Rose, op. cit. **12:** 424, 1909.
*Lemaireocereus gummosus* Britt. & Rose, op. cit. 425.

Erect or more often ascending-sprawling plant 1–3 m. tall, forming thickets 10 m. or more in diameter; branches 5–8 cm. in diameter, dark green; ribs 8 or sometimes 9, low and obtuse; areoles about 2 cm. apart; spines stout, radials 8–12, terete or somewhat flattened, 8–15 mm. long, centrals 3–6, stouter and strongly flattened, one about 4 cm. long, considerably surpassing others, all of them gray or reddish gray with slightly darker tips; flowers borne one at an areole, 10–15 cm. long, limb 6–8 cm. in diameter; perianth segments 1–2.5 cm. long, purplish; tube slender, bearing a few scales 1–2 mm. long with tufts of feltlike hairs in axils, ovary with felted and spine-bearing areoles rather closely arranged at anthesis; fruit ovoid to globose, 6–8 cm. in diameter, scarlet with purple flesh that is sharply but pleasantly acid and of pleasing flavor; seeds about 2.5 mm. long, finely rugose and pitted.

Rocky hillsides, along arroyos, and on desert alluvial plains, Lower Sonoran Zone and barely entering the Upper Sonoran Zone, near the coast about 5 miles north of Ensenada south to the Cape Region of Baja California and on most of the islands in the Gulf including Tiburón Island and on the adjacent mainland at Punta Sargento, Sonora. Flowering period extended throughout much of the year, but heaviest in July–Aug.

## 11. CARNEGIEA Britt. & Rose, Jour. N.Y. Bot. Gard. 9: 187, 1908

A large columnar cactus with erect, stout, many-ribbed stems and branches, and crowded areoles bearing spines and tufts of brown felt. Flowers borne singly at uppermost areoles, funnelform-campanulate, diurnal, tube cylindrical; scales on tube felted in axils; inner perianth segments white, waxy, short, spreading or reflexed. Ovary oblong. Stamens numerous; stigma lobes narrowly linear, slightly exceeding stamens. Fruit ovoid to ellipsoid, sparingly spinose, fleshy, containing a red edible pulp filled with small, black, shining seeds. Embryo sharply curved; cotyledons incumbent.

**1. Carnegiea gigantea** (Engelm.) Britt. & Rose, Jour. N.Y. Bot. Gard. **9:** 188, *fig.* 32, 1908

*Cereus giganteus* Engelm., in Emory, Notes Mil. Rec. 159, 1848.
*Pilocereus engelmannii* Lemaire, Ill. Hortic. **9:** Misc. 97, 1862.
*Pilocereus giganteus* Rümpler, in Förster, Handb. Cact. ed. 2, 662, 1885.

Stem simple and upright or with 1 to several lateral branches, up to 16 m. high, branches 3–6.5 dm. in diameter; ribs 12–25, obtuse, 1–3 cm. high, varying in width with water supply; areoles 2–4 cm. apart on older growth, closely crowded at apex of stem, densely brown-felted; spines at top of flowering plant acicular, yellowish brown, porrect; those on sterile branches or older parts of plant broader, stouter, subulate, central stouter than radials, up to 7–8 cm. long, usually dark brown to black; flowers nocturnal, 10–12 cm. long, 5–8 cm. in diameter when fully expanded; tube 1–1.5 cm. long, green; throat 2.5–3.5 cm. long, expanded; perianth segments waxy-white; filaments white; style 5–6.5 cm. long, cream-colored to white; ovary bearing scales with felted axils; fruit red to purple, 6–10 cm. long, splitting irregularly; seeds about 0.75 mm. in diameter.

On gravelly slopes, rocky ridges, and outwash fans, Lower Sonoran Zone, from northern Arizona and along the Colorado River in Riverside and Imperial Counties, California, to northern Sonora. Flowering chiefly in May but sparingly in August following summer rains.

## 12. RATHBUNIA Britt. & Rose, Contr. U.S. Nat. Herb. 12: 414, 1909

Slenderly branched, shrubby, or bushy cactus with curving or arching, older branches with ascending to erect offshoots and younger branches, juvenile plants sprawling-prostrate with very weak, slender spines. Ribs 4–8. Areoles distinctly separated, flowering and nonflowering ones alike. Spines subulate. Flowers diurnal, solitary, narrowly tubular; tube bearing distant long scales; limb slightly to markedly oblique. Perianth segments short, spreading or recurved. Filaments closely appressed to style, exserted about 1 cm. beyond corolla. Style exserted 5–15 mm. beyond stamens. Ovary scaly with felt and some spines in axils of scales. Fruit globose, spiny but eventually smooth and spineless. Seeds compressed, minutely pitted, basal hilum large and oblique.

**1. Rathbunia alamosensis** (J. M. Coulter) Britt. & Rose, Contr. U.S. Nat. Herb. **12:** 415, 1909

*Cereus alamosensis* J. M. Coulter, Contr. U.S. Nat. Herb. **3:** 406, 1896.
*Cereus sonorensis* Rünge, Monats. Kakteenk. **11:** 135, 1901.
*Rathbunia sonorensis* Britt. & Rose, Contr. U.S. Nat. Herb. **12:** 415, 1909.
*Cereus pseudosonorensis* Gürke, Monats. Kakteenk. **20:** 147, 1910.

Stems columnar as soon as plant passes the flaccid, prostrate, juvenile stage, with stem 6–8 cm. in diameter and erect to ascending to 1–2 m., then curving and arching over so tip rests on the ground and usually strikes root, other erect branches given off at intervals on arching stems, a colony often to 10 m. or more across and impenetrable; ribs 5–8, obtuse; radial spines 11–18, straight, whitish or light gray, spreading; centrals 1–4, stouter than radials, 3–5 cm. long, one or two porrect, others ascending; flowers scarlet, usually 7–10 cm. long, perianth segments 1–1.5 cm. long; tube 1.5–2 cm. long, scales on tube bearing only felt in axils, those on ovary also bearing 1–6 axillary spines; stigma lobes 5–6, creamy white; fruit globose, bright red, 3–4.5 cm.

in diameter, bearing clusters of 5–6 slender whitish spines; seeds about 1.5–2 mm. long.

In alluvial and sandy soil of desert plains, flats, and bajadas, Lower Sonoran Zone, central Sonora south into Sinaloa and Tepic. March–May.

The juvenile stage scarcely resembles the adult plants, for the stems are slender, flaccid, completely prostrate and sprawling. They grow over each other in a profuse tangle and resemble a pile of woolly caterpillars. These young stems are 1–2 cm. in diameter, the joints 4–15 cm. long, and the spines are slender, weak, almost bristlelike, and of a light gray color. The cortex and epidermis are pale yellow in sharp contrast to the dark green of the mature plants. These bristly weak stems root copiously along the lower side. These seedlings or juvenile stages are found only rarely in the field and were recognized only when a small clump of them was found organically attached to the adult form, where a few of the stems had assumed the coarser characteristics of the mature cactus.

## 13. LOPHOCEREUS Britt. & Rose, Contr. U.S. Nat. Herb. 12: 426, 1909

Stout columnar cacti with simple or few-branched stems or sometimes much branched at or near base. Ribs coarse, few. Areoles on older, lower part of plant bearing only a few short-subulate spines, upper flowering ones with many slender, bristlelike spines and dense mat of feltlike wool. Flowers 1 to several at an areole, funnelform, tube short and narrow, corollas beginning to open late in evening and often closed again by daylight of next morning, odorless; outer perianth segments greenish, inner ones pink. Stamens short, included. Fruit globose, red, naked or with a few small spines in axils of lower scales, dehiscing irregularly. Seeds small and numerous, black, shining; hilum depressed.

Ribs 5–10; flowers remaining open until nearly noon, white to pale flesh-pink:
- Mature plants 2–4 m. tall; areoles 7–10 mm. in diameter, each bearing a conspicuous gland; ribs mostly 5–9 .......................................... 1. *L. schottii* var. *schottii*
- Mature plants to 8 m. tall; areoles 3–5 mm. in diameter, gland inconspicuous; ribs at tips of flowering branches 7–10 ............................ 1*a*. *L. schottii* var. *australis*

Ribs 10–15; flowers wholly nocturnal, closed soon after daybreak, deep pink ... 2. *L. gatesii*

### 1. Lophocereus schottii (Engelm.) Britt. & Rose, Contr. U.S. Nat. Herb. 12: 427, 1909, var. schottii

*Cereus schottii* Engelm., Proc. Amer. Acad. **3:** 288, 1856.
*Pilocereus schottii* Lemaire, Rev. Hort. **1862:** 428, 1862.
*Cereus sargentianus* Orcutt, Gard. & Forest **4:** 436, 1891.
*Lophocereus sargentianus* Britt. & Rose, Contr. U.S. Nat. Herb. **12:** 427, 1909.
*Lophocereus schottii* var. *sargentianus* Borg, Cacti 154, 1937.

Erect plant to 4 m. tall, usually colonial and branching near base to form dense clumps as much as 10–15 m. in diameter; branches 8–12 cm. in diameter, grayish; ribs mostly 5–7, rarely 8 or 9, these broad at base and sharply angular, 1.5–2.5 cm. high, separated by broad intervals, gray-green and often distinctly glaucous on young stems, particularly near tips; bristles of flowering areoles 25 or more, acicular, often twisted or somewhat flattened, straight, 2–6 cm. long, gray, set in a circular rounded mound made up of numerous small tufts of yellowish to gray felt, the whole areole 7–10 mm. in diameter, bearing a blunt ovoid gland at base; lower nonflowering areoles 4–6 mm. in diameter, not felted, each bearing 3–7 short-subulate radial spines and 1–3 centrals, all with bulbous bases, these spines 5–10 mm. long, grayish; flowers 3–4 cm. long, white to pale flesh-pink, remaining open until nearly noon, tube slender and about as long as flaring part of corolla, outer segments minutely granular-puberulent on outside, finely ciliolate along margins; style, stigma lobes, and filaments white or faintly tinged with rose, included; fruit globose or obovoid, 2–3 cm. in diameter, naked or nearly so; seeds black and shining, 2.5–3 mm. long, slightly lenticular-obovoid, the raphe prominent and the hilum brown and excavated.

On alluvial plains and rocky hillsides, Lower Sonoran Zone, southern Arizona southward throughout the drier parts of Sonora and Baja California to the Cape Region. April–Aug.

**1*a*. Lophocereus schottii** var. **australis** (K. Brandegee) Borg, Cacti 154, 1937

*Cereus schottii* var. *australis* K. Brandegee, Zoe **5:** 4, 1900.
*Lophocereus australis* Britt. & Rose, Contr. U.S. Nat. Herb. **12:** 427, 1909.

Similar to var. *schottii* but the plants branching less freely, branches taller (to 8 m.) and more slender (5–8 cm. in diameter); areoles small, mainly 3–5 mm. in diameter, bearing an inconspicuous basal gland or this sometimes lacking; flowers deep pink and strictly nocturnal, closing soon after sunup; seeds barely 2 mm. long.

Sandy and silty llanos, gently sloping hillsides, and coastal bluffs and slopes, Lower Sonoran Zone, southwestern Sonora and southern Baja California.

**Lophocereus schottii** forma **monstrosus** Gates, Cactus & Succ. Jour. **3:** 38, 1931, with stems bearing broad, rounded protuberances and bosses roughly arranged in poorly defined "ribs," the stem often 15 or 20 cm. in diameter; areoles few and spineless; and flowers and fruit unknown; occurs sporadically with var. *schottii* in Baja California.

**2. Lophocereus gatesii** M. E. Jones, Cactus & Succ. Jour. **5:** 546, fig., 1934

An erect, freely branching plant 1.5–2.5 m. tall and forming dense clumps 2–3 m. in diameter; stems curving outward strongly at base, then erect, 6–9 cm. in diameter, pale olive green, somewhat pruinose; ribs 10–15, 10–12 mm. high, rather sharply angled; areoles below flowering part 1–1.5 cm. apart, 5–6 mm. in diameter, densely short-felted; radial spines 8–10, acicular, gray and spreading, 2–6 cm. long; central spines slenderly subulate, gray to gray-brown, 1–1.5 cm. long; areoles on flowering portion of branches orbicular, 6–9 mm. in diameter, more conspicuously brown- to gray-lanate, bearing 15–20 gray acicular spines 4–8 cm. long; flowers nocturnal, 2.5–3 cm. long and about as wide when fully open; outer perianth segments oblong-acuminate, brownish without and with coral-pink margins, 1–1.5 cm. long, inner ones deep coral-pink; stamens numerous, pale pink, included, anthers white; style pink; fruits globose, naked, greenish tinged with purple, 1.5–2.5 cm. in diameter.

Sandy and silty soil in bottoms of canyons and on coastal plains, Lower Sonoran Zone, the Magdalena Plain, Baja California. March–Aug.

## 14. MYRTILLOCACTUS Console, Boll. R. Ort. Bot. Palermo 1: 8, 1897

Large plants with short trunks and stout, much-branched tops forming enormous, candelabra-like growth, the stems few-ribbed, erect or nearly so, bearing only one type of areole and all areoles containing similar spines. Flowers small, diurnal, several borne at an areole, tubes short, perianth segments spreading widely. Ovary small, bearing a few minute scales with woolly axils, without spines. Fruit globose, small, juicy and pleasantly acid. Seeds small, black, hilum basal.

**1. Myrtillocactus cochal** (Orcutt) Britt. & Rose, Contr. U.S. Nat. Herb. **12:** 427, 1909

*Cereus cochal* Orcutt, W. Amer. Sci. **6:** 29, 1889.
*Cereus geometrizans* var. *cochal* K. Brandegee, Zoe **5:** 4, 1900.

Plants 1–4 m. tall with many erect, short, stocky branches from the sturdy woody trunk, the latter 2–4 dm. in diameter, grayish, and with few or no spines; ribs on branches 6–8, obtusely angled or rounded, separated by shallow broad intervals; areoles lightly felted and 1–3 cm. apart on ribs, each bearing 5 stout, subulate, radial spines 8–12 mm. long and a single spreading central spine 12–20 mm. long, each flowering areole producing several flowers; flowers about 2.5 cm. long and broad, perianth segments pale green tinged with purple, oblong, 2.5–3.5 mm. wide, 10–15 mm. long; stamens about 1.5 cm. long, filaments white; stigma lobes 5–6, white; fruit globose, scarlet, often tinged with yellow or brown on outer surface, 12–20 mm. in diameter, slightly acid but pleasantly flavored and edible; seeds obovoid, 1.2–1.5 mm. long, dull black, rather coarsely pitted.

Hillsides, and mesas, Lower and Upper Sonoran Zones, in vicinity of San Carlos Canyon 15 miles south of Ensenada, Baja California, south to the Cape Region. Flowers almost throughout the year.

## 15. ECHINOCEREUS Engelm., in Wisliz., Mem. Tour North. Mex. 91, 1848*

Plants small, rarely 0.5 m. high, sometimes simple but usually branching from base to form many-stemmed clumps; stems soft and fleshy, cylindric, erect or strongly ascending, with several to many continuous, more or less tuberculate ribs. Spines acicular, never hooked. Flowers diurnal, arising from ruptures in epidermis just above lateral spiniferous areoles, hypanthium forming a distinct and sometimes long tube above ovary. Fruit fleshy, thin-skinned, covered with spines that are detached easily when fruit is mature; seeds black, tuberculate.

Flowers yellow:
  Plants densely cespitose, forming clumps to 1 m. in diameter; flowers (including ovary) 3–4 cm. long; radial spines 8–10; ribs 8–10 ........................ 1. *E. maritimus*
  Plants solitary or of 2–3 stems; flowers to 10 cm. long, perianth segments 4–5 cm. long; radial spines 15–24; ribs 15–20 ................. *6a. E. pectinatus* var. *neomexicana*
Flowers not yellow:
  Flowers white; plants simple or sparsely cespitose, robust, with 21–25 ribs; San Estéban and San Lorenzo Islands, Gulf of California ........................... 2. *E. grandis*
  Flowers not white:
    Spines all less than 15 mm. long, centrals and radials similar in size and form:
      Plant densely cespitose; Cape Region of Baja California ............... 3. *E. sciurus*
      Plant simple or branching sparsely; not from peninsular Baja California:
        Mature plants over 3 dm. tall and with stems to 8 cm. in diameter; spines golden yellow aging brown; San Pedro Nolasco Island .......... 4. *E. websterianus*
        Mature plants usually less than 3 dm. tall and 5 cm. in diameter; Sonora:
          Perianth segments broadly oblanceolate or spatulate, erosely dentate ........... ............................................................ 5. *E. scopulorum*

* Text contributed by Dr. George E. Lindsay.

Perianth segments narrowly lanceolate, entire ..... 6. *E. pectinatus* var. ***bristolii***
Spines or some of them more than 15 mm. long, centrals and radials dissimilar in size or form:
Principal central spine distinctly flattened through its entire length .. 7. *E.* ***brandegeei***
Principal central spine not distinctly flattened through its entire length:
Flowers less than 2 cm. long; plants forming very dense, compact clumps ........ ........................................... 8. *E.* ***barthelowanus***
Flowers more than 3 cm. long:
Flowers rose-pink with orange throat; central Baja California .. 9. *E.* ***ferreirianus***
Flowers lavender to magenta:
Central spines 2–6, of approximately equal size or at least well developed, often flattened at the base:
Spines, centrals and radials, not uniformly golden yellow ................ .............................. 10. *E. engelmannii* var. ***engelmannii***
Spines, centrals and radials, all golden yellow .......................... ................................. 10*a*. *E. engelmannii* var. ***nicholii***
Central spine 1, often with 1 or 2 short, rudimentary, auxiliary spines in some of the areoles, never flattened at base:
Ribs less than 13; rudimentary central spines if present gray ............. ...................................... 11. *E. fendleri* var. ***robustus***
Ribs 13 or more; rudimentary central spines if present straw-colored, reddish, or white ...................... 11*a*. *E. fendleri* var. ***boyce-thompsonii***

**1. Echinocereus maritimus** (M. E. Jones) K. Schum., Gesamtb. Kakteen 273, 1898

*Cereus maritimus* M. E. Jones, Amer. Nat. **17:** 973, 1883.
*Cereus flaviflorus* Engelm., in J. M. Coult., Contr. U.S. Nat. Herb. **3:** 391, 1896.
*Echinocereus flaviflorus* Hildm. ex K. Schum., Gesamtb. Kakteen 264, 1898.
*Echinocereus hancockii* Dawson, Desert Pl. Life **21:** 89, *fig. 1,* 1949.

Cespitose, forming dense clumps to 1 m. in diameter and 30 cm. high; stems globose to cylindric, to 20 cm. long; ribs 8–10; spines yellowish or horn color fading to gray; central spines 4, stout, angled, to 6 cm. long; radial spines about 10, spreading, to 5 cm. long; flowers observed from April through November, yellow, 3–4 cm. long including ovary, usually not opening broadly because of interlaced spines of the plant; inner perianth segments oblanceolate, rounded at apex; fruit globular, red, spine-covered, about 3 cm. in diameter.

Usually in sedimentary hills and flats, Upper and Lower Sonoran Zones, west coast of Baja California and adjacent islands from Ensenada south to below Vizcaíno Bay. The plants blossom occasionally over a long flowering season.

The population on Natividad Island is sparsely spined and small, while a very robust and heavily armed form, which was described as *E. hancockii,* occurs at San Hipolito Bay. Giant forms occur also on Cedros Island, and the species is variable throughout its range.

**2. Echinocereus grandis** Britt. & Rose, Cactaceae **3:** 18, *fig. 18*; *pl. 3*, *fig. 2*, 1922

Simple or forming small clusters; stems to 5 dm. tall and 12 cm. in diameter; ribs 21–25, low; areoles large, longer than broad, about 1 cm. apart; spines white or cream-colored, short and stiff, centrals 8–12, radials 15–25; flowers white, 5–6 cm. long and not opening broadly, with short tube; ovary and tube densely covered with pale yellow spines and white hairs; outer perianth segments white with green midstripe, inner ones narrow, 1.5 cm. long, white with green bases; filaments green; style white; stigma lobes green; fruit densely spiny.

Rocky hillsides and washes, Lower Sonoran Zone, San Estéban and San Lorenzo Islands, Gulf of California. Reported from San Pedro Nolasco Island, but collections there were probably sterile *E. websterianus.* April–May.

**3. Echinocereus sciurus** (K. Brandegee) Britt. & Rose, Cactaceae **3:** 22, *pl. 4, fig. 1,* 1922

*Cereus sciurus* K. Brandegee, Zoe **5:** 192, 1904.

Cespitose, forming dense clumps to 60 cm. broad; stems slender and to 20 cm. long; ribs 12–17, low, tuberculate; areoles small, approximate, circular, woolly in youth; spines horn-colored or yellowish, with brown tips, acicular, about 18, centrals and radials indistinguishable but those in center of areole shorter; flowers 7 cm. long and 9 cm. broad when fully expanded; ovary and tube covered with numerous slender, brown-tipped spines; inner perianth segments in 2–4 rows, light rose; filaments cream; style cream; stigma lobes 10–12, obtuse, green; seeds 1 mm. long, tuberculate.

Sandy slopes, rocky hillsides and washes, Lower Sonoran and Tropical Zones, vicinity of La Paz to Cabo San Lucas, Baja California. April–July.

**4. Echinocereus websterianus** Lindsay, Cactus & Succ. Jour. **19:** 153, *figs. 102–103,* 1947

Cespitose, stems to 6 dm. tall, to 8 cm. in diameter; ribs 14–18, 5 mm. high, slightly tuberculate; areoles approximate; spines golden yellow aging to brown; central spines 6–8, short-acicular, to 1 cm. long, radials 14–18, spreading, about 1 cm. long; flowers 6 cm. long, 3 cm. broad; ovary and tube tuberculate, with circular areoles bearing 8–12 white acicular spines and many white cobwebby hairs; outer perianth segments oblanceolate, apiculate, lavender-pink with olive midstripe, inner ones oblanceolate, erose at tip, distal one-third lavender-pink, central one-third clear white, basal portion shining-green; filaments green; style cream; stigma lobes 11, olive-green.

In soil enriched by guano on cliffs and steep slopes, Lower Sonoran Zone, San Pedro Nolasco Island, Gulf of California. April–July.

**5. Echinocereus scopulorum** Britt. & Rose, Cactaceae **3:** 30, *fig. 34,* 1922

Simple, cylindric, to 40 cm. tall; ribs 13 or more, low, tuberculate; areoles circular; spines highly colored, pink or brown with darker tips; central spines 3–6, similar to radials, which are spreading; flowers from near top of plant, to 9 cm. broad, funnelform, with a tube 2 cm. long; inner perianth segments 4 cm. long, rose or purplish rose, paler on the outside, usually in 2 rows, oblanceolate to spatulate, erosely dentate, acute; stamens greenish; style stout, much longer than stamens; stigma lobes 12, linear.

Rocky areas along the coast, Lower Sonoran and Tropical Zones, from San Pedro and San Carlos Bays, Sonora, to Topolobampo, Sinaloa. May–July.

**6. Echinocereus pectinatus** (Scheidw.) Engelm. var. **bristolii** Marshall, Saguaroland Bull. **10:** 81, 1956

*Echinocereus bristolii* Marshall, Cactus & Succ. Jour. **9:** 160, *fig.,* 1938.

Simple or sparsely cespitose, cylindric, 20 cm. long, 4–5 cm. in diameter; ribs 15–16, tuberculate; areoles elongate, woolly in youth; spines white, centrals with red tips; central spines 3, two upper divergent, lower deflexed, 5–15 mm. long, stiff; radial spines about 20, acicular, spreading, lower ones to 8 mm. long, upper ones shorter and bristlelike; flowers violet-pink; ovary 12 mm. long, globose, covered with spines; tube

15 mm. long; inner perianth segments about 40, 45 mm. long, 6 mm. wide, narrowly lanceolate, entire, violet-pink; filaments numerous, pale green; stigma white; stigma lobes 11, green, 7 mm. long; fruit globose, red, spiny.

Steep slopes and rocky cliffs, Lower and Upper Sonoran Zones, southern Sonora near Alamos, south into Sinaloa. May.

Marshall considered this species to be very near *E. scopulorum,* from which it differs in having more numerous and narrower perianth segments, more central spines, and other minor characters. Marshall also placed *E. floresii* Backeb. in the synonymy of *E. pectinatus* var. *bristolii.*

**6a. Echinocereus pectinatus** var. **neomexicanus** (J. M. Coulter) L. Benson, Proc. Calif. Acad. IV, **25**: 256, 1944

*Echinocereus dasyacanthus* Engelm., in Wisliz., Mem. Tour North. Mex. 100, 1848.
*Cereus dasyacanthus* Engelm., in A. Gray, Pl. Fendl. 50, 1849.
*Cereus ctenoides* Engelm., Proc. Amer. Acad. **3**: 279, 1856.
*Cereus dasyacanthus* var. *neo-mexicanus* J. M. Coulter, Contr. U.S. Nat. Herb. **3**: 384, 1896.
*Echinocereus spinosissimus* Walton, Cact. Jour. **2**: 162, 1899.

Stems 1–3 dm. tall, solitary, cylindrical; ribs 15–20, about 2–3 cm. high; areoles elliptical, 1.5–2.5 mm. long, 3–5 mm. apart, spines practically hiding stems; central spines 5–8, spreading in several directions, 3–10 mm. long, pink in youth, gray in age; radial spines 15–25, appressed to spreading or lower deflexed, 10–15 mm. long; flowers yellow, to 10 cm. long, perianth segments 4–5 cm. long, oblong, acute to rounded; fruit globose or essentially so, purplish, 2.5–3.5 cm. in diameter, spiny.

Desert grasslands and open oak savannah, mostly Upper Sonoran Zone but occasionally in the Lower Sonoran Zone, Pima County, Arizona, east disjunctly to Texas and south into northern Sonora and Chihuahua. June–Aug.

Britton and Rose (Cactaceae **3**: 19, 1922) placed *E. spinosissimus* Walton and *E. rubescens* Dams in the synonymy of this variety (which they gave specific rank as *E. dasyacanthus* Engelm.), but these plants are too poorly known to us to permit a satisfactory disposition of the epithets. Marshall and Bock (Cactaceae 119, 1946) suggested that *E. rubescens* might have to be merged with *E. dasyacanthus.*

**7. Echinocereus brandegeei** (J. M. Coulter) K. Schum., Gesamtb. Kakteen 290, 1898

*Cereus brandegei* J. M. Coulter, Contr. U.S. Nat. Herb. **3**: 389, 1896.
*Cereus sanborgianus* J. M. Coulter, op. cit. 391.
*Cereus mamillatus* Engelm., in J. M. Coulter, op. cit. 405.
*Echinocereus sanborgianus* K. Schum., Gesamtb. Kakteen 274, 1898.
*Echinocereus mamillatus* Britt. & Rose, Cactaceae **3**: 41, *fig. 54,* 1922.

Cespitose, forming large open clumps to 2 m. wide; stems clavate, sometimes to 1 m. long and 6 cm. in diameter, longer ones procumbent; ribs about 10, very strongly tubercled; areoles circular; spines yellow to reddish or gray; central spines usually 4, very stout, distinctly flattened throughout their length, erect or porrect, to 8–10 cm. long; radial spines about 12, spreading, acicular; flowers purplish, about 5 cm. long and broad, opening widely; closely set areoles on ovary and tube bearing long white wool and pale acicular spines; inner perianth segments lanceolate, light purple to magenta with darker base; stamens dark red or purple; style and stigma lobes cream; fruit globular, 3 cm. in diameter, spiny; seeds black, tuberculate.

Sandy plains and rocky hilltops, Lower Sonoran Zone, central Baja California south to the Cape Region. July–Oct.

The typical form, from near El Arco, is a heavy, gray-spined plant. The species varies considerably, with an attractive, golden-spined form near Bahía de la Concepción, a population of very robust plants west of Comondú, and much smaller and more delicate types from the Cape District, La Paz, and islands of the lower gulf. *Cereus mamillatus*, which Coulter described from Engelmann's manuscript notes of a fragment of a plant collected by William Gabb near Mulegé, is probably referable to this species.

### 8. **Echinocereus barthelowanus** Britt. & Rose, Cactaceae **3:** 41, 1922

Cespitose, forming large compact clusters to 8 dm. in diameter; stems cylindric, 1–3 dm. long, to 5 cm. in diameter; ribs about 10, tuberculate; areoles approximate, white-felted in youth; spines acicular and hiding plant, pinkish when young, fading to gray or brown with darker brown tips; spines not clearly differentiated into radials and centrals but centermost about 4, to 7 cm. long, surrounded by about 14 shorter radiating spines; flowers only 10–12 mm. long, red; perianth segments oblong, 3–4 mm. long; ovary minute, strongly tubercled, densely covered with white or pinkish-tipped spines half as long as flower.

Low hills and sandy washes, Lower Sonoran Zone, Magdalena Island, Baja California; possibly on adjacent Santa Margarita Island also. May–July.

### 9. **Echinocereus ferreirianus** Gates, Saguaroland Bull. **7:** 8, 1953

Cespitose, forming open clumps to 3 dm. tall and 4–5 dm. in diameter; branches columnar, 6–8 cm. in diameter, tapering toward a rounded apex, green or sometimes brown or red or purple; ribs about 11–12, low, obtuse, tuberculate; areoles circular, gray-felted, 1–1.5 cm. apart; spines gray, brown, or purple, acicular; radial spines 9–13, about 1 cm. long, appressed, spreading; central spines usually 3, to 5 cm. long, lower porrect or deflexed, upper ones ascending, shorter; flowers funnelform, 6 cm. long, 4 cm. in diameter; ovary ovate, covered with areoles bearing white wool and 6–9 acicular spines 2–10 mm. long; tube 2–2.5 cm. long, spiny; inner perianth segments about 33, oblanceolate, apiculate, entire, outer surface pale purple, inner surface rose-pink abruptly changing to orange on lower third; filaments light purple at base and golden above; style white to cream or very light green; stigma lobes 8–10, white to light green, linear; fruit globose-ellipsoid, 4 cm. long, 3.5 cm. in diameter, olive-green, sometimes tinged purple, covered with thin soft skin bearing areoles with 6–12 acicular spines; seeds black, punctate, more than 1 mm. long.

Rocky mountain slopes, low hills, and cliffs, Lower Sonoran Zone, on the islands and mainland of the peninsula in the vicinity of Bahía de los Angeles, Baja California. May–June(?).

### 10. **Echinocereus engelmannii** (Parry ex Engelm.) Rümpler, in Förster, Handb. Cact. ed. 2, 805, 1885, var. **engelmannii**

*Cereus engelmannii* Parry ex Engelm., Amer. Jour. Sci. II, **14:** 338, 1852.
*Cereus engelmanni* var. *chrysocentrus* Engelm., Proc. Amer. Acad. **3:** 283, 1856.
*Cereus engelmanni* var. *variegatus* Engelm., loc. cit.
*Echinocereus engelmanni* var. *chrysocentrus* Rümpler, in Förster, Handb. Cact. ed. 2, 806, 1885.
*Echinocereus engelmanni* var. *variegatus* Rümpler, in Förster, loc. cit.
*Echinocereus chrysocentrus* Orcutt, Cactography 4, 1926.

Cespitose, forming uncrowded clumps to 1 m. in diameter and to 3 dm. high; stems cylindrical, firm, with 10–13 ribs that are weakly tuberculate; spines acicular, white to brown, yellow, red, or variegated, spines in one areole often different colors; central spines 2–6, all large and well developed, 3–9 cm. long, stout but somewhat flexible, curving or twisted, usually somewhat flattened at base; radial spines 6–12, usually straight but sometimes twisted; flowers magenta, 4–8 cm. in diameter, funnelform; tube about 12 mm. long above ovary; perianth segments oblong, 3–4 cm. long, acuminate; areoles bearing felt and bristles; fruit red, spiny, spherical to ovoid, about 3 cm. long.

Flats, hillsides, and outwash fans, Lower and Upper Sonoran Zones, throughout the Sonoran Desert areas of California and Arizona south to Guaymas, Sonora, and southern Baja California. Feb.–May.

This species is variable in all characters. It grows from sea level to 5,000 feet elevation.

**10*a*. Echinocereus engelmannii** var. **nicholii** L. Benson, Proc. Calif. Acad. IV, **25:** 258, 1944

Differs from the typical variety in having all yellow spines and lavender perianth segments.

Desert habitats, Lower Sonoran Zone, between Ajo and Tucson in southern Arizona south into adjacent northern Sonora.

This golden-spined variety represents a rather homogeneous population. The *engelmannii* complex is very plastic, however, and other variants of possibly equal stature are not recognized.

**Echinocereus decumbens** Clover & Jotter, Bull. Torrey Club **68:** 417, *fig. 7,* 1941, which was combined as a variety of *E. engelmannii* by Lyman Benson, was described from specimens in Marble Canyon, Arizona, and also reported from the Sonoran Desert at Palm Canyon in the Kofa Mountains, Yuma County, Arizona. It is known only from sterile material.

**11. Echinocereus fendleri** (Engelm.) Rümpler var. **robustus** (Peebles) L. Benson, Proc. Calif. Acad. IV, **25:** 259, 1944

*Echinocereus rectispinus* var. *robustus* Peebles, Amer. Jour. Bot. **25:** 675, 1938.
*Echinocereus robustus* Peebles, Jour. Wash. Acad. **30:** 219, 1940.

Cespitose, forming clumps of 5–20 stems which are 2–5 dm. long and usually 8–10 cm. in diameter; ribs 8–13, firm; spines pale gray tipped with brown or black; central spine straight, porrect, 2.5–7.5 cm. long, in some areoles accompanied by small accessory centrals, rather flexible, slender; flowers 5–7 cm. in diameter; floral tube about 12 mm. long; perianth segments magenta to purplish; fruit ovoid, about 3 cm. long, spiny.

Rocky slopes of the Sonoran and Chihuahuan Deserts and lower part of the desert grasslands, Lower and Upper Sonoran Zones, from eastern Pinal and eastern Pima Counties to the upper Gila River valley and lower parts of Cochise County, Arizona, south into northern Sonora. April–May.

This is the only *Echinocereus* in the desert about Tucson.

**11*a*. Echinocereus fendleri** var. **boyce-thompsonii** (Orcutt) L. Benson, Proc. Calif. Acad. IV, **25:** 260, 1944

*Echinocereus boyce-thompsoni* Orcutt, Cactography 4, 1926.

Cespitose, forming clumps of to 10 stems, these 1–2.5 dm. long, 5–7.5 cm. in diameter; ribs firm, 12–18; spines straw-colored, red, or white, straight, flexible, very slender; central spine 2.5–4 cm. long, accompanied in at least some areoles by 1 or 2 accessory centrals, erect or deflexed; flowers usually 6–6.5 cm. in diameter.

Desert grasslands and open forest and chaparral, upper margins of the Lower Sonoran and Upper Sonoran Zones, from the vicinity of Superior, Arizona, south and east in Coconino, Yavapai, Gila, Graham, and Pinal Counties. March–May.

Some forms of this variety seem to be transitional to *E. engelmannii.*

## 16. FEROCACTUS Britt. & Rose, Cactaceae 3: 123, 1922*

Usually simple but sometimes branched; stems ovoid to cylindric, often large, with thick, prominent, and sometimes tuberculate ribs. Areoles large, tomentose or woolly in youth, often divided into a superior floriferous and an inferior spiniferous section. Spines large and strong, usually in 2 or 3 distinct series, produced from lower part of areole; nectariferous gland spines produced between spine fascicle and flower bud in young areoles and persisting as dwarf spines or spurs. Flowers broadly funnelform to campanulate, with scales on ovary and tube intergrading with outer and inner perianth segments; stamens numerous, inserted in a short tube; style thick, divided at tip into several to many stigma lobes. Fruit a single loculed berry with thick, semi-fleshy or fleshy walls, and usually dehiscing through a basal pore. Seeds with a thin, pitted or sculptured testa; embryo curved; hypocotyl large.

Spines all similar in shape and size:
- Spines 22–25; Angel de la Guarda Island, Gulf of California ........ 1. *F. johnstonianus*
- Spines 6–10; plants from various other islands in the Gulf of California:
  - Stem massive-columnar; mature specimens over 1 m. tall and 5 dm. in diameter; juvenile specimens usually depressed-globular, broader than tall .. 2. *F. diguetii* var. *diguetii*
  - Stem subcolumnar; mature specimens less than 1 m. tall and 5 dm. in diameter; juvenile specimens globular and not flattened ............ 2*a*. *F. diguetii* var. *carmenensis*

Spines dissimilar in shape and size:
- Mature plants usually less than 4 dm. tall:
  - Plant somewhat conical, always taller than broad; from southern Baja California:
    - Flowers orange or red; principal central spine usually strongly curved at tip ........ .................................... 3. *F. townsendianus* var. *townsendianus*
    - Flowers yellow; principal central spine straight or only slightly curved at tip ....... ....................................... 3*a*. *F. townsendianus* var. *santa-maria*
  - Plant not conical, often as broad as tall; northwestern Baja California:
    - Flowers orchid; inner perianth segments spatulate or subovate, to 2 cm. long, 6–8 mm. wide ............................................... 4. *F. fordii* var. *fordii*
    - Flowers red or orange; inner perianth segments linear-lanceolate, to 4 cm. long, 4–5 mm. wide .................................... 4*a*. *F. fordii* var. *grandiflorus*
- Mature plants more than 4 dm. tall:
  - Areoles without marginal bristles or setaceous spines; all radial spines similar in character to central spines but smaller:
    - Central spines 4, cruciform:
      - Spines usually gray; principal central spine usually only slightly flattened but strongly hooked; flowers over 5 cm. long; outer perianth segments broad, spatulate ................................ 5. *F. peninsulae* var. *peninsulae*
      - Spines usually red; principal central spine usually much flattened and not strongly hooked; flowers less than 5 cm. long; outer perianth segments narrow, lanceolate, acuminate .......................... 5*a*. *F. peninsulae* var. *viscainensis*

* Text contributed by Dr. George E. Lindsay.

Central spine 1:
Central spines always straight, 25 cm. long; central Baja California . . . . . . . . . . . . . . . . . . . . . . . . . . . . . . . . . . . . . . . . . . . . . . . . . . . . . . . . . . . 6. *F. rectispinus*
Central spine usually hooked at least in young plants, less than 9 cm. long; Sonora and Arizona . . . . . . . . . . . . . . . . . . . . . . . . . . . . . . . . . . . . . . . . . . . . . 7. *F. covillei*
Areoles with marginal bristles or setaceous spines unlike central spines:
Four centermost spines neither twisted nor tortuous, lower one usually hooked:
Flowers red or sometimes yellow, blossoming in midsummer; fruit maturing following winter; areoles usually with setaceous radial spines . . . . . . . . . . . . . . . . . . . . . . . . . . . . . . . . . . . . . . . . . . . . . . . . . . . . . . . . . . . . 8. *F. wislizenii* var. *wislizenii*
Flowers yellow, blossoming in spring; fruit maturing by early summer; no setaceous radial spines . . . . . . . . . . . . . . . . . . . . . . . . . . . 8*a*. *F. wislizenii* var. *tiburonensis*
Four centermost spines or some of them twisted and tortuous (except rarely in *F. gracilis* var. *coloratus*):
Central spines 4:
Flowers red; fruit 4–7.5 cm. long, extending well above tops of spines; Smith Island group, Bahía de los Angeles, Baja California . . . . . . . . . . 9. *F. gatesii*
Flowers yellow; fruit less than 4 cm. long, not extending well above tops of spines:
Central spines more or less tortuous and spreading:
Plants often over 6 dm. tall; spines usually yellow, red, or brown . . . . . . . . . . . . . . . . . . . . . . . . . . . . . . . . . . . . . . . . . . 10. *F. acanthodes* var. *acanthodes*
Plants under 6 dm. tall; body broad with flat apex; spines gray . . . . . . . . . . . . . . . . . . . . . . . . . . . . . . . . . . . . . . . . . 10*a*. *F. acanthodes* var. *tortulospinus*
Central spines more or less appressed and recurved . . . . . . . . . . . . . . . . . . . . . . . . . . . . . . . . . . . . . . . . . . . . . . . . . . . . . . . . . . 10*b*. *F. acanthodes* var. *lecontei*
Central spines more than 4:
Central spines golden yellow or rarely bright red, diamond-shaped in cross section, twisted; Cedros Island . . . . . . . . . . . . . . . . . . . . . . 11. *F. chrysacanthus*
Central spines red, principal one distinctly flattened rather than diamond-shaped, only principal one twisted; Rosario to Miller's Landing, Baja California:
Plants often over 1 m. tall; widest central spine usually less than 5 mm. wide . . . . . . . . . . . . . . . . . . . . . . . . . . . . . . . . . . . . . . . . . . . . 12. *F. gracilis* var. *gracilis*
Plants less than 1 m. tall; widest central spine more than 6 mm. wide . . . . . . . . . . . . . . . . . . . . . . . . . . . . . . . . . . . . . . . . . . . 12*a*. *F. gracilis* var. *coloratus*

## 1. Ferocactus johnstonianus Britt. & Rose, Cactaceae 4: 287, 1923

*Echinocactus johnstonianus* Fosberg, Amer. Midl. Nat. **27**: 257, 1942.

Stem simple, globular to short-cylindric, to over 1 m. tall and 3.5 dm. broad; ribs 24–31, slightly tuberculate; spines golden yellow aging to brown, about 22–25 per areole, to 6 cm. long, all about the same, not differentiated into central and radial series, subulate, slightly recurved outward; flowers yellow, funnelform, to 5 cm. long, 3.5 cm. wide, with wide, orbicular, imbricate, ovary scales; inner perianth segments yellowish, to 2 cm. long and 7 mm. wide, margins erose, tips short-acuminate; filaments yellow below, reddish above; style 2 cm. long, distal 1 cm. divided into 8–13 flesh-colored stigma lobes; fruit when dried 3 cm. long, 1.5 cm. wide, or 5–6 cm. long including withered perianth; seeds shining-black, irregularly angled, averaging 2.5 mm. long, 1.5 mm. wide, 1.25 mm. thick, covered with regular oval depressions.

Rocky cliffs, hillsides, and alluvial plains, Lower Sonoran Zone, on Isla Angel de la Guarda, Gulf of California. March–July.

**2. Ferocactus diguetii** (Weber) Britt. & Rose, Cactaceae **3:** 131, *pl. 11, fig. 2; pl. 12, fig. 3,* 1922, var. **diguetii**

*Echinocactus digueti* Weber, Bull. Mus. Hist. Nat. Paris **4:** 100, *fig. 1,* 1898.

Stem simple, massive-columnar, to 4 m. tall, 6 dm. thick; ribs 25–35, to 3 cm. high, vertical, straight or becoming sinuate in moribund specimens; areoles elliptic, to 2 cm. long, 1 cm. wide, bearing tan tomentum which turns gray and erodes away; spines clear yellow or occasionally reddish brown, usually 7–8 but sometimes only 4 and in some juvenile specimens to 10, not differentiated into central and radial series, subulate, very slightly curved outward; flowers red, funnelform, 4 cm. long and wide; floral tube V-shaped in long section, imbricate fimbriate scales on ovary; inner perianth segments 2 cm. long, 8 mm. wide, dark red with yellow margin, oblanceolate; stamens yellowish below, tinged with red above; style red or orange, 17 mm. long, with longitudinal ridges continuing into stigma lobes, these 5–10 mm. long, yellow; fruit 3 cm. long, 2 cm. wide when dry, usually dehiscing through basal pore; seeds glossy-brown, irregular, 1.5 mm. long, 1 mm. wide, with small oval pits rather than polygonal sculpturing; testa smooth in a band around a large deep hilum.

Rocky hills and gravel slopes, Lower Sonoran Zone, on islands in the Gulf of California including Santa Catalina, Monserrate, Dansante, San Diego, and Cerralvo. Also reported from Isla Coronados near Loreto but probably now extinct there. March–May.

This is the largest, most spectacular *Ferocactus,* and specimens on Santa Catalina Island are sometimes 12 feet tall.

**2*a*. Ferocactus diguetii** var. **carmenensis** Lindsay, Cactus & Succ. Jour. **27:** 167, *fig. 156,* 1955

Differs from var. *diguetii* in size, never over 1 m. tall and 4 dm. in diameter, usually much smaller. Juvenile plants tend to be globular rather than flattened, with somewhat heavier spines than the typical variety.

Carmen Island, Gulf of California.

**3. Ferocactus townsendianus** Britt. & Rose, Cactaceae **3:** 127, *fig. 133; pl. 12, fig. 1,* 1922, var. **townsendianus**

Stem simple, short-cylindric, more or less conical or constricted toward apex, to 5 dm. tall, 3 dm. wide; ribs about 16, rather tall, slightly tubercled, usually spiraled; areoles oval, bearing short tan tomentum in youth; spines gray or brown; central spines usually 4, cruciform, annulate, upper three ascending, straight or slightly recurved, lower longer, flattened, with curved or hooked tip, to 10 cm. long, 3.5 mm. wide; radial spines 14–16, radiating, setaceous to subulate, three lower ones heavier and somewhat resembling centrals; flowers large, funnelform, 5–6 cm. long, nearly as broad; scales on ovary fleshy, intergrading with red outer perianth segments; inner perianth segments yellow with red midstripe and base, oblong-lanceolate, with serrulate-undulate margins and acute tip, 32 mm. long, 7 mm. wide; filaments red, 5–15 mm. long; style 30 mm. long, 4 mm. wide, grooved, tip divided into about 20 pink or greenish brown stigma lobes; fruit globular, yellow, about 2.5 cm. long, 2 cm. wide, or about 4.5 cm. long including persistent perianth; seeds reddish brown, rounded, 1.5 mm. long, 1 mm. wide, with oval or oblong pits and small, oval, brown hilum.

Hills and coastal plains, Lower Sonoran Zone, on the mainland of Baja California south of the 26th parallel and on San José, Santa Margarita, and Magdalena Islands. May–Aug.

This variable species is closely related to *F. peninsulae* but is separated by its characteristic conical shape, smaller size, and smaller seed.

**3a. Ferocactus townsendianus** var. **santa-maria** (Britt. & Rose) Lindsay, Cactus & Succ. Jour. **27:** 170, 1955

*Ferocactus santa-maria* Britt. & Rose, Cactaceae **3:** 131, 1922.
*Echinocactus santa-maria* Rose ex Schick., Monats. Kakteenk. **29:** 13, 1919. (Nomen subnudum.)

Differs from var. *townsendianus* in having clear yellow flowers, a globular rather than conical shape, larger reniform seed with polygonal reticulations rather than oval or oblong pits, and somewhat straighter spines.

Rocky hillsides on Magdalena Island near Santa María Bay and on coastal bluffs at Cabo San Lucas, Baja California. June–Aug.

The morphology of the flowers and the same number and arrangement of spines indicates a very close relationship between the two varieties.

**4. Ferocactus fordii** (Orcutt) Britt. & Rose, Cactaceae **3:** 126, *fig. 132*, 1922, var. **fordii**

*Echinocactus fordii* Orcutt, Rev. Cact. **1:** 56, 1899.

Simple or sometimes cespitose, depressed-globose to short-cylindric, to 5 dm. tall, 2.5 dm. wide, with a flattened or even invaginated base; ribs about 21, tuberculate; areoles tomentose, oval with a truncate floriferous part at upper edge; spines gray; centrals 4, cruciform, flattened or angled, to 5 cm. long, 3 mm. wide, three upper ones straight and ascending, lower one porrect and curved or hooked at tip; radial spines about 17; flowers rose-purple or orchid, funnelform, about 3 cm. long and broad; inner perianth segments about 32, spatulate to subobovate, 2 cm. long, 6–8 mm. wide, somewhat fringed, more or less aristate, midstripe deep orchid, margins lighter orchid; filaments yellow at base, red at top; style reddish, 2 cm. long, terminal 4 or 5 mm. divided into 9–11 yellow stigma lobes; fruit oval, pink to yellow; seeds black, plump but somewhat angled, with fine reticulate sculpturing, 2 mm. long, nearly 2 mm. broad, 1.5 mm. thick.

Coastal plains, silty flats, and low hills, lower margins of Upper Sonoran Zone and Lower Sonoran Zone, western Baja California from near Rancho San Antonio south to Cape San Eugenio, and on Isla San Martín. March–Aug.

**4a. Ferocactus fordii** var. **grandiflorus** Lindsay, Cactus & Succ. Jour. **27:** 164, *fig. 154*, 1955

Differs from var. *fordii* in having a taller habit, to almost 1 m. high in exceptional specimens, and in the flowers, which are red or orange rather than purple, and to 6 cm. long, with linear-lanceolate inner perianth segments to 4 cm. long and only 4 or 5 mm. wide.

West coast of Baja California from Cape San Eugenio to below Abreojos Point, and on Natividad Island. April–May.

**5. Ferocactus peninsulae** (Engelm. ex Weber) Britt. & Rose, Cactaceae **3:** 133, *fig. 140*, 1922, var. **peninsulae**

*Echinocactus peninsulae* Engelm. ex Weber, Bull. Mus. Hist. Nat. Paris **1:** 320, 1895.
*Ferocactus horridus* Britt. & Rose, Cactaceae **3:** 128, 1922.

Stem simple, ovoid becoming clavate, to 2.5 m. tall, 4 dm. wide; ribs 12–20, prominent; areoles 20 mm. long, 8 mm. wide, constricted between flower and spine-bearing section; spines grayish red with yellow tips; central spines 4, cruciform, annulate, straight, only lower one flattened, hooked at the tip, 4–15 cm. long and 6 mm. wide; radial spines variable but usually about 11, annulate, subulate, straight or sometimes twisted and setaceous; flowers funnelform, 5–6 cm. long; perianth segments golden yellow with red midstripe; stamens orange, appressed against style; fruit globular, yellow, to 3 cm. long, 2–2.5 cm. wide, with fleshy yellow scales, seeds reddish brown or black, about 2 mm. long, 1.5 mm. wide, sculptured with oval or elongated reticulations.

Hillsides, outwash bajadas, and desert plains, Lower Sonoran Zone, central Baja California from Bahía de los Angeles and Sierra San Borjas to the Cape Region, where it is replaced by *F. townsendianus*. April–May.

*Ferocactus peninsulae* intergrades with *F. rectispinus,* a yellow-flowered species with long, straight spines and without fine radial spines. *Ferocactus peninsulae* is closely related to *F. wislizenii* or at least to the variant of *F. wislizenii* that occurs in central Sonora, but differs in having a clavate stem, fewer ribs and fewer spines, and in other minor characters.

**5*a*. Ferocactus peninsulae** var. **viscainensis** (Gates) Lindsay, Cactus & Succ. Jour. **27:** 169, 1955

*Ferocactus viscainensis* Gates, op. cit. **4:** 324, *fig.,* 1933.

Differs from var. *peninsulae* in having shorter, stiffer, more tubular flowers; shorter, less differentiated spines which somewhat resemble those of *F. gracilis* var. *coloratus*; and other inconstant and intangible characters.

Gravelly and silty flats and slopes, Lower Sonoran Zone, between San Andrés and Calmallí, Baja California. July–Aug.

**6. Ferocactus rectispinus** (Engelm., in J. M. Coulter) Britt. & Rose, Cactaceae **3:** 134, *fig. 142,* 1922

*Echinocactus emoryi rectispinus* Engelm., in J. M. Coulter, Contr. U.S. Nat. Herb. **3:** 362, 1896.
*Echinocactus rectospinus* Britt. & Rose, Jour. N.Y. Bot. Gard. **12:** 269, 1911.

Stem simple, globose to cylindric, to 1.5 m. tall, 4.5 dm. wide; ribs about 21, tuberculate, high, sometimes spiraled; areoles large, 3 cm. long, 1 cm. wide, with short, light gray tomentum; areoles separated about 1 cm. in youth becoming approximate in age, with truncate upper portion; spines all straight, terete, annulate, reddish or blotched red and yellow; central spine 1, porrect, 9–25 cm. long, 2.5 mm. wide; radial spines 7–9, like central but only 1.5–6 cm. long; flowers clear light yellow, 6 cm. long, with lemon-yellow, lanceolate, acuminate, inner perianth segments; fruit yellow, fleshy, 3.5 cm. long, 2.5 cm. wide, basal pore not opening; seeds black, about 2 mm. long, 1.5 mm. wide, pitted with small, round or oval depressions, not reticulate, with small, round, terminal hilum.

Rocky hills, Lower Sonoran Zone, between San Ignacio and Comondú, Baja California. Aug.–Sept.

This species has a wide altitudinal range, from sea level at Bahía de la Concepción to between 4,000 and 5,000 feet in the Sierra de las Palmas. It intergrades or hybridizes with *F. peninsulae*. Specimens near Comondú have spines over 26 cm. long, probably the record spine-length in the *Echinocactanae*.

**7. Ferocactus covillei** Britt. & Rose, Cactaceae **3:** 132, *figs. 138–139*, 1922
*Echinocactus covillei* Berger, Kakteen 238, 1929.

Stem simple, globular to cylindric, to 2.5 m. tall, 6 dm. in diameter; ribs 15–32, to 5 cm. high and broad, distinctly tuberculate in youth but not in mature specimens; areoles large, flowering ones sometimes 4 cm. long, 17 mm. wide, spine-bearing portion nearly 3 cm. long; young areoles bearing dense yellow tomentum; spines variable, usually pinkish gray, very robust, annulate, never setaceous; central spine 1, porrect, terete or sometimes flattened dorsiventrally, hooked or sometimes straight, 4–10 cm. long, in some specimens 4 mm. in diameter near base; radial spines 6–8, erect-spreading, straight or slightly curved; flowers to 7 cm. long, 5 cm. broad, mahogany, red, orange, or yellow, with broad scales on ovary intergrading into outer perianth segments; inner perianth segments linear-oblong, acuminate, often serrate, mahogany-red to lemon-yellow; fruit obovoid to oblong, to 5 cm. long and 3 cm. broad, yellow, fleshy ovary walls covered with thick cordate scales; seeds dark brown or black, rounded, 2 mm. long, 1.5 mm. wide, pitted with fine, circular or oval depressions, sometimes slightly crested on side opposite hilum.

Rocky hills and gravelly flats, Lower Sonoran Zone, south-central Arizona to south of Guaymas, Sonora. June–Aug.

In Arizona mostly between 1,500 and 3,000 feet elevation in Pima County, but at Guaymas, Sonora, it occurs at sea level. In Arizona the central spines are hooked, but in the southern part of its range mature specimens have all straight spines. Young specimens in all areas have hooked central spines.

**8. Ferocactus wislizenii** (Engelm.) Britt. & Rose, Cactaceae **3:** 127, *pl. 1*; *pl. 12, fig.* 2, 1922, var. **wislizenii**

*Echinocactus wislizeni* Engelm., in Wisliz., Mem. Tour North. Mexico 96, 1848.
*Echinocactus emoryi* Engelm., in Emory, Notes Mil. Rec. 157, 1848.
*Echinocactus wislizeni* var. *decipiens* Engelm., in Rothrock, Wheeler Rep. 128, 1878.
*Echinocereus emoryi* Rümpler, in Förster, Handb. Cact. ed. 2, 804, 1885.
*Echinocactus wislizeni* var. *albispina* Toumey, Gard. & Forest **8:** 154, 1895.
*Echinocactus falconeri* Orcutt, W. Amer. Sci. **12:** 162, 1902.
*Echinocactus arizonicus* Kunze, Monats. Kakteenk. **19:** 149, 1909.
*Echinocactus wislizeni* var. *phoeniceus* Kunze, Torreya **13:** 75, 1913.
*Ferocactus emoryi* Orcutt, Cactography 5, 1926.
*Ferocactus arizonicus* Orcutt, op. cit. 6.
*Ferocactus falconeri* Orcutt, loc. cit.
*Ferocactus phoeniceus* Orcutt, loc. cit.
*Echinocactus hertrichii* Weinberg, Desert **1:** 40, *figs.*, 1929 (as *hertichii*).
*Ferocactus wislizeni* var. *phoeniceus* Y. Ito, Cacti **1952:** 105, 1952.
*Ferocactus wislizeni* var. *albispinus* Y. Ito, loc. cit.

Stem simple or occasionally sparsely branched from base, globular to columnar, to 3 m. tall, 6 dm. wide; ribs 20–30, not strongly tuberculate, about 3 cm. tall; areoles large, oval to elliptic, brown-tomentose in youth, spaced 2–3 cm. apart in flowering portion, later becoming approximate; spines variable in size, number, and color; 4 central spines cruciform, ashy gray to white or red, upper three usually similar, terete, straight, about 5 cm. long, lower strongest, flattened laterally, often hooked at tip, often 8–10 cm. long, 5 mm. wide; radial spines variable, from setaceous to acicular, usually about 12, to 5 cm. long; flowers appearing in midsummer, red to yellow, funnelform, 4–5 cm. long and broad; inner perianth segments ovate to lanceolate, about 2 cm. long, 5–10 mm. wide, red with yellow margins or sometimes pure yellow, serrulate, tips mucronate; stamens numerous, filaments red, 5–10 mm. long; style yellow, stigma lobes about 20, yellow; fruit yellow, narrow-ovoid, to 5 cm. long, 3 cm. wide, fleshy; seeds reddish brown to almost black, 2.25–2.50 mm. long, 1.75 mm. wide, rounded or sometimes angled, testa covered with shallow intricate sculpturing, depressions wrinkled.

Rocky hillsides, grassy mesas, and alluvial flats, Lower and lower parts of Upper Sonoran Zones, southern Arizona to southwestern Texas, south into northwestern Chihuahua, northern and western Sonora, and northwestern Sinaloa. July–Sept.

This species varies a great deal through its broad geographical and altitudinal range, and probably more than two varieties could be recognized. There is evidence that *F. wislizenii* var. *wislizenii* hybridizes with *F. acanthodes* var. *lecontei.*

**8*a*. Ferocactus wislizenii** var. **tiburonensis** Lindsay, Cactus & Succ. Jour. **27:** 166, *fig. 155,* 1955

Differs from var. *wislizenii* in flowering in April and May rather than middle to late summer, in the absence of setaceous radial spines and the absence of distinctly flattened, hooked, lower central spines, and in the less conspicuous secondary figuring in the depressions in the seed coat.

Known only from Tiburón Island, Gulf of California.

**9. Ferocactus gatesii** Lindsay, Cactus & Succ. Jour. **27:** 150, *figs. 146–147,* 1955

Stem simple, globular to subcolumnar, to 1.5 m. tall, 3 dm. in diameter; ribs 30–32, acute, tuberculate; areoles oval, 1–1.5 cm. long; spines pink, yellow, or horn-colored, aging to gray; central spines 4, erect-spreading, flattened laterally, annulate, to 7 cm. long, 3 mm. wide, lower longest and sometimes curved at tip; radial spines about 16, radiating, upper and lower ones resembling centrals but those at side twisted, setaceous; flowers large, red, funnelform, to 6 cm. long and broad, produced in a circle around apex of plant; outer perianth segments to 2 cm. long and 1 cm. wide, obtuse; inner ones variable, red with yellow margins, to 3 cm. long, 1 cm. wide, tip obtuse to acuminate or apiculate; innermost series of segments narrower, oblong-lanceolate with undulate-cuspidate, ciliate or slightly lacerate margins and apiculate-attenuate tips; stamens numerous, filaments red, 5–15 mm. long; style yellow, 2 cm. long, 2.5 mm. wide, stigma lobes about 16; fruit very large, elongate; ovary to 7.5 cm. long, 2.5 cm. wide, or including persistent withered perianth to 10 cm. long and extending well above top of plant; seeds large, black, to 2.5 mm. long, 1.75 mm. wide.

Among granite rocks, Lower Sonoran Zone, on barren islets of the Smith Island group at the north side of the entrance of Bahía de los Angeles, Baja California. April–June.

**10. Ferocactus acanthodes** (Lemaire) Britt. & Rose, Cactaceae **3:** 129, *figs. 134–137; pl. 15, figs. 1–2,* 1922, var. **acanthodes**

*Echinocactus acanthodes* Lemaire, Cact. Gen. Nov. Sp. 106, 1839.
*Echinocactus viridescens, β.? cylindraceus* Engelm., Amer. Jour. Sci. II, **14:** 338, 1852.
*Echinocactus cylindraceus* Engelm., Proc. Amer. Acad. **3:** 275, 1856.
*Ferocactus rostii* Britt. & Rose, Cactaceae **3:** 146, *fig. 153b,* 1922.
*Ferocactus cylindraceus* Orcutt, Cactography 5, 1926.
*Echinocactus rostii* Berger, Kakteen 238, 1929.
*Echinocactus acanthodes* var. *rostii* Munz, Man. S. Calif. 328, 1935.
*Ferocactus acanthodes* var. *rostii* Marshall & Bock, Cactaceae 148, 1941.

Stem simple or cespitose, globose to cylindric, to 3 m. tall, 4 dm. wide; ribs 20–30, vertical, about 2 cm. high, with transverse creases between areoles; areoles large, oval, 15 mm. long, 10 mm. broad, becoming contiguous, youngest bearing brown wool to 15 mm. long; spines variable but typically flexuous, spreading, annulate, from white to red or often yellow or brown; central spines 4, flattened, twisted, 5–15 cm. long, lower one longer and broader, sometimes curved or hooked at tip; radial spines variable, usually several larger ones above and below centrals with radiating setaceous lateral spines or bristles, total number of both types 15–25; flowers yellow or sometimes tinged with red, broad-funnelform, 4–6 cm. long and wide; filaments pink or yellow; style about 25 mm. long and 2.5 mm. thick, ridged; stigma lobes about 17; fruit globular, yellow, to 3 cm. long, 2 cm. wide, opening by basal pore; seeds dark reddish brown or black, plump, rounded-angular, about 2.5 mm. long, 2 mm. wide, 1.5 mm. thick; testa reticulate.

Rocky slopes and outwash fans, Lower Sonoran Zone, San Diego, Imperial, Riverside, and San Bernardino Counties, California; northeastern Baja California; Yuma County, Arizona; and northwestern Sonora. March–June.

This is a plastic polymorphic complex with diverse, though intergrading, geographical types. Three taxa, *F. rostii, F. lecontei,* and *F. tortulospinus,* which were once considered to be distinct species, have recently been combined in *F. acanthodes.* Here *F. lecontei* and *F. tortulospinus* are considered to be varieties of *F. acanthodes,* but the holotype specimen of *F. rostii* appears to be inseparable from the typical variety.

**10*a*. Ferocactus acanthodes** var. **tortulospinus** (Gates) Lindsay, Cactus & Succ. Jour. **27**: 168, 1955

*Ferocactus tortulospinus* Gates, Cactus & Succ. Jour. **4**: 343, *fig.*, 1933.

Differs from the typical variety in its depressed stout stem and the constant, dull grayish-red color of its spines. Always simple, and not over 6 dm. tall and 4 dm. wide. Flowers and fruits are indistinguishable from those of *F. acanthodes.*

Rocky hills in a small section in the arid interior of Baja California, near 29° 35′ N., 114° 40′ W. The type locality is low hills ten miles north of Laguna Seca Chapala. Probably the unusual habit, consistent color and type of armament, and disjunct range of *F. tortulospinus* are sufficient to maintain it as a separate variety of *F. acanthodes.*

**10*b*. Ferocactus acanthodes** var. **lecontei** (Engelm.) Lindsay, Cactus & Succ. Jour. **27**: 169, 1955

*Echinocactus lecontei* Engelm., Proc. Amer. Acad. **3**: 274, 1856.
*Echinocactus wislizeni lecontei* Engelm., in Rothrock, Wheeler Rep. 128, 1878.
*Ferocactus lecontei* Britt. & Rose, Cactaceae **3**: 129, 1922.

Differs from var. *acanthodes* in having more appressed spines, a more clavate body, and smaller seeds. The spines are not as twisted, and the upper and lower of the principal central spines are usually wider and recurved against the body of the plant. None of the characters which separate the two varieties is absolutely constant.

Rocky hills and slopes, Lower and Upper Sonoran Zones, in eastern Riverside and San Bernardino Counties, California, southern Nevada, and all of Arizona except the northeastern section. It occupies the northern and eastern portions of the range of *F. acanthodes,* often at high altitudes, particularly in the Mojave Desert in California and the Charleston Mountain region of Nevada. April–May.

**11. Ferocactus chrysacanthus** (Orcutt) Britt. & Rose, Cactaceae **3**: 127, 1922

*Echinocactus chrysacanthus* Orcutt, Rev. Cact. **1**: 56, 1899.

Stem simple or rarely cespitose with up to 6 heads, to 1 m. tall, 3 dm. broad; ribs about 21, tuberculate, 2 cm. tall; areoles oval, to 2 cm. long, 1 cm. wide, bearing light gray tomentum; spines bright yellow or red, in some specimens gray; central spines about 10, to 5 cm. long, flattened, annulate, more or less twisted, one sometimes curved or hooked at tip; radial spines 4–12 or more, white, radiating, sometimes bristle-like but usually acicular, intergrading with heavier central type spines at lower side of areole; flowers yellow or orange, 4.5 cm. long, to 4 cm. broad, campanulate; scales on ovary and tube red edged with yellow, intergrading into outer perianth segments, which are yellow with red midstripe; inner perianth segments satiny-yellow or sometimes with red midstripe, about 25 mm. long, 8 mm. wide; filaments red, 5–15 mm. long, style pink, 15 mm. long, with about 11 pink stigma lobes; fruit yellow, to 3 cm. long, 1.5 cm. wide, fleshy, persistent on plant, sometimes opening by basal pore; seeds large, 2 mm. long, 1.5 mm. wide, 0.5 mm. thick, dull black; testa sculptured with oval or angular depressions.

Rocky hillsides, Lower Sonoran Zone, along the east and south sides of Cedros Island; one specimen was found on West San Benito Island. Reported from the adjacent coast of Baja California, possibly in error. June–July.

**12. Ferocactus gracilis** Gates, Cactus & Succ. Jour. **4:** 323, *fig.*, 1933, var. **gracilis**

Stem simple, globose to cylindric, to 3 m. tall, 3 dm. wide; ribs about 24, slightly tuberculate and expanded under areoles, 2 cm. high, deep green; areoles oval to elliptic, to 2 cm. long, 6 mm. wide, with light gray tomentum in youth; spines in 2 series, central spines 7–13, annulate, red with yellow tips; four principal ones cruciform, upper and lower flattened, lateral two subulate, lower longest, to 6 cm. long, 5 mm. wide, rather convolute and curved or hooked at tip; radial spines white, flattened-setaceous, twisted, lower three acicular to subulate; flowers red, funnelform, about 4 cm. long, 3.5 cm. wide, with widely placed scales on ovary intergrading into obtuse, red, outer perianth segments; inner perianth segments linear-lanceolate, 25 mm. long, 5–7 mm. wide, midstripe lavender-red, margins yellow, serrulate; filaments numerous, fine, appressed against style; style 22 mm. long, yellow and red above; stigma lobes 9; fruit yellow, oblong, about 2.5 cm. long, not opening by basal pore; seeds black, shiny, 1.75–2 mm. long, 1.10–1.35 mm. wide, rounded, not angled.

Rocky hillsides, outwash fans, silt flats, and gravelly plains, Lower Sonoran Zone, southern end of the Sierra San Pedro Mártir south to below Punta Prieta, Baja California. June–Aug.

**12*a*. Ferocactus gracilis** var. **coloratus** (Gates) Lindsay, Cactus & Succ. Jour. **27:** 169, 1955

*Ferocactus coloratus* Gates, op. cit. **4:** 344, *fig.*, 1933.

Differs from var. *gracilis* in having shorter stems and shorter spines with upper and lower of four principal centrals at least 6 mm. and often over 1 cm. wide; flowers similar to those of var. *gracilis* but with less red pigment and less broadly expanded; seeds more angular, irregular, and slightly larger than those of *gracilis*.

Known only from the area between Punta Prieta and Miller's Landing, Baja California. April–June. Cultivated plants flowered in July.

The population of barrel cacti in which var. *coloratus* occurs is variable, and the specimens which have been collected to represent *coloratus* are usually the extreme examples. It is possible that the population represents a hybrid swarm containing genetic material from

*F. gracilis* var. *gracilis, F. peninsulae* var. *peninsulae,* which occurs in the adjacent Sierra San Borjas, and *F. peninsulae* var. *viscainensis,* which occurs in the Vizcaíno Desert to the south.

## 17. ECHINOMASTUS Britt. & Rose, Cactaceae 3: 147, 1922

Plants small, globose to short-cylindric, with low, somewhat spiraled ribs divided into low but usually distinct tubercles. Areoles bearing several spreading, often intricately intertangled, acicular spines; central spines 1 to several or absent, straight or slightly curved but not hooked, when present usually stoutish. Flowers borne at base of short woolly groove nearly buried by spine cluster in young areoles near apex of plant, pink, magenta or yellow; scales on hypanthium erose, thin. Fruit oblong, scaly, at length dry, dehiscing by a basal pore or short vertical fissure; scales and their axils naked. Seeds black, muricate; hilum ventral, depressed.

Areoles circular or nearly so; central spines 4–9, all alike, spreading at about equal angles; ribs weakly tuberculate:
  Flowers magenta or deep pink .......................... 1. *E. johnsonii* var. *johnsonii*
  Flowers yellow ...................................... 1a. *E. johnsonii* var. *lutescens*
Areoles distinctly elliptical; central spines 1–2, lower much shorter than upper if two present, larger (or solitary) central spine erect, straight; ribs markedly tuberculate .......... .................................................... 2. *E. erectocentrus*

### 1. Echinomastus johnsonii (Parry) E. M. Baxter, Calif. Cactus 75, *fig.*, 1935, var. johnsonii

*Echinocactus johnsoni* Parry, in Engelm., Bot. King Expl. 117, 1871.
*Echinocactus johnsoni* var. *octocentrus* J. M. Coulter, Contr. U.S. Nat. Herb. 3: 374, 1896.
*Ferocactus johnsonii* Britt. & Rose, Cactaceae 3: 141, *fig. 149,* 1922.

Simple oval plant 8–15 cm. tall, 6–10 cm. in diameter; ribs 17–21, narrow, 3–6 mm. high, undulate-tuberculate, almost completely hidden by interlocking spines; areoles 1.5–2 cm. apart vertically, with a short, narrow, woolly groove running from upper margin of areole to axil of tubercle; apex of plant devoid of spines but a heavily felted-over, circular patch 1–1.5 cm. in diameter at apex; radial spines 9–14, grayish to yellowish, tinged with red toward apex, often becoming darker in age, 1–2 cm. long, radiate-spreading; central spines 4–9, stouter, darker red, 2–3.5 cm. long, straight, distinctly bulbous at base, divaricately spreading; flowers 4–6 cm. long, nearly as wide, petals pink to deep rose or magenta and with a silvery sheen; scales of ovary obtuse, membranous on margins; fruit oblong, 10–15 mm. long, nearly naked; seeds finely reticulate-pitted.

Rocky hillsides in lime-impregnated soil, Lower Sonoran Zone, Inyo County, California, and adjacent Nevada south into western Yuma County, Arizona. March–April.

### 1*a*. Echinomastus johnsonii var. lutescens (Parish) Wiggins, comb. nov.

*Echinocactus johnsoni* var. *lutescens* Parish, Bull. S. Calif. Acad. 25: 83, 1926.
*Echinomastus arizonicus* Hester, Cactus & Succ. Jour. 5: 504, *fig.*, 1934.

Similar to var. *johnsonii* in size, spine arrangement, and color but flowers canary yellow with greenish bases and lower midribs of perianth segments or centers of flowers sometimes pale maroon.

Among limestone boulders and ledges, Lower Sonoran Zone, Yuma County, Arizona. April–May.

**Echinomastus intertextus** (Engelm.) Britt. & Rose, Cactaceae **3:** 149, *fig. 156,* 1922, differs from *E. johnsonii* in having 3 of the 4 central spines turned upward and distinguishable from radials only with difficulty, other one much shorter and porrect, entire arrangement of spines closely interwoven; flowers only 2–2.5 cm. long, outer perianth segments white-margined.

Occasional on rocky hillsides, Upper Sonoran Zone, southeastern Arizona to Texas but possibly straggling into the uppermost edge of the Sonoran Desert along its eastern border in Arizona and Sonora.

**2. Echinomastus erectocentrus** (J. M. Coulter) Britt. & Rose, Cactaceae **3:** 148, *fig. 154,* 1922

*Echinocactus erectocentrus* J. M. Coulter, Contr. U.S. Nat. Herb. **3:** 376, 1896.

Plants ovoid to short-cylindrical, 6–15 cm. tall, 4–10 cm. in diameter, pale bluish green but surface usually partly hidden by heavy armament; ribs 19–21, low, usually slightly spirally oblique, with closely set, low tubercles 2–5 mm. high and with a narrow groove along upper side; areoles broadly elliptical, 1.5–2 mm. wide, 2.5–3.5 mm. long; radial spines 11–17 (usually 13–14), terete, straight, pectinately appressed, 7–21 mm. long, ashy gray at base, grading into pink, dull red, or purplish toward tips; central spines usually solitary but sometimes 2, upper erect or essentially so, 1.5–3 cm. long, bulbous at base, if two centrals present lower one porrect, much shorter than upper; flowers diurnal, rose-pink, 3–5 cm. high, 3–4 mm. in diameter, perianth segments abruptly short-apiculate, often somewhat erosulate, 6–8 mm. wide; scales on ovary cordate, 3–4 mm. wide, about as long, finely erosulate, acute at apex; stamens 8–10 mm. long, included; style exceeding stamens 5–8 mm., greenish; stigmas 6–8, pink to pale red; mature fruit not seen.

Rocky hillsides, Lower Sonoran and arid parts of Upper Sonoran Zones, Pinal, Cochise, and Pima Counties, Arizona, south at least to Puerto Peñasco in northwestern Sonora. March–April.

## 18. ECHINOCACTUS Link & Otto, Verh. Ver. Beförd. Gartenb. 3: 420, 1827

Single or cespitose plant with globose or cylindrical stems clothed with dense mat of wool or naked at apex. Ribs few to many. Areoles large, very spiny; spines annulate. Flowers borne on crown of plant, yellow or pink, medium in size; flower tube covered with imbricate, persistent, pungent scales. Ovary clothed with narrower scales having axillary mats of wool. Fruit densely white-woolly, dry, thin-walled, oblong. Seeds black, smooth, shining or papillose; hilum subbasal.

Plants cespitose at maturity; ribs 10–21; spines hairy when young; flowers yellow .................................................................. 1. *E. polycephalus*
Plants solitary; ribs 6–8; spines glabrous in youth; flowers pink ..... 2. *E. horizonthalonius*

**1. Echinocactus polycephalus** Engelm. & Bigelow, in Engelm., Proc. Amer. Acad. **3:** 276, 1856

*Echinocactus polycephalus* var. *flavispinus* Haage Jr., Monats. Kakteenk. **9:** 43, 1899.

Solitary when young but soon forming clumps of 3–60 heads, each globular to oblong-ovoid, 1–3(7) dm. high; ribs 10–21, 2–3 cm. high, tuberculately irregular; areoles 10–12 mm. in diameter; radial spines 8–10, unequal, 2.5–5 cm. long, woolly and reddish when young, glabrate and gray to black in age, subulate, flattened; central spines 3–5, stouter than radial ones, more or less annulate, curved but not hooked, 3–10 cm. long; flowers 5–6 cm. long, yellow; perianth segments linear-oblong, entire; scales on flower tube and ovary small, obscured by wool; fruit 1.5–2.5 cm. long, dehiscing by a basal pore; seeds angulate, dull black, 3–4 mm. long.

Rocky hillsides and gravelly slopes, Lower Sonoran Zone, deserts of San Bernardino County, California, to southern Utah, Arizona, and northern Sonora. Feb.–March.

**2. Echinocactus horizonthalonius** Lemaire, Cactus Gen. Nov. Sp. 19, 1839

*Echinocactus equitans* Scheidweiler, Bull. Acad. Brux. **6**[1]**:** 88, 1839.
*Echinocactus horizonthalonius* var. *curvispinus* Salm-Dyck, Cact. Hort. Dyck. **1849:** 156, 1850.
*Echinocactus horizonthalonius* var. *centrispinus* Engelm., Proc. Amer. Acad. **3:** 276, 1856.
?*Echinocactus parryi* Engelm., loc. cit.
*Echinocactus laticostatus* Engelm. & Bigelow, Pacif. R. Rep. **4:** 32, 1857.
*Echinocactus horizonthalonius* var. *obscurispinus* R. Meyer, Monats. Kakteenk. **21:** 181, 1911.

Solitary globose, depressed-globose, or rarely short-cylindrical plants 1–2.5 dm. tall; ribs usually 8(7–13), broad and obtuse, frequently spirally arranged; areoles circular or essentially so, 5–7 mm. in diameter, 6–12 mm. apart on each rib; central spine solitary, curvingly deflexed, 2–3.5 cm. long, dorsiventrally flattened, weakly annulate, reddish to gray, glabrous; radial spines 6–9, confined to sides and top of areole, those of adjacent areoles interlocking, appressed, more slender and more markedly annulate than central spine, 2–4 cm. long, straight or curved; flowers in nascent areoles, pink to rose, 5–7 cm. long, about as broad, densely long-woolly on the ovary and hypanthium; outer perianth segments lance-linear, attenuate-aristate at apex; inner perianth segments oblong to narrowly oblanceolate, 5–9 mm. wide, often erosulate around distal margin; stamens many, filaments very delicate, 8–12 mm. long; fruit densely long-woolly, 2.5–3.5 cm. long, lower half reddish at maturity; seeds blackish, finely papillose, about 2 mm. long.

Limestone ledges and hillsides, Lower and Upper Sonoran Zones, Pima County, Arizona, to Texas and adjacent northern Mexico. May–June.

## 19. SCLEROCACTUS Britt. & Rose, Cactaceae 3: 212, 1922

Simple or rarely cespitose, globose to short-cylindrical cactus with undulate, tuberculate, prominent ribs and intertangled spines of 3 kinds. Areoles bearing short, terete, white, acicular radial spines, maroon to bright red; longer, terete, strongly hooked central spines; and angulate, flattened, somewhat tortuous, white, unhooked spines in upper part of areole. Flowers borne above and adjacent to areoles near apex of plant, subcampanulate. Ovary oblong, with tufts of short wool in axils of scattered scales. Fruit pyriform, dehiscing by a basal pore, nearly naked. Seeds tuberculate, large; embryo curved, endosperm abundant.

**1. Sclerocactus polyancistrus** (Engelm. & Bigelow) Britt. & Rose, Cactaceae **3:** 213, *pl. 23, figs. 3–4,* 1922

*Echinocactus polyancistrus* Engelm. & Bigelow, in Engelm., Proc. Amer. Acad. **3:** 272, 1856.

Simple, globose to oblong plant up to 4 dm. high; ribs 12–17, obtuse, 1–1.5 cm. high, strongly undulate; radial spines 15–20, terete, 1–2.5 cm. long; hooked central spines 6–8, spreading, 3–6 cm. long, borne on lower two-thirds of areole, upper third bearing 2–4 (usually 3) flattened, erect, white spines 4–12 cm. long; flowers magenta, 4–6 cm. long; inner perianth segments oblong, 3–4 cm. long; tube 2–3 mm. long; stamens numerous, half as long as perianth; style 1–1.5 cm. longer than stamens; stigma lobes greenish; fruit becoming dry, nearly scaleless, 3–4 cm. long; seeds black, 4 mm. long, hilum sublateral.

Occasional on gravelly slopes and mesas, Lower Sonoran Zone, Mojave Desert, California, into southern Nevada and Utah and northwestern Arizona. May.

## 20. COCHEMIEA Walton, Cactus Jour. 2: 50, 1899

Usually in cespitose clusters from a central root; stems cylindrical and covered with closely crowded, spirally arranged, ungrooved tubercles with clear juice. Areoles terminal, producing both radial and central spines, centrals 1–12, one or more hooked or all (if more than one) straight, radials straight. Flowers in axils of upper mature tubercles, narrowly tubular, somewhat curved and slightly oblique; perianth segments in 2–3 series, red. Stamens and style red, exserted. Ovary naked. Fruit indehiscent, globose to obovoid, with a large terminal scar. Seeds small, black, finely reticulate-pitted.

Central spines 1–4; axils of tubercles woolly but rarely setose:
- Spines all straight, almost entirely hiding tubercles; flowers 4–5 cm. long; filaments yellow .......... 1. *C. halei*
- Spines of 2 kinds, one or more hooked, others straight, not hiding tubercles; flowers about 3 cm. long; filaments red:
  - Central spines normally solitary, hooked; radial spines 7–9; tubercles somewhat flattened .......... 2. *C. poselgeri*
  - Central spines 2–4 or sometimes solitary; radial spines 10–15; tubercles terete:
    - Radial spines white with black tips, 1–3.5 cm. long; central spines 1–4 .......... 3. *C. setispina*
    - Radial spines reddish brown, 8–12 mm. long; central spines 4 ....... 4. *C. maritima*

Central spines 8–11; axils of tubercles setose as well as woolly (known positively only from Cedros Island) .......... 5. *C. pondii*

### 1. Cochemiea halei (Brandegee) Walton, Cactus Jour. 2: 50, 1899

*Mamillaria halei* Brandegee, Proc. Calif. Acad. II, 2: 161, *pl. 6,* 1889.
*Cactus halei* J. M. Coulter, Contr. U.S. Nat. Herb. 3: 106, 1894.

Cespitose with 1–10 or more stems 2–5 dm. tall and 5–8 cm. in diameter from a tough subfleshy root; stems nearly erect and crowded together; tubercles rounded, about 1 cm. long, crowded together and nearly hidden by interlacing straight spines, woolly but without setae in axils; radial spines 10–20, rigid, straight, 6–15 mm. long; central spines 3–4, all straight, to 25 mm. long; all spines, both central and radial, reddish brown to yellowish or grayish with slightly darker tips; flowers 4–5 cm. long, scarlet; filaments yellow, anthers scarlet, exserted; stigmas scarlet; fruit clavate, bright red, 12–15 mm. long; seeds black, minutely pitted-reticulate, appearing smooth without a lens.

On rocky hillsides or silty soil, Lower Sonoran Zone, Magdalena and Santa Margarita Islands and the adjacent mainland, Baja California. Said to be very abundant on the two islands. The perianth segments are subpersistent on the fruits after withering. March–April.

**2. Cochemiea poselgeri** (Hildmann) Britt. & Rose, Cactaceae **4:** 22, *fig.* 23; *pl.* 2, *fig.* 3, 1923

*Mammillaria poselgeri* Hildmann, Garten-Zeit. **4:** 559, 1885.
*Mamillaria roseana* Brandegee, Zoe **2:** 19, 1891.
*Mammillaria radiana* Quehl, Monats. Kakteenk. **2:** 104, 1892.
*Cactus roseanus* J. M. Coulter, Contr. U.S. Nat. Herb. **3:** 105, 1894.
*Cochemiea rosiana* Walton, Cactus Jour. **2:** 51, 1899.

Stems several to many from a central root, 2.5–4.5 cm. in diameter, erect, spreading and prostrate or decumbent, or even pendent over rocks and sides of cliffs, as much as 2 m. long; tubercles ovoid, 8–12 mm. long, 10–14 mm. apart, somewhat flattened at right angles to long axis of stem; areoles and axils lanate, axils rarely slightly setose also; older tubercles becoming naked, grayish, younger ones green or faintly glaucous; radial spines 7–11, reddish brown or eventually grayish, 9–16 mm. long, all straight and radiately spreading; central spine solitary, hooked, stouter than radials, to 3.5 cm. long; flowers in upper axils about 3 cm. long, scarlet; stamens and styles scarlet; fruit broadly obovoid to subglobose, 6–8 mm. in diameter, scarlet; seeds black, minutely pitted.

Rocky cliffs, canyon walls, on steep rocky slopes, and on silt flats, Lower Sonoran Zone, interior of central Baja California from the vicinity of Canípole and Comondú and along the Gulf coast from San Nicolás Bay southward. Near Comondú pendent stems 1–2 m. long resemble giant slender caterpillars. July–Aug.

**3. Cochemiea setispina** (J. M. Coulter) Walton, Cactus Jour. **2:** 51, 1899

*Cactus setispinus* J. M. Coulter, Contr. U.S. Nat. Herb. **3:** 106, 1894.
*Mamillaria setispina* Engelm. ex K. Brandegee, Erythea **5:** 117, 1897.

Few to many cespitose stems 3–6 cm. in diameter, 2–3.5 dm. tall; tubercles short-ovoid, terete or essentially so in cross section, their axils lanate but not setose; radial spines 10–12, acicular, radiately spreading, 10–35 mm. long, all straight, white with black or dark gray tips, closely interlacing; central spines 1–4, stouter than radials, 2–5 cm. long, one of them, usually the lowest and longest, strongly hooked, hook usually turning upward; flowers unknown; fruit obovoid, about 3 cm. long, scarlet; seeds black and finely pitted.

Rocky hillsides and canyons, Lower Sonoran Zone, interior desert ranges, vicinity of San Borjas Mission east of Punta Prieta to Calmallí, on Isla Angel de la Guarda, and possibly more widely distributed in these imperfectly known desert ranges. July–Aug.

**4. Cochemiea maritima** Lindsay, Cactus & Succ. Jour. **8:** 144, 1937

*Cochemiea maritima* Shurly, Cact. Jour. [British] **4:** 12, 1935. Nomen nudum.

Cespitose in clusters to 1 m. in diameter or more; stems 1–5 dm. tall, erect or procumbent; tubercles conic or slightly flattened laterally, glaucous, becoming corky and gray in age; areoles and axils of tubercles lanate when young; radial spines 10–15, acicular, 8–12 mm. long, reddish brown; central spines 4, reddish brown, upper three ascendingly spreading, 1–2 cm. long, straight, lowermost porrect, 2–5 cm. long, hooked, stouter than others; flowers from upper axils, about 3 cm. long, scarlet, upper segments flaring and tips reflexed; style nearly twice as long as corolla, scarlet; stamens exserted half as far as style, filaments and anthers red; fruit obovoid to globose, scarlet; seeds about 0.5 mm. long, black, minutely pitted.

Rocky maritime cliffs and slopes along the coast near Punta Blanca at least to Punta Rosalia, west of Punta Prieta, Baja California. March–April.

**5. Cochemiea pondii** (Greene) Walton, Cactus Jour. **2:** 51, 1899

*Mamillaria pondii* Greene, Pittonia **1:** 268, 1889.
*Cactus pondii* J. M. Coulter, Contr. U.S. Nat. Herb. **3:** 102, 1894.

Stems 1 to several from a stout root, cylindrical, 7–30 cm. tall, densely covered with an interlacing armament of spines which nearly hides underlying stem; axils of tubercles lanate-setose, areoles white-tomentose; radial spines white or grayish, 15–30, slender and spreading; central spines 8–11, longer and stouter than radials, 2–3 cm. long, one or two of them hooked; flowers slender, tubular, 4–5 cm. long, bright scarlet; fruit purplish red, ovoid to obovoid, 15–18 mm. long.

Rocky cliffs and hillsides, Lower Sonoran Zone, Cedros Island. Possibly also on the adjacent mainland, an area nearly unknown botanically. April–June.

## 21. CORYPHANTHA (Engelm.) Lemaire, Cactées 32, 1868

Solitary or cespitose plants with globose to cylindric stems bearing conspicuous, spirally arranged tubercles. Tubercles mammillate, narrowly grooved from apex to base along adaxial side when mature. Flowers borne near top of plant in axils of young tubercles, comparatively large and showy; perianth segments withering-persistent. Ovary naked or sparsely scaly in some species. Fruit ovoid to oblong or cylindrical, greenish or yellowish, slightly juicy. Seeds brownish, smooth to finely reticulate or granulose; hilum subbasal; embryo curved.

Flowers yellow to salmon; central spines 1–4:
- Radial spines 6–15; central spine solitary, usually hooked or strongly curved at tip; fruit 4–6 cm. long; tubercles 15–25 mm. long; seeds about 3 mm. long .. 1. *C. robustispina*
- Radial spines 20–30; central spines 2–4, straight or slightly curved but not hooked; fruit 18–20 mm. long; tubercles 8–12 mm. long; seeds 2 mm. long .......... 2. *C. deserti*

Flowers pink, rose, or lavender; central spines 5–14:
- Plants solitary or 1–5-branched, never forming large mounds; central spines 8–14, stouter than acicular radials; fruit green, clavate ......................... 3. *C. alversonii*
- Plants with several to many crowded, spherical to ovoid stems forming mounds to 6 dm. wide and high; central spines 5–6, similar to radials; fruit ellipsoidal, green with a purplish tinge ................................................ 4. *C. aggregata*

**1. Coryphantha robustispina** (Schott) Britt. & Rose, Cactaceae **4:** 33, 1923

*Mamillaria robustispina* Schott, in Engelm., Proc. Amer. Acad. **3:** 265, 1856.
*Cactus robustispinus* Kuntze, Rev. Gen. Pl. **1:** 261, 1891.
*Mamillaria brownii* Toumey, Bot. Gaz. **22:** 253, 1896.
*Cactus brownii* Toumey, loc. cit.

Small, ovoid, spherical or short-cylindrical plant or sometimes with a few clustered stems, each stem 5–10 cm. in diameter, 5–40 cm. tall; tubercles truncate-conical, 15–30 mm. long, 10–15 mm. in diameter at base, crowded only at base of stem or plant, often with large glands in groove; areoles broadly elliptical; radial spines 6–15, spreading almost at right angles to axis of tubercle, straight, rigid, stout, 20–25 mm. long, grayish or yellowish, rarely with purplish or black tips; central spine solitary, nearly at right angles to radials, abruptly hooked or strongly curved at tip, yellowish when young, becoming gray in age, usually dark purplish to nearly black toward tip, 2.5–4 cm. long; flowers yellow to salmon, 3–4 cm. in diameter when fully open;

perianth segments entire or faintly erosulate; fruits cylindrical, 4–6 cm. long; seeds 3–3.5 mm. long, shining, reddish brown.

Valley floors, outwash slopes, mesas, and hillsides, Lower and Upper Sonoran Zones, Pima, Santa Cruz, and Cochise Counties, Arizona, south into northern Sonora. June–July.

**2. Coryphantha deserti** (Engelm.) Britt. & Rose, Cactaceae **4:** 46, *figs. 44–46*, 1923

*Mamillaria deserti* Engelm., in Brewer & Wats., Bot. Calif. **2:** 449, 1880.
*Cactus radiosus* var. *deserti* J. M. Coulter, Contr. U.S. Nat. Herb. **3:** 121, 1894.
*Mamillaria radiosa* var. *deserti* Schumann, Gesamtb. Kakteen 481, 1898.

Simple or very rarely 1–3-branched, globose, ovoid, to short-cylindric plant 5–25 cm. high, 6–9 cm. in diameter, densely covered with interlocking spines; tubercles slender, crowded, 8–12 mm. long, 5–6 mm. wide at base, gray-green, obscured by spines; areoles broadly elliptical, 2–2.5 mm. wide, 3–4 mm. long; radial spines 20–30, acicular, unequal, 9–16 mm. long, spreading, white or ashy-gray, sometimes black-tipped; central spines 2–4, stout, terete, 5–15 mm. long, somewhat spreading, those at top of plant black or purple-black on upper half, fading through red to white at base, whole spine light at base of old plants; flowers straw-colored to burnt orange, 2.5–3 cm. wide when fully expanded, outer perianth segments linear-lanceolate, 2–3 mm. wide, acute to attenuate at apex, ciliate along margins, sometimes tinged with pink toward tips; fruit green, oblong, 18–20 mm. long, walls thin; seeds obliquely obovate, curved, minutely reticulate-pitted, reddish brown, 2–2.2 mm. long.

Infrequent on rocky hillsides and mesas, Lower Sonoran Zone, Inyo County, California, to the southern edge of the Mojave Desert in eastern San Bernardino County, east through southern Nevada into southern Utah. May–June.

**3. Coryphantha alversonii** (J. M. Coulter) Orcutt, Cactography 3, 1926

*Cactus radiosus* var. *alversoni* J. M. Coulter, Contr. U.S. Nat. Herb. **3:** 122, 1894.
*Mamillaria alversonii* Zeiss., Monats. Kakteenk. **5:** 70, 1895.
*Mamillaria radiosa* var. *alversonii* Schumann, Gesamtb. Kakteen 481, 1898.

Stems simple or infrequently 1–5-branched at base; stems oblong to short-cylindric, 1–2 dm. high, 5–8(10) cm. in diameter; tubercles short and thick, 8–9 mm. high, 5–6 mm. in diameter at base; radial spines 20–35, acicular, 1–2 cm. long, white or slightly tipped with black or dark brown, closely interlocking-spreading, nearly concealing stem; central spines 8–14, much stouter, terete, unequal, 1–2.5 cm. long, divaricately spreading, white to ash-yellow at base, shading through deep maroon to black from about middle toward tips; flowers 2.5–3 cm. long, about 1.5 cm. broad when fully expanded; perianth segments with rose-colored midveins, shading into light purple toward margins; outer perianth segments ciliate-margined; style and stigma lobes white; fruit clavate, green, naked or nearly so, 2–2.5 cm. long; seeds brown, minutely tuberculate.

Infrequent on rocky mesas and in desert mountain canyons, Lower Sonoran Zone, Morengo Valley to Indio, California, and in the mountains between the Mojave and Colorado Deserts. May.

**4. Coryphantha aggregata** (Engelm.) Britt. & Rose, Cactaceae **4:** 47, *fig. 47*; *pl. 4*, 1923

*Mamillaria aggregata* Engelm., in Emory, Notes Mil. Rec. 157, 1848.
*Cereus aggregatus* J. M. Coulter, Contr. U.S. Nat. Herb. **3:** 396, 1896, as to name.
*Echinocereus aggregatus* Rydb., Bull. Torrey Club **33:** 146, 1906, as to name.

Low-growing, ovoid, solitary plant or more commonly much branched and producing several to many heads in mounds to 6 dm. high and about as wide, each branch 5–8 cm. long, 2.5–3.5 cm. in diameter; tubercles cylindrical, 4–6 mm. in diameter, 5–8 mm. long, hidden by interlacing spines; radial spines 15–20, whitish with slightly brownish tips, 12–25 mm. long, slightly flattened, those of contiguous areoles closely interwoven; central spines 5–6, straight, white with brownish tips, upper ones pointing upward, 15–20 mm. long, lower ones 6–10 mm. long, spreading at different angles, slender, terete; flowers lavender to purplish or sometimes pink, 3.5–5 cm. in diameter when fully open, perianth segments oblanceolate, abruptly acuminate, 2–6 mm. wide, outer ones ciliate-margined; fruit ellipsoidal, 10–18 mm. in diameter, 15–25 mm. long, green with a purplish tinge; seeds dark reddish brown, about 2 mm. long.

Along washes, on grassy flats and hillsides, Lower Sonoran Zone to lower edge of Arid Transition Zone, eastern Pima and Santa Cruz Counties, Arizona, into adjacent New Mexico; south into northern Sonora. May–June.

A solitary specimen without flowers or fruit was collected on desert hills a few miles south of Sásabe in northern Sonora that resembles *C. pectinata*, but the plant has more numerous radial spines than normally found in the areoles of *C. pectinata*, so accurate determination awaits collection of more adequate material.

## 22. BARTSCHELLA Britt. & Rose, Cactaceae 4: 57, 1923

Cespitose with short-oblong to globose heads closely crowded in flat or slightly rounded clumps. Tubercles comparatively large, slightly coalescent with those adjacent, terete in cross section, grooveless, with clear juice. Radial spines straight, central spines usually hooked. Flowers large, borne near top of plant, lavender to pale purple. Stamens and style included. Fruit obovoid to clavate, circumscissile. Seeds black, dull, slightly pitted, base cylindrical, hilum triangular, faintly depressed.

**1. Bartschella schumannii** (Hildmann) Britt. & Rose, Cactaceae **4:** 58, *fig. 55; pl. 7, fig. 6,* 1923

*Mamillaria schumannii* Hildmann, Monats. Kakteenk. **1:** 125, *pl. 8,* 1891.
*Mamillaria venusta* K. Brandegee, Zoe **5:** 8, 1900.

At first solitary but becoming cespitose with as many as 40 heads from one root; heads ovoid, 3–5 cm. in diameter, 2.5–7 cm. tall; tubercles 5–7 mm. high, nearly as broad at base, green or tinged with purplish or glaucous, tip often slightly concave, areoles and axils at first lanate, soon naked; radial spines 8–15, slenderly subulate, 6–15 mm. long, white or pale yellow, tinged with deep red toward tips; central usually solitary but sometimes 2–4, porrect, 8–15 mm. long, hooked or if more than one, straight ones much shorter than hooked ones; flowers rose, about 4 cm. in diameter, lobes 10–12, lanceolate, spreading; stamens shorter than petals; stigma lobes 5–6, green; fruit clavate, 1.5–2 cm. long, scarlet, circumscissile 4–6 mm. above base, the persistent basal cup remaining almost hidden by tubercles and spines; seeds about 1 mm. long, cylindrical base about one-third as long as main body, slightly constricted at junction of two parts, surface dull black, shallowly pitted.

On granitic detritus of hillsides and among disintegrating rocks, Lower Sonoran Zone, vicinity of La Paz south to the tip of the Cape Region, Baja California. Quite common in the vicinity of San José del Cabo. May; Sept.

The basal portions of the fruits partially buried among the tubercles are quite numerous, so it is almost impossible to cut longitudinally through a stem without splitting several of the deep narrow cups packed with seeds. Some of these cups are as much as 9 mm. deep and only 2–3 mm. in diameter.

## 23. MAMMILLARIA Haw., Syn. Pl. Succ. 177, 1812

Solitary or few-branched perennial cactus with globose, short-cylindrical or sometimes elongated-cylindrical stems with watery or milky juice and tubercles arranged in spirals. Tubercles terete or angled, rarely grooved on upper side, naked or more commonly bearing wool or bristles or both in axils. Areoles crowning tubercles, central spine or spines like radials or quite different; if different, 1 or more of centrals usually hooked at tip. Flowers axillary to old tubercles, funnelform to campanulate, rather small. Perianth segments mostly narrow, entire, erose, or serrulate toward tips. Stamens many, inserted in tube, usually included. Style equaling or slightly exceeding stamens; stigma lobes mostly linear. Fruits obovoid to clavate, naked, scarlet, orange, or rarely white. Seeds small, brownish, black, or greenish, punctate, granulose, or sometimes slightly rugulose; hilum basal or subbasal.

Tubercles with milky sap; seeds brown:
  Central spines absent ........................................ 1. *M. peninsularis*
  Central spine or spines present:
    Radial spines less than 10 per areole:
      Tubercles sharply angled from base to tip ........................ 2. *M. bocensis*
      Tubercles obscurely angled or terete to tips:
        Central spine solitary:
          Outer perianth segments ciliate:
            Central spines 4–6 mm. long; radial spines 1–5 mm. long; seeds dull brownish red ........................................ 3. *M. dawsonii*
            Central spines 25–30 mm. long; radial spines 8–13 mm. long; seeds dark brown ........................................ 4. *M. gatesii*
          Outer perianth segments entire ........................ 5. *M. baxteriana*
        Central spines 1–4 ........................................ 6. *M. brandegeei*
    Radial spines 10–20 per areole:
      Radial spines 5–10 mm. long; central spines 1–7, 12–25 mm. long:
        Bristles regularly present in axils of tubercles; radial spines chalky white ........................................ 7. *M. evermanniana*
        Bristles lacking in axils of tubercles; radial spines horn-colored, yellowish, or brownish or, if white, dark brown to black at tips:
          Margins of outer perianth segments ciliate to serrulate, pinkish cream; filaments pink to cream ........................................ 8. *M. johnstonii*
          Margins of outer perianth segments entire, bright green or greenish yellow; filaments white:
            Tubercles gray-green, in 8 and 13 spirals; central spines 4–7, dark brown ........................................ 9. *M. arida*
            Tubercles blue-green, in 13 and 34 spirals; central spines 1–3, purple-black ........................................ 10. *M. lewisiana*
      Radial spines 4–7 mm. long; central spine 1 or rarely 2, 10–12 mm. long ........................................ 11. *M. tesopacensis*
Tubercles with watery sap; seeds chiefly black (a few brown):
◆ Central spines or one of them practically always hooked:
  ■ Radical spines 10–20:
    ● Bristles present in at least some of the axils:
      Central spine solitary ........................................ 12. *M. capensis*

Central spines usually 2 or more:
Inner perianth segments linear, widely spreading, white; filaments white ........ 13. *M. angelensis*
Inner perianth segments lanceolate or lance-ovate, slightly spreading, pinkish, rosy, or purplish at least along central stripe; filaments usually pinkish:
Stigma lobes orange or reddish; spines closely interlacing, obscuring tubercles:
Axillary bristles less than 10; tubercles in 5 and 8 spirals ... 14. *M. fraileana*
Axillary bristles 15–20; tubercles in 8 and 13 spirals .... 15. *M. verhaertiana*
Stigma lobes greenish; spines less closely interlaced, rarely obscuring tubercles:
Outer perianth segments entire; filaments white or rarely faintly pinkish; axillary bristles 5–15 .................................. 16. *M. dioica*
Outer perianth segments serrate or ciliate; filaments pink; axillary bristles 1–3 or in *M. swinglei* often lacking:
Seeds obovoid; tubercles blue-green, 3–4 mm. wide at base; axillary spines 1–3, consistently present; Baja California ........... 17. *M. armillata*
Seeds globose; tubercles dark green, 5–7 mm. wide at base; axillary spines often lacking; Sonora ............................ 18. *M. swinglei*

● Bristles absent from axils of tubercles:
Stigma lobes dark red:
Flowers 1–2 cm. long; tubercles 10–15 mm. long, 8–10 mm. wide at base; plant rarely cespitose ........................................ 19. *M. mainae*
Flowers 2–3 cm. long; tubercles 5–6 mm. long, about 5 mm. wide at base; plant usually cespitose .................................... 20. *M. fasciculata*
Stigma lobes yellow to green:
Central spines 20–30 mm. long; seeds larger, to 1.7 mm. long; tubercles to 2 cm. long ............................................... 21. *M. wilcoxii*
Central spines 7–12 mm. long; seeds 1 mm. long or less; tubercles 1 cm. long or less:
Seeds obpyriform; filaments white to pinkish:
Tubercles in 5 and 8 spirals; one central spine always hooked; roots usually more or less tuberous:
Roots strongly tuberous; central spines 4; filaments white to pink ....... .......................... 22. *M. blossfeldiana* var. *blossfeldiana*
Roots partially fibrous; central spines 3; filaments purplish ............ ............................ 22a. *M. blossfeldiana* var. *shurliana*
Tubercles in 8 and 13 spirals; central spines sometimes straight; roots wholly fibrous ...................................... 23. *M. guirocobensis*
Seeds globose; filaments cream, purplish, or tan (seeds obpyriform but filaments purplish, see 22*a*):
Outer perianth segments entire, midstripes pink to brownish; axils of tubercles scantily woolly or sometimes naked:
Stigma lobes olive-green; lower central spine hooked .................. ............................... 24. *M. goodridgei* var. *goodridgei*
Stigma lobes deep red; central spines all straight ...................... .............................. 24*a*. *M. goodridgei* var. *rectispina*
Outer perianth segments ciliate or serrulate, midstripes purplish or brownish green; axils of tubercles consistently naked:
Tubercles 8–9 mm. long; radial spines 10–15; filaments purplish ......... .......................................... 25. *M. sheldonii*
Tubercles 4–5 mm. long; radial spines 15–20; filaments tan above, white toward base ............................... 26. *M. hutchisoniana*

■ Radial spines usually more than 20 per areole:
Seeds 1.2–1.5 mm. long, bearing a corky, aril-like base; stigma lobes cream; radial spines 40–60 per areole ................................. 27. *M. tetrancistra*

Seeds 0.8–1 mm. long, naked; stigma lobes greenish; radial spines 20–35 per areole:
Outer perianth segments entire; tubercles in 5 and 8 spirals, blue-green:
Seeds globose, not constricted; fruit 9–10 mm. long; inner perianth segments white-margined .................................... 28. *M. insularis*
Seeds obovoid, constricted just above hilum; fruit 2–3 cm. long; inner perianth segments pinkish throughout ........................... 29. *M. boolii*
Outer perianth segments ciliate; tubercles in 8 and 13 or in 13 and 21 spirals, gray-green:
Axils of tubercles bearing 10–20 white bristles about equaling tubercles; seeds very small, about 0.5 mm. long ...................... 30. *M. phitauiana*
Axils of tubercles naked or woolly but not bristly; seeds about 1 mm. long:
Axils totally naked; filaments deep pink; radial spines 20–30, 6–12 mm. long, white to yellow, tips dark brown ................... 31. *M. microcarpa*
Axils woolly; filaments rose to magenta; radial spines 25–35, 3–6 mm. long, snow-white ................................... 32. *M. oliviae*

◆ Central spines nearly always straight, not hooked:
Radial spines usually less than 25 per areole:
Flowers 3–4 cm. long; inner perianth segments and filaments carmine ............ ............................................. 33. *M. mazatlanensis*
Flowers 1–2 cm. long; inner perianth segments white, yellow, cream, or pink:
Central spine solitary, sometimes an occasional one on a plant hooked; flowers very small, 10–12 mm. long; radial spines pale yellowish or light horn-colored ..... ............................................. 34. *M. cerralboa*
Central spines 2–6, always straight; flowers 12–20 mm. long; radial spines white, pinkish, or pale brown:
Plants profusely branching, often with over 100 stems; outer perianth segments with olive-green midstripe, margins entire; central spines white with reddish brown tips ...................................... 35. *M. multidigitata*
Plants simple, or if cespitose having about 20 stems or less; outer perianth segments with pinkish or brownish midstripes, margins serrulate or erose; central spines pink, red, or brownish with darker tips:
Axils of tubercles scantily woolly; seeds pyriform .............. 36. *M. inaiae*
Axils of tubercles densely woolly; seeds subglobose .......... 37. *M. albicans*
Radial spines usually 25–50 per areole:
Stigma lobes linear, equal; axils of tubercles densely woolly, and with several fine bristles 2–3 mm. long; radial spines 25–30 per areole ........ 38. *M. neopalmeri*
Stigma lobes revolute-undulate, unequal; axils of tubercles scantily woolly or naked; radial spines 35–50 .................................... 39. *M. radiaissima*

**1. Mammillaria peninsularis** (Britt. & Rose) Orcutt, Cactography 8, 1926

*Neomammillaria peninsularis* Britt. & Rose, Cactaceae **4:** 85, 1923.

Plants solitary or sometimes in clusters, turbinate, deeply seated in ground, 4–12 cm. in diameter, flattened top often nearly flush with soil; tubercles in 5 and 8 spirals, bluish green, 4–5 mm. wide at base, 5–10 mm. long, sharply 4-angled, keeled ventrally; juice milky; axils woolly in youth, naked in age; areoles slightly woolly in youth; radial spines 4–8, subulate to acicular, 3–8 mm. long, pale yellow with brown tips, spreading to ascending; central spine none or sometimes one similar to radials; flowers about 15 mm. high; outer perianth segments linear-lanceolate, acuminate, 2–5 mm. long, green at base, midstripe brownish green, margins pale green and serrate; inner perianth segments greenish yellow with a reddish brown midstripe toward apex, tips acuminate or emarginate-apiculate; filaments greenish white; stigma lobes bright green; mature fruits unknown; immature fruits clavate, about equaling tubercles, surrounded by dense tufts of yellowish wool 4–6 mm. high.

Gravelly flats and hillsides, Lower Sonoran and arid Tropical Zones, southern Baja California. This attractive *Mammillaria* may not enter the Sonoran Desert, but its presence near the southern limit makes its inclusion desirable. It should be sought near Rancho Arroyo Seco. April–May.

**2. Mammillaria bocensis** Craig, Mam. Handbook 56, *fig. 38*, 1945

Plant depressed-globose to short-cylindrical, to 8.5 cm. wide, slightly taller, from a strong taproot and some fibrous roots, armament interlacing slightly or not at all; tubercles pyramidal and somewhat angled, keeled ventrally, 8–12 mm. broad at base, 10–13 mm. long; juice milky; axils scantily woolly but without bristles; radial spines 6–8, acicular, straight, chalky white with brownish to black tips or older spines becoming reddish or grayish, 5–14 mm. long, slightly ascending; central spine 1, acicular to slenderly subulate, straight or slightly curved, 8–12 mm. long, slightly swollen at base, reddish brown with dark brown to blackish tip, porrect; flowers 1.5–2 cm. long; outer perianth segments pale greenish yellow at base, with reddish or brownish green midstripe and pale pink to light green, ciliate margins, lanceolate, acuminate; inner perianth segments greenish cream, pinkish at margins; filaments white to pale pink; stigma lobes pale yellow-green; fruit scarlet, clavate, about 2.5 cm. long, nearly half as broad; seeds pyriform, light brown, granular, hilum laterally subbasal.

In sandy alluvial soil, mostly under shrubs, Lower Sonoran Zone, near the coast in southern Sonora. This species may not enter the Sonoran Desert, but its known occurrence so near our area makes its inclusion desirable. May–June.

**3. Mammillaria dawsonii** (Houghton) Craig, Mam. Handbook 67, *fig. 48*, 1945
*Neomammillaria dawsonii* Houghton, Cactus & Succ. Jour. **7**: 88, *fig.*, 1935.

Plant simple, depressed-globose, 3–5 cm. broad, deeply seated and extending only 1.5–2.5 cm. above soil; tubercles in 5 and 8 spirals, conical and slightly angular, 4–6 mm. long, about 4 mm. broad at base, dull, light green; juice milky; axils slightly woolly and devoid of bristles; radial spines 6–10, acicular, horn-colored with darker tips, 1–5 mm. long, spreading almost at right angles to axis of tubercles, interlacing only on youngest tubercles; central spine 1, acicular, straight or slightly curved, light brown basally, darker at tip, porrect, about 6 mm. long; flowers 10–12 mm. long; outer perianth segments lanceolate, acute, reddish brown with pale greenish yellow, serrate margins; inner perianth segments greenish yellow with brownish green midstripes; filaments white to cream; stigma lobes greenish yellow; fruit clavate, pinkish above, fading to white below, about 15 mm. long; seeds about 1 mm. long or less, dull red.

In shade of shrubs, Lower Sonoran Zone, western coast of central Baja California. April.

**4. Mammillaria gatesii** M. E. Jones, Cactus & Succ. Jour. **8**: 99, *fig.*, 1937

Simple or slightly cespitose plant with a globose to short-cylindric body 10–15 cm. wide, 15–20 cm. high, tubercles conic, 9–15 mm. long, about 15 mm. broad at base, terete in cross section, arranged in 8 and 13 spirals; axils white-woolly, devoid of bristles; radial spines 6–8, acicular, straight and smooth, 8–14 mm. long, yellowish white with dark brown or blackish tips, eventually dull whitish, upper two shorter than others; central spine 1, stout-acicular to slenderly subulate, straight, 2.5–3 cm. long, reddish brown or purplish in youth, later horn-colored, porrect; flowers about

1.5 cm. long; outer perianth segments ovate-lanceolate, greenish yellow with purplish midstripe, margins usually ciliate; inner perianth segments golden to greenish yellow, lacerate on margins; filaments pale yellow; stigma lobes light green; fruit about 1.5 cm. long, clavate; seeds about 1 mm. long, dark brown, minutely pitted.

Near the sea coast, Lower Sonoran and arid Tropical Zones, vicinity of San José del Cabo to Todos Santos, southern Baja California. March–April.

**5. Mammillaria baxteriana** (H. E. Gates) Boed., in Backeb. & Knuth, Kaktus-ABC 398, 1935

*Neomammillaria baxteriana* H. E. Gates, Cactus & Succ. Jour. **6:** 3, *fig.*, 1934.
*Neomammillaria marshalliana* H. E. Gates, op. cit. 4, *fig.*
*Neomammillaria pacifica* H. E. Gates, op. cit. 5, *fig.*
*Mammillaria marshalliana* Boed., in Backeb. & Knuth, Kaktus-ABC 398, 1935.
*Mammillaria pacifica* Boed., in Backeb. & Knuth, loc. cit.

Simple or sometimes slightly cespitose plant with depressed-globose body about 15 cm. wide and high; tubercles in 13 and 21 spirals, conical, 10–13 mm. long, somewhat quadrangular and 6–9 mm. wide at base; juice milky; axils woolly, without bristles; radial spines 7–13, acicular, straight, 10–15 mm. long, white or with tips brownish in youth; central spine 1, stout-acicular, straight, stiff, smooth, white with a brown tip, later becoming chalky, porrect; flowers 12–15 mm. long; outer perianth segments ovate-lanceolate, acute, brownish purple along midstripe, yellow and entire along margins; inner perianth segments greenish yellow with rose-purple midstripe; filaments pale yellow; stigma lobes greenish yellow; fruit red, clavate, about 2 cm. long; seeds obpyriform, dull reddish brown, about 1 mm. long.

Rolling hills, granitic cliffs, and coastal slopes, Lower Sonoran and Tropical Zones, Cape Region, southern Baja California, and on Isla Espíritu Santo in the Gulf of California. April.

**6. Mammillaria brandegeei** (J. M. Coulter) K. Brandegee, Erythea **5:** 116, 1897

*Cactus brandegei* J. M. Coulter, Contr. U.S. Nat. Herb. **3:** 96, 1894.
*Cactus gabbii* J. M. Coulter, op. cit. 109.
*Mammillaria gabbii* Engelm. ex J. M. Coulter, loc. cit., as a synonym.
*Mammillaria gabbii* Engelm. ex K. Brandegee, Erythea **5:** 116, 1897.
*Mammillaria brandegeei* var. *gabbii* Craig, Mam. Handbook 71, *fig. 51,* 1945.

Simple or cespitose plant to 10 cm. high with a globose to cylindrical body and dark green, conical tubercles 6–10 mm. long and quadrangular at base, arranged in 13 and 21 spirals; axils densely white-woolly, rarely with a single bristle; juice milky; radial spines 8–15, slenderly acicular, straight, smooth, whitish with yellowish or brownish tips, 7–10 mm. long; central spines 1–4, stout-acicular, straight to slightly curved, reddish brown toward base, darker toward tips, 10–20 mm. long, porrect; flowers small, 8–10 mm. long; outer perianth segments ovate, acute, midstripe brown, margins bright green, ciliate; inner perianth segments greenish yellow with a reddish midline, lanceolate; filaments white; stigma lobes green; fruit clavate, about 15 mm. long; seeds curved-pyriform, dull brown, about 1 mm. long.

Hillsides and mesas, chiefly Lower Sonoran Zone, vicinity of Hamilton's Ranch to Calmallí, Baja California. April.

**7. Mammillaria evermanniana** (Britt. & Rose) Orcutt, Cactography 7, 1926

*Neomammillaria evermanniana* Britt. & Rose, Cactaceae **4:** 97, *fig. 95,* 1923.

Simple to cespitose, globose to cylindric plant 5–7 cm. broad; tubercles light green, conic, terete in cross section, 8–9 mm. long, about 6 mm. broad at base, in 13 and 21 spirals; juice milky; axils densely white-woolly and with a few slender bristles 5–8 mm. long; radial spines 12–15, acicular, straight, subflexuous, chalky white or lower ones reddish tan at tip, 5–8 mm. long; central spines 2–4, acicular, straight or slightly curved, 12–15 mm. long, reddish brown in youth, later chalky white with brown tips, divergent; flowers narrow, 12–15 mm. long; outer perianth segments lanceolate, acute, brownish magenta with narrow, pink, serrate margins; inner perianth segments reddish pink to purplish white; stigma lobes olive-green; fruit clavate, red, about 10 mm. long; seeds light brown, curved-pyriform, about 1 mm. long, minutely rugulose.

Arid plains and hillsides, Lower Sonoran Zone, Gulf coast of Baja California from vicinity of Loreto south and on the adjacent islands. April–May.

**8. Mammillaria johnstonii** (Britt. & Rose) Orcutt, Cactography 7, 1926

*Neomammillaria johnstonii* Britt. & Rose, Cactaceae **4:** 80, *fig. 72b,* 1923.
*Mammillaria johnstonii* var. *typica* Craig, Mam. Handbook 99, *fig. 79,* 1945.
*Mammillaria johnstonii* var. *sancarlensis* Craig, loc. cit. *fig. 80.*
*Mammillaria johnstonii* var. *guaymensis* Craig, loc. cit. *fig. 81.*

Simple or rarely somewhat cespitose at base, body globose, 15–20 cm. tall; tubercles in 13 and 21 spirals, bluish green, 10–13 mm. long, 6–9 mm. broad at base, quadrangular clear to tip; juice milky; axils scantily woolly, without bristles; radial spines 10–18, acicular, straight, 6–9 mm. long, white to horn-colored with brownish to black tips, slightly ascending; central spines usually 2 but sometimes to 6, subulate, straight or sometimes tortuous, 10–25 mm. long, light purplish to black, upper erect, others divergent; flowers 1.5–2 cm. long; outer perianth segments ovate-lanceolate, acute to obtuse, reddish to purplish brown with pale green to pinkish green margins, margins ciliate to serrate; inner perianth-segments linear-lanceolate, pink to brownish with white to cream margins; filaments pink above, cream below; stigma lobes olive-green; fruit red, clavate, about 2.5 cm. long; seeds curved-pyriform, 1–1.2 mm. long, dull light brown.

Near coast, Lower Sonoran Zone, southwestern Sonora, mainly in the vicinity of Guaymas. April.

**Mammillaria macdougalii** Rose, in Bailey, Cyclop. Hort. **4:** 1982, 1916, differs from *M. johnstonii* in having tubercles that are nearly terete at tips, their axils with long white wool, radial spines to 20 mm. long, flowers 3–3.5 cm. long, and filaments pale green.

Hillsides, Upper Sonoran Zone, southeastern Arizona in Pima, Santa Cruz, and Cochise Counties, but it may occur as a straggler along the upper margins of the Sonoran Desert in that region and in adjacent northern Sonora. March–April.

**9. Mammillaria arida** Rose, Monats. Kakteenk. **23:** 181, 1913

*Neomammillaria arida* Britt. & Rose, Cactaceae **4:** 73, *pl. 7, fig. 3,* 1923.

Simple globose plant deeply embedded in soil, 5–6 cm. broad across somewhat flattened top; tubercles in 8 and 13 spirals, dull gray-green, nearly terete in cross section near apex, quadrangular at base; juice milky; axils scantily woolly, without bristles; radial spines about 15, acicular, straight, 6–10 mm. long, pale yellowish with darker tips; central spines 4–7, stout-acicular, straight, 12–16 mm. long, dark brown, porrect; flowers about 1 cm. long, twice as wide; outer perianth segments narrowly lanceolate, reddish brown to purplish with bright green, entire margins; inner perianth segments cream to light yellow tinged with green and midstripe reddish beneath; filaments white; stigma lobes greenish to yellowish; fruit clavate, about 1.5 cm. long; seeds brown, about 1 mm. long.

Arid places, Lower Sonoran Zone, Pichilinque Island and vicinity of La Paz, Baja California. June–July.

**10. Mammillaria lewisiana** H. E. Gates, Cactus & Succ. Jour. **27**: 185, *fig. 169*, 1955

Plants simple, globose to depressed-globose above ground, tapering gradually to a fleshy tapering taproot bearing numerous fibrous roots at its lower end, aerial part of stem 8–12 cm. in diameter, 5–8 cm. tall; tubercles in 13 and 34 spirals, diamond-shaped in cross section, 10–14 mm. wide at base, 8–10 mm. long, becoming flaccid in age, bluish green, with thick milky sap; axils of tubercles bearing a small tuft of short white wool, otherwise naked; areoles circular, 1–2 mm. in diameter, bearing very short, white to tan wool in youth; radial spines 10–13, very slenderly acicular, 6–10 mm. long, light gray with pale brown tips; central spines 1–3, uppermost (or solitary) one curved upward, sometimes strongly so, purple-black, 12–20 mm. long, slender, not hooked; lower ones acicular, 6–8 mm. long, straight to somewhat flexuous, pointing downward, purple-black; flowers campanulate, about 2 cm. long, 1 cm. wide, tube naked, yellow tinged with green; outer perianth segments lanceolate, 2–3 mm. wide, 4–10 mm. long, greenish yellow with a faint brown midstripe on outer surface, acute and entire at apex; inner perianth segments 12–15 mm. long, otherwise similar to outer ones; filaments white; fruits broadly clavate, 4–5 mm. in diameter, 10–15 mm. long, white to pink; seeds about 0.8 mm. long, one-half as wide, reddish brown, faintly rugulose.

Along washes and gentle desert slopes, Lower Sonoran Zone, Vizcaíno Desert, central Baja California. April?

**11. Mammillaria tesopacensis** Craig, Mam. Handbook 104, *fig. 86*, 1945

Simple, globose to cylindrical plant 8–13 cm. in diameter, 12–20 cm. tall, from fibrous roots; tubercles in 13 to 21 spirals, blue-green to gray-green, conic-pyramidal, 7–8 mm. wide at base, 10–12 mm. long, obtusely angled at base, bearing scanty tufts of wool in axils when young, later naked; juice milky; areoles 1.5–2 mm. in diameter, scantily woolly when young, soon naked; radial spines 10–15, slenderly acicular, straight, smooth, stiff, 4–7 mm. long, reddish tan near base and blackish near tips in youth, eventually becoming ash-brown, spreading to appressed, interlocked only at apex of plant; central spine 1 or rarely 2, straight, stout-acicular, smooth, 10–12 mm. long, colored like radicals, not hooked; flowers broadly funnelform, about 2 cm. wide and long; outer perianth segments broadly lanceolate, dull brownish red along midvein, margins cream with tinge of green, ciliate, obtuse at apex; inner perianth segments linear-lanceolate, 2–3 mm. wide, acute at apex, cream with pinkish midvein, margins entire to serrate or ciliate; stamens with pink filaments; anthers yellow; fruit short-clavate to obovoid, 8–10 mm. in diameter, 15–20 mm. long, bright red; seeds about 1 mm. long, brownish, faintly rugose.

Hillsides, Lower Sonoran Zone, southern Sonora. May.

**12. Mammillaria capensis** (H. E. Gates) Craig, Mam. Handbook 156, *fig. 137*, 1945

*Neomammillaria capensis* H. E. Gates, Cactus & Succ. Jour. **4**: 372, *fig.*, 1933.
*Ebnerella capensis* F. Buxb., Oest. Bot. Zeitschr. **98**: 89, 1951.
*Chilita capensis* F. Buxb., Sukkulentenk. **5**: 17, 1954.
*Mammillaria capensis* var. *pallida* Remski, Bot. Gaz. **116**: 165, 1954 (incorrectly attributed to H. E. Gates). Nomen nudum.

More or less cespitose plant; shoots cylindric, 3–5 cm. in diameter, to 25 cm. long; tubercles ovoid-cylindric, 4–5 mm. long, about as wide at base, olive-green, closely crowded in 5 and 8 spirals; juice watery; axils naked or with 1–3 fine short bristles; radial spines about 13, acicular, stiff, 8–15 mm. long, whitish at base, then reddish brown and finally black at tips; central spine 1, acicular, strongly hooked at apex, 15–20 mm. long, colored like radials; flowers narrowly funnelform, about 2 cm. long; outer perianth segments oblong, obtuse, greenish pink with pale pink margins; inner perianth segments pale pink with reddish or brownish midstripe, oblong, obtuse; stigma lobes greenish yellow; fruit clavate, about 2 cm. long, orange to scarlet; seeds globose, black and glossy, 1.3 mm. long, basal hilum extended and set off from rest of seed by a constriction.

Sandy or gravelly soil, Lower Sonoran Zone, southern Baja California, from vicinity of La Paz south. April–May.

**13. Mammillaria angelensis** Craig, Mam. Handbook 165, *fig. 146,* 1945

*Ebnerella angelensis* F. Buxb., Oest. Bot. Zeitschr. **98:** 89, 1951.
*Chilita angelensis* F. Buxb., Sukkulentenk. **5:** 18, 1954.

Simple columnar plant 10–15 cm. tall, 5–6 cm. in diameter; tubercles conical, terete in cross section, slightly keeled ventrally, 8–9 mm. long, 6–7 mm. wide at base, olive-green, in 8 and 13 spirals; juice watery; axils densely white-woolly and with 15–20 tortuous white bristles about 1 cm. long; radial spines about 16, acicular, straight, white or some tan at tip, 5–10 mm. long, nearly horizontal except in youth when ascending; central spines 4 or sometimes 3, 8–14 mm. long, lower one hooked and longer than others, all dull purplish brown or lighter at base, hooked one porrect, others spreading and not easily distinguished from radials; flowers about 2 cm. long, 3 cm. wide; outer perianth segments linear, obtuse and short-ciliate at apex, cream-white with a pink midstripe; filaments white; stigma lobes yellowish olive-green; fruit clavate; seeds black, minutely pitted.

Rocky hillsides and gravelly slopes, Lower Sonoran Zone, eastern coast of Baja California in the vicinity of Bahía de los Angeles and on Isla Angel de la Guarda. March–April.

**14. Mammillaria fraileana** (Britt. & Rose) Boed., Mam. Vergl. Schlüss. 30, 1933

*Neomammillaria fraileana* Britt. & Rose, Cactaceae **4:** 157, 1923.
*Chilita mainae* Orcutt, Cactography 2, 1926.
*Ebnerella fraileana* F. Buxb., Oest. Bot. Zeitschr. **98:** 89, 1951.

Simple or somewhat cespitose plant with cylindrical stems 3–4 cm. in diameter, to 15 cm. long; tubercles conic, 5–6 mm. long, 3–4 mm. broad at base, in 5 and 8 spirals; juice watery; axils bearing 1–8 bristles 1–3 mm. long or sometimes naked, not woolly; radial spines about 12, acicular, straight, reddish brown on young tubercles, becoming sordidly chalky in age, 8–10 mm. long; central spines 3–4, acicular, lowest hooked, 10–18 mm. long, others straight, 8–12 mm. long, all reddish brown in young areoles, later becoming more or less chalky; flowers about 20 mm. long and wide; outer perianth segments lanceolate, purplish brown with lighter margins; inner perianth segments pinkish, acuminate; filaments pink; fruit pinkish, clavate, about 1.5 cm. long; seeds globose, constricted between body and basal hilum.

Rocky hillsides, Lower Sonoran Zone, Pichilinque, Cerralvo, and Santa Catalina Islands off the Gulf coast of southern Baja California and in the vicinity of La Paz on the peninsula. April–May.

**15. Mammillaria verhaertiana** Boed., Monats. Kakteenk. **22:** 152, 1912

*Neomammillaria verhaertiana* Britt. & Rose, Cactaceae **4:** 164, 1923.
*Chilita verhaertiana* Orcutt, Cactography 2, 1926.
*Ebnerella verhaertiana* F. Buxb., Oest. Bot. Zeitschr. **98:** 90, 1951.

Plant simple, short-cylindric, 5–10 cm. tall, 3–4 cm. in diameter; tubercles conical, weakly keeled ventrally, 7–10 mm. long, 6–8 mm. wide, in 8 and 13 spirals, bright green, obscured by close armament of interlacing spines; axils white-woolly and bearing 15–20 slender white bristles 3–6 mm. long; radial spines 15–20, acicular, straight, dark yellowish brown on young growth, becoming nearly white, upper ones 6–8 mm. long, lower ones 10–16 mm. long; central spines 4–6, upper ones straight, acicular, yellowish with brownish tips, lower one hooked, 8–12 mm. long; flowers funnelform, about 2 cm. long and broad; outer perianth segments oblong to lanceolate, acute to obtuse, olive-green with yellowish white, serrate margins; inner perianth segments with a pale pink midstripe, margins white, entire; filaments rose to whitish; stigma lobes orange-yellow; fruit orange-red; seeds black, minutely pitted.

Gravelly slopes and rocky hillsides, Lower Sonoran Zone, Isla Angel de la Guarda and vicinity of Bahía de los Angeles, Baja California. March–April.

**16. Mammillaria dioica** K. Brandegee, Erythea **5:** 115, 1897

*Mammillaria fordii* Orcutt, W. Amer. Sci. **10:** 83, 1900.
*Neomammillaria dioica* Britt. & Rose, Cactaceae **4:** 158, 1923.
*Chilita fordii* Orcutt, Cactography 2, 1926.
*Mammillaria incerta* Parish, in Jepson, Fl. Calif. **2:** 549, 1936.
*Ebnerella dioica* F. Buxb., Oest. Bot. Zeitschr. **98:** 89, 1951.
*Chilita dioica* F. Buxb., Sukkulentenk. **5:** 17, 1954.

Solitary or cespitose plant with cylindric body 4–25 (sometimes to over 30) cm. tall; tubercles short-cylindrical, bluish green, in 8 and 13 spirals, somewhat quadrangular and 3–4 mm. wide at base, 4–8 mm. long; juice watery; axils densely white-woolly, with several (5–15) bristles nearly as long as tubercles; radial spines 10–20, white, or sometimes tips rose-colored, purplish, brown or black, 5–8 mm. long, spreading; central spines 1–4, brown, lower one stouter and hooked; flowers yellowish white with a pink or purplish midrib, 10–20 mm. long, incompletely dioecious; outer perianth segments reddish to purple with very narrow, yellow margins, lanceolate, acute to acuminate, entire; inner perianth segments usually notched at apex; filaments white; stigma lobes light green to brownish green; fruit scarlet, clavate, 10–25 mm. long; seeds black, shining, minutely pitted, obpyriform, hilum slightly lateral.

Grassy hillsides and gravelly slopes, Lower and Upper Sonoran Zones, San Diego County, California, south to the Cape Region, Baja California, and east to the desert. April–July.

**17. Mammillaria armillata** K. Brandegee, Zoe **5:** 7, 1900

*Neomammillaria armillata* Britt. & Rose, Cactaceae **4:** 157, *pl. 14, fig. 4; pl. 15, figs. 2–4,* 1923.
*Chilita armillata* Orcutt, Cactography 2, 1926.
*Neomammillaria lapacena* H. E. Gates, Desert Pl. Life **3:** 117, *fig.,* 1932.
*Ebnerella armillata* F. Buxb., Oest. Bot. Zeitschr. **98:** 89, 1951.

Simple or infrequently cespitose plant 4–5 cm. in diameter, to 30 cm. tall; tubercles conical, 3–4 mm. wide, quadrangular at base, 4–5 mm. long, bluish green, arranged in 5 and 8 spirals; juice watery; axils scantily woolly and with 1–3 slender bristles; radial spines 9–15, slenderly acicular, straight, stiff, white or yellowish toward base, dark brown at tip, 7–12 mm. long; central spines 1–4, stout-acicular, 10–20 mm. long, yellowish gray to dark brown, upper ones ascending and straight, lower one or two longer than others (to 22 mm.), hooked, porrect; spines form alternating encircling bands of light and dark color; flowers 10–20 mm. long, about 20 mm. wide; outer perianth segments lanceolate, acute, greenish, margins serrate; inner perianth segments white to pink; filaments pink; stigma lobes rose; fruit clavate, red, 15–30 mm. long; seeds obpyriform, dull black, constricted slightly just above hilum, minutely pitted.

Arid slopes, hillsides, and mesas, Lower Sonoran and arid Tropical Zones, southern Baja California. April–May.

**18. Mammillaria swinglei** (Britt. & Rose) Boed., Mam. Vergl. Schlüss. 33, 1933

*Neomammillaria swinglei* Britt. & Rose, Cactaceae **4:** 158, 1923.
*Chilita swinglei* Orcutt, Cactography 2, 1926 (as *swingleri*).
*Ebnerella swinglei* F. Buxb., Oest. Bot. Zeitschr. **98:** 90, 1951.

Simple or sparingly cespitose, cylindrical plant 3–5 cm. in diameter, 10–25 cm. tall, with dark green, conic-cylindrical tubercles 8–10 mm. long and 5–7 mm. broad at base, arranged in 8 and 13 spirals; axils naked or with a few slender bristles; juice watery; radial spines 11–18, stout-acicular, straight, stiff, dull gray to whitish with dark tips, 7–14 mm. long, mostly horizontally interlaced and nearly hiding tubercles; central spines 1–4, stout-acicular to subulate, 8–15 mm. long, lower sometimes straight, usually hooked, longer than upper ones, dark purplish brown to black near tips; flowers broadly campanulate, to 3 cm. wide; outer perianth segments lanceolate, purplish brown with white to greenish, serrate margins; margins of inner perianth segments entire; filaments pink; stigma lobes yellowish green; fruit dark red, clavate, 1.5–2 cm. long; seeds globose, black, and glossy, about 1 mm. in diameter, minutely pitted.

Arid hills and coastal slopes, Lower Sonoran Zone, vicinity of Guaymas, Sonora. Johnston reported it very common among volcanic rocks on the islands in Guaymas Bay. April.

**19. Mammillaria mainae** K. Brandegee, Zoe **5:** 31, 1900

*Neomammillaria mainae* Britt. & Rose, Cactaceae **4:** 154, *fig. 172,* 1923.
*Chilita mainae* Orcutt, Cactography 2, 1926.
*Ebnerella mainae* F. Buxb., Oest. Bot. Zeitschr. **98:** 89, 1951.

Simple or rarely cespitose, hemispherical to conic-globose plant 5–12 cm. tall, 4–7 cm. in diameter; tubercles cylindric-conic and often slightly curved, 8–10 mm. wide at base, 10–15 mm. long, bluish green to gray-green, terete in cross section, in 8 and 13 spirals; axils naked; radial spines 10–15, acicular, straight, 6–10 mm. long, scarcely at all interlaced except on youngest growth, yellowish with brown tips, often becoming chalky; central spines 1–3, yellowish with dark tips, one hooked, others if present straight, 15–20 mm. long; flowers 10–20 mm. long, to 25 mm. wide, rosy; outer perianth segments linear-lanceolate, obtuse, brownish green to reddish green with pinkish cream, ciliate margins; inner perianth segments paler; filaments purplish to rose-red; stigma lobes rose-red to purplish; fruit obovoid, red, 6–8 mm. long, 3–5 mm. wide; seeds obpyriform, dull black, about 1 mm. long, minutely pitted.

Desert plains and hillsides, usually in gravelly soil, Lower Sonoran Zone (sometimes in lower Upper Sonoran Zone), southwestern Arizona south to Sinaloa. June–July.

**20. Mammillaria fasciculata** Engelm., in Emory, Notes Mil. Rec. 157, *fig. 8,* 1848

*Cactus fasciculatus* Kuntze, Rev. Gen. Pl. **1:** 259, 1891.
*Mammillaria thornberi* Orcutt, W. Amer. Sci. **12:** 161, 1902.
*Neomammillaria fasciculata* Britt. & Rose, Cactaceae **4:** 162, *fig. 180,* 1923.
*Chilita thornberi* Orcutt, Cactography 2, 1926.
*Ebnerella fasciculata* F. Buxb., Oest. Bot. Zeitschr. **98:** 89, 1951.

Closely cespitose, with several to many cylindrical shoots in clumps to 3 dm. wide, individual heads 5–8 cm. wide, to 20 cm. long; tubercles rather flabby, purplish green, conic to cylindrical, about 5 mm. wide and long, in 5 and 8 spirals; axils naked; juice watery; radial spines 13–20, slenderly acicular, straight, semiflexuous, white with brown to blackish tips, 5–8 mm. long, practically obscuring tubercles; central spine usually 1, sometimes 2 or 3, stiff, hooked, 12–18 mm. long, brownish or blackish, porrect; flowers broadly funnelform, 2 cm. wide, 3 cm. long; outer perianth segments broadly lanceolate, greenish at base, brownish purple above, acute to obtuse, margins serrate; inner perianth segments carmine to purplish with whitish margins; filaments rose below, purplish above; stigma lobes dark red; fruit clavate, about 8–10 mm. long, scarlet; seeds globose, glossy-black, minutely pitted.

Mostly in shade of shrubs, flood plains or bottomland flats, Lower Sonoran Zone, Maricopa County, Arizona, south into Sonora. May–June.

**Mammillaria yaquensis** Craig, Mam. Handbook 320, 1945, which can scarcely, if at all, be distinguished from *M. fasciculata* on spine characters, probably is a mere form of the latter. *Mammillaria yaquensis* was obtained from the Yaqui Valley in southern Sonora.

**21. Mammillaria wilcoxii** Toumey, in K. Schumann, Gesamtb. Kakteen 545, 1898

*Neomammillaria wilcoxii* Britt. & Rose, Cactaceae **4:** 153, *pl. 13, fig. 1,* 1923.
*Chilita wilcoxi* Orcutt, Cactography 2, 1926.
*Mammillaria wrightii* var. *wilcoxii* W. T. Marshall, Des. Bot. Gard. Ariz. **1:** 100, 1950.
*Ebnerella wilcoxii* F. Buxb., Oest. Bot. Zeitschr. **98:** 90, 1951.

Simple, or rarely branching at base, body depressed-globose to short-cylindrical, to 1 dm. tall; tubercles widely separated, conical-cylindric, terete in cross section, about 5 mm. broad at base, 1–2 cm. long, dark green, in 5 and 8 spirals or in 8 and 13 spirals; axils naked; juice watery; radial spines 14–30 (rarely over 22), bristlelike to slenderly acicular, 10–15 mm. long, spreading tangential to outline of plant and nearly hiding tubercles, white with brownish tips; central spines 1–3, dark amber to reddish brown with whitish bases, one or more hooked, tapering toward tip and sometimes slightly flattened, porrect; flowers about 3.5 cm. long, 4 cm. wide; outer perianth segments linear-lanceolate, light cream at base, brownish green with pale green to cream, ciliate margins; inner perianth segments cream with pink to purplish midstripes, entire; filaments white; stigma lobes light green; fruit pink to purplish, obovoid, about 2.5 cm. long; seeds obpyriform, 1.5–1.7 mm. long, about 1 mm. broad, glossy-black, minutely pitted.

Desert grassland, upper margin of Lower Sonoran and low Upper Sonoran Zones, Pima and Santa Cruz Counties, Arizona, to the Huachuca Mountains and south into Sonora. Mostly above the limits of the Sonoran Desert. May–June.

**22. Mammillaria blossfeldiana** Boed., Monats. Deutsch. Kakt. Ges. **3:** 209, *fig.,* 1931, var. **blossfeldiana**

*Neomammillaria blossfeldiana* H. E. Gates, Cactus & Succ. Jour. **4:** 371, *fig.,* 1933.
*Ebnerella blossfeldiana* F. Buxb., Oest. Bot. Zeitschr. **98:** 89, 1951.
*Chilita blossfeldiana* F. Buxb., Sukkulentenk. **5:** 17, 1954.

Simple, globose to cylindrical plant 4–5 cm. in diameter, 10–15 cm. tall, with short-conic, grayish or purplish green tubercles 5–6 mm. broad at base and 5–8 mm. long, arranged in 5 and 8 spirals; axils scantily woolly; juice watery; radial spines 15–20, slenderly acicular, stiff, 5–7 mm. long, grayish white toward base, tips dark brown to black, spreading tangentially; central spines usually 4, lowest one 9–10 mm. long, hooked, three upper ones acicular, straight, 6–8 mm. long, all whitish at base, purple to black at tip; flowers about 2 cm. long, 3–3.5 cm. broad; outer perianth segments linear-lanceolate, brownish or reddish, with greenish white, entire margins and brownish green base; inner perianth segments carmine with darker midstripe and white margins; filaments whitish to pale pink; stigma lobes olive-green, or turning yellowish in age; fruit clavate, about 2 cm. long, orange-red; seeds obpyriform, slightly less than 1 mm. long, faintly and minutely pitted, glossy-black.

Arid slopes and alluvial plains, Lower Sonoran Zone, western coast of Baja California in the vicinity of Santa Rosalia Bay. April.

**22*a*. Mammillaria blossfeldiana** var. **shurliana** (H. E. Gates) Wiggins, comb. nov.

*Mammillaria blossfeldiana* var. *shurliana* H. E. Gates, Cactus & Succ. Jour. **13:** 78, 1941. (Nomen nudum.)

*Mammillaria shurliana* H. E. Gates, Cactus & Succ. Jour. Gr. Brit. **18:** 30, 1956.

Differing from var. *blossfeldiana* only in usually having 3 instead of 4 central spines, reddish purple instead of white or pinkish filaments, smoother seeds, and more fibrous roots.

Desert alluvium and mesas, Lower Sonoran Zone, Mesquital Rancho, vicinity of Calmallí, and on Isla Guadalupe, Baja California.

**23. Mammillaria guirocobensis** Craig, Mam. Handbook 220, *fig. 201,* 1945

*Ebnerella guirocobensis* F. Buxb., Oest. Bot. Zeitschr. **98:** 89, 1951.

Simple or sparingly cespitose, cylindrical plant 3–5 cm. in diameter, 5–11 cm. tall; tubercles conic-cylindrical, 6–7 mm. long, nearly as wide, quadrangular at base, olive-gray-green, weakly keeled ventrally, rather crowded in 8 and 13 spirals; axils naked or rarely with a few bristles; juice watery; radial spines 18–20, acicular, straight, 5–8 mm. long, yellowish to chalky; central spines 1–3, upper two straight, about 6 mm. long, lower one either straight, 2–3 mm. long, or hooked and 8–10 mm. long, reddish brown; flowers about 2 cm. long and wide; outer perianth segments linear-lanceolate, light greenish or reddish brown with pale greenish or greenish white, ciliate margins; inner perianth segments whitish, with reddish brown midline and pale green border, acute, margins entire; filaments purplish pink; stigma lobes olive-green; fruit clavate, scarlet, about 12 mm. long, half as wide; seeds obovoid, glossy-black, about 1 mm. long, minutely pitted.

Mostly foothills in gravelly soil, Sonoran Zones, southern Sonora and adjacent Chihuahua and Sinaloa. Entering the Sonoran Desert in the vicinity of Quiriego, Sonora. March–April.

**24. Mammillaria goodridgei** Scheer, in Salm-Dyck, Cact. Hort. Dyck **1849:** 91, 1850 (as *M. goodrichii*), var. **goodridgei**

*Mammillaria goodridgii* Scheer, in Seem., Bot. Herald 286, 1856.

*Cactus goodridgii* Kuntze, Rev. Gen. Pl. **1:** 260, 1891.

*Neomammillaria goodridgei* Britt. & Rose, Cactaceae **4:** 158, 1923 (not as to *pl. 15, fig. 5*).

*Chilita goodridgei* Orcutt, Cactography 2, 1926.

*Ebnerella goodridgei* F. Buxb., Oest. Bot. Zeitschr. **98:** 89, 1951.

Simple or cespitose, globose to cylindrical plant 3–4 cm. in diameter, 6–10 cm. tall; tubercles conic-cylindric, 3–5 mm. long, about as wide, obscurely quadrangular below, crowded in 8 and 13 spirals, hidden by spines; axils scantily woolly or naked; radial spines 11–15, rigid, acicular, 4–7 mm. long, white with darker tips; central spines 3–4, upper two or three straight, acicular, white, 6–8 mm. long, lower one strongly hooked, 8–10 mm. long, porrect; flowers broadly funnelform, about 1.5 cm. long, 2.5 cm. broad; outer perianth segments oblong, obtuse, entire, deep pink to brownish red with greenish midstripe and light pink margins; inner perianth segments pink to rose-red with white to cream margins, about 20 mm. long; filaments cream; stigma lobes greenish; fruit clavate, 15–20 mm. long, scarlet; seeds globose, black, minutely pitted.

Arid soil and rocky places, Lower Sonoran Zone, Cedros Island off Baja California. March–April.

**24a. Mammillaria goodridgei** var. **rectispina** Dawson, in Lindsay & Dawson, Cactus & Succ. Jour. **24:** 80, *fig. 46,* 1952

Differs from var. *goodridgei* in having straight instead of hooked central spines, and deep red instead of olive-green stigma lobes.

With var. *goodridgei* at elevations of 600 to 800 feet, Cedros Island. March–April.

**25. Mammillaria sheldonii** (Britt. & Rose) Boed., Mam. Vergl. Schlüss. 30, 1933

*Neomammillaria sheldonii* Britt. & Rose, Cactaceae **4:** 156, 1923 (not as to *fig. 175*).
*Chilita sheldoni* Orcutt, Cactography 2, 1926.
*Ebnerella sheldonii* F. Buxb., Oest. Bot. Zeitschr. **98:** 90, 1951.

Simple or with several heads from base of the plant, this cylindrical, 4–6 cm. in diameter, 5–25 cm. tall; tubercles conical, 8–9 mm. long, 5–6 mm. wide, quadrangular at base, dull green, in 8 and 13 spirals; juice watery; axils naked; radial spines 10–15, acicular, straight, 6–9 mm. long, whitish at base, reddish brown above with purplish tips; central spines 1–3, upper ones (if porrect) straight, 8–9 mm. long, hooked one to 12 mm. long, dark reddish brown to purplish; flowers broadly funnelform, about 2 cm. high, 3 cm. broad; outer perianth segments linear-lanceolate, olive-green at base, brownish green with pinkish green margins above, acute, ciliate; inner perianth segments pink to purplish with white margins, to 7 mm. wide; filaments light purple; stigma lobes light olive-green; fruit clavate, pale scarlet, 25–30 mm. long; seeds globose, slightly constricted above base, about 1 mm. long, glossy-black and slightly pitted.

Gravelly slopes, hillsides, and plains, Lower Sonoran Zone, from vicinity of Puerto Libertad to Hermosillo, Guaymas, and the Yaqui Valley, Sonora. March–April.

**26. Mammillaria hutchisoniana** (H. E. Gates) Boed., in Backeb. & Knuth, Kaktus-ABC 387, 1935

*Neomammillaria hutchisoniana* H. E. Gates, Cactus & Succ. Jour. **6:** 4, *fig.,* 1934.
*Neomammillaria bullardiana* H. E. Gates, loc. cit.
*Mammillaria bullardiana* Boed., in Backeb. & Knuth, Kaktus-ABC 387, 1935.
*Ebnerella hutchisoniana* F. Buxb., Oest. Bot. Zeitschr. **98:** 89, 1951 (as *hutchinsoniana*).
*Ebnerella bullardiana* F. Buxb., loc. cit.
*Chilita hutchisoniana* F. Buxb., Sukkulentenk. **5:** 17, 1954 (as *hutchinsoniana*).

Usually cespitose plant with globose to cylindrical stems 4–6 cm. in diameter, 10–15 cm. tall; tubercles 4–5 mm. tall and broad, in 8 and 13 spirals; axils naked; juice watery; radial spines 15–20, acicular, straight, 5–8 mm. long, purplish to black in youth, soon whitish; central spines usually 3, two upper ones straight, stout-acicular, about 6–8 mm. long, lower one slightly longer, strongly hooked, all white with purplish tips; flowers broadly campanulate, about 2 cm. long, 3 cm. wide, segments often somewhat recurved; outer perianth segments broadly lanceolate, obtuse, purplish or pinkish purple with wide, pale greenish white, shallowly serrate margins; inner perianth segments pinkish cream with a darker pink to purplish midstripe, about 7 mm. broad, entire; filaments white, tannish above; stigma lobes golden brown; fruit ovoid, scarlet, about 2 cm. long, nearly half as wide; seeds globose, slightly less than 1 mm. long, black, finely pitted.

Sandy or gravelly deserts, Lower Sonoran Zone, Isla Natividad and Sierra San Borja, central Baja California, south to vicinity of La Paz. April.

**27. Mammillaria tetrancistra** Engelm., Amer. Jour. Sci. II, **14:** 337, 1852

*Mammillaria phellosperma* Engelm., Proc. Amer. Acad. **3:** 262, 1856.
*Cactus phellospermus* Kuntze, Rev. Gen. Pl. **1:** 261, 1891.
*Cactus tetrancistrus* J. M. Coulter, Contr. U.S. Nat. Herb. **3:** 104, 1894.
*Phellosperma tetrancistra* Britt. & Rose, Cactaceae **4:** 60, *figs. 57–58,* 1923.
*Neomammillaria tetrancistra* Fosberg, Bull. S. Calif. Acad. **30:** 57, 1931 (as *tetrancista*).

Stems oblong to cylindrical, 5–35 cm. high, densely spiny; root carrotlike or branched; tubercles commonly elongated, with clear juice; radial spines 30–60, acicular, white or slightly darkened toward tip, spreading, forming a dense white covering and obscuring tubercles; central spines 4 (rarely only 1), from one to all of them hooked, 1–2.5 cm. long; flowers 2.5–3.5 cm. long, rose, orchid or light purple; tube naked, slender, greenish at base; scales at apex of tube and outer perianth segments ciliate-margined; style and stigma lobes cream-colored; fruit 1.5–3.5 cm. long, depressed-umbilicate at apex; seeds 1.8–2.2 mm. long, dull black, body somewhat rugulose, base corky.

Gravelly or stony slopes and mesas, Lower Sonoran Zone, southern Nevada and Utah to northern Baja California and western Arizona. April.

**28. Mammillaria insularis** H. E. Gates, Cactus & Succ. Jour. **10:** 25, *fig.,* 1938

*Ebnerella insularis* F. Buxb., Oest. Bot. Zeitschr. **98:** 89, 1951.
*Chilita insularis* F. Buxb., Sukkulentenk. **5:** 17, 1954.

Cespitose plant with several depressed-globose heads 5–6 cm. in diameter in a compact clump; tubercles bluish green, bluntly conical, 6–7 mm. long, and equally wide at angular base, in 5 and 8 spirals; axils naked or scantily woolly; juice watery; radial spines 20–30, acicular, 4–5 mm. long, white, closely interlaced; central spine solitary, slightly swollen at base, black, brownish or yellowish, about 10 mm. long, strongly hooked; flowers broadly funnelform, about 15 mm. long and 20–25 mm. wide; outer perianth segments lanceolate, acute, light green or paler along entire margins; inner perianth segments light pink with white margins; stigma lobes greenish yellow; fruit clavate, orange-red, 9–10 mm. long, 2–3 mm. in diameter; seeds globose, dull black, with a flattened, reddish brown hilum, lightly pitted, about 0.8 mm. in diameter.

In rock crevices and along gypsum deposits, Lower Sonoran Zone, Smith Island, Bahía de los Angeles, and Isla San Marcos near Santa Rosalia, Gulf of California. Aug.

**29. Mammillaria boolii** Lindsay, Cactus & Succ. Jour. **25:** 48, 1953

*Chilita boolii* F. Buxb., Sukkulentenk. **5:** 17, 1954.

Solitary or few-branched, cespitose plant with branches arising from axils of tubercles near base of main stem, solitary plants becoming truncate and nearly level with surface of soil when dried out, when supplied with moisture stems spherical or nearly so, 2.5–3 cm. in diameter, 3–4 cm. tall; roots fibrous; juice watery; tubercles in 5 and 8 spirals, terete or lower ones somewhat flattened, 5–8 mm. in diameter at base, 8–10 mm. long, bluntly conical, glaucous-green, their axils sparsely lanate in youth, naked in age; areoles circular, 2–3 mm. in diameter, lanate among bases of spines in youth, later naked; radial spines 19–23, nearly equal, 12–15 mm. long, acicular, spreading almost parallel to surface of plant, pale yellowish to white; central spine solitary, 15–20 mm. long, porrect, rather stout and strongly hooked at apex, yellow to stramineous at base, reddish to brownish toward tip; flowers pink to lavender-pink, about 2.5 cm. long and wide when fully open; outer perianth segments oblanceolate, acute to apiculate, entire, midvein olive-green on outer surface; inner perianth segments similar in shape and size, but pink to lavender-pink throughout; fruit slenderly clavate, 4–5 mm. in diameter, 2–3 cm. long, orange; seeds dull black, finely pitted, constricted just above hilum.

In crevices in volcanic rocks, Lower Sonoran Zone, near Guaymas, Sonora. April.

**30. Mammillaria phitauiana** (E. M. Baxter) Werd., in Backeb., Neue Kakteen 96, 1931

*Neomammillaria phitauiana* E. M. Baxter, Cactus & Succ. Jour. **2:** 472, *2 figs.*, 1931.
*Ebnerella phitauiana* F. Buxb., Oest. Bot. Zeitschr. **98:** 90, 1951.
*Chilita phitauiana* F. Buxb., Sukkulentenk. **5:** 18, 1954.

Plant simple or often 2–7-branched near base; stem or stems short-cylindrical, 4–6 cm. in diameter, 12–18 cm. tall, from slender fibrous roots; tubercles in 8 and 13 spiral rows, conic-cylindrical, 6–7 mm. in diameter at base, 5–8 mm. long, somewhat flattened dorsoventrally and slightly keeled; sap watery; axils bearing as many as 20 persistent white bristles equaling to slightly exceeding tubercles; radial spines 18–24, slenderly acicular, 4–12 mm. long, banded with white and pale brown toward base, darker brown bands toward tips, parallel to surface of plant or slightly spreading, interlocked; central spines 4, upper three straight, lower usually hooked but sometimes straight, all smooth and slightly bulbous at base, 8–10 mm. long, colored much as radials in white to dark bands, porrect; flowers narrowly funnelform to nearly tubular, 12–15 mm. long; outer perianth segments pinkish with darker midvein, lanceolate, acute at apex, serrulate; inner perianth segments narrow, with white, nearly entire margins; fruit subglobose to clavate, 10–12 mm. long, red; seeds black, pyriform, about 0.5 mm. long.

Rocky hillsides, Lower Sonoran Zone, western flanks of the Sierra de la Laguna east of Todos Santos, southern Baja California. April.

**31. Mammillaria microcarpa** Engelm., in Emory, Notes Mil. Rec. 157, *fig. 3,* 1848

*Mammillaria grahami* Engelm., Proc. Amer. Acad. **3:** 262, 1856.
*Cactus grahami* Kuntze, Rev. Gen. Pl. **1:** 260, 1891.
*Mammillaria grahamii* var. *arizonica* Quehl, Monats. Kakteenk. **6:** 44, 1896.
*Coryphantha grahami* Rydb., Fl. Rocky Mts. 581, 1917.
*Neomammillaria microcarpa* Britt. & Rose, Cactaceae **4:** 155, *figs. 170, 173–174,* 1923.
*Neomammillaria milleri* Britt. & Rose, op. cit. 156, *fig. 184a.*
*Chilita grahami* Orcutt, Cactography 2, 1926.
*Chilita milleri* Orcutt, loc. cit.
*Mammillaria milleri* Boed., Mam. Vergl. Schlüss. 30, 1933.
*Ebnerella microcarpa* F. Buxb., Oest. Bot. Zeitschr. **98:** 89, 1951.
*Chilita microcarpa* F. Buxb., Sukkulentenk. **5:** 12, 1954.

Cespitose or simple plant 4–8 cm. high, with small tubercles becoming corky in age; axils of tubercles naked; radial spines 15–30, spreading, white or with dark tips,

6–10 mm. long, usually nearly hiding flesh of plant; central spines 1–3, dark brown to purplish black, if more than one the lower stouter and hooked, 12–18 mm. long; flowers near top of plant, 2–2.5 cm. long, broadly funnelform; outer perianth segments ovate, obtuse; inner perianth segments obovate, acuminate, with purplish midribs and whitish margins; style purple; stigma lobes green; fruit clavate, scarlet, 2–2.5 cm. long; seeds black, shining, globose, about 1 mm. in diameter.

Sandy plains, gravelly slopes, and rocky ridges, Lower Sonoran Zone, Santa Rosa Mountains, Colorado Desert, to southern Utah, southwestern Texas, and the desert areas of central Baja California, Sonora, and Chihuahua, Mexico. Feb.–July.

**32. Mammillaria oliviae** Orcutt, W. Amer. Sci. **12:** 163, 1902

*Neomammillaria oliviae* Britt. & Rose, Cactaceae **4:** 135, *fig. 147,* 1923.
*Chilita oliviae* Orcutt, Cactography 2, 1926.
*Ebnerella oliviae* F. Buxb., Oest. Bot. Zeitschr. **98:** 90, 1951.

Simple or cespitose plant with globose to short-cylindric heads 5–7 cm. in diameter, 5–10 cm. long; tubercles dull gray-green, quadrangular and about 5 mm. wide at base, 6–10 mm. long; juice watery; axils scantily woolly; radial spines 25–35, acicular, straight, practically snow-white, 3–6 mm. long; central spines 1–4 (usually 1–3), lower one usually about 3 mm. long, subulate, sometimes to 12 mm. long, slender, flexuous and hooked, upper ones 4–9 mm. long, acicular, straight, stiff, all white or lower one with brownish tip; flowers funnelform, 2–3 cm. wide, slightly shorter; outer perianth segments linear-lanceolate, greenish near base, reddish brown with paler ciliate margins; inner perianth segments rose with paler entire margins; filaments rose to magenta; stigma lobes olive-green; fruit clavate, 2–2.5 cm. long, scarlet; seeds globose to obovate, glossy-black, about 0.9 mm. long.

Desert slopes and lower foothills, Sonoran Zones, Pima and Cochise Counties, Arizona, south into central Sonora. March–April, and July.

**33. Mammillaria mazatlanensis** K. Schumann, Monats. Kakteenk. **11:** 154, 1901

*Mammillaria littoralis* K. Brandegee, Monats. Kakteenk. **17:** 80, 1907.
*Neomammillaria mazatlanensis* Britt. & Rose, Cactaceae **4:** 138, *fig. 151,* 1923.
*Chilita mazatlanensis* Orcutt, Cactography 2, 1926.
*Ebnerella mazatlanensis* F. Buxb., Oest. Bot. Zeitschr. **98:** 89, 1951.

Crowded cespitose plant with globose to cylindrical heads 3–4 cm. in diameter, to 12 cm. long; tubercles rather remote, broadly conical, 9–10 mm. wide at base, 4–8 mm. long, gray-green, arranged in 5 and 8 spirals; axils naked or with 1–2 short bristles; juice watery; radial spines 13–15, slender-acicular, straight, stiff, 5–10 mm. long, creamy white; central spines 3–4, acicular, all straight, stiff, 8–15 mm. long, cream-colored at base, reddish brown above, spreading to porrect; flowers funnelform, 3–4 cm. long, very showy; outer perianth segments lanceolate to spatulate, obtuse, with a brownish red midstripe, margins white or nearly so, serrate; inner perianth segments lanceolate, carmine, acute, serrate near apex; filaments carmine; stigma lobes bright green; fruit clavate, 15–20 mm. long, brownish to reddish yellow; seeds irregularly subglobose, glossy-black, about 1 mm. long.

Arid soil along coastal plains, Lower Sonoran Zone, southern Sonora and Sinaloa. March and July.

**Mammillaria mazatlanensis** var. **monocentra** Craig, Mam. Handbook 242, *fig. 219,* 1945, differs in having a single central spine about 7 mm. long. The flowers are unknown.

Arid habitats, Lower Sonoran Zone, Yaqui Valley, Sonora.

**34. Mammillaria cerralboa** (Britt. & Rose) Orcutt, Cactography 7, 1926

*Neomammillaria cerralboa* Britt. & Rose, Cactaceae **4:** 116, *fig. 121a,* 1923.

Simple or somewhat cespitose plant with 2–6 cylindrical branches 4–6 cm. in diameter, 8–15 cm. tall, from several robust fibrous roots to 3 mm. in diameter; tubercles closely crowded, in fairly low-pitched spirals, obtusely conical, terete in cross section, 3–4 mm. in diameter, 5–6 mm. long, yellowish green, with watery sap; areoles circular, 1.5–2 mm. in diameter, short-woolly in youth; axils sparsely lanate in youth, soon naked; radial spines 8–10, slenderly acicular but stiff, straight, 7–10 mm. long, yellowish, ascending-spreading about equally from areole; central spine solitary, 1.5–2 cm. long, acicular but nearly twice as stout as radials, yellowish to pale reddish brown, straight or an occasional one abruptly slightly hooked; flowers narrowly funnelform, 10–12 mm. long, borne about 3 cm. below apex of stem, color in field uncertain; fruit clavate, red, 15–20 mm. long, perianth persistent; seeds black, obovoid, 1–1.3 mm. long, hilum obliquely truncate, surface punctate-pitted.

Rocky hillsides and canyon walls, Lower Sonoran Zone, Cerralvo Island, Gulf of California, Baja California. March–April.

**35. Mammillaria multidigitata** W. T. Marshall ex Lindsay, Cactus & Succ. Jour. **19:** 152, *figs. 99–101,* 1947

*Mammillaria multidigitata* W. T. Marshall, Cactus & Succ. Jour. **12:** 5, *fig.,* 1940. (Nomen provisorum.)

Many-stemmed, cespitose plant forming masses with over 100 stems or less than 12, individual stems 2–5 cm. in diameter, 5–20 (usually 5–8) cm. tall; juice watery; tubercles crowded, obtusely conical, about 3 mm. wide at base, 4–6 mm. long, green, obscured by spines except on older basal parts of plant; axils of tubercles slightly woolly; radial spines 15–25, slenderly acicular, straight, 6–8 mm. long, white, spreading and closely interlocking; central spines 2–4, usually 4, straight and acicular, spreading, 6–8 mm. long, white with reddish brown tips; flowers funnelform, white, 12–15 mm. long; outer perianth segments oblong, obtuse, entire, an olive-green midstripe on outer surface; inner perianth segments similar but lacking green midstripe; filaments white with a greenish tinge; anthers yellow; fruit clavate, bright red, 12–15 mm. long, 3–4 mm. in diameter; seeds black, about 1 mm. long.

Steep rocky slopes covered with extensive deposits of guano, Lower Sonoran Zone, San Pedro Nolasco Island, Sonora. April–May.

**36. Mammillaria inaiae** Craig, Cactus & Succ. Jour. **10:** 111, *fig.,* 1939

*Ebnerella inaiae* F. Buxb., Oest. Bot. Zeitschr. **98:** 89, 1951.
*Chilita inaiae* F. Buxb., Sukkulentenk. **5:** 15, 1954.

Simple or infrequently cespitose, cylindrical plant 5–6 cm. in diameter, 10–20 cm. tall; tubercles light green, 4–5 mm. wide at base, 6–8 mm. long, slightly grooved ventrally just below areole, arranged in 13 and 21 spirals; axils scantily woolly and with occasional bristles in youth; juice watery; radial spines 17–24, acicular, straight, stiff, 4–6 mm. long, white clear to tip; central spines 2 or less commonly 3, subulate, straight or slightly curved near apex but not hooked, 8–9 mm. long, reddish, purplish or nearly black, especially near tip, upper one ascending, lower porrect; flowers 15–18 mm. long, about 2 cm. wide; outer perianth segments oblanceolate, acute, reddish brown with pinkish white, serrate margins; inner perianth segments white to cream suffused with pink or pinkish brown along midstripe, margins entire; filaments light green below, pink toward tips; stigma lobes light green; fruit ovoid, 10–12 mm. long, about 5 mm. in diameter; seeds pyriform, glossy-black, about 0.8 mm. long.

Coastal slopes and low hills, Lower Sonoran Zone, vicinity of San Carlos Bay near Guaymas, Sonora. July.

**37. Mammillaria albicans** (Britt. & Rose) Berger, Kakteen 308, 1929

*Neomammillaria albicans* Britt. & Rose, Cactaceae **4:** 138, *fig. 152,* 1923.
*Neomammillaria slevinii* Britt. & Rose, op. cit. 139, *fig. 153.*
*Chilita albicans* Orcutt, Cactography 2, 1926.
*Chilita slevinii* Orcutt, loc. cit.
*Mammillaria slevinii* Boed., Mam. Vergl. Schlüss. 44, 1933.

Simple or cespitose plant with cylindrical stem 3–6 cm. in diameter, 10–20 cm. long, often in clumps of 10–15 branches; tubercles pale green, broadly conical, 5–6 mm. broad at base, 4–5 mm. long, closely crowded and obscured by spines; axils densely woolly; juice watery; areole densely white-woolly in youth; radial spines 14–18, slender, acicular, straight, stiff, 5–8 mm. long, closely interlaced, white, pinkish or pale brown, slightly darker to nearly black at tips; central spines 3–4 (sometimes 6), acicular, straight, stiff, 8–10 mm. long, pinkish to brownish with darker tips, none hooked; flowers about 2 cm. broad; outer perianth segments pinkish; stigma lobes white or faintly greenish; fruit clavate, red, 10–18 mm. long; seeds subglobose, dull black.

Arid soil and rocky hillsides, Lower Sonoran Zone, islands along the Gulf coast of Baja California between La Paz and Loreto. July–Aug.

**38. Mammillaria neopalmeri** Craig, Mam. Handbook 267, *fig. 242,* 1945

*Cactus palmeri* J. M. Coulter, Contr. U.S. Nat. Herb. **3:** 108, 1894.
*Mammillaria dioica* var. *insularis* K. Brandegee, Erythea **5:** 115, 1897.
*Neomammillaria palmeri* Britt. & Rose, Cactaceae **4:** 140, *pl. 14, fig.* 7, 1923.
*Chilita palmeri* Orcutt, Cactography 2, 1926.
*Mammillaria palmeri* Boed., Mam. Vergl. Schlüss. 43, 1933. Not *M. palmeri* Jacobi, 1856.

Plants crowded-cespitose in clumps 1–2 dm. tall, 1–2.5 dm. in diameter, with as many as 30 stems 6–7 cm. in diameter, these subglobose to short-cylindrical; roots fibrous; tubercles in 8 and 13 spirals, bluntly conical, crowded and quadrangular at base, about 4 mm. high, 5–6 mm. wide at base, more or less glaucous; sap watery; axils densely matted with white wool and to 12–15 slender tortuous bristles 2–3 mm. long; areoles oval, about 1 mm. wide, 1.5 mm. long, lanose in youth, eventually naked; radial spines 25–30, slenderly acicular, somewhat spreading, 5–6 mm. long, white throughout, straight, smooth and rigid, intricately interlaced; central spines 3–5 (usually 4), 7–8 mm. long, usually all straight, acicular, upper erect, barely above plane of radials, lower porrect, laterals divergent-ascending, usually brownish with darker brown tips, lower central sometimes hooked; flowers funnelform, 10–12 mm. long, 9–10 mm. wide; outer perianth segments oblong, entire, obtuse, tan with flush of rose along midrib; inner perianth segments paler to nearly white with greenish midline, 3–4 mm. wide, obtuse, entire; stamens with white filaments and orange-yellow anthers; stigma lobes olive-green; fruit clavate, scarlet, 4–5 mm. in diameter, 10–25 mm. long; seeds black and glossy, minutely but distinctly pitted.

On guano-covered hill slopes and ridges, Lower Sonoran Zone, San Benito Islands, Baja California. Feb.–April.

**39. Mammillaria radiaissima** Lindsay, in Craig, Mam. Handbook 292, *figs. 263–264,* 1945

Cespitose, depressed-globose plant to 5 cm. in diameter and high; tubercles rather flabby, conical, 5–6 mm. wide at base, 9–10 mm. long; areoles white-woolly in youth; axils naked or scantily woolly; juice watery; radial spines 35–50, slenderly acicular, straight, 10–18 mm. long, white, obscuring tubercles; central spines 1–8, slenderly acicular and similar to radials, 4–6 mm. long, pale brown, porrect; flowers 1.5 cm. wide, 3 cm. long; outer perianth segments lance-spatulate, yellow tinged with red, acute, margins entire; inner perianth segments broadly lanceolate, acute, citron-yellow tinged with red along midstripe, margins entire; filaments white; stigma lobes revolute-wavy or convoluted, irregular and somewhat unequal; fruits and seeds not seen.

Rocky slopes, Lower Sonoran Zone, near Escondido Bay about 110 miles north of La Paz, Baja California. Flowering period in native habitat unknown.

## Family 99. LYTHRACEAE. Loosestrife Family

Annual or perennial herbs, or in tropics often shrubs or trees, with simple alternate, opposite, or verticellate leaves. Flowers perfect, axillary or whorled. Calyx tubular, 4–6-toothed and often with accessory teeth or toothlike appendages in sinuses. Petals 4–6, inserted on calyx tube. Stamens 4–12, inserted near base of calyx tube, included or exserted, often unequal. Ovary superior, enclosed by but free from calyx, 1–4-celled; style 1, slender; stigma 2–8-lobed. Seeds small (in ours), without endosperm.

Calyx regular, not gibbous at base; seeds minute, less than 0.5 mm. long, often flattened on one side but not lenticular; stems angular:
- Leaves opposite, sessile, base auriculate; capsule globose, rupturing irregularly; stamens 4 .......... 1. *Ammania*
- Leaves alternate, petiolate, or if sessile at least not auriculate; capsule lanceolate or narrowly ovoid, dehiscing regularly; stamens 6 or 12 .......... 2. *Lythrum*

Calyx slightly irregular and oblique at apex, gibbous at base; seeds 1–1.2 mm. long, lenticular; stems terete .......... 3. *Cuphea*

### 1. AMMANIA L., Sp. Pl. 119, 1753

Annual herbs with 4-angled stems and opposite, sessile, auriculate leaves. Flowers in 1–5-flowered, axillary cymes, purplish. Calyx campanulate in flower, subglobose to globose in fruit; tube 8-ribbed; calyx teeth 4, usually alternating with 4 minute accessory teeth situated in sinuses. Petals 4, small and deciduous, purple, or sometimes lacking. Stamens 4, inserted below middle of calyx tube. Capsule globose, rupturing irregularly. Seeds minute, obovoid, faveolate-reticulate.

**1. Ammania coccinea** Rottb., Pl. Hort. Havn. Descr. 7, 1773

*Ammania latifolia* Torr. & Gray, Fl. N. Amer. **1:** 480, 1840. Not *A. latifolia* L., 1753.

Sparingly to profusely branched herb 2–15 dm. tall, erect, branches ascending; leaves nearly horizontal, broadly linear, 2–6 mm. wide, 2–6.5 cm. long, acute at apex, truncate-auriculate at base, green and glabrous, entire; calyx urceolate-campanulate, 2.5–3 mm. long at anthesis, ribs 8, conspicuous, dark green, tissue between ribs pale green or whitish; calyx teeth spreading, about 0.5 mm. long, alternating accessory ones minute; petals pale purplish, blade obovate, exserted 1.5–2.5 mm., rounded, fugacious; stamens and style about equaling petals, filaments very slender; capsule globose, closely crowded in axils, 3–5 mm. in diameter, outer wall chartaceous, inner one deep red, the two separating readily; seeds numerous, closely crowded on thick central placenta, obovoid, about 0.2–0.3 mm. long, reticulate, often shining as though varnished, somewhat angled.

About the margins of temporary pools and in low ground, Sonoran Zones, California to New Jersey and south to South America. Fairly common along the Colorado River from Needles south. The whole plant often turns red at maturity. Reported from Comondú, Baja California, by Brandegee. May–Sept.

**Ammania auriculata** Willd. var. **arenaria** (H. B. K.) Koehne, Bot. Jahrb. **1:** 245, 1881, occurs in Sonora a short distance east of the desert, but no specimens are known from the Sonoran Desert.

## 2. LYTHRUM L., Sp. Pl. 446, 1753

Slender herbs with 4–5-angled stems, alternate (in ours) or opposite, sessile leaves and small, axillary, purplish or whitish flowers. Calyx tube cylindrical or nearly so, 8–12-ribbed; teeth 4–6, erect, smaller than greenish accessory appendages, these spreading in flower but erect in fruit. Petals 4–6, inconspicuous. Stamens 4–6. Capsule cylindrical or oblong, 2-celled. Seeds numerous, minute.

Stamens 6; plants perennial .......................................... 1. *L. californicum*
Stamens 12; plants annual .............................................. 2. *L. bryantii*

### 1. **Lythrum californicum** Torr. & Gray, Fl. N. Amer. **1:** 482, 1840

*Lythrum sanfordi* Greene, Pittonia **2:** 12, 1889.

Erect perennial herb to 2 m. tall with slender, paniculately branching stems and exfoliating papery bark; leaves linear-lanceolate or lowest ovate-oblong, 1–8 mm. wide, 1–6.5 cm. long, gradually narrowing to each acute end, margins faintly emarginate and sparsely appressed-puberulent; inflorescence usually a spiciform slender raceme 5–30 cm. long; pedicels 0.2–1 mm. long; calyx about 5 mm. long at anthesis, 5–8 mm. long in fruit, 12-ribbed, teeth erect, minute, appendages 1–1.5 mm. long, spreading in flower, erect in fruit, greenish; petals 4, broadly obovate, exserted 4–6 mm., bright purple; stamens 6, inserted near base of calyx tube, equal, exserted 2–3 mm. beyond calyx tube; style filiform, 7–10 mm. long; fruiting calyx urceolate-cylindrical; seeds minute, narrowly ellipsoid to linear-lanceolate in outline, one and one-half times to twice as long as broad.

Damp soil in low places and along streams or about seepy places, Lower Sonoran Zone to Transition Zone, central California into northern Baja California, thence east to Nevada and Texas and south into Mexico. Rare in the desert proper but collected along the western edge of the Colorado Desert and in Pima County, Arizona. May–Sept.

### 2. **Lythrum bryantii** Brandegee, Proc. Calif. Acad. II, **2:** 155, 1889

Annual herb with slender, erect to ascending stems 1–3 dm. tall, these simple or branching at or near base; leaves narrowly to broadly elliptic to broadly oblong or sometimes lanceolate, 2–6 mm. broad, 7–20 mm. long, obtuse or rounded at apex, broadly cuneate at base, thin and plane, glabrous; flowers solitary in axils, often slightly reflexed, slender pedicels 1–4 mm. long, bracteate at about middle; calyx vasiform-cylindric, 5–6 mm. long at anthesis, 12-ribbed, teeth and appendages nearly equal; petals 4, blades broadly obovate, exserted 4–6 mm., bright purple; stamens 12, inserted near base of calyx tube, six short ones 1.5–2 mm. long and included, six longer ones exserted 1–3 mm., opposite the calyx teeth; fruiting calyx often spreading to slightly deflexed; mature seeds unknown, immature ones ovoid, apparently slightly flattened.

Along washes and on moist flats, Lower Sonoran Zone, western Baja California from Comondú south on the Magdalena Plain. Dec.–March.

The petals on this species are slightly broader and more showy than those of *L. californicum*.

### 3. CUPHEA P. Br., Nat. Hist. Jamaica 216, 1756

Annual or perennial herbs or frutescent plants with viscid terete stems and branches and opposite, verticillate, or rarely alternate leaves. Flowers solitary or in few- to many-flowered cymes. Calyx tube many-ribbed, base obliquely gibbous, mouth more or less oblique; teeth 6, with accessory alternating appendages or these lacking. Petals usually 6, dorsal pair usually larger. Stamens 11, two dorsal usually shorter than unequal lower nine. Style filiform; stigma capitate, slightly bilobed. Capsule oblong, included in calyx, dehiscing laterally.

**1. Cuphea wrightii** A. Gray, Pl. Wright. **2:** 56, 1853
*Parsonia wrightii* Kearney, Trans. N.Y. Acad. **14:** 37, 1894.

Annual, 1–4 dm. tall, erect or with ascending branches, sparsely hirsute with simple and gland-tipped, spreading, purplish or reddish hairs; leaves opposite; petioles slender, 3–20 mm. long; leaf blades ovate to lance-ovate, rounded to obtuse at base, acute at apex, 3–15 mm. broad, 1–3.5 cm. long, plane, sparsely appressed-pubescent; flowers solitary or 2–4 in axillary cymes, forming racemiform terminal inflorescence to 1.5 dm. long; upper leaves reduced to small purplish bracts; pedicels 1–6 mm. long; calyx 4–6 mm. long at anthesis, purplish, about 12-ribbed, sparsely hirsute, teeth deltoid, minute; petals 2–2.5 mm. long, pale purplish; capsule ovoid, 5–7 mm. long, thinly chartaceous; seeds about 1 mm. long, lenticular.

Canyons and hillsides, mainly Upper Sonoran and Transition Zones, southern Arizona to Panama. Barely reaching the upper edges of the Sonoran Desert in the Baboquivari Mountains and near Cumpas, Sonora. Aug.–Oct.

## Family 100. COMBRETACEAE. Combretum Family

Trees, shrubs, or scandent plants with opposite or alternate, entire, estipulate leaves, sometimes armed with spines. Inflorescence spicate, racemose, or in heads, bracteate. Flowers perfect or polygamodioecious. Calyx tube adnate to ovary, valvately 4–5-lobed. Petals 4 or 5 or none, small and inconspicuous. Stamens as many or twice as many as petals, inserted on calyx. Ovary 1-celled. Style simple; stigma capitate, entire. Fruit coriaceous or drupaceous, 1-seeded, indehiscent.

Flowers in globose, conelike heads; drupes flattened and scalelike; stamens 5; leaves alternate; petals none .................................................. 1. *Conocarpus*
Flowers in racemes; drupes ellipsoid, not scalelike; stamens 10; leaves opposite; petals 5 ....
............................................................ 2. *Laguncularia*

### 1. CONOCARPUS L., Sp. Pl. 176, 1753

A shrub or sometimes a tree to 20 m. tall with alternate, subfleshy, short-petiolate, entire leaves bearing 2 basal glands. Flowers perfect and minute in globose heads arranged racemosely or paniculately. Calyx tube compressed-urceolate, 5-lobed, limb deciduous. Petals none. Stamens 5. Ovary 1-celled, 2-ovuled; style simple, short, with a simple stigma. Fruit a small, flattened, 1-seeded drupe that becomes leathery and scale-like, those of a head remaining closely crowded in an imbricate pattern in the globose head.

**1. Conocarpus erecta** L., Sp. Pl. 176, 1753

Characters of the genus; bark thin, dark brown, irregularly ridged and scaly; leaf blades elliptic to obovate, 2–10 cm. long, 1–3 cm. wide, acute at both ends or obtuse or emarginate at apex, glabrous or sericeous, leathery, bright green, 2 small glands at juncture of petiole and blade; heads 5–8 mm. in diameter at anthesis, flowers greenish; calyx about 2 mm. long; stamens exserted 1–2 mm.; fruiting heads conelike, 8–15 mm. long, brownish, individual drupes about 3 mm. long.

In coastal swamps and along esteros, islands in Gulf of California and along eastern shores of southern Baja California; Sinaloa to tropical America and western Africa. Sept.–Dec.

## 2. LAGUNCULARIA Gaertn. f., Fruct. **3**: 209, *pl. 217*, 1807

Shrubs or trees with opposite, petiolate, coriaceous, entire leaves bearing a pair of glands at base of blade. Flowers in axillary and terminal spikes, polygamodioecious. Calyx tube turbinate, adnate to ovary; limb shallowly urceolate and weakly 5-toothed; teeth persistent. Petals 5, caducous. Stamens 10, in 2 series. Ovary 1-celled, crowned by an epigynous fleshy disk. Stigma 2-lobed. Fruit coriaceous, narrowly obovoid, somewhat ribbed and angled, 1-seeded.

**1. Laguncularia racemosa** (L.) Gaertn. f., Fruct. **3**: 209, *pl. 217*, 1807

*Conocarpus racemosa* L., Syst. Nat. ed. 10, 930, 1759.

Characters of the genus; leaf blades oval to oblong, 1–4 cm. wide, 2–12 cm. long, rounded at apex, obtuse or sometimes acute at base, this bearing 2 conspicuous glands; flowers perfect or polygamous, in lax spikes; calyx sericeous, 2–3 mm. long at anthesis; petals 5, suborbicular, about equaling calyx lobes; drupe leathery, sericeous, 10-ribbed, 10–15 mm. long, often reddish, 1-seeded.

Growing with *Rhizophora* and *Avicennia* along esteros and shores of bays, Lower Sonoran Zone to Tropical Zone, along the coast from Magdalena Bay and Bahía de la Concepción south to tropical South America; also in the West Indies, Florida, along the coast of Mexico and South America, and in western Africa. The wood is hard but rather brittle. On the Sonoran coast it extends northward to San Carlos Bay. July–Oct.

# Family 101. RHIZOPHORACEAE. Mangrove Family

Trees or shrubs with opposite, stipulate, or rarely alternate and estipulate leaves; blades leathery and simple, entire or serrate. Flowers perfect, axillary or cymose. Calyx tube more or less adnate to ovary, divided above into 3–14 valvate persistent lobes. Petals usually same number as sepals, inserted on calyx tube. Stamens usually twice to three or four times as many as petals (twice as many in ours). Style simple; stigma capitate or lobed. Fruit fleshy or leathery, usually indehiscent, 1–5-celled, each locule 1–2-seeded.

## 1. RHIZOPHORA L., Sp. Pl. 443, 1753

Tree or shrub, with numerous "prop roots" descending from branches, intricately branched and forming dense, virtually impenetrable thickets. Leaves opposite, petiolate, leathery, entire, glabrous. Stipules caducous. Inflorescences axillary, few-flowered. Calyx tube cupulate, 4-parted. Petals 4, inserted on a fleshy disk, entire. Stamens 8–12; filaments short. Ovary 2-celled; stigma bidentate. Embryo exalbuminous, radical clavate, perforating the apex of capsule while latter is still on tree.

### 1. Rhizophora mangle L., Sp. Pl. 443, 1753

A shrub (in our area) or sometimes a tree to 25 m. tall; bark thin, grayish or brownish, shallowly furrowed or nearly smooth on younger stems; leaves opposite, blades elliptic to obovate, 5–15 cm. long, glabrous and dark green, obtuse or abruptly acute at apex, base more or less decurrent on petiole; calyx leathery, 4-lobed, lobes lanceolate, 8–12 mm. long; petals 4, yellowish white, linear, about equaling calyx lobes, pubescent with crisped white hairs; stamens 8, 5–6 mm. long, anthers equaling or slightly exceeding fleshy filaments; fruit baccate, conical, 2–2.5 cm. long, brown and leathery; seed usually germinating and radical pushing through capsule as a clavate seedling root and growing downward 2–3 dm. before it falls.

Along the muddy shores of bays and esteros of the coast, Lower Sonoran and Tropical Zones, from San Jorge at the upper end of Magdalena Bay, from Mulegé and Guaymas and Tamaulipas south on both coasts of Mexico; a small colony on Smith Island, Bahía de los Angeles, Baja California. Widely distributed along tropical shores of both North and South America. March–Nov.

## Family 102. ONAGRACEAE. Evening-Primrose Family*

Herbs or shrubs with simple, entire to pinnatifid, alternate or opposite, estipulate leaves (except in *Jussiaea*). Flowers perfect, mostly 4-merous, in terminal racemes or spikes or axillary. Hypanthium adnate to, and usually prolonged more or less beyond apex of, ovary, often lined with an annular disk. Sepals 4 (or 5 in *Jussiaea*), usually equal. Petals 4 (or 5 in *Jussiaea*), inserted at summit of hypanthium, often under margin of disk; anthers versatile or basifixed. Ovary inferior, usually 4- (sometimes 2- or 5-) celled; style slender; stigma capitate, discoid or 4-lobed. Fruit a capsule or nutlike or embedded in the smaller stems. Seeds usually numerous, small.

Sepals persistent, all of them divided clear to ovary; seeds covered by a layer of capsule wall ........ 1. *Jussiaea*
Sepals deciduous, borne on a short to long calyx tube, not divided to ovary (one sepal divided to ovary in *Lopezia*):
  Seeds bearing a tuft of hairs at one end or subterminally winged; corollas and sepals scarlet:
    Seeds bearing a tuft of fine hairs; hypanthium 2–3 cm. long with 8 lobelike appendages within; anthers not mucronate ........ 2. *Zauschneria*
    Seeds minutely winged subterminally, not hairy; hypanthium 12–15 mm. long, without internal appendages; anthers mucronate ........ 3. *Xylonagra*
  Seeds naked, neither hairy nor winged (finely fimbriate along upper margin in *Clarkia*); corollas and sepals not scarlet (or if corolla scarlet capsules indehiscent):
  ◆ Stamens 8; sepals regular and equal; disk in hypanthium annular:
      Fruit a dehiscent capsule:
        Anthers basifixed; seeds minutely fimbriate along upper margin ....... 4. *Clarkia*

* Contributed by John Hunter Thomas.

Anthers versatile; seeds not fimbriate .......................... 5. *Oenothera*
Fruit indehiscent, nutlike, or embedded in swollen branch:
Ovary 4-celled (sometimes imperfectly so); plant herbaceous; hypanthium shorter than the sepals; capsule not embedded in stem .................... 6. *Gaura*
Ovary 2-celled; plant a low shrub; hypanthium longer than sepals; capsules embedded in swollen stem .................................... 7. *Burragea*
♦ Stamens 2, one sterile; sepals slightly unequal, posterior 3 united along basal third; disk semicircular, lining base of coalesced posterior sepals ................ 8. *Lopezia*

## 1. JUSSIAEA L., Sp. Pl. 388, 1753

Herbs, shrubs or small trees with alternate, simple, stipulate leaves and white to yellow axillary flowers. Flowers regular. Calyx tube not prolonged above ovary. Sepals 4–6, persistent. Petals 4–6, caducous. Stamens twice as many as petals, inserted in 2 series under margin of epigynous disk. Ovary 4–6-celled, many-ovuled; style simple; stigma capitate, shallowly lobed. Capsule indehiscent or tardily and imperfectly dehiscent. Seeds naked, often covered by a layer of capsule wall.

Plants puberulent with spreading hairs; flowers 4-merous; seeds multiseriate and free in each locule ............................................ 1. *J. suffruticosa* var. *octofila*
Plants glabrous; flowers 5–6-merous; seeds uniseriate and enclosed in endocarp in each locule ................................................... 2. *J. repens* var. *peploides*

### 1. **Jussiaea suffruticosa** L. var. **octofila** (DC.) Munz, Darwiniana **4:** 239, 1942

*Jussiaea octofila* DC., Prod. **3:** 57, 1828.
(For additional synonymy see Munz, Darwiniana **4:** 239–240, 1942.)

An herb or erect shrub to 2.5 m. tall, pubescent throughout with fine, spreading hairs, branching freely or only above; leaves lanceolate or ovate-lanceolate, 8–15 mm. wide, 2–6 cm. long, subsessile or very short petiolate, acute to acuminate at apex, subentire, several brownish or reddish marginal glands usually along each edge, with an obvious submarginal vein; pedicels to 2 cm. long at anthesis, elongating slightly in fruit; bracteoles setaceous; hypanthium 10–15 mm. long, 4-angled, pubescent; sepals 4, ovate, 3–5 mm. wide, 7–10 mm. long; petals 4, deep yellow, 1–2 cm. long, claw nearly obsolete; anthers nearly equaling filaments; capsules clavate-cylindric or cylindric, 2.5–5 mm. in diameter, 2.5–5 cm. long, obtusely quadrangular; seeds obovoid, reddish brown, lustrous.

In low, moist ground, Texas and Arkansas to Florida south through the West Indies and Mexico to Venezuela and Peru. It occurs within the Sonoran Desert at Guaymas, Sonora. Jan.–June; Sept.–Nov.

### 2. **Jussiaea repens** L. var. **peploides** (H. B. K.) Griseb., Cat. Pl. Cubens. 107, 1866

*Jussiaea peploides* H. B. K., Nov. Gen. & Sp. **6:** 97, 1823.
(For additional synonymy see Munz, Darwiniana **4:** 272, 1942.)

Perennial herb with creeping or floating stems that root abundantly at nodes, glabrous or essentially so; petioles 1–2.5 cm. long; leaf blades oblong to spatulate-oblong, or oblanceolate, 5–20 mm. wide, 1–4 (occasionally –6) cm. long, obtuse to acute at apex, gradually narrowed toward base; stipules deltoid-squamate, 0.5–1.5 mm. long; pedicels slender, to 1.5 mm. long and ascending at anthesis, sharply deflexed in fruit; sepals 5, linear-lanceolate, 4–9 mm. long, 1.5–2.5 mm. wide, acute; petals 5, yellow, narrowly to broadly obovate and often emarginate, 8–12 mm. long; capsules 2–2.5

mm. in diameter, 1–2.5 cm. long, rather prominently ribbed; fruiting pedicels to 3 cm. long.

In and about margins of ponds or quiet pools, Tropical to Transition Zones, southern Oregon to Texas, the West Indies and south into South America. Collected at Horcasitas, Sonora, near the upper limits of the Sonoran Desert. In Baja California it occurs in the foothills of the Sierra San Pedro Mártir and in the Cape Region but has not been collected within the limits of the desert proper. April–Sept.

## 2. ZAUSCHNERIA Presl, Rel. Haenk. 2: 28, 1835

Decumbent to erect perennials, more or less woody at least at base, with shredding papery bark and opposite or alternate, nearly or quite sessile, often fasciculate leaves. Flowers conspicuous, in terminal spikes or spiciform racemes. Hypanthium scarlet, globose at base, tubular above, bearing 4 erect and 4 reflexed appendages near summit within. Sepals 4. Petals 4, scarlet. Stamens 8, four short and four longer ones alternating. Ovary 4-celled, quadrangular in cross section; stigma shallowly 4-lobed. Capsule imperfectly 4-celled. Seeds many, obovoid, silky-comose at apex.

Plants suffrutescent at base; leaves linear to lanceolate, usually less than 6 mm. wide ....... .............................................. 1. *Z. californica* subsp. *californica*
Plants entirely herbaceous; leaves oval to broadly lanceolate, 7–15 mm. wide ............ .............................................. 1a. *Z. californica* subsp. *latifolia*

### 1. **Zauschneria californica** Presl, Rel. Haenk. 2: 28, *pl. 52,* 1835, subsp. **californica**

(For synonymy see Keck, Carnegie Inst. Wash. Pub. No. **520**: 220, 1940.)

Suffrutescent plants 3–9 dm. tall, usually much branched and frequently more or less decumbent, herbage gray- or green-pilose and often glandular; leaves linear-lanceolate to linear-oblong, 1.5–6 mm. wide, 0.5–4 cm. long, entire or slightly denticulate, upper mostly alternate and lower opposite, lateral veins rarely evident; hypanthium 2–3 cm. long; sepals erect, lanceolate, 8–10 mm. long; petals 2-cleft, 8–12 mm. long; capsule 1.5–2 cm. long, straight or slightly curved, sessile or nearly so; seeds about 1.5 mm. long, coma 5–7 mm. long.

Arid slopes and flats, mainly Upper Sonoran Zone, Lake County, California, south into northern Baja California. It occurs in the vicinity of Palm Springs on the western margin of the Colorado Desert but the plants there approach subsp. *latifolia* in width of the leaves. The stems are suffrutescent at the base and the upper leaves are alternate in the Palm Canyon specimens. June–Dec.

### 1*a*. **Zauschneria californica** subsp. **latifolia** (Hook.) Keck, Carnegie Inst. Wash. Pub. No. **520**: 220, 1940

(For synonymy see Keck, loc. cit.)

Differing from subsp. *californica* in having leaves chiefly opposite throughout plant; herbaceous instead of suffrutescent stems; lateral veins of leaves distinctly visible; and leaf blades 7–15 mm. wide, oval to broadly lanceolate in outline.

Dry ridges and arid slopes, mostly Upper Sonoran and Arid Transition Zones, southern Oregon to Trinity County, California, along the western slopes of the Sierra Nevada to the Sierra San Pedro Mártir, and east to southwestern New Mexico, northern Sonora and into Nevada. A specimen that appears to belong here was collected near Cataviñá, Baja California, and another at Cottonwood Springs at the northern edge of the Colorado Desert. July–Nov.

## 3. **XYLONAGRA** Donn. Smith & Rose, Contr. U.S. Nat. Herb. **16:** 294, 1913

A low brushy shrub with numerous short ascending branches. Leaves alternate, solitary or fascicled; tips glandular. Inflorescence terminal, racemose. Flowers pedicelled, ascending or spreading, scarlet. Hypanthium a short tube that abruptly broadens into a long funnelform throat. Sepals 4, ovate to triangular. Petals 4. Stamens 8; filaments short; anthers mucronate. Style exserted slightly beyond stamens; stigma capitate. Ovary 4-celled. Capsule loculicidally dehiscent. Seeds few and uniseriate in each cell, each bearing a small subterminal wing. Embryo straight, oblanceolate; cotyledons three times as long as radical.

**1. Xylonagra arborea** Donn. Smith & Rose subsp. **wigginsii** Munz, Aliso **4:** 500, 1960

*Oenothera arborea* Kell., Proc. Calif. Acad. **2:** 32, 1859, in part.
*Hauya californica* S. Wats., Proc. Amer. Acad. **20:** 366, 1885, in part.
*Hauya arborea* Curran, Proc. Calif. Acad. II, **1:** 253, 1888, in part.
*Xylonagra arborea* Donn. Smith & Rose, Contr. U.S. Nat. Herb. **16:** 294, 1913, in part.

Shrub to 1.7 m. tall with slender, light-barked branches that are reddish and minutely cinereous-pubescent when young; leaves fascicled or solitary; petioles 0.5–8 mm. long, slender; leaf blades ovate-lanceolate, 3–9 mm. wide, to 2.5 cm. long, entire, acute to obtuse at the apex and tipped with a yellowish, orange, or reddish gland, sparsely cinereous-strigulose on both surfaces; inflorescence often 1–3-flowered; pedicels 3–5 mm. long; hypanthium 2–3 cm. long, scarlet; sepals 8–13 mm. long; petals 10–18 mm. long, erect; stamens equaling or slightly exceeding petals; style puberulent; capsule woody, 10–15 mm. long, rich brown; seeds 5–10 in each cell, body 2.5–3 mm. long, wing increasing length to 4–5 mm.

Desert hills and coastal plains, Lower Sonoran Zone, Cedros Island and on the mainland of Baja California from Playa Santa Catarina to San Bartolomé Bay, possibly somewhat farther south. Feb.–Oct.

Most of the plants from Cedros Island are referable to *X. arborea* subsp. *arborea* and differ from *X. arborea* subsp. *wigginsii* in having smaller flowers and gland-tipped hairs on the flowers.

## 4. **CLARKIA** Pursh, Fl. Amer. Sept. **1:** 260, *pl. 11*, 1814

Annual herbs with erect or ascending, simple or branched, terete stems, often with thin exfoliating epidermis below. Leaves alternate, linear to spatulate, solitary or fascicled. Inflorescence spicate or paniculate with erect or nodding buds. Hypanthium obconic to narrowly funnelform. Sepals 4, distinct or partially united at tips and turned to one side instead of reflexed. Petals 4, white to purple, obovate, erose, entire, or lobed. Stamens 8, four opposite the petals shorter than others; anthers equal or nearly so, sparsely hairy. Stigma lobes ovoid to linear, short. Capsule 4-angled, terete, or heavily 8-ribbed, sessile to long-pedicelled, dehiscing loculicidally. Seeds several to many, brown, finely fimbriate along upper margins.

Sepals usually distinct in anthesis; buds erect; petals lavender to purple; capsule 2–3 mm. thick, strigose-puberulent ...................... 1. *C. purpurea* subsp. *quadrivulnera*
Sepals usually united at tips and turned to one side in anthesis; buds nodding; petals white, pink or pale rose; capsule 1–1.5 mm. thick, glabrous or very sparsely puberulent ...... ........................................................ 2. *C. epilobioides*

**1. Clarkia purpurea** (Curt.) Nels. & Macbr. subsp. **quadrivulnera** (Dougl.) Lewis & Lewis, Univ. Calif. Pub. Bot. **20:** 305, 1955

(For full synonymy see Lewis & Lewis, loc. cit.)

Erect plant 1–8 dm. tall with simple or ascendingly branched stems; leaf blades sessile or short-pedicelled, lanceolate to spatulate, 2–8 mm. wide, 1–5 cm. long, acute to rounded at apex, sparsely puberulent to glabrate, slightly paler beneath than above; inflorescence to 5 dm. long; hypanthium 2–6 mm. high, bearing a ring of hairs about one-third of way up from base; sepals lanceolate, 5–10 mm. long, usually free and reflexed in anthesis, sometimes more or less united at tips; petals lavender to purple, often with a dark purple spot at base when light-colored, rounded or truncate and usually somewhat erose at apex, 5–20 mm. long, clawless; shorter stamens one-third to one-half as long as others; anthers 2–4 mm. long; capsules 2–3 mm. thick, 1–3.5 cm. long, 8-ribbed, sparsely to densely strigose-puberulent; seeds about 1 mm. long, fimbriate crest about one-fifth as long as body.

Grassy hillsides and plains, chiefly Upper Sonoran Zone but entering the higher parts of the Lower Sonoran Zone, Washington to northern Baja California and east into central Arizona. May–July.

**2. Clarkia epilobioides** (Nutt.) Nels. & Macbr., Bot. Gaz. **65:** 60, 1918

*Oenothera epilobioides* Nutt., in Torr. & Gray, Fl. N. Amer. **1:** 511, 1840.
*Sphaerostigma epilobioides* Walp., Rep. **2:** 78, 1843.
*Godetia epilobioides* S. Wats., Bot. Calif. **1:** 231, 1876.

Erect, usually branched sparingly from about middle but sometimes profusely branched whole length, 1–7 dm. tall; petioles slender, 1–6 mm. long, or leaves sessile; leaf blades linear, lanceolate or spatulate, 2–6 mm. wide, 1–4 cm. long, acute to rounded at apex, gradually or abruptly narrowed to base, glabrous or minutely puberulent on margins; buds nodding; hypanthium 1–1.5 mm. long, inner ring of hairs near summit; sepals united and turned to one side in anthesis, 3–5 mm. long; petals obovate, 4–8 mm. wide, 6–12 mm. long, usually white faintly tinged with pink, sometimes pale lavender, a basal spot present or absent, short-clawed; filaments 2–6 mm. long, shorter ones in any particular flower about two-thirds as long as longer ones; anthers 1–1.5 mm. long; capsule linear, 1–1.5 mm. in diameter, 1.5–3 cm. long, sparsely strigillose to glabrate at maturity; seeds about 0.5 mm. long, apical crest minute.

Mainly on shaded hillsides, Upper Sonoran Zone, but entering the Lower Sonoran Zone in Baja California and Arizona; central California to northern Baja California (Cataviñá) and central Arizona. April–May.

## 5. OENOTHERA L., Sp. Pl. 346, 1753

Annual to perennial, acaulescent or caulescent herbs with basal or alternate cauline leaves or both. Inflorescence spicate, racemose or cymose. Hypanthium extending beyond ovary and this portion deciduous. Sepals 4, reflexed in anthesis. Petals 4. Stamens 8, equal or in 2 unequal series; anthers chiefly versatile. Stigma of 4 linear lobes, discoid or capitate. Capsules thin-walled to hard and woody, straight, curved or spirally coiled, 4-celled, 4-valved, dehiscent. Seeds numerous, naked. Embryo straight or nearly so.

Stigmas with 4 linear lobes; flowers vespertine:

Plants with well-developed, usually leafy stems, flowers borne on stems (see also *O. primiveris* var. *caulescens*):

Flowers yellow; buds erect; seeds biseriate in each locule:

Plants prostrate or decumbent; cauline leaves obovate; capsules not narrowed upward .................................... 1. *O. drummondii* var. *thalassaphila*

Plants erect; cauline leaves lanceolate; capsule gradually tapering toward apex ...... ............................................ 2. *O. hookeri* subsp. *venusta*

Flowers white to rose; buds nodding; seeds uniseriate or in several rows in each locule:

Capsule clavate, distinctly broader near apex than below; seeds clustered on slender funiculi, in more than 2 rows in each locule:

Petals 10–15 mm. long; plants decumbent or ascending; hypanthium 10–20 mm. long; petals white or pink ............................ 3. *O. kunthiana*

Petals 5–10 mm. long; plants erect or ascending; hypanthium 4–8 mm. long; petals rose to red-violet ........................................... 4. *O. rosea*

Capsule cylindrical, tapering slightly toward apex; seeds uniseriate in each locule:

Plants annual; basal leaf blades 2–8 cm. long, rhombic; capsule woody:

Hypanthium pubescent with spreading hairs 1–1.5 mm. long, none appressed:

Petals 2.5–4 cm. long; capsules 4–5 cm. long; leaf blades subentire to coarsely dentate ............................................ 5. *O. deltoides*

Petals 1.5–2.5 cm. long; capsules 2–4 cm. long; leaf blades pinnately divided in lower half ...................................... 6. *O. wigginsii*

Hypanthium pubescent with appressed hairs or with both appressed and spreading ones:

Capsules slender, 1.5–2.5 mm. thick at base; hypanthium pubescent with both appressed and spreading hairs ......... 7a. *O. californica* subsp. *arizonica*

Capsules heavier, 2.5–5 mm. thick at base; hypanthium appressed-pubescent without spreading hairs ............................. 5. *O. deltoides*

Plants perennial (or sometimes biennial); basal leaves smaller, narrower; capsules not woody .............................. 7. *O. californica* subsp. *californica*

Plants usually acaulescent, flowers borne at bases of leaves in the rosette:

Capsules much longer than broad, not winged; borders of seeds turned inward to form a groove over the raphe:

Flowers white, aging pink; plants perennial; hypanthium glabrous within; anthers 10–14 mm. long:

Capsules pedicelled, cylindrical-ovoid, 3–4 cm. long ........................... ...................................... 8. *O. caespitosa* var. *marginata*

Capsules sessile, ovoid, 1–2 cm. long (Baja California) ....................... .................................. 8*a*. *O. caespitosa* var. *brandegeei*

Flowers yellow, aging red; plants annual; hypanthium minutely puberulent within in lower half; anthers 8–10 mm. long:

Plants acaulescent; stems occasionally to 1 dm. tall ......................... ...................................... 9. *O. primiveris* var. *primiveris*

Plants caulescent; stems 1–3 dm. high .......... 9*a*. *O. primiveris* var. *caulescens*

Capsules short, as broad as long, broadly 4-winged; seeds narrowly winged but wings not forming a groove ......................................... 10. *O. triloba*

Stigmas capitate, discoid or with very short, ovoid lobes; flowers diurnal:

♦ Hypanthium lined with a lobed disk; plants erect, simple to sparingly branched:

Plants shrubby; leaves mostly oblong-lanceolate, not deeply pinnatifid ............... ......................................................... 11. *O. crassifolia*

Plants annual herbs; leaves deeply pinnatifid:
Stigmas globose; longer stamens two-thirds as long as petals:
Anthers all alike; hypanthium 2–3 mm. long; capsules present in lowest axils as well as higher up; capsules somewhat contorted, not deflexed .................................................. 12. *O. angelorum*
Anthers of shorter stamens unlike those of larger ones; hypanthium about 1 mm. long; capsules borne only in upper part of plant; capsules straight, sharply deflexed ............................................ 13. *O. leptocarpa*
Stigmas cylindrical; longer stamens one-third as long as petals .................................................. 14. *O. sceptrostigma*

♦ Hypanthium not lined with a disk; plants decumbent to erect, often much branched or acaulescent:
Capsules sessile or virtually so:
Flowers borne in terminal spikes; petals white or pale yellow, drying pinkish:
Capsules linear or nearly so, slightly broader at base than above, refracted or spreading, mostly over 2.5 cm. long:
Petals 5–7 mm. long; capsules commonly refracted; style 10–13 mm. long; sepals 5–6 mm. long ........................................ 15. *O. refracta*
Petals 2.5–3 mm. long; capsules divaricately spreading; styles 3.5–4 mm. long, scarcely equaling petals; sepals 1.5–2.5 mm. long ... 16. *O. chamaenerioides*
Capsules subfusiform, broadest in middle, mostly appressed-ascending with slightly spreading tips, 1.5–2.5 cm. long:
Base of capsules 1–1.5 mm. thick; petals longer than wide ....................................................... 17. *O. decorticans* var. *desertorum*
Base of capsules 3 mm. thick; petals no longer than wide ....................................................... 17*a*. *O. decorticans* var. *condensata*
Flowers borne in axils of foliage leaves; petals yellow, often drying greenish:
Sepals 1–7 mm. long; petals 1.5–7 mm. long; anthers 0.5–1 mm. long:
Petals 3–7 mm. long; herbage cinereous with close fine pubescence or glabrate:
Herbage cinereous; stems spreading .......... 18. *O. micrantha* var. *exfoliata*
Herbage glabrate, pubescence spreading, sparse; stems erect ................................................ 18*a*. *O. micrantha* var. *ignota*
Petals 1.5–3 mm. long; herbage villous or hirsutulous but green ................................................ 18*b*. *O. micrantha* var. *jonesii*
Sepals 7–10 mm. long; petals 8–15 mm. long; anthers 2.5–3 mm. long ........................................................ 19. *O. bistorta* var. *hallii*
Capsules distinctly pedicelled (or very shortly pedicelled in *O. cardiophylla*):
Leaves orbicular-cordate, not pinnatifid, cauline; pedicels often less than 5 mm. long; pollen shed in tetrads:
Hypanthium 18–40 mm. long; style 30–60 mm. long; petals 8–20 mm. long ........................................................ 20. *O. arenaria*
Hypanthium 4.5–14 mm. long; style 8–23 mm. long; petals 3–12 mm. long:
Leaves cordate to cordate-orbicular, about as long as broad; petals usually over 4.5 mm. long; pubescence usually villous ........................................................ 21. *O. cardiophylla* subsp. *cardiophylla*
Leaves ovate-acuminate, longer than broad; petals under 4.5 mm. long; pubescence mostly glandular ............ 21*a*. *O. cardiophylla* subsp. *cedroensis*
Leaves usually basal, ovate, oblong, or lanceolate, usually pinnatifid; pedicels usually over 5 mm. long; pollen grains shed individually:
■ Capsules linear, usually under 2 mm. in diameter, 2–9 cm. long:
Petals usually under 6 mm. long, often fading to red; buds pendulous before anthesis ............................. 22*a*. *O. brevipes* subsp. *arizonica*

Petals usually over 6 mm. long, not fading to red; buds not pendulous before anthesis:
Sepals with caudate projections below tips; plants stout, subsucculent, villous .......................................... 22. *O. brevipes* subsp. *brevipes*
Sepals lacking projections; plants usually slender, not subsucculent, strigose .................................... 22*b*. *O. brevipes* subsp. *pallidula*
Capsules clavate, usually over 2 mm. in diameter, 1–2.5 cm. long:
Plants with spreading, villous pubescence below; petals usually yellow:
Petals not fading after anthesis ......... 23*a*. *O. claviformis* subsp. *peirsonii*
Petals fading to brick-red after anthesis:
Petals 1.5–2 mm. long; style 5–7 mm. long (Baja California) .............. ................................ 23*b*. *O. claviformis* subsp. *wigginsii*
Petals 3–5 mm. long; style 7–12 mm. long (Arizona and Sonora) ....... ................................ 23*c*. *O. claviformis* subsp. *rubescens*
Plants glabrous or short-appressed-pubescent below, not spreading-villous; petals yellow or white:
Petals yellow ............................ 23*d*. *O. claviformis* subsp. *yumae*
Petals white:
Leaves, stems, and often the inflorescence glandular (Arizona and adjacent Sonora) ......................... 23*e*. *O. claviformis* subsp. *peeblesii*
Leaves and stems rarely glandular, inflorescence never glandular (mainly California and northern Baja California) .......................... ................................ 23. *O. claviformis* subsp. *aurantiaca*

**1. Oenothera drummondii** Hook. var. **thalassaphila** (Brandegee) Munz, Amer. Jour. Bot. **22:** 651, 1935

*Oenothera thalassaphila* Brandegee, Univ. Calif. Pub. Bot. **10:** 185, 1922.

Perennial with prostrate or decumbent suffrutescent stems and densely canescent-puberulent herbage, branches 1–5 dm. long; lower leaves obovate to oblanceolate, entire to remotely sinuate-dentate, 6–15 mm. wide, 1.5–7.5 cm. long, acute to rounded at apex, gradually or abruptly narrowed to base; petioles 0.5–3 cm. long; cauline leaves obovate, oblong-ovate or oblanceolate, subsessile, 0.8–4 cm. long, entire or with 1–3 low, rounded teeth on one or both margins near base; flowers solitary in axils, showy; buds erect; hypanthium 1.5–4 cm. long, densely but minutely appressed-puberulent outside, pubescent near base within; sepals lance-linear, 1.5–3 cm. long, usually reflexed in pairs at anthesis, free tips 1–3 mm. long; petals yellow, drying rose or pink, broadly obovate, 2–3 cm. long; stamens glabrous, filaments slightly flattened, 12–20 mm. long; anthers linear, 7–8 mm. long; style slender, about equaling petals; stigma-lobes 3–6 mm. long; capsules sessile, spreading, cylindrical, 2.5–4 cm. long, 2–3.5 mm. thick, straight or somewhat curved, minutely appressed-puberulent; seeds fusiform, 1–1.2 mm. long, in 2 rows in each locule.

Sandy soil, Lower Sonoran Zone, Cape Region, Baja California, extending north along the coast a few miles above Todos Santos. Said to occur near Rancho Conejo on the Pacific side of the peninsula in the latitude of La Paz. March.

**2. Oenothera hookeri** Torr. & Gray subsp. **venusta** (Bartlett) Munz, El Aliso **2:** 21, *pl. 3, figs. D–G,* 1949

*Oenothera venusta* Bartlett, Rhodora **16:** 36, 1914.

Plants to 2.5 m. tall, freely branching, pubescence grayish, hairs long, spreading or appressed, often glandular in upper parts; basal leaves elliptic-oblanceolate, sinuateserrulate, strigose, blades 15–20 cm. long, 2.5–3.5 (5.5) cm. wide; petioles 5–15 cm. long; cauline leaves numerous, lanceolate, wavy margined, 5–15 cm. long, 1.2–2.5 cm. wide; inflorescence to 8 dm. long; hypanthium green, 4.5 cm. long, pilose and glandular; sepals green, 3–4.5 cm. long, pilose to appressed-villous, glandular, lacking papillae at base of long hairs, tips 3–5 mm. long; petals pale yellow, 3–4.5 cm. long, usually equally broad; stigma lobes 5–7 mm. long; capsule 2–4 cm. long, about 5 mm. thick at base, pubescent; seeds 1.2–1.5 mm. long.

Occurring on the western base of the Sierra Nevada and in the coastal drainage of southern California and northern Baja California. It has been collected at Vallecitos Creek Canyon and at Jacumba in San Diego County and may occasionally be found within the western boundry of the Colorado Desert. June–Oct.

**3. Oenothera kunthiana** (Spach) Munz, Amer. Jour. Bot. **19**: 759, 1932

*Hartmannia kunthiana* Spach, Nouv. Ann. Mus. Paris **4**: 362, 1835.
(For full synonymy see Munz, Amer. Jour. Bot. **19**: 759, 1932.)

Perennial herb (sometimes blooming the first season) with numerous ascending, slender stems 1–6 dm. long from a slender taproot; herbage sparsely crisped-pubescent with spreading hairs and younger parts commonly appressed-canescent; basal leaves oblanceolate, 1–2 cm. wide, 3–10 cm. long, mostly sinuate-pinnatifid into ovate to lanceolate, lateral lobes 1–8 mm. long, terminal lobe rhomboid-ovate, 1–3 cm. long; cauline leaves 1–3 cm. long, similarly lobed or sinuate-dentate to subentire; flowers few, solitary in upper axils, sessile or nearly so; hypanthium 1–2 cm. long, strigillose, puberulent within at base; sepals lance-linear, 7–10 mm. long, commonly coherent in pairs or 3 together at anthesis; petals white or pinkish, 1–1.5 cm. long; style equaling or surpassing stamens; stigma lobes 3–5 mm. long; capsule obovoid, body 4–5 mm. thick, 9–12 mm. long, 4-ribbed and 4-winged, these about 1 mm. wide, pedicels 5–30 mm. long in fruit; seeds asymmetrically obovoid, 1 mm. long, brown.

Hillsides, plains and along arroyos, Sonoran Zones, Texas to Sonora and central Mexico. In the Sonoran Desert occurring near Hermosillo, Tesopaco and Cedros, Sonora. March.

**4. Oenothera rosea** Soland., in Ait., Hort. Kew. **2**: 3, 1789

(For synonymy see Munz, Amer. Jour. Bot. **19**: 762, 1932.)

Perennial with several slender, simple or branched stems 1–5 dm. long, ascending or erect from a lignescent caudex, finely strigillose throughout, occasionally with a few additional spreading hairs on lower stems; leaves distant, oblanceolate to narrowly obovate, 2–5 cm. long, entire to sinuate-dentate or pinnatifid, upper ones reduced; flowers in slender racemes; hypanthium 4–8 mm. long, slender, strigillose-canescent outside, puberulent within; sepals ovate-acuminate, 5–8 mm. long, free tips 0.6–1 mm. long; petals rose to red-violet, 5–10 mm. long, broadly obovate; stamens about as long as petals, subequal; anthers 2.5–4 mm. long; style barely equaling stamens; stigma lobes 1.5–2 mm. long; body of capsule obovoid, 8–10 mm. long, 3–4 mm. in diameter (exclusive of 1 mm. wide wings or these narrower); pedicel-like stipe 5–20 mm. long, hollow, ribbed; seeds brown, asymmetrically obovoid, about 0.6 mm. long.

Plains, stream banks and hillsides, Sonoran Zones, central Arizona (Sacaton) to Texas and south to Bolivia and Peru. April–Aug.; Nov.

**Oenothera speciosa** Nutt., Jour. Acad. Phila. **2**: 119, 1821, differs chiefly in having larger more deeply pinnatifid leaves, more cinereous herbage and nodding instead of erect buds. It has escaped from cultivation in many parts of the world and has been collected within the limits of the Sonoran Desert in southern Baja California and in Arizona. Its natural range is from Kansas and Missouri to Texas and south into Durango and San Luis Potosí, Mexico. March–Aug.

**5. Oenothera deltoides** Torr., in Frem., Rep. 315, 1845

*Oenothera trichocalyx* f. *albicauloidea* Lévl., Mon. Onoth. 346, 1909.
*Oenothera trichocalyx* f. *mollis* Lévl., loc. cit.
*Oenothera trichocalyx* f. *refracta* Lévl., loc. cit.
*Oenothera trichocalyx* var. *cineracea* Jepson, Man. Fl. Pl. Calif. 681, 1925.
*Oenothera deltoides* var. *typica* Munz, Amer. Jour. Bot. **18:** 313, 1931.
*Oenothera deltoides* var. *cineracea* Munz, op. cit. 316, 1931.

Coarse annuals 2–10 dm. tall with a simple, erect, central stem and often several decumbent basal branches 1–10 dm. long, whole plant pale green, glabrous or strigillose to spreading-pubescent on lower parts, often spreading-pubescent above; lower leaf blades rhombic-obovate to oblanceolate, 5–30 mm. wide, 2–8 cm. long, subentire to coarsely dentate, petioles about as long; cauline leaves gradually reduced to sessile smaller ones; flowers axillary in upper half of plant, buds nodding; hypanthium 2–4 cm. long, slender; sepals linear-lanceolate, acuminate, 8–30 mm. long, 2.5–4.5 mm. wide, distinct and reflexed in anthesis; petals white, aging pink, 2.5–4 cm. long; anthers 3–9 mm. long; capsules 4–5 cm. long, 2–5 mm. in diameter at base; seeds obovoid, 1.5–2 mm. long.

Desert slopes and plains, Lower Sonoran Zone, Mojave and Colorado Deserts into Arizona and northern Baja California on the eastern side of the mountain axis to the vicinity of Punta Prieta, and at least to Punta Peñasca, Sonora. Feb.–May.

**6. Oenothera wigginsii** Klein, Aliso **5:** 180, 1962

Winter annuals, 15 cm. tall, often with several decumbent-ascending basal branches 5–27 cm. long, the foliage dark green, woolly-pubescent when young; rosette leaves with broadly oblanceolate to rhombic-ovate blades, obtuse to acute, pinnately divided in the lower half, 6–20 cm. long, petioles 3–10 cm. long; cauline leaves reduced and clustered near apices of branches, densely pilose; flowers vespertine, subterminal; hypanthium 2–3 cm. long, slender; sepals lanceolate, often reddish, 10–15 mm. long, distinct and separately reflexed at anthesis; petals white, fading to pink, 1.5–2.5 cm. long, 2–3 cm. wide; anthers about 5 mm. long; capsule straight, tapered above, 2–4 cm. long; seeds lanceolate to ovate in outline, 1.5–2.5 mm. long.

Sand hills, Upper Sonoran Zone, along the Pacific Coast from near Cape Colnett to Punta Baja, northwestern Baja California. Feb.–May.

**7. Oenothera californica** S. Wats., Bot. Calif. **1:** 223, 1876, subsp. **californica**

*Oenothera albicaulis* var. *californica* S. Wats., Proc. Amer. Acad. **8:** 582, 1873.
*Anogra californica* Small, Bull. Torrey Club **23:** 176, 1896.
*Oenothera albicaulis* race *transiens* var. *melanosperma* Lévl., Mon. Onoth. 342, 1909.
*Oenothera pallida* var. *californica* Jepson, Man. Fl. Pl. Calif. 681, 1925.

Perennial with coarse, decumbent to ascending stems 1–6 dm. long, these appressed-pubescent and with some spreading hairs throughout, somewhat cinereous; leaves glabrous to hoary-canescent or villous, basal ones spatulate to oblanceolate, cauline oblong to lanceolate, all subentire to deeply and regularly sinuate-dentate, 5–15 mm. broad, 1–6 cm. long, acute to obtuse at apex, sessile to petiolate with petioles 1–3 cm. long; buds stout; hypanthium 2–4 cm. long, glabrous to villous; sepals lance-linear, 15–20 mm. long, adherent or distinct in anthesis; petals broadly obovate, 2–3 cm. long, rounded to emarginate; anthers 5–10 mm. long; style glabrous; capsules terete, divaricate and usually somewhat curved upward, 2–5 cm. long, 2.5–3 mm. in diameter at base; seeds about 1.5 mm. long.

Interior hills and valleys, chiefly Upper Sonoran and Lower Transition Zones, but barely entering the edge of the desert, Kern and Mono Counties, California, and southwestern Nevada to northern California. Rare along the western margin of the Colorado Desert (near Jacumba, San Diego County; eastern base of Sierra Pinal, southwest of Mexicali). Feb.–Sept.

**7a. Oenothera californica** subsp. **arizonica** (Munz) Klein, Aliso **5:** 179, 1962

*Oenothera deltoides* var. *arizonica* Munz, Amer. Jour. Bot. **18:** 315, 1931.
*Oenothera deltoides* var. *arizonica* f. *floccosa* Munz, loc. cit.

Pubescence on upper parts of plant of very short, closely appressed hairs and spreading ones 1.5–2 mm. long interspersed with short ones; buds with a purple spot at base of each spreading hair; petals 10–25 mm. long; capsule slender, cylindrical, 35–60 mm. long.

In desert areas, Lower Sonoran Zone, central and southwestern Arizona. March–May.

**8. Oenothera caespitosa** Nutt. var. **marginata** (Nutt.) Munz, Amer. Jour. Bot. **18:** 733, 1931

*Oenothera marginata* Nutt. ex Hook. & Arn., Bot. Beechey 342, 1838.
*Pachylophus hirsutus* Rydb., Bull. Torrey Club **31:** 571, 1904, in part.
*Pachylophus cylindrocarpus* A. Nels., Bot. Gaz. **47:** 429, 1909.

Plants perennial, caespitose, with a thick taproot, 1–3 dm. tall, villous-hirsute throughout; leaf blades oblanceolate, 3–12 cm. long, 1–2 cm. wide, sinuate-pinnatifid, petioles winged, about equaling blade; hypanthium 5–8 cm. long, villous without, glabrous within; sepals 2.5–3.5 cm. long; petals obcordate, white, aging pink, 2.5–4 cm. long; capsules 3–4 cm. long, linear-cylindric, not recurved, low tuberculate on angles, on pedicels 1–4 cm. long; seeds obovoid, about 3 mm. long, furrowed along raphe.

Ranging from California and Washington east to Colorado and Arizona. Occurring in the Upper Sonoran and Arid Transition Zones in Yuma and Maricopa Counties, Arizona. April–Aug.

**8a. Oenothera caespitosa** var. **brandegeei** Munz, Amer. Jour. Bot. **18:** 732, 1931

Acaulescent perennial plant with sparsely villous herbage; leaves oblanceolate, 3–12 cm. long, 1–2 cm. wide, lateral part pinnatifid into narrowly lanceolate, ascendingly directed lobes, terminal part ovate to rhombic-ovate; hypanthium 3–4 cm. long, villous without, glabrous within; sepals 8–12 mm. long; petals white, aging pink, 10–12 mm. long; capsule 1–2 cm. long, lance-ovoid, coriaceous, sessile, with rounded tubercles on angles and some on faces; seeds nearly 3 mm. long, obovoid, conspicuously furrowed along raphe.

Arid, stony ridges and hillsides, Lower Sonoran Zone, vicinity of Bahía de los Angeles and adjacent central Baja California.

**9. Oenothera primiveris** A. Gray, Pl. Wright. **2:** 58, 1853, var. **primiveris**

(For synonymy see Munz, Amer. Jour. Bot. **18:** 735–736, 1931.)

Acaulescent annual from a rather heavy taproot, cespitose, occasionally with stems to 1 dm. long; herbage villous throughout or leaf blades sometimes glabrate on upper or both surfaces; leaf blades oblanceolate, 3–30 mm. wide, 1–12 cm. long, rarely subentire, usually deeply pinnatifid, lobes ovate to lanceolate and these again toothed or lobed, all lobes rounded or obtuse; petioles one-half to as long as blades; hypanthium 2–6 cm. long, puberulent on lower half within; sepals linear-lanceolate, 3–5 mm. wide, 12–28 mm. long, usually reflexed in pairs at anthesis; petals yellow, aging orange-red, obovate, 2–3 cm. long, emarginate; stamens about two-thirds as long as petals, glabrous; anthers 8–10 mm. long; style equaling petals; capsule square in cross section with a heavy median rib down each face and somewhat reticulate, gradually tapering to apex, 6–8 mm. thick at base, 2–3.5 cm. long; seeds in 2 rows in each locule, 2.5–3 mm. long, a narrow groove over raphe.

Dry hills and plains, Lower Sonoran Zone, Mojave and Colorado Deserts to Nevada, Utah, western Texas and south into Sonora and Baja California at least to vicinity of Pozo Alemán. Feb.–May.

**9*a*. Oenothera primiveris** var. **caulescens** Munz, Leafl. West. Bot. **4:** 239, 1946

Differing from the preceding in being caulescent; plants 1–3 dm. tall.

Ranging with the species. In southwestern Arizona it appears to replace the species in part. March–April.

**10. Oenothera triloba** Nutt., Jour. Acad. Phila. **2:** 118, 1821

(For synonymy see Munz, Amer. Jour. Bot. **17:** 359–360, 1930.)

Winter annual or biennial, usually acaulescent but sometimes with decumbent to ascending stems to 2 dm. long; herbage glabrate to finely strigillose and often with calyces, ovaries, and veins and margins of leaves glandular-pubescent; leaves mostly crowded, basal, blades oblanceolate, 1–4 cm. wide, 3–20 cm. long, runcinate-pinnatifid almost to midrib, lobes acute to acuminate and dentate or serrate, linear to lanceolate, divergent or recurved, gradually smaller toward base of leaf, terminal lobe ovate to lanceolate, acute, considerably larger than laterals; petioles shorter than blades; flowers often nearly hidden by leaves, vespertine; hypanthium 2–10 cm. long; sepals linear-lanceolate, 2–4 mm. wide, 10–18 mm. long, with free tips 4–6 mm. long; petals delicate, pale yellow, orbicular-obovate, 1–2 cm. long, emarginate; stamens two-thirds as long as petals; anthers 5–9 mm. long; capsule subwoody, 5–10 mm. wide at base, 10–20 mm. long, with 4 reticulate wings 1–3 mm. wide; seeds many, shallowly concave and weakly carinate ventrally, summit narrowly winged.

Arid soils, Sonoran Zones, Kentucky and Tennessee to Oklahoma and Texas, and in central Baja California. March–May.

**11. Oenothera crassifolia** Greene, Bull. Calif. Acad. **1:** 188, 1885

Shrubby or suffrutescent plant to 1.5 m. tall, occasionally to 2 cm. thick at base, considerably branched above and often with several stems from base, glabrous and slightly glaucous throughout, branches quite brittle when young; leaves thickish, sessile or nearly so, oblong-ovate to lanceolate-oblong, entire or more commonly sinuately toothed, acute to obtuse at apex, 5–20 mm. wide, 1–4(6) cm. long, upper ones gradually reduced and usually entire; flowers axillary; calyx tube obconic, 2.5–3 mm. long lined with a shallowly lobed, glabrous, orange-colored disk; sepals lanceolate, 7–10 mm. long, acute at apex; petals pale yellow, streaked with red near base, usually drying orange or reddish, 10–15 mm. long and wide; filaments 4–5 mm. long, glabrous; anthers linear, 3–5 mm. long, sparsely and minutely pubescent; style 10–12 mm. long, pubescent below; stigma globose, about 1 mm. in diameter; capsule linear, nearly straight, slightly curved or coiled almost into a circle, 1–2 mm. in diameter, 2–3.5 cm. long; seeds linear-obovoid, 1–1.3 mm. long, brown with minute purplish dots.

Sandy coastal slopes and dunes, Sonoran Zones, vicinity of San Quintín, south along the western coast of Baja California at least to Scammon's Lagoon. Feb.–May.

**12. Oenothera angelorum** S. Wats., Proc. Amer. Acad. **24:** 49, 1889

Annual or possibly biennial under favorable conditions, erect and often with several ascending branches 3–6 dm. long arising at or near base, these slender, glabrous and weakly glaucescent throughout; leaf blades lanceolate to oblanceolate, glabrate to finely pubescent, mainly basal, deeply and fairly regularly pinnatifid almost to midrib, 1–2 cm. wide, 4–10 cm. long (upper ones gradually reduced), lateral lobes linear to ovate-deltoid, spreading or ascending; flowers present even in lower axils, scattered above in pseudoracemose fashion axillary to reduced leaves; hypanthium obconic, 2–3 mm. long, lined with an indistinct disk and pubescent on lower part within, finely strigillose without; sepals linear-lanceolate to oblong-lanceolate, 6–10 mm. long, 2.5–4 mm. wide, often with scattered, reddish purple dots, finely and sparsely strigillose to glabrous; petals obovate, 8–15 mm. long and wide, yellow, drying rose; stamens in two lengths, shorter ones one-half, larger ones two-thirds, as long as petals; anthers all alike, 2.5–3 mm. long, sparsely pubescent; stigma globose; capsules cylindrical or slightly enlarged at base, more or less contorted, 1–2 mm. in diameter, 2–5 cm. long, bearing a beak 3–6 mm. long; seeds 1.5–2 mm. long, brownish with purple spots.

Sandy plains, Lower Sonoran Zone, central Baja California in the vicinity of Bahía de los Angeles, Punta Prieta, and Calmallí. Feb.–April.

**13. Oenothera leptocarpa** Greene, Pittonia **1:** 302, 1889

*Eulobus californicus* Nutt., in Torr. & Gray, Fl. N. Amer. **1:** 515, 1840.
*Oenothera californica* Greene, Pittonia **1:** 290, 1889. Not *O. californica* S. Wats., 1876.
*Oenothera eulobus* Lévl., Mon. Onoth. 235, 1905.
*Oenothera crassifolia* var. *leptocarpa* Lévl., op. cit. 209.

Erect coarse-stemmed annual with few to several stiffish, erect or ascending, glabrous or glabrate, slightly glaucescent branches to 1.5 m. tall; leaves mostly in basal rosettes, lanceolate in outline, 1–2.5 cm. wide, 5–15(25) cm. long, dentate to pinnatifid with narrow and entire to broader and again toothed or lobed lateral lobes; basal leaves early deciduous; flowers solitary in upper axils; hypanthium obconical, 1–1.5 mm. long, pubescent and lined with an orange, lobed disk within, strigillose to pubescent without; sepals lanceolate, 5–8 mm. long, glabrous or finely and sparsely pubescent, reflexed in anthesis; petals yellow to pale orange, often with reddish spots near base within, rhombic-obovate, 5–14 mm. long; longer stamens with linear anthers, 1.5–2 mm. long, shorter ones with subglobose anthers about 0.5 mm. long; stigma globose, 1–2 mm. in diameter; capsules squarish in cross section, straight, usually strongly deflexed, about 1–1.5 mm. in diameter, 3–12 cm. long, inconspicuously short-beaked; seeds about 1 mm. long, brown with purple flecks.

Hillsides and plains, Sonoran Zones, Kern County, California, to western Arizona, central Sonora and central Baja California. Usually on open hillsides or plains, rarely among shrubs. The basal leaves are scarcely ever present when collected in flower and young fruit. Feb.–July.

**14. Oenothera sceptrostigma** Brandegee, Proc. Calif. Acad. II, **2:** 156, 1889

Annual with a simple stem 6–12 cm. high or more commonly several spreadingly ascending to decumbent stems 1–3.5 dm. long, sparsely and minutely appressed-puberulent, older stems becoming subglabrate and somewhat shining; leaves basal and cauline, lanceolate in outline, 0.6–1.5 cm. broad, 2.5–14 cm. long, deeply pinnatifid almost to midrib into narrow, secondarily toothed or lobed segments 1–8 mm. long, lateral lobes spreading or very slightly deflexed, acute to obtuse at tips, 0.5–1.5 mm. wide; flowers axillary on stem and often somewhat crowded about base of plant; hypanthium obconic, 4–5 mm. long, minutely strigillose without, glabrous within, disk poorly developed; sepals lance-linear, 7–10 mm. long, reddish, sparsely puberulent; petals bright yellow drying rose or pink, 10–15 mm. long, nearly as wide; longer stamens about 5 mm. long, shorter ones about 3 mm. long; anthers all alike, 2.5–3 mm. long, glabrous; stigma cylindrical, 2.5–3 mm. long; capsules cylindrical, straight or somewhat curved but not coiled, about 0.6–0.8 mm. in diameter, 2.5–5 cm. long, narrow beak 5–6 mm. long; seeds obovoid, about 1 mm. long, dull brown, minutely cellular-granular.

Sandy plains, Lower Sonoran Zone, Calmallí, Vizcaíno Desert, and south on the Magdalena Plain to vicinity of El Refúgio (latitude 24°40′ N.), Baja California. Nov.–April.

**15. Oenothera refracta** S. Wats., Proc. Amer. Acad. **17:** 373, 1882

*Sphaerostigma refractum* Small, Bull. Torrey Club **23:** 192, 1896.
*Oenothera deserti* M. E. Jones, Contr. West. Bot. **12:** 15, 1908.
*Sphaerostigma deserti* Heller, Muhlenbergia **9:** 68, 1913.

Erect to spreadingly ascending branched annual 1.5–4 dm. tall, commonly sparsely strigillose and glandular-puberulent throughout; stems slender, usually reddish in youth and becoming white with exfoliating, papery epidermis in age; leaves oblong-linear, lanceolate or oblanceolate, 3–8 mm. wide, 2–5 cm. long, entire to shallowly and sparingly denticulate, often tinged with red, tip callous, upper subsessile, lower ones gradually narrowed to petioles to 2 cm. long; inflorescence racemose to paniculate; hypanthium narrowly funnelform, 4–6 mm. long, finely puberulent within, nearly or quite glabrous without; sepals lance-oblong, 5–6 mm. long; petals broadly obovate, 5–7 mm. long, pale yellow or whitish; stamens about equaling petals; anthers 2–2.5 mm. long; capsule linear, refracted to spreading, 0.5–0.8 mm. in diameter, 3–5 cm. long, curved, straight or rarely coiled; seeds linear-obovoid, about 1 mm. long, pale buff.

Desert hillsides and plains, Lower Sonoran Zone, southern Utah to Inyo and Kern Counties, California, and south into Arizona, the Colorado Desert and northeastern Baja California. March–June.

**16. Oenothera chamaenerioides** A. Gray, Pl. Wright. **2:** 58, 1853

*Sphaerostigma chamaenerioides* Small, Bull. Torrey Club **23:** 189, 1896.
*Sphaerostigma erythra* Davidson, Bull. S. Calif. Acad. **1:** 118, *pl. 9,* 1902.
*Oenothera erythra* J. F. Macbride, Contr. Gray Herb. No. **65:** 41, 1922.

Erect annual branching from base and above, 1–5 dm. tall, stems slender, reddish, often glandular-puberulent on lower portions, strigillose above; leaves both basal and cauline, thin, bright green or sometimes suffused with red, ovate-lanceolate, lanceolate or narrowly oblanceolate, entire, 3–9 mm. wide, 1–6 cm. long, upper subsessile, lower on petioles to one-half as long as blade; inflorescence racemose, to 2 dm. long in fruit; hypanthium narrowly funnelform, 1.5–3 mm. long, glabrous without, puberulent within; sepals narrowly ovate, 1.5–2.5 mm. long, sparsely and minutely puberulent to glabrous; petals barely exceeding sepals, spatulate, white but usually drying pinkish or rose; stamens nearly equal, 2.5–3 mm. long, including 1 mm. long anthers; capsule linear, terete in cross section, about 0.5–0.8 mm. in diameter, 2.5–5 cm. long, divaricate-spreading, glabrous, truncate and beakless at apex; seeds linear, 0.8–1 mm. long, pale gray, ends obliquely truncate.

Arid hills and deserts, Lower Sonoran Zone, southern Nevada and Utah to the Colorado Desert, western Texas, central Sonora, and central Baja California. (It has been collected as far south as Torres, Sonora.) Feb.–June.

**17. Oenothera decorticans** (Hook. & Arn.) Greene var. **desertorum** Munz, Bot. Gaz. **85:** 246, 1928

Erect annual 1–5 dm. tall with a simple stem or more commonly branched from near base and more or less above, epidermis shining white or sometimes slightly reddish and readily exfoliating in papery shreds and sheets, finely puberulent, and somewhat glandular in inflorescence, glabrate below; leaves near base of stems oblanceolate to lance-ovate, entire or minutely and remotely denticulate, acute at apex, gradually narrowing to petiole, 5–15 (rarely to 30) mm. wide, 1.5–8 cm. long, petioles to 2 cm. long, sometimes longer in very vigorous plants; upper leaves reduced to lanceolate or narrowly elliptic foliose bracts to 1.5 cm. long; inflorescence compact spikes to 1.5 dm. long; hypanthium narrowly funnelform, 3.5–5 mm. long, glabrous without, puberulent within; sepals ovate to lance-ovate, 3.5–5 mm. long, minutely and sparsely puberulent; petals 4–5.5 mm. long, white, broadly obovate, but distinctly longer than broad; subequal stamens about equaling petals; anthers 1–2 mm. long, glabrous; capsules slender, 1–1.5 mm. in diameter at base, contorted so the tip often turns downward, 1.5–2 cm. long, gradually narrowed to apex; seeds cinerous, about 1 mm. long.

Desert plains, washes and hillsides, Lower Sonoran Zone, southern Nevada through the Owens Valley region, the western part of the Mojave and Colorado Deserts where it occurs sparingly near Palm Springs. March–Aug.

**17*a*. Oenothera decorticans** var. **condensata** Munz, Bot. Gaz. **85:** 247, 1928

Differing from var. *desertorum* in being somewhat lower with stouter stems usually not over 15–18 cm. tall, epidermis pure white and rarely tinged with red; petals 4–5 mm. long and almost or fully as wide instead of broadly spatulate; capsules quite woody and heavy, about 3 mm. in diameter at base, quadrangular in cross section with a lower, supplementary ridge running longitudinally along each lateral face, capsules often very closely crowded on fruiting stems.

Desert sandy areas, Lower Sonoran Zone, Colorado Desert, through the eastern part of the Mojave Desert and east to southwestern Utah, southwestern Arizona and adjacent northwestern Sonora. It probably occurs in northeastern Baja California also, but I have seen no specimens from that area. Feb.–June.

**18. Oenothera micrantha** Hornem ex Spreng. var. **exfoliata** (A. Nels.) Munz, Amer. Jour. Bot. **19:** 778, 1932

*Sphaerostigma micranthum exfoliatum* A. Nels., Bot. Gaz. **40:** 59, 1905.
*Sphaerostigma pallidum* Abrams, Bull. Torrey Club **32:** 539, 1905.
*Oenothera abramsi* J. F. Macbride, Contr. Gray Herb. No. **65:** 41, 1922.
*Oenthera micrantha* var. *abramsii* Jepson, Man. Fl. Pl. Calif. 684, 1925.
*Oenothera hirta* var. *exfoliata* Munz, Bot. Gaz. **85:** 264, 1928.

Annual with several ascending stems, these and leaves minutely but closely cinereous-puberulent with closely appressed hairs, 0.5–6 dm. tall and leafy throughout; basal leaves linear-lanceolate to oblanceolate, entire or virtually so, 3–15 mm. broad, 1.5–10 cm. long, gradually narrowing to petioles usually shorter than blades; lower cauline leaves somewhat broader and shorter; upper cauline leaves often ovate-clasping, more or less undulate, much reduced; flowers axillary to nearly all leaves, even basal ones; hypanthium 1.5–5 mm. long, cinereous without; sepals 4–6 mm. long; petals yellow, 3–6 mm. long, often emarginate; longer stamens about two-thirds as long as petals; capsules quadrangular, curved or contorted, 1–2 mm. in diameter, 1–2.5 cm. long, attenuate-beaked; seeds dark brown to dull black, about 1 mm. long.

Desert regions, Lower Sonoran Zone, Colorado and Mojave Deserts east into western Arizona and south into northeastern Baja California. March–May.

**18*a*. Oenothera micrantha** var. **ignota** Jepson, Man. Fl. Pl. Calif. 684, 1925

*Oenothera hirta* var. *ignota* Munz, Bot. Gaz. **85:** 263, 1928.

Differing from var. *exfoliata* in having stems and herbage sparsely invested with spreading, rather coarse, white hairs but usually becoming glabrate in time, not appressed-cinereous or pallid; stems often simple and erect instead of several and ascending.

This variety occurs chiefly in the interior hills and valleys, in the Upper Sonoran Zone, from Madera County, California, into Baja California. However, it enters the margins of the Sonoran Desert in the Borrego Valley and along the western edge of the Colorado Desert. Specimens that appear to belong here, or perhaps represent intergrades with var. *jonesii*, occur near El Mármol and in the vicinity of Calmallí, Baja California. It is possible that the immature plants from these two localities represent a separate entity but they present no characters by which I can segregate them. Feb.–Aug.

**18*b*. Oenothera micrantha** var. **jonesii** (Lévl.) Munz, Amer. Jour. Bot. **19:** 778, 1932

*Oenothera hirtella* Greene, Fl. Fran. 215, 1891.
*Oenothera hirta* var. *jonesii* Lévl., Mon. Onoth. 213, 1905.
*Sphaerostigma arenicolum* A. Nels, Bot. Gaz. **40:** 58, 1905.
*Sphaerostigma micranthum jonesi* A. Nels., op. cit. 59.

This variety differs from var. *exfoliata* in virtually the same manner as does var. *ignota* but the herbage does not become glabrate, the spreading coarse hairs persisting; it further differs from both of the above varieties in having smaller flowers with petals only 2–3 mm. long.

Lower interior valleys, Upper Sonoran Zone, Mendocino and Amador Counties, California, south into northern Baja California where it barely enters the upper margins of the Sonoran Desert near the southern end of the Sierra San Pedro Mártir. It also occurs as an infrequent straggler along the western margin of the Colorado Desert, where it intergrades with both of the other varieties mentioned above. March–July.

**19. Oenothera bistorta** Nutt. ex Torr. & Gray var. **hallii** (Davidson) Jepson, Man. Fl. Pl. Calif. 685, 1925

*Sphaerostigma hallii* Davidson, Muhlenbergia **3:** 107, 1907.

Annual with 1–several erect or ascending, moderately branched pallid stems, these and leaves closely pubescent with minute, appressed hairs; leaves linear-oblanceolate to linear-lanceolate, 2–6 mm. wide, 1–7 cm. long, entire or denticulate, cauline leaves often secund, broader and shorter than basal ones, subsessile to subcordate-clasping; flowers axillary; hypanthium obconic, 3–5 mm. long, villous without, puberulent within; sepals broadly lanceolate, 7–10 mm. long, short-villous; petals yellow, often drying greenish, obovate to obcordate, 8–15 mm. long, often with a dark spot near base; longer stamens about two-thirds as long as petals; anthers 2.5–3 mm. long; capsules curved to contorted, 1.5–2.5 mm. in diameter, 1–4 cm. long, puberulent, beak 1–3 mm. long; seeds brown, obovoid, about 1 mm. long.

Desert plains and washes, Lower Sonoran Zone, western Colorado Desert, probably entering into northern Baja California.

The variety **veitchiana** Hook., Bot. Mag. **84:** *pl. 5078,* 1858, differing only in possessing green, sparsely villous instead of cinereous herbage, occurs in the coastal drainage of southern California and may extend into the northern limit of the Sonoran Desert in Baja California. Occasional intergrades between the two are known. March–May.

**20. Oenothera arenaria** (A. Nels.) Raven, Univ. Calif. Pub. Bot. **34:** 76, 1962

*Chylisma arenaria* A. Nels., Amer. Jour. Bot. **21:** 575, 1934.
*Oenothera cardiophylla* var. *splendens* Munz & Jhtn., Bull. Torrey Club **49:** 354, 1923.
*Oenothera cardiophylla* var. *longituba* Jepson, Man. Fl. Pl. Calif. 686, 1925.

Perennial from a woody taproot, to 1.8 m. tall, or an annual to 2.5 dm. tall, villous; leaves cordate-deltoid, coarsely dentate or doubly dentate, reduced upward, blades to 6 cm. long and 6 cm. wide; petioles to 6 cm. long; inflorescence nodding, sometimes with glandular hairs; hypanthium 18–40 mm. long, villous without; sepals narrowly deltoid-lanceolate, 8–15 mm. long, to 5.5 mm. wide at base; petals yellow, broadly obovate, 8–20 mm. long, 7–28 mm. wide; stamens about half as long as petals, the anthers 5–8 mm. long, ciliate; capsules 30–44 mm. long, ascending, villous; pedicels 2–5 mm. long; seeds 0.5–0.7 mm. long.

Sandy washes, rocky areas, and granitic slopes, Lower Sonoran Zone, vicinity of Needles south along the Colorado River to the Tinajas Altas Range of Arizona and Sonora and west to the northern end of the Salton Sea. Dec.–April.

**21. Oenothera cardiophylla** Torr., Pacif. R. Rep. **5:** 360, 1858, subsp. **cardiophylla**

*Oenothera cardiophylla* var. *petiolaris* M. E. Jones, Proc. Calif. Acad. II, **5:** 682, 1895.
*Chylismia cardiophylla* Small, Bull. Torrey Club **23:** 193, 1896.
*Oenothera cardiophylla* var. *typica* Munz, Amer. Jour. Bot. **15:** 226, 1928.

Annual or sometimes apparently a short-lived perennial, 1–7 dm. tall with erect, coarse, freely branching stems and minutely and closely puberulent herbage; leaf blades cordate-orbicular, about as long as broad, 0.8–5 cm. wide, 1–6 cm. long, irregularly dentate, obtuse to rounded at apex, finely cinereous to glabrate; petioles 1–7 cm. long, cinereous; inflorescences of terminal racemes or some flowers solitary in upper axils; hypanthium obconic to funnelform, 4.5–12 mm. long, finely puberulent within and without; sepals ovate, 5–9 mm. long, acute to slightly attenuate; petals yellow, aging deep rose to red, broadly obovate, 3–12 mm. long, 2–12 mm. wide; longer stamens about two-thirds as long as petals; style barely exceeding petals; capsules broad-cylindrical, 2–6 cm. long, about 3 mm. in diameter, slightly curved, ascending, sometimes somewhat secund; seeds broadly obovoid, about 0.5 mm. long, pale brown.

Arid slopes, washes and plains, Lower Sonoran Zone, Inyo County, California, south through the Mojave and Colorado Deserts to central Baja California and Cedros Island. Feb.–June.

**21*a*. Oenothera cardiophylla** subsp. **cedroensis** (Greene) Raven, Univ. Calif. Pub. Bot. **34:** 79, 1962

*Oenothera cedroensis* Greene, Bull. Calif. Acad. **1:** 187, 1885.

Differing from subsp. *cardiophylla* in the following: leaves ovate-acuminate, longer than broad; pubescence predominantly glandular; petals 3–4.5 mm. long.

Arroyos and borders of washes, Lower Sonoran Zone, Cedros Island and the adjacent mainland of Baja California, west to Bahía de los Angeles and south to the Vizcaíno Desert.

**22. Oenothera brevipes** A. Gray, in Torr., Pacif. R. Rep. **4:** 87, 1857, subsp. **brevipes**

*Chylisma brevipes* Small, Bull. Torrey Club **23:** 194, 1896.
*Oenothera brevipes* var. *typica* Munz, Amer. Jour. Bot. **15:** 228, 1928.

Annual with 1 to several coarse stems 1–4 dm. tall from a rather heavy taproot, spreading-villous to glabrate throughout; leaves mainly basal but some cauline, larger ones 1–4 cm. broad, 5–15 cm. long, blades ovate to oblong-cordate and repandly denticulate or rachis below terminal part with several ovate to lanceolate, lateral lobes, frequently with veins conspicuously reddish or reddish brown beneath, lamina between veins pale green beneath, darker above; petioles to 6 cm. long; upper leaves reduced to denticulate bracts; inflorescence a terminal raceme or sometimes paniculate; hypanthium obconic, 4–8 mm. long, pubescent within at least in lower part, villous without; sepals lance-ovate, 6–10 mm. long, with free tips 0.5–3 mm. long; petals broadly obovate, 6–18 mm. long, 7–25 mm. wide, yellow; stamens about one-half as long as petals; anthers 4–6 mm. long; capsules 2–3 mm. in diameter, 2–9 cm. long, spreading; pedicels 5–20 mm. long; seeds dark straw-colored, obovoid, 1–1.5 mm. long.

Desert washes and plains, Lower Sonoran Zone, southern Esmeralda County, Nevada, south to northern Imperial County, California, and from southwestern Utah south to northern Yuma County, Arizona. March–June.

**22*a*. Oenothera brevipes** subsp. **arizonica** Raven, Univ. Calif. Pub. Bot. **34:** 84, 1962

Differing from subsp. *brevipes* in the following: often branched above, stems villous; terminal leaflets lance-ovate; hypanthium 3–5 mm. long; sepals villous to strigose, free caudate part terminal or lacking; petals 3–8 mm. long, 4–7 mm. wide; anthers 2.5–5 mm. long; capsules 18–60 mm. long; pedicels 2.5–5 mm. long.

Rocky slopes and flats, Lower Sonoran Zone, eastern Imperial County, California, and Yuma County, Arizona. March–April.

**22*b*. Oenothera brevipes** subsp. **pallidula** (Munz) Raven, Univ. Calif. Pub. Bot. **34:** 83, 1962

*Oenothera brevipes* var. *pallidula* Munz, Amer. Jour. Bot. **15:** 229, 1928.
*Oenothera pallidula* Munz, Leafl. West. Bot. **2:** 88, 1938.

Differing from subsp. *brevipes* in the following: usually branching above, stems usually strigose; terminal leaflets lance-ovate to ovate; inflorescence a dense terminal raceme; hypanthium obconic, 4–5 mm. long; sepals strigose, lacking projections, or these reduced; petals obovate, 7–12 mm. long; anthers 4–6 mm. long; capsule 20–42 mm. long; pedicels 2–10 mm. long.

Sonoran Zones, ranging from Inyo County, California, east through southern

Nevada to southern Utah and south along the Colorado River to southeastern California and western Arizona. April–July.

**23. Oenothera claviformis** Torr. & Frem. subsp. **aurantiaca** (S. Wats.) Raven, Univ. Calif. Pub. Bot. **34:** 100, 1962

*Oenothera scapoidea* var. *aurantiaca* S. Wats., Proc. Amer. Acad. **8:** 595, 1873, in part.
*Chylismia scapoidea aurantiaca* Davids. & Moxley, Fl. S. Calif. 254, 1923.
*Oenothera clavaeformis* var. *aurantiaca* Munz, Amer. Jour. Bot. **15:** 237, 1928.

Annual with 1 to several erect or ascending, mostly unbranched stems from the base, 0.5–5 dm. tall, finely strigillose or sometimes glabrate in age; leaves chiefly basal, simple and irregularly dentate to deeply pinnatifid, 0.6–3 cm. wide, 2–8 cm. long, sometimes with several discrete pinnae along the upper part of petiole, the blades appressed-pubescent, veins often reddish beneath; cauline leaves gradually reduced to bracts 5–12 mm. long; inflorescence in racemes to 2.5 dm. long, only a few flowers opening at a time; pedicels slender, 6–25 mm. long; hypanthium 3–5 mm. long, strigillose or sometimes glabrate, pubescent within; sepals lanceolate-ovate, 4–5 mm. long, tips 0.5–0.8 mm. long, free; petals obovate to nearly orbicular, 2.5–8 mm. long, white, rarely purple-dotted at the base; filaments 2–5 mm. long, pubescent with spreading white hairs; anthers 2.5–4.5 mm. long; capsule clavate, over 2 mm. in diameter, 12–30 mm. long (usually 15–18 mm. long), somewhat curved, ascending; pedicels 8–25 mm. long; seeds obovoid, pale buff to light brown, rounded or slightly angled, 0.8–1.2 mm. long.

Desert washes and flats, Lower Sonoran Zone, Death Valley region south into the Colorado Desert and east to southern Nevada. March–June.

**23*a*. Oenothera claviformis** subsp. **peirsonii** (Munz) Raven, Univ. Calif. Pub. Bot. **34:** 105, 1962

*Oenothera clavaeformis* var. *peirsonii* Munz, Amer. Jour. Bot. **15:** 238, 1928.

Differing from subsp. *aurantiaca* in having bright yellow petals that do not fade after anthesis and spreading-villous pubescence below.

Arid plains and washes, Lower Sonoran Zone, western Colorado Desert south to the vicinity of San Felipe, Baja California. Feb.–May.

**23*b*. Oenothera claviformis** subsp. **wigginsii** Raven, Univ. Calif. Pub. Bot. **34:** 103, 1962

Differing from subsp. *aurantiaca* in having spreading-villous pubescence below and yellow petals, 1.5–2 mm. long, which fade to brick-red after anthesis.

Lower Sonoran Zone, central Baja California from the vicinity of San Fernando to Bahía de los Angeles. Feb.–March.

**23*c*. Oenothera claviformis** subsp. **rubescens** Raven, Univ. Calif. Pub. Bot. **34:** 103, 1962

Differing from subsp. *aurantiaca* in having spreading-villous pubescence below, yellow petals, 3–5 mm. long, which fade to brick-red after anthesis.

Sandy flats, Lower Sonoran Zone, southwestern Arizona south to the vicinity of Puerto Peñasco. March–April.

**23*d*. Oenothera claviformis** subsp. **yumae** Raven, Univ. Calif. Pub. Bot. **34:** 104, 1962

Differing from subsp. *aurantiaca* in having yellow petals.

Dry dunes and sandy flats, Lower Sonoran Zone, from southwestern Imperial County, California, to the vicinity of Puerto Peñasco, Sonora. Feb.–April.

**23*e*. Oenothera claviformis** subsp. **peeblesii** (Munz) Raven, Univ. Calif. Pub. Bot. **34:** 101, 1962

*Oenothera clavaeformis* var. *peeblesii* Munz, Leafl. West. Bot. **2:** 158, 1939.

Differing from subsp. *aurantiaca* in having glandular pubescent leaves, stems, and often inflorescences.

Desert washes and flats, Lower Sonoran Zone, central Arizona in Yavapai, Pinal, Pima, Maricopa, Cochise, and eastern Yuma counties and adjacent Sonora. March–May.

## 6. GAURA L., Sp. Pl. 347, 1753

Annual or perennial herbs or rarely suffrutescent at base. Leaves simple, subentire to sinuate or dentate. Flowers spicate or racemose, 3–4-merous, usually irregular. Hypanthium obconic to cylindrical. Stamens 8 (or 6), a scale usually at base of each filament. Style filiform, deflexed. Stigma more or less 4-lobed. Ovary 4- (3-) celled or only partially so owing to obsolete partitions, each cell 1–2-ovuled. Capsule woody, 3–4-sided, indehiscent or 3–4-cleft apically, 1–4-seeded.

Petals 1–2 mm. long; sepals 1.5–2 mm. long; capsule sessile or essentially so; plant annual or biennial .................................... 1. *G. parviflora* var. *lachnocarpa*

Petals about 6 mm. long; sepals 9–10 mm. long; capsule narrowed to a stipelike base; plant perennial .......................................... 2. *G. coccinea* var. *arizonica*

**1. Gaura parviflora** Dougl. ex Hook var. **lachnocarpa** Weatherby, Rhodora **27:** 14, 1925

*Gaura australis* Griseb., Goett. Abh. **24:** 132, 1879.

Biennial or winter annual 2–30 dm. tall, usually freely and widely branched above, soft-villous throughout with spreading white hairs and having some shorter glandular ones; cauline leaves sessile or nearly so, lanceolate to lance-ovate, acute to acuminate, 1–2.5 cm. wide, 3–10 cm. long, entire to remotely sinuate-denticulate; caducous bracts 2–5 mm. long, linear-lanceolate; inflorescence 1–3 dm. long, tips usually nodding; hypanthium 1.5–2.5 mm. long, short-pubescent with divergent hairs; sepals 1.5–2 mm. long, oblong-lanceolate, separate and reflexed in anthesis, puberulent; petals 1–2 mm. long, erect, spatulate; stamens scarcely equaling sepals, without basal scales; anthers about 1 mm. long; capsule sessile, about 2 mm. in diameter, 6–10 mm. long, narrowly fusiform, 4-angled above, 1–2-seeded, these brown.

Along streams or arroyos, Sonoran Zones, Utah to Missouri and south through Mexico and in Argentina. Within the Sonoran Desert it is common from central Arizona through Sonora, and from Mulegé into the Cape Region in Baja California. Feb.–July.

**2. Gaura coccinea** (Nutt.) Pursh var. **arizonica** Munz, Bull. Torrey Club **65:** 225, 1938

Perennial herb with several stems 3–6 dm. tall and numerous ascending, canescent-strigillose, densely leafy branches; leaves lanceolate or lance-ovate, 3–8 mm. wide, 2–5 cm. long, margins undulate; spikes erect, at least in fruit, tip often nodding in bud,

1–3 dm. long; floral bracts lance-linear, 4–12 mm. long; hypanthium narrow, 6–7(10) mm. long, cinereous-strigillose; sepals 9–10 mm. long, cinereous-strigillose; petals about 6 mm. long; anthers 4–5 mm. long; capsule 1.5–2 mm. in diameter, 10–12 mm. long, fusiform, usually cinereous-strigillose, beak acuminate, abruptly narrowed to a stipelike base.

Hillsides, canyons and plains, Sonoran Zones, northern and central Arizona to New Mexico and Coahuila. Occurring between Casa Grande and Tucson, Arizona. April–May.

## 7. BURRAGEA Donn. Smith & Rose, Contr. U.S. Nat. Herb. 16: 297, 1913

A low shrub with thin, reddish, exfoliating bark and alternate, subsessile, somewhat thickish, crowded leaves. Inflorescence of short terminal spikes. Hypanthium slender, longer than the 4 bright red sepals, partially closed at mouth by an annular disk. Petals 4, rose-red. Stamens 8, alternating 4 long and 4 short ones. Style slender; stigma capitate, shallowly bilobed. Ovary embedded in swollen branch, 2-celled. Capsule obscurely diamond-shaped, 2-seeded, persistently embedded in stem, closely crowded against adjacent fruits.

Plants glandular-pubescent; leaves linear to narrowly oblanceolate ........................................................ 1. *B. fruticulosa* subsp. *fruticulosa*
Plants glabrous; leaves oblanceolate ...................... 1a. *B. fruticulosa* subsp. *glabra*

### 1. **Burragea fruticulosa** (Benth.) Donn. Smith & Rose, Contr. U.S. Nat. Herb. **16:** 298, 1913, subsp. **fruticulosa**

*Gaura? fruticulosa* Benth., Bot. Sulph. 15, 1844.
*Gongylocarpus frutescens* Curran, Proc. Calif. Acad. II, **1:** 231, 1888.
*Gongylocarpus fruticulosus* Brandegee, op. cit. II, **2:** 158, 1889.
*Burragea frutescens* Donn. Smith & Rose, Contr. U.S. Nat. Herb. **16:** 298, 1913.
*Burragea fruticulosa* var. *frutescens* Munz, Aliso **4:** 502, 1960.

A copiously branched shrub 3–6 dm. tall; young shoots reddish to purplish, puberulent with intermingled, glandular and nonglandular hairs; leaves solitary but internodes mostly very short and leaves closely spaced, linear to narrowly oblanceolate, 2–6 mm. wide, 2–4 cm. long, acute at apex, gradually narrowing toward base, finely glandular-puberulent, entire; inflorescence several- to many-flowered, 2–10 cm. long; hypanthium 1.5–3 cm. long, 1–1.5 mm. in diameter; sepals lanceolate, 8–10 mm. long, acuminate, separately deflexed at anthesis, minutely and sparsely strigillose near base within; petals obovate, short-clawed, 7–10 mm. long; stamens subequal, 5–9 mm. long; style filiform, glabrous, equaling or exceeding stamens; capsule woody, diamond-shaped by compression within woody stem, terminal face flush with swollen stem, this 4–6 mm. in diameter in fruit and 2–10 cm. long, capsules indehiscent or very tardily separating; seeds dark brown, 3–4 mm. long, angled.

Arid slopes, Lower Sonoran Zone, Santa Margarita and Magdalena Islands and along the shores of Magdalena Bay, Baja California. Oct.–May.

### *1a.* **Burragea fruticulosa** subsp. **glabra** (J. H. Thomas) Munz, Aliso **4:** 502, 1960

*Burragea glabra* J. H. Thomas, Madroño **10:** 164, 1950.

Plants low shrubs, freely branched, branches often stunted, young ones purplish, aging reddish brown, bark shredding, glabrous; leaves numerous on ends of twigs, oblanceolate, 5–8 mm. wide, 2–3.5 mm. long, subsessile, glabrous; hypanthium narrow, 2–2.5 cm. long; sepals lanceolate, about 1 cm. long, reflexed in anthesis, tips acute, free in bud; petals orbicular to ovate, about 1 cm. long, slightly clawed; stamens subequal, 5–9 mm. long; style equaling or exceeding stamens; capsule woody, embedded in swollen stem, this 4–7 mm. in diameter, 2–8 cm. long in fruit, tardily dehiscent; seeds 3–4 mm. long.

Known only from Santa María Bay, Baja California. March.

### 8. **LOPEZIA** Cav., Ic. **1:** 12, 1791

Erect herbs with petiolate, dentate, alternate leaves or lower ones opposite. Flowers small, in racemes or in subcorymbose clusters, pedicellate. Calyx tube subglobose, very short. Sepals 4, linear, deciduous, somewhat unequal, posterior 3 united at base. Petals 4, slightly irregular, posterior ones narrower than other pair, short- to long-clawed. Stamens 2, posterior bearing an anther, other sterile and somewhat petaloid; anther linear-oblong. Ovary 4-celled; style short and filiform; stigma truncate-capitate. Capsule globose to short-cylindric, coriaceous, 4-valved. Seeds many, obovoid, in several series in each cell, minutely granulate.

**1. Lopezia clavata** Brandegee, Proc. Calif. II, **2:** 157, *pl. 4,* 1889

Erect annual herbs 2–15 dm. tall, simple or branched above, glabrous or leaf blades sparsely hirsute and ciliate with coarse white hairs; leaves ovate, 1–4 cm. wide, 2.5–8 cm. long, acuminate at apex, acute or obtuse at base, thin, upper ones alternate, lower ones opposite or nearly so; uppermost leaves reduced to sessile bracts 1 cm. or less in length; petioles slender, to 2 cm. long; flowers in terminal, open racemes; pedicels subfiliform, 5–10 mm. long at anthesis, elongating to nearly twice as long in fruit; posterior 3 sepals 3 mm. long, lanceolate-ovate, united about one-third of their length and this united area covered by a yellow, subfleshy disk, anterior sepal slightly longer, free; petals 4, anterior pair broadly oblanceolate, 4–5 mm. long, shortly but distinctly clawed, posterior pair 3–4 mm. long, sessile on margin of semicircular disk; posterior stamen nearly 2 mm. high, antheriferous, anterior one petaloid and somewhat embracing pistil, about as long as fertile stamen; capsule clavate-cylindrical, 5–8 mm. long, 1–1.5 mm. in diameter, very sparsely pubescent to glabrate, 4-valvate at summit; seeds broadly obovoid, brown, 0.6–0.8 mm. long, arranged in 2 rows in each cell.

Canyons and shaded hillsides, Sonoran Zones, vicinity of Comondú to the Cape Region, Baja California. In the Sierra de la Laguna in the Cape Region it occurs at an elevation of about 5,000 feet. Oct.–Feb.

## Family 103. **HALORAGACEAE.** WATERMILFOIL FAMILY

Herbaceous, perennial, aquatic plants with stems partially or wholly immersed. Leaves in whorls, simple or much dissected. Flowers perfect or unisexual, small, sessile in axils of leaves or of bracts. Calyx tube adherent to ovary; sepals minute or lacking. Petals usually 4 or wanting. Stamens 1–8. Ovary inferior, 1–4-celled. Fruit indehiscent, often splitting into separate nutlets.

### 1. **MYRIOPHYLLUM** L., Sp. Pl. 992, 1753

Leaves entire, dentate-serrate or pectinately pinnatifid, in ours the divisions of submerged leaves linear. Flowers in ours axillary. Petals 2–4, sessile. Stamens 2–8. Ovary 4-celled; stigmas 4, recurved and plumose.

**1. Myriophyllum brasiliense** Camb., in A. St. Hil., Fl. Bras. Merid. **2:** 252, 1829
*Myriophyllum proserpinacoides* Gill. ex Hook. & Arn., Bot. Misc. **3:** 313, 1833.

Stems simple or branching, 3–5 mm. in diameter, 2–5 dm. long; leaves in whorls of 3–6, 2–3.5 cm. long, with 8–15 linear segments 3–8 mm. long on each side of midrib; flowers about 1 mm. long, on pedicels of equal length; plumose stigmas equaling hypanthium at anthesis; fruit deeply 4-grooved, about 3 mm. long.

In ponds, Lower Sonoran Zone to Transition Zones, at only a few known localities in eastern Mojave Desert, California; near Sacaton, Arizona; and several localities in Mexico. Native of South America and possibly transported to the Mexican and United States localities by migrating water fowl. May–Aug.

## Family 104. ARALIACEAE. Ginseng Family

Perennial herbs, shrubs or trees with alternate, verticillate or rarely opposite leaves. Inflorescences various, often of clustered umbels. Flowers perfect or polygamodioecious. Sepals minute or absent, 5 when present. Petals 5, valvate or somewhat valvate, inserted on margin of hypanthium. Stamens 5, borne on epigynous disk; anthers introrse. Ovary inferior, 1- to several-celled; styles as many as locules in ovary; each cell 1-ovuled. Fruit a berry or a drupe. Seeds flat or triangular; endosperm abundant; embryo small.

### 1. ARALIA L., Sp. Pl. 273, 1753

Characters of the family. Leaves pinnately (in ours) or ternately decompound; petioles sheathing. Flowers white or greenish, racemose, corymbose, or in paniculate umbels. Sepals 5 or wanting. Petals with inflexed tips. Ovary 5-celled. Fruit a berry.

Leaflets densely puberulent beneath; pedicels glabrous; sepals not ciliolate ............... 1. *A. pubescens*

Leaflets glabrous beneath or sparsely puberulent along midrib only; pedicels puberulent; sepals minutely erose-ciliolate ................................ 2. *A. scopulorum*

**1. Aralia pubescens** DC., Cat. Hort. Monsp. 80, 1813

Shrub or small tree, usually 3–6 m. tall, with thick corky bark; young twigs finely striate, finely short-puberulent with curled or crisped hairs; petioles slender, 5–12 cm. long, puberulent; leaflets 5–11, commonly 7, ovate to elliptic-ovate, 1–4.5 cm. wide, 3–9 cm. long, acute to acuminate, rounded or obtuse at base, serrate with each tooth tipped with a small, yellowish, sessile gland, finely puberulent on both surfaces or sometimes only beneath; peduncles 2–10 cm. long, sometimes having a solitary flower in axil of a small ovate bract 1–2 cm. below the umbel; pedicels slender, glabrous, 9–12 mm. long at anthesis, to 18 mm. long in fruit; sepals deltoid, about 0.4 mm. long; petals greenish white, 2–2.5 mm. long; styles slender, distinct, erect or slightly spreading, nearly straight; berry ellipsoidal to subglobose, 4–4.5 mm. broad, 4–5.5 mm. long, 5-seeded.

Shaded hillsides and canyons, mainly Upper Sonoran Zone, central Sonora (hills west of Cumpas) south to Oaxaca. Aug.–Nov.

**2. Aralia scopulorum** Brandegee, Proc. Calif. Acad. II, **2:** 165, *pl. 8,* 1889

Openly branched shrub or small tree 1–7 m. tall, with light-colored, smooth, corky bark; twigs, petioles, and leaflets glabrous; stipules 3–4 mm. long, minutely ciliolate along free margin; petioles 3–7 cm. long; leaflets 3–7, narrowly to broadly ovate, 1.2–3.2 cm. wide, 2.5–6 cm. long, acuminate at apex, obtuse, rounded, or subcordate at base, glabrous or rarely with a few scattered hairs along midrib beneath, margins finely serrate, teeth curved lightly inward and each blunted by a yellowish, sessile, apical gland; peduncles more or less puberulent, 1–6 cm. long; pedicels sparsely and minutely puberulent, 8–11 mm. long at anthesis; sepals deltoid, about 0.3 mm. long, margins irregularly and minutely ciliolate; petals broadly ovate, greenish, 2.5–3 mm. long, deflexed in anthesis; fruit 5–6-seeded, 5–6 mm. long, distinctly grooved and ridged.

Deep shaded canyons, Lower Sonoran Zone, central Baja California from mountains southeast of Calmallí south into the Sierra de la Giganta east of Comondú. July–Feb.

## Family 105. GARRYACEAE. Silk-Tassel Family

Evergreen shrubs or small trees with erect or strongly ascending, somewhat quadrangular twigs and opposite, simple, entire, short-petiolate, leathery leaves. Flowers dioecious, apetalous, borne in slender pendent aments; staminate flowers in threes in axils of connate and decussate bracts, their sepals 4, linear to lance-oblong, valvate in bud; pistillate flowers solitary in each axil, their sepals obsolete or 2, very small and bractlike, inserted near apex of ovary. Stamens 4, alternating with sepals. Ovary inferior, 1-celled; ovules 2, pendulous; styles 2, persistent, linear-lanceolate, stigmatic on inner surface. Fruit berrylike, rupturing irregularly, containing a bitter pulp that surrounds 1–2 seeds, dries black, and separates from brittle epicarp. Seeds smooth, endosperm horny; embryo small.

### 1. GARRYA Dougl., in Lindl., Bot. Reg. **20:** *pl. 1686,* 1834

Characters of the family.

Aments densely spiciform, flowers closely crowded; bracts markedly connate, deltoid, about as wide as long; leaves distinctly to densely pubescent to tomentose beneath; fruits more or less puberulent at maturity:
- Leaves densely tomentulose beneath with tangled curly hairs, glabrous and light green above . . . . . . . . 1. *G. veatchii*
- Leaves sparsely to moderately appressed-strigose beneath with forward-pointing, straight or slightly curved hairs, gray-glaucous above:
  - Upper surfaces of leaves yellow-green, not glaucous . . . . . 2. *G. flavescens* var. *flavescens*
  - Upper surfaces of leaves glaucous or bluish . . . . . . . . 2a. *G. flavescens* var. *pallida*

Aments lax, flowers less crowded; bracts only slightly connate at very base or wholly distinct, those of pistillate aments linear-lanceolate, leaflike, several times as long as wide; leaves glabrous or glabrate; fruits glabrous at maturity:
- Leaf margins minutely callose-denticulate, not revolute; upper surface of leaves dull, often slightly pulverulent in age; mature leaves one and one-half to two and one-half times as long as wide . . . . . . . . 3. *G. wrightii*
- Leaf margins entire, thin, faintly revolute; upper surface of leaves glossy, yellow-green; mature leaves three or four times as long as wide . . . . . . . . 4. *G. salicifolia*

**1. Garrya veatchii** Kell., Proc. Calif. Acad. **5:** 40, 1873

*Garrya flavescens* var. *palmeri* S. Wats., Bot. Calif. **1:** 276, 1876.
*Garrya veatchii* var. *palmeri* Eastw., Bot. Gaz. **36:** 458, 1903.
*Garrya veatchii* var. *undulata* Eastw., loc. cit.

Compact, erect, rounded shrub 1–3 m. tall; young twigs densely white-tomentose, later glabrate, and with reddish brown bark; petioles 3–6 mm. long, 1.5–2.5 mm. broad, densely short-tomentose; leaf blades ovate, ovate-elliptic or lance-ovate, 1–3 cm. wide, 2.5–6.5 cm. long, coriaceous, plane, slightly undulate or subrevolute, acute to obtuse at base, acute to rounded and often apiculate at apex, densely short-tomentose with crinkled, intertangled, white hairs beneath, glabrate, bright green and shining above, veins fairly prominent but not markedly raised above surface; staminate aments 3–7 cm. long, solitary or in fascicles, simple; bracts floccose, deltoid, upper ones 2.5–4 mm. long, lowermost to 7 mm. long and often with foliaceous tips; pistillate aments 2.5–6 cm. long, compact; fruits globose to ovoid-globose, 6–8 mm. in diameter, brownish or purplish, densely tomentose and with longer, slightly crinkled hairs intermingled with and extending beyond tomentum, at length fruit pubescent in irregular patches but not wholly glabrate; seeds broadly obovoid, about 3–4 mm. long, dull black.

In chaparral on slopes and along intermittent streams, mainly Upper Sonoran Zone but straggling into upper margins of Lower Sonoran Zone, Santa Barbara County, California, south to Cedros Island, Baja California. Dec.–May.

**2. Garrya flavescens** S. Wats., Amer. Nat. **7:** 301, 1873, var. **flavescens**

*Garrya mollis* Greene, Leaflets Bot. Obs. **2:** 86, 1910.

Compact rounded shrub 1–3 m. tall with cinereous-silky young branches and glabrate, brownish, smooth bark in age; petioles to 1.5 cm. long, strigose with white hairs in youth, eventually often glabrate; leaf blades broadly elliptic, broadly lanceolate-ovate, to oval, stiffly coriaceous, broadly cuneate to acute at base, usually acute (sometimes rounded), minutely apiculate at apex, 1–3.5 cm. wide, 3–8 (rarely –25) cm. long, yellow-green and glabrous to glabrate above, cinereous-strigose with straight or slightly curved, upward-pointing, appressed hairs beneath, these not dense enough to obscure venation; staminate aments unbranched, 3–5 cm. long, their bracts broadly ovate, attenuate at apex, 4–7 mm. long, connate into a deep cup, tips of lowermost sometimes foliaceous; pistillate aments compact, 2.5–6 cm. long, their bracts densely sericeous-pubescent, tips of lowermost often slightly recurved; fruits broadly obovoid, 6–8 mm. in diameter, silky-sericeous with straight appressed hairs; seeds broadly ellipsoidal, 3.5–4 mm. long, 2.8–3 mm. in diameter, dull black, smooth.

Arid slopes and ridges, Upper Sonoran and Transition Zones or occasional along upper margins of the Lower Sonoran Zone, southern Nevada and southern Utah south to western Texas and northern Mexico. Jan.–April.

**2*a*. Garrya flavescens** var. **pallida** (Eastw.) Bacigalupi ex Ewan, Bull. Torrey Club **64:** 519, 1937

*Garrya pallida* Eastw., Proc. Calif. Acad. III, **2:** 287, 1902.

Differing from var. *flavescens* only in the presence of a bluish or whitish bloom on the upper side of the leaves.

Chaparral slopes and ridges, mainly Upper Sonoran and Arid Transition Zones but entering the margins of the Lower Sonoran Zone in canyons and on north-facing, shady slopes, Santa Barbara and Fresno Counties, California, south to northern Baja California; apparently intergrading with var. *flavescens* in Inyo County and in several areas in the Mojave Desert. Feb.–April.

**3. Garrya wrightii** Torr., Pacif. R. Rep. **4:** 136, 1857

An erect, compact or rather openly branched shrub to 3 m. tall, with sparsely appressed-pubescent young twigs and reddish brown, glabrate bark on older stems; petioles mostly 3–8 mm. long, sparsely strigose at first but soon glabrate; leaf blades elliptic-ovate to suborbicular, acute to obtuse at base, acute and apiculate to rounded and shallowly emarginate at apex, 1–3.5 cm. wide, 2.5–5.5 cm. long, dull yellowish green, coriaceous, minutely and irregularly callous-denticulate along margins, very sparingly appressed-strigose when very young, soon wholly glabrate on both surfaces, rarely turning black in drying, reticulate venation obvious above; aments loose, their bracts only faintly connate at very base, narrowly lance-ovate to linear-oblanceolate, 5–12 mm. long, midvein prominent; fruits globose, 5–7 mm. in diameter, sparsely strigose when very young, glabrate by maturity, purplish or blackish, persistent bases of styles rather conspicuous; seeds broadly obovoid, 3–3.5 mm. long, dull black, often faintly rugose.

On arid slopes, upper parts of Lower Sonoran Zone upward into Arid Transition Zone, Coconino and Greenlee Counties, Arizona, south into northern Sonora and east to western Texas. March–Aug.

**4. Garrya salicifolia** Eastw., Bot. Gaz. **36**: 463, 1903

Slenderly branched shrub to 5.5 m. tall, often with many branches from base, with very sparsely strigose, quadrangular twigs and dark gray, irregularly checkered bark on older stems; petioles slender, 3–9 mm. long, at first sparsely short-strigose, soon glabrate; leaf blades narrowly elliptical, 10–18 mm. wide, 3–6 cm. long, broadly cuneate at base, acute and short-apiculate at apex, bright green, sparingly puberulent with a few scattered, simple, appressed hairs when young, soon wholly glabrate on both surfaces, often turning black during drying, venation prominent, margins entire but faintly revolute on many plants; aments loose, 2.5–5 cm. long, nodes to 2 cm. apart; bracts lance-elliptical to oblanceolate, 10–18 mm. long or the lowermost leaflike and to 25 mm. long; fruit slightly depressed-globose, 6–8 mm. in diameter, 5–6 mm. high, greenish brown to nearly black, glabrous.

Forested canyons and margins of meadows, mostly in pine-oak, Upper Sonoran or Arid Transition Zone but straggling into Lower Sonoran Zone in shaded canyons, Sierra de la Laguna, southern Baja California. Aug.–Dec.

## Family 106. **UMBELLIFERAE.** Parsley Family*

Herbs with alternate or basal, usually compound leaves with usually sheathing petioles. Perfect or staminate flowers usually in simple or compound umbels. Calyx lobes 5, free or united, or obsolete. Petals 5. Stamens 5. Ovary inferior, 2-celled. Fruit a schizocarp of 2 one-seeded mericarps.

Inflorescence a dense bracteate head .................................. 15. *Eryngium*
Inflorescence a distinct umbel, more or less spreading, never capitate:
  Leaves simple or the blade absent; umbels simple or proliferous:
    Ovary and fruit covered with stellate hairs; foliage more or less stellate-pubescent .... ........................................................ 2. *Bowlesia*
    Ovary and fruit glabrous; foliage glabrous:
      Leaves with a definite ovate to orbicular blade .................... 1. *Hydrocotyle*
      Leaves reduced to hollow, cylindric, jointed phyllodes .............. 11. *Lilaeopsis*

* Text contributed by Mildred E. Mathias, University of California at Los Angeles, and Lincoln Constance, University of California, Berkeley, California.

Leaves variously compound; umbels compound (or rarely simple by reduction):
Ovary and fruit armed with bristles, callous teeth, or papillae:
Stems and leaves variously hispid; involucre foliaceous; bractlets of involucel foliaceous, sometimes pinnately divided; fruit armed with barbed or hooked bristles:
Fruit flattened laterally, bristles hooked, not barbed; stylopodium conic; calyx teeth prominent .......... 6. *Caucalis*
Fruit flattened dorsally, bristles barbed at tip; stylopodium absent; calyx teeth absent .......... 7. *Daucus*
Stems and leaves glabrous or somewhat roughened, never hispid; involucre mostly absent; bractlets of the involucel linear or filiform and entire, or absent; fruit armed with short bristles, callous teeth, or papillae:
Involucel absent; umbels sessile; fruit papillate .......... 3. *Apiastrum*
Involucel present; umbels pedunculate; fruit not papillate:
Fruit covered with short bristles; seed face more or less sulcate ... 4. *Spermolepis*
Fruit covered with callous teeth; seed face concave .......... 5. *Ammoselinum*
Ovary and fruit not armed, sometimes pubescent:
Ribs of the fruit not prominently winged; ovary and fruit terete in cross section or somewhat flattened laterally:
Petals conspicuously unequal; fruit subglobose .......... 10. *Coriandrum*
Petals equal; fruit orbicular to oblong:
Flowers yellow; plants with aniselike odor .......... 8. *Foeniculum*
Flowers white; plants without aniselike odor .......... 9. *Apium*
Ribs of the fruit (some or all of them) prominently winged; fruit more or less flattened dorsally:
Plants annual or biennial, with dill-like odor .......... 13. *Anethum*
Plants perennial, without dill-like odor:
Lateral ribs winged, wings thin; dorsal ribs absent or filiform .... 14. *Lomatium*
Lateral and dorsal ribs winged; wings slightly enlarged at base in cross section .......... 12. *Cymopterus*

## 1. HYDROCOTYLE L., Sp. Pl. 234, 1753

Plants glabrous, perennial, with slender, floating or creeping stems or rootstocks. Leaf blades simple, round, peltate or nonpeltate; stipules present. Flowers in a simple axillary umbel or an interrupted spike; calyx teeth minute; corolla white, greenish or yellow; stylopodium depressed to conspicuously conic. Fruit transversely ovate to orbicular, 1–3 mm. long, strongly flattened laterally; ribs slender or obsolete; oil-bearing cells conspicuous to obsolete; seed face plane to convex.

Leaves peltate, entire to shallowly lobed:
Inflorescence a simple umbel .......... 1. *H. umbellata*
Inflorescence an interrupted spike .......... 2. *H. verticillata*
Leaves nonpeltate, lobed about to middle .......... 3. *H. ranunculoides*

### 1. Hydrocotyle umbellata L., Sp. Pl. 234, 1753

Leaves orbicular-peltate, 5–75 mm. in diameter, crenate or crenately lobed; petioles mostly slender, 0.5–40.5 cm. long; peduncles usually longer than leaves, 1.5–35 cm. long; umbels simple, many-flowered; pedicels 2–25 mm. long, spreading and reflexed; fruit 1–2 mm. long, 2–3 mm. broad.

Plants creeping on wet ground or floating, Lower and Upper Sonoran Zones, Riverside County, California. March–July.

**2. Hydrocotyle verticillata** Thunb., Diss. Hydroc. 2, 1798
*Hydrocotyle cuneata* Coult. & Rose, Contr. U.S. Nat. Herb. **7:** 28, 1900.

Leaves orbicular-peltate, 5–60 mm. in diameter, 7–14-veined, shallowly 8–13-lobed, lobes crenate; petioles slender, 0.5–26 cm. long; inflorescence an axillary, simple, interrupted, once- or twice-bifurcate spike 1.5–17 cm. long, with 2–7 few-flowered verticils, flowers sessile or subsessile; fruit 1–3 mm. long, 2–4 mm. broad.

Plants on wet ground or aquatic, Lower Sonoran Zone, Maricopa and Gila Counties, Arizona. The variety *triradiata* (A. Rich) Fernald is distinguished from the species by the usually unbranched inflorescence and the pedicellate fruit. It has been collected in Pima County, Arizona; northern Sonora. April–Sept.

**3. Hydrocotyle ranunculoides** L. f., Suppl. Syst. Veg. 177, 1781

Leaves not peltate, roundish-reniform with a sinus at base, 5–80 mm. long, 5–80 mm. broad, 5–6-lobed about to middle, lobes crenate or lobulate; petioles mostly slender, 1–34.5 cm. long; peduncles shorter than leaves, 4–60 mm. long; umbels simple, 5–10-flowered; pedicels 1–3 mm. long, spreading and ascending; fruit 1–3 mm. long, 2–3 mm. broad.

Plants creeping on wet ground or floating, Lower Sonoran Zone, Pima County, Arizona. May–Aug.

## 2. BOWLESIA Ruiz & Pav., Fl. Peruv. Prod. 44, *pl. 34,* 1794

Plants annual, stellate-pubescent, prostrate or suberect, dichotomously branching. Leaf blades simple, suborbicular, palmately 5–7-lobed; stipules present. Peduncles axillary, shorter than leaves; umbels simple, few-flowered; involucral bracts small. Calyx teeth prominent; corolla greenish white; stylopodium depressed-conic. Fruit broadly ovate, 1–1.5 mm. long, stellate-pubescent, turgid, narrowed at commissure; ribs and oil tubes obsolete; seed face plane or convex.

**1. Bowlesia incana** Ruiz & Pav., Fl. Peruv. **3:** 28, *pl. 268,* 1802
*Bowlesia septentrionalis* Coult. & Rose, Contr. U.S. Nat. Herb. **7:** 31, *fig. 3,* 1900.

Plants 1.2–5 dm. long or high; leaf blades 4–30 mm. long, 5–45 mm. broad, lobes entire to dentate; petioles 0.5–7 cm. long; peduncles 1–21 mm. long; umbels 2–6-flowered; fruit sessile or subsessile.

Common, usually among bushes, Lower Sonoran Zone, Pima, Pinal, and probably Yuma Counties, Arizona; Baja California; Sonora. Feb.–May.

## 3. APIASTRUM Nutt. ex Torr. & Gray, Fl. N. Amer. 1: 643, 1840

Plants annual, slender, glabrous; stems sometimes simple, usually di- or trichotomously branching. Leaves 2–3-ternate, segments subfiliform to linear. Umbels compound, sessile in axils or opposite upper leaves; involucre and involucel none. Calyx teeth obsolete; corolla white; stylopodium minute and depressed. Fruit ovate or cordate, 1–1.5 mm. long, papillate-roughened; ribs inconspicuous; oil tubes solitary in intervals, 2 on commissure; seed face concave to sulcate.

**1. Apiastrum angustifolium** Nutt. ex Torr. & Gray, Fl. N. Amer. **1:** 644, 1840

Plants 0.5–5 dm. high; leaves orbicular, blades 1–5 cm. long, some or all opposite, ultimate divisions linear-filiform to oblong, obtuse, 5–25 mm. long, 0.5–2 mm. broad;

petioles 2–4 cm. long; peduncles of lower umbels up to 8 cm. long; rays 2–3, unequal, 1–5 cm. long or obsolete; pedicels 3–4, unequal, 15 mm. long to obsolete.

Sandy soils, Lower and Upper Sonoran Zones, foothills of northern Coast Ranges and Sierra Nevada south into Baja California; eastern Pima County, Arizona. March–May.

## 4. SPERMOLEPIS Raf., Neogen. 2, 1825

Plants annual, caulescent, alternately branching, glabrous. Leaves ternately decompound; segments filiform. Peduncles axillary and terminal, exceeding leaves; umbels compound; involucre absent; bractlets of involucel few, filiform, shorter than pedicels, glabrous or callous-toothed. Calyx teeth obsolete; corolla white; stylopodium low-conic. Fruit ovoid, 1.5–2 mm. long, laterally compressed, covered with short echinate bristles; oil tubes 1–3 in intervals, 2 on commissure; seed face more or less sulcate.

**1. Spermolepis echinata** (Nutt.) Heller, Contr. Herb. Frankl. & Marsh. **1:** 73, 1895
*Leptocaulis echinatus* Nutt. ex DC., Coll. Mém. **5:** 39, 1829.

Low and often spreading, 0.5–4 dm. high; leaves ovate, blades 7–25 mm. long, 7–20 mm. broad, ultimate divisions mucronulate, 2–18 mm. long, 0.5–1 mm. broad; petioles 3–20 mm. long; peduncles 8–65 mm. long; rays 5–14, suberect, unequal, 1–15 mm. long; pedicels 1–6, up to 7 mm. long, central umbellets 1-flowered, sessile.

Sandy plains and washes, Lower Sonoran Zone to Arid Transition Zone, Pima, Maricopa, and Pinal Counties, Arizona; Sonora; San Diego County, California; Baja California. Feb.–May.

## 5. AMMOSELINUM Torr. & Gray, Pacif. R. Rep. 2: 165, 1855

Plants annual, caulescent, branching, more or less roughened. Leaves ternate, then bipinnate, segments linear. Peduncles axillary and terminal, up to 4 cm. long, or none; umbels compound; involucre mostly absent; bractlets of involucel few, linear, acute, somewhat callous-toothed, about equaling pedicels. Calyx teeth obsolete; corolla white; stylopodium low-conic. Fruit oblong-ovate, 3–5 mm. long, somewhat constricted toward apex, subcordate at base, laterally compressed, covered with callous teeth; ribs corky, acute to rounded; oil tubes 3 in intervals, 2 on commissure; seed face concave.

**1. Ammoselinum giganteum** Coult. & Rose, Contr. U.S. Nat. Herb **7:** 89, 1900
*Ammoselinum occidentale* Munz & Jhtn., Bull. Torrey Club **52:** 224, 1925.

Plants 1–2 dm. high; leaves obovate, blades 15–25 mm. long, 2–3 cm. broad, ultimate divisions obtuse, mucronulate, 4–13 mm. long, about 1 mm. broad; petioles 3–8 mm. long, auriculate-sheathing throughout; rays 4–8, unequal, 22 mm. long to obsolete; pedicels 1–10, unequal, 8 mm. long to obsolete.

Desert mesas, Lower Sonoran Zone, Pima, Pinal, and Maricopa Counties, Arizona, south into Coahuila. March.

## 6. CAUCALIS L., Sp. Pl. 240, 1753

Plants annual, caulescent, branching, hispid. Leaves pinnately decompound, segments short, linear to filiform. Peduncles axillary and terminal; umbels compound (or occasionally simple by reduction); involucre of foliaceous bracts shorter than rays;

bractlets of involucel foliaceous, entire or pinnately divided, mostly shorter than fruits. Calyx teeth prominent; corolla white; stylopodium conic. Fruit ovate-oblong, 3–7 mm. long, laterally compressed; carpels with 5 filiform, bristly, prominent ribs and 4 prominent, winged, secondary ribs bearing hooked bristles; oil tubes solitary in intervals, 2 on commissure; seed face deeply sulcate.

**1. Caucalis microcarpa** Hook. & Arn., Bot. Beechey 348, 1839

Slender, 8–40 cm. high, more or less hispid throughout or sometimes glabrescent; leaves oblong to ovate, blades 2–6 cm. long, 2–5 cm. broad, pinnately decompound; leaflets 2–8 mm. long, up to 2 mm. broad; rays 1–9, unequal, 1–8 cm. long; pedicels 2–9, unequal, 15 mm. long to obsolete.

Sandy washes, plains, and rocky hillsides, Lower and Upper Sonoran Zones, eastern British Columbia to Idaho, Utah, Arizona, and Baja California. April–June.

## 7. DAUCUS L., Sp. Pl. 242, 1753

Plants annual, caulescent, variously hispid. Leaves 3–4-pinnatisect; leaflets small, linear. Peduncles terminal; umbels compound; involucre foliaceous, equaling or exceeding rays, pinnately decompound into short, linear or lanceolate segments; bractlets of involucel linear, about equaling pedicels. Calyx teeth obsolete; corolla white; stylopodium none. Fruit oblong, 3–5 mm. long, somewhat flattened dorsally; primary ribs slender, bristly, secondary ribs with a single row of prominent barbed bristles; oil tubes solitary in intervals, 2 on commissure; seed face plane to concave.

**1. Daucus pusillus** Michx., Fl. Bor. Amer. **1:** 164, 1803

Plants 3–90 cm. high; stems usually solitary, simple or few-branched above, retrorsely papillate-hispid; leaves oblong, blades 3–10.5 cm. long, 1.5–7 cm. broad, ultimate divisions 1–5 mm. long, 0.5–1 mm. broad, acute; petioles 4–15 cm. long; peduncles 1.1–4.5 dm. long, retrorsely papillate-hispid; rays few to numerous, unequal, 0.4–4 cm. long, compact in fruit; pedicels unequal, 2–9 mm. long.

Grassy valleys and hillsides, Lower Sonoran Zone to Transition Zone, British Columbia to Baja California east to North and South Carolina; in Pima, Maricopa, and Yuma Counties, Arizona. March–June.

## 8. FOENICULUM Adans., Fam. Pl. 2: 101, 1763

Plants perennial, caulescent, branching, glabrous, with a strong odor of anise; stem tall. Leaves large, pinnately decompound; segments filiform, glaucous. Peduncles axillary and terminal, exceeding leaves; umbels compound; involucre and involucel none. Calyx teeth obsolete; corolla yellow; stylopodium conic. Fruit oblong, 3.5–4 mm. long, laterally compressed; ribs slender, acute; the oil tubes solitary in intervals, 2 on commissure; seed face slightly concave.

**1. Foeniculum vulgare** Mill., Gard. Dict. ed. 8, *Foeniculum,* no. 1, 1768

Plants stout, 9–21 dm. high; leaves ovate to deltoid, blades 3 dm. long, 4 dm. broad, dark green, ultimate divisions 4–40 mm. long, about 0.5 mm. broad; petioles 7–14 cm. long, wholly and broadly sheathing; peduncles 15–65 mm. long; rays 15–40, spreading-ascending in flower, ascending to suberect in fruit, somewhat unequal, 1–6.5 cm. long; pedicels several, 2–10 mm. long, subequal.

Naturalized from Europe, occasionally Lower Sonoran Zone, commoner in the Upper Sonoran Zone, Arizona, Sonora, and Baja California; widely scattered in various parts of the United States. May–Sept.

## 9. APIUM L., Sp. Pl. 264, 1753

Plants annual or biennial, glabrous, branched above. Leaves pinnate to ternate-pinnately decompound; umbels compound (or simple by reduction), axillary and terminal, sessile to short-pedunculate; involucre and involucel none. Calyx teeth absent or inconspicuous; corolla white; stylopodium short-conic. Fruit ovate to suborbicular, 1–3 mm. long, laterally compressed, somewhat constricted at commissure; ribs obtuse, conspicuous; oil tubes solitary in intervals, 2 on commissure; seed face more or less plane.

Plants annual; leaves pinnately or ternate-pinnately decompound, segments linear to filiform ........................................................ 1. *A. leptophyllum*

Plants biennial; leaves pinnate, segments ovate to suborbicular or cuneate .............. ............................................................ 2. *A. graveolens*

**1. Apium leptophyllum** (Pers.) F. Muell. ex Benth. & Muell., Fl. Austr. **3:** 372, 1866

*Apium ammi* Urban, in Mart., Fl. Bras. **11**$^1$**:** 341, *pl. 91, fig. 1,* 1879.

Plants annual, 0.5–6 dm. high; leaves oblong-ovate, blades 3.5–10 cm. long, 3.5–8 cm. broad, 3–4-pinnately decompound, leaflets linear to filiform, 1.5–7 mm. long, 0.5–1 mm. broad; petioles 2.5–11 cm. long, sheath white-scarious-margined; cauline leaves ternate-pinnately decompound, leaflets 4–35 mm. long; peduncles absent or up to 2 cm. long; rays 3–5, 1–2.2 cm. long; pedicels 2–8 mm. long; fruit 1.2–3 mm. long, 1.5–2 mm. broad.

Widespread weed, Humboldt County, California, south to Baja California; Pima, Cochise, and Maricopa Counties, Arizona; southeastern United States, West Indies, South America, and Europe. March–Sept.

**2. Apium graveolens** L., Sp. Pl. 264, 1753

*Celeri graveolens* Britt., in Britt. & Brown, Ill. Fl. ed. 2, **2:** 660, *fig. 3179,* 1913.

Plants perennial, 5–15 dm. high; leaves oblong to obovate, blades 7–18 cm. long, 3.5–8 cm. broad, pinnate with few pairs of ovate to suborbicular leaflets, 2–4.5 cm. long, 2–5 cm. broad, usually 3-lobed to near middle; petioles 3–26 cm. long; cauline leaves frequently cuneate, sometimes laciniate; peduncles absent or up to 1.2 cm. long; rays 7–16, 0.7–2.5 cm. long; pedicels 1–6 mm. long; fruit about 1.5 mm. long, 1.5–2 mm. broad.

Escaped from cultivation in Maricopa and Pinal Counties, Arizona; more or less naturalized in many parts of North America. May–Oct.

## 10. CORIANDRUM L., Sp. Pl. 256, 1753

Plants annual, caulescent, glabrous. Leaves biternate to ternate-pinnate, leaflets mostly cuneate, doubly lobed at apex, upper leaflets linear. Peduncles axillary or terminal; umbels compound; involucre absent; bractlets of involucel few, inconspicuous, lanceolate. Calyx teeth acute, unequal; corolla white or roseate, petals conspicuously unequal; stylopodium conic. Fruit subglobose, 1.5–5 mm. in diameter, not constricted at commissure; ribs filiform; oil tubes wanting; seed face concave.

**1. Coriandrum sativum** L., Sp. Pl. 256, 1753

Plants 2–7 dm. high; basal leaves ovate, blades 3–15 cm. long, 2–10 cm. broad, simple and ternately or pinnately lobed, or pinnate, leaflets flabelliform, 1–2 cm. long, 5–10 mm. broad, variously toothed or incised; petioles 2–15 cm. long; cauline leaves pinnately dissected, ultimate divisions of upper leaves linear to filiform, 2–15 mm. long, 0.5–1.5 mm. broad, obtuse, entire; peduncles 3–10 cm. long or occasionally abortive; rays 2–8, 1–2.5 cm. long; pedicels 2–5 mm. long.

Sparingly naturalized in Pima County, Arizona; Baja California; and various other places in North America. May–July.

## 11. LILAEOPSIS Greene, Pittonia 2: 192, 1891

Plants perennial, glabrous; stems creeping and rooting in the mud. Leaves reduced to hollow cylindric phyllodes, transversely septate, elongate when growing in water. Peduncles shorter than leaves; umbels simple; involucre of few small bracts. Calyx teeth minute; corolla white; stylopodium depressed. Fruit subglobose or slightly compressed laterally, 2–2.5 mm. long; lateral ribs very thick and corky, dorsal ribs filiform; oil tubes solitary in intervals, 2 on commissure; seed face somewhat convex.

**1. Lilaeopsis recurva** A. W. Hill, Jour. Linn. Soc. **47:** 535, *pl. 19, figs. 5–6,* 1927
*Lilaeopsis schaffneriana* Coult. & Rose, Bot. Gaz. **24:** 48, 1897, as to Arizona specimens.

Leaves linear to filiform, terete, 2–24 cm. long, 0.5–3 mm. broad; peduncles 1.5–6 cm. long, shorter than leaves; umbels 6–8-flowered; pedicels slender, recurved, pendulous, 5–15 mm. long; fruit with all ribs rounded.

Known only from southern Arizona, the type collection being from Santa Cruz River Valley near Tucson, Pima County. Rarely collected.

## 12. CYMOPTERUS Raf., Jour. Phys. 89: 100, 1819

Plants perennial from long taproots, low, acaulescent or subacaulescent, glabrous or pubescent. Leaves 2–4-pinnatisect, segments narrow, short. Peduncles shorter than to exceeding leaves; umbels compound; involucre present; bractlets of involucel usually conspicuous, usually dimidiate, foliaceous to scarious. Calyx teeth inconspicuous; corolla white, yellow or purple; stylopodium absent. Fruit ovate to oblong, 4–18 mm. long, more or less compressed dorsally; lateral wings present, dorsal wings usually present; oil tubes small, 1–9 in intervals, 2–12 on commissure; seed face slightly to deeply concave.

Bractlets white or whitish, few-nerved; pedicels 3–12 mm. long ............ 1. *C. bulbosus*
Bractlets purple or greenish white, conspicuously many-nerved; pedicels less than 1 mm. long to obsolete ............ 2. *C. multinervatus*

**1. Cymopterus bulbosus** A. Nels., Bull. Torrey Club **26:** 241, 1899
*Phellopterus bulbosus* Coult. & Rose, Contr. U.S. Nat. Herb. **7:** 168, 1900.

Plants with a pseudoscape, 5–35 cm. high, glabrous; leaves ovate-oblong, blades 1–9 cm. long, 0.6–7 cm. broad, pinnate to bipinnate, pallid, somewhat fleshy, leaflets entire to pinnately lobed, lobes obtuse, linear, 1–10 mm. long, 1–5 mm. broad, confluent; petioles 1.5–9 cm. long; peduncles usually several and exceeding leaves, 1.5–26 cm. long; involucre conspicuous, varying from a low sheath to conspicuous, connate, whitish, 1–3-nerved bracts; involucel similar to involucre, equaling or exceeding flowers; fertile rays 3–8, spreading, 5–50 mm. long; pedicels 3–12 mm. long; flowers purplish; fruit ovoid-oblong to oblong, 7–17 mm. long, 4–12 mm. broad, wings narrow or slightly enlarged at base, narrower to broader than body; oil tubes 1–9 in intervals, 3–10 on commissure.

Alkaline or heavy adobe soils, Lower Sonoran Zone to Arid Transition Zone, southwestern Wyoming to eastern Arizona and western Texas. April–June.

**2. Cymopterus multinervatus** (Coult. & Rose) Tidestrom, Proc. Biol. Soc. Wash. **48:** 41, 1935

*Phellopterus multinervatus* Coult. & Rose, Contr. U.S. Nat. Herb. **7:** 169, 1900.

Plants with a pseudoscape, 4–20 cm. high; leaves ovate-oblong, blades 1–8.5 cm. long, 0.5–6 cm. broad, pinnate, bipinnate or occasionally ternate-pinnate, somewhat fleshy, pallid, leaflets entire to pinnately lobed, lobes acute or obtuse, mucronulate, 0.5–6 mm. long, 0.5–2 mm. broad, usually confluent, glabrate; involucre a low scarious sheath or of 1 or 2 conspicuous, nerved bracts, or a conspicuous, purplish, connate cup with irregular, many-nerved lobes; involucel of conspicuous, ovate to ovate-oblong, subconnate, many-nerved, greenish or purple bractlets, usually with a whitish margin, about equaling flowers; umbels compact, fertile rays usually 1–5, glabrous, 5–25 mm. long, inner rays shorter and umbellate usually sterile; pedicels less than 1 mm. long; flowers purplish; fruit ovoid to ovoid-oblong, 8–17 mm. long, 8–17 mm. broad, wings long and slender, slightly enlarged at base, twice to thrice as broad as body; oil tubes 3–9 in intervals, 5–12 on commissure.

Gravelly slopes and ridges, rocky outcrops, and bajadas, upper margins of Lower Sonoran Zone into Arid Transition Zone, Maricopa, Pima, Yuma, and probably Pinal Counties, Arizona; northern Sonora; southeastern California; and north to southern Idaho. March–May.

## 13. ANETHUM L., Sp. Pl. 263, 1753

Plants annual or biennial from slender subfusiform roots, caulescent, branching, glabrous, with a strong odor of dill. Leaves quadripinnatisect, segments filiform, distinct. Peduncles axillary and terminal, exceeding leaves; umbels compound; involucre and involucel mostly absent. Calyx teeth obsolete; corolla yellow; stylopodium conic, margin crenulate. Fruit ovate, about 4 mm. long, compressed dorsally; lateral wings much narrower than body of fruit; dorsal wings obsolete; oil tubes solitary in intervals, 2 on commissure; seed face plane.

**1. Anethum graveolens** L., Sp. Pl. 263, 1753

Plants 4–17 dm. high; leaves oblong to obovate, blades 13–35 cm. long, 11–20 cm. broad, leaflets 4–20 mm. long, less than 0.5 mm. broad; upper cauline leaves greatly reduced; rays 10–45, spreading, 3–10 cm. long; pedicels 20–50, 6–10 mm. long.

Sparingly escaped in Pima and Pinal Counties, Arizona, near Tucson and Sacaton.

## 14. LOMATIUM Raf., Jour. Phys. 89: 101, 1819

Plants perennial, acaulescent or short-acaulescent, glabrous or pubescent. Leaves ternately or pinnately decompound, segments filiform to ovate. Peduncles equaling or exceeding leaves; umbels compound; involucre mostly absent; bractlets of involucel filiform to obovate, foliaceous to subscarious, distinct to connate, rarely none. Calyx teeth small; corolla greenish white, yellow or purple; stylopodium absent. Fruit ovate to linear, 4.5–15 mm. long, compressed dorsally, lateral wings present, dorsal ribs absent or filiform, wings thin; oil tubes small, 1 to several in intervals, 2 to several on commissure; seed face slightly concave.

**1. Lomatium nevadense** (S. Wats.) Coult. & Rose, Contr. U.S. Nat. Herb. **7:** 220, *fig. 60,* 1900

*Peucedanum nevadense* S. Wats., Proc. Amer. Acad. **11:** 143, 1876.

Plants 1–4.5 dm. high, pubescent; leaves oblong to obovate, blade 5–6 cm. long, tripinnate, leaflets crowded, oblong, 2–3 mm. long, about 1 mm. broad, apiculate; peduncles exceeding leaves; involucel of conspicuous, linear and distinct or obovate and connate, scarious-margined bractlets, about equaling flowers; rays 8–22, spreading, 1–2.5 cm. long, unequal; pedicels 3–10 mm. long, umbellets about 20-flowered; flowers white; fruit ovate to oblong-obovate, 6–8 mm. long, 4–6 mm. broad, puberulent, wings narrower than body.

Rocky slopes and mesas, upper margins of Lower Sonoran Zone into Arid Transition Zone, Gila, Maricopa, and Pinal Counties, Arizona; western Oregon south to Sonora and New Mexico. March–May.

The variety *parishii* (Coult. & Rose) Jepson, is distinguished from var. *nevadense* by its glabrous ovaries and fruits. It occurs in Maricopa and Pima Counties, Arizona.

## 15. ERYNGIUM L., Sp. Pl. 232, 1753

Plants perennial, caulescent, glabrous; stems simple or branched. Leaf blades lanceolate to oblanceolate and reticulate-veined, or linear and parallel-veined, entire to pinnatifid. Peduncles exceeding leaves; inflorescence a dense bracteate head; bracts linear to ovate-lanceolate, entire to spinulose-serrate; floral bractlets usually entire. Calyx lobes lanceolate to ovate, obtuse or acute; corolla white or blue; stylopodium absent. Fruit ovoid, flattened laterally, covered with hyaline scales or tubercles; ribs obsolete; oil tubes several, inconspicuous; seed face plane.

Leaf venation parallel; heads 15–25 mm. long without a coma; bracts shorter than heads ........................................................ 1. *E. sparganophyllum*

Leaf venation reticulate; heads 5–15 mm. long, comate; bracts about equaling heads ........................................................ 2. *E. nasturtiifolium*

**1. Eryngium sparganophyllum** Hemsl., in Hook., Ic. Pl. **26:** *pl. 2508,* 1897

Slender, caulescent, glabrous perennials, 4–12 dm. high; stems solitary to several, erect; basal leaves numerous, linear to linear-lanceolate, 1–9 dm. long, 5–15 mm. broad, broadest at base, attenuate at apex, entire or obsoletely spinulose-dentate with teeth 2 mm. long or less, axillary spines absent, venation parallel; cauline leaves few and reduced; inflorescence branching, heads ovoid or ovoid-oblong, 15–25 mm. long, 10–15 mm. broad; bracts 8–12, much shorter than heads; bractlets 5 mm. long, slightly exceeding fruit; coma wanting; fruit ovoid, 3–4 mm. long.

Wet seeps and marshy ground, Lower Sonoran and Upper Sonoran Zones, Pima County, Arizona, east into New Mexico and south into Mexico. March–June.

**2. Eryngium nasturtiifolium** Juss., in Delar. f., Eryng. 46, *pl. 17,* 1808

Prostrate, ascending or erect, glabrous biennials or perennials 1–3 dm. high or long; stems several, leafy, branched from base; basal leaves spatulate or oblanceolate,

3–10 cm. long, 1–3 cm. broad, cuneate at base, coarsely dentate to runcinate-pinnatifid, teeth or lobes mucronulate or spinose, venation reticulate; cauline leaves like basal, spinose-dentate or spinose-pinnatifid; inflorescence branching, heads numerous, small, nearly sessile, ovoid to ovoid-cylindric, 5–15 mm. long, 4–8 mm. broad; bracts 5–9, about equaling heads; bractlets 4–9 mm. long, exceeding fruit; coma of several short bractlets 3–7 mm. long, resembling bracts; fruit ovoid-globose, 2–3 mm. long.

Margins of vernal pools, on moist playas, and wet sand bordering streams and ponds, Lower Sonoran Zone, the Magdalena Plain to the Cape Region, Baja California, and southern Sonora into Sinaloa. March–April.

## Family 107. **LENNOACEAE.** Lennoa Family

Fleshy root parasite without chlorophyll; stems brownish and succulent; leaves reduced to scales. Flowers perfect, regular, small, in densely crowded heads or spikes. Calyx deeply parted; segments narrow, hairy or plumose. Corolla tubular, 5–9-lobed. Stamens as many as and alternating with lobes of corolla, inserted on throat; anthers 2-celled, longitudinally dehiscent. Ovary superior. Pistil 1, of 6–14 united carpels, each carpel divided into 2 cells by a dorsal partition, each cell 1-ovuled; style 1, simple; stigma subcapitate to peltate, somewhat crenate. Fruit a drupelike capsule, irregularly circumscissile into 12–28 nutlets. Embryo globose, not differentiated into cotyledons and radicle.

Flowers in a dense terminal spike; sepals densely short-puberulent; corolla lobes spreading 4–6 mm. in anthesis, puberulent without .............................. 1. *Pholisma*
Flowers in a peltate terminal disk; sepals plumose; corolla lobes spreading 2.5–3.5 mm. in anthesis, glabrous ............................................ 2. *Ammobroma*

### 1. **PHOLISMA** Nutt. ex Hook., Ic. Pl. **7:** *pl. 626,* 1844

Fleshy herbaceous plants without chlorophyll, parasitic on roots of other flowering plants. Stems simple or sparingly branched, bearing small, bractlike scales. Inflorescence a simple or branched spike. Bractlets none. Calyx of 5–7 linear, puberulent, distinct lobes. Corolla regular, narrowly funnelform, lobes 5–7, rounded, nearly truncate, faintly prickly on margins. Stamens 5–7, adnate to about middle of corolla tube, included. Ovary subglobose, of 6–10 carpels, each carpel 2-celled by a dorsal false partition. Stigma capitate, crenately 6–10-lobed. Fruit subglobose, tardily dehiscing into small, ovoid-flattened nutlets.

**1. Pholisma arenarium** Nutt. ex Hook., Ic. Pl. **7:** *pl. 626,* 1844

*Lennoa arenaria* Fourn., Bull. Soc. Bot. Fr. **16:** 11, 1869.
*Pholisma depressum* Greene, Bull. Calif. Acad. **1:** 198, 1885.
*Pholisma paniculatum* Templeton, Bull. S. Calif. Acad. **37:** 98, *pl. 25–26,* 1939.

Fleshy, simple or sparingly branched plant 1–2 dm. tall, brownish; part exposed above sand glandular-puberulent; bracts on buried part of stem lance-linear to broadly lanceolate, 1–2 cm. long, 3–8 mm. broad at base, glabrous or essentially so, upper exposed bracts often ovate-deltoid, 5–10 mm. long, glandular-puberulent; spike compact, simple or somewhat branched, 1–8 cm. broad, 1.5–10 cm. long; calyx lobes linear to narrowly clavate, 7–8 mm. long, slightly shorter than corolla, glandular-puberulent with short, crisped, stoutish hairs; corolla funnelform, cylindrical tube only 1–2 mm. long, flaring throat purplish except along white margins of lobes, 7–9 mm. long, 4–6 mm. broad, broad rounded lobes nearly truncate, emarginate and slightly sinuate;

fruit depressed, subglobose, 2.5–3 mm. in diameter, enclosed within base of withering corolla and tardily breaking into 12–20 nutlets.

In sandy soil, usually in dunes, Sonoran Zones, San Luis Obispo County to San Diego and in the Mojave and Colorado Deserts, extending into northern Baja California. Parasitic on the roots of such shrubs as *Haplopappus, Chrysothamnus, Eriodictyon, Eriogonum, Hymenoclea,* and probably others. April–July.

## 2. AMMOBROMA Torr. ex A. Gray, Pl. Thurb. 327, 1854

Fleshy parts devoid of chlorophyll and parasitic on roots of other flowering plants. Stems mainly buried in sand, simple, scaly-bracted, terminal part expanded into a discoid receptacle bearing numerous small, closely crowded flowers. Calyx lobes 6–9, narrowly linear, distinct, plumose-hairy. Corolla tubular-funnelform, with 6–9 lobes, these very short, nearly truncate, plicate. Stamens 6–9, adnate to corolla throat and alternate with lobes, included. Ovary of 6–10 carpels, each falsely 2-celled by a false partition; stigma capitate, faintly lobed or crenate on margins. Fruit globose, dividing into 12–20 nutlets at maturity.

### 1. Ammobroma sonorae Torr. ex A. Gray, Pl. Thurb. 327, 1854

Fleshy plant with a stem 1–2.5 cm. in diameter, 1.5–2.5 dm. long, often buried in sand almost to disklike inflorescence; lower scales linear-spatulate, 1.5–2.5 cm. long, glabrous, white or suffused with pale purple or violet, upper scales (those above sand) lance-linear, 8–15 mm. long, densely puberulent with rather coarse, crisped, tawny hairs, adjacent stem similarly clothed; inflorescence a flattened disk 4–15 cm. in diameter, central part somewhat concave, very densely covered with pale lilac to nearly white flowers, those at margin of disk sessile, those in center short-pedicellate; calyx lobes 6–10, narrowly linear, 7–9 mm. long, pale lilac and sparsely puberulent at base, densely plumose-pubescent with tangled, pale tan hairs upward; corolla 8–9 mm. long, tubular in basal 5–5.5 mm., limb abruptly and narrowly funnelform, 4 mm. long, 2.5–3.5 mm. wide, glabrous, suffused with pale lilac, lobes 6–10, plicate, very short, whole corolla nearly truncate at apex, glabrous; stamens as many as corolla lobes, included; fruit breaking up into ovoid, laterally flattened nutlets about 1 mm. long, surface of nutlets finely and obscurely reticulate, brownish.

In sandy soil, Lower Sonoran Zone, parasitic on shrubby plants, particularly on *Coldenia*; between Yuma and Imperial Valley south into northeastern Baja California. May.

# Family 108. THEOPHRASTACEAE. Theophrasta Family

Shrubs or small trees with entire or serrate, estipulate leaves arranged alternately or pseudoverticillately. Inflorescence racemose or umbellate. Flowers perfect, regular. Calyx lobes 5, distinct or connate at base, imbricate, stiffish, rounded at apex. Corolla sympetalous, funnelform or limb rotate, fleshy. Stamens 5, alternating with 5 staminodia. Style simple; stigma entire or very faintly lobed. Fruit baccate or a drupe.

## 1. JACQUINIA L. ex Jacq., Stirp. Amer. 53, 1763

Shrubs or small trees with coriaceous leaves. Flowers 5-merous. Sepals free or slightly connate at base, ovate to suborbicular, margins ciliate to finely crenate. Corolla lobes imbricate, spreading in anthesis, entire or emarginate. Staminodia petal-like but smaller than corolla lobes. Ovary ellipsoid or clavate, glabrous, many-ovuled. Fruit ovoid. Seeds few, ovoid, immersed in mucilaginous material, with a cartilaginous endosperm.

**1. Jacquinia pungens** A. Gray, Pl. Thurb. 325, 1854

*Jacquinia donnell-smithii* Mez, Pflanzenreich $4^{236a}$: 39, 1903.

Shrub or small tree 1–6 m. tall, with light gray bark and a dense crown of dark green, pungently tipped leaves, more or less tufted at ends of branchlets; young twigs sparsely arachnoid-puberulent; leaf blades lance-elliptic, 4–8 mm. wide, 2–5 cm. long, plane and dark green above, narrowly revolute and paler beneath, numerous minute pores in lower surface, only midrib obvious, pungently pointed with a brownish spine 1–2.5 mm. long, cuneately narrowed to subsessile base; racemes few-flowered, about equaling subtending leaves; pedicels 3–5 mm. long; sepals suborbicular, about 3 mm. long, margin thin; corolla yellow, tube 4–5 mm. long, limb 6–8 mm. broad; fruit ovoid, about 1.5 cm. broad, 1.8–2.2 cm. long, pungently tipped by persistent style; seeds ovoid-lenticular, 3–4 mm. wide, 5–7 mm. long, reddish brown, embedded in a gelatinous or mucilaginous matrix.

Gravelly plains and arid hillsides, Lower Sonoran Zone, central Sonora from vicinity of Carbo south to Chiapas and Vera Cruz; Guatemala. Also on Tiburón Island, Sonora. May.

## Family 109. **PRIMULACEAE.** Primrose Family

Herbs with simple leaves and perfect regular flowers in terminal racemes or umbels or solitary in axils. Calyx 4–8 (usually –5) -toothed, -lobed or -parted. Corolla sympetalous, often deeply 4–8-lobed or -parted and appearing choripetalous. Stamens inserted near base of corolla tube, opposite lobes. Ovary superior, or base adnate to calyx tube, 1-celled, several- to many-ovuled; placenta central. Style simple; stigma entire. Fruit a capsule, longitudinally valvular or circumscissile. Seeds several to many, small, minutely pitted.

Flowers in umbels; corolla minute, constricted at throat; leaves all basal ...... 1. *Androsace*
Flowers not in umbels; corolla open at throat; at least some cauline leaves present on lower part of stem, or all cauline:
  Ovary superior; flowers solitary in axils; capsules circumscissile; corolla deeply parted, lobes rotate ...................................................... 2. *Anagallis*
  Ovary adnate to calyx on lower half; flowers in racemes; capsules longitudinally valvate; corolla lobes shorter than tube or barely equaling it .................... 3. *Samolus*

### 1. **ANDROSACE** L., Sp. Pl. 141, 1753

Annual or perennial herbs, more or less stoloniferous and cespitose. Leaves all in a basal rosette or rarely cauline. Flowers perfect, regular, borne in umbels or rarely solitary at apex of slender, erect or ascending scapes. Calyx campanulate to subglobose, 5-lobed or -parted about to middle or somewhat deeper. Corolla infundibuliform or hypocrateriform, with a short tube, lobes 5, spreading. Stamens inserted in corolla tube, included. Style short; stigma capitate. Ovary globose, several- to many-ovuled. Fruit dehiscing from apex to base by 5 valves. Seeds few (in ours) or sometimes numerous.

Calyx lobes erect or nearly so in fruit, 1.2–1.5 mm. long, reddish in flower ........... ........................................... 1. *A. occidentalis* var. *occidentalis*
Calyx lobes spreading to slightly recurved in fruit, 2–2.5 mm. long and green in flower ........................................... 1a. *A. occidentalis* var. *arizonica*

**1. Androsace occidentalis** Pursh, Fl. Amer. Sept. **1**: 137, 1814, var. **occidentalis**
*Androsace platysepala* Woot. & Standl., Bull. Torrey Club **34**: 519, 1907.
*Amadea occidentalis* Lunell, Amer. Midl. Nat. **4**: 504, 1916.

A small, delicate, annual herb 1 dm. tall or less, sparsely puberulent throughout; leaves in a basal rosette, blades narrowly obovate to oblanceolate or oblong, 1.5–4 mm. wide, 4–15 mm. long, obtuse to short-acuminate at apex, narrowed to a short petiole or subsessile, entire or slightly denticulate; scapes several, ascending to erect, 1.5–5 cm. long; bracts subtending umbel ovate or elliptic, usually somewhat rhombic, 2–5 mm. long, foliaceous, green; umbels 2–15-flowered (usually 2–8-flowered); pedicels slender, 5–20 mm. long, outer ones curved-ascending, inner erect or nearly so; calyx tube 1.5–2 mm. long in flower, green or often reddish, lobes narrowly ovate or narrowly triangular, 1.2–1.5 mm. long in flower, green or reddish, erect or ascending in fruit; corolla minute, limb about 2 mm. broad, tube about equaling calyx tube; capsule about 3 mm. in diameter; seeds dark brown, 0.6–0.8 mm. long, minutely granular-papillate.

On grassy hillsides and along streams and washes, Lower Sonoran to Transition Zones, Santa Cruz and Pima Counties, Arizona, to Texas, Illinois, and southern Canada. Feb.–April.

**1*a*. Androsace occidentalis** var. **arizonica** (A. Gray) St. John, Victoria Mus. Mem. No. **126**: 54, 1922
*Androsace arizonica* A. Gray, Proc, Amer. Acad. **17**: 221, 1882.
*Primula arizonica* Derganc, Allg. Bot. Zeit. **10**: 110, 1904.
*Androsace occidentalis* subsp. *arizonica* G. T. Robbins, Amer. Midl. Nat. **32**: 153, 1944.

Differs from var. *occidentalis* in having slightly longer, greener sepals 2–2.5 mm. long that are spreading or even slightly reflexed in fruit; said to have fewer scapes, but this character doesn't seem to be correlated with the shape and size of the sepals.

Commoner at lower altitudes in Arizona and New Mexico than var. *occidentalis,* but the two occur together and intergrade. Near Ajo and Sells, Pima County, and in the Tucson Mountains. Feb.–April.

## 2. ANAGALLIS L., Sp. Pl. 148, 1753

Low to decumbent annuals with opposite or less commonly ternate, entire leaves. Flowers axillary, pedicellate. Calyx deeply 5-cleft; segments nearly distinct. Corolla deeply 5-parted; tube reduced to a scant ring; lobes rounded, convolute in bud, rotate in anthesis. Stamens 5; filaments hirsute. Capsule ovoid to globose, circumscissile. Seeds numerous.

**1. Anagallis arvensis** L., Sp. Pl. 148, 1753

Stems prostrate or decumbent, 1–6 dm. long, often forming a fairly compact mat but not rooting at nodes; leaves opposite, sessile, ovate, acute, 6–14 mm. wide, 10–20 mm. long, rounded, truncate or subcordate at base; pedicels slender, 1.5–4 cm. long, surpassing subtending leaves; flowers remaining closed on cloudy days; sepals lanceolate, 3–4.5 mm. long, acuminate, margins scarious toward base; corolla salmon-colored, 8–10 mm. broad, petals minutely glandular-ciliate distally, filaments hirsute with pale bluish hairs; capsule 4–5 mm. in diameter, papery, style persistent; seeds dark brown, about 0.8 mm. long, minutely pitted.

An introduced weed occurring sparingly in the Lower Sonoran Zone about fields and along roadsides; widely distributed in North America and occurring in the Sonoran Desert in the Salt River Valley drainage in Arizona, in Sonora and northern Baja California. April–Sept.

## 3. SAMOLUS L., Sp. Pl. 171, 1753

Herbaceous perennial or sometimes frutescent, with alternate or basal, linear, oblong or spatulate leaves. Flowers in terminal racemes or corymbs. Pedicels bracteate at base or about middle or bractless. Calyx tube adnate to ovary for about lower half; lobes 5, persistent. Corolla subcampanulate, 5-lobed. Stamens 5, inserted in tube or throat of corolla, alternating with staminodes which are inserted just below sinuses of corolla. Ovary globose; style short; stigma capitate. Capsule ovoid to globose, 5-valved. Seeds many.

Pedicels bractless; corollas 3–4 mm. long; leaves mostly basal; inflorescence leafless ........................................................................ 1. *S. ebracteatus*
Pedicels with a small bract near middle; corollas 1–2 mm. long; stems leafy, a few small leaves at base of branches of inflorescence .......................... 2. *S. floribundus*

### 1. Samolus ebracteatus H. B. K., Nov. Gen. & Sp. 2: 223, *pl. 129,* 1817

*Samodia ebracteata* Bando, Ann. Sci. Nat. II, **20**: 350, 1843.

Perennial with 1 to several erect or ascending stems 1–5 dm. tall from a sturdy taproot; leaves mainly basal or on lower part of stem, broadly spatulate to oblanceolate-spatulate or sometimes suborbicular, 1–4 cm. wide, 3–9 cm. long, entire, cuneately to abruptly narrowed at base to a broad petiole 5–25 mm. long, obtuse or rounded at apex, glabrous; flowering scapes 1–4 dm. long or more, leafless; pedicels spreading to ascending, stiffish, 5–15 mm. long, bractless, sparsely and minutely glandular-puberulent but soon glabrate; calyx tube 1–1.5 mm. long, lobes deltoid-ovate, about equaling tube; corolla 3–4 mm. long, tube short-cylindric, lobes about 1 mm. long and broad, rounded, spreading; capsule globose, about 3 mm. in diameter, 5-valved; seeds dark brown, minutely faveolate-pitted, about 0.2 mm. long.

Along streams and in moist meadows, Lower Sonoran to Transition Zones, Texas to Florida, the mainland of Mexico, the West Indies, in southern Nevada, and at San Marcos Island, at Loreto, and in the Cape Region, Baja California. Entering the Lower Sonoran Zone as a straggler along streams. April–Aug.

### 2. Samolus floribundus H. B. K., Nov. Gen. & Sp. 2: 181, 1817

*Samolus americanus* Spreng., Syst. **1**: 702, 1825.
*Samolus valerandi* var. *americanus* A. Gray, Man. ed. 2, 274, 1856.
*Samolus valerandi* var. β *floribundus* Knuth, Pflanzenreich $4^{237}$: 338, 1905.

Slender perennial with 1 to several (oftenest solitary) stems 1–4 dm. tall from a slender taproot; basal leaves suborbicular-obovate to oblong-spatulate, 8–20 mm. wide, to 6 cm. long, obtuse, rounded, or nearly truncate, narrowed to a short broad petiole 2–10 mm. long; cauline leaves similar but usually gradually smaller and gradually narrower, uppermost elliptic, 1–2.5 cm. long, entire; pedicels slender, spreading, each bearing a minute bractlet 0.5 mm. long or less near middle, younger pedicels sparingly glandular-viscid, 7–15 mm. long; calyx 1.5–2 mm. high at anthesis, deltoid lobes constituting about two-fifths of total length; corolla white, minute, 1.5–2 mm. long; capsule about 2 mm. in diameter.

Along streams and on seepy hillsides, mainly Upper Sonoran and Transition Zones, through much of temperate North and South America and extending into the tropics. Straggling a short distance into the Sonoran Desert along the streams in southern Arizona and Baja California. April–Aug.

## Family 110. **PLUMBAGINACEAE.** Plumbago Family

Perennial herbs or suffrutescent plants with basal or alternate, simple, entire leaves. Flowers perfect, regular, 5-merous, in panicles of spikelike racemes, in globose heads, or racemose. Calyx tubular to funnelform. Corolla sympetalous, tubular or barely united at base. Stamens opposite corolla lobes and adnate to tube or base of claw or nearly free. Ovary superior, 1-celled, 1-ovuled, often 5-angled near apex; styles 5. Fruit usually included within calyx, indehiscent, circumscissile or valvate. Seed 1, pendulous.

Plants with long, slender, scandent stems; leaves distributed along stem, not in basal tufts; calyx ribs stipitate-glandular; capsule 5–7 mm. long .................... 1. *Plumbago*
Plants acaulescent; leaves in basal tufts or rosettes; calyx ribs (in ours) glabrous; utricle 3–3.5 mm. long ................................................ 2. *Limonium*

### 1. **PLUMBAGO** L., Sp. Pl. 151, 1753

Perennial herb or suffrutescent plant with slender, often subscandent branches and petiolate or auriculate-amplexicaulate leaves. Flowers in paniculate spikes. Calyx tubular, stipitate-glandular, 5-lobed, plicate and with scarious or subscarious sinuses above. Corolla salverform; tube slender; limb 5-lobed, spreading or rotate. Stamens inserted at base of corolla and free. Capsule membranous, circumscissile near the base.

**1. Plumbago scandens** L., Sp. Pl. ed. 2, 215, 1762

*Plumbago mexicana* H. B. K., Nov. Gen. & Sp. **2:** 221, 1817.

A decumbent to scandent, straggly shrub or suffrutescent plant with slender weak stems to 3 m. long; herbage glabrous except in stipitate-glandular inflorescence; leaves elliptic or lance-elliptic, oblong or ovate, acute or acuminate, to 4 cm. wide, 10 cm. long or less, acute at base, narrowed to a short petiole 3–18 mm. long; bracts lance-ovate, attenuate, 3–6 mm. long; flowers sessile or subsessile in many-flowered spikes paniculately arranged; calyx 8–12 mm. long at anthesis, 5 green nerves closely beset with conspicuous stipitate glands; corolla white, 2–3.5 cm. long, tube very slender, lobes 6–10 mm. long; anthers barely exserted; capsule ovoid, 5–7 mm. long, membranous.

In canyons and on shaded brushy hillsides or along fence-rows and stone walls, Sonoran Zones, Pima County, Arizona, in southern Florida, and thence south into tropical America. In Baja California from Loreto south and in all except the driest parts of the desert in Sonora. Common in our area but nowhere abundant. May–Sept.

## 2. LIMONIUM Mill., Gard. Dict. abr. ed. 4, 1754

Perennial, acaulescent herb with leaves in a basal tuft or rosette; root crown often branched one or more times. Flowering scapes one to several, branching, erect or ascending. Flowers small, paniculate-cymose, in few-flowered, bracteolate clusters arranged in secund spikes. Calyx campanulate to tubular, usually 10-ribbed; limb scarious, 5-toothed. Corolla gamopetalous only at base, lobes resembling distinct petals. Stamens 5, adnate to bases of corolla lobes. Styles 5, distinct, stigmatic along inner surface. Ovary superior, 1-celled, 1-ovuled. Fruit an utricle, thin-walled.

**1. Limonium californicum** (Boiss.) Heller var. **mexicanum** (S. F. Blake) Munz, Aliso **4:** 96, 1958

*Limonium mexicanum* S. F. Blake, Rhodora **18:** 59, *pl. 118, fig. C1–5*, 1916.
*Limonium commune* var. *mexicanum* Jepson, Fl. Calif. **3:** 77, 1939.

Caudex reddish, sublignescent, often bearing several crowns; leaves oblong-obovate, obtuse, rounded or retuse at apex, 1.5–7 cm. wide, 5–20 cm. long, thick, coriaceous, tapering to reddish petioles nearly equaling blade, both glabrous; scapes 1.5–5 dm. tall, solid, paniculately branched above, densely floriferous at tips of branches; secund spikes 1–3.5 cm. long; bracts broadly obovate to suborbicular, abruptly acute or apiculate, scarious along margins, 3–4 mm. long, glabrous; calyx narrowly obconic, 4.5–5.5 mm. long, glabrous, 5 ribs prominent, others faint; corolla lavender to pale blue-violet, slightly exceeding calyx; fruit broadly spindle-shaped, 3–3.5 mm. long, 0.8–1 mm. wide, dark reddish brown, faintly and minutely reticulate.

Coastal salt marshes, Lower and Upper Sonoran Zones, Moro Bay in San Luis Obispo County, California, and San Diego County south to Scammon's Lagoon, Baja California. June–Nov.

# Family 111. SAPOTACEAE. Sapote Family

Trees or shrubs with alternate, coriaceous, entire leaves and glabrous or pubescent twigs and herbage. Flowers solitary or in axillary glomerules, sessile or pedicellate, regular and perfect or rarely polygamodioecious. Calyx of 4–6 nearly distinct lobes or more or less coalesced, sometimes in 2 distinct series. Corolla sympetalous; tube short, campanulate, urceolate or infrequently cylindrical; limb of similar equal lobes, imbricate in bud. Stamens inserted in 1 or 2 series in tube of corolla, often alternating with linear or subpetaloid staminodia. Anthers basifixed or versatile, introrsely or laterally dehiscent. Ovary superior, 2–5- or sometimes many-celled, each cell 1-ovuled. Style simple, glabrous. Fruit indehiscent; pericarp fleshy or leathery. Embryo straight; endosperm present.

Branchlets often spine-tipped; 2 lateral appendages at base of each corolla lobe; endosperm scanty or none .......... 1. *Bumelia*
Branchlets unarmed; appendage lacking; endosperm abundant .......... 2. *Sideroxylon*

## 1. BUMELIA Swartz, Prod. Veg. Ind. Occ. 49, 1788

Shrubs or trees to 8 m. tall or more (in ours), frequently armed with stiff spinose branches. Leaves alternate or pseudo-opposite, estipulate, persistent or deciduous; lateral veins often obscure. Flowers in fasciculate clusters, mostly 5-merous. Sepals 5, unequal. Corolla sympetalous, 5-lobed, with 2 lateral appendages at base of each lobe or appendages lacking. Stamens 5, inserted at base of corolla, alternating with 5 petaloid acuminate staminodia. Ovary superior, 5-celled, each cell 1-ovuled, more or less pubescent; style slender.

Corolla 4–5 mm. long; leaves 1–3 cm. long, finely but distinctly puberulent on both surfaces, glaucous .................................................... 1. *B. occidentalis*
Corolla 6–7.5 mm. long; leaves 2.5–4.5 cm. long, glabrous or nearly so above, green ..... .................................................... 2. *B. peninsularis*

**1. Bumelia occidentalis** Hemsl., Biol. Centr. Amer. Bot. **2:** 298, 1881

*Bumelia fragrans* Brandegee, Zoe **5:** 106, 1901. Not *B. fragrans* Ridley, 1890.
*Bumelia brandegei* Blake, Contr. Gray Herb. **52:** 76, 1917.
*Bumelia cuneifolia* M. E. Jones, Contr. West. Bot. **18:** 63, 1933.

A rather dense, much-branching shrub or small tree to 6 m. tall with stiff, divaricate, spinose branches; leaves oblanceolate or some obovate, 2–10 or rarely to 20 mm. wide, 8–30 mm. long, cuneately tapering to base, rounded, obtuse or infrequently emarginate at apex, finely puberulent on both surfaces (slightly more densely so beneath than above) and glaucous, veins inconspicuous or obscure; flowers in several-flowered axillary fascicles; pedicels 5–20 mm. long, finely appressed-puberulent; sepals 2.5–3 mm. long, rounded, finely strigose-cinereous; corolla yellow, tube short, 1–2 mm. long, lobes rounded, 2–3 mm. long, appendages lance-ovate, 1–1.5 mm. long, 1 or 2 often lacking; staminodia 3–4 mm. long, lanceolate, somewhat plicate; fruit ellipsoid to subglobose, 1–1.5 cm. long, black, slightly glaucous, mostly 1-seeded.

Along arroyos and on alluvial plains, Lower Sonoran Zone, vicinity of Comondú and south to the Cape Region in Baja California, and from Montezuma and vicinity of San Pedro de las Cuevas south at least to Guaymas, Sonora. Feb.–March.

**2. Bumelia peninsularis** Brandegee, Zoe **5:** 107, 1901

A rounded, stiff-branched shrub 3–4 m. tall with short stiff spines tipping branchlets; twigs densely appressed-pubescent in youth, later glabrate; leaves ovate-elliptic to oblong-elliptic, 1–2.5 cm. wide, 2.5–4.5 cm. long, rounded at apex, abruptly tapered to base, rather densely pubescent along midrib beneath, more sparsely so on blade; petioles 3–8 mm. long; flowers in few-flowered fascicles; pedicels 4–8 mm. long, glabrous or essentially so; sepals rounded, 4–5 mm. long, glabrous or with a few scattered hairs; corolla tube 2–2.5 mm. long, lobes to 5 mm. long, nearly erect; staminodia 3–3.5 mm. long, petaloid; anthers barely exserted; ovary glabrous to short-pubescent; fruit broadly ellipsoid, 12–15 mm. long, blue-black.

Along canyons and arroyos, Lower Sonoran and Tropical Zones, southern Baja California. Dec.–Feb.

## 2. SIDEROXYLON L., Sp. Pl. 192, 1753

Shrubs or trees with coriaceous, usually long-petiolate, estipulate leaves and small flowers fasciculate at nodes of branchlets. Sepals 5, ovate to orbicular, usually obtuse or rounded, subequal. Corolla tubular-campanulate, 5 lobes imbricate in bud. Stamens 5, inserted at base of and opposite corolla lobes. Ovary usually 5-celled. Fruit ovoid to globose, mostly 1-seeded through abortion of other ovules. Seeds basal; endosperm fleshy.

**1. Sideroxylon leucophyllum** S. Wats., Proc. Amer. Acad. **24:** 59, 1889

Shrub or small tree to 3 m. tall with a sturdy but soon branched trunk to 3 dm. in diameter; leaves on vigorous branches oblong, 1.5–3.5 cm. wide, 4–10 cm. long, densely but finely tomentose on both surfaces, obtuse to acute at base, obtuse to rounded at apex, on short stout petioles to 1.5 cm. long; leaves on basal branches 1.5–3 cm. long, 6–8 mm. wide; flowers in dense crowded fascicles in axils of leaves; sepals 4–5 mm. long, suborbicular, densely cinereous-tomentose with hairs slightly longer than those of leaves; corolla greenish yellow, equaling or slightly exceeding sepals, lobes about 1–2 mm. long and wide; fruit ovoid to globose, tomentose in youth, eventually glabrate, about 1.5–2 cm. in diameter.

On rocky hillsides and sometimes along washes, Lower Sonoran Zone, Tajo Canyon, Sierra Juárez (32° 16′ N.), south along the eastern slopes of the desert ranges to Santa María Canyon, about 65 miles south of Bahía de los Angeles, and on adjacent islands in the Gulf of California. May.

## Family 112. EBENACEAE. Persimmon Family

Shrubs or trees with alternate, entire, estipulate, deciduous or persistent, often leathery leaves. Flowers small, white or greenish, solitary, or staminate cymose, regular, perfect or dioecious. Calyx 3–6-lobed, persistent and often accrescent in fruit. Corolla urceolate, campanulate or salverform, 3–7-lobed, lobes imbricate. Stamens usually two to four times as numerous as corolla lobes, inserted on a hypogynous disk or on base of corolla tube; filaments free or connate in pairs; anthers introrse, longitudinally dehiscent. Ovary superior, 2–16-celled; ovules 1–2 in each cell and pendulous. Style simple or divided. Fruit usually a succulent berry. Seeds frequently flattened; endosperm present.

Calyx 3-lobed; ovary 3–6-celled, 1–6-seeded .................................... 1. *Maba*
Calyx 3–7-lobed; ovary 2–16-celled, 1–16-seeded .......................... 2. *Diospyros*

### 1. MABA Forst., Char. Gen. Pl. 121, *pl. 61*, 1776

Shrubs or small trees with hard, dark-colored wood and small scattered leaves. Flowers dioecious, rarely monoecious or perfect, axillary. Calyx campanulate, accrescent in fruit, 3-lobed. Corolla campanulate or tubular, in ours 3-lobed. Staminate flowers with 3 to many stamens (in ours usually 9) and a small rudimentary ovary. Pistillate flowers usually with 1 to several staminodia. Ovary 3–6-celled, 1–6-ovuled; style branches 3. Fruit baccate, succulent, globose or ovoid. Seeds 1–6, with ruminate endosperm.

**1. Maba intricata** (A. Gray) Hiern, Trans. Cambr. Phil. Soc. **12:** 126, 1873
*Macreightia intricata* A. Gray, Proc. Amer. Acad. **5:** 163, 1861.

Dense, rounded or flat-topped shrub 1–3 m. tall and often nearly as broad; bark pale gray, or reddish brown on young twigs; leaves coriaceous, gray-green, obovate to oblong-obovate, 5–10 mm. wide, 1–2 cm. long, rounded or slightly emarginate at apex, broadly cuneate to obtuse at base, minutely puberulent on both surfaces with very fine, appressed, pale tawny or rufous hairs and closely stippled beneath with minute pellucid dots; calyx about 4 mm. long at anthesis, lobes 3, obovate, 2 mm. wide, rounded or obtuse at apex, finely puberulent with pale brownish hairs; corolla fleshy, tube about 3 mm. long, lobes ovate, acute at apex, 3–4 mm. long, densely strigose with closely appressed, tawny hairs without, inner surface glabrous; stamens mostly 6, inserted at base of tube, filaments 0.2–0.4 mm. long, anthers lance-ovate, attenuate, about 1.4 mm. long, no rudimentary ovary apparent in staminate flowers; fruiting calyx lobes 6–8 mm. long, heavily coriaceous; berry globose, 8–12 mm. in diameter, sparsely puberulent or eventually glabrate.

Rocky slopes, cliffs, and canyon walls, mainly Arid Tropical Zone, Isla San Josef and Isla Cerralvo; Cape Region, Baja California. July–Aug.

The description of the staminate flowers was drawn from specimens not yet in anthesis, so the measurements offered are tentative. The anthers contained apparently normal pollen grains.

## 2. DIOSPYROS L., Sp. Pl. 1057, 1753

Shrubs or trees, sometimes of considerable size, with alternate or rarely subopposite, deciduous or persistent leaves. Flowers solitary or cymose, axillary, dioecious or rarely perfect. Calyx 3–7-lobed, usually 4–5-lobed, pubescent and accrescent in fruit. Corolla urceolate, tubular or campanulate, minutely pubescent without, mostly 5-lobed. Staminate flowers bearing 4–16 stamens; anthers oblong-linear to lance-oblong; rudimentary ovary present or lacking. Pistillate flowers usually with sterile stamens or antherless staminodia. Ovary ovoid to globose, pubescent or glabrous, usually with twice as many locules as styles, mostly 2–8-celled (in ours to 16-celled). Fruit globose or conical-ovoid. Seeds oblong, somewhat flattened to lenticular; endosperm ruminate.

Sepals acute to short-acuminate, broadest at or below middle, more or less puberulent; leaves tomentulose or markedly glaucous in age:
- Leaves, twigs and calyces finely tomentose .............. 1. *D. californica* var. *californica*
- Leaves, twigs and calyces glabrate ........................ 1a. *D. californica* var. *tonsa*

Sepals rounded, broadest near apex, at length glabrate; leaves glabrate and shining in age (at least above), not glaucous .................................... 2. *D. sonorae*

**1. Diospyros californica** (Brandegee) I. M. Johnston, Proc. Calif. Acad. IV, **12:** 1124, 1924, var. **californica**

*Diospyros texana* var. *californica* Brandegee, Zoe **5:** 164, 1903.

Shrub or small tree with dark gray bark and rather stiff, knotty branchlets; young twigs densely tomentulose with brownish or rusty, closely intertangled, curved hairs; petioles short, 2–5 mm. long; leaf blades broadly elliptic to obovate, 2–3.5 cm. wide, 4–6 cm. long, acute or rounded at base, mostly obtuse, rounded or sometimes emarginate at apex, densely rusty-tomentulose beneath, sparsely so and gray-green above, margins often somewhat undulate or slightly sinuous; pedicels stoutish, 3–10 mm. long, tomentulose; calyx 5-parted almost to base, sepals ovate to ovate-acuminate, 4–6 mm. wide, 10–18 mm. long in fruit, finely and densely puberulent both inside and out, tardily glabrescent without; fruit 1.5–3 cm. in diameter, at first tomentulose, eventually glabrous and shining; seeds smooth, dull brown, flattened, 8–10 mm. wide, 10–13 mm. long, about 3 mm. thick and distinctly but shallowly grooved around dorsal surface.

Along arroyos, in rocky canyons, and on outwash slopes, Lower Sonoran and Arid Tropical Zones, southern Baja California. July–Aug.

**1*a*. Diospyros californica** var. **tonsa** I. M. Johnston, Proc. Calif. Acad. IV, **12:** 1124, 1924

Differing from var. *californica* in having broadly elliptic leaves 5–9 cm. long, which become glabrous and heavily glaucous in age; branchlets at first puberulent but very soon glabrous; otherwise as in the species.

Known only from the type locality, San Bernardo, in the Cape Region, Baja California.

**2. Diospyros sonorae** Standley, Contr. U.S. Nat. Herb. **18:** 120, 1916

Small to large tree with a dense, symmetrical, rounded crown; branchlets slender, grayish, densely puberulent when very young but soon glabrate; leaves persistent, oblong to oblanceolate-oblong, 2–4.5 cm. wide, 5–13 cm. long, rounded or sometimes shallowly emarginate at apex, rounded, obtuse or broadly cuneate at base, coriaceous, pale gray-green, minutely puberulent beneath, glabrous above, midrib impressed above, prominent beneath; petioles stout, 2–3 or rarely to 6 mm. long, minutely puberulent; flowers not seen; fruit solitary or 2–3 in axillary fascicles on stout pedicels 2–3 mm. in diameter, 6–15 mm. long; fruiting calyx deeply 5-lobed, 3–4 cm. broad, lobes lance-oblong, 5–10 mm. broad, obtuse, rounded or infrequently acute at apex, finely puberulent but eventually glabrate; fruit globose or depressed-globose, 2.5–3 cm. in diameter, fulvous-puberulent in youth, later glabrate, mostly 8-celled; seeds 7–8 mm. broad, 10–12 mm. long, dark brown, inner edge acute, much thinner than outer.

Sandy valley floors and adjacent slopes, Lower Sonoran and Arid Tropical Zones, southern Sonora near Alamos and south into Sinaloa. Also growing in the outskirts of Hermosillo but probably cultivated there. July–Aug.

## Family 113. **OLEACEAE.** Olive Family

Herbs, shrubs, or trees with simple or pinnately compound, estipulate leaves arranged alternately or opposite, or in some the lower leaves opposite, upper alternate. Flowers regular, mostly small, perfect or unisexual. Calyx small, 4–15-toothed or -lobed or rarely lacking. Corolla sympetalous, choripetalous, or wanting; petals or corolla lobes 4–6 when present. Stamens 2 or 4. Ovary superior, 2-celled. Style simple; stigma mostly capitate. Fruit a drupe, samara, or capsule, 1–4-seeded.

Leaves pinnately compound (in ours); fruit a samara; plant a tree ........... 1. *Fraxinus*
Leaves simple; fruit not a samara; plants shrubs or perennial herbs:
  Flowers small, greenish or whitish, with 1–2 petals or these lacking; fruit a drupe; plants large woody shrubs ........................................................ 2. *Forestiera*
  Flowers larger, yellow, sympetalous, with 4–6 lobes; fruit a didymous circumscissile capsule; plants small shrubs or suffrutescent .............................. 3. *Menodora*

### 1. **FRAXINUS** L., Sp. Pl. 1057, 1753

Shrubs or trees with opposite, usually odd-pinnate but sometimes simple leaves. Flowers in racemes, panicles, or fascicles, usually dioecious but polygamous in some, small, greenish or white. Calyx 4-parted or -lobed. Petals lacking or 2–4. Stamens usually 2, with large anthers. Fruit a 1-seeded samara with a long, flat, terminal wing.

Twigs quadrangular in youth; wing extending to base of fruit; leaves 1-foliolate or less commonly 3-foliolate, blades broadly ovate to orbicular, nearly truncate at base ........... ........................................................ 1. *F. anomala*

Twigs terete or essentially so; wing of fruit subterminal, not extending to base of fruit; leaves 3–5-foliolate or rarely 7-foliolate; leaflets elliptic, ovate, or narrowly obovate, cuneate to rounded at base:

Leaves, twigs, and pedicels densely puberulent:

Leaflets all sessile or essentially so ........................ 2. *F. velutina* var. *velutina*

Leaflets all or mostly on definite stalks 3–10 mm. long .... 2a. *F. velutina* var. *toumeyi*

Leaves, twigs, and pedicels glabrous or essentially so; leaflets definitely stalked, sometimes with a few hairs along midrib:

Leaflets thin, not noticeably coriaceous .................... 2b. *F. velutina* var. *glabra*

Leaflets thick, coriaceous ................................ 2c. *F. velutina* var. *coriacea*

**1. Fraxinus anomala** Torr. ex S. Wats., Bot. King Expl. 283, 1871

*Fraxinus anomala* var. *triphylla* M. E. Jones, Proc. Calif. Acad. II, **5:** 707, 1895.

Low spreading shrub or a small tree 2–6 m. tall, with glabrous, distinctly quadrangular young twigs, older twigs obtusely quadrangular to nearly terete, their bark tawny, brownish, or eventually gray and faintly checked; leaves simple or on some twigs 2–3-foliolate, petioles slender, 1–4 cm. long, glabrous, channelled along upper side; blades of simple leaves broadly ovate to suborbicular, 2–5.5 cm. wide, 2–6 cm. long, glabrous, distinctly paler beneath than above, broadly and abruptly cuneate to subtruncate at base, obtuse, rounded, or bluntly apiculate at apex, inconspicuously serrulate to entire along margins; trifoliolate leaves with lateral leaflets similar to but considerably smaller than terminal one; inflorescences 4–6 cm. long, branches glabrous; some flowers perfect, others pistillate, calyces cup-shaped, 1–1.3 mm. long, about as wide, thin, with 4 low, broadly deltoid teeth; stamens 2.5–3.5 mm. long, anthers slightly longer than filaments; styles slender, 1.8–2.2 mm. long; samaras obovate, 8–10 mm. wide, 15–22 mm. long, wing obvious to base of fruit, often emarginate at apex, glabrous.

Along bottoms and shaded lower slopes of canyons (in our area) and on mountain sides, mainly Upper Sonoran and Arid Transition Zones, Utah, Colorado, and New Mexico to the Panamint and Providence Mountains in California, and in Apache, Yavapai, and northwestern Gila Counties, Arizona. April.

Both simple and trifoliolate leaves occur on an occasional shrub, hence the varietal recognition of the trifoliolate variant is unwarranted.

**Fraxinus lowellii** Sarg., in Rehd., Proc. Amer. Acad. **53:** 211, 1917 (*F. anomala* var. *lowellii* Little, Jour. Wash. Acad. **42:** 370, 1952), is similar to the arborescent examples of *F. anomala* but regularly with 3–7 leaflets, these ovate to oblong-lanceolate, cuneate at the base and more markedly crenate-serrate along the margins.

Along streams in Coconino, Mohave, Gila, and Maricopa Counties, Arizona, mostly in Upper Sonoran and Arid Transition Zones, but possibly straggling to upper edge of the desert in shaded canyons. March–May.

**2. Fraxinus velutina** Torr., in Emory, Notes Mil. Rec. 149, 1848, var. **velutina**

*Fraxinus pistaciaefolia* Torr., Pacif. R. Rep. **4:** 128, 1857.

A slender tree 5–10 (or rarely –15) m. tall with spreading stout branches and a rounded crown; bark gray or faintly tinged with red, deeply fissured into interrupted ridges which scale off in thin flakes; twigs terete, closely but finely puberulent with short, spreading hairs when young or in some varieties glabrous or nearly so; leaves 8–25 cm. long, usually less than 15 cm. long, petiole grooved on upper side, densely puberulent; leaflets 3–5 or rarely 7, elliptic, ovate, or narrowly obovate, acute at apex, rounded to cuneate at base, finely serrate above middle or nearly entire, finely tomentose on both sides in youth, glabrous above in age, 2–6 cm. long, 1–2 cm. wide, sessile or on short petiolules to 3 mm. long or terminal leaflet on a petiolule to 15 mm. long; flowers dioecious, on slender puberulent pedicels in elongating panicles; calyx cup-shaped, 0.5–1 mm. high; fruits oblong-obovate to elliptic (including wing), body terete or nearly so in cross section, 1.5–3 mm. wide, 10–16 mm. long, wing terminal, narrowly obovate or elliptic, rounded, acute, or slightly emarginate at apex, 3–5.5 mm. wide, extending 10–16 mm. beyond body of fruit, decurrent about halfway down sides of latter.

Chiefly mountain canyons, Upper Sonoran Zone but infrequently following the course of the stream down into the Lower Sonoran Zone, western margins of the Colorado Desert and in Inyo County, California, and from Arizona to Texas, south into northern Mexico. March–May.

**2*a*. Fraxinus velutina** var. **toumeyi** (Britt.) Rehd., Proc. Amer. Acad. **53:** 204, 1917

*Fraxinus toumeyi* Britt., N. Amer. Trees 803, *fig. 732,* 1908.

Differing from var. *velutina* in having leaves that average somewhat larger, leaflets distinctly petiolulate on slender petiolules 4–10 mm. long, and fruits often somewhat larger with broader, more rounded wings than in plants with sessile leaflets.

Mainly mountain canyons, Upper Sonoran Zone to lower Transition Zone, but planted as a shade tree and occasionally escaping about Tucson and Phoenix, Arizona. Apparently occurring sparingly in northern Sonora also.

**2*b*. Fraxinus velutina** var. **glabra** Thornber ex Rehd., Proc. Amer. Acad. **53:** 207, 1917

Differing from var. *toumeyi* in having glabrous foliage and twigs, or leaflets bearing a narrow line of pubescence along sides of midrib beneath when young.

Similar habitats to those occupied by the species, and occurring in Arizona and northern Sonora.

**2*c*. Fraxinus velutina** var. **coriacea** (S. Wats.) Rehd., Proc. Amer. Acad. **53:** 206, 1917

Similar to var. *glabra* but with thicker, more coriaceous, strongly veined leaflets.

Southern Utah to southern California and Arizona, probably also in northern Sonora. A specimen that may belong here was collected in the foothills near the southwest flank of the Sierra San Pedro Mártir in Baja California.

**Fraxinus trifoliata** (Torr.) Lewis & Epling, Amer. Midl. Nat. **24:** 743, 1940, occurs on the chaparral-covered slopes in the Upper Sonoran Zone in southern California and south to the western slopes of the Sierra San Pedro Mártir, but it has not been found within the limits of the desert proper.

**Hesperelaea palmeri** A. Gray, Proc. Amer. Acad. **11:** 83, 1876, a small tree with perfect yellow flowers in terminal panicles and drupaceous fruits, was described from Guadalupe Island, Baja California, but it has never been found on the mainland, and apparently is now extinct. Its taxonomic relationships are puzzling.

## 2. FORESTIERA Poir., in Lam., Encycl. Suppl. 2: 664, 1812

Large shrubs or small trees with many slender to rigid branches and simple, entire, or serrulate, opposite or fasciculate leaves that are persistent or deciduous and often markedly punctate beneath; margins often revolute. Flowers small, greenish, apetalous or nearly so, dioecious or polygamous, in lateral racemes or fascicles. Calyx minute and 4-lobed or none. Corolla of 1–2 small deciduous petals or lacking. Stamens 2–4. Fruit a small drupe, globose, obovoid or elliptic, symmetrically curved; flesh usually bluish or blackish.

Leaves puberulent on both surfaces, not conspicuously punctate; drupes asymmetrically curved, about twice as long as wide .......................... 1. *F. phillyreoides*
Leaves glabrous or soon glabrate, distinctly punctate beneath; fruit symmetrical, ellipsoidal or ovoid, less than twice as long as wide:
  Leaf blades linear to narrowly oblong-spatulate, 1–3.5 mm. wide, mostly obtuse or rounded at apex, margins entire ..................................... 2. *F. angustifolia*
  Leaf blades ovate, obovate or oblanceolate, 6–11 mm. wide, usually acute at apex, margins crenulate to serrate ......................................... 3. *F. neomexicana*

**1. Forestiera phillyreoides** (Benth.) Torr., Bot. Mex. Bound. 167, 1859

*Piptolepis phillyreoides* Benth., Pl. Hartw. 29, 1840.
*Forestiera porulosa* var. β *phillyreoides* A. Gray, Proc. Amer. Acad. **4:** 365, 1860.
*Forestiera shrevei* Standley, Field Mus. Bot. Ser. **17:** 205, 1937.

A much-branched shrub 2–3.5 m. tall with rather stiffish, divaricate twigs with finely striate, pale to dark gray bark; leaves oblong to oblong-oblanceolate, or narrowly obovate, 4–10 mm. broad, 1.5–2.5 cm. long, rounded, minutely emarginate or slightly apiculate at apex, broadly cuneate at base, finely puberulent with slightly intertangled, simple hairs on both surfaces, or becoming glabrate above, not noticeably punctate, margins narrowly revolute; scales of flower buds 6–10 pairs, pale yellow, broadly ovate, about 1 mm. long, acute, faintly keeled, margins ciliolate-puberulent; flowers 2–6 in a fascicle; stamens usually 4, anthers dark purple; pedicels slender, 4–7 mm. long, glabrous; drupe ellipsoid, 2.5–3.5 mm. in diameter, 5–8 mm. long, purplish black, usually asymmetrically curved, seed striate.

Arid slopes and canyons, Lower and Upper Sonoran Zones, southern Arizona from Yuma County to Pima and Maricopa Counties south into Mexico. Dec.–March.

**2. Forestiera angustifolia** Torr., Bot. Mex. Bound. 168, 1859

*Forestiera porulosa* var. γ? *angustifolia* A. Gray, Proc. Amer. Acad. **4:** 365, 1860.

Intricately branched shrub 1.5–5 m. tall with slender but rather stiff branches, dark gray bark and glabrous foliage; leaves linear to narrowly oblong-spatulate, acute or obtuse at apex, cuneate at base, 1–3.5 mm. wide, 1.5–3 cm. long, entire, plane or margins sometimes faintly revolute, conspicuously punctate beneath, bright green; scales of flower buds 6–8 pairs, pale yellow, angled on back, glabrous except for ciliolate-puberulent margins, obtuse at apex, 0.5–0.8 mm. long, slightly wider than long; stamens 2–3 mm. long; fruit on pedicels 2.5–5 mm. long, ovoid to broadly elliptic or obovoid, 6–8 mm. long, black or purplish, symmetrical, seed longitudinally striate, 5–7 mm. long.

Canyons and hillsides, Sonoran Zones, central Sonora (Cumpas, Sierra Babiso), Coahuila, Nuevo León, Tamaulipas, and western Texas.

The material from Sonora seems to belong here but it may be distinct. No flowers are available. The leaves of the specimens from Cumpas are 1–1.5 mm. broad, 1.5–2.5 cm. long, quite glabrous, but conspicuously punctate beneath. Feb.–March.

**Forestiera macrocarpa** Brandegee, Zoe 4: 404, 1894, is a glabrous shrub with elliptic or oblong-ovate leaves 2–3 cm. long and dark blue, oblong fruit 12–15 mm. long. It is known only from the Cape Region in Baja California but may occur in the mountains immediately north of La Paz.

**3. Forestiera neomexicana** A. Gray, Proc. Amer. Acad. **12:** 63, 1876

*Forestiera acuminata* var. β *parvifolia* A. Gray, op. cit. **4:** 364, 1860.
*Adelia parvifolia* Coville, Contr. U.S. Nat. Herb. **4:** 148, 1893.

Glabrous, more or less spiny, divaricately branching shrub 1–4 m. tall with yellowish to light gray bark; leaves often in fascicles, ovate, obovate, or oblanceolate-oblong, 6–11 mm. wide, 1.5–3 cm. long, cuneate at base, acute or sometimes obtuse at apex, shallowly crenulate or serrate, distinctly punctate-dotted beneath, sparsely so above, margins sometimes faintly revolute; petioles 1–3 mm. long; flowers polygamodioecious, in clusters, staminate sessile, pistillate pedicellate; scales of flower buds obovate to nearly orbicular, 1.5–2.5 mm. long, yellowish, ciliolate near base, otherwise glabrous; anthers yellow, 0.5 mm. long, nearly globose; filaments about 2 mm. long; drupes ovoid to broadly ellipsoidal, blue-black with a faint to pronounced, glaucous bloom, 6–8 mm. long; style slender, 1–1.5 mm. long, deciduous.

Valleys, mountain sides, and canyons, Lower and Upper Sonoran Zones, Alameda County, California, to the western margins of the Colorado Desert, east to Utah, southwestern Colorado, New Mexico, and western Texas. March–May.

## 3. MENODORA H. & B., Pl. Aequin. 2: 98, 1809

Perennial herbs or small shrubs woody at least at base, with simple (in ours), entire or lobed, sessile or subsessile, chiefly alternate leaves (or lower ones opposite). Flowers perfect, showy, solitary or corymbose; corolla yellow, sympetalous, rotate-campanulate to salverform; limb 4- to 6-lobed. Stamens 2. Capsules thin-walled, didymous, circumscissile, each cell commonly containing 2 or 4 seeds.

Herbage and calyces usually scabrous-puberulent; calyx lobes 7 or more; leaves of inflorescence foliaceous though small:
- Plants with simple or slightly branched stems, mostly herbaceous:
  - Leaves chiefly linear or linear-lanceolate; corolla lobes slightly mucronate; anther connectives prolonged into a rounded protuberance ......... 1. *M. scabra* var. *scabra*
  - Leaves chiefly ovate, oblong-elliptic, or oblong-lanceolate; corolla lobes distinctly mucronate; connectives prolonged into cuneate protuberance ... 1*a*. *M. scabra* var. *laevis*
- Plants more profusely branched; stems woody at base .... 1*b*. *M. scabra* var. *ramosissima*

Herbage and calyces glabrous or essentially so; calyx lobes 5–6 (rarely with 1–5 additional very small, intermediate ones); leaves of inflorescence reduced to subulate bracts ...... ........................................................ 2. *M. scoparia*

**1. Menodora scabra** A. Gray, Amer. Jour. Sci. II, **14:** 44, 1852, var. **scabra**

Plants herbaceous from a heavy woody rootstock or woody at base, growth of season 1–4 dm. long; stems, leaves, and calyces minutely but distinctly scabrous-puberulent, at least along margins of leaves and sepals, stems sometimes subglabrate; leaves subsessile, linear-oblong to elliptic-oblong, acute at apex, cuneate at base, 1.5–6 mm. wide, 1–3 cm. long, margins faintly revolute or plane; inflorescence corymbose; pedicels 1–5 mm. long, minutely scabrous; calyx lobes 7–13, linear, 3–5 mm. long at anthesis, only slightly longer in fruit, acute to apiculate at apex; corolla yellow, tube about equaling calyx lobes, lobes broadly obovate, 3.5–6 mm. wide, 6–8 mm. long, rounded; stamens 6–10 mm. long, somewhat exserted; fruit 6–9 mm. high, to 1.5 cm. broad, capsule circumscissile, chartaceous, shining; seeds 5–6 mm. high, concavo-convex, curved, sordid-yellow, foveolate-reticulate, weakly keeled or angled on inner face.

Arid slopes and mesas, Lower and Upper Sonoran Zones, southern Utah and Colorado to southeastern California, Arizona, Texas, and adjacent Mexico; Sonora and Baja California. March–Sept.

**1*a*. Menodora scabra** var. **laevis** (Woot. & Standl.) Steyermark, Ann. Mo. Bot. Gard. **19:** 137, 1932

*Menodora laevis* Woot. & Standl., Contr. U.S. Nat. Herb. **16:** 158, 1913.

Differing from var. *scabra* mainly in having ovate to elliptic-oblong or oblong-lanceolate, distinctly mucronate leaves 1–5 mm. broad and 2–17 mm. long; distinctly mucronate corolla lobes; and connectives of anthers prolonged into cuneate or subcuneate, pointed protuberances instead of these being rounded.

Arid regions, Lower and Upper Sonoran Zones, southeastern California to southwestern Texas and northern Chihuahua. Within our area found near Tucson.

**1*b*. Menodora scabra** var. **ramosissima** Steyermark, Ann. Mo. Bot. Gard. **19:** 139, 1932

Differing from the two entities above mainly in being more densely paniculately branched, suffruticose; in having somewhat larger leaves (these 1–9 mm. broad and to 4 cm. long); prolongation of connective short-cylindric; seeds more irregularly pitted.

Arid mesas, slopes, and ridges, Lower and Upper Sonoran Zones, central Arizona to Texas, Chihuahua, and Nuevo León.

The varieties listed above intergrade with each other. They are difficult to differentiate.

**2. Menodora scoparia** Engelm., in A. Gray, Bot. Calif. **1:** 471, 1876

*Menodora scabra* var. *glabrescens* A. Gray, in S. Wats., Cat. Pl. Wheeler Exp. 15, 1874.

A paniculately branched, erect perennial to 9 dm. tall, with several to many slender, slightly angled, glabrous stems; leaves oblong-obovate to oblanceolate, 1–4 mm. wide, 1–3 cm. long, gradually reduced above to linear-subulate bracts, acute to obtuse at apex, short-petiolate below, upper sessile, glabrous or very sparingly scaberulent; pedicels slender, 3–10 mm. long; calyx 3–6 mm. long, lobes usually 5–6, sometimes with smaller intermediate ones, linear-lanceolate, 1.5–4 mm. long; corolla 9–10 mm. long, yellow, tube 3–5 mm. long, somewhat pilose at throat within, lobes obovate, obtuse, 5–8 mm. long; capsule 4–6 mm. high, 8–10 mm. broad; seeds 4 in each cell, 4–5 mm. long, 2.5–3 mm. wide, foveolate-reticulate.

Arid hills and mesas, Lower and Upper Sonoran Zones, entering the desert only along its upper margins, southeastern San Diego County, California, to Arizona, Baja California, Coahuila, and Nuevo León. April–Sept.

## Family 114. LOGANIACEAE. Logania Family

Shrubs, trees, herbs, or scandent vines with entire or dentate, opposite, estipulate leaves usually having petioles of a pair of leaves united by a stipular line. Flowers perfect, regular, variable as to size and arrangement. Calyx tubular or campanulate, 4–5-lobed. Corolla sympetalous, tubular, funnelform or salverform with a 4–5-lobed limb. Stamens as many as and alternate with corolla lobes inserted in throat or deeper in corolla tube. Style simple; stigma capitate or lobed. Fruit a capsule (in ours), a berry, or a drupe. Seeds various.

## 1. BUDDLEIA L., Sp. Pl. 112, 1753

Shrubs or small trees usually invested with a tomentum of stellate hairs. Leaves ample, entire or dentate. Flowers small, in dense cymes or heads, these solitary in axils or spicate or paniculate. Calyx campanulate, 4-toothed. Corolla small, short-salverform, with 4 imbricate lobes rotately spreading in anthesis. Stamens 4. Fruit a septicidally bivalvate capsule. Seeds small, several to many, usually somewhat angled by mutual compression.

Smaller leaves near tips of branches and of inflorescence entire; flowering heads in spikes or lower ones solitary on axillary peduncles, 1–1.8 cm. in diameter in anthesis; calyces 4–4.5 mm. long at anthesis:

Flower heads all sessile, interruptedly arranged on spikes to 3 dm. long; corollas nearly uniformly pilose within throat; corolla lobes about as long as wide .. 1. *B. sessiliflora*

Flower heads chiefly solitary on axillary peduncles 5–20 mm. long; corollas with a dense ring of coarse hairs at apex of throat, sparsely pilose with shorter, finer hairs on lower part and in tube; corolla lobes 1.5 mm. long, 2 mm. wide .......... 2. *B. wrightii*

Smaller leaves as well as larger ones distinctly dentate or denticulate; flower heads in broad panicles, 5–8 mm. in diameter in anthesis; calyces about 2.5 mm. long at anthesis .... ........................................................ 3. *B. crotonoides*

### 1. Buddleia sessiliflora H. B. K., Nov. Gen. & Sp. **2**: 345, *pl. 182*, 1817

*Buddleia verticillata* H. B. K., op. cit. 346, *pl. 184*.
*Buddleia pseudoverticillata* Mart. & Gal., Bull. Acad. Brux. **12**$^2$: 24, 1845.
*Buddleia melliodora* Kunth, Ind. Sem. Hort. Berol. 10, 1845.
*Buddleia barbata* Kunth, op. cit. 12, 1847.
*Buddleia pringlei* A. Gray, Proc. Amer. Acad. **19**: 86, 1883.
*Buddleia simplex* Kränzlein, Ann. Nat. Hofm. Wien **26**: 396, 1912.

A shrub or somewhat arborescent, 1–5 m. tall; branches sparsely puberulent to closely but finely tomentulose with tawny hairs; leaves narrowly lanceolate, elliptic, or ovate-rhombic, 1–4 cm. broad, 4–15 cm. long, dark green and sparsely arachnoid-lanate to densely tomentulose above, paler and more hairy beneath, veins rather prominent beneath, acute or slightly attenuate at apex, cuneately narrowed to a narrowly winged petiole to 2 cm. long, margins serrate on lower larger leaves, entire on smaller ones in and near inflorescence; heads sessile, 1–1.8 cm. in diameter, forming interrupted terminal and axillary spikes to 3 dm. long; calyx 4–4.5 mm. high, lobes deltoid, 1.5 mm. long and wide, densely tomentulose; corolla 5–6 mm. long, tomentulose without, greenish yellow, fragrant, fairly uniformly pilose within throat, lobes 4, about 2 mm. long; filaments 1 mm. long, sparsely pilose; anthers 1.2–1.5 mm. long; capsule obovoid, about 3–4 mm. long; seeds pale brown, 0.4–0.6 mm. long.

Along streams and in canyons, Lower Sonoran Zone to Tropical Zone, Arizona to Coahuila, San Luis Potosí, and Oaxaca. Near Ímuris in Sonora. Feb.–May.

### 2. Buddleia wrightii Robinson, Bot. Gaz. **16**: 341, 1891

Slenderly branched shrub 2–3 m. tall with sparsely arachnoid-tomentose to glabrate branches; leaves narrowly lanceolate or lance-elliptic to ovate-lanceolate, 1–9 cm. broad, 3–26 cm. long, acuminate to attenuate at apex, narrowly cuneate to abruptly decurrent at base, finely and sparsely arachnoid-tomentulose on both surfaces, or eventually glabrate above, margins serrate-dentate on larger leaves, entire on smaller subterminal ones and those in inflorescence; petioles slender, 1–4 cm. long; flowers in subglobose heads 1–1.5 cm. in diameter, heads usually solitary in each axil, on peduncles 5–20 mm. long; calyx 4–4.5 mm. high, lobes 1.5 mm. high and wide, densely tomentulose without; corolla 5.5–6 mm. long, densely pilose with a ring of coarse whitish hairs around upper edge of throat, sparsely puberulent with finer hairs below and in tube, corolla lobes 1.5 mm. long, about 2 mm. wide; filaments 0.5–0.8 mm. long, sparsely pilose; capsule ovoid, slightly surpassing calyx.

Along streams, Lower Sonoran and Tropical Zones, Sonora to Tepic. In Sonora it occurs as far north as Cumetal on the Río de los Alisos. Flowering in spring.

**3. Buddleia crotonoides** A. Gray, Proc. Amer. Acad. **5:** 165, 1861

Sturdy shrub 2–5 m. tall with densely tomentose, terete twigs that eventually become glabrate; leaves lanceolate to broadly ovate-lanceolate or deltoid-ovate, 2–7 cm. broad, 6–15 cm. long, acute to slightly attenuate at apex, acute to truncate or subcordate at base, crenulate, thick, densely tomentose on both surfaces in youth, later glabrate or nearly so above, veins prominent beneath, reticulate near margins, blade sometimes decurrent along comparatively short, stoutish petiole, latter 1–2 cm. long or shorter on upper leaves; inflorescence broadly pyramidal, paniculate, to 3 dm. long or more, nearly as wide; heads globose, 5–8 mm. in diameter, sessile on branches of panicle, each branch an interrupted spike of heads, whole inflorescence densely tomentose; calyx broadly campanulate, 2.5 mm. tall, lobes deltoid, equaling tube, densely tomentose; corolla about 4 mm. long, puberulent without, lobes 4, 1.5 mm. long, 1.5–2 mm. wide, each lobe bearing a dense, brush-like tuft of white hairs across its base on inner surface, corolla tube pilose with fewer shorter hairs; capsules narrowly ovoid, 2.5–3 mm. high, tawny-tomentulose; seeds dark brown, about 0.5 mm. long.

Canyon floors and along washes, Lower Sonoran and Arid Tropical Zones, southern Baja California. Flowering after rains.

## Family 115. GENTIANACEAE. Gentian Family

Glabrous herbs with bitter colorless juice and simple, entire (3-foliolate in *Menyanthes*), opposite, and mostly sessile leaves. Flowers perfect, regular. Calyx 4–5-lobed or -parted. Corollas 4–5-lobed, usually withering-persistent; lobes usually convolute in bud. Stamens as many as corolla lobes and inserted in tube or throat. Ovary superior, 1-celled; placentae 2, parietal. Style 1 or lacking; stigmas 2. Fruit a septicidal, 2-valved capsule. Seeds minute; embryo straight; endosperm present.

Corolla salverform, pink; anthers spirally twisted after dehiscence; corolla lobes 1.2 cm. long or less ........................................................ 1. *Centaurium*

Corolla campanulate, blue or lavender; anthers straight or slightly recurved after dehiscence, not spirally twisted; corolla lobes 1.8 cm. long or more ................ 2. *Eustoma*

### 1. CENTAURIUM Hill, Brit. Herb. 62, 1756

Ours annual, glabrous herbs with sessile, often clasping, opposite leaves. Flowers small to moderate in size, 4–5-merous, white, pink, or pale red, in cymose panicles or sometimes solitary and terminal, pedicellate or subsessile. Calyx tube short-cylindrical; lobes linear to linear-lanceolate, acute or acuminate, frequently appressed to corolla tube, keeled and scarious-margined. Corollas tubular-salverform; lobes usually distinctly shorter than tube. Stamens inserted in corolla tube near its mouth, alternate with corolla lobes; filaments narrowly winged; anthers ovate to oblong, usually spirally twisted after dehiscence. Ovary superior, 1-celled, fusiform. Style slender;

stigma with 2 obcuneate short lobes. Capsule 2-valved; placentation marginal. Seeds spherical to oblong, minute, reticulate.

Corolla lobes 7–12 mm. long, only slightly shorter than tube; stems usually with 6 pairs of leaves or more; anthers linear:
  Pedicels all equaling calyx, usually longer than whole flower; calyx shorter than or barely equaling corolla tube .......... 1. *C. calycosum*
  Pedicels often shorter than calyx, some flowers subsessile; calyx usually equaling corolla throat .......... 2. *C. venustum*
Corolla lobes 2.5–5 mm. long, much shorter than tube; stems rarely with more than 6 pairs of leaves; anthers ovate:
  Basal leaves not conspicuously wider than cauline leaves, not forming a rosette; cauline leaves usually strongly ascending .......... 3. *C. exaltatum*
  Basal leaves conspicuously broader than cauline ones and forming a rosette; cauline leaves spreading or weakly ascending .......... 4. *C. nudicaule*

**1. Centaurium calycosum** (Buckl.) Fernald, Rhodora **10:** 54, 1908

*Erythraea calycosa* Buckl., Proc. Acad. Phila. **1862:** 7, 1863.
*Erythraea calycosa* var. *arizonica* A. Gray, Syn. Fl. N. Amer. **2**[1]**:** 113, 1878.
*Centaurium arizonicum* Heller, Muhlenbergia **4:** 86, 1908.

Erect, glabrous, simple or more frequently much-branched annuals to 5 or 6 dm. tall; leaves obovate, oblong, or lanceolate, 2–10 mm. wide, 2.5–6 cm. long, sessile or some gradually narrowed to a short, narrowly winged petiole 2–7 mm. long, acute to apiculate or sometimes rounded and mucronulate, upper ones smaller and narrower than lower; flowers about 2 cm. long, on pedicels mostly 1–3 cm. long, pedicels usually longer than calyx; calyx 8–10 mm. long, about 1 mm. wide at base, lobes lance-subulate, scarious-margined, erect; corolla tube slender, 10–13 mm. long, usually equaling or, in lowland plants, surpassing calyx lobes; corolla lobes 7–11 mm. long, to 5 mm. wide, obovate, acute or apiculate, pink, throat yellow; stamens exserted 3–6 mm. beyond throat; style 1–2 mm. longer than stamens, 2 stigma lobes abruptly and broadly funnelform; capsules 8–10 mm. long, membranous, transparent; seeds subglobose, slightly angulate, 0.3 mm. in diameter, microscopically favose-reticulate, dull black or ridges of testa nearly transparent.

In moist soil in early spring, Lower Sonoran to Arid Transition Zones, western Arizona, southern Nevada, and Utah to Texas and northern Mexico. March–Nov.

**2. Centaurium venustum** (A. Gray) Robinson, Proc. Amer. Acad. **45:** 397, 1910

*Erythraea chironioides* Torr., Bot. Mex. Bound. 156, *pl. 42*, 1859, in part.
*Erythraea venusta* A. Gray, Bot. Calif. **1:** 479, 1876.

Very similar to former species but averaging smaller, usually 1–2.5 dm. tall and less profusely branched; calyx lobes reaching nearly or fully to top of corolla throat; corolla 15–25 mm. long, lobes rose to vermilion, elliptic to oblong, 8–12 mm. long, three-fifths as long as, to equaling tube, throat white with 5 red spots or crescents just below each lobe; stigma, capsules, and seeds as in *C. calycosum*.

Moist or eventually dry flats and slopes, Lower Sonoran to Arid Transition Zones, northern Sacramento Valley and adjacent foothills, south through the Sierra Nevada to northern Baja California. Sparingly present in both the Mojave and Colorado Deserts. May–Aug.

This plant is doubtfully distinct from *C. calycosum,* but extensive field observations and careful garden work should be undertaken before definitive conclusions are attempted. Inasmuch as no nomenclatorial changes are involved, it is here retained in specific rank pending the critical study needed to determine its status.

**3. Centaurium exaltatum** (Griseb.) Wight, in Piper, Contr. U.S. Nat. Herb. **11:** 449, 1906

*Cicendia exaltata* Griseb., in Hook., Fl. Bor. Amer. **2:** 69, *pl. 157A*, 1837.
*Erythraea douglasii* A. Gray, Bot. Calif. **1:** 480, 1876.
*Erythraea exaltata* Coville, Contr. U.S. Nat. Herb. **4:** 150, 1893.

Stems slender and strict, 1–4 dm. tall; leaves erect and nearly appressed to stem or strongly ascending, blades oblong or upper ones lance-linear, 8–30 mm. long, 1–7 mm. wide, acute, and often apiculate; flowers in loose cymes, pedicels slender, 1–6 cm. long; calyx lobes lance-linear, 6–8 mm. long; corolla tube 8–10 mm. long, lobes 4–5, pink, 2.5–5 mm. long, oblong, obtuse, or rounded; stamens exserted about length of anthers, these oval, 2–2.5 mm. long; capsules 8–10 mm. long; seeds ovoid, dull black, reticulate-pitted, 0.3–0.4 mm. long.

Mainly moist soil along streams and in seepy places in meadows and on hillsides, Sonoran Zones, foothills and mountains in southern California north to eastern Washington and Idaho, east to Utah, infrequently into northern Baja California, and possibly as a rare plant in western Arizona. Chiefly above the limits of the Sonoran Desert but collected on the western margin of the San Felipe Desert, Baja California, at about 2,500 feet altitude. April–May.

Specimens indistinguishable from this species were collected along stream banks (at three points) between La Paz and San José del Cabo, in the Cape Region of Baja California. The species probably occurs rather generally down the east side of the peninsula.

**4. Centaurium nudicaule** (Engelm.) Robinson, Proc. Amer. Acad. **45:** 397, 1910

*Erythraea nudicaulis* Engelm., in A. Gray, Proc. Amer. Acad. **17:** 222, 1882.

Slender plants 0.5–3.5 dm. tall; basal leaves broadly obovate to oblong-spatulate, 4–13 mm. wide, 8–26 mm. long, obtuse or rounded at apex, subsessile; cauline leaves narrowly spatulate to linear-lanceolate, 0.5–2.5 mm. wide, to 18 mm. long, erect and appressed to stems or strongly ascending, much shorter than internodes they subtend; flowers few, on slender pedicels 2–8 cm. long; calyx lobes linear-subulate, 6–10 mm. long, about equaling corolla tube and throat; corollas 8–14 mm. long, lobes usually 5, oblong or narrowly oblanceolate, 4–5 mm. long, pinkish; anthers and stamens about as in *C. exaltatum*; capsules 7–8 mm. long; seeds about 0.3 mm. in diameter, reticulate-pitted, dull black.

Along streams, mainly Upper Sonoran Zone, Pima County, Arizona, and the Cape Region, Baja California. April–May.

Two specimens from the Cape Region have smaller calyces and corollas (5–6 mm. and 10–11 mm. long, respectively) than material from Arizona. No qualitative differences are apparent.

## 2. EUSTOMA Salisb., Parad. Lond. *pl. 34*, 1805

Glaucous (in ours) perennial herbs with erect or ascending, leafy stems to 7 dm. tall. Leaves opposite, sessile or amplexicaul, oblong to oblanceolate. Flowers 5–6-merous except for the bicarpellate, 2-valved capsule, axillary and terminal, long-pedicelled. Calyx lobes keeled, long-acuminate. Corolla deeply campanulate, blue or white; lobes more or less erose-denticulate. Stamens inserted on corolla throat; anthers straight or recurved in age. Capsule oval to ellipsoid, obtuse, 2-valved. Seeds small, numerous, faveolate-pitted.

**1. Eustoma exaltatum** (L.) Griseb., in A. DC., Prod. **9:** 51, 1845

*Gentiana exaltata* L., Sp. Pl. ed. 2, 331, 1762.
*Eustoma silenifolium* Salisb., Parad. Lond. *pl. 34*, 1805.

Erect plants 3–8 dm. tall; stems ascendingly branching from about middle upward, glabrous and somewhat glaucous throughout; herbage blackened in drying; leaf blades elliptic to oblong or sometimes obovate, 0.5–3 cm. wide, 2.5–9 cm. long, acute, obtuse, or rounded, thickish, sessile and at least lower and middle ones subcordately clasping; flowers showy, 2.5–3.5 cm. long, blue or lavender, on pedicels 2.5–8 cm. long; calyx lobes lance-attenuate, 10–15 mm. long, keeled and scarious-margined toward base; corolla tube 10–12 mm. long, lobes narrowly obovate to oblong, 1.8–2.5 cm. long, mostly about twice as long as tube, minutely erose-denticulate and often apiculate at apex; stamens about one-third as long as corolla lobes; capsule broadly elliptic, 10–15 mm. long, usually obtuse, drying shiny-black and irregularly but finely wrinkled; seeds about 0.4 mm. in diameter, faveolate-pitted, black or dark brown.

In moist places along streams, in meadows or in sandy or alkaline ground, Lower and Upper Sonoran Zones, foothills and lower mountains of southern California, in suitable habitats to Florida; south through Baja California and the mainland of Mexico to Venezuela. Occurring in the Sonoran Desert along the western margin of the Colorado Desert and in Sonora. Rather rare in Arizona. June–Sept.

**Swertia parryi** Kuntze, Rev. Gen. Pl. **2:** 430, 1891, approaches the margins of the Sonoran Desert and ranges from southern California into Arizona and south almost to the southern end of the Sierra San Pedro Mártir, but always, as far as is known, in the Upper Sonoran Zone above the limits of the desert.

## Family 116. **APOCYNACEAE.** Dogbane Family

Perennial, herbaceous or slightly woody plants or shrubby or arborescent in some, often with milky acrid juice. Leaves simple, entire, opposite or alternate, sessile or petiolate, estipulate. Flowers perfect, regular. Calyx lobes 5, distinct to base or essentially so. Corolla cylindrical-campanulate, tubular-salverform, or funnelform; lobes 5, convolute in bud. Stamens 5, inserted separately on corolla; anthers often connivent and somewhat adherent to stigma, bearing appendages apically, or basally, or both or none. Pistils 2, separate; stigmas often united in flower, superior or nearly so. Fruit a pair of erect or divergent follicles or a drupe. Seeds 1 to many, often comose. Embryo straight.

Follicles 1–4 cm. in diameter; seeds flattened, membranously winged ......... 1. *Plumeria*
Follicles less than 1 cm. in diameter, or fruit a drupe; seeds cylindrical or ellipsoid, not strongly flattened, either naked or comose, not winged:
  Seeds comose at each end, longitudinally ridged; corolla tube shorter than lobes ........ ........................................................ 2. *Haplophyton*
  Seeds comose at one end only or naked, pitted or wrinkled but not strongly and regularly ridged; corolla tube longer than lobes:
    Anthers without appendages; seeds naked, not comose; leaves alternate:
      Plants herbaceous; fruit a follicle; stigma (in ours) bearing 2 terminal blunt apiculations and a reflexed membranous appendage ...................... 3. *Amsonia*

Plants shrubby or arborescent; fruit a translucent white or pinkish drupe; stigma cylindrical, unappendaged .......................................... 4. *Vallesia*
Anthers with basal appendages; seeds apically comose; leaves opposite or verticillate:
Corolla 2–6.5 cm. long, without appendages within; stamens inserted high in corolla tube; ovary superior:
Plants erect; corolla slenderly funnelform, 3–6.5 cm. long ...... 5. *Macrosiphonia*
Plants mostly scandent; corolla salverform, 2–2.5 cm. long ........ 6. *Mandevilla*
Corolla (in ours) 3–4 mm. long, with 5 basal appendages opposite corolla lobes; stamens inserted at base of corolla tube; ovary partly inferior ...... 7. *Apocynum*

## 1. PLUMERIA L., Sp. Pl. 209, 1753

Shrubs or trees with thick, glabrous or pubescent branches, alternate, petiolate, ample leaves, and copious milky sap. Flowers large, showy, in terminal, twice- or thrice-branched cymes. Calyx 5-lobed and eglandular, lobes broadly obtuse and somewhat imbricate in bud. Corolla salverform or funnelform, with a slender tube flaring into a broader throat; lobes 5, oblong. Stamens 5, inserted near base of tube, included. Anthers oblong, obtuse, without appendages. Ovary of 2 distinct carpels, each many-ovuled. Fruit of 2 divergent, ellipsoid to oblong or linear, more or less spindleform, leathery follicles. Seeds oblong to lanceolate, compressed, membranously winged.

**1. Plumeria acutifolia** Poir., in Lam., Encycl. Suppl. **2:** 667, 1812

*Plumeria mexicana* Lodd., Bot. Cab. **11:** *pl. 1024*, 1825.
*Plumeria lambertiana* Lindl., Bot. Reg. **16:** *pl. 1378*, 1830.
*Plumeria gouani* D. Don ex G. Don, Gen. Hist. Pl. **4:** 94, 1838.

A shrub or small tree to 8 or 9 m. tall with finely puberulent but eventually glabrate, fleshy branches; leaves ample, elliptic, oblong or broadly oblanceolate, 5–10 cm. broad, to 40 cm. long, acute to acuminate at apex, narrowly to broadly cuneate at base, dark green and glabrous above, paler and sparsely pubescent along numerous parallel lateral veins beneath; petioles stoutish to slender, 1–3.5 cm. long; flowers white, fragrant; peduncles stout, subfleshy, to 10 cm. long; cymes several-flowered; pedicels slender, about 1 cm. long; calyx lobes deltoid-ovate, 1–1.5 mm. long; corolla tube about 2 cm. long, glabrous; corolla throat 4–5 mm. long; corolla lobes 1.5–2 cm. long, oblong; follicles spindleform, 1–4 cm. in diameter, 10–25 cm. long, leathery, acute or attenuate at apex; seeds 7–9 mm. long, or with wings to 2 cm. long, 4–5 mm. wide, tip of wing erosulate.

Along arroyos and canyons or lower slopes, Lower Sonoran and Tropical Zones, southern Baja California and central Sonora (Divisadero) to Chihuahua, south into the American tropics, and introduced into the tropics of the Old World. March–Sept.

## 2. HAPLOPHYTON A. DC., Prod. 8: 412, 1844

A much-branched perennial plant with erect stems, woody below. Leaves alternate or sometimes a few opposite or subopposite, bright green, lance-ovate to lanceolate, thin. Flowers in showy terminal clusters of 2–3 or solitary. Calyx 5-parted into linear-subulate lobes. Corolla salverform, yellow; lobes broad, equaling or exceeding tube. Stamens inserted about middle of slender corolla tube, included. Anthers connivent around, but not attached to, stigma. Ovary of 2 distinct carpels, each carpel many-ovuled. Styles filiform. Fruit of 2 slender, terete, erect or slightly spreading follicles, or 1 sometimes abortive. Seeds linear, black, bearing a deciduous tuft of silky white hairs at each end.

Leaves 2–5 cm. long; seeds with continuous longitudinal ridges, 8–10 mm. long, brown ........................................................ 1. *H. cimicidum* var. *cimicidum*
Leaves mostly less than 3 cm. long; seeds with broken instead of continuous ridges, 6–7.5 mm. long, black .................................... 1a. *H. cimicidum* var. *crooksii*

## 1. Haplophyton cimicidum A. DC., Prod. 8: 412, 1844, var. cimicidum

Plants woody only near base; stems slender, erect or erect-ascending, 3–6 dm. tall, finely strigose, eventually glabrate, bright green; petioles slender, 1–3 mm. long; leaf blades lanceolate to ovate-lanceolate, 7–16 mm. wide, 2–5 cm. long, rounded or obtuse at base, acute or short-acuminate at apex, entire, scaberulous-puberulent on both surfaces; pedicels 5–20 mm. long; calyx lobes linear-subulate, 5–6 mm. long, puberulent with subappressed stiffish hairs; corolla tube 7–8 mm. long, sparsely puberulent without, pubescence arranged more or less in lines; corolla lobes 10–12 mm. long, 8–9 mm. wide, rounded, sparsely puberulent on outer surface; follicles 7–10 cm. long, terete, finely strigose; seeds 8–10 mm. long, brown to nearly black, sometimes tinged with green, longitudinally ridged, ridges smooth and not interrupted by transverse grooves.

Rocky hillsides and arroyos, Lower Sonoran Zone to Upper Sonoran Zone, central Sonora, south to Guatemala and east to Cuba. July–Oct.

## 1*a*. Haplophyton cimicidum var. crooksii L. Benson, Torreya 42: 9, 1942

*Haplophyton crooksii* L. Benson, Amer. Jour. Bot. **30**: 630, 1943.

Differs from var. *cimicidum* only in having smaller leaves, and seeds that are slightly smaller and ridges with longitudinal costae that are more or less interrupted by transverse grooves or depressions on these ridges.

Arid hillsides and canyons, Lower Sonoran Zone, southern Arizona, northern Sonora, and east to Texas and Chihuahua. June–Oct.

## 3. AMSONIA Walt., Fl. Car. 98, 1788

Herbaceous, mostly erect, glabrous or pubescent perennials with milky juice. Leaves alternate or subverticillate, sessile or petiolate, entire, membranous to subfleshy. Inflorescence a terminal thyrse or corymbiform cyme. Calyx deeply 5-parted into acuminate or subulate lobes. Corolla salverform, pubescent within; tube cylindrical, flaring into a short narrow throat, this often somewhat constricted at mouth; corolla lobes ovate to lanceolate, spreading to ascending. Stamens 5, inserted on and adnate to about middle of corolla tube, included; anthers ovate to ovate-lanceolate, without appendages. Carpels 2, united at apex in flower but later distinct. Stigma capitate or 2-lobed, surrounded by a papillose mass and bearing a membranous reflexed appendage. Fruit of 2 follicles, these cylindrical, torose or unconstricted. Seeds few to many, cylindrical to spindleform, straight or slightly curved, unappendaged.

Corolla tube and throat 2–4 cm. long; leaves filiform to linear-lanceolate or rarely narrowly lanceolate; corolla lobes 10–15 mm. long ........................ 1. *A. longiflora*
Corolla tube and throat 18 mm. long or less; leaves ovate to lanceolate; corolla lobes 8 mm. long or less:
  Follicles more or less constricted between seeds; corolla tube 5–12 mm. long:
    Follicles usually strongly torose, constrictions encircling follicles; corolla lobes spreading; herbage glabrous or tomentose with subappressed hairs:
      Herbage glabrous; seeds 8–10 mm. long ........................ 2. *A. brevifolia*
      Herbage tomentose; seeds 10–13 mm. long ...................... 3. *A. tomentosa*

Follicles slightly compressed between seeds in only one plane, not strongly torose; corolla lobes erect or ascending; pubescence of spreading hairs ........ 4. *A. kearneyana*

Follicles not constricted between seeds; corolla tube 10–18 mm. long:

Calyx lobes sparingly short-ciliate with stoutish hairs or glabrous; corolla lobes 3–4 mm. long; leaves linear-lanceolate, 1 cm. wide or less .................... 5. *A. palmeri*

Calyx lobes copiously long-ciliate with slender spreading hairs; corolla lobes 5–6 mm. long; leaves ovate-lanceolate, to 1.5 cm. wide ...... 6. *A. hirtella* var. *pogonosepala*

**1. Amsonia longiflora** Torr., Bot. Mex. Bound. 159, 1859

*Amsonia grandiflora* Alexander, Torreya **34:** 116, 1934.

Herbaceous perennial 3.5–6 dm. tall from a thickened woody root; stems ascending or erect, usually clustered at base; herbage glabrous; leaves linear-lanceolate to filiform or rarely narrowly lanceolate, 1–6 mm. broad, 2.5–7 cm. long, sessile, gradually attenuate toward each end; inflorescence rather loose, of 5–10 flowers each; pedicels slender, 2–5 mm. long; calyx 6–8 mm. long, the lobes subulate; corolla trumpet-shaped, tube 2–4 cm. long, constricted at mouth, lobes 10–15 mm. long, oblong-lanceolate, spreading; stigma bearing 2 obtuse terminal apiculations; follicles slender, 7–9 cm. long, not constricted between seeds, acuminate at apex, sessile; seeds 5–15, elliptic-oblong, truncate at each end, 5–10 mm. long, brownish, wrinkled or pitted.

Chiefly in the Upper Sonoran Zone, from southern Arizona to Texas and south into Durango. Barely entering the Sonoran Desert between Tucson and Nogales. April–May.

**2. Amsonia brevifolia** A. Gray, Proc. Amer. Acad. **12:** 64, 1876

Herbaceous glabrous perennial 1.5–4 dm. tall from a thickened fibrous root with sparingly branched, erect to ascending, clustered stems; leaves numerous, alternate, ovate, ovate-oblong, or oblong-lanceolate, 5–23 mm. wide, 2–6 cm. long, acute to acuminate at apex, acuminate to obtuse or rarely rounded at base, glossy-green, distinctly veined beneath, subsessile to short-petiolate; inflorescence rather crowded; pedicels 5–15 mm. long, or nearly lacking in some flowers; calyx 2–4 mm. long with bluish subulate lobes; corollas white, tinged with blue or lavender on tube, throat often yellowish; corolla tube 7–12 mm. long, constricted at mouth; corolla lobes ovate to ovate-oblong, 4–6 mm. long, spreading; stigma obtusely apiculate with 2 blunt lobes; follicles 5–8 cm. long, torose, glabrous, 3–10-seeded, terminal segment attenuate; seeds elliptic, truncate at one or both ends, tapering gradually toward one end, more abruptly so at other, or sometimes tapering very little, 8–10 mm. long, 3–4.5 mm. wide, variously wrinkled, pitted or ruminate, particularly near ends, yellowish brown, dull.

Along washes and open rocky ground, mainly Lower Sonoran Zone, northern edge of the Colorado Desert, Riverside County, California, to southern Nevada, southern Utah, and barely entering northern Arizona in Mohave County. April–June.

**3. Amsonia tomentosa** Torr. & Frem., in Frem., Second Rep. 316, 1845

*Amsonia brevifolia* var. *tomentosa* Jepson, Man. Fl. Pl. Calif. 768, 1925.

Densely tomentose, herbaceous perennial 3–4 dm. tall from a woody rootstock; branches usually crowded and clustered at base, erect or ascending; leaves alternate, numerous, ovate-oblong to oblong-lanceolate, 1–1.5 cm. wide, 2–4 cm. long, acute at both ends, uppermost conspicuously attenuate at apex; inflorescences small and dense; pedicels 5–20 mm. long, or sometimes nearly lacking, closely white-tomentose; calyx

2–3 mm. long, lobes linear to subulate; corolla salverform, tube slender below, flaring above and markedly constricted at mouth, 5–8 (or rarely –10) mm. long, lobes ovate to oblong, 4–7 mm. long, white or often suffused with sordid-purple; stigma apiculate with 2 obtuse lobes; follicles 4–8 cm. long, 3–5 mm. in diameter, articulate into constricted segments, finely tomentose, 3–7-seeded, tapering toward each end; seeds elliptical, 10–13 mm. long, tapering toward ends and sometimes somewhat arcuate, 3–4 mm. in diameter, reddish brown.

Gravelly washes and sandy slopes, mostly Lower Sonoran Zone, Chuckwalla Mountains, Colorado Desert, to southern Nevada, southwestern Utah, and Mohave County, Arizona. April–June; rarely into Oct.

**4. Amsonia kearneyana** Woodson, Ann. Mo. Bot. Gard. **15:** 415, 1928

An herbaceous perennial 4–8 dm. tall with erect or ascending, sparingly branched stems; stems and herbage pilose with more or less spreading hairs; leaves alternate to subverticillate, lanceolate to oblong-lanceolate, 7–15 mm. wide, 3.5–7 cm. long, subsessile or petiolate, crenate at base, attenuate at apex, sparsely pilose but green on both surfaces, more densely pilose on margins and along some veins beneath; inflorescence many-flowered, dense; pedicels 1 mm. long or less; calyx 3–5 mm. long, densely pilose-ciliate lobes subulate-aristate; corolla tube 10–12 mm. long, slightly constricted at mouth, tinged with purple below; corolla lobes ovate to oblong, 3–5 mm. long, erect or ascending, stigma with 2 apiculations; follicles to 7 cm. long, slightly compressed, shallowly impressed on faces between seeds, 4–6 mm. wide, glabrous, short-acuminate at apex; seeds cylindrical, truncate or obliquely flattened at ends, 6–10 mm. long, 3–4 mm. in diameter, reddish brown, ruminately wrinkled and reticulate, especially on ends.

Known only from the western slopes of the Baboquivari Mountains, western Pima County, Arizona, near the upper margin of the desert. March–April.

**5. Amsonia palmeri** A. Gray, Proc. Amer. Acad. **12:** 64, 1876

Herbaceous, essentially glabrous perennial 3–6 dm. tall, clustered stems sparingly branched above, branches erect or ascending; leaves numerous, linear-lanceolate to narrowly oblong-lanceolate, 2.5–10 mm. broad, 2.5–7.5 cm. long, gradually narrowing toward each end, glabrous; inflorescence loosely few-flowered, surpassing solitary leaves, pedicels 1–3 mm. long or some flowers sessile; calyx lobes subulate, 3–4 mm. long, sparsely ciliate with rather stoutish hairs 0.5–1 mm. long, or glabrate; corolla tube 10–18 mm. long, greenish or purplish toward base, flaring slightly about one-half its length, glabrous; corolla lobes ovate-oblong, 3–4 mm. long, spreading to erect, white; follicles straight, 4–10 cm. long, glabrous, sessile, unconstricted, 5–10-seeded; seeds oblong-cylindrical, 4–8 mm. long, truncate at one or both ends, often obliquely so, ruminately pitted and wrinkled, reddish chocolate-brown.

Rocky slopes and washes, Lower and low Upper Sonoran Zones, Mohave County, Arizona. March–June; occasionally to Sept.

**6. Amsonia hirtella** Standley var. **pogonosepala** (Woodson) Wiggins, Contr. Dudley Herb. **4:** 21, 1950

*Amsonia pogonosepala* Woodson, Ann. Mo. Bot. Gard. **15:** 412, 1928.

Herbaceous, nearly glabrous perennial from a heavy woody root, stems clustered, erect to ascending, 5–8 dm. tall; leaves numerous, alternate, glabrous, mostly ovate-lanceolate, 6–15 mm. broad, 3.5–7 cm. long, acute to acuminate at both ends, on petioles 1–3 mm. long, upper leaves often lanceolate, subsessile; inflorescence loosely many-flowered; pedicels 1–4 mm. long, glabrous; calyx lobes subulate, 3–5 mm. long, conspicuously ciliate with long, slender, white hairs; corolla tube 12–15 mm. long, constricted at mouth, purplish toward base, narrow throat about twice the diameter of slender tube but of equal length; corolla lobes white, ovate to oblong, 5–6 mm. long, spreading or at first ascending; follicles 1.2–8 cm. long, sessile, acuminate at apex, unconstricted, straight, glabrous, 4–12-seeded; seeds oblong-cylindrical, 5–10 mm. long, ends more or less truncate, ruminately pitted and wrinkled, dull reddish brown.

Rocky hillsides and canyons, Upper and Lower Sonoran Zones, central and northern Arizona. March–April.

### 4. **VALLESIA** Ruiz & Pavon, Fl. Peruv. Prod. 28, *pl.* 5, 1794

Shrubs or small trees with short-petiolate, alternate, entire, persistent leaves and smooth, rather thin bark. Flowers small, in cymes on stoutish peduncles opposite the leaves. Calyx deeply 5-parted into ovate lobes, eglandular. Corolla salverform; tube cylindrical but enlarged into a short throat just below the 5 lobes. Stamens 5, inserted near apex of corolla tube, included; anthers lanceolate to cordate, not appendaged. Ovary 2-carpellate, 2–4-ovuled; style filiform; stigma cylindrical, pubescent. Fruit a 1–2-seeded drupe. Seeds oblong, compressed; embryo curved.

Corolla, including lobes, 5–7 mm. long; corolla lobes 1.2–1.5 mm. long; calyx lobes 0.5 mm. long .......... 1. *V. glabra*
Corolla, including lobes, 10–16 mm. long; corolla lobes 3.5–6 mm. long; calyx lobes 0.8–1.8 mm. long:
  Leaves and twigs glabrous; corolla lobes 5–6 mm. long; calyx lobes 0.8–1 mm. long; leaves dark green on both surfaces .......... 2. *V. baileyana*
  Leaves and twigs densely puberulent; corolla lobes 3.5–4 mm. long; calyx lobes 1.5–1.8 mm. long; leaves paler beneath than above .......... 3. *V. laciniata*

**1. Vallesia glabra** (Cav.) Link, Enum. Hort. Berol. **1:** 207, 1821

*Rauwolfia glabra* Cav., Ic. **3:** 50, *pl.* 297, 1794.
*Vallesia cymbaefolia* C. G. Ortega, Hort. Mat. 58, 1798.
*Vallesia dichotoma* Ruiz & Pavon, Fl. Peruv. **2:** 26, *pl.* 151, 1799.
*Rauwolfia oppositiflora* Sessé & Moc., Pl. Nov. Hisp. 32, 1888.

Slenderly branched shrub or small tree to about 6 m. tall, with glabrous or sparsely puberulent twigs and foliage; petioles 3–5 mm. long; leaf blades narrowly lanceolate to oblong-lanceolate, 6–20 mm. wide, 2.5–7 cm. long, broadly cuneate to rounded at base, acute or slightly acuminate at apex, slightly fleshy or leathery, veins inconspicuous; cymes few-flowered, shorter than subtending leaves; pedicels 2–5 mm. long; calyx lobes ovate, about 0.5 mm. long; corolla 5–7 mm. long, tube 4–4.5 mm. long, slightly smaller and then constricted above, lobes lanceolate-ovate, 1.5 mm. long; fruits white, oblong, 8–10 mm. long, flesh thin, translucent; seeds 2.5–3 mm. in diameter.

Canyons, plains, and lower hillsides, Lower Sonoran and Tropical Zones, central Baja California and northwestern Sonora to southern Mexico, the West Indies, southern Florida, and South America. Oct.–May; flowering sparingly during other months also.

**2. Vallesia baileyana** Woodson, Ann. Mo. Bot. Gard. **24:** 14, 1937

A glabrous shrub; leaves petiolate, oblong-lanceolate, 12–20 mm. wide, 6–9 cm.

long, acute at apex, obtuse to rounded at base, blade thin but firm; petioles 7–9 mm. long; stipules persistent, minutely denticulate; inflorescences subumbellate and more or less dichotomously branching, shorter than subtending leaves and each bearing 8–20 whitish flowers; pedicels 3–4 mm. long, each subtended by a deciduous, minute, ovate-lanceolate bract; calyx teeth narrowly ovate, acute, 0.8–1 mm. long; corollas salverform, tube 9–10 mm. long, about 1 mm. in diameter at base, throat slightly inflated and bearing the stamens, lobes oblong-ovate, rounded, 5–6 mm. long, spreading; anthers ovoid, about 1 mm. long; stigma subcapitate and obscurely bilobed.

Known only from the type locality, near Guaymas, Sonora. March.

### 3. Vallesia laciniata Brandegee, Proc. Calif. Acad. II, 2: 182, 1889

Erect symmetrical shrub 1–2 m. tall with leafy, rather crowded branches; bark light gray, gray-green or gray-brown, twigs densely but finely puberulent with minute spreading hairs; stipules narrowly ovate, 1.5–2.8 mm. long, acute to short-acuminate; petioles stoutish, 5–9 mm. long, densely puberulent; leaf blades lance-elliptic to broadly lanceolate, 1.5–2.3 cm. broad, 3–9 cm. long, acute to obtuse at apex, rounded to broadly cuneate at base, finely and densely puberulent on both surfaces, eventually sparsely puberulent to subglabrate and dark green above, paler and veins prominent beneath; cymes rather dense, several- to many-flowered; peduncles and pedicels densely puberulent; calyx lobes broadly ovate, abruptly acuminate, 1.5–1.8 mm. long; corolla 10–12 mm. long, tube 7–8 mm. long, yellowish, glabrous, about 1 mm. in diameter below, abruptly swollen to 2.5–3 mm. in diameter, then constricted to about 1.2 mm. just below the lance-ovate lobes, these 3.5–4 mm. long; fruit a fleshy white to pink drupe, oblong, 10–12 mm. long; seed solitary owing to abortion of several ovules.

Rocky soil in bottoms of arroyos, Lower Sonoran Zone, central Baja California. Mainly at Comondú, where it occurs in close proximity to *V. glabra*. Dec.–April.

## 5. MACROSIPHONIA Muell. Arg., in Mart., Fl. Bras. 6[1]: 137, 1860

Low erect plants with fruticose or suffrutescent stems. Leaves opposite or verticillate, short-petiolate or sessile, margin entire or undulate-crispate. Flowers large and showy, short-pedicellate, terminal or lateral. Calyx deeply 5-lobed, glandular within at base, lobes linear or lance-linear. Corolla slenderly funnelform, tube elongate and gradually enlarged above. Corolla lobes 5, broad, contorted in bud. Stamens inserted near apex of tube, included within throat. Anthers oblong or sagittate, bearing obtuse appendages at base, connivent and lightly adherent to large stigma. Ovary 2-carpellate, these distinct. Fruit of 2 straight or curved, terete, erect follicles. Seeds oblong, bearing a tardily deciduous tuft of terminal silky hairs.

Leaves ovate or broadly elliptic, longer than broad; flowers puberulent without ........ 1. *M. brachysiphon*

Leaves orbicular to reniform-orbicular, as broad as long; flowers glabrous without ........ 2. *M. hesperia*

### 1. Macrosiphonia brachysiphon (Torr.) A. Gray, Syn. Fl. N. Amer. 2[1]: 83, 1878

*Echites brachysiphon* Torr., Bot. Mex. Bound. 158, 1859.
*Macrosiphonia dulcis* A. Nels., Amer. Jour Bot. 21: 577, 1934.

Erect suffrutescent plants branching from base, with slender stems 1.5–6 dm. tall, finely puberulent; bark reddish brown or eventually grayish; leaves broadly elliptic, ovate-elliptic or oblong, 5–13 mm. broad, 8–33 mm. long, rounded at base, acute to obtuse at apex, entire, finely puberulent to glabrate and dark green above, paler, densely puberulent and conspicuously reticulately veined beneath, margins plane or somewhat crispate, undulate; petioles 1–2 mm. long, puberulent; flowers 3–6.5 cm. long, white, fragant; calyx tube 6–10 mm. long, densely but finely puberulent, lobes 5–7 mm. long, linear-oblong, acute or obtuse at apex, often slightly tinged with red; corolla tube 1.5–2 mm. in diameter, gradually flaring to a throat 4–5 mm. in diameter, latter about equaling tube in length, both puberulent without; corolla lobes 1–2 cm. long, nearly as broad as long; follicles brownish, puberulent, 6–8 (or 10) cm. long, mostly about 2.5–3 mm. in diameter, straight or slightly curved; seeds 5–7 mm. long, linear-terete, blunt at base, truncate and bearing a tuft of yellowish silky hairs 10–14 mm. long at apex.

Arid slopes and mesas, Sonoran Zones, Cochise, Pima, and Santa Cruz Counties, Arizona, to New Mexico and northern and central Sonora. July–Sept.

**2. Macrosiphonia hesperia** I. M. Johnston, Proc. Calif. Acad. IV, **12:** 1125, 1924

A much-branched shrub 5–10 dm. tall with numerous loosely tufted stems; younger twigs and leaves densely pubescent with fine brownish hairs; petioles 3–4 mm. long; leaves opposite, broadly ovate to orbicular, 18–24 mm. broad, 2–3 cm. long, dark green and velvety-hirsute above, finely white-tomentose beneath, obtuse to rounded at base, obtuse and short-mucronate at apex; flower 6–7 cm. long, solitary or in pairs, terminal on pedicels 4–17 mm. long; calyx about 8 mm. long, lobes oblong to spatulate, brownish-hirsute outside, glabrous and each lobe with 8–10 pectinate glands at its base inside; corolla 6–7 cm. long, glabrous without, tube 1–1.5 mm. in diameter, 4–5 cm. long, throat cylindrical, 3–4 mm. wide, 6–8 mm. long; corolla lobes obovate, about 1 cm. wide, 1–1.3 cm. long; follicles 10–12 cm. long, terete but somewhat torose, finely canescent-puberulent; seeds linear to oblong, 5–8 mm. long, striate, each bearing a coma about as long as the seed.

Rocky cliffs and steep rocky slopes, Lower Sonoran Zone, along the east coast and on the adjacent islands of Baja California from the vicinity of Loreto south. Aug.–Nov.

## 6. MANDEVILLA Lindl., Bot. Reg. 24: *pl.* 7, 1840

Suffruticose or suffrutescent herbs, shrubs or lianas with latex-containing stems of various types. Leaves opposite or verticillate with several inconspicuous glands near base or along midrib or rarely glandless; petioles usually girdling stems in an appendiculate stipular ring. Inflorescence usually lateral but sometimes terminal or subterminal, bracteate. Calyx 5-parted, bearing 5 to many squamellae within base of equal or subequal lobes. Corolla salverform, infundibuliform or tubular-salverform; tube straight or slightly glabrous, 5-lobed. Stamens 5, inserted near apex of corolla tube; anthers connivent and agglutinated to stigma, each one bearing a sagittate peltate connective. Carpels 2, united at apex in flower but later distinct, many-ovuled; placenta axial. Nectaries 2–5 at base of follicles, or rarely absent. Follicles 2, terete, slender. Seeds many, apically comose.

**1. Mandevilla foliosa** (Muell. Arg.) Hemsl., Biol. Centr. Amer. Bot. **2:** 316, 1881

*Amblyanthera foliosa* Muell. Arg., Linnaea **30:** 427, 1860.
*Laseguea foliosa* Miers, Apoc. So. Amer. 253, 1878.
*Trachelospermum stans* A. Gray, Proc. Amer. Acad. **21:** 394, 1886.
*Secondatia stans* Standley, Contr. U.S. Nat. Herb. **23:** 1165, 1924.

Erect or ascending, suffrutescent herbs with slender terete stems 1–1.5 m. tall, densely but minutely puberulent when young, later glabrate, bearing irregularly scattered, rather conspicuous, pustule-like lenticels, each encircled with a narrow reddish band; leaves opposite, ovate-elliptic, 2–7 cm. wide, 6–15 cm. long, acute to acuminate at apex, rounded and obscurely cordate at base, thin, bright green, sparsely and minutely pilose and sparingly glandular near base of midrib on upper surface, slightly more densely puberulent beneath; petioles 5–10 mm. long; nodal appendages minute; inflorescence lateral, shorter than subtending leaves, 3–12-flowered; bracts lanceolate, 3–10 mm. long, scarious; calyx lobes deltoid-lanceolate, acuminate, 4–8 mm. long; corolla salverform, glabrous without, tube 10–15 mm. long, straight, lobes oblong-obovate, 7–10 mm. long, oblique, spreading; stamens inserted about middle of corolla tube; anthers broadly auriculate; nectaries 5, shorter than ovary; follicles falcate, somewhat moniliform, 8–12 cm. long, minutely puberulent or finally glabrate; seeds 9–10 mm. long, each bearing a pale yellowish coma 1.5–2 cm. long.

Canyons, rocky mesas, and lava fields, Sonoran and Tropical Zones, east central Sonora (Sierra Batuc) and Chihuahua to Mexico, Michoacán, and Morelos. June–Aug.

## 7. APOCYNUM L., Sp. Pl. 213, 1753

Herbaceous perennials with milky, latex-containing juice, from thickened rhizomes. Stems erect or ascending, or rarely prostrate, pubescent or glabrous; cortex containing long tough fibers. Leaves opposite, petiolate or sessile, entire, thin, glabrous or pubescent, bearing 2–6 small pectinate glands at base of petioles. Inflorescence cymose, glabrous or pubescent and bracteate or ebracteate. Calyx 5-lobed; tube adnate to receptacle, eglandular within. Corolla sympetalous, campanulate to cylindrical, 5-lobed or -parted, glabrous, bearing a sagittate appendage opposite base of each lobe. Stamens 5, alternate with corolla lobes; anthers with a sagittate connective. Carpels 2, distinct, but united temporarily at apex in flower, many-ovuled. Follicles 2, terete, divaricate. Seeds terete, apically comose.

**1. Apocynum sibiricum** Jacq. var. **salignum** (Greene) Fernald, Rhodora **37:** 328, 1935

*Apocynum salignum* Greene, Pittonia **5:** 64, 1902.
*Apocynum hypericifolium* var. *salignum* Beg. & Bel., Atti R. Acad. Lincei V, **9:** 115, 1913.
(For synonymy see Woodson, Ann. Mo. Bot. Gard. **17:** 141–142, 1930.)

Stems 2–5 dm. tall, glabrous throughout; leaves sessile or essentially so, ovate to lanceolate, cordate or obtuse at base, obtuse to rounded at apex, to 4 cm. wide, 10 cm. long or less, often mucronate, bright green above, paler or glaucous beneath; bracts semifoliaceous, conspicuous; calyx lobes 1.5–2 mm. long; corolla tubular-cylindrical, 3–4 mm. long, about half as broad as long; follicles 4–10 cm. long, straight; seeds 3.5–4 mm. long, each bearing a coma 15–20 mm. long.

Along streams and ditches, on meadows and hillsides, mainly Upper Sonoran and Transition Zones, British Columbia to Manitoba and Saskatchewan south to Texas, New Mexico, Arizona, and southern California; barely straggling into the northeastern extremity of the Sonoran Desert region near Whitewater and Twentynine Palms. June–July.

**Apocynum suksdorfii** Greene var. **angustifolium** (Wooton) Woodson, Ann. Mo. Bot. Gard. **17:** 119, 1930, has been reported from Pima County, Arizona, at elevations between

3,000 and 7,000 feet by Kearney, but I have seen no specimens from the Sonoran Desert that could be referred to this species.

**Nerium oleander** L., Sp. Pl. 209, 1753, a glabrous to sparsely puberulent shrub or small tree with short-petiolate, ternate, linear-oblong to linear-oblanceolate, entire, coriaceous leaves and white or pink, fragrant flowers in terminal showy cymes, is cultivated extensively in the warm-temperate and subtropical parts of the world and has become established as an escape in a few places within the limits of the Sonoran Desert. It is planted abundantly in Arizona, Sonora, southern California, and occasionally in Baja California.

## Family 117. ASCLEPIADACEAE. MILKWEED FAMILY*

Plants perennial, herbaceous, suffrutescent or rarely somewhat woody throughout; stems frequently twining or scandent. Leaves simple, mostly opposite or in whorls, usually petiolate, entire. Flowers perfect, regular, 5-merous (except for 2 carpels), mostly in umbelliform cymes, extremely specialized in structure. Stamens and style coherent into a gynostegium which is adnate to base of corolla. A corona (crown) of separate or united, fleshy, laminate or hoodlike appendages usually present between corolla and anthers, adnate to column, corolla, or both. Anthers winged and scariously tipped with a small hyaline scale; pollen in narrowly pyriform masses called pollinia, in ours 1 pollinium in each anther cell, the pollinia of adjacent anthers attached in pairs by transverse filaments or translator arms connected to a small clip or corpusculum, pendulous, horizontal, or erectly ascending from attachment of translators. Stigma conical or depressed, entire or shallowly lobed or angled. Ovary superior, of 2 pistils which are attached apically by a common stigma but otherwise distinct. Fruit a pair of follicles or one abortive. Seeds mostly obovoid, strongly flattened, concavo-convex, usually bearing a terminal coma of silky hairs or fine bristles.

The generic alignment used in this treatment of the *Asclepiadaceae* does not follow entirely that generally accepted by American botanists during the past half-century or more. Nor is it in complete accordance with Woodson's ultraconservative treatment published in the *Annals of the Missouri Botanical Garden,* Volume 28, 1941. I am retaining, in full generic status, *Astephanus* and *Metastelma,* both of which Woodson reduced to synonymy. I have accepted, with some reluctance, his reduction of *Rothrockia, Mellichampia,* and *Himantostemma* to synonymy under *Matelea,* since it seems to me that Woodson presents a better case for the merging of these into the larger complex than he does for the reduction of the genera retained here but reduced to synonymy by him.

Corona none; stamineal column and anthers unappendaged ................ 1. *Astephanus*
Corona present; stamineal column or anthers, or both, appendaged:
  Stems erect or ascending, not twining; corona consisting of 5 hoods, these appendaged within by a hornlike crest ........................................... 2. *Asclepias*
  Stems twining or scandent; corona various but lobes not hoodlike and without internal appendages:
    Pollinia strictly pendulous, uniformly fertile to attachment of translators:
      Corona double, consisting of an outer ring adnate to corolla, and 5 turgid vesicles attached to stamen column within; corolla broadly rotate ...... 3. *Sarcostemma*
      Corona single, adnate to corolla; corolla campanulate-rotate:
        Corolla lobes somewhat imbricate, glabrous or sparsely puberulent within ........ ..................................................... 4. *Cynanchum*

---

* Manuscript checked and partially revised by Richard W. Holm.

Corolla lobes valvate in bud, usually densely white-pilose within (except *M. californicum*) .......................................... 5. *Metastelma*

Pollinia strictly erect or horizontal, not pendulous, sterile near attachment of translators or, in *Marsdenia*, uniformly fertile:

Fruit fusiform, more or less attenuate toward one or both ends; pollinia horizontal or nearly so, a sterile hyaline portion at or near attachment to translators:

Anthers without dorsal appendages; fruits smooth or muricate but not angled ..................................................... 6. *Matelea*

Anthers with more or less laminate dorsal appendages; fruits strongly angled, somewhat ridged longitudinally .................................. 7. *Gonolobus*

Fruit ellipsoid in outline, not attenuate at either end; pollinia strictly erect from attachment of translators, uniformly fertile ..................... 8. *Marsdenia*

## 1. ASTEPHANUS R. Br., Mem. Wern. Soc. 1: 54, 1809

Suffruticose, shrubby or vinelike plants with opposite leaves or these greatly reduced; branches sometimes spinescent. Flowers small or minute, in axillary umbelliform cymes. Calyx 5-parted; lobes acute. Corolla urceolate or subcampanulate, shallowly 5-parted or 5-lobed; lobes narrow, glabrous or papillose-puberulent within. Corona none. Stamens inserted near base of corolla; filaments connate into a short tube; anthers tipped with a small inflexed membrane. Pollinia pendulous. Stigma obtuse, conical or rostrate. Follicles subfleshy, terete, acuminate. Seeds rough-granulate.

**1. Astephanus utahensis** Engelm., Amer. Nat. **9:** 349, 1875

*Cynanchum utahense* Woodson, Ann. Mo. Bot. Gard. **28:** 215, 1941.

A slender glabrous vine with numerous branches 1–10 dm. long from a woody perennial root; leaves narrowly linear, 0.5–1.5 mm. wide, 2–4 cm. long, subsessile, acute; cymes on peduncles 0.5–2.5 cm. long, mostly 3–10-flowered; pedicels slender, 2–5 mm. long at anthesis; calyx lobes lanceolate, 1.5–2 mm. long, minutely and sparsely strigose with fine white hairs; corolla yellow, about 2.5 mm. high, ovate lobes turned inward and slightly hooded, glabrous without, finely puberulent within; follicles 4–6 cm. long, 5–6 mm. in diameter, spindle-shaped, glabrous; seeds few, 7–8 mm. long, each bearing a coma of silky tawny hairs 12–15 mm. long.

Scrambling through shrubs or semiprostrate on arid slopes and plains, mainly Lower Sonoran Zone, southern Nevada and southwestern Utah to the Mojave Desert and along the western margin of the Colorado Desert in southern California. Occurring in northwestern Arizona and possibly in northern Baja California. April–June.

## 2. ASCLEPIAS L., Sp. Pl. 214, 1753

Perennial or annual caulescent herbs with alternate, opposite, scattered or whorled leaves, these entire or slightly crispate-erose, lanate or glabrate. Flowers greenish, cream-colored or purple-tinged, pedicellate, in pedunculate umbelliform inflorescences. Calyx small, deeply 5-lobed, usually bearing 2 glands at each sinus. Corolla rotate, reflexed, or 5 lobes ascending in anthesis. Column often short and scarcely evident. Hoods mostly erect or ascending, involute-concave, shorter than, equaling or surpassing anthers, usually somewhat pitcher-shaped, pendulous and saccate at base, entire to 2–3-toothed at apex, with or without crests or horns, these included or exserted. Anther wings usually angulate at or below middle, corneous. Pollinia pendulous. Stigma 5-angled, flat-topped. Follicles 2, one often abortive, naked or echinate with soft spreading processes. Seeds flattened, concavo-convex, narrowly winged; coma silky, present in all of ours.

Stems leafless except in early youth, reedlike; leaves very early deciduous:
  Hoods barely equaling or shorter than anther head, not dilated at apex:
    Corolla lobes greenish brown, 5–6 mm. long; hoods about one-half to two-thirds as long as slender column .......... 1. *A. albicans*
    Corolla lobes honey-yellow, 7–8 mm. long; hoods equaling stoutish column .......... 2. *A. masonii*
  Hoods distinctly longer than anther head, dilated at apex:
    Follicles 5–12 cm. long; seeds 5–6 mm. long; coma 2.5–3 cm. long ....... 3. *A. subulata*
    Follicles 3.5–4 cm. long; seeds about 2 mm. long; coma 10–12 mm. long .......... 4. *A. leptopus*
Stems leafy, not reed-like; leaves persistent:
  Hoods distinctly stipitate:
    Leaves opposite, not in threes, fours, or whorls; hoods 3–4 mm. high; microphyllous axillary branches none:
      Corolla lobes red-orange, 7–8 mm. long; plants perennial; leaves 1–3 cm. wide .......... 5. *A. curassavica*
      Corolla lobes white to pink, 4–5 mm. long; plants annual; leaves 2–15 mm. wide .......... 6. *A. angustifolia*
    Leaves mostly in threes or fours; hoods 1–1.5 mm. high; microphyllous branches often present in upper leaf axils:
      Leaves linear to lanceolate-oblong, to 2.5 cm. wide; inflorescences usually in pairs or clusters .......... 7. *A. fascicularis*
      Leaves linear, 1–4 mm. wide; inflorescences usually solitary at each flowering node .......... 8. *A. subverticillata*
  Hoods sessile or essentially so, not noticeably stipitate:
    Corolla lobes reflexed-rotate at anthesis; hoods not laterally coalesced at base; horns not wholly included within hoods:
      Hoods two to three times as long as columns:
        Corolla lobes 7–15 mm. long:
          Plants densely white-tomentose; leaves to 14 cm. broad; corollas rose-purple; horns much shorter than hoods .......... 9. *A. speciosa*
          Plants puberulent, pilose, or glabrate but scarcely tomentose; leaves rarely to 6 cm. wide; corollas greenish, yellowish, orange, or reddish, not purple; horns equaling or exceeding hoods:
            Hoods 7–10 mm. long, their apices truncate and slightly curved outward; anther head 3–5 mm. broad; stems often decumbent:
              Horns almost wholly adnate to hoods; follicles 3–5 cm. long; hoods not attenuate toward base .......... 10. *A. nyctaginifolia*
              Horns free and strongly reflexed at tips; follicles 7–9 cm. long; hoods obovate-flabellate, strongly attenuate toward base .......... 11. *A. oenotheroides*
            Hoods 4–5 mm. long, erect, their apices not curved outward; anther head 1.5–2 mm. wide; stems erect or ascending, not decumbent .......... 12. *A. tuberosa* subsp. *terminalis*
        Corolla lobes 4–5 mm. long:
          Leaves broadly ovate to suborbicular, 1–4 cm. wide; herbage arachnoid-lanate; hoods truncate, 2.5–3 mm. long .......... 13. *A. nummularia*
          Leaves filiform to acicular, 1–4 mm. wide; herbage inconspicuously puberulent to glabrate; hoods attenuate-acuminate, 4–6 mm. long ...... 14. *A. macrotis*
      Hoods shorter than or about equaling column:
        Leaves broadly ovate to ovate-lanceolate, to 10 cm. wide and 20 cm. long, margins erose; stems arachnoid-lanate; follicles 6–12 cm. long .......... 15. *A. erosa*

Leaves acicular, linear, or lanceolate, 1–9 mm. wide, their margins not erose; stems puberulent but not lanate; follicles 3.5–7 cm. long:

Stems erect or ascending, to 1.5 m. tall; corollas greenish or white; leaves narrowly linear, about 1 mm. in diameter, subterete ............. 16. *A. linaria*

Stems decumbent to ascending, 0.5–4 dm. long; corollas pinkish or purplish; leaves linear-lanceolate to lanceolate, 2–9 mm. wide, flat:

Hoods 3–3.5 mm. long; leaves 1.5–3.5 cm. long, eventually glabrate except on margins ......................................... 17. *A. involucrata*

Hoods 1–1.5 mm. long; leaves 5–11 cm. long, arachnoid-puberulent ......... ............................................... 18. *A. brachystephana*

Corolla lobes curving upward at anthesis; hoods laterally coalesced at base; horns wholly included within hoods ......................... 19. *A. asperula* subsp. *asperula*

**1. Asclepias albicans** S. Wats., Proc. Amer. Acad. **24:** 59, 1889

A perennial shrubby plant with 1 to several stems 1–3.5 m. tall, these waxy-white, 4–8 mm. in diameter; leaves in threes, narrowly linear, 1–3 cm. long, glabrous or glabrate and glaucous, early deciduous; peduncles 1–3 cm. long; umbels terminal or in a terminal panicle, few- to several-flowered; pedicels pubescent, about 5–12 mm. long; corolla white, tinged with purple to greenish brown, 5–6 mm. long, hoods yellowish, saccate, subtruncate and cleft along top, about half as long as stamineal column, bearing a slender curved horn barely exserted at tips; follicles spindle-shaped, attenuate at each end, erect, smooth, waxy-white, 8–10 mm. in diameter, 8–12 cm. long.

Rocky hillsides, washes, and mesas, Lower Sonoran Zone, southeastern Mojave Desert, throughout the Colorado Desert in favorable habitats, east into southwestern Arizona (Maricopa and Yuma Counties), and into adjacent northwestern Sonora, western Sinaloa, and the desert areas of Baja California to the Cape Region. March–June.

**2. Asclepias masonii** Woodson, Ann. Mo. Bot. Gard. **41:** 145, *fig. 72,* 1954

Erect perennial with several slender glaucescent stems bearing a few sessile linear leaves about 1 mm. in diameter and 2–5 cm. long near tips of young growth, but these slender leaves early caducous, leaving stems naked; inflorescences lateral from uppermost nodes on short peduncles 1–3 cm. long, many-flowered; peduncles and pedicels minutely pilose, latter 1–1.5 cm. long; calyx lobes ovate-lanceolate, 2.5–3 mm. long, puberulent; corolla pale yellowish, lobes 7–8 mm. long, deflexed-rotate; gynostegium short-stipitate and stoutish, cream; column obconic, about 1.5 mm. tall and 2.5 mm. wide; hoods erect, ovate, 7–8 mm. long; horns strongly adnate, broadly falciform and well exserted from hoods, introrsely arching over anther head, this truncate, 4–4.5 mm. in diameter, about 2.5 mm. high; follicles narrowly fusiform and apiculate, about 1.5 cm. in diameter, 10–13 cm. long, pilosulous in youth, glabrate at maturity, pendent from peduncles; seeds 4–5 mm. long, coma yellowish or greenish, about 2.5 cm. long.

Arid hills and sandy plains, Lower Sonoran Zone, vicinity of Magdalena Bay, Baja California. March.

**3. Asclepias subulata** Decne., in A. DC., Prod. **8:** 571, 1844

Stems in erect or ascending clusters, straight and rushlike, 1–1.5 (or 2) m. tall from a woody root and base; herbage glabrous, usually greenish; leaves linear, 1.5–4 cm. long, deciduous very early; inflorescences in terminal panicles or sometimes a single terminal one, 5–15-flowered; peduncles 1–2 cm. long; pedicels slender, 10–15 mm. long; corollas dull yellowish or whitish yellow, lobes reflexed, 6–8 mm. long, acute or obtuse; hoods ascending and elongated, about twice as long as stamineal column, apex nearly scimitar-shaped in outline and with edges folded toward each other, lower part narrower; horn a thin, winglike structure attached near middle of hood, its sharp point scarcely exserted between folds and pointing horizontally toward center of ring of hoods; follicles slenderly spindle-shaped, long-acuminate at each end, 5–12 cm. long, about 5–8 mm. in diameter; seeds 5–6 mm. long; coma 2.5–3 cm. long.

Arid washes and less frequently on rocky hillsides or silty plains, chiefly Lower Sonoran Zone, western edge of the Colorado Desert and eastern Mojave Desert to southern Nevada, western Arizona, and south into Sinaloa and the southern part of Baja California. Flowering sporadically throughout the year.

**4. Asclepias leptopus** I. M. Johnston, Proc. Calif. Acad. IV, **12:** 1127, 1924

Much-branched, many-stemmed plant 4–6 dm. tall from a suffrutescent base; stems slender and soon nearly or quite leafless, glabrous, green or faintly glaucescent near nodes; leaves opposite, filiform, 1.5 mm. wide or less, 4–7 cm. long, margins revolute, deciduous; inflorescences in terminal panicles, 3–8-flowered; peduncles to 2 cm. long or some cymes sessile; pedicels slender, 9–15 mm. long, puberulent; calyx lobes linear-oblong, acute, about 2 mm. long; corolla lobes oblong, about 2 mm. wide, 5 mm. long, white faintly tinged with green, strongly reflexed; hoods oblong, about 3 mm. high, one and one-half to two times as long as anther column, entire, orange with a median maroon line on back; horns adnate to hood for about two-thirds the length of latter, falcate, incurved and arching over stamen column, exserted about 1.5–2 mm.; follicles 3.5–4 cm. long, about 3.5 mm. in diameter, attenuate at apex, smooth and glabrous, ascending; seeds about 2 mm. long, light brown; coma about 1 cm. long.

On cliffs and rocky slopes, Lower Sonoran Zone, southern Sonora (near Guaymas), southwestern Chihuahua, and northern Sinaloa. July–Dec.

**5. Asclepias curassavica** L., Sp. Pl. 215, 1753
*Asclepias bicolor* Moench, Meth. 717, 1794.

Herbaceous or slightly woody at base; stems erect, to 2 m. tall, at first sparsely strigose but soon glabrate or nearly so; leaves opposite, lance-linear to lance-elliptic, 1–3 cm. wide, to 10 or 15 cm. long, acute, acuminate or attenuate at each end, sparsely puberulent, soon glabrate, slightly paler beneath than above; peduncles terminal and in upper axils, to 6 cm. long; cymes few- to several-flowered; pedicels slender, 1–1.5 cm. long; calyx lobes narrowly lanceolate, 3–3.5 mm. long, puberulent, reddish; corolla lobes bright red-orange, lance-oblong, 7–8 mm. long, nearly glabrous, reflexed, obtuse to acute at apex; corona and anthers on a slender column 2–2.5 mm. high; hoods yellow, 3.5–4 mm. high, saccate at base, distinctly stipitate, obliquely acute and slightly spreading at apex, exceeding anthers 1–1.5 mm.; horns yellowish, slender, exserted 2–2.5 mm. and curving inward over anthers; corona scales broadly ovate, 1–1.5 mm. long, inwardly arched over column; follicles erect, fusiform, 6–9 mm. in diameter, 7–9 cm. long, purplish, often glaucescent, glabrous or nearly so; seeds about 5 mm. long, dark brown, with a lighter brown, narrow wing; coma about 1.5 cm. long.

Margins of fields and along ditches and streams, Lower Sonoran to Tropical Zones, vicinity of Comondú, Baja California, and Sinaloa south to tropical South America; widely distributed in the warmer regions of both hemispheres. Occurring also in Florida and Louisiana. Flowering sporadically throughout the year.

**6. Asclepias angustifolia** Schweig., Enum. Pl. Hort. Bot. Regiomont. 13, 1812
(For synonymy see Woodson, Ann. Mo. Bot. Gard. **41:** 62, 1954.)

Somewhat suffrutescent annuals 1.5–8 dm. tall with simple to several times branching stems with minutely puberulent lines below each node but eventually glabrate; leaves opposite, short-petiolate, linear to linear-lanceolate, 2–15 mm. wide, 3–15 cm. long, acuminate at apex, attenuately narrowed at base, moderately revolute to plane, weakly pilosulous beneath and glabrous above or wholly glabrous; inflorescences solitary or few at upper nodes, each one few- to several-flowered; peduncles 1–2 cm. long; pedicels erect, 10–15 mm. long; corollas white to pink, reflexed-rotate lobes 4–5 mm. long; gynostegium white, stipitate, column about 1 mm. high and broad; hoods cuculate, oblong, 3–4 mm. long, distinctly stipitate; horns basal, narrowly acicular, exceeding hoods and arching inward over anther head, this cylindrical, about 2 mm. broad and high; follicles narrowly fusiform, 5–8 mm. in diameter, 6–10 cm. long, smooth and glabrous or nearly so, erect on erect pedicels; seeds 5–6 mm. long, with shining white coma 1.5–3 cm. long.

Arroyos, canyons, and along stream beds, Lower and Upper Sonoran Zones, southern Arizona south to southern Mexico. Flowering more or less throughout the year.

**7. Asclepias fascicularis** Decne., in A. DC., Prod. **8:** 569, 1844
*Asclepias macrophylla* Nutt., Jour. Acad. Phila. II, **1:** 180, 1847.
*Asclepias macrophylla* var. *comosa* Dur. & Hilg., Pacif. R. Rep. **5**$^3$**:** 10, *pl. 8,* 1857.
*Asclepias fasciculata* Hemsl., Biol. Centr. Amer. Bot. **2:** 824, 1881 (error).

Erect herbaceous perennials 4–10 dm. tall with small axillary branches; stems more or less puberulent in lines below nodes or glabrate in age; leaves mostly in threes and fours on flowering stems, linear to oblong-lanceolate, 1–25 mm. wide, 3–12 cm. long, acuminate to obtuse at apex and base, thin, glabrous, to weakly pilosulous beneath; petioles 1–3 mm. long; leaves on dwarf branches opposite and much reduced; inflorescences several- to many-flowered, paired or clustered in uppermost axils; peduncles 2–4 cm. long; pedicels slender, 7–15 mm. long; calyx lobes narrowly deltoid, 1–1.5 mm. long, pilosulous to glabrate; corolla reflexed-rotate, grayish pink to nearly white, with lobes 3–4 mm. long; gynostegium gray-pink to white, with a cylindrical column about 1 mm. long; hoods 1–1.5 mm. long, ovate, cuculate, short-stipitate, narrowly acicular horns basal, inwardly arcuate, about 1.5 mm. long; follicles narrowly fusiform, 5–12 cm. long, 7–10 mm. in diameter, smooth and glabrous, erect on erect or ascending pedicels; seeds oval, 6–7 mm. long, the silky-white coma 1.5–2 cm. long.

In moist to very dry soil of plains, hillsides, and outwash slopes, upper margins of Lower Sonoran Zone to the drier ridges of the Transition Zone, eastern Washington and Idaho to Utah and south to northern Baja California. Chiefly above and north of the Sonoran Desert. May–Oct.

**8. Asclepias subverticillata** (A. Gray) Vail, Bull. Torrey Club **25:** 178, 1898
*Asclepias verticillata* var. *subverticillata* A. Gray, Proc. Amer. Acad. **12:** 71, 1876.

Perennial, erect or ascendingly branched herbs 1.5–12 dm. tall from woody rootstocks, with dwarf, microphyllous, sterile branches usually in many axils, more or less puberulent in lines below nodes; leaves usually 3 or 5 at each node but sometimes paired, on petioles 1–2 mm. long, blades linear, 1–4 mm. wide, 2–13 cm. long, glabrous to scantily pilosulous; inflorescences several- to many-flowered, usually solitary at upper nodes, on slender peduncles 1.5–3 cm. long; pedicels 5–8 mm. long, erect in fruit; calyx lobes narrowly deltoid, 1.5–2 mm. long, weakly pilosulous to glabrate; corolla white or sometimes suffused with greenish purple, reflexed-rotate lobes 3–5 mm. long; gynostegium white, narrowly stipitate, column about 1 mm. high, slightly narrower; hoods cucullate, oval, 1.5 mm. long, short-stipitate, with a narrowly acicular, basal horn moderately overarching anther head, this cylindrical, about 1.5 mm. broad and long; follicles narrowly fusiform, 6–8 mm. in diameter, 5–9 cm. long, smooth and glabrous or scantily pilosulous; seeds 7–8 mm. long, with a white coma about 2 cm. long.

Sandy and rocky outwash slopes, flood plains, and flats, Lower Sonoran to Transition Zones, Utah and Colorado to Arizona, Texas, and central Mexico; disjunctly present in Idaho. June–Aug.

**9. Asclepias speciosa** Torr., Ann. Lyc. N.Y. **2:** 218, 1828

Herbaceous perennials with simple stout stems 6–10 dm. tall; herbage densely white-tomentose; leaves opposite, short-petiolate, broadly ovate, oval, narrowly oblong or ovate-lanceolate, 3–14 cm. broad, 6–20 cm. long, broadly obtuse to rounded at apex, obtuse, rounded or occasionally shallowly cordate at base, firm, densely and persistently white-tomentose beneath, less densely tomentose and usually glabrate in age above; petioles stoutish, 5–15 mm. long; inflorescences few, solitary at some upper nodes, each several- to many-flowered, densely white-tomentose; peduncles 1–10 cm. long, stout; pedicels 2–3 cm. long, tomentose; flowers relatively large, showy; calyx lobes lanceolate, 5–6 mm. long, densely tomentose; corolla lobes 10–15 mm. long, rose-purple, tomentose without, glabrous or nearly so within; gynostegium rose to pinkish cream, subsessile, column broadly obconic, 1 mm. long, 3 mm. in diameter; hoods narrowly ovate-lanceolate, widely spreading-ascending, gradually attenuate upward, 10–14 mm. long; horns adnate toward base, acicular-falciform, sharply introrse, much shorter than hoods; anther head 3 mm. long, 4–4.5 mm. in diameter; follicles erect on deflexed pedicels, fusiform, 2–3 cm. in diameter, 9–12 cm. long, spiny to smooth, white-tomentose; seeds 6–9 mm. long; coma 3–4 cm. long.

Hillsides, plains, arroyos, roadsides, and margins of fields, Lower Sonoran to Canadian Zones, British Columbia to Minnesota and south to Arizona and Texas. May–Sept.

**10. Asclepias nyctaginifolia** A. Gray, Proc. Amer. Acad. **12:** 69, 1876

*Podostemma nyctaginifolium* Greene, Pittonia **3:** 237, 1897.

Stems several from base, 1–3 dm. long, decumbent to ascending, sparsely puberulent to glabrate; petioles 1–2 cm. long; leaf blades ovate to broadly oblong, 5–11 cm. long, 2–3.5 cm. wide, rounded, obtuse, or subtruncate at base, acute or obtuse at apex, green, glabrous or sparsely puberulent along veins; peduncles very short, cymes often subsessile; pedicels slender, 1–1.5 cm. long, spreadingly puberulent; calyx lobes lance-ovate, 3–4 mm. long, puberulent; corolla lobes oblong-lanceolate, 10–11 mm. long, thin, greenish or greenish yellow, reflexed; hoods narrowly oblong, 9–10 mm. long, erect, sides closely appressed, crispate near apex, narrower above than below, twice to two and one-half times as long as anther column; horn yellow, winglike, attached near apex of hood and drawn to a slender point, exserted 1–1.5 mm., erect; hyaline corona scales ovate, longer than broad, inflexed; follicles narrowly ovoid, acute to rounded at base, abruptly attenuate toward apex, 3–5 cm. long.

Arid slopes and gravelly plains, mostly Upper Sonoran Zone, eastern Mojave Desert to Santa Cruz and Pima Counties, Arizona; southwestern New Mexico; and northern Sonora (near Tajitos, northwest of Caborca). May–Aug.

**11. Asclepias oenotheroides** Cham. & Sch., Linnaea **5:** 123, 1830
(For synonymy see Woodson, Ann. Mo. Bot. Gard. **41:** 156, 1954.

Herbaceous perennial with clustered, simple to branching stems 0.5–4.5 dm. long, these decumbent to ascending, minutely puberulent; leaves opposite, rhombic-ovate to oblong-lanceolate, 1–6 cm. wide, 4–12 cm. long, acute to obtuse at apex, acute to broadly obtuse and abruptly cuneate at base, pale green, minutely puberulent beneath; petioles 1–2 cm. long; inflorescences solitary at a few upper nodes, subsessile to short-pedunculate, each several-flowered; peduncles to 2 cm. long, puberulent; petioles slender, 1.5–2.5 cm. long, minutely puberulent; calyx lobes ovate-lanceolate, 3–4 mm. long, finely puberulent; corolla strongly reflexed, greenish white to yellow, lobes 8–14 mm. long; gynostegium short-stipitate, pale greenish cream, with a broadly obconic column 1.5 mm. high and 2.5 mm. broad; hoods narrowly obovate-flabellate, 7–10 mm. long, narrowed from above middle downward, tip broadly flared and minutely erose to repand, fully twice as high as anther head; horns adnate almost to tip of hoods, tip free, falciform, strongly incurved, a smaller posterior appendage often present; anther head truncate-conic, 3 mm. high, 4–5 mm. in diameter; follicles erect on deflexed pedicels, fusiform to ovoid, short-apiculate, 1.5–2 cm. in diameter, 7–9 cm. long, smooth, pilosulous to glabrate; seeds 6–8 mm. long, coma tawny, 2–2.5 cm. long.

Heavy rocky soil on mesas, hills, and plains, Upper Sonoran to Tropical Zones, New Mexico to Texas and south through Mexico to Nicaragua. Feb.–Oct.

**12. Asclepias tuberosa** L., subsp. **terminalis** Woodson, Proc. Nat. Acad. Sci. **39:** 79, 1953

Conspicuously hirsutulous to hispid, perennial herbs from deep woody rootstocks, with stoutish stems in clusters from the crown, these 2–9 dm. tall, mostly unbranched below inflorescences; leaves usually crowded, opposite, short-petioled, narrowly lanceolate to broadly oblanceolate, acuminate to rounded at apex, mostly obtuse to truncate at base, 3–30 mm. wide, 3–11 cm. long, thin but firm, sometimes irregularly crisped-revolute, hirsutulous on both surfaces but more densely so beneath than above; inflorescences of 1 to several subterminal and terminal branches bearing few to many umbelliform cymes in axils of somewhat reduced leaves; calyx lobes lance-deltoid, 2–3 mm. long, minutely hirsutulous; corolla lobes 7–8 mm. long, orange, reddish or yellow, reflexed-rotate; gynostegium usually orange, less commonly yellow, with a narrowly obconic column about 2 mm. high and 1.5 mm. in diameter, hoods lanceolate, cuculate, 4–5 mm. long; horns basal, narrowly acicular, slightly exceeding hoods, inwardly arcuate over cylindrical anther head, this about 2 mm. broad and high; follicles narrowly fusiform, 1–1.5 cm. in diameter, 8–15 cm. long, smooth, pilosulous, erect on deflexed pedicels; seeds 5–7 mm. long, broadly oval; coma silvery, 3–4 cm. long.

Arroyos, prairies, hillsides, and oak and pine forests, upper margins of Lower Sonoran Zone into Transition and possibly Canadian Zones, Minnesota to Michigan, south to Utah, Arizona, Texas, and northern Mexico. June–Sept.

**13. Asclepias nummularia** Torr., Bot. Mex. Bound. 163, *pl. 45, fig. A,* 1859

A small plant with 1 to several decumbent stems 2–6 cm. long from a woody rootstock; herbage arachnoid-lanate, cinereous; leaves 2–4 pairs closely crowded near apex of short stem, broadly ovate to suborbicular, 1–4 cm. wide, 1.5–4 cm. long, apiculate or short-acuminate at apex, on a short broad petiole 1–4 mm. long; peduncles erect, 1–7 cm. long; cymes 15–20-flowered; pedicels slender, 1–1.5 cm. long, sparsely arachnoid-puberulent; calyx lobes ovate, 1–1.5 mm. long; corolla lobes greenish white, oblong-ovate, 4–4.5 mm. long, strongly reflexed; hoods obliquely truncate on inner side, spreading-ascending, entire, about 3 mm. long and about twice as long as anther column; horns narrowly falcate, exserted from about middle of each hood and overarching anther column; corona scales strongly inflexed, broadly ovate to subreniform; follicles slenderly ovate-attenuate, 5–6 cm. long, erect or ascending, minutely lanate.

Arid slopes and mesas, Lower Sonoran Zone to margins of Arid Transition Zone, Pima, Santa Cruz, and Cochise Counties, Arizona, to western Texas and south to San Luis Potosí. March–Aug.

**14. Asclepias macrotis** Torr., Bot. Mex. Bound. 164, *pl. 45, fig. B,* 1859

Much branched, suffrutescent perennial with slender stems 1–3 dm. tall, inconspicuously puberulent in lines decurrent from nodes in youth but later glabrate; leaves opposite, sessile, filiform to acicular, 1–4 mm. wide, 3.5–8 cm. long, revolute and glabrous; inflorescences solitary at upper nodes, each few- to several-flowered; peduncles 2–10 mm. long, slender; pedicels 8–10 mm. long, very slender, deflexed in fruit; calyx lobes ovate, 2–3 mm. long; corolla pale greenish yellow often slightly tinged with purple on outer surface, lobes 4–5 mm. long, rotate-reflexed; gynostegium subsessile, cream to yellowish, column about 2 mm. wide, 0.5 mm. long; hoods ovate-lanceolate, narrowly attenuate-acuminate, 4–6 mm. long, ascending-spreading; horns strongly adnate to and much shorter than hoods, ascending and only slightly introrse, tips usually pilosulous; anther head conic, about 1 mm. high, 2 mm. broad; follicles erect from deflexed pedicels, narrowly fusiform, 5–7 mm. in diameter, 4–7 cm. long, smooth, pilosulous to glabrate; seeds 5–6 mm. long, with white to tawny coma 2.5–3.5 cm. long.

Arid hills, mesas, and bajadas, Lower and Upper Sonoran Zones, southern Colorado and western Oklahoma to Arizona and Coahuila. May–Oct.

**Asclepias lemmonii** A. Gray, Proc. Amer. Acad. **19:** 85, 1883, is related to *A. macrotis* but has broadly oval, broadly cordate, amplexicaul leaves 9–11 cm. wide; flowers with corolla lobes 10–13 mm. long; and densely setose-pilose follicles.

Growing on limestone slopes and in canyons in oak and pine woods from Arizona to Jalisco, may occur occasionally along the upper limits of the Sonoran Desert in Pima County, Arizona, and in adjacent Sonora. June–Sept.

**15. Asclepias erosa** Torr., Bot. Mex. Bound. 162, 1859

*Asclepias leucophylla* var. *obtusa* A. Gray, Bot. Calif. **1:** 620, 1876.
*Asclepias erosa* var. *obtusa* A. Gray, Syn. Fl. N. Amer. **2:** 94, 1878.
*Asclepias demissa* Greene, Leaflets Bot. Obs. **2:** 231, 1912.

Stems erect, stoutish, 6–15 dm. tall; herbage finely arachnoid-lanate with white slender hairs when young, usually glabrate in age; leaves opposite, ample, ovate-lanceolate to broadly ovate, acute to acuminate at apex, sessile, rounded to subcordate at base, to 10 cm. broad and 20 cm. long, upper ones much reduced, margins somewhat cartilaginous, finely and irregularly erose, midrib heavy and conspicuous; peduncles stoutish or slender, 2–12 cm. long, many-flowered; pedicels slender, about 2–3 cm. long; flowers pale cream or tinged with green; calyx lobes lanceolate, 3–4 mm. long, finely lanate at base, sparsely so toward apex; corolla lobes oblong, 6–7 mm. long, acute or abruptly short-acuminate; hoods orbicular-obovate, 3.5–4 mm. high, truncate or rounded at apex, erect to slightly spreading and about equaling anther column; horns broadly falcate, strongly exserted from hood and overarching stamen column, conspicuous; corona scales ovate, incurved; follicles oblong-ovate, 2–2.5 cm. in diameter, 6–12 cm. long, attenuate at apex, erect from tips of deflexed pedicels, peduncles often erect or ascending; seeds 10–12 mm. long, granular and lanate-puberulent.

Arid hills and desert plains, Lower and Upper Sonoran Zones, Inyo and Kern Counties, California, to southern Nevada, southern Utah, western Arizona, south into northern Sonora and Baja California. April–Oct.

**Asclepias eriocarpa** Benth., Pl. Hartw. 323, 1849, differs from *A. erosa* in having leaves with entire margins; hoods nearly as broad as long and shorter than anther head; and horns wholly adnate and longer than hoods.

Flats, rocky hillsides, woods and streambanks of the Upper Sonoran and Transition Zones, in California and northern Baja California but not as yet collected within the limits of the Sonoran Desert.

**Asclepias latifolia** Raf., Atl. Jour. 146, 1832, has leaves comparatively broader in relation to their length than those of *A. erosa,* peduncles obsolete to barely 1.5 cm. long; outer margins of hoods lower than inner and often retuse; and horns somewhat exceeding instead of barely equaling hoods.

Prairies, plains, and badlands from Nebraska to Inyo County, California, south into Arizona and Texas. This may reach the upper edges of the Sonoran Desert in Arizona but no specimens from within our area have been seen.

## 16. **Asclepias linaria** Cav., Ic. **1:** 42, *pl.* 57, 1791

Stems sometimes herbaceous but oftener woody, at least at base, finely whitish-pubescent, to 1.5 m. tall, lower parts conspicuously tuberculate where leaves have fallen leaving warty leaf scars; leaves spirally approximate, narrowly linear, 0.5–1 mm. broad, tightly revolute and subterete, 2–6 cm. long, scattered but moderately crowded, forming tufts near apices of branches, sparsely puberulent but eventually glabrate; cymes axillary, 10–30-flowered; peduncles 1–2.5 cm. long; bractlets linear-subulate, 5–6 mm. long; pedicels 10–12 mm. long, slender, finely cinereous-strigulose; calyx lobes subulate, 2.5–3 mm. long; corolla lobes lance-ovate, 3–4 mm. long, abruptly acute to acuminate, glabrous or essentially so, greenish or white; stamineal column elevated 0.5–0.75 mm. above corolla lobes; hoods spreading-ascending to nearly erect, 2.5–3 mm. long, ovate in outline, slightly saccate at base, margins strongly inflexed, greenish white; horns barely exserted about 0.5 mm., erect or nearly so; follicles ovoid-attenuate, 1–1.2 cm. broad, 3.5–6 cm. long, glabrous; seeds 4–5 mm. long, granular.

Rocky hillsides, canyons, and arroyos, Lower Sonoran to Tropical Zones, scattered in southern Arizona and south throughout most of Mexico. April–Nov.

## 17. **Asclepias involucrata** Engelm., in Torr., Bot. Mex. Bound. 163, 1859

A small plant with 1 to several decumbent to ascending stems 0.5–1.5 dm. long from a deep-seated root; herbage lightly cinereous-puberulent with short, spreading, crispate hairs, leaves eventually glabrate except on ciliate margins; leaves lanceolate, 2–6 mm. wide, 1.5–3.5 cm. long, acute to rounded at base, acute to obtuse at apex, on petioles 1–3 mm. long; cymes sessile, subtended by 2–6 leaves, 4–15-flowered; pedicels slender, 1–1.5 cm. long, sparsely puberulent; calyx lobes lance-ovate, about 1.5 mm. long, puberulent; corolla lobes purplish or pinkish, ovate, 4–4.5 mm. long, acute, glabrous or essentially so; hoods ovate, acute, ascending, 3–3.5 mm. long, longer than anther column but only slightly exceeding it because of their spreading tips, margins in nearly same horizontal plane at top of disk; horns slightly exserted and curved falcately inward from about middle of hoods; corona scales broadly ovate, inflexed; follicles ovate, acute but scarcely attenuate at apex, 5–7 cm. long.

Arid plains and hillsides, chiefly Upper Sonoran and Arid Transition Zones, Santa Cruz and Pima Counties, Arizona, to southern Utah and Colorado, western New Mexico, and south to Durango, Mexico. March–July.

**18. Asclepias brachystephana** Engelm., in Torr., Bot. Mex. Bound. 163, 1859

Stems usually several from a somewhat woody root, erect, ascending or procumbent, 1–4 dm. long, finely arachnoid-puberulent and in youth often slightly cinereous; leaves opposite, linear-lanceolate, 3–9 mm. wide at acute to obtuse base, 5–11 cm. long, acute and often apiculate at apex, finely arachnoid-puberulent, midvein conspicuous beneath; peduncles axillary, 5–20 mm. long, few-flowered; pedicels slender, 9–15 mm. long, minutely lanate-cinereous; bracts subulate, 2–3 mm. long, purplish; calyx lobes ovate, 1.5–2 mm. long, acuminate; corolla lobes purplish, ovate, 4–5 mm. long, sparsely puberulent without, glabrous and greenish-tinged within; hoods yellow, about 1.5 mm. high, erect, apex truncate but with an angulate erect tooth applied to stamineal column; horn whitish, erect, barely exserted 0.5–0.75 mm. above lip of hood; anther column about 3 mm. high, conspicuously surpassing hoods; corona scales broadly ovate, nearly 1 mm. long, slightly inflexed; follicles ovate-attenuate, 4–6 cm. long, rounded at base, arachnoid-lanate, ascending from tips of deflexed peduncles and pedicels.

Arid plains and hillsides, mainly Upper Sonoran Zone, Pima County, Arizona, to Texas, south through Sonora, Chihuahua, and Coahuila to Zacatecas. April–Sept.

**19. Asclepias asperula** (Decne.) Woodson, Ann. Mo. Bot. Gard. **41:** 193, 1954, subsp. **asperula**

*Asclepias capricornu* subsp. *occidentalis* Woodson, Ann. Mo. Bot. Gard. **32:** 371, 1945.
*Anantherix decumbens* Nutt., Trans. Amer. Phil. Soc. II, **5:** 203, 1837.
*Asclepiodora decumbens* A. Gray, Proc. Amer. Acad. **12:** 66, 1876, in part.

Perennial with scabrous-puberulent, simple, strongly angled stems 2–5 dm. long, solitary or several from a stout root, procumbent to erect; leaf blades linear to lanceolate, to 3 cm. broad, 4–15 cm. long, acute to broadly rounded at base, tapering to acute or attenuate apex, thick and coriaceous, scabrous-puberulent along veins and margins; peduncles rather stout, to 2 dm. long but usually shorter, many-flowered; pedicels 1.5–2 cm. long, sparsely and unevenly scaberulous-puberulent; flowers depressed-globose in bud; calyx lobes narrowly lance-triangular, 4–5 mm. long; corolla lobes curvingly ascending, greenish, 7–9 mm. long, abruptly acute at apex, finely and sparsely puberulent without; hoods forming a ring around column, united at base, scimitar-shaped and curving upward and inward, purplish, obtuse at apex, with a narrow, winglike crest attached near apex and included between infolded margins of nearly tubular hood; stamen column about equaling hoods, 3.5–4 mm. broad; hyaline corona scales triangular-ovate, inflexed; follicles erect (but pedicel deflexed), 1.5–1.8 cm. in diameter, 7–9.5 cm. long, rounded at base, tapering toward apex, glabrous and smooth; seeds obovate, 6–7 mm. long, light brown, granular, wing nearly 1 mm. broad.

Arid plains and along arroyos, mainly Upper Sonoran Zone, Colorado and Utah to the eastern Mojave Desert and south in Mexico to Tamaulipas and Nuevo León.

Barely crossing the upper margins of the Sonoran Desert near Ímuris, Sonora, and near Tucson, Arizona. March–Aug.

## 3. SARCOSTEMMA R. Br., Mem. Wern. Soc. 1: 50, 1809*

Herbaceous or fruticose, scandent perennials with petiolate opposite leaves. Flowers regular, usually fragrant, white to purplish, in pedunculate umbelliform cymes. Calyx 5-parted, small. Corolla rotate to campanulate, 5-lobed; lobes often somewhat contorted. Corona consisting of an outer ring which is short, adnate to corolla or gynostegium or both, and 5 turgid vesicles attached at base of stamen column. Pollinia pendulous. Follicles terete, smooth, slender or turgid, few- to many-seeded. Seeds flattened, comose at one end.

Inflorescence corymbiform, obviously compound, peduncle not equaling subjacent internode in thickness; leaves usually densely white-tomentose beneath .......... 1. *S. pannosum*
Inflorescence umbelliform, pedicels all inserted at nearly same point; leaves never densely white-tomentose beneath:
  Peduncle equaling or exceeding subjacent internode in thickness, sometimes bearing 2 cymes; leaves elliptic to lance-elliptic, not cordate or hastate at base .... 2. *S. clausum*
  Peduncle not equaling subjacent internode in thickness, usually bearing but 1 cyme; leaves usually hastate or cordate at base, very narrowly lanceolate or linear when not so:
    Leaves very long, narrowly lanceolate, margin usually undulate-crispate; column exceeding corona vesicles in length .............................. 3. *S. crispum*
    Leaves lanceolate to linear, margin not undulate-crispate; column much shorter than corona vesicles:
      Leaves broadly lanceolate to linear, base cordate or hastate (rarely obtuse); corona ring free from corona vesicles:
        Leaves broadly lanceolate, three to four times as long as broad, cordate at base; flowers white ..................... 4. *S. cynanchoides* subsp. *cynanchoides*
        Leaves narrowly lanceolate, more than three to four times as long as broad, hastate (rarely truncate or obtuse); flowers purple ................................ ................................. 4a. *S. cynanchoides* subsp. *hartwegii*
      Leaves narrowly lanceolate to linear or oblanceolate, base obtuse to cuneate; corona ring adnate to base of corona vesicles:
        Leaves narrowly lanceolate to linear; stems and leaves densely canescent; corolla lobes 4–5 mm. long ..................................... 5. *S. hirtellum*
        Leaves oblanceolate; stems and leaves puberulent; corolla lobes 4.5–6 mm. long ..................................................... 6. *S. arenarium*

### 1. Sarcostemma pannosum Decne., in A. DC., Prod. 8: 540, 1844

*Philibertia pavoni* Hemsl., Biol. Centr. Amer. Bot. **2:** 319, 1881.
*Philibertia pannosa* Hemsl., op. cit. 320.
*Philibertia tomentella* T. S. Brandegee, Univ. Calif. Pub. Bot. **4:** 90, 1910.
*Funastrum pannosum* Schlecht., Rep. Spec. Nov. **13:** 286, 1914.
*Funastrum pavonii* Schlecht., loc. cit.
*Funastrum tomentellum* Schlecht., op. cit. 287.

Slender vine with herbaceous or frutescent stems climbing to 5 m. or more; stems and herbage finely puberulent in youth, stems glabrate in age; petioles slender, cinereous, 0.5–2.5 cm. long; leaf blades broadly obovate-elliptic, 2–6 cm. wide, 3–10 cm.

* Text contributed by Richard W. Holm.

long, apex mostly obtuse or rounded but abruptly cuspidate or acuminate, base usually cordate with a narrow sinus, finely puberulent and bright green above, cinereous-tomentulose and with conspicuously white veins beneath; peduncles 4–10 cm. long, many-flowered; corolla white, 1.5–2.5 cm. in diameter; calyx lobes lanceolate-attenuate, 3–5 mm. long, densely puberulent; corolla lobes obtuse to subacute, puberulent without; corona ring 1–1.5 mm. high, corona vesicles obovoid, 2.5 mm. long; follicles 1.5–2.5 cm. in diameter, rounded at base, somewhat attenuate at apex, 4–7 cm. long or slightly more, glabrous or essentially so at maturity.

Hillsides, grassy slopes, and canyons, Upper Sonoran, Lower Sonoran, and Tropical Zones, southern Baja California, central Sonora, southwestern Chihuahua, and northern Sinaloa; Tamaulipas and Jalisco southeasterly to central Oaxaca. July–Nov.

**2. Sarcostemma clausum** (Jacq.) Sch., in Roem. & Sch., Syst. Veg. **6:** 114, 1820
(For nearly 60 synonyms see R. Holm, Ann. Mo. Bot. Gard. **37:** 511–512, 1950.)

A vine climbing vigorously to 8 m. over shrubs and trees; stems slender, glabrous or nearly so, sometimes minutely puberulent at nodes; leaves narrowly to broadly elliptic, 8–38 mm. broad, 3–8 cm. long, acute or obtuse and abruptly short-acuminate at apex, acute, obtuse, rounded or very rarely subcordate at base, leathery, finely puberulent with soft tangled hairs at first, eventually nearly or wholly glabrate, usually with a small cluster of minute, bristlelike glands at base of midrib on upper surface; peduncles slender, to 15 cm. long or more; flowers many; pedicels 1–2 cm. long, subglabrous; calyx lobes oblong-lanceolate, acute, 3–3.5 mm. long, densely and finely puberulent; corolla white, 16–18 mm. in diameter, lobes 7–8 mm. long, oblong, fimbriate-ciliate and densely puberulent on outer surfaces, sparsely so with shorter hairs within; corona ring about 1 mm. high, attached at upper margin to vesicles of stamineal column; corona vesicles about 2 mm. high; follicles 1.5–2 cm. in diameter at widest part, 4–8 cm. long, obtuse at base, attenuate at apex, puberulent but eventually glabrate, minutely striate; seeds reddish brown, 3–5 mm. long, granular on both surfaces, a slender vein down middle on inner surface, margin raggedly erose-incised around apical part; coma 2–2.5 cm. long, silvery white.

Along streams and in canyons, Upper and Lower Sonoran and Tropical Zones, Cape Region of southern Baja California, central Sonora, east to Florida, and south into South America. Flowering more or less throughout the year.

**3. Sarcostemma crispum** Benth., Pl. Hartw. 291, 1848
*Sarcostemma heterophyllum* Engelm., in Torr., Pacif. R. Rep. **5** (App.): 362, 1857.
*Sarcostemma undulatum* Torr., Bot. Mex. Bound. 161, 1859.
*Philibertia undulata* A. Gray, Proc. Amer. Acad. **12:** 65, 1876.
*Philibertia crispa* Hemsl., Biol. Centr. Amer. Bot. **2:** 318, 1881.
*Philibertella crispa* Vail, Bull. Torrey Club **24:** 306, 1897.
*Funastrum crispum* Schlechter, Rep. Spec. Nov. **13:** 284, 1914.

Slender vine climbing and scrambling to height of 2 m. or more; young stems minutely cinereous-puberulent but soon glabrate or nearly so; leaves linear, linear-lanceolate or hastate-lanceolate, 4–8 cm. long, 2–8 mm. broad just above hastate base, this often 10–15 mm. wide, margins undulate-crispate to entire, minutely puberulent, midvein often conspicuously white above; petioles 3–5 mm. long; peduncles 5–20 mm. long, 5–10-flowered; calyx lobes linear-lanceolate, about 3 mm. long, finely puberulent; corolla dull greenish purple, lobes 8–9 mm. long, oblong, obtuse, finely puberulent without, glabrous within, obscurely undulate; corona ring saucer-shaped, about 1 mm.

deep, free from stamineal column; corona vesicles oblong-ovoid, constricted at their insertion on column; follicles 10–16 cm. long, spindle-shaped, attenuate at each end, minutely and persistently puberulent; seeds about 4 mm. long, minutely granular-tuberculate, margins slightly revolute but entire.

Arid slopes and canyons, Upper and Lower Sonoran Zones to Tropical Zone, Colorado through Texas, and central Arizona to central Mexico and northeastern Baja California. May–Aug.

**4. Sarcostemma cynanchoides** Decne., in A. DC., Prod. **8:** 540, 1844, subsp. **cynanchoides**

*Philibertia cynanchoides* A. Gray, Proc. Amer. Acad. **12:** 64, 1876.
*Philibertella cynanchoides* Vail, Bull. Torrey Club **24:** 307, 1897.
*Funastrum cynanchoides* Schlechter, Rep. Spec. Nov. **13:** 284, 1914.

Slender herbaceous stems to 2 m. long from a woody root and short caudex; herbage glabrous or nearly so; petioles slender, 1–2 cm. long; leaf blades ovate or hastate-ovate, usually cordate at base, abruptly short-acuminate at apex, 1–3 cm. wide, 2–5.5 cm. long, thin, slightly paler beneath than above; umbels many-flowered on peduncles 2–5 cm. long; pedicels 1–2 cm. long, puberulent; flowers white, about 1 cm. broad; calyx lobes ovate, acute, puberulent without; corolla lobes acute, minutely puberulent beneath; corona ring very shallow; corona vesicles suboblongoid, widest above middle; follicles 3–7 cm. long, obtuse or rounded at base, attenuate at apex, about 1.5 cm. in diameter, finely striate; seeds obovate, about 6 mm. long, granulate or tuberculate, reddish brown; coma 3–3.5 cm. long.

Scrambling over bushes along streams and arroyos, Upper Sonoran to Tropical Zones, Maricopa, Cochise, and Pima Counties, Arizona, to southern Utah, western Texas, and south at least to Coahuila and Nuevo León, Mexico. May–Sept.

***4a.* Sarcostemma cynanchoides** subsp. **hartwegii** (Vail) R. Holm, Ann. Mo. Bot. Gard. **37:** 530, 1950

*Sarcostemma linearis* Decne., in Benth., Pl. Hartw. 25, 1840. Not *S. linearis* Spreng., 1820.
*Sarcostemma heterophyllum* Engelm., sensu Torr., Pacif. R. Rep. **5** (App.): 372, 1857, as to specimens cited, not as to type.
*Philibertia linearis* A. Gray, Proc. Amer. Acad. **12:** 64, 1876.
*Philibertella hartwegii* Vail, Bull. Torrey Club **24:** 308, 1897, based on *S. linearis* Decne.
*Funastrum hartwegii* Schlecht., Rep. Spec. Nov. **13:** 285, 1914.
*Funastrum heterophyllum* Standley, Contr. U.S. Nat. Herb. **23:** 1170, 1924, as to description, not as to type.

Stems very numerous, from a woody root, very slender, herbaceous above, 1–3 m. long or more, somewhat glaucescent; leaves linear to lanceolate, sometimes auriculate-lobed or even cordate-hastate at base, 1–5 (or sometimes –10) mm. wide, 2.5–6 cm. long, short-petiolate, glabrous to puberulent; peduncles slender, 1–5 cm. long, few- to many-flowered; pedicels 5–12 mm. long, sparsely puberulent with spreading hairs; calyx lobes ovate, 1–1.5 mm. long, puberulent; corollas purplish, 8–10 mm. broad, lobes acute to slightly acuminate, sparsely puberulent to subglabrous without, ciliate on margins, glabrous within; corona ring rectanguloid, widest below middle, 0.5–0.8 mm. high, free from vesicles; corona vesicles 1–1.5 mm. high; follicles slender, 6–9 mm. in diameter, attenuate at each end, 7–11 cm. long; seed 5–6 mm. long, minutely granulate on convex side, more coarsely so on concave face, wing narrow, forced down into concave face, entire or essentially so.

Along arroyos and in arid valleys, Sonoran Zones, Los Angeles County, California, to southern Utah, western Texas, and south to Coahuila, Querétaro, Jalisco, and Baja California. Feb.–Sept.

**5. Sarcostemma hirtellum** (A. Gray) R. Holm, Ann. Mo. Bot. Gard. **37:** 532, *fig. 14,* 1950

*Sarcostemma heterophyllum* var. *hirtellum* A. Gray, Bot. Calif. **1:** 478, 1876.
*Philibertia linearis* var. *hirtella* A. Gray, Proc. Amer. Acad. **12:** 64, 1876.
*Philibertella hirtella* Vail, Bull. Torrey Club **24:** 309, 1897.
*Philabertia hirtella* Parish, Muhlenbergia **3:** 126, 1907.
*Funastrum hirtellum* Schlechter, Rep. Spec. Nov. **13:** 286, 1914.

Canescent slender vine with numerous stems 2–20 dm. long from a woody base; stems and herbage closely puberulent with short, stiffish, spreading hairs with a sudoral odor when bruised; leaves linear or narrowly lance-linear, acute at both ends, 1–2.5 mm. wide, 1.5–2.5 cm. long; peduncles 0.5–5 cm. long, few- to several-flowered; pedicels slender, 5–9 mm. long; calyx lobes lance-ovate, about 2 mm. long; corollas greenish or greenish yellow, 8–9 mm. broad, lobes narrowly ovate, acute, sparsely puberulent without, glabrous within; corona ring thin, adnate to corona vesicles, barely 1.5 mm. high; follicles paired, lance-attenuate, acute or rounded at base, 4–5 mm. in diameter, 5–6 cm. long, attenuate at apex, densely puberulent, cinereous; seeds 5–6 mm. long, concavo-convex, granular-tuberculate, more coarsely so on concave face than on other, wing very narrow, entire; coma about 2 cm. long.

Along desert washes, Lower Sonoran Zone, Colorado and eastern Mojave Deserts to southern Nevada and western Arizona. March–May.

**6. Sarcostemma arenarium** Decne., in Benth., Bot. Sulph. 34, 1844

Stems twining or trailing, very thick at base, puberulent above; leaves more or less caducous, narrowly oblanceolate, obtuse, cuneate, 3–8 mm. wide, 3.5 cm. long, coriaceous, somewhat revolute when dried; peduncles 2.2–3.5 cm. long, few- to several-flowered; pedicels 1–1.3 cm. long, densely pilose; calyx lobes narrowly lanceolate, involute, pilose without; corolla lobes ovate, obtuse, 4.5–6 mm. long, pilose without; corona ring prominent, somewhat revolute, adnate to base of corona vesicles which are spherical to obovoid, 1–1.5 mm. long; follicles usually solitary, rounded at base, attenuate at apex, 9 mm. in diameter, 7 mm. long, puberulent; seeds 3 mm. wide, 7 mm. long; coma about 3 cm. long.

Sandy desert flats and arroyos of the Pacific coastal strip, Lower Sonoran Zone, central Baja California south to Magdalena Island. Flowering sporadically.

## 4. CYNANCHUM L., Sp. Pl. 212, 1753

Plants herbaceous or suffrutescent, more or less scandent, glabrous or pubescent. Leaves opposite, sessile or petiolate, usually ample. Flowers small, in umbelliform cymes, yellowish green. Calyx lobes 5, often glandular within, acute to obtuse at apex. Corolla campanulate-rotate; lobes somewhat contorted. Corona shallowly lobate, adnate to anther column, of 5 separate or united scales, occasionally with internal processes or rarely (not in ours) wholly lacking. Pollinia pendulous, their faces uniformly flattened or rounded, fertile to attachment of arms of translators. Stigma conical, entire to lightly but distinctly bilobed. Follicles slender, mostly smooth. Seeds flattened, slightly concavo-convex; coma silky, copious.

Leaves broadly ovate-cordate; central lobe of each corona segment long-acuminate, equaling corolla lobes, much longer than anther column; pedicels 5–15 mm. long ............ ............................................................ 1. *C. sinaloense*
Leaves linear-lanceolate, to filiform; corona segments merely acute, half as long as anther column; pedicels 1–4 mm. long:
Corolla 8 mm. long, outer surfaces of lobes brownish; plants to 2 m. tall ............... ........................................................ 2. *C. peninsulare*
Corolla 3–5 mm. long, lobes yellowish; plants 1 m. tall or less ............ 3. *C. palmeri*

**1. Cynanchum sinaloense** (Brandegee) Woodson, Ann. Mo. Bot. Gard. **28:** 210, 1941

*Roulinia sinaloensis* Brandegee, Zoe **5:** 243, 1908.
*Mellichampia sinaloensis* Kearney & Peebles, Jour. Wash. Acad. **29:** 488, 1939.

A slender scrambling vine to 5 m. long or more, with opposite ample leaves and glabrous herbage; petioles slender, 2–3.4 cm. long; leaf blades lance-ovate to broadly ovate-cordate, 1.5–5 cm. broad, 3–8.5 cm. long, bright green above, paler beneath, acute or more frequently acuminate at apex, sinus broad or basal lobes rounded-auriculate and overlapping; peduncles 1–3.5 cm. long; cymes 3–10-flowered; pedicels slender, 5–15 mm. long, sparsely puberulent; calyx lobes linear-subulate, 4–5 mm. long, sparsely puberulent near base; corolla cream-yellow, 7–8 mm. long, lobes oblong, rotate, margins somewhat involute, arachnoid-puberulent within, tips slightly recurved; corona scales 3-lobed, two lateral lobes appearing like short wings toward base of attenuate central one, this 5–6 mm. long, about three times as long as stamen column; stigma conical, slightly 5-rayed; follicles 9–10 cm. long, 1.5–2 cm. broad, broadly spindle-shaped, slightly narrowed toward base, attenuate at apex, glabrous, finely striate, brownish.

Along arroyos, rocky canyons, and outwash slopes, Lower Sonoran Zone, southern Arizona south to Sinaloa and probably to Michoacán. July–Sept.

**2. Cynanchum peninsulare** S. F. Blake, Contr. Gray Herb. **52:** 83, 1917

*Cynanchum palmeri* var. *peninsulare* I. M. Johnston, Proc. Calif. Acad. IV, **12:** 1129, 1924.

Scandent or shrubby plant to 2 m. tall, with numerous stems from a woody base, these glabrous except at sparsely pubescent nodes; leaves opposite, linear, 1–2 mm. wide, 3–8.5 cm. long, sessile and plane or nearly so, appressed-pubescent along margins, otherwise glabrous; cymes 2–7-flowered, subsessile, peduncles about 1 mm. long, pubescent; pedicels 1–2 mm. long; calyx lobes narrowly lanceolate, acuminate, about 1.5 mm. long, appressed-pubescent; corolla greenish yellow but erect, linear-lanceolate, obtuse lobes brownish on outside, 7–8 mm. long; corona segments oblong-elliptic, entire, broadly rounded, about 2 mm. long, 1 mm. wide, adnate to anther column at base; appendages of anthers appressed to stigmatic rostrum; stigma conical, acuminate, minutely bifid at apex, about equaling corona; follicles fusiform, slightly narrowed at base, attenuate toward apex, glabrous and somewhat glaucescent, 9–10 cm. long; seeds about 5–6 mm. long; coma shining, white, about 1.5 cm. long.

Sandy soil near the coast, Lower Sonoran and Subtropical Zones, western coast of southern Baja California. Nov.–Feb.

**3. Cynanchum palmeri** (S. Wats.) S. F. Blake, Contr. Gray Herb. **52:** 83, 1917

*Pattalias palmeri* S. Wats., Proc. Amer. Acad. **24:** 60, 1889.

Several slender scandent stems to 1 m. long from a thick, woody, branching rootstalk, lower parts of stems with pale, irregularly furrowed, corky bark; leaves linear,

2–5 cm. long, sessile, acute, glabrous or margins puberulent; flowers yellow, dull but not brownish, in 2–6-flowered, subsessile cymes; pedicels 1–4 mm. long, puberulent; calyx lobes 1.5–2 mm. long, about half as long as corolla; corolla lobes ovate or oblong, 3.5–4.5 mm. long, lobes of corona ovate-lanceolate, acute, about 0.8 mm. long or slightly less, attached to column at base of anthers; stigma column conical; follicles 6–8 mm. in diameter, 9–10 cm. long, narrowed toward base and long-attenuate toward apex, smooth and glabrous.

On talus slopes and sandy or gravelly soil near the Gulf Coast, Lower Sonoran Zone, central Baja California near Mulegé and on Isla San Marcos. Flowering following rains.

## 5. METASTELMA R. Br., Mem. Wern. Soc. 1: 52, 1809

Slender twining vines with small, narrow, opposite leaves. Flowers small, in umbelliform cymes. Calyx lobes equaling or exceeding corolla tube. Corolla small, rotate-campanulate, lobes sparsely to densely pubescent within except for naked median line. Corona adnate to base of corolla tube, its scales narrow, erect or curved inward over anthers; anthers borne on stout stalks shorter than or exceeding length of anther sacs, tipped with a membranous incurved appendage. Gynostegium equaling or slightly shorter than anthers; stigma flat to subconic. Follicles slender.

Corolla lobes densely villous within; leaves acute or obtuse but not cordate at base; stigma flat or slightly rounded:
- Flowers exclusive of pedicel 1.5–3 mm. long; appendages of anthers broader than long:
  - Calyces puberulent, lobes twice as long as tube; corona scale inflexed at apex; leaves linear, 3–6 mm. wide .................................... 1. *M. minutiflorum*
  - Calyces glabrous, lobes equaling or shorter than tube; corona scale erect or slightly spreading; leaves elliptic, 5–10 mm. wide ...................... 2. *M. latifolium*
- Flowers exclusive of pedicel 3–5 mm. long; appendages of anthers longer than broad:
  - Calyx lobes lance-attenuate, three-fourths as long as petals, much exceeding gynostegium; corolla lobes narrowly linear .............................. 3. *M. watsonianum*
  - Calyx lobes ovate or lanceolate, one-fourth to one-third as long as petals, shorter than gynostegium; corolla lobes oblong to lance-ovate:
    - Corona scale barely equaling anthers or slightly shorter; stems and calyces puberulent ...................................................... 4. *M. arizonicum*
    - Corona scale conspicuously longer than anthers, two-thirds as long as petals; stems and calyces glabrous ........................................ 5. *M. pringlei*

Corolla lobes minutely puberulent within; leaves truncate-rounded to subcordate at base; stigma subconic ............................................ 6. *M. californicum*

### 1. Metastelma minutiflorum Wiggins, Contr. Dudley Herb. 3: 71, *pl. 17, figs. 1–9,* 1940

Twining vine 1–2 m. high, suffrutescent at base; stems slender, finely puberulent along a single line from one node to next above; petioles puberulent along upper surface, slender, 3–6 mm. long; leaf blades linear to lance-linear, 3–6 mm. wide, 1.5–4 cm. long, dark green and sparsely puberulent along midvein above, lighter and glabrous beneath, margins slightly revolute; flowers 2–3 mm. long, in sessile clusters or on peduncles 1–1.5 mm. long; pedicels 1.5 mm. long or less, puberulent; calyx puberulent, lobes ovate, acute, less than half as long as corolla, fleshy, tips incurved; corolla tube broadly campanulate, lobes ovate, spreading, densely pubescent within; gynostegium sessile, about 1 mm. high or less; anther appendage broadly ovate, corona scales lanceolate, attached at base of anthers, slightly exceeding gynostegium, tips curving inward; retinaculum dark brown, linear; follicle slender, 4–5.5 cm. long, glabrous; seeds 5–6 mm. long, dark brown.

Rocky hillsides, Lower Sonoran Zone, central to southern Sonora, and northern Sinaloa. Aug.–Oct.

**2. Metastelma latifolium** Rose, Contr. U.S. Nat. Herb. **1:** 106, 1891

Climbing vine from a woody base with slender, glabrous, twining stems; leaves oblong to lance-elliptic, 3–10 mm. wide, 10–25 mm. long, rounded to acute at base, obtuse and abruptly apiculate at apex, bright green above, lighter beneath, glabrous or with a few scattered minute hairs along faintly revolute margins and on midrib, especially above; petiole slender, 1.5–3 mm. long, a sparse line of puberulence along upper side; flowers 1.5–2 mm. long in several-flowered umbelliform cymes on axillary peduncles 1–6 mm. long; pedicels slender, 3–4 mm. long, glabrous; calyx lobes ovate, acute, about equaling tube, sparsely puberulent; corolla white, lobes oblong, obtuse, densely villous on inner surface; column very short, almost lacking; anthers nearly as broad as high, apical appendage obtuse, nearly twice as broad as long; corona scales lanceolate, not stalked nor constricted at base, about equaling gynostegium; follicles unknown.

Collected at Alamos in 1890 (*Palmer 665*) and not known otherwise. This slender inconspicuous vine should be sought along the margins of the desert in southern Sonora. Probably flowering after summer rains.

**3. Metastelma watsonianum** Standley, Contr. U.S. Nat. Herb. **23:** 1175, 1924

*Metastelma albiflora* S. Wats., Proc. Amer. Acad. **24:** 60, 1889. Not *M. albiflorum* Griseb., 1861.

Woody vine with glabrous stems except for a puberulent line extending from above one or both of each pair of leaves to node above; leaves oblong-linear or oblong, 1.5–4 mm. wide, 1–2 cm. long, rounded to acute at base, acute to obtuse and mucronate at apex, dark green above, lighter beneath, glabrous; petioles slender, 1.5–3 mm. long; flowers about 4 mm. high, in few-flowered cymes on very short peduncles; pedicels stout, 1 mm. long or less; calyx lobes linear or lance-attenuate, three-fourths as long as petals, sparsely puberulent; corolla lobes narrowly linear, white, densely villous within; gynostegium short-stalked, stalk equaling or exceeding anthers; corona scales linear, equaling or slightly exceeding anthers; anther appendages broadly ovate, slightly longer than broad; follicles fusiform, glabrous, 3–4 cm. long.

Exposed cliffs and hillsides in the vicinity of Guaymas, Sonora, the type locality, and from near Divisadero, east central Sonora. Aug.–Feb. The long calyx lobes make this species easy to recognize among the Metastelmas of the Sonoran Desert.

A specimen from Apache Gap, Superstition Mountains, Arizona (*Abrams 13004*), seems to belong here. It has rather narrow leaves and the corona scales are slightly longer than those of the Guaymas plants, but these differences seem to be within the expected range of variation of the species.

**4. Metastelma arizonicum** A. Gray, Proc. Amer. Acad. **19:** 85, 1883

Slender-branched, woody vine 0.5–2.5 m. high with strongly revolute, linear leaves; stems glabrous except for a narrow line of puberulence running from immediately above one or both leaves of each pair to next node above; leaves 1–2 mm. wide, 5–18 mm. long, acute at each end, shining and dark green above, tightly revolute, glabrous or essentially so; petioles 1–2 mm. long; flowers 3–4 mm. long, on short axillary peduncles; calyx lobes lanceolate, 1–1.25 mm. long, sparsely puberulent, fleshy; corolla lobes oblong, obtuse, white, densely villous within; gynostegium short-stipitate; corona scales lanceolate, rounded to a minute stalk at base, adnate to very base of gynostegium just above corolla tube, slightly exceeding anthers but not appendages; anther appendages broadly ovate, slightly longer than broad, erect or slightly inflexed; fruit fusiform, 3.5–5.5 cm. long, 4–5 mm. wide, glabrous; seeds brown, 4–5 mm. long.

Growing among shrubs on rocky hillsides, Lower Sonoran Zone, southern Arizona in Pinal, Maricopa, and Pima Counties. Flowering throughout the year.

**5. Metastelma pringlei** A. Gray, Proc. Amer. Acad. **21:** 396, 1886

A slender twining vine with woody base and light green stems that are glabrous except for 1 or 2 lines of puberulence from node to node; leaves linear-lanceolate to lance-elliptic, 2–5 mm. wide, 8–22 mm. long, rounded at base, acute to obtuse and minutely apiculate at apex, dark green above, lighter beneath, glabrous, margins faintly and narrowly revolute; petioles 3–6 mm. long, sparsely and minutely puberulent along upper side; flowers white, in few-flowered, short-pedunculate cymes; pedicels 2–4 mm. long, glabrous or with a line of puberulence along one side; calyx lobes linear, acute, about one-third as long as corolla lobes, these lance-oblong, obtuse, densely villous within; corona scales linear-attenuate, one and one-half times to twice as long as gynostegium, erect; anther appendages broadly ovate, acute to subapiculate at apex, slightly longer than broad; follicles slender, 4–5 cm. long.

Chihuahua, Coahuila, and Durango and reported somewhat uncertainly from Baja California. A specimen from Guaymas (*Palmer 626*, 1887) referred here as a possible variety by Gray (in herb.) is *M. watsonianum* Standley. Flowering after rains.

**6. Metastelma californicum** Benth., Bot. Sulph. 33, *pl. 18*, 1844

A slender twining vine from a woody base; stems green or purplish, to 1.5 or 2 m. long, glabrous except for 1 or 2 lines of puberulence from node to node; leaves elliptic to ovate-oblong, rounded to subcordate at base, obtuse and abruptly mucronate at apex, 2–8 mm. wide, 10–28 mm. long, glabrous or essentially so, margins faintly to strongly revolute; petioles 3–5 mm. long, finely puberulent on upper side; flowers in subsessile cymes, white; pedicels slender, 3–5 mm. long; calyx lobes narrowly ovate, acute to abruptly apiculate, 0.5 mm. long, glabrous, rotate-campanulate; corolla lobes ovate or oblong, 2.5–3 mm. long, finely puberulent along margins; gynostegium nearly sessile, corona scales lanceolate, erect, about equaling or slightly exceeding anthers; appendages broadly ovate, slightly longer than broad, rounded at apex, inflexed; stigma depressed-conical; follicles 4–6 mm. in diameter, 2–4.5 cm. long, glabrous.

Rocky hillsides and gravelly washes, Lower Sonoran Zone, vicinity of Punta Prieta to the Cape Region, Baja California. Aug.–March.

One specimen in the Dudley Herbarium that is undoubtedly this species is labeled "Orendain, Jalisco, *Marcus E. Jones 27310*." It may possibly represent a case of mixed labels, for it arrived in a shipment that was mostly from Baja California. The Jaliscan record is to be viewed with suspicion.

## 6. MATELEA Aubl., Hist. Pl. Guiane Franc. 1: 277, *pl. 109*, 1775

Scandent vines or sometimes suffruticose and short-stemmed. Leaves mostly petiolate, opposite, cordate or base acute to truncate. Flowers in umbelliform or racemiform cymes, small or large, green, yellowish, brownish, purplish or nearly black. Calyx 5-parted; lobes much shorter than to equaling corollas, often glandular within. Corollas rotate to campanulate, shallowly to deeply lobed; lobes spreading, ascending, or erect, often markedly reticulate-veined. Corona more or less annular, variously toothed, lobed, or appendaged. Anthers relatively simple, without dorsal appendages. Pollinia horizontal or mainly so, one or both faces somewhat excavated, a hyaline sterile portion near attachment of translator arms. Follicles smooth, muricate, or with rather long, spinelike projections. Seeds flattened, comose.

Follicles sparsely covered with spinelike projections:
  Corolla lobes covered with linear-spatulate hairs on inner surface; corona of 15 scales ........ 1. *M. pringlei*
  Corolla lobes glabrous or short-pubescent within; corona of 5 scales:
    Leaves broadly ovate-cordate to suborbicular-cordate; follicles abruptly expanded and truncate at base, 12–18 mm. in diameter ........ 2. *M. altatensis*
    Leaves narrowly lanceolate-ovate, acute to truncate at base; follicles acute or slightly rounded at base, not truncate, 2.5–3.5 cm. in diameter ........ 3. *M. caudata*
Follicles smooth, without spinelike processes or projections:
  Leaves rarely over 10 mm. long; follicles 5–6 cm. long; stems 1–4 dm. long, plant often somewhat bushy; flowers mostly solitary or in pairs ........ 4. *M. parvifolia*
  Leaves 2–8 cm. long; follicles 8–15 cm. long; stems scandent, to 2 m. long or more, plant not bushy; flowers in 5–15-flowered umbels:
    Stigma produced with a 5-dentate column; corolla disk rotate, saucer-shaped, much shallower than height of corona and anther column:
      Corolla greenish yellow; corona lobes bidentate or with lateral processes 1.5 mm. long:
        Corona lobes bidentate, lateral teeth 0.2–0.4 mm. long; corolla lobes 6–10 mm. long, not papillose within ........ 5. *M. cordifolia*
        Corona lobes with lateral caudate processes about 1.5 mm. long; corolla lobes 15–18 mm. long, papillose within ........ 6. *M. umbellata*
      Corolla purple-black; corona lobes with exterior filiform appendages 3–4 mm. long ........ 7. *M. fruticosa*
    Stigma conical, not produced into a column; corolla broadly campanulate, cup three to four times as deep as height of anther column ........ 8. *M. chrysantha*

### 1. **Matelea pringlei** (A. Gray) Woodson, Ann. Mo. Bot. Gard. **28:** 223, 1941

*Himantostemma pringlei* A. Gray, Proc. Amer. Acad. **20:** 294, 1885.

A scrambling or scandent vine from a rather limber woody caudex bearing pale yellowish, slightly fissured, thin, corky bark; herbaceous seasonal stems minutely but closely cinereous-puberulent with short spreading hairs; petioles slender, 8–14 mm. long, puberulent; leaf blades narrowly to broadly ovate-cordate, 1–2.5 cm. broad, 1–3.2 cm. long, rounded to acuminate at apex, basal sinus narrow to broad and basal lobes broadly rounded, green and sparsely puberulent above, paler and more densely so beneath; flowers usually in pairs, on a short peduncle 1–5 mm. long; pedicels slender, 8–10 mm. long, puberulent; calyx lobes linear-lanceolate, 3–3.5 mm. long; corolla disk rotate, 5–6 mm. in diameter, lobes lanceolate, 7–8 mm. long, brownish purple, covered near base within with numerous linear-spatulate hairs 2.5–3.4 mm. long, a few such hairs near margins nearly to tips, lobes strongly reflexed shortly after anthesis; corona of 10 erect filiform scales 4–6 mm. long and 5 shorter ones; follicles turgidly fusiform, 1.5–2.2 cm. in diameter, 8–10 cm. long, more or less attenuate at each end, distantly covered with spreading, spinelike appendages 3–10 mm. long; seeds 5–6 mm. long, granular, wing slightly erose-dentate; coma 3–4 cm. long, silvery-silky.

Sandy washes and rocky hillsides, Lower Sonoran Zone, northwestern Sonora and in Baja California from the vicinity of Calmallí to the tip of the Cape Region. It may occur more or less continuously around the head of the Gulf of California, but no definite record of its occurrence there has been found. Aug.–March.

**2. Matelea altatensis** (Brandegee) Woodson, Ann. Mo. Bot. Gard. **28:** 236, 1941
*Gonolobus altatensis* Brandegee, Zoe **5:** 244, 1908.

Scandent vine, climbing to 6 m. or more, base woody; stems slender, spreadingly hirsute and also minutely puberulent; leaves opposite; petioles slender, 2–5 cm. long, hirsute; leaf blades ovate-cordate to suborbicular-cordate, 3–8 cm. broad, 4–9 cm. long, acuminate at apex, sinus usually nearly closed by broad, rounded, basal lobes, sparsely hirsute, particularly along veins, on both surfaces; peduncles 2–6 cm. long, puberulent and hirsute; inflorescences axillary, of 2–12-flowered, subumbelliform-racemiform cymes; pedicels slender, 1.5–3 cm. long, hirsute; calyx lobes narrowly ovate-lanceolate, 4–5 mm. long, 1.5–2 mm. wide, hirsute without, glabrous within; corolla greenish yellow, rotate, 1.5–2 cm. in diameter, parted nearly to middle into 5 broadly deltoid-rounded lobes, finely but conspicuously reticulate-veined; corona of 5 fleshy rounded processes attached to corolla and surrounding base of column, extending nearly to anthers, bearing 2 hyaline scales near base within, which are joined to the anther column but with free tips, curving outward and then inward in a narrow-mouthed U; anthers opening transversely, wings acute, corneus, scales terminal, hyaline, deltoid, curving inward over the flat stigma; follicles fusiform, attenuate toward each end, abruptly expanded into a truncate, marginally muricate ring 6–8 mm. broad over apex of pedicel, body of fruit 12–18 mm. in diameter, 7–11 cm. long, bearing scattered, spinelike processes 2–8 mm. long over whole fruit, these sometimes arranged in 2–5 obscure rings.

Growing over shrubs and trees along arroyos and on desert llanos, Lower Sonoran Zone, central Sonora and south into Sinaloa. July–Sept.

**3. Matelea caudata** (A. Gray) Woodson, Ann. Mo. Bot. Gard. **28:** 233, 1941
*Gonolobus caudatus* A. Gray, Proc. Amer. Acad. **21:** 399, 1886.
*Vincetoxicum caudatum* Standley, Contr. U.S. Nat. Herb. **23:** 1190, 1924.

Subshrubby plant with slender, sparsely puberulent stems 1–6 dm. long; branches ascending or arching-spreading; woody base ensheathed in light yellow or tawny, corky, irregularly fissured bark; leaves opposite; petioles slender, 5–10 mm. long, sparingly puberulent; leaf blades caudately lanceolate-attenuate, acute to subtruncate at base, 2.4 cm. long, to about 1 cm. wide near base, glabrous or nearly so, margins finely crispate-undulate; peduncles stout, about 1 cm. long, usually 1-flowered; calyx lobes linear-lanceolate; corolla rotate, brownish, somewhat fleshy, lobes lanceolate-subulate, 4–5 mm. long, glabrous; corona a fleshy subcrateriform ring around base of anther column, entire or undulate; fruit ellipsoidal, slightly narrowed toward each end, 2.5–3.5 cm. in diameter, 8–9 cm. long, sparsely muricate with stout, spinelike processes 3–8 mm. long; seeds tawny to light brown, 7–8 mm. long; coma about 3 cm. long, silky, slightly yellowish.

Arid gravelly flats, Lower and Upper Sonoran Zones, central Sonora to southern Chihuahua and south to Puebla and Oaxaca. Aug.–Sept.

**4. Matelea parvifolia** (Torr.) Woodson, Ann. Mo. Bot. Gard. **28:** 230, 1941

*Gonolobus* (?) *parvifolius* Torr., Bot. Mex. Bound. 166, 1859.
*Lachnostoma hastulatum* A. Gray, Proc. Amer. Acad. **11:** 87, 1876.
*Gonolobus hastulatus* A. Gray, op. cit. **12:** 78, 1876.
*Vincetoxicum hastulatum* Heller, Muhlenbergia **1:** 2, 1900.
*Vincetoxicum parvifolium* Heller, loc. cit.
*Gonolobus californicus* Jepson, Man. Fl. Pl. Calif. 771, 1925.

Low twining herbs from a woody rootstock and base; numerous slender branches ascending or twining and scandent, spreadingly hirsutulous, 1–4 dm. long; herbage minutely puberulent with spreading hairs; petioles slender, 3–8 mm. long; leaf blades narrowly deltoid-sagittate to cordate-sagittate, tapering-acuminate, 5–10 mm. long, 2–4(6) mm. wide; flowers solitary or in pairs in axils, without peduncles, subsessile, greenish, 4–5 mm. long; calyx lobes narrowly lanceolate, about 1.5 mm. long; corolla lobes oblong to lanceolate, rotate, pubescent within, a short, downwardly pointing process in each sinus; corona a corolla-like, 5-lobed ring united to anther column by 5 thin vertical membranes, nearly equaling column; anthers with hyaline scarious scales inflexed over stigma; follicles fusiform, 5–6 cm. long, smooth or sparingly tuberculate.

Arid valleys, plains, mesas, and arroyos, Lower Sonoran Zone, and arid parts of Upper Sonoran Zone, discontinuous from eastern Mojave Desert, central and western Colorado Desert, to western Texas and south to central Baja California. March–Oct.

**5. Matelea cordifolia** (A. Gray) Woodson, Ann. Mo. Bot. Gard. **28:** 222, 1941

*Rothrockia cordifolia* A. Gray, Proc. Amer. Acad. **20:** 295, 1885.

Suffrutescent plant with scandent, hirsute and puberulent stems, scrambling to 3–4 m. from a woody base; leaves opposite, petioles slender, 1.5–4 cm. long, spreadingly hirsute and with very short, stiffish hairs mingled with longer ones; leaf blades broadly ovate-cordate, 2–8 cm. long, 1–6 cm. broad, narrow sinus nearly closed by broad rounded lobes, apex acute to acuminate, bright green and sparsely puberulent on upper surface, paler and more densely puberulent, especially along veins beneath; peduncles 1–4 cm. long; inflorescences umbelliform to racemiform, each flower subtended by 1 to several lance-subulate bracts 1.5–2.5 mm. long; pedicels 1–2.5 cm. long, slender, hirsute; calyx lobes lanceolate, 3–5 mm. long, hirsutulous, slightly exceeding sinuses of corolla; corolla rotate, disk nearly flat, 5 lobes oblong-spatulate, greenish white, 6–10 mm. long, glabrous, obtuse at apex or rarely acute, corona of 5 fleshy segments arranged in a ring around base of anther column, each segment minutely toothed centrally; anthers horizontal or nearly so; stigma produced into a column nearly twice as high as corona and anthers, apex peltate-conical, erosely and shallowly 5-lobed; follicles fusiform, 12–15 mm. in diameter, 10–12 cm. long, brownish, finely striate, glabrous, often irregularly mottled with light and dark green before maturity; seeds obovate, 5–7 mm. long, dark brown, granulate-tuberculate; coma 3–4 cm. long.

Mostly along arroyos and on sandy plains, Lower Sonoran Zone, northwestern and central Sonora south at least to the Río Mayo region and from central Baja California to the Cape Region. Aug.–Oct.

**6. Matelea umbellata** (Brandegee) Woodson, Ann. Mo. Bot. Gard. **28:** 223, 1941

*Rothrockia umbellata* Brandegee, Zoe **5:** 165, 1903.

Twining, hirsute, slender stems 1–1.5 m. long; leaves opposite; petioles 2–4 cm. long, lightly hirsute; leaf blades ovate-cordate, gradually acuminate at apex, 2–4 cm.

wide, 6–8 cm. long, dark green and sparsely pubescent on upper surface, paler and slightly more densely hirsute beneath; flowers in axillary, umbelliform or racemiform cymes, these usually 1–6-flowered; peduncles 2–3 cm. long, hirsute; pedicels 1–2 cm. long, each subtended by narrowly lanceolate bracts 2–3 mm. long; corolla yellowish green, broadly rotate, nearly 3.5 cm. in diameter, lobes ovate-oblong, 15–18 mm. long, whitish papillose-puberulent on upper surface; corona 5-lobed, thick and fleshy, especially about lower part near base, each lobe furnished with a thickened band that is somewhat free and 2-lobed near middle, with rounded lateral lobes, a caudate process 1.5 mm. long from sides; stigma produced into a short column with a globose apex; follicles fusiform, glabrous, 12–15 cm. long, smooth.

Lower slopes of the mountains and along arroyos, Lower Sonoran and Arid Tropical Zones, southern Baja California. Sept.–Nov.

**7. Matelea fruticosa** (Brandegee) Woodson, Ann. Mo. Bot. Gard. **28:** 222, 1941
*Rothrockia fruticosa* Brandegee, Zoe **5:** 165, 1903.

Stems fruticose below, slender and herbaceous above, hirsute with subappressed stiff hairs; leaves opposite; petioles 3–5 cm. long, hirsute; leaf blades ovate-cordate, 2.5–5 cm. wide, 3–9 cm. long, acute to gradually acuminate at apex, sinus narrow or nearly closed by auriculately rounded basal lobes, dark green and strigulose above, pale and papillose-puberulent on lamina, hirsute on veins beneath; peduncles to 5 cm. long, few-flowered; pedicels slender, 1–3.5 cm. long; calyx lobes lanceolate, 4.5–6 mm. long, purplish without, hirsute near base and marginally; corolla purple-black, broadly rotate, lobes oblong-lanceolate, 9–12 mm. long, 3–4 mm. broad, minutely and sparsely puberulent without, glabrous within; corona segments fleshy, main portion about 1 mm. high but bearing a filiform erect appendage about 4 mm. long from outer median angle; anthers barely exceeding body of corona lobes; stigma column about 1 mm. longer than anthers, purplish; follicles fusiform, 9–12 cm. long, attenuate at each end.

Hillsides, washes, and plains, Lower Sonoran and Arid Tropical Zones, southern Baja California. Sept.–Nov.

The star-shaped, purple-black flowers are very striking.

**8. Matelea chrysantha** (Greenm.) Woodson, Ann. Mo. Bot. Gard. **28:** 222, 1941
*Gonolobus chrysanthus* Greenm., Proc. Amer. Acad. **32:** 299, 1897.
*Vincetoxicum chrysanthum* Standley, Contr. U.S. Nat. Herb. **23:** 1190, 1924.

Scandent plant growing to 5–6 m. or more, basal parts woody; stems slender, sparsely hirsute and with minute papillose puberulence between their bases; leaves opposite; petioles 2–3 cm. long, pubescent in same manner as stems but long hairs fewer; leaf blades ovate to oblong-ovate, cordate at base, gradually acuminate at apex, sinus usually fairly broad but basal lobes rounded-auriculate and approaching each other at their tips, lamina sparsely subappressed-hirsute above, papillose-granular and slightly paler beneath; peduncles to 4 cm. long, few-flowered; pedicels 1.5–2 cm. long; calyx lobes ovate-lanceolate, 4–4.5 mm. long; corolla broadly campanulate-urceolate, 4–5 mm. deep, lobes oblong-lanceolate, greenish yellow, 5–6 mm. long, ascending or somewhat rotate; corona annular, shallowly 5-lobed, fleshy and closely applied to stamen column; anthers surpassing corona 0.5–1 mm.; stigma depressed-conical, barely surpassing anthers; follicles slenderly fusiform, attenuate at each end, 10–15 cm. long, glabrous and smooth; calyx lobes subpersistent, glabrate, 1.2–2 mm. wide, 6–8 mm. long in fruit.

Growing among shrubs and along rocky canyons, Lower Sonoran to Tropical Zones, east-central Sonora (near San Pedro de las Cuevas) to Morelos, Oaxaca, and Michoacán. Aug.–Oct.

## 7. GONOLOBUS Michx., Fl. Bor. Amer. 1: 119, 1803

Perennial, twining, herbaceous vines with thin, cordate, opposite leaves. Herbage puberulent to subglabrate. Inflorescences axillary, of few-flowered, pedunculate, umbelliform cymes. Sepals free, ascending. Corolla campanulate, bearing a small coralline ring in throat, or this absent in some; lobes 5, erect or essentially so, entire, slightly contorted. Corona fleshy, erect, often adnate to corolla tube much of its length. Stamens inserted at base of corolla; filaments united into a tube; each anther tipped with an erect or slightly inflexed, more or less laminate, membranous appendage; anther wings small, corneous. Pollinia oblong-elliptic, horizontal to slightly ascending or descending, one or both faces somewhat excavate, a sterile hyaline area near attachment of translators. Stigma conic; disk 5-angled. Follicles lance-ovoid, angled or narrowly winged and finely striate-ridged longitudinally. Seeds compressed, copiously comose.

### 1. Gonolobus arizonicus (A. Gray) Woodson, Ann. Mo. Bot. Gard. 28: 243, 1941

*Lachnostoma arizonicum* A. Gray, Proc. Amer. Acad. **20:** 296, 1885.

Stems slender and twining, often climbing to heights of 2 m. or more, white-hispid with slender, simple, spreading hairs; internodes 3–20 cm. long; petioles slender, hispidulous, 0.5–0.8 mm. in diameter, 1.5–4 cm. long; leaf blades thin, lance-ovate, acute at apex, broadly cordate at base, 1.5–3 cm. wide, 3–6 cm. long, sparsely to moderately hispidulous, dark green above, paler and rather prominently veined beneath; peduncles slender, 1–2 cm. long, 1–7-flowered; pedicels about equaling peduncles, hispidulous; sepals narrowly ovate-lanceolate, 1.2–1.5 mm. wide, 4–5 mm. long, ascending and closely applied to corolla tube; corolla 9–11 mm. long, about as wide in full flower, lobes oblong, 5–6 mm. long, cream with greenish veins, apex rounded, puberulent without and within on portions not in contact with adjacent petals while in bud, those parts in contact in bud glabrous; faucal ring absent; corona of 5 slightly fleshy, straight, erect, minutely emarginate appendages 1.5–2 mm. long, each inserted beneath a sinus of corolla and adnate to tube except at distal 0.2–0.3 mm.; stamen column 3–3.5 mm. high; anthers about 1 mm. high, 1.3–1.5 mm. wide, nearly as deep horizontally, truncate and capped by corneous, short, blunt wings, each bearing a thin membranous appendage from inner brownish ridge of wing, appendages erect or inflexed; pollinia horizontal or essentially so, translators broad and ribbonlike, hyaline except at tips where attached to corpusculum, here brownish; stigma disk pentagonal, about 3 mm. in diameter, greenish; follicles 5–8 cm. long, 1.5–2 cm. in diameter, bearing 4–5 thin, slightly undulant, longitudinal wings 2–3 mm. high, surface between wings finely striate longitudinally, minutely and sparsely puberulent when young, glabrate in age; seeds reddish brown, 5–6 mm. long; coma of silvery white hairs 2–2.5 cm. long.

Along streams in canyons, Lower Sonoran and Upper Sonoran Zones, Sycamore Canyon, Santa Cruz County, Arizona, and the Rincon, Santa Catalina, and Baboquivari Mountains, Arizona. May–Sept.

## 8. MARSDENIA R. Br., Mem. Wern. Soc. 1: 28, 1809

Scandent plants usually more or less woody near base; bark often corky and somewhat winged or angulate on older stems. Leaves petiolate, ample. Flowers small or medium in size, in umbelliform cymes. Calyx lobes 5, glandular within. Corolla campanulate, usually pubescent within throat; lobes spreading or ascending, contorted; sinuses sometimes appendaged. Corona scales 5, adnate to stamen tube. Stigma rostrate or depressed. Follicles thick, leathery, smooth. Seeds flattened; coma copious, silky and silvery.

### 1. Marsdenia edulis S. Wats., Proc. Amer. Acad. 24: 61, 1889

A vine climbing to 8 m. or more; lower stems woody, 3–5 cm. in diameter, and bearing a corky, irregularly ridged bark 1–1.5 cm. thick; petioles slender, 1–1.5 cm. long; leaf blades ovate to lance-ovate, acuminate at apex, rounded or obtuse at base, to 5 cm. broad, 4–15 cm. long, glabrous; peduncles stout, usually less than 1 cm. long; cymes few- to many-flowered, axillary; calyx lobes broadly ovate, obtuse, about 3 mm. long, ciliate-margined; corolla white or faintly tinged with brown within, 5–6 mm. long, minutely ciliate, otherwise glabrous, throat of corolla filled with slender stiffish hairs, these obscuring stamen column and corona; follicles 6–12 cm. long, 2–3.5 cm. in diameter, rounded at base, usually obtuse at apex; seeds dark brown, 8–10 mm. long, granular-tuberculate.

Growing over shrubs and trees along arroyos and on rocky hillsides, Lower Sonoran Zone, from Sierra Picú, Sonora, south into Sinaloa. Flowering sporadically following rains; young fruit collected. March–Nov.

## Family 118. CONVOLVULACEAE. Morning-Glory Family

Annual or perennial herbs or vines, or rarely shrubs or trees (parasitic in *Cuscuta*). Leaves alternate, estipulate, blades various or leaves reduced to functionless scales. Flowers perfect, regular, solitary or in cymes. Sepals 4–5, distinct or more or less united at base, often imbricated, persistent, equal or unequal, in 1 or 2 whorls. Corolla sympetalous, usually convolute in bud; limb entire, 5-angled, or shallowly 5-lobed, often involute after flowering. Stamens 5, alternate with corolla lobes; filaments equal or unequal, commonly dilated at base and pubescent, inserted in and adnate to corolla tube; anthers 2-celled, erect or incumbent, straight or spirally twisted. Ovary superior, 2–5-celled, 2–4 ovules per cell, seated in a fleshy disk. Styles 2 or united into 1; stigmas simple or sometimes 2-lobed, linear, oblong, or capitate. Fruit a capsule, 2–4- or 5-celled. Seeds comparatively large; coat coriaceous or membranous. Embryo curved; cotyledons foliaceous, usually folded. Endosperm mucilaginous.

Plants parasitic, virtually devoid of chorophyll, stems yellow, orange, or reddish, attached to hosts by knob-like haustoria; flowers minute, white, 6 mm. long or less; leaves reduced to minute scales .......... 1. *Cuscuta*

Plants autophytic, green; flowers mostly larger and often showy; leaves functional, mostly ample:

  Styles 2 and distinct nearly to base, each with 2 stigma lobes; stigma lobes linear, terete .......... 2. *Evolvulus*

  Style 1 or if 2-lobed each lobe entire; stigma lobes capitate, globose, flabelliform, or oblong, not terete-linear:

    Corolla imbricate in bud; leaves sessile or nearly so .......... 3. *Cressa*

Corolla plicate-convolute in bud; leaves petiolate:
Plants small herbs or suffrutescent plants, not twining vines; corolla campanulate-rotate, tube very short .................................. 4. *Jacquemontia*
Plants twining or sometimes creeping or trailing vines or rarely trees or shrubs; corollas funnelform, deeply campanulate or salverform, tube evident:
Stigmas 2, ovate to linear-spatulate, sessile or style very short ...... 5. *Convolvulus*
Stigmas entire or of 2 globose sessile lobes:
Anthers spirally twisted at or shortly after anthesis:
Corollas white or yellow, campanulate-funnelform; filaments usually glandular at base; capsule longitudinally valvate .................... 6. *Merremia*
Corollas scarlet, salverform; filaments not glandular; capsule with a caplike, thickened lid or operculum, dehiscing circumscissilely ..... 7. *Operculina*
Anthers straight, not spirally twisted:
Corollas funnelform or broadly campanulate; stamens and style well included ........................................................ 8. *Ipomoea*
Corollas salverform or tubular-salverform; stamens and style somewhat exserted or barely equaling corolla tube:
Bracts small, inconspicuous, not enclosing flowers; corolla limb thrice as wide as corolla tube or more:
Corollas white or purple; limb 3–10 cm. broad; corolla tube slender-cylindric ........................................... 9. *Calonyction*
Corollas scarlet or tinged with yellow; limb about 1 cm. broad; corolla tube narrowly funnelform ....................... 10. *Quamoclit*
Bracts large and conspicuous, more or less enclosing flower; corolla limb only slightly wider than tube .......................... 11. *Exogonium*

## 1. CUSCUTA L., Sp. Pl. 124, 1753

Herbaceous parasites with slender, often filiform, yellowish or orange-colored, leafless, much-branching stems nearly devoid of chlorophyll, twining about herbaceous or woody hosts and obtaining food by means of knob-like, deeply penetrating haustoria. Leaves reduced to small functionless scales. Flowers small, in loose to dense, cymose clusters, chiefly sympetalous, 4–5-parted. Calyx campanulate to turbinate or rarely cylindrical, shallowly to deeply parted; lobes acuminate to orbicular, entire or denticulate. Corolla campanulate to cylindrical; lobes shorter than, equaling, or longer than tube. Stamens inserted in throat of corolla, alternating with corolla lobes; filaments of uniform diameter or subulate. Scalelike, more or less fringed appendages at base of corolla opposite each stamen or absent in some species. Styles distinct (in ours) or united; stigmas capitate or elongated. Ovary 2-celled, each cell 2-ovuled. Fruit a capsule, circumscissile or rupturing irregularly. Seeds 1–4; embryo filiform or with a knob at one end, without visible cotyledons.

Capsules obviously circumscissile:
Styles subulate; calyx lobes orbicular, denticulate ......................... 1. *C. erosa*
Styles about same diameter throughout; calyx lobes narrower, usually entire:
Perianth segments obtuse:
Corolla lobes about equaling tube; flowers short; calyx lobes deltoid ............ ........................................................ 2. *C. applanata*
Corolla lobes shorter than tube; flowers long; calyx lobes ovate:
Flowers about 3 mm. long; scales nearly reaching filaments, bridged at or above middle of corolla tube .................................. 3. *C. americana*

Flowers 5–7 mm. long; scales shorter than tube, bridged below middle of corolla tube:
Calyx lobes not markedly overlapping; corolla tube furrowed exterior to stamen insertions; stigma one and one-half to two times as wide as styles .......... .................................... 4. *C. corymbosa* var. *grandiflora*
Calyx lobes strongly overlapping; corollas not furrowed exterior to stamen insertions; stigmas three to four times as thick as styles; anthers nearly sessile ................................................ 5. *C. macrocephala*
Perianth segments acute to acuminate:
Calyx lobes about equaling or exceeding corolla tube:
Scales fringed only above middle; calyx lobes overlapping each other ........... ................................................ 6. *C. odontolepis*
Scales fringed nearly to base, or reduced to mere wings in *C. gracillima* var. *esquamata*:
Stamens shorter than corolla lobes; corollas not glandular:
Corolla lobes ovate-acuminate to lance-attenuate; calyx lobes lanceolate; flower clusters rather loose ............................. 7. *C. desmouliniana*
Corolla lobes ovate to lanceolate but merely acute; calyx lobes deltoid to ovate; flowers in very dense clusters:
Flowers 4–6 mm. long ....................... 8. *C. umbellata* var. *reflexa*
Flowers 2 mm. long ......................... 8*a*. *C. umbellata* var. *dubia*
Stamens mostly distinctly longer than corolla lobes; corollas somewhat glandular .................................... 9. *C. gracillima* var. *esquamata*
Calyx lobes much shorter than corolla tubes:
Scales fringed nearly to base, about one-third as long as corolla tube; calyx lobes tuberculately keeled; calyx squarish at base ............... 10. *C. tuberculata*
Scales fringed mostly above middle, about half as long as corolla tube; calyx lobes not tuberculate; calyx tapering to base:
Flowers 3–4.5 mm. long; corolla papillate near base, not grooved below sinuses ................................................ 11. *C. leptantha*
Flowers 4–5 mm. long; corolla glabrous, distinctly plicate-grooved below each sinus ........................................ 12. *C. polyanthemos*
Capsules not circumscissile:
Capsule depressed-globose:
Calyx lobes equaling or slightly surpassing corolla tube, entire; flowers not fleshy, smooth; tips of corolla lobes not inflexed; flowers 2–3 mm. long ............. ................................................ 13. *C. campestris*
Calyx lobes one-half to two-thirds as long as corolla tube, faintly erosulate; flowers fleshy, slightly papillate, tips of corolla lobes inflexed; flowers 4–5 mm. long .... .................................... 14. *C. indecora* var. *neuropetala*
Capsules globose or ovoid, but not depressed, often somewhat conical:
Scales absent or represented by a narrow ridge or slight wing; calyx lobes one-fourth to one-half as long as corolla tube:
Capsules ovoid, apiculate; corolla somewhat granulate, broadly campanulate ...... .................................... 15*b*. *C. californica* var. *apiculata*
Capsules globose, not apiculate; corolla narrowly campanulate to cylindrical:
Flowers papillose-puberulent .................. 15*a*. *C. californica* var. *papillosa*
Flowers glabrous:
Styles equaling or exceeding ovary; calyx lobes acuminate; flowers short-pedicellate ................................ 15. *C. californica* var. *californica*
Styles much shorter than ovary; calyx lobes obtuse or rounded; flowers sessile ................................................ 16. *C. occidentalis*
Scales well developed; calyx lobes about equaling or surpassing corolla tube:
Calyx lobes orbicular, denticulate; flowers virtually sessile ........ 17. *C. denticulata*

Calyx lobes deltoid to lanceolate, entire; flowers short-pedicellate:
  Scales equaling corolla tube; corolla lobes faintly denticulate; plants parasitic on desert trees .......................................... 18. *C. veatchii*
  Scales one-half to two-thirds as long as corolla tube; corolla lobes not denticulate; plants parasitic mainly, if not entirely, on plants of saline soils .... 19. *C. salina*

**1. Cuscuta erosa** Yuncker, Univ. Ill. Biol. Monogr. **6:** 116, *pl. 2, fig. 8*; *pl. 10, fig. 61*, 1921

Stems of medium diameter; flowers about 3 mm. long, glabrous, pentamerous, on pedicels 3–6 mm. long, in closely clustered, cymose panicles, somewhat tinged with reddish brown; calyx lobes orbicular or broadly obtuse, overlapping, fleshy in middle and often with a dorsal apiculation or midrib somewhat knobby and tuberculate, margins membranous and finely denticulate nearly or quite to base, lobes nearly distinct and usually slightly shorter than corolla tube, sometimes equaling it; corolla campanulate, lobes upright or spreading, about as long as or slightly shorter than tube, oblong, obtuse, often with a horn-like, short apiculation on outside of each lobe near its apex; stamens shorter than corolla lobes; anthers about equaling subulate filaments; scales broad, somewhat shorter than corolla tube, finely fringed, bridged at about their middle; styles narrowly subulate, longer than ovary; capsule globose, circumscissile, frequently 1-seeded, usually capped by withered corolla; seeds about 1.5 mm. in diameter, globose or ovate, if more than one in capsule then somewhat angular; hilum a narrow transverse line or small dot.

Parasitic on *Hymenoclea, Ambrosia, Ipomoea, Merremia,* and numerous other herbaceous plants, Lower Sonoran Zone, southern Arizona and south into Sonora. Sept.–Oct.

**2. Cuscuta applanata** Engelm., Trans. Acad. St. Louis **1:** 479, 1859
*Cuscuta alata* Brandegee, Univ. Calif. Pub. Bot. **3:** 388, 1909.

Stems medium to coarse, bright orange; flowers pentamerous, 2–3 mm. long, somewhat fleshy to membranous, on pedicels shorter than flowers or subsessile, glabrous, in dense cymose panicles; calyx campanulate, slightly shorter than or barely equaling corolla tube, lobes triangular-ovate, obtuse, irregularly keeled along midnerve and below sinuses, entire; corolla campanulate, lobes ovate to oblong, obtuse or sometimes acutish, nearly as long as or equaling corolla tube, spreading, margin slightly uneven but scarcely erose; stamens somewhat shorter than corolla lobes, anthers ovate, slightly cordate, about equal to filaments, these not subulate; scales spatulate, fringed toward apex with medium length segments, longer than corolla tube, visible above rim of corolla, bridged about one-fourth to one-third of way up from base; styles exserted, as long as or longer than ovary; stigmas capitate; capsule depressed-globose, somewhat 4-lobed around seeds, circumscissile, surmounted by withering corolla, 1–4-seeded; seeds brown, 1.4 mm. long; hilum short, transverse or oblique.

Parasitic on various hosts, Sonoran Zones to Tropical Zone, northern Arizona to Texas and south to San Luis Potosí and Zacatecas. July–Oct.

**3. Cuscuta americana** L., Sp. Pl. 124, 1753
(For extensive synonymy see Mem. Torrey Club **18:** 220, 1932.)

Stems medium to coarse; flowers glabrous, 2–3 mm. long, closely crowded into compact clusters, usually turning dark brown in drying, pentamerous, subsessile;

calyx tubular, nearly or quite as long as corolla tube, lobes ovate-orbicular, usually much broader than long, obtuse, overlapping; corolla cylindrical, nearly included within calyx, lobes upright to slightly spreading, lobes becoming fully exserted when corolla is borne upward by growing capsule; scales triangular or somewhat oblong, fringed, shorter than corolla tube or rarely equaling it, bridged about middle or above; anthers subsessile, an apiculation frequently borne on connective; capsule ovoid, usually 1-seeded; seeds about 1.5 mm. long, ovoid; hilum oblique or a rounded spot.

Parasitic on various hosts, Lower Sonoran to Tropical Zones, from the vicinity of Guaymas, Sonora, south and east throughout central Mexico and the West Indies to the Bahamas; southern Florida. July–Nov.

**4. Cuscuta corymbosa** Ruiz & Pav. var. **grandiflora** Engelm., Trans. Acad. St. Louis **1:** 483, 1859

*Cuscuta popayanensis* H. B. K., Nov. Gen. & Sp. **3:** 123, 1819. Not *C. popayanensis* Poeppig, 1859.
*Cuscuta cymosa* Willd., Sch., Syst. Veg. **6:** 205, 1820.
*Cuscuta inclusa* Choisy, Mém. Soc. Genève **9:** 275, *pl. 2, fig. 2,* 1841.
*Cuscuta patens* Benth., Bot. Sulph. 35, 1844.
*Cuscuta laxiflora* Benth., op. cit. 138.
? *Cuscuta corymbosa* var. β *patens* Progel, in Mart., Fl. Bras. **7:** 377, 1871.

Stems medium; flowers glabrous, 5–7 mm. long, pentamerous, subsessile on pedicels much shorter than flowers, forming corymbose panicles; calyx membranous, rather loosely investing corolla, angled, lobes short and obtuse, slightly overlapping, usually reaching to slightly above middle of corolla tube; corolla broadly tubular, almost globular in many flowers, furrowed along stamen attachments in lower part, and bulging between furrows, lobes about one-fourth as long as corolla tube, obtuse, upright to spreading; anthers sessile or essentially so; scales narrow, with short scattered processes, adnate most of their length and set out on a ridge somewhat away from corolla tube in most specimens, usually bridged below middle; capsule globose, surrounded by withered corolla; seeds about 1.5 mm. long, compressed, with an oblique or transverse hilum.

Lower Sonoran to Tropical Zones, the Magdalena Plain, Baja California, to Tamaulipas and Central America. Aug.–Nov.

**5. Cuscuta macrocephala** Schaffner ex Yuncker, Univ. Ill. Biol. Monogr. **6:** 126, *pl. 4, fig. 19a–e; pl. 12, fig. 108,* 1921

Stems relatively coarse, fully 1 mm. in diameter; flowers 5–6 mm. long, glabrous, pentamerous, on pedicels 1–7 mm. long, in cymose clusters that are either loose or densely crowded; calyx 4–5 mm. deep, 3–4 mm. wide, fleshy, lobes rounded, 1.5–2 mm. long, greatly overlapped at margins, nearly equaling corolla, slightly angled below sinuses; corolla tube cylindrical or nearly so, about equaling calyx or slightly shorter, glabrous, lobes short, broadly ovate, obtuse to rounded, usually upright or slightly spreading, about 0.6–1 mm. long, overlapping; anthers subsessile, about 1 mm. long; scales deltoid-oblong, shorter than corolla tube, shallowly but finely fringed to base or nearly so, bridged at or a trifle below middle; styles slender, much longer than ovary, stigmas three to four times as wide as diameter of styles, usually somewhat knobbed or tuberculate; capsule circumscissile, globose, about 3 mm. in diameter; seeds 1.5–2 mm. long, obtusely angled, reddish brown, 1–4 in each capsule.

Parasitic on a number of herbaceous and shrubby plants, Lower Sonoran to

Tropical Zones, southern Baja California, and Sinaloa to Tamaulipas. Flowering throughout the year following rains.

**6. Cuscuta odontolepis** Engelm., Trans. Acad. St. Louis **1:** 486, 1859

*Cuscuta odontolepis* var. *typica* Yuncker, Univ. Ill. Biol. Monogr. **6:** 129, *pl. 2, fig. 7a–d; pl. 12, fig. 116; pl. 13, fig. 153,* 1921.

Stems slender; flowers glistening-white, 3–5 mm. long, pentamerous, in dense, cymose, paniculate clusters on short pedicels or subsessile, each usually subtended by an acute ovate bract; calyx equaling or shorter than corolla tube, papillose-verrucose especially toward base, lobes broadly ovate-deltoid, acute, more or less overlapping; corolla cylindrical to narrowly campanulate, lobes shorter than or about equaling corolla tube, ovate, acute, upright, spreading, or at length reflexed, a hornlike apiculation near apex on dorsal side on some flowers, lacking on others; stamens shorter than corolla lobes, anthers oblong, about equaling filaments; scales large, oblong to spatulate, dentate or fringed about upper part only, usually slightly shorter than corolla tube but sometimes reaching filaments; ovary globose, much exceeded in length by slender styles, these exserted in fruit; capsule globose, thickened and slightly tuberculate or rugose at apex around styles, usually 4-seeded, circumscissile; seeds about 1.2 mm. long, light brown, angled; hilum oblique, linear.

On various hosts, including *Amaranthus, Euphorbia,* and *Atriplex,* Lower Sonoran to Tropical Zones, Pima County, Arizona, to central Mexico. Aug.–Oct.

**7. Cuscuta desmouliniana** Yuncker, Univ. Ill. Biol. Monogr. **6:** 130, *pl. 5, fig. 25; pl. 10, fig. 67,* 1921

*Cuscuta desmouliniana* var. *typica* Yuncker, op. cit. 131.
*Cuscuta desmouliniana* var. *attenuiloba* Yuncker, loc. cit., *pl. 5, fig. 25f–h.*

Stems slender and very profuse; flowers about 2 mm. long (sometimes to 2.5 mm. long), pentamerous or rarely tetramerous, slightly papillate, on pedicels about as long as flowers or slightly longer; calyx narrowly campanulate, lobes triangular or lanceolate, acute to acuminate, longer than corolla tube, tips more or less spreading; corolla campanulate, lobes longer than tube, lanceolate-acuminate, ascending to spreading, or reflexed in fruit; stamens about two-thirds as long as corolla lobes, anthers shorter than filaments; scales oblong, sparingly fringed with short processes nearly full length, about three-fourths as long as corolla tube; styles slender, longer than globose ovary; stigmas capitate; capsule circumscissile, globose, thin, usually 3–4-seeded; seeds ovoid, about 1 mm. long, somewhat compressed and angled; hilum short, oblong, oblique.

On *Euphorbia* and some other low herbaceous plants, Lower Sonoran Zone, central to southern Sonora. Aug.–Oct.

**8. Cuscuta umbellata** H. B. K. var. **reflexa** (Coulter) Yuncker, Univ. Ill. Biol. Monogr. **6:** 132, *pl. 12, fig. 114,* 1921

*Cuscuta californica* var. *reflexa* Coulter, Contr. U.S. Nat. Herb. **1:** 45, 1890.

Stems slender but profuse; flowers glabrous or sometimes slightly puberulent, 4–6 mm. long, pentamerous, on pedicels 2–8 mm. long, forming dense compound cymes; calyx turbinate, lobes as long as or sometimes longer than campanulate corolla, acute to acuminate; corolla lobes about equaling corolla tube, oblong to lanceolate, acute to acuminate, usually reflexed; stamens shorter than corolla lobes, anthers oblong to oval, shorter than or about equaling filaments; scales spatulate, profusely fringed near apex, firmly attached to tube below; styles longer than globose ovary; capsule depressed-globose, with a ring of low, broad, rounded tubercles about intrastylar aperture, circumscissile, invested by withering corolla in fruit; seeds about 1 mm. long, somewhat angled, yellowish; hilum linear, transverse.

Growing on various hosts, mostly herbaceous, Lower Sonoran to Tropical and Upper Sonoran Zones, southern Arizona to Texas and south into Sonora, Baja California, and Sinaloa. June–Dec.

**8*a*. Cuscuta umbellata** var. **dubia** Yuncker, Univ. Ill. Biol. Monogr. **6:** 133, *pl.* 2, *fig.* 9*f–g*, 1921

Stems very slender; flowers only about 2 mm. long; calyx lobes acutely triangular and forming angles at sinuses; corolla lobes about equaling tube, spreading.

Known only from the type locality, beach near Guaymas, Sonora.

**9. Cuscuta gracillima** Engelm. var. **esquamata** Yuncker, Univ. Ill. Biol. Monogr. **6:** 133, 1921

Stems very slender and often disappearing between flower clusters before fruit matures; flowers about 2 mm. long, smooth and glabrous or minutely papillate, frequently tetramerous, on pedicels usually longer than flowers; calyx turbinate, lobes triangular-lanceolate, acute to acuminate, longer than corolla tube; corolla campanulate, somewhat glandular, lobes about as long as tube or slightly longer, triangular to lanceolate, acute; stamens slightly shorter than to a trifle longer than corolla lobes, anthers oblong or oval, shorter than filaments; scales reduced to a few short processes or lacking; styles capillary, much longer than ovary; capsule depressed-globose, irregularly and somewhat tardily circumscissile; seeds 1–4, about 1 mm. long, ovate; hilum oblique, oblong.

Known only from the vicinity of the type locality, El Taste, southern Baja California. Flowering period variable with rainy seasons.

**10. Cuscuta tuberculata** Brandegee, Univ. Calif. Pub. Bot. **3:** 389, 1909

Stems slender; flowers in umbellate-racemose clusters, 2.5–4 mm. long, pentamerous, on pedicels shorter or longer than flowers; calyx nearly truncate at base, short-cylindrical, lobes about one-half to three-fourths as long as corolla tube, tringular to lanceolate, acute or subacuminate, often keeled to give calyx an angular appearance; corolla cylindrical, lobes equaling or slightly shorter than corolla tube, upright, triangular-acute or triangular-lanceolate, basal part of corolla, especially part enclosed by calyx, minutely papillate; stamens shorter than corolla lobes, oblong anthers equaling or shorter than slender filaments; scales ovate to broadly spatulate, reaching about to or slightly beyond middle of corolla tube, fringed to base with short processes, bridged at about middle or slightly below; ovary globose-conical, with a thickened ring around base of styles; styles much longer than ovary, exserted and usually about equaling corolla lobes; capsule globose, slightly umbonate apically, circumscissile, bearing withered corolla; seeds 1–4, about 0.8 mm. long, ovoid, compressed, pale reddish brown, minutely papillate.

Parasitic on *Boerhaavia* and *Amaranthus* and probably other herbaceous hosts, Lower Sonoran Zone, southern Arizona into central Sonora, west central Baja California, and south toward the Cape Region. Aug.–Nov.

**11. Cuscuta leptantha** Engelm., Trans. Acad. St. Louis **1:** 489, 1859

*Cuscuta palmeri* S. Wats., Proc. Amer. Acad. **24:** 64, 1889.
*Cuscuta leptantha* var. *typica* Yuncker, Univ. Ill. Biol. Monogr. **6:** 135, *pl. 6, fig. 34a–e*, 1921.
*Cuscuta leptantha* var. *palmeri* Yuncker, op. cit. 136, *pl. 6, fig. 34f; pl. 11, fig. 91*, 1921.

Stems very slender and frequently nearly absent by time flowers are present; flowers in umbellate clusters, 3–4.5 mm. long, mostly tetramerous, on pedicels equaling or longer than flowers; calyx shallowly campanulate, finely papillose, lobes triangular-ovate, acute, slightly fleshy but not keeled, reaching almost to middle of corolla tube; corolla cylindrical, fleshy and papillose on basal part, lobes triangular to lanceolate, acute, nearly as long as corolla tube, upright to reflexed; stamens about half as long as corolla lobes, ovate anthers about two-thirds as long as filaments; scales oblong, truncate to broadly and roundly deltoid, reaching a little beyond middle of corolla tube, fringed nearly or quite to base with short processes, bridged at about one-quarter to one-half their length; styles slender, much longer than ovary, exserted; capsule globose, slightly umbonate, circumscissile, 2–4-seeded, capped with drying corolla; seeds about 0.8 mm. long, ovoid, oblique, somewhat angled or sides rounded, flattened on one or two faces; hilum oblique or transverse.

Parasitic on several herbaceous hosts, Lower Sonoran Zone, western Texas to Sonora, central Baja California, and south to central Mexico. Occurring at Bahía de los Angeles, Baja California. Nov.–May.

**12. Cuscuta polyanthemos** Schaffner ex Yuncker, Univ. Ill. Biol. Monogr. **6:** 136, *pl. 6, fig. 31; pl. 9, fig. 92*, 1921

Stems very slender but profuse, disappearing early; flowers in umbellate clusters, pentamerous or tetramerous, 4–5 mm. long, on pedicels usually two or three times as long as flowers; calyx short-campanulate, lobes triangular, acute, about equaling calyx tube and reaching less than halfway to top of corolla tube, often even shorter; corolla cylindrical, or slightly tapering toward base, 4–5 mm. long, lobes narrowly triangular to lanceolate, acute, slightly spreading to reflexed, about half as long as corolla tube, basal part of tube somewhat plicate-furrowed exterior to stamen attachments; stamens about half as long as corolla lobes, anthers oblong-oval, shorter than filaments, latter slightly subulate; scales oblong, sparingly fringed with short close processes, reaching about to middle of corolla tube or less, bridged about middle or slightly higher; ovary globose, slightly pointed or umbonate; styles several times as long as ovary; capsule globose, slightly umbonate, lightly glandular, circumscissile; seeds 1–4, about 1 mm. long, oval, more or less angled; hilum oblique, oblong.

Parasitic on *Euphorbia* and other hosts, Lower Sonoran Zone, southern Sonora into Sinaloa. March, and probably through the summer.

**13. Cuscuta campestris** Yuncker, Mem. Torrey Club **18:** 138, *fig. 14*, 1932.

*Cuscuta pentagona* γ *calycina* Engelm., Amer. Jour. Sci. **45:** 76, (Oct.) 1843.
*Cuscuta arvensis* var. δ *calycina* Engelm., Trans. Acad. St. Louis **1:** 495, 1859.
? *Cuscuta arvensis* var. *capsici* Deg. & Lin., Zeitsch. Pflanzenkrank. **17:** 269, 1907.
*Cuscuta pentagona* var. *subulata* Yuncker, Amer. Jour. Bot. **10:** 3, 1923, in part.

Stems medium; flowers 2–3 mm. long, in compact globular clusters, 5-parted, often somewhat glandular, on pedicels shorter than flowers or some sessile; calyx broadly and shallowly campanulate or bowl-shaped, broader than long, lobes oval or nearly orbicular, sometimes broader than long, often overlapping when young, about equaling short corolla tube; corolla broadly and shallowly campanulate, lobes broadly triangular, acute, tips often inflexed, about equaling tube; stamens shorter than corolla lobes, anthers nearly or quite as long as slender filaments; scales ovate to oblong, copiously fringed with rather long processes, slightly exserted or about equaling corolla tube, bridged below middle; ovary globose, nearly equaling styles; capsule depressed-globose, usually splitting corolla, this often remaining attached at base of capsule; seeds ovate, about 1.2–1.5 mm. long.

Parasitic on a large number of hosts, among them *Artemisia, Ambrosia, Aster, Cirsium, Ipomoea,* and *Xanthium,* in several life zones from southern California to Virginia and south to South America. Also present in various parts of all the other continents. Parasitic on several crop plants and the seed apparently disseminated with the seeds of the crops. April–Dec.

**14. Cuscuta indecora** var. **neuropetala** (Engelm.) A. S. Hitchcock, Contr. U.S. Nat. Herb. **3:** 549, 1896

*Cuscuta neuropetala* Engelm., Lond. Jour. Bot. **2:** 198, (June) 1843.
*Cuscuta neuropetala* γ *littoralis* Engelm., Bost. Jour. Nat. Hist. **5:** 223, 1845.
*Cuscuta pulcherrima* Scheele, Linnaea **21:** 750, 1848.
*Cuscuta decora* var. β *pulcherrima* Engelm., Trans. Acad. St. Louis **1:** 502, 1859.
*Cuscuta decora* var. γ *subnuda* Engelm., loc. cit.
*Cuscuta indecora* var. *portoricensis* Urban, Symb. Antill. **4:** 502, 1910.
*Cuscuta indecora* var. *subnuda* Yuncker, Amer. Jour. Bot. **10:** 9, 1923.

Stems medium to coarse; flowers 4–5 mm. long, fleshy and translucent-white, more or less papillose-hispid, in loose or compact clusters on pedicels mostly longer than flowers; calyx shallowly bowl-shaped, lobes broadly deltoid-ovate, equaling or longer than calyx tube, about half as long as corolla tube to equaling it; corolla broadly campanulate, lobes usually erect but sometimes spreading, broadly triangular, acute, usually with inflexed tips, usually shorter than corolla tube or about equaling it; stamens one-half to three-fourths as long as corolla lobes; anthers broadly ovate, about as long as filaments; scales oblong-spatulate, equaling or slightly surpassing corolla tube, deeply fringed about apex and more shallowly so nearly to base, bridged near or slightly below middle; styles slender, equaling or slightly longer than globose, faintly umbonate ovary; capsule globose, thickened and slightly umbonate at apex, 2–4-seeded; seeds about 1.6 mm. long, finely scurfy.

Parasitic on a number of native hosts as well as on several crop plants, Idaho to Michigan and south to the West Indies, California, and central Mexico. Occurring near Hermosillo, Sonora, and near Tucson, Arizona. July–Nov.

**15. Cuscuta californica** Choisy, Mém. Soc. Genève **9:** 279, 1841, var. **californica**

*Cuscuta acuminata* Nutt. ex Engelm., Trans. Acad. St. Louis **1:** 498, 1859.
*Cuscuta californica* var. β *graciliflora* Engelm., op. cit. 499.
*Cuscuta californica* var. γ *longiloba* Engelm., loc. cit.

Stems medium; flowers in loose, cymose-paniculate clusters, 3–5 mm. long, on short pedicels or subsessile; calyx campanulate, slightly shorter than to slightly exceeding corolla tube, lobes triangular to lanceolate, acute to acuminate, overlapping slightly near base, fleshy at base; corolla campanulate-cylindrical, often shallowly furrowed over stamen-attachments and saccate between them, lobes narrowly lanceolate, acute to slightly acuminate, reflexed, equaling to longer than tube (connivent in bud to form a sharp point); stamens shorter than lobes, ovate to oblong anthers about equaling or slightly shorter than faintly subulate filaments; scales absent or represented by faint ridges below each stamen, these ridges dilated and coalesced at base of corolla tube;

styles much longer than globose ovary; capsule depressed-globose to globose, not circumscissile, rupturing irregularly; seeds about 1.4 mm. long, reddish brown.

Parasitic on a number of hosts, among them *Franseria, Eriogonum, Adenostoma, Abronia, Asclepias* and others, mostly Upper Sonoran Zone, but entering the margins of the Sonoran Desert in eastern San Diego County; Washington to northern Baja California. March–Aug.

**15*a*. Cuscuta californica** var. **papillosa** Yuncker, Univ. Ill. Biol. Monogr. **6:** 152, *pl. 11, fig. 76,* 1921

Differing from var. *californica* only in having pedicels and flowers densely papillose-puberulent. From the San Bernardino Mountains to Indio and the Borrego Valley, Imperial and San Diego Counties, California.

**15*b*. Cuscuta californica** var. **apiculata** Engelm., Trans. Acad. St. Louis **1:** 499, 1859

Similar to var. *californica* except that corolla is granulate with low rounded tubercles and ovary is ovoid-pointed. Known only from the type locality, along the Colorado River, eastern California. Possibly not entering our area.

**16. Cuscuta occidentalis** Millsp., in Millsp. & Nutt., Field Mus. Bot. Ser. **5:** 204, 1923

*Cuscuta californica* var. δ *breviflora* Engelm., Trans. Acad. St. Louis **1:** 499, 1859.

Stems medium; flowers mostly sessile in compact clusters, about 3 mm. long, often glandular; calyx shallowly and broadly campanulate, lobes broadly deltoid, acute or abruptly acuminate, slightly shorter than to slightly longer than corolla tube, fleshy and thickened at base; corolla globose-campanulate, somewhat saccate between stamen attachments, lobes lanceolate, acuminate, spreading, nearly as long as corolla tube; stamens much shorter than corolla lobes, filaments about equaling anthers and faintly subulate; scales none; styles about as long as to slightly longer than globose ovary; capsule globose, encased in withered corolla; seeds about 1.5 mm. long, oval, flattened on one or two surfaces.

Parasitic on various hosts, Sonoran Zones and occasionally into the Transition Zone, Washington and Oregon east to Colorado and south to Nevada and central Baja California. March–Aug.

**17. Cuscuta denticulata** Engelm., Amer. Nat. **9:** 348, 1875

Stems very slender but very profuse, disappearing early; flowers about 2 mm. long, subsessile, in short, few-flowered, more or less bracteate clusters; bracts 1–3, ovate-lanceolate, acute; calyx rounded-campanulate, lobes orbicular or obtuse, deeply divided and overlapping, denticulate around membranous margin, nearly enclosing corolla; corolla narrowly campanulate, urceolate in fruit, lobes ovate to oval, slightly overlapping, about equaling tube, faintly denticulate near apex or entire; stamens about half as long as rounded corolla lobes, filaments and anthers about equal; scales oblong, about reaching filaments or sometimes reaching anthers, rounded at apex, denticulate, bridged about middle or a trifle below; styles much shorter than ovoid ovary, stigmas small, globose; capsule conical, mostly 1-seeded, bearing withered corolla; seeds about 1 mm. long, globose-ovoid, rounded, hilum small, embryo with a terminal rounded knob.

Parasitic on *Larrea* and a number of composites, mainly Lower Sonoran Zone, Colorado and Mojave Deserts to southern Utah and western Arizona. In our area it appears chiefly on *Larrea.* March–Oct.

**18. Cuscuta veatchii** Brandegee, Proc. Calif. Acad. II, **2:** 189, 1889

*Cuscuta veatchii* var. *typica* Yuncker, Univ. Ill. Biol. Monogr. **6:** 159, *pl. 8, fig. 48a–e; pl. 10, fig. 60,* 1921.

Slender to medium-stemmed; flowers solitary or more commonly in 2–10-flowered clusters, about 2–2.5 mm. long, glabrous, on pedicels 0.5–1 mm. long; calyx thin, lobes ovate-deltoid, acute, microscopically denticulate, equaling or slightly longer than corolla tube, somewhat overlapping each other; corolla campanulate, lobes ovate, acute, about equaling tube, spreading, minutely denticulate along edges; scales nearly or quite reaching filaments, rather copiously fringed with medium length processes, attached to corolla tube about to middle of scale; filaments about equaling or slightly shorter than oval anthers; styles slender, one-half as long as to equaling globose-ovoid ovary; stigmas capitate; capsule globose-ovoid, about 1.5 mm. high, withered corolla persistent; seeds usually 1, sometimes 2 in each capsule, if only one then subglobose, light brown; embryo with a large globose knob at one end.

Parasitic most frequently on *Bursera* and *Pachycormus veatchii,* Lower Sonoran Zone, Baja California from the San Felipe Desert to Vizcaíno Desert. March–April.

The seedlings often develop in the tangled masses of the previous year's stems still clinging to the branches of the host tree instead of starting in the ground and then attaching themselves to the host plant. Haustoria develop on seedling stems less than 5 mm. long, usually attacking the very young leaves of the host, later penetrating the seasonal twigs.

**19. Cuscuta salina** Engelm., in Brewer & Wats., Bot. Calif. **1:** 536, 1876

*Cuscuta californica* var. ε ? *squamigera* Engelm., Trans. Acad. St. Louis **1:** 499, 1859.
*Cuscuta subinclusa* var. β *abbreviata* Engelm., op. cit. 500
*Cuscuta squamigera* Piper, Contr. U.S. Nat. Herb. **11:** 455, 1906.
*Cuscuta salina* var. *squamigera* Yuncker, Univ. Ill. Biol. Monogr. **6:** 161, *pl. 12, fig. 126,* 1921.

Stems slender, profuse; flowers in umbellate-cymose clusters, mostly 5-parted, 2–3 mm. long, on pedicels shorter than flowers, or some of them sessile; calyx campanulate, lobes ovate-lanceolate, acute to acuminate, nearly to fully as long as corolla tube; corolla cylindro-campanulate, lobes lanceolate, sometimes granulate, acute to acuminate, slightly spreading to upright, or widely spreading in age, about as long as corolla tube, sometimes more or less overlapping; stamens about one-half or one-third as long as corolla lobes, anthers oval, equaling or slightly exceeding length of faintly subulate filaments; scales oblong, narrow, about one-half to two-thirds as long as corolla-tube, fringed nearly to base with very short processes, firmly attached to tube nearly full length of scales, bridged below middle; styles slender, shorter than to about equaling globose, slightly pointed ovary; capsule globose to ovoid, narrowly umbonate, mostly 1-seeded, invested by or capped by withering corolla; seeds about 1.5 mm. long, ovoid, somewhat rostrate.

Parasitic on plants growing in saline soils, Lower Sonoran to Transition or possibly Canadian Zones, but at low elevations, British Columbia to Baja California and east to Utah, Arizona, and Sonora. Not common in the Sonoran Desert but occasional along the coast to southern Baja California. April–Nov.

## 2. EVOLVULUS L., Sp. Pl. ed. 2, 391, 1762

Annual or perennial herbs or small shrubs. Stems prostrate to erect. Leaves simple, entire, petiolate or sessile. Flowers perfect, pedicellate or sessile in axils, or on slender, 1- to few-flowered peduncles. Sepals 5, distinct, equal or subequal, obtuse to acuminate. Corolla small, rotate, funnelform or salverform, blue, white or yellow; limb obscurely 5-lobed. Stamens 5, included or equaling corolla; anthers oblong or ovate. Ovary ovoid to globose, 2-celled, 4-ovuled. Styles 2, separate to base or nearly so, each bearing 2 stigma lobes; stigmas terete, elongate. Capsule ovoid or globose, 1–4-seeded. Seeds smooth or slightly roughened.

Flowers on distinct slender peduncles, not sessile:
  Corolla 4–8 mm. in diameter .......................... 1. *E. alsinoides* var. *acapulcensis*
  Corolla 10–15 mm. in diameter:
    Herbage appressed-villous .......................... 2. *E. arizonicus* var. *arizonicus*
    Herbage spreading-villous ............................ 2a. *E. arizonicus* var. *laetus*
Flowers sessile or short-pedicellate in axils of leaves, not pedunculate:
  Stems and leaves patently villous on both surfaces; sepals lanceolate or narrowly lanceolate ............................................................ 3. *E. pilosus*
  Stems and leaves appressed-sericeo-villous on lower surface; sepals oblong-lanceolate .... ................................................ 4. *E. sericeus* var. *discolor*

### 1. **Evolvulus alsinoides** L. var. **acapulcensis** (Willd.) v. Ooststr., Mededeel. Bot. Mus. & Herb. Rijks Univ. Utrecht **14:** 34, 1934

*Evolvulus acapulcensis* Willd., in Sch., Syst. Veg. **6:** 199, 1820.

Perennial herb with simple or sparingly branched stems 1–3 dm. high, appressed-pilose and often with some spreading hairs throughout; leaves lance-elliptic to lance-linear, 2–7 mm. wide, 12–32 mm. long, short-petiolate, acute at both ends or apiculate at apex, green; peduncles filiform, equaling to somewhat exceeding leaves, 1–3-flowered; bracteoles subulate-linear, 1–2 mm. long; pedicels 3–3.5 mm. long; sepals lanceolate, 1.5–2 mm. long; corolla light blue, 4–8 mm. in diameter; capsule globose, 2.5 mm. in diameter, 2-valved; seeds ovoid, light brown, smooth, 1–1.25 mm. high.

Rocky hillsides and plains, Lower Sonoran to Tropical Zones, southern Arizona to Texas and south through Sonora, Chihuahua, and Baja California into southern Mexico. March–Dec.

### 2. **Evolvulus arizonicus** A. Gray, Syn. Fl. N. Amer. **2**[1]**:** 218, 1878, var. **arizonicus**

Perennial from a slender woody root; stems several, erect or ascending, simple, 1–4 dm. high; herbage closely appressed-sericeous throughout; leaves lance-linear to lanceolate, 1.5–5 mm. wide, 10–25 mm. long, uppermost much smaller, acute at both ends, sessile or on petioles 1–1.5 mm. long; peduncles filiform, 1.5–3 cm. long, ascendingly spreading, much longer than subtending leaves, 1–3-flowered; bracteoles subulate, about 2 mm. long; pedicels 5–8 mm. long, filiform; sepals lance-acuminate, 2–2.5 mm. long; corolla rotate, 10–15 mm. broad, light blue; capsule broadly ovoid to subglobose, 2.5 mm. wide, 3 mm. high; seeds obovoid, smooth, brownish, microscopically papillose.

Plains, hillsides, and mountains, Upper and Lower Sonoran Zones, Graham and Yavapai Counties, Arizona, south into northern Mexico; also in Argentina. March–Oct.

**2a. Evolvulus arizonicus** var. **laetus** (A. Gray) v. Ooststr., Mededeel. Bot. Mus. & Herb. Rijks Univ. Utrecht **14:** 76, 1934

*Evolvulus laetus* A. Gray, Proc. Amer. Acad. **17:** 228, 1882.

Similar to var. *arizonicus* but with brownish pubescence, this more spreading and heavier than in that variety. Stems erect, simple, 1–2.5 dm. high; leaves narrowly lance-elliptic, sessile, acute at apex, 2–3 mm. wide, 10–18 mm. long, equally pubescent on both surfaces; peduncles slender, about one and one-half times to twice as long as subtending leaves; bracteoles linear, 1.5–2 mm. long; pedicels 4–6 mm. long; sepals lanceolate, acute, about 2 mm. long; corolla 10–15 mm. wide, light blue; capsule globose, 2.5 mm. high.

Southwestern Arizona and adjacent Sonora, mostly in the Upper Sonoran Zone of the foothills and lower parts of the mountains, but entering the margins of the Sonoran Desert in northern Sonora.

**3. Evolvulus pilosus** Nutt., Gen. **1:** 174, 1818

*Evolvulus argenteus* Pursh, Fl. Amer. Sept. **1:** 187, 1814. Not *E. argenteus* R. Br., 1810.
*Evolvulus nuttallianus* Sch., Syst. Veg. **6:** 198, 1820.
*Evolvulus mollis* Small, Bull. N.Y. Bot. Gard. **1:** 285, 1899.

Low perennial from woody roots and base, densely pubescent throughout with rusty, brown or grayish spreading hairs; stems several, erect or ascending, 8–15 cm. long, simple or sparingly branched; leaves crowded, erect or ascending, elliptic, narrowly oblong or oblanceolate, sessile or nearly so, acute at each end or some obtuse at apex, 1.5–5 mm. wide, 8–20 mm. long; flowers solitary in axils full length of stems, short-pedicellate; pedicels reflexed in fruit; sepals narrowly lanceolate, 4–5 mm. long; corolla blue to purple, 8–12 mm. wide; filaments twice as long as anthers; capsule ovoid, 3–4 mm. long, stramineous, 1–2-seeded; seeds 2–3 mm. long, brown.

Ranging from Montana and North Dakota to Arizona, Texas, Arkansas, and Tennessee. In Arizona it is found mostly above the desert regions but is included here on the basis of a specimen in the United States National Herbarium from "Tumacacori and vicin." *D. Griffiths 3950,* March–April, 1903. March–Sept.

**4. Evolvulus sericeus** Benth. var. **discolor** (Benth.) A. Gray, Syn. Fl. N. Amer. ed. 2, **2**[1]**:** 436, 1886

*Evolvulus discolor* Benth., Pl. Hartw. 6, 1839.
*Evolvulus holosericeus* β *obtusatus* Choisy, in A. DC., Prod. **9:** 444, 1845.
*Evolvulus oreophilus* Greene, Leaflets Bot. Obs. **1:** 151, 1905.
*Evolvulus wilcoxiana* House, Bull. Torrey Club **33:** 315, 1906.

Perennial from woody root, with prostrate to ascending, simple stems 5–20 cm. long; stems and undersurfaces of leaves sericeous-villous with moderately appressed, shining, brownish hairs; leaves somewhat distichous, sessile or nearly so, elliptic, oblong or ovate-oblong, acute to obtusish at apex, acute to rounded at base, 4–8 mm. wide, 7–25 mm. long, usually glabrous on upper surface; flowers solitary in leaf-axils, short-pedicellate; bractlets 2–5 mm. long, linear; sepals ovate-oblong, acute, 4–5 mm. long, sericeous-villous; corolla white to pale blue, 10–15 mm. wide; filaments three times as long as anthers; capsule globose, about equaling sepals; seeds 1–4, brown, smooth.

Rocky hillsides, Lower and Upper Sonoran Zones, Arizona to Texas and south into Mexico. March–Sept.

Reported from California by van Ooststroom on basis of specimen collected "California, Colorado Desert, San Diego County, Santa Catalina Mission, Sept. 1889, C. R. Orcutt." It is mostly in the mountains above the desert.

## 3. CRESSA L., Sp. Pl. 223, 1753

Small, silky-canescent, perennial herbs with erect to prostrate or decumbent, non-twining stems and alternate, sessile or subsessile, entire, more or less crowded leaves. Flowers small, solitary in upper axils, pedicellate. Sepals 5, erect or nearly so. Corolla imbricate in bud, white; tube about equaling sepals, oblong-campanulate; lobes 5, spreading or later reflexed. Stamens 5, filaments subulate, exserted slightly from throat of corolla. Ovary 2-celled, each 2-ovuled. Styles 2, distinct; stigmas capitate. Capsule usually 1-seeded by abortion of other ovules.

**1. Cressa truxillensis** H. B. K., Nov. Gen. & Sp. **3:** 119, 1819

*Cressa cretica* δ *truxillensis* Choisy, in A. DC., Prod. **9:** 440, 1845.

Plants usually much-branched from base, from a deep-seated perennial but slender rootstock, rarely rhizomatous; stems procumbent to erect, 0.5–2 dm. long, tufted; herbage finely silky-canescent throughout; leaves sessile or nearly so, oblong-ovate, to narrowly elliptical near tips of stems, 4–7 (rarely 10) mm. long, 1.5–4.5 mm. wide; pedicels 1–3 mm. long, each bearing 2 narrowly elliptic or lanceolate bractlets immediately below calyx; calyx lobes broadly ovate, obtuse at apex, 3–3.5 mm. long, somewhat imbricated; corolla limb 4–6 mm. in diameter, withering-persistent in fruit, lobes ovate-deltoid, 2–2.5 mm. long, glabrous within, silky-pilose without; capsule ovoid, 4–6 mm. high, long-hairy near summit, often 1-seeded.

Usually in alkaline or saline soil, Sonoran and Tropical Zones, Sacramento Valley, California, to southern Utah, Texas, and south to tropical America. Generally distributed in warm regions around the world. Very abundant at scattered localities on playas and near the coast throughout our area. March–Sept. in northern part of the Sonoran Desert, practically the year around farther south.

## 4. JACQUEMONTIA Choisy, Mém. Soc. Genève 6: 476, 1833

Usually scandent but some low, erect shrubs or prostrate, pubescent throughout or glabrous. Leaves alternate, entire, often cordate, or rarely lobed or toothed, petiolate. Flowers small to medium, blue or white, in axillary cymes or capitate, each subtended by 2 foliaceous bractlets. Sepals 5, subequal or outer two larger than inner three. Corolla narrowly to broadly campanulate; limb plicate, 5-angulate. Stamens 5, inserted near base of corolla tube, equal or unequal, shorter than corolla; filaments usually slightly dilated at base. Ovary 2-celled, 4-ovuled. Styles filiform; stigma lobes 2, ovate to oblong. Capsule globose or broadly ovoid, 2-celled, 4- or 8-valved. Seeds usually glabrous or glabrate.

Plants annual, herbaceous; corolla 7–8 mm. long; sepals all alike ............. 1. *J. palmeri*
Plants perennial, woody at base; corollas 12–25 mm. long; outer sepals slightly to markedly longer and wider than inner ones:
  Peduncles glabrous; outer sepals only slightly larger than inner ones ........ 2. *J. albida*
  Peduncles puberulent to tomentose; outer sepals much larger than inner ones:
    Sepals sparsely pilosulous to subglabrate, acute, outer ones accrescent in fruit; cymes open; corollas lavender or whitish .................................. 3. *J. pringlei*

Sepals densely pilosulous, acuminate, outer ones not markedly accrescent in fruit; cymes usually dense; corollas blue:

Sepals long-acuminate; herbage moderately tomentulose; cymes 3–7-flowered ....... .......................................... 4. *J. abutiloides* var. *abutiloides*

Sepals short-acuminate; herbage densely tomentose; cymes 1–3-flowered ........... ...................................... 4a. *J. abutiloides* var. *eastwoodiana*

**1. Jacquemontia palmeri** S. Wats., Proc. Amer. Acad. **24**: 63, 1889

*Jacquemontia palmeri* var. *varians* Brandegee, Zoe **5**: 170, 1903.

Erect or twining annual herbs 3–10 dm. tall with slender stems and sparsely pubescent herbage; petioles slender, 5–12 mm. long; leaf blades narrowly lance-ovate to broadly ovate, 1–2.5 cm. broad, 2–4.5 cm. long, acute to acuminate at apex, rounded to shallowly cordate at base, thin; peduncles slender, 1.5–6 cm. long, usually exceeding subtending leaves, 1–6-flowered, flowers sometimes subracemosely arranged; bracts filiform, 1–3 mm. long, inconspicuous; pedicels 2–10 mm. long, puberulent; sepals virtually alike, ovate-lanceolate, acuminate, 1.5–2 mm. broad, 4.5–5.5 mm. long at anthesis, not accrescent, sparsely pilosulous; corolla 7–8 mm. long, blue, limb often 12–15 mm. broad; stigma lobes broadly ovoid to nearly orbicular; capsule globose, 4–5 mm. in diameter; seeds dark brown to black, 2–2.5 mm. long, minutely granular in a fine, irregularly reticulate pattern.

Along arroyos, on outwash slopes, and valley floors, Lower Sonoran Zone and probably into the lower margins of the Upper Sonoran Zone, southwestern Pima County, Arizona, south throughout Sonora and in southern Baja California. Aug.–Nov.

**2. Jacquemontia albida** Wigg. & Roll., Contr. Dudley Herb. **3**: 277, *pl. 64*, 1943

Scrambling slender vine with twining, slender, glabrous stems; petioles slender, 1–3.5 cm. long, glabrous or sparsely pubescent with simple hairs; leaf blades ovate, 1.5–4 cm. wide, 3–6 cm. long, acuminate at apex, cordate at base, bright green, glabrous; peduncles slender, 5–15 cm. long, mostly dichotomously branched near tip, glabrous; cymes 2–5-flowered; pedicels slender, 5–10 mm. long, glabrous; bracts and bracteoles narrowly lanceolate, 2–5 mm. long, acute to acuminate; sepals ovate, acute to subcuspidate, glabrous, outer slightly broader than inner ones, 3–4 mm. broad, 4–6 mm. long; corolla infundibuliform, white, 2–2.5 cm. long, faintly lobed; stamens included, unequal, anthers pale yellow, about 1.5 mm. long; stigma lobes nearly orbicular; capsule oblong-ovoid, 5–6 mm. long, blunt at apex, glabrous; seeds dull black, glabrous or sparsely granular with minute, golden-brown papillae.

Along arroyos, growing over shrubs and trees, Lower Sonoran Zone, from the region just southwest of Hermosillo to southern Sonora. Aug.–Oct.

**3. Jacquemontia pringlei** A. Gray, Proc. Amer. Acad. **17**: 227, 1882

*Jacquemontia pringlei* var. *glabrescens* A. Gray, op. cit. **21**: 402, 1886.

Slender twining vines climbing to 3 m. or sometimes a prostrate ground vine, stems densely but finely pubescent with a tangled mixture of simple and few-rayed, stellate trichomes or subglabrate, woody near base; petioles slender, 0.5–4 cm. long, pubescent; leaf blades narrowly to broadly ovate-cordate, 1–3.5 cm. broad, 1.5–6.5 cm. long, acuminate at apex, pubescent or less commonly glabrate; peduncles slender, 3–10 cm. long, bearing open, few- to several-flowered cymes; bracts and bracteoles linear, 3–6 mm. long; pedicels to 1 cm. long, slender; outer sepals ovate, acuminate,

4–5 mm. wide, 5–8 mm. long at anthesis, accrescent and 9–13 mm. long in fruit, sparsely pubescent to glabrate; corollas infundibuliform, 1.2–2.5 cm. long, pale lavender or white, very shallowly 5-angulate; stamens unequal, 7–12 mm. long; stigma lobes distinctly oblong; capsule ovoid, 5–6 mm. high, glabrous, acute to briefly acuminate at apex; seeds dull dark brown, minutely papillose in an irregular reticulum.

Rocky canyons and hillsides, Lower Sonoran to arid parts of Tropical Zone, Pima County, Arizona, to Oaxaca and Chihuahua; in El Salvador. Aug.–Oct.

**4. Jacquemontia abutiloides** Benth., Bot. Sulph. 34, 1844, var. **abutiloides**

A scrambling and twining perennial vine with slender branches 5–20 dm. long; younger stems and herbage closely tomentose with crowded, several-rayed, stellate hairs that are nearly white when young, turning tawny to golden-brown with age; stems eventually glabrate and bark turning gray; internodes 2–8 cm. long; petioles 3–30 mm. long, stoutish to slender, densely stellate-tomentose; leaf blades broadly ovate, acute and abruptly acuminate to rounded at apex, distinctly cordate with a broad sinus at base, 1–3.5 cm. wide, 1.5–6 cm. long on vigorously growing stems, markedly reduced on short lateral branches from older wood, harshly tomentose with white to reddish brown hairs on both surfaces, pinnately 5–7-veined, veins conspicuous beneath; flowering peduncles 2–10 cm. long, mostly 3–7-flowered, flowers crowded; bracts linear, 4–7 mm. long, stellate-puberulent; calyces 6–10 mm. long, angular in bud, stellate-puberulent, lobes narrowly to broadly ovate, acute to attenuate, 3.5–4.2 mm. wide, puberulent within and without; corolla broadly funnelform-rotate, clear blue, 2–3.5 cm. in diameter, glabrous; stamens 6–9 mm. long, filaments slender, glabrous; anthers straight, about 2.5 mm. long, white; styles about equaling stamens to one and one-half times as long; stigma lobes broadly ovoid, 0.8–1 mm. long, divergently spreading; capsules ovoid, acuminate at apex, glabrous, 4–5 mm. in diameter, 5–6 mm. high; seeds broadly obovoid to subglobose, 3–3.5 mm. long, dark brown to black, dull, minutely granulose-papillate and slightly reticulately ridged transversely on back.

Rocky hillsides, canyons, and arroyos, Lower Sonoran and Tropical Zones, central Baja California to the Cape Region. Reported from Sonora also, but no specimens from that side of the Gulf assignable to this species have been seen. Sept.–June and probably flowering occasionally during the rest of the year.

**4*a*. Jacquemontia abutiloides** var. **eastwoodiana** (I. M. Johnston) Wiggins, Contr. Dudley Herb. **4:** 21, 1950

*Jacquemontia eastwoodiana* I. M. Johnston, Proc. Calif. Acad. IV, **12:** 1133, 1924.

Differing from var. *abutiloides* in having shorter, less pronouncedly acuminate sepals, more densely tomentose herbage, and mostly 1–3-flowered cymes on generally shorter peduncles.

Rocky situations, Lower Sonoran Zone, mainly along the Gulf of California side of the peninsula from Isla Tortuga south to the Cape Region. The two expressions of the species appear to intergrade completely. Dec.–May.

## 5. CONVOLVULUS L., Sp. Pl. 153, 1753

Twining or trailing, herbaceous, mostly perennial vines with alternate petiolate leaves. Flowers borne singly or in few-flowered inflorescences on short to long peduncles; each flower normally subtended by a pair of small to conspicuous bracts, these remote or immediately below calyx. Sepals 5, subequal or outer ones sometimes broader than inner, acuminate to obtuse. Corolla broadly funnelform; limb 5-angled or rarely 5-lobed, plicate in bud, white, cream, pink, blue or purple. Stamens 5, inserted near base of corolla, equal or unequal, included; filaments filiform or somewhat dilated at base; anthers oblong, straight. Ovary 2-celled, 4-ovuled. Styles slender; stigmas 2, ovate to linear-spatulate, sessile or on short branches of style. Capsule ovoid to globose, 4-valved or rupturing irregularly. Seeds glabrous or puberulent; cotyledons usually emarginate and plicate.

Bracts larger than sepals, closely subtending flowers and obscuring sepals; herbage glabrous; corolla 3–3.5 cm. long .............................. 1. *C. aridus* subsp. *longilobus*

Bracts reduced to scales much shorter than sepals, attached several times their length below calyx; herbage pubescent; corolla 1.5–2 cm. long:

- Leaves oblong-sagittate to ovate, very obtuse; sepals suborbicular, 3–5 mm. long, not cordate at base; herbage sparsely and unevenly pubescent with spreading hairs ........ ............................................................ 2. *C. arvensis*
- Leaves hastate, central lobe linear to narrowly lanceolate, acute; sepals oblong, 6–9 mm. long, subcordate at base; herbage uniformly appressed-sericeous ...... 3. *C. incanus*

### 1. **Convolvulus aridus** Greene subsp. **longilobus** Abrams, Contr. Dudley Herb. **3:** 358, *pl. 68, fig. 4,* 1946

Twining or trailing plants scrambling to a height of 3 m. over shrubs and rocks; stems and leaves minutely tomentulose but green throughout; leaves strongly hastate, 3–4 cm. long, central lobe triangular-lanceolate to narrowly lanceolate, basal lobes about one-half as long as central lobe, frequently sharply and deeply 2-toothed or 2-parted; peduncles slender, 8–10 cm. long, mostly 1-flowered, those on small lateral branches shorter; bracts 11–14 mm. long, broadly ovate to broadly oblong, thin, often reddish purple, faintly undulate-crisped along margin, obtuse to rounded at base, obtuse, rounded or subtruncate and often apiculate at apex, equaling or slightly exceeding sepals, these ovate, acute, 8–10 mm. long, closely invested by bracts; corolla creamy or white, 3–3.5 cm. long, slightly tinged with reddish purple along 5 midveins; capsules ovoid, apiculate, 8–9 mm. long.

Mostly in chaparral, Upper Sonoran Zone, from Orange and Riverside Counties, California, south into Baja California, where it enters the Lower Sonoran Zone south of the Sierra San Pedro Mártir at Cataviñá. March–July.

### 2. **Convolvulus arvensis** L., Sp. Pl. 153, 1753

Stems prostrate, slender, 2–12 dm. long, often forming dense mats, arising from very deep-seated perennial roots; herbage minutely villous-pubescent or sometimes nearly glabrous except for margins of leaves; petioles 3–20 mm. long; leaf blades elliptic, oblong-sagittate, or sometimes ovate, 1–4 cm. long, to 3 cm. wide, usually obtuse or at times retuse, rarely acute, basal lobes triangular to lance-deltoid, usually equaling about one-fourth width of blades or less in length, acute, base of blade subcordate to truncate; peduncles 1–3-flowered, 1–4.5 cm. long; bracts narrowly spatulate or subulate, attached near or slightly above middle of peduncle, 2–5 mm. long; calyx broadly cupulate, lobes broadly oblong to suborbicular, 3–5 mm. long, obtuse or rounded, margins hyaline, obscurely gibbous, green; corolla white with 5 narrowly lanceolate, purplish stripes and often a faint wash of purple between them on outside, 1.5–2 cm. long, open-funnelform and not angled; capsule ovoid, apiculate, 5–8 mm. high, 4–7 mm. in diameter; seeds about 4 m. long, black, dull, minutely reticulate and tuberculate, sparsely and unevenly puberulent.

An introduced weed from Europe and occurring from British Columbia to northern Baja California and east across the continent; uncommon in the desert areas but occurring in irrigated ground in Arizona. April–July.

### 3. Convolvulus incanus Vahl, Symb. Bot. 3: 23, 1794

Stems herbaceous, scrambling over bushes to a height of 1–2 m. from a woody, deep-seated, perennial root; herbage finely and closely sericeous-pubescent, or hairs on younger parts often rusty or tawny; petioles to 1.5 cm. long but mostly 5–8 mm. long; leaf blades hastate or less commonly ovate-sagittate, 1.5–4 cm. long, central lobe usually linear, oblong, or linear-lanceolate, mostly 1.5–4 mm. wide but to 10 mm. wide or slightly more, basal lobes usually several-toothed or shallowly lobed; peduncles 1–6 cm. long, 1–3-flowered; bracts subulate, 1–2.5 mm. long, attached 5–15 mm. below calyx; calyx campanulate, sepals oblong, 6–9 mm. long, obtuse to faintly retuse at apex, appressed-sericeous, subcordate at base in age, margins membranous; corolla 1.5–2 cm. long, pink, usually turning purplish or bluish in drying, rather conspicuously sericeous on midveins without, 5-angled; capsules subglobose, 7–8 mm. in diameter, weakly apiculate; seeds dull black, microscopically granulate, 4–4.5 mm. high.

Bushy slopes and dry mesas, Sonoran Zones (mainly Upper Sonoran), Pima County, Arizona, to Colorado, Nebraska, Texas, and south into Mexico. Barely reaching the edge of the Sonoran Desert in Pima County. April–Oct.

## 6. MERREMIA Dennst., Schlüssel Hort. Ind. Malobaricus 23, 1818

Twining vines or rarely small erect shrubs, with glabrous to densely pubescent herbage and entire, lobed or divided leaves. Flowers solitary or in few-flowered clusters in axils. Bracts linear to lanceolate. Sepals subequal or outer larger than inner, glabrous, pubescent or glandular-pubescent. Corolla campanulate, fairly large. Stamens equal or nearly so; filaments usually glandular at base; anthers spirally twisted after dehiscence. Style filiform. Stigma globose or of 2 globose lobes, included. Ovary 2–3-celled, 4–6-ovuled. Capsule dehiscent by 4–6 longitudinal valves or rarely irregularly. Seeds usually 4–6, rarely solitary, glabrous to densely pubescent.

Pedicels and calyces glabrous, without glandular hairs; calyx lobes 1–4 cm. long in flower; corollas large, 3–8 cm. long; leaflets entire or faintly undulate-margined:
- Corollas yellow, about 5 cm. long; sepals in fruit 3–3.5 cm. long; leaflets 1–4 cm. long ........ 1. *M. aurea*
- Corollas white, 6–8 cm. long; sepals in fruit 3.5–4 cm. long; leaflets 3–16 cm. long ........ 2. *M. palmeri*

Pedicels and calyces with some glandular hairs; calyx lobes 5–9 mm. long in flower; corollas 1.5–2.5 cm. long; leaflets serrate ........ 3. *M. quinquefolia*

### 1. Merremia aurea (Kell.) O'Donell, Lilloa 6: 494, 1941

*Aniseia aurea* Kell., Proc. Calif. Acad. **5:** 83, 1873.
*Ipomoea aurea* Kell. ex Curran, Bull. Calif. Acad. **1:** 143, 1885.
*Operculina aurea* House, Muhlenbergia **5:** 68, 1909.

A twining vine woody at base and climbing over shrubs and trees to 8 m. or more, glabrous or with a few yellowish hairs; petioles 10–25 mm. long; leaflets 5, oblong, ovate-lanceolate or oblanceolate, obtuse to acuminate at apex, frequently widest above middle, 0.5–2 cm. broad, 1–4 cm. long, mostly entire but sometimes some leaflets lobate near base, glabrous; peduncles 2–4 cm. long, slightly expanded apically, 1-flowered; sepals oblong to obovate, subequal, outer 1–2.5 cm. long in flower, obtuse or rounded, glabrous; corolla bright yellow, about 5 cm. long, 6–8 cm. in diameter; filaments 12–15 mm. long; anthers 8–10 mm. long, tightly twisted spirally after anthesis; capsule bilocular, usually 4-seeded, broadly obovoid to subglobose, 2–2.5 cm. long, glabrous; seeds dull black, 8–11 mm. long, minutely and closely puberulent.

Clambering over trees and shrubs on hillsides and along arroyos, Lower Sonoran and arid parts of Tropical Zones, from vicinity of Loreto and La Purísima south to the Cape Region, Baja California. Flowering practically the year around.

**2. Merremia palmeri** (S. Wats.) Hallier f., Jahrb. Hamburg. Wiss. Anst. **16,** Beiheft **3:** 38, 1899

*Ipomoea palmeri* S. Wats., Proc. Amer. Acad. **24:** 63, 1889.
*Operculina palmeri* House, Bull. Torrey Club **33:** 502, 1906.

A rampant climber with long slender stems to 5 m. or more in length, glabrous throughout; petioles slender, 1.5–5 cm. long; leaves 5-foliolate, leaflets sessile or on short petiolules to 3 mm. long, lanceolate, entire, slenderly acuminate at apex or sometimes at both ends, oftener merely acute to obtuse at base and widest below middle, 6–30 mm. wide, 3–16 cm. long, bright green above, paler beneath; flowers solitary; peduncles 3–10 cm. long, slightly swollen at juncture of more slender pedicel, this 1–3 cm. long; bracts lanceolate, 2–4 mm. long, caducous; sepals unequal, exterior ones about 1.5 cm. wide, 2.5 cm. broad in flower, oblong or obtuse or rounded at apex, to 3.5–4 cm. long and chartaceous in fruit, inner ones 5–8 mm. shorter than outer; corolla white, 6–8 cm. long, nearly as broad; stamens subequal; anthers 8–10 mm. long, tightly twisted spirally after dehiscence; capsule broadly ovoid, 3–3.5 cm. long, abruptly acute at apex, glabrous; seeds dark brown to black, about 1 cm. long, minutely puberulent.

Climbing over shrubs and trees along arroyos, plains, and outwash slopes, Lower Sonoran and arid ridges in Tropical Zones, from the vicinity of Ures and Llano, Sonora, south into Sinaloa. Aug.–Jan.

**3. Merremia quinquefolia** (L.) Hallier f., Bot. Jahrb. **16:** 552, 1893

*Ipomoea quinquefolia* L., Sp. Pl. 162, 1753.
(For additional synonymy see O'Donell, Lilloa **6:** 529, 1941.)

Twining or prostrate vine with slender, somewhat striate stems, these glabrous or sparsely hirsute with scattered, large, yellowish hairs; leaves 5-foliolate or rarely 3-foliolate; petioles 1–5 cm. long, glabrous or hirsute with scattered yellow hairs; leaflets subsessile, elliptic to narrowly lanceolate, acuminate and often mucronate (in our specimens), 3–5 cm. long, 3–6 mm. wide, serrate; peduncles surpassing leaves, 2–7 cm. long, slender, glandular-puberulent, each bearing a lax cyme of 3–10 flowers or rarely a single flower; bracts 0.5–2.5 mm. long, narrowly ovate, caducous; pedicels 5–15 mm. long, slender, glandular-puberulent; sepals slightly unequal, oblong, obtuse, glabrous and eglandular without, 5–9 mm. long, margin scarious, outer shorter than inner ones; corolla white, broadly campanulate, 15–20 (or sometimes 25) mm. long, nearly as broad; stamens subequal, filaments 5–7 mm. long, glandular-pubescent at base; capsule subglobose, 7–8 mm. in diameter, glabrous; seeds 2–3 mm. long, pubescent.

Mainly in tropical regions, scrambling over low shrubs or prostrate, southern Baja California, Sinaloa, and Tamaulipas south to Peru. Sept.–March.

## 7. OPERCULINA Silva Manso, Enum. Subst. Bras. 16, 1836

Twining vines with herbaceous or woody, perennial stems and entire, lobed or divided leaves; stems or peduncles or both often more or less winged. Calyx pear-shaped, somewhat constricted above. Sepals closely imbricated, subequal or unequal, chartaceous or scarious, at least along margins, usually enlarging in fruit. Corollas campanulate to subsalverform, showy. Stamens (in ours) included. Anthers spirally twisted. Ovary 2-celled, 4-ovuled. Style slender or filiform. Stigma globose or of 2 subglobose lobes. Capsule differentiated into an upper thicker portion forming an operculum-like cap, and a thinner lower portion dehiscing regularly or irregularly around base of cap. Seeds large, brown or black, mostly smooth.

### 1. Operculina rubicunda House, Bull. Torrey Club **33:** 498, 1906

*Ipomoea alata* Rose, Contr. U.S. Nat. Herb. **1:** 108, *pl. 10,* 1891. Not *I. alata* R. Br., 1810, nor G. Don, 1838; not *Operculina alata* (Ham.) Urban, 1902.

A twining vine with milky juice and slender, wingless, lightly striate stems several meters long; petioles very slender, 1–2 cm. long, glabrous; leaf blades lanceolate-cordate to triangular-ovate, acute to acuminate at apex, shallowly cordate at base, 1.5–3.5 cm. wide, 3–6 cm. long, thin, slightly paler beneath than above; peduncles 3–7 cm. long, usually surpassing subtending leaves, bearing 2 membranous wings 0.5–2.5 mm. wide, 8–20 mm. long below middle, mostly 1-flowered; pedicel broadly clavate, 8–15 mm. long, slightly constricted at apex, glabrous, often tinged with red; sepals acute, about 2 cm. long, some 3–5 mm. shorter than longer ones, often reddish, membranous; corollas subsalverform, scarlet, tube 6–10 mm. in diameter, about 5 cm. long, limb 2.5–3 cm. broad; anthers about equaling tube; capsule globose or nearly so, about 1.5 cm. in diameter, usually 4-seeded; seeds 6–8 mm. long, black, glabrous.

Along arroyos, canyons, and hillsides, upper edge of Lower Sonoran Zone to Tropical Zone, east central Sonora near Mátape south to Michoacán. Aug.–Nov.

## 8. IPOMOEA L., Sp. Pl. 159, 1753

Annual or perennial, twining, trailing or creeping vines, or rarely erect shrubs or trees with ample leaves and showy, axillary, solitary or cymose flowers on short to long peduncles. Sepals 5, equal, subequal or unequal, margins often thin, membranous or scarious. Corolla funnelform or campanulate; limb entire, 5-angled or 5-lobed, more or less plaited, particularly in bud. Stamens 5, equal or unequal, included; filaments filiform or dilated at base; anthers ovate to linear. Ovary ovoid to globose, 2–4-celled, 4–6-ovuled. Styles filiform, included; stigmas 1 or 2, capitate to globose. Capsule ovoid to globose, 2–4-valved, 2–6-seeded. Seeds usually angled on inner face or margins, glabrous, scanty, puberulent, short-hirsute or silky-comose on angles.

Plants more or less twining, scandent, or prostrate-creeping vines; flowers appearing with or after leaves:
 Leaf blades oblong-lanceolate, cuneate to rounded at base, to 25 cm. long; corolla white with a pink to lavender or purple throat .......................... 1. *I. longifolia*
 Leaf blades ovate, cordate, or deeply lobed or divided, not lance-oblong; corolla usually pink, lavender, blue, or purple on limb, sometimes whitish in throat:
 ♦ Leaf blades pedately 3–9-divided or -parted:
  Sepals 2–2.5 cm. long, ovate-acuminate; leaves velvety-strigose beneath; corollas 6–8 mm. long ............................................ 2. *I. heterophylla*

Sepals 4–15 mm. long, oblong to lance-linear; leaves glabrous or essentially so beneath;
Leaf segments mostly lanceolate, 4–8 mm. wide or basal ones wider; sepals lance-subulate, densely hispid on lower one-half to three-fourths .................................................................. 3. *I. barbatisepala*
Leaf segments filiform to linear-oblong or rarely linear-oblanceolate, rarely over 3 mm. wide, mostly less; sepals oblong-ovate or linear-lanceolate, glabrous:
Corolla 8–12 mm. long; sepals oblong to ovate, 4–6 mm. long, acute to obtuse and apiculate .......................................... 4. *I. costellata*
Corolla 2.5–3.5 cm. long; sepals linear-lanceolate to narrowly ovate-lanceolate, 8-12 mm. long, attenuate at apex ...................... 5. *I. leptotoma*

♦ Leaf blades entire or shallowly 3-lobed, not deeply divided or parted, if lobed not pedately so:
Pedicels somewhat thicker than peduncles; outer sepals muricate-warty in lower one-half:
Corollas 8–10 cm. long; pedicels 3–5 cm. long; sepals 7–8 mm. long .................................................................. 6. *I. pedicellaris*
Corollas 1.5–1.8 cm. long; pedicels 10–12 mm. long; sepals 4–5 mm. long .................................................................. 7. *I. peninsularis*
Pedicels not markedly thicker than peduncles, usually thinner, at least at base; sepals not muricate-warty:
Leaf blades orbicular, orbicular-cordate, or slightly obcordate, leathery-fleshy; plants creeping, mostly along strand and on sand dunes; stems subsucculent, often 1–1.5 cm. in diameter .................................. 8. *I. pes-caprae*
Leaf blades not obcordate, rarely orbicular, not fleshy; plants twining or ascending, not strand plants; stems thinner, rarely over 3 mm. in diameter except at very base:
Peduncles 10–30 cm. long, 5–15-flowered; stems and petioles frequently muricate-warty .................................................. 9. *I. perlonga*
Peduncles 1–7 cm. long, 1–5-flowered; stems and petioles not muricate-warty:
Leaf blades cordate, but not lobed nor angulate:
Peduncles viscid; capsules oblong-ovate, about twice as long as broad .................................................................. 10. *I. viscosa*
Peduncles not viscid; capsules mostly ovoid to subglobose, little if any longer than broad:
Foliage and stems cinereous-pubescent with forked and stellate hairs; sepals obtuse to rounded at apex, not apiculate ........... 11. *I. scopulorum*
Foliage glabrous or pubescent with simple hairs only, not cinereous; sepals acute, or apiculate if obtuse:
Bracts subfoliaceous, 12–20 mm. long, immediately below 1–5-flowered, headlike cluster; pedicels 1–3 mm. long ............ 12. *I. iodantha*
Bracts smaller, subulate; pedicels 2.5–12 mm. long or more, flowers not crowded into headlike clusters:
Corolla 5–7 cm. long or more, purple; sepals lance-elliptic, 12–15 mm. long, densely pilose at base ..................... 13. *I. purpurea*
Corolla 4–5 mm. long, yellow; sepals lance-ovate, 3–3.5 mm. long, sparsely pilose full length .................. 14. *I. minutiflora*
Leaf blades or some of them distinctly 3-lobed or angulate and obscurely toothed near base:
Roots tuberous; plants perennial; sepals very unequal, glabrous; leaves 3–5-angulate and obscurely toothed toward base .............. 15. *I. jicama*
Roots not tuberous; plants annual; sepals subequal, more or less pubescent; leaves 3-lobed but not strongly angulate nor toothed:
Sepals ciliate or sparsely ciliate on backs, scarious-margined, tips apiculate; corollas 1.5–2 cm. long .............................. 16. *I. triloba*

Sepals densely hirsute at base, not scarious-margined, tips attenuate or acute; corollas 2.5–4 cm. long:

Sepals lanceolate, acute and erect at apex, not attenuate nor spreading; capsule ovoid .................................. 17. *I. hirsutula*

Sepals ovate, attenuate into slender-spreading process at tips; capsule subglobose .................................. 18. *I. hederacea*

Plants arborescent; flowers appearing before leaves .................... 19. *I. arborescens*

## 1. **Ipomoea longifolia** Benth., Pl. Hartw. 16, 1839

*Convolvulus shumardianus* Torr., Bot. Expl. Red River 291, 1852.
*Convolvulus queretarensis* Sessé & Moc., La Naturaleza [Pl. Nov. Hisp.] II, **1:** app. 24, 1887.
*Ipomoea carletonii* Holz., Contr. U.S. Nat. Herb. **1:** 211, *pl. 17,* 1892.

A widely creeping perennial from a heavy, deep-seated root, with stout, coarsely striate stems 5–10 mm. in diameter, 1–10 m. long, glabrous; leaves erect or ascending in 2–3 distinct ranks from prostrate stem; petioles slender, 6–20 mm. long; leaf blades narrowly lanceolate, 1–3 cm. wide, 7–25 cm. long, entire, obtuse to rounded (or infrequently cuneate) at base, acute to obtuse at apex, rather leathery, glabrous, midrib broad and conspicuous, lateral veins inconspicuous; peduncles to 10 cm. long or more, 1-flowered, jointed at or above middle; bracts lance-subulate, 3–5 mm. long, caducous; pedicels 1–2 cm. long; outer sepals 12–15 mm. long, broadly oblong, obtuse or rounded at apex, margins thin; inner sepals similar but to 2 cm. long; corolla 8–12 cm. long, broadly funnelform, white faintly tinged with purple on limb, deep purple in throat, limb 6–8 cm. in diameter; filaments densely villous near base; capsule ovoid, firmly coriaceous, 2–2.5 cm. long; seeds oblong, minutely hairy on angles.

Mostly on grassy flats and gentle slopes, Upper Sonoran to Tropical Zones, southern Arizona to Oklahoma and Texas and south into Sonora, Querétaro, and Durango. July–Oct.

## 2. **Ipomoea heterophylla** C. G. Ortega, Hort. Mat. 9, 1797

*Ipomoea ortegae* Poir., in Lam., Encycl. Suppl. **4:** 633, 1816.
*Ipomoea willdenowii* Roem. & Sch., Syst. Veg. **4:** 211, 1819.
*Convolvulus heterophyllus* Spreng., Syst. **1:** 592, 1825.
*Batatas heterophylla* G. Don, Gen. Hist. Pl. **4:** 261, 1838.
*Batatas ? willdenowii* G. Don, loc. cit.
*Convolvulus willdenowii* Steudel, Nom. ed. 2, **1:** 412, 1840.
*Pharbitis heterophylla* Choisy, in A. DC., Prod. **9:** 344, 1845.

Slender twining vine with sparsely strigose-villous stems and foliage; petioles slender, 2–3 cm. long; leaf blades deeply 5-lobed, lobes lanceolate, 7–15 mm. wide at broadest point, tapering toward each end, attenuate at apex, 2–6 cm. long, entire, sparsely silky-strigose above, densely so beneath; peduncles stoutish, 5–12 mm. long, strigose, 1-flowered; bracts linear-subulate, 5–9 mm. long, about 1 mm. wide, deciduous; pedicels equaling to one and one-half times as long as peduncles; sepals equal, lance-ovate and attenuate at apex, 10–12 mm. wide, 2–2.5 cm. long, silky-strigose, nearly glabrous within; corolla narrowly funnelform, 6–8 cm. long, bluish, limb 4–5 cm. broad; capsule depressed-ovoid, 12–15 mm. broad, about 10 mm. high, firm; seeds purplish brown, 7–8 mm. long, minutely faveolate-pitted and puberulent on angles.

Plains and mesas, Sonoran Zones, Baboquivari Mountains and Cochise County, Arizona, to Texas and adjacent Mexico. Barely entering the Sonoran Desert on the western slopes of the Baboquivari Mountains, Pima County. Aug.–Oct.

**3. Ipomoea barbatisepala** A. Gray, Syn. Fl. N. Amer. 2[1]: 212, 1878

A twining vine with slender stems 0.2–1.5 m. long (or possibly more), glabrous throughout except for hispid-barbate sepals; roots not tuberous; petioles slender, 1.5–5 cm. long; leaf blades pedately 5-parted, basal pair of lobes often again deeply lobed on lower side near base, then leaf 7-veined, other lobes linear-lanceolate to narrowly lanceolate, 4–8 mm. wide at widest part, 1.5–6 cm. long, attenuate apically, somewhat cuneate at base; peduncles 2–3 cm. long, slender; bracts filiform-subulate, lax, 3–5 mm. long, subpersistent, often present until fruit begins to ripen; pedicels about 1 cm. long, often sharply deflexed from tip of peduncles; sepals narrowly lance-subulate, 10–15 mm. long, equal, thickly beset from about sinuses to within 3–4 mm. of apex, with porrect, stiff, tawny or rusty hairs; corolla narrowly funnelform, about 2 cm. long, limb about 1 cm. broad, purplish pink (sometimes white) gradually fading to a whitish tube; capsules broadly ovoid to oblong-ovate, 6–9 mm. high, coriaceous, glabrous; seeds brownish, 5–7 mm. long, minutely scurfy or glabrous.

Climbing over herbs and shrubs, canyons, hillsides, and talus slopes, Sonoran Zones, Pima, Santa Cruz, and Cochise Counties, Arizona, to western Texas and south through Sonora to Oaxaca. Aug.–Oct.

**4. Ipomoea costellata** Torr., Bot. Mex. Bound. 149, 1859

A slender, glabrous, much-branched, annual, twining vine or spreading-ascending herb with branches 1–12 dm. long or possibly longer but usually less than 5 dm. long; roots slender, filiform; petioles slender, 0.5–2.5 cm. long; leaf blades 5–9-divided into filiform, linear or narrowly oblanceolate segments, these 0.5–3 mm. wide, to 3.5 cm. long, three central segments simple, entire, outer pair usually again divaricately 2–3-lobed from near base; peduncles slender, 1–2.5 cm. long; bracts linear, 1.5–2.5 mm. long, persistent; pedicels 4–10 mm. long; sepals subequal, oblong or oblong-ovate, 4–6 mm. long, obtuse and minutely apiculate, margins papery-membranous, inner slightly longer than outer, widening somewhat in fruit; corolla 8–12 mm. long, funnelform but with a narrow limb 5–6 mm. broad, lavender or purplish, tube paler; capsules globose, faintly apiculate at apex, 5–6 mm. in diameter; seeds obovoid, dark brown, minutely faveolate-pitted, sublustrous.

Arid plains, hillsides, and washes, Sonoran Zones, Yavapai County to Pima County, Arizona, east to Texas, and south to central Mexico and Baja California. Aug.–Oct.

**5. Ipomoea leptotoma** Torr., Bot. Mex. Bound. 150, 1859

*Ipomoea radiatifolia* Kell., Calif. Acad. **7:** 163, 1877.
*Pharbitis leptoloma* Peter, in Engler & Prantl, Nat. Pflanzenf. **4**[3a]: 32, 1897.
*Ipomoea divergens* House, Muhlenbergia **3:** 40, 1907.

Herbaceous annual with slender, twining or ascending, moderately branching stems 1–8 dm. long, these inconspicuously lined longitudinally with 8–10 very low, narrow ridges, glabrous or nearly so; petioles slender, 2–5 cm. long, glabrous, very sparsely hirsute, or sometimes with a few scattered conical processes 0.5–1 mm. long; leaf blades pedately 3–5-parted into linear lobes 1–3 (rarely 4) mm. wide, 2–8 cm. long, basal pair of lobes usually again 3-parted near base, lower two lobes of this trio shorter than upper one, leaf segments glabrous or sparsely and irregularly hirsute; peduncles 3–7 cm. long, slender, 1–3-flowered; bracts linear-subulate, 2–5 mm. long, persistent; sepals narrowly ovate-lanceolate to linear-lanceolate, attenuate at apex, 8–12 mm. long, margins slightly scarious, glabrous; corolla funnelform, 2.5–3.5 cm. long, bluish to purple or rarely white, tube and lower throat usually whitish or pale cream; capsule ovoid, about 4 mm. wide, 5–6 mm. long, obtuse or rounded at apex; seeds oblong, 3–3.5 mm. long, dark brown, minutely puberulent on angles below and around circular hilum.

Hillsides, arroyos, and grassy plains, Sonoran Zones, southern Arizona and New Mexico south into Oaxaca. Aug.–Dec.

**Ipomoea leptotoma f. wootonii** (Kelso) Wiggins, Contr. Dudley Herb. **4:** 21, 1950, (*Ipomoea lepotoma* var. *wootonii* Kelso, Rhodora **39:** 151, 1937) differs from typical representatives of the species only in bearing simple spreading hairs on the stems, leaves, and sometimes on the calyces. The degree of hairiness passes through a gradual transition from the completely glabrous condition to noticeable but relatively sparsely arranged, spreading hairs. This scarcely warrants varietal rank.

Range and habitat of the species.

### 6. Ipomoea pedicellaris Benth., Bot. Sulph. 135, 1844

Slender, twining vine climbing to 8–10 m., often covering shrubs and small trees with a continuous blanket of foliage and flowers; stems and herbage glabrous or essentially so; petioles slender, 2.5–7 cm. long or more; leaf blades cordate, 3–6 cm. wide, 3–8 cm. long, sinus broad and open, 1–1.5 cm. deep, apex acuminate, dark green above, paler beneath, veins inconspicuous; peduncles 1.5–3.5 cm. long, 1–5-flowered; bracts minute, caducous; pedicels stout, subfleshy, two to three times as thick as peduncle, 3–5 cm. long, glabrous; sepals oblong or oblong-ovate, 7–8 mm. long, outer ones subfleshy, muricate-warty at least on lower half, apiculate, inner ones thinner, rounded at apex, margins nearly scarious; corolla 8–10 cm. long, funnelform, purple, limb 6–8 cm. broad; capsule oblong, 7–8 mm. in diameter, 15–18 mm. long, glabrous; seeds dull brownish black, ovoid, 8–9 mm. long, finely puberulent with short yellowish hairs.

Plains, valleys, and hillsides, Lower Sonoran to Tropical Zones, central Sonora to Guerrero. Aug.–Nov.

### 7. Ipomoea peninsularis Brandegee, Zoe 5: 168, 1903

Slender, loosely twining vine from a tuberous root; herbage glabrous or sparsely puberulent; petioles slender, 1–5 cm. long; leaf blades cordate-ovate, 2.5–4 cm. wide, 4–6 cm. long, acuminate at the apex, thin, sinus closed to narrowly deltoid, veins very slender, lamina thin, glabrous or sparsely puberulent along bases of veins; peduncles subfiliform to slender, 3–6 cm. long, 1–3-flowered; bracts minute, about 1 mm. long; pedicels 10–12 mm. long; sepals ovate, subequal, 4–5 mm. long, slightly tuberculate along midrib near base, scarious along upper margin, apiculate at apex; corollas funnelform, 1.5–1.8 cm. long, pale violet, limb 12 mm. broad; capsules ovoid-apiculate, 7–9 mm. long; seeds dark brown, 5–6 mm. long, minutely puberulent, angulate in cross section, ovoid.

Hillsides and along watercourses, Lower Sonoran and Tropical Zones, southern Baja California. Oct.–Feb.

### 8. Ipomoea pes-caprae (L.) R. Br., in Tuckey, Narr. Exped. Zaire 477, 1818

*Convolvulus pes-caprae* L., Sp. Pl. 159, 1753.
*Convolvulus brasiliensis* L., loc. cit.
*Ipomoea biloba* Forsk., Fl. Aeg.-Arab. 44, 1775.
*Ipomoea brasiliensis* Sweet, Hort. Sup. Lond. 35, 1818.
(For additional synonymy see Blumea **3:** 532–539, 1940.)

Perennial creeping plant with stout stems 1–10 m. long and milky, astringent juice; branches subsucculent, often to 1 cm. in diameter; herbage glabrous; petioles stout to slender, 1–7 cm. long; leaf blades orbicular-cordate, orbicular, broadly ovate or obcordate, base obtuse or rounded or cordate, 2.5–8 cm. broad, 5–10 cm. long, leathery, glabrous, veins rather conspicuous beneath; peduncles stoutish, 1–10 cm. long, 1- to several-flowered; bracts lanceolate-ovate, 3–5 mm. long, caducous; pedicels 1–3 cm. long, slender in flower but usually greatly thickened in fruit; sepals unequal, outermost broadly oblong, 4–5 mm. long, inner successively longer, 8–10 mm. long, margins thin but not scarious, backs minutely and faintly granulate or smooth; corollas broadly funnelform, 2.5–3 cm. long, purplish, limb about as broad; capsules ovate, about 2 cm. long, 12–15 mm. wide, valves nearly woody; seeds angulate along inner edge, rounded on back, 8–10 mm. high, densely pubescent with brownish spreading hairs 0.5–1 mm. long.

Along the upper beach and on coastal sand dunes, Lower Sonoran to Tropical Zones, circumtropical. In the Sonoran Desert area it extends from the vicinity of Magdalena Bay south. Dec.–April.

**9. Ipomoea perlonga** Robinson, Proc. Amer. Acad. **29:** 319, 1894

A vigorously growing vine twining and scrambling over shrubs and trees to 5 m. or more; stems often stoutish, frequently these and petioles beset with conical processes 0.5–2 mm. long; herbage minutely silky-strigose to glabrous; petioles slender, 5–12 cm. long; leaf blades ovate-cordate to orbicular-cordate, 6–15 cm. wide, to 20 cm. long, dark green above, paler and sometimes silvery-strigillose or glabrous beneath, sinus often closed by overlapping rounded lobes, apex deltoid-acuminate, veins moderately conspicuous beneath; peduncles often surpassing leaves, to 30 cm. long or more, 5–15-flowered; bracts linear-lanceolate, 3–7 mm. long, caducous; pedicels 1–1.5 cm. long, glabrous in our material; sepals equal or essentially so, ovate, acute, 5–6 mm. long, 3–4.5 mm. wide, mucronulate, margins thin; corolla light blue, about 4 cm. long, minutely puberulent without on plicae; capsules ovoid, apiculate, 10–12 mm. high, on deflexed pedicels, valves thin but firm, finely reticulate; seeds ovoid-angulate, 5–6 mm. high, dull brownish black, minutely scurfy.

Along arroyos and on gentle slopes, Lower Sonoran to Tropical Zones, central Sonora and Baja California to Central America. Sept.–Nov.

**10. Ipomoea viscosa** Wiggins, Contr. Dudley Herb. **4:** 21, *pl. 2, figs. 1–4,* 1950

A slender vine with trailing stems 1–2 m. long, 1.5–2 mm. in diameter, striate, terete in cross section, glabrous or very sparsely hirsute immediately adjacent to petioles and peduncles; petioles slender, 1–5 cm. long, striate, shallowly canaliculate on upper surface, sparsely hirsute at base or glabrous throughout; leaf blades narrowly ovate-cordate or -subhastate, 2–4 cm. broad, 3–6 cm. long, sinus broad, 5–10 mm. deep, gradually acuminate toward apex, thin, light green, glabrous, 5–7-veined at base; peduncles slender, surpassing leaves, 3–7 cm. long, sparsely hirsute at base, very viscid from a point about 5–8 mm. above base almost to apex, 1–5-flowered, often forking and bearing a solitary flower in fork, two branches 2–4 cm. long, viscid; bracts ovate, 1–1.5 mm. long, attached immediately beneath sepals and similar to them, persistent; pedicels stoutish, 7–15 mm. long, about twice as thick at apex as at base, glabrous and nonviscid, striate; sepals subequal, broadly ovate, 4–5.5 mm. long, 2.5–3.5 mm. wide, obtuse to rounded at membranous apex, minutely mucronulate, glabrous; corolla blue, funnelform, 13–18 mm. long, limb nearly as broad, slightly angulate; stamens included, 8–11 mm. long, filaments sparsely hairy at base; anthers ovate, about 1 mm. long; stigma capitate, shallowly 2-lobed; capsule ovoid to oblong-ovoid, mucronate, 10–12 mm. long, 4.5–6 mm. in diameter, 2-celled, glabrous, valves thin but firm, minutely striate-reticulate; seeds narrowly ovoid, 5–6 mm. long, about 3 mm. wide, mottled dull black and light brown, minutely puberulent with tufts of dark and pale brown hairs corresponding to mottled pattern of seed coat.

Mostly in shade of shrubs or clambering near and through them, on desert plains and along stream banks, Lower Sonoran Zone, central Sonora into Sinaloa. Aug.–Oct.

### 11. Ipomoea scopulorum Brandegee, Zoe **5:** 169, 1903

Perennial, scrambling or twining, somewhat woody vine with stoutish stems 2.5–3.5 mm. in diameter, 1–2 m. long, sparsely to densely pubescent throughout with a mixture of simple, forked, and few-rayed, stellate hairs; petioles slender, 2–5 cm. long, moderately densely pubescent; leaf blades cordate-ovate, 4–6 cm. broad, to 8 cm. long, sinus shallow and broad, or some leaves nearly truncate at base, acute to short-acuminate at apex, entire, stellate-puberulent on both sides; peduncles stoutish, 2–4 cm. long, 1–4-flowered; bracts subulate, 1–3 mm. long, caducous; pedicels 2–3 cm. long; sepals unequal, 7–10 mm. long, 4–6 mm. wide, broadly oblong to broadly ovate, rounded or obtuse at apex, slightly scarious on upper margins; corolla funnelform-rotate, tube 4–5 cm. long, limb 6–7 cm. broad, slightly angulate, angles apiculate; capsule ovoid, 8–10 mm. wide, 10–14 mm. high, leathery; seeds brownish, 6–8 mm. high, silky-comose with white or tawny hairs 8–10 mm. long on 2 dorsilateral angles, otherwise glabrous.

Among rocks, Lower Sonoran and Tropical Zones, southern Baja California and in the vicinity of Mazatlán. Sept.–Dec.

### 12. Ipomoea iodantha Brandegee, Univ. Calif. Pub. Bot. **4:** 383, 1913

Stems slender, sparsely hirsute with long, slender, spreading hairs or glabrate, strongly twining; roots tuberous; petioles 8–25 mm. long, sparsely hirsute; leaf blades cordate, abruptly short-acuminate at apex, sinus 1.5–2 cm. deep, about one-half as wide or less, margins entire, smoothly rounded or curved, not lobate; peduncles 2–3 cm. long, 1- to several-flowered, sparsely hirsute to glabrous; bracts lanceolate, 12–20 mm. long, pilose, subfoliaceous; pedicels 1–3 mm. long, glabrous or nearly so; sepals subequal, 15–20 mm. long, 2.5–3 mm. wide, narrowly lanceolate, basal 4–5 mm. densely hirsute with slender, spreading, simple hairs 3–5 mm. long, remainder glabrous, acute to attenuate-acuminate at apex, thin but not membranous; corolla funnelform, 2.5–3.2 cm. long, purplish or bluish, limb about 2 cm. broad, tube pale cream or whitish; capsule ovoid, 7–8 mm. high.

Along arroyos and on hillsides, Lower Sonoran to Tropical Zones, southern Baja California and in Sinaloa. Not definitely reported from the Sonoran Desert proper but to be expected in the Sierra de la Giganta north of La Paz. Sept.–Jan.

### 13. Ipomoea purpurea (L.) Lam., Tab. Encycl. **1:** 466, 1793

*Convolvulus purpureus* L., Sp. Pl. ed. 2, 219, 1762.
(For synonymy see Ann. N.Y. Acad. Sci. **18:** 199, 1908.)

Annual twining vine often climbing 2–5 m. over shrubs; stems, leaves, and calyces sparsely to densely hirsute or infrequently glabrous; petioles slender, 5–10 cm. long;

leaf blades broadly ovate, cordate, acute or commonly acuminate at apex, deeply cordate at base, 4–10 cm. broad, slightly longer, dark green above, paler beneath; peduncles slender, 2–7 cm. long, 1–5-flowered; bracts linear-subulate or filiform, 3–5 mm. long, persistent; pedicels 1–2 cm. long, often deflexed in fruit; sepals subequal, lanceolate to oblong-lanceolate, 1.5–4.5 mm. wide, 12–15 mm. long, acute at apex, densely hirsute and somewhat thickened toward base, thin and sparsely puberulent on upper two-thirds; corollas funnelform, blue, purple, pink, variegated or rarely white, 5–7 cm. long, limb 4–5 cm. wide; capsules depressed-globose, usually 3-celled, 7–9 mm. in diameter, glabrous; seeds dark brown, 5–6 mm. long, smooth and glabrous, dull.

Naturalized or escaped from cultivation in many parts of temperate North America, occurring spasmodically as an escape in waste places along irrigation ditches and streams, southern Arizona southward. A native of tropical America. May–Nov.

**14. Ipomoea minutiflora** (Mart. & Gal.) House, Muhlenbergia **5:** 71, 1909

*Convolvulus minutiflorus* Mart. & Gal., Bull. Acad. Brux. **12**²: 262, 1845.
*Ipomoea filipes* Benth., in Meissn., in Mart., Fl. Bras. **7:** 274, 1869.

Slender, weakly twining or scandent annual with several stems 2–6 dm. long from a light taproot; stems and herbage sparsely hirsute with slender, simple, spreading to subappressed hairs nearly confined to minute angles of stems and petioles but scattered on lamina of leaves; petioles slender, 0.5–5 cm. long; leaf blades broadly ovate-cordate, 1.5–4 cm. broad, often slightly wider than long, sinus broad and open, apex slightly deltoid-acuminate; peduncles subfiliform, 1–4 cm. long, 1–3-flowered; bracts subulate, about 1 mm. long, persistent; pedicels slender, 2.5–6 mm. long, curved or deflexed in fruit; sepals equal, lance-ovate, 3–3.5 mm. long, sparsely hirsute with slender spreading hairs; corolla deep yellow, 4–5 mm. long, campanulate-funnelform; capsules broadly ovoid to globose, 3–4 mm. in diameter; seeds 2.5–3 mm. long, glabrous or minutely scurfy, dark brown.

Along arroyos and stream banks and in open forests, margins of Lower Sonoran to Tropical Zones, central Sonora south to tropical South America. Sept.–Nov.

**15. Ipomoea jicama** Brandegee, Proc. Calif. Acad. II, **2:** 188, 1889

*Ipomoea odorata* Eastw., Leafl. West. Bot. **3:** 257, 1943.

Stems slender, woody at base, strongly twining, epidermis yellowish, under a lens minutely but very distinctly and regularly fenestrate-reticulate, glabrous throughout; roots tuberous, fleshy; petioles slender, 5–30 mm. long; leaf blades broadly ovate-cordate to angulate-pentagonal, shallowly and coarsely angulate-toothed along lower half, acuminate at apex, shallowly and broadly cordate at base, 1.5–3.5 cm. wide, 2–4 cm. long, very thin; peduncles rather stout, mostly 5–8 mm. long, apparently 1-flowered; bracts subulate, 1–1.5 mm. long, caducous; pedicels slightly more slender than peduncle, 1–3 cm. long, nearly twice as thick at apex as at base; sepals very unequal, 8–20 mm. long, outermost shortest, lance-ovate, acute and bearing a terminal apiculation 1–1.5 mm. long, thin, but margins not membranous, inner ones slightly wider, more markedly membranous, with more slender, slightly longer apiculations; corolla broadly funnelform, pinkish to purplish, 6–7 cm. long, limb 5–6 cm. broad; capsules depressed-ovoid, 7–9 mm. in diameter, 6–8 mm. high, tipped by persistent base of style; seeds broadly obovoid, 5–6 mm. high, somewhat flattened on inner face, dark brown, minutely scurfy or reticulate-alveolate, puberulent near hilum.

Among volcanic rocks on mesas and along streams and on arid plains, Lower Sonoran Zone, central Baja California south to the Cape Region. Sept.–Dec.

**16. Ipomoea triloba** L., Sp. Pl. 161, 1753

(For synonymy see Ann. N.Y. Acad. Sci. **18:** 253, 1908.)

Annual with slender, twining or ascending stems, often climbing several meters, sparsely hirsute to glabrate; petioles slender, mostly shorter than blades, latter ovate-cordate or hastately 3-lobed, 4–10 cm. long, acuminate at apex, sparsely hirsute or glabrous; sinus broad, comparatively shallow; peduncles commonly longer than leaves, to 1 dm. long, 1–5-flowered; bracts linear-lanceolate, 2–3 mm. long, caducous; pedicels 6–12 mm. long; sepals nearly equal, oblong, lanceolate, or sometimes outer ones suborbicular, 6–9 mm. long, apiculate at apex, ciliate or sparsely hirsute on back; corolla 1.5–2 cm. long, pink, lavender, or purple, limb 10–15 mm. broad; capsules subglobose, 6–8 mm. in diameter, sparsely pilose with slender spreading hairs or glabrous; seeds dark brown, glabrous or sparsely puberulent on angles and near hilum.

Roadsides, grassy areas, and along stream banks, Lower Sonoran to Tropical Zones, Pima County, Arizona, and southern Florida south into tropical America. Sept.–Feb.

**17. Ipomoea hirsutula** Jacq. f., Eclog. Pl. **1:** 63, *pl. 44,* 1811–16

(For synonymy see Ann. N.Y. Acad. Sci. **18:** 199, 1908.)

Annual twining vine with slender, hirsutulous to glabrate stems and sparsely hirsute to glabrate foliage; petioles slender, much exceeding blades, to 12 cm. long or more; leaf blades ovate-cordate or hastately 3-lobed, 2.5–5 cm. wide, 3–7 cm. long, short-acuminate at apex, sinus broad and shallow or base sometimes truncate, when 3-lobed middle lobe ovate-acuminate, lateral ones asymmetrically ovate, lower side more strongly curved than upper; peduncles 2–6 cm. long, 1–3-flowered; bracts linear-subulate, 3–4 mm. long, deciduous; pedicels 1–2 cm. long, slightly broader at apex than at base; sepals narrowly ovate-lanceolate to lanceolate, 12–15 mm. long, somewhat thickened and densely hirsute at base, more sparsely so with appressed hairs above, acute, erect; corollas funnelform, 2.5–4 cm. long, purple with lighter plicae, limb 1.5–3 cm. in diameter; capsules ovoid, 6–7 mm. high; seeds ovoid, 4–5 mm. high, dark brown, glabrous, smooth and sublustrous.

A wayside weed and along arroyos and ditches, Lower Sonoran to Tropical Zones, southern Arizona to tropical South America. Not common in our area. The first leaves produced are ovate-cordate and not lobed, the later ones deeply 3-lobed. Aug.–Jan.

**18. Ipomoea hederacea** (L.) Jacq., Ic. Pl. Rar. **1:** *pl. 36,* 1781–86

*Convolvulus hederaceus* L., Sp. Pl. 154, 1753.
*Ipomoea desertorum* House, Ann. N.Y. Acad. Sci. **18:** 203, 1908.
(For additional synonymy see House, op. cit. 202–203.)

A vigorously growing annual with slender twining stems climbing to 2 m. or more, hirsute or retrorsely hirsutulous throughout; petioles slender, 1.5–6 cm. long; leaf blades ovate-orbicular in outline, deeply 3-lobed, to 10 cm. long, shallowly to deeply cordate at base, lobes ovate, acute to short-acuminate, entire or the lateral ones sometimes shallowly repandly dentate; peduncles 5–15 mm. long, 1–3-flowered; bracts linear-lanceolate, 4–8 mm. long, persistent; sepals subequal, ovate with a caudate-attenuate terminal appendage, 1.5–3 cm. long, basal broader part hirsute with spreading hairs 3–4 mm. long, caudate-attenuate part glabrous or ciliate-margined, spreading or recurved at tips; corollas funnelform, white to pale blue or purple, 2.5–4 cm. long; capsule depressed-globose, usually 3-celled, 9–10 mm. in diameter, glabrous; seeds ovoid, angulate, 5–6 mm. high, black, with irregular patches of closely appressed, glistening, grayish hairs.

A native of tropical and subtropical America, naturalized from Arizona to Florida and north to Maine and Nebraska, in our area occurring in southern Arizona and central Sonora. May–Nov.

### 19. **Ipomoea arborescens** (Humb. & Bonpl.) G. Don, Gen. Hist. Pl. **4:** 267, 1838

*Convolvulus arborescens* Humb. & Bonpl. ex Willd., Enum. Hort. Berol. 204, 1809.
*Argyreia ?oblonga* Benth., Bot. Sulph. 133, 1844.
*Convolvulus quahutzehuatl* Sessé & Moc., La Naturaleza [Pl. Nov. Hisp.] II, **1:** app. 23, 1887.
*Ipomoea arborescens* var. *glabrata* Gentry, Carnegie Inst. Wash. Pub. No. **527:** 213, 1942. Not *I. murcodes* var. *glabrata* A. Gray, 1891.
*Ipomoea arborescens* var. *pachylutea* Gentry, loc. cit.

A large tree to 12 m. tall with smooth white bark and stoutish puberulent twigs; leaves ovate, rounded, obtuse, acute or attenuate-acute at apex, truncate to shallowly and broadly cordate at base, 3–8 cm. broad, 8–20 cm. long, sparsely pubescent on both sides, slightly more densely so beneath than above, about 8–15 lateral veins on each side, these narrow and inconspicuous; petioles slender, 2–7 cm. long, more densely pubescent than blades; flowers appearing before leaves in few- to many-flowered panicles, usually only one flower of a cluster opening at a time; peduncles 3–6 mm. long, puberulent; bracts small, caducous; pedicels 1–2 cm. long, densely puberulent; sepals broadly ovate, subequal, 6–10 mm. long, obtuse or rounded, closely but finely puberulent without, strigose-silky within; corolla white, fleshy, 4–5 cm. long and wide, tube tinged with greenish or cream; capsule narrowly ovate, 15–22 mm. long, firmly leathery, 4-valved; seeds dark reddish brown, 8–10 mm. long, bearing a fringe of white or yellowish, silky hairs 10–15 mm. long on upper two-thirds of two exterior angles.

Gravelly or rocky hillsides, Lower Sonoran to Tropical Zones, central Sonora to Morelos, Vera Cruz, and El Salvador. Nov.–March.

This species is either very polymorphic, or there are several entities occurring within the range as delimited above. At present, however, I have insufficient data to warrant recognizing these entities as nomenclatorial units. Inasmuch as some of the variants occur well outside the limits of the Sonoran Desert, I prefer to treat the whole complex as a single, polymorphic species until such time as intensive study, supplemented by field observations, may provide more reliable data for treating the arborescent species of Mexican Ipomoeas.

## 9. **CALONYCTION** Choisy, Mém. Soc. Genève **6:** 441, 1833

Twining vines, usually with glabrous stems and foliage. Leaves entire or lobed, often cordate. Flowers large, white or purple, solitary or in few-flowered cymes borne by axillary peduncles. Sepals 5, unequal or subequal. Corolla salverform, tube long and comparatively slender, limb broad, shallowly 5-angled. Stamens slightly exserted or barely included. Stigma capitate, bilobed. Ovary and capsule 2-celled, 4-seeded. Capsule dehiscing longitudinally. Seeds large, dark brown or black, glabrous.

Corolla purple, its tube 3–5 cm. long; seeds glabrous .................. 1. *C. muricatum*
Corolla white, its tube 8–12 cm. long; seeds minutely puberulent:
Stems often armed with short recurved prickles; outer sepals tipped with slender but sturdy, hornlike appendages; leaves entire ....................... 2. *C. aculeatum*
Stems unarmed; outer sepals without appendages; leaves often coarsely and irregularly dentate toward base ............................................. 3. *C. tastense*

**1. Calonyction muricatum** (L.) G. Don, Gen. Hist. Pl. **4:** 264, 1838

*Convolvulus muricatus* L., Mant. 44, 1767.
*Ipomoea muricata* Jacq., Pl. Hort. Schönbr. **3:** 40, *pl. 323,* 1798.
*Ipomoea spinulosa* Brandegee, Zoe **5:** 169, 1903.

A glabrous vigorous vine climbing to 8–10 m.; stems usually armed with few to many short, stout, recurved prickles 1–4 mm. long; juice milky; petioles to 15 cm. long or more; leaf blades broadly cordate-ovate to orbicular-cordate, 5–16 cm. long, abruptly acuminate, deeply cordate at base, slightly paler beneath than above; peduncles stout, 1–6 (rarely to 15) cm. long, 1–3- (rarely 5-) flowered; pedicels conspicuously thickened above, 2–2.5 cm. long, sometimes nearly 1 cm. in diameter immediately below apical constriction; outer sepals narrowly lanceolate, 1–1.8 cm. long, tips subulate-cuspidate, inner ones broadly oblong-ovate, only slightly longer than outer but cusp shorter; corolla pale lavender near edges of limb, shading to deep purple in throat, tube usually whitish, 3–5 cm. long, limb spreading 3.5–4.5 cm.; stamens about equaling throat; capsules broadly ovoid, 1.5–2 cm. broad, 2–2.5 cm. long, mucronate at apex, glabrous; seeds 10–14 mm. long, dark brown, glabrous.

Hillsides, canyons, and arroyos, Lower Sonoran Zone to Tropical Zone, southern Baja California, central Sonora, and Chihuahua south to tropical South America. Sept.–Jan.

**2. Calonyction aculeatum** (L.) House, Bull. Torrey Club **31:** 590, 1904

*Convolvulus aculeatus* L., Sp. Pl. 155, 1753.
*Ipomoea alba* L., op. cit. 161.
*Ipomoea bona-nox* L., Sp. Pl. ed. 2, 228, 1762.
*Calonyction speciosum* Choisy, Mém. Soc. Genève **6:** 441, 1833.
*Calonyction bona-nox* Bojer, Hort. Maurit. 228, 1837.
*Ipomoea aculeata* Kuntze, Rev. Gen. Pl. **2:** 442, 1891.

Herbaceous or suffrutescent vine climbing over shrubs and trees several meters above ground, glabrous; stems striate and often armed with short, stout, recurved prickles 1–3 mm. long; petioles slender, straight or somewhat tortuous, 3–6 cm. long; leaf blades ovate-cordate to suborbicular-cordate, 3–10 cm. broad, 4–15 cm. long, acuminate, broadly cordate at base, entire or obscurely 3–5-lobed, paler beneath than above, glabrous; peduncles stout to slender, 1–10 cm. long, 1- to few-flowered; pedicels slightly thickened in youth, 1–2 cm. long, greatly thickened in age; outer sepals oblong, midrib extended as a firm, slender to stout, hornlike appendage, 1–1.5 cm. long, thinner part slightly shorter than appendage, inner sepals broadly oblong, obtuse or rounded, 1–1.5 cm. long; corolla white, tube 8–12 cm. long, limb 7–10 cm. broad; capsule broadly ovoid, 2–2.5 cm. broad, 2.5–3.5 cm. long, acuminate at apex, glabrous; seeds dark brown to black, shining, minutely puberulent.

Along arroyos, canyons, and on hillsides and valleys, Lower Sonoran to Tropical Zones, southern Baja California and Sinaloa to tropical South America. Widely distributed in the tropics of both hemispheres. Nov.–Feb.

**3. Calonyction tastense** (Brandegee) House, Bull. Torrey Club **33:** 318, 1906

*Ipomoea tastensis* Brandegee, Zoe **5:** 168, 1903.

A woody vine with branching, glabrous, unarmed stems climbing over trees and shrubs to 10 m. or more; petioles slender, 3–7 cm. long, straight or "curved as if used in climbing"; leaf blades glabrous, cordate-ovate, long-acuminate, 2–6 cm. wide, 4–8 cm. long, sometimes entire but usually bearing a few coarse, deltoid to deltoid-acuminate teeth 2–5 mm. long along outer margin near base, sinus broad, rounded; peduncles short and stoutish, 1–3 cm. long, mostly 1-flowered; pedicels striate, about 2 cm. long, nearly three times as thick at apex as at base; sepals oblong to oblong-lanceolate, 1.5–2.5 cm. long, outermost shortest, abruptly short-acuminate to long-acuminate, 3.5–6 mm. wide, thin, margins scarious; corolla white, narrowly funnelform tube 8–12 cm. long, limb 8–10 cm. broad, lobes shallow, abruptly acuminate; stamens inserted near base of throat, exserted about 1 cm.; capsule nearly globose, 1.5–2 cm. in diameter; seeds about 1 cm. long, minutely puberulent.

Rocky hillsides and canyons, Lower Sonoran and arid parts of Tropical Zones, southern Baja California. Sept.–Jan.

## 10. QUAMOCLIT Moench, Meth. 453, 1794

Annual or perennial, twining vines with alternate, entire, lobed, or pinnately parted leaves. Flowers axillary, solitary or in few- to several-flowered cymes. Sepals 5, equal or subequal, membranous or herbaceous. Corolla (in ours) scarlet, narrowly salverform, limb narrow, and much shorter than tube, shallowly 5-lobed. Stamens 5, somewhat exserted; filaments filiform. Anthers linear, not twisted. Ovary 2-celled or falsely 4-celled, 4-ovuled. Style slender, exserted; stigma capitate. Capsule subglobose to ovoid, usually 4-seeded. Seeds dark brown to black, glabrous or sparsely puberulent in irregular patches.

Leaves entire or 3–5-lobed; sepals conspicuously acuminate; capsules subglobose, 7–8 mm. in diameter:
- Leaves entire, cordate or subsagittate ..................... 1. *Q. coccinea* var. *coccinea*
- Leaves deeply 3–5-lobed ............................. 1*a*. *Q. coccinea* var. *hederifolia*

Leaves pinnately divided into filiform or narrowly linear segments; sepals obtuse or short-mucronate; capsules ovoid, 7–10 mm. long ........................ 2. *Q. pinnata*

**1. Quamoclit coccinea** (L.) Moench, Meth. 453, 1794, var. **coccinea**

*Ipomoea coccinea* L., Sp. Pl. 160, 1753.

(About a dozen specific epithets have been applied to this attractive plant under *Ipomoea, Convolvulus,* and *Quamoclit.* A monographic study of the genus, impossible here, would be necessary to properly dispose of all of them, so only the name-carrying synonym is listed. The same practice is used in dealing with the two following entities.)

A freely climbing, rather slender, glabrous annual with showy, bright red flowers; stems often several meters long; petioles slender, 2.5–10 cm. long; leaf blades ovate to deltoid-cordate, 3–8 cm. wide, to 10 cm. long, entire or somewhat sagittately lobed basally, acute to acuminate, truncate to broadly and shallowly cordate at base; peduncles slender, to 2.5 dm. long, much surpassing leaves, 1- to several-flowered; pedicels 3–15 mm. long; sepals oblong, obtuse, about 2 mm. wide (outer nearly as broad as long), 4–5 mm. long, an infraterminal, subterete, horn-like appendage 4–8 mm. long borne by each one, tips of these curved inward against corolla tube; corolla scarlet, sometimes orange-tipped, tube 2–4 cm. long, limb cup-shaped, 1–2 cm. broad, very shallowly angulate; stamens barely exserted; capsule subglobose, 7–8 mm. in diameter, mucronate at apex, glabrous; seeds dark brown, 5–6 mm. long, glabrous or irregularly puberulent.

Climbing over shrubs, rocks, and trees along arroyos, in plains, hillsides, and mesas, Lower Sonoran into Upper Sonoran and Tropical Zones, Coconino and Cochise Counties, Arizona, to Missouri, Pennsylvania, and Florida and south into tropical America. In Baja California from near San Matias Pass south. Aug.–Dec.

**1*a*. Quamoclit coccinea** var. **hederifolia** (L.) House, Ann. N.Y. Acad. Sci. **18:** 262, 1908

*Ipomoea hederifolia* L., Sp. Pl. ed. 2, 229, 1762.
*Ipomoea coccinea* var. *hederifolia* A. Gray, Syn. Fl. N. Amer. **2**[1]**:** 209, 1878.

Similar to var. *coccinea* in all respects except shape and depth of lobing of leaves, these in var. *hederifolia* deeply 3-lobed, or sometimes 5-lobed, each lobe usually narrower at base than at middle, long-acuminate, two lateral lobes often obscurely lobed on outer margin so general shape of whole blade is somewhat sagittate.

Similar habitats to those occupied by the species, Lower Sonoran to Tropical Zones, Arizona to Texas, Baja California, and tropical America.

**2. Quamoclit pinnata** (Desr.) Bojer, Hort. Maurit. 224, 1837

*Ipomoea quamoclit* L., Sp. Pl. 159, 1753.
*Convolvulus pennatus* Desr., in Lam., Encycl. **3:** 567, 1791–92.
*Quamoclit quamoclit* Britt., in Britt. & Brown, Ill. Fl. **3:** 22, *fig. 2943*, 1898.

Slender-stemmed, twining vine several meters long; stems glabrous; leaves broadly ovate in outline, 2–4 cm. wide, 3–8 cm. long, deeply divided into 8–16 pairs of linear to subfiliform divisions, segments entire and glabrous or margins sparsely beset with distant, forwardly pointing, stoutish hairs; petioles linear, 5–25 mm. long, resembling a division of leaf blade; peduncles 3–15 cm. long, 1- to several- (mostly 1–2-) flowered; pedicels slenderly clavate, 1.5–2.5 cm. long, faintly striate; sepals broadly oblong or the shorter broadly ovate, 3–6 mm. long, obtuse or rounded at apex, an infraterminal mucronulation sometimes present, margins subscarious; corolla scarlet, tube 2.5–3 cm. long, limb 1 cm. broad or less, lobes triangular or broadly deltoid-ovate; stamens barely exserted; capsule broadly ovoid, 7–10 mm. long, 5–6 mm. wide, obtuse at apex, glabrous; seeds dark brown, 4–5 mm. long, bearing irregularly convoluted lines of golden brown, minute hairs.

Over and among shrubs and brush, waste places, arroyos, and wooded hillsides, Lower Sonoran to Carolinian and Tropical Zones, Kansas and Virginia south to Florida, Texas, Sonora, and tropical America. Occurring in our area near Mátape, about 40 miles east of Hermosillo, Sonora, and south. Sept.–Jan.

## 11. EXOGONIUM Choisy, Mém. Soc. Genève 6: 443, 1833

Vigorously climbing, woody vines with usually entire, ample, alternate, petiolate leaves, these often deciduous before flowers open. Flowers solitary or in axillary cymes or racemes, each flower subtended by a bract, this frequently large and brightly colored. Sepals equal, subequal, or very unequal. Corolla tubular-salverform, tube long and slender, limb very narrow. Stamens inserted in corolla tube, somewhat exserted. Style exserted; stigma capitate, shallowly bilobed. Ovary 2-celled, 4-ovuled. Seeds usually hairy.

**1. Exogonium bracteatum** (Cav.) Choisy ex G. Don, Gen. Hist. Pl. **4:** 264, 1838

*Ipomoea bracteata* Cav., Ic. **5:** 51, *pl. 477,* 1799.
*Ipomoea spicata* H. B. K., Nov. Gen. & Sp. **3:** 112, 1819.
*Ipomoea cincta* Roem. & Sch., Syst. Veg. **4:** 254, 1819.
*Convolvulus obvallatus* Spreng., Syst. **1:** 595, 1825.
*Exogonium olivae* Bárcena, Viaje Cav. Cacahuam. 29, 1874.
*Convolvulus bractiflorus* Sessé & Moc., La Naturaleza [Pl. Nov. Hisp.] II, **1:** app. 23, 1887.

A glabrous woody vine to 10 m. tall or more from a large, watery, sweet, fleshy root; petioles slender, 1.5–2 cm. long; leaf blades ovate to ovate-cordate, 2.4–4 cm. broad, 3–9 cm. long, light green, acute to long-acuminate at apex, truncate to cordate at base, entire, usually deciduous by time flowers open; racemes 5–20 cm. long; bracts reniform or broadly ovate-cordate, 2–3.5 cm. long, acute to obtuse, usually infolded to enclose flower, prominently veined, pinkish, bright reddish, or magenta (or green when very young), more or less imbricated and reflexed; sepals broadly ovate, 4–5.5 mm. long, thin, rounded or obtuse at apex, margins subscarious; corollas 2.5–3.5 cm. long, red; stamens exserted 3–8 mm.; capsule narrowly ovoid, 6–9 mm. long, short-acuminate, glabrous; seeds 5–6 mm. long, dull black, minutely puberulent.

Climbing over shrubs and trees, Lower Sonoran to Tropical Zones, southern Baja California and Chihuahua to Morelos and Oaxaca. Dec.–May.

## Family 119. POLEMONIACEAE. Phlox Family

Annual or perennial herbs, shrubs, or vines. Leaves opposite or alternate, simple, cleft, lobed, or compound. Inflorescence paniculate, glomerate, a flat-topped cyme or compact heads, less commonly flowers solitary. Calyx membranous, chartaceous or herbaceous, 5-lobed, 5-toothed, or of 5 nearly distinct sepals; sinuses often hyaline part way to base. Corolla sympetalous, campanulate, funnelform, or salverform, 5-lobed; tube from short to very long; throat usually evident. Stamens 5, equally or unequally inserted in corolla tube or throat; anthers oval to sagittate, basifixed or versatile. Ovary mostly 3-celled; style simple; stigma lobes usually 3. Fruit a capsule, usually dehiscing loculicidally. Seeds 1 to many in each cell.

Leaves (in ours) simple, merely toothed or entire:
- Calyx slightly zygomorphic; corolla distinctly bilabiate; seeds small, ovoid, 1–several in each cell, mucilaginous when wet; leaves alternate .................... 1. *Loeselia*
- Calyx regular; corolla regular, not bilabiate; seeds larger, lenticular, unchanged when wet; leaves opposite .................................................... 2. *Phlox*

Leaves lobed, divided, pinnatifid or bipinnatifid (uppermost sometimes simple and entire):
- Cauline leaves alternate; leaf blades linear and entire, toothed, pinnatifid or pinnately dissected (palmately dissected in *Leptodactylon*):
  - Stems more or less scapose; upper cauline leaves markedly reduced in comparison with lower ones .................................................... 3. *Gilia*
  - Stems leafy throughout, or leaves more numerous near tips of branches, not scapose above:
    - Veins in corolla throat and lobes rarely anastomosing; stems leafy throughout; corolla salverform, but with broad tube ................................ 4. *Ipomopsis*
    - Veins in corolla usually anastomosing freely; stems more densely leafy at tips of branches, lower parts often leafless; corolla funnelform, often bilabiate:
      - Calyx zygomorphic; leaf blades pinnately dissected or linear; anthers sagittate .... ........................................................ 5. *Eriastrum*
      - Calyx regular, spiny; leaf blades pinnately toothed, teeth bristle-tipped; anthers oval ............................................... 6. *Langloisia*

Cauline leaves opposite (sometimes alternate in *Leptodactylon*):
 Plants rounded shrubs to 2 m. tall; leaf divisions acerose, rigid, pungently tipped ...... .......................................................... 7. *Leptodactylon*
 Plants annual herbs 3 dm. tall or less; leaf divisions rarely rigid, not pungently tipped .......................................................... 8. *Linanthus*

## 1. LOESELIA L., Sp. Pl. 628, 1753

Annual or perennial herbs or shrubs with alternate simple, entire to toothed leaves. Flowers in terminal, few-flowered cymes, solitary in axils or in axillary clusters; bracts small and barely noticeable or imbricated and conspicuous. Calyx slightly zygomorphic, herbaceous or with membranous sinuses, lobes fused nearly to apices, cuspidate. Corolla narrowly funnelform; limb 5-lobed but more deeply cleft on sides and more or less bilabiate. Stamens inserted equally near base of throat, exserted slightly beyond corolla throat but often barely equaling lobes; filaments glabrous. Ovary glabrous, ovoid, each cell 1–2- or several-ovuled. Style glabrous, exserted. Capsule about equaling calyx, often enveloped by bracts. Seeds small, narrowly to broadly winged, mucilaginous when wet.

Leaves lanceolate; flowers not obscured by bracts; corollas 15 mm. long; calyces bearing stipitate-glandular hairs .......................................... 1. *L. glandulosa*
Leaves ovate; flowers somewhat obscured by foliaceous setose-ciliate bracts; corollas about 10 mm. long; calyces not glandular .......................................... 2. *L. ciliata*

### 1. **Loeselia glandulosa** (Cav.) G. Don, Gen. Hist. Pl. **4:** 248, 1838

*Hoitzia glandulosa* Cav., Ic. **4:** 45, *pl.* 367, 1797.
(For additional synonymy see Contr. U.S. Nat. Herb. **23:** 1211, 1924.)

Stems slender, 1–5 dm. tall from a woody base, minutely floccose-puberulent throughout, younger stems often tinged with red or pink; leaves alternate, lanceolate, 3–25 mm. wide, 1.5–6 cm. long, acute at apex, cuneate and decurrent on short petiole (or sessile) at base, serrate on margins with each tooth mucronate, sparsely puberulent with simple and a few glandular hairs; flowers in 1–3-flowered axillary cymes; peduncles, bracts, pedicels, and calyces puberulent with simple and glandular, spreading hairs; bracts linear-lanceolate, 5–6 mm. long; calyces 6–7 mm. long, teeth 1–1.5 mm. long, cuspidate; corollas pink-lavender, narrowly funnelform-salverform, about 1.5 cm. long, lobes nearly or quite equaling throat and tube, rounded and entire; stamens and styles about equaling corolla lobes; capsule narrowly ovoid, about equaling calyx tube.

Rocky hillsides and canyon walls, Lower Sonoran to Tropical Zones, central Sonora to Chihuahua, Vera Cruz, and Chiapas. Sept.–April.

### 2. **Loeselia ciliata** L., Sp. Pl. 628, 1753

*Hoitzia aristata* H. B. K., Nov. Gen. & Sp. **3:** 128, 1818.
*Hoitzia loeselia* Spreng., Syst. **1:** 626, 1825.
*Loeselia involucrata* G. Don, Gen. Hist. Pl. **4:** 248, 1838.
*Hoitzia lupulina* Hook. & Arn., Bot. Beechey 441, 1841.

Slender, erect or ascending annual or short-lived perennial 1–3 dm. tall, only slightly woody at base or entirely herbaceous; stems very slender, 0.5–1 mm. in diameter, sparsely puberulent when young but soon glabrate, smooth, reddish or yellowish; leaves opposite or alternate; petioles 1–8 mm. long, puberulent, often narrowly winged above by decurrent blade; leaf blade ovate to ovate-lanceolate, 6–20 mm. wide,

12–35 mm. long, obtusely to abruptly acute-cuneate and somewhat decurrent on petiole at base, acute or obtuse at apex, serrate with each serration cuspidate, sparsely puberulent on both surfaces; flowers on slender axillary peduncles 0.5–3.5 cm. long; bracts foliaceous, ovate, 5–9 mm. wide, 7–12 mm. long, margins setose-ciliate but not serrate, nearly hiding flowers; calyx narrowly campanulate, about 5 mm. high, veins green, sinuses hyaline to white-scarious, teeth lance-attenuate, 1–1.5 mm. long; corolla lavender, about 10 mm. long, lobes equaling tube, puberulent near base; stamens and style about equaling lobes.

Arroyos, hillsides, mesas, and valleys in the open or among shrubs, Lower Sonoran to Tropical Zones, southern Baja California and from southern Sonora and Chihuahua to Vera Cruz and Guerrero, Mexico. Oct.–May.

## 2. PHLOX L., Sp. Pl. 151, 1753

Erect or diffuse, perennial or rarely annual herbs with entire, opposite or infrequently alternate leaves. Flowers perfect, regular, in terminal, corymbiform or paniculate cymes or sometimes solitary in the axils. Calyx tubular or tubular-campanulate, 5-ribbed, 5-toothed, or 5-lobed, these acute or acuminate; margin usually scarious. Corolla salverform with a slender tube; lobes obovate, orbicular, or obcordate. Stamens 5, unequally inserted in corolla tube, included; anthers ovate. Style slender, 3-lobed. Ovary ovoid, 3-celled, 3-valved. Capsule thin. Seeds ovoid, wingless or very narrowly winged, mucilaginous or unchanged in water.

**1. Phlox tenuifolia** E. Nels., Rev. W. N. Amer. Phloxes 27, 1899
*Phlox gilioides* A. Nels., Univ. Wyo. Pub. Sci. **1:** 127, 1926.

Slender tufted perennial, suffrutescent, 3–7.5 dm. tall, or in shrubs stems slender, over 1 m. tall, glabrous; leaves opposite below, alternate near inflorescence, linear to linear-lanceolate, 1–3 mm. broad, 1.5–5 cm. long, sessile, acute or mucronulately tipped, glabrous or with a few crinkled hairs at base or margins scaberulous-ciliate; inflorescence cymose; pedicels slender, 1–2 cm. long, sparsely to moderately stipitate-glandular; flowers fragrant; calyces 9–12 mm. long, villous on tube with spreading, thin-walled, simple hairs and a few glandular ones, lobes linear-subulate, erect or somewhat spreading, acicular and sparsely pubescent at apex; corollas narrowly funnelform, 1.5–2.5 cm. long, yellowish tube about equaling oblong-obovate, white to lavender lobes, these rounded at apex; capsules ovoid, 4–5 mm. long, obtuse at apex.

Canyons and mountain sides, Lower Sonoran Zone to lower edge of Transition Zone, central Arizona. Barely entering the Sonoran Desert in the lower part of Sabino Canyon, Pima County, and in eastern Maricopa County, Arizona. March–May, rarely again in Aug.–Oct.

**Phlox austromontana** Coville, Contr. U.S. Nat. Herb. **4:** 151, 1893, a lower, more compact, gray-green plant with rose or pink flowers and calyces with finer, less glandular pubescence, ranges into northern Arizona and has been reported from the Harquahala Mountains in Maricopa or Pima County, and may straggle into the upper edges of the Sonoran Desert along canyons and watercourses. May–Aug.

## 3. GILIA Ruiz & Pav., Fl. Peruv. Prod. 25, *pl. 4,* 1794*

Erect or spreading annual or perennial herbs branched from the base or above. Leaves alternate, entire or variously toothed, cleft or pinnately lobed. Inflorescence

*Contributed by Alva Day.

cymose, loosely paniculate to subglomerate, the flowers borne terminally or on pedicels mostly opposite a leaf or the inflorescence a congested, headlike, leafy-bracted cyme. Calyx segments equal, flanked on margins by hyaline tissue which is joined in the lower one-quarter to three-quarters to form a tube that is distended or ruptured by the growing capsule. Corolla funnelform to salverform or campanulate, tube very short to very long and often widening above to form a throat; lobes narrowly to broadly ovate or oblong, entire or denticulate. Stamens inserted in the sinuses of the corolla lobes or equally to unequally inserted in the tube. Capsule ovoid to subglobose or oblong-ovoid, dehiscing by splitting between the 3 valves. Seeds irregularly angled, 2–30 or more in each cell, some species mucilaginous when wet.

Stamens inserted in corolla tube:
- Leaves or leaf lobes linear:
  - Corolla 8–35 mm. long, pink:
    - Plants annual herbs; upper leaves entire; corolla 8–13 mm. long, lobes 4–5 mm. wide .......... 1. *G. palmeri* subsp. *palmeri*
    - Plants suffrutescent perennials; upper leaves with pair of lobes near base; corolla 25–35 mm. long, lobes 11–14 mm. wide .......... 1*a*. *G. palmeri* subsp. *spectabilis*
  - Corolla 3.5–6.2 mm. long, yellow .......... 2. *G. filiformis*
- Leaves ovate, coarsely serrate .......... 3. *G. latifolia*

Stamens inserted in sinuses of corolla lobes:
- Pubescence of lower leaves arachnoid, trichomes long, fine, and often densely tangled or woolly:
  - Stems at base glabrous and glaucous; leaves strap-shaped, sinuately toothed .......... 4. *G. sinuata*
  - Stems at base lightly to densely arachnoid-pubescent, not glaucous; leaves variously dissected or linear, but not as above:
    - Upper corolla throat purple, or yellow with purple spots; calyx glandular-pubescent .......... 5. *G. minor*
    - Upper corolla throat light violet, white or yellow, and not as above; calyx glabrous or arachnoid-pubescent, rarely sparsely glandular:
      - Capsules broad at base, narrowing and elongate above; calyx, if not green, coarsely red-blotched .......... 6. *G. flavocincta*
      - Capsules globular or ovoidal, not elongate; calyx, if not green, minutely red-dotted:
        - Flowers 8–10.5 mm. long; corolla throat full, limb pink .......... 7. *G. exilis*
        - Flowers 4–8 mm. long; corolla throat briefly flaring; limb white or pale blue .......... 8. *G. mexicana*
- Pubescence of lower leaves not arachnoid, trichomes coarse:
  - Calyx pubescent; capsules broadly ovoidal or globular:
    - Corolla tube longer than, to twice as long as calyx; trichomes of lower leaves more or less straight, translucent .......... 9. *G. scopulorum*
    - Corolla tube about equal to calyx; trichomes of lower leaves geniculate, white .......... 10. *G. stellata*
  - Calyx glabrous; capsules narrowly ovoidal .......... 11. *G. australis*

**1. Gilia palmeri** S. Wats., Proc. Amer. Acad. **24:** 61, 1889, subsp. **palmeri**

Annual herb 10–60 cm. tall with one to several slender, leafy stems from a slender taproot; stems stoutish, with short internodes, glabrous or upper part puberulent; leaves linear, 0.8–3 cm. long, 1 mm. wide, cuspidate at tip, entire, uppermost shorter and bractlike, axils of leaves filled with white hairs, blades glabrous or puberulent; inflorescence a simple raceme or loose panicle, flowers borne singly on puberulent pedicels 5–15 mm. long; calyx 4–5 mm. long, herbaceous parts linear, acuminate, lobes equal to or a little shorter than tube, puberulent, exceeded in length by mature capsule but becoming distended and usually splitting in sinuses with dehiscence of capsule; corolla 8–13 mm. long, violet to deep pink, tube shorter than calyx, lobes oval and denticulate, 4–5 mm. broad, 2–3 times as long as corolla tube; stamens equally inserted low in tube and included; capsule 6.5–8.6 mm. long, dehiscing by splitting from top to bottom between the 3 valves; seeds about 8 in each cell, mucilaginous when wet.

Rocky and sandy plains and hills, Lower Sonoran Zone, along the eastern coastal slopes, Baja California. Dec.–April.

**1*a*. Gilia palmeri** subsp. **spectabilis** A. Day, subsp. nov.

Suffrutescent herb; leaves 3–5 cm. long, upper with a pair of lobes at or near base; pedicels 2.5–4.5 cm. long; calyx 9–11 mm. long; corolla 2.5–3.5 cm. long, lobes broadly expanded (11–14 mm. wide), seven times as long as corolla tube.

Known only from the type locality where it occurs in abundance.

Type: Hills of Sierra del Volcán, 4 miles east of El Mármol, Baja California, *Wiggins* 7570 (Dudley Herbarium No. 263981), collected Feb. 13, 1935.

Herba suffrutescens, foliis 3–5 cm. longis; folia superiora ad basem jugum unum lobarum ferenda; pedicellus 2.5–4.5 cm. longus; calyx 9–11 mm. longus; corolla 2.5–3.5 cm. longa, lobis amplis fatulis 11–14 mm. latis quam tuba septies longioribus.

**2. Gilia filiformis** Parry ex A. Gray, Proc. Amer. Acad. **10:** 75, 1874

*Tintinabulum filiforme* Rydb., Fl. Rocky Mts. 698, 1917.

Erect annual, 3–18 cm. high; stem simple and glabrous at base, paniculately slender-branched and lightly glandular-pubescent above; leaves alternate, entire, narrowly linear to filiform, glabrous to lightly glandular-pubescent, 0.5–3 cm. long, lower leaves a little longer than upper leaves but not crowded or forming a basal rosette; flowers borne almost throughout plant on slender, glandular pedicels, each opposite a leaf; calyx 2.3–4 mm. long, glabrous, lobes three to four times as long as shallow tube, acuminate, hyaline-margined almost to acute tips; corolla campanulate, yellow, 3.5–6.2 mm. long, one and one-third to two times as long as calyx, lobes narrowly oblong, 1–1.5 mm. wide, slightly wavy-margined at apex; stamens inserted just above base of tube, unequal in length, longest 2.5–3 mm. long; capsule ovoid, 2.5–3 mm. long, dehiscing by splitting from top to bottom; seeds 6–9 in each cell, not mucilaginous when wet.

Sandy canyons and washes, Lower Sonoran Zone, Inyo County, California, south in the Mojave Desert, east to southern Utah and central Mohave County, Arizona, probably overlapping into the northern limits of the Sonoran Desert. March–May.

**3. Gilia latifolia** S. Wats., Amer. Nat. **9:** 347, 1875

*Navarretia latifolia* Kuntze, Rev. Gen. Pl. 433, 1891.

Erect annual, 5–20 cm. high; stems simple or with several branches from near base, often arising from a cluster of basal leaves; stems and leaves glandular-hairy; leaves ovate or oblong with broad blade and irregularly serrate with teeth deltoid and cuspidate, lower leaves 5–12 cm. long, 1–5 cm. wide with petioles 1–5 cm. long, upper leaves smaller and in inflorescence bractlike, simple or toothed; inflorescence open-paniculate to somewhat glomerate, flowers borne singly on glandular-pubescent pedicels 2–10 mm. long; calyx 3.5–5.5 mm. long, later equaling or exceeding capsule, glandular-pubescent, lobes as long as tube, acuminate; corolla narrow-funnelform, 6–11 mm.

long, bright pink within, pale or buff outside, lobes narrow and three-fourths as long as corolla tube; stamens subequally inserted on middle of corolla tube, unequal in length and included; capsule 5–7 mm. long, ovoid to somewhat elongate, dehiscing by splitting from top and valves recurving; seeds very small, 60–100 in each capsule, not mucilaginous when wet.

Desert plains and washes, Lower Sonoran Zone, southeastern California from Inyo County to Imperial County and east to southwestern Utah and western Arizona in Mohave and Yuma Counties. March–May.

**4. Gilia sinuata** Dougl., in A. DC., Prod. **9:** 313, 1845

*Gilia inconspicua* var. *sinuata* A. Gray, Bot. Calif. **1:** 498, 1876.
*Gilia arenaria* subsp. *leptantha* var. *borealis* Brand, Pflanzenreich $4^{250}$: 1907.
*Gilia inconspicua* subsp. *eu-inconspicua* var. *variegata* Brand, op. cit. 105.
*Gilia inconspicua* subsp. *sinuata* Brand, op. cit. 105.
*Gilia tenuiflora* var. *sinuata* Jepson, Fl. Calif. **3:** 179, 1943.

Erect annual 5–30 cm. tall, branched from near base or above or simple; stems glabrous and glaucous at base, glandular above; lower leaves in a rosette, sinuately toothed or with acute-tipped lobes, sometimes bipinnate, lightly to densely woolly-pubescent, rarely glabrous, cauline leaves shorter, sessile with strap-shaped to ovate blades, sinuately toothed with teeth shorter than width of rachis, uppermost leaves minute and bractlike; inflorescence paniculate; flowers borne singly on pedicels that vary in length on single plant, (2–) to 5–15 mm. long, sometimes subglomerate; calyx 2.5–5.2 mm. long, glandular-dotted, lobes one-half to one-third as long as tube; corolla funnelform, 6–12–8 mm. long, tube usually longer than calyx, yellow or violet, throat yellow, lobes oval or narrower and tapering to a point, violet; stamens inserted in sinuses and usually shorter than corolla lobes; pollen white; capsule ovoid, 4–5.5 mm. long, dehiscing by splitting from top and valves recurving or splitting from top to bottom; seeds 4–9 in each cell, mucilaginous when wet .

Sandy slopes and washes, Lower Sonoran to Arid Transition Zones, east of the Cascades and Sierra Nevada in Washington, Oregon, and California, east to Idaho, Utah, and eastern Arizona, and mountains of interior southern California. March–June.

**5. Gilia minor** A. Grant & V. Grant, Aliso **3:** 254, 1956

Annual herb 6–25 cm. tall with 3–8 stems from a basal cluster of leaves, branching above, cobwebby-pubescent below middle, sparsely glandular above; basal leaves pinnately or rarely somewhat bipinnately lobed, 1.5–4 cm. long, cobwebby-pubescent, rachis 0.7–1.2 mm. wide, lobes 1.4–5 mm. long; cauline leaves progressively reduced and less dissected upward on stem, uppermost bractlike, entire or with 1 pair of lobes at base; inflorescence strict, 1–2 flowers axillary to a bract; pedicels unequal, 1–11 mm. long, lengthening in fruit; calyx 2–3.2 mm. long at anthesis, sparsely to densely gland-dotted, enlarging but not over three-fourths as long as capsule at maturity; corolla 4.2–7.8 mm. long, tube short, included, upper tube and throat dark violet or throat yellow and veined with dark violet, or yellow with dark violet spot just below each lobe; corolla lobes pale violet; stamens barely exserted from orifice; capsule oblong-ovoidal, 5–7 mm. long.

Sandy flats and washes, Lower Sonoran Zone, San Benito County to Los Angeles County; Kern Valley in southern Sierra Nevada to Inyo County, east into western Arizona in the vicinity of Wickenburg and Wenden. March–April.

**6. Gilia flavocincta** A. Nels., Amer. Jour. Bot. **21:** 577, 1934

*Gilia ophthalmoides* subsp. *flavocincta* A. & V. Grant, Aliso **3:** 262, 1956.
*Gilia ophthalmoides* subsp. *australis* A. & V. Grant, op. cit. 263.

Annual herb with 1 to many slender, commonly flexuous stems 10–45 cm. tall, sparsely arachnoid-pubescent below, lightly glandular to glabrous above, leafy below middle; basal leaves in a loose cluster, pinnately or bipinnately lobed, lightly arachnoid-puberulent, rachis 0.4–2 mm. wide, lobes narrowly linear; cauline leaves similar to basal to middle of stem; inflorescence often congested in flower, expanding in fruit, 1–8 flowers axillary to a bract; pedicels 1.5–23 mm. long, glandular-pubescent to glabrous, elongating in fruit; calyx 2.8–5 mm. long, sparsely arachnoid-pubescent to glabrous; corolla 7.5–27 mm. long, slenderly funnelform, tube 0.8–1 mm. wide, orifice 1.8–7 mm. wide; lobes 2–6 mm. long; stamens subequal, slightly exserted; style equaling or exceeding stamens; capsule ovoidal, 4.6–6.7 mm. long, broad at base, narrowing toward apex.

Sandy canyons, mountain slopes, and arroyos, Lower Sonoran and lower margins of Transition Zones, Yuma County to Maricopa and southern Yavapai Counties, Arizona, to southeastern Arizona, southern New Mexico, and northern Sonora, Mexico. Feb.–May.

**7. Gilia exilis** (A. Gray) Abrams, Fl. Los Angeles, ed. 3, 289, 1917

*Gilia latiflora* var. *exilis* A. Gray, Syn. Fl. N. Amer. ed. 2, **2**[1]**:** 411, 1886.
*Gilia arenaria* subsp. *exilis* Brand, Pflanzenreich **4**[250]**:** 104, 1907.
*Gilia arenaria* var. *abramsii* Brand, Ann. Conserv. & Jard. Bot. Genève **15-16:** 330, 1913.
*Gilia lineata* Davidson, Bull S. Calif. Acad. **22:** 71, 1923.
*Gilia tenuiflora* var. *exilis* Jepson, Fl. Calif. **3:** 178, 1943.
*Gilia abramsii* Mason & A. Grant, Madroño **9:** 216, 1948.
*Gilia abramsii* subsp. *integrifolia* Mason & A. Grant, loc. cit.
*Gilia latiflora* subsp. *exilis* Mason & A. Grant, op. cit. 219.
*Gilia ochroleuca* subsp. *exilis* A. & V. Grant, Aliso **3:** 222, 1956.

Erect annual 15–30 cm. tall; stems usually several from base, glabrous or woolly-pubescent below, becoming glandular or partly glabrous on upper branches of inflorescence; lower leaves numerous, in loose cluster, moderately arachnoid-pubescent, simple, pinnately or rarely bipinnately lobed, 2–6 cm. long, rachis and lobes narrowly linear, 1–1.5 mm. wide, lobes 4–15 mm. long; cauline leaves much shorter, digitately lobed to entire; inflorescence full, divaricately branched with 1-flowered filiform pedicels 4–27 mm. long, glandular just beneath flowers; calyx 2–3.2 mm. long in flower, subequal to or exceeding capsule at maturity, lightly woolly-pubescent, lobes about one-half as long as tube; corolla 8–10.5 mm. long, funnelform, yellow in lower throat, violet above, lobes pinkish, tube slightly exserted from calyx, slightly longer than throat, this broadly expanded; stamens inserted in sinuses, shorter than corolla lobes; capsule ovoid, 3.8–4.5 mm. long, dehiscent to base; seeds about 4 in each cell, mucilaginous when wet.

Sandy areas near creeks, mainly Upper Sonoran and Transition Zones, coastal mountains of San Bernardino, Riverside, and San Diego Counties, California, barely reaching Sonoran Desert in Riverside County. April–June.

**8. Gilia mexicana** A. & V. Grant, Aliso **3:** 255, 1956.

Erect annual with 1–several leafy stems 10–33 cm. tall from a basal rosette or tuft of leaves; stems slender, somewhat flexuous, gray-arachnoid below middle, sparsely pubescent with minute glandular hairs on inflorescence; basal leaves pinnately or bipinnately lobed, rachis and lobes narrow, 0.8–1.3 mm. wide, primary lobes 4.5–6.5 mm. long; lower cauline leaves similar to basal ones, one-third as long as to equaling internodes near base, progressively smaller upward; uppermost leaves bractlike, linear, entire or with small pair of basal lobes; inflorescence paniculate, 1–2 flowers axillary to a bract; pedicels unequal, 1–4 mm. long at anthesis, 5–12 mm. long in fruit; calyx 2.5–4 mm. long, glabrous or sparsely glandular or arachnoid-pubescent, lobes acute; corolla 4.5–8 mm. long, tube usually included, throat sometimes included in calyx or corolla twice as long as calyx; corolla tube and throat white with yellow spots; lobes white to pale blue with violet dots along midvein; stamens short, equaling or slightly surpassing orifice; capsule narrowly ovoid, small, 3.5–5 mm. long.

Hillsides, arroyos, and plains, Lower and Upper Sonoran Zones, Pima and Pinal Counties, Arizona, to New Mexico, south into Chihuahua. Barely entering the margins of the Sonoran Desert. March–April.

**9. Gilia scopulorum** M. E. Jones, Bull. Torrey Club **8:** 70, 1881

*Gilia scopulorum* var. *typica* Brand, Pflanzenreich **4**[250]: 109, 1907.
*Gilia scopulorum* var. *covillei* Brand, loc. cit.

Erect annual, 10–30 cm. tall; stems simple or paniculately branched from near base and above, pubescent with straight, translucent hairs, often gland-tipped, becoming shorter and more glandular in inflorescence; lower leaves 3–9 cm. long, 5–20 mm. broad, coarsely toothed or incised or pinnately divided, lobes pinnatifid, ultimate divisions often acute, pubescence as on stems; upper leaves sessile, much shorter, with fewer divisions, uppermost 2–5 mm. long, 3-toothed; inflorescence paniculately branched with numerous flowers borne singly on glandular pedicels of varying lengths, mostly longer than 1 cm.; calyx glandular, 3–4.3 mm. long, enlarging with maturing capsule, calyx lobes acerose, one-half as long as calyx tube; corolla funnelform, 10–14.5 mm. long, tube white, yellow or pale violet, slender, exceeding calyx more than twice, throat white or yellow, lobes pink, pale violet or white, narrow, 1.4–3.4 mm. long; stamens inserted in sinuses of corolla lobes, shorter than corolla lobes; capsule ovoid, subglobular, 4.5–5.5 mm. long, dehiscing by splitting from top to bottom between the 3 valves; seeds 4–6 in each cell, mucilaginous when wet.

Desert washes and rocky slopes, Lower Sonoran Zone, Death Valley and eastern Mojave Desert in California and east to southern Utah; in the Sonoran Desert from southeastern California to Mohave and Yuma Counties in Arizona. March–May.

**10. Gilia stellata** Heller, Muhlenbergia **2:** 117, 1906

*Gilia tenuiflora* var. *newloniana* Jepson, Fl. Calif. **3:** 179, 1943.

Erect annual, 10–40 cm. tall, stems stout, simple or branched from near base and above, lower stems and leaves faintly hoary with short, white, geniculate hairs, upper stems and leaves bearing shorter, glandular hairs; basal leaves 2–10 cm. long, bipinnate, rachis narrow and straight, primary lobes lying at right angles to rachis, rounded in outline, with 2–7 apiculate teeth radiating on all sides; upper leaves much shorter and less dissected, uppermost leaves 2–5 mm. long and 3–5-toothed; inflorescence paniculate with flowers borne singly on glandular pedicels 2–30 mm. long and of varying lengths on same plant; calyx 3.2–4.2 mm. long, later becoming as long and broad as capsule by lengthening and by distention of hyaline regions, pubescent with glandular or geniculate hairs, lobes acuminate, about half as long as tube; corolla funnelform, 6.3–10.4 mm. long, pale blue-violet to white, usually with a ring of purple spots in throat, tube about as long as calyx, throat subequal to longer than lobes;

stamens inserted in sinuses of corolla lobes and shorter than corolla lobes; capsule broad, almost globular, 5–7 mm. long, dehiscing by splitting between 3 valves from top to about middle; seeds 5–10 in each cell, mucilaginous when wet.

Desert slopes and washes, Lower Sonoran Zone, Death Valley and eastern Mojave Desert in California, east to southern Utah; in the Sonoran Desert from southeastern California to Maricopa and Pinal Counties in Arizona and south to Sonora and Baja California, Mexico. March–May.

**11. Gilia australis** (Mason & A. Grant) V. Grant & A. Grant, Aliso **3:** 87, 1954
*Gilia splendens* subsp. *australis* Mason & A. Grant, Madroño **9:** 213, 1948.

Erect annual, 10–40 cm. tall, stems branched from near base or above, pubescent with coarse translucent hairs below but becoming increasingly glandular above; basal and lower cauline leaves 2–6 cm. long, conspicuously longer than upper cauline leaves, bi- to tripinnate with narrow, stemlike rachis, primary lobes numerous but not crowded, finely toothed or lobed, pubescent with coarse translucent hairs, upper cauline leaves like lower except shorter and with fewer lobes, leaves of inflorescence minute and often entire; inflorescence paniculately branched with flowers borne singly on slender, glandular pedicels of varying lengths; calyx 2–4 mm. long, glabrous, lobes about half as long as tube, acuminate and often recurved; corolla 5–11 mm. long, violet to white with yellow spots in throat and outer surface of lobes usually streaked with purple, throat subequal to longer than tube; stamens inserted in sinuses and shorter than corolla lobes; capsules ovoid to narrowly elongate, 1–2 times as long as mature calyx, 4–6 mm. long, dehiscing by splitting from top and valves recurving; seeds 14–30 in each cell, mucilaginous when wet.

Sandy slopes and washes, mainly Upper Sonoran and Transition Zones, southern California in the San Bernardino and San Jacinto Ranges, descending to the northwestern edge of the Colorado Desert in Riverside County, south in San Diego County to northern Baja California. April–June.

## 4. IPOMOPSIS Michx., Fl. Bor. Amer. 1: 141, 1803

Annual, biennial or perennial herbs or shrubs with a basal rosette of leaves (obscure in some) and stems leafy to apex; herbage often villous or stipitate-glandular. Leaves pinnatifid; tips of segments bearing horny mucros. Inflorescence loosely racemose to congested cymes; each flower subtended by a bract. Calyx 5-lobed or 5-toothed, villous, glandular or glabrous; lobes separated by broad hyaline sinuses. Corolla salverform or tubular, white, violet, red, yellow or pink; veins in tube and limb rarely anastomosing. Stamens unequally inserted in tube or in sinuses of corolla, unequal in length. Capsules promptly or tardily dehiscent; each cell 1- to several-seeded. Seeds slenderly elliptic to fusiform, often curved, less commonly spheroidal, rounded or angular.

Plants perennial herbs or rounded shrubs; stamens long, exserted 8–12 mm. beyond corolla tube; corolla slightly zygomorphic:
  Plants rounded shrubs 5–15 dm. tall; corollas pink to white, tube 2.5–3 cm. long; calyx 10–14 mm. long . . . . . . . . . . 1. *I. gloriosa*
  Plants erect herbs 1–3 dm. tall; corolla scarlet, tube 1–1.5 cm. long; calyx 5–6 mm. long . . . . . . . . . . 2. *I. tenuifolia*
Plants annual herbs; stamens included or exserted less than 5 mm.; corolla regular:
  Capsule 8–15 mm. long; corolla blue-violet to nearly white, 3.5–5 cm. long, tube 4–6 times as long as corolla lobes . . . . . . . . . . 3. *I. longiflora*

Capsule 3–5 mm. long; corolla white tinged with pink, 4–6.5 mm. long, tube 2–3 times as long as corolla lobes:
Inflorescence a congested, terminal cyme; seeds 2–3 in each cell of capsule .......... 4. *I. polycladon*
Inflorescence diffuse, the flowers borne almost throughout the plant; seeds 10–30 in each cell of capsule .......... 5. *I. sonorae*

**1. Ipomopsis gloriosa** (Brandegee) A. Grant in V. Grant, Aliso **3:** 357, 1956
*Gilia gloriosa* Brandegee, Proc. Calif. Acad. II. **2:** 184, *pl.* 9, 1889.
*Leptodactylon gloriosum* Wherry, Amer. Midl. Nat. **34:** 383, 1945.

Dense rounded shrub 8–15 dm. tall and broad with glandular-pubescent herbage and divaricately spreading, pinnately 3–5-lobed, acerose rigid leaves 1.2–2 cm. long, when 3-lobed lateral lobes about one-third of way up from base, 2–6 mm. long; fasciculate leaves 7–12 mm. long, simple, entire; bark on young twigs yellowish, grayish and striate on older ones; flowers solitary on stoutish pedicels 2–5 mm. long in upper axils; calyx cylindric, 10–14 mm. long, densely glandular, sinuses scarcely membranous; lobes linear, pungent, erect, shorter than tube; corollas funnelform, pale pink to nearly white, tube 2.5–3 cm. long; corolla lobes 10–12 mm. long, spreading, oval and entire; stamens inserted on lower one-third of tube, exserted about 10 mm.; capsule subcylindric, 8–10 mm. long, many-seeded, 3-valved at apex.

Rocky and gravelly benches and along arroyos, Lower Sonoran Zone, Santa Rosalia Bay to Miller's Landing, Baja California. May–July.

**2. Ipomopsis tenuifolia** (A. Gray) V. Grant, Aliso **3:** 357, 1956
*Loeselia tenuifolia* A. Gray, Proc. Amer. Acad. **11:** 68, 1876.
*Loeselia guttata* A. Gray, op. cit. **20:** 302, 1885.
*Gilia tenuifolia* A. Gray, Syn. Fl. N. Amer. ed. 2, **2**[1]**:** 411, 1886.
*Gilia truncata* Davidson, Bull. S. Calif. Acad. **22:** 72, *pl. 19*, 1923.

An erect perennial with numerous slender stems from a woody base; branches 1–3 dm. tall, with pale yellowish bark below; herbage puberulent; leaves narrowly linear, 15–25 mm. long, cuspidate, entire, or lower ones pinnately parted into 1–3 remote pairs of short subulate or linear lobes 1–4 mm. long; flowers in terminal, rather loose, 2–6-flowered clusters or sometimes solitary; calyces 5–6 mm. long, tube 4–5 mm. long, lobes cuspidate; corollas scarlet, tubular, tube and throat about 15 mm. long, lobes 3-dentate, 5–6 mm. long, nearly regular; stamens and style conspicuously exserted, 5–8 mm. longer than corolla lobes; capsule ovoid, about 6 mm. long, distending and finally rupturing pinkish-membranous tissue below sinuses of calyx; seeds 6–16 in each capsule.

Arid ridges and rocky canyon sides, Lower Sonoran Zone, western margin of the Colorado Desert and south to the eastern flanks of the Sierra San Pedro Mártir, Baja California. March–Oct.

**3. Ipomopsis longiflora** (Torr.) V. Grant, Aliso **3:** 361, 1956
*Cantua longiflora* Torr., Ann. Lyc. N.Y. **2:** 221, 1828.
*Gilia longiflora* G. Don, Gen. Hist. Pl. **4:** 245, 1838.

Erect, usually much branched, bushy annual, 10–50 cm. tall or taller, a basal rosette ephemeral or lacking, stems leafy; herbage lightly to densely pubescent in lower part of plant with crinkled and appressed somewhat coarse hairs, minutely puberulent on upper branches; leaves pinnately lobed with 1-3 pairs of lobes or upper leaves simple, rachis and lobes narrowly linear, less than 1 mm. wide, acerose, pubescent with hairs like those on stems; inflorescence paniculately much branched with numerous flowers borne singly on pedicels 0.5–3 cm. long; calyx 7–8.5 mm. long, growing with capsules and sometimes exceeding them, midrib regions of calyx narrowly linear, minutely puberulent, lobes acerose, one-third as long as tube, pubescent on inside with crinkled hairs; corolla pale blue-violet to almost white, 3.5–5 cm. long, tube elongate (four to six times as long as lobes), narrow, only slightly broader at top than at base, abruptly expanding; lobes rotate, oval with a short cusp at tip of each; stamens unequally inserted in upper part of tube, unequal, barely exserted; capsule 8–15 mm. long, narrow, dehiscing from top toward base, valves recurved; seeds 7–9 in each cell, mucilaginous when wet.

Sandy plains, canyons and river valleys, mostly Upper Sonoran and Transition Zones, Colorado to Mexico and east to Texas; mostly above the Sonoran Desert but descending to its margins in Pima County, Arizona. April–Oct.

**4. Ipomopsis polycladon** (Torr.) V. Grant, Aliso **3:** 361, 1956
*Gilia polycladon* Torr., Bot. Mex. Bound. 146, 1859.

Small annual, erect with 2–6 stems arising from a basal rosette of leaves; stems 5–16 cm. long, appearing naked below due to elongated internodes, pubescent near base and lightly above with coarse, white, crinkled hairs, these intermixed with minute hairs on upper parts of plant; leaves 1–3 cm. long, pubescent with both minute and coarse hairs as on stems, once-pinnate with linear rachis or upper leaves pinnately toothed and with broader rachis, lobes or teeth short (to 3.5 mm. long) cuspidate; inflorescence headlike leafy-bracted cymes borne at ends of branches; pedicels very short, densely puberulent, each subtended by a pinnately toothed, spine-tipped bract; calyx 3–4.5 mm. long, later equaling or exceeding capsule, lobes spine-tipped, about as long as tube, pubescent with coarse hairs on inside; corolla tubular-funnelform, white tinged with pink, 4–6 mm. long, exceeding, but less than twice as long as calyx, lobes one-third to one-half as long as corolla tube; stamens very short, inserted in tube, included; capsule ovoid, 4 mm. long, dehiscing from top to bottom; seeds 2–3 in each cell, mucilaginous when wet.

Gravelly slopes, Lower and Upper Sonoran Zones, eastern Oregon to southeastern California and northern Mohave County, Arizona, east to Utah and south to Texas and Mexico. April–May.

**5. Ipomopsis sonorae** (Rose) A. Grant, in V. Grant, Aliso **3:** 361, 1956
*Gilia sonorae* Rose, in Vasey & Rose, Contr. U.S. Nat. Herb. **1:** 90, 1890.

Small annual with 4–12 stems spreading outward and upward from base; stems 5–22 cm. long, leafy but sometimes with longer internodes in lower parts than in inflorescence, appearing somewhat naked below and more leafy above, minutely puberulent; leaves once-pinnate, 5–20 mm. long, upper cauline leaves almost as long as lower, rachis linear, 0.4–0.8 mm. wide, lobes in 2–5 pairs, linear, equal in width to rachis and 0.5–5 mm. long, tapering abruptly to a short cusp, minutely puberulent; calyx 2–4 mm. long, growing as long as capsule, pubescent like leaves with a few coarse hairs near base intermixed with minute hairs, lobes about half as long as tube, tapering abruptly to a short cusp; corolla white tinged with pink, 4.5–6.5 mm. long, less than twice as long as calyx, lobes rounded, slightly tapering toward apex, one-half to one-third as long as tube; stamens unequally exserted about middle of corolla tube, unequal and longest exserted slightly from tube; capsule 3–5 mm. long, ovoid, dehiscing from top to bottom between valves; seeds small, 10–30 in each cell, not mucilaginous when wet.

Valleys in the foothills, Lower Sonoran Zone, western Sonora and northwestern Sinaloa, Mexico. March–April.

### 5. ERIASTRUM Woot. & Standl., Contr. U.S. Nat. Herb. 16: 160, 1913

Erect, simple to paniculately or corymbosely branched, annual or perennial herbs with linear to pinnately toothed or dissected leaves and puberulent to lanate herbage. Flowers sessile in bracteate heads or rarely solitary or pedicellate; heads usually densely arachnoid-woolly. Calyx deeply cleft into 5 linear, subequal or unequal lobes; sinuses hyaline-membranous. Corolla white, blue, yellow, rarely pink or bicolored, salverform or funnelform. Stamens 5, inserted at base of corolla throat (in ours) or less commonly just below sinuses between lobes. Anthers sagittate, elliptic or cordate. Capsule ellipsoid or obovoid. Seeds 1 to several in each cell, mostly mucilaginous when wetted.

Corolla lobes equaling or longer than tube; stamens inserted at base of throat:
  Stamens very unequal in length; leaves usually pinnately parted:
    Corolla noticeably bilabiate; leaves with 1–3 pairs of lateral lobes .................. 1. *E. eremicum* subsp. *eremicum*
    Corolla only slightly bilabiate; leaves with 1 pair of lateral lobes or upper ones simple .......... 1a. *E. eremicum* subsp. *yageri*
  Stamens equal in length or nearly so; leaves entire .. 2. *E. sapphirinum* subsp. *ambiguum*
Corolla lobes distinctly shorter than tube; stamens inserted midway in throat:
  Corolla lobes not apiculate; heads moderately lanate:
    Corolla 7–9 mm. long, nearly regular; anthers 0.5 mm. long .......... 3. *E. diffusum* subsp. *diffusum*
    Corolla 10–12 mm. long, slightly irregular; anthers 0.7–1 mm. long .......... 3a. *E. diffusum* subsp. *jonesii*
  Corolla lobes apiculate; heads densely lanate .......... 3b. *E. diffusum* subsp. *harwoodii*

**1. Eriastrum eremicum** (Jepson) H. L. Mason, Madroño **8:** 78, 1945, subsp. **eremicum**

*Navarretia densifolia* var. *jacumbana* Brand, Ann. Conserv. & Jard. Bot. Genève **15-16:** 340, 1913.
*Hugelia eremica* Jepson, Man. Fl. Pl. Calif. 793, 1925.
*Gilia eremica* Craig, Bull. Torrey Club **61:** 416, 1934.
*Gilia eremica* var. *typica* Craig, op. cit. 417.
*Gilia eremica* var. *zionis* Craig, op. cit. 418.

Erect to spreading-ascending annuals 0.5–2 dm. tall, usually divaricately branched from base, stems and leaves sparingly floccose to glabrate; leaves 0.5–3 cm. long, pinnatifid into 3–7 linear lobes or in small specimens rarely entire, lobes or apex cuspidate, usually sparingly floccose at least near base; flowers congested in hemispheric heads 0.5–1 cm. wide at ends of branches and embedded in a mass of white wool; bracts mostly pedately 3-lobed, central lobe to 10 mm. long, lateral 2–4 mm. long, tips of all more or less recurved; calyx 4–5 mm. long, linear subequal lobes cuspidately tipped, sinuses membranous; corolla tube and throat 7–9 mm. long, lobes 5–6 mm. long, usually more or less bilabiately arranged, violet; stamens very unequal, inserted at base of throat, longest exserted; anthers sagittate; pistil exserted; capsules ovoid to oblong, 2–2.5 mm. wide, 3–6 mm. long, each cell several-seeded.

Desert washes, plains, hillsides, and mesas, mostly Lower Sonoran Zone, eastern San Diego County, California, to southern Nevada, southern Utah, and southern Arizona south into northeastern Baja California. March–July.

**1*a*. Eriastrum eremicum** subsp. **yageri** (M. E. Jones) H. L. Mason, Madroño **8:** 78, 1945

*Gilia virgata* var. *yageri* M. E. Jones, Contr. West. Bot. **13:** 2, 1910.
*Gilia eremica* var. *yageri* Craig, Bull. Torrey Club **61:** 420, 1934, as to name, not as to lectotype.
*Gilia eremica* var. *arizonica* Craig, op. cit. 419.

Differing from subsp. *eremicum* in having leaves that are simple and entire or some of them with 1 or rarely 2 pairs of lateral lobes, and corollas that are essentially regular, not markedly bilabiate.

Desert flats, arroyos, hillsides, and mesas, Lower Sonoran Zone, in western Arizona.

**2. Eriastrum sapphirinum** (Eastw.) H. L. Mason subsp. **ambiguum** (M. E. Jones) H. L. Mason, Madroño **8:** 80, 1945

*Gilia floccosa* var. *ambigua* M. E. Jones, Contr. West. Bot. **13:** 2, 1910.
*Gilia virgata* var. *ambigua* Craig, Bull. Torrey Club **61:** 412, 1934.
*Hugelia virgata* var. *ambigua* Jepson, Fl. Calif. **3:** 165, 1943.

Erect, diffusely paniculately branched annual 2–6 dm. tall with distinctly viscid herbage, glandular, puberulent, or sometimes lightly floccose in flowering heads; leaves linear, entire, 1–3 cm. long, cuspidate at tip, rarely with a single pair of short lobes near base; heads compact, few-flowered, or reduced to a solitary flower in depauperate plants; bracts 3–7-lobed, not membranously margined, usually shorter than calyx; calyx 5–6 mm. long; corolla funnelform, white, pale yellow, or pale blue, 10–12 mm. long, tube over twice as long as throat; stamens inserted at base of throat, 2–3 mm. long; capsules about 4 mm. long, narrowly ovoid, each cell several-seeded.

Desert slopes, Lower Sonoran Zone, chiefly in the lower mountains along the western and southern edges of the Mojave Desert, but entering the northwestern part of the Colorado Desert in Riverside and eastern San Diego Counties, California. April–Sept.

**3. Eriastrum diffusum** (A. Gray) H. L. Mason, Madroño **8:** 76, 1945, subsp. **diffusum**

*Gilia filifolia* var. *diffusa* A. Gray, Proc. Amer. Acad. **8:** 272, 1870.
*Navarretia filifolia* var. *diffusa* Brand, Pflanzenreich $4^{250}$: 167, 1907.
*Welwitschia diffusa* Rydb., Fl. Rocky Mts. 688, 1917.
*Welwitschia filifolia diffusa* Tidestrom, Proc. Biol. Soc. Wash. **48:** 42, 1935.
*Hugelia diffusa* Jepson, Fl. Calif. **3:** 167, 1943.

Erect to ascending, diffusely branched annual 5–15 cm. tall, more or less floccose-puberulent throughout; leaves linear to 3–5-parted, 5–20 mm. long, lobes linear, cuspidate-tipped; flowers in lanate-covered heads; bracts similar to leaves, 3–7-lobed, to about 1 cm. long, lobes slightly recurved; calyx 5–7 mm. long, unequal acerose lobes nearly obscured by wool, membranes reaching about two-thirds of distance to tips of lobes; corollas nearly regular, 7–9 mm. long, tube 4–5 mm. long, lobes about one-half as long as tube proper, throat 0.5–1.5 mm. long; stamens 1–2 mm. long, equal or unequal, inserted at middle of throat; anthers 0.5 mm. long; capsule 3–4 mm. long, each cell 2–3-seeded.

Desert regions, Lower Sonoran Zone, southern Nevada and Inyo County, California, to southern Utah, Texas, and south into northern Chihuahua, Sonora, and Baja California. March–May.

**3*a*. Eriastrum diffusum** subsp. **jonesii** H. L. Mason, Madroño **8:** 77, 1945

*Gilia eremica* var. *yageri* Craig, Bull. Torrey Club **61:** 420, 1934, as to lectotype only.

Differing from subsp. *diffusum* in having corollas 10–12 mm. long, stamens 2–3 mm. long, and anthers 0.7–1 mm. long.

Desert plains, mesas, and arroyos, Lower Sonoran Zone, Pima and Gila Counties, Arizona, south into Sonora. March–May.

**3*b*. Eriastrum diffusum** subsp. **harwoodii** (Craig) H. L. Mason, Madroño **8:** 77, 1945

*Gilia filifolia* var. *harwoodii* Craig, Bull. Torrey Club **61:** 424, 1934.
*Hugelia diffusa* var. *harwoodii* Jepson, Fl. Calif. **3:** 167, 1943.

Differing from subsp. *diffusum* in being more conspicuously white-lanate on the herbage and having apiculate corolla lobes.

Arid plains, Lower Sonoran Zone, along or near the Colorado River in the eastern Mojave and Colorado Deserts, San Bernardino and Riverside Counties, California. March–May.

## 6. LANGLOISIA Greene, Pittonia 3: 30, 1896

Low, rigidly branching annuals with alternate, linear, spatulate or cuneate-ovate leaves, pinnatifidly toothed, lower teeth reduced to long bristles, upper ones bristle-tipped. Herbage floccose-puberulent to glabrate. Flowers in bracteate terminal heads; bracts with bristle-tipped teeth like the leaves. Calyx tubular; lobes equal, conspicuously bristle-tipped, splitting to base in fruit. Corolla tubular-funnelform, more or less bilabiate with 3 lobes in upper lip, 2 in lower, or 5 lobes equal or essentially so and not bilabiate. Stamens inserted in corolla throat, equaling or shorter than corolla lobes. Capsule obtusely triangular in cross section, 2–9-seeded. Seeds somewhat mucilaginous when wetted.

Leaves markedly dilated at apex; some lateral bristles on lower part of leaves 2–3-forked; corolla essentially regular; corolla lobes broadly rounded ........... 1. *L. setosissima*
Leaves linear to spatulate, not dilated; lateral bristles all simple; corolla distinctly bilabiate; corolla lobes acute or nearly truncate:
  Corolla lobes acute, entire, one-fourth to one-half as long as tube; calyx 3–5 mm. long (exclusive of terminal bristles); stamens barely equaling or slightly shorter than shortest corolla lobe ........................................................ 2. *L. schottii*
  Corolla lobes nearly truncate, more or less tridentate, nearly as long as tube; calyx 5–7 mm. long; stamens usually equaling longer corolla lobes ............. 3. *L. matthewsii*

**1. Langloisia setosissima** (Torr. & Gray) Greene, Pittonia **3:** 30, 1896

*Navarretia setosissima* Torr. & Gray ex Torr., Ives Rep. 22, 1860.
*Gilia setosissima* A. Gray, Proc. Amer. Acad. **8:** 271, 1870.
*Loeselia setosissima* Peter, in Engler & Prantl, Nat. Pflanzenf. $4^{3a}$: 54, 1893.
*Langloisia setosissima* var. *campyloclados* Brand, Pflanzenreich $4^{250}$: 171, 1907.

Low or depressed, many-branched plant 2–7 cm. high, 3–15 cm. wide, branches horizontally spreading to ascending; herbage sparsely floccose-puberulent to glabrate, stems often reddish-tinged; leaves 10–30 mm. long (basal longest and narrowest), mostly 3–7 mm. wide at widest part, cuneately and abruptly dilated upward from slender petiole, tridentate at apex and usually with 1–2 pairs of teeth below, all teeth bristle-tipped, lower rachis and petiole bearing some 2–3-forked bristles as well as simple ones; bracts foliaceous, 3-toothed and bristle-tipped; calyx lobes 5–6 mm. long, each tipped by a stiff bristle 3–4 mm. long, these commonly equaling corolla tube; corolla lavender or bluish, essentially regular, 12–18 mm. long, lobes rounded, 2–3 mm. wide, 4–5.5 mm. long; stamens reaching about to middle of corolla lobes; capsule ovate to oblong, 5–8 mm. long, acute at apex, each cell 3–9-seeded.

Arid mesas, gravelly hills, and along arroyos, Lower Sonoran Zone, throughout the Colorado Desert and eastern Mojave Desert into Inyo County, California; Nevada; Idaho; east into Mohave, Maricopa, and Yuma Counties, Arizona, and probably into adjacent northern Baja California and Sonora. Feb.–June.

**Langloisia punctata** (A. Gray ex Cov.) Goodding, Bot. Gaz. **37**: 58, 1904, has leaves and habit similar to *L. setosissima* but flowers larger, lobes spotted with purple dots and nearly as long as corolla tube. This species is common in the Mojave Desert and extends east into Nevada and northern Arizona, but it has not been collected, apparently, within the Sonoran Desert. The nearest approach known to me is represented by specimens collected "near Virginia Dale [Mine], Colorado Desert, Riverside County" by H. M. Hall (*6037*) on May 12, 1905. There are two errors on the labels of these specimens. Virginia Dale Mine is in San Bernardino County and is located on the southern edge of the Mojave Desert instead of in the Colorado Desert. It is possible that *L. punctata* does range beyond the Eagle Mountains into the northern fringe of the Colorado Desert. March–April.

### 2. Langloisia schottii (Torr.) Greene, Pittonia 3: 30, 1896

*Navarretia schottii* Torr., Bot. Mex. Bound. 145, 1859.
*Gilia setosissima* var. *exigua* A. Gray, Proc. Amer. Acad. **8**: 271, 1870.
*Gilia schottii* S. Wats., Bot. King Expl. 267, 1871.
*Loeselia schottii* A. Gray, Bot. Calif. **2**: 466, 1880.

A low tufted annual 2–12 cm. tall and often with several horizontal ascending branches 2–10 cm. long, but more commonly with short divaricate branches 1–5 cm. long; herbage floccose-pubescent to glabrate; leaves mostly crowded toward tips of branches, linear to spatulate-lingulate, 3–8 mm. wide, 12–20 mm. long, or basal ones 4–5 cm. long, regularly toothed as though shallowly pinnatifid, teeth bristle-tipped or replaced by simple, unforked bristles 1–5 mm. long; foliaceous bracts often somewhat cuneate-dilated and 3-toothed at apex, teeth and lower margin bristly; calyx lobes 3–5 mm. long, tipped with stiff, erect bristles 2–4 mm. long, equal, sparsely to moderately arachnoid-villous; corolla bilabiate, pink, pale lavender, or rarely white or yellowish, 4–8 mm. long, upper 3-lobed lip often purple-spotted, lower 2-lobed lip sometimes spotted at base, lobes divergent, one-fourth to one-half as long as corolla tube, acute; stamens as long as or slightly shorter than shortest corolla lobes; capsule ovoid, 3–4 mm. long, cells 2–6-seeded.

Sandy flats, washes, valleys, and mesas, Lower and less commonly in the Upper Sonoran Zones, central San Diego County, California, east throughout the Colorado Desert and eastern Mojave Desert to southern Utah, and to Mohave and Yuma Counties, Arizona; south into northwestern Sonora and adjacent Baja California. March–May.

### 3. Langloisia matthewsii (A. Gray) Greene, Pittonia 3: 30, 1896

*Loeselia matthewsii* A. Gray, Bot. Calif. **2**: 466, 1880.
*Gilia matthewsii* A. Gray, Syn. Fl. N. Amer. ed. 2, **2**[1]: 409, 1886.
*Navarretia matthewsii* Coville, Contr. U.S. Nat. Herb. **4**: 153, 1893.

A low subprostrate plant or main axis erect and surrounded by horizontal to spreading branches 2–15 cm. long; herbage arachnoid-pubescent with rather coarse, white, crinkled hairs; leaves mainly crowded toward ends of branches, linear to spatulate, 0.5–3.5 cm. long, regularly but shallowly pinnatifid, lobes bristle-tipped (or lower lobes wholly replaced by bristles), bristles simple, frequently somewhat recurved; bracts linear, 5–10 mm. long, bristle-margined; calyx 5–7 mm. long, each lobe tipped with a simple, erect, stiff spine 3–4 mm. long; corolla distinctly bilabiate, white, pale blue or pink, 15–20 mm. long, lobes oblong, nearly or quite truncate and tridentate or subentire at apex, each of the 3 upper lobes with a reddish arch beneath a white spot, 2 lower lobes smaller and usually unmarked; stamens equaling or slightly surpassing longest corolla lobes; capsule ovate, 3–5 mm. long, each cell 2–6-seeded; seeds yellowish, about 1 mm. long.

Sandy washes, mesas, valleys, and plains, Sonoran Zones, Mount Diablo Range and in the southern Sierra Nevada from Kern and Inyo Counties, California, into southern Nevada, western Arizona, and south through the Mojave and Colorado Deserts into northern Sonora and Baja California. March–June.

## 7. LEPTODACTYLON Hook. & Arn., Bot. Beechey 369, 1838

Suffrutescent or shrubby plants of straggly or compact habit, 1–20 dm. tall. Leaves alternate to subopposite (sometimes both arrangements on same plant), palmately parted into acicular, pungently tipped lobes, usually with others forming compact axillary fascicles. Inflorescence cymose or glomerulate at tips of branches or less commonly solitary in axils, sessile or nearly so. Calyx lobes 5, entire, pungent, equal or subequal; sinuses membranous about two-thirds of their length. Corolla salverform or narrowly funnelform, showy; lobes rotately spreading. Stamens inserted on throat or in tube, included or exserted. Style included; stigmas 3–4-lobed. Capsule cylindric or nearly so, 3–4-celled, each locule few- to many-seeded.

Lateral lobes of leaves usually one-half to two-thirds as long as central one; herbage densely glandular; corollas 10–12 mm. long ................................ 1. *L. veatchii*

Lateral lobes of leaves one-fourth to one-third as long as central one or completely lacking; herbage arachnoid-tomentose to glabrate, scarcely at all glandular; corolla 15–20 mm. long ............................................ 2. *L. pungens* subsp. *hallii*

### 1. Leptodactylon veatchii (Parry ex Greene) Wherry, Amer. Midl. Nat. 34: 384, 1945

*Gilia veatchii* Parry ex Greene, Bull. Calif. Acad. **1**: 198, 1885.

Compact rounded shrub 2–4 dm. tall and wide with very numerous, densely leafy branches; herbage densely glandular-puberulent, older parts turning dark brown or blackish; leaves opposite or subopposite on sterile branches, often alternate on flowering ones, palmately 3-lobed (rarely 5-lobed), spreading central lobe 10–14 mm. long, lateral ones mostly 6–10 mm. long, departing near base of central rachis; flowers terminal; calyx 7–9 mm. long, lobes slightly unequal; corolla 10–12 mm. long, ochroleucous within, purplish without, lobes broadly obovate, rotate, 5–7 mm. long.

Among lava boulders, Cedros Island. Possibly on the adjacent mainland but known definitely only from the island. March–April.

### 2. Leptodactylon pungens (Torr.) Rydb. subsp. hallii (Parish) H. L. Mason, in Abrams, Ill. Fl. Pacif. States 3: 455, 1951

*Gilia hallii* Parish, Erythea **7**: 94, 1899.
*Gilia tenuiloba* Parish, op. cit. 95.
*Gilia pungens* var. *hallii* Milliken, Univ. Calif. Pub. Bot. **2**: 42, 1904.
*Leptodactylon hallii* Heller, Muhlenbergia **1**: 146, 1906.
*Leptodactylon tenuilobum* Heller, loc. cit.
*Gilia pungens* subsp. *eu-pungens* var. *tenuiloba* Brand, Pflanzenreich $4^{250}$: 128, 1907.
*Gilia pungens* subsp. *hallii* Brand, loc. cit.
*Leptodactylon pungens* var. *tenuilobum* Jepson, Man. Fl. Pl. Calif. 807, 1925.
*Leptodactylon pungens* var. *hallii* Jepson, loc. cit.

Shrubby plants 1–8 dm. tall with erect, spreading or decumbent branches, crowded leaves and arachnoid, sparsely glandular herbage; leaves palmately 3–5-cleft, central lobes 5–14 mm. long, lateral ones 1–4 mm. long or sometimes obsolete, all lobes pungent, acerose, rigid; calyx 7–9 mm. long, lobes acerose, exceeding hyaline pseudotube 1–2.5 mm.; corollas 15–20 mm. long, narrowly funnelform, pink, lilac or yellowish; stamens included; capsule cylindric, many-seeded.

Rocky hillsides and arid slopes, upper margins of Lower Sonoran into the Transition Zones, eastern Washington and Idaho south mainly east of the Sierra Nevada to Arizona and southern California. Entering the upper fringe of the Sonoran Desert near Mountain Spring, eastern San Diego County, California. April–Aug.

## 8. LINANTHUS Benth., Bot. Reg. 19: under *pl. 1622,* 1833

Erect, ascending or spreading annuals or perennials; stems often dichotomously branched. Leaves opposite or rarely alternate, entire or palmately parted. Inflorescence open-cymose to congested in heads terminating branches, solitary flowers often in forks of cyme. Calyx deeply cleft, with or without a hyaline membrane uniting lobes for varying distances above base; distinct sepals sometimes hyaline-margined. Corolla campanulate, funnelform or nearly salverform; tube very short to very long. Stamens inserted equally on throat or tube, glabrous or puberulent at base. Style slender; stigma lobes exserted or included. Capsule ellipsoid to oblong-cylindrical, each cell 1- to several-seeded. Seeds producing mucilage or spiracles or unchanged when wetted.

Flowers not in heads, on slender pedicels or in axillary cymules or scattered:
- Inflorescence cymose but not dichotomous; flowers not vespertine:
  - Flowers sessile or subsessile in glomerules; corolla cream-yellow, 5–8 mm. long ........ 1. *L. lemmonii*
  - Flowers pedicellate, in open cymes; corolla golden-yellow, 6–12 mm. long ........ 2. *L. aureus*
- Inflorescence regularly dichotomous; flowers vespertine:
  - Corolla white, often tinged with brownish on backs of lobes; calyx glabrous; testa of seed more or less bladdery-hyaline:
    - Filaments hairy at base; corolla 2–3 cm. broad; leaves palmately divided ........ 3. *L. dichotomus*
    - Filaments glabrous; corolla 10–15 mm. broad; leaves simple or infrequently with 2–3 lobes ........ 4. *L. bigelovii*
  - Corolla yellow; calyx pubescent; testa of seed not bladdery nor hyaline .... 5. *L. jonesii*

Flowers aggregated into heads or geminate in small compact clusters:
- Corolla white, 3–8 mm. long; calyx lobes distinct essentially to base:
  - Leaves linear or oblong-linear, entire; corolla 3–4 mm. long; disk at base of corolla 5-lobed ........ 6. *L. maculatus*
  - Leaves 3-lobed, lobes linear; corolla 5–8 mm. long; disk entire .......... 7. *L. demissus*
- Corolla pink, 10–25 mm. long; calyx lobes united by a hyaline membrane along lower portion:
  - Calyx lobes hyaline-margined to tip; flowers sessile in 1–3-flowered glomerules; corolla lobes entire; filaments glabrous ........ 8. *L. bellus*
  - Calyx lobes naked at tip; flowers short-pedicellate (or less commonly subsessile); corolla lobes minutely erose-denticulate; filaments puberulent ........ 9. *L. dianthiflorus*

### 1. Linanthus lemmonii (A. Gray) Greene, Pittonia 2: 257, 1892

*Gilia lemmonii* A. Gray, Syn. Fl. N. Amer. $2^1$: 394, 1878.
*Navarretia lemmonii* Kuntze, Rev. Gen. Pl. 2: 433, 1891.

Delicate annuals with spreading, ascending or erect, slender stems 3–15 cm. long; stems often reddish, these and herbage puberulent, sometimes densely so and canescent; leaves 3–5-cleft into linear divisions 2–5 mm. long, tips acute and mucronulate; flowers in terminal and axillary cymules, subsessile; calyx 4–5 mm. long, united into a tube to about middle, lobes linear, slightly flaring, equal, minutely cuspidate; corolla short-funnelform, 5–8 mm. long, cream-colored or pale yellow, tube 1.5–2.5 mm. long, included in calyx, throat and lobes about equal, a hairy ring at summit of tube; stamens inserted on throat immediately beneath sinuses, about one-half as long as corolla lobes; anthers yellow, ovate, 0.5–0.8 mm. long; stigmas exserted; capsules narrowly oblong, 3.5–5 mm. long, each cell several-seeded.

Sandy plateaus, mesas, and valleys, mainly Upper Sonoran and Arid Transition Zones, from the San Bernardino Valley south into eastern San Diego County, where it straggles into the upper margins of the Colorado Desert near Jacumba and the Mountain Springs grade. Reported also from northern Baja California. April–July.

**2. Linanthus aureus** (Nutt.) Greene, Pittonia **2:** 257, 1892

*Gilia aurea* Nutt., Jour. Acad. Phila. **1:** 155, 1848.
*Leptosiphon aureus* Benth. ex Vilm.-Andr., Fl. Pl. Terre ed. 2, 470, 1866.
*Navarretia aurea* Kuntze, Rev. Gen. Pl. **2:** 433, 1891.
*Dactylophyllum aureum* Heller, Muhlenbergia **2:** 231, 1906.

Erect, or, less commonly, spreading, annual 3–18 cm. tall; branches horizontal, in spreading plants to 20 cm. long, stems and herbage glabrous, stipitate-glandular or somewhat hirsute, upper internodes elongate, wiry; leaves cleft into 3 linear segments 3–6 cm. long, mucronulate; flowers in open cymes; pedicels slender, 0.5–2.5 cm. long; calyx 5–6 mm. long, glabrous, hispid or stipitate-glandular, usually short-villous toward tips of lobes, membranous tube about two-thirds as long as calyx; corolla 6–12 mm. long, campanulate to short-funnelform, pale to golden-yellow, throat often spotted with purple, lobes nearly or quite equaling throat, tube about 2 mm. long, included in calyx tube with a hairy ring at its summit; stamens inserted just above middle of throat, scarcely surpassing orifice; stigma lobes about one-third as long as slender style, exserted; capsule oblong, 4–5 mm. long, several-seeded.

Arid slopes and washes, Lower and Upper Sonoran Zones, Ventura and Inyo Counties, California, southward into northern Baja California and eastward to southern Nevada and New Mexico. Occurring in our area mainly on the mountain slopes impinging upon the lower deserts. March–June.

**3. Linanthus dichotomus** Benth., Bot. Reg. **19:** under *pl. 1622,* 1833

*Gilia linanthus* Steudel, Nom. ed. 2, **1:** 684, 1840.
*Gilia dichotoma* Benth., in A. DC., Prod. **9:** 314, 1845.
*Navarretia dichotoma* Kuntze, Rev. Gen. Pl. **2:** 433, 1891.
*Gilia dichotoma* var. *uniflora* Brand, Pflanzenreich **4**$^{250}$**:** 144, 1907.
*Gilia dichotoma* var. *integra* M. E. Jones, Contr. West. Bot. **12:** 53, 1908.

Erect, simple or sparingly dichotomously branched annual 5–20 cm. tall; stems usually glabrous, reddish and somewhat glaucous, internodes elongate; leaves palmately 3–5-parted into linear lobes 5–20 mm. long, or lower leaves sometimes simple and entire, glabrous, acute or mucronulate at apex; flowers in dichotomous cymes, vespertine, solitary in forks on pedicels 1–10 mm. long; calyx urceolate, 8–10 mm. long, free tips of lobes extending 2–3 mm. beyond the membranous tube, slightly flaring; corolla funnelform, 2–3 cm. long, tube 5–7 mm. long, included in calyx, throat 1.5–3 mm. long; corolla lobes broadly obovate, entire or erosulate, white within, usually with a purplish or brownish wash on portion of backs that is exposed in bud; stamens inserted low in corolla tube, included; filaments villous on dilated base; stigma included below anthers; capsule oblong, 7–10 mm. long, each cell several-seeded; seeds with a membranous coat.

Hillsides, valleys, and sandy slopes and washes, Upper and less commonly Lower Sonoran Zones, along the western margins of the Mojave Desert north in the Coast Ranges to the San Francisco Bay area and the foothills of the Sierra Nevada to Inyo and Tulare Counties, California, east into southern Nevada and Arizona. Barely entering the Sonoran Desert in Mohave County, Arizona. March–May.

**4. Linanthus bigelovii** (A. Gray) Greene, Pittonia **2:** 253, 1892

*Gilia dichotoma* var. *parviflora* Torr., Bot. Mex. Bound. 147, 1859.
*Gilia bigelovii* A. Gray, Proc. Amer. Acad. **8:** 265, 1870.
*Navarretia parviflora* Kuntze, Rev. Gen. Pl. **2:** 433, 1891.

An erect annual 1–3 dm. tall, simple or dichotomously branched, glabrous throughout, internodes elongated and wiry; leaves linear and entire or 2–3-cleft into linear segments 1–3 cm. long; flowers sessile or short-pedicellate in forks of cymes, vespertine; calyx 7–12 mm. long, cylindrical, hyaline membrane between linear lobes forming a tube two-thirds to three-fourths as long as lobes and truncate across top, tips of lobes erect or slightly flaring, mucronulate; corolla narrowly funnelform, 10–15 mm. long, tube 5–8 mm. long, throat scarcely distinguishable from tube, lobes obovate, rounded, 4–5 mm. long, white within, washed with purplish or brownish on backs; stamens 2–3 mm. long, inserted on upper half of tube, glabrous; stigma and stamens included; capsule cylindrical-oblong, several-seeded; seeds ellipsoid, coat hyaline and bladdery.

Grassy or sandy hillsides, slopes, washes, and desert flats, Lower and Upper Sonoran Zones, Colorado and Mojave Deserts north in the inner Coast Ranges to Stanislaus County, California, east to southern Nevada and western Texas, and south into northern Sonora and northern Baja California. March–May.

**5. Linanthus jonesii** (A. Gray) Greene, Pittonia **2:** 254, 1892

*Gilia jonesii* A. Gray, Syn. Fl. N. Amer. ed. 2, $2^1$: 407, 1886.
*Navarretia jonesii* Kuntze, Rev. Gen. Pl. **2:** 433, 1891.
*Gilia bigelovii* var. *jonesii* Brand, Pflanzenreich $4^{250}$: 144, 1907.
*Linanthus bigelovii* var. *jonesii* Jepson & Mason, in Jepson, Man. Fl. Pl. Calif. 802, 1925.

Erect annual 2–10 cm. tall, stems slender to subfiliform, simple or sparingly branched dichotomously, glabrous throughout except on pedicels and calyces; leaves filiform, 1–2.2 cm. long, entire; flowers sessile to short-pedicellate at ends of branches and in axils of cyme, pedicels stipitate-glandular, at least near apex; calyx cylindrical, 5–6 mm. long, stipitate-glandular and sometimes sparsely villous also, slender linear lobes extending 1–1.5 mm. beyond hyaline membranous portion of tube; corolla tubular-funnelform, 8–10 mm. long, yellow or rarely white, lobes obovate, about 3 mm. long; stamens 2 mm. long, inserted in corolla tube, these and stigma included, filaments glabrous; capsule ellipsoid-oblong, 4–5 mm. long, each cell several-seeded; seeds reniform or subreniform, constricted at hilum, testa not hyaline nor bladdery.

Flats and sandy slopes, Lower Sonoran Zone, eastern Mojave and Colorado Deserts, east into Arizona and south into northwestern Sonora and at least to the vicinity of Calmallí, Baja California. March–May.

This and *L. bigelovii* are similar in general appearance, but the glandular-puberulent pedicels and calyces, coupled with the constricted seeds with closely applied seed coats, readily separate *L. jonesii* from *L. bigelovii.*

**6. Linanthus maculatus** (Parish) Milliken, Univ. Calif. Pub. Bot. **2:** 55, 1904
*Gilia maculatus* Parish, Bull. Torrey Club **19:** 93, 1892.

Small annual 1.5–3 cm. tall, more or less branched and sparsely leafy, sparsely villous throughout; leaf blades oblong-linear, 2–5 mm. long, entire, cuspidate, puberulent above and ciliate-margined, glabrous beneath; flowers subsessile in lower forks and crowded into small, several-flowered, headlike clusters at ends of branches; calyx narrowly campanulate, 2–2.5 mm. long, lobes distinct nearly to base, slightly unequal, sparsely puberulent to glabrous on backs, ciliate-margined, tips minutely and abruptly cuspidate; corolla campanulate, white, 3–4 mm. long, barely surpassing calyx, throat and tube almost indistinguishable, lobes about one-third as long as combined throat and tube, broader than long; stamens inserted on summit of tube, included in throat; capsule ovoid, 1.5–2 mm. long, several-seeded; seeds unchanged when moistened.

Sandy washes and flats, Lower Sonoran Zone, northwestern corner of the Colorado Desert in the vicinity of Palm Springs and Whitewater, and to the vicinity of Twentynine Palms, San Bernardino County, California. March–May.

**7. Linanthus demissus** (Torr.) Greene, Pittonia **2:** 257, 1892
*Gilia dactylophyllum* Torr., in Ives, Rep. 22, 1860. (Nomen confusum.)
*Gilia demissa* A. Gray, Proc. Amer. Acad. **8:** 263, 1870.
*Navarretia demissa* Kuntze, Rev. Gen. Pl. **2:** 433, 1891.
*Linanthus dactylophyllum* Rydb., Fl. Rocky Mts. 698, 1917.

Erect, ascending or widely spreading annual with dichotomously branched stems 2.5–10 cm. long, moderately villous throughout or glabrate, stems often reddish-tinged; leaves palmately 3-cleft into linear lobes 3–10 mm. long or less commonly entire or 5-cleft, opposite at base, alternate near tips of flowering branches; flowers short-pedicellate or subsessile in compact, headlike cymes; sepals distinct to base or nearly so, lanceolate, hyaline-margined nearly to tip, 3–4 mm. long, minutely cuspidate, finely and sparsely puberulent to glabrate; corolla campanulate, 5–6 (rarely 8) mm. long, tube very short, lobes ovate, about equaling combined tube and throat, white with 2 dark lines at base of each lobe; stamens inserted on base of throat, about 2 mm. long, included, filaments glabrous; stigma slightly exceeding stamens; capsule ellipsoidal, 2–3 mm. long, several-seeded.

Open desert, arid hills, Lower Sonoran Zone, southern Mojave Desert north to Inyo County, California, and east into Nevada, southern Utah, and Arizona; occasional in the Sonoran Desert in Maricopa and northern Yuma Counties, Arizona. March–May. Not known from the Colorado Desert.

**8. Linanthus bellus** (A. Gray) Greene, Pittonia **2:** 256, 1892
*Gilia bella* A. Gray, Proc. Amer. Acad. **20:** 301, 1885.
*Navarretia bella* Kuntze, Rev. Gen. Pl. **2:** 433, 1891.
*Linanthus peirsonii* H. L. Mason, Madroño **2:** 23, 1931.

Annual with several to many spreading-ascending, wiry, glabrous, reddish stems 2–6 cm. long, bearing 1 or 2 whorls of leaves on each stem; leaves opposite, usually 3-cleft but rarely simple or as much as 7-cleft, divisions 1–4 mm. long, 0.5–1 mm. thick, sparsely villous; flowers sessile, in 1–3-flowered, bracteate glomerules at end of each branch; calyx about 4 mm. long, tubular part 1–1.5 mm. long, lobes oblong, ovate, hyaline-margined to glabrous tip, puberulent and each with a purple spot toward base; corolla short-funnelform, 10–15 mm. long, lilac to pink, throat golden-yellow and with a purple spot below each lobe, these broadly obovate, subtruncate, margins usually entire, sometimes very faintly erosulate-denticulate, lobes equaling or exceeding combined throat and very short tube; stamens glabrous, inserted near base of throat; capsules ellipsoidal, 3–4 mm. long, several-seeded.

Sandy or gravelly, arid soil on hillsides, washes, and desert plains, Sonoran Zones, southwestern Colorado Desert and adjacent flanks of the mountains south to the southern end of the Sierra San Pedro Mártir. March–May.

**9. Linanthus dianthiflorus** (Benth.) Greene, Pittonia **2:** 254, 1892

*Fenzlia dianthiflora* Benth., Bot. Reg. **19:** under *pl. 1622,* 1833.
*Gilia dianthoides* Endl., Atakta Bot. *pl. 29,* 1833.
*Gilia fenzlia* Steudel, Nom. ed. 2, **1:** 683, 1840.
*Gilia dianthiflora* Steudel ex. Benth., in A. DC., Prod. **9:** 314, 1845.
*Fenzlia speciosa* Nutt., Jour. Acad. Phila. **1:** 157, 1848.
*Fenzlia concinna* Nutt., loc. cit.
*Navarretia dianthiflora* Kuntze, Rev. Gen. Pl. **2:** 433, 1891.
*Gilia dianthoides alba* Orcutt, W. Amer. Sci. **7:** 132, 1891.

Annual with erect, decumbent or prostrate stems 3–15 cm. long, slender, reddish- or purplish-tinged stems simple or diffusely branched, glabrous to puberulent or rarely floccose about nodes; leaves simple, entire, filiform, 5–30 mm. long, glabrous or puberulent near base, particularly along margin, acute at apex; flowers in few-flowered, leafy-bracteate cymes or solitary on slender pedicels to 5 mm. long or subsessile, pedicels when present puberulent; calyx 8–10 mm. long, tubular, cleft about half the length, linear lobes without a hyaline margin, puberulent within; corolla short-funnelform, 10–25 mm. long, tube 2–5 mm. long, throat yellow, 4–5 mm. long, bearing stamens and a ring of puberulence at its base, corolla lobes pink or white, broadly obovate, longer than combined tube and throat, delicately thin margins erose-denticulate; stamens equaling throat, puberulent at base; stigma lobes about equaling style, exserted; capsule ellipsoidal, 5–8 mm. long, many-seeded; seeds frequently narrowly winged on angles.

Grassy and sandy slopes, valleys, and washes, Sonoran Zones, western Colorado Desert to the coast in Santa Barbara County, California, and south in Baja California at least to the latitude of San Quintín. Feb.–May.

## Family 120. **HYDROPHYLLACEAE.** Waterleaf Family

Annual, biennial or perennial herbs or shrubs with alternate or opposite, simple, divided or compound leaves, and cymose or solitary, axillary flowers. Flowers perfect, regular or essentially so. Calyx lobes usually 5, similar or unequal, with or without auricles, often accrescent in fruit. Corolla sympetalous, lobes usually 5, nearly or quite equal, longer or shorter than corolla tube, usually with a pair of scales at base of each filament. Stamens mostly 5, hypogynous, exserted or included. Pistil 1, compound, consisting of 2 united carpels; style deeply 2-parted to entire. Stigmas 2, capitate or rarely subulate. Fruit a loculicidal capsule dehiscent by 2 valves or both loculicidally and septicidally dehiscent by 4 valves or irregularly dehiscent. Capsule 1-celled or imperfectly 2-celled by intrusion of parietal placentae. Ovules few to numerous. Seeds 1 to over 100. Endosperm present; cotyledons 2, entire.

Ovary unilocular, placentae nearly filling and forming lining of capsule:
  Herbage neither viscid nor scented; ovules borne only on front of placentae; seeds reticulate or alveolate ........................................ 1. *Pholistoma*
  Herbage viscid and scented; ovules borne on both faces of placentae; seeds smooth or corrugated ........................................ 2. *Eucrypta*
Ovary partially or completely 2-celled by intrusion of narrow parietal placentae:
  Calyx lobes similar or nearly so, or if differing in size and shape plants annual and usually with toothed or divided leaves:
    Stamens equally inserted, if unequal in length, seeds strongly corrugated:
      Corolla blue or violet to white, deciduous, or if marcescent seeds not transversely corrugated; pedicels rarely equaling flowers ........................ 3. *Phacelia*
      Corolla yellow, withering-persistent and enclosing mature capsule; pedicels equaling or longer than pendulous flowers; seeds foveolate ............ 4. *Emmenanthe*
    Stamens unequal or unequally inserted or both:
      Annual or perennial herbs; leaves not coriaceous; capsule membranous or only slightly cartilaginous; seeds usually 6 to many; flowers in reduced cymes or solitary .......................................................... 5. *Nama*
      Shrubs, woody throughout; leaves coriaceous; capsule cartilaginous; seeds usually 2–12; flowers in open panicles or cymes ........................ 6. *Eriodictyon*
  Calyx lobes dimorphic, outer conspicuously enlarged, cordate, and strongly venose in fruit, inner lobes linear; plants perennial with entire leaves ................ 7. *Tricardia*

## 1. PHOLISTOMA Lilja, Bot. Notiser 40, 1839

Green or glaucous, succulent, annual herbs with reclining or scrambling, angled, brittle stems having uncinate hairs on angles. Herbage hispidulous to glabrate. Lowermost leaves opposite, others alternate, ovate, ovate-lanceolate, to oblong, cordate to truncate at base, pinnately divided into various-shaped, entire or few-toothed lobes, usually markedly bristly on veins below and on margins. Flowers solitary in axils or in few-flowered, raceme-like cymes. Calyx broadly campanulate, 5-cleft nearly to base; sinuses with sepal-like auricles or naked. Corollas semirotate to bowl-shaped, white, blue or violet, lobes exceeding the tube; a pair of scales at base of each filament. Stamens exceeding corolla tube. Styles short. Capsules globose, loculicidally dehiscent by 2 valves. Seeds 1–6, globose, regularly reticulate-faveolate.

Foliage green; calyx with conspicuous auricles, enveloping mature capsule; capsule 4–8 mm. in diameter:
  Cauline leaves auriculate-clasping; corolla scales broad and conspicuous ........................................ 1. *P. auritum* var. *arizonicum*
  Cauline leaves narrowly winged but only slightly connate on petioles; corolla scales narrow and inconspicuous ........................................ 2. *P. racemosum*
Foliage glaucous; calyx without auricles, stellate-rotate under mature capsule; capsule 2–4 mm. in diameter ........................................ 3. *P. membranaceum*

### 1. **Pholistoma auritum** (Lindl.) Lilja var. **arizonicum** (M. E. Jones) Constance, Bull. Torrey Club **66:** 348, 1939

*Nemophila arizonica* M. E. Jones, Contr. West. Bot. **12:** 50, 1908.
*Nemophila aurita* var. *arizonica* Brand, Pflanzenreich $4^{251}$: 45, 1913.

Stems slender, weak and scrambling, sometimes 1 m. long, uncinate-bristly on angles; herbage bright green, sparsely hispid throughout; leaves oblong to ovate-lanceolate in outline, 1.5–8 cm. long, 0.5–3.5 cm. broad, with 2–5 lobes on each side, these oblong, ovate, or nearly deltoid, usually retrorse, obtuse at tips; petioles moderately winged (wings 1–5 mm. wide), auriculate-clasping; flowers solitary or in 1–4-flowered cymes; pedicels 1–2.5 cm. long, sparsely uncinate-hispid; calyx lobes lanceolate to ovate-lanceolate, 1.5–2 mm. wide, 5–7 mm. long, sparsely hispid; auricles barely 1 mm. long; corolla 8–12 mm. broad, blue or violet; corolla scales triangular, purple, margins fimbriate; styles 2.5–3 mm. long; capsule 4–5 mm. in diameter, hispid with subulate white hairs; seeds 2.5–3 mm. in diameter, dark brown.

Canyons and under bushes and trees on rocky hillsides and plains, Sonoran Zones, Pinal, Maricopa, and Pima Counties, Arizona, and said to occur in "southeastern California," but I have seen no specimens from west of the Colorado River. Feb.–April.

**2. Pholistoma racemosum** (Nutt.) Constance, Bull. Torrey Club **66:** 349, 1939

*Nemophila racemosa* Nutt. ex A. Gray, Proc. Amer. Acad. **10:** 315, 1875.
*Viticella racemosa* J. F. Macbride, Contr. Gray Herb. **59:** 30, 1919.
*Nemophila erodifolia* Millsp. ex Millsp. & Nutt., Field Mus. Bot. Ser. **5:** 205, 1923.

Scrambling or reclining plant with glaucous stems and herbage; stems 3–8 dm. long, sparsely hispidulous to glabrate; lower leaves ovate to deltoid-ovate in outline, 4–10 cm. long, 2–6 cm. wide, obtuse at apex, subcordate to truncate at base, with 5–9 oblong or ovate, obtuse to acute, entire or 3–5-toothed lobes on each side, sparsely hispidulous above, more densely so beneath; petioles narrowly winged but not auriculate, 1–2 cm. long; flowers in few-flowered cymes or less commonly solitary; calyx lobes 2–3 mm. long, 1–1.5 mm. wide, linear-lanceolate to ovate-lanceolate, auricles similar in shape, 1–1.5 mm. long; corolla 6–10 mm. broad, white or blue, lobes obovate, often puberulent toward apex; corolla scales triangular but narrow, or obsolete, upper portion often free and minutely fimbriate; style about 2.5 mm. long; capsule 5–8 mm. in diameter, enclosed by accrescent calyx; seeds mostly 4–6, sometimes only 1, 1–2 mm. in diameter.

Canyons, rocky hillsides, mesas, and under trees and shrubs on gravelly hillsides, chiefly Upper Sonoran Zone but entering the Lower Sonoran Zone in the southern part of its range; Santa Barbara County and adjacent islands south near the coast to the southern foothills of the Sierra San Pedro Mártir and infrequently in favorable locations to central Baja California (Rancho Mesquital). Feb.–May.

**3. Pholistoma membranaceum** (Benth.) Constance, Bull. Torrey Club **66:** 350, 1939

*Ellisia membranacea* Benth., Trans. Linn. Soc. **17:** 274, 1835.
*Macrocalyx membranaceus* Kuntze, Rev. Gen. Pl. **2:** 434, 1891.
*Nemophila membranacea* Greene, Man. Bay Reg. 252, 1894.
*Ellisia membranacea* var. *hastifolia* Brand, Pflanzenreich $4^{251}$: 38, 1913.
*Nyctelea membranacea* Tidestrom, Proc. Biol. Soc. Wash. **48:** 42, 1935.

Stems and foliage glaucous; stems glabrous, 0.5–6 dm. long, slender; lower leaves oblong to oval in outline, 3–12 cm. long, 1–5 cm. broad, obtuse at apex, subcordate to truncate at base, with 5–11 oblong, obtuse, rather distant lobes on each side, these entire or with a single broad lateral tooth, sparsely hispidulous on both surfaces; petioles narrowly winged but not auriculate; cymes 2–10-flowered or flowers rarely solitary; calyx lobes 2–3 mm. long, 1–2 mm. wide, oval, obtuse, auricles lacking; corolla 4–10 mm. broad, white and usually with a purple spot on each lobe, these ovate; corolla scales triangular, very small and inconspicuous, entire; style 1.5–2 mm. long; capsule 2–4 mm. in diameter, calyx stellate-spreading under globose capsule; seeds 1–2, 2–3 mm. in diameter.

Hillsides, under shrubs, in canyons, and on rocky slopes, Sonoran Zones, foothills of Coast Ranges and Sierra Nevada from Contra Costa and Mariposa Counties, California, south into Baja California and east into the desert ranges of Inyo and San Bernardino Counties; occurring in the Sonoran Desert along the western and northern margins of the Colorado Desert and along the eastern escarpment of the Sierra San Pedro Mártir. March–June.

## 2. EUCRYPTA Nutt., Jour. Acad. Phila. II, 1: 158, 1848

In our area delicate annual herbs with strong-scented, glandular-viscid foliage and diffusely branching stems. Lowest leaves opposite, petiolate; petioles narrowly winged; blades oblong to ovate, truncate at base, pinnately divided into oblong-lanceolate or spatulate lobes which are again 1–2-pinnatifid, thin, more or less hispid on both surfaces; upper leaves reduced, less intricately divided to entire, sessile and often auriculate-clasping. Flowers usually 4–15 in solitary, geminate, or paniculate, loose, naked or few-bracteate cymes. Pedicels filiform. Calyx shallowly campanulate, divided one-half to two-thirds of way to base into ovate, oblong or spatulate obtuse lobes, these more or less hispid on both surfaces; sinuses naked. Corolla campanulate, white, yellowish, bluish or purplish; lobes shorter than tube (in ours), equaling or slightly shorter than calyx. Corolla scales minute, triangular or none and a minute fold between each pair of filaments. Stamens about equaling tube. Capsules globose or ovoid, hispid. Ovules borne on both front and back of each placenta. Seeds 5–15, inner ones lenticular and smooth and the outer terete and corrugated, or all seeds alike, corrugated.

Mature calyx stellate-rotate beneath the capsule; seeds dimorphic; sepals rarely stipitate-glandular ............................... 1. *E. chrysanthemifolia* var. *bipinnatifida*

Mature calyx campanulate, the lobes erect and enclosing the capsule; seeds all alike; sepals commonly stipitate-glandular ....................................... 2. *E. micrantha*

### 1. **Eucrypta chrysanthemifolia** (Benth.) Greene var. **bipinnatifida** (Torr.) Constance, Lloydia **1**: 147, 1939

*Phacelia micrantha* var. ? *bipinnatifida* Torr., Ives Rep. 21, 1860.
*Ellisia torreyi* A. Gray, Proc. Amer. Acad. **20**: 302, 1885.
*Macrocalyx bipinnatifidus* Coville, Contr. U.S. Nat. Herb. **4**: 157, 1893.
*Eucryta torreyi* Heller, Cat. N. Amer. Pl. 7, 1898.
*Ellisia torreyi* var. *bipinnatifida* Brand, Pflanzenreich $4^{251}$: 42, 1913.
*Nyctelea torreyi* Tidestrom, Proc. Biol. Soc. Wash. **48**: 42, 1935.

Weak and rather diffuse plants with slender stems 1–3 dm. long; lower leaves oblong to broadly ovate, 2–4.5 cm. long, 1–2.5 cm. wide, pinnately divided into 7–9 oblong lobes which are usually again shallowly pinnatifid and with rounded or obtuse segments; flowers 4–8 in much-branched cymes; pedicels slender, recurved in fruit; calyx lobes 1–2 mm. long, 1–1.5 mm. wide, deltoid or ovate, obtuse; corolla 2–3 mm. broad, white or bluish, equaling or slightly exceeding calyx; corolla scales minute, denticulate; style about 0.8 mm. long, barely exserted beyond calyx or included; mature capsule 2–3 mm. in diameter; seeds 6–8, inner ones lenticular-orbicular, smooth, about 1 mm. in diameter, outer ones oblong-ovoid, terete, 0.8–1 mm. long, corrugated, dark brown.

Often under trees and shrubs, hillsides and arid slopes and plains, Lower Sonoran Zone, Mojave and Colorado Deserts to southeastern Nevada, west central Arizona (as far east as Pima County), and south into northern Sonora and to Guadalupe Island and Comondú, Baja California. Feb.–April.

**2. Eucrypta micrantha** (Torr.) Heller, Muhlenbergia **2:** 163, 1906

*Phacelia micrantha* Torr., Bot. Mex. Bound. 144, 1859.
*Macrocalyx micranthus* Coville, Contr. U.S. Nat. Herb. **4:** 157, 1893.
*Phacelia pinetorum* M. E. Jones, Zoe **4:** 279, 1893.
*Ellisia micrantha* Brand, Pflanzenreich **4**[251]: 42, 1913.
*Nyctelea micrantha* Woot. & Standl., Contr. U.S. Nat. Herb. **19:** 535, 1915.
*Eucrypta pinetorum* Rydb., Fl. Rocky Mts. 701, 1917.
*Nyctelea pinetorum* Tidestrom, Contr. U.S. Nat. Herb. **25:** 442, 1925.

Weak diffuse herb with stems to 3 dm. long, these often stipitate-glandular and sparsely to moderately hispid; lower leaves oblong to oval, 10–30 mm. long, 5–20 mm. wide, pinnatifid into 7–9 oblong to spatulate, straight or falcate segments, these entire or few-toothed; petioles about equaling blades, slender; upper leaves similar in shape but much reduced, lobed, toothed, or entire, short broad petioles auriculate-clasping; inflorescences 4–12-flowered; pedicels usually erect in fruit, these and calyces often stipitate-glandular and villous; calyx lobes oblong to spatulate, 1.5–2 mm. long, 1–1.5 mm. wide at anthesis, elongating somewhat in fruit; corolla 2–4 mm. broad, purplish, blue or less commonly white with a yellow tube, lobes oblanceolate, barely exceeding calyx; corolla scales none, a V-shaped fold between adjacent stamens; style 1–1.5 mm. long; capsules 2–3 mm. in diameter, enclosed by elongated, erect or slightly spreading calyx lobes; seeds 7–15, all alike, oblong and terete but becoming curved and vermiform, about 0.8 mm. long, transversely corrugated.

Rocky hillsides, desert slopes and plains, often under shrubs and trees, Lower and Upper Sonoran Zones; Inyo County, California, south through the Mojave and Colorado Deserts to central Baja California and east to Utah, Texas, and northern Sonora. Feb.–May.

## 3. PHACELIA Juss., Gen. 129, 1789

Annual, biennial or perennial herbs from taproots or woody crowns; herbage variously pubescent and often more or less glandular. Leaves chiefly alternate, at least basal, petiolate, entire or pinnately lobed, pinnatifid or dissected. Flowers in simple or branched terminal cymes, these lax and only vaguely scorpioid or more compact and strongly scorpioid, few- to many-flowered. Calyx 5-cleft nearly to base; lobes usually equal. Corolla tubular to rotate, variously lobed, shorter than to several times as long as calyx. Stamens equally inserted on corolla tube, subequal to unequal, included or exserted, a pair of scales often present at base of filament and adnate or partially free from corolla tube, or the scales absent or obsolete. Style bifid. Capsule 1- or 2-celled at maturity, loculicidally dehiscent. Ovules 2 to many on each placenta. Seeds terete in cross section or cymbiform, with a prominent median ridge separating two deep excavations on ventral side.

◆ Seeds foveolate, pitted or papillate, not strongly transversely corrugated, or if corrugated excavated on each side of a prominent ridge ventrally:
■ Corolla scales present, adnate to corolla tube, sometimes greatly reduced:
Seeds terete in cross section, not excavated on inner side; herbage usually not malodorous:
Leaves variously pinnately toothed or lobed or pinnatifid, often finely decompound, their segments again toothed, lobed, or pinnatifid, ovate to oblong in outline, thin, not subfleshy:
Plants perennial, often somewhat woody at base; calyx lobes spatulate .......... 1. *P. ramosissima*

Plants annual, entirely herbaceous; calyx lobes linear-oblong or narrowly ovate:
Calyx lobes considerably enlarged, to 10 mm. long in fruit; corollas lavender:
Corolla 4–7 mm. long, 3–4 mm. broad; capsules hirsute but without pustules at bases of hairs; inflorescence heavily glandular, moderately hispid .... 2. *P. cryptantha*
Corolla 8–12 mm. long, about as wide; capsules hispid with pustules at bases of hairs; inflorescence densely hispid, scantily glandular .... 3. *P. cicutaria* var. *hispida*
Calyx lobes not conspicuously enlarged, not over 8 mm. long in fruit; corollas bluish, purple, or white:
Corollas 6–12 mm. long, broadly campanulate, as wide as long, bluish, purplish, or rarely cream-colored; stems not conspicuously weak or slender:
Calyx lobes linear or lance-subulate, about 1 mm. wide or less; capsule ovoid, hairy only at apex .... 4. *P. tanacetifolia*
Calyx lobes oblanceolate to obovate, to 3 mm. broad; capsule globose, hairy to base .... 5. *P. distans*
Corollas 3–5 mm. long, tubular-campanulate or funnelform-campanulate, longer than wide, white or sordid; stems weak and very slender:
Capsules globose, shorter than calyx lobes; stamens wholly included .... 6. *P. umbrosa*
Capsules ovoid, nearly equaling calyx lobes; stamens equaling corolla lobes .... 7. *P. pauciflora*
Leaves entire or shallowly lobed, orbicular or nearly so, thick and subfleshy .... 8. *P. rotundifolia*
Seeds cymbiform, excavated on each side of a prominent ridge on inner surface; herbage usually malodorous:
Pedicels subfiliform, 3–6 mm. long; calyx lobes distinctly scarious and subglabrate toward tips:
Calyx lobes 5–6 mm. long in fruit, oblanceolate; seeds transversely corrugated as well as foveolate on back .... 9. *P. pedicellata*
Calyx lobes 8–10 mm. long in fruit, obovate; seeds foveolate but not corrugated on back .... 10. *P. scariosa*
Pedicels thicker, less than 2.5 mm. long at anthesis; calyx lobes not markedly scarious:
Seeds transversely corrugated as well as foveolate on back, 2.5–3.5 mm. long; capsule ovoid or ovoid-subglobose; corollas bluish, violet, purplish or yellowish:
Stamens and style distinctly exserted; corollas violet, purple, or yellowish; seeds 3–3.5 mm. long:
Corolla 6–10 mm. long, violet or purple .... 11. *P. crenulata* var. *crenulata*
Corolla 3–4 mm. long, yellowish .... 11*a*. *P. crenulata* var. *minutiflora*
Stamens and style included; corollas blue or whitish; seeds 2.5–3 mm. long .... 12. *P. coerulea*
Seeds foveolate but not corrugate on back, 2–2.5 mm. long; capsule globose; corollas sordid-white .... 13. *P. popei* var. *arizonica*
■Corolla scales absent, or represented by a dilation of basal part of stamen filament and free from corolla:
Corolla subrotate, 7–20 mm. long (or deep), tube about as long as limb; corollas sordid-white, or violet with white spots below lobes:
Pedicels 1–3 cm. long in fruit; corolla violet with 5 white spots below lobes; seeds 50–60, 0.6–0.8 mm. long:
Corolla 10–20 mm. long; basal pads on filaments quadrate, slightly toothed; stems erect .... 14. *P. parryi* var. *parryi*
Corolla 10–12 mm. long; basal pads ovate, not toothed; stems mostly spreading .... 14*a*. *P. parryi* var. *celata*

Pedicels 0.5–1 cm. long in fruit; corolla sordid-white; seeds 8–15, about 1–1.5 mm. long .................................................... 15. *P. longipes*

Corolla tubular-campanulate or campanulate-funnelform, 15–40 mm. long, tube longer than lobes; corollas violet or deep blue:

Corolla violet, tubular-campanulate, constricted at throat, lobes 3–8 mm. long ...... ........................................................ 16. *P. minor*

Corolla deep blue, campanulate-funnelform, not constricted at throat, lobes 8–10 mm. long ................................................ 17. *P. campanularia*

◆ Seeds strongly corrugated transversely; ovules numerous:

Leaves broadly ovate to orbicular, entire to crenate-dentate; calyx lobes shorter than mature capsule; seeds 50 or more; style divided about to middle:

Stamens wholly included, filaments more or less pilose toward base; seeds 1–1.2 mm. long ...................................................... 18. *P. pachyphylla*

Stamens equaling tube or barely exserted, filaments glabrous; seeds 0.6–0.8 mm. long ........................................................ 19. *P. neglecta*

Leaves ovate to linear-oblong in outline, pinnately lobed to bipinnatifid; calyx lobes equaling or exceeding capsules; seeds 20 or less; styles divided about one-fourth their lengths or less:

Corollas 2–4.5 mm. long, shorter than to equaling calyx lobes at anthesis:

Calyx lobes linear or linear-oblanceolate, tips somewhat spreading in fruit .......... ...................................................... 20. *P. ivesiana*

Calyx lobes spatulate, tips erect .................................. 21. *P. affinis*

Corollas 8–15 mm. long, at least twice as long as calyx lobes at anthesis:

Seeds 15–20, about 0.5 mm. long; corolla scales usually obsolete; corolla limb white to lavender ............................................ 22. *P. brachyloba*

Seeds 8–12, 1–1.2 mm. long; corolla scales linear-lanceolate; corolla limb blue ...... ...................................................... 23. *P. fremontii*

## 1. **Phacelia ramosissima** Dougl. ex Lehm., Stirp. Pug. **2:** 21, 1830

*Phacelia decumbens* Greene, Pittonia **5:** 17, 1902.
(For additional synonymy see Jepson, Fl. Calif. **3:** 259, 1943.)

Several-stemmed, erect or ascending plant 5–12 dm. tall from a perennial root, finely cinereous-pubescent and hirsute or hispid throughout, glandular-viscid in inflorescence; leaves broadly ovate to oblong, 5–10 cm. long, 2–6 cm. wide, pinnate, narrowly ovate to oblong lobes toothed, incised or sometimes twice pinnatifid, having appearance of distinct leaflets; petioles slender, 1–3 cm. long; upper leaves similar in shape and lobing but smaller and sessile or nearly so; cymes dense but scattered, 2–8 cm. long, erect in fruit; pedicels rarely over 1 mm. long even in fruit; calyx lobes spatulate, 5–8 mm. long, 1–2.5 mm. wide at widest part, densely glandular-hispid, slightly enlarged in fruit; corolla campanulate, pale sordid-blue or sordidly white, 5–8 mm. long, 6–10 mm. broad, ovate lobes 2–3 mm. long, spreading; stamens 7–10 mm. long, somewhat exserted, glabrous; corolla scales ovate, attenuate toward apex, adnate at base of filaments only; style 7–10 mm. long, glabrous, cleft about two-thirds of way to base; capsule ovoid, 3–4 mm. long, hispid toward apex, slightly less pubescent at base, acute; seeds usually 2–4, ovoid to oblong, 2.5–3 mm. long, dark brown, uniformly finely foveolate.

Rocky or brushy slopes and along watercourses, Lower Sonoran to Transition Zones, coastal areas from northern Baja California to Washington, the Sierra Nevada, and east to Utah and Arizona; entering the Sonoran Desert in central Arizona (Maricopa County) and probably at other points. March–May.

**2. Phacelia cryptantha** Greene, Pittonia **5:** 21, 1902

*Phacelia hispida* var. *brachyantha* Coville, Contr. U.S. Nat. Herb. **4:** 158, 1893.
*Phacelia eremica* Jepson, Man. Fl. Pl. Calif. 823, 1925.
*Phacelia cryptantha* var. *derivata* Voss, Bull. S. Calif. Acad. **33:** 174, 1935.

Annual 1–5 dm. tall from a slender taproot, short-glandular-pubescent and hispid throughout; inflorescence most heavily glandular; stems much branched from base, branches ascending to erect, forming a rounded clump; leaves oblong or ovate, 3–9 cm. long, 1.5–5 cm. wide, pinnate or pinnatifid, divisions ovate to oblong, remote, toothed; petioles slender, usually distinctly shorter than blade; cymes simple or few-branched, lax, many-flowered, branches 2–10 cm. long, these erect in fruit; calyx lobes linear-oblanceolate with a very narrow claw and dilated at apex, 4–7 mm. long, 0.5–1 mm. wide, enlarging to about 1 cm. long in fruit, densely hispid; corolla campanulate-funnelform, 4–7 mm. long, 3–4 mm. wide across limb, lobes lavender, 2–3 mm. long; stamens 3–4 mm. long, glabrous, included; corolla scales inserted in upper part of tube, adnate full length or with an acute free tip; style 3–4 mm. long, hispid toward base, cleft about to middle or slightly below; capsule globose, about 2 mm. long, hispid; seeds usually 4, dark brown, 1.5–2.5 mm. long, foveolate in an irregular pattern.

Moist hillsides and canyons of arid ranges, Lower and Upper Sonoran Zones, western margin of the Colorado Desert to Inyo County, California; east into Nevada and to Pinal and Maricopa Counties, Arizona. April–June.

**3. Phacelia cicutaria** Greene var. **hispida** (A. Gray) J. T. Howell, Leafl. West. Bot. **3:** 120, 1942

*Phacelia ramosissima* var. *hispida* A. Gray, Proc. Amer. Acad. **10:** 319, 1875.
*Phacelia hispida* A. Gray, Syn. Fl. N. Amer. **2**$^{1}$**:** 161, 1878. Not *P. hispida* Buckl., 1862.
*Phacelia eximia* Eastw., Bull. Torrey Club **32:** 204, 1905.
*Phacelia hispida* var. *heliophila* J. F. Macbride, Contr. Gray Herb. **49:** 29, 1917.

Annual with stems usually widely branched, 2–6 dm. tall, densely hispid and hispidulous throughout or somewhat glandular; lower leaves oblong-ovate, 4–10 cm. long, 2.5–5 cm. broad, incised to pinnate, divisions again incised; inflorescence paniculately branched, lax, cymes 10–15 cm. long; calyx lobes conspicuously dilated at apex, 6–8 mm. long, 0.5–1 mm. wide, densely bristly-hispid, often elongating to about 10 mm. in fruit; pedicels slender, 1–3 mm. long; corolla lavender to sordid-yellow with a wash of lavender, broadly campanulate, 8–12 mm. long, about as broad, lobes 4–5 mm. long, obovate; stamens equaling corolla lobes or slightly exserted, glabrous; scales conspicuous, adnate full length; style about equaling or slightly exceeding stamens; capsule subglobose, about 3 mm. long, conspicuously hispid with stout white hairs; seeds usually 4, about 1.5–2 mm. long, coarsely foveolate.

Rocky hillsides, under shrubs, and along washes, Lower and Upper Sonoran Zones, mainly in the chaparral areas from Santa Barbara County south into northern Baja California but entering the Sonoran Desert along the western margins of the Colorado Desert and in adjacent Baja California, south discontinuously to Miller's Landing. March–May.

**4. Phacelia tanacetifolia** Benth., Trans. Linn. Soc. **17:** 280, 1835

*Phacelia tanacetifolia* f. *staminea* Brand, Univ. Calif. Pub. Bot. **4:** 216, 1912.
*Phacelia tanacetifolia* subvar. β *tenuisecta* Brand, Pflanzenreich **4**$^{251}$**:** 91, 1913.

Erect, rather stout, sparingly branched annual to 1.5 m. tall, bristly-hispid with scattered spreading hairs throughout; leaves thinly hispidulous; racemes densely bristly-hirsute; leaves ovate in outline, to 2.5 dm. long, once or twice pinnately divided or parted or pinnatifid, often very finely divided, ultimate segments 5–20 mm. long and again toothed or pinnatifid; cymes branched, compact, individual branches to 1 dm. long, often considerably surpassing leaves; calyx lobes linear or lance-subulate, 4–5 mm. long; corolla open-campanulate, 6–8 mm. long, blue, lobes broadly ovate or suborbicular, about one-fourth as long as tube, spreading; stamens exserted nearly length of corolla; corolla scales usually wholly adnate, tapering toward apex, the pairs connivent in front of filament by dilated middle part; capsule ovoid to subglobose, 3–4 mm. long, hairy only at apex, glabrous below; seeds usually 2, sometimes 3 or 4, about 3 mm. long, strongly transversely rugose.

Sandy, rocky, or clay foothills and plains and valleys, Lower Sonoran to Arid Transition Zones, Central Valley, California, and Mojave Desert south through the coastal ranges into northern Baja California and the northwestern corner of the Colorado Desert; introduced into western Arizona with grain seed. March–May.

**5. Phacelia distans** Benth., Bot. Sulph. 36, 1844

*Phacelia tanacetifolia* var. *tenuifolia* Torr., Bot. Mex. Bound. 143, 1859.
*Phacelia scabrella* Greene, Pittonia **1**: 35, 1887.
*Phacelia arthuri* Greene, op. cit. 224, 1888.
*Phacelia leptostachya* Greene, Erythea **2**: 190, 1894.
*Phacelia commixta* Greene, Pittonia **5**: 21, 1902.
*Phacelia distans* subvar. β *ammophila* Greene ex Brand, Univ. Calif. Pub. Bot. **4**: 216, 1912.
*Phacelia distans* var. *australis* Brand, loc. cit.

Slender to robust annual 2–12 dm. tall from a sturdy taproot, hispidulous to scabrous throughout; petioles usually shorter than blades; lower leaves oblong-oval to broadly ovate, 4–12 cm. long, 2.5–9 cm. broad, 1–2-pinnate, with oblong or lanceolate divisions, these toothed, incised, or pinnatifid, appressed-hirsute; flowers many, short-pedicellate, in scattered, simple to few-branched cymes 2–12 cm. long, erect in fruit; calyx lobes narrowly oblanceolate to obovate, often unequal, 4–6 mm. long, to 3 mm. wide, densely pubescent with stiff spreading hairs exceeding smaller softer ones; corolla cream, bluish, or purplish, broadly campanulate, deciduous, 6–12 mm. long, about as broad, lobes 3–5 mm. long, obovate; stamens 8–13 mm. long, glabrous; scales ovate, free at apex; style 7–12 mm. long, glabrous, cleft about three-fourths its length; capsule globose, 2–3 mm. in diameter, hispid; seeds 2–4, 2–3 mm. long, coarsely foveolate, brown.

Hillsides, valleys, sand dunes, and along arroyos, Lower Sonoran to Arid Transition Zones, central California to central Baja California and east to Nevada and Arizona. Feb.–May.

**6. Phacelia umbrosa** Greene, Erythea **2**: 191, 1894

*Phacelia hispida* var. *umbrosa* Brand, Pflanzenreich **4**$^{251}$: 88, 1913.

Annual with weak branching stems to 6 dm. long, sparingly hispidulous throughout but not glandular or viscid, sometimes glabrate; leaves ovate in outline, 3–7.5 cm. long, thin, pinnate or pinnately cleft into 3–5 divisions, these segments ovate, shallowly incised, lobed or crenate, 12–35 mm. long, the lower ones with petiolules; cymes dense but slender, simple or few-branched, elongated in fruit; calyx lobes linear, acute at apex, 4–5 mm. long, hispid; corolla tubular-campanulate, 3–5 mm. long, white, barely equaling or shorter than calyx lobes; stamens included; corolla scales united below, dilated toward apex; capsule globose, hispidulous; seeds usually 4, densely papillate.

Arid slopes and desert flats, Lower Sonoran Zone, southwestern part of the Colorado Desert, south into Baja California. A poorly known plant which may not be distinct from *P. distans.* March–May.

### 7. **Phacelia pauciflora** S. Wats., Proc. Amer. Acad. **24:** 61, 1889

Slender, weak-stemmed, simple or somewhat branching annual 1–3 dm. tall, apparently shade-loving; herbage sparsely hispid with stiff, spreading or slightly deflexed, whitish hairs; leaves narrowly ovate to oblong in outline, 1–3 cm. wide, 2–7 cm. long, pinnately divided and segments of lower ones again pinnatifid; petioles slender, to 2.5 cm. long; cymes short, rather compact, few-flowered, 1–3 cm. long, simple or geminately branched; pedicels subfiliform, 1–2 mm. long; calyx densely hispid without, especially on lower half, glabrous within, lobes linear-oblanceolate, 4–5 mm. long at anthesis, to 8 mm. long in fruit; corolla campanulate-funnelform, 4–5 mm. long, white, rounded lobes about 2 mm. wide, one-half as long; stamens barely equaling corolla lobes, slightly unequal, filaments glabrous; corolla scales very narrow (about 0.2 mm. wide), about 1.5 mm. long, united at base over filament, truncate at apex; style 3 mm. long, cleft to middle; capsule ovoid, 3–5 mm. long, hispid, especially on upper third; seeds 8–12, foveolate.

Sandy washes and rocky hillsides, Lower Sonoran Zone, east central Baja California from Bahía de los Angeles to Barríl. Probably more widespread than indicated by this range. Feb.–March.

### 8. **Phacelia rotundifolia** Torr. ex S. Wats., Bot. King Expl. 253, 1871

Annual with erect or ascending stems branching near base or simple, to 3 dm. tall, glandular-hirsute and hirsutulous throughout; leaves broadly ovate to orbicular, 5–20 mm. long and broad, usually subcordate at base, rounded at apex, coarsely toothed or rarely shallowly lobed, slender petioles usually longer than or equaling leaf blade; cymes lax, few-flowered, 1–2 cm. long, erect in fruit; pedicels slender, to 5 mm. long in fruit, ascending or spreading; calyx lobes linear or narrowly linear-oblanceolate, 2–4 mm. long, 0.5–1 mm. broad, tips slightly spreading, glandular-hirsute, to 6 mm. long in fruit; corolla tubular-campanulate, 4–5 mm. long, 2.5–3 mm. broad, violet to white, usually with a yellowish tube, at least in lower part, the lobes 1–2 mm. long; stamens 2–3 mm. long, with glabrous filaments; corolla scales narrow, about 1 mm. long, free from or attached to base of filaments; style 1.5–2 mm. long, cleft one-fourth to one-third its length, sparsely pubescent toward base; capsule oblong, about 4 mm. long, acute, puberulent; seeds numerous, ovoid, about 0.5 mm. long or less, coarsely foveolate.

Gravelly slopes and desert mesas, Sonoran Zones, Colorado and Mojave Deserts to southern Utah, and in Mohave and northern Yuma Counties, Arizona. March–May.

### 9. **Phacelia pedicellata** A. Gray, Syn. Fl. N. Amer. **2**[1]**:** 160, 1878

A glandular-hirsute annual 1.5–4 dm. tall with erect or ascending stems usually branched above base; leaves ovate, 6–12 cm. long, 4–10 cm. wide, pinnate or pinnatifid, lobes often petiolulate, irregularly lobed or dentate and somewhat undulate-crisped, thickish, dark green; petioles stoutish, usually equaling or much shorter than blades; glandular exudate malodorous; cymes compact, many-flowered, 3–5 cm. long, arranged in paniculately branched terminal inflorescence, erect in fruit, with slender pedicels 3–6 mm. long and sharply deflexed at tips; calyx lobes oblanceolate, 3–4 mm. long, 1–1.5 mm. wide, glandular-hirsute, scarious and 5–6 mm. long in fruit; corolla broadly campanulate, 4–6 mm. long, about as broad, blue, purple, or rarely white or pale lilac, obovate lobes 1–1.5 mm. long; stamens 6–8 mm. long, exserted, glabrous; corolla scales lunate to rectangular, somewhat auriculate at base, adnate full length; style 6–8 mm. long, glabrous, cleft about to middle; capsule ovoid, 3–4 mm. long; seeds usually 4, 2.5–3 mm. long, thickish, inner surface excavate on each side of a prominent median ridge, foveolate and transversely corrugated on back.

Desert talus slopes, washes, and canyons, mainly Lower Sonoran Zone, Colorado and Mojave Deserts east into western Arizona and south along the desert slopes at least to the southern end of the Sierra San Pedro Mártir, Baja California. March–May.

**10. Phacelia scariosa** Brandegee, Proc. Calif. Acad. II, **2:** 185, 1889

Slender to sturdy annual 1–4 dm. tall or more with erect or ascending stems branching mostly from above base, hirsutulous with fine spreading hairs and somewhat glandular throughout, inflorescences somewhat more densely glandular; leaves ovate in outline, 2–8 cm. long, 1.5–6 cm. wide, pinnately 3–5-lobed, at least lower lobes petiolulate, all of them again more or less lobed or dentate, terminal lobe of each blade much larger than others; petioles 1–5 cm. long; cymes lax, erect or ascending, usually more or less branched, individual racemes to 5 cm. long; pedicels subfiliform, 3–5 mm. long at anthesis, deflexed and very little if any longer in fruit; calyx lobes broadly obovate, 3–3.5 mm. long at anthesis, rounded at apex, becoming scarious, 6–10 mm. long and subglabrate at apex in fruit, then very conspicuous; corollas broadly campanulate, 4–5 mm. long, 5–6 mm. wide, light blue with a white throat, obovate to suborbicular lobes slightly spreading, about as long as throat; stamens exserted 2–3 mm., glabrous; corolla scales triangular, nearly 2 mm. long, undulate, narrow apex ending in a minute tooth, central portion curling inward toward and often overlapping filament; capsule broadly ovoid, 2–2.5 mm. wide, 3–3.5 mm. long, puberulent and sparsely glandular; seeds usually 4, cymbiform-rounded, 2–2.6 mm. long, regularly foveolate on back, inner side deeply excavate on each side of a prominent median ridge, this and slightly inrolled margins transversely and pectinately corrugated, these ridges again foveolate, dark cinnamon-brown.

Rocky hillsides, arroyos, and sandy canyons, Lower Sonoran Zone, central Baja California (San Ignacio) southward to the Cape Region, on the adjacent islands, and at Guaymas, Sonora. This is one of the most distinctive phacelias in the Sonoran Desert. Nov.–April.

**11. Phacelia crenulata** Torr. ex S. Wats., Bot. King Expl. 251, 1871, var. **crenulata**

*Phacelia ambigua* M. E. Jones, Contr. West. Bot. **12:** 52, 1908.
*Phacelia crenulata* var. *vulgaris* Brand, Pflanzenreich **4**$^{251}$**:** 78, 1913.
*Phacelia crenulata* var. *ambigua* J. F. Macbride, Contr. Gray Herb. **49:** 25, 1917.
*Phacelia crenulata* var. *funerea* Voss ex Munz, Man. S. Calif. 409, 1935.

Annual or biennial with 1 to several stems 1.5–15 dm. tall from a strong taproot, glandular-pubescent and more or less hirsute or hispid throughout; lower leaves oblong to ovate, 3–12 cm. long, 1.5–5 cm. wide, pinnately lobed, lobes oval, crenate to again lobed; upper leaves similar to lower but progressively smaller and lobes less distinct or even reduced to coarse teeth; flowers many, short-pedicellate in corymbose or paniculate clusters of cymes 5–10 cm. long, these erect in fruit; calyx lobes oblanceolate to ovate, 4–6 mm. long, 1–1.5 mm. wide at anthesis, somewhat larger in fruit, glandular-hispid; corolla violet to blue-purple, broadly campanulate, 6–10 mm. long, about as broad, obovate lobes crenulate, 3–5 mm. long; stamens 10–14 mm. long, glabrous, exserted; scales lunate to rectangular, adnate; style 12–15 mm. long, exserted, cleft about two-thirds its length; capsule globose-ovoid, 3–4 mm. long, sparsely hirsute toward apex; seeds usually 4, oblong-ovoid, 3–3.5 mm. long, inner surface excavate on each side of a prominent ridge, foveolate and transversely corrugated on outer side.

Rocky slopes, usually in the open, Lower and Upper Sonoran Zones, Colorado and Mojave Deserts, California, east to southern Utah, throughout most of Arizona, and into New Mexico, south into northwestern Sonora and Baja California. Feb.–June.

**11*a*. Phacelia crenulata** var. **minutiflora** (Voss) Jepson, Fl. Calif. **3:** 266, 1943

*Phacelia minutiflora* Voss ex Munz, Man. S. Calif. 409, 1935.

Smaller in all its parts; corollas yellowish white, 3–4 mm. long; stamens exserted only 1–2 mm. beyond corolla.

Colorado Desert in eastern San Diego County, southern Imperial County, and adjacent Baja California. Feb.–June.

**12. Phacelia coerulea** Greene, Bull. Torrey Club **8:** 122, 1881

*Phacelia invenusta* A. Gray, Proc. Amer. Acad. **20:** 303, 1885.

A simple to considerably branched annual 1–4 dm. tall from a slender taproot, glandular-hispid throughout, stems erect or essentially so; lower leaves oblong-ovate to ovate, 3–8 cm. long, 1–2.5 cm. wide, pinnately lobed or divided, on slender petioles 1–5 cm. long; upper leaves progressively smaller, uppermost sinuate instead of divided; flowers few to many, subsessile in loosely paniculate cymes 5–10 cm. long, these spreading in fruit; calyx lobes 2–3 mm. long, 1–1.5 mm. wide, lanceolate to narrowly ovate, glandular-hispid; corolla blue to white, broadly campanulate, soon deciduous, 3–4 mm. long and broad, obovate lobes 1–2 mm. long, only slightly exceeding calyx lobes; stamens 2–3 mm. long, glabrous, included; scales lunate, often auriculate at apex; style 2–3 mm. long, cleft about two-thirds its length, included; capsule ovoid, 2–3 mm. long, sparsely hispid along dorsal suture; seeds usually 4, ovoid-cymbiform, 2.5–3 mm. long, thickish, a prominent corrugated ridge running length of inner side, foveolate and transversely corrugated on back, dark brown.

Rocky slopes, Lower and Upper Sonoran Zones, Yuma, Maricopa, and Mohave Counties, Arizona, east to Texas and south into Sonora and Chihuahua. Reported from "southeastern California," but I have seen no material from the Colorado Desert. March–June.

**13. Phacelia popei** Torr. & Gray var. **arizonica** (A. Gray) Voss, Bull. Torrey Club **64:** 94, 1937

*Phacelia arizonica* A. Gray, Syn. Fl. N. Amer. $2^1$: 394, 1878.

Annual with several procumbent to ascending stems 0.5–3 dm. long, villous and hirsutulous throughout, inflorescence somewhat glandular; leaves ovate, oblong, or oblong-oblanceolate in outline, 8–20 mm. wide, 3–8 cm. long, pinnatifid, lobes entire, dentate or again pinnatifid, mostly obtuse; petioles slender, 5–18 mm. long; cymes compact, usually branched, 1–3 cm. long in flower, erect and to 10 cm. long in fruit; pedicels slender, 1 mm. long or less at anthesis, to 4 mm. long and ascending in fruit; calyx lobes lance-elliptic, about 2 mm. long in anthesis, about 4 mm. long in fruit;

corollas campanulate, 4–5 mm. long, about as broad, white or pale lavender, lobes obovate, 1.6–2 mm. wide, slightly longer, margins irregularly denticulate; stamens 6–8 mm. long, glabrous; corolla scales broadly ovate or elliptic, 0.6–0.8 mm. long, rounded at both ends, adnate along sides to both corolla tube and filament, free at apex only, where it is somewhat rounded, glabrous; style 6–7 mm. long, pubescent below cleft; capsule globose, about 3 mm. in diameter, sparsely hirsutulous; seeds usually 4, 2–2.5 mm. long, dark brown, deeply excavated on each side of a prominent median ridge on ventral surface, back rounded, foveolate on all surfaces, reticulations arranged in transverse lines on back.

Rocky hillsides and adobe flats, Sonoran Zones, Santa Cruz and Pima Counties, Arizona, east into New Mexico and southward into northern Sonora. Feb.–May.

Occurring in the Sonoran Desert near Altar, Sásabe, and Magdalena, Sonora.

**14. Phacelia parryi** Torr., Bot. Mex. Bound. 144, 1859, var. **parryi**

Annual 1.5–6.5 dm. tall with simple or sparingly branched stems from a slender taproot, glandular-hispid throughout; lower leaves oblong-ovate to broadly ovate, 2–5 cm. long, 1–4 cm. wide, truncate to cordate at base, obtuse at apex, coarsely dentate, doubly dentate, or sometimes lobed, usually surpassed by petiole; upper leaves similar but smaller and shorter-petiolate; cymes lax, few-flowered, simple or few-branched, 1–3 dm. long; pedicels slender, 1–3 cm. long and spreading-horizontal or slightly recurved in fruit; calyx lobes linear-oblong, 5–8 mm. long, about 1 mm. broad, slightly enlarged in fruit; corolla open-campanulate or oftener subrotate, 1–2 cm. long, 1.5–3 cm. wide, violet with 5 white or pale yellowish spots below lobes in tube, lobes obovate, nearly equaling tube; stamens 1–2 cm. long, about equaling to slightly exceeding corolla, filaments pilose, with a quadrate, dilated, slightly toothed basal pad; style 1–2 cm. long, pubescent, cleft one-third to one-half its length; capsule ovoid-oblong, 6–10 mm. long, acuminate, sparsely hairy at apex, slightly exceeding calyx; seeds 50–60, ovoid, angled, 0.6–0.8 mm. long, coarsely foveolate, dark brown.

Arid slopes, especially in disturbed soil or on burns, Lower Sonoran to Arid Transition Zones, southern Monterey County, California, to the southern flanks of the Sierra San Pedro Mártir, and east to the margins of the Colorado and San Felipe Deserts. March–May.

**14*a*. Phacelia parryi** var. **celata** Jepson & Hoover ex Jepson, Fl. Calif. **3:** 276, 1943

Similar in general appearance to var. *parryi* but rarely over 2 dm. high; stems usually several and broadly spreading; corolla 10–12 mm. long, slightly broader; filaments dilated at base into an ovate pad without squarish angles.

Gravelly flats and washes, vicinity of Whitewater and the Santa Rosa Mountains, eastern San Diego and adjacent Riverside Counties, California. March–May.

**15. Phacelia longipes** Torr. ex A. Gray, Proc. Amer. Acad. **10:** 322, 1875

*Phacelia cordifolia* S. Wats. ex Brand, Pflanzenreich $4^{251}$: 72, 1913.

Erect or ascending annual 1–4 dm. tall, glandular-hirsute throughout; stems branching from base or above, some prostrate; leaves broadly ovate to suborbicular, 1.5–4 cm. long, 1–2.5 cm. wide, cordate at base, obtuse at apex, crenate to crenate-dentate, blades exceeded by petioles; cymes lax, few-flowered, 0.5–2 dm. long, simple or few-branched, erect in fruit, pedicels 0.5–1 cm. long in fruit; calyx lobes linear-oblong, 3–6 mm. long, 0.5–1 mm. wide, slightly enlarged in fruit; corolla subrotate to open-campanulate, sordid-white, 7–12 mm. long, about as broad, usually lobed about to middle, lobes obovate; stamens slightly pilose near base, 1–1.5 cm. long, dilation glabrous; capsule ovoid-oblong, 5–6 mm. long, acuminate, sparsely pubescent; seeds 8–15, ovoid-oblong, 1–1.5 mm. long, dark brown, coarsely foveolate.

Dry slopes and along washes, upper edges of the Lower Sonoran to Arid Transition Zones, northwestern corner of the Colorado Desert north to Santa Barbara County. Rare in our area and barely entering it above Whitewater. April–July.

**16. Phacelia minor** (Harv.) Thell., Ber. Bayer. Bot. Ges. **14:** 79, 1914

*Whitlavia grandiflora* Harv., Lond. Jour. Bot. **5:** 312, *pl. 11*, 1846. Not *Phacelia grandiflora* (Benth.) A. Gray, 1875.
*Whitlavia minor* Harv., loc. cit., *pl. 12*.
*Phacelia whitlavia* A. Gray, Proc. Amer. Acad. **10:** 322, 1875.
*Phacelia whitlavia* var. *jonesii* Brand, Pflanzenreich $4^{251}$: 71, 1913.

Annual 2–6.5 dm. tall with simple or branching, glandular-hispid stems from a stoutish taproot; lower leaves oblong-ovate to broadly ovate, 2–7 cm. long, 1.5–6 cm. wide, truncate to cordate at base, obtuse at apex, coarsely and usually doubly dentate or serrate, on petioles equaling to longer than blades; upper leaves with shorter petioles and gradually somewhat reduced; inflorescence few- to many-flowered, lax, open, simple to few-branched cymes 1–4 dm. long, these erect in fruit, fruiting pedicels curved or spreading horizontally, 1–4 cm. long; calyx lobes 5–10 mm. long, about 1 mm. broad, linear-oblong, somewhat enlarged in fruit; corolla tubular-campanulate and slightly constricted at throat, violet or very rarely white, 1.5–4 cm. long, 1–3 cm. wide, obovate lobes 3–8 mm. long; stamens 2–4.5 cm. long, slightly exserted, filaments bearing an oblong, dilated, pubescent pad at base; style equaling to slightly shorter than stamens, pubescent, cleft about one-third of way to base; capsule ovoid, 8–12 mm. long, acuminate; seeds numerous, ovoid, angled, about 1 mm. long, coarsely foveolate, dark brown.

Arid slopes, particularly on burned areas, and desert slopes of the coastal mountains, Lower and Upper Sonoran Zones, northern Los Angeles County south into northern Baja California and along the northwestern edge of the Colorado Desert; mostly on the coastal side of the divide. Feb.–June.

**17. Phacelia campanularia** A. Gray, Syn. Fl. N. Amer. $2^1$: 164, 1878

*Phacelia minor* var. *campanularia* Jepson, Man. Fl. Pl. Calif. 819, 1925.

A glandular-hispid annual 1–5 dm. tall from a tough taproot; stems several and usually branching from base; leaf blades suborbicular to ovate-oblong, 1.5–7.5 cm. long, 1–5 cm. broad, truncate to cordate at base, obtuse or rounded at apex, crenate to dentate or serrate, on petioles 1–20 cm. long; racemes loose, several- to many-flowered; pedicels slender, spreading in fruit, 5–25 mm. long; calyx lobes 4–12 mm. long, 0.5–2 mm. wide, linear-oblong, acute or obtuse at apex, somewhat accrescent in fruit; corolla funnelform-campanulate, 1.5–4 cm. long, 1–4 cm. broad, deep blue or rarely white, lobes 8–10 mm. long, considerably broader, obovate; stamens 2–4.5 cm. long and therefore somewhat exserted, filaments with a quadrate, glabrous or pubescent dilation at base; anthers 1–1.5 mm. long; styles 1.5–3.5 cm. long, cleft about one-fourth to one-third of its length, pubescent; capsule ovoid-oblong, 8–12 mm. long, acuminate, puberulent and somewhat glandular; seeds numerous, ovoid-oblong, 1–1.5 mm. long, brownish, coarsely foveolate.

Sandy washes, rocky hillsides, and occasionally sandy plains, Lower and Upper Sonoran Zones, Mojave and Colorado Deserts and the arid mountains immediately adjacent. Not known east of the Colorado River. March–June.

**18. Phacelia pachyphylla** A. Gray, Proc. Amer. Acad. **19:** 88, 1883

Rather stoutish, subfleshy annual 1–2 dm. or much less tall, with few to several spreading to ascending branches, glandular-hirsutulous throughout; leaf blades ovate to orbicular, thickish, cordate at base, 2–2.5 cm. in diameter, or smaller in dwarf plants, entire or crenate, petioles usually longer than blades; flowers subsessile in compact terminal cymes, cymes to 5 cm. long in fruit, usually exceeding foliage; calyx lobes oblong, 2–3 mm. long, up to 5 mm. long in fruit, densely glandular-pubescent; corolla pale purple to pink, narrowly campanulate, 5–7 mm. long, suborbicular lobes 1.5–2 mm. long, scales narrowly linear, 1–2 mm. long; stamens 2.5–4 mm. long, included, pubescent; style 2–3 mm. long, divided about to middle; capsule ovoid to subglobose, about 5 mm. long, glandular-puberulent; seeds numerous, oblong-ovoid, 1–1.25 mm. long, brown, transversely corrugated.

Gravelly desert mesas and alkaline plains, Lower Sonoran Zone, central and eastern Mojave Desert south into the Colorado Desert and northeastern Baja California; possibly in Arizona. April–May.

**19. Phacelia neglecta** M. E. Jones, Contr. West. Bot. **12:** 50, 1908

Diminutive compact annual 1–10 cm. tall with subfleshy thickish leaves, petioles, and stems, densely hirsute but sparingly glandular throughout; petioles 5–25 mm. long; leaf blades broadly ovate to suborbicular or subreniform (then broader than long), 5–25 mm. long, reniform ones to 3 cm. broad, finely to coarsely crenate, rounded to truncate at base, obtuse to broadly rounded at apex; inflorescence very compact and nearly hidden by foliage at anthesis; flowers sessile or essentially so; calyx lobes lance-linear, 2.5–3 mm. long in anthesis, to 4.5 mm. long in fruit; corolla campanulate-tubular, white, 3.5–4.5 mm. long, barely exceeding calyx, lobes suborbicular, 1.5–2 mm. wide, about as long; stamens barely equaling corolla tube, filaments glabrous; corolla scales linear, slightly sinuate, extending from base of tube upward two-thirds to three-fourths of way to corolla sinuses, adnate; styles about 2 mm. long, puberulent toward base, cleft nearly to middle; capsule subglobose, 4–4.5 mm. in diameter, puberulent toward apex; seeds numerous, ovoid, 0.6–0.8 mm. long, reddish brown, transversely corrugated, corrugations again transversely ridged.

Gravelly washes and alkaline plains, Lower Sonoran Zone, central Colorado and Mojave Deserts, Imperial County to Inyo County, California, east into Pima County, Arizona. March–April.

**20. Phacelia ivesiana** Torr., Ives Rep. 21, 1860

*Phacelia campestris* A. Nels., Bull. Torrey Club **26:** 242, 1899.

Small annual 0.5–3 dm. tall with numerous prostrate to ascending branches, hirsutulous and slightly glandular throughout; lower leaves linear-oblong, 1.5–4 cm. long, 5–15 mm. wide, pinnately lobed into obtuse and entire segments; petioles slender, usually longer than blades; flowers few, short-pedicellate in lax, simple or few-branched, short cymes, only slightly exceeding foliage, to 6 cm. long and erect in fruit; calyx lobes linear to linear-oblanceolate, unequal, 3–5 mm. long, about 0.5 mm. wide, hirsute-ciliate, to 6 mm. long and slightly exceeding capsule in fruit, tips often somewhat spreading; corolla white to purplish, broadly funnelform, 2–4 mm. long, about as broad, lobes 0.5–1 mm. long, oval, whole corolla about equaling or slightly shorter than calyx; stamens 1–2 mm. long, glabrous, included; scales very narrow to obsolete; styles 1–1.5 mm. long, divided about one-fourth their length, included, glabrous; capsule ovoid-oblong, 3–5 mm. long, obtuse, hispidulous near apex; seeds 10–15, brown, 1–1.25 mm. long, coarsely corrugated on back.

Sandy arid soil, on hillsides, plains, and along washes, Lower and Upper Sonoran Zones, Colorado and Mojave Deserts, California, to Wyoming, Utah, and New Mexico. March–June.

**21. Phacelia affinis** A. Gray, Syn. Fl. N. Amer. ed. 2, **2**[1]**:** 417, 1886

*Phacelia rugulosa* Lemmon ex Greene, Pittonia **1:** 175, 1888.

A slender annual 1–3 dm. tall with few to several erect or ascending branches, puberulent to hispidulous and glandular-puberulent throughout; basal leaves narrowly oblong, pinnatifid to pinnately lobed, 2–6 cm. long, 0.5–2 cm. wide, divisions obtuse and few-toothed; petioles usually shorter than blades, slender; upper leaves similar but short-petiolate to subsessile; flowers few to many, subsessile to short-pedicellate in simple or few-branched cymes, these to 15 cm. long, exceeding foliage, erect; calyx lobes spatulate, 4–5 mm. long, about 1 mm. broad, hirsute-ciliate and glandular-hirsutulous, to 6–8 mm. long in fruit and exceeding capsules; corolla narrowly campanulate, deciduous, 2.5–4.5 mm. long, 2–3 mm. wide, pale purplish to nearly white, lobes oval, about 1 mm. long, about equaling calyx; stamens unequal, about 2 mm. long, included, glabrous; scales inconspicuous or obsolete; style about 2 mm. long, cleft about one-fourth its length, included; capsule oblong, obtuse, 4–5 mm. long, hispid near apex; seeds 15–20, oblong-ovoid, about 1 mm. long, brown, reticulate and transversely corrugated on back.

Arid slopes, hillsides, and along washes, Lower Sonoran Zone, western margin of the Colorado Desert to southern Nevada, Utah, and Pima County, Arizona, south in Baja California to the vicinity of San Ignacio. Probably also in northern Sonora. March–May.

**22. Phacelia brachyloba** (Benth.) A. Gray, Proc. Amer. Acad. **10:** 324, 1875

*Eutoca brachyloba* Benth., Trans. Linn. Soc. **17:** 277, 1835.
*Phacelia cooperae* A. Gray, Proc. Amer. Acad. **15:** 49, 1879.
*Phacelia orcuttiana* A. Gray, op. cit. **19:** 88, 1883.
*Phacelia leucantha* Lemmon ex Greene, Pittonia **1:** 175, 1888.

Annual with 1 to several erect or ascending branches 1–5 dm. long from base, hispidulous throughout, inflorescence more or less, usually densely, glandular; lower leaves narrowly oblong to oblong-ovate, 2.5–8 cm. long, 1–2.5 cm. wide, pinnatifid or pinnately lobed into obtuse, few-toothed to entire divisions; petioles slender, shorter than blades; upper leaves similar to basal ones but smaller and less deeply lobed, uppermost often toothed only; flowers short-pedicellate in many-flowered, paniculately arranged cymes to 2 dm. long, erect in fruit; calyx lobes spatulate, 4–5 mm. long, 1–1.5 mm. wide, unequal, glandular-hispidulous and hirsute-ciliate, little if at all enlarged in fruit and about equaling or slightly exceeding capsule; corolla white to lavender with a yellow throat, broadly funnelform to campanulate, 8–12 mm. long, about as broad, lobes orbicular, about 3 mm. long; stamens about 4 mm. long, unequal, glabrous; scales mostly obsolete; style 3–4 mm. long, pubescent toward base; capsule oblong-ovoid, 4–5 mm. long, thinly hispidulous, obtuse; seeds 15–20, ovoid, about 0.5 mm. long, foveolate, and transversely corrugated, brown.

Sandy soil, on hillsides and along washes, Lower Sonoran to Arid Transition Zones, western margins of the Colorado and Mojave Deserts west to the coastal slopes and north to Monterey County, California. Barely entering the Sonoran Desert area along its northwestern margin. May–June.

**23. Phacelia fremontii** Torr., Ives Rep. 21, 1860

*Phacelia brannani* Kell., Proc. Calif. Acad. **7**: 90, 1877.
*Phacelia hallii* Brand, Jahresb. Kgl. Gymnas. Sorau Beilage 8, 1911.

An ascending annual with few to several stems 1–3 dm. tall, stems and leaves puberulent to hirsutulous, inflorescence glandular with stipitate, black or dark brown glands and somewhat short-villous; leaves oblong to ovate-oblong, 2–6 cm. long, 0.5–2 cm. wide, pinnatifid with entire, toothed, or again lobed divisions, blades longer than petioles; cymes few-branched or simple, few-flowered, erect in fruit; pedicels short, rarely over 1.5 mm. long even in fruit; calyx lobes spatulate-linear, 3–6 mm. long, 0.5–2 mm. wide, to 8 mm. long in fruit; corolla campanulate or broadly funnelform, 8–15 mm. long, 10–18 mm. broad, bright blue, pale blue, or sometimes nearly white, tube yellow, orbicular lobes 2–5 mm. long; stamens 3–8 mm. long, somewhat unequal, with glabrous filaments; corolla scales linear-lanceolate, wholly adnate to corolla tube but free from filaments or attached to them only at base; style 3–5 mm. long, cleft about one-fifth its length, pubescent toward base; capsule ovoid to oblong, 3–5 long, hispid and obtuse at apex; seeds few (8–12), ovoid, 1–1.2 mm. long, dark brown, transversely corrugated.

Arid flats, along washes, and often under shrubs, Lower Sonoran to Arid Transition Zones, mainly in the Mojave Desert and north in the Tehachapi and inner Coast Ranges to southwestern Fresno County, California; east from the Mojave Desert into southern Nevada, Utah, and northern Arizona; barely entering the Sonoran Desert in Mohave County, Arizona. March–June.

## 4. EMMENANTHE Benth., Trans. Linn. Soc. 17: 281, 1835

Erect, hirsute, glandular-viscid annuals with pinnately cleft or divided leaves or these rarely only dentate. Flowers in loose, somewhat scorpioid, terminal cymes, slenderly pedicellate, pendulous at or soon after anthesis. Sepals 5, subequal, distinct to base or nearly so. Corolla campanulate, shallowly 5-lobed, cream-colored or yellow, withering-persistent, devoid of scales. Stamens 5, inserted near base of corolla, included. Ovary 2-celled; placentae wing-like and borne at right angles on inner edge of partitioned halves. Style slender, deciduous; stigma lobes short. Capsule oblong, compressed, 2-celled or imperfectly 4-celled, each cell 4- to several-seeded. Seeds oval, flattened, reticulate-foveolate in rather regular lines.

**1. Emmenanthe penduliflora** Benth., Trans. Linn. Soc. **17**: 281, 1835

Simple to much-branched plants 1–5 dm. tall, with heavy-scented, stipitate-glandular puberulence throughout; leaves oblong, 5–20 mm. broad, 3–8 cm. long, pinnatifid with narrowly deltoid to oblong, entire or dentate lobes, decurrent at base to a short, winged or clasping petiole; cymes usually several to many; pedicels filiform, 6–15 mm. long or longer in fruit, recurved, villous and stipitate-glandular; sepals ovate-lanceolate, 6–10 mm. long, 2–3 mm. wide; corolla 8–12 mm. long, 5–10 mm. broad, orbicular lobes 1–2 mm. long, sparsely puberulent along and below midvein of each lobe; style 1.5–2 mm. long, lobes reflexed, or curled into 2 minute circles; capsule 8–10 mm. long, thin-walled, sparsely villous and glandular; seeds dark brown, 1.2–1.8 mm. long, plano-convex or sometimes concavo-convex, conspicuously foveolate-reticulate on both sides.

Gravelly or rocky soil, especially common on burns, mainly Upper Sonoran and Arid Transition Zones but occasionally along upper margins of Lower Sonoran Zone, from central Coast Ranges and Sierra Nevada south into northern Baja California and east to Nevada and Arizona, where it enters the Sonoran Desert near the mouth of Sabino Canyon, Pima County, Arizona. March–May.

## 5. NAMA L., Syst. Nat. ed. 10, 950, 1759

Herbaceous or less commonly suffrutescent, erect, ascending or prostrate annuals or perennials with alternate entire leaves. Flowers solitary in axils of leaves or in short, lateral or terminal cymes. Calyx divided nearly or entirely to base; lobes 5, linear, lanceolate, or spatulate. Corolla tubular to campanulate-funnelform, finely puberulent on outside; lobes rounded and usually spreading. Stamens subequal or unequal, usually included; filament bases somewhat dilated, adnate portion with a free wing or margin or this lacking, glabrous with one exception. Styles 2, free to base or nearly so, or connate part or full length. Ovary superior, or partially inferior by adnation to basal part of calyx, many-ovuled, 1-celled but falsely 2-celled by ingrowth of placentae, usually pubescent. Capsule ovoid to oblong, loculicidally dehiscent, sometimes septicidally dehiscent as well. Seeds numerous, variously pitted, reticulate or alveolate, sometimes angular, transversely corrugated or wrinkled as well as pitted.

Ovary partially inferior; calyx tubular and adnate to ovary in basal part, lobes indurated and tips recurved in fruit; styles connate to stigma .................... 1. *N. stenocarpum*
Ovary wholly superior; calyx lobes erect, herbaceous in fruit; styles distinct or connate only in basal one-fourth:
  Petioles decurrent on stems; leaf blades broadly obovate, 5–35 mm. broad; pedicels commonly to 1.5 (sometimes to 5) cm. long in fruit .................. 2. *N. jamaicense*
  Petioles not decurrent on stems; leaf blades linear to oblanceolate, 1–8 mm. wide; pedicels rarely 5 mm. long:
    Plants generally erect or with ascending branches, not matted; seeds yellowish or yellow-brown:
      Herbage hispid with stiff harsh hairs, sometimes with softer ones intermingled; leaves linear-oblanceolate; dilated part of filament less than twice as wide as free part; stems not strikingly slender, about 1.5–2 mm. in diameter:
        Leaf blades only slightly revolute or wholly plane, not cinereous:
          Calyx lobes linear or linear-lanceolate; pubescence of harsh stiff hairs only ...... ...................................... 3. *N. hispidum* var. *spathulatum*
          Calyx lobes linear-spatulate; pubescence a mixture of harsh, soft-silky, and glandular hairs ........................... 3*a*. *N. hispidum* var. *sonorae*
        Leaf blades strongly revolute, cinereous with soft, somewhat appressed hairs among harsh ones .............................. 3*b*. *N. hispidum* var. *revolutum*
      Herbage villous-hirsute with silky, somewhat appressed hairs, not strongly hispid; leaves obovate-spatulate; dilated part of filaments twice as wide as free part; stems very slender, 0.5–1 mm. in diameter ......................... 4. *N. coulteri*
    Plants prostrate, often forming dense mats; seeds dark brown:
      Corolla broadly funnelform, 8–15 mm. long, 6–12 mm. wide; style 3.5–5 mm. long; calyx lobes linear-lanceolate, 5–8 mm. long:
        Leaf blades 1–7 mm. wide; calyx not conspicuously cinereous, only moderately pilose:
          Leaf blades 2–7 mm. wide, 1–2.5 cm. long ...... 5. *N. demissum* var. *demissum*
          Leaf blades 1–3 mm. wide, 1–4 cm. long .......... 5*a*. *N. demissum* var. *deserti*
        Leaf blades 0.5–2 mm. wide; calyx conspicuously silky-gray-pilose .............. .......................................... 5*b*. *N. demissum* var. *lineare*
      Corolla tubular-funnelform, 4–5 mm. long, 2–3 mm. wide; style 1–1.5 mm. long; calyx lobes linear-spatulate, 3–4 mm. long .................... 6. *N. pusillum*

**1. Nama stenocarpum** A. Gray, Proc. Amer. Acad. **10:** 331, 1875

*Marilaunidium stenocarpum* Kuntze, Rev. Gen. Pl. **2:** 434, 1891.
*Conanthus stenocarpus* Heller, Cat. N. Amer. Pl. 6, 1898.
*Nama humifusum* Brand, Jahresb. Kgl. Gymnas. Sorau Beilage 9, 1911.

Diffuse leafy annuals with ascending or procumbent stems from a slender taproot, stems 1–3 dm. long, hirsute throughout; leaves oblanceolate, 1.5–3.5 cm. long, 4–10 mm. wide, tapering toward base into a short petiole, or sessile or slightly auriculate, margins undulate; flowers in compact axillary clusters of 2–5 flowers, leafy-bracteate, and in terminal cymes; calyx lobes linear-spatulate, 4–6 mm. long in flower, somewhat accrescent and indurate in fruit, basal tubular part closely adherent to partially inferior ovary; corolla funnelform, 5–7 mm. long, 3–4 mm. wide, ovate lobes about 1 mm. long; stamens unequal, equally inserted low in corolla tube, dilated part linear, tip of free part acutely angled, each wing about equaling diameter of free part of filament; style 1.5–2 mm. long, divided about one-third of its length; capsule linear-oblong, 5–6 mm. long, loculicidally dehiscent; seeds very numerous (often over 100), irregularly ovoid, somewhat flattened, 0.2–0.3 mm. long, light brown, finely reticulate.

Muddy ground about ponds and lakes and on river banks, Lower and Upper Sonoran Zones, southern California to Texas and adjacent northern Mexico. March–May.

**2. Nama jamaicense** L., Syst. Nat. ed. 10, 950, 1759

(For synonymy see Amer. Jour. Bot. **20:** 520, 1933.)

A moderately slender, prostrate, ascending, or procumbent annual with rather simple branches 1–5 dm. long, strigose-hirsute throughout; leaves broadly ovate, suborbicular, ovate-spatulate, or obovate-spatulate, 1.5–8 cm. long, 0.5–3.5 cm. wide, narrowed to distinct petioles about one-fourth as long as blades; petioles decurrent on stems and narrowly winged below leaf blades; flowers mostly solitary or in 2–3-flowered axillary clusters, subsessile to long-pedicellate, pedicels to 1.5 or rarely 5 cm. long in fruit; calyx lobes linear to narrowly spatulate, 4–5 mm. long in flower, to 10 mm. long in fruit and closely appressed to ovary and later to capsule but not adnate; corolla subtubular-funnelform, 6–7 mm. long, 2.5–3.5 mm. wide; stamens unequally inserted on corolla tube near base, dilated base free; style 1.5–2 mm. long, distinct, to connate one-half its length; capsule 5–8 mm. long, linear-oblong; seeds 50–70, light brown, about 0.5 mm. long, finely alveolate, rounded and scarcely angled.

Along watercourses and in moist meadows, Sonoran to Tropical Zones, southern Sonora (Tesopaco; Alamos) to Tamaulipas and San Luis Potosí, south to Central America; also in southern Florida, the West Indies, and in Argentina, where it is probably introduced. Dec.–April.

**3. Nama hispidum** A. Gray var. **spathulatum** (Torr.) C. L. Hitchcock, Amer. Jour. Bot. **20:** 524, 1933

*Nama biflorum* var. ? *spathulatum* Torr., Pacif. R. Rep. **7:** 17, 1858.

Annual herb with several branches from base, these ascending to erect, 5–30 cm. long; herbage densely hispid throughout with stiffish white hairs, from pustulate bases; leaves 1–7 cm. long, 1–8 mm. wide, mostly oblong-spatulate, plane or somewhat revolute, tapering gradually to base; flowers solitary or in 3–5-flowered terminal cymes; calyx lobes linear-lanceolate, 4–7 mm. long, hispid; corolla tubular-campanulate, pale purplish to lavender, with a yellowish throat, 8–15 mm. long, limb often nearly as wide; stamens about one-half as long as corolla, unequally inserted 1–4 mm.

from base of corolla tube, adnate portion only slightly wider than free portion, often marked by a faint free ridge; styles 2–4 mm. long, distinct; capsule 20–100-seeded, 3–4.5 mm. long, oblong, seeds about 0.5 mm. long, yellow to light brown, minutely alveolate-reticulate, markings regular.

Desert plains and arroyos, less commonly on the hillsides, Lower Sonoran Zone, eastern Colorado Desert, Imperial County, California, to Texas and south to Coahuila, Nuevo León, Sinaloa, and northern Baja California. Feb.–June.

**3*a*. Nama hispidum** var. **sonorae** C. L. Hitchcock, Amer. Jour. Bot. **20:** 525, 1933

Branches rarely over 12 cm. long, sparingly villous-hirsute and stipitate-glandular, particularly on undersurface of leaves and on inflorescences; sepals linear-spatulate, about 6 mm. long.

Arid parts of Sonora from the vicinity of Altar south. Feb.–June.

**3*b*. Nama hispidum** var. **revolutum** Jepson, Man. Fl. Pl. Calif. 832, 1925

Differing from the above in having soft-hirsute pubescence of finer hairs intermingled with the harsh stiff ones, and in the narrow linear leaves that are very strongly revolute.

Arid desert flats, in sandy or heavy soil, Lower Sonoran Zone, eastern Riverside and Imperial Counties, California, and adjacent Arizona south into central Sonora and northern Baja California. March–May.

**4. Nama coulteri** A. Gray, Proc. Amer. Acad. **8:** 283, 1870

*Nama hispidum* var. *coulteri* Brand, Pflanzenreich $4^{251}$: 154, 1913, in part.
*Nama parvifolium* var. *brevistylum* Brand, op. cit. 155.
*Marilaunidium coulteri* Kuntze, Rev. Gen. Pl. **2:** 434, 1891.
*Conanthus coulteri* Heller, Cat. N. Amer. Pl. 6, 1898.

A low annual with ascending branches 5–30 cm. long, slender and wiry in appearance, reddish brown, mostly 0.5–1 mm. in diameter; leaves mostly near tips of branches and at upper forks; herbage short-villous with rather soft hairs throughout; leaves obovate, obovate-spatulate to spatulate, 3–8 mm. wide, 1–3 cm. long, obtuse or rounded at apex, gradually narrowed to base; flowers mostly solitary in axils or in 2–4-flowered terminal cymes; calyx lobes linear-spatulate, 4–7 mm. long at anthesis, to 9 mm. long in fruit; corolla lavender with yellowish throat, 7–10 mm. long, 7–9 mm. wide; stamens about one-half as long as corolla, unequally inserted 1–2 mm. above base of corolla tube, adnate dilated part of filaments twice as wide as free portion, its margins free; styles 2–4 mm. long; capsules oblong, 4–5 mm. long, 20–80-seeded; seeds yellowish, about 0.4 mm. long, finely alveolate-reticulate.

Sandy desert flats, Lower Sonoran Zone, central Baja California south to the Cape Region, and central Sonora south into Sinaloa. Nov.–May.

**5. Nama demissum** A. Gray, Proc. Amer. Acad. **8:** 283, 1870, var. **demissum**

*Marilaunidium demissum* Kuntze, Rev. Gen. Pl. **2:** 434, 1891.
*Conanthus demissus* Heller, Cat. N. Amer. Pl. 6, 1898.

Prostrate, diffusely branching annuals with villous-hirsute and strigose branches 3–15 cm. long and often forming a dense mat, leaves virtually confined to compact clusters at tips of branches; leaf blades obovate to spatulate or linear-spatulate, 2–7 mm. wide, 1–2.5 cm. long, gradually narrowed to petiole which about equals blade, strigose and hirsute; flowers subsessile in terminal, few-flowered cymes and in axils of branches;

calyx lobes narrowly linear-lanceolate, villous-hirsute, usually conspicuously cinereous, 5–8 mm. long; corolla campanulate, 9–15 mm. long, 6–12 mm. wide, oval lobes 2–3 mm. long, purple, throat paler or yellowish; stamens subequally inserted, dilated adnate portion only slightly wider than free terete part; styles 3–5 mm. long, distinct; capsule linear-oblong, 3–4 mm. long, 10–18-seeded; seeds ovoid, about 0.5 mm. long, dark brown, shallowly and regularly reticulate-pitted.

Sandy soil on desert slopes and along washes, Lower and Upper Sonoran Zones, Colorado and Mojave Deserts and adjacent hills to Utah, Arizona, and northern Baja California. Feb.–June.

**5*a*. Nama demissum** var. **deserti** Brand, Pflanzenreich $4^{251}$: 159, 1913

Differing from var. *demissum* only in having longer, narrower, linear to narrowly spatulate leaves 1–3 mm. wide and 1–4 cm. long.

Arid sandy soil, Lower Sonoran Zone, Colorado and Mojave Deserts east into southern Nevada and western Arizona. Feb.–May.

**5*b*. Nama demissum** var. **lineare** C. L. Hitchcock, Amer. Jour. Bot. **20:** 531, 1933

Differs from var. *demissum* only in extremely narrow, linear leaves averaging about 1 mm. in width, and in a more densely silky-pilose, gray pubescence on calyces.

Sandy desert plains, Lower Sonoran Zone, vicinity of Punta Prieta, central Baja California, south to the southern edge of the Vizcaíno Desert. Jan.–April.

**6. Nama pusillum** Lemmon ex A. Gray, Proc. Amer. Acad. **20:** 305, 1885
*Conanthus pusillus* Heller, Cat. N. Amer. Pl. 6, 1898.

Diffusely branched, mostly matted, prostrate annual with dichotomously branched stems 1.5–10 cm. long (branches sometimes ascending); herbage densely short-hirsute with grayish hairs throughout; leaf blades ovate, 2–3 mm. long, 1.5–3 mm. wide, obtuse, entire, tapering to a slender petiole nearly equaling blade; flowers sessile or subsessile in axils, solitary; calyx lobes linear to linear-oblanceolate, 3–4 mm. long, densely hirsute with spreading hairs; corolla pink, nearly salverform, 4–5 mm. long, 2–3 mm. wide, oval lobes about 0.5 mm. long; stamens unequally inserted, about one-half as long as corolla, free portion adnate to bases of filament and bifid in 2 divergent ridges that form an inverted "V" below free portion; style 1–1.5 mm. long, bifid to base; capsule ovoid, 2.5–3 mm. long, 20–30-seeded; seeds ovoid, angular, about 0.5 mm. long, finely transversely ridged in a regular pattern.

Desert plains and hillsides and along arroyos, Lower and Upper Sonoran Zones, mainly Mojave Desert but barely entering the northeastern part of the Colorado Desert in Riverside County, California. April–May.

## 6. ERIODICTYON Benth., Bot. Sulph. 35, 1844

Shrubs with resinous aromatic leaves and young branches. Leaves alternate, coriaceous or firm, usually dentate and petiolate but sometimes sessile, veins pinnate and finely reticulate beneath, pubescent to densely tomentose, usually more or less glutinous-resinous. Flowers in terminal panicles of scorpioid cymes. Calyx lobes 5, distinct, narrowly linear, equal or essentially so. Corolla campanulate to funnelform; tube naked within, mostly pubescent without; lobes 5, rounded, margins often undulate. Stamens 5, subequal to unequal, usually unequally inserted at or below middle of corolla tube, included or slightly exserted; filaments sparsely hairy. Ovary 2-celled by meeting of thin placentae or only incompletely 2-celled. Styles 2, distinct. Capsule loculicidally and septicidally dehiscent, 2–12-seeded.

Leaves densely white-tomentose on both surfaces; corollas 10–16 mm. long, broadly funnelform ........................................ 1. *E. crassifolium*
Leaves green and glabrous or sparsely pubescent above, tomentose beneath; corollas 4–10 mm. long, campanulate:
  Calyx sparsely hirsutulous; corolla white; leaves narrowly linear, 2–5 mm. broad ........ ........................................ 2. *E. angustifolium*
  Calyx densely hirsutulous to tomentose; corolla lavender or less commonly white; leaves linear-lanceolate to ovate-lanceolate, 5–20 mm. wide ................ 3. *E. lanatum*

**1. Eriodictyon crassifolium** Benth., Bot. Sulph. 35, 1844
*Eriodictyon crassifolium* subsp. *grayanum* Brand, Pflanzenreich **4**[251]: 139, 1913.
*Eriodictyon crassifolium* var. *typicum* Brand, loc. cit.

Grayish shrub to 2 (or sometimes 4) m. tall with densely cinereous-tomentose herbage and twigs; leaves ovate to oblong, 3–12 cm. long, 1.5–3 cm. wide, thick and rigid, crenate, equally tomentose on both sides; petioles short, 3–8 mm. long; inflorescence rounded or flat-topped, to 15 cm. broad; calyx densely villous, cinereous, sepals linear, 3.5–5 mm. long; corolla broadly funnelform, 10–16 mm. long, pale lavender to rose, densely pubescent without; stamens unequally inserted below middle of corolla tube, hairy and narrowly winged on adherent part below insertion, barely equaling corolla tube; capsule ovoid, 2.5–3 mm. long, densely pubescent, 6–12-seeded; seeds finely striate-reticulate, brownish.

Arroyos, washes, and rocky mountain sides, mainly Upper Sonoran Zone, southwestern Kern County to San Diego County. It enters the Lower Sonoran Zone in the vicinity of Palm Springs, on the western margin of the Colorado Desert. April–June.

**2. Eriodictyon angustifolium** Nutt., Jour. Acad. Phila. II, **1**: 181, 1848

Small shrubs 5–12 dm. tall with nearly or quite glabrous twigs; leaves narrowly linear, 2–5 (rarely –7) mm. broad, 3–10 cm. long, entire or less commonly remotely dentate, green and glabrous, but resinous above, white-tomentose between veins beneath, acute at apex, narrowed toward short-petiolate base; panicles 4–15 cm. long, rather lax; calyx lobes sparsely hirsutulous to subglabrate, 2–3 mm. long, green; corolla campanulate-funnelform, 4–7 mm. long, white, limb about as wide; stamens equaling corolla tube or slightly exserted; capsules ovoid, about 2 mm. long.

Arid slopes and washes, Lower Sonoran Zone to Arid Transition Zone, eastern Mojave Desert in the New York Mountains, east into Arizona, Nevada, and southern Utah and south along the western margin of the Colorado and San Felipe Deserts to the southern end of the Sierra San Pedro Mártir, Baja California. April–Aug.

**3. Eriodictyon lanatum** (Brand) Abrams, in Abrams & Smiley, Bot. Gaz. **60**: 126, 1915
*Eriodictyon californicum* var. *lanatum* Brand, Pflanzenreich **4**[251]: 142, 1913.

Shrubs 5–15 dm. tall with persistently tomentose branchlets; leaves linear-lanceolate to ovate-lanceolate, 5–20 mm. wide, 4–9 cm. long, undulate or serrate, green and sparsely pubescent to glabrate and glutinous above, densely white-tomentose beneath, particularly between veins, these often tawny, margins often slightly revolute, acute at apex, tapering gradually to a petiole 5–20 mm. long; cymes 2–20 cm. broad, rounded or flat-topped, open in fruit; calyx lobes lance-linear, 2–3 mm. long, somewhat glutinous and densely hirsute; corolla funnelform-campanulate, purplish or less commonly white, 6–10 mm. long, limb 3–5 mm. broad; corolla lobes ovate, about 2 mm. long; stamens and style included; capsule ovoid, 2–2.5 mm. long, hirsutulous; seeds 1–1.5 mm. long.

Mostly in chaparral on rocky hillsides and along arroyos, chiefly Upper Sonoran Zone, Riverside County, California, south into northern Baja California but barely entering the western edge of the Colorado Desert near Mountain Spring, San Diego County. April–June.

### 7. **TRICARDIA** Torr. ex S. Wats., Bot. King Expl. 258, 1871

Perennial herbs of low stature from a stout taproot, with petiolate leaves in a basal rosette and a few alternate ones sessile or subsessile on stems. Flowers pedicellate in loose, raceme-like, terminal cymes. Calyx divided nearly to base; 3 outer sepals cordate, becoming conspicuously larger, venose and somewhat scarious in fruit, 2 inner ones linear and much smaller. Corolla broadly campanulate and slightly narrowed at throat, shallowly lobed, shorter than or barely equaling calyx, purplish and deciduous. Stamens unequal but equally inserted on the lower part of corolla tube, included. Corolla scales linear, free from stamens. Style 2-cleft, included. Ovules 4 on each placenta, pendulous. Capsule 1-celled, oblong, acute, with scarious walls. Seeds 4–8, dark brown, minutely reticulate-alveolate.

**1. Tricardia watsonii** Torr. ex S. Wats., Bot. King Expl. 258, *pl. 24,* 1871

Root woody, with a sturdy rootstock 3–10 cm. long above branching roots, upper part of this clothed with persistent bases of leaves of previous seasons; stems usually several from apex of rootstock, 1–3 dm. tall, finely silky-villous throughout when young but soon glabrate; basal leaves 3–5 cm. long, 1–2.5 cm. wide, lanceolate to ovate-lanceolate, cuneate at base and gradually tapering to narrow petiole, this 1–3 cm. long, sparsely lanate-villous; cauline leaves progressively reduced and becoming sessile, upper ones broadly ovate, 1–1.5 cm. long, all of them entire to obscurely crenate, sparsely villous to glabrate in age; cymes mostly simple, to 15 cm. long; pedicels slender, 2–5 mm. long, gracefully recurved in fruit; calyx lobes 5–6 mm. long in flower, 15–25 mm. long in fruit, outer ones orbicular to broadly ovate-cordate, rounded or obtuse at apex; corolla 5–6 mm. long, 6–8 mm. broad, lobes 2–3 mm. long, ovate to orbicular; capsule 7–9 mm. long; seeds dark brown, 3–4 mm. long.

Arid rocky canyons in the desert ranges, Sonoran Zones, western margin of the Colorado and Mojave Deserts east into southern Nevada and Utah and northwestern Arizona. Not common in our area but occurring at Palm Springs, Riverside County, California. April–July.

## Family 121. **BORAGINACEAE.** Borage Family

Herbaceous, suffrutescent, or shrubby plants, usually with hispid, somewhat bristly herbage. Leaves simple, entire, mostly alternate. Flowers perfect, regular, sympetalous, small or moderate in size, solitary or in glomerate-racemose or spicate cymes, cymes frequently secund and scorpioidly coiled, usually bracteate. Calyx mostly deeply 5-lobed, often somewhat irregular. Corolla cylindrical with a rotate limb, 5-lobed, often with saccate appendages or folds partly closing throat. Stamens 5, inserted in corolla tube alternate with lobes. Ovary superior, usually 4-ovulate, entire or 2–4-lobed, usually becoming bony at maturity, commonly breaking into four 1-seeded nutlets in fruit or some nutlets abortive. Style seated at apex of fruit or on receptacle or an elongation of this (gynobase) between nutlets. Endosperm scanty or none.

Style terminal on ovary, not inserted between nutlets:
Fruit drupaceous; shrubs, small trees, or scandent perennial vines:
Style twice bifid into 4 (sometimes 6) clavate or spatulate lobes; calyx tubular or narrowly campanulate, lobes not valvate in bud .......................... 1. *Cordia*
Style once bifid into 2 lobes; calyx globose-urceolate or campanulate and valvate in bud, or 5-lobed nearly to base:
Calyx globose-urceolate or campanulate, lobes valvate in bud; stigmas terminal on lobes of style .............................................. 2. *Bourreria*
Calyx 5-lobed nearly to base, not valvate; stigmatic surface forming a ring at base of bifid part of style ......................................... 4. *Tournefortia*
Fruit dry; herbs or suffrutescent plants:
Style 2-cleft; stigmas 2, distinct at tips of style lobes; flowers often in forks of branches .. ........................................................ 3. *Coldenia*
Style entire or slightly bidentate at apex; stigmatic tissue forming a ring below a sterile apical portion:
Sterile tip of style glabrous or minutely puberulent; corolla (in ours) not obviously constricted at throat ...................................... 5. *Heliotropium*
Sterile tip of the style bearing a tuft of bristles; corolla abruptly constricted at throat .. ........................................................ 6. *Euploca*
Style borne on receptacle or on a gynobase between nutlets:
Mature calyces not burlike or markedly irregular:
Nutlets stellately spreading, markedly compressed dorsiventrally ......... 7. *Pectocarya*
Nutlets erect or weakly divergent, not strongly compressed:
Gynobase conical or pyramidal, usually one-half as high as nutlets or more; nutlets often with a triangular depression on inner face and attached to gynobase therein:
Nutlets emarginate, somewhat tumid, margins with barbed appendages ......... ........................................................ 8. *Lappula*
Nutlets acute or obtuse but not emarginate, not tumid or beset with barbed appendages:
Gynobase usually slightly surpassing nutlets; stigma sessile; plants with filiform stems ................................................ 9. *Eremocarya*
Gynobase shorter than nutlets; stems often slender but scarcely filiform:
Corollas white; cotyledons entire; appendages or folds usually partly closing throat:
Nutlets without a ventral groove, attached by caruncular scar at basal end of a ventral keel .................................. 10. *Plagiobothrys*
Nutlets with a ventral groove, this usually forked at base; attachment in triangular slit at fork .............................. 11. *Cryptantha*
Corollas orange or yellow; cotyledons 2-lobed; throat unappendaged ........ ................................................... 12. *Amsinckia*
Gynobase flat or slightly convex, much shorter than nutlets; attachment surface on nutlets flat or nearly so ................................. 13. *Antiphytum*
Mature calyces markedly irregular, with stout appendages near base of the lobes, thus burlike ........................................... 14. *Harpagonella*

## 1. CORDIA L., Sp. Pl. 190, 1753

Shrubs or small trees, often with pubescent herbage and twigs. Leaves alternate or uncommonly subopposite, petiolate, entire or dentate. Flowers sessile or pedicellate, in cymes, racemes, spikes, or heads, small or large and conspicuous, white or yellowish. Calyx tubular or campanulate, smooth or striate-costate; lobes 3–5, short or somewhat attenuate; calyx sometimes enlarging considerably in fruit, either exceeding or shorter than fruit. Corolla infundibuliform, hypocrateriform, or campanulate; lobes usually 4–6 and shallow. Stamens usually 4–6 (same number as calyx lobes), equally or unequally attached to corolla tube, exserted or included. Ovary 4-celled; style elongate, divided into 4–6 clavate stigmas. Fruit drupaceous, usually included within enlarging calyx. Seeds 1–4, without endosperm.

Inflorescence slenderly spicate; flowers small; corollas 3–4 mm. long; calyx shallowly campanulate, not strongly costate; leaves lanceolate or linear-lanceolate .. 1. *C. brevispicata*
Inflorescence cymose; flowers large, the corollas 1.5–3 cm. long; calyx tubular or narrowly campanulate, 10-costate; leaves ovate, obovate, or broadly oval:
Leaves 6–12 cm. long, entire; glabrous or essentially so; calyx 10–14 mm. long at anthesis ........................................................ 2. *C. sonorae*
Leaves 4 cm. long or less (usually less than 2 cm. long), strigose, finely dentate; calyx 5–7 mm. long at anthesis ........................................ 3. *C. parvifolia*

### 1. Cordia brevispicata Mart. & Gal., Bull. Acad. Brux. 11[2]: 331, 1844

*Cordia palmeri* S. Wats., Proc. Amer. Acad. **24:** 62, 1889.
*Cordia socorrensis* Brandegee, Erythea **7:** 5, 1899.

A rather straggly shrub 1–3 m. tall with puberulent and slightly glandular twigs and leaves (glandular exudate with an unpleasant musty ordor); bark on older stems light gray or rarely to nearly black with light buff, subcircular lenticels; petioles 2–6 mm. long, densely puberulent; leaf blades linear-lanceolate to oblong, 3–18 mm. wide, 2–6.5 cm. long, margins crenatulate and slightly revolute, dark green, scabrous, and slightly rugose above, lower surface paler, densely pubescent and prominently veined, acute to obtuse at apex, cuneate at base; inflorescence spicate, 1–4 cm. long, 6–9 mm. in diameter, on a peduncle about as long, flowers closely crowded; calyx broadly campanulate, 2.5–3.5 mm. long, nearly as wide, minutely glandular-puberulent, 4–5-lobed, lobes 1–1.5 mm. long, triangular; corolla white, 4–5 mm. long, limb spreading, 3.5–5 mm. wide, shallowly 4–5-lobed; drupes bright red, 4–5 mm. in diameter, only lower half enclosed by bowl-like calyx.

Gravelly or rocky canyons and arroyos, Lower Sonoran Zone, vicinity of Comondú and Agua Verde Bay, Baja California, south to the Cape Region and the Revillagigedo Islands and near Guaymas, Sonora. Oct.–April.

### 2. Cordia sonorae Rose, Contr. U.S. Nat. Herb. 1: 106, *pl.* 9, 1891

A small symmetrical tree to 7 m. tall with a sturdy trunk to 15 cm. in diameter and dark gray, slightly fissured or nearly smooth bark; young twigs densely puberulent to tomentulose but soon glabrate; petioles 1–1.5 cm. long, finely but densely puberulent on upper side; leaf blades elliptic to oblong, 1.5–4.5 cm. wide, 5–11 cm. long, mostly obtuse or abruptly and slightly acuminate at apex, broadly cuneate to rounded at base, entire, coriaceous, scantily scaberulous to glabrous and shining above, paler and essentially glabrous beneath; flowers in dense cymes 5–15 cm. broad, branchlets tomentulose with sordid or greenish brown hairs; pedicels stoutish, 2–5 mm.

long; calyx cylindrical, 3.5–4 mm. in diameter, 10–14 mm. long, strongly 10-costate, densely pubescent, teeth deltoid, attenuate, 2–2.5 mm. long; corolla white, turning purplish with age, 2–3 cm. long, limb 2.5–3.5 cm. wide, sparsely villous along median veins without; stamens 5–7, about equaling funnelform throat; fruits not seen.

Hillsides and along arroyos and canyons, Lower Sonoran and Tropical Zones, vicinity of Guaymas, Sonora, southward into Guerrero. March–May.

**3. Cordia parvifolia** A. DC., Prod. **9:** 498, 1845

*Cordia greggii* Torr., Bot. Mex. Bound. 135, 1859.
*Cordia greggii* var. (?) *palmeri* S. Wats., Proc. Amer. Acad. **24:** 61, 1889.
*Cordia watsoni* Rose, in Vasey & Rose, Contr. U.S. Nat. Herb. **1:** 89, 1890.

An open shrub 1–3 m. tall with dark gray, essentially smooth bark flecked with small, pale gray lenticels; twigs somewhat divaricate, densely strigose but eventually glabrate; leaves on young twigs and in fascicles on stubby lateral "spur" branches, mostly obovate or ovate, 3–15 (usually less than 10) mm. wide, 1–3 cm. long, on slender puberulent petioles 2–10 mm. long, blades acute to rounded at apex, broadly cuneate at base, dentate, scabrous with subappressed white hairs from pustulate bases, more densely so beneath, veins impressed above and conspicuous beneath; cymes few-flowered, headlike; calyx narrowly campanulate, 5–7 mm. long, 10-costate, densely puberulent, hairs grayish on basal portion of calyx, grading into dark brown ones near apex and on lance-subulate teeth; corolla white, 1.5–2 cm. long, turning purplish or brownish with age, limb to 3 cm. broad, thin; fruits enclosed within enlarging calyx, 6–9 mm. in diameter.

Alluvial flats, canyons, arroyos, and gently sloping hillsides, Lower Sonoran, Upper Sonoran, and Tropical Zones, northwestern Sonora near Altar to Coahuila, Durango, and Zacatecas, and in central Baja California. Feb.–Nov.

## 2. BOURRERIA P. Br., Nat. Hist. Jamaica 168, 1756

Shrubs or trees with glabrous or pubescent twigs and herbage, and alternate, petiolate, entire leaves. Flowers white, rather large, in terminal, corymblike cymes, these few- to many-flowered. Calyx globose-urceolate to campanulate, 2–5-lobed or dentate, lobes valvate. Corolla salverform, 5-lobed; tube short; limb broadly spreading; lobes imbricate in bud. Stamens 5, attached in corolla tube, included or exserted. Ovary 4-locular. Style bifid at apex; stigmas capitate or clavate. Fruit a drupe, containing 4 bony nutlets.

**1. Bourreria sonorae** S. Wats., Proc. Amer. Acad. **24:** 62, 1889

Openly growing shrub or rarely a small tree 3–6 m. tall with reddish brown bark on younger twigs, gray and lightly striate on older branches, at first minutely but closely pubescent with intertangled curved hairs; leaves rather crowded on numerous short spur branches, more remote on terminal shoots, the petioles 5–8 mm. long or less; leaf blades oblong-oblanceolate to obovate, 8–25 mm. wide, 1.5–4.5 cm. long, harshly scabrous above, subvelvety-pubescent and conspicuously veined beneath; cymes few-flowered; calyx lobes lance-triangular, 3–5 mm. long, acute to acuminate, puberulent; corollas 10–12 mm. long and wide; drupe black, globose, 6–10 mm. in diameter, styles persistent, 10–15 mm. long; stigmas capitate.

Rocky canyons and along arroyos, Lower Sonoran Zone, from the vicinity of Bahía de la Concepción south to the Cape Region, on the adjacent islands in the Gulf, and in the vicinity of Guaymas, Sonora. Nov.–Jan.

## 3. COLDENIA L., Sp. Pl. 125, 1753

Low herbs or shrubs, often with diffuse, prostrate or procumbent branches. Leaves small, alternate, entire or less commonly dentate or somewhate lobed, often crisped. Flowers small, axillary or in terminal heads, sessile or short-pedicellate. Calyx 4–5-parted; sepals linear to lanceolate. Corolla tubular-rotate; tube short; limb 4–5-lobed; interior of the throat naked or bearing as many scales as corolla lobes. Stamens 4–5, inserted in tube, included; filaments very short. Ovary ovoid, shallowly 4-lobed, 4-celled, or 2-celled and falsely 4-celled. Styles 2, slender. Fruit drupaceous, globose, usually containing 4 bony nutlets, 4-lobed or smooth.

Fruit unlobed until completely mature; the style terminal on globose ovary; flowers solitary in leaf axils or forks of stems:
- Corolla white, 6–7 mm. long .......................... 1. *C. canescens* var. *canescens*
- Corolla blue, 8–12 mm. long .......................... 1a. *C. canescens* var. *pulchella*

Fruit and ovary distinctly 2–4-lobed; styles borne between apices of nutlets; flowers in dense clusters at dichotomies of stems and branches:
- Corollas 3–4 mm. long; leaves 2–6 mm. long, markedly cuspidate; corollas without appendages in throats; sepals joined about half their length; nutlets granulate .................................................................. 2. *C. cuspidata*
- Corollas 4.5–6 mm. long; leaves 5–10 mm. long, acute to obtuse at apex but not strikingly cuspidate; corollas appendaged in throats; sepals distinct full length; nutlets not granulate:
  - Leaf blades with 2–3 pairs of shallowly impressed veins; nutlets globose or nearly so; calyx glabrous or slightly puberulent within ...................... 3. *C. palmeri*
  - Leaf blades with 4–6 pairs of deeply impressed veins; nutlets ovoid-elongate; calyx long-villous within .................................................. 4. *C. plicata*

### 1. Coldenia canescens DC., in A. DC., Prod. 9: 559, 1845, var. canescens

*Stegnocarpus canescens* Torr., in Torr. & Gray, Pacif. R. Rep. **2:** 169, *pl.* 7, 1855.
*Coldenia canescens* var. *subnuda* I. M. Johnston, Proc. Calif. Acad. IV, **12:** 1137, 1924.

Low, prostrate or procumbent perennials with gnarled, rough-barked stems 5–25 cm. long, often forming a rather dense mat 2–4 dm. in diameter; leaves rather crowded, tomentose-cinereous and with scattered, spreading, simple, stiff hairs extending 0.5–1.5 mm. beyond tomentum; petioles 0.5–3 mm. long; leaf blades ovate to oblong, 5–10 mm. long, acute at apex, rounded or acute at base, entire, margins revolute, veins not impressed and usually completely obscured by tomentum; flowers mostly solitary (rarely 2–3-flowered clusters) in axils of leaves and forks of branches; sepals narrowly lanceolate or linear-subulate, about 3 mm. long, tomentose and sparsely hirsute; corolla white, 6–7 mm. long, limb rotate, lobes about 1.5 mm. long, rounded, puberulent on outer surfaces; filaments triangular, connate at base; style cleft about one-fourth its length; fruit globose, smooth, not lobed, breaking at maturity into 4 ovate nutlets about 2–2.5 mm. high, with smooth rounded backs and obscurely rugose, plane, lateral surfaces.

Sandy or gravelly soils on desert plains, mesas, and along arroyos, Lower Sonoran Zone, eastern Mojave Desert and Colorado Desert to Texas, south into adjacent northern Mexico. Feb.–Sept.

### 1*a*. Coldenia canescens var. pulchella I. M. Johnston, Jour. Arnold Arb. 20: 379, 1939

Differs from var. *canescens* in having corollas 8–12 mm. long, limb nearly as wide and blue, bluish, or lavender instead of white.

Rocky slopes, mesas, and banks of arroyos, Lower Sonoran Zone, Chocolate Mountains, Imperial County, California, east into western Arizona near Quartzite and the Kofa Mountains.

**2. Coldenia cuspidata** I. M. Johnston, Proc. Calif. Acad. IV, **12:** 1137, 1924

Prostrate to procumbent perennial with woody stems and rootstocks, usually closely appressed to ground but sometimes to 2 dm. tall; branches brittle, densely villous-tomentose when young, older ones with a rough, brown papery bark by exfoliation of tomentum; leaves in crowded fascicles; petioles dilated, triangular at base, 1–2 mm. long, densely villous on margins; leaf blades ovate to lanceolate-ovate, 1–4 mm. wide, 2–6 mm. long, rounded at base, cuspidate-acute at apex, entire, strongly revolute, midrib and 2–3 pairs of veins obscured by pubescence of appressed and spreading hairs; flowers in few flowered axillary clusters; calyx sessile, crowded among leaves, sepals narrowly oblanceolate, about 2 mm. long, obtuse to acutish, joined together in about lower half, densely hirsute-villous; corolla salverform, 3–4 mm. long, pale bluish, unappendaged in throat, lobes rounded, wider than long; stamens slightly unequal; style 2-parted, lobes 1–3 mm. long; ovules usually 2, one often abortive; nutlets 1–1.3 mm. long, ovate-oblong, minutely granulate in lineate patterns.

Sandy washes, mesas, plains, and on talus slopes, Lower Sonoran Zone, Baja California from the vicinity of Santa Rosalia to the Magdalena Plain and in Sonora between Hermosillo and Tastiota. Aug.–April.

**3. Coldenia palmeri** A. Gray, Proc. Amer. Acad. **8:** 292, 1870

*Coldenia angelica* S. Wats., Proc. Amer. Acad. **24:** 62, 1889.
*Coldenia brevicalyx* S. Wats., loc. cit.
*Tiquiliopsis palmeri* Rydb., Fl. Rocky Mts. 711, 1917, (as *Triquiliopsis*).

Low, spreading to prostrate perennial with white-barked, hirsutulous-pubescent, dichotomously branched stems 1–4 dm. long, leaves cinereous-strigillose and sparsely hispidulous, particularly beneath and along revolute margins; petioles linear, terete, or slightly flattened, not dilated at base, hirsutulous; leaf blades ovate or rhombic, 3–6 mm. wide, 5–10 mm. long, acute to truncate at base, acute at apex, with 2–3 pairs of deeply impressed veins and midrib; flowers in dense axillary clusters; sepals linear-subulate, 2.5–3 mm. long, short-hispidulous outside, glabrous or minutely and sparsely puberulent within; corolla tubular-rotate, lavender or pale blue, tube 3–4 mm. long, limb 4–5 mm. in diameter, lobes ovate, glabrous or essentially so; filaments dilated into narrow plates just below point of insertion; nutlets globose or subglobose, smooth and shining, about 1 mm. in diameter.

Sandy flats, mesas, and arroyos, Lower Sonoran Zone, western Colorado Desert to western Arizona, north along Colorado River to vicinity of Needles and south into northern Baja California and northwestern Sonora at least to Punta Peñasca. March–Oct.

**4. Coldenia plicata** (Torr.) Coville, Contr. U.S. Nat. Herb. **4:** 163, 1893

*Tiquilia brevifolia* var. *plicata* Torr., Bot. Mex. Bound. 136, 1859.
*Coldenia palmeri* of S. Wats., and recent authors, not A. Gray, 1870.

A spreading to decumbent perennial with freely branching, dichotomous stems forming a dense mat 1–2 dm. tall and to 4–5 dm. in diameter; branches slender, moderately puberulent to glabrate, bark eventually papery and exfoliating, pale reddish tan to silvery; petioles slender, terete, not dilated, 1–5 mm. long, hirsutulous;

leaf blades broadly elliptic, ovate, or obovate, 3–5 mm. wide, 4–8 (or sometimes 10) mm. long, with 4–6 pairs of deeply impressed lateral veins, sharply but narrowly revolute, densely puberulent with grayish strigose hairs; flowers in small crowded clusters in axils of leaves and branches, usually subtended by a whorl of leaves; sepals subulate, 2–2.5 mm. long, hirsute-villous, particularly within; corolla bluish or pinkish, 6–7 mm. long, limb about 3 mm. in diameter, tube minutely glandular-puberulent between sepals and base of corolla lobes; filaments continued below point of attachment as narrow plates; style cleft about one-half its length or slightly more; nutlets ovoid, slightly elongated, smooth and shining, about 1 mm. high, 1–3 usually aborting.

Sandy plains and washes or on dunes, Lower Sonoran Zone, Colorado Desert, eastern Mojave Desert, western Arizona, and south in northern Baja California to the vicinity of San Quintín. April–Oct.

**Coldenia hispidissima** (Torr.) A. Gray, Proc. Amer. Acad. **5**: 340, 1862, occurs in northern Arizona, but no specimens from the Sonoran Desert have been seen. It has linear to lanceolate leaf blades without obvious nerves and is usually conspicuously and pungently setose; base of petiole is expanded, indurate, and villous; ovate nutlets are finely tuberculate on back and angled on inner face. It may occur along the northern fringe of our area in Arizona.

## 4. TOURNEFORTIA L., Sp. Pl. 140, 1753

Shrubs, small trees, or scandent vines with alternate entire leaves, and small flowers in dichotomous scorpioid cymes, spikes, or racemes. Calyx 5-parted into linear, lanceolate, or oblong sepals. Corolla salverform or narrowly infundibuliform; tube cylindrical; throat naked, lobes 5, variously shaped and arranged in bud. Stamens 5, inserted in tube, included. Ovary 4-celled, 4-ovuled. Fruit a drupe, containing 4 nutlets. Seeds pendulous or oblique; endosperm fleshy.

### 1. Tournefortia hartwegiana Steudel, Nom. ed. 2, 2: 693, 1841

*Tournefortia capitata* Mart. & Gal., Bull. Acad. Brux. **11²**: 332, 1844.

An erect, rather slenderly branched shrub 1–4 m. tall with scabrous-pubescent herbage, twigs, and inflorescences; leaves elliptic-oblong, lanceolate, or oblanceolate, 1–4 cm. broad, 5–13 cm. long, acuminate at apex, attenuate-cuneate at base, scabrous above, pubescent to scaberulous beneath, margins often slightly revolute and veins prominent beneath; peduncles and branches of cymes usually surpassing subtending leaves; spikes of cymes 2–4 cm. long, scorpioid in youth; flowers subsessile, fragrant; calyx lobes linear or subulate, 3–6 mm. long, canescent; corolla tube cylindrical, 5–8 mm. long, canescent-pubescent without; corolla limb white, 5–7 mm. broad, lobes rounded or obtuse, naked within; drupes 5–7 mm. in diameter, waxy-white; nutlets about 4 mm. long.

In canyons and on alluvial plains or bottoms, Lower Sonoran to Tropical Zones, central Baja California and southern Sonora to Tamaulipas, Oaxaca, and San Luis Potosí. Nov.–April.

**Tournefortia volubilis** L., Sp. Pl. 140, 1753, a suberect to scandent shrub with grayish-sericeous leaves and long, slender, lax spikes, occurs in the Cape Region of Baja California and in the subtropical forests near Alamos in southern Sonora, but it has not been reported from the Sonoran Desert. It occurs sparingly as a straggler in canyons and on ridges of the Sierra de la Giganta near Loreto.

## 5. HELIOTROPIUM L., Sp. Pl. 130, 1753

Herbaceous, suffrutescent, or rarely shrubby plants with small to large leaves. Inflorescence scorpioid spikes or racemes or flowers solitary along stems. Calyx 5-toothed or -lobed, persistent or deciduous, sometimes slightly accrescent in fruit. Corolla tubular with a spreading limb, 5-lobed or -toothed, white, yellow, or blue. Stamens 5, inserted in corolla tube, included. Ovary 4-celled; style terminal; stigma sessile or on a distinct slender style, apex bearing a conic or cylindrical, sterile appendage above receptive region of stigma. Fruit dry, lobed or unlobed, breaking into 2–4 bony, 1–2-seeded nutlets.

Plants subsucculent; leaves and stems glabrous and glaucous:
  Corollas inconspicuous, 1.2–5 mm. in diameter, white or faintly purplish; calyx lobes shorter than fruit, appressed ................. 1. *H. curassavicum* var. *curassavicum*
  Corollas larger, 3–7 mm. in diameter, turning deep purple in throat; calyx lobes longer than fruit, erect ............................ 1a. *H. curassavicum* var. *oculatum*
Plants not succulent; leaves and stems obviously puberulent or pubescent, not glaucous:
  Spikes naked; calyx lobes distinctly shorter than mature nutlets:
    Nutlets scurfy with blunt, vesiclelike cells; stigma sessile or essentially so ............ ............................................... 2. *H. angiospermum*
    Nutlets pubescent with slender hairs, not scurfy; stigmas on evident styles:
      Leaves broadly ovate, 2–7 cm. broad; corolla 5–6 mm. long; plants erect or ascending ............................................... 3. *H. macrostachyum*
      Leaves oblanceolate-elliptic, 2–10 mm. broad; corolla 1.5–2 mm. long; plants often procumbent ........................................ 4. *H. procumbens*
  Spikes more or less bracteate, although not uniformly so; calyx lobes equaling or considerably surpassing nutlets at maturity:
    Nutlets glabrous or essentially so; corollas 12–15 mm. in diameter ..... 5. *H. wigginsii*
    Nutlets distinctly puberulent or strigose-pubescent; corollas 5–6 mm. in diameter:
      Lateral veins obvious, distinctly impressed above; spikes uniformly bracteate, each bract 1–2 mm. long; pubescence on stems reflexed .......... 6. *H. greenmanii*
      Lateral veins obscure or lacking, not impressed; spikes irregularly leafy-bracteate, bracts 1–6 mm. long; pubescence on stems ascendingly appressed:
        Leaves linear, strongly revolute; plants perennial; corollas 4–5 mm. long ........ ............................................... 7. *H. limbatum*
        Leaves broadly elliptic to oblanceolate; plants annual; corollas 1–1.5 mm. long ... ............................................... 8. *H. fruticosum*

### 1. Heliotropium curassavicum L., Sp. Pl. 130, 1753, var. curassavicum
(For synonymy see Contr. Gray Herb. **81:** 14, 1928.)

Stems several to many from a perennial root, 1–10 dm. long, these and leaves subfleshy and usually more or less glaucous; leaf blades lance-elliptic, oblanceolate, or obovate, 3–10 mm. wide, to 6 cm. long, gradually narrowed to a subpetiolate base, acute to rounded at apex, fleshy, glabrous and glaucous, often purplish in age; spikes geminate or sometimes in threes or fours, tightly scorpioid at tip in youth, naked, 3–7 cm. long; calyx subglobose, lobes narrowly ovate, 1–2 mm. long at anthesis, equal, glabrous and glaucous, rather closely applied to fruit in age; corolla minute, inconspicuous, 1–2.5 mm. long, limb about as wide, white, without a purple "eye"; stigma minute, glabrous, about 0.2 mm. long; fruit depressed-globose, 1.5–2 mm. high, 2–2.5 mm. wide, nutlets ovoid, rounded and smooth or faintly rugulose on back.

In marshy places, along stream beds, on alkaline flats, and along the sea coast, Upper Sonoran to Tropical Zones, southern Arizona to southernmost Texas and

Florida, and south into South America. Flowering practically throughout the year. Replaced in our area almost entirely by the following variety.

**1*a*. Heliotropium curassavicum** var. **oculatum** (Heller) I. M. Johnston ex Tidestrom, Proc. Biol. Soc. Wash. **48:** 42, 1935

*Heliotropium oculatum* Heller, Muhlenbergia **1:** 58, 1904.
*Heliotropium spathulatum* subsp. *oculatum* Ewan, Bull. S. Calif. Acad. **41:** 56, 1942.

Similar to var. *curassavicum* but with more conspicuous corollas 3–6 mm. long, limb 3–5 (or 7) mm. wide, throat at first yellowish, later becoming distinctly purplish; sepals 2–3 mm. long in flower and exceeding fruit at maturity; fruit to 3 mm. wide.

Valleys and in saline moist soils, mainly Lower Sonoran Zone, southwestern Utah to central California and south throughout Baja California and Sonora; north to Washington. Flowering throughout the year.

**2. Heliotropium angiospermum** Murr., Prodr. Stirp. Göttingen 217, 1770

*Heliotropium parviflorum* L., Mant. 201, 1771.
(For additional synonymy see Contr. Gray Herb. **81:** 10, 1928, but add *H. humile* Lam., Tab. Encycl. **1:** 393, 1792.)

Erect, ascending, or procumbent herbs from perennial roots; stems rather slender, 2–6 dm. long or more, sparsely pubescent with a few spreading and many appressed white hairs 0.5–2 mm. long, bases of larger ones slightly pustulate; leaves broadly elliptic to broadly ovate, on petioles 3–15 mm. long, blades dark green and very sparsely pubescent above, paler and more densely pubescent along veins beneath, acute at apex, obtuse to broadly cuneate at base; spikes solitary or geminate, tightly scorpioid at apex in youth, later straight or slightly arcuate, 5–15 cm. long, naked; sepals oblong to narrowly ovate, about 1 mm. long at anthesis, 1–1.5 mm. long in fruit, sparsely strigose, spreading and shorter than fruits at maturity; corolla white, about 1.5 mm. long, sparsely strigose without, limb 1–1.5 mm. in diameter; fruits broadly depressed-ovoid, 2–2.5 mm. in diameter, 1.5–2 mm. high, nutlets scurfy with small, white, vesicle-like cells, firmly attached in pairs, each with a low dorsal ridge; stigmas sessile or essentially so.

Valleys, arroyos, along streams, and on shaded hillsides, Lower Sonoran to Tropical Zones, central Sonora and central Baja California throughout Mexico, southern Texas, southern Florida, and the West Indies to northern Chile and eastern Bolivia. Aug.–April.

**3. Heliotropium macrostachyum** (DC.) Hemsl., Biol. Centr. Amer. Bot. **2:** 375, 1882

*Heliophytum macrostachyum* DC., in A. DC., Prod. **9:** 556, 1845.

Perennial herb from a slender woody rootstock; stems erect, 3–6 dm. tall, simple or sparingly branched, hirsute with spreading, slightly tawny hairs 1.5–2 mm. long, and finely puberulent with shorter, gland-tipped hairs among spreading ones; petioles slender, 2–5 cm. long, pubescent; leaf blades broadly ovate, 2–7 cm. broad, 4–10 cm. long, acute to subcordate at base, abruptly acute or short-acuminate at apex, slightly sinuate-crenate on margins, puberulent beneath especially along veins, puberulent and hirsute above, usually darkening in drying; spikes usually geminate, 1–1.5 dm. long, tightly scorpioid at apex, fruits 5–10 mm. apart in age; sepals narrowly lanceolate, 1.5–2 mm. long, glandular-puberulent and sparsely hirsute; corolla white, tube 5–6 mm. long, limb 4–5 mm. broad; fruit depressed-globose, 3–3.5 mm. wide, 2–2.5 mm. high, distinctly puberulent.

On hillsides and along arroyos under shrubs, margins of Lower Sonoran to Tropical Zones, vicinity of Ures, Sonora, south into southern and eastern Mexico. Sept.–March.

A rocky hillside 18 miles northeast of Ures along the road to Baviácora is the most northwesterly known locality for the species.

### 4. **Heliotropium procumbens** Mill., Gard. Dict. ed. 8, no. 10, 1768

*Heliotropium inundatum* Swartz, Prod. Veg. Ind. Occ. 40, 1788.
(For additional synonymy see Contr. Gray Herb. **81:** 52, 1928.)

Herbaceous perennial with several prostrate to erect stems 1–4 dm. long from a woody rootstock, branches and herbage sparsely to densely strigose-canescent with slender white hairs from white pustulate bases; leaves elliptic, oblanceolate, or linear-lanceolate, 2–10 mm. wide, 1–4 cm. long, acute or abruptly short-acuminate at apex, gradually narrowed toward base, lower ones petiolate, entire, veins inconspicuous; spikes solitary to ternate, usually geminate, naked, 2–8 cm. long, tightly scorpioid at tip when young, later straight or arcuate; calyx lobes linear-lanceolate, about 1 mm. long in flower, strigulose, to 1.5 or 2 mm. long and abaxial one slightly longer than others in fruit; corolla white, tube barely exceeding sepals, strigose on exterior of tube and along midribs of lobes of limb, latter 1–1.25 mm. in diameter, inconspicuous; fruit nearly globose, slightly depressed and tip drawn to a low blunt point, nutlets densely strigose, about 1 mm. tall; style short, slightly tapering stigma extending about 0.2–0.4 mm. beyond tips of nutlets.

In saline soil or along arroyos and margins of pools, Lower Sonoran to Tropical Zones, southern Texas and east-central Baja California south into South America. Aug.–May.

Plants from Agua Verde Bay collected by I. M. Johnston are floccose-pubescent with tangled, somewhat spreading hairs and appear to be annual. At least, they flowered the first season. The flowers and fruits are indistinguishable from those of the perennial plants from San José del Cabo and La Paz.

### 5. **Heliotropium wigginsii** I. M. Johnston, Jour. Arnold Arb. **29:** 232, 1948

Annual herb 2–5 dm. tall with erect or ascending, moderately branched stems sparsely pubescent with ascendingly appressed, whitish, stiff hairs; leaves oblong to narrowly oblanceolate, 2.5–7 mm. wide, 1.5–4 cm. long, acute or short-acuminate apically, gradually narrowed to the cuneate base or short-petiolate, dark green, appressed-scaberulous with ascending hairs above, densely appressed-pubescent and canescent beneath, margins narrowly revolute, midrib slightly impressed above, prominent beneath, lateral veins scarcely visible; racemes loose, interruptedly and irregularly leafy-bracteate, often naked toward tips, 5–15 cm. long, scorpioid in youth; pedicels 0.5–1.5 mm. long; calyx 3–3.5 mm. long, narrowly funnelform in flower, lobes lance-linear, equal, coarsely strigose; corolla white, plicate, appressed-pubescent on tube and on exposed plicae without, tube 3–4 mm. long, limb about 15 mm. in diameter, a small lobe occupying base of each sinus between main lobes and surmounting inwardly folded plicae; throat of tube slightly constricted and densely puberulent with short, thickish, transversely septate hairs; stamens inserted about middle of tube; anthers narrowly ovate, 1–1.2 mm. long, tips more or less connivent; style and stigma about 2 mm. long, 5-lobed annulus 0.6–0.8 mm. in diameter, set about 1–1.4 mm. above nutlets; fruit 1.6–2 mm. high, 2.5–3.2 mm. in diameter, distinctly 4-lobed, each nutlet somewhat rugosely ridged on back, glabrous, readily separating from receptacle.

About margins of summer rain pools and on desert plains and mesas, Lower Sonoran Zone, central to southern Sonora and in central Baja California. Aug.–Oct.

**6. Heliotropium greenmanii** Wiggins, Contr. Dudley Herb. **4:** 22, 1950

*Heliotropium mexicanum* Greenm., Proc. Amer. Acad. **33:** 484, 1898. Not *H. mexicanum* Sessé & Moc., 1888.

Perennial herb, suffrutescent at base, with several to many ascending branches 1–5 dm. long from a woody rootstock; stems puberulent and subcanescent with somewhat reflexed hairs and a few intermingled spreading ones; leaves oblanceolate, 3–9 mm. wide, 1.5–3.5 cm. long, obtuse to abruptly short-acuminate at apex, tapering gradually toward base, on petioles 0.5–1.5 mm. long or subsessile, midrib and lateral veins slightly impressed above, upper surface finely and moderately puberulent with an admixture of spreading and appressed hairs, more densely and slightly more coarsely pubescent beneath, especially along midrib, margin narrowly revolute; spikes solitary or geminate, mostly pedunculate, 2–10 cm. long, bracteate; bracts narrowly lanceolate, 1–2 mm. long; calyx lobes lance-linear, 2.5–3 mm. long at anthesis, slightly unequal, to 4 mm. long in fruit, finely strigose-canescent; corolla 5–6 mm. long, somewhat plicate, puberulent on tube and exposed sections of plicae, limb 5–6 mm. broad, white; fruit about 1–1.5 mm. in diameter, completely obscured by erect calyx lobes.

Hillsides and open slopes, Lower and Upper Sonoran Zones, near Batúc, Sonora, and east into southern Chihuahua and south to Oaxaca. Sept.–March.

**7. Heliotropium limbatum** Benth., Pl. Hartw. 20, 1839

Low rounded plant with several to many decumbent to ascending, herbaceous branches 5–25 cm. long from a woody perennial rootstock, this branching into several parts 1–4 cm. below surface of soil; herbage scaberulous-canescent throughout; leaves linear, 1–2.5 mm. wide, 6–20 mm. long, acute at apex, gradually narrowed toward base into a slender petiole 1–5 mm. long, margins narrowly but tightly revolute, lower ones crowded into dense tufts; inflorescence a scorpioid spiciform raceme terminal on each branch, 1–3 cm. long, bracteate; bracts similar to leaves but reduced, 3–6 mm. long; calyx about 2–2.5 mm. long at anthesis, 3–4.5 mm. long in fruit, lance-ovate lobes distinct nearly to base, densely strigose-canescent; corolla white with a faint yellow eye, about 5 mm. long, strigose-puberulent without, limb 3–4.5 mm. broad; fruit depressed-globose, 1.5–2 mm. in diameter, shallowly lobed, enclosed in ascendingly spreading calyx lobes.

On gravelly or rocky hillsides, Lower Sonoran to Tropical Zones, vicinity of Ures, Sonora, south into central Mexico. July–Oct.

**8. Heliotropium fruticosum** L., Syst. Nat. ed. 10, 913, 1759

*Heliotropium phyllostachyum* Torr., Bot. Mex. Bound. 137, 1859.

(For additional synonymy see Contr. Gray Herb. **81:** 66, 1928, but exclude *H. humile* Lam., Tab. Encycl. **1:** 393, 1792.)

Annual herbs sparingly branched from near or above base, branches spreading-ascending, 3–15 cm. long; herbage sparsely strigose with slender whitish hairs, midribs of leaves bearing heavier hairs with slightly pustulate bases; leaves elliptic to oblanceolate, 1.5–6 mm. wide, 7–20 mm. long, acute to rounded at apex, broadly cuneate at base, lower ones short-petiolate, dark green above, paler beneath, margin very narrowly but tightly revolute; flowers axillary or borne in a leafy-bracteate spike with some flowers devoid of subtending bracts; sepals narrowly ovate-lanceolate, unequal, 1–1.5 mm. long at anthesis, strigose and somewhat pustulate, slightly elongated in fruit, adaxial sepal 1–1.5 mm. long, abaxial one nearly or fully twice as long, others intermediate; corollas inconspicuous, white, 1–1.5 mm. long, limb about as wide, finely strigulose without; fruit depressed-globose, 1.2–1.5 mm. high, 1.5–2 mm. broad, nutlets rounded on back, brownish, puberulent with fine white hairs, lateral faces each bearing a central pit; style and stigma together 0.6–0.8 mm. long, separated by a 5-lobed collar, stigma about two-thirds as long as style.

Dry hillsides and valley floors, Upper and Lower Sonoran and Tropical Zones, southern Arizona south through Mexico and Central America to Colombia, Venezuela, and Brazil and east throughout the West Indies. In our area, July–Nov.

## 6. EUPLOCA Nutt., Trans. Amer. Phil. Soc. II, 5: 189, 1837

Small annuals with ascending or erect branches and hirsute or hirsutulous herbage. Flowers solitary in axils of leaves. Calyx deeply 5-lobed or -toothed. Corolla white; tube slightly dilated near apex, then abruptly constricted; limb spreading, shallowly lobed. Stamens 5, inserted low in corolla throat, included; anthers ovate-lanceolate. Style somewhat elongated; stigma annular-peltate, caplike, entire or obscurely 2-lobed, surmounted by a tuft of bristles. Fruit of 2 globose lobes, each lobe splitting into 2 hemispherical, 1-seeded nutlets. Endosperm present.

**1. Euploca convolvulacea** Nutt. subsp. **californica** (Greene) Abrams, Ill. Fl. Pacif. States **3:** 534, 1951

*Heliotropium californicum* Greene, Bull. Calif. Acad. **1:** 202, 1885.
*Euploca aurea* Rose & Standley, Contr. U.S. Nat. Herb. **16:** 16, *pl. 11*, 1912.
*Heliotropium convolvulaceum* var. *californicum* I. M. Johnston, Contr. Arnold Arb. **3:** 83, 1932.
*Euploca albiflora* var. *californica* Jepson & Hoover, in Jepson, Fl. Calif. **3:** 299, 1943.

An ascendingly branched, annual herb with branches to 3 dm. tall, pungently hispid-bristly throughout, especially on angles of stems and leaf margins, hairs with pustulate bases; petioles 3–10 mm. long; leaf blades ovate to lance-oblong, 5–15 mm. wide, 1–3 cm. long, acute or short-acuminate at apex, broadly cuneate to rounded at base; flowers fragrant, on pedicels 1.4 mm. long; calyx lobes linear-lanceolate, hispid, 4–5 mm. long; corolla tube 5–7 mm. long, distinctly swollen in upper one-half, then abruptly constricted again, densely hirsute with stiff appressed hairs, these extending to edge of limb along main veins; corolla limb white, about 10 mm. broad, bearing 5 small triangular apiculations at the tips of veins but not lobed; fruit about 3 mm. in diameter, each lobe hispid, dark brown.

Sandy flats, plains, and hillsides, Lower Sonoran Zone, near the Colorado River in the eastern Mojave and Colorado Deserts, east into adjacent Arizona; south into northern Sonora. Feb.–Oct.

## 7. PECTOCARYA DC. ex Meisn., Pl. Vasc. Gen. Comm. 279, 1840

Spreading to prostrate, slender, annual herbs, more or less strigose-pubescent throughout. Leaves narrowly linear, forming a sparse rosette in youth, later cauline leaves distant. Flowers in leafy-bracted, false racemes, small, white. Pedicels short, deflexed at maturity. Sepals 5, equal or unequal, spreading in fruit. Corolla tubular-rotate to tubular-funnelform, the tube short; lobes 5, ascending, short, somewhat imbricate; throat partly closed by small appendages. Stamens inserted in corolla tube, included. Style short. Stigmas capitate. Nutlets 4, linear to obovate, commonly paired and divaricate, usually margined with a thin wing, this usually bearing a fringe of uncinate hairs, often a denser tuft of hairs at distal end of nutlets. Gynobase broadly pyramidal, low.

Nutlet body linear to oblong; margins of nutlet lacerate, undulate, or bearing uncinate bristles; sepals not conspicuously setose:
  Nutlets all alike or essentially so; sepals equal; pedicel basal, not appearing lateral:
    Margins of nutlets pectinately lacerate or dentate almost entire length, usually also uncinately bristly at distal end:
      Nutlets conspicuously recurved; marginal teeth gradually merging into slender uncinate hairs .......... 1. *P. recurvata*
      Nutlets straight or nearly so; marginal teeth broader, stout, narrowing abruptly into hairs:
        Margins of nutlets narrow and inconspicuous, teeth distinct or essentially so .......... 2. *P. linearis* var. *ferocula*
        Margins of nutlets broad and conspicuous, teeth confluent at base .. 3. *P. platycarpa*
    Margins of nutlets entire or undulate along sides, armed only at distal end with a tuft of uncinate hairs .......... 4. *P. penicillata*
  Nutlets heteromorphic, one pair in each tetrad broadly margined, other pair narrowly margined or wing lacking; sepals distinctly unequal; pedicels appearing to be attached laterally to receptacle:
    Nutlets of each fruit in 2 planes, one pair inwardly curved, other pair recurving; nutlets oblong when viewed from above .......... 5. *P. heterocarpa*
    Nutlets of each fruit spreading in 1 plane; nutlets elliptic-ovate when viewed from above .......... 6. *P. peninsularis*
Nutlet body obovate to rhomboidal; margins entire, not lacerate into separate appendages, but bearing slender uncinate hairs on and near edges of membranous wings; sepals conspicuously setose .......... 7. *P. setosa*

**1. Pectocarya recurvata** I. M. Johnston, Contr. Arnold Arb. **3:** 97, 1932

Plants slender, prostrate to ascending, 5–20 cm. long, minutely cinereous-strigulose; leaves filiform-linear to narrowly linear-oblanceolate, 0.6–2 mm. broad, 1–3 cm. long, rather harshly strigose; flowers axillary to leaves from very base to apex of individual branches, minute; sepals lance-linear, 1.5–2 mm. long, slightly longer in fruit, strigose; corolla short-salverform, about 2 mm. long, lobes ascending; nutlets all alike, linear, about 0.8 mm. broad, 2.5–3.2 mm. long, distinctly and often strongly reflexed, often forming a semicircle, margins deeply dissected to form a series of fairly distinct, short teeth, each gradually tapering into and terminated by a slender, elongated, uncinate hair equaling or surpassing tooth, these well-spaced appendages and their terminal hairs pectinately arranged and exceeding width of body of nutlet, armament at tip not appreciably denser than that at sides.

Arid slopes, plains, mesas, and washes, Lower Sonoran Zone, southern Nevada and Inyo County, California, south through the Mojave and Colorado Deserts to northern Baja California and east into Arizona and northern Sonora, crossing to the Pacific side of the peninsula at San Quintín, Baja California. Feb.–May.

**2. Pectocarya linearis** DC. var. **ferocula** I. M. Johnston, Contr. Arnold Arb. **3:** 95, 1932

A low, prostrate or procumbent annual with slender, strigulose, pale green stems 2–15 cm. long; leaves filiform-linear or narrowly linear, to 2 mm. wide, 0.5–4 cm. long, strigose and sparsely hispid, bases of spreading hairs usually pustulate; flowers small, borne almost to base of each branch; pedicels short, 0.2–1 mm. long; calyx symmetrical, sepals equal, 1–1.5 mm. long and erect at anthesis, spreading in fruit, strigulose; corolla white, about 2 mm. long; nutlets all alike, spreading, tips slightly recurved, body linear-oblong or sometimes spatulate-oblong, 0.5–1 mm. wide, 2–3.5 mm. long, margin comparatively inconspicuous, divided into short, crowded, broadly deltoid teeth, each of these abruptly terminated by an uncinate hair which equals or exceeds length of deltoid marginal division, divisions and hairs at tips of nutlet only slightly more crowded than on sides, body of nutlets strigulose on back, sparsely and minutely puberulent to glabrate within.

Arid slopes and gravelly places, mainly Upper Sonoran Zone, Santa Barbara County, California, and the adjacent Channel Islands south into central Baja California and in Argentina; entering the Sonoran Desert in eastern San Diego County. Feb.–May.

According to I. M. Johnston, var. *linearis* does not occur in North America.

**3. Pectocarya platycarpa** Munz & Jhtn., in I. M. Johnston, Contr. Gray Herb. **81:** 81, 1928

*Pectocarya gracilis* var. *platycarpa* Munz & Jhtn., in I. M. Johnston, Contr. Gray Herb. **70:** 36, 1924.

Plants annual, stems slender but stiffish and prostrate to spreading-ascending, 1–2 dm. long, cinereous-strigulose; leaves linear to linear-oblanceolate, 0.5–1.8 mm. broad, 1–3.5 cm. long, strigose; pedicels slender, 1–2 mm. long; sepals 2.5–3 mm. long, linear or lance-linear; corolla white, about 2 mm. long; nutlets usually heteromorphic, body linear or oblong or spatulate-oblong, 0.6–1 mm. wide, 2.5–3 mm. long, 3 of them usually conspicuously margined with a broad, deeply lacerate wing that is glabrous within, teeth coarse, broadly deltoid, pallid, tipped with short uncinate bristles shorter than width of supporting margin, fourth nutlet, if different, with a narrower, more dissected wing and a closely puberulent body.

Arid gravelly benches, hillsides, and mesas, Lower Sonoran Zone, Colorado and Mojave Deserts to southern Utah, Sonora, and Baja California. Feb.–April.

**4. Pectocarya penicillata** (Hook. & Arn.) A. DC., Prod. **10:** 120, 1846

*Cynoglossum penicillatum* Hook. & Arn., Bot. Beechey 371, 1839.
*Pectocarya linearis* var. *penicillata* M. E. Jones, Proc. Calif. Acad. II, **5:** 709, 1895.
*Pectocarya miser* A. Nels., Bot. Gaz. **37:** 278, 1904.
*Pectocarya penicillata* var. *genuina* I. M. Johnston, Contr. Gray Herb. **70:** 37, 1924.

Stems prostrate to procumbent, several from base, slender, 5–25 cm. long, cinereous-strigose; leaves filiform, to linear or sometimes spatulate, 0.3–2 mm. wide, 1–3 cm. long, strigose; flowers small; sepals elliptic-lanceolate or linear-lanceolate, 1.5–2 mm. long at anthesis, to 2.5 mm. long in fruit, strigose; corolla white, about 2 mm. long; nutlets all alike, oblong, divaricate, straight or very slightly inflexed at tip, 0.5–0.8 mm. broad, 1.6–2.5 mm. long, cartilaginous margin upturned to incurved along length of body, unarmed on this portion, nearly entire, or undulate, rounded distal end bearing a tuft of crowded uncinate bristles, back of body and margins puberulent, inner face more finely and less densely so.

Sandy or gravelly soil, mainly Upper Sonoran Zone, British Columbia to northern

Baja California and east to Wyoming, but barely entering the margins of the Colorado and Mojave Deserts. Feb.–May.

**5. Pectocarya heterocarpa** (I. M. Johnston) I. M. Johnston, Jour. Arnold Arb. **20:** 399, 1939

*Pectocarya penicillata* var. *heterocarpa* I. M. Johnston, Contr. Gray Herb. **70:** 37, 1924.

Similar to *P. penicillata* in habit and size, but stems slightly more robust and strigulose with shorter, finer hairs than in that species; leaves linear to narrowly oblanceolate, 0.5–1.2 mm. wide, 5–25 mm. long, strigulose; flowers small, about 2 mm. long; sepals as in *P. penicillata*; nutlets heteromorphic, consisting of 2 broadly margined nutlets, these divaricate and somewhat inflexed, margins lacerate-toothed and deltoid teeth tipped with uncinate hairs; other 2 nutlets unmargined and somewhat reflexed, all bearing a tuft of uncinate hairs distally, the fruits at bases of branches usually of 4 unmargined reflexed nutlets.

Arid, gravelly, and sandy slopes, valleys, and washes, Lower Sonoran Zone, Colorado and Mojave Deserts to Inyo County, California, east to Utah and Arizona and south into northern Sonora and Baja California. Feb.–April.

**6. Pectocarya peninsularis** I. M. Johnston, Jour. Arnold Arb. **20:** 401, 1939

A slender annual herb with several strigose to hispidulous, prostrate stems 5–15 cm. long from a slender root; leaves narrowly linear, about 0.5–1 mm. wide, 1–4 cm. long, strigulose; flowers dimorphous; pedicels short, appearing to be attached to side of receptacle; sepals unequal, adaxial one shortest, similar to those in *P. heterocarpa*; flowers at base of plant cleistogamous; nutlets strongly heteromorphic, about 1 mm. wide, 2 mm. long, 2 nutlets very narrowly winged or wingless, other 2 conspicuously margined, wing lacerate into deltoid teeth tipped with uncinate hairs, all nutlets in essentially same plane; nutlets about base of branchlets somewhat narrower than upper ones, 0.5–0.8 mm. wide, tips of the nutlets bearing a tuft of uncinate hairs, the backs lightly strigose.

Arid sandy plains and washes, Lower Sonoran Zone, vicinity of San Quintín south in Baja California to Calmallí or beyond. Feb.–April.

*Pectocarya peninsularis* is very closely related to *P. heterocarpa,* but the characters that separate the 2 species seem to be reasonably constant.

**7. Pectocarya setosa** A. Gray, Proc. Amer. Acad. **12:** 81, 1876

*Gruvelia setosa* Rydb., Bull. Torrey Club **40:** 479, 1913.
*Pectocarya setosa* var. *genuina* I. M. Johnston, Contr. Gray Herb. **70:** 38, 1924.
*Pectocarya setosa* var. *aptera* I. M. Johnston, loc. cit.
*Pectocarya setosa* var. *holoptera* I. M. Johnston, op. cit. 39.

Stems erect or ascendingly and profusely branched from near base, 1–2.5 dm. tall, minutely strigose-pubescent and hispid throughout; leaves linear to linear-spatulate or sometimes narrowly oblanceolate, 1–3.5 mm. wide, 5–30 mm. long, bases of hairs conspicuously white-pustulate; flowers mostly on upper parts of branches; sepals linear or lance-linear, 2–3 mm. long, surpassing nutlets, each armed with several stout, stiff, divergent, tawny or rusty bristles; corolla white, about 2 mm. long; nutlets in 2 divergent pairs, broadly obovate, 2–3 or rarely all 4 of them conspicuously broad-margined with a thin entire wing, usually 1–2 (rarely all 4) wingless, body cuneate-obovate, about 2 mm. long, margin and upper face set with slender uncinate hairs, outer face puberulent, all 4 nutlets in same plane.

Sandy arroyos, slopes, and plains, Lower and Upper Sonoran Zones, inner South Coast Ranges from San Joaquin County to San Luis Obispo County, Mojave and Colorado Deserts north; east of the Sierra Nevada to eastern Washington; east into southern Nevada and central Arizona; south into northern Baja California. March–June.

## 8. LAPPULA Moench, Meth. 416, 1794

Annual herbs with canescent-villous to glabrate stems and herbage and alternate, usually narrow leaves. Flowers small, white or blue, borne in bracteate racemes. Pedicels usually short, often thickening somewhat in fruit. Calyx 5-parted into narrowly ovate, nearly distinct lobes. Corolla with a short tube, usually with small appendages in short throat; limb small, 5-lobed. Stamens 5, inserted in throat, included. Ovary of 4 ovoid nutlets, these burlike, armed with terminally barbed prickles arranged along outer margins or over whole back of each nutlet.

Margins of nutlets strongly inflated, inwardly curved; prickles appearing to be seated upon margin .............................................. 1. *L. texana* var. *coronata*
Margins of nutlets thinner, appearing to be made up of confluent bases of prickles, not swollen ............................................ 2. *L. redowskii* var. *desertorum*

**1. Lappula texana** (Scheele) Britt. var. **coronata** (Greene) Nels. & Macbr., Bot. Gaz. **61:** 41, 1916

*Lappula coronata* Greene, Pittonia **4:** 94, 1899.
*Echinospermum coronatum* K. Schumann, in Just, Bot. Jahresb. **27**[1]**:** 522, 1901.
*Lappula heterosperma* var. *homosperma* A. Nels., Bot. Gaz. **34:** 29, 1902.
*Lappula texana* var. *homosperma* Nels. & Macbr., op. cit. **61:** 41, 1916.

Slender herb 1–2 dm. tall, more or less branched from base or strict and simple, stems and foliage moderately hirsutulous throughout; leaves linear to lance-linear, 1–5 mm. broad, 1–2.5 cm. long, short-petiolate or upper subsessile, acute to obtuse at apex, entire; racemes dense in bud and early flower but elongating in fruit, often to 10 cm. long or more; calyx lobes linear, 1.5–2 mm. long at anthesis, hirsutulous; corolla white or pale blue, barely surpassing calyx lobes, limb about 1.5 mm. broad; nutlets homomorphous, about 3 mm. high, margin strongly inflated and incurved-cupulate, appendages terete or essentially so, seeming to stand upon curved inflated margin, 0.5–1 mm. long, back of nutlet smooth or finely papillate within cupulate pit.

Arid mesas, hillsides, and along washes, Lower and Upper Sonoran Zones, southern Arizona and adjacent Sonora. April–June.

**2. Lappula redowskii** (Hornem.) Greene var. **desertorum** (Greene) I. M. Johnston, Contr. Arnold Arb. **3:** 93, 1932

*Lappula desertorum* Greene, Pittonia **4:** 95, 1899.
*Lappula collina* Greene, op. cit. 96.
(For additional synonymy see Contr. Arnold Arb. **3:** 93, 1932.)

An erect herb 1–6 dm. tall or more, usually branching above middle, hirsutulous and somewhat cinereous throughout; basal leaves oblanceolate, to 1 cm. broad, 5–10 cm. long, blades gradually narrowing to petiole and often equaled by it; upper leaves gradually reduced, uppermost lanceolate, sessile or nearly so, 1–3 cm. long or less; racemes dense in flower, elongate in fruit; calyx lobes linear, about 1.5 mm. long at anthesis, twice as long in fruit; corolla white or pale blue, about 2 mm. long, limb 1–1.5 mm. broad; nutlets broadly ovoid, 2.5–3 mm. tall, about as wide, usually homomorphous, margins thin and apparently composed of fused bases of flattened, narrowly deltoid prickles, rarely with one nutlet bearing completely distinct appendages; outer face minutely tuberculate or papillate.

Arid, sandy plains and gentle slopes, Sonoran Zones, southeastern Washington to Wyoming and Montana, south to Arizona, New Mexico, and Sonora; also in Patagonia. March–Sept.

## 9. EREMOCARYA Greene, Pittonia 1: 58, 1887

A profusely branching annual herb with very slender, ascending, nearly leafless stems and leaves mostly in a basal rosette. Roots and often lower part of stems containing a purple stain. Racemes dense in flower, spiciform, evenly leafy-bracteate beneath each flower. Calyx small, 5-parted to base. Corolla minute, white. Ovules and nutlets 4, similar in size and shape. Gynobase slenderly columnar, scarcely distinguishable from rather broad style. Style dilated and broader than stigma in fruit, persistent.

Corolla limb 0.5–1.2 mm. broad; appendages in throat obscure ........................ 1. *E. micrantha* var. *micrantha*

Corolla limb 1.5–2.5 mm. broad; appendages in throat more pronounced ........................ 1*a*. *E. micrantha* var. *lepida*

### 1. Eremocarya micrantha (Torr.) Greene, Pittonia 1: 59, 1887, var. micrantha

*Eritrichium micranthum* Torr., Bot. Mex. Bound. 141, 1859.
*Krynitzkia micrantha* A. Gray, Proc. Amer. Acad. **20**: 275, 1885.
*Cryptantha micrantha* I. M. Johnston, Contr. Gray Herb. **68**: 56, 1923.
*Eremocarya abramsiana* Brand, Pflanzenreich $4^{252}$: 77, 1931.

Stems slender, several from base, dichotomously much branched, erect or curvingly ascending, to 1.5 dm. tall, finely pubescent with closely appressed, fine, white hairs, each branch ending in a crowded, terminal, leafy-bracteate, spikelike raceme; basal leaves linear, 0.8–1.5 mm. wide, 3–7 mm. long, canescent-strigose and sometimes short-villous, obtuse or rounded at apex, upper ones scarcely reduced, exceeding subtended flowers; spikes solitary or geminate, dense, 1–4 cm. long, leafy-bracteate throughout; calyces 1.5–2.5 mm. long in fruit, biseriate and somewhat asymmetrical, lobes linear or narrowly linear-lanceolate, erect, hirsute; corolla 0.5–1.2 mm. wide, inconspicuous, appendages in throat weakly developed; nutlets usually 4, alike or slightly heteromorphous, 1–1.3 mm. long, narrowly ovate, smooth or minutely tuberculate, dark gray or brownish, abaxial often slightly largest and most persistent; ventral groove narrow, running full length of nutlet, barely broadened at base; gynobase subulate, surpassing nutlets and nearly equaling calyx lobes.

Arid, desert, and semidesert soils, Sonoran Zones, southeastern Oregon to Utah, south to northern Baja California and east through northern Sonora to western Texas. Feb.–June.

### 1*a*. Eremocarya micrantha var. lepida (A. Gray) J. F. Macbride, Proc. Amer. Acad. 51: 545, 1916

*Eritrichium micranthum* var. *lepidum* A. Gray, Syn. Fl. Amer. $2^1$: 193, 1878.
*Krynitzkia micrantha* var. *lepida* A. Gray, Proc. Amer. Acad. **20**: 275, 1885.
*Eremocarya lepida* Greene, Pittonia **1**: 59, 1887.
*Cryptantha micrantha* var. *lepida* I. M. Johnston, Contr. Gray Herb. **68**: 57, 1923.
*Cryptantha micrantha* subsp. *lepida* Mathew & Raven, Madroño **16**: 171, 1962.

Differing from var. *micrantha* in having a larger corolla, the limb being 1.5–2.5 mm. broad; stems usually coarser; appendages in throat of corolla somewhat more strongly developed.

Sandy soil, Lower Sonoran to Transition Zones, southern Sierra Nevada from Tulare County to Mono County; western San Bernardino County south into northern Baja California; entering the Colorado Desert along its western margin.

## 10. PLAGIOBOTHRYS Fisch. & Mey., Ind. Sem. Hort. Petrop. 2: 46, 1835

Slender annuals with soft-pubescent herbage, young pubescence often rusty or tawny, especially on calyces. Leaves alternate or in a basal rosette, sessile or lower ones narrowed abruptly just above broad base to a petiole, then gradually widening to narrow blade. Flowers in more or less scorpioid spikes or spikelike racemes. Spikes loose and elongated or short and dense. Pedicels short, persistent. Calyx lobes 5, ovate to lanceolate, distinct to base or nearly so, usually densely pubescent. Corolla white, short, barely surpassing calyx lobes, 5-lobed; small crests or processes in throat of orifice. Nutlets 4, or fewer by abortion, ovoid, strongly incurved, keeled dorsally and from near middle to apex ventrally, margins often keeled or angled. Nutlets attached to gynobase by a ventral, raised and rounded scar (caruncle) situated at base or near middle of nutlet; caruncle leaving a depression on gynobase when shed.

Nutlets tessellate with blocklike sections ................................ 1. *P. jonesii*
Nutlets not tessellate, these either wrinkled, ridged, or tuberculate or with a combination of these features on back:
  Calyx circumscissile; leaves containing purple dye, especially in midrib; nutlets attached by level scar ................................................ 2. *P. arizonicus*
  Calyx not circumscissile; leaves green, not containing purple dye; nutlets attached by elevated scar or stipe:
    Stipe about equaling body of nutlet; nutlets usually united in pairs ...... 3. *P. pringlei*
    Stipe much shorter than body of nutlet; nutlets separate ........................... ........................................ 4. *P. californicum* var. *fulvescens*

### 1. Plagiobothrys jonesii A. Gray, Syn. Fl. N. Amer. ed. 2, $2^1$: 430, 1886

*Sonnea jonesii* Greene, Pittonia 1: 23, 1887.

Simple or several-branched annual 1–3 dm. tall with ascending branches; stems and leaves finely pubescent with short appressed hairs but also sparsely bristly-hirsute with spreading hairs 1–2 mm. long, latter from white basal pustules; leaves narrowly lanceolate to oblanceolate, 4–8 (or sometimes 10) mm. wide, 2–9 cm. long (including the petioles on basal leaves), mostly obtuse at apex, the pustules conspicuous; spikes dense in flower, elongating to 5–15 cm. in fruit, leafy-bracteate toward base; calyx lobes lance-linear, 4–6 mm. long at anthesis, 6–11 mm. long in fruit, equal or nearly so, conspicuously hispid, especially on midrib, distinct to base, persistent; corolla 4–5 mm. long, 1–1.5 mm. broad, white, soon deciduous; nutlets usually 2–3, broadly ovate, 2.5–3 mm. long, strongly keeled dorsally, weakly keeled laterally, tips incurved, surface tessellate with closely crowded, several-sided, tuberculate or papillate blocks or flattened tubercles; caruncle elongated, extending along ventral keel.

Gravelly outwash fans and desert arroyos, mainly Lower Sonoran Zone, Inyo County east into southern Nevada and south into the eastern Mojave Desert and adjacent Arizona. Feb.–April.

**2. Plagiobothrys arizonicus** (A. Gray) Greene ex A. Gray, Proc. Amer. Acad. **20:** 284, 1885

*Eritrichium canescens* var. *arizonicum* A. Gray, Proc. Amer. Acad. **17:** 227, 1882.

Annual herb with 1 to several stems from base, these usually simple, erect, ascending or sometimes slightly decumbent, 1–3 dm. long; herbage hispid with slender spreading hairs 1–2.5 mm. long and sparsely puberulent with tangled, mostly appressed, delicate hairs among bases of spreading hairs, the stems, roots, and leaves, particularly midribs, containing a purplish dye; leaves lanceolate to linear-oblanceolate, 1.5–5 (or 8) mm. broad, 1–6 cm. long, basal ones gradually narrowed to slender base, acute to obtuse at apex, strigose and with some spreading hairs, not noticeably pustulate; spikes compact at anthesis, elongated and lax in fruit, then to 15 cm. long, naked or few-bracteate toward base; calyx lobes ovate, distinct about one-half way to base, densely tawny-hirsute; calyx 3–4 mm. long in fruit, at length circumscissile, lobes connivent, equal; corolla about 2 mm. long, 1.5–2 mm. broad; nutlets usually 2, sometimes only 1, or 3 or 4, ovoid, short-acute, 1.5–2.5 mm. long, more or less transversely rugulose, dorsal and lateral keels often somewhat reticulate; caruncle sunken below crest of ventral keel in a transverse groove.

Arid sandy hillsides and plains, Lower and Upper Sonoran Zones, central San Joaquin Valley and Inyo County to Arizona, northern Sonora, and eastern San Diego County, California. Feb.–May.

**3. Plagiobothrys pringlei** Greene, Pittonia **1:** 21, 1887

*Echidocarya arizonica* A. Gray, Proc. Amer. Acad. **11:** 89, 1876. Not *Eritrichium canescens* var. *arizonicum* A. Gray, 1882.

A prostrate or decumbent annual with several rather coarse stems 1–2.5 dm. long from slender taproot; herbage hispid throughout; leaves not developing a distinct rosette, lowermost commonly opposite, upper ones alternate, blades linear to linear-oblong, 2–3(5) cm. long, acute or obtuse at apex; flowers borne throughout length of stems, even in axils of basal leaves; calyx lobes narrowly linear-lanceolate, 1.5–2 mm. long, somewhat spreading in fruit; corolla white, minute; nutlets 4, ovoid, finely rugulose and somewhat finely muriculate, keeled both dorsally and ventrally, ventral keel produced into a slender stipe nearly as long as body of nutlet, nutlets commonly united about to middle in pairs.

Arid plains and hillsides, Lower Sonoran and lower part of the Upper Sonoran Zones, Maricopa County to Pima County, Arizona, and south into northern Sonora. Feb.–March.

**4. Plagiobothrys californicus** (A. Gray) Greene var. **fulvescens** I. M. Johnston, Contr. Gray Herb. **68:** 74, 1923

Prostrate or decumbent annual with several slender, often very leafy stems 1–4 dm. long from slender taproot, bearing axillary flowers almost to base; herbage hispid-fulvescent throughout; leaf blades linear to ovate-oblong, 1–4 cm. long, hispid on both surfaces; racemes spikelike, extending beyond leaves and in age remotely flowered, only a few flowers borne near base of stems, lower part of raceme bracteate, terminal part naked; calyx lobes lance-linear, 1.5–2 mm. long, moderately hispid with slightly rusty or tawny hairs; corollas white, barely equaling calyx lobes, limb 1.5–2.5 mm. broad; nutlets mostly 4, separate, ovate, 1.5–1.8 mm. long, delicately rugose transversely and keeled toward tip dorsally, often muriculate or tuberculate between and on thin ridges, ventral keel thin, drawn down to a broadish, nearly rounded stipe about one-fifth as long as body of nutlet.

Along creek banks, meadows, and hillsides, mostly Upper Sonoran Zone to Transition Zone, but descending into the Lower Sonoran Zone between Rosario and San Augustín, Baja California; northern and central Arizona and along western margins of the Mojave and Colorado Deserts to the southern flanks of the Sierra San Pedro Mártir, Baja California. Feb.–April.

## 11. **CRYPTANTHA*** Lehm. ex Fisch. & Mey., Ind. Sem. Hort. Petrop. **2**: 35, 1836

Annual herbs with hispid-bristly herbage. Leaves simple, entire, linear to lanceolate or spatulate, lower pairs opposite, upper ones alternate. Inflorescence fasciculate or in scorpioid spikes or racemes, bracteate. Calyx 5-lobed to base or nearly so, often slightly accrescent in fruit, usually deciduous. Corolla white, small or minute, tube cylindrical, equaling or surpassing calyx lobes; throat short, more or less closed below with 5 small intruded appendages; corolla lobes 5, spreading-ascending or rotate. Stamens 5, inserted below middle of corolla tube, included; anthers oblong, about equaling filaments. Style simple, very short and surpassed by nutlets, or elongated and longer than nutlets, attached to apex of gynobase well below tips of nutlets. Ovules 4 or less commonly 2. Fruit of 1–4 uniovulate nutlets, these ovate to lanceolate, smooth or tuberculate or slightly prickly, often more or less heteromorphous within same flower, attached to subulate or pyramidal gynobase (an elongation of receptacle) through an elongated, median, ventral groove that often forks near base; nutlets erect or tips slightly connivent.

Surface of nutlets papillate, tuberculate or muricate, or sometimes when heteromorphous odd nutlet smooth:
  Nutlets bent; gynobase pyramidal .......................................... 1. *C. pusilla*
  Nutlets straight; gynobase subulate:
    Margin of nutlets prominently winged or with a knifelike edge:
      Plants usually short-lived perennials (sometimes annual) with many diffuse branches; inflorescence racemose .......................................... 2. *C. racemosa*
      Plants annuals; branches ascending; inflorescence loosely or densely scorpioid:
        Nutlets 4:
          Margins of nutlets merely knifelike or very narrowly winged:
            Fruiting calyx 3–6 mm. long:
              Nutlets markedly heteromorphous; fruiting calyx 3 mm. long, pedicels 1–2 mm. long .......................................... 3. *C. fastigiata*
              Nutlets homomorphous; fruiting calyx 4–6 mm. long, sessile ... 4. *C. costata*
            Fruiting calyx 2 mm. or less long .......................... 5. *C. angelica*
          Margins of nutlets conspicuously winged, wings one-third to more than one-half width of nutlet:
            Calyx broadly ovate in fruit; pubescence soft, with or without coarse spreading bristles intermixed; wings of nutlet unlike body of nutlet in color, wings of at least some nutlets crenate or erose:
              Nutlets heteromorphous ................. 6. *C. pterocarya* var. *pterocarya*
              Nutlets homomorphous:
                Corolla 0.5–1.5 mm. broad ............ 6a. *C. pterocarya* var. *cycloptera*
                Corolla 4–7 mm. broad ................................ 7. *C. oxygona*

* Key to the species constructed by Roxana S. Ferris.

Calyx narrowly ovate in fruit, pubescence all coarse; wings of nutlet like body of fruit in color, always entire ........................ 8. *C. holoptera*
Nutlets 1 or sometimes 2 .................................... 9. *C. utahensis*
Margin of nutlets rounded or merely angled:
Nutlets heteromorphous (if more than one):
Odd nutlet smooth and shining, others roughened (when more than one):
Ovules 2; nutlets 1 or 2:
Calyx not conspicuously pilose ............... 10. *C. maritima* var. *maritima*
Calyx conspicuously pilose .................... 10*a*. *C. maritima* var. *pilosa*
Ovules 4; nutlets 1–4 ..................... 10*b*. *C. maritima* var. *cedrosensis*
Odd nutlet differing only in size, all similarly roughened:
Fruiting calyx deflexed, adaxial lobe most hirsute ......... 11. *C. echinosepala*
Fruiting calyx erect or with spreading tips, lobes equally pubescent or most hirsute on abaxial side:
Mature calyx not gibbous, not appressed to rachis; stems erect and branching; racemes densely flowered ......................... 12. *C. angustifolia*
Mature calyx gibbous and strongly appressed to flattened rachis; stems conspicuously sinuous and brittle ..................... 13. *C. dumetorum*
Nutlets all similar (if more than one):
Nutlets 1, rarely 2 ............................................ 14. *C. decipiens*
Nutlets normally 4:
Nutlets 0.5–0.7 mm. long; fruiting calyx 1–2 mm. long:
Corolla conspicuous, 2–3 mm. broad .................. 15. *C. grayi* var. *grayi*
Corolla inconspicuous, 0.5–1 mm. broad:
Nutlets very sparsely tuberculate .............. 15*a*. *C. grayi* var. *nesiotica*
Nutlets densely tuberculate ............. 15*b*. *C. grayi* var. *cryptochaeta*
Nutlets 1.5–2.5 mm. long; fruiting calyx 4–10 mm. long (2–3.5 in *C. muricata*):
Nutlets triangular-ovate; fruiting calyx broad at base, short lobes mostly erect and not connivent ................... 16. *C. muricata* var. *denticulata*
Nutlets lanceolate to lanceolate-ovate; fruiting calyx ovate-oblong to lanceolate, long linear lobes connivent above nutlets and spreading or recurved above:
Spikes of inflorescence solitary, bracteate at base ........... 17. *C. pondii*
Spikes of inflorescence geminate or ternate (when solitary, bractless):
Lower part of calyx conspicuously tufted with long, white, villous hairs; stems and leaves with only spreading pubescence:
Corolla inconspicuous ................ 18. *C. barbigera* var. *barbigera*
Corolla conspicuous .............. 18*a*. *C. barbigera* var. *fergusoniae*
Lower part of calyx not conspicuously tufted with long, white, villous hairs; stems and leaves with appressed and spreading pubescence:
Stems slender, flexuous .............. 19. *C. nevadensis* var. *nevadensis*
Stems erect ......................... 19*a*. *C. nevadensis* var. *rigida*
Surface of nutlets smooth; nutlet 1:
Adaxial nutlet developing ........................... 10. *C. maritima* var. *maritima*
Abaxial nutlet developing:
Fruiting calyx 6–10 mm. long, bristles spreading, tawny .............. 20. *C. ganderi*
Fruiting calyx 2–5 mm. long, bristles deflexed and uncinate, white-encrusted ........ ........................................................ 21. *C. flaccida*

**1. Cryptantha pusilla** (Torr. & Gray) Greene, Pittonia **1:** 115, 1887

*Eritrichium pusillum* Torr. & Gray, Pacif. R. Rep. **2:** 171, 1857.

Plants annual with many prostrate-ascending, slender branches 1–2 dm. long, pubescence rather densely strigose and more sparsely spreading-hirsute; leaves crowded below, less abundant above, narrowly oblanceolate, 1.5–2.5 cm. long, little reduced above, densely hirsute, hairs from pustulate bases with some finer hairs intermixed; corolla small, 1.5–2 mm. long; calyx sessile or nearly so, early deciduous; calyx lobes equal, hirsute, lance-ovate to oblong-lanceolate, midrib but little thickened, 2–2.5 mm. long; nutlets 4, homomorphous, strongly bent above base, about 1 mm. long, glossy, light-colored with pale low tuberculations, margins knifelike, groove open or closed, areola shallow, triangular; style much surpassing nutlets; gynobase narrowly pyramidal, equaling nutlets.

Sandy and rocky soils, mainly Upper Sonoran Zone; southern Arizona and Sonora to Texas and Durango. March–June.

## 2. **Cryptantha racemosa** (S. Wats. ex A. Gray) Greene, Pittonia **1:** 115, 1887

*Eritrichium racemosum* S. Wats. ex A. Gray, Proc. Amer. Acad. **17:** 226, 1882.
*Krynitzkia ramosissima* A. Gray, op. cit. **20:** 277, 1885.
*Krynitzkia racemosa* Greene, Bull. Calif. Acad. **1:** 208, 1885.
*Cryptantha suffruticosa* Piper, Proc. Biol. Soc. Wash. **32:** 42, 1919.
*Cryptantha racemosa* var. *lignosa* I. M. Johnston, Univ. Calif. Pub. Bot. **7:** 445, 1922.
*Johnstonella racemosa* Brand ex I. M. Johnston, Contr. Gray Herb. **74:** 32, 1925.
*Johnstonella racemosa* var. *lignosa* Brand ex I. M. Johnston, loc. cit.

Annual or short-lived perennial with strongly suffruticose base, 1–10 dm. tall, stems solitary with numerous ascending branches or profusely branched and forming a rounded, rather dense growth; younger parts green, faintly strigose and markedly hirsute, older parts brownish and bark exfoliating; leaves oblanceolate, basal ones 2–12 mm. broad, 2.5–6 cm. long, upper ones 1.5–2.5 mm. broad, 1.5–4 cm. long, acute at apex, hirsute, hairs from pustulate bases; pedicels 1–4 mm. long, often somewhat recurved; calyx lobes lance-linear, hirsute, 1–1.5 mm. long in flower, about twice as long in fruit; corolla minute, barely surpassing calyx, limb 1–1.5 mm. broad; nutlets 4, heteromorphous, ovate, acute with tips curved slightly outward, closed groove broadening to a triangular basal areola, odd nutlet 1–2 mm. long, finely tuberculate or muricate or both; smaller nutlets 0.8–1.5 mm. long, narrowly winged dark brown with paler tuberculations; style nearly equaling calyx lobes; gynobase nearly equaling smaller nutlets.

Arid, sandy or rocky soil, Lower Sonoran Zone, southern Nevada and adjacent California into western Arizona, the Mojave and Colorado Deserts, and south through Baja California to Carmen Island. Feb.–Sept.

## 3. **Cryptantha fastigiata** I. M. Johnston, Jour. Arnold Arb. **20:** 388, 1939

Herbaceous or somewhat suffruticose plant 1–10 dm. tall with erect or ascending, strict to profusely branched stems, sparsely clothed with stiff, spreading, and variously appressed, rigid hairs, spreading hairs to 2 mm. long and conspicuously pustulate at base; leaves numerous, narrowly to linearly oblanceolate, 1–3 mm. wide (or rarely wider), mostly 2–5 cm. long, gradually reduced toward tips of stems, basal ones gradually narrowed to a slender petiole and to 10 cm. long, sparsely hirsute with stiff pustulate hairs 1–1.5 mm. long; spikes scorpioid, solitary or geminate, rather lax, minutely bracteate; calyx lobes lance-linear, about 2 mm. long in flower, densely hispid, about 3 mm. long in fruit; corolla white, tube about 1 mm. long, limb 3–4 mm. broad; nutlets 4, heteromorphous, ovate-triangular, strongly convex on back, dark brown with pale tubercles and pale, sharply angled margins, 3 shorter ones 1–1.5 mm. long, abaxial one 1.5–1.9 mm. long; stoutish subulate style 0.5–1 mm. longer than nutlets.

Sandy arroyos and plains and rocky hillsides, Lower Sonoran Zone, on the islands of the Gulf of California from Isla Angel de la Guarda to Carmen Island and the adjacent peninsula. March–April.

**4. Cryptantha costata** Brandegee, Bot. Gaz. **27:** 453, 1899

*Cryptantha seorsa* J. F. Macbride, Contr. Gray Herb. **48:** 46, 1916.

A rather coarse plant with stiffish branches 1–2 dm. long, younger stems densely villous-strigose and with some intermingled spreading hairs, canescent, older parts of stems becoming brownish and less pubescent; leaves linear to lanceolate, 1.5–4 mm. wide, 1–3 cm. long, acute at apex, villous-strigose, sparsely hispid and pustulate; spikes 1–5 cm. long, rather stiff, solitary or geminate, sparsely leafy-bracted toward base; calyx lobes linear-lanceolate, about 2–3 mm. long in flower, 4–6 mm. long in fruit, with tips somewhat connivent, hirsute on midrib, strigose on margins, subequal; corolla white, 2–2.5 mm. long, inconspicuous; nutlets 4, essentially alike, abaxial one slightly larger than others, 1.6–1.8 mm. long, triangular-oblong-ovate, rugulose or faintly verrucose and strongly curved on back, face flat or slightly convex, smooth, margins narrowly but sharply winged; groove shallow, dilated to a narrow triangular pit below; style surpassing nutlets 0.1–0.5 mm.

Sandy washes, hillsides, and desert flats, Lower Sonoran Zone, eastern Mojave Desert and Colorado Desert from Borrego Valley east to the Colorado River and Yuma County, Arizona. Probably also in Baja California and Sonora. Feb.–March.

**5. Cryptantha angelica** I. M. Johnston, Proc. Calif. Acad. IV, **12:** 1143, 1924

Erect and sparsely branched to much-branched, rounded plant 1–2.5 dm. tall, main branches from base of central stem spreading-ascending, usually with additional tertiary branches near tips, brown and subglabrate to glabrous at base, green and canescent-hispidulous above; leaves linear, 0.3–2 mm. wide, 8–30 mm. long, densely pustulate and hispidulous below, sparsely hispid and sparingly pustulate above, basal leaves withering long before upper ones; inflorescence of numerous scorpioid spikes 1–3 cm. long; calyx lobes linear, about 1 mm. long, spreadingly hispidulous, to 2 mm. long in fruit; corolla inconspicuous, about 1 mm. long, lobes about 0.25 mm. long, rounded; tube equaling or slightly shorter than calyx; style about 0.5 mm. longer than nutlets; nutlets 4, abaxial one larger than others, all nutlets narrowly ovoid, sharp-margined, dark, but with paler tubercles, larger nutlet about 0.7 mm. long, others 0.5–0.6 mm. long; lower one-third of groove of a nutlet usually dilated into a shallow triangular pit.

Sandy washes and flats, Lower Sonoran Zone, Isla Angel de la Guarda and shores of Bahía de los Angeles and near San Ignacio Lagoon, Baja California. Nov.–March.

**6. Cryptantha pterocarya** (Torr.) Greene, Pittonia **1:** 120, 1887, var. **pterocarya**

*Eritrichium pterocaryum* Torr., Bot. Mex. Bound. 142, 1859.
*Eritrichium pterocaryum* var. *pectinatum* A. Gray, Proc. Amer. Acad. **10:** 61, 1874.
*Krynitzkia pterocarya* A. Gray, op. cit. **20:** 276, 1885.
*Krynitzkia pterocarya* var. *pectinatum* A. Gray, loc. cit.
*Cryptantha pterocarya* var. *purpusii* Jepson, Man. Fl. Pl. Calif. 849, 1925.

Erect, copiously ascendingly branched herb 1–5 dm. tall, herbage finely puberulent with appressed and a few spreading hairs; leaves linear to oblanceolate-linear, 1–5 mm. wide, 1–4 cm. long, obtuse at apex, strigose to hispid, coarsely pustulate beneath, finely so above; inflorescence geminate or ternate, inconspicuously bracteate below or naked, branches 2–12 cm. long; calyx 2–3 mm. long at anthesis, 3–5 mm. long in fruit, densely puberulent with appressed and some spreading hairs, symmetrical, tardily deciduous, calyx lobes ovate-lanceolate, slightly longer than nutlets, connivent; corolla small and inconspicuous, 0.5–1.5 mm. broad, white; nutlets 4, the adaxial one wingless, other three distinctly winged; body of nutlet lanceolate, 2–3 mm. long, muricate or verrucose on back, winged margin entire or crenate; groove open or closed, dilated into an open areola below; gynobase about two-thirds as high as nutlets, slender; style surpassing body of nutlets but shorter than wings.

Arid, sandy or gravelly soil, Lower and Upper Sonoran Zones, eastern Washington and adjacent Idaho to southern California and Arizona; occasionally in northwestern Sonora. Feb.–June.

**6*a*. Cryptantha pterocarya** var. **cycloptera** (Greene) J. F. Macbride, Contr. Gray Herb. **48:** 44, 1916

*Krynitzkia cycloptera* Greene, Bull. Calif. Acad. **1:** 207, 1885.
*Cryptantha cycloptera* Greene, Pittonia **1:** 120, 1887.

Differing from var. *pterocarya* only in having all 4 nutlets broadly winged.

Sandy plains and washes, mainly Lower Sonoran Zone, Inyo County to eastern San Diego County, California; east into southern Nevada, southern Utah, Colorado, and western Texas and south to central Baja California (Calmallí).

**7. Cryptantha oxygona** (A. Gray) Greene, Pittonia **1:** 120, 1887

*Eritrichium oxygonum* A. Gray, Proc. Amer. Acad. **19:** 89, 1883.
*Krynitzkia oxygona* A. Gray, op. cit. **20:** 277, 1885.

Moderately or sparsely branched herb 1–4 dm. tall with strigose and villous-hispid stems and foliage; leaves linear to lance-linear, 1–3 mm. broad, 1–6 cm. long, obtuse at apex, inconspicuously but densely pustulate, gradually reduced toward geminate or ternate, scorpioid spikes; spikes 1–6 cm. long, usually dense and 1–3 cm. long, naked; calyx lobes lanceolate, 2–2.5 mm. long at anthesis, to 4 mm. long in fruit, somewhat connivent at tips, silky-strigose on margins, sparsely hispid on thickened midrib; corolla white, conspicuous, limb 4–7 mm. broad; nutlets 4, homomorphous, body oblong-ovate, 2–3 mm. long, nearly equaling calyx lobes, muricate or tuberculate on low, slightly rounded back, margins of nutlets knifelike or narrowly winged; groove closed or less commonly open, dilated to a triangular areola at base; gynobase two-thirds as high as nutlets, subulate; style surpassing nutlets.

Arid grassy hillsides and plains, Lower Sonoran and lower parts of Upper Sonoran Zones, San Joaquin Valley to the northeastern edge of the Colorado Desert and to southwestern Nevada. April–May.

**8. Cryptantha holoptera** (A. Gray) J. F. Macbride, Contr. Gray Herb. **48:** 44, 1916

*Eritrichium holopterum* A. Gray, Proc. Amer. Acad. **12:** 81, 1876.
*Krynitzkia holoptera* A. Gray, op. cit. **20:** 276, 1885.
*Oreocarya holoptera* Greene, Pittonia **1:** 58, 1887.

Erect annual with coarse stems and branches 1–6 dm. tall, rarely slightly suffruticose toward base; herbage hirsute with stiff hairs from pustulate bases intermingled with finer appressed hairs; leaves lance-linear to oblanceolate, lowermost 3–12 mm. wide, 3–6 cm. long, upper ones lance-linear, 1–3 mm. wide, 1–3.5 cm. long, coarsely pustulate and hirsute beneath, more sparsely so above, acute to obtuse at apex; spikes 2–10 cm. long, sparsely bracteate, at first dense, later lax; pedicels to 1.5 mm. long, ascending to recurved; calyx lobes to 3.5 mm. long in fruit, spreadingly hirsute, subequal; corolla minute, shorter than calyx, limb about 1 mm. broad; nutlets 4, homomorphous, oblong-ovate to triangular-ovate, 1.5–2.5 mm. long, dark brown with pale tuberculations on convex back, margins narrowly to broadly winged, inner face smooth; groove open or closed above, dilated to a narrowly triangular areola below, pale gray to white; gynobase about equaling nutlets; style surpassing nutlets but not equaling calyx lobes.

Arid desert slopes, plains, and washes, Lower Sonoran Zone, eastern Mojave Desert, the Colorado Desert, and adjacent western Arizona. Collected at Signal Mountain, Imperial County, California. Feb.–April.

**9. Cryptantha utahensis** (A. Gray) Greene, Pittonia **1:** 120, 1887

*Eritrichium holopterum* var. *submolle* A. Gray, Proc. Amer. Acad. **13:** 374, 1878.
*Krynitzkia utahensis* A. Gray, Syn. Fl. N. Amer. ed. 2, **2**[1]**:** 427, 1886.
*Cryptantha submollis* Coville, Contr. U.S. Nat. Herb. **4:** 166, 1893.

An erect or ascendingly branched herb 1–4 dm. tall; stems finely but closely strigose or somewhat short-hirsute; leaves linear to linear-oblanceolate, 1–4 mm. wide, 1–7 cm. long, mostly rounded at apex, gradually narrowed toward base, short-hirsute and pustulate, particularly beneath, much reduced toward inflorescence; spikes solitary or more commonly geminate, 8–50 mm. long, dense and naked in flower, elongating somewhat in fruit; calyx densely strigose with grayish hairs, lobes lanceolate, 1–1.5 mm. long at anthesis, about 2–3 mm. long in fruit, unequal, connivent; corolla white, 2–3 mm. broad, tube about equaling calyx lobes; ovules 4, 2 or 3 usually abortive; nutlets 1 or less commonly 2, abaxial one usually functional, 1.8–2.5 mm. long, 1–1.5 mm. wide, pale gray, broadly lanceolate to narrowly ovate in outline, granulate or muricate-papillate or rarely spinulose on slightly convex or nearly flat back, margins sharply angled or almost winged; groove narrow but open, dilated to a small areola below; gynobase about two-thirds as high as nutlet; style slightly shorter than nutlet.

Arid slopes and washes, Lower Sonoran Zone, Inyo County, California, to southern Utah, northwestern Arizona, and south to the northwestern part of the Colorado Desert. Feb.–April.

**10. Cryptantha maritima** (Greene) Greene, Pittonia **1:** 117, 1887, var. **maritima**

*Krynitzkia ramosissima* Greene, Bull. Calif. Acad. **1:** 203, (Aug.) 1885. Not *K. ramosissima* A. Gray, (Jan.) 1885.
*Krynitzkia maritima* Greene, op. cit. 204.

Loosely branched, ascending herb 1–3 dm. tall with commonly reddish, strigose or hirsute stems; leaves linear to linear-lanceolate, 1–4 mm. wide, 1–3.5 cm. long, acute at apex, narrowed at base, usually hirsute and coarsely pustulate; spikes solitary or geminate, 1–6 cm. long, often crowded or glomerate, unevenly bracteate throughout; fruiting calyx 1.8–3 mm. long, stiffly ascending, subsessile and slightly asymmetrical, lobes lance-linear, stiffish, margins of 3 abaxial lobes loosely or appressed pilose, midrib thickened and sparsely hirsute, adaxial lobes slightly shorter, less hirsute; corolla 1.5–2 mm. long, limb 0.5–1 mm. broad; nutlets usually 1 or 2, oblong-lanceolate, 1–2 mm. long, minutely tuberculate, larger one (if 2 or more are present) firmly attached next to adaxial calyx lobe, groove narrow and closed throughout or dilated at base into a small areola; gynobase subulate, one-half to two-thirds as long as nutlet; style about equaling shorter nutlet or slightly surpassed by larger one.

Arid plains, washes, and hillsides, Lower Sonoran Zone and occasionally in the

Upper Sonoran Zone, southern Nevada through the Mojave and Colorado Deserts to Magdalena Island, Baja California, and to the Coast and coastal islands in southern California. Feb.–April.

**10*a*. Cryptantha maritima** var. **pilosa** I. M. Johnston, Univ. Calif. Pub. Bot. **7:** 445, 1922

Differing from var. *maritima* in having the calyx conspicuously pilose with long white hairs.

Habitat and range essentially that of var. *maritima,* but the plants of this variety less numerous than the typical ones.

**10*b*. Cryptantha maritima** var. **cedrosensis** (Greene) I. M. Johnston, Contr. Gray Herb. **74:** 48, 1925

*Krynitzkia cedrosensis* Greene, Bull. Calif. Acad. **1:** 204, 1885.
*Cryptantha cedrosensis* Greene, Pittonia **1:** 117, 1887.

Ovules 4, nutlets 1–4; otherwise as in var. *maritima.*

This variety has been reported only from Cedros Island, but it might reasonably be expected to occur on the adjacent mainland of Baja California.

**11. Cryptantha echinosepala** J. F. Macbride, Contr. Gray Herb. **56:** 57, 1918

Slender, rather loosely branched herb 1–3 dm. tall with reddish (usually), strigose, and somewhat hirsute stems; leaves linear to linear-lanceolate, 2–5 mm. wide, 1.5–4.5 cm. long, pustulate and short-hirsute beneath, minutely strigose or subglabrate above; spikes geminate or sometimes solitary, naked or rarely with a single basal or subbasal bract, 2–5 cm. long; fruiting calyx ovate, 2–3 mm. long, spreading or reflexed, asymmetrical, lobes lance-linear, very unequal, strigose and hirsute along midrib, adaxial lobe longest and most densely hirsute; corolla inconspicuous, limb about 1 mm. wide, tube shorter than or barely equaling calyx; nutlets usually 4, heteromorphous, pale gray, with a narrow groove scarcely dilated below; odd nutlet next to adaxial calyx lobe, ovate, 1.2–1.5 mm. long, minutely muriculate-tuberculate, margin angulate; consimilar nutlets smaller, about 1 mm. long, lance-ovate, minutely tuberculate; gynobase shorter than 3 similar nutlets; style about equaling odd nutlet.

Sandy draws and rocky hillsides, Lower Sonoran Zone, near Bahía de la Concepción and south in Baja California. Feb.–April.

**12. Cryptantha angustifolia** (Torr.) Greene, Pittonia **1:** 112, 1887

*Eritrichium angustifolium* Torr., Pacif. R. Rep. **5:** 363, 1858.
*Krynitzkia angustifolia* A. Gray, Proc. Amer. Acad. **20:** 272, 1885.

A much-branched herb 8–25 cm. tall, usually with many slender, ascending or spreading-ascending, brownish stems sparsely hispidulous with slender white hairs, epidermis eventually exfoliating in irregular strips and shreds; leaves linear, usually 1 mm. wide or less, 5–30 mm. long, hispidulous with whitish hairs from pustulate bases; inflorescence of numerous short scorpioid spikes, these elongating in fruit; calyx lobes lance-linear, 1–1.5 mm. long at anthesis, hispid with stiff spreading hairs; corolla white, about 1.5 mm. long, limb 1.5–2 mm. broad; nutlets 4, heteromorphous, all ovoid, acute, brownish or pale gray, minutely papillate, abaxial nutlet 1–1.4 mm. long, others 0.6–0.8 mm. long, lateral angles rounded or rather sharp, ventral groove narrow above, broadened to a narrowly triangular, shallow pit in lower one-third to one-half.

In gravelly or rocky soil on hillsides, along washes, and on disturbed soil, Lower and Upper Sonoran Zones, Inyo County to Utah and western Texas, south through the eastern Mojave and Colorado Deserts into central Baja California and northern Sonora. Feb.–June.

**13. Cryptantha dumetorum** (Greene ex A. Gray) Greene, Pittonia **1:** 112, 1887

*Krynitzkia dumetorum* Greene ex A. Gray, Proc. Amer. Acad. **20:** 272, 1885.
*Cryptantha intermedia* var. *dumetorum* Jepson, Man. Fl. Pl. Calif. 849, 1925.

A laxly branched, straggly, brittle herb with stems often scrambling through bushes or sprawling, 1–6 dm. long, densely to sparsely strigose with stiffish stout hairs; leaves lanceolate, 2–4 (or rarely to 8) mm. broad, 1.5–4 cm. long, closely pustulate below, more sparsely so above, hairs more or less appressed or marginal ones strongly spreading; spikes geminate or solitary, lax and usually remotely flowered, to 10 cm. long, naked or with foliaceous bracts near base, rachis flattened; abaxial calyx lobes (3) lanceolate, strigose, midribs deflexed-hirsute, 2 adaxial lobes partly united, hirsute only on two outer margins; corolla limb about 1 mm. wide, inconspicuous; nutlets 4, granulate and muriculate, adaxial one 2–3 mm. long, distorting calyx, groove broad and open, other 3 alike, 1.5–2 mm. long, lanceolate, groove closed or very narrow; gynobase narrow, shorter than smaller nutlets; style barely equaling or slightly shorter than nutlets.

Arid slopes, often under shrubs, Lower Sonoran Zone, Mojave Desert east into southern Nevada, barely entering Colorado Desert at Whitewater. March–May.

**14. Cryptantha decipiens** (M. E. Jones) Heller, Muhlenbergia **8:** 48, 1912

*Krynitzkia decipiens* M. E. Jones, Contr. West. Bot. **13:** 6, 1910.

An openly branched herb 1–4 dm. tall with ascendingly spreading, slender branches, these and leaves strigose and short-hispid; leaves linear, 1–3 mm. broad, 1–3 cm. long, obtuse at apex, minutely pustulate and sparsely hispid among strigose hairs; the spikes geminate, ternate, or rarely solitary, 4–10 cm. long, remotely or closely flowered, slender and devoid of bracts; calyx lobes lance-linear, 2.5–7 mm. long, unequal, connivent above nutlets, with tips usually spreading or recurved, thickened midrib usually hirsute, margins strigose, abaxial lobe longest and most hirsute; corolla 0.8–1 mm. wide; ovules 4; nutlets 1 or rarely 2, those next to abaxial calyx lobe developing, ovate to ovate-lanceolate, 1.5–2.4 mm. long, brownish or rarely grayish, granulate, muriculate or low-papillate, rounded and convex on back, angles rounded, groove dilated below into a triangular areola; gynobase about one-half as long as nutlet; style surpassed by nutlet.

Sandy desert slopes, arroyos, and plains, Lower Sonoran Zone, vicinity of Palm Springs, Riverside County, to western Arizona and through the eastern Mojave Desert to southern Nevada. March–May.

**15. Cryptantha grayi** (Vasey & Rose) J. F. Macbride, Contr. Gray Herb. **48:** 43, 1916, var. **grayi**

*Krynitzkia grayi* Vasey & Rose, Proc. U.S. Nat. Mus. **11:** 536, 1889.

A slender annual 5–28 cm. tall with several strict or spreadingly branched stems, villous-hispid or appressed-pubescent; leaves linear, 1–2 mm. wide, 1–4 cm. long, densely pustulate-hispid beneath, sparsely so above; spikes mostly geminate, densely flowered, 1–4 cm. long, ebracteate, lax in fruit; calyx lobes lance-linear, rigid, tips slightly connivent, short-hirsute on midrib, villous-hispid on margins, 1–2 mm. long in fruit, adaxial lobe usually pubescent but not hirsute; corolla white, tube 0.8–1.5 mm. long, limb 2–3 mm. in diameter; nutlets 4, homomorphous, ovate, apex acute, 0.5–0.7 mm. long, triangular-ovate, dark brown with lighter, minute, low tuberculations, margins rounded or slightly angled; groove closed above, broadened to a narrowly triangular, shallow areola; gynobase subulate, shorter than or about equaling nutlets; style markedly surpassing nutlets.

Arid, sandy or rocky places, Lower Sonoran Zone, southern Vizcaíno Desert and the Magdalena Plain, Baja California. Feb.–April.

**15*a*. Cryptantha grayi** var. **nesiotica** I. M. Johnston, Proc. Calif. Acad. IV, **12:** 1146, 1924

Corolla limb 0.5–1 mm. broad; nutlets very sparsely tuberculate or smooth, or slightly wrinkled; otherwise as in var. *grayi.*

On sand dunes, Lower Sonoran Zone, islands off the eastern shore of Baja California from Coronados Island to Isla Espíritu Santo.

**15*b*. Cryptantha grayi** var. **cryptochaeta** (J. F. Macbride) I. M. Johnston, Proc. Calif. Acad. IV, **12:** 1145, 1924

*Cryptantha micromeres* var. *cryptochaeta* J. F. Macbride, Contr. Gray Herb. **48:** 46, 1916.
*Cryptantha filiformifolia* J. F. Macbride, op. cit. 45.

Corolla limb 0.5–1 mm. broad; nutlets densely but minutely tuberculate; otherwise as in var. *grayi.*

Sandy washes and arid plains, Lower Sonoran Zone, Cape Region, Baja California, and in Sonora from the vicinity of Torres south to Alamos.

**16. Cryptantha muricata** (Hook. & Arn.) Nels. & Macbr. var. **denticulata** (Greene) I. M. Johnston, Contr. Gray Herb. **74:** 71, 1925

*Krynitzkia denticulata* Greene, Bull. Calif. Acad. **1:** 205, 1885.
*Cryptantha denticulata* Greene, Pittonia **1:** 114, 1887.
*Cryptantha densiflora* Nels. & Kenn., Proc. Biol. Soc. Wash. **19:** 156, 1906.

Sparsely and loosely long-branched, erect annual, villous-hirsute throughout; leaves narrowly oblanceolate; spikes few, naked; corolla minute, 1–2 mm. broad; fruiting calyx sessile, 2–3.5 mm. long, ovate in outline, lobes connivent, midribs prominent, hirsute mixed with few to many spreading bristles; nutlets 4, homomorphous, 1.5–2 mm. long, triangular-ovate, somewhat truncate at base, margins acute, thickened, obtusely angled ventrally, verrucose to tuberculate, groove closed or narrowed above, spreading below into a small areola; style surpassing nutlets; gynobase elongate, over one-half length of nutlets.

Rocky slopes and canyons, Sonoran Zone, eastern California from Lake Tahoe south to the mountains of southern California, east to Nevada, southern Utah, and western Arizona. April–May.

Rare in the Sonoran Desert. (Tucson, Pima County, according to Kearney and Peebles.)

**17. Cryptantha pondii** Greene, Pittonia **1:** 291, 1889

*Cryptantha bartolomaei* Greene, Pittonia **2:** 232, 1892.

A loosely branched or nearly strict plant 0.5–3 dm. tall with strigose and sparsely short-hispid herbage; leaves linear to linear-lanceolate, 1–3 mm. broad, 1–2(5) cm. long, obtuse or rarely acute at apex, strigose and somewhat hispid, minutely pustulate; spikes geminate to ternate, 1–2 cm. long, sparingly leafy-bracteate near base; calyx lobes lance-linear, to 2–3 mm. long in fruit, connivent above but tips somewhat spreading, margins finely white-villous, midribs thickened and unarmed, abaxial lobe longer than rest, thus calyx asymmetrical; corolla tube about 2 mm. long, limb 1.5–3 mm. in diameter, hemispherical appendages inserted in funnelform throat; nutlets 4, all alike, 1.2–1.7 mm. long, ovate-lanceolate, acute, smooth to minutely tuberculate above middle, more or less truncate at base, back flattened, margins angled, groove narrow to nearly closed, divaricately forked at base; gynobase subulate, nearly two-thirds to three-fourths as long as nutlets; style distinctly surpassing nutlets.

Sandy plains and washes, Lower Sonoran Zone, western central Baja California. Feb.–April.

**18. Cryptantha barbigera** (A. Gray) Greene, Pittonia **1:** 114, 1887, var. **barbigera**

*Eritrichium barbigerum* A. Gray, Syn. Fl. N. Amer. **2**[1]**:** 194, 1878.
*Krynitzkia barbigera* A. Gray, Proc. Amer. Acad. **20:** 273, 1885.
*Krynitzkia mixta* M. E. Jones, Contr. West. Bot. **13:** 6, 1910.

Erect or ascendingly branched herb 1–4 dm. tall with 1 to several very bristly, hispid stems, scarcely any appressed hairs on stems or leaves; leaves lance-linear to oblong, 3–7 (or to 12) mm. broad, 1–7 cm. long, obtuse at apex, bristly-hirsute, inconspicuously pustulate; spikes usually geminate, sometimes ternate or solitary, ebracteate, to 15 cm. long, rather densely flowered; calyx lobes linear-lanceolate to lanceolate, 4–10 mm. long, connivent above but tips recurved, margins conspicuously villous with white hairs, midrib hirsute, abaxial lobe longest; corolla inconspicuous, limb 1–2 mm. broad; nutlets 1–4, lance-ovate, all alike if more than one, 1.5–2.5 mm. long, strongly verrucose on rounded or obscurely angled edges and convex back, groove narrow or broad, but dilated into a triangular basal areola; gynobase slender, about three-fourths as high as nutlets; style equaling or slightly surpassing nutlets.

Desert sands, arroyos, and hillsides, Lower Sonoran Zone, Colorado and Mojave Deserts to southern Utah, central Arizona, and western middle Baja California. Feb.–June.

**18*a*. Cryptantha barbigera** var. **fergusonae** J. F. Macbride, Contr. Gray Herb. **56:** 59, 1918

*Cryptantha fergusonae* Brand, Rep. Spec. Nov. **24:** 58, 1927.

Differing from var. *barbigera* in having larger corollas with limbs 5–6 mm. in diameter, and groove on adaxial face of nutlet dilated nearly to apex.

Known only from the vicinity of Palm Springs, California.

**19. Cryptantha nevadensis** Nels. & Kenn., Proc. Biol. Soc. Wash. **19:** 157, 1906, var. **nevadensis**

*Krynitzkia barbigera* var. *inops* Brandegee, Zoe **5:** 228, 1906.
*Cryptantha arenicola* Heller, Muhlenbergia **2:** 242, 1906.
*Cryptantha leptophylla* Rydb., Bull. Torrey Club **36:** 678, 1909.

Slender herb 1–5 dm. tall with erect or flexuous, strigose stems sometimes sparsely hirsute with spreading hairs in addition to appressed ones, stems solitary to several, strict or branched; leaves linear to linear-oblanceolate, acute to obtuse at apex, 1–5 mm. broad, 1–4 cm. long, somewhat pustulate, mostly hispid with appressed or subappressed hairs; spikes geminate or ternate, naked or sparsely bracteate at base, congested or elongated, to 15 cm. long; calyx lobes linear to lanceolate, to 5–12 mm. long in fruit, connivent but with tips recurving, villous on margins, hirsute on midrib;

corolla inconspicuous, limb 1–2 mm. broad; nutlets 4, all alike, ovate to lanceolate, 2.5–2.9 mm. long, convex on back, verrucose, muriculate near tips, margins slightly angled; groove open or closed, dilated into a small areola at base; gynobase slender, to three-fourths as long as nutlets; style nearly or quite equaling nutlets.

Desert slopes, plains, arroyos, and mouths of canyons, Lower and Upper Sonoran Zones, Colorado and Mojave Deserts to southern Utah, Arizona, northern Sonora, and central Baja California. Feb.–May.

**19*a*. Cryptantha nevadensis** var. **rigida** I. M. Johnston, Contr. Gray Herb. **74:** 68, 1925

*Cryptantha intermedia* var. *rigida* Brand, Pflanzenreich $4^{252}$: 70, 1931.

The calyx lobes more slender and less spreading; plant more erect and less hirsute; nutlets acutely ovate-oblong, about 2 mm. long, verrucose but not muricate toward tips; otherwise as in var. *nevadensis.*

Arid desert regions, Lower Sonoran Zone, western Mojave Desert and adjacent hills, east into western Arizona (near Wickenburg).

**20. Cryptantha ganderi** I. M. Johnston, Jour. Arnold Arb. **20:** 386, 1939

Annual herb 1–4 dm. tall branching from base or slightly above it, stems mainly strigose but with a few somewhat spreading, slender hairs; leaves linear or narrowly oblong, 1–2.5 mm. wide, 1.5–5 cm. long, obtuse or rarely acute at apex, basal ones gradually narrowed to base, upper abruptly rounded basally, strigose on upper surface and near margins beneath, pustulate and with scattered spreading hairs beneath; spikes solitary, densely scorpioid at apex or when young, soon lax, calyces 0.5–2 cm. apart, ebracteate or with 1–2 basal leafy bracts; calyx lobes linear or linear-lanceolate, 3–4 mm. long and nearly equal at anthesis, 6–10 mm. long and unequal in fruit, densely hispid with spreading hairs 2–2.5 mm. long on lower one-third to one-half, densely villous within, sparsely so above, somewhat connivent but tips spreading, abaxial one longest; corolla inconspicuous, about 2.5 mm. long, 1 mm. broad; ovules 4, usually only abaxial one maturing; nutlet 1, lanceolate-acuminate, 1.5–2 mm. long, light gray, smooth or very slightly rugulose, shining, sometimes minutely spotted, groove narrow but dilated to form a triangular areola at base; gynobase 1–1.5 mm. long; style shorter than nutlet.

Arid washes and gentle slopes, Lower Sonoran Zone, Borrego Valley, eastern San Diego County, south to central Baja California. Feb.–April.

**21. Cryptantha flaccida** (Dougl. ex Lehm.) Greene, Pittonia **1:** 115, 1887

*Myosotis flaccida* Dougl. ex Lehm., Stirp. Pug. **2:** 22, 1830.
*Eritrichium oxycaryum* A. Gray, Proc. Amer. Acad. **10:** 58, 1874.
*Krynitzkia oxycarya* A. Gray, op. cit., **20:** 269, 1885.
*Cryptanthe multicaule* Howell, Fl. N.W. Amer. 487, 1901.
*Cryptantha howellii* A. Nels., Bot. Gaz. **34:** 30, 1902.

Simple or ascendingly branched annual 1–4.5 dm. tall with pale gray-green, strigose stems and herbage; leaves linear, filiform, or linear-oblanceolate, 1–2 mm. wide, 2–6 cm. long, closely strigose, ascending and rather firm; spikes solitary to 5-branched, 4–16 cm. long, usually rather stiff, ebracteate; calyx lobes linear-lanceolate, 2–5 mm. long in fruit, connivent above with tips usually spreading, strigose to ciliate on margins, thickened midrib hirsute with encrusted, arcuate to uncinate, stiff hairs, abaxial lobe longest and most densely hirsute, fruiting calyx asymmetrical, closely appressed to stem; corolla 1–4 mm. broad; ovules 4, all except abaxial one abortive; nutlet 1, lanceolate-ovate to rostrate-acuminate, subterete or slightly compressed, smooth or very finely granulate; groove very narrow, margins often overlapping, dilated to form a small areola at base; gynobase minute, rarely over 0.5 mm. tall; style one-third to one-half as high as nutlet.

Grassy hillsides or in openings in the forest or chaparral, mainly Upper Sonoran and Transition Zones, eastern Washington and adjacent Idaho south to eastern San Diego County, California. March–April.

Doubtfully included on the basis of a specimen collected by T. S. Brandegee in "the Colorado Desert."

## 12. **AMSINCKIA** Lehm., Del. Sem. Hort. Hamb. 7, 1831

Erect annual with rough-hispid stems and foliage; hairs usually pustulate-swollen and often hardened or granular at base. Leaves alternate, lanceolate, linear-lanceolate or oblanceolate, lower ones petiolate, upper sessile or essentially so. Inflorescence of elongated scorpioid spikes. Calyx lobes lanceolate, usually 5, sometimes 3 or 4 through fusion of 2 or 3 lobes. Corolla yellow or orange, funnelform or salverform; tube cylindrical; throat sometimes with folds or hairs or both but without crests. Stamens 5, inserted in corolla tube, included; anthers ovate, more or less connivent. Ovary deeply 4-lobed, forming four 1-seeded nutlets; these tuberculate or transversely ridged on back or smooth, attached to gynobase basally or suprabasally. Cotyledons fleshy, deeply 2-lobed.

Corolla tube 10-nerved below stamens; nutlets tuberculate or dorsal keel and margins somewhat echinate; calyx lobes 5, equal, entire:
- Nutlets tuberculate but not dentate on margins or dorsal keel ........................................................ 1. *A. intermedia* var. *intermedia*
- Nutlets with thin, interrupted-dentate or echinate rugae on margins and keel .................................................. 1*a*. *A. intermedia* var. *echinata*

Corolla tube 20-nerved below stamens; nutlets tessellate with low rounded rugae; calyx lobes 3–5, unequal, larger ones 2–3-dentate at apex ........................ 2. *A. tessellata*

### 1. **Amsinckia intermedia** Fisch. & Mey., Ind. Sem. Hort. Petrop. **2**: 26, 1836, var. **intermedia**

(For synonymy see Jepson, Fl. Calif. **3**: 324–325, 1943.)

Erect and slender, often strict plant 3–15 dm. tall, or under favorable conditions frequently and widely branched; stems, herbage, and inflorescence bristly-hispid with spreading hairs; basal leaves narrowly oblanceolate or oblong, entire, to 2 dm. long, gradually narrowed to a slender petiole 1–6 cm. long; upper leaves gradually reduced to linear-lanceolate bracts 1 cm. long or less, intermediate ones usually lanceolate, sessile or subsessile; spike leafy-bracted at base, 5–30 cm. long or more, tip continuing to produce flowers after basal ones have matured nutlets; calyx lobes linear-lanceolate, reddish-hispid, 3–5 mm. long in flower, elongating to 6–10 mm. in fruit; corolla dark yellow to orange, 7–12 mm. long, rotate limb 3–6 mm. wide, glabrous without, short throat narrow, 10-nerved below stamens; nutlets ovoid, incurved, dorsally keeled, scabrous-rugose, tuberculate in intervals between ridges, usually grayish, 1.5–3 mm. long.

Grassy hillsides, valleys, and along washes, Lower Sonoran to Transition Zones, British Columbia to Idaho, Arizona and central Baja California. Uncommon in the drier parts of the desert but entering the Colorado Desert along its western margin and in favorable localities in Baja California. Feb.–June.

**1*a*. Amsinckia intermedia** var. **echinata** (A. Gray) Wiggins, Contr. Dudley Herb. **4:** 22, 1950

*Amsinckia echinata* A. Gray, Proc. Amer. Acad. **10:** 54, 1874.
*Benthamia echinata* Druce, Rep. Bot. Exch. Cl. Brit. Isles **4:** 298, 1916.
*Amsinckia menziesii* var. *echinata* Brand, Pflanzenreich $4^{252}$: 211, 1931, as to Fort Mohave specimen.

Differing from var. *intermedia* in having margins, dorsal keel, and sometimes transverse rugae of nutlets very thin and irregularly toothed; sometimes these are little more than laterally flattened tubercles instead of ridges on backs of nutlets; corolla tubes 10-nerved below stamens.

Arid plains and arroyos, Lower Sonoran Zone, Arizona and possibly in the eastern parts of the Colorado Desert. Strikingly sculptured nutlets of this type have been examined on specimens from the vicinity of Tucson, Arizona. March–May.

**2. Amsinckia tessellata** A. Gray, Proc. Amer. Acad. **10:** 54, 1874

(For synonymy see Jepson, Fl. Calif. **3:** 322, 1943.)

Rather stout, often profusely branched annual 3–8 dm. tall with coarsely hispid stems and herbage, hairs conspicuously pustulate at base; leaves linear, lanceolate, oblong or narrowly ovate, lower ones gradually narrowed to a short petiole, upper sessile, 2–10 cm. long, conspicuously spreading-hispid; spikes 1–5 cm. long in flower, elongating to 20 cm. or more in fruit, flowering tips dense, later rather lax; calyx lobes 3–5, often of two narrow ones and one broader, this one 2–3-dentate at apex, 5–8 mm. long in flower to 12 mm. long in fruit, sparsely hispid; corolla yellow or pale orange, 8–12 mm. long, 20-nerved below stamens, limb 2–4 mm. broad; nutlets broadly ovoid, erect or slightly incurved, 2.5–3.2 mm. long, back flattened or slightly rounded, closely tessellate and usually with some transverse, low, rounded rugae and a few small rounded tubercles, keel low, rounded.

Grassy slopes and valley floors, Lower and Upper Sonoran Zones, east of the Cascades from Washington and Idaho south through the hills and valleys of Utah, Nevada, and California to Arizona, the Colorado and Mojave Deserts, and to about 29° 30′ N., in Baja California. Feb.–June.

## 13. ANTIPHYTUM A. DC. ex Meisn., Pl. Vasc. Gen. Comm. 280, 1840

Suffrutescent or fruticose perennials with strigose-villous or canescent herbage. Leaves all opposite or partly opposite and partly alternate or completely alternate, linear to broadly lanceolate. Inflorescence usually a 2-branched cyme. Flowers short-pedicellate. Calyx 5-divided or -parted into linear to lanceolate segments. Corolla white, yellow, or blue; tube short; throat bearing occluding squamae; corolla limb 5-lobed, rotate or subrotate. Stamens 5, inserted in tube, included. Ovary 4-lobed; nutlets distinct, attached to a flat or pyramidal gynobase, short-stipitate or attached by a scar.

**1. Antiphytum peninsulare** (Rose) I. M. Johnston, Contr. Gray Herb. **68:** 51, 1923

*Krynitzkia peninsularis* Rose, in Vasey & Rose, Contr. U.S. Nat. Herb. **1:** 85, 1890.
*Amblynotopsis peninsularis* J. F. Macbride, Contr. Gray Herb. **48:** 41, 1916.

Perennial herb 1–4 dm. tall from a woody base and rootstock; stems and herbage hispid-strigose with stiff whitish hairs with pustulate bases; leaves mainly alternate but with a few opposite, linear to narrowly lanceolate, 0.5–2 mm. broad, 8–25 mm. long, lower part of stem often bearing fascicles of short linear leaves on stubby dwarf branches; branches of cymes 2–4 cm. long, flowers rather crowded in bud, more distant (3–10 mm. apart) after anthesis; pedicels 1–5 mm. long, hirsute; calyx lobes slightly unequal, linear-subulate, 2–3 mm. long at anthesis, to 6 mm. long in fruit; corolla white, tube 4–5 mm. long, limb 5–6 mm. broad, glabrous without; nutlets ovoid, about 2 mm. long, erect, grayish, strongly papillate-tuberculate, attached by a triangular ventral scar near base; style barely surpassing nutlets.

In canyons and washes, Lower Sonoran Zone, eastern coastal area of Baja California in the vicinity of Santa Rosalia. Feb.–Sept.

## 14. HARPAGONELLA A. Gray, Proc. Amer. Acad. 11: 88, 1876

Small annual herb simulating *Pectocarya* in general habit and habitat, with appressed-pubescent herbage. Leaves linear or lance-linear, basal petiolate, upper sessile or nearly so. Flowers sparsely arranged in or near axils of upper leaves, small. Calyx lobes unequal, 3 basal distinct or essentially so, upper pair united most of their length to form a burlike envelope around nutlets and bearing 5–9 soft glochidiate spines on their backs, inequality among lobes accentuated in fruit. Corolla white, subrotate-salverform, only slightly exceeding calyx. Nutlets usually 2, ovate to oblong, basally attached, one usually hirsutulous and sterile, other glabrous and fertile, completely enclosed by united, spine-bearing sepals.

Body of fruit 1.5–2 mm. in diameter; spines 1.5–3 mm. long; branches and stems rarely over 6 cm. long .......... 1. *H. palmeri* var. *palmeri*

Body of fruit 2–3 mm. in diameter; spines to 4 mm. long; branches and stems often to 15 cm. long .......... 1*a*. *H. palmeri* var. *arizonica*

### 1. Harpagonella palmeri A. Gray, Proc. Amer. Acad. 11: 88, 1876 var. palmeri

Stems erect to ascending or more commonly with several slender branches 3–15 cm. long from or near base, then ascending or procumbent; herbage thinly strigose-hirsute with slender white hairs; leaf blades linear to linear-lanceolate, 0.5–2.5 mm. wide, 1–3 cm. long, only lower gradually narrowed to a slender petiole; flowers crowded into a short, faintly scorpioid spike in bud, internodes elongating after anthesis so fruits are distant and mostly axillary; 3 free calyx lobes lance-linear, about 1 mm. long at anthesis, strigose; corolla barely 2 mm. long, rotate limb about 1.5 mm. in diameter; fruit burlike, body 1.5–2 mm. long, strigose, spines 1.5–3 mm. long, bearing hooked spines on upper two-thirds or three-fourths; nutlets about 1 mm. long, fertile one glabrous.

In sandy or clay soil, Lower and Upper Sonoran Zones, Riverside County and Santa Catalina Island, California, south to central Baja California. Feb.–April.

### 1*a*. Harpagonella palmeri var. arizonica I. M. Johnston, Contr. Gray Herb. 73: 75, 1924

Differing from var. *palmeri* only in having slightly more robust, larger stems, larger fruits and somewhat thicker and longer spines on fruiting sepals.

Sandy soil on flats and gentle slopes, Lower Sonoran Zone, Pima and Maricopa Counties, Arizona. March–April.

## Family 122. **VERBENACEAE.** Vervain Family*

Herbs, shrubs, woody vines, or trees with mostly tetragonal branches and twigs, not prominently nodose or articulate, often annulate. Leaves mostly opposite, sometimes whorled, alternate, or scattered, thick- or thin-textured, deciduous, estipulate, sessile or petiolate, simple or sometimes palmately compound or 1-foliolate; blades entire or variously dentate, incised, or cleft. Inflorescence axillary or terminal, of cymes, racemes, spikes, panicles, thyrsi, heads, or false umbels, sometimes involucrate, axillary ones mostly solitary. Flowers sessile or pedicellate, perfect or imperfect, hypogynous, sometimes diclinous or polygamous, rarely polygamodioecious, large or small, usually irregular. Calyx gamosepalous, campanulate, tubular, or salver-shaped, persistent, usually accrescent, mostly 4-lobed or -toothed or sometimes rim subentire. Corolla regular or irregular, gamopetalous, funnel- or salver-shaped, usually with a well-developed tube; limb 4- or 5- (rarely 7- to many-) parted, often more or less 2-lipped. Stamens mostly 4 and didynamous or reduced by abortion to 2, sometimes 4 or 5 and equal, inserted on corolla tube; staminodes sometimes present. Gynoecium of 2 (rarely 4 or 5) united carpels, 1 sometimes aborted. Ovary mostly compound, sessile or sometimes subtended by a disk, mostly somewhat 4-lobed, at first 2- or 5-celled but almost invariably soon becoming 4–10-celled through formation of false partitions, never with a free central placenta or columella; axial placentae lobes bearing each 1 ovule so that cells not subsequently divided by cross-walls contain each 2 ovules, while cells later divided by partitions contain each 1 ovule. Ovules anatropous and basal or hemianatropous and lateral. Fruit usually a dry schizocarp or drupaceous, with a thin and dry, or fleshy exocarp and more or less hard endocarp, 2–4-celled and indehiscent when ripe or dehiscent into 2 (rarely 4 to 10) 1- or 2-celled pyrenes. Seeds plainly testate; embryo never viviparous; radicle short and inferior; cotyledons 2, flat, more or less thickened and parallel.

Inflorescence spicate or racemiform, indeterminate and centripetal:
- Inflorescence spicate, often subcapitate during anthesis and elongating after anthesis; flowers sessile or subsessile:
  - Fruit composed of 4 one-seeded nutlets or cocci ........................ 1. *Verbena*
  - Fruit composed of only 2 pyrenes (or by abortion only 1):
    - Pyrenes 1-celled and 1-seeded:
      - Anthers appendaged ........................................ 2. *Burroughsia*
      - Anthers not appendaged:
        - Calyx very small, inconspicuous, thin-membranous, usually hidden by subtending bracts or bractlets:
          - Calyx rim truncate or shallowly toothed; corolla 4- or 5-lobed; fruit drupaceous, usually with a fleshy and juicy exocarp and a hard endocarp, either 2-celled or separating into 2 one-celled pyrenes ..................... 3. *Lantana*
          - Calyx rim 2- or 4-cleft or conspicuously toothed; corolla 4-lobed; fruit small, dry, with a hard and thin or papery exocarp, separating into 2 one-celled pyrenes:
            - Spikes elongate at anthesis, with scattered, often distant, flowers ... 4. *Aloysia*
            - Spikes dense and congested at anthesis, often subcapitate, with closely imbricate flowers:

---

* Text contributed by Dr. Harold N. Moldenke.

Mostly herbaceous, with trailing or ascending stems, rooting at nodes, sometimes somewhat woody; spikes greatly elongating after anthesis; bractlets always small, mostly cuneate-obovate, not 4-ranked; hairs malpighiaceous .......................................... 5. *Phyla*

Erect bushes, subshrubs, shrubs, or trees; spikes not usually elongating after anthesis; bractlets often very large, mostly ovate or lanceolate, often more or less 4-ranked; hairs not malpighiaceous ..... 6. *Lippia*

Calyx large, elongate-tubular, conspicuous, not hidden by subtending bracts .... ...................................................... 7. *Bouchea*

Pyrenes 2-celled and 2-seeded (or by abortion 1-celled and 1-seeded) ....... 8. *Priva*

Inflorescence racemiform and elongate, or reduced and axillary; flowers pedicellate:

Drupes composed of 4 pyrenes, each 2-celled and 2-seeded; calyx mostly obvolute and rostrate after anthesis ........................................... 9. *Duranta*

Drupes composed of 2 pyrenes, each 2-celled and 2-seeded (or only 1-seeded by abortion); calyx not obvolute or rostrate after anthesis ................... 10. *Citharexylum*

Inflorescence cymose, determinate and centrifugal:

Drupes composed of one 4-celled pyrene; leaves mostly palmately compound or 1-foliolate ............................................................ 11. *Vitex*

Drupes composed of 4 (or through abortion less numerous) 1-celled pyrenes; leaves always simple ........................................................ 12. *Tetraclea*

## 1. VERBENA L., Sp. Pl. 18, 1753

Herbs, or rarely woody at base, annual or perennial; stems and branches procumbent, ascending, or erect, glabrous or variously pubescent. Leaves mostly opposite, dentate (rarely entire) or variously lobed, incised, or pinnatifid. Inflorescence spicate; spikes terminal, usually densely many-flowered, often flat-topped and pseudo-umbellate, sometimes greatly elongate with scattered flowers, very rarely also axillary. Flowers small or medium-sized, each solitary in axil of a usually narrow bractlet. Calyx usually tubular, 5-angled, 5-ribbed, unequally 5-toothed. Corolla salverform or funnelform; tube straight or curved, often slightly ampliate at apex; limb flat, weakly 2-lipped, lobes 5, usually rather elongate, obtuse, rounded, or emarginate at apex, 2 posterior ones outermost and anterior one innermost in bud. Stamens 4, didynamous, inserted in upper half of corolla tube, included; anthers ovate, with parallel or slightly divergent thecae, connective unappendaged or provided with a glandular appendage. Style single, usually short, shortly 2-lobed at apex, posterior lobe smooth and nonstigmatiferous, anterior lobe broader, papillose, and stigmatic. Ovary 2-carpellate, 4-lobed, completely 4-celled, even at time of anthesis, with 1 ovule in each cell. Fruit mostly enclosed by mature calyx, dry, schizocarpous, readily separating on maturity into 4 one-seeded, linear or linear-oblong, crustaceous nutlets (cocci).

Sterile style lobe usually not protruding beyond stigmatic surface; ovary entire or very shallowly depressed; schizocarp commonly not constricted along lines of cleavage; anthers unappendaged; corolla mostly small, often not showy; calyx rarely more than twice as long as schizocarp, not contorted beyond it (Sect. *Verbenaca* Schau.):

Spikes panicled at apices of stems and branches, subtended chiefly by inconspicuous bracts; floral bractlets not prominent:

Leaves 1–2-pinnatifid or 3–5-cleft or deeply incised, at least those of middle and lower portions of stem:

Schizocarp a little longer than broad; leaves all deeply cleft or incised............ ...................................................... 1. *V. menthifolia*

Schizocarp about twice as long as broad; leaves diverse, the basal incised-dentate, middle stem leaves 1–2-pinnatifid, upper sparingly toothed or entire....2. *V. halei*

Leaves merely serrate or crenate:
Leaves subsessile or short-petiolate, with an attenuate base, not scabrous ............ 3. *V. carolina*
Leaves obviously petiolate, with a rounded-cuneate base, very harshly scabrous ....... 4. *V. scabra*
Spikes solitary or in threes at apices of stems and branches, or panicled and subtended by leafy bracts at base; floral bractlets often conspicuous:
Leaves predominantly entire, mostly linear, lower ones with a few salient teeth ....... 5. *V. perennis*
Leaves not predominantly entire:
Leaf blades serrate, serrate-dentate, or shallowly incised; spikes with more or less contiguous fruits:
Plants coarse, more or less densely hirsute-pubescent or hirsute-villous; leaf blades elongate-elliptic to ovate-orbicular; spikes stout at anthesis, 7–10 mm. in diameter ............................................ 6. *V. macdougalii*
Plants more slender, hirtellous; leaf blades lanceolate-elliptic to spatulate; spikes slender at anthesis, 5–6 mm. in diameter:
Leaves hirtellous, not canescent, coarsely serrate; rachis angulate; bractlets lance-acuminate; corolla limb 3–4 mm. broad .................. 7. *V. orcuttiana*
Leaves hirtellous and canescent, shallowly incised; rachis scarcely angulate; bractlets ovate-acuminate; corolla limb 6–8 mm. broad .................... 8. *V. neomexicana* var. *hirtella*
Leaf blades deeply incised-dentate, pinnatifid, or 3-cleft:
Leaves more or less strongly plicate, conspicuously whitish-veined near margins ............ 9. *V. plicata*
Leaves not strongly plicate nor whitish-veined:
Spikes not essentially or conspicuously bracteose; flowers readily seen; corolla limb 4–10 mm. broad:
Leaf blades, at least the lower ones, oblong-ovate or obtusely elliptic-ovate, not narrowly elongate, usually 3-cleft with segments incised-dentate; spikes dense at anthesis ................................ 10. *V. lasiostachys*
Leaf blades, at least lower ones, oblong-lanceolate to spatulate, narrowly elongate, usually incised-pinnatifid or incised-dentate; spikes slender at all times:
Plants canescent-hirtellous; leaf blades merely dentate or somewhat shallowly incised; bractlets broadly ovate ........ 8. *V. neomexicana* var. *hirtella*
Plants more or less hirsute; leaf blades pinnately cleft; bractlets lanceolate ................................ 8*a*. *V. neomexicana* var. *xylopoda*
Spikes conspicuously bracteose with many foliaceous bracts; flowers inconspicuous, hidden by bracts; corolla limb 2.5–3 mm. broad .......... 11. *V. bracteata*
Sterile style lobe protruding well beyond stigmatic surface; ovary in its late stages definitely (although shallowly) lobed at distal end; schizocarp constricted along lines of cleavage; connective of upper anthers mostly appendaged; corolla mostly large and showy; calyx usually more than twice as long as schizocarp and constricted or contorted beyond it (Sect. *Glandularia* Schau.):
Nutlets suggesting a tendency toward a beak (a slight protrusion horizontal to axis of schizocarp); commissural face practically reaching tip of nutlet ...... 12. *V. delticola*
Nutlets without a beak or tendency toward one; commissural face definitely not reaching tip of nutlet:
Leaf blades shallowly lobed, incised, or toothed ..... 13*a*. *V. gooddingii* var. *nepetifolia*
Leaf blades 3-cleft, incised-pinnatifid, or bipinnatifid:
Corolla tube protruding well beyond calyx at anthesis:
Corolla tube approximately twice as long as calyx at anthesis .................. 14. *V. elegans* var. *asperata*

Corolla tube one-third to one-half longer than calyx at anthesis:
Plants decumbent-ascending or prostrate, densely hirsute; calyx somewhat glandular, hispidulous-hirsute, scarcely viscid:
Plants with decumbent-ascending, loose habit; leaf margin slightly revolute .............................................. 15. *V. ciliata* var. *ciliata*
Plants with prostrate compact habit; leaf margin strongly revolute .......... ........................................ 15*a*. *V. ciliata* var. *pubera*
Plants ascending-erect, more or less hirsute; calyx densely glandular-hirsutulous and somewhat viscid-pubescent ......................... 16. *V. wrightii*
Corolla tube only slightly longer than calyx at anthesis:
Calyx teeth acute-subulate, short (less than half as long as calyx tube):
Bractlets very short, 2–3 mm. long; leaf blades abundantly pinnatifid-incised; calyx sparsely and irregularly short-pubescent, not glandular ............ ............................................... 17. *V. bajacalifornica*
Bractlets longer, 4–10 mm. long; leaf blades 3-cleft, divisions incised; calyx densely hirsutulous or villous-hirsute, more or less glandular:
Calyx villous-hirsute with long, white, spreading hairs; bractlets broadly lanceolate or ovate, 6–10 mm. long, acuminate, very densely long-ciliate on margins; corolla 9–12 mm. long ........ 13. *V. gooddingii* var. *gooddingii*
Calyx glandular-pubescent and sparsely hirsutulous; bractlets narrowly lanceolate, about 4 mm. long, not acuminate, only sparsely long-ciliate; corolla 7–8 mm. long ....................................... 18. *V. shrevei*
Calyx teeth subulate-setaceous, elongate (approximately half as long as calyx tube) ...................................................... 19. *V. setacea*

**1. Verbena menthifolia** Benth., Pl. Hartw. 21, 1839
*Verbena setosa* Mart. & Gal., Bull. Acad. Brux. **11**$^2$: 321, 1844.

Stems decumbent or ascending, branched, sparsely and minutely hispidulous; leaves ovate, tapering at base into a margined petiole, 3–6 cm. long, deeply cleft or subincised with divisions remotely serrate-dentate, strigillose on both surfaces, somewhat pustulate above; spikes panicled, slender, elongate, compact only at apex; bractlets ovate-lanceolate, acuminate, variable in length, usually shorter than calyx, ciliate and sparsely strigillose; calyx 2.5–3 mm. long, strigillose, sparsely (if at all) glandular, its teeth minute; corolla blue or light purple, its tube only slightly longer than calyx, its limb about 6 mm. wide, lobes more or less truncate; cocci trigonous with convex backs, 2–2.5 mm. long, striate, raised-reticulate above, commissural faces muricate.

In bottomlands, ravines, streamsides, and along roadsides, Sonoran Zones, western edge of the Colorado Desert to Texas and south into northern Baja California and Oaxaca.

**2. Verbena halei** Small, Bull. Torrey Club **25**: 617, 1898
*Verbena leucanthemifolia* Greene, Pittonia **5**: 135, 1903.

Stems usually several from a woody base, erect, ascendingly branched, glabrous or strigillose above; leaves 3–10 cm. long, strigillose on both surfaces, diverse in outline (basal and lower stem leaves oblong to ovate) tapering into a petiole approximately as long as blade, irregularly dentate or incised, middle stem leaves 1–2-pinnatifid with shorter petioles, and uppermost sparingly dentate or entire; spikes paniculately disposed, slender and elongate; bractlets about half as long as calyx, appressed, ciliate; fruiting calyx 3–3.5 mm. long, strigillose, subtruncate, broad veins terminating in unequal subulate teeth; corolla varying from lavender or pale lavender to blue, bluish lavender, pale blue, or purple, tube scarcely longer than calyx, limb 6–7 mm. wide, lobes retuse; cocci trigonous, approximately 2.5 mm. long, usually prominently striate, raised-reticulate at apex, commissural faces muricate.

In sandy or calcareous meadows, prairies, fields, and woodlands, on rocky hillsides and along roadsides, Sonoran Zones, from Alabama to Oklahoma and from south-central Arizona to Querétaro.

### 3. **Verbena carolina** L., Syst. Nat. ed. 10, 852, 1759

*Verbena caroliniana* L. ex Schau., in A. DC., Prod. **11:** 546, 1847, as a synonym.
*Verbena caroliniana* Spreng. ex Schau., loc. cit., as a synonym.
*Verbena carolinensis* Dill. ex Schau., loc. cit., as a synonym.
*Verbena pauciflora* Mart. & Gal. ex Walp., Rep. **6:** 687, 1847.
*Verbena paucifolia* Turcz., Bull. Soc. Nat. Mosc. **36**$^{2}$**:** 196, 1863.
(For additional synonymy see Ann. Mo. Bot. Gard. **20:** 268, 1933.)

Stems erect or ascending, to 1.5 m. tall, usually solitary, branched, hirsute-hispid; leaves subsessile or narrowed into a short petiole; leaf blades lanceolate-oblong or somewhat elliptic, 3–12 cm. long, 1.5–3 cm. wide, obtusish or acute at apex, coarsely crenate-serrate, strigose and somewhat pustulate above, hirsute-strigose especially along midrib and veins beneath; spikes paniculately disposed, slender, open in fruit; bractlets ovate, varying in length, usually about half as long as calyx or shorter, acuminate-subulate, somewhat spreading, ciliate-margined; corolla varying from light purple, lavender, or blue to white; fruiting calyx approximately 2 mm. long, hispidulous, obtuse and mucronate lobes connivent; corolla tube scarcely exserted; corolla limb inconspicuous, about 2 mm. broad; mature schizocarp enclosed by calyx; nutlets trigonous with a convex back, hardly 1.5 mm. long, faintly striate, commissural faces meeting at right angles and almost smooth.

In diverse soils and habitats, Lower Sonoran to Hudsonian Zones, from Pima County, Arizona, through most of Mexico including southern Baja California to El Salvador; ascending to 10,500 feet.

### 4. **Verbena scabra** Vahl, Eclog. Amer. **2:** 2, 1798

Stems to 1 m. or more tall, erect, solitary, simple or branched, hispidulous; leaves petiolate; leaf blades ovate to elongate-ovate, 3–13 cm. long, 2.5–5 cm. wide, acute or obtusish at apex, serrate-dentate along margins, very scabrous and commonly strigillose above, less scabrous and somewhat paler beneath, also hispidulous along veins; spikes paniculately disposed, slender, pedunculate, copiously and closely flowered; bractlets ovate-acuminate, about half as long as calyx, hispidulous; fruiting calyx 2.5–3 mm. long, hispidulous, ovoid, with somewhat unequal lobes acutely connivent, diverging from rachis of spike by 45° or more; corolla blue, tube scarcely if at all longer than calyx, limb about 2 mm. wide, lobes obtuse; anthers glandless; stigmatic surface midway between 2 almost equal, obtusish, sterile lobes; cocci trigonous, 1–1.3 mm. long, faintly striate, reticulate above, commissural faces extending to tip of schizocarp, meeting sharply at right angles, muriculate.

In mostly rich soil of low ground of marshes and swamps, and along the edges of streams and lakes, Lower Sonoran to Tropical Zones, from North Carolina to Florida and the Greater Antilles, west to southern California and south to Baja California, Coahuila, and Nuevo León. (Tucson, Pima County, Arizona.)

### 5. **Verbena perennis** Woot., Bull. Torrey Club **25:** 262, 1898

Stems several from a woody base, divaricately ascending-erect, more or less strictly branched, glabrate or often finely glandular and slightly hispidulous with short, stiff, antrorse hairs; leaves predominantly linear, 1–4 cm. long, entire or pinnately few-lobed, erect-ascending, sparsely hispidulous, margins revolute; spikes terminal, pedunculate, slender-filiform, elongate; bractlets ovate, 1.5–3 mm. long, acute at apex, hispidulous, ciliate-margined; calyx 4–5 mm. long, more abundantly pubescent along nerves, its lobes subequal, short, acute; corolla deep blue, its tube slightly longer than calyx, pubescent, its limb 5–7 mm. wide, lobes repand; fruit more or less remote; schizocarp about 3 mm. long, strongly constricted along lines of cleavage; cocci subcylindric, reticulate-scrobiculate except at base, commissural faces smooth or slightly scabrous and not extending to tip of nutlet.

On rough grassy slopes, gravelly or rocky hillsides, flats, canyons, and low rocky ridges, Sonoran to Transition Zones, western Texas to Pima County, Arizona, and to Coahuila.

### 6. **Verbena macdougalii** Heller, Bull. Torrey Club **26:** 588, 1899

Stems 3–8 dm. tall, stout, simple or occasionally branched, obtusely 4-angled, cinereous-green, hirsute-pubescent; leaves short-petiolate or narrowed into a subpetiolar base; leaf blades oblong-elliptic or elongate-ovate, 6–10 cm. long, coarsely and irregularly serrate-dentate, hirtellous, rugose and minutely pustulate above, densely pilose-pubescent and prominently veined beneath; spikes solitary or sometimes several, short-pedunculate, thick, 7–10 mm. in diameter at anthesis, comparatively dense both at anthesis and in fruit; bractlets lanceolate-subulate, for most part noticeably longer than calyx, pubescent, ciliate-margined; calyx 4–5 mm. long, rather densely pubescent, lobes very obtuse, terminating in short subulate teeth; corolla deep purple, tube scarcely protruding beyond calyx, limb about 6 mm. wide; cocci trigonous with convex dorsal surface, about 2.5 mm. long, raised-reticulate toward distal end, strongly or faintly striate below, commissural faces reaching tip of coccus, muriculate or almost smooth.

In open places, mostly above the Lower Sonoran Zone, from southern Wyoming to western Texas, New Mexico, and Arizona; entering the Sonoran Desert at Mountain Park west of Tucson, Pima County, Arizona.

### 7. **Verbena orcuttiana** Perry, Ann. Mo. Bot. Gard. **20:** 284, 1933

Stems several from a common base, 4-angled in cross section, glabrous or very sparsely hirtellous, branching; leaves rather obscurely petiolate, petiole broadly margined, 1–2 cm. long; leaf blades lanceolate-elliptic to spatulate, 4–6 cm. long, tapering into margined petiole, coarsely serrate, rugose and appressed-pubescent above, more densely spreading-pubescent or hirtellous beneath and prominently veined; spikes pedunculate, solitary or somewhat panicled, strict, mostly dense, elongate; rachis more or less angulate; bractlets lanceolate-acuminate, shorter than calyx, midrib and margins more or less decurrent along rachis, sparsely ciliate-margined; calyx about 4 mm. long, appressed-pubescent and finely glandular, its teeth short-subulate or acuminate, more or less connivent above schizocarp; corolla tube about as long as calyx; corolla limb 3–4 mm. broad; nutlets trigonous, about 2 mm. long, raised-reticulate at apex, striate toward base, commissural faces extending to tip of nutlet, muriculate-scabrous.

On tablelands and mountains, dry flats, and the margins of dry pools, Sonoran Zones, northeastern Baja California.

**8. Verbena neomexicana** (A. Gray) Small var. **hirtella** Perry, Ann. Mo. Bot. Gard. **20:** 298, 1933

Plants slender, densely canescent-hirtellous throughout; stems upright, branched; leaves sessile, 1–5 cm. long, oblong or linear, varying to oblong-lanceolate or spatulate at base of plant, more or less shallowly incised with rather few, divergent, linear or ovate lobes or teeth, densely canescent-hirtellous on both surfaces, usually also somewhat scabrous and finely pustulate above, veins quite prominent beneath, margins slightly revolute; spikes solitary or tending to be panicled, usually short-pedunculate; bracts usually broadly ovate-acuminate, commonly about as long as calyx; calyx about 3 mm. long, densely canescent-hirtellous, sometimes slightly glandular, teeth short and subulate; corolla blue or blue-purple, tube scarcely longer than calyx, limb about 8 mm. wide; nutlets trigonous, with a convex back, about 2 mm. long, very shallowly reticulate-scrobiculate on upper half, longitudinally striate below, commissural faces extending to tip of nutlet.

In dry arroyos, plains and rocky hillsides, often in shaded situations, Lower Sonoran to Arid Transition Zones, Gila County, Arizona, to Texas, Nuevo León, and at least to vicinity of El Mármol, Baja California.

**8*a*. Verbena neomexicana** var. **xylopoda** Perry, Ann. Mo. Bot. Gard. **20:** 297, 1933

This variety differs from previous one in being often coarser in growth, more or less hirsute rather than hirtellous, more glandular; leaf blades copiously pinnately cleft; bractlets usually more narrowly lanceolate; calyx about 4 mm. long and glandular-hirsute, with acuminate teeth; and corolla limb 6–10 mm. wide; corolla showy, varying from blue or dark blue to blue-lavender, blue-violet, or purple.

In sandy soil on hillsides, mountain valleys, dry plains, open, dry rocky slopes, and along moist streamsides, Lower Sonoran to Arid Transition Zones, southern California to Texas, San Luis Potosí, Sonora, and northern Baja California. Introduced in Florida.

**9. Verbena plicata** Greene, Pittonia **5:** 135, 1903

Coarse herb, with stems decumbent or ascending, branched, hirtellous; lower leaf blades elliptic-ovate, 1–4 cm. long, narrowed at base into a margined petiole of approximately same length, broadly obtuse at apex, plicate, coarsely incised-dentate, often 3-lobed, more or less canescent, rugose and somewhat appressed-hirsute above, hirsute beneath and prominently marked (particularly near margins) with whitish veins; upper leaves similar but smaller, often appearing spatulate; spikes terminal on stems and branches, ordinarily not compact; bractlets green, herbaceous, ovate-lanceolate, usually surpassing calyx, at times barely equaling it, 5–7 mm. long, 1.5–4 mm. wide, gradually attenuate from base to apex, acute at apex, hirsute; calyx 3.5–4 mm. long, more or less glandular-hirsute, lobes very obtuse or subtruncate, terminating abruptly in subulate teeth; corolla varying from blue, bluish lavender, or lavender to purplish blue or purple, tube scarcely longer than calyx, limb 4–6 mm. wide, anterior lobe retuse; nutlets cylindric-trigonous, 2–2.5 mm. long, shallowly scrobiculate above, changing to indefinitely striate toward base, commissural faces finely muricate-scabrous to practically smooth, not reaching tip of nutlet.

In open sandy ground, flats, prairies, and ravines, along roadsides, and frequent on slopes of gravel hills and cliffs, Lower and Upper Sonoran Zones, from the vicinity of Tucson, Arizona, to Texas and Nuevo León.

**10. Verbena lasiostachys** Link, Enum. Hort. Berol. **2:** 122, 1822

*Verbena prostrata* R. Br., in Ait., Hort. Kew. ed. 2, **4:** 41, 1812. Not *V. prostrata* Savi, 1802.

Stems at first erect or ascending, at length spreading and diffusely branched, sparsely villous; leaves ovate or oblong-ovate, with a cuneate base tapering into a margined petiole, 5–10 cm. long, 1–4.5 cm. wide, commonly 3-cleft, coarsely and irregularly serrate-dentate with apiculate teeth or incised, veins impressed above, more or less conspicuous beneath, both surfaces pilose to sparsely villous; spikes solitary or more often loosely paniculate, dense before anthesis, becoming elongated and open in fruit; bractlets lanceolate-subulate, not longer than calyx, villous and at times finely glandular; calyx 4–5 mm. long, villous or glandular-hirsute, subtruncate, distal end connivent above schizocarp, teeth subulate; corolla white or pale blue, its tube a little longer than calyx, very sparsely pubescent outside or apparently glabrous, its limb 3–5 mm. broad; nutlets oblong-trigonous with convex back, 2–2.5 mm. long, slightly thicker at distal end, raised-reticulate above, striate below, often with striae fading out toward base, commissural faces more or less muricate, approximately reaching tip of nutlet.

In sandy or granitic soil of waste places, often in moist places along streams and canyons, mainly Upper Sonoran and Arid Transition Zones, northern Baja California to Oregon but barely entering the Sonoran Desert in eastern San Diego County, California. Introduced in New York.

**11. Verbena bracteata** Lag. & Rodr., Anal. Cienc. Nat. Madrid **4:** 260, 1801

*Verbena prostrata* Savi, Mem. Soc. Ital. **9:** 349, *pl. 7,* 1802.
*Verbena bracteosa* Michx., Fl. Bor. Amer. **2:** 13, 1803.
(For additional synonymy see Flora of Texas **3:** 33, 1942.)

Stems usually several from a common base, diffusely branched, decumbent or ascending, rarely erect, coarsely hirsute; leaves 1–6 cm. long, pinnately incised or usually 3-lobed (lateral lobes narrow and divaricate, middle lobe large, cuneate-obovate, incised-dentate or cleft), narrowed at base into a short, margined petiole, hirsute on both surfaces, midrib and larger veins slightly prominent beneath; spikes terminal, sessile, comparatively thick, conspicuously bracteose; bracts much longer than calyx, spreading-ascending, recurved in age, coarsely hirsute, foliaceous, lowermost often incised and quite leaflike, upper linear-lanceolate, acute to acuminate at apex, entire; calyx 3–4 mm. long, hirsute particularly along nerves, its lobes very short, connivent over schizocarp; corolla varying from bluish or pale violet to lavender, lilac, or purplish, tube protruding slightly beyond calyx, very finely pubescent outside throat, limb 2.5–3 mm. wide; nutlets trigonous, approximately 2 mm. long, sharply raised-reticulate above, striate below, commissural faces reaching to distal end of nutlets, muricately scabrous.

In low and newly cleared ground, grassy places, waste soil, and along roadsides and lake margins, from Minnesota and Illinois to Virginia and Florida, west to British Columbia and California, south to Nuevo León and Baja California; introduced on ballast at seaports from Maine and Ontario south along the Atlantic seaboard. In our area occurring at Yuma, in San Bernardino County, at Blue Lake, Imperial County, California, possibly in Pinal and Yavapai Counties, Arizona, and in Baja California.

**12. Verbena delticola** Small, in Perry, Ann. Mo. Bot. Gard. **20:** 314, 1933

Stems decumbent to ascending, branched, more or less hirsute; leaves ovate-deltoid, 3–7 cm. long with truncate-cuneate base narrowed into margined petiole, obtusish or subacute at apex, coarsely serrate-dentate, often 3-lobed, usually thin-textured, sparsely appressed-hirsute on both surfaces, trichomes often with minute bulbous bases on upper surface; spikes pedunculate, fascicle-like during anthesis, becoming elongate in fruit; bractlets linear-attenuate, shorter than or equaling calyx; flowers odorless or with a faint odor; calyx 7–8 mm. long, sparsely glandular, hirsute, lobes slender, subulate, unequal; corolla showy, varying from red, deep pink, rose-pink, or bright magenta-pink to deep pinkish lavender, lavender-blue, violet, red-purple, or deep purple, tube 1–1.5 cm. long, protruding well beyond calyx, pubescent or glabrate outside, limb 7–10 mm. wide, lobes emarginate; nutlets about 3 mm. long, subcylindric, reticulate from apex to definitely broadened base, commissural faces practically reaching tip of nutlets, muricately scabrous, suggesting a tendency toward a beak at apex.

In sandy, adobe, clay, loam, or alluvial soils, or even limestone or rocky areas in diverse situations, Sonoran Zones, central Sonora to Texas and south to Puebla and Vera Cruz, Mexico. Often in cultivated ground in fields and orchards.

**13. Verbena gooddingii** Briq., Ann. Conserv. & Jard. Bot. Genève **10:** 103, 1907, var. **gooddingii**

*Verbena verna* var. *fissa* A. Nels., Amer. Jour. Bot. **18:** 437, 1931.
*Verbena bipinnatifida* var. *gooddingii* Jepson, Fl. Calif. $3^2$: 382, 1943.

Perennial; stems usually several from a common base, erect or decumbent-ascending, cinereous-green, branched, densely pilose or somewhat villous, often glandular; leaves 3–5 cm. long, tapering at base into a short, margined petiole, 3-cleft, divisions coarsely toothed or incised, cinereous-green, more or less villous-hirsute on both surfaces, margins slightly (if at all) revolute, midrib and veins slightly prominent beneath; spikes pedunculate, fascicle-like in anthesis, somewhat elongate in fruit; bractlets usually a little shorter than calyx, lanceolate, acuminate at apex, villous-hirsute, long-ciliate along margins; calyx 8.5–11 mm. long, villous-hirsute, more or less glandular, lobes slender, subulate, unequal; corolla varying from pink, pale or deep lavender to bluish lavender, blue-violet, or blue in age, tube very little longer than calyx, pubescent outside, its limb 8–12 mm. wide, segments retuse; nutlets 3–3.5 mm. long, subcylindric, reticulate-scrobiculate or alveolate except at striate base, commissural faces retrorsely hispidulous, making a white, linear-oblong band which shows up white against brown nutlet.

In sandy, limestone, and granitic soil on dry rocky slopes or gravelly hillsides, dry sandy washes, canyon floors, ravines, roadbanks, gravel bars, and grasslands, Lower Sonoran to Arid Transition Zones, southern Utah and Nevada to southern California, Sonora, and northern Baja California.

**13*a*. Verbena gooddingii** var. **nepetifolia** Tidestrom, Proc. Biol. Soc. Wash. **38:** 15, 1925

*Verbena arizonica* Briq., Ann. Conserv. & Jard. Bot. Genève **10:** 102, 1907. Not *V. arizonica* A. Gray, 1883.
*Verbena verna* A. Nels., Amer. Jour. Bot. **18:** 436, 1931.
*Verbena bipinnatifida* var. *nepetifolia* Jepson, Fl. Calif. $3^2$: 383, 1943.

Differs from var. *gooddingii* in having its leaves broadly ovate, merely coarsely and unevenly dentate, at times shallowly lobed or incised, abruptly narrowed into a cuneate base, not as decidedly 3-cleft or pinnately lobed, and pubescence varying from strongly pilose to densely villous; corolla varies from pink or lavender to blue in age, drying white, and is fragrant.

In granitic soil, dry and rocky slopes, red clay hillsides, the sides of ravines, on banks of small streams, and along arid roadsides, often associated with *Prosopis* and *Simmondsia*, Lower and Upper Sonoran Zones, from Arizona and southern California to central Baja California and Nuevo León.

**14. Verbena elegans** H. B. K. var. **asperata** Perry, Ann. Mo. Bot. Gard. **20:** 319, 1933

Perennial herb; stems decumbent-ascending, more or less branched, hirsute-hispid; leaves lanceolate-ovate, 2–5 cm. long, 1–3.5 cm. wide, cuneate base narrowed into a margined petiole, variously incised-pinnatifid, trifid, cleft, or lobed, somewhat appressed-hirsute on both surfaces, dark green above, lighter beneath, often with much abbreviated branches or fascicles of smaller leaves in their axils; spikes terminal, pedunculate, dense, many-flowered, fascicle-like at anthesis; bractlets mostly half to two-thirds as long as calyx, lanceolate, attenuate, glandular-hirsute, ciliate-margined; calyx 8–10 mm. long, glandular-hispidulous-hirsute, teeth subulate and unequal, larger about 2 mm. long, often whole calyx or at least teeth purple; corolla varying from bright pink or rose, deep rose or light red to bright purple or rhodamine-purple, tube 15–18 mm. long, finely but sparsely pubescent outside, throat gibbous, limb 10–12 mm. wide; nutlets subcylindric, 3.5–4 mm. long, reticulate-scrobiculate, chiefly on upper half, lower part striate, commissural face muriculate.

On limestone or even bare sandstone outcroppings in mountains and canyons, in various soils and situations but characteristic of arroyo margins, ascending to 8,000 feet, chiefly Upper Sonoran and Transition Zones, Sonora to Texas, Hidalgo, and Zacatecas but entering the Sonoran Desert near Hermosillo and Magdalena, Sonora.

**15. Verbena ciliata** Benth., Pl. Hartw. 21, 1839, var. **ciliata**

Spreading annual or perennial; stems several from a common base, decumbent, branched, hirsute-hispidulous; leaves short-petiolate or subsessile, subpinnatifid or commonly trifid with divisions more or less deeply incised, ultimate segments linear-oblong, with slightly revolute margins, hirsute-strigillose on both surfaces; spikes subsessile or short-pedunculate, more or less compact even in fruit; bractlets a little shorter than calyx, lanceolate, subulate or acuminate at apex, hirsute-ciliate; flowers with a faint lilac fragrance; calyx 7–8 mm. long, somewhat glandular, hirsute-hispidulous, its lobes short, subulate, unequal; corolla showy, varying from red, pinkish lavender, or lavender to pink, lilac-, blue-, violet-, or reddish-purple, tube 10–12 mm. long, pubescent outside, limb 6–8 mm. wide, lobes retuse; nutlets 2.5–3 mm. long, reticulate-scrobiculate above, striate toward base, commissural faces muricately scabrous.

On flats, sandy plains, dry limestone hills, mesas, and rocky slopes, mainly above the Sonoran Desert (ascending to nearly 12,000 feet) from Arizona to Oklahoma and Texas and south to Oaxaca, with isolated records from Tennessee and South Dakota. Entering the margin of the Sonoran Desert near Tucson and Aguila, Arizona.

**15a. Verbena ciliata** var. **pubera** (Greene) Perry, Ann. Mo. Bot. Gard. **20:** 332, 1933

*Verbena pubera* Greene, Pittonia **5:** 136, 1903.

Differs from var. *ciliata* in its more or less prostrate compact habit and all segments of leaves, as well as central portions, narrowly oblong or linear, their margins strongly revolute.

In cindery soil, on sandy clay flats, sandy and grassy plains, and washes, mainly Upper Sonoran Zone, southern California to western Texas. Possibly entering the Sonoran Desert along the eastern margin in Arizona.

**16. Verbena wrightii** A. Gray, Syn. Fl. N. Amer. **2**[1]**:** 337, 1878

Stems usually several from a common base, decumbent-ascending to erect, branched, sparsely hispid-hirsute; leaves subsessile or contracted into a short, margined petiole; leaf blades 2–4 cm. long and about as wide, bipinnatifid or trifid, with divisions more or less deeply incised, ultimate segments narrowly lanceolate, hirtellous or hirsute on both surfaces; spikes short-pedunculate, more or less compact even in fruit; bractlets shorter than calyx, lanceolate, acute or acuminate at apex, hirsute-ciliate; fruiting calyx 7–9 mm. long, densely glandular, somewhat viscid-pubescent, hirsute on nerves, lobes short, subulate, unequal; flowers fragrant; corolla varying from pink, deep pink-lavender, or lavender to rose, rose-lavender, rose-purple, magenta or purplish, tube 11–12 mm. long, pubescent around throat, limb 6–8 mm. wide, lobes retuse; nutlets 2.5–3 mm. long, reticulate-scrobiculate above, striate toward base, commissural faces muricately scabrous.

In moist rich loam among boulders, loose sandy or rocky soil, or even dry, light adobe clay or calcareous soil, Lower Sonoran to lower parts of Arid Transition Zones (to 7,000 feet), western Arizona near Wickenburg to Colorado, Oklahoma, and Texas and south to San Luis Potosí.

**17. Verbena bajacalifornica** Moldenke, Phytologia **2:** 22, 1941

Annual herb; stems erect, simple or sparsely branched, 8–15 cm. long, obtusely tetragonal, sparsely hirsutulous with mostly scattered, stiff, whitish, nonglandular hairs about 1 mm. long, glabrescent in age, sometimes decumbent at base and rooting from lower nodes; petioles distinct, very slender, 1–10 mm. long, densely or sparsely hirsutulous with stiff, white, nonglandular hairs like stems, slightly margined; leaf blades chartaceous, uniformly green on both surfaces, ovate in outline, 8–27 mm. long, 4–18 mm. wide, sparsely hirsutulous-pilose with rather short and subappressed whitish hairs on both surfaces, more densely so along midrib and larger veins beneath, abundantly pinnatifid-incised, sometimes obscurely 3-parted with divisions again abundantly pinnatifid-incised, lobes rounded, subrevolute along margins; spikes erect, long-pedunculate, 5–15 or more cm. long; peduncles slender, obtusely tetragonal, 2–6.5 cm. long, sparsely hirsutulous-pilose with rather scattered, nonglandular, whitish hairs; rachis densely many-flowered, more densely hirsutulous, not glandular, flowers close together and densely imbricate before, during, and even after anthesis, or 2 or 3 lowermost to 5 mm. apart in fruit; bractlets very small, lanceolate, 2–3 mm. long, about half as long as calyx, attenuate at apex, glabrate except for long-ciliate margin; calyx 4–5 mm. long, irregularly short-pubescent with whitish spreading hairs, obscurely (if at all) glandular; corolla 7–8 mm. long, slightly projecting from calyx, tube slightly puberulent at apex outside, limb about 3 mm. wide.

Known only from the type locality, 18 miles north of El Refúgio, Baja California.

**18. Verbena shrevei** Moldenke, Phytologia **2:** 26, 1941

Annual herb; stems decumbent at base, slender, obtusely tetragonal, more or less spreading-pubescent with whitish, often glandular hairs, often many-branched with erect or ascending branches, which are usually somewhat more densely spreading-pubescent; leaves petiolate or uppermost subsessile; petioles very slender, 1–10 mm. long, more or less winged, densely hirsutulous or spreading-pubescnt; leaf blades chartaceous, rather uniformly green on both surfaces, elliptic in outline, 1–2 cm. long, 9–17 mm. wide, rather densely strigose on both surfaces, usually more or less 3-parted, divisions sparingly incised-lobed, lobes rounded at apex; spikes elongating to 10 cm. or more, densely many-flowered, rachis elongating even during anthesis and thus separating individual flowers by 4–13 mm. toward base of spike; peduncles (1–4 cm. long) and rachis slender, obtusely tetragonal, densely spreading-pubescent, or hirsutulous with whitish, often glandular hairs; bractlets lanceolate, about 4 mm. long, shorter than calyx, attenuate at apex, densely glandular-pubescent on back, sparsely long-ciliate with longer, stiff, white, nonglandular hairs on margins, not acuminate; calyx 5–6 mm. long, rather densely glandular-pubescent and also more or less scattered white-hirsutulous; corolla small, 7–8 mm. long, slightly projecting from calyx, tube minutely puberulent at apex outside, limb about 4 mm. wide.

Known only from the type locality, 19 miles northeast of Comondú, Baja California.

**19. Verbena setacea** Perry, Ann. Mo. Bot. Gard. **20:** 339, *pl. 15,* 1933

Stems decumbent-ascending, soft-pubescent; leaves 3–5 cm. long, with cuneate-truncate base narrowed into a margined petiole, trifid with segments coarsely dentate, lateral lobes small, venation somewhat prominent beneath, softly villous-pubescent on both surfaces; spikes short-pedunculate; bractlets linear-lanceolate, about three-fourths as long as calyx, subulate-setaceous, soft-pubescent, ciliate-margined; calyx about 6 mm. long, villous-pubescent, teeth about 2 mm. long, half as long as tube, subulate-setaceous; corolla tube about 7 mm. long, pubescent outside; corolla limb 6–7 mm. broad; anthers not glandular; mature nutlets not known.

Known only from the type locality, among rocks, Calmallí, Baja California.

## 2. BURROUGHSIA Moldenke, Phytologia 1: 411, 1940

Somewhat woody perennials or shrubby, much branched. Stems short, somewhat ligneous, covered with a light-colored, cracked cortex; branches numerous, slender, flexuous, ascending, often fascicled. Leaves simple, opposite or ternate, sessile, small, usually lanceolate in outline, pinnatifid or crenate, often plicate and prominently veined beneath. Inflorescence axillary, spicate or capitate; peduncles solitary, two to four times as long as subtending leaves; heads hemispheric or oblong and few-flowered, or spikes oblong and dense. Bractlets ovate or ovate-lanceolate, distinct, acute, pubescent, ciliate-margined. Calyx much exceeding bractlets, 2-parted, persistent, united nearly to top and truncate or minutely 2-toothed. Corolla purplish or white, much exserted, hypocrateriform; tube cylindric; limb bilabiate, upper lip 2-lobed, lower 3-cleft, all lobes crenate or crisped, ventral lobe somewhat larger than others. Stamens 4, didynamous; filaments longer than anthers, ventral ones with slender erect appendages springing from connective and exceeding anthers; anther cells somewhat separate and oblique. Pistil very short; stigma strictly simple, oblique, disklike. Ovules 1 in each cell.

**1. Burroughsia fastigiata** (Brandegee) Moldenke, Phytologia **1:** 412, 1940

*Lippia fastigiata* Brandegee, Proc. Calif. Acad. II, **2:** 196, 1889.

Densely branched shrub, 3–6 dm. tall, with mintlike odor, cinereous throughout; branches numerous, decussate-opposite but so crowded as to appear fascicled, densely cinereous-tomentose, less so in age; leaves sessile, numerous, usually decussate-opposite, sometimes ternate, erect or ascending, usually so close as to be imbricate, cuneate-lanceolate, 4–20 mm. long, 2–6 mm. wide, usually only about 5 mm. long and 2 mm. wide, rather coarsely crenate-dentate above middle with obtuse antrorse teeth, revolute-margined, densely cinereous-tomentose on both surfaces, not glandular, midrib and other venation invisible because of dense tomentum; inflorescence axillary, capitate or subcapitate, 1 or 2 per node, rather few toward tips of branches, rather long-pedunculate; peduncles solitary, slender or comparatively stout, ascending-divergent, 5–23 mm. long, usually two to four times as long as subtending leaves, obtusely tetragonal, densely cinereous-tomentose; heads hemispheric to oblong, dense but rather few-flowered; bractlets ovate, distinct, about 2 mm. long and 1 mm. wide, attenuate-acute at apex, tomentellous, rather less densely so than rest of plant; calyx much exceeding bractlets, about 4 mm. long, very densely cinereous-tomentose, 2-lobed, lobes united nearly to top and truncate; corolla white, pinkish, light lavender, or purple, about 5 mm. long, slender cylindric tube equaling or very slightly exceeding calyx, upper part puberulent on outside, all lobes very short; anther cells on a dilated connective; style about 0.5 mm. long; stigma attached to front of upper half of style.

On dry playas, plains, and in flatlands and sandy valleys, Lower Sonoran Zone, vicinity of Calmallí to the Magdalena Plain, Baja California.

## 3. LANTANA L., Sp. Pl. 626, 1753

Erect herbs or shrubs, sometimes subscandent or scandent, usually more or less scabrous and hirsute-pubescent or tomentose with simple hairs. Leaves opposite or ternate, dentate, often rugose. Inflorescence dense cylindric spikes or heads, usually axillary, pedunculate. Flowers red, yellow, blue, or white, sessile in axils of solitary bractlets, which are oblong, lanceolate, or ovate, often acuminate at apex and spreading or subimbricate. Calyx small, membranous, truncate and entire or sinuate-dentate. Corolla tube cylindric, slender, equal in diameter throughout or slightly ampliate above, limb spreading, regularly or obscurely 2-lipped, 4- or 5-fid, lobes broadly obtuse or retuse at apex. Stamens 4, didynamous, inserted at about middle of corolla tube, included; anthers ovate, with parallel thecae. Ovary 1-carpellate, 2-celled, each cell 1-ovulate. Style usually short; stigma rather thick, oblique or sublateral. Ovules basal and erect or attached laterally near base of each cell. Fruit drupaceous; exocarp usually more or less fleshy (rarely dry); endocarp hard, 2-celled or splitting into two 1-celled pyrenes. Seeds without endosperm.

Bractlets narrow-lanceolate or oblong, usually not pronouncedly imbricate, 1–2 mm. wide; corolla mostly orange, yellow, or red; heads large, mostly 2–3 cm. wide .... 1. *L. horrida*

Bractlets broadly ovate, pronouncedly imbricate, outermost 4–8 mm. wide; corolla mostly white, pink, or purplish; heads small, mostly 1–1.5 cm. wide:

Leaf blades usually very densely canescent-tomentose or velutinous beneath; peduncles usually short, 1–7.5 cm. long; bractlets mostly subacute or obtuse at apex, densely short-pubescent or velutinous; upper leaf surface more or less bullate, marginal teeth small and rounded ........................................ 2. *L. velutina*

Leaf blades appressed-strigose beneath; peduncles usually elongate, usually 5–14 cm. long; bractlets mostly acuminate at apex, appressed-strigose; upper leaf surface not bullate, marginal teeth coarse or appressed, acute:

Leaf blades sparsely strigose beneath with distinct, straight, stiff hairs, marginal teeth narrow, regularly attenuate, and very sharp, often greatly appressed ........... ........................................................ 3. *L. achyranthifolia*

Leaf blades densely strigose-pubescent beneath with massed soft hairs which are not notably straight or stiff, marginal teeth usually broadly triangular, very abruptly acute ................................................ 4. *L. macropoda*

**1. Lantana horrida** H. B. K., Nov. Gen. & Sp. **2:** 211, 1818

*Lantana rubra* Berland., in Teran & Berland., Mem. Comision Limites 15, 1832.
(For additional synonymy see Moldenke, in Lundell, Flora of Texas **3:** 48–49, 1942.)

Shrub, 0.5–2 m. tall, or perennial herb, much branched; bark gray or brown; roots woody; stems and branches slender, spreading, obtusely tetragonal, varying from very densely armed with stout recurved prickles to unarmed, and from abundantly hirsute with spreading whitish hairs to glabrous and scabrellous or smooth to touch, younger ones usually hispidulous or hirtellous; nodes more or less annulate; principal internodes 2–6.5 cm. long; leaves decussate-opposite; petioles 2–12 mm. long; blades chartaceous or submembranous, bright or dark green on both surfaces, ovate or subrotund-ovate, 2.5–9 (usually 3–5) cm. long, 1.5–6.5 (usually 2–4) cm. wide, acute or obtuse at apex, usually truncate or subtruncate at base, very coarsely serrate from apex almost to base with large, somewhat irregular, widely spreading, acute or obtuse teeth, short-strigose and scabrous above, hirtellous or hispidulous on venation beneath to merely puberulent, short-pubescent, or subglabrate, often subbullate above in drying, slender midrib, 4–6 pairs of secondaries, and veinlet-reticulation often impressed above and sharply prominent beneath, sometimes flat or veinlets even indiscernible; inflorescence axillary, 2 per node; peduncles slender, 3–9 cm. long, usually slightly surpassing subtending leaves, hispidulous-hirsute to subglabrate; bractlets narrow-lanceolate or oblong, 4–9 mm. long, 1–2 mm. wide, acute at apex, rather sparsely or densely strigillose, innermost usually half as long as corolla, outermost occasionally larger and forming an involucre almost as long as corolla tube; heads large, hemispheric, 2–3 cm. wide, densely many-flowered, not elongating; flowers showy, with a faint odor; corolla bright or deep orange or red and yellow, orange-red, or dark yellow, tube 7–10 mm. long, densely pubescent outside, limb 5–9 mm. wide; fruit black or dark blue, edible.

In fields, thickets, and scrub, on gravelly hills, and outwash plains, often growing in large patches, Lower Sonoran to Arid Transition Zones, from Mississippi to New Mexico south to Sonora and San Luis Potosí.

**2. Lantana velutina** Mart. & Gal., Bull. Acad. Brux. **11²:** 325, 1844

*Lantana odorata* var. *berlandieri* Torr., Bot. Mex. Bound. 128, 1859.
*Camara velutina* Kuntze, Rev. Gen. Pl. **2:** 504, 1891.
*Lantana involucrata* var. *velutina* Standley, Field Mus. Bot. Ser. **11:** 172, 1936.

Small, diffusely branched shrub to 2 m. tall, with an odor like rhubarb; stems unarmed, smooth; branches and branchlets slender, more or less densely short-pubescent or tomentose with spreading or subappressed hairs, less so in age; petioles slender, 1–9 mm. long, densely tomentose, often alate; leaf blades chartaceous, ovate or elliptic-ovate, 1–4 cm. long, 7–28 mm. wide, mostly acute or subacute at apex, rarely short-acuminate or rounded, cuneately narrowed into petiole at base, regularly crenate along margins, with small obtuse and antrorse teeth, mostly reticulate-rugose or bullate and densely pubescent or velutinous above, usually very densely and softly canescent-tomentose or velutinous beneath; peduncles slender, elongate, 1–7.5 cm. long, more or less densely short-pubescent with spreading or more rarely subappressed hairs; heads hemispheric or slightly ovate-elongate in age; bractlets large, conspicuous, membranous, closely imbricate, broadly ovate, lowermost to 12 mm. long and 8 mm. wide, subacute or obtuse at apex, broadly rounded and sessile at base, venose, densely appressed-pubescent or velutinous, shorter than corolla or lowermost equaling or surpassing corolla; corolla white, lavender or pinkish in age, yellow in throat, fragrant, tube about 6 mm. long, densely strigose-pubescent outside, lobes densely strigose-pubescent outside, glabrate or minutely appressed-puberulent within; fruit pink, sweet, edible.

In waste places, on mesas, slopes of foothills, alluvial plains, sandy bottoms, and llanos, Upper Sonoran to Tropical Zones, from Sonora and Baja California to Panama.

**3. Lantana achyranthifolia** Desf., Cat. Pl. Paris ed. 3, 392, 1829

*Lippia purpurea* Jacq. f., Eclog. Pl. **1:** 126, *pl. 85,* 1816.
*Lantana purpurea* Benth. & Hook., Gen. Pl. **2:** 1142, 1876. Not *L. purpurea* Hornem., 1815.
*Lippia fimbriata* Rusby, Mem. Torrey Club **4:** 244, 1895.
*Lantana macropodioides* Greenm., Field Mus. Bot. Ser. **2:** 339, 1912.

Suffrutescent, 0.5–2 m. tall; stems unarmed; branches slender, elongate, ascending, light yellowish green, densely appressed-strigose with straight, white, antrorse, stiff hairs, sparsely so in age; internodes 3–11 cm. long; leaves opposite or ternate; petioles slender, 1–10 mm. long, densely strigose; blades chartaceous, narrowly lanceolate to broadly ovate, 2.5–8 cm. long, 1.5–5.3 cm. wide, acute or acuminate (rarely obtuse) at apex, acuminate or rarely only acute at base, coarsely and sharply serrate with usually narrow and regularly attenuate teeth, more or less scabrous-strigose above with closely appressed, antrorse, whitish or yellowish hairs similarly appressed-strigose beneath but stiff, hairs rather scattered, denser on large venation, resinous-punctulate beneath, slender midrib and 2–5 pairs of ascending secondaries often slightly impressed above, very prominent beneath; inflorescence axillary, abundant, showy; peduncles slender, 4–14 cm. long, erect or ascending, 2 or 3 per node, densely appressed-strigose; heads hemispheric or ovoid-cylindric, 1–2 cm. long and wide, densely many-flowered, not greatly elongating after anthesis; bractlets broadly ovate, pronouncedly imbricate, foliaceous, lowest to 12 mm. long and 8 mm. wide, attenuate-acuminate at apex, truncate at base, sessile, entire, appressed-strigose like leaves but hairs somewhat shorter, often more or less resinous-punctulate, often red; calyx globose, about 1 mm. long, sericeous, margin truncate and setose-fimbriate or ciliate; corolla equaling or slightly surpassing bracts or somewhat shorter than lowest bracts, opening yellow, quickly turning white with a yellow or orange center, fading violet, blue or purple in age, tube about 5 mm. long, strongly curved, very densely puberulent or short-pubescent outside, lower third narrow-cylindric, mouth slightly contracted, larger lip fan-shaped, about 2 mm. long and 3 mm. wide, three or four times as large as lobes on smaller lip, margins slightly erose; ovary broadly oblong, light brown, glabrous; stigma lateral, oblong-expanded, about half as long as the style; fruit globular, crustaceous, entirely enclosed in fruiting calyx, with a contracted, flattened, 2-winged base.

In sandy alluvial soils in canyons, valleys, fields, arroyo margins, and meadows, mostly in moister areas than those in the Sonoran Desert, ascending to 6,200 feet, from Sonora and Chihuahua to Tamaulipas and south to Guatemala; also from Colombia and Venezuela south to Peru and Argentina.

**4. Lantana macropoda** Torr., Bot. Mex. Bound. 127, 1859

Aromatic shrub, 1.5 m. tall or less; stems slender, tetragonal, simple or branching, gray, more or less strigose with appressed whitish hairs, glabrescent in age; branches slender, tetragonal, densely white-strigose, especially on younger parts; nodes annulate; principal internodes 0.5–5 cm. long; leaves decussate-opposite; petioles slender, 2–10 mm. long, densely white-strigose with antrorse hairs; blades thin-chartaceous, ovate or lanceolate, lighter green beneath, 5–35 mm. long, 2.5–20 (mostly 6–15) mm. wide, sharply acute at apex, cuneately attenuate to gradually cuneate-acute at base, sharply serrate from apex almost to base with rather coarse, antrorse, often rather spreading teeth, densely white-strigose above, more densely strigose-pubescent with whitish or yellowish hairs and densely glandular-punctate beneath; midrib flat or subimpressed above, subprominent beneath; secondaries 3–5 per side, obscure or subimpressed above, prominulous beneath, extending to the sinuses between the teeth but not into the abruptly acute teeth themselves; inflorescence axillary, 1 or 2 per node; peduncles slender, tetragonal, often sulcate, 1–10 cm. long, white-strigose; heads densely many-flowered, at first hemispheric, later elongating to 2 cm., 1–1.5 cm. wide; bractlets broadly ovate, lowermost to 7 mm. long and 4 mm. wide, long-acuminate at apex, densely white-strigose, abundantly glandular-punctate, closely imbricate; corolla white or pink, varying to light lavender or purple, often with a yellow "eye," tube 4–5 mm. long, limb about 3 mm. wide; fruit white, translucent.

In clay, clay-loam, and black soil on gravelly hills, dry rocky slopes, mesas, flats, and limestone bluffs, chiefly Lower Sonoran Zone, Arizona to Texas south to Sonora and San Luis Potosí.

## 4. ALOYSIA Ortega ex Pers., Syn. Pl. 2: 139, 1806

Usually sweet-aromatic, occasionally polygamodioecious shrubs. Leaves decussate-opposite or whorled, simple, estipulate, deciduous, blades entire or toothed, not pinnatifid or lobed. Inflorescence centripetal (racemose), in form of axillary, usually numerous, slender, mostly elongate, loosely flowered spikes or racemes. Flowers perfect or imperfect, hypogynous, each subtended by a narrow, inconspicuous or conspicuous bractlet. Calyx gamosepalous, tubular-campanulate, or sometimes 2-parted, usually spreading-hirsute or -villous, not inflated at maturity, its tube angled, not flattened, its limb 4-lobed, lobes slender, nearly equal. Corolla gamopetalous, zygomorphic, hypocrateriform, 2-lipped, upper lip and lobes of lower lip nearly equal. Stamens 4, didynamous, usually inserted at about middle of corolla tube, included, sometimes absent; anthers 2-celled, unappendaged, thecae parallel. Pistil 1, included. Ovary 2-celled, each cell 1-ovulate; ovules erect. Fruit small, dry, composed of 2 thin-walled nutlets. Seeds without endosperm.

Spikes very slender, 4–9 mm. in diameter; bractlets small and very inconspicuous, 1–2 mm. long, 1 mm. or less wide; leaves small, 2–27 mm. long, 2–13 mm. wide; calyx 4-lobed:
  Leaf blades ovate or subrotund, many-toothed, densely tomentose beneath ........... 1. *A. wrightii*
  Leaf blades oblong, elliptic, obovate, or obovate-oblong, entire or few-toothed, densely puberulent or strigillose-hispidulous beneath:
    Leaf blades narrowly oblong or elliptic, entire, appressed-puberulent beneath .......... 2. *L. lycioides* var. *lycioides*

Leaf blades more broadly elliptic, obovate, or oblong-elliptic, larger ones usually 2–8-toothed, densely strigillose beneath and shortly hispidulous along venation ....... ............................................ 2a. *A. lycioides* var. *schulziae*

Spikes very thick, about 1.5 cm. in diameter; bractlets large and very conspicuous, 7–8 mm. long, about 3 mm. wide; leaves 2–8 cm. long, to 2.6 cm. wide; calyx 2-parted ......... ............................................................ 3. *A. barbata*

## 1. **Aloysia wrightii** (A. Gray) Heller, Muhlenbergia **1:** 147, 1906

*Lippia wrightii* A. Gray, Amer. Jour. Sci. II, **16:** 98, 1853.

Diffuse branching shrub, to 1.5 m. tall and about as wide-spreading; stems slender or to about 2.5 cm. in diameter, brittle, wood bright yellow, thin membranous bark readily peeling off in long strips; branches and twigs slender, numerous, sharply tetragonal, densely grayish-puberulent, margined on angles, margins eventually peeling off; nodes not annulate; principal internodes 4–42 mm. long, usually much abbreviated; leaves opposite, numerous, often crowded, strong-scented; petioles very slender, 1–4 mm. long or obsolete, densely appressed-puberulent with sordid, grayish or yellowish hairs; blades usually small, ovate or subrotund, 2–15 mm. long, 2–13 mm. wide, rarely slightly larger, rounded at apex, regularly crenulate or crenate-serrate, with small rounded teeth, varying from acute or obtuse to rounded or truncate at base, subbullate and strigillose-scabrous above, very densely tomentellous with short, appressed, sordid, grayish or yellowish hairs, more or less resinous-glandular, slender midrib, secondaries, and abundant veinlet-reticulation usually deeply impressed above and hidden by tomentum beneath; inflorescence axillary, 2 per node, surpassing the subtending leaves, very numerous; spikes slender, 1–4 cm. long, 4–5 mm. in diameter, densely many-flowered; peduncles very slender or subfiliform, 1–10 mm. long, densely puberulent; rachis slender, or subfiliform, densely puberulent; bractlets lanceolate, 1.5–2 mm. long, inconspicuous, long-acuminate at apex, densely puberulent; calyx tubular, about 2 mm. long, very densely spreading-hirsute with sordid gray hairs, its rim 4-lobed, with very small, equal lobes; corolla white, tube slightly longer than calyx, glabrous inside, limb about 2 mm. wide, puberulent outside.

On various substrata and exposures common in the Upper Sonoran Zone and ascending to Transition Zone, from southern Nevada to Texas and southeastern California south to Durango and Zacatecas.

## 2. **Aloysia lycioides** Cham., Linnaea **7:** 237, 1832, var. **lycioides**

*Verbena gratissima* Gill. & Hook. ex Hook., Bot. Misc. **1:** 160, 1830.
*Lippia lycioides* Steudel, Nom. ed. 2, **2:** 54, 1841.
*Aloysia floribunda* Mart. & Gal., Bull. Acad. Brux. **11**$^{2}$**:** 320, 1844.
"*Lippia ligustrina*" and "*Aloysia ligustrina*" of American authors.

Slender shrub, 2–3 m. tall, greatly branched, often with a spread of 2 m.; roots woody; bark pale; branches and branchlets slender, gray, subtetragonal, stiff, pulverulent-puberulent, incanous, often spinescent at tips; wood yellow; nodes not annulate; principal internodes 0.5–3 cm. long; leaves opposite, often with a fascicle of smaller ones in their axils, crowded; petioles subfiliform, 1–3 mm. long, and incanous-pulverulent or puberulent, or obsolete; blades chartaceous or submembranous, narrowly oblong or elliptic, varying to lanceolate-oblong, 3–27 mm. long, 2–8 mm. wide, obtuse or acute (sometimes mucronulate or emarginate) at apex, entire, attenuate into petiole at base, minutely strigillose and scabrellous-pustulate above, very densely puberulent-pulverulent, incanous, and glandular-punctate beneath; midrib very slender, flat or subimpressed above, subprominent beneath; secondaries and veinlet-reticulation rarely visible; inflorescences 1 or 2 per node, greatly surpassing leaves, 2–7 cm. long, rather densely many-flowered, slender, usually 5–9 mm. in diameter; peduncles subfiliform, 5–15 mm. long, densely puberulent; rachis subfiliform, densely short-pubescent; bractlets very small and inconspicuous, lanceolate, 1–1.5 mm. long, usually less than 1 mm. wide, acuminate at apex, caducous, densely puberulent; flowers strongly odorous, with fragrance of vanilla; calyx tubular-campanulate, 2.5–3 mm. long, 4-costate, densely white-hirsute with stiff spreading hairs, glandular beneath hairs, rim 4-lobed, lobes linear-subulate, unequal; corolla white or tinged with violet, tube slightly longer than calyx, glabrous outside, limb about 3 mm. wide, loosely villous in throat.

In rocky creek beds, dry rocky arroyos, and mesquite valleys, Lower Sonoran to Arid Transition Zones, Sonora to New Mexico, Texas, and south to Oaxaca; also from Brazil and Paraguay to Uruguay and Argentina.

**2*a*. Aloysia lycioides** var. **schulzae** (Standley) Moldenke, Phytologia **2:** 464, 1948

*Lippia ligustrina* var. *schulzii* Standley, Field. Mus. Bot. Ser. **4:** 256, 1929.
*Aloysia ligustrina* var. *schulzii* Moldenke, Phytologia **1:** 95, 1934.

This variety differs from var. *lycioides* in its leaf blades being broader, elliptic or obovate to oblong-elliptic, 5–18 mm. long, 5–7 mm. wide, or larger, obtuse or acute at apex and base, entire or larger ones with 1–4 rather large and salient teeth on each side, tuberculate or strigillose and scabrous above, densely strigillose beneath with longer and more distinct, easily visible hairs, often hispidulous on larger venation with short, ascending, and rather rigid hairs, usually revolute-margined, midrib and secondaries usually impressed above.

In sandy or alluvial soil, arroyos, rocky or gravelly hillsides, low moist spots in deserts, and rocky valleys, Lower Sonoran Zone, central Sonora to Texas and Nuevo León.

**3. Aloysia barbata** (Brandegee) Moldenke, Phytologia **1:** 416, 1940

*Lippia barbata* Brandegee, Proc. Calif. Acad. II, **2:** 196, 1889.
*Lippia montana* Brandegee, op. cit. **3:** 163, 1891.

Small, intricately or openly branched shrub, 1–3.5 m. tall, more or less rough-pubescent throughout; bark gray; stems and branches obtusely tetragonal, densely short-pubescent or puberulent, often with much longer, stiffer, white, antrorse, and substrigose hairs interspersed, much less pubescent or glabrate in age, brownish, sometimes slightly flattened at nodes; principal internodes mostly short, 5–25 mm. long; leaves opposite, usually clustered at ends of short branches, odorous; petioles slender, 3–10 mm. long on mature leaves, slightly margined above, sparsely villous, often resinous-dotted like twigs; blades thick, ovate, lanceolate-ovate, oblong- or round-ovate, 1.5–8 cm. long, 15–26 mm. wide, acute, obtusish or rounded at apex, short-cuneate to acute or rounded at base, serrate-dentate or crenate, teeth sometimes bristle-tipped or appressed and blunt, rugose and very scabrous above, venation usually impressed above, prominent beneath; inflorescence racemose or appearing subcapitate, short-pedunculate; peduncles slender, 5–25 mm. long, densely spreading-pubescent or hirsutulous, resinous-punctate; racemes spikelike, 1.5–2 cm. in diameter, 1–3 cm. long, lax; bractlets herbaceous, large, conspicuous, ovate to broadly ovate to oblong or obovate, lowest 7–9 mm. long, about 3 mm. wide, those of staminate flowers about half as long, all 3–5-nerved, acuminate at apex, densely sericeous-pubescent and resinous-punctate on outer surface, densely puberulent on inner surface; flowers polygamodioecious or dioecious; pedicels about 1 mm. long, glandular-puberulent; calyx 2-parted,

1.5–2 mm. long, very densely villous-hirsute on outer surface with erect, rather soft, white hair and densely resinous-granular, segments weakly navicular, acute or obtusely 2-dentate at apex; corolla 7–8 mm. long, minutely puberulent or glabrate outside, puberulent within, limb 4 mm. wide, larger lobes about 3 mm. wide and 2 mm. long, all erose-undulate margined, puberulent on outer and glabrous on inner surface; stamens inserted practically at mouth of corolla tube; style capillary, about 3 mm. long, glabrous; stigma lateral, oblique, about 0.7 mm. long; ovary small, subglobose or turbinate, about 0.6 mm. long and wide, glabrate, situated on a small disk; fruits globose, about 2 mm. long, conspicuously white-hirsute; nutlets easily separable.

On shady canyon slopes and washes in the foothills, Lower Sonoran Zone, central Baja California.

## 5. **PHYLA** Lour., Fl. Cochinch. 66, 1790

Perennial procumbent or creeping herbs with trailing or ascending stems, rarely suffrutescent to shrubby, subglabrate or appressed-strigose with malpighiaceous hairs. Leaves decussate-opposite, usually variously dentate (except at base), flat or pinnately plicatulate above. Inflorescence spicate, axillary; spikes cylindric, densely many-flowered, usually greatly elongate in fruit, solitary, paired or in threes in leaf axils. Flowers small, sessile, borne singly in axils of small, cuneate-obovate or flabelliform, closely imbricate bractlets, not at all 4-ranked; otherwise with characters as in *Lippia.*

Plants usually creeping and widespreading, open in growth, often with ascending branches, usually green and only finely strigillose; leaves mostly large, cuneate-spatulate, spatulate, or obovate, thick-textured, usually abundantly toothed .................. 1. *P. incisa*

Plants usually densely matted, more or less densely canescent-strigose throughout; leaves mostly very small and cuneiform, usually thin-textured and few-toothed, sometimes entire or subentire ........................................ 2. *P. nodiflora* var. *rosea*

### 1. **Phyla incisa** Small, Fl. SE. U.S. 1012, 1337, 1903

(For synonymy see Moldenke, in Lundell, Flora of Texas **3:** 57, 1942.)

Stems mostly prostrate, often rooting and swollen at nodes, slender, obtusely tetragonal, simple or branched, often purplish, densely or sparsely appressed-strigillose with small white hairs; branches often numerous, varying from decumbent to ascending or erect, mostly very slender, tetragonal, abundantly appressed-strigose with small white hairs; principal internodes 1.8–5 cm. long; leaves often with a cluster of smaller ones in their axils, variable in shape and size; petioles usually obsolete, if distinguishable only 1–3 mm. long; blades firmly chartaceous, often thick-textured, uniformly green on both surfaces, narrowly oblong or cuneiform to broadly obovate, 1–5 cm. long, 2–15 mm. wide, acute to obtuse or rounded at apex, sparsely or abundantly dentate above middle, cuneate, rather sparsely or densely appressed-strigillose with small, inconspicuous, white hairs on both surfaces; inflorescences 1 or 2 per node, erect; peduncles slender, tetragonal, 2–9 cm. long, appressed-strigillose; heads at first globose, later cylindric and elongating to 3 cm. long, 5–7 mm. wide; bractlets obovate, 2–3 mm. long, acute at apex, abundantly white-strigillose; corolla white, with a yellow center, tube 2–2.5 mm. long, limb about 2 mm. wide; fruit dry, obovoid.

In open ground, clay or sandy flats, along seashores and watercourses, from Colorado, Oklahoma, and Missouri to Vera Cruz and Mexico, D.F., west to southern California and Baja California.

**2. Phyla nodiflora** (Bauh.) Greene var. **rosea** (D. Don) Moldenke, Phytologia **2:** 22, 1941

*Zappania nodiflora* var. *rosea* D. Don, in Sweet, Brit. Flow. Gard. II, **3:** *pl. 225,* 1834.
*Lippia uncinulifera* Nees ex Walp., Nova Acta Acad. Leop.-Carol. **19** (suppl. 1): 337, 1843.
*Lippia filiformis* Schrad. ex Schau., in A. DC., Prod. **11:** 585, 1847, as a synonym.
*Lippia nodiflora* Cham. ex Schau., loc. cit., as a synonym.
*Lippia litoralis* Phil., Anal. Univ. Chile **90:** 624, 1896.
*Lippia nodiflora* var. *pusilla* Briq., Arkiv Bot. **2** (10): 19, 1904.
*Lippia nodiflora* subsp. *canescens* f. *pseudo-sarmentosa* Reyn., Bull. Soc. Bot. Fr. **62:** 205, 1915.
*Phyla nodiflora* var. *pusilla* Moldenke, Phytologia **1:** 170, 1935.

Stems prostrate, usually densely matted, creeping, rooting at nodes; branches slender, numerous, short, procumbent, densely and conspicuously canescent-strigose; petioles slender, 2–8 mm. long or rarely obsolete, sometimes cuneate-winged and appearing as though a part of blade; blades mostly very thin-textured and small, cuneiform to oblanceolate or elliptic, occasionally narrowly oblong, usually 5–28 mm. long, 2–9 mm. wide, usually few-toothed toward the apex, sometimes entire or subentire, usually densely and conspicuously canescent-strigose on both surfaces, acute or obtuse at apex, long-cuneate at base; peduncles slender, 1 or 2 per node, usually 1–5 cm. long or less, rarely elongate to 10 cm., densely canescent-strigose; spikes at first globose, cylindric and elongate to 3 cm. in age and 6–9 mm. in diameter; bractlets obovate or subrhomboid-cuneate, subequaling corolla tube, mucronate-acuminate at apex, densely canescent-strigose; calyx hyaline-membranous, flattened, about equaling corolla tube, deeply 2-cleft, slightly 2-carinate, keels puberulent, lobes lanceolate; corolla rose-purple or white, often tinged with lavender or with a yellow "eye," slightly surpassing bractlets, limb exigous, slightly strigillose outside.

In dry, sandy or adobe soil in nearly every type of site as a naturalized introduction from California to Texas and in Florida and south to Guerrero and Vera Cruz. A native of temperate South America. Common in the Imperial Valley region.

## 6. LIPPIA L., Sp. Pl. 633, 1753

Erect bushes, shrubs, or subshrubs, glabrous or variously pubescent with simple hairs or often hirsute or tomentose. Leaves decussate-opposite or ternate, rarely alternate, entire, toothed, or lobed, flat or rugose above. Inflorescence spicate, solitary or fascicled and axillary, or aggregate into terminal corymbs or panicles, spikes mostly contracted into heads or cylindric. Flowers small, sessile, borne singly in axils of large, ovate or lanceolate bractlets, these more or less 4-ranked. Calyx small, membranous, ovoid-campanulate or compressed, often 2-carinate or 2-alate, its rim 2- or 4-fid, or 4-dentate. Corolla tube cylindric, straight or incurved, slender, slightly exserted from calyx or rarely elongate, equal in diameter throughout or ampliate above; corolla limb oblique, spreading, somewhat 2-lipped, 4-parted, lobes broad, often retuse at apex, posterior one entire, emarginate, or bifid to about middle, lateral ones exterior, anterior one often larger. Stamens 4, didynamous, inserted at about middle of corolla tube, included or slightly exserted; anthers ovate, unappendaged, cells parallel. Ovary 2-celled, each cell 1-ovuled. Style often short; stigma rather incrassate, oblique or recurved. Ovules basal and erect or affixed laterally near base. Fruit small, dry, included by calyx and sometimes adnate to it, dividing into 2 pyrenes at maturity, pericarp papery and hard, exocarp membranous and rarely distinct from pyrenes. Seeds without endosperm.

Spikes 4–9 mm. long, rarely to 13 mm. long .................... 1. *L. palmeri* var. *palmeri*
Spikes 1–2.5 cm. long, rarely shorter .......................... 1a. *L. palmeri* var. *spicata*

**1. Lippia palmeri** S. Wats., Proc. Amer. Acad. **24:** 67, 1889, var. **palmeri**

A low, rounded, bushy, much-branched, and twiggy shrub, 0.5–2 m. tall; branches very slender, short, irregular, tetragonal, appressed-pubescent; branchlets and twigs very numerous; nodes annulate; principal internodes 0.5–4 cm. long; leaves richly aromatic, with a strong, sagelike odor; petioles filiform, 1–5 mm. long, appressed-pubescent; blades ovate or rounded-ovate to elliptic, 1–2.5 cm. long, 3–10 mm. wide, obtuse or rounded to subacute at apex, usually decurrent into petiole at base, subcrenate or coarsely crenate, scaberulous or rugose above and finely substrigose, more densely appressed-pubescent and resinous-punctate beneath, slender midrib and about 3 pairs of short secondaries often impressed above and prominent beneath; peduncles solitary or in pairs in the leaf axils, very short and inconspicuous, 1–3 mm. long, substrigose; spikes capitate, 4–13 (usually only 4–9) mm. long, often few-flowered and loose; bractlets often pink, decussate, lower united to middle, upper distinct, ovate, 3–4 mm. long, acute or acuminate, densely appressed-pubescent, more or less resinous-dotted; calyx thin, not carinate; corolla white or cream-colored, turning yellow, pink or rose, salverform, its tube about 4 mm. long, densely appressed-pubescent and more or less resinous-dotted outside, limb about 3 mm. wide, lobes puberulent toward base outside, otherwise glabrous.

In gravelly or basaltic, clay soils and rocky sand or sandy bottom soil, on dry open hills and mesas, rocky benches, and the drier lower slopes of hills, Lower Sonoran Zone, Magdalena Island and the adjacent peninsula, Baja California, to central Sonora and Sinaloa.

**1*a*. Lippia palmeri** var. **spicata** Rose, in Vasey & Rose, Contr. U.S. Nat. Herb. **1:** 75, 1890

Differs from var. *palmeri* in having its spikes 1–2.5 cm. long, more compact and conspicuous.

In habitats similar to those of var. *palmeri,* Lower Sonoran Zone, in Baja California, Tiburón Island, and Sonora.

## 7. BOUCHEA Cham., Linnaea 7: 252, 1832

Annual or perennial densely pubescent or glabrous herbs or low shrubs. Leaves decussate-opposite or ternate, usually petiolate, sometimes sessile; blades usually serrate or serrate-crenate, rarely incised, dissected, or entire. Inflorescence racemose, rarely spicate, terminal (seldom axillary), elongate, loosely to densely many-flowered, bracteolate, indeterminate and centripetal. Flowers solitary in axils of bractlets, usually pedicellate; bractlets subulate or lanceolate, sometimes enlarged and leaflike. Calyx persistent, tubular, 5-ribbed, ribs terminating in 5 more or less unequal teeth. Corolla gamopetalous, infundibular, or hypocrateriform, zygomorphic; tube cylindric, straight and erect or curvate; limb oblique, spreading, unequally 5-lobed, 2 posterior lobes shorter than 3 anterior lobes. Stamens 4, didynamous, included; posterior pair inserted at about middle of corolla tube and anterior pair inserted somewhat above middle; filaments short; anthers ovate to subcordate, 2-celled. Pistil single, compound. Style filiform; stigma 2-lobed, anterior lobe club-shaped, posterior lobe toothlike, on a level with anterior pair of stamens. Ovary 2-celled, oblong, each cell 1-ovulate. Fruit dry, schizocarpous, linear, beaked, included in calyx or exserted, separating into 2 cocci at maturity; cocci completely separate or coherent at base, dorsal surface more or less reticulate-ridged, commissural surface plane, furrowed or ridged, sometimes a little roughened.

Leaf blades pinnately dissected nearly to midrib .......................... 1. *B. dissecta*
Leaf blades merely dentate or crenate .................... 2. *B. prismatica* var. *brevirostra*

**1. Bouchea dissecta** S. Wats., Proc. Amer. Acad. **24:** 68, 1889
(For synonymy see Fedde, Rep. Spec. Nov. **49:** 124, 1940.)

Erect annual herb, to 6.5 dm. tall, usually less than 55 cm. tall; stems slender, acutely or obtusely tetragonal (subterete at the very base), often sulcate, densely to minutely canescent-puberulent or glabrate below; branches few, divaricate, opposite, slender, tetragonal, straw-colored or purplish, canescent-puberulent; nodes annulate; principal internodes 2–7 cm. long; leaves opposite, often with short branchlets in their axils; petioles slender, 3–23 mm. long, minutely puberulent; blades thin-chartaceous, light green on both surfaces, ovate, 2–7 cm. long, 1–4.5 cm. wide, deeply dissected or pinnately cleft with 8–11 divisions, each division reaching almost to midrib and entire or again 1–3-toothed, -lobed or -dissected, minutely strigillose or puberulent on both surfaces or glabrate; midrib and veinlets slender and delicate; racemes slender and elongate, terminal or occasionally terminating short axillary branchlets, spikelike, 10–30 cm. long, less than 1 cm. wide, loosely many-flowered; peduncles short, slender, tetragonal, puberulent; rachis slender, straight or slightly flexuous, tetragonal, puberulent; bractlets linear or subulate, 1.5–3 mm. long; pedicels to 1.5 mm. long in fruit; flowers closely appressed to rachis, slightly overlapping, short-pedicellate or subsessile; calyx narrow, thin-textured, slightly curved, 7–8 mm. long, about 1.25 mm. wide, 5-costate, 2-lipped, slightly pubescent or glabrous, rim short-toothed or 5-mucronulate; corolla pink or white, tube about 8 mm. long, glabrous, lobes spreading, 2.5–3.75 mm. long, slightly venose; cocci curved, 10–12 mm. long, beak 3.5–4 mm. long, blunt at apex, narrowed at middle, glabrous.

In shade on llanos, in pebbly, slightly red soil and fields, and on rocky ridges, ledges, and hill slopes, Lower Sonoran to Tropical Zones, central Sonora to Guerrero.

**2. Bouchea prismatica** (L.) Kuntze var. **brevirostra** Grenz., Ann. Mo. Bot. Gard. **13:** 80, *pl. 9, figs. 9–12; pl. 12, fig. 30,* 1926
*Lomake brachiata* Raf., Aut. Bot. 73, 1840.
*Stachytarpha bifurca* Benth. ex Schau., in A. DC., Prod. **11:** 572, 1847, as a synonym.
(For additional synonymy see Fedde, Rep. Spec. Nov. **49:** 104, 1940.)

Erect or sprawling herb, 10–40 cm. tall; stems and numerous ascending branches slender, obtusely tetragonal, medullose, minutely puberulent throughout, often more densely so and canescent on younger parts, often sulcate; nodes annulate; leaves opposite, often with several small ones on short branchlets in their axils; petioles slender, 11–25 mm. long, short-pubescent or puberulent; blades thin-chartaceous, somewhat darker above than beneath, mostly elliptic, 1–4 cm. long, 8–30 mm. wide, often falcate, obtuse or rounded at apex, coarsely but regularly serrate, acute or obtuse and entire at base, rather densely short-pubescent on both surfaces when immature, sparsely so or glabrate in age; spikes terminal and terminating long or short, axillary branches which often far surpass terminal spike, erect, 5–15 cm. long, about 1 cm. wide (or less) in anthesis, many-flowered; peduncles slender, short, puberulent; rachis slender, angled, often slightly flexuous, minutely puberulent; flowers somewhat spreading in anthesis, closely appressed to rachis in bud and fruit, overlapping; bractlets lanceolate-subulate, 4–6 mm. long, puberulent; pedicels 1 mm. long or less; calyx narrow, slightly curved, thin-textured, 5–7.5 mm. long, about 1.25 mm. wide, minutely appressed-puberulent, slightly plicate, translucent between ribs, rim 5-mucronulate; corolla from deep or light purple to rose-red, red-violet, light violet, lilac, lavender, pink or blue, tube 8–10 mm. long, glabrous, lobes spreading, suborbicular, 1–2.5 mm. long and wide, rounded at apex; cocci about 7.5 mm. long, beak only about 0.5 mm. long, somewhat curved, obtuse or emarginate at apex, often with persistent style base attached.

In fields, clearings, llanos, and waste places, on open stony hillsides, along roadsides, and among shrubbery in sunny barrancas, Sonoran Zones, from western Texas and southern Arizona to El Salvador; also in St. Thomas, Barbados, Curaçao, and Colombia.

## 8. **PRIVA** Adans., Fam. Pl. **2:** 505, 1763

Herbaceous caulescent perennials, harshly pubescent throughout. Leaves decussate-opposite or subopposite, thin-membranous. Stems, branches, and branchlets more or less tetragonal, often decumbent. Inflorescence terminal and axillary, indeterminate, racemiform, often subspicate in anthesis, narrow, pedunculate, erect or subflexuous. Flowers small, solitary in axil of a small bractlet, arranged in a spirally alternate or pseudosecund manner on elongate rachis, never whorled. Calyx tubular in anthesis, 5-ribbed, slightly 5-plicate, slightly zygomorphic or nearly actinomorphic, terminating in 4 short, equal or subsequal teeth, persistent, enlarging with and investing fruit. Corolla more or less zygomorphic, hypocrateriform or infundibular, surpassing calyx; tube straight or somewhat curved, more or less ampliate at apex; limb spreading, oblique, more or less bilabiate, abaxial lip 3-lobed with 1 large central and 2 medium-sized lateral lobes, axial lip 2-lobed, with usually very small lobes. Fertile stamens 4, didynamous, upper pair better developed and usually inserted slightly above middle of corolla tube, lower pair inserted approximately at middle of tube, all included or equaling tube; filaments slender; anthers erect, ovate or oblong, 2-celled, dorsifixed at or below middle, introrse; connective usually conspicuously thickened and sagittate, unappendaged; thecae parallel or divergent at base, opening by a lateral longitudinal slit; fifth (posterior) stamen reduced to a minute staminode or absent. Style usually equaling lower stamens, 2-lobed at apex, anterior lobe longer and recurved or erect, stigmatiferous at apex, posterior lobe reduced, very minute and toothlike, not stigmatiferous. Ovary 4-celled or (by abortion) 2-celled; ovules basal, erect, anatropous, solitary in each cell. Fruit a dry, often woody schizocarp included by fruiting calyx, composed of 2 usually similar, 2- (or by abortion 1-) celled cocci, which separate easily on maturity; pericarp hard, dorsal surface echinate, scrobiculate or ridged, commissural surface excavated, concave or plane. Seeds without endosperm.

### 1. **Priva lappulacea** (L.) Pers., Syn. Pl. **2:** 139, 1806

(For synonymy see Moldenke, in Lundell, Flora of Texas **3:** 71, 1942.)

Perennial herb, to 1 m. tall; stems spreading and decumbent or procumbent; branches usually acutely tetragonal, often sulcate, pilose-pubescent with curved or uncinate hairs, glabrescent in age below; larger nodes usually annulate with a band of longer hairs; petioles slender, 8–30 mm. long, more or less pilose with scattered whitish hairs; blades ovate, 1.4–14.5 cm. long, 0.9–8.5 cm. wide, acute or acuminate at apex, uniformly and coarsely serrate, mostly subtruncate or subcordate (or acute when young) at base, pilose or strigose above with scattered, whitish, bulbous-based hairs, pilose and puberulent beneath with scattered hairs; inflorescence terminal, 4.5–21 cm. long, 5–15 mm. wide, flowers many, loosely alternate; peduncles slender, 0.8–5.8 cm. long, puberulent-pilose, uppermost leaves at base of peduncle often much reduced; bractlets usually surpassing pedicels at anthesis; calyx oblong-campanulate, 2–3.1 mm. long, about 2.3 mm. wide, densely short-tomentose with short uncinate hairs and interspersed with longer straight hairs, obscurely 5-ribbed only toward apex, its rim 5-apiculate; corolla usually blue, pink, violet or purple, occasionally lavender or white, with a few scattered hairs on both surfaces, tube about 3.6 mm. long on axial and 3.3 mm. on abaxial side; fruiting calyx broadly ovate, thin-membranous, conspicuously inflated, enclosing fruit and short-beaked at apex, 5–7 mm. long, 3–4.5 mm. wide, densely hispidulous with uncinate whitish hairs; cocci glabrous, each about 3 mm. long and 2 mm. wide, dorsal surface echinate with 2 parallel, longitudinal, marginal rows of short, straight, sharp spines.

In thickets and weedy flats, along small arroyos, and on canyon floors; common as a weed in clearings and along streets, Sonoran to Tropical Zones, from Florida and Texas through the West Indies, Mexico, and Central America to Peru, Brazil, and Bolivia; naturalized in Java. Occurring in the Sonoran Desert near Mátape, about 40 miles northeast of Hermosillo, Sonora.

## 9. **DURANTA** L., Sp. Pl. 637, 1753

Glabrescent or pubescent, often spinose, shrubs and trees. Leaves decussate-opposite or whorled, entire or dentate. Inflorescence indeterminate, racemiform, elongate, mostly terminal, rarely axillary. Flowers small, pedicellate, each borne in axil of a small bractlet. Calyx tubular or subcampanulate, truncate, 5-costate and subplicate, each rib terminating in a short subulate tooth, posterior tooth smallest. Corolla hypocrateriform, its tube cylindric, straight or curved above, exserted from calyx; limb spreading, regular or oblique, 5-parted, mostly pubescent at mouth; lobes rounded, usually unequal, mostly pubescent on inner surface. Stamens 4, didynamous, included, inserted at or above middle of corolla tube; filaments very short; anthers sagittate, dorsifixed, erect. Ovary more or less completely 8-celled, composed of four 2-celled carpels, each cell 1-ovulate. Style terminal, shorter than or equaling lower stamens; stigma obliquely subcapitate, shortly and unequally 4-lobed. Fruiting calyx accrescent, flask-shaped, usually surpassing and closely appressed to fruit but not coalesced with it, usually coarctate-rostrate at apex. Fruit drupaceous, mostly completely included by mature calyx; exocarp fleshy; endocarp hard; pyrenes 4, each 2-celled and 2-seeded.

### 1. **Duranta repens** L., Sp. Pl. 637, 1753

(For synonymy see Moldenke, in Lundell, Flora of Texas **3:** 76, 1942.)

An extremely variable and polymorphic shrub or small tree to 6 m. tall; branches slender, often drooping or trailing, unarmed or spiny; branchlets tetragonal; leaves numerous, ovate-elliptic, oval or obovate, 1.5–5 cm. long, obtuse to apiculate or acuminate at apex, entire or serrate above middle, cuneate at base, glabrate on both surfaces; racemes terminal and axillary, 5–15 cm. long, loosely many-flowered, erect or usually recurved, often paniculate; bractlets minute, occasionally subfoliaceous; pedicels 1–5 mm. long; calyx tubular, 3–4 mm. long, angled, its teeth triangular at base, subulate at apex, shorter than tube; corolla blue, lilac or lavender, sometimes purple, tube surpassing calyx, limb 7–9 mm. wide; fruit yellow, globular, 7–11 mm. in diameter, completely enclosed by accrescent yellowish calyx, which is prolonged into a curved beak.

Found throughout subtropical and tropical America south to Argentina; also introduced into the Old World tropics and widely cultivated. In our area El Taste, Baja California.

## 10. CITHAREXYLUM L., Sp. Pl. 625, 1753

Trees or shrubs; branches and branchlets tetragonal, sometimes spiny; leaf scars mostly large, corky, and elevated. Leaves decussate-opposite or whorled, deciduous, estipulate, entire or dentate, usually bearing a pair of prominent glands at base of blade. Inflorescence indeterminate, racemiform or spicate, axillary and terminal, mostly elongate and many-flowered, rarely reduced to only a few flowers, erect or nutant, mostly simple, occasionally sparsely branched. Flowers small, each subtended by a tiny inconspicuous bractlet. Calyx tubular or cyathiform, thin, accrescent, its rim truncate or 5-toothed or -lobed. Corolla infundibular or hypocrateriform, mostly yellow or white; tube narrow-cylindric and regular; limb spreading, 5-parted, with broad, slightly irregular lobes, the 2 hindermost outside in bud, mostly more or less pubescent in throat. Stamens 4, didynamous, inserted at or above middle of corolla tube, included, a fifth stamen represented by a very rudimentary staminode; filaments very short; anthers ovate or sagittate, introrse, erect, with 2 parallel thecae opening by longitudinal slits and a thickened connective which often surpasses thecae. Style terminal, included, often thickened toward apex; stigma very shortly bifid. Ovary perfectly or imperfectly 4-celled, composed of 2 carpels, each cell containing 1 lateral anatropous ovule. Fruiting calyx conspicuously enlarged and indurated, cupuliform or patelliform, shorter than fruit. Fruit drupaceous, with juicy exocarp and hard endocarp, with two 2-celled and 2-seeded pyrenes, which are often separated by a median fissure.

Leaf blades narrowly oblong or elliptic, entire .............................. 1. *C. shrevei*
Leaf blades flabelliform, dentate .................................. 2. *C. flabellifolium*

### 1. Citharexylum shrevei Moldenke, Phytologia **1:** 415, 1940

Shrub; branches and branchlets slender, gray, subterete; twigs extremely slender and tenuous, white or very light gray, minutely puberulent with very short, white hair, obtusely tetragonal; nodes not annulate; principal internodes 4–25 mm. long; leaves opposite; petioles very slender and less than 1 mm. long or obsolete; blades chartaceous, gray-green on both surfaces, narrowly oblong or elliptic, 4–22 mm. long, 2.5–5.5 mm. wide, bluntly acute or obtuse at apex, entire and usually more or less revolute along margins, acute or subcuneate at base, strigillose-roughened on both surfaces with minute, whitish, bulbous-based hairs; midrib very slender, very slightly prominulous beneath; secondaries about 3 per side, very slender, usually indiscernible above and obscure beneath; vein and veinlet reticulation indiscernible; inflorescence and fruit not known.

Known only from the original collection made at Picu Pass, Sonora.

### 2. Citharexylum flabellifolium S. Wats., Proc. Amer. Acad. **24:** 67, 1889

Stiff, intricately branched, compact shrub, 1–3 m. tall and as broad; branches and branchlets medium-slender, widely ascending, subterete, gray, medullose, densely canescent-puberulent, glabrate in age, 4-canaliculate; twigs and young shoots light gray, acutely tetragonal, densely canescent-pubescent, often spine-tipped; nodes not annulate; principal internodes 4–20 mm. long; leaf scars small, on short, slender sterigmata; leaves opposite, but mostly clustered on extremely short, spur- or knoblike, lateral twigs; petioles slender, 2–4 mm. long, densely canescent-pubescent; blades thin-chartaceous, dark green and uniformly colored on both surfaces, nigrescent in drying, flabelliform, 6–16 mm. long, 7–17 mm. wide, rounded or subtruncate at apex, acutely narrowed to cuneate base, regularly dentate almost to middle with broad, more or less apiculate teeth, densely or sparsely puberulent on both surfaces especially along larger venation beneath; midrib slender, plane above, prominulous beneath; secondaries slender, about 3 pairs, ascending, not at all arcuate, each ending in a tooth-apiculation; veinlet reticulation indiscernible; racemes terminal or on short leafy spurs, 2.5–9 cm. long, 1–1.5 cm. wide, very loosely 5–15-flowered, flowers distant, alternate or subopposite, often reflexed, each subtended by a comparatively large bractlet; peduncles 5–11 mm. long, canescent-pubescent; rachis similar to twigs, canescent-pubescent; pedicels very slender, 1–2 mm. long, elongated to 4 mm. in fruit, canescent-pubescent; bractlets lanceolate, 3–6 mm. long, about 1.5 mm. wide, pubescent; prophylla linear, 1–3 mm. long, often several subtending each pedicel, pubescent; corolla pale purple or dark violet, about 12 mm. long, sweet-scented, yellowish tube shorter than calyx, tomentose within, broadly expanded limb with nearly equal, rounded lobes; fruiting calyx light, thin, slightly obvolute so as to include immature fruit, later splitting irregularly to base as fruit matures, sparsely puberulent in 5 parallel lines leading to 5 apiculations, splitting between these lines; fruit oblong-subglobose, 4–6 mm. long and wide, 2-lobed, glabrous, not fleshy, dull black, rugate-wrinkled in drying, juice staining a persistent black.

In coarse calcareous soil in lowlands, mesquite-grassland, gulches, and mountain ravines and about the summits of ocean bluffs, ascending to at least 500 feet, Lower Sonoran Zone, Baja California, Carmen Island, and Sonora.

## 11. VITEX L., Sp. Pl. 638, 1753

Trees or shrubs, glabrous to tomentose or villous throughout. Leaves decussate-opposite, palmately compound, mostly 3–7-foliolate, leaflets chartaceous or membranous, mostly petiolulate, entire or dentate, rarely incised. Inflorescence cymose; cymes short and dense or loosely divaricate, sessile or pedunculate in leaf axils or aggregate in terminal, racemiform or thyrsoid or laxly diffuse panicles or more rarely contracted into heads. Flowers perfect, more or less zygomorphic. Bractlets and prophylla small, mostly linear, sometimes longer than calyx. Calyx campanulate or rarely tubular-infundibular, 5-dentate or 5-fid or rarely 3-fid, teeth mostly slightly unequal. Corolla white, blue, violet, or yellowish, hypocrateriform, zygomorphic, its tube short or rarely elongate, cylindric, straight or slightly incurved; limb oblique, spreading, more or less bilabiate, upper lip bifid, lower lip trifid, 2 posterior lobes exterior and usually shorter, lateral lobes larger, anterior lobe inmost and largest, entire or emarginate. Stamens 4, didynamous, inserted in corolla tube, often exserted; anthers 2-celled, thecae distinct, subparallel, divergent or arcuate, attached near their apex, dehiscing by longitudinal slits. Pistil bicarpellary; style filiform, shortly bifid at apex; branches acute. Ovary at first imperfectly 2-celled, during anthesis usually 4-celled, cells 1-ovulate; ovules attached laterally at or above middle of cell. Fruiting calyx often accrescent, usually patelliform or very shallowly cupuliform, rarely subincluding fruit. Fruit drupaceous, more or less fleshy; endocarp hard, often horny, sometimes very much incrassate in relation to cells, 4-celled. Seeds obovate or oblong, erect, without endosperm.

Leaflets mostly 3; petioles to 5 cm. long, mostly pubescent or villosulous-tomentose ........ ........................................................... 1. *V. mollis*
Leaflets mostly 5; petioles to 10 cm. long, mostly merely puberulent ...... 2. *V. pyramidata*

**1. Vitex mollis** H. B. K., Nov. Gen. & Sp. **2:** 245, 1817
*Vitex lasiophylla* Benth., Bot. Sulph. 155, 1844.
*Cornutia ternata* Sessé & Moc., Pl. Nov. Hisp. II, **1:** app. 103, 1889.

Shrub or tree to 18 m. tall; trunk to 1.2 m. in diameter; branchlets obtusely tetragonal or subterete, gray-brown or buff, short-pubescent or puberulent, becoming subglabrate in age; twigs densely villous-tomentose when young, merely densely pubescent in age, pubescence cinereous, albidous, or flavescent, gray or brownish in age; principal internodes 1–6 cm. long; leaves 1–3-foliolate (rarely 5-foliolate); petioles 0.3–5 cm. long, densely villosulous-tomentose when young, becoming sparsely pubescent in age; leaflets usually unequal, 2 lateral ones (if present) much reduced or occasionally subequal, all short-petiolulate on densely pubescent, canaliculate and margined petiolules 1–9 mm. long, lateral petiolules usually much shorter or obsolete; blades rather firmly chartaceous, central one oblong, elliptic, or obovate, 2–14.5 cm. long, 1.3–5.8 cm. wide, obtuse or rounded at apex, often emarginate, rarely subacute, entire (or rarely serrate above middle on shoots), obtuse or acute at base, sometimes cuneate or slightly inequilateral, densely velutinous-villous or tomentulose, pubescence before and during anthesis flavescent, after anthesis cinereous, or brownish; inflorescence axillary, cymose, 1.5–6.5 cm. long, 3–15-flowered; peduncles 1–3.5 cm. long, flattened; pedicels obsolete or to 2 mm. long; bractlets linear, 2–5 mm. long, densely tomentose or pubescent; calyx 2–4 mm. long and wide, densely villous, its rim 5-toothed; corolla blue or white with a violet lip, or upper lip pale lilac and lower lip deeper lilac, purple streaked in throat, about 1 cm. long, densely pubescent on outer surface, limb about 1 cm. wide, densely puberulent within; fruiting calyx incrassate, about 6 mm. long and 10 mm. wide, pubescent; fruit black or bluish black, nearly spherical, 10–20 mm. in diameter, edible.

On dry or partly wooded hillsides, stream banks, in valleys and arroyos, along roadsides, and in the short-tree forest and lower margin of the oak forest, Sonoran to Tropical Zones, from central Baja California, Sonora, and Chihuahua to Oaxaca.

**2. Vitex pyramidata** Robinson, Proc. Amer. Acad. **29:** 321, 1894
*Cornutia pentaphylla* Sessé & Moc., Pl. Nov. Hisp. II, **1:** app. 103, 1889. Not *Vitex pentaphylla* Merr., 1909.

Tall shrub or sometimes a large tree to 15 m. tall, densely and intricately branched; branchlets dichotomous, usually densely puberulent with sordid-gray hairs; nodes annulate; principal internodes 1–3.5 cm. long; leaves 5-foliolate (rarely 3-foliolate); petioles 4.5–10 cm. long, disciform at apex, puberulent with minute, sparse, grayish or sordid hairs; leaflets usually unequal, 2 lowermost decidedly smaller than 3 central ones, all long-petiolulate on puberulent, distinctly canaliculate and margined petiolules 2–22 mm. long; blades thin-chartaceous or submembranous at anthesis, subcoriaceous or coriaceous in age, the central one oblong or elliptic or ovate, 3.5–18 cm. long, 1.7–6.7 cm. wide, rather abruptly acute or rounded at apex, entire, broadly rounded or subcordate at base, sparsely pulverulent-puberulent above, becoming subglabrate except for the midrib and secondaries, densely short-pubescent and resinous-granular beneath with cinereous or sordid pubescence, varying to densely grayish-tomentulose; inflorescence axillary, thyrsoid-paniculate, often paired in uppermost axils, 10–26 cm. long, 3–16 cm. wide, mostly composed of 1–4 pairs of lateral, long-stalked panicles and a terminal one, each panicle composed of 3–9 rather long-stipitate, many-flowered cymes, densely puberulent throughout; peduncles 4–13 cm. long, tetragonal-flattened, puberulent; pedicels 1–2 mm. long, densely puberulent; large and foliaceous bracts often present, 3–5-foliolate, long-stipitate, resembling leaves in color and pubescence; bractlets linear, oblong, lanceolate, or spatulate, 2–4 mm. long, puberulent; prophylla linear, about 1 mm. long, puberulent; calyx about 2 mm. long, minutely puberulent, its rim 5-toothed; corolla lavender or bright blue, with a white throat, about 12 mm. long, lightly puberulent outside, its limb very strongly bilabiate, about 1 cm. wide, glabrous within except for a pubescent band on larger lip leading into throat, glabrous outside except for ciliolate margins, segments rounded, ovate, upper suberect, lowest largest; stamens exserted, recurved at summit; fruiting calyx patelliform, incrassate, about 1 cm. in diameter, glabrous; fruit black, nearly spherical, about 10 mm. in diameter, edible.

On rocky hillsides, prairies, and basaltic mesas, in arroyos, and at edges of craters, Lower Sonoran to Tropical Zones, Sonora to Yucatán. Employed by the natives for food, fuel, and construction. When burned, the ash is blue.

## 12. TETRACLEA A. Gray, Amer. Jour. Sci. II, 16: 98, 1853

Low erect perennial herbs or subshrubs, often suffrutescent only at base. Leaves simple, deciduous, oblong or ovate, dentate or entire, those in inflorescence similar but smaller. Inflorescence axillary, cymose; cymes pedunculate, mostly 3-flowered. Flowers short-pedicellate. Calyx broadly campanulate, 10-nerved, deeply 5-parted, accrescent, persistent, lobes subequal. Corolla white or tinged with red; tube exserted, narrow-cylindric; limb spreading, 5-parted; lobes obovate-oblong, subequal. Stamens 4, didynamous, long-exserted, incurved-ascending, anterior pair longer; anthers 2-celled, thecae parallel. Disk present, equal, short. Ovary shortly 4-lobed at apex. Style bifid at apex; branches subulate and subequal. Pyrenes obovoid, hard, reticulate-rugose, attached to beyond middle by a broad keeled areole, fallen ones not leaving a gynobase. Seeds attached laterally.

### 1. **Tetraclea coulteri** A. Gray, Amer. Jour. Sci. II, **16:** 98, 1853

(For synonymy see Moldenke, in Lundell, Flora of Texas **3:** 86, 1942.)

Herbaceous perennial, with several branching, obscurely 4-angled, ascending or spreading stems, about 4 dm. tall, from a woody root; branchlets very slender, gray, obtusely tetragonal, densely puberulent with appressed, strigose, whitish hairs; nodes not annulate, or swollen; principal internodes 9–32 mm. long; leaves decussate-opposite; petioles slender, 4–10 mm. long, flattened, distinctly margined from apex to base, appressed-puberulent throughout, not ampliate at base; blades thin-chartaceous or submembranous, uniformly bright green on both surfaces, ovate, 1.6–3.5 cm. long, 6–18 mm. wide, sharply acute and mucronulate at apex, irregularly and coarsely dentate with 2 or 3 teeth on each margin or lower ones entire, acute or acuminate at base and prolonged into petiole, finely puberulent with whitish substrigose hairs on both surfaces; midrib plane above, prominulous beneath; secondaries 1–3 per side, ascending, slightly arcuate, plane and mostly obscure above, prominulous beneath, not conspicuously anastomosing; vein and veinlet reticulation very sparse, obscure or indiscernible on both surfaces; cymes mostly 3- (sometimes reduced to 2- or 1-) flowered; peduncles slender, 3–8 mm. long, densely puberulent; corolla cream-colored, tinged with red outside, lobes elliptic-obovate, entire; pyrenes pyriform, very strongly and coarsely reticulated, finely pubescent, commissural scar somewhat rugose.

In sandy clay soil, dry open ground, and broad, damp, river beds, chiefly in the foothills, Sonoran Zones, southwestern Pima County, Arizona, to Texas and south to Puebla.

## Family 123. **AVICENNIACEAE.** Black-Mangrove Family*

Shrubs or trees of maritime regions, mostly inhabiting saline or brackish, coastal, mangrove lagoons. Branches, branchlets, and twigs commonly terete, prominently nodose, and articulate. Leaves opposite, thick-textured, persistent, petiolate; estipulate; blade entire. Inflorescence axillary or terminal, determinate and centrifugal (cymose), spicate or subcapitate; axillary inflorescences mostly paired. Flowers sessile, perfect, hypogynous, small. Calyx composed of 5 nearly separate sepals, segments ovate and plainly imbricate, subtended by a pseudo-involucre composed of a scalelike bractlet and 2 alternate, scalelike prophylla, which are slightly shorter than calyx and are imbricate with each other and with calyx segments. Corolla actinomorphic, gamopetalous at base, campanulate-rotate, 4-parted. Stamens 4, inserted in corolla tube, equal or subdidynamous. Gynoecium composed of 2 united carpels. Ovary compound, but with a free, central, often more or less 4-winged placenta; ovules 4, pendent, orthotropous. Fruit a compressed oblique capsule, with a juicy and somewhat fleshy, usually tomentellous exocarp, dehiscent by 2 valves, by abortion regularly only 1-seeded. Seeds without a testa; embryo viviparous; radicle hairy; colyledons 2, folded lengthwise.

### 1. **AVICENNIA** L., Sp. Pl. 110, 1753

Characters of the family.

**1. Avicennia germinans** (L.) L., Sp. Pl. ed. 3, 891, 1764

*Bontia germinans* L., Syst. Nat. ed. 10, 1122, 1759.
*Avicennia nitida* Jacq., Enum. Pl. Carib. 25, 1760.
*Avicennia tomentosa* Jacq., loc. cit.

Tree to 20 m. tall; branches and twigs slender, brownish, glabrous, often shiny; petioles slender 2–27 mm. long, mealy or glabrous; leaf blades chartaceous or subcoriaceous, dark green or gray-green, shiny above, whitish furfuraceous to brunnescent or nigrescent on both surfaces, oblong, lanceolate, elliptic or obovate, 4.5–15 cm. long, 1.8–4.4 cm. wide, acute to obtuse at apex, acute or acuminate to cuneate at base; spikes 1.5–6.5 cm. long, 1–1.5 cm. wide at anthesis; flowers usually opposite, 1–15 per spike, sessile, usually closely crowded; calyx lobes ovate, 3–5 mm. long, 2–3 mm. wide, densely appressed-pubescent on outer surface, glabrous within; corolla white or yellow, 12–20 mm. long, parted to about middle, tube equaling or shorter than calyx, practically glabrous, lobes 2–2.5 mm. long, oblong or subquadrate, spreading, densely

---

* Text contributed by Harold N. Moldenke.

cinereous-puberulent outside with appressed hairs, velutinous-tomentose within; stamens slightly exserted; fruit obpyriform to ovate, asymmetric, 12–20 mm. long, 7–12 mm. wide, apiculate at apex in youth, densely white- or gray-pulverulent, often more or less appressed-strigose at apex.

Mangrove lagoons, from southern Florida and southeastern Texas, Bermuda, the Bahamas, and both coasts of Mexico, through the West Indies to the coasts of Brazil and Peru; also on the Galápagos and other islands off the coast of tropical and subtropical America. In our area from the vicinity of Mulegé, Magdalena Bay, and Guaymas south along the coast.

## Family 124. **LABIATAE.** Mint Family

Aromatic punctate-glandular herbs, shrubs, or rarely trees, with mostly quadrangular stems and opposite, simple, estipulate leaves. Flowers variously clustered, often in interrupted spicate, racemose, or cymose arrangements, typically irregular and bilabiate but nearly regular in some genera. Calyx regular or 2-lipped, usually 5-toothed or 5-lobed, persistent. Corolla usually 2-lipped; upper lip entire or more commonly 2-lobed; lower lip 3-lobed. Stamens inserted in corolla tube alternate with lobes, mostly 4 and didynamous, sometimes reduced to 2; staminodia present or absent; anthers 2-celled and introrse, or confluently 1-celled, or sometimes 1-celled by abortion of one cell. Ovary 4-lobed or 4-parted, superior, each division 1-ovuled; ovules anatropous; style arising from central depression between ovary lobes; stigma branches 2. Fruit of 4 nutlets. Seed erect (transverse in *Scutellaria*); endosperm scanty or none; embryo small, usually straight.

Ovary 4-lobed; style not basal; nutlets almost completely united laterally:
  Upper lip of corolla much smaller than lower; stamens equaling or shorter than corolla lobes . . . . . . . . 1. *Teucrium*
  Upper lip of corolla about same length as lower; stamens slightly to much exserted and equaling or exceeding corolla lobes . . . . . . . . 2. *Trichostema*
Ovary deeply 4-parted; style basally inserted; nutlets almost distinct, attached basally:
  Calyx 2-lipped with each lip entire:
    Plants shrubby; calyx inflated and bladder-like in fruit; inflorescence spicate or cymose . . . . . . . . 3. *Salazaria*
    Plants herbaceous; calyx saccate or helmet-like on upper side, not inflated; flowers solitary in axils . . . . . . . . 4. *Scutellaria*
  Calyx equally 5- or 10-toothed, or if bilabiate, lips 2- or 3-toothed:
    Calyx teeth 10, recurved and hooked in fruit; stamens and style included in corolla tube . . . . . . . . 5. *Marrubium*
    Calyx teeth 5, not recurved in fruit; stamens and style exserted beyond corolla tube and often exceeding lobes:
      Stamens ascending, not declined nor enveloped in lower lip:
        Corolla strongly 2-lipped, lips unequal:
          Fertile stamens 4; staminodia none:
            Corolla tube without a pilose ring within near base; calyx teeth attenuate, not mucronate nor spine-tipped; plants annual . . . . . . . . 6. *Lamium*
            Corolla tube with a pilose ring within near base; calyx teeth mucronate or spine-tipped; plants perennial . . . . . . . . 7. *Stachys*
          Fertile stamens 2; staminodia present or wanting:
            Upper corolla lip more or less galeate, distinctly concave:
              Anther connectives elongated, articulate with filaments, pollen sacs distinctly separated, or lower one wanting; calyx 2-lipped . . . . . . . . 8. *Salvia*

Anther connectives not elongated, pollen sacs approximate; calyx equally 5-toothed, not 2-lipped .................................. 9. *Monarda*

Upper corolla lip plane or essentially so, usually erect:

Staminodia represented by scales, or lacking; nutlets ovoid .... 10. *Hedeoma*

Staminodia of short filaments; nutlets oblong ............ 11. *Poliomintha*

Corolla nearly regular, lobes almost or quite equal:

Flowers in dense terminal heads; calyx tube minutely 15-nerved; stigma branches unequal ............................................ 12. *Monardella*

Flowers in axillary whorls or interrupted terminal spikes; calyx tube 10-nerved; stigma branches equal .................................. 13. *Mentha*

Stamens declined, more or less enveloped by lower lip of corolla:

Calyx teeth erect, not spreading; corollas strongly 2-lipped, lower lip often saccate-drooping ............................................ 14. *Hyptis*

Calyx teeth stellately spreading in fruit; corolla weakly bilabiate, lower lip not saccate-drooping .................................. 15. *Asterohyptis*

## 1. TEUCRIUM L., Gen. Pl. ed. 5, 247, 1754

Annual or perennial herbs or suffrutescent plants with entire, dentate, or incised, opposite leaves. Flowers in terminal racemes or axillary to reduced upper leaves. Calyx tubular or campanulate, rarely inflated, equally or subequally 5-toothed or 5-lobed and 10-nerved, often more or less saccate. Corolla tube usually included, naked to densely bearded within, limb bilabiate but upper lip very short, lower lip conspicuous and spreading, with small lateral lobes. Stamens 4, didynamous, inserted near apex of corolla tube, exserted; anthers bilocular, cells divergent or divaricate. Style 2-lobed; lobes subequal, subulate. Nutlets obovate, reticulate-roughened.

Leaves densely pubescent beneath, finely serrate or faintly biserrate; flowers in terminal bracteate racemes; calyx tube campanulate and somewhat saccate, teeth shorter than tube ........................................ 1. *T. canadense* var. *angustatum*

Leaves glabrous or essentially so, entire, remotely few-toothed or deeply incised; flowers axillary to leaves, not in dense racemes; calyx tube narrowly campanulate to tubular, not saccate, teeth equaling or exceeding tube:

Pedicels 1–8 mm. long; corolla 7–15 mm. long, pale blue with maroon to dark purple spots at base of lobes; plants annual, densely leafy, mostly procumbent ................ ........................................... 2. *T. cubense* subsp. *depressum*

Pedicels 1.5–4 cm. long; corolla 15–22 mm. long, white with lavender or purplish veins and streaks; plants perennial, sparsely and remotely leafy, erect or nearly so ........... ..................................................... 3. *T. glandulosum*

### 1. **Teucrium canadense** L. var. **angustatum** A. Gray, Syn. Fl. N. Amer. **2**[1]: 349, 1878

*Teucrium mexicanum* Sessé & Moc., La Naturaleza **2**[1]: app. 92, 1889.
*Teucrium nashii* Kearney, Bull. Torrey Club **21**: 484, 1894.

Erect perennial herbs to 1 m. tall from running rootstocks; stems distinctly quadrangular, pubescent with fine, curved, downwardly directed, appressed hairs; leaves usually exceeding internodes, narrowly elliptic or lance-elliptic, 0.1–2 cm. broad, 3.5–8 cm. long, cuneately narrowed toward base to a petiole 5–12 mm. long, acute or slightly acuminate at apex, serrate or obscurely biserrate, dark green and sparsely puberulent, with closely appressed, forwardly directed hairs on upper surface, densely appressed-puberulent with silvery hairs beneath; inflorescence a terminal raceme 1–2 dm. long; bracts lance-linear, 5–12 mm. long; pedicels 1–4 mm. long, minutely puberulent; calyx narrowly campanulate, somewhat saccate, finely puberulent but eglandular, 5–6 mm. long at anthesis, broadly deltoid, slightly unequal teeth 2–2.5 mm. long; corolla 10–15 mm. long, pale bluish, lower lobe 4–6 mm. long; nutlets broadly obovate, 3–4 mm. tall, slightly roughened.

Mainly along stream beds and ditches, Tropical, Lower Sonoran, and Arid Transition Zones, southern Arizona to Florida and south into Mexico. May–Sept.

**2. Teucrium cubense** Jacq. subsp. **depressum** (Small) McClintock & Epling, Brittonia **5:** 505, 1946

*Teucrium depressum* Small, Bull. N.Y. Bot. Gard. **1:** 288, 1899.
*Teucrium cubense* var. *densum* Jepson, Man. Fl. Pl. Calif. 861, 1925.

Annual herbs with several to many branches 1–5 dm. long, these usually procumbent-ascending or central ones erect, glabrous except in inflorescences where pubescent with coarse, curly or crinkled, white hairs, margins of leaves sometimes ciliate-pubescent, stems only faintly quadrangular; leaves near base and middle of stems 1.5–3.5 cm. long, rather broadly obovate in outline, 3–5-lobed nearly or quite to midrib, upper ones usually 3-parted nearly to base, lobes linear-oblanceolate, 1–3.5 mm. wide, often irregularly and shallowly toothed toward apex; pedicels slender, sparsely pubescent, 1–8 mm. long; calyx 5–6 mm. long, teeth lance-subulate, about equal, 2–3.5 mm. long; corolla 7–15 mm. long, pale blue with maroon to purplish spots at base of lobes, pubescent without, tube scarcely equaling to slightly exceeding calyx lobes; nutlets about 2.5 mm. high, reticulately ridged and spreading-puberulent.

Mainly in sandy or silty soil, llanos, arroyos, and stream banks, mainly Lower Sonoran Zone, northern Colorado Desert in the vicinity of Hayfields east to Texas and south into northeastern Baja California, along the coastal plain through Sonora into Sinaloa, and through Chihuahua to Nuevo León. March–May.

**3. Teucrium glandulosum** Kell., Proc. Calif. Acad. **2:** 23, 1860

Perennial herbs to 1.5 m. tall, with sparsely leaved, quadrangular, glabrous stems from a woody base; leaves chiefly lanceolate, entire, 2.5–12 mm. broad, 1.5–6 cm. long, sessile or nearly so, usually much shorter than internodes, sometimes lower ones irregularly 3-lobed or remotely 3–5-toothed on one or both margins; flowers axillary to reduced leaves, thus tips of branches nearly racemose; pedicels 15–40 mm. long, straight or arcuate; calyces campanulate at anthesis, 8–11 mm. long, lobes lance-ovate, 4–8 mm. long, equaling to markedly exceeding tube; corolla 15–22 mm. long, white, streaked with violet or lavender, pubescent with curled hairs along midveins outside, densely bearded within, lateral lobes of lower lip nearly one-half as long as central one; stamens about 1.5 cm. long, filaments bearded below middle; nutlets oval, 2.5–3 mm. long, lightly grooved longitudinally, densely puberulent on upper third.

Along streams, about seeps, and at margins of ponds, Lower Sonoran and lower margins of Upper Sonoran Zones, Whipple Mountains in the eastern Mojave Desert, the Castle Dome Mountains in Yuma County, Arizona, and from the southwestern flanks of the Sierra San Pedro Mártir to the Magdalena Plain and Cedros Island, Baja California. Nov.–June.

## 2. TRICHOSTEMA L., Sp. Pl. 598, 1753

Annual herbs or suffrutescent perennials with linear, elliptic, or ovate, usually entire, petiolate or subsessile leaves and glandular inflorescences, herbage pungently odoriferous. Flowers in 1- to many-flowered cymes or in racemes. Calyx regularly and deeply 5-lobed or irregular owing to a partial fusion of upper 3 teeth. Corolla blue, lavender or pink, 5-lobed, anterior lobe longer than 4 upper ones, pubescent without, glabrous within. Stamens 4, more or less didynamous, inserted near bottom of throat, ascending, arched, or nearly straight; anther cells usually divergent at maturity. Styles about equaling stamens; stigma lobes slightly unequal. Nutlets joined along lower one-third, rugose-reticulate, usually puberulent to hirtellous.

Annual herbs; stamens straight or nearly so, only slightly exserted ........ 1. *T. brachiatum*
Perennial shrubby plants to 12 dm. tall; stamens strongly arched, conspicuously exserted .... ............................................................ 2. *T. parishii*

## 1. Trichostema brachiatum L., Sp. Pl. 598, 1753

*Isanthus coeruleus* Michx., Fl. Bor. Amer. **2:** 4, *pl. 30,* 1803.
*Isanthus pumilus* Raf., New Fl. **2:** 34, 1837.
*Isanthus pubescens* Raf., op. cit. 35.
*Isanthus multiflorus* Raf., loc. cit.
*Isanthus brachiatus* B.S.P., Prel. Cat. N.Y. 44, 1888.
*Isanthus brachiatus* var. *linearis* Fassett, Rhodora **35:** 388, 1933.

Annual herb 1–4 dm. tall with simple to freely branching stems, hirtellous to pilose throughout with spreading or downwardly curved but not appressed hairs and with admixtures of capitate-glandular hairs of varying density but with very few of these glandular hairs on specimens in our area; leaves elliptic or rarely nearly linear, 4–16 mm. wide, 2–5 cm. long, acute at apex, gradually narrowed at base to petioles 1–10 mm. long, entire to irregularly serrate, both surfaces hirtellous with upwardly arching, nonglandular hairs and with a scattering of gland-tipped hairs; cymes 1–3-flowered, axillary, peduncles to 2 cm. long; bracts 1.5–3 mm. long; calyx tube 1–2 mm. long, lobes 1.5–4 mm. long at anthesis, to 5 mm. long in fruit; corolla tube straight, 1.5–3.5 mm. long, included within calyx, lower lip of corolla 1.3–2.2 mm. long, slightly longer than 4 upper lobes; stamens straight or essentially so, exserted only 1–3 mm.; nutlets 2–3 mm. long, about one-half to two-thirds as broad, hirtellous and glandular-puberulent, alveolate with rather prominent ridges.

Along streams and valleys, chiefly of the Mississippi Valley, the Appalachian area, and into New England, but apparently indigenous specimens occurring in Gila and Navajo Counties, Arizona, according to Lewis. The species may not descend into the Sonoran Desert but can be expected along the Gila River.

## 2. Trichostema parishii Vasey, in Coult., Bot. Gaz. 6: 173, 1881

*Trichostema lanatum* var. *denudatum* A. Gray, Syn. Fl. N. Amer. ed. 2, $2^1$: 459 (suppl.), 1886.

A rounded, freely branched shrub or subshrub 4–12 dm. tall, hirtellous throughout with downwardly appressed hairs; leaves linear to linear-lanceolate, 1.5–6 mm. broad, 2–8 cm. long, usually exceeding internodes and often bearing fascicles of smaller leaves in their axils, acute or rarely obtuse at apex, sessile or short-petiolate at base, dark green and sparsely hispidulous on upper surface, densely lanate-cinereous beneath, margins entire, lightly revolute; cymes axillary to reduced leaves on long, terminal, raceme-like inflorescence, some of lower cymes again racemiform; peduncles densely gray-pubescent, 5–20 mm. long; pedicels 1–10 mm. long, lanate with white, pinkish, lavender or bluish hairs; calyces 3–5 mm. long, ovate to lance-ovate teeth acute to short-acuminate, subequal, about as long as calyx tube; corolla pubescent to lanate without, tube straight, 4–7 mm. long, lower lip 5–9 mm. long; stamens 15–25 mm. long, strongly arched; nutlets broadly obovoid, 2–4 mm. long, irregularly reticulate, more or less pubescent.

On arid ridges and outwash fans, mainly Upper Sonoran Zone, Los Angeles County, California, south to western flanks of the Sierra San Pedro Mártir, barely entering the Sonoran Desert along its western margin in Imperial County near Mountain Springs, and in eastern San Diego County. March–July.

## 3. SALAZARIA Torr., Bot. Mex. Bound. 133, 1859

A brittle, subspinose, straggly subshrub from a woody base with herbaceous, gray-green, glandular-puberulent stems that die back to persistent base each year. Leaves ovate to oval, entire, usually obtuse and shorter than internodes. Flowers in few-flowered cymes in axils of upper leaves or bracts. Calyx tubular or urceolate at anthesis, equally 2-lipped, inflated into a papery, reticulately veined, purplish bladder in fruit. Corolla violet on lower lip, white or cream on upper, tubular, with a short limb; upper lip and lateral lobes more or less adherent to form a short galea surrounding stamens and style. Stamens 4, didynamous, anterior pair longer. Nutlets depressed-obovoid, tuberculately roughened.

**1. Salazaria mexicana** Torr., Bot. Mex. Bound. 133, *pl. 39,* 1859

Rounded to straggly plant to 1 m. tall or less, growth open and usually sparsely leaved; leaves 3–8 mm. wide, 5–25 mm. long, thickish and rather leathery; pedicels 1–5 mm. long; calyx 5–6 mm. long at anthesis, 1.5–2 cm. long and nearly as wide, with a small terminal ostiole 3–4 mm. in diameter in fruit; corolla 10–18 mm. long, pubescent with short, spreading, pale lavender to whitish hairs without; nutlets 3–4 mm. long.

Desert hillsides, mesas, and along arroyos, Lower and Upper Sonoran Zones, Mojave and Colorado Deserts to Utah, western Texas, and south into northern Mexico, extending at least to central Baja California. March–June.

## 4. SCUTELLARIA L., Sp. Pl. 598, 1753

Annual or perennial herbs with 1 to several stems from slender rhizomes or from a woody caudex; herbage slightly or not at all aromatic. Leaves petiolate, entire or shallowly toothed. Flowers axillary near tip of stem or racemose or spicate. Calyx campanulate, somewhat gibbous, 2-lipped; upper lip with a crestlike projection on back, often deciduous; lower lip persistent. Corolla exserted, bilabiate, dilated above throat; upper lip more or less galeate, entire or emarginate, arched; lower lip spreading or deflexed; lateral lobes joined to upper lip. Stamens 4, in pairs, included within upper lip of corolla; upper pair of anthers 2-celled, lower 1-celled. Style unequally 2-lobed. Nutlets 4, subglobose or depressed, smooth to tuberculate.

Herbage distinctly cinereous; stems from a woody caudex; corollas densely pilose within .... .............................................. 1. *S. potosina* subsp. *platyphylla*
Herbage green, not cinereous; stems from slender rhizomes, sometimes bearing tubers; corollas globose or sparingly pubescent within:
  Pubescence on stems and leaves of straight hairs; corollas 12–16 mm. long; upper corolla lip shorter than lower:
    Rhizomes regularly bearing tubers; nutlets muriculate; corolla blue ..... 2. *S. tuberosa*
    Rhizomes without tubers; nutlets rugose; corolla white ........................... ........................................ 3. *S. bolanderi* subsp. *austromontana*
  Pubescence on stems and leaves of fine curved hairs; corollas 20–28 mm. long; upper corolla lip longer than lower ........................................... 4. *S. austiniae*

**1. Scutellaria potosina** Brandegee subsp. **platyphylla** Epling, Univ. Calif. Pub. Bot. **20:** 64, 1942

Stems few to many from a branching woody caudex, erect or slightly decumbent at base, 10–40 cm. tall, densely puberulent, somewhat cinereous; leaves ovate to ovate-lanceolate, 5–12 mm. wide, 10–25 mm. long, obtuse at apex, broadly rounded to subcuneate at base, entire or faintly and remotely crenate, puberulent and closely gland-dotted on both surfaces; petioles 1–4 mm. long; flowers few in upper axils, ascending; pedicels 1–2 mm. long, hirsutulous; calyx 3–4 mm. long, nearly as broad, scale broad and flaring, densely glandular and sparsely hirsute; corolla blue-violet, 12–19 mm. long, hirsutulous without, lanate within; nutlets black, about 1 mm. long, roughly tuberculate.

Hillsides and along streams, mostly Upper Sonoran Zone, Cochise, Gila, and Maricopa Counties, Arizona, to Texas and south into northern Mexico. Mainly at altitudes above the Sonoran Desert. March–July.

**2. Scutellaria tuberosa** Benth., Lab. Gen. & Sp. 441, 1834

*Scutellaria pilosiuscula* Nutt. ex Benth., in A. DC., Prod. **12:** 429, 1848.
*Scutellaria tuberosa* subsp. *australis* Epling, Madroño **5:** 54, 1939.

Stems erect or rarely trailing, 5–30 cm. long, hirsute to nearly glabrous, arising from creeping, tuber-bearing, slender rhizomes; tubers 5–20 mm. long, 2–5 mm. in diameter; leaves thin, ovate, 5–18 mm. wide, 8–25 mm. long, obtuse to rounded at apex, broadly rounded to truncate at base, remotely crenate to entire, on slender hirsute petioles 5–25 mm. long; flowers few, solitary in upper axils on slender, hirsute-glandular petioles 1–3 mm. long; calyx broadly campanulate, 3–4 mm. long, purplish at tip, moderately to densely hirsute; corolla blue, 12–16 mm. long, 5–6 mm. wide at throat, tube curved slightly upward, moderately hirsutulous without, upper lip 2–3 mm. long, entire, lower lip shallowly 3-lobed, 4–6 mm. long, its middle lobe emarginate; nutlets black, 1–1.5 mm. long, strongly muriculate.

Woods and chaparral, mainly Transition and Upper Sonoran Zones, Alameda and Santa Cruz Counties, California, south to foothills of Sierra San Pedro Mártir, Baja California. Rarely reaching margin of Colorado Desert as a straggler. March–July.

**3. Scutellaria bolanderi** A. Gray subsp. **austromontana** Epling, Madroño **5:** 58, 1939

Stems slender, 1 to many from a slender rhizome, erect to decumbent, 1–4.5 dm. long, sparsely short-pubescent with spreading, moderately curved hairs; leaves thin, spreading, ovate to ovate-oblong, 8–15 mm. wide, 15–38 mm. long, obtuse to rounded at apex, rounded to subcordate at base, remotely crenate-serrate or uppermost entire, sparsely pubescent; petioles 1–3 mm. long; flowers few in upper axils; pedicels slender, 1–2 mm. long, hirsute and somewhat glandular; calyx 3–4 mm. long, pale green, hirsute; corolla white, 12–15 mm. long, sparsely hirsute without, slender tube curved upward at base and expanding abruptly into broad throat; upper lip 2–3 mm. long, lower lip shallowly 3-lobed, 4–6 mm. long; nutlets dark brown, 1 mm. long, rugose, with a low median dorsal ridge.

Meadows and stream banks, Upper and Lower Sonoran Zones, southern Mojave Desert to eastern San Diego County, California. Reaching margins of Sonoran Desert rarely as a straggler along arroyos. May–Aug.

**4. Scutellaria austiniae** Eastw., Bull. Torrey Club **30:** 493, 1903

*Scutellaria linearifolia* Eastw., loc. cit.

*Scutellaria angustifolia* var. *austiniae* Leonard, Contr. U.S. Nat. Herb. **22:** 726, 1927.

Stems few to many, 10–30 cm. tall, from slender woody rhizomes devoid of tubers; herbage eglandular, puberulent with minute curved hairs; leaves ascending, oblong-lanceolate to oblong-linear or lower ones ovate-elliptic, 3–15 mm. wide, 15–35 mm. long, entire or essentially so, obtuse to rounded at apex, cuneately narrowed at base to a short petiole 1–6 mm. long; flowers in upper axils; pedicels 2–4.5 mm. long; calyx 3–4 mm. long, at anthesis to 8 mm. long in fruit, minutely puberulent, often purplish or pinkish on upper lip; corolla blue-purple, 20–28 mm. long, sparsely to moderately glandular-puberulent on outside, erect, slender tube curved upward, limb 10–12 mm. wide, upper lip strongly galeate, distinctly longer than lower; nutlets dark brown to nearly black, 1.8–2 mm. long, distinctly tuberculate on back and sides.

Gravelly stream banks and rocky hillsides, mostly Arid Transition Zone, San Jacinto and Santa Rosa Mountains, California, to Shasta County, California. Rarely a straggler along streams at upper margin of Colorado Desert. May–Aug.

## 5. MARRUBIUM L., Sp. Pl. 582, 1753

Perennial herb with pubescent to lanate herbage and rugose, dentate to incised leaves. Flowers in compact multiflowered verticels forming interrupted spikes; bracteoles subulate or obsolete. Flowers small, white or pale lavender. Calyx tubular, in ours equally 10-toothed, each tooth tipped by a small, recurvingly hooked spine or bristle. Corolla tube included; upper lip erect, medianly notched, laterally reflexed; lower one 3-toothed. Stamens 4, didynamous. Nutlets ovoid to oblong, obtuse or truncate at apex, black, somewhat reticulate-roughened.

**1. Marrubium vulgare** L., Sp. Pl. 583, 1753

Usually a rounded, compact, perennial herb to 6 dm. tall and 2–10 dm. broad from a tough taproot; stems white-lanate, quadrangular except on old sublignescent growth; herbage with a characteristic odor and bitter taste; leaves broadly ovate, oval, suborbicular or subreniform, 1–5 cm. wide, 1–4(6) cm. long, dentate, acute to rounded or truncate at apex, acute to cordate at base, conspicuously rugose, pubescent and gray-green above, cinereous-lanate beneath, lower ones on petioles shorter than to equaling blades, subsessile toward tips of spikes; calyx 3–5 mm. long at anthesis, slightly larger and lanate within in fruit; corolla white, 2–4 mm. longer than calyx; nutlets oblong, about 2 mm. long.

A weed of roadsides, fields, fence rows, and pastures widely distributed as an introduced weed in much of North America; comparatively rare in the desert regions but occurring within the Sonoran Desert near Tempe, Phoenix, and Tucson, Arizona, and in southern Sonora, mainly along streams and irrigating ditches. March–Sept.

## 6. LAMIUM L., Sp. Pl. 579, 1753

Annual or perennial herbs with toothed or incised, cordate leaves and small flowers in axillary and terminal, verticillate clusters. Calyx tubular-campanulate, 5-ribbed, 5-toothed; teeth lanceolate, subequal or upper longer than lower. Corolla exceeding calyx; tube dilated above, 2-lipped; upper lip erect, entire, somewhat cuculate; the lower lip spreading, 3-lobed; middle lobe emarginate, often with a lateral appendage.

Stamens 4, didynamous; anterior pair longer; anthers 2-celled, with divaricate, often hirsute sacs. Style 2-cleft at apex; ovary 4-celled. Nutlets smooth or tuberculate.

**1. Lamium amplexicaule** L., Sp. Pl. 579, 1753

Annual or biennial, with sparsely pubescent herbage and decumbent to ascending stems 5–35 cm. long, these often branching at base and from some axils; leaves broadly ovate to suborbicular, truncate to cordate at base, obtuse to rounded at apex, coarsely crenate, dark green above, slightly paler beneath, lower 5–10 mm. wide, slenderly petiolate, upper sessile and often clasping, 20–25 mm. wide, nearly as long; cymules few-flowered; calyx hispidulous, 4–5 mm. long, teeth equaling or slightly longer than broad tube; corolla purple to lavender, 10–16 mm. long, tube slender, sparsely hirsute below, becoming more densely pubescent upward, upper lip closely bearded with lavender to purplish hairs, lower lip with very small lateral lobes, its middle lobe spotted with white and deep purple; nutlets obovoid-oblong, rounded on back, trigonous in cross section, narrowly grooved down ventral midline, pale brown, with paler numerous tubercles; hilum oblique.

Occasional in cultivated fields and lawns in Arizona; introduced from Europe and widely distributed from British Columbia to southern California and east to the Atlantic coast. March–Oct.

## 7. STACHYS L., Sp. Pl. 580, 1753

Annual or perennial herbs with erect, ascending, or prostrate stems and elongate internodes. Leaves petiolate or upper subsessile; blades ovate, oblong, or less commonly deltoid, glabrous or more commonly pubescent. Flowers in remote whorls or glomerules. Calyx narrowly turbinate, tubular or narrowly campanulate, broadly campanulate to urceolate in fruit, short-pedicellate; teeth deltoid to lanceolate, usually narrowed to a slender apical spine or bristle. Corolla strongly bilabiate; upper lip bifid, lower trifid; tube cylindrical, glabrous within except for a pilose ring toward base. Stamens 4, didynamous, inserted at about middle of corolla tube, bearded toward base, usually exserted from corolla tube. Style bifid; 2 lobes slightly unequal. Nutlets obovoid, apically truncate, lightly verrucose to smooth.

Corolla bright red; tube 15–20 mm. long; calyx tube 5–8 mm. long at anthesis .......... 1. *S. coccinea*

Corolla purplish or lavender; tube 7–15 mm. long; calyx tube 2.5–5 mm. long at anthesis:

- Plant perennial; corolla tube puberulent within but hardly annulate; calyx teeth usually distinctly shorter than tube .......... 2. *S. bigelovii*
- Plant annual; corolla tube annulate-pilose near base within; calyx teeth equaling or slightly exceeding tube .......... 3. *S. tenerrima*

**1. Stachys coccinea** Jacq., Pl. Hort. Schönb. **3:** 18, *pl. 284,* 1798

*Stachys cardinalis* Kunze, Bot. Zeit. **2:** 645, 1844.
*Stachys oaxacana* Fernald, Proc. Amer. Acad. **35:** 564, 1900.

A rather robust annual 3–7 dm. tall with quadrangular, sparsely to densely puberulent, erect or ascending stems; internodes elongate, to 15 cm. long; petioles slender, to 7 cm. long on lower leaves, 1–5 mm. long on upper leaves or these subsessile; leaf blades elliptic to broadly ovate or deltoid-ovate, 2–8 cm. long, 0.5–4.5 cm. broad, obtuse to short-acuminate at apex, broadly cuneate to subcordate at base, margin crenate to crenate-serrate, lamina sparsely to densely puberulent on both surfaces, few to many subsessile glands intermingled with pubescence; verticils about 6-flowered; calyx tube 5–8 mm. long, 3–3.5 mm. in diameter, puberulent without and within, teeth deltoid, acute to subulate at apex, 2–3.5 mm. long; corolla bright red, tube 15–20 mm. long, densely puberulent without, pilose-annulate near base within, lower lip 9–14 mm. long, upper nearly orbicular, 6–8 mm. in diameter; stamens barely equaling upper lip; nutlets obovoid, about 3 mm. high, black, minutely and sparsely puberulent, truncate or obtuse at apex.

Shaded canyons and along watercourses, Lower and Upper Sonoran Zones, Arizona to Texas, south to central Mexico, and from central Baja California to the Cape Region. Jan.–Sept.

**2. Stachys bigelovii** A. Gray, Proc. Amer. Acad. **8:** 371, 1872

A slender, rather lax, perennial herb but often blooming first year; stems simple or sparingly branched, 4–8 dm. tall, erect or sometimes prostrate, sparsely puberulent with short fine hairs or nearly glabrous; petioles slender, 1–5 cm. long or uppermost leaves subsessile; leaf blades deltoid-ovate or upper ovate to elliptic, 1–4 cm. broad, 2–6 cm. long, obtuse at apex, broadly cuneate to subcordate at base, sparsely hirtellous to glabrate on both surfaces, margin obtusely serrate or dentate; verticils in interrupted spicate terminal inflorescences, mostly 6-flowered; pedicels slender, 1–2 mm. long; calyx turbinate, tube 3–4 mm. long at anthesis, sparsely hirtellous and rather closely glandular-dotted without, sparsely but distinctly puberulent within, at least near apex; calyx teeth deltoid-acuminate, 2–2.5 mm. long and spreadingly curved at or soon after anthesis; corolla purplish, tube 9–15 mm. long, densely puberulent without, puberulent but scarcely annulate within; lower corolla lip 7–8 mm. long, upper 3–4 mm. long; stamens exserted 1–3 mm. from corolla tube; nutlets broadly obovoid, about 2 mm. high, about as wide, dark brown, glabrous, faintly reticulate.

Along streams and washes and on hillside seeps, Lower and Upper Sonoran Zones and into the Tropical Zone, Texas to Zacatecas and San Luis Potosí and in southern Baja California, where it occurs in the Sonoran Desert near El Refúgio. March–Sept.

**3. Stachys tenerrima** Epling, Rep. Spec. Nov. Beih. **80:** 32, 1934

Rather slender annual herbs 3–5 dm. tall with sparsely puberulent to subglabrate stems and foliage; petioles slender, 0.5–4 cm. long; leaf blades narrowly to broadly ovate, 1–3.5 cm. wide, 3–6 cm. long, obtuse to acute at apex, rounded to shallowly subcordate at base, margins coarsely serrate, serrations obtuse to rounded; verticils few, remote on slender terminal branches to 15 cm. long; pedicels subfiliform, 1.5–4 mm. long; calyx campanulate, tube 2.5–3 mm. long, sparsely and minutely puberulent to subglabrate, glabrous or essentially so within; calyx teeth deltoid, tips filiform-subulate, equaling or slightly exceeding tube; corolla purplish or pale lavender, tube 7–8 mm. long, sparsely puberulent without, annulate-puberulent near base within; lower corolla lip 6–7 mm. long, upper one 4–5 mm. long; stamens exserted 2–3 mm.; nutlets dark brown, obovoid, about 2 mm. long, glabrous, minutely pebbled, stretching calyx into a broadly campanulate cup with spreading teeth.

Streamsides and mountain canyons, Lower Sonoran and Tropical Zones, southern Baja California. Oct.–Jan.

## 8. SALVIA L., Sp. Pl. 23, 1753

Aromatic herbs or shrubs with punctate-glandular leaves and flowers arranged in interrupted spikes or sometimes in racemes or panicles. Calyx ovoid, cylindrical, or campanulate, usually 2-lipped; upper lip entire or 3-toothed; lower 2-toothed or 2-cleft. Corolla strongly bilabiate; upper lip more or less arched and concave, entire, emarginate, or 2-lobed; lower lip 3-lobed, spreading or drooping. Stamens 2, inserted in throat of corolla; anther connective elongate, articulated with filament, often somewhat fleshy at lower end, bearing a functional anther cell at apex of upper arm, on some at lower end also, and in others lower arm wholly suppressed or sterile. Style slender, often surpassing stamens, 2-cleft at summit. Ovary deeply 4-parted; nutlets attached basally. Mature nutlets ovoid to short-oblong, smooth, in some developing mucilage when wet.

Plants herbaceous, annual or if perennial only faintly woody at base:
- Lower arm of connective bearing a functional anther theca; middle lobe of lower lip of corolla fimbriate:
  - Basal leaves thistle-like, spinose-toothed; corollas lavender, 20–30 mm. long; nutlets 3–4 mm. long . . . . . . . . 1. *S. carduacea*
  - Basal leaves not thistle-like, unarmed; corolla blue, 10–13 mm. long; nutlets 2–2.2 mm. long . . . . . . . . 2. *S. columbariae*
- Lower arm of connective without a functional anther theca; corolla lobes sometimes ciliate or erose, but not fimbriate:
  - Plants strictly herbaceous, annual; corolla tubes not over 4.5 mm. long; calyx not over 5.5 mm. long at anthesis:
    - Inflorescence of several- to many-flowered verticils, verticils interruptedly arranged; bracts broadly ovate to orbicular:
      - Lower arms of connective and base of filaments densely hirsute; calyx densely hirsute on veins to tips of teeth . . . . . . . . 3. *S. galinsogifolia*
      - Lower arms of connectives fleshy, glabrous, filaments glabrous or sparsely hirsute; calyx hirsute at base, nearly glabrous near tips of teeth . . . . . . . . 4. *S. hyptoides*
    - Inflorescence more or less racemose, flowers solitary or 2–10 in each verticil; verticils lax; bracts lanceolate to ovate, often acuminate, frequently caducous:
      - Calyx campanulate, only slightly longer in fruit than at anthesis; leaves cinereous beneath . . . . . . . . 5. *S. riparia*
      - Calyx tubular to cylindrical at anthesis, enlarging markedly in fruit; leaves not cinereous beneath:
        - Stems and leaves moderately puberulent with crispate, white hairs; calyx 3.5–5 mm. long in fruit . . . . . . . . 6. *S. misella*
        - Stems and leaves hispid or glandular-hirtellous, at least on veins and margins; calyces 7–10 mm. long in fruit:
          - Stamens included within obviously galeate upper lip of corolla; nutlets 2–3 mm. long; inflorescences 2–15 cm. long:
            - Bracts and calyx teeth setose-tipped, not markedly glandular; leaf blades ovate, crenate-serrate, not incised . . . . . . . . 7. *S. setosa*
            - Bracts and calyx teeth acute, not setose, strongly glandular; leaf blades oblong-elliptic to linear-lanceolate, irregularly incised-serrate . . . . . . . . 8. *S. subincisa*
          - Stamens clearly exserted beyond upper corolla lip; nutlets 1.1–1.6 mm. long; inflorescences 5–25 cm. long . . . . . . . . 9. *S. tiliifolia*
  - Plants perennial, sometimes suffrutescent at base; corolla tubes 6–20 mm. long; calyx 5.5–10 mm. long at anthesis:
    - Corolla blue, tube 6–6.5 mm. long; leaf blades rhomboid-ovate, 2–8 cm. long; calyx hirtellous to nearly glabrous, 5.5–7 mm. long . . . . . . . . 10. *S. seemannii*
    - Corolla purple or lavender, tube 15–20 mm. long; leaf blades broadly lanceolate to ovate, 4–15 cm. long; calyx cinereous, 7–10 mm. long:
      - Calyx cylindrical; corolla lavender; twigs soft-tomentulose; lower corolla lip 7–9 mm. long . . . . . . . . 11. *S. peninsularis*

Calyx narrowly campanulate; corolla purple; twigs cinereous-puberulent; lower corolla lip 10–15 mm. long .................................. 12. *S. marci*

Plants shrubby, distinctly woody at least at base:

Calyx teeth deltoid to ovate, obtuse to acute or short-apiculate, not spinose-tipped or aristate:

Corolla lobes blue; nutlets rarely over 2 mm. long:

Calyx flaring and only slightly oblique at orifice; stamens shorter than hoodlike upper lip of corolla:

Leaf blades lanceolate to lance-ovate, 5–15 mm. wide; lower corolla lip 3–3.4 mm. long, several-toothed; lower arm of connective lanate ........... 13. *S. parryi*

Leaf blades lance-ovate to deltoid-ovate, 10–25 mm. wide; lower corolla lip 7–9 mm. long, not toothed; lower arm of connective glabrous ............ 14. *S. similis*

Calyx cylindrical to narrowly campanulate, not markedly flaring, orifice often strongly oblique; stamens usually equaling or surpassing upper corolla lip:

Inflorescence capitate; bracts membranous-chartaceous, white; connective shorter than filaments, lower arm obsolete ....................... 15. *S. mohavensis*

Inflorescence of several verticils, interruptedly arranged; bracts herbaceous, green or purplish; lower arm of connective present, or if absent bracts not white:

Calyx strongly punctate-glandular, glands not obscured by fine puberulence:

Leaf blades acute to short-acuminate at apices, dark green on upper surfaces, not coriaceous; corolla with coarse blue hairs on outside of upper lip ................................................................ 16. *S. palmeri*

Leaf blades obtuse to rounded at apices, tomentose-cinereous on upper surface, coriaceous; corolla glabrous or short-puberulent with fine hairs ......................................................... 17. *S. platycheila*

Calyx densely puberulent or tomentose, glands obscured by hairs:

Twigs lanate-canescent in youth; leaf blades sessile, 2.5 cm. long or less, teeth pungently spinose-tipped ........................... 18. *S. californica*

Twigs cinereous-puberulent or scurfy in youth, not lanate; leaves petiolate, blades usually over 2.5 cm. long, teeth not spinose-tipped:

Nutlets glabrous; leaf blades broadly ovate; filaments and connectives straight or slightly curved:

Bracts and calyces hirsutulous, mostly green; pubescence on twigs of simple minute hairs ................................. 19. *S. pinguifolia*

Bracts and calyces lanate, purplish; pubescence on twigs of branched hairs ............................................... 20. *S. cedrosensis*

Nutlets densely puberulent; leaf blades lance-ovate to elliptic-lanceolate; filaments and connectives strongly arcuate .............. 21. *S. goldmanii*

Corolla lobes white or lavender; nutlets 2–3 mm. long:

Twigs and petioles floccose-lanate; corolla lavender with small red spots on lower lip; upper corolla lip 3.5–4 mm. long; nutlets 2–2.4 mm. long; inflorescences of interrupted verticils .................................... 22. *S. chionopeplica*

Twigs, petioles, and leaves minutely tomentose-canescent; corolla white with faint lines of lavender on lower lip; upper corolla lip 1 mm. long; nutlets 2.5–3 mm. long; inflorescences thyrsoid-paniculate ......................... 23. *S. apiana*

Calyx teeth spinose-tipped to aristate:

Leaves petiolate, not leathery nor spinose at tips of teeth; lower arm of connective obsolete or nearly so, not bearing an anther theca:

Corolla white; calyx teeth aristate, to 6 mm. long; nutlets 3.5–4 mm. long .......................................................... 24. *S. vaseyi*

Corolla blue, lavender, violet, or rose; calyx teeth setose-tipped, but not aristate, rarely over 2 mm. long; nutlets 2–2.5 mm. long:

Upper lip of corolla 2–3 mm. long; stamens about equaling upper lip of corolla; leaf blades rugose, but scarcely bullate on upper surface:

Calyx 4–5 mm. long, purplish; leaf blades oblanceolate, 1–3 cm. long; corolla blue ........................................................ 25. *S. munzii*
Calyx 6–8 mm. long, greenish; leaf blades oblong-lanceolate, 2–6 cm. long; corolla blue-lavender .......................................... 26. *S. mellifera*
Upper lip of corolla 4–6 mm. long; stamens exserted 4–6 mm.; leaf blades markedly bullate on upper surface .............................. 27. *S. eremostachya*
Leaves sessile, leathery, teeth spine-tipped; lower arm of connective bearing a rudimentary or functional anther theca ............................ 28. *S. greatae*

**1. Salvia carduacea** Benth., Lab. Gen. & Sp. 302, 1833
*Salvia gossypina* Benth., Pl. Hartw. 330, 1849.

Annual, with 1 to several erect, stout, scapelike stems 1–6 dm. tall, and white-woolly, thistlelike herbage; leaves basal, sinuate-pinnatifid, 3–35 cm. long, to 10 cm. wide, spinose-toothed; flowers 1 to several in dense, headlike, verticillate clusters 2–6 cm. broad, each head subtended by several lanceolate, pectinate-toothed bracts 2–4 cm. long; calyx 10–15 mm. long, densely long-woolly, each lobe tipped with a slender brownish spine 3–5 mm. long; corollas lavender, 20–30 mm. long, upper lip cleft and erose-denticulate, lower with a fan-shaped middle lobe with fimbriate margins and smaller lateral lobes; filaments slender, short, upper arm of connective about equaling lower lip and bearing an anther cell, lower arm much shorter and bearing a fertile anther cell; nutlets broadly obovoid, 3–4 mm. long, 2–2.5 mm. wide, smooth and sublustrous, pale brown.

Sandy and gravelly desert slopes, arroyos, and flats, Lower and Upper Sonoran Zones, Contra Costa County, California, to northern Baja California and east to the western parts of the Mojave and Colorado Deserts. March–June.

**2. Salvia columbariae** Benth., Lab. Gen. & Sp. 302, 1833
*Pycnosphace columbariae* Rydb., Fl. Rocky Mts. 747, 1917.

Annual herbs with 1 to several erect, naked, peduncle-like stems 1–6 dm. tall and basal leaves, or with 1–4 nodes above base also bearing somewhat reduced leaves, whole plant more or less cinereous with short recurved hairs, often somewhat purplish; blades of basal leaves 5–15 cm. long, 1–2-pinnatifid into toothed or irregularly incised divisions, cinereous-tomentose; petioles about equaling blade; stem leaves similar, but reduced; flowers in capitate verticils 2–4 cm. in diameter, subtended by suborbicular, green to purplish, awn-tipped bracts 6–14 mm. long, then sparsely ciliate along margins; calyx 8–10 mm. long, upper lip of oblique orifice about three times as long as lower, teeth of latter minutely spinescent; corolla blue, 10–13 mm. long, upper lip erose-denticulate and shallowly cleft, erect, lower with short lateral lobes and a larger, fan-shaped middle lobe fimbriate along its margins; filaments very short, both arms of connective bearing anther cells; nutlets narrowly obovoid, rounded on back, flattened dorsiventrally, 2–2.2 mm. long, dark buff with irregular flecks of dark brown, smooth and shining.

Sandy, gravelly, or rarely clay soil on hillsides and plains, Lower and Upper Sonoran Zones, Sacramento Valley, California, to southern Nevada, Arizona, Sonora, and central Baja California. March–July.

**3. Salvia galinsogifolia** Fernald, Proc. Amer. Acad. **35:** 498, 1900

Annual, with sharply quadrangular, appressed-hirsutulous stems 2–6 dm. tall, internodes slender, 3–10 cm. long; petioles slender, 1–5 cm. long; leaf blades broadly ovate, truncate to rounded at base, acute to acuminate at apex, 2.5–4 cm. wide, 3–6 cm. long, sparsely hirsute along veins, otherwise glabrous, dark green above, paler beneath, margins crenate-serrate; flowers in globose verticils 8–15 mm. in diameter, 1–15 cm. apart on slender scapes; bracts broadly ovate to ovate-orbicular, abruptly short-acuminate, 5–8 mm. long, sparsely hirsutulous to glabrate, pinkish to purplish, somewhat punctate-glandular; calyx 3–3.5 mm. long, strongly hirsute on veins and with slightly shorter hairs on outside between angles, thus older glomerules grayish; corollas bluish, tube 3–3.5 mm. long, upper lip 1–1.5 mm. long, erect and nearly straight but strongly concave-convex, pubescent on outside, regularly ciliate-margined with gland-tipped hairs; lower lip 3.5–4 mm. long, its middle lobe flabelliform, emarginate, erosulate-denticulate and sinuate along its margins, pubescent without; filaments of stamens slender, 0.6–0.8 mm. long, glabrous, slightly arcuate, upper arm of connective 0.2–0.3 mm. long, bearing a fertile anther, lower arm 1.0–1.7 mm. long, lanceolate, sterile, hairy, connivent with that of other stamens and two lower arms held together by intertangled hairs.

Shaded slopes, meadows, and stream banks, Lower and Upper Sonoran Zones, southern Sonora south to Guerrero, east to Chihuahua and Durango. Nov.–March.

**4. Salvia hyptoides** Mart. & Gal., Bull. Acad. Brux. **11:** 74, 1844

*Salvia elsholtzioides* Benth., Bot. Sulph. 152, *pl. 50,* 1844.
*Salvia hyptoides* var. *subspicata* Fernald, Proc. Amer. Acad. **35:** 498, 1900.

Annual herbs 20–60 cm. tall with slender quadrangular stems appressed-hirsutulous, internodes 3–15 cm. long; petioles slender, 1–5 cm. long, appressed-hirsutulous; leaf blades lanceolate-ovate to deltoid-ovate, 1.5–4 cm. wide, 3–6 cm. long, truncate-rounded at base, acute to acuminate at apex, crenate-dentate, dark green and sparsely hirtellous above, paler beneath, punctate-glandular on both surfaces; scapes slender, often 2–3 dm. long; bracts broadly ovate-cordate to nearly orbicular, 6–10 mm. wide and long, erose-denticulate, glandular-punctate, thin, often pinkish or purplish, glomerules globose, 10–15 mm. in diameter, lower ones 1–10 cm. apart, upper closer together; calyx 3.5–4.5 mm. long, purplish, densely hispid at base, hairs fewer and shorter on adaxial side and on teeth, these about 1 mm. long; corolla tube 2.5–3.5 mm. long, upper lip about 1 mm. long, lower 2.5–4.5 mm. long, both lips sparsely puberulent without; filaments slender, 0.4–0.6 mm. long, arcuate; lower connective arm lance-deltoid, 1.4–1.6 mm. long, sterile, fleshy; upper arm 0.4 mm. long, strongly arcuate; nutlets lenticular-orbicular, 0.8 mm. long and wide, dark brown, smooth and shining.

Clay slopes and rocky hillsides, Tropical and Lower Sonoran Zones, southern Sonora to Michoacán, south to Venezuela and Colombia. Sept.–Feb.

**5. Salvia riparia** H. B. K., Nov. Gen. & Sp. **2:** 300, 1817

*Salvia privoides* Benth., Bot. Sulph. 150, 1844.
*Salvia viscosa* Sessé & Moc., Pl. Nov. Hisp. II, **1:** app. 8, 1887.
*Salvia lateriflora* Fernald, Proc. Amer. Acad. **35:** 494, 1900.

Diffusely branched and often procumbent annual herb 3–5 dm. tall with slender, crispate-puberulent or rarely glabrate branches; petioles slender, to 1.5 cm. long, gradually broadening upward; leaf blades ovate, acute at apex, broadly cuneate to subtruncate at base, 8–20 mm. wide, 1.5–3.5 cm. long, crenate-serrate, subcinereous to dark green on upper surface, usually ashy beneath, sparsely and indistinctly gland-dotted; verticils few-flowered, 1–5 cm. apart; bracts 3–4 mm. long, ovate-acuminate;

spikes 1–2 dm. long; pedicels 0.5–3 mm. long; calyx 3–3.5 mm. long, campanulate, hirsute, with gland-tipped, capitate hairs, lips divergent, upper rounded or truncate, 1–3-mucronate, lower teeth spinose-tipped; corolla blue, tube 2.5–3 mm. long, upper lip about 2 mm. long, galeate, lower lip 3 mm. long, with irregular whitish stripes in throat; filament nearly straight, connective arms one and one-half to two times as long as filament, lower one doubled back on itself and bearing a flattened callosity at tip.

Beds and banks of arroyos, moist meadows, and hillsides, Lower Sonoran and Tropical Zones, Sonora and Baja California to Tamaulipas, south to Peru. Nov.–March.

### 6. **Salvia misella** H. B. K., Nov. Gen. & Sp. **2:** 290, 1818

*Salvia obscura* Benth., Lab. Gen. & Sp. 245, 1833.

Much-branched annual with procumbent to erect, slender branches 3–4 dm. long, herbage sparsely to moderately puberulent with crisped whitish hairs, older stems and leaves often glabrate; petioles slender, 4–10 mm. long; leaf blades broadly ovate to rhomboidal, 1–2 cm. wide, 1.5–4 cm. long, cuneate to subtruncate at base, acute at apex, crenate-serrate, dark green and sparsely pubescent with appressed hairs, slightly more densely hispidulous beneath; verticils 1–12-flowered, subtended by broadly ovate, abruptly acuminate bracts 2–4 mm. long, forming racemes 1–3 dm. long; pedicels 1–2 mm. long; calyx 2–2.5 mm. long at anthesis, ovoid-cylindrical, hispid with gland-tipped, spreading hairs, upper lobe rounded, acuminate, lower lip of 2 distinct, rounded-deltoid, acuminate teeth 1–1.5 mm. long, calyx 3.5–5 mm. long in fruit; corolla blue, tube 2.5 mm. long, upper lip 1.5 mm. long, galeate, straight; lower lip 2.5 mm. long; nutlets broadly ellipsoidal, 1.2 mm. wide, 1.8 mm. long, dark buff mottled with brown, smooth and sublustrous.

Arroyos, canyons, and shaded hillsides, Lower Sonoran to Transition Zones, southern Baja California and Tamaulipas, Mexico, south to Yucatán, east to Cuba. Aug.–Feb.

### 7. **Salvia setosa** Fernald, Proc. Amer. Acad. **35:** 493, 1900

*Salvia occidentalis* var. *garberi* Chapm., Bot. Gaz. **3:** 11, 1878.

Annual with few to many ascending slender branches 2–4 dm. tall; herbage sparsely hispid with curved, rather coarse hairs to glabrate; petioles 5–30 mm. long, gradually widening upward; leaf blades narrowly to broadly ovate, 1–4 cm. wide, 2–7 cm. long, acute, broadly cuneate at base, crenate-serrate, dark green, sparsely to moderately hispid or eventually glabrate on both surfaces; flowers 1–3 in each verticil, forming racemes 2–12 cm. long; bracts linear-lanceolate to ovate-acuminate, 3–6 mm. long, setose-tipped; pedicels stoutish, 1–2 mm. long; calyx 4–5 mm. long at anthesis, 7–10 mm. long in fruit; hispid with simple, stiffish, white hairs and with sessile globose glands between bases of hairs, upper lip rounded, usually with 3 but sometimes with only 1 spinose tip; lower lip of 2 deltoid-rounded, spinose-tipped teeth 2–3 mm. long; corolla blue, tube 3 mm. long, upper lip puberulent, 2 mm. long, shallowly hooded, lower lip 3–3.5 mm. long, middle lobe broadly retuse-obcordate; stamens included, filament 0.8–1 mm. long, lower arm of connective appearing doubled back upon itself, 1.5–1.8 mm. long when folded, papillate-glandular at its tip, upper arm 1.4–1.6 mm. long, slightly curved; nutlets obovoid, 2–2.2 mm. long, brownish, smooth and sublustrous.

Grassy slopes, flats, and shaded hillsides, Lower Sonoran and Tropical Zones, southern Baja California and Sonora to Florida, south into Oaxaca and Nayarit, Mexico. Aug.–Dec.

**8. Salvia subincisa** Benth., Pl. Hartw. 20, 1839

Erect annual herb with several to many slender branches 3–5 dm. tall, internodes 3–10 cm. long, quadrangular in cross section, closely glandular-hirtellous with capitate hairs; petioles slender, glandular-hirtellous, 1–1.5 cm. long, broadening gradually above to cuneate base of leaf blades, these oblong-elliptic to linear-lanceolate, 3–15 mm. wide, 1.5–6 cm. long, acute to attenuate at apex, irregularly incised-serrate, sparsely hirtellous and minutely glandular, slightly paler beneath than above; flowers usually solitary but sometimes in 2–3-flowered verticils 1–3 cm. apart in a raceme 3–15 cm. long; bracts broadly ovate, 2–3.5 mm. long, ciliate-glandular, caducous; calyx 4.5–5.5 mm. long in flower, 7–8.5 mm. long and flaring in fruit, markedly glandular-hispid; corolla tube 4–4.5 mm. long, upper lip 2–2.5 mm. long, bluish, minutely and sparsely hirtellous without; lower lip 7–8 mm. long, slightly hirtellous beneath; upper and lower arms of connective about equal, upper barely included in upper lip, lower arm with a conspicuous glandular lateral callosity; nutlets turgidly ellipsoidal, 2.5–3 mm. long, 1.8–2 mm. wide, smooth but dull, dark buff or pale brown.

Along canyons and on rocky shaded slopes and margins of meadows, Lower and Upper Sonoran Zones, southeastern Arizona and New Mexico south to Sonora, Chihuahua, Durango, and Aquascalientes. Aug.–Sept.

**9. Salvia tiliifolia** Vahl, Symb. Bot. **3:** 7, 1794

*Salvia fimbriata* Kunth, in H. B. K., Nov. Gen. & Sp. **2:** 199, *pl. 149,* 1818.

Erect, sparsely to much-branched annual herb 3–12 dm. tall with quadrangular, crispate-hirtellous stems; petioles slender, 1–6 cm. long, appressed-hirtellous; leaf blades thin, broadly ovate, acute to slightly attenuate at apex, broadly cuneate, rounded, or subtruncate at base, crenate-serrate, hispid-pilose on margins and veins, sparsely hirtellous elsewhere; inflorescences 5–25 cm. long, of 3–12-flowered verticils 1–2 cm. apart; bracts lanceolate, thin, 2–4 mm. long, caducous; pedicels slender, 2–5 mm. long, hirtellous, curved and fruiting calyces often pendent-reflexed; calyx 4.5–5.5 mm. long in flower, to 9 mm. long in fruit, sparsely appressed-hispidulous on prominent veins, lower teeth 1.5–2 mm. long; corolla blue, tube 3.5–4 mm. long, upper lip 1–2 mm. long, sparsely hirtellous on back near hooded tip, lower lip 2–2.5 mm. long, nearly glabrous; stamens barely exserted; nutlets broadly ellipsoidal, moderately compressed dorsiventrally, 1–1.2 mm. wide, 1.4–1.6 mm. long, minutely mottled with pale buff and dark brown, smooth and lustrous.

Fence rows, meadows, shaded hillsides, and plains, Lower Sonoran and Tropical Zones, Sonora and Chihuahua south to northern South America. Aug.–Nov.

**10. Salvia seemannii** Fernald, Proc. Amer. Acad. **35:** 516, 1900

Perennial herb or occasionally suffrutescent plant to 2 m. tall with numerous ascending, distinctly quadrangular, sparsely hirsute to glabrous stems; petioles slender, 1–5 cm. long, strongly grooved, hirtellous; leaf blades rhomboidal-ovate, 1–4 cm. wide, 2–8 cm. long, acute to short-acuminate at apex, narrowly to broadly cuneate at base, coarsely serrate, thin, dark green, and hirtellous to glabrate above, paler and inconspicuously glandular-punctate beneath; inflorescence of few-flowered verticils forming interrupted panicles 1–4.5 dm. long; bracts lanceolate, 2–3 mm. long; pedicels slender, 1–4 mm. long; calyx 5.5–7 mm. long, sparsely hirtellous to nearly glabrous, upper lip entire, rounded-acute, lower lip with 2 teeth, these acute, 1.5 mm. long; corolla blue, tube 6–6.5 mm. long, markedly saccate on lower side of back, upper lip bluntly galeate, 4 mm. long, bearded with purplish hairs 0.3–0.8 mm. long on outside, lower lip 6–7 mm. long, glabrous; stamens included, filaments 1.5 mm. long, slightly curved; upper arm of connective 2 mm. long, nearly straight, lower arm 2.5–3 mm. long, bearing a small appendage near its juncture with filament, semihastate in outline; nutlets broadly ellipsoidal, 0.7 mm. wide, 1.4–1.6 mm. long, dark buff to light brown, smooth, sublustrous.

Stream banks, margins of moist meadows, and canyons, Lower Sonoran and Tropical Zones, Sonora and Chihuahua to Durango and Sinaloa. Sept.–Nov.

### 11. **Salvia peninsularis** Brandegee, Zoe **5:** 108, 1901

Perennial herb with numerous erect or ascending branches 6–12 dm. tall, twigs and leaves soft-tomentulous or cinereous-puberulent; petioles 1.5–3 cm. long; leaf blades broadly ovate to deltoid-ovate, 2.5–6 cm. wide, 5–12 cm. long, truncate to broadly and shallowly subcordate at base, gradually tapering to acute to short-attenuate apex, sharply serrate, dark green above, slightly paler beneath; flowers pedicellate, in few-flowered verticils spicately arranged, inflorescence 10–30 cm. long; bracts small, caducous; pedicels 2–5 mm. long, slender; calyx 7–9 mm. long, cinereous-puberulent, upper lip 5-veined, minutely 3-toothed, lower lip of 2 equal deltoid teeth 1.5–1.8 mm. long; corolla lavender, tube 15–18 mm. long, slightly curved; upper lip galeate, 6–8 mm. long, pubescent with purplish hairs on back, lower lip 7–9 mm. long, glabrous or nearly so; stamens included in upper lip or barely exserted, filament 2–3 mm. long, nearly straight, upper arm of connective 5–6 mm. long, erect, straight or nearly so, lower arm 6–7 mm. long, extending downward in corolla tube about to apex of calyx, sterile; upper part of style and back of stigma branches heavily bearded.

Foothills and mountain canyons, Lower Sonoran Zone, central Baja California. Sept.–Dec.

### 12. **Salvia marci** Epling, in M. E. Jones, Contr. West. Bot. **18:** 52, 1933

Perennial herbaceous plant with many erect, obtusely quadrangular, cinereous-puberulent stems 5–12 dm. tall, lowermost branches sometimes decumbent and bearing upright lateral shoots 2–5 dm. long; petioles 1–3 cm. long, cinereous-puberulent with fine appressed hairs; leaf blades lance-ovate, 2–4 cm. wide, 4–15 cm. long, rounded at base, gradually attenuate toward tips, serrate-dentate, finely puberulent but green on upper surface, ashy and densely appressed-puberulent beneath, glandular-punctate on both surfaces, veins distinct; flowers in 6–12-flowered verticils spicately arranged, spikes 10–30 cm. long, smaller ones in axils of upper leaves on vigorous shoots; bracts small, caducous; calyx narrowly campanulate, 8–10 mm. long, minutely appressed-puberulent, cinereous, punctate-glandular, teeth deltoid, 1–1.5 mm. long; corolla purple, tube 18–20 mm. long, upper lip 12–18 mm. long, strongly galeate, soft-pubescent with long purplish hairs; lower lip 10–15 mm. long, nearly glabrous; nutlets ellipsoidal, slightly compressed dorsiventrally, 1.8–2 mm. long, 1 mm. wide, pale buff with faint, pale brown speckling and lines.

Canyon beds and rocky slopes, Lower Sonoran Zone, Sierra de la Giganta, Baja California. Oct.–Dec.

**13. Salvia parryi** A. Gray, Proc. Amer. Acad. **8:** 369, 1872

*Salvia confinis* Fernald, Proc. Amer. Acad. **35:** 523, 1900.

A small shrub 4–12 dm. tall with slender, minutely and closely canescent branches, older branches glabrate, striate-fissured, light brown; leaves lanceolate to lance-ovate, 5–15 mm. wide, 1–5 cm. long, rounded to broadly cuneate at base, acute to obtuse at apex, finely and regularly crenate-serrate, moderately to densely tomentose, and greenish or canescent on upper surface, more densely tomentose with minute dendritic trichomes and ashy beneath; petioles puberulent-tomentose, 2–10 mm. long; flowers blue, in verticillate heads in an interrupted spikelike raceme 10–15 cm. long, each head 8–15 mm. in diameter; bracts lance-ovate to broadly ovate, subcordate at base, acute to slightly attenuate at apex, 5–10 mm. long, often purplish, usually reflexed in age; calyx flaringly campanulate, 5–6 mm. long, 3–4 mm. in diameter at mouth, strongly striate-veined, floccose-lanate with much-branched hairs, canescent; corollas 6–7 mm. long, dark blue, lanate without, upper lip entire, boat-shaped, 3 mm. long, lower lip fan-shaped, 3–3.4 mm. long, shallowly 6–9-toothed, central lobe scarcely longer than lateral ones; stamens slightly shorter than and included within upper lip; lower arm of each connective lanate, with a wedge-shaped callosity two to three times as thick as shaft, triangular-acute at the tip, included within throat.

Gravelly and sandy soil, Lower and Upper Sonoran Zones, Santa Cruz, Cochise, and Pima Counties, Arizona, southwestern New Mexico, and south in Sonora at least to the vicinity of Magdalena. April–Sept.

**14. Salvia similis** Brandegee, Zoe **5:** 108, 1901

Suffrutescent plant to 2 m. tall with slender branches, cinereous with closely appressed, minute, white hairs or becoming glabrate with age; petioles slender, 1–3 cm. long; leaf blades lance-ovate to deltoid-ovate, 1–2.5 cm. wide, 2–8 cm. long, acute to attenuate at apex, cuneate to truncate at base, crenate-serrate, hirtellous-cinereous in youth, dark green and closely punctate-glandular in age; verticils few-flowered, arranged in spiciform racemes 5–15 cm. long; bracts linear to lanceolate, 5–15 mm. long, hirtellous; pedicels 1–4 mm. long, slender; calyx 5.5–7 mm. long, flaring upward, distinctly veined, densely hirtellous, punctate-glandular; corolla pale to dark blue, with white markings at base of lower lip, tube 5–6.5 mm. long, markedly saccate about middle of lower side; upper lip 3.5–4 mm. long, hooded, hirtellous; lower lip 7–9 mm. long, middle lobe not emarginate; stamens included, connective arms about equal, or lower slightly longer; style strongly bearded; nutlets 2–2.4 mm. long, pale brown with irregular dots and lines of dark brown, smooth, dull.

Margins of arroyos, rocky hillsides, and shaded canyons, Lower Sonoran Zone, Vizcaíno Desert south to the Cape Region, Baja California. Sept.–April.

**15. Salvia mohavensis** Greene, Pittonia **2:** 235, 1892

*Audibertia capitata* A. Gray, Proc. Amer. Acad. **7:** 387, 1868. Not *Salvia capitata* Schlecht., 1853.
*Audibertiella capitata* Briq., Bull. Herb. Boiss. **2:** 73, 1894.
*Ramona capitata* Briq., op. cit. 440.

Low, compact, much-branched shrub 4–10 dm. tall with pale brownish or dark buff, puberulent twigs and striate-flaky, gray bark on older stems, leafy parts of twigs usually 15 cm. long or less; herbage gland-dotted throughout; petioles 5–10 mm. long; leaf blades broadly ovate to oblong-ovate, 3–8 mm. wide, 5–18 mm. long, acute to subcordate at base, acute to rounded at apex, finely crenate, rugose and glandular above, minutely hispidulous and strongly reticulate-veined beneath; flowers in subglobose terminal heads 1–2 cm. in diameter, subtended by membranous, whitish, ovate-elliptic to linear-lanceolate bracts 10–15 mm. long, margins of bracts tomentulose; calyx 8–10 mm. long, pubescent and strongly punctate-glandular; corolla light blue to lavender, 15–20 mm. long, scabrous, upper lip 3–3.5 mm. long, bifid to middle, lower lip plane or nearly so, middle lobe entire, 4–5 mm. long; stamens exserted, connective shorter than filament, lower arm obsolete.

Arroyos, canyons, and rocky slopes, Lower Sonoran Zone, Mojave and Colorado Deserts, California, to western Arizona, and northwestern Sonora, Mexico. April–July.

**16. Salvia palmeri** A. Gray, Proc. Amer. Acad. **21**: 408, 1886

A slenderly branched shrub 6–10 dm. tall, with cinereously pubescent twigs and dark brown, nearly smooth bark on older branches; petioles slender, 1–2 cm. long, cinereous and closely sublanate with simple and dendritically branched hairs intermingled; leaf blades ovate, 1.5–4 cm. wide, 3–7 cm. long, broadly cuneate to rounded at base, acute and short-attenuate at apex, finely serrate marginally, dark green and minutely puberulent above, ashy, thickly sublanate and punctate-glandular beneath, veins distinct but only moderately prominent; flowers pedicellate, in verticils of 3–6, arranged in a spiciform raceme 3–7 cm. long; pedicels slender, 1–3 mm. long, minutely tomentulose; calyx narrowly campanulate, 5–6 mm. long, markedly ribbed, closely tomentulose and punctate-glandular, teeth 1.5–2 mm. long, deltoid, apiculate; corolla pale blue, 10–14 mm. long, tube 6–7 mm. long, upper lip bluntly galeate, 5–6 mm. long, pubescent with rather coarse, bluish hairs on back; lower lip 3-lobed, about as long as upper, middle lobe orbicular-retuse, glabrous within, sparsely hirsute beneath; stamens included within upper lip, filament nearly straight, 2 mm. long, connective arms essentially equal, 2–2.3 mm. long, lower one paralleling filament, sterile, with a small flange at its apex; nutlets broadly ellipsoidal, 1.5–1.8 mm. long, dark brown with pale buff arch over apex, smooth, glabrous, dull.

Canyon slopes and hillsides, Lower and Upper Sonoran Zones, central Sonora and Chihuahua. Sept.–Oct.

**17. Salvia platycheila** A. Gray, Proc. Amer. Acad. **8**: 292, 1870

An intricately branching, brittle shrub 1–2 m. tall with rather coarse branches and canescent to glabrate twigs; leaves coriaceous, densely appressed-tomentulose in youth, less so in age, petioles stoutish, 5–15 mm. long, blades broadly ovate to deltoid-ovate, 2–5 cm. long, 1.5–3.5 cm. wide, obtuse to rounded at apex, truncate, rounded, or rarely subcordate at base, subentire to shallowly crenulate; flowers in sparse verticils 1–5 cm. apart; bracts linear-lanceolate, 5–10 mm. long, purplish; calyx 8–10 mm. long, minutely but densely appressed-puberulent, canescent or purplish, finely punctate-glandular, subequal lips obtuse, upper 5–7-veined and curved in fruit; corolla blue or purplish, tube 6–7 mm. long, somewhat ventricose, upper lip about 5 mm. long, lower slightly longer.

Canyons and dry rocky ridges, Lower Sonoran Zone, on islands along the eastern shore of Baja California from Isla Santa Catalina southward. Sept.–Feb.

**18. Salvia californica** Brandegee, Proc. Calif. Acad. II, **2**: 197, 1889

Erect, much-branched shrub 1–2 m. tall, to 3 m. in diameter, branches erect or ascending; young twigs densely lanate-canescent, at length glabrate and revealing pale stramineous bark, this finally slightly striate and dull gray; leaves elliptic to ovate, 5–15 mm. wide, 10–25 mm. long, sessile, rounded to obtuse at base, acute at apex, irregularly toothed, each tooth pungently apiculate, densely to moderately floccose-lanate on both surfaces, rather prominently reticulate-veined beneath in mature leaves; flowers in interrupted, spikelike racemes of verticillate heads 1–2 cm. in diameter, bracts similar to leaves but progressively reduced from bottom toward top of inflorescences; calyx narrowly campanulate, 5–6 mm. long, striate at base, floccose-lanate, triangular teeth 0.6–1.2 mm. long, apiculate; corollas blue, 9–12 mm. long, bifid upper lip 2–2.5 mm. long, middle lobe of lower lip 4–5 mm. long, slightly erose or faintly toothed apically, fully twice as long as lateral lobes; stamens about equaling lower lip; nutlets obovoid, somewhat flattened dorsiventrally, about 2 mm. long, usually slightly apiculate at apex, pale brown to gray-brown, smooth and dull.

Sandy arroyos and plains, Lower Sonoran Zone, Bahía de los Angeles south to vicinity of El Arco and west to Miller's Landing, Baja California. May–Dec.

**19. Salvia pinguifolia** (Fernald) Woot. & Standl., Contr. U.S. Nat. Herb. **16:** 169, 1913

*Salvia ballotaeflora* var. *eulaliae* Fernald, Proc. Amer. Acad. **35:** 522, 1900.
*Salvia ballotaeflora* var. *pinguifolia* Fernald, op. cit. 523, 1900.

Low shrub 0.5–1.5 m. tall with cinereous-puberulent twigs and leaves, hairs simple and minute; twigs slender, internodes 1–3 cm. long; petioles 4–15 mm. long; leaf blades broadly deltoid-ovate, 2–5 cm. long and about as wide, obtuse to short-acuminate at apex, subcordate to broadly cuneate at base, margins coarsely crenate-serrate, upper surfaces soft-hirsutulous, pale green, lower surface ashy to silvery and densely but minutely tomentulose; bracts lanceolate, deciduous, 4–6 mm. long; flowers in lax glomerules, whole inflorescence 3–8 cm. long; calyx 6–7 mm. long at anthesis, 9–10 mm. long in fruit, cinereous-hirsutulous, lower lip 3.5–5 mm. long; corollas bluish, tube 4–5 mm. long, upper lip 4.5–5 mm. long, lower one usually slightly longer, both lips puberulent without; stamens about equaling lips or slightly exserted, filaments and connectives nearly straight.

Mountainsides and canyons, mostly Upper Sonoran Zone, southeastern Arizona to western Texas and south into Chihuahua. Aug.–Dec.

**20. Salvia cedrosensis** Greene, Bull. Calif. Acad. **1:** 212, 1885

A low, slenderly branched shrub 1 m. tall or less, twigs distinctly quadrangular and minutely but closely puberulent-canescent with glistening white, branched hairs; leaves on slender petioles 2–10 mm. long, blades broadly ovate to lance-ovate, 4–25 mm. wide, 1–4.5 cm. long, truncate, subcordate, or broadly cuneate at base, acute to obtusely rounded at apex, serrate-dentate, puberulent-canescent to glabrate, glandular-punctate on both surfaces; racemes of interrupted verticillate heads or whorls 1.5–2 cm. in diameter, each raceme 5–15 cm. long; bracts minute or lowermost pair to 7 mm. long, foliaceous; calyx broadly tubular, 6–8 mm. long, usually purplish, strongly striate-lined, somewhat lanate, teeth 1–2 mm. long, triangular; corollas dark blue, 14–18 mm. long, pubescent on outer surface of both lips, tube glabrous; upper lip entire, boat-shaped, 3.5–4 mm. long, straight; lower lip slightly curved downward, 5–5.5 mm. long, margins of central lobe slightly wavy but hardly erose; lower arm of connective equaling or slightly exceeding upper arm, about 4 mm. long, straight or faintly curved, glabrous or minutely granular-papillate.

Washes and rocky hillsides, Lower Sonoran Zone, Cedros Island and vicinity of Magdalena Bay, Baja California. Jan.–March.

**21. Salvia goldmanii** Fernald, Proc. Amer. Acad. **35:** 527, 1900

A straggly shrub 1–2 m. high with brownish, scurfy-pubescent, quadrangular twigs, hairs dendritically few-branched; leaves ovate-lanceolate to elliptic-lanceolate, 6–25 mm. wide, 2–6 cm. long, obtuse to broadly cuneate at base, acute to apiculate at apex, serrulate-denticulate, puberulent but green on upper surface, densely short-lanate, cinereous, and distinctly veined beneath; petioles 1–1.5 cm. long; upper leaves and those in axillary fascicles much reduced; inflorescence of 5–12 few-flowered verticils 0.5–3 cm. apart, only lowest subtended by a pair of foliaceous bracts about 1 cm. long; calyx tubular-campanulate, floccose-scurfy, 6–7 mm. long, upper lip entire, teeth of lower lip broadly deltoid-apiculate, 1–1.5 mm. long; corolla 10–12 mm. long, tube only slightly exceeding calyx, sublanate with dendritically much-branched white hairs on outside, upper lip 4–5 mm. long, hoodlike; lower lip 6–7 mm. long, 3-lobed, middle lobe somewhat erose and undulate-margined; lower arm of anther connective lanceolate, twice as long as strongly curved, stoutish filament, upper arm equaling or slightly exceeding lower arm, moderately arcuate; nutlets densely puberulent.

Canyons, stream banks, and rocky slopes, mostly Upper Sonoran Zone but straggling into the upper margins of the Lower Sonoran Zone, southern Sonora and Chihuahua, south into Sinaloa. Aug.–Nev.

**22. Salvia chionopeplica** Epling, Ann. Mo. Bot. Gard. **27:** 260, *pl. 29, figs. 3–4,* 1940

A much-branched shrub 6–15 dm. tall with floccose-canescent twigs and young leaves, older branches with dark gray, slightly striate, thin bark; petioles slender, floccose, 5–15 mm. long; leaf blades ovate to oblong-lanceolate, 5–15 mm. wide, 1.5–5 cm. long, obtuse to rounded at apex, narrowly to broadly cuneate at base, margins crenulate, upper surface strongly bullate, lower markedly reticulate-venose, both surfaces more or less floccose-snowy with dendritically branched white hairs; flowers in dense globose glomerules 1–3.5 cm. in diameter, 1–5 such glomerules on each flowering branch, strongly floccose-lanate; bracts foliaceous, to 3.5 cm. long; calyx 9–10 mm. long in flower, 12–13 mm. long in fruit, densely floccose, orifice oblique, faintly toothed, lower teeth 0.3–0.4 mm. long; corolla lavender with small spots of red, slender tube 12–13 mm. long, with a median pilose ring inside, upper lip 3.5–4 mm. long, lower 5–6 mm. long, its middle lobe broadly obcuneate and slightly deflexed; stamens exserted and exceeding lips 2–4 mm., filament proper nearly equaling upper lip of corolla, lower arm of connective minute, about 0.6 mm. long, upper arm arcuate, 3–3.5 mm. long; nutlets oblong, 2–2.4 mm. long, dark brown.

Rocky hillsides and outwash slopes, Lower and Upper Sonoran Zones, foothills along western slopes of Sierra San Pedro Mártir from vicinity of San José del San Telmo (Meling's Ranch) south to San Fernando, Baja California. March–May.

**23. Salvia apiana** Jepson, Muhlenbergia **3:** 144, 1908

*Audibertia polystachya* Benth., Lab. Gen. & Sp. 314, 1833. Not *Salvia polystachya* Mart. & Gal., 1844.
*Ramona polystachya* Greene, Pittonia **2:** 235, 1892.
*Audibertiella polystachya* Briq., Bull. Herb. Boiss. **2:** 73, 1894.
*Salvia californica* Jepson, Fl. W. Mid. Calif. 460, 1901. Not *S. californica* Brandegee, 1889.

Shrubby plants 1–3 m. tall with woody base, herbaceous upper branches, and densely and minutely tomentose-canescent, evergreen leaves, these mostly at base of erect seasonal branches; petioles 1–2 cm. long; leaf blades oblong-lanceolate, 5–10 cm. long, acute at apex, cuneate to rounded at base, evenly crenulate, hoary-tomentose on both surfaces; inflorescences thyrsoid panicles to 10 cm. broad, 20–60 cm. long; bracts and bractlets 3–20 mm. long, lanceolate to narrowly oblanceolate, reflexed in age; calyx 5–6 mm. long, upper lip truncate to shallowly 3-toothed, lower lip shorter, with 2 triangular teeth; corolla white to pale lavender, 12–16 mm. long, upper lip 1 mm. long or less, lower prolonged, 3–5 mm. long, its middle lobe unguiculate-rounded apically, erosulate; stamens well exserted, divaricate; styles slender, exserted 12–20 mm.; nutlets oblong, 2.5–3 mm. long, pale brown, dull, faintly reticulate-lined with darker brown.

Sandy washes, rocky hillsides and benches, Lower and Upper Sonoran Zones, Santa Barbara County, California, south to within 35 miles of Punta Prieta, Baja California. Mostly above the limits of the Sonoran Desert. March–July.

Hybrids between this species and *S. mellifera* have been found sporadically almost throughout the range of *S. apiana*.

**24. Salvia vaseyi** (Porter) Parish, Muhlenbergia **3:** 126, 1907

*Audibertia vaseyi* Porter, Bot. Gaz. **6:** 207, 1881.
*Audibertiella vaseyi* Briq., Bull. Herb. Boiss. **2:** 73, 1894.
*Ramona vaseyi* Briq., op. cit. 440.

A much-branched, rounded shrub 0.5–1.5 m. tall with long, wandlike flowering branches with leaves crowded at base, young stems cinereous, minutely and thinly hirtellous, dotted with sessile globose glands; petioles 1–2 cm. long, cinereous; leaf blades oblong-ovate to oblong-lanceolate, 0.5–3 cm. wide, 1–5.5 cm. long, acute to rounded at apex, truncate to shallowly subcordate at base, finely crenate, densely strigose-tomentulose and glandular-dotted on both surfaces; verticils few- to many-flowered, subglobose, 2–3.5 cm. in diameter, distantly arranged in an interrupted spike; bracts narrowly ovate, aristate-spinulose, 1–2 cm. long; calyx 7–8 mm. long in flower, to 12 mm. long in fruit (including aristate teeth), awnlike teeth to 6 mm. long; corolla white, 15–22 mm. long, upper lip suborbicular, retuse, sparsely puberulent on outer surface, lower lip 7–10 mm. long, glabrous, central lobe subreniform; style and stamens exserted beyond length of lower lip; lower connective arms obsolete or essentially so, not prolonged below articulation with filament; nutlets ellipsoidal, somewhat dorsiventrally compressed, 2 mm. wide, 3.5–4 mm. long, pale brown, smooth and shining.

Gravelly washes and rocky slopes, Lower Sonoran Zone, western edge of the Colorado Desert from southern San Bernardino County, California, to northern Baja California. April–June.

**25. Salvia munzii** Epling, Madroño **3:** 169, 1935

*Salvia mellifera* var. *jonesii* Munz, Bull. S. Calif. Acad. **26:** 24, *pl. 4, figs. 20–21,* 1927.
*Salvia mellifera* subsp. *jonesii* Abrams, Ill. Fl. Pacif. States **3:** 641, 1951.

Low compact shrub to 1 m. tall with slender, purplish or reddish twigs and dark gray, striate bark on older stems; leaves oblanceolate, 3–8 mm. wide, 1–3 cm. long, serrate, rugose and dark green above, reticulate beneath, punctate-glandular and cinereous-pubescent on both surfaces, blades narrowing gradually to a short petiole 1–6 mm. long or subsessile; glomerules spicately arranged, 1–5 cm. apart, 1–1.5 cm. in diameter, bracts foliaceous, or broadly ovate in upper verticils; calyx 4–5 mm. long, orifice oblique, upper lip slightly bicuspidate, lower of 2 distinct, spinulose-tipped teeth 1.5–2 mm. long, purplish, puberulent and punctate-glandular; corolla clear blue, 8–10 mm. long, arcuately curved, glabrous, upper lip about 2 mm. long, lower 4–5 mm. long, middle lobe nearly orbicular, undulate-margined; filaments 1.8–2 mm. long, straight or nearly so, upper arm of connective 0.6–1 mm. long, essentially straight, lower arm obsolete; nutlets obovoid, 1.8–2 mm. long, dark brown, smooth but dull.

Clay and rocky hillsides, and along arroyos and canyons, mainly Upper Sonoran Zone but occasionally in Lower Sonoran Zone, southern San Diego County, California, south to vicinity of Rosario, Baja California. Nov.–April.

**26. Salvia mellifera** Greene, Pittonia **2:** 236, 1892

*Audibertia stachyoides* Benth., Lab. Gen. & Sp. 313, 1833. Not *Salvia stachyoides* Kunth, 1817.
*Audibertiella stachyoides* Briq., Bull. Herb. Boiss. **2:** 73, 1894.
*Ramona stachyoides* Briq., op. cit. 440.

Leafy, much-branching shrub 1–2.5 m. tall with cinereous-tomentose to glabrate twigs and punctate-glandular herbage; petioles to 12 mm. long or leaves subsessile; leaf blades oblong-lanceolate, 5–20 mm. wide, 2–6 cm. long, acute to rounded at apex, crenulate, rugose and dark green above, cinereous and tomentulose beneath; verticils 1–6 cm. apart, to 2 cm. in diameter; bracts ovate to oblong, 5–10 mm. long, cuspidate, greenish to purplish, densely hispidulous; calyx 6–8 mm. long, hispidulous, upper lip tricuspidate, lower of 2 distinct, spine-tipped teeth about 1.5 mm. long; corolla pale blue-lavender, 10–12 mm. long, tube with a narrow ring of hairs within, upper lip 2–3 mm. long, shallowly retuse, lower lip 4–5 mm. long, middle and lateral lobes nearly equal; stamens slightly exserted, upper arm of connective about equaling filament, lower arm a short spur, sterile.

Brushy slopes, hillsides, and along arroyos, mainly Upper Sonoran Zone but entering upper parts of Lower Sonoran Zone, Contra Costa and Stanislaus Counties, California, south to northern Baja California. Feb.–Aug.

**27. Salvia eremostachya** Jepson, Man. Fl. Pl. Calif. 870, 1925

Low, much-branched shrub 2–12 dm. tall, twigs canescent with spreading glandular hairs; petioles stoutish, 3–8 mm. long, narrowly margined; leaf blades oblong to linear, 4–6 mm. wide, 1.5–3.5 cm. long, truncate to slightly subcordate at base, obtuse to rounded at apex, crenulate and usually revolute along margins, scabrous-hispid on both surfaces, bullate above, strongly reticulate-veined beneath; flowers in 1–3 whorls 1–1.5 cm. in diameter on each flowering shoot; bracts round-ovate, short-acuminate, thin, glandular and sparsely hispid, to 10 mm. long; calyx 8–11 mm. long, with an entire or 2–3-spinose upper lip 4–5 mm. long, lower lip of 2 distinct lobes 4.5–5 mm. long; corolla violet to rose, tube 14–17 mm. long, somewhat curved to nearly straight, bearing a hairy ring at about middle within, lobes of upper lip 4–6 mm. long, entire to erose; lower lip with middle lobe rounded but more or less cleft, margins irregularly erose; filaments and connectives about equal, exserted 4–6 mm. beyond lower lip, lower arm of connective obsolete.

Rocky desert slopes and canyons, Lower Sonoran Zone, Santa Rosa Mountains, San Diego County, California, south into northern Baja California. April–Nov.

**28. Salvia greatae** Brandegee, Zoe **5:** 229, 1906

A much-branched shrub 1–1.5 m. tall with white, moderately hirsute twigs and light gray, flaky bark on older branches; leaves sessile or short-petiolate, blades thick and leathery, ovate to ovate-lanceolate, pinnatifid-toothed, with 2–3 pairs of stiff, spine-like teeth and attenuate into a prominent terminal spine, 1–2.5 cm. wide, 2–4 cm. long, including spines, prominently veined, thinly tomentose with branched slender hairs; verticils 6–10-flowered, 1.5–2 cm. in diameter, 3–4 cm. apart, each glomerule subtended by a pair of bracts resembling leaves and by several smaller inner ones; calyx 7–8 mm. long, tomentose, its upper lip tipped with a spine 3–4 mm. long and 2 smaller ones near its base, lower lip bifid into 2 linear-lanceolate, spinulose-tipped lobes 1.5–2 mm. long; corolla pale lavender, 14–15 mm. long, its upper lip 3 mm. long, bilobed, lower lip 3-lobed, slightly longer, its middle lobe irregularly erose-fimbriate; lower arm of anther connective about one-half as long as upper arm, bearing an anther cell near its tip.

Desert washes, Lower Sonoran Zone, Orocopia Mountains, Riverside County, California. April.

## 9. MONARDA L., Sp. Pl. 22, 1753

Perennial or annual, aromatic herbs with erect quadrate stems and dentate or serrate, usually petiolate leaves, herbage copiously punctate-glandular. Flowers in capitate clusters and often in upper axils, showy, subtended by foliaceous or colored bracts and bractlets. Calyx tubular, 15-veined, subequally 5-toothed, usually villous in throat. Corolla puberulent or glandular without, glabrous within, 2-lipped; upper lip erect, more or less hooded, entire to emarginate; lower lip spreading, 3-lobed, middle lobe largest. Fertile stamens 2, exserted or included within hood of upper corolla lip; anthers linear, with divaricate sacs, upper pair rudimentary or lacking. Styles 2-lobed at apex; ovary deeply 4-parted. Nutlets ovoid, smooth.

### 1. **Monarda austromontana** Epling, Madroño **3:** 28, 1935

Erect, slender, annual plant 1.5–8 dm. tall, stems usually simple but sometimes branching in some or most of axils; herbage cinereous, densely punctate-glandular, puberulence fine, closely appressed; petioles 3–20 mm. long, faintly winged; leaves narrowly elliptical, oblong, or oblanceolate, 8–15 mm. wide, 3–7 cm. long, gradually tapering to petiole, acute to obtuse at apex, shallowly and distantly subserrate or sometimes nearly entire, appressed-puberulent on both sides or becoming glabrate above, margins sparingly and irregularly bristly; flowering glomerules 2–3 cm. in diameter, 1–6 cm. apart, subtended by lanceolate bracts 1–3 cm. long, these bristly-ciliate and aristate, usually reflexed at maturity, green, whitish, or tinged with purple; calyx tubes 4.5–6.5 mm. long, sparsely puberulent to glabrate, teeth 1.5–4 mm. long, aristate; corolla white to rose, usually immaculate, tube 9–13 mm. long, abruptly dilated to a throat 3.5–6 mm. long, upper lip 5–7 mm. long with a notch 1–1.5 mm. deep, lower lip 4–5 mm. long; stamens included within upper lip; nutlets 1.2–1.4 mm. long, broadly ovoid, pale brown, smooth but dull.

Slopes; mesas, and mountain meadows, usually with grasses, brushland upward into pine forests, rarely straggling to upper edges of Sonoran Desert along washes or in cultivated fields; southeastern Arizona and adjacent New Mexico south into Durango, Mexico. July–Oct.

## 10. HEDEOMA Pers., Syn. Pl. 2: 131, 1807

Annual or perennial herbs or somewhat suffrutescent plants with strongly aromatic, glandular-punctate twigs and herbage. Leaves small, sessile or short-petioled, entire or shallowly toothed. Flowers small, in few-flowered cymules or solitary in axils of upper leaves. Calyx tubular, 13-nerved, 5-toothed, faintly to distinctly 2-lipped; upper 3 teeth united below middle; lower pair distinct and slightly exceeding upper. Corolla blue or purple, 2-lipped; upper lip erect, entire, emarginate or 2-lobed, straight and nearly plane; lower lip 3-lobed, spreading. Anther-bearing stamens 2, partially over-arched by upper lip, 2 sterile stamens reduced to scales or lacking. Style glabrous, 2-lobed at apex. Nutlets ovoid, smooth.

Corolla tube glabrous within or nearly so; stems 25–50 cm. long, erect and rather stiff ........ 1. *H. oblongifolia*

Corolla tube hirtellous and often with a dense ring of hairs near middle within; stems mostly less than 25 cm. tall, slender, wiry and often decumbent at base:

- Leaves strongly dentate, 10–25 mm. long; a ring of erect, stiff, white hairs in calyx throat; corollas 9–10 mm. long ........ 2. *H. dentata*
- Leaves minutely dentate to subentire, rarely over 15 mm. long; hairs not conspicuous in throat of calyx; corollas 5–7 mm. long:
  - Leaves 5–10 mm. long; calyx tube 2.5–3 mm. long; corolla about equaling calyx teeth ........ 3. *H. nana* subsp. *nana*
  - Leaves 10–15 mm. long; calyx tube 3.5–4.5 mm. long; corolla slightly exceeding calyx teeth ........ 3*a*. *H. nana* subsp. *macrocalyx*

## 1. Hedeoma oblongifolia (A. Gray) Heller, Muhlenbergia 1: 4, 1900

*Hedeoma piperita* var. *oblongifolia* A. Gray, Proc. Amer. Acad. 8: 367, 1872.
*Hedeoma thymoides* var. *oblongifolia* A. Gray, Syn. Fl. N. Amer. $2^1$: 362, 1878.

Suffrutescent plants 2.5–5 dm. tall, branches usually numerous, erect or ascending, internodes 1–5 cm. long, rather closely pilose with curled or crisped hairs 0.5–1 mm. long; leaves ovate to oblong, 2–8 mm. wide, 5–15 mm. long, subsessile or on petioles to 2 mm. long, cuneate to obtuse at base, acute to minutely apiculate at apex, faintly and distantly serrulate or upper ones entire, hispidulous with curved grayish hairs, veins moderately prominent beneath; cymules several-flowered; pedicels slender, 1–4 mm. long, hispidulous; calyx tube 4–5 mm. long, slightly arcuate and somewhat swollen near base in fruit, hispidulous on prominent nerves, teeth reddish or purplish, linear-subulate, upper 3 about 1.5–2 mm. long, lower pair 2–3 mm. long; corollas purplish, tube about equaling calyx tube, hirsutulous without, nearly glabrous within, lobes about equaling or slightly surpassing calyx teeth; nutlets 0.8–1 mm. long, cinnamon brown, dull.

Hillsides and mountain slopes, upper margins of the Lower Sonoran to the Upper Sonoran Zones, Mohave and Pima Counties, Arizona, into New Mexico and adjacent northern Mexico. March–Oct.

## 2. Hedeoma dentata Torr., Bot. Mex. Bound. 130, 1859

Suffrutescent plants with several erect or ascending branches 3–6 dm. long from a slender rootstock; herbage pilose to nearly glabrous; internodes 5–30 mm. long; leaves elliptic to rhomboidal, 3–9 mm. wide, 10–25 mm. long, broadly cuneate at base and gradually narrowing to petiole, 2–5-dentate on each side, or uppermost sometimes subentire, thin, nearly scabrous, hairs short, slightly curved and stiff, punctate glands numerous; cymules several-flowered; pedicels very slender, 2–4 mm. long; calyx tube slenderly cylindrical, 1–1.3 mm. wide, 4–5 mm. long, finely ribbed and short-hispidulous, a ring of erect white hairs at apex of tube, swollen in lateral one-fourth to one-third; calyx teeth subulate, upper three 1–1.2 mm. long, somewhat spreading, lower pair straight, erect, 1.6–2 mm. long, all of them ciliate-hispid marginally;

corollas pale blue, puberulent without, tube 6–7 mm. long, upper lip 1.5 mm. long, lower lip 3 mm. long; stamens equaling upper lip, anthers deep purple or reddish; nutlets narrowly obovoid, about 0.8 mm. long, pale brown, smooth and dull.

In rock crevices, on steep rocky slopes, and occasionally along washes, mostly Upper Sonoran and Arid Transition Zones, southern Arizona south into northern Mexico. April–Oct.

**3. Hedeoma nana** (Torr.) Briq., in Engler & Prantl, Nat. Pflanzenf. **4**[3a]**:** 294, 1897, subsp. **nana**

*Hedeoma dentata* var. *nana* Torr., Bot. Mex. Bound. 130, 1859.
*Hedeoma thymoides* A. Gray, Syn. Fl. N. Amer. **2**[1]**:** 362, 1878. Not *H. thymoides* Pers., 1807.
*Hedeoma nana* subsp. *typica* W. S. Stewart, Rep. Spec. Nov. Beih. **115:** 29, 1939.

Low suffrutescent plants with several to many slightly decumbent to ascending stems 10–25 cm. long, herbage moderately closely beset with white, downwardly curving, simple hairs 0.5–1 mm. long, a tinge of red or lavender often present on various parts; petioles slender, 1–3.5 mm. long; leaf blades ovate to elliptic, sometimes rhomboidal-ovate, 2–6 mm. wide, 5–10 mm. long, denticulate to subentire, obtuse to acute at base, acute at apex, dark green above, paler and somewhat veiny beneath, finely punctate; cymules few-flowered; pedicels 1–4.5 mm. long; calyx tube mostly 2.5–3 mm. long, finely ribbed, distinctly swollen in basal one-half at maturity, hispidulous; calyx teeth reddish, upper three ovate-subulate, 1 mm. long, slightly curved outward, lower pair narrowly subulate, 2 mm. long, curved inward; corolla blue to purple, often with darker spots and streaks in lower lip, tube about equaling calyx, hispidulous without; nutlets almost plano-convex, 1.2–1.4 mm. long, pale brown to deep buff, smooth, dull.

Rock crevices, rocky slopes, and exposed ridges, upper margin of the Lower Sonoran to Arid Transition Zones, eastern Mojave Desert, California, to Colorado, Arizona, Texas, and northern Mexico. Barely entering the Sonoran Desert in Maricopa and Pima Counties, Arizona. March–Oct.

**3*a*. Hedeoma nana** subsp. **macrocalyx** W. S. Stewart, Rep. Spec. Nov. Beih. **115:** 29, 1939

Differing from subspecies *nana* in having basal leaves more persistent and 10–15 mm. long, often purplish beneath; calyx tube 3.5–4.5 mm. long; corolla tube 5–6 mm. long, thus limb distinctly exceeding calyx.

Occurring with var. *nana* in Mohave, Coconino, Yavapai, Pinal, and Pima Counties, Arizona. March–Oct.

**Hedeoma drummondii** Benth., Lab. Gen. & Sp. 368, 1834, differs from all the above in having the calyx teeth connivent at maturity and nearly or quite closing the tube, and in having entire, narrowly elliptic leaves. It may occur as a straggler along the upper margins of our area, but no specimens from within the limits of the Sonoran Desert have been seen.

## 11. POLIOMINTHA A. Gray, Proc. Amer. Acad. 8: 295, 1873

Suffrutescent or a small shrub, with hoary-canescent twigs and herbage. Leaves small, linear to oblong, entire. Flowers in small axillary clusters near tips of branches. Calyx tubular, 15-veined; tube shaggy-pilose to puberulent; throat strongly annulate-villous; teeth 5, subequal, more or less connivent. Corolla tubular, straight or slightly arcuate, pilose-annulate within with ascending hairs; limb bilabiate; upper lip straight and nearly plane, emarginate; lower lip spreading, deeply 3-lobed; middle lobe often emarginate. Fertile stamens 2 (the lower pair), with ascending parallel filaments, tips often incurved; sterile stamens represented by short filaments. Stigma lobes slightly unequal in thickness, somewhat curved. Nutlets oblong, smooth.

Calyx tube shaggy-villous without; corolla white with purplish flecks to pale blue, tube calyx tube .................................................... 1. *M. spicata*
Calyx tube densely but minutely pubescent; corolla yellow-orange, tube 18–23 mm. long; plants suffrutescent ............................................ 2. *P. conjunctrix*

**1. Poliomintha incana** (Torr.) A. Gray, Proc. Amer. Acad. **8:** 296, 1870
*Hedeoma incana* Torr., Bot. Mex. Bound. 130, 1859.

Small shrub 3–8 dm. tall with many slender, erect or ascending branches; older branches more or less ridged, with dark gray, scaly, thin bark; twigs of season quadrangular, cinereous-puberulent, hairs fine and clearly appressed, punctate dots numerous but minute; leaves sessile or very short-petiolate, narrowly oblong to linear, 1–4 mm. wide, 7–24 mm. long, acute at base, acute to rounded at apex, cinereous-puberulent, punctate-glandular; inflorescence 3–15 cm. long; pedicels 0.5–1.5 mm. long; calyx tube cylindrical, 5–6.5 mm. long, markedly striate-ridged, pilose with spreading white hairs 1.5–3 mm. long, teeth subequal, 2–2.5 mm. long, purplish near tips, less hairy than tube; corolla white flecked with numerous specks of lavender or often uniformly pale blue, puberulent without, tube 6–10 mm. long, upper lip straight or slightly curved upward, 2.5–3 mm. long, lower lip spreading, 3–4.5 mm. long; stamens about equaling upper lip; nutlets oblong-obovoid, 2.8–3 mm. long, light brown, smooth but dull.

Flats, plains, eroded slopes, and sandy washes, upper margins of Lower Sonoran and Upper Sonoran Zones, southern Mojave Desert to southern Utah, northern Arizona, and east to Texas. Rarely entering the Sonoran Desert in Arizona. May–Aug.

**2. Poliomintha conjunctrix** Epling & Wiggins, Contr. Dudley Herb. **3:** 85, *pl. 21*, 1940

A perennial herb with a lignescent base and slender, wandlike branches to 6 dm. tall; twigs and herbage smoothly covered with minute, densely crowded, ashy hairs and somewhat punctate-dotted, internodes usually longer than leaves; petioles about 1 mm. long; leaf blades narrowly elliptic to oblong, 1–5 mm. wide, 1–2 cm. long, obtuse at apex, entire, often reflexed; inflorescence 5–20 cm. long, axillary cymules mostly 3-flowered, on peduncles 2–3 mm. long; pedicels 1–2 mm. long, cinereous; calyx 13–15-veined, cylindrical, 5.5–6.5 mm. long, subequal teeth 0.8–1.2 mm. long, subulate and more or less connivent; corolla orange or yellow-orange, tube 18–23 mm. long, puberulent on outer surface, with a ring of retrorse hairs just above base within, upper lip about 5 mm. long, lower lip 6–7 mm. long, narrowly 3-lobed about to middle; stamens exserted 7–8 mm. beyond corolla tube; mature nutlets unknown, immature ones dark brown, smooth.

Rolling hills and along arroyos, Lower Sonoran Zone, southern flanks of the Sierra San Pedro Mártir, Baja California. May–June.

## 12. MONARDELLA Benth., Lab. Gen. & Sp. 331, 1834

Aromatic, annual or perennial herbs, sometimes woody at base, with entire to serrate, simple leaves. Flowers in terminal heads subtended by broad, often colored, membranous to herbaceous bracts. Calyx tubular, nearly equally 5-toothed, mostly 15-nerved; teeth narrowly triangular, erect, often ciliate-margined. Corolla white, rose, lavender, or purple; upper lip 2-lobed, erect; lower horizontal to declined, 3-lobed, the lobes linear-oblong. Stamens 4, all fertile, exserted, with divergent anther cells. Style unequally 2-cleft at apex. Nutlets broadly oblong, smooth, dull to sub-lustrous.

Bracts broadly ovate; herbage minutely appressed-puberulent; corollas rose:
  Plants annual; heads 10–15 mm. in diameter; calyx 5–6 mm. long .................... ........................................ 1. *M. lanceolata* var. *microcephala*
  Plants perennial; heads 2–3 cm. in diameter; calyx 8–9 mm. long ......... 2. *M. linoides*
Bracts elliptic-lanceolate, herbage hispid; corollas white .................. 3. *M. arizonica*

**1. Monardella lanceolata** A. Gray var. **microcephala** A. Gray, Syn. Fl. N. Amer. ed. 2, **2**[1]**:** 459, 1886

Annual herbs with erect divaricately branched stems 2–4.5 dm. tall, herbage minutely puberulent and moderately to copiously gland-dotted, often tinged with red or purplish; petioles slender, 2–10 mm. long; leaf blades linear-oblong to lanceolate, entire or the largest remotely serrulate, 1.5–6 mm. wide, 1.5–4 cm. long, cuneate or gradually narrowed toward base, acute at apex; heads 10–15 mm. in diameter; bracts ovate, 2.5–4 mm. wide, 4–6 mm. long, veiny, puberulent, with coarser hairs on veins, sometimes faintly tinged with rose; calyx 5–6 mm. long, puberulent, narrowly ovate-triangular teeth 1–1.2 mm. long, hispid-ciliate on margins; corolla pale to deep rose, 9–10 mm. long, slender tube barely exserted, puberulent, lobes linear, about 4 mm. long, cupulate at tips.

Rocky slopes, washes, and grassy areas, Upper Sonoran Zone and occasionally in the Lower Sonoran Zone, eastern San Diego County, California, south to the Sierra San Pedro Mártir, Baja California. June–Sept.

**2. Monardella linoides** A. Gray, Proc. Amer. Acad. **11:** 101, 1876

*Monardella anemonoides* Greene, Pittonia **5:** 86, 1902.
*Madronella linoides* Greene, Leaflets Bot. Obs. **1:** 169, 1906.

Stems several to many from a woody base, erect or ascending, 2–5 dm. tall, silvery-puberulent with closely crowded, retrorsely appressed, microscopic hairs; petioles 5 mm. long or less, silvery; leaves linear-lanceolate, 1.5–4 mm. wide, 1–2.5 cm. long, acute at apex, narrowed to a cuneate base, entire, minutely puberulent-silvery on both surfaces; heads 2–3 cm. in diameter, terminal on branches; bracts broadly ovate to ovate-oblong, 10–15 mm. long, membranous, distinctly veined, white or tinged with rose, puberulent and minutely gland-dotted, margins soft-ciliate; calyx 8–9 mm. long, prominently veined, puberulent to short-pubescent, teeth lanceolate, 1.5–2 mm. long, hirsute-pubescent, erect; corollas 12–15 mm. long, pale rose, retrorsely hispidulous without; stamens exserted 2–5 mm.

Dry slopes and washes, mostly Upper Sonoran Zone, straggling to upper margins of Lower Sonoran Zone, southern Sierra Nevada, California, to desert slopes of eastern San Diego County. June–Aug.

**3. Monardella arizonica** Epling, in Kearney & Peebles, Jour. Wash. Acad. **29:** 489, 1939

Suffrutescent plant with several to many slightly decumbent to erect branches 1–3.5 dm. tall; herbage hispidulous and with numerous short-stipitate, glandular hairs, moderately canescent; main leaves lanceolate, 2–4 mm. wide, 10–16 mm. long, obtuse to acute at base, acute at apex, entire, margins slightly revolute; numerous smaller leaves in axillary fascicles; flowering heads terminal and on branches 1–5 cm. long from upper axils, 10–18 mm. in diameter; bracts elliptic-lanceolate, 1.5–2 mm. wide, 7–9 mm. long, hispidulous and glandular-puberulent, thin but strongly veined, tinged with red or lavender on distal one-third to one-half; calyx tube 4.5–5 mm. long, hispidulous, teeth subequal, narrowly triangular to triangular-subulate, 1–1.5 mm. long; corolla white, tube 6–8 mm. long, short-hispidulous without, lobes linear, 3–4 mm. long, faintly tinged with rose, minutely hispidulous on outside, glabrous within; stamens exceeding corolla-lobes 2–3 mm., filaments sparsely villous; nutlets oblong, plano-convex in cross section, 1.8–2 mm. long, pale brown with darker brown specks and blotches, sublustrous.

Rocky outcrops, steep slopes, ridges, ledges, and canyons, Lower and Upper Sonoran Zones, Mohave, Yavapai, Maricopa, Pima, and Yuma Counties, Arizona. March–Oct.

**Monardella nana** A. Gray subsp. **leptosiphon** (Torr.) Abrams, Muhlenbergia **8:** 31, 1912, with slender stems from a woody rootstock, herbage villous-pubescent, and corollas pale yellow or cream with tubes 20–25 mm. long and lobes 10–12 mm. long, occurs in open coniferous forest and adjacent chaparral from Palomar Mountain, San Diego County, California, south into northern Baja California. It may rarely occur as a straggler along washes at the upper edges of the Sonoran Desert, but no specimens from within our area have been seen. June–Aug.

## 13. MENTHA L., Sp. Pl. 576, 1753

Perennial herbs with erect or diffuse stems, sessile or petioled, punctate leaves, and pungently fragrant or odorous herbage. Flowers small, in axillary whorls, or in congested or interrupted terminal spikes. Calyx campanulate to cylindric, 10-nerved, regular to slightly 2-lobed, 5-toothed. Corolla tube shorter than calyx, bilabiate; upper lip entire to emarginate; lower 3-lobed. Stamens 4, equal, erect, included or exserted; filaments glabrous; anther sacs oblong to linear, parallel. Ovary deeply 4-parted; style 2-cleft at apex. Nutlets normally 4, ovoid, smooth.

Leaves sessile or essentially so; bracts equaling or surpassing calyx; calyx lobes about equaling calyx tube ........ 1. *M. spicata*
Leaves petiolate; bracts rarely equaling calyx; calyx lobes shorter than tube .... 2. *M. piperita*

### 1. Mentha spicata L., Sp. Pl. 576, 1753

*Mentha viridis* L., Sp. Pl. ed. 2, 804, 1763.

Stolons present; stems erect and sparingly branched to somewhat decumbent and much branched, 3–12 dm. tall, often reddish to purplish, glabrous to sparingly hirsute; leaves elliptic, oblong-lanceolate, or ovate-lanceolate, 5–25 mm. wide, 1.5–6 cm. long, lower ones usually largest, sharply serrate, acute to acuminate at apex, obtuse to subcordate at base, strongly veined, dark green but slightly lighter beneath than above, sparingly hirsute beneath; spikes terminal, leafless, 2–8 cm. long, 1 cm. or less in diameter; bracts lance-subulate, ciliate, equaling or slightly surpassing calyx; calyx about 2 mm. long, densely glandular, teeth about equaling tube, markedly ciliate-margined; corolla pale lavender, about twice as long as calyx, glabrous; stamens barely exceeding corolla tube; styles well exserted, about 2 mm. longer than stamens.

Moist stream banks, occasional in irrigated fields, and along ditches in southern

Arizona; extensively naturalized in temperate North America; introduced from Europe. June–Oct.

**2. Mentha piperita** L., Sp. Pl. 576, 1753

Perennial from slender, leafless stolons or sometimes exposed parts leafy-bracted; stems slender, erect or decumbent, simple or branched, 3–8 dm. long, glabrous, often purplish; petioles slender, 3–8 mm. long; leaf blades ovate to lanceolate, 1–2 cm. wide, 1.5–4 cm. long, acute to obtuse at apex, rounded at base, dark green, firm, regularly serrate, glabrous or sparingly hirsute beneath; spikes of interrupted whorls about 1 cm. in diameter, 2–12 cm. long in fruit; bracts narrowly lance-acuminate, naked or sparsely ciliate, rarely equaling calyx lobes; calyx tube narrowly campanulate, about 2 mm. long, closely gland-dotted, teeth about 1 mm. long, sparsely ciliate, calyx otherwise glabrous; corolla white to rose-purple, lobes 1–1.4 mm. long; stamens nearly equaling corolla lobes, styles somewhat longer.

Occasional along irrigation ditches and wet places about springs; naturalized from Europe and widely distributed in North America; rare in the Sonoran Desert. July–Oct.

## 14. HYPTIS Jacq., Collect. 1: 101, 1786

Annual or perennial herbs or small to large shrubs, with opposite, usually toothed leaves, and bilabiate flowers borne in headlike clusters in axils of reduced leaves, often forming verticils in an interrupted cymose arrangement. Calyx straight or oblique, sessile or pedicellate; tube ovoid, campanulate, or cylindrical, 5-lobed or 5-toothed, often markedly ridged beneath each acute to awn-tipped lobe. Corolla 2-lipped; upper lip emarginate, erect or spreading; lower lip saccate, drooping or nearly horizontal; tube cylindrical to infundibuliform. Stamens 4, inserted in corolla tube, declined and resting on lower lip; filaments mostly hairy, each anther-bearing; anthers confluently 2-celled. Style glabrous; stigma arms equal or unequal. Nutlets ovoid to subglobose, often somewhat angled marginally, smooth or slightly reticulate-roughened.

Plants coarse annual herbs, sparsely pubescent to subglabrous; calyx tube 8–10 mm. long in fruit, acicular teeth 3–6 mm. long; nutlets 3–5 mm. long, 2.5–3 mm. wide .......... .......... 1. *H. suaveolens*

Plants shrubs or suffrutescent perennials, usually cinereous to lanate, at least on young branches; calyx tube rarely over 5 mm. long in fruit, teeth mostly less than 3 mm. long:

- Flower clusters on distinct peduncles 0.5–4 cm. long:
  - Leaves dark green, sparsely pubescent on upper surface; calyx teeth distinctly less pubescent than tube .......... 2. *H. decipiens*
  - Leaves lanate-puberulent on both surfaces; calyx teeth nearly as woolly as tube .......... 3. *H. laniflora*
- Flower clusters sessile or on short peduncles less than 1 cm. long:
  - Leaves thick, densely cinereous to lanate with dendritically branched hairs; plants distinctly shrubby, often to 5 m. tall; calyx lobes acicular, 1.5–5 mm. long:
    - Leaf blades oblong-lanceolate to elliptic, often arcuately curved downward; corolla tube densely white-lanate without .......... 4. *H. tephrodes*
    - Leaf blades ovate to suborbicular, not arcuately curved; corolla tube sparsely pubescent without:
      - Pedicels slender, 1–2.5 mm. long, or flowers sessile; nutlets smooth .... 5. *H. albida*

Pedicels stoutish, 1–6 mm. long, flowers almost never sessile; nutlets minutely granular-roughened:
Leaves mostly broadly ovate to suborbicular, upper surface lanate or cinereous, not green and glabrate ........................ 6. *H. emoryi* var. *emoryi*
Leaves mostly narrowly ovate, sparsely clothed with dendritic hairs, often green above:
Leaf blades 1–2 cm. wide, cinereous or lanate beneath .................... ........................................ 6a. *H. emoryi* var. *palmeri*
Leaf blades 3–3.5 cm. wide, usually greenish on both surfaces ............. .................................... 6b. *H. emoryi* var. *amplifolia*
Leaves thin, dark green, sparsely scaberulous-puberulent with simple curved hairs and punctate-glandular beneath; plants suffrutescent, rarely over 1.2 m. tall; calyx lobes broadly deltoid, 0.6–0.8 mm. long ............................. 7. *H. collina*

**1. Hyptis suaveolens** (L.) Poit., Ann. Mus. Paris **7:** 472, *pl. 29, fig. 2,* 1806

*Ballota suaveolens* L., Syst. ed. 10, 1100, 1759.
*Hyptis plumieri* Poit., Ann. Mus. Paris **7:** 473, 1806.
*Mesosphaerum suaveolens* Kuntze, Rev. Gen. Pl. **2:** 525, 1891.
*Hyptis congesta* Leonard, Jour. Wash. Acad. **17:** 70, 1927.

Erect, moderately branching, annual herb 0.3–3 m. tall, stem obtusely quadrangular, strongly grooved between angles and hollow; internodes to 15 cm. long, sparsely villous and with many sessile or short-stalked, minute glands interspersed among scattered hairs, base of plant often densely pilose with spreading unbranched hairs 1–3 mm. long; leaves narrowly to broadly ovate, 1–5 cm. wide, 3–10 cm. long, rounded, subcordate, or occasionally broadly cuneate at base, usually acute at apex, margins simply or doubly serrate-dentate, bright green and sparsely hirsute to subglabrous above, paler and usually minutely hirsute beneath, stalked glands usually numerous; petioles usually about one-half as long as blade or less but sometimes equaling blade; inflorescence of few-flowered, pedunculate cymules, peduncle about equaling calyces or slightly shorter; flowering calyces 4–5 mm. long, turbocampanulate, mouth slightly oblique, lobes subulate, slightly dilated at base, about equaling tube at anthesis, tube to 1 cm. long in fruit, strongly 10-nerved and transversely reticulate-veined on internerve faces, sparsely hirsute and copiously glandular, somewhat shining, dark brown, teeth somewhat spreading; corolla blue to purple, tube 4–6 mm. long, slightly hirsute but strongly glandular toward apex without and on lobes, lower lip 1.5–2 mm. long; stamens barely exceeding throat, filaments hairy; nutlets 3–5 mm. long, to 3 mm. wide, squarrose to emarginate at apex, dark brown to black, finely granulate on back, bearing a few white hairs on inner face; hilum rather large, nearly white.

Openings in the forest, clearings, roadsides, and abandoned fields, mostly Tropical Zone but reaching southern margins of the Sonoran Desert, widely distributed in the American tropics and north to southern Sonora. Sept.–Jan.

**2. Hyptis decipiens** M. E. Jones, Contr. West. Bot. **18:** 53, 1933

An ascending, many-branched shrub 0.5–1.5 m. tall, branches slender, moderately to rather closely floccose-tomentulose with dendritic hairs when young but soon glabrate and bark tawny to light brown; internodes 1–6 cm. long; petioles 5–20 mm. long, minutely scurfy-puberulent with stoutish hairs bearing many minute, lateral, simple branches near tip of each erect hair; leaf blades ovate, lower ones 2–6 cm. wide, 3.5–8 cm. long, obtuse to rounded or less commonly acute at apex, rounded to subtruncate at base, margins irregularly dentate-crenate, teeth again minutely few-toothed, main crenulations 2–3 mm. high, acute to minutely apiculate; upper surface of leaf blades dark green, sparsely beset with stoutish, erect hairs bearing lateral, much shorter hairs near tips, lower surface slightly paler, more densely clothed with peculiar, many-branched, dendritic hairs; leaves of inflorescence reduced, uppermost cuneate at base; inflorescence 1–4 dm. long, with numerous tomentose glomerules on slender peduncles 5–15 mm. long; pedicels very slender, 3–5 mm. long, tomentulose; calyces narrowly campanulate, 3.5–4.2 mm. long, tomentose, lobes lance-attenuate, 2–2.5 mm. long, less densely tomentose than tube, reddish or purplish under investiture; corollas lavender, tube 3.5–4 mm. long, lobes 2–3 mm. long, sparsely to moderately pubescent without with dendritic white hairs; stamens about equaling lower lip or slightly shorter, filaments densely hirsute with spreading, white, mainly simple hairs; style equaling or slightly exceeding upper lip, upwardly curved; mature seeds unknown.

Canyons and adjacent, gently sloping hillsides, Lower Sonoran and arid sites in Tropical Zones, southern Baja California, mostly confined to the area south of La Paz. Oct.–March.

**3. Hyptis laniflora** Benth., Bot. Sulph. 42, *pl. 20,* 1844
*Mesosphaerum laniflorum* Kuntze, Rev. Gen. Pl. **2:** 526, 1891.

A weak, openly branched shrub 2–5 m. tall, branches often arching, younger twigs cinereous with a close covering of minute, dentritically branched, whitish hairs; older branches glabrate; bark reddish brown; internodes 2–5 cm. long; leaves coriaceous, rhomboid-ovate, broadly ovate, or suborbicular, 1–3 cm. wide, 2–6 cm. long, rounded, subcordate, or cuneate at base, acute to rounded at apex, margins irregularly crenate-dentate, teeth 1.5–3 mm. long, both surfaces lanate-puberulent in youth, upper or both surfaces glabrate or lower surface remaining cinereous in age; petioles slender, 3–15 mm. long; inflorescence paniculate-pyramidal; cymules loose to compact, globose, 1–2 cm. in diameter; peduncles slender, 1–4 cm. long, divaricate; pedicels filiform, 2–6 mm. long; calyx tube 1.5–2.5 mm. long at anthesis, 3.5–4 mm. long in fruit, densely clothed in silvery-white wool; calyx lobes lanceolate-subulate, 2–3.5 mm. long, attenuate, reddish or purplish under the wool; corollas blue-purple, 6–8 mm. long, tube 4–5 mm. long, sparsely puberulent above; lower lip 1.5–2 mm. long; stamens barely exserted, not quite equaling upper lip, filaments hairy; nutlets broad, 1.8–2.3 mm. long, 1.2–1.5 mm. wide, dull reddish brown, smooth but with 3 faint lines on outer surface; hilum broad, conspicuously white.

Sandy plains, arroyos, and rocky canyons, Lower Sonoran and Tropical Zones, Comondú south to Cabo San Lucas, Baja California. Sept.–May.

**Hyptis laniflora** var. **insularis** (Standley & Goldm.) I. M. Johnston, Proc. Calif. Acad. IV, **12:** 1150, 1924 [*Mesophaerum insularis* Standley & Goldm., Contr. U.S. Nat. Herb. **13:** 375, 1911; *Hyptis insularis* Standley, op. cit. **23:** 1276, 1924], differs from var. *laniflora* in having more markedly orbicular leaves more persistently cinereous on both surfaces, and comparatively shorter, broader calyx lobes. Islands in the Gulf of California off the coast of southern Baja California. Poorly known.

**4. Hyptis tephrodes** A. Gray, Proc. Amer. Acad. **5:** 164, 1861
*Mesosphaerum tephrodes* Kuntze, Rev. Gen. Pl. **2:** 527, 1891.

A robust, ascendingly branched shrub 2–5 dm. tall; herbage densely tomentose with very fine, closely intertangled, white, several-rayed, stellate-dendritic hairs; older stems and branches becoming subglabrous; bark reddish brown; internodes 1–5 cm. long; leaves oblong-lanceolate or rarely broadly elliptic-ovate, 8–25 mm. wide, 2.5–8 cm. long, often arcuately curved downward, obtuse or less commonly acute at apex, broadly cuneate to rounded at base, margins shallowly crenate-dentate with teeth about 1 mm. long, upper surfaces cinereous-tomentose with very fine, stellate-dendritic hairs, lower surface paler and more heavily tomentose, prominently veined; petioles 3–8 mm. long, tomentose; inflorescence pyramidal, of several spreading-ascending branches, each bearing 5–15 globose cymules in interrupted spikes, each lateral branch and some of basal cymules subtended by a slightly reduced pair of leaves; cymules dense, 15–50-flowered, 1.5–2 cm. in diameter; pedicels slender, 1–3 mm. long or some flowers nearly sessile, lanate with tangled, many-branched, white hairs; calyx turbinate, tube 1–2 mm. long, densely white-lanate without; calyx teeth acicular to nearly filiform, 2–2.5 mm. long, reddish or slightly lavender, sparsely lanate; corollas blue to violet, 6–8 mm. long, tube densely white-lanate, lobes sparsely lanate without, glabrous within; stamens slightly exserted, often about equaling lateral lobes of lower lip, filaments sparsely hirsute with crinkly hairs; style exserted about 2 mm. beyond corolla lobes; nutlets dark brown, glabrous, dull, 1.8–2.2 mm. long, 0.8–1 mm. wide, basal funicular groove prominent, lighter-colored.

Sandy arroyos, plains, or among rocks along mountain streams, Lower Sonoran and Tropical Zones, southern Baja California from the vicinity of La Paz southward; Sinaloa and Nayarit. Aug.–March.

**5. Hyptis albida** H. B. K., Nov. Gen. & Sp. **2:** 319, 1817

*Mesosphaerum albidum* Kuntze, Rev. Gen. Pl. **2:** 526, 1891.

Openly and ascendingly branched shrub 3–5 m. tall; younger twigs and herbage cinereous, floccose-tomentose with fine, dendritically branched, white hairs; older branches subglabrate; bark reddish brown, finely and obscurely striate; internodes 1–6 cm. long; leaves ovate-oblong to lance-ovate, 8–32 mm. wide, 1–6 cm. long, acute to obtuse at apex, truncate, rounded or broadly cuneate at base, margins crenate-serrate, teeth 0.5–1 mm. high, upper surface cinereous to green, closely scurfy-tomentulose to subglabrous, lower surface paler and more densely lanate with minute, many-branched hairs, strongly rugulose-veined; leaves in inflorescence somewhat reduced and usually narrower than lower ones; inflorescence interrupted-spicate, or pyramidal owing to lateral branchlets bearing interrupted, globose, sessile cymules, these loose to compact, mostly about 2 cm. in diameter; calyces sessile or on pedicels to 2.5 mm. long, narrowly turbinate to nearly cylindrical, tube 1.5–2 mm. long at anthesis, 2.5–3 mm. long at maturity, densely lanate with intertangled, dendritic, whitish hairs; calyx teeth acicular, 1.5–2.5 mm. long, lavender, less densely lanate than tube; corollas deep lilac, tube 3–4 mm. long, usually about equaling calyx teeth, sparsely puberulent with small dendritic hairs without, lobes puberulent without, glabrous within, upper lip about twice as long as lower lip; stamens slightly exserted, nearly equaling lower lip, filaments moderately hairy, lavender anthers slightly hairy at base; style exserted 1.5–2.5 mm.; nutlets oblong, 0.8–1 mm. wide, 1.5–2 mm. long, brownish, smooth but dull; hilum narrow.

On rocky hillsides, gravelly bajadas, abandoned fields, and old clearings, Lower Sonoran and Tropical Zones, northern Sonora (near Llano) to San Luis Potosí, Michoacán, and Guerrero. Flowering in our area from Sept.–March.

**6. Hyptis emoryi** Torr., in Ives, Rep. 20, 1860, var. **emoryi**

*Mesosphaerum emoryi* Kuntze, Rev. Gen. Pl. **2:** 526, 1891.

A straggly-branched shrub 0.5–3 m. tall, younger branches cinereous with fine, closely crowded, dendritically branched hairs forming a more or less scurfy layer of white or gray wool, older ones becoming glabrate and displaying pale gray, longitudinally fissured bark; internodes 0.5–8 cm. long; leaves lance-ovate to broadly ovate or occasionally suborbicular, acute to rounded at apex, broadly cuneate to truncate or rarely subcordate at base, 8–25 mm. wide, 1.5–4.5 cm. long, margins serrate-dentate, with some of larger teeth again minutely toothed, densely scurfy-lanate with dendritic hairs in youth, only slightly less densely so in age; petioles stoutish, 3–8 mm. long, densely scurfy-lanate; inflorescence of globose cymules on short peduncles 1 cm. long or less in axils of reduced leaves, or sessile, sometimes congested in a nearly cylindrical spike, each cymule 1–2 cm. in diameter; pedicels usually rather stoutish, 1–6 mm. long, loosely scurfy-lanate; calyx tube turbinate, 1.5–2.5 mm. long at anthesis, 3–5 mm. long in fruit, densely lanate or scurfy with closely crowded, relatively short, several-branched hairs; calyx teeth subulate to acicular, 1.2–3 mm. long at anthesis, to 4.5 mm. long in fruit, usually purplish or lavender, less densely lanate than tube; corolla pale blue-lavender, tube 3–4.4 mm. long, exposed part and lobes sparsely and minutely scurfy; stamens and style about equaling upper lip, filaments hirsute; nutlets pale brown, oblong, 1.5–2 mm. long, about 1 mm. wide, minutely granular-roughened.

Along desert washes, in canyons, and on rocky hillsides, mainly Lower Sonoran Zone, eastern Mojave Desert, eastern San Diego County, California, east into Arizona and south into Sonora and at least to the Magdalena Bay area in Baja California. Sept.–May.

**6*a*. Hyptis emoryi** var. **palmeri** (S. Wats.) I. M. Johnston, Proc. Calif. Acad. IV, **12:** 1149, 1924

*Hyptis palmeri* S. Wats., Proc. Amer. Acad. **24:** 68, 1889.
*Mesosphaerum palmeri* Goldm., Contr. U.S. Nat. Herb. **16:** 363, 1916.

Differing from var. *emoryi* in leaves generally somewhat narrower (1–2 cm. wide, to 5 cm. long) with green upper surface sparsely clothed with small dendritic hairs or nearly glabrous, paler and somewhat cinereous-lanate beneath.

Rocky hillsides, plains, and arroyos, north-central Sonora south at least to Guaymas and on the islands and mainland in the vicinity of Bahía de los Angeles in Baja California. Aug.–March.

This is a weak variety and intergrades with var. *emoryi.*

**6*b*. Hyptis emoryi** var. **amplifolia** I. M. Johnston, Proc. Calif. Acad. IV, **12:** 1149, 1924

Very similar to var. *palmeri* but leaves larger, mostly 3–3.5 cm. wide, 6–8.5 cm. long, green above and usually greenish beneath.

Rocky canyons, slopes, and arroyos, mainly along the eastern side of the peninsula from the vicinity of Loreto south in Baja California. Sept.–April.

It seems to intergrade with, or at least approach, *H. laniflora* in several characteristics.

**7. Hyptis collina** Brandegee, Proc. Calif. Acad. II, **3:** 164, 1891

Suffrutescent or shrubby plant 3–12 dm. tall, branches slender, numerous, and ascending, distinctly quadrangular and finely but sparsely scaberulous-puberulent near tips, glabrous and reddish below, internodes 3–10 cm. long; leaves ovate, thin, 8–15 mm. wide, 1.5–2.5 cm. long, acute to rounded at apex, broadly cuneate to subtruncate at base, crenate-dentate, teeth 1–1.5 mm. long, larger ones often again minutely denticulate, dark green, glabrous and sparsely punctate-glandular above, sparsely scaberulous-puberulent with simple, forwardly curved hairs and more densely punctate-glandular beneath; petioles slender, 3–10 mm. long; inflorescence racemose, to 4 dm. long, 3–10-flowered glomerules axillary to greatly reduced leaves; peduncles 4–5 mm. long, slender, bracts subtending glomerules linear, 1.5–2 mm. long; flowers sessile; calyx turbinate, 3–3.2 mm. long, 10-nerved and intricately reticulate between nerves, dark green, closely set with many short, simple, gland-tipped, brownish hairs, teeth narrowly deltoid, 0.6–0.8 mm. long, equaled or usually slightly surpassed by many erect, silky-white, simple hairs fringing inner margin of calyx throat; corolla bluish purple to lavender, tube paler, 5–6 mm. long, moderately puberulent with simple spreading hairs on outside, lateral lobes of lip scarcely differing from central one, glabrous within, dotted with oil globules on outside; stamens barely equaling or slightly exceeding throat, their filaments pubescent; nutlets oblong, dull black, 0.7–0.8 mm. wide, 1.5–1.8 mm. long, slightly and unequally quadrangular in cross section.

Along arroyos and on rocky ridges, chiefly arid parts of Tropical Zone, southern Baja California, possibly not reaching the Sonoran Desert. Sept.–Feb.

## 15. ASTEROHYPTIS Epling, Bull. Torrey Club **60:** 17, 1933

Suffruticose or shrubby plants with divaricately ascending branches. Leaves lanceolate to ovate, short-petiolate, more or less pubescent and gland-dotted. Inflorescence of dense, globose cymules in axils of reduced leaves in interrupted spikes or more congested and moniliform in cylindrical spikes paniculately arranged. Calyx tube campanulate, 10-nerved, truncate, hispid to naked; teeth subulate, stellate-spreading. Corolla tube cylindrical, moderately ampliate above, hirtellous within; lobes rounded, subequal and only slightly bilabiate; lower lip nearly entire and scarcely concave, not contracted nor callose at base. Stamens 4, didynamous, posterior pair shorter, included, anterior slightly exserted. Style slightly exserted; stigma lobes obovate, plane; gynobase shorter than ovules. Nutlets ovate, slightly complanate, minutely punctate-rugulose.

Calyx teeth 2–4 mm. long at maturity, strongly spreading; calyx tube glabrous within; glomerules 8–12 mm. in diameter at maturity ........................ 1. *A. stellulata*
Calyx teeth 0.4–0.6 mm. long at maturity, ascendingly spreading; calyx tube bearing ring of silvery hairs in throat; glomerules 5–7 mm. in diameter at maturity ........................................................ 2. *A. seemannii*

### 1. Asterohyptis stellulata (Benth.) Epling, Bull. Torrey Club **60:** 17, *fig. 1*, 1933

*Hyptis pubescens* Benth., Lab. Gen. & Sp. 129, 1833.
*Hyptis stellulata* Benth., loc. cit.
*Mesosphaerum pubescens* Kuntze, Rev. Gen. Pl. **2:** 527, 1891.
*Mesosphaerum stellulatum* Kuntze, loc. cit.

Aromatic shrub 2–3 m. tall with rather slender, divaricately ascending branches with obtuse angles and pronounced interangle channels on quadrangular internodes, stems and herbage moderately short-hispid with simple, slightly ascending hairs; internodes 2–8 cm. long; leaves ovate to lance-ovate, 1–3.5 cm. wide, 3–8 cm. long, acute at apex, narrowly rounded to rounded-truncate at base, crenate-serrate with teeth to 1.5 mm. high, dark green and sparsely hirtellous above, paler and hirsutulous to subtomentose beneath; petioles to 1.5 cm. long, hirtellous; inflorescences of globose sessile cymules in interrupted spikes on lateral and terminal branches, or rarely congested into cylindrical spikes, cymules 8–12 mm. in diameter at maturity; calyx tubes turbinate-campanulate, 1.5–2 mm. long at anthesis, 2.5–3.5 mm. long in fruit, strongly 10-nerved, hispidulous and closely dotted with globose brownish glands, teeth rigidly acicular, 1.5–2.5 mm. long at anthesis, nearly twice as long and divaricately spreading in fruit, intercostal areas of tube reticulately veined on upper one-third to one-half; corolla white or faintly flushed with pale lavender, tube 1.5–3 mm. long, nearly glabrous, lobes nearly equal, spreading, ciliate along margins and sparsely hirsutulous on outer surface; stamens inserted very high in throat of corolla, filaments about as long as anthers; nutlets dark dull brown, oblong-lenticular, 1–1.2 mm. long, about 0.8 mm. wide, hilum rather broad, only slightly paler than rest of nutlet.

Canyons and rocky hillsides, Lower Sonoran and Tropical Zones, southern Sonora to Durango, Vera Cruz, and Oaxaca. Sept.–March.

Scarcely entering the desert proper but to be expected as a straggler in canyons near the southern limits of the Sonoran Desert.

**2. Asterohyptis seemannii** (A. Gray) Epling, Bull. Torrey Club **60:** 20, 1933

*Hyptis seemanni* A. Gray, Proc. Amer. Acad. **21:** 407, 1886.
*Hyptis seemanni* var. *stenophylla* B. L. Robinson, Proc. Bost. Soc. Nat. Hist. **31:** 267, 1904.

Suffrutescent or shrubby plant 1–2 m. tall with subterete, somewhat grooved stems, hispid-villous herbage, and internodes 1.5–5 cm. long; leaves oblong-lanceolate, 0.5–2 cm. wide, 3–10 cm. long, acute to somewhat attenuate at apex, rounded to cuneate at base, serrate with teeth about 0.5 mm. high, dark green and moderately hirsute with soft, forward-pointing hairs above, slightly paler and more densely hirsute beneath, copiously gland-dotted among hairs; petioles 5–10 mm. long, short-hispidulous; inflorescence paniculate with several to many slender moniliform branches, cymules sessile, subglobose, 5–7 mm. in diameter at maturity; calyx tube 1.2–1.5 mm. long at anthesis, to 2 mm. long at maturity, closely short-hispid, bearing a dense ring of ascending, silvery-white hairs 0.4–0.6 mm. long in throat; calyx teeth about 0.4 mm. long, ascending, not appreciably elongating in fruit; corolla white, tube about 2 mm. long; nutlets dull brown, about 1 mm. long.

Canyons and grassy slopes, mostly in palm-oak belt, southern Sonora into Sinaloa and east into Chihuahua. Sept.–Jan.

## Family 125. **SOLANACEAE.** Potato Family

Herbs or shrubs with alternate leaves (upper leaves opposite in *Petunia*). Flowers perfect, regular, solitary, umbellate, cymose, or paniculate, axillary or terminal. Calyx rotate, campanulate, or tubular, 5-toothed or 5-cleft (sometimes 4-toothed), usually persistent and in some inflated in fruit. Corolla tubular, campanulate, funnelform, or rotate, 5-lobed; limb valvate or plicate in bud. Stamens 5, inserted in tube, alternate with lobes; anthers ovoid to linear, dehiscing longitudinally or by terminal pores or slits, often connivent. Ovary superior, 2-celled, several- to many-ovuled; style 1, slender or stout; stigmas capitate, entire or slightly 2-lobed. Fruit a berry or capsule. Seeds several to many, ovoid, subspherical, or strongly compressed. Embryo subperipheral, straight or strongly curved. Endosperm scanty but present.

Seeds moderately to strongly compressed laterally; embryo strongly curved:
- Plants shrubby; branches spinose:
  - Berry reddish or fruit greenish and bony; corolla tubular, urceolate or funnelform, not rotate ........ 1. *Lycium*
  - Berry black, yellow, or sometimes greenish but scarcely bony; corolla rotate ........ 8. *Solanum*
- Plants herbaceous; branches usually not spinose:
  - Calyx enlarged or inflated and investing berry in fruit:
    - Corolla 5 mm. long or less, urceolate to subglobose; stamens wholly included ........ 2. *Margaranthus*
    - Corolla rotate to campanulate, mostly 1 cm. in diameter or more; stamens rarely wholly included:
      - Berry partially enclosed in fruiting calyx, upper one-third to one-fourth exposed in fruit; calyx herbaceous; tomentose pads near base of corolla alternating with filaments ........ 3. *Chamaesaracha*
      - Berry wholly enclosed by calyx in fruit; calyx more or less papery; tomentose pads lacking on corolla:
        - Stigmas slightly bilobed; herbage glabrous, pubescent, or glandular but not scurfy ........ 4. *Physalis*
        - Stigmas capitate; herbage granular-scurfy ........ 5. *Quincula*
  - Calyx not conspicuously enlarged in fruit or if so, never enveloping berry:
    - Fruit a berry, juicy, fleshy, or leathery; calyx rotate to campanulate, not long-tubular or circumscissile near base; corolla less than 5 cm. long (or broad):
      - Stems sharply 4-angled; calyx saucer-shaped and markedly reticulate-veined in fruit; corolla greenish ........ 6. *Saracha*
      - Stems not markedly 4-angled; calyx herbaceous, not conspicuously veined in fruit; corolla white, yellow, lavender, or purple but not greenish:
        - Anthers dehiscing longitudinally; seeds pungently peppery (in ours); fruit bright red or sometimes yellow ........ 7. *Capsicum*
        - Anthers opening through a terminal pore, a subterminal slit, or rarely longitudinally; seeds not peppery; fruit not red ........ 8. *Solanum*
    - Fruit capsular, bearing heavy prickles, often dehiscing irregularly; calyx tubular or prismatic, circumscissile near base; corolla 5–25 cm. long, funnelform ........ 9. *Datura*

Seeds not strongly compressed; embryo straight or nearly so:
- Flowers over 1 cm. long, racemose or paniculate, in ours white or yellow; plant erect or ascending; seeds ovoid to reniform ........ 10. *Nicotiana*
- Flowers 5–7 mm. long, solitary in axils, purplish; plants prostrate or decumbent; seeds nearly spherical ........ 11. *Petunia*

## 1. LYCIUM L., Sp. Pl. 191, 1753

Shrubs or small trees with mostly rigid, spinose branches and alternate, fasciculate, glabrous to pubescent, more or less succulent leaves. Flowers borne singly or in groups of 2–6, short-pedicellate. Calyx tubular to campanulate, 4–6-lobed, slightly accrescent or somewhat ruptured by enlarging fruit. Corolla white, greenish, or tinged with purple, tubular to funnelform or narrowly campanulate with 4–6 rotate or reflexed lobes. Stamens 4–6, included to exserted; filaments subequal or unequal, adnate for different distances to tube of corolla, often pubescent near base of free portion; anthers oval, versatile. Style slender. Stigma shortly bilobed. Ovary bicarpellary, 2-celled, 2- to many-seeded, fleshy to dry and indurated. Seeds mostly somewhat flattened, minutely pitted. Embryo coiled; cotyledons slender.

Pericarp somewhat indurate; fruit 1–few-seeded:
- Fruit constricted near apex or base; leaves plane, not terete; flowers 1 cm. long or more, corolla lobes one-half as long as tube or less:
  - Fertile seeds in both upper and lower compartments of each carpel; constriction near summit; calyx tube subtruncate at base ............................ 1. *L. cooperi*
  - Fertile seeds in upper compartments only; fruit constricted near base; calyx tube rounded or obtuse at base, not truncate:
    - Corolla glabrous without; filaments adnate to lower half of corolla tube; stamens equaling corolla tube; style shorter than stamens ............... 2. *L. macrodon*
    - Corolla pubescent without; filaments adnate to basal two-fifths of corolla tube; stamens exserted 2–4 mm.; style longer than stamens ......... 3. *L. dispermum*
- Fruit not constricted; leaves terete, ovoid; flowers 4–6 mm. long, corolla lobes one-half to equaling tube:
  - Calyx slightly angular but not compressed, lobes subequal; ovary 1-ovulate in each cell .................................................... 4. *L. californicum*
  - Calyx strongly compressed laterally, carinate, lateral lobes obsolescent; ovary 2–3-ovulate in each cell .................................................. 5. *L. carinatum*

Pericarp not indurate; fruit juicy, several- to many-seeded:
- Leaves glaucous, glabrous; corolla 15–26 mm. long; berry glaucous, 8–10 mm. in diameter ........................................................ 6. *L. pallidum*
- Leaves not conspicuously glaucous, often pubescent; corolla usually not over 15 mm. long; berry not glaucous, mostly under 8 mm. in diameter:
  - Corolla tube slender, 0.8–2.5(–3) mm. wide at apex; leaves linear to linear-spatulate, rarely over 3 mm. wide (to 5 mm. in *L. andersonii*):
    - Filaments bearing only a few scattered hairs on basal part of free portion; corolla tube 1.5–2 times as wide at top as at base:
      - Leaves 3–16 mm. long:
        - Corolla 10–14 mm. long; flowers mostly 5-merous ........................................................................ 7. *L. andersonii* var. *andersonii*
        - Corolla 5–8 mm. long; flowers mostly 4-merous ......................................................................... 7*a*. *L. andersonii* var. *wrightii*
      - Leaves 20–35 mm. long ...................... 7*b*. *L. andersonii* var. *deserticola*
    - Filaments densely pilose on basal 1 mm. of free portion; corolla tube 2–4 times as wide at top as at base:
      - Calyx 1–2 mm. long; stamens equaling corolla lobes; branches slender, arcuate:
        - Style exserted less than 4 mm:
          - Corolla 9–13 mm. long ................... 8. *L. berlandieri* var. *peninsulare*
          - Corolla 4–6 mm. long .................... 8*a*. *L. berlandieri* f. *parviflorum*
        - Style exserted 4–7 mm. ..................... 8*b*. *L. berlandieri* var. *longistylum*
      - Calyx 2.5–3 mm. long; stamens about as long as corolla tube; branches thick, knotty, ascending, not arcuate .................................. 9. *L. densifolium*
  - Corolla tube broader, 3 mm. wide or more at apex; leaves obovate to oblanceolate-spatulate, mostly over 3 mm. wide (rarely 1 mm. in *L. parishii*):
    - Leaves and calyx essentially glabrous:
      - Corolla lobes rotate, densely lanate-ciliate; style exserted 2–3 mm.; stamens about equaling corolla lobes ..................................... 10. *L. torreyi*
      - Corolla lobes reflexed, glabrous; style barely equaling corolla tube; stamens surpassing corolla lobes 1–2 mm. ......................... 11. *L. megacarpum*
    - Leaves and calyx densely glandular-puberulent:
      - Calyx tube subcylindrical; calyx lobes one-third as long as tube or shorter:
        - Free portion of filaments nearly glabrous; stamens unequal, longest scarcely exserted:

Corollas glabrous without; herbage not viscid-glandular:
Leaves 3–6 mm. broad, 10–25 mm. long ..... 12. *L. fremontii* var. *fremontii*
Leaves 7–15 mm. broad, 15–30 mm. long ............................ 12*a*. *L. fremontii* var. *congestum*
Corollas glandular-pubescent without; herbage viscid-glandular ............................................ 12*b*. *L. fremontii* var. *viscidum*
Free portion of filaments densely pubescent at base; stamens subequal, exserted 2–5 mm. ............................................ 13. *L. exsertum*
Calyx tube campanulate; calyx lobes one-third as long as, to equaling tube:
Corolla tube broadly campanulate, not markedly constricted above ovary; calyx lobes deltoid to lance-subulate, unequal, acute ........... 14. *L. brevipes*
Corolla tube narrowly funnelform, conspicuously constricted above ovary; calyx lobes broadly ovate-elliptic, equal, acute to obtuse .......... 15. *L. parishii*

**1. Lycium cooperi** A. Gray, Proc. Amer. Acad. **7:** 388, 1868

*Lycium cooperi* var. *pubiflora* A. Gray, Syn. Fl. N. Amer. $2^1$: 238, 1878.

A stout, intricately branched, spiny shrub 1–2.5 m. high; bark lustrous-tan to silvery when young, finely striate and light gray in age; branchlets densely pubescent in youth, later glabrate; spines heavy, straight, 3–12 mm. long; leaves 3–10 in a fascicle, oblanceolate to obovate-spatulate, 1–3 cm. long, 4–10 mm. wide, obtuse, rounded, or truncate at apex, tapering to a short petiole, finely puberulent to hispidulous; flowers 1–3 in a fascicle; pedicels stoutish, 5–12 mm. long, glandular-hispidulous; calyx oblong-campanulate, apex narrower than base, glandular-hispidulous, tube 4–8 mm. long, 3–4 mm. wide at apex at anthesis, lobes 4 or usually 5, triangular to ovate-lanceolate, 2–6 mm. long; corolla greenish white, tinged with purple, cylindrical tube 8–15 mm. long, glabrous to puberulent outside, lobes 4 or usually 5, ovate, rotate to somewhat reflexed, margins revolute; stamens 4–5, about equaling tube; filaments adnate about to middle of corolla tube, densely pilose; stigma equaling or slightly surpassing stamens; fruit ovoid, 5–10 mm. high, indurated, with 2 lateral constrictions parallel to partition about one-third of way down from top of fruit, 1 seed in each upper, smaller compartment, 1–8 in each lower one.

Arid mesas and sandy slopes, mostly higher parts of the Lower Sonoran Zone; Esmeralda County, Nevada, southwestern Utah, Inyo County, California, and Mojave Desert south into northern Arizona and the Colorado Desert. March–April.

**2. Lycium macrodon** A. Gray, Proc. Amer. Acad. **6:** 45, 1862

A spiny, intricately branched, pubescent shrub 1–3 m. tall with slender, gray to light brown branches and blunt spines 2–5 mm. long; twigs gray-tomentose; leaves linear-spatulate to oblanceolate, 0.6–5.5 cm. long, 2–8 mm. wide, acute to rounded at apex, tapering gradually to a petiole 1–5 mm. long, minutely puberulent to glabrate; calyx campanulate, tube 3–4 mm. long, puberulent to glabrate, lobes lanceolate, 3–8 mm. long, acute, sinuses rounded; corolla tube greenish white, 9–12 mm. long, 2.5–3 mm. in diameter at summit, glabrous without, lobes 5, ovate, 2–4 mm. long, rotate, margins sparsely ciliolate and revolute; stamens equaling or slightly exceeding corolla tube, filaments adnate to lower half of corolla tube, densely pilose on adnate part; fruit with indurated pericarp, constricted about two-fifths of way above base, 1–2 seeds in each upper compartment, 1 to several abortive ovules in lower parts.

Along washes and sandy flats, Lower Sonoran Zone, southern Arizona to southern Sonora. Feb.–May.

**3. Lycium dispermum** Wiggins, Contr. Dudley Herb. **1:** 204, *pl. 19, figs. 5–10,* 1934

A stiff shrub 1–2 m. tall with slender, crooked, spiny branches; twigs scantily glandular-puberulent but soon glabrate; spines 3–10 mm. long; leaves obovate to oblong-lanceolate, 2–5 mm. wide, 5–30 mm. long, acute to rounded at apex, tapering to a petiole 1–2 mm. long, bright green, glabrous, veins prominent, whitish; petioles slender, minutely puberulent, 5–7 mm. long; calyx campanulate, 6–9 mm. long, minutely glandular-puberulent; calyx lobes 5, linear-lanceolate, 1.5–2 times as long as tube, acute; corolla tube greenish white, 9–12 mm. long, 2.5–3 mm. in diameter at base, 3–4.5 mm. in diameter at summit, sparsely puberulent without, lobes 5, pale lilac with darker veins, ovate, 4–5 mm. long, rotate to reflexed, margins revolute and closely ciliolate; stamens exserted 2–4 mm.; filaments subequal, adnate to lower two-fifths of corolla tube, pilose from near base of adnate part upward 3–4 mm. on free part; style 1–4 mm. longer than stamens; fruit bicarpellary, pericarp indurated, reticulate, constricted about one-fifth of way above base, a single seed in upper part of each carpel, part below constriction fleshy and containing 1–5 abortive ovules.

Known only from the type locality, Tacimaroa, on the Yaqui River near Corral, Sonora. Collected in flower in March.

**4. Lycium californicum** Nutt. ex A. Gray, Bot. Calif. **1:** 542, 1876

*Lycium californicum* var. *arizonicum* A. Gray, Syn. Fl. N. Amer. ed. 2, $2^1$: 437, 1886.

A compact shrub 0.3–2 m. tall with knotty, tangled, bluntly spinose branches and grayish bark; twigs glabrous or minutely puberulent and glabrate; leaves sessile or essentially so, elliptic-terete, succulent, 2–12 mm. long, 1–3 mm. wide, glabrous or temporarily puberulent; calyx campanulate, 2–2.5 mm. long, 1–2 mm. wide, 2–4- or sometimes 5-lobed, obscurely angled, puberulent, lobes deltoid, acute, about one-third as long as tube, ciliolate; corolla broadly short-funnelform, white or tinged with purple, tube 2–3 mm. long, about 2 mm. in diameter at apex, 4-lobed, lobes ovate, rotate or reflexed, margins revolute and sparsely ciliolate; stamens 4, exserted 1–1.5 mm.; filaments adnate to lower three-fifths of corolla tube, densely pilose on basal part of free portion and adjacent corolla tube but glabrous on adnate portion; style equaling or slightly exceeding stamens; berry ovoid, 2–4 mm. long, bicarpellate, each carpel containing 1 seed, pericarp indurated at maturity.

Mostly near the coast but occasionally in alkaline soil inland, Upper and Lower Sonoran Zones, Santa Monica Mountains, California, to southwestern Arizona and south at least to Puerto Libertad, Sonora, and central Baja California. Feb.–July.

**5. Lycium carinatum** S. Wats., Proc. Amer. Acad. **24:** 65, 1889

A low, compact, rounded shrub 0.3–1 m. high with spinose, intertangled, stiff branches, puberulent twigs, and grayish bark; leaves terete, ovoid to short-spatulate, succulent, minutely and densely puberulent at first but soon glabrate, light green, slightly glaucous, sessile; flowers small, 5–7 mm. long; pedicels short, about equaling calyx tube; calyx campanulate, 2–3.5 mm. long, glabrous except for ciliolate margins of lobes, strongly compressed and carinate, 4-lobed with 2 large and 2 short minute lobes or these small lateral ones obsolescent, longer lobes about one-third to one-half as long as tube; corolla tube 1.5–2 times as long as calyx tube, glabrous without, white with faint tinge of lilac on 4 ovate, often emarginate lobes, these about one-half to two-thirds as long as tube, rotate; stamens 4, equaling corolla lobes, thus exserted 2–2.5 mm. when corolla lobes are spreading; filaments adnate about halfway up corolla tube, densely pilose on basal part of free portion, glabrous on adnate part; fruit ovoid, bicarpellate, each carpel bearing 2–3 ovules.

Saline flats and near the coast, vicinity of Guaymas. Feb.–March.

**6. Lycium pallidum** Miers, Ann. Mag. Nat. Hist. II, **14:** 131, 1854

A spreading, much-branched, spiny shrub 1–2 m. tall with glabrous to sparsely pubescent, somewhat flexuous branches; bark smooth and lustrous when young, whitish; spines 5–10 mm. long, slender; leaves ovate to oblong-spatulate, 3–15 mm. wide, 1–4 cm. long, acute or sometimes rounded at apex, tapering to a short petiole, glaucous, glabrous or nearly so; pedicels slender, 6–12 mm. long; calyx shallowly campanulate, 5–8 mm. long, 5-lobed, lobes ovate to lanceolate, acute, 3–5 mm. long, glabrous and glaucous; corolla tube narrowly funnelform, 12–20 mm. long, 5–6 mm. broad at throat, greenish, tinged with purple, glabrous, 5 lobes oval to rhombic, 3–6 mm. long, margins sparsely ciliolate; stamens exserted 3–12 mm., filaments adnate almost to middle of corolla tube, pilose below; style about equaling stamens; berry bright red, or blue-glaucescent, ovoid, 8–10 mm. in diameter, 20–50-seeded.

Arid, sandy or rocky soil, Upper Sonoran Zone and higher parts of Lower Sonoran Zone, northern Arizona, southern Colorado, and western Texas south into Sonora, San Luis Potosí, and Zacatecas; mostly above the limits of the Sonoran Desert. April–June.

**7. Lycium andersonii** A. Gray, Proc. Amer. Acad. **7:** 388, 1868, var. **andersonii**

*Lycium andersoni* var. *pubescens* S. Wats., Proc. Amer. Acad. **24:** 65, 1889.

A rounded shrub 0.5–3 m. tall with spinose rigid branches and flexuous, light-barked twigs; spines numerous, slender, 5–20 mm. long; leaves linear to linear-spatulate, 1–2 (3.5) mm. wide, 3–16 mm. long, rounded to acute at apex, gradually tapering to base, glabrous or slightly scurfy; pedicels 3–9 mm. long, filiform; calyx shallowly campanulate, glabrous to sparsely puberulent, 1–2.5 mm. long, 4–5-toothed, teeth about one-fourth as long as tube, sparsely ciliolate; corolla narrowly tubular-funnelform, 10–14 mm. long, 1.5–3 mm. wide at top, sordid-lavender, 4–5-lobed, lobes ovate, 1.5–2.5 mm. long, ciliolate; stamens equaling corolla tube or exserted 2–3 mm.; filaments adnate to basal one-third of corolla tube, sparsely pilose on lower part of free portion; style about equaling stamens; fruit ellipsoid or ovoid, 3–8 mm. long, 2–5 mm. wide, juicy, bright orange-red at maturity; seeds numerous.

Arid washes and stony or sandy slopes, mostly Lower Sonoran Zone, Colorado and Mojave Deserts, southern Nevada, and southern Utah to New Mexico and Arizona south through Sonora and Baja California to northern Sinaloa. Also in Cuyama Valley, Santa Barbara County, California. Feb.–May.

***7a.* Lycium andersonii** var. **wrightii** A. Gray, Bot. Calif. **1:** 543, 1876

Differs from var. *andersonii* in having shorter, 4-lobed flowers only 5–8 mm. long and smaller, elliptic-spatulate leaves 3–8 mm. long.

Southern Arizona south to the Río Yaqui, Sonora. Feb.–April; Aug.–Sept.

***7b.* Lycium andersonii** var. **deserticola** (C. L. Hitchcock) Jepson, Fl. Calif. **3:** 461, 1943

*Lycium andersonii* f. *deserticola* C. L. Hitchcock, Ann. Mo. Bot. Gard. **19:** 280, 1932.

Differs from var. *andersonii* in having more slender leaves 20–35 mm. long.

Sandy washes and flats, occasionally on rocky hillsides, Lower Sonoran Zone,

Colorado Desert, California, to Arizona and south into Sonora and Baja California. Feb.–May.

**8. Lycium berlandieri** Dunal var. **peninsulare** (Brandegee) C. L. Hitchcock, Ann. Mo. Bot. Gard. **19:** 249, 1932

*Lycium peninsulare* Brandegee, Univ. Calif. Pub. Bot. **6:** 359, 1916.
*Lycium parvifolium* var. *peninsulare* I. M. Johnston, Proc. Calif. Acad. IV, **12:** 1154, 1924.

An open, spreadingly branched shrub 0.6–2.5 m. high with slender, densely leafy branches bearing few slender, spine-tipped laterals; leaves linear to elliptic-spatulate, 1.2–5 mm. wide, 10–30 mm. long, glabrous or sometimes sparsely and minutely puberulent; pedicels slender, 3–20 mm. long; calyx cup-shaped, 1–2 mm. long, 3–5-lobed, glabrous save for a tuft of hair at tip of each lobe, lobes one-third as long as tube; corolla pale lavender, tube 6–8 mm. long, constricted just above ovary, glabrous without, 4–5-lobed, lobes 3–5 mm. long, reflexed; stamens strongly exserted, equaling corolla lobes; filaments adnate to lower one-third of corolla tube, pilose on basal one-fourth to one-third of free portion; style about equaling stamens; fruit ellipsoid, 4–4.5 mm. in diameter, orange-red, juicy; seeds 8–30, 1.5–2 mm. long, yellowish.

Arid ravines and hillsides, Lower Sonoran Zone, southern Sonora and the Cape Region of Baja California. March–Sept.

**8*a*. Lycium berlandieri** f. **parviflorum** (A. Gray) C. L. Hitchcock, Ann. Mo. Bot. Gard. **19:** 246, 1932

*Lycium parviflorum* A. Gray, Proc. Amer. Acad. **6:** 48, 1862.

Differs from the above in having smaller flowers only 4–6 mm. long, lobes of which are one-third to one-half as long as tube.

Appearing with the species in southern Arizona, Sonora, Coahuila, and San Luis Potosí.

**8*b*. Lycium berlandieri** var. **longistylum** C. L. Hitchcock, Ann. Mo. Bot. Gard. **19:** 248, 1932

Similar to the above but slightly less densely leafy, and with the style exserted 4–7 mm.

Foothills and desert mountains of Pima and Yuma Counties, Arizona; Lower Sonoran Zone.

**9. Lycium densifolium** Wiggins, Contr. Dudley Herb. **1:** 198, *pl. 18, figs. 7–9,* 1934

An intricately branched, nearly glabrous shrub 1–1.5 m. high with knotty, densely leafed, rigid, spinose branches; leaves 3–10 in a fascicle, linear-spatulate, 1–2.5 mm. wide, 5–16 mm. long, fleshy, acute or rarely rounded at apex, sparsely glandular-puberulent but soon glabrate; pedicels slender, 4–6 mm. long; calyx subcylindrical, 1.5 mm. wide, 2.5–3 mm. long, subequally 4-lobed, lobes 0.5–1.25 mm. long, sparsely ciliolate; corolla white, faintly veined with pale lavender, tube 4–5 mm. long, 2–2.5 mm. wide at top; corolla lobes 4, rounded, spreading, glabrous, 1–1.5 mm. long; stamens unequal, longest slightly exserted, others included; filaments adnate to lower one-fourth of corolla tube, pilose on basal 1 mm. of free portion; style about equaling longer stamens; berry globose to ellipsoid, 3–4 mm. in diameter, orange-red, many-seeded.

Gravelly mesas, Lower Sonoran Zone, from the vicinity of Cape Colnett south to central Baja California. Feb.–May.

**10. Lycium torreyi** A. Gray, Proc. Amer. Acad. **6:** 47, 1862
*Lycium torreyi* var. *filiforme* M. E. Jones, Proc. Calif. Acad. II, **5:** 714, 1895.

A spreading shrub 1.3–5 m. tall with heavy spines 5–10 mm. long, terete, leafy, glabrate branches, and elliptic to spatulate, short-petioled, glabrous, pallid leaves 1–5 cm. long and 3–15 mm. broad; pedicels slender, 5–20 mm. long; calyx shallowly to deeply crateriform, 2.5–4.5 mm. long, 1.5–2.5 mm. in diameter at anthesis, sparsely puberulent but soon glabrous, 5-lobed, lobes deltoid, acute, one-fourth to one-half as long as tube; corolla 8–15 mm. long, greenish or white, tinged with lavender, tube narrowly funnelform, 3–4 mm. in diameter at top, glabrous without, (4)5 lobes 2–4 mm. long, rounded to ovate, rotate, margins densely ciliate-lanate; stamens about equaling lobes; filaments adnate to lower half of corolla tube, densely pilose on basal 1.5 mm. of free portion, adjacent corolla tube also pilose; style exserted 2–3 mm.; fruit ovoid, 6–10 mm. long, bright red, 8–30-seeded.

Southern Mojave Desert, Colorado Desert, southern Nevada, southwestern Utah, to western Texas, Chihuahua, Sonora, and Hidalgo, Mexico. Common along the banks of the Colorado River near Blythe. March–May.

**11. Lycium megacarpum** Wiggins, Contr. Dudley Herb. **3:** 73, *pl. 19, figs. 1–3,* 1940

A low, often prostrate, rounded, intricately branched, rigid, spinescent, glabrous shrub 0.3–1.5 m. high; bark dark gray to nearly black, finely and irregularly fissured, eventually slightly scaly; spines straight, slender, 5–15 mm. long; young twigs tawny to greenish, glabrous; leaves in fascicles of 2–6 on knobby, lateral, dwarf branches and borne alternately on young growth, oblanceolate to obovate-spatulate, 3–8 mm. wide, 5–30 mm. long, rounded, obtuse, or acute at apex, tapering gradually to cuneate base, fleshy, glabrous, deep green with purplish midvein, lateral veins scarcely apparent; flowers borne singly or 2–3 on a dwarf branch; pedicels stoutish, 3–7 mm. long, glabrous; calyx tubular-campanulate, 3.5–5 mm. long, 2–3 mm. in diameter at anthesis, subequally 5-lobed, lobes deltoid to ovate, acute, 0.7–2 mm. long, calyx often somewhat 2-lipped; corolla tubular-funnelfoım, tube 9–12 mm. long, contracted just above ovary, then flaring gradually to throat, this 3–4 mm. wide; corolla lobes broader than long, rounded, 1.5–2.0 mm. long, strongly reflexed, margins revolute, glabrous, eciliate; stamens subequal, exserted 3–6 mm., filaments adnate to basal two-fifths to one-half of corolla tube, moderately pilose on most of adnate part and on the free part almost to apex of corolla tube, adjacent corolla tube pilose about halfway from apex of adnation to throat; style barely equaling corolla tube; stigma crateriform; berries bright tomato-red, 9–12 mm. long, nearly as wide, 2-celled, many-seeded.

Rocky hillsides and along arroyos, Lower Sonoran Zone, from near San Augustín, Baja California, south to the Vizcaíno Desert. Feb.–May.

**12. Lycium fremontii** A. Gray, Proc. Amer. Acad. **6:** 46, 1862, var. **fremontii**
*Lycium gracilipes* A. Gray, Proc. Amer. Acad. **12:** 81, 1876.

A stout, intricately branched, compact shrub 1–3 m. tall, with sparingly armed branches bearing spines 10–15 mm. long and densely glandular-puberulent foliage; leaves spatulate, 8–25 mm. long, 3–9 mm. wide, acute to rounded at apex; pedicels 4–25 mm. long, densely glandular-puberulent; calyx tubular, 2–3 mm. in diameter, 4–8 mm. long at anthesis, 5 lobes 0.8–1.2 mm. long; corolla tubular to funnelform, white to lavender, with purplish veins, tube 8–15 mm. long, 3–3.5 mm. wide at apex, glabrous without, lobes 5, rotate or spreading, 2–6 mm. long, rounded, margins sparsely ciliolate or glabrous; stamens unequal, longest rarely equaling corolla lobes; filaments adnate to basal two-fifths of corolla tube, glabrous or sparingly pilose at base of free portion, adjacent corolla tube glabrous or sparsely pilose; style about equaling corolla tube; fruit ovoid, 5–9 mm. long, bright red, 40–60-seeded.

Washes and rocky hillsides, Lower Sonoran Zone, Colorado Desert, southwestern Arizona, and south in Sonora to the Río Yaqui, and to the Cape Region in Baja California. Feb.–May; Aug.–Dec.

**12*a*. Lycium fremontii** var. **congestum** C. L. Hitchcock, Ann. Mo. Bot. Gard. **19:** 295, 1932

*Lycium umbellatum* Rose, in Vasey & Rose, Contr. U.S. Nat. Herb. **1:** 74, 1890. Not *L. umbellatum* Ruiz & Pavon, 1799.

Differs from var. *fremontii* in having larger leaves and flowers, and in being more densely pubescent. Central to southern Sonora, and in the Cape Region, Baja California.

**12*b*. Lycium fremontii** subsp. **viscidum** Wiggins, Contr. Dudley Herb. **1:** 203, *pl. 19, figs. 1–4,* 1934

Herbage densely glandular-puberulent and quite viscid; corolla tube 3–3.5 mm. wide at base, 3.5–4.5 mm. wide at throat, glandular-puberulent without and on lower surface of lobes; fruit broadly ovoid, 10–12 mm. long.

Sandy slopes and washes, Santa Maria Plains, south to the vicinity of Cataviñá, Baja California.

**13. Lycium exsertum** A. Gray, Proc. Amer. Acad. **20:** 305, 1885

*Lycium fremonti* var. β? *bigelovii* A. Gray, Proc. Amer. Acad. **6:** 47, 1862.
*Lycium retusum* Rob. & Fern., op. cit. **30:** 120, 1894.

An openly branched, sparingly armed shrub 1–4 m. high with densely pubescent twigs and dark gray or brown bark; leaves spatulate to obovate, 3–7 mm. wide, 5–20 mm. long, obtuse or rounded at apex, densely glandular-puberulent; pedicels 3–6 mm. long, mostly recurved and flowers pendent; calyx tubular-campanulate, 3–6 mm. long, densely glandular-puberulent, lobes 5, deltoid, acute, 1–1.5 mm. long; corolla whitish, greenish or tinged with purple and brown, funnelform, slightly narrowed just above ovary, 8–12 mm. long, glabrous without, lobes broadly ovate or obovate, spreading, about 2 mm. long; stamens exserted 2–5 mm., filaments adnate to basal fifth of corolla tube, densely pilose on lower two-thirds of free part; style often sparsely pilose near stigma; berry ovoid, 6–8 mm. long, 20–35-seeded.

Along washes, Lower Sonoran Zone, southwestern Arizona and northern Baja California to Sonora and northern Sinaloa. Jan.–April.

**14. Lycium brevipes** Benth., Bot. Sulph. 40, 1844

*Lycium richii* A. Gray, Proc. Amer. Acad. **6:** 46, 1862.
*Lycium palmeri* A. Gray, op. cit. **8:** 292, 1870.
*Lycium cedrosense* Greene, Pittonia **1:** 268, 1889.

An intricately branched, spinose, ascending shrub 1–4 m. tall with tan to dark gray bark, pubescent twigs, and fleshy, puberulent, more or less glandular foliage; leaves elliptic, obovate, or spatulate, 5–25 mm. long, 3–10 mm. broad, densely puberulent and somewhat glandular when young, often glabrate in age, acute to rounded at apex; pedicels 1–6 (or 10) mm. long, puberulent; calyx campanulate, 2.5–6 mm. long, 1.5–3 mm. in diameter, 2–4-lobed, lobes from half as long as to equaling tube but longer ones of any one calyx more than half as long as tube, puberulent and usually glandular; corolla tubular to narrowly campanulate, white to lilac with darker stripes on lobes and in throat, glabrous, mostly 4-lobed, tube 4–8 mm. long, lobes broadly subreniform or oval, from one-third to as long as tube, rotate, margins slightly revolute and rarely ciliolate; stamens unequal, from slightly shorter than tube to equaling corolla lobes; filaments adnate to lower half of corolla tube, densely pilose on lower part of free portion and sometimes on adnate part, adjacent corolla tube also pilose; stigma about equaling corolla lobes; fruit globose to obovoid, bright red, 4–6 mm. wide, 5–8 mm. long, many-seeded.

Sandy washes, gravelly slopes, and coastal flats and bluffs, lower part of the Upper Sonora Zone and in the Lower Sonoran Zone, Colorado Desert and San Clemente Island, California, south to the Cape Region, Baja California, and Sinaloa. Feb.–April.

**15. Lycium parishii** A. Gray, Proc. Amer. Acad. **20:** 305, 1885

*Lycium pringlei* A. Gray, loc. cit.

An intricately branched, spinose shrub 1–3.5 m. tall with glandular-puberulent foliage and silvery to dark gray bark; leaves obovate-spatulate, 1–5 mm. broad, 3–12 mm. long, obtuse or rounded at apex; pedicels 5–10 mm. long, cinereous-puberulent; calyx campanulate, tube about 2–2.5 mm. long, acute at base, densely puberulent, 5 lobes equal, ovate, elliptic, or sometimes lance-ovate, equaling tube, densely puberulent; corolla narrowly funnelform, 6–10 mm. long, tube 1–1.5 mm. wide at base, 3–3.5 mm. wide at throat, contracted just above ovary, glabrous without, corolla lobes 5, rotate, ovate, 2–3 mm. long, margins finely ciliolate; stamens slightly exserted, about equaling corolla lobes; filaments adnate to lower half of corolla tube, densely pilose on basal one-fourth of free portion; style equaling stamens; fruit ovoid, bright red, 4–7 mm. long, 7–15-seeded.

Sandy plains and washes, Lower Sonoran Zone, near San Bernardino to San Diego County, California, east to southern Arizona and adjacent Sonora. March–April.

## 2. MARGARANTHUS Schlecht., Ind. Sem. Hort. Halens. 1, *pl. 1*, 1838

Small annual herbs closely resembling *Physalis* in habit, with branched leafy stems. Leaves simple, slenderly petiolate; blades entire to sinuate. Flowers solitary in upper axils, small, short-pedicellate. Calyx small, campanulate at anthesis, 5-toothed, inflated-globose with a contracted mouth in fruit. Corolla urceolate to subglobose, somewhat ventricose and 5-sulcate above short tube; orifice truncate and nearly entire to minutely 5-toothed. Stamens included, inserted near the bottom of tube; filaments short; anthers oblong, erect, longitudinally dehiscent. Basal disk minute. Ovary 2-celled; style filiform; stigma entire or faintly bilobed; ovules numerous. Fruit baccate, globose, included in inflated calyx. Seeds flattened; embryo curved; cotyledons nearly terete.

**1. Margaranthus solanaceus** Schlecht., Ind. Sem. Hort. Halens. 1, *pl. 1,* 1838

*Margaranthus tenuis* Miers, Ann. Mag. Nat. Hist. II, **5:** 250, 1850.

Erect or ascendingly branched herbs 2–6 dm. tall, branching sympodially so each leaf appears to have a branch and a flower axillary to it, a narrow band of simple, short, curved hairs extending from node to node and along adaxial side of each petiole, stems and leaves otherwise glabrous or essentially so; petioles slender, 1–3 cm. long; leaf blades elliptic to elliptic-ovate, 8–25 mm. wide, 2–5 cm. long, broadly cuneate at base, acute to slightly attenuate at apex, margin entire or sometimes sinuate; pedicels 2–4 mm. long at anthesis, slender, coarsely puberulent with subappressed white hairs, usually reflexed in fruit; calyces 2–3 mm. long and tubular-subcampanulate at anthesis, puberulent, broadly deltoid teeth 0.4–0.6 mm. long, erect; corolla tube about equaling calyx, inflated upper part 2–3 mm. in diameter, purplish at anthesis, bearing a few scattered minute hairs exteriorly near minute incurved teeth; fruiting calyx globose-ovoid, 10–12 mm. long, finely reticulate-veined, greenish; berries 5–6 mm. in diameter, glabrous; seeds about 2 mm. long, strongly flattened laterally.

Canyons, hillsides, and on streamside alluvium, in shaded areas, mostly Upper Sonoran and Transition Zones but occurring occasionally at the upper edges of the Sonoran Desert, in Cochise, Santa Cruz, and Pima Counties, Arizona, east to Texas and south into Mexico. Aug.–Nov.

## 3. CHAMAESARACHA A. Gray, Bot. Calif. 1: 540, 1876

Low perennial herbs with entire to pinnatifid leaves tapering to narrowly margined petioles. Flowers solitary or in few-flowered fascicles. Calyx campanulate, 5-toothed or 5-lobed, slightly enlarged and partially surrounding fruit at maturity. Corolla rotate, white or cream, faintly flushed with purple; throat tomentose within. Stamens inserted near base of corolla, about equaling plicate corolla limb; anthers oblong. Stigmas obscurely 2-lobed. Berries globose, on recurved pedicels. Seeds reniform, flattened, finely rugose-favose.

Pubescence on herbage and calyces of coarse, dentritic to few-rayed, stellate hairs; leaves mostly pinnatifid, short-petiolate to sessile ........................ 1. *C. coronopus*
Pubescence on herbage and calyces of simple hairs intermingled with gland-tipped ones, thus more or less viscid; leaves entire to shallowly pinnatifid, distinctly petiolate ....... ............................................................ 2. *C. conoides*

### 1. Chamaesaracha coronopus (Dunal) A. Gray, Bot. Calif. 1: 540, 1876

*Solanum coronopus* Dunal, in A. DC., Prod. **13**[1]: 64, 1852.
*Withania? coronopus* Torr., Bot. Mex. Bound. 155, 1859.
*Saracha coronopus* A. Gray, Proc. Amer. Acad. **10**: 62, 1874.

Rootstocks slender and subligneous; stems herbaceous, ascending to procumbent, 1–5 dm. long, slightly angular in cross section, longitudinally ridged or subalate, ridges more pubescent than intervening channels; herbage, including calyces, sparsely to moderately scurfy with short, coarse, dendritic to stellately branched hairs, branches stubby, white, older parts eventually subglabrate in some plants, longer simple hairs sparsely scattered among stellate ones on some specimens; leaves oblong-linear to elliptic-oblanceolate, usually pinnatifid with teeth 1–2 mm. long from a midlamella 2–5 mm. wide but vigorously growing plants often with leaves shallowly pinnatifid or only undulate, 5–15 mm. wide, 4–6 cm. long, acute, obtuse, or rounded at apex, gradually narrowed to subsessile base; flowers solitary in axils; pedicels slender, to 4 cm. long and often deflexed and sigmoidly curved in fruit; calyx 3–4 mm. deep at anthesis, densely pubescent-scurfy, teeth broadly to narrowly deltoid, 1–1.5 mm. long; corolla rotate or shallowly cup-shaped, 12–18 mm. in diameter, greenish yellow, more or less blotched with purple, densely puberulent around base of filaments, tips of lobes sparsely pubescent; stamens erect, filaments 2–3.5 mm. long, glabrous, anthers about 2 mm. long, yellowish; calyx accrescent in fruit and nearly or quite equaling dryish fruit, orifice broad; berry globose, 5–8 mm. in diameter, glabrous, dry or nearly so; seeds brownish, laterally compressed, weakly reniform, 1.8–2 mm. wide, 2.5–3 mm. long, about 1 mm. thick, testa strongly reticulate-foveolate.

Along ditches, roadsides, or pasture land and hillsides, Lower Sonoran Zone to Arid Transition Zone, throughout much of Arizona into southern Utah and east to Kansas and Texas. April–Sept.

**2. Chamaesaracha conoides** (Moric.) Britt., Mem. Torrey Club **5:** 287, 1894

*Solanum conoides* Moric. ex Dunal, A. DC., Prod. **13**[1]**:** 64, 1852.
*Withania? sordida* Dunal, in A. DC., op. cit. 456.
*Saracha sordida* A. Gray, Proc. Amer. Acad. **10:** 62, 1874.
*Chamaesaracha sordida* A. Gray, Syn. Fl. N. Amer. **2**[1]**:** 232, 1878.

Copiously branched, glandular-pubescent plant with ascending or slightly decumbent stems 1–4 dm. long, viscid and often discolored with adhering soil and debris; leaves numerous, lance-ovate to broadly ovate, blades 4–18 mm. wide, 1–3.5 cm. long, rounded to acute at apex, cuneately narrowed to narrowly winged petiole, this to 1.5 cm. long, margin entire, undulate, or occasionally pinnatifid; flowering pedicels 1–2.5 cm. long, glandular-hirsute; calyx 3–4 mm. deep, teeth 1–1.5 mm. long, ascending or erect, lanceolate, triangular; corolla rotate, 10–15 mm. in diameter, greenish yellow to purplish, densely pubescent in throat; stamens 3–5 mm. long, erect, filaments glabrous; anthers yellowish, 2 mm. long; berries dry, globose to depressed-globose, 5–6 mm. in diameter, glabrous, pale yellow to whitish; seeds subreniform, 2–2.5 mm. long, yellowish brown, faveolate-reticulate.

Roadsides, grassy plains, rocky hillsides, and mesas, mainly Upper Sonoran Zone, but reaching the upper edge of the Sonoran Desert in Pima County, Arizona; southeastern Arizona to Colorado, Kansas, Texas, and northern Mexico. April–Aug.

## 4. PHYSALIS L., Sp. Pl. 182, 1753

Annual or perennial herbs with entire to sinuate-dentate leaves. Flowers solitary from axils of leaves or in some species in 2–5-flowered, axillary fascicles. Calyx campanulate, 5-toothed, enlarged and inflated in fruit, 5–10-angled, 10-ribbed, reticulate, enclosing berry. Corolla obscurely 5-lobed, openly campanulate to campanulate-rotate, plicate in bud, yellow, whitish, or purple, often with center of different color, this usually darker than rest of corolla. Stamens 5, inserted near base of corolla tube; anthers oval or oblong, dehiscing longitudinally. Style slender; stigma faintly bilobed. Seeds numerous, reniform, flattened, finely pitted.

Anthers purplish, bluish, or greenish, not clear yellow; plants annual:
- Pedicels shorter than fruiting calyces; anthers broadly ovate or elliptic, less than twice as long as broad:
  - Plant subglabrous, sometimes sparsely pubescent on buds and young leaves; corolla 10–15 mm. wide ........................................ 1. *P. ixocarpa*
  - Plant distinctly pubescent, at least on stems, peduncles, and petioles; leaf blades sometimes subglabrous; corolla 4–10 mm. wide:
    - Calyx lobes equaling or exceeding tube at anthesis; leaves acute or obtuse, not acuminate, finely but densely pubescent:
      - Leaf blades strongly sinuate-dentate, obtuse at apex, rounded or obtuse at base; corolla 5–6 mm. wide ................................ 2. *P. neo-mexicana*

Leaf blades weakly sinuate-dentate or subentire, acute at apex, often subcordate at base; corolla 6–10 mm. wide .............................. 3. *P. pubescens*

Calyx lobes much shorter than tube at anthesis; leaves abruptly acuminate, thin, subglabrous except along veins and margins .................... 4. *P. leptophylla*

Pedicels exceeding fruiting calyces; anthers linear or oblong-elliptic, two to four times as long as broad (except in No. 7):

Fruiting calyx 3–4 cm. long; leaf blades three-fourths as wide as long .............................................................................. 5. *P. filipendula*

Fruiting calyx 1.5–2.5 cm. long; leaf blades two to five times as long as broad:

Corolla 12–20 mm. in diameter; limb rotate, whitish with a yellow center; anthers 3–4.5 mm. long; fruiting calyx lobes narrowly deltoid, acuminate ............................................................................ 6. *P. acutifolia*

Corolla 5–6 mm. in diameter, subcampanulate, yellow; anthers 1.5–2.5 mm. long; fruiting calyx lobes broadly deltoid, acute ..................... 7. *P. lancifolia*

Anthers yellow, not blue, greenish, or purple; plants perennial (except Nos. 8 and 9):

Plants annual:

Pedicels shorter than fruiting calyces; fruiting calyces 4–6 cm. long; calyx lobes exceeding tube at anthesis ........................................ 8. *P. nicandroides*

Pedicels exceeding fruiting calyces; fruiting calyx 2–2.5 cm. long; calyx lobes shorter than tube at anthesis ............................................ 9. *P. greenei*

Plants perennial:

Pedicels shorter than fruiting calyces:

Leaf blades lanceolate, three to five times as long as broad, cuneate at base; plant essentially glabrous ........................................ 10. *P. longifolia*

Leaf blades broadly ovate, about as broad as long, base rounded, truncate or subcordate; plant distinctly pubescent:

Pubescence of stellate or branched hairs; calyx lobes narrowly deltoid, equaling or exceeding tube at anthesis ................................. 11. *P. fendleri*

Pubescence of simple hairs; calyx lobes broadly deltoid, shorter than tube at anthesis:

Pedicels shorter than flowers at anthesis; foliage somewhat canescent; calyx lobes two-thirds to four-fifths as long as tube ................ 12. *P. hederifolia*

Pedicels longer than flowers at anthesis; foliage green; calyx lobes one-fourth to one-half as long as tube:

Corolla subrotate:

Leaves thick, ovate to subcordate .......... 13. *P. crassifolia* var. *crassifolia*

Leaves thin, strongly cordate .......... 13*a*. *P. crassifolia* var. *cardiophylla*

Corolla narrowly funnelform .......... 13*b*. *P. crassifolia* var. *infundibularis*

Pedicels longer than fruiting calyces:

Anthers oval, 1.5–2 mm. long; leaves lanceolate or lance-ovate, 2.5–4 times as long as broad; plant glabrous ....................................... 14. *P. glabra*

Anthers oblong, 2–3.5 mm. long; leaves ovate, cordate, or deltoid, only slightly if at all longer than broad:

Corolla yellow or whitish, not purple:

Leaf blades entire or nearly so; corolla not drying bluish:

Fruiting calyx muriculate at angles:

Corolla 12–15 mm. broad, greenish-yellow, with a darker throat; calyx lobes shorter than tube .............................. 15. *P. muriculata*

Corolla 20–25 mm. broad, clear yellow; calyx lobes equaling or slightly exceeding tube ........................................ 16. *P. flava*

Fruiting calyx puberulent but not muriculate:

Corolla subrotate:

Leaves thick, ovate to subcordate ........ 13. *P. crassifolia* var. *crassifolia*

Leaves thin, strongly cordate ........ 13*a*. *P. crassifolia* var. *cardiophylla*
Corolla funnelform ................. 13*b*. *P. crassifolia* var. *infundibularis*
Leaf blades sinuate-dentate; corolla drying bluish or purplish:
Leaves 1.5–4 cm. long, basal cordate; pedicels about equaling petioles ........ ....................................... 17. *P. versicolor* var. *versicolor*
Leaves about 1 cm. long; pedicels exceeding leaves ....................... ................................... 17*a*. *P. versicolor* var. *microphylla*
Corolla deep purple, with a yellow throat ...................... 18. *P. purpurea*

**1. Physalis ixocarpa** Brot. ex Hornem., Hort. Hafn. Suppl. 26, 1819

*Physalis aequata* Jacq. f. ex Nees, Linnaea **6:** 470, 1831.

Erect to spreading annual 3–10 dm. high, glabrous or young leaves and calyces sparsely pubescent; stems slender, much-branched, angulate; petioles 1.5–3.5 cm. long; leaf blades ovate to elliptic, shallowly sinuate-dentate, 1–3.5 cm. wide, 3–6 cm. long, base cuneate and somewhat asymmetrical, apex acute to short-acuminate; pedicels 3–4 mm. long in flower, about 5–6 mm. long in fruit, shorter than fruiting calyx; calyx campanulate, sparsely pubescent, 3–4.5 mm. long with deltoid lobes shorter than tube at anthesis; corolla 10–15 mm. in diameter, bright yellow with purple center; anthers tinged with purple or green; fruiting calyx ovoid, 1.5–2 cm. long, obscurely 10-angled, sparingly hairy; berry purple.

A native of "Mexico" but frequently cultivated and occasionally escaping along ditch banks and hedge rows. June–Oct.

**2. Physalis neo-mexicana** Rydb., Mem. Torrey Club **4:** 325, 1896

Stout erect annual with an obtusely angled stem; foliage and stems minutely but densely short-puberulent, only slightly glandular; leaves 3–5 cm. long, broadly ovate to suborbicular, obtuse at apex, rounded or obtuse at base, sinuately crenate; petioles equaling or slightly shorter than blades; pedicels short and stout, rarely over 2 cm. long in fruit; calyx tubular-campanulate, 3–6 mm. long, lobes lance-deltoid, acute but not acuminate, equaling or slightly exceeding tube at anthesis; corolla subrotate-campanulate, 5–6 mm. in diameter, yellow with a darker center; anthers ovate, about 1.5 mm. long, purplish or greenish; fruiting calyx ovoid, 2.5–3.5 cm. long, moderately sunken at base, strongly 10-angled, shining, reticulate-veined, lobes lance-deltoid, acuminate, nearly twice as long as broad.

Mostly in the Upper Sonoran and Transition Zones, New Mexico, southern Colorado, and eastern Arizona; introduced in southern California at Elsinore. Aug.–Oct.

**3. Physalis pubescens** L., Sp. Pl. 183, 1753

A diffusely branching, slender annual that is finely pubescent throughout with spreading hairs; petioles 6–25 mm. long, blades 1.5–6 cm. long, 1–4 cm. wide, broadly ovate, rounded to subcordate and often asymmetrical at base, acute to apiculate at apex, shallowly sinuate-crenate or nearly entire; pedicels slender, densely pubescent, equaling or slightly exceeding petioles; calyx tubular-campanulate, 4–5 mm. long at anthesis; lobes about equaling tube at anthesis; corolla rotate or campanulate, 6–10 mm. wide, light yellow with darker center, sometimes tinged with green on outside; anthers ovate, 1–1.5 mm. long, greenish to purplish; fruiting calyx membranous, scarcely angled, ovoid-pyramidal, somewhat retuse at base, lobes acuminate at apex.

Sandy soil along streams and washes, Lower Sonoran and Tropical Zones, from southern California to Pennsylvania and Florida; south into Mexico and the Galápagos Islands. Occurring in the Sonoran Desert near Yuma and on the delta of the Colorado River in both Baja California and Sonora. Sept.–Oct.

**4. Physalis leptophylla** Rob. & Greenm., Proc. Amer. Acad. **29:** 389, 1894

A much-branched, erect or ascending annual 3–6 dm. high, with finely villous stems, petioles, and calyces; leaves ovate, 1–2 cm. wide, 1.5–4 cm. long, obtuse or cordate at base, rarely asymmetrical, entire, apex abruptly long-acuminate, thin, bright green, often nearly glabrous except on ciliate margins and along main veins; flowers small, usually nodding, on pedicels 2–4 mm. long; calyx densely glandular-puberulent, about 3 mm. long, teeth ovate, shorter than tube; corolla campanulate, 4–6 mm. long, whitish or light yellow with darker spot at center; anthers ovate, about 1 mm. long, greenish; fruiting calyx 1.5 cm. long, thin, ovoid, rounded and slightly retuse at base, lobes 2–3.5 mm. long.

Not definitely known from the Sonoran Desert proper but occurring at Alamos, Sonora, and south into Sinaloa. July–Oct.

**5. Physalis filipendula** Brandegee, Univ. Calif. Pub. Bot. **10:** 187, 1922

A slender glabrous annual with striate stems branching above; leaves ovate, 4–8 cm. wide, 5–12 cm. long, sinuate-dentate, acute at apex, gradually narrowing at base to a slender petiole nearly as long as blade; pedicels 8–12 mm. long in anthesis, 4–6 cm. long in fruit, sparsely furfuraceous; calyx campanulate with deltoid-acuminate lobes equaling or exceeding tube, sparsely furfuraceous at anthesis, later glabrate; corolla light yellow, subrotate, 10–13 mm. in diameter; anthers linear, about 3 mm. long, purplish or bluish; fruiting calyx ovate, acuminate, 2–2.5 cm. wide, 3–4 cm. long.

Sandy soil along washes and ditches, Lower Sonoran Zone, southern Baja California. Flowers following adequate rain.

**6. Physalis acutifolia** (Miers) Sandw., Kew Bull. **14:** 232, 1960

*Physalis wrightii* A. Gray, Proc. Amer. Acad. **10:** 63, 1874.
*Chamaesaracha physaloides* Greene, Bull. Torrey Club **9:** 122, 1882.

Erect or ascending annual 1–10 dm. high with strongly angled, much-branched stems and sparingly pubescent to subglabrous foliage; petioles slender, 1.5–5 cm. long; leaf blades lanceolate, 6–35 mm. wide, 2.5–8 cm. long, deeply sinuate-toothed, cuneate at base, acute, somewhat attenuate at apex, margins finely ciliate; pedicels slender, 5–20 mm. long, finely but closely puberulent at anthesis, often subglabrate and 2–4 cm. long in fruit; calyx campanulate, scarcely angular, 3–5 mm. long with narrowly deltoid lobes about equaling tube at anthesis; corolla rotate, 12–20 mm. in diameter, whitish or light yellow with deeper yellow center; anthers greenish, linear, 3–4.5 mm. long, on slender filaments about as long; fruiting calyx globose-ovoid, 1.5–2.5 cm. long, obscurely 10-angled, lobes acuminate.

Roadsides, fields, and ditches, mostly Lower Sonoran Zone, Colorado Desert, California, to western Texas and south through Sonora into Sinaloa. April–Sept.

**7. Physalis lancifolia** Nees, Linnaea **6:** 473, 1831

Erect robust annual 5–8 dm. high with ascending, angled, glabrous branches and subglabrous foliage; petioles 1–3 cm. long; leaf blades lanceolate, 6–15 or rarely 20 mm. wide, 3.5–7 cm. long, attenuate at ends, entire to shallowly sinuate-toothed, dark green, subglabrous or with a few stiff short hairs on veins; pedicels filiform, 1.5–3 cm. long at anthesis; calyx puberulent, tubular-campanulate, about 2 mm. in diameter, about 3–4 mm. high at anthesis, lobes deltoid, half as long as tube; corolla yellow, narrowly campanulate, 5–6 mm. long, and about as wide at anthesis; anthers ovate, 1.5–2.5 mm. long, greenish or purplish; fruiting calyx broadly ovoid, 2–2.5 cm. long, on pedicels equaling or exceeding calyx, lobes broadly deltoid, not acuminate.

Sandy soil in fields, roadsides, and ditch banks, Lower Sonoran Zone, Imperial Valley, California, to Texas and south to central Mexico. March–June.

### 8. **Physalis nicandroides** Schlecht., Linnaea **19:** 311, 1847

A robust, erect, much-branched annual 8–12 dm. high with strongly angled, clammy, glandular-villous stems; petioles 3–10 cm. long, villous; leaf blades broadly ovate to suborbicular, 3–7 cm. wide, 4.5–15 cm. long, thin, dark green above, slightly paler beneath, villous on veins, sparsely hispidulous with short stout hairs elsewhere, shallowly but sharply and irregularly dentate, rounded, truncatish to subcordate at base, short-acuminate at apex; pedicels slender, villous, 2–3 cm. long at anthesis; calyx broadly campanulate, villous, tube 2–3 mm. long, linear-lanceolate lobes 4–5 mm. long at anthesis; corolla rotate, 10–12 mm. in diameter, yellow, somewhat villous on outside; anthers yellow, 3–4 mm. long, on filaments about as long; fruiting calyx broadly cordate-ovate, sharply angled, chartaceous, 4–6 cm. long, cordate at base, lobes lance-attenuate, 10–15 mm. long.

Roadsides, waste ground, and hedgerows, Lower Sonoran Zone, southern Sonora, Baja California, and south to southern Mexico. Feb.–April; Aug.–Nov.

### 9. **Physalis greenei** Vasey & Rose, Contr. U.S. Nat. Herb. **1:** 18, 1890

*Physalis pedunculata* Greene, Pittonia **1:** 268, 1889. Not *P. pedunculata* Mart. & Gal., 1845.

Erect spreading annual 1–4 dm. tall with slender, angular, somewhat flexuous, puberulent branches and finely puberulent foliage; petioles slender, 1–2.5 cm. long; leaf blades ovate, sinuate-dentate, acute at apex, broadly cuneate to subcordate at base, 1.5–3 cm. long, 1.5–2.5 cm. wide; pedicels 1–3.5 cm. long at anthesis, to 5 cm. long in fruit; calyx broadly tubular-campanulate, 5–6 mm. long, 3.5–4 mm. broad at anthesis, hispidulous; corolla 12–15 mm. broad, rotate, greenish yellow; fruiting calyx ovoid, rounded to slightly sunken at base, acute to acuminate at apex, 1.5–2.5 cm. long, moderately hispidulous on low angles, on a pedicel 2–3.5 cm. long.

Sandy soil, Upper and Lower Sonoran Zones, San Diego County south near coast to the vicinity of Rosario and San Augustín, Baja California; Cedros Island. March–July.

### 10. **Physalis longifolia** Nutt., Trans. Amer. Phil. Soc. II, **5:** 193, 1836

*Physalis pumila* var. ?*sonorae* Torr., Bot. Mex. Bound. 153, 1859.
*Physalis lanceolata* var. *laevigata* A. Gray, Proc. Amer. Acad. **10:** 68, 1874.

Stout erect perennial 0.5–1 m. tall with angulate-striate, glabrous stems and lance-elliptic, glabrous or subglabrous, entire leaves 1–2.2 cm. wide and 3–9 cm. long, acute, cuneate and somewhat decurrent on petioles at base; petioles 1–4 cm. long; pedicels 8–12 mm. long at anthesis, scarcely longer in fruit; calyx campanulate, truncatish at base, sparsely puberulent toward base, lobes lanceolate, 2.5–3 mm. wide, 4–5 mm. long, equaling or slightly exceeding limb; corolla campanulate-rotate, 12–20 mm. wide, about 10–15 mm. long, yellow with darker center; anthers 2–3 mm. long, yellow; fruiting calyx ovoid, 2–3.2 cm. long, glabrous or essentially so, distinctly veined; berry yellow.

Plains and stream banks, Wyoming and South Dakota to Iowa and south disjunctly through Alabama, and Arizona into Mexico. April–Aug.

**11. Physalis fendleri** A. Gray, Proc. Amer. Acad. **10:** 66, 1874

Compact perennial 2–4 dm. high from a deep-seated rootstock, and branching from base, with ascending, slightly striate branches, puberulent throughout with fine, branching, and forked hairs; petioles 6–25 mm. long, narrowly winged; leaf blades broadly ovate to deltoid, 1–3.5 cm. wide, 1.2–4 cm. long, green to cinereous, sinuate-dentate; pedicels slender, 4–10 mm. long at anthesis, 10–20 mm. long in fruit; calyx campanulate, 5–7 mm. high, lobes narrowly deltoid, about equaling or slightly exceeding tube at anthesis; corolla campanulate-subrotate, 8–10 mm. in diameter, greenish yellow with brownish center; anthers 3 mm. long, yellow; fruiting calyx narrowly ovoid, 2–3 cm. long, obscurely angled; berry yellow.

Sandy soil, Upper and Lower Sonoran Zones, eastern Mojave Desert to Utah, Colorado, Arizona, and New Mexico. Mostly above the Sonoran Desert. March–Aug.

**Physalis fendleri** var. **cordifolia** A. Gray, Syn. Fl. N. Amer. **2**[1]: 395, 1878, differs from the above in having somewhat larger, cordate leaves with sinuate instead of dentate margins. Huachuca Mountains and Kingman, Arizona. Might occur as a rare straggler.

**12. Physalis hederifolia** A. Gray, Proc. Amer. Acad. **10:** 65, 1874

*Physalis palmeri* A. Gray, Syn. Fl. N. Amer. **2**[1]: 235, 1878.
*Physalis digitalifolia* Britton, Mem. Torrey Club **5:** 288, 1894.

Erect or rarely decumbent, cinereous-puberulent perennial 3–5 dm. high; petioles 5–35 mm. long; leaf blades ovate-deltoid to cordate-subreniform, 1–3 cm. wide, 1.5–5 cm. long, coarsely sinuate-dentate, acute to rounded at apex, broadly cuneate to cordate at base; pedicels 5–15 mm. long at anthesis, recurved and flowers nodding; calyx tubular-campanulate, 6–8 mm. long, viscid-puberulent, lobes lance-deltoid, two-thirds to about four-fifths as long as tube at anthesis; corolla campanulate-rotate, 12–15 mm. wide, yellow with greenish throat and center; anthers 2.5–3 mm. long, yellow; fruiting calyces ovoid, obtusely 10-angled, strongly reticulate-veined, 2–3 cm. long, usually two to three times as long as pedicels; berry yellow.

Sandy plains, arroyos, and desert canyons, Lower Sonoran Zone to Arid Transition Zone, southeastern California to Texas and south into northern Mexico. June–Sept.

**13. Physalis crassifolia** Benth., Bot. Sulph. 40, 1844, var. **crassifolia**

Spreadingly branched, compact perennial 3–6 dm. high, 3–15 dm. wide, with finely viscid-puberulent foliage and stems; petioles slender, equaling or somewhat exceeding blade; leaf blade broadly ovate, deltoid, or ovate-cordate, usually small, 1–2.5 cm. wide, 1.5–3.5 cm. long, entire or margins shallowly sinuate, acute to rounded at apex, green, somewhat viscid, finely puberulent; pedicels slender, 8–15 or rarely to 30 mm. long at anthesis, scarcely longer in fruit; calyces campanulate, subtruncate at base, 3–5 mm. long, lobes short, broadly deltoid, 1–1.5 mm. long; corolla campanulate-subrotate, 10–15 mm. broad, dull yellow, tube slightly exceeding calyx, with 5 narrow pubescent bands from base of tube to tips of lobes; anthers about 3 mm. long, yellow; fruiting calyx ovoid, 1.5–2.5 cm. long, obscurely angled, mouth usually open; berry greenish.

Sandy and rocky places, mainly Lower Sonoran Zone, Colorado and Mojave Deserts, California, to Utah and Texas, south to central Sonora and the Cape Region of Baja California. Feb.–June; Aug.–Dec.

**13*a*. Physalis crassifolia** var. **cardiophylla** (Torr.) A. Gray, Syn. Fl. N. Amer. **2**[1]: 235, 1878

*Physalis cardiophylla* Torr., Bot. Mex. Bound. 153, 1859.

Differs from var. *crassifolia* in having thinner, more deeply cordate leaves. It occurs mostly toward the northern limits of the species.

**13*b*. Physalis crassifolia** var. **infundibularis** I. M. Johnston, Proc. Calif. Acad. IV, **12:** 1156, 1924

Differs from var. *crassifolia* only in having a funnelform corolla as long as or longer than wide.

Northeastern part of Baja California and adjacent islands; San Luis Gonzales Bay; Isla de la Guarda; San Estéban Island.

**14. Physalis glabra** Benth., Bot. Sulph. 39, 1844

*Physalis hastata* Rydb., Mem. Torrey Club **4:** 363, 1896.

Openly branched, scrambling or decumbent, glabrous perennial with slender, slightly angulate stems and small distant leaves; petioles 8–40 mm. long, slender; leaf blades narrowly lanceolate to narrowly ovate, more or less asymmetrical at base, sometimes subhastate, 4–15 mm. wide, 1–5 cm. long, entire or slightly undulate-margined, conspicuously veined; pedicels filiform, 15–25 mm. long at anthesis; calyx broadly tubular-campanulate, 4–5 mm. long and nearly as broad at anthesis, lobes broadly deltoid, 1–1.5 mm. long, puberulent along margins; corolla 8–10 mm. wide, dull yellow with a greenish center, 5 triangular tufts of white hairs on throat between bases of short filaments; anthers oval, 1.5–2 mm. long, yellow; fruiting calyx ovoid, 1.5–2 cm. long, about two-thirds as wide, obscurely 10-angled, glabrous or nearly so, exceeded by pedicel; berry apparently greenish.

Sandy plains, arroyos, and dunes, Lower Sonoran and Tropical Zones, southern Baja California, mainly the Cape Region. Aug.–Jan.

**15. Physalis muriculata** Greene, Bull. Calif. Acad. **1:** 209, 1885

A glandular-pubescent perennial 1–3 dm. high; branches striate-angulate, slender, internodes 2–5 cm. long; petioles 1–3 cm. long, slender, puberulent; leaf blades 1.5–4 cm. wide, 2–5 cm. long, broadly ovate, entire, truncate to subcordate at base, acute to obtusely rounded at apex, thin, deep green; pedicels slender, 1.5–3 cm. long at anthesis; calyx broadly campanulate, 4–5 mm. high and nearly as wide at anthesis, lobes deltoid, acute, 1.5–2.5 mm. long; corolla campanulate-rotate, 12–15 mm. wide, greenish yellow with a brownish throat; anthers about 2.5 mm. long; fruiting calyx broadly ovoid, 2.5–3 cm. long, nearly as wide, distinctly angled and strongly reticulate, angles muriculate; pedicels 3–5 cm. long in fruit, glandular-pubescent and muriculate; berry yellow, very viscid; seeds granular-tuberculate, yellowish brown, 1.5 mm. long.

Among volcanic rocks, Lower Sonoran Zone, along the coastal slopes from San Quintín to Punta Prieta, Baja California. Feb.–May.

**16. Physalis flava** Wiggins, Contr. Dudley Herb. **3:** 73, *pl. 19, figs. 7–8,* 1940

A much-branched, rounded perennial 3–5 dm. high with robust, striate-angled, leafy branches moderately glandular-pubescent throughout; petioles narrowly winged, 1.5–3 cm. long; leaf blades ovate-lanceolate to broadly ovate, 1.2–3 cm. wide, 1.5–4 cm. long, entire or shallowly sinuate-angled, broadly cuneate to shallowly cordate at base, acute at apex, thin, green, sparsely pubescent; pedicels slender, 2–4 cm. long at anthesis, short-villous; calyx campanulate, 3.5–4.5 mm. wide, 6–8 mm. long at anthesis, sparsely puberulent with more densely puberulent angles, lobes narrowly lance-deltoid, slightly unequal, equaling tube (3–4 mm. long), margins densely puberulent; corolla funnelform, 1.5–2 cm. long, 2–2.5 cm. broad, clear yellow, without a dark center, shallowly 5-lobed, sparsely puberulent on outside of each lobe, margin finely but distinctly ciliate, a ring of tangled white hairs at about middle of outside surface of tube; stamens about equaling corolla tube, filaments glabrous; anthers oblong, 3 mm. long, yellow; style equaling or slightly exceeding stamens, faintly greenish or purplish; stigma capitate, faintly bilobed; fruiting calyx (immature) ovoid, 2.5–3 cm. long, about 1.5–2 cm. wide, sparsely puberulent and somewhat muriculate on angles near base.

Desert areas, among granite boulders, Lower Sonoran Zone, vicinity of San Quintín to Cataviñá, Baja California. Feb.–May.

**17. Physalis versicolor** Rydb., Bull. Torrey Club **22:** 307, 1895, var. **versicolor**

A much-branched perennial with spreading to scrambling branches 2–7 dm. long, finely puberulent in youth, finally glabrate; branches somewhat zigzag, distinctly striate-angled; petioles slender, equaling to somewhat exceeding blades, these thin, ovate to ovate-cordate or lower cordate-reniform, 1.5–4 cm. long, rounded to subcordate and often asymmetrical at base, mostly acute at apex, sinuate-dentate, teeth acute to rounded; calyx hemispherical-campanulate, 3–4 mm. high, lobes deltoid, about one-fourth as long as tube, all densely glandular-puberulent; corolla subrotate, 8–10 mm. wide, light yellow with brownish spots at center, turning bluish in drying; anthers 2–2.5 mm. long, yellow; fruiting calyx thin, cylindrical-ovoid, finely reticulate, 2–2.5 cm. long, slightly sunken at base, teeth not connivent; pedicels slightly exceeding calyces.

Rocky and sandy places, Lower Sonoran Zone, southern Arizona and New Mexico to northern Sinaloa. Aug.–Sept.

**17*a*. Physalis versicolor** var. **microphylla** Rydb., Bull. Torrey Club **22:** 307, 1895

Differs from var. *versicolor* in having deltoid, coarsely toothed leaves only about 1 cm. long; pedicels nearly twice as long as blades; white corollas with sordid yellow center; and a nearly spherical fruiting calyx 1.5 cm. long.

Vicinity of Guaymas; San Francisco Island; San Diego Island; Cerralvo Island.

**18. Physalis purpurea** Wiggins, Contr. Dudley Herb. **3:** 74, *pl. 19, figs. 9–10,* 1940

Erect herbaceous perennial about 3–5 dm. high branching from base, with finely puberulent, striate-angled stems, and bright green leaves with puberulent veins and subglabrous lamina; petioles slender, 2–5 cm. long, sparsely puberulent, faintly winged; leaf blades deltoid-ovate, 2–3 cm. wide, 2.5–4.5 cm. long, coarsely but shallowly sinuate-dentate or angled, rounded to subtruncate and sometimes oblique at base, acute to obtuse at apex, veins prominulous beneath; pedicels slender, 2–3.5 cm. long at anthesis, densely puberulent; calyx campanulate, truncatish and moderately hispidulous at base, 3–5 mm. wide at top, 5–7 mm. long at anthesis, lobes deltoid, about 2–2.5 mm. long, sparsely puberulent on outer surface, densely puberulent on margins; corolla campanulate-rotate, deep purple with yellow tube and throat, 8–10 mm. high, 12–15 mm. wide, sparsely puberulent beneath each lobe on outer surface;

stamens exserted about length of anthers; anthers oblong, about 3 mm. long, yellow; style equaling to slightly surpassing stamens; fruiting calyx (immature) ovoid, moderately 10-angled, finely and sparsely puberulent, 1.5 cm. long, on a pedicel over twice as long.

In heavy clay soil among rocks, Lower Sonoran Zone, near Guaymas, Sonora. Feb.–April.

## 5. QUINCULA Raf., Atl. Jour. 145, 1832

A low, diffuse, granular-scurfy, perennial herb from a heavy rootstock. Leaves elliptic to oblanceolate, sinuate to pinnatifid, subsucculent. Flowers borne in pairs from axils of leaves or less commonly solitary or 3–5-fasciculate. Calyx broadly campanulate, with 5 deltoid teeth, enlarging to enclose berry and becoming 5-angled, reticulate in fruit, lobes connivent in fruit. Corolla rotate, pentagonal, violet to purple; stamens 5, free, erect; anthers yellow, dehiscing by a longitudinal slit. Style about equaling anthers; stigma capitate. Seeds few, reniform, flattened, rugose-tuberculate, margins thickened.

**1. Quincula lobata** (Torr.) Raf., Atl. Jour. 145, 1832
*Physalis lobata* Torr., Ann. Lyc. N.Y. **2:** 226, 1827.
*Physalis sabeana* Buckley, Proc. Acad. Phila. **1862:** 6, 1863.

Low-spreading or prostrate, branches 1–3 dm. long, sparsely scurfy-puberulent throughout with rounded, white, unicellular, pustulate hairs; leaf blades elliptic, oblanceolate, spatulate, or oblong, 5–30 mm. wide, 1–7 cm. long, sinuate-dentate to pinnatifid with rounded lobes, acute to rounded at apex, cuneate at base, tapering gradually into petiole; pedicels 5–20 mm. long at anthesis; calyx 3.5–6 mm. long at anthesis, rounded at base, lobes deltoid, half as long as tube or less; corolla violet or purple, 1.5–2.5 cm. in diameter; stamens 3–5 mm. high; anthers oval, 1.5–2.5 mm. long; fruiting calyx broadly ovate, 1.5–2.5 cm. long, nearly as wide, sharply 5-angled, sunken at base, on pedicels about equaling or somewhat exceeding them.

Sandy washes and dry, open ground, Upper and Lower Sonoran Zones, Arizona to Colorado, Kansas, and Texas and south into northern Mexico. March–Oct.

## 6. SARACHA Ruiz & Pav., Fl. Peruv. Prod. 31, *pl. 34,* 1794

Diffusely branched, large, perennial herbs from an elongated tuberous root; stems sharply 4-angled. Leaves large, thin, long-petioled, entire or essentially so. Flowers in axillary umbels, rarely solitary; pedicels recurved in fruit. Calyx broadly campanulate, shallowly 5-lobed, enlarging and saucer-shaped in fruit. Corolla subrotate to broadly campanulate, greenish, plicate in bud. Stamens 5, inserted near base of corolla tube; filaments slender. Anthers erect, longitudinally dehiscent. Ovary 2-celled, many-ovuled. Style slightly thickened toward apex; stigma emarginate or shallowly 2-lobed. Berry juicy, subglobose, dark purple at maturity. Seeds many, compressed; embryo strongly curved.

**1. Saracha procumbens** (Cav.) Ruiz & Pav., Fl. Peruv. **2:** 43, *pl. 180,* 1799
*Atropa procumbens* Cav., Ic. **1:** 53, *pl. 72,* 1791.
*Saracha umbellata* DC., Cat. Hort. Monsp. 142, 1813.
*Bellinia procumbens* Roem. & Sch., Syst. Veg. **4:** 687, 1819.

A large, coarse, perennial herb to 1 m. high with procumbent to ascending, quadrangular stems and ample, very thin leaves from a fleshy vertical root; herbage sparsely hispid with short coarse hairs; petioles slender, 3–8 cm. long, narrowly but distinctly winged, blade ovate, acute to acuminate at apex, broadly cuneate to truncate at base, 1–12 cm. wide, 3–20 cm. long, entire or nearly so; flowers on very short peduncles, in pairs or several-flowered umbels; pedicels slender, 1–3 cm. long at anthesis; calyx 5–6 mm. wide at anthesis, to 2.5 cm. wide with broadly ovate-deltoid, spreading lobes in fruit; corolla 12–15 mm. broad, greenish; berries 7–10 mm. in diameter, glabrous, dark purple and more or less juicy when ripe; seeds numerous, 1.5 mm. in diameter.

Hillsides, often in shelter of rocks or under trees, Upper Sonoran to Tropical Zones, southern Arizona to Central and South America. Rare in the Sonoran Desert. Aug.–Sept.

## 7. **CAPSICUM** L., Sp. Pl. 188, 1753

Annual or perennial herbs or plants woody at base, stems erect, ascending, or some procumbent, profusely branched. Leaves entire to repand, petiolate. Flowers usually solitary but sometimes in 2–3-flowered clusters in axils. Calyx short, broadly campanulate to saucer-like, truncate or shallowly 5-lobed or 5-toothed. Corolla subrotate, with a 5–6-toothed limb plicate in bud and a very short tube, white, slightly greenish, or purplish. Stamens usually 5; filaments slender, inserted in corolla tube; anthers shorter than filaments to about equaling them, usually connivent, dehiscing longitudinally. Ovary 2–3-celled; style slender, exceeding anthers; stigma slightly clavate or dilated at apex. Fruit a berry, red, yellow or sometimes white, globose to slenderly oblong, in some strains considerably inflated. Seeds compressed, smooth to lightly reticulate-rugose, mild to very pungent. Embryo curved to subspiral. Endosperm fleshy but meager.

### 1. **Capsicum annuum** L. var. **baccatum** (L.) Kuntze, Rev. Gen. Pl. **2:** 449, 1891

*Capsicum baccatum* L., Mant. **1:** 46, 1767.
*Capsicum frutescens* var. *baccatum* Irish, Rep. Mo. Bot. Gard. **9:** 99, 1898.
(Many other epithets have been applied to this taxon.)

A profusely branching annual or perennial with slender, geniculate, sparsely puberulent to glabrate stems, often scrambling through shrubs to a height of 2 m. or forming a compact rounded plant to 1.5 m. in diameter, bark tawny on basal branches of perennial plants; petioles slender, 1–2.5 cm. long, narrowly alate, leaf blades ovate to lanceolate-ovate, 1–4 cm. wide, 2–6 cm. long, acute to slightly attenuate at apex, rounded to cuneate at base, entire or rarely slightly repand, mostly sparsely puberulent along veins beneath; flowers solitary or appearing clustered in fruit; pedicels 1–2 cm. long in flower, to 4 cm. long in fruit, slender, slightly enlarged upward; calyx 1.5–2 mm. long at anthesis, minutely 5-toothed, sparingly hispid to glabrate; corolla white, 6–9 mm. in diameter, puberulent along margins; anthers about 2 mm. long, often greenish; berries globose to short-ellipsoidal, 6–8 mm. in diameter, red in our material; seeds 2.5–3 mm. high, broadly subreniform, yellowish.

Common along fence rows, arroyos, ditches, and in shrubs, Lower Sonoran to Tropical Zones, southern Arizona to Texas and south to South America. Aug.–Oct.

## 8. **SOLANUM** L., Sp. Pl. 184, 1753

Herbs or shrubs with glabrous, pubescent, or tomentose stems and foliage, sometimes climbing and some armed with stiff prickles or spines. Leaves simple, entire, lobed, or intricately parted or divided, petiolate. Flowers in cymes, umbels, panicles, or racemes. Calyx campanulate to rotate, 5-toothed or -cleft, in some enlarging in fruit. Corolla rotate, 5-angled or -lobed; tube very short; limb plaited in bud. Stamens 5, inserted on corolla tube; filaments short; anthers oblong, acute to acuminate or attenuate upward, connivent around style, dehiscent by a terminal pore, by a short, introrsely oriented, subterminal slit, or longitudinally. Ovary 2-celled; stigmas small, capitate or obscurely bilobed. Fruit a globose berry, fleshy or leathery, juicy or dry when ripe. Seeds numerous, more or less flattened; embryo annular.

Plants unarmed; pubescence of simple, gland-tipped, or sparingly branched trichomes, or absent:
  Corollas 3–10 mm. in diameter; plants annual (except *S. nodiflorum,* which often flowers first season of growth):
    Berries purple or black at maturity; sepals more or less reflexed in fruit .......................... 1. *S. nodiflorum*
    Berries whitish, yellowish, brownish, or greenish at maturity; sepals curved upward and closely applied to fruit:
      Pedicels strongly deflexed in fruit; berry milky-translucent apically when ripe; seeds strongly verrucose on sides; calyx lobes not enlarging in fruit .... 2. *S. deflexum*
      Pedicels not markedly deflexed in fruit; berry yellowish or greenish; seeds minutely granular-tessellate, not verrucose; calyx lobes enlarging noticeably in fruit:
        Leaves ovate, shallowly sinuate-toothed; berries yellowish or brownish; herbage, especially inflorescences, markedly glandular-viscid ......... 3. *S. sarachoides*
        Leaves often elliptic, pinnatifid or deeply lobed; berries greenish; herbage scaberulous-pubescent with appressed hairs, not glandular-viscid ..... 4. *S. triflorum*
  Corollas (7–) 10–25 mm. in diameter; plants perennial (but often flowering first season in *S. furcatum*):
    Berries black or purple at maturity; corolla white:
      Peduncles strongly deflexed at maturity; berries containing 10–15 granules around base, 5–6 mm. in diameter .................................. 5. *S. furcatum*
      Peduncles not markedly deflexed at maturity; berries devoid of granular concretions, 6–9 mm. in diameter ...................................... 6. *S. douglasii*
    Berries greenish or red at maturity; corollas lavender to purple:
      Fruits greenish at maturity; leaves entire or hastately lobed at base, not compound; seeds 1.5–2.5 mm. long:
        Glandular and dendritic hairs numerous; leaves mostly 2–4 cm. long or more, not attenuate at base ................................ 7. *S. xanti* var. *xanti*
        Glandular and dendritic hairs few to nearly absent; plants nearly glabrous; leaves mostly less than 3 cm. long, attenuate at both ends .......................... 7a. *S. xanti* var. *glabrescens*
      Fruits red; leaves compound, of 3–7 leaflets; seeds 3–3.5 mm. long .......................... 8. *S. seaforthianum*
Plants moderately to strongly armed with stiff spines; pubescence mainly of closely crowded, stellate trichomes:
  Plants annual herbs; leaves bipinnatifid, pinnatifid, or deeply pinnately lobed; seeds moderately to strongly rugosely ridged across flattened or rounded back, dull black:
    Spines all terete; gland-tipped hairs on stems and pedicels few or none; longest anther 9–12 mm. long .................................. 9. *S. rostratum*
    Spines on stems and petioles often flattened laterally, elliptical at base; gland-tipped hairs numerous among stellate ones on stems and pedicels; longer anthers 5–6 mm. long .................................. 10. *S. grayi*
  Plants shrubby, or perennial, more or less suffrutescent herbs; leaves not pinnatifid (except rarely in *S. carolinense*); seeds finely reticulate-faveolate, not transversely ridged dorsally, usually yellowish or brownish:

Stamens markedly unequal, longer anthers 12–16 mm. long; armament differing on fruiting and nonfruiting calyces, fruiting ones strongly spinose, nonfruiting merely pubescent ........................................................ 11. *S. amazonium*
Stamens subequal, longer anthers about 10 mm. long or less; armament not differing on fruiting and nonfruiting calyces:
Plants herbaceous to suffrutescent, from spreading rhizomes; pedicels not markedly expanded at apex in fruit:
Seeds bright yellow, 2.2–2.5 mm. long; leaf blades repand to pinnatifid, 2–7 cm. wide ........................................................ 12. *S. carolinense*
Seeds brownish, 3.5–4 mm. long; leaf blades entire to repand, 0.4–3.5 cm. wide .. ........................................................ 13. *S. elaeagnifolium*
Plants shrubby, not rhizomatous; pedicels distinctly expanded toward apex in fruit:
Corolla white; inflorescence with gland-tipped, unbranched hairs intermingled with stellate trichomes; seeds 2–2.5 mm. long, yellowish; spines on branches strongly flattened, elliptical at base ........................ 14. *S. madrense*
Corolla blue to purple; inflorescence not glandular-hispid; seeds 3–4.5 mm. long, brownish to olivaceous; spines chiefly terete ................ 15. *S. hindsianum*

**1. Solanum nodiflorum** Jacq., Ic. Pl. Rar. **2:** 288, *pl. 326,* 1788–89

*Solanum chenopodioides* Lam., Illustr. **2:** 18, 1793, in part.
*Solanum nigrum* var. *dillenii* A. Gray, Syn. Fl. N. Amer. **2**[1]**:** 228, 1878.
*Solanum nigrum* var. *nodiflorum* A. Gray, loc. cit.
*Solanum photeinocarpum* Nakamura & Odashima, Jour. Soc. Trop. Agr. Taiwan 8: 54, *fig. 2,* 1936.
*Solanum nigrum* sensu most Western authors. Not *S. nigrum* L.

Perennial subwoody plant to 1.5 m. tall with straggly stems, but often flowering first year and misinterpreted as annual; herbage glabrous or subglabrous to subscabrous on angle of stems, with scattered, subappressed, simple hairs on younger growth; leaves ovate to lance-ovate, 0.5–7 cm. wide, 3–12 cm. long (usually not over 3.5 mm. wide or 8 cm. long), entire to sinuately toothed, cuneate to truncate at base, base gradually decurrent along upper part of petiole, this from one-fourth as long as blade to equaling it; peduncles slender, 1–3.5 cm. long, usually exceeding pedicels, these 5–12 mm. long at anthesis, deflexed in fruit; sepals 1–2.5 mm. long, thickish, usually sparsely puberulent, spreading to reflexed in fruit; corollas white to faintly tinged with purple, 4–6 mm. in diameter, lobes broadly lance-ovate, about 2 mm. long, glabrous within, sparsely puberulent along midrib and on distal one-half without; anthers yellow, 1.5–2.4 mm. long, equal or essentially so; filaments 0.6–1 mm. long, sparsely hairy, particularly along margins; berry globose, 5–8 mm. in diameter, black or purple, shiny, without granules or containing 1–4 small ones; seeds 30–60, pale cream, 1.5–1.8 mm. in diameter, minutely granular with low rounded tubercles.

Margins of fields, along arroyos and stream banks, and on shaded hillsides, Lower Sonoran to Tropical Zones and into the Upper Sonoran and Transition Zones to the north, Washington south throughout the Pacific states and Mexico, widely dispersed in the tropics of both the New and the Old World areas. Flowering virtually throughout the year under favorable moisture conditions.

**2. Solanum deflexum** Greenm., Proc. Amer. Acad. **32:** 301, 1897

*Salpichroa wrightii* A. Gray, Syn. Fl. N. Amer. **2**[1]**:** 231, 1878.

Annual herb 1–4 dm. tall with slender, often slightly zigzag stems and moderately to sparsely hirsutulous herbage; petioles 3–20 mm. long, shallowly grooved on upper surface, minutely winged from near base to base of blade, hirsutulous with simple hairs of various lengths; leaf blades thin, lance-ovate, entire, plane to slightly undulate, 8–25 mm. wide, 1.5–6 cm. long, rounded to broadly subcuneate at base, acute at apex, bright green and sparsely hirsute above, paler and slightly more densely pubescent beneath, veins nearly white; flowers usually 2 per umbel but sometimes solitary; peduncles slender, 1–8 mm. long or lacking; pedicels exceeding peduncles, very slender, to 1.5 cm. long, ascending at anthesis, strongly deflexed immediately thereafter; calyx 2–3 mm. long at anthesis, hirsutulous, lobes about equaling tube, ovate or lanceolate; corollas white, 8–10 mm. in diameter, lobes short and broad, glabrous within, puberulent without, especially along midribs of lobes; stamens yellow, equal, about 3 mm. long, on glabrous filaments 0.5–1 mm. long; berries globose, 6–8 mm. in diameter, milky-white and translucent in terminal half, purplish-tinged below, seeds confined to basal part of berry; seeds strongly flattened, ovoid-reniform, about 3 mm. long, yellowish, lustrous, strongly verrucose in radiating lines from center toward margins, and minutely papillose between and on rugae.

Mostly in shade of shrubs along arroyos and on brushy hillsides, Lower Sonoran to Tropical Zones, southern Arizona south to Central America. July–Oct.

### 3. **Solanum sarachoides** Sendt., in Mart., Fl. Bras. **10:** 18, *pl. 1,* 1846

*Solanum gracile* sensu A. Gray, Syn. Fl. N. Amer. **2**[1]**:** 228, 1878. Not *S. gracile* Link, 1827–33.
*Solanum villosum* of various Western authors. Not *S. villosum* Mill., 1768.

Decumbent or ascendingly branched annual with stems 1–5 dm. long, herbage viscid-villous throughout; leaves ovate, 1–3(4) cm. wide, 2.5–7 cm. long, gradually or abruptly narrowing at base and somewhat decurrent on petiole, acute to obtuse at apex, sinuately but shallowly toothed; peduncles usually exceeding pedicels, 5–10 or rarely –20 mm. long; pedicels 3–5 mm. long at anthesis, markedly swollen toward apex in fruit; calyx 2–2.5 mm. long at anthesis, villous, enlarging and becoming somewhat scarious and partially enclosing berry at maturity, its lobes then 3–5 mm. long; corolla white, 3–5 mm. in diameter, lobes narrowly triangular, short-villous on outer surface near tips; anthers erect, 2–2.5 mm. long, yellow, filaments glabrous, 0.4–0.6 mm. long; berries globose, 6–7 mm. in diameter, yellowish to brownish, somewhat lustrous; seeds yellowish, ovoid-reniform, moderately flattened, 2–2.2 mm. long, minutely granular or tessellate in concentric lines, often slightly lustrous.

Cultivated fields, roadsides, and neglected agricultural lands, introduced from South America and sparingly established from British Columbia to southern California and Arizona. April–Oct.

### 4. **Solanum triflorum** Nutt., Gen. **1:** 128, 1818

Annual herb with prostrate to decumbent branches 1–4 dm. long, herbage sparsely scaberulous-pubescent with appressed hairs; leaves ovate to elliptic, pinnatifid or deeply lobed, 1–2 cm. wide, to 4 cm. long, individual lobes or teeth lanceolate, 1.5–3 mm. wide, 3–7 mm. long, acute; petioles 1–1.5 cm. long, blade decurrent as a narrow wing about halfway to base; peduncles stoutish, 5–15 mm. long, 1–6- but oftener 3-flowered; pedicels 2–5 mm. long at anthesis, to 12 mm. long and distinctly swollen toward apex in fruit; calyx 2.5–3 mm. long, with lance-ovate lobes equaling or slightly exceeding tube at anthesis, enlarging in fruit and then 3–4 mm. wide, 3–5 mm. long, more or less clasping base of berry; corolla white or slightly tinged with green, 6–8 mm. in diameter, lobes ovate-attenuate, minutely puberulent on outside and on margins near tips; anthers all alike, 2.5–3 mm. long, erect and straight, on glabrous filaments about 1 mm. long; berries globose, 6–10 mm. in diameter, greenish at maturity, smooth and sublustrous; seeds pale yellowish or tawny, obovoid, moderately flattened but with rounded margins, 2.5–2.8 mm. long, funicular notch scarcely noticeable, microscopically and uniformly tuberculate-tessellate, dull.

Roadsides, stream beds, and sandy soil, upper margins of Lower Sonoran to Transition Zones, northern margin of the Sonoran Desert in Arizona north to Washington, Alberta, and North Dakota, east to Kansas and New Mexico; east of the Sierra Nevada in California. May–Sept.

**5. Solanum furcatum** Dunal, in Poir., Encycl. Suppl. **3:** 750, 1814

*Solanum chenopodioides* Lam., Illustr. **2:** 18, 1793, in part.
*Solanum rancaguense* Dunal, in A. DC., Prod. **13¹:** 150, 1852.

Perennial herb (often flowering first year) with slender straggling stems to 1 m. long, usually reclining or scrambling over other plants; herbage glabrous to sparsely puberulent on younger parts and on angles of stems; leaves ovate to elliptic-ovate, 2–4.5 cm. wide, 4–8 cm. long, shallowly and broadly sinuate-toothed, angulate-dentate, or subentire, acute to slightly acuminate at apex, narrowly decurrent on upper part of petiole, glabrous or with a few scattered simple hairs along veins beneath; petioles slender, 1–4 cm. long; peduncles slender, 1.5–4 cm. long, usually dichotomously forked near apex, deflexed at maturity, several-flowered, glabrous to sparsely pubescent with simple, mostly antrorse hairs; pedicels slender, 6–10 mm. long, glabrous to sparsely puberulent; calyx campanulate, 2–3 mm. deep, narrowly deltoid lobes equaling tube; corollas white or faintly lavender, rotate, 10–22 mm. in diameter, sinus between lobes shallow; filaments slightly unequal, 1–1.5 mm. long, pubescent marginally; anthers erect, yellow, 3–3.5 mm. long; berries globose, black or dark purple, 5–6 mm. in diameter, deciduous when ripe, containing 10–15 reniform to globose granules roughly ringing base of berry; seeds 12–25(40), ovoid-reniform, pale yellow, about 1.4 mm. wide, 1.6–1.8 mm. long, flattened, minutely reticulate-pitted.

Stream banks, shrubby areas, and margins of fields, Lower Sonoran to Humid Transition Zones, as an introduced weed from Peru, scattered along the Pacific Coast from Oregon to central Baja California. March–Oct.

**6. Solanum douglasii** Dunal, in A. DC., Prod. **13¹:** 48, 1852

*Solanum umbelliferum* var. *trachycladum* Torr., Pacif. R. Rep. **7:** 17, 1858.
*Solanum nigrum* var. *douglasii* A. Gray, Syn. Fl. N. Amer. **2¹:** 228, 1878.
*Solanum arizonicum* Parish, Proc. Calif. Acad. III, **2:** 165, 1901.

Herbaceous to slightly shrubby perennial 0.6–2 m. tall with sharply angled to narrowly winged stems and puberulent to subglabrate herbage, angles of stems distinctly roughened by heavy conical bases of antrorse simple hairs; leaves ovate, 1–6 cm. wide, 2–18 cm. long, coarsely sinuate-dentate, acute to short-acuminate at apex, cuneate to subtruncate at base, green and sparsely puberulent on both surfaces; petioles 1–2.5 cm. long, slightly winged near blade; peduncles 1–3 cm. long, several-flowered; pedicels slender, 5–12 mm. long; calyx 2–3 mm. long at anthesis, lobes lance-oblong to broadly ovate, 1.5–2 mm. long, appressed-puberulent; corolla white with greenish spots at center, 7–20 mm. in diameter, rotate or ovate to lance-deltoid lobes spreading-ascending to slightly reflexed, 6–9 mm. long, acute, glabrous within, finely puberulent along margins and midrib without; anthers erect, yellow, about 3 mm. long, subequal, almost three times as long as villous filaments; berry black, globose, smooth and glabrous, 6–9 mm. in diameter; seeds numerous, about 1.3 mm. in diameter, pale yellow, minutely reticulate-pitted, funicular notch shallow.

On partially shaded slopes, along canyons, stream banks, and margins of clearings and meadows, mainly Upper Sonoran Zone but in favorable habitats in the Lower Sonoran Zone, coastal central California south to the Cape Region of Baja California, east into Arizona, Tamaulipas, and the Federal District. Flowering more or less throughout the year.

**7. Solanum xanti** A. Gray, Proc. Amer. Acad. **11:** 90, 1876, var. **xanti**

*Solanum cupuliferum* Greene, Erythea **3:** 72, 1895.

Suffrutescent perennial 2–15 dm. tall, herbage short-villous with an admixture of simple unbranched hairs, 2–5-rayed, dendritic trichomes, and shorter, gland-tipped hairs throughout plant and persistent, or nearly glabrous in one variety; leaves ovate, lance-ovate, or oblong-ovate, subentire to hastately lobed at base, 2–4(10) cm. long, blades slightly decurrent on petioles, latter to 1.5 cm. long; inflorescences umbelliform cymes; peduncles 5–12 mm. long, densely glandular-villous, 4–9-flowered; pedicels slender, 5–15 mm. long, villous with a larger proportion of hairs dendritic than in pubescence of twigs and leaves; calyx 4–6 mm. deep, broadly deltoid lobes one-third as long as or equaling broadly campanulate tube, calyx enlarging to 12 mm. or less in diameter in fruit; corolla dark lavender to purple, 1.5–2.5 cm. in diameter, finely puberulent outside, glabrous within, lobes broadly deltoid, rarely as long as wide; stamens equal or essentially so, filaments 1.5–2 mm. long, glabrous; anthers yellow, erect, 3.2–4.8 mm. long; berries globose, 6–12 mm. in diameter, greenish, smooth and glabrous, dull to sublustrous; seeds broadly ovate-lenticular, 1.5–2.5 mm. long, 1.4–2 mm. wide, yellowish to pale brown, minutely reticulate.

Mountain sides, outwash bajadas, and at margins of openings in chaparral, mostly Upper Sonoran to Transition Zones, but straggling into the upper edges of the Lower Sonoran Zone in eastern San Diego County, California, and in Arizona and Baja California; Mendocino and Amador Counties, California, south to northern Baja California, east into Arizona. Feb.–Aug.

***7a.*** **Solanum xanti** var. **glabrescens** Parish, Proc. Calif. Acad. III, **2:** 169, 1901

Similar to var. *xantii* but subglabrous with fewer, more scattered, simple hairs which are shorter than those in var. *xantii* and almost devoid of glandular and dendritic hairs; leaves 1–3 cm. long or less, often 6 mm. wide or less, usually attenuate at each end; flowers and fruit as in nominate variety.

Foothills and lower mountains, mostly Upper Sonoran Zone, Monterey County, California, to northern Baja California, reaching the margins of the Sonoran Desert near the southern end of the Sierra San Pedro Mártir in Baja California. March–Sept.

**8. Solanum seaforthianum** Andr., Bot. Rep. **8:** *pl. 504,* 1808

Scrambling or climbing vine with slender glabrous stems 1–5 m. long; petioles 1–5 cm. long, fine golden hairs along upper side in youth but soon glabrate; leaf blade decompound with 3–7 lobes or lowermost lobes distinct and petiolulate, segments elliptical to lance-ovate, 0.5–3 cm. wide, 1–10 cm. long, terminal one largest, often somewhat attenuate at apex, lamellae thin, dark green and glabrous above, finely puberulent especially along veins and paler beneath; uppermost leaves sometimes simple, ovate-lanceolate; inflorescence paniculate, 5–10 cm. long, nearly as wide at base; pedicels slender, to 1.5 cm. long, usually equaling or surpassing stouter peduncles; calyx about 3 mm. wide, with broadly deltoid teeth 0.5–1 mm. long, glabrous;

corollas purple, deeply lobed, 1.5–2.5 cm. in diameter, lance-elliptic lobes glabrous except on major veins and margins; stamens subequal, anthers 3–3.5 mm. long, rather stout, erect, opening by terminal pores, on filaments 1.5–3 mm. long; fruits globose, 8–15 mm. in diameter, shining, red; seeds numerous, 3–3.5 mm. long, slightly reniform, laterally compressed, shallowly reticulate-punctate.

Roadsides, around dwellings and abandoned fields, less commonly in arroyos and on hillsides, Lower Sonoran to Tropical Zones, southern Baja California to Tamaulipas and south to Central America, northern South America, and widely distributed in the West Indies. Dec.–June.

**9. Solanum rostratum** Dunal, Hist. Sol. 234, *pl. 24*, 1813

*Solanum heterandrum* Pursh, Fl. Amer. Sept. 156, *pl. 7*, 1814.
*Androcera lobata* Nutt., Gen. **1:** 129, 1818.
*Androcera rostrata* Rydb., Bull. Torrey Club **33:** 150, 1906.

Stout annual herb with erect, moderately branched stems 4–8 dm. tall, entire plant densely stellate-pubescent and well armed with yellow subacute spines 3–12 mm. long on stems, petioles, main veins of leaves, and calyces, spines terete in cross section; petioles rather stout, 0.5–5 cm. long; leaf blades oval to ovate in outline, irregularly pinnately 3–7-lobed or once- to twice-pinnatifid, to 5 cm. wide and 12 cm. long, lobes oblong to suborbicular, obtuse to rounded; flowers in few-flowered racemes; pedicels 5–10 mm. long, 1.5–2 mm. in diameter; calyces 5–7 mm. long at anthesis, lobes lanceolate, about equaling tube, rapidly accrescent and 10–14 mm. long in fruit, loosely enclosing berry, spines spreading-ascending, those along midrib largest; corollas yellow, lobes lanceolate, 3–4 mm. wide, 10–12 mm. long, glabrous within, markedly stellate-hirsute on median portion of back; filaments about 1 mm. long, glabrous; 4 stamens subequal, 5–6 mm. long, straight or nearly so, fifth 9–12 mm. long, introrsely curved in distal third; berry globose, 9–10 mm. in diameter, brownish, smooth and glabrous; seeds numerous, obovoid-reniform, moderately flattened laterally, 2.6–3.2 mm. long, 2–2.3 mm. wide, coarsely rugose-netted on back and less obviously so on lateral faces, faveolate-reticulate in a fine meshwork throughout, dull black.

Arid slopes and plains, Transition to Tropical Zones, South Dakota to Texas, Arizona, and central Mexico; adventive in California and Oregon, and from Ontario to Florida. A pest on overgrazed range land. March–Sept.

Mainly above the Sonoran Desert but entering it in Pima County, Arizona, and possibly elsewhere.

**10. Solanum grayi** Rose, Contr. U.S. Nat. Herb. **1:** 108, 1891

Weedy, erect, ascendingly or spreadingly branched, very spiny, annual herb 3–18 dm. tall, stems and leaves bearing scattered simple hairs and numerous gland-tipped hairs among yellowish spines, latter 3–15 mm. long, slender but stiffly rigid, spreading; leaves broadly ovate in outline, bipinnatifid, 3–8 cm. wide, 5–12 cm. long, divisions again pinnatifid or irregularly toothed, dark green and bearing a few scattered golden spines along ribs above, paler and with scattered, 4–6-rayed, stellate trichomes beneath; petioles spiny, to 3 cm. long; inflorescences few-flowered, racemose, 5–12 cm. long at anthesis; pedicels 2–5 mm. long, slender in flower, to 1 cm. long and stouter in fruit, spinose; calyx 3–4 mm. long at anthesis, densely viscid-glandular with tack-shaped hairs among spines, lobes linear-lanceolate, about equaling cup, firm and accrescent, to 18 mm. long in fruit; corolla bright yellow to white, 8–15 mm. in diameter, lobes ovate-deltoid, 2–3 mm. wide, 4–8 mm. long, glabrous within, hispidulous without; anthers yellow, tinged with purple, unequal, 3–4 about 4 mm. long, other 1 or 2 5–6 mm. long and strongly curved near tip; fruits ovoid to subglobose, 8–12 mm. in diameter, smooth and glabrous, reddish; seeds broadly ovoid-reniform, 2.6–3.2 mm. long, moderately flattened, transversely ridged across rounded back, finely reticulate-granulose, dull black.

Fields, pastures, stream banks, and shaded fence rows, Lower Sonoran to Tropical Zones, vicinity of Altar, Sonora, southeast at least to Colima. Sept.–Jan.

**Solanum sisymbriifolium** Lam., Illustr. **2:** 25, 1793, differs from *S. grayi* in having larger, less finely dissected leaves, flowers 2.5–3 cm. in diameter, and stouter, less numerous spines on stems and leaves.

Reported from the vicinity of Superior, Pinal County, Arizona, where it is adventive from tropical America. It has not been detected as an established plant within the confines of the Sonoran Desert but may be expected as a casual introduction in agricultural areas.

## 11. Solanum amazonium Ker, Bot. Reg. 1: *pl. 71*, 1815

*Solanum obtusifolium* Schlecht., Linnaea **5:** 113, 1830. Not *S. obtusifolium* Humb. & Bonpl. ex Dunal, 1813.
*Solanum verae-crucis* Steudel, Nom. ed. 2, **2:** 607, 1841.

An open to straggly shrub to 1 m. tall or sometimes only frutescent and smaller; twigs densely tomentose with fine stellate hairs, investiture pale gray to golden, persistent; scattered, golden-brown, rigid, spreading spines 1–15 mm. long irregularly spaced on vigorously growing branchlets; petioles slender, 1–5 cm. long, stellate-tomentulose; leaf blades ovate to ovate-lanceolate in outline, 1.5–10 cm. wide, 2.5–18 cm. long, sinuately margined to 3–5-lobed one-third to one-half the way to midrib, lobes acute to rounded, dark green and sparsely stellate-puberulent with very fine hairs above, more densely so and paler to tawny-tomentulose beneath, petioles and lower surface of blade often spinose; racemes 4–10 cm. long, several-flowered; pedicels slender, 8–15 mm. long, stellate-tomentulose and spinose; calyx cup 3–4 mm. deep, nearly as wide, those of abortive flowers unarmed or essentially so, distinctly and rather closely set with golden spines 3–6 mm. long in basal fruiting flower; calyx lobes linear-attenuate, sometimes dilated at base, 5–10 mm. long on unarmed calyces, to 2.5 cm. or more in length and strongly spinose on armed fruiting flowers, these often deltoid-ovoid at base, attenuate-linear above; corolla rotate, 2.5–4 cm. in diameter, deep purple with a yellowish central star, glabrous within, lobes narrowly to broadly deltoid, somewhat irregular, stellate-tomentulose and somewhat streaked with white along midveins on outer surface; stamens unequal, 2 of them 6–8 mm. long, nearly straight, other 3 strongly curved, 12–16 mm. long, yellow or often tinged with purple, dehiscing by apical pores; fruits globose, 1.5–2 cm. in diameter, many-seeded, dehiscing irregularly; seeds flattened, about 3 mm. long, 2–2.5 mm. wide, finely and shallowly faveolate-reticulate, tawny-gray or sordid.

Canyons, rocky hillsides, valley floors, and as a roadside weed, Lower Sonoran to Tropical Zones, central Sonora and Chihuahua to Oaxaca, Campeche, Zacatecas, and Yucatán. Aug.–May.

## 12. Solanum carolinense L., Sp. Pl. 187, 1753

A perennial herb 3–12 dm. tall from widely spreading rhizomes; stems erect or ascending, green, sparsely and finely stellate-pubescent, armed with scattered, yellowish, spreading spines 2–5 mm. long on stems, petioles, midribs and sometimes on main lateral veins of leaf blades; leaves ovate to oblong, repand, lobed, or occasionally pinnatifid, 2–7 cm. wide, 3.5–15 cm. long, lobes obtuse or acute, both surfaces stellate-pubescent with yellowish, more or less appressed hairs; petioles 3–20 mm. long, stellate-hirsute; flowers in racemose cymes, surpassing leaves at anthesis and then appearing terminal, but distinctly lateral in fruit; pedicels 5–15 mm. long, pubescent but not spinose; calyx lobes lance-acuminate, 5–7 mm. long, slightly spinose near base or bearing only stellate hairs; corolla 2–3.5 cm. in diameter, pale blue to nearly white, sparsely stellate-puberulent to glabrate within, densely stellate-puberulent along and near midveins of lobes without, lobes ovate-lanceolate, 6–7 mm. long; anthers subequal, erect, 5–9 mm. long; fruits globose to slightly depressed-globose, 9–20 mm. in diameter, glabrous, orange-yellow; seeds obovoid-lenticular, bright yellow, about 2.5 mm. long, minutely and shallowly reticulate-pitted, glossy.

A wayside weed and occasionally in fields, mostly Upper Sonoran to Carolinian Zones, Arizona and New Mexico north to Oregon and east to Texas, Florida, and North Carolina. Adventive with us and uncommon. May-Aug.

**13. Solanum elaeagnifolium** Cav., Ic. **3:** 22, *pl. 243*, 1794

*Solanum leprosum* C. G. Ortega, Hort. Matr. 115, 1800.
*Solanum flavidum* Torr., Ann. Lyc. N.Y. **2:** 227, 1828.
*Solanum texense* Engelm. & Gray, Bost. Jour. Nat. Hist. **5:** 227, 1845.
*Solanum roemerianum* Scheele, Linnaea **21:** 767, 1848.

Perennial herb or suffrutescent plant to 1 m. tall from tough creeping rhizomes; foliage and stems canescent with finely stellate pubescence; stems, petioles, and midribs of leaves sparsely to densely prickly with slender yellowish spines 1–5 mm. long or rarely plants nearly unarmed; petioles shallowly canaliculate on upper surface, 3–20 mm. long; leaf blades linear, oblong or lanceolate, 4–25 or rarely –35 mm. wide, 3–10 cm. long, broadly cuneate to rounded at base, obtuse to acute at apex, repand-dentate or smaller ones entire, veins rather prominent; flowers cymose; peduncles, pedicels, and calyces usually prickly with straight yellow spines to 4 mm. long; calyx lobes ovate to lance-linear, to 1 cm. long; corollas 2–3 cm. in diameter, violet or blue, lobes deltoid, 4–7 mm. wide, glabrous within, stellate-puberulent on outer surfaces exposed in bud; anthers 7–9 mm. long, subequal, yellow; berries globose, 9–14 mm. in diameter, yellow to brownish, smooth and glabrous; seeds ovoid-reniform, strongly flattened, 3.5–4 mm. long, pale brown, smooth and sublustrous.

Sandy plains, arroyos, and outwash slopes, Upper and Lower Sonoran Zones, throughout most of Arizona, introduced in California, and east to Colorado, Kansas, Arkansas, and Texas, south throughout much of Mexico and Central America to Argentina. April–Oct.

**14. Solanum madrense** Fernald, Proc. Amer. Acad. **35:** 558, 1900

Erect, shrubby plant 1–2 m. tall, usually from a single trunk, young branches loosely stellate-tomentose and almost scurfy with short-rayed, grayish or ochraceous hairs and sparingly armed with straight or slightly curved prickles 2–6 mm. long, these usually flattened parallel to branch, thus narrowly elliptical in basal cross section, yellowish or pale brownish; petioles stoutish, 1–3.5 cm. long, densely stellate-tomentose, usually armed with 1 to several spines similar to but smaller than those on branches; leaf blades broadly ovate to lance-ovate, 4–13 cm. wide, 5–15 cm. long, bluntly and shallowly and angulately sinuate-lobed (2–3 lobes on each side), acute or short-acuminate at apex, unequally subcordate at base, dark olive-green and scaberulous-stellate above, paler and tomentulose and sometimes sparsely armed beneath;

inflorescence racemose, extra-axillary, 3–6 cm. long in anthesis, often to twice as long in fruit, rachis and pedicels sparsely stellate-puberulent and closely invested with gland-tipped, simple hairs between stellate trichomes; pedicels slender, 1–1.5 cm. long in flower, stouter and strongly swollen toward apex in fruit; calyx cup about 3 mm. deep and broad, tomentulose, lobes lance-linear, 2–2.5 mm. long at anthesis, to 9 mm. long and 3 mm. wide in fruit; corolla white, deeply 5-lobed, 2.5–3 cm. in diameter, lobes 5–6 mm. wide, 8–10 mm. long, stellate-scaberulous on back, puberulent with fine, simple or few-rayed hairs near tip, especially along distal margins; anthers subequal, slender, 8–10 mm. long; berries globose, 8–15 mm. in diameter, black in dried specimens, smooth and dull; seeds broadly elliptic, strongly flattened, 2–2.5 mm. long, only faintly lustrous, microscopically granular-tuberculate, yellowish.

Canyons and arroyos, mostly in the oak forests but straggling into the upper margins of the Lower Sonoran Zone along arroyos in southern Sonora, east into adjacent Chihuahua and south to Colima and Guerrero. Flowering following both spring and summer rains.

**15. Solanum hindsianum** Benth., Bot. Sulph. 39, 1844

An openly branching shrub 1–3 m. tall, usually armed with scattered spines 3–12 mm. long, bark on older branches dark gray or slightly mottled, slightly checked or fissured longitudinally; younger twigs tawny to cinereous with many-rayed, fine, closely interwoven, stellate hairs; leaves ovate to oblong, rounded to subtruncate or occasionally subcordate at base, acute to obtuse at apex, entire, 5–20 mm. wide, 2–6 cm. long, prominently pinnately veined, closely invested with fine, felt-like, stellate tomentum, cinereous to tawny in color; petioles stoutish, 3–15 mm. long, densely stellate-tomentose; flowers solitary in upper axils, appearing subumbelliform in bud; pedicels 5–10 mm. long at anthesis, stellate-puberulent, becoming dilated above in fruit and then often 15–20 mm. long, 3–4 mm. in diameter at apex; calyx cup 4–6 mm. in diameter, 3–5 mm. deep, markedly 5-angled below lobes, these lanceolate to nearly subulate, 6–10 mm. long, slightly unequal, densely stellate-tomentulose; corollas rotate, blue to purple, 2.5–5 cm. in diameter, lobes twice as broad as long, bearing a deeper colored midstripe from center of flower to tip of each lobe, glabrous within, each lobe with a lanceolate strip of stellate pubescence 3–6 mm. wide along underside of its midrib, space between these lines glabrous or nearly so; anthers subequal, 6–10 mm. long, on glabrous filaments 2–3 mm. long, erect or nearly so; fruit globose, 1–2 cm. in diameter, smooth and glabrous, pale green with darker green stripes, lustrous; seeds numerous, obovoid-reniform, strongly flattened, 3–4.5 mm. long, minutely reticulate-netted, dull brown to olivaceous.

On rocky hillsides, outwash bajadas, along arroyos and canyons, and on sandy plains, Lower Sonoran Zone and margins of the Upper Sonoran Zone, near San Quintín, Baja California, and northwestern Sonora south to the Cape Region and southern Sonora. Aug.–April, probably flowering sporadically during rest of the year.

**Lycopersicon esculentum** Mill., Gard. Dict. ed 8, no. 2, 1768, the common garden tomato, occurs as a waif in many localities in the Sonoran Desert when rainfall is sufficient, and may persist for several years in the most favorable habitats, as along arroyos and irrigation ditches. It is scarcely naturalized, however, in our area.

## 9. DATURA L., Sp. Pl. 179, 1753

Coarse, erect or ascendingly branched, rank-smelling herbs (in ours) with alternate, short-petioled, entire, sinuate or lobed blades. Flowers large and showy, solitary in forks of branches, short-pedunculate, white or purple-tinged or yellowish, heavily fragrant. Calyx cylindric or prismatic, 5-toothed, usually circumscissile near base, lower part persistent below capsule as a spreading or reflexed collar. Corolla funnelform, convolute-plicate in bud. Stamens included; filaments filiform; anthers linear, longitudinally dehiscent. Stigma bilobed. Capsule ovoid to globose, spiny, apically 4-valved or dehiscing irregularly, falsely 4-celled. Seeds many, subreniform to oblong, strongly compressed. Embryo small, strongly curved.

Plants perennial; corollas 15–25 cm. long; seeds buff to pale brown, with a cordlike margin, sides smooth; capsule irregularly dehiscent ............................ 1. *D. inoxia*
Plants annual; corollas 6–15 cm. long; seeds black at maturity, coarsely verrucose-warty, margin not corded; capsule apically 4-valved ........................ 2. *D. discolor*

**1. Datura inoxia** Mill., Gard. Dict. ed. 8, no. 5, 1768

*Datura meteloides* DC. ex Dunal, in A. DC., Prod. **13**[1]: 544, 1852.

Erect, widely branching, perennial herb 5–18 dm. tall with cinereous-puberulent herbage; leaves broadly ovate, usually asymmetric at base, 4–15 cm. long, sinuately and irregularly repand or less commonly entire, petioles one-fourth as long as or sometimes equaling blade; calyx tube 6–10 cm. long, 5 teeth ovate, 1–3 cm. long, attenuate at apex, basal persistent part usually rotate, 3.5–5 cm. in diameter, finely and sparsely puberulent on both surfaces; corolla white, usually flushed with lavender or purple, 15–25 cm. long, limb 10–20 cm. wide, bearing 5 slender teeth 1–2 cm. long; anthers white or sometimes pale lavender before dehiscing, 12–15 mm. long; capsules globose, 2.5–4 cm. in diameter, irregularly dehiscent, nodding, puberulent, sometimes glabrate in age, closely prickly, prickles 5–12 mm. long; seeds cream to buff, ovoid-reniform and strongly flattened, 4–5 mm. long, smooth on sides, margin strongly cord-like.

Sandy washes, fields, and rocky hillsides, Lower and Upper Sonoran Zones, lower Sacramento Valley, California, to Texas and south to South America. Throughout the year, but mainly March–Oct.

**2. Datura discolor** Bernh., Linnaea **8**: Litt. Ber. 138, 1833

*Datura thomasii* Torr., Pacif. R. Rep. **5**: 362, 1858.

Erect annual 1–8 dm. tall with finely puberulent, mostly bright green but sometimes slightly cinereous herbage; petioles 1.5–7 cm. long; leaf blades ovate, 3–15 cm. long, sinuately and angularly few-toothed to entire, acute at apex, broadly tapered to subtruncate at base; calyx prismatic, 4–8 cm. long, lobes ovate-attenuate, 8–15 mm. long, persistent base mostly rotate or somewhat reflexed and 1.5–4 cm. in diameter in fruit; corolla white (or rarely pale yellow) with a purplish flush in throat, mostly 8–15 cm. long, limb 5–8 cm. in diameter, corolla teeth narrowly subulate, 4–7 mm. long; anthers 7–9 mm. long, white; capsules nodding, globose, 2.5–3 cm. in diameter (exclusive of spines), glandular-puberulent, spines slender, 1–2 cm. long, puberulent; seeds black at maturity, rugulose and minutely pitted.

Sandy flats, arroyos, margins of cienagas and playas, Lower and Upper Sonoran Zones, Riverside and Imperial Counties, California, to Baja California, Arizona, Sonora, and the West Indies. March–Oct., or spasmodically throughout the year following rains.

## 10. NICOTIANA L., Sp. Pl. 180, 1753

Narcotic-poisonous, heavy-scented, usually viscid-puberulent, annual or perennial herbs or shrubs. Leaves alternate, entire or sometimes repand, petiolate or sessile. Inflorescences few- to many-flowered, racemose or paniculate, terminal. Calyx tubular-campanulate or ovoid, 5-toothed or 5-cleft, persistent. Corolla funnelform, salverform, or tubular; limb shallowly 5-lobed, usually spreading. Stamens 5; filaments filiform, inserted near base of corolla tube. Style slender; stigma capitate. Ovary 2-celled or rarely 4-celled. Capsule ovoid, acute, 2- or 4-valved at summit. Seeds numerous, small, ovoid to reniform, minutely reticulate-punctate.

Shrubs; flowers yellow; leaves glacuous, eglandular ........................ 1. *N. glauca*
Herbs; flowers white or greenish white; leaves glandular and pubescent:
  Leaves auriculate-clasping at base; plants biennial or perennial:
    Calyx lobes equaling capsule; corolla slightly constricted at orifice ................................................................ 2. *N. trigonophylla*
    Calyx lobes exceeding capsule; corolla not constricted at orifice ......... 3. *N. palmeri*
  Leaves not auriculate-clasping; plants annual:
    Calyx lobes deltoid, shorter than tube; mature calyx somewhat pock-marked ................................................................ 4. *N. attenuata*
    Calyx lobes linear or lanceolate, some of them at least equaling or exceeding calyx tube; mature calyx not pock-marked:
      Plants 1–3 dm. high; lower leaves lanceolate, 2–4 cm. long or less; pedicels shorter than mature capsules; seeds dull .......................... 5. *N. greeneana*
      Plants 2–15 dm. high; lower leaves oblong, ovate, obovate, or spatulate, 3.5–30 cm. long; pedicels mostly equaling or exceeding mature capsules; seeds somewhat shining:
        Corolla tube slightly constricted at orifice, limb 8–10 mm. wide; leaves 2–8 cm. long ................................................... 6. *N. clevelandii*
        Corolla tube not constricted at orifice, the limb 15–18 mm. wide; lower leaves 10–30 cm. long ............................................ 7. *N. mexicana*

### 1. Nicotiana glauca R. Graham, Edinb. New Phil. Jour. **1828:** 175, 1828

Glabrous shrub or small tree 8 m. high or less, with ample, long-petiolate, glaucous, somewhat leathery, ovate to lanceolate-oblong leaves 5–18 cm. long, blades acute or obtuse at apex, acute to subcordate at base, entire or slightly repand; flowers greenish yellow, in lax terminal panicles; calyx tubular-campanulate, 8–12 mm. long, 5-dentate, teeth much shorter than tube; corolla tubular, 3.0–4.5 cm. long, minutely villosulous, limb narrow; capsule ovoid, acute, 10–13 mm. long, 4-valved at summit; seeds reddish brown, slightly shining, finely favose-reticulate.

A native of South America, established and spontaneous from central California to Texas and south throughout Mexico; mostly in the Lower and Upper Sonoran Zones. Flowering throughout the year.

### 2. Nicotiana trigonophylla Dunal, in A. DC., Prod. **13**[1]**:** 562, 1852

*Nicotiana ipomopsiflora* Dunal, in A. DC., Prod. **13**[1]**:** 559, 1852.
*Nicotiana multiflora* Torr., Pacif. R. Rep. **5:** 362, 1858.
*Nicotiana glandulosa* Buckl., Proc. Acad. Phila. **1862:** 6, 1863.

Erect, simple or branched, biennial or perennial herb 2–9 dm. tall, glandular-pubescent throughout; lower leaves petiolate, oblanceolate or spatulate, 1–4 cm. wide, 5–15 cm. long, tapering gradually to more or less winged petiole; upper leaves lanceolate to oblong-ovate, sessile, auriculate, smaller; inflorescence laxly paniculate-racemose; pedicels 5–10 mm. long; calyx campanulate-ovoid, 3–4 mm. wide, 7–10 mm. long, teeth lance-triangular, 3–5 mm. long, about equaling mature capsule; corolla white, tubular, 12–18 mm. long, slightly constricted at orifice, limb 8–10 mm. broad; capsules 2-valved, each valve bilobed at apex; seeds numerous, dark reddish brown, somewhat shining.

Sandy areas and along washes, mostly Lower Sonoran Zone, Mojave and Colorado Deserts, California, to Nevada, Texas, Nayarit, and southern Baja California. Flowering throughout the year.

**3. Nicotiana palmeri** A. Gray, Syn. Fl. N. Amer. **2**[1]: 242, 1878
*Nicotiana trigonophylla* var. *palmeri* M. E. Jones, Contr. West. Bot. **12**: 52, 1908.

Simple or more often branched perennial 5–12 dm. high, glandular-pubescent throughout; old stems white, tomentulose; lower leaves oblanceolate, 1.5–4 cm. wide, 5–15 cm. long, sessile or with a winged petiole, acute to short-acuminate, margins often undulate; upper leaves lance-elliptic, sessile, auriculate, smaller; inflorescence racemose-paniculate; calyx ovoid-campanulate, tube 3–5 mm. wide, 5–8 mm. long, teeth subequal, triangular-lanceolate, equaling to once and a half as long as tube, exceeding mature capsule; corolla narrowly funnelform, white, faintly tinged with green near throat, glandular-puberulent without, tube 18–20 mm. long, 4–6 mm. wide at unconstricted throat, limb 10–14 mm. wide, remaining expanded through day; capsule ovoid, acute, 8–10 mm. long; seeds shining.

Rocky hillsides, cliffs, and washes, Lower Sonoran Zone, western Arizona from the Bill Williams River south to northern Sonora and Baja California. Jan.–April.

**4. Nicotiana attenuata** Torr. ex S. Wats., Bot. King Expl. 276, *pl. 27, figs. 1–2,* 1871

Erect, simple or branching annual 3–10 dm. high, glandular-pubescent throughout, or somewhat glabrate in age; leaves petiolate, lower 2–5 cm. wide, 8–15 cm. long, ovate to broadly lanceolate, upper narrower and sometimes sessile, mostly acute at both ends; inflorescence racemose or paniculate-racemose; calyx ovoid-campanulate, 6–8 mm. high at anthesis, pock-marked when mature, teeth deltoid, subequal, shorter than tube; corolla narrowly funnelform, white, tube 2–2.5 cm. long, essentially glabrous except near apex, limb 8–10 mm. broad; corolla lobes obtuse, 2–3 mm. long; capsule 8–10 mm. long, 4-valved at apex; seeds 0.5–0.7 mm. long, dull.

Washes, sandy slopes, roadsides, and fields, mostly Arid Transition and Upper Sonoran Zones, rare in the desert, British Columbia to Montana, south to Baja California and Sonora, east to New Mexico, and Texas. May–Oct.

**5. Nicotiana greeneana** Rose, in Vasey & Rose, Contr. U.S. Nat. Herb. **1**: 18, 1890

Low, glandular-pubescent, branching annual 1–3 dm. high; main branches often decumbent; leaves lanceolate, petiolate, 3–10 mm. wide, 2–4 cm. long, thin, sparingly pubescent, acute to acuminate at apex, acute to rounded at base; inflorescence of few-flowered, bracteate, terminal racemes 5–12 cm. long; pedicels 1–3 mm. long; calyx narrowly ovoid-campanulate, 2–3 mm. wide, 5–8 mm. long (including unequal, linear-lanceolate lobes), longer lobes exceeding tube; corolla slender, narrowly tubular-funnelform, 10–12 mm. long, white, tinged with green, limb 4–6 mm. wide; capsules broadly ovoid, 4-valved, 5–6 mm. high; seeds about 0.6 mm. long, dull gray-brown, minutely reticulate-ridged and rounded-tuberculate.

On volcanic hillsides among rocks, Lower Sonoran Zone, Cedros Island and on the adjacent mainland of Baja California. Feb–May. (Hardly distinct from *N. clevelandii.*)

**6. Nicotiana clevelandii** A. Gray, Syn. Fl. N. Amer. **2**[1]: 242, 1878

Viscid-pubescent, erect annual 2–6 dm. high, simple to freely branched; leaves ovate to lanceolate, acute, lower usually short-petiolate, upper sessile, 3–8 cm. long, petioles narrowly winged; inflorescence racemose, comparatively few-flowered; calyx campanulate, tube about 3–4 mm. long, 2–3 mm. wide, teeth linear, unequal, 1.5–5 mm. long, longer exceeding tube; corolla 1.5–2 cm. long, whitish, tinged with violet, glabrous or sparsely puberulent without near summit, limb 8–10 mm. broad; capsule 5–8 mm. high, 4-valved; seeds reddish brown, punctate-reticulate, shining.

Sandy washes and hillsides, Lower and Upper Sonoran Zones, Santa Barbara County to the Colorado Desert, California, to southeastern Arizona and southern Baja California. Feb.–April.

**7. Nicotiana mexicana** Schlecht., Ind. Sem. Hort. Halens. 4, 1840

Erect, simple or branching, glandular-villous annual 3–12 dm. high; stems densely rusty-villous above, nearly glabrate at base; lower leaves ovate to ovate-lanceolate, 3–12 cm. wide, 8–25(or 30) cm. long, short-petiolate, acute to short-cuneate at apex, acute or broadly cuneate at base, villosulous on veins, margins entire; upper leaves mostly sessile, lance-acuminate; inflorescence racemose; pedicels slender, 5–10 mm. long, villous-glandular; calyx narrowly campanulate, tube 2–3 mm. wide, 4–5 mm. long and slightly constricted at throat, lobes linear, unequal, 4–12 mm. long, 0.5 mm. wide, viscid-pubescent, erect or slightly spreading; corolla white, faintly veined with green, narrowly funnelform, 1.5 mm. wide at summit of calyx tube, 3–4 mm. wide at unconstricted throat, 16–20 mm. long, limb spreading, 15–18 mm. wide.

Widely distributed in Mexico, mostly above the desert; an occasional straggler in the Sonoran Desert, as near Torres, Sonora. Feb.–May. (Doubtfully distinct from *N. tabacum* L.)

**Nicotiana plumbaginifolia** Viv., Elench. Pl. Hort. Bot. Dinegro 26, *pl. 5*, 1802, with lanceolate, somewhat auriculate-clasping, undulate-margined leaves; subequal linear calyx lobes 6–8 mm. long; and slender, tubular-funnelform corolla 2.5–3 cm. long; occurs in Sinaloa and foothills in east-central Sonora.

**Nicotiana rustica** L., Sp. Pl. 180, 1753, is widely established in Mexico and may appear as a naturalized adventive in moist plains in southern Sonora and Baja California. Its corolla is broader, nearly campanulate.

## 11. PETUNIA Juss., Ann. Mus. Paris 2: 215, 1803

Annual or perennial, viscid herbs with alternate to subopposite leaves. Flowers terminal or solitary in axils, minute to large and showy. Calyx 10-nerved, deeply 5-lobed. Corolla funnelform, white, violet, or purple. Stamens 5, of three different lengths, inserted near base of corolla tube. Stigma capitate, obscurely bilobed. Fruit an ovoid capsule, septicidally bivalved. Seeds small, subspherical.

**1. Petunia parviflora** Juss., Ann. Mus. Paris **2**: 216, *pl. 47, fig. 1*, 1803

Prostrate annual herbs with diffusely branched stems 0.2–4 dm. long and glandular-viscid herbage; leaves linear, elliptic, or oblong-oblanceolate, 4–18 mm. long, 1–4.5 mm. wide, gradually narrowed to a short petiole or nearly sessile, 1-nerved, acute to rounded at apex, margins entire to slightly undulate; flowers solitary in axils; pedicels slender, 2–5 mm. long at anthesis, elongating to 1–2 cm. in fruit; calyx lobes linear-spatulate to linear-lanceolate, 0.5–1 mm. wide, 2–3 mm. long in flower, 5–8 mm. long in fruit, cup 1–2 mm. deep; corolla 5–7 mm. long, pale blue to purplish, often paler to nearly white on one side, lobes rounded but apiculate, more or less spreading; capsule broadly ovoid, 3–4 mm. long, firm, yellowish or whitish, glabrous; seeds about 0.6 mm. long, amber to pale brown, favose-reticulate.

Sandy flats, arroyos, sand bars, and stream banks, Lower and Upper Sonoran Zones, Sacramento Valley and central Coast Ranges of California to Texas and Florida, south throughout much of Mexico and Central America into South America. Flowering sporadically throughout the year and profusely Dec.–Sept. in much of our area.

## Family 126. **SCROPHULARIACEAE.** Figwort Family

Annual or perennial herbs or a few shrubby, partially parasitic in some genera. Leaves alternate or opposite, simple, entire to pinnately dissected, sessile or petiolate. Flowers variously arranged, perfect, nearly regular to strikingly irregular. Calyx campanulate to tubular or of 4–5 distinct sepals, sometimes reduced to 1 or 2 distinct, leaflike or bractlike lobes, when tubular or campanulate 4–5-toothed or 4–5-lobed. Corolla tubular, campanulate, rotate or galeate; throat often ventricose; base of tube spurred, saccate or gibbous in some genera. Stamens 2, 4, or 5, commonly 4 and didynamous, a fifth stamen then often represented by a scale, filament, or gland; anthers 1–2-celled, cells variously arranged. Style 1; stigma entire, emarginate or bilobed. Ovary superior, nearly or completely 2-celled; ovules many. Fruit a capsule, loculicidally or septicidally dehiscent or both. Seeds small, numerous, variously sculptured or smooth.

Corolla spurred, saccate, or gibbous on lower side near base; capsule opening by terminal or subterminal pores or rupturing irregularly:
- Corolla tube with a slender spur at base .................................... 1. *Linaria*
- Corolla tube saccate or gibbous but not spurred:
  - Fertile stamens 2; seeds with an inrolled wing, distinctly cup-shaped ...... 2. *Mohavea*
  - Fertile stamens 4; seeds not cup-shaped:
    - Plants herbaceous; leaves opposite; corollas (in ours) not red:
      - Capsule asymmetrical, dehiscing by subterminal pores or rupturing very irregularly ...................................................... 3. *Antirrhinum*
      - Capsule symmetrical, dehiscing fairly regularly by a longitudinal or diagonal rupture ................................................ 4. *Asarina*
    - Plants shrubby or suffrutescent; leaves opposite and ternate; corollas deep red or scarlet ................................................ 5. *Galvezia*

Corolla tubular or funnelform; throat often ventricose but not spurred, saccate, nor gibbous; capsule dehiscing longitudinally and regularly:
- Upper corolla lobes overlapping externally in bud, lower ones inside and covered:
  - Flowers in cymose or paniculate clusters; stigmas capitate or pointed, not markedly 2-lobed or 2-lipped:
    - Capsule subglobose; fifth stamen minute or absent; seeds embedded in long hairs inside capsule ................................................ 6. *Russelia*
    - Capsule ovoid or oblong; fifth stamen represented by a sterile filament or scale:
      - Corolla 4–8 mm. long; sterile filament a broad scale ............ 7. *Scrophularia*
      - Corolla 10–30 mm. long; sterile stamen represented by a slender filament ........ ................................................ 8. *Penstemon*

Flowers in racemes or axillary; stigmas flattened except in *Scoparia,* usually markedly bilobed:
Calyx tubular or campanulate, prismatic, teeth rarely longer than tube; anther cells divergent or divaricate .......................................... 9. *Mimulus*
Calyx of sepals distinct nearly to base, not prismatic nor strongly angled; anther cells parallel:
Corolla campanulate, tubular, or salverform, lobes equaling or shorter than tube; stamens and style included; pedicels usually 1–2-bracteolate:
Anther cells separated on short arms of connective; sepals equal or nearly so ... ........................................................ 10. *Stemodia*
Anther cells proximate, not noticeably separated; sepals usually unequal or no more than subequal:
Corolla campanulate, nearly regular, white; stigma only minutely emarginate ........................................................ 12. *Bacopa*
Corolla tubular or cylindrical, markedly bilabiate, blue or yellow; stigma deeply bilobed:
Bracteoles at apex of pedicel; corollas blue or purplish; style curved or inflexed .............................................. 11. *Conobea*
Bracteoles at base of pedicel; corollas yellowish; style straight ........... ............................................... 13. *Mecardonia*
Corolla rotate, lobes longer than tube; stamens and style exserted; pedicels not bracteolate ................................................... 14. *Scoparia*
Upper corolla lobes interior, overlapped by overlying lower lobes in bud:
Stamens 2; corolla nearly regular, not galeate; capsule compressed or nearly globose ............................................................ 15. *Veronica*
Stamens 4; corolla irregular; capsule ovoid to lanceolate, not compressed except in *Dicranostegia*:
Bracts more or less lobed or dissected, or if entire lower corolla lip reduced to small teeth 1–3 mm. long; calyx tubular or of a distinct, leaflike lobe:
Calyx deeply cleft before and behind, lateral lobes entire to bilobed; lower corolla lip reduced to 3 rudimentary teeth ........................... 16. *Castilleja*
Calyx subequally 4-lobed or of a single leaflike lobe; lower corolla lip nearly equaling upper lip or of 3 saccate or plane lobes somewhat shorter than upper lips but not rudimentary:
Corolla flattened dorsiventrally and expanded laterally; calyx of a single (sometimes 2 and distinct) leaflike lobe ...................... 17. *Dicranostegia*
Corolla not flattened dorsiventrally; calyx tubular, equally or subequally 4-lobed or 4-toothed:
Lower lip of corolla 3-saccate; corolla purplish or crimson, upper lip strongly galeate ........................................... 18. *Orthocarpus*
Lower lip of corolla of 3 spreading, nearly plane lobes; corolla white with pinkish tips, upper lip scarcely galeate .................. 19. *Clevelandia*
Bracts merely dentate; calyx campanulate ...................... 20. *Lamourouxia*

## 1. LINARIA Mill., Gard. Dict. abr. ed. 4, 1754

Erect or diffuse, annual or perennial herbs with narrow sessile leaves and glabrous herbage. Flowers in terminal racemes, subtended by short bracts. Sepals 5, distinct. Corolla yellow, bluish, or violet, bearing a basal ventral spur, 2-lipped, lower one often folded upward near its base to form a palate. Stamens 4, didynamous; anthers distinct, glabrous. Stigmas united, slender, scarcely capitate. Capsule globose to cylindrical, rupturing by a transverse, irregular, subterminal tear. Seeds small, variously ridged, tuberculate or narrowly winged.

**1. Linaria texana** Scheele, Linnaea **21:** 761, 1848

*Linaria canadensis* var. *texana* Pennell, Proc. Acad. Phila. **73:** 502, 1921.

Slender, erect, annual or biennial, with 1 to several branches 1–8 dm. long from base, stems and leaves glabrous, dark green, rather densely leafy below, pedunculate-racemose above; leaves of sterile stems crowded, ovate, elliptic or linear, entire, obtuse to acute at both ends, 1–3 mm. wide, 3–20 mm. long, sessile or petioles 1–2 mm. long; leaves on fertile stems linear to linear-oblong, opposite or ternate at base of stem, alternate above, entire, acute or obtuse, sessile, 1–3 mm. wide, 5–30 mm. long; raceme spikelike, 5–30 cm. long, on peduncles 2–15 cm. long; pedicels filiform, 2–10 mm. long, glabrous or finely and usually sparsely glandular-puberulent; bracts subulate, 0.5–2 mm. long; calyx lobes lanceolate, 2–3 mm. long; corolla pale blue with darker veins, 9–12 mm. long (exclusive of spur), glabrate without, upper lip 4–5 mm. long, reflexed, oblong lobes 2–3 mm. long, lower lip deflexed-spreading; palate prominent, 2-ridged, densely pubescent; stamens glabrous, capsule globose to subcylindrical, 2.5–4 mm. long, irregularly dehiscent by 2 subterminal pores; styles persistent; seeds subconic-cylindrical, often angled, truncate at one or both ends, dark gray or black, minutely tuberculate over whole surface.

Along arroyos, stream banks, and on grassy hillsides or in open chaparral, Upper and Lower Sonoran Zones, Oregon to northern Baja California east to Colorado, South Carolina, Georgia, and Sonora; throughout Mexico and Central America to South America. March–June in our area.

## 2. MOHAVEA A. Gray, Pacif. R. Rep. 4: 122, 1857

Annual with lanceolate leaves. Flowers solitary in axils on short pedicels, but internodes so short that inflorescence is spikelike. Sepals distinct. Corolla yellow, much exceeding calyx, with a short tube and 2 fan-shaped lips; upper lip distinctly 2-lobed; lower shallowly 3-lobed, its palate small but strongly inflated and with a central hairy crest and a depression on each side of crest. Stamens 2, connivent over tip of style, included within arched ridge of upper lip. Capsule thin-chartaceous, rupturing irregularly at apex.

**1. Mohavea confertiflora** (Benth.) Heller, Muhlenbergia **8:** 48, 1912

*Antirrhinum confertiflorum* Benth., in A. DC., Prod. **10:** 592, 1846.
*Mohavea viscida* A. Gray, Pacif. R. Rep. **4:** 122, 1857.

Erect, 0.6–3 dm. high, simple or sparingly branched from near base, viscid-pubescent; all save the lower leaves alternate, ovate to lanceolate, cuneate at base, short-petiolate, acute to acuminate at apex, 5–20 mm. wide, 1.5–6.5 cm. long; sepals lanceolate, 10–12 mm. long; corolla pale cream-yellow, both lips penciled with fine purple dots in vertical lines, erose-denticulate, a purple splotch on each side of hairy, bright yellow crest of lower lip; style persistent; capsule ovoid, 10–14 mm. long; seeds 2–2.5 mm. long, brown to black, with a flattened, narrowly margined crest on back and an erose-denticulate, involute wing about 0.8 mm. wide forming a cavity on ventral side.

On rocky desert slopes, arroyos and mesas, Lower Sonoran Zone, southeastern Nevada and San Bernardino County, California, to eastern Arizona and south to central Baja California, Mexico. March–May.

## 3. ANTIRRHINUM L., Sp. Pl. 612, 1753

Annual or perennial herbs with erect or scandent stems and often with tortile branches, petioles, or pedicels. Leaves opposite or whorled, commonly with upper ones alternate, entire, petiolate or sessile. Flowers in upper axils or in terminal racemes, small and inconspicuous to showy. Calyx 5-parted or deeply 5-lobed. Corolla saccate or gibbous on lower side near base of tube, 2-lipped, folds of upper portion of tube (palate) nearly or quite closing throat. Functional stamens 4, inserted near base of tube; filaments often dilated near apex, sterile stamen usually represented by a gland. Style deciduous or persistent. Capsule ovoid, dehiscing irregularly or by apertures near apex. Seeds several to many, ovoid, subglobose, or cup-shaped.

Plants robust, lateral branches nearly equaling main one, not at all climbing or tortile; seeds with a conspicuous, thin-corky, cup-shaped wing larger than body .... 1. *A. cyathiferum*

Plants slender, lateral branches much smaller than central axis, often more or less tortile or peduncles tortile; seeds tuberculate or ridged, but not with a cup-shaped wing:

- Leaves broadly ovate; plant glandular-pubescent throughout; seeds with 10–15 deeply cristate, narrow ridges, or some of them sometimes subentire; capsule 6–8 mm. long at maturity .................................................. 2. *A. nuttallianum*
- Leaves linear to narrowly oblanceolate; plants glabrous except at very base and in inflorescence; seeds papillate or with 4–6 nearly entire, corky wings with rolled margins; capsule 4–6 mm. long at maturity:
  - Sepals narrowly ovate-lanceolate, sparsely puberulent mainly near base with very fine, tack-shaped hairs; seeds papillate ...................... 3. *A. kingii* var. *watsonii*
  - Sepals linear-lanceolate, pubescent to tips with coarser tack-shaped hairs; seeds longitudinally ribbed with 4–6 corky, rolled-margined ribs, not strongly papillate ..... ........................................................ 4. *A. costatum*

### 1. Antirrhinum cyathiferum Benth., Bot. Sulph. 40, *pl. 19*, 1844

*Antirrhinum chytrospermum* A. Gray, Proc. Amer. Acad. **12:** 81, 1876.

Erect, moderately coarse annuals 5–50 cm. tall, commonly branching sparsely to profusely from base, secondary branches nearly as robust as main leader, whole plant viscid-glandular-pubescent; leaves opposite at base, alternate above, broadly ovate, entire, obtuse to acute at apex, rounded or broadly cuneate at base, 3–15 mm. wide, 5–25(30) mm. long, winged petiole 5–13 mm. long; flowers solitary in axils, from near base of stems to apex; pedicels slender, 1–10 mm. long, strongly recurved after anthesis; calyx campanulate, somewhat oblique, glandular-pubescent, lobes subequal, lanceolate, about 4–5 mm. long at anthesis, 6–7 mm. long in fruit; corolla bluish, 10–13 mm. long, glandular-puberulent on outside, very faintly saccate at base, tube and throat together 5–6 mm. long, 2.5–3 mm. wide, upper lip about 4 mm. long, suberect, somewhat arched, glabrous within, bilobed about to middle, lower lip 4.5–5 mm. long, spreading, palate yellowish and glandular-villous, lip 3-lobed about to middle; fertile stamens 7–8 mm. long, filaments not dilated, strongly geniculate; fifth stamen a rudimentary gland or scale; style persistent, glandular-puberulent on lower half; capsule globose, thin-walled, 7–8 mm. in diameter, sparsely puberulent, dehiscing irregularly; seeds pale-stramineous, body 0.8–1 mm. long, with a thin-corky, cup-shaped wing on concave side, wing one and one-half to two times as large as body.

In sandy soil on plains, along arroyos, and on desert hillsides, Lower Sonoran Zone, Maricopa and Pima Counties, Arizona, into northwestern Sonora (Pinacate Mountain; Tiburón Island), and Baja California from the San Felipe Desert to the Cape Region. Oct.–May.

**2. Antirrhinum nuttallianum** Benth., in A. DC., Prod. **10:** 592, 1846

*Antirrhinum nuttallianum* var. *effusum* A. Gray, Bot. Calif. **1:** 622, 1876.
*Antirrhinum subsessile* A. Gray, Bot. Gaz. **9:** 53, 1884.
*Antirrhinum nuttallianum* var. *subsessile* Jepson, Man. Fl. Pl. Calif. 899, 1925.

Annual, or apparently at times biennial, with an erect, simple, or sparingly branched plant with ascending branches, often scrambling through adjacent plants by tortile, slender, nearly horizontal, lateral branches to a height of 12 dm., glandular-pubescent throughout; leaves mostly alternate, ovate to subcordate-ovate, 2–20 mm. wide, 5–40 mm. long, reduced to ovate bracts in inflorescence, petioles glandular-pubescent, 1–10 mm. long; inflorescence 5–30 cm. long, a lax simple raceme or profusely paniculately branching; pedicels 2–20 mm. long, ascending, spreading, or tortile, slender to capillary; calyx oblique, 3–5 mm. long, lobes ovate to lanceolate, subequal, dorsal one longest, only slightly enlarged in fruit; corolla violet-blue, 10–12 mm. long, with a lavender tube and yellow, more or less reticulate palate, glandular-puberulent without, tube 4.5–6 mm. long, slightly gibbous at base, glabrous within except for two lines of hairs below palate; upper lip 4–5 mm. long, erect or nearly so, lobes about 1.5 mm. long, lower lip 5–6 mm. long, erect, palate with tack-shaped, yellow hairs in throat; stamens in pairs, filaments somewhat dilated upward, geniculate, glabrous except at joints, fifth stamen rudimentary; style equaling stamens, glandular-pubescent except at glabrous stigma; capsule ovoid-cylindrical, 6–8 mm. long, 3.5–4.5 mm. in diameter, dehiscent by 2 definite subapical pores; seeds 0.6–0.8 mm. long, subcylindrical, dark brown to nearly black, cristate or irregularly papillate-cristate.

Grassy hillsides, along arroyos, margins of vernal pools, and maritime cliffs, Sonoran Zones, from the Santa Barbara Islands, Ventura County, and Los Angeles County, California, south along both sides of the coastal mountains into Baja California to the southern flanks of the Sierra San Pedro Mártir, and in the foothills near Tucson, Arizona. Feb.–Sept.

**3. Antirrhinum kingii** S. Wats. var. **watsonii** (Vasey & Rose) Munz, Proc. Calif. Acad. IV, **15:** 367, 1926

*Antirrhinum watsoni* Vasey & Rose, Proc. U.S. Nat. Mus. **11:** 533, 1889.

Erect simple herbs 1–5 dm. tall, strict or sparingly branched, stems finely and sparsely puberulent at very base, glabrous or essentially so above; leaves linear, narrowly oblong, or linear-oblanceolate, 1–2.5 mm. wide, 1.5–5 cm. long, acute or obtusish at apex, gradually narrowed toward base, sessile or short-petiolate, entire, glabrous or sparsely arachnoid-puberulent on upper surface just above axil in some specimens, reduced toward tip of branches; flowers axillary, on slender pedicels 2–5 mm. long at anthesis, lengthening to 1–2 cm. in fruit and sometimes somewhat tortile, glabrous; sepals lanceolate to narrowly ovate, 3–3.5 mm. long at anthesis, slightly larger (to 4.5 mm.) in fruit, sparsely and inconspicuously puberulent with very fine, gland-tipped hairs, these mostly toward base of calyx; corolla bluish, 6–7 mm. long, sparingly glandular-puberulent without, densely bearded on folds of pale yellowish-white palate within; capsule asymmetrically ovoid, 4–5 mm. long, stramineous, sparsely glandular-puberulent with tack-shaped, slender hairs, dehiscing by irregular subterminal pores; seeds cylindrical, 0.6–0.8 mm. long, 0.4–0.5 mm. in diameter, dark brown to nearly black, dull, closely beset with blunt papillae about 0.1 mm. long.

Sandy or rocky slopes and washes, Lower Sonoran Zone, from vicinity of Rosario to Miller's Landing, east to the Gulf and on Cedros Island, Baja California. Feb.–April.

**4. Antirrhinum costatum** Wiggins, Contr. Dudley Herb. **4:** 22, *pl. 2, figs. 5–12,* 1950

Annual, erect, sparingly branched herb 1.5–7 dm. tall with sparsely villous basal parts of stem, inflorescence, and bases of small leaves, otherwise glabrous; leaves linear, 15–55 mm. long, glabrous or with a few gland-tipped hairs on upper surface near base; flowers axillary, on slender pedicels 2–5 mm. long, these about twice as long in fruit, glandular-hirsute; sepals linear-lanceolate, 2.5–4 mm. long, pubescent with spreading, gland-tipped hairs; corolla purplish or bluish, with yellow or whitish throat, 7–8 mm. long, upper lip 3–3.5 mm. long, erect to slightly reflexed, with rounded lobes, lower lip folded upward to nearly close throat, tip then turned abruptly downward, three subequal lobes rounded, sulcus in palate densely glandular-hirsute; stamens equaling throat, puberulent at base; sterile stamen represented by a minute callosity; capsule asymmetrically ovoid, 5–6 mm. long, glandular-puberulent, dehiscing by subapical pores; seeds tawny to dull black, 0.6–0.8 mm. long, with 4–6 corky ridges.

Among boulders and under shrubs, Lower Sonoran Zone, central Sonora, Mexico. Feb.–March.

## 4. ASARINA Mill., Gard. Dict. ed. 8, 1768

Annual or perennial herbs, often climbing-scandent by tortile petioles and peduncles, glabrous or pubescent. Leaves alternate or lower ones opposite, dentate to hastately or angularly lobed. Flowers axillary, rosaceous to violet or less commonly yellowish. Sepals 5, linear to lanceolate. Corolla slightly gibbous at base, spreadingly but shallowly bilabiate, upper lip 2-lobed, lower 3-lobed and basally biplicate, throat rather broad, hairy on plicae. Stamens 4, didynamous, included; anthers with locules divergent basally, confluent at apex. Capsule globose, locules equal or slightly unequal, usually dehiscing regularly. Seeds several, tuberculate, rugose or slightly winged.

Plants not prostrate; stems vinelike, usually climbing by tortile pedicels and petioles; herbage glabrous except at base of plant; leaves ovate, lanceolate, hastate or 3–5-lobed:
- Corolla purplish, carmine, or rarely white, glabrous without, 25–30 mm. long; leaves hastate or 3–5-lobed; perennial .......... 1. *A. antirrhiniflora*
- Corolla bright yellow, glandular-puberulent without, 11–13 mm. long; leaves lance-linear above, ovate at base of plant; annual .......... 2. *A. filipes*

Plants prostrate or decumbent; stems stiff, stout, not climbing; pedicels and petioles not tortile; herbage densely glandular-villous throughout; leaves broadly ovate-cordate to reniform:
- Leaf blades shallowly dentate, teeth again serrate; fruiting pedicels strongly contorted; corolla 25–28 mm. long; stamens exserted .......... 3. *A. flaviflora*
- Leaf blades palmately 5–7-lobed, lobes entire; fruiting pedicels straight or merely curved; corolla 20–22 mm. long; stamens included .......... 4. *A. acerifolia*

**1. Asarina antirrhiniflora** (Humb. & Bonpl. ex Willd.) Pennell, Proc. Acad. Phila. **99:** 174, 1947

*Maurandya antirrhiniflora* Humb. & Bonpl. ex Willd., Hort. Berol. **2:** *pl. 83,* 1806.
(For additional synonymy see Proc. Calif. Acad. IV, **15:** 383, 1926.)

Scandent perennial herb climbing to 2 m. or more by tortuous petioles and pedicels, stems slender, vine-like, glabrous except for sparse tomentum at base of stems; leaves thin, green, triangular, hastate or 3–5-lobed, 5-25 mm. long, often nearly as wide, often cordate at base, central lobe acuminate, lateral lobes mucronate; petioles 5–25 mm. long; flowers solitary in axils; pedicels subfiliform, 1–2 cm. long, glabrous; calyx narrowly campanulate, 5-parted almost to base, sepals lance-linear or narrowly lanceolate, 10–13 mm. long at anthesis, to 15 mm. long with connivent tips in fruit; corolla purplish, carmine, or rarely white, 2.5–3 cm. long, glabrous without, throat broadly ampliate, whitish but tinged with blue, about 1 cm. wide, with 2 prominent pubescent pleats fusing toward tip to form a conspicuous yellow palate, upper lobes 5–6 mm. long, lower about 6 mm. long; filaments glabrous at base, slightly geniculate, densely pubescent and clavate above geniculae; capsule ovoid-globose, 7–8 mm. long, surpassed by slightly connivent sepals, dehiscence subterminal, irregular; seeds 1–1.2 mm. long, brown, with cristate or tuberculate, curly ridges.

Mainly in lime soils, Lower and Upper Sonoran Zones, from the Providence Mountains, San Bernardino County, California, through Arizona and New Mexico to Texas and south to central Mexico. April–Sept.

**2. Asarina filipes** (A. Gray) Pennell, Proc. Acad. Phila. **99:** 174, 1947

*Antirrhinum filipes* A. Gray, Proc. Amer. Acad. **7:** 376, 1868.
*Antirrhinum cooperi* A. Gray, loc. cit.

Annual climbing herb with filiform stems 3–10 dm. long and much branched below; herbage glabrous except at very base; lower leaves ovate, entire, 5–20 mm. long, obtuse to rounded at apex, upper leaves lanceolate or linear, to 3 cm. long, then again reduced near tips of branches, petioles 5–15 mm. long; flowers solitary in axils on tortile filiform pedicels 3–8 cm. long; calyx lobes lanceolate, 0.8–1 mm. wide, 3.5–4.5 mm. long, sparsely and minutely puberulent with fine, tack-shaped, glandular hairs; corolla bright yellow, 11–13 mm. long, tube saccate at base, 5–6 mm. long, pubescent within from base of tube to base of palate, minutely and sparsely glandular-puberulent without, upper lip 5–6 mm. long, erect, lobes about 2 mm. long, broadly rounded-ovate, lower lip 5–6 mm. long, erect but with deflexed lobes, these about 2.5 mm. long; palate prominent, coarsely hairy, flecked with dark spots; stamens 6–8 mm. long, glabrous except on genicula; capsule globose, 3–5 mm. in diameter, minutely glandular-puberulent, dehiscing by irregular subterminal slits; seeds 0.8–1 mm. long, pale, longitudinally ridged with corky, irregular, narrow wings, these often minutely tuberculate.

Twining about desert shrubs on mountain sides, desert slopes, and along washes, chiefly Lower Sonoran Zone but extending into lower parts of Upper Sonoran Zone, eastern Mojave Desert east to southern Utah, south into western Arizona and to the western margins of the Colorado Desert; probably in adjacent Baja California and Sonora. March–May.

**Asarina stricta** (Hook. & Arn.) Pennell, Proc. Acad. Phila. **99:** 175, 1947, with blue flowers, slightly shorter pedicels, stems stouter at base, and gibbous corolla tube, occurs in chaparral of the Upper Sonoran Zone from Marin County, California, southward into northern Baja California. It may occur sporadically along the upper margins of the western Colorado Desert and adjacent Baja California but has not been reported from the area.

**3. Asarina flaviflora** (I. M. Johnston) Pennell, Proc. Acad. Phila. **99:** 175, 1947

*Maurandya flaviflora* I. M. Johnston, Proc. Calif. Acad. IV, **12:** 1162, 1924.

Probably perennial, with prostrate stems forming dense mats 2–5 dm. in diameter, herbage oily-villous throughout; leaves broadly cordate to reniform, 20–25 mm. long, 25–40 mm. wide, coarsely toothed and teeth again serrate; petioles slender, straight or curved, 1–3 cm. long; flowers solitary in axils; pedicels slender, 20–25 mm. long, becoming contorted and to 10 cm. long in fruit; calyx 11–12 mm. long in flower, segments 7–9 mm. long, foliaceous in anthesis, slightly accrescent and firmer, lobes ovate in fruit; corolla pale yellow, cylindrical, 25–28 mm. long, glabrous without and within except on ampliate lower part of throat, this part pubescent with flat yellow hairs; corolla lobes broadly ovate to orbicular, anterior lobes erect, posterior ones arched; stamens 4, exserted 4–6 mm., filaments flattened; capsule broadly ovoid, slightly compressed laterally, 8–10 mm. broad, apices of locules somewhat crest-like; seeds brown, 1.5–2 mm. long, bearing coarse, longitudinal, corky ridges.

Known only from the type locality, Las Animas Bay, Baja California. In flower and fruit in early May.

**4. Asarina acerifolia** (Pennell) Pennell, Proc. Acad. Phila. **99:** 174, 1947
*Maurandya acerifolia* Pennell, Jour. Wash. Acad. **19:** 69, 1929.

Prostrate or decumbent perennial forming loose, matlike masses with densely glandular-villous herbage; stems brittle, becoming thick, white-villous near base; petioles slender, 1.5–2 cm. long, straight or slightly curved; leaf blades broadly cordate to reniform, palmately 5–7-lobed, about 1.5 cm. long, 2–2.5 cm. wide, lobes entire, triangular, rounded or acute; flowers solitary in axils, pale yellow; pedicels 10–20 mm. long at anthesis, becoming thicker and somewhat recurved but not contorted in fruit; sepals triangular-ovate, about 6 mm. long at anthesis, slightly accrescent in fruit; corolla 20–22 mm. long, cylindric, nearly glabrous without, finely pubescent at base and on bases of filaments within, 2 infolding ridges posteriorly near apex of tube but not forming a palate, these densely pubescent with flat yellow hairs; lobes uniformly spreading, ovate-orbicular, 4–5 mm. long; stamens didynamous, included; capsule globose, 7–8 mm. long, each cell rupturing subdistally and irregularly; seeds 1–1.5 mm. long, gray to blackish, with low corky longitudinal ridges.

Canyon walls, upper edge of the Lower Sonoran Zone, in side canyons of the Salt River, eastern Maricopa County, Arizona. April–May.

## 5. GALVEZIA Domb., in Juss., Gen. 119, 1789

Perennial herb or (in ours) suffrutescent plants with slender, elongated, glabrous to minutely glandular-pubescent branches and alternate, opposite, or ternately whorled, entire leaves. Flowers axillary, upper ones often subracemose. Pedicels slender, often strongly inflexed after anthesis. Calyx 5-parted into narrowly lanceolate or lance-ovate, subequal segments, glabrous or glandular-pubescent. Corolla tubular, bilabiate, slightly saccate at base, bright red; throat open; upper lip erect or nearly so, minutely bilobed; lower lip 3-lobed, spreading. Stamens 4, more or less didynamous, two-thirds as long as to equaling corolla tube. Anther locules confluent at apex. Style subclavate. Ovary ovoid, glabrous or puberulent, many-ovuled. Capsule ovoid; locules equal, dehiscing by an irregular pore near apex of each locule. Seeds many, minute, truncate at ends, minutely winged or costate with thin cristate ridges or sometimes somewhat echinate.

Leaves small, usually considerably less than 1 cm. long ............ 1. *G. juncea* var. *juncea*
Leaves larger, usually 1–2.5 cm. long ........................ 1*a*. *G. juncea* var. *pubescens*

**1. Galvezia juncea** (Benth.) Ball, Jour. Linn. Soc. **22:** 154, 1886, var. **juncea**

*Maurandia juncea* Benth., Bot. Sulph. 41, 1844.
*Saccularia veatchii* Kell., Proc. Calif. Acad. **2:** 17, 1860.
*Antirrhinum junceum* A. Gray, Proc. Amer. Acad. **7:** 377, 1868.
*Galvesia juncea* var. *typica* Munz, Proc. Calif. Acad. IV, **15:** 376, 1926.

Erect or spreading, much-branched shrubs with slender stems to 2 m. long; nodes 2–10 cm. long, glabrous and glaucous or densely glandular-puberulent; leaves opposite or verticillate in threes, narrowly lance-elliptic to broadly obovate or suborbicular, 1–3 mm. wide, 3–10 mm. long, acute to rounded at apex, glabrous to densely glandular-puberulent; flowers in pairs or threes; pedicels slender, 1–3 cm. long; calyx herbaceous, glabrous or glandular-puberulent, sepals lanceolate to oblong-ovate, subequal, 2.5–5 mm. long in flower, to 7 mm. long in fruit; corolla bright red, 2.5–3 cm. long, glandular-pubescent without, slightly saccate basally on anterior side, tube 15–25 mm. long, 3–5 mm. in diameter, tawny-pubescent within, throat scarcely larger than tube, lower lip reflexed, 3-lobed, 4–8 mm. long, lobes rounded, upper lip erect or nearly so, about as long as lower one, slightly 2-lobed; fertile stamens 4, subequal to strongly didynamous, longer pair about equaling corolla, or all of them one-half to two-thirds as long as corolla, filaments dilated and glabrous at base, pubescent above slight genicula; anthers confluent at tips and divergent; stigma slightly flattened, entire or minutely bilobed; capsule ovoid, 8–10 mm. long, 5–7 mm. in diameter, glabrous or sparsely glandular-pubescent with slender spreading hairs, dehiscing by 2 subterminal, rounded, irregular pores; seeds dark brown to black, dull, 0.8–1.2 mm. long, truncate at ends, longitudinally ridged-cristate.

On hillsides, along arroyos, and occasionally on plains, mainly Lower Sonoran Zone, from near the coast about halfway between Ensenada and San Diego south to the Cape Region, on the islands along the eastern shores of Baja California, and on Cedros Island. Feb.–Oct.

**1*a*. Galvezia juncea** var. **pubescens** (Brandegee) I. M. Johnston, Proc. Calif. Acad. IV, **12:** 1161, 1924

*Galvesia glabrata* Brandegee, Zoe **5:** 167, 1903.
*Galvesia speciosa* var. *pubescens* Brandegee, loc. cit.
*Galvezia rupicola* Brandegee, Univ. Calif. Pub. Bot. **6:** 360, 1916.
*Galvezia juncea* var. *foliosa* I. M. Johnston, Proc. Calif. Acad. IV, **12:** 1161, 1924.

Differing from var. *juncea* only in having larger leaves, these often 1–2.5 cm. long and to 8 mm. wide, and in having capsules broader in proportion to length, often subglobose and tending to be somewhat pendulous; the plants may be glabrous or sparsely pubescent on stems, calyces, and young capsules.

Rocky hillsides and along arroyos, Lower Sonoran Zone, San Lorenzo and San Pedro Nolasco Islands, and in the vicinity of Rancho Mesquital, south into the Cape Region, Baja California.

## 6. RUSSELIA Jacq., Enum. Pl. Carib. 6, 1760

Shrubs or suffrutescent plants with strongly angulate stems and opposite or verticillate leaves usually copiously furnished with resinous glands. Inflorescence cymose, bracteate. Calyx 5-toothed or -parted; segments strongly imbricate. Corolla red; tube cylindrical; limb small, shallowly 5-lobed, faintly bilabiate; lobes rounded. Stamens 4, didynamous, included; anthers with locules divaricate at base, confluent at apex; sterile stamen minute or absent. Style slender, very slightly dilated at apex.

Capsule subglobose, septicidally dehiscent by 2 valves. Seeds many, more or less pubescent or embedded in long hairs within capsule.

**1. Russelia polyedra** Zucc., Abh. Akad. Münch. **2:** 328, 1832–36

*Russelia retrorsa* Greene, Pittonia **1:** 176, 1888.

Suffrutescent plants 0.5–2.5 m. tall with numerous slender, ascending, 6-angulate stems, these sparsely to densely pubescent between and often on angles; leaves in pairs or sometimes ternate, broadly ovate, 1–6 cm. long, 8–25 mm. broad, broadly cuneate to rounded at base, acute at apex, dentate, sparsely short-hispid along veins and on young blades, dotted with glistening, transparent, resinous glands, particularly beneath; flowers 1–5 on each peduncle, latter shorter than subtending leaves, 1–3 mm. long, hispidulous; pedicels slender, 1–3 mm. long, puberulent; calyx 5-parted to base, sparsely hispidulous and gland-dotted, sepals broadly ovate, 0.8–1 mm. wide, abruptly attenuate-subulate, 2–2.5 mm. long, slender tip often outwardly curved; corolla bright red, 12–15 mm. long, tube slightly funnelform, nearly or quite glabrous without, moderately hirsute with coarse hairs full length within; limb faintly bilabiate, 5 lobes rounded, 1–1.5 mm. long, about as wide; longer pair of stamens and style nearly equaling tube; capsule subglobose, 3–5 mm. long, glabrous; seeds embedded in dense mass of coarse, pale brown hairs.

Rocky hillsides, canyons, and occasionally on gentle slopes, Lower Sonoran to Tropical Zones, central Sonora and southern Baja California to Tamaulipas, Chiapas, and Vera Cruz. April–Nov.

## 7. SCROPHULARIA L., Sp. Pl. 619, 1753

Erect perennial herbs with quadrangular stems, strong-scented herbage, and paniculate inflorescences. Leaves opposite, petiolate, toothed. Sepals 5, distinct almost to base. Corolla greenish purple to deep maroon; tube short, cylindrical or subglobose; upper lip flat, horizontally spreading; lower lip with lateral lobes vertical or nearly so, central one deflexed to recurved. Stamens 4, faintly didynamous, included; anther cells divergent; vestigial sterile filament scalelike or lacking. Style slender; stigma pointed. Capsule ovoid or subglobose, septicidally dehiscent. Seeds numerous, more or less tumid, shallowly furrowed.

Capsules globose-conic; sepals broadly rounded to orbicular; corolla 5–8 mm. long; leaf blades triangular-ovate, irregularly and often deeply dentate or laciniate ............ ............................................... 1. *S. californica* subsp. *floribunda*

Capsules narrowly ovoid-conic; sepals ovate but acute; corolla 4–6 mm. long; leaf blades lanceolate, more regularly and usually shallowly dentate .............. 2. *S. parviflora*

**1. Scrophularia californica** Cham. & Sch. subsp. **floribunda** (Greene) Shaw, Aliso **5:** 176, 1962

*Scrophularia californica* var. *floribunda* Greene, Man. Bay Reg. 273, 1894.

(For additional synonymy see Aliso **5:** 176, 1962.)

Erect or ascendingly branched perennial 6–12 dm. tall with minutely glandular-pubescent herbage, pubescence densest in inflorescence; petioles 1–6 cm. long, flattened, margins almost narrowly winged; leaf blades ovate to oblong-lanceolate or triangular-ovate, 3–10 cm. wide, 4–15 cm. long, acute at apex, truncate or cordate at base, irregularly and usually doubly dentate or incised; panicles 5–10 cm. wide, to 5 dm. long, branches spreading; pedicels slender, 3–10 mm. long; sepals broadly rounded to nearly orbicular, 2–3 mm. long, margins thin; corolla 5–8 mm. long, dorsally reddish to garnet-brown, base of tube whitish or greenish, ventral lobes greenish yellow, median lobe strongly recurved; upper filament clavate to obovate; capsule globose-conic, 4–7 mm. long, rather abruptly apiculate; seeds dark brown or black, broadly ovoid, about 0.5 mm. long, finely rugulose transversely and microscopically striate longitudinally.

Hillsides, along washes, and in openings in chaparral, mainly Upper Sonoran Zone but entering the desert mountains at upper edge of the Lower Sonoran Zone, Inner Coast Ranges, and western slopes of Sierra Nevada south into northern Baja California. March–Sept.

**2. Scrophularia parviflora** Woot. & Standl., Contr. U.S. Nat. Herb. **16:** 173, 1913

*Scrophularia glabrata* Davidson, Bull. S. Calif. Acad. **1:** 26, *fig. 3,* 1902. Not *S. glabrata* Ait., 1789.
*Scrophularia davidsonii* Pennell, Notulae Naturae **43:** 8, 1940.

Erect or ascendingly branched herb 5–12 dm. tall with rather slender, minutely puberulent to glabrate stems; petioles slender, only faintly flattened, 0.5–5 cm. long; leaf blades finely and sparsely glandular-puberulent, broadly lanceolate to ovate-lanceolate, 1–3 cm. broad, 2–8 cm. long (or basal ones somewhat larger), acute at apex, cuneate to subcordate at base, regularly once-dentate, teeth 0.5–5 mm. long; panicles loose, spreading 1–4 dm. long; pedicels slender, 1–1.5 cm. long; sepals ovate, 1–1.5 mm. wide, 2–2.5 mm. long, acute or obtuse at apex; corolla brownish red, 4–6 mm. long, lobes broadly rounded; capsule narrowly ovoid-conic, 3–3.5 mm. in diameter, 5–6 mm. long, gradually narrowed apically to slender style; seeds dull black, ovoid, 0.5–0.8 mm. long, transversely rugulose with low broad ridges, these shallowly pitted.

In rich soil mostly in forested areas, chiefly Upper Sonoran Zone to Arid Transition Zone, Mohave and Coconino to Pima and Cochise Counties, Arizona, and southwestern New Mexico, reaching edge of Lower Sonoran Zone in Pima County. July–Aug.

## 8. PENSTEMON Mitch., Nov. Gen. Pl. Virg. 36, 1769

Perennial herbs or shrubs with opposite leaves, basal often petiolate, upper ones sessile and sometimes connate-perfoliate. Flowers showy, paniculate. Calyx 5-parted almost to base; sepals more or less imbricate, margins often submembranous. Corolla tubular and usually ventricose, 2-lipped; upper lip 2-lobed, erect, spreading or reflexed; lower lip 3-lobed or -cleft. Fertile stamens 4, didynamous, inserted near base of corolla tube, more or less arched; sterile filament (staminode) inserted on upper side of corolla at union of tube and throat, curving downward and extending toward orifice. Anthers 2-celled, these often confluent. Style slender; stigma narrowly capitate. Capsules 2-celled, septicidal and in some more or less loculicidal as well. Seeds many, angled.

Corolla sulphur-yellow; filaments strongly pubescent at base; plants shrubby, to 2 m. tall .................................... 1. *P. antirrhinoides* subsp. *microphyllus*
Corolla not sulphur-yellow; filaments glabrous (viscid-puberulent in *P. pseudospectabilis*); plants herbaceous or slightly woody at base:
Cauline leaves linear, 0.5–1 mm. wide; corolla 10–12 mm. long, funnelform, blue-purple throughout ................................................ 2. *P. thurberi*

Cauline leaves linear-lanceolate or broader, 5 mm. wide or more; corolla 13–30 mm. long, if funnelform not blue-purple throughout:

Leaves (or at least some of them) distinctly dentate or serrate; corolla not scarlet:

Corolla rose, crimson, or purplish, tube twice as long as calyx or more, throat 4–9 mm. broad at orifice; stems to 1 m. tall:

Leaves prominently serrate; corolla pink to rose-purple, with prominent lines in throat:

Pedicels and calyces glandular-puberulent .................................. ......................... 3. *P. pseudospectabilis* subsp. *pseudospectabilis*

Pedicels and calyces glabrous ....... 3*a*. *P. pseudospectabilis* subsp. *connatifolius*

Leaves often entire on upper part of plant, lower ones irregularly and sparsely dentate; corolla crimson or red-purple, faintly lined or lineless in throat:

Upper leaves distinct; anther sacs glabrous:

Staminode usually bearded; upper leaves deltoid-lanceolate to cordate ....... .................................. 4. *P. clevelandii* subsp. *clevelandii*

Staminode glabrous; upper leaves cuneate-obovate ......................... .................................. 4*a*. *P. clevelandii* subsp. *angelicus*

Upper leaves connate-perfoliate; anther sacs ciliolate-denticulate ............. ......................................... 4*b*. *P. clevelandii* subsp. *connatus*

Corolla white, tinged with purple and with purplish guide lines, tube about as long as calyx, throat 15–20 mm. wide at orifice; stems to 2.5 m. tall ............. ........................................................ 5. *P. eximius*

Leaves all entire; corolla scarlet or rose:

Staminode conspicuously bearded; corolla rose, 13–21 mm. long, glandular-puberulent without ......................................................... 6. *P. parryi*

Staminode glabrous or sparsely bearded; corolla scarlet, 18–33 mm. long, glabrous without and within:

Corolla tubular, scarcely at all ampliate; stamens included; anthers glabrous:

Cauline leaves lanceolate, 0.5–3 cm. wide; staminode glabrous; corolla 25–35 mm. long ........................................... 7. *P. centranthifolius*

Cauline leaves subulate-lanceolate, 2–5(–8) mm. wide; staminode glabrous or sparsely bearded; corolla 18–26 mm. long ............... 8. *P. subulatus*

Corolla moderately ampliate above; stamens slightly exserted; anthers minutely puberulent:

Stems and leaves glabrous ........................ 9. *P. eatonii* subsp. *eatonii*

Stems and leaves puberulent:

Stamens included or barely exserted ........... 9*a*. *P. eatonii* subsp. *undosus*

Stamens long-exserted ....................... 9*b*. *P. eatonii* subsp. *exsertus*

**1. Penstemon antirrhinoides** Benth. subsp. **microphyllus** (A. Gray) Keck, in Abrams, Ill. Fl. Pacif. States **3:** 770, 1951

*Penstemon microphyllus* A. Gray, Pacif. R. Rep. **4:** 119, 1857.
*Penstemon plummerae* Abrams, Bull. Torrey Club **33:** 445, 1906.
*Penstemon antirrhinoides* var. *microphyllus* Munz & Johnston, Bull. Torrey Club **49:** 43, 1922.

A spreading, moderately branched shrub 0.5–2 m. tall with yellowish gray-green, canescent herbage and finely puberulent, cinereous twigs; leaves linear, oblanceolate, or elliptic, 2–8 mm. wide, 5–22 mm. long, gradually narrowing to a petiole 1–5 mm. long, acute or sometimes obtuse at apex, entire; panicles leafy, 5–30 cm. long; peduncles 1–3-flowered; pedicels 1–2 mm. long; sepals lanceolate or lance-oblong, 5–8 mm. long, acute or slightly acuminate, somewhat imbricate, glandular-puberulent; corolla bright yellow, lined with red without, 8–10 mm. broad at throat, 15–20 mm. long, throat flaring abruptly from tube which about equals calyx; upper lip 9–12 mm.

long, erect to arcuate, boat-shaped and low-keeled; sterile filament densely bearded on lower two-thirds; anther sacs 1.2–1.3 mm. long; capsules narrowly ovoid-acuminate, 7–10 mm. long, glabrous.

Rocky hillsides and outwash slopes of desert ranges, upper Lower Sonoran and lower Upper Sonoran Zones, western margins of the Colorado Desert to central Arizona and south to the Sierra San Pedro Mártir, Baja California. April–June.

**2. Penstemon thurberi** Torr., Pacif. R. Rep. **7:** 15, 1858

*Leiostemon thurberi* Greene, Leaflets Bot. Obs. **1:** 223, 1906.
*Penstemon scoparius* A. Nels., Wyo. Univ. Pub. Bot. **1:** 132, 1926.

Several- to many-stemmed, strongly suffrutescent plant 2–6 dm. tall with erect slender branches, glabrous throughout; leaves linear, erect or ascending, about 0.5–1 mm. broad, 1.5–5 cm. long, margins more or less involute in drying; panicles slender, to 2 dm. long, 2–5 cm. wide; peduncles 1–3-flowered; sepals ovate-acuminate, 2.5–3 mm. long, often reddish, margins submembranous; corolla bluish purple or rarely pinkish, 10–12 mm. long, funnelform, throat 5–6 mm. broad at orifice, straight but ventricose, glabrous except at base of lower lip, where it is sparsely bearded; upper lip moderately reflexed-spreading, all lobes rounded, 3–4 mm. long; stamens didynamous, longer pair slightly exserted, shorter included; sterile filament inserted at about middle of corolla, equaling throat, glabrous; capsules ovoid, 4.5–7 mm. long, acute; seeds dark brown or blackish, 1.5–2 mm. long, minutely foveolate-reticulate.

Sandy or rocky slopes and plains, Lower and Upper Sonoran Zones, western margin of Colorado Desert east into New Mexico, south to northeastern flanks of the Sierra San Pedro Mártir. April–June.

**3. Penstemon pseudospectabilis** M. E. Jones, Contr. West. Bot. **12:** 66, 1908, subsp. **pseudospectabilis**

*Penstemon pseudospectabilis* subsp. *typicus* Keck, Amer. Midl. Nat. **18:** 805, 1937.

Slender to moderately robust herb 5–10 dm. tall, glabrous and glaucous throughout except in inflorescence; leaves thin to coriaceous, basal ones lanceolate-ovate to broadly ovate, 1–4 cm. broad, to 8 cm. long, finely and sharply serrate-dentate, teeth often caudate, on petioles one-fourth as long as to equaling blade; cauline leaves ovate-lanceolate, sessile; upper leaves connate-perfoliate, forming disks to 6 cm. wide, to 12 cm. long; inflorescence 1–5 dm. long or more, 4–8 cm. wide, peduncles and pedicels erect or nearly so; cymules usually 2–4-flowered; pedicels and sepals sparsely to moderately glandular-puberulent; sepals ovate, short-acuminate, 3.5–5 mm. long; corolla pink to rose-purple with a yellowish tube and dark lines in throat, tubular-funnelform, 20–26 mm. long, throat ventricose, viscid-puberulent near orifice, lobes about equal, rounded, 3–4 mm. long, spreading or reflexed; filaments viscid-puberulent toward base; staminode slender, usually glabrous or sparsely bearded near tip; capsules broadly ovoid, 4–6 mm. in diameter, 7–11 mm. long; seeds foveolate-reticulate, 1.5–2 mm. long, blackish.

Arid mountain slopes, Lower Sonoran Zone, southeastern Mojave Desert to western Pima County, Arizona, and central Imperial County, California. March–April.

**3*a*. Penstemon pseudospectabilis** subsp. **connatifolius** (A. Nels.) Keck, Amer. Midl. Nat. **18:** 807, 1937

*Penstemon connatifolius* A. Nels., Amer. Jour. Bot. **18:** 437, 1931.
*Penstemon crideri* A. Nels., op. cit. **23:** 270, 1936.

Similar to subsp. *pseudospectabilis* except that pedicels and calyces are glabrous.

Mountain slopes and along washes, Lower and Upper Sonoran Zones, Mohave and Cochise Counties, Arizona, eastward into New Mexico. Barely entering the fringe of the Sonoran Desert along the western flanks of the Santa Catalina Mountains. April–June; Oct.–Nov.

**4. Penstemon clevelandii** A. Gray, Proc. Amer. Acad. **11:** 94, 1876, subsp. **clevelandii**

*Penstemon clevelandii* subsp. *typicus* Keck, Amer. Midl. Nat. **18:** 810, 1937.

Erect herbaceous plant with 1 to several stems 3–7 dm. tall from a stout woody root; herbage glabrous or glandular-puberulent; leaves deep green to glaucous, entire to sharply dentate, basal ones ovate, 1–3 cm. wide, 3–7 cm. long, on petioles 1–4 cm. long, upper cauline leaves deltoid-lanceolate to cordate, distinct, sessile; inflorescence 1–3 dm. long, 3–6 cm. wide, cymules 2–8-flowered, peduncles and pedicels each 5–12 mm. long, suberect or slightly divergent; calyx dark reddish, sepals ovate to suborbicular, obtuse to acuminate, 4–6 mm. long, 2–3.5 mm. wide, sparsely glandular-puberulent; corolla 17–25 mm. long, 5–8 mm. wide at throat, crimson or red-purple, concolorous in throat, tube 5–9 mm. long, shorter than gradually ampliate throat, glandular-puberulent within, lobes subequal, quadrate, rotately spreading; anther sacs explanate, glabrous; fertile filaments glabrous, staminode 8.5–11 mm. long, more or less bearded; capsules broadly ovoid, 6–8 mm. long, 5–7 mm. wide; seeds angular, 1.5–2 mm. long, dull black.

Arid canyon sides bordering the deserts, mainly Lower Sonoran Zone, eastern San Diego County, California, south to central Baja California. Feb.–May. On the eastern escarpment of the Sierra San Pedro Mártir this subspecies ascends to an elevation of approximately 4,000 feet above sea level.

***4a.* Penstemon clevelandii** subsp. **angelicus** (I. M. Johnston) Keck, Amer. Midl. Nat. **18:** 811, 1937

*Penstemon clevelandii* var. *angelicus* I. M. Johnston, Proc. Calif. Acad. IV, **12:** 1165, 1924.

Differing from subsp. *clevelandii* in having the upper leaves cuneate-obovate, the inflorescences subracemose instead of thyrsoid, and staminode glabrous.

Known only from the type locality, Palm Canyon on Isla Angel de la Guarda, Baja California.

***4b.* Penstemon clevelandii** subsp. **connatus** (Munz & Johnston) Keck, Amer. Midl. Nat. **18:** 811, 1937

*Penstemon clevelandii* var. *connatus* Munz & Johnston, Bull. Torrey Club **49:** 357, 1923.

Differing from subsp. *clevelandii* in having upper cauline leaves conspicuously connate-perfoliate, pedicels and peduncles glabrous, and anther sacs denticulate-ciliolate but not explanate.

Along canyons, in rocky or sandy soil, Lower Sonoran Zone, western margin of the Colorado Desert, Riverside County, California.

**5. Penstemon eximius** Keck, Amer. Midl. Nat. **18:** 798, 1937

Robust, erect, perennial herb to 2.5 m. tall with 1 to several stems from a heavy woody root; leaves and stems glabrous, pedicels, peduncles, and calyces glandular-puberulent; leaves coriaceous, green, sharply dentate, teeth narrow and somewhat caudate, blades of lower leaves ovate-lanceolate, 3–5 cm. wide, 6–15 cm. long, on stoutish petioles to 5 cm. long, upper ones lanceolate, cordate-clasping, those at base of inflorescence lance-triangular, 1–3 cm. long, glandular-puberulent; thyrse 4–10 cm. broad, to 8 dm. long; peduncles 2–4-flowered, erect or ascending, 1–2 cm. long; sepals ovate-oblong, 2.5–3.5 mm. wide, 4–7 mm. long, obtuse, margins submembranous; corolla white suffused with rose-purple, 25–30 mm. long, 15–20 mm. wide at mouth, viscid-puberulent, tube abruptly ampliate into a wide throat, this over twice as long as tube, lower lip villous near base within, lobes broad, rounded, 5–6 mm. long; fertile stamens glabrous, included; staminode 9–10 mm. long, bearded, glandular-puberulent at base; capsules ovoid, acute or acuminate, 6–8 mm. in diameter, 10–15 mm. long; seeds about 2 mm. long, truncate, seed coat loose, finely foveolate-reticulate.

Sandy washes and rocky hillsides and canyons, Lower Sonoran Zone, eastern escarpment of the Sierra Juárez and Sierra San Pedro Mártir, thence west around the southern tip of the Sierra San Pedro Mártir and south to Miller's Landing and the vicinity of Calmallí, Baja California. April–June.

**6. Penstemon parryi** (A. Gray) A. Gray, Syn. Fl. N. Amer. **2**[1]**:** 264, 1878

*Penstemon puniceus* var. *parryi* A. Gray, in Torr., Bot. Mex. Bound. 113, 1859.
*Penstemon shantzii* A. Nels., Amer. Jour. Bot. **23**: 270, 1936.

Herbaceous, stems 1 to several from a tough root, erect or ascending, 3–12 dm. tall, glabrous and glaucous throughout; basal leaves elliptic, spatulate or oblanceolate, 1–2.5 cm. wide, 4–15 cm. long, gradually narrowing to a somewhat winged petiole 1–4 cm. long, upper ones narrowly lanceolate to lanceolate-oblong, 1–2.5 cm. wide, 3–12 cm. long, obtusish at tip, auriculate-clasping at base; inflorescence virgate, 1–5 dm. long; peduncles few- to many-flowered, erect or divergent, to 2.5 cm. long, glabrous; calyx 3–5 mm. high, ovate sepals glabrous or rarely glandular-puberulent, acute or short-acuminate; corolla rose, cerise, or rose-magenta, 13–21 mm. long, tube 3–5 mm. long, expanding rather abruptly into moderately ventricose throat, limb spreading, glandular-puberulent without and on anterior portion within, pilose at base of lower lip; anther sacs glabrous, peltately explanate; staminode dilated apically and retrorsely bearded with stiff yellow hairs; capsules ovoid, 4–6 mm. long.

Hillsides, outwash slopes, and canyons leading down to the desert, Lower and low Upper Sonoran Zones, Pinal County, Arizona, south to southern Sonora. March–May.

**7. Penstemon centranthifolius** Benth., Scroph. Indicae 7, 1835

*Chelone centranthifolia* Benth., Trans. Hort. Soc. Lond. II, **1**: 481, 1835.

Erect herbaceous plant with 1 to several stems 3–12 dm. tall from a tough root; herbage glabrous and glaucous, often strikingly purplish or reddish; basal leaves spatulate, petiolate, cauline ones linear-lanceolate to ovate-lanceolate, sessile to auriculate-clasping, acuminate to rounded and more or less apiculate, 0.5–3 cm. wide, 4–10 cm. long, ascending; thyrse virgate, about half as long as stems; peduncles erect, 1–5-flowered, 1–1.5 cm. long; sepals ovate to suborbicular, abruptly acute or apiculate, 3–6 mm. long, margins broadly scarious, often erose; corolla 25–35 mm. long, 4.5–6 mm. broad at mouth, tubular, scarlet or basal part yellowish, lobes scarcely spreading, rounded, 2–3 mm. long, glabrous outside and within; stamens and staminode glabrous, included; anther sacs peltately explanate; capsules broadly ovoid, 7–11 mm. long.

Mainly in coastal regions on grassy hillsides or on openings in the chaparral, chiefly Upper Sonoran Zone from Lake County, California, south to the Sierra San Pedro Mártir, Baja California. Barely entering the uppermost edge of the Lower Sonoran Zone on the eastern flanks of the Sierra San Pedro Mártir. Feb.–July.

**Penstemon cerrosensis** Kell., Proc. Calif. Acad. **2:** 19, 1860, with reflexed basal leaves and glandular-pubescent pedicels, calyces, and corollas, occurs on Cedros Island and may appear on mainland adjacent, but has not as yet been reported from our area.

### 8. Penstemon subulatus M. E. Jones, Contr. West. Bot. 12: 63, 1908

Slender herbaceous plant with few to several stems 2–7 dm. tall, glabrous throughout; leaves green or faintly glaucous, basal ones narrowly linear-oblanceolate to broadly oblanceolate or rarely elliptic, 3–15 mm. wide, to 10 cm. long, gradually tapering to a narrowly winged petiole 1–3 cm. long, often long-attenuate at apex; cauline leaves linear-lanceolate to linear-subulate, 2–5(8) mm. wide, to 8 cm. long, sessile but not auriculate-clasping; inflorescence an elongated thyrse 1–4 dm. long; sepals ovate, 3–5 mm. long, abruptly acuminate or subulate, glabrous, or minutely scurfy with small sessile glands on and near margins; corolla scarlet, tubular, 18–26 mm. long, 3–5 mm. wide at mouth, lobes similar, 2–2.5 mm. long, rounded, moderately spreading, glabrous without and within; shorter stamens about three-fourths as long as, longer pair nearly equaling, tube; anthers peltately explanate; staminode slender, glabrous or uncommonly bearing a few scattered hairs near apex; capsules ovoid, 6–8 mm. long, 4–5 mm. wide; seeds truncate-cylindric, sharply and irregularly angled, 1.5–2.5 mm. long, minutely foveolate-reticulate coat rather loose.

Stony hillsides, mesas, and rocky canyons, Lower and Upper Sonoran Zones, central Mohave County to western Graham County, Arizona. March–May.

### 9. Penstemon eatonii A. Gray, Proc. Amer. Acad. 8: 395, 1872, subsp. eatonii

Erect, few-stemmed herb 3–8 dm. tall, glabrous throughout, with striking suffusion of red or red-purple on stems and inflorescences; basal leaves narrowly oblanceolate to oblong-spatulate, 1–2.5 cm. wide, 4–12 cm. long, obtuse or rounded or less commonly acute at apex, tapering to a short or long, slightly winged petiole 1–6 cm. long; cauline leaves lanceolate, to 2 cm. wide, 15 cm. long, sessile, uppermost somewhat auriculate-clasping, acute at apex, entire; inflorescence slender, 2–4 dm. long, 4–6 cm. broad, short peduncles (2–10 mm. long) and slender pedicels erect or ascending, flowers spreading or slightly deflexed; sepals narrowly ovate-acuminate, 1.5–2 mm. wide, 4–7 mm. long, margins scarious and faintly and shallowly erosulate; corollas narrowly funnelform, scarlet, 25–30 mm. long, 6–8 mm. wide at mouth, glabrous without and within, lobes spreading, broadly rounded, 4–5 mm. long; longer stamens equaling lobes; anther sacs splitting only about one-half their length, ciliolate-denticulate along suture; staminode glabrous.

Desert mountain slopes and plains, Lower Sonoran to Arid Transition Zones, southern Mojave Desert and northernmost edge of Colorado Desert east to Colorado, south in Arizona to northern Gila County. March–June.

### 9*a*. Penstemon eatonii subsp. undosus (M. E. Jones) Keck, in Kearney & Peebles, Jour. Wash. Acad. 29: 491, 1939

*Penstemon eatonii* var. *undosus* M. E. Jones, Proc. Calif. Acad. II, **5:** 715, 1895.
*Penstemon coccinatus* Rydb., Bull. Torrey Club **36:** 691, 1909.

Differing from subsp. *eatonii* in having glandular-puberulent stems, leaves, and inflorescences and included or barely exserted stamens.

In habitats similar to those occupied by the typical form, eastern Mojave Desert to southern Utah, south in Arizona to Gila County.

**9*b*. Penstemon eatonii** subsp. **exsertus** (A. Nels.) Keck, in Kearney & Peebles, Jour. Wash. Acad. **29:** 491, 1939

*Penstemon exsertus* A. Nels., Amer. Jour. Bot. **18:** 438, 1931.
*Penstemon amplus* A. Nels., op. cit. **25:** 115, 1938.

Different from subsp. *eatonii* in having glandular-puberulent herbage and from both it and subsp. *undosus* in having long-exserted stamens.

Along the valley slopes and adjacent canyons bordering the Salt River, in Pinal and Gila Counties, Arizona. Mainly above the Sonoran Desert.

## 9. MIMULUS L., Sp. Pl. 634, 1753

Annual or perennial herbs or suffrutescent to shrubby plants with opposite, sessile or petiolate, entire, toothed, or lobed leaves. Flowers axillary or in terminal racemes. Calyx tubular or campanulate, mostly 5-angled and -toothed, usually plicate, persistent and often accrescent; teeth equal or unequal. Corolla bilabiate or subequally lobed; tube cylindrical or funnelform, expanding into throat; upper lip 2-lobed, exterior in bud, erect or spreading in anthesis; lower lip usually with 2 pubescent patches on floor of throat. Stamens 4, didynamous; anther cells confluent at apex, more or less divaricate basally. Style glabrous or pubescent; stigma bilobed or peltate-funnelform. Capsule ovoid, oblong, or linear, membranous to cartilaginous, loculicidally dehiscent. Seeds numerous, oval, smooth, striate or reticulate.

Calyx 1.8–2.5 cm. long in flower; corolla scarlet or carmine; stamens slightly but distinctly exserted:
  Corolla tube equaling or barely exceeding calyx ......... 1. *M. cardinalis* var. *cardinalis*
  Corolla tube 10–25 mm. longer than calyx .......... 1*a*. *M. cardinalis* var. *verbenaceus*
Calyx 5–15 mm. long in flower; corolla yellow or red-purple; stamens included except in *M. aridus,* plant then shrubby:
  Calyces strongly angled or prismatic; all 4 stamens fertile:
    Pedicels usually longer than calyces at anthesis; capsule usually dehiscent to base along both sutures:
      Calyx teeth equal or nearly so; plant viscid-villous, somewhat slimy .............................................................. 2. *M. floribundus*
      Calyx teeth unequal, posterior one largest; plant glabrous, puberulent, or slightly pubescent but not viscid-slimy:
        Plants perennial; style glabrous; calyx 5–8 mm. long at anthesis:
          Leaves broadly ovate .......................... 3. *M. glabratus* var. *glabratus*
          Leaves mostly subcordate ..................... 3*a*. *M. glabratus* var. *fremontii*
        Plants annual or sometimes perennial; style puberulent (sometimes glabrous in *M. guttatus*); calyx 6–18 mm. long at anthesis:
          Upper tooth on calyx 1.2–1.6 times as long as others; seeds longitudinally striate; corolla to 4 cm. long .................................... 4. *M. guttatus*
          Upper tooth on calyx 2–3 times as long as others; seeds reticulate; corolla to 2.5 cm. long ........................................... 5. *M. nasutus*
    Pedicels shorter than or barely exceeding calyces at anthesis; capsule usually dehiscent only part way to base along upper suture:
      Plants annual herbs; calyx distinctly inflated in fruit:

Corolla rose-red or reddish purple; plants 3–25 cm. tall, robust, usually with several ascending branches from base:
Calyx teeth broadly triangular, equal or nearly so; corolla rose-red with darker markings on upper side of throat; anthers glabrous ....... 6. *M. fremontii*
Calyx teeth subulate, unequal; corolla red-purple with yellowish throat; anthers hispid ............................................ 7. *M. bigelovii*
Corolla yellow; plants 1–8 dm. tall, rarely branched from base .... 8. *M. brevipes*
Plants suffrutescent; capsules not inflated in fruit .................... 9. *M. aridus*
Calyces very faintly angled, somewhat saccate; 2 stamens often sterile .... 10. *M. pilosus*

**1. Mimulus cardinalis** Dougl. ex Benth., Scroph. Indicae 28, 1835, var. **cardinalis**
(For synonymy see Ann. Mo. Bot. Gard. **11:** 138, 1924.)

A vigorous perennial with stems 2.5–5 dm. tall from a creeping rootstock, villous and more or less viscid throughout; leaves obovate to oblong, 1–4 cm. wide, 2–12 cm. long, sessile and somewhat auriculate, 3–5-veined from base, irregularly or evenly serrate, sometimes doubly so; pedicels stoutish, 3–15 cm. long; calyx tubular to oblong-prismatic, 18–25 mm. long, often dotted and suffused with red, particularly along ribs, teeth broadly ovate, acute or short-acuminate, ciliate-margined, 4–5 mm. long; corolla bright scarlet or pale reddish yellow, broadly cylindrical, 4–5 cm. long, tube only slightly exserted from calyx, upper lip erect or with lobes slightly reflexed, middle lobe of lower lip spreading, lateral pair reflexed; stamens distinctly exserted, partly enclosed by upper lip, filaments glabrous; anthers densely villous with soft white hairs; capsule oblong, acuminate, 1–1.5 cm. long; seeds oblong, finely reticulate.

Along streams and on shaded moist banks, chiefly Upper Sonoran and lower Transition Zones, Oregon to Utah and south into Arizona and Baja California. May–Sept.

**1*a*. Mimulus cardinalis** var. **verbenaceus** (Greene) Kearney & Peebles, Jour. Wash. Acad. **29:** 491, 1939
*Mimulus verbenaceus* Greene, Leaflets Bot. Obs. **2:** 2, 1909.
*Mimulus lugens* Greene, op. cit. 3.

Differing from var. *cardinalis* in having a longer corolla tube that is exserted 10–25 mm. beyond calyx; slender pedicels to 15 cm. long; and longitudinally wrinkled seeds.

Along streams and on moist shaded banks, Lower and Upper Sonoran Zones, northern Arizona south into Chihuahua and Sinaloa. Barely entering the Sonoran Desert along the flanks of the Santa Catalina Mountains, Pima County, Arizona, and near Baviácora, Sonora. March–Sept.

**2. Mimulus floribundus** Dougl. ex Lindl., Bot. Reg. **13:** *pl. 1125,* 1828
(For synonymy see Ann. Mo. Bot. Gard. **11:** 216, 1924.)

Slender annuals 1–5 dm. tall with viscid-villous, more or less slimy stems and foliage, hairs white and glistening; leaves remote, ovate to ovate-lanceolate, 1.5–5 cm. long, 5–30 mm. wide, acute at apex, truncate or subcordate or rarely broadly cuneate at base, thin, dentate with short broad teeth, on slender or faintly winged petioles 3–18 mm. long or sometimes a few of them sessile; pedicels filiform, 8–20 mm. long, often spreading-ascending at maturity; calyx cylindric-campanulate, 4.5–7 mm. long, often suffused or spotted with red, to 10 mm. long in fruit, teeth triangular to broadly lanceolate, 1–1.5 mm. long, about equal; corolla yellow with a few minute maroon flecks in throat, 7–12 mm. long, tube about equaling calyx, throat ampliate, short, lobes mostly erect, unequal, rounded, lower lip sometimes spreading; style and stamens glabrous, included; capsule oblong, chartaceous, 4–5 mm. long; seeds about 0.3 mm. long, elliptic-oblong, faintly and irregularly wrinkled longitudinally.

Moist banks and along stream beds, chiefly Upper Sonoran to Transition Zones, British Columbia to Wyoming and south to Arizona, California, and northern Mexico. Barely entering the Sonoran Desert along the eastern escarpment and southern tip of the Sierra San Pedro Mártir, and in Pima County, Arizona. In our area flowering Feb.–May.

**3. Mimulus glabratus** H. B. K., Nov. Gen. & Sp. **2:** 370, 1817, var. **glabratus**

*Mimulus andicolus* H. B. K., loc. cit.
*Mimulus jamesii* var. *texensis* A. Gray, Syn. Fl. N. Amer. **2**[1]**:** 277, 1878.
*Mimulus tener* Phil., Anal. Mus. Nac. Chile **1:** 70, 1891.

A low, nearly or quite glabrous perennial in aquatic or semiaquatic habitats, often with creeping stolons and rooting freely at nodes, upright branches hollow, weak, usually less than 2 dm. tall; leaves broadly ovate, 1.5–5 cm. long, 0.8–3 cm. broad, irregularly dentate, 3–5-nerved, infrequently slightly lobed at base; petioles equaling or shorter than blades, often slightly winged, uppermost leaves subcordate, sessile; flowers axillary toward tips of branches, on slender to filiform pedicels 1–3.5 cm. long; calyx campanulate, 5–8 mm. long, to 12 mm. long in fruit, often tinged with red, teeth broadly triangular, lower ones 1–1.5 mm. long, upper ones 2–3 mm. long; corolla 7–15 mm. long, yellow, heavily bearded within on floor of throat, middle lobe of lower lip markedly longer than lateral ones; style glabrous; capsule oblong, 7–8 mm. long, rounded at apex and constricted at base; seeds oval, minutely striate longitudinally.

In muddy places or shallow water along streams and about ponds, Lower Sonoran to Transition Zones, Oklahoma, Arizona, and Texas south into South America. Represented in the Sonoran Desert near Sacaton, Arizona, and in the Sierra de la Giganta, Baja California. Flowering nearly throughout the year.

**3*a*. Mimulus glabratus** var. **fremontii** (Benth.) Grant, Ann. Mo. Bot. Gard. **11:** 190, 1924

*Mimulus jamesii* β *fremontii* Benth., in A. DC., Prod. **10:** 371, 1846.
(For additional synonymy see Ann. Mo. Bot. Gard. **11:** 190, 1924.)

Differing from var. *glabratus* in having leaves mostly suborbicular or subcordate or nearly reniform and with less markedly dentate margins; calyx and pedicels often sparsely pubescent.

In shallow water and muddy places, Manitoba to Michigan and Montana, thence south into northern Mexico; reaching the margins of the Sonoran Desert in Arizona and Sonora, where it is commoner than var. *glabratus.*

**4. Mimulus guttatus** DC., Cat. Hort. Monsp. 127, 1813

(For synonymy see Ann. Mo. Bot. Gard. **11:** 157–159, 1924.)

A polymorphous, glabrous to pubescent annual (or under favorable conditions perennial) herb with creeping rootstocks and stolons rooting freely at nodes; stems fistulous, more or less quadrangular, to 6 dm. tall; leaves mostly rounded-ovate, ovate-oblong, or oblong-lanceolate, to 15 cm. long and 8 cm. wide, coarsely and irregularly dentate and often more or less irregularly lobed near base of blade, petioles of basal leaves usually surpassing blades, uppermost leaves sessile, broadly ovate to suborbicular; inflorescence usually racemose, to 2 dm. long or more or dwarfed to 1 to few flowers; pedicels stoutish, about one and one-half times as long as calyx at anthesis; calyx campanulate, often spotted or washed with red, 8–17 mm. long at anthesis, much inflated, papery, and 1–2.5 cm. long in fruit, teeth broadly triangular, upper tooth longer and narrower than rest; corolla yellow, usually spotted with red in throat, 1–4 cm. long, densely hairy on floor of throat, upper lip much shorter than lower one; style glabrous or puberulent; capsule broadly oblong, 8–15 mm. long, chartaceous, constricted or short-stipitate at base; seeds oblong-elliptic, pale brown, finely striate longitudinally.

In moist situations, Montana to Alaska, south to New Mexico, Chihuahua, and northern Baja California. Barely entering the margins of the Sonoran Desert along streams and about seeps in shaded canyons. Flowering through most of the year in our area.

**5. Mimulus nasutus** Greene, Bull. Calif. Acad. **1:** 112, 1885

(For synonymy see Ann. Mo. Bot. Gard. **11:** 177, 1924.)

Annual herb 1–6 dm. tall with puberulent to nearly glabrous or rarely pubescent herbage and often with several branches from base; stems quadrangular, fistulous when robust, frequently narrowly winged on angles; leaves broadly ovate to oblong, to 12 cm. long, coarsely and irregularly dentate or lobed, often with additional small lobes on petiole, petioles of lower leaves clasping, uppermost leaves sessile; inflorescence racemose; pedicels glabrous or nearly so, 8–15 mm. long at anthesis, to 3 cm. long in fruit; calyx narrowly campanulate, 7–10 mm. long at anthesis, 12–18 mm. long in fruit, often appressed-puberulent and with tufts of white short hairs in sinuses between lance-triangular teeth, upper tooth two to three times as long as others, lateral teeth closely infolded over capsule shortly after corolla falls; corolla yellow, dotted and splotched with brown or reddish brown in throat, 7–23 mm. long, densely bearded on floor of throat; style puberulent; capsule oblong, 6–10 mm. long, short-stipitate; seeds finely reticulate in longitudinal lines.

In wet soil on shaded banks, along streams, and in canyons, mainly Upper Sonoran to Canadian Zones, British Columbia to Idaho and south into Chihuahua, Sonora, and Baja California. Barely touching the margins of the Sonoran Desert in shaded canyons in Arizona, and occurring in the canyon in the vicinity of Santa Agueda near Santa Rosalia, Baja California. March–May; Sept.

**6. Mimulus fremontii** (Benth.) A. Gray, Proc. Amer. Acad. **11:** 96, 1876

*Eunanus fremonti* Benth., in A. DC., Prod. **10:** 374, 1846.

A low annual 4–20 cm. tall, usually with several branches from near base; stems and sometimes leaves more or less tinged with red, glandular-pubescent and with an admixture of white shaggy hairs or rarely nearly glabrous, foliage with a strong mephitic odor; leaves oblong, spatulate, or oblanceolate, 2–8 mm. wide, 1–3 cm. long, acute to rounded at apex, tapering to a cuneate sessile base or to a very short petiole above, lower ones usually petiolate, obovate, blades irregularly dentate toward apex; flowers numerous in upper axils; pedicels 1–2 mm. long at anthesis; calyx broadly campanulate, slightly constricted at throat, 8–10 mm. long, teeth nearly equal, broadly triangular, acute to obtuse, 1.5–2 mm. long; corolla 2–2.5 cm. long, rose-red with darker lines and spots in roof of throat, 2 yellow ridges spotted with red on its floor, corolla tube barely equaling calyx, throat ampliate, 13–18 mm. in diameter, rounded lobes somewhat unequal; stamens included, glabrous or filaments of longer pair glandular-pubescent; anthers glabrous; capsule lanceolate, attenuate, often slightly curved, 7–10 mm. long; seeds oblong, apiculate at each end, finely reticulate.

Arid sandy slopes and valley floors, Lower and Upper Sonoran Zones, eastern Monterey County, California, south through the Inner Coast Ranges to the western Mojave Desert and northern Baja California; entering the Sonoran Desert in eastern San Diego County and on the eastern flanks of the Sierra Pinal, Baja California. March–June.

### 7. **Mimulus bigelovii** (A. Gray) A. Gray, Proc. Amer. Acad. **11:** 96, 1876

*Eunanus bigelovii* A. Gray, Pacif. R. Rep. **4:** 121, 1857.

Annual with 1 to several branches 3–25 cm. long from near base, glandular-pubescent throughout, calyces and lower surfaces of leaves often villous with slender tawny hairs; stems and foliage often reddish; leaves thin, elliptic to broadly obovate, 5–12 mm. broad, 15–35 mm. long, acute at apex, gradually narrowed to a slightly winged petiole or a broadly cuneate, subsessile base, margins entire or denticulate toward apex; flowers usually crowded near tips of branches; pedicels 1.5–4 mm. long; calyces oblong-campanulate, 8–12 mm. long, reddish or purplish, throat oblique, teeth subulate from a triangular base, unequal, lower teeth 1–1.5 mm. long, upper one 2–3 mm. long, its tip often slightly spreading; corolla reddish purple, 15–30 mm. long, narrow tube included, throat narrowly funnelform, with a large yellow patch on its floor, abruptly expanded into a spreading limb 15–20 mm. wide, subequal lobes broadly rounded; anthers somewhat hispid; capsule lanceolate, equaling or slightly exceeding calyx tube, sometimes surpassing longest calyx tooth; seeds oblong, 0.8–1 mm. long, rounded at each end, minutely papillate.

Arid plains, washes, and hillsides, Lower and Upper Sonoran Zones, Kern and Los Angeles Counties, California, south into northern Baja California and east into southern Nevada and northwestern Arizona. March–May. This is the commonest *Mimulus* in the Sonoran Desert, occurring frequently throughout the Colorado Desert.

### 8. **Mimulus brevipes** Benth., Scroph. Indicae 28, 1835

*Eunanus brevipes* Greene, Bull. Calif. Acad. **1:** 105, 1885.

An erect annual 1–8 dm. tall with simple or branched stems and glandular-pubescent foliage, stems, and inflorescences; basal leaves oblong or obovate, 5–8 cm. long, 1–4 cm. wide, irregularly dentate or rarely entire; cauline leaves remote, linear-lanceolate, 1.5–5 mm. broad, 2–6 cm. long, sessile, finely dentate or entire; pedicels 2–5 (rarely -13) mm. long; calyx broadly tubular-campanulate, 10–18 mm. long at anthesis, throat strongly oblique and sometimes slightly constricted, teeth triangular to ovate-lanceolate, unequal, lower ones 2.5–5 mm. long, upper one to 8 mm. long, slightly curved upward; corolla yellow, 2–5 cm. long, strongly bilabiate, broadly funnelform, tube proper barely equaling lower side of calyx, expanding abruptly into broad campanulate throat, this ventricose on lower side; corolla lobes spreading, nearly equal, forming a limb 2–4 cm. broad; stamens included, glabrous; capsule ovate-oblong, 9–12 mm. long, 3.5–4.5 mm. in diameter, splitting full length along upper suture, irregularly along lower one; seeds numerous, oblong, apiculate at each end, minutely papillate.

On hillsides, mesas, and ridges, especially common following burns, mainly Upper Sonoran Zone, Santa Barbara County, California, south to the southern flanks of the Sierra San Pedro Mártir, Baja California, where it barely enters the fringe of the Sonoran Desert proper. March–May.

**9. Mimulus aridus** (Abrams) Grant, Ann. Mo. Bot. Gard. **11:** 336, 1924

*Diplacus aridus* Abrams, Bull. Torrey Club **32:** 540, 1905.

A low suffrutescent perennial 2–5 dm. tall with yellowish, somewhat scaly bark and glabrous glutinous herbage; leaves numerous and crowded, oblong, narrowly oblanceolate, or obovate, 4–10 mm. wide, 2–4 cm. long, entire or irregularly denticulate, short-petiolate or sessile, margins often narrowly but tightly revolute, mostly acute at apex, narrowly cuneate at base; flowers numerous in upper axils; pedicels stoutish, 5–6 mm. long, very glutinous; calyx narrowly funnelform, 3–3.5 cm. long, teeth broadly triangular to subulate, unequal, lower teeth 3–5 mm. long, upper one 7–9 mm. long; corolla pale buff to yellow, 4.5–5.5 cm. long, glabrous tube included, expanding into a funnelform throat 1.5–2 cm. long, limb bilabiate, 2–2.5 cm. broad, lobes short, rounded, slightly emarginate; stamens and style exserted, latter sparsely puberulent; capsule narrowly ovoid to oblong, 12–15 mm. long; seeds oval, apiculate at each end, minutely muriculate.

Arid hillsides, Lower and Upper Sonoran Zones, southeastern part of San Diego County, where it barely enters the Sonoran Desert. March–May.

**10. Mimulus pilosus** (Benth.) S. Wats., Bot. King Expl. 225, 1871

*Herpestis pilosa* Benth., Comp. Bot. Mag. **2:** 57, 1836.
*Mimulus exilis* Dur. & Hilg., Pacif. R. Rep. **5:** 12, *pl. 12,* 1855.
*Mimetanthe pilosa* Greene, Bull. Calif. Acad. **1:** 181, 1885.

Annual villous herb with erect or ascending branches 1–4 dm. tall, usually profusely branching from near base; leaves oblong to oblanceolate, 3–15 mm. wide, 1–10 cm. long, gradually narrowed to a sessile base or to a short winged petiole in a few cases, often arachnoid-villous; flowers solitary in axils of most leaves; pedicels 5–15 mm. long at anthesis, becoming as much as 35 mm. long in fruit, slender, villous; calyx 6–7 mm. long, scarcely at all angled, slightly constricted at oblique throat, lobes unequal, shortest about as long as calyx tube, uppermost tooth about twice as long, triangular-lanceolate to nearly subulate; corolla yellow, 4–9 mm. long, tubular-cylindrical with 2 slightly hairy, ventral ridges flecked with maroon, lobes rounded and spreading, limb 3–5 mm. in diameter; filaments minutely pubescent toward tips; anthers glabrous; capsule ovoid, 4–7 mm. long, dehiscing through terminal one-third to one-half of placenta; seeds elliptic, pale brown or yellowish, about 0.3 mm. long, apiculate at base, rounded to truncate at apex, finely reticulate in longitudinal lines.

Stream banks, arroyos, and gravelly bars, upper margins of the Lower Sonoran Zone into the Transition Zone, Washington to the central Sierra San Pedro Mártir, Baja California, and east into Utah and Arizona. April–Sept.

## 10. STEMODIA L., Syst. Nat. ed. 10, 1118, 1759

Annual or perennial herbs or sometimes suffrutescent. Leaves opposite or verticillate. Flowers solitary in axils of leaves or uppermost in foliate or bracteate, spiciform inflorescences, pedicellate; pedicels often 1–2-bracteolate just beneath sepals. Calyx 5-parted to base; sepals narrow, equal or posterior one slightly largest. Corolla tubular-infundibuliform, bilabiate; upper lobe suberect, entire or emarginate; lower one 3-lobed and spreading. Stamens 4, didynamous, included; anthers of 2 distinct, parallel, often slightly stipitate pollen sacs. Style slender, dilated and often 2-lobed at apex. Capsule ovoid, septicidally and often loculicidally dehiscent. Seeds numerous, small, striate or minutely reticulate-pitted.

Plants perennial; larger leaves subsessile and auriculate-clasping; upper flowering stems spiciform; upper leaves reduced to bracts ............................ 1. *S. durantifolia*
Plants annual; leaves all petiolate, not clasping; flowering stems not spiciform; upper leaves scarcely reduced:
  Corollas 8–10 mm. long; sepals 6–8 mm. long; leaf blades with teeth 1–2 mm. long; plant slightly viscid-glandular ............................................ 2. *S. palmeri*
  Corollas 6–7 mm. long; sepals 3.5–5 mm. long; leaf blades with fine teeth 0.5 mm. long; plant very viscid ...................................................... 3. *S. pusilla*

## 1. **Stemodia durantifolia** (L.) Swartz, Obs. Bot. 240, 1791

*Capraria durantifolia* L., Syst. Nat. ed. 10, 1116, 1759.
*Stemodia arizonica* Pennell, Notulae Naturae No. **43:** 3, 1940.

Rather weedy perennial herbs with erect or slightly procumbent, terete or obscurely angled stems 1–5 dm. tall; herbage sparsely to moderately glandular-puberulent, usually darkening in drying; decumbent stems rooting at nodes; lower leaves broadly oblanceolate or oblong-oblanceolate, 7–15 mm. broad, 2–4.5 cm. long, acute or obtuse at apex, slightly auriculate and clasping at base, serrate from apex about two-thirds of way to base; upper leaves (in inflorescence) lance-elliptic, 1.5–3 mm. wide, 0.5–1.5 cm. long, serrulate, subpetiolate; pedicels slender, 1–6 mm. long, puberulent with tack-shaped, glandular hairs; bracteoles linear, 2–3.5 mm. long, 1–2 immediately below calyx; sepals linear-lanceolate, subequal, 4–5 mm. long, glandular-puberulent; corolla 8–10 mm. long, bluish or purplish, tube yellowish, 4–8 mm. long; capsule broadly ovoid, 4–5 mm. long, 2- or 4-valved; seeds dark brown, about 0.3 mm. long, smooth or faintly and very shallowly striate.

Along streams, ditches, and margins of pools, Upper Sonoran to Tropical Zones, southern California and southern Arizona to Texas and south into tropical America. In our area flowering almost throughout the year.

## 2. **Stemodia palmeri** A. Gray, Proc. Amer. Acad. **21:** 403, 1886

Erect, strict to moderately branched annual 1–4 dm. tall with slender, terete, finely and rather sparsely villous stems, many hairs on younger parts gland-tipped; leaves on slender petioles 3–12 mm. long, blade narrowly to broadly ovate, 4–10 mm. wide, 7–18 mm. long, cuneate at base, rather remotely dentate with 2–6 teeth on each side, then sometimes again faintly 1–2-toothed, midrib rather prominent beneath, lateral veins delicate, surfaces puberulent, a few slender long hairs to 1.5 mm. long intermingled with short ones; flowers solitary in axils of only slightly reduced leaves; pedicels subfiliform, 5–12 mm. long at anthesis, often to 2 cm. long in fruit; sepals linear, subequal, 6–8 mm. long, tip of posterior one slightly spreading or reflexed; corolla purplish, 8–10 mm. long, yellowish tube 6–8 mm. long, glabrous, upper lip erect, faintly emarginate, lower lip 3-lobed, each lobe rounded, 1–1.5 mm. long; capsule narrowly ovoid, 3.5–5 mm. long, apex arcuate; seeds obovoid, pale brown, smooth or nearly so, about 0.3–0.4 mm. long.

Sandy meadows, arroyos, and stream sides, Lower and Upper Sonoran Zones, southern Sonora and Durango south into Sinaloa. Nov.–May.

**3. Stemodia pusilla** Benth., Bot. Sulph. 144, 1844

Slender annual herb with 1 to several erect or ascending stems 0.5–3 dm. long, minutely villous and viscid-glandular throughout; stems terete; leaves petiolate, petioles slender, slightly margined toward blade, 2–12 mm. long; leaf blades elliptic to broadly ovate, 2–11 mm. broad, 3–15 mm. long, cuneate basally, acute or obtuse at apex, usually finely dentate with teeth 0.5 mm. long or less, often arachnoid-pubescent and viscid, especially beneath; pedicels filiform, 2–6 mm. long at anthesis, only slightly longer in fruit; sepals linear, 3.5–5 mm. long, posterior one slightly longer than other and faintly dilated at apex; corolla blue, purplish, or whitish, with a yellow tube, 6–7 mm. long, lobes spreading only slightly or nearly erect, upper lip entire or barely emarginate; capsule ovoid, 3–3.5 mm. long, straight; seeds 0.2–0.3 mm. long, brownish, nearly or quite smooth, dull.

Wooded hillsides, canyon walls, and along arroyos, Lower Sonoran Zone to Tropical Zone, Jalisco and Nayarit, and probably southern Baja California. Nov.–Feb.

A specimen from Miraflores, Baja California, is assigned here with some hesitancy, as it has corollas similar to those of *S. palmeri* in size and shape, but all other characters like those of *S. pusilla.*

## 11. CONOBEA Aubl., Hist. Pl. Guiane Franc. 2: 639, *pl. 258*, 1775

Erect annual, or procumbent perennial, glabrous or pubescent herbs with opposite, crenulate-serrate or incised leaves. Flowers usually small, solitary or in pairs in axils, bibracteolate at apex of pedicels, blue or white. Sepals 5, narrowly lanceolate, equal, subequal, or unequal. Corolla tubular, with spreading bilabiate limb; lobes subequal. Stamens 4, didynamous, included; filaments filiform; anthers with parallel locules. Style curved or inflexed at apex; stigma shallowly bilobed. Ovules numerous. Capsule ovoid-globose to oblong or linear, septicidally dehiscent. Seeds many, ovoid or oblong, more or less striate.

Plants annual, roots not indurated; sepals unequal, shortest one-half to two-thirds as long as longest; midribs of sepals inconspicuous ........................ 1. *C. intermedia*
Plants biennial or perennial, roots often considerably indurated; sepals nearly equal; midribs of sepals conspicuous ........................................... 2. *C. polystachya*

**1. Conobea intermedia** A. Gray, in Torr., Bot. Mex. Bound. 117, 1859
*Schistophragma intermedia* Pennell, Notulae Naturae No. **43:** 2, 1940.

Annual herbs branching from base and from several of lower nodes, 0.5–1.5 dm. tall, sparsely to moderately glandular-pubescent with slender spreading hairs, strong-scented; stems slender, 0.5–1.2 mm. in diameter, quadrangular; leaves ovate in outline, 0.5–1.5 cm. long, deeply incised into 3–5 narrow, entire or faintly 2–3-toothed lobes 0.5–2 mm. wide, blade narrowly to broadly cuneate at base, tapering to a narrowly winged petiole nearly or quite equaling blade; flowers blue, axillary; calyx tubular-campanulate, sepals ovate-lanceolate, acute, margins scarious or submembranous, sparsely to moderately glandular-pubescent, midrib inconspicuous, not raised, shorter sepals 1.8–2.5 mm. long, longest (adaxial) one 3–3.8 mm. long at anthesis; corolla 6–7 mm. long, minutely and sparsely puberulent without, tube about 2.5–3 mm. long, yellowish toward base; corolla limb 4–5.5 mm. broad, lobes rounded or truncate; longer stamens about equaling throat; capsule lance-ovoid, 5–6 mm. long, about 1.5 mm. in diameter, tapering toward slightly attenuate apex; seeds oblong, 0.5–0.7 mm. long, purplish black, spirally striate.

Hillsides, outwash slopes, and along arroyos, Lower Sonoran to Arid Transition Zones, Cochise and Pima Counties, Arizona, to New Mexico and south into Sonora and Baja California. Sept.–Oct.

**2. Conabea polystachya** (Brandegee) Minod, Bull. Soc. Genéve II, **10:** 226, 1918

*Stemodia polystachya* Brandegee, Proc. Calif. Acad. II, **2:** 191, 1889.

Perennial, prostrate or decumbent herbs from a slender but indurated root, branches 1–2 dm. long, larger ones to 3 mm. in diameter, quadrangular, densely glandular-pubescent; leaves broadly ovate in outline, sometimes nearly flabelliform, glandular-pubescent with slender spreading hairs, 5–12 mm. wide, 6–10 mm. long, deeply incised into 3–7 oblong or obovate, entire to 3-toothed lobes, base of blade broadly cuneate or truncate; petioles 5–10 mm. long; flowers solitary in axils; pedicels filiform, 1–4 mm. long, glandular-villous; sepals linear-lanceolate, subequal, 4–5 mm. long, glandular-villous, slightly attenuate, midnerve rather prominent; corolla blue or purplish, 6.5–8 mm. long, minutely and sparsely puberulent without, rounded or truncate lobes 1.5–2 mm. long; capsules lance-subulate, 6.5–9 mm. long, 1.5–2.2 mm. in diameter at base, glabrous; seeds oblong-obovoid, about 0.8 mm. long, purplish, spirally striate, ridges low and evenly rounded.

In crevices in rocky canyon walls and hillsides, Lower Sonoran Zone, central Baja California and south to the Cape Region. Brandegee himself reduced the species to synonymy under *Conobea intermedia,* but the perennial prostrate habit, the slightly larger flowers and seeds, the subequal sepals with conspicuous midribs, and the denser villous indument seem to me to warrant its retention as a definite entity. Feb.–April.

## 12. BACOPA Aubl., Hist. Pl. Guiane Franc. 1: 128, *pl. 49*, 1775

Decumbent or erect, glabrous plants of muddy or wet habitats. Leaves opposite, entire or serrate. Flowers solitary in axils, pedicellate, usually bibracteolate. Sepals 5, imbricate, outer much broader and somewhat longer than inner ones. Corolla (in ours) campanulate-tubular, faintly bilabiate; lobes 5, nearly equal, spreading. Stamens 4, subequal or didynamous, included; anther cells distinct, subparallel. Style slender, slightly dilated and only faintly bilobate at apex. Ovules numerous. Capsule ovoid, often bisulcate, dehiscing septicidally, valves entire or bilobed, becoming free from placental tissue. Seeds many, ovoid to truncate-obovoid, smooth or longitudinally ridged.

**1. Bacopa monnieria** (L.) Wettst., in Engler & Prantl, Nat. Pflanzenf. $\mathbf{4}^{3b}$: 77, 1891

*Lysimachia monnieri* L., Cent. Pl. **2:** 9, 1756.
*Gratiola monniera* L., Amoen. Acad. **4:** 306, 1759.
*Bramia radica* Lam., Encycl. **1:** 459, 1785.
*Monnieria cuneifolia* Michx., Fl. Bor. Amer. **2:** 22, 1803.
*Habershamia cuneifolia* Raf., Aut. Bot. 44, 1840.
*Bramia monniera* Drake, Pl. Polyn. Franc. 142, 1893.
*Bramia monnieri* Pennell, Proc. Acad. Phila. **71:** 243, 1920.
*Bacopa monnieria* var. *cuneifolia* Fernald, Rhodora **37:** 440, 1935.
*Bacopa monnieri* Pennell, Proc. Acad. Phila. **98:** 94, 1946.
(For a discussion of the variants "*monnieri*" and "*monnieria*" see Rhodora **37:** 440, 1935; and Proc. Acad. Phila. **98:** 94, footnote 11, 1946.)

Stems widely creeping, 1–5 dm. long or more, 1.5–3.5 mm. in diameter, rooting at most of nodes; leaves cuneate-obovate, 3–9 mm. broad, 5–16 mm. long, rounded at apex, petioles broad, 1–2 mm. long or subsessile; pedicels slender, 0.6–2 cm. long at anthesis, often slightly elongating in fruit; bracteoles lance-linear or subulate, 1–3 mm. long, appressed or tips slightly spreading; outermost sepals broadly ovate, 2.5–4 mm. wide, 4–6 mm. long, inner narrowly lance-ovate, about 1 mm. wide, 4–5 mm. long; corolla white, campanulate, 6–8 mm. long, nearly as wide across limb, glabrous within and without; stigma often only minutely emarginate; capsule ovoid, included within calyx; seeds oblong, about 0.4–0.5 mm. long, one-third or one-half as wide, light brown, longitudinally ridged.

In moist situations, in warmer parts of the world; within the Sonoran Desert, occurring along streams and about ponds, from central Sonora south and in southern Baja California. Flowering year around.

## 13. **MECARDONIA** Ruiz & Pav., Fl. Peruv. Prod. 95, 1794

Perennial herbs with 4-angled, erect or diffuse or rarely creeping stems and opposite, toothed or serrate leaves. Flowers solitary in axils of leaves, pedicels inconspicuously bibracteolate at base. Sepals 5, distinct to base or essentially so, unequal, upper one broadest and often longest. Corolla cylindrical or cylindric-campanulate, white, purple, or yellow, bilabiate; upper lip shallowly notched; lower 3-lobed. Stamens 4, more or less didynamous, adnate to corolla tube almost or quite to throat. Style slender but short; stigma 2-lobed. Capsule ovoid, septicidally dehiscent, valves 2-cleft. Seeds numerous, small.

Plants annual, erect; longest sepal 5–6 mm. long, posterior one about 1.5 mm. broad, faintly serrulate or entire; corolla 6–8 mm. long .............................. 1. *M. exilis*

Plants perennial, procumbent; longest sepal 7–10 mm. long, posterior one 3–4 mm. broad, distinctly serrulate; corolla 8–9 mm. long ....................... 2. *M. vandellioides*

### 1. **Mecardonia exilis** (Brandegee) Pennell, Proc. Acad. Phila. **98:** 87, 1946

*Herpestis exilis* Brandegee, Proc. Calif. Acad. II, **2:** 191, 1889.

Erect annual about 1 dm. tall, with 1 to several erect or ascending branches from base, not branching above, stems, leaves, and calyces glabrous, but leaves and sepals minutely and inconspicuously punctate-glandular; leaves oblong-oblanceolate, 3–7 mm. broad, 7–18 mm. long, finely and shallowly serrulate from about middle or a few of lower ones entire, acute at apex, gradually tapering toward base to a slightly margined petiole 1–5 mm. long or upper ones sessile; flowers solitary in axils, on slender pedicels 1.8–3.2 cm. long, each subtended at base by 2 narrowly oblanceolate bracteoles 4–7 mm. long; sepals subequal, posterior one ovate-lanceolate, 5–6 mm. long, only slightly broader and longer than lower, inner two lance-linear, about 0.5–1 mm. shorter than longest; corolla yellow, veined with purple, 6–8 mm. long, upper lip entire or faintly emarginate, erect or curved upward slightly, about 0.8–1 mm. longer than 3-lobed lower lip, latter slightly spreading, corolla-throat densely glandular-pilose within at base of upper lip, glabrous below; capsule narrowly ovoid, 5–6 mm. long, glabrous; seeds oblong, about 0.3 mm. long, dark brown or black, minutely reticulate-pitted.

Along washes and on arid plains and playas, Lower Sonoran Zone, Magdalena Plain, Baja California. March–April.

**2. Mecardonia vandellioides** (H. B. K.) Pennell, Proc. Acad. Phila. **98:** 87, 1946
*Herpestis vandellioides* H. B. K., Nov. Gen. & Sp. **2:** 367, 1817.
*Pagesia vandellioides* Pennell, Notulae Naturae No. **43:** 2, 1940.

More or less procumbent, perennial herb with slender quadrangular stems 0.5–2.5 dm. long, root stout and woody, herbage glabrous throughout; leaves oblanceolate, elliptic or ovate, 2–10 (rarely –15) mm. wide, 5–23 mm. long, acute or obtuse at apex, gradually or abruptly narrowing at base to a slightly winged petiole 1.5–6 mm. long, blade serrate from near base to apex, inconspicuously veined, minutely dotted with punctate glands; flowers solitary in axils on pedicels 1.5–3 cm. long; bracteoles at base of petioles narrowly lanceolate to oblanceolate, 1–2.5 mm. long; posterior sepal ovate, 3–4 mm. wide, to 10 mm. long, at least in fruit, usually minutely and remotely serrulate, acute at apex; inner sepals linear-lanceolate, they and anterior outer ones nearly as long as posterior sepals; corolla yellowish with darker veining, 8–9 mm. long, glabrous without, upper lip emarginate to shallowly 2-lobed, lower deeply 3-lobed, throat coarsely villous at base of upper lip, glabrous below; capsule ovoid, 7–8 mm. long, glabrous; seeds black, cylindric-oblong, minutely reticulate-pitted, 0.4–0.5 mm. long.

Canyon floors and along streams and arroyos, Lower Sonoran to Tropical Zones, occasionally into the Upper Sonoran Zone, Cochise, Santa Cruz, and Pima Counties, Arizona, south into Mexico. March–May; Sept.

## 14. SCOPARIA L., Sp. Pl. 116, 1753

Herbaceous or fruticose plants with opposite or whorled leaves and glabrous or pilose herbage. Leaves punctate-glandular. Flowers small, yellowish white or pale blue, axillary. Calyx 4–5-parted to base, sepals ovate or lanceolate. Corolla subequally 4-lobed; tube much shorter than rotate imbricate lobes; throat densely villous. Stamens 4, subequal; pollen sacs distinct, parallel or divergent. Style slightly clavate toward apex; stigma truncate, emarginate. Capsule ovoid to globose, septicidally dehiscent; valves membranous, entire, inflexed margins separating from placenta. Seeds numerous, obovoid, somewhat angled; testa scrobiculate or reticulate-pitted.

**1. Scoparia dulcis** L., Sp. Pl. 116, 1753
*Gratiola micrantha* Nutt., Amer. Jour. Sci. **5:** 287, 1822.
*Scoparia grandiflora* Nash, Bull. Torrey Club **23:** 105, 1896.

A much-branched annual or short-lived perennial (?), weedy plant to 6 dm. tall or more, with nearly glabrous stems, leaves, and inflorescences, a few fine hairs often on upper side of bases of petioles and on stems immediately below nodes; leaves lance-elliptic or narrowly oblanceolate, 2–12 mm. broad, 1.5–5 cm. long, narrowly cuneate at base, short-petiolate, acute and often apiculate at apex, serrate or serrulate with slender incurved teeth from apex about three-fourths of way to base, finely punctate-glandular; flowers 1–3 in axils of verticillate reduced leaves; pedicels filiform, 3–8 mm. long; sepals broadly ovate, 2–2.5 mm. long, about 1 mm. wide; corolla white or nearly so, tube less than 0.5 mm. long, lobes rotately spreading, about 3 mm. long, rounded, densely villous at base with tangled hairs 0.5–1.5 mm. long; capsule subglobose-ovoid, about 3 mm. high, glabrous; seeds dark brown, truncate, reticulate-pitted in longitudinal rows.

Wayside weed and along streams and ditches, mainly Tropical Zone, coast of the Gulf of Mexico from Florida west to Texas, in the West Indies, and throughout much of tropical America. Occurring in southern Baja California and to be expected in damp places in the southern half of the Sonoran Desert. Nov.–April.

## 15. VERONICA L., Sp. Pl. 9, 1753

Annual or perennial herbs or rarely lignescent plants with opposite or less commonly verticillate leaves or upper ones sometimes alternate, often reduced above. Flowers in terminal, racemiform or axillary inflorescences, ebracteolate. Calyx 4–5-parted, with imbricate sepals. Corolla tube short, rarely equaling calyx; limb 4–5-lobed, spreading; lateral lobes outside in bud slightly wider than anterior and posterior lobes. Stamens 2, exserted; anther sacs divergent or parallel, confluent at apex. Capsule compressed or nearly subglobose, bisulcate, loculicidally dehiscent. Seeds few to many, ovate to globose, smooth to rugulose.

Leaves linear to oblong-spatulate, three to five times as long as wide; pedicels shorter than sepals; seeds 0.3–0.4 mm. long, smooth or faintly pitted .................................................. 1. *V. peregrina* subsp. *xalapensis*

Leaves broadly ovate, not over one and one-half times as long as broad; pedicels several times as long as sepals; seeds 1.5–2 mm. long, transversely rugulose ........... 2. *V. persica*

### 1. **Veronica peregrina** L. subsp. **xalapensis** (H. B. K.) Pennell, Torreya **19:** 167, 1919

*Veronica xalapensis* H. B. K., Nov. Gen. & Sp. **2:** 389, 1817.
*Veronica sherwoodii* M. E. Peck, Torreya **28:** 56, 1928.

Erect annual 0.5–4 dm. tall, often ascendingly branched from lower nodes, finely and moderately glandular-puberulent throughout, or older stems and leaves glabrate; leaves linear to oblong-spatulate, 1–5 mm. wide, 6–26 mm. long, entire or remotely and shallowly serrate, sessile or narrowed to a short margined petiole; flowers axillary to reduced leaves, forming terminal racemiform inflorescences to 2 dm. long; pedicels 1–1.5 mm. long; sepals linear-lanceolate to narrowly ovate, 3–4 mm. long, glabrous or sparsely glandular-puberulent; corollas whitish with purplish or bluish veins, rounded lobes rotate, 1.5–2 mm. long; capsule obcordate, distinctly flattened, 3–4 mm. high and wide; seeds numerous, 0.3–0.4 mm. long, pale brown, smooth or slightly lined or pitted.

A widespread weedy plant, particularly along streams, ditches, on seepy hillsides, and margins of ponds, Tropical to Transition Zones, British Columbia to Newfoundland south, mainly west of the Mississippi River, to central America. Occurring in Arizona, Sonora, and Baja California. March–April in our area.

### 2. **Veronica persica** Poir., in Lam., Encycl. **8:** 542, 1808

*Veronica precox* Raf., Atl. Jour. 79, 1832.
*Veronica diffusa* Raf., New Fl. **4:** 38, 1838.

Decumbent-creeping plants or some stems upright, 1–4 dm. long, stems and herbage irregularly and sparsely arachnoid-pubescent with a mixture of fine and coarse hairs in patches and lines; leaves mostly alternate above base of stems, on short, faintly margined petioles 1–8 mm. long; leaf blades broadly ovate, 4–12 mm. broad, 5–15 mm. long, truncate at base, obtuse or rounded at apex, dentate-serrate with 3–6 teeth on each side; flowers axillary to upper leaves, these only slightly reduced; pedicels slender, 1–3 cm. long, tips more or less arcuate-recurved in fruit; sepals broadly ovate or elliptic, 4–5 mm. long, to 3.5 mm. wide, rather prominently veined, ciliate-margined; corolla with dark bluish posterior lobes, anterior ones paler, nearly orbicular, 4–5 mm. long, slightly hairy at base; stamens about 4 mm. long; capsules broadly obcordate, tumid, about 4 mm. high, 5–8 mm. wide; seeds 1.5–2 mm. long, deeply concave-excavated on inner face, rounded and transversely rugulose or pitted on back.

Established as a weed in lawns, parkways, and roadsides at various places throughout the United States and southern Canada, where it was introduced from Eurasia. Occurring within our area at Tempe, Maricopa County, Arizona.

**Veronica americana** Schwein., in Benth., in A. DC., Prod. **10:** 468, 1846, with flowers in axillary racemes and growing in aquatic or semiaquatic habitats, occurs in Arizona slightly north of our area, but has not been reported, as yet, from within the Sonoran Desert.

## 16. CASTILLEJA Mutis ex L. f., Suppl. Syst. Veg. 47, 1781

Erect or ascending, annual or perennial herbs or sometimes suffrutescent at base, with alternate leaves and spikelike racemes. Bracteoles none. Bracts entire or variously lobed or divided, usually colored, at least at tip. Calyx of 2 or 4 lobes, these unequally separated, usually lateral sinuses much shallower than posterior and anterior ones. Corolla greenish or tinged with red or yellow along margins of galea, 2-lipped; upper (galea) united to tip into a narrow, hoodlike structure that usually includes or enfolds stamens and style; lower lip short, often rudimentary. Stamens 4, didynamous; anther cells attached unequally. Capsule ovoid to cylindric-ovoid, turgid, glabrous, loculicidally dehiscent. Seeds many, cylindrical to irregularly angled; testa loose and reticulate, thin.

Plants annual; bracts exceeding flowers:
  Leaves lance-linear, entire; corolla 25–35 mm. long .................... 1. *C. stenantha*
  Leaves deeply dissected into 3–9 linear lobes; corolla 15–18 mm. long .... 2. *C. bryantii*
Plants perennial; bracts barely equaling or much shorter than flowers:
  Stems sparsely hirsute or glabrate, not lanate-pubescent; inflorescence markedly glandular:
    Leaves (at least some of them) 2–3-parted; galea strongly pubescent dorsally; lower lip 2–4 mm. long; calyx cleft more deeply before than behind ........... 3. *C. affinis*
    Leaves all entire, lance-linear; galea finely puberulent dorsally; lower lip 1–2 mm. long; calyx cleft about equally before and behind .................... 4. *C. californica*
  Stems densely pubescent, often lanate; inflorescence usually not conspicuously glandular:
    Corolla 18–25 mm. long; older parts of stems densely clothed with axillary fascicles of small leaves 3–8 mm. long .................................... 5. *C. foliolosa*
    Corolla 25–40 mm. long; older stems rarely with axillary fascicles of reduced leaves:
      Calyx equally cleft before and behind, lobes 2-toothed to depth of 5–6 mm.; pubescence of tangled, mostly branched hairs; corolla 2.5–3 cm. long ........ 6. *C. integra*
      Calyx much more deeply cleft before than behind, lobes entire or faintly 2-toothed; pubescence mainly of coarse unbranched hairs; corolla 2.5–4 cm. long .. 7. *C. laxa*

### 1. Castilleja stenantha A. Gray, Syn. Fl. N. Amer. $2^1$: 295, 1878

Erect annual with slender, sparsely pubescent to glabrate, often purplish stems 2–10 dm. tall, some gland-tipped hairs often interspersed among spreading glandless ones; stems simple or branched; leaves lanceolate-attenuate, entire, 3–12 mm. wide at base, 2–10 cm. long, sparsely to densely scaberulous, especially on midrib and margins; bracts leaf-like, erect and scarlet-tipped at apex of spike but becoming green and spreading by time subtended flower opens, 2–3 cm. long at anthesis; calyx green, 22–35 mm. long, cleft before and behind one-half to two-thirds its length, lobes 2-toothed or -cleft to depth of 0.5–4 mm., teeth acute or attenuate; corolla greenish, yellowish, or suffused with red, 25–35 mm. long, galea 15–20 mm. long, slender, dull yellow and finely puberulent dorsally, narrow margins thin, ciliate; lower lip 2–3 mm. long, yellow, spreading at right angles to corolla tube; capsule narrowly ovoid, 10–12 mm. long; seeds elliptic, about 1 mm. long, light gray, testa foveolate in longitudinal lines.

Moist soil along streams and in marshy places, Lower Sonoran to Transition Zones, in the foothills and lower mountains, Lake and Fresno Counties, California, south to the western flanks of the Sierra San Pedro Mártir, Baja California. Barely reaching the Sonoran Desert in the vicinity of Valle Trinidad, between the Sierra Juárez and the Sierra San Pedro Mártir. Feb.–Aug.

**2. Castilleja bryantii** Brandegee, Proc. Calif. Acad. II, **2:** 192, 1889

Slender annual 1.5–6 dm. tall, strict or branching from near base, sparsely hirsute with spreading hairs 0.5–1 mm. long, slender stems often reddish or purplish; leaves from lower 2–3 nodes linear, entire, 1–1.5 mm. wide, 1–3 cm. long, those above divided into 3–9 linear lobes 3–20 mm. long, central rachis 1–2.5 mm. broad, sparsely pubescent; spikes slender, 2–6 cm. long in flower, much elongated (to 2.5 dm.) in fruit; floral bracts 1.5–2.5 cm. long, deeply dissected into 3–7 linear lobes, their tips usually reddish or yellowish; calyx 16–20 mm. long, slender, cleft equally in front and behind about halfway to base, each lobe cleft one-third to one-half its length into linear segments 0.5–1.5 mm. wide, red-tipped or yellow-tipped, segments on upper side slightly wider than lower pair; corolla 15–18 mm. long, pale yellow to reddish, tube about twice as long as short slender galea, lower lip with 3 short incurved lobes; stamens 2, 3, or 4 in same plant; capsule oblong-ovoid, 6–9 mm. long; seeds about 0.5 mm. long, dark gray, thin testa foveolate-reticulate, rather close.

Rocky hillsides and along arroyos, Lower and Upper Sonoran Zones, central Baja California south into the Cape Region. March–April.

**3. Castilleja affinis** Hook. & Arn., Bot. Beechey 154, 1833

Erect or ascendingly branched perennial with sparsely hirsute, reddish stems and leaves, stiffer scabrous hairs often intermingled with and exceeded by softer ones; stems 4–6 dm. tall, slightly woody at base; leaves lanceolate to narrowly oblong, 3–5 cm. long, entire or with 1–2 pairs of slender spreading lobes 2–15 mm. long, 0.5–1 mm. wide, main leaves often subtending small fascicles of reduced leaves 2–5 mm. long; spikes to 10 cm. long; bracts 2–3 cm. long, usually bearing 1 pair of narrow lobes 2–6 mm. long near bright red tips, pubescent or scaberulous; calyx 2–3 cm. long, cleft about to middle in front, usually slightly less deeply so behind, lobes cleft laterally 4–10 mm. into obtuse or acute, linear or oblong teeth, upper half of calyx bright red; corolla slender, 25–35 mm. long, galea 15–20 mm. long, attenuate to apex, densely pubescent with ascendingly appressed hairs dorsally, thin margins bright red or yellowish, lower lip thickened, dark green, 2–4 mm. long, usually well exserted beyond calyx tube; capsules 10–15 mm. long, acute or short-acuminate; seeds cylindric-truncate, 1.5–2.2 mm. long, light brown, loose testa deeply and coarsely foveolate-reticulate.

Sandy or rocky slopes, often in the chaparral, Upper Sonoran and Transition Zones, Del Norte and Shasta Counties, California, south into northern Baja California, where it enters the Sonoran Desert between Rosario and El Mármol. Jan.–June.

**4. Castilleja californica** Abrams, Bull. S. Calif. Acad. **1:** 68, 1902

Erect or ascendingly branched perennials 3–8 dm. tall, pilose with straight glandless hairs on lower stems and leaves, inflorescence often moderately glandular in addition to nonglandular hairs; leaves linear or narrowly lanceolate, 1–2.5 mm. wide, 1–3 cm. long, entire or rarely a few of upper ones with 1–2 pairs of slender lobes, lower ones usually subtending fascicles of much reduced leaves 2–4 mm. long; bracts lanceolate, about 2 cm. long, entire or with a pair of slender lobes near apex, tips bright red; calyx 12–27 mm. long, red-tipped, equally cleft before and behind about to middle or slightly below, each lobe with 2 lanceolate teeth 3–5 mm. long; corolla 15–45 mm. long, slender, arcuate, galea about equal to or slightly longer than tube, finely puberulent dorsally, margins thin and reddish, lower lip dark green or brown, 1–2 mm. long, usually exserted beyond calyx tube; capsule ovoid-acuminate, 10–15 mm. long; seeds angled or somewhat pyramidal, 1.5–2 mm. long, pale buff, deeply foveolate-reticulate.

Mostly in chaparral or open forest, Upper Sonoran and Transition Zones, coastal slopes of the mountains from San Luis Obispo County California, into northern Baja California; barely entering the Colorado Desert in the vicinity of Jacumba and Mountain Springs, San Diego County, and in Imperial County, California. Feb.–July.

**5. Castilleja foliolosa** Hook. & Arn., Bot. Beechey 154, 1833
*Castilleja clementis* Eastw., Bull. Torrey Club **32:** 212, 1905.

Erect perennial with 1 to several stems from a heavy woody root, white-tomentose throughout with finely branched, tangled hairs; stems woody at base, forming clumps of herbaceous stems 2–6 dm. tall, older branches bearing crowded fascicles of small woolly leaves 3–8 mm. long in axils of primary leaves; primary leaves linear or linear-lanceolate, 1–2.5 mm. wide, 12–25 mm. long, entire, acute or obtuse at apex, floccose-tomentose; spikes usually short, 1–3 cm. long in flower, elongating to 10 cm. in fruit; bracts red-tipped, usually 3–5-lobed, 12–18 mm. long, central lobes 2–3 mm. broad, lateral ones about 1 mm. broad; calyx red-tipped, a yellow band between red tip and green base, 16–21 mm. long, equally cleft before and behind about two-fifths its length, lobes rounded-truncate or shallowly emarginate, lanate; corolla 18–25 mm. long, galea 7–15 mm. long, minutely pubescent dorsally, its margins thin, pale red, lower lip dark green, about 2 mm. long, hidden within calyx tube; capsule ovoid, acute, 8–15 mm. long; seeds ovoid, 1.5–1.8 mm. long, testa whitish, deeply foveolate-reticulate, loose.

Sandy margins of arroyos, rocky hillsides, or gravelly outwash fans, mostly Upper Sonoran and Transition Zones, Coast Ranges and Sierra Nevada from Humboldt and Tuolumne Counties, California, south to the Sierra San Pedro Mártir, Baja California. Reaching the western margins of the Colorado and San Felipe Deserts. Jan.–July.

**6. Castilleja integra** A. Gray, in Torr., Bot. Mex. Bound. 119, 1859

An erect or ascending perennial with 1 to several stems 1–4 dm. long from a woody root; stems wholly herbaceous and densely arachnoid-pubescent to white-lanate; hairs fine, simple or few-branched; leaves linear, oblong, or linear-lanceolate, 1–4 mm. wide, 2–8 cm. long, often somewhat arcuate, moderately to densely pubescent, often glabrate or subglabrate on upper surface in age; spikes 3–12 cm. long, often 5 cm. in diameter, elongating in fruit; bracts lanceolate to ovate-lanceolate, 2–4 cm. long, entire or sometimes with 1–3 lobes, bright red at apex; calyx 2–2.5 cm. long, about equally cleft two-fifths its length before and behind, 2 lobes bifid about 5–6 mm., teeth lance-oblong, acute or obtuse, red-tipped; corolla 2.5–3 cm. long; galea 10–15 mm. long, usually shorter than tube, greenish red and finely puberulent on back, margins thin, bright red; lower lip short, 1.5–2 mm. long, greenish teeth curved upward and inward; capsule ovoid, 10–14 mm. long; seeds cylindrical, truncate at ends, about 1.5 mm. long, one-half as wide, thin brownish testa finely foveolate.

Rocky hillsides, upper margins of the Lower Sonoran Zone into the Transition Zone, mostly among oaks and pines, Pima County, Arizona, into Colorado, New Mexico, and adjacent northern Chihuahua and Sonora. Barely entering the Sonoran Desert. A single, rather meager specimen from near Laguna Chapala Seca, Baja California, seems to belong here, but the determination is questionable. March–Sept.

### 7. **Castilleja laxa** A. Gray, in Torr., Bot. Mex. Bound. 119, 1859

Ascendingly branched perennial with several slender stems 1–5 dm. tall from a woody rootstock, stems canescent with a mixture of short and long hairs, or somewhat lanate, some glandular pubescence in inflorescence; leaves linear or oblong, 1.5–5 mm. wide, 1.5–4 cm. long, entire, acute or obtuse, more or less scabrous in youth, often glabrate above in age, often subtending small fascicles of leaves 3–6 mm. long; spikes rather broad, 3–4 cm. long, to 6 cm. broad; bracts lanceolate or ovate-lanceolate, 1.5–2 cm. long, red-tipped at first, later green; calyx 20–26 mm. long, cleft about to middle in front, about one-half as deeply behind, lobes broadly rounded, truncate, entire or slightly emarginate, brightly red-tipped; corolla 25–40 mm. long, galea about equaling tube, densely puberulent dorsally near tip, sparsely so below, margins bright red, lower lip 1.5–3 mm. long, greenish or brownish; capsules broadly ovoid, 6–9 mm. long; seeds truncate-cylindrical, about 1 mm. long, pale buff, finely foveolate-reticulate.

Rocky hillsides, under oaks and pines and along canyons, mainly Upper Sonoran and Transition Zones, southern Arizona and northern Sonora. In the Sonoran Desert near Tucson, Arizona. April–Oct.

## 17. **DICRANOSTEGIA** (A. Gray) Pennell, Proc. Acad. Phila. **99**: 189, 1947

Erect, ascending, or procumbent, summer-blooming annual with rather brittle, paniculately branching stems and yellow roots. Leaves alternate, entire to dissected. Flowers in spikes, head, or solitary in axils of leaves, each subtended by an entire, parted, or dissected bract. Calyx bifid into 2 linear-attenuate, laterally placed lobes (each comprised of 2 wholly united sepals). Corolla cylindrical with upper part expanded laterally to form a flattened, beak-like structure, bilabiate; upper lip enclosing pistil and stamens; lower one equaling upper or nearly so, obscurely 3-crenulate or entire, somewhat saccate. Stamens (in ours) 4, didynamous; anthers 1–2-celled, one cell attached below other when two are present. Capsule lanceolate or ovoid, more or less flattened. Seeds irregularly angled, reticulate-pitted.

### 1. **Dicranostegia orcuttiana** (A. Gray) Pennell, Proc. Acad. Phila. **99**: 190, 1947

*Cordylanthus orcuttianus* A. Gray, Proc. Amer. Acad. **19**: 95, 1883.
*Adenostegia orcuttiana* Greene, Pittonia **2**: 181, 1891.

Divaricately branched, erect or decumbent annual 1–4.5 dm. tall with stoutish, densely hirsute, often reddish stems, pubescence sparse in age; leaves 2–6 cm. long, deeply dissected into linear divisions 1 mm. or less wide, central rachis to 3.5 mm.

wide near base, this and segments 1-nerved, densely to sparsely hispid with white hairs to 1 mm. long; spikes dense, 2–6 cm. long, elongating to nearly twice that length in fruit; floral bracts oblong, 20–25 mm. long, rounded at tip, with 2–3 pairs of linear lobes 0.5–1 mm. wide and 3–10 mm. long borne along sides, setose-hispid along margins and on veins; calyx lobes 2, each about 6 mm. long, linear-acuminate, pubescent on margins with soft hairs intermingling with stiffly spreading ones; corolla about equaling bracts, tube about 10 mm. long, throat 5–7 mm. long, soft-pubescent, white, spotted with purple, galea greenish yellow; stamens 4, anthers 1-celled or with a vestige of second cell, upper cell sterile, lower functional; capsule slightly asymmetrical, 6–7 mm. long; seeds 1–1.5 mm. long, coat loose, deeply reticulate-pitted, ridges very thin, brownish.

On grassy hillsides and along arroyos, Upper and Lower Sonoran Zones, near Tijuana, Baja California, south to the vicinity of Rancho Mesquital and Miller's landing. March–Sept.

## 18. ORTHOCARPUS Nutt., Gen. 2: 56, 1818

Annual herbs with glabrous or pubescent, erect or procumbent stems and alternate or rarely opposite leaves, latter usually dilated at base and dissected above, upper ones often colored at tips. Flowers in terminal, spiciform, bracteate racemes or solitary in axils of leaves. Calyx tubular-campanulate, more or less membranous at base, 4-lobed or -toothed. Corolla tube slender, limb strongly bilabiate, upper lip erect, galeate, margins more or less inflexed; lower lip much shorter and 3-plicate or -saccate, terminating in 3 erect teeth. Stamens 4, didynamous, ascending into galea; anthers 1–2-celled, if two present one terminal on filament, other attached medianly a short distance below terminal cell. Style filiform; stigma entire. Capsule ovate to oblong, loculicidally dehiscent. Seeds few to many, reticulate-foveolate.

**1. Orthocarpus purpurascens** Benth., Scroph. Indicae 13, 1835

*Orthocarpus purpurascens* var. *palmeri* A. Gray, Syn. Fl. N. Amer. **2**[1]: 300, 1878.
*Orthocarpus purpurascens* var. *typicus* Keck, Proc. Calif. Acad. IV, **16**: 541, 1927.

Stems simple to diffusely branched from near base and closely ascending, 1–4 dm. tall, villous-pubescent throughout; leaves sessile, 1–5 cm. long, parted into few to many linear or filiform divisions 1–12 mm. long or lower ones entire, linear, villous-pubescent with shining, white, spreading hairs; spikes 2–20 cm. long, dense; bracts 10–20 mm. long, central portion lanceolate, bearing 2–4 pairs of linear or filiform divisions palmately or pectinate-ascendingly disposed, each lobe or at least upper ones with crimson to purple, spatulate-dilated tips, conspicuously pilose with shining white hairs at base; calyces 12–20 mm. long, 4-lobed about to middle or slightly below, lobes resembling those of bracts in shape and color; corolla 12–30 mm. long, purplish or crimson, lower lip purple, crimson, pink, yellow, or white but usually purple tipped with yellow, 3–5 mm. wide, 3–7 mm. long, shallowly to deeply saccate with 3 minute teeth about 1 mm. long; upper lip galeate, 2–3 mm. longer than lower lip, abruptly hooked, crimson or purple, densely pubescent dorsally and tipped with a brush of soft purplish hairs; stamens enclosed by galea; anthers 1.5–2 mm. long, sparsely puberulent; capsule ovoid, 8–15 mm. long; seeds ovate to elliptic, 0.5–0.8 mm. long, loose coat reticulate-foveolate.

Grassy valley floors and hillsides, Lower and Upper Sonoran Zones, upper Sacramento Valley and adjacent foothills south to the vicinity of Rosario and San Augustín, Baja California; east into Arizona and northwestern Sonora. March–May.

## 19. CLEVELANDIA Greene, Bull. Calif. Acad. 1: 182, 1885

Small annual with erect or ascending, slender stems and alternate, deeply dissected leaves. Inflorescence slender, bracteate, spiciform racemes terminating branches. Calyx tubular-campanulate, 4-cleft. Corolla slender; narrow tube about equaling gradually flaring throat; limb shallowly bilabiate; upper lip short, spreading, slightly bifid; the lower lip with 3 spreading plane lobes, not at all saccate nor plicate. Stamens 4, didynamous, included; anthers with 2 unequal cells. Style filiform; stigma entire. Capsule oblong, loculicidally bivalved. Seeds numerous, small, minutely but evenly foveolate-reticulate; testa rather close.

### 1. **Clevelandia beldingii** (Greene) Greene, Bull. Calif. Acad. 1: 182, 1885

*Orthocarpus beldingi* Greene, op. cit. 123.

Erect and strict or branching moderately from base upward; stems slender, moderately hirsute throughout, 1–4 dm. tall; leaves 1–2.5 cm. long, central rachis 0.5–1.5 mm. broad, with 1–3 pairs of linear lobes 1–5 mm. long or upper ones entire, hirsutulous and with numerous gland-tipped hairs intermingled with nonglandular ones; spikes 2–10 cm. long, about 1.5 cm. in diameter, inflorescence quite glandular-puberulent; bracts 7–10 mm. long, mostly slenderly 3-lobed, median one about twice as long as spreading laterals, glandular-puberulent; calyx tube 6–8 mm. long, 4 linear lobes about equaling it, erect or irregularly spreading; corolla 11–15 mm. long, white with a tinge of pink toward apex, faintly arcuate, mouth about twice as broad as diameter of corolla tube, upper throat faintly plicate and minutely puberulent on outer surface; capsules 5–6 mm. long, about 2 mm. in diameter, acute; seeds pale gray, elliptical, 0.4–0.6 mm. long, minutely reticulate-foveolate.

Sandy banks and beds of arroyos and on outwash fans, Lower Sonoran Zone, southern Baja California from the southern Sierra de la Giganta southward. Flowering nearly throughout the year when rainfall permits growth.

## 20. LAMOUROUXIA H. B. K., Nov. Gen. & Sp. 2: 335, *pls. 167–169,* 1817

Erect, scandent, or decumbent, annual or perennial herbs with opposite leaves. Flowers showy, variously arranged. Calyx campanulate, 4-lobed, often somewhat compressed. Corolla tube elongated, compressed or ventricose; limb distinctly bilabiate; upper lip erect, more or less galeate, entire or slightly emarginate; lower lip shorter, often twice plicate, divided into 3 spreading oblong lobes. Stamens 4, didynamous, enclosed within galea. Style entire; stigma obtuse. Capsule ovoid, loculicidally dehiscent into 2 entire valves. Seeds numerous, oblong; testa loose or closely appressed, reticulate-foveolate.

### 1. **Lamourouxia viscosa** H. B. K., Nov. Gen. & Sp. 2: 338, 1817

*Lamourouxia cordata* Schlecht., Linnaea 5: 103, 1830.
*Lamourouxia coccinea* A. Gray, Proc. Amer. Acad. 21: 404, 1886.

A rather sturdy herb to 1.5 m. tall from a stoutish woody root, densely viscid-pubescent throughout with intermingled, tack-shaped, glandular hairs and slender nonglandular ones; leaves elliptic to lanceolate, sessile, to 3.5 cm. wide, 3–10 cm. long, acute at apex, obtuse to subcordate at base, sharply dentate, teeth acuminate, prominently veined, usually turning dark brown in drying; inflorescence of terminal bracteate racemes 1–2 dm. long; bracts broadly ovate to suborbicular, 1.5 cm. long or usually less, dentate; pedicels 1–5 mm. long, densely glandular-pubescent; calyx 6–7 mm. long, ovate-acuminate lobes about equaling tube; corolla bright red, 3.5–4.5 cm. long, densely pubescent without, straight, sinuses between upper and lower lips 1–1.5 cm. deep; capsules ovoid-acuminate, 7–10 mm. long, 5–6 mm. in diameter; seeds about 1 mm. long, narrowly oblong, loose testa deeply reticulate-foveolate.

Hillsides and canyons, margin of Lower Sonoran Zone into Tropical Zone, Sierra Batúc, Sonora, south into central and southern Mexico. Sept.–Dec.

## Family 127. **BIGNONIACEAE.** Bignonia Family

Trees, shrubs or woody vines with opposite or ternate, simple or compound leaves, terminal leaflet often modified into a tendril. Flowers usually large and showy, perfect, somewhat irregular. Calyx tubular or campanulate, often closed in bud, truncate, 5-toothed or spathaceous after anthesis. Corolla sympetalous, straight or more commonly somewhat curved, faintly bilabiate or lobes only slightly unequal, rounded and rotate or spreading. Stamens mostly 4, inserted near apex of corolla tube or at base of throat, usually included. Fruit baccate or a loculicidal or septicidal capsule. Seeds (in ours) flattened and membranously winged.

Plants scandent vines; stems often with aerial roots; terminal leaflets often replaced by a trifid tendril; each branch often ending in a sharp hook . . . . . . . . . . . . . . . . . . . . . 1. *Bignonia*
Plants erect shrubs or trees; aerial roots none; leaflets all photosynthetic, not replaced by tendrils:
  Flowers white, pale pink or faintly purplish; leaves simple, entire; fringes of seed-wings finely dissected into long, silky hairs or fimbriae . . . . . . . . . . . . . . . . . . . . 2. *Chilopsis*
  Flowers yellow or deep pink; leaves pinnately or digitately compound; seed-wings entire or erose, not hairy-margined:
    Plants trees to 20 m. tall; leaves digitately compound; flowers deep pink .. 3. *Tabebuia*
    Plants shrubs or small trees, rarely to 8 m. tall; leaves pinnately compound; flowers bright yellow . . . . . . . . . . . . . . . . . . . . . . . . . . . . . . . . . . . . . . . . . . . . . 4. *Tecoma*

### 1. **BIGNONIA** L., Sp. Pl. 622, 1753

Scandent or climbing vines with stems woody at base, often herbaceous above. Leaves opposite, bifoliolate, at least some of them with tendrils, rarely simple; leaflets usually petiolulate, entire or less commonly dentate. Flowers in terminal panicles or axillary cymes. Calyx campanulate or tubular, truncate or oblique at apex or variously toothed. Corolla tube often elongate, expanding above into a funnelform throat, straight or curved; limb distinctly or obscurely bilabiate; upper lip 2-lobed; lower 3-toothed or 3-parted, all lobes broad or rarely oblong, spreading. Stamens 4, didynamous, included or slightly exserted; anthers glabrous, oblong or linear locules divaricate, or rarely parallel. A soft, fleshy disk present. Capsule linear, compressed parallel to partition between two locules, septicidally dehiscent. Seeds thin, usually with a hyaline wing, arranged in 1 or 2 series in each cell.

**1. Bignonia unguis-cati** L., Sp. Pl. 623, 1753

*Bignonia pubescens* L., Sp. Pl. ed. 2, 870, 1763.
*?Doxantha mexicana* Miers, Proc. Hort. Soc. Lond. **3:** 190, 1863.
*Bignonia acutistipula* Schlecht., Linnaea **26:** 375, 1854.
*Bignonia lanuginosa* Hemsl., Biol. Centr. Amer. Bot. **2:** 491, 1882.
*Bignonia californica* Brandegee, Zoe **5:** 170, 1903.

A woody vine climbing to 15 m. or more; stems often with aerial roots; bark pale tan or grayish, smooth; leaves trifoliolate, terminal leaflet often replaced by a trifid tendril; branchlets often bearing a terminal, stout hook; petioles 5–12 mm. long, glabrous; leaflets ovate or ovate-lanceolate, 1–3 cm. wide, 3–8 cm. long, rounded or obscurely subcordate at base, acute to acuminate at apex, entire or irregularly and faintly dentate, leathery, dark green and glabrous above, paler and sparsely puberulent beneath; flowers in axillary cymes or sometimes solitary; pedicels 1–2.5 cm. long; calyx broadly and obliquely campanulate, 10–15 mm. long at anthesis, faintly and broadly 4-lobed; corolla yellow, 4–5 cm. long, limb somewhat bilabiate, 3–4.5 cm. broad, lobes broad, rounded, 1–1.5 cm. long, 1.5–2 cm. wide; capsule strongly flattened, 10–14 mm. broad, to 4 dm. long, dark brown, dull and finely pustulate, inner surface often lustrous; seeds about 5–6 mm. long, 10–12 mm. broad, hyaline wing to 2.5 cm. long.

Climbing over trees, Lower Sonoran to Tropical Zones, southern Sonora and Cape Region, Baja California, throughout Mexico and Central America into northern South America and the West Indies. Flowering after rains.

## 2. **CHILOPSIS** D. Don, Edinb. Phil. Jour. **9**: 261, 1823

An erect shrub with several slender branches arising from the base. Leaves linear to lanceolate, entire. Flowers in racemes or narrow thyrsoid, terminal panicles. Calyx ovoid-campanulate, inflated, deeply 2-lipped; upper lip 3-toothed; lower bidentate. Corolla funnelform-ampliate; tube proper about equaling calyx; throat gradually expanding; limb 5-lobed and bilabiate, lobes undulate-crispate and often erose. Stamens 4, didynamous; filaments incurved, included; anthers glabrous, locules linear, divaricate. Ovary sessile; ovules many. Capsule linear, subterete, loculicidally dehiscent into 2 concave valves. Seeds flattened; wings finely divided into hairlike segments.

### 1. **Chilopsis linearis** (Cav.) Sweet, Hort. Brit. 283, 1827

*Bignonia? linearis* Cav., Ic. **3**: 35, *pl. 269,* 1794.
*Chilopsis saligna* D. Don, Edinb. Phil. Jour. **9**: 261, 1823.
*Chilopsis glutinosa* Engelm., in Wisliz., Mem. Tour North. Mexico 45, 1848.
*Chilopsis linearis* var. *originaria* F. R. Fosberg, Madroño **3**: 365, 1936.
*Chilopsis linearis* var. *glutinosa* F. R. Fosberg, loc. cit.
*Chilopsis linearis* var. *arcuata* F. R. Fosberg, op. cit. 366.

Slender shrub or small tree to 8 m. tall, with a short, stout trunk and several to many erect or ascending, slender branches; older bark irregularly ridged and scaly, that on younger growth smooth or somewhat striate, thin; leaves alternate and opposite, linear, 2–6 mm. broad, 10–30 cm. long, attenuate toward each end, glabrous or minutely glandular-puberulent, often slightly to markedly arcuate; petioles 1–3 mm. long; calyx 8–12 mm. long, papery, distinctly inflated in bud, lips ascending to spreading in flower, glabrous or arachnoid-lanate; corolla pinkish or pale purplish, 2.5–3.5 cm. long, only weakly bilabiate, lobes erose-crispate, about as broad as long; capsules linear, 10–30 cm. long, 4–6 mm. in diameter, terete or nearly so, glabrous, dark brown; body of seeds about 1 cm. long, 2.5–3.5 mm. wide, tipped at each end by silky hair-like processes 1–1.5 cm. long.

Along arroyos and outwash fans, Lower and Upper Sonoran Zones, Mojave and Colorado Deserts and the San Jacinto Valley, Riverside County, California, to Texas, Durango, Zacatecas and Tamaulipas. April–Aug.

## 3. **TABEBUIA** Gomes, Obs. Bot. **2:** 7, *pl.* 2, 1803

Shrubs or trees with glabrous or pubescent twigs and foliage. Leaves opposite, simple to digitately 5–7-foliolate, leaflets petiolulate. Flowers ample, in terminal cymes or racemes, pedicellate. Calyx tubular, variously cleft or toothed after anthesis. Corolla tubular-ampliate, straight or slightly curved; limb slightly bilabiate, with 5 subequal, rounded, spreading lobes. Stamens 4, didynamous, included; anthers glabrous, with linear or oblong, divergent to divaricate locules. Ovary sessile, many-ovuled. Capsule oblong to linear, subterete, loculicidally dehiscent. Seeds in 1–3 series in each cell, flattened, membranously winged.

### 1. **Tabebuia palmeri** Rose, Contr. U.S. Nat. Herb. **1:** 109, *pl. 11*, 1891

An openly branched tree 5–20 m. tall with a rounded to umbrella-like crown and a straight, cylindrical trunk with gray, nearly smooth bark; petioles to 10 cm. long or more, glabrous; leaflets 5, elliptic, elliptic-ovate or sometimes slightly obovate, rounded to broadly cuneate at base, acuminate-attenuate at apex, to 8 cm. broad and 20 cm. long, entire or faintly crenulate toward apex, glabrous except for small tufts of fine white hairs in axils of veins along midribs, blackening in drying; petiolules 1–4 cm. long; flowers in crowded terminal cymes, appearing well before leaves; pedicels stout, 1–5 mm. long, densely puberulent with fine, branched, whitish hairs; calyx 6–7 mm. long, lanate-puberulent, teeth broadly ovate, 1–2 mm. long, fully as wide; corollas pink with darker veins, 6–8.5 cm. long, tube narrow, 1–1.5 cm. long, broadening into a slightly curved, narrowly funnelform throat, sparsely puberulent without and within, lobes 1.5–2 cm. wide, 2–2.5 cm. long, rounded, margins somewhat undulate-crispate; stamens about 2 cm. long; capsules 2.5–3.5 cm. long, 10–15 mm. in diameter.

On forested or brushy hillsides, Lower Sonoran and Tropical Zones, southern Sonora to Guerrero. Oct.–March.

**Tabebuia chrysantha** (Jacq.) Nicholson, Dict. Gard. **4:** 1, 1887, is a tree similar in habit but with scaly bark, yellow corollas, and densely rusty-pubescent calyces and pedicels. It occurs in the vicinity of Alamos, Sonora, and south into northern South America, but has not been reported definitely from within the limits of the Sonoran Desert.

## 4. **TECOMA** Juss., Gen. 139, 1789

Erect shrubs or small trees, or less commonly twining or scandent vines, with glabrous or pubescent leaves and twigs. Leaves opposite, pinnate or simple, leaflets often dentate. Inflorescences racemose or paniculate. Calyx tubular-campanulate, subequally 5-dentate. Corolla elongate-tubular or funnelform, straight or curved, ampliate or subventricose toward mouth; limb slightly bilabiate, 5 lobes only slightly dissimilar, more or less spreading. Stamens 4, didynamous, inserted below middle of tube, included or exserted; anthers glabrous or sparsely pilose, linear locules parallel at first, later divergent. Disk annular. Capsule linear or narrowly elliptic, straight or curved, more or less compressed parallel to partition, loculicidally dehiscent. Seeds hyaline-winged, compressed.

Leaflets 1–2.5 cm. wide, shallowly serrate ........................ 1. *T. stans* var. *stans*
Leaflets mostly 1 cm. wide or less, saliently incised-dentate ...... 1*a*. *T. stans* var. *angustata*

**1. Tecoma stans** (L.) Juss., Gen. 139, 1789, var. **stans**

*Bignonia stans* L., Sp. Pl. ed. 2, 871, 1763.
*Tecoma sambucifolia* H. B. K., Nov. Gen. & Sp. **3:** 143, 1818.
*Tecoma mollis* H. B. K., op. cit. 144.
*Tecoma incisa* Sweet, Hort. Brit. 284, 1827.
*Tecoma stans* γ *velutina* A. DC., Prod. **9:** 224, 1845.
*Tecoma stans* β *apiifolia* Hort. ex A. DC., loc. cit.
*Stenolobium molle* Seem., Jour. Bot. Brit. & For. **1:** 90, 1863.
*Stenolobium stans* Seem., op. cit. 88.
*Stenolobium sambucifolium* Seem., op. cit. 90.
*Stenolobium tronadora* Loesener, Rep. Spec. Nov. **16:** 210, 1919.
*Stenolobium quinquejugum* Loesener, op. cit. 211.

Shrub or small tree to 5–8 m. tall with slender, ascending branches and thin, smooth, buff to brownish bark on younger branches, dark gray and slightly fissured bark on older ones; leaves opposite; petioles 2–6 cm. long, glabrous or sparsely puberulent; leaflets 3–9 (mostly 5–7), lanceolate to ovate-lanceolate, usually 1.5–3 cm. broad, 4–12 cm. long, acute to attenuate at apex, rounded to cuneate at base, margin subentire to dentate with teeth to 2 mm. long, glabrous on both surfaces or puberulent beneath; inflorescence 1–2 dm. long; bracts lanceolate, 3–5 mm. long, glabrous or sparsely puberulent; pedicels 5–12 mm. long; calyx tube 4–6 mm. long, lobes ovate-attenuate, 1.5–2.5 mm. long; corolla bright yellow, 4–5 cm. long, tube narrow, about 1 cm. long, expanding abruptly into a campanulate-funnelform throat, limb slightly bilabiate, lobes 1–1.5 cm. long and broad, rounded, spreading; fruit linear, 7–20 cm. long, about 1 cm. in diameter, flattened, straight or slightly curved, body of the seeds 7–9 mm. long, yellowish, papery-white wings increasing size to 8–10 mm. wide, 2–2.5 cm. long.

Sandy valleys, along arroyos and at the bases of outwash fans, Lower Sonoran to Tropical Zones, southern Sonora and the Cape Region, Baja California, south into tropical South America; West Indies. Flowering after rains.

**1*a*. Tecoma stans** var. **angustatum** Rehd., Mitt. Deutsch. Dendr. Ges. **1915:** 227, 1915

*Stenolobium incisum* Rose & Standl. ex Woot. & Standl., Contr. U.S. Nat. Herb. **16:** 174, 1913.
*Tecoma incisa* I. M. Johnston, Jour. Arnold Arb. **21:** 264, 1940. Not *T. incisa* Sweet, 1827.

Differing from var. *stans* in being a small shrub 1–2.5 m. tall; in having a slightly higher average number of leaflets per leaf (usually 7 and sometimes to 13); narrower leaflets usually less than 1.5 cm. broad; and more deeply incised-serrate-dentate leaflet-margins, teeth often 4–5 mm. long.

Sandy or gravelly valley floors, arroyos, outwash fans and hillsides, mainly Lower Sonoran Zone but ranging into the Upper Sonoran Zone, southern Arizona to Texas and south into the interior of Sonora and Chihuahua; not closely approaching the coast. Flowering after rains.

## Family 128. **MARTYNIACEAE.** Unicorn-Plant Family

Annual or perennial herbs with estipulate leaves and perfect, irregular, sympetalous flowers. Leaves petiolate, lower opposite, upper often alternate. Calyx 4–5-cleft or 4–5-parted, or divided to base only on lower side. Corolla with a 5-lobed and slightly 2-lipped limb and an oblique, often decurved tube; lobes subequal, upper 2 exterior in bud. Anther-bearing stamens 2 or 4, posterior pair sometimes sterile; anthers 2-celled, longitudinally dehiscent. Ovary superior, 1-celled, often appearing 2-celled by intrusion of two placentae; ovules few to many, anatropous; style slender; stigma 2-lobed. Fruit in ours a loculicidal capsule with a fleshy exocarp separating in age from woody sculptured endocarp. Seeds often compressed, angular; endosperm none; cotyledons large.

### 1. PROBOSCIDEA Keller, in Schmidel, Icon. (ed. Keller) 49, 1762

Stoutish, annual or perennial herbs from a fibrous root system or from a fleshy, large, fusiform, yellowish root. Herbage glandular or viscid-pubescent, strong-scented. Leaf blades broad. Flowers in loose terminal racemes. Calyx spathe-like, shallowly 5-lobed but split to base on lower side, membranous, deciduous. Corolla campanulate-funnelform, declined; limb oblique, 5-lobed. Capsule 2-valved, crested below and sometimes above also, ending in a prominent incurved beak, falsely 2-celled by extension of placentae; exocarp fleshy and separating from woody endocarp. Seeds many, tuberculate.

Plants perennial, root fusiform, fleshy, 5–10 cm. in diameter; bracts ovate to orbicular; fruit crested along both dorsal and ventral sutures .................... 1. *P. altheifolia*

Plants annual, roots mostly fibrous, if fleshy less than 2 cm. in diameter; bracts linear, oblong, or narrowly ovate; fruit crested along dorsal suture only:

- Corollas yellow with brownish, orange or purplish lines in throat; leaf blades usually slightly wider than long; bracts markedly falcate .................. 2. *P. arenaria*
- Corollas purplish, pinkish, or if yellow the lobes with strong blotches of purple, throat usually yellow or orange; leaf blades longer than broad; bracts straight to slightly falcate:
  - Calyx 2–2.5 cm. long; bracts 8–12 mm. long; corolla reddish purple to white veined with deeper purple, 4–5 cm. long; rudiment of fifth stamen present ............ 3. *P. louisianica*
  - Calyx 1–1.5 cm. long; bracts 5–8 mm. long; corolla dull yellow, brownish or orange in throat, purplish on limb, 2.5–3.8 cm. long; fifth stamen lacking:
    - Leaf blades glandular on both surfaces; body of fruit 5–7 cm. long; corolla 2.5–3.5 cm. long ............................................ 4. *P. parviflora*
    - Leaf blades glabrous or nearly so except on veins beneath; body of fruit 4–5 cm. long; corolla 3–3.8 cm. long .................................. 5. *P. sinaloensis*

**1. Proboscidea altheifolia** (Benth.) Decne., Ann. Sci. Nat. V, **3:** 324, 1865

*Martynia altheaefolia* Benth., Bot. Sulph. 37, 1844.
*Martynia palmeri* S. Wats., Proc. Amer. Acad. **24:** 66, 1889.

Coarse, viscid-puberulent perennial 3–5 dm. tall from a fusiform, fleshy, yellowish root 3–15 cm. in diameter, apex of root usually 10–30 cm. below surface of soil; petioles 3–10 cm. long, densely glandular-pubescent; leaf blades broadly ovate to reniform, shallowly 3–5-lobed, 1–6 cm. long, 1.5–8 cm. wide; racemes several-flowered, to 20 cm. long; bracts broadly ovate to orbicular, 5–6 mm. long; calyx 10–15 mm. long, viscid-glandular, lobes obtuse to rounded at apex; flowers fragrant; corolla 2.5–3.5 cm. long, limb 3–4 cm. in diameter, clear yellow with reddish or deeper yellow lines and specks in upper throat, tube brownish yellow outside, clear yellow streaked with deeper yellow, brownish, or purple inside; longer pair of stamens about 2 cm. long, shorter ones 12–16 mm. long; body of fruit slender, 5–6 cm. long, 1–2 cm. in diameter, strongly crested along dorsal suture, distal teeth often prolonged into a slender horn nearly as long as main one, ventral suture lightly crested or nearly smooth; seeds rhomboidal-oblong, 6–8 mm. long, dark gray to dull black, warty with radiately arranged ridges on low bosses.

Sandy soil, Lower Sonoran Zone, Colorado Desert, California, into western Arizona, south into Sinaloa and southern Baja California. May–Aug. in the Colorado Desert; flowering practically throughout the year in the southern parts of its range.

**2. Proboscidea arenaria** (Engelm.) Decne., Ann. Sci. Nat. V, **3:** 326, 1865

*Martynia arenaria* Engelm., in Wisliz., Mem. Tour North. Mexico 100, 1848.

Spreading annual from a slender taproot and fibrous lateral rootlets; branches often somewhat decumbent, 20–60 cm. long, usually 30–40 cm. high; herbage viscid-pubescent; lower leaves opposite, upper alternate; petioles slender, 5–15 cm. long; blades broadly ovate to reniform, shallowly 5–7-lobed, 2.5–4.5 cm. long, 3.5–5.5 cm. wide, cordate at base, margins slightly undulate-dentate, viscid-puberulent on both surfaces; racemes 5–20 cm. long, 5–10-flowered, slender pedicels 3–6.5 cm. long, abruptly deflexed in fruit; bracts narrowly oblong, often slightly falcate, 5–10 mm. long; calyx 10–15 mm. long, lobes acute to obtuse, membranous, divided about halfway to base; corolla 3.5–4 cm. long, yellow with brownish flecks within, lightly washed with brown on outside of tube and throat, limb spreading, about 4 cm. broad, throat ventricose; stamens 4, 2–2.5 cm. long, filaments pilose near base; fruit slenderly cymbiform, body 12–16 mm. in diameter, 4.5–6 cm. long, strongly crested with irregularly winged and toothed processes 5–7 mm. long along dorsal suture, terminal process prolonged into a slender horn 6–8 cm. long, primary horns stiffly curved or hooked, 8–12 cm. long; seeds ashy-gray, 7–9 mm. long, warty and irregularly angled by compression.

Sandy flats, mesas, and gentle gravelly slopes, Lower Sonoran Zone, southern Arizona to western Texas and south into northern Mexico. July–Oct.

**3. Proboscidea louisianica** (Mill.) Thell., Mem. Soc. Sci. Cherbourg **38:** 480, 1912

*Proboscidea jussieui* Keller, in Schmidel, Icon. (ed. Keller) 49, *pls. 12–13,* 1762. (Not legitimate binomial.)
*Martynia louisiana* Mill., Gard. Dict. ed. 8, no. 3, 1768.
*Martynia louisianica* Mill., op. cit. in corrigenda.
*Martynia proboscidea* Gloxin, Obs. 14, 1785.
*Martynia alternifolia* Lam., Encycl. **2:** 112, 1786.
*Proboscidea louisiana* Woot. & Standl., Contr. U.S. Nat. Herb. **19:** 602, 1915.
*Martynia jussieui* J. T. Howell, Leafl. West. Bot. **1:** 40, 1933.

Annual from a short taproot and fibrous lateral roots, stems decumbent to ascending, 3–10 dm. long; herbage slimy-viscid-pubescent throughout; leaves opposite to alternate above, petioles 6–25 cm. long, to 6 mm. in diameter; leaf blades broadly ovate to suborbicular, obtuse to rounded at apex, cordate at base, 6–20 cm. wide, usually slightly shorter than wide, mostly sinuate-margined; racemes open, several-flowered, 10–20 cm. long, elongating to twice as long in fruit; pedicels 1–3 cm. long in flower, rarely over 5 cm. long and deflexed in fruit; bracts 8–12 mm. long, lanceolate to oblong; calyx 2–2.5 cm. long, lobes obtuse to acute, divided about one-third of distance to base; corolla 4–5 cm. long, reddish purple to whitish veined with deep purple, tube often yellowish with purple lines or blotches, limb 4–5 cm. wide, spreading; fertile stamens 2–2.5 cm. long, rudiment of fifth stamen usually evident; body of fruit 2–2.5 cm. in diameter, 4–6 cm. long, curved horns 8–12 cm. long, upper suture crested, lower smooth; seeds 5–6 mm. wide, 8–9 mm. long, dull black, warty, faveolate-pitted.

Mostly in cultivated areas in moist soil, Lower and Upper Sonoran Zones, Sacramento Valley south to northern Mexico, east to the Atlantic coast; rare in the Sonoran Desert. March–Oct.

**4. Proboscidea parviflora** (Woot.) Woot. & Standl., Contr. U.S. Nat. Herb. **19:** 602, 1915

*Martynia parviflora* Woot., Bull. Torrey Club **25:** 453, 1898.

Densely viscid-pubescent annual with spreading-decumbent branches to 1 m. long, 3–4 dm. high; upper leaves alternate; petioles slender, 5–16 cm. long; leaf blades broadly ovate, 2–10 cm. wide, 2.5–12 cm. long, sometimes nearly twice as large, entire or shallowly and obtusely 3–7-lobed, obtuse to rounded at apex, truncate to cordate at base; raceme open, few-flowered; pedicels 1–2 cm. long at anthesis, 3–5 cm. long and deflexed in fruit; bracts broadly ovate to oblong, 5–7 mm. long, thin, often reddish; calyx 10–15 mm. long, lobes 3–6 mm. long, ciliate-pilose marginally, surfaces glandular-pilose; corolla 2.5–3.5 cm. long, slightly ventricose, viscid-glandular, throat dull yellow with purple along upper side and spreading to upper lobes or over whole limb, limb flaring to slightly reflexed, 2–3 cm. in diameter; rudiment of fifth stamen very short or lacking; body of fruit about 2 cm. in diameter, 5–7 cm. long, endocarp roughly sculptured, horns about twice as long as body, crest of slender, irregularly dentate teeth 2–3 mm. wide, 5–7 mm. long, along upper suture; seeds obovoid, about 5 mm. wide, 8–10 mm. long, somewhat angled, dull black, strongly warty-tuberculate.

Sandy or gravelly soil, Lower Sonoran Zone, Arizona to Texas, south in Mexico to San Luis Potosí, Durango, and Sinaloa. March–Oct.

**5. Proboscidea sinaloensis** Van Eseltine, N.Y. Agric. Exp. Stat. Tech. Bull. No. **149:** 24, *fig.* 9, 1929

*Martynia sinaloensis* Wigg. & Roll., Contr. Dudley Herb. **3:** 277, 1943.

Viscid-pubescent annual from a short slender taproot and numerous fibrous laterals; stem erect, 3–4 dm. tall, lateral spreading branches somewhat decumbent, to 6 dm. long; petioles 5–12 cm. long, densely glandular-pilose; leaf blades ovate, 4–9 cm. wide, 5–15 cm. long, moderately to strongly 3–5-lobed, terminal lobe deltoid, larger than lateral ones, lower surface glandular on main veins beneath, otherwise glabrous or essentially so; raceme 10–25 cm. long, often 8–15-flowered; pedicels slender, 2–5 cm. long at anthesis, 5–8 cm. long and deflexed in fruit; bracts ovate to lanceolate, usually somewhat falcate, 6–8 mm. long, glandular-pilose; calyx 10–15 mm. long, glandular-pilose, uppermost lobe lance-deltoid, 6–9 mm. long, lateral ones about one-half as long; corolla 3–3.8 cm. long, tube lightly ventricose, purplish with a yellow area on abaxial distal part of throat and narrow yellow stripes along sides of tube, limb spreading, 2.5–3.5 cm. wide, lavender to purple, carmine spots within; stamens didynamous, 2 cm. long, rudiment of fifth stamen absent; body of fruit about 2 cm. in diameter, 4–5 cm. long, crested along upper suture with irregular narrow teeth, terminal teeth prolonged into slender horns 1.5–3 cm. long, main horns 8–10 cm. long; seeds obovoid, 5–7 mm. wide, 8–10 mm. long, 2–3 mm. thick, dull black, warty with low broad tubercles, these shallowly foveolate-pitted.

Arroyos and valley floors, Lower Sonoran Zone, central Sonora into Sinaloa. Aug.–Oct.

## Family 129. **OROBANCHACEAE.** Broomrape Family

Root parasites with yellowish or purplish, erect, simple or branched stems, leaves reduced to alternate appressed scales, and without chlorophyll. Flowers perfect, irregular, sympetalous, solitary or in pedunculate bracteate racemes or spikes, and usually 2-bracteate. Calyx free from ovary, 4–5-toothed or 4–5-cleft, or split almost or completely to base along one or both sides. Corolla irregular; limb 5-lobed and 2-lipped. Stamens 4, didynamous, inserted on corolla tube; rudiment of fifth stamen sometimes present; filaments slender; anthers 2-celled, with parallel equal sacs. Ovary superior, 1-celled, with 4 parietal placentae; ovules many. Style slender; stigma discoid or 2–4-lobed. Capsule 2–4-valved. Seeds many, small, wrinkled, striate, or reticulated; embryo minute, with scarcely differentiated colyledons.

## 1. OROBANCHE L., Sp. Pl. 632, 1753

Characters of family. Flowers yellowish or purplish, with or without bracts. Calyx 4–5-cleft into acute to acuminate lobes. Corolla with an erect, 2-lobed upper lip, and spreading, 3-lobed lower lip. Anther sacs usually mucronate at base. Ovary ovoid. Stigma peltate, funnelform, entire or bilobed.

Flowers solitary on long slender pedicels; floral bracts none ............ 1. *O. fasciculata*
Flowers in spikes or racemes, corymbs or panicles; 2 floral bracts subtending each flower:
  Stigma bilobed; palatal folds lacking or weakly developed; corolla lobes plainly short-toothed at apex ............................................ 2. O. *multicaulis*
  Stigma peltate-crateriform; palatal folds obvious in throat; corolla lobes acute to rounded but not apiculate-toothed:
    Lower flowers on pedicels to 5–18 mm. long; inflorescence openly corymbose; calyx 14–16 mm. long .................................... 3. *O. grayana* var. *feudgei*
    Lower flowers as well as upper sessile or nearly so; inflorescence spicate; calyx 5–12 mm. long:
      Corolla lobes acute or short-pointed; anthers usually pubescent, at least after dehiscence ............................................ 4. *O. cooperi*
      Corolla lobes broad and rounded at apex; anthers glabrous ..................... ........................................ 5. *O. multiflora* var. *arenosa*

### 1. Orobanche fasciculata Nutt., Gen. 2: 59, 1818

*Loxanthes fasciculatus* Raf., Neogen. 3, 1825.
*Anoplanthus fasciculatus* Walp., Rep. **3**: 480, 1844–45.
*Aphyllon fasciculatum* Torr. & Gray ex A. Gray, Man. ed. 2, 281, 1856.
*Thalesia fasciculata* Britt., Mem. Torrey Club **5**: 298, 1894.

Stems fleshy, mostly subterranean but extending 2–6 cm. above surface, simple or several-branched, erect, bearing lance-ovate scales 5–10 mm. long and several to many 1-flowered pedicels 3–15 cm. long, glandular-puberulent throughout; calyx campanulate, 5–9 mm. long, narrowly to broadly triangular-subulate lobes about as long as tube, attenuate at apex; corolla 18–32 mm. long, yellow suffused with purple or stramineous throughout, tube straight or slightly curved downward, lobes 4–6 mm. long, rounded, ciliate along margins; stamens inserted about one-fourth to one-third of length of tube above base, filaments broadened at base, glabrous; anthers 1.2–1.6 mm. long, cream to white, glabrous, short-apiculate at base of each locule; capsule broadly ovoid, 7–10 mm. long, smooth and glabrous; seeds 0.2–0.3 mm. long, flattened, minutely faveolate-pitted, stramineous to dark brown.

Parasitic on roots of numerous plants, mostly Upper Sonoran and Transition Zones, British Columbia and Washington to northern Baja California, east to Michigan and Texas. April–Sept. Barely reaching the upper edges of the Sonoran Desert.

### 2. Orobanche multicaulis Brandegee, Univ. Calif. Pub. Bot. 6: 361, 1916

Stems erect or base sharply bent, simple or few-branched, 10–25 cm. tall, mostly

7–10 mm. in diameter, moderately puberulent, glandular; cauline bracts ovate to broadly lanceolate, 5–6 mm. long, rounded, or more commonly acute to short-acuminate at apex; floral bracts narrowly lanceolate, 8–10 mm. long, densely glandular-puberulent; inflorescence spiciform, 4–12 cm. long, densely glandular-puberulent, lower flowers on pedicels to 2.5 cm. long, upper sessile or nearly so; calyx shallowly cup-shaped, 7–11 mm. long, equally 5-parted to middle or slightly deeper, lobes lanceolate to subulate, dark purplish brown to nearly black, densely glandular-puberulent, elongating somewhat (to 15 mm.) in fruit; corolla 18–25 mm. long, yellowish or whitish on tube, purplish on lobes and sometimes on tube, sparsely villous near base of tube, gradually more densely so upward, lobes closely puberulent, subequal, obtuse to acute and distinctly short-toothed or apiculate at apex; lower lip with poorly developed palatal folds or none; anthers white, 1.4–1.6 mm. long, glabrous except along lips of thecae after anthesis, these villous-ciliate with crinkled slender hairs; stigma bilobed, about 1.5 mm. wide, margins of lobes short-lanate; capsules ovoid, 8–10 mm. long.

Parasitic on roots of *Franseria, Acacia,* and other plants, Lower Sonoran Zone, southern Arizona to central Sonora and northern Magdalena Plain, Baja California. March–Oct.

**3. Orobanche grayana** G. Beck var. **feudgei** Munz, Bull. Torrey Club **57:** 616, *pl. 38, fig. 8,* 1930

Stems erect, stout, 10–25 cm. tall, simple or branching moderately at or near base, glandular-puberulent throughout; cauline bracts ovate, obtuse, 5–6 mm. long; inflorescence corymbose to subpaniculate, densely many-flowered, 5–10 cm. long; pedicels slender, 5–18 mm. long; calyx 14–16 mm. long; corolla purplish brown with yellow markings on rather prominent palatal folds, upper lip 10–12 mm. long, its lobes 2–3 mm. long, lanceolate, lower lip 9–12 mm. long, with lanceolate lobes divided nearly to throat; anthers markedly hairy; stigma peltate; capsule ovoid, 8–10 mm. long.

Parasitic on *Artemisia,* mostly Upper Sonoran and Transition Zones, straggling into upper edge of Lower Sonoran Zone, Kern County, California, to Inyo County, south to northern Baja California. April–July.

**4. Orobanche cooperi** (A. Gray) Heller, Cat. N. Amer. Pl. 7, 1898

*Aphyllon cooperi* A. Gray, Proc. Amer. Acad. **20:** 307, 1885.
*Orobanche ludoviciana* var. *cooperi* Beck, Bibl. Bot. **4:** 81, 1890.
*Mycorrhiza cooperi* Rydb., Bull. Torrey Club **36:** 695, 1909.
*Orobanche ludoviciana* var. *latiloba* Munz, op. cit. **57:** 621, *pl. 39, fig. 18,* 1930.

Plants stout and fleshy, 1–4.5 dm. tall, simple or branching sparingly at base, viscid-puberulent throughout; stems brownish to purplish; cauline bracts 5–10 mm. long, obtuse to rounded; inflorescence 5–20 cm. long, spicate and densely flowered, or branching into 2–3 axillary spikes; lowest flowers on slender pedicels slightly longer than scales, upper sessile, or all sessile; calyx densely puberulent, 5–10 mm. long, lobes lance-attenuate, three to five times as long as shallow cup; floral bracts 3–8 mm. long, acute to obtuse; corolla 1.5–3 cm. long, purplish within, prominent palatal folds yellow, puberulent and gray-purple without, lips 4–8 mm. long, upper lip erect, shallowly cleft at apex, lobes broadly to narrowly acute, erect to reflexed; lower lip deeply cleft, lobes spreading; anthers sparsely villous; stigma peltate, margins more or less crenulate.

Sandy desert flats and washes, parasitic on *Franseria, Hymenoclea,* and other shrubs, Lower Sonoran Zone, Death Valley region, California, to southern Utah, south to New Mexico, Sonora, and Baja California. Feb.–May.

**5. Orobanche multiflora** Nutt. var. **arenosa** (Suksd.) Munz, Bull. Torrey Club **57:** 623, *pl. 39, fig. 21,* 1930
*Aphyllon arenosum* Suksd., Allg. Bot. Zeitschr. **17:** 27, 1906.

Plants erect, stoutish, 5–20 cm. tall, finely glandular-puberulent with grayish hairs; cauline bracts ovate to ovate-lanceolate, 5–10 mm. long, often obtuse and abruptly acute-apiculate; inflorescence spicate, 4–10 cm. long, rather dense, only lowest flowers short-pedicellate, or even these sessile; floral bracts linear-lanceolate, 4–9 mm. long; calyx irregularly cleft into linear-lanceolate lobes 8–12 mm. long, tube half as long or less; corolla 15–20 mm. long, yellowish or tinged with pink or purple, densely glandular-puberulent without, upper lip about 5 mm. long, shallowly cleft into rounded lobes, lower lip 3–4 mm. long, divided to base into narrow, obtuse to rounded lobes; anthers white, glabrous; capsules 7–10 mm. long.

Sandy soil, parasitic mainly on *Compositae*, chiefly Upper Sonoran and Transition Zones, straggling into upper edges of Lower Sonoran Zone, Idaho to Utah, south to southern California and Santa Catalina Mountains, Arizona. March–July.

**Orobanche bulbosa** Beck, Bibl. Bot. **4:** 83, *pl. 1, fig. 7,* 1890, with very fleshy, bulbous stems, flowers in a compact thyrsoid panicle, and nearly obsolete palatal folds, occurs in chaparral from Solano and El Dorado Counties, California, south to the southwestern flanks of the Sierra San Pedro Mártir, Baja California. It may occur as an occasional waif along the upper margins of the Sonoran Desert.

## Family 130. **ACANTHACEAE.** Acanthus Family*

Herbs or shrubs. Leaves simple, estipulate. Flowers irregular to nearly regular, perfect. Calyx persistent; segments 5, or occasionally 4, imbricate. Corolla gamopetalous; limb 5-lobed or 2-lipped. Stamens 4, didynamous, or 2 only. Staminodes sometimes present. Anther sacs 2 or 1, longitudinally dehiscent. Ovary superior, 2-celled; ovules 2–10 in each cavity. Styles filiform, simple. Stigmas 1 or 2. Fruit a capsule, 2-celled, 2-valved. Seeds usually flat and rounded, borne on retinacula (papilliform or, more often, hook-shaped projections from placenta), testa smooth or roughened, often developing mucilage when moistened.

Stamens 4; flowers white, pink or purple:
- Anthers 1-celled; corolla strongly irregular, 2-lipped; rigid, often spinescent shrubs with whitish stems .................... 1. *Berginia*
- Anthers 2-celled; corolla only slightly irregular, 5-lobed:
  - Anther sacs mucronate or aristulate at base .................... 2. *Dyschoriste*
  - Anther sacs muticous at base .................... 3. *Ruellia*

Stamens 2:
- Peduncles bearing many lanceolate coriaceous bracts; seeds borne on papilliform retinacula; flowers blue .................... 4. *Elytraria*
- Peduncles, if present, without bracts; seeds borne on hooklike retinacula:
  - Bracts large, conspicuously 4-ranked; flowers borne in dense spikes; flowers white or yellow .................... 5. *Tetramerium*

---

* Text, excluding that for the genus *Carlowrightia,* contributed by E. C. Leonard, United States National Herbarium, Smithsonian Institution. Published by permission of the Secretary of the Smithsonian Institution.

Bracts small or, if large, not conspicuously 4-ranked; flowers variously arranged:
Inner bracts posteriorly connate to near tip, forming a capsulelike involucre, much larger than outer bracts; flowers white or yellow .................. 6. *Henrya*
Inner bracts narrow, not at all connate, usually shorter than outer bracts:
Stems conspicuously 6-angled; flowers subtended by 2 conspicuous cordate bracts; placenta separating from capsule walls at maturity; flowers purple .......... ........................................................ 7. *Dicliptera*
Stems terete or subquadrangular; placenta remaining attached to capsule walls at maturity:
Anther cells parallel, equal, both inserted at same height on filament; shrubs:
Corolla very open, 1.5 cm. long or less, lobes about twice as long as tube ..... ................................................... 8. *Carlowrightia*
Corolla tubular-funnelform, more than 2 cm. long, lobes equaling or shorter than tube; flowers red, rarely yellow or orange ......... 9. *Anisacanthus*
Anther cells more or less divergent, unequal, inserted at different heights on filament:
Corolla tube very slender and elongate, cylindrical, terete; corolla white or yellowish; suffrutescent herb ......................... 10. *Siphonoglossa*
Corolla tube more or less ampliate; corolla red, purple or scarlet, or rarely white in *Justicia*; shrubs:
Anther cells muticous at base or lower cell sometimes minutely apiculate ... .................................................. 11. *Jacobinia*
Anther cells calcarate at base:
Corolla tube short, tube usually broadly ampliate at throat; flowers sessile .................................................. 12. *Justicia*
Corolla tube elongate, only slightly ampliate at throat; flowers pedicellate ................................................ 13. *Beloperone*

## 1. **BERGINIA** Harv. ex Benth. & Hook., Gen. Pl. **2:** 1096, 1876

Low shrubs with whitish branches. Leaves sessile or nearly so, entire. Flowers small, sessile in axils of small bracts, forming terminal, more or less interrupted spikes. Bractlets narrow, rigid, resembling calyx lobes. Calyx 5-parted; segments linear-lanceolate, acute, rigid. Corolla white or pink; tube short, ampliate above; limb 2-lipped. Stamens 4, anthers glabrous. Ovules 2 in each cell.

Leaves scaberulo-puberulous:
Inflorescence puberulous, hairs eglandular ................... 1. *B. virgata* var. *virgata*
Inflorescence glandular-pubescent ..................... 1*a*. *B. virgata* var. *glandulifera*
Leaves densely hirtellous ................................................ 2. *B. palmeri*

### 1. **Berginia virgata** Harv. ex Benth. & Hook., Gen. Pl. **2:** 1097, 1876, var. **virgata**

*Pringleophytum lanceolatum* A. Gray, Proc. Amer. Acad. **20:** 293, 1885.

A bushy shrub to 2 m. high; branches whitish, tips scaberulous; petioles short; leaf blades oblong to oblong-lanceolate or linear-lanceolate, 1–4 cm. long, to 14 mm. wide, obtuse or acute, cuneate at base, scaberulo-puberulous; bracts oblong-elliptic; flowers borne in spikes to 13 cm. long; calyx 5–6 mm. long, minutely puberulous, ciliolate, hairs usually glandular, segments linear-oblong, acute, those of lowermost flowers abruptly so, all more or less aristulate; corollas about 12 mm. long, puberulous, upper lip whitish, lower salmon-pink; capsules clavate, 6–7 mm. long, about 5 mm. broad, retrorsely puberulous; seeds flat, hirtellous.

Gravelly slopes, Lower Sonoran Zone, Baja California and Sonora. Feb.–May.

**1a. Berginia virgata** var. **glandulifera** Leonard & Morton, in Wiggins, Contr. Dudley Herb. **4:** 24, 1950

Leaves of branchlets oval, obtuse, to rounded, or occasionally emarginate, narrowed or rounded at base; bracts subtending flowers suborbicular, 4–6 mm. long, 3.5–5 mm. wide, obtuse, these and calyx glandular-pubescent; similar in other respects to typical form of species.

Arroyos and rocky hillsides, Lower Sonoran Zone, vicinity of Comondú to the Cape Region. Nov.–April.

*Johnston 3820,* collected at Puerto Ballandra, Carmen Island, Baja California, may be referred to this variety. The inflorescence of this specimen is glandular-pubescent, but the leaflets and bracts are oblong-lanceolate, agreeing in this respect with the typical form of the species.

**2. Berginia palmeri** Rose, Contr. U.S. Nat. Herb. **1:** 86, 1890

Shrub to 50 cm. high; tips of branches pilosulous; leaves ovate to oval, 7–15 mm. long, to 8 mm. wide, obtuse or rounded at apex, hirtellous; spikes about 6 cm. long or less; calyx densely glandular-pilosulous; corolla 8–10 mm. long, pink; seeds flat, puberulous.

Rocky washes and canyons, Lower Sonoran Zone, known from the vicinity of Santa Rosalia and on Carmen Island, Baja California. Feb.–April.

## 2. DYSCHORISTE Nees, in Wall., Pl. Asiat. Rar. 3: 75, 1832

Herb. Leaf blades entire, petioled, or subsessile. Flowers solitary or clustered in the axils. Calyx very slender; segments 5, usually subulate. Corolla funnelform, purple, 5-lobed, slightly 2-lipped. Stamens 4, didynamous. Anthers 2-celled, cells oblong, parallel or slightly divergent, sharply mucronate at base. Capsule included in persistent calyx, oblong-linear, glabrous, 2–4-seeded, separating with difficulty at maturity into 2 valves. Seeds 1 or 2 in each valve, flattened, suborbicular, mucilaginous when moistened.

**1. Dyschoriste decumbens** (A. Gray) Kuntze, Rev. Gen. Pl. **2:** 486, 1891

*Calophanes decumbens* A. Gray, Syn. Fl. N. Amer. ed. 1, **2**[1]**:** 325, 1878.

*Calophanes oblongifolius* sensu Torr., Bot. Mex. Bound. 122, 1859. Not *C. oblongifolius* D. Don, 1833.

Puberulous herbs; stems branched from base, usually spreading from a ligneous base, occasionally erect; leaves spatulate to oblanceolate, 2–4 cm. long, 0.5–1.1 cm. wide, usually obtuse and sometimes mucronate at apex, attenuate at base; flowers few, borne in foliose, bracteate, axillary clusters; calyx 15–20 mm. long, at maturity as much as 10 mm. longer than capsule, 5-parted, tube 5–7 mm. long, lobes subulate-setaceous, barely twice as long as tube; corolla purple, 18–20 mm. long, lower half slender, slightly ampliate at base, enlarged to 7 or 8 mm. at throat, limb about 15 mm. broad, lobes rounded; capsules about 1 cm. long, 4-seeded.

Dry soil, Lower Sonoran Zone, western Texas to Arizona and the plateau region of northern Mexico; lower mountains in the Cape Region, Baja California. Oct.–May.

## 3. RUELLIA L., Sp. Pl. 634, 1753

Perennial herbs or shrubs. Leaves petioled, entire or undulate. Flowers usually large and showy, solitary, or clustered in axils, or borne in terminal cymose panicles.

Calyx usually 5-parted, segments often narrow. Corolla white or purple (usually mauve), funnelform or salverform, sometimes saccate, lower portion of the tube usually narrow; the limb of 5 obtuse or rounded, spreading lobes. Stamens 4, didynamous; anther sacs blunt at base or minutely apiculate. Capsules oblong or clavate. Seeds flat, mucilaginous when moistened.

Shrubs:
  Branchlets, leaves, and inflorescence glabrous, glutinous .............. 1. *R. peninsularis*
  Branchlets, leaves, and inflorescence pubescent:
    Pubescence eglandular; flowers white .............................. 2. *R. leucantha*
    Pubescence glandular; flowers purple or violet:
      Leaf blades orbicular, cordate at base ............................... 3. *R. cordata*
      Leaf blades ovate to lanceolate, narrowed or rounded at base ....... 4. *R. californica*
Herbs:
  Ovary and capsules glabrous (except calluses at tip):
    Leaf blades ovate to suborbicular, obtuse or rounded at apex ......... 5. *R. intermedia*
    Leaf blades lanceolate to oblong-elliptic, acute to obtuse at apex ... 6. *R. malacosperma*
  Ovary and capsules glandular-puberulous:
    Leaves relatively small, seldom more than 4 cm. wide; inflorescence glandular-puberulous ........................................ 7. *R. nudiflora* var. *nudiflora*
    Leaves much larger, at least some of them more than 5 cm. wide; inflorescence densely glandular-pilosulous ............................ 7a. *R. nudiflora* var. *glabrata*

**1. Ruellia peninsularis** (Rose) I. M. Johnston, Proc. Calif. Acad. IV, **12:** 1172, 1924
*Calophanes peninsularis* Rose, Contr. U.S. Nat. Herb. **1:** 75, 1890.

A compact shrub to 1.3 m. high; stems much branched, older portions gray or whitish, tips and leaves glabrous, glutinous; petioles 3–8 mm. long; leaf blades ovate to oblong-ovate, 1–3 cm. long, 0.5–2 cm. wide, acute or acuminate at apex, narrowed or rounded at base, entire; flowers solitary, borne on peduncles 1–2 mm. long; bracts subulate, 3–4 mm. long, deciduous; pedicels 1 mm. long; calyx 5–6 mm. long, segments linear-subulate, appressed-hirtellous within, ciliolate; corolla purple, 3–5 cm. long, minutely and sparingly pubescent, lower portion of tube slender, about 2 mm. in diameter, enlarged from below middle to about 1 cm. at throat, limb 2–3 cm. broad, lobes suborbicular, about 1 cm. in diameter; filaments 6 and 10 mm. long; anthers linear-oblong, 4 mm. long, 2 mm. broad, basal lobes often unequal and slightly mucronulate; capsules 15–20 mm. long, 4 mm. broad, pointed at apex, narrowed to a slender solid base 6–8 mm. long and 2 mm. broad, sparingly glandular-puberulous, 4-seeded; seeds flat, suborbicular, about 3 mm. in diameter, appressed-pubescent when dry, mucilaginous-pubescent when moistened.

Dry gravelly slopes, Lower Sonoran Zone, southern Baja California and in the vicinity of Guaymas, Sonora. Oct.–April.

**2. Ruellia leucantha** Brandegee, Zoe **5:** 109, 1901

A shrub, forming rounded clumps to 80 cm. high; stems much branched, obtusely quadrangular, gray, tips, together with leaves and calyx, densely pubescent, hairs mixed with minute glands; petioles about 5 mm. long; leaf blades ovate to oblong-ovate, 3–6 cm. long, 1–4 cm. wide, acute or short-acuminate at apex, narrowed at base and decurrent on petiole, entire, ciliate; flowers solitary, axillary, usually crowded at ends of branches; bracts narrowly oblanceolate, 1.5 cm. long; pedicels about 1 mm. long; calyx 15–20 mm. long, tube 2 mm. long, segments linear-attenuate; corolla 5–6 cm. long, white, minutely pubescent, bearing small glands, slender portion of tube about 2.5 cm. long and 2–2.5 mm. in diameter, obconic part about 1.5 cm. broad at throat, limb 3–4 cm. broad, lobes rounded, 1.5 cm. in diameter; filaments 7 and 9 mm. long, glabrous; anthers oblong, 5.5 mm. long, 2 mm. broad, apiculate at tip by extension of connective, basal lobes minutely mucronulate; styles about 5 cm. long, sparingly puberulous; capsule pyriform, 1–1.5 cm. long, 5–6 mm. broad, short-pointed at apex, narrowed at base (solid portion about 2 mm. long), puberulous, hairs intermixed with glands; seeds flat, suborbicular, 4–5 mm. broad, pubescent, mucilaginous when moistened.

Common on dry slopes, Lower Sonoran to Tropical Zones, Bahía de la Concepción to the Cape Region, Baja California, and near Guadalupe, Sinaloa. Oct.–May.

### 3. Ruellia cordata Brandegee, Zoe **5:** 173, 1903

A low shrub; stems glandular-pubescent; petioles to 3 mm. long (uppermost leaves sessile); leaf blades cordate, to 12 mm. long and 15 mm. wide, obtuse, glandular-pubescent; flowers solitary, borne on straight, stout, axillary peduncles to 2 cm. long; bracts foliaceous, orbicular, subsessile, 6–8 mm. broad, usually wider than long; calyx lobes 10–15 mm. long, broadly to narrowly spatulate, obtuse or acute, glandular-pubescent and ciliate; corollas light purple, up to 4 cm. long, tube exceeding calyx lobes; ovary 8-ovuled.

Known only from the type locality, Comondú, Baja California.

### 4. Ruellia californica (Rose) I. M. Johnston, Proc. Calif. Acad. IV, **12:** 1171, 1924

*Calophanes californica* Rose, Contr. U.S. Nat. Herb. **1:** 85, 1890.

Shrub to 1.3 m. high; branches numerous, older parts glabrous, whitish, tips glandular-hirtellous or glandular-puberulous; petioles 2–4 mm. long, glandular-pubescent; leaf blades ovate to lanceolate, 1–3 cm. long, 0.5–2 cm. wide, acute or acuminate at apex, narrowed or rounded at base, entire, both surfaces glandular-hirtellous, sometimes glabrate; flowers borne on short axillary peduncles 2–8 mm. long, usually solitary; bractlets linear, 2–5 mm. long, glandular-puberulous; calyx 1–1.5 cm. long, glandular-puberulous, segments linear-lanceolate, 1 mm. wide at base; corolla purple or violet, 4.5–5.5 cm. long, sparingly puberulous, funnelform, 1–1.5 cm. in diameter at throat, limb 2–3 cm. broad, lobes oval, emarginate; capsules 12–15 mm. long, 4 mm. broad, pointed, rather abruptly narrowed to a solid stipitate base 6 mm. long and 2 mm. broad, sparingly glandular-puberulous, 4-seeded; seeds suborbicular, 3 mm. in diameter, brown.

On dry rocky slopes and along arroyos, sometimes in low thickets, Lower Sonoran Zone, a common and often abundant shrub endemic in the Sonoran Desert. Aug.–May.

The sticky leaves and odor suggest green tobacco.

### 5. Ruellia intermedia Leonard, Jour. Wash. Acad. **17:** 512, 1927

Herbs from thick-fibrous roots; stems 5–40 cm. high, erect or ascending, branched, quadrangular, puberulous, a few longer hairs sometimes intermixed; leaf blades ovate to suborbicular, 2–7 cm. long and 1.5–4 cm. wide, obtuse or rounded at apex, rounded or narrowed at base and decurrent on petiole (those of axillary shoots smaller), thin, entire or undulate, both surfaces more or less puberulous especially along costa and veins, sometimes beset with a few additional long white hairs; petioles 5–20 mm. long;

flowers borne in axillary cymes with a flower in forks; peduncles to 10 cm. long (usually not exceeding 4 cm.), puberulous; calyx 10–22 mm. long, puberulous, short hairs often mixed with longer spreading ones, some of them gland-tipped, tube 1–2 mm. long, segments linear, about 1 mm. wide at base, often widely spreading at maturity, corollas 4–6 cm. long, blue or purple, puberulous, slender portion of tube about 2.5 cm. long and 2–3 mm. broad, expanded portion funnelform, slightly oblique, 1–1.5 cm. broad at throat, limb 2–4 cm. broad, lobes ovate, 10–15 mm. long, 10–14 mm. wide, undulate; filaments 6 and 12 mm. long, glabrous; anthers oblong, 3.5 mm. long, style 3–3.5 cm. long, sparingly and minutely pilose; capsules 18 mm. long, 4–5 mm. broad, pointed at apex, gradually narrowed to a solid stipitate base about 2 mm. long, glabrous except for pilose cuneate calluses near tip, 10–12-seeded; seeds lenticular, 2–2.5 mm. broad, mucilaginous-pilose when moistened.

Fields and waste places, Lower Sonoran and Tropical Zones, western coastal states of Mexico from Sonora to Guerrero. Aug.–Oct.

**6. Ruellia malacosperma** Greenm., Proc. Amer. Acad. **34:** 572, 1899

Herb; stems to 60 cm. high, glabrous or sparingly pilose; petioles 0.5–3 cm. long, glabrous or pilose; leaf blades lanceolate to oblong-elliptic, to 12 cm. long and 3.5 cm. wide, gradually narrowed to an acute or obtuse tip, narrowed at base, undulate, glabrous or costa and margins pilose; inflorescence axillary, once or twice dichotomous, peduncles slender, 3–6 cm. long, glabrous; calyx 1.2–2 cm. long, more or less glandular-puberulous or sometimes glabrate, segments linear-subulate; corolla mauve, finely pubescent, to 6 cm. long, narrow portion of tube 1–1.5 cm. long, expanded part funnelform or campanulate, slightly oblique, 1–1.5 cm. broad at throat, limb 3–4.5 cm. broad, lobes rounded, 1–1.5 cm. wide; capsules 2–3 cm. long, cylindric, glabrous except for apical calluses (these puberulous), usually 20-seeded.

Lowlands of Mexico and Central America. Oct.–April.

**7. Ruellia nudiflora** (Engelm. & Gray) Urban, Symb. Antill. **7:** 382, 1912, var. **nudiflora**

*Dipteracanthus nudiflora* Engelm. & Gray, Bost. Jour. Nat. Hist. **5:** 229, 1845.

*Cryphiacanthus viscosus* Oerst., Nat. For. Kjøbenhavn Vid. Medd. **1854:** 128, 1855. Not *Ruellia viscosa* H. B. K., 1818.

Roots clustered, thick-fibrous; stems erect or ascending, 5–30 cm. high, usually branched, obscurely quadrangular, puberulous, occasionally beset with longer spreading, additional hairs, sometimes glabrate; petioles 0.5–4 cm. long, puberulous and pilose; leaf blades ovate or oblong, 2–12 cm. long, 1.5–7 cm. wide, obtuse, narrowed or rarely rounded at base and decurrent on petiole, thickish, undulate-crisped, sparingly puberulous or pilosulous, sometimes glabrate; inflorescence a terminal compound cyme, branches ascending, with a pediceled flower in each fork or, when peduncles are very short, an interrupted spike of verticillasters, or occasionally a single verticillastrate head; peduncles to 5 cm. long but usually much shorter, puberulous, more or less glandular; bracts linear, 3–10 mm. long; pedicels 3–5 mm. long, glandular-puberulous; calyx 12–18 mm. long, glandular-puberulous, tube very short, segments linear-subulate; corollas purple, finely pubescent, 3–6 cm. long, slender part of tube to 3 cm. long and 1–2 mm. in diameter, expanded part funnelform or narrowly campanulate, 1–1.5 cm. broad at throat, limb 1.5–3.5 cm. broad, lobes suborbicular, 10–15 mm. broad, erose; stamens 9 and 11 mm. long, glabrous; anthers 4.5 mm. long; ovary puberulous; style 3.5 cm. long, sparingly pilosulous; capsule 15 mm. long, 4 mm. broad, pointed at tip, short-stipitate at base, glandular-puberulous, 8–16-seeded; seeds lenticular, 3.5 mm. long, 3 mm. broad, mucilaginous-pilose when moist.

Open places, fields and roadsides, Lower Sonoran and Tropical Zones, Arizona to Texas and throughout much of Mexico. April–Aug.

**7a. Ruellia nudiflora** var. **glabrata** Leonard, Jour. Wash. Acad. **17:** 518, 1927

*Ruellia tuberosa* var. *occidentalis* A. Gray, Syn. Fl. N. Amer. ed. 1, **2**[1]**:** 325, 1878, in part.

Large herbs; stems erect, branched, tips glandular-puberulous, lower part sparingly pilose, cystoliths prominent; petioles 1–3 cm. long, glabrous or sparingly pilose; leaf blades ovate to oblong-ovate, 6–19 cm. long, 3–10 cm. wide, rounded or obtuse at apex, rounded then narrowed at base and decurrent on petiole, undulate or crisped, glabrous or costa and veins sparingly pilose, cystoliths prominent; branches of inflorescence, pedicels, and calyces glandular-pilosulous; corollas 2.5–4 cm. long, slender part of tube about 10 mm. long and 2 mm. in diameter, upper expanded part funnelform, 1.5 cm. broad at throat, limb 2–3 cm. broad, lobes rounded, 10–15 mm. wide, entire or undulate; stamens 6 and 12 mm. long; anthers 3.5–4 mm. long; style about 2 cm. long, sparingly puberulous; capsules 15 mm. long, 3–6 mm. broad, glandular-puberulous.

Swampy meadows, swales, and thickets, Lower Sonoran to Tropical Zones, Arizona, New Mexico, Texas, Mexico and Guatemala. April–Oct.

## 4. ELYTRARIA Michx., Fl. Bor. Amer. 1: 8, 1803

Caulescent or acaulescent herbs. Leaves alternate or sometimes subopposite, basal or crowded at ends of branches. Flowers borne in dense peduncled spikes, both spike and peduncle bearing imbricate coriaceous bracts. Calyx scarious, segments narrow, entire or toothed at apex. Corolla white or blue; tube slender; limb 2-lipped; upper lip 2-lobed; lower one 3-lobed. Stamens 2, barely exserted. Anthers 2-celled; sacs equal, parallel, sometimes awn-tipped at base. Staminodes usually wanting. Ovary 2-celled. Ovules 6–10 in each cavity. Capsules narrow, contracted at base, acute at apex.

**1. Elytraria imbricata** (Vahl) Pers., Syn. Pl. **1:** 23, 1805

*Justicia imbricata* Vahl, Eclog. Amer. **1:** 1, 1796.
*Verbena squamosa* Jacq., Pl. Hort. Schönbr. **1:** 3, *pl. 5,* 1797.
*Elytraria tridentata* Vahl, Enum. **1:** 107, 1804.
*Elytraria scorpioides* Roem. & Sch., Syst. Veg. **1:** 128, 1822.
*Elytraria squamosa* Lindau, Anal. Inst. Fisico-Geogr. Costa Rica **8:** 299, 1895.

Acaulescent, or leaves crowded at tip of a glabrous or sparingly pilosulous stem to 30 cm. long; leaf blades ovate to oblong or obovate, rarely linear-lanceolate, usually 3–12 cm. long, 1.5–4 cm. wide, blunt or acutish at apex, narrowed at base to a slender winged petiole, both surfaces appressed-pilose or glabrate, margins undulate; scapes numerous, axillary, usually 5–24 cm. long, simple or branched (sometimes leafy at tip), covered by tightly appressed, ovate to subulate scales; spikes 1 to several, to 6 cm. long; bracts oblong to elliptic, 3–6 mm. long, 1–2 mm. wide, firm, awn-tipped and bearing near apex a pair of triangular or rhombic hyaline teeth; bractlets 3 mm. long; calyx segments thin, upper bidentate; corolla blue, 5–8 mm. long; capsule glabrous.

Dry rocky slopes, banks, or ledges, Lower Sonoran to Tropical Zones, southern Arizona to Texas, Mexico, Central America, the West Indies, northern and western South America, and in India and the Philippine Islands. Sept.–May.

A widely distributed species, extremely variable in its habit of growth. In certain parts of Sonora it is known as Cordonsillo.

## 5. TETRAMERIUM Nees, in Benth., Bot. Sulph. 147, 1844

Herbs or small shrubs. Stems erect or spreading-decumbent, occasionally quadrangular, smooth, whitish, bark exfoliating, tips of branches pubescent in 2 lines. Leaves sessile or petiolate, rotund-ovate to linear, entire, both surfaces pubescent to glabrous. Inflorescence terminal and axillary, spicate, more or less imbricate; spikes 4-angled. Bracts single, herbaceous, opposite, entire, sessile to short-petiolate, loose or imbricate, cordate to linear, 3–5-nerved, pubescent to glabrous, ciliate, more or less mucronate, erect or recurved. Bractlets 1 to several, herbaceous, sessile or short-petiolate, filiform to ovate-acuminate, pubescent to glabrous, ciliate. Flowers usually solitary, borne in axils of bracts. Calyx 4–5-parted, cleft nearly to base; segments usually equal, filiform to ovate-acuminate, pubescent to glabrate, ciliate. Corolla tubular-infundibular, bilabiate; anterior lip 3-lobed, lobes oblanceolate to oblong, nearly equal; posterior lip usually subemarginate; tube straight or slightly curved, one-half to one-third length of corolla, expanding slightly at base, abruptly dilated at throat. Stamens 2, included; anther cells converging slightly toward apex, basal lobe of one slightly longer than that of other. Capsules obovate, apiculate, contracted below into a solid base, more or less hispid, usually 4-seeded. Seeds tuberculate or muriculate.

Bracts oblong-ovate, rounded to subtruncate at tip, inconspicuously ciliate .. 1. *T. fruticosum*
Bracts lanceolate to ovate-lanceolate, acute to acuminate at apex, hispid-ciliate ............. ........................................................... 2. *T. hispidum*

### 1. Tetramerium fruticosum Brandegee, Zoe 5: 171, 1903

Herbs or small shrubs, to 30 cm. high; stems freely branched, branches slender, minutely puberulous, with scattered longer hairs intermixed, or older branchlets inconspicuously pubescent in 2 lines or glabrous; leaves lanceolate to ovate, 1–3 cm. long and 0.5–1.4 cm. wide, gradually acuminate, mucronulate, rounded or obtuse at base, at first puberulous, short hairs intermixed with longer ones, glabrescent; inflorescence more or less loosely spicate, spikes 1–6 cm. long, internodes usually 2 mm. long; bracts oblong-ovate, 7–15 mm. long, 4–6 mm. wide, rounded to subtruncate, mucronulate, cuneate and subpetiolate at base, closely glandular-puberulous, with a few scattered longer hairs intermixed, glabrescent, inconspicuously ciliate; bractlets linear-lanceolate, 7–10 mm. long, 0.5–0.8 mm. wide, pilose; calyx segments 5, lanceolate, 5 mm. long, pilose; corolla 10–12 mm. long, straw-colored, middle lobe of anterior lip obovate, 5–6 mm. long, lateral lobes oblanceolate, 6–7 mm. long, 3 mm. wide, posterior lip obovate, 6 mm. long, about 2.7 mm. wide, narrowed to base, subemarginate; capsule 5–6 mm. long, about 2 mm. broad, glabrous, constricted solid basal part about 2 mm. long.

Along arroyos and under trees, Lower Sonoran and Tropical Zones, central Baja California south to the Cape Region and in central to southern Sonora. Oct.–May.

### 2. Tetramerium hispidum Nees, in A. DC., Prod. 11: 468, 1847

*Tetramerium nervosum* var. *hispidum* Torr., Bot. Mex. Bound. 125, 1859.
*Dianthera sonorae* S. Wats., Proc. Amer. Acad. 24: 67, 1889.

Herb to 30 cm. high; stems terete, branched, brittle, pilosulous or glabrous; leaf blades lanceolate to ovate-lanceolate, 1–7 cm. long, 0.5–2.5 cm. wide, obtuse at apex, rounded to cuneate at base, pilosulous; petioles to 8 mm. long, slender, pilose; flowers borne in terminal and lateral spikes to 9 cm. long and about 8 mm. in diameter, rachis glabrous or sparingly pilose; bracts lanceolate to ovate-lanceolate, 7–15 mm. long, about 4 mm. wide, acute, ending in a spine 0.5 mm. long, ciliate, 3- or 5-nerved, closely imbricate; bractlets narrowly lance-aristate, 2.5 mm. long, 0.5 mm. wide, tip pilose; calyx segments similar to bractlets but slightly longer; corolla 1 cm. long, white to deep yellow with occasional purplish markings, glabrous, tube slender, lips 5 mm. long, upper lip entire, lower 3-lobed, lobes elliptic, 3 mm. long; anther lobes parallel, slightly unequal, about 1 mm. long; capsule 4.5 mm. long, pubescent or glabrate; seeds papillose.

Dry open ground, Lower Sonoran and Tropical Zones, Texas, New Mexico, and Arizona south to the Cape Region, Baja California, and Costa Rica. Sept.–June.

## 6. **HENRYA** Nees, in Benth., Bot. Sulph. 148, *pl. 49,* 1844

Fruticose to suffruticose perennials. Stems erect, spreading, or decumbent, branches terete or quadrangular (sometimes winged), evenly pubescent or hairs disposed in 2 lines. Leaves petiolate, entire, pubescent to glabrous. Inflorescences terminal and axillary, spicate; bracts lanceolate-ovate to linear-pubescent to glabrate. Flowers solitary, sessile, or short-pedicellate, subtended by 2 involucral bracts, these connate from base nearly to apex on posterior side, pubescent to glabrous, mucronate at or near apex. Calyx lobes 5, small, lanceolate-ovate to filiform, pubescent or glabrous. Corolla tubular-infundibuliform, bilabiate; tube straight or slightly curved, slightly enlarged near base; posterior lip obovate-oblong; anterior lip 3-lobed; lobes nearly equal. Stamens 2, exserted, extending beyond corolla lips, cells 2, nearly parallel, obtuse. Capsules obovate, apiculate, compressed at base, glabrous, 2-seeded. Seeds disciform, hispid.

Leaf blades ovate to lanceolate, usually acuminate at apex .................. 1. *H. insularis*
Leaf blades ovate to obovate, rounded to merely acute at apex ............. 2. *H. brevifolia*

### 1. **Henrya insularis** Nees, in Benth., Bot. Sulph. 148, *pl. 49,* 1844

*Henrya scorpioides* sensu Nees, in A. DC., Prod. **11:** 491, 1847, in part.
*Tetramerium scorpioides* sensu Hemsl., Biol. Centr. Amer. Bot. **2:** 526, 1882, in part.

A branched shrub to 1 m. high; stems evenly glandular-pilose, becoming pubescent in 2 lines or finally glabrate; petioles to 2.5 cm. long, glandular-pilose to glabrous; leaf blades ovate to oblong or lanceolate, to 8 cm. long, 4 cm. wide, usually acuminate at apex, acute to rounded at base, usually 3–8-veined on either side of costa, glandular-pilose to glabrate; spikes to 25 cm. long, becoming rather loose; cauline bracts oblanceolate, to 3 mm. long and 1.5 mm. wide, mucronulate, glandular-pilose; involucral bracts oblanceolate, 7–9 mm. long, about 2 mm. broad, acute, glandular-pilose, mucro erect directly on apical margin, 0.4 mm. long; calyx segments linear-lanceolate, about 1 mm. long, glandular-puberulent; corolla 10 mm. long, glabrous, white or yellow, with purple markings on lower lip; capsule 7 mm. long, 1.8 mm. broad, glabrous.

On rocky slopes, Lower Sonoran to Tropical Zones, southern Sonora and the Cape Region, Baja California, south to Panama; rare in the Sonoran Desert. Feb.–May.

### 2. **Henrya brevifolia** Happ, Ann. Mo. Bot. Gard. **24:** 547, 1937

*Henrya costata* var. *glandulosa* Brandegee, Zoe **5:** 171, 1903.

Branching shrub; stems glandular-pilose, becoming pubescent in 2 lines or glabrate; petioles to 0.5 cm. long, glandular-pilose to glabrous; leaf blades ovate to obovate, to 5 cm. long and 1.5 cm. wide, rounded to acute at apex, cuspidate, rounded at base, usually 3–5-veined on either side of costa, glandular-pilose, glabrescent; spikes to 12 cm. long, rather loose; cauline bracts oblanceolate, 3–5 mm. long, about 1 mm. wide, mucronulate, glandular-hispid; involucral bracts oblanceolate, 6–8 mm. long, about 2 mm. broad, acute, glandular-hispid with additional long simple hairs, mucro directly on apical margin, erect, 0.3 mm. long; calyx segments linear-lanceolate, 0.8 mm. long, glandular-puberulous; corolla 10–12 mm. long, pale yellow, glabrous; capsule 6 mm. long, about 1.5 mm. broad, slightly puberulous near apex.

Along arroyos and in partial shade on hillsides, Lower and Upper Sonoran Zones, central Sonora east into northeastern Mexico. March–May.

## 7. DICLIPTERA Juss., Ann. Mus. Paris **9:** 267, 1807

Herbs. Leaves entire, mostly ovate to lanceolate, petioled. Inflorescence cymose, flowers 1–3, subtended by an involucre of 2 conspicuous bracts; involucres 3–5, umbellate on a short or elongate, axillary peduncle. Calyx 5-parted, hyaline. Corolla narrow, slightly ampliate; limb 2-lipped. Stamens 2; anther sacs often unequal, superposed, longer one sometimes calcarate at base. Staminodes none. Capsules ovate or suborbicular; placenta separating elastically from walls. Seeds 2 or 4.

### 1. Dicliptera resupinata (Vahl) Juss., Ann. Mus. Paris **9:** 268, 1807

*Justicia resupinata* Vahl, Enum. **1:** 114, 1804.
*Dicliptera torreyi* A. Gray, Proc. Amer. Acad. **20:** 309, 1885.
*Diapedium torreyi* Woot. & Standl., Contr. U.S. Nat. Herb. **19:** 598, 1915.

Annual or perennial, suffrutescent herbs to 60 cm. high; stems branching, glabrous or sparingly puberulous about nodes, hairs curved; petioles to 2 cm. long; leaf blades lanceolate to lance-oblong or ovate, to 8 cm. long, acuminate (tip blunt), narrowed at base, glabrous or nearly so, inconspicuously ciliolate; common peduncles short or elongate (4 cm.), bibracteate at apex (bracts small, linear-oblanceolate), 3–5-branched, branches to 3 cm. long; involucral bracts cordate, deltoid-subcordate, or rarely round-obovate, to 7 mm. long and 8 mm. wide, rounded or obtuse at apex, often emarginate, sometimes subapiculate, very flat, veiny, glabrous or nearly so; corolla purple, with darker purple dots, about 1.5 cm. long, lips obovate; capsules about 5 mm. long, glabrous, flat; seeds flat, about 2 mm. broad, sparingly and minutely hispid.

Dry wooded slopes, Mexico and southwestern Arizona. With age the plants are usually leafless and the bracts often become papery and white. Sept.–May.

Some botanists recognize part of the material here described as *D. resupinata* as an additional species, *D. pseudoverticillaris* A. Gray, characterized by basally truncate or abruptly cuneate, partly united bracts and by sessile or subsessile cymes, in contrast to the broadly cordate or slightly cuneate bracts and the much longer peduncled cymes (surpassing the leaves) attributed to true *D. resupinata,* but the characters are highly variable, and examination of a large series of specimens shows no adequate basis for segregation.

## 8. CARLOWRIGHTIA A. Gray, Proc. Amer. Acad. **13:** 364, 1878

Herbaceous to suffrutescent, straggling undershrubs with slender branches and entire, linear to ovate, opposite leaves. Flowers few, comparatively small, in loose spikes, racemes or panicles. Calyx 5-parted or 5-cleft. Corolla lobes 4, imbricate in bud, white or cream streaked with purple or pale purple throughout; tube slender, scarcely dilated at throat. Stamens 2; anther cells equal, inserted at same level. Capsules flattened, clavate, apiculate to acuminate at apex, distinctly stipitate. Seeds 4, flattened, variously papillate or sculptured on faces.

Inflorescence racemose-paniculate; corollas purple; filaments short-pilose along one side with purple hairs; leaves narrowly linear .............................. 1. *C. linearifolia*

Inflorescence of simple or interrupted spikes, these sometimes forming a leafy panicle; corollas white to cream, with spots of yellow and streaks of purple but never uniformly purple; filaments glabrous (or sparingly bearded below in *C. pectinata*); leaves narrowly lanceolate to broadly ovate (or if linear the flowers not as in *C. linearifolia*):

- Leaves linear to narrowly lanceolate, four to ten times as long as broad:
  - Mature capsules sparsely hirsutulous; outer pair of seeds much thinner than inner, margins coarsely dentate-winged, inner ones coarsely papillate but wingless; rachis of inflorescence subglabrous; sepals 1.2–1.6 mm. long .......... 2. *C. pectinata*
  - Mature capsules glabrous; outer pair of seeds closely but finely papillose, inner sparsely so except on margins, neither pair winged; rachis of inflorescences puberulent to hirsute, often glandular; sepals 2.5–3.5 mm. long ................ 3. *C. arizonica*
- Leaves ovate-lanceolate, ovate, or ovate-cordate, rarely over two and one-half times as long as broad:
  - Capsules narrowly clavate, terminally subtruncate, bicorniculate at summit, 1.5–2 mm. wide; seeds clavate-oblong, minutely papillate only at larger end; leaf blades thin, dark green, sparsely appressed-pubescent above, paler beneath ..... 4. *C. sinaloensis*
  - Capsules broadly clavate, body abruptly narrowed to stipe, apiculate at apex, 4–5 mm. wide; seeds cordate, sparsely to closely papillate over most of surface; leaf blades thick, usually canescent, moderately to densely puberulent:
    - Branches of inflorescences and leaves minutely puberulent; gland-tipped hairs few or lacking on rachises ........................................ 5. *C. cordifolia*
    - Branches of inflorescences and leaves pubescent or more coarsely puberulent; coarse, gland-tipped hairs conspicuous on rachises of inflorescences .... 6. *C. californica*

## 1. **Carlowrightia linearifolia** (Torr.) A. Gray, Proc. Amer. Acad. **13:** 364, 1878

*Schaueria linearifolia* Torr., Bot. Mex. Bound. 123, 1859.

Herbaceous to suffrutescent plant with slender, often decumbent stems 2–5 dm. long, these minutely puberulent or becoming glabrate and subshining in age; leaves linear, 0.5–1.2 mm. wide, 1–3.5 cm. long, entire, gradually narrowed at base, acute to obtuse, minutely puberulent with short, straight hairs, midvein impressed beneath; lateral flowering branches 1–3 dm. long in axils of upper 6–12 pairs of leaves; inflorescence racemose on slender, apically bibracteate, puberulent peduncles 5–12 mm. long, each bearing 1–2 flowers; sepals linear-subulate, 2.8–3.2 mm. long, minutely puberulent; corolla purple, glabrous, tube slender, 2–2.5 mm. long, flaring slightly, lobes obtusely rounded at apex, 4–6 mm. long; filaments dark purple, about equaling corolla, short-pilose along one side on lower part, usually glabrous above; anthers sagittate; style filiform, 6–8 mm. long; capsule 10–12 mm. long, glabrous, body broadly ovate, 3–4 mm. wide, apiculation about 1 mm. long, stipe 5–6 mm. long; seeds orbicular-cordate, thin, sparsely sprinkled with low rounded papillae, margins scarcely more papillate than faces.

On mesas and along washes, Lower Sonoran Zone to the lower margin of the Upper Sonoran Zone, southeastern Arizona east to western Texas. Aug.–Sept.

**2. Carlowrightia pectinata** Brandegee, Proc. Calif. Acad. II, **3:** 160, 1891
*Carlowrightia ?fimbriata* Brandegee, op. cit. 161.

Perennial, herbaceous to suffrutescent plant with numerous slender geniculate stems 2–10 dm. long, branches often rooting at nodes; older bark white and shreddy; leaves linear to narrowly lanceolate-acuminate, 1–7 mm. wide, 1.5–9 cm. long, narrower ones often revolute, glabrous or sparsely pubescent with coarse, short, appressed hairs, blade tapering gradually to sessile base, this decurrent 2–3 mm. below node; inflorescence paniculate-spicate, flowers sessile, scattered on slender branches; bracts about 2 mm. long, linear-subulate; bractlets 0.5–1 mm. long; sepals linear, about equal, 1.2–1.6 mm. long, subglabrous to minutely puberulent; corollas 7–8 mm. long, white or buff, streaked with pale to dark purple, tube 1–1.5 mm. long, lobes sparsely short-hirsute on outside, more densely so along midveins; filaments sparsely bearded below, purplish, nearly equaling corolla lobes; anthers often muticous; capsule clavate, 7–9 mm. long, sparsely puberulent, apiculation about 0.2–0.3 mm. long, body equaling or slightly shorter than stipe, 3–3.5 mm. wide; inner pair of seeds minutely muriculate on outer surface, more coarsely so on inner face, outer pair smooth on faces, with one edge curling over edge of inner seed, other edge and radicular angle strongly pectinate-winged, teeth smooth to fimbriate distally.

Shaded hillsides, streamsides, and in canyons, Lower Sonoran and Tropical Zones, Cape Region of Baja California and in southern Sonora, south into Sinaloa. Feb.–Nov.

**3. Carlowrightia arizonica** A. Gray, Proc. Amer. Acad. **13:** 364, 1878

Bushy or straggly, suffrutescent plant 2–6 dm. tall, with slender, brittle, gray-green, minutely puberulent branches, older bark irregularly striate-fissured and buff to gray-brown; leaves linear-lanceolate to ovate-lanceolate, 2–5 mm. wide, 1–3 cm. long, short-petiolate, rounded at base, acute, apiculate, or acuminate at apex, sparsely puberulent to subglabrous, only midrib obvious; inflorescence openly paniculate-spicate, floriferous branches 3–12 cm. long, glandular-puberulent and sparsely hirsute; flowers sessile; bracts and bractlets subulate, 1.5 mm. or less in length; sepals 2.5–3.5 mm. long, subulate, puberulent; corollas cream-colored, striped with lavender or purple, tube 3–4 mm. long, oblong-oblanceolate lobes 6–8 mm. long, sparsely short-hirsute without; capsules 10–12 mm. long, body slightly shorter than slender stipe, glabrous; seeds tawny, 3–3.5 mm. long, outer ones uniformly papillate-granulose with small low bosses, inner ones sparsely so to nearly smooth.

Dry stony slopes, hillsides, and margins of washes, Lower Sonoran Zone and margins of Upper Sonoran Zone, southern Arizona and adjacent northern Mexico. March–May.

**4. Carlowrightia sinaloensis** Leonard & Gentry, Brittonia **6:** 327, 1948

Suffrutescent, sparsely and finely pubescent plant 3–10 dm. tall, with pale yellow to whitish bark on older branches; petioles slender, 1.5–3 cm. long; leaf blades ovate to ovate-lanceolate, 2–4.3 cm. wide, 4–9 cm. long, acuminate, dark green above, paler beneath, sparingly pilose above, more densely so beneath, ciliate-margined; inflorescences lax, spikes 2–5 cm. long, with puberulent, sparingly glandular rachises; bracts subulate, 2–3 mm. long, sparsely glandular-puberulent; bractlets 1.5–2 mm. long; sepals subulate, 2–2.5 mm. long, minutely and sparingly puberulent; corolla white, streaked with pale purple, tube very slender, 1.5–2 mm. long, lobes 5 mm. long, puberulent on outer surfaces; stamens about one-half to two-thirds as long as corolla lobes; anthers oval, 1.5 mm. long; capsules narrowly clavate, 11–13 mm. long, 1.5–2 mm. wide through body, with a lateral spur about 0.5 mm. long at upper angle of each valve, sparsely puberulent to subglabrate; seeds bluntly oblanceolate, about 2 mm. long, faintly granulose or with a few scattered minute papillae near distal end, wingless, dark brown to nearly black.

Mostly in shade of larger shubs and trees, rocky hillsides and along arroyos and canyons, Lower Sonoran Zone and lower parts of Upper Sonoran Zone, southern Sonora to Sinaloa. Oct.–Nov.

### 5. **Carlowrightia cordifolia** A. Gray, Proc. Amer. Acad. **21**: 406, 1886

Straggly, herbaceous to suffrutescent plants 1 m. or less tall, with finely puberulent herbage; petioles 5–25 mm. long; leaf blades lanceolate-cordate to broadly ovate-cordate, 1–5 cm. wide, 1.5–7 cm. long, shallowly cordate to subtruncate at base, short-acuminate or apiculate at apex, 4–7 pairs of lateral veins apparent, dark green above, slightly paler beneath; inflorescences open panicles of slender, sparingly floriferous spikes 1.5–10 cm. long; bracts lance-subulate, 2.5–3 mm. long, 3-nerved; bractlets similar to and only slightly shorter than bracts; sepals linear-subulate, equalling bracts, minutely puberulent; corollas pale yellow or white, striped with purple, tube 3–4 mm. long; corolla lobes 10–12 mm. long, glabrous or essentially so, upper one with a slender claw 3–4 mm. long; filaments 8–10 mm. long, glabrous; anthers oval, purplish brown, 0.8–1 mm. long; capsules 12–13 mm. long, body 4–5 mm. wide, 5–7 mm. long, glabrous; seeds 3.5–4.2 mm. long, outer ones smooth or with a few scattered papillae, inner ones more densely and prominently papillate, often lavender to purplish.

Rocky hillsides, desert flats, and along canyons, Lower and Upper Sonoran Zones, central Sonora south into Sinaloa, east into Chihuahua, and in southern Baja California. Aug.–Nov.

This species shows considerable variation in leaf size and in density of pubescence.

### 6. **Carlowrightia californica** Brandegee, Zoe **5**: 172, 1903

Diffuse suffrutescent plant 2–8 dm. tall, with sparingly puberulent to glabrate branches and herbage; leaves short-petiolate or uppermost reduced, subsessile; petioles to 1.5 cm. long, puberulent and often with a few scattered, longer, spreading, simple hairs; leaf blades lance-ovate to ovate, 0.5–2.5 cm. wide, 1–5 cm. long, rounded to subcordate at base, acute to apiculate at apex, firm, dark green above, slightly paler beneath, netted venation obvious; inflorescences open panicles of slender spikes 3–20 cm. long, rachises closely puberulent with simple and gland-tipped hairs; bracts linear-subulate, 3 mm. long, glandular-puberulent and sparingly hirsute; bractlets similar, slightly shorter; sepals subulate-attenuate, 2.5–3.5 mm. long, glandular-puberulent; corollas 12–15 mm. long, nearly as wide, tube 4–5 mm. long, lobes 6–8 mm. long, cream to nearly white, with faint pink to purplish veins, upper one with a yellow central spot, glabrous within, sparsely hirsute without; filaments glabrous, 7–9 mm. long; capsules 11–13 mm. long, body 4–5 mm. wide, strongly apiculate, glabrous; seeds 4–4.5 mm. long, outer ones smooth except for a few minute papillae around marginal rim, inner ones sparsely to moderately papillose on both surfaces or sometimes only on one surface.

On gravelly flats, arroyos, and rocky outwash slopes and hillsides, Lower Sonoran Zone, central Baja California south to the Cape Region; possibly also in Sinaloa. Flowering after rains during any month of the year.

## 9. ANISACANTHUS Nees, Linnaea 16: 307, 1842

Branched shrubs. Stems terete, tips usually pubescent, older portions covered with brown or gray bark, exfoliating in thin strips. Leaves petiolate, entire. Flowers red, borne in contracted, subsessile, bracteate, axillary cymes. Bracts foliose, usually small. Calyx 5-parted; segments narrow, subequal. Corolla tube slender, elongate, straight or slightly incurved; limb bilabiate; posterior lip entire; anterior one 3-parted. Stamens 2; anther cells equal. Ovules 2 in each cell. Capsules glabrous. Seeds discoid, more or less tuberculate.

**1. Anisacanthus thurberi** (Torr.) A. Gray, Syn. Fl. N. Amer. ed. 1, **2**[1]: 328, 1878
*Drejera thurberi* Torr., Bot. Mex. Bound. 124, 1859.
*Drejera puberula* Torr., op. cit. 123.

Shrub to 1.5 m. high; stems sparingly pubescent in 2 lines; petioles 1–5 mm. long; leaf blades lanceolate to oblong, to 5 cm. long, 1.5 cm. wide, pilosulous to glabrate; bracts and bractlets oblanceolate to linear, acute, to 10 mm. long, 3 mm. wide, puberulous, often slightly pilose on costa and margin; pedicels 2–5 mm. long, glandular; calyx about 1 cm. long, glandular-puberulous, segments linear-attenuate; corollas to 4 cm. long, tube tubular-funnelform, lobes equaling or shorter than tube, upper lip slightly emarginate; capsules 15–18 mm. long, about 6 mm. in diameter; seeds 2, about 5 mm. in diameter, slightly tuberculate.

Gravelly or rocky ravines, Lower Sonoran Zone and margins of Upper Sonoran Zone, central Arizona and southwestern New Mexico; throughout Sonora. Oct.–March.

## 10. SIPHONOGLOSSA Oerst., Nat. For. Kjøbenhavn Vid. Medd. 1854: 159, 1855

Suffrutescent herbs. Stems diffuse, branched. Leaves entire. Flowers paired, axillary, subsessile. Calyx segments 4. Corolla white or yellowish white, 2-lipped; tube slender, elongate, terete; upper lip entire or 2-parted; lower one spreading, 3-lobed. Stamens 2, included or at least surpassed by upper lip; anthers 2-celled; sacs unequally inserted. Staminodes none. Ovary 2-celled, with 2 ovules in each cavity. Capsules contracted into a solid, stipelike base.

**1. Siphonoglossa longiflora** (Torr.) A. Gray, Syn. Fl. N. Amer. ed. 1, **2**[1]: 328, 1878
*Adhatoda ?longiflora* Torr., Bot. Mex. Bound. 125, 1859.

Herbs to 35 cm. high; stems branched, cinereous-puberulous, erect or ascending; leaf blades lanceolate, 2–6 cm. long, to 15 mm. wide, blunt at tip, cuneate at base, undulate, short-petioled, glabrous or costa slightly puberlous; calyx 5–7 mm. long, segments subulate-linear, about as long as subtending bracts; corolla white or yellowish white, 3–4 cm. long, tube slenderly cylindric, about twice as long as lips, upper lip notched, lower one 3-lobed, each lobe about as large as upper lip; capsules 8–9 mm. long, glabrous; seeds dark brown, cordate, flattened, muricate.

Canyons and mountain sides, Lower and Upper Sonoran Zones, southern Arizona and western Mexico. May–Oct.

## 11. JACOBINIA Moric., Pl. Nouv. Amer. 156, 1846

Shrubs. Leaves entire, petioled. Flowers sessile or short-pedicellate, axillary, cymose. Calyx 5-parted; segments narrow. Corolla red; tube usually slender and elongate, straight or curved; limb 2-lipped; upper lip narrow, erect, entire or minutely

2-lobed; lower lip 3-lobed. Stamens 2; anther cells more or less unequally attached, muticous or lower all minutely apiculate. Staminodes none. Capsules clavate, 4-seeded.

Stems and leaves pubescent .......................................... 1. *J. ovata* var. *ovata*
Stems and leaves glabrous or essentially so .................... 1a. *J. ovata* var. *subglabra*

**1. Jacobinia ovata** A. Gray, Proc. Amer. Acad. **21:** 405, 1886, var. **ovata**

Shrubs to 2 m. high; branches grayish, pubescence a mixture of minute hairs and longer ones (to 1 mm. long), or lower stems glabrate; petioles 4–10 mm. long; leaf blades ovate to oblong-ovate, to 6 cm. long and 3 cm. wide, short-acuminate to obtuse or acutish, acute or rounded at base and decurrent on petiole, densely to sparingly pubescent, usually more so beneath than above, hairs beneath often longer and denser on costa and veins; flowers sessile or short-pedicellate, crowded in axils of uppermost leaves; bracts and bractlets narrowly linear-lanceolate, pubescent, pinnate-veined, bracts 1.5 mm. wide, bractlets 0.75 mm. wide, slightly shorter and more strongly ciliate than bracts; calyx 8 mm. long, pubescent, tube 1.5 mm. long, segments linear, 1–1.25 mm. wide, acuminate, ciliate, costa strong, lateral veins faint; corolla pubescent, 2.5–3 cm. long, tube slender, lips about 15 mm. long, upper lip erect, triangular-ovate, about 5 mm. wide near base, bidentate at apex, lower lip 3-lobed, lobes oblong, 8 mm. long, 3 mm. wide, obtuse; stamens exserted but not exceeding upper lip, basal anther lobes slightly spreading; ovary glabrous; style subglabrous; capsules glabrous, about 1 cm. long, narrowed to a solid base about 2 mm. long; seeds subcordate, flattened, reddish, turning gray at maturity, glabrous, rugose.

Dry rocky ravines, Lower Sonoran to Tropical Zones, southern Arizona, Sonora, Sinaloa, and western Chihuahua. Sept.–April.

**1*a*. Jacobinia ovata** var. **subglabra** S. Wats., Proc. Amer. Acad. **24:** 67, 1889

Branches glabrous; leaf blades ovate, obtuse, rounded, or subtruncate at base, glabrous or basal portion of costa sparingly pubescent. Similar in other respects to var. *ovata.*

Ravines in the vicinity of Guaymas, Sonora, Mexico. March–April.

## 12. JUSTICIA L., Sp. Pl. 15, 1753

Shrubs. Leaves petiolate, usually ovate to lanceolate, entire. Flowers mostly in secund spikes, rarely fasciculate or solitary. Bracts small. Calyx segments narrow, nearly equal or one much smaller than others. Corollas white to purple, red, or scarlet, sometimes with white or purple markings; tube longer or shorter than limb; throat ampliate; upper lip 2-lobed or notched; lower one 3-lobed. Stamens 2; anther cells 2, unequally attached, at least the lower more or less calcarate. Ovules 2 in each cavity. Capsules oblong to obovate.

Calyx 4-parted:
  Inflorescence glandular-pilose; corolla white or light mauve .............. 1. *J. salviiflora*
  Inflorescence glandular-puberulous; corolla rose or reddish purple:
    Branchlets closely white-tomentose ........................ 2. *J. insolita* var. *insolita*
    Branchlets pubescent with spreading hairs ................ 2a. *J. insolita* var. *tastensis*
Calyx 5-parted:
  Calyx segments unequal, posterior one much shorter and narrower than others; bracts shorter than calyx ........................................................ 3. *J. hians*
  Calyx segments similar; bracts longer than calyx:
    Bracts and bractlets spatulate, relatively large, obtuse .................... 4. *J. palmeri*
    Bracts and bractlets linear-lanceolate or oblanceolate, small ............ 5. *J. mexicana*

**1. Justicia salviiflora** H. B. K., Nov. Gen. & Sp. **2:** 233, 1817
*Justicia paniculata* Rose, Contr. U.S. Nat. Herb. **1:** 348, 1895.

Small shrub to 1 m. high; branches numerous, glandular-pilose or older stems glabrate, whitish; leaves petiolate, lance-oblong to oblong-ovate, to 8 cm. long and 4.5 cm. wide, acute to acuminate, narrowed at base, pubescent beneath, glabrescent above; flowers mostly borne in few-flowered, secund, glandular-pilose spikes; bracts and bractlets small, filiform, to about 3 mm. long; calyx deeply 4-parted, segments filiform, about 8 mm. long; corolla 8–15 mm. long, tube somewhat swollen, white or light mauve, upper lip erect, rostrate, bidentate, lower one 3-lobed, spreading; stamens barely exserted, anther lobes oblique, hairy on back, separated by a broad connective broader above, lower cell spurred; capsule about 15 mm. long, puberulous, 4-seeded; seeds about 3 mm. broad, densely short-pubescent.

Canyon slopes, Lower and Upper Sonoran Zones, throughout western Mexico. Feb.–April.

**2. Justicia insolita** Brandegee, Proc. Calif. Acad. II, **2:** 195, 1889, var. **insolita**

A small dense shrub to 50 cm. high, branches short, numerous, stiff, spinose, minutely and closely white-tomentose; leaves oblong-ovate to lanceolate, to 25 mm. long and 10 mm. wide, obtuse or rounded at apex, sometimes emarginate, narrowed at base, minutely pubescent to almost glabrous, short-petioled; flowers scattered, sessile, borne at or near ends of branches; bracts and bractlets linear, 2–4 mm. long, glandular-puberulent; calyx 7 mm. long, deeply 4-parted, glandular-puberulous, segments linear-acuminate; corollas rose-purple, about 20 mm. long, deeply 2-lipped, tube very short and broad, 2-sulcate at base in front, throat rugose-veined, lips 2–4 times longer than tube, upper one galeate, emarginate, lower 3-parted, lobes oblong-ovate; stamens exserted but shorter than lips, anther cells parallel but not even, widely separated by a broad connective, upper cell muticous, lower one spurred; ovary 4-ovuled; capsule to 18 mm. long, narrowed to a sterile, solid, stipitate base 8-10 mm. long, puberulous, some hairs glandular; seeds flat, oblique at base, densely covered with retrorsely barbed bristles.

Along arroyos and on rocky cliffs, Lower Sonoran and Tropical Zones, southern Baja California. Oct.–April.

**2*a*. Justicia insolita** var. **tastensis** Brandegee, Zoe **5:** 173, 1903

A shrub to 2 m. high, branches numerous, slender, hirsute or older glabrate; leaf blades ovate to oval or oblong, to 3 cm. long and 1.5 cm. wide, acute to obtuse, short-petioled, sparingly to densely pubescent; flowers in loose secund spikes; corolla bright reddish purple, having whitish, featherlike markings on lower lip.

Dry arroyos, southern Baja California. Dec.–May.

**3. Justicia hians** (Brandegee) Brandegee, Univ. Calif. Pub. Bot. **6:** 194, 1915
*Beloperone hians* Brandegee, Proc. Calif. Acad. II, **2:** 194, 1889.
*Justicia wigginsii* Leonard, Jour. Wash. Acad. **32:** 342, 1942.

Shrubs; branches numerous, striate, closely white-puberulous or older parts glabrate, gray; petioles slender, to 5 mm. long; leaf blades ovate to elliptic or lanceolate, to 2 cm. long and 1 cm. wide, rounded to acute at apex, narrowed at base, inconspicuously puberulous; flowers borne in loose secund spikes, these puberulous with glandular and eglandular hairs intermixed; bracts in threes, middle one 2.5 mm. long, subcucullate, lateral ones lanceolate, 3 mm. long, about 0.75 mm. wide near base; calyx deeply 5-parted, posterior segment subulate, about 3 mm. long, others linear-lanceolate, about 6 mm. long and 1.5 mm. wide, rather firm, indistinctly striate-veined; corolla 3 cm. long, red, finely but sparingly pubescent, tube about 1 cm. long, somewhat saccate, upper lip straight, erect, 2-lobed at apex, lower lip somewhat spreading, 3-lobed, lobes oblong, about 7 mm. long, rounded at apex, middle lobe about 4 mm. wide at base, lateral ones about 2 mm. wide; stamens about as long as upper lip, anther cells unequal, one 2 mm. long, other 2.5 mm. long, spreading at base; capsule 11 mm. long, sparingly glandular-puberulous; seeds flattened, 2.5 mm. broad, strongly tuberculate, minutely papillose.

Rocky banks, Lower Sonoran Zone, central to southern Baja California. Oct.–April.

**4. Justicia palmeri** Rose, Contr. U.S. Nat. Herb. **1:** 75, 1890

Branched shrubs to 2 m. high; branches minutely cinereous-tomentulose; leaves short-petiolate, ovate to lanceolate, to 7 cm. long and 3 cm. wide, obtusely acuminate, obtuse or acute at base, glabrous or minutely and sparingly pubescent; flowers fasciculate in axils or borne in short, leafy-bracted spikes; bracts in threes, middle one elliptic to spatulate, to 15 mm. long and 7 mm. wide, rounded or obtuse and minutely apiculate at apex, narrowed at base, lateral ones about half as large; calyx about 4 mm. long, deeply 5-parted, segments lance-acuminate, minutely pubescent; corollas scarlet, 3–3.5 cm. long, deeply 2-lipped, tube 13–15 mm. long, upper lip minutely 2-lobed, lower one 3-parted, lobes oblong, 6–8 mm. long; anther cells parallel, unequally inserted, lower one shortly spurred; capsules glabrous, 10–12 mm. long, stipe about 5 mm. long; seeds 4, flattened, brown, rugose.

Gravelly slopes, Lower Sonoran Zone, central to southern Baja California. Aug.–March.

**5. Justicia mexicana** Rose, Contr. U.S. Nat. Herb. **1:** 348, 1895

A slender shrub to 2 m. high; branches pubescent or older glabrate; petioles slender, to 1.5 cm. long; leaf blades lanceolate to ovate, to 10 cm. long and 5 cm. wide, acute to acuminate at apex, acute or obtuse at base, sparingly pubescent to glabrate; flowers borne in terminal or axillary clusters; bracts in threes, middle one narrowly oblanceolate, 10–12 mm. long, 1.5–3.5 mm. wide, lateral ones linear-subulate, all ciliate and both hirsute and puberulous; calyx deeply 5-parted, about 7 mm. long, segments narrowly lanceolate, ciliate and both pilose and puberulous; corollas scarlet, about 2.5 cm. long, finely pubescent, tube 1.5–2 cm. long, lower portion slender, about 1.5 mm. in diameter, gradually enlarged from near base to 5 mm. wide at throat, upper lip oblong, entire or slightly notched, erect, lower lip spreading, 3-lobed, lobes oblong, rounded; stamens exserted beyond tube, anther cells unequally inserted, lower one slightly spurred; capsules about 12 mm. long, glabrous, stipe slender, 5–6 mm. long; seeds reddish, glabrous.

Rocky slopes, Lower Sonoran and Tropical Zones, southern Sonora and Sinaloa. Sept.–April.

## 13. BELOPERONE Nees, in Wall., Pl. Asiat. Rar. 3: 76, 102, 1832

Herbs or shrubs. Leaves usually elliptic or ovate, entire, occasionally crenate or undulate. Flowers solitary in axils of bracts, or racemose; bracts small, or large and conspicuous. Calyx 5-parted; segments narrow. Corollas red or purple; tube slender, elongate, scarcely ampliate above; limb deeply 2-lipped; upper lip narrow, erect, entire or minutely 2-lobed; lower lip 3-lobed and more or less spreading. Stamens 2; anther cells unequal, usually superposed, at least lower one calcarate at base. Staminodes none. Capsules clavate, 4-seeded.

Flowers borne in terminal, small-bracted racemes; calyx 7–9 mm. long ...... 1. *B. californica*
Flowers borne in axils of large, leaflike bracts; calyx 12–18 mm. long ....... 2. *B. purpusii*

**1. Beloperone californica** Benth., Bot. Sulph. 38, 1844

Shrub to 2 m. high, often leafless; branches whitish-tomentulose; petioles to 3 cm. long; leaf blades oblong-ovate to rounded-ovate, 1–6.5 cm. long, to 4 cm. wide, obtuse or acute, rounded or cordate at base, pubescent, entire or sometimes a little crenate; flowers borne in racemes 2.5–5 cm. long, these forming a pyramidal panicle; pedicels 1 mm. long, these and peduncles glandular-puberulous; bracts linear-subulate, glandular-puberulous; calyx 7–9 mm. long, glandular-puberulous, segments linear-lanceolate, 5–9 mm. long, 1.5–2 mm. wide at or near base, 3-veined; corollas red, to 3.5 cm. long, lips 10 mm. long, posterior one erect, triangular-ovate, emarginate at apex, anterior lip spatulate, 3-lobed, lobes about 2 mm. long, middle one 2.5 mm. wide, shallowly emarginate, lateral ones somewhat narrower, oblique at tip; capsule about 2 cm. long (narrow, solid, basal portion about 15 mm. long), pubescent; seeds gray, 3 mm. in diameter, 2 mm. thick, smooth.

Dry gravelly, sandy, or rocky slopes, Lower Sonoran to Tropical Zones, southern Arizona and California south through Baja California, Sonora, and Sinaloa. Aug.–May.

**2. Beloperone purpusii** Brandegee, Zoe **5**: 172, 1903

Spreading leafy shrubs; stems densely soft-puberulous and villous, longer hairs intermixed with minute ones; petioles to 2 cm. long (usually much less), both puberulous and villous; leaf blades ovate, to 9 cm. long and 5 cm. wide, acute to obtuse and apiculate at apex, obtuse, rounded, or subcordate at base, entire or undulate, densely and softly villous; flowers solitary, borne in axils of large, opposite, leaflike bracts, these broadly cordate, to 5 cm. wide and long, gradually intergrading into stem leaves, softly villous; peduncles to 2 cm. long, villous; bractlets linear-subulate, to 5 mm. long, villous; calyx 12–18 mm. long, deeply divided, segments nearly equal, lanceolate, to 3 mm. wide, villous, long hairs intermixed with minute ones, more or less viscid; corolla to 4 cm. long, pubescent, red or purplish, tube about 5 mm. in diameter at mouth, lips about 2 cm. long, posterior one erect, minutely 2-lobed at apex, lower lip spreading, cuneate, 1 cm. wide near tip, 3-lobed, lobes rounded, about 4 mm. long; capsules 2 cm. long, villous; seeds ash-colored, mottled when mature, nearly spherical, sides rugose.

Banks of ravines, Lower Sonoran and Tropical Zones, southern Baja California; barely reaching the margins of the Sonoran Desert. Nov.–April.

## Family 131. PLANTAGINACEAE. Plantain Family

Annual or perennial, acaulescent, short-stemmed or rarely stoloniferous herbs. Leaves basal (in ours) or in caulescent species alternate or opposite, estipulate, with apparently parallel venation. Inflorescence long-scaped, terminal spikes or heads.

Flowers small, perfect, polygamous, or monoecious, bracteate. Calyx 4-parted, persistent, lobes imbricate. Corolla 4-lobed, scarious to membranous, usually marcescent. Stamens 4 or less, commonly 2, inserted in tube or throat of corolla; filaments very slender; anthers versatile, 2-celled, longitudinally dehiscent. Ovary superior, 1–2-celled or falsely 3–4-celled. Style filiform, simple, stigmatic longitudinally. Fruit a pyxis, circumscissile, or nutlet. Seeds 1 to several in each cavity; endosperm fleshy; cotyledons slender.

## 1. PLANTAGO L., Sp. Pl. 112, 1753

Acaulescent or rarely leafy-stemmed; scapes arising from axils of alternate leaves. Calyx lobes equal, or in some two larger than other pair. Corolla salverform, with cylindrical tube or this constricted at throat; limb spreading in anthesis. Fruit a pyxis, usually 2-celled; seeds mostly with flattened or concave inner faces, more or less mucilaginous when moistened.

Plants perennial:
- Seeds 6–18, 1–1.2 mm. long; leaf blades mostly not over twice as long as wide; sepals 2–2.5 mm. long .......... 1. *P. major*
- Seeds 2 or 3, 1.6–2.5 mm. long; leaf blades usually three to five times as long as wide; sepals 3–3.5 mm. long:
  - Flowers perfect; seeds 2, shining, deeply excavate on inner face; bracts equaling or exceeding sepals; corolla lobes rotate-spreading .......... 2. *P. lanceolata*
  - Flowers polygamodioecious; seeds 3, dull, flat on inner face; bracts shorter than sepals; corolla lobes of fertile flowers connivent, forming a beak in fruit .......... 3. *P. hirtella* var. *galeottiana*

Plants annual (or rarely biennial):
- Flowers perfect; corolla lobes rotate-spreading, not connivent into a beak; seeds brown to nearly black:
  - Seeds 6–14, fusiform, nearly terete in cross section, 1–1.4 mm. long; bracts rather fleshy, glabrous or nearly so .......... 4. *P. heterophylla*
  - Seeds 2, concavo-convex, about 2 mm. long; bracts not fleshy, sparsely to densely villous:
    - Bracts orbicular to broadly ovate, rounded or obtuse at apex, 2–3 mm. long, closely applied to calyx:
      - Sepals usually tinged with brown; corolla lobes brown at base, usually brown-striped to middle or above; herbage moderately villous .......... 5. *P. insularis* var. *insularis*
      - Sepals green, rarely brownish; corolla lobes white to rusty, usually not brown-spotted or striped; herbage densely villous .......... 5*a*. *P. insularis* var. *fastigiata*
    - Bracts linear to linear-subulate, 5–10 mm. long, more or less divergent-ascending .......... 6. *P. purshii* var. *oblonga*
- Flowers polygamodioecious; corolla lobes erect and closely connivent in fertile flowers; seeds golden yellow or wine red:
  - Seeds 2, wine red, flat to faintly excavate on inner face; corolla tube coarsely short-puberulent outside near apex; sepals acute to apiculate, coarsely but evenly hirsute .......... 7. *P. rhodosperma*
  - Seeds 2–4, golden yellow, deeply excavate on inner face; corolla tube glabrous throughout; sepals ovate to oblong, not acuminate, only midrib hirsute .......... 8. *P. virginica*

### 1. Plantago major L., Sp. Pl. 112, 1753

Perennial, sparsely puberulent to glabrous, acaulescent herb with horizontally spreading leaves in a basal rosette; petioles broad and thin, 2–25 cm. long, channeled on upper side; leaf blades ovate, mostly broadly cuneate to subtruncate or rarely subcordate at base, obtuse to rounded at apex, 2–12 cm. wide, 5–25 cm. long, mostly 5-veined, margin entire to undulate or irregularly toothed; scapes erect or somewhat decumbent, 6–40 cm. long, sometimes bearing foliose, ovate to elliptic bracts 2–3.5 cm. long at base of particularly large spikes; spikes dense, 5–6 mm. in diameter, 5–35 cm. long at anthesis, to 10 mm. in diameter in fruit; bracts ovate to lanceolate-ovate, 2–4 mm. long or lowest larger; sepals broadly ovate, 2–2.5 mm. long; corolla lobes 1–1.5 mm. long, reflexed, acute to slightly apiculate, margins often involute; capsule broadly ovoid, about 2–2.5 mm. in diameter, 3–4 mm. long, brownish, circumscissile about middle, 6–18-seeded; seeds red-brown, 1–1.2 mm. long, irregularly angled, minutely foveolate-striate, a distinct, centrally placed, attachment scar on one face.

Roadsides, fields, ditch banks, and waste places, widely distributed in Temperate and Subtropical Zones, introduced from Europe; in agricultural areas and about towns throughout our area. March–Dec.

**2. Plantago lanceolata** L., Sp. Pl. 113, 1753

*Arnoglossum lanceolatum* S. F. Gray, Nat. Arr. Brit. Pl. **2:** 293, 1821.
*Plantago flexuosa* Gaud. ex Rapin, Mém. Soc. Linn. Paris **6:** 456, 1827.

Perennial herb from a short rootstock, conspicuously brown-villous at bases of leaves; leaves narrowly lanceolate-oblong to oblanceolate, mostly ascending to erect, acute to acuminate at apex, petiole elongate, narrowly to moderately winged, blade 0.5–2.5 (3.5) cm. wide, 5–25 cm. long, entire to remotely dentate, nearly glabrous above, sparsely to densely pubescent beneath, most copiously so along 3–5 nearly parallel veins; scapes slender, 1–6 dm. tall, spikes densely flowered, ovoid in youth, finally elongate-cylindrical, 5–9 mm. in diameter, 1.5–10 cm. long, tip often attenuate in bud; flowers perfect; bracts ovate, abruptly long-acuminate, hyaline marginally and conspicuously surpassing calyx, 4–7 mm. long, those at base of spike slightly longer than upper ones; sepals broadly ovate, 3–3.5 mm. long, with a green midrib and scarious margins, tips often emarginate, abaxial ones often united at base or nearly to tip; corolla brownish, glabrous, lobes ovate, spreading, 2–2.5 mm. long; capsule ellipsoid-oblong, about 3.5 mm. long, dehiscing near base; seeds 2, deeply excavated on hilar face, brown and shining.

Fields, roadsides, and ditch banks, introduced from Europe and widely distributed throughout the Temperate Zone of the continent; less common in the desert. March–Oct.

**3. Plantago hirtella** H. B. K. var. **galeottiana** (Decne.) Pilger, Pflanzenreich **$4^{269}$:** 257, 1937

*Plantago galeottiana* Decne., in A. DC., Prod. **$13^{1}$:** 726, 1852.
*Plantago virginica* var. *maxima* A. Gray, Bot. Calif. **1:** 611, 1876.
*Plantago subnuda* Pilger, Notizblatt **5:** 260, 1912.
*Plantago durvillei* subsp. *subnuda* Pilger, Pflanzenreich **$4^{269}$:** 234, 1937.

Perennial herbs with short thick caudex clothed with intermingled villous hairs and persistent leaf bases; leaves elliptic to broadly obovate, 1.5–8 cm. wide, 5–22 cm. long, narrowed below to a winged petiole less than one-half as long as blade, veins 5–7, prominent, moderately pilose beneath, more densely so on adaxial surface, margins entire to sparingly and shallowly denticulate; scapes erect, 10–40 cm. tall, densely villous with septate flattened white hairs near apex, sparsely so and often reddish near base; spikes polygamous, 5–35 cm. long, 5–12 mm. in diameter, sparsely flowered below, densely so above, bracts ovate to lance-ovate, 2–2.5 mm. long, carinate, pubescent on back, margins scarious and ciliate; sepals obtuse, 3–3.5 mm. long, midrib rather thick and pubescent, margins scarious and only slightly puberulent to glabrous; corolla lobes of staminate or perfect flowers spreading, about 2 mm. long, narrowly obovate, of pistillate flowers erect, nearly enclosing style; stamens 4, anthers white; capsule ovoid, about 3.5 mm. long; seeds usually 3, flattened on one side, 1.6–2 mm. long, greenish brown, minutely tuberculate.

Moist ground along ditches and in fields, mostly Upper Sonoran Zone to Humid Transition Zone, Washington to the Cape Region, east to Arizona, New Mexico, and central Mexico. Flowering nearly throughout the year.

**4. Plantago heterophylla** Nutt., Trans. Amer. Phil. Soc. II, **5:** 177, 1837

*Plantago perpusilla* Decne., in A. DC., Prod. **13**[1]**:** 697, 1852.
*Plantago californica* Greene, Bull. Calif. Acad. **1:** 123, 1885.

Small annual herb, herbage puberulent with short, somewhat tangled to spreading, whitish hairs; leaves spreading, linear and entire or with a few short, narrow, spreading teeth, blades in ours 1–5 cm. long; scapes several, slender, arcuately spreading-ascending, 2–10 cm. long, sparsely puberulent to glabrate; spikes loosely flowered, 5–40 mm. long, 3–4 mm. in diameter; bracts broadly ovate, obtuse, 1.8–2.2 mm. long, moderately fleshy and becoming wrinkled in drying, margins with a wide, scarious, white wing, glabrous or essentially so; sepals broadly oblong, obtuse to rounded at apex, 2–2.4 mm. long, fleshy part slightly narrower than in bracts, scarious wing white, 0.4–0.8 mm. wide; corolla lobes about 0.6 mm. long, narrowly lanceolate, ascending or sometimes spreading in fruit; stamens 2; capsules ovoid, 3–3.5 mm. long, stramineous but with a reddish tinge, dehiscing about one-third of way from base; seeds 6–14, narrowly ellipsoidal, about 1.4 mm. long, dark brown to nearly black, comparatively coarsely foveolate in longitudinal lines and faintly transversely ridged, slightly angled, nearly terete in cross section.

In saline soils, Lower Sonoran Zone, Sacramento Valley, California, to southern California, Arizona, and Sonora, east to Virginia and Florida. March–May.

**5. Plantago insularis** Eastw., Proc. Calif. Acad. III, **1:** 112, 1898, var. **insularis**

*Plantago brunnea* Morris, Bull. Torrey Club **27:** 115, 1900.
*Plantago fastigiata* var. *brunnea* Pilger, Pflanzenreich **4**[269]**:** 374, 1837.

Small annual with spreading to ascending, slender scapes 2–30 cm. tall, canescent with silky hairs, these especially prominent just below spikes; leaves narrowly to broadly linear-lanceolate, acuminate, entire or sparingly callous-toothed, 2–10 mm. wide, 1.5–15 cm. long, gradually narrowed to villous, narrowly winged petiole, 1–3-nerved, usually arachnoid-villous; spikes oblong-cylindrical, 1–2.5 cm. long, 6–10 mm. in diameter, or in depauperate plants spheroidal, 3–6 mm. in diameter; bracts broadly ovate to orbicular, about 3 mm. long, midrib usually brownish, densely villous to puberulent, margins broadly scarious; sepals similar to bracts, about 3 mm. long, degree of villosity on midrib greatly variable; corolla lobes about 2 mm. long, faintly rhomboidal, obtuse but faintly acuminate, usually distinctly brownish at base and often with brown stripes nearly to apex, spreading both in anthesis and fruit; stamens 4, well exserted; fruit broadly ovoid, about equaling calyx, wall of pyxis thin, often reddish, dehiscing just below middle, wall of lower part usually thinner than that of deciduous part; seeds 2, pale reddish brown to nearly orange, deeply excavate on inner face, smooth and shining, incurled edges around pit often nearly vitreous.

Sandy soils, Upper and Lower Sonoran Zones, Santa Barbara County, California, and on coastal islands south to Vizcaíno Desert and Guadalupe Island, Mexico. Jan.–May.

**5*a*. Plantago insularis** var. **fastigiata** (Morris) Jepson, Man. Fl. Pl. Calif. 956, 1925

*Plantago minima* A. M. Cunningham, Proc. Indiana Acad. **1896**: 202, 1897. Not *P. minima* DC., 1805.
*Plantago fastigiata* Morris, Bull. Torrey Club **27**: 116, 1900.
*Plantago scariosa* Morris, op. cit. 117.
*Plantago gooddingii* Nels. & Kenn., Muhlenbergia **3**: 142, 1908.
*Plantago insularis* var. *scariosa* Jepson, Man. Fl. Pl. Calif. 956, 1925.

Differing from var. *insularis* mainly in being more densely villous, in having green instead of brownish midribs on bracts and sepals, and less brown on corolla lobes; otherwise as in var. *insularis*.

Desert slopes, washes, and flats, Lower Sonoran Zone, Death Valley region, California, south into Baja California, east to Utah, Arizona, and Sonora. Oct.–May in Baja California; mostly Jan.–May in Arizona and Sonora.

**6. Plantago purshii** var. **oblonga** (Morris) Shinners, Field & Lab. **18**: 117, 1950

*Plantago picta* Morris, Bull. Torrey Club **28**: 118, 1901. Not *P. picta* Colenso, 1890.
*Plantago oblonga* Morris, op. cit. 119.
*Plantago ignota* Morris, loc. cit.
*Plantago xerodea* Morris, op. cit. **36**: 515, 1909.
*Plantago spinulosa* var. *oblonga* Poe, Bull. Torrey Club **55**: 411, 1928.
*Plantago purshii* var. *picta* Pilger, Pflanzenreich $\mathbf{4}^{269}$: 369, 1937.

Annual herbs with erect or ascending scapes 2–15 cm. tall, villous to white-arachnoid; leaves narrowly linear-oblanceolate, 1–5 mm. wide, 3–14 cm. long, villous, usually only midrib obvious; spikes oblong to long-cylindrical, 5–8 mm. in diameter, 2–15 cm. long, flowers closely crowded; bracts linear to linear-subulate, 5–10 mm. long, lower ones often two to three times as long as calyx lobes, hirsute to villous, green, more or less divergent; calyx lobes oval to obovate, 2–3 mm. long, green, densely silky-villous; corolla lobes broadly ovate, 1.5–2.5 mm. long, white with a brownish basal spot, spreading; stamens only slightly exceeding corolla, small; capsule ovoid, 3–4 mm. long, dehiscing at or slightly above middle; seeds 2, about 2 mm. long, dark brown, minutely pitted, rather dull, flattened and slightly excavated on inner face, margin of excavation slightly winged.

Arid hills and desert slopes and flats, Lower and Upper Sonoran Zones, Mojave and Colorado Deserts south into northeastern Baja California, east to New Mexico and Sonora. Feb.–May.

**7. Plantago rhodosperma** Decne., in A. DC., Prod. $\mathbf{13}^{1}$: 722, 1852

*Plantago virginica* var. *pectinata* Kuntze, Rev. Gen. Pl. **2**: 532, 1891.
*Plantago rubra* A. M. Cunningham, Proc. Indiana Acad. **1896**: 204, 1897.

Annual or winter biennial from a slender taproot; leaves few to many, rosulate, elliptic to oblanceolate, 5–25 mm. wide, 3–15 cm. long, entire to sparingly pectinately toothed, sparsely to moderately pubescent on both faces, more densely hirsute on veins with coarse, multicellular, white hairs; scapes usually several, 2–25 cm. long, erect or arcuately ascending, hirsute; spikes cylindrical, 5–9 mm. in diameter, 1–10 cm. long; bracts lanceolate, 2–2.5 mm. long, acute, prominently keeled, hirsute-ciliate on back and margin of narrow scarious wings; flowers polygamodioecious, sepals elliptic-oblong, 2.5–3 mm. long, moderately keeled, apiculate or acute at apex, coarsely and sparsely hirsute, often purplish; corolla lobes erect and connivent over capsule in pistillate or perfect flowers, more or less spreading in staminate flowers, lance-ovate, 3–4 mm. long, acute to apiculate at apex, auriculate at base, white to pale stramineous; corolla tubes puberulent with short, coarse, blunt hairs outside near apex; capsules 3–3.5 mm. long, ovoid, dehiscing at middle or slightly below, 2-seeded, often reddish or brownish; seeds about 2.5 mm. long, nearly flat or only faintly excavate on hilar face, dark wine-red, minutely and shallowly reticulate-faveolate, margin very narrowly winged.

Coastal bluffs and arid mesas, Lower and Upper Sonoran Zones, Orange County, California, south into northern Baja California, east to Kansas, Texas, and northern Mexico; mostly above the Sonoran Desert. March–June.

**8. Plantago virginica** L., Sp. Pl. 113, 1753

*Plantago caroliniana* Walt., Fl. Car. 85, 1788.
*Plantago connivens* Moench, Meth. Suppl. 174, 1802.
*Plantago purpurascens* Nutt. ex Rapin, Mém. Soc. Linn. Paris **6:** 454, 1827.
*Plantago missouriensis* Steudel, Flora **32:** 409, 1849.

Annual herb from a slender taproot; herbage sparsely to moderately villous with coarse septate hairs, usually soon glabrate; leaves spreading to ascending, oblanceolate to spatulate or less commonly obovate to elliptic, 3–5-nerved, 0.5–2.5 cm. wide, 1.5–15 cm. long, entire to repand-dentate, gradually narrowed to slender, faintly winged petiole, this about one-half as long as blade; scapes erect or ascending, 5–22 cm. tall, villous to glabrate; spikes dense or sometimes interrupted near base, cylindrical, 5–7 mm. in diameter, 5–20 cm. long; flowers dioecious or polygamodioecious; bracts linear-lanceolate to lanceolate, 2–2.3 mm. long, sparsely villous with coarse septate hairs along midrib; calyx lobes ovate to oblong, unequal in width, 2–2.5 mm. long, brownish midrib prominent, sparsely villous, broad margins scarious, white, glabrous or with coarse cilia along edge; corolla lobes spreading in staminate flowers, erect and closely connivent in pistillate flowers, ovate, minutely apiculate, 1.8–2 mm. long, white to faintly rusty; stamens exserted 2–2.4 mm.; fruit ovoid to oblong-ovoid, 2–2.5 mm. long, 2–4-seeded; seeds 1.8–2.2 mm. long, moderately excavated-concave on hilar face, golden yellow to pale orange, minutely and shallowly pitted, sublustrous.

Along streams, ditches, and cienegas, Lower Sonoran to Transition Zones, Arizona to the Atlantic Coast, south into northern Mexico; introduced sparingly in California and Oregon. March–June.

## Family 132. **RUBIACEAE.** Madder Family

Herbs, shrubs, or trees with simple, opposite or verticillate, stipulate leaves, stipules often connate and fringed. Flowers regular and symmetrical or nearly so, perfect but often dimorphous or sometimes polygamous. Calyx tube adnate to ovary; limb variously lobed or divided, persistent or deciduous. Corolla sympetalous, 4–5-lobed, rotate, campanulate, salverform, or funnelform, often valvate in bud. Stamens as many as corolla lobes and alternate with them, inserted in tube or throat; filaments usually short; anthers linear to oblong, included or exserted. Ovary 2–4-celled (in ours), inferior or sometimes upper one-third to one-half free from calyx tube; each cell 1- to many-seeded; style simple or bilobed, short or elongate. Fruit a capsule, berry, drupe, drupelet, or nutlet; seeds free within the carpel or embedded in juicy pulp. Embryo small, straight or curved; endosperm fleshy or horny.

Ovary partially superior, i.e., calyx tube adnate to lower one-half to two-thirds of ovary .... .............................................................. 1. *Houstonia*

Ovary wholly inferior:

Seeds broadly winged; fruit an ellipsoidal capsule 1.5–3 cm. long, 6-costate .. 2. *Hintonia*

Seeds wingless; fruit not a capsule, or if so then less than 5 mm. long, not costate:

Plants shrubs or small trees; fruits 2- to many-seeded:

Inflorescences globose, many-flowered heads; fruit dry, obpyramidal, splitting from base upward into 2–4 one-seeded nutlets; stamens 4 .......... 3. *Cephalanthus*

Inflorescences of solitary flowers or in few-flowered fascicles; fruit berrylike or a drupe, seeds embedded in a juicy pulp, pericarp leathery to subwoody; stamens usually 5:

Branches rigid, more or less spinescent; fruit baccate, several- to many-seeded; corolla lobes contorted in bud .................................. 4. *Randia*

Branches flexuous, not spinescent; fruit a small drupe, 2-seeded; corolla lobes not contorted in bud ........................................ 5. *Chiococca*

Plants annual or perennial herbs or sometimes suffrutescent; fruits 2–4-seeded, dry except in some species of *Galium*:

Ovary 3–4-celled; stigmas 3–4; fruit splitting into 3–4 one-seeded nutlets ........... ........................................................ 6. *Richardia*

Ovary 2-celled; stigmas 2, bilobed, or capitate; fruits various but not as above:

Flowers in involucrate heads or axillary clusters; pedicels not jointed to calyx; stigmas 2-branched or -lobed:

Fruits separating from partition longitudinally or obliquely, not circumscissile; seeds not 4-lobed:

Carpels separating parallel to partition, each carpel indehiscent; seeds not grooved longitudinally on adaxial face:

Carpel walls papery, fragile; leaves flat, not striate; flower heads subtended by 4 leafy bracts, terminal on branchlets ................... 7. *Crusea*

Carpel walls firm; leaves linear to lanceolate, longitudinally striate; flowers in axillary clusters ..................................... 8. *Diodia*

Carpels separating obliquely at base, partition sometimes adhering to one carpel and closing it, other carpel open on adaxial side; seeds longitudinally grooved adaxially ........................................ 9. *Staelia*

Fruits circumscissile; seeds distinctly 4-lobed ................. 10. *Mitracarpus*

Flowers in axillary or terminal cymes or panicles; pedicels jointed with the calyx; stigmas capitate ........................................... 11. *Galium*

## 1. HOUSTONIA L., Sp. Pl. 105, 1753

Erect or prostrate, glabrous or pubescent herbs or low shrubs with opposite, stipulate, sessile or petiolate leaves, sometimes with fascicles of smaller leaves in their axils; stipules persistent, entire, dentate or laciniate. Flowers axillary or cymose, often dimorphous; pedicels recurved or erect in fruit; hypanthium globose to turbinate, compressed or terete; calyx lobes 4, remote. Corolla funnelform to salverform, 4-lobed, valvate in bud. Stamens 4, inserted in corolla throat; anthers dorsifixed. Ovary 2-celled, partially inferior; style branches 2, linear; ovules few to numerous in each cell. Capsule partially superior, loculicidally dehiscent. Seeds concavo-convex; endosperm horny; embryo straight, clavate.

Plants annual; stems filiform, 1 mm. or less in diameter; capsules almost wholly inferior:

Pedicels sharply recurved in fruit; hypanthium minutely papillose:

Branches prostrate; internodes shorter than leaves; leaves mostly sessile ............. ............................................ 1. *H. prostrata* var. *prostrata*

Branches erect; internodes mostly equaling or exceeding leaves; leaves mostly short-petiolate .......................................... 1*a.* *H. prostrata* var. *parvula*

Pedicels erect or ascending in fruit; hypanthium not papillose:

Calyx lobes attenuate, acute, exceeding hypanthium at anthesis; corolla 7–9 mm. long .. .......................................... 2. *H. asperuloides* var. *asperuloides*

Calyx lobes obtusish, equaling or shorter than hypanthium at anthesis; corolla 3–6 mm. long:

Capsule turbinate; flowers mostly pedicellate; corolla purplish, glabrous .......... .......................................... 2*a.* *H. asperuloides* var. *brandegeana*

Capsule subglobose; flowers mostly sessile; corolla white, minutely puberulent outside .......................................... 3. *H. arenaria*

Plants perennial; stems stouter, usually 2.5 mm. or more in diameter at base; capsules about three-fourths inferior:

Stems angled; plants shrubby throughout; leaves often fascicled in axils:

Plants stout, bushy; internodes 5–10 mm. long; capsule 2 mm. wide .. 4. *H. mucronata*

Plants slender, branches spreading; internodes 1–2 cm. long; capsules 1.5 mm. wide or less .......................................... 5. *H. gracilenta*

Stems terete; plants woody only at base; leaves not fascicled .......... 6. *H. brevipes*

**1. Houstonia prostrata** Brandegee, Zoe **5:** 105, 1901, var. **prostrata**

*Hedyotis vegrandis* W. H. Lewis, Rhodora **63:** 222, 1961.

Prostrate, much-branched annual forming dense mats 5–15 cm. in diameter; branches slender, minutely and sparsely papillose, internodes mostly shorter than leaves; stipules about 0.5 mm. long, sparsely ciliolate; leaves sessile, linear to linear-lanceolate, 0.6–2.5 mm. wide, 5–12 mm. long, acute to obtuse, deep green, and sparsely papillose above, paler, narrowly revolute, and more densely papillose beneath; flowers axillary, on stout pedicels 0.5–1 mm. long, these recurved in fruit; hypanthium glabrous but minutely papillose, 0.6–0.8 mm. long; calyx lobes lanceolate-oblong, acute, about as long as hypanthium; corolla funnelform, white tinged with purple, 1–1.5 mm. long, exterior glabrous; capsule 2.5–3 mm. broad, minutely white-papillose, subdidymous, nearly three-fourths inferior; seeds 1.2–1.5 mm. long, black, cymbiform.

Sandy washes and stream banks, Lower Sonoran and Tropical Zones, southern Baja California. Aug.–Jan.

**1*a*. Houstonia prostrata** var. **parvula** (Brandegee) Wiggins, comb. nov.

*Houstonia parvula* Brandegee, Zoe **5:** 221, 1905.
*Hedyotis sinaloae* W. H. Lewis, Rhodora **63:** 222, 1961.

Differing from var. *prostrata* in having internodes, at least lower ones, equaling or exceeding leaves by as much as half their length, short-petiolate leaves, and erect or ascending branches, thus presenting an open instead of a prostrate, matlike habit; plants 5–15 cm. tall.

Sandy or silty washes and llanos, Lower Sonoran Zone, central Sonora to Sinaloa. Aug.–Jan.

**2. Houstonia asperuloides** (Benth.) A. Gray, Proc. Amer. Acad. **5:** 158, 1861, var. **asperuloides**

*Hedyotis asperuloides* Benth., Bot. Sulph. 19, *pl. 13,* 1844.

Erect or ascending, diffusely branched annual 1–3 dm. high with slender, subangulate, glabrous to sparsely puberulent stems, internodes exceeding leaves; stipules 0.5–1 mm. long, 2–6-setiferous; leaves sessile, linear, 5–25 mm. long, 1 mm. wide or less, glabrous to sparsely puberulent, mostly acute at apex; flowers axillary and in terminal cymes, on filiform pedicels 5–25 mm. long; hypanthium narrowly turbinate, 1 mm. long or less at anthesis, glabrous; calyx lobes lance-linear, acute, 1–1.5 mm. long at anthesis, often horizontally spreading in fruit; corolla funnelform, 7–9 mm. long, minutely puberulent outside, lobes and upper part of tube pink, lobes ovate, obtuse, nearly or quite as long as tube; stamens slightly exserted; capsules turbinate, 2–3 mm. long, 1–1.5 or rarely 2 mm. wide, almost wholly inferior, retuse to slightly rounded at apex; seeds peltate, concavo-convex, brownish, minutely scrobiculate.

Sandy plains and dunes, Lower Sonoran Zone, Comondú south to the Cape Region, Baja California. Sept.–May.

**2*a*. Houstonia asperuloides** var. **brandegeana** (Rose) Wiggins, comb. nov.
*Houstonia brandegeana* Rose, in Vasey & Rose, Contr. U.S. Nat. Herb. **1:** 70, 1890.
*Hedyotis asperuloides* f. *brandegeana* W. H. Lewis, Rhodora **63:** 221, 1961.

Leaves 5–8 mm. long, puberulent on upper surface, glabrous beneath; pedicels 5–30 mm. long; calyx lobes ovate to oblong, obtuse or abruptly acute at apex, often shorter than hypanthium; corolla purplish, 4–6 mm. long, glabrous outside; otherwise as in var. *asperuloides* and intergrading with it.

Sandy plains and washes, Lower Sonoran and Tropical Zones, southern Baja California. Sept.–March.

**3. Houstonia arenaria** Rose, in Vasey & Rose, Contr. U.S. Nat. Herb. **1:** 70, 1890
*Hedyotis arenaria* W. H. Lewis, Rhodora **63:** 221, 1961.

Erect or ascending annual 1–4 dm. high with slender, subangulate, glabrous stems; stipules minute, scarious, laciniate-setiferous; leaves sessile or essentially so, linear to lance-linear, 10–35 mm. long, 1–3 (rarely 5) mm. wide, acute at apex, gradually attenuate to base, glabrous; flowers mostly sessile-recurved on lax cymes, terminal ones pedicellate; hypanthium broadly turbinate, 0.4–0.6 mm. long at anthesis, glabrous; calyx lobes oblong, obtuse or abruptly acute at apex, equaling or usually exceeding hypanthium, 1–1.7 mm. long in fruit; corolla narrowly funnelform, white, turning pink when dry, 3–5 mm. long, usually minutely and sparsely puberulent outside; corolla lobes oblong, obtuse, or rounded, about half as long as slender tube, glabrous within; anthers barely exserted; capsule subglobose, 2–2.5 mm. high and wide, lightly costate beneath spreading calyx lobes, almost wholly inferior; seeds oblong, concavo-convex, brownish, minutely pitted, 0.3–0.5 mm. long.

Sandy places, Lower Sonoran and Subtropical Zones, southern Baja California. Sept.–April.

**4. Houstonia mucronata** (Benth.) B. L. Robinson, Proc. Amer. Acad. **45:** 401, 1910
*Hedyotis mucronata* Benth., Bot. Sulph. 19, 1844.
*Houstonia fruticosa* Rose, Contr. U.S. Nat. Herb. **1:** 132, 1892.

Erect glabrous shrub 2–9 dm. high with crowded angled branches and leaves in dense axillary fascicles; bark gray-yellow; stipules 1 mm. long or less, glandular-setiferous, lobes short-triangular; leaves sessile, linear, terete, 3–15 mm. long, acute, mucronulate at apices; flowers in few-flowered, terminal cymes, sessile or on pedicels 2–10 mm. long; hypanthium about 1 mm. long at anthesis; calyx lobes 1.5–2 mm. long, lanceolate, acute to acuminate, thick, bearing a stipitate gland in each sinus; corolla white tinged with rose outside, 10–12 mm. long, salverform, tube slender, lobes 2.5–3 mm. long, ovate, acute; capsule subglobose, 2–2.5 mm. long, about 2 mm. wide, two-thirds to three-fourths inferior, rounded or truncate at apex; seeds 0.5–0.8 mm. long, oblong, peltate, dark brown, minutely scaly.

Dunes and beaches, Lower Sonoran Zone, Magdalena Island and along the coast of the Gulf of California from Loreto to eastern coast of the Cape Region and on the adjacent islands, Baja California. Aug.–March.

**5. Houstonia gracilenta** I. M. Johnston, Proc. Calif. Acad. IV, **12:** 1174, 1924

*Hedyotis gracilenta* W. H. Lewis, Rhodora **63:** 222, 1961.

Compact decumbent perennial 1–2 dm. high and up to 8 dm. broad with intricately branched, slender, angulate, glabrous to slightly glandular stems; bark on old stems roughened, gray to brown; internodes 1–2 cm. long; leaves opposite and in axillary fascicles, sessile, filiform, 8–10 cm. long, acute, flattened, coriaceous; stipules 1 mm. long or less, triangular, bearing 1–2 gland-tipped setae; flowers in loose, few-flowered cymes; hypanthium 0.5–0.7 mm. high at anthesis, glabrous; calyx lobes about 1.5 mm. long at anthesis; corolla 10–12 mm. long, salverform, pink, tube 5–6 mm. long; throat cylindrical, about 3 mm. long, pink with darker rose lines; corolla lobes oblong, acute, 2 mm. long; style and stamens included; capsule oblong-globose, 2–2.5 mm. long, 1.2–1.8 mm. wide, about three-fourths inferior.

Known only from the type locality, Lower Sonoran Zone, rocky hillsides on San Diego Island, Gulf of California, about halfway between Loreto and La Paz. Collected in flower during May.

**6. Houstonia brevipes** Rose, in Vasey & Rose, Contr. U.S. Nat. Herb. **1:** 83, 1890

*Hedyotis brevipes* W. H. Lewis, Rhodora **63:** 221, 1961.

Erect, diffusely branched perennial with slender, glabrous, glaucous, terete branches from a stout taproot; stipules minute, 1–4-setiferous; leaves sessile, linear, flat, 12–35 mm. long, 0.5–1 mm. wide, acuminate; inflorescence a loose cyme, flowers sessile to pedicellate on pedicels to 10 mm. long; hypanthium 0.5–0.8 mm. long, glabrous; calyx lobes lanceolate, acute, somewhat fleshy, glabrous, exceeding hypanthium at anthesis, equaling or slightly surpassing capsule in fruit; corolla salverform, 10–16 mm. long, glabrous, pinkish, lobes oblong-ovate, mostly obtuse, 3–4 mm. long, tube slender; capsule subglobose, 2–2.5 mm. in diameter, rounded at apex, about two-thirds inferior; seeds concavo-convex, oblong, black.

Sandy washes and plains, Lower Sonoran Zone, from the vicinity of Santa Rosalia south to Loreto and on adjacent islands in the Gulf, Baja California. Oct.–April.

## 2. HINTONIA Bullock, Hook. Ic. Pl. 33: *pl. 3295,* p. 1, 1935

Erect shrub or small tree with opposite petiolate leaves and light gray, copiously lenticellate bark. Stipules small, ovate, acute, puberulent or glabrous. Leaves elliptic to oblong-lanceolate, minutely puberulent to glabrate. Flowers large. Hypanthium turbinate, 6–8-costate. Calyx lobes 6, narrowly lanceolate, finally deciduous. Corolla infundibuliform, white, plicate-valvate in bud, villous within. Stamens inserted at base of tube, included or slightly exserted, filaments straight. Anthers linear, basifixed. Disk conspicuous. Ovary 2-celled, inferior; style filiform, equaling stamens. Capsules ellipsoid, somewhat compressed laterally, 6–8-costate, lenticellate, septicidally dehiscent. Seeds compressed, numerous, broadly winged.

**1. Hintonia latiflora** (Sessé & Moc.) Bullock, Hook. Ic. Pl. **33:** *pl. 3295,* p. 4, 1935

*Coutarea latiflora* Sessé & Moc. ex DC., Prod. **4:** 350, 1830.
*Portlandia pterosperma* S. Wats., Proc. Amer. Acad. **24:** 52, 1889.
*Coutarea pterosperma* Standley, N. Amer. Fl. **32:** 127, 1921.

Shrub or small tree 1–6 m. high with grayish or brownish, terete, glabrous, conspicuously lenticellate branches; stipules deltoid, acute, 2–4 mm. long, acute or glandular-mucronate, puberulent to glabrate; petioles slender, 4–20 mm. long; leaf blades ovate to ovate-oblong, 1.5–6 cm. wide, 3–10 cm. long, rounded to acute and decurrent at base, obtuse, acute, or abruptly acute at apex, dark green, glabrous or sparsely puberulent, lateral veins moderately prominent; flowers solitary in axils on pedicels 1–2.5 cm. long, with small subulate bracteoles at or below middle; hypanthium 5–8 mm. long; calyx lobes 6, linear to subulate, 10–20 mm. long; corolla white, 6–8 cm. long, 6-lobate; stamens included; capsule oval, 1.5–2 cm. wide, slightly compressed, 6-costate, finely tuberculate and prominently lenticellate, brownish; seeds oval to orbicular, 5–7 mm. long, dark brown.

Canyons and hillsides, Lower Sonoran Zone and in arid subtropical thorn forests, vicinity of Guaymas east to Chihuahua and south into Colima. July–Aug.

## 3. CEPHALANTHUS L., Sp. Pl. 95, 1753

Slenderly branched shrub or small tree with opposite or verticillate, short-petiolate, entire leaves. Inflorescences in terminal or axillary, globose heads. Flowers small, bracteolate, white or yellowish; calyx tube narrowly obpyramidal, with 4 small, obtuse-deltoid lobes; corolla tubular-funnelform, with 4 short, erect to spreading lobes. Stamens 4, inserted in corolla throat; filaments short; anthers oblong, bicuspidate at base. Ovary 2-celled; ovules 1 in each cell, pendulous; style exserted, with a capitate stigma. Fruit obpyramidal, 1–2-seeded, dry. Endosperm cartilaginous; cotyledons narrowly oblong.

**1. Cephalanthus occidentalis** L. var. **californicus** Benth., Pl. Hartw. 314, 1849

Shrub or small tree 1–8 m. tall, with opposite or verticillate leaves and branchlets, glabrous to pubescent; petioles 6–10 mm. long; leaf blades ovate to elliptic, acute to acuminate at apex, rounded to subtruncate at base, entire, somewhat leathery, 2–4 cm. wide, 5–12 cm. long, glabrous, dark green above, pale and distinctly veined beneath; peduncles 2–7 cm. long; heads globose, 2–3 cm. in diameter, receptacle puberulent; calyx sessile, greenish, about 2–2.5 mm. long, persisting in fruit, lobes 0.4–0.6 mm. long; corolla white, 6–10 mm. long, lobes oblong, obtuse, often black-tipped; style filiform, twice as long as corolla or slightly longer; fruit about 4 mm. long; seeds flattened, margins acute.

Margins of streams in canyons, Lower and Upper Sonoran Zones, occasional in the Great Valley, California, east into Arizona, grading into the variety *occidentalis* which extends to the Atlantic seaboard. June–Sept.

Barely reaching margins of the Sonoran Desert in the Catalina Mountains, Pima County, Arizona.

## 4. RANDIA L., Sp. Pl. 1192, 1753

Shrubs or trees with ascending or spreading, rigid or flexible branches, these often bearing axillary or supra-axillary or terminal spines. Leaves opposite, sessile or petiolate, thin or coriaceous, glabrous or pubescent. Stipules small, intrapetiolar. Flowers small or large, axillary or terminal, fasciculate or solitary, perfect or unisexual; hypanthium ovate to turbinate. Calyx truncate or lobed; lobes sometimes elongated or foliaceous. Corolla funnelform to campanulate, glabrous or pubescent; lobes contorted in bud. Stamens 5, inserted in throat of corolla. Ovary mostly 2-celled. Fruit baccate, globose to oval, with a thick hard pericarp or this sometimes soft and thin, smooth or tuberculate. Seeds horizontal, flattened, immersed in pulpy placentae; cotyledons orbicular.

Plants unarmed; fruit obpyriform, 6–10 cm. long ........................ 1. *R. laevigata*
Plants armed; fruit globose, 0.8–9 cm. in diameter:
 Fruit puberulent, at least when young; leaves puberulent to pilose on both surfaces; spines in clusters of 4 at ends of branches:
  Leaves distinctly acute or apiculate at apex; fruit tuberculate or costate; corolla tube 20–32 mm. long:
   Fruit spinose-tuberculate, 4.5–10 cm. long and wide; leaf blades 3–8.5 cm. long; corolla densely pubescent without ........................ 2. *R. echinocarpa*
   Fruit somewhat costate, 1.5–3.5 cm. long; leaf blades 6–22 cm. long; corolla glabrous to sparsely pubescent without ................................ 3. *R. armata*
  Leaves mostly obtuse, rounded or abruptly subacute at apex; fruit smooth; corolla tube about 2 cm. long (unknown in *R. sonorensis*):
   Stipules 6–10 mm. long, mostly glabrous outside; leaf blades 2.5–8.5 cm. long; fruit 3–4.5 cm. in diameter .................................... 4. *R. megacarpa*
   Stipules 1–1.5 mm. long, densely puberulent outside; leaf blades 8–33 mm. long; fruit 2–3.5 cm. in diameter .................................. 5. *R. sonorensis*
 Fruit glabrous; leaves glabrous or sparsely puberulent along veins beneath and ciliolate on margins; spines in pairs:
  Branches divaricately spreading; fruit 1.2–2.5 cm. in diameter; seeds numerous ...... ........................................................ 6. *R. thurberi*
  Branches virgate-ascending; fruit 5–9 mm. in diameter; seeds 2–6 ...... 7. *R. obcordata*

### 1. **Randia laevigata** Standley, Contr. U.S. Nat. Herb. **20:** 201, 1919

An unarmed shrub 1.5–3 m. high with ascending open branches and large elliptic, obovate-oblong, or rhombic-ovate, glabrous leaves 10–25 cm. long and 5–12 cm. wide; stipules connate at base, deltoid-ovate, acute to cuspidate-mucronate, 8–10 mm. long, glabrous to puberulent without, glabrous within; leaves sessile to short-petiolate, long-attenuate to base, acute or acuminate at apex, chartaceous, bright green, lustrous above, puberulent in youth, soon glabrate, 10–13 lateral nerves along each side conspicuous beneath; calyx and hypanthium glabrous, calyx lobes triangular-subulate, 2–3 mm. long; corolla not seen; fruit subglobose to obpyriform, 6–10 cm. long, 2–3.5 cm. in diameter, glabrous, umbonate, borne on a terminal or axillary peduncle 1.5–3 cm. long; seeds numerous, oval to suborbicular, 6–10 mm. long, yellowish brown.

Rocky hillsides and on canyon walls, Lower Sonoran and Subtropical Zones, central Sonora to Durango and Nayarit. July–Aug.

### 2. **Randia echinocarpa** Sessé & Moc., in DC., Prod. **4:** 385, 1830

*Solena echinocarpa* D. Dietr., Syn. Pl. **1:** 799, 1839.
*Genipa echinocarpa* A. Gray, Proc. Amer. Acad. **21:** 380, 1886.
*Basanacantha echinocarpa* Bullock, Hook. Ic. Pl. **33:** *pl. 3298,* 1935.

A rigidly branching shrub 2–3 m. high with stout, divergent, gray branchlets and bearing 4 divergent spines 1–3 cm. long and a cluster of leaves at tips; stipules ovate, obtuse to acuminate, 1–2 mm. long, pilose to glabrous without, pubescent within; leaves sessile or subsessile, blades oval, obovate, or rhombic-ovate, 3–8.5 cm. long, 1.5–5 cm. wide, from obtuse to acuminate at base and long-attenuate to rounded at apex, densely puberulent on both surfaces, at least when young; flowers terminal, solitary or clustered, perfect or polygamodioecious; calyx lobes linear, 1.5–4 mm. long, sericeous; corolla densely appressed-pubescent without, tube 25–32 mm. long, lobes oval to broadly ovate, 12–23 mm. long, glabrous within; calyx lobes of pistillate flowers 1 cm. long; fruit subglobose, 4.5–10 cm. in diameter, pubescent, bearing irregular flattened tubercles 0.5–3 cm. long; seeds numerous, 6–10 mm. broad.

Dry hillsides, Lower and Upper Sonoran Zones, central Sonora to Chihuahua, Guerrero, and Vera Cruz. July–Aug.

**3. Randia armata** (Swartz) DC., Prod. **4:** 387, 1830

*Gardenia armata* Swartz, Prod. Veg. Ind. Occ. 51, 1788.
*Basanacantha armata* Jackson & Hook. f., Ind. Kew. 277, 1893.

Shrub 1–6 m. high with gray or brown bark and slender, divaricate, glabrate branches bearing 4 stout spines 5–25 mm. long at their apices; stipules broadly ovate, 3–8 mm. long, mucronate, thin, imbricate, glabrous; petioles 5–20 mm. long, puberulent but soon glabrate; leaf blades ovate, oblong-ovate, or obovate, 6–22 cm. long, 2–11 cm. wide, acute to abruptly acuminate at both ends, glabrous or sparsely puberulent, bright green above, paler and puberulent along nerves beneath, veins prominent; flowers dioecious, in clusters of 2–8 at ends of branchlets; calyx and hypanthium glabrous or puberulent; calyx lobes lanceolate-linear or oblong, 4–9 mm. long; corolla white or yellowish, glabrous or sparsely puberulent outside, tube 2–2.5 cm. long, 5 lobes obovate, 10–12 mm. long, obtuse; fruit oval to subglobose, 15–35 mm. long, 10–25 mm. in diameter, umbonate, smooth or obscurely costate, glabrous or sparsely puberulent; seeds numerous, 4–6 mm. broad, suborbicular.

Dry slopes and ridges at low elevations, mostly in Subtropical and Tropical Zones, southern Baja California to Oaxaca and South America. July-Aug.

**4. Randia megacarpa** Brandegee, Zoe **5:** 237, 1908

A shrub 3–6 m. high with gray or brownish branches and stout divaricate branchlets bearing 4 stout spines 5–15 mm. long and a tuft of leaves at their tips; stipules imbricate, ovate, 5–10 mm. long, acute or rarely obtuse, glabrous outside, white-pilose within; petioles 2–12 mm. long; leaf blades oval to obovate, 2.5–8.5 cm. long, 1.5–7.5 cm. wide, acute and short-decurrent to truncate at base, obtuse to rounded and apiculate at apex, short-pilose above, paler and densely tomentose beneath, nerves prominent; flowers perfect, sessile, white, terminal; hypanthium 5–6 mm. long, sericeous; calyx lobes 5, linear to linear-spatulate, 2.5–3.5 mm. long; corolla pilose, tube 18–20 mm. long, lobes oval to oblong, 13–18 mm. long, obtuse; fruit globose, 3–5 cm. in diameter, umbonate, short-stipitate at base, smooth, densely puberulent when young; seeds oval to orbicular, 7–10 mm. long.

Canyons and thickets near watercourses, central Baja California to the Cape Region, and on some southern islands in the Gulf of California. July–Aug.

**5. Randia sonorensis** Wiggins, Contr. Dudley Herb. **3:** 75, *pl. 19, figs. 4–6,* 1940

An erect to ascendingly branched shrub 2–3 m. tall; slender, divaricate, young twigs rather ridged, pubescent with rusty-brown, spreading hairs, internodes 1–2 cm.

long; bark on older branches glabrate and finely fissured; terminal twigs each bearing 4 subulate, straight, spreading spines 6–15 mm. long; stipules 1–1.5 mm. long, broadly ovate, abruptly acuminate, densely puberulent on outer surface; leaves in terminal fascicles, or in pairs on rapidly growing twigs, petioles 3–15 mm. long, narrowly wing-margined, densely puberulent, leaf blades elliptic to ovate-elliptic or suborbicular, 6–22 mm. wide, 8–33 mm. long, cuneate at base, acute, obtuse, or rounded at apex, densely puberulent, with longer, densely crowded hairs on veins; flowers unknown; fruit globose, 2–3.5 cm. in diameter, finely puberulent; seeds numerous, 6–8 mm. long, strongly flattened, dark brown to nearly black, embedded in sooty to black pulp.

Rocky hillsides, Lower Sonoran Zone, between Ures and Baviácora, Sonora. July?

### 6. **Randia thurberi** S. Wats., Proc. Amer. Acad. **24:** 53, 1889

A compact shrub 1–2.5 m. high with stout, divaricate, reddish brown to grayish, appressed-pilose branches bearing numerous pairs of divergent spines 8–15 mm. long; stipules ovate-deltoid, 1–1.5 mm. long, thin, acute to short-acuminate, appressed-pilose without and at base within; leaves mostly in clusters near tip of short lateral spurs; petioles 7–15 mm. long or nearly obsolete; leaf blades oval, orbicular, cuneate-orbicular, spatulate, or cuneate-obovate, 6–45 mm. long, 5–30 mm. wide, cuneate, rounded, obtuse, or short-decurrent on petiole at base, obtuse to retuse, sometimes apiculate at apex, chartaceous to subcoriaceous, glabrous and slightly glaucous or sparsely puberulent along veins beneath, nerves inconspicuous, 2–3 on each side; flowers unknown except for 5 linear to oblanceolate calyx lobes 1.5–5 mm. long remaining on solitary, sessile, terminal fruits, these globose, 12–25 mm. in diameter, greenish yellow, usually somewhat mottled, smooth, glabrous; seeds numerous, suborbicular, 6 mm. long, dark brown.

Dry gravelly hillsides and washes, Lower Sonoran Zone, northern Sonora to Sinaloa. July?

### 7. **Randia obcordata** S. Wats., Proc. Amer. Acad. **24:** 53, 1889

An erect, virgately branched shrub 2–4 m. high; branches stout, light gray or less commonly brownish, numerous stubby, wart-like, lateral spurs 3–15 mm. apart and bearing crowded tufts of subsessile leaves; few to many pairs of stout, divergent, straight or slightly curved spines 4–10 mm. long borne at nodes of larger shoots; stipules minute, 1 mm. long or less, imbricate, finely puberulent to glabrate; leaf blades flabellate, broadly cuneate-obovate to obcordate, 4–20 mm. long, nearly as broad, decurrent at base along nearly obsolete, ciliolate petiole, retuse, rounded, or truncate at apex, glabrous, light green to somewhat glaucous, the lateral veins inconspicuous; flowers solitary, sessile; hypanthium turbinate, truncate at apex, glabrous; calyx lobes 4–5, narrowly triangular, 0.5–0.75 mm. long; corolla 5–6 mm. long, glabrous without, tube slender, 3–3.5 mm. long, sparsely puberulent on lower half within; corolla lobes 1.2–1.8 mm. long and wide, rounded to subemarginate with a minute apiculation in notch, prominently nerved; anthers alternate with corolla lobes, sessile, inserted in throat, ovate, acute, 1.5 mm. long; fruit globose, 5–9 mm. in diameter, black, lustrous, glabrous, sometimes rugulose in drying, pericarp thick and hard; seeds 2–5, oval to suborbicular, brownish, 4–5 mm. long.

Dry plains, hillsides, and along arroyos, Lower Sonoran Zone, central Sonora and southern Baja California to Colima. July.

## 5. CHIOCOCCA P. Br., Nat. Hist. Jamaica 164, 1756

Shrubs or small trees with terete branchlets and opposite, petiolate, membranous to coriaceous, entire leaves. Stipules persistent. Flowers small, white, in axillary, simple, or paniculate racemes; hypanthium ovoid to turbinate, pubescent or glabrous, usually laterally compressed; calyx lobes 5, persistent; corolla funnelform, lobes valvate. Stamens 5, included or exserted, inserted at base of corolla tube, filaments pubescent, coalesced at base; anthers basifixed. Ovary 2-celled, inferior; style filiform; stigma entire or bilobed; ovules solitary in each cell, compressed. Fruit a small coriaceous drupe, orbicular or essentially so and usually compressed at right angles to septum. Seeds laterally compressed; endosperm fleshy; embryo straight.

Leaves glabrous, lustrous above; corolla 6–8 mm. long, glabrous outside ........ 1. *C. alba*
Leaves puberulent, not prominently lustrous; corolla 4–6 mm. long, sparsely puberulent on outer surface:
  Calyx lobes deltoid, 0.6–1 mm. long, erect in fruit; hypanthium glabrous at anthesis; corolla lobes nearly equaling the tube ............ 2. *C. pubescens* var. *peninsularis*
  Calyx lobes lanceolate to oblong, 1.8–2 mm. long, ascending-spreading in fruit; hypanthium densely puberulent; corolla lobes about half as long as the tube .. 3. *C. petrina*

### 1. Chiococca alba (L.) Hitchcock, Rep. Mo. Bot. Gard. 4: 94, 1893

*Lonicera alba* L., Sp. Pl. 175, 1753.
*Chiococca racemosa* L., Syst. Nat. ed. 10, 917, 1759.
*Chiococca macrocarpa* Mart. & Gal., Bull. Acad. Brux. **11**[1]: 230, 1844.

Shrub or small tree 2–6 m. high with spreading to scandent, slender, gray to brownish branches and green, glabrous or sparsely puberulent branchlets; stipules often broader than long, abruptly mucronate, 1–2 mm. long; petioles 2–12 mm. long; leaf blades variable, ovate, lanceolate, elliptic, or rounded-oval, 2–10 cm. long, 1–5 cm. wide, acute to short-acuminate at apex, obtuse to rounded and short-decurrent at base, coriaceous, glabrous, dark green and lustrous above, paler beneath, margins plane to slightly revolute; inflorescence racemose to paniculate, peduncle exceeding subtending petiole and sometimes surpassing leaf blade, glabrous, few- to many-flowered; calyx and hypanthium 2–2.5 mm. high, glabrous or rarely obscurely puberulent; calyx lobes deltoid or deltoid-subulate, acute, 0.6–1 mm. long; corolla 6–8 mm. long, white, throat 2–3 mm. wide, the lobes triangular, one-third to one-half as long as tube; stamens included; fruit white, orbicular, compressed, 4–6 (or rarely –8) mm. long, glabrous, persistent calyx lobes erect; seeds 3–4 mm. long, puncticulate.

In moist places, Tropical Zone, Florida to Texas, to the mountains of the Cape Region, Baja California, and in the West Indies and South America. March-July.

### 2. Chiococca pubescens Standley var. peninsularis Wiggins, Contr. Dudley Herb. 4: 24, 1950

A shrub with slender, terete, grayish branches, internodes from one-half to one-fourth as long as leaves, and densely puberulent branchlets; stipules broadly deltoid-triangular, 1–2 mm. high, as broad or broader than high, abruptly acuminate, sparsely puberulent, margins ciliate; petioles 2–5 mm. long, densely short-pilose to tomentulose; leaf blades 7–18 mm. wide, 16–30 mm. long, broadly ovate or elliptic, acute, apiculate, or rarely obtusish at apex, acute to obtuse and short-decurrent at base, closely and finely short-pilose to tomentulose beneath, sparsely puberulent to glabrate and dark green above, midrib prominent, lateral veins inconspicuous, margins narrowly revolute; inflorescence racemose or slightly paniculate, few-flowered, densely tomentulose peduncles often surpassing subtending leaves; bracteoles filiform, about 2 mm. long; pedicels slender, 2–6 mm. long, tomentulose at base, glabrous toward apex; hypanthium and calyx 2–2.5 mm. long, glabrous; calyx lobes narrowly deltoid to ovate, abruptly mucronulate, margins faintly ciliolate, erect in fruit; fruit orbicular, compressed, 4–5 mm. high, whitish, glabrous; seeds reddish brown, minutely and closely tuberculate.

Hillsides and along canyons, Lower Sonoran and Subtropical Zones, southern Baja California. July–Aug.

**3. Chiococca petrina** Wiggins, Contr. Dudley Herb. **3:** 76, *pl. 20, figs. 5–13,* 1940

A dense, ascendingly branched shrub 1–2 m. high, older branches dark gray or brownish, terete, younger ones green, minutely puberulent, internodes mostly shorter than leaves; stipules triangular-cuspidate, 1 mm. long; petioles short, 1–2 mm. long, puberulent; leaf blades ovate, ovate-elliptic, or oval, 2–10 mm. wide, 6–20 mm. long, acute and apiculate at apex, acute to rounded at base, leathery, green and shining though sparsely short-puberulent, becoming glabrate, midnerve and laterals inconspicuous above, slightly paler, more densely puberulent, and midvein prominent beneath, margins subrevolute; racemes few-flowered, shorter than or about equaling leaves, short-peduncled, bracts minute; calyx and hypanthium 3 mm. long at anthesis, densely short-puberulent, hairs white; calyx lobes narrowly lanceolate to oblong, about 2 mm. long, more or less spreading; corolla tube 4–5 mm. long, 2 mm. wide at throat, sparsely puberulent on outside, lobes triangular-ovate, obtuse, 1.5 mm. long, spreading or subreflexed; stamens barely equaling corolla tube; style exserted 1–2 mm.; stigma bilobate; fruit green, 3–3.5 mm. long, compressed, densely puberulent; seeds finely and uniformly favose-reticulate, 2.5 mm. long.

Rocky hillsides and canyons, Lower and Upper Sonoran Zones, central Sonora to Chihuahua and southern Sonora. July–Aug.

## 6. RICHARDIA L., Sp. Pl. 330, 1753

Annual or perennial herbs with terete, often striate, decumbent to erect stems and pubescent herbage. Leaves opposite, flat, entire; stipules sheathing, bristle-fringed. Flowers generally perfect, in terminal involucrate clusters on axillary peduncles. Hypanthium turbinate to subglobose. Sepals 4–8, moderately persistent. Corolla funnelform, white or pink, glabrous in throat; lobes 3–6, valvate in bud. Stamens as many as corolla lobes, adnate to throat of corolla tube. Disk small. Ovary 3–4-celled; styles filiform, united; stigmas 3–4. Ovule solitary in each cavity, amphitropous. Fruit of 3–4 carpels, crowned by persistent sepals. Seeds with 2 grooves on lower face. Endosperm horny.

**1. Richardia scabra** L., Sp. Pl. 330, 1753

Annual; stems erect and sparingly branched or prostrate to decumbent and more diffusely branched, branches 1–5 dm. long, striate to quadrangular; herbage hispid with straight stiffish hairs; stipules connate, bowl-shaped, 1–1.5 mm. high, bristle-margined; petioles of lower leaves 5–15 mm. long, broadly winged, gradually widening to blade, this elliptic, oblong, oblanceolate, or lanceolate, 1–2 cm. wide, 2.5–5 cm.

long, acute to rounded at apex, coarsely hispid; upper leaves smaller, with shorter petioles; flower clusters depressed-globose, subtended by 2 unequal pairs of foliaceous bracts, inner pair slightly to markedly smaller than outer, these to 2 cm. wide, to 3.5 cm. long; hypanthium 1–1.5 mm. long; sepals lanceolate to ovate-lanceolate, 1.5–2 mm. long, hispid; corolla white to faint pink, 4–6 mm. long, lobes 4–6, narrowly triangular to lance-triangular, more or less ciliate, 2–2.5 mm. long; hypanthium falling from fruit at maturity, carpels separating, narrowly obovoid, 2.5–3 mm. long, pale brown or gray-brown, rounded and coarsely muricate on back, longitudinally sulcate along inner face; seeds yellowish, smooth.

A weed along irrigation ditches and in fields, margins of Lower Sonoran Zone into Tropical Zone, southern Sonora and Baja California south to Peru. Sept.–Feb.

## 7. **CRUSEA** Cham. & Schlecht., Linnaea **5:** 165, 1830

Annual or perennial herbs with opposite, flat-bladed leaves and short-sheathing stipules. Flowers perfect, numerous, in compact heads or clusters subtended by 4 foliaceous bracts. Hypanthium ovoid to obovoid, more or less flattened. Sepals 4, persistent, sometimes with extra teeth. Corolla funnelform, pink to white; tube slender; lobes 4, valvate, spreading to rotate at anthesis. Stamens 4, adnate to the corolla tube to its mouth or nearly so; filaments bent; anthers exserted. Ovary 2-celled; styles partially to wholly united, very slender. Ovule solitary in each ovary cell, amphitropous. Fruit didymous, the carpels separating from each other but indehiscent, their walls very thin. Seeds oblong.

### 1. **Crusea subulata** (Pavon) A. Gray, Proc. Amer. Acad. **19:** 78, 1883

*Spermacoce subulata* Pavon ex DC., Prod. **4:** 543, 1830.
*Borreria subulata* DC., Prod. **4:** 543, 1830.
*Diodia subulata* Standley, Field Mus. Bot. Ser. **8:** 386, 1931.

Erect, strict, or slightly branching, annual herb 1–2.5 dm. tall, stems slender, quadrangular, hispid along angles, glabrous or sparsely puberulent between them; stipules connate-vasiform, 1.5–3 mm. deep, 3–4 mm. wide, white-hyaline, bearing several slender white bristles 1.5–3 mm. long along rim; leaf blades linear-lanceolate, linear-oblanceolate, or subulate, 0.8–3 mm. wide, 1–2 cm. long, rather tightly revolute to flat, acute and setose-apiculate at apex, minutely puncticulate-dotted on upper surface, sparsely scabrous-hispid along margins; flowers terminal on axillary peduncles, some of which do not elongate, thus flowers clustered in axils of leaves; hypanthium 1.5–2 mm. high; sepals winged, larger broadly subulate, 1.4–2 mm. long, pungently tipped, glabrous on back, with a few stiff short setae along margins, smaller ones scarcely more than coarse setae; corolla white to pink, 4–6 mm. long, tube about 0.6 mm. in diameter, lobes oblong, spreading, 1.5–2 mm. long; anthers 0.6–0.8 mm. long, exserted 1–2 mm.; carpels 2–2.2 mm. long, narrowly obovoid, shallowly depressed-concave at apex, longitudinal scar along inner face white, rather conspicuous, back strigose with rather coarse, short hairs forming distinct lines, carpel wall tan or yellowish, firm; seed pale yellow-brown, closely enclosed in carpel, smooth, strongly grooved along inner face.

Open hillsides, among trees, or under bushes, upper margins of Lower Sonoran Zone to Arid Transition Zone, Pima, Santa Cruz, and Cochise Counties, Arizona, to southern New Mexico, south into northern Mexico. Barely reaching the margins of the Sonoran Desert. Aug.–Oct.

## 8. **DIODIA** L., Sp. Pl. 104, 1753

Annual or perennial herbs with erect, ascending, decumbent or procumbent stems. Leaves opposite, entire, sessile, usually distinctly striate longitudinally; stipules sheathing, fringed or lacerate along margins. Flowers small, perfect, in axillary glomerules. Hypanthium obovoid to obconic, persistent. Sepals 2–4, sometimes with additional smaller teeth. Corolla funnelform to salverform, white, pink, or lilac; corolla lobes 4, short, valvate in bud. Stamens usually 4, adnate to throat of corolla tube. Disk none. Ovary usually 2-celled; styles united; stigma bilobed. Ovules solitary in each locule. Fruit oblong to globose, the 2 carpels with firm walls, separating from each other but indehiscent. Seeds free from carpel wall, oblong to elliptic.

### 1. **Diodia teres** Walt. var. **angustata** A. Gray, Syn. Fl. N. Amer. **1**[2]**:** 35, 1884

Erect, sparingly branched to simple-stemmed annual 1–3 dm. tall, with sparsely puberulent to glabrous herbage; stipules connate, shallowly bowl-shaped, chartaceous, 1.5–2 mm. deep, 2.5–3.5 mm. broad, margins beset with slender bristles 2–6 mm. long; leaves linear to narrowly lanceolate, 0.5–4 mm. wide, 1.5–3 cm. long, plane or margins slightly revolute, midrib impressed above, prominent beneath, apex acute to short-acuminate, broadly cuneate to rounded at base; flowers several in axils of leaves; hypanthium about 1 mm. long at anthesis, hispid; sepals slightly unequal, lance-subulate to narrowly triangular, 1–2 mm. long, ciliolate marginally, cuspidate at apex; corolla white, pale pink, or lavender, 4–5 mm. long, tube very slender, villous within throat, lobes narrowly triangular, 1.5–2 mm. long, sparsely hispid along margins near tips; anthers barely exserted; carpels separating from each other, bearing persistent sepals at maturity, body obovoid, about 3 mm. long, 2 mm. wide, tawny-brown, sparingly hispid-scaberulous on rounded back, inner face nearly flat, smooth, with a low, rounded, longitudinal ridge near base, walls firm, closely enclosing seed, this 2–2.5 mm. long, yellowish, rounded on back, thin-grooved on flat inner face.

Moist meadows, swales, and grassy slopes, Lower and Upper Sonoran Zones, southern Arizona to Sinaloa and southern Baja California. Aug.–Sept., and in Mexico often Jan.–April.

## 9. **STAELIA** Cham. & Schlecht., Linnaea **3:** 364, *pl. 3, fig. 3,* 1828

Herbaceous or suffruticose plants with erect, and simple or diffusely branching, procumbent to ascending stems. Stipules connate into a bristle-margined cup. Leaves sessile or short-petiolate, subsetaceous to linear or oblong. Flowers small, in axillary clusters, sessile, perfect. Hypanthium ovoid to subglobose. Sepals 2 with smaller accessory teeth, or 4 and subequal, persistent. Corolla funnelform, glabrous within, often pubescent without, 4-lobed, lobes spreading, valvate in bud. Stamens 4, inserted in throat; anthers slightly exserted. Disk none. Ovary 2-celled, 2-ovuled; style slender, with 2 slender branches. Fruit membranous to subcrustaceous, carpels separating obliquely from persistent or subpersistent partition, or this sometimes adhering to and falling with a carpel. Seeds ovoid to oblong, flat and longitudinally grooved adaxially, granulate-tessellated on back. Endosperm fleshy; embryo slender, with elliptical cotyledons.

### 1. **Staelia scabra** (Presl) Standley, Field Mus. Bot. Ser. **8:** 385, 1931

*Diphragmus scaber* Presl, Bot. Bemerk. 81, 1844.
*Spermacoce asperifolia* Mart. & Gal., Bull. Acad. Brux. **11**[1]**:** 132, 1844.
*Borreria asperifolia* Robinson, Proc. Amer. Acad. **45:** 409, 1910.

Sparingly to much-branched, procumbent, short-lived, perennial herb with stems 0.5–3.5 dm. long and scaberulous-hispid herbage; stipules connate into cups 1–2 mm. deep, 3–5 mm. broad, marginal bristles 2–2.5 mm. long; leaves elliptic to lance-ovate, 4–10 mm. wide, 1.5–4.5 cm. long, acute at apex, rounded to subtruncate at base, margins faintly revolute, blade dark green above, paler beneath, midrib and veins prominent; hypanthium about 2 mm. long, scaberulous; sepals mostly 4, lance-deltoid, about 2 mm. long; corolla white, 6–8 mm. long, lance-ovate lobes and upper part of tube closely hispid with coarse, stiff, white hairs; stamens exserted 1–2.5 mm.; fruit subglobose, 3–4 mm. long, subtruncate at apex, scaberulous-hispid, valves separating from persistent septum and from basal ring of hypanthium, or sometimes septum adhering to one carpel and falling with it, sepals persistent, usually 3 attached to one carpel, only 1 to the other carpel; seeds dark brown to black, 1.2–1.4 mm. wide, 2–2.3 mm. long, deeply grooved longitudinally and flat on adaxial face, rounded and closely but narrowly tessellate transversely on abaxial surface.

Trail sides, along ditches and arroyos, mostly in sandy soil, Lower Sonoran and Tropical Zones, southern Sonora and Cape Region of Baja California, Mexico, south in the American tropics. Jan.–Nov.

## 10. MITRACARPUS Zucc., in Roem. & Sch., Syst. Mant. 3: 210, 1827

Small, annual or suffrutescent perennial with sparingly branched, erect or ascending, quadrangular stems, opposite leaves, and connate stipules forming a shallow cup. Leaves sessile or short-petiolate. Flowers in dense, terminal and axillary clusters or heads, terminal one usually largest and often subtended by 4 leaves. Hypanthium ovoid to ovoid-campanulate, lobes in 2 pairs unequal in size, color, and texture, larger pair usually at least twice as large as smaller. Corolla tubular-salverform, white, equaling or surpassing sepals. Stamens 4, with short filaments inserted near apex of corolla tube; anthers ovoid to linear, slightly exserted; style 2-lobed. Capsule 2-celled, 2-seeded, circumscissile at or near middle, sepals persistent on caplike, distal part. Seeds peltately attached to placenta, distinctly 4-lobed, 4-grooved on sides.

Plants perennial; leaves 3–10 cm. long; larger sepals 4–6 mm. long ...... 1. *M. schizangius*
Plants annual; leaves mostly less than 3 cm. long; larger sepals 3.5 mm. long or less:
  Corolla tube barely equaling larger sepals ........................ 2. *M. breviflorus*
  Corolla tube one and one-half to two times as long as larger sepals:
    Larger sepals 1.6–2 mm. long; stems and young leaves glabrous or glabrate; corolla puberulent without ........................................ 3. *M. linearis*
    Larger sepals 3–3.5 mm. long; stems and leaves hirsute or sometimes subglabrate in age; corolla minutely granular without ........................ 4. *M. hirtus*

### 1. Mitracarpus schizangius DC., Prod. 4: 572, 1830

Suffrutescent perennial from a woody rootstock, stems ascending or slightly decumbent, 2–5 dm. long, slightly silvery-sublustrous, sparsely hirsute; stipules connate into a cup 3–4 mm. deep, 3–6 mm. wide, minutely foveolate, white or silvery, marginal bristles to 4 mm. long, acicular, smooth; leaf blades narrowly to broadly lanceolate or less commonly narrowly elliptic, 3–6(12) mm. wide, 3–10 cm. long, dark green above, paler beneath, sparsely puberulent on both surfaces, scaberulous along midvein beneath and slightly revolute margin, acute to short-attenuate at apex; fascicles of smaller, lance-linear leaves frequently in axils; heads 8–15 mm. in diameter, terminal usually larger than axillary; hypanthium 1.5–2 mm. long, hispidulous to base or nearly so; larger pair of sepals green, narrowly subulate, 4–6 mm. long, sparsely setose along margins, shorter pair more abruptly attenuate, 1.5–2.2 mm. long; corolla white, tube

5–6 mm. long, glabrous, lobes narrowly ovate, 1–1.5 mm. long; anthers linear, 1–1.2 mm. long, exserted 1.5–2 mm.; seeds pale brown, 1.2–1.3 mm. long, finely foveolate, sublustrous, 4-lobed.

Rocky slopes and sandy arroyos, margins of Lower Sonoran and Tropical Zones, southern Baja California to southern Mexico. Sept.–May.

## 2. **Mitracarpus breviflorus** A. Gray, Pl. Wright. **2:** 68, 1853

Erect, simple or sparingly branched annual 1–2.5 dm. tall, with sparsely hispidulous to glabrate stems and leaves and 1 to several axillary clusters of flowers subtended by a pair of leaves, terminal head often subtended by 4 leaves, heads 4–8 mm. in diameter; stipules connate into a white-hyaline cup 1.5–2 mm. deep, 2–4 mm. in diameter, margin ciliate-bristly; leaf blades narrowly oblanceolate, 2–6 mm. wide, 15–25 mm. long, short-petiolate or sessile, scaberulous along midvein beneath and slightly revolute margins, upper surface usually glabrate; hypanthium about 1.5 mm. long at anthesis, hyaline-chartaceous at base, sparsely puberulent above line of dehiscence, this slightly below middle; sepals 4, in unequal pairs, larger ones green, stiff, 1.5–2.2 mm. long, ciliolate-margined, smaller ones white, erose-ciliolate, 0.4–0.6 mm. long; corolla waxy-white, tube 1–1.2 mm. long, glabrous, lobes one-fourth as long; anthers ovoid-elliptic, about 0.3 mm. long; seeds pale buff, 1 mm. long, foveolate-reticulate, prominently 4-lobed.

Hillsides, canyon floors, and under shrubs, Lower Sonoran and Upper Sonoran Zones, southern Arizona to Texas and adjacent Mexico. Aug.–Oct.

## 3. **Mitracarpus linearis** Benth., Bot. Sulph. 20, 1844

Slender annual 1–3 dm. high, with simple or moderately branching stems and glabrous herbage or young branchlets sparingly puberulent but soon glabrate; stipules connate into a subhyaline chartaceous cup 1.5–2 mm. deep, 2–4 mm. wide, foveolate, glabrous or sparsely hispid near apex, marginal bristles fine, weak, 0.2–0.8 mm. long; leaf blades linear to lance-elliptic or narrowly oblanceolate, 1–7(10) mm. wide, 1–3(4) cm. long, dark green above, paler beneath, acute, margins narrowly but tightly revolute; heads 8–15 mm. in diameter, terminal larger than axillary ones; hypanthium about 1.5–2 mm. long, hyaline and glabrous below line of dehiscence, sparingly hirsute above; sepals 4, one pair green, lance-subulate, 1.6–2 mm. long, ciliolate-margined, other pair white or nearly so, 0.8–1.2 mm. long, both pairs persistent on lid of capsule; corolla waxy-white, 3–4 mm. long, distinctly exceeding sepals, minutely granular and sparsely puberulent without, ovate segments spreading, 0.8–1.2 mm. long; anthers exserted about their own length; seeds 1–1.2 mm. long, pale olive-green, sublustrous, shallowly and finely foveolate, lobing prominent, grooves darker than adjacent ridges.

Usually in partial shade beside rocks or under shrubs and trees, Lower Sonoran and Tropical Zones, southern Baja California. Nov.–April.

## 4. **Mitracarpus hirtus** (L.) DC., Prod. **4:** 572, 1830

*Spermacoce hirta* L., Sp. Pl. ed. 2, 148, 1762.
*Spermacoce villosa* Swartz, Prod. Veg. Ind. Occ. 29, 1788.
*Spermacoce diffusa* Willd. ex Roem. & Sch., Syst. Veg. **3:** 531, 1818.
*Mitracarpus villosus* Cham. & Schlecht., Linnaea **3:** 363, 1828.
*Mitracarpus simplex* Rusby, Mem. N.Y. Bot. Gard. **7:** 381, 1927.

Erect or ascending, annual herb with 1 to several stems 1–3 dm. long, herbage scabrous to hirsute throughout or stems sometimes glabrate in age, stems and branches obscurely quadrangular or subterete, but with 4 narrow ridges; stipules broadly cupped, 1–2.5 mm. deep, about twice to four times as wide, white, foveolate-reticulate, moderately hirsute, margins ciliate with lance-subulate bristles 2–3.5 mm. long, bristles also sparsely ciliate-margined; leaves elliptic to oblong, 3–20 mm. wide, 2–6 cm. long, bright green above, slightly paler beneath, sparingly hirsute, scaberulous, or nearly glabrous except along margins, acute, short-petiolate; flowers numerous in crowded globose heads at nodes, sessile; hypanthium ovoid, about 2 mm. long, glabrous below, sparsely scaberulous above line of dehiscence; sepals 4, two green, broadly lanceolate, 3–3.5 mm. long, other pair narrower, nearly hyaline, 2–2.3 mm. long, all persistent and more or less puberulent or scaberulous; corollas waxy-white, about equaling longer sepals, glabrous or slightly granular outside; anthers broadly ellipsoidal, 0.4–0.6 mm. long, barely exserted; fruit about 2 mm. long, walls chartaceous, dehiscing slightly below middle; seeds distinctly cruciately 4-lobed on inner face but lobing visible from opposite face, yellow-brown, minutely and shallowly foveolate-pitted, 1–1.2 mm. long.

Hillsides, rocky banks, stream sides, and along irrigation ditches, Lower Sonoran to Tropical Zones, southern Baja California and Sonora to northern South America. Sept.–March.

## 11. GALIUM L., Sp. Pl. 105, 1753

Annual or perennial plants with slender, 4-angled stems and branches, and opposite or whorled leaves. Flowers perfect, polygamo-monoecious, or -dioecious, small, in axillary or terminal cymes or panicles, pedicels jointed with calyx. Calyx tube ovoid to globose, with a minutely toothed limb or this lacking. Corolla rotate to shallowly campanulate, usually 4-lobed, lobes involute and inflexed in bud, spreading in flower, often acuminate. Stamens 4 or rarely 3, with small anthers and short filaments. Ovary inferior, 2-celled, 1 ovule in each cell. Styles 2, short; stigmas capitate. Fruit didymous, fleshy or dry, glabrous or more frequently hispid, separating into 2 indehiscent carpels. Seeds concavo-convex or spherical; endosperm horny; embryo curved; cotyledons foliaceous.

Flowers not individually subtended by leaflike bracts; midrib and margins of leaves usually not conspicuously white:
- Plants annual; fruits with hooked hairs:
  - Corolla lobes acute; leaves in whorls of 6–8, to 5 cm. long; inflorescence not markedly proliferous .......... 1. *G. aparine*
  - Corolla lobes obtuse to rounded; leaves in whorls of 4, 4–8 mm. long; branches of inflorescence proliferating markedly .......... 2. *G. proliferum*
- Plants perennial; fruits glabrous or clothed with straight hairs:
  - Fruit dry at maturity, not markedly pulpy, densely hispid:
    - Corollas reddish or deep maroon; corolla lobes acuminate; flowers polygamo-monoecious .......... 3. *G. wrightii* var. *rothrockii*
    - Corollas greenish white or greenish yellow; corolla lobes acute; flowers dioecious:
      - Leaves linear; corollas greenish white, 1.5–2.5 mm. in diameter .......... 4. *G. angustifolium* var. *diffusum*
      - Leaves elliptic to lanceolate; corollas greenish yellow, 2.5–3.5 mm. in diameter .......... 5. *G. stellatum* var. *eremicum*
  - Fruit pulpy and berrylike, sparsely hirsute to glabrous at maturity:
    - Stems strict, procumbent to scrambling, 5–20 dm. long; fruit hirsute in youth, nearly glabrous at maturity; leaves ovate to elliptic-lanceolate, not rigid .......... 6. *G. pubens* var. *grande*

Stems cespitose, erect, matted, 3–15 cm. long; fruit glabrous; leaves linear-subulate, rigid and spine-tipped .................................... 7. *G. andrewsii*

Flowers individually subtended by 4 leaflike bracts; midribs and margins of leaves white, nearly vitreous ........................................... 8. *G. microphyllum*

## 1. **Galium aparine** L., Sp. Pl. 108, 1753

*Galium vaillantii* DC., in Lam. & DC., Fl. Franc. ed. 3, **4:** 263, 1805.
*Galium agreste* α *echinospermum* Wallr., Sched. 59, 1822.
*Galium aparine* β *minor* Hook., Fl. Bor. Amer. **1:** 290, 1833.
*Galium spurium* var. *echinospermum* Hayak, Fl. Steierm. **2:** 393, 1913.

Slender, scrambling or prostrate annual from a slender root, stems 1–15 dm. long, sometimes stiffer and erect in dry habitats, retrorsely hispid-scabrous on angles; leaves in whorls of 6–8 (rarely only 4), linear to lanceolate or oblanceolate, 1–6 mm. wide, to 5 cm. long, cuspidate at apex, retrorsely hispid on midribs and margins; flowers in 1–3-flowered cymes or becoming cymose-paniculate in mature plants; corolla lobes acute, 0.4–0.8 mm. long; fruiting pedicels straight, 5–20 mm. long, divaricately spreading, glabrous or sparsely hispid; fruits 1.5–3.5 mm. in diameter, covered with white hooked bristles 0.4–0.6 mm. long.

In widely varying habitats, mostly in canyons and intertwined in shrubs in our area, widely distributed in North America, South America, Europe, and Asia; barely entering the margins of the Sonoran Desert. Feb.–Aug.

## 2. **Galium proliferum** A. Gray, Pl. Wright. 67, 1853

*Galium virgatum* var. *diffusum* A. Gray, Pl. Wright. 80, 1852.
*Galium proliferum* var. *subnudum* Greenm., Proc. Amer. Acad. **33:** 461, 1898.

Simple to moderately branched annuals with decumbent to erect stems 1–4 dm. tall, angles usually scabrous but sometimes glabrous; leaves in whorls of 4, oblong to narrowly ovate, 1–4 mm. wide, 4–8 mm. long, short-petiolate to sessile, abruptly short-apiculate at apex, scaberulous along margins; internodes to 6 cm. long; flowers nearly sessile between 2 leaflike bracts at tips of short-pedunculate branches, these usually proliferating; pedicels stout, recurved, shorter than fruit; corollas white to pale yellow, 2.5–3 mm. in diameter, lobes broadly ovate, about 0.5 mm. long, rounded to obtuse; fruit, exclusive of hairs, 2.5–3 mm. in diameter, closely beset with hooked hairs about 1 mm. long.

Rocky slopes, often under larger shrubs, Lower Sonoran and Upper Sonoran Zones, Providence Mountains, San Bernardino County, California, to western Texas and northern Mexico. March–April.

## 3. **Galium wrightii** A. Gray var. **rothrockii** (A. Gray) Ehrendf., in Ferris, Contr. Dudley Herb. **5:** 99, 1958

*Galium rothrockii* A. Gray, Proc. Amer. Acad. **17:** 203, 1882.

Slender-stemmed perennial with several mostly erect, nearly glabrous branches 2–5 dm. long from a woody base; leaves in whorls of 4, rigid, narrowly linear, acute, 5–10 mm. long, sparsely scabrous-prickly on slightly revolute margins; inflorescence diffusely branched, terminal panicles 1–3 dm. long; flowers polygamo-monoecious, dark red to purplish brown, corollas 1.5–2.5 mm. in diameter, lobes attenuate-acuminate; styles and style branches very slender; fruit dry, purplish, 1.5–2 mm. in diameter, including straight, spreading, sparse bristles.

Rocky slopes and canyons, mostly Arid Transition Zone but straggling to upper edge of Lower Sonoran Zone, Clark Mountains, San Bernardino County, California, to New Mexico and northern Mexico. May–Aug.

**4. Galium angustifolium** Nutt. var. **diffusum** Hilend & Howell, Leafl. West. Bot. **1:** 134, 1934

Lax, diffusely branched, shrubby perennial with stems 5–10 dm. long with nearly glabrous branches; leaves 2–4 at a node, 0.5–1.5 mm. wide, 3–12 mm. long, scaberulous along margins and midrib; flowers dioecious, very numerous in broad diffuse panicles 0.5–3 dm. long; bracts mostly in pairs, lance-elliptic, 1–2 mm. long; pedicels slender, 1–2.5 mm. long at anthesis, to 5 mm. long in fruit, mostly glabrous; corolla greenish white, 1.5–2.5 mm. in diameter, segments acute to apiculate; fruit dry, 1.5–2 mm. long and broad, densely hispid with coarse, simple, white hairs to 1 mm. long.

Arid desert ridges and slopes, Lower Sonoran and margins of Upper Sonoran Zones, desert slopes of the San Bernardino Mountains, California, south along the western edge of Colorado Desert into northern Baja California. April–June.

**5. Galium stellatum** Kell. subsp. **eremicum** (Hilend & Howell) Ehrendf., in Ferris, Contr. Dudley Herb. **5:** 7, 1956

*Galium acutissimum* A. Gray, Proc. Amer. Acad. **7:** 350, 1868.
*Galium stellatum* var. *eremicum* Hilend & Howell, Leafl. West. Bot. **1:** 137, 1934.

Low dioecious shrub 2–5 dm. tall, with intricately branched, basally woody stems; bark pale gray and shreddy; younger branches strongly 4-angled, scabrid along angles, densely puberulent with short, spreading, simple hairs; leaves in whorls of 4, elliptic to lance-linear, 3–8 mm. long, rigidly divergent, acuminate-cuspidate, pale gray-green, scaberulous-puberulent, white midvein prominent; inflorescence of dense to diffuse panicles, many-flowered, leafy with reduced leaves; pedicels straight, 2–3 mm. long in fruit; corolla greenish yellow, 2.5–3.5 mm. in diameter, lobes ovate, acute, sparsely hirsutulous; fruit dry, 2.5–3 mm. wide, including crowded, spreading, white hairs.

Dry desert slopes and arroyos, mostly Lower Sonoran Zone, southern Nevada to Utah, south to northern Baja California and Sonora. March–May.

**6. Galium pubens** A. Gray var. **grande** (McClatchie) Jepson, Man. Fl. Pl. Calif. 961, 1925

*Galium grande* McClatchie, Erythea **2:** 124, 1894.

Perennial with strict, scrambling or procumbent stems 5–20 dm. long, some spreading branches above; stems and leaves cinereous-pubescent; leaves in whorls of 4, narrowly ovate to elliptic-lanceolate, 1–12 mm. wide, 6–22 mm. long, margins narrowly revolute, paler beneath than above; flowers dioecious, staminate inflorescence paniculate, with few-flowered cymules on short pedicels; pistillate flowers solitary or in axillary pairs, greenish to purplish, 2.5–3.5 mm. in diameter, lobes acute to short-acuminate; mature fruit fleshy, purplish, 2.5–3 mm. in diameter, hispidulous in youth, eventually glabrous or essentially so, black in dried specimens.

Hillsides and canyons, mostly Upper Sonoran and Arid Transition Zones, straggling into upper edges of Lower Sonoran Zone along western side of San Felipe Desert; Tule River in southern Sierra Nevada, California, to northern Baja California. April–May.

**7. Galium andrewsii** A. Gray, Proc. Amer. Acad. **6:** 537, 1865

Densely cespitose, usually matted perennials with much-branched, densely leafy stems 3–15 cm. tall from slender woody roots; stems glabrous or scabrous on angles; leaves crowded, in whorls of 4, linear-subulate, shining gray-green, 4–10 mm. long, rigid, spinulose-tipped, often scabrous-spinulose along margins; staminate flowers in small clusters or singly on capillary pedicels equaling or surpassing leaves; pistillate flowers solitary on short axillary pedicels shorter than leaves; corolla greenish white, 2–2.5 mm. in diameter, lobes acute; fruit 2–3 mm. in diameter, glabrous, berrylike and turning black when ripe or in dried specimens.

Arid rocky ridges, mostly in chaparral and under pines, Upper Sonoran and Arid Transition Zones but barely entering the Lower Sonoran Zone along the western margin of the Colorado Desert, the Sierra Nevada and Coast Ranges from Glenn County, California, to northern Baja California. May–Aug.

**8. Galium microphyllum** A. Gray, Pl. Wright. **1:** 80, 1852

*Relbunium microphyllum* Hemsl., Biol. Centr. Amer. Bot. **2:** 63, 1881.
*Galium nitens* M. E. Jones, Contr. West. Bot. **15:** 153, 1929.

Slenderly branched, often cespitose perennial 1–4 dm. tall or bright green, sparsely hirsute to glabrous stems decumbent and matted; internodes 4-angled, angles whitish and minutely papillate, shining; leaves in whorls of 4, linear to elliptic-linear or oblong, 1–2.5 mm. wide, 5–15 mm. long, acute and apiculate at apex, margins and midribs distinctly whitish, minutely granular to subpapillate; flowers solitary in axils on peduncles 2–3 mm. long, each flower subtended by 4 leaflike bracts 2–3.5 mm. long, sessile in cup of involucre; corolla greenish yellow, 2–2.5 mm. in diameter, broadly ovate lobes obtuse, sparsely hirsutulous without; fruit 2.5–3 mm. broad, bearing a few short blunt tubercles or processes at first, later nearly smooth, purplish when ripe.

Rocky hillsides and arroyos, mostly Upper Sonoran and Arid Transition Zones but straggling into margins of Lower Sonoran Zone, Gila County, Arizona, to Texas and northern Mexico; Cape Region, Baja California. March–Oct.

## Family 133. **CAPRIFOLIACEAE.** Honeysuckle Family

Vines, shrubs, trees, or perennial herbs with opposite, simple or pinnately compound leaves; stipules none, minute, or nectariferous glands. Flowers perfect, regular or irregular, cymose, thyrsoid, spicate, capitate, or in involucrate pairs. Calyx adnate to the ovary, 3–5-toothed or -lobed. Corolla 5-lobed, 2-lipped in *Lonicera*, the limb rotate, or ascending. Stamens 5 or less commonly 4, alternating with corolla lobes, inserted on tube; anthers 2-celled, versatile, dehiscing lengthwise. Ovary inferior, 1–6-celled; style slender; stigma capitate or 2–5-lobed; ovules 1 to several in each cell. Fruit a berry, drupe, or capsule.

Corolla rotate or urceolate; flowers in flat compound cymes; shrubs or small trees; leaves pinnately compound .......... 1. *Sambucus*
Corolla tubular, 2-lipped; flowers spicate, capitate, or in pairs and subtended by bracts; woody vines; leaves simple, usually entire .......... 2. *Lonicera*

### 1. **SAMBUCUS** L., Sp. Pl. 269, 1753

Shrubs or small trees with pinnate leaves and serrate petiolulate leaflets. Flowers small, numerous, white, yellowish, or pinkish, in compound, depressed or thyrsoid cymes. Calyx tube ovoid to turbinate, 3–5-toothed. Corolla rotate, regular, mostly 5-lobed, the lobes imbricate or valvate in bud. Stamens 5, inserted at base of corolla; filaments slender, usually spreading. Ovary 3–5-celled; styles 3-parted; ovules 1 in each cell. Fruit a berrylike drupe, with 3–5 one-seeded nutlets embedded in the juicy pulp.

**1. Sambucus mexicana** Presl ex DC., Prod. **4:** 322, 1830
(For long synonymy see Ferris, Ill. Fl. Pacif. States **4:** 44, 1960.)

Shrubs or trees 3–10 m. tall with 1 to several trunks, furrowed, dark gray bark and suckers or rapidly growing branches with large pith and conspicuous lenticels on smooth young bark; leaflets 5–7, 2–6 cm. long, oval, ovate-lanceolate, or oblong, abruptly acuminate to cuspidate at apex, cuneate to rounded at base, petiolulate to subsessile, serrate, firm, glabrous to densely tomentulose, green on both surfaces; cymes flat-topped, 5–20 cm. broad; flowers pale yellow to creamy white, 5–7 mm. in diameter; filaments about 2 mm. long; fruit globose, 5–6 mm. in diameter, dark blue to nearly black but bearing a dense white bloom when ripe.

Along arroyos, streams, and occasionally on open flats, Lower and Upper Sonoran Zones, Sacramento Valley, California, to Arizona, Texas, and northern Mexico. April–Oct.

## 2. LONICERA L., Sp. Pl. 173, 1753

Erect shrubs or, in ours, woody climbers with slender stems, shreddy bark, and simple opposite leaves. Flowers spicate, capitate, or in pairs subtended by involucrate bracts. Calyx tube ovoid to subglobose, 5-toothed. Corolla tubular, funnelform, or campanulate, irregular, often gibbous at base, limb 5-lobed and 2-lipped. Stamens 5, inserted on corolla tube, exserted in ours. Ovary 2–3-celled; ovules several in each cell; style slender; stigma capitate. Fruit a berry, 2–3-celled or rarely 1-celled, few-seeded. Seeds ovoid to oblong; embryo terete; endosperm fleshy.

Leaves whitish beneath . . . . . . . . . . . . . . . . . . . . . . . . . . . . . . . . 1. *L. subspicata* var. *johnstonii*
Leaves green on both sides, not whitish beneath . . . . . . . . . . . 1*a*. *L. subspicata* var. *denudata*

**1. Lonicera subspicata** Hook. & Arn. var. **johnstonii** Keck, Bull. S. Calif. Acad. **25:** 67, 1926
*Lonicera johnstonii* McMinn, Ill. Man. Calif. Shrubs 541, 1939.

Basal stems woody and stout, upper branches loosely climbing over shrubs, young twigs reddish to pale brown, glandular-puberulent; leaves leathery, ovate-oblong to suborbicular, 1–3.5 cm. long, 0.5–2.5 cm. wide, dark green and sparsely glandular-puberulent above, whitish and more densely hairy beneath, margins slightly revolute; petioles stoutish, 1–5 mm. long; flowers in spikes 2–6 cm. long, with several few-flowered whorls; corolla cream-colored to yellow tinged with pink, 10–14 mm. long, short-pubescent, lips of limb more or less recurved; stamens equaling corolla lobes, filaments sparingly pilose near base; style pilose below; berries yellow or reddish when ripe, 5–7 mm. long; seeds 2.5–3 mm. long, shallowly pitted, pale brown.

On brushy hillsides and dry slopes, mostly Upper Sonoran Zone but straggling into Lower Sonoran Zone, San Benito County, California, to northern Sierra San Pedro Mártir, Baja California. May–June.

**1*a*. Lonicera subspicata** var. **denudata** Rehd., Rep. Mo. Bot. Gard. **14:** 176, 1903
*Lonicera denudata* Davids. & Moxley, Fl. S. Calif. 344, 1923.

Differing from var. *johnstonii* in having less orbicular leaves 5–15 mm. wide, their blades shining green on both sides and less densely puberulent.

Brushy hillsides, mainly Upper Sonoran Zone but reaching edge of Lower Sonoran Zone, southern San Diego County, California, to the southern Sierra San Pedro Mártir in Baja California. May–Oct.

## Family 134. VALERIANACEAE. Valerian Family

Annual or perennial, strong-scented herbs with opposite or basal, simple or pinnately divided, estipulate leaves. Flowers small, perfect or polygamous, slightly irregular, in corymbose, paniculate, or capitate cymes, or rarely monochasial. Calyx limb absent or inconspicuous in flower, sometimes developing into an awned or plumose pappus in fruit. Corolla 5-lobed, with a slender, often spurred or gibbous base, the limb somewhat zygomorphic. Ovary inferior, 1–3-celled; ovule solitary, pendulous. Embryo straight. Endosperm none.

### 1. PLECTRITIS DC., Prod. 4: 631, 1830

Annual glabrous herbs with opposite, ovate to oblong leaves. Inflorescence thyrsiform, capitate, verticillate, or spiciform; bracts linear, fused into a many-parted structure. Calyx reduced to an inconspicuous ring. Corolla campanulate to funnelform, white to pink, bilabiate to nearly regular; tube gibbous to distinctly spurred. Stamens 3, included to exserted. Ovary 1-celled; stigma 2–3-parted. Fruit dorsally keeled, with inrolled or spreading wings or these obsolete, the ventral wall with glands forming 2 rows or 2 prominent mammillae. Cotyledons incumbent or accumbent.

**1. Plectritis ciliosa** (Greene) Jepson subsp. **insignis** (Suksd.) Morey, Contr. Dudley Herb. **5**: 121, 1959

*Aligera insignis* Suksd., Deutsch. Bot. Monats. **15**: 146, 1897.
*Aligera rubens* Suksd., loc. cit.
*Aligera patelliformis* Suksd., W. Amer. Sci. **12**: 53, 1901.
*Plectritis davyana* Jepson, Fl. W. Mid. Calif. 475, 1901.
*Aligera intermedia* Suksd., Werdenda **1**: 44, 1927.
*Aligera glabrior* Suksd., loc. cit.
*Aligera barbata* Suksd., loc. cit.
*Plectritis californica* var. *rubens* Dyal, Amer. Midl. Nat. **42**: 495, 1949.

Slender plants 0.5–5.5 dm. tall, with obovate to oblong-fusiform, petiolate leaves 2–6 cm. long below, reduced and sessile above; inflorescence verticillate-spiciform; corolla 1.5–3.5 mm. long, deep pink, with darker spots at the base of lower lobes, tube narrowly funnelform, spur slender, about equaling tube and exceeding ovary, tip not distended, limb distinctly bilabiate; stamens about equaling upper corolla lip; fruit pale straw-yellow to brown, 3–4 mm. long, pubescent with stout, cylindrical, stiff hairs, keel rounded, with a dorsal groove, and brushlike rows of hairs; wings usually inwardly curled, connivent at base, spreading above; cotyledons incumbent.

In partial to deep shade, Upper Sonoran Zone and margins of Lower Sonoran Zone, Napa County, California, to Washington and south to vicinity of Tucson, Arizona, and nothern Baja California. Feb.–May.

## Family 135. CUCURBITACEAE. Gourd Family*

Annual or perennial, trailing or climbing, monoecious or dioecious herbaceous vines. Stems with tendrils, usually ribbed or grooved. Leaves alternate, estipulate, usually petiolate, entire to palmately lobed or dissected, usually emarginate. Flowers minute to large, axillary; calyx and corolla adnate, cylindrical to campanulate or salverform, usually 5-merous; staminate racemose, paniculate, or umbellate. Stamens 1 to 5,

* Text contributed by Howard Scott Gentry.

inserted in bottom, on sides, or on rim of calyx tube; anthers straight, curved, or folded, 1- or 2-celled, free or united to form an androecium, longitudinally dehiscent. Pistillate flowers solitary, geminate, umbellate, or rarely paniculate, adnate calyx and corolla forming a hypanthium. Ovary inferior, 1- to 5-celled; placenta parietal or axillary; ovules 1-many, horizontal, ascending or pendulous. Fruits very small to very large, fusiform to globose, smooth or echinate, dehiscent or indehiscent, rarely operculate. Seeds commonly flattened-ovoid.

Ovules and seeds horizontal, rarely ascending, usually numerous; fruits medium to large; stamens normally 3; roots commonly enlarged or tuberous:
- Stamens complexly folded, inserted at base of corolla tube; anthers connate (except in *Luffa*):
  - Staminate flowers racemose; fruits with a spongelike network of fibers within, dryish:
    - Fruits ribbed or striped; stamens distinct; seeds horizontal, compressed .... 1. *Luffa*
    - Fruits not ribbed or striped; stamens connate; seeds ascending, globose .... 2. *Marah*
  - Staminate flowers solitary or fasciculate; fruits without a spongelike network of fibers, succulent or fleshy:
    - Large vines, mostly perennial from large roots; fruits gourds, smooth, globose; staminate flowers solitary .......................................... 3. *Cucurbita*
    - Small annuals; fruits smaller, echinate or smooth, ovoid; staminate flowers usually in fascicles ................................................ 4. *Cucumis*
- Stamens merely curved or straight, inserted on sides or at top of calyx tube; anthers wholly distinct or partly connate:
  - Style surrounded at base by an annular disk; fruits greenish, not ripening red or yellowish, long-pedunculate:
    - Receptacle and calyx tube cylindric; anthers sessile, dorsifixed ...... 5. *Apodanthera*
    - Receptacle and calyx tube campanulate; anthers subsessile, basifixed ... 6. *Melothria*
  - Style not surrounded at base by annular disk; fruits bright red or yellowish when ripe:
    - Plants dioecious; roots of single tubers; seeds with rounded apex .. 7. *Maximowiczia*
    - Plants monoecious; roots of several tubers; seeds emarginate at apex ... 8. *Tumamoca*

Ovules and seeds erect, ascending, or pendulous, 1 to several; fruits small to medium, echinate or smooth; roots rarely enlarged; stamens 1–5:
- Anthers not fused to form a horizontal ring around a capitate androecium:
  - Ovules or seeds normally 2 or more per cell, erect or ascending:
    - Fruits smooth; staminate flowers large (2–4 cm. long), usually solitary; filaments distinct below ...................................... 9. *Schizocarpum*
    - Fruits echinate (except sometimes in *Vaseyanthus*); staminate flowers small (1 cm. wide or less), racemose or paniculate; filaments united:
      - Fruits obconic to globose to oblong, indurate; body 15 mm. long or less, 2- to 3-seeded or 1-seeded by abortion; stamens 3–5 .............. 10. *Vaseyanthus*
      - Fruits oblong to ovoid, 15 mm. or more long, normally 4–16-seeded; stamens 5 .... ................................................ 11. *Echinopepon*
  - Ovules or seeds one, pendulous or inverted from apex or side of ovary; fruits small (5–10 mm. long):
    - Pistillate flowers each enveloped in a pair of dentate bracts; corolla lobes bifurcate; fruit smooth; anthers 5 .................................. 12. *Sicyosperma*
    - Pistillate flowers not enveloped in bracts; corolla lobes entire; fruits echinate; anthers 3:
      - Fruits solitary or in pairs, oblique, long-rostrate; plants perennial .. 13. *Brandegea*
      - Fruits capitate in congested, sessile umbels, not oblique, not rostrate; plants annual ...................................................... 14. *Sicyos*
- Anthers fused, forming a continuous horizontal ring around a capitate androecium ..... ........................................................ 15. *Cyclanthera*

## 1. LUFFA Cav., Icon. 1: 7, *pl.* 9, 1791

Rather coarse annual, monoecious vines with fibrous fruits. Leaves usually lobate. Male flowers racemose, long-pedunculate. Calyx shallowly campanulate, 5-lobate. Corolla rotate. Stamens 3, distinct, inserted in bottom of calyx tube; anthers linear, flexuous, connivent. Pistillate flowers solitary, short-pedunculate; hypanthium fusiform, 5-lobed. Corolla rotate. Stigma of 3-forked or deeply split lobes. Fruit fibrous, 3-celled, many-seeded, large, smooth or coarsely prickle-covered, ribbed or striped, cylindric to oblong-ovoid; placentae 3. Seeds horizontal, oblong, compressed.

**1. Luffa operculata** (L.) Cogn., var. **intermedia** Cogn., in Rose, Contr. U.S. Nat. Herb. **1**: 330, 1895

Large climbing vine forming dense masses of foliage; stems slender, narrowly grooved, sparsely hispidulous; tendrils bifid, strongly petiolate, hispidulous; petioles 4–6 cm. long; leaf blades 4–6 cm. long, 5–7 cm. wide, broadly cordate with a wide sinus, 5-lobate with open sinus between lobes, these broadly ovate to spatulate, constricted below, denticulate, apiculate, muricately hispid above and more densely so beneath; staminate racemes several- to many-flowered, bracteate, 5–10 cm. long; calyx hispid, lobes triangular-ovate, acuminate; corolla about 12–15 mm. broad, pale yellow; petals ovate, mucronate, pubescent, about twice as long as sepals; fruits short-pedunculate, oblong-ovoid, 5–7 cm. long, ribbed longitudinally, ribs sparsely and coarsely prickly, sparsely white-pubescent, glabrate in age, dry; seeds oblong or depressed-ovoid, faintly spotted, with 2 oblique bumps above hilum on each face.

Margins of summer pools and water courses, Lower Sonoran to Tropical Zones, vicinity of Hermosillo, Sonora, south along the coastal plain to Nayarit. June–Sept.

## 2. MARAH Kell., Proc. Calif. Acad. 1: 38, 1854

Perennial, trailing or climbing, monoecious vines with large globose or fusiform roots. Stems coarse with slender lateral branches. Leaves shallowly to deeply 5- to 7-lobate. Tendrils simple or bifid. Staminate inflorescences racemose, simple or compound. Calyx and corolla adnate, calyx shallowly campanulate. Corolla 5-lobed; lobes triangular to ovate, spreading. Stamens normally 3, rarely 4, forming a capitate androecium with a short column of united filaments. Anthers linear, folded, upper ends unequal. Pistillate flowers solitary in same axils as staminate, usually similar to but larger than staminate. Ovary ovoid to globose, 2- to 4-celled. Ovules 2–4 in each cell, ascending. Fruits large, ovoid to globose, strongly echinate. Seeds large, globose to ovoid, scarcely compressed.

Fruit 5–6 cm. in diameter, 4- to 12-seeded; ovary not broader than long; leaves scabrous above with conic trichomes .......................................... 1. *M. macrocarpa*
Fruit 2–3 cm. in diameter, 2- to 4-seeded; ovary broader than long; leaves slightly scabrous above with scattered, low, orbicular trichomes ......................... 2. *M. gilensis*

**1. Marah macrocarpa** (Greene) Greene, Leaflets Bot. Obs. **2**: 36, 1910
*Echinocystis macrocarpa* Greene, Bull. Calif. Acad. **1**: 188, 1885.
*Micrampelis macrocarpa* Greene, Pittonia **2**: 129, 1890.

Leafy vines, perennial from an enlarged root; main stems coarse, glabrous, strongly ribbed, lateral branches slender, sparsely pubescent at nodes only, occasionally pustulate along ridges; leaves thin, somewhat scabrous with conic trichomes above, smooth and sparsely pubescent or glabrous below, shallowly or deeply 5- to 9-lobed, lobes triangular-ovate to linear-lanceolate, rounded to acute, mucronate or aristate, margins sometimes coarsely toothed; tendrils simple or bifid and petiolate, sparsely hispid or glabrous; petioles mostly shorter than leaf blades, stout, striate, sparsely hispid; staminate flowers in simple or compound racemes to 20 cm. long; corollas spreading, 8–15 mm. broad, sparsely pubescent outside, minutely stipitate-glandular within; pistillate flowers similar to staminate but larger; ovary ovoid; fruits large, 5–6 cm. in diameter, subglobose to broadly ovoid, densely long-echinate, 4-celled, many-seeded; seeds large, obovoid or subglobose, 12–16 mm. long, smooth, light yellowish or gray.

Among shrubs and rocks of canyons, washes, and bajadas, Upper and Lower Sonoran Zones, western edge of the Colorado Desert and south to the Cape Region, Baja California. Feb.–April.

**2. Marah gilensis** (Greene) Greene, Leaflets Bot. Obs. **2:** 36, 1910

*Megarrhiza gilensis* Greene, Bull. Torrey Club **8:** 97, 1881.
*Echinocystis gilensis* Greene, Bull. Calif. Acad. **1:** 189, 1885.
*Micrampelis gilensis* Britt., Trans. N.Y. Acad. **8:** 67, 1889.

Perennial from a large tuberous root; stems essentially glabrous, slender, striate; tendrils small, slender, simple or bifid, glabrous; leaves thin, glabrous or nearly so below, somewhat scabrous above with scattered orbicular trichomes, blades 3–8 cm. long, 3- to 7-lobate, lobes triangular-ovate to lanceolate, acute, broadly emarginate; petioles 1–4 cm. long, striate, glabrous; staminate flowers in simple or compound racemes; corolla white, spreading, 6–10 mm. broad, sparsely pubescent outside, minutely stipitate-glandular within; stamens 3, deeply flexed; ovary glandular-pubescent, densely echinate, broader than long, solitary, beak short; fruits 2–3 cm. long and broad, 2- to 4-celled, 2- to 4-seeded, somewhat succulent when green, becoming dry, dehiscing irregularly, densely echinate, globose; seeds somewhat flattened, ovoid to globose.

Canyons and hillside thickets, Lower and Upper Sonoran Zones, Catalina and Ajo Mountains, Arizona, east into New Mexico. Feb.–April.

### 3. CUCURBITA L., Sp. Pl. 1010, 1753

Coarse, scabrous, monoecious, juicy and often strong-scented perennial herbs from thickened tuberous roots and with prostrate, trailing, or climbing stems. Stems sometimes rooting at nodes, angled or striate, with hairlike, scaly, or prickly trichomes. Tendril sessile or petiolate, often with hold-fast terminal glands. Leaves entire, lobed or divided, divisions commonly 5, and often variously again toothed or lobed, palmately veined, veins large and conspicuous beneath. Flowers solitary, gamosepalous and gamopetalous; staminate commonly more numerous than pistillate. Calyx cylindric to campanulate; lobes narrow, distinct. Corolla large, campanulate; 5 lobes yellowish, conspicuous. Anthers 3, linear, connate and complexly folded into a terete and somewhat twisted anthericum on short, distinct filaments. Pistil single with 3 pairs of divergent stigmatic lobes. Ovary 3- to 5-celled, constricted near top. Fruit a gourd with numerous horizontal, flattened seeds.

Tendrils long-petiolate; leaves longer than broad, triangular-ovate, coarse, usually not lobed; calyx lobes at anthesis about equaling calyx tube .................. 1. *C. foetidissima*
Tendrils sessile or short-petiolate; leaves broader than long, broadly ovate, cordate, or orbicular in outline, variously lobed; calyx lobes at anthesis much shorter than calyx tube:

Blades lobed less than halfway to base, lobes broad and rounded .......... 2. *C. cordata*
Blades lobed to middle or beyond, lobes linear to lanceolate:
Leaves pedately divided nearly to base, lobes linear-ovate to linear-lanceolate and often variously again lobed; staminate calyx cylindric at anthesis:
Leaf lobes rounded at tip, to 6 cm. long, numerous; petioles 1.5–5 cm. long ......... ........................................................ 3. *C. cylindrata*
Leaf lobes acuminate, to 12 cm. long, mostly 5; petioles 3–8 cm. long .... 4. *C. digitata*
Leaves palmately lobed or divided, lobes lanceolate-acuminate to triangular-lanceolate, entire to irregularly, sparingly and obscurely lobed again:
Leaf blades to 8–9 cm. long; upper 3 lobes lanceolate-acuminate, 2 or more times as long as broad; veins often conspicuously white-pubescent above ... 5. *C. palmata*
Leaf blades to 15 cm. long; upper 3 lobes triangular-acutish to obtuse, mostly less than twice as long as broad; veins not markedly white-pubescent ..... 6. *C. californica*

## 1. Cucurbita foetidissima H. B. K., Nov. Gen. & Sp. **2:** 123, *figs. 162, 163,* 1817

*Cucumis perennis* James, in S. H. Long, Exped. Rocky Mts. **2:** 20, 1820.
*Cucurbita perennis* A. Gray, Bost. Jour. Nat. Hist. **6:** 193, 1850.
*Pepo foetidissima* Britt., in Britt. & Brown, Illus. Fl. N. U.S. and Canada ed. 2, **3:** 291, 1913.

A coarse, large-leaved prostrate vine with a large, deeply penetrating tuberous root, long-lived; stem radiate, forming a dense cover 30–40 cm. high over several square meters, scabrous with curved, dimorphic, hairlike or conic trichomes; leaves triangular-lanceolate to quadrangular-lanceolate, blades commonly 10–20 cm. long, evenly and progressively smaller toward the tips of stems, densely and coarsely white-pubescent above, scabrous beneath with conical trichomes along veins; petioles thick, scabrous, commonly one-half as long as blade; tendrils thick, long-petiolate, branches short and capitately coiled; peduncles short; staminate calyx lobes subulate, about equaling calyx tube; corollas bearing several kinds of hairs, coarser trichomes prominent on veins; fruits depressed-globose, 6–8 cm. high, green with conspicuous, whitish stripes, white-mottled; seeds small, oblong-ovate, 12 mm. long, 6–7 mm. broad, the margins obtuse.

In sandy alluvium of waste lands and fallow fields, Lower Sonoran to Carolinian Zones, Nebraska and Missouri to California and south into northern Mexico; entering the Sonoran Desert in Arizona and edges of the Colorado Desert. May–Aug.

## 2. Cucurbita cordata S. Wats., Proc. Amer. Acad. **24:** 50, 1889

Perennial prostrate vine with gray-canescent foliage; stems slender, sharply angled, retrorsely scabrous; tendrils small, capitately coiled, tipped with holdfast glands; leaf blades orbicular to reniform in outline, cleft one-third of way to base into 3–5 main lobes, sublobes forming an irregularly crenate margin, closely appressed-pubescent and papillate, veins more coarsely scabrous-hispid below; petioles about as long as blades, stout, bearing reflexed trichomes; staminate flowers about 7 cm. long; corolla pubescent-hirsute; calyx lobes very slender and short; fruit depressed-globose, about 8 cm. thick, 7.5 cm. high, green with 5 narrow stripes of lighter color and secondary stripes of intermediate blotches; seeds plump, ovate, 10–11 mm. long, 7–8 mm. broad, obtuse.

Known only from the type locality, "sandy plains near Los Angeles Bay," where it was collected by Edward Palmer in 1887.

**3. Cucurbita cylindrata** L. H. Bailey, Gentes Herb. **6:** 304, *fig. 156*, 1943

Prostrate vine with cinereous foliage, perennial from a complex system of several fusiform tap roots 3–6 dm. long, and many horizontal branches; stems slender, sometimes rooting at nodes, older striate with rounded ridges, muricately pilose-hirsute; tendrils epetiolate, 3–5-branched from base, terminally capitate with holdfast glands; leaves deeply 5-lobed, each lobe with 3–5 sublobes, latter linear to oblong, rounded at tip, hoary with muriculate and more slender trichomes, paler below and on veins above, blades 2–6 cm. long, 3–7 cm. wide; petioles mostly shorter than blades; flowers 8–10 cm. long; staminate calyces cylindric, tube about 2.5 cm. long, lobes subulate, 6–8 mm. long, muriculate, hispid; corolla densely pubescent, veiny, lobes acute; pistillate calyx less than 1 cm. long; fruits depressed-globose, 8 cm. wide, 7 cm. high, green with 10 whitish narrow stripes; seeds small, thick, irregularly ovate, 9–10 mm. long, with very obtuse margins.

Low to medium elevations, Lower Sonoran Zone, vicinity of Calmallí to Loreto, Baja California. March-May.

**4. Cucurbita digitata** A. Gray, Pl. Wright. **2:** 60, 1853

Perennial prostrate vine with a deep, tuberous root; branches slender and distantly running, rarely climbing; stems slender, glabrous, ribbed, whitish-pustulate with flat, oval trichomes on angles; tendrils very short-petiolate, 3- to 5-parted, branches gland-tipped; leaves 5-cleft nearly to base of blade, lobes 4–10 cm. long, linear-lanceolate to linear-oblanceolate, variably sublobed, green, bearing conic trichomes above and below, sometimes paler below; petioles stout, conspicuously ribbed, mostly shorter than lobes, muricate and hispid; flowers 4–6 cm. long; calyx cylindric to narrowly campanulate, sparsely hispid, tube 2.5–3 cm. long, lobes subulate, 3–5 mm. long; corolla sparsely hispid; fruit globose, slightly depressed, dark vivid green with 10 narrow stripes and variably speckled; seeds ovate, obtusely pointed, 10–11 mm. long, 7–8 mm. wide, edges thick and obtuse.

Mostly in sandy alluvium of valleys and along washes, Lower Sonoran Zone, southeastern California, southern Arizona, southern New Mexico, and northwestern Sonora. June–Oct.

**5. Cucurbita palmata** S. Wats., Proc. Amer. Acad. **11:** 137, 1876

A grayish green, perennial, prostrate vine with a deep fusiform root; stems strongly ribbed, rather densely pilose and muriculate when young, muriculate only on angles in age; tendrils nonpetiolate, mostly 3-branched, retrorsely hispid, gland-tipped, small, weak; leaf blades palmately cleft to middle or deeper, 3–9 cm. long and about as wide or slightly narrower, conically appressed-hispid above, densely so along veins, appressed hispid below and muriculate, lobes lanceolate, acuminate or acute, occasionally irregularly sublobed; petioles on mature leaves equaling or exceeding blades, much shorter than blades on young or depauperate leaves, stout, strongly or obscurely ribbed, mostly densely muriculate and hispid; staminate flowers 5–7 cm. long, pubescent, on peduncles 3–5 cm. long; calyx narrowly campanulate, tube 3–3.5 cm. long, lobes 4–6 mm. long, subulate; ovary pubescent; fruits depressed-globose, dull light green, narrowly and faintly 10-striped; seeds ovate to oblong, white, 9–12 mm. long, 6–8 mm. broad.

Sandy soils, Lower Sonoran Zone, Mojave and Colorado Deserts, California, east into Arizona and south into northern Baja California. April–Sept.

**6. Cucurbita californica** Torr. ex S. Wats., Proc. Amer. Acad. **11:** 138, 1876

Perennial, prostrate vine with coarse stems and leaves, and a thick tuberous root; stems prominently angled, sordidly appressed-hispid; tendrils 3- to 5-parted on very short petioles, weak, hispid; leaf blades palmately 5-cleft to about middle, lobes broadly triangular-lanceolate, variably sublobed or entire, short white-hispid above, muriculate, pustulate, and ephemerally hispid below; petioles stout, mostly shorter than blades, ribbed, strongly muricate, hispid; flowers 6–8 cm. long, hispid and coarsely floccose-pubescent; calyx narrowly campanulate, tube of staminate flowers 2–2.5 cm. long, lobes subulate, 8–10 mm. long; peduncles 5–8 cm. long; fruits globose, faintly 10-striped and mottled, 7–8.5 cm. in diameter; seeds 13–14 mm. long, 8–9 mm. broad, thick, dull white.

Arid plains and desert slopes, Lower Sonoran Zone, Sacramento and San Joaquin Valleys, California, east into southern Nevada and Arizona, south to the Colorado Desert. April–Sept.

## 4. CUCUMIS L., Sp. Pl. 1011, 1753

Monoecious, annual vines with yellow flowers, simple tendrils, and entire or palmately 3- to 7-lobed, petiolate leaves. Staminate flowers fasciculate or rarely solitary in axils of leaves. Calyx tube campanulate to turbinate, 5-lobed. Corolla campanulate, 5-lobed. Stamens 3; anthers 1- or 2-celled, linear, erect, flexuous, basifixed, united, stalked, connective produced beyond anther cells. Pistillate flowers in axils separate from those bearing staminate flowers, pedunculate. Calyx and corolla 5-lobed. Style short; stigmas 3, obtuse. Ovary narrowly ovoid to oblong, with 3 to 5 placentae. Ovules many, horizontal. Fruits ovoid, oblong, or globose, smooth or echinate, succulent or pulpy, indehiscent, many-seeded. Seeds flattened, oblong to oval, smooth.

**1. Cucumis dipsaceus** Ehrenb. ex Spach, Hist. Veg. Phan. **6:** 211, 1838

Small scabrous vines with long-petiolate subentire leaves and yellowish, densely prickled fruits; stems coarse, prostrate or trailing, striate, scabrous with stiff erect pointed trichomes on ridges; tendrils simple, loosely coiled, scabrous; leaves orbicular to subreniform, cordate, emarginate sinus narrow, scarcely lobed, margin undulate and coarsely dentate, coarsely hispid above, more densely so and muriculate below, blades 5–8 cm. long; petioles distinctly longer than blades, prickly; staminate flowers few, fasciculate, 12–15 mm. long; calyx prickly, teeth subulate, longer than tube; corolla yellow, campanulate; pistillate flowers solitary, similar to staminate; fruits densely echinate with yellow prickles, ovoid, 4–5 cm. long, prickles 4–5 mm. long, fruiting peduncles stout; seeds numerous, whitish, elliptic to narrowly ovoid, about 5 mm. long.

Along ditches and arroyos, Lower Sonoran and Tropical Zones, introduced from La Purísima to the Cape Region, Baja California. March-May.

## 5. APODANTHERA Arn., Hook. Jour. Bot. 3: 274, 1841

Perennial dioecious or monoecious vines with thick roots and trailing or climbing branches, and short-petiolate, weak, simple or divided tendrils. Leaves long-petiolate, denticulate, deeply lobed, or entire, harshly pubescent. Staminate flowers racemose, bracteate. Calyx tubular, lobes filiform. Stamens 2–3; anthers sessile, dorsifixed, somewhat curved, usually connate, at least in middle. Pistillate flowers solitary, long-pedunculate. Style surrounded at base by an annular disk. Fruits ovoid to oblong, 4–7 cm. long, longitudinally ribbed or striped, smooth or nearly so, many-seeded. Seeds horizontal, ovate, flattened.

Leaf blades deeply 5-lobed, sinuses open, cordate; calyx lobes about equaling calyx tube .... ........................................................................ 1. *A. palmeri*
Leaf blades entire to undulate-margined, not lobed, reniform; calyx lobes less than half as long as calyx tube .................................................. 2. *A. undulata*

### 1. **Apodanthera palmeri** S. Wats., Proc. Amer. Acad. **24:** 50, 1889

Perennial prostrate vines with large, long, white roots, dissected leaves and thin-skinned, edible fruits; stems slender, narrowly striate, hispid-pubescent to glabrate; tendrils in opposite pairs, bifid, short-petiolate; leaves long-petiolate, palmately divided halfway to base or deeper, sinuses forming loops, lobes broadly ovate, denticulate, constricted at base, acute, with sparse, conic trichomes above, paler and bearing slender, simple hairs and conic trichomes below; staminate calyx tube cylindrical, glandular-hispid, 10–15 mm. long, lobes subulate, about equaling tube, spreading; corolla about twice as long as calyx lobes, spreading, pubescent; female flowers 2.5–4 cm. long, pubescent, peduncles usually equaling flowers; fruits ellipsoidal, 3–4 cm. long, striped with dull green and white, many-seeded, pericarp thin; seeds whitish or dull yellowish, ovate, flattened, about 8 mm. long.

Sandy alluvial plains, Lower Sonoran Zone, vicinity of Carbo, Sonora, south at least to Guaymas, Sonora. July–Sept.

### 2. **Apodanthera undulata** A. Gray, Pl. Wright. **2:** 60, 1853

Coarse perennial vine with gray to green foliage; stems thick, multistriate, hispid and sparsely muriculate; tendrils short-petiolate, mostly bifid, short, sparsely pilose; petioles 4–7 cm. long; leaf blades 6–9 cm. long, 7–10 cm. wide, reniform, entire, evenly and remotely denticulate or markedly undulate-margined and sparsely hispid and often muriculate above, whitish and densely hispid below, usually finely muriculate between veins and coarsely muricate on veins below; inflorescence with filiform bracts 5–20 mm. long; staminate calyx tube slender, 2–2.5 cm. long, lobes subulate, less than half as long as tube; corolla lobed to calyx rim, lobes about equaling calyx tube, pubescent; fruit ovoid, longitudinally ribbed, 4–5 cm. long, pubescent, on peduncles 4–7 cm. long.

Arid plains and mesas, Lower Sonoran to Arid Transition Zones, southern Arizona to Texas and adjacent northern Mexico. June–Sept.

## 6. **MELOTHRIA** L., Sp. Pl. 35, 1753

Annual or perennial, mesophytic, mostly monoecious vines with slender stems, and small to medium, smooth pepos. Leaves subentire to strongly lobed. Staminate flowers racemose, rarely corymbose or solitary. Calyx campanulate, lobes small. Corolla 5-lobed, lobes entire. Stamens 3, distinct, their short filaments inserted on sides of corolla tube; anthers distinct or rarely coherent, oblong to suborbicular, straight or slightly curved, one of them 1-celled, other two 2-celled, or occasionally all of them 2-celled; connective truncate or apiculate. Pistillate flowers solitary, fasciculate, or umbellate. Hypanthium campanulate. Staminodia 3. Ovary fusiform to oblong or globose, constricted below hypanthium; placenta 3. Style short, surrounded at base by an annular disk. Stigma 3-lobed or rarely only 2-lobed. Fruit baccate, oblong or globose or ovoid, many-seeded, unarmed and glabrate. Seeds horizontal, smooth or finely sculptured.

**1. Melothria donnell-smithii** Cogn., in Coult., Bot. Gaz. **16:** 9, 1891

Small climbing or trailing, leafy, monoecious vine; stems slender, grooved and ribbed, grooves whitish, ridges green, those on older stems callous-margined, papillate-hispid; tendrils small, simple; leaves cordate to angular-cordate in outline, occasionally deeply 3- to 5-lobate, lobes acute to obtuse, mucronate, broadly triangular to lanceolate, margins dentate to undulate, blades rather thick, 5–8 cm. long and wide, scabrous on both surfaces with sparse to dense, translucent trichomes, basal sinuses narrow or open, 1 cm. deep or more; petioles 3–6 cm. long, striate, papillate, hispid; staminate peduncles slender, single or paired, 1.5–4 cm. long, hispidulous; pedicels about 1 cm. long, congested at apex of peduncles; buds puberulent, densely so on corolla; receptacle glabrate; corolla 7–9 mm. wide, spreading, lobes puberulent, shallowly emarginate; pistillate flowers solitary; fruiting peduncles slender, 2–6 cm. long; fruits oblong to ovoid, faintly and irregularly striped, 3.5–5 cm. long, smooth, many-seeded; seeds pale white to yellowish, 4–5 mm. long, 2.5–3 mm. wide.

Open subtropical scrub and along washes, Lower Sonoran to Tropical Zones, central Sonora to Guatemala. June–Sept.

## 7. MAXIMOWICZIA Cogn., in DC., Monogr. Phan. 3: 726, 1881

Dioecious vines with dissected leaves, perennial from thickened rootstock extending slightly above ground. Stems slender, glabrous, with short internodes. Tendrils simple, small. Leaves deeply 3- to 5-dissected, lobes variously again lobed, linear to ovate, glabrous or scabrid-pubescent. Staminate flowers racemose, normally 5-merous. Calyx tubular to narrowly campanulate; lobes acute. Corolla narrowly campanulate; lobes emarginate to bifurcate, margins wavy, villous. Stamens 3, distinct, borne on orifice of tube; anthers 3, with 5 cells, sessile, slightly curved. Pistillate flowers solitary. Corolla rotate, tube and lobes more slender than those in staminate flowers. Stigma 3-lobed. Ovary fusiform. Fruit a globose to ovoid berry, several- to many-seeded, red to yellowish when ripe. Seeds horizontal, irregularly ovoid, flattened or rounded on lateral faces, rounded to pointed at apex, the margins raised.

Petals narrowed at base, apex widened; leaves glabrous or scabrous with sparse, cystolithic, flat trichomes:
- Lobes of leaves narrow, 2–6 mm. wide; fruits globose, about as wide as long; pericarp thin .......... 1. *M. lindheimeri* var. *tenuisecta*
- Lobes of leaves broad, 10–40 mm. wide; fruits ovoid, longer than wide, pericarp thick:
  - Leaves distinctly glaucous, glabrous; seeds not rugose ...... 2. *M. sonorae* var. *sonorae*
  - Leaves not glaucous, markedly hispid; seeds rugose ... 2*a*. *M. sonorae* var. *peninsularis*

Petals not narrowed at base, apex not widened; leaves scabrous-pubescent with simple trichomes; pericarp of fruit thin .......... 3. *M. insularis*

**1. Maximowiczia lindheimeri** Cogn. var. **tenuisecta** (A. Gray) Cogn., in DC., Monogr. Phan. **3:** 728, 1881

*Sicydium lindheimeri* var. *tenuisecta* A. Gray, Pl. Wright. **1:** 75, 1852.
*Ibervillea tenuisecta* Small, Fl. SE. U.S. 1136, 1903.

Stems slender, glabrous, glaucous, finely striate; tendrils very small, jointed 1–3 mm. above enlarged base, glabrous; leaves pedately 3- to 5-parted, irregularly sublobed, sublobes linear, 3–6 mm. wide, coarsely toothed toward apices, glabrous above, pustulate hispid below; petioles 1–2 cm. long, glabrous; staminate flowers racemose, campanulate, glabrous externally; corolla lobes 3–4 mm. long, spreading, yellow, base narrower than broadened, emarginate or undulate apex; pistillate flowers solitary, small; calyx glabrous; corolla puberulent, short-pedunculate; hypanthium glabrous; fruit globose, 15–30 mm. long, reddish when ripe; seeds 6–12, ovoid, 5 mm. long, margin flat, lateral faces transversely corky-pleated near margin, emarginate at narrowed base, pointed to rounded at tip.

Desert flats and grassy plains, Lower Sonoran and Subtropical Zones, southeastern Arizona to eastern Texas, south into Sonora, Mexico. June–Sept.

**2. Maximowiczia sonorae** S. Wats., Proc. Amer. Acad. **24:** 51, 1889, var. **sonorae**
*Ibervillea sonorae* Greene, Erythea **3:** 75, 1895.

Climbing glaucous vine from a conspicuous, perennial, tuberous bottle-shaped root, neck and part of body above ground; stems slender, glabrous, with short internodes; tendrils glabrous, nearly sessile; leaves orbicular to broadly ovate in outline, deeply dissected into three primary lobes, each lobe narrowed toward base and with 7–9 sublobes, each sublobe 1–4 cm. wide, coarsely and deeply or shallowly toothed toward apex, glaucous and glabrous or sometimes sparsely white-pustulate beneath, whole blade 4–10 cm. long; staminate flowers in racemes about 15 cm. long, distal buds opening first; calyx tubular, glabrous; corolla campanulate, villous without and within, lobes 4–5 mm. long, narrowed at base, spreading divergently from emarginate apex, margins wavy; pistillate flowers on peduncles 2–3 cm. long, solitary, rotate, lobes 7–8 mm. long, tube 1–2 mm. in diameter; ovary ovoid, glabrous; fruit ovoid, 3–5 cm. long, at first greenish with faint stripes of whitish dots, ripening reddish; seeds many, dull black, somewhat flattened, ovate, 6–8 mm. long, rounded at apex, narrowed at base, margin distinct, lateral faces felt-like.

Foothill valleys and desert plains, Lower Sonoran Zone, northern Sonora south to the Río Mayo. July–Sept.

**2*a*. Maximowiczia sonorae** var. **peninsularis** I. M. Johnston, Proc. Calif. Acad. IV, **12:** 1178, 1924

Differs from var. *sonorae* in the nonglaucous, hispid leaves; thinner rind of fruit, and in more rugose seeds; perennial stems longer than in var. *sonorae.*

Along washes and on mesas and plains, Lower Sonoran Zone, central Baja California to the Cape Region, Baja California. Aug.–Nov.

**3. Maximowiczia insularis** Brandegee, Univ. Calif. Pub. Bot. **6:** 361, 1916

*Maximowiczia sonorae* var. *brevicaulis* I. M. Johnston, Proc. Calif. Acad. IV, **12:** 1179, 1924.

Climbing vines; rootstocks swollen, with persistent woody shoots 5–15 dm. long; stems somewhat flexuous, oldest reddish, older whitish, exfoliating, glabrous, glaucous, newer ones green with sparse ascending pubescence; tendrils simple, fine; leaves narrowly and deeply dissected, primary lobes 4–7, narrowed toward base, broadly spatulate or ovate, crenate or dentate, or more frequently sublobed, cuneate to spatulate, apices truncate or coarsely toothed, minutely muricate above, coarsely muricate-hispid beneath, whole blade about 2 cm. in diameter; petioles 1–2.5 cm. long; staminate flowers rotate, about 1 cm. long, light greenish; calyx glabrous; corolla densely hispid within and without, lobes about 4 mm. long, linear, somewhat narrower across emarginate apex than across base; pistillate flowers similar to staminate, solitary; fruits globose, 2.5–3 cm. long, orange-red when ripe, sometimes glaucous, few- to many-seeded; seeds ovoid, 5–7 mm. long, slightly narrowed toward truncate base, rounded at apex, uniformly pale yellowish brown, margin distinct, smoothly wrinkled, lateral faces rounded, tuberculate.

Along arroyos and valley floors, Lower Sonoran Zone, northern Vizcaíno Desert and south along the outer coast through the Magdalena Plain, Baja California and on the adjacent islands. Flowering both in spring and summer, following rains.

## 8. TUMAMOCA Rose, Contr. U.S. Nat. Herb. 16: 21, 1912

Xerophytic dioecious vines with dissected leaves, from perennial tuberous roots. Stems woody to somewhat succulent at base, annual growth above slender; internodes short. Tendrils small, simple. Leaves deeply lobed or dissected. Staminate flowers in few-flowered racemes, tubular, 5-merous; corolla lobes spreading. Calyx lobes minute. Stamens 3, borne on corolla tube below orifice. Pistillate flowers solitary, similar to staminate, withering-persistent. Fruit globose, several-seeded, berrylike, red or yellow at maturity. Seeds horizontal, ovoid, truncate at apex.

### 1. **Tumamoca macdougalii** Rose, Contr. U.S. Nat. Herb. **16:** 21, *pl. 17*, 1912

Roots tuberous, each about 5–15 cm. long, united into a woody crown with a short, persistent woody stem 10–15 cm. long; annual stems slender, weakly striate, glabrous or nearly so, somewhat flexuous, nodes inconspicuous; leaves pedately parted, lobes 3, spreading, 2–4 cm. long, narrowly linear to linear-lanceolate or ovate, variously and irregularly sublobed, obtuse to rounded at tip, mucronate, glabrous beneath, pustulate-scabrous above; hypanthium narrowly tubular, about 1 cm. long; corolla pale yellow or greenish yellow, glabrous, lobes of the pistillate flowers linear-lanceolate, longer than those of staminate, one-third to one-half as long as tube; ripe fruit conspicuously yellow to red, succulent, 2–several-seeded; seeds 7–8 mm. long, quadrate above narrowed base, tuberculate-rugose, truncate at apex.

Along gullies and sandy washes of hills and valleys, Lower Sonoran Zone, vicinity of Tucson, Arizona, south into central Sonora. June–Sept.

## 9. SCHIZOCARPUM Schrad., Ind. Sem. Hort. Gött. 1830

Monoecious vines, ours annual; stems slender, climbing. Tendrils branched. Leaves entire to lobate, thin, petiolate. Staminate flowers solitary or racemose. Calyx tubular, the lobes narrow, distinct. Corolla yellow, campanulate, large, 5-merous. Stamens 3; filaments distinct below, united above with connate, folded anthers; connectives produced beyond anthers. Pistillate flowers short-pedunculate, in same axils as staminate; calyx and corolla adnate. Ovary fusiform, with 3 placentae. Ovules 12 to many. Fruits smooth, tapering at one or both ends, ovoid, indehiscent or tardily and irregularly bursting. Seeds ascending, borne in double vertical rows along each side of each placenta, ovoid, flattened.

### 1. **Schizocarpum palmeri** Cogn. & Rose, Contr. U.S. Nat. Herb. **1:** 100, 1890

Slender climbing annual with thin leaves; stems striate, pilose with long, white hairs, sometimes minutely glandular-puberulent or merely puberulent; tendrils long-petiolate, unequally bifid, pubescent and often with minute clavate glands; leaves cordate, entire to shallowly or deeply trilobate, blades 4–8 cm. long, nearly as broad, obtuse to acute, mucronate, sparsely pubescent to glabrous above, pubescent and muriculate beneath, basal sinus broad and open, margins denticulate; petioles 2.5–6 cm. long, conspicuously long-pilose and often clavate-glandular; staminate flowers solitary or rarely in few-flowered racemes, pubescent and densely clavate-glandular on young peduncles and calyces; sepals linear to lanceolate, hispidulous; corolla 4–5 cm. long, hispid or puberulent, lobes lanceolate, acute, spreading; pistillate flowers similar; ovary densely hispid and clavate-glandular; seeds about 30 in each locule.

On coastal plains and valleys, Lower Sonoran Zone, central Sonora into Sinaloa. June–Sept.

## 10. VASEYANTHUS Cogn., Zoe 1: 368, 1891

Small, slender, herbaceous vines with palmately lobed or dissected leaves, and small, globose, smooth or echinate fruits. Stems sulcate, pubescent or glabrous, internodes relatively short. Tendrils petiolate, usually bifid. Leaves petiolate, commonly hispid-pubescent with various types of trichomes, often scabrous. Flowers small, 5-merous, monoecious. Staminate flowers in simple or compound racemes. Calyx tube usually cupulate to salverform, lobes minute. Corolla cupulate, limb spreading. Stamens 3–5; filaments united into a column, this bearing 3–5 anthers with linear to horseshoe-shaped thecae; pollen grains large, smooth, depressed-globose, obscurely 5- to 6-colpate. Pistillate flowers small, pedunculate, solitary, in same axils as staminate. Calyx and corolla adnate. Ovary conic to ovoid-oblong, sometimes oblique, 2-celled or rarely 3-celled, long-rostrate. Style short; stigma thickened, discoid. Ovules solitary in each cell, erect or ascending, or sometimes horizontal, attached near base of cell, upper ovule often abortive. Fruit narrowly ovoid to globose, thickly long-rostrate; rostrum persistent or deciduous; body dry, indurate, vesicular, indehiscent or rarely dehiscent, armed or smooth, 1- to 2-seeded. Seeds erect or ascending, obovate, clavate, or ovoid, somewhat compressed.

Mature fruits 12–20 mm. in diameter; prickles 4–7 mm. long; stamens 5; leaf lobes ovate, rounded to acute, often mucronate ............................... 1. *V. brandegei*
Mature fruits less than 1 cm. in diameter; prickles 2–3 mm. long or lacking; stamens 3–5; leaf lobes aristate to mucronate, acute:
Fruits echinate, globose:
Beak deciduous or slightly persistent on mature fruits; leaf lobes mostly triangular, acute, thickish (peninsular and insular) .................. 2. *V. insularis* var. *insularis*
Beak conspicuously persistent on mature fruits; leaf lobes usually acuminate, thin, aristate (mainland and San Pedro Nolasco Island) ...... 2*a*. *V. insularis* var. *palmeri*
Fruits smooth or slightly echinate, globose to oblong (peninsular and insular) ..........
.................................................. 2*b*. *V. insularis* var. *inermis*

### 1. Vaseyanthus brandegei (Cogn.) Rose, Contr. U.S. Nat. Herb. 5: 119, 1897

*Echinocystis brandegei* Cogn., Proc. Calif. Acad. II, 3: 59, 1890.
*Vaseyanthus insularis* var. *brandegei* I. M. Johnston, Proc. Calif. Acad. IV, 12: 1182, 1924.

Coarse herbaceous vines with deeply lobed leaves and strongly echinate, globose, indurate fruits; stems coarse, strongly ribbed, hispid but finally glabrate; tendrils bifid, stoutly long-pedunculate, sparsely hispid near base; leaves thick, orbicular in outline, 2–5 cm. wide, densely hispid on both surfaces and somewhat muriculate above, 5–7-lobed, with open, rounded sinuses, lobes lanceolate, ovate, or spatulate, mostly rounded but sometimes acute to mucronate at apex; petioles hispid with curved hairs; staminate flowers in compound racemes, peduncles mostly shorter than leaves, numerous; pedicels glabrous, persistent; corolla white, 5–6 mm. broad; stamens 5; ovary densely echinate, ovoid, with a slender beak falling at maturity; fruit globose, 12–20 mm. in diameter, 2-celled, 1–2-seeded, strongly echinate, sparsely pilose; seeds compressed, dark brown, broadly oval, narrowed at the base.

Upper strand and washes near the coast, Lower Sonoran Zone, southern Baja California. Nov.–April.

**2. Vaseyanthus insularis** (S. Wats.) Rose, Contr. U.S. Nat. Herb. **5:** 120, 1897, var. **insularis**

*Echinopepon insularis* S. Wats., Proc. Amer. Acad. **24:** 51, 1889.
*Vaseyanthus insularis* var. *originalis* Gentry, Madroño **10:** 149, 1950.

Annual or short-lived perennial; stems slender, striate, puberulent or pubescent with curled hairs, or sometimes glabrous; leaves cordate-orbicular, 2–7 cm. in diameter, sparsely to densely hispid on both surfaces, shallowly to deeply 5- to 7-lobed, lobes triangular to lanceolate, denticulate, acute to acuminate, mucronate to aristate, basal sinus broadly open; petioles equaling or much exceeding blades; staminate flowers in racemes shorter than or exceeding leaves; calyx shallowly campanulate, teeth green, minute; corolla 5-cleft, 4–5 mm. broad; stamens 3, 4, or 5; pistillate flowers short-pedunculate, slightly larger than staminate; body of fruit globose to oblong, 7–9 mm. in diameter, indurate, spongy or vesicular, covered with stiff prickles 1–3 mm. long, long-conic beak smooth, tardily deciduous; seeds 1–2, erect, smooth, oblong-obovate, slightly compressed, 4–5 mm. long, base broad and flat.

Arid slopes and upper beaches, Lower Sonoran Zone, coast of Gulf of California from Bahía San Francisquito south to La Paz, and on several adjacent islands. Nov.–April.

**2*a*. Vaseyanthus insularis** var. **palmeri** (S. Wats.) Gentry, Madroño **10:** 149, 1950

*Echinopepon palmeri* S. Wats., Proc. Amer. Acad. **24:** 52, 1889.
*Brandegea palmeri* Rose, Contr. U.S. Nat. Herb. **5:** 120, 1897.

Stems sparsely puberulent to glabrous; petioles slender; leaves thin, sparsely to somewhat densely pubescent, rarely scabrous, commonly divided beyond middle of blade, lobes lanceolate, often saliently toothed, acute to acuminate, aristate; body of fruit globose, pericarp sparsely puberulent; prickles 2–3 mm. long, glabrous or glabrate; beak narrow, acute, persistent.

Sandy and gravelly washes and on upper beaches, Lower Sonoran Zone, vicinity of Guaymas, Sonora, and on San Pedro Nolasco Island. Nov.–April.

**2*b*. Vaseyanthus insularis** var. **inermis** I. M. Johnston, Proc. Calif. Acad. IV, **12:** 1182, 1924

*Vaseyanthus rosei* Cogn., Zoe **1:** 368, *pl. 11*, 1891.

Stems and leaves scabrous-pubescent; leaves small, dried blades mostly less than 3 cm. wide, broader than long, rarely deeply 5–7-lobed, mostly shallowly 5–9-lobed, lobes broad, angular, obtuse to acute, mucronate to aristate; body of fruit unarmed, obliquely globose or oblong, with a persistent, attenuate beak, 1-seeded when globose, often 2-seeded if oblong, tardily dehiscent around middle.

Sandy and gravelly areas, Lower Sonoran Zone, Gulf of California coast of Baja California from Bahía de los Angeles to the Cape Region, and on several islands in the northern part of the Gulf to Isla Tortuga. Nov.–April.

## 11. ECHINOPEPON Naud., Ann. Sc. Nat. V, **6:** 17, 1866

Slender annual herbs with long-pedunculate, bifid tendrils and lobate, deeply lobed, or dissected, thin, petiolate leaves. Staminate flowers in racemes or panicles in axils of leaves; flowers caducous from pedicels, these persistent on rachis. Corollas small, rotate, white. Stamens 4 or more commonly 5; anthers horseshoe-shaped, forming a capitate androecium surmounting fused filaments. Pistillate flowers solitary in same axils as staminate, short-pedunculate. Calyx adnate to corolla; teeth minute. Corolla rotate, usually slightly larger than those of staminate flowers. Stigma short-stipitate, capitate. Ovary 2-celled, 4–10-seeded, ovoid, conspicuously echinate, short- to long-pedunculate, pubescent or glabrous, beak caducous or persistent, often forming a calyptra and operculate when mature. Seeds quadrangular in outline above narrowed base, tenpin-shaped, gray to brown, tuberculate, keeled, and variously patterned.

Peduncles greatly elongate in fruit (to 15 cm. long), sinuous, coiled like a slender spring; seeds dark brown .................................... 1. *E. cirrhopedunculatus*
Peduncles not greatly elongate, not springlike, less than 8 cm. long; seeds whitish to light brown:
  Fruit stipitate-glandular, 2.5–3 cm. long (exclusive of beak); prickles 1–2 cm. long ...... ........................................................ 2. *E. wrightii*
  Fruit not stipitate-glandular, less than 2 cm. long; prickles 1 cm. long or less:
    Peduncles 5–8 cm. long; prickles to 1 cm. long; leaves deeply dissected; nodes conspicuously floccose ............................................ 3. *E. peninsularis*
    Peduncles 3 cm. long or less; prickles to 5 mm. long; leaves shallowly lobed; nodes not conspicuously floccose ........................................ 4. *E. minimus*

### 1. **Echinopepon cirrhopedunculatus** Rose, Contr. U.S. Nat. Herb. **1:** 100, *pl. 4,* 1890

Slender annual vine; stems finely striate, internodes long, glabrous except at nodes; tendrils relatively short-petiolate, unequally bifid; leaves thin, membranous, muriculate-hispid above, punctate and glabrous below, cordate, mostly 5-lobed, longer than broad, center lobe distinctly longer than laterals, triangular-ovate to lanceolate, acute, aristate, 2–6 cm. long, lateral lobes triangular-ovate, obtuse, mucronate; petioles glabrous, about equaling blades; staminate inflorescence racemose to paniculate, 8–15 cm. long; corollas 5–6 mm. long; pistillate flowers solitary on slender peduncles; fruiting peduncles 8–16 cm. long or more, springy like slender coiled springs; fruits 1.5–2 cm. long (excluding beak), glabrous, echinate except on beak, operculately dehiscent; seeds about 5 mm. long, flattened slightly, quadrate above narrowed, flattened base, apically truncate, rugose and tuberculate, dark brown.

Grassy plains and alluvial bottom lands, Lower Sonoran and Subtropical Zones, central Sonora to central Sinaloa. July–Nov.

### 2. **Echinopepon wrightii** (A. Gray) S. Wats., Bull. Torrey Club **14:** 158, 1887

*Elaterium wrightii* A. Gray, Pl. Wright. **2:** 61, 1853.
*Echinocystis wrightii* Cogn., Mém. Cour. Acad. Belg. **28:** 88, 1878.

Coarse annual vines with undulate to lobate leaves and large, long-spined fruits; stems conspicuously ribbed, with spreading glandular-pubescent vestiture, not floccose at nodes; tendrils strongly ribbed, mostly trifid, pubescent; leaves orbicular, deeply and broadly notched at base, dentate, undulate, or 3–5-lobed, lobes acute, mucronate, blade

5–8 cm. wide, finely hispid on both surfaces; petioles sulcate, equaling or longer than blades; staminate flowers in simple or compound racemes; calyx pubescent; corolla 6–8 mm. wide; pistillate flowers glandular-pubescent without, glandular-punctate within; ovary glandular-pubescent; fruit obovate, tapering to base, 2–2.5 cm. long (excluding beak), stipitate-glandular, 4–8-seeded, its prickles slender, 1–2 cm. long; seeds 6–7 mm. long, compressed, quadrangular above narrowed base, rugose and tuberculately ridged, pale gray to light brown.

Alluvial plains and gentle slopes, Lower and Upper Sonoran Zones, southeastern Arizona, south through much of Sonora. July–Oct.

**3. Echinopepon peninsularis** Gentry, Hancock Pac. Exped. **13**²: 170, *pl. 3, figs. 7–9,* 1949

*Echinopepon minimus* var. *peninsularis* Stocking, Madroño **13:** 97, 1955.

Annual vine with deeply lobed leaves, large flowers, and long-pedunculate fruits; stems slender, finely striate, glabrous except for a conspicuous ring of hyaline trichomes at each node; tendrils long-petiolate, bifid but arms unequal, glabrous; leaf blades thin, broadly ovate to orbicular, 3–6 cm. long, broadly notched at base, finely muriculate-hispid above, glabrous below, usually deeply 5–7-lobed, sinuses rounded and open, lobes ovate to lanceolate, serrulate or denticulate distally, obtuse to acuminate; staminate inflorescence elongate, peduncles 9–12 cm. long, glandular-pubescent at tip and on pedicels; corolla white, spreading, 12–15 mm. broad, petals faintly striate and minutely erose along margins; pistillate flowers solitary, similar to staminate; ovary 5–6-seeded, ovules obliquely ascending slightly above horizontal; fruits on peduncles 6–8 cm. long, narrowly ovoid, 12–15 mm. long, sparsely pubescent; mature seeds unknown.

Washes and alluvial flats, Lower Sonoran and Subtropical Zones, southern Baja California. Dec.–May.

**4. Echinopepon minimus** (Kell.) S. Wats., Proc. Amer. Acad. **24:** 52, 1889

*Marah minima* Kell., Proc. Calif. Acad. **2:** 18, 1860.
*Echinocystis minima* Brandegee, Proc. Calif. Acad. II, **3:** 138, 1891.
*Elaterium minimum* S. Wats., Proc. Amer. Acad. **12:** 252, 1877.

Small scabrous annual vine with mostly trilobed leaves and small, short-prickled fruits; stems very slender, obscurely sulcate, glabrous except at sparsely floccose nodes; tendrils small, glabrous, two arms usually unequal; leaf blades broadly ovate in outline, mostly broadly 3-lobed, rarely 5-lobed, blades 2–5 cm. long, hispid and scabrous above, glabrous and puncticulate beneath, lobes obtuse to acute, serrulate to denticulate distally; petioles commonly shorter than blades; staminate racemes 3–6 cm. long, glandular-stipitate near tip, otherwise glabrous; flowers caducous, corolla white, spreading, about 1 cm. long; stamens 5; body of fruit ovoid to obovoid, 1–1.5 cm. long, sparsely pubescent, prickles mostly 5 mm. long or less, their bases broadly flattened; peduncles rarely exceeding fruit in length; seeds 6–8, 4–5 mm. long, narrow, flattened base nearly equaling quadrate, upper part, muriculate, light gray with transverse brownish markings.

Sandy slopes, arroyos, and canyon beds, Lower Sonoran and Subtropical Zones, central Baja California to the Cape Region and on the islands from Cedros Island south. Nov.–April.

## 12. **SICYOSPERMA** A. Gray, Pl. Wright. **2:** 62, 1853

Monecious annual with slender, climbing stems and thin, angulate to lobed leaves. Flowers small; staminate in short-pedunculate, compound racemes; the pistillate solitary or in short racemes. Staminate corollas salverform, 5-lobed; lobes spreading, notched. Stamens forming a capitate androecium attached in bottom of flower cup; anthers 5, connate, erect, 1-celled. Pistillate flowers in same axils as staminate; corolla like that of staminate flower. Stigma short-styled, 3-lobed, spreading. Ovary ovoid, 1-celled, 1-ovuled, pendulous. Fruits 1-seeded, unarmed, smooth, dry, indehiscent.

**1. Sicyosperma gracile** A. Gray, Pl. Wright. **2:** 62, 1853

Stems slender, twining, smooth, narrowly grooved, or striate, glabrous or sparsely pubescent; tendrils slender, petiolate, bifid; leaf blades mostly broadly triangular to shallowly 3-lobed, thin, conic-hispid on both surfaces, margins entire to denticulate; petioles shorter than blades, finely to coarsely hispid; staminate flowers in short racemes or panicles; pedicels very slender, long; corolla white, 2–3 mm. wide, petals bifid at apex, bearing fine, clavate glands marginally; pistillate flowers solitary or in few-flowered, bracteate racemes, minute flowers hidden among cordate, sessile, incised bracts; fruits smooth, indehiscent, enclosed in loosely enveloping bracts, whitish, about 5 mm. long, closely investing solitary seed.

Canyons and adjacent slopes, Lower and Upper Sonoran Zones, southern Arizona and northern Sonora. Aug.–Sept.

## 13. **BRANDEGEA** Cogn., Proc. Calif. Acad. II, **3:** 58, 1890

Small perennial monoecious vines with minute flowers and inconspicuous fruits. Leaves 3–5-lobed, scabrous. Tendrils simple. Staminate flowers in racemes or panicles. Calyx shallowly cup-shaped, 5-lobed. Corolla rotate, 5-lobed; segments triangular-ovate, acute. Stamens 3; anthers horseshoe-shaped, distinct but sessile on a short column formed by fused filaments. Pistillate flowers usually in pairs at nodes, similar to staminate. Ovary oblique, long-rostrate. Style very short; stigma hemispherical. Ovule 1, erect. Fruit small, oblique, slightly compressed laterally, smooth or sparsely echinate, irregularly dehiscent; pericarp thin. Seed narrowly triangular-obovate to clavate, muriculate or warty.

**1. Brandegea bigelovii** (S. Wats.) Cogn., Proc. Calif. Acad. II, **3:** 58, 1890

*Elaterium bigelovii* S. Wats., Proc. Amer. Acad. **12:** 252, 1877.
*Echinocystis parviflora* S. Wats., op. cit. **17:** 373, 1882.
*Cyclanthera monosperma* Brandegee, Proc. Calif. Acad. II, **2:** 159, 1889.
*Brandegea monosperma* Cogn., op. cit., **3:** 59, 1890.

Small, slender-stemmed, climbing vine, perennial from thick roots; stems grooved, glabrous or nearly so, sparsely scaly-pustulate on angles; tendrils slender, simple; petioles commonly 1–2 cm. long; leaves 3–5-lobed, blades 1–6 cm. long, hastate to quadrangular in outline, deeply cordate, lobes linear, triangular, or ovate, mucronate, glabrous and punctate beneath, pustulate-scabrous above; staminate flowers 3–4 mm. broad, white; pistillate flowers rarely more than 2 at a node; fruit small, 8–12 mm. long; seeds subcylindric-clavate, 4–5 mm. long, muriculate.

Arroyos, canyons, hillsides and plains, Lower Sonoran Zone, southeastern California and adjacent Arizona to central Sonora and Baja California. Oct.–May.

## 14. SICYOS L., Sp. Pl. 1013, 1753

Annual monoecious climbing vines with capitate clusters of small fruits, and angled or striate stems. Tendrils with 3–5 branches, petiolate. Leaves subentire to deeply lobed, petiolate. Staminate inflorescence racemose or subumbellate, usually long-pedunculate. Calyx shallowly cup-shaped to rotate, 5-lobed and minute. Corolla salverform; limb spreading, 5-lobed. Stamens 3; connate filaments inserted in bottom of corolla tube; anthers connate or free. Pistillate flowers in sessile clusters or in short-pedunculate umbel-like heads, borne in same axils as staminate flowers. Style about equaling corolla tube; stigma broad, bilobed, flattened. Ovary 1-celled. Ovule solitary, pendulous. Fruit dry, ovoid, echinate with retrorsely barbulate prickles. Seeds reticulate.

Leaves shallowly lobed; fruits normally 8–12 in each cluster; prickles reduced or lacking on lateral faces of fruit except at the base .............................. 1. *S. sinaloae*
Leaves deeply lobed; fruits normally 15–20 in each cluster; prickles present over entire fruit ........................................................ 2. *S. peninsularis*

### 1. **Sicyos sinaloae** Brandegee, Univ. Calif. Pub. Bot. **6**: 370, 1917

Scandent, glandular-puberulent herb; stems 2–3 m. long, multistriate, long-floccose, coarsely to finely hispid and stipitate-glandular, glandular trichomes longer at nodes; tendrils stout-petiolate; leaves shallowly 5–7-lobed, 5–8 cm. long, 6–9 cm. wide, hispid and glandular-floccose on both surfaces, more coarsely hispid on veins, deeply notched with a broad, rounded basal sinus, lobes triangular, acute, margins serrulate to denticulate; petioles 2–4 cm. long, coarsely long-hispid; staminate flowers long-pedunculate, clustered at tip of peduncle, stipitate-glandular; corolla whitish, 4–5 mm. wide; pistillate flowers 10–15, umbellate on peduncles 1–1.5 cm. long; corolla rotate, lobes lanceolate; ovary setose; fruits in clusters 2–2.5 cm. in diameter, densely echinate with slender yellowish prickles on margins and toward base, these nearly lacking on upper part of lateral faces, prickles 3–5 mm. long, deciduous; mature fruit blackish, glabrate, shrunken to outline of reticulate seed.

Alluvial bottom lands and coastal plains, Lower Sonoran and Subtropical Zones, central Sonora south into Sinaloa. Aug.–Nov.

### 2. **Sicyos peninsularis** Brandegee, Zoe **5**: 159, 1903

Stems deeply grooved, pubescent, with stipitate-glandular hairs intermingled with non-glandular trichomes, nodes hyaline-floccose; tendrils stout, with 4–5 branches; leaves deeply triangular-lobed, cordate in outline, 8–12 cm. broad, densely short-hispid beneath, less densely pubescent above, upper 3 lobes acuminate, deeply notched and basal sinus outlined by veins along half its margins, distal margins of lobes serrulate to dentate; petioles 2–4 cm. long, densely glandular-pubescent; staminate flowers on peduncles 15–20 cm. long, densely glandular-pubescent; corolla 5 mm. in diameter, white; stamineal column about 2 mm. long; pistillate flowers 10–30, crowded in compact heads; peduncle 1–1.5 cm. long; fruits 15–20 per head, densely setose, cuneate, 10–12 mm. long, about 6 mm. wide, prickles 2–3 mm. long.

Lower mountain slopes and canyons, Lower Sonoran and Subtropical Zones, southern Baja California, mainly south of the Sonoran Desert. July–Oct.

## 15. CYCLANTHERA Schrader, Ind. Sem. Hort. Gött., 1831

Small, monoecious annual vines with dissected or lobed leaves. Staminate flowers racemose or paniculate, minute, 5-merous, cupulate to salverform. Calyx lobes spreading. Stamens forming a short-stalked androecium with a single anther forming a ring around its apex. Pistillate flowers solitary or rarely geminate in same axils as staminate. Corolla limb rotate, tube very slender, often curved. Ovary obliquely globose to ovoid, rostrate, echinate, bursting irregularly. Seeds ovoid, flattened.

### 1. Cyclanthera dissecta (Torr. & Gray) Arn., Hook. Jour. Bot. 3: 280, 1841

*Discanthera dissecta* Torr. & Gray, Fl. N. Amer. 1: 697, 1840.

Slender, glabrous, climbing annual; stems angled, elongated; leaves pedately compound, blade 4–7 cm. long, usually 3–7-foliolate, leaflets cuneate, ovate to lanceolate, acute to acuminate, dentate or incised along margins; staminate flowers racemose or paniculate on a long peduncle; sepals triangular or ovate-triangular, 1–2 mm. long; corolla 5–6 mm. broad; fruit slightly oblique, 2.5–3 cm. long, slender-echinate.

Along arroyos, canyons and in alluvial bottom land, mainly Upper Sonoran to Carolinian Zones, mountains of southeastern Arizona to Kansas, Louisiana, and northern Mexico. June–Nov.

## Family 136. CAMPANULACEAE. Bluebell Family

Ours herbs, or slightly suffrutescent plants with milky or acrid juice, and alternate, simple or sometimes pinnatisect, petiolate to sessile, estipulate leaves. Inflorescence spicate, racemose, paniculate, or of solitary flowers in axils of leaves. Flowers perfect, regular to irregular. Calyx tube adnate to ovary most of its length to nearly free; limb regular, slightly irregular, or strongly irregular, 5-lobed or -parted; lobes valvate or imbricate in bud, usually persistent. Corolla actinomorphic to strongly zygomorphic, sympetalous, 5-parted or -lobed, white, blue, purple, or red, epigynous. Stamens 5, inserted at base of the corolla tube, free or connate through part of their length into a stamineal tube surrounding style; anthers 5, free or connate into a tube, all alike or dimorphic. Style slender; stigma 5-lobed, 5-parted, or capitate and entire. Fruit a capsule in ours, opening by slits, pores, or valves. Seeds numerous, minute, smooth or sculptured.

Corollas regular; stamens with filaments and anthers free:
- Flowers of 2 kinds, early ones cleistogamous, later opening normally; corolla lobes much longer than corolla tube; capsule opening by subterminal pores or slits below sepals between ribs .......... 1. *Triodanis*
- Flowers all alike and opening normally, none cleistogamous; corolla lobes shorter than corolla tube; capsule opening apically, dehiscing around base of style .... 2. *Githopsis*

Corollas irregular; stamens with filaments connate into a tube through part of their length; anthers free or connate into a tube:
- Anthers free; stems and branches slender and wiry, usually more or less zigzag; appendages usually present on filaments of 2 of the 5 stamens; corollas white or nearly so, less than 1 cm. long .......... 3. *Nemacladus*
- Anthers connate into a tube; stems and branches more robust, not wiry nor zigzag; appendages lacking on stamens; corollas blue, purple or red (rarely white in albino forms), over 1 cm. long:
  - Hypanthium prolonged into a short blunt spur at base; 2 sepals diverging at right angles from hypanthium tube, markedly separated from other 3 ......... 4. *Heterotoma*
  - Hypanthium nearly regular, not prolonged into a spur; sepals ascending or slightly spreading, departing from hypanthium cap at about same level ........ 5. *Lobelia*

## 1. TRIODANIS Raf., New Fl. 4: 67, 1837

Annual, erect or procumbent herbs; stems simple or with ascending lateral branches with 4–5 subalate angles continuous with or decurrent from leaf bases. Leaves sessile, often equaling internodes. Inflorescence spiciform, often densely so; bracts imbricate; 1–8 flowers in each axil; flowers sessile or nearly so. Lower flowers cleistogamous, upper ones open. Calyx lobes green, 5 in open flowers, 3–5 in cleistogamous ones, equal or slightly unequal. Corollas blue, funnelform, 5-parted; tube short. Stamens 5; filaments flattened and ciliate below; anthers linear, exceeding filaments. Hypanthium short-cylindrical to elliptic. Ovary inferior, 2–3-celled. Capsule opening by terminal or subterminal pores. Seeds very numerous, minute.

Pores in capsule linear, 0.2–0.4 mm. wide; seeds with low tubercles in longitudinal lines ........................................................... 1. *T. holzingeri*

Pores in capsule round, oval, or broadly elliptical, 0.5–1.5 mm. wide; seeds smooth or slightly roughened by minute, irregularly spaced processes:

Open flowers borne at several upper nodes; seeds smooth to muriculate on surface; bracts as broad as or broader than long; pores near middle of capsule ...... 2. *T. perfoliata*

Open flower usually solitary, terminal, others cleistogamous; seeds smooth and polished; bracts usually longer than broad; pores at or near apex of capsule ...... 3. *T. biflora*

### 1. Triodanis holzingeri McVaugh, Wrightia 1: 45, 1945

Slender erect herb 2.5–6 dm. tall, strict or sparingly branched, hispid at base, retrorsely hispidulous to scabrous above; basal leaves sessile to short-petiolate, ovate to oblanceolate, obtuse or rounded at apex, but with a short mucro, 5–10 mm. wide, 16–24 mm. long, 5–8-toothed along each side; upper leaves ovate, acute or mucronulate; bracts sessile-auriculate, 5–15 mm. wide, 7–20 mm. long, crenate; flowers in upper one-third of infloresence usually open, lower ones cleistogamous, 1–3 at a node, usually only uppermost producing fruits; calyx lobes deltoid to attenuate, 4–6 mm. long; open corollas bluish purple, glabrous or hirsute along midveins of lobes, these 7–9 mm. long, tube 1–2 mm. long; filaments 2 mm. long or less, dilated below; capsules ellipsoidal to oblong, angular, 0.7–2.5 mm. in diameter, 6–12 mm. long, usually sulcate; pores linear, near middle of capsule; seeds chestnut-brown, lustrous, low-tuberculate in longitudinal lines.

Canyons and along sandy washes, Lower Sonoran to Arid Transition Zones in our area, southeastern Arizona to Texas, Wyoming, and Tennessee. March–May in Arizona.

### 2. Triodanis perfoliata (L.) Nieuwl., Amer. Midl. Nat. 3: 192, 1914

*Campanula perfoliata* L., Sp. Pl. 169, 1753.
*Specularia perfoliata* A. DC., Monogr. Camp. 351, 1830.
(For additional synonymy see Wrightia 1: 34, 1945.)

Erect and strict to decumbent and sparingly branched herbs 1–6 dm. tall, with pilose, hispid, or scabrous stems; leaves sessile, ovate-cordate, or lowermost elliptical to obovate and petiolate, obtuse to rounded at apex, coarsely crenate with 3–6 teeth on each side or sometimes nearly entire, to 25 mm. wide and 35 mm. long; bracts reniform to ovate, sessile, cordate-clasping, tip obtuse to acute and mucronate, often crenate; flowers in upper axils open, those below cleistogamous; calyx lobes in open flowers 5–8.5 mm. long, those of cleistogamous flowers 1–4.5 mm. long, triangular-lanceolate with subulate tips; corollas of closed flowers minute, less than 1 mm. long; corollas of open flowers blue-lavender, tube 2–3.5 mm. long, lobes 8–12 mm. long; capsule obovoid to oblong, 4–8 mm. long in cleistogamous flowers, slightly larger in open flowers, pores broadly elliptic to rounded, valves 1.5–3 mm. long, lightly curled after opening; seeds ellipsoidal, brown, 0.5–0.6 mm. long, moderately lustrous, smooth to muriculate in irregular patterns.

Hillsides, in shade of rocks, and under shrubs, Lower Sonoran to Transition and Carolinian Zones, almost throughout the United States but rare to lacking in deserts and very wet soil, south to northern South America. March–Aug.

**3. Triodanis biflora** (Ruiz & Pav.) Greene, Man. Bay Reg. 230, 1894

*Campanula biflora* Ruiz & Pav., Fl. Peruv. **2:** 55, *pl. 200,* 1799.
*Specularia biflora* Fisch. & Mey., Ind. Sem. Hort. Petrop. **1:** 17, 1835.
(For additional synonymy see Wrightia **1:** 31, 1945.)

Slender, erect to decumbent herbs 1–4.5 dm. tall, sparsely to densely hirsute below, retrorsely hispid to scabrous above; leaves broadly ovate to elliptic, lower obtuse and short-petiolate, upper acute and sessile, crenate, 6–12 mm. wide, 1–3 cm. long; bracts ovate, sessile and moderately clasping, acute and mucronulate at apex, ascending and often concealing capsules, 6–16 mm. wide, 8–35 mm. long, crenate on upper two-thirds; most nodes bearing flowers, terminal and 2–3 lateral ones open, those below cleistogamous; calyx lobes triangular-ovate, acute to mucronate, 3.5–7 mm. long, those of cleistogamous flowers smaller; corollas of closed flowers 0.5 mm. long, open corollas blue to purple, with tubes 1–2 mm. long, lobes 5–9 mm. long, 2–3 mm. wide; capsule glabrous to hispid, clavate to ovoid or obovoid, 4.5–9 mm. long, pores oval to circular, near apex; seeds chestnut-brown, smooth and highly polished, 0.5–0.6 mm. long, nearly as wide.

Mostly on disturbed, rather dry soils, Lower Sonoran to Transition Zones, Oregon to Baja California, east across the southern United States to Virginia; Mexico to South America. April–May; into June in Arizona.

## 2. GITHOPSIS Nutt., Trans. Amer. Phil. Soc. II, 8: 258, 1843

Small annual herbs with angled stems, cuneate-obovate, sessile leaves, and flowers borne solitary in axils. Calyx with strongly 10-ribbed tube and 5 foliaceous, long, narrow lobes. Corolla 5-lobed, campanulate-tubular. Stamens 5, inserted at base of corolla tube; short filaments dilated at base; anthers lance-linear. Ovary inferior, 3-celled; stigma 3-lobed. Capsule obconic to clavate, strongly striate-ribbed, crowned with rigid persistent calyx lobes, opening by a terminal circular orifice by falling of style. Seeds numerous, small.

Flowers 12–25 mm. long; capsule 2.5–4 mm. in diameter at summit; corolla 3–10 mm. long ........................................................ 1. *G. specularioides*
Flowers 3–6 mm. long; capsules 1–1.5 mm. wide at summit; corolla 1.5–3 mm. long ...... ........................................................ 2. *G. filicaulis*

**1. Githopsis specularioides** Nutt., Trans. Amer. Phil. Soc. II, **8:** 258, 1843

*Githopsis specularioides* β *hirsuta* Nutt., loc. cit.
*Githopsis calycina* Benth., Pl. Hartw. 321, 1849.
*Githopsis latifolius* Eastw., Proc. Calif. Acad. IV, **20:** 154, 1931.
*Githopsis specularioides* subsp. *candida* Ewan, Rhodora **41:** 308, 1939.

Stems simple and erect to profusely branched and spreading-decumbent, 3–20 cm.

long, grayish or light green, persistently hirsute to glabrate, hairs spreading to deflexed and varying greatly in size and stiffness, most prominent on angles of stems; leaves ovate, oblong, or broadly linear, sessile, shallowly and remotely to markedly toothed, 3–20 mm. long, numerous on young plants, deciduous in age; calyx tube strongly ribbed, 2.5–4 mm. wide at apex, 7–12 mm. long in fruit, calyx lobes lance-linear, acute at apex, 6–13 mm. long, slightly to markedly spreading in fruit; corolla blue, usually slightly shorter than calyx lobes, tube longer than lobes; capsule 10–12 mm. long, brownish, strongly striate-ribbed; seeds narrowly ellipsoidal, reddish brown, sublustrous, 0.6–0.8 mm. long.

In heavy clay soil around swales or in shelter of rocks, Lower Sonoran to Transition Zones, Washington south through the Coast Ranges and Sierra Nevada to northern Baja California; Guadalupe Island, Mexico. March–June.

**2. Githopsis filicaulis** Ewan, Rhodora **41:** 312, 1939

Very slender, lax annual with almost filiform stems to 1 dm. long; leaves ovate-oblong, glabrate, 1–3 mm. long, their margins serrulate; flowers sessile, calyx tube only lightly striate, 2.5–5 mm. long, lobes about as long or slightly shorter; corolla shorter than calyx lobes; seeds narrowly elliptical, 0.6–0.9 mm. long, reddish and lustrous.

Known only from two localities, Mission Canyon near San Diego, California, and at Vallecito, Baja California. Included doubtfully in our area because Orcutt did not clearly indicate at which "Vallecito" in northern Baja California he obtained his collection. April–May.

## 3. NEMACLADUS Nutt., Trans. Amer. Phil. Soc. II, 8: 254, 1843

Small annual herbs with wiry to filiform, ascending, mostly zigzag, often diffusely branched stems, and leaves in a basal rosette. Cauline leaves reduced and bractlike, entire. Inflorescence racemose on all branches, often nearly to base of plant. Pedicels slender, spreading or spreading-ascending, usually bowed and often turned upward abruptly near tip to place flower erect. Hypanthium adherent to lower part of ovary but lobes free, usually glabrous. Sepals ovate, triangular, or lanceolate, blunt at apex, weakly irregular or all about alike. Corolla irregular, bilabiate; lower lip bilobed; upper 3-lobed; corolla tube short, usually much shorter than lobes. Stamen filaments free at base and sometimes distally also, but fused into a tube surrounding style for at least part of their length; two uppermost generally bearing an appendage near base; appendages consisting of a small stipe tipped with few to several transparent, cylindrical, glassy cells, these stellately spreading to reflexed; anthers elliptic to subglobose, usually obtuse but mucronulate at apex. Style about equaling stamineal column; stigma capitate. An elliptical to orbicular, sessile to short-stipitate gland inserted on roof of ovary alternating with bases of upper three sepals, opposite upper corolla lobes. Capsules 2-celled, loculicidally dehiscent. Seeds globose to ellipsoidal or cylindrical, ridged, furrowed, or pitted.

Capsules narrowly ovoid to ovoid-elliptical; seeds with 8–10 undulate longitudinal ridges and 10–12 shallow pits in each ridge; appendages attached near middle of upper filaments:

Corolla 4–8 mm. long; filament tube 3.5–7 mm. long; bracts puberulent on adaxial surface only . . . . . . . . . . . . . . . . . . . . . . . 1. *N. longiflorus* var. *longiflorus*

Corolla 3–3.5 mm. long; filament tube 2.2–2.8 mm. long; bracts usually puberulent on both surfaces . . . . . . . . . . . . . . . . . . . . . 1*a*. *N. longiflorus* var. *breviflorus*

Capsules broadly ovoid to hemispherical; seeds variously sculptured, but not as above; appendages attached near base of filaments or lacking:

Seeds coarsely alveolate-pitted, without prominent longitudinal ridges; pedicels distinctly sigmoid-curved; corolla tube about equaling calyx lobes .......... 2. *N. sigmoideus*

Seeds longitudinally ridged, ridges straight or zigzag-undulate; pedicels curved upward near tips or nearly straight but not markedly sigmoid; corolla tube much shorter than calyx lobes, limb parted almost to base:

Ridges on seeds straight, separated by narrow, deeply impressed grooves, each ridge with transverse lines; stems and branches green to purplish but without silvery-gray sheen; glands finely papillate distally:

Calyx lobes 0.8–2.5 mm. long at anthesis; capsules 2–3.5 mm. long; glands sessile or nearly so; terminal cells of appendages exceeding stipe and deflexed, or appendages lacking:

Pedicels curved upward at tip; appendages lacking; calyx lobes 1.5–2.5 mm. long .................................. 3. *N. glanduliferus* var. *glanduliferus*

Pedicels straight or nearly so; appendages present; calyx lobes 0.8–1.5 mm. long at anthesis .............................. 3*a*. *N. glanduliferus* var. *orientalis*

Calyx lobes 2.5–3 mm. long at anthesis; capsules 3.5–4 mm. long; glands on stipes as long as diameter of glands; terminal cells of appendages stellately spreading, shorter than stipes, not deflexed ............. 3*b*. *N. glanduliferus* var. *australis*

Ridges on seeds undulate or zigzag, each ridge with 10–12 shallow pits; stems and at least lower parts of branches with a silvery-gray sheen; glands smooth distally:

Basal leaves entire or shallowly toothed; pedicels slender but not capillary; stipe of appendages shorter than terminal cells, these strongly reflexed; glands sessile ... ........................................ 4. *N. rubescens* var. *rubescens*

Basal leaves pinnatifid; pedicels capillary; stipe longer than terminal cells, these stellately spreading; glands short-stipitate .............. 4*a*. *N. rubescens* var. *tenuis*

**1. Nemacladus longiflorus** A. Gray, Proc. Amer. Acad. **12:** 60, 1876, var. **longiflorus**

Ascending zigzag branches 5–10(20) cm. tall, minutely and sparsely puberulent throughout or only at base; basal leaves obovate to oblanceolate, 1.5–3 mm. wide, 5–12 mm. long, entire to finely crenate, grayish, mostly densely puberulent with coarse short hairs on both surfaces; cauline leaves narrower, 2–3 mm. long; pedicels widely spreading to obliquely ascending, usually curved upward abruptly near tip so flower is erect, 6–10 mm. long in flower, to 25 mm. long in fruit, slender; bracts ovate to lanceolate, 1–2 mm. long, ciliate marginally; hypanthium free nearly to base, 1–1.5 mm. long at anthesis, lobes three to four times as long as cup-shaped, adnate part and closely appressed to corolla tube at anthesis, 1–2.5 mm. long in fruit; corolla 4–8 mm. long, white with a faint pink tinge or with pink lines and flecks, with a yellow spot at base of each upper lobe within, these bearded at base; appendages minute and delicate, attached below middle of filament-tube, this 3.5–7.5 mm. long; capsule ovoid to fusiform, 1.5–2 mm. in diameter, 3–4.5 mm. long; seeds ellipsoidal, about 0.4 mm. long, longitudinally marked with 8–10 undulate or zigzag ridges and 10–12 rows of faint pits on each ridge.

Sandy hillsides, arroyos, and sand bars, Lower and Upper Sonoran Zones, San Gabriel Mountains, Los Angeles County, California, into western Arizona and south in northern Baja California to the southern Sierra San Pedro Mártir. March–June.

**1*a*. Nemacladus longiflorus** var. **breviflorus** McVaugh, Amer. Midl. Nat. **22:** 526, 1939

Differing from var. *longiflorus* in having corollas only 3–3.5 mm. long and corolla tube 2–2.5 mm. long, therefore corolla lobes much smaller than in var. *longiflorus*; filament tube only 2.2–2.8 mm. long; bracts pubescent on both surfaces instead of adaxial surface only (as in var. *longiflorus*).

Gravelly hillsides and sandy desert slopes, Lower and Upper Sonoran Zones, western margin of Mojave Desert in the Little San Bernardino Mountains to south-central Arizona and southern San Felipe Desert, Baja California. April–June.

## 2. **Nemacladus sigmoideus** G. T. Robbins, Aliso **4:** 144, 1958

Delicate annuals 5–20 cm. tall with slender, dichotomously branched or sometimes simple stems; herbage usually reddish brown, glabrous to sparsely puberulent; flowering branches weakly zigzag; basal leaves rhomboid-ovate to elliptic, 2–8 mm. long, with short to elongated petioles, margins entire to crenate, sparsely puberulent with very coarse, short, white hairs on one or both surfaces, more densely puberulent marginally; bracts lance-ovate to lance-oblong, 1–2 mm. long, usually conduplicate-folded and embracing base of subtended pedicel, glabrous or sparsely puberulent, mucronulate; pedicels filiform, slightly to strongly sigmoid, abruptly turned up at tip, glabrous; hypanthium broadly campanulate, cup 0.6–0.8 mm. deep at anthesis, sepals lance-elliptic, slightly spreading, 0.8–1.2 mm. long, acute, glabrous; corolla campanulate, 2.5–3 mm. long, short tube about equaling calyx lobes, sparsely hirsute to glabrous at mouth, corolla lobes white with tinge of pink or yellow at tips and bases, upper three lobes slightly exceeding widely spreading lower pair; filament tube abruptly curved near apex, bearing a few moniliform hairs just beneath anthers, these 0.2–0.3 mm. long, quadrate; stamineal appendages near base of filaments, each bearing a few rod-like hairs; capsule half-inferior, about equaling calyx at maturity; seeds reddish brown, broadly ellipsoidal to narrowly obovoid, 0.3–0.4 mm. in diameter, 0.5–0.6 mm. long, coarsely alveolate-reticulate.

Sandy washes, flats, and gravelly outwash slopes, Lower and Upper Sonoran Zones, Mojave Desert and adjacent mountain ranges east into western Nevada and Arizona, south into northern Baja California. April–June.

## 3. **Nemacladus glanduliferus** Jepson, Man. Fl. Pl. Calif. 975, 1925, var. **glanduliferus**

Diffusely ascendingly branched annuals 5–25 cm. tall with coarse, white, short hairs forming sparse puberulence on basal leaves and lower parts of stems and in axils of some bracts, or nearly glabrous throughout, branches usually brownish or purplish, dull to faintly lustrous; leaves in basal rosette few to as many as 20, elliptical to oblanceolate, 1–5 mm. wide, 3–20 mm. long, obscurely toothed to distinctly coarsely pinnatifid, usually puberulent only along margins of relatively wide petioles and on adaxial side of blade or sometimes subglabrate; bracts linear to lanceolate, 2–5 mm. long, obtuse at tip, flat or only faintly folded at base, often spreading or reflexed, glabrous or sparsely ciliate on margins; pedicels slender, but hardly capillary, smooth and glabrous, curving upward near tip, 5–9 mm. long at anthesis; hypanthium broadly campanulate, bowl 0.5–0.8 mm. long, glabrous; calyx lobes lanceolate or lance-ovate, 1.2–2.5 mm. long, acute; corolla broadly campanulate, lobes white, tinged with pink, spreading, 2–3 mm. long; filament tube 1.6–2.4 mm. long, glabrous, without appendages, curved at tip; anthers 0.2–0.35 mm. long; capsule about half inferior, hemispherical to ovoid, 2–3.5 mm. long and in diameter, apex acute; seeds cylindrical to ellipsoidal, 0.4–0.6

mm. long, bearing 6–8 longitudinal ridges separated by sharply impressed lines, each ridge transversely lined.

Sandy or gravelly slopes, flats, and arroyos, mainly Lower Sonoran Zone, southern Mojave Desert into northern Baja California. March–May.

**3*a*. Nemacladus glanduliferus** var. **orientalis** McVaugh, Amer. Midl. Nat. **22:** 540, 1939

Habit of var. *glanduliferus* but pedicels forming more acute angles (less than 60°) with stem, ascending and only slightly if at all curved upward at tip, and usually ciliate on upper side near base; calyx lobes 0.8–1.5 mm. long; filament tube 1.2–1.8 mm. long, bearing appendages with terminal clavate, sharply reflexed cells about 0.4 mm. long; capsule 1.5–2 mm. in diameter, slightly longer than broad.

Sandy and gravelly flats, arroyos, and hillsides, Lower Sonoran Zone, Inyo and Los Angeles Counties, California, to Arizona, Sonora, and south in Baja California to the Vizcaíno Desert, chiefly west of the mountains. March–May.

**3*b*. Nemacladus glanduliferus** var. **australis** (Munz) McVaugh, Amer. Midl. Nat. **22:** 540, 1939

*Nemacladus rigidus* var. *australis* Munz, Amer. Jour. Bot. **11:** 242, *pl. 10,* 1924.

Strict and scarcely to diffusely and abundantly branched, delicate annuals 5–20 cm. tall, often fully as wide as high; stems moderately zigzag, slender, green or turning reddish in age; basal leaves oblanceolate, 1–8 mm. wide, 6–25 mm. long, coarsely but shallowly few-toothed; bracts linear, 3–5 mm. long, glabrous or sparsely puberulent on upper surface; pedicels slender, 5–8 mm. long, spreading but curved upward near tip; hypanthium broadly campanulate, bowl 0.4–0.7 mm. deep; sepals lance-linear, 2.5–3 mm. long, acute, spreading-ascending, glabrous; corolla white with faint tinge of pink on outside, 4–6 mm. in diameter, lanceolate lobes spreading, sparsely hirsute on inner surface and margins, glabrous outside; glands deep yellow, 0.5–0.6 mm. in diameter, definitely short-stipitate, closely tufted-papillose on distal surface; appendages inserted near base of stamens, about 0.4 mm. long, bearing 3–5 cylindrical, vitreous, transparent rods apically, these spreading to extending in same direction as appendages, not at all reflexed; capsules 3.5–4 mm. long; seeds as in var. *glanduliferus*.

Sandy washes and canyons, Lower Sonoran Zone, near Puertocitos, Baja California, west to vicinity of Rosario, and south to Bahía de los Angeles. March–May.

**4. Nemacladus rubescens** Greene, Bull. Calif. Acad. **1:** 197, 1885, var. **rubescens**

*Nemacladus adenophorus* Parish, Bull. S. Calif. Acad. **2:** 28, 1903.
*Nemacladus rigidus* var. *rubescens* Munz, Amer. Jour. Bot. **11:** 245, 1924, in part.

Moderately to much-branched annual 5–25 cm. tall, with zigzag branches and a basal rosette of oblanceolate to narrowly obovate leaves 3–7 mm. wide, 7–27 mm. long, these entire to shallowly toothed; stems and branches usually with a silvery-gray sheen, slightly pubescent near base or less commonly wholly glabrous; bracts narrowly ovate to elliptic, about 1 mm. long, blunt at apex, faintly conduplicate, glabrous; pedicels slender, ascending-spreading, slightly curved at base or apex or both, 4–10 mm. long; hypanthium broadly campanulate at anthesis, cup-shaped in fruit, lobes oblong, about 1 mm. long, glabrous; corolla 2–3 mm. long, white to pale yellow flecked with purplish or brownish specks near ends of lobes, tube about 0.5 mm. long or less, lobes ciliate-margined; filament tube slender, 2–3 mm. long, nearly straight; appendages bearing 3–5 reflexed clavate rods 0.5–0.7 mm. long and considerably exceeding stipe; anthers

0.5–0.7 mm. long; capsule ovoid, slightly pointed, 1.5–2 mm. in diameter and high; seeds broadly ellipsoidal, 0.5–0.7 mm. long, amber, with zigzag or irregularly spaced, transverse ridges.

In sandy or gravelly soils of washes and bajadas, Lower Sonoran Zone, Kern County, California, to southern Nevada, south to Arizona, the Colorado Desert, and northeastern Baja California. April–May.

### 4*a*. **Nemacladus rubescens** var. **tenuis** McVaugh, Amer. Midl. Nat. **22:** 536, 1939

Branches usually forked several times, slender, weakly zigzag to nearly straight; rosette leaves deeply pinnatifid, blades 1.5–4 mm. wide, 1–2 cm. long, hirsute on both surfaces; pedicels wide-spreading, nearly capillary, bowed at middle, abruptly turned upward at tip, 10–18 mm. long in fruit; bracts and corolla as in var. *rubescens*; anthers 0.4–0.5 mm. long; stamineal appendages with stipes 0.5–1 mm. long, each bearing 4–12 spreading, transparent, glistening, rodlike cells 0.2–0.4 mm. long; seeds as in var. *rubescens*.

Sandy washes and canyon beds, Lower Sonoran Zone, Inyo County, California, south through the Mojave and Colorado Deserts to northern Baja California. March–May.

**Nemacladus pinnatifidus** Greene, Bull. Calif. Acad. **1:** 197, 1885; **Nemacladus ramosissimus** Nutt., Trans. Amer. Phil. Soc. II, **8:** 254, 1843; and **Nemacladus gracilis** Eastw., Bull. Torrey Club **30:** 500, 1903, all occur in sandy and gravelly soil in the chaparral zones above and west of the Mojave and Colorado Deserts, and both *N. pinnatifidus* and *N. ramosissimus* occur in the foothills of the Sierra San Pedro Mártir at least halfway toward the southern end of that range. None of them has been found within the Sonoran Desert but may occur sporadically as stragglers along washes in years of high rainfall.

## 4. **HETEROTOMA** Zucc., Flora **15**[2]**:** Beibl. 100, 1832

Annual or perennial herbs with erect, decumbent, or prostrate-creeping stems and simple, toothed, or incised leaves with pinnate or palmate veins. Inflorescence secund-racemose at top of a naked stem. Pedicels slender, spreading-ascending, usually ebracteolate. Flowers inverted in anthesis. Hypanthium asymmetrically 1-sided, lower part reduced and produced into a conic to linear spur, bearing 2 short calyx lobes opposite shorter stamens, other 3 calyx lobes longer. Corolla strongly zygomorphic, tube prolonged into a spur on lower side, cleft two-thirds of its length dorsally. Filaments connate into a tube distally, distinct at base, two lower prolonged into corolla spur; two smaller anthers densely tufted-hirsute at tip, larger three curved apically. Capsule enclosed by hypanthium for two-thirds of its dorsal length, bilocular, dehiscent by apical loculicidal valves. Seeds minute, ellipsoidal, smooth and shining.

### 1. **Heterotoma aurita** Brandegee, Proc. Calif. Acad. II, **3:** 149, 1891

*Lobelia cotensis* M. E. Jones, Contr. West. Bot. **15:** 152, 1929.
*Lobelia amabilis* M. E. Jones, op. cit. **18:** 68, 1933.

Erect annual 10–75 cm. tall with simple to ascendingly branched stems and sparingly white-pilose leaves and lower stems; slender, naked, flowering branches usually glabrous; leaves petiolate, blades lanceolate-ovate to broadly ovate, 1–3 cm. wide, 1.5–6 cm. long, broadly cuneate to truncate or cordate at base, rounded to acute at apex, irregularly and coarsely sharp-serrate, white hairs more numerous on petioles, margins, and lower surface than on upper surface, petioles about equaling blades of lower leaves, shorter on upper ones; naked stem equaling or exceeding leaf-bearing portion; bracts linear, 2–5 mm. long; pedicels slender, 1–3 cm. long at anthesis; hypanthium tube 4–6 mm. long, split nearly to base along lower side, with a conical spur 1–2 mm. long just beneath cleft; sepals linear to linear-lanceolate, 2–3 mm. long, upper three erect or ascending, lower pair sharply turned outward at right angles to upper three; corolla blue-purple with whitish to greenish spot at base of lower lip, tube 8–10 mm. long, produced downward into a blunt spur 1.5–2.5 mm. long within hypanthium spur, lower lobes 3–4 mm. wide, 6–9 mm. long, spreading at right angles to tube, upper pair linear, 1–1.5 mm. wide, 5–6 mm. long, crossing over each other near base; stamineal tube, including anthers, about equaling corolla tube; anther tube 1.5–2 mm. long, grayish blue, shorter two anthers terminally bearded, longer three hirsute on backs; capsules ellipsoidal, 4–5 mm. long, membranous; seeds turgidly ellipsoidal, 0.4–0.5 mm. long, often slightly flattened on one side, dark amber, smooth and shining.

On shaded banks, under rocks, and among shrubs in damp soil, Lower Sonoran and Tropical Zones, mostly in mountain canyons, Cape Region and on Isla Partida and probably other islands in the southern part of the Gulf of California. Dec.–May.

**Heterotoma cordifolia** (Hook. & Arn.) McVaugh, Bull. Torrey Club **67:** 143, 1940, and **Heterotoma goldmanii** Fernald, Proc. Amer. Acad. **36:** 504, 1901, occur in Sinaloa and southern Sonora, but neither has been reported from the Sonoran Desert.

## 5. LOBELIA L., Sp. Pl. 929, 1753

Annual or perennial herbs with entire, serrate, or dentate leaves and glabrous to pubescent herbage. Leaves sessile to petiolate, their margins beset with minute, yellow to brown callosities. Inflorescence racemose, spicate, or flowers inserted in axils of upper leaves. Calyx tube adnate to ovary, obovoid to hemispherical; calyx lobes slender. Corolla tubular-salverform, irregular, dorsal fissure extending nearly to base of tube in most species; tube often fenestrate. Stamens mostly free from corolla. Anther tube with opening partially closed by incurved tips of longer anthers; two shorter anthers usually tufted at tips. Corollas, stamens, and style withering-persistent. Fruit capsular, dehiscing by apical valves. Seeds small, variously marked and sculptured.

Corolla lobes blue, 5–8 mm. long; hypanthium 2.5–3 mm. in diameter at anthesis; anther tube 2–2.5 mm. long . . . . . . . . . . . . . . . . . . . . . . . . 1. *L. dunnii*
Corolla lobes red, 10–20 mm. long; hypanthium 3–6 mm. in diameter at anthesis; anther tube 3.5–9 mm. long:
  Pedicels equaling or exceeding length of corollas in fruit; anther tubes yellowish, 5–9 mm. long; calyx lobes deltoid-ovate to lance-triangular, 2–5 mm. long at anthesis . . . . . . . . . . . . . . . . 2. *L. laxiflora* var. *angustifolia*
  Pedicels usually shorter than flowers (5–15 mm. long); anther tubes bluish, 3.5–4.5 mm. long; calyx lobes lance-linear to narrowly subulate, 8–15 mm. long at anthesis . . . . . . . . . . . . . . . . 3. *L. cardinalis* subsp. *graminea*

### 1. Lobelia dunnii Greene, Pittonia 1: 297, 1889

*Palmerella debilis* A. Gray, Proc. Amer. Acad. **11:** 80, 1876. Not *Lobelia debilis* L. f., 1781.
*Laurentia debilis* McVaugh, Bull. Torrey Club **67:** 144, 1940.

Slender perennial herb 20–50 cm. tall with sessile leaves and glabrous herbage or inflorescence sometimes slightly puberulent; stems erect and simple or sometimes with ascending lateral branches; leaf blades linear to linear-lanceolate, 2–8 mm. wide, 4–12 cm. long, or the basal ones slightly larger, minutely and remotely serrulate or upper ones entire; inflorescence mostly subcapitate, 2–12 cm. tall, bearing 2–25 flowers; pedicels ascending or faintly spreading, 2–9 mm. long, glabrous; bracts linear, to 1.5 cm. long; hypanthium narrowly campanulate, 3–4.5 mm. long at anthesis, usually glabrous; sepals lance-acicular, 5–8 mm. long, curving outward slightly above middle; corolla 15–30 mm. long, lobes blue, tube white or faintly washed with blue, sparsely and finely hirsute, splitting from base upward about halfway on dorsal and lateral sides in age, densely hirsute within; corolla lobes deep blue, abruptly spreading, 5–8 mm. long, two upper narrower and slightly shorter than lower three; filaments distinct at base, fused into a tube near apex, pubescent, about equaling corolla tube; anther tube 2–2.5 mm. long, glabrous or bearing some white or whitish, stiff hairs; capsule 5–6 mm. long, walls papery; seeds lenticular-ellipsoidal, 0.5 mm. long, smooth and shining, light brown.

Sandy soil in canyons or on seepy walls, Lower and Upper Sonoran Zones, along western margins of San Felipe Desert, northern Baja California. Sept.–Oct.

**2. Lobelia laxiflora** H. B. K. var. **angustifolia** A. DC., in DC., Prod. **7:** 383, 1839

*Lobelia loretensis* M. E. Jones, Contr. West. Bot. **18:** 68, 1933.
(For full synonymy see N. Amer. Fl. **32A:** 97, 1942.)

Long-lived herbaceous perennial or sometimes suffrutescent or shrubby plant with few to many erect or ascending stems from a stout woody root; stems to 3 m. but usually 1–1.5 m. tall, glabrous below, glabrous to short-pubescent above; lower leaves often deciduous, cauline ones 5–15 mm. wide, 5–15 cm. long, from thin and nearly membranous to thicker and rugose, subentire to slightly sinuate in outline, mostly finely and regularly serrate, narrowly lanceolate to linear-lanceolate in outline; inflorescence racemose, 5–40 cm. long, pedicels appressed to stem at base or often their full length, 2.5–7 cm. long, often reddish or purplish; bracts leafy, lanceolate, 2–5 cm. long; hypanthium broadly bowl-shaped, 3–5 mm. deep, 4–6 mm. wide at anthesis, sometimes glabrous but usually finely and moderately puberulent; sepals deltoid-ovate to lance-triangular, 2–5 mm. long at anthesis, only slightly longer in fruit, persistent; corolla red, 2.5–3.5 cm. long, tube 16–26 mm. long, fenestrate in lower 5–10 mm., minutely puberulent, lobes 10–20 mm. long, in some forms yellowish; filament tube 15–30 mm. long, slightly deflexed, yellow, pink, or reddish at base, filaments connate in upper one-half or two-thirds; anther tube 5–9 mm. long, yellowish, two smaller anthers densely tufted with stiff white hairs at apex, larger three sparsely hirsute to glabrous toward base, more densely hirsute distally; fruiting hypanthium 7–8 mm. in diameter, 3–4 mm. high, brownish; seeds ellipsoidal-lenticular to ovoid, amber to pale brown, smooth but with faint longitudinal lines, 0.4–0.6 mm. long, slightly pointed at one end.

Among rocks and under trees, along streams and seepy areas, Upper Sonoran Zone to Tropical Zone, southern Arizona to San Luis Potosí, Oaxaca, and the southern half of Baja California. Mainly April–Oct., but flowering sparsely throughout rest of year.

**3. Lobelia cardinalis** L. subsp. **graminea** (Lam.) McVaugh, in Woodson & Schery, Ann. Mo. Bot. Gard. **27:** 347, 1940

*Lobelia graminea* Lam., Encycl. **3:** 583, 1791.
*Lobelia splendens* Willd., Enum. Hort. Berol. *pl. 86,* 1809.
(For added synonymy see N. Amer. Fl. **32A:** 81, 1942.)

Erect or ascendingly branched, perennial herbs with fibrous roots and giving rise to few to many offshoots; stems 4–20 dm. tall, glabrous to puberulent, often purplish;

cauline leaves linear to narrowly lanceolate, 4–15 mm. wide, 5–15 cm. long, thin, narrowed at base, finely and regularly toothed, glabrous or essentially so; inflorescence 5–50 cm. long, stems peduncle-like and naked below flowers; bracts rarely leafy, usually lance-linear, 5–20 mm. long, entire or minutely toothed; pedicels slender, 4–10 mm. long, ascending, puberulent with short coarse hairs; hypanthium broadly bowl-shaped, 2–3 mm. deep, 3–5 mm. wide at anthesis, short-puberulent to subglabrate; sepals lance-linear to narrowly subulate, 8–15 mm. long; corolla deep crimson (or rarely pink to nearly white), glabrous or sparsely puberulent, tube 13–22 mm. long, fenestrate near base, lobes of lower lip spreading-deflexed, 3–3.5 mm. wide, 12–20 mm. long, upper lobes linear, erect, equaling lower; filament tube usually surpassing corolla tube by 4–6 mm., red, sparsely hirsute below; anther tube 3.5–4.5 mm. long, bluish, shorter pair of anthers bearing dense terminal tufts of white hairs, longer pair moderately hirsute on distal one-fourth; capsule 5–7 mm. high, 8–10 mm. in diameter, distinctly 10-veined; seeds light brown, 1–1.2 mm. long, coat deeply foveolate-pitted.

Boggy cienegas, stream banks, and shaded canyons, Lower Sonoran to Arid Transition Zones, southern California to southern Nevada, Texas, south to Oaxaca. July–Oct.

The material in our area has been further segregated by McVaugh as var. *pseudosplendens* of subsp. *graminea,* and the range of var. *graminea* given as extending south to Panama. July–Oct.

## Family 137. **GOODENIACEAE.** Goodenia Family

Perennial herbs or small shrubs with watery juice and alternate or rarely opposite, or basal, simple, estipulate leaves. Inflorescence cymose, racemose, or less commonly capitate, paniculate or a single flower. Flowers perfect, usually zygomorphic. Calyx adnate to ovary, 5-lobed, or consisting of a narrow ring of tissue. Corolla 5-lobed, with a short or long tube, sometimes bilabiate or 1-lipped, valvate or induplicate in bud. Stamens 5, inserted on ovary alternate with corolla lobes, free or adnate to basal part of corolla tube, distinct or anthers connate into a cylinder around style; anthers 2-celled, introrse, dehiscing longitudinally. Ovary inferior (in ours), sometimes partially or almost wholly superior, 1–2-celled, of 2 carpels. Ovules 1–many, on a basal or axile placenta. Style 1, slender; stigma simple or 2–3-branched, surrounded by a shallow cup. Fruit a capsule, berry, drupe, or nut. Embryo straight. Endosperm fleshy.

### 1. **SCAEVOLA** L., Mant. **2:** 145, 1771

Erect or scandent herbs or shrubs with alternate leaves. Corolla (in ours) only slightly irregular; lobes nearly equal. Anthers free. Ovary 2-celled, one ovule in each cell. Style curved to sharply bent near apex; stigma entire or slightly bilobed, included in a shallow cup. Fruit a succulent, indehiscent drupe in ours, usually 1-seeded by abortion of one ovule.

**1. Scaevola plumieri** (L.) Vahl, Symb. Bot. **2:** 36, 1791

*Lobelia plumieri* L., Sp. Pl. 929, 1753.
*Scaevola lobelia* Murr., Syst. ed. 13, 178, 1774.
(For additional synonymy see Das Pflanzenr. IV, **277:** 120, 1912.)

An erect shrub 6–12 dm. tall, with terete, stoutish, freely branching stems, glabrous except for tufts of slender white hairs in axils of leaves and above leaf-scar after dehiscence of old leaves; leaf blades fleshy, obovate-elliptical to oblong-obovate, obtuse to rounded at apex, 3–8 cm. long, 1.5–3.5 cm. wide, gradually narrowed below to broad, winged petiole, this 5–12 mm. long, margins of blade entire, veins scarcely evident; cymes 1–3-flowered, on peduncles 1–5 cm. long; bracts linear-lanceolate, acute, 3–7 mm. long; calyx truncate or very shallowly and obscurely 5-lobed, collar about 1 mm. high; corolla white, 2–2.5 cm. long, glabrous outside, hirsute within tube with multibranched hairs, lobes oblong, short-acuminate, margins slightly undulate and lacerate-fringed near base, 3–4 mm. wide, 10–14 mm. long; style greenish, about 1.5 cm. long, hirsute on lower one-third to one-half and at top, glabrous between; stigma cup 2–2.5 mm. in diameter, hirsute, turned sharply to one side; fruits broadly ellipsoidal to subglobose, 1–1.8 cm. long, black or purplish, smooth and lustrous.

Sandy soil and upper beaches, Lower Sonoran and Tropical Zones, southern Baja California and Sinaloa; Central America; West Indies and tropics of South America, Africa, and Ceylon. Jan.–May, in our area.

## Family 138. **COMPOSITAE.** Sunflower Family

Herbs, shrubs, or small trees with alternate, opposite, or basal leaves; blades ample or in some reduced to mere scales. Flowers in heads borne on enlarged summit of a short to long peduncle (a receptacle) and surrounded by an involucre of few to many bracts. Involucral bracts herbaceous, coriaceous, chartaceous or scarious, free and distinct or more or less united, in 1 to several series, equal or unequal, strongly to not at all imbricated. Receptacle naked or with hairs, bristles or chaffy bracts, flattened, convex, conical, spheroidal, or subcylindric. Flowers few (sometimes only 1) to many in each head, one or more types in each head. Calyx reduced to scales; hairs, bristles, or paleae at apex of inferior ovary and becoming the pappus, or entirely lacking. Corollas of disk flowers tubular and regularly or (in *Chaenactis*) slightly irregularly 4–5-toothed or (in the *Vernonieae*) deeply bilabiate, with one lip 2-, the other 3-, lobed or toothed. Ray flowers marginal in 1–3 series; limb ligulate and apically toothed, usually pistillate or, in *Cichoreae,* all flowers ligulate and perfect, marginal and inner ones thin, differing only in size of ligule. Ray flowers, in heads containing both ray and disk flowers, usually pistillate and fertile (sometimes perfect or neutral); disk flowers perfect and all fertile, all sterile or both conditions occurring among disk flowers in different parts of same head. Plants sometimes dioecious or monoecious. Stamens as many as corolla lobes (usually 5) and alternate with them; filaments distinct, inserted at or near base of corolla tube (or well above base of tube in *Vernonieae*); anthers united laterally to form a tube (free or nearly so in *Ambrosieae*), caudate, sagittate or rounded at base, with or without appendages of varying shapes and sizes at tips, usually dehiscing along inner side. Styles divided above into 2 distinct branches which bear stigmatic lines along all or a portion of inner face, often appendaged terminally and rarely stigmatic to summit, this usually acute or elongated in some tribes, truncate or obtuse in others, usually penicillate or hispidulous. Ovary inferior, 1-celled, bearing a single, anatropous, erect ovule, becoming an achene in fruit; achene enclosed in a burlike structure formed from indurated involucral bracts of pistillate head in *Ambrosieae,* variously invested with hairs or glabrous, striate, winged or smooth (sometimes beaked in *Madieae* and *Cichoreae*).

Filaments of stamens inserted well above base of corolla tube (rays none; corolla never yellow; anthers sagittate at base) ...................... Tribe 1. *Vernonieae*, p. 1447
Filaments of stamens inserted at or very near base of corolla tube:
Corollas of disk flowers regular or nearly so, not deeply bilabiate; heads never of ligulate flowers only (if rays present, some regular, disk flowers present in same head):
Style branches thickened toward apex; stigmatic lines only on lower one-half of style branches (flowers never yellow) ................. Tribe 2. *Eupatorieae*, p. 1447
Style branches not thickened upward; stigmatic lines approaching or reaching summit of style branches (flowers often yellow):
Anthers not caudate nor sagittate at base; rays often present; involucral bracts rarely white-scarious or bristly:
Heads usually unisexual; fruit often a bur; anthers nearly or quite distinct; rays none ..................................... Tribe 6. *Ambrosieae*, p. 1563
Heads rarely unisexual; fruit not burlike; anthers always united full length; rays often present:
Receptacles chaffy-bracted, at least between ray and disk flowers, often over whole surface:
Involucral bracts not enfolding ray achenes, usually in several well-imbricated series; receptacle chaffy over whole surface; ray achenes not differing markedly from disk achenes ........... Tribe 5. *Heliantheae*, p. 1515
Involucral bracts each enfolding a ray achene, in a single series; receptacle often with bracts only between ray and disk flowers; ray achenes markedly different from disk achenes .................... Tribe 7. *Madieae*, p. 1579
Receptacles naked or nearly so (sometimes with a few hairs or bristles in *Anthemideae* and *Helenieae*):
Pappus of awns, paleae, bristles or wanting, rarely of silky capillary hairs; outer involucral bracts if shorter than inner rarely calyculate; main involucral bracts usually not black-tipped:
Style branches of fertile flowers distinctly flattened, with an acute to obtuse terminal appendage; herbage usually not strong-scented (involucral bracts well imbricated in several series except in *Erigeron* and *Conyza*) ...................................... Tribe 3. *Astereae*, p. 1466
Style branches not markedly flattened, truncate or less commonly short-appendaged; herbage often strong-scented:
Involucral bracts poorly or not at all imbricated, herbaceous; pappus if present not a reduced ring or crown ...... Tribe 8. *Helenieae*, p. 1581
Involucral bracts well imbricated in several series, scarious or thin-chartaceous, at least marginally; pappus a low ring or crown or wanting ... ................................ Tribe 9. *Anthemideae*, p. 1630
Pappus of silky capillary hairs; outer involucral bracts usually calyculate; main bracts often black-tipped ................ Tribe 10. *Senecioneae*, p. 1636
Anthers caudate or sagittate at base; ray flowers none; involucral bracts either white-scarious or bristly:
Involucral bracts dry, white-scarious; receptacle not bristly; anthers unappendaged terminally; plants not bristly; heads small, less than 1 cm. in diameter ....... ........................................... Tribe 4. *Inuleae*, p. 1506
Involucral bracts herbaceous or coriaceous, often bristle-margined; receptacle densely bristly; anthers long-appendaged terminally; plants bristly or spiny; heads larger, 1–8 cm. in diameter, rarely smaller ... Tribe 11. *Cynareae*, p. 1644
Corollas deeply bilabiate or all ligulate, only one type of corolla present in any head:
Juice never milky; ligulate flowers none; corollas always bilabiate .................. ............................................. Tribe 12. *Mutisieae*, p. 1649
Juice milky; corolla all ligulate, never bilabiate ........ Tribe 13. *Cichorieae*, p. 1653

## Tribe 1. VERNONIEAE. Vernonia Tribe

Herbs, shrubs, or small trees with alternate or rarely opposite leaves. Heads discoid with white, red, purple or bluish flowers, these never yellow. Involucres cylindric to hemispherical; bracts graduated, dry or indurated, tips rarely herbaceous. Flowers slenderly tubular, in ours regular, perfect and fertile. Receptacles naked. Anthers sagittate or caudate at base. Style branches elongate, acute, hispidulous full length exteriorly. Achenes ribbed. Pappus biseriate, outer paleaceous or setose, shorter than inner capillary or setose bristles, or pappus sometimes lacking.

### 1. VERNONIA Schreb., Gen. Pl. ed. 8, 2: 541, 1791

Characters of the tribe. Leaves alternate. Flowers purple or purplish. Pappus of capillary bristles only.

**1. Vernonia palmeri** Rose, Contr. U.S. Nat. Herb. **1:** 101, 1891
*Eremosis palmeri* Gleason, Bull. N.Y. Bot. Gard. **4:** 233, 1906.
*Vernonia chacalana* S. F. Blake, Contr. Gray Herb. **52:** 19, 1917.

A freely branching shrub to 5 m. tall, branchlets and leaves sericeous-pubescent in youth, later glabrate; petioles 3–18 mm. long; leaf blades lanceolate, elliptic-lanceolate or rarely oblanceolate, 1–5.5 cm. wide, 3–18 cm. long, acute apically, acute to obtuse at base, entire to remotely and shallowly serrulate, sparsely pubescent and resinous-dotted above, densely pubescent and resinous-dotted beneath, finally glabrate; heads short-pedunculate, in rounded to pyramidal inflorescences to 2 dm. long; involucres cylindric to narrowly campanulate, 5–6 mm. high, bracts chartaceous, outer ovate or triangular and obtuse, densely puberulent and ciliolate-margined, inner progressively longer, thinner, less pubescent and more acute, midvein green, rest stramineous, resinous near tip; flowers 3–5 in each head, about 3.5 mm. long, purplish; achenes about 2 mm. long, hirsutulous; pappus white, bristles capillary, outer 1–1.5 mm. long, inner about 4 mm. long.

On mountain slopes, open hillsides, and in forests, chiefly Upper Sonoran Zone, southern Sonora to Durango and Tepíc. Barely reaching the edge of the Sonoran Desert. Feb.–May.

## Tribe 2. EUPATORIEAE. Eupatory Tribe

Herbs, shrubs, or (in tropics) small trees with opposite or alternate, usually divided or deeply lobed leaves. Heads discoid. Involucral bracts in few series, usually distinctly imbricated, herbaceous. Receptacle flat, convex or rarely spherical, naked or sparsely bristly or hairy. Flowers perfect and fertile. Corollas tubular, regular, white, bluish or purple, never yellow. Anthers rounded at base, not caudate nor sagittate. Style branches semicylindric, elongated, more or less thickened upward, truncate or obtuse at apex, stigmatic only on lower one-half or less. Achenes prismatic-linear to obovate, 4–5-angled. Pappus of bristles, hairs, awns or paleae.

Pappus of a low paleaceous crown one-fifth as long as achene .............. 2. *Ageratum*
Pappus longer, some paleae or bristles one-half as long as achene or more:
  Pappus in 2 series, an inner of bristles, an outer alternating one of hyaline squamellae:
    Plants perennial, mostly suffrutescent; leaves petiolate ............... 3. *Hofmeisteria*
    Plants annual; leaves sessile, linear ....................................... 4. *Malperia*

Pappus in a single series of bristles or hairs, squamellae absent:
Pappus bristles finely plumose; heads in virgate leafless panicles; plants annual ........ ........................................................ 5. *Carminatia*
Pappus bristles smooth or finely barbellate, not plumose; heads in corymbose, cymose or thyrsoid panicles, not virgate; plants perennial (except in *Brickellia diffusa*):
Achenes 5-nerved and 5-angled .................................. 6. *Eupatorium*
Achenes mostly 10-nerved or 10-ribbed, sometimes 10–20-nerved ....... 7. *Brickellia*

## 2. AGERATUM L., Sp. Pl. 839, 1753

Herbs or shrubs with slender stems and mostly opposite, ovate, oblong or lanceolate leaves. Heads in terminal corymbose clusters on a long, slender, virtually naked peduncle, discoid, many-flowered. Involucres campanulate; bracts linear, unequal, imbricated in 2–3 series. Receptacle plane to convex, naked or chaffy-bracteate. Corollas white to purplish, equal and regular, limbs 5-parted, longer than or about equaling tubes. Anthers obtuse at base, terminally short-appendaged. Style branches obtuse, elongate, well exserted. Achenes 5-angled. Pappus of short, obtuse to aristate paleae in a single series or these coalesced to form a dentate cupule at top of achene.

**1. Ageratum salicifolium** Hemsl., Biol. Centr. Amer. Bot. **2:** 83, 1881
*Ageratum strictum* Hemsl., loc. cit.
*Carelia salicifolia* Kuntze, Rev. Gen. Pl. **1:** 325, 1891.
*Carelia stricta* Kuntze, loc. cit.

Herbaceous to fruticose perennial to 1.3 m. tall with slender, finely puberulent and faintly scaberulous branches and opposite leaves on slender petioles 5–12 mm. long; leaf blades narrowly lance-ovate to narrowly lanceolate, 0.5–1.8 cm. wide, 4–7 cm. long, narrowly to broadly cuneate and somewhat decurrent at base, attenuate apically, coarsely and remotely serrate to subentire, bright green and puberulent above, paler, densely puberulent, stipitate-glandular and conspicuously 3-veined beneath; upper leaves reduced, narrow and often entire; peduncles to 1.5 dm. long; involucres about 4 mm. long, 4–5 mm. wide, bracts subulate to lance-linear, acuminate to attenuate with a strong mid-nerve, puberulent with short, spreading hairs and sparsely gland-dotted; corollas about 3 mm. long, whitish, sparsely glandular; style branches pale purplish, exserted 1.5–2.5 mm.; achenes narrowly clavate, 2–2.5 mm. long, brownish, glabrous; pappus a low crown 0.3–0.5 mm. high, yellowish or whitish, very shallowly 5-toothed to subtruncate.

In canyons and along shaded water courses in the lower mountains and hills, upper edge of the Lower Sonoran Zone into the Upper Sonoran and Tropical Zones, central Sonora to Morelos, Mexico. Sept.–Oct.

## 3. HOFMEISTERIA Walp., Rep. 6: 106, 1846–47

Herbaceous, suffrutescent or truly shrubby, rather densely crowded, much-branched, perennial plants with opposite or alternate leaves or both arrangements usually on a single plant. Petioles usually equaling or surpassing blades; leaf blades dentate, lobed, or once- to thrice-pinnatisect, often fleshy or subfleshy. Heads cylindrical to broadly campanulate, solitary on long peduncles or paniculately arranged. Involucral bracts several-seriate, strongly graduate, imbricated, narrowly acuminate or attenuate, dryish, mostly 3-nerved. Receptacle naked. Corollas white, pinkish, or lavender, subfiliform, tube many times as long as minute lobes. Achenes linear or prismatic, 2–5-ribbed. Pappus of 2–15 finely barbellate bristles longer than achenes, and about same number of shorter, often erose or laciniate, hyaline squamellae, these sometimes awn-tipped.

Heads cylindric or narrowly turbinate, in cymes or corymbs; leaves entire, toothed or crenate, rarely deeply lobed; involucral bracts about 18–25, tips often spreading:
- Leaf blades lanceolate to lance-ovate, usually cuneate at base; petioles 2.5–10 times as long as blade .......... 1. *H. pluriseta*
- Leaf blades broadly ovate to orbicular or orbicular-reniform, truncate to cordate at base; petioles 1–2.5 times as long as blade:
  - Pappus bristles 8–15; squamellae 8–10, mostly lance-attenuate, often awned .......... 2. *H. laphamioides* var. *laphamioides*
  - Pappus bristles 4–6, usually 5; squamellae usually same number as and alternating with bristles, obtuse to truncate and erosulate or finely laciniate, not awned:
    - Leaf blades crenate or shallowly dentate, not divided; ovate to orbicular in outline .......... 2*a*. *H. laphamioides* var. *pauciseta*
    - Leaf blades distinctly bi- to tripinnatifid, narrowly ovate to oblong in outline .......... 3. *H. gentryi*

Heads campanulate, solitary on long peduncles; leaves usually deeply divided or pinnatifid; involucral bracts about 50 or more, subrigid, tips erect:
- Pappus bristles 5(–7); squamellae narrowly lance-linear, not erosulate above; lobes of leaves terete or nearly so; plants glabrous and glaucous .......... 4. *H. crassifolia*
- Pappus bristles mostly 2–3; squamellae oblong or broadly lance-ovate, more or less erosulate at blunt apex; plants essentially glabrous to densely pubescent, not glaucous:
  - Heads 10–11 mm. high; corolla lobes triangular; leaf blades lobed halfway to base or pinnatifid, divisions oblong, not filiform; pappus bristles 2 or rarely 3; paleae oblong, minutely erosulate but not laciniate:
    - Leaf blades ovate in outline, deeply pinnatifid, lobes oblong and some again toothed:
      - Herbage glabrous or only slightly puberulent on youngest growth and tips of peduncles; leaf blades divided into broad, rather coarse segments .......... 5. *H. fasciculata* var. *fasciculata*
      - Herbage densely glandular-pubescent; leaf blades divided into narrow fine divisions .......... 5*a*. *H. fasciculata* var. *pubescens*
    - Leaf blades suborbicular in outline, usually 3-lobed about to middle, lobes often again crenately toothed .......... 5*b*. *H. fasciculata* var. *xantii*
  - Heads 7–8 mm. high; corolla lobes roundish; leaf blades bipinnately 3–9-parted into filiform segments; pappus bristles usually 3; paleae laciniate .......... 6. *H. filifolia*

## 1. **Hofmeisteria pluriseta** A. Gray, in Torr., Pacif. R. Rep. **4:** 96, *pl.* 9, 1857

A much-branched, compact to straggly shrub 2–3.5 dm. tall, branches slender, more or less flexuous and finely striate; herbage finely glandular-puberulent throughout except on glabrate, pale brownish, thin bark on older stems; petioles subfiliform, 2–6.5 cm. long; leaf blades ovate to lanceolate in outline, mostly 1.5–5 mm. wide, 2.5–10 mm. long but occasionally to 15 mm. wide and 2.5 cm. long, entire or more frequently with 1–(3) pairs of spreading triangular teeth at or near base, blade always less than one-half as long as petiole; heads cylindric-turbinate, 6–9 mm. long, in loose corymbs or cymes; involucral bracts lance-linear, acuminate, strongly striate, 0.5–1 mm. wide, 2.5–6 mm. long, exposed parts hispid-glandular, tips of outer bracts usually spreading; corollas white, very slenderly tubular, about 4–5 mm. long; achenes narrowly ellipsoid to oblong, 3–4 mm. long, faintly 4-ribbed, strigulose; pappus of 10–12 slender bristles 4–5 mm. long, irregularly alternating with nearly as many narrowly lanceolate, acute to attenuate squamellae 1–2 mm. long.

Among rocks in steep washes and on faces of cliffs, Lower Sonoran Zone, southern Nevada through Inyo County, California, and south into central Baja California and northern Sonora. Feb.–May, and sometimes in the fall.

**2. Hofmeisteria laphamioides** Rose, Contr. U.S. Nat. Herb. **1:** 79, 1890, var. **laphamioides**

*Hofmeisteria pluriseta* var. *laphamioides* I. M. Johnston, Proc. Calif. Acad. IV, **12:** 1186, 1924.

A dense, rounded, much-branched shrub 0.5–1.5 m. tall with glandular-puberulent herbage and striate branchlets; leaves mostly opposite below, alternate toward inflorescence; petioles 5–10 (20) mm. long; leaf blades broadly ovate to orbicular, 4–28 mm. wide and long, crenate to dentate, shallowly or not at all lobed, thickish; heads cylindric-turbinate or narrowly campanulate, in cymes or corymbs, 8–10 mm. long; involucral bracts in 5–6 series, lance-linear, 2–7 mm. long, 1–3-nerved, puberulent on exposed parts, margins scarious, tips of outer ones spreading to slightly reflexed; corollas white, filiform-tubular, about 5 mm. long; achenes 2.5–3 mm. long, narrowly obpyramidal, slightly 4-angled, strigulose; pappus bristles tawny, 10–12, sometimes slightly dilated and hyaline at base, with 4–12 short, lance-attenuate, often awned squamellae alternating with main bristles.

Arid talus slopes, rocky hillsides and faces of cliffs, Lower Sonoran Zone, northwestern Sonora, east central Baja California and on the adjacent islands in the Gulf of California. March–April.

**2*a*. Hofmeisteria laphamioides** var. **pauciseta** (I. M. Johnston) S. F. Blake, Contr. U.S. Nat. Herb. **23:** 1430, 1926

*Hofmeisteria pluriseta* var. *pauciseta* I. M. Johnston, Proc. Calif. Acad. IV, **12:** 1187, 1924.

Differing from var. *laphamioides* in having only 4–8 pappus setae alternating with 4–8 oblong, obtuse squamellae that are erosulate and awnless.

East central Baja California and on San Pedro Nolasco and Espíritu Santo Islands in the Gulf of California. March–April.

**3. Hofmeisteria gentryi** Wiggins, Contr. Dudley Herb. **4:** 25, 1950

A small, rounded shrub 4–6 dm. tall with striate, gray-barked branches, densely hispidulous young growth and brittle leaves rather crowded toward tips of branches; petioles slender, 12–15 mm. long, hispidulous; leaf blades broadly ovate in outline, 15–22 mm. wide, 15–25 mm. long, bi- and tripinnately lobed or divided, lobes again irregularly toothed, mostly 5–7 mm. wide, 6–10 mm. long, densely hispidulous-glandular; heads 1–2 on terminal peduncles 2–6 cm. long, turbinate at anthesis but soon subhemispherical; involucral bracts in 4–6 series, lance-linear, 2–7 mm. long, attenuate-acuminate, mostly 3-nerved, glandular-puberulent on exposed parts, margins of inner ones hyaline, only lowest spreading or deflexed at tips; corollas white, subfiliform, 5–6 mm. long; style branches purplish, exserted 2–2.5 mm.; achenes narrowly obpyramidal, about 3 mm. long, dark brown, hispidulous with stiffish, blunt, ascending hairs; pappus of 4–6 scaberulous bristles 5–6 mm. long, some of them slightly hyaline-dilated at base, alternating with 4–6 oblong squamellae 1–1.5 mm. long, these with obtuse to truncate, erosulate tips.

On basaltic cliffs, from sea level to 3,000 feet, Lower Sonoran Zone, Sierra de la Giganta above Puerto Escondido, Baja California. Type collected in April.

**4. Hofmeisteria crassifolia** S. Wats., Proc. Amer. Acad. **24:** 53, 1889

Dense, suffrutescent, rounded plant 6–25 cm. high and fully as wide, with numerous brittle stems from a woody caudex, glabrous and more or less glaucous throughout; leaves crowded toward tips of branches, alternate, 1–5 cm. long, more or less triangular in outline, once or twice ternately parted into fleshy, linear, nearly terete lobes 1–12 mm. long, petiole 1–3.5 cm. long, about as thick as larger lobes; heads broadly campanulate to subhemispherical, solitary on scapes 5–18 cm. long, these glabrous and shining; involucral bracts 60–100, in 6–9 series, narrowly lance-ovate to lance-linear, 2–8 mm. long, 1–3-nerved, attenuate-acuminate at apex, stramineous, glabrous except along ciliolate margins; corollas subfiliform-tubular, 4–4.5 mm. long, slightly expanded above into narrowly funnelform pinkish limb, lobes narrowly triangular; achenes dull black, narrowly obpyramidal, about 2–2.5 mm. long, slightly flattened and 4-angled, glabrous; pappus of 5–6 slender bristles nearly equaling corollas and about as many shorter squamellae dissected into shorter, unequal bristles 1–2.5 mm. long.

On rocky hillsides, promontories and faces of cliffs, Lower Sonoran Zone, near Guaymas, Sonora, and on San Pedro Nolasco Island in the Gulf of California. Probably March–May; specimens were collected in April.

**5. Hofmeisteria fasciculata** (Benth.) Walp., Rep. **6:** 106, 1846–47, var. **fasciculata**
*Helogyne fasciculata* Benth., Bot. Sulph. 20, *pl. 14,* 1844.

Suffruticose or frutescent plants with several to many branches from a woody caudex forming dense to straggly plants 2–6 dm. tall; herbage essentially glabrous or tips of peduncles, outer involucral bracts and youngest leaves sparsely puberulent; leaves opposite below, alternate above; petioles slender, 1–4 cm. long, mostly 2 mm. or less in width; leaf blades deltoid to suborbicular in outline, 1.5–4 cm. long and wide, ternately parted or lobed, lobes again irregularly lobed or dissected into oblong to triangular segments; peduncles erect, to 2 dm. long, or with a few reduced, bractlike leaves scattered from base almost to involucre; heads broadly campanulate, 8–11 mm. tall at anthesis; involucral bracts 75–100, lance-linear, 1.5–8 mm. long, subulate, 3-nerved, glabrous or sparsely glandular-puberulent, slightly brownish-tinged; corollas pale lavender, 3.5–4.5 mm. long, very slender; achenes black, about 3 mm. long, glabrous or sparsely strigulose on 4 angles; pappus of 2 or less commonly 3 slender, faintly barbellate bristles about equaling corollas, alternating with 2–4 obtuse to truncate, oblong, finely erose-laciniate squamellae about 1 mm. long.

Arroyos, flats, slopes and crevices in cliffs, Lower Sonoran Zone, vicinity of Bahía de los Angeles to Magdalena Bay, Baja California, and on adjacent islands along both coasts. Jan.–Aug.

**5*a*. Hofmeisteria fasiculata** var. **pubescens** (S. Wats.) Robinson, Proc. Amer. Acad. **47:** 192, 1911
*Hofmeisteria pubescens* S. Wats., op. cit. **24:** 54, 1889.

Usually somewhat taller (to 1.2 m.) than var. *fasciculata* and densely glandular-puberulent; leaf lobes often blunter but more finely divided than in var. *fasciculata.*

Cliffs and steep rocky slopes, Lower Sonoran Zone, vicinity of Mulegé south at least to Isla Catalina. March–May.

**5*b*. Hofmeisteria fasciculata** var. **xantii** A. Gray, in Brewer & Wats., Bot. Calif. **1:** 299, 1876

*Hofmeisteria fasciculata* var. *grayi* Brandegee, Zoe **5:** 160, 1903.

Leaf blades mostly suborbicular in outline, mostly 3-lobed about halfway to middle; lobes coarsely repand-toothed or crenate; heads somewhat larger, often to 1.5 cm. in diameter.

Mostly in crevices of maritime cliffs, Lower Sonoran and Subtropical Zones, southern Baja California, chiefly in the Cape Region. March–April.

**6. Hofmeisteria filifolia** I. M. Johnston, Proc. Calif. Acad. IV, **12:** 1185, 1924

Herbaceous perennial with numerous slender, slightly flexuous, striate stems forming dense clumps 1–6 dm. high; herbage sparsely to densely glandular-puberulent throughout; leaves alternate or basal opposite, 4–7 cm. long, 2–6 cm. wide, once- to twice-pinnately parted into linear or subfiliform lobes 0.5–1 mm. wide, 2–20 mm. long, lobes of lowermost leaves sometimes slightly dilated distally; peduncles slender, 1.5–9 cm. long, usually bearing 1 to several scattered setaceous bracts 1–3 mm. long; heads narrowly campanulate, solitary, 7–8 mm. long, 4–5 mm. wide; involucral bracts numerous, lance-linear, subulate, well imbricated, 3-nerved, glandular-puberulent, tips erect; corollas pale pink, subfiliform, 3–5.4 mm. long, rounded lobes broader than long; achenes black, about 3 mm. long, marginal angles strigulose; pappus of 3–4 (usually 3) antrorsely barbed bristles about equaling corolla and 3 broadly oblong, hyaline, terminally erose-laciniate squamellae about 1 mm. long.

Known only from the type locality, on Isla Angel de la Guarda, Baja California. March–May.

## 4. MALPERIA S. Wats., Proc. Amer. Acad. 24: 54, 1889

Erect, slender annual moderately branched above. Leaves alternate or lowermost opposite, narrow, mainly sessile. Heads few, loosely cymose, turbinate. Involucral bracts in several series, linear to lance-linear, unequal and closely imbricated, thin, several-nerved. Corollas slender; tube subfiliform, many times as long as narrowly triangular-ovate lobes. Style branches filiform, obtuse, clavately thickened toward apex and exserted at maturity. Receptacle flat, naked. Achenes slenderly prismatic, 5-angled. Pappus of 3 minutely hispidulous bristles equaling corollas and 3–4 alternating, minute, truncate, erosulate squamellae.

**1. Malperia tenuis** S. Wats., Proc. Amer. Acad. **24:** 54, 1889

*Hofmeisteria tenuis* I. M. Johnston, Proc. Calif. Acad. IV, **12:** 1188, 1924.

Plant slender, 1–3 dm. tall, stems and branches shining, glabrous or minutely and sparsely puberulent with a few scattered, deciduous, whitish hairs; leaves narrowly linear, 0.3–1.5 mm. wide, 1.5–5 cm. long, entire, sessile; involucres 7–10 mm. high, bracts narrowly ovate-lanceolate (lowermost) to lance-linear, greenish tinged with purplish and puberulent on backs, scarious and white along margins; corollas faintly tinged with pink; achenes about 3 mm. long, dull black, very sparsely and minutely appressed-puberulent on angles; pappus bristles 4–5 mm. long, squamellae mostly 0.5 mm. long or less.

Along washes and on arid slopes, Lower Sonoran Zone, southwestern Imperial County, California, south along the eastern side of the main ranges at least to Barríl, Baja California. Feb.–May.

## 5. CARMINATIA Moc. ex DC., Prod. 7: 267, 1838

Erect or ascendingly and sparingly branched herbs with opposite, deltoid-ovate, thin, slender-petiolate, toothed leaves. Heads discoid, cylindrical, in terminal, leafless, virgate panicles. Involucres graduated, of slender, few-nerved, thin, acute to attenuate bracts. Corollas white or faintly tinged with pink or lavender, very slender. Style branches filiform, well exserted. Achenes linear-prismatic, slightly 5-angled. Pappus bristles about 10, about equaling the achene, finely silky-plumose.

### 1. Carminatia tenuiflora DC., Prod. 7: 267, 1838

*Carminatia tenuifolia* Steudel, Nom. ed. 2, 1: 299, 1840.

Plants usually erect, often strict, 1–12 dm. tall, more or less glandular-puberulent throughout or eventually nearly glabrate; petioles slender, 1–9 cm. long; leaf blades to 7 cm. long, to 6 cm. wide, usually truncatish at base, acute to acuminate at apex, coarsely repand-dentate to sinuous-dentate; panicles to 3 dm. long or more on large plants, or reduced to 2–3 heads on depauperate specimens; peduncles 2–6 mm. long; involucres 7–13 mm. long, bracts linear-subulate or outer lance-linear, mostly 3-nerved, greenish and shining, margins pale-scarious, nearly or quite glabrous; corollas 6–7 mm. long; achenes about 5 mm. long, black, dull to slightly shining; pappus bristles silvery, 5–7 mm. long, silky-plumose.

Mostly rich soils in canyon bottoms, Tropical Zone to Upper Sonoran Zone, southern Arizona to western Texas southward to southern Baja California and Central America. Collected in the Sonoran Desert between Baviácora and Cumpas, Sonora. Sept.–Nov.

## 6. EUPATORIUM L., Sp. Pl. 836, 1753

Herbs, shrubs or small trees with opposite or rarely alternate, membranous to coriaceous, filiform to orbicular leaves. Flowers perfect and fertile. Heads discoid, 3–100-flowered, usually in thyrsoid or corymbose panicles. Corollas very slender, only slightly ampliate above, red, purple, blue, lavender, white, or rarely yellowish or greenish. Anthers appendaged at tip, entire at base. Style branches long and slender, well exserted at anthesis, uniformly threadlike or more commonly more or less clavate toward tips, often colored. Achenes columnar to prismatic-obovoid, 5-nerved or 5-angled, glabrous to densely hispidulous. Pappus bristles numerous, capillary, white, sordid, stramineous, roseate, purplish or tawny, subequal or less frequently outermost much shorter than inner, usually minutely barbellate.

Stems strongly quadrangular; leaves obovate-spatulate, sessile, to 30 cm. long .............................................................. 1. *E. quadrangulare*

Stems not markedly angular; leaves lanceolate, sagittate, ovate or deltoid, always broader below middle than above it, petiolate, rarely over 12 cm. long:

  Involucral bracts linear, 1 mm. wide or less, usually distinctly 1–3-veined to tip, chartaceous; leaf blades lanceolate, ovate or deltoid, if sagittately lobed at base then as broad as long:

    Leaf blades broadly ovate, deltoid-ovate or broadly deltoid-hastate, nearly or fully as wide as long; involucres mostly 3–8 mm. wide:

      Stems and petioles of mature leaves glabrous or essentially so; involucral bracts linear-subulate and attenuate at apex; leaf blades subcordate at base ..... 2. *E. purpusii*

Stems and petioles minutely but closely and distinctly puberulent; involucral bracts linear-oblong to linear-lanceolate, acute or obtuse but not attenuate at apex; leaf blades broadly cuneate to truncate at base:

Achenes black, sparsely hispidulous on angles; pappus bristles not crinkly at base; leaf blades deltoid-ovate, rounded at base, paler beneath than above, margins crenulate-serrate to subentire .............................. 3. *E. collinum*

Achenes tawny, densely strigose-hispid over whole surface; pappus bristles crinkly wavy at base; blades of larger leaves deltoid-hastate, smaller ones often elliptic, not paler beneath than above, margins entire:

Leaves and branchlets green, glabrous or puberulent ........................ .................................. 4. *E. peninsulare* var. *peninsulare*

Leaves and branchlets grayish tomentulose .... 4*a*. *E. peninsulare* var. *epipolium*

Leaf blades lance-ovate to lanceolate, 1.5–5 times as long as broad; involucres mostly 1.5–2.5 mm. wide at anthesis:

Heads 3–6-flowered; involucral bracts not strongly graduate .. 5. *E. solidaginifolium*

Heads 10–25-flowered; involucral bracts distinctly graduate in 4–9 series:

Leaves scabrous and prominently veined beneath, 5–12 cm. long, margins ascendingly serrate; heads 7–9 mm. long; involucral bracts obtuse, densely puberulent, often purplish ..................................... 6. *E. bertholdii*

Leaves hardly scaberulous, inconspicuously veined, 1.5–5 cm. long, margins spreadingly dentate; heads 4–6 mm. long; involucral bracts acute, sparsely puberulent to subglabrate, green .............................. 7. *E. pycnocephalum*

Involucral bracts spatulate, upper portion ovate, 1.2–2.8 mm. wide, outer ones not visibly veined toward tip, subfoliaceous; leaf blades lance-sagittate ........ 8. *E. sagittatum*

## 1. **Eupatorium quadrangulare** DC., Prod. **5:** 150, 1836

Stout perennial with a woody base and herbaceous, sharply quadrangular, hollow stems; younger herbage and inflorescence puberulent to pubescent or tomentulose, older parts usually glabrate or nearly so; leaves opposite, spatulate or ovate but with a broadly winged petiolar part, 3–12 cm. wide, 6–30 cm. long, biauriculate at base, appearing sessile, serrate with apiculate teeth; heads usually 8–10-flowered, borne in densely crowded terminal clusters, narrowly campanulate or turbinate, 7–8 mm. high; involucral bracts well imbricated, stramineous, ovate-oblong, 1–6 mm. long, obtuse to rounded, striate, puberulent to glabrate, margins scarious; flowers fragrant; corollas white or faintly tinged with pink or lavender, about 5 mm. long; achenes slenderly prismatic, about 3 mm. long, obtusely 5-angled, glabrous; pappus bristles capillary, tawny, 3–4 mm. long.

Canyons and along water courses, Upper Sonoran to Tropical Zones, southern Sonora, where it barely enters the Sonoran Desert, to Durango, Vera Cruz, Guatemala and El Salvador. Oct.–March.

## 2. **Eupatorium purpusii** Brandegee, Erythea **7:** 3, 1899

An erect, moderately branched perennial herb 4–6 dm. tall, glabrous except on upper part of stems, peduncles and youngest leaves, which are slightly puberulent; stems reddish or purplish; leaves opposite, thin, broadly ovate to deltoid-ovate, 1.5–6 cm. wide, to 7 cm. long, truncate to subcordate at base, acute to slightly acuminate at apex, crenate-serrate, glabrous; petioles slender, reddish, 3–6 cm. long; heads in corymbose clusters on peduncles 2–14 mm. long, broadly hemispheric-campanulate, 7–8 mm. high and broad, about 75-flowered; involucral bracts subequal, linear-acuminate, 3–5 mm. long, puberulent, 2–3-ribbed; corollas slenderly tubular in basal one-third of their length, slightly ampliate above, about 3 mm. long, white or slightly purplish; achenes linear, about 2 mm. long, dull black, finely but sharply 5-angled, glabrous; pappus bristles finely capillary, about equaling corolla, deciduous.

In shaded canyons, usually along a watercourse, Lower Sonoran Zone, in mountains east of Calmallí south to the Cape Region, Baja California. Feb.–April.

**3. Eupatorium collinum** DC., Prod. **5:** 164, 1836

*Eupatorium nigrescens* Hook. & Arn., Bot. Beechey 297, 1840.
*Krystenia collina* Greene, Leaflets Bot. Obs. **1:** 9, 1903.

A shrub 1–5 m. tall, with rather slender, terete stems with brownish to grayish bark and minutely puberulent branchlets; leaves deltoid-ovate, 2.5–7 cm. wide, 5–10 cm. long, acuminate at apex, broadly cuneate to truncate and rarely subcordate at base, firmly membranous, green on both sides but somewhat paler beneath than above, obscurely puberulent to spreading-pubescent on veins beneath, sometimes glabrate except on veins, crenate, serrate or less commonly subentire on margins, minutely glandular-punctate beneath; petioles slender, 1–3 cm. long; heads in round-topped corymbs and fastigiately grouped on axillary branchlets, 6–8 mm. high, about 25–50-flowered; involucres narrowly turbinate-campanulate, 2.5–3 mm. wide, 5–6 mm. high, bracts narrowly ovate to lance-oblong, 1.5–5 mm. long, mostly obtuse, striate-nerved, puberulent; corollas white, slender, about 4 mm. long; achenes about 2.5 mm. long, dull black, 4–5-angled, glabrous except on angles; pappus bristles capillary, rusty or tawny, 3.5–4 mm. long, persistent.

Canyon slopes and abandoned fields, Lower Sonoran to Tropical Zones, central Sonora and Chihuahua to Central America. Sept.–April.

**4. Eupatorium peninsulare** Brandegee, Erythea **7:** 4, 1899, var. **peninsulare**

A moderately branched shrub 1–2 m. tall with pale gray bark on older stems and puberulent to glabrate, yellowish branchlets, these faintly striate; leaves opposite, mostly halberd-shaped or hastate, but a few ovate to obovate, latter 1.5–3 cm. wide, 3–5 cm. long, others 5–7 cm. wide, 6–12 cm. long, strongly 3-nerved, sparsely scaberulous to glabrous, entire to slightly crenate-serrate, bright green on both sides; petioles 1–3 cm. long; heads about 10–15-flowered, in terminal panicles and in smaller axillary groups, 8–10 mm. long, each peduncle subtended by a lance-linear bract 1–2.5 mm. long; involucres campanulate, 6–7 mm. long, bracts ovate to lance-linear, 1.5–5 mm. long, acute to rounded, puberulent to glabrate, strongly 3–5-nerved, scarious margins arachnoid-ciliolate; corollas white or faintly tinged with lavender on lobes, about 5 mm. long, slender; achenes linear-clavate, about 4 mm. long, 5-ridged, densely hispidulous with ascending stiff hairs, tipped with a persistent style base about 0.3 mm. long; pappus bristles about 4 mm. long, antrorsely barbellate, often wavy-crinkled at base, persistent, faintly enlarged at tip.

In shaded canyons, Lower Sonoran Zone to Subtropical Zone, Baja California in the Sierra de la Giganta near Loreto south to the tip of the peninsula.

**4*a*. Eupatorium peninsulare** var. **epipolium** B. L. Robinson, Contr. Gray Herb. **73:** 17, 1924

Like var. *peninsulare* in all respects except that branchlets and leaves are grayish-tomentulose. Southern Baja California. Nov.–March.

**5. Eupatorium solidaginifolium** A. Gray, Pl. Wright. **1:** 87, 1852

A low, slenderly branched, rather straggly shrub 3–6 dm. tall with sparsely scaberulous to subglabrate branchlets and leaves, branches of the inflorescence usually finely puberulent; leaves narrowly ovate to lanceolate, 1–2 cm. wide, 2.5–5.5 cm. long, rounded to subtruncate at base, attenuate at apex, subentire, with a few minute scattered serrations irregularly spaced, sparsely hispidulous along veins beneath; petioles 3–10 mm. long; heads 3–6-flowered, usually in closely aggregated groups of 2–5 at tips of thyrsoid panicles, each head cylindrical to narrowly campanulate, about 4 mm. long, 1–2.2 mm. wide at anthesis; involucral bracts 8–12, lance-linear, 3-nerved, acute to slightly acuminate, greenish, sparsely ciliolate along margins and sometimes puberulent along nerves; corollas white, 2.5–2.8 mm. long; style branches clavate, exserted 1–1.5 mm.; achenes light brown, 1.5 mm. high, angled and rather densely strigulose; pappus bristles about equaling corollas, white to slightly greenish tinged.

Mainly on limestone outcrops or lime soil, Lower and Upper Sonoran Zones, southern Arizona to western Texas and south into Sonora, Chihuahua, Durango and Coahuila. May–Oct.

### 6. **Eupatorium bertholdii** Sch.-Bip., in Seem., Bot. Herald 299, 1856

A slender, upright shrub 1–1.5 m. tall, with brownish, puberulent to pubescent, lignescent branchlets; leaves ovate to lance-ovate, 1–5 cm. wide, 4.5–13 cm. long, acute to acuminate at apex, acute to rounded at base, shallowly serrate to entire, rather prominently 3-veined from base, scabrous on both surfaces, paler beneath than above, finely punctate-glandular and reticulate-veined beneath; petioles 2–12 mm. long; heads 10–13-flowered, 7–9 mm. long, sessile to short pedunculate, in rather compact panicles; involucres cylindric, 2–2.8 mm. wide, 5–6 mm. long, bracts well graduated, suborbicular to oblong, obtuse or rounded and usually brownish or purplish at tips, stramineous below, 3–5-nerved, glabrous to sparsely and minutely puberulent; corollas about 4 mm. long, white or purplish tinged; achenes prismatic, faintly 4–5-nerved and angled, glabrous, about 1.5 mm. long; pappus bristles tawny to sordid, 3–3.5 mm. long.

Mostly in the oak belt, Upper Sonoran Zone, but straggling into the upper margins of the Sonoran Desert along canyons, central Sonora (between Baviácora and Cumpas) south to Tepíc. Oct.–Jan.

### 7. **Eupatorium pycnocephalum** Less., Linnaea **6:** 404, 1831

*Eupatorium diversifolium* Schrad., Ind. Sem. Hort. Gött. **1829:** 2, 1829. (Nomen nudum.)
*Eupatorium virgatum* Schrad., loc. cit. (Nomen nudum.)
*Eupatorium schiedeanum* Schrad., op. cit. **1832:** 3, 1832.
*Eupatorium sonorae* A. Gray, Pl. Wright. **2:** 74, 1853.

A subshrubby, weedy plant 3–8 dm. tall with slender, striate-puberulent, mostly herbaceous stems and sparsely puberulent, opposite leaves; petioles slender, 0.5–3 cm. long; leaf blades deltoid-ovate, broadly cuneate to subtruncate at base, acute to acuminate at apex, 0.6–2 cm. wide, 1.5–5 cm. long, irregularly crenate-serrate; heads 4–6 mm. long, nearly as wide, about 25-flowered, aggregated in dense rounded corymbs disposed in leafy-bracted panicles; involucres broadly campanulate to nearly hemispheric, about 4 mm. high, outer bracts ovate and acute to acuminate, puberulent, 0.5–1.2 mm. long, middle and inner ones elliptic to oblong, rounded or obtuse at tip and nearly or quite glabrous, striate-nerved; corollas pale purple or rarely white, about 2 mm. long; achenes prismatic, 1–1.4 mm. long, faintly angled and sparsely puberulent on angles; pappus bristles whitish, slightly longer than achenes.

Mountain sides and canyons, Lower Sonoran to Tropical Zones, central Arizona south to the American tropics. Sept.–Dec.

**8. Eupatorium sagittatum** A. Gray, Pl. Wright. **1:** 88, 1852

An intricately branched, slender-stemmed, somewhat straggly and often scandent shrub 1–2 m. tall with minutely puberulent herbage; leaves opposite, ovate to lanceolate and sagittate, 4–12 mm. wide just above spreading basal lobes, 1–3 cm. long, lobes 1.5–6 mm. long, margins entire, lower surface minutely but distinctly punctate-glandular; heads 8–12 mm. long, about 35-flowered, disposed in open, long-branched corymbs, peduncles 0.3–2 cm. long; involucres turbinate, 6–8 mm. long, bracts broadly ovate to oblong, 1.2–2.8 mm. wide, outer ones all broadly ovate and bright green at tip, this portion slightly spreading, densely puberulent, and more or less punctate-glandular, without visible nerves, lower part stramineous and 3–5-nerved; corollas about 5 mm. long, gradually ampliate from base, lilac, glabrous; achenes linear-clavate, about 4 mm. long, 4–5-angled, dark brown, glabrous except for a few minute, antrorsely appressed hairs on light-colored angles; pappus bristles tawny to rusty, about equaling corollas, persistent.

Rocky hillsides, coastal benches and canyons, Lower Sonoran Zone, vicinity of Hermosillo, Sonora, south into Sinaloa and near Todos Santos, Baja California. Flowering more or less throughout the year when moisture is available.

## 7. BRICKELLIA Ell., Bot. S.C. & Ga. 2: 290, 1822–24

Suffruticose or fruticose perennials or rarely annual herbs with erect, ascending or scrambling branches. Leaves opposite, alternate or rarely verticillate, sessile or petiolate. Heads usually paniculate, less commonly in racemes, cymes or corymbs, discoid. Involucres cylindrical to campanulate; bracts numerous, more or less striate and imbricate in several series, inner linear, others gradually becoming shorter and broader toward base of involucre, outermost usually ovate. Receptacle usually naked and flat. Corollas slenderly tubular, rarely forming a discernible throat; lobes 5, deltoid to oblong-lanceolate, one-tenth to one-fifth as long as tube. Anthers appendaged at tip, rounded at base. Style enlarged into a globose or cylindrical part and often pubescent near base; style-branches clavate. Achenes prismatic-linear, lightly to strongly 5–20-ribbed, usually hispidulous. Pappus of 10–80 slender capillary-like equal bristles, these smooth, barbellate or rarely subplumose.

Heads 6–24-flowered:
- Leaves very small, 3–12 mm. long or less; heads 10–12-flowered or if more, branchlets subspinescent:
  - Branchlets subspinescent, stiff; leaves spatulate; heads 12–13 mm. long, about 18–20-flowered .......... 1. *B. frutescens*
  - Branchlets rarely spinescent, often flexuous; leaves ovate to suborbicular; heads 8–11 mm. long, 8–12-flowered:
    - Tips of outer involucral bracts squarrose-spreading; pubescence stiffish; herbage scabrous; achenes 3.5 mm. long .......... 2. *B. scabra*
    - Tips of outer involucral bracts appressed-erect, not squarrose; pubescence softer; herbage not markedly scabrous; achenes 2.3 mm. long .......... 10. *B. desertorum*
- Leaves larger, usually much over 12 mm. long (smaller in *desertorum*); heads mostly 15–24-flowered:
  - Leaves sessile or subsessile, oblong to narrowly elliptic, conspicuously reticulate-veined; corollas pale yellow with faintly purplish lobes .......... 3. *B. venosa*
  - Leaves short- to long-petiolate, blade moderately veined, ovate, hastate or rhombic, not linear-oblong; corollas white, ochroleucous or purplish, not clear yellow:

Outer bracts all gradually smaller than inner ones; leaves rarely over 7 cm. long:
Leaves cordate and sometimes amplexicaul; heads in pseudo-racemes ............ 4. *B. betonicifolia*
Leaves cuneate, truncate or rounded at base; heads not racemosely disposed:
Style branches one-half as long as corolla tubes, conspicuously exserted and often strongly recurved; leaves hastate, 3–10 cm. long .............. 5. *B. hastata*
Style branches less than one-fourth as long as corolla tubes, slightly exserted, usually ascending; leaves rarely over 5 cm. long, often much less:
Leaves coarsely hastately toothed at base; achenes 3.5 mm. long:
Branches, leaves, and outer bracts more or less puberulent; corollas 8–8.2 mm. long; leaves not fleshy .......................... 6. *B. coulteri*
Branches, leaves, and outer bracts glabrous; corollas 4.5–7 mm. long; leaves fleshy ............................................ 7. *B. glabrata*
Leaves finely or coarsely toothed, but not hastate; achenes 2–3 mm. long:
Achenes 1.7–2.3 mm. long; leaves irregularly or coarsely toothed; corollas 4–6 mm. long:
Peduncles filiform, 8–18 mm. long; corollas 4–4.3 mm. long .. 8. *B. diffusa*
Peduncles stouter, often less than 8 mm. long; corollas 4.5–6 mm. long:
Leaves broadly cuneate at base, coarsely and acutely dentate, 1–4 cm. long; heads 16–17-flowered, pedunculate ........ 9. *B. baccharidea*
Leaves truncate at base, irregularly cuneate-serrate, 3–9 mm. long; heads 8–12-flowered, subsessile ...................... 10. *B. desertorum*
Achenes 3 mm. long; leaves regularly and rather finely toothed; corollas 7.5 mm. long ...................................... 11. *B. californica*
Outer bracts very unequal, 1 to several of them much longer than some of those interior to them; leaves to 13 cm. long ........................ 12. *B. floribunda*
Heads 28–60-flowered:
Corollas 7–9 mm. long; heads 10–16 mm. long; achenes 3.5 mm. long:
Leaves sessile or subsessile, ovate-acuminate, spiny-toothed; heads 75–80 per cent as wide as long:
Outer involucral bracts entire; branches and peduncles glandular-puberulent with very short hairs ....................................... 13. *B. atractyloides*
Outer involucral bracts dimorphic, some having 1 to several small lateral teeth; branches and peduncles glandular-puberulent with longer, spreading hairs:
Teeth on 1–2 outer bracts ............................ 14. *B. arguta* var. *arguta*
Teeth on several outer bracts .................... 14*a*. *B. arguta* var. *odontolepis*
Leaves petiolate, ovate to elliptic-oblong, teeth not spiny-tipped; heads rarely over 65 per cent as wide as long:
Heads corymbosely, paniculately or cymosely arranged; leaves crenate-dentate or crenate-serrate, acute or obtuse at tip:
Achenes 3 mm. long; heads 9–10 mm. long; leaves crenate-dentate, rounded at tips ......................................................... 15. *B. brandegei*
Achenes 3.6–5 mm. long; heads 14–16 mm. long; leaves more coarsely crenate-serrate, acute or obtuse at tips:
Heads about 30-flowered; tips of outer bracts often retuse-recurved; leaves 3–8 cm. long .......................................... 16. *B. peninsularis*
Heads 45–50-flowered; tips of outer bracts closely appressed-ascending; leaves 1.5–3 cm. long:
Leaves reniform-cordate; achenes 3.5 mm. long .......... 17. *B. macromera*
Leaves oblong-spatulate to oblanceolate; achenes 4.5–5 mm. long ........... 18. *B. vollmeri*
Heads in terminal, subumbellate panicles; leaves entire or faintly undulate-dentate, slightly acuminate at tips .............................. 19. *B. rhomboidea*
Corollas 12–14 mm. long; heads 2.5 cm. long; achenes 9–10 mm. long ..... 20. *B. incana*

**1. Brickellia frutescens** A. Gray, Proc. Amer. Acad. **17:** 207, 1882

Intricately branched, cinereous-tomentose, suffruticose plant with slender branches to 1 m. long and shorter, lateral, subspinescent, rather rigid branchlets with gray-puberulent, slightly sulcate bark; leaves spatulate-oblong, obtuse, 1–3.2 mm. wide, 3–12 mm. long, entire, pale green and tomentulose on both surfaces; peduncles terminating lateral branches, 1–3 cm. long, curving-ascending; heads 5–6 mm. wide, 12–13 mm. long; involucral bracts linear-oblong, usually 4-nerved, rounded and short-mucronulate at apex, lower ones shorter and broader, acute or more or less squarrose; corollas slenderly tubular, purplish, 6.5–7.5 mm. long, glabrous, lobes ovate, acute, about one-sixth as long as tube; achenes 3.3–3.5 mm. long, hispidulous along ridges; pappus bristles about equaling corolla tube.

Desert hillsides and arroyos, Lower Sonoran Zone and arid hillsides in Upper Sonoran Zone, western margin of the Colorado Desert south into Baja California on both flanks of the Sierra San Pedro Mártir and in southern Nevada. March–June.

**2. Brickellia scabra** (A. Gray) Coulter & Nels., Man. Bot. Rocky Mts. 487, 1909

*Brickellia microphylla* var. *scabra* A. Gray, Proc. Amer. Acad. **11:** 74, 1876.
*Coleosanthus scaber* Greene, Pittonia **3:** 100, 1896.

Fruticose perennial 4–8 dm. tall; stems slender, leafy, ascending, glandular-villous with spreading hairs and simple, stiff, scabrous ones; leaves small, those of stems suborbicular to broadly ovate, 8–12 mm. long, rather coarsely dentate, those on lateral branchlets gradually reduced in size, scabrous-puberulent; flowering branchlets slenderly elongated, ascending to erect, bearing 1–2 heads; flowering heads 10–11 mm. long, 10–12-flowered, usually terminal but collectively forming a broad lax panicle; involucres cylindrical, with about 30 bracts, inner ones lance-linear, acute, 3–4-striate, outer ones successively shorter and comparatively broader, tips spreading, scabrous-puberulent; corollas ochroleucous, outer ones purplish-tinged, about 5–6 mm. long; achenes 3.5 mm. long, finely but distinctly ribbed; pappus bristles white, short-barbellate.

Arroyos, canyons and hillsides, upper margins of the Lower Sonoran to Transition Zones, Yuma County, Arizona, to southern Wyoming, Nevada, and New Mexico. Mainly above the Sonoran Desert but reaching it in the mountains of Yuma County. July–Oct.

**3. Brickellia venosa** (Woot. & Standl.) B. L. Robinson, Mem. Gray Herb. **1:** 50, 1917

*Coleosanthus venosus* Woot. & Standl., Contr. U.S. Nat. Herb. **16:** 177, 1913.

Fruticose plant 6–8 dm. tall with spreading, canescent-puberulent to tomentulose branches and opposite, spreading, narrowly oblong to linear leaves 2–9 mm. wide, 2.5–5.5 cm. long, obtuse at apex, broadly cuneate at base, entire both at base and apex but rounded-dentate between, reticulate-veined beneath; panicles lax, their branches leafless or nearly so, sparsely puberulent; heads about 24-flowered, 6–6.5 mm. in diameter, 10–11 mm. high on slender erect peduncles equaling or slightly longer than heads; involucres cylindro-subcampanulate with about 25 bracts in 6–7 imbricate series, often purple-tinged, lower ones broader than long, nearly orbicular, lightly pubescent and abruptly mucronulate and often retuse at apex, upper ones successively longer, innermost lanceolate, acute and mucronate, 3–5-nerved, pubescent; receptacle glabrous; corollas 6–6.2 mm. long, pale yellow, with slightly purplish lobes; achenes sericeous-puberulent; pappus bristles about 40, nearly equaling cylindrical corolla tube.

Rocky slopes, canyon walls and mesas, particularly on limestone, Lower and Upper Sonoran Zones, southern Arizona and New Mexico south into Sonora and Chihuahua. Sept.–Oct.

**4. Brickellia betonicifolia** A. Gray, Pl. Wright. **2:** 72, 1853

An erect, perennial herb with 1 to several stems 3–9 dm. tall; upper parts spreadingly villous; leaves opposite, broadly ovate to oblong, 1.5–3.2 cm. wide, 3–7 cm. long, crenate-serrate, acute to obtuse at apex, cordate and sometimes amplexicaul at base, sparsely pubescent on both surfaces; petioles 3–6 mm. long; heads 12–15 mm. high, about 14–16-flowered, erect on slender peduncles 1–2.5 cm. long and more or less racemosely disposed in a subcylindrical inflorescence 1–3.5 dm. long; peduncles spreadingly villous, without glandular hairs; outer involucral bracts ovate, acute to acuminate, 1–1.5 mm. long, innermost lance-linear to linear-subulate, to 9 mm. long, 3–5-nerved and striate, greenish, glabrous to puberulent, scarious margins ciliolate; corollas purplish, about 8 mm. long; achenes 3.5 mm. long, appressed-sericeous; pappus bristles white, equaling corollas, finely barbellate.

Grasslands and brushy hillsides, mainly Upper Sonoran and Transition Zones, mountains of central Arizona to Texas and south into northern Mexico. Barely entering the Sonoran Desert near Magdalena, Sonora. June–Oct.

**5. Brickellia hastata** Benth., Bot. Sulph. 21, 1844

*Bulbostylis ?hastata* Walp., Rep. **4:** 279, 1848.
*Coleosanthus hastatus* Kuntze, Rev. Gen. Pl. **1:** 328, 1891.

Fruticose plant with minutely cinereous-tomentulose stems and branches, 1 m. tall or less; leaves opposite, oblong-lanceolate, hastately lobed at base and coarsely toothed or upper ones entire, 1–6 cm. broad (at base), 3.5–10 cm. long, glabrous or nearly so above, obscurely reticulate, sparsely tomentulose and slightly paler beneath, punctate-glandular; branches terminated by corymbosely branching inflorescences of 7–15 heads, these 12–13-flowered, 8–11 mm. long; involucral bracts lance-ovate to lance-linear, outer much wider and shorter than inner ones, mucronate, 3–4-striate, more or less pubescent, substramineous; corollas 7–7.5 mm. long, slightly expanded above, ovate-deltoid acute lobes granulate-glandular without, about one-eighth as long as glabrous cylindrical tube; style branches narrowly clavate, long exserted; achenes 3–3.5 mm. long, hirsutulous on ribs; pappus bristles about equaling corolla tube.

Arid plains and rocky or sandy slopes, Lower Sonoran Zone, Comondú south to the Cape Region and on Santa Margarita and Magdalena Islands, Baja California. Feb.–April.

**6. Brickellia coulteri** A. Gray, Pl. Wright. **1:** 86, 1852

*Coleosanthus coulteri* Kuntze, Rev. Gen. Pl. **1:** 328, 1891.

Fruticose perennial with oppositely branched, ascendingly spreading, rather brittle stems 3–10 dm. tall, these glandular-puberulent or puberulent with crisped, nonglandular, yellow-gray hairs; leaves opposite, deltoid to rhombic-ovate, hastate, sharply 1–3-toothed, and truncate to subcordate at base, acute at apex, 1–4 cm. long, sparsely puberulent with crisped hairs; heads laxly cymose-paniculate on slender peduncles, each about 17–18-flowered; involucral bracts about 20, narrowly lanceolate to linear, 4–5-striate, acute to acuminate, tips of exterior ones rather lax, apiculate to mucronate, dorsally short-puberulent near tips; corollas 8–8.2 mm. long, yellowish or greenish white tinged with purple, lobes ovate, obtusish, about one-eighth as long as cylindrical tube; style branches barely exserted; achenes 3.5 mm. long, yellow-brown to yellow-gray, hispidulous; pappus bristles silky, nearly equaling corolla tube.

Arid rocky slopes and canyons, Lower and Upper Sonoran Zones, central and southwestern Arizona to central Mexico and to the Cape Region, Baja California. Sept.–May.

B. L. Robinson (Mem. Gray Herb. **1:** 61, 1917) recognized two forms—(*a*) **coulteri** (as *"typica"*), with glandular-puberulent branches and peduncles; and (*b*) **eglandulescens,** with grayish, crispate, nonglandular pubescence on these parts.

**7. Brickellia glabrata** (Rose) B. L. Robinson, Mem. Gray Herb. **1:** 63, 1917

*Brickellia brachiata* Vasey & Rose, Contr. U.S. Nat. Herb. **1:** 83, 1890. Not A. Gray, 1886.
*Brickellia brachiata* var. *glabrata* Rose, op. cit. 132, 1892.

Glabrous, oppositely branched fruticose plant usually less than 1 m. tall, with terete, yellow-gray, slightly shining stems and branches; leaves deltoid-ovate to rhombic-oblong, coarsely 2–4-dentate on each side, obtuse to rounded at base, acute to obtuse at apex, 1–2.5 cm. long, nearly or fully as broad basally, glabrous and somewhat fleshy, often turning black in drying; petioles 4–5 mm. long; heads about 5 mm. in diameter, 12–15 mm. long, disposed in cymose panicles on slender, divergent peduncles 1.5–2.5 cm. long, about 15–16-flowered; involucral bracts 30–32, oblong or lowermost broadly oval, 4–5-striate, rounded to acute and mucronate at apex, glabrous; corollas about 7 mm. long, greenish white; style branches about two to three times as long as corolla lobes, exserted about their own length; achenes 3.5 mm. long, subsericeous; pappus bristles yellowish.

Rocky canyons and hillsides, Lower Sonoran Zone, central Baja California. Oct.–March.

**8. Brickellia diffusa** (Vahl) A. Gray, Pl. Wright. **1:** 86, 1852

*Eupatorium diffusum* Vahl, Symb. Bot. **3:** 94, 1794.
*Eupatorium capillare* Desv., in Hamilt., Prod. Pl. Ind. Occ. 51, 1825.
(For additional synonymy see Mem. Gray Herb. **1:** 25, 1917.)

An erect, paniculately branched, sparsely puberulent to glabrous annual to 2 m. tall; branchlets very slender and ascending; leaves deltoid-ovate to suborbicular, 1.5–4 cm. wide, 3–6 cm. long, very thin, acuminate at apex, cuneate to subtruncate at base, serrate to crenate-dentate, bearing minute punctate glands beneath; petioles slender, 1–4 cm. long; peduncles filiform, ascending, 6–18 mm. long; heads 6–8-flowered, nearly cylindrical, about 8 mm. long; involucral bracts lance-subulate, outer 1–2 mm. long, inner 6–8 mm. long, 1–3-nerved, subscarious on margins, greenish; corollas greenish white, 4–4.3 mm. long; achenes 1.7–2 mm. long, striate, yellowish gray and minutely strigulose with very fine, silky, short hairs; pappus bristles 20–25, delicately capillary, silvery white.

On brushy hillsides and in canyons, Lower and Upper Sonoran Zones to Tropical Zones, southern Sonora, where it barely enters the edge of the desert, to Chihuahua and south to tropical South America. Apparently flowering whenever moisture conditions permit growth.

**9. Brickellia baccharidea** A. Gray, Pl. Wright. **1:** 87, 1852

*Coleosanthus baccharideus* Kuntze, Rev. Gen. Pl. **1:** 328, 1891.

Fruticose plant with slender, flexuous, gray-brown stems and thin exfoliating bark;

branches ascending, leafy, slender branchlets puberulent; leaves mostly alternate, rhombic-ovate, coriaceous, coarsely laciniate-dentate except at broadly cuneate, entire base, 1–4 cm. long, about one-half to two-thirds as wide, green, subglabrous, strongly reticulate and often vernicose; heads 16–17-flowered, on short peduncles one-half as long as to equaling heads; involucral bracts 5–9 mm. long, brownish stramineous, 3–4-striate, narrowly oblong or lower ones nearly ovate, all rounded or obtuse, firm, often resinous but subglabrous; achenes 2–2.3 mm. long, strongly ribbed, gray-villosulous; pappus bristles white, about 20, scaberulous.

On rocky hillsides and slopes, often on limestone, upper margins of the Lower Sonoran Zone into the lower parts of the Transition Zone, southeastern Arizona to western Texas and south into adjacent Sonora. Barely entering the Lower Sonoran Zone near Tucson, Arizona, and near Baviácora, Sonora. Oct.–May.

**10. Brickellia desertorum** Coville, Proc. Biol. Soc. Wash. **7:** 68, 1892

*Coleosanthus desertorum* Coville, Contr. U.S. Nat. Herb. **4:** 119, 1893.
*Brickellia californica* var. *desertorum* Parish ex Hall, Univ. Calif. Pub. Bot. **3:** 33, 1907.

Intricately branched, fruticose plant to 1 m. tall, with light gray, fibrous bark on main stem; branches slenderly flexuous, sparsely puberulent, occasionally indurated and subspinose; leaves opposite or alternate, ovate, 3–13 mm. long and wide, irregularly crenate-serrate, densely cinereous-tomentulose; petioles 1–3 mm. long; heads subsessile in small glomerules in upper axils, each head 8–12-flowered, about 2.5 mm. in diameter, 8–9 mm. long; involucral bracts about 21, greenish stramineous or brown-stramineous, 3–4-nerved, all acute or obtusish, but not mucronate, slightly puberulent dorsally; corollas whitish, 4.5–5 mm. long; achenes 2.3 mm. long, villosulous; pappus bristles 12–15, capillary, white and smooth.

Desert hillsides, canyons and outwash slopes, Lower Sonoran Zone, western margin of Colorado Desert, California, to southern Nevada and western Arizona. Sept.–March.

**11. Brickellia californica** (Torr. & Gray) A. Gray, Pl. Fendl. 64, 1849

(For synonymy and segregation of varieties see Mem. Gray Herb. **1:** 68–71, 1917.)

Fruticose plant 6–10 dm. tall with terete, crisply puberulent stems and ascending branches at first whitish, later purplish, brownish or grayish; leaves numerous, usually alternate, deltoid-ovate, dentate or crenate-serrate, usually obtuse, base subtruncate or broadly cordate, 3-veined from base, puberulent with crisped hairs, concolorous, 1–5 cm. long and wide; petioles 4–20 mm. long; heads numerous, small, solitary in axils or often aggregated into cymes, secund racemes or short spikes, each head 8–18-flowered, about 3 mm. in diameter, 12–14 mm. long; receptacle glabrous; involucral bracts narrowly oblong, subacute to rounded at tips, 3–6-striate, greenish stramineous or purple-tinged, outermost ones oval, rounded at tips; corollas 7–7.5 mm. long, greenish white often tinged with purple; achenes about 3 mm. long; pappus bristles white or slightly purplish brown, scaberulous.

Rocky hillsides, mesas, canyons, arroyos and outwash fans, Lower Sonoran to Transition Zones, California and Idaho to Colorado, Texas, Chihuahua, Sonora and Baja California. July–Oct.

**12. Brickellia floribunda** A. Gray, Pl. Wright. **2:** 73, 1853

*Coleosanthus floribundus* Kuntze, Rev. Gen. Pl. **1:** 328, 1891.
*Brickellia floribunda* var. *amblyolepis* Gray ex Pringle, Pl. Mex. 2, 1885.

Suffrutescent puberulent plants 1–1.5 m. tall with upper parts of branches glandular-viscid; stems terete, pale green or gray-brown, younger parts often spotted with purple; leaves alternate, deltoid-ovate or rhomboid-ovate, acute, irregularly few-toothed, strongly 3-veined, 3–9 cm. wide, 4–13 cm. long, more or less glandular-puberulent on both surfaces; heads in alternately branched, ovoid, many-headed panicles 2–4 dm. in diameter; heads about 15-flowered, 4–4.5 mm. in diameter, 9–10 mm. long; involucral bracts about 22, oblong, rounded and slightly apiculate at apex, 3–4-striate, inner ones lance-linear, glabrous, stramineous, outer ones ovate-lanceolate, acute, subherbaceous, glandular-hirtellous, often one or two more or less elongated and nearly as long as inner; corollas pale ochroleucous or greenish white; achenes blackish, 2 mm. long, villous with grayish nonglandular hairs; pappus bristles about 35, white, finely barbellate.

Mainly in rich soil in canyon bottoms, Lower and Upper Sonoran Zones, southern Arizona to New Mexico, Chihuahua and Sonora. Sept.–Oct.

**13. Brickellia atractyloides** A. Gray, Proc. Amer. Acad. **8**: 290, 1870

*Coleosanthus atractylodes* Kuntze, Rev. Gen. Pl. **1**: 328, 1891.
*Coleosanthus venulosus* A. Nels., Bot. Gaz. **37**: 262, 1904.

Much-branched, fruticose perennial with rather stiff, flexuous, grayish branches shedding their bark in longitudinal shreds; young branchlets ascending, quite leafy; leaves usually alternate, ovate-acuminate and subsessile, stiff-coriaceous, sharply few-toothed, green and reticulate-veined on both surfaces; heads few to many, when numerous disposed in loose corymbs, each 10–13 mm. in diameter, 13–16 mm. long, about 50-flowered; involucre campanulate, with about 30 bracts, inner and median ones subequal, outermost ovate or lanceolate-oblong, acuminate, finely many-nerved; receptacle glabrous; corollas white or yellowish, 7 mm. long; achenes 4 mm. long, dark gray, hispidulous on angles; pappus bristles about 20, white, nearly smooth.

Rocky hillsides, outwash fans and canyons, Lower Sonoran Zone, southern Nevada and Utah to Yuma County, Arizona, and eastern California. March–Sept.

**14. Brickellia arguta** B. L. Robinson, Mem. Gray Herb. **1**: 102, 1917, var. **arguta**

*Coleosanthus atractyloides* Coville, Contr. U.S. Nat. Herb. **4**: 119, 1893. Not Kuntze, 1891.
*Brickellia atractyloides* Hall, Univ. Calif. Pub. Bot. **3**: 29, 1907. Not A. Gray, 1870.

Stiffly much-branched, fruticose perennial, branchlets and peduncles pubescent with spreading hairs, hairs somewhat enlarged at base, often gland-tipped; leaves opposite or alternate, ovate-acuminate, subsessile, sharply dentate-serrate, veins pubescent, intervein areas more or less covered with globose sessile glands; involucral bracts 4–5-striate, outer ovate to ovate-lanceolate, attenuate, sometimes part of them bearing 1 or 2 lateral teeth, inner ones lance-linear, short-acuminate; flowers and achenes with characters of *B. atractyloides.*

Desert mountain slopes and washes, Lower Sonoran Zone, western Mojave and Colorado Deserts north into Inyo County, California, and south into northern Baja California. March–May.

**14*a*. Brickellia arguta** var. **odontolepis** B. L. Robinson, Mem. Gray Herb. **1**: 103, 1917

Differing from var. *arguta* in having more conspicuous lateral teeth on a larger number of outer involucral bracts.

Colorado Desert, California, and south into the San Felipe Desert, northeastern Baja California.

**15. Brickellia brandegei** B. L. Robinson, Mem. Gray Herb. **1:** 106, 1917

Divaricately and oppositely branched, densely glandular-puberulent, fruticose perennial with brownish terete stems 1–2 m. long; leaves opposite, broadly ovate, rather finely crenate-dentate, 12–23 mm. wide, 14–25 mm. long, rounded at apex, broadly cordate, strongly 3-veined from base, subcoriaceous, short-pubescent with grayish hairs; petioles 3–4 mm. long; heads in terminal, oppositely spreading, few-headed panicles, each head about 6 mm. in diameter, 9–10 mm. long, about 28-flowered; peduncles slender, 12–22 mm. long, sharply curved downward in bud, straight in flower and fruit; involucres campanulate, bracts 4–5-striate, inner ones narrowly oblong, outer ones ovate to ovate-oblong, acute at tips; corolla yellowish, about 7 mm. long, narrowly tubular, achenes about 3 mm. long, hirsutulous with dusty gray, slightly spreading hairs; pappus bristles silky, white.

Along washes, among other shrubs and on hillsides, Lower Sonoran Zone, Sierra de la Giganta to the Cape Region, Baja California. Sept.–Jan.

**16. Brickellia peninsularis** Brandegee, Zoe **5:** 160, 1903

Fruticose plant with scabrous, terete, whitish glandular-pubescent stems to 2.5 m. tall; leaves opposite, ovate to ovate-oblong, obtuse at apex, 2–4.5 cm. wide, 3–8 cm. long, crenate-serrate, truncate to cordate at base, green and glandular-puberulent above, slightly paler and pubescent to tomentose beneath; petioles slender, 1–2.5 cm. long; heads about 30-flowered, solitary or corymbose-paniculate on slender peduncles 3–12 cm. long, each head 8–9 mm. in diameter, 13–14 mm. long, broadest near base; involucral bracts about 24, inner lance-linear ones acute and subglabrous, outer gradually shorter, 4–7-striate, oblong-lanceolate, acuminate, glandular-puberulent, tips usually squarrose-recurved; corollas 7–8 mm. long, narrowly tubular; achenes 4.5 mm. long, strigose-sericeous; pappus bristles about 50, white, minutely barbellate.

Rocky and gravelly hillsides and washes, Lower Sonoran Zone, Cape Region, Baja California and on Socorro Island. Oct.–April.

**17. Brickellia macromera** B. L. Robinson, Mem. Gray Herb. **1:** 107, 1917

Fruticose perennial with erect stems 6 dm. tall or more, and widely spreading branches densely hirtellous with crisped hairs; internodes to 13 cm. long, bark longitudinally fissured; leaves opposite or subopposite, reniform-ovate, cordate, rounded at apex, 2–4 cm. wide, 1.5–3 cm. long, crenate-serrate, slightly gray-pubescent on both surfaces, minutely punctate-glandular beneath; petioles 2–5 mm. long; heads few, in subumbellate cymes at ends of branches, each about 45–50-flowered and 1.5 cm. long; involucral bracts slender, 4–6-striate, outermost ovate-oblong, acute, green, intermediate and inner ones oblong, submembranous, rose-tinged, rounded at tips; corollas slenderly tubular, white, 8–9 mm. long; achenes about 3.6 mm. long, reddish brown but obscured by grayish subappressed hairs; pappus bristles about 30, delicately capillary-filiform, smooth or essentially so.

Rocky canyons, hillsides, and bottoms of washes, Lower Sonoran Zone, eastern side of Baja California in the vicinity of Bahía de la Concepcíon, south into the Sierra de la Giganta and near Comondú. Oct.–April.

**18. Brickellia vollmeri** Wiggins, Contr. Dudley Herb. **3:** 286, *pl. 65, figs. 6–17,* 1943

Rounded shrub 6–10 dm. tall with gray striate bark; branchlets stramineous, glandular-puberulent; leaves alternate, oblong-spatulate to oblanceolate, 3–6 mm. wide,

10–25 mm. long, irregularly serrate to entire, cuneate at base, acute to obtuse at apex, 3-veined, glandular-scaberulous, fascicles of 2–5 small entire leaves often in axils of longer ones; petioles 2–5 mm. long, glandular-puberulent; heads loosely cymose, 2–6 on a flowering branch, about 50-flowered, narrowly campanulate, 7–10 mm. wide, 13–16 mm. high, on peduncles 1–4 cm. long; involucral bracts in 5–7 series, outer ones greenish, herbaceous, strongly 3-nerved, densely glandular-puberulent, acute, entire; middle and inner bracts linear, 6–10 mm. long, 1–1.5 mm. wide, chartaceous, whitish with 3 conspicuous green nerves, erosulate to emarginate at apex, sparsely puberulent on backs; corollas slender, 7–9 mm. long, glabrous, yellowish tinged with purple, lobes narrowly deltoid, acute; achenes columnar, 4.5–5 mm. long, 10–15-ribbed, sparsely scaberulous toward apex; pappus of 30–40 slender, smooth, white to sordid bristles 6–8 mm. long.

On rocky hillsides, upper edge of the Lower Sonoran Zone, along the eastern escarpment of the Sierra San Pedro Mártir adjoining the San Felipe Desert. March–May.

**19. Brickellia rhomboidea** Greene, Pittonia **2:** 103, 1890

*Brickellia floribunda* var. S. Wats., Proc. Amer. Acad. **24:** 54, 1889.
*Coleosanthus rhomboideus* Greene, Erythea **1:** 54, 1893.

Fruticose plant with whitish or pale brown, flexuous stems and branches and puberulent herbage; leaves opposite; petioles 6–15 mm. long, slender; leaf blades broadly rhomboid-ovate to subdeltoid-ovate, 2.5–5.5 cm. wide, 3–5 cm. long, coarsely undulate-dentate or entire, rounded, subtruncate or faintly cordate at base, acute to short-acuminate at apex; heads in subumbellate panicles at apices of branches, each about 10 mm. long and 45–50-flowered; peduncles filiform, flexuous, 1–3 cm. long, glandular-puberulent; involucres campanulate-cylindrical, with about 35 bracts in several imbricate series, 4–5-striate, inner ones linear-oblong with roundish tips, outer gradually shorter, acute, dorsally puberulent; corollas white but yellowish toward limb, about 7 mm. long, lobes faintly glandular exteriorly; achenes about 3 mm. long, finely villous; pappus bristles slender, about 30, white.

Rocky hillsides, Lower Sonoran Zone, vicinity of Guaymas, Sonora, and on the islands in the harbor. July–Oct. (?).

**20. Brickellia incana** A. Gray, Proc. Amer. Acad. **7:** 350, 1868

*Coleosanthus incanus* Kuntze, Rev. Gen. Pl. **1:** 328, 1891.

Greatly branched, incanus-tomentose perennial from a woody base; stems and branches erect or ascending, terete, white-tomentose when young, later glabrate and bark shredding longitudinally; leaves alternate, sessile, ovate to orbicular-ovate, coarsely serrate on lower two-thirds or uppermost entire, 1–2 cm. long, at first tomentulose but eventually glabrous or essentially so; heads terminal on branches, solitary, about 2 cm. in diameter, 2.5 cm. long, about 60-flowered; involucres turbinate-campanulate, bracts ovate-oblong, imbricate in several series, acute at apex and dorsally white-tomentose, 3–5-nerved; corollas slenderly tubular, about 14 mm. long, lobes 0.8–1 mm. long, deltoid-ovate; achenes sericeous, about 1 cm. long; pappus bristles 40–45, white.

Sandy washes and canyon bottoms, Lower Sonoran Zone, Colorado and Mojave Deserts to southern Nevada and western Arizona. May–Oct.

## Tribe 3. **ASTEREAE.** Aster Tribe

Annual or perennial herbs or small to moderate-sized shrubs. Herbage usually scentless or essentially so but occasionally resinous or gummy. Leaves alternate. Heads small to medium-sized, radiate or discoid, bracts usually well imbricated. Receptacle naked, alveolate; alveolae sometimes breaking into scales or palea-like segments. Rays present or absent, yellow, white, pink, lavender, or purple. Disk flowers perfect in all except *Baccharis* and *Archibaccharis*. Anthers not caudate. Style branches usually ovate to lanceolate, sometimes truncate. Achenes angled, striate, or ribbed, sometimes slightly compressed. Pappus of bristles, awns, paleae, or squamellae or sometimes reduced to a minute crown or none.

Plants not dioecious, heads containing at least some perfect flowers, these either fertile or sterile:
  Heads radiate, ligulate flowers pistillate or neuter:
    Rays yellow:
      Ray flowers about equaling or slightly surpassing disk corollas, inconspicuous, sometimes lacking:
        Pappus none ............................................. 8. *Gymnosperma*
        Pappus present, at least on disk achenes:
          Ray achenes devoid of pappus; ray flowers 12–19; ligules about equaling tubes ... ............................................. 9. *Xanthocephalum*
          Ray achenes with pappus of about 10 paleae about one-half as long as those of disk achenes; ray flowers 1–11; ligules usually shorter than their tubes ..... ............................................. 10. *Gutierrezia*
      Ray flowers exceeding disk corollas, usually moderately showy to conspicuous:
        Involucral bracts strongly chartaceous or coriaceous, their margins erose-fimbriate; heads (in ours) spherical ............................ 11. *Acamptopappus*
        Involucral bracts mostly herbaceous, at least at tips, their margins not fimbriate; heads not spherical:
          Ray achenes epappose; leaves 5–10 cm. long:
            Leaves lanceolate, tapering toward each end; pappus of outer disk flowers less developed than that of inner ones; ligules not longer than tubes of ray flowers ....................................... 9. *Xanthocephalum*
            Leaves ovate-lanceolate to oblong, mostly truncate to cordate at base; pappus of all disk achenes alike; ligules considerably longer than tubes of ray flowers ............................................ 12. *Heterotheca*
          Ray achenes pappose; leaves smaller, rarely over 4 cm. long, usually less than 3 cm. long:
            Pappus in 2 series, outer one of short paleae or squamellae, inner of capillary bristles; achenes somewhat compressed ................. 13. *Chrysopsis*
            Pappus not in 2 distinct series, of capillary bristles only; achenes 4–5-angled or striate but not compressed .......................... 14. *Haplopappus*
    Rays white, pink, lavender or purplish, not yellow:
      Pappus reduced to a short crown or of squamellae and awns less than 1 mm. long:
        Appendages of style branches broadly ovate, obtuse, much shorter than stigmatic portion; ray flowers 30–40 ............................ 17. *Aphanostephus*
        Appendages of style branches linear, acute, three to four times as long as stigmatic portion; ray flowers 2–20 ................................ 18. *Greenella*
      Pappus of 1 to many bristles, these usually nearly or quite equaling achenes:
        Ray achenes epappose or with fewer bristles than disk achenes:
          Involucre hemispherical, bracts graduate and well imbricated, glandular along midrib; tips of style branches acute ........................ 19. *Psilactis*

Involucre campanulate, bracts equal, densely pubescent but not gland-dotted; tips of style branches obtuse or truncate ....................... 20. *Monoptilon*
Ray achenes bearing as many pappus bristles as disk achenes:
Rays mostly oblong, usually 2 mm. wide or more; involucral bracts strongly graduate; style appendages usually lanceolate:
Plants perennial, with rhizomes or with fibrous roots; if annuals, leaves entire or essentially so .......................................... 21. *Aster*
Plants annual, biennial or short-lived perennials with a taproot; leaves spinulose-tipped and entire, or spinulose-dentate to pinnatifid with spinose tips, or dissected, segments spinose-tipped ............... 22. *Machaeranthera*
Rays mostly linear to filiform, rarely 2 mm. wide; involucral bracts subequal, usually not strongly graduate; style branches usually obtuse, sometimes lanceolate and acute .......................................... 23. *Erigeron*
Heads discoid; ray flowers lacking:
Involucral bracts erose-fimbriate ............................. 11. *Acamptopappus*
Involucral bracts not erose-fimbriate:
Leaves and young branches conspicuously punctate-resinous:
Involucral bracts not in distinct vertical ranks ................. 14. *Haplopappus*
Involucral bracts in distinct vertical ranks .................. 15. *Chrysothamnus*
Leaves and young branches not conspicuously punctate-resinous:
Plants shrubby, woody nearly to tips of branches:
Leaves linear, 1.5–4 cm. long; achenes 4–5 mm. long; bracts in 5 vertical ranks .. ............................................ 15. *Chrysothamnus*
Leaves scalelike, 1–2 mm. long; achenes 2.5–3 mm. long; bracts not in vertical ranks ................................................ 21. *Aster*
Plants herbaceous, woody only at very base or not at all:
Marginal florets perfect; corollas slightly irregular but 5-toothed .. 16. *Lessingia*
Marginal florets pistillate, 2–3-denticulate but not ligulate .......... 24. *Conyza*
Plants dioecious or polygamodioecious:
Plants dioecious; heads strictly pistillate or staminate ..................... 25. *Baccharis*
Plants polygamodioecious; pistillate heads with 1–15 hermaphroditic sterile flowers in center, rest of florets pistillate .............................. 26. *Archibaccharis*

## 8. GYMNOSPERMA Less., Syn. Comp. 194, 1832

Suffrutescent to shrubby plants with alternate, glabrous, glandular-glutinous, linear to oblong-lanceolate, entire leaves and corymbose-paniculate cymes of small, inconspicuously radiate heads. Involucres narrowly campanulate, often slightly constricted at apex, graduate. Involucral bracts chartaceous, yellowish but with slightly greenish tips, obovate to lance-linear, obtuse or innermost sometimes acute. Disk corollas very slender at base, abruptly ampliate above into a 5-lobed limb several times as broad as tube. Ray flowers narrow, inconspicuous, pistillate. Achenes oblong, glabrous to sparsely puberulent, 4–5-ridged or angled. Pappus none.

### 1. Gymnosperma corymbosum DC., Prod. 5: 312, 1836

*Selloa glutinosa* Spreng., Nov. Prov. Hal. 36, 1819.
*Gymnosperma multiflorum* DC., Prod. 5: 312, 1836.
*Gymnosperma scoparium* DC., loc. cit.

A profusely branched, leafy shrub to 12 dm. tall with essentially glabrous, or more or less glutinous-viscid branchlets and foliage; leaves alternate, sessile, linear to narrowly lance-elliptic, acute to attenuate at both ends, entire, densely punctate-glandular on both surfaces, 2–6 mm. wide, 2–8 cm. long, 3-nerved but laterals inconspicuous;

heads small, cymose-paniculate, mostly 4–5 mm. long, radiate; involucres ovoid to narrowly campanulate, often slightly constricted at apex, bracts graduate and imbricated, yellowish, chartaceous, acute to obtuse or rounded, margins hyaline-scarious; disk corollas 4–8, yellow, about 2 mm. long, very slender below, abruptly ampliate into a limb about 0.7 mm. wide; ray corollas only slightly longer than those of disk, inconspicuous, 0.2–0.4 mm. wide, yellow; achenes oblong, about 1 mm. long, sparsely and minutely puberulent to glabrous, 4–5-ribbed; pappus none.

Rocky hillsides and canyons, Lower and Upper Sonoran Zones, central Arizona to Texas and south to Central America. March–Dec.

## 9. XANTHOCEPHALUM Willd., Gesel. Nat. Fr. Berl. Mag. 1: 140, 1807

Annual or perennial herbs or sometimes suffruticose, erect or ascendingly branched plants. Leaves alternate, narrow, entire or dentate. Heads small to medium in size, radiate, solitary at tips of branches or loosely corymbose. Involucres hemispherical to broadly campanulate, bracts imbricate in several series, graduate, appressed to disk or tips somewhat spreading. Receptacle flat, foveolate, naked. Ray flowers pistillate; ligules yellow, spreading but small and inconspicuous. Disk flowers regular; throat narrowly ampliate, 5-lobed. Anthers entire at base. Style branches flattened; appendages triangular to lanceolate. Achenes compressed, smooth or striate. Pappus of several more or less united awns or reduced to a mere crown on disk achenes, lacking on ray achenes.

**1. Xanthocephalum gymnospermoides** (A. Gray) Benth. & Hook. ex Rothr., in Wheeler's Rep. **6:** 140, 1878

*Gutierrezia ?gymnospermoides* A. Gray, Pl. Wright. **2:** 79, 1853.

Stout herbaceous annual to 1.5 m. tall, glutinous and sometimes sparingly tomentose in youth, very leafy to apex; leaves lanceolate to oblong-lanceolate, 5–20 mm. wide, 5–10 cm. long, tapering toward each end, entire or sometimes sparingly denticulate, obscurely pinnately veined; heads numerous in corymbose cymes, radiate; pedicels stipitate-glandular; involucres hemispherical, 6–8 mm. high, bracts linear to narrowly oblong, subequal, greenish and loose to slightly spreading at tips; ray flowers 12–19, ligules yellow, 4–5 mm. long, about equaling their tubes; disk flowers 40–60, golden yellow; achenes oblong, compressed, those of rays glabrous, devoid of pappus; disk achenes sparingly strigillose, with a very short, toothed, coroniform pappus on outer ones, inner usually with 4–8 longer, stiffish, setiform paleae in addition to crown.

Along arroyos and about saline swales, Lower and Upper Sonoran Zones, Pima, Santa Cruz, Pinal, and Cochise Counties, Arizona, into New Mexico and northern Sonora and Chihuahua. Aug.–Oct.

**Xanthocephalum wrightii** (A. Gray) A. Gray, Proc. Amer. Acad. **8:** 632, 1873 (*Gutierrezia wrightii* A. Gray, Pl. Wright. **2:** 78, 1853), has linear leaves, only 2–4 mm. wide; heads solitary at tips of branches; rays 30–50; achenes glabrous; pappus a short undulate crown.

This species occurs well above the Lower Sonoran Zone in Arizona but approaches the margins of the desert in the vicinity of Alamos, Sonora. It may occur as a straggler along streams in the southern parts of our region.

## 10. GUTIERREZIA Lag., Gen. & Sp. Pl. 30, 1816

Herbaceous to suffruticose, much-branched, low plants with resinous, more or less glutinous, nearly glabrous herbage. Leaves alternate, linear-filiform to oblanceolate, punctate-dotted, 1-nerved or rarely 3-nerved, entire. Heads small, turbinate-campanulate, radiate or discoid in rather many-headed terminal cymes or heads solitary at tips of branches. Bracts graduate in few series, coriaceous, yellowish and often green at tips. Ray flowers yellow, 1–12 or wanting, when present about equaling or slightly surpassing disk, inconspicuous. Disk flowers yellow, minute. Achenes angled or striate, turbinate to obconic, scarious-paleaceous squamellae equaling or shorter than achenes; paleae of ray flowers about one-half as long as those of disk flowers.

Heads containing only 2–3 flowers; involucres narrowly cylindrical, about 1.5 mm. in diameter; achenes of disk flowers abortive .......................... 1. *G. microcephala*
Heads containing 4–15 flowers; involucres campanulate or turbinate, 1.5–6 mm. in diameter; achenes of both disk and ray flowers fertile:
 Involucres turbinate; involucral bracts narrow, several times as long as wide, margins not membranous; achenes mostly 1.8–2 mm. long:
  Heads clustered at ends of branches; flowers 4–5; plants to 90 cm. tall .. 2. *G. sarothrae*
  Heads solitary at ends of branches; flowers 10–15; plants low, rarely over 50 cm. tall ... ........................................................ 3. *S. bracteata*
 Involucres hemispheric to campanulate; involucral bracts as broad as long, margins membranous; achenes 1.3–1.5 mm. long ............................ 4. *G. serotina*

### 1. Gutierrezia microsephala (DC.) A. Gray, Mem. Amer. Acad. II, 4: 74, 1849

*Brachyris microcephala* DC., Prod. **5**: 313, 1836.
(For synonymy see Contr. Gray Herb. **188**: 57, 1960.)

A globose, ascendingly branched shrub 2–10 dm. tall with yellowish green, slightly fistulous, glabrous branches about 1 mm. in diameter from a woody base 1–2 cm. in diameter; cauline leaves narrowly lanceolate, 2–4 mm. wide, 2–5 cm. long, resinous-glutinous, with fascicles of smaller, narrower leaves in their axils, all leaves entire; stems corymbosely branched above and bearing numerous narrowly turbinate heads 1–1.5 mm. in diameter, 3–4 mm. long, at tips of slender peduncles 1–8 mm. long; involucral bracts 5–8, lanceolate, with moderately greenish-herbaceous tips; ligulate flowers usually 1–2, rarely to 4, 3–4 mm. long; disk flowers 1–3, sparsely glandular-punctate, 2–3 mm. long; achenes of tubular flowers abortive, of ligulate flowers 2 mm. long, silky-strigulose; pappus of 6–10 squamellae 2–3 mm. long on abortive achenes, about one-half as long in fertile achenes.

Hillsides and rocky ridges, mostly Upper Sonoran Zone, southeastern California from Inyo to Riverside Counties, to southern Utah, central Colorado and western Texas, south in Mexico to Coahuila, Chihuahua and Sonora. July–Oct.

### 2. Gutierrezia sarothrae (Pursh) Britt. & Rusby, Trans. N.Y. Acad. 7: 10, 1887

*Solidago sarothrae* Pursh, Fl. Amer. Sept. 540, 1814.
(For additional synonymy see Contr. Gray Herb. **188**: 52–53, 1960.)

Bushy suffrutescent plant with numerous slender stems 3–9 dm. tall from a woody base; herbage scaberulous and resinous-punctate; leaves of two kinds, primary ones 2–7 cm. long, 1–3 mm. wide, ascending-spreading, acute; secondary ones in fascicles in axils of primary leaves and branchlets, 1–2 mm. wide, to 2 cm. long, glabrous to tomentose, resinous-punctate; heads corymbosely arranged, numerous, usually in fascicles of

3–10 at tips of branchlets, sometimes solitary, each head oblong-turbinate, 3–10 mm. long, 1.5–5 mm. in diameter; bracts ovate to oblong, narrow, acute, slightly green-herbaceous at tips, otherwise yellow and subcoriaceous, margins thin; ligulate flowers 4–5(7), 2–5 mm. long; tubular flowers 4–5, these 2–3 mm. long; achenes terete, about 1.8–2 mm. long, silky-strigulose; pappus of 8–10 scarious squamellae about equal to achenes in tubular flowers, one-half to two-thirds as long in ligulate flowers.

Arid hillsides and plains, mainly Upper Sonoran Zone but entering higher parts of the Lower Sonoran Zone, coast ranges in southern California, to Texas, Baja California, Sonora, and Chihuahua, and east of the Sierra Nevada to Manitoba, Minnesota and Kansas. May–Oct.

**3. Gutierrezia bracteata** Abrams, Bull. Torrey Club **34:** 265, 1907

*Gutierrezia californica* var. *bracteata* Hall, Univ. Calif. Pub. Bot. **3:** 36, 1907.

A small suffrutescent, paniculately branched, shrub 2–5 dm. tall, with slender, flexible, greenish branches from a woody root crown; leaves filiform, 1–2 mm. wide, 3–5 cm. long, short-pubescent to glabrous, much smaller leaves in axils of some branches; heads in open-paniculate inflorescences, usually solitary but occasionally 2 at tips of branchlets; peduncles 0.5–5 cm. long; involucres turbinate to conical, 5–6.5 mm. long, 2–3.5 mm. in diameter; involucral bracts narrowly lanceolate, 1–2 mm. wide, to 4 mm. long, carinate or strongly convex, swollen and green at tip, tightly appressed-imbricated; ligulate flowers usually 5 (3–6), 6–8 mm. long; tubular flowers usually 4(2–6), 4–6 mm. long, achenes cylindrical, 1.8–2 mm. long, strigulose; pappus scales 10–12, about equaling achenes in tubular flowers, 0.5–1.5 mm. long in ligulate flowers.

Hillsides, mesas, and outwash slopes, Lower Sonoran and lower margins of Upper Sonoran Zones, inner Coast Ranges from Yolo County, California, south to central Baja California. Feb.–Oct.

**4. Gutierrezia serotina** Greene, Pittonia **4:** 57, 1899

*Gutierrezia polyantha* A. Nels., Amer. Jour. Bot. **25:** 17, 1938.

Low hemispheric shrub 1–5 dm. tall, with green to gray branches, nearly glabrous, young branches striate, leafy to tips; leaves singly arranged along branches or in fascicles on lower parts, linear, to 1 mm. in diameter, 1–2 cm. long, glabrous; heads mostly solitary, in corymbose inflorescences, sessile or short-pedunculate; involucres hemispheric to turbinate-campanulate, 2–3 mm. in diameter, 2–4 mm. long; involucral bracts broadly ovate to orbicular, in 2–3 loose series, glabrous, margins membranous, slightly thickened and greenish or brownish at tip; ligulate flowers usually 7 (4–8), 4–5 mm. long; tubular flowers usually 8 (5–10), salverform, 3–4 mm. long; achenes 1.3–1.5 mm. long, pubescent with hairs in parallel rows; pappus scales 10–12, 1–2 mm. long in tubular flowers, 0.5–1.5 mm. long in ligulate flowers.

Plains and lower mountain slopes, Lower Sonoran Zone, central Arizona, into northern Sonora and northwestern Chihuahua. March–Oct.

## 11. ACAMPTOPAPPUS A. Gray, Proc. Amer. Acad. 8: 634, 1873

Low, rounded shrubs with numerous, erect or ascending, brittle branchlets, grayish bark and small, alternate, entire leaves. Heads radiate or discoid, solitary or subcorymbosely disposed, spherical to hemispherical, 12- to many-flowered. Involucral bracts broadly oblong, obtuse, imbricated in 3–5 series, coriaceous to chartaceous, margins erose-fimbriate and scarious. Receptacle slightly convex, alveolate and slightly fimbriolate through breakdown of alveolae. Flowers yellowish, cylindro-funnelform, 5-lobed. Style appendages subulate, thickish. Achenes turbinate, 5-nerved beneath dense sericeous-villous covering of hairs. Pappus of 30–40 setiform to narrowly paleaceous, silvery awns and bristles.

**1. Acamptopappus sphaerocephalus** (Harv. & Gray) A. Gray, Proc. Amer. Acad. **8:** 634, 1873

*Haplopappus sphaerocephalus* Harv. & Gray, Mem. Amer. Acad. II, **4:** 76, 1849.

A low, much-branching shrub to 5 dm. tall with glistening whitish bark on young branches, gray-shreddy bark on older ones; branchlets pale green, striate, glabrous; leaves linear to linear-spatulate, 1–3 mm. wide, 5–18 mm. long, acute at apex, gradually narrowing toward base, glabrous or with minutely hispid-ciliolate margins, 1-nerved; heads depressed-globose, 7–12 mm. in diameter, 6–10 mm. high (all from the Colorado Desert examined, discoid); involucral bracts broadly obovate to broadly oblong, 2–3 mm. wide, coriaceous except for margins, these chartaceous and conspicuously laciniate- or fimbriate-erose, glabrous, a crescent-shaped greenish spot following curve of the distal end just below apex; corollas yellowish, 4–5 mm. long, funnelform, limb about 3 times as wide as tube, lance-ovate lobes about 1 mm. long; achenes 3–3.5 mm. long, densely hirsute with stiffish, ascending hairs; pappus paleae narrowly subulate to linear-spatulate, some of them slightly dilated and turned abruptly inward at apex, slightly exceeding corolla, often breaking off just above base to leave a narrow fringed crown attached to achene.

Rocky washes and arid hillsides, Lower Sonoran Zone, western Colorado Desert to the Mojave Desert and east to northern Arizona and southern Utah. March–June.

## 12. HETEROTHECA Cass., Bull. Philom. 137, 1817

Erect, leafy herbs with alternate, usually toothed leaves and medium to large heads corymbosely to paniculately arranged. Heads radiate, many-flowered, both disk and ray flowers fertile. Involucres hemispherical to broadly campanulate, of numerous narrow, rather rigid, imbricated bracts in several series. Receptacle plane, alveolate. Ray flowers pistillate. Disk flowers perfect, slightly and gradually ampliate into a 5-lobed limb. Style appendages narrowly lanceolate or rarely short-triangular. Achenes triangular-compressed, obtuse at apex, silky-hirsute, at least in disk flowers. Pappus of disk flowers in 2 series, outer of numerous short, stout bristles or setalike squamellae, inner of long, scaberulous-capillary, rusty to brownish bristles. Pappus lacking on ray achenes.

Ray achenes glabrous; style branches lanceolate, two-thirds as long as stigmatic portion; involucre canescent; upper leaves cordate-clasping .................. 1. *H. subaxillaris*

Ray achenes hirsute at least on angles; style appendages narrowly ovate, not over one-half as long as stigmatic portion; involucre glandular, not canescent; upper leaves sessile but not cordate-clasping ............................................ 2. *H. grandiflora*

**1. Heterotheca subaxillaris** (Lam.) Britt. & Rusby, Trans. N.Y. Acad. **7:** 10, 1887

*Inula subaxillaris* Lam., Encycl. **3:** 259, 1789.
*Inula scabra* Pursh, Fl. Amer. Sept. **2:** 531, 1814.
*Heterotheca lamarckii* Cass., Dict. Sci. Nat. **21:** 131, 1821.
*Chrysopsis scabra* Ell., Bot. S.C. & Ga. **2:** 339, 1824.
*Heterotheca scabra* DC., Prod. **5:** 317, 1836.

Erect annual to biennial herb 4–15 (20) dm. tall with striate and short-hirsute stems, moderately branched above; leaves ovate-lanceolate to somewhat lyrate in outline, 0.8–3.5 cm. wide, 2.5–10 cm. long, at least upper cordate-clasping, saliently serrate-dentate, acute to obtuse and apiculate at apex, short hirsute-scabrous on both sides; heads corymbosely or broadly paniculately arranged, 6–8 mm. high, 10–18 mm. wide at anthesis; involucral bracts in several series, lance-linear to subulate, rather rigid, outer ones acute and apiculate, innermost attenuate and bearing a scraggly brush of spreading and ascending hairs on terminal part, pale greenish to stramineous below, tips often brownish or reddish; ray flowers 30–50, ligules about 5 mm. long, 1 mm. wide, yellowish but soon turning brownish; disk flowers very slender, corollas 4–5 mm. long, gradually and slightly ampliate upward; achenes of ray flowers about 3 mm. long, glabrous, epappose; achenes of disk flowers about as long as ray achenes but more slender, densely silky-villous; pappus bristles reddish-brown, longer inner ones about 15 mm. long.

Along streams, ditches, and fence-rows and in waste ground, Lower Sonoran to Upper Carolinian Zones, central Arizona to Oklahoma, Florida, and Delaware, south into central Mexico; sparingly introduced in southern California. Entering the Sonoran Desert near Tucson, and in Sonora. Aug.–Nov.

**2. Heterotheca grandiflora** Nutt., Trans. Amer. Phil. Soc. II, **7:** 315, 1841

*Heterotheca floribunda* Benth., Bot. Sulph. 24, 1844.

Erect biennial weedy plants 1–2 m. tall, with 1 to several stems from root-crown, each corymbosely or paniculately branching at summit, conspicuously striate and moderately clothed with spreading, stiffish, white hairs 1–2 mm. long and more finely glandular-puberulent between bases of long ones; leaves ovate to elliptic or oblong, 1–3.5 cm. wide, 4–8 cm. long, coarsely serrate, villous-hispid and somewhat glandular, lower ones on petioles 1–4 cm. long, obtuse to acute at each end, upper ones sessile, often with stipule-like auricular lobes at base but not cordate; inflorescence densely glandular, ill-smelling; heads several to many, 1–2 cm. broad; involucral bracts linear, acute to slightly attenuate, rather closely glandular-puberulent on greenish rounded back; ray flowers yellow, ligules 1–1.5 mm. wide, 4–6 mm. long; disk corollas slenderly tubular-funnelform, gradually ampliate upward, about 4 mm. long; style appendages narrowly ovate, one-half as long as stigmatic part or less; ray achenes 3–3.5 mm. long, 1.5 mm. wide, angles strongly rounded-carinate and minutely puberulent, interangle faces usually glabrous; disk achenes strongly flattened, 3.5–4 mm. long, hispidulous; inner pappus bristles reddish, about 5 mm. long.

A common weed along ditches, roadsides and in waste places, Lower and Upper Sonoran Zones, southern California to central Arizona and northern Baja California, north to Monterey and Butte Counties, California. Not common in the Sonoran Desert. Aug.–Oct.

## 13. CHRYSOPSIS Nutt., Gen. 2: 150, 1818

Perennial or rarely annual herbs with villous to glabrous stems and herbage and alternate, sessile, entire to few-toothed leaves. Heads small to moderate size, solitary at tips of branches or laxly corymbose in arrangement, radiate. Ray flowers pistillate, yellow. Disk flowers perfect, usually fertile. Involucre campanulate to hemispheric, narrow bracts in several series, imbricated. Receptacle alveolate. Disk corollas tubular, slightly ampliate near apex, 5-toothed. Style branches slender; appendages linear to subulate. Achenes oblong-linear to obovoid, more or less compressed. Pappus in 2 series, inner of numerous elongated, slightly unequal, capillary, faintly barbellate bristles, outer ones much shorter, narrowly paleaceous.

Herbage canescent; involucres densely puberulent, long white hairs obscuring few glands or these wanting; leaves oblong ........................... 1. *C. villosa* var. *echioides*
Herbage green; involucres sparsely hispid and closely glandular; leaves linear-oblanceolate .. ............................................................... 2. *C. hispida*

**1. Chrysopsis villosa** Nutt. var. **echioides** (Benth.) A. Gray, Syn. Fl. N. Amer. **1**[2]**:** 123, 1884

*Chrysopsis echioides* Benth., Bot. Sulph. 25, 1844.

Erect or ascending plants 2–10 dm. tall with canescent-villous stems and leaves; hairs slender, spreading, 1–2.5 mm. long, more or less tangled or appressed on some parts; leaves oblong to lance-oblong, 2.5–6 mm. wide, 1–3 cm. long, or upper ones somewhat smaller, sessile or lowermost petiolate, acute-mucronate at apex, densely villous on both surfaces, margins entire; inflorescences corymbose or paniculate; heads 8–10 mm. high, 1–2 cm. wide; involucral bracts lance-linear to linear-subulate, attenuate at apex, strigose-canescent, white hairs obscuring scattered glands; ligules of ray flowers 1–1.5 mm. wide, 4–7 mm. long; disk corollas 5–6 mm. long, scarcely 1 mm. in diameter at throat; ray and disk achenes alike, compressed, about 3 mm. long, densely strigose; pappus bristles tawny to rufous, about 7 mm. long.

Arid grassy slopes and in openings in chaparral, mainly Upper Sonoran Zone, extending sparingly into the Lower Sonoran and Arid Transition Zones, central California to northern Baja California, east into Arizona. May–Oct.

**2. Chrysopsis hispida** (Hook.) DC., Prod. **7:** 279, 1838

*Diplopappus hispidus* Hook., Fl. Bor. Amer. **2:** 22, 1834.

Differing from previous species mainly in having more scattered, spreading, stiffish hairs with numerous short-glandular hairs closely crowded between them; herbage greenish instead of canescent; leaves linear-oblanceolate, 1–3 mm. wide, 5–20 mm. long, villous to hispid; peduncles glandular-puberulent and more or less leafy-bracted; involucral bracts 6–7 mm. high, minutely but densely glandular; outer row of pappus squamellae whitish.

Rocky slopes and washes, mostly Upper Sonoran Zone, but barely entering the margin of the Sonoran Desert near Whitewater in Riverside County, California. April–Oct.

The taxonomy of this genus is badly in need of revision. The characters that have been used to separate species and varieties appear to be unreliable and the true identity of many of the populations represented in herbaria is in doubt.

## 14. HAPLOPAPPUS Cass., Dict. Sci. Nat. 56: 168, 1828*

Annual and perennial herbs and shrubs, often glandular or resinous. Leaves alternate, entire, toothed, lobed or rarely dissected. Heads small to large, solitary or cymosely, corymbosely, or paniculately disposed. Involucres obconic to hemispheric with numerous, nearly equal to very unequal, herbaceous to chartaceous or coriaceous bracts, loosely or closely imbricated in several series, their tips usually greenish. Receptacle

---

* The key to the species is partially adapted from Hall, "The Genus Haplopappus; a Phylogenetic Study in the Compositae." Carnegie Institution of Washington, Publication No. 389. 1928.

alveolate, walls of these sometimes breaking down to form bractlike scales. Ray flowers 1–100 or rarely absent; ligules usually yellow, sometimes purplish. Disk flowers several to numerous, perfect, usually fertile; corolla subcylindric to funnelform, 5-toothed or 5-lobed. Anthers oblong, obtuse. Style branches exserted, ovate to subulate; appendages either ovate and shorter than stigmatic portion, linear and equaling, to exceeding stigmatic portion, or unreceptive part not distinguishable from receptive part. Achenes subcylindric, subprismatic, turbinate or fusiform, 4–5-angled or striate or smooth, glabrous to pubescent. Pappus of numerous capillary, unequal, scabrous bristles in 1 to several series.

Involucres broadly campanulate to hemispheric:
  Achenes 2–3 mm. long, turbinate; teeth or lobes of leaves bristle-tipped; pappus bristles sordid to brown (Sect. *Blepharadon*):
    Plants annual, stems herbaceous; involucral bracts eglandular ........... 1. *H. gracilis*
    Plants perennial, base of stems woody; involucral bracts more or less glandular:
      Stems nearly leafless, upper leaves reduced and remote; ligules 5–6 mm. long; leaves glabrous to cinereous and glabrate; achenes 2.5–3 mm. long ..... 2. *H. junceus*
      Stems leafy at least to peduncles; ligules 6–10 mm. long; leaves cinereous, scabrous or tomentulose, rarely glabrous; achenes 2–2.5 mm. long:
        Bracts strongly squarrose above middle; leaves thickish, obtusely dentate to subentire ............................................. 3. *H. arenarius*
        Bracts not squarrose; leaves thinner, spinulose-dentate to pinnatifid:
          Peduncles obvious, nearly naked:
            Herbage markedly scabrous or cinereous and glandular; involucres to 8–12 mm. wide; lobes or teeth of leaves oblong or broader ...................... .................................... 4. *H. spinulosus* subsp. *scabrellus*
            Herbage glandular, broadly scabrous or cinereous; involucres to 20 mm. wide; lobes or teeth of leaves narrowly linear .............................. ................................ 4*a*. *H. spinulosus* subsp. *gooddingii*
          Peduncles short, leafy or bracted almost or quite to heads:
            Larger leaves conspicuously incised; herbage densely glandular ............ ................................ 4*b*. *H. spinulosus* subsp. *incisifolius*
            Larger leaves toothed or pinnatifid, not incised; herbage cinereous and obscurely glandular .................... 4*c*. *H. spinulosus* subsp. *australis*
  Achenes 4 mm. long; leaves entire, not bristle-tipped; pappus bristles silky-silvery, soft, not sordid nor brown (Sect. *Stenotopsis*) ......... 5. *H. linearifolius* subsp. *interior*
Involucres turbinate, subcylindric or narrowly campanulate:
  Disk corollas abruptly ampliate at throat; style appendages ovate, less than one-half as long as stigmatic part (Sect. *Isocoma*):
    Involucres 5–8 mm. high; corolla lobes of disk flowers ovate, 0.5–1 mm. long; outer bracts greenish or brownish, but not dark-tipped nor subterminally resin-dotted:
      Bracts greenish, without a resin pocket near tip; stems brownish, not white and shining:
        Leaves entire, 1–4 mm. wide; involucres 6–8 mm. high ........................ ......................................... 6. *H. venetus* subsp. *furfuraceus*
        Leaves 3–4-dentate at tips, 5–10 mm. wide (including lobes); involucres 5–6 mm. high .................................. 6*a*. *H. venetus* subsp. *tridentatus*
      Bracts with conspicuous, wartlike resin pockets near tips; stems white and shining:
        Leaves entire; flowers 6–13; involucres 5–6.5 mm. high ....................... ....................................... 7. *H. acradenius* subsp. *acradenius*
        Leaves denticulate to dentate; flowers 15–25; involucres 6–8 mm. high .......... .................................... 7*a*. *H. acradenius* subsp. *eremophilus*

Involucres 4–5.5 mm. high; corolla lobes of disk flowers lanceolate to linear-lanceolate, 1–2 mm. long; outer bracts with conspicuous dark or obscurely thickened, resin-dotted tips:
Leaves pinnately parted into linear lobes 2–10 mm. long, middle rachis 1–1.5 mm. wide; outer bracts conspicuously dark-tipped ................ 8. *H. tenuisectus*
Leaves entire or shallowly dentate, main blade 1.5–5 mm. wide, not lobed; outer bracts obscurely resin-dotted but not conspicuously darkened at tip ........... ................................................ 9. *H. heterophyllus*
Disk corollas gradually ampliate above; style appendages oblong to linear or rarely ovate, one-half as long as to exceeding stigmatic part (Sect. *Ericameria*):
Heads in panicles, racemose panicles, or solitary at tips of branches:
Ray flowers present:
Achenes puberulent only at summit; outer involucral bracts acute; ray flowers 5–10; style appendages 1.5–2 times as long as stigmatic part ........ 10. *H. pinifolius*
Achenes densely villous throughout; outer involucral bracts more or less obtuse; ray flowers 1–6; style appendages about equaling stigmatic part .............. ...................................... 11. *H. palmeri* subsp. *pachylepis*
Ray flowers lacking:
Flowers 10–12; heads few, large; involucre 4–6 mm. broad, of 16–22 bracts; receptacle naked ......................................... 12. *H. propinquus*
Flowers 4–6; heads numerous, small; involucres 2–2.5 mm. broad, of 8–12 bracts; receptacle chaffy ...................................... 13. *H. sonoriensis*
Heads arranged in regular cymes:
Leaves filiform or linear to oblanceolate; pappus about equaling corolla:
Involucral bracts 12–20; style appendages linear-filiform; disk flowers 9–16; pappus brownish ............................................ 14. *H. laricifolius*
Involucral bracts 10–12; style appendages ovate; disk flowers 5–8; pappus dull white ................................................ 15. *H. cooperi*
Leaves obovate or spatulate; pappus distinctly shorter than corolla:
Achenes 2–2.5 mm. long; leaves entire, to 2 cm. long; style appendages apparent, slightly shorter than stigmatic part .......... 16. *H. cuneatus* var. *spathulatus*
Achenes 1–1.5 mm. long; leaves dentate with spinulose-tipped teeth, 1 cm. long or less; style branches not visibly differentiated into appendages and stigmatic part ............................................ 17. *H. vernicosus*

## 1. **Haplopappus gracilis** (Nutt.) A. Gray, Mem. Amer. Acad. II, **4:** 76, 1849

*Dieteria gracilis* Nutt., Jour. Acad. Phila. II, **1:** 177, 1848.
*Aster dieteria* Kuntze, Rev. Gen. Pl. **1:** 315, 1891.
*Eriocarpum gracile* Greene, Erythea **2:** 109, 1894.
*Sideranthus gracilis* A. Nels., Bot. Gaz. **37:** 266, 1904.

Erect, strict to moderately branched annual 0.5–3 dm. tall, branching from base or throughout; stems sparsely striate, strigose; leaves linear to lance-spatulate, 1–3 mm. wide, 1–3 cm. long, serrate-dentate to pinnately or bipinnately divided, each tooth or lobe tipped with a slender, white bristle, white-strigose, midrib prominent, lateral veins obscured; uppermost leaves reduced, mostly entire but with marginal bristles; heads solitary or cymosely arranged; involucres subhemispherical, 6–8 mm. high, 8–10 mm. broad, numerous bracts regularly imbricated in 5–6 series, linear-lanceolate, attenuate into a whitish awn, herbaceous in middle, bright green toward apex to just below bristle, cinereous to strigose, margins white-scarious; receptacle bearing numerous linear white scales about 1 mm. long; ray flowers 18–28, ligules 7–8 mm. long, yellow; disk flowers 4.5–5.5 mm. long, tube abruptly expanding into throat; achenes about 2.5 mm. long, canescent; pappus bristles whitish to brownish, unequal, largest nearly equaling corolla and slightly dilated below.

Rocky hillsides, mesas, and gravelly plains, Lower and Upper Sonoran Zones, eastern borders of California to Utah and Colorado, south into Texas, central Baja California, Sonora, Chihuahua, Coahuila, and Durango, Mexico. Flowers following rains. April–Oct.

### 2. **Haplopappus junceus** Greene, Bull. Calif. Acad. **1:** 190, 1885

*Eriocarpum junceum* Greene, Erythea **2:** 108, 1894.
*Sideranthus viridis* Rose & Standl., Contr. U.S. Nat. Herb. **16:** 19, *pl. 15*, 1912.
*Sideranthus junceus* David. & Mox., Fl. So. Calif. 377, 1923.

Perennial herb with slender wiry stems 3–10 dm. tall; branches striate, cinereous to subtomentose on young shoots, soon nearly glabrous or slightly scabrous below, slightly glandular upward; leaves few, scattered, linear and entire to pinnatifid, 1–4 mm. wide, 1–2 cm. long, glabrous to cinereous and glabrate, each tooth or lobe bristle-tipped; upper leaves reduced, 5 mm. long or less, entire, bristle-tipped; heads terminating bracteate branches, hemispheric, 5–8 mm. high, 10–12 mm. broad, bracts numerous, linear, acute to acuminate, bristle-tipped, puberulent and slightly glandular, upper third and midrib green, white toward base; ray flowers 15–25, ligules 5–6 mm. long; disk corollas 5–6.3 mm. long; style branches 1.2–2 mm. long, appendages narrowly triangular, shorter than stigmatic part; achenes 2.5–3 mm. long, villous; pappus pale brown, unequal.

Arid ridges, mesas, mountain slopes, and desert plains, Lower and Upper Sonoran Zones, San Diego County, California, into adjacent northern Baja California and northwestern Sonora. June–Oct.

### 3. **Haplopappus arenarius** Benth., Bot. Sulph. 24, 1844

Compact subshrub 1.5–10 dm. tall with much-branched, villous and glandular stems; leaves oblanceolate-spatulate to obovate-spatulate, 3–12 mm. wide, 1–3.5 cm. long, entire to dentate, densely and coarsely glandular-pubescent, thickish; upper leaves narrower; heads mostly solitary at tips of densely leafy branchlets 2–10 cm. long; involucres broadly hemispheric, 6–8 mm. high, 12–18 mm. broad, numerous bracts imbricated in 5–6 series, linear, acute to acuminate, thick and herbaceous except along subscarious margins, often strongly squarrose above middle, finely and densely stipitate-glandular; ray flowers 30–50 with ligules 6–10 mm. long; disk corollas 5–6.5 mm. long, abruptly expanded into tubular throat, glabrous; style appendages broadly ovate, blunt, much shorter than stigmatic part; achenes cylindric-turbinate, 2–2.3 mm. long, striate, densely appressed-strigose; pappus bristles unequal, longest about equaling corolla, brown and fragile.

On sand dunes and coastal slopes, Lower Sonoran Zone, southern Baja California from La Paz south; possibly on adjacent islands. May–Oct.

### 4. **Haplopappus spinulosus** (Pursh) DC. subsp. **scabrellus** (Greene) Hall, Carnegie Inst. Wash. Pub. No. **389:** 74, 1928

*Eriocarpum scabrellum* Greene, Erythea **2:** 108, 1894.
*Sideranthus viscidus* Woot. & Standl., Contr. U.S. Nat. Herb. **16:** 180, 1913.
*Haplopappus spinulosus* var. *scabrellus* S. F. Blake, Contr. Gray Herb. **52:** 24, 1917.

Perennial herbs 2–6 dm. tall from a woody caudex, stems erect to divergent, leafy except toward apex; herbage gray-scabrous with a slight admixture of glandular hairs particularly toward heads; leaves oblong, 2–6 mm. wide, 1–2.4 cm. long, dentate or entire and bristle-tipped, teeth also bristle-tipped when present; heads terminating long peduncles with remote bracts on upper parts; involucres 6–8 mm. high, 8–12 mm. broad, glandular-scabrid; bracts imbricated in 4–6 series, linear, acute to attenuate, usually bristle-tipped, erect and appressed when young, spreading and reflexed in age, green and glandular to tomentulose in median portion, thin and chartaceous along margins; ray flowers 15–50, ligules 8.5–10 mm. long; disk corollas 4.5–6 mm. long, glabrous or rarely puberulent; style appendages broadly ovate, about one-half as long as stigmatic part; achenes 2–2.5 mm. long, striate under dense appressed pubescence; receptacle alveolate, not chaffy; pappus bristles unequal, one-half as long as to equaling disk corollas, brownish.

Arid plains and mesas, Lower Sonoran and lower, arid parts of Upper Sonoran Zones, central Arizona to southern Colorado, Texas, and Baja California, south into Sonora, Hidalgo, Chihuahua, Coahuila, and Nuevo León. March–Oct.

*4a.* **Haplopappus spinulosus** subsp. **gooddingii** (A. Nels.) Hall, Carnegie Inst. Wash. Pub. No. **389:** 75, 1928

*Sideranthus gooddingii* A. Nels., Bot. Gaz. **37:** 266, 1904.
*Haplopappus gooddingii* Munz & Jhtn., Bull. Torrey Club **49:** 44, 1922.
*Haplopappus spinulosus* var. *gooddingii* S. F. Blake, Contr. U.S. Nat. Herb. **25:** 543, 1925.

Differing from subsp. *scabrellus* in having rather bright green herbage; sparsely to densely stipitate-glandular pubescence (or rarely slightly cinereous); leaves linear with few teeth or narrowly linear short lobes to pinnatifid with slender, divergent lobes; and densely stipitate-glandular to almost scabrid involucral bracts, these 6–7 mm. long.

Arid plains, mesas and hillsides, Lower Sonoran Zone and arid parts of lower edge of Upper Sonoran Zone, southern Nevada to adjacent Arizona, eastern Mojave and Colorado Deserts, California, south to the southern slopes of the Sierra San Pedro Mártir, Baja California. March–June; sometimes flowering also in the autumn.

*4b.* **Haplopappus spinulosus** subsp. **incisifolius** (I. M. Johnston) Hall, Carnegie Inst. Wash. Pub. No. **389:** 75, 1928

*Haplopappus arenarius* var. *incisifolius* I. M. Johnston, Proc. Calif. Acad. IV, **12:** 1190, 1924.

Stems woody, spreading and much branched to form a rounded shrub, leafy nearly to heads; herbage minutely stipitate-glandular or rarely with traces of a fine tomentum; leaves oblong-oblanceolate in outline, 5–10 mm. wide, incisely serrate to bipinnatifid, teeth and lobes bristle-tipped; involucres 6–7 mm. high, 12–20 mm. wide at anthesis, bracts stipitate-glandular, often squarrose in age, greenish along median line and at tips; pappus rusty-brown.

Coastal slopes, sand dunes and beaches, Lower Sonoran Zone, islands along eastern coast of Baja California from South San Lorenzo Island southward. March–July.

A specimen from near Puerto Libertad, Sonora, approaches this subspecies and may belong here.

**Haplopappus arenarius** var. **rossii** I. M. Johnston, Proc. Calif. Acad. IV, **12:** 1191, 1924, is relegated tentatively to this subspecies as a "minor variation" by Hall. It was collected on San Marcos Island and is imperfectly known.

*4c.* **Haplopappus spinulosus** subsp. **australis** (Greene) Hall, Carnegie Inst. Wash. Pub. No. **389:** 77, 1928

*Eriocarpum australe* Greene, Erythea **2:** 108, 1894.
*Sideranthus australis* Rydb., Bull. Torrey Club **27:** 621, 1900.
*Sideranthus puberulus* Rydb., op. cit. 622.

Stems herbaceous; herbage scabrid to cinereous but not tomentulose, very obscurely glandular, glands mostly hidden by short but spreading pubescence; heads terminating peduncle-like branches that are leafy almost or quite to heads; involucres 5–6 mm. high, puberulent and minutely glandular.

Arid ridges, slopes, and plains, Lower and Upper Sonoran Zones, southern Idaho to Sonora, Texas, and Oklahoma. April–Aug.

**5. Haplopappus linearifolius** DC. subsp. **interior** (Coville) Hall, Carnegie Inst. Wash. Pub. No. **389:** 158, 1928

*Haplopappus interior* Coville, Proc. Biol. Soc. Wash. **7:** 65, 1892.
*Stenotus interior* Greene, Erythea **2:** 72, 1894.
*Haplopappus linearifolius* var. *interior* M. E. Jones, Proc. Calif. Acad. II, **5:** 697, 1895.
*Stenotopsis interior* Rydb., Bull. Torrey Club **27:** 617, 1900.
*Stenotus linearifolius interior* Hall, Univ. Calif. Pub. Bot. **3:** 48, 1907.
*Stenotopsis linearifolia* var. *interior* J. F. Macbride, Contr. Gray Herb. **49:** 59, 1917.

A much-branched, leafy shrub 4–15 dm. tall with rounded outline and brittle woody branches; twigs striate, resinous, usually puberulent near tips; leaves linear to linear-oblanceolate, narrowed very gradually toward base, 1–3 mm. wide, 1–2.5 cm. long, glabrous or margins sparsely scaberulous, conspicuously punctate-resinous, thickish, sometimes subterete; heads numerous, solitary but cymosely arranged on naked peduncles 2–8 cm. long; involucres broadly hemispheric, 8–10 mm. high, 10–18 mm. broad, with linear to broadly lanceolate bracts in 2 series, acute to acuminate, middle part slightly herbaceous and scaberulous to stipitate-glandular, margins scarious and glabrous; ray flowers 12–18, ligules 9–21 mm. long, 4–5 mm. wide, yellow; disk corollas tubular, gradually expanded upward, 6.5–10 mm. long; achenes nearly linear, about 4 mm. high, silky-pubescent; pappus bristles copious, silky, silvery-white, equaling or slightly exceeding disk corollas.

Arid interior plains and desert slopes and mesas, Lower Sonoran and lower edges of Upper Sonoran Zones, western margins of the Colorado and Mojave Deserts to southwestern Colorado and northern Baja California. March–June.

**6. Haplopappus venetus** (H. B. K.) S. F. Blake subsp. **furfuraceus** (Greene) Hall, Carnegie Inst. Wash. Pub. No. **389:** 226, 1928

*Bigelovia furfuracea* Greene, Bull. Calif. Acad. **1:** 87, 1885.
*Haplopappus fasciculatus* Vasey & Rose, Proc. U.S. Nat. Mus. **11:** 530, 1888.

A spreadingly much-branched shrub, often somewhat decumbent, 1–4 dm. tall, often 10 dm. broad, with glabrous but viscid-resinous leaves, resinous exudate often forming a distinct scurf on leaves and young stems; leaves linear to linear-spatulate, acute to obtuse, entire or with a few minute subapical teeth, blade 1–4 mm. wide, 1–2.5 or rarely to 4 cm. long, thickish and rather stiff; heads few in close, rounded, terminal, simple to compound cymes, involucres 6–8 mm. high with obtuse but usually mucronate (at least inner) bracts, these resinous-viscid, green along midvein and at tips, with narrow, scarious, erosulate margins; ray flowers none; disk flowers 15–30, corolla ventricose, 5.5–7 mm. long; style appendages ovate, less than one-half as long as stigmatic part; achenes 4-angled, about 2 mm. long, silky-strigose; pappus bristles slightly unequal, pale brown, longest equaling corollas.

Coastal slopes and islands off the west coast of Baja California, Lower and Upper Sonoran Zones, from near Ensenada south to Magdalena Island, Baja California; on

Santa Catalina and San Clemente Islands off the coast of southern California. Feb.–Oct.

**6a. Haplopappus venetus** subsp. **tridentatus** (Greene) Hall, Carnegie Inst. Wash. Pub. No. **389:** 225, 1928

*Bigelovia tridentata* Greene, Bull. Torrey Club **10:** 126, 1883.
*Isocoma tridentata* Greene, Erythea **2:** 111, 1894.
*Haplopappus tridentatus* S. F. Blake, Contr. U.S. Nat. Herb. **23:** 1493, 1926.

Differing from last mainly in having more densely crowded, erect or strongly ascending main stems; leaves mostly with 3–4 prominent lobes or teeth near apices, larger of which are 5–10 mm. wide near tips; heads more numerous in loose leafy cymes; involucres 5–6 mm. high with more or less squarrose, nearly erect bracts when mature, and achenes 2.3–2.6 mm. long.

According to Hall, known only from Cedros Island but included because the area on the mainland of the peninsula immediately adjacent to Cedros Island has been very incompletely explored and because of a specimen in the Dudley Herbarium collected by A. W. Anthony in 1896 and labeled, "Islands off the coast of Lower California and on adjacent mainland."

**7. Haplopappus acradenius** S. F. Blake, in Tidestrom, Contr. U.S. Nat. Herb. **25:** 546, 1925, subsp. **acradenius**

*Bigelovia acradenia* Greene, Bull. Torrey Club **10:** 126, 1883.
*Aster acradenius* Kuntze, Rev. Gen. Pl. **1:** 317, 1891.
*Isocoma acradenia* Greene, Erythea **2:** 111, 1894.
*Isocoma veneta* var. *acradenia* Hall, Univ. Calif. Pub. Bot. **3:** 64, 1907.
*Haplopappus acradenius* subsp. *typicus Hall,* Carnegie Inst. Wash. Pub. No. **389:** 233, 1928.

An erect shrub 3–10 dm. tall with white to stramineous bark and brittle, striate, glabrous or granular herbage; leaves narrowly oblanceolate to spatulate-oblanceolate, 2–3 cm. long, 3–8 mm. broad, subsessile but gradually narrowed to base, entire or nearly so, glabrous or scaberulous; smaller leaves in axillary fascicles; heads small, nearly sessile, in clusters on peduncle-like branchlets about 1 cm. long; involucres 5–6.5 mm. high, bracts imbricated in 3–4 series, oblong, chartaceous at base, acute to mucronate, bearing a conspicuous, subepidermal, resin pocket near tip; ray flowers none; disk corollas 6–13, about 5.5–6.5 mm. long, abruptly ventricose, hirtellous to glabrous; style appendages about one-half as long as stigmatic part; achenes 3–4 mm. long, 5-ribbed, silvery-villous; pappus bristles yellowish, numerous, longest equaling or slightly surpassing corollas.

Desert hillsides and plains, Lower Sonoran Zone, Mojave Desert, California, east into southern Nevada and Maricopa and Pinal Counties, Arizona. Aug.–Oct.

**7a. Haplopappus acradenius** subsp. **eremophilus** (Greene) Hall, Carnegie Inst. Wash. Pub. No. **389:** 233, 1928

*Isocoma eremophila* Greene, Leaflets Bot. Obs. **1:** 171, 1906.

Differing from last mainly in having dentate or denticulate leaves 1.5–5 cm. long, their blades glabrous to scaberulous; heads larger and from sessile to distinctly pedunculate; involucres 6–8 mm. high; disk flowers 15–25 in each head.

Desert slopes and washes, Lower Sonoran Zone, Colorado Desert, California; probably extending into northeastern Baja California. Mainly Oct.–Dec. but sometimes flowering in May and June.

**8. Haplopappus tenuisectus** (Greene) S. F. Blake ex Benson, Amer. Jour. Bot. **27:** 188, 1940

*Linosyris coronopifolia* A. Gray, Pl. Wright. **1:** 96, 1852. Not *Haplopappus coronopifolius* DC., 1836.
*Bigelovia coronopifolia* A. Gray, Proc. Amer. Acad. **8:** 638, 1873.
*Aster coronopifolius* Kuntze, Rev. Gen. Pl. 317, 1891.
*Isocoma coronopifolia* Greene, Erythea **2:** 111, 1894.
*Isocoma tenuisecta* Greene, Leaflets Bot. Obs. **1:** 169, 1906.
*Isocoma fruticosa* Rose & Standl., Contr. U.S. Nat. Herb. **16:** 18, *pl. 13,* 1912.
*Haplopappus fruticosus* S. F. Blake, op. cit. **23:** 1493, 1926.

An erect subshrub with numerous erect to ascending branches forming a compact rounded shrub 3–8 dm. tall; branchlets slender, yellowish, minutely puberulent to glabrate; leaves broadly oblong in outline, 0.5–1.5 cm. wide, 1–4.2 cm. long, pinnately parted into 2–4 pairs of linear acute lobes 1–1.5 mm. wide, to 9 mm. long, central rachis about as wide as or very slightly wider than lobes, cinereous-puberulent and moderately dotted with punctate-resinous glands; heads in compact glomerules cymosely arranged at tips of branches; involucres turbinate-campanulate, 3–3.5 mm. wide, 4–5.5 mm. long at anthesis, bracts narrowly lance-ovate to linear in 3–5 series, stramineous and chartaceous below, greenish, resin-dotted and puberulent medianly near tip, margins scarious and erosulate; ray flowers lacking; disk flowers 6–12 or rarely to 15, about 5 mm. long, narrowly tubular at base, rather abruptly ampliate slightly above middle, lobes linear, 1–1.3 mm. long, glabrous; achenes mostly 2.5–3 mm. long, moderately strigose-silky; pappus bristles pale brownish, longer ones about equaling corollas.

Arid plains, hillsides and mesas, mainly Lower Sonoran Zone, Pima and Cochise Counties, Arizona, and in southwestern Texas south into Sonora and Chihuahua. Aug.–Oct.

**9. Haplopappus heterophyllus** (A. Gray) S. F. Blake, in Tidestrom, Contr. U.S. Nat. Herb. **25:** 546, 1925

*Linosyris heterophylla* A. Gray, Pl. Wright. **1:** 95, 1852.
*Linosyris wrightii* A. Gray, loc. cit.
*Linosyris hirtella* A. Gray, loc. cit.
*Bigelovia wrightii* A. Gray, Proc. Amer. Acad. **8:** 639, 1873.
*Isocoma heterophylla* Greene, Erythea **2:** 111, 1894.
*Isocoma hirtella* Heller, Muhlenbergia **1:** 6, 1900.
*Isocoma wrightii* Rydb., Bull. Torrey Club **33:** 152, 1906.
*Isocoma limitanea* Rose & Standl., Contr. U.S. Nat. Herb. **16:** 18, *pl. 14,* 1912.
*Isocoma oxylepis* Woot. & Standl., op. cit. 180, 1913.
*Haplopappus pluriflorus* Hall, Carnegie Inst. Wash. Pub. No. **389:** 237, 1928, in part.

Suffrutescent to suffruticose plants 3–6 dm. tall with hirtellous to subglabrous branchlets, peduncles, and foliage; leaves linear to linear-oblanceolate, 1.5–5 mm. wide, 2–4(6) cm. long, entire, sparsely short-dentate or lowermost sometimes laciniate-dentate, densely punctate-resinous, usually hispidulous along margins, rather thick and stiff; heads numerous, sessile or short-pedicellate in rounded to flattened terminal cymes; involucres turbinate-campanulate, 4–5 mm. high, 2.5–3.5 mm. in diameter at anthesis, bracts oblong to lance-oblong or lowermost ovate, obtuse to acute, stramineous below, obscurely greenish toward tips, minutely puberulent to glabrate, margins scarious and erosulate; ray flowers none; disk flowers 7–15, about 5–6.5 mm. long, abruptly ampliate above, throat and narrowly lanceolate lobes together about 1.5–1.8

mm. long; achenes about 2 mm. long, silky-strigose; pappus bristles about 4–4.5 mm. long, stramineous.

Along washes, mesas, and on sandy plains, Lower and Upper Sonoran Zones, southwestern Arizona to Colorado, western Texas, Chihuahua, Sonora, and the arid parts of northern Baja California. June–Sept.

**10. Haplopappus pinifolius** A. Gray, Proc. Amer. Acad. **8:** 636, 1873

*Aster pityphyllus* Kuntze, Rev. Gen. Pl. **1:** 316, 1891.
*Chrysoma pinifolia* Greene, Erythea **3:** 12, 1895.
*Ericameria pinifolia* Hall, Univ. Calif. Pub. Bot. **3:** 54, 1907.

A rather stout, open shrub 0.5–3 m. tall with greenish to pale brown bark on young growth, dark gray to nearly black and longitudinally fissured in age; branchlets fastigiately crowded, striate, faintly or not at all punctate-resinous, glabrous or essentially so; leaves linear to subfiliform, 0.5–1 mm. wide, 15–35 mm. long, thickish, 1-nerved, often nearly terete, acute, slightly resinous-punctate, usually with fascicles of very small leaves in axils of main ones; heads solitary in spring, cymosely paniculate to racemosely disposed during autumn flowering season; involucres of vernal heads 8–15 mm. broad, 7–10 mm. high, subtended by bractlike clusters of leaves, those of autumnal heads 5–7 mm. broad and high, less conspicuously leafy-bracted; bracts 20–26, in 3–4 series, lance-ovate to lanceolate, acute, yellowish with a pale brownish midrib and scarious-erosulate margins; ray flowers 5–10 in autumnal heads, 15–30 in vernal ones, ligules 1.5–2 mm. wide, 3.5–5 mm. long; disk corollas 6–7.5 mm. long, puberulent on tube, gradually ampliate; style appendages linear, usually longer than stigmatic portion; achenes subcylindric, 4.5–5 mm. long, striate or ribbed, glabrate to pubescent at summit in age; pappus bristles tawny to reddish, copious, exceeding disk corollas.

Mainly coastal slopes on arid ridges, Upper Sonoran Zone, but reaching the western margins of the Colorado Desert in San Diego County, California, ranging from Los Angeles County south into San Diego County and adjacent northern Baja California. March–June and Sept.–Nov.

**11. Haplopappus palmeri** A. Gray subsp. **pachylepis** Hall, Carnegie Inst. Wash. Pub. No. **389:** 267, 1928

A brushy plant to 1.5 m. tall with a sturdy, woody, main stem and numerous crowded, erect, resinous-punctate branchlets; main leaves about 1 mm. in diameter, 1–3 cm. long, subterete, strongly revolute, margins meeting, glabrous to sparsely puberulent when young, usually with stout, fascicled leaves in axils of main ones; heads many in pyramidal panicles or sometimes racemosely disposed; peduncles 1–20 mm. long, scaly-bracteate; involucres obconic to nearly cylindric, 3.5–4.5 mm. broad, 5–7 mm. high, of 16–24 bracts imbricated in 4–5 series, these oblong-lanceolate or outer shorter, chartaceous, midrib thickened, yellow-green, resinous, borders white-scarious and ciliolate to erosulate near tip; ray flowers 1–6 with ligules 1.2–1.5 mm. wide, 4–6 mm. long; disk flowers 5–15, corollas 5–8 mm. long, gradually ampliate, sparsely puberulent; style appendages slightly shorter than to slightly exceeding stigmatic part; achenes 3–4 mm. long, 5-angled, appressed-villous; pappus bristles brownish, soft, slightly exceeding disk corollas.

Mainly on arid brushy ridges, Upper Sonoran Zone, Los Angeles County to San Diego County, California, but occasionally extending eastward to the margins of the Mojave and Colorado Deserts. March through Dec., but mostly Sept.–Dec.

**12. Haplopappus propinquus** S. F. Blake, Contr. U.S. Nat. Herb. **23:** 1490, 1926

*Bigelovia brachylepis* A. Gray, Bot. Calif. **1:** 614, 1876.
*Aster brachylepis* Kuntze, Rev. Gen. Pl. 317, 1891.
*Chrysoma brachylepis* Greene, Erythea **3:** 12, 1895.
*Ericameria brachylepis* Hall, Univ. Calif. Pub. Bot. **3:** 56, 1907.
*Haplopappus brachylepis* Hall, op. cit. **7:** 273, 1919. Not *H. brachylepis* Phil., 1894.

A much-branched shrub to 2 m. tall with brittle, erect or ascending branches and gray to brownish bark; twigs densely leafy, somewhat striate, glabrous, closely and conspicuously resinous-punctate; leaves linear, nearly terete, acute to obtuse, about 1 mm. in diameter or less, 0.5–2 cm. long, strongly revolute so margins meet to form a thin groove, densely resinous-punctate, many with fascicles of 2 to several smaller leaves 3–6 mm. long in their axils; heads narrowly racemose to paniculately disposed, often numerous; involucres turbinate, 4–6 mm. high, nearly or fully as wide, of 16–22 bracts in 3 or sometimes 4 loosely imbricated series, lowermost ovate, others lance-linear to oblong-lanceolate, acute to obtuse, brownish-resinous along midline, scarious margined, glabrous or essentially so except along ciliolate margins near tips; ray flowers none; disk flowers 10–12, corolla slender, 6–7 mm. long with spreading lanceolate lobes 1 mm. long; style appendages linear, equaling to exceeding stigmatic part; receptacle naked; achenes 3–4 mm. long, densely silky-strigose; pappus bristles soft, white to sordid, about equaling disk corollas.

Brushy hillsides and along washes, chiefly Upper Sonoran Zone, San Diego County, California, south into northern Baja California, entering the Sonoran Desert in eastern San Diego County. April–May and Oct.–Dec.

**13. Haplopappus sonoriensis** (A. Gray) S. F. Blake, Contr. U.S. Nat. Herb. **23:** 1490, 1926

*Ericameria diffusa* Benth., Bot. Sulph. 23, 1844. Not *Haplopappus diffusus* DC., 1836.
*Solidago diffusa* A. Gray, Proc. Amer. Acad. **5:** 159, 1861.
*Linosyris sonoriensis* A. Gray, op. cit. **8:** 291, 1870.
*Bigelovia diffusa* A. Gray, op. cit. 640, 1873.
*Aster sonoriensis* Kuntze, Rev. Gen. Pl. 317, 1891.
*Chrysoma diffusa* Greene, Erythea **3:** 10, 1895.

Diffusely branched shrub to 2.5 m. tall with numerous slender, erect or ascending stems; bark greenish to light brown in youth, turning light gray and longitudinally fissured in age; branchlets very leafy, striate, punctate-resinous but glabrous; leaves filiform, nearly terete, mostly obtuse but with a short terminal mucro, about 0.5 mm. in diameter, 5–30 mm. long, grooved by meeting of strongly revolute margins, closely resinous-punctate; heads in oblong to rounded paniculate clusters, numerous; involucres narrowly turbinate, 2–2.5 mm. wide, 4–4.5 mm. long, of 8–12 lanceolate to oblong-linear bracts loosely imbricated in about 3 series, all bracts greenish and thickened just below apex, margins scarious, white to stramineous; receptacle bearing 1 to several chaffy awns 0.5–1 mm. long; ray flowers none or rarely one; disk flowers 4–6, their corollas 4–5 mm. long, puberulent on tube, lobes narrowly lanceolate, 1.2–1.5 mm. long; style appendages ovate, one-half as long as stigmatic part or less; achenes 1.5–2 mm. long, silky-strigose; pappus bristles sordid to brownish, equaling to slightly exceeding corollas.

Sandy plains, arroyos and coastal dunes, Lower Sonoran Zone, from the vicinity of Kino Bay, Sonora, and Bahía de los Angeles, Baja California, south along the coastal slopes to southern Baja California and Sinaloa. March–May and Aug.–Dec.

**14. Haplopappus laricifolius** A. Gray, Pl. Wright. **2:** 80, 1853

*Aster laricifolius* Kuntze, Rev. Gen. Pl. **1:** 318, 1891.
*Chrysoma laricifolia* Greene, Erythea **3:** 11, 1895.
*Bigelovia nelsonii* Fernald, Proc. Amer. Acad. **36:** 505, 1901.
*Ericameria nelsonii* S. F. Blake, Contr. Gray Herb. **52:** 26, 1917.

Fastigiately branched, rather compact, erect, leafy shrub to 1 m. tall with brittle woody branches; older bark dark gray, younger yellowish; branchlets striate below leaf bases, conspicuously resinous-punctate but glabrous; leaves broadly linear, acute to short-acerose, 1–1.8 mm. wide, 1–2 cm. long, subfleshy, conspicuously impressed-punctate-resinous, rather densely crowded on younger branchlets; heads in irregular, much-bracted cymes; involucres turbinate-campanulate, 3–5 mm. wide, about as high at anthesis, spreading considerably in fruit, bracts imbricated in 4–5 series, linear to lance-linear, acute to slightly acuminate, brownish yellow in median part, white-scarious marginally, glabrous or rarely faintly hirtellous along margins; ray flowers 3–6 or sometimes wanting in part of heads, ligules 1.5–2 mm. wide, 4–5 mm. long, barely exceeding disk corollas; disk flowers 9–16, corollas 5–6 mm. long, gradually ampliate, lobes lanceolate, acute, about 1 mm. long; achenes 3.5–4 mm. long, densely white-strigose; pappus light brown, about equaling disk corollas.

Arid slopes, mesas, and canyons, Lower and Upper Sonoran Zones, eastern Mojave Desert, California; Pima and Mohave Counties, Arizona, to western Texas and south into Sonora and Chihuahua. Aug.–Nov.

**15. Haplopappus cooperi** (A. Gray) Hall, Carnegie Inst. Wash. Pub. No. **389:** 275, 1928

*Bigelovia cooperi* A. Gray, Proc. Amer. Acad. **8:** 640, 1873.
(For additional synonymy see Carnegie Inst. Wash. Pub. No. **389:** 276–78, 1928.)

A rounded, much-branched, leafy shrub to 2 m. tall with brown bark turning gray or nearly black in age, longitudinally fissured and somewhat shreddy; twigs striate and resinous-punctate, minutely puberulent; leaves linear to linear-spatulate, about 1 mm. wide or less, 5–20 mm. long, minutely puberulent and more or less resinous-punctate, acute at apex, larger ones usually with smaller ones fascicled in their axils; heads in rounded terminal cymes on peduncles 3–15 mm. long; involucres narrowly campanulate, 3–4 mm. broad, 4–5 mm. high, of 10–12 loosely imbricated, ovate to oblong bracts in 3–4 series, all thin, more or less puberulent and greenish along mid-nerve, white-scarious along margins; ray flowers 1–2 or wanting in some heads, ligules about 2 mm. wide, 5–6 mm. long when present; disk flowers 5–8, corollas 4–5 mm. long, gradually ampliate, puberulent to subglabrous; corolla lobes widely spreading, 1–1.5 mm. long; style appendages ovate, much shorter than stigmatic portion; achenes 3–3.5 mm. long, obtusely 4-angled, white-villous; pappus bristles dull white, slightly shorter than disk corollas.

Along washes and desert slopes, upper parts of Lower Sonoran Zone and in arid, lower parts of Upper Sonoran Zone, Inyo County, California, to northern Baja California and southern Nevada. April–June.

**16. Haplopappus cuneatus** A. Gray var. **spathulatus** (A. Gray) S. F. Blake, Contr. U.S. Nat. Herb. **23:** 1489, 1926

*Bigelovia spathulata* A. Gray, Proc. Amer. Acad. **11:** 74, 1876.
*Chrysoma cuneata* var. *spathulata* Greene, Erythea **3:** 11, 1895.
*Ericameria cuneata* var. *spathulata* Hall, Univ. Calif. Pub. Bot. **3:** 52, 1907.

A low, rounded or flat-topped shrub 1–10 dm. tall with greenish to brown bark and leafy, faintly striate, glabrous twigs, only faintly or not at all punctate-glandular; leaves oblanceolate-spatulate, entire or faintly undulate, often slightly retuse at summit, 3–8 mm. wide, 5–20 mm. long, 1-nerved, thickish, obscurely if at all resinous-punctate, glabrous; heads solitary or in small, compact, few-headed cymes; peduncles 2–8 mm. long; involucres turbinate, 3–4 mm. broad, 5–7 mm. high, of 20–30 bracts imbricated in 4–5 series, outer bracts lanceolate, acute, inner ones linear-oblong and obtuse, all chartaceous, glabrous and moderately resinous, not carinate, margins scarious; ray flowers none; disk flowers usually 15–20, corollas slenderly tubular, only slightly ampliate above, 5.5–7 mm. long, more or less puberulent on tube, lance-attenuate lobes erect, 0.7–1 mm. long; style appendages only slightly shorter than stigmatic portions; achenes about 2–2.5 mm. long, silky-pubescent; pappus bristles shorter than disk corollas, scanty, brownish.

On rocky hillsides, canyon walls, and rocky outwash slopes, Lower and Upper Sonoran Zones, bordering the desert areas from southern California into northern Baja California. Sept.–Dec.

**17. Haplopappus vernicosus** Brandegee, Proc. Calif. Acad. II, **2:** 168, 1889

A low shrub 3–5 dm. tall with densely crowded, leafy, brittle branches; bark brown on young branches, dark gray on older parts, fissured and slightly shreddy; twigs sharply angled, glabrous, resinous and granular; leaves spreading to ascendingly curved, cuneate-obovate to cuneate-oblanceolate, 4–5 mm. wide, 6–9 mm. long, each with 2–7 broadly triangular, rigidly mucronate-tipped teeth, 1-nerved, closely impressed-punctate glands obscured by copious exudate of resin, often with minute leaves fascicled in axils of main ones; heads in small glomerules (of 2–4 heads) at tips of branches, often equaled by subtending leaves; involucres cylindric-turbinate, 3–4 mm. wide, 5–6 mm. high, of about 20 oblong, obtuse bracts imbricated in 4–5 series, inner ones thin and stramineous, outer ones greenish along midrib and resinous-punctate just below tip, margins ciliolate; ray flowers 3–5, ligules about 1 mm. wide, 2 mm. long; disk flowers 4–7, corollas cylindric, 4.5–5 mm. long, scarcely ampliate above, minutely puberulent; style branches lanceolate, not visibly divided into stigmatic and nonreceptive parts; achenes 1–1.5 mm. long, silky-strigose; pappus bristles sparse, brownish, shorter than corollas.

Arid hillsides and along dry watercourses, Lower Sonoran Zone, between Rosario and San Augustín, Baja California. Poorly known and apparently rare. May.

## 15. CHRYSOTHAMNUS Nutt., Trans. Amer. Phil. Soc. II, 7: 323, 1840

Much-branched shrubs and subshrubs with alternate slender leaves tapering gradually to base, these entire or obscurely serrulate. Heads small, variously disposed, discoid. Involucre cylindric; bracts well imbricated in fairly distinct vertical ranks, coriaceous to chartaceous, midrib prominent. Disk flowers 4–20; corolla tubular-funnelform, gradually expanding into rather narrow throat; limb 5-lobed or 5-cleft. Style appendages lance-attenuate, subulate or filiform, slightly shorter to considerably longer than stigmatic portion, exserted. Achenes slender, terete or faintly flattened or angled, glabrous to pubescent. Pappus of many soft, dull white to reddish, slender bristles.

Branchlets minutely but densely tomentulose; leaves flattish, tomentulose, not dotted with impressed, resinous glands ........................ 1. *C. nauseosus* subsp. *gnaphalodes*

Branchlets closely dotted with impressed resinous glands; leaves terete or nearly so, puberulent to glabrous but resinous-punctate:

Involucral bracts stramineous, without an apical greenish, resinous spot; style appendages to twice as long as stigmatic parts; leaves 1–3 cm. long ........... 2. *C. paniculatus*

Involucral bracts stramineous below but with a conspicuous greenish, resinous, subapical spot; style appendages shorter to only slightly longer than stigmatic parts; leaves 0.5–2 cm. long ............................................ 3. *C. teretifolius*

**1. Chrysothamnus nauseosus** (Pall.) Britt. subsp. **gnaphalodes** (Greene) Hall & Clements, Carnegie Inst. Wash. Pub. No. **326:** 211, 1923

*Chrysothamnus speciosus* var. *gnaphalodes* Greene, Erythea **3:** 110, 1895.
*Chrysothamnus gnaphalodes* Greene, Pittonia **4:** 42, 1899.

Rounded shrub to 2 m. tall with erect or ascending branches, densely tomentose twigs, and leaves devoid of resinous-punctate glands; branchlets yellowish green to gray, obscurely striate, erect; tomentum fine, closely packed; leaves linear, 1 mm. wide or less, 1.5–4 cm. long or upper ones less than 3 cm. long, often recurved, flat or nearly so; heads in rounded terminal cymes; involucres 6–8 mm. high; bracts lance-linear, slightly keeled, minutely tomentulose to glabrate, in 5 distinct vertical rows; corollas 7–8 mm. long, tube sparsely puberulent to glabrous, about 4 mm. long; style appendages slightly shorter than stigmatic part; achenes 4–5 mm. long, densely silky-strigillose; pappus bristles 5–6 mm. long, minutely barbellate, faintly tawny.

Arid slopes and along washes, Lower and Upper Sonoran Zones, margin of Colorado Desert near Banning to Mono County, California, southern Nevada and northern Arizona. June–Nov. Rare in the Sonoran Desert.

**2. Chrysothamnus paniculatus** (A. Gray) Hall, Univ. Calif. Pub. Bot. **3:** 58, 1907

*Linosyris viscidiflora* var. *paniculata* A. Gray, Bot. Mex. Bound. 80, 1859.
*Bigelovia paniculata* A. Gray, Proc. Amer. Acad. **8:** 644, 1873.
*Ericameria paniculata* Rydb., Fl. Rocky Mts. 853, 1917.

A much-branched, rounded shrub 6–20 dm. tall, with numerous, erect, leafy, brittle, resinous and viscid, glabrous, green twigs; resinous-punctate glands usually very conspicuous; leaves nearly terete, mostly 0.5 mm. or less in diameter, 1–3 cm. long, glabrous, closely and conspicuously resinous-punctate, resin sometimes forming a pale, scurfy bloom; heads many in dense terminal panicles; involucres turbinate, 5–6 mm. high, 3–5 mm. wide, of 13–17 obtuse-oblong, glabrous, stramineous bracts without terminal spots, in 5 poorly defined vertical rows; ray flowers none; disk flowers 5–8, tubular-funnelform, 5–6 mm. long, gradually ampliate, lobes erect or nearly so, lance-ovate, 0.8–1 mm. long; style appendages lance-linear, equaling to twice as long as stigmatic part; achenes 3–4 mm. long, villous with appressed to slightly spreading silky hairs; pappus bristles white to faintly tawny, about equaling corollas.

Dry rocky slopes and arroyos, Lower and drier parts of Upper Sonoran Zones, Colorado and Mojave Deserts, California, to southwestern Utah and northern Arizona. Mainly Oct.–Dec., but flowering also in May and June.

**3. Chrysothamnus teretifolius** (Dur. & Hilg.) Hall, Univ. Calif. Pub. Bot. **3:** 57, 1907

*Linosyris teretifolia* Dur. & Hilg., Jour. Acad. Phila. II, **3:** 41, 1855.
*Bigelovia teretifolia* A. Gray, Proc. Amer. Acad. **8:** 644, 1873.

A rounded or flattish, densely leafy shrub to 1 or sometimes 2 m. tall with dark gray, somewhat fissured bark and brittle, erect, glabrous but densely resinous-punctate, striate branchlets; leaves linear-terete, 0.5–2 cm. long, obtuse, densely resinous-punctate, 1 mm. in diameter or less; heads moderately numerous in short terminal spikes; involucres turbinate, 6–8 mm. high; bracts 16–20 in 5 clearly defined vertical ranks, oblong, inner ones abruptly acute, chartaceous and stramineous, faintly spotted at tip, outer ones obtuse, stramineous below, dark brownish and thickened near apex; ray flowers none; disk flowers 5–6, corollas tubular-funnelform, 6–6.5 mm. long, gradually and moderately ampliate above, glabrous, erect lobes about 1 mm. long; style branches ovate to short-lanceolate, shorter than or barely exceeding stigmatic part; achenes 4–5 mm. long, densely strigulose, pappus bristles white to tawny, about equaling corollas.

Rocky and gravelly slopes, higher parts of the Lower Sonoran Zone and arid, lower margins of the Upper Sonoran Zone, Mono and Inyo Counties, California, to adjacent southern Nevada, northwestern Arizona, and southwesterly to the northern margins of the Colorado Desert. Sept.–Dec.

## 16. LESSINGIA Cham., Linnaea **4:** 203, 1829

Annual summer and autumn herbs of erect to depressed habit with tomentose to glabrate, usually glandular herbage and alternate, entire to serrate or deeply lobed leaves, basal ones rosulate and petiolate, upper ones sessile. Heads discoid, small, 3–40-flowered, in paniculate, spiciform or glomerulate inflorescences. Involucres narrowly turbinate to narrowly campanulate; bracts green-tipped, loosely imbricate in several series. Receptacle convex. Corollas 5-lobed, outer slightly irregular and somewhat reflexed, inner tubular to funnelform. Style branches flattened; appendages truncate to subulate, sometimes with a terminal cusp. Achenes turbinate, rounded or compressed, silky-villous. Pappus of unequal bristles, often in 2 series (in ours), united into clusters or distinct, or reduced to 5 paleaceous awns.

Plants erect, to 5 dm. tall, paniculately and openly branched; cauline leaves usually glabrate, usually glandular over whole surface ........... 1. *L. glandulifera* var. *glandulifera*
Plants spreadingly branched, compact, rarely over 1.5 dm. tall; cauline leaves densely tomentose, glandular only on margins ........... 1*a*. *L. glandulifera* var. *tomentosa*

### 1. Lessingia glandulifera A. Gray, Proc. Amer. Acad. 17: 207, 1882, var. glandulifera

*Lessingia germanorum* var. *vallicola* J. T. Howell, Univ. Calif. Pub. Bot. **16:** 19, 1929.
*Lessingia germanorum* var. *tenuipes* J. T. Howell, op. cit. 20.
*Lessingia germanorum* var. *glandulifera* J. T. Howell, op. cit. 22.

Annual 1–7.5 dm. tall with numerous slender branches loosely tomentose in youth, becoming glabrate in age; lowest leaves oblanceolate, petiolate, acute to obtuse, 3–10 mm. wide, 1.5–6 cm. long, pinnately lobed to toothed, tomentulose to tardily and partly glabrate; cauline leaves oblong to oblanceolate, 0.5–6 mm. wide, 1.5–2.5 cm. long, entire to shallowly lobed, heavily and conspicuously glandular marginally, sparsely glandular on upper surface, often without glands beneath; heads solitary at tips of branches, 17–33-flowered; involucre broadly turbinate, 5–8 mm. high; bracts closely imbricate, in about 6 series, elliptic to narrowly linear-oblanceolate, greenish toward tips, conspicuously stipitate-glandular marginally and on upper one-third to one-half; corollas yellow, with a brownish or purplish band in throat, outer corollas palmately cleft; style appendages with a prominent cusp or sometimes subulate; achenes 2–3 mm. long; pappus of 30–40 light brown, distinct bristles 1.5–3 mm. long in 1–2 series.

On gravelly or rocky slopes and grassy hillsides in openings in the brush, mostly Upper Sonoran Zone, but straggling into Lower Sonoran Zone, San Joaquin County, California, to cismontane southern California and northern Baja California, and in the foothills of the Sierra Nevada from Placer County south. May–Nov.

**1*a*. Lessingia glandulifera** var. **tomentosa** (Greene) Ferris, Contr. Dudley Herb. **5:** 101, 1958

*Lessingia tomentosa* Greene, Leaflets Bot. Obs. **2:** 32, 1910.
*Lessingia germanorum* var. *tomentosa* J. T. Howell, Univ. Calif. Pub. Bot. **16:** 24, 1929.

Differing from var. *glandulifera* mainly in being a lower, more compact plant rarely exceeding 1.5 dm. in height; being densely tomentose throughout except on involucres, these 5–6 mm. high with marginal glands; style branches nearly twice as long (2–2.5 mm.); slightly more numerous pappus bristles.

Desert slopes and arroyos, Lower Sonoran Zone, southwestern part of the Colorado Desert, San Diego County, California, and adjacent northern Baja California. May–Nov.

## 17. APHANOSTEPHUS DC., Prod. 5: 310, 1836

Annual (in ours) or perennial herbs with alternate, linear to spatulate or obovate, entire, toothed or pinnatifid leaves. Heads small to medium; solitary at tips of branches, radiate. Ray flowers in 1–2 series, pistillate and fertile. Disk flowers perfect, fertile. Involucres hemispheric, linear bracts imbricate in few series. Ligules white or purple tinged. Disk corollas yellow, slightly ampliate above, 5-toothed. Style branches flattened, appendages broadly ovate, obtuse. Achenes 4–5-angled, about 10-ribbed. Pappus reduced to a short, ciliate-fringed crown.

**1. Aphanostephus humilis** (Benth.) A. Gray, Pl. Wright. **1:** 93, 1852

*Leucopsidium humile* Benth., Pl. Hartw. 18, 1839.
*Aphanostephus arizonicus* A. Gray, Proc. Amer. Acad. **16:** 81, 1880.

Low annual herbs usually with several ascending branches 5–35 cm. long from a rosette of oblanceolate, pinnatifid leaves 6–10 mm. wide, 3–6 cm. long, or these soon dying and leaving base of stems naked, herbage cinereous-puberulous; cauline leaves smaller, entire to pinnatifid, teeth or lobes acute and often faintly apiculate; heads solitary at tips of branches, peduncles 1–6 cm. long, bearing several reduced, bract-like leaves scattered below head; involucres 5–7 mm. high, 8–12 mm. wide, bracts linear, acute to attenuate, greenish and puberulent along midrib, scarious and white to rosaceous along margins; ray flowers 30–40, ligules about 1 mm. wide or less, 4–6 mm. long, white or often tinged with purple; disk corollas very narrowly tubular, about 2 mm. long, yellow; achenes broadly obconic, about 1 mm. long, brownish, minutely and sparsely appressed-puberulent along low, rounded ribs; pappus crown barely discernible, microscopically erosulate-ciliolate.

Plains and along arroyos and riverbanks, mainly Lower Sonoran Zone, south central Arizona to Texas and south into Chihuahua, Coahuila, and Nuevo León, Mexico. March–Oct.

## 18. GREENELLA A. Gray, Proc. Amer. Acad. 16: 81, 1880

Low, slender winter annuals or short-lived perennials with several to numerous erect or ascending branches, glabrous herbage and linear to lanceolate, entire, alternate leaves. Heads small, on slender naked peduncles terminating branches, scattered or slightly corymbosely disposed, radiate or discoid. Involucres broadly campanulate to hemispheric; bracts oblong, coriaceous along midnerve, scarious on margins, obtuse, imbricated in few series. Ligules, when present, white. Disk corollas tubular-funnelform, 5-lobed. Style branches exserted, linear, flattened, appendages linear, about four times as long as stigmatic part. Achenes oblong-turbinate, about 8-nerved, canescent-puberulent. Pappus a small, dissected, hyaline crown much shorter than achene, or of lanceolate to linear squamellae.

Plants annual, simply branched; heads broadly campanulate to hemispheric, 5–7 mm. in diameter, usually broader than long; rays 10–20 .................... 1. *G. arizonica*
Plants perennial, intricately branched; heads cylindrical, 2.5–4 mm. long, 1–1.8 mm. in diameter, distinctly longer than broad; rays 2–5 .................... 2. *G. ramulosa*

## 1. **Greenella arizonica** A. Gray, Proc. Amer. Acad. **16:** 82, 1880

Erect or ascending annual 1–3 dm. tall with 1 to many branches from base; herbage glabrous, pale green and faintly glaucous; branchlets faintly but distinctly ribbed, low ribs nearly white; leaves linear to lance-linear, 0.3–3 mm. broad, 1.5–5 cm. long, sessile, acute and slightly mucronate at apex, minutely scaberulous on margins, peduncles slender, 1–5 cm. long, often with a bractlike leaf 2–5 mm. long just below head; involucres 3–4 mm. high, bracts oblong or lowermost narrowly ovate, 0.5–1 mm. wide, conspicuously green on distal one-fourth to one-third, margins hyaline-membranous, faintly erosulate toward apex; ray flowers about 10–20, ligules white, 1–1.2 mm. wide, 3.5–5 mm. long; disk flowers numerous, corollas greenish yellow, slenderly tubular-funnelform, about 3 mm. long, spreading lobes about 0.2 mm. long; achenes of ray and disk flowers alike, broadly obturbinate, about 1.5 mm. long, rather strongly 4–5-angled, densely canescent-puberulent with very short, spreading hairs; pappus of a minute, dissected, hyaline crown about 0.2 mm. long or less.

Gravelly plains and low hills, chiefly Lower Sonoran Zone, eastern Yuma County to Cochise County, Arizona, south into northern Sonora. March–May.

## 2. **Greenella ramulosa** Greene, Pittonia **1:** 302, 1889

An intricately branched, suffrutescent perennial 1.5–3 dm. tall, often nearly or fully as wide, brittle branchlets rather crowded and compactly interlaced, minutely glandular-viscid, rather markedly angled; leaves linear-elliptic to narrowly oblanceolate, 0.4–3.5 mm. wide, 3–25 mm. long, acute at apex, spinulose-ciliolate, 1-nerved, impressed-punctate, lower ones early deciduous, upper reduced to mere bracts; heads solitary at tips of branches, short-pedunculate, 2.5–4 mm. high, few-rayed; involucral bracts oblong, obtuse, glandular-granular along greenish tip and median part, thin-margined, 2–3-seriate and arranged in distinct vertical rows; ray flowers 2–5 (usually 3), white, changing to rose purple with age, ligules about 3 mm. long; disk flowers about 6, white, about 1.5 mm. long; achenes oblong-turbinate, about 1.5 mm. long, 10-striate; pappus of ray achenes of about 15 lanceolate hyaline squamellae 0.5 mm. long, that of disk achenes of about twice as many linear-lanceolate, paleaceous awns slightly over 1 mm. long.

Arid coastal slopes and plains, Lower Sonoran Zone, western side of Baja California in the vicinity of San Bartólome Bay. March–June.

## 19. PSILACTIS A. Gray, Mem. Amer. Acad. II, 4: 71, 1849*

Slender, moderately branching annuals, biennials, or short-lived perennials with erect or ascending stems. Herbage glabrous to canescent. Leaves alternate, entire to pinnatifid. Heads small, radiate; rays white or purplish, especially beneath; disk flowers yellow, both ray and disk flowers fertile. Involucres hemispheric; bracts narrow, imbricated in few series, graduate, herbaceous at or near tip and usually along glandular-dotted midnerve, otherwise chartaceous or coriaceous, margins of at least inner ones somewhat scarious. Ligules spreading. Disk corollas tubular; limb only slightly ampliate, regularly 5-toothed. Style branches flattened; short appendages acute. Achenes more or less compressed, slightly striate, minutely puberulent. Ray achenes without pappus. Disk achenes with a pappus of capillary bristles about equaling achenes.

Leaves toothed to pinnatifid, teeth and lobes spinose-tipped; stems glandular throughout and usually more or less pilose; plant much branched from base:
  Ray flowers 50–80; ligules blue-purple, 5–6 mm. long; disk achenes 1.2–1.5 mm. long; phyllaries in 2–3 series .......... 1. *P. coulteri*
  Ray flowers 30–40; ligules pink-purple, 6–8 mm. long; disk achenes 2.0–2.4 mm. long; phyllaries in 3–5 series .......... 2. *P. crispa*
Leaves entire or rarely toothed, acute but not spinose-tipped; stems eglandular and often glabrous or sometimes hirsute below inflorescence; plant branching chiefly above .......... 3. *P. asteroides*

### 1. **Psilactis coulteri** A. Gray, Mem. Amer. Acad. II, 4: 72, 1849

Plants 1–5 dm. tall, usually much branched from base, moderately to rather densely stipitate-glandular throughout, often with longer nonglandular hairs at least on lower parts of stems; basal leaves oblanceolate, petiolate, 5–25 mm. wide, 2.5–6 cm. long, coarsely serrate-dentate to pinnatifid, teeth and lobes spinose-tipped; upper leaves lance-elliptic to oblanceolate, 2–8 mm. wide, to about 3 cm. long or less, mostly only saliently serrate-dentate or subpinnatifid; heads terminating branchlets, somewhat corymbosely arranged; involucres hemispheric, 4–6 mm. high, 5–8 mm. wide at anthesis, bracts lance-ovate to lance-linear, green toward tips and outer ones along midrib also, densely glandular; ray flowers 50–80, blue-purple, mostly in 2 series; ligules 5–6 mm. long; disk corollas very slender, about 4.5 mm. long, glabrous; achenes terete-turbinate, finely striate, minutely silky-strigulose, 1.2–1.5 mm. long, truncate at apex, those of ray flowers epappose, disk achenes with a pappus of capillary silky bristles 1.5–2 mm. long.

River bottoms, sandy plains, and roadsides, Lower Sonoran Zone, southern Nevada into Inyo and San Bernardino Counties, California, and south through Arizona into northeastern Sonora. Apparently not in the Colorado Desert. Feb.–Oct.

### 2. **Psilactis crispa** Brandegee, Proc. Calif. Acad. II, 2: 169, 1889

Annual or biennial herb 3–4 dm. tall from a sturdy taproot, with stout, suffruticose, leafy, ascending stems and densely glandular-puberulent herbage; leaves sessile, linear-oblong, 5–20 mm. wide, 2–3 cm. long, or upper ones smaller, irregularly incised-dentate, crispate; heads 1–4 on short lateral shoots of upper stems; peduncles 4–15 mm.

* R. C. Jackson (Amer. Jour. Bot. 49: 676, 1962) has suggested that *Psilactis* may have to be merged with *Machaeranthera*, and gives several arguments for such a realignment. *Psilactis coulteri* A. Gray, is frequently misidentified as *Machaeranthera tanacetifolia* (H. B. K.) Nees.

long, stipitate-glandular; involucres hemispheric, 6–8 mm. tall, 10–15 mm. wide, erect at anthesis, more or less reflexed in age; phyllaries 30–55, imbricate in 3–5 series, lance-linear, 1.5–2 mm. wide, 3–8 mm. long, densely glandular-mealy, greenish toward moderately spreading tips, scarious along margins, white-chartaceous on lower part; ray flowers 30–40, pistillate and fertile, ligules 0.8–1 mm. wide, 6–8 mm. long, pinkish to purplish, tube 3–4 mm. long; disk corollas yellow, 4–6 mm. long, lobes erect, 0.3–0.6 mm. long; disk achenes fusiform, 2–2.4 mm. long, sparsely appressed-pubescent; pappus bristles about 30, in one whorl, 2.8–3.4 mm. long; ray achenes without pappus, slightly shorter than disk achenes.

Along arroyos, on alkaline plains and mesas, Lower Sonoran Zone, central Baja California. April–Oct.

**3. Psilactis asteroides** A. Gray, Mem. Amer. Acad. II, **4:** 72, 1849
*Aster boltoniae* Greene, Pittonia **3:** 248, 1897.
*Psilactis lepta* Shinners, Field & Lab. **17:** 53, 1959.

Erect annual with slender striate stems 3–6 dm. tall and subglabrate or sparsely hirsute herbage almost devoid of glands except in inflorescence; leaves linear to oblong, 1–5 mm. wide, 1–4 cm. long, entire or slightly toothed, not spinulose-tipped at ends of leaves or teeth, often sparsely and finely villous along midvein and margins; heads radiate, solitary at tips of branches on sparsely branched peduncles 1–4 cm. long; involucres hemispheric, 3–5 mm. high, to 8 mm. broad at anthesis; bracts linear, green except on margins and toward base of inner ones, minutely but closely glandular-puberulent; ligules 4–5 mm. long, purplish; disk corollas slenderly tubular-funnelform, about 3 mm. long; achenes broadly fusiform to obturbinate, terete in cross section, 1.5–2 mm. long, striate, minutely silky-strigulose, those of ray flowers epappose, of disk flowers with about 30 silky capillary, slightly rufescent hairs about equaling or slightly exceeding achenes.

River bottoms, about ponds and ditches, and along roadsides, Lower and Upper Sonoran Zones, Pinal and Navajo Counties, Arizona, to Texas, south into Mexico. Uncommon in the Sonoran Desert. June–Oct.

## 20. MONOPTILON Torr. & Gray, Bost. Jour. Nat. Hist. 5: 106, 1845

Low desert annuals with prostrate to decumbent, radiating branches bearing a terminal radiate head at upturned tip of each branch. Herbage hispid and somewhat glandular. Leaves alternate, small, oblong to broadly oblanceolate. Heads radiate. Involucre campanulate, of equal, linear, slightly imbricated bracts. Receptacle flat, naked. Rays rather conspicuous, white or tinged with purple or rose. Disk corollas yellow, narrowly tubular-funnelform. Style branches flattened, slender, tips obtuse or truncate. Achenes oblong-obovate, slightly compressed, nerved along margins, pubescent. Pappus of 1–12 bristles, with or without added paleae.

Pappus a scarious cup and a single plumose bristle about equaling disk corolla; disk corolla puberulent below .......... 1. *M. bellidiforme*
Pappus of several nonplumose bristles, alternating with shorter paleae; disk corolla nearly glabrous .......... 2. *M. bellioides*

**1. Monoptilon bellidiforme** Torr. & Gray, Bost. Jour. Nat. Hist. **5:** 106, *pl. 13,* 1845

Stems spreading or central one erect, 1–8 cm. long; herbage villous-hirsute and often somewhat cinereous; leaves spatulate to linear-spatulate or narrowly obovate,

4–20 mm. long, 0.5–2.5 mm. wide; involucres about 5 mm. high, bracts linear, acute to attenuate, densely villous-hirsute, tips often purplish; rays 20–30 or more, ligule oblong, about 5 mm. long; ray corollas 3.5–4 mm. long, sparsely long-villous with slender jointed hairs on tube; achenes obovoid, about 1.5 mm. long, finely strigulose; pappus a low crown of paleae about 0.2 mm. long or less and a solitary bristle about equaling disk corolla, finely plumose at tip.

On desert plains and gentle slopes, Lower Sonoran Zone, Mojave Desert and northern borders of the Colorado Desert, east into northwestern Arizona and southern Utah. March–May.

**2. Monoptilon bellioides** (A. Gray) Hall, Univ. Calif. Pub. Bot. **3:** 75, 1907

*Eremiastrum bellioides* A. Gray, Mem. Amer. Acad. II, **5:** 321, 1854.
*Eremiastrum orcuttii* S. Wats., Proc. Amer. Acad. **25:** 132, 1890.
*Eremiastrum bellioides orcuttii* Coville, Contr. U.S. Nat. Herb. **4:** 125, 1893.

Very similar to *M. bellidiforme* in habit and general appearance but differing in following characters: tubes of disk corollas glabrous or very sparsely puberulent; pappus of 3–12 oblong to cuneate, more or less setose-pinnatifid paleae and 1–12 slender, nonplumose bristles of slightly unequal length.

In sandy soil along arroyos, desert plains, and on desert mesas, Lower Sonoran Zone, Inyo County, California, to southern Utah and central Arizona, south into northern Sonora and northeastern Baja California. March–May. Much commoner than the last.

## 21. ASTER L., Sp. Pl. 872, 1753

Perennial herbs or rarely annuals, or biennial, suffrutescent or shrubby plants from rhizomes, horizontal woody caudices or taproots with many fibrous roots. Leaves alternate, entire, sometimes toothed. Inflorescence corymbose, racemose, or paniculate; heads small to large; radiate or less commonly discoid; rarely solitary at tips of branches. Involucres campanulate to hemispheric; bracts imbricated in several series, graduate, scarious along margins, usually herbaceous along midrib or at tip or both. Receptacle plane or convex. Ray flowers pistillate, fertile; ligules purplish or white, spreading or rarely erect, often showy, sometimes lacking. Disk flowers perfect and fertile, regular, yellow or less often white or purplish; limb slightly ampliate, 5-toothed. Style branches flattened; appendages short to elongate, triangular to narrowly lanceolate, papillose or hirtellous. Achenes compressed, hairy or glabrous, sometimes ribbed. Pappus of numerous capillary bristles or sometimes with narrow, slightly paleaceous ones, in 2–3 irregular series, subequal or outer shorter than inner.

Plants shrubby or suffrutescent, woody at least at base; leaves not grasslike, 5 cm. long or less:
  Ray flowers none; leaves mostly minute, 1–2 mm. long, lowermost sometimes to 12 mm. long, 2.5 mm. wide .......... 1. *A. intricatus*
  Ray flowers present; leaves 1–5 cm. long, 2–15 mm. wide:
    Disk corollas 7–8 mm. long; rays lilac, 8–10 mm. long; achenes 2–3 mm. long, strigillose .......... 2. *A. frutescens*
    Disk corollas about 3 mm. long; rays white, 2–3 mm. long; achenes 2 mm. long, glabrous .......... 3. *A. spinosus*
Plants herbaceous throughout; leaves linear, grass-like, to 15 cm. long:
  Disk corollas 4–5 mm. long; involucral bracts densely glandular-dotted; ligules 3–4 mm. long; plants perennial .......... 4. *A. pauciflorus*
  Disk corollas 2.8–3 mm. long; involucral bracts not markedly glandular; ligules 1.5–3 mm. long; plants annual .......... 5. *A. exilis*

**1. Aster intricatus** (A. Gray) S. F. Blake, Jour. Wash. Acad. **27:** 378, 1937

*Linosyris carnosa* A. Gray, Pl. Wright. **2:** 80, 1853.
*Aster carnosus* A. Gray, in Hemsl., Biol. Centr. Amer. Bot. **2:** 120, 1881. Not *A. carnosus* Gilib., 1781.
*Bigelovia intricata* A. Gray, Proc. Amer. Acad. **17:** 208, 1882.
*Leucosyris carnosa* Greene, Fl. Fran. 384, 1897.

A very intricately branched, rounded shrub 2–6 dm. tall with reduced, scalelike leaves and glabrous herbage; branches pale green to whitish, smooth, somewhat striate; leaves 1–2 mm. long, or rarely lowermost linear, to 12 mm. long, 1.5–2.5 mm. wide; heads numerous, terminating branchlets, discoid; involucres 4–7 mm. high and wide, turbinate-campanulate; bracts lance-subulate or outermost narrowly ovate, acute to acuminate, midribs slightly greenish, margins thin, minutely erosulate-ciliolate; disk corollas slender, 7–8 mm. long, gradually and narrowly ampliate upward, glabrous; achenes slender, 2.5–3 mm. long, minutely strigose; pappus bristles capillary, 5–6 mm. long, shining white to faintly sordid.

In saline soils, Lower Sonoran Zone, Los Angeles, San Bernardino, and Inyo Counties, California, to Pima and Pinal countries, Arizona, in southern Nevada, and in Mariposa and Madera Counties, California. May–Oct.

**2. Aster frutescens** S. Wats., Proc. Amer. Acad. **24:** 55, 1889

An erect or ascendingly branched shrub about 1 m. tall or less with pale yellowish to whitish bark and glandular-hirtellous younger parts; leaves broadly ovate to broadly oblong, 5–15 mm. wide, 1–2.5 cm. long, repandly spiny-toothed, harshly scaberulous, sessile, prominently veined; heads solitary at tips of branchlets but more or less cymosely arranged, radiate; involucres 7–10 mm. high, 10–15 mm. wide, bracts lanceolate-attenuate, graduate, 0.6–1 mm. wide, moderately glandular-puberulent, whitish or yellowish and chartaceous below, green-herbaceous at least toward tips, tack-shaped glands particularly prominent on upper margins; rays lilac, ligules about 1 cm. long; disk corollas slender, 7–8 mm. long, gradually and narrowly ampliate upward, glabrous; achenes 2–3 mm. long, silky-strigillose; pappus bristles silvery to tawny, 6–7 mm. long.

Rocky hillsides and benches, Lower Sonoran Zone, eastern central Baja California, particularly in the vicinity of Isla Angel de la Guarda and on the adjacent islands. Collected in flower in July on Pond Island.

**3. Aster spinosus** Benth., Pl. Hartw. 20, 1839

*Leucosyris spinosa* Greene, Pittonia **3:** 244, 1897.

A much-branched, glabrous, perennial herb to 3 m. tall with pale green, strongly striate, reedy, nearly leafless stems and usually with stiff, spine-tipped, axillary branchlets 3–25 mm. long; leaves on rapidly growing young shoots linear to spatulate, 1–5 cm. long, flaccid, deciduous, those on upper branches scalelike, 1–5 mm. long, deciduous; heads numerous, solitary at tips of branches; involucres turbinate-campanulate, 5–10 mm. wide at anthesis, bracts ovate-lanceolate to linear, acute to slightly attenuate at apex, glabrous, thin, midrib slightly herbaceous, margins scarious and very minutely erosulate; ray flowers few and inconspicuous, white, ligules about 3 mm. long, drying brown; disk corollas slender, about 3 mm. long, only faintly ampliate upward; achenes slightly flattened, about 2 mm. long, glabrous, faintly 4–5-ribbed; pappus bristles capillary, whitish, 4–5 mm. long.

Along ditch banks, streams, and in low moist soil, Lower Sonoran Zone, eastern

San Diego County, California, to southern Nevada, southern Utah, Texas, and south to Central America. May–Dec.

**4. Aster pauciflorus** Nutt., Gen. **2:** 154, 1818

*Aster thermalis* M. E. Jones, Proc. Calif. Acad. II, **5:** 694, 1895.
*Aster hydrophilus* Greene ex Woot. & Standl., Contr. U.S. Nat. Herb. **16:** 187, 1913.

An erect, moderately branched perennial herb 1.5–8 dm. tall with green slender stems and herbage sparsely to closely dotted with tack-shaped, gland-tipped hairs, eglandular hairs lacking; leaves narrowly linear, grasslike, 2–5 mm. wide, 3–12 cm. long, entire, acute at apex, sessile or lowermost narrowed to a narrowly winged petiole 1–6 cm. long; heads cymosely or paniculately arranged at tips of branches, radiate; involucres 8–12 mm. wide, 6–8 mm. high, bracts lance-linear, herbaceous throughout except along very narrow scarious margin, acute to short-acuminate, densely glandular-dotted; rays rather numerous, pale purple, 0.5–1 mm. wide, ligule about 3–4 mm. long; disk corollas very slender, 4–5 mm. long, gradually and narrowly ampliate upward, pink-tinged at apex; achenes about 2 mm. long, finely and rather sparsely puberulent; pappus bristles capillary, 4–5 mm. long.

Along river banks and valley floors, Sonoran Zones, Pima and Cochise Counties, Arizona, to Saskatchewan and Texas, south into Mexico. May–Sept.

**5. Aster exilis** Ell., Bot. S. C. & Ga. **2:** 344, 1824

An erect, openly branched, glabrous annual to 1.5 m. tall, with dark green, somewhat angular-striate stems; leaves linear, lanceolate, or less commonly narrowly oblanceolate, 2–12 mm. wide, 2.5–15 cm. long, acute, entire, lowermost gradually narrowed to slightly winged petioles 1–5 cm. long; uppermost leaves reduced to linear or subulate bracts 2–7 mm. long; heads in open diffuse panicles, numerous, involucres turbinate-campanulate, 5–6 mm. high, bracts linear, green, with a pale midrib and hyaline, virtually entire margins; rays numerous, light pink to pale purple, ligules 1.5–3 mm. long, about 0.5 mm. wide; disk corollas slender, about 3 mm. long, white except for pink-tinged, minute, triangular lobes; achenes about 1.5–2 mm. long, faintly 4–5-angled and ribbed, sparsely and minutely silky-strigulose; pappus bristles finely capillary, 3–4 mm. long, faintly rusty.

Moist soil along ditches and streams, Sonoran and Tropical Zones, Sacramento Valley, California, south into Baja California, east through Arizona to Arkansas, South Carolina, and Florida, south into tropical America. April–Nov.

## 22. MACHAERANTHERA Nees, Gen. & Sp. Aster 224, 1832

Annual, biennial, or perennial herbs or shrubs from a definite tap root and often with a branching caudex. Leaves alternate, pinnatifid, pinnately lobed, dentate, or entire, but segments, teeth and tips of entire leaves usually spinulose-tipped, sometimes only acute. Heads solitary to numerous, small to large. Involucre with several series of narrow to broad phyllaries; these herbaceous or greenish above middle or near tip, coriaceous or chartaceous toward base, loose, squarrose, or appressed. Receptacle naked but usually distinctly alveolate. Ray flowers well developed, usually pistillate and fertile; ligules bluish, purplish, or rarely white; ray flowers rarely lacking. Disk flowers usually numerous, perfect; anthers not caudate; style branches flattened; stigmatic lines submarginal; appendages acute to obtuse, varying in length. Achenes turbinate to linear-oblong, glabrous to villous, several-nerved or smooth. Pappus of numerous unequal barbellate, white to brownish bristles.

Lower leaves once- or twice-pinnatifid:
  Involucres 3–5 mm. high; green tips of bracts rhombic to rhombic-lanceolate, mostly shorter than chartaceous, whitish part .......................... 1. *M. parviflora*
  Involucres 6–12 mm. high; green tips of bracts subulate to lance-triangular, usually equaling or exceeding chartaceous, whitish part:
    Tips of involucral bracts erect or nearly so, triangular or lance-triangular, slightly if at all narrower than base .......................................... 2. *M. tagetina*
    Tips of involucral bracts spreading, narrowly subulate to linear-subulate, distinctly narrower than base ........................................ 3. *M. tanacetifolia*
Leaves entire, serrate or toothed, not pinnatifid:
  Leaves strongly spiny-toothed; heads (including rays) 3.5–8 cm. broad; involucres 1.5–2 cm. high; plants woody at base:
    Stems and leaves essentially glabrous; pubescence of achenes coarse and long, grading into shorter pappus-bristles .................................... 4. *M. orcuttii*
    Stems and leaves more or less pubescent and glandular; pubescence of achenes much shorter than pappus bristles:
      Leaves and stems puberulent with tack-shaped, glandular hairs and short, nonglandular ones; branches leafy nearly or quite to heads .................. 5. *M. cognata*
      Leaves or stems or both bearing few to many slender, nonglandular hairs in addition to glandular ones, often cinereous to sparsely sublanate in youth; peduncles naked or nearly so, 5–15 cm. long ................................ 6. *M. tortifolia*
  Leaves entire or toothed, but teeth not strongly spinulose-tipped; heads usually less than 3 cm. broad; involucres 5–10 mm. high; plants herbaceous:
    Bracts and peduncles canescent-puberulent, not markedly glandular; achenes 2.5–3 mm. long, minutely puberulent:
      Herbaceous tips of involucral bracts triangular, shorter than chartaceous, light-colored base, appressed; rays few .................................. 7. *M. canescens*
      Herbaceous tips of involucral bracts subulate, equaling or exceeding chartaceous basal part, usually spreading; rays many .......................... 8. *M. tephrodes*
    Bracts and peduncles densely covered with tack-shaped, gland-tipped hairs, non-glandular hairs virtually lacking; achenes 2–2.5 mm. long, villous ..... 9. *M. bigelovii*

## 1. **Machaeranthera parviflora** A. Gray, Pl. Wright. **1:** 90, 1852

*Machaeranthera tanacetifolia* var. *pygmaea* A. Gray, op. cit. **2:** 74, 1853.
*Aster parviflorus* A. Gray, in Brew. & Wats., Bot. Calif. **1:** 322, 1876. Not *A. parviflorus* Nees, 1818.
*Aster tanacetifolius* var. *pygmaeus* A. Gray, Syn. Fl. N. Amer. **1**[2]**:** 206, 1884.
*Machaeranthera pygmaea* Woot. & Standl., Contr. U.S. Nat. Herb. **16:** 189, 1913.

Low, often much-branched annual 1–3 dm. tall with slender branches and sparsely glandular-puberulent herbage; lower leaves once-pinnatifid, 8–25 mm. long, lobes about 0.5 mm. wide, to 3 mm. long; uppermost leaves entire or faintly dentate; heads numerous, solitary at tips of branchlets; involucres 3–5 mm. high, 5–7 mm. broad, bracts well imbricated, their greenish tips rhombic, shorter than chartaceous part, glandular-puberulent, mostly appressed; rays 18–30, purplish or bluish to nearly white, ligules 3–5 mm. long; disk flowers slender, corollas 3–3.5 mm. long, glabrous; achenes slenderly clavate, about 1.5 mm. long, strigulose-villous; pappus bristles capillary, 2.5–3 mm. long.

Arid mesas, arroyos, and plains, Lower and Upper Sonoran Zones, Pima and Maricopa Counties, Arizona, to New Mexico, southern Utah, and northern Sonora. Feb.–Sept.

**2. Machaeranthera tagetina** Greene, Pittonia **4:** 71, 1899
*Machaeranthera tanacetifolia* var. humilis A. Gray, Pl. Wright. **2:** 74, 1853.
*Machaeranthera humilis* Standley, Muhlenbergia **5:** 48, 1909.
*Aster tagetinus* S. F. Blake, in Tidestrom, Contr. U.S. Nat. Herb. **25:** 563, 1925.

Annual, or possibly biennial under favorable conditions, stems slender, moderately branched, 1–4 dm. long, puberulent with both eglandular and gland-tipped hairs, slightly cinereous; lower leaves pinnatifid, 1.5–2.5 cm. long, 3–7 lobes acute, spinulose-tipped, 0.5–3 mm. broad, 2–6 mm. long; upper leaves reduced and denticulate, all hirsutulous and glandular-puberulent; stems leafy to heads, these 10–15 mm. wide, bracts lance-linear, 3–8 mm. long, green herbaceous part triangular to lance-triangular, nearly as wide as chartaceous part, densely glandular; rays 10–20, ligules 2–2.5 mm. wide, 6–7 mm. long, purplish; disk corollas 6–6.5 mm. long, slender, slightly and gradually ampliate, glabrous; achenes about 4 mm. long, silky-villous; pappus bristles slender, 5–6 mm. long, whitish.

Arid mesas, plains, and roadsides, Lower and Upper Sonoran Zones, Pima and Mohave Counties, Arizona, to New Mexico and southern Utah. April–Oct.

**3. Machaeranthera tanacetifolia** (H. B. K.) Nees, Gen. & Sp. Aster. 225, 1832
*Aster tanacetifolius* H. B. K., Nov. Gen. & Sp. **4:** 95, 1820.
*Machaeranthera parthenium* Greene, Pittonia **4:** 99, 1899.

A weedy annual 1.5–3.5 dm. tall, considerably branched above, less profusely from base; herbage rather closely beset with gland-tipped and a few eglandular hairs; lower leaves once- or rarely twice-pinnatifid, 1.5–3 cm. long, lobes 1–3 mm. broad, 1.5–5 mm. long, acute to obtusish, sometimes slightly apiculate; uppermost leaves linear, entire or faintly denticulate near base; peduncles slender, 1–5 cm. long, bearing 1–3 small linear bracts; involucres 6–8 mm. high; bracts linear-subulate, green tips more or less spreading, usually abruptly narrowed just above chartaceous base, finely glandular-puberulent; rays 12–20, purplish or blue, ligules 4–5 mm. long; disk corollas about 5 mm. long, gradually and narrowly ampliate; achenes about 2 mm. long, silky-villous with ascending hairs; pappus bristles 4–5 mm. long, whitish.

Arid hillsides, plains, roadsides, and waste ground, Lower Sonoran to Arid Transition Zones, Alberta to South Dakota and south to Arizona, New Mexico, Texas, and northern Mexico. May–Nov.

**4. Machaeranthera orcuttii** (Vasey & Rose) Cronq. & Keck, Brittonia **9:** 239, 1957
*Aster orcuttii* Vasey & Rose, Bot. Gaz. **16:** 113, *pl. 11*, 1891.
*Xylorrhiza orcuttii* Greene, Pittonia **3:** 48, 1896.

Densely leafy, rounded shrub 4–8 dm. tall with pearly white, somewhat shining bark and essentially glabrous herbage; leaves obovate to oblong, 6–18 mm. wide, 2–5 cm. long, serrately spinulose-toothed, acute to obtuse at apex, rather prominently 3-veined, glabrous or sparingly ciliate-glandular along margin near base, leathery, upper ones sessile, lower gradually narrowed to a cuneate base; heads terminal on leafy branches, solitary, radiate; involucres 1.5–2 cm. high, to 3 cm. wide, bracts linear-lanceolate, attenuate tips green, ascending, margins of thinner yellowish or stramineous basal part erosulate to ciliolate or glandular or both; rays 3–4 mm. wide, 1.5–2 cm. long, lavender to mauve or purple; disk corollas 7–8 mm. long, sparsely puberulent just at apex of tube, ampliate slightly above middle; achenes 3–4 mm. long, densely strigose; pappus bristles coarse, 7–8 mm. long, larger ones slightly paleaceous and puberulent along margins.

Rocky canyons and washes, Lower Sonoran Zone, western margins of the Colorado Desert and probably in northern Baja California. March–June.

**5. Machaeranthera cognata** (Hall) Cronq. & Keck, Brittonia **9:** 239, 1957

*Aster cognatus* Hall, Univ. Calif. Pub. Bot. **6:** 173, 1915.

Suffrutescent plants 4–8 dm. tall with pearly white bark; stems sparsely glandular-puberulent above or when young lower parts soon glabrate; leaves oblong-ovate, 1–2 cm. broad, 2–4.5 cm. long, acute to obtuse at apex, subcordate and sessile to narrowly cuneate and slightly petiolate at base, rather prominently veined, sparsely glandular-puberulent, spinulose-toothed; heads solitary at tips of leafy branches, peduncles densely glandular-puberulent, 1–5 cm. long; involucres about 1.5 cm. high, 2–2.5 cm. wide, bracts linear to linear-lanceolate, attenuate, outer ones green, glandular-puberulent, to 25 mm. long, inner ones lance-acuminate, 10–15 mm. long, green only at tips, yellowish and erosulate-margined below; rays 25–30, ligules 1.5–2 cm. long, bluish or violet; disk corollas 8–10 mm. long, glabrous; achenes about 4 mm. long, densely villous-strigose; pappus bristles stiffish, some of them subpaleaceous toward base, scaberulous, about equaling disk corollas.

Rocky hillsides, Lower Sonoran Zone, northwestern part of Colorado Desert, California. March–June.

**6. Machaeranthera tortifolia** (Torr. & Gray) Cronq. & Keck, Brittonia **9:** 239, 1957

*Haplopappus tortifolius* Torr. & Gray, Bost. Jour. Nat. Hist. **5:** 109, 1845.
*Aster tortifolius* A. Gray, Proc. Amer. Acad. **7:** 353, 1868. Not *A. tortifolius* Michx., 1803.
*Aster mohavensis* Coville, Contr. U.S. Nat. Herb. **4:** 126, 1893. Not *A. mohavensis* Kuntze, 1891.
*Xylorrhiza tortifolia* Greene, Pittonia **3:** 48, 1896.
*Xylorrhiza lanceolata* Rydb., Bull. Torrey Club **37:** 146, 1910. Not *Aster lanceolata* Willd., 1803.
*Aster abatus* S. F. Blake, in Tidestrom, Contr. U.S. Nat. Herb. **25:** 562, 1925.

Subshrubby plant 5–10 dm. tall from a woody base; bark pale yellow to whitish, herbage sparsely tomentose when young, often merely scabrous-cinereous in age; leaves oblong to linear, 6–18 mm. wide, 2.5–6 cm. long, serrately spinose-toothed to incised or uppermost rarely entire, rather prominently veined and scabrous with tack-shaped, glandular hairs in age; peduncles to 2 dm. long, glandular-puberulent and scaberulous; heads solitary; involucres about 15 mm. high, 18–24 mm. wide, bracts attenuately lance-linear, tips nearly filiform, outer ones greenish, densely glandular-puberulent, inner ones stramineous, margins erosulate; rays 40–60, lavender, ligules 2–3 mm. wide, 1.5–2 cm. long; disk corollas 7–8 mm. long, glabrous, achenes 3.5–4.5 mm. long, moderately strigose; pappus bristles stiffish, scaberulous, 5–6 mm. long, slightly tawny.

Rocky hillsides and mesas, Lower Sonoran Zone, Inyo County to Imperial County, California, east into northwestern Arizona, southern Nevada, and southwestern Utah. March–May.

**7. Machaeranthera canescens** (Pursh) A. Gray, Pl. Wright. **1:** 89, 1852

*Aster canescens* Pursh, Fl. Amer. Sept. 547, 1814.
*Dietaria canescens* Nutt., Trans. Amer. Phil. Soc. II, **7:** 300, 1840.
*Machaeranthera oxylepis* Greene, Pittonia **4:** 25, 1899.
*Machaeranthera verna* A. Nels., Bot. Gaz. **37:** 267, 1904.
*Machaeranthera scoparia* Greene, Leaflets Bot. Obs. **2:** 227, 1912.

Erect, openly branching, perennial herb 1 m. tall or less, with canescent-puberulent stems and herbage, some glandular hairs intermingled with grayish eglandular ones; leaves narrowly lanceolate, 1–7 mm. wide, to 5 cm. long, sharply dentate, scaberulous, each tooth mucronate, upper leaves subsessile, lower ones narrowed to a short petiole; heads solitary at tips of branchlets, somewhat paniculately or cymosely arranged, radiate; involucres hemispheric, 6–8 mm. high, to 1.5 cm. wide, bracts lance-linear, mostly erect, greenish herbaceous tips usually triangular and considerably shorter than chartaceous basal part, puberulent and somewhat glandular; rays few, violet to purple or rarely wanting, ligules about 5 mm. long; disk corollas about 5 mm. long; achenes 2.5–3 mm. long, minutely pubescent; pappus bristles capillary, about 5 mm. long, faintly rusty to silvery.

Roadsides, openings in chaparral, and in pine forests, Lower Sonoran to Transition Zones, British Columbia to Colorado and south to Arizona and southern California. Occurring very sparingly in the Sonoran Desert, where its place is taken largely by the following. Aug.–Nov.

### 8. **Machaeranthera tephrodes** Greene, Pittonia **4:** 24, 1899

*Dietaria asteroides* Torr., in Emory, Notes Mil. Rec. 141, 1848. Not *Aster asteroides* MacM., 1892.
*Machaeranthera canescens* var. *latifolia* A. Gray, Pl. Wright. **2:** 75, 1853.
*Aster canescens* var. *latifolius* A. Gray, Syn. Fl. N. Amer. **1**$^{2}$**:** 206, 1884. Not *A. latifolius* Mill., 1768.
*Aster canescens* var. *tephrodes* A. Gray, loc. cit.
*Machaeranthera asteroides* Greene, Pittonia **3:** 63, 1896.
*Aster tephrodes* S. F. Blake, in Tidestrom, Contr. U.S. Nat. Herb. **25:** 563, 1925.

Very similar to last, and differing from it chiefly in having slightly larger heads; involucres 10–11 mm. high; involucral bracts densely pubescent with appressed, whitish, eglandular hairs and only a few scattered glandular hairs or these lacking, green tips of bracts lanceolate or subulate, usually longer than chartaceous basal part and often spreading to somewhat reflexed; peduncles and leaves puberulent to pubescent with eglandular hairs, glands few or lacking; corollas, achenes, and pappus about as in *A. canescens.*

Mainly along valley bottoms in alluvial soil or along washes, Lower and Upper Sonoran Zones, western Riverside County and eastern San Diego County, California, to southern Nevada and New Mexico, south into northern Baja California and Sonora. March–Oct.

### 9. **Machaeranthera bigelovii** (A. Gray) Greene, Pittonia **3:** 63, 1896

*Aster bigelovii* A. Gray, Pacif. R. Rep. **4:** 97, *pl. 11,* 1857.
*Aster townshendii* Hook. f., Bot. Mag. *pl. 6430,* 1879.
*Machaeranthera aspera* Greene, Pittonia **3:** 62, 1896.

Erect, moderately branched, perennial herb 1–8 dm. tall with dark green, slightly scaberulous herbage closely beset with tack-shaped, gland-tipped hairs, eglandular hairs lacking or essentially so; leaves lanceolate to lance-ovate, 3–12 mm. wide, 1.5–6 cm. long, rather remotely but sharply dentate, teeth acute to mucronate; heads terminating branchlets, radiate, showy; involucres 5–7 mm. high, bracts linear to subulate, outer ones green nearly full length, inner ones green only along midvein near tip, glandular-puberulent throughout, tips often recurved or spreading; rays 20–35, violet to purple, ligules 4–6 mm. long; disk corollas 4–5 mm. long, scarcely ampliate; achenes 2–2.5 mm. long, sparsely villous with fine, slightly appressed hairs; pappus bristles 4–6 mm. long, silvery to faintly sordid or pinkish.

Rocky slopes, canyons, and roadsides, mainly Upper Sonoran and Transition Zones, but occasionally along the upper margins of the desert in Pima, Pinal, and Mohave Counties, Arizona, into Nevada, Colorado, and New Mexico at moderate altitudes. March–Oct.

## 23. ERIGERON L., Sp. Pl. 863, 1753

Perennial or less commonly annual or biennial herbs with alternate leaves and solitary, few, or numerous heads on naked or slightly bracteate peduncles. Involucres campanulate to hemispheric. Phyllaries subequal to unequal and strongly imbricate, herbaceous or more or less scarious, particularly toward tip and along margins. Receptacle naked, flat or slightly convex. Pistillate flowers numerous, usually bearing evident but narrow ligules equaling or surpassing pappus bristles; ligules lavender, blue, pink, or purple. Disk flowers usually numerous, hermaphroditic, tubular, 5-toothed. Style appendages short, lanceolate and acute to broadly triangular and obtuse. Achenes small, usually 2-nerved. Pappus of fragile, capillary bristles and often with a few shorter outer setae or squamellae.

Plants biennial or annual; disk corollas 2–3 mm. long:
- Eglandular hairs on stems mostly 1 mm. long or less; plants more or less stoloniferous; cauline leaves usually entire, mostly 5 mm. or less wide; peduncles and stems rarely glandular ........ 1. *E. divergens* var. *cinereum*
- Eglandular hairs on stems 1–3 mm. long; plants not stoloniferous; lower and middle cauline leaves coarsely lobed, often 1.5 cm. wide; peduncles glandular upward; stems often glandular between eglandular leaves ........ 2. *E. lobatus*

Plants perennial; disk corollas 2.7–6.5 mm. long:
- Stems hispid or hispidulous; lower leaves persistent:
  - Pappus bristles about 12 or fewer; leaves pinnately parted into linear or oblong lobes; ligules 50–150; disk corollas 2.7–4 mm. long .. 3. *E. delphinifolius* var. *oreophilus*
  - Pappus bristles 20–35; leaves entire; ligules 15–65; disk corollas 4–6.5 mm. long ........ 4. *E. foliosus*
- Stems glabrous, striate, reed-like; leaves soon deciduous below ........ 5. *E. oxyphyllus*

### 1. Erigeron divergens Torr. & Gray var. cinereum A. Gray, Pl. Wright. 1: 91, 1852

*Erigeron cinereum* A. Gray, Mem. Amer. Acad. II, **4:** 68, 1849. Not *E. cinereus* Hook. & Arn., 1836. (For additional synonymy see Brittonia **6:** 261, 1947.)

Annual or biennial plants 1–5 dm. tall, herbage and stems puberulent to pilose with spreading, white, eglandular hairs to 0.2–1 mm. long; lower leaves oblanceolate to obovate, 3–14 mm. wide, to 7 cm. long, entire to coarsely few-lobed or pinnatifid, terminal lobe acute to rounded, all others usually acute, blades often equaled or slightly exceeded by slender petioles; upper leaves narrowly oblanceolate to linear, entire, 1–3 cm. long; heads few, terminating branchlets on peduncles 1–6 cm. long; involucres 8–12 mm. broad, 5–7 mm. high, bracts lance-linear, hispidulous, midrib green, slightly thickened, tips and somewhat scarious margins sometimes faintly tinged with lavender; ray flowers numerous, lavender-purple, ligules 5–6 mm. long; disk corollas slender, 2.5–3 mm. long, lobes bright yellow, rest paler or white; achenes plump, about 1.3 mm. long, strigulose along two faint costae; longer pappus bristles 2–2.5 mm. long, finely capillary, shorter ones paleaceous, fimbriate, 0.2–0.4 mm. long.

Plains, roadsides, valley floors, and rocky hillsides, Lower Sonoran Zone to lower edge of Canadian Zone, British Columbia to South Dakota and south to Texas, central Baja California, Sonora, Chihuahua, and Coahuila. Feb.–Oct.

### 2. Erigeron lobatus A. Nels., Amer. Jour. Bot. 21: 580, 1934

Erect biennial or short-lived perennial 2–4 dm. tall with 1 to several slender stems from somewhat woody rootstock, moderately branched above; stems, leaves, and peduncles hirsute with spreading white hairs 1–3 mm. long and with intermingled puberulence of short, tack-shaped, gland-tipped hairs 0.2 mm. long or less; lower and often the middle leaves oblanceolate to obovate in outline, to 1 cm. wide, 3–6 cm. long, pinnatifid, pinnately lobed or dentate, lobes or teeth acute, often faintly apiculate; upper leaves linear to oblanceolate, 1–5 mm. wide, to 4 cm. long, entire; peduncles slender, 3–10 cm. long, densely hirsute toward apex; heads solitary at ends of branchlets but often somewhat cymosely arranged; involucres hemispheric, 5–8 mm. in diameter, 3–4 mm. high, bracts linear to linear-oblanceolate, acute to attenuate, 0.6–1 mm. wide, hirsute and glandular-puberulent, eglandular hairs slender, white, spreading; ray flowers purplish or lavender, ligules 5–6.5 mm. long; disk corollas 2–2.5 mm. long, tube short (about 0.5 mm. long), expanded abruptly into a tubular throat, glabrous, shining white except lobes and upper part of throat; achenes about 1 mm. long, yellowish, faintly compressed and 2-ribbed or smooth, glabrous or sparsely and minutely strigulose; outer pappus bristles slightly paleaceous, about 0.2 mm. long, inner ones slender, 1–1.5 mm. long.

Sandy plains, arroyos, and hillsides, Lower Sonoran Zone and lower margin of Upper Sonoran Zone, Mohave, Maricopa, Yuma, Gila, Pinal, and Pima Counties, Arizona, and northern Sonora. Feb.–May.

**3. Erigeron delphinifolius** Willd. var. **oreophilus** (Greenm.) Cronq., Brittonia **6:** 264, 1947

*Erigeron oreophilus* Greenm., Proc. Amer. Acad. **41:** 257, 1905.

Short-lived perennial 1.5–8 dm. tall with coarsely hirsute stems and leaves, hairs spreading to upwardly curved, usually with smaller, gland-tipped hairs between bases of scattered eglandular ones; leaves obovate to broadly triangular in outline, deeply pinnatifid, to 2.8 cm. broad, 1.2–4.5 cm. long, lobes narrow, mostly linear, coarsely hirsute and usually stipitate-glandular; heads few to many, corymbosely disposed; disk 8–12 mm. wide; involucre 4–6 mm. high, sparsely strigose to coarsely hirsute and granular- to stipitate-glandular; phyllaries linear, somewhat attenuate, imbricate, herbaceous; ligules 50–150, white, 5–13 mm. long; disk corollas 2.7–4 mm. long; achenes 2-nerved, sparsely strigulose; pappus bristles 5–12, deciduous, outer paleae inconspicuous, setaceous.

Rocky hillsides, canyon walls, and cliffs, Lower and Upper Sonoran Zones, central and southern Arizona into New Mexico and south into northern Mexico. Aug.–Nov.

**4. Erigeron foliosus** Nutt., Trans. Amer. Phil. Soc. II, **7:** 309, 1840

*Erigeron douglasii* Torr. & Gray, Fl. N. Amer. **2:** 177, 1841.
*Erigeron mariposanus* Congdon, Erythea **7:** 185, 1900.
*Erigeron striatus* Greene, Bull. S. Calif. Acad. **1:** 39, 1902.

Erect perennial herb 1–10 dm. tall from a slender to stout taproot and root crown, with strigose to glabrate stems and hirsute-strigose leaves, blades of these narrowly oblanceolate to linear-oblanceolate, 1.5–6 mm. wide, 1–8 cm. long, margins usually with a few conical-based, stiff hairs; heads few to several, corymbiformly disposed or less commonly solitary; involucre 4–7 mm. high, 8–12 mm. wide; phyllaries linear to lance-linear, greenish, especially along midrib, pubescent or glandular or both, acute to slightly attenuate at apex; ray flowers 15–65, ligules bluish, 5–15 mm. long, 1–2.7 mm. wide; disk corollas 4–6.5 mm. long, tube to 1.5 mm. long; achenes about 1.5 mm. long, 2-nerved, sparsely strigulose to glabrous; inner pappus bristles 20–35, sordid, outer of a few inconspicuous paleae.

Hillsides and along washes, mainly Upper Sonoran and Transition Zones but rarely straggling into the upper margins of the Lower Sonoran Zone, Humboldt and El Dorado Counties, California, south to San Diego County and probably into northern Baja California. May–July.

### 5. Erigeron oxyphyllus Greene, Erythea 3: 20, 1895

Perennial herbs 5–10 dm. tall from a branched stoutish caudex, stems glabrous and striate-angled; leaves thin, moderately hirsute or strigose in youth, later with sparsely ciliate-hirsute margins, otherwise glabrous; lower leaves oblanceolate, to 15 mm. wide, 5–10 cm. long, acute and mucronate at apex; upper leaves linear, to 3 mm. wide, 6 cm. long; all leaves withering-deciduous rather early in season, stems then reed-like; heads cymosely disposed on bracteate ascending branchlets; involucre 7–13 mm. wide, 4.5–6 mm. high; involucral bracts sharply acute to acuminate, glabrous, with brownish midrib, margins scarious on inner ones; ray flowers 20–45, ligules lavender to bluish, 1–1.8 mm. wide, 5–8 mm. long, disk corollas 4–5.3 mm. long, tube about one-fifth of total length; achenes about 1 mm. long, 2-nerved, villous-hirsutulous; pappus bristles in 2 series, the inner of slender, slightly unequal capillary bristles, outer of slender inconspicuous setae.

Arid rocky slopes, Lower and arid lower parts of Upper Sonoran Zones, Mohave, Maricopa, and Pinal Counties, Arizona. Feb.–May.

## 24. CONYZA L., Sp. Pl. 861, 1753; emend. Less., Syn. Compos. 203, 1832

Annual or perennial herbs with alternate, entire, dentate, or dissected leaves and small, numerous, discoid heads corymbosely arranged. Involucres campanulate; bracts linear to lanceolate, imbricated in 2 to several series. Marginal florets pistillate, filiform-tubular, 2–3-denticulate at apex but not ligulate or ligule reduced and shorter than pappus bristles. Disk corollas tubular, few, regularly 5-dentate. Style branches flattened, usually slender; appendages lanceolate, short to elongate. Achenes small, slightly flattened. Pappus of capillary bristles arranged in 1 or 2 rows; when biseriate, outer ones very short.

Lower and middle cauline leaves cuneate-obovate, coarsely toothed to pinnatifid or parted:
- Leaves toothed to coarsely pinnatifid; achenes strigillose and sometimes faintly glandular .. .......... 1. *C. coulteri*
- Leaves once- or twice-pinnately parted, lobes mostly linear; achenes glabrous, puncticulate in fine lines .......... 2. *C. sophiifolia*

Leaves all linear to linear-lanceolate and entire, or sometimes lowermost remotely and shallowly few-toothed:
- Heads 2–4 mm. in diameter, 3–4 mm. high; stems and involucres sparsely hispidulous to glabrous .......... 3. *C. canadensis*
- Heads 6–8 mm. high and wide; stems and involucres densely hispidulous and hispid ..... .......... 4. *C. bonariensis*

### 1. Conyza coulteri A. Gray, Proc. Amer. Acad. 7: 355, 1868

*Erigeron discoidea* Kell., Proc. Calif. Acad. **5:** 55, 1873.
*Conyzella coulteri* Greene, Fl. Fran. 386, 1897.
*Eschenbachia coulteri* Rydb., Bull. Torrey Club **33:** 154, 1906.

Weedy annuals 2–8 dm. tall with striate stems and hispidulous stems and herbage, sometimes also somewhat finely glandular; leaves oblanceolate to linear-lanceolate, subsessile, to 1.5 cm. wide, 2–7 cm. long, rather coarsely serrate-dentate to pinnatifid or uppermost entire; heads numerous, small, somewhat crowded into racemose panicles or corymbosely disposed; involucres 3–4 mm. broad, 3.5–5 mm. high, bracts lance-linear, green, attenuate at apex, villosulose with fine, white, rather tangled hairs and often sparsely to densely gland-dotted; marginal flowers pistillate, tubular, very slender, a little shorter than pappus bristles; disk corollas slenderly tubular, perfect, 2.5 mm. long; disk achenes about 1 mm. long, minutely hispidulous to nearly glabrate, sometimes slightly glandular as well; marginal achenes 0.5–0.8 mm. long; pappus bristles capillary, about 3 mm. long, shining, white.

Along arroyos, on plains, in fields, and on valley floors, Lower Sonoran to Transition Zones (sometimes into lower edge of Canadian Zone), southern California to Colorado, Texas, southern Baja California, and central Mexico. April–Oct.

**2. Conyza sophiifolia** H. B. K., Nov. Gen. & Sp. **4:** 72, *pl. 326,* 1820

*Conyza coulteri* var. *tenuisecta* A. Gray, Syn. Fl. N. Amer. **1**[2]**:** 221, 1884.
*Eschenbachia tenuisecta* Woot. & Standl., Contr. U.S. Nat. Herb. **16:** 186, 1913.

Differs from *C. coulteri* in being slightly less densely villous and more glandular; in having leaves only 1–3.5 cm. long and once- or twice-pinnately parted into linear lobes; heads smaller, fruiting head only about 3 mm. high; achenes glabrous and puncticulate in fine lines.

Arid hills, mesas, and plains, mostly Upper Sonoran to Transition Zones, rarely straggling into Lower Sonoran Zone along watercourses, southeastern Arizona into adjacent New Mexico and south into Mexico; occurring in southern Baja California near the margins of the desert. Aug.–Oct.

**3. Conyza canadensis** (L.) Cronq., Bull. Torrey Club **70:** 632, 1943

*Erigeron canadensis* L., Sp. Pl. 863, 1753.
*Conyzella canadensis* Rupr., Mém. Acad. St. Pétersb. VII, **14:** 51, 1869.
*Leptilon canadense* Britt. & Brown, Ill. Fl. **3:** 391, 1898.
*Marsea canadensis* Badillo, Bol. Soc. Venez. Cienc. Nat. **10:** 256, 1946.

Erect, sparingly to moderately branching annual 1.5–10 dm. tall, with sparingly to densely hispid herbage and lightly striate stems; leaves linear-oblong to linear-oblanceolate, 1.5–10 mm. wide, 2–10 cm. long, entire or remotely few-toothed near apex in lowermost petiolate ones, others entire, acute to attenuate, those on sparsely pubescent plants with hairs confined to margins; heads numerous, small, arranged in narrow panicles 1–3 dm. long or sometimes more; involucres narrowly campanulate to nearly tubular at anthesis, 2–3.5 mm. wide, 3–4 mm. high, bracts linear to lance-subulate, herbaceous midrib green and slightly thickened, margins subscarious, tips acute to attenuate; ray flowers inconspicuous, ligules usually just about equaling pappus bristles; disk corollas nearly white, very slender, about 2 mm. long; achenes about 1.2 mm. long, plump, faintly flattened, glabrous or very sparsely and minutely strigulose; pappus bristles capillary, 2–2.5 mm. long.

Roadsides, wasteland, stream banks, and dry beds, arid cultivated fields, Lower Sonoran and Tropical to Transition Zones, nearly throughout temperate North America and south into the West Indies and South America. Introduced into the Old World as a weed. March–Oct.

**4. Conyza bonariensis** (L.) Cronq., Bull. Torrey Club **70:** 632, 1943

*Erigeron bonariense* L., Sp. Pl. 863, 1753.
(For additional synonymy see Abrams & Ferris, Ill. Fl. Pacif. States **4:** 382, 1960.)

A weedy annual 1–4 dm. tall with 1 to several stems from base and corymbosely branching above, herbage hirsute and slightly scabrous, eglandular or essentially so; leaves numerous, linear, 1.5–4(8) mm. wide, 1.5–5 cm. long, remotely and shallowly dentate to entire; heads 6–8 mm. in diameter, numerous, corymbosely arranged; peduncles 3–20 mm. long; involucral bracts lance-subulate to linear-subulate, 4–6 mm. long, greenish, hirsutulous, only margins slightly scarious; ray flowers inconspicuous, ligules barely equaling or slightly surpassing pappus, 3–4.5 mm. long; disk corollas 3–3.5 mm. long, slender tube about 1.2 mm. long, lobes narrowly lance-triangular, about 0.2–0.3 mm. long; achenes brownish, about 1.2 mm. long, faintly compressed, minutely and sparsely strigulose to glabrate; pappus bristles capillary, 3–4 mm. long, tawny or shining white in youth.

Along irrigation ditches, river bottoms, and occasionally as a wayside weed, Tropical and Upper Sonoran to Carolinian Zones, southern California to southern Arizona and in the southeastern United States; widely distributed in the warmer parts of both Eastern and Western Hemispheres. June–Sept.

## 25. BACCHARIS L., Sp. Pl. 860, 1753

Shrubs or less commonly perennial herbs with a woody base and alternate, entire or toothed leaves. Herbage glabrous, granular-scurfy, glutinous or puberulent. Plant dioecious. Heads small or of moderate size, discoid, disposed in cymes or panicles or rarely solitary. Involucres campanulate to cylindric-campanulate, phyllaries graduate, dry or with a greenish midrib, scarious along margins. Receptacle flat or somewhat convex, naked or uncommonly with a few paleae among flowers. Staminate heads with hermaphroditic but sterile flowers; tubular corollas deeply 5-toothed, teeth spreading, recurved or coiling downwardly; pappus of narrowly clavate to paleaceous bristles, these often contorted below and subplumose near tips, usually barely equaling corollas. Pistillate heads with very slender, tubular-filiform corollas; teeth minute, not at all spreading. Achenes small, 5–10-nerved, light-colored; pappus of soft capillary bristles in 1 to several series, much longer than those of staminate flowers.

Plants perennial herbs, woody only at base; herbage finely but definitely puberulent with short, spreading, often glandular hairs; leaves 12 mm. long or less . . 1. *B. brachyphylla*

Plants shrubby or subshrubby, branches woody; herbage glabrous, glutinous, or granular-scurfy but not puberulent; leaves, at least on vigorous branches, mostly more than 12 mm. long:

- Achenes 8–10-ribbed; leaves mostly less than 5 cm. long, not strikingly willowlike:
  - Plants broomlike, branches numerous, erect, strongly striate-angled; leaves entire or finely few-toothed, usually deciduous before flowering time:
    - Leaves linear to linear-lanceolate, acute; receptacles alveolate but naked; staminate heads 3.5–7 mm. long; pistillate pappus 10–12 mm. long . . . . . . 2. *B. sarothroides*
    - Leaves cuneate-obovate, usually obtuse at apex; receptacles usually with a few paleaceous bracts among flowers; staminate heads 2.5–4 mm. long; pistillate pappus 2–4 mm. long . . . . . . . . . . . . . . . . . . . . . . . . . . . . . . . . . . . . 3. *B. sergiloides*
  - Plants more open, rarely broomlike, branches less strongly striate-angled; leaves usually coarsely toothed, not deciduous before flowering time . . . . . . . . . . . . . 4. *B. emoryi*
- Achenes 5-ribbed; leaves 5–12 cm. long, willowlike:
  - Leaves finely and pungently toothed or serrate; involucres 2–2.5 mm. long; pappus bristles of pistillate flowers slightly rusty . . . . . . . . . . . . . . . . . . . . . . . 5. *B. thesioides*
  - Leaves remotely toothed with blunt teeth or entire; involucres 3–5 mm. long; pappus bristles of pistillate flowers silvery:
    - Heads in terminal paniculate cymes or corymbs; leaves usually denticulate and plainly 3-veined . . . . . . . . . . . . . . . . . . . . . . . . . . . . . . . . . . . . . . . . . . . 6. *B. glutinosa*

Heads in small cymose panicles on short lateral branches; leaves usually entire, obscurely 3-veined ............................................ 7. *B. viminea*

**1. Baccharis brachyphylla** A. Gray, Pl. Wright. **2:** 83, 1853

A much-branched perennial herb 2–10 dm. tall, roots and base woody, slender, lightly striate branches pale green or slightly cinereous, puberulent with short spreading hairs, many of these gland-tipped; leaves linear, entire, acute, 0.5–1.5 mm. wide, 5–12 mm. long, upper ones reduced to scalelike bracts often less than 5 mm. long; heads numerous, paniculately disposed on upper branchlets; involucres campanulate, 3–5 mm. high; phyllaries ovate to oblong-lanceolate, pale greenish and puberulent along midrib, margins scarious and often irregularly erosulate, especially toward tips; staminate corollas barely equaling or slightly surpassing involucre, pappus about equaling corollas; pistillate corollas about 3 mm. long, styles exserted 1–2 mm., achenes brownish, about 2 mm. long, puberulent between 5 ribs; pappus bristles of pistillate flowers silky, rusty or tawny, 6–8 mm. long.

Mesas, arroyos, and rocky gulches and hillsides, Lower Sonoran Zone, eastern San Diego and Riverside Counties, California, into Arizona, south into Sonora and central Baja California. May–Nov.

**2. Baccharis sarothroides** A. Gray, Proc. Amer. Acad. **17:** 211, 1882

*Baccharis arizonica* Eastw., Proc. Calif. Acad. IV, **20:** 155, 1931.

A densely fastigiately branched, glutinous shrub 1–4.5 m. tall with conspicuously striate-angled branchlets often crowded into broomlike masses; leaves on young, rapidly growing branches linear to lance-linear, 1.5–3.5 mm. wide, 1–2.5 cm. long, acutish at both ends, rather leathery, entire, soon deciduous, those of branchlets reduced to minute bracts 1–2.5 mm. long; heads numerous, solitary or in few-headed glomerules at tips of branchlets, staminate 3.5–7 mm. long, pistillate about 1 cm. long at anthesis; phyllaries in several series, narrowly ovate to oblong, or oblong-lanceolate, somewhat granular-glutinous on exposed portions, margins scarious, only faintly erosulate toward tips; receptacles flattish, alveolate but naked; staminate corollas 3.5–4 mm. long, about equaled by narrowly clavate pappus bristles, clavate portion slightly scaberulous, lower part glabrous, often twisted; pistillate corollas nearly filiform, about 3 mm. long; achenes about 2 mm. long, cylindric, prominently 8–10-ribbed, glabrous; pappus bristles of pistillate flowers rusty to brownish, 10–12 mm. long.

Hillsides, bottom lands, and outwash slopes on gravelly, rocky, or less commonly saline soil, Lower and Upper Sonoran Zones, southern California to southwestern New Mexico, northern Sonora, and northern Baja California. Flowering almost throughout the year.

**3. Baccharis sergiloides** A. Gray, in Torr., Bot. Mex. Bound. 83, 1859

A much-branched, suffrutescent, glutinous plant 1–2 m. tall with strongly striate-angled branches usually crowded into broom-like tufts; leaves on vigorous branches cuneate-obovate, 5–15 mm. wide, 1–3.5 cm. long, entire or sparingly and obscurely toothed, leathery, veins usually prominent, usually deciduous, those of fastigiate flowering branchlets chiefly bract-like, 2 mm. long or less; heads numerous, in small glomerules or solitary at tips of branchlets, arranged in naked panicles; staminate heads 2.5–4 mm. high, pistillate to 5 or 5.5 mm. long; phyllaries ovate to lance-ovate, or innermost of pistillate heads lance-linear, minutely granulose, margins scarious and faintly erosulate-ciliolate; corollas about 2 mm. long; receptacles often somewhat conical, alveolate and usually with a few paleae intermingled with flowers, at least in pistillate heads; achenes cylindric, 1–1.2 mm. long, shining and glabrous, 10–ribbed; pappus of pistillate flowers 2–4 mm. long, finely capillary, white or slightly sordid.

Along washes, in canyons, and on mountain sides, Lower and Upper Sonoran Zones, western margins of the Colorado Desert to Inyo County, California, east to southern Utah and south into northern Mexico. Flowering practically throughout the year.

**4. Baccharis emoryi** A. Gray, in Torr., Bot. Mex. Bound. 83, 1859

A glutinous glabrous shrub to 5 m. tall with strongly striate-angled branches, these sometimes fastigiate and broomlike but usually more open; leaves oblong-elliptic to broadly oblanceolate or less commonly linear-oblong, 2.5–10 or sometimes to 25 mm. wide, 1.5–6 cm. long, usually coarsely few-toothed, but uppermost narrower ones often entire, acute to obtuse at apex, cuneate at base, veins 3 from base, moderately prominent; heads solitary or in small glomerules at tips of branchlets, campanulate, staminate 6–7.5 mm. long at anthesis, pistillate 10–14 mm. long, narrower than staminate; phyllaries ovate to lance-ovate in staminate heads, ovate to linear in pistillate, well imbricated in several series, median part slightly thickened, margins scarious, markedly erose-ciliate toward tips, especially on longer innermost phyllaries of pistillate heads; staminate corollas 3.5–4 mm. long, lobes slender and recurved or coiled, anthers exserted nearly their full length, pappus bristles of staminate flowers barely equaling corollas, subplumose and somewhat paleaceous toward tips; pistillate corollas capillary-cylindric, about 5 mm. long, lobes microscopic; achenes about 1.5 mm. long, stramineous, finely but distinctly ribbed; pappus of pistillate flowers 10–12 mm. long, silky, white or faintly tawny.

Along arroyos and streams, Lower and Upper Sonoran Zones, Los Angeles to San Diego County, California, east to southern Utah and Texas, south into northern Mexico. Sept.–Dec. in our area.

**5. Baccharis thesioides** H. B. K., Nov. Gen. & Sp. **4:** 61, 1820

*Baccharis ptarmicaefolia* DC., Prod. **5:** 419, 1836.
*Baccharis sulcata* DC., loc. cit.

An openly branched, slightly glutinous shrub 1–2 m. tall with moderately striate-sulcate branches; leaves linear-oblanceolate to narrowly oblong, 1.5–6.5 mm. wide at broadest part, this usually above middle, 1.5–8.5 cm. long, acute at apex, narrowly cuneate toward base, regularly dentate-serrate with fine spinulose teeth 0.2–0.6 mm. long, finely granulose-glutinous on both surfaces, lateral veins obscure or apparently lacking; heads small, 4–6 mm. high, numerous in rounded panicles; involucres campanulate; phyllaries graduate and imbricate in 4–6 series, ovate to lance-linear, acute or obtuse, green and granular along midrib, white-scarious and slightly erosulate on margins; pappus of staminate florets barely equaling involucre, exserted anthers 1–2 mm. longer; pistillate florets filiform, about 3 mm. long; achenes 1.5–2 mm. long, stramineous, 5-ribbed; pappus of pistillate flowers slightly rusty, 4–6 mm. long.

Mountain sides and canyons, mostly in the Upper Sonoran Zone but straggling into upper margins of the Lower Sonoran Zone in the vicinity of Cumpas and Ures, Sonora; Arizona and New Mexico south to Jalisco and Mexico, D.F. Aug.–Nov.

**Baccharis alamosana** S. F. Blake, Contr. U.S. Nat. Herb. **22:** 595, 1924. This species is very similar in appearance to *B. thesioides* but has lanceolate instead of oblanceolate leaves that are usually less than 4 cm. long; minute heads with involucres only 2–2.5 mm. high; and pappus 2.5 mm. long.

Known only from the vicinity of Alamos, in southern Sonora, but its sporadic occurrence along the upper margins of the Sonoran Desert in southern Sonora is not unlikely.

### 6. **Baccharis glutinosa** Pers., Syn. Pl. **2:** 425, 1807

*Molina viscosa* Ruiz & Pav., Syst. Veg. Peruv. Chil. 207, 1798. Not *Baccharis viscosa* Lam., 1783.
*Baccharis farinosa* Spreng., in Ersch & Grub., Allgem. Encycl. **7:** 27, 1821.
*Baccharis coerulescens* DC., Prod. **5:** 402, 1836.
*Baccharis alamani* DC., loc. cit.
*Baccharis longifolia* DC., loc. cit.

An erect, openly branched shrub 2–5 m. tall with somewhat striate, glabrous, but minutely and sparsely granular-glutinous, leafy branches; leaves linear to lanceolate, 4–18 mm. wide, 5–12 cm. long, acuminate at apex, narrowly and gradually cuneate at base, usually remotely and shallowly toothed but sometimes entire, short-petiolate, larger ones distinctly 3-veined, moderately glutinous; heads numerous in rounded terminal panicles; involucres broadly campanulate, 3–5 mm. long, phyllaries ovate to lance-linear, stramineous, margins hyaline and finely erosulate; lobes of staminate flowers slightly exceeding pappus bristles; pistillate corollas much shorter than pappus, the style branches exserted 1–1.5 mm.; achenes about 1 mm. long, with 5 low ribs, glabrous; pappus bristles silky, 4–6 mm. long.

Along arroyos, stream banks, ditches, and margins of reservoirs and ponds, Lower and Upper Sonoran Zones, central and southern California to Colorado and Texas, south into South America. March–Dec. in our area.

### 7. **Baccharis viminea** DC., Prod. **5:** 400, 1836

A loosely branched shrub 1–4 m. tall very similar in appearance to *B. glutinosa* but differing from that species mainly in having slightly less glutinous herbage, entire or nearly entire leaves 2.5–8 cm. long, their blades faintly 3-veined or 1-veined; heads in corymbose clusters at tips of numerous, short, lateral branches disposed along several decimeters of main branches; otherwise as in *B. glutinosa.*

Along watercourses, ditches, and about ponds, Lower and Upper Sonoran Zones, southern California to southern Utah, south into Arizona and northern Baja California. March–Oct. with the main flowering periods April–May and in Sept.

## 26. **ARCHIBACCHARIS** Heering, Jahrb. Hamburg, Wiss. Aust. **21:** Beih. **3:** 40, 1904

Plants with the general habits and vegetative characters of *Baccharis.* Heads polygamodioecious. Staminate heads containing only hermaphrodite sterile flowers with tubular-ampliate corollas whose 5-lobed limbs are spreading to slightly recurved; pappus bristles barely equaling or slightly shorter than corollas, often somewhat dilated and paleaceous apically. Pistillate heads with 1 to several hermaphrodite sterile flowers occupying center of head and surrounded by pistillate flowers with capillary corollas and longer pappus bristles.

### 1. **Archibaccharis peninsularis** S. F. Blake, Jour. Wash. Acad. **33:** 267, 1943

A low bush with slender, spreading to subprocumbent stems 3–6 dm. long and scaberulous-puberulent branchlets and herbage; leaves numerous, obovate to rhombic-obovate, 1–2.5 cm. broad, 2.5–5.5 cm. long, acute to obtusely rounded and coarsely toothed apically, broadly cuneate at base, scaberulous beneath, glabrate and glutinous on upper surface; petioles 3–10 mm. long, expanding gradually into cuneate base of blade; heads corymbosely arranged at ends of branches, campanulate, about 3 mm. long, phyllaries ovate to oblong, midrib stramineous, hyaline margins ciliolate; staminate corollas purplish, about 1 mm. longer than inner phyllaries; pistillate flowers not seen.

Known only from the type locality in a small canyon in the Sierra de la Giganta near Loreto, Baja California. In flower in Dec.

## Tribe 4. **INULEAE.** Everlasting Tribe

Annual or perennial herbs, mostly with white-woolly herbage. Leaves alternate or rarely opposite, entire or more or less dentate. Heads small, discoid. Involucres spherical to narrowly campanulate; bracts frequently white and at least the margins scarious. Pistillate flowers with very slender, usually filiform corollas. Sterile flowers perfect or staminate. Anthers caudate at base. Style branches slender, stigmatic to apex, without appendages. Pappus of numerous capillary hairs or none.

Marginal flowers pistillate; pappus bristles equal or subequal if present; corollas glabrous or minutely and sparingly puberulent:
- Involucral bracts dry but herbaceous, not scarious; plants robust herbs or shrubs, to 2.5 m. tall .......... 27. *Pluchea*
- Involucral bracts wholly or partially scarious; plants smaller herbs, rarely over 1 m. tall, usually much less:
  - Receptacles chaffy-bracted; central perfect flowers few, often sterile; involucral bracts scarious only apically or marginally:
    - Fertile flowers devoid of pappus; central flowers sterile:
      - Fruiting bracts saclike or boat-shaped, each enclosing an achene; pappus bristles usually present in sterile flowers .......... 28. *Stylocline*
      - Fruiting bracts open, merely subtending achenes; pappus none .......... 29. *Evax*
    - Fertile flowers of 2 kinds, outer without pappus and enclosed in their bracts, inner without bracts, with pappus of capillary bristles; central flowers often fertile .......... 30. *Filago*
  - Receptacles naked; central perfect flowers numerous, usually fertile; involucral bracts wholly scarious .......... 31. *Gnaphalium*

Marginal flowers perfect, fertile, as are central ones; pappus bristles of 2 kinds—long, slender ones and more numerous, shorter, somewhat connate ones; corollas densely puberulent .......... 32. *Pelucha*

## 27. **PLUCHEA** Cass., Bull. Soc. Philom. Paris **1817**: 31, 1817

Annual or perennial herbs or shrubs with alternate leaves and numerous discoid heads in terminal corymbose, cymose, or paniculate inflorescences. Flowers numerous, purplish. Involucres campanulate; bracts chartaceous to subscarious, graduated and imbricated. Receptacle flattish to somewhat convex, naked. Marginal flowers pistillate and fertile, with tubular filiform corollas; central ones hermaphroditic but sometimes sterile, their corollas 5-lobed, or 5-cleft. Anthers caudate. Achenes small, 3–5-ribbed or grooved. Pappus of capillary bristles in a single series.

Stems winged by decurrent leaf bases; leaves serrate-denticulate, very densely viscid-glandular .......... 1. *P. adnata* var. *parvifolia*

Stems not winged; leaves entire or coarsely and remotely toothed, silky or moderately viscid-glandular:

Pappus bristles not dilate at apex; leaves glandular-puberulent, not silky-sericeous, 1–10 cm. wide:

Plants stout annual herbs; phyllaries reddish or purplish; leaves 1–4 cm. wide .................................................................. 2. *P. purpurascens*

Plants shrubby; phyllaries stramineous; leaves 2.5–10 cm. wide .......... 3. *P. odorata*

Pappus bristles of perfect flowers dilated at apex; leaves sericeous, silky, 2–7 mm. wide .................................................................. 4. *P. sericea*

**1. Pluchea adnata** (Humb. & Bonpl.) Mohr var. **parvifolia** (A. Gray) Blake, Contr. U.S. Nat. Herb. **23:** 1510, 1926

*Pluchea subdecurrens* var. *parvifolia* A. Gray, Proc. Amer. Acad. **5:** 160, 1861.

Suffrutescent plants 1–2 m. tall with densely viscid-glandular herbage; stems winged with decurrent bases of leaves, wings 0.5–2.5 mm. wide; leaves sessile, lanceolate to lance-linear, 3–12 mm. wide, 2.5–8 cm. long, rather finely and regularly serrate-denticulate, each tooth gland-tipped, glandular-puberulent on both surfaces, veins prominulous beneath; heads in rounded corymbose inflorescences 4–10 cm. broad, each one 5–8 mm. high, phyllaries narrowly ovate to lance-ovate, purple-tipped, glandular-puberulent; pappus bristles of perfect flowers slightly clavate-dilated toward tips; achenes 1 mm. long or less; pappus bristles of pistillate flowers capillary, not dilated, 4–5 mm. long.

In moist soil along streams and about ponds, Lower Sonoran and Tropical Zones, central to southern Baja California. March–May.

**2. Pluchea purpurascens** (Sw.) DC., Prod. **5:** 452, 1836

*Conyza purpurascens* Sw., Prod. Veg. Ind. Occ. 112, 1788.

*Pluchea camphorata* of California authors, not DC., 1836.

An erect robust annual 3–10 dm., or under very favorable conditions sometimes to 5 m., tall with soft-puberulent, markedly glandular herbage; leaves ample, lanceolate to oblong-ovate, 1–4 cm. wide, 5–15 cm. long, acute at each end, glandular-dentate, veins prominulous beneath; petioles 0.5–2 cm. long or upper leaves sessile; heads numerous in flat- or round-topped, corymbose inflorescences, each head broadly campanulate, 5–7 mm. high, 6–8 mm. wide at anthesis; phyllaries broadly ovate to linear-lanceolate or oblong, acute to short-acuminate at apex, glandular-puberulent except on subscarious, often ciliolate margins, purplish or reddish; pistillate corollas capillary, about 3 mm. long, about equaling pappus bristles; hermaphrodite corollas 4–4.5 mm. long, ovate lobes puberulent without, slightly surpassing pappus; achenes about 1 mm. long, faintly ribbed and minutely puberulent, pale brown.

In moist saline soil, Lower and Upper Sonoran Zones to Carolinian and Tropical Zones, central California to the Atlantic Coast south into Mexico. In our area, Aug.–Oct.

**3. Pluchea odorata** (L.) Cass., Dict. Sci. Nat. **42:** 3, 1826

*Conyza odorata* L., Syst. Nat. ed. 10, **2:** 1213, 1759.

A coarsely branched shrub to 2.5 m. tall with sordid-tomentose or less commonly cinereous-tomentose branchlets, glabrate on older parts; leaves elliptic, lance-elliptic or oblong-ovate, 2.5–10 cm. wide, 6–20 cm. long (uppermost sometimes smaller), entire or remotely repand-dentate, acute to obtuse at apex, usually broadly cuneate at base, pale and densely sordid-tomentulose beneath, darker green and subglabrate above; petioles stoutish, to 2.5 cm. long; panicles or corymbs 7–25 cm. wide; heads broadly campanulate, 5–7 mm. high, outer phyllaries ovate, inner lance-oblong, stramineous and subchartaceous, glandular-puberulent, broadly spreading in maturity; achenes slender, 0.6–0.8 mm. long; pappus bristles all alike, silky-capillary, 3–5 mm. long.

Moist banks of arroyos, in canyons, and along margins of marshy plains, Lower Sonoran and Tropical Zones, southern Baja California and Sinaloa to northern South America, the West Indies and in southern Florida. Flowering nearly throughout the year.

**4. Pluchea sericea** (Nutt.) Coville, Contr. U.S. Nat. Herb. **4:** 128, 1893

*Polypappus sericeus* Nutt., Jour. Acad. Phila. II, **1:** 178, 1848.
*Tessaria borealis* Torr. & Gray ex A. Gray, Mem. Amer. Acad. II, **4:** 75, 1849.
*Pluchea borealis* A. Gray, Proc. Amer. Acad. **17:** 212, 1882.
*Berthelotia sericea* Rydb., Bull. Torrey Club **33:** 154, 1906.

Slender, willow-like shrub 1–5 m. tall with erect, considerably branched, silvery-silky branchlets and leaves, pubescence closely appressed; leaves alternate, rather crowded, lance-linear to narrowly elliptic, sessile, acute at apex, cuneately tapering to base, 2–7 mm. wide, 1–4.5 cm. long, entire, 1-nerved, eglandular; heads in small corymbose or paniculate glomerules at tips of branchlets; involucres broadly campanulate, about 5 mm. high and wide, bracts broadly ovate (outermost) to oblong, purplish and puberulent on back, villous-ciliate along margins; hermaphrodite corollas 4.5–5.5 mm. long, glabrous, lobes one-half as long as combined tube and throat, some pappus bristles clavate-dilated; pistillate corollas filiform, about as long as hermaphrodite ones; achenes about 1 mm. long, glabrous; pappus bristles of pistillate flowers capillary, not dilated, 4–6 mm. long.

Along streams and arroyos, sometimes in saline soil, mainly Lower Sonoran Zone but into Upper Sonoran Zone, Santa Barbara and Inyo Counties, California, to southern Utah and Texas, south into Chihuahua, northern Sonora, and central Baja California. Feb.–May, less commonly Sept.–Oct.

## 28. STYLOCLINE Nutt., Trans. Amer. Phil. Soc. II, 7: 338, 1841

Low, erect or spreading, woolly annuals with small, ovoid or globose heads in small terminal clusters. Leaves alternate, entire. Receptacle columnar, bearing 3–5 sterile perfect flowers at its apex subtended by as many slightly curved or plane hyaline bracts. Pistillate flowers constituting major part of head, each completely enclosed by lower or median portion of its woolly bract. Pappus none in fertile flowers, sometimes a few bristles on sterile central ones. Achenes brown, straight or slightly curved.

Hyaline part of bracts of fertile flowers broadly ovate to suborbicular, extending nearly or quite to base of bract .......... 1. *S. gnaphalioides*
Hyaline part of bracts of fertile flowers lanceolate, forming about half of total length of bract:
  Bracts of fertile flowers (including tip) 3–4.5 mm. long; leaves lance-linear, acute, callus-tipped, involucral ones surpassing heads; achenes 1–1.4 mm. long .. 2. *S. micropoides*
  Bracts of fertile flowers 2–2.4 mm. long; leaves broadly spatulate to oblong, obtuse, not callus-tipped; achenes 0.6–0.8 mm. long .......... 3. *S. sonorensis*

**1. Stylocline gnaphalioides** Nutt., Trans. Amer. Phil. Soc. II, **7:** 338, 1841

*Stylocline arizonica* Coville, Proc. Biol. Soc. Wash. **7:** 79, 1892.

Diffusely branched from base or depauperate plants simple, erect, 3–20 cm. tall; leaves broadly linear to spatulate, 1–2.5 mm. wide, 5–12 mm. long, obtuse, floccose; fertile pistillate flowers numerous, their bracts broadly ovate to suborbicular, 2–2.5 mm. long, plane or slightly convex on outer surface, central part on inner side forming a sac completely enfolding achene, this part firm, greenish on outside and densely long-woolly, broad margins extending clear to base, hyaline, glabrous; achenes devoid of pappus, brownish, slightly curved but flower terminal; sterile flowers almost equaling their bracts; abortive achene bearing a few caducous bristles or naked.

Sandy plains and washes, Lower and drier parts of Upper Sonoran Zones, Lake, Marin, Fresno, and Inyo Counties to the western margin of the Colorado Desert, cismontane southern California, and south to central Baja California. March–May.

**2. Stylocline micropoides** A. Gray, Pl. Wright. **2:** 84, 1853

Branched at or near base; branches ascending or diffuse, 2–15 cm. high; cauline leaves linear, acute or mucronulate, 5–10 mm. long, those subtending heads lanceolate-linear, acute or usually cuspidate, 5–12 mm. long; pistillate fertile flowers numerous, enclosed in oblong woolly bracts 1.8–2.5 mm. long without lateral wings but apical half forming an oblong-ovate, hyaline, glabrous tip; sterile flowers naked or subtended by oblong, glabrous, hyaline paleae, abortive achenes often bearing 3–4 slender bristles.

Sandy washes, Lower Sonoran Zone, southern San Joaquin Valley and southern Nevada to the western edge of the Colorado Desert and from Inyo County and the Mojave Desert to New Mexico and central Sonora. March–April.

**3. Stylocline sonorensis** Wiggins, Contr. Dudley Herb. **4:** 26, 1950

Low floccose annual 2–5 cm. high with erect-ascending or spreading-ascending branches; basal and cauline leaves elliptic to broadly oblong, 1–2 mm. wide, 2.5–5 mm. long, densely white-floccose, leaves subtending heads broadly oblong to spatulate, 2–3 mm. wide, 5–9 mm. long, obtuse or abruptly and minutely apiculate, woolly on both surfaces, not noticeably surpassing clustered heads; heads closely clustered, ovoid to subglobose, 2.2–3 mm. high, woolly but with hyaline tips of bracts extending beyond the wool; sterile basal bracts apparently none; bracts of fertile flowers 2–2.4 mm. long (including terminal hyaline part), densely woolly on back of firm basal part enclosing achene, hyaline part lance-elliptic, 2–2.5 times as long as broad, acute, sometimes extending to base of bract as narrow wing 0.1–0.2 mm. wide, glabrous, median stripe pale straw-colored; achene dark stramineous to pale yellow-brown, 0.6–0.8 mm. long, smooth and glistening, slightly flattened laterally.

Grassy hillsides in gravelly soil, upper margins of Lower Sonoran Zone and lower parts of Upper Sonoran Zone, northern Sonora. April.

## 29. EVAX Gaertn., Fruct. 2: 393, *pl. 165,* 1791

Low, densely woolly annuals with entire, alternate leaves and small discoid heads encircled by bractlike leaves. Receptacle hemispheric to short-columnar. Pistillate fertile flowers numerous toward exterior of head, each subtended by an oblong, nearly plane, scarious bract. Perfect sterile flowers central, few, subtended by open or slightly involute scarious bracts. Achenes narrowly obovoid, slightly compressed, smooth or minutely papillose. Pappus none.

**1. Evax multicaulis** DC., Prod. **5:** 459, 1836

*Micropus minimus* DC., op. cit. 461.
*Filaginopsis multicaulis* Torr. & Gray, Fl. N. Amer. **2:** 263, 1842.
*Diaperia multicaulis* Benth. & Hook., Gen. Pl. **2:** 298, 1873.

Diffusely branched from base with leafy, decumbent to ascending branches 0.5–1.5 dm. long, these bearing small, densely crowded, terminal glomerules of heads nearly hidden by bractlike leaves; cauline leaves broadly spatulate, 1.5–3.5 mm. wide, 5–12 mm. long, sessile, arachnoid-tomentose; heads subglobose, 2–3 mm. high; bracts of fertile flowers oblong, scarious below, densely inflexed-woolly at apex, those of sterile flowers woolly farther down back; achenes 1–1.2 mm. long, smooth, yellowish.

Sandy soil, Lower Sonoran Zone, from vicinity of Tucson, Arizona, to western Texas, southern Oklahoma, and northern Sonora. March–April.

## 30. FILAGO L., Sp. Pl. 927, 1753

Small, herbaceous, annual plants with slender, depressed to erect stems, alternate entire leaves, and canescent to lanate herbage. Heads small, discoid, in terminal or axillary glomerules; outer flowers pistillate, outermost of these subtended by boat-shaped, open, paleaceous bracts, epappose, more centrally placed ones without subtending bracts but with a pappus of capillary bristles; innermost flowers hermaphrodite, usually without subtending bracts, pappose, usually fertile. Outermost involucral bracts dorsally lanate, margins more or less scarious. Receptacle plane to convex, deeply alveolate. Pistillate flowers tubular-filiform, minutely 2–3-dentate. Perfect flowers slenderly tubular, slightly ampliate and 4–5-toothed above. Anthers slightly sagittate basally. Achenes small, subterete to somewhat compressed, ribless, often minutely papillose. Pappus of inner flowers of slender, scaberulous bristles in 1–2 series.

Flowers within inner whorl of bracts 4–7, all perfect; leaves conspicuously surpassing subtended heads; internodes slender, nearly naked, often purplish or brownish .......... 1. *F. arizonica*

Flowers within inner whorl of bracts 10–25, only central 4–5 perfect, others pistillate; leaves equaling or barely surpassing subtended heads; internodes less slender, usually lanate, greenish:

- Achenes all smooth; plants diffuse to ascending, heads mostly terminal; hyaline appendage of outer bracts about equaling body .......... 2. *F. depressa*
- Achenes of pappose flowers papillose; plants usually erect, heads both terminal and axillary; hyaline appendage of outer bracts about one-half as long as body .......... 3. *F. californica*

### 1. Filago arizonica A. Gray, Proc. Amer. Acad. 8: 652, 1873

Diffuse to erect herb with slender stems 2–15 cm. long, internodes slender, purplish, usually 0.5 mm. in diameter or less, 0.5–3 cm. long, at first sublanate but later glabrescent; heads in small, compact, axillary and terminal glomerules, subtended by linear, oblong or narrowly oblanceolate leaves 1–2.2 mm. wide, 3–12 mm. long, acute at both ends, sessile, finely lanate-canescent, longer ones conspicuously surpassing glomerules; heads ovoid, 2.5–3.5 mm. high; outer bracts boat-shaped, silky-lanate without, glabrous and shiny within, margins hyaline-scarious, each one except outermost 3–6 subtending a pistillate epappose flower, back green beneath tangled wool, hyaline tip less than one-half as long as body; inner bracts paleaceous, oblong, only slightly boat-shaped, glabrous or nearly so, white to stramineous; central flowers 4–7, perfect, glabrous, about 1.2–1.4 mm. long; achenes smooth, about 0.6–0.8 mm. long; pappus bristles scaberulous, white, about 1.5 mm. long.

Gravelly slopes and plains, Lower Sonoran Zone, Pinal, Maricopa, and Pima Counties, Arizona, to the Mojave and Colorado Deserts, the Channel Islands off southern California, and northern Baja California and Sonora. March–April.

**2. Filago depressa** A. Gray, Proc. Amer. Acad. **19:** 3, 1883

Depressed annual with slender, more or less lanate, prostrate or decumbent stems 2–12 cm. long; leaves basal and cauline, oblong to lance-ovate, 1–3 mm. wide, 2–8 mm. long, sessile, about equaling or barely surpassing glomerules of heads in their axils; internodes short or only slightly elongated, rarely over 12 mm. long, usually less, mostly lanate, greenish or tardily brownish; glomerules dense, disposed from base to tips of stems; heads ovoid, 2.5–3.5 mm. long; oblong outer bracts lanate on back, hyaline tip about as long as body, or its margin extending along full length of bract; inner bracts narrower, hyaline tips extending slightly beyond wool of outer bracts; flowers immediately interior to paleaceous bracts pistillate, fertile, pappose; innermost 4–8 flowers perfect, pappose, fertile; achenes all smooth, greenish brown, slightly compressed; pappus bristles white or silvery, minutely scaberulous.

Open plains and slopes, Lower Sonoran Zone, Inyo County to Imperial and eastern San Diego Counties, California, east to Pima County, Arizona. March–April.

**3. Filago californica** Nutt., Trans. Amer. Phil. Soc. II, **7:** 405, 1841
*Oglifa californica* Rydb., Fl. Rocky Mts. 914, 1917.

Erect or ascending annual herb 5–35 cm. tall with leafy lanate stems; leaves rather numerous at base of stems, narrowly oblanceolate, 1–2.5 mm. wide, 8–15 mm. long, acute at apex, narrowed to a slender petiole, at first lanate but glabrate in age, upper leaves subtending axillary and terminal glomerules linear-oblong, mostly less than 5 mm. long, moderately lanate; heads ovoid-turbinate, 3–4 mm. long; outermost bracts ovate to lance-ovate, sterile, lanate dorsally, those immediately interior to them lance-ovate, shallowly boat-shaped, lanate dorsally, tipped with a hyaline appendage about one-half as long as body, each subtending filiform, pistillate, fertile, epappose flowers; innermost bracts lance-linear or oblong, hyaline-paleaceous, nearly flat, tips extending 0.2–0.3 mm. beyond lanate parts of outer bracts; flowers immediately interior to hyaline bracts pistillate, pappose, fertile; innermost 4–8 flowers perfect, pappose, fertile; achenes of outer epappose flowers smooth, those of flowers with pappus minutely papillose.

Open slopes and plains, Lower and Upper Sonoran Zones, upper Sacramento Valley to northern Baja California, east from Inyo County, California, to southern Utah, northern Sonora, and central Arizona. March–May.

## 31. GNAPHALIUM L., Sp. Pl. 850, 1753

Annual or perennial herbs with alternate leaves and woolly herbage. Leaves entire, sessile, often decurrent, rarely petiolate. Heads small, discoid, in cymose glomerules, corymbs, panicles, or spikes. Receptacles flat or convex, naked. Involucres ovoid to campanulate, of imbricated, scarious, chaffy bracts, these greenish, brownish, roseate or white, often shining, bases often buried in dense wool. Outer flowers pistillate, filiform, minutely 3–4-dentate; central flowers perfect, tubular, slightly ampliate and shallowly 5-toothed apically. Anthers sagittate. Style branches terete, truncate to capitate. Achenes oblong to obovoid, terete or slightly compressed, smooth, papillose or roughened. Pappus bristles capillary, in a single series, falling separately, or connate at base and falling together.

Pappus bristles connate in a ring at base:
Involucral bracts brownish or purplish at tips; heads spicately arranged; leaves spatulate, distinctly greener above than beneath .......................... 1. *G. purpureum*

Involucral bracts slightly yellowish or greenish, not purple-tipped; heads in terminal and axillary glomerules; leaves mostly linear to oblanceolate ........ 2. *G. pedunculosum*

Pappus bristles distinct, falling separately:

Involucral bracts mainly embedded in loose wool, only tips showing; glomerules of heads leafy, leaves exceeding heads .................................... 3. *G. palustre*

Involucral bracts woolly only at base, conspicuously chaffy-scarious; glomerules only slightly leafy or naked, rarely exceeded by leaves:

Leaves equally woolly on both surfaces or essentially so:

Heads in broadly branched, corymbose panicles; involucral bracts shining-white; leaves not noticeably decurrent ...................................... 4. *G. wrightii*

Heads in dense, subglobose, terminal glomerules; involucral bracts stramineous or yellowish; leaves distinctly decurrent .......................... 5. *G. chilense*

Leaves densely woolly beneath, greenish above, distinctly bicolored:

Involucral bracts dull pearly white, not shining; stems densely white-tomentose; lower leaves linear to linear-oblanceolate, attenuate .............. 6. *G. leucocephalum*

Involucral bracts white or faintly greenish or yellowish, distinctly shining; stems arachnoid-tomentose; lower leaves broadly oblanceolate to oblong-lanceolate, obtuse to acute but not attenuate:

Leaves submembranous, lower ones 10–18 mm. wide; heads in compact terminal glomerules, these only slightly leafy ...................... 7. *G. oxyphyllum*

Leaves thicker, lower ones 3–8 mm. wide; heads in openly branched, pedunculate glomerules, these conspicuously leafy ......................... 8. *G. sonorae*

## 1. **Gnaphalium purpureum** L., Sp. Pl. 854, 1753

*Gnaphalium ustulatum* Nutt., Trans. Amer. Phil. Soc. II, **7:** 404, 1841.

*Gnaphalium pannosum* Gandoger, Bull. Soc. Bot. Fr. **65:** 42, 1918. Not *G. pannosum* Schultz-Bip. 1845 or A. Gray, 1883.

Biennial or short-lived perennial, often flowering first season, usually considerably branched from base and with ascending to erect stems 1–3.5 dm. high, these closely woolly-canescent; leaves broadly spatulate, 4–15 mm. wide, 2–6 cm. long, finely and compactly tomentose beneath, greenish and glabrate above, mostly obtuse to rounded and often minutely apiculate at apex, gradually narrowed to a petiole-like base; heads spicately arranged, inflorescence 1–10 cm. long; heads 3.5–5 mm. high, outer bracts brownish or purplish, at least at tips, obtuse, shining, innermost bracts oblong, thinner and paler than outer, often acutish or minutely erosulate and apiculate at apex; pistillate flowers about 1.5–1.8 mm. long, very slender; pappus bristles 7–15, connate in a funnelform ring at base, falling together; achenes straw-colored, about 0.5 mm. long, faintly roughened.

Open plains and slopes, ditch banks and waysides, Lower Sonoran to Transition and Carolinian Zones, southern California to British Columbia and Arizona; Kansas to Maine. March–May.

**Gnaphalium spathulatum** Lam., Encycl. **2:** 759, 1786. This weedy plant, which is said to be strictly annual and with greenish instead of brownish or purplish bracts, occurs in the cismontane parts of southern California and may reach the western margins of the Sonoran Desert.

## 2. **Gnaphalium pedunculosum** I. M. Johnston, Contr. Gray Herb. **68:** 99, 1923

Annual herbs with simple or slightly branched, erect stems 1–3 dm. tall, these and lower sides of leaves floccose-woolly; leaves oblong to narrowly oblong-oblanceolate or less commonly spatulate, 2–5 mm. wide, 1–3.5 cm. long, entire, lanate on both sides and concolorous or slightly greener above than below, obtuse but often minutely apiculate; heads 3–4 mm. broad, 4–5 mm. high, in dense terminal and axillary, subglobose glomerules; outer involucral bracts lance-ovate, embedded in wool at base, greenish or slightly stramineous on scarious shining part; inner bracts oblong to lance-linear, about 3 mm. long, faintly stramineous, scarious; capillary pistillate flowers 30–50, about 2 mm. long; perfect flowers 4–5, about 2.5 mm. long; pappus bristles 10–12, connate in a ring at base; achenes ellipsoidal, terete, greenish yellow, about 0.5 mm. long, faintly and sparsely papillose.

Sandy washes and plains, Lower and Upper Sonoran Zones, Durango to San Luis Potosí and southern Baja California. March–May. The Baja California specimens are assigned to this species with hesitation but there seem to be no constant characters separating them from the mainland material.

**3. Gnaphalium palustre** Nutt., Trans. Amer. Phil. Soc. II, **7:** 403, 1841

*Gnaphalium palustre* var. *nanum* Jepson, Man. Fl. Pl. Calif. 1067, 1925.

A low annual, commonly much branched at base with erect or ascending stems 1–2.5 dm. long; herbage loosely floccose-lanate; wool persistent on stems but often more or less deciduous from spatulate to linear-spatulate leaves, these to 5 mm. wide and 3 cm. long but usually smaller, uppermost ones subtending and exceeding heads, usually lanceolate to oblong; heads in dense subglobose glomerules at tips of branchlets, each 2–3 mm. high; involucral bracts deeply embedded in loose wool, only scarious tips showing, scarious part usually obtuse, often denticulate; flowers and pappus bristles about equaling longer involucral bracts; pappus bristles distinct or virtually so, falling separately or in small groups of 2–4, minutely scaberulous.

Along streams, grassy plains and valley floors, Lower Sonoran to Transition Zones, central Baja California to British Columbia, east to Alberta, Wyoming, and New Mexico. March–Oct.

**4. Gnaphalium wrightii** A. Gray, Proc. Amer. Acad. **17:** 214, 1882

*Gnaphalium viridulum* I. M. Johnston, Contr. Gray Herb. **70:** 86, 1924.

Erect or ascending, short-lived, perennial herb much branched at base, densely and closely white-woolly; leaves crowded and more or less matted at base, these oblong to oblanceolate or somewhat spatulate, to 1 cm. wide, 2–4.5 cm. long; cauline leaves oblong to oblong-oblanceolate, 1.5–5 mm. wide, 1.5–3 cm. long, densely tomentose on both sides, rarely slightly greenish on upper surface, acute and often apiculate at apex, sessile and not at all or only slightly decurrent at base; inflorescence of numerous, small, ovoid-turbinate heads arranged in an open corymbose panicle of few-headed glomerules, each head 4–5.5 mm. long; bracts well imbricated in 3–5 series, white and shining, outermost woolly at base only, mostly obtuse, sometimes apiculate; pistillate flowers capillary, about 3 mm. long, about equaled by pappus bristles; perfect flowers about same length; achenes ovoidal, 0.5–0.7 mm. long, pale brown, smooth, dull; pappus bristles distinct.

Arid rocky slopes, upper margins of Lower Sonoran into the Transition Zones, Mohave, Cochise, Pinal, Pima, and Santa Cruz Counties, Arizona, to western Texas and northern Mexico. Aug.–Oct.

**5. Gnaphalium chilense** Spreng., Syst. **3:** 480, 1826

(For synonymy see Abrams & Ferris, Ill. Fl. Pacif. States **4:** 469, 1960.)

Erect or ascending annual or biennial with 1–several stems 2–6 dm. tall from base; herbage closely covered with greenish or yellowish tomentum; leaves linear-lanceolate to spatulate, 2–8 cm. long, 3–15 mm. wide, acute to obtuse at apex, slightly but distinctly decurrent at base, lower ones narrowed to a winged petiole, upper strictly sessile; heads few, usually in a single dense glomerule at apex of each stem, less commonly glomerules arranged in a panicle; heads broadly turbinate-campanulate, about 5–6 mm. long, fully as wide at anthesis, base embedded in loose wool; bracts ovate to oblong, well imbricated, scarious, white or slightly yellowish or greenish, obtuse and commonly somewhat erosulate at apex; flowers 2.5–3 mm. long, pistillate capillary, perfect ones slightly more robust; pappus bristles distinct, 5–6 mm. long; achenes brownish, about 0.5 mm. long, smooth.

Along streams, ditches, arroyos, and in rather moist situations, Lower Sonoran to Transition Zones, Washington to Montana south to Texas, Arizona, southern California, and northern Baja California. May–Oct.

**6. Gnaphalium leucocephalum** A. Gray, Pl. Wright. **2:** 99, 1853

Erect perennial with 1 to several erect stems 3–8 dm. tall; stems leafy, persistently tomentose with dense, finely packed wool; leaves narrowly linear-attenuate or lowermost linear-oblanceolate, 1.5–6 mm. wide, 2–8 cm. long, green and only puberulent-glandular above, persistently white-tomentose beneath, sessile; inflorescence a corymbose panicle, often 2 dm. broad; heads many, turbinate, 6–7 mm. high, 7–8 mm. wide at anthesis, bracts well imbricated in 4–6 series, broadly ovate to oblong, dull pearly white, obtuse and more or less erosulate at apex, only bases of outermost embedded in loose wool; flowers about 4 mm. long, pistillate barely equaling capillary pappus bristles, perfect ones 0.1–0.3 mm. longer than pistillate or equaling them; achenes about 0.8 mm. long, pale brown or stramineous, smooth, dull; pappus bristles distinct, falling separately.

Sandy washes, disturbed areas, and dry hillsides, upper parts of the Lower Sonoran to lower margin of the Transition Zones, Cochise, Pima, and Maricopa Counties, Arizona, to cismontane southern California and northern Sonora. July–Oct.

**7. Gnaphalium oxyphyllum** DC., Prod. **6:** 225, 1837

Erect or ascending biennial or short-lived perennial with 1 to several stems 3–7 dm. tall branching from base, these loosely floccose-tomentose in youth but wool soon deciduous; leaves numerous, rather crowded, oblong-linear, oblanceolate, or spatulate, basal ones 1–2 cm. wide, 5–12 cm. long, narrowed to a petiole-like, slightly decurrent base, submembranous, finely arachnoid-lanate with closely appressed, white hairs beneath, glabrate and pale green on upper surfaces, acute to apiculate-acuminate at apex, basal ones soon withering, uppermost smaller, linear to oblong; heads few to many, in small, fairly compact glomerules, corymbosely or paniculately arranged at tips of branches, each head ovoid, 3–4.5 mm. long at anthesis; bracts faintly greenish white, shining, imbricated in 3–5 series, ovate to oblong, acute to obtuse, often erosulate at apex, only outermost smaller ones slightly arachnoid-lanate at base; flowers barely 2 mm. long, very slender; pappus bristles barely equaling longest bracts, distinct, falling separately or in groups of 2–3, silvery or slightly tawny.

Shady woods and along streams or canyons, Lower Sonoran to Tropical Zones, southern Sonora and central Baja California south into central Mexico. March–Dec.

**8. Gnaphalium sonorae** I. M. Johnston, Contr. Gray Herb. **68:** 99, 1923

Erect perennial with 1 to several loosely tomentose stems 2–5 dm. tall, much branched from base into broad, corymbose panicles at apex; leaves entire, persistently tomentose beneath, lightly so or subglabrate and green on upper surface, acute at apex, sessile by a broad base, lowermost oblong-oblanceolate, 3–8 mm. wide, 2–4 cm. long, upper ones linear-oblong, 2–4 mm. wide, to 2.5 cm. long; heads in corymbosely arranged glomerules, each glomerule usually subtended by 1 to several linear to oblong leaves 1–2 cm. long; each head about 3 mm. wide, 4–5 mm. long, loosely lanate at base; bracts white, imbricated in about 3–4 series, outer oblong-ovate, acute, inner ones lance-oblong, acute to rounded; flowers 30–50, about 3.5 mm. long; perfect flowers 5–6; achenes oblong, 0.5–0.6 mm. long, smooth; pappus bristles distinct, falling separately.

Dry hillsides and along arroyos, Lower Sonoran Zone, central Sonora in the vicinity of Hermosillo. July–Oct.

## 32. PELUCHA S. Wats., Proc. Amer. Acad. 24: 55, 1889

Heavy-scented, much-branched shrub with alternate, subfleshy, toothed or lobed leaves and thinly gray-tomentose herbage. Heads in terminal cymose panicles, short-pedunculate, discoid. Involucres campanulate; bracts subequal, in 2 series. Flowers all perfect, tubular. Anthers sagittate-caudate. Achenes oblong-turbinate, densely silky-pubescent. Pappus of 10–15 antrorsely barbellate bristles and 25–30 shorter, irregular, somewhat connate bristles.

### 1. Pelucha trifida S. Wats., Proc. Amer. Acad. 24: 55, 1889

Irregularly branched shrub 4–12 dm. tall, tomentum thin, grayish, soon deciduous; stems glabrate and light brown, sparsely and irregularly fissured; leaves oblanceolate in outline, 5–20 mm. long, subfleshy, bluntly 3-toothed or 3-cleft near apex, lobes to 4 mm. long, 1–2.5 mm. wide, central rachis about as wide, ill-smelling; heads 8–10 mm. high; involucres 4–5 mm. high, bracts herbaceous, lanceolate to oblong, obtuse or acute to slightly acuminate at apex, uniformly and sparsely tomentose, 1-nerved; flowers about 20, tubular, corollas 6–7 mm. long, densely puberulent with irregularly spreading to subappressed brownish hairs on outer surface, ovate teeth about 1 mm. long; anthers exserted 2–3 mm.; achenes 3 mm. long, pubescence white, but turning brown on herbarium sheets; longer pappus bristles 3–4 mm. long, shorter ones mostly 1.5 mm. long or less.

Rocky ridges and canyon walls, Lower Sonoran Zone, Isla Angel de la Guarda, Isla San Pedro Mártir, and on the peninsula back of Las Ánimas Bay, Baja California. March–May.

## Tribe 5. HELIANTHEAE. Sunflower Tribe

Annual or perennial herbs or shrubs, usually with yellow flowers (some with white, orange, or roseate corollas) and often with resinous juice. Heads discoid or radiate, disk flowers perfect but either fertile or sterile; ray flowers pistillate or neutral. Receptacle flat to slightly convex, beset with chaffy bracts (or naked or nearly so in *Lagascea, Coulterella, Elvira,* and *Milleria*), each subtending a flower and sometimes closely investing it. Anthers obtuse or rarely minutely sagittate or bidentate at base, but never caudate. Style branches ovate, obtuse, elongate or rarely truncate at apex; stigmatic lines well marked. Achenes thick and tumid to greatly flattened, obovoid to linear, sometimes narrowly to markedly winged, wing then often continuing into a pair of persistent awns. Pappus of awns, paleae, squamellae, or stiff bristles but never of capillary hairs, sometimes lacking.

Involucres tubular, calyxlike, each subtended by a small or moderate-sized bract:
  Involucres herbaceous, not corky-winged; style branches attenuate-acute .... 33. *Lagascea*
  Involucres corky-winged and permanently investing achene; style branches obtuse ...... .......................................................... 34. *Coulterella*
Involucres not tubular nor calyxlike, bracts distinct or essentially so:
  Chaffy receptacular bracts greatly reduced or wanting; heads few-flowered, usually with less than 10 flowers:
    Involucre of 1 large bract opposing 2–3 smaller ones; heads greatly flattened parallel to opposing bracts; ligule of fertile flower entire or nearly so; heads arranged in racemes or panicles ................................................ 35. *Elvira*
    Involucre of subequal bracts, these not opposed; head not flattened; ligule of fertile flower small but noticeable, 3-toothed; heads in glomerules or fascicles .......... ........................................................ 36. *Milleria*
  Chaffy receptacular bracts not reduced, often equaling involucres; heads usually more than 10-flowered:
    Disk flowers perfect but all of them sterile, their ovules abortive:
      Corollas of ray flowers deciduous before achenes fall, not leathery nor indurated; corolla tube obvious:
        Achenes of ray flowers free from adjacent disk flowers, falling separately:
          Corollas white; anthers green; heads in terminal corymbs; filaments pubescent .. .................................................... 37. *Guardiola*
          Corollas and anthers yellow; heads not corymbosely arranged; filaments glabrous or nearly so ........................................38. *Melampodium*
        Achenes of ray flowers attached to two adjacent disk flowers by a callous thread and falling with them:
          Heads erect; achenes glandular-puberulent and more or less arachnoid-villous, keeled on inner face .................................. 39. *Parthenium*
          Heads nodding; achenes glabrous, keel lacking ................ 40. *Parthenice*
      Corollas of ray flowers persistent and falling with achenes, leathery or indurated; corolla tube reduced to mere ring .......................... 41. *Tragoceros*
    Disk flowers fertile (part of them sometimes sterile in *Sclerocarpus* and *Coreocarpus*):
      Corollas of ray flowers firm, persisting until fall of achenes:
        Achenes of disk flowers strongly compressed or markedly angled; all leaves entire:
          Leaves sessile; involucral bracts in 3–7 series, outer and inner very unequal; ray achenes markedly triquetrous ............................... 42. *Zinnia*
          Leaves petiolate (or uppermost sometimes sessile); involucral bracts in 2–3 series, outer nearly as long as inner; ray achenes 4-angled .......... 43. *Sanvitalia*
        Achenes of disk flowers terete or obtusely angled; larger leaves dentate or serrate ...................................................... 44. *Heliopsis*
      Corollas of ray flowers herbaceous, deciduous before achenes fall:
        Rays very numerous, usually less than 3 mm. long; disk achenes markedly corky-margined; ray achenes often abortive ......................... 45. *Eclipta*
        Rays fewer or lacking, usually over 5 mm. long if present; disk achenes not corky-margined:
          Achenes broadly obovoid, tumid, persistently invested in indurated, rugose, receptacular bracts; 1–5 of outer involucral bracts much longer than others ... ............................................... 46. *Sclerocarpus*
          Achenes not tumidly obovoid, free from their subtending bracts; outer involucral bracts not conspicuously longer than inner ones:
            Ray flowers neutral or wanting:
              Pappus of awns, paleae, bristles or wanting, but never plumose:
                Elements of pappus never rigid, retrorsely barbed awns, if hairy pubescence soft:

Peduncles much enlarged upward, fistulose; heads large, 5–15 cm. in diameter ........................................ 47. *Tithonia*

Peduncles not markedly enlarged upward, solid, or if fistulose heads much smaller, usually less than 3 cm. in diameter:

Achenes angular or only slightly compressed; pappus not pubescent; anthers entire or faintly bidentate at base:

Pappus paleae 1–3, often accompanied by several squamellae, all of these persistent:

Heads bearing 10–100 flowers or more in each; involucres not constricted at apex; involucral bracts usually indurated at base ........................................ 48. *Viguiera*

Heads with 1–8 flowers in each; involucres urceolate, somewhat constricted at apex; involucral bracts not indurated toward base ........................................ 49. *Alvordia*

Pappus paleae 1–2, caducous, additional squamellae rarely present ........................................ 50. *Helianthus*

Achenes strongly compressed laterally; pappus paleae pubescent, continuous with lateral margins of achenes; anthers minutely sagittate or bidentate:

Plants perennial herbs or shrubs; achenes without a crown ........ ........................................ 51. *Encelia*

Plants annual herbs; achenes with a prominent crown in addition to awns ........................................ 52. *Geraea*

Elements of pappus 2–8 rigid, retrorsely barbed awns firmly attached to achene ........................................ 58. *Bidens*

Pappus of plumose paleae or awns; rays none .................. 59. *Bebbia*

Ray flowers present and fertile:

Heads 1–8-flowered; involucres narrowly cylindric and often urceolate:

Achenes quadrangular, slightly compressed; pappus of awns and squamellae but awns not retrorsely barbed .................... 49. *Alvordia*

Achenes linear to narrowly cuneate, markedly compressed; pappus of 2–8 retrorsely barbed, persistent awns ..................... 58. *Bidens*

Heads 15-flowered or more; involucres broadly cylindric, hemispheric, campanulate or obturbinate, not urceolate:

Pappus paleae not retrorsely barbed:

Achenes cuneate, strongly flattened but not winged on margins:

Herbs; disk corollas 6–8 mm. long .................... 52. *Geraea*

Shrubs; disk corollas (in ours) only 4 mm. long ....... 53. *Zexmenia*

Achenes obovate to oblong, margins (of at least some of them) winged:

Ray flowers (in ours) yellow; wings of achenes entire:

Involucral bracts imbricated in 2–3 series, not connate at base; leaves entire to dentate but not dissected; herbage usually pubescent .................................. 54. *Verbesina*

Involucral bracts in 2 different series, bracts of each series somewhat connate at base; outer bracts much narrower than inner ones; leaves dissected into narrow divisions; herbage glabrous or nearly so .............................. 55. *Coreopsis*

Ray flowers white (lined with rose or yellow beneath); wings of achenes often crenate ...................... 56. *Coreocarpus*

Pappus paleae moderately to strongly armed with retrorsely pointed barbs:

Ray achenes with narrowly inrolled marginal wings, epappose; disk achenes slightly beaked ...................... 57. *Heterosperma*

Ray achenes not inrolled along margins; disk achenes beakless ....... ........................................ 58. *Bidens*

## 33. LAGASCEA Cav., Anal. Cienc. Nat. Madrid 6: 331, 1803

Annual or perennial herbs or shrubs. Leaves usually opposite, ovate, petiolate, often acuminate apically. Heads small, 1–2-flowered, crowded in dense, campanulate to sub globose, terminal glomerules, these clusters subtended by leaflike bracts. Involucre calyxlike, tubular, 4–6-toothed. Flowers perfect. Corolla tubular at base; throat more broadly cylindrical; limb 5-toothed, exserted beyond tubular involucre. Style branches long, attenuate. Achenes columnar or slightly clavate. Pappus of 2 to several short scales or reduced to mere rudiments.

### 1. **Lagascea decipiens** Hemsl., Diagn. Pl. Mex. 33, 1879

*Nocca decipiens* Kuntze, Rev. Gen. Pl. **1:** 354, 1891, as *Noccaea.*
*Calhounia nelsonae* A. Nels., Wyo. Univ. Pub. Bot. **1:** 55, 1924.

A loosely branching shrub 1–4 m. tall with slender, faintly striate branches moderately to densely hirsutulous with spreading whitish hairs 0.2–0.6 mm. long; bark brownish or grayish; leaves ovate-acuminate or ovate-attenuate, 1–4.5 cm. wide, 4–10 cm. long, obtuse, rounded, or broadly cuneate at base, entire to shallowly serrate, strigose-hirsute and dotted with minute, golden, globose glands on both surfaces, dark green above, slightly paler beneath, rather prominently 3-veined from base; petioles 3–9 mm. long, hirsutulous; glomerules 2.5–3 cm. in diameter, globose or nearly so; involucres 8–10 mm. long, 5–6-toothed, slightly sericeous-hirsute, teeth 1–1.8 mm. long, lance-subulate; corollas yellow, puberulent without, tube and throat 4–5 mm. long, lobes narrowly ovate to oblong, 1.5–2 mm. long, spreading; achene about 5 mm. long, silky-hirsute near apex with erect, bristle-like hairs, glabrous below.

Canyons, along arroyos, and occasionally on desert plains, Lower Sonoran, Upper Sonoran, and Tropical Zones, southern Arizona to Chihuahua and Jalisco. Feb.–April and Sept.–Oct.

## 34. COULTERELLA Vasey & Rose, Contr. U.S. Nat. Herb. 1: 71, *pl. 1,* 1890

Compact, glabrous shrubs with opposite, fleshy, sessile leaves. Heads small, in terminal and subterminal glomerules. Involucres tubular, 3–4-toothed at apex, 1-flowered or rarely 2-flowered, corky-winged and permanently investing achene, subtended by a minute bract. Corolla tubular, 5-toothed. Anthers exserted, sagittate. Style branches elongated, obtuse. Achenes terete, linear-cuneate. Pappus a minute crown or none.

### 1. **Coulterella capitata** Vasey & Rose, Contr. U.S. Nat. Herb. 1: 71, *pl. 1,* 1890

Much branched, low and spreading to moderately ascending to 1 m. tall, usually slightly wider than tall, with pungently aromatic herbage; leaves fleshy, 2–4 mm. thick, 5–20 mm. wide, 1–3.5 cm. long, irregularly and sparsely toothed, sessile; heads in glomerules 1–2.5 cm. in diameter, glomerules corymbosely arranged, each head subtended by an ovate bract 3–6 mm. long; involucres 5–6 mm. long, fleshy, 3–4-winged and 3–4-toothed; corollas lemon-yellow, lobes linear-oblong, about 2 mm. long; anthers 2–2.5 mm. long, almost wholly exserted; involucre in fruit 6–8 mm. long, spongy; achenes striate, glabrous, 3.5–3.8 mm. long, bearing a short annular apical cup.

Usually near the beach, Lower Sonoran Zone, interruptedly from Isla Monserrate, Gulf of California, south to the vicinity of La Paz and on some of the adjacent islands. Oct.–May.

## 35. ELVIRA Cass., Dict. Sci. Nat. 30: 67, 1824

Annual or short-lived perennial herbs with opposite, lightly dentate leaves and small heads borne in axils of leaves and at tips of branches in fascicles or glomerules. Heads radiate with 1–3 pistillate, fertile ray flowers and 1–4 perfect but sterile disk flowers. Involucral bracts 2–4, herbaceous, usually larger, much broader one opposite 2–3 smaller bracts, one or more of smaller bracts often more or less 2–3-lobed. Receptacle naked. Ray corollas with slender tubes, limb small, nearly entire. Disk corollas with a very short tube and a narrowly campanulate throat regularly 5-toothed at apex. Styles of sterile disk flowers entire, not divided. Achenes obovate, compressed, smooth and glabrous, enfolded by subtending bracts.

**1. Elvira biflora** (L.) DC., Prod. **5:** 503, 1836

*Milleria biflora* L., Sp. Pl. 919, 1753.
*Elvira martyni* Cass., Dict. Sci. Nat. **30:** 68, 1824.

Herbage slightly scaberulous; leaves lance-ovate, 2–6 cm. long, shallowly and remotely serrulate; involucral bracts orbicular-cordate, 4–5 mm. long, to 9 mm. wide, reticulately veined; each head 2-flowered, one pistillate and fertile, other perfect but sterile.

Canyons and along watercourses, Lower Sonoran to Tropical Zones, southern Sonora to Central America. This plant has a superficial resemblance to some of the Acanthaceae and can be overlooked easily. It is a weedy herb which has been collected near Alamos and may occur within the southern limits of the Sonoran Desert.

## 36. MILLERIA L., Sp. Pl. 919, 1753

Erect, often robust, annual herbs with opposite, membranous, dentate, petiolate leaves. Heads small, paniculately disposed. Heads bearing one fertile, pistillate ray flower and 1–5 tubular, perfect, sterile disk flowers. Receptacle small, naked. Involucre ovoid, of several slightly unequal membranous bracts, one of these broadly concave-cucullate, 2–3 smaller ones somewhat attenuate. Ligule of pistillate flower yellow, somewhat exserted and spreading, minutely 3-dentate. Perfect flowers regular; throat campanulate; limb exceeding short tube, 5-lobed apically. Anthers slightly auriculate at base, sagittate. Achenes ovoid, compressed parallel to radius of involucre, closely enfolded by membranous involucral bract; pericarp crustaceous. Pappus none.

**1. Milleria quinquiflora** L., Sp. Pl. 919, 1753

*Milleria dichotoma* Cav., Ic. **1:** 58, *pl. 82*, 1791.
*Milleria glandulosa* DC., Prod. **5:** 503, 1836.
*Milleria maculata* Mill., Gard. Dict. ed. 8, no. 2, 1768.

Erect herbs to 1.5 m. tall with angular-striate to angular-costate stems, often considerably branched above; younger herbage sparsely scaberulous-pubescent; lower leaves broadly ovate, to 15 cm. wide, 18–20 cm. long, coarsely but shallowly serrate-dentate, acute to acuminate at apex, broadly cuneate to subtruncate at base and gradually narrowing to a winged petiole, this one-half to two-thirds as long as blade; inflorescence dichotomously branching into a cymose panicle, branches glandular-pubescent with tack-shaped, glandular hairs; heads 4–5 mm. high at anthesis; ligule of ray flower 3–5 mm. long; achenes 6–7 mm. long, strongly ridged diagonally, completely enfolded by accrescent membranous bracts.

Along ditches and stream banks, Lower Sonoran to Tropical Zones, northern Sonora to Central America. Sept.–Oct.

## 37. **GUARDIOLA** Cerv., in H. & B., Pl. Aequin. **1:** 143, *pl. 41,* 1808

Suffrutescent or herbaceous plants with opposite leaves and small, terminal, solitary to cymose-paniculately arranged heads. Involucres cylindrical to narrowly campanulate; bracts few, subequal, multinerved, subherbaceous but with thin submembranous margins. Receptacle bearing paleaceous bracts. Rays 1–5, small, pistillate, fertile; disk flowers 3–8; corollas tubular with a very short, abruptly ampliate throat and 5 relatively long lobes, sterile. Anthers entire at base; filaments pubescent. Achenes oblong, compressed, smooth to multistriate. Pappus none.

**1. Guardiola platyphylla** A. Gray, Pl. Wright. **2:** 91, 1853

Erect or often somewhat straggly, suffrutescent plants with one to many slender, striate, glabrous stems 1–2.2 m. tall from a woody base; upper branches often numerous and more or less glaucescent; leaves broadly ovate to deltoid-ovate, 1.5–4.5 cm. broad, 2–5 cm. long, acute to obtuse or rounded at tip, truncate to shallowly cordate at base, spinulose-dentate, leathery and conspicuously netted-veined, glaucous, glabrous, sessile or on short, stout petioles 1–4 mm. long; heads terminally arranged in pyramidal panicles or cymose-paniculate, usually 3–10 heads in a cluster; involucres cylindric, 7–9 mm. long at anthesis, outer bracts oblong, obtuse, or rounded at apex, greenish, inner ones acute, hyaline-scarious; ray flowers 1–5, usually 5, ligules 4–5 mm. long; disk corollas 6–7 mm. long, tube very slender, glabrous, abruptly ampliate into a throat about 0.2–0.4 mm. long and slightly wider, lobes 1.5–2 mm. long; anthers partially exserted; achenes 4–5 mm. long, brownish.

Canyon bottoms and tree-covered slopes, Lower and Upper Sonoran Zones, Pima and Gila Counties, Arizona, south into southern Sonora and southwestern Chihuahua. Feb.–Oct.

## 38. **MELAMPODIUM** L., Sp. Pl. 921, 1753

Annual or perennial herbs or suffrutescent plants with glabrous, scabrous, or hirsute, punctate-glandular herbage and frequently with dichotomously branching stems. Leaves opposite, entire or dentate. Heads small or of moderate size, radiate at apices of dichotomously branching stems or pedunculate in upper axils. Corollas yellow or rarely white. Ray flowers in a single external series, pistillate and fertile. Disk flowers perfect but sterile. Involucral bracts in 2 series; outer herbaceous and often connate at base; inner ones each closely enfolding a ray achene and in fruit often expanding toward apex to form a hood, this with or without a slender awn or appendage. Receptacle convex or conical, beset with paleaceous bracts subtending sterile disk flowers. Achenes obovoid, smooth, closely invested by subtending bract, this barely equaling achene or surpassing it, variously ridged, knobbed, tuberculate or muticous, in some exterior apical portion drawn into a slender curved, recurved, coiled, or spiral awnlike process.

Fruiting bracts exceeding invested achenes and forming a hood or cup:
  Plants woody, at least at base; rays white, conspicuous, 9–13 mm. long; hood of fruiting bracts not awned ........................................ 1. *M. leucanthum*

Plants herbaceous; rays yellow, 2–6 mm. long; hood of fruiting bracts tipped with a recurvingly coiled, slender awn:

Ligules 4–6 mm. long; peduncles surpassing subtending leaves; fruiting bracts strongly tuberculate-muticous along dorsal margins .............. 2. *M. appendiculatum*

Ligules 1–2 mm. long; peduncles mostly shorter than subtending leaves or heads subsessile; fruiting bracts merely ridged dorsally and laterally, not tuberculate or only weakly so .............................................. 3. *M. longicorne*

Fruiting bracts barely equaling, or shorter than, enclosed achenes, not developing into a hood or cup:

Plants annual; rays 1–4 mm. long; disk flowers slightly surpassing subtending bracts:

Heads pedunculate; peduncles usually surpassing subtending leaves; ligules 3–4 mm. long, well exceeding involucral bracts ........................ 4. *M. cupulatum*

Heads sessile or subsessile among several subtending leaves; ligules 1–2 mm. long, barely exceeding bracts .............................................. 5. *M. hispidum*

Plants perennial, somewhat woody at base; rays 6–10 mm. long; disk flowers often surpassed by their subtending receptacular bracts ..................... 6. *M. sinuatum*

### 1. **Melampodium leucanthum** Torr. & Gray, Fl. N. Amer. **2:** 271, 1842

Suffrutescent plant with few to numerous slender branches 1–4 dm. long from a stoutish woody root, strigose and somewhat cinereous throughout, hairs forwardly appressed, white and stiffish; lower parts of stems woody with brownish or gray-brown, somewhat fissured bark; leaves linear to oblong-oblanceolate, 2–5 mm. wide, 1.5–4 cm. long, gradually narrowed below, acute or obtusish at apex, entire or remotely repand-dentate, midrib prominent beneath; peduncles slender, to 10 cm. long or more, usually well surpassing leaves; heads 1.8–3 cm. wide in anthesis, showy; involucre 7–12 mm. in diameter, lobes usually 5, 4–5 mm. wide and long, rounded or obtuse, appressed cinereous-strigose, ligules white, strongly veined with purple, 9–13 mm. long, firm, sparsely puberulent along veins on outside; corollas of disk flowers about 2–2.5 mm. long, puberulent without; receptacular bracts oblong-oblanceolate, erosulate or ciliate at apex, chaffy to papery; fruit 4–5 mm. high, investing bract transversely rugulose, ridges short and irregularly arranged, terminal portion projected into an oblique hood about one-half as long as body, orifice toward disk flowers, its margins rolled inward, whole structure more or less arcuately curved.

Dry slopes and mesas, particularly in limestone, Lower and Upper Sonoran Zones, central and southeastern Arizona to Kansas, Texas, Chihuahua, and northeastern Sonora. March–Oct.

### 2. **Melampodium appendiculatum** Robinson, Proc. Amer. Acad. **36:** 457, 1901

Slender erect annual 3–6 dm. tall with sparsely pubescent herbage and stems branching (often dichotomously) almost from base; leaves linear to oblong, 3–10 mm. wide, 2.5–5 cm. long, acute to attenuate at apex, only slightly narrowed basally to a subauriculate sessile base, entire to obsoletely serrate, stoutish hairs forwardly subappressed, surface of blade somewhat scabrous; peduncles slender, erect, 2–7 cm. long; involucre saucer-shaped, gamophyllous, 5 lobes broadly deltoid, rounded or abruptly short-apiculate, hispid, rather prominently veined, margin hyaline-scarious; ray flowers yellow, 8–10, ligules oblong, 4–6 mm. long; disk corollas 2–2.5 mm. long, puberulent on outside, narrow tube abruptly expanding into a narrowly funnelform throat, lobes about 0.5 mm. long, spreading but soon reflexed; involucral bracts scarious, oblong-oblanceolate, 2.5–3 mm. long, slightly constricted near tip and then expanded abruptly into a fan-shaped, erosulate, apical portion about one-fourth as long as body;

body of fruit about 3 mm. long, faintly ridged longitudinally, hood scarcely broader than body, oblique, outer margin drawn into a slender, awn-like appendage 2–3 mm. long, recurvingly coiled and appressed-hirsutulous, margins of hood slightly tuberulate or toothed below awn.

Margins of meadows, stream banks, outwash slopes, and hillsides, Lower and Upper Sonoran Zones, the foothill area in Sonora and east into southern Chihuahua. The two minor varieties proposed by Robinson are of doubtful value. Sept.–Nov.

**3. Melampodium longicorne** A. Gray, Mem. Amer. Acad. **5:** 321, 1854

Erect annual 1–4 dm. tall moderately branched from near base; stems and branches spreading-hispidulous; leaves oblanceolate to oblong-spatulate, 5–15 mm. wide (exclusive of projecting lobes), 1.5–5 cm. long, entire or with 1–2 pairs of abruptly spreading lobes 3–9 mm. long, when only one pair present lobes located slightly below middle of blade, cuneate toward sessile base, acute at apex, blade sparsely scabro-hispidulous; heads sessile or subsessile at tips of branchlets, subtended by leaves 3–6 times as long as heads; outer involucral bracts distinct, obovate, abruptly short-acuminate, sparsely hispid on midrib and along margins, 3.5–4 mm. long; ray flowers inconspicuous, ligules 1–2 mm. long or less, barely equaling or usually shorter than outer involucral bracts; disk flowers few, often only 3–5, inconspicuous; fruit about 4 mm. long, lightly but distinctly ridged, 3 low ridges on each lateral face and one fainter dorsal one, outer angles each bearing 1–3 low conical tuberculations; cup oblique, outer margin drawn into a recurvingly coiled, externally puberulent awn; achene black, shining, longitudinally striate with low broad ridges.

Canyons, hillsides, and stream banks, mainly Upper Sonoran Zone but straggling to the margins of the desert, Cochise and Santa Cruz Counties, Arizona, to northern Sonora and Chihuahua. March–Oct.

**4. Melampodium cupulatum** A. Gray, Proc. Amer. Acad. **8:** 291, 1870

Erect annual 1.5–5 dm. tall branching almost from base, with scabro-puberulent herbage, stems and branches somewhat striate, often reddish; leaves lance-linear to lanceolate, 3–15 mm. wide, 2–10 cm. long, acute at apex, cuneately narrowed to base, this sessile or subsessile, margins entire to shallowly repand, lighter beneath than above; peduncles slender, 1–5, or in fruit to 10 cm. long, rather densely hispidulous just beneath heads; involucres broadly cup-shaped or deeply saucer-shaped, gamophyllous, 5 lobes broadly ovate, obtuse except for an abrupt acumination, moderately hispidulous, about 3 mm. broad; ray flowers usually 10–15, ligules yellow, oblong, 3–4 mm. long, sparsely puberulent and distinctly gland-dotted on outer surface; disk flowers numerous, about 2 mm. long, each subtended by a scarious oblong bract with a flabelliform erosulate apex; fruit dark brown, 3–3.5 mm. long, bearing several low longitudinal ridges and several conical-muticous tubercles along angles and about summit, smaller minute tuberulations on sides and back betwen low ridges, involucre barely overcapping achene and not forming a cup or hood, without a cusp or awn.

Along stream courses, on open plains, mesas, and hillsides, Lower Sonoran and Tropical Zones, central Sonora, south into Sinaloa and Nayarit, and in central and southern Baja California. Sept.–March.

**5. Melampodium hispidum** H. B. K., Nov. Gen. & Sp. **4:** 273, *pl. 399,* 1820

*Melampodium coronopifolium* Schultz-Bip. ex Hemsl., Biol. Centr. Amer. Bot. **2:** 145, 1881. (Nomen nudum.)

Erect annual 1–4 dm. tall, strict or branching almost from base, sparsely hispidulous throughout, stems more or less angled and striate, often reddish; leaves lance-linear to broadly elliptic-lanceolate, 5–12 mm. wide, 2–4.5 cm. long, entire or nearly so, sparsely scabro-hispidulous, paler beneath than above, acute at apex, cuneately narrowed to subsessile base; heads short-pedunculate or sessile, markedly surpassed by subtending leaves; involucres deeply saucer-shaped or broadly cup-shaped, about 5 mm. in diameter, gamophyllous, 5 lobes broadly ovate to deltoid-ovate, 2.5–3 mm. wide, sparsely hispidulous; ligules about 1–2 mm. long, rarely equaling outer involucral bracts; disk flowers few, often not over 3–8, slightly exserted beyond scarious, apically erosulate, receptacular bracts; fruit 3–3.5 mm. long, longitudinally ridged and strongly roughened with conical tuberculations along outer angles and less pronouncedly warty and transversely ridged on sides.

Hillsides and along watercourses, mostly Upper Sonoran Zone, but extending into the margins of the desert in canyons and along streams, southern Arizona to southern Mexico. Aug.–Sept.

**6. Melampodium sinuatum** Brandegee, Proc. Calif. Acad. II, **3:** 144, 1891

Perennial herbs 1.5–4 dm. tall from a slender woody root; stems slender, retrorsely hispid-scaberulous, often reddish; leaves firm, oblong-oblanceolate, lance-ovate or oblong-elliptic, acute at apex, slightly narrowed to a subauriculate sessile base, hirsutulous-scaberulous, especially on margins, obviously pinnately veined beneath, irregularly and shallowly repand-sinuate or smaller ones sometimes entire; peduncles slender, 2–12 cm. long; involucres broadly cup-shaped to hemispherical, 6–8 mm. broad, gamophyllous, moderately to densely pubescent with rather coarse, spreading hairs, and dotted with aromatic resinous glands, 5 lobes rounded, 2.5–3 mm. wide, 1.5–2 mm. long; ray flowers 9–14, ligules pale yellow, 6–10 mm. long, glabrous within, finely puberulent and gland-dotted on outer surface, persistent; disk corollas yellow, 1–1.3 mm. long, the narrow, short tube nearly as long as and gradually expanding into funnelform throat, lobes narrowly ovate, puberulent without, reflexed; receptacular bracts membranous, yellowish, erosulate at slightly flabelliform apex, glabrous; fruit about 3–3.5 mm. long, dull brown, finely ridged along angles and mid-dorsal line, muticous-tuberculate along angles and toward summit, sparsely puberulent.

Hillsides and along arroyos, Lower Sonoran and Tropical Zones, southern Baja California. Aug.–Nov.

## 39. PARTHENIUM L., Sp. Pl. 988, 1753

Suffrutescent plants or shrubs with alternate, entire to pinnatisect, glabrous to tomentose leaves. Heads small, radiate, in terminal panicles or lax corymbose inflorescences. Involucres broadly campanulate or hemispherical, bracts imbricate in 2 to several series, obtuse, subequal or outer ones successively shorter. Receptacle convex or conical with membranous paleae subtending perfect flowers. Ray flowers white, pistillate and fertile, ligules broad and short, bidentate. Disk flowers white, tube expanding into a small, 5-toothed, ampliate throat; anthers entire at base; achenes abortive. Achenes of pistillate flowers dorsally compressed, keeled along inner face, bearing 2–3 short hairy paleae at apex.

Leaves 1.5–4 cm. long, deeply lobed or pinnatifid; plants low shrubs about 0.5 m. tall or less ........ 1. *P. incanum*

Leaves 8–30 cm. long, entire or shallowly sinuate-repand; plants tall herbs 1–4 m. tall ........ 2. *P. stramonium*

**1. Parthenium incanum** H. B. K., Nov. Gen. & Sp. **4:** 260, *pl. 391*, 1820
*Parthenium ramosissimum* DC., Prod. **5:** 532, 1836.

Much-branched, compact shrub 3–5 dm. tall with finely tomentose herbage and light gray, striate and eventually slightly fissured bark; leaves ovate in outline, 0.8–2.5 cm. wide, 1.5–4 cm. long, somewhat irregularly lobed or pinnatifid about to middle of blade, densely tomentose, veiny and cinereous beneath, pale green, less densely tomentulose and moderately veined on upper surface, lobes mostly obtuse to rounded, 5–10 mm. long, entire or larger ones again shallowly and sparingly toothed; petioles 2–5 mm. long, densely tomentulose; heads in terminal cymose panicles, on naked or nearly naked branches 2–12 cm. long; involucres broadly cup-shaped, 3–4.5 mm. wide at anthesis, bracts graduate, ovate to suborbicular, tomentulose; ray flowers with ligules about 1–1.5 mm. long, only slightly exceeding involucres, mostly erect or only weakly spreading; sterile disk flowers about 2 mm. high; achenes 2–2.5 mm. long, dorsiventrally flattened, attached to subtending bract and to enfolded paleae of 2 sterile disk flowers immediately anterior to it by slender callous threads, these ultimately separating from achene at base but remaining attached to it above, thin bracts thus falling with achene, all these structures and achene glandular-puberulent and sparsely arachnoid-tomentulose; pappus paleae 2–3, minute.

Arid mesas and plains, Lower and Upper Sonoran Zones, Coconino, Mohave, and Yavapai Counties, Arizona, to Western Texas and south into Sonora, Coahuila, and Chihuahua. June–Oct.

**Parthenium argentatum** A. Gray, in Torr., Bot. Mex. Bound. 86, 1859, with lance-ovate to lanceolate, acute to acuminate, entire or slightly lobed, silvery-canescent leaves ("Guayule"), has been cultivated for its rubber in Arizona south of Tucson and escaped sparingly in the region.

**2. Parthenium stramonium** Greene, Pittonia **4:** 240, 1901
*Parthenium arctium* Bartlett, Proc. Amer. Acad. **44:** 635, 1909.

An upright, open, often slightly branched shrub 1–4 m. tall with cinereous-tomentulose branchlets and light gray, nearly smooth bark on older larger stems; leaf blades triangular-ovate to triangular-lanceolate, 2–10 cm. wide, 8–30 cm. long, truncate to subcordate and often asymmetrical at base, acute at apex, entire to weakly repand or crenulate, glabrescent and green above, tomentulose-velvety and strongly veined beneath; petiole to 5 cm. long, faintly winged above; heads small, numerous, in broadly cymose panicles; involucres hemispheric, 3–4.5 mm. broad, bracts graduate, obtuse to rounded at apex, sparsely to densely arachnoid-tomentulose, margins scarious; ligules 0.5–1 mm. long, fully as broad; disk flowers about 1.5 mm. long, puberulent; achenes similar to those of *P. incanum*.

Along rocky canyons and on outwash slopes, Lower and Upper Sonoran Zones, Sonora and Chihuahua. March–Sept.

## 40. PARTHENICE A. Gray, Pl. Wright. 2: 85, 1853

Erect herb with canescent to glabrate stems and alternate, petiolate, ample, finely tomentulose leaves. Heads small, disposed in lax panicles, discoid. Involucre hemispherical, bracts imbricate in several series, obtuse, outer ones subherbaceous, inner larger and membranous, subtending pistillate flowers. Outer whorl of flowers pistillate and fertile with tubular corollas, inner ones perfect, sterile, tubular-funnelform, ampliate limb 5-dentate. Anthers obtuse and entire at base, each bearing a straight, muticous, apical appendage. Achenes of pistillate flowers obovoid, dorsiventrally compressed, smooth and glabrous, attached by a callous thread to bases of paleae of 2 adjacent disk flowers and to a plane or slightly concave, larger, subtending bract and falling with them and paleae of disk flowers.

**1. Parthenice mollis** A. Gray, Pl. Wright. **2:** 85, 1853

Moderately to profusely branched herb to 2 m. tall; leaves ample, blades lance-ovate to broadly ovate-cordate, 1–15 cm. broad, 3–20 cm. long, upper smaller ones entire, larger coarsely and irregularly dentate, bright green and sparsely puberulent to glabrate above, paler and more densely puberulent beneath; petioles to 10 cm. long, blades often decurrent on upper portion; panicles 1–4 dm. long; involucres 4–5 mm. in diameter at anthesis; bracts obtuse or rounded, thin, densely short-puberulent and somewhat scurfy; corollas of sterile disk flowers 1–1.3 mm. long, tube very short, abruptly expanding into narrowly funnelform throat, lobes minute, densely arachnoid-puberulent or scurfy-villous, hairs of adjacent corollas intertangled; corollas of pistillate flowers tubular, about 0.3 mm. long or less; achenes 1–1.3 mm. long, compressed, attached to and falling with paleae of 2 disk flowers immediately adjacent.

Along watercourses and on hillsides and mesas, Lower and Upper Sonoran Zones, Arizona to Colorado, New Mexico, and northern Mexico to central Baja California and southern Sonora. Feb.–Sept.

## 41. TRAGOCEROS H. B. K., Nov. Gen. & Sp. 4: 248, *pl. 385,* 1820

Erect annual herbs with entire opposite leaves and pubescent herbage. Heads pedunculate at tips of terminal or axillary branchlets or sometimes sessile in axils. Heads radiate, ray flowers pistillate and fertile; disk flowers perfect but sterile. Involucres ovoid to turbinate; phyllaries graduate, imbricated in a few series. Receptacle plane, chaffy-bracted. Ray corollas ligulate; lamina spreading or recurved, firm, persistent. Disk corollas tubular, regular; throat and limb gradually ampliate, 5-toothed. Anthers entire or subentire at base. Style of disk flowers not divided. Achene of ray flowers oblong, somewhat compressed, epappose.

**1. Tragoceros zinnioides** H. B. K., Nov. Gen. & Sp. **4:** 249, *pl. 385,* 1820
*Tragoceros mocinianus* A. Gray, Proc. Amer. Acad. **21:** 388, 1886.

Erect, strict or much-branched herb 1.5–2.5 dm. tall, branching from base or only near top; stems slender, yellowish to slightly reddish, hirsute with minute subappressed hairs intermingled with spreading ones; leaves opposite, lance-linear to oblong-lanceolate, 4–12 mm. wide, 1.5–4 cm. long, acute to slightly attenuate at apex, rounded, broadly cuneate, or truncate at base, entire, sparsely hispid with subappressed to spreading hairs and dark green upper surface, paler, punctate-glandular and slightly more densely hirsute beneath, 1-nerved; heads sessile, terminating branchlets, each subtended by a pair of leaves; involucres cylindro-turbinate or cylindro-campanulate, 6–7 mm. high, phyllaries chartaceous and stramineous, glabrous, somewhat shining, rounded at apex, each with several fine nerves, outer ones broadly ovate, inner obovate; ray corollas firm to subrigid, ascendingly curved, white tinged with pale green, 5–8 mm. long, with 2 cord-like, submarginal veins, each vein ending in a minute tooth at anthesis, later thinner tissue between them splitting to produce 2 curved, ascending or slightly recurved, stiffish, hirsutulous awns that persist on achenes; achenes 5–6 mm. long, strongly compressed, with 3 low but sharp ridges on outer face, a single median one on inner face, puberulent on both faces, ridges on outer face minutely tuberculate.

Roadsides and on grassy, often shaded slopes, Lower Sonoran to Tropical Zones, vicinity of Ures, Sonora, south to central Mexico. Aug.–Sept.

## 42. ZINNIA L., Syst. ed. 10, 1221, 1759

Annual or perennial herbs or low shrubs with opposite, entire leaves. Heads radiate, pedunculate. Involucres cylindric to hemispheric; phyllaries oblong to obovate, graduated, dry-chartaceous or coriaceous, closely imbricated in 3 to many series, rounded or obtuse, often somewhat colored, usually erose to ciliate at apex. Receptacle conical to columnar, bearing a scarious-chaffy bract at base of, and enfolding each flower, bracts somewhat carinate, obtuse and often erosulate at apex. Ray flowers pistillate, fertile; ligules oval to oblong or suborbicular, 1–3-toothed or -lobed, white, red, yellow, or purple, tubed or tubeless, persistent. Disk corollas tubular, slightly zygomorphic, 5-toothed; limb scarcely ampliate, often velvety within; disk flowers fertile. Anthers truncate to free-sagittate at base; appendages apical, sterile, lance-triangular. Style branches acute to truncate at apex, faintly if at all appendaged. Achens of ray flowers compressed laterally or triangular, glabrous to ciliate-margined and pubescent, awnless, 2-toothed, or sometimes with a single awn from summit of inner angle. Achenes of disk flowers compressed to triangular in cross section, usually 3-dentate, 1–3 awns present or absent.

Plants perennial; leaves linear, 2–2.5 mm. wide; ligules 3–7, usually white or yellow, but sometimes reddish:
- Rays white with green nerves beneath; to 10 mm. long; leaves 1-nerved or faintly 3-nerved at base only; disk achenes 2.5–3.5 mm. long; receptacular bracts yellow throughout ........ 1. *Z. acerosa*
- Rays yellow on both sides, to 18 mm. long; leaves 3-nerved at least half way to apex; disk achenes 4–5 mm. long; receptacular bracts yellowish below, red-tipped ........ 2. *Z. grandiflora*

Plants annual; leaves lance-linear to ovate, to 3 cm. wide; ligules 6–15, often red ........ 3. *Z. peruviana*

### 1. Zinnia acerosa (DC.) A. Gray, Pl. Wright. 1: 105, 1850

*Diplothrix acerosa* DC., Prod. **5**: 611, 1836.
*Zinnia pumila* A. Gray, Mem. Amer. Acad. II, **4**: 81, 1849.
*Crassina pumila* Kuntze, Rev. Gen. Pl. **1**: 331, 1891.

A much-branched low rounded or flat-topped subshrub 1–2.5 dm. tall, with slender, cinereous-puberulent branches, irregularly scaly grayish bark and linear to oblanceolate-linear, acerose leaves 0.3–2.5 mm. wide; leaves numerous, often longer than internodes, 5–20 mm. long, cinereous-puberulent, golden punctate glands interspersed among hairs; peduncles 5–35 mm. long; involucres campanulate, 5–7 mm. long, 3–5 mm. wide, phyllaries suborbicular to oblong, greenish and tomentulose at first, later glabrate and stramineous but remaining dull; ray flowers mostly 5–7, ligules suborbicular to broadly oblong, 7–10 mm. long, white or faintly yellowish, strongly green-veined beneath, persistent, strongly reflexed in fruit; disk corollas 5–6 mm. long, often tinged with purple; disk achenes 2.5–3.5 mm. long, striate, strigose or only upwardly ciliate, pappus usually of 2–3 unequal awns or these much reduced; ray achenes oblanceolate, 3-angled in cross section, truncate at apex, 2.6–4 mm. long, awns fused to and continuous with main corolla nerves, or awns 3 and free, or solitary at inner angle at summit; receptacular chaffy bracts uniformly yellow, erose at apex.

Arid rocky slopes and mesas, Lower and Upper Sonoran Zones, southern Arizona to Sonora, Texas, Chihuahua, Zacatecas, Nuevo León, and San Luis Potosí. March–Nov.

**2. Zinnia grandiflora** Nutt., Trans. Amer. Phil. Soc. II, **7**: 348, 1841

*Zinnia grandiflora* f. *viridiflora* Horr & McGregor, Trans. Kans. Acad. **55**: 174, 1952.

Subshrub to 2.5 dm. tall, with greenish, striate, more or less strigose stems; leaves linear, 3-nerved at least half way to apex, 2–2.2 mm. wide, to 2.5 cm. long, strigose-scabrous; peduncles to 1.5 cm. long; heads narrowly campanulate, to cylindric, 5–8 mm. tall; phyllaries oblong, obtuse, often with a dark band at or just below summit, subsquarrose, erose-ciliate, often reddish at tips; rays 3–6, yellow on both sides, ovate to suborbicular, 10–18 mm. long, glabrous on upper surface, sparsely pubescent to glabrous beneath; disk flowers 18–24, red or green, to 10 mm. long, glabrous without; disk achenes oblanceolate, compressed or triquetrous, 4–5 mm. long, awns 1–4 or lacking, mostly 2 and unequal; ray achenes oblanceolate-linear, triquetrous, awns adnate to ray or free at tips, awn of inner angle minute or none; receptacular chaffy bracts obtuse, erose, usually reddish at tips.

Foothill and mountain slopes, upper edge of Lower Sonoran into arid Transition Zones, northern and eastern Arizona to Texas, Sonora, Coahuila, Durango, and Zacatecas. April–Nov.

**3. Zinnia peruviana** (L.) L., Syst. ed. 10, 1221, 1759

*Chrysogonum peruvianum* L., Sp. Pl. 920, 1753.
*Zinnia multiflora* L., Sp. Pl. ed. 2, 1269, 1763.
(For additional synonymy see Torres, Brittonia **15**: 12, 1963.)

An erect, simply branched annual 1–5 or rarely to 9 dm. tall; stems strigose or hirsute, striate, green in youth, eventually yellow or purplish; leaves lance-ovate, 10–25 mm. wide, 2.5–8 cm. long, sessile, rounded to subcordate at base, acute at apex, sparsely puberulent-scabrous and dotted with yellow, punctate glands on both surfaces, blades mostly 3-nerved; heads short-pedunculate at tips of few branches, a subtending pair of leaves often surpassing peduncles; involucres 12–18 mm. high, campanulate; phyllaries erect, spreading at maturity, obovate, glabrous, often shining, reddish tinged and slightly swollen at tips; ray flowers 6–15, ligules oblanceolate, 4–6 mm. wide, 8–25 mm. long, ciliate-margined, pubescent at throat, glandular on both surfaces, mostly glabrous above throat on upper surface, puberulent-scabrous beneath, dark red (in ours) or sometimes greenish, strongly 2-ribbed; disk flowers 12–50, lobes yellow inside, purplish to nearly black and glabrous outside, tube yellow below; disk achenes oblanceolate-linear, triquetrous or compressed, striate, sparsely ciliate, 7–9 mm. long, 1-awned at apex; ray achenes linear-oblanceolate to narrowly elliptic, 9–10 mm. long, striate, sericeous, awnless or with two tufts of short hairs at upper angles; receptacular chaffy bracts obtuse, erose, red to purple at tips, yellow below, exceeded by disk flowers.

Grassy slopes, plains and in partial shade on rocky hillsides, Upper Sonoran to Tropical Zones, Arizona to Florida, the West Indies, and South America. April–Oct. in our area.

## 43. SANVITALIA Gualt., in Lam., Jour. Nat. Hist. 2: 176, *pl. 33*, 1792

Annual or perennial herbs with simple or much-branched stems and opposite entire leaves. Heads small to medium-sized, pedunculate at ends of branchlets, radiate.

Involucres hemispherical to broadly campanulate; phyllaries slightly unequal, obscurely imbricate in 2–3 series. Receptacle plane or convex, bearing at base of each flower a folded, membranous-scarious bract, callous-cuspidate at tip. Ray flowers pistillate, fertile; ligules sessile and persistent on achenes. Disk flowers tubular, regularly 5-toothed; limb only slightly ampliate. Style branches truncate or tipped with a short appendage. Ray achenes narrowly 4-sulcate, pappus of 3 short awns or tubercles. Disk achenes 4-angled or triquetrous, epappose or essentially so.

### 1. **Sanvitalia abertii** A. Gray, Mem. Amer. Acad. II, **4:** 87, 1849

Annual herb 1–4 dm. tall, strict, or slender branches emerging from all of axils except lowermost; branchlets at first puberulent with ascendingly curved hairs, soon glabrate, often reddish-tinged in age; leaves lance-linear, 2–8 mm. broad, 2.5–8 cm. long, usually obviously 3-veined, narrowly cuneate at base, acute at apex, sparsely but sharply scaberulous on both surfaces, more densely so on margins; petioles 2–12 mm. long, blade gradually decurrent on upper part of petiole; involucres broadly campanulate, 5–6 mm. long, as wide or slightly wider; phyllaries linear to oblong, obtuse or acute at apex, rather strongly ribbed, broader ones tipped with a greenish callus, glabrous; ray flowers 5–10, ligules ovate, 1–5 mm. long, sessile, erect or slightly spreading, whitish or faintly greenish; receptacle conical; disk corollas about 2 mm. long or less; receptacular bracts very strongly ridged medially, surpassing flowers; ray achenes 3–4 mm. long, whitish, with 5 broad low ridges, not tuberculate, 3 terminal awns stout, conical, about 0.6 mm. long, somewhat spreading; disk achenes 2.5–3 mm. long, strongly 4-ribbed, ribs sinuate-interrupted or tuberculate, chalky-gray, epappose.

Arid mesas and rocky slopes, upper margins of Lower Sonoran Zone and into the Upper Sonoran and Transition Zones, eastern San Bernardino County, California, and northwestern Arizona to Texas, Chihuahua, and Sonora. July–Oct.

## 44. **HELIOPSIS** Pers., Syn. Pl. **2:** 473, 1807

Glabrous, villous or scabrous perennial herbs with opposite petiolate leaves and rather large radiate heads on long, scapelike peduncles, these often somewhat fleshy or enlarged just beneath heads. Involucres broadly campanulate to hemispheric; bracts narrow, unequal, 2–3-seriate, outer bracts herbaceous. Receptacles conical or elongate, beset with stiff paleaceous bracts subtending disk flowers. Ray flowers several, ligules spreading, firm and persistent, pale yellow, pistillate, fertile. Disk corollas regular; short tube only slightly ampliate to form cylindric throat, 5-toothed at apex. Style branches obtuse, short-appendaged, hirsutulous. Achenes plump, obtusely 3–4-angled; pappus none.

Axils of leaves sparingly pubescent to glabrous; ray achenes smooth or slightly rugulose .... ............................................... 1. *H. parvifolia* var. *parvifolia*
Axils of leaves densely tomentulose; ray achenes distinctly tuberculate .................. ...............................................1a. *H. parvifolia* var. *rubra*

### 1. **Heliopsis parvifolia** A. Gray, Pl. Wright. **2:** 86, 1853, var. **parvifolia**

An erect or ascending perennial herb with one to many stems 2–10 dm. tall from a woody rootstock; stems slender, slightly striate, sparsely puberulent with crispate hairs and often with 2 distinct lines of dense, slightly longer hairs leading from base of petiole to next lower node; leaves opposite; petioles slender, flattened, 1–5 cm. long, puberulent, especially along faintly winged margins; leaf blades triangular-ovate to lanceolate, 0.5–6 cm. broad, 2–10 cm. long, usually shallowly to sharply toothed or smaller ones entire, sometimes slightly hastate, sparsely short-puberulent and slightly scaberulous, mostly 3-veined; peduncles slender, to 3 dm. long, usually somewhat swollen just below heads; phyllaries obovate to oblong, several-nerved, puberulent, outer ones distinctly herbaceous, exceeding inner ones, 2.5–5 mm. long, obtuse; ligules 9–10 mm. wide, 2–3.2 cm. long, pale yellow beneath, slightly darker above, firm and persistent; disk corollas greenish yellow, 3–3.5 mm. long, glabrous; receptacular bracts coriaceous, linear-lanceolate, about 1–1.5 mm. longer than achenes, finely erosulate on margins; disk achenes about 4 mm. long, smooth to faintly rugulose, dark brown; ray achenes curved, greenish, smooth or rugulose.

Along washes and on canyon walls and slopes, Lower Sonoran to Transition Zones, southern Arizona to southwestern Texas and south into Chihuahua, Sonora and Durango. July–Oct.

**1*a*. Heliopsis parvifolia** var. **rubra** (T. R. Fisher) Wiggins, stat. nov.

*Heliopsis rubra* T. R. Fisher, Madroño **12:** 152, *fig. 2–4,* 1954.

Differing from var. *parvifolia* in having conspicuous concentration of white hairs in axils of leaves and lateral branches; in slightly to distinctly more scabrous leaf blades; in slightly darker colored tips of receptacular bracts in heads; and in irregular, coarse tuberculation on ray achenes and some disk achenes.

Canyon walls and along rocky water courses, Lower Sonoran Zone, central Baja California south to the Cape Region. Oct.–April.

## 45. ECLIPTA L., Mant. 157, 1771

Annual or perennial herbs with diffusely branching, decumbent to erect, more or less strigose stems and opposite, entire to dentate leaves. Heads small, radiate, pedunculate, apical or in axils of leaves, solitary or in few-headed clusters. Involucres hemispheric to broadly campanulate; bracts indistinctly biseriate, herbaceous, subequal or inner ones shorter than outer. Receptacle plane to convex, bearing a few slender paleae among inner ray flowers. Ray flowers narrow, entire to bidentate, pistillate and fertile. Disk flowers perfect and fertile, tubular; limb scarcely ampliate, shallowly 4–5-toothed. Style branches flattened; appendages very short, obtuse or triangular. Achenes of ray flowers narrowly triquetrous, often sterile. Disk achenes slightly compressed laterally, 4-angled, glabrous, more or less corky along angles, tuberculate or transversely rugose between them, epappose or with 2 short paleaceous scales.

**1. Eclipta alba** (L.) Hassk., Pl. Jav. Rar. 528, 1848

*Verbesina alba* L., Sp. Pl. 902, 1753.
*Verbesina conyzoides* Trew, Pl. Rar. 8, *pl. 6,* 1763.
*Eclipta erecta* L., Mant. 286, 1771.

Decumbent to suberect, annual or biennial herb with striate, ascendingly strigulose stems 1–8 dm. long; leaves opposite, lanceolate, oblong-lanceolate, or narrowly elliptic, 3–14 mm. wide, 1.5–6.5 cm. long, remotely serrulate, acute to obtuse at apex, sessile or short-petiolate, strigose-pubescent, hairs on under surface from conspicuous pustulate bases; heads on peduncles 0.4–6 cm. long, terminal and from axils of many leaves; involucres hemispheric, 5–9 mm. in diameter, phyllaries lanceolate, about 1.5–2.2 mm.

wide, 4–6 mm. long, slightly attenuate, 3–5-nerved, strigose-pubescent; rays white, numerous, 1.5–3 mm. long, numerous or sometimes almost wanting; disk corollas greenish, 1.5–2 mm. long, 4-toothed; achenes 1.5–2 mm. long, tuberculate on faces, those of disk corollas compressed-tetragonal in cross section, more or less corky-margined when mature, ray achenes trigonous, corollas deciduous some time before achenes are shed.

Moist ground along streams, shores of ponds, and along irrigation ditches, Upper Sonoran to Tropical Zones, Arizona to Nebraska, northern California, and Massachusetts and south into South America. March–Nov.

## 46. SCLEROCARPUS Jacq., Ic. Pl. Rar. *pl. 176,* 1782

Annual or perennial herbs with strigose to villous stems and alternate or opposite, dentate to entire leaves. Heads radiate, small to medium in size, pedunculate. Involucre campanulate, bracts more or less herbaceous, in 2 series, inner ones subtending ray flowers, 1–5 of outer ones obviously longer than others and somewhat spreading-ascending. Receptacle convex to conical, receptacular bracts subtending and enclosing disk flowers and becoming indurated, falling with mature achenes. Ray flowers neutral, ligules yellow, entire or denate. Disk flowers perfect, innermost usually sterile, outer fertile; disk corollas tubular, limb slightly ampliate and 3–5-parted or -toothed. Style branches elongated, appendages obtuse or acute, subclavate, slightly pubescent. Achenes tumidly obovoid, attenuate toward base, invested by rugose, indurated receptacular bract.

Involucral bracts all alike, narrowly spatulate, 0.5–3 mm. wide, spreadingly hispid with stiffish hairs 1.5–2.5 mm. long; receptacular bracts of sterile disk flowers linear, long-hispid at tips .................................................. 1. *S. spathulatus*

Involucral bracts dimorphic, outer ovate, 2.5–6 mm. wide, appressed-strigose with hairs 0.5–1 mm. long; receptacular bracts of sterile disk flowers oblong, short-hispid at tips .................................................. 2. *S. divaricatus*

### 1. Sclerocarpus spathulatus Rose, Contr. U.S. Nat. Herb. **1:** 103, 1891

An erect, branching, often rather straggly, annual herb to 1.5 m. tall, branches more or less striate and sparsely strigose with ascending, closely appressed hairs; leaves opposite below, alternate above lower 2–6 nodes; petioles slender, to 4 cm. long, flattened, sparsely scabrous-strigose; leaf blades ovate to lanceolate, 8–32 mm. wide, 2–8 cm. long or basal ones somewhat larger, scaberulous above, strigose beneath, broadly cuneate at base, acute at apex, serrate-dentate; heads on terminal peduncles to 1.5 or 2 dm. long; involucres about 1 cm. broad, bracts spatulate, 8–11 mm. long, 0.5–3 mm. wide at broadest part, hispid with rather coarse, white, spreading hairs 1.5–2.5 mm. long; ray flowers 8–15, yellow, very slender tube 3–5 mm. long, strigose, ligules 6–8 mm. wide, 10–12 mm. long, sparsely strigose on lower side; disk flowers slenderly tubular, greenish or greenish yellow, 7–10 mm. long, only slightly ampliate above, strigose with coarse hairs, those along margins of lance-linear lobes black and conspicuous under a lens; receptacular bracts subtending sterile, central disk flowers, linear-spatulate, about 5–6 mm. long, bearing a tuft of stiffish tangled hairs at tips; achenes obovoid, 4–5 mm. long, completely enclosed in black tuberculate bracts, puberulent.

Along arroyos, roadsides, canyons, and on flats and valley floors, often in shade of trees or shrubs, Lower Sonoran Zone, central Sonora southward into Sinaloa, Mexico. Sept.–Nov.

**2. Sclerocarpus divaricatus** (Benth.) Benth. & Hook. f. ex Hemsl., Biol. Centr. Amer. Bot. **2:** 164, 1881

*Gymnopsis divaricatus* Benth., Bot. Sulph. 116, 1844.
*Sclerocarpus triunfonis* M. E. Jones, Contr. West. Bot. **18:** 77, 1933.

Weedy annual herbs to 1 m. tall or more with slender, shallowly striate, sparsely strigose stems and branches; petioles slender, 1–4 cm. long; leaf blades lanceolate to lance-ovate, 2–5 cm. long, coarsely serrate-dentate, broadly cuneate to subtruncate at base, acute to attenuate at apex, strigose with forwardly appressed hairs on both sides; peduncles slender, 2–10 cm. long, terminating stems and branches; heads 2–2.5 cm. broad, involucres about 1 cm. broad; involucral bracts herbaceous, ovate to oblong-ovate, 2.5–6 mm. wide, 5–10 mm. long, appressed-strigose with fine hairs 0.5–1 mm. long; receptacular bracts 1.5–2.5 mm. long at anthesis, whitish and sparsely hirsute at tip; ray flowers bright orange within, paler yellow without, ligules 5–6 mm. wide, 8–10 mm. long, strigose on lower surface; disk corollas 8–10 mm. long, yellow to orange with black-margined lobes, puberulent without; in fruit, receptacular bracts oblong, 5–6 mm. long, coriaceous, sparsely hirsutulous at tips; achenes obovoid, about 3 mm. long, grayish or stramineous, only slightly tuberculate or rugose.

Shaded slopes, valleys, and along streams, margin of Lower Sonoran into Tropical Zones, southern Baja California, Sinaloa, and Yucatan south into Central America. Aug.–Oct.

## 47. TITHONIA Desf. ex J. F. Gmel., Syst. Nat. 1259, 1791

Herbaceous or shrubby plants with upper leaves alternate, lower sometimes opposite, linear-lanceolate to broadly ovate, triple-nerved, petiolate. Heads medium to large, radiate, rays uniseriate, neutral, disk flowers numerous, fertile. Involucres hemispheric to broadly campanulate, bracts 2–5-seriate, unequal to subequal, oval to lanceolate, indurate, ribbed below, herbaceous or membranous at tips. Receptacle convex, chaffy with rigid, striate, acute to aristate-acuminate, concave, persistent paleae. Ray flowers 8–20; ligules oval to oblong. Disk corollas with cylindrical or funnelform throat and short narrow tube, 5-toothed. Anthers cordate-sagittate; appendages ovate. Style branches dorsally hispid above with lanceolate to linear-lanceolate, acuminate, hispid appendages. Achenes oblong, thickened to subquadrangular in cross section, glabrous or pubescent. Pappus of 1–2 paleaceous awns and 4–12 squamellae (these free or united), or pappus lacking.

Plants annual; ligules orange, 7–12 mm. long; stems sparsely hispid-pilose; pappus awn solitary, 5.5–8 mm. long ........ 1. *T. thurberi*
Plants shrubby; ligules yellow, 20–35 mm. long; stems densely and canescently pilose-tomentose; pappus awns 1–2, 2–3 mm. long, or lacking ........ 2. *T. fruticosa*

**1. Tithonia thurberi** A. Gray, Proc. Amer. Acad. **8:** 655, 1873

*Tithonia palmeri* Rose, Contr. U.S. Nat. Herb. **1:** 104, 1891.
*Urbanisol thurberi* Kuntze, Rev. Gen. Pl. **1:** 371, 1891.

Slender, moderately branched annual 0.5–2 m. tall with whitish, striate stems sparsely to densely hispid-pilose with spreading, tuberculate-based, rather coarse hairs interspersed with finer ones, eventually glabrate; lower leaves opposite, upper ones alternate, blades ovate to broadly triangular-ovate, acute to acuminate, 3–27 cm. wide, 5–28 cm. long, cuneately decurrent along petiole from a cuneate to broadly cordate base, crenate-dentate with mucronulate teeth, deep green and hispidulous with incurved hairs above, paler green and sparsely granular or hispidulous and hispid-pilose along veins beneath; petioles 2–10 cm. long, margin irregularly dentate on larger leaves; peduncles fistulose, striate, 10–30 cm. long, spreadingly hispid-pilose and finely hispidulous; heads 2.5–3.5 cm. wide; involucres of graduated bracts in 3 series, 1–2 cm. tall, outer bracts lanceolate, oblong-ovate, or obovate, acute to acuminate, hispid-pilose and sparsely hispidulous, indurated below, tip herbaceous; innermost bracts broadest, submembranous, obtuse to mucronulate at tip, only granular or strigulose; rays 5–10 (usually about 8), orange-yellow, oval-oblong, 4–6 mm. wide, 7–12 mm. long; disk corollas 6–6.2 mm. long, hispidulous on lobes, glandular-pilosulous below, tube about 1 mm. long, throat cylindrical; receptacular bracts 1–1.5 cm. long, oblong, abruptly acuminate at tip with a small tooth on each side just below tip, striate-ribbed, smooth; achenes oblong-obovate, 8.5–9 mm. long; pappus awn solitary on outer angle of achene, linear-subulate, 5.5–8 mm. long; squamellae 4–8, unequal, 4 mm. long or less, lanceolate.

In rich soil along banks of streams, ditches, and margins of reservoirs, usually in partial shade, Lower and Upper Sonoran Zones, Pima and Santa Cruz Counties, Arizona, south to central Sonora. Aug.–Sept.

**2. Tithonia fruticosa** Canby & Rose, Contr. U.S. Nat. Herb. **1:** 104, *pl.* 5, 1891

A robust shrub 3–4 m. tall; woody stems grayish, to 10 cm. in diameter; younger branches densely canescent-pilose-tomentose with spreading hairs; lower leaves opposite, upper alternate, blades ovate to lanceolate, acuminate to attenuate and often falcate, crenate-serrate, cuneately narrowed to winged petiole, 2–14 cm. wide, 6–30 cm. long, 3-nerved and conspicuously veined beneath, canescent-pilose on both sides, more densely so beneath than above, gland-dotted; petioles 3–5 cm. long, thin-margined but not auriculate; heads terminal and axillary, 7–9.5 cm. wide; peduncles shorter than subtending leaves, fistulose and striate; disk 2–2.5 cm. high, 2.5–4.5 cm. thick; involucral bracts in 4 series, strongly graduated, 2–2.8 cm. long, broadly oval to obovate, 8–12 mm. wide, rounded or infrequently obtuse at tip, obscurely indurate at base, inner ones sparsely pubescent, outer ones densely canescent-pilosulous with appressed, tuberculate-based hairs; rays 14–20, yellow, narrowly elliptic, 2–3.5 cm. long; disk corollas 8–9 mm. long, hispidulous without, at least below; receptacular bracts lance-ovate, abruptly acute, 12–14 mm. long, glabrous to hispidulous above; achenes 4–5 mm. long, appressed-pilose; pappus squamellae connate into a laciniate-fimbriate paleaceous crown 1–1.5 mm. high, awns lacking or of teeth about 2–3 mm. long.

Along watercourses in shaded canyons, Lower Sonoran and Tropical Zones, southern Sonora to Chihuahua, Durango, and Sinaloa. March–April.

**Tithonia auriculata** (Brandegee) S. F. Blake, Contr. Gray Herb. **54:** 9, 1918 (*Gymnolomia auriculata* Brandegee, Zoe **5:** 223, 1905), an herbaceous, many-stemmed perennial with linear-lanceolate, cordate-auriculate, subsessile leaves; heads 3.8–6 cm. broad; ligules 14–24 mm. long; and epappose, glabrous achenes 3.5 mm. long, has been collected in the Alamos region in southern Sonora by H. S. Gentry and may occur as a straggler along streams in the southernmost part of our area nearby, but no specimens from that territory have come to my attention.

## 48. VIGUIERA H. B. K., Nov. Gen. & Sp. 4: 224, *pl. 379*, 1820

Annual or perennial herbs, frutescent plants or shrubs, with leaves opposite at least below. Heads small to large, radiate, corollas usually yellow, sometimes with whitish rays or purplish disk. Involucres subcampanulate to subcylindric; phyllaries 2- to several-seriate, subequal or graduated, usually indurated and ribbed at base and herbaceous toward apex. Receptacle slightly convex to low-conical, a chartaceous or scarious, persistent palea usually subtending and embracing each disk achene. Rays 5–32; ligules usually 2–3, denticulate apically. Disk corollas tubular-funnelform, 5-toothed; tube shorter than throat. Anthers faintly sagittate; apical appendages ovate. Style branches with acute to acuminate, terminal, sterile appendages. Ray achenes sterile. Disk achenes often more or less quadrangular. Pappus of 1–3 awns and several squamellae, or none.

Plants annual; leaves linear to linear-lanceolate, five to ten times as long as wide, margins more or less revolute; pappus none .................................. 1. *V. ciliata*

Plants perennial; leaves lanceolate to broadly ovate or deltoid-ovate, rarely more than five times as long as broad, usually proportionately wider, margins usually not revolute; pappus present:

- Involucral bracts pale green or canescent, not blackish, narrowly ovate to linear, not oblong-ovate and rounded:
  - Plants distinctly shrubby:
    - Involucres 4–10 mm. wide; achenes 2.8–3.5 mm. long; ligules 5–15 mm. long; leaves rarely tomentose:
      - Leaf blades lance-ovate, laciniately toothed, often sinuately margined; receptacular paleae 7–8 mm. long .................................... 2. *V. laciniata*
      - Leaf blades ovate or deltoid-ovate, not sinuately margined, shallowly toothed or entire; receptacular paleae 4–5 mm. long:
        - Involucres 4–6 mm. wide; leaf blades ovate, entire, 5–12 mm. long; leaves alternate ............................................. 3. *V. microphylla*
        - Involucres 6–10 mm. wide; leaf blades ovate to deltoid-ovate, usually dentate, mostly over 2 cm. long; leaves opposite:
          - Pubescence of spreading or slightly curved hairs, these usually harsh; leaves rarely cinereous beneath; margins dentate:
            - Heads usually several, cymosely disposed; leaf blades 1.5–6.5 cm. long; bases of hairs not strongly pustulate-tuberculate .................................................................. 4. *V. deltoidea* var. *deltoidea*
            - Heads usually 1–2 on long peduncles; leaf blades rarely over 3 cm. long; bases of hairs strongly pustulate-tuberculate ..... 4*a*. *V. deltoidea* var. *parishii*
          - Pubescence mostly softer, appressed, often of tangled hairs, rarely harsh; leaves cinereous, at least beneath, mostly entire ................................................................ 4*b*. *V. deltoidea* var. *chenopodina*
    - Involucres 10–20 mm. broad; achenes 4–4.8 mm. long; ligules 8–25 mm. long; leaves distinctly tomentose or softly pilosulous:
      - Upper surface of leaves harshly scabrous with stiff, slightly curved hairs; achenes 4 mm. long; pappus squamellae 0.5–1.5 mm. long, erosulate ............................................................ 4*c*. *V. deltoidea* var. *tastensis*
      - Upper surface of leaves silky-pubescent to subtomentose; achenes 4.5–4.8 mm. long; pappus squamellae 0.6–0.8 mm. long, lacerate .............. 5. *V. tomentosa*
  - Plants suffrutescent, frutescent or herbaceous:
    - Leaves crowded near ground; heads on tall, scapelike branches; ligules 20–25 mm. long; herbage densely lanate-tomentose ........................ 6. *V. lanata*

Leaves not basal, distributed nearly to inflorescences; ligules 6–20 mm. long; herbage sparsely and more harshly pubescent, not lanate:

Disk corollas 6 mm. long; receptacular paleae 8–10 mm. long; achenes 6–6.5 mm. long; leaf blades usually cordate at base .................... 7. *V. cordifolia*

Disk corollas 3–4 mm. long; receptacular bracts 4–5.5 mm. long; achenes 2.5–3.8 mm. long; leaf blades truncate to cuneate at base:

Leaf blades incised-serrate with 5–10 pairs of coarse, often again denticulate teeth; achenes 2.8–3 mm.. long .............................. 8. *V. subincisa*

Leaf blades serrate to entire, if toothed teeth never denticulate on their margins; achenes 3.5–3.8 mm. long:

Leaf blades ovate to broadly lanceolate, to 8 cm. wide .................... .......................................... 9. *V. dentata* var. *dentata*

Leaf blades narrowly lanceolate, 5–23 mm. wide .......................... .................................... 9*a*. *V. dentata* var. *lancifolia*

Involucral bracts dark green to nearly black, broadly ovate or oblong-ovate, rounded at apex ................................................... 10. *V. purisimae*

## 1. Viguiera ciliata (Rob. & Greenm.) S. F. Blake, Contr. Gray Herb. **54:** 113, 1918

*Heliomeris multiflora* var. *hispida* A. Gray, Pl. Wright. **2:** 87, 1853.
*Gymnolomia hispida* Rob. & Greenm., Proc. Bost. Soc. Nat. Hist. **29:** 93, 1899. Not *V. hispida* Baker, 1884.
*Gymnolomia hispida* var. *ciliata* Rob. & Greenm., loc. cit.
*Gymnolomia ciliata* Rydb., Bull. Torrey Club **37:** 328, 1910.

Slender annual or possibly sometimes biennial with one to several simple or slightly branching stems to about 1 m. tall from a slender taproot, stems sparsely strigose to densely hispid-pilose with spreading hairs, often reddish; leaves mostly alternate, linear to linear-lanceolate, 1–5 mm. wide, 3–8.5 cm. long, attenuate to acuminate at each end, more or less revolute, sparsely tuberculate-strigillose and hispid-ciliate along margins, 1-veined; petioles short, to 2 mm. long, hispid; heads few to many, 1.5–4 cm. wide, on peduncles to 7 cm. long, terminating upper branches; involucres 6–7 mm. high, 8–14 mm. wide, its bracts biseriate, 6–12 mm. long, lanceolate to linear-lanceolate, acuminate to attenuate, more or less callous-mucronate, 1–3-ribbed, sparsely ciliate-hispid along margins and sometimes on back, especially below; rays 12–14, ligules 6–12 mm. long; disk corollas about 3 mm. long, sparsely puberulent below; receptacular paleae 4.5–5 mm. long, abruptly cuspidate, hispid-pilose toward apex; achenes blackish, glabrous, about 2 mm. long; pappus none.

Along canyons, streams, ditches, and in fields, Lower and Upper Sonoran Zones, southern Utah to New Mexico, Arizona, and central or southern Sonora. Sept.–Oct.

## 2. Viguiera laciniata A. Gray, in Torr., Bot. Mex. Bound. 89, 1859

Rounded, rather straggly shrub to 1.5 m. tall with slender, brownish, puberulent and somewhat scurfy branchlets and dark gray, slightly fissured bark on older stems; petioles slender, 5–9 mm. long, pale; leaf blades lanceolate to lance-ovate, to 1.5 cm. wide, 1.5–4.5 cm. long, laciniately repand-lobed, green and harshly tuberculate-hispid on both surfaces, conspicuously reticulate-veined beneath; heads 1.5–3 cm. in diameter, in cymose panicles of 3–15 heads; involucres hemispheric, about 1 cm. in diameter; phyllaries lance-subulate, 4–10 mm. long, hispidulous; ligules yellow, 4–5 mm. wide, 10–15 mm. long, puberulent beneath, especially along veins; disk corollas tubular-funnelform, about 5 mm. long, sparsely puberulent or nearly glabrous, tube proper 1–1.2 mm. long; receptacular paleae 7–8 mm. long, oblong, acute, puberulent on back;

achenes black, about 3 mm. long, compressed and 4-angled, sparsely strigillose with slightly reddish hairs; pappus squamellae about 0.3 mm. long, laciniate-erose, reddish or tawny.

Along dry washes, on hillsides and rocky mesas, Lower and Upper Sonoran Zones, San Diego County, California, south to central Baja California. Feb.–June.

**3. Viguiera microphylla** Vasey & Rose, Proc. U.S. Nat. Mus. **11:** 535, 1889

A slenderly branched, open shrub to 6 dm. tall with minutely strigillose, faintly striate stems and small alternate leaves on short petioles 1–2.5 mm. long; leaf blades ovate, 3–8 mm. wide, 5–12 mm. long, broadly cuneate to truncate at base, mostly obtuse to rounded at apex, entire, canescent-strigillose on both surfaces; heads corymbosely or paniculately disposed, 2–4 on each flowering branch, peduncles slender, 1.5–6 cm. long; involucres campanulate, 4–6 mm. broad at anthesis, phyllaries ovate to oblong-lanceolate, obtusish, strigillose-canescent; ligules yellow, 2.5–3.5 mm. wide, 5–6 mm. long, sparsely puberulent beneath; disk corollas about 4 mm. long, somewhat puberulent without, lobes broadly spreading or slightly reflexed; receptacular paleae oblong, erosulate and acute at apex; achenes about 3 mm. long, densely strigillose with reddish brown, silky hairs; main pappus paleae about 2 mm. long, subulate, squamellae about one-third as long but about same shape.

Sandy plains and washes, Lower Sonoran Zone, west-central Baja California. Feb.–May.

**4. Viguiera deltoidea** A. Gray, Proc. Amer. Acad. **5:** 161, 1861, var. **deltoidea**

*Viguiera deltoidea* var. *genuina* S. F. Blake, Contr. Gray Herb. **54:** 90, 1918.
*Viguiera triangularis* M. E. Jones, Contr. West. Bot. **18:** 75, 1933.

An open, more or less straggly shrub 1–3 m. tall with slender, strigillose to hispid stems and peduncles; leaves petiolate, petioles 3–12 mm. long, strigillose, blades deltoid-ovate, 1.5–6.5 cm. long, 0.5–4.5 cm. wide, usually dentate, truncate, subcordate or less commonly broadly cuneate at base, acute to obtuse at apex, harshly to rather softly pubescent on both surfaces, hairs spreading or curved but rarely appressed, moderately reticulate beneath; heads several at ends of branching, naked, terminal peduncles; involucres campanulate, 6–10 mm. wide, phyllaries lance-linear, 3–6 mm. long, acute, indurated at base, puberulent, dotted with minute, spherical, golden glands; ligules yellow, 3–7 mm. wide, 6–15 mm. long, puberulent along several veins beneath, glabrous within; disk corollas yellow, 4–5 mm. long, narrowly funnelform-tubular, puberulent without, ovate-deltoid lobes spreading; receptacular bracts paleaceous, 4–5 mm. long, acute to muticous at apex; achenes about 3 mm. long, somewhat compressed, oblong, distinctly ridged, strigose with brownish, forwardly appressed hairs; pappus squamellae unequal, 0.5–1 mm. long, erosulate at apex, paleae lance-subulate, attenuate, slightly longer.

Plains, arroyos, hillsides, and mesas, chiefly Lower Sonoran Zone, central Baja California to the Cape Region. Nov.–April.

**4*a*. Viguiera deltoidea** var. **parishii** (Greene) Vasey & Rose, Contr. U.S. Nat. Herb. **1:** 72, 1890

*Viguiera parishii* Greene, Bull. Torrey Club **9:** 15, 1882.

Leaves somewhat smaller, more strongly reticulate than in var. *deltoidea,* hairs on leaves and stems with heavy conical bases, conspicuously spreading; heads mostly solitary, long-peduncled, sometimes 2–3 on a peduncle.

Arid mesas, hillsides, margins of arroyos and plains, Lower Sonoran Zone, southern Arizona and Inyo County, California, south into northern Sonora and Baja California. March–June and Sept.–Nov.

**4b. Viguiera deltoidea** var. **chenopodina** (Greene) S. F. Blake, Contr. Gray Herb. **54:** 91, 1918

*Viguiera chenopodina* Greene, Leaflets Bot. Obs. **2:** 154, 1911.

Differing from var. *deltoidea* chiefly in having entire, deltoid-ovate to rhombic-ovate leaves 1.5–2.5 cm. long and canescent-strigillose pubescence at least on under side of leaf blades.

Dry washes, sandy flats and mesas, Lower Sonoran Zone, Vizcaíno Desert and the Magdalena Plain, and on Isla Carmen, Baja California. Feb.–May.

**4c. Viguiera deltoidea** var. **tastensis** Brandegee, Zoe **5:** 161, 1903

Shrub to 6 m. tall, conspicuously different from all other varieties of *V. deltoidea* in having deltoid-ovate leaf blades to 6 cm. wide, 5–10 cm. long, on petioles 1.5–2 cm. long; lower surfaces of leaves densely and softly hispidulous-pilosulous; heads in corymbose clusters 4.5–5 cm. wide, involucres 1.5–2 cm. wide; ligules 6–10 mm. wide, 1.5–2.5 cm. long, densely pubescent near base on lower surface; disk corollas 5 mm. long, tube proper about 1 mm. long; receptacular bracts 8–9 mm. long, puberulent along main vein dorsally, acute but not erosulate; achenes black, 4 mm. long, strigillose-hispid with upwardly appressed, silky hairs; main pappus paleae 3–4 mm. long, linear-subulate; squamellae 0.5–1.5 mm. long, erosulate.

Margins of arroyos and rocky hillsides, Lower Sonoran and Tropical Zones, southern Baja California. March–Nov.

**5. Viguiera tomentosa** A. Gray, Proc. Amer. Acad. **5:** 161, 1861

Shrubby, to 3 m. tall with slender, pilosulose, straggly or drooping branches, nodes often more densely pilose than internodes; leaves opposite except in inflorescences, ovate to triangular-ovate, acuminate at apex, truncate to shallowly cordate at base, serrulate to subentire, canescent above, densely soft-tomentose, canescent beneath, 2–8 cm. wide, 4–10 cm. long; petioles stoutish, densely canescent-pilosulous, 5–16 mm. long; heads corymbosely arranged, 3–4 cm. wide on peduncles 1–5 cm. long, these densely pubescent; involucres 10–16 mm. wide, bracts biseriate, 5–10 mm. long, their bodies ovate to oblong-ovate, apex linear-lanceolate, callous-mucronate at tip, densely pilose-tomentose; rays about 8, 5–10 mm. wide, 8–20 mm. long, bright yellow; disk corollas pilose on tubes, about 5.5 mm. long, dark brownish yellow; receptacular paleae 7–9.5 mm. long, nearly glabrous; achenes 4.5–4.8 mm. long, appressed-silky-pilose; awns slender, 3.8 mm. long, 4–6 squamellae 0.6–0.8 mm. long, lacerate, distinct.

Valley floors, arroyos, and rocky hillsides, Lower Sonoran and Tropical Zones, southern Baja California from the vicinity of Loreto south. Sept.–March.

**6. Viguiera lanata** (Kell.) A. Gray, Proc. Amer. Acad. **17:** 218, 1882

*Bahiopsis lanata* Kell., Proc. Calif. Acad. **2:** 35, 1863.

Frutescent plants 2–6 dm. tall with densely pannose-tomentose leaves, stems, peduncles, and phyllaries; leaves crowded at or near base of several- to many-stemmed plant; petioles stoutish, 0.5–3 cm. long; leaf blades oval, broadly ovate or upper narrowly ovate, 1–4 cm. wide, 2–5.5 cm. long, entire or obscurely dentate, reticulate-veined, broadly cuneate to truncate at base; heads corymbosely disposed at tips of paniculately branching, nearly leafless, scapose stems; involucres about 1.5 cm. in diameter, deeply saucer-shaped; phyllaries 6–9 mm. long, linear to lance-oblong, acute or abruptly obtusish; ligules yellow, 7–10 mm. wide, about 2.5 cm. long, puberulent beneath; disk corollas 4–5 mm. long, densely hispidulous on short tube, slightly less so on narrowly funnelform throat; receptacular paleae scarious-chartaceous, oblong, acute, about 8 mm. long, hispidulous along midvein dorsally; achenes 4–4.5 mm. long, dull black, strigillose with rufous hairs, 4-angled; pappus squamellae about 1 mm. high, erose, tawny.

Known only from Cedros and Natividad Islands but to be expected along the adjacent headland of the peninsula. March–May.

**7. Viguiera cordifolia** A. Gray, Pl. Wright. **1:** 107, 1852

Herbaceous perennial with 1 to several stems about 1 mm. tall from a heavy root, stems finely granular and moderately beset with stiff, coarse, pustulate-based, roughened hairs; leaves opposite, or rarely alternate above; petioles stout, 0.3–1 cm. long; leaf blades ovate, ovate-deltoid or uppermost often lanceolate, acute to acuminate or mucronate at apex, cuneate-rounded to shallowly cordate at base, sharply serrate or upper ones entire, 1.5–7 cm. wide, 2–10 cm. long, green and rough with curved, tuberculate-based, roughish hairs and somewhat granular-glandular, strongly 3-veined from base; heads numerous in terminal panicles, 2.5–4.5 cm. wide; peduncles 1.5–15 cm. long, hispidulous-scabrous; involucres 10–18 mm. wide, 7.5–15 mm. high, bracts lance-subulate and attenuate to narrowly oblong or spatulate-oblong, mucronate or acuminate, herbaceous to strongly indurated, hispid-scabrous with stiff, upwardly curved, whitish hairs along midrib, some of them to 1.5 mm. long; rays 6–8, ligules 3–9 mm. wide, 9–20 mm. long; disk corollas about 6 mm. long, gland-dotted, puberulent at base and on teeth; receptacular paleae 8–10 mm. long, acuminate or cuspidate; achenes 6–6.5 mm. long, appressed-pilose; awns about 5 mm. long, 4–6 squamellae 0.8–2 mm. long, distinct or united at base.

Canyons, rocky slopes, and plains, mostly Upper Sonoran and Transition Zones, but occasionally in the Lower Sonoran Zone, southern and eastern Arizona to western Texas, south into adjacent northern Mexico. June–July and Sept.–Oct.

**8. Viguiera subincisa** Benth., Bot. Sulph. 27, 1844

Suffrutescent plant 3–6 dm. tall, stems minutely and sparsely strigillose in and just below inflorescences, glabrous below; leaves opposite near base, alternate above, blades ovate, 2.5–4.5 cm. wide, 4–6.5 cm. long, obtuse to subacuminate at apex, cuneate to subtruncate at base, incisely serrate with 5–10 pairs of coarse triangular teeth, these often again more finely dentate, slightly carnose, minutely and sparsely strigillose on both surfaces; petioles 1–4.5 cm. long; heads 10–26 in narrow, long-peduncled, terminal, cymose panicles 1.5–3.5 cm. in diameter; involucres 10–12 mm. in diameter, phyllaries 10–12, ovate-oblong, in 2 series, about 3.5 mm. long, subequal, constricted near middle, densely strigillose, indurated and ribbed at base; rays 6–8, ligules 3–6 mm. wide, 6–16 mm. long; disk corollas about 3.5 mm. long, puberulent without; receptacular paleae 4.5–5.5 mm. long, mucronulate and puberulent dorsally at tip; achenes 2.8–3 mm. long, silky-villous; awns 2.5–2.8 mm. long; squamellae usually 4, quadrate, more or less united, about 1 mm. long.

On sandy or rocky soil, Lower Sonoran Zone, vicinity of Magdalena Bay, Baja California. Jan.–May.

**9. Viguiera dentata** (Cav.) Spreng., Syst. **3:** 615, 1826, var. **dentata**

*Helianthus dentatus* Cav., Ic. **3:** 10, *pl.* 220, 1795.
*Viguiera brevipes* DC., Prod. **5:** 579, 1836.
(For additional synonymy see Contr. Gray Herb. **54:** 83, 1918.)

Perennial herb with numerous slender branches 1–2 mm. tall, these sparsely strigillose to subglabrate; leaves opposite below, alternate toward inflorescences, ovate-rhombic to broadly lanceolate, 1–8 cm. wide, 3–13 cm. long, cuneate to truncate at base and tapering into petiole, acute to long-acuminate at apex, serrate to entire, bright green above, paler beneath, sparsely strigillose to densely pilose on both surfaces, bases of hairs pustulate; petioles to 5.5 cm. long, often narrowly winged toward blade; heads numerous, 2–3.7 cm. wide, cymosely paniculate, on slender, naked or minutely bracteolate peduncles 3–14 cm. long; fruiting disks 1–1.4 cm. broad, 5–13 mm. high; involucral bracts 3-seriate, graduate to subequal, phyllaries with ovate indurated bases and costae and linear to oblong, herbaceous apices, appressed-canescent to hispidulous; rays 10–12, yellow, ligules 7–15 mm. long, 3–7.5 mm. wide; disk corollas puberulent, 3–4 mm. long; achenes obovate-oblong, appressed-pubescent, 3.5–3.8 mm. long; awns slender, 2.2–2.8 mm. long, 4 squamellae quadrate, free, about 0.8 mm. long, fimbriate at apex.

Arid slopes and canyons, in fields, along ditches, and occasionally in woods, Lower and Upper Sonoran Zones, southern Arizona to Texas and south to Chiapas and Oaxaca. June–Oct.

**9*a*. Viguiera dentata** var. **lancifolia** S. F. Blake, Contr. Gray Herb. **54:** 86, 1918

Differing from var. *dentata* in having narrowly lanceolate leaves 5–23 mm. wide, 4–15 cm. long, acutely cuneate at base, acuminate at apex, mostly sparsely strigose-pilose and bright green.

Arid canyons, washes, and rocky hillsides, Lower Sonoran Zone, Pima County, Arizona, to central Sonora. March–Oct.

**10. Viguiera purisimae** Brandegee, Proc. Calif. Acad. II, **2:** 173, 1889

An openly branched shrub or suffrutescent perennial to 12 dm. tall; branches ashy-gray, obscurely striate and harshly scabrous with heavy, conical-based, abruptly curved and appressed hairs; leaves opposite, petioles to 3 cm. long, appressed-scabrous; leaf blades obovate to elliptic, 1–4 cm. wide, 2–6 cm. long, rounded or obtuse and abruptly apiculate at apex, broadly to narrowly cuneate at base, strongly 3-veined from base, dark green and scaberulous above, paler and only slightly more densely puberulent beneath; heads in corymbose panicles of 2–8 heads; involucres hemispheric, 1–1.8 cm. broad, 8–10 mm. deep, phyllaries ovate, apiculate, graduate and imbricate, dark green to blackish, sparsely puberulent on back, densely hirsute-ciliate on margins; ligules 10–12 mm. long, only faintly puberulent beneath; disk corollas very heavily veined along margins of lobes and below sinuses; achenes dull black, 2.5–3 mm. long, glabrous, crowned with a minute ring of reddish squamellae 1 mm. long and 2 awns 2.5–4.5 mm. long.

Arroyos and bottomland, Lower Sonoran Zone, west-central Baja California. Nov.–March.

## 49. ALVORDIA Brandegee, Proc. Calif. Acad. II, 2: 174, 1889

Suffrutescent herbs or small shrubs with opposite, entire or subentire leaves and small heads in pedunculate glomerules in upper axils and at tips of branches. Heads sessile or nearly so, radiate or discoid. Involucres narrowly campanulate to urceolate, somewhat contracted near tip. Phyllaries graduate, in few series, moderately imbricate. Receptacular bracts chaffy, embracing achenes. Ray flowers fertile or absent. Disk flowers perfect, fertile; corollas tubular-funnelform, gradually ampliate upward, 5-lobed. Achenes oblong, quadrangular and slightly compressed, truncate at apex. Pappus of 2–4 paleaceous awns and two or three times as many short paleaceous squamellae.

Rays present, 1–3 in each head; phyllaries dark green, strigose dorsally; longer awns of pappus equaling or exceeding achenes:
 Achenes of disk flowers 3.5–4 mm. long, as long as pappus awns; leaves lanceolate to ovate-lanceolate, acute at apex .................... 1. *A. glomerata* var. *glomerata*
 Achenes of disk flowers 2.5–2.8 mm. long, shorter than pappus awns; leaves oval, ovate, or obovate, usually obtuse at apex .................. *1a. A. glomerata* var. *insularis*
Rays absent, or minute and inconspicuous when rarely present; phyllaries stramineous to yellowish, glabrous dorsally, marginally ciliolate toward tips; awns of pappus rarely as long as achenes:
 Leaves ovate, one and one-half to three times as long as wide; inflorescence sparingly branched, 1.5–2.5 cm. in diameter; disk flowers 2 (rarely 3) in each head ........ ............................................................ 2. *A. brandegeei*
 Leaves narrowly lanceolate, three to six times as long as wide; inflorescence moderately branched, 3–4 cm. in diameter; disk flower 1 (rarely 2) in each head ........... ............................................................ 3. *A. fruticosa*

## 1. **Alvordia glomerata** Brandegee, Proc. Calif. Acad. II, **2:** 174, 1889, var. **glomerata**

A rounded shrub 1–2 m. tall and nearly as broad with slender, shallowly striate, strigillose branches, these at length glabrate, often flesh-colored; petioles 3–12 mm. long, strigillose-scaberulous; leaf blades ovate to lance-ovate, 7–20 mm. broad, 2–6.5 cm. long, cuneate at base, acute to slightly attenuate at apex, entire or faintly crenate-dentate, distinctly 3-nerved, scabrous-strigillose on both surfaces, dark green; heads in dense glomerules 1.5–2.2 cm. broad at tips of branches; involucres 6–7 mm. long, narrowly campanulate; phyllaries ovate to oblong-ovate, 1.2–3 mm. broad, basal ones 1.5–2 mm. long, densely strigillose dorsally, dark green or somewhat cinereous, margins densely puberulent-ciliate on upper one-half or more; ray flowers usually 1–3, ligules 5–6 mm. long, erect to spreading, yellow; disk flowers 4–6 per head, 5–6 mm. long, glabrous; receptacular bracts closely enfolding achenes, acute; achenes of ray flowers linear, sterile, 3–4 mm. long, pale, with 3–6 lanceolate paleae 0.4–1 mm. long; disk achenes 3.5–4 mm. long, narrowly obpyramidal, black, strigillose, truncate above, with 2–4 lanceolate paleae 3.5–4 mm. long and 6–12 shorter ones about half as long.

Rocky hillsides, mesas, and outwash slopes, Lower Sonoran Zone, central Baja California from vicinity of Calmallí south through the Sierra de la Giganta and on Isla Partida. Dec.–April.

## *1a.* **Alvordia glomerata** var. **insularis** Carter, Proc. Calif. Acad. IV, **30:** 161, 1964

Differing from var. *glomerata* in having leaves slightly less than twice as long as wide, often obtuse but abruptly cuspidate at apex; involucres 5–7 mm. long; outer phyllaries only slightly strigillose; ligules of ray flowers 4–5 mm. long; achenes of disk flowers smaller, only 2.5–2.8 mm. long.

Rocky arroyos and adjacent hillsides, Lower Sonoran Zone, islands off the east coast of Baja California from Isla San Marcos to Isla Santa Catalina. March–April; Nov.

**2. Alvordia brandegeei** Carter, Proc. Calif. Acad. IV, **30:** 163, 1964

*Alvordia fruticosa* sensu S. F. Blake, Contr. Gray Herb. **52:** 42, 1917. Not *A. fruticosa* Brandegee, Erythea **7:** 5, 1889.

An openly branched shrub 1–2 m. tall with slender, slightly striate, finely cinereous-puberulent, eventually glabrate branches; cortex tawny, gray and slightly exfoliating in age; leaves opposite; petioles 2–7 mm. long, finely strigillose or some hairs longer and spreading; leaf blades broadly ovate, 5–22 mm. wide, 1–5 cm. long, broadly cuneate to rounded at base, acute or abruptly acuminate at apex, entire or faintly crenate, strongly 3-nerved, scaberulous on both surfaces, hairs from white pustulate bases; inflorescences 1.5–2.5 cm. in diameter, glomerules 1–5 at tips of branches; heads 2–3 mm. in diameter, 5.4–6.2 mm. long; phyllaries narrowly ovate to oblong, mostly obtuse, sparsely ciliolate along upper margins, 1–5.5 mm. long, yellowish, glabrous and shining dorsally, midnerve rather prominent; ray flowers usually none (a single, inconspicuous ray flower rarely present); disk flowers 2–3, glabrous, corollas about 5 mm. long; achenes 2.4–3 mm. long, somewhat flattened, puberulent; pappus of 15–20 lanceolate, subequal paleae 1.5–2.5 mm. long.

Hillsides and outwash slopes, Lower Sonoran and Tropical Zones, southern Baja California. Sept.–May.

**3. Alvordia fruticosa** Brandegee, Erythea **7:** 5, 1889

*Alvordia angusta* S. F. Blake, Contr. Gray Herb. **52:** 42, 1917.
*Alvordia fruticosa* var. *angusta* Wiggins, Contr. Dudley Herb. **4:** 26, 1950.

Openly branched shrub 1–2 m. tall with cinereous-puberulent branches eventually glabrate; leaves lanceolate to linear-lanceolate, 6–32 mm. wide, 3–11.5 cm. long, acute at apex; inflorescences subglobose, 3–4 cm. in diameter, of numerous heads in short-pedunculate glomerules; involucres 4–7 mm. long, 1.2–2.2 mm. in diameter, each usually containing 1, or rarely 2, disk flowers; ray flowers none; phyllaries 3–6, glabrous dorsally, ciliolate marginally, outer ones lightly keeled; disk corolla 5–6 mm. long, lobes 1.2–2 mm. long; achene enfolded in receptacular bract, 2.5–3 mm. long; pappus awns 1–2.5 mm. long.

Sandy beaches and brushy mountain slopes, Lower Sonoran and Tropical Zones, southern Baja California. March–April; Nov.–Jan.

## 50. HELIANTHUS L., Sp. Pl. 904, 1753

Herbs or shrubs with alternate or opposite leaves, scabrous-pubescent to strigose-canescent herbage and small to large, radiate, pedunculate heads. Involucres hemisphercial to broadly saucerlike with herbaceous or less commonly membranous bracts imbricated in 2–several series. Receptacle plane to convex, each disk flower subtended by a firm to rigid, paleaceous bract. Ray flowers several to many, sterile; ligules spreading. Disk flowers tubular, gradually ampliate upward with a globose swelling near base, shallowly 5-dentate. Anthers entire at base or minutely bilobed. Style branches elongated; sterile appendages short, pubescent. Achenes oblong to obovate, more or less 4-angled and somewhat compressed. Pappus of 2 awnlike paleae, sometimes with a few additional minute, caducous squamellae.

Plants annual; roots neither tuberous or woody:
Involucral bracts linear or lanceolate, gradually tapering to tips; pappus paleae very short, often not reaching swollen part of disk corollas; leaves lanceolate, not cordate:
Involucral bracts lanceolate, equaling or only slightly surpassing disk corollas; hairs fine, subappressed, not coarsely pustulate at base:
Leaves not canescent; achenes sparsely hispid with hairs less than 0.4 mm. long, or subglabrate .................................. 1. *H. petiolaris* var. *petiolaris*
Leaves distinctly canescent; achenes hispid with silky, white to rufous hairs 0.2–1.2 mm. long .................................. 1a. *H. petiolaris* var. *canescens*
Involucral bracts linear, one and one-half to two times as long as disk flowers, weakly imbricated; hairs coarse, conspicuously spreading from heavy, white-pustulate bases .................................................. 2. *H. anomalus*
Involucral bracts broadly ovate, abruptly attenuate into a slender tail; pappus paleae one-half to two-thirds as long as disk corollas; leaves broadly ovate, lower ones cordate ............................................. 3. *H. annuus* subsp. *lenticularis*
Plants perennial from a tuberous or woody root; base of stems woody in *H. niveus*:
Root tuberous; plant herbaceous; herbage not canescent ............. 4. *H. californicus*
Root woody; plant suffrutescent to shrubby; herbage canescent with closely appressed, white hairs ........................................................ 5. *H. niveus*

**1. Helianthus petiolaris** Nutt., Jour. Acad. Phila. **2:** 115, 1821, var. **petiolaris**

(For synonymy see Abrams & Ferris, Ill. Fl. Pacif. States **4:** 114, 1960.)

Erect, strict to moderately branched annual 3–10 dm. tall or more with scaberulous but distinctly green stems and leaves; petioles slender, 1–5 cm. long, sparsely scaberulous, often tinged with red; leaf blades narrowly lanceolate to ovate-lanceolate, 0.5–5 cm. broad, to 15 cm. long, cuneate to subtruncate at base, acute at apex, scaberulous and dark green above, paler and more densely pubescent beneath, margins entire to serrate or sinuate-serrate; peduncles slender, to 15 cm. long or more, sparsely puberulent below, densely so just beneath heads; involucres 1–1.8 cm. wide, phyllaries lanceolate to lance-ovate, 6–10 mm. long, slightly acuminate, equaling or barely surpassing disk corollas, scaberulous dorsally; ligules 1.5–2 cm. long; disk corollas 4–5 mm. long, nearly glabrous, lobes dark brown to nearly black; receptacular bracts rather thin, puberulent dorsally, entire and acuminate to obscurely tridentate at apex, inner ones often black-tipped; achenes 4.5–5 mm. long, striate and faintly mottled, sparsely silky-puberulent below, more densely so above; pappus paleae about 1 mm. long, thin, caducous.

Hillsides and valley floors, upper edge of Lower Sonoran to the Transition Zones, San Bernardino Mountains, in southern California, to British Columbia, Missouri, and Texas. March–Oct.

**1*a*. Helianthus petiolaris** var. **canescens** A. Gray, Smiths. Contr. **3**[5]**:** 108, 1852

*Gymnolomia encelioides* A. Gray, Proc. Amer. Acad. **19:** 4, 1883.
*Helianthus petiolaris canus* Britt., Mem. Torrey Club **5:** 334, 1894.
*Helianthus canus* Woot. & Standl., Contr. U.S. Nat. Herb. **16:** 190, 1913.

Differing from var. *petiolaris* mainly in having densely strigose-canescent herbage, shorter leaf blades 2–5 cm. long, shorter involucral bracts 5–8 mm. long, smaller achenes 3.5–4.5 mm. long, and involucral chaffy scales merely puberulent on middle cusp.

Sandy or stony soil, mainly Lower Sonoran Zone, western margin of the Colorado Desert, San Diego County, California, to Texas and south into northern Mexico. March–June.

**2. Helianthus anomalous** S. F. Blake, Jour. Wash. Acad. **21:** 333, 1931

Erect, strict to much-branched, annual herb 2–8 dm. tall, hispid with conspicuously spreading, coarse, white hairs from heavy pustulate bases on stems, peduncles, leaves, and involucres; petioles 0.5–2.5 cm. long; leaf blades lance-elliptic to linear-lanceolate and rhombic-ovate, 5–15 mm. wide, 2–4.5 cm. long, narrowly to broadly cuneate at base, acute at apex, spreadingly hispid and with smaller subappressed hairs intermingled with spreading ones, margins entire or obscurely serrate; peduncles slender, to 20 cm. long; disks 1.5–2 cm. in diameter; phyllaries linear or linear-lanceolate, 1–2 mm. wide, 12–22 mm. long, conspicuously hispid, extending beyond disk corollas and between ray flowers; ligules pale yellow, about 1.5 cm. long; disk corollas about 5 mm. long, minutely puberulent below, lobes brownish; receptacular bracts 5–7 mm. long, puberulent-scaberulous dorsally, 3-dentate at apex; achenes about 4 mm. long, finely striate, mottled and minutely puberulent, particularly toward apex; pappus paleae minute and early caducous.

Hillsides, plains, and on outwash fans, Lower Sonoran to Transition Zones, Mohave County, Arizona, to southern Utah. June–Sept.

**3. Helianthus annuus** L. subsp. **lenticularis** (Dougl. ex Lindl.) Cockerell, Bot. Gaz. **45:** 338, 1908

*Helianthus lenticularis* Dougl. ex Lindl., Bot. Reg. **15:** *pl. 1265*, 1829.
*Helianthus aridus* Rydb., Bull. Torrey Club **32:** 127, 1905, in part.

Erect, simple to moderately branched annual 3–30 dm. tall, rough-hispid with rather coarse, spreading to subappressed hairs on stems, branches, and leaves; petioles slender on upper leaves, broader near base of plant, often one-half as long as to equaling blades; leaf blades deltoid-ovate, 2–18 cm. broad, to 25 cm. long, finely to coarsely serrate or uppermost sometimes nearly entire, rounded, subtruncate, cordate or broadly cuneate at base, acute or abruptly acuminate at apex, ascendingly scabrous-hispidulous; peduncles 2–20 cm. long or more, densely hispid-scabrous; involucres 2–5 cm. broad, bracts broadly ovate, abruptly narrowed to a slender, attenuate-acuminate, tail-like tip, scaberulous on back, densely white-ciliate on margins, 4–8 mm. wide, to 2 cm. long; rays 15–30, 2–3 cm. long; receptacular bracts stiff, 8–12 mm. long, scaberulous dorsally, 3-toothed at apex; disk corollas 8–10 mm. long; puberulent; achenes 4–5 mm. long, minutely puberulent.

A common roadside weed in waste places and along arroyos, Sonoran Zones, Saskatchewan to Texas west to California and south into northern Mexico; established as an introduced weed in many parts of the United States. March–Oct.

**4. Helianthus californicus** DC., Prod. **5:** 589, 1836

A slender perennial 1–3 m. tall from a somewhat woody, tuberlike root, stems glabrous or nearly so, often reddish, shallowly striate and more or less shining; petioles comparatively short, 1–1.5 cm. long, nearly glabrous except along ciliolate margins; leaf blades lanceolate to linear-lanceolate or lowermost sometimes ovate, upper ones 1–2.5 cm. broad, 10–20 cm. long, usually entire, cuneate at base, scabrous with short, stout, forwardly pointing hairs on both surfaces; peduncles short, 1–5 cm. long, scabrous; heads in terminal cymose panicles; involucres 10–12 mm. high, about 2 cm. broad; involucral bracts lance-linear, imbricated, outer ones tapering gradually into tail-like, spreading tips, 1.5–2 cm. long, inner ones about equaling disk flowers, all scaberulous-puberulent; ray flowers 15–20, 2–2.5 cm. long; receptacular bracts obtuse but usually abruptly apiculate, puberulent dorsally; disk corollas 5–6 mm. long; achenes about 5 mm. long, glabrous; pappus paleae 2–3, lanceolate.

In springy places or along streams, upper margins of the Lower Sonoran to Transition Zones, northern Baja California to central California, barely touching the Sonoran Desert near Neji and in the vicinity of Valle Trinidad, northern Baja California. July–Oct.

**5. Helianthus niveus** (Benth.) Brandegee, Proc. Calif. Acad. II, **2:** 173, 1889

*Encelia nivea* Benth., Bot. Sulph. 27, 1844.
(For synonymy see Contr. U.S. Nat. Herb. **23:** 1547, 1926.)

Suffrutescent (sometimes flowering first season), somewhat decumbent plant 3–6 dm. tall with canescent-strigillose stems and leaves; petioles about equaling blades, these ovate, lance-ovate, or oblong, obtuse to acuminate at apex, cuneate to subcordate at base, 0.5–3 cm. wide, 1.5–5 cm. long, entire to serrulate, usually densely canescent-strigillose on both surfaces but sometimes greenish above; heads on short to long peduncles (1–20 cm.), disks 1.5–2.5 cm. broad, about 10 mm. high; involucral bracts lanceolate to lance-linear, densely hirsute with spreading to appressed, whitish hairs, 1–2 mm. wide, 5–10 mm. long; rays 12–22, pale yellow, 1–3 cm. long; receptacular bracts 6–8 mm. long, obscurely tridentate and puberulent at apex, glabrous or nearly so below; disk corollas purplish, about 5 mm. long, minutely but densely puberulent below; achenes about 4 mm. long, striate and finely flecked with dark brown or purplish on a pale gray background, sparsely silky-strigillose; pappus paleae caducous, barely equaling silky hairs at apex of achene.

Sandy soil, mainly Lower Sonoran Zone, Colorado Desert, Riverside County, California, south into central Baja California and Sonora. Feb.–May; sometimes in late fall.

**Helianthus similis** (Brandegee) S. F. Blake, Contr. Gray Herb. **54:** 189, 1918 (*Viguiera similis* Brandegee, Zoe **5:** 260, 1908), a thinly tomentose, suffruticose plant about 1 m. tall or less with cordate, acuminate-attenuate, serrate leaves, is known definitely only from the Cape Region in Baja California, but it may occur along the southern margin of the desert west and southwest of La Paz.

## 51. ENCELIA Adans., Fam. Pl. 2: 128, 1763

Shrubs, suffrutescent plants, or perennial herbs with alternate leaves and mostly pubescent herbage. Heads small or medium, radiate or rarely discoid, solitary to paniculately arranged. Involucres hemispheric to broadly campanulate; bracts more or less imbricate in 2–3 series, subequal or outer ones shorter than inner, ovate-lanceolate to linear-lanceolate. Receptacle convex, bearing soft scarious paleae enfolding achenes and deciduous with them. Rays entire to 3-lobed, yellow, neutral when present, rarely absent. Disk corollas cylindric-funnelform; limb 5-toothed and puberulent; tube very short. Style branches obtuse, pubescent on outer side. Achenes strongly compressed, oblong to obovate, villous on margins, glabrous or partly to wholly villous-pubescent on sides. Pappus of 2 slender pubescent awns or wanting.

Leaves entire or toothed but not lobed:
  Heads several–many, paniculately arranged; branches of inflorescence glabrous or essentially so:
    Disk corollas yellow .................................. 1. *E. farinosa* var. *farinosa*
    Disk corollas purplish or blackish:
      Leaves white-farinose; involucral bracts puberulent ..............................
      ............................................ 1a. *E. farinosa* var. *phenicodonta*

Leaves soon glabrate, green; involucral bracts nearly glabrous .................... ............................................ *1b. E. farinosa* var. *radians*
Heads solitary or very few in an inflorescence; peduncles pubescent:
Disk corollas yellow; plants usually distinctly shrubby to base of peduncles:
Pubescence of stems and peduncles scabrous; rays rarely present; inner involucral bracts linear-lanceolate .................................... 2. *E. frutescens*
Pubescence of stems and peduncles softer, not scabrous; rays always present; all involucral bracts ovate-lanceolate ................. 3. *E. virginensis* subsp. *actonii*
Disk corollas purple or brownish purple; plants usually shrubby only at or near base, herbaceous above:
Involucres 10–13 mm. high; achenes pubescent on middle of each face:
Leaf blades cordate or truncate basally; achenes 4.5 mm. long; involucral bracts glandular on back ........................................ 4. *E. palmeri*
Leaf blades cuneate to subtruncate basally; achenes 5.5–6 mm. long; involucral bracts tomentose or scaberulous on back:
Achenes awnless; involucral bracts densely tomentulose dorsally ............. ...................................... 5. *E. californica* var. *californica*
Achenes with 1–2 slender awns; involucral bracts scaberulous dorsally ......... ..................................... 5a. *E. californica* var. *asperifolia*
Involucres 5–6 mm. high; achenes glabrous on faces, villous marginally ............ ....................................................... 6. *E. halimifolia*
Leaves laciniately lobed:
Rachis and lobes of leaves 1–2 mm. wide, glabrous or essentially so; involucral bracts puberulent only along margins .................................... 7. *E. ventorum*
Rachis of leaves 3–8 mm. wide, finely but densely puberulent; involucral bracts generally puberulent dorsally ............................................ 8. *E. laciniata*

**1. Encelia farinosa** A. Gray ex Torr., in Emory, Notes Mil. Rec. 143, 1848, var. **farinosa**

A rounded, much-branched plant to about 1.5 m. tall from a woody base, stems and branches white-mealy in youth, later glabrate and exuding an aromatic brownish resin; leaves mostly toward ends of branches, lanceolate to broadly ovate, 2–5 cm. wide, 3–10 cm. long, acute to obtuse at apex, entire to repandly dentate or undulate, densely white-farinose or sometimes subglabrate in age; petioles 1–4.5 cm. long, narrowly margined; heads in nearly naked panicles at ends of glabrate, whitish yellow branches extending to 2.5 dm. beyond leaves; involucres 1–1.5 cm. in diameter, bracts imbricated in 3–4 series, outer linear, inner gradually broader at base but longer, sparsely and loosely puberulent along margins and midveins, usually glabrate at maturity, 3.5–7 mm. long, slightly thickened at blunt tips; rays 10–12, oblong, 7–12 mm. long, conspicuous; disk corollas 3.5–4.5 mm. long, yellow, glandular below; receptacular bracts 6–7 mm. long, entire or laterally 1-toothed; achenes 4–4.5 mm. long, obovate and emarginate, glabrous between margins and main body, latter and margins villous with brownish silky hairs; pappus none.

Desert washes, slopes, mesas, and plains, Lower Sonoran Zone, Colorado and Mojave Deserts to southern Nevada, Arizona, and south to Sinaloa and southern Baja California. Nov.–May.

**1*a*. Encelia farinosa** var. **phenicodonta** (S. F. Blake) I. M. Johnston, Proc. Calif. Acad. IV, **12:** 1198, 1924

*Encelia farinosa* f. *phenicodonta* S. F. Blake, Proc. Amer. Acad. **49:** 362, 1913.

Differing from var. *farinosa* only in having purplish disk flowers.

Mojave and Colorado Deserts, California, into Arizona and south throughout the arid parts of Sonora and Baja California. Sept.–May.

**1*b*. Encelia farinosa** var. **radians** Brandegee ex S. F. Blake, Proc. Amer. Acad. **49:** 362, 1913

Differing from var. *farinosa* in having leaves soon glabrate, essentially glabrous involucres and purplish disk corollas.

Desert bajadas and open, rocky hillsides, Lower Sonoran and Tropical Zones, southern California, chiefly in the Cape Region, barely entering the Sonoran Desert near La Paz. March–June.

**2. Encelia frutescens** (A. Gray) A. Gray, Proc. Amer. Acad. **8:** 657, 1873

*Simsia frutescens* A. Gray, in Torr., Bot. Mex. Bound. 89, 1859.
*Encelia frutescens* f. *radiata* Hall, Univ. Calif. Pub. Bot. **3:** 135, 1907.
*Encelia frutescens* f. *ovata* Hall, loc. cit.

A much-branched, low, usually rounded shrub 1–1.6 m. tall with white stems and scabrous herbage, branches slender, ending in monocephalous peduncles to 2.5 dm. long; leaves ovate to oblong, obtuse to acute at apex, broadly cuneate to truncate at base, 6–16 mm. wide, 10–30 mm. long, dark green, sparsely scabrous above with stout hairs from pustulate bases, densely so beneath, margins entire; involucres 6–10 mm. high, bracts imbricate and unequal, hispid-scabrous on backs and margins, sometimes somewhat glandular, outermost ovate-acuminate, inner linear-lanceolate; disk 1–2.5 cm. broad; rays mostly lacking but about 8–9 mm. long when rarely present; disk corollas 5–6 mm. long, short tube glandular-puberulent; receptacular paleae 8–10 mm. long, glandular-puberulent dorsally; achenes black, 6.5–8 mm. long, 2.5–3.2 mm. wide, white-villous along margins and with some silky hairs along middle of each face; pappus none or of 1–2 weak, villous awns 1–1.2 mm. long.

Stony slopes and mesas, Lower Sonoran Zone, Mojave and Colorado Deserts into Arizona and south into northern Sonora and Baja California. March–Oct.

**3. Encelia virginensis** A. Nels. subsp. **actonii** (Elmer) Keck, Aliso **4:** 101, 1958

*Encelia actonii* Elmer, Bot. Gaz. **39:** 47, 1905.
*Encelia frutescens* f. *actonii* Hall, Univ. Calif. Pub. Bot. **3:** 135, 1907.
*Encelia frutescens* var. *actonii* S. F. Blake, Proc. Amer. Acad. **49:** 365, 1913.

Much-branched, shrubby-based perennial to 1 m. tall, stems and leaves with spreading simple hairs and stipitate glands, at length becoming glabrous on old parts; leaves ovate to elliptic, 1.5–4.5 cm. long, white-cinereous with dense, closely appressed hairs; peduncles 2–4 dm. long; heads radiate; involucres with densely pubescent bracts nearly devoid of glands; rays 1.5–3 cm. long; disk yellow; achenes 4–5 mm. long, glabrous on sides, long-villous along thin margins.

Dry gravelly or sandy slopes, mesas and outwash slopes, Lower and Upper Sonoran Zones, Mono and Inyo Counties, California, into southern Nevada and western Arizona, south to eastern San Diego County. Feb.–July.

**4. Encelia palmeri** Vasey & Rose, Proc. U.S. Nat. Mus. **11:** 535, 1889

A much-branched, rounded, suffrutescent plant about 1 m. tall with rather sharply angled, whitish stems and branches bearing a dense hispid pubescence; leaves broadly ovate, 10–35 mm. wide, 1.5–4 cm. long, cordate or less commonly truncate at base, acute, obtuse or rounded at apex, hispid-canescent to subvelutinous on both sides, entire or dentate with bluntly triangular teeth; petioles 3–10 mm. long, bases broadened;

heads several in a hispid-glandular, paniculate inflorescence; disk 1–2 cm. broad; involucre about 1 cm. high, bracts linear-lanceolate to lanceolate, somewhat unequal in 3 series, densely villous marginally, glandular on back, moderately pubescent with closely appressed hairs within; rays 15–20, ligules broadly oval, about 10 mm. long, puberulent on tube and lower surfaces; disk corollas brownish purple, subglabrous, about 5 mm. long; receptacular bracts 7–8 mm. long, dorsally glandular-puberulent near tips; achenes 4.5 mm. long, villous marginally, sparsely villous on midline of sides.

Hillsides, mesas, and sandy areas, Lower Sonoran Zone, Vizcaíno Desert to the Cape Region, Baja California. Feb.–June.

**5. Encelia californica** Nutt., Trans. Amer. Phil. Soc. II, **7:** 357, 1841, var. **californica**

A much-branched, rounded to moderately spreading, suffrutescent plant 0.6–3.6 m. tall (usually 1 m. tall or less); branches and peduncles canescent with fine, incurved to spreading, softish hairs; petioles 3–30 mm. long; leaf blades ovate to lanceolate, 0.4–3 cm. wide, 1–6 cm. long, acute at apex, cuneate to rounded at base, entire to repand-dentate, appressed-pubescent but distinctly green; peduncles nearly naked, terminal and from upper axils; disk 1.5–2.5 cm. in diameter; involucres 10–13 mm. high, lanceolate bracts densely tomentulose, in 2–3 series; rays 14–20, puberulent on tube and undersurface, ligules 1.5–3 cm. long; disk corollas purplish, 5 mm. long, puberulent on teeth, otherwise glabrous; receptacular bracts 7.5–10 mm. long, glandular-puberulent dorsally near tips; achenes 5.5–6 mm. long, villous on margins and along midlines of faces; pappus none.

Coastal slopes, mesas, and hillsides, mostly Upper Sonoran Zone but entering margins of the Lower Sonoran Zone near the southern limits of its range, Santa Barbara County, California, to vicinity of Rosario, Baja California, and inland to Riverside, California. Also on the Channel Islands and in western Arizona. Jan.–June.

**5a. Encelia californica** var. **asperifolia** S. F. Blake, Proc. Amer. Acad. **49:** 368, 1913

Differing from var. *californica* mainly in having more scabrous leaves, stems and peduncles; slightly shorter, more scabrous involucres; and achenes occasionally with 1–2 slender aristate pappus paleae.

Sandy washes, rocky hillsides, and desert plains, Lower Sonoran Zone, vicinity of Rosario and San Fernando, Baja California, south at least to the Vizcaíno Desert. Feb.–June.

**6. Encelia halimifolia** Cav., Ic. **3:** 6, *pl. 210,* 1795

*Pallasia grandiflora* Willd., Sp. Pl. **3:** 2261, 1804.
*Encelia conspersa* Benth., Bot. Sulph. 26, 1844.

Diffusely branched, suffrutescent plant 6–15 dm. tall with slightly canescent branchlets, leaves and peduncles, hairs fine and incurved; leaves ovate to oblong, acute to obtuse at apex, cuneate to truncate at base, 8–20 mm. wide, 2–3.5 cm. long, entire to slightly repand-dentate, sparsely to moderately pubescent with incurved, often scaberulous hairs but not markedly canescent; petioles 3.5–10 mm. long; peduncles terminal and axillary, simple or 2–3-branched, nearly naked; involucres 5–6 mm. high, 10–14 mm. in diameter, bracts in 2–3 series, unequal, linear-lanceolate to narrowly ovate, ciliate on margins, somewhat glandular dorsally; rays 10–12, ligules broadly oval, 7–10 mm. long, puberulent on tube and back; disk corollas purplish, 4.5–5 mm. long, glabrous; receptacular bracts 6–7.5 mm. long, glandular-pubescent on back; achenes 3–4 mm. long, villous on margins, sparsely villous along midline near apex or glabrous laterally; pappus none.

Sandy plains and low hills, Lower Sonoran Zone, west-central to southern Sonora, and in the vicinity of Magdalena Bay, Baja California. Aug.–Oct.

### 7. **Encelia ventorum** Brandegee, Proc. Calif. Acad. II, **2:** 175, 1889

A much-branched, suffrutescent plant 0.5–1.5 m. tall with glandular branches; leaves crowded near tips of branches, fleshy, 3–7 cm. long, central rachis linear, 1–2 mm. wide, and bearing 1–7 alternate linear lobes 1–2 mm. wide and 5–20 mm. long, mostly above middle, glabrous or essentially so but shining from exudate of impressed glands; peduncles solitary at tips of branches, monocephalous, 0.5–2 cm. long; involucres 7–10 mm. high, 1–1.6 cm. broad, bracts unequal, lance-ovate to lanceolate, glandular-dotted and glutinous dorsally, ciliate marginally, becoming reflexed and rather rigid in age; rays 8–10, truncate, 7–8 mm. long, villous on tube; disk corollas 4.5–5 mm. long, purplish, puberulent without, tube short, throat cylindric-funnelform; receptacular bracts 7–11 mm. long, 2–3-nerved, greenish, glandular-puberulent on midrib; achenes 5.5–8 mm. long, oblong, truncate, villous along narrow margins and at apex, glabrous on sides.

Sandy slopes, dunes, and upper margins of the beaches, Lower Sonoran Zone, along west coast of Baja California from about the latitude of Punta Prieta south to the vicinity of Magdalena Bay. Feb.–July.

### 8. **Encelia laciniata** Vasey & Rose, Proc. U.S. Nat. Mus. **11:** 535, 1889

Suffrutescent, much-branched, more or less glandular-pubescent plant 0.5–1 m. tall, younger parts usually closely beset with stiffish slender hairs; leaves ovate to obovate in outline, acute or more commonly obtuse at apex, unequally and irregularly laciniately lobed, central rachis 3–8 mm. wide between lobes, 2.5–5.5 cm. long, whole leaf 1–2.5 cm. broad, finely but densely puberulent on both surfaces; peduncles terminal and from upper axils, monocephalous; heads erect in flower, nodding in fruit, 10–12 mm. high, 1–2 cm. broad; involucral bracts lanceolate, loosely disposed in 2–3 series, tomentulose and ciliate, glandular between hairs, midrib part often bearing only sessile glands; rays 8–12, oval, nearly entire, 6–7 mm. long, tube puberulent; disk corollas about 5 mm. long, purplish, glandular-puberulent at base and on backs of teeth; receptacular bracts 8–10 mm. long, glandular-pubescent toward tip; achenes 5–6.5 mm. long, obovate, emarginate apically, margins densely villous, apex sparsely so.

Sandy soil near the sea, Lower Sonoran Zone, San Gregorio to Lagoon Head, Baja California, and on Ascension Island. Feb.–May.

**Encelia stenophylla** Greene, Bull. Torrey Club **10:** 41, 1883, a suffrutescent, more or less glutinous plant with linear leaves 1–2 mm. wide, 4–8 cm. long, crowded toward base of stems; heads in corymbose panicles at tips of stems, 7–12 mm. wide, 7–8 mm. high; involucral bracts lance-ovate, 3–4 mm. long, resinous and sparsely ciliate; rays 5–6, resin-dotted on tube and back, ligules oval, 5–5.5 mm. long; disk corollas 4–5 mm. long, resin-dotted; achenes 5 mm. long, villous over whole surface; pappus of 2 slender awns about equaling disk corollas.

This species occurs on Cedros Island but has not as yet been found on the mainland of Baja California. Flowering in the spring.

## 52. GERAEA Torr. & Gray, Amer. Jour. Sci. II, 3: 275, 1847

Annuals or perennials with erect, more or less branched stems, alternate dentate leaves and glandular-pubescent herbage. Heads paniculately arranged or sometimes solitary, radiate or discoid; flowers yellow. Involucres hemispherical; bracts linear to oblong, in 2–3 series. Receptacles flat or nearly so, bearing slightly scarious, conduplicate, deciduous paleae that fall with achenes. Rays, when present, cuneate, neutral. Disk corollas with cylindrical tubes and ampliate throats; limb 5-dentate and puberulent. Disk achenes narrowly cuneate, strongly compressed, villous on edges, margins produced into 2 strong awns decurrent into a conspicuous crown.

Heads radiate; plants annual; involucral bracts densely canescent-ciliate ..... 1. *G. canescens*
Heads discoid; plants perennial; involucral bracts densely glandular but not canescent ..... ........ 2. *G. viscida*

### 1. Geraea canescens Torr. & Gray, Amer. Jour. Sci. II, 3: 275, 1847

*Simsia canescens* A. Gray, Mem. Amer. Acad. IV, 1: 85, 1849.
*Encelia eriocephala* A. Gray, Proc. Amer. Acad. 8: 657, 1873.
*Encelia eriocephala* var. *paniculata* A. Gray, Syn. Fl. N. Amer. $1^2$: 282, 1884.
*Geraea canescens* var. *paniculata* S. F. Blake, Proc. Amer. Acad. 49: 357, 1913.

Erect, simple to moderately branched annual 1–6 dm. tall, canescently hirsute and stipitate-glandular throughout or stems and peduncles eventually subglabrate; lower leaves on narrowly winged petioles 0.5–2.5 cm. long, upper ones subsessile; leaf blades ovate, obovate, or lance-oblong, 0.5–4 cm. wide, 1–7 cm. long, acute at apex, cuneate at base, strongly 3-nerved, more or less dentate above middle or smaller ones entire; heads few, solitary to somewhat paniculate, to 5 cm. in diameter; involucres 8–12 mm. high, bracts unequal, linear-lanceolate and acute or attenuate, 7–10 mm. long, densely ciliate with long white hairs along margins except at tip, densely glandular on back; rays 10–15, golden yellow, 10–20 mm. long, 6–10 mm. wide; disk corollas 5–6 mm. long; receptacular paleae 8–9 mm. long, fimbriate-ciliate; achenes 6–7 mm. long, narrowly cuneate, silky-villous, black except for white margins; crown yellowish, thick; paleae continuous with margins, half as long as achene, denticulate.

Common in sandy and gravelly soil, Lower Sonoran Zone, Colorado and Mojave Deserts to Inyo County, California, east to southern Utah and south into Arizona, Sonora, and northern Baja California. Jan.–June.

### 2. Geraea viscida (A. Gray) S. F. Blake, Proc. Amer. Acad. 49: 357, 1913

*Encelia viscida* A. Gray, Proc. Amer. Acad. 11: 78, 1876.

An erect, sparingly branched, sparsely hirsute-villous and densely glandular-viscid, perennial herb to 1 m. tall from a slender lignescent taproot; stems pale green to nearly white, leafy, spreading hairs often to 3 mm. long; leaves sessile, ovate to ovate-oblong, 1–3 (or rarely to 5) cm. wide, 3–10 cm. long, cordate or auriculate at base, acute, obtuse or rounded at apex, irregularly and coarsely dentate to sinuate on margins, midrib prominent; heads discoid, solitary or in few-headed panicles, terminating branchlets or peduncles from upper axils, 1.5–4 cm. in diameter; involucres hemispherical, bracts oblong, 2–5 mm. broad, 8–15 mm. long, obtuse or rather abruptly acute at apex, densely glandular-pubescent, margins ciliate with hairs 1–2 mm. long; disk corollas 7–8 mm. long, somewhat glandular-puberulent without; receptacular paleae 11–14 mm. long, thin, glandular-puberulent and acute to attenuate toward tips; achenes 7–9 mm. long, body blackish, margins and crown white; pappus awns 2, 3.5–5 mm. long.

Gravelly slopes and sandy flats, mainly Upper Sonoran Zone but reaching the upper margins of the Lower Sonoran Zone, along the eastern flanks of the Sierra Pinal and Sierra San Pedro Mártir in Baja California; eastern San Diego County, California. April–Sept.

## 53. ZEXMENIA Llav. & Lex., Nov. Veg. Desc. fasc. 1: 13, 1824

Annual or perennial herbs or shrubs with pubescent herbage and opposite, sessile to short-petiolate, more or less serrate leaves. Peduncles solitary or heads borne in corymbose or umbelliform clusters. Heads medium-sized, radiate; flowers yellow. Involucres subglobose, campanulate, or short-cylindric; bracts in 1–5 series. Ligules usually conspicuous, pistillate. Disk corollas numerous, usually surpassing involucres. Achenes fusiform to cuneate, smooth and glabrous or glabrate, those of ray flowers trigonous and 3-awned, those of disk flowers strongly flattened and 2-awned. Pappus of 2–3 awns continuous with body of achene and often several very short intermediate squamellae.

### 1. Zexmenia fruticosa Rose, Contr. U.S. Nat. Herb. 1: 103, 1891

A much-branched shrub 1–1.5 m. tall or more with slender, ascending, striate and strigillose branches; petioles 2–7 mm. long, scaberulous; leaf blades ovate to lance-ovate, 5–28 mm. wide, 2–6 cm. long, rounded to obtuse at base, acute to slightly attenuate at apex, dark green and sparsely scaberulous above, paler and more densely scaberulous to hispid beneath, rather prominently 3-veined, margins serrulate-dentate; peduncles slender, 1–5 cm. long; heads usually solitary, sometimes in cymes of 3–4, about 1 cm. high; involucres broadly campanulate, bracts 1.5–3 mm. wide, 5–8 mm. long, acute or obtuse and short-acuminate, strigillose and slightly canescent; ray flowers 5–10; disk corollas yellow, about 4 mm. long, nearly glabrous; receptacular bracts lanceolate to oblong, outer obtuse, inner attenuate, subscarious; achenes about 4 mm. long, those of ray flowers slightly curved, ciliate-angled, with 3 bristles, those of disk flowers straight, with 2 bristles 2–2.5 mm. long, these faintly barbellate.

Wooded hillsides and rocky canyons, margins of Lower Sonoran into Upper Sonoran Zones, central to southern Sonora. Barely entering the Sonoran Desert south of Divisadero. Aug.–Dec.

## 54. VERBESINA L., Sp. Pl. 901, 1853

Herbs or shrubs with alternate or opposite, linear to broadly ovate, mostly dentate leaves. Heads radiate or discoid, solitary to numerous in narrow to broad panicles. Flowers yellow or less commonly white or orange. Involucres hemispheric to broadly campanulate; bracts somewhat imbricate and in 2–3 series, unequal and more or less herbaceous. Receptacular paleae concave, enfolding outer edge of disk achenes. Rays usually fertile. Disk flowers tubular, only slightly ampliate upward. Achenes strongly compressed, winged along both margins, oblong to obovate. Pappus of 2 awns continuous with marginal wings or very rarely lacking.

Plants annual; leaves more or less canescent on both surfaces .......... 1. *V. encelioides* var. *exauriculata*

Plants perennial, usually shrubby; leaves green at least on upper surface:

  Achenes 8–12 mm. long, glabrous or nearly so; involucres 1.5–3 cm. broad:

    Heads solitary; ligules 10–15 mm. long; leaves strongly scabrous, lower ones sessile .......... 2. *V. rothrockii*

Heads few in terminal panicles; ligules 12–20 mm. long; leaves nearly glabrous, lower ones short-petiolate ........................................... 3. *V. dissita*
Achenes 4–8 mm. long, often pubescent; involucres mostly 10 mm. broad or less:
Cymes hidden by leaves; involucres 3–4 mm. high; rays orange-yellow ............... ........................................................ 4. *V. oligocephala*
Cymes not hidden by leaves; involucres 5–12 mm. high; rays yellow or none:
Leaves auriculate-clasping; involucral bracts somewhat squarrose at tips ........... ........................................................ 5. *V. palmeri*
Leaves narrowly petiolate, not auriculate-clasping; involucral bracts not squarrose:
Leaves scabrous-hispidulous; phyllaries oblong-ovate, acute to obtuse; rays consistently present ..................................... 6. *V. peninsularis*
Leaves appressed-hirsute; phyllaries lanceolate to linear, acute to acuminate; rays often lacking ............................................... 7. *V. erosa*

**1. Verbesina encelioides** (Cav.) Benth. & Hook. var. **exauriculata** Robins. & Greenm., Proc. Amer. Acad. **34:** 544, 1899

*Encelia albescens* A. Gray, Proc. Amer. Acad. **8:** 658, 1873.
*Verbesina exauriculata* Cockerell, Nature **66:** 607, 1902.
*Ximenesia exauriculata* Rydb., Bull. Torrey Club **33:** 154, 1906.
*Ximenesia encelioides exauriculata* F. C. Gates, Trans. Kansas Acad. Sci. **42:** 138, 1939.

Annual herb 2–10 dm. tall with canescent-puberulent stems and foliage or leaves green above in age; petioles to 3 cm. long, wingless, without basal auriculate appendages; leaf blades ovate to lance-ovate, saliently dentate, 3–10 cm. long, 1–5 cm. wide, broadly cuneate to obtuse basally, mostly acute at apex, appressed-pubescent-canescent beneath, sparsely scaberulous and soon green above; heads usually solitary on naked peduncles 1–2 dm. long; involucres 1–1.8 cm. broad, about 8–10 mm. high, phyllaries lance-linear, 1–2 mm. wide, canescent, acute to acuminate; rays yellow, 10–15 mm. long; disk flowers yellow, 4–5 mm. long, glabrous or essentially so on limb, puberulent on narrow tube, this expanded abruptly into cylindro-funnelform throat; achenes 4–5 mm. long, body black, 1.5–2 mm. wide, sparsely villous to glabrate, wings stramineous to nearly white, slightly corky, 0.6–1 mm. wide, fimbriate-ciliate marginally; pappus awns none.

Plains, foothills, arroyos, and margins of fields and roadsides, Lower and Upper Sonoran and Carolinian Zones, Great Valley of California to Montana, Kansas, and Texas, south into Mexico. April–Nov.

**2. Verbesina rothrockii** Robins. & Greenm., Proc. Amer. Acad. **34:** 541, 1899

*Actinomeris wrightii* A. Gray, Pl. Wright. **2:** 89, 1853.
*Verbesina wrightii* A. Gray, Proc. Amer. Acad. **19:** 12, 1883. Not *V. wrightii* Griseb., 1866.

Herbaceous perennial from a slender woody root; stems slender, simple or sparingly branched, 3–7 dm. tall, striate, hirsutulous with slender, whitish, spreading hairs; leaves opposite below, uppermost alternate, ovate to lance-ovate, 1–3 cm. broad, 2–6.5 cm. long, sessile and slightly auriculate-clasping, irregularly serrate-dentate, acute at apex, pustulate-scabrous on both surfaces, green above, slightly canescent beneath; peduncles to 2.5 dm. long; involucres 1.5–3 cm. broad, 10–12 mm. high; phyllaries in 2–3 series, oblong-ovate to oblong or obovate-oblong, 1.5–4 mm. wide, acute to obtuse, green or faintly cinereous, sparsely scaberulous; rays 1–1.5 cm. long; disk flowers 5–6 mm. long; puberulent below; achenes 10–12 mm. long, body dark gray, glabrous, 5–6.5 mm. wide (including corky, stramineous or pale yellow wing); pappus paleae 1–2 mm. long.

Rocky hillsides and outwash slopes, Lower Sonoran to Transition Zones, southern

Arizona and New Mexico into Coahuila. Barely touching the Sonoran Desert in the foothills of the Santa Rita Mountains in Pima County, Arizona. May–Sept.

**3. Verbesina dissita** A. Gray, Proc. Amer. Acad. **20:** 299, 1885

Suffrutescent perennial from a woody root; stems green, sparsely strigillose to nearly glabrous, 6–10 dm. tall, remotely leafy; leaves opposite below, alternate above, ovate, 4–8 cm. long, entire to serrate, acute at apex, sessile and auriculate-clasping or lowermost short-petiolate, thin, slightly scabrous but distinctly green on both surfaces; heads several in terminal panicles on peduncles 2–15 cm. long; involucres 10–12 mm. high, 1.5–2 cm. broad, bracts lance-linear, oblong, or narrowly obovate, acute, obtuse, or rounded at apex, appressed-pubescent and somewhat cinereous; rays yellow, 1.2–2 cm. long; disk flowers 5–6 mm. long, short tube puberulent and abruptly expanded into a cylindro-funnelform throat; achenes 8–10 mm. long, body black or dark gray, glabrous or nearly so, thin corky wings slightly broader above than below.

Hillsides and canyons, chiefly Upper Sonoran Zone, San Bernardino and Orange Counties, California, to northern Baja California. Possibly not entering the Sonoran Desert but to be expected in the southern foothills of the Sierra San Pedro Mártir. March–July.

**4. Verbesina oligocephala** I. M. Johnston, Proc. Calif. Acad. IV, **12:** 1200, 1924

Ascendingly branched shrub 5–12 dm. tall with white-tomentose younger branchlets and grayish or cinereous older stems; leaves opposite, on narrowly winged petioles 5–8 mm. long; leaf blades ovate to lance-ovate, acute at apex, rounded to cuneate at base and narrowed into wings of petiole, 4–7 cm. long, light green, scabrous, hairs with pustulate bases; heads small, in few-headed terminal corymbose cymes more or less hidden by leaves; peduncles 4–11 mm. long, densely short-villous; involucres campanulate, 4–8 mm. wide, 3–4 mm. high, bracts ovate-oblong, biseriate, tips rounded and somewhat recurved; rays 10–12, ligules orange-yellow, 5–6 mm. long, pubescent tube 2–2.5 mm. long; disk corollas 5–6 mm. long; receptacular paleae 5–6 mm. long, oblong-linear, puberulent dorsally near tip; achenes cuneate-oblong, 5–6 mm. long, silky-villous, especially above; wings narrow; pappus of 2 ciliate awns nearly as long as achenes.

Canyons and hillsides, Lower Sonoran Zone, eastern slopes of Sierra de la Giganta, Baja California. March–June.

**5. Verbesina palmeri** S. Wats., Proc. Amer. Acad. **24:** 56, 1889

Suffruticose plant 5–13 dm. tall, forming rather compact, rounded clumps, hispid-scabrous throughout with fragile, readily deciduous, spreading hairs from conspicuous pustulate bases; leaves opposite or uppermost sometimes alternate; petioles comparatively short, 5–15 mm. long, broadly winged and auriculate-clasping; leaf blades deltoid-ovate or rhombic-ovate, 2.5–5 cm. wide, 3–8 cm. long, coarsely dentate to nearly entire, acute at apex, scabrous but green on both surfaces; heads fairly numerous in loosely arranged, cymose panicles on pedicels 1.5–6 cm. long; involucres 5–6 mm. high; phyllaries biseriate, inner ones linear, thin and erect, acute to acuminate, outer ones spatulate-oblong, obtuse at apex, herbaceous, slightly squarrose; rays 8–11, ligules 6–10 mm. long; disk flowers about 4 mm. long; achenes obovate, body glandular-scaberulous to villous, thin wings ascendingly ciliate; awns 2, erect or slightly spreading, about equaling disk corollas.

Arid rocky canyons, Lower Sonoran Zone, along or near the eastern coast of Baja California in the vicinity of Bahía de los Angeles. Apparently flowering at almost any season following favorable rainfall. The type was collected in November or December.

**6. Verbesina peninsularis** S. F. Blake, Proc. Biol. Soc. Wash. **37:** 58, 1924
*Verbesina cayucensis* M. E. Jones, Contr. West. Bot. **18:** 76, 1933.

Suffrutescent or fruticose perennial 1 m. tall or more with finely hispidulous-pilosulous, faintly striate, pale green stems; leaves opposite below, uppermost alternate; leaf blades triangular-ovate to lanceolate, 1–6 cm. wide, 5–12 cm. long, coarsely and spreadingly toothed, often somewhat hastate-lobed at base, acuminate at apex, broadly cuneately decurrent at base to short, faintly winged petioles 2–7 mm. long, rather prominently 5-veined from base, sparingly hispidulous and green on both surfaces, hairs on upper surface from pustulate bases; heads 4–12 in each loose cymose panicle, peduncles 1–8 cm. long, densely hispidulous; involucres 8–12 mm. in diameter, 7–8 mm. high, phyllaries ovate-oblong, acute to obtuse, 1.5–3 mm. wide, herbaceous, erect, finely subappressed-hispidulous; rays yellow, 7–10 mm. long; disk corollas 3–3.5 mm. long, yellow; anthers dark purple, well exserted; achenes oblong, 4–5 mm. long, sparingly silky-villous, thin wing about 0.5 mm. broad, finely ciliate; pappus paleae 2, erect, minutely puberulent, about equaling disk corollas.

Rocky hillsides, canyons, and mesas, Lower Sonoran Zone, eastern Baja California from between San Ignacio and Santa Rosalia to the vicinity of Loreto. Sept.–Nov.

**7. Verbesina erosa** Brandegee, Proc. Calif. Acad. II, **3:** 146, 1891

Many-stemmed, suffrutescent plant to 1.5 m. tall from a woody root; stems hirsutulous to spreading-hirsute; leaves opposite below, alternate above, ovate to lance-ovate, 3–7 cm. broad, 6–12 cm. long, sharply serrate or uppermost entire, acuminate at apex, broadly cuneate to rounded but not decurrent at base, subsessile or lower ones on winged petioles 5–18 mm. long, scabrous on upper surface, hirsute to ascendingly pilose and often canescent beneath; heads few in loosely corymbose panicles; involucres hemispherical, 6–8 mm. high, bracts 2–3-seriate, ovate-lanceolate to linear, acute to acuminate, hirsute; rays present or lacking in some heads, ligules 8–10 mm. long when present; achenes obovate, body usually glabrous, bordered with a thin, irregularly lacerate-ciliate wing often broader than body or ray achenes often wingless; awns 2, slightly spreading and about equaling achene.

Deep canyons and mountainsides, Lower to Upper Sonoran Zones, Sierra San Francisquito to the Cape Region, Baja California. Flowering time uncertain.

**Verbesina pustulata** M. E. Jones, Contr. West. Bot. **18:** 77, 1933, is closely related to *V. erosa* but with evenly serrate leaves truncate-rounded at base and griseous-hirsute beneath; pappus awns unequal, one scarcely one-half as long as the other.

Described from the higher parts of the Sierra de la Laguna, Baja California, but it may occur in canyons in the Sierra de la Giganta.

## 55. COREOPSIS L., Sp. Pl. 907, 1753

Annual or perennial herbs or sometimes fruticose. Leaves opposite or rarely alternate, often chiefly basal, entire, dentate or variously parted or dissected. Heads medium to large, pedunculate, solitary or corymbose-paniculate, radiate. Involucres hemispheric to obturbinate; bracts in 2 series, bracts of each series somewhat connate at base; inner membranous, brownish or yellowish; outer narrower, green, thicker, and usually shorter than inner. Receptacle flat, chaffy-bracteate. Rays and disk corollas yellow (in ours). Achenes flat or meniscoidal, oblong to orbicular, margins often winged, smooth or ciliate. Pappus of 2 bristles, scales or teeth, one from each angle of achene, or pappus none.

Ray and disk achenes different; disk achenes ciliate, their pappus of 2 paleae; ray achenes glabrous and epappose .......................................... 1. *C. bigelovii*
Ray and disk achenes alike, nonciliate; pappus reduced to a cupule ...... 2. *C. californica*

**1. Coreopsis bigelovii** (A. Gray) Hall, Univ. Calif. Pub. Bot. **3:** 141, 1907

*Pugiopappus bigelovii* A. Gray, in Torr., Pacif. R. Rep. **4:** 104, 1857.
*Pugiopappus breweri* A. Gray, Proc. Amer. Acad. **8:** 660, 1873.
*Leptosyne bigelovii* A. Gray, Syn. Fl. N. Amer. **1**$^{2}$**:** 300, 1884.

Erect glabrous annual with several scapose stems 1–6 dm. tall, each stem monocephalous and somewhat leafy toward base, naked or with 1–2 small bracts above; leaves alternate but basally crowded, blade ovate, 2–8 cm. long, once- or twice-pinnate into linear, callous-tipped, obtuse lobes 1–2 mm. wide and 5–30 mm. long; heads erect, 1–1.5 cm. high, 2–4.5 cm. in diameter (including rays); involucres subcylindric, abruptly truncate at base; outer bracts 4–7, linear, 4–12 mm. long, somewhat spreading; inner bracts 6–8, ovate, 3–6 mm. wide, 6–10 mm. long, yellowish green, scarious-margined, acute; rays 5–10, ligule golden yellow; disk corollas yellow, 1.5–2 mm. wide, 3.5–4 mm. long; ray achenes 3–5 mm. long, obovate to oblong, glabrous and epappose; disk achenes 1–1.5 mm. wide, 4–6 mm. long, body glabrous except for a median row of ascending hairs, margins silky-villous with ascending cilia; pappus of 2 persistent lanceolate paleae 2–3 mm. long.

Desert and coastal slopes and plains, Lower and Upper Sonoran Zones, southern Monterey and Tulare Counties to Inyo, San Bernardino, Riverside, and San Diego Counties, California, barely reaching the Sonoran Desert in eastern San Diego County. March–June.

**2. Coreopsis californica** (Nutt.) H. K. Sharsmith, Madroño **4:** 217, 1938

*Leptosyne californica* Nutt., Trans. Amer. Phil. Soc. II, **7:** 363, 1841.
*Leptosyne newberryi* A. Gray, Proc. Amer. Acad. **7:** 358, 1868.

Erect, virtually glabrous annual with several scapose stems 0.5–4.5 dm. tall, each monocephalous, leafy near base, slightly reddish; leaves near base erect or strongly ascending, linear-filiform, about 0.5 mm. in diameter, 2–15 cm. long, mostly entire but sometimes with 1–3 small filiform pinnae 1–6 mm. long, glabrous except for a few widely scattered, capitate hairs; heads 6–10 mm. high, 10–35 mm. wide; involucres ellipsoid-tubular, rounded at base; outer bracts 2–7, pale green or sometimes flecked with red, bearded and gibbous at base, narrowly lanceolate, about 1 mm. wide, 4–7 mm. long, obtuse, callous tip reddish and spreading; inner bracts 5–10, lanceolate to obovate, 2–4 mm. wide, 6–10 mm. long, thin, scarious-margined, constricted at middle, upper half spreading at anthesis; rays 5–12, ligule yellow, 5–15 mm. long; disk corollas 3–4 mm. long; receptacular paleae 4–5.5 mm. long; achenes all alike, obovate, 2.5–5 mm. long, dull, microscopically papillate, villous; marginal wing corky; apex of achene forming a short cupule, pappus paleae none.

Arid plains, washes, and hillsides, Lower and Upper Sonoran Zones, southeastern Santa Barbara County, California, to Pinal County, Arizona, and south into northern Baja California. Feb.–May.

Several species of *Coreopsis* have been cultivated as ornamentals and sometimes escape to nearby fields and roadsides. Both *C. atkinsoniana* Dougl. and *C. tinctoria* Nutt. are so cultivated and either or both might occur along irrigation ditches and roadsides in our area.

## 56. COREOCARPUS Benth., Bot. Sulph. 28, *pl. 16,* 1844

Annual herbs or suffrutescent perennials from a woody root. Leaves opposite, once- to thrice-pinnatifid or pinnatisect. Heads small, pedunculate, radiate; rays white with roseate or purplish lines on backs or yellow with or without darker lines. Involucres hemispheric to turbinate, of 5–12 subequal, submembranous, ovate to ovate-oblong bracts in 2 series, outer ones obtuse, inner acuminate. Receptacle flat with membranous narrow paleae subtending disk florets. Ray flowers pistillate, fertile. Disk flowers perfect but sometimes sterile, tubular, regular, yellow, slightly ampliate above, with a ring of hairs at base of throat. Style branches with hispid, subulate appendages. Anthers entire at base. Achenes obcompressed, with a crustaceous, entire to deeply cut wing along each margin, and with 2 retrorsely hispid slender awns or these lacking; awns often present on outer achenes, lacking on inner of same head.

Plants herbaceous annuals:
  Ultimate division of leaves ovate, deltoid or broadly oblong, not linear; awns usually present, at least on ray achenes; ligules of ray flowers white with roseate to purple lines on back .................................. 1. *C. parthenioides* var. *parthenioides*
  Ultimate divisions of leaves linear or essentially so; awns less commonly present on achenes; ligules bright yellow, with brownish or purplish lines on back ........................................................ 1*a*. *C. parthenioides* var. *heterocarpus*
Plants suffrutescent perennials:
  Wings of achenes toothed or lobed, thinnish; body of achenes 2.5–3 mm. long; ligules white to pale yellow tinged with rose:
    Achenes, or at least some in each head, 2-awned; herbage glabrous or essentially so; involucral bracts 5–6 mm. long:
      Lobes of leaves 1–4.5 mm. wide, not filiform ...... 2. *C. arizonicus* var. *arizonicus*
      Lobes of leaves filiform, less than 1 mm. wide ...... 2*a*. *C. arizonicus* var. *filiformis*
    Achenes devoid of apical awns; herbage sparsely hispid; involucral bracts 3–4 mm. long:
      Ligules 2.5–3 mm. long; involucres turbinate; leaf blades 10–18 mm. long ........................................................................ 3. *C. johnstonii*
      Ligules 5–8 mm. long; involucres campanulate; leaf blades 20–35 mm. long ........................................................................ 4. *C. sonoranus*
  Wings of achenes thick, entire to obscurely crenulate along margins; body of achenes 4–4.5 mm. long; ligules bright yellow ........................... 5. *C. dissectus*

### 1. Coreocarpus parthenioides Benth., Bot. Sulph. 28, *pl. 16,* 1844, var. parthenioides

*Leptosyne parthenioides* A. Gray, Syn. Fl. N. Amer. **1**²: 301, 1884.
*Coreocarpus shrevei* Sherff, Bot. Gaz. **97**: 185, 1935.
*Coreocarpus shrevei* var. *latilobus* Sherff, op. cit. 187.
*Coreocarpus robustior* Sherff, loc. cit.

Slender herbs 0.5–4 dm. tall, stems simple or branching from base or above; herbage glabrous or essentially so; petioles as long as or shorter than ovate to deltoid-oblong leaf blades, these 2–5 cm. long, bipinnatifid into primary deltoid lobes 1–2 cm. long and nearly as wide, these with 3–4 lobes or broad divisions or sometimes nearly entire; heads on slender peduncles 1–6 cm. long, somewhat ternately arranged, 1–2.2 cm. broad, including rays; involucre hemispheric to broadly campanulate, bracts

3–5 mm. long, outer obtuse and abruptly apiculate, dark-lined; rays oblong, to 7 mm. long, white above, strongly purple-lined and roseate beneath; disk flowers about 2 mm. long, bright yellow, glabrous; achenes black, body 2–2.5 mm. long (sometimes to 4 mm. long), wing 0.2–0.4 mm. wide, deeply crenulate to distinctly toothed or lobed, those of ray flowers usually bearing 2 retrorsely hispid awns 0.5–1 mm. long, those of disk flowers with or without awns.

Sandy plains, clay soil, and rocky hillsides, Lower Sonoran Zone, vicinity of El Mármol to the Magdalena Plain, Baja California, and near Guaymas, Sonora. Feb.–April.

**1*a*. Coreocarpus parthenioides** var. **heterocarpus** (A. Gray) S. F. Blake, Proc. Amer. Acad. **49**: 343, 1913

*Coreocarpus heterocarpus* A. Gray, Proc. Amer. Acad. **5**: 162, 1861.
*Leptosyne heterocarpa* A. Gray, Syn. Fl. N. Amer. **1**[2]: 301, 1884.
*Coreocarpus involutus* Greene, Pittonia **1**: 290, 1889.

Similar to var. *parthenioides* but segments of leaves narrowly linear to subfiliform, margins of achenes strongly incurved and frequently dissected into discrete lobes, and awns less uniformly present.

Arid plains and hillsides, Lower Sonoran Zone, Natividad Island, Bahía de los Angeles, and Lagoon Head south to the Cape Region, Baja California. Dec.–April.

This is an extremely variable species, very sensitive, apparently, to fluctuations in rainfall.

**2. Coreocarpus arizonicus** (A. Gray) S. F. Blake, Proc. Amer. Acad. **49**: 344, 1913, var. **arizonicus**

*Leptosyne arizonica* A. Gray, Proc. Amer. Acad. **17**: 218, 1882.
*Coreopsis arizonica* Hoffm., in Engler & Prantl, Nat. Pflanzenf. **4**[5]: 243, *fig. 118S*, 1890.

A much-branched, suffrutescent plant with slender, striate, glabrous stems 2–6 dm. tall; petioles slender, 1–3 cm. long; leaf blades to 10 cm. long, broadly ovate to deltoid in outline, pinnately divided into 3–5 ovate divisions and then again divided into several linear lobes 1.5–3 mm. wide, or uppermost leaves merely pinnately 3–5-parted into linear lobes, ultimate segments acute to attenuate-acuminate; heads in few-headed cymes; involucres campanulate, bracts broadly ovate to oblong, mostly 5–6 mm. long, obtuse and abruptly acuminate, or inner ones acute, strongly purple-lined, narrowly scarious-margined; ligules pale yellow, 3–5 mm. long; disk flowers abruptly ampliate, 2–2.5 mm. long, anthers exserted 1–1.5 mm.; body of achene about 3 mm. long; wing incised into broadly cuneate lobes; 2 retrorsely spinulose awns present at apex of some achenes, lacking on others.

Stream banks and rocky slopes, upper margin of Lower into the Upper Sonoran Zones, Pima and Santa Cruz Counties, Arizona, south into northern Sonora and Chihuahua. Feb.–May and Sept.–Oct.

**2*a*. Coreocarpus arizonicus** var. **filiformis** (Greenm.) S. F. Blake, Proc. Amer. Acad. **49**: 344, 1913

*Leptosyne arizonica* var. *filiformis* Greenm., Proc. Amer. Acad. **40**: 44, 1904.

Differing from var. *arizonicus* in having filiform leaf division 1 to 6 cm. long.

Southwestern Chihuahua and on Isla San Pedro Nolasco off the coast of Sonora. Feb.–May.

**3. Coreocarpus johnstonii** Sherff, Bot. Gaz. **97:** 604, 1936

An erect suffrutescent plant to 3 dm. tall from a woody root, trichotomously branched and glabrous or sparsely hispidulous, branchlets divaricate and somewhat fleshy; petioles 3–8 mm. long, glabrous or sometimes hispidulous; leaf blades 10–18 mm. long, bipinnatifid into subfleshy, linear-oblong or ovate segments; heads numerous, more or less congested in cymose clusters at tips of branches, 6–9 mm. broad, 5–6 mm. high at anthesis; involucres turbinate with 5–7 subequal ovate bracts about 3 mm. long, glabrous; ligules oblong, 2.5–3 mm. long, yellowish white, entire or nearly so; disk flowers slender, about 4 mm. long; outer achenes 2.5–3 mm. long, 1–1.5 mm. wide, narrow subscarious wings subentire to distinctly incised into separate teeth, inner ones narrower and very narrowly winged; terminal awns none.

Known only from the type locality, San Pedro Bay, near Guaymas, Sonora. July–March.

**4. Coreocarpus sonoranus** Sherff, Bot. Gaz. **97:** 604, 1936

Low, much-branched shrub 3–6 dm. tall and nearly or fully as wide; bark on older parts tawny, striate and somewhat fissured, younger branches sparsely puberulent or hispidulous; petioles slender, 1–1.8 cm. long; leaf blades ovate in outline, 2–3.5 cm. long, bipinnatifid into slender, linear or oblong segments 1–3 mm. wide, 3–10 mm. long; heads laxly disposed in few-headed cymes on slender naked branches 1–5 cm. long, 14–18 mm. wide, 5–6 mm. tall; involucres campanulate with oblong-ovate bracts 3–4 mm. long, these abruptly acuminate at apices; rays 4–6, ligules white-rosaceous, 5–8 mm. long; disk flowers yellow, about 3 mm. long; achenes about 3 mm. long, body of outer 1–1.2 mm. wide, blackish, pale wings subcontinuous or separated into discrete teeth about 1 mm. long; inner achenes narrower; terminal awns lacking.

Near beach on rocky cliffs and hillsides, Lower Sonoran Zone, in the vicinity of Guaymas, Sonora. July–March.

**5. Coreocarpus dissectus** (Benth.) S. F. Blake, Proc. Amer. Acad. **49:** 344, 1913

*Acoma dissecta* Benth., Bot. Sulph. 29, *pl. 17,* 1844.
*Leptosyne dissecta* A. Gray, Syn. Fl. N. Amer. $\mathbf{1}^2$: 301, 1884, as to synonym.
*Coreocarpus dissectus* var. *longilobus* S. F. Blake, Proc. Amer. Acad. **49:** 345, 1913.

Low procumbent shrub or suffrutescent plant 3–5 dm. tall and often as broad or broader, stems slender, finely striate, reddish and glabrous; petioles slender, 1–3.5 cm. long; leaf blades broadly ovate in outline, 3–8 cm. long, often nearly as broad, 1–2-pinnatifid into linear to subfiliform segments 0.5–2 mm. wide, 1–3 cm. long, acute, dark green and glabrous; heads turbinate, cymosely arranged at tips of slender naked branches 3–10 cm. long; involucres 4–5 mm. in diameter, 5–6 mm. high, bracts broadly ovate to oblong, obtuse and acuminate to acute, only moderately dark-lined; rays yellow, 3–4.5 mm. long, obtuse, minutely 2–3-toothed apically; disk flowers 3.5–4 mm. long, tube about 1 mm. long and sparsely puberulent, cylindrical throat dark-lined, glabrous; ray achenes 4–4.5 mm. long, 1.5–2.2 mm. broad, wings somewhat inrolled, entire to crenulate; disk achenes scarcely 1 mm. wide, nearly wingless or narrow wing dissected into discrete teeth or lobes, body pappillose with corky excrescences; awns lacking.

Canyons and rocky hillsides and cliffs, Lower Sonoran Zone, central Baja California and the adjacent islands to the Cape Region. Feb.–April.

## 57. HETEROSPERMA Cav., Ic. 3: 34, *pl. 267,* 1795

Annual herbs with opposite, dentate or incised or pinnatifid leaves and glabrous to villous herbage. Heads small, terminal or axillary on long slender peduncles. Involucre campanulate; bracts few in 2 series, exterior ones herbaceous, often ciliate-margined, narrower than inner, latter subconnate at base, striate, each subtending a ligulate flower. Receptacle plane; membranous paleae subtending disk flowers, deciduous. Ray flowers yellow, ligulate. Disk flowers tubular, regular, yellow; limb more or less ampliate, shallowly 5-toothed. Appendages on style branches acute to subulate. Ray achenes strongly obcompressed, margin somewhat inrolled and narrowly winged, inner face excavate-concave, smooth dorsally, epappose. Disk achenes narrower, linear, wingless or nearly so, slightly beaked apically, armed with 2 retrorsely barbed, ascending or slightly spreading awns.

Leaf segments narrowly oblanceolate; receptacular bracts nearly one and one-half times as long as involucral bracts; ligules of ray flowers minute, 1–2 mm. long ................ ........................................................ 1. *H. pinnatum*
Leaf segments linear; receptacular bracts scarcely longer than involucral bracts; ligules of ray flowers larger, 5–8 mm. long .................................... 2. *H. xanti*

### 1. Heterosperma pinnatum Cav., Ic. 3: 34, *pl. 267,* 1795

*Heterosperma tagetinum* A. Gray, Mem. Amer. Acad. II, **4:** 87, 1849.

Erect annuals 3–6 dm. tall, somewhat paniculately branched above, herbage moderately hispid to glabrate, stems striate, densely hispidulous in 2 lines from midlines between bases of adjoining leaves to nodes below; leaves broadly ovate in outline, 1.5–5.5 cm. long, pinnately divided into 5–9 (mostly 5–7) linear segments about 1–2.5 mm. wide, 0.5–2.5 cm. long, central one usually exceeding laterals, all acute; peduncles in upper axils and terminating branches 1–7 cm. long; involucres narrowly campanulate to subcylindrical, 2–3.5 mm. wide, 5–6 mm. long at anthesis, outer bracts usually 3–5, linear to linear-oblong, green and herbaceous, margins ciliate-hispid; inner involucral bracts broadly ovate-oblong, 2–2.5 mm. wide, 4–5 mm. long, obtusish but abruptly apiculate at apex, submembranous, stramineous and conspicuously lined with 10–15 darker nerves, glabrous; ligules of ray flowers suborbicular to broadly elliptic, mostly 1–1.5 mm. long, inconspicuous; disk corollas minute, barely exceeding or only equaling subtending receptacular bracts; ray achenes strongly obcompressed, 3.5–5 mm. long, about 2 mm. wide, reddish brown, margins inrolled and rugulose, inner part more or less papillose with low tubercles, epappose; disk achenes oblong to linear, progressively longer from outer to innermost ones, longitudinally striate or slightly ridged, outer angles antrorsely barbellate, outer ones sometimes epappose but at least longest (to 14 mm. long) inner ones bearing 2 retrorsely barbed awns 1–2 mm. long, these deciduous.

Along streams, ditches, and in moist canyons, upper margins of Lower Sonoran into the Transition Zones, northern Arizona to Texas and south to Guatemala. Barely entering the Sonoran Desert near Cumpas, Sonora. July–Oct.

### 2. Heterosperma xanti A. Gray, Proc. Amer. Acad. 5: 162, 1861

(For synonymy see N. Amer. Fl. II, **2:** 63, 1955.)

Annual herb 1–2.5 dm. tall with slender, simple, and erect or spreading-ascending stems branched at base, somewhat hispidulous in youth but soon glabrate; leaves opposite, petioles 4–16 mm. long, glabrous or margins slightly ciliate near base; leaf blades pinnately divided, 1–2.5 cm. long, segments somewhat fleshy, obtuse at apex, 0.5–1.5 mm. wide, to 1.5 cm. long, glabrous or midrib and margins minutely setose-ciliate; peduncles slender, to 10 cm. long; heads to 1.8 cm. broad (including spreading rays); involucre subhemispheric, 5–6 mm. in diameter, 4–5 mm. high at anthesis, 6–8 exterior bracts green, narrowly linear, 2–3.5 mm. long, margins not ciliate; inner bracts ovate-oblong to broadly lanceolate, purplish, acuminate or slightly attenuate, glabrous except for a small fringe of short spreading hairs sometimes along margins of acuminate portion; ray flowers 6–8, yellow, oblong to elliptic-ovate ligules 5–8 mm. long; disk corollas 1.8–2 mm. long, yellow; achenes purplish, blackish, or outer ones often brown, those of ray flowers about 3 mm. long, dorsally convex, sparsely and lightly tuberculate, margins incurved, inner face more or less ridged and tuberculate; achenes of disk flowers linear, more or less tetragonal, 6–9 mm. long, body glabrous, bearing 2 (or 2–4) slender, straight, moderately spreading, retrorsely barbed awns 1–3 mm. long, these tardily deciduous or subpersistent.

Sandy slopes and banks of streams, chiefly Upper Sonoran Zone but straggling into Lower Sonoran Zone along watercourses, southern Baja California. Apparently flowering in either spring or late summer following rains, flowering specimens having been collected in March, September, and October.

## 58. **BIDENS** L., Sp. Pl. 831, 1753

Annual or perennial herbs or less commonly somewhat suffrutescent plants with procumbent to erect, terete or angled stems and opposite, entire to thrice-pinnately dissected leaves. Heads small to moderate in size, discoid or radiate. Involucres campanulate to hemispherical; bracts in 2 series, those of outer series distinct, usually herbaceous, interior ones submembranous and often hyaline toward margins. Ray flowers obscure to conspicuous or lacking. Disk flowers few to many, tubular, regularly and shallowly 4–5-toothed, only slightly ampliate above. Anthers entire or slightly sagittate at base. Appendages on style branches short and acute or longer and subulate. Achenes dorsiventrally compressed or 3–4-angled, cuneate to linear, usually bearing 2–8 retrorsely barbed, stiff awns apically, bodies glabrous to sparsely hispidulous.

Plants perennial, although often flowering first season; leaves not divided or, uncommonly, lower ones pinnately lobed or parted; heads 2–7 cm. broad at anthesis; rays 1.5–3 cm. long, conspicuous:
- Achenes linear, body 12–18 mm. long, usually 4-aristate; outer involucral bracts glabrous or sparsely puberulent at base, at least not ciliate-margined ...... 1. *B. sambucifolia*
- Achenes cuneate or linear-cuneate, body 4–9 mm. long, usually 2-aristate; outer involucral bracts ciliate-margined:
  - Leaves sessile, not even lower ones divided; outer involucral bracts 6–8 ..... 2. *B. laevis*
  - Leaves petiolate, lower ones sometimes divided or parted; outer involucral bracts 8–17 .............................................................. 3. *B. aurea*

Plants annual; leaves pinnately once- to thrice-parted; heads 4–13 mm. in diameter at anthesis; rays small, less than 1 cm. long, inconspicuous or lacking (except in *B. pilosa* var. *radiata*):
- Heads discoid; outer involucral bracts ciliate-margined:
  - Outer involucral bracts 4–6, only 1–2.5 mm. long; foliage and stems glabrous ....... ........................................................ 4. *B. leptocephala*
  - Outer involucral bracts 7–9, 4–5 mm. long; foliage and stems hispidulous in youth but soon glabrate ........................................ 5. *B. pilosa* var. *pilosa*

Heads radiate, although obscurely so; outer involucral bracts nonciliate marginally except in *B. pilosa*:
Ligules 7–15 mm. long; outer involucral bracts ciliate-margined ........................................................................ 5a. *B. pilosa* var. *radiata*
Ligules 2–6 mm. long, inconspicuous; outer involucral bracts glabrous or puberulent basally but not ciliate-margined:
Leaves usually once-pinnate; rays 4–6 mm. long; body of achenes 6–12 mm. long ........................................................ 6. *B. duranginensis*
Leaves twice- and thrice-pinnately lobed or divided; rays 1–2 mm. long; body of achenes 10–18 mm. long ........................................ 7. *B. bipinnata*

**1. Bidens sambucifolia** Cav., Ic. **3:** 15, *pl.* 229, 1795

*Coreopsis sambucifolia* Cav., Descr. 225, 1802.
*Bidens alamosana* Rose, Contr. U.S. Nat. Herb. **1:** 104, *pl.* 6, 1891.

Perennial herb to 3 m. tall with glabrous foliage and lanceolate to ovate-lanceolate leaves 8–12 cm. long on petioles 2–3 cm. long, blade pinnately divided into 3–5 lanceolate lobes, these serrate, more or less connate at base by decurrent wings on rachis; heads corymbosely arranged but solitary on peduncles, 2.5–3.5 cm. broad (including rays), 10–15 mm. high at anthesis; peduncles 4–15 cm. long; involucres minutely pubescent to glabrate, outer 10–12 bracts linear or linear-lanceolate, 4–6 mm. long, inner ones broader and thinner, strongly lined; ray flowers 9–12, oblong-obovate ligules 12–15 mm. long, yellow to orange; disk flowers slenderly tubular, 2.5–3 mm. long; achenes linear, acutely quadrangular in cross section, blackish, body 12–18 mm. long, usually crowned with 4, but sometimes with 3 or 5, retrorsely barbed awns 2–3.5 mm. long, one sometimes shorter than others.

Along streams and ditches in moist soil, Lower Sonoran to Tropical Zones, southern Sonora into Sinaloa. March–Dec.

**2. Bidens laevis** (L.) B. S. P., Prel. Cat. N.Y. 29, 1888

*Helianthus laevis* L., Sp. Pl. 906, 1753.
(For synonymy see N. Amer. Fl. II, **2:** 96, 1955.)

Annual or perennial (depending upon conditions) herb 3–10 dm. tall from a procumbent base or wholly erect; herbage glabrous; leaves sessile, linear-lanceolate, lanceolate, or ovate-lanceolate, 5–15 cm. long, acute to acuminate at apex, regularly serrate marginally; heads ligulate, 3–7 cm. broad at anthesis, 8–11 mm. high; involucres somewhat hispid at base, the 6–8 exterior bracts linear-lanceolate, acute to obtuse at apex, sparsely setose-ciliate along margins; ray flowers 7–8, yellow, ligules obovate-lanceolate, 1.5–3 cm. long; achenes narrowly cuneate, plane or 3–4-angled, angles or margins retrorsely hooked, angles often more or less tuberculate between hooks, body of achenes 6–9 mm. long, bearing 2–4 retrorsely barbed awns 3–5 mm. long.

Usually in water of slowly flowing streams or in very wet soil nearby, Lower Sonoran to Transition and Tropical Zones, California to Florida and south to Argentina. Established as a weed in the Hawaiian Islands. Sept.–Feb. in our area.

**3. Bidens aurea** (Ait.) Sherff, Bot. Gaz. **59:** 313, 1915

*Coreopsis aurea* Ait., Hort. Kew. **3:** 252, 1789.
*Bidens heterophylla* C. G. Ortega, Hort. Mat. 99, *pl.* 12, 1798.
(For additional synonymy see N. Amer. Fl. II, **2:** 100, 1955.)

Erect perennial herb but often flowering first season, with tetragonous, simply branched stems to 1 m. tall; herbage glabrous or sometimes pubescent on young growth;

leaves thin, linear, lanceolate, or elliptic-oblong, 0.5–1.2 cm. wide, to 10 cm. long, sharply serrate-dentate, undivided or less commonly lower ones 3–5-parted, on petioles 0.2–4 cm. long; heads radiate, 2–5 cm. broad, about 1 cm. high; outer involucral bracts 8–17, linear, acute and somewhat indurated at apex, hispid-ciliate on margins but glabrous on face, 3–6 mm. long, 1 mm. wide or less; inner bracts thinner, slightly wider, but about as long; ray flowers 5–6, yellow, ligules 1–3 cm. long, conspicuous, obovate; achenes linear-cuneate, slightly quadrangular in cross section, black or dark gray, often more or less papillose toward base, body 4–7 mm. long, tipped with 2 yellowish, retrorsely barbed awns 1.5–2.7 mm. long.

In moist soil near streams, springs, and ditches, Upper Sonoran to Tropical Zones, southern Arizona to Guatemala. Rare in the desert proper. July–Oct.

**4. Bidens leptocephala** Sherff, Bot. Gaz. **64:** 22, 1917

An annual herb with moderately branched, quadrangular, striate, slender stems 1–5 dm. tall; petioles sparsely ciliate-hispid, connate at base, 0.3–4 cm. long; leaf blades 1.5–5.5 cm. wide, 2–10 cm. long, once- or twice-pinnately divided into linear to ovate segments, hispid to glabrate, margins and lower surfaces of veins more or less hispid-ciliate; heads obscurely radiate or sometimes apparently discoid, 4–8 mm. wide, 3–5 mm. high at anthesis, on slender peduncles 2–8 cm. long; exterior involucral bracts 4–6, linear, marginally ciliate, 1–2.5 mm. long, inner ones about one-half as long again, lanceolate, often puberulent at apices; ray flowers usually 2–3, ligules about 2.5 mm. long, one-half as wide, pale yellow to nearly white; achenes 5–13, linear, somewhat quadrangular, outer with body 6–8 mm. long, hispidulous, inner ones longer, body 9–14 mm. long, hispidulous above, glabrous below, all bearing 2 retrorsely barbed awns 1–3 mm. long.

In moist soil, usually in shaded places, Lower Sonoran to Transition Zones, central Arizona to Chihuahua, Sonora, and Baja California. Aug.–Oct.

**5. Bidens pilosa** L., Sp. Pl. 832, 1753, var. **pilosa**

(For synonymy see N. Amer. Fl. II, **2:** 107, 1955.)

Erect, much-branched, annual herb 3–18 dm. tall with quadrangular, often sparsely pilose stems; petioles 1–6.5 cm. long, pilose-ciliate marginally; leaf blades ovate in outline, 5–20 cm. long, pinnately divided into 3–7 ovate acute segments 1.5–7 cm. long, these serrate, appressed-pilose to subglabrate, terminal segment oblong-ovate or lanceolate, acute to acuminate at apex, decurrent at base; heads discoid, 5–7 mm. high, 7–8 mm. wide at anthesis, 25–40-flowered on slender peduncles 1–10 cm. long; involucres pubescent at base, outer bracts 7–9, linear to linear-spatulate, 4–5 mm. long, ciliate-margined, somewhat indurated-apiculate at apex; inner bracts lanceolate, slightly longer; achenes linear, straight or outer ones slightly incurved, obcompressed to tetragonal, glabrous below, tuberculate and strigose above, body 4–16 mm. long, inner achenes longest, each bearing 2–3, or rarely to 5, retrorsely barbed awns 2–4 mm. long.

In moist soil along streams, ditches, ponds, and springs, Upper Sonoran to Tropical Zones, California to Florida and south throughout Mexico and the West Indies to Argentina; in Africa, Asia, Australia, New Zealand, and many of the islands of the Pacific and Indian Oceans. Flowering in the spring or summer after rains.

**5*a*. Bidens pilosa** var. **radiata** Schultz-Bip., in Barker-Webb & Berthel., Hist. Nat. Canar. Isl. III, **2**$^{2}$**:** 242, 1836–50

(For synonymy see N. Amer. Fl. II, **2:** 113, 1955.)

Plants somewhat smaller and segments of leaves usually ovate; heads radiate, 1.5–3 cm. wide at anthesis; ray flowers 5–6, whitish, yellow, or purplish, 7–15 mm. long; otherwise as in var. *pilosa*.

In moist situations, Upper Sonoran to Tropical Zones, with much the same distribution as, but commoner than, var. *pilosa* in our area. Aug.–Sept.

**6. Bidens durangensis** Sherff, Bot. Gaz. **70:** 90, *pl. 11, figs. g–m,* 1930
*Bidens anthriscoides* var. *angustiloba* DC., Prod. **5:** 601, 1836.

A glabrate annual herb 5–9 dm. tall with slightly quadrangular, striate stems minutely pubescent in youth; petioles 2.5–4 cm. long, connate and ciliate-margined near base; leaf blades 10–12 cm. long, pinnately divided into 3–5 ovate to ovate-lanceolate, serrate or dentate segments; heads numerous, radiate with inconspicuous ray flowers, 8–13 mm. wide, 4–7 mm. tall at anthesis, on slender peduncles 3–8 cm. long; involucres hispid at base, outer bracts 7–8, linear, glabrous above base, not ciliate-margined, 2–3 mm. long, tips somewhat indurated; inner bracts narrowly lanceolate, about one and one-half times as long as outer ones; ray flowers 3–6, pale yellow or nearly white, narrowly ovate ligules 4–6 mm. long; achenes linear, dusky, glabrous or sparsely strigillose above, body 6–12 mm. long, bearing 2–4 retrorsely barbed awns 1.5–2.5 mm. long.

Moist situations, mostly Upper Sonoran Zone, central Sonora and Chihuahua to central Mexico, reaching the margins of the Sonoran Desert between Ures and Cumpas, Sonora. Sept.–Oct.

**7. Bidens bipinnata** L., Sp. Pl. 832, 1753
(For synonymy see N. Amer. Fl. II, **2:** 107, 1955.)

Erect, glabrous or minutely hispidulous annual 3–17 dm. tall, stem somewhat quadrangular and finely striate; petioles slender, 2–5 cm. long; leaf blades 4–20 cm. long, usually 2–3-pinnately divided into lanceolate or deltoid-lanceolate segments, these cuneate at base, acute to obtuse at apex, entire or nearly so marginally; heads small, 4–6 mm. wide, 5–7 mm. high at anthesis; involucres puberulent at base, outer bracts 7–10, linear, 3–5 mm. long, acute; inner bracts linear-lanceolate, about 4–7 mm. long; ray flowers pale yellow to nearly white, ligules inconspicuous, about 2 mm. long; inner achenes linear, quadrangular, black or dusky, body 10–18 mm. long, attenuate upward, often sparsely hispidulous above, glabrous below, body of 2–4 of outer achenes thicker, minutely tuberculate-hispid, 7–12 mm. long, both types bearing 2, or rarely 3 or 4, retrorsely barbed awns 2–4 mm. long.

Moist ground, Tropical to Transition Zones, widely distributed in warm temperate regions of both the Old and New Worlds, barely reaching the margins of the Sonoran Desert. July–Nov. in our area.

None of the species of *Bidens* included here is abundant in the Sonoran Desert, occurring only where streams, springs, irrigation ditches, or ponds provide sufficient water to permit their growth.

## 59. BEBBIA Greene, Bull. Calif. Acad. 1: 179, 1885

Much-branched, somewhat brittle, suffrutescent plants with few, comparatively small, mostly opposite leaves. Heads solitary on ends of branches but often more or less cymosely arranged, discoid, all flowers perfect and fertile. Involucres campanulate; bracts striate, appressed, imbricate in several ranks. Receptacle chaffy-bracteate;

bracts lanceolate, persistent, partially enclosing achenes they subtend. Corollas yellow, 5-toothed. Achenes turbinate, obtusely trigonous, slightly obcompressed. Pappus bristles slender, strongly plumose or of persistent awns or bristles.

Stems nearly naked; leaves linear or narrowly oblong, 0.5–3 mm. wide; inner involucral bracts stramineous, dullish:
  Herbage glabrous or nearly so, not scabrous ..................... 1. *B. juncea* var. *juncea*
  Herbage scabrous-hispid ................................... 1a. *B. juncea* var. *aspera*
Stems moderately leafy; leaves lanceolate to lance-ovate, to 15 mm. wide; inner involucral bracts roseate, lustrous ........................... 1b. *B. juncea* var. *atriplicifolia*

## 1. **Bebbia juncea** (Benth.) Greene, Bull. Calif. Acad. **1**: 180, 1885, var. **juncea**

*Carphephorus junceus* Benth., Bot. Sulph. 21, 1844.
*Bebbia filifolia* M. E. Jones, Contr. West. Bot. **18**: 80, 1933.

Intricately branched, compact, suffrutescent plant of rounded outline 8–12 dm. tall or stems scrambling up through other shrubs; herbage glabrous or essentially so, pale green and somewhat glaucous; leaves filiform to linear, 0.5–3 mm. wide, 1–3 cm. long, entire or remotely and obscurely toothed, readily deciduous, leaving stems naked and reedlike; peduncles 1–10 cm. long; heads 10–18 mm. wide, 9–12 mm. high at anthesis; involucral bracts ovate to lance-linear, 1–6 mm. long, outer ones herbaceous and minutely villosulous, inner ones stramineous, submembranous, glabrous except for a fine marginal ciliation; corollas yellow or sometimes whitish, tubular, 4–6.5 mm. long, scarcely at all ampliate, teeth minute; achenes 3–3.5 mm. long, densely strigose with ascending, fine but stiffish hairs; pappus of 12–18 plumose bristles 4–7 mm. long, slightly tawny.

Arid slopes, mesas, outwash fans, and rocky or sandy arroyos, Lower Sonoran Zone and lower arid parts of Upper Sonoran Zone, Orange and San Diego Counties, California, to southern Nevada, central Arizona, and south into Sonora and Baja California. Flowering almost throughout the year.

## 1*a*. **Bebbia juncea** var. **aspera** Greene, Bull. Calif. Acad. **1**: 180, 1885

*Bebbia aspera* A. Nels., Bot. Gaz. **37**: 273, 1904.

Differing from var. *juncea* only in having stems, peduncles, and leaves moderately to closely invested with short, stout, upwardly curved hairs from pustulate bases, thus making the herbage rough-scaberulous.

The range of var. *juncea* but less common in Baja California than in the Colorado Desert and Arizona. Flowering nearly throughout the year.

## 1*b*. **Bebbia juncea** var. **atriplicifolia** (A. Gray) I. M. Johnston, Proc. Calif. Acad. IV, **12**: 1197, 1924

*Carphephorus atriplicifolius* A. Gray, Proc. Amer. Acad. **5**: 159, 1861.
*Bebbia atriplicifolia* Greene, Bull. Calif. Acad. **1**: 181, 1885.

Plants forming broader, more flat-topped masses than var. *juncea*; herbage scaberulous or hispidulous to glabrate; leaves lanceolate to lance-ovate, to 1.5 cm. wide, 2–4 cm. long, coarsely and irregularly repand-dentate, broadly cuneate to subhastate at base, acute apically; heads usually more crowded than in var. *juncea* and paniculately arranged; inner involucral bracts usually roseate, lustrous, outer ones subtomentose.

Rocky slopes, hillsides, talus slopes, and canyons, mainly Lower Sonoran Zone, vicinity of Comondú and Loreto to San José del Cabo, Baja California. Flowering almost throughout the year.

## Tribe 6. **AMBROSIEAE.** Ragweed Tribe

Coarse monoecious weeds or shrubs with small, greenish or white flowers in numerous heads. Leaves alternate or rarely lower ones opposite. Flowers unisexual, staminate heads in racemes or spikes or rarely staminate flowers in same head as pistillate ones. Receptacles of staminate or of bisexual heads with chaffy bracts. Rays none. Corollas of staminate flowers tubular, regular; anthers distinct or nearly so, not caudate. Pistillate flowers few, often in enclosing fused bracts of involucre, leaving a narrow orifice at apex. Corollas of pistillate flowers tubular or often reduced to a ringlike rudiment or none. Fruit often a more or less prickly bur. Pappus none.

Staminate flowers in center of bisexual heads; pistillate flowers marginal; fruits not burlike; involucres not enfolding pistillate flowers:
- Pistillate flowers not enclosed by broad paleae; achenes not strongly flattened or winged .......... 60. *Iva*
- Pistillate flowers subtended by broad paleae, slightly enfolded by them; achenes strongly flattened, margins winged and toothed .......... 61. *Dicoria*

Staminate flowers in heads separate from pistillate flowers; pistillate flowers wholly enclosed by bur- or nutlike involucres:
- Involucral bracts of staminate heads united, at least at base, forming a cup; receptacle of staminate heads flat or slightly convex; spines on fruits not hooked:
  - Pistillate heads bearing several scarious, transversely oriented wings; leaves linear or subfiliform .......... 62. *Hymenoclea*
  - Pistillate heads bearing tubercles or spines, not winged; leaves or their divisions broader:
    - Involucres tuberculate, unarmed or bearing a few short spines in a single series; pistillate heads 1-flowered .......... 63. *Ambrosia*
    - Involucres armed with distinct spines (in series except in *Franseria bryantii*); pistillate heads 2–5-flowered .......... 64. *Franseria*
- Involucral bracts of staminate heads distinct; receptacle of staminate heads conical to elongate; spines on fruits more or less hooked .......... 65. *Xanthium*

## 60. **IVA** L., Sp. Pl. 988, 1753

Annual or perennial herbs or low shrubs with lowermost leaves opposite, those of inflorescence and sometimes upper stems alternate, entire to pinnatifid. Heads numerous in spikes, racemes, or panicles, ebracteate or each subtended by a leaflike bract, often nodding. Involucres hemispheric to turbinate, of 3–9 distinct or somewhat united bracts, these in a single series or slightly imbricated. Staminate flowers 3–50, subtended by filiform to spatulate paleae; corollas funnelform, glabrous or glandular-puberulent, acutely 5-lobed or 5-toothed. Anthers oblong; tips acute, incurved. Pistillate flowers 1–8, marginal, corollas vestigial or wanting. Style thick; stigma capitate or slightly peltate. Achenes cuneate to obovate, slightly compressed dorsiventrally.

Plants perennial, shrubby; achenes 1.9–2.5 mm. long; staminate flowers 8–25 per head:
- Leaves simple, entire; phyllaries rounded, entire at apex; achenes resin-dotted, glabrous .......... 1. *I. hayesiana*
- Leaves pinnatifid with filiform lobes; phyllaries irregularly toothed at apex, villous; achenes villous, not resin-dotted .......... 2. *I. acerosa*

Plants annual; achenes 1.2–1.5 mm. long; staminate flowers over 25 per head .......... 3. *I. ambrosiifolia*

**1. Iva hayesiana** A. Gray, Proc. Amer. Acad. **11:** 78, 1875

A fruticose perennial with numerous spreading-ascending branches to 1 m. long, puberulent and hispidulous throughout; leaves of stems chiefly opposite, oblanceolate to oblong-spatulate, entire, obtuse at apex, hirsutulous and glandular-punctate, 3–6 cm. long on a short stoutish petiole; leaves in inflorescence alternate, linear-oblong, 1–2.5 cm. long; heads numerous, in leafy racemes; involucres hemispheric, 5–6 mm. broad, of 5–6 distinct, rounded-obovate, slightly imbricate bracts; paleae spatulate; staminate flowers 12–20, corolla trumpet-shaped; anther tips acute; pistillate flowers 3–6 (usually 5); achenes lenticular-obovate, 2–2.2 mm. long, glandular-muricate.

Mostly along watercourses or margins of salt marshes, Lower and Upper Sonoran Zones, southern California south along and near the coast to central Baja California and on the adjacent islands. May–Oct.

**2. Iva acerosa** (Nutt.) Jackson, Univ. Kan. Sci. Bull. **41:** 828, 1960

*Oxytenia acerosa* Nutt., Jour. Acad. Phila. II, **1:** 172, 1848.

A shrub 1–2 m. tall with erect, striate, grayish-strigose branches; leaves pinnatifid into 3–7 linear-filiform segments 1 mm. wide, 5–10 cm. long, or uppermost leaves entire, filiform, leaves often deciduous, branches then reedlike; heads numerous; involucres hemispherical, 4–5 mm. wide, 3–4 mm. high; phyllaries 5, distinct, ovate-acuminate, canescent dorsally; staminate flowers 9–22, corollas about 2.5 mm. long; paleae of staminate flowers spatulate, villous; pistillate flowers usually 5, bearing a fleshy disk at base of style; paleae of pistillate flowers elliptic, broadly oblanceolate, irregularly toothed at apex, long-villous, sometimes reduced to a mere vestige; achenes slightly compressed, brownish, 2–2.5 mm. long, villous with hairs equalling style.

Mainly on saline soil, mesas and plains, Lower and Upper Sonoran Zones, southeastern California, southern Nevada, and northern Arizona to southern Utah, southwestern Colorado and New Mexico. July–Sept.

**3. Iva ambrosiifolia** (A. Gray) A. Gray, Syn. Fl. N. Amer. **1²:** 246, 1884

*Euphrosyne ambrosiaefolia* A. Gray, Pl. Wright. **1:** 102, 1852.

*Cyclochaena ambrosiaefolia* Benth. & Hook. ex B. D. Jackson, Ind. Kew. **1:** 678, 1895.

Annual herb 3–6 dm. tall with terete hispid stems and alternate leaves on petioles 3–5 cm. long; leaf blades ovate in outline, 5-8 cm. long, bipinnatifid with confluent segments, ultimate ones obovate and toothed, hispid on both surfaces; heads many in leafy branched panicles, slenderly pedunculate; involucres hemispheric, 4–5.5 mm. broad, nearly as deep; bracts 5, obovate to oblanceolate, acute, hispidulous dorsally, faintly scarious at margins; staminate flowers many, 1.5–1.8 mm. long, glabrous; paleae of pistillate flowers cuneate-obovate, truncate at apex, concave; those of staminate flowers filiform with clavate tips; pistillate corollas obsolete or a faint vestige; achenes 2–2.5 mm. long, cuneate, slightly flattened, muricate in age, margins acute.

Along arroyos and streams, Lower and Upper Sonoran Zones, Pima County, Arizona, to western Texas and adjacent northern Mexico. May–Oct.

## 61. DICORIA Torr. & Gray ex A. Gray, in Torr., Bot. Mex. Bound. 86, 1859

Small erect herbs with hispid-canescent stems and foliage. Leaves opposite below, alternate above, petiolate, closely and finely canescent-tomentulose. Heads small, arranged in loose panicles of nearly leafless spikes. Involucres of 5 distinct, herbaceous, outer bracts and 1–2 larger, scarious, inner ones subtending fertile flowers, accrescent in fruit or these usually absent in staminate heads. Staminate flowers 6–12; corolla regular, 5-dentate; filaments monodelphous. Pistillate flowers 1–2, without corollas; style branches linear. Achenes planoconvex, lacerate-pectinate along thin margins. Pappus of several small squamellae.

Mature achenes solitary, surpassing subtending bracts; mature stems strigose-canescent; upper leaves lanceolate-oblong .......................................... 1. *D. brandegei*
Mature achenes usually 2, shorter than subtending bracts; stems rough-hispid, sometimes canescent but hairs spreading; upper leaves rounded or broadly ovate, or obtuse at apex:
Achenes 4–5 mm. long; scarious fruiting bracts 6–8 mm. long in fruit .... 2. *D. canescens*
Achenes 5.5–6.5 mm. long; scarious fruiting bracts 10–15 mm. long in fruit ........... ........................................................... 3. *D. clarkae*

## 1. **Dicoria brandegei** A. Gray, Proc. Amer. Acad. **11:** 76, 1876

Cinereous erect herb 2–10 dm. tall with strigose-canescent stems and herbage; leaves mostly alternate, narrowly ovate-lanceolate to lanceolate, 2.5–4 cm. long, 5–12 mm. wide, entire to shallowly dentate, finely and densely strigose-canescent; heads numerous in slightly leafy panicles; involucres 3–4 mm. wide, outer bracts 5, elliptic, finely strigose-canescent, obtuse to rounded at apex; inner phyllaries subtending pistillate flowers elliptic to oblong, thin but herbaceous, hispidulous, slightly shorter than mature achenes, those subtending staminate flowers filiform with clavate apices; staminate flowers 6–12, funnelform corollas sparsely pilose; achene usually solitary in each head or sometimes none, flat or planoconvex, oblong, 5–6 mm. long, hispidulous on faces, wing cleft into lobes that are again faintly toothed laterally and apically.

Along dry watercourses on plains and hillsides, Lower and Upper Sonoran Zones, southern Utah, southwestern Colorado, northern Arizona, and rarely in the eastern parts of the Mojave and Colorado Deserts, California. June–Dec.

## 2. **Dicoria canescens** A. Gray, in Torr., Bot. Mex. Bound. 87, 1859

*Dicoria oblongifolia* Rydb., N. Amer. Fl. **33:** 12, 1922.
*Dicoria hispidula* Rydb., loc. cit.

Annual herb 3–12 dm. tall with much-branched stems, these striate, harshly hispidulous with coarse, pustulate-based, spreading hairs, canescent in youth, usually greenish in age; leaves opposite below, alternate toward inflorescences, petiolate, blades lanceolate to broadly ovate or suborbicular, particularly above, densely white-hirsute on both surfaces, lower to 3 cm. broad, 4–5.5 cm. long; heads numerous in sparsely leafy panicles; involucres hemispherical, 3–4 mm. broad, outer bracts 5, linear to elliptic or narrowly obovate, 2–2.5 mm. long, rounded or obtuse at apex, hispidulous; fruiting bracts obovate to suborbicular, membranous, glandular-puberulent, 6–8 mm. long at maturity, only slightly concave or plane, not forming a hood, glandular-erose on margins; receptacular paleae subtending staminate flowers slenderly clavate, about equaling flowers; staminate flowers 5–7, with sparingly pilose funnelform corollas; pistillate flowers 2; achenes 4–5 mm. long, slightly pilose, scarious or chartaceous wings dentate, body black, dull.

Desert washes, plains, and gentle slopes, Lower Sonoran Zone, southwestern Utah to the Mojave and Colorado Deserts, California, south into Arizona and Sonora. June–Dec.

**3. Dicoria clarkae** Kenn., Muhlenbergia **4:** 2, 1908

Stoutish herbaceous annuals 5–10 dm. tall with spreading, harshly hispid and slightly canescent branches; leaves alternate or lowermost ones opposite; petioles to 1.5 or 2 cm. long; leaf blades ovate, 2–4 cm. wide, 3–6 cm. long, entire to sinuate-dentate, strongly hispid with harsh spreading hairs; heads numerous in rather dense, practically leafless panicles; involucres 3–4.5 mm. broad, 5 outer bracts oblong to ovate-lanceolate, grayish-pubescent, tips obtusish to rounded; receptacular bracts subtending pistillate flowers, scarious, nearly orbicular, strongly hooded, reticulate-veined and densely glandular-puberulent, 10–15 mm. long, more or less erose on margins; bracts subtending staminate flowers linear; staminate flowers 6–8, funnelform corollas 2.5–3 mm. long, pilose; pistillate flowers 2–5, usually 3; achenes lanceolate, 5.5–6.5 mm. long, 2–3 mm. wide, sparsely puberulent, glandular-dotted and slightly carinate on both surfaces, margin 0.8–1.2 mm. wide, irregularly toothed.

Sandy desert sinks and plains, Lower Sonoran Zone, southwestern Nevada and adjacent California, sparingly along the Colorado River about to Yuma. May–Jan.

## 62. HYMENOCLEA Torr. & Gray, in A. Gray, Mem. Amer. Acad. II, 4: 79, 1849

Open, slenderly branched, rounded shrubs with slender linear leaves and more or less resinous-viscid herbage. Flowers monoecious; staminate in nodding heads with saucer-shaped, 4–8-lobed involucres. Staminate corollas funnelform, hyaline, 5-toothed, pubescent, with oblong anthers inwardly curving at tips and filaments distinct above. Pistillate heads erect, 1-flowered; involucre fusiform with 5–12 flat, membranous, bladelike wings inserted transversely at about middle of body, this produced into a truncate slender beak. Achene wholly enclosed within winged involucre.

Wings of fruit spirally arranged or in several series; diameter of winged fruit 8–10 mm. . . . . . . . . . . 1. *H. salsola*
Wings of fruit usually in a single equatorial whorl; diameter of winged fruit 3–6 mm.:
  Fruit, including beak, 6–8 mm. long; wings mostly 5, flabelliform-obovate, nearly or quite as wide as long . . . . . . . . . . 2. *H. pentalepis*
  Fruit, including beak, 4–6 mm. long; wings 5–12, oblanceolate to narrowly obovate, mostly distinctly longer than broad . . . . . . . . . . 3. *H. monogyra*

**1. Hymenoclea salsola** Torr. & Gray, in A. Gray, Mem. Amer. Acad. II, **4:** 79, 1849

A rounded, often straggly shrub 1–2.5 m. tall with pale stramineous bark and slender, striate, sparsely and finely puberulent branches; leaves filiform, 5–35 mm. long, about 0.5 mm. wide, subterete, minutely pilosulous; staminate heads above pistillate heads or intermingled with them, sessile; staminate involucres 15–20-flowered, 3–4 mm. wide, 5–7 lobes rounded, crenate, finely pilosulous; receptacular paleae broadly spatulate, pinnately veined; staminate corollas funnelform, puberulent; pistillate heads subtended by 1 to several linear bractlets; body of fruit about 6 mm. long, with 7–12 wings in several series or in an imperfect spiral; beak about 2 mm. long; wings reniform, narrowed to a petiole-like base, with a deep pit in axil of each.

Along washes, in canyons, and on the open desert, Lower and Upper Sonoran Zones, Mojave and Colorado Deserts to southern Utah, central Arizona, and into adjacent Sonora and Baja California. March–May.

**2. Hymenoclea pentalepis** Rydb., N. Amer. Fl. **33:** 14, 1922

*Hymenoclea hemidioica* A. Nels., Amer. Jour. Bot. **25:** 117, 1938.

A slenderly branched, rounded or straggly shrub 1–1.5 m. tall, branches green, striate, more or less resinous-glutinous, glabrous; leaves few, filiform, 1–5 cm. long, terete or nearly so, glandular-punctate and glutinous; staminate heads fewer than and mixed with pistillate ones, involucres turbinate, 3–3.5 mm. wide, about as high, with 6–7 ovate obtuse lobes, glandular-puberulent and often rather resinous; receptacular paleae of staminate flowers oblanceolate, shorter than funnelform, puberulent, staminate corollas; tips of anthers acute, incurved; pistillate heads numerous, often crowded, subtended by 1 to several short, leaflike bracts and a single obovate hyaline scale; fusiform body of pistillate involucre 5–6 mm. long, usually with 5 broadly obovate-flabelliform, short-cuspidate or erosulate wings inserted in a single median whorl, or occasionally with an extra wing about halfway between main whorl and apex of slender conical beak, latter 1.5–2.2 mm. long; wings 2–2.5 mm. long, to 3 mm. wide.

Along washes, on open desert, and rocky hillsides, Lower Sonoran Zone, Mohave, Graham, Maricopa, and Pima Counties, Arizona, to Imperial County, California, south into northeastern Sonora and central Baja California. Feb.–May.

**3. Hymenoclea monogyra** Torr. & Gray in A. Gray, Mem. Amer. Acad. II, **4:** 79, 1849

Openly branched, rounded or straggly shrub 1–4 m. tall with prominently striate-ridged, slender branchlets, these at first puberulent, later glabrate and resinous-glutinous; leaves filiform, nearly terete, 2–10 cm. long or lower ones with several filiform divisions 1–3 cm. long, somewhat grooved and finely tomentulose beneath; heads very numerous in small, densely crowded, axillary clusters; staminate heads usually intermingled with pistillate ones, with turbinate, puberulent, 5–6-lobed involucres 3–4 mm. broad, ovate lobes entire or indistinctly toothed; receptacular paleae of staminate heads linear-spatulate; staminate corollas membranous, funnelform, about 2 mm. long, minutely puberulent, with triangular lobes; pistillate heads very numerous, each subtended by an ovate, scale-like bract; fruit 4–5 mm. long, body narrowly fusiform, with a single whorl of 5–12 scarious, narrowly elliptic to narrowly obovate wings 1.5–2 mm. long inserted at about middle of body; beak about 2 mm. long; entire fruit sparsely granular-dotted or smooth in age.

Mostly in sandy soil of desert plains or along washes, Lower Sonoran Zone and arid places in the lower parts of the Upper Sonoran Zone, San Diego and San Bernardino Counties, California, to western Texas and south into Chihuahua, Sonora, Baja California, and Sinaloa. Sept.–Oct.

## 63. AMBROSIA L., Sp. Pl. 987, 1753

Annual or perennial herbs with opposite or alternate, usually lobed or dissected leaves and more or less pubescent herbage. Staminate heads above pistillate ones in spikes or racemes, usually nodding, their involucres saucer-shaped to turbinate, with 5–12 lobes. Receptacular paleae of staminate heads filiform with more or less dilated or clavate tips. Staminate corollas funnelform, glabrous to puberulent, regularly 5-toothed, filaments monodelphous; anthers with incurved, cuspidate to acuminate tips; pistils abortive. Pistillate heads less numerous, erect, involucres nut-like, usually with tubercles or erect spines near truncate apex; beak 3–4-toothed; pistillate flower solitary in each involucre, devoid of a corolla.

Leaves entire or 3–5-cleft; body of fruit glabrous; plant 1–5 m. tall, annual .... 1. *A. aptera*
Leaves once- to thrice-pinnatifid into fine or medium segments; body of fruit puberulent; plants 0.1–1 m. tall, perennial:
Fruits armed with 4–7 conical spines or tubercles around upper angles; leaves once-pinnatifid; plants 3–10 dm. tall, hirsute:
Hairs on stems ascending or appressed ............ 2. *A. psilostachya* var. *psilostachya*
Hairs on stems widely spreading ................. 2a. *A. psilostachya* var. *californica*
Fruits unarmed; leaves twice- or thrice-pinnatifid; plants 1–2 dm. tall, silky-canescent ... ............................................................ 3. *A. pumila*

**1. Ambrosia aptera** DC., Prod. **5:** 527, 1836

*Ambrosia trifida* var. *texana* Scheele, Linnaea **22:** 156, 1849.

Annual weed 1–5 m. tall with angled, striate and often scaberulous stems; lower leaves ovate to orbicular in outline, 3–5-cleft, 3–12 cm. broad, to 25 cm. long, upper ones somewhat smaller, simple or deeply 3-cleft, shallowly serrate-dentate or denticulate, dark green above, paler beneath, scabrous on both sides with short, conical, forwardly curved hairs; petioles 3–15 cm. long, scabrous, wingless; staminate heads numerous in terminal racemes or panicles; staminate involucres saucer-shaped, 3–3.5 mm. broad, with 6–8 low rounded lobes, sparsely scabrous; receptacular paleae minute; staminate corollas broadly funnelform, glabrous, about 1.5 mm. high; pistillate heads in small clusters in lower parts of racemes, subtended by linear to elliptic leaves 0.5–2 mm. broad, 5–15 mm. long; body of fruit obovoid, 4–4.5 mm. long, with 4–8 obtuse longitudinal ridges, each of these tipped with a small conical tubercle, intervals often pitted; beak broadly conical, about 1 mm. long or less.

Bottomland along ditches, fields, and roadsides, Lower Sonoran to Transition and Carolinian Zones, Colorado to Illinois and south to Texas, Arizona, and adjacent Sonora and Chihuahua. July–Oct.

**2. Ambrosia psilostachya** DC., Prod. **5:** 526, 1836, var. **psilostachya**

*Ambrosia coronopifolia* Torr. & Gray, Fl. N. Amer. **2:** 291, 1842.
*Ambrosia lindheimeriana* Scheele, Linnaea **22:** 156, 1849.
*Ambrosia psilostachya* var. *lindheimeriana* Blankenship, Rep. Mo. Bot. Gard. **18:** 173, 1907.

Perennial herb 3–10 dm. tall from a creeping rootstalk, branched above; branches striate, hirsute with ascending, more or less appressed hairs; leaves lanceolate in outline, subsessile, pinnatifid, to 2 cm. wide, 3–12 cm. long, glandular-granuliferous and hirsutulous, sometimes scaberulous, dark green above, often canescent beneath, divisions linear-lanceolate to lance-oblong, acute to acuminate, entire or few-toothed or rarely again pinnatifid; staminate heads numerous in terminal spiciform racemes; staminate involucres obliquely and broadly obconic, 2–2.5 mm. broad, margin barely crenate, sides hirsutulous and glandular-granuliferous; pistillate heads usually solitary in axils just below racemes, sometimes in 2–3-headed clusters; body of the fruit about 2.5 mm. long, obovoid, somewhat rugulose, hirsutulous, tipped with a broadly conic beak 0.5–0.8 mm. long and with 4–7 short acute tubercles at upper ends of obtuse angles.

Plains, along washes, in fields, and on hillsides and roadsides, Lower Sonoran to Transition Zones, Washington to Saskatchewan, in Illinois, and south into California, Arizona, Texas, and northern Mexico. July–Oct.

**2*a*. Ambrosia psilostachya** var. **californica** (Rydb.) S. F. Blake, in Tidestrom, Contr. U.S. Nat. Herb. **25:** 581, 1925

*Ambrosia californica* Rydb., N. Amer. Fl. **33:** 20, 1922.

Differing from var. *psilostachya* in having widely spreading hairs on stems, staminate involucres 3–4 mm. wide, and often slightly larger fruits.

A common weed in waste places along roads and ditches, Lower Sonoran to Transition Zones, northern California to northern Baja California, east to Nevada, Colorado, and Arizona. Commoner in southern California than the typical variety, or possibly completely replacing it there. July–Oct.

**3. Ambrosia pumila** (Nutt.) A. Gray, Proc. Amer. Acad. **17:** 217, 1882

*Franseria pumila* Nutt., Trans. Amer. Phil. Soc. II, **7:** 344, 1840.
*Hemiambrosia heterocephala* Delpino, "Studi Lign. Anem. 61. 1871."

Small perennial herb 1–2 dm. tall from a deep-seated taproot; stems and herbage finely silky-canescent; leaves chiefly alternate, lance-oblong to broadly ovate in outline, 1–2.2 cm. wide, 3–6 cm. long, bi- or tripinnatifid into numerous linear or linear-oblong divisions 0.5–1 mm. wide, 1–3 mm. long, finely silky-canescent; staminate heads comparatively few in terminal racemes 2–8 cm. long; staminate involucres cup-shaped, 3–4 m. wide, crenately 5–7-toothed on margin, teeth low and rounded, canescent; staminate corollas about 1.5 mm. long, glabrous to minutely glandular-puberulent, slightly surpassing spatulate, pubescent-tipped receptacular paleae; pistillate heads in small clusters or solitary in axils below racemes, puberulent; body of fruit obovoid, about 2 mm. long, puberulent but devoid of tubercles on upper angles.

Margins of drying vernal pools, on clay hillsides and roadsides, Upper and Lower Sonoran Zones, vicinity of San Diego, California, and south at least to Calmallí, Baja California. April–June.

## 64. FRANSERIA Cav., Ic. 2: 78, 1793

Annual or perennial herbs or shrubs (usually low) with alternate or opposite, mostly petiolate, entire to much-dissected leaves. Plants monoecious. Staminate heads usually nodding, in racemes or spikes at tips of branches; involucre saucer-shaped to turbinate, crenate, toothed or lobed; receptacular paleae filiform to spatulate, usually puberulent to villous. Staminate corollas funnelform, 5-toothed. Filaments monodelphous; anthers distinct, usually with tips somewhat chartaceous, cuspidate, inwardly curved. Pistils of staminate flowers with a peltate abortive stigma. Pistillate heads burlike or nutlike, erect or spreading; involucre of several bracts fused together, with several to many straight or hooked spines and tipped with 1 to several 2-toothed or 2-lipped beaks, outer of two teeth longer than and usually curved over inner one.

Spines on fruit 4–7 (mostly 5), shining-white, 1.5–3 cm. long; pistillate heads solitary at base of staminate spikes .......... 1. *F. bryantii*
Spines on fruit usually 10 or more, brownish or yellowish; pistillate heads usually in small clusters at base of staminate racemes:
  Plants herbaceous to rootstock; beaks very oblique:
    Fruit 6–10 mm. long; spines straight, not hooked at apex; plants annual; stems hirsute with some spreading, pustulate-based hairs .......... 2. *F. acanthicarpa*
    Fruit 2–3 mm. long; spines hooked at tip; plant perennial; stems strigose but not hirsute .......... 3. *F. confertiflora*
  Plants shrubby, at least at base; beaks only slightly oblique:
    Pistillate heads nearly always 1-flowered; fruit 1-beaked:
      Fruit floccose-woolly or arachnoid-villous with long hairs; spines 10–18, subulate, 3–5 mm. long .......... 4. *F. eriocentra*
      Fruit glabrous; spines 5–8, conic, about 0.5 mm. long .......... 5. *F. acuminata*

Pistillate heads 2–8-flowered (usually 2–3-flowered); fruit with as many beaks as pistillate flowers:
- Spines 6–60; fruit less than 10 mm. long:
  - Spines conical, not flattened or channeled:
    - Stems tomentose and glandular in youth, finally glabrate; spines with a pit above at base; staminate involucres 7–8 mm. wide:
      - Fruit hirsute as well as glandular; leaf segments 2–4 mm. wide .................................................. 6. *F. camphorata* var. *camphorata*
      - Fruit merely glandular; leaf segments usually less than 2 mm. wide .......................................... 6a. *F. camphorata* var. *leptophylla*
    - Stems hispid; spines subulate-caudate, not pitted above near base; staminate involucres 3.5–4.5 mm. wide ............................ 7. *F. hispida*
  - Spines subulate and flattened, or terete and channeled on upper side:
    - Leaves deeply pinnatifid or bipinnatifid; bark often white:
      - Spines subulate, flattened, not hooked; young branches and leaves distinctly cinereous ............................................ 8. *F. dumosa*
      - Spines subterete, channeled on upper side, hooked at apex; gray branchlets and leaves only faintly cinereous or green ................ 9. *F. magdalenae*
    - Leaves not pinnatifid, usually lobed or toothed or sometimes entire; bark usually grayish, brownish or stramineous:
      - Spines not hooked, more or less flattened at base; leaf blades dentate, serrate or entire:
        - Leaf blades mostly ovate or deltoid-ovate; fruit densely lanate between spines, not apparently glandular ...................... 10. *F. chenopodifolia*
        - Leaf blades mostly lanceolate or narrowly lance-ovate; fruit sparsely villous and distinctly glandular between spines and on their bases ...................................................... 11. *F. deltoidea*
      - Spines moderately to strongly hooked, terete but channeled on upper side; leaf blades 3–5-lobed:
        - Leaf blades triplinerved from base; cordate or subcordate:
          - Spines 15–18; fruit densely stipitate-glandular; leaf blades with spreading hairs beneath ................................ 12. *F. cordifolia*
          - Spines 25–30; fruit sparsely pilosulous, obscurely glandular; leaf blades with appressed hairs ................................ 13. *F. divaricata*
        - Leaf blades pinnately veined and sometimes cleft, not cordate at base:
          - Spines of fruits broadly flattened and pitted at base, faintly hooked or straight; leaf blades deltoid-lanceolate, caudate-acuminate at apex ........................................................ 14. *F. flexuosa*
          - Spines of fruit nearly terete, faintly channeled above, strongly hooked; leaf blades ovate or lance-ovate in outline, not caudate-acuminate:
            - Leaf blades 4–8 cm. long, sinuately toothed but not strongly lobed; fruits 5–6 mm. long ........................ 15. *F. sanctae-gertrudis*
            - Leaf blades 8–22 cm. long, strongly 3–7-lobed and coarsely serrate-dentate; fruits 8–9 mm. long ..................... 16. *F. arborescens*
- Spines 100 or more; fruits usually more than 1 cm. long:
  - Leaves petiolate, to 18 cm. long; staminate involucres 5–8 mm. broad, lobes not caudate-acuminate .................................. 17. *F. ambrosioides*
  - Leaves sessile, 3–8 cm. long; staminate involucres 12–20 mm. broad, lobes caudate-acuminate, 4–6 mm. long ................................ 18. *F. ilicifolia*

### 1. **Franseria bryantii** Curran, Proc. Calif. Acad. II, **1:** 232, 1888

*Acanthambrosia bryantii* Rydb., N. Amer. Fl. **33:** 22, 1922.
*Ambrosia bryantii* Payne, Madroño **16:** 234, 1962.

A low, spreadingly branched shrub 2–5 dm. tall with gray to stramineous bark and nearly glabrous, whitish, striate young branches; leaves ovate to oblong-lanceolate in outline, 1–2.5 cm. wide, 2–3.5 cm. long, bi- or tripinnatifid into slender, linear or oblong segments 1–6 mm. long, their margins slightly to strongly revolute, sparingly strigillose to glabrate, dark green; staminate heads subsessile at tips of short, terminal and axillary racemes; involucres campanulate, about 2–2.5 mm. in diameter, 2.5–3.5 mm. high, irregularly and shallowly lobed, glandular-granuliferous on sides, hispidulous-ciliolate on margins of lobes; staminate flowers exceeding involucres 1–1.5 mm.; pistillate heads solitary, axillary, subsessile; body of fruit 5–6 mm. long, bearing a solitary subulate beak about 3 mm. long, armed with 4–7 terete, subulate, whitish spines 1.5–3 cm. long, mature fruit and spines glabrous, more or less glistening.

Hillsides, plains, mesas, and arroyos, Lower Sonoran Zone, central to southern Baja California. Jan.–April.

**2. Franseria acanthicarpa** (Hook.) Coville, Contr. U.S. Nat. Herb. **4:** 129, 1893

*Ambrosia acanthicarpa* Hook., Fl. Bor. Amer. **1:** 309, 1833.
*Franseria hookeriana* Nutt., Trans. Amer. Phil. Soc. II, **7:** 345, 1840.
*Gaertneria hookeriana* Kuntze, Rev. Gen. Pl. **1:** 339, 1891.
*Gaertneria acanthicarpa* Britt., Mem. Torrey Club **5:** 332, 1894.
*Franseria palmeri* Rydb., N. Amer. Fl. **33:** 25, 1922.

A spreading or rounded, annual herb 3–6 dm. high and often somewhat wider; stems several to numerous, obviously striate and hirsute with few to many spreading white hairs from pustulate bases, often cinereous to canescent; leaves petiolate, bipinnatifid almost to midrib or sometimes only pinnatifid; petioles 0.5–5 cm. long; leaf blades ovate in outline, 1–4.5 cm. wide, 5–12 cm. long, strigose and hirsute on both sides, divisions linear to oblong, acute to rounded at apex; heads numerous in moderately to densely leafy panicles of racemes, racemes of staminate ones sometimes 10–12 cm. long; peduncles of staminate heads 1.5–4 mm. long, ascending; involucres saucer-shaped, 2.5–4 mm. wide, dark green with upper lobes dark-ribbed, sparsely hispidulous to glabrate, lobes broadly ovate, about 1 mm. long and wide, obtuse or rounded; pistillate heads mostly in small axillary clusters below staminate racemes, 1-flowered; fruit ellipsoidal, 6–10 mm. long, minutely glandular-granuliferous, often somewhat reticulate-veined, hirsutulous to glabrate, beak subulate, 0.5–1.4 mm. long; spines 8–20, lanceolate, 3–4 mm. long, flattened and more or less connate at base, often broadly and shallowly grooved above near base, straight, not hooked.

Sandy soil, roadsides, fields, and waste places, upper margins of Lower Sonoran into Transition and possibly Canadian Zones, southern California to Texas, Missouri, Alberta, and Saskatchewan. Occurring sporadically along the fringe of the Sonoran Desert near Palm Springs and Cabazon, Riverside County, California. March–Dec.

**3. Franseria confertiflora** (DC.) Rydb., N. Amer. Fl. **33:** 28, 1922

*Ambrosia fruticosa* DC., Prod. **5:** 525, 1836. Not *A. fruticosa* Medic., 1775.
*Ambrosia confertiflora* DC., op. cit. 526.
*Ambrosia longistylis* A. Gray, Mem. Amer. Acad. II, **4:** 79, 1849. Not *A. longistylis* Nutt., 1840.
*Franseria tenuifolia* Harv. & Gray, in A. Gray, op. cit. 80.
*Franseria tenuifolia* var. *tripinnatifida* A. Gray, Bost. Jour. Nat. Hist. **6:** 227, 1850.
*Xanthidium tenuifolium* Delpino, "Studi Lign. Anem. 62. 1871."
*Gaertneria tenuifolia* Kuntze, Rev. Gen. Pl. **1:** 339, 1891.
*Franseria incana* Rydb., N. Amer. Fl. **33:** 30, 1922.

Herbaceous perennial 3–20 dm. tall with appressed-hispid, very leafy stems and short-petioled, bi- or tripinnately dissected leaves; leaf blades 5–15 cm. long, strigosehispidulous on both sides, segments linear, oblong or narrowly ovate, acute, terminal ones elongated; heads small, numerous, in terminal panicles; staminate heads on reflexed peduncles 1–2 mm. long; involucres turbinate, 1–1.5 mm. high, 1.5–2 mm. wide, hirsutulous and glandular-puberulent, irregularly 8–10-lobed, these broadly deltoid, ciliolate; receptacular paleae filiform-spatulate; corollas pale yellow, puberulent; pistillate heads disposed singly or in small clusters near base of racemes, 1–2-flowered; fruit ovoid, glandular-puberulent and sparsely villous, 2–3 mm. long with a single conic beak incurved at tip; spines 6–20, dilated and somewhat grooved or pitted toward base, hooked at apex, 0.6–0.8 (or rarely 1) mm. long.

Hillside slopes, mesas, and sometimes a weed in fields and along roadsides, chiefly Lower Sonoran Zone but also in Upper Sonoran and Carolinian Zones, southern Colorado and Oklahoma to southern California, Texas, and south into northern Mexico. March–Oct.

**4. Franseria eriocentra** A. Gray, Proc. Amer. Acad. **7:** 355, 1868

*Gaertneria eriocentra* Kuntze, Rev. Gen. Pl. **1:** 339, 1891.

An upright or ascendingly spreading shrub to 1 m. tall, branchlets cinereous-strigillose when young but soon glabrate and pale gray or stramineous bark becoming shallowly and irregularly fissured; leaves lanceolate to oblanceolate in outline, short-petiolate to subsessile, 5–18 mm. wide, 2–4 cm. long, pinnately lobed or incised, cuneate at base, acute at apex, finely gray-tomentulose on both surfaces but soon subglabrate or minutely and sparsely puberulent and green above; heads in panicles, staminate heads subsessile or short-pedunculate; involucres saucer-shaped, 2.5–3.5 mm. in diameter, tomentulose and glandular-puberulent, 5–7 lobes lance-triangular; receptacular paleae about equaling corollas; corollas puberulent, narrowly funnelform, yellow with purple-tipped lobes; pistillate heads sessile, subtended by several lanceolate bractlets 1.5–2.5 mm. long, 1-flowered; body of fruit fusiform, 7–8 mm. long, beak narrowly conic, not hooked; spines 10–18, subulate, straight or slightly curved and 3–5 mm. long, slightly flattened, upper 6–8 in a fairly uniform whorl, rest scattered below, whole fruit glandular-puberulent and long-villous with very fine, silky, white hairs.

Washes and rocky hillsides, Lower Sonoran to Transition Zones, Providence and New York Mountains, San Bernardino County, California, east to southern Utah and south into Mohave, Gila, Maricopa, and Pinal Counties, Arizona. Barely entering the northern fringe of the Sonoran Desert. April–May.

**5. Franseria acuminata** Brandegee, Proc. Calif. Acad. II, **2:** 171, 1889

A suffrutescent, nearly glabrous plant 3–6 dm. tall, much branched above; leaves chiefly alternate, 3–5 cm. wide, 5–8 cm. long, ovate-lanceolate to lanceolate in outline, pinnately divided nearly to midrib into 3–8 pairs of unequal, lanceolate and more or less acuminate lobes, terminal lobe longer than laterals and long-acuminate, blades bright green and very sparsely strigose; heads arranged in racemose-paniculate inflorescences; staminate heads many, their involucres turbinate, glabrous, with 5–8 triangular mucronate lobes; staminate flowers 10–20 in each head, subtended by minute, slightly clavate paleae; pistillate heads 1–2-flowered, arranged in small clusters in upper axils; fruit ovoid, 4.5–5 mm. long, glabrous, with 5–8 very short, incurved spines 0.5 mm. long or less arranged in 2 series; beak short, conical.

Along sandy arroyos and gravelly mesas, Lower Sonoran Zone, near the ocean along the western side of the peninsula, central Baja California. March–June.

**6. Franseria camphorata** Greene, Bull. Calif. Acad. **1:** 192, 1885, var. **camphorata**

A low perennial 3–6 dm. tall, woody only at base, stems tomentose and glandular-viscid when young, soon glabrate or subglabrate and reddish; petioles 1–3 cm. long, sparsely tomentulose to glabrate; leaf blades ovate to triangular in outline, 1.5–3.5 cm. wide, 3–8 cm. long, interruptedly bipinnatifid, white-tomentose beneath, more or less floccose but green above, rachis narrowly winged, usually with small triangular lobes between main segments, final divisions lanceolate, usually finely and somewhat irregularly toothed; staminate heads on terminal racemes, or these sometimes paniculate; peduncles of staminate heads 3–5 mm. long, spreading to recurved; involucres saucer-shaped, 5–8 mm. broad, glandular-puberulent, teeth 8–10, broadly deltoid; staminate corollas 2–2.5 mm. long; pistillate heads below staminate, subtended by linear bracts 3–5 mm. long, 2–3-flowered; fruit 6–7 mm. long, broadly obovoid to subglobose, finely glandular-granuliferous and sometimes sparsely hispid; heads 2–3, broadly conic, about 1.5 mm. long and as broad at base; spines 6–10, subconic, straight or nearly so, usually with a pit in upper axil.

Rocky hillsides, mesas, and desert plains, Lower Sonoran Zone, northern central Baja California and on Guadalupe Island. Jan.–May.

**6*a*. Franseria camphorata** var. **leptophylla** A. Gray, Proc. Amer. Acad. **22:** 309, 1887

*Franseria leptophylla* Rydb., N. Amer. Fl. **33:** 32, 1922.

Differing from var. *camphorata* in having the leaves more deeply divided into narrower lobes; fruit 5–6 mm. long and bearing 8–12 slightly more slender spines.

Arid hillsides and along washes, Lower Sonoran Zone, west-central Sonora and southern flanks of the Sierra San Pedro Mártir to central Baja California. March–May.

**7. Franseria hispida** Benth., Bot. Sulph. 25, 1844

*Gaertneria hispida* Kuntze, Rev. Gen. Pl. **1:** 339, 1891.

A low shrubby plant 1–2.5 dm. tall with erect, white-hispid stems; leaves triangular in outline, about 5–7 cm. wide and long, bipinnately divided with small ovate lobes which are often again divided into smaller divisions between larger segments, rachis winged, blades green, glandular-pruinose and hispid with slender, white, spreading hairs, ultimate divisions round-toothed; racemes dense and short; staminate involucres 3.5–4.5 mm. in diameter, with 6 ovate, acute, almost distinct lobes, glandular-puberulent without; paleae of staminate receptacle linear, slightly dilated at apex, viscid hirsutulous, particularly toward tips; pistillate involucres ovate, 2–4-flowered, 2–4-celled, viscid-glandular; fruit about 4 mm. long, armed with 6–9 subulate straight spines about 1 mm. long, arranged in 2 series; beaks 2–4, conic, acute, 0.8–1 mm. long.

Sandy and rocky soil, Lower Sonoran Zone, vicinity of Magdalena Bay, Baja California. Nov.–March.

**8. Franseria dumosa** A. Gray ex Torr., in Frém., Rep. 316, 1845

*Franseria albicaulis* Torr., Pl. Frém. 16, 1853.
*Franseria dumosa* var. *albicaulis* A. Gray, in Torr., Bot. Mex. Bound. 87, 1859.
*Gaertneria dumosa* Kuntze, Rev. Gen. Pl. **1:** 339, 1891.

A much-branched, rounded, shrubby perennial 2–6 dm. tall; branches stiff, more or less spinose, densely strigillose-canescent when young, eventually glabrate; bark gray and slightly striate-fissured; leaves ovate in outline, 0.8–2 cm. long, pinnately to bipinnately divided into short, rounded or obtuse lobes 0.5–2 mm. wide, 0.5–3 mm. long, often having a lacelike appearance, densely grayish-tomentose on both sides; petioles 2–20 mm. long; inflorescences racemose or spiciform, staminate and pistillate heads often intermingled throughout length of racemes; staminate heads on peduncles 0.2–3 mm. long; involucres broadly saucer-shaped, 4–5 mm. wide, strigillose-canescent, lobes 5–8, broadly triangular-ovate; nearly as long as wide; receptacular paleae linear, with dilated rhombic-spatulate tips, finely villous; corollas puberulent, yellow; pistillate heads 2-flowered; fruit 4–5.5 mm. long, subglobose, moderately glandular-puberulent, beaks 2, conic-subulate, straight, 1–1.5 mm. long; spines 30–40, narrowly subulate, flattened toward base, 1.5–2.2 mm. long, tips not hooked.

Arid plains, mesas, outwash slopes, and arroyos, Lower Sonoran Zone, Mojave and Colorado Deserts, California, through southern Nevada into southern Utah, south into northern Sonora and along the eastern side of Baja California to the Sierra de la Giganta and the southern part of the Magdalena Plain. Occasionally ascending to 6,000 feet on dry ridges. Feb.–Dec.

**9. Franseria magdalenae** Brandegee, Proc. Calif. Acad. II, **2:** 170, 1889
*Franseria intricata* Rydb., N. Amer. Fl. **33:** 33, 1922.

Rounded, rather divaricately branched, stiffish shrub 2–6 dm. tall with strigose-puberulent branchlets or young branchlets sometimes cinereous-tomentulose; leaves ovate to oblong-ovate in outline, 3–7 cm. long, 1–4.5 cm. wide, bipinnatifid into lanceolate to linear, dentate, divaricate divisions 1–4 mm. wide, strigose and green above, moderately to densely tomentose beneath; inflorescences racemose or pistillate heads sometimes aggregated in clusters at base of raceme; rachis of raceme pale, sparsely hirsute with spreading slender hairs; peduncles slender, 3–10 mm. long, ascending-spreading but sometimes abruptly reflexed at very tip; involucres shallowly saucer-shaped, 5–7 mm. in diameter, with 5–10 broadly ovate or deltoid, abruptly apiculate lobes, sparsely to densely floccose or strigose; receptacular paleae filiform, puberulent; staminate flowers 25–40, lobes purplish; pistillate heads usually 2-flowered; fruit ovoid, 4–5 mm. long, beaks 2, conic-subulate, about 1.5 mm. long; spines 20–40, slenderly subulate, not flattened, 2.5–3 mm. long, sharply hooked at apex.

Sandy plains, arroyos, and clay hillsides, Lower Sonoran Zone, vicinity of Punta Prieta and Bahía de los Angeles, south to Magdalena Island, Baja California. Nov.–March.

**10. Franseria chenopodifolia** Benth., Bot. Sulph. 26, 1844
*Gaertneria chenopodifolia* Abrams, Bull. N.Y. Bot. Gard. **6:** 461, 1910.
*Franseria lancifolia* Rydb., N. Amer. Fl. **33:** 36, 1922.

A low, fairly compact, rounded shrub 2–4 dm. tall, often slightly wider than high, branches and leaves white-tomentose when young, former eventually subglabrate; petioles slender to stoutish, 2–12 mm. long, tomentulose; leaf blades broadly ovate to deltoid, acute to obtuse at apex, cuneate to truncate at base, 0.5–3 cm. wide (usually 0.5–2), 1–3 (or rarely 5) cm. long, dentate or rarely obscurely 3-lobed, densely white-tomentose and strongly veined beneath, slightly tomentose or in age glabrate above; inflorescence spiciform or paniculate; staminate heads crowded, subsessile or short-peduncled; involucres rotate, 6–7 mm. in diameter, shallowly 8–10-lobed, densely tomentulose; receptacular paleae linear-clavate, villous; staminate corollas 1.2–1.6 mm. long, slightly purplish, puberulent; pistillate heads clustered at base of inflorescence, 2-flowered; fruit subglobose, 5–6 mm. long, arachnoid-villous, armed with 20–30 subulate spines 2–3 mm. long arranged in 4–5 series, spines strongly flattened near their bases and some of them more or less connate, not hooked; beaks 2, straight, about equaling spines.

Hillsides, slopes, mesas, and arid ridges, Lower and lower parts of Upper Sonoran Zones, San Diego County, California, south to southern Baja California. Feb.–May.

**11. Franseria deltoidea** Torr., Pl. Frém. 15, 1853

*Gaertneria deltoidea* Kuntze, Rev. Gen. Pl. **1**: 339, 1891.

A shrubby perennial with numerous ascending to erect branches from a woody base; branches densely and minutely tomentulose with grayish or yellowish hairs with intermingled resinous glands, eventually glabrate and viscid; leaves numerous, petiolate, finely and closely tomentulose on both surfaces in youth, often glabrate above in age; petioles to 1.5 cm. long, slightly winged above; leaf blades deltoid to rhombic-lanceolate, broadly cuneate to truncate at base, acute to attenuate at apex, 0.4–1.5 cm. wide, 1.5–3.2 cm. long, crenulate-dentate; heads in terminal racemes or panicles, staminate ones on peduncles 0.5–3 mm. long; involucres saucer-shaped, 3–5 mm. broad, lobes 8–10, broadly triangular, densely tomentulose in youth, later glabrate and conspicuously veined; staminate corollas 2–2.5 mm. long, minutely puberulent; receptacular paleae linear, slightly dilated at apex, slightly shorter than corollas; pistillate heads usually in small clusters at bases of racemes, subtended by small, leaflike bracts, 2–3-flowered; fruit globose or broadly ellipsoidal, 5–6 mm. long, arachnoid-villous and glandular-puberulent, bearing 2–3 subulate, somewhat flattened beaks 1.5–2 mm. long; spines 15–30, distinctly flattened, glandular-puberulent, straight or nearly so, broad and more or less connate at base, rather abruptly narrowed to an attenuate slender tip, not hooked, lowermost series much broader than upper ones.

Hillsides, mesas, arroyos, and plains, Lower Sonoran Zone, Maricopa, Pinal, and Pima Counties, Arizona, south to the vicinity of Santa Ana and Puerto Libertad, Sonora. Dec.–April.

**12. Franseria cordifolia** A. Gray, Syn. Fl. N. Amer. **1**$^{2}$: 445, 1884

*Gaertneria cordifolia* Kuntze, Rev. Gen. Pl. **1**: 339, 1891.
*Franseria malvacea* Rydb., N. Amer. Fl. **33**: 34, 1922.

Shrubby perennial with several slender ascending branches from a woody base; branches minutely but densely white-tomentose when young, soon sparingly floccose or glabrate, striate; petioles slender, nearly or quite equaling blades; leaf blades ovate-cordate in outline, 1–5 cm. broad, 1.5–6.5 cm. long, moderately 3–7-lobed and rather coarsely dentate, densely cinereous-tomentulose beneath, sparsely so to glabrate and deep green and often scaberulous above; heads in terminal racemes to 1.5 dm. long; staminate involucres saucer-shaped, 4–5 mm. in diameter, densely puberulent in youth, later subglabrate, lobes much broader than long; receptacular paleae linear-clavate, short-villous with curved or crinkled hairs; staminate corollas finely puberulent on brownish lobes; pistillate heads solitary or in small (or sometimes large) clusters at base of raceme, sessile or on short axillary branches to 2.5 cm. long, 2-flowered; fruit ellipsoidal, 6–8 mm. long, densely glandular-puberulent, beaks conical, united below, somewhat hooked; spines 8–15, usually in 2–3 series, conical at base, 1–1.5 mm. long, somewhat hooked at tips.

Locally abundant in canyons, arroyos, and on rocky slopes, Lower Sonoran Zone, Maricopa and Pima Counties, Arizona, south into Sonora, Sinaloa, and San Luis Potosí. Jan.–April.

**13. Franseria divaricata** Brandegee, Proc. Calif. Acad. II, **2:** 171, 1889

Divaricately branching shrub 3–6 dm. tall with canescent-puberulent, somewhat glandular twigs; leaves short-petiolate; petioles 1–9 mm. long; leaf blades broadly ovate to suborbicular-ovate in outline, 3–5-lobed almost or quite to middle, lobes cuneate-obovate to broadly oblong, dentate or doubly dentate on their margins, minutely canescent-puberulent in youth, greener in age; heads in racemes terminating branches; staminate heads short-peduncled; 20–30-flowered; staminate involucres rotate or shallowly saucer-shaped, about 5 mm. broad, shallowly 5–10-lobed, tomentulose or puberulent in youth, glabrate in age; receptacular paleae filiform-spatulate, villous at tips; staminate corollas closely puberulent; pistillate heads in small clusters below staminate, mostly 2-flowered, minutely glandular-puberulent; fruit broadly obovoid to subglobose, 5–8 mm. long; beaks 2, stoutish, conical-subulate, about 1.5 mm. long; spines 25–30, subulate or conic-subulate, glandular-puberulent, 1.5–2 mm. long, hooked at apex, grooved on upper side near base but not appreciably flattened.

Sandy plains, arroyos, and gravelly slopes, Lower Sonoran Zone, along the Pacific side of central Baja California. Feb.–May.

**14. Franseria flexuosa** A. Gray, Proc. Amer. Acad. **20:** 298, 1885

A low, much-branched shrub 2–4 dm. tall with flexuous twigs canescent-puberulent when young, soon glabrate, purple-tinged and finely striate; petioles short, 2–6 mm. long, with decurrent base of blade forming a narrow wing along upper part; leaf blades deltoid-lanceolate, somewhat hastate, to 3 cm. wide from tip to tip of lower, hastately arranged lobes, 2–5 cm. long, coriaceous, sinuately lobed or toothed, caudate-acuminate at apex, broadly cuneate to subtruncate and slightly decurrent at base, conspicuously pinnately and reticulately veined beneath, teeth saliently spine-tipped, canescent-puberulent on both surfaces when young, in age often subglabrate and minutely punctate-glandular; heads in narrow panicles or racemose; staminate involucres rotate, 5–6 mm. in diameter, villous with slender tangled hairs, 5–7 lobes ovate, acute or apiculate; corollas greenish on lobes, villous; pistillate heads 2–3-flowered, subtended by ovate bractlets 3–4 mm. long; fruit subglobose, 5–7 mm. long, villous and glandular-puberulent, beaks 2–3, subulate; spines lance-subulate, pronouncedly flattened, channeled on upper surface and with a conspicuous pit in axil of each, straight or slightly curved, not hooked at apex.

Sandy canyons and rocky hillsides, Lower Sonoran Zone, along the desert slopes of the Sierra Juárez and Sierra San Pedro Mártir, Baja California. March–May.

**15. Franseria sanctae-gertrudis** Rydb., N. Amer. Fl. **33:** 35, 1922

A slenderly branched shrub 1–3 m. tall with pale gray to nearly white stems; twigs canescent-hispidulous in youth; leaves short-petiolate, alternate, blades ovate to ovate-lanceolate, 4–8 cm. long, 2–5 cm. wide, coarsely and sinuately toothed but not divided, somewhat acuminate at apex, hispidulous but green on both surfaces, slightly paler beneath than above; heads paniculately arranged, staminate ones above pistillate, their rotate involucres 5 mm. broad, finely pilose, lobes 5–7, triangular, acute; paleae of staminate receptacles filiform to subulate, sparsely hirsute; staminate corollas puberulent; style branches thick, stigmas peltate and penicillate; pistillate heads 2-flowered; fruit about 6 mm. long, bearing 12–15 subulate, pilose, hooked spines in 3 series; beaks 2, terete, spreading, hooked at tips.

Rocky hillsides, Lower Sonoran Zone, along the mountains east of Punta Prieta in central Baja California. A poorly known species. Probably flowering Feb.–May.

**16. Franseria arborescens** Brandegee, Zoe **5**: 162, 1903

*Franseria carduacea* Greene, Leaflets Bot. Obs. **2**: 156, 1911.
*Franseria acerifolia* M. E. Jones, Contr. West. Bot. **18**: 78, 1933.

Shrub or small tree to 5 m. tall with strongly striate, canescent-puberulent or tomentulose branches, these later glabrate; leaves alternate; petioles 2–3 cm. long, tomentulose, often somewhat scurfy; leaf blades ovate to lance-ovate in outline, 5–8 cm. wide, 8–22 cm. long, pinnately deeply cleft into 3–7 coarsely serrate lobes, cinereous-puberulent and strongly veined beneath, darker green and puberulent to scaberulous above, lobes and apex usually acute to acuminate; heads conspicuously paniculate, staminate heads above pistillate; involucres rotate, 5–6 mm. broad, deeply cleft into 6–12 lanceolate to linear-lanceolate, hispidulous lobes; receptacular paleae subulate, puberulent, shorter than puberulent tubular corollas, these about 1.5 mm. long; pistillate heads 2–3-flowered, subtended by linear bracts 3–6 mm. long; fruit elliptic-fusiform, hirsutulous, 8–9 mm. high, bearing 2 elongate-conic, slightly hooked beaks; spines 7–12, in about 3 series, subulate, hooked at tip, rather closely pubescent.

Canyons and along washes, Lower Sonoran to Tropical Zones, vicinity of Loreto to the Cape Region, Baja California. Sept.–Nov.

**17. Franseria ambrosioides** Cav., Ic. **2**: 79, *pl. 200,* 1793

*Xanthidium ambrosioides* Delpino, "Studi Lign. Anem. 63. 1871."
*Gaertneria ambrosioides* Kuntze, Rev. Gen. Pl. **1**: 339, 1891.

A subshrubby perennial 1–2 m. tall with grayish-hirsute and glandular-puberulent branches; leaves simple, petiolate, petioles 1–3 cm. long, hirsute; leaf blades lanceolate or lance-ovate, 1.5–4 or rarely to 6 cm. wide, 4–18 cm. long, truncate, cordate or broadly to narrowly cuneate at base, irregularly and rather coarsely dentate, thickish, obviously reticulate-veined beneath, green but somewhat grayish-hirsutulous, acute to acuminate or attenuate at apex; heads numerous in terminal or axillary, simple or branched racemes; staminate heads numerous, above pistillate ones; staminate involucres saucer-shaped, 5–8 mm. broad, with 7–12 acutely triangular lobes, hirsutulous; receptacular bracts of staminate heads filiform, about one-half as long as corollas; staminate corollas tubular, about 1.5 mm. long, membranous, puberulent and somewhat granuliferous; pistillate involucres subtended by 1 to few lanceolate, leaflike bracts 3–10 mm. long; fruit ellipsoidal, 10–15 mm. long, armed with numerous, spreading, faintly flattened, slender spines 1–3 mm. long, these glandular-granuliferous and hooked or coiled at tip; beaks subulate, 3–4 mm. long, hooked or coiled.

Sandy washes and canyons, Lower Sonoran Zone, Yavapai and Pinal Counties, Arizona, south into Durango, Sinaloa, and southern Baja California. Feb.–May.

**18. Franseria ilicifolia** A. Gray, Proc. Amer. Acad. **11**: 77, 1876

*Gaertneria ilicifolia* Kuntze, Rev. Gen. Pl. **1**: 339, 1891.

A moderately branched, densely leafy shrub 3–12 dm. tall with hirsute and glandular-hispidulous branches; leaves sessile, often slightly clasping, 2–4.5 cm. wide, 3–8 cm. long, ovate to ovate-lanceolate in outline, spinulosely coarse-dentate, dark green on both surfaces, prominently reticulate-veined, hispidulous and scaberulous; inflorescences compact, racemes to 15–20 cm. long, often 2–2.5 cm. in diameter; staminate heads on more or less recurved or spreading peduncles 5–10 mm. long; involucres saucer-shaped, 1.2–2 cm. broad, lobes 7–12, ovate-lanceolate and caudate-acuminate, 4–6 mm. long, surpassing disk, hirsutulous; staminate corollas campanulate, densely glandular-puberulent; pistillate involucres usually 2-flowered; fruit 10–12 mm. long, broadly ellipsoidal, densely glandular-puberulent; beaks 2, 3–4 mm. long; spines numerous, terete, glandular-puberulent and hispidulous, 3–5.5 mm. long, hooked at tip.

Along washes, sandy plains, and outwash fans, Lower Sonoran Zone, Colorado Desert, California, southwestern Arizona, and south at least to Bahía de los Angeles, Baja California. Feb.–May.

## 65. XANTHIUM L., Sp. Pl. 987, 1753

Annual, often much-branched herbs with slender to robust stems and mostly pubescent herbage. Leaves alternate, petiolate, lobed, incised or toothed, often beset with numerous punctate resinous glands having a strong odor. Heads monoecious, staminate ones hemispheric or cup-shaped, borne above pistillate ones, their bracts free. Pistillate involucre burlike, 2-celled, 2-beaked, covered with slender to robust, stiff (or weak), terminally hooked prickles or spines; each cell normally 1-seeded.

Leaves lanceolate to lance-ovate, densely tomentulose-canescent beneath; axils of leaves armed with 3-forked slender yellowish spines; prickles on fruit slender, weak .. 1. *X. spinosum*
Leaves broadly ovate to deltoid, scabrous but not tomentose-canescent; axils of leaves unarmed; prickles on fruit stouter, stiff .......................... 2. *X. strumarium*

### 1. Xanthium spinosum L., Sp. Pl. 987, 1753

*Xanthium xanthocarpon* Wallr., Beitr. Bot. **1**²: 241, 1844.
*Acanthoxanthium spinosum* Fourr., Ann. Soc. Linn. Lyon II, **17**: 110, 1869.

A considerably branched annual with erect to ascending branches 3–12 dm. tall, branches slightly striate, sparsely pubescent on internodes, rather densely so at nodes; leaves lanceolate to ovate-lanceolate, 0.5–3 cm. wide, 4–12 cm. long, acute to acuminate at apex, cuneate to rounded or subtruncate at base, 2–4-lobed with acute lobes to 2.5 cm. long or upper ones entire, densely canescent-tomentulose beneath and on midvein above, on slender petioles to 2 cm. long; a rigid, 3-parted spine with straight branches to 3 cm. long axillary to each leaf, these yellowish, shining; fruit oblong-cylindric or oblong-ellipsoidal, 9–14 mm. long, yellowish green, minutely puberulent, beset with numerous slender, rather weak, hooked prickles 2–3.5 mm. long, beaks very short or virtually lacking.

Margins of fields, roadsides, waste ground, and along ditches as a weed nearly throughout the warm temperate regions of the world. In our area occurring in Arizona, Imperial Valley, California, Sonora, and in northern Baja California. July–Oct.

### 2. Xanthium strumarium L., Sp. Pl. 987, 1753

(For extensive synonymy and notes on numerous poorly differentiated varieties see Hitchcock et al., Vasc. Pl. Pacif. N.W. **5**: 334, 1955.)

A moderately to much-branched, robust annual to 1.5 m. tall, with coarse foliage and scabrous to short-pilose herbage throughout; leaf blades deltoid-ovate to suborbicular, shallowly 3–5-lobed, truncate to deeply cordate at base, 7–20 cm. long, scaberulous on both surfaces of blade, petioles about equaling blades, hispid to subglabrous; heads in several to many short-pedunculate, axillary inflorescences; fruiting bur broadly cylindric to ovoid, usually 1–3.5 cm. long, with 2 straight to hooked stoutish beaks at apex, covered with numerous hooked prickles over rest of body; prickles glabrate to puberulent, brownish or reddish when ripe, 2.5–7 mm. long (usually 2.5–4 mm. long), bur between prickles usually puberulent.

Common weed in fields, along roadsides, and in waste places, especially in swales where water stands and along ditches, Lower Sonoran to Transition and Carolinian Zones, virtually cosmopolitan in United States and southern Canada, less common in our area; Imperial Valley and agricultural areas in Baja California and Sonora; Arizona. Feb.–Oct.

## Tribe 7. MADIEAE. Tarweed Tribe

Annual or occasionally perennial herbs or suffrutescent plants with glandular, often heavy-scented herbage. Leaves alternate or less commonly opposite. Heads radiate. Involucral bracts equal, in a single series, each enclosing, partially or wholly, one of the ray achenes. Receptacular bracts usually forming a ring or cup separating the ray from the disk flowers but in some also scattered among disk flowers. Ray flowers pistillate and fertile, often epappose. Disk flowers perfect, fertile or sterile. Anthers obtuse at base, not caudate. Disk achenes often pubescent, their pappus paleaceous or less commonly none.

Ray achenes only half-enclosed by subtending involucral bracts; disk achenes uniformly abortive ........................................................ 66. *Hemizonia*

Ray achenes wholly enclosed by infolded edges of involucral bracts; some disk flowers usually fertile, others in same head abortive ................................ 67. *Layia*

## 66. HEMIZONIA DC., Prod. 5: 692, 1836

Annual, biennial, or short-lived perennial herbs with alternate leaves, herbage often more or less glandular and aromatic. Heads numerous, small. Involucres subcylindric, ovoid, or narrowly campanulate, bracts in 1 series, with only lower part enfolding ray achenes, upper portion flat. Receptacular bracts subscarious, deciduous. Ray flowers pistillate, fertile; ligules usually inconspicuous. Disk flowers perfect, sterile, tubular, with a narrow to broadly campanulate limb. Ray achenes obcompressed, back broad and rounded, rather turgid, each partially enclosed by lower part of its subtending involucral bract, epappose. Disk achenes abortive, with or without pappus bristles.

Leaves not spinose-tipped; disk achenes with lacerate pappus paleae; receptacular bracts forming circle between ray and disk flowers ........................ 1. *H. kelloggii*

Leaves pungently spinose-tipped; disk achenes epappose; receptacular bracts throughout receptacle, not forming a circle between ray and disk flowers ........................ ................................................ 2. *H. pungens* subsp. *laevis*

### 1. Hemizonia kelloggii Greene, Bull. Torrey Club 10: 41, (April) 1883

*Hemizonia wrightii* A. Gray, Proc. Amer. Acad. **19:** 17, (October) 1883.
*Deinandra wrightii* Greene, Fl. Fran. 424, 1897.

Erect, diffusely and widely branched annual 2–8 dm. tall, upper branches paniculately ascending-spreading, rather densely hirsute with spreading hairs below, sparsely hirsute and glandular above or eglandular hairs lacking in and near inflorescences; lower leaves narrowly oblanceolate, 1–4 mm. wide, 2–6 cm. long, subentire, few-toothed, or few-lobed, upper ones gradually reduced to oblong green bracts 2–5 mm. long; heads solitary at tips of numerous branchlets, short-pedunculate; involucres 4–4.5 mm. high, bracts densely stipitate-glandular, rounded on back, free apical portion 0.5–1 mm. long; ray flowers yellow, ligules 3–6 mm. long; disk corollas 2.5–3 mm. long, throat campanulate; ray achenes black, 2–2.2 mm. long, rugulose dorsally and on sides, beak set at inner apical angle, erect; pappus of sterile disk achenes lance-subulate, 1–1.5 mm. long, more or less finely lacerate toward tips, united into a short tube at base or totally free.

On arid slopes, sandy washes, and mesas, Lower and Upper Sonoran Zones, Los Angeles and San Bernardino Counties, California, south into northern Baja California and east to Imperial Valley; introduced in the vicinity of Tucson, Arizona. May–July.

**2. Hemizonia pungens** (Hook. & Arn.) Torr. & Gray subsp. **laevis** Keck, Madroño **3:** 14, 1935

A much-branched annual 2–6 dm. tall, often fully as broad, stems yellowish, sparsely to densely hirsute with spreading white hairs to 2 mm. long; herbage scarcely at all glandular; basal leaves forming a rosette 1–2.5 dm. in diameter, each leaf oblong in outline, irregularly incisely lobed to a rachis 2–4 mm. wide, lobes narrowly ovate to oblong-spatulate, 2–28 mm. long, larger ones often again lobed; cauline leaves oblong, 1–5 mm. wide, 0.5–2.5 cm. long, entire to remotely and irregularly lobed, each leaf or lobe pungently bristle-tipped, sparsely hirsute-ciliate, fascicles of smaller entire leaves in axils of larger ones, uppermost leaves crowded and almost involucrate, those immediately beneath heads sparsely setose-ciliate; heads solitary or in small loose cymes; involucres hemispherical, 4–5 mm. high, 5–7 mm. wide; involucral bracts narrowly ovate, sparsely and minutely puberulent; rays 20–40, yellow ligules 2–3 mm. long, spreading, shallowly to markedly bifid; disk flowers numerous, corollas about 2.5 mm. long; receptacle flat or slightly convex, bearing narrowly oblanceolate, scarious, sparsely villous bracts throughout; ray achenes black, nearly smooth, 1–1.3 mm. long, beak only a slight protuberance; pappus none on either ray or disk achenes.

Low alkaline swales or flats, mainly Upper Sonoran Zone, Los Angeles and San Bernardino Counties, California, south into San Diego County and barely reaching the margins of the Sonoran Desert in western Imperial County. May–Aug., or rarely to Oct.

**Hemizonia pungens** (Hook & Arn.) Torr. & Gray, has been reported as a casual weed in Pima County, Arizona, but I have seen no specimens from that area and do not know which subspecies is represented by the Arizona material.

## 67. LAYIA Hook. & Arn., Bot. Beechey 148, 1833

Annual herbs with alternate leaves and hirsute, often glandular herbage. Heads radiate, ray flowers pistillate, fertile; disk flowers perfect and fertile or sterile. Involucre broadly campanulate to hemispherical, subequal bracts in a single series, flattened on back below, thin margins infolded to partially enclose ray achenes. Ray flowers 8–20, white, yellow, or yellow with white tips. Receptacle flat, a circle of thin bracts separating the ray and disk flowers; additional receptacular bracts often present interior to ring. Ray achenes obcompressed, epappose, usually glabrous. Disk achenes usually pubescent, most of them sterile, usually bearing 5–20 pappus bristles or awns.

Rays white; pappus bristles with tangled woolly hairs at base ............ 1. *L. glandulosa*
Rays yellow with white tips; pappus bristles naked or lacking ..........................
........................................... 2. *L. platyglossa* subsp. *campestris*

**1. Layia glandulosa** (Hook.) Hook. & Arn., Bot. Beechey 358, 1840
*Blepharipappus glandulosus* Hook., Fl. Bor. Amer. **1:** 316, 1834.

Erect and strict or more commonly more or less branched annual 1–4.5 dm. tall, moderately hispid throughout, with tack-shaped blackish glands intermingled with eglandular hairs, particularly toward heads; basal leaves narrowly oblong to linear, 1–15 mm. wide, 2.5–5 cm. long, mostly remotely toothed to pinnatifid, upper ones smaller and entire; peduncles slender, 1–15 cm. long, usually markedly stipitate-glandular just beneath hemispheric involucres, these 7–10 mm. high, 1–1.5 cm. wide, bracts lance-linear, 7–10 mm. long, acute at apex, hirsute and stipitate-glandular; rays showy, white or tinged with pink, 1–2 cm. long, rather deeply 3-toothed; disk corollas about 4 mm. long; achenes about 4 mm. long, those of the ray flowers terete, glabrous, epappose; disk achenes slightly thicker, black, sparsely to moderately strigillose with silky ascending hairs, pappus of 10–12 silvery white, lanceolate or lance-attenuate paleae 3–4 mm. long, with a tangle of silky shining wool at base, particularly inside.

Open grassy hillsides and valley floors, mainly Upper Sonoran Zone, British Columbia to northern Baja California and from Idaho to New Mexico and Arizona. Reaching the margins of the Sonoran Desert along the western edge of the Colorado Desert and in Pima and Maricopa Counties, Arizona. March–June.

**2. Layia platyglossa** (Fisch. & Mey.) A. Gray subsp. **campestris** Keck, Aliso **4:** 106, 1958

*Madaroglossa elegans* Nutt., Trans. Amer. Phil. Soc. II, **7:** 393, 1841.
*Layia elegans* Torr. & Gray, Fl. N. Amer. **2:** 394, 1843.
*Layia platyglossa* var. *breviseta* A. Gray, Proc. Amer. Acad. **9:** 193, 1874.
*Blepharipappus elegans* Greene, Pittonia **2:** 246, 1892.

Simple or commonly branching annual with erect to diffuse branches 2–8 dm. tall; herbage hirsute and sparsely to densely stipitate-glandular, gray-green; basal leaves oblong, remotely toothed to pinnatifid, 2–7 cm. long, upper ones gradually reduced, those just below heads scarcely more than lance-linear bracts, entire, all of them moderately hirsute; peduncles 1–10 cm. long or more, thickened toward apex; involucres broadly campanulate to hemispheric, 6–12 mm. high, 10–15 mm. wide, bracts linear, green, 1–1.5 mm. wide, acute, hirsute and stipitate-glandular on back; ray flowers 5–15, ligules yellow with white tips or less commonly pinkish on under surface, 6–15 mm. long, 5–10 mm. wide, 3-toothed; disk flowers 30–100, corolla 4–6 mm. long; disk achenes slightly flattened, striate, 2.8–5 mm. long, densely strigillose with silky hairs; pappus of 18–32 stout, awn-like bristles slightly dilated toward base, upwardly scaberulous but not plumose or woolly, very rarely lacking; ray achenes 3–3.8 mm. long, glabrous, without pappus bristles.

Open grassy hillsides, fallow fields, and valley floors, Lower and Upper Sonoran Zones, vicinity of San Francisco Bay, sparsely north to Butte County, south through the Coast Ranges mostly west of the Great Valley to central Baja California. March–June.

## Tribe 8. **HELENIEAE.** Sneezeweed Tribe

Herbaceous or less commonly suffruticose plants with alternate or opposite leaves, or lower opposite, upper alternate. Herbage often heavy-scented. Heads radiate or discoid. Involucres cylindric, turbinate, hemispheric, or rotate; bracts in 1–5 series. Receptacles convex to cylindric or globose, naked or in some genera sparsely beset with bristles, narrow paleae, or hairs. Anthers mostly rounded at base. Style branches elongate or short, acute to truncate and dilated. Achenes linear to obpyramidal, compressed, angled or terete. Pappus of paleae, awns, bristles, or squamellae or a combination of some of these, or wholly lacking.

Involucral bracts comparatively broad, imbricated in 3–5 series; plants fleshy, growing in saline soils . . . . . . . . . . . . . . . . . . . . . . . . . . . . . . . . . . . . . . . . . . . . . . . . . . . . . . . 68. *Jaumea*
Involucral bracts not markedly broad, subequal in 1–2 (–3) series; plants usually not conspicuously fleshy, rarely in saline habitats:
Ray corollas marcescent and persistent, falling with achenes:
Achenes not heavily callous-thickened at apex; ligules broad as or broader than long, mostly 4–8; pappus of lacerate or ciliate paleae or squamellae; plants perennial . . . . . . . . . . . . . . . . . . . . . . . . . . . . . . . . . . . . . . . . . . . . . . . . . . . . . . . . . . . . . 69. *Psilostrophe*
Achenes strongly callous-thickened apically; ligules longer than broad, numerous except in *Baileya pauciradiata* which is an annual; pappus none . . . . . . . . . . . . 70. *Baileya*
Ray corollas soon deciduous from achenes, or flowers all tubular, not marcescent or persistent:
Achenes flattened but not angled, ribs or veins only marginal:
Pappus a small crown of squamellae and usually of 1–2 additional barbellate bristles; margins of achenes more or less callous and ciliate . . . . . . . . . . . . . . . . 71. *Perityle*
Pappus of 1–2 bristles, without squamellae, or achenes rarely epappose; margins of achenes not callous, glabrous or essentially so . . . . . . . . . . . . . . . . . . 72. *Laphamia*
Achenes distinctly 3–5-angled, or if angles obscure then obconic and with distinct nerves on faces:
Involucral bracts devoid of conspicuous oil glands:
Receptacle strongly conical, cylindrical, or subglobose:
Bracts in 2–3 series, distinct:
Leaves opposite, at least at base; receptacles naked; bracts often partially enclosing marginal achenes:
Heads radiate; herbage not markedly aromatic, usually pubescent; all leaves usually opposite . . . . . . . . . . . . . . . . . . . . . . . . . . . . . . . . . . . . . . . . 73. *Baeria*
Heads discoid; herbage pungently aromatic, glabrous; upper leaves usually alternate . . . . . . . . . . . . . . . . . . . . . . . . . . . . . . . . . . . . . . 81. *Amblyopappus*
Leaves alternate or basal; receptacles with few to many bristles or paleae among disk flowers; bracts not enclosing marginal achenes:
Stems more or less winged by decurrent leaf bases, leafy to summit; bracts linear to subulate; heads subglobose . . . . . . . . . . . . . . . . . . . . 86. *Helenium*
Stems not winged, usually leafy only at or near base; bracts ovate, oblong, or lanceolate; heads strongly convex but scarcely globose . . . . 87. *Gaillardia*
Bracts in a single series, slightly united toward base . . . . . . . . . . . 80. *Eriophyllum*
Receptacle flat or slightly convex:
Pappus none or reduced to a low callous crown:
Heads 1–2-flowered (in ours); achenes 10-ribbed, only faintly compressed; leaves opposite, sessile . . . . . . . . . . . . . . . . . . . . . . . . . . . . . . . . . . . . . . . . . 74. *Flaveria*
Heads several- to many-flowered; achenes 4-angled; leaves (at least the upper) alternate, petiolate:
Plants suffruticose perennials; achenes with a low callous ring; anthers short-caudate . . . . . . . . . . . . . . . . . . . . . . . . . . . . . . . . . . . . . . . . . . . . 75. *Amauria*
Plants annual; achenes usually with pappus of bristles but these sometimes lacking; anthers obtuse at base . . . . . . . . . . . . . . . . . . . . . 80. *Eriophyllum*
Pappus of bristles, awns, or squamellae:
Squamellae of pappus dissected into plumose bristles, united at base and falling together . . . . . . . . . . . . . . . . . . . . . . . . . . . . . . . . . . . . . . . . . . 76. *Syntrichopappus*
Squamellae, paleae, or bristles of pappus not plumose, separate and falling singly:
Style appendages acute or conical:
Margins and tips of bracts membranous, white, roseate, or purplish; involucre hemispherical . . . . . . . . . . . . . . . . . . . . . . . . 77. *Hymenopappus*

Margins of bracts herbaceous, usually greenish; involucres mostly turbinate or rotate:

Heads radiate; receptacle papillose or setose; pappus paleae with mid-nerve usually excurrent as an awn:

Involucre turbinate; pappus paleae 12–20; receptacle papillose, slightly convex .................................... 78. *Hymenothrix*

Involucre rotate; pappus paleae 6–10; receptacle strongly convex to conical ...................................... 87. *Gaillardia*

Heads discoid; receptacle naked (except rarely slightly bristly in *Chaenactis*); pappus paleae acute to obtuse but not awned:

Corollas all regular and of equal size; achenes 9–18 mm. long ................................................ 79. *Palafoxia*

Corollas at margin of disk slightly irregular, larger than central ones; achenes 4–6 mm. long ......................... 82. *Chaenactis*

Style appendages obtuse, truncate, or rounded, often dilated:

Bracts in a single series, somewhat united at base ....... 80. *Eriophyllum*

Bracts in 2–3 series, distinct:

Heads discoid:

Marginal corollas irregular; involucres mostly turbinate ........................................................ 82. *Chaenactis*

Marginal corollas regular; involucres hemispheric or rotate .................................................. 84. *Trichoptilium*

Heads radiate:

Ligules linear; herbage conspicuously punctate-glandular; throat of disk corollas campanulate ........................... 83. *Bahia*

Ligules broadly oblong; herbage silky-villous, not punctate-glandular; throat of disk corollas funnelform to tubular .... 85. *Hymenoxys*

Involucral bracts usually with 1 to several conspicuous oil glands; herbage pungently and often disagreeably aromatic:

Style branches of perfect flowers elongate; leaves not bristly-ciliate at base, mostly alternate:

Pappus in 2 series, the inner of 5 lanceolate paleae, outer of numerous slender bristles ................................................ 88. *Nicolletia*

Pappus not of 2 distinct series:

Rays present; leaves usually pinnate:

Involucral bracts united nearly to apex; pappus squamellae 3–10 ................................................ 89. *Tagetes*

Involucral bracts distinct to base; pappus squamellae usually 10–20 or more ................................................ 90. *Dyssodia*

Rays none; leaves linear or pinnate:

Leaves pinnate; achenes glabrous ........................... 90. *Dyssodia*

Leaves entire; achenes puberulent ....................... 91. *Porophyllum*

Style branches of perfect flowers short, often dilated at apex, truncate or rounded; leaves bristly-ciliate at base, opposite ............................ 92. *Pectis*

## 68. JAUMEA Pers., Syn. Pl. 2: 397, 1807

Perennial herbs or somewhat shrubby plants with narrow, entire, opposite leaves. Heads solitary, radiate; peduncles short to very long. Involucre campanulate-turbinate with broad, rather firm bracts in 3–4 irregular imbricated series, striated, rounded to ovate. Receptacle conic, naked but alveolate. Ray flowers in a single series, pistillate and fertile; ligules broad, 3- to many-nerved, 3-toothed, yellow. Disk flowers numerous, glabrous; tube short and slightly dilated at base and throat or cylindrical and scarcely dilated at either end; lobes slenderly ovate-lanceolate. Achenes linear-oblong, angled or ridged, glabrous or pilose. Pappus represented by a callous margin or of 15–20 plumose bristles.

Peduncles only slightly exceeding subtending leaves; involucre 1–1.5 cm. high; achenes glabrous, 10-ridged; plant herbaceous .................................... 1. *J. carnosa*
Peduncles 1–2.5 dm. high, much exceeding leaves; involucre 2–2.5 cm. high; achenes pilose, 4–5-angled; plant subshrubby .................................... 2. *J. peduncularis*

**1. Jaumea carnosa** (Less.) A. Gray, in Torr., Bot. Wilkes Exp. 360, 1874
*Coinogyne carnosa* Less., Linnaea **6:** 521, 1831.

Subsucculent herb from a creeping rootstock with erect or procumbent stems 1–3 dm. long, glabrous, stems more or less striate; leaves fleshy, opposite, linear-oblanceolate to oblanceolate-spatulate, 2–5 mm. wide, 2–4 cm. long, rounded or obtuse but usually apiculate at apex, sessile and connate at base; involucre 1–1.5 cm. high, about 1 cm. broad, bracts broadly ovate to suborbicular, obtuse or rounded at tips, margins thin, backs finely striate, glabrous; ligules yellow, about 1 mm. wide, 5–7 mm. long, spreading or ascending, strongly 3-nerved; disk corollas with short tubes slightly enlarged at base and slightly flaring at throat, lobes ovate; achenes oblong, about 3 mm. long, 10-ridged; pappus reduced to a callous margin.

On salt marshes and along streams near the ocean, Upper and Lower Sonoran Zones, Puget Sound, Washington, south along the coast into northern Baja California. April–Sept.

**2. Jaumea peduncularis** (Hook. & Arn.) Benth. & Hook. ex Oliver, Fl. Trop. Afr. **3:** 395, 1877
*Chaethymenia peduncularis* Hook. & Arn., Bot. Beechey 298, *pl. 62,* 1838 or 1839.

A much-branched, somewhat shrubby plant 1 m. tall or less with striate, glabrous, somewhat purplish stems and opposite, narrowly lanceolate, glabrous leaves 5–18 mm. wide, 5–15 cm. long, tapering to each end, sessile, midrib prominent beneath, slightly impressed above; peduncles 1–2.5 dm. long, finely striate, purplish; involucres 2–2.5 cm. high, nearly as broad, bracts broadly ovate, striate, acute to obtuse at apex, purplish, outermost one-third to one-half as long as inner ones; ligules 12–15, about 1 cm. long, 6–7 mm. broad, many-nerved, 3-lobed at apex; disk corollas 7–9 mm. long, equaled or slightly surpassed by scabrous-plumose, brownish pappus bristles; achenes 7–10 mm. long, brownish, 4–5-angled, densely puberulent with short, appressed, stiffish, tawny hairs; pappus 7–12 mm. long.

Along streams, Lower Sonoran and Tropical Zones, southern Sonora to Jalisco and Tepíc. Oct.–Jan.

## 69. PSILOSTROPHE DC., Prod. 7: 261, 1838

Perennial herbs or small shrubs with tomentose herbage. Leaves alternate, entire or scantily and shallowly lobed. Heads corymbosely arranged on short peduncles, radiate. Involucres cylindrical, campanulate or turbinate, of a single series of woolly bracts, sometimes 1–4 scarious bracts within herbaceous ones and 1–2 small calyculate ones below. Receptacle naked. Ray flowers pistillate; ligules broad, 3-lobed, yellow, becoming papery and persistent in age. Disk flowers perfect, fertile, with a short tube and cylindrical throat; corolla limb of 5 short, regular, glandular lobes. Anthers lanceolate at tips, obtuse at base. Style branches truncate at tips. Achenes linear, obtusely angled, glabrous to villous. Pappus squamellate.

**1. Psilostrophe cooperi** (A. Gray) Greene, Pittonia **2:** 176, 1891
*Riddellia cooperi* A. Gray, Proc. Amer. Acad. **7:** 358, 1868.

A perennial suffrutescent plant forming few- to many-stemmed clumps 3–6 dm. tall, stems and leaves densely white-lanate when young; leaves linear, 1–4 mm. broad, 2–10 cm. long, becoming green and nearly glabrous above in age, midrib then rather conspicuous; peduncles 2–10 cm. long; involucres campanulate, 4–6 mm. wide, 7–9 mm. long; bracts 10–20, oblong, densely lanate without, acute at apex; ligules 4–8, broadly ovate, 8–15 mm. long, one-half to three-fourths as broad, pale yellow, coarsely 3-toothed, papery-persistent; disk flowers exserted 3–4 mm. beyond involucre at anthesis, sparsely arachnoid-villous and glandular-granuliferous without, strongly granuliferous on margins of lobes; achenes 5–7 mm. long, glabrous; pappus of oblong to spatulate, entire or erose squamellae one-half as long as to nearly equaling achenes.

Along washes, on gravelly slopes and rocky hillsides, mainly Lower Sonoran Zone, Mojave and Colorado Deserts, California, to southern Nevada, southern Utah, and south through the arid western half of Arizona into northern Sonora and Baja California. March–Sept.

## 70. BAILEYA Harv. & Gray ex Torr., in Emory, Notes Mil. Rec. 144, 1848

Annual or perennial, floccose herbs with alternate, pinnatifid to entire leaves and radiate heads solitary on long peduncles. Involucre hemispheric, bracts woolly, in 2 series, nearly equal. Receptacle flat to convex, naked but alveolate. Ray flowers 5 to many, pistillate and fertile; ligules ovate to oblong, 7-nerved, 3-toothed, scarious-papery in age and persistent. Disk flowers numerous, perfect and fertile, with short corolla tubes and tubular-funnelform throats, limb 5-toothed, glandular on margins of teeth. Anthers linear. Style branches short, truncate. Achenes oblong, slightly angled, ribbed or striate, truncate at apex, those of the ray flowers callous-thickened at apex. Pappus none.

Heads cymosely disposed toward tips of branches; rays 4–6, indistinctly 3-toothed; disk 5–6 mm. in diameter at anthesis .................................... 1. *B. pauciradiata*
Heads mostly solitary at tips of stems or major branches; rays 20–50, distinctly 3-toothed; disk 7–15 mm. in diameter at anthesis:
  Stems leafy nearly to apex; peduncles 10 cm. long or less; some ribs on achenes stronger than others ................................................ 2. *B. pleniradiata*
  Stems leafy only below middle; peduncles 10–30 cm. long; achenes equally striate-ribbed ......................................................... 3. *B. multiradiata*

**1. Baileya pauciradiata** Harv. & Gray, in A. Gray, Mem. Amer. Acad. II, **4:** 105, 1849

A floccose annual 1–3.5 dm. tall with simple erect stems or ascendingly branched; lower leaves 5–12 cm. long, oblanceolate in outline, slightly to markedly pinnatifid with linear to oblong divisions 5–15 (or 25) mm. long, densely but loosely villous; upper leaves linear, mostly entire; heads corymbiform-cymose on peduncles 2–5 cm. long; involucre 5–6 mm. high, about as broad, of 8–10 floccose bracts; ligules 4–6, yellow, oval, 6–8 mm. long, 5–6 mm. wide, shallowly 3-toothed at truncate apex; disk flowers 10–20, corollas about 3 mm. long, glandular-granuliferous without; achenes linear to slightly clavate, 4.5–5 mm. long, striately ribbed and indistinctly angled, hispidulous-scabrous and slightly glandular, truncate at apex.

Sandy desert plains and outwash slopes, Lower Sonoran Zone, Colorado Desert, California, adjacent southwestern Arizona in Yuma and Pima Counties, and south into northern Sonora and Baja California. March–May; less common following summer rain in Sept.–Oct.

**2. Baileya pleniradiata** Harv. & Gray, in A. Gray, Mem. Amer. Acad. II, **4:** 105, 1849

*Baileya multiradiata pleniradiata* Coville, Contr. U.S. Nat. Herb. **4:** 133, 1893.
*Baileya nervosa* M. E. Jones, Contr. West. Bot. **8:** 34, 1898.
*Baileya pleniradiata* var. *perennis* A. Nels., Bot. Gaz. **47:** 431, 1909.
*Baileya perennis* Rydb., N. Amer. Fl. **34:** 10, 1914.

A floccose annual with several to many loosely woolly, ascending stems 1.5–3 dm. long branching from base; lower leaves oblanceolate in outline but more or less pinnatifid into linear or oblong divisions, whole leaf 4–8 cm. long; upper leaves linear to narrowly spatulate, to 3.5 cm. long, mostly entire; heads on terminal peduncles 3–10 cm. long; involucre hemispheric, 6–8 mm. high, 7–10 mm. broad, of 20–30 linear floccose bracts; ligules 20–40, oval-obovate, about 6 mm. wide, 8–10 mm. long, shallowly 3-toothed at apex; disk flowers 40–50, corollas 3.5–4 mm. long, somewhat granuliferous without; achenes 3.5–4 mm. long, truncate at apex, tapering toward base, distinctly ribbed and angled, ribs of angles strongest.

Arid plains, mesas, and along arroyos, Lower and Upper Sonoran Zones, Colorado and Mojave Deserts, California, to southern Utah, western Texas, and northern Mexico. March–Nov.

**3. Baileya multiradiata** Harv. & Gray ex Torr., in Emory, Notes Mil. Rec. 144, 1848

*Baileya multiradiata* var. *nudicaulis* A. Gray, Syn. Fl. N. Amer. **1**$^2$: 318, 1884.
*Baileya pleniradiata* var. *multiradiata* Kearney, Trans. N.Y. Acad. **14:** 42, 1894.

Biennial or sometimes a short-lived perennial with floccose stems and leaves, stems branching at base, decumbent to ascending, 2–4 dm. long, leafy only on lower portion; basal leaves 3–5 cm. long, spatulate, 3-lobed and crenate early in season, later ones deeply 3-cleft and lobed, all densely white-floccose; lower stem leaves similar or strongly pinnatifid with obovate divisions; upper cauline leaves linear to spatulate, entire, 2–4 cm. long; peduncles 1–3 dm. long; involucres 7–8 mm. high, 10–15 mm. broad, lanate; bracts 20–35, lance-linear; ligules 25–50, cuneate-oblong, 8-12 mm. long, broadly 3-toothed, orange at anthesis, fading to pale yellow in age; disk flowers numerous, granuliferous, 3.5–4.5 mm. long; achenes cylindrical-truncate, 3–4 mm. long, evenly striate, glandular-granuliferous.

Arid plains, arroyos, and outwash slopes, mainly Lower Sonoran Zone, Arizona to southern Utah, western Texas, southern California, Baja California, Sonora, and Chihuahua. March–Oct.

## 71. PERITYLE Benth., Bot. Sulph. 23, 1844*

Annual or perennial herbs, shrubs or subshrubs with glabrous to tomentose herbage and alternate or opposite, petiolate leaves. Leaf blades entire to deeply dissected.

* Adapted from Mary Louise Everly, "A Taxonomic Study of the Genus Perityle and Related Genera," Contr. Dudley Herb. **3:** 375–96, 1947.

Heads solitary or corymbosely arranged at tips of branches, radiate or sometimes discoid only. Involucre hemispheric to turbinate; bracts more or less cucullate, 1–2-ribbed, often spreading at maturity. Receptacle naked, flat. Ray flowers pistillate, fertile; ligules white or yellow, with a slender, glandular-puberulent throat. Disk flowers several to many, yellow, sometimes becoming reddish in age; tube slender, glandular-puberulent; throat funnelform to cylindrical; limb shallowly 4–5-toothed; styles exserted. Achenes oblong to elliptic, flattened, margins slightly ciliate to villous, often calloused. Pappus a crown of erose to laciniate squamellae and a slender barbellate awn at one or both of the apical angles of achene, or awns completely lacking.

Awn of pappus only one or completely lacking:
  Squamellae much reduced, inconspicuous:
    Involucre 6–7 mm. high, 10–12 mm. broad; rays 6–7 mm. long; herbage densely glandular ........ 1. *P. crassifolia* var. *robusta*
    Involucre 4–5 mm. high, 5–7 mm. broad; rays 8–9 mm. long; herbage obscurely glandular ........ 2. *P. incompta*
  Squamellae prominent:
    Plants perennial; herbaceous branches from a woody base:
      Rays white; awn minute, about twice as long as marginal cilia of achene ........ 3. *P. spilanthoides*
      Rays yellow; awn half as long as disk corolla or more:
        Stems puberulent to glandular-puberulent, not viscid-villous; involucres 7–10 mm. wide; rays 6–8 mm. long ........ 4. *P. leptoglossa*
        Stems viscid-villous; involucre 10–15 mm. wide; rays 8–12 mm. long ........ 5. *P. palmeri*
    Plants annual, from a slender taproot:
      Margins of achenes not calloused; rays 2–3 mm. long:
        Rays white; bracts oblanceolate ........ 6. *P. emoryi*
        Rays yellow; bracts lanceolate ........ 7. *P. aurea*
      Margins of achenes distinctly calloused; rays 3–8 mm. long:
        Rays 6–8 mm. long; leaf blades deeply lobed ........ 8. *P. lobata*
        Rays 3–4 mm. long; leaf blades crenate, coarsely toothed or obscurely lobed ........ 9. *P. californica*
Pappus of squamellae and 2 short slender awns:
  Plant perennial, somewhat canescent; leaf blades coarsely dentate, not lobed or incised ........ 3. *P. spilanthoides*
  Plant annual, glandular-puberulent, not canescent; leaf blades incised or deeply lobed ........ 10. *P. microglossa*

## 1. **Perityle crassifolia** Brandegee var. **robusta** (Rydb.) Everly, Contr. Dudley Herb. **3:** 382, 1947

*Perityle robusta* Rydb., N. Amer. Fl. **34:** 16, 1914.

Perennial from somewhat woody base, 3–5 dm. tall, stems striate, densely glandular-puberulent; leaves mostly alternate; petioles subfleshy, about 1 cm. long; leaf blades reniform to cordate-ovate, thick and crisped, 0.5–1.5 cm. long, deeply lobed or some of them strongly crenate, densely glandular-puberulent; peduncles 1–4 cm. long; involucres 6–7 mm. high, 10–12 mm. broad; heads radiate, rays white, spatulate, 6–7 mm. long; disk corollas yellow, glandular-puberulent, about 2 mm. long, throat about equaling corolla tube; achenes slightly curved, pubescent to glabrous on flattened faces, slightly calloused and densely ciliate along margins; pappus a minute crown of lacerate squamellae and a single, retrorsely barbed awn about three-fourths as long as corolla.

In rocky places, often on the faces of cliffs, Lower Sonoran Zone, along the Gulf coast of Baja California from the vicinity of Agua Verde Bay south to the Cape Region, and on the adjacent islands in the Gulf of California. Oct.–May.

**2. Perityle incompta** Brandegee, Univ. Calif. Pub. Bot. **6:** 503, 1919
*Perityle macromeres* S. F. Blake, Proc. Biol. Soc. Wash. **37:** 59, 1924.

Annual herb 4–6 dm. tall, with stoutish, striated, sparsely villous, obscurely glandular stems, slightly more densely villous near nodes than on internodes; leaves opposite below, alternate above; petioles 1–3 cm. long; leaf blades ovate-suborbicular in outline, 1–4 cm. long, 2–5.5 cm. wide, lower leaves deeply 3-lobed, lateral lobes again with 2 lobes on lower sides, terminal lobe again deeply 3–5-lobed, thinnish, sparsely villous; upper leaves less deeply lobed or coarsely dentate; heads radiate; involucre 4–5 mm. high, 5–7 mm. broad; bracts oblanceolate, glandular-puberulent; rays about 13, white, 8–9 mm. long, 2.5–3 mm. wide; disk flowers yellow, tube 1.3–1.6 mm. long, throat 1–2 mm. long, funnelform-campanulate; achenes linear-ovate, about 3 mm. long, flattened, densely ciliate and narrowly calloused on margins, glabrous on faces; pappus a minute crown of squamellae and a single awn about two-thirds as long as the disk corolla, awn retrorsely barbed above middle.

In sandy soils along arroyos and on the desert plains, Lower Sonoran Zone, from the vicinity of San Ignacio south to the Cape Region, and on the islands off the west coast of southern Baja California. Nov.–June.

**3. Perityle spilanthoides** (Schultz-Bip.) Rydb., N. Amer. Fl. **34:** 17, 1914
*Galinsogeopsis spilanthoides* Schultz-Bip., in Seem., Bot. Herald 307, 1856.
*Pericome spilanthoides* Benth. & Hook. ex Hemsl., Biol. Centr. Amer. Bot. **2:** 215, 1881.
*Perityle microcephala* A. Gray, Proc. Amer. Acad. **21:** 391, 1886.
*Perityle saxosa* Brandegee, Zoe **5:** 225, 1905.

A much-branched perennial with grayish-puberulent stems 2–5 dm. tall; leaves opposite, puberulent; petioles 1–2 cm. long; leaf blades deltoid, acuminate, coarsely dentate, cuneate to truncate at base, thin, 2–5 cm. long; heads corymbosely arranged, radiate; peduncles 1–6 cm. long, puberulent; involucre about 5 mm. broad, 4 mm. high; bracts oblanceolate, densely puberulent, acute tips ciliate; ray flowers 10–12, white, rays 2–4 mm. long, 1.5–2 mm. broad, throat about 1 mm. long; disk corollas 1.5–2 mm. long, tube glandular-puberulent, almost equaling campanulate throat; achenes oblong, about 2 mm. long, puberulent on flattened faces, slightly calloused and ciliate on margins; pappus a minute crown of squamellae and 1 or 2 short awns, these unequal when two are present.

Rocky slopes and cliffs, Sonoran Zones, southeastern Arizona, eastern Sonora to Chihuahua, Durango, and Sinaloa. Barely entering the eastern limits of the Sonoran Desert. Sept.–Nov.

**4. Perityle leptoglossa** Harv. & Gray, in A. Gray, Mem. Amer. Acad. II, **4:** 77, 1849
*Leptopharynx leptoglossa* Rydb., N. Amer. Fl. **34:** 22, 1914.

A somewhat branching perennial 1–3 dm. high from a woody base; stems grayish-puberulent; leaves mainly alternate; petioles 1–3 cm. long; leaf blades ovate to deltoid, 3–5-veined, thickish, 1.5–2 cm. long and as wide at base, puberulent, crenate or doubly crenate to slightly 3-lobed, lobes obtuse to acute; peduncles 1–3 cm. long; heads radiate; involucre turbinate, about 7 mm. high, 7–10 mm. wide; bracts oblanceolate, boat-shaped at base, puberulent, acute tips ciliate; ray flowers about 20, ligules yellow, 6–8 mm. long, tube about 3 mm. long, glandular-puberulent; disk corollas yellow, throat 3–4 mm. long, tube densely glandular-puberulent, about 1.5 mm. long; achenes oblong, 3 mm. long, glabrate on flattened faces, densely villous-ciliate and callous along margins; pappus a crown of fimbriate squamellae and a solitary scabrous awn almost equaling disk corolla or this awn sometimes lacking.

Under ledges and around rocks on hillsides and canyons, Lower Sonoran Zone, central Sonora in the vicinity of Hermosillo, Torres, and Ures. Aug.–May.

**5. Perityle palmeri** S. Wats., Proc. Amer. Acad. **24:** 57, 1889

*Leptopharynx cordifolia* Rydb., N. Amer. Fl. **34:** 22, 1914.
*Leptopharynx palmeri* Rydb., op. cit. p. 23.
*Perityle cordifolia* S. F. Blake, Contr. U.S. Nat. Herb. **23:** 1602, 1926.

Perennial 1–2 dm. tall from a woody base; leaves mostly opposite, a few upper ones alternate; stems viscid-villous; petioles 1–4 cm. long; leaf blades thin, broadly cordate, 3–5-veined, 1–4 cm. broad and long, villous and glandular-puberulent on both surfaces, doubly dentate or sometimes sharply incised into 7–8 divisions; peduncles 4–7 cm. long; involucre hemispheric, 9–10 mm. high, 10–15 mm. broad, bracts oblanceolate-villous and glandular-puberulent, tips acute, ciliate; ray flowers about 30, ligules bright yellow, 8–12 mm. long; disk corollas glandular-puberulent, tube about 2 mm. long, throat 3–3.5 mm. long; achenes oblong, about 2.5 mm. long, nearly glabrous on flattened faces, broad callous margin villous-ciliate; pappus a crown of well-developed squamellae and a single scabrous awn about equaling disk corolla.

Rocky slopes and canyon sides, Lower Sonoran Zone, vicinity of Guaymas, Sonora, and south into Sinaloa. Oct.–March.

**6. Perityle emoryi** Torr., in Emory, Notes Mil. Rec. 142, 1848

*Perityle nuda* Torr., Pacif. R. Rep. **4:** 100, 1857.
*Perityle emoryi* var. *nuda* A. Gray, Bot. Calif. **1:** 397, 1876.
*Perityle californica* var. *nuda* A. Gray, Syn. Fl. N. Amer. **1**$^{2}$**:** 321, 1884.
(For additional synonymy see Contr. Dudley Herb. **3:** 388, 1947.)

A much-branched, winter annual 0.5–4 dm. tall, with sparingly hirsute and glandular-puberulent to nearly glabrous stems; leaves chiefly alternate or some of lower ones opposite; petiole 1–4 cm. long or upper leaves subsessile; leaf blades ovate, triangular-ovate, suborbicular or cordate, hirsute to glandular-puberulent, 1–4 cm. long, deeply toothed, palmately lobed or lobes serrate to coarsely dentate; peduncles slender, 1–5 cm. long; involucre of 2 rows of bracts, about 5 mm. high, 6–8 mm. broad; bracts oblanceolate, tips ciliate; ray flowers 10–12, ligules white, broadly rectangular, 2–3 mm. long, 2–2.5 mm. wide, slender tube 1.5–2 mm. long, glandular-puberulent; disk corollas 2–3 mm. long, throat and tube about equal, latter glandular-puberulent; achenes 2–3 mm. long, flattened faces glabrous to puberulent, margins thin, not at all calloused, short-hirsute to villous-ciliate; pappus a crown of small lanceolate squamellae and a weakly barbellate awn about equaling disk corolla or awn lacking.

Rocky hillsides, cliffs, and canyon walls, chiefly Lower Sonoran Zone, southern Nevada, southern California, and Arizona, to Sinaloa and southern Baja California. Throughout the year, but mainly Nov.–May.

**7. Perityle aurea** Rose, Contr. U.S. Nat. Herb. **1:** 84, 1890

A low, much-branched, and spreading annual with stems 1–4 dm. long, somewhat pubescent and glandular throughout; leaves both alternate and opposite; petioles short, stoutish; leaf blades cordate, reniform or suborbicular, 1–3 cm. long, 1–2 cm. wide, irregularly lobed and serrate, lower leaves sometimes broader than long, upper ones reduced, all of them glandular-puberulent; involucres about 6 mm. broad, 4–5 mm. high; bracts lanceolate, densely glandular-puberulent, tips acute, ciliate; ray flowers yellow, ligules 2–3 mm. long; disk corollas yellow, tube 1.5–2 mm. long, throat slenderly campanulate, about 1 mm. long, glandular-puberulent; achenes about 2 mm. long, flattened faces glabrous or sparingly villous, margins thin, not calloused, ciliate; pappus a crown of united squamellae and a short awn about equaling disk corolla.

Among rocks and in crevices in cliffs, Lower Sonoran Zone, along the Gulf in the central part of Baja California from Tres Virgenes (north of Santa Rosalia) to Canipolé and on San Marcos Island. Oct.–May.

**8. Perityle lobata** (Rydb.) I. M. Johnston, Proc. Calif. Acad. IV, **12:** 1205, 1924

*Leptopharynx lobata* Rydb., N. Amer. Fl. **34:** 23, 1914.

A low, sometimes trailing annual with viscid-villous stems 1–2 dm. long; petioles slender, 2–3 cm. long; leaves opposite, blades orbicular to ovate-cordate, 1–4 cm. wide, 4–5 cm. long, palmately 5-lobed, with lobes irregularly toothed and sinuses rounded; peduncles 4–5 cm. long; heads radiate; involucre 7–8 mm. high and about as wide; bracts oblanceolate, 2-veined, viscid-villous, tips acute, ciliate; ray flowers 8–12, ligules yellow, 6–8 mm. long, about 2 mm. wide, tube about 3 mm. long, glandular-puberulent; disk corollas 4.5–5 mm. long, tube about 1.5 mm. long, glandular-puberulent, throat 3–4 mm. long, somewhat campanulate; style appendages subulate, hispidulous; achenes oblong, flattened faces glabrous, margins calloused and villous-ciliate; pappus a crown of lanceolate squamellae and a single, rather robust, barbellate awn about equaling disk corolla.

Canyon floors and along steep slopes, Lower Sonoran Zone, central Baja California. Dec.–April.

**9. Perityle californica** Benth., Bot. Sulph. 23, 1844

*Perityle plumigera* Harv. & Gray, Mem. Amer. Acad. II, **4:** 77, 1849.
*Perityle deltoidea* S. Wats., Proc. Amer. Acad. **24:** 57, 1889.

A much-branched annual with glandular-puberulent stems 1–3 dm. long; petioles slender, 1–1.5 or 3 cm. long; leaves opposite or alternate, blades ovate to suborbicular, upper ones oblong-lanceolate, glandular-puberulent, 2–5 cm. long, margin coarsely toothed or somewhat lobed; peduncles 1–5 cm. long, glandular-puberulent; involucre hemispheric, 5–8 mm. broad, 4–6 mm. high; bracts oblanceolate, glandular-puberulent, tips ciliate; ray flowers 10–15, ligules light yellow, 3–4 mm. long; disk corollas yellow, glandular-puberulent, 2–3 mm. long; achenes about 2 mm. long, faces glabrous to villous, margins calloused and short-hirsute to villous-ciliate; pappus a crown of erose lanceolate squamellae and a solitary subplumose awn almost as long as disk corolla.

In gravelly soil or about rocks on hillsides and at the bases of cliffs, Lower Sonoran Zone, Baja California from the vicinity of Bahía de los Angeles and Calmallí south to the Cape Region, and from central Sonora south into Sinaloa. Oct.–April.

**10. Perityle microglossa** Benth., Bot. Sulph. 119, 1844

*Perityle acmella* Harv. & Gray, Mem. Amer. Acad. II, **4:** 77, 1849.
*Perityle microglossa* var. *effusa* A. Gray, Syn. Fl. N. Amer. **1**[2]**:** 322, 1884.
*Perityle effusa* Rose, Contr. U.S. Nat. Herb. **1:** 104, 1891.
*Perityle urticifolia* Rydb., N. Amer. Fl. **34:** 15, 1914.

A much-branched, or less commonly a strict, annual 2–5 dm. tall, glabrate to glandular-puberulent below, puberulent to glandular-puberulent above; lower leaves usually opposite, upper alternate; petioles 1–3 cm. long; leaf blades ovate to cordate, 3-veined, lower usually incised-dentate, upper often deeply 3–5-lobed, sometimes hastately so; heads radiate; involucres 3–5 mm. high, 4–7 mm. broad; bracts oblanceolate, densely glandular-puberulent, tips ciliate; ray flowers 6–12, ligules 2–5 mm. long, light yellow to white, throat 1–2 mm. long, glandular-puberulent; disk corollas yellow, about 2 mm. long, glandular-puberulent; achenes oblong, slightly hirsute to glabrous on flattened faces, 1.5–2 mm. long, narrowly calloused margin hirsute-ciliate; pappus a crown of lanceolate squamellae and 2 short, unequal, barbellate awns considerably shorter than disk corollas.

Margins of arroyos and about rocks on hillsides, Lower Sonoran to Tropical Zones, southern Arizona to Nicaragua and in the Cape Region of Baja California. Nov.–June.

## 72. LAPHAMIA A. Gray, Pl. Wright. 1: 99, 1852

Low suffruticose perennials with opposite leaves (or uppermost alternate) and small to medium-sized, radiate or discoid heads. Involucres campanulate to hemispheric; bracts linear, herbaceous, subequal and in 1–2 series, often more or less carinate but not surrounding achenes. Receptacles flat or faintly convex, naked. Ray flowers, when present, pistillate and fertile with oval, obovate, or rectangular ligules. Disk flowers few; tube distinctly glandular-puberulent; throat subcylindric to campanulate. Style branches of disk flowers with subulate hirsutulous appendages. Achenes compressed but not ciliate-margined, sometimes margin calloused. Pappus of 1–2 awnlike bristles or lacking.

Heads discoid; stems villous or puberulent; disk corollas 3 mm. long:
  Leaf blades dentate but not dissected or parted; disk flowers 30–50; heads 6–7 mm. wide at anthesis .......... 1. *L. palmeri*
  Leaf blades twice-pedately dissected into oblong lobes; disk flowers about 20; heads 4–5 mm. wide at anthesis .......... 2. *L. dissecta*
Heads radiate; stems glabrous; disk corollas 4.5–5 mm. long:
  Rays yellow; leaf segments filiform to linear, 5–15 mm. long, whole leaf 3–6 cm. long .......... 3. *L. saxicola*
  Rays white; leaf segments obovate, 2–6 mm. long, or lower leaves merely dentate, whole blade about 1 cm. long .......... 4. *L. gilensis*

### 1. Laphamia palmeri A. Gray, Proc. Amer. Acad. 13: 372, 1878

*Monothrix palmeri* Rydb., N. Amer. Fl. 34: 21, 1914.

Erect or ascendingly branched, rather compact, suffruticose plant 1–2.5 dm. tall with scaberulous-puberulent and somewhat glandular herbage; petioles 2–5 mm. long; leaf blades broadly ovate, deltoid or subcordate, 7–12 mm. wide, 9–15 mm. long, sharply and irregularly sinuate-dentate; heads in leafy corymbs of 2–10 heads or sometimes solitary; peduncles 2–25 mm. long; involucres turbinate-campanulate, 4–5 mm. high, 6–7 mm. wide at anthesis, bracts linear-oblanceolate, acute at apex, somewhat boat-shaped at base, back pilosulous and stipitate-glandular; rays none; disk corollas 30–50, pale yellow or cream colored, about 3 mm. long, cylindro-campanulate throat longer than short tube; achenes oblong, about 2.5 mm. long, a fine callous ridge along each margin, finely and sparsely puberulent; pappus of a single slender awn 2–2.5 mm. long.

On rocky hillsides and in crevices on cliffs, Lower and Upper Sonoran Zones, northwestern Arizona and southern Utah. June–Aug. Probably rarely entering the Sonoran Desert.

**2. Laphamia dissecta** Torr., in A. Gray, Pl. Wright. **2:** 81, 1853

*Laphamia lemmonii* A. Gray, Proc. Amer. Acad. **16:** 101, 1880.
*Laphamia lemmonii* var. *pedata* A. Gray, Syn. Fl. N. Amer. **1**$^{2}$**:** 319, 1884.
*Perityle dissecta* A. Gray, op. cit. 320.
*Leptopharynx dissecta* Rydb., N. Amer. Fl. **34:** 24, 1914.
*Leptopharynx lemmonii* Rydb., loc. cit.

A low perennial 1–2 dm. tall from a stout woody caudex; stems several, erect, moderately branched or simple, villous and glandular-puberulent; leaves chiefly alternate, rather closely arranged; petioles 3–7 mm. long, glandular-villous; leaf blades cordate, suborbicular or subreniform, to 12 mm. broad, 4–8 mm. long, twice pedately dissected into oblong to oblong-linear lobes, these obtuse to rounded at apex, rarely over 2 mm. long; heads solitary at ends of branches on stoutish peduncles 1–5 mm. long; involucres broadly campanulate, 4–5 mm. broad, 5–6 mm. high at anthesis; bracts linear, moderately boat-shaped, acute at apex, villous and somewhat glandular; rays lacking; disk corollas about 20, 3 mm. long, tube shorter than abruptly ampliate throat; achenes oblong, 2 mm. long, finely and sparsely puberulent, a faint callous border near apex; pappus of a single slender bristle about as long as achene or often wanting.

Crevices of cliffs, mainly Upper Sonoran Zone but entering the Lower Sonoran Zone in narrow canyons, Pima County, Arizona, to southwestern Texas and adjacent northern Mexico. May–Oct.

**3. Laphamia saxicola** Eastw., Proc. Calif. Acad. IV, **20:** 159, 1931

A slender, much-branched, suffruticose plant 1–3 dm. tall from a woody caudex, branches frequently hanging pendant on face of cliffs; stems finely striate, glabrous except just below heads; leaves broadly ovate in outline, 3–6 cm. long, bipinnately dissected into linear or subfiliform segments 0.5–1.5 mm. wide, 5–15 mm. long, their tips mostly acute; petioles 5–10 mm. long, channeled above; peduncles 0.5–2.5 cm. long, glandular-puberulent; heads solitary at ends of branches but loosely corymbosely arranged, about 1 cm. tall, 12–15 mm. broad at anthesis; involucral bracts linear, 5–6 mm. long, strongly boat-shaped at base, rather abruptly narrowed to a short-attenuate, acute tip, glandular-puberulent; ray flowers yellow, ligules 3–4 mm. long; disk flowers 4.5–5 mm. long, slender tube nearly as long as cylindrical throat; achenes oblong, dark brown to nearly black, 2.5–3 mm. long, callous-margined, faces sparsely strigillose with ascending hairs, callous ridges more densely so; pappus bristle exceeding body of achene, to 5 mm. long.

Growing in crevices in cliffs, Lower and Upper Sonoran Zones, Gila and Maricopa Countries, Arizona. Collected only in May but probably flowering later.

**4. Laphamia gilensis** M. E. Jones, Zoe **2:** 15, 1891

*Leptopharynx gilensis* Rydb., N. Amer. Fl. **34:** 24, 1914.
*Laphamia arizonica* Eastw., Proc. Calif. Acad. IV, **20:** 159, 1931.
*Laphamia dura* A. Nels., Amer. Jour. Bot. **23:** 266, 1936.

A very slender-stemmed, suffruticose plant with branches 1–3 dm. long from a comparatively heavy, woody caudex, branches striate, glabrous and erect; leaves mostly opposite, 3-parted, on petioles 1–2 cm. long, blades about 1 cm. long, 3-lobed into entire or denticulate, ovate to obovate lobes 2–6 mm. long (terminal one much larger than lateral pair) or lower leaves barely dentate; heads corymbosely arranged but solitary at ends of branches, 6–8 mm. high, 9–15 mm. wide at anthesis; involucral bracts linear to oblanceolate, boat-shaped at base, attenuate-acute and densely glandular-puberulent at apex; rays white, ligules about 5 mm. long; disk corollas about 5 mm. long, densely glandular, slender tube about 1.5 mm. long; achenes 3–3.5 mm. long, sparsely puberulent, callous margins slightly more densely puberulent than faces; pappus bristle nearly equaling disk corolla.

Crevices in cliffs and in rocky canyons, Lower Sonoran Zone, Pinal, Pima, and Maricopa Counties, Arizona. April–Oct.

## 73. BAERIA Fisch. & Mey., Ind. Sem. Hort. Petrop. **2:** 29, (Jan.) 1836

Small annuals with opposite, sessile, linear to spatulate, entire to laciniate-pinnatifid leaves and glabrous to pubescent herbage. Heads pedunculate. Involucres hemispheric to cylindric, of 3–15 herbaceous, nearly plane to lightly carinate bracts, these distinct or slightly united below, sometimes partially embracing an achene. Receptacle hemispheric to oblong-conic, muricate or deeply pitted. Ray flowers usually present but ligules often minute, yellow. Disk corollas few to many, with a narrow tube and narrowly campanulate throat; yellow limb 5-lobed. Style branches truncate to ovate and often with a terminal apiculation. Achenes 4-angled or flattened, narrowly cuneate to clavate-linear. Pappus of 2 to several awns or squamellae or both, or wanting.

Receptacles subulate; heads apparently discoid, rays minute; involucres subcylindric ....... 1. *B. microglossa*
Receptacle conic to subglobose; heads obviously radiate; involucres hemispheric or campanulate:
  Herbage only slightly glandular; pappus of 2–5 lanceolate-attenuate awns ....... 2. *B. chrysostoma* subsp. *gracilis*
  Herbage distinctly glandular-pubescent; pappus of 6–12 squamellae, some or all of them truncate and erosulate, awned or awnless ....... 3. *B. californica*

### 1. Baeria microglossa (DC.) Greene, Fl. Fran. 438, 1897

*Burrielia microglossa* DC., Prod. **5:** 664, 1836.
*Lasthenia microglossa* Greene, Man. Bay Reg. 205, 1894.
*Pentachaeta laxa* Elmer, Bot. Gaz. **41:** 318, 1906.

Slender, much-branched annual to 2 dm. tall with weak slender stems and sparsely soft-pubescent herbage, hairs subappressed; leaves linear, entire, 0.4–3 mm. wide, 1–3 cm. long; peduncles slender, 1–4 cm. long; involucre subcylindric, 5–8 mm. high, with only 3–4 oblong and acute bracts; receptacle subulate; ray flowers 1–3, pistillate and fertile, their ligules minute, shorter than style branches or completely lacking; disk flowers 8–15, slenderly cylindric-campanulate; style branches with short-ovate, sterile tips; achenes sublinear, slightly narrower below than above, slightly compressed, sparsely hirsutulous with upwardly subappressed bristles; pappus of 2–4 flattened squamellae narrowly attenuate into subulate awns or these often lacking on the ray achenes.

Infrequent in the shade of rocks, Lower Sonoran and lower parts of Upper Sonoran Zones, western middle California to the western margins of the Mojave and Colorado Deserts. Barely entering our region at Palm Springs and vicinity. March–May.

**2. Baeria chrysostoma** Fisch. & Mey. subsp. **gracilis** (DC.) Ferris, Contr. Dudley Herb. **5:** 100, 1958

*Burrielia gracilis* DC., Prod. **5:** 664, 1836.

(For additional synonyms see Abrams & Ferris, Ill. Fl. Pacif. States **4:** 215, 1960.)

Slender annuals with simple erect stems to 1.5 dm. tall or under favorable conditions more robust, branching from base, with stems to 2.5 dm. long, sparsely strigillose to moderately hirsute throughout, sparsely beset with scattered, tack-shaped glands or glandless; leaves filiform to linear, mostly less than 2 mm. wide, 1–5 cm. long, acute, entire; peduncles slender, 1–15 cm. long; involucre broadly hemispheric, 2.5–4 mm. high, 5–9 mm. wide; bracts 9–12, lance-ovate, 1–2 mm. wide, 4–6 mm. long, acute at apex, distinct to base, sparsely strigose and sparsely gland-dotted; ligules 1–3.5 mm. long, to 1.5 mm. wide; disk flowers about 2.5 mm. high, sparsely glandular-puberulent without; achenes linear-clavate, 2–2.5 mm. long, sparsely pubescent with short, upwardly pointing hairs, obscurely 4-angled, black; pappus of 2–5 lance-attenuate awns, basal part slightly cucullate and narrowing rapidly to awn, margins entire.

Sandy plains and gentle slopes, Lower and Upper Sonoran Zones, central California to the borders of the desert in Baja California and east into Arizona. March–May.

**3. Baeria californica** (Hook.) Chambers, Contr. Dudley Herb. **5:** 68, 1957

*Hymenoxys californica* Hook., Bot. Mag. **67:** *pl. 3838,* 1840.

(For synonymy see Abrams & Ferris, Ill. Fl. Pacif. States **4:** 218–19, 1960.)

Simple to diffusely branched, erect annual to 3 dm. tall, minutely glandular with tack-shaped glands throughout and usually also loosely villous at nodes, toward heads, and less commonly on leaves; leaves once- or sometimes twice-pinnatifid into linear or filiform divisions 3–30 mm. long, lower leaves to 8 cm. long, upper ones smaller and sometimes entire; peduncles 1–10 cm. long, glandular and villous; involucre hemispheric, 3–5 mm. high, to 12 mm. wide; bracts 8–12, broadly ovate, acute, 4–5 mm. long, villous-ciliate on margins and with a prominent midrib, partially enclosing subtended achenes and eventually falling with them; receptacle narrowly conical, deeply pitted, glabrous to hirsute; ray flowers 8–13; ligules oblong, 3–10 mm. long; disk corollas 2–2.2 mm. long, glandular without; achenes about 2 mm. long, weakly compressed, 1–2-nerved on each face, scaberulous; pappus of 6–12 squamellae, rounded, acute or truncate at apex, entire or erosulate on upper part, tapering into or terminated by a slender awn nearly equaling disk corolla, or awnless.

Sandy to clay soils, grassy slopes or mesas, Lower and Upper Sonoran Zones, coastal slopes from southern California south at least to Playa Santa Catarina, Baja California. March-May.

## 74. FLAVERIA Juss., Gen. 186, 1789

Annual or perennial herbs with opposite, narrow, entire or dentate, sessile leaves that are often connate at base. Heads many, narrow, short-pedicellate to sessile, in dense cymes or glomerules on solitary or paniculately arranged peduncles. Involucres cylindric; bracts few, equal or subequal. Receptacle naked or sparsely setose. Ray flowers solitary or lacking, if present pistillate and fertile. Disk flowers 1–15, perfect and fertile; corolla tube longer than campanulate throat, or throat funnelform and then exceeding glandular or puberulent tube; corolla limb 5-lobed. Style branches of disk flowers obtuse, reflexed. Achenes oblong, 10-ribbed, weakly compressed; pappus wanting in ours.

**1. Flaveria trinervia** (Spreng.) C. Mohr, Contr. U.S. Nat. Herb. **6:** 810, 1901
*Oedera trinervia* Spreng., Bot. Gart. Halle 63, 1800.
(For additional synonymy see N. Amer. Fl. **34:** 142–43, 1915.)

A divaricately and more or less dichotomously branched, annual, robust herb to 12 dm. tall, stems and branches somewhat ribbed and angled, often tinged with red, unevenly puberulent but eventually glabrate; leaves connate at base, lanceolate to elliptic, 5–20 mm. wide, 3–9 cm. long, shallowly serrate, strongly 3-nerved, sparsely puberulent, paler beneath than above, lower on petioles to 1.5 cm. long; cymes congested into head-like clusters, 1–2.5 cm. broad, subtended by 2–3 pairs of sessile leaves; heads usually 1–2-flowered, flowers in 1-flowered heads either ligulate and pistillate or regular and perfect; involucral bracts 1–2, thin, pale green, if 2, second shorter than inner one; corollas of ligulate flowers 1.5 mm. long, with a 3-dentate ligule, those of disk flowers 2 mm. long with a villous tube and campanulate throat; achenes 2–2.5 mm. long, oblong-clavate, black or dark gray, ribs distinct, acute, but low.

In moist soil along ditches and roadside drains, Lower Sonoran to Tropical Zones, Arizona to Alabama and south into the West Indies and Tropical South America. Probably adventive in Arizona. May–Nov.

## 75. AMAURIA Benth., Bot. Sulph. 31, 1844

Low, suffruticose, glandular-pubescent perennials with petiolate, opposite lower leaves and alternate upper leaves. Heads radiate, corymbose. Involucre hemispheric; bracts biseriate, equal, partially enfolding ray achenes below. Receptacle pitted. Ligulate ray flowers pistillate with smooth style branches. Disk flowers numerous, perfect, with hispid style branches and caudate anthers. Achenes slightly flattened, linear, 4-angled. Pappus reduced to a short callous crown.

Achenes pubescent on angles only; ligules 7–8 mm. long ................ 1. *A. rotundifolia*
Achenes pubescent on faces as well as on angles; ligules 5–6 mm. long .... 2. *A. brandegeana*

**1. Amauria rotundifolia** Benth., Bot. Sulph. 31, 1844
*Perityle fitchii* A. Gray, in Torr., Pacif. R. Rep. **4:** 100, 1857.
*Laphamia peninsularis* Greene, Bull. Calif. Acad. **1:** 8, 1884.
*Perityle rotundifolia* Brandegee, Zoe **4:** 210, 1893.

Erect or ascending, with moderately branched, densely glandular-pubescent stems 1–2 dm. high; petioles 1–5 cm. long; blades ovate to orbicular-cordate, doubly and irregularly dentate to shallowly 5–15-lobed and dentate, equaling petioles, glandular on both surfaces; heads few on peduncles 1–5 cm. long; involucres 6–7 mm. high, 8–12 mm. wide; bracts oblong to oblong-lanceolate, acute, glandular-puberulent, strongly 2-veined, inner with scarious margins; ray flowers 16–20, with white ligules 7–8 mm. long, tubes densely stipitate-glandular; disk flowers glandular-puberulent on tube, style tips acute; achenes 2.5–3 mm. long, black, hispidulous with stout golden hairs on angles, faces glabrous.

Rocky hillsides, slopes, and dry canyon walls, Lower Sonoran Zone, vicinity of San Quintín to Loreto and on Isla San Martín, Baja California. Oct.–April.

**2. Amauria brandegeana** (Rose) Rydb., N. Amer. Fl. **34:** 30, 1914
*Perityle brandegeana* Rose, Bot. Gaz. **15:** 114, 1890.

Erect or ascending perennial (often flowering first season) 0.6–3 dm. high; stems glabrous below; petioles 0.5–3.5 cm. long; blades broadly ovate- to orbicular-cordate, 5–35 mm. long, shallowly to deeply 3–7-lobed and finely to coarsely dentate, glandular-puberulent on both sides; heads on glandular peduncles 1–4 cm. long; involucres 4–5.5 mm. high; bracts oblanceolate, acute, 2-ribbed, glandular-puberulent, margins scarious and ciliate; rays white, 5–6 mm. long; disk flowers 3 mm. long, densely glandular on tube, this equaling throat; style tips obtuse; achenes black, 3–4 mm. long, sparingly hispidulous on faces, densely so on angles, many of those of disk flowers abortive.

Sandy and dry saline flats, Lower Sonoran Zone, along western side of Baja California from vicinity of Pozo Alemán to southern Magdalena Plain. Oct.–April.

## 76. SYNTRICHOPAPPUS A. Gray, in Torr., Pacif. R. Rep. 4: 106, 1857

Low, floccose, much-branched annuals with alternate, linear to spatulate, entire or 3-lobed leaves. Heads solitary on short terminal peduncles, radiate. Involucre campanulate, of 5–8 oblong concave bracts partly enfolding ray achenes. Receptacle flat, naked. Ray flowers as many as involucral bracts, pistillate and fertile, with yellow (in ours), or white tinged with purplish, or rose ligules. Disk flowers 15–25, corolla tube much shorter than narrowly funnelform throat. Anthers linear-lanceolate at tips. Style branches with lanceolate flattened tips. Achenes linear to obpyramidal, pubescent, 5-ribbed. Pappus (in ours) of a ring of squamellae dissected into barbellate bristles and falling together.

### 1. **Syntrichopappus fremontii** A. Gray, in Torr., Pacif. R. Rep. **4:** 106, 1857

Stems usually several from base, 3–12 cm. long, erect or ascending; leaves oblong-spatulate, 1–4 mm. wide, 7–20 mm. long, rounded or obtuse and entire or shallowly 2–3-toothed or 2–3-lobed at or near apex, cuneate at base, densely floccose but glabrate in age; peduncles 5–15 mm. long; involucre campanulate, 3–5 mm. high and broad, bracts scarious-margined and acute to short-acuminate; ray flowers mostly 5 with golden-yellow ligules about 2 mm. wide, 3–4 mm. long; disk corollas 3–4 mm. long, glandular and puberulent with loose curly hairs on tube and throat without, limb sparsely glandular, longer than very short tube; achenes about 3 mm. long, clavate, angled and puberulent; pappus dissected into 30–40 silvery-white, barbellate bristles 3–4 mm. long, entire ring of pappus falling intact.

Rocky slopes, desert plains and mesas, Lower and arid parts of Upper Sonoran Zones, Mojave and Colorado Deserts to southern Nevada and Utah and adjacent Arizona. March–June.

## 77. HYMENOPAPPUS L'Hér., Diss. 1788

Annual, biennial, or perennial herbs usually with more or less tomentose herbage. Leaves alternate, pinnatifid or lowermost sometimes entire. Heads usually discoid or rarely radiate, corymbosely disposed. Involucres hemispheric to campanulate, bracts in 2–3 series, outer herbaceous, inner more or less petaloid, rather broad, yellowish or whitish and scarious. Disk flowers many, perfect and fertile, with a slender, glandular-puberulent tube and 5-lobed, campanulate throat; lobes ovate, acute. Appendages on style branches conic to subulate. Achenes obpyramidal, 4–5-angled, striate and hirsutulous. Pappus of 12–16 membranous squamellae; midrib present or absent, and rarely slightly excurrent (not developed in ours).

**1. Hymenopappus filifolius** Hook. var. **lugens** (Greene) Jepson, Man. Fl. Pl. Calif. 1128, 1925

*Hymenopappus lugens* Greene, Pittonia **4:** 43, 1899.
(For additional synonymy see Turner, Rhodora **58:** 230, 1956.)

Scapose or subscapose perennial with 1 to several stems 1–4.5 dm. tall from a woody, more or less cespitose caudex, stems striate and sparsely to densely floccose; leaves oblanceolate to lanceolate in outline, bipinnatifid into linear-filiform divisions 1 mm. or less wide, 3–22 mm. long, punctate-glandular and floccose, midvein deeply impressed on upper surface of each segment; heads 1–4, corymbosely arranged on peduncles 3–10 cm. long; involucre hemispheric, 5–6 mm. high, 8–12 mm. wide; bracts obovate to elliptic, 3–3.5 mm. wide, densely floccose on lower part, outer ones purplish at scarious tips and margins, inner ones similarly marked with yellow; corollas greenish yellow, tube 1–1.5 mm. long; corolla throat tubular-campanulate, 3–3.5 mm. long, about three to four times as long as short ovate lobes; achenes about 5 mm. long, silky-hirsute; pappus squamellae oblong-obovate, obtuse, usually erosulate, midnerve yellowish, rather prominent.

Rocky hillsides, mesas and outwash slopes, margins of Lower Sonoran to Transition Zones, Mojave Desert and adjacent mountains to southern Utah, Colorado, western New Mexico, and northern Baja California, entering the Colorado Desert near Whitewater and in Baja California. Chiefly above our area. May–Sept.

## 78. HYMENOTHRIX A. Gray, Mem. Amer. Acad. II, 4: 102, 1849

Annual or perennial herbs with alternate leaves and lightly pubescent herbage. Leaves petiolate, bi- or triternately dissected into linear or oblong divisions, these entire or again cleft. Heads in corymbose or corymbose-paniculate clusters, radiate or discoid. Involucre turbinate; bracts narrow, in 2 series of 4–8 each. Receptacles papillose. Ray flowers 8–10 or more, pistillate and fertile when present. Disk flowers many, perfect and fertile, with slender glandular tubes, throat about equaling tube. Anthers bidentate at base, with ovate appendages at tip. Style branches with conic to subulate tips. Achenes narrowly obpyramidal, 4- or 5-angled. Pappus of 12–20 hyaline lanceolate paleae; costa excurrent into a barbellate awn.

Plants puberulent throughout; heads radiate; corollas yellow; throat of disk corolla about equaling tube .................................................... 1. *H. wislizenii*
Plants glabrous above; heads discoid; corollas white or faintly purplish; throat of disk corollas much shorter than tube ........................................... 2. *H. wrightii*

**1. Hymenothrix wislizenii** A. Gray, Mem. Amer. Acad. II, **4:** 102, 1849

*Hymenopappus wislizenii* var. *setiformis* M. E. Jones, Contr. West. Bot. **12:** 47, 1908.

Erect annual herb to 12 dm. tall, branching profusely above, stem and ascending branches striate and puberulent; leaves biternate or triternate, 5–12 cm. long, divisions linear to oblong, acute or apiculate at apices, puberulent; heads in many-headed, corymbose-paniculate clusters; peduncles slender, to 1 cm. long; involucres turbinate, 3–5 mm. high and broad; bracts narrowly lanceolate-oblong, 1 mm. broad or less, puberulent, tips scarious or nearly so; ligules 3–4 mm. long, yellow, shallowly 3-toothed; disk corollas yellow, about 5 mm. long, slender tube about 2 mm. long, granuliferous, abruptly expanded to the cylindrical throat, this of equal length but about twice as wide as tube; corolla lobes triangular, about 0.8 mm. long, acute; achenes black, 3–4 mm. long, sparingly hirsutulous, truncate at apex, narrowing sharply to base; pappus of 10–15 narrowly lanceolate, subscarious squamellae with a strong midnerve excurrent into a barbellate awn, whole scale 4–5 mm. long.

Desert flats, along washes, canyons, and rocky hillsides, Lower and Upper Sonoran Zones, Arizona to New Mexico, Chihuahua, and central Sonora, particularly in the lower mountains and foothills. July–Oct.

**2. Hymenothrix wrightii** A. Gray, Pl. Wright. **2:** 97, 1853

*Hymenopappus wrightii* Hall, Univ. Calif. Pub. Bot. **3:** 179, 1907.
*Trichymenia wrightii* Rydb., N. Amer. Fl. **34:** 56, 1914.
*Hymenopappus wrightii* var. *viscidulus* Jepson, Man. Fl. Pl. Calif. 1128, 1925.

Perennial herb (but often flowering first year) with slender, striate, ascendingly branched stems to 1 m. tall, densely hirsute at base, progressively less hairy upward, terminal one-third to one-half of stems and branches glabrous; lower leaves 5–10 cm. long, triternately divided into linear or filiform divisions 3–25 mm. long, upper leaves biternately divided or uppermost simple or 2–3-lobed; heads paniculate-corymbose, ultimate peduncles slender, 1 cm. long or less; involucre turbinate-campanulate, 6–7 mm. high, 8–12 mm. broad, bracts lance-linear, 3–6 mm. long, minutely and sparsely puberulent, tips scarious and pale brownish or yellowish; corollas 4–5 mm. long, lobes and granuliferous tube each about 2 mm. long, throat scarcely discernible; achenes 4–5 mm. long, about 1 mm. wide at truncate apex, with low sharp ridges but these obscured by closely appressed, stiffish hairs; pappus paleae narrowly lance-attenuate, 4–5 mm. long, midrib excurrent as a short awn.

Gravelly and rocky hillsides, mainly Upper Sonoran Zone, but entering the upper margins of the Sonoran Desert along the eastern escarpment of the Sierra Pinal, northern Baja California; southern California to New Mexico and south into central Sonora and northern Baja California. Sept.–Nov.

## 79. PALAFOXIA Lag., Gen. & Sp. Pl. 26, 1816

Herbaceous annuals or perennials from a sturdy, sometimes woody taproot with strigose to hispid, densely glandular to eglandular stems and herbage. Leaves alternate or lower ones opposite, entire. Heads oblong or turbinate, in cymose or corymbose, terminal clusters, discoid. Involucral bracts in 2–3 series, herbaceous, subequal. Receptacle flat, naked. Corollas regular; tube and lobes shorter than narrow throat. Tip of stamen tube slightly exserted; anthers rounded at base. Style branches filiform, hispidulous. Achenes linear to obpyramidal, quadrangular in cross section, pubescent. Pappus of several hyaline scarious scales, often reduced on marginal achenes.

Plants erect; longer pappus scales of inner florets lance-linear, acerose midrib exceeding hyaline margins:
  Leaves 6 mm. broad or less; nonglandular hairs on leaves coarse, curved, not appressed; achenes 9–14 mm. long; peduncles glandular-pubescent .... 1. *P. linearis* var. *linearis*
  Leaves 8–12 mm. broad; hairs on leaves fine, usually closely appressed; achenes 15–18 mm. long; peduncles pubescent but scarcely at all glandular ... 1a. *P. linearis* var. *gigantea*
Plants suffrutescent, decumbent; longer pappus scales of inner florets oblong, obtuse, midrib usually shorter than hyaline margins .............................. 2. *P. leucophylla*

**1. Palafoxia linearis** (Cav.) Lag., Gen. & Sp. Pl. 26, 1816, var. **linearis**

*Ageratum lineare* Cav., Ic. **3:** 3, *pl. 205,* 1795.
*Stevia linearis* Cav., Praelect. No. 464, 1797.
*Stevia lavandulaefolia* Schlecht., in Suppl. to Willd. Enum. Hort. Berol. 57, 1814.
*Paleolaria carnea* Cass., Bull. Soc. Phil. 198, 1816.

Herbaceous annual or rarely short-lived perennial with divaricately branching stems to 8 dm. tall, scabrous to rough-hispid throughout and glandular above; leaves linear to linear-lanceolate, 2–6 mm. wide, 3–6 cm. long, entire, gradually narrowing toward each end, thickish, 1-nerved; peduncles 2–12 cm. long, sparsely scabrous and usually densely stipitate-glandular; heads 2–3 cm. long; involucre oblong to narrowly obconic; bracts linear-oblong, 10–15 mm. long, acute to obtuse, scabrous and glandular; florets 10–18 per head; corollas 8–10 mm. long, throat about 5 mm. long, glabrous; achenes 9–14 mm. long, linear, straight, more or less hispidulous; pappus scales 3–8, unequal, larger ones lance-linear, 7–10 mm. long, shorter ones obtuse, with included ribs or reduced to small callose midribs and hyaline margins.

Desert sandy soil, Lower Sonoran Zone, southern Nevada, southwestern Arizona, Mojave and Colorado Deserts, California, and south into central Sonora and Baja California. Feb.–Nov.

**1*a*. Palafoxia linearis** var. **gigantea** M. E. Jones, Contr. West. Bot. **18:** 79, 1933

*Palafoxia linearis* var. *arenicola* A. Nels., Amer. Jour. Bot. **23:** 265, 1936.

Erect, 0.6–2 m. tall, herbage green, scaberulous with fine, rather closely appressed hairs; leaves 8–12 mm. broad, to 8 cm. long, distinctly 3-veined; peduncles nearly eglandular; florets about 25 per head; achenes 15–18 mm. long, sometimes cinereous-hispid.

In sand dunes, Lower Sonoran Zone, between Holtville and Yuma, California. Feb.–Nov.

**2. Palafoxia leucophylla** A. Gray, Proc. Amer. Acad. **8:** 291, 1870

*Palafoxia arenaria* Brandegee, Proc. Calif. Acad. II, **2:** 178, 1889.
*Palafoxia linearis* var. *leucophylla* I. M. Johnston, Proc. Calif. Acad. IV, **12:** 1202, 1924.

Shrubby, with numerous decumbent stems 3–10 dm. tall, densely hispid-scabrous throughout, moderately stipitate-glandular on stems and lower leaves and progressively much more densely so above, or rarely densely glandular throughout; leaves oblong, 3–6 mm. wide, 2.5–6 cm. long, gradually narrowed to base, obtuse or rounded at apex; peduncles mainly 2–3 cm. long, heads in cymose clusters of 2–6, about 2 cm. long; involucre narrowly obconic, bracts 9–14 mm. long, acute or obtuse; corolla 8–9 mm. long, lobes 1.5–2 mm. long, roseate; achenes linear, straight, 9–12 mm. long, finely hispidulous; pappus scales 8, larger ones 6–7 mm. long, hyaline margins obtuse at apex and surpassing midrib.

Sand dunes and sandy areas, Lower Sonoran Zone, Barríl on the Gulf side and from Boca de las Animas on the Pacific side of Baja California south to the Cape Region. Also reported from Coahuila. Jan.–May.

## 80. ERIOPHYLLUM Lag., Gen. & Sp. Pl. 28, 1816

Annual (in ours) or perennial herbs or suffrutescent plants with tomentose or floccose herbage or wool tardily and partially deciduous. Leaves alternate except on seedlings and young shoots, entire or more commonly toothed, lobed, or divided. Heads radiate or rarely discoid. Involucres campanulate to hemispherical with bracts in a single series, firm, distinct or somewhat united toward base. Receptacle slightly to strongly convex, foveolate, naked or rarely sparsely chaffy. Ligulate flowers pistillate and fertile. Disk flowers few to many, regular, perfect, mostly fertile; tube glandular or hirsute or both, usually much shorter than campanulate to oblong throat. Anthers obtuse at base; apical appendages acute. Style tips obtuse or rounded. Achenes prismatic, slender, 4-angled, rarely flattened. Pappus of firm, hyaline or opaque, nerveless, usually erosulate paleae, or rarely lacking.

Heads solitary on naked peduncles; receptacle strongly convex:
  Pappus of awnless erosulate paleae 0.5 mm. long or less; achenes 2–3 mm. long; ligules yellow; pappus squamellae not cleft into several bristles:
    Involucral bracts acute to short-acuminate; peduncles equaling or usually surpassing subtending leaves .......... 1. *E. ambiguum*
    Involucral bracts obtuse; peduncles barely equaling subtending leaves:
      Pappus squamellae shorter than corolla tube; ligules yellow; achenes glabrate .......... 2. *E. wallacei* subsp. *wallacei*
      Pappus squamellae longer than corolla tube; ligules white; achenes strigose .......... 2a. *E. wallacei* subsp. *australe*
  Pappus of 4–5 lance-subulate awns 2–3 mm. long alternating with short squamellae 1 mm. long; achenes 3–3.5 mm. long; ligules white or rose-tinted .......... 3. *E. lanosum*
Heads sessile in leafy-bracted clusters; receptacle nearly flat; pappus paleae cleft into several bristle-like divisions .......... 4. *E. pringlei*

**1. Eriophyllum ambiguum** (A. Gray) A. Gray, Proc. Amer. Acad. **19:** 26, 1883

*Lasthenia ambigua* A. Gray, Proc. Amer. Acad. **6:** 547, 1865.
*Bahia ambigua* A. Gray, Bot. Calif. **1:** 382, 1876.
*Bahia parviflora* A. Gray, loc. cit.
*Eriophyllum paleaceum* Brandegee, Bot. Gaz. **27:** 450, 1899.
*Eriophyllum parviflorum* Rydb., N. Amer. Fl. **34:** 86, 1915.
*Eriophyllum parishii* Hall ex Constance, Madroño **2:** 114, 1934, as a synonym.

Floccose, sparingly branched annual 0.5–2 dm. tall or less; leaves narrowly oblanceolate to obovate-spatulate, 1–2.2 cm. long, entire or obscurely undulate-repand near apex; peduncles 1–4 cm. long, slender; involucres broadly campanulate, 4–5 mm. wide, 6–7 mm. high, bracts densely floccose, lance-elliptic to lanceolate, short-acuminate to acute, often more or less united toward base, margins thin; ray flowers 6–9, with yellow ligules 2–7 mm. long, 1.5–2.5 mm. wide; disk corollas 2–2.5 mm. long, densely granular and glandular-hispid, tube much shorter than throat and lobes; achenes about 2.5–3 mm. long, sparsely hirsute or inner ones glabrous; pappus of 2–4 minute lanceolate squamellae or none.

Grassy hillsides and valleys or along arroyos and sandy washes, Lower and drier parts of Upper Sonoran Zones, southern San Joaquin Valley south into Riverside County and probably into northern Baja California along the desert sides of the ranges. March–May.

**2. Eriophyllum wallacei** (A. Gray) A. Gray, Proc. Amer. Acad. **19:** 25, 1883, subsp. **wallacei**

*Bahia wallacei* A. Gray, in Torr., Pacif. R. Rep. **4:** 105, 1857.
*Actinolepis wallacei* A. Gray, Proc. Amer. Acad. **9:** 198, 1874.
*Antheropeas wallacei* Rydb., N. Amer. Fl. **34:** 98, 1915.
(For additional synonymy see Univ. Calif. Pub. Bot. **18:** 118, 1937.)

Lanate annual with erect or ascending branches 1.5–10 cm. long branching from base; leaves obovate to broadly spatulate, 3–15 mm. long, entire, permanently floccose; peduncles short, often barely surpassing subtending leaves, heads rather crowded; involucres turbinate to campanulate, 4.5–5 mm. high, 5–6 mm. broad, with 8–10 narrowly oblanceolate, obtusish bracts, densely lanate without; receptacle strongly convex; ray flowers 8–10, with yellow ligules about 3 mm. wide, 4–5 mm. long, obscurely

2–3-toothed; disk corollas glandular-puberulent, 2–2.5 mm. long; tips of style branches conic; achenes linear-obpyramidal, about 2 mm. long, obtusely angled, sparsely appressed-hirsute to glabrous, blackish; pappus of 6–8 oblong, faintly erose, opaque, whitish squamellae about 0.5 mm. long or sometimes reduced to a mere crown or rarely wanting.

Arid plains and mesas, mainly Lower Sonoran Zone, but entering arid parts of the Upper Sonoran Zone adjacent to the Colorado and Mojave Deserts, California, to southern Utah, western Arizona, northern Baja California and Sonora. March–June.

**2*a*. Eriophyllum wallacei** subsp. **australe** (Rydb.) Wiggins, Contr. Dudley Herb. **3:** 304, 1944

*Antheropeas australe* Rydb., N. Amer. Fl. **34:** 98, 1915.

Differing from subsp. *wallacei* in having slightly narrower leaves, white instead of yellow ray flowers, achenes from sparsely to densely strigose, awnlike squamellae equaling or slightly exceeding corollas of disk flowers.

Hillsides and gravelly slopes, arid parts of Upper Sonoran Zone and upper margin of Lower Sonoran Zone, Riverside County, California, to the southern flanks of the Sierra San Pedro Mártir, Baja California. March–May.

**3. Eriophyllum lanosum** (A. Gray) A. Gray, Proc. Amer. Acad. **19:** 25, 1883

*Burrielia lanosa* A. Gray, in Torr., Pacif. R. Rep. **4:** 107, 1857.
*Actinolepis lanosa* A. Gray, Proc. Amer. Acad. **9:** 198, 1874.
*Antheropeas lanosum* Rydb., N. Amer. Fl. **34:** 98, 1915.
*Antheropeas tenuifolium* Rydb., op. cit. 99.

Loosely floccose, spreadingly branched annual with slender stems 5–15 cm. long, these erect, ascending or the lowermost often decumbent; leaves linear to narrowly oblanceolate, entire, 1–3 mm. wide, 5–20 mm. long, acute to apiculate at apex, gradually narrowing toward base; peduncles slender, 1–6 cm. long; involucres campanulate, 5–8 mm. wide, 5–6 mm. high; bracts 8–11, oblanceolate, short-acuminate, 1–1.5 mm. wide, floccose; ray flowers 8–10, with white ligules 4–6 mm. long, 3–4 mm. wide; disk corollas yellow, 2.5–3 mm. long, sparsely glandular-puberulent, tube about equaling throat and limb; achenes linear-obpyramidal, 3–3.5 mm. long, sparsely strigose, black; pappus of 4–5 slender, lance-subulate, scaberulous awns about equaling corollas and about as many or a few more short, obtuse, whitish squamellae about 1 mm. long.

Arid mesas and gravelly slopes and washes; Lower Sonoran Zone, Colorado and Mojave Deserts to southern Utah and south into Arizona and to central Baja California. March–April.

**4. Eriophyllum pringlei** A. Gray, Proc. Amer. Acad. **19:** 25, 1883

*Actinolepis pringlei* Greene, Fl. Fran. 441, 1897.

Much-branched, floccose annuals with decumbent to erect stems 1–5 cm. long; leaves narrowly spatulate, 2–10 mm. long, 3-lobed or 3-toothed at apex or rarely entire, long-villous; heads clustered at ends of leafy-bracted, ascending branches, densely white-tomentose; involucres campanulate, 5–6 mm. high and nearly as broad, of 6–8 elliptic, slightly boat-shaped bracts, these obtuse but usually short-apiculate at apex; rays none; disk corollas yellow, about 2.5 mm. long, slightly glandular tube about one and one-half times as long as narrowly campanulate throat; achenes narrowly obpyramidal, about 2 mm. long, brown-hairy; pappus of 10–12 lanceolate silvery squamellae, each cleft into several bristle-like divisions, about equaling throat of disk flowers.

Sandy desert plains, mesas, and gravelly slopes, Lower Sonoran Zone, Mojave and Colorado Deserts, California, into southern Nevada, western and southern Arizona, and northern Baja California. March–May.

## 81. AMBLYOPAPPUS Hook. & Arn., in Hook., Jour. Bot. 3: 321, 1841

Low annual herbs with slender, erect or sharply ascending branches, alternate leaves, and aromatic herbage. Lower leaves pinnately 3–5-divided into linear segments, upper ones entire. Heads small, discoid. Involucres subglobose, of 4–6 broad bracts in 2 equal series; bracts concave and more or less enclosing achenes of marginal flowers. Receptacle conic. Marginal flowers pistillate, with tubular, briefly 2–3-toothed, slightly irregular corollas. Central flowers, perfect, short-tubed; throat campanulate; limb 5-lobed. Achenes obpyramidal, bluntly angled, glandular-puberulent and hirsute. Pappus of 8–12 squamellae, these obtuse.

**1. Amblyopappus pusillus** Hook. & Arn., in Hook., Jour. Bot. **3:** 321, 1841

*Aromia tenuifolia* Nutt., Trans. Amer. Phil. Soc. II, **7:** 396, 1841.
*Infantea chilensis* Remy, in Gay, Fl. Chil. **4:** 259, 1849.

Upright, many-stemmed annual to 3.5 dm. tall, glandular-dotted throughout but glands falling from older parts, penetratingly aromatic; leaves 1–3.5 cm. long, lobes and blades of upper ones 1 mm. broad or less; heads subglobose, truncate, 3–3.5 mm. high, 4–5 mm. wide; bracts broadly obovate, rounded or obtuse, glandular-granuliferous, bluntly low-keeled where they enfold marginal achenes; corollas bright yellow, barely 1 mm. long; achenes black, about 2 mm. long, sparsely appressed-hirsute with yellowish hairs, faintly puberulent; pappus squamellae rounded and slightly erosulate, 0.3–0.5 mm. long, white or faintly purplish.

Grassy slopes and bluffs, Lower and Upper Sonoran Zones, Ventura County, California, and adjacent Channel Islands south at least to Rosario, Baja California, and in Chile. April–June.

## 82. CHAENACTIS DC., Prod. 5: 659, 1836

Small to robust, annual or biennial herbs or low perennials with alternate, entire to 4-times pinnate, or irregularly dissected leaves and glabrous to pubescent or glandular-pubescent herbage. Heads pedunculate, solitary or corymbose, hemispheric, campanulate or turbinate, 10–50-flowered. Involucral bracts subequal, slightly imbricate, in 1–4 series, linear, and acuminate to oblong and obtuse. Receptacles flat, alveolate, naked or occasionally with a few bristles. Corollas tubular-funnelform, white, yellowish, or pinkish, 5-toothed, marginal ones slightly irregular, puberulent to glabrate. Achenes clavate, terete or compressed laterally, brownish to black, often strigose. Pappus of 4–20 hyaline, linear to obovate paleae with erosulate tips or margins or sometimes obsolete, an outer ring of reduced squamellae sometimes present at base of larger paleae.

Pappus well developed; glandular hairs viscid:
  Corollas 9–12 mm. long, twice as long as achenes; anthers included ..... 1. *C. macrantha*
  Corollas 4–8 mm. long, equaling or slightly exceeding achenes; anthers partially exserted at anthesis:
    Pappus of 8 or more paleae 3.5–6 mm. long, in 2 series; achenes laterally compressed ........ 2. *C. lacera*

Pappus of 4–5 paleae with or without reduced outer squamellae; achenes terete or nearly so:
  Receptacles bearing few to many pungently tipped, linear bracts; involucral bracts acuminate, tips terete, reddish:
    Heads 9–10 mm. long, hemispherical; florets many .................................................................... 3. *C. carphoclinia* var. *carphoclinia*
    Heads 6–8 mm. high, turbinate-cylindrical; florets few .................................................................... 4*a*. *C. carphoclinia* var. *attenuata*
  Receptacles naked; involucral bracts acute or obtuse, tips not terete nor reddish:
    Pappus exceeding florets in bud; herbage glabrate; leaf segments terete and more or less fleshy ............................................ 4. *C. fremontii*
    Pappus much shorter than florets in bud; herbage pubescent; leaf segments flattened, not obviously fleshy (sometimes slightly so in *C. furcata*):
      Corollas white or faintly cream; leaf segments 2–12 mm. long:
        Involucral bracts often lobed or toothed, not glandular; corollas 7–8 mm. long .................................................. 5. *C. furcata*
        Involucral bracts entire, hispid and glandular; corollas 4–5 mm. long:
          Heads 10–12 mm. high; corolla tube abruptly expanded near base to cylindrical throat; rachis of leaves 4–7 mm. wide ............. 6. *C. latifolia*
          Heads 8–10 mm. high; corolla tube gradually expanding into narrowly funnelform throat; rachis 1–2 mm. wide ..... 7. *C. stevioides* var. *stevioides*
      Corollas usually lemon yellow to deep yellow; leaf segments to 20 mm. long:
        Achenes densely strigulose; corollas white to lemon yellow .................................................................... 7*a*. *C. stevioides* var. *thornberi*
        Achenes sparsely strigulose; corollas cream to deep yellow .................................................................... 8. *C. glabriuscula* var. *denudata*
Pappus lacking or only a few rudimentary squamellae present; glands on leaves mealy, not viscid .................................................. 9. *C. artemisiifolia*

**1. Chaenactis macrantha** D. C. Eaton, Bot. King Expl. 171, 1871

A moderately branched, leafy annual 0.5–3.5 dm. tall, usually with several branches from base; herbage at first densely arachnoid-lanate, later subglabrate, at least on lower parts of stems; leaves narrowly ovate in outline, 2–6 cm. long, 0.5–1.5 cm. wide, irregularly bipinnate into linear acute segments 1–7 mm. long, more or less persistently lanate; peduncles 0.5–3 cm. long, densely to sparsely lanate or rarely glabrate; heads narrowly campanulate, 15–22 mm. long; involucral bracts linear-lanceolate, 1–1.5 mm. wide, 9–12 mm. long, acute or faintly acuminate, moderately lanate; corollas white, 9–12 mm. long, slender, narrow tube expanding abruptly into throat and limb, densely puberulent without; anthers mostly included; achenes clavate, 5–6 mm. long, dark brown to black, strigose; pappus usually of 4 ovate acute paleae nearly equaling corollas and 4 short, obovate, slightly erose paleae exterior to and alternating with longer paleae.

Arid plains and mountain slopes, Upper and Lower Sonoran Zones, southern Idaho to central Utah, southward along the eastern side of the Sierra Nevada to the Mojave Desert and the vicinity of Tucson, Arizona. April–May.

**2. Chaenactis lacera** Greene, Pittonia **1:** 291, 1889

Erect or ascendingly branched, leafy annuals 1–3 dm. tall with herbage moderately pubescent with rather coarse, flaccid, clammy hairs, glandular hairs present only on peduncles and involucres; leaves 2–6 cm. long, irregularly and rather distantly bipinnately dissected, segments 0.5–2 mm. wide, 2–15 mm. long, whole blade rather skeleton-like in appearance, petioles on lower leaves 1–2.5 cm. long; peduncles stoutish to slender, 1–3 cm. long or longer and bearing several linear, leaf-like bracts, closely glandular-puberulent with tack-shaped hairs; involucres broadly campanulate to hemispherical, 8–10 mm. high, nearly as broad to slightly wider; involucral bracts linear-oblong, acute, 1.5–2.2 mm. wide, 7–10 mm. long, glandular-puberulent and bearing a few white, clammy hairs; corollas white tinged with reddish purple, 5–6 mm. long, gradually and uniformly expanded from base to limb, sparsely glandular-puberulent without; achenes clavate, 5–6 mm. long, often slightly curved and laterally compressed, dark brown, strigillose, marginal ones somewhat thicker and shorter than central ones; pappus paleae 8–12, broadly lanceolate, unequal, 3.5–6 mm. long, upper margins and apex more or less lacerate-erose.

Desert plains and coastal dunes, Lower Sonoran Zone, Miller's Landing south near the coast at least to Lagoon Head and on Natividad Island, Baja California. March–April.

**3. Chaenactis carphoclinia** A. Gray, Bot. Mex. Bound. **2:** 94, 1859, var. **carphoclinia**

*Chaenactis paleolifera* A. Nels., Bot. Gaz. **47:** 434, 1909.

Annual (or sometimes biennial) with slender, often wiry, slightly zigzag stems 1–3.5 dm. tall, ascendingly spreading branches and 1–3-times pinnate leaves 1–4.5 cm. long; herbage minutely and sparsely puberulent to glabrate, stems often reddish; leaf segments filiform to linear, 1–10 mm. long; peduncles slender, 1–5 cm. long, faintly glandular-puberulent but eventually nearly glabrous; heads campanulate, 9–10 mm. high, 6–12 mm. wide at anthesis; involucral bracts lance-linear, slenderly acuminate at apex, 6–8 mm. long, conspicuously surpassing florets in bud, finely glandular-puberulent, green; receptacular bracts glabrous except at extreme tips, pungently setaceous, reddish at tips, distinctly surpassing florets in bud; corollas white, 4–5 mm. long, minutely hispidulous; achenes 4–5 mm. long, narrowly clavate, black, strigose with whitish or silvery hairs; pappus of 4 truncate to acuminate, somewhat lacerate paleae 2–2.5 mm. long.

Arid slopes and arroyos, Lower and Upper Sonoran Zones, Mojave and Colorado Deserts to southern Nevada and south into Arizona, northern Sonora, and Baja California. Feb.–June.

***3a.* Chaenactis carphoclinia** var. **attenuata** (A. Gray) M. E. Jones, Proc. Calif. Acad. II, **5:** 699, 1895

*Chaenactis attenuata* A. Gray, Proc. Amer. Acad. **10:** 73, 1874.
*Chaenactis peirsonii* Jepson, Madroño **1:** 259, 1929.

Similar to var. *carphoclinia* but more delicate, smaller, and more slender; heads 3–5 mm. wide, few-flowered; pappus paleae 1–1.5 mm. long, truncate; receptacular bracts sometimes reduced or wanting.

Range similar to that of var. *carphoclinia* but apparently not extending south of the Gila River Valley.

**4. Chaenactis fremontii** A. Gray, Proc. Amer. Acad. **19:** 30, 1883

Erect or ascendingly branched (or rarely the basal branches procumbent), nearly glabrous annual 1–3 dm. tall; leaves 1–8 cm. long, pinnately divided into 3–7 linear lobes, these 0.5–1.5 mm. wide, 7–28 mm. long, glabrous or essentially so, petiole to

2 cm. long; stems often reddish; peduncles terminating scantily leaved branches, 1–6 cm. long, scantily to densely and minutely puberulent toward apex or sometimes glabrous; involucres broadly campanulate, 10–12 mm. long, nearly as wide, bracts lance-linear, 1–2 mm. wide, acute or slightly attenuate at apex, sparingly glandular-puberulent on upper part, usually subglabrous below; corollas white, 5–6 mm. long, central ones trumpet-shaped, marginal ones irregularly palmatifid, scantily puberulent to glabrous; achenes clavate, 5–6 mm. long, dark brown to nearly black, sparsely to closely strigose; pappus of 4–5 unequal hyaline paleae 2–8 mm. long, usually 2 short blunt ones 2–4 mm. long and 2 longer lance-attenuate to nearly awned paleae, shorter ones slightly erosulate.

Sandy or gravelly soil, Lower Sonoran Zone, southern Nevada to southern Utah, Mojave and Colorado Deserts, south into Arizona and probably northern Baja California. March–June.

**5. Chaenactis furcata** Stockw., Contr. Dudley Herb. **3:** 126, *pl. 35, figs. 4–6,* 1940

A spreadingly branched annual with several scapelike stems 5–15 cm. long arising from a basal rosette of green, irregularly and remotely pinnate-divided leaves, these 4–8 cm. long, segments 2–12 mm. long, linear-obtuse, entire or sometimes again branched or forked, glabrous or sparsely arachnoid-puberulent at bases of petioles; branches nearly leafless, each terminating in a single, broadly campanulate head 10–14 mm. long, densely glandular-puberulent just below involucres; involucral bracts 7–10 mm. long, in 2 series, narrowly lance-linear, acute to slightly acuminate, moderately glandular-pubescent, some of outer ones slightly lobed or shallowly furcate near apex; corollas white, faintly tinged with purple, 7–8 mm. long, sparsely stipitate-glandular on outer surfaces of lobes; achenes 5–6 mm. long, finely strigose; pappus of 4–5 oblanceolate, minutely erosulate paleae one-half to three-fourths as long as corollas.

Known only from the type locality, arid desert hillsides, Lower Sonoran Zone, about twenty-one miles west of Bahía de los Angeles, Baja California. Collected in Feb.

**6. Chaenactis latifolia** Stockw., Contr. Dudley Herb. **3:** 128, *pl. 36, figs. 4–6,* 1940

Leafy-stemmed, somewhat succulent annual 1.5–3.5 dm. tall; leaves 3–7 cm. long, pinnately or sometimes bipinnately dissected into narrow segments 5–10 mm. long, 1–2 mm. wide, petiole winged, about equal to blade, sparsely hirsute and stipitate-glandular; involucres 9–10 mm. high, broadly campanulate; phyllaries oblong, acute, hispid-glandular on back; corollas cream-colored, central ones 5–6 mm. long, expanded near base, glandular-puberulent on outside, marginal corollas more widely expanded, more densely pubescent on outside, hairs somewhat arachnoid-lanate; achenes clavate, 4.5–5 mm. long, coarsely strigose; pappus paleae 4, lanceolate, acute, 3.5–4 mm. long.

Arid hillsides and outwash slopes, Upper and Lower Sonoran Zones, eastern San Diego County, California, into adjacent northern Baja California. March–July.

**7. Chaenactis stevioides** Hook. & Arn., Bot. Beechey 353, 1840, var. **stevioides**

*Chaenactis brachypappa* A. Gray, Proc. Amer. Acad. **8:** 390, 1872.
*Chaenactis stevioides* var. *brachypappa* Hall, Univ. Calif. Pub. Bot. **3:** 194, 1907.
*Chaenactis gillespiei* Stockw., Contr. Dudley Herb. **3:** 123, *pl. 32, figs. 5–8,* 1940.

Annual 1–3 dm. tall with one to several ascending or erect, leafy stems and pinnate or bipinnate, petiolate leaves 2–5 cm. long; petioles to 3 cm. long, often arachnoid-tomentulose when young, later glabrate; leaf blades oblong-ovate in outline, linear segments 2–10 mm. long, obtuse at apices, hispidulous to glabrate; peduncles slender, 0.5–6 cm. long, glandular and finely puberulent; heads campanulate, 8–15 mm. long at anthesis; involucral bracts linear-oblanceolate, 5–8 mm. long, acute at apex, glandular-hispidulous, green; corollas white, 4–6 mm. long, very slender and expanding abruptly into limb, puberulent without; achenes slenderly clavate, dark reddish brown to nearly black, 4–6 mm. long, strigillose; pappus of 4 lanceolate to oblong, hyaline paleae, often attenuate and slightly erosulate toward apex, about equaling corolla tubes or slightly shorter, those of a particular achene subequal.

Open hillsides, plains, and arroyo beds, Lower and Upper Sonoran Zones, southern Idaho to Colorado, central New Mexico, Sonora, and northern Baja California. March–June.

**7*a*. Chaenactis stevioides** var. **thornberi** Stockw., Contr. Dudley Herb. **3:** 128, 1940

A larger, more robust plant; leaf segments linear-obtuse, to 1.5 cm. long; corollas cream, lemon, or yellow.

Arid outwash slopes, hillsides, and plains, Lower Sonoran Zone, central Arizona south into northern Sonora. Feb.–April.

**8. Chaenactis glabriuscula** DC. var. **denudata** (Nutt.) Munz, Man. S. Calif. Bot. 567, 1935

*Chaenactis denudata* Nutt., Proc. Acad. Phila. II, **1:** 177, 1847.

Annual 1–2.5 dm. tall with a dense rosette of basal leaves and scantily leaved, scapose stems; leaves 2–7 cm. long, ovate in outline, pinnate into flat linear divisions 3–20 mm. long, densely puberulent with fine tangled hair but sometimes subglabrate in age; peduncles, plus scapose, nearly naked stems, 1 dm. long or slightly more, puberulent or tardily subglabrate; heads broadly turbinate, 10–12 mm. high; involucral bracts lanceolate, 7–8 mm. long, at first lanate, later sparsely so and somewhat glandular; corollas yellow, 4–5 mm. long, finely puberulent and somewhat glandular; achenes 4–5 mm. long, strigillose; pappus of 4 subequal, linear-oblong, slightly erosulate paleae nearly equaling corollas.

Arid mountain slopes and along washes, Lower and Upper Sonoran Zones, Los Angeles, San Bernardino, and Riverside Counties, California, entering the Colorado Desert near Cabazon, Riverside County. March–May.

**9. Chaenactis artemisiifolia** (Harv. & Gray ex A. Gray) A. Gray, Proc. Amer. Acad. **10:** 74, 1874

*Acarphaea artemisiaefolia* Harv. & Gray ex A. Gray, Mem. Amer. Acad. II, **4:** 98, 1849.
*Acicarphaea artemisiaefolia* Walp., Ann. Bot. Syst. **2:** 877, 1852.

Erect robust annual 1.5–20 dm. tall with spreading-ascending branches, glandular-puberulent to tardily glabrate stems and peduncles, and petiolate, deeply 2- to 4-times pinnate or pinnatifid leaves with narrow delicate divisions, leaf blades 3.5–20 cm. long, ultimate segments 2–12 mm. long, minutely glandular-puberulent; petioles 1–3.5 cm. long; peduncles 0.5–8 cm. long, densely glandular-puberulent; heads 10–15 mm. long, 12–20 mm. wide at anthesis; involucral bracts linear-oblong, 8–10 mm. long, acute to slightly attenuate at apex, hispid and slightly glandular-puberulent, green; corollas white, 5–7 mm. long, finely stipitate-glandular without, outer ones slightly surpassing involucral bracts; achenes 4–6 mm. long, dark brown to black, somewhat laterally compressed, finely and sparsely strigillose with golden brown hairs, marginal achenes rather abruptly or gradually curved near base; pappus lacking or rarely of a few rudimentary squamellae.

Open hillsides or gaps in chaparral, particularly abundant on burned-over areas, Upper Sonoran Zone and western margins of Lower Sonoran Zone, Santa Barbara County, California, to the southern flanks of the Sierra San Pedro Mártir where it barely enters the Sonoran Desert. March–June.

## 83. BAHIA Lag., Gen. & Sp. Pl. 30, 1816

Annual or perennial herbs or shrubs with leaves alternate or opposite or both, conspicuously punctate-glandular, more or less dissected. Heads usually radiate, corymbosely disposed. Involucres turbinate to hemispheric; bracts oblanceolate, herbaceous, in 2–3 series, thin. Ray flowers pistillate, fertile; ligules linear. Disk flowers perfect and fertile; tube densely glandular; throat campanulate. Achenes narrowly obpyramidal, more or less hirsutulous below. Pappus of 8–10 squamellae with either a callous base or with a midrib, sometimes produced into an awn-like tip.

Plants perennial; leaves simple to deeply 3-lobed; achenes 4.5–5.2 mm. long; squamellae not produced as awns .................................. 1. *B. absinthifolia* var. *dealbata*

Plants annual; leaves biternately divided; achenes 3–3.2 mm. long; squamellae, at least of inner achenes, excurrent beyond tips as short awns .................... 2. *B. biternata*

### 1. Bahia absinthifolia Benth. var. dealbata (A. Gray) A. Gray, Pl. Wright. 1: 121, 1852

*Bahia dealbata* A. Gray, Mem. Amer. Acad. II, **4:** 99, 1849.
*Picradeniopsis dealbata* Woot. & Standl., Contr. U.S. Nat. Herb. **16:** 192, 1913.

Perennial 1–3 dm. tall from a woody base, much-branched from base, seasonal stems and leaves white-tomentose; leaves opposite, short-petiolate or subsessile, blade 1–6 cm. long, entire to 3-cleft into linear or lanceolate, entire or toothed segments, lowermost leaves lanceolate and entire; peduncles terminal, 2–6 cm. long; involucre hemispheric, 5.5–7 mm. high, 12–13 mm. broad; bracts 12–15, broadly oblanceolate, somewhat acuminate at apex, tomentose; ligules linear, 18–20 mm. long, yellow; disk corollas yellow with glandular-dotted tubes about 2 mm. long, throat campanulate, about 1 mm. long; achenes 4.5–5.2 mm. long, narrowly obpyramidal, 4-angled, distinctly striate and villous on faces; pappus of obovate truncate squamellae about 1.5 mm. long, dorsally thickened about two-thirds their length.

Arid plains and slopes, rocky hillsides, and along arroyos, Lower Sonoran to lower edge of Arid Transition Zones, Pima and Santa Cruz Counties, Arizona, to western Texas, northern Chihuahua, and northern Sonora. Very abundant in caliche soils in southern Arizona and northern Sonora. April–Oct.

### 2. Bahia biternata A. Gray, Pl. Wright. 2: 95, 1853

*Schkuhria biternata* A. Gray, Proc. Amer. Acad. **9:** 199, 1874.

Annual 1–3 dm. tall with cinereous-puberulent slender stems and foliage; leaves alternate, biternately dissected into obtuse, linear, entire or 2–3-lobed segments, on petioles much shorter than blade; peduncles 4–5 cm. long; involucre about 5 mm. high, hemispheric; bracts 10–12, obovate, obtuse at apex, densely puberulent; ray flowers 8–10, ligules bright yellow, cuneate-obovate, 6–7 mm. long, 3–4 mm. wide; disk flowers glandular, tube 1.5–2 mm. long, throat narrowly campanulate; achenes 3–3.2 mm. long, 4-angled, narrowly obpyramidal, strongly striate and hirsute; pappus of

12–14 obovate squamellae, midribs, particularly of inner achenes, frequently excurrent beyond thinner portion.

Along arroyos and watercourses and on brushy or forested hillsides, mainly Upper Sonoran Zone but in Lower Sonoran Zone along streams, from near Sacaton, Arizona, to western Texas and south into Sonora. May–Oct.

## 84. TRICHOPTILIUM A. Gray, in Torr., Bot. Mex. Bound. 97, 1859

Low floccose annuals with slender spreading stems and incised-dentate leaves oppositely arranged below, alternate above. Heads discoid on slender peduncles terminating branches. Involucre broadly hemispheric. Bracts about 20 in 2 subequal series, those of inner whorl thinner and narrower than outer. Receptacle flat and naked. Ray flowers none. Disk flowers perfect, yellow; corolla tube very short; throat funnelform; lobes short, spreading, outer ones of marginal flowers slightly enlarged. Anther tips oblong-lanceolate. Style branches obtuse. Achenes obpyramidal, villous. Pappus of 5 lanceolate squamellae, each dissected into a number of unequal, slender, finely barbellate bristles.

**1. Trichoptilium incisum** (A. Gray) A. Gray, in Torr., Bot. Mex. Bound. 97, 1859
*Psathyrotes incisa* A. Gray, Mem. Amer. Acad. II, **5:** 322, 1854.

Stems 3–15 cm. long, villous-floccose; leaves oblanceolate to spatulate, 1–5 cm. long, 3–9 mm. wide, acute, incised-dentate, narrowed into a marginal petiole, densely white-floccose; peduncles 1–9 cm. long, densely glandular-puberulent, especially toward apex, scarcely at all floccose except at very base; involucres 6–7 mm. high, 7–12 mm. wide; involucral bracts lanceolate to oblanceolate, outer acute or slightly attenuate, inner ones obtuse or rounded and often reddish at tip, all glandular-puberulent and floccose-villous; flowers 30–50, corollas about 4 mm. long, puberulent, inner ones with lobes 1–1.5 mm. long, outer with lobes 2.5–3.5 mm. long; achenes 1 mm. broad, 3 mm. long, 5-angled, hirsute-strigose; pappus golden, 4–5 mm. long, bases of each squamella thickened into a slight callosity.

Sandy or gravelly slopes, mesas, and along arroyos, Lower Sonoran Zone, Colorado and Mojave Deserts, California, into southern Nevada, western Arizona, and south into northern Baja California and Sonora. Feb.–May.

## 85. HYMENOXYS Cass., Dict. Sci. Nat. 55: 278, 1828

Annual or perennial herbs with caulescent or acaulescent scapose habit and more or less silky-villous herbage. Leaves simple and entire to pinnately lobed or divided, alternate, crowded at base, or scattered along stem or both. Heads radiate or discoid, on slender peduncles. Involucres hemispheric; bracts in 2 subequal series, densely villous. Receptacles naked, convex. Ray flowers pistillate, fertile; ligules broadly oblong to obovate, usually puberulent beneath. Disk flowers many, perfect, fertile; tube very short; throat narrowly funnelform to nearly tubular. Anther tips broadly triangular-ovate. Style appendages dilated-truncate. Achenes obpyramidal, pentagonal, hirsute. Pappus of 5–7 hyaline squamellae; midrib often produced into an awn.

Plants perennial, acaulescent; leaves entire; bracts obtuse . . . . . . . . 1. *H. acaulis* var. *arizonica*
Plants annual, caulescent; leaves pinnately to bipinnately divided into linear lobes; bracts acute . . . . . . . . . . . . . . . . . . . . . . . . . . . . . . . . . . . . . . . . . . . . . . . . 2. *H. odorata*

**1. Hymenoxys acaulis** (Pursh) Parker var. **arizonica** (Greene) Parker, Madroño **10:** 159, 1950

*Tetraneuris arizonica* Greene, Pittonia **3:** 266, 1898.
*Tetraneuris pilosa* Greene, in Rydb., Fl. Colo. 379, 1906.
*Actinea acaulis* var. *arizonica* S. F. Blake ex Munz, Man. S. Calif. Bot. 570, 1935.

Cespitose acaulescent perennial with a woody branched caudex bearing remnants of leaves from previous season and densely crowded clusters of linear-oblanceolate leaves 2–5 mm. wide, 3–7 cm. long, these sparsely silky-villous in youth, at length glabrate; scapes slender, 5–20 cm. long, silky-villous when young, later glabrate; involucres 7–8 mm. high, 10–15 mm. broad at anthesis; involucral bracts ovate to elliptic, obtuse, densely silky-villous, margins commonly reddish; ligules obovate, 5–6 mm. wide, 8–10 mm. long, bright yellow, distinctly 2–3-toothed at apex; disk corollas about 4 mm. long, tube much shorter than throat; achenes obpyramidal, 3–3.5 mm. long, densely hirsutulous; pappus of broadly obovate squamellae about 2 mm. long, each abruptly contracted into a short slender awn.

Dry rocky slopes and mesas, Lower Sonoran to drier parts of Transition Zones, Chuckwalla Mountains, Imperial County, and eastern San Bernardino and Inyo Counties, California, east into Nevada, Utah, Arizona, and New Mexico. April–Dec.

**2. Hymenoxys odorata** DC., Prod. **5:** 661, 1836

*Actinea odorata* Kuntze, Rev. Gen. Pl. **1:** 303, 1891.
(For additional synonymy see N. Amer. Fl. **34:** 117–18, 1915.)

Moderately to much-branched annual with ascending-spreading stems 1–3 dm. tall, these striate, puberulent, often purplish toward base; leaves 4–6 cm. long, pinnately to bipinnately divided into linear to filiform divisions, inconspicuously punctate-glandular and sparsely hirsutulous; heads corymbose-cymose on striate peduncles 2–8 cm. long; involucres broadly turbinate, 4–8 mm. high, 5–10 mm. broad, hirsute with crinkled white hairs; outer bracts 6–10, lanceolate, acute or short-acuminate, with dark green tips and margins, thick and coarse on back, slightly united at base, 3–5 mm. long; inner bracts slightly longer and narrower with narrower margins; ray flowers 10–18; ligules 2–4 mm. wide, 3–8 mm. long; disk corollas about 4 mm. long, sparsely to densely glandular, tube about 1 mm. long, gradually narrowing to throat; achenes 2.5–3 mm. long, hirsute; squamellae ovate, about 1.5–2 mm. long, acuminate, slightly awn-tipped.

Mainly in moist alluvial soil along streams, Lower and Upper Sonoran to Tropical Zones, Colorado and Kansas to vicinity of Yuma, Arizona, Texas, south to Tamaulipas and Coahuila. Jan.–July.

## 86. HELENIUM L., Sp. Pl. 886, 1753

Annual or perennial herbs with alternate, narrow, more or less decurrent leaves producing wings below their insertion on stems; herbage punctate-glandular. Heads solitary or in leafy-bracted corymbs, mostly radiate. Involucres flat and rotate, at length usually reflexed. Bracts in 2–3 series, linear to subulate, subequal or inner ones shorter. Receptacle convex to conic, naked or with a few bracts between ray flowers and disk flowers. Rays pistillate and fertile, neutral and sterile, or sometimes lacking; ligules usually large and conspicuous. Disk flowers many, perfect and fertile, brown, yellow or purplish; tube very short; throat deeply campanulate. Achenes obpyramidal, 4–5-angled, ribbed, pubescent on ribs and angles or rarely glabrous. Pappus of 4–8 thin squamellae.

Heads discoid; squamellae of pappus obtuse, not awn-tipped, without an obvious midrib ... ........................................................................ 1. *H. thurberi*

Heads radiate; squamellae of pappus acute or acuminate, tipped by a short slender awn, midrib obvious:

Leaves entire or dentate, not laciniate; herbage puberulent but distinctly green; squamellae about 1 mm. long ............................................ 2. *H. puberulum*

Leaves, or at least lower ones, laciniately divided; herbage cinereous-puberulent; squamellae 1.5–1.75 mm. long ...................................... 3. *H. laciniatum*

**1. Helenium thurberi** A. Gray, Proc. Amer. Acad. **19:** 32, 1883

*Heleniastrum thurberi* Kuntze, Rev. Gen. Pl. **1:** 342, 1891.

Ascending, much-branched annual 4–10 dm. tall with slender, puberulent, obviously winged stems and branches; lower leaves oblanceolate, 5–15 cm. long, more or less puberulent and subentire to sharply dentate, narrowing to a slightly winged petiole; upper leaves narrowly lanceolate to linear-spatulate, 2–7 cm. long, entire; peduncles slender, 3–10 cm. long, striate; heads discoid, broadly ovoid to subglobose, more or less corymbose, 5–12 mm. in diameter; involucral bracts linear, 3–7 mm. long, granuliferous and slightly hirsutulous; corollas greenish yellow, faintly tinged with purplish or brownish at tips of lobes, about 1 mm. long, tube virtually lacking, throat campanulate, granuliferous; achenes barely 1 mm. long, hirsutulous along low ridges; pappus squamellae broadly ovate to suborbicular, obtuse or rounded, ribless and awnless, about 0.2 mm. long, subhyaline.

Marshy places along streams, ditches and about ponds, Lower and Upper Sonoran and Tropical Zones, southern Arizona to Chihuahua and Tepíc and in central Baja California. April–Oct.

**2. Helenium puberulum** DC., Prod. **5:** 667, 1836

(For synonymy see N. Amer. Fl. **34:** 124–25, 1915.)

Erect or ascending, sparingly to much-branched annual 3–12 dm. tall with angled, irregularly puberulent stems; lower leaves broadly oblanceolate, 2–4 cm. wide, 1–3 dm. long, entire to denticulate, minutely puberulent, gradually narrowing to sessile, strongly decurrent base, wings extending entire distance to next node below; upper leaves linear-lanceolate, long-attenuate, usually much shorter than decurrent wings, to 10 cm. long or more, entire; peduncles slender, markedly striate, 5–30 cm. long; bracts subulate, 4–10 mm. long, hirsutulous, strongly reflexed at anthesis; ray flowers pistillate, with yellow ligules 4–6 mm. long, 2–5 mm. wide; disk flowers brownish yellow to purplish, head spherical or nearly so, 10–20 mm. in diameter, corollas 2.5–3 mm. long, tube 0.2–0.4 mm. long, throat abruptly widening to three to four times width of tube, cylindrical, glandular-granuliferous, broadly triangular lobes about as long as tube; achenes about 1.5 mm. long, finely ribbed, hirsutulous on angles, brownish; pappus of 4–8 lance-ovate, hyaline squamellae about 1 mm. long, with short terminal awns.

Along stream banks, ditches, and about temporary and permanent ponds, Lower and mainly Upper Sonoran Zones, central California south into northern Baja California, where it barely enters the edge of the Sonoran Desert along the eastern escarpment of the Sierra San Pedro Mártir. July–Nov.

**3. Helenium laciniatum** A. Gray, Proc. Amer. Acad. **9:** 203, 1874

*Heleniastrum laciniatum* Kuntze, Rev. Gen. Pl. **1:** 342, 1891.

Annual herb 1–3 m. tall with much-branched, cinereous-puberulent stems and foliage; leaves linear to lanceolate, 2–5 cm. long, at least lower ones laciniate-pinnatifid, weakly decurrent; peduncles 3–8 cm. long; involucral bracts linear-subulate, surpassing reflexed ligules, these 4–6 mm. long, yellow; head of disk flowers globose, 8–12 mm. in diameter; ray flowers pistillate; corollas of disk flowers 2 mm. long, with a minute tube; achenes 1–1.2 mm. long; pappus squamellae ovate, 1.5–1.75 mm. long, ovate, abruptly contracted into an awnlike tip.

Along streams and about ponds, probably mainly Upper Sonoran Zone, imperfectly known. Originally reported from southern California and Arizona, thence southward into Sonora, but more recently excluded from manuals of the California flora and doubtfully listed from Arizona. Probably in the foothills skirting the eastern margins of the Sonoran Desert in Sonora. Aug.–Dec.

## 87. GAILLARDIA Foug., Mém. Acad. Sci. Paris **1786:** 5, *pl. 1–2,* 1788

Annual, biennial, or perennial, scapose or caulescent herbs or rarely suffruticose at base. Leaves alternate or basal, entire, toothed, or pinnatifid. Heads pedunculate, discoid or radiate. Involucre saucer-shaped to rotate at anthesis with ovate, oblong, or lanceolate bracts in 2–3 series, at least upper half of bracts reflexed in fruit. Receptacle convex to subglobose, usually with soft and short-conic to elongate-subulate or bristle-like fimbrillae among disk flowers. Ray flowers usually neutral, sometimes pistillate and fertile, or wanting; ligules, when present, broadly cuneate or flabelliform, yellow or purple or both, deeply 3-toothed or 3-cleft. Disk flowers perfect and fertile, with very short, glabrous tubes abruptly enlarged to trumpet-shaped, cylindric, or campanulate, often puberulent throat; lobes from ovate and obtuse to lanceolate and acuminate, often bearing glandular moniliform hairs. Anthers auriculate at base. Appendages of style branches from short and glabrous to filiform and hispidulous. Achenes broadly obpyramidal, more or less appressed-hirsute, about twice as long as broad.

Plants annual; disk flowers yellow; pappus squamellae 2–2.5 mm. long, shorter than disk corollas:
  Pappus squamellae retuse at apex, not awned .............. 1. *G. arizonica* var. *arizonica*
  Pappus squamellae short-awned .......................... *1a.* *G. arizonica* var. *pringlei*
Plants perennial; disk flowers purple; pappus squamellae 5–6 mm. long, surpassing disk corollas .................................................... 2. *G. pinnatifida*

### 1. Gaillardia arizonica A. Gray, Syn. Fl. N. Amer. 1²: 353, 1884, var. arizonica

*Gaillardia pedunculata* A. Nels., Bot. Gaz. **47:** 432, 1909.

Scapose or subscapose annual with several erect stems 1–2.5 dm. tall from base; leaves mostly at base, 4–6 cm. long, pinnately lobed with rounded or oblong, crenate divisions, sparsely to moderately villous; peduncles erect, 5–15 cm. long; involucres 1–2 cm. in diameter, bracts linear-lanceolate, acute, 7–8 mm. long, villous with thin-walled hairs that become flattened in drying; receptacle somewhat chaffy with thin but stiffish, subulate scales; ray flowers neutral, ligules 4–5 mm. wide, 10–15 mm. long, deeply 3-lobed, yellow; disk 15–18 mm. broad, yellow; disk corollas about 5 mm. long, glabrous tube about 1 mm. long or less, throat sparsely puberulent, narrowly funnelform; corolla lobes of disk flowers triangular, glandular-puberulent; achenes 3–4 mm. long, densely hirsute with spreading-ascending hairs; pappus of 8–10 obovate squamellae about 2.5 mm. long, rounded or slightly retuse at apex, midrib lacking or obscure, awn none.

Arid plains and mesas, Lower and Upper Sonoran Zones, southern Nevada and southern Utah south to Mohave, Maricopa, and Pima Counties, Arizona. Gradually replaced by the following variety southward. May–Sept.

**1*a*. Gaillardia arizonica** var. **pringlei** (Rydb.) S. F. Blake, in Tidestrom, Contr. U.S. Nat. Herb. **25:** 598, 1925

*Gaillardia pringlei* Rydb., N. Amer. Fl. **34:** 139, 1915.
*Gaillardia crinita* Rydb., loc. cit.

Differing from var. *arizonica* by squamellae of pappus having prominent midribs excurrent as a short awn.

Mesas and plains, Lower and Upper Sonoran Zones, Coconino County, Arizona, south into northern Sonora. March–May.

**2. Gaillardia pinnatifida** Torr., Ann. Lyc. N.Y. **2:** 214, 1828

*Gaillardia multiceps* Greene, Bull. Torrey Club **24:** 512, 1897.
*Gaillardia mearnsii* Rydb., op. cit. **37:** 443, 1910.
*Gaillardia linearis* Rydb., N. Amer. Fl. **34:** 137, 1915.
*Gaillardia crassa* Rydb., op. cit. 138.
*Gaillardia globosa* A. Nels., Wyo. Univ. Pub. Bot. **1:** 135, 1926.

Cespitose perennial with a much-branched, low caudex, erect stems 1–3.5 dm. tall, leafy toward base, herbage more or less strigose-canescent; leaves narrowly oblanceolate in outline, 3–8 cm. long, to 2.5 cm. broad, pinnatifid with linear to oblong, acute to obtuse divisions 2–10 mm. long, or earliest merely lobed or toothed, hirsute and punctate-glandular; peduncles slender, 1–2.5 dm. long; involucres 1–2 cm. broad, bracts lance-subulate, lance-acuminate, or attenuate, moderately villous with long, weak, white hairs that flatten in drying, punctate-glandular; ray flowers yellow, ligules 8–12 mm. long, lightly veined with purple, deeply 3-lobed; disk corollas about 5 mm. long, short tube and much longer throat yellow, sparsely pubescent, lobes purple and densely glandular-hirsute; style branches purplish; achenes about 3 mm. long, densely silky-hirsute with spreading-ascending hairs; pappus of several narrowly lance-attenuate squamellae 5–6 mm. long, narrowing to a slender awn surpassing disk corolla.

Mesas, plains, and openings in brush and in pine forests, upper margins of Lower Sonoran to lower part of Transition Zones, Utah and Colorado to Texas, Arizona, and northern Mexico. May–Oct.

## 88. NICOLLETIA A. Gray, in Torr., Frém. Rep. 315, 1845

Perennial (sometimes flowering first year), glabrous herbs from slender creeping rootstocks. Leaves alternate, pinnatifid, with narrow, gland-tipped segments. Heads radiate, subsessile or pedunculate; rays white, purplish, or pinkish, often darker on undersurface. Involucre campanulate or turbinate with 8–12 scarious-margined, oblong bracts each bearing 1 to few conspicuous oil glands near apex. Ligules few, fertile. Disk flowers many, perfect and fertile. Achenes narrowly clavate, hirsute with short, ascending, white or slightly rusty hairs. Pappus in 2 series, outer of numerous capillary bristles, inner of 5 lanceolate paleae tipped with slender scabrous awns.

Leaf segments 1.5–3 mm. wide; pappus paleae entire, gradually narrowing to awn; involucres 15–18 mm. high. . . . . . . . . . . . . . . . 1. *N. occidentalis*
Leaf segments 1 mm. wide or less; pappus paleae 3-cleft at the apex, central awn forming central tooth; involucres 10–13 mm. high . . . . . . . . . . . . . . . . 2. *N. trifida*

**1. Nicolletia occidentalis** A. Gray, in Torr., Frém. Rep. 316, 1845

Stem stout with ascending, striate, glabrous and glaucous branches 1–3.5 dm. high; leaves fleshy, 1.5–8 cm. long, pinnatifid with 2–5 pairs of linear segments, these and rachis 1.5–3 mm. wide, each awn-tipped and bearing an oil gland near tip; heads subsessile on short peduncles to 5 mm. long; involucre 8–12 mm. broad, 15–18 mm. high; main bracts oblong, 8–12, with a single gland just below abruptly acute apex and sometimes with 1–3 additional ones below near margins; accessory bracts 1–6, ovate, acute, one-fourth to half as long as involucres; rays pinkish or white with pinkish backs, mostly ascending, 4–5 mm. long; disk flowers yellow, 7–8 mm. long, tube 1.5–2 mm. long, lobes glandular-puberulent, ovate-lanceolate; achenes black, 7–8 mm. long; paleae narrowly lanceolate, tawny, 7–8 mm. long, gradually narrowing to short awn; bristles slightly shorter.

Sandy soils of the western Mojave Desert and from the vicinity of San Felipe, San Diego County. April–June.

**2. Nicolletia trifida** Rydb., N. Amer. Fl. **34:** 180, 1915

A spreadingly branched, bushy herb 1–3.5 dm. high with glabrous, striate, lightly glaucous stems (plants erect and sparingly branched when flowering first year); leaves 2–5 cm. long, pinnately divided into 3 or sometimes 5, narrowly linear, awn-tipped segments about 1 mm. wide, each segment bearing an oil gland just beneath awn; peduncles stoutish, 5–35 mm. long; involucres turbinate, 5–7 mm. wide, 10–13 mm. high, main bracts 8–10, linear-lanceolate, short-acuminate, each bearing an oil gland just below apex; accessory bracts 1–3 (rarely 5), ovate to lanceolate, 1–2 mm. long; ligules 8–14 mm. long, entire or shallowly 3-toothed, white with a broad, brownish red, median stripe on underside; disk corollas 6–8 mm. long, anthers exserted 2–3 mm.; achenes black, 6–7 mm. long; pappus paleae 4–6 mm. long, trifid at apex, middle lobe with a scaberulous awn about 1 mm. long; bristles 3–5 mm. long.

Sandy and gravelly flats and hillsides, Lower Sonoran Zone, foothills on the southwestern flank of the Sierra San Pedro Mártir, south in Baja California to the Cape Region. Nov.–May.

## 89. TAGETES L., Sp. Pl. 887, 1753

Annual or perennial herbs with opposite leaves (or upper sometimes alternate) conspicuously dotted with punctate, ill-scented glands. Heads usually radiate, terminating ends of naked branches or in leafy cymes. Involucres fusiform, campanulate, or cylindric, bracts in a single series, narrow, united to near apex and bearing a row of glands along each margin. Receptacle plane to convex, favose. Ray flowers few (if present), pistillate and fertile, their ligules usually yellow or orange. Disk flowers several, perfect and fertile, corolla tube narrowly cylindric, short, gradually ampliate into narrowly funnelform throat. Achenes clavate-elongate, angled, terete, or flattened. Pappus of 3–10 unequal squamellae, often somewhat united, 1–5 of them linear, subulate or bristlelike and longer than rest which usually are rounded to truncate at apex.

Ligules 5–8, oblong, 8–10 mm. long; involucres 7–8 mm. broad; leaflets 1.5–6 cm. long, 5–12 mm. wide . . . . . . . . . . . . . . . . . . . . . . . . 1. *T. lacera*

Ligules 1–5, oval to orbicular, 1.5–2 mm. long; involucres 1.5–5 mm. broad; leaflets or divisions less than 1.5 cm. long, 0.5–5 mm. wide:

Leaves pinnatifid or bipinnatifid into filiform divisions 0.5–1.5 mm. wide, divisions not dentate; involucres 6.5–8 mm. long; achenes strongly striate; pappus bristles not over 3.5 mm. long .................................................. 2. *T. filifolia*

Leaves pinnately compound, leaflets to 3.5–4 mm. wide, serrate-dentate; involucres 15–19 mm. long; achenes scarcely striate; longer pappus bristles 5–8 mm. long ............ .................................................. 3. *T. subulata*

## 1. **Tagetes lacera** Brandegee, Zoe **1:** 314, 1890

A glabrous annual or sometimes persisting as a short-lived perennial, with 1 to several erect, terete, faintly striate stems to 1 m. tall; leaves mostly opposite but uppermost alternate, 7–15 cm. long, rachis slender, narrowly margined and more or less toothed, each tooth usually bearing an orbicular or elliptic gland at its base and terminating in a slender seta to 1.5 mm. long or more; leaflets 5–11, lanceolate, acute to subcuneate at base, acute to attenuate at apex, 5–12 mm. wide, 1.5–6 cm. long, rather sharply serrate, dark green above, paler beneath, conspicuously gland-dotted on both surfaces; peduncles slender basally, gradually thickened upward, 5–16 cm. long; involucres cylindro-campanulate, 7–8 mm. broad, 12–15 mm. high and usually slightly narrowed at orifice in anthesis; involucral bracts fused except for 1–2 mm. long, deltoid, free tips, each bract with 2 rows of 3–4 submarginal, elliptic to oblong glands and a single suborbicular, subapical gland; ray flowers 5–8, oblong, 2–2.5 mm. wide, 8–10 mm. long, retuse at apex, yellow; disk corollas 7–8 mm. long, puberulent, greenish yellow; achenes black, 7–8 mm. long, minutely puberulent with upwardly appressed, fine, short hairs; pappus usually of 2 (sometimes 1 or to 4) subulate, awn-tipped, minutely retrorsely barbed squamellae 5–6 mm. long and several minute, obovate, awnless squamellae about 0.5–0.8 mm. long, latter more or less lacerate.

Mostly in shade of trees and shrubs, Upper Sonoran and upper margins of Lower Sonoran and Tropical Zones, southern Baja California. Sept.–Nov.

## 2. **Tagetes filifolia** Lag., Gen. & Sp. Pl. 28, 1816

*Tagetes multifida* DC., Prod. **5:** 646, 1836.
*Tagetes congesta* Hook. & Arn., Bot. Beechey 299, 1837.

Slender annual 0.8–4.5 dm. tall with glabrous or minutely and sparsely puberulent stems and leaves and strictly simple to paniculately branching stems; leaves opposite or rarely uppermost alternate, 1–3.5 cm. long, pinnatifid or bipinnatifid into linear-filiform divisions 0.2–1 mm. wide; heads numerous, except on depauperate plants, paniculately arranged; peduncles slender, 1–2 cm. long or less; involucres fusiform, 6.5–8 mm. long, 1.5–2.5 mm. broad, rather prominently ribbed, terete or essentially so; bracts 5, each abruptly cuspidate into an awnlike tip about 0.3–0.5 mm. long, bearing many small orbicular glands irregularly arranged in 2–5 poorly defined rows from about one-fourth of way from base to summit, more crowded above than below; ray flowers 1–3 or less commonly lacking, ligules yellow, oval, about 1.5 mm. long, inconspicuous, erect; disk flowers several, yellow corollas about 3 mm. long, glabrous; achenes dull black, 4.5–5 mm. long, strongly striate, sparsely strigillose along some ribs with golden brown, fine hairs; pappus of 2 bristle-like, ascendingly hispidulous awns 2.5–3.2 mm. long, other 3 paleaceous, oblong, rounded at apex, 0.6–1.2 mm. long, usually slightly unequal.

In sandy soil along arroyos and streams and on mountain sides, Lower Sonoran and Tropical Zones, southern Baja California to San Luis Potosí and south into Central America. Aug.–Oct.

**3. Tagetes subulata** Cerv., in Llave & Lex., Nov. Veg. Desc. **1:** 31, 1824
*Tagetes multiseta* DC., Prod. **5:** 645, 1836.
?*Tagetes oligocephala* DC., loc. cit.

An erect, glabrous, paniculately much-branched annual 2–5 dm. tall; leaves opposite below, alternate above, 1.5–3 cm. long, ovate to oblong in outline, pinnate, with 5–13 leaflets 1 cm. long or less, each of these pectinately dissected into several linear, bristle-tipped divisions or teeth, bearing several rather large, conspicuous glands; heads usually numerous in leafy panicles; peduncles slender, 2.5–7 cm. long; involucres cylindro-fusiform, 15–19 mm. long, 3.5–5 mm. broad, narrowed to tip; bracts 4–5 with deltoid-ovate tips, bearing 2 lateral irregular rows of 2–4 linear, dark-colored glands and a single, small, oval one in free terminal part; ray flowers 3–5, inconspicuous, ligules 1.5–2 mm. long, nearly or fully as wide, scarcely spreading; disk flowers 4–9, their corollas 7–8 mm. long, tube and throat about equal; achenes dull black, 4–5 mm. long, faintly 4–5-angled and not noticeably striate, sparsely and minutely strigillose with yellowish hairs; pappus of 1–2 lance-subulate paleae 9–10 mm. long, other 3–4 linear, acute to truncate, 5–8 mm. long, tawny or slightly ferruginous.

Along watercourses, mountain sides and in gravelly soil among shrubs, Lower and Upper Sonoran to Tropical Zones, southern Baja California and southern Sonora to northern South America. Sept.–Feb.

## 90. DYSSODIA Cav., Anal. Cienc. Nat. Madrid 6: 334, 1803

Annual or perennial herbs or suffruticose, with opposite or attenuate, entire to pinnatisect leaves bearing punctate translucent glands, these usually marginal or submarginal, strong scented. Heads solitary at tips of branches or in corymbs or cymes, pedunculate or subsessile, small to large, radiate. Involucres campanulate to hemispheric, usually calyculate, bracts glandular, linear, usually in 2 series, more or less united at base or in some almost to apex. Ray flowers yellow, orange, or rarely white, pistillate and fertile. Disk corollas yellow, slender; tube cylindric, often imperfectly differentiated from tubular-funnelform throat. Achenes cuneate-oblong to linear, more or less ribbed and angled, often hispidulous on angles. Pappus of 10–20 squamellae or paleae, these often dissected into numerous bristles or tipped with 1–3 bristles.

Plants annual:
- Heads discoid; accessory bracts setose-ciliate, setae 1–2.5 mm. long; leaves setose-toothed; heads narrowly campanulate, 10–12 mm. high .................... 1. *D. porophylla*
- Heads radiate; bracts not setose-ciliate or cilia less than 1 mm. long; leaves not setose; involucres hemispheric, rarely over 5–8 mm. high:
  - Involucral bracts linear, lance-linear, or linear-oblanceolate, four to eight times as long as broad; rays narrow, 1–3 mm. wide:
    - Ligules white; all pappus squamellae dissected into 5–7 bristles; achenes 3–3.5 mm. long; stems spreading, not over 1 dm. long .................... 2. *D. concinna*
    - Ligules yellow; outer pappus squamellae merely erose; achenes 2 mm. long; stems erect, 1.5–5 dm. long .................... 3. *D. anomala*
  - Involucral bracts suborbicular to broadly oval, not over twice as long as broad; rays broad, 4–7 mm. wide:
    - Leaves pinnately dissected into 3–7 linear lobes 0.4–1 mm. wide; stems erect .................... 4. *D. anthemidifolia*
    - Leaves entire or obscurely and broadly lobed; stems prostrate or procumbent-spreading .................... 5. *D. littoralis*

Plants perennial:
  Ligules orange; heads large; involucres 12–20 mm. broad, 12–15 mm. high:
    Leaves chiefly opposite, 3–5 leaflets mostly suborbicular; pappus squamellae 10–12 mm. long . . . . . . . . . . 6. *D. speciosa*
    Leaves chiefly alternate, simple or divisions lanceolate or subulate; pappus squamellae 6–8 mm. long:
      Achenes glabrous, 6–7 mm. long; leaves lanceolate to ovate, saliently dentate but not divided; principal bracts 20–30 . . . . . . . . . . 7. *D. cooperi*
      Achenes hirsutulous, 5–6 mm. long; leaves pinnately dissected into 3–5 subulate or lanceolate lobes; principal bracts 14–20 . . . . . . . . . . 8. *D. porophylloides*
  Ligules yellow; heads small; involucres 3–6 mm. wide and high:
    Leaves linear-filiform, acerose, entire; pappus squamellae each divided into 3–5 bristles . . . . . . . . . . 9. *D. acerosa*
    Leaves pinnatifid into 3–7 stiff, filiform, spinulose divisions; some pappus squamellae bearing only a single bristle or merely erosulate:
      Inner pappus squamellae usually 3-awned, outer one a single short awn; achenes hispidulous on angles . . . . . . . . . . 10. *D. thurberi*
      Inner pappus squamellae 1-awned, outer obtuse, erosulate, awnless; achenes glabrous . . . . . . . . . . 11. *D. pentachaeta*

### 1. **Dyssodia porophylla** (Cav.) Cav., Anal. Cienc. Nat. Madrid **6:** 334, 1803

*Pteronia porophyllum* Cav., Ic. **3:** 13, *pl. 225,* 1795.
(For additional synonymy see N. Amer. Fl. **34:** 163, 1915, under *Lebetina porophyllum* Rydb.)

Erect branched annual 3–15 dm. tall with strongly striate, glabrous stems and branches and alternate, pinnately parted leaves 2–5 cm. long, 1–2.5 cm. broad, lobes or divisions obovate to oblong, saliently dentate, at least some teeth tipped with slender setae 1–4 mm. long, elliptic glands marginal in sinuses and nearly terminal in larger lobes; heads terminating branches but corymbosely disposed; involucres broadly turbinate, 10–12 mm. high, about as broad at anthesis; accessory bracts 15–20, pectinately divided into 3–6 pairs of bristle-like segments 1–2.5 mm. long, subterminal gland 1–1.5 mm. long and broader than rachis just below gland; main bracts about 20, linear, abruptly acuminate, strongly ridged, bearing a single subterminal gland and usually 2 submarginal ones about halfway to base, terminal horn about 1 mm. long; rays lacking or if present barely equaling disk corollas, latter yellow, about 5 mm. long, throat about equaling slender tube; achenes black, striate but glabrous, 4.5–5 mm. long; pappus squamellae tawny, 6–7 mm. long, all dissected into 5–10 slender bristles.

On gravelly ridges, grassy flats, and along arroyos, Lower Sonoran into Upper Sonoran and Tropical Zones, central Sonora to Vera Cruz and Guatemala. Barely entering the margins of the Sonoran Desert. Aug.–Jan.

### 2. **Dyssodia concinna** (A. Gray) Robinson, Proc. Amer. Acad. **49:** 507, 1913

*Hymenatherum concinnum* A. Gray, Syn. Fl. N. Amer. **1**$^2$**:** 446, 1884.
*Boeberastrum concinnum* Rydb., N. Amer. Fl. **34:** 162, 1915.

A small depressed annual with numerous spreading-ascending branches 1–10 cm. long from base, glabrous or sparsely puberulent and somewhat glaucous; leaves alternate, 1–2 cm. long, pinnately parted into linear to subfiliform, obtuse divisions 1–5 mm. long; heads clustered at summits of leafy branches, subsessile or on very short peduncles; involucres short-campanulate to subhemispheric, 5–6 mm. high, 3–4 mm. broad; bracts 12–15, united almost to apex, dotted with 2 rows of orbicular glands, scarious-margined on free, ovate upper portion; ray flowers 10–12, with white ligules 4–5 mm. long; disk corollas yellow, 4–5 mm. long, glabrous, tube about as long as narrowly trumpet-shaped throat; achenes clavate-linear, about 3.5 mm. long, sparsely hairy on low angles; pappus of 10 squamellae, each dissected into 5–7 bristles 2–2.5 mm. long.

Rocky mesas and gravelly outwash slopes, Lower Sonoran Zone, Pima County, Arizona, south into southern Sonora. May–June.

**3. Dyssodia anomala** (Canby & Rose) Robinson, Proc. Amer. Acad. **49:** 506, 1913

*Hymenatherum anomalum* Canby & Rose, Contr. U.S. Nat. Herb. **1:** 105, 1891.
*Thymophylla anomala* Rydb., N. Amer. Fl. **34:** 173, 1915.

A slender, moderately branched, erect or ascending annual 1.5–5 dm. tall with glabrous or sparsely and minutely puberulent herbage, stems striate-angled, green; leaves opposite and alternate, 2–5 cm. long, pinnatifid into 3–9 filiform or narrowly linear segments 1–3 cm. long, sparingly and inconspicuously dotted with oblong, pale orange glands, segment tips acerose-apiculate; peduncles slender, 3–13 cm. long; involucres hemispherical, 5–8 mm. wide, 4–5 mm. high; bracts equal, in 2 series of 6–8 each, oblanceolate, acute to short-acuminate, free except at base, glabrous, except along scarious ciliolate margins, with 1–5 oblong glands; ray flowers 8–15, with oblong yellow ligules 1.5–2 mm. wide and 7–10 mm. long; disk corollas yellow, about 3 mm. long, minutely granular-dotted, narrowly trumpet-shaped throat longer than slender tube, lobes narrowly ovate, about 0.4 mm. long; achenes about 2 mm. long, clavate-truncate, angled and striate, minutely and sparsely strigose with rufous hairs; pappus of 2 series and of 8–10 hyaline or scarious, oblanceolate, erosulate squamellae, inner with midvein excurrent on a slender awn, outer squamellae merely erosulate, scales, including awns, 1.5–4 mm. long.

Grassy hillsides and mesas, chiefly Upper Sonoran Zone but occasionally in the Lower Sonoran Zone, lower mountains and foothills from upper reaches of Río Bavispe in northeastern Sonora south to the vicinity of Alamos, Sonora. Sept.–Nov.

**4. Dyssodia anthemidifolia** Benth., Bot. Sulph. 29, 1844

*Boeberastum anthemidifolium* Rydb., N. Amer. Fl. **34:** 162, 1915.

Erect to decumbent, sparsely to moderately branched, glabrous annual 1–3 dm. tall, stems slightly striate; lower leaves opposite, upper alternate, 2–5 cm. long, pinnatifid into 3–9, linear, fleshy, mucronate lobes 2–12 mm. long, each division bearing a rounded gland just below apex; peduncles terminating branches, 1–10 cm. long, slightly expanded just below receptacle and bearing 1 to several small linear bracts remote from head; involucres hemispherical, 5–7 mm. high, 8–12 mm. broad; bracts 8–10, in 2 subequal series, broadly oval, 2–3 mm. wide, 5–7 mm. long, rounded and submembranous at apex, bearing 3–5 orbicular to elliptic, orange-yellow glands; ray flowers bright yellow, about 8, with broadly oval ligules 5–7 mm. wide, 6–8 mm. long, obtuse or retuse at apex; receptacle strongly convex, hispid with stiffish hairs 0.5–0.8 mm. long; disk corollas yellow, 3–4 mm. long, glabrous, short tube broadening gradually to narrowly trumpet-shaped throat, narrowly triangular-ovate lobes about 0.8–1 mm. long; achenes black, narrowly obpyramidal-truncate, 2–3 mm. long, appressed-hirsutulous on angles; pappus about equaling achenes, of 10–15 squamellae, each divided above middle into 4–10 slender barbellate bristles.

Arid plains, along arroyos, and sandy gentle slopes, Lower Sonoran Zone, central Baja California south to the Cape Region. Feb.–April.

**5. Dyssodia littoralis** Brandegee, Zoe **5:** 163, 1903

*Boeberastrum literale* Rydb., N. Amer. Fl. **34:** 162, 1915.

A branched, subfleshy, procumbently spreading annual with stems 5–18 cm. long and glabrous herbage; leaves alternate, obovate to spatulate, 4–8 mm. wide, 1.5–5 cm. long, fleshy, obtuse, rounded or emarginate at apex, entire, irregularly and obscurely repand-dentate or sometimes shallowly and broadly lobed, elliptic glands between margins and midrib, or submarginal on lobed leaves; peduncles stout, 1–2.5 cm. long, slightly broadened just below heads, strongly striate; involucres hemispheric, 1.5–2 cm. broad, bracts broadly elliptic-oblong, obtuse, 6–7 mm. long, 3–4 mm. wide, scarious-margined, conspicuously dotted with 5–11 orbicular glands; ligules 6–10 mm. long, 4–5 mm. wide; disk corollas yellow, about 4 mm. long, finely puberulent, tube 1–1.2 mm. long; achenes black, narrowly obpyramidal, 2.5–3 mm. long, hirsutulous; pappus squamellae 15–20, 3–4 mm. long, each dissected into 5–10 slender bristles.

Sandy soil, usually near the beach, Lower Sonoran and Tropical Zones, western coast of southern Baja California. Feb.–May.

**6. Dyssodia speciosa** A. Gray, Proc. Amer. Acad. **5:** 163, 1861

*Lebetina speciosa* A. Nels., Bot. Gaz. **47:** 435, 1909.

*Clomenocoma speciosa* Rydb., N. Amer. Fl. **34:** 165, 1915.

A fruticose perennial with divaricately branching, glabrous, brittle stems 2–8 dm. tall; leaves opposite, 2–5 cm. long, 3–5-foliolate, leaflets 8–15 mm. long, rounded in outline, coarsely incised-dentate or 3–5-cleft with incised lobes, glabrous to sparsely puberulent, petiolate; peduncles naked or nearly so, 5–20 cm. long, somewhat expanded upward just below receptacles; involucres campanulate-turbinate, 12–15 mm. high, 15–20 mm. broad; main bracts 17–20, linear-lanceolate with caudate-attenuate tips, usually bearing a single oblong gland near apex and 2–4 marginal ones below; secondary bracts subulate, more or less recurved, each bearing a single conspicuous subterminal gland; bristles of receptacle less than 1 mm. long, subulate to filiform; ray flowers 12–16 with bright orange, oblong ligules 10–16 mm. long; disk corollas 10–11 mm. long, yellow, with lanceolate lobes about 1.5 mm. long; achenes 6–7 mm. long, striate, scaberulous; pappus of about 10 squamellae 10–12 mm. long, each dissected into 7–9 slender bristles.

Arid plains, gentle slopes, mesas, and along arroyos, often growing up through shrubs, Lower Sonoran Zone, southern Baja California, mainly in the Cape Region but extending north to the southern edge of the Magdalena Plain. Flowering throughout the year if moisture conditions are favorable.

**7. Dyssodia cooperi** A. Gray, Proc. Amer. Acad. **9:** 201, 1874

*Lebetina cooperi* A. Nels., Bot. Gaz. **47:** 435, 1909.

*Clomenocoma cooperi* Rydb., N. Amer. Fl. **34:** 166, 1915.

*Clomenocoma laciniata* Rydb., loc. cit.

A low perennial with several erect, puberulent or finely glabrate and straw-colored branches 3–7 dm. tall from a woody base; leaves thick, alternate, simple, sessile, 1–2 cm. long, ovate to lanceolate, saliently dentate with spine-tipped teeth, acute at apex, scabrous-puberulent but glandless; peduncles slender, 3–5 cm. long; involucres broadly turbinate, about 15 mm. high and broad; main bracts 20–30, attenuately linear-subulate, some without glands, those with glands bearing 1–2 lateral ones about midway between base and apex, several near tip; lesser bracts subulate, about half as long as main ones, thickened at base, bearing a single apical or subapical gland; bristles of receptacle subulate, about 1 mm. long; ligules orange or orange-yellow, oblong, 8–10 mm. long; disk corollas nearly 1 cm. long, lobes lanceolate, about 1 mm. long; achenes clavate, 6–7 mm. long, striate but glabrous; squamellae of pappus 10–15, each dissected into 7–9 bristles and about 6–8 mm. long.

On dry mesas, outwash slopes, and in canyons and arroyos, Lower Sonoran Zone, southern Nevada into southeastern California and western Arizona; probably also in adjacent Sonora and Baja California. March–Oct.

**8. Dyssodia porophylloides** A. Gray, Mem. Amer. Acad. II, **5:** 322, 1854

*Lebetina porophylloides* A. Nels., Bot. Gaz. **47:** 435, 1909.
*Clomenocoma porophylloides* Rydb., N. Amer. Fl. **34:** 166, 1915.

Perennial herb 1–5 dm. tall from a thick woody base; herbage with a strong disagreeable odor; stems erect or ascending, glabrous, somewhat angled; leaves opposite below, alternate above, thick, pinnatifid into 3–5 subulate to lanceolate, entire or spinulose-serrate divisions 5–15 mm. long, glandless; peduncles 3–13 cm. long, often with 1–3 minute subulate bracts remote from head; heads 1.5–2 cm. long at anthesis; involucres turbinate, 9–12 mm. high, nearly as broad; main bracts 14–20, linear, acute to short-mucronate and ciliate at apices with several conspicuous, oblong to rounded glands arranged subapically and near margins; accessory bracts subulate, 2–5 mm. long, usually with a single gland; ligules elliptic, 5–6 mm. long, orange; disk corollas 8–9 mm. long, very slender, narrowly lanceolate lobes often purple-tinged, about 1.5 mm. long; achenes brownish, 5–6 mm. long, finely striate, strigillose; pappus of 8–12 squamellae with thickened bases, 7–8 mm. long, each dissected into 5–11 barbellate bristles.

Rocky slopes, outwash fans, arroyos, and mesas, Lower and Upper Sonoran (chiefly Lower) Zones, northern Arizona, southeastern California, and adjacent northern Baja California and Sonora. March–Oct.

**9. Dyssodia acerosa** DC., Prod. **5:** 641, 1836

*Aciphyllaea acerosa* A. Gray, Mem. Amer. Acad. II, **4:** 91, 1849.
*Hymenatherum acerosum* A. Gray, Pl. Wright. **1:** 115, 1852.
*Dyssodia fusca* A. Nels., Bot. Gaz. **47:** 436, 1909.

A suffruticose perennial with many much-branched, ascending-spreading to erect stems 1–2 dm. tall from a woody caudex; bark grayish, irregularly striate and shreddy; leaves opposite or alternate, linear to filiform, acerose, 5–20 mm. long, usually with fascicles of smaller ones in their axils, sparsely scaberulous and with scattered, orbicular to elliptical glands; heads subsessile, radiate; involucres campanulate-turbinate, 5–7 mm. high, 3–5 mm. wide; main bracts 8–14, linear, united almost to acute or obtuse apices, puberulent, conspicuously gland-dotted in 2 rows; ray flowers pistillate and fertile, with yellow ligules 3–4 mm. long, about 2 mm. wide; disk corollas slender, yellow, about 4 mm. long, narrowly triangular lobes about 0.5 mm. long, slender tube about equaling cylindrical throat; achenes about 3 mm. long, finely striate, obscurely angled and sparsely strigillose; pappus bristles 3–4 mm. long, yellowish.

Arid mesas and rocky slopes, Lower and Upper Sonoran Zones, southern Nevada to Texas and south to central Mexico. April–Oct.

**10. Dyssodia thurberi** (A. Gray) Robinson, Proc. Amer. Acad. **49:** 508, 1913

*Hymenatherum thurberi* A. Gray, Proc. Amer. Acad. **19:** 41, 1883.
*Dyssodia cupulata* A. Nels., Bot. Gaz. **47:** 435, 1909.
*Thymophylla thurberi* Woot. & Standl., Contr. U.S. Nat. Herb. **16:** 191, 1913.

Pubescent, slightly canescent perennial with slender, much-branched, leafy stems 1–2 dm. tall from a suffruticose base; leaves mostly opposite or a few upper ones alternate, sessile, 1–3 cm. long, pinnatifid into 5–7 filiform, spinulose-tipped divisions 2–12 mm. long; peduncles slender, 2–8 cm. long, puberulent to glabrate; involucres turbinate-campanulate, 4–5 mm. high, about as broad, naked or with 1–3 minute, accessory, subulate bracts at base; main bracts in 2 series, united almost to apex, outer ones linear, acute, with free ciliolate margins decurrent nearly to base, with 5–9 glands in 2 rows; inner bracts oblong-spatulate with rounded-triangular, subscarious, erosulate-ciliolate tips and 3–5 subapical glands; ligules yellow, 1–1.5 mm. wide, 3–3.5 mm. long; disk corollas 2.5 mm. long with a very short tube, lobes triangular-ovate, glabrous; achenes 2–3 mm. long, strigillose on low angles; pappus of 10 lanceolate, awn-tipped squamellae 2–3 mm. long, inner usually 3-awned, outer 1-awned.

Dry mesas and rocky slopes, mainly Lower Sonoran Zone, northwestern Arizona and southern Nevada to Texas and southwestern California and south into northern Sonora, Chihuahua, and Coahuila. April–Oct.

**11. Dyssodia pentachaeta** (DC.) Robinson, Proc. Amer. Acad. **49:** 507, 1913

*Hymenatherum pentachaetum* DC., Prod. **5:** 642, 1836.
*Thymophylla pentachaeta* Small, Fl. SE. U.S. 1295, 1903.
*Thymophylla gracilis* Rydb., N. Amer. Fl. **34:** 176, 1915.

Low, profusely branched, densely puberulent perennial 0.5–1.8 dm. tall from a suffruticose base; leaves mainly opposite, 1–2 cm. long, crowded, pinnatifid into 5–7 stiff, spinose-tipped, filiform or linear divisions 2–8 mm. long, pilose-puberulent and somewhat canescent; peduncles slender, 3–7 cm. long, puberulent to glabrate; involucres turbinate-campanulate, 4–6 mm. high, nearly as broad, usually with 1–3 subulate basal bracts 0.5–1.5 mm. long; main bracts in 2 subequal series, united nearly to apices, outer narrowly linear, with free, scarious, ciliate margins decurrent almost to base with submarginal glands; inner bracts with broadly triangular, erose, scarious tips and 2–5 glands; ligules yellow, about 3 mm. long, 1.5–2 mm. wide; disk corollas about 2.5 mm. long, tube shorter than narrowly trumpet-shaped throat; achenes black, linear-clavate, 2.5–3 mm. long, glabrous; pappus in 2 series, inner of 5 lanceolate, awn-tipped squamellae nearly 3 mm. long, 5 outer oblong-spatulate, awnless, about 1 mm. long.

Arid slopes and mesas, Lower and Upper Sonoran Zones, northern and south-central Arizona to Texas and south at least to Nuevo León, Mexico. April–Aug.

## 91. POROPHYLLUM (Vaill.) Adans., Fam. Pl. 2: 122, 1763

Annual or perennial herbs or small shrubs with alternate or opposite leaves or both arrangements on same plant; blades from filiform and entire to pinnatifid or broad and sinuate, petiolate or sessile, with pellucid impressed, heavily scented glands along margins and sometimes scattered over surface. Heads discoid, solitary on ends of branches or in dense cymes or corymbs. Peduncles short and stout to long and slender, sometimes somewhat expanded toward apex. Involucres cylindric to broadly campanulate. Bracts 5–10, linear, oblong, or narrowly obovate, distinct or slightly connate at base, sometimes slightly gibbous, usually with 2 rows of submarginal glands and 1 or more near apex. Receptacle naked. Flowers all perfect and fertile. Corollas yellowish or purplish or white tinged with purple; tube slender and much longer or shorter than narrowly funnelform throat. Anthers rounded at base. Style branches filiform-subulate, hirtellous. Achenes slender, often tapering to both ends, many-striate, often hispidulous. Pappus of numerous fine barbellate to hispidulous bristles.

Plants annual; involucres 18–22 mm. high; achenes 10–13 mm. long; throat and lobes of corollas deep purple .......... 1. *P. macrocephalum*
Plants perennial; involucres 7–14 mm. high; achenes 3–10 mm. long; corollas white, pale purple, or yellowish, sometimes streaked with deep purple but not uniformly so:
  Heads solitary at ends of branches:
    Corolla tube shorter than slender throat; corollas yellow or ochroleucous:
      Peduncles 2–4 cm. long; achenes 7–8 mm. long; involucre oblong, 10–12 mm. high .......... 2. *P. maritimum*
      Peduncles 1 cm. long or less; achenes 3–3.5 mm. long; involucres broadly campanulate, 7–8 mm. high:
        Leaves linear, 5–25 mm. long, terete, entire; corollas 7 mm. long .......... 3. *P. crassifolium*
        Leaves cuneate, 5–12 mm. long, tridentate at apex; corollas 5–6 mm. long .......... 4. *P. tridentatum*
    Corolla tube longer than throat; corollas whitish or purplish, not yellow or ochroleucous .......... 5. *P. gracile*
  Heads in dense cymes or corymbs:
    Corollas white or faintly purplish; pappus 6–7 mm. long, not purple-tinged:
      Stems green; leaf blades 2–5 mm. wide, sessile; flowers 25–30; peduncles 5–20 mm. long .......... 6. *P. pausodynum*
      Stems shining; leaf blades 4–8 mm. wide, distinctly petiolate; flowers 12–15; peduncles 5–8 mm. long .......... 7. *P. brachypodum*
    Corollas ochroleucous; pappus 5–6 mm. long, often tinged with purple:
      Flowers 5–7; leaves less than 2 mm. wide; involucres 3–3.5 mm. broad .......... 8. *P. confertum* var. *confertum*
      Flowers 12–15; leaves 2–4 mm. wide; involucres 5–6 mm. broad .......... 8*a*. *P. confertum* var. *ochroleucum*

### 1. **Porophyllum macrocephalum** DC., Prod. **5:** 648, 1836

*Kleinia glandulosa* Moc. & Sessé, in DC., loc. cit., as a synonym.

Erect glabrous annual to 1 m. tall, moderately branched above; leaves sometimes opposite below, alternate above, on slender petioles 1–2 cm. long, blades broadly oval to obovate, 1–3 cm. long, 0.5–1.5 cm. wide, thin, pale green, more or less repand, with conspicuous, oblong or elliptic glands along margins, one in each sinus; heads solitary at ends of branches; peduncles stoutish, slightly clavate-dilated above, 2–10 cm. long; involucres cylindric in youth, 5–6 mm. broad, 18–22 mm. high, in age turbinate and 8–15 mm. broad; bracts usually 5, linear, acute or obtuse, margins subhyaline-scarious, glands several, linear, in 2 rows; corollas 10–13 mm. long, slender tube pale, much longer than and gradually expanding into narrowly funnelform, dark purple throat; corolla lobes lanceolate, 1–1.5 mm. long; achenes dark brown to nearly black, 12–13 mm. long, angled, striate and hispidulous; pappus of numerous fine, minutely barbellate, sordid-brown bristles 8–10 mm. long.

Rocky canyons, slopes, and outwash fans, upper parts of Lower Sonoran and arid parts of Upper Sonoran Zones, southern Arizona south through Mexico into Central America and northern South America. Aug.–Oct. in our area.

### 2. **Porophyllum maritimum** Brandegee, Zoe **5:** 162, 1903

A suffrutescent perennial 1–2 dm. tall with several to many spreading-ascending branches from woody base, branches striate and glaucous, glabrous; leaves sessile and terete, rather markedly fleshy, 1.5–3 cm. long, 1–2 mm. in diameter, abruptly acuminate, bearing 2–4 punctate glands; heads solitary on terminal and axillary peduncles 2–4 cm. long; involucres oblong-cylindric, 10–12 mm. high, conspicuously glaucous, 5 bracts oblong, obtuse, purple-tinged, bearing 2 lines of linear to oblong, submarginal glands, reflexed after shedding achenes; flowers 12–20, corollas 6.5–7 mm. long, ochroleucous with purplish streaks in throat, this longer than tube proper; achenes fusiform, 7–8 mm. long, finely hispidulous with yellowish hairs; pappus of hispidulous, yellowish or rose-tinged bristles 6–7 mm. long.

Chiefly near the coast on rocky slopes and cliffs, Lower Sonoran and Tropical Zones, southern Baja California. Jan.–April and following summer rains.

### 3. **Porophyllum crassifolium** S. Wats., Proc. Amer. Acad. **24**: 57, 1889

*Porophyllum leptophyllum* I. M. Johnston, Proc. Calif. Acad. IV, **12**: 1210, 1924.
*Porophyllum tridentatum* var. *crassifolium* I. M. Johnston, op. cit. 1211.

A low, ill-smelling undershrub with much-branched, brittle, glabrous and slightly glaucous, and often purplish stems 1–3 dm. long; leaves very fleshy, linear to narrowly elliptic, 5–25 mm. long, sessile, acute at each end and tipped with a gland and often short-apiculate at apex, a few rounded glands scattered below; heads solitary at ends of short branches on stoutish peduncles 1 cm. long or less; involucres broadly campanulate, 7–8 mm. high, fully as broad; bracts 5–8, oval to broadly oblong, 2.5–3 mm. wide, subscarious and obtusish at apex, rounded and green on backs, bearing 3–7 rounded to oblong glands subapically and near margins; corollas yellow, about 7 mm. long, sparsely granuliferous, tube usually much shorter than narrowly cylindro-funnelform throat, lobes narrowly lance-ovate, about 1 mm. long; achenes 3 mm. long, sparsely puberulent to glabrate; pappus of numerous hispidulous tawny bristles about equaling corollas.

Rocky hillsides and canyon walls, Lower Sonoran Zone, along the eastern coast of central Baja California and on the adjacent islands in the Gulf of California. Feb.–April.

**Porophyllum coloratum** (H.B.K.) DC., Prod. **5**: 650, 1836, a slender annual with petiolate, lance-linear to oblanceolate leaves, slender stems, and dark purple involucres 8–10 mm. high, occurs in the mountains immediately adjacent to the Sonoran Desert in southern Sonora and may enter our area as a straggler.

### 4. **Porophyllum tridentatum** Benth., Bot. Sulph. 30, 1844

A low, profusely branched, brittle, rather closely leafy undershrub 1 dm. tall or less, zigzag stems often decumbent and older parts bearing conspicuous leaf scars; leaves alternate, sessile, fleshy, cuneate-obovate in outline, 5–12 mm. long, 1.5–4 mm. wide, saliently 3-cleft at apex, a conspicuous gland just below terminal apiculation on each lobe; heads solitary on peduncles 2–10 mm. long at tips of branches; involucres campanulate, 7–8 mm. high, 4–5 mm. broad at anthesis, spreading to hemispheric as achenes mature; involucral bracts 5, oblong or inner 1 or 2 elliptic, 1.5–3 mm. wide, glabrous or finely and sparsely puberulent, purplish and somewhat glaucous, acute to obtuse at apex, 1–3 conspicuous glands near apex, margins of inner broader bract scarious; flowers 30–40; corollas yellowish, 5–6 mm. long, tube 1.5–2 mm. long, throat cylindro-funnelform; corolla lobes ovate, 0.3–0.5 mm. long, minutely papillose-puberulent along margins; achenes black, sparsely hispidulous, 3–3.5 mm. long; pappus bristles yellowish or tawny, 5–6 mm. long.

Coastal areas, on sandy soil and about rocky outcrops, Lower Sonoran Zone, along western side of the peninsula along the Magdalena Plain and the adjacent islands, Baja California. Nov.–April.

**5. Porophyllum gracile** Benth., Bot. Sulph. 29, 1844

*Porophyllum caesium* Greene, Leaflets Bot. Obs. **2:** 155, 1911.
*Porophyllum leucospermum* Greene, loc. cit.
*Porophyllum junciforme* Greene, op. cit. 154.
*Porophyllum vaseyi* Greene, loc. cit.
*Porophyllum pinifolium* Rydb., N. Amer. Fl. **34:** 192, 1916.
*Porophyllum putidum* A. Nels., Amer. Jour. Bot. **18:** 440, 1931.
*Porophyllum nodosum* M. E. Jones, Contr. West. Bot. **18:** 82, 1933.

Slender, moderately branched, suffrutescent perennial 2–6 dm. tall with angled or slightly sulcate, faintly glaucous or quite green or purplish stems and scattered, filiform, sparsely glandular, thin leaves 0.5–1 mm. wide and 2–5.5 cm. long; heads solitary at ends of branches on peduncles 1–3 cm. long and slightly dilated upward; involucres cylindro-campanulate, 8–14 mm. high; bracts 5, linear to narrowly oblong, obtuse or less commonly acute at apex, rounded on back and slightly gibbous at base, purplish and somewhat glaucous, margins subhyaline-scarious, at least above, glands few, usually in 2 rows, oblong or linear; corollas 7–8 mm. long, white or faintly purplish with darker purple streaks, distinctly puberulent, tube shorter than narrow throat; achenes 8–10 mm. long, brownish, tapering to both ends, hispidulous with stiff, ascendingly spreading hairs about 0.2–0.3 mm. long; pappus of tawny, stramineous or rose-colored, very slender bristles 5–8 mm. long.

Rocky slopes, outwash fans, canyons, and arroyos, mainly Lower Sonoran Zone but also in dry situations in Upper Sonoran Zone, Colorado and Mojave Deserts, California, to southern Nevada, western and central Arizona, and south into Sonora and the Cape Region in Baja California. March–Oct.

**6. Porophyllum pausodynum** Robins. & Greenm., Proc. Amer. Acad. **32:** 32, 1896

*Porophyllum seemannii* S. Wats., Proc. Amer. Acad. **24:** 57, 1889. Not *P. seemannii* Schultz-Bip., 1857.

Erect shrub to 1 m. tall with angled, sulcate, green and slightly glaucous branches that at length turn reddish; leaves alternate, sessile or subsessile, blades linear to linear-oblanceolate, somewhat succulent, 2–5 mm. wide, 4–6 cm. long, somewhat narrowed at each end but often obtuse or rounded at apex, bearing 2–3 narrowly ellipsoidal glands along each margin and a single apical one; heads in dense cymes, each on a peduncle 5–20 mm. long; involucres narrowly campanulate, 9–10 mm. high, 3–4.5 mm. wide; bracts 5, oblong, obtuse, pale green, with 2 rows of linear glands about halfway between median line and margins, margins of innermost bract scarious; flowers 25–30; corollas white, 7–8 mm. long, slender puberulent tube much longer than cylindro-funnelform throat, lance-ovate lobes 0.6–0.8 mm. long; achenes 6–7 mm. long, light brown or rusty, sparsely hispidulous, distinctly striate-ribbed; pappus bristles stramineous, scaberulous, 6–7 mm. long.

On rocky slopes and basaltic cliffs, Lower Sonoran Zone, vicinity of Guaymas, Sonora. Aug.–Oct.

**7. Porophyllum brachypodum** Robinson, Proc. Amer. Acad. **35:** 341, 1900

Differing from the last chiefly in having more shiny, strict branches; distinctly petiolate, lanceolate to linear leaves 4–8 mm. wide, only 3–4 cm. long; 12–15 instead of 25–30 flowers in each head; and peduncles only 5–8 mm. long instead of 5–20 mm. long as in *P. pausodynum*.

Rocky hillsides and cliffs in the vicinity of Guaymas, Sonora, the type locality. This may not be distinct from *P. pausodynum*.

**8. Porophyllum confertum** Greene, Leaflets Bot. Obs. **2:** 155, 1911, var. **confertum**

A suffrutescent perennial 5–25 dm. tall with terete, striate, glabrous, ascending branches and alternate subsessile leaves; leaf blades linear, 1–2 mm. wide, 4–6 cm. long, sparingly glandular; heads short-pedunculate in dense corymbs; involucres narrowly cylindro-campanulate, 3–3.5 mm. wide, 9–10 mm. high; involucral bracts 5, linear-obtuse at apex, green and glabrous, with 2 rows of submarginal oblong glands; flowers 5–7 in each head, pale yellow; corollas 6.5–7 mm. long, slender tube about twice as long as cylindro-funnelform throat; corolla lobes lance-ovate; achenes slender, about 7 mm. long, tapering somewhat to apex, dark brown to black, minutely puberulent; pappus bristles about 5 mm. long, hispidulous, stramineous or somewhat tinged with purple.

In gravelly washes, Lower Sonoran Zone, Cerralvo Island, just off the coast of Baja California near La Paz and on the adjacent peninsula. Flowering almost throughout the year.

**8*a*. Porophyllum confertum** var. **ochroleucum** (Rydb.) I. M. Johnston, Proc. Calif. Acad. IV, **12:** 1209, 1924

*Porophyllum ochroleucum* Rydb., N. Amer. Fl. **34:** 189, 1916.

Differing from var. *confertum* mainly in having slightly broader leaves (2–4 mm. wide) and broader involucres (5–6 mm. wide) containing 12–15 flowers per head.

Gravelly washes, Lower Sonoran Zone, southern Baja California. Flowering after rains.

## 92. PECTIS L., Syst. Nat. ed. 10, 1221, 1759

Low, branching, annual or perennial herbs with opposite, glandular-dotted, mostly entire leaves bearing several pairs of subbasal or lateral marginal bristles, and with strongly aromatic herbage. Heads mostly small, solitary or in small cymes, usually radiate. Involucres cylindric, oblong, campanulate, or turbinate, with 3–12 free, glandular-dotted bracts in a single series, these usually gibbous basally and partially enclosing ray achenes. Receptacle flat to low-convex, naked. Ray flowers few, usually same number as involucral bracts, pistillate and fertile; ligules yellow, often tinged with red or purple. Disk flowers few, perfect; corollas yellow, with a short tube gradually broadening into a narrowly funnelform throat, 5-lobed or slightly bilabiate. Styles hispidulous; branches obtuse, short, without appendages. Achenes linear, terete or slightly angled, glabrate to pubescent. Pappus of few to many awns, bristles, squamellae, or less often reduced to a small crown.

Pappus of squamellae, ray achenes often with 1–2 additional awns; heads sessile or nearly so, equaled or surpassed by subtending leaves:
- Bracts 5–10; disk flowers 4–15; involucres 4–7.5 mm. high:
  - Ray achenes with 1–2 awns in addition to squamellae; glands on leaves marginal; bracts 8–10 .......... 1. *P. angustifolia*
  - Ray achenes with 2 narrowly lanceolate squamellae and 1–3 smaller bristles; glands on leaves scattered; bracts 5 .......... 2. *P. prostrata*
- Bracts 3; disk flowers 4–6; involucres 8–10 mm. high .......... 3. *P. cylindrica*

Pappus of bristles, awns, or reduced to a minute crown, or some with additional outer squamellae or ray achenes with squamellae only; heads pedunculate, slightly to greatly exceeding subtending leaves:
- Plants annual; ray flowers 4–10; pappus of disk achenes 4 mm. long, or shorter:
  - Involucral bracts 7–10; ray flowers 7–10; pappus bristles of ray achenes 12–20:
    - Pappus bristles faintly barbellate, but not plumose; peduncles to 8 cm. long, much longer than leaves; involucral bracts glandless or nearly so ........ 4. *P. palmeri*
    - Pappus bristles short-plumose; peduncles 1–3 cm. long, only slightly longer or shorter than leaves; involucral bracts with 3–7 conspicuous glands ........ 5. *P. papposa*
  - Involucral bracts 4–6; ray flowers 4–6; pappus bristles 2–6 or represented by a mere crown:
    - Pappus bristles or awns slender, not dilated at base; involucral bracts glandless or essentially so; branchlets and peduncles subfiliform .................. 6. *P. filipes*
    - Pappus bristles stoutish, somewhat dilated at base; involucral bracts gland-dotted; branchlets and peduncles usually slender but not subfiliform:
      - Involucral bracts conspicuously ciliate; pappus of disk flowers of 2–5, of rays 3–6, retrorsely barbed awns ...................................... 7. *P. coulteri*
      - Involucral bracts not ciliate; pappus of 2–3 subequal awns, or that of disk flowers a minutely squamellate crown:
        - Awns retrorsely barbed; bracts and peduncles puberulent:
          - Leaves narrowly linear, setose-ciliate only near base; involucre 3–4 mm. high .. ................................................ 8. *P. ambigua*
          - Leaves oblanceolate, setose-dentate almost to apex; involucres 4–5 mm. high ... ................................................9. *P. multiseta*
        - Awns smooth; bracts and peduncles glabrous:
          - Ligules 4–4.5 mm. long, rotate; glands on bracts 2–3 pairs, submarginal, about halfway between margins and midline; leaves linear, 0.5–1.5 mm. wide; disk corollas glandless ................................ 10. *P. vollmeri*
          - Ligules 1–1.2 mm. long, erect; glands on bracts 6–8 pairs, virtually marginal; leaves lanceolate, to 5 mm. wide; disk corollas with a dark subapical gland on 1–5 of teeth ........................................ 11. *P. linifolia*
- Plants perennial; ray flowers 12–15; pappus bristles 4–6 mm. long ........ 12. *P. longipes*

## 1. **Pectis angustifolia** Torr., Ann. Lyc. N.Y. **2:** 214, 1828

*Helioreos angustifolius* Raf., Atl. Jour. 145, 1832.
*Pectidopsis angustifolia* DC., Prod. **5:** 98, 1836.
*Pectis fastigiata* A. Gray, Mem. Amer. Acad. II, **4:** 62, 1849.
*Pectis angustifolia* var. *subaristata* A. Gray, Pl. Wright. **1:** 82, 1852.
*Pectis papposa* var. *sessilis* M. E. Jones, Contr. West. Bot. **12:** 46, 1908.

A diffuse, spreading-ascending, much-branched annual 0.5–2 dm. tall; stems striate, glabrous; leaves linear, 0.6–1.5 or rarely to 2 mm. broad, 1–4 cm. long, subfleshy, glabrous, slightly to strongly dilated with scarious whitish margins at base, this dilated part bearing 3–5 pairs of subulate, bristle-tipped lobes or setae, marginal glands broadly ellipsoidal, conspicuous; heads fastigiate at ends of branches, 6–8 mm. high at anthesis; involucres turbinate, 4–5 mm. high, 2.5–3.5 mm. broad at anthesis; involucral bracts 8–10, narrowly linear, strongly indurate and gibbous at base, margins strongly involute except at apex, back rounded, glabrous, each bract bearing a conspicuous, ellipsoidal, terminal gland and 1–2 smaller ones slightly lower; ray flowers 8–10, ligules yellow, 0.5–0.8 mm. wide, 3–4 mm. long; disk flowers 10–15, corollas about 3 mm. long, tube sparsely and minutely puberulent to glabrate, about equaling gradually ampliate throat; corolla lobes lanceolate; achenes black, 3–4 mm. long, bearing a few scattered, appressed, silky hairs; pappus a crown of 4–5 squamellae about 0.2 mm. long and sometimes with 1 to several additional slender awns on achenes of ray flowers and sometimes on both ray and disk achenes.

Arid mesas and sandy or gravelly plains, Lower and Upper Sonoran Zones, Navajo, Coconino, and Gila Counties, Arizona, to Utah, Colorado, Nebraska, Texas, and northern Mexico. Barely reaching the margins of the Sonoran Desert. July–Oct.

**2. Pectis prostrata** Cav., Ic. **4:** 12, *pl. 324,* 1797

*Chthonia prostrata* Cass., Dict. Sc. Nat. **9:** 173, 1817.
*Pectis costata* Ser. & Merc. ex DC., Prod. **5:** 100, 1836.

A diffuse annual branching from the base with branches 4–20 cm. long and decumbent, or in depauperate plants with few, ascending or erect branches, stems densely short-pilose in decurrent lines immediately below insertion of leaves; leaves linear to lance-oblong, 2–3.5 mm. wide, 1.5–3 cm. long, obtuse to acute at apex, with 2–5 pairs of slender bristles near base, entire, faintly scaberulous along margins and midribs, dotted with numerous scattered glands; heads axillary and in terminal 2–9-flowered glomerules, surpassed by subtending leaves; involucres oblong, 2–3 mm. wide, 5–7.5 mm. high; bracts 5, oblong-linear, rounded to truncate or faintly emarginate at apex, thick rounded keel running from base to apex, sparsely puberulent, margins scarious near apex, bearing several oblong glands subapically; ligules of 5 ray flowers oblong, about 3 mm. long; disk corollas 4–15, about 2 mm. long; achenes 3–4 mm. long, strigillose; pappus of disk achenes of 5 unequal lanceolate squamellae about equaling corollas, that of ray achenes of 2 narrowly lanceolate squamellae and 1–3 slender bristles.

Sandy plains and arid mountain slopes, Lower and Upper Sonoran into the Tropical Zones, southern Arizona to western Texas and south to northern South America; also in Florida, Cuba, and Jamaica. Flowering following summer rains.

**3. Pectis cylindrica** (Fernald) Rydb., N. Amer. Fl. **34:** 198, 1916

*Pectis prostrata* var. *cylindrica* Fernald, Proc. Amer. Acad. **33:** 68, 1897.

A diffuse, decumbent, many-branched annual with stems 1–3 dm. long, branches pilose along 2 lines running from point of coalescence between bases of oppositely arranged leaves to node below; leaves linear to linear-oblanceolate, 1.5–3 mm. broad, 1.5–2 (or rarely 3.2) cm. long, entire, obtuse to rounded at apex, ciliate with 2–5 pairs of stiffish spreading bristles 0.5–2.5 mm. long at base, glabrous, dotted with numerous small scattered glands; heads axillary and usually solitary, although rather crowded near ends of branches before internodes elongate, subsessile; involucres cylindric, 2–2.5 or rarely 3 mm. broad, 9–10 mm. long, of 3 striate glabrous bracts, these slightly rounded-keeled to near obtuse, rounded, retuse or faintly apiculate apex, pale green or yellowish, sparingly gland-dotted; ray flowers usually 3, oblong ligules about 2 mm. long, yellow; disk flowers 4–6, corollas 2–2.5 mm. long, very slender; achenes 4–4.5 mm. long, linear-clavate or linear-fusiform, black, finely hirsutulous with fine, silvery-appressed hairs; pappus of 4–6, slightly unequal, narrowly lanceolate squamellae 2–2.8 mm. long, each of these acuminately tapering into a slender awn.

Along gravelly washes and on sandy plains, Lower and into arid parts of Upper Sonoran Zones, Yavapai, Pinal, and Cochise Counties, Arizona, to New Mexico and south through Sonora to northern Sinaloa. May–Sept.

**4. Pectis palmeri** S. Wats., Proc. Amer. Acad. **24:** 58, 1889

*Pectis mearnsii* Rydb., N. Amer. Fl. **34:** 209, 1916.

Diffusely branched annual 1–2 dm. tall with ascending-spreading, dichotomously branching, glabrous stems; leaves linear, 2–4 mm. wide, 1–4 cm. long, glands marginal, conspicuous, 1–2 pairs of bristles at or near base; heads solitary in forks; peduncles slender, 1–8 cm. long; involucres broadly turbinate, 5–6 mm. high, nearly as broad; bracts 8–10, strongly keeled below and slightly gibbous, glandless or nearly so, bearing a small apical tuft of soft short hairs; ray flowers 8–10, ligules 1.5–2.5 mm. wide, 4–6 mm. long, bright yellow; disk flowers 25–40, corollas about 5 mm. long; achenes linear-clavate, black, finely strigose; pappus of disk achenes of 15–25 brownish scabrous bristles 2–3 mm. long, of ray achenes of 2–3 bristles slightly dilated at base.

Plains, mesas, and hillsides, Lower Sonoran Zone, southern Arizona south into southern Sonora and in Baja California. Feb.–Oct.

**5. Pectis papposa** Harv. & Gray, in A. Gray, Mem. Amer. Acad. II, **4:** 62, 1849

*Pectis tenella* Rothr., Wheeler Rep. **6:** 171, 1878. Not *P. tenella* DC., 1836.
*Pectis palmeri* S. Wats., Proc. Amer. Acad. **24:** 58, 1889, in part.

A yellowish green annual with slender, spreading-ascending or procumbent, dichotomously branched, glabrous stems 1–3 dm. long; leaves filiform or narrowly linear, 1–6 cm. long, 1–2 mm. wide or less, with 2–5 pairs of bristles near base and conspicuous elliptical marginal glands; heads subfastigiate in leafy cymes; peduncles 1–3 cm. long, usually shorter than subtending leaves; involucres turbinate, 3–5 mm. broad, 4.5–6 mm. high; bracts 7–9, narrowly linear and strongly involute, strongly keeled and gibbous at base, obtuse and scarious-margined at apex, irregularly dotted with 3–7 conspicuous glands, these somewhat concentrated near apex; ray flowers 7–9, with yellow ligules 1.5–2 mm. wide, 4–6 mm. long; disk flowers 10–15, their corollas slender, 4–5 mm. long, their lobes lanceolate; achenes linear-clavate, black, 4–5 mm. long, sparsely strigillose; pappus of disk achenes of 12–20 sparsely short-plumose or barbellate bristles 3–4 mm. long or rarely reduced to a crown, that of ray achenes a short crown of united squamellae, 1 or 2 of which are sometimes produced into a slender barbellate awn.

Sandy or gravelly plains and mesas, Lower Sonoran Zone, southeastern California to New Mexico and south into Sonora and Baja California. June–Oct., or earlier in Baja California.

**6. Pectis filipes** Harv. & Gray, in A. Gray, Mem. Amer. Acad. II, **4:** 62, 1849

*Pectis filipes* var. *subnuda* Fernald, Proc. Amer. Acad. **33:** 76, 1897.

Annual, with slender to subfiliform, moderately branched, ascending stems 1–3 (rarely to 5) dm. long, these minutely puberulent, glabrate or glabrous; leaves linear, 0.5–3 mm. wide, 1–6 cm. long, glabrous to faintly scabrous, often subrevolute, with marginal glands and 1–5 pairs of basal and subbasal bristles; heads in loose leafy cymes; peduncles very slender to subfiliform, 1–8 cm. long; involucres narrowly turbinate, 2–4 mm. wide, 4–6 mm. high; involucral bracts narrowly oblanceolate, obtuse, lightly keeled and faintly gibbous-callous below, glabrous or nearly so, often glandless, margin thin and scarious but not ciliolate; ray flowers usually 5, with ligules 1–1.5 mm. wide, 5–6 mm. long, pale yellow but drying reddish in age; disk flowers 5–6, about 3 mm. long, throat only slightly broader than tube, glabrous or essentially so; achenes linear, 4–5 mm. long, black, minutely strigillose; pappus of 1–4 slender awns 2.8–4 mm. long and often with a few minute squamellae between their slightly dilated bases; awns minutely barbellate with ascending hairs, or nearly smooth.

Mesas, rocky slopes, and sandy plains, Lower Sonoran to lower margins of Transition Zones, southern Arizona to western Texas, western Chihuahua, and northern Sonora. July–Oct.

### 7. **Pectis coulteri** Harv. & Gray, in A. Gray, Mem. Amer. Acad. II, **4:** 62, 1849

Spreadingly branched, prostrate or decumbent annual with slender, sparsely and minutely puberulent stems 5–25 cm. long; leaves linear, 0.5–2 mm. wide, 1–3 cm. long, divided near base into 2–5 pairs of linear-subulate lobes 1–3 mm. long, each tipped with a slender bristle, orbicular glands in a single row along each margin; heads in leafy cymes; peduncles slender, 2–35 mm. long; involucres cylindro-turbinate, 3–5 mm. wide, 5–7 mm. high; bracts 5–6, linear-oblong, callose at base, margins subscarious, purplish, ciliolate with tangled hairs, irregularly dotted with elliptical glands; ray flowers 5–6; ligules 2–3 mm. wide, 5–7 mm. long, 3 median veins purplish beneath; disk flowers 7–15, with slender glabrous corollas about 3 mm. long; achenes linear, 4–5 mm. long, minutely strigillose; pappus of 3–6 stout, subulate, spreading awns 1–1.5 mm. long on ray achenes, of 2–5 (usually 2) slightly more slender, ascending ones on disk achenes.

Open plains, Lower Sonoran Zone, southern Arizona to southern Sonora. Feb.–May.

### 8. **Pectis ambigua** Fernald, Proc. Amer. Acad. **33:** 84, 1897

*Pectis multiseta* Vasey & Rose, Contr. U.S. Nat. Herb. **1:** 72, 1890. Not *P. multiseta* Benth., 1844.

Annual; stems several, diffusely branching from base, 5–20 cm. long, mostly procumbent or a few ascending to erect, sparsely and minutely puberulent, often reddish; leaves linear, linear-oblanceolate, or rarely linear-lanceolate, 4–15 mm. long, entire, bearing 1–6 pairs of tawny bristles below middle, dotted irregularly with scattered, suborbicular, yellowish glands, acute at apex; heads in leafy cymes or some lower ones solitary in axils of leaves; peduncles 2–22 mm. long, persistent; involucre oblong-campanulate, 3–4 mm. high, 2–3 mm. broad; bracts 5, linear-oblong to narrowly obovate, rounded at apex, puberulent, strongly keeled and dotted with linear to oblong-elliptic glands; ray flowers 4–5; ligules yellow, 4–6 mm. long, faintly 3-toothed; disk flowers 5–10, corollas about 2.5 mm. long, glabrous or essentially so; achenes about 3 mm. long, black, sparsely appressed-hispidulous with fine whitish hairs, ray achenes slightly longer than those of disk flowers; pappus of disk achenes a low crown of minute, whitish or brownish squamellae about 0.1 mm. high, that of ray achenes of 2–3 unequal, narrowly subulate, retrorsely scaberulous, spreading awns 0.5–1 mm. long.

Arid plains, mesas, and along washes, Lower Sonoran Zone, central Baja California south to the Cape Region. March–May.

### 9. **Pectis multiseta** Benth., Bot. Sulph. 20, 1844

Rather diffuse annual with much-branched, minutely puberulent, reddish or purplish stems 1–3 dm. long; leaves lance-elliptic to lanceolate, 5–25 mm. long, 1–4 mm. wide, serrulate, each tooth and apex bristle-tipped, lamina dotted with scattered glands of various sizes; peduncles slender, 1–3.5 cm. long; involucres turbinate, 2.5–3.5 mm. wide, 4–5 mm. high, 5 bracts obovate-cuneate, obtuse to rounded, strongly keeled below, ciliolate, bearing several elongated glands, these often one-half as long as bract, purplish; ray flowers 4–5, with yellow ligules 2.5–3 mm. wide, 4–6 mm. long, often purple-veined beneath; disk flowers 15–20, corollas glabrous or nearly so, about 3 mm. long; achenes 3–4 mm. long, linear, puberulent; pappus of 1–2 stout, subulate, retrorsely scabrous, spreading awns about 1 mm. long on disk achenes and usually 3 on ray achenes or sometimes a crown of small squamellae.

Sandy plains, gravelly slopes, and mesas, Lower Sonoran Zone, southern Baja California. Jan.–April.

**10. Pectis vollmeri** Wiggins, Contr. Dudley Herb. **4:** 27, *pl. 3, figs. 1–10,* 1950

Erect or ascending annual 1–3.5 dm. tall, branching dichotomously or trichotomously, herbage glabrous; stems slender, reddish and lustrous; leaves linear to very narrowly lanceolate, 1–3.5 cm. long, 0.5–1.5 mm. wide, faintly revolute, 1-veined, with 5–12 small glands along each margin, 1–3 pairs of bristles near base; heads on slender peduncles 3–8 cm. long; involucres turbinate, 6–7 mm. long, 3–3.5 mm. wide at anthesis; involucral bracts 5, oblong, obtuse and sparsely ciliolate at apex, keeled below, margins scarious, central ridge bearing subterminal impressed gland and 3–6 narrowly elliptical submarginal ones along its sides; ray flowers 5, ligules yellow, oval, 2–2.6 mm. wide, 4–4.5 mm. long, rotately spreading, faintly purplish on under surface; disk flowers 8–15, corollas about 4 mm. long, gradually ampliate upward, tube sparsely and minutely puberulent, teeth about 0.5 mm. long, glandless; achenes all alike, linear, 4–4.5 mm. long, smooth, dark reddish brown to black, slightly dilated at base.

Rocky hillsides and arroyos, Lower Sonoran Zone, eastern slopes of the Sierra de la Giganta from vicinity of Santa Rosalia at least to southern end of Bahía de la Concepción, Baja California. Oct.–March.

**11. Pectis linifolia** L., Syst. Nat. ed. 10, 1221, 1759

*Verbesina linifolia* L., op. cit. 1226.
*Pectis punctata* Jacq., Enum. Pl. Carib. 28, 1760.
*Pectidium punctatum* Less., Linnaea **6:** 707, 1831.
*Tetracanthus linearifolius* A. Rich., in Sagra, Hist. Cuba **11:** 60, 1853.
*Pectis punctata* S. Wats., Proc. Amer. Acad. **24:** 58, 1889.
*Pectis linifolia* var. *marginalis* Fernald, Proc. Amer. Acad. **33:** 85, 1897.

Erect slender annual 1–10 dm. tall, slightly to moderately branched dichotomously, branches glabrous and somewhat angled; leaves linear to lanceolate, mostly plane, 1–5 mm. wide, 2–6 cm. long, dark green above, glaucous beneath, sparsely and inconspicuously glandular marginally, often with a few smaller scattered ones as well, 1–2 pairs of bristles borne at base or these lacking; peduncles slender, 1–3 cm. long; involucre 2–3 mm. broad, 5–6 mm. high, glabrous or minutely puberulent, bracts 4–5, linear, at length involute, obtuse at apex, with nearly marginal lines of 6–9 linear black glands along each edge; ray flowers 4–5, with faintly purplish-tinged ligules about 1 mm. long and nearly erect, dotted with 1–3 glands near tip; disk flowers 2–5, corollas 2–3 mm. long, purplish, glabrous, 1–5 teeth gland-dotted at apex; achenes dull black, 4–7 mm. long, minutely striate, puberulent to glabrous; pappus of 2–3 stoutish, divaricate, smooth, purplish awns 1.5–2.5 mm. long.

Mostly in shade, on rocky slopes, along ditches, and watercourses, margins of the Lower Sonoran to Transition and Tropical Zones, southern Arizona to northern South America and the West Indies. Growing in shade along irrigation ditch near Altar, Sonora. Aug.–Oct.

**12. Pectis longipes** A. Gray, Pl. Wright. **2:** 69, 1853

A perennial herb with a slender cespitose caudex and several to many decumbent branches 0.5–2.5 dm. long, these closely leafy below, glabrous, ascending at tips, striate and often reddish-tinged; leaves linear to narrowly linear-oblanceolate, 1–3 mm. wide, 2–4 cm. long, thickish, mucronate at apex, faintly to strongly revolute, bearing 2–3 pairs of subulate setae 1.5–4 mm. long at base and 5–11 conspicuous glands along each margin, a shallow indentation marking each gland; heads solitary at ends of branches and in upper axils, on slender peduncles 5–15 cm. long; involucres turbinate-campanulate, 6–8 mm. high and about as broad; bracts 12–15, linear, keeled and slightly gibbous below, plane and acute at apex, bearing a conspicuous, subapical, elliptical gland, margins scarious; ray flowers 12–15, ligules yellow, 1.5–2 mm. wide, 6–9 mm. long; disk flowers about 40–75, their corollas 4–5 mm. long, gradually ampliate upward; achenes 4–5 mm. long, dark brown to black, moderately appressed-villous; pappus of ray achenes of 2 slender setae 3–5 mm. long and sometimes a few additional squamellae; pappus of disk achenes of 20–40 unequal, scabrous, capillary bristles 2–6 mm. long.

Rocky slopes, mesas, and gravelly flats, Lower and Upper Sonoran Zones, Pima County, Arizona, to western Texas and south into adjacent Mexico. Barely reaching the margins of the Sonoran Desert. April–Sept.

**Pectis imberbis** A. Gray, Pl. Wright. **2:** 70, 1853, a perennial species with sulcate stems to 1 m. tall; thick linear leaves 5–7 cm. long; involucres 6–8 mm. tall, bearing submarginal lines of black linear glands and a terminal tuft of short fine hairs; gland-dotted corolla lobes; and pappus of stout subulate awns and short, somewhat toothed squamellae, occurs at elevations slightly above those of the Sonoran Desert at a number of localities, if not continuously, along the eastern border of our area and may be expected as a straggler in the foothill areas of Sonora.

## Tribe 9. **ANTHEMIDEAE.** Mayweed Tribe

Aromatic or strong-scented annuals or perennials. Leaves alternate, often dissected. Bracts imbricated, commonly dry-scarious at least along margins. Receptacle flat to narrowly conical, naked or with bristlelike or chaffy bracts or hirsute. Rays present or none. Anthers not caudate. Style branches obtuse or truncate, often fimbriate or erose. Pappus of a short scarious crown, or none.

Receptacles distinctly conical, two to four times as long as broad:
- Heads radiate, ligules conspicuous; receptacle with narrow chaffy bracts and a few hairs on upper part .................... 93. *Anthemis*
- Heads discoid; receptacle naked throughout .................... 94. *Matricaria*

Receptacles flat to convex but never higher than broad:
- Ray flowers ligulate and conspicuous .................... 95. *Chrysanthemum*
- Ray flowers minute and heads appearing discoid, or their corollas completely suppressed:
  - Heads solitary on peduncles several to many times as long as heads; marginal flowers devoid of corollas; achenes borne on small but distinct pedicels ........ 96. *Cotula*
  - Heads in dense racemes or panicles, on short peduncles usually shorter than heads; marginal flowers with minute but distinct corollas; achenes not pedicelled .................... 97. *Artemisia*

## 93. **ANTHEMIS** L., Sp. Pl. 893, 1753

Annual or perennial herbs with alternate, pinnatifid or bipinnatifid leaves. Heads solitary at ends of branches, usually radiate. Involucres saucer-shaped, with thin, nearly equal bracts in 2–3 series. Receptacle narrowly conic to hemispheric, chaffy-bracted, paleae membranous or stiffish and bristle-like, usually persistent. Ray flowers 10–15, pistillate and fertile or neutral; ligules white. Disk flowers perfect and fertile, with cylindrical tubes and short, funnelform, 5-toothed throat. Anther tips obtuse. Style branches truncate and fimbriate at apex. Achenes obovoid to (in ours) cylindrical, narrowly ribbed, often glandular-tuberculate. Pappus none.

**1. Anthemis cotula** L., Sp. Pl. 894, 1753

*Maruta cotula* DC., Prod. **6:** 13, 1837.

(For additional synonymy see N. Amer. Fl. **34:** 229, 1916, under *Maruta cotula* [L.] DC.)

A strong-scented, leafy annual 2–6 dm. tall with erect or ascending, much-branched, striate and sparsely to densely pubescent stems; leaves narrowly to broadly ovate in outline, 1.5–7 cm. long, sparsely pubescent with slender spreading hairs, bipinnatifid into numerous linear to filiform, entire or further incised, cuspidate divisions; peduncles slender, 1–7 cm. long, sparsely to moderately hispidulous with fine, pustulate-based, white hairs; involucres 3–4 mm. high, 6–8 mm. broad, bracts elliptical, acute to obtuse, subscarious on margins and toward tips, sparsely villous with fine, rather tangled hairs on backs; ligules 2–2.5 mm. wide, 6–9 mm. long; disk corollas pale greenish yellow, 1.5–2 mm. long, minutely and sparsely glandular-granuliferous on lower part of tube, ovate-deltoid lobes about 0.5 mm. long, tube about equaling throat; achenes pale brown, obovoid, about 1 mm long, lightly but distinctly 10-ribbed, sparsely glandular-tuberculate near apex.

Margins of fields, roadsides and waste plains, Lower Sonoran to Canadian Zones, naturalized from Eurasia and widely distributed almost throughout the United States and southern Canada and adjacent northern Mexico. March–Oct. in our area.

## 94. MATRICARIA L., Sp. Pl. 891, 1753

Annual or perennial herbs, often with sweet-scented herbage, usually glabrous. Leaves alternate, bi- or tripinnatifid into filiform to linear divisions. Heads corymbosely arranged or solitary at ends of branches, radiate or (in ours) discoid. Involucre saucer-shaped to hemispherical, with narrow, oblong or ovate, rather thin bracts with scarious margins in 2–4 series, subequal, somewhat imbricate. Receptacle naked, conical. Rays, when present, pistillate and fertile, white. Disk flowers numerous, perfect and fertile; corollas cylindro-campanulate. Achenes oblong, small, ribbed on inner side. Pappus a low crown or sometimes almost lacking.

**1. Matricaria matricarioides** (Less.) Porter, Mem. Torrey Club **5:** 341, 1894

(For synonymy see N. Amer. Fl. **34:** 232, 1916, under *Chamomilla suaveolens* [Pursh] Rydb.)

A much-branched, leafy annual with ascending stems 1–3 dm. tall; herbage sweet-scented, glabrous; leaves 2–4 cm. long, twice- or thrice-pinnatifid into short, linear-filiform divisions 0.5–6 mm. long; heads numerous, terminating leafy branches, discoid; involucres deeply saucer-shaped, about 3 mm. high, 5–8 mm. in diameter; bracts in 2–3 series, elliptic, rounded at apex, subequal, margins whitish and thin-scarious; receptacle narrowly conical, naked; disk corollas about 1.5 mm. long, greenish yellow, ill-defined tube only slightly longer than campanulate throat, with 4 broadly ovate lobes; achenes pale greenish brown, somewhat oblique, smooth on round back and between 4 ribs running full length of inner side; pappus a minute crown of squamellae or absent.

Roadsides, river bottoms, and waste land, Lower Sonoran to Canadian Zones,

Alaska to Montana, Arizona, and Baja California. Occurring sparingly in the desert regions. Spring and summer.

## 95. CHRYSANTHEMUM L., Sp. Pl. 887, 1753

Annuals with alternate, dentate to pinnatifid leaves and radiate heads solitary at ends of branches. Involucres hemispherical, bracts imbricate in 3–4 series, broad and thin, margins more or less papery. Receptacle naked, convex. Ray flowers 10–20, pistillate and fertile, ligules conspicuous, yellow. Disk flowers perfect, fertile, with yellow glabrous corollas. Corolla tube cylindrical, about equaling narrowly funnelform throat. Anthers with ovate tips. Styles of the ray flowers with short, oblong, truncate or obtuse branches; those of disk flowers included, tips truncate to slightly peltate and slightly fimbriate at apices. Achenes of ray flowers short and obpyramidal, 2–3-angled and narrowly winged; those of the disk flowers cylindric to prismatic, 8–10-ribbed. Pappus none.

### 1. Chrysanthemum coronarium L., Sp. Pl. 890, 1753

*Matricaria coronaria* Desr., in Lam., Encyl. **3:** 737, 1791.
*Pyrethrum breviradiatum* Ledeb., Mém. Acad. St. Pétersb. **5:** 577, 1815.
*Pinardia coronaria* Less., Syn. Comp. 255, 1832.
*Glebionis coronaria* Cass. ex Spach, Hist. Veg. **10:** 181, 1841.

A leafy annual with stems 3–6 dm. tall; herbage glabrous, stems and branches striate-grooved; leaves usually bipinnatifid, 3–6 cm. long, obovate in outline, cauline ones sessile and auriculate basally, primary divisions oblanceolate, secondary ones lance-linear; heads on peduncles 5–18 cm. long; involucres hemispherical, 8–10 mm. high, 12–20 mm. wide; outer bracts ovate, inner elliptic to oblong and much longer, all of them yellow with light brown, scarious margins; ray flowers about 15–20; ligules pale yellow, 4–5 mm. wide, 10–15 mm. long; disk corollas about 5 mm. long; ray achenes as broad as long, 3-angled, middle angle slightly winged; disk achenes 4-angled, with inner angle tipped with a short spur.

Sparingly introduced along ditches, about dwellings, along roadsides, and in waste ground, near Chandler and Tucson, Arizona; native in Europe. Flowering in the spring.

## 96. COTULA L., Sp. Pl. 891, 1753

Annual or perennial herbs, usually with a strong odor. Leaves alternate, usually more or less toothed, lobed, or dissected. Heads solitary, discoid. Involucres rotate to saucer-shaped, of numerous narrow subequal bracts in 1–3 series. Receptacle flat to convex, more or less papillose. Marginal flowers pistillate and fertile, without corollas; achenes borne on a distinct pedicel. Style branches oblong, obtuse, truncate tips fimbriate; base of style persistent on apex of achene. Disk flowers perfect and usually fertile but sometimes sterile; disk corollas tubular, throat very short, slightly campanulate, 4-lobed. Tips of anthers ovate. Achenes compressed, margined or winged, wings of marginal achenes broader than those of central flowers. Pappus none.

### 1. Cotula coronopifolia L., Sp. Pl. 892, 1753

*Cotula montevidensis* Spreng., Syst. **3:** 497, 1826.
*Lancisia coronopifolia* Rydb., N. Amer. Fl. **34:** 286, 1916.

A glabrous, decumbent or creeping, succulent-stemmed annual with branches 1–4 dm. long, these ascending or erect at the tips, rooting at nodes on parts of stems in contact with soil; leaves moderately succulent, sessile, broad base clasping or entirely sheathing stem and forming a cup 2–5 mm. deep, blades laciniately pinnatifid or toothed, 1–2.5 cm. wide, to 8 cm. long, exceeding internodes, lobes to 4 mm. wide, 1–2 cm. long, or basal and uppermost leaves linear to oblong-spatulate and entire; peduncles slender, 1–5 cm. long; involucres rotate, 7–12 mm. broad, bracts oblong to lanceolate-oblong, subequal, obtuse, 3–4 mm. long, tips slightly scarious; marginal pistillate flowers in a single series, without corollas, each achene borne on a narrowly winged pedicel 0.5–1 mm. long in fruit, achenes emarginate at each end, oval, laterally winged, glandular-hispidulous; disk flowers with yellow tubular corollas about 1.5 mm. long, glandular-granuliferous, shallowly 4-lobed; disk achenes about 1 mm. long, oblong, with very narrow wings, on stalks 0.2–0.4 mm. long or less.

Chiefly in saline soil in damp places, Lower Sonoran to Transition Zones, thoroughly naturalized from northern Baja California to British Columbia, barely reaching the margins of the Sonoran Desert below Jacumba in eastern San Diego County, California. Feb.–Sept.

## 97. ARTEMISIA L., Sp. Pl. 845, 1753

Annual and perennial herbs and shrubs with more or less aromatic herbage. Leaves alternate, entire to deeply lobed or dissected. Heads small, nodding or erect in paniculate or racemose-paniculate inflorescences, with minute ray flowers (these none in *A. tridentata* and its section), only slightly smaller than the disk flowers. Involucres ovoid, campanulate or hemispherical, the bracts imbricated in 2–4 series, at least inner ones with scarious margins. Receptacle flat to strongly convex, naked or pubescent. Ray flowers pistillate and fertile or lacking, 2–4-toothed, with a 2-cleft, exserted style. Disk flowers perfect, fertile or abortive. Style branches truncate, erose to fimbriate at the apex. Achenes ellipsoidal, obovoid, or prismatic, angled, striate, or smooth. Pappus none.

Plants woody except at tips of seasonal branches; lobes of leaves filiform; achenes angled ........ 1. *A. californica*
Plants herbaceous or only slightly woody at extreme base; lobes or blades of leaves not filiform; achenes ribbed or ellipsoidal, not angled:
 Disk flowers fertile; leaves white-tomentose beneath and often above:
  Main leaves entire, toothed, or shallowly lobed, not glandular:
   Panicles compact, nearly or quite leafless; involucres 3–4 mm. high ........ 2. *A. ludoviciana* subsp. *ludoviciana*
   Panicles loosely open and leafy; involucres 2.5–3 mm. high:
    Leaves tomentose only beneath, usually bright green above, lobes 3–10 cm. long, margins not revolute ........ 2*a*. *A. ludoviciana* subsp. *sulcata*
    Leaves tomentose on both surfaces, lobes 1–2 cm. long, margins often narrowly revolute ........ 2*b*. *A. ludoviciana* subsp. *albula*
  Main leaves, or many of them, bipinnately dissected almost to midrib, more or less glandular ........ 2*c*. *A. ludoviciana* subsp. *mexicana*
 Disk flowers sterile, their achenes abortive; leaves green and nearly or quite glabrous on both surfaces ........ 3. *A. dracunculus*

**1. Artemisia californica** Less., Linnaea **6:** 523, 1831

*Artemisia fischeriana* Bess., Nouv. Mém. Soc. Nat. Mosc. **3:** 21, 1834.
*Artemisia fischeriana* var. *vegetior* Bess., op. cit. 88.
*Artemisia foliosa* Nutt., Trans. Amer. Phil. Soc. II, **7:** 397, 1841.
*Artemisia abrotanoides* Nutt., op. cit. 399.
*Crossostephium californicum* Rydb., N. Amer. Fl. **34:** 243, 1916.
*Crossostephium foliosum* Rydb., loc. cit.

A rounded, freely branched, leafy shrub to 2 m. tall with a woody trunk and main branches, fibrous bark and a pungent odor; twigs gray to white with a dense canescent pubescence; leaves sessile, 2–5 (or rarely 10) cm. long, pinnately or ternately divided into filiform segments 0.5–1 mm. broad, minutely but densely cinereous, usually with few to several less divided or entire leaves fascicled in axils of main ones; inflorescences leafy-bracteate, racemose-paniculate, 1–3 dm. long to 1 dm. broad; heads radiate, peduncled and nodding; involucres hemispheric, 2.5–4 mm. high, 3–5 mm. broad, with 12–20 bracts, inner ones with scarious margins, outer wholly herbaceous, all canescent on exposed parts, obtuse; ray flowers 6–10 (in ours) with a 4-toothed, nearly regular, tubular corolla about 1 mm. long; disk flowers 15–30, corollas campanulate, 5-toothed, 1.5–2 mm. long; styles of disk corollas barely exserted, those of ray flowers well exserted; achenes oblong-turbinate, with a slight, more or less undulate rim, 5-angled.

On exposed, often rocky slopes of hills and lower mountains, mainly Upper Sonoran Zone, from Marin County, California, south to the southern flanks of the Sierra San Pedro Mártir, Baja California. Barely entering the fringe of the Sonoran Desert near Palm Springs, Riverside County, and in the San Felipe Valley, San Diego County. April–Oct.

**2. Artemisia ludoviciana** Nutt., Gen. **2:** 143, 1818, subsp. **ludoviciana**

*Artemisia ludoviciana* subsp. *typica* Keck, Carnegie Inst. Wash. Pub. No. **520:** 330, 1940.
(For additional synonymy see Proc. Calif. Acad. IV, **25:** 441–42, 1946.)

A slender, moderately branched, erect or ascending, perennial herb 3–10 dm. tall; leaves numerous and rather crowded, linear, lanceolate, elliptic, or oblanceolate, entire to few-toothed or subapically lobed, 3–10 cm. long, densely and persistently white-tomentose on both sides or some merely floccose-villous and green above, then often glabrate above in age, lobes usually entire; inflorescence a compact panicle 1.5–5 cm. broad, 1–2.5 dm. long; heads erect or nodding, 3–5 mm. high, often in small glomerules; involucres ovoid to campanulate, 2–3 mm. wide, 3–4 mm. high, obtuse bracts 7–13, usually densely tomentose but sometimes glabrate; ray flowers 5–12; disk flowers 6–21; achenes ellipsoidal, smooth, not ribbed or angled, about 1 mm. long.

Hillsides, mesas, and flat plains, mainly Upper Sonoran and Transition Zones, San Bernardino and Riverside Counties, California, northeast of the Sierra Nevada to Washington and Alberta, east to Virginia and New England. Barely touching the margin of the Sonoran Desert on the northwestern edge of the Colorado Desert in Riverside County, California. July–Sept.

**2*a*. Artemisia ludoviciana** subsp. **sulcata** (Rydb.) Keck, Proc. Calif. Acad. IV, **25:** 444, 1946

*Artemisia sulcata* Rydb., N. Amer. Fl. **34:** 270, 1916.

Differing morphologically from subsp. *ludoviciana* chiefly in having more or less sulcate stems; leaves bright green and glabrous or nearly so on upper surface, and involucres slightly smaller (2.5–3 mm. high).

Plateaus and mountain slopes, mostly Upper Sonoran and Transition Zones, Gila and Pima Counties, Arizona, into southern Sonora and Chihuahua. Barely entering the Sonoran Desert at Fort Lowell, Arizona, and east of Hermosillo, Sonora. Aug.–Oct.

**2*b*. Artemisia ludoviciana** subsp. **albula** (Woot.) Keck, Proc. Calif. Acad. IV, **25:** 446, 1946

*Artemisia microcephala* Woot., Bull. Torrey Club **25:** 455, 1898. Not *A. microcephala* Hillebr., 1888.
*Artemisia albula* Woot., Contr. U.S. Nat. Herb. **16:** 193, 1913.

Differing from subsp. *ludoviciana* in having leafy-bracteate panicles; small, obovate to elliptic, shallowly lobed to lanceolate and entire leaves, the lobes rarely over 1–2 cm. long; and slightly smaller (2.5–3 mm. long) heads.

Dry hills and flats, Lower Sonoran to Transition Zones, Inyo, Riverside, and eastern San Diego Counties, California, into adjacent Baja California, east to southern Colorado, western Texas, Sonora, and Chihuahua. Occurring only along the fringes of the Sonoran Desert. May–Oct.

**2*c*. Artemisia ludoviciana** subsp. **mexicana** (Willd.) Keck, Proc. Calif. Acad. IV, **25:** 452, 1946

*Artemisia mexicana* Willd., in Spreng., Syst. **3:** 490, 1826.
*Artemisia vulgaris* subsp. *mexicana* Hall & Clem., Carnegie Inst. Wash. Pub. No. **326:** 80, 1923.
(For additional synonymy see Proc. Calif. Acad. IV, **25:** 452, 1946.)

Differing from the other subspp. of *A. ludoviciana* included here in having more intricately dissected, slenderly lobed, basal leaves that are white-tomentose beneath, green and more or less glandular on upper surface; and slightly smaller heads (2.5–3.5 mm. long).

Mountain sides and mesas, Lower Sonoran to Canadian Zones, Colorado to Missouri, Arizona, and Sonora, south to Guatemala. Entering the Sonoran Desert between Cumpas and Baviácora, Sonora. Aug.–Nov.

**3. Artemisia dracunculus** L., Sp. Pl. 849, 1753

(For synonymy see Abrams & Ferris, Ill. Fl. Pacif. States **4:** 410, 1960.)

Erect, ascendingly branched, perennial herb 3–12 dm. tall from a somewhat creeping rootstock; stems and main branches glabrous, markedly striate, often reddish; lowermost leaves 3-cleft into lance-linear divisions 1–5 cm. long or more, others lance-linear, entire, 1–4 mm. wide, 1.5–8 cm. long, gradually tapered to base, sessile, acute at apex, glabrous, green on both surfaces or rarely slightly villous; inflorescences dense, leafy, racemose panicles to 5 dm. long, heads numerous, on very short, slender peduncles, nodding; involucres campanulate, 2–2.8 mm. long, bracts acute to obtuse, slightly scarious-margined, glabrous; disk flowers barely exceeding involucres, perfect but their achenes aborting; ray flowers 6–30, corolla 0.5–1 mm. long, cleft on one side of minute tube; achenes of ray flowers ellipsoid, ribless, glabrous.

Dry mesas, hillsides, and along arroyos, Lower Sonoran to Transition and probably Canadian Zones, northern Baja California to Manitoba and east to Texas; occurring along the fringes of the Sonoran Desert, usually along watercourses. Aug.–Oct.

Several other species of *Artemisia* occur regularly in the mountains adjacent to the Sonoran Desert but none of them has as yet been reported from the desert proper. Among these are *A. douglasiana* Bess., *A. palmeri* A. Gray, and *A. tridentata* Nutt. The first two of these might be expected only along the western margin of the Colorado and San Felipe Deserts in southern California and northern Baja California, since their known range is along the coastal slopes west of our area. *Artemisia tridentata* occurs both west of the Colorado Desert and in the mountains of central and eastern Arizona so might occur along the eastern fringe of the Sonoran Desert as well as on its western margins.

## Tribe 10. **SENECIONEAE.** Groundsel Tribe

Herbs or suffruticose plants or woody shrubs. Leaves alternate or basal or rarely opposite. Involucres with bracts in 1–2, or rarely in 3–4 series, equal or nearly so, not imbricated. Heads radiate or discoid. Flowers (in ours) all yellow. Anthers not caudate at base. Style branches hispidulous their full length. Pappus bristles slender, numerous, falling separately.

Heads discoid; corolla tubes either much shorter or much longer than throat; pappus often tawny or rufous; involucral bracts in 2–4 series:
  Plants shrubby; leaves spatulate, linear or reduced to scales; style branches obtuse or acute at apex:
    Leaves, except earliest ones, reduced to small ovate scales; herbage not punctate-glandular; corolla tube much longer than throat; bracts obtuse, in 3–4 series .................................................................. 98. *Lepidospartum*
    Leaves linear, terete, not reduced to scales; herbage punctate-glandular; corolla tube much shorter than throat; bracts subulate, in 2 series ........... 99. *Peucephyllum*
  Plants herbaceous; leaves suborbicular; style branches truncate at apex ... 100. *Psathyrotes*
Heads radiate (ligules minute in some); corolla tubes of disk flowers usually about equaling throats; pappus silvery-white; involucral bracts usually in a single, or sometimes in 2 series ........................................................ 101. *Senecio*

## 98. **LEPIDOSPARTUM** A. Gray, Proc. Amer. Acad. **19:** 50, 1883

A rigid, broomlike shrub with alternate leaves, earliest of which are spatulate, with blades about 1 cm. long, others reduced to small ovate scales. Heads discoid. Receptacle somewhat convex, naked. Bracts chartaceous, oblong, obtuse at apex, regularly imbricated in 3–4 series. Corolla tube slender, much longer than short campanulate throat; lobes spreading, narrowly lance-ovate, exceeding throat. Anthers distinctly exserted, lanceolate at apices, sagittate. Style branches flattish, acute at tips. Achenes linear-terete, lightly 8–10-nerved. Pappus of many capillary, minutely scaberulous bristles.

### 1. **Lepidospartum squamatum** (A. Gray) A. Gray, Proc. Amer. Acad. **19:** 50, 1883

*Linosyris squamata* A. Gray, Proc. Amer. Acad. **8:** 290, 1870.
(For additional synonymy see Abrams & Ferris, Ill. Fl. Pacif. States **4:** 464, 1960.)

A much-branched, strictly ascending shrub to 1.5 m. tall with numerous slender branches sparingly puberulent when young, soon glabrate except for small tufts of tangled silky hairs in axils of scalelike leaves; leaves narrowly ovate, acuminate, 1.5–2.5 mm. long, inserted 2–10 mm. apart; heads numerous, 8–12 mm. long, panicles forming broad corymbs at tips of larger branches; peduncles 2–20 mm. long, bearing a few scale-like bracts or leaves; involucres turbinate-campanulate, 6–7 mm. long, 4–6 mm. wide; bracts 2–6 mm. long, lower ones acute, inner mostly obtuse, with a strong mid-nerve, sparingly arachnoid-puberulent to glabrate; corollas about 6 mm. long, tube very slender, 3–3.5 mm. long, throat campanulate, about equaling acute, narrowly lance-ovate lobes; achenes obpyramidal, 1.5–2 mm. long, glabrous; pappus bristles 6–7 mm. long, silky, minutely barbellate, white to rufous.

Dry stream beds, sandy washes, and brushy hillsides, mainly Upper Sonoran Zone but entering the edge of the Lower Sonoran Zone along the western margin of the Colorado Desert; Monterey and San Benito Counties, California, south to northern Baja California. April–Dec.

## 99. PEUCEPHYLLUM A. Gray, in Torr., Bot. Mex. Bound. 74, 1859

Much-branched shrub with ascending branches and glandular-punctate, alternate, linear leaves. Heads solitary at tips of branches, subsessile, discoid. Involucre campanulate, of about 12 linear-subulate bracts in 2 series, outer subterete, inner flattened and faintly scarious-margined. Receptacle flat, naked, alveolate. Flowers 10–18, all perfect and fertile. Corollas with an elongate cylindrical throat and very short tube and lobes. Anther tips obtuse. Style branches linear, obtuse at apex, flattened within, papillose-puberulent. Achenes oblong-obconic, faintly 10-ribbed, densely hirsute-strigose with tawny or rufous hairs. Pappus of numerous scaberulous, rufous bristles.

### 1. Peucephyllum schottii A. Gray, in Torr., Bot. Mex. Bound. 74, 1859

*Psathyrotes schottii* A. Gray, Proc. Amer. Acad. **9:** 206, 1874.
*Inyonia dysodioides* M. E. Jones, Contr. West. Bot. **8:** 42, 1898.
*Peucephyllum schottii* var. *latisetum* I. M. Johnston, Proc. Calif. Acad. IV, **12:** 1212, 1924.

A much-branched shrub 1–3 m. tall; leaves subterete, filiform, 1–2 cm. long, about 1 mm. in diameter, closely punctate-glandular and viscid; involucre 8–10 mm. wide and high; bracts 0.6–1 mm. wide, attenuately linear-subulate, punctate-glandular; corollas 6–8 mm. long, yellow, triangular lobes about 0.5 mm. long, glandular-granuliferous; achenes about 3.5 mm. long; pappus bristles tawny to rufous, 3–5 mm. long.

Dry rocky slopes and along arroyos, mainly Lower Sonoran Zone, Colorado Desert, California, to southwestern Utah, southern Nevada, western Arizona, and central Baja California. March–June.

## 100. PSATHYROTES A. Gray, Pl. Wright. 2: 100, 1853

Low, divaricately much-branched annual or perennial herbs with alternate, petiolate, suborbicular leaves. Heads discoid, pedunculate, solitary in forks of branches. Involucre turbinate, of 2 series of about 5 bracts each, outer herbaceous, inner scarious at least along margins. Receptacle flat or slightly convex, naked and markedly alveolate. Flowers all perfect and fertile; tube short; throat cylindrical-elongate; corolla lobes short, rounded-ovate, villous on outside. Anther tips lanceolate. Style branches truncate, glabrous or nearly so. Achenes obconic, more or less 10-ribbed but these usually obscured by dense ascending pubescence. Pappus of numerous, minutely scaberulous, reddish-fuscous bristles about as long as achene.

Plants annual; outer bracts linear to spatulate; involucres 5–6 mm. broad ....... 1. *P. annua*
Plants perennial; outer bracts broadly ovate to obovate-spatulate; involucres 8–10 mm. broad ........................................................................ 2. *P. ramosissima*

### 1. Psathyrotes annua (Nutt.) A. Gray, Pl. Wright. 2: 100, 1853

*Bulbostylis annua* Nutt., Jour. Acad. Phila. II, **1:** 179, 1848.

A low, rounded, winter annual 3–15 cm. high, with densely scurfy-tomentose leaves and pale yellowish, floccose stems branching divaricately from base; leaf blades rhombic, suborbicular, or sometimes reniform, 5–15 mm. wide, 4–12 mm. long, shallowly sinuate-dentate, thickish, truncate to subcuneate at base; heads solitary in or slightly above forks of branches on stoutish peduncles 3–10 mm. long; involucre 5–6 mm. high and broad; outer bracts linear to lance-linear, tips acute to spatulate, densely tomentulose, inner narrowly lanceolate, obtuse at apex, scarious below, puberulent at tips; corollas 3.5–4.5 mm. long, tube about 0.5 mm. long, throat nearly cylindrical, only slightly broader than tube, lobes about as long as tube, villous without; achenes obconic, 2–2.5 mm. long, densely hirsute with tawny or rufous hairs; pappus bristles 2–3 mm. long.

Rocky or gravelly hillsides and mesas, Lower Sonoran Zone, Colorado and Mojave Deserts, California, to southern Utah and southward into northern Sonora and Baja California. March–Aug.

**2. Psathyrotes ramosissima** (Torr.) A. Gray, Proc. Amer. Acad. **7:** 363, 1868
*Tetradymia ramosissima* Torr., in Emory, Notes Mil. Rec. 145, 1848.

A low perennial of rounded outline, with divaricately branched stems 5–20 cm. long, these and leaves more or less villous as well as scurfy; petioles 5–15 mm. long; leaf blades suborbicular to reniform, 5–25 mm. wide, 4–20 mm. long, coarsely and shallowly sinuate-dentate, subcuneate to shallowly cordate at base; peduncles usually 5 mm. long or less; involucres 6–7 mm. high, 8–10 mm. wide; outer bracts 5, spatulate-obovate to broadly obovate, villous-tomentose, tips often spreading; inner bracts lanceolate, less tomentose and nearly scarious along margins; corollas about 5 mm. long, ovate lobes densely villous without; achenes obconic, about 2.5 mm. long, densely hispid with ascending-spreading, rufous hairs 1.5–2 mm. long; pappus bristles rufous, 2.5–3 mm. long.

In sandy or gravelly soil of plains, mesas, or hillsides, Lower Sonoran Zone, Colorado Desert, California, to southwestern Utah, western Arizona, and northeastern Baja California. Flowering throughout the year.

## 101. SENECIO L., Sp. Pl. 866, 1753

Annual or perennial herbs, suffruticose plants, shrubs, or scandent or climbing perennial vines or rarely arborescent, with alternate, entire to intricately divided, pinnately or palmately veined leaves. Heads radiate or discoid. Involucres cylindrical to hemispherical, usually subtended by calyculate, irregularly arranged bracteoles, main bracts uniseriate or subbiseriate by overlapping of margins. Ray flowers, when present, in a single whorl, fertile. Disk flowers perfect, their corollas slenderly tubular to abruptly ampliate into a funnelform throat, 5-toothed. Anthers obtuse to faintly sagittate at base. Style branches rounded-obtuse to truncate, narrowly appendaged. Achenes terete or nearly so, usually ribbed, glabrous or strigillose. Pappus of numerous silky hairs or setae, these falling separately.

Annual herbs:
- Involucres narrowly campanulate to cylindric, twice as long as broad; ligules 0.5–2 mm. long, barely equaling disk, inconspicuous:
  - Disk corollas about 5 mm. long; herbage glabrous; leaves oblong to ovate .................................................................... 1. *S. mohavensis*
  - Disk corollas 3–3.5 mm. long; herbage sparsely pubescent, at least in inflorescence; leaves linear to oblanceolate .......................................... 2. *S. aphanactis*
- Involucres cup-shaped to hemispheric, often broader than long; ligules 10–15 mm. long, conspicuous:
  - Leaves not obviously fleshy; lower leaves dentate to lobed .................................................................... 3. *S. californicus* var. *californicus*

Leaves distinctly thickened and somewhat fleshy; lower leaves entire or nearly so ...... ........................................ 3a. *S. californicus* var. *amophilus*

Perennial herbs or shrubs:

Involucral bracts linear or linear-lanceolate, acute to attenuate; leaves coarsely dentate to lyrately lobed, not willowlike:

Leaf blades or their lobes linear, revolute; achenes 4–5.5 mm. long:

Herbage white-tomentose, at least in youth; plants distinctly woody at base:

Upper leaves linear and entire; involucres not obviously calyculate, few bracteoles more than one-fourth as long as main bracts ................. 4. *S. longilobus*

Upper leaves usually pinnately lobed; involucres strongly calyculate, bracteoles to three-fourths as long as main bracts ............. 5. *S. douglasii* var. *douglasii*

Herbage glabrous or essentially so; plants herbaceous almost to base ............... ........................................ 5a. *S. douglasii* var. *monoensis*

Leaf blades entire, coarsely toothed or variously lobed, lobes not linear or revolute; achenes 2–3.5 mm. long:

Stems herbaceous to base or only slightly suffruticose; leaves dentate to once-pinnatifid; throat of disk corolla rarely over twice as wide as tube:

Herbage more or less tomentose or leaves tardily glabrate on upper surface; involucral bracts pubescent at base .......................... 6. *S. neo-mexicanus*

Herbage glabrous or tufts of wool in axils of leaves; involucral bracts essentially glabrous:

Achenes glabrous; leaves oblanceolate to oblong-spatulate .......... 7. *S. stygius*

Achenes strigillose; leaves lanceolate to oblong-spatulate ......... 8. *S. lemmonii*

Stems distinctly woody well above base; leaves bi- to tripinnatifid; throat of disk corollas 3 times as wide as slender tube ......................... 9. *S. cedroensis*

Involucral bracts broadly oblong, obtuse to rounded at apex; leaves finely serrate to entire, willowlike; shrub to 2.5 m. tall ................................. 10. *S. salignus*

### 1. **Senecio mohavensis** A. Gray, Syn. Fl. N. Amer. 1²: 446, 1884

Erect glabrous annual 1–4.5 dm. tall, branching freely above or from base or in drought-dwarfed plants simple; stems and leaves often purple-tinged; leaves alternate, submembranous, ovate, oblong, obovate, or obovate-spatulate, 1–4 cm. wide, 2–7 cm. long, irregularly dentate or occasionally laciniate-dentate, lower ones with a narrowly winged petiole shorter than to equaling blade, upper ones sessile and amplexicaul-auriculate; heads in terminal corymbose cymes; peduncles slender, 0.5–3 cm. long; heads 8–9 mm. high at anthesis, discoid or with greatly reduced, inconspicuous ligules; involucres narrowly campanulate to nearly cylindrical, main bracts 10–13, linear, acute at apex, dark green and distinctly 1-ribbed, with narrow, subscarious, hyaline or white margins and a slightly brownish tip; basal bractlets 0.5–1.5 mm. long, inconspicuous; disk flowers 18–20; disk corollas slender, about 5 mm. long, gradually ampliate throat about 1 mm. long, teeth minute, about 0.2 mm. long; ray flowers 4–5 mm. long when present, erect; achenes 2.5–3 mm. long, terete, linear, closely and finely puberulent with ascending, slightly spreading, light gray hairs; pappus bristles silky, numerous, 4–5 mm. long.

Along arroyos, in shade of rocks, and at bases of cliffs, Lower Sonoran Zone, Inyo County, California, southward through the Mojave and Colorado Deserts to central Baja California, east into western Arizona and northern Sonora. Feb.–May.

### 2. **Senecio aphanactis** Greene, Pittonia 1: 220, 1888

An erect, sparingly branched annual 0.5–3 dm. tall with glabrous herbage or sparsely tomentulose in inflorescences, penetratingly aromatic; stems often reddish toward base; leaves linear to oblanceolate, 1–12 mm. broad, 1–4 cm. long, rarely entire, more commonly coarsely dentate or rather deeply pinnately lobed, segments acute to obtuse or rounded at apex, glabrous or faintly tomentulose when young; lower leaves narrowed to slender petioles 1 cm. long or less, upper ones sessile; heads radiate, 1 to several, terminal, on short peduncles 1–12 mm. long; involucres slightly flask-shaped, 5–6 mm. high, narrowly lance-linear bracts yellowish green, scarious-margined, obtuse, each head subtended by 1–3 minute bracteoles or these lacking; ray corollas yellow, ligules 0.5–1 mm. long, barely exceeding involucre and slightly spreading, inconspicuous; disk corollas 3–3.5 mm. long, barely equaling involucre; achenes 2–2.5 mm. long, subcylindric or slightly clavate, densely but minutely puberulent with ascending, faintly spreading, whitish hairs; pappus bristles shining white, about 3 mm. long.

Open hillsides and dry sandy soil, Lower and Upper Sonoran Zones, San Francisco Bay Region, California, south to the vicinity of Laguna Chapala Seca, Baja California. Feb.–March.

**3. Senecio californicus** DC., Prod. **6:** 426, 1837, var. **californicus**

*Senecio californicus* var. *laxior* DC., loc. cit.
*Senecio coronopus* Nutt., Trans. Amer. Phil. Soc. II, **7:** 413, 1841.

An erect, simple or moderately branched, glabrous annual 1–5 dm. tall with stems and lower leaves often tinged with red; leaves oblong-spatulate, narrowly oblanceolate or lanceolate, 3–20 mm. wide, 2–7 cm. long, entire, coarsely and remotely dentate, or subpinnatifid into linear-oblong lobes to 1 cm. long, upper leaves sessile and auriculate-clasping, lower ones usually narrowed gradually to a narrowly winged petiole, midrib rather prominent beneath; heads few to several in loose cymes, 8–10 mm. high, 10–18 mm. wide, radiate; involucres broadly cup-shaped or hemispheric, of 18–20 linear-oblanceolate bracts, these acute and usually brown-tipped, a few bracteoles 0.5–1.2 mm. long below and exterior to main ones; ray flowers 12–18, ligules yellow, 10–15 mm. long, 2.5–3.5 mm. wide, spreading and conspicuous; disk flowers numerous, corollas 6–7 mm. long, glabrous, tube very slender, about equaling slender, gradually ampliate throat, small teeth about 0.5 mm. long; achenes black, terete, rather broadly clavate-cylindrical, striately about 10-ribbed, abruptly flaring at apex, strigillose-cinereous on ribs; pappus bristles silky, glossy white, 6–7.5 mm. long.

Sandy soil, sand dunes and canyon floors, mainly Upper Sonoran Zone but entering the higher margins of the Lower Sonoran Zone, Monterey County, California, south into the northern one-fourth of Baja California. March–May.

**3*a*. Senecio californicus** var. **ammophilus** (Greene) Greenm., Ann. Mo. Bot. Gard. **2:** 590, 1915

*Senecio ammophilus* Greene, Bull. Calif. Acad. **1:** 193, 1885.

Differing from var. *californicus* in having thick, somewhat succulent leaves, lower usually being oblanceolate and nearly or quite entire.

On sand dunes and the higher margins of the beach, Lower Sonoran Zone, vicinity of San Quintín at least to Rosario, Baja California. Feb.–May.

**4. Senecio longilobus** Benth., Pl. Hartw. 18, 1839

*Senecio filifolius* Nutt., Trans. Amer. Phil. Soc. II, **7:** 414, 1841. Not *S. filifolius* Berg., 1767.
*Senecio orthophyllus* Greene, Leaflets Bot. Obs. **1:** 221, 1906.

Much-branched, suffruticose plant to 1.5 m. tall, very leafy, white-tomentose throughout or sometimes tardily glabrate; branches rather robust, striate but striations usually hidden by dense tomentum; leaves linear and entire to pinnately divided into few to several linear divisions to 3.5 cm. long, 1–2 mm. wide, whole leaf to 10 cm. long, larger ones usually with several smaller entire ones fascicled in axils, densely tomentose or occasionally subglabrate in age; heads few to several in corymbose cymes, radiate, 10–15 mm. high; involucres broadly turbinate or turbinate-campanulate, not calyculate, bracts linear, 8–9 mm. long, acute to attenuate at apex, lanate to subglabrate on heavy midpart, margins subscarious and glabrous; ray flowers 8–13, ligules yellow, 2–3 mm. wide, 10–12 mm. long; disk flowers 40–50, slenderly tubular, 8–9 mm. long, slightly ampliate throat about one-third as long as tube; achenes about 4 mm. long, densely canescent-hirtellous with ascendingly subappressed hairs; pappus silvery, 7–8 mm. long.

On stony mesas, outwash slopes, along arroyos, and on sandy soil, Lower Sonoran to Transition Zones, Pima County, Arizona, to southern Utah, Colorado, Texas, and northern Mexico. Entering the Sonoran Desert near Tucson, Arizona, and near Magdalena, Sonora. April–Nov.

### 5. **Senecio douglasii** DC., Prod. **6:** 429, 1837, var. **douglasii**

Suffruticose or shrubby plant to 1.5 m. tall with numerous ascending striate stems and white-tomentose to glabrate or glabrous, gray-green, pungently aromatic herbage; leaves 3–12 cm. long, pinnately divided into 3–9 narrowly linear, revolute lobes 1–1.5 mm. wide, 1–6 cm. long, or upper leaves entire, usually sparsely arachnoid-tomentulose beneath, even in age; heads 10–12 mm. high, in broad corymbose cymes; involucres broadly turbinate, calyculate, main bracts usually 21, linear, attenuate-acute at apex, rather strongly ribbed, 7–9 mm. long, sparsely to densely tomentulose; secondary bracteoles several, 2–5 mm. long; ray flowers 10–12, ligules oblong, 2.5–3 mm. wide, 7–11 mm. long, yellow; disk flowers 50–60, slender, 8–9 mm. long, slender tube about equaling tubular-funnelform throat, teeth about 0.8–1 mm. long; achenes terete, 4–5 mm. long, truncate at each end, weak ribs obscured by fine but close, upwardly appressed puberulence; pappus bristles silvery, 7–9 mm. long.

Along washes and stream beds, Lower Sonoran to Transition Zones, western central California to northern Baja California and east to the western margins of the Mojave and Colorado Deserts. April–Aug.

### 5*a*. **Senecio douglasii** var. **monoensis** (Greene) Jepson, Man. Fl. Pl. Calif. 1149, 1925

*Senecio monoensis* Greene, Leaflets Bot. Obs. **1:** 221, 1906.
*Senecio lathyroides* Greene, op. cit. **2:** 21, 1909.
*Senecio filicifolius* Greenm., Ann. Mo. Bot. Gard. **1:** 274, 1914.
*Senecio pectinatus* A. Nels., Wyo. Univ. Pub. Bot. **1:** 141, 1926.

Differing from var. *douglasii* in being consistently glabrous throughout or with a few arachnoid hairs at bases of involucres and on under sides of leaves; slightly smaller, more slender habit; less shrubby; leaves more finely dissected; bracteoles slightly shorter.

Sandy or rocky mountain slopes, mesas, and arroyos, Lower Sonoran to Transition Zones, Mono and Inyo Counties, California, to southern Utah and south into Sonora and Baja California. March–Oct.

**Senecio millelobatus** Rydb., Bull. Torrey Club **27:** 171, 1900 (*Senecio tampicanus* A. Gray, Pl. Wright. **1:** 125, 1852), an herbaceous perennial with glabrous to slightly tomentulose herbage, leaves 1.5–10 cm. long, pinnately divided into numerous obovate to cuneate divisions, and radiate heads 7–9 mm. high, occurs in northern Arizona to Texas and south into northern Mexico. It normally occurs north of and at a higher elevation than in the Sonoran Desert but it occurs along the Agua Fria River in Maricopa County, Arizona, at an elevation of about 1,000 feet, presumably a straggler. May–June.

**6. Senecio neo-mexicanus** A. Gray, Proc. Amer. Acad. **19:** 55, 1883

*Senecio toumeyi* Greene, Pittonia **3:** 349, 1898.
*Senecio mutabilis* Greene, op. cit. **4:** 113, 1900.
*Senecio willowensis* Greenm., Monogr. Senecio I. Teil 24, 1901. (Name only)
*Senecio blumeri* Greene, Leaflets Bot. Obs. **2:** 20, 1909.
*Senecio encelia* Greene, op. cit. 22.
*Senecio papagonius* A. Nels., Wyo. Univ. Pub. Bot. **3:** 110, 1937.

Erect herbaceous perennial with 1 to several erect or ascending branches 1.5–6 dm. tall from a stoutish rootstock; herbage more or less white-tomentose throughout or sometimes stems and upper sides of leaves glabrate; stems finely striate under tomentum, often reddish-tinged; basal and lower leaves on slender petioles 1–6.5 cm. long, blades obovate to oblanceolate, 0.5–3 cm. wide, 2–10 cm. long, subentire to slightly lyrate, margins usually irregularly dentate, thick and firm in texture, often tinged with purple; upper leaves sessile, oblong to lanceolate, dentate, gradually reduced toward inflorescence, usually persistently tomentose; heads numerous on corymbosely arranged inflorescences, each head 10–12 mm. high, radiate; involucres campanulate, sparingly calyculate, main bracts usually 21, lance-linear, 6–8 mm. long, acute at apex, scarious-margined, tomentose toward base but usually glabrate above; bracteoles 0.5–2 mm. long, subulate; ray flowers 10–13, yellow, ligules 7–10 mm. long, spreading and conspicuous; disk flowers numerous, slender, 5–6 mm. long, tube gradually expanding into very narrow, slender throat which about equals tube; achenes 2–2.5 mm. long, brownish, strongly 5-nerved and usually with a smaller rib in each interval between main ones, latter sparsely and minutely hirtellous with rather stiff, short hairs; pappus bristles numerous, silky, 6–10 mm. long.

Mountain slopes and canyons, mostly Upper Sonoran Zone to Canadian Zone, Pima County, Arizona, into Colorado and New Mexico, occasionally straggling into the higher parts of the Sonoran Desert along the canyons of the Santa Catalina Mountains, Pima County, Arizona. April–Aug.

**7. Senecio stygius** Greene, Leaflets Bot. Obs. **2:** 21, 1909

*Senecio prolixus* Greenm., Ann. Mo. Bot. Gard. **1:** 264, 1914.

A nearly glabrous, herbaceous perennial with 1 to several erect stems 1–5.5 dm. tall from a simple or branching rootstock; stems terete, finely striate, glabrous or minutely white-tomentulose or puberulent toward base and in axils of leaves; basal and lower leaves petiolate, 1–6 cm. long, blades oblanceolate to oblong-spatulate in outline, dentate or lyrately pinnatifid into oblong to cuneate-oblong, coarsely dentate lobes which, in larger leaves, are separated by a deep rounded sinus wider than adjacent lobes; upper leaves subsessile, oblanceolate-spatulate, coarsely dentate, glabrous or with sparse tomentum in their axils; heads radiate, 8–10 mm. high, in loose corymbose cymes 1–2.5 dm. broad; involucres campanulate, glabrous, slightly calyculate, main bracts lance-linear, 5–7 mm. long, acute to slightly attenuate at apex, thinnish, 1–3-nerved, margins scarious; ray flowers 10–15, yellow, ligules oblong, 5–9 mm. long, spreading; disk flowers numerous, corollas about 6 mm. long, throat about twice as broad as slender glabrous tube, teeth 0.5–0.8 mm. long, nearly erect or weakly spreading; achenes 2.5–3 mm. long, yellowish, glabrous, 4–8-ribbed, alternate ribs much heavier than intervening ones; pappus bristles very fine, 4–6 mm. long, silvery.

About springs and under shrubs, Lower Sonoran to Canadian Zones, Mono and eastern San Bernardino Counties, California, to southern Nevada and northwestern Arizona; occurring in the Sonoran Desert at Wickenburg, Mohave County, Arizona. April–Aug.

**8. Senecio lemmonii** A. Gray, Proc. Amer. Acad. **17:** 220, 1882

*Senecio decorticans* A. Nels., Amer. Jour. Bot. **23:** 266, 1936.

An erect or ascendingly branched, perennial herb 1.5–7 dm. tall or often suffruticose at base, glabrous throughout or with tufts of fine white tomentum in axils of leaves; stems smooth, brownish with a thin, more or less deciduous cortex; leaves narrowly lanceolate or oblong, 0.5–2.5 cm. wide, 3–12 cm. long, remotely, coarsely, and irregularly salient-dentate or uppermost entire, lower ones narrowed to a petiole nearly equaling blade, upper ones sessile and auriculate-amplexicaul, submembranous, midrib rather prominent; heads few to many in corymbose cymes, radiate, 10–12 mm. high; involucres campanulate, calyculate, bracts linear, abruptly narrowed to a slightly attenuate or acute apex, strongly 1-ribbed, glabrous or essentially so, margins pale, subscarious; ray flowers 12–18, ligules 7–9 mm. long, 1.5–2.5 mm. wide, spreading but soon more or less curled downwardly; achenes 2.5–3 mm. long, subcylindric, outer ones slightly curved, weakly 8–10-ribbed and minutely puberulent with short, stiffish, ascending, grayish hairs.

Dry hillsides and rocky slopes, mostly among shrubs, chiefly Lower Sonoran Zone, southern Arizona and northern Baja California. Feb.–May.

**9. Senecio cedrosensis** Greene, Bull. Calif. Acad. **1:** 194, 1885

A suffruticose plant with a rather low, rounded habit and distinctly woody basal branches, flowering branches extending above leafy part, often to 6–8 dm. tall; bark on older branches light gray; younger branches striate, yellow-green or tinged with red, sparingly crisp-hirsute with slender jointed hairs or soon glabrate; leaves elliptic to lanceolate or spatulate in outline, 0.5–5 cm. broad, 1.5–10 cm. long, bi- or tripinnatisect into short, irregularly dentate or incised divisions to 5 mm. long, dark green, minutely puberulent to sparsely arachnoid-pubescent beneath, a tuft of closely packed wool in axils of narrowly winged petioles; heads radiate, 8–10 mm. high, in corymbose cymes; involucres broadly campanulate, slightly calyculate, linear bracts about 21, linear, 5–6 mm. long, attenuate and brown-tipped at apex, glabrous or sparsely arachnoid-puberulent at base; ray flowers 11–15, ligules 6–8 mm. long, about 2 mm. wide, yellow; disk flowers numerous, corollas 5–6 mm. long, exceeding involucre 2–4 mm., throat funnelform, about three times as broad at apex as slender tube, this about as long as throat, teeth triangular-ovate, 0.5–0.7 mm. long; achenes terete, 3–3.5 mm. long, truncate, finely hirtellous with grayish hairs; pappus white, 7–8 mm. long.

Among rocks and on lava flows, Lower Sonoran Zone, vicinity of San Quintín, Baja California, south to Cedros Island and the adjacent mainland. Feb.–May.

**10. Senecio salignus** DC., Prod. **6:** 430, 1837

*Senecio vernus* DC., loc. cit.
*Cineraria verna* Mairet, in DC., loc. cit., as synonym.
*Senecio axillaris* Klatt, Abh. Naturf. Ges. Halle **15:** 333, 1882.

A woody, moderately branched shrub to 2.5 m. tall, glabrous throughout; bark reddish brown on younger branches, tawny to gray on old growth, youngest twigs green and finely striate; leaves narrowly lanceolate or narrowly elliptic, 5–10 mm. wide, to 10 cm. long, tapering gradually to each end, usually distinctly serrate, teeth slightly callus-tipped, dark green on upper surface, slightly paler beneath, petioles gradually merging with blade; inflorescences corymbose cymes to 10 cm. broad; heads radiate, 10–12 mm. high; involucres campanulate, bracts 7–9, broadly oblong or oblanceolate, 2–2.8 mm. wide, 5–6 mm. long, obtuse or rounded at tip, somewhat indurated at base, thin, margins scarious; ray flowers 2–8, or sometimes lacking, ligules pale yellow, 5–7 mm. long but rather lightly rolled downward, about 2 mm. wide; disk flowers 15–25, slender, 7–8 mm. long, abruptly expanded at base of tube, gradually and slightly ampliate upward to narrow throat which equals tube, teeth narrowly triangular-lanceolate, about 1.5 mm. long, downwardly curled; achenes brownish, 2.5–3 mm. long, weakly 8–10-ribbed, ascendingly and sparsely strigillose; pappus bristles 5–6 mm. long, white.

Chiefly in moist rich soil along streams, Lower and Upper Sonoran Zones, southern Arizona south to Guatemala. Feb.–May.

## Tribe 11. **CYNAREAE.** Thistle Tribe

Small to large herbs with alternate, usually prickly leaves. Heads moderate to large in size, discoid. Involucral bracts imbricated in several series, usually prolonged into a spine or bristle or margins membranous. Receptacle flat or nearly so, densely hairy or bristly. Corollas tubular; lobes linear. Anthers long-caudate at base, tips long-appendaged. Pappus bristly or plumose, less commonly paleaceous.

Achenes attached to receptacle by bases, not notched; leaves prickly-margined; heads large, usually over 2 cm. long and broad; corollas not yellow:
  Pappus bristles plumose; leaves not conspicuously blotched with white ..... 102. *Cirsium*
  Pappus bristles narrowly subulate, barbellate but not plumose; leaves conspicuously blotched with white ............................................ 103. *Silybum*
Achenes attached to receptacle in a lateral notch just above base; leaves not prickly margined; heads smaller, rarely 2 cm. in diameter; flowers (in ours) yellow ....... 104. *Centaurea*

## 102. **CIRSIUM** Adans., Fam. Pl. **2**: 116, 1763

Biennial or perennial herbs with alternate, often woolly leaves usually armed with spines on margins and spiny or prickly involucres. Heads discoid, medium to large, pedunculate or sessile. Involucres campanulate, subcylindric, or hemispheric, bracts imbricated in several to many series, at least some of them spine-tipped. Receptacles flat or convex, densely and persistently bristly. Corolla tube slender; throat tubular, usually equalled or surpassed by linear lobes. Anthers sagittate at base. Style branches linear to filiform; style often with a ring of hairs below stigmatic branches. Achenes more or less 4-angled, or sometimes many-ribbed, glabrous. Pappus bristles numerous, in several series, connate in a ring at base, often plumose.

Plants wholly glabrous; inner involucral bracts acute but not spinose .... 1. *C. trachylomum*
Plants more or less tomentose or arachnoid-pubescent; inner bracts spine-tipped, except in *C. arizonicum*:
  Outer involucral bracts spreading or reflexed:
    Cauline leaves strongly decurrent, often for one-third their length; middle involucral bracts with broadly ovate bodies, abruptly tipped with slender acicular spines less than 1 cm. long ............................................ 2. *C. mohavense*

Cauline leaves weakly decurrent less than one-third their length; middle involucral bracts lance-ovate, gradually narrowed to subulate spines 1–2 cm. long .......... ........................................... 3. *C. neomexicanum*

Outer involucral bracts ascending, not reflexed nor strongly spreading:

Corollas bright red to carmine; corolla lobes longer than throat; involucral bracts sparsely arachnoid-lanate ................................ 4. *C. arizonicum*

Corollas white, pink, or lavender; corolla lobes shorter than throat; involucral bracts densely arachnoid-lanate, at least in youth .................... 5. *C. californicum*

## 1. **Cirsium trachylomum** S. F. Blake, Jour. Wash. Acad. **25:** 322, 1935

Erect, moderately branching herb to 2 m. tall with strongly striate, glabrous, fistulous stems; leaves lanceolate to lance-oblong in outline, lower ones petiolate, to 3 dm. long, irregularly lobed and marginally spinose, some yellowish spines 1–2 cm. long; upper leaves sessile, slightly auriculate-clasping, pale green, thick and coriaceous, glabrous or sparsely and ephemerally arachnoid when young; peduncles 3–15 cm. long, bearing 1 to several bracteate leaves 1–3.5 cm. long immediately below heads; involucres campanulate, about 3 cm. high and nearly as wide at anthesis, outermost, lower bracts ovate-lanceolate, 3–3.5 mm. wide, 6–10 mm. long, tapering gradually to stiff, yellowish, terminal spine, tips slightly spreading, margins finely ciliolate-scaberulous, back glabrous, middle bracts gradually longer and less spinose, innermost narrowly lanceolate and acute to attenuate at apex but not spine-tipped; corollas bright red, 2.5–3 cm. long, tube 7–9 mm. long, lobes 11–13 mm. long, slightly longer than narrow throat; achenes dark brown, slightly lustrous, 5–6 mm. long; pappus bristles white to sordid, 1.5–2.2 cm. long, finely plumose.

Along watercourses in boulder-strewn canyons and at margins of meadows, mainly Upper Sonoran and Transition Zones, Sierra San Pedro Mártir, entering margins of the Sonoran Desert near San Matías Pass, at the northern end of the Sierra San Pedro Mártir. Collected in September and October but probably beginning to flower in July or August.

## 2. **Cirsium mohavense** (Greene) Petrak, Bot. Tidsskr. **31:** 68, 1911

*Carduus mohavensis* Greene, Proc. Acad. Phila. **1892:** 361, 1893.

Strict to moderately branched biennial 1–1.5 m. tall, densely arachnoid-lanate throughout when young, stems, upper surfaces of leaves, and involucres glabrate or sparsely lanate in age; leaves narrowly lanceolate to oblong in outline, 1–10 cm. broad (including spines), 3–35 cm. long, more or less lobed, lobes not overlapping but often again slightly lobed or irregularly toothed, bearing numerous yellow spines 2–20 mm. long, middle cauline leaves decurrent on stem for one-third their length or more, lowermost petiolate and not decurrent; peduncles slender, to 2.5 dm. long, naked except for a few remote foliar bracts scattered well below heads or sometimes crowded rather closely beneath involucres when heads are borne in panicles; involucres ovoid to oblong in bud, hemispheric in flower, 1.5–2 cm. high at anthesis, bracts ovate to ovate-lanceolate, body usually equalled by slender, yellow, at length deflexed, terminal spines, margins entire; corollas reddish, linear acute lobes 6–7 mm. long, equaling throat; anther tips attenuate-subulate, nearly or quite equaling corolla lobes; pappus bristles of outer achenes scabrous, of inner ones plumose to near tip, terminal part scabrous.

In alkaline meadows and about desert seeps, Lower Sonoran Zone, mainly Mojave Desert region but entering the Sonoran Desert along its northwestern border. March–June.

**3. Cirsium neomexicanum** A. Gray, Pl. Wright. **2:** 101, 1853
*Cirsium arcuum* A. Nels., Amer. Jour. Bot. **25:** 118, 1938.

Robust, erect or ascending, moderately to much-branched biennial to 2.5 m. tall with stems, foliage, and involucres densely arachnoid-lanate, at least in youth; leaves oblong or oblong-lanceolate, to 6 cm. broad, 1–4.5 dm. long, reduced to spinose bracts 1–2 cm. long above, irregularly and deeply lobed and lobes often more or less overlapping on larger leaves, margins divaricately spinose with yellow acicular spines 0.5–2 cm. long, lower only slightly decurrent; petioles of lower leaves narrowly winged and spiny, upper leaves sessile; heads solitary or in panicles, peduncles to 3 dm. long or much shorter; involucres urceolate-campanulate to hemispheric, 2.5–3.5 cm. high at anthesis, densely arachnoid-lanate in youth, sparsely so in age; involucral bracts lance-ovate, each tipped with a lance-subulate spine 0.8–2 cm. long, these abruptly deflexed or spreading, yellowish; corollas white, pale pink, or faintly lavender, about 2.5 cm. long, slender tube 10–13 mm. long, throat 4–6 mm. long, exceeded by linear, acute lobes; achenes 5–7 mm. long, buff, strongly 4-angled, smooth and shining; pappus bristles 1.5–2 cm. long, those of outer achenes scabrous, faintly dilated at apex, those of inner achenes plumose.

Mesas, plains, and foothills, Lower Sonoran to Transition Zones, southern California to Colorado, New Mexico, Arizona, and northern Mexico. March–Sept.

**4. Cirsium arizonicum** (A. Gray) Petrak, Bot. Tidsskr. **31:** 68, 1911
*Cnicus arizonicus* A. Gray, Proc. Amer. Acad. **10:** 44, 1874.

Erect, strict to moderately branched, biennial or short-lived perennial 1–1.5 m. tall or less with densely arachnoid-lanate stems and leaves in youth, upper surfaces of leaves and involucral bracts glabrate or nearly so in age; basal leaves narrowly oblong, to 4 dm. long, on narrowly and irregularly winged petioles to 1 dm. long, blades deeply and irregularly lobed, lobes often again lobed or coarsely toothed and spiny-margined, upper surface dark green and nearly glabrate, lower persistently white-lanate; upper leaves gradually reduced, sessile, lance-ovate in outline; heads in panicles, slenderly short-pedunculate, 3–4.5 cm. high at anthesis; involucres campanulate, bracts narrowly lance-ovate to lanceolate, outer ones spine-tipped, upper part of body bent inward sharply, then tip ascending-spreading, inner ones thin and acute but scarcely spinose, all of them ascending, not reflexed or spreading, nearly glabrate in age; corollas crimson to carmine, about 3 cm. long, linear lobes 13–15 mm. long, twice as long as slender throat and slightly longer than tube; achenes 6–7 mm. long, dark brown, smooth and sublustrous, strongly compressed; pappus bristles about 2 cm. long, plumose.

Rocky slopes and in openings in chaparral, mainly Upper Sonoran to Transition Zones, southern Utah to northern Sonora, entering the margins of the Sonoran Desert near Tucson, Arizona, and between Nogales and Magdalena, Sonora. May–Oct.

**5. Cirsium californicum** A. Gray var. **bernardinum** (Greene) Petrak, Beih. Bot. Centralbl. **35:** 477, 1917
*Carduus bernardinus* Greene, Proc. Acad. Phila. **1892:** 361, 1893.

A tall, usually paniculately branched biennial 0.5–2.5 m. high, densely leafy toward base, covered with a close white tomentum of very fine tangled hairs in youth, wool somewhat deciduous with age; leaves narrowly oblanceolate to oblong in outline, 1–3 (or sometimes 4) dm. long, 2–4 cm. wide, sinuately lobed to deeply pinnatifid, moderately spinose, persistently tomentose beneath, glabrate and dark green above in age, lower ones petiolate, those from about middle of stem upward sessile and somewhat auriculate-clasping, uppermost much reduced; heads solitary on peduncles 1–4 dm. long; involucres ovoid in bud, hemispheric after anthesis, 2–3 cm. high, about as wide or slightly broader; outer involucral bracts with lanceolate to ovate, coriaceous bodies gradually narrowed to a subulate yellowish spine 1–3 mm. long, ascending or only slightly spreading, densely arachnoid-lanate on lower portion; inner bracts narrowly lanceolate, straight, tips herbaceous and often more or less cuspidate but not bristle-tipped; corollas white, cream-colored, or rarely pale purple, about 2 cm. long, linear corolla lobes slightly shorter than slender throat; achenes strongly compressed, dark brown, smooth and sublustrous, 5–6 mm. long; pappus bristles silvery, about 1.5 cm. long, plumose.

Dry sandy soil and on outwash fans, mainly Upper Sonoran Zone, Kern County, California, south to northern Baja California. Barely entering the margins of the Sonoran Desert at Palm Springs and along the eastern flank of Sierra San Pedro Mártir. June–Aug.

**Cynara scolymus** L., Sp. Pl. 827, 1753, the cultivated artichoke, a native of Europe, with very large pinnatifid, arachnoid-lanate leaves and brilliant blue-purple flowers in heads often 10 cm. broad, escapes from cultivation occasionally and has appeared along roadsides near Phoenix and Tempe, Arizona.

## 103. SILYBUM Vaill. ex Adans., Fam. Pl. 2: 116, 1763

Erect glabrous herbs with alternate ample leaves irregularly splotched with white along veins, sinuately lobed and again dentate; teeth spine-tipped. Heads large, pedunculate, solitary. Involucres broadly subglobose; bracts imbricated in several series, broadened at base and fimbriate-spinose, tip lance-subulate and terminated by a stiff, straight, spreading spine. Receptacle flat or nearly so, densely set with stiff bristles. Corolla tube slender, slightly ampliate into slender throat, equalled or exceeded by 5 linear lobes. Filaments of stamens connate into a tube at base or nearly to middle, glabrous; anthers sagittate at base. Style pubescent near base. Achenes obovate-oblong, compressed, glabrous. Pappus of numerous subpaleaceous unequal setae, these united into a short tube at base and deciduous as a unit.

### 1. Silybum marianum (L.) Gaertn., Fruct. 2: 378, *pl. 168,* 1802

*Carduus marianus* L., Sp. Pl. 823, 1753.
*Mariana mariana* Hill, Hort. Kew. 61, 1768 or 1769.

A robust, erect, simple or considerably branched annual, nearly glabrous herb 6–15 dm. tall; leaves oblong-lanceolate or oblong-oblanceolate, lower ones to 2.5 dm. wide, 4–7 dm. long, upper much smaller, moderately or shallowly sinuate-lobed, beset marginally with stiffish yellow or whitish spines 0.5–2 cm. long, lobes obtuse or rounded, tip usually acute, blade heavily splotched and streaked with white, especially along veins; peduncles to 4 dm. long; heads subglobose to hemispheric, 3–4 cm. high, 5–8 cm. broad; bracts of involucre stifly spinose-margined, lower part gradually constricted at apex, spreading, spine-tipped part then abruptly dilated to a width about equaling appressed body, terminal spine often to 4 cm. long; corollas reddish purple, subfiliform tube 1.5–2 cm. long, slender throat 3–6 mm. long, about equaled or slightly exceeded by linear acute lobes; achenes 5–7 mm. long, 3–4 mm. wide, 1–1.5

mm. thick, smooth and glabrous, mottled light and dark brown; pappus bristles about 1.5 cm. long, creamy white, slightly paleaceous and united into a distinct ring at base.

An occasional roadside weed and in waste ground, Lower Sonoran to Transition Zones, introduced from Europe and established in numerous localities from coast to coast in the United States and southern Canada; entering the Sonoran Desert sparingly in Maricopa County, Arizona. May–Sept.

## 104. CENTAUREA L., Sp. Pl. 909, 1753

Annual or perennial herbs with entire to pinnatifid leaves and small to large heads. Involucres ovoid to globose; bracts imbricated in several to many series, dentate or fimbriate marginally. Receptacle flat, bristly. Corollas white, purplish or yellow, marginal ones usually somewhat larger than central ones and neutral, former perfect and fertile, or in some species all flowers perfect and fertile. Corolla tube slender; limb 5-parted or 5-lobed. Anthers sagittate at base. Style branches short and obtuse, often somewhat connate. Achenes obovoid to oblong, compressed or 4-angled, obliquely or laterally attached to receptacle. Pappus of bristles, scales, or less commonly lacking.

Spines of involucral bracts purplish, less than 1 cm. long; hairs on leaves and stems more or less spreading, not arachnoid .......................................... 1. *C. melitensis*

Spines of involucral bracts yellow, to 2.5 cm. long; hairs on leaves and stems arachnoid-matted .......................................................... 2. *C. solstitialis*

### 1. Centaurea melitensis L., Sp. Pl. 917, 1753

Strict and slightly branched to much-branched and rounded annual 3–12 dm. tall, grayish-pubescent with slender to stoutish, spreading or slightly tangled hairs; basal leaves oblanceolate in outline, 2.5–5 cm. broad, 1–1.8 dm. long, lyrately pinnatifid into several pairs of broadly oblong to spatulate, rounded lobes, terminal lobe suborbicular, sinuses between lateral lobes about equaling or slightly narrower than lobes; cauline leaves narrowly oblanceolate, entire or remotely and faintly dentate, their bases decurrent almost to next node below; heads ovoid to subglobose, 1–1.5 cm. high, solitary on slender, moderately leafy branches or 2–5 short-pedunculate in terminal clusters; involucral bracts with broadly ovate bodies 2.5–4 mm. wide, each bearing a slender, purplish, terminal spine 5–10 mm. long, this usually with 2–5 pairs of shorter, laterally spreading, marginal spines at its base; corollas yellow, 8–10 mm. long, all alike, throat exceeding lobes; achenes 2.5–3 mm. long, stramineous to grayish, smooth and glabrous, flattened, delicately lined with light gray; pappus bristles somewhat unequal, 1–2 mm. long, persistent.

Waste ground, fallow fields, roadsides, and in rocky ground, Lower Sonoran to lower margins of Transition Zones, an introduced European weed established extensively in the western United States and sparingly eastward. Occurring in the Sonoran Desert in Pima County, Arizona, and Imperial County, California. May–July.

### 2. Centaurea solstitialis L., Sp. Pl. 917, 1753

An erect, strict to much-branched annual 3–6 dm. tall, often with many branches from base and these rounded, cottony-pubescent with very fine, matted, white hairs; basal leaves oblanceolate in outline, to 15 cm. long, pinnatifid into oblong acute lobes usually much narrower than intervening sinuses, lobes progressively smaller toward petiole; cauline leaves linear to narrowly lanceolate, reduced greatly toward inflorescences; heads broadly ovoid to ovoid-globose, about 2 cm. high at anthesis; involucres 10–14 mm. high, 8–11 mm. wide (exclusive of spines); bracts ovate to oblong, yellowish, each tipped with a rigid, spreading, yellow spine 5–25 mm. long and often bearing 1–3 pairs of small bristles 1–2.5 mm. long at base of main spine; corollas yellow, all alike, linear-subulate lobes 4–5 mm. long, about equaling tubular throat; achenes about 2 mm. long, truncate at summit, grayish or mottled with flecks of purple or brown; pappus bristles unequal, 1.5–4 mm. long, persistent.

In waste places, roadsides, grain fields, and along fence-rows, Lower and Upper Sonoran Zones, a European weed occasionally established in widely scattered localities in the United States and a serious pest in the Great Valley of California. Occurring sparingly in Yuma County, Arizona, and Imperial and eastern San Diego Counties, California. July–Sept.

## Tribe 12. **MUTISIEAE.** Mutisia Tribe

Herbs, shrubs, vines or small trees with alternate leaves and (in ours) heads in corymbose, cymose or paniculate inflorescences or less commonly solitary. Heads homogamous, corollas regular and deeply lobed or some or all deeply bilobed, or the heads heterogamous with ligulate ray flowers. Involucres nearly tubular, with many phyllaries in several well-imbricated series. Receptacle naked. Anthers caudate at base and with elongated apical appendages. Style branches short and broad, not appendaged. Achenes cylindrical to fusiform, slender. Pappus of simple or plumose bristles or narrow paleae.

Heads in glomerules (rarely single); corollas equally 5-lobed; pappus of stiff, narrowly lanceolate to acicular paleae .......................................... 105. *Gochnatia*
Heads in corymbose, cymose, or paniculate clusters or rarely solitary; corollas distinctly bilobed; pappus of soft, slender bristles:
  Involucral bracts in 3 to many series; corollas white, pink, bluish or purple, never yellow ....................................................... 106. *Perezia*
  Involucral bracts in 2 series; corollas pale to deep yellow .................... 107. *Trixis*

## **105. GOCHNATIA** H. B. K., Nov. Gen. & Sp. **4:** 19, *pl. 309,* 1820

Shrubs or approaching arborescent size in some with alternate, entire or toothed, mostly petiolate leaves. Heads usually in glomerules, discoid. Involucral bracts in several series, graduate and well imbricated, more or less indurated. Corollas narrowly tubular, equally 5-lobed, lobes linear, spreading or ascendingly spreading. Anthers long-caudate at base. Achenes cylindrical, pubescent. Pappus of many slender stiffish bristles.

### **1. Gochnatia arborescens** Brandegee, Zoe **5:** 163, 1903

Arborescent, 3–8 m. tall with a stout trunk to 4 dm. in diameter with fissured, dark gray bark; branchlets stiffish, striate, moderately puberulent in youth but later glabrate and with reddish brown bark; petioles 3–5 mm. long, puberulent; leaf blades broadly ovate, obtuse to rounded or abruptly subcuneate at base, obtuse and often abruptly acuminate at apex, 1.5–7 cm. wide, 2.5–10 cm. long, thick and rather coriaceous, sparsely and minutely puberulent, finely gland-dotted, prominently veined beneath; heads cylindrical, 5–8 mm. in diameter, 15–18 mm. high, 12–20-flowered; involucral bracts ovate to lance-ovate, 1.5–15 mm. long, in about 10 series, slightly striate, stramineous, shining, sparsely puberulent to glabrous, thin margins ciliolate; corollas about 1 cm. long, lobes curling downward; achenes subcylindrical, 6–8 mm. long, hispidulous with ascending subappressed hairs; pappus bristles stiffish, about 1 cm. long, scaberulous toward base, but appearing smooth.

Along arroyos and shallow canyons, Lower Sonoran and Subtropical Zones, southern Baja California, mostly in the Cape Region. March–May.

## 106. PEREZIA Lag., Amen. Nat. 1: 31, 1811

Perennial or rarely annual herbs with entire to pinnately parted, alternate leaves, stems leafy or scapiform and leaves mainly in a basal rosette. Root woody, caudex often densely lanate with felty masses of rust-colored wool. Heads solitary to paniculate, corymbose or cymose, mostly many-flowered. Involucral bracts in 3 to many series, imbricated, linear-subulate to ovate, outer often reddish or purplish. Flowers perfect and fertile, white, pink, bluish or purplish; corolla limb more or less irregular, 2 lobes forming upper lip, 3 lower ones coalesced to form lower lip. Anthers tailed at base, tipped with elongate appendages. Stigma lobes flattened, truncate. Achenes cylindrical to fusiform. Pappus of several to many light-colored or white, soft to stiffish bristles, these free or sometimes fused at base.

Plants dwarf, usually less than 3 dm. tall; heads solitary, about 15–25-flowered; leaves suborbicular, 2–5 cm. long, nearly as wide . . . . . . . . . . . . . . . . . . . . . . . . . . . . . . . . . 1. *P. nana*

Plants not dwarf, usually 5–15 dm. tall; heads several to many, corymbosely or paniculately arranged, about 10-flowered; leaves ovate to lanceolate, to 30 cm. long:

Achenes densely stipitate-glandular but not hirsutulous; involucral bracts 1.5 mm. wide or less, widest toward base:

Involucral bracts acute to obtuse, not markedly acuminate or attenuate, not stipitate-glandular; heads 8–11-flowered, 10–15 mm. high at anthesis . . . . . . . . . 2. *P. wrightii*

Involucral bracts acuminate to attenuate, stipitate-glandular dorsally; heads 4–6-flowered, 18–22 mm. high at anthesis . . . . . . . . . . . . . . . . . . . . . . . . . . . . . . . . . . . . 3. *P. thurberi*

Achenes stipitate-glandular and coarsely hirsutulous as well; wider involucral bracts to 2.5 mm. broad, usually widest toward apex, acuminate . . . . . . . . . . . . . . 4. *P. palmeri*

### 1. Perezia nana A. Gray, Pl. Wright. 1: 125, 1852

Low perennial herb 0.5–3 dm. tall from a woody, platform-like rootstock 1–5 cm. below soil, densely covered with a thick brownish tomentum and bearing several to many tough woody roots 1–2 mm. in diameter on its lower side; stems erect or ascending, simple or moderately branched, finely scabrous with simple and gland-tipped hairs; leaves sessile or short-petioled, leathery, pale green, obovate-suborbicular or suborbicular, 2–5 cm. long and nearly or quite as wide, coarsely and unequally spinulose-dentate, scaberulous, veins conspicuous on both surfaces; heads solitary at ends of branches on stoutish peduncles 4–10 mm. long or subsessile; involucre campanulate, about 1.5 cm. high, 9–12 mm. wide; bracts broadly ovate and abruptly attenuate to lance-linear and acute to apiculate, in 4–5 series, inner ones narrowest, often purplish, margins slightly scarious and lanate-ciliate below, backs scaberulous-puberulent on exposed parts; flowers fragrant; corollas 10–14 mm. long, glabrous, lobes lanceolate, 3–4 mm. long; achenes linear, 5–6 mm. long, strongly ribbed, puberulent with short, strongly spreading hairs; pappus of numerous slender silky hairs 10–15 mm. long, these silvery white to slightly tawny.

Mesas, arid plains, and desert slopes, usually under shrubs, Lower and Upper Sonoran Zones, western Pima County, Arizona, to western Texas and south into adjacent northern Mexico. March–June.

**2. Perezia wrightii** A. Gray, Pl. Wright. **1:** 127, 1852

*Perezia wrightii* var. *subpuberula* A. Gray, op. cit. 128.
*Perezia arizonica* A. Gray, in Brewer & Wats., Bot. Calif. **1:** 422, 1876.
*Perezia schaffneri* A. Gray ex Urbina, Cat. de Pl. Mex. 197, 1897.

Erect, more or less branching perennial 3–12 dm. tall with subglabrous to glandular-scabrid stems; leaves oblong-lanceolate to elliptic-oblong or lance-ovate, sessile, main ones 2–5.5 cm. wide, 6–13 cm. long, upper ones reduced, coarsely and irregularly dentate, more or less cordate- or sagittate-amplexicaul, thin to rather rigid, minutely scabrid above, more copiously so and glandular beneath; heads in subglomerate cymules, 8–11-flowered, 10–15 mm. high at anthesis; involucral bracts linear- to lance-oblong, obtuse to acute or rarely some slightly acuminate, 0.6–1.3 mm. wide, 2–7 mm. long, in 2–3 series, ciliate on margins, glabrous to puberulent but not glandular; corollas pink-lavender, sparsely to densely stipitate-glandular; achenes linear-fusiform, 5–6 mm. long, densely glandular-puberulent but not hirsutulous; pappus white or nearly so, biseriate, stiffish, only moderately coalesced into a ring at base.

On rocky hillsides and gravelly or sandy soils, Upper and Lower Sonoran Zones, southern Utah to western Texas and central Arizona south into central Mexico. Jan.–June; and sometimes in the autumn.

**3. Perezia thurberi** A. Gray, Mem. Amer. Acad. II, **5:** 324, 1854

Perennial, with 1 to several erect stems to 12 dm. tall from a woody rootstock; stems and leaves scabrous and puberulent with gland-tipped hairs; leaves sessile, lower ones somewhat rounded-auriculate, broadly ovate to obovate, 2–10 cm. broad, 3.5–20 cm. long, coarsely and irregularly spinulose-dentate, dark green above, slightly paler beneath, veins prominent on both surfaces; heads numerous in panicles to 15 cm. wide and nearly as high, each head 4–6-flowered, 2–3 mm. in diameter in bud, 18–22 mm. long at anthesis; involucres narrowly campanulate, about 1 cm. long; bracts in 4–5 series, lance-linear, acuminate, stipitate-glandular, rounded on back, 1-nerved, scarious margins very narrow; corollas white or faintly tinged with purple, 10–12 mm. long, main segments 5–6 mm. long; anthers bright purple at tips; achenes linear, 4–5 mm. long, brownish, densely covered with brownish, tack-shaped or tubular, glandular hairs; pappus bristles about 1 cm. long, white, finely barbellate, fully twice as thick at base as at apex.

Rocky slopes and canyons, mainly Upper Sonoran and Transition Zones, but entering the Sonoran Desert near Ures, Sonora, where it grows at an elevation of approximately 2,700 feet above sea level. Southern Arizona into southwestern New Mexico and south into central Mexico. June–Nov.

**Perezia microcephala** (DC.) A. Gray, Pl. Wright. **1:** 127, 1852, with slightly smaller heads and involucral bracts all acute or abruptly acuminate instead of attenuate, superficially resembles *P. thurberi* and may possibly enter the margins of the Sonoran Desert along its western margin in Imperial County, California, and adjacent northern Baja California, but to the present time it has not been found below the Upper Sonoran Zone.

**4. Perezia palmeri** S. Wats., Proc. Amer. Acad. **24:** 58, 1889

Erect to ascendingly or spreadingly branched perennial to 1.5 m. tall with slender, subglabrous to scaberulous, somewhat flexuous stems; leaves sessile, thin-chartaceous, oblong-lanceolate and usually shallowly cordate-clasping, 2.5–5 cm. wide, 7–15 cm. long or lower ones probably larger, upper ones attenuate-acuminate, acutely pungent-dentate and denticulate; floral leaves linear-lanceolate, 2–6 cm. long; heads 13–18 mm.

high, about 10-flowered, in diffuse panicles; peduncles 1–3 cm. long, densely glandular-puberulent; involucres campanulate; involucral bracts 2–12 mm. long, outer broadly ovate, abruptly and sharply acuminate, to 2.5 mm. wide, inner ones spatulate-oblong, abruptly acuminate, broadest about middle, 1.5–2 mm. wide, all bracts glandular-puberulent dorsally; corollas "lilac," very sparsely puberulent toward base of tube; achenes subcylindric-fusiform, 5–6 mm. long, dark brown, densely stipitate-glandular and coarsely hirsutulous; pappus bristles shining white to faintly tawny, 8–12 mm. long, strongly coalesced at base to form a conspicuous ring, biseriate.

Arid slopes among rocks and in chaparral-like brush, chiefly Lower Sonoran Zone, vicinity of Bahía de los Angeles to the Sierra de la Giganta above Loreto, Baja California. Dec.–March.

**Perezia pinetorum** Brandegee, Zoe **5**: 105, 1901, resembles *P. palmeri* but has longer, narrower, attenuate phyllaries and more densely puberulent stems. The two may not be specifically distinct. *P. pinetorum* was described from the Sierra de la Laguna in the Cape Region of Baja California and may occur within our area.

## 107. TRIXIS P. Browne, Hist. Jamaic. 312, 1756

Shrubs or subshrubs with alternate, ovate-oblong or less commonly linear, entire or dentate leaves and loosely paniculate to densely corymbose or rarely subsolitary heads. Involucral bracts in 2 series; outer ones unequal, linear to ovate, more or less foliaceous; inner series of 5–10 equal linear-oblong scales. Corollas yellow, strongly bilabiate, lower (outer) lip spreading, tridentate or subentire, upper lip of 2 erect, nearly distinct, linear teeth. Anthers sagittate and bicaudate at base. Achenes slenderly cylindrical to slightly flask-shaped, finely papillose, more or less expanded into a disklike summit. Pappus of numerous white to tawny, roughish bristles.

Outer involucral bracts shorter than inner, not markedly foliaceous; heads 1.5–2 cm. long, 12–14-flowered; at least some leaves short-petiolate:
  Leaves densely glandular-dotted beneath; inner involucral bracts corky-thickened at base ........................................................ 1. *T. californica*
  Leaves silvery-sericeous beneath, not gland-dotted; inner involucral bracts not noticeably thickened basally ........................................ 2. *T. peninsularis*
Outer involucral bracts usually longer than inner, somewhat foliaceous; heads 2–2.5 cm. long, usually 18–20-flowered; leaves sessile, appressed-villous beneath ........ 3. *T. longifolia*

### 1. Trixis californica Kell., Proc. Calif. Acad. 2: 182, *fig. 53,* 1862

*Trixis suffruticosa* S. Wats., Bot. Calif. **2**: 459, 1880.
*Trixis angustifolia* var. *latiuscula* A. Gray, Syn. Fl. N. Amer. **1**$^2$: 410, 1884.

A profusely branched shrub about 1 m. tall or less with slender, brittle, appressed-puberulent branchlets, these soon glabrate and with light yellowish gray bark; leaves lanceolate to linear-lanceolate, 5–20 mm. broad, 2–10 cm. long, acute to acuminate at apex, cuneate to obtuse at base, sessile to short-petiolate, entire to mucronulate-dentate, sparsely appressed-puberulent on both surfaces or sometimes glabrous above, densely glandular with small spherical glands beneath; heads in corymbose panicles, 1.5–2 cm. high, usually 12–14-flowered; inner involucral bracts 8–10, linear-oblong to narrowly lanceolate, 9–15 mm. long, somewhat chaffy above, thickened and slightly keeled at base, acute to short-acuminate at apex, sparsely to moderately gland-dotted; outer bracts shorter, elliptic, lance-linear or narrowly obovate, gland-dotted and with a few simple nonglandular hairs intermixed; corollas about 10 mm. long, sparsely puberulent at tips of lobes without; achenes 5–8 mm. long, densely glandular with short stout hairs; pappus about twice as long as achenes.

Mainly on rocky slopes and ridges, Lower and arid parts of Upper Sonoran Zones, western margin of the Colorado Desert to western Texas and south into Baja California, Sonora, Chihuahua, Coahuila, and San Luis Potosí. Feb.–June, and sometimes flowering again in the autumn.

**2. Trixis peninsularis** S. F. Blake, Contr. U.S. Nat. Herb. **22:** 654, 1924

A rounded, much-branched shrub about 1 m. tall with numerous branchlets, sericeo-tomentose at first, later glabrate, bark then light brown; leaves lanceolate to lance-elliptic, 1–2 cm. broad, 3–12 cm. long, acute to acuminate at apex, cuneately narrowed to a petiole 3–8 mm. long or sometimes subsessile, entire or denticulate, usually revolute-margined, moderately pubescent but soon glabrate above, softly sericeous-pilose with appressed hairs beneath; heads rather crowded in cymose panicles, 1.5–1.8 cm. high, usually 12–14-flowered; outer involucral bracts 4, elliptic to oblong, mostly 6–9 mm. long, puberulent and somewhat glandular, inner ones usually 8, lance-linear to linear, 11–12 mm. long, appressed-pubescent and with some intermingled glands, not noticeably thickened or keeled at base; corollas with tubes 6–7.5 mm. long, lips nearly as long; achenes slenderly subcylindrical, 6–8 mm. long, papillose-hispidulous with short, stout, 1-celled hairs; pappus bristles straw-colored, about 1 cm. long.

Arid hillsides, Lower Sonoran to Tropical Zones, mainly in the Cape Region, Baja California. Jan.–March.

**3. Trixis longifolia** D. Don, Trans. Linn. Soc. **16:** 191, 1833

*Perdicium longifolium* Sessé & Moc. ex D. Don, op. cit. 192, 1833, as a synonym.
*Trixis obvallata* Hook. & Arn., Bot. Beechey 300, *pl. 65,* 1840.
*Trixis conferta* Benth., Pl. Hartw. 289, 1848.

A much-branched shrub with numerous erect, striate, densely leafy stems, branchlets closely invested with very fine, subappressed, tawny hairs; leaves crowded, subsessile, lanceolate or upper ones ovate, 5–20 mm. wide, 3–8 cm. long, narrowly cuneate to obtuse at base, acute, acuminate or somewhat attenuate at apex, entire to shallowly and obscurely dentate, sparsely appressed-puberulent above, more densely so with longer hairs beneath, midribs often somewhat tawny beneath; heads crowded in corymbose or cymose clusters, 2–2.5 cm. tall, usually about 18–20-flowered; outer involucral bracts lance-ovate, 18–20 mm. long, foliaceous, usually equaling or surpassing inner ones, gland-dotted and appressed-villous; inner bracts linear to narrowly oblong, acute to short-attenuate, mostly 12–15 mm. long, invested like outer ones; corollas bright yellow, about 15 mm. long; achenes cylindrical, 6–8 mm. long, scaberulous and stipitate-glandular; pappus bristles tawny, 12–15 mm. long.

Rocky hillsides, Lower Sonoran to Tropical Zones, southern Sonora to Colima. Feb.–June.

## Tribe 13. **CICHORIEAE.** Chicory Tribe

Herbs (in ours), low shrubs, or in the tropics rarely trees with milky juice and alternate or basal leaves. Heads solitary on naked scapes or sessile in axils or terminating leafy or bracteate branches. Receptacles naked or bearing chaffy or bristle-like bracts. Involucral bracts in few series, main ones often equal or nearly so, smaller calyculate ones usually at base of head. Flowers perfect, all with ligulate, truncate,

5-toothed corollas, usually twisted into a caplike mass following anthesis and remaining on achenes for some time. Anthers sagittate or auriculate at base, short-appendaged at tips. Style branches semi-cylindric, often attenuate toward apex, stigmatic full length on inner side. Achenes often excavate at base, apex truncate and cupulate or projected into a slender beak. Pappus bristles capillary, paleaceous or none.

Flowers bright blue; plant 1–2 m. tall; heads often in several-headed, axillary clusters .......... 108. *Cichorium*
Flowers white, yellow, pink or pale lavender, not bright blue; plants rarely over 0.5 m. tall; heads solitary on slender scapes or short to long peduncles, not clustered:
  Pappus paleaceous or none, never capillary or plumose:
    Pappus none; leaves of basal rosette suborbicular to broadly oblong, flat on ground .......... 109. *Atrichoseris*
    Pappus present; leaves lanceolate, oblong or linear, rarely flat on ground .......... 110. *Microseris*
  Pappus of barbellate or plumose bristles, capillary or nearly so:
    Bristles of pappus plumose:
      Outer involucral bracts orbicular, with very broad, scarious margins; receptacle with stiffish, linear to lanceolate, scarious bracts; leaves all basal ...... 111. *Anisocoma*
      Outer involucral bracts linear to lanceolate, herbaceous or only narrowly scarious-margined; receptacle naked or with a few fine hairs; leaves both basal and cauline:
        Heads 0.8–2 cm. broad at anthesis; achenes truncate at apex, beakless .......... 112. *Stephanomeria*
        Heads 2.5–4 cm. broad at anthesis; achenes narrowed upward into a stout or slender beak .......... 113. *Rafinesquia*
    Bristles of pappus capillary, smooth or minutely barbellate but never plumose:
      Heads usually nodding in bud; pappus bristles falling in small groups, 1–5 often subpersistent on achene after fall of rest .......... 114. *Malacothrix*
      Heads usually erect in bud; pappus bristles falling separately or in a ring or all persistent, but not behaving as in *Malacothrix*:
        Receptacle clothed with bristles .......... 115. *Calycoseris*
        Receptacles naked:
          Achenes distinctly flattened; involucres turbinate or cylindro-urceolate in bud:
            Involucres campanulate or hemispheric in flower, mostly 6–10 mm. in diameter at base in bud; achenes truncate, beakless .......... 116. *Sonchus*
            Involucres cylindric in flower, usually less than 5 mm. in diameter at base in bud; achenes beaked .......... 117. *Lactuca*
          Achenes terete or somewhat 4–5-angled, not markedly flattened; involucres ovoid to broadly cylindric in bud:
            Ligules pink or rosy; achenes beakless; pappus bristles deciduous .......... 118. *Lygodesmia*
            Ligules yellow, orange, or purplish; achenes beaked; pappus bristles persistent and achenes expanding into globose head at maturity ...... 119. *Agoseris*

## 108. CICHORIUM L., Sp. Pl. 813, 1753

Erect, more or less branching, perennial herbs with basal and alternate leaves, cauline leaves much reduced. Heads many, moderate in size, solitary or in several-headed clusters. Involucres narrowly campanulate; bracts herbaceous, in 2 series, inner erect, outer spreading at tips. Receptacle flat, naked or weakly fimbrillate. Rays 5-denticulate at truncate apex. Anthers sagittate. Style slender; branches obtuse. Achenes 5-ribbed, beakless, truncate at apex. Pappus of 2–3 short, obtuse paleae.

**1. Cichorium intybus** L., Sp. Pl. 813, 1753

*Cichorium intybus* β *divaricatum* DC., Prod. **7**: 84, 1838.

Perennial to 2 m. tall from a deep-seated taproot, stems much-branched, somewhat striate and sparingly hispid, green or slightly glaucous; basal leaves spreading, 1–2.5 dm. long, runcinate-pinnatifid, lobed, or sometimes only dentate, oblanceolate to spatulate in outline, narrowed to faintly winged petioles about one-half as long as blade or less; cauline leaves oblong to lanceolate, much reduced upward, auriculate-clasping at base, entire to saliently lobed; heads numerous, usually in several-headed, sessile clusters scattered along stems; involucres cylindric or narrowly campanulate, 4–5 mm. in diameter, 8–10 mm. long, bracts lance-ovate to narrowly lanceolate, their margins sparsely ciliate with rather coarse, gland-tipped, whitish hairs; ligules blue, 1–1.5 cm. long, 3–4.5 mm. wide, 5 teeth narrowly triangular; achenes broadly obovoid, 2–2.5 mm. long, glabrous and striate, truncate at apex, maturing quickly and shed immediately; pappus of 15–20 minute, rounded or obtuse scales about 0.2–0.3 mm. long.

Roadsides, along irrigation ditches and in waste ground, introduced from Europe and widely distributed from coast to coast in the United States and Canada southward into Texas, northern Sonora and Baja California. Feb.–Oct.

## 109. ATRICHOSERIS A. Gray, Syn. Fl. N. Amer. $1^2$: 410, 1884

Scapose winter annual with glabrous herbage and scape branched above to support several to many heads. Leaves basal, obovate, often spotted, spinulose-dentate. Involucres of 12–15 equal, lance-linear bracts in 2 series, their margins scarious. Receptacle plane or nearly so, scrobiculate. Ligules white. Achenes oblong; ribs corky-thickened. Pappus none.

**1. Atrichoseris platyphylla** (A. Gray) A. Gray, Syn. Fl. N. Amer. $1^2$: 410, 1884

*Malacothrix platyphylla* A. Gray, Proc. Amer. Acad. **9**: 214, 1874.

Characters of genus; leaves 3–10 cm. long, to 2.5 cm. wide, often turning purplish; scapes 2–7 dm. tall, whitish or pale rose, bearing an ovate-lanceolate bract 2–5 mm. long at base of each branch; involucres 5–8 mm. long, acute or obtusish at apex; ligules 1–1.5 cm. long, white, flowers fragrant; achenes 3.5–4.2 mm. long, pale stramineous to nearly white, truncate and with a small pit in apical end, minutely glandular-puberulent with fine, tack-shaped hairs.

In sandy washes, canyons and rocky hillsides, Lower Sonoran Zone, Death Valley region to northeastern Baja California and east to southern Utah, and to Mohave and Yuma Counties, Arizona. March–May.

## 110. MICROSERIS D. Don, Phil. Mag. 11: 388, 1832

Herbaceous, mostly acaulescent or short-stemmed plants with glabrous or slightly puberulent herbage. Leaves basal, entire to pinnatifid, lobes often more or less falcate. Heads solitary at tips of an erect, slender peduncle. Involucres cylindric or nearly so, calyculate with small outer bracts, main ones linear or linear-lanceolate, subequal, loosely imbricated. Ligules short and inconspicuous, yellow. Achenes turbinate, cylindric or fusiform, usually truncate. Pappus of 5–10 paleae, each elongated and scabrous or slightly plumose toward apex.

**1. Microceris linearifolia** (Nutt.) Schultz-Bip., Pollichia **22–24:** 308, 1866

*Uropappus linearifolius* Nutt., Trans. Amer. Phil. Soc. II, **7:** 425, 1841.
(For additional synonymy see Contr. Dudley Herb. **4:** 283, 1955.)

Characters of genus; scapes 1–6 dm. tall, erect, often several from a single root; leaves linear, 1–5 mm. wide, 6–15 cm. long, entire or with 2 to several pairs of remote, salient lobes to 1.5 cm. long, moderately villous with more or less tangled, white hairs; peduncles often swollen and fistulous just beneath heads, latter 1.5–2.2 cm. long, erect in bud, outer involucral bracts about 1 cm. long, acute, inner twice as long and slightly attenuate; ligules barely surpassing involucre, ephemeral, pale yellow but drying reddish or purplish; achenes fusiform, black, 10–12 mm. long, attenuate into 2–3 mm. long beak, low ribs antrorsely scaberulous; pappus-paleae 10–14 mm. long, narrowly lanceolate, bifid at apex with a slender, faintly scaberulous awn 2–4 mm. long projecting from between attenuate teeth.

Mesas, plains, outwash slopes and arroyos, Lower Sonoran to Transition Zones, Washington to the southern flanks of the Sierra San Pedro Mártir, Baja California, and east into Idaho, Nevada and New Mexico. March–June.

**Microseris heterocarpa** (Nutt.) K. Chambers, Contr. Dudley Herb. **4:** 286, 1955, with heads nodding in bud; outermost involucral bracts rarely as much as one-fourth as long as main ones; achenes stramineous to violet-gray or dark gray, sometimes flecked with black, occurs from the Sacramento Valley, California, to northwestern Baja California and on Guadalupe Island. It may grow disjunctly along the west coast of the peninsula south of Rosario, an area poorly explored botanically.

## 111. ANISOCOMA Torr. & Gray, Bost. Jour. Nat. Hist. 5: 111, *pl. 13, figs. 7–11,* 1845

Annual with leaves in a basal rosette and 1 to several scapes from a strong taproot, each scape bearing a solitary, yellow-flowered head. Involucre cylindric or narrowly campanulate; bracts with broad, scarious margins and a green or purplish midrib, outer ones orbicular scales, successive inner ones longer, well imbricated and comparatively narrower, innermost linear and acute. Receptacle flat, bearing linear, scarious bracts. Ligules truncate at apex, minutely toothed, outer ones often roseate on lower surface. Achenes linear-turbinate, truncate, 10–15-nerved, puberulent. Pappus silvery white, of 10–15 plumose bristles, in 2 series, outer ones sometimes merely scaberulous instead of plumose.

**1. Anisocoma acaulis** Torr. & Gray, Bost. Jour. Nat. Hist. **5:** 111, *pl. 13, figs. 7–11,* 1845

*Pterostephanus runcinatus* Kell., Proc. Calif. Acad. **3:** 21, *fig. 4,* 1863.

Leaves numerous, 6–12 mm. wide, 3–5 cm. long, oblong in outline, pinnately lobed and larger lobes coarsely dentate, floccose-lanate beneath but wool soon deciduous in irregular patches; scapes glabrous, 0.5–2 dm. tall; involucres 2–3 cm. high at maturity, bracts often tinged with red along edge and flecked with a few small red spots; heads 4–6 cm. wide in anthesis; ligules 1.5–2 cm. long, about 4 mm. wide, 5-dentate at truncate apex; achenes about 5 mm. long, ascendingly hispidulous, a shallow hyaline cup about 0.2 mm. deep crowning slightly narrowed, truncate apex; pappus bristles 2–2.5 cm. long, faintly dilated at base, plumose only on upper one-half or two-thirds.

On sandy desert soil and rocky mountain slopes, Lower Sonoran to Transition

Zones, Mojave and Colorado Deserts and the contiguous mountains east into Nevada and northwestern Arizona. March–June.

## 112. STEPHANOMERIA Nutt., Trans. Amer. Phil. Soc. II, 7: 427, 1841

Annual or perennial herbs with simple or paniculately branched stems and upper leaves reduced to herbaceous bracts. Lower leaves linear to oblong, entire to pinnatifid. Heads sessile or short-pedunculate, few-flowered. Involucral bracts linear to lanceolate. Ligules equal, white, cream or rose-pink, opening early in morning and usually closing before noon. Receptacle flat, naked or alveolate. Achenes linear, oblong or slightly turbinate, strongly 5-angled and often transversely rugose, sometimes sharply grooved between angles, truncate. Pappus bristles 4–20, plumose at least toward apex, distinct or weakly coalesced basally in small groups.

Pappus bristles finely plumose to base, lateral capillary hairs intricately intertangled; leaves linear, but remotely and runcinately lobed .......................... 1. *S. tenuifolia*
Pappus bristles plumose only above basal one-fourth or one-third, lateral capillary hairs usually less tangled; leaves oblong or if linear not so strongly runcinate-lobed:
  Plants perennial; pappus bristles brownish or tawny, lateral capillary hairs mostly 1.5–2.5 mm. long:
    Stems mostly divaricate and stiffish; leaves oblong, runcinately lobed ................................................................ 2. *S. pauciflora* var. *pauciflora*
    Stems mostly slender, flexuous and ascending; leaves filiform-linear, entire or essentially so ........................................... 2a. *S. pauciflora* var. *myrioclada*
  Plants annual or rarely biennial; pappus bristles silvery white, lateral capillary hairs rarely over 0.5–1 mm. long:
    Pappus bristles 10–20, usually joined basally into clusters of 3–5, weakly or not at all dilated at base ......................................... 3. *S. exigua* var. *exigua*
    Pappus bristles 4–6, all separate and distinctly dilated at base ........................................................................ 3a. *S. exigua* var. *pentachaeta*

### 1. Stephanomeria tenuifolia (Torr.) Hall, Univ. Calif. Pub. Bot. 3: 256, 1907

*Prenanthes ?tenuifolia* Torr., Ann. Lyc. N.Y. **2**: 210, 1828.
*Ptiloria tenuifolia* Raf., Atl. Jour. 145, 1832.
*Lygodesmia minor* Hook., Fl. Bor. Amer. **1**: 295, *pl. 103, fig. A,* 1834.
*Stephanomeria minor* Nutt., Trans. Amer. Phil. Soc. II, **7**: 427, 1841.

Perennial, glabrous, pale green herbs 2–6 dm. tall from a slender, woody rootstock, stems branching paniculately from near base into slender, ascending, nearly straight branchlets; basal leaves narrowly oblong in outline, 4–8 cm. long, main blade 1–3 mm. broad, runcinate, remote lobes to 5 mm. long, upper leaves linear-oblanceolate, rarely over 1 mm. broad, entire or essentially so; heads terminal on branchlets, involucres 8–10 mm. long at anthesis, about 5–6-flowered; ligules exceeding involucre 4–6 mm., slightly pinkish; achenes striate, smooth, about 3 mm. long; pappus bristles silvery white, 15–25, 5–7 mm. long, very finely plumose to base, not dilated, lateral hairs intricately intertangled.

Arid hillsides, dry washes, roadsides, open plains and in clearings in pine forests, mainly Upper Sonoran to Transition Zones, Washington to Montana south into southern California, southern Colorado and central Arizona. Entering the margin of the Sonoran Desert near Tucson, Arizona. July–Aug.

**2. Stephanomeria pauciflora** (Torr.) A. Nels., in Coult. & Nels., New Man. Rocky Mts. 588, 1909, var. **pauciflora**

*Prenanthes ?pauciflora* Torr., Ann. Lyc. N.Y. **2:** 210, 1828.
*Ptiloria pauciflora* Raf., Atl. Jour. 145, 1832.
*?Stephanomeria runcinata* Nutt., Trans. Amer. Phil. Soc. II, **7:** 428, 1841.
*Ptiloria runcinata* Davids. & Moxley, Fl. S. Calif. 355, 1923.

A perennial, rounded herb with intricately much-branched stems 3–6 dm. tall from a woody rootstock, base of stems sometimes slightly woody; herbage glabrous, pale green; basal leaves narrowly oblong, lanceolate or oblanceolate in outline, 3–7 cm. long, remotely runcinate-pinnatifid, rarely over 1.5 cm. wide, including lobes; upper leaves reduced, linear to subfiliform, entire, often more or less revolute, uppermost mere scales so branchlets form an open, nearly leafless system; heads solitary on axillary and terminal, short peduncles; involucres 8–10 mm. long at anthesis, bracts acute or slightly callus-tipped; flowers mostly 3–5, ligules flesh-colored, 5–6 mm. long; achenes 3–4 mm. long, striate between angles, minutely roughened but rarely rugose; pappus bristles 12–25, sordid to tawny or faintly rufous, 6–7 mm. long, plumose above basal one-fourth, lateral capillary hairs 1.5–2.5 mm. long, bases slightly dilated.

Along washes, gravelly bajadas, plains and arid mesas, Lower Sonoran to Transition Zones, Colorado and Mojave Deserts, California, to Kansas, Texas, northern Sonora, and central Baja California. Flowering almost throughout the year.

**2*a*. Stephanomeria pauciflora** var. **myrioclada** (D. C. Eaton) Munz, Man. S. Calif. 589, 1935

*Stephanomeria myrioclada* D. C. Eaton, Bot. King Expl. 198, *pl. 20, figs. 1–4,* 1871.
*Ptiloria myrioclada* Greene, Pittonia **2:** 130, 1890.
*Ptiloria tenuifolia myrioclada* S. F. Blake, in Tidestrom, Contr. U.S. Nat. Herb. **25:** 623, 1925.
*Stephanomeria runcinata* var. *myrioclada* Jepson, Man. Fl. Pl. Calif. 998, 1925.

Differing from var. *pauciflora* only in having linear-filiform, mainly entire leaves and involucres only 5–6 mm. high.

Desert plains and washes, Lower Sonoran Zone, eastern Mojave Desert into southern Nevada and northwestern Arizona and south into northern Sonora. A vaguely marked variety. Flowering following rains.

**3. Stephanomeria exigua** Nutt., Trans. Amer. Phil. Soc. II, **7:** 428, 1841, var. **exigua**

*Ptiloria exigua* Greene, Pittonia **2:** 132, 1890.

Erect or ascendingly branched annual (or sometimes biennial) herb with 1 to many branches from base or simple below and paniculately branched above, 1.5–6 dm. tall, pale green, glabrous below, often sparsely glandular-puberulent with tack-shaped, glandular hairs above; lower leaves narrowly oblong, auriculately clasping to petiolate, remotely lobed, 2–5 cm. long, uppermost reduced to minute herbaceous bracts; heads on short axillary and terminal peduncles, scattered or in panicles; involucres of 5–6 main bracts, 6–7 mm. high, smaller, basal bracts 1–2 mm. long; flowers usually 5, ligules 5–7 mm. long, rose but soon sordid; achenes stramineous, 3–3.5 mm. long, strongly 5-angled, narrowly but sharply grooved midway between angles, adjacent faces tuberculate-rugose in 2 rows, one on either side of groove; pappus bristles 10–18, silvery white, 4–6 mm. long, usually united by their slightly dilated bases into 4–5 clusters, plumose on upper one-half to two-thirds, lateral hairs rather stiff and less than 0.5 mm. long.

Sandy washes, hillsides, dunes and mesas, Lower and Upper Sonoran Zones or sometimes into the Transition Zone, mainly in the desert areas but approaching or reaching the coast in southern California; northern Baja California to Monterey County, California, and east to Wyoming and New Mexico. April–Sept.

**3*a*. Stephanomeria exigua** var. **pentachaeta** (D. C. Eaton) Hall, Univ. Calif. Pub. Bot. **3:** 260, 1907

*Stephanomeria pentachaeta* D. C. Eaton, Bot. King Expl. 199, *pl. 20, figs. 8–10,* 1871.
*Ptiloria pentachaeta* Greene, Pittonia **2:** 133, 1890.
*Ptiloria exigua* var. *pentachaeta* Davids. & Moxley, Fl. S. Calif. 355, 1923.

Differing from var. *exigua* in having only 4–6 completely distinct, basally dilated pappus bristles; upper branches somewhat less commonly glandular-pubescent.

Desert washes, mesas and hillsides, Lower Sonoran Zone, western Colorado Desert, southern Nevada and near Tucson, Arizona. March–June.

## 113. RAFINESQUIA Nutt., Trans. Amer. Phil. Soc. II, 7: 429, 1841

Glabrous, more or less branching annuals with stems often stoutish and fistulous. Leaves alternate, toothed or more frequently pinnatifid or upper ones entire. Heads in panicles, rather large, solitary at tips of branchlets. Involucres cylindro-conical at anthesis; main bracts 7–15, equal, lanceolate to lance-acuminate, scarious-margined; outer secondary bracts much shorter and very unequal. Ligules white or outer ones roseate on back, inner much shorter than outer ones. Receptacle flat, naked. Achenes subfusiform, tapering upward into stout or slender beak. Pappus bristles 10–15, capillary, plumose.

Body of achenes tapering gradually into beak, this shorter than body, stoutish; pappus bristles silvery white, narrowly subulate, 10–15 mm. long; ligules 15–20 mm. long .................... 1. *R. neomexicana*

Body of achenes tapering abruptly into slender beak, this longer than body; pappus bristles dull white, sordid or tawny, capillary, 6–9 mm. long; ligules 5–7 mm. long .................... 2. *R. californica*

**1. Rafinesquia neomexicana** A. Gray, Pl. Wright. **2:** 103, 1853

*Nemoseris neomexicana* Greene, Pittonia **2:** 193, 1891.

Simple to considerably branched, glabrous plant 1.5–5 dm. tall with rather weak, often zigzag, fistulous, purplish stems, these frequently growing upward through shrubs; leaves to 15 cm. long, 1–3 cm. wide, deeply and often runcinately pinnatifid, lobes to 2 cm. long, acute; blades sessile and auriculate-clasping or short-petioled; heads solitary and scattered at tips of branchlets; involucres about 2 cm. long, main bracts lance-linear, with purplish midribs; ligules white within, rose-tinged without, outer ones 15–20 mm. long; achenes 12–15 mm. long, body tapering only slightly into beak, nearly twice as long as latter, minutely and closely papillate or puberulent, pale to dark gray; pappus bristles 10–15 mm. long, silvery white, about three times as wide as thick at base, terminal one-fourth to one-third often smooth and not plumose.

Arid plains, mesas and gentle slopes, Lower Sonoran and Upper Sonoran Zones, Inyo County, California, south through the Mojave and Colorado Deserts into central Baja California and east to southern Utah, western Texas and northern Sonora. March–May.

**2. Rafinesquia californica** Nutt., Trans. Amer. Phil. Soc. II, **7:** 429, 1841
*Nemoseris californica* Greene, Pittonia **2:** 193, 1891.

Robust herb 2–15 dm. tall, stem to 2 cm. in diameter, usually much-branched above, glabrous; leaves oblong or oblanceolate in outline, denticulate to coarsely and shallowly or more finely and runcinately pinnatifid, or upper ones nearly entire, lowermost to 2 dm. long and petiolate, middle and upper ones sessile and auriculate-clasping; heads paniculately to cymosely arranged at tips of branchlets; involucres 15–18 mm. long, to 12 mm. broad, main bracts 10–20, linear to lance-acuminate, green or faintly purple-tinged, scarious border very narrow, outer bracts one-half as long or less and subulate; ligules white, 5–7 mm. long; achenes usually brown, weakly 5-ribbed, minutely puberulent, body 5–6 mm. long, tapering rather abruptly to a slender, glabrous beak 6–8 mm. long; pappus bristles dull white, sordid or tawny, finely capillary to base, plumose to apex, 6–9 mm. long.

Among shrubs and at margins of open areas, mainly Upper Sonoran Zone but straggling into the upper margins of the Lower Sonoran Zone, Humboldt County, California, to the southern flanks of the Sierra San Pedro Mártir, Baja California, and east to southern Utah and central Arizona. Rare in the Sonoran Desert and entering it only along its upper margins. March–May.

## 114. MALACOTHRIX DC., Prod. 7: 192, 1838

Annual or perennial herbs or rarely suffrutescent, with mainly basal, dentate to pinnatisect leaves. Flowering stems scapose or sometimes leafy. Heads small to medium in size, solitary or in panicles, always pedunculate, usually nodding in bud. Involucre narrowly cylindric to hemispheric; bracts strongly graduated and imbricate or nearly subequal with much smaller ones forming a calyculate base, usually narrowly to broadly scarious-margined. Receptacles flat or slightly convex, naked or bristly. Flowers numerous, with white, yellow or rarely pinkish ligules. Achenes truncate-columnar, 10–15-ribbed, terete or 4–5 ribs more highly developed than others, borders or angles then often denticulate. Pappus of soft or scabrous, silky bristles, often more or less united at base and falling in a ring, or 1 to several separate and persistent, a crown of minute teeth also present on apex of achene in some species.

Involucral bracts strongly graduated, ovate to orbicular, each phyllary 3–4 mm. broad, with a green or purplish center line and broad scarious margins; stem leaves conspicuous, cordate-clasping .......... 1. *M. coulteri*

Involucral bracts scarcely graduated except for basal, calyculate ones, all linear to lanceolate, each phyllary 1–1.5 mm. wide, margins narrowly subscarious; stem leaves when present less conspicuous, not cordate-clasping:

Flowering branches naked or nearly so, simple or sparingly branched; leaf segments narrowly linear to filiform, rarely if at all dentate; involucres usually lanate at base, at least when young:

Scapes simple, leafless; leaves persistently arachnoid-lanate; involucres 10–15 mm. high .......... 2. *M. californica*

Scapes, or some of them, sparingly branched and slightly leafy; leaves glabrous or glabrate; involucres 5–12 mm. high .......... 3. *M. glabrata*

Flowering branches more or less leafy, branched so heads are paniculately arranged; leaf-segments ovate to oblong, entire or toothed; involucres glabrous:

Persistent pappus bristles 1–4, or none; apical crown of achenes of 14–25 small, scarious teeth; blades of basal leaves dentate to pinnatifid, usually less than 2 cm. wide:

Achenes pentagonal in cross section, 5 ribs more prominent than other 10:
  Persistent pappus bristles none; teeth on crown of achene 20–25, 0.05–0.1 mm. high; achenes 2.2–2.6 mm. long .............................. 4. *M. sonchoides*
  Persistent pappus bristles 1, rarely 2; teeth on crowns of achenes 14–18, 0.1–0.2 mm. high; achenes 1.4–1.8 mm. long:
    Achenes stramineous to pale brown; cauline leaves often toothed; plants usually extensively branched only in upper half of main stems .... 5. *M. clevelandii*
    Achenes dark purplish brown; cauline leaves mostly entire; plants often extensively branched from base ................................. 6. *M. similis*
Achenes round in cross section, all 15 ribs fine, of equal width and height:
  Ligules surpassing involucre only 1–2 mm. at anthesis, not noticeably purplish beneath; achenes 0.3–0.45 mm. in diameter; teeth of crown obvious, extending above rim of achene summit:
    Persistent pappus bristles usually 2, rarely 3 or 4; margins of basal leaves irregularly and doubly dentate .................................. 7. *M. sonorae*
    Persistent pappus bristle 1, rarely 2; margins of basal leaves once dentate ....... ................................................ 8. *M. stebbinsii*
  Ligules surpassing involucres 4–6 mm. at anthesis, distinctly purplish beneath; achenes 0.5–0.65 mm. in diameter; teeth of crown minute, usually hidden behind rim of achene ...................................... 9. *M. fendleri*
Persistent pappus bristles mostly 5–7; apical crown of achenes distinctly 5-toothed with broad, deltoid teeth; blades of basal leaves runcinately lobed, mostly 2–3.5 cm. wide ............................................................ 10. *M. xanti*

**1. Malacothrix coulteri** A. Gray, Pl. Fendl. 113, 1849
*Malacolepis coulteri* Heller, Muhlenbergia **2:** 147, 1906.

Annual herb 1–6 dm. tall with simple stems paniculately branched above or branching from base, plant then rounded in outline; stems and leaves glabrous but more or less glaucescent; basal leaves lance-linear, oblanceolate, spatulate or ovate in outline, 1–3.5 cm. broad, 3–10 cm. long, sinuately pinnatifid, subfleshy, slightly narrowed to a slender petiole equaling blade or only slightly narrowed to a broad, sessile base; cauline leaves ovate, lanceolate-ovate or oblong-spatulate, 0.5–2 cm. wide at base, to 6 cm. long, less deeply pinnatifid or subentire, broad base auriculate-clasping, terminal lobe often elongated; flowering branches paniculate, heads several to many, subglobose in bud; involucres 12–18 mm. high, often slightly broader, bracts suborbicular to broadly oblong, silvery-scarious but with a thicker, linear, purplish or green central portion, regularly imbricated in about 5 ranks, innermost lance-linear; ligules pale yellow, linear, 8–12 mm. long, drying reddish; achenes 2–2.5 mm. high, 15-ribbed and slightly 4–5-angled, rib forming each angle slightly higher than others; summit of achenes minutely denticulate by extension of ribs; pappus bristles capillary and weakly joined at base to form a ring, 1–2 slightly more robust bristles persistent, 3–5 mm. long.

Mainly sandy or gravelly soil, Lower Sonoran and arid lower parts of Upper Sonoran Zones, lower San Joaquin Valley at least to Cataviñá, Baja California, and east to southern Utah, and to Pinal and Pima Counties, Arizona. March–May.

**2. Malacothrix californica** DC., Prod. **7:** 192, 1838

Annual with several erect or strongly ascending leafless scapes 1–4 dm. long; leaves all basal, forming a dense rosette, blades lanceolate to oblong in outline, 0.5–2 dm. long, pinnatifid into narrowly linear or filiform lobes 0.5–2.5 mm. wide, to 3 cm. long or terminal ones often 5–6 cm. long, these lobes entire or rarely only faintly and sparsely denticulate, herbage conspicuously woolly with long, soft, intertangled hairs when young, some of them persisting into maturity; heads solitary; involucres 10–15 mm. high, 1.5–2.2 cm. wide at anthesis; bracts lance-linear to subulate in 3–4 ranks, midribs and tips usually strongly marked with reddish or purplish, margins scarious, outer bracts densely woolly, inner subglabrous; ligules yellow on upper surface, purplish or reddish beneath, 12–18 mm. long; receptacle bearing delicate bristles or these sometimes lacking; achenes slenderly fusiform to nearly linear, 2.5–3 mm. long, delicately striate, truncate and minutely denticulate at apex; pappus bristles silvery white, 6–10 mm. long, 1–2 of them usually persistent.

Mainly in sandy or gravelly soil, upper margins of Lower Sonoran into Transition Zones, Sacramento Valley, California, to northern Baja California and the adjacent coastal islands, east to the western margins of the Mojave, Colorado, and San Felipe Deserts. March–June.

**3. Malacothrix glabrata** (A. Gray) A. Gray, Syn. Fl. N. Amer. **1**[2]**:** 422, 1884

*Malacothrix californica* var. *glabrata* A. Gray ex D. C. Eaton, Bot. King Expl. 201, 1871.

Ascendingly moderately branched, wholly glabrous plant except slightly pubescent on youngest leaves in some plants, and on smallest outer involucral bracts; stems 1–4 dm. tall, bearing 1–2 cauline leaves near base, usually with 2–3 or rarely only 1 large head at tips of upper branches; basal leaves oblong-oblanceolate, 2–8 cm. long, sinuately lobed to remotely pinnatifid into linear segments, glabrous; involucres 5–12 mm. high, glabrous or lowest bracts sparsely arachnoid-pubescent; ligules yellow, often tinged with purple or lavender on lower surface; achenes cylindric, 1.6–2 mm. long, finely and evenly 15-ribbed or nerved, glabrous, truncate and with a low ring of tissue at apex, bearing a white-scarious, denticulate ring inside the broader ring of tissue; pappus bristles white, 5–6 mm. long, 1 or sometimes 2 persistent or semi-persistent, the rest deciduous, sometimes all deciduous.

Sandy and gravelly plains, rocky hillsides and mesas, Lower Sonoran and arid parts of Arid Transition Zones, southern Idaho to central Baja California, and northern Sonora; often very abundant. March–June.

**4. Malacothrix sonchoides** (Nutt.) Torr. & Gray, Fl. N. Amer. **2:** 486, 1843

*Leptoseris sonchoides* Nutt., Trans. Amer. Phil. Soc. II, **7:** 439, 1841.

Erect and sparingly branched to much-branched annual with spreading-ascending, glabrous, sparsely leafy stems 0.6–3.5 dm. long; leaves mainly basal, oblong or spatulate in outline, 0.5–2.5 cm. wide, to 10 cm. long, pinnatifid, lobes again irregularly and spreadingly dentate, thickish, glabrous, often slightly glaucous; cauline leaves remote, reduced but even smaller ones usually dentate; heads paniculately disposed on bracteate peduncles to 4 cm. long; involucres broadly campanulate, about 8 mm. high, basal calyculate bracts ovate to elliptic, 2–3 mm. long, main ones linear-oblong, greenish, scarious-margined, scarcely imbricate and nearly equal; ligules deep yellow, 10–12 mm. long; achenes about 2.5 mm. long, 15-ribbed, with 5 ribs heavier and slightly higher than others, apex truncate and rimmed with a thin, whitish crown minutely denticulate along its upper margin; pappus bristles silky, white, 4–6 mm. long, all usually deciduous.

Sandy arroyos and plains, outwash slopes and rocky hillsides and mesas, Lower Sonoran to Transition Zones, central Mojave Desert to Idaho, Nebraska, and Arizona; barely entering the northern margins of the Sonoran Desert. March–June.

**5. Malacothrix clevelandii** A. Gray, in Brewer & Wats., Bot. Calif. **1:** 433, 1876

Diffusely branched, slender-stemmed annual 1–6 dm. tall with numerous heads paniculately arranged at tips of branches; herbage glabrous throughout, stems and lower leaves often reddish-tinged; basal leaves linear to narrowly lanceolate, 2–7 cm. long, to 1.5 or 2 cm. wide, saliently lobed to pinnatifid or less commonly with linear segments; cauline leaves remote, linear to lanceolate, entire or nearly so, gradually reduced upward to linear bracts 5–10 mm. long; involucres cylindro-campanulate, 4–8 mm. tall, 2–5 mm. broad; bracts linear, 0.4–0.7 mm. wide, equal and scarcely imbricated, green with purple-dotted tips, margins narrow, scarious; basal bracts calyculate, ovate, 1–2 mm. long; ligules pale yellow or rarely white, exceeding involucre only 1–3 mm.; achenes linear-fusiform, truncate at apex, 1.4–1.8 mm. long, finely 15-ribbed, 5 ribs more prominent than rest, buff to pale brown; crown truncate, bearing 14–17 minute whitish, scarious teeth and 1 persistent seta; deciduous pappus bristles silvery white, 5–6 mm. long.

Along arroyos and streams, on hillside burns and less commonly on grassy hillsides, Lower and Upper Sonoran Zones, Tehama County, California, to northwestern Baja California; barely entering the Sonoran Desert on the flanks of the Santa Rosa Mountains, Riverside County, California. March–May.

**6. Malacothrix similis** Davis & Raven, Madroño **16:** 262, 1962

Annual glabrous herb 5–35 cm. tall, usually considerably branched from base, stems 1–11; basal leaves linear-lanceolate, entire, dentate, lobed or pinnatifid, 2–6 cm. long, 1–2 cm. wide, rachis oblong; cauline leaves reduced, entire or nearly so; heads narrowly campanulate, 6–10 mm. high, 3–6 mm. broad, involucral bracts as in *M. clevelandii*; ligules yellow, exceeding involucres 1–2.5 mm. in anthesis; achenes truncate-fusiform, 1.4–1.7 mm. long, 0.25–0.3 mm. wide, finely 15-ribbed, 5 ribs more prominent than other 10, body of achene pentagonal in cross section, usually dark brown to purplish brown, rarely stramineous; apex of achene flared, bordered by ring of 15–18 white-scarious teeth 0.1–0.2 mm. high, irregular; persistent pappus bristle 1.

Grassy plains and gentle slopes, Lower and Upper Sonoran Zones, Todos Santos Islands near Ensenada, Baja California, to Cedros Island; Santa Barbara and Ventura Counties, California, and Channel Islands. March–May.

**7. Malacothrix sonorae** Davis & Raven, Madroño **16:** 264, 1962

Annual herb 1–3.5 dm. tall, usually unbranched at base, but sometimes with 6–10 stems from base, all branched considerably toward apex; herbage glabrous throughout; basal leaves lanceolate to oblanceolate, irregularly and doubly dentate, rachis broadest near apex; heads campanulate, 6–9 mm. high, 4–6.6 mm. wide; bracts narrow, nearly scarious at margins, outermost reduced, 1–2 mm. long; ligules yellow, exceeding involucre 1–2 mm. at anthesis, soon withering-deciduous; achenes cylindrical, 1.7–2 mm. long, 0.3–0.4 mm. in diameter, attenuate toward base, straight or slightly curved, stramineous to gray-brown, finely 15-ribbed, all ribs equal in width and height; body of achene round in transverse section, slightly flaring-expanded at summit; crown of 16–18 white-scarious teeth, these straight, narrow, regularly or only slightly irregularly arranged; persistent pappus bristles usually 2, sometimes 1, 3, or very rarely 4; deciduous pappus bristles silvery white, 4–5 mm. long.

Open slopes on hills and lower mountainsides, Lower Sonoran Zone, Pima County, Arizona, into northern Sonora. March–May.

**8. Malacothrix stebbinsii** Davis & Raven, Madroño **16:** 265, 1962

Annual herb 1–6 dm. tall, usually simple and unbranched at base, but occasionally with 3–9 stems, all branched generously toward apex or if depauperate, simple; herbage glabrous throughout; basal leaves lanceolate to oblanceolate, simply dentate or rarely pinnatifid, 2–8 cm. long, 5–18 mm. wide; cauline leaves smaller, often entire; heads campanulate, 7–10 mm. high, 3.5–8 mm. in diameter; involucral bracts often tinged with purplish or red, the outermost 1–2.5 mm. long; ligules yellow or rarely white, 1–2 mm. longer than involucre; achenes narrowly columnar-fusiform, 1.7–2.3 mm. long, 0.3–0.45 mm. in diameter, straight or faintly curved, gray-brown to stramineous, finely 15-ribbed, all ribs equal in height and width; body of achene terete, not angular, slightly flaring-expanded at apex; crown of 14–17 white-scarious teeth 0.1–0.2 mm. tall, irregular, lance-linear; persistent pappus bristle 1, rarely 2; deciduous pappus bristles silvery white, 4–7 mm. long.

Margins of boulders, shaded canyons, and open hillsides, Lower and Upper Sonoran Zones, Washoe County, Nevada, and Inyo County, California, to northwestern part of the Colorado Desert and through foothills in Arizona into northern Sonora, Mexico. April–May.

**9. Malacothrix fendleri** A. Gray, Pl. Wright. **2:** 104, 1853

Sparingly to moderately branched annual with spreading to ascending stems 0.5–3 dm. long, stems, upper leaves and involucres glabrous; basal leaves oblong-oblanceolate to spatulate in outline, 2–8 cm. long, sinuately lobed to remotely pinnatifid, lobes or teeth deltoid to oblong, mostly less than 1 cm. long, leaf blades sparsely arachnoid to glabrous, petioles often with a dense mat of wool on upper surface near base; cauline leaves few, smaller, lobed to entire, glabrous or essentially so; heads fairly numerous, paniculately arranged or some solitary at tips of unbranched but sparsely leafy branches; involucres broadly campanulate, 7–8 mm. tall, basal calyculate bracts ovate, 1.5–3 mm. long, main ones equal, scarcely imbricated, 1–1.5 mm. wide, purplish at tip and often along midrib, margins narrowly scarious; ligules yellow, washed with purple or lavender on underside, 7–10 mm. long; achenes cylindric, 1.8–2 mm. long, dark brown, finely and evenly 15-ribbed or striate, glabrous, truncate at apex and bearing a terminal annular crown with a whitish denticulate ring set near its base inside, this not equaling truncate ring; pappus bristles white, 5–6 mm. long, 1–2 of them persistent or all of them deciduous.

Sandy plains, mesas and rocky hillsides, upper margins of Lower Sonoran to lower parts of Transition Zones, eastern and southern Arizona into northern Sonora and east to Texas. March–June.

**10. Malacothrix xanti** A. Gray, Proc. Amer. Acad. **9:** 213, 1874

Glabrous annual with 1 to several erect or strongly ascending, paniculately branching stems to 5 dm. tall from a basal rosette 1.5–3.5 dm. in diameter; basal leaves oblanceolate in outline, 8–18 cm. long, 1.5–7 cm. wide, runcinately lobed, lobes often again coarsely and saliently dentate with deltoid to lance-subulate teeth to 1 cm. long; petioles rather slender, narrowly winged, 1–4 cm. long, gradually or abruptly widening into blade, latter submembranous, often purple-tinged; cauline leaves lance-linear to linear, entire or somewhat toothed, to 5 cm. long; peduncles slender, 1–5 cm. long, bracteate; involucres cylindro-campanulate, 7–9 mm. long, basal calyculate bracts narrowly ovate to ovate-lanceolate, 1–2 mm. long, main ones linear, about 1 mm. wide, acute, green with purplish tips, scarious margins very narrow; ligules linear-spatulate, 8–10 mm. long, pink or pale lavender; achenes 2–2.5 mm. long, buff-colored, 15-ribbed, 5 ribs heavier and higher than others and terminating apically in a broadly deltoid or rounded minute tooth, teeth forming a 5-toothed cupulate rim around apex of achene; pappus bristles silky, white, about 4 mm. long, minutely barbellate, 3–5 of them persistent.

Along arroyos, on rocky hillsides and mesas, Lower Sonoran Zone, vicinity of Purísima and Comondú south to the Cape Region, Baja California. Sept.–May.

## 115. CALYCOSERIS A. Gray, Pl. Wright. 2: 104, *pl. 14,* 1853

Ascendingly much-branched desert annuals glabrous at base but usually sparsely to densely dotted with tack-shaped glandular hairs above. Leaves pinnatisect into a few linear to filiform, spreading lobes. Heads moderate-sized, solitary at tips of branches. Involucres calyculate, campanulate at anthesis; main bracts lance-linear, equal, narrowly scarious-margined, outer calyculate ones much shorter and loosely spreading. Flowers numerous; ligules bright yellow or white and turning purplish. Receptacle flat, bearing finely capillary bristles among flowers. Achenes 5–6-ribbed, fusiform, tapering above into a short beak expanded at apex into a small denticulate cup. Pappus bristles shining white, numerous, united at base and deciduous as a ring.

Ligules bright yellow; achenes light brown or gray, deeply sulcate between ribs, minutely papillate .................................................. 1. *C. parryi*

Ligules white, turning purple in age; achenes dark brown, shallowly sulcate between ribs, tuberculate .................................................. 2. *C. wrightii*

### 1. **Calycoseris parryi** A. Gray, in Torr., Bot. Mex. Bound. 106, 1859

Plant simple and erect to much-branched with spreading-ascending branches, 0.5–3 dm. tall; lower leaves pinnatisect into linear divisions 0.5–2 mm. wide, to 2 cm. long, whole blade to 10 cm. long, central rachis 5 mm. wide or less, glabrous or essentially so; upper leaves linear and entire or with 1–3 pairs of subbasal lobes; upper one-half of stems, branches, and involucres more or less glandular with reddish, brownish, or black, tack-shaped glands; involucres 9–15 mm. long at anthesis, main bracts to 2.5 mm. wide, acute or slightly acuminate; ligules bright yellow, showy, about 3 mm. wide, 10–15 mm. long; achenes 6–7 mm. long over all, slender beak one-half to two-thirds as long as fusiform, deeply sulcate body, latter pale brown or gray, very minutely and uniformly papillate but not tuberculate; pappus bristles 7–9 mm. long.

In sandy soil and along washes, Lower Sonoran Zone, Mojave and Colorado Deserts, California, to southern Utah and western Arizona. March–May.

### 2. **Calycoseris wrightii** A. Gray, Pl. Wright. 2: 104, *pl. 14,* 1853

*Calycoseris wrightii* var. *californica* Brandegee, Zoe **5**: 155, 1903.

Similar to the previous species in habit, character of leaves, peduncles and involucres but glandular hairs on peduncles and involucres paler in color; ligules white with pinkish or pinkish-brown spots or streaks on under side and turning purplish or reddish in drying; achenes about 6 mm. long, including beak, this usually 1–1.5 mm. long, body of achene dark brown, rather shallowly sulcate between ribs and bearing a row of low, rounded tubercles along each side of sulcus; pappus bristles 7–8 mm. long.

Sandy plains, rocky mesas and slopes, Lower and into Upper Sonoran Zones, Colorado and Mojave Deserts to southern Nevada, southern Utah, western Texas and northwestern Sonora. Often growing up through shrubs. March–May.

## 116. SONCHUS L., Sp. Pl. 793, 1753

Coarse, leafy-stemmed, more or less succulent annual herbs with more or less glaucous, nearly glabrous herbage, tack-shaped glandular hairs often on peduncles and involucres. Leaves subentire to pinnatifid, often spinulose-toothed, middle cauline ones frequently auriculate-clasping. Heads cymosely or umbellately arranged. Involucres turbinate in bud, campanulate at anthesis, with few thin bracts, basal ones calyculate, chief ones often callous-ribbed in age. Receptacles plane, naked. Flowers several to many; ligules pale yellow. Achenes obcompressed, striate with low ribs, truncate and beakless. Pappus of many very soft, fine, silky white hairs, these usually falling together.

Achenes 3-ribbed on each face, smooth between ribs, thin-margined; leaf auricles rounded .......... 1. *S. asper*

Achenes striate, rugulose or papillate-rugulose between ribs, edges not markedly thin; auricles acute:

- Leaf blades serrate or with lobes 0.5–3 cm. wide; achenes about twice as broad as thick; involucres more or less arachnoid-lanate at base in bud .......... 2. *S. oleraceus*
- Leaf blades pinnately parted into narrow lobes mostly less than 5 mm. wide, to 2 cm. long; achenes only slightly wider than thick; involucres glabrous in bud ... 3. *S. tenerrimus*

### 1. Sonchus asper (L.) Hill, Herb. Brit. 1: 47, 1769

*Sonchus oleraceus* var. *asper* L., Sp. Pl. 794, 1753.

A simple or scantily branched annual 0.3–2.5 m. tall; herbage glabrous or commonly upper stems, peduncles and involucres beset with tack-shaped, glandular hairs; basal leaves oblanceolate to spatulate in outline, to 3 dm. long, blades lyrately or runcinately pinnatifid into broadly ovate to oblong lobes and these saliently dentate with spinulose teeth or sometimes blades only dentate; petiole often equaling blade; cauline leaves similar but usually sessile, auricles to 2 cm. long, rounded and saliently toothed; heads urceolate-turbinate in bud, on peduncles 1–10 cm. long; involucres 10–16 mm. long and campanulate or cylindric in anthesis; main bracts lance-linear, acute to attenuate, thickened along midrib toward base in age, glabrous or sparsely glandular-pubescent, outer bracts ovate, more commonly glandular but rarely thickened; ligules 3–6 mm. long, quickly withering; achenes ovoid, 2–2.5 mm. long, about 1 mm. wide, strongly compressed, each face 3-ribbed and smooth in intervals, lateral margins very thin; pappus hairs 6–10 mm. long, often much tangled, thereby holding several achenes together in a cluster.

At scattered localities along roadsides, in fields and waste ground, Lower Sonoran into Canadian Zones, widely distributed as a European weed introduced almost throughout North America. Entering the Sonoran Desert in Imperial Valley, California, and in the agricultural regions in central and southern Arizona. April–Aug.

### 2. Sonchus oleraceus L., Sp. Pl. 794, 1753

A sparingly branched or simple herb to 2.5 m. tall with coarse, lyrately or runcinately pinnatifid leaves, terminal lobe of which is often much larger than other and broadly triangular, margins all more or less dentate; basal leaves 1–3 dm. long, petiole usually equaling blade; auricles of cauline leaves lance-ovate to lance-attenuate, acute; peduncles frequently bearing tack-shaped glands; heads often umbellate, involucres 8–16 mm. long at anthesis, bracts similar to those of *S. asper*; ligules 4–7 mm. long, pale yellow; achenes obovoid, moderately compressed, 2.5–3 mm. long, pale brown, striately ribbed and minutely rugulose on ribs and in intervals, margins not markedly thin; pappus hairs copius, to 12 mm. long.

Fields, roadsides and waste land, Lower Sonoran to Canadian and Carolinian Zones, nearly throughout temperate North America, where it has become established as a weed; a native of Europe. Entering the Sonoran Desert in the agricultural regions of Maricopa and Pima Counties, Arizona, and in eastern San Diego, Riverside, and central Imperial Counties, California. April–Sept.

**3. Sonchus tenerrimus** L., Sp. Pl. 794, 1753

A more slender but very leafy herb 3–10 dm. tall with short-pedunculate heads and glabrous herbage; leaves oblong in outline, 2–3 cm. wide, 5–15 cm. long, deeply pinnatifid into linear or narrowly lanceolate lobes to 3 cm. long, these entire or spinulose-denticulate, spreading or somewhat runcinate; auricles narrow, lanceolate to ovate, acute to attenuate; heads campanulate at anthesis, chief bracts lance-linear, acute to attenuate, 8–12 mm. long, midrib heavily thickened in age; outer ligules about 5 mm. long, turning reddish in drying; achenes narrowly obovoid, 2.5–3 mm. long, weakly compressed, striate-ribbed and distinctly but finely rugulose both on ribs and in intervals, margins rounded; pappus hairs 8–10 mm. long.

Fairly well established in sandy and rocky soil from San Quintín south to the vicinity of Magdalena Bay, Baja California, and on the adjacent islands off the west coast of the peninsula; less common northward into coastal southern California. A native of Europe. March–June.

## 117. LACTUCA L., Sp. Pl. 795, 1753

Leafy annual or perennial herbs with glabrous or sparsely pubescent herbage. Leaves from linear and entire to oblong or spatulate and pinnatifid, sessile or petiolate. Heads small to medium in size, paniculately arranged. Involucres cylindrical or cylindro-urceolate, slender; bracts more or less graduate, in 2 to several series. Ligules yellow or bluish, rather small. Achenes strongly compressed, gradually or abruptly tapering above into a short to long beak, usually finely sculptured. Pappus hairs numerous, soft and silky, silvery white.

**1. Lactuca serriola** L., Cent. Pl. **2:** 29, 1756

*Lactuca scariola* L., Sp. Pl. ed. 2, 1119, 1763.

Erect, simple or few-stemmed annual with slender to robust, smooth and glabrous to hirsute-prickly stems 2–20 dm. tall, paniculately branched above; leaves 4–22 cm. long, elliptic, oblong, or oblanceolate in outline, merely denticulate to deeply pinnatifid with denticulate margins, sessile and auriculate-clasping or basal ones petiolate, more or less soft-prickly along midrib beneath; auricles deltoid-ovate and acute to attenuate; paniculate inflorescence many-headed, 1–4 dm. tall; involucres 10–13 mm. high, innermost bracts linear, scarious-margined, tips often slightly recurved at anthesis; ligules pale yellow, only slightly exceeding longer bracts; body of achene fusiform, about 3 mm. long, dark olive green, strongly flattened, striate, hirsutulous above, tapering rather abruptly into filiform beak, this 3–4.5 mm. long, expanded into a flat disk 0.3–0.4 mm. in diameter at tip; pappus hairs silvery white, 5–7 mm. long.

Roadsides, fence-rows and in waste ground, Lower Sonoran to upper parts of Transition Zones, abundant throughout temperate North America, occurring sparingly along the margins of the Sonoran Desert. Naturalized from Europe. May–Sept.

Besides the typical form with pinnatifid leaves, there is another (**L. serriola** f. **integrifolia** Bogenhard) with merely denticulate, unlobed leaf blades. The latter occurs with, but is less common than, the typical form in our area.

A single specimen of a member of the Cichorieae was collected at La Purísima by Brandegee and identified by him as *Lactuca intybacea* Jacq. However, the achenes of this specimen are not those of the genus *Lactuca* as I understand it, but resemble those of some species of *Sonchus*. Nothing like Brandegee's specimen has been collected in the southern half of Baja California since he obtained the material, and his collection may represent a casual introduction that has not persisted. Until close comparison of the specimen from La Purísima can be made with a large suite of Old World specimens or the plant is found in Baja California again, it seems inadvisable to perpetuate a "record" of *L. intybacea* Jacq. from the peninsula. That species is imperfectly known, but originally was reported from Nicaragua.

## 118. LYGODESMIA D. Don, Edinb. New Phil. Jour. 6: 311, 1829

Annual or perennial herbs with glabrous or puberulent, rushlike stems and narrow, entire or dentate leaves. Heads small to large, erect in bud, scattered or terminal, few-flowered. Involucres (in ours) cylindro-ovoid; bracts few, equal, basal ones forming a small calyculus. Ligules whitish to rose-colored, small. Achenes linear-prismatic to subcylindric, beakless, broadly few-ribbed, truncate at each end. Pappus bristles numerous, capillary, not plumose.

### 1. **Lygodesmia exigua** (A. Gray) A. Gray, Proc. Amer. Acad. **9:** 217, 1874

*Prenanthes exigua* A. Gray, Pl. Wright. **2:** 105, 1853.
*Stephanomeria minima* M. E. Jones, Contr. West. Bot. **17:** 31, 1930.

A diffusely branched glabrous annual 1–5 dm. tall with rather slender, spreading-ascending stems; basal leaves oblanceolate, 2–3.5 cm. long, entire or more commonly dentate or pinnatifid into acute lobes; cauline leaves reduced to herbaceous bracts 1.5 cm. long or less; heads solitary on slender peduncles 5–25 mm. long; involucres about 5 mm. high at anthesis, main bracts usually 4–5, green except for narrow, scarious-hyaline margins; ligules white to pinkish, exceeding involucres 2–3 mm., withering quickly; achenes whitish, about 3 mm. long; ribs broad, flat, separated by 4–5 narrow grooves, almost microscopically papillate; pappus bristles shining white, 2–2.5 mm. long.

Open hillsides, mesas and desert plains, Lower Sonoran to lower parts of Transition Zones, Mojave and Colorado Deserts, California, to southern Nevada, Colorado, New Mexico, and western Texas. March–June.

## 119. AGOSERIS Raf., Fl. Ludov. 58, 1817

Annual or more frequently perennial herbs from stout, deep-seated taproots. Stems naked or nearly so, scapose, each bearing a solitary, rather large head. Leaves chiefly basal, oblong, oblanceolate, lanceolate or linear, sometimes a few reduced, scattered ones on stem. Involucres campanulate to subcylindric, bracts graduate and imbricate, outer ovate to oblong, inner linear to lanceolate. Corollas yellow (in ours), orange or purple, ligules usually showy. Achenes terete, oblong or fusiform, 10-ribbed, narrowed slightly above or tapering to a distinct beak. Pappus of numerous fine, white bristles or hairs inserted on expanded disk at apex of beak. Bracts of involucre reflexed and achenes expanded into a globose head in fruit.

**1. Agoseris heterophylla** (Nutt.) Greene, Pittonia **2:** 178, 1891

*Macrorhynchus heterophyllus* Nutt., Trans. Amer. Phil. Soc. II, **7:** 430, 1841.
*Troximon chilense* A. Gray, Proc. Amer. Acad. **9:** 216, 1874.
*Troximon heterophyllum* Greene, Bull. Torrey Club **10:** 88, 1883.

Annual, from a slender taproot; leaves oblong, spatulate or linear, entire, denticulate or sinuate-pinnatifid, 0.2–3 cm. wide, 5–15 cm. long, sparsely villous or glabrous, lobes on pinnatifid blades ovate to oblong, spreading or ascending and usually nearly as wide as long; scapes slender, 5–40 cm. tall, often several from a single root, glabrous or very sparsely villous; involucres 10–18 mm. high, often nearly as broad, bracts lance-acuminate, inner ones glabrous and hyaline-margined, outer ones shorter and arachnoid-villous to glabrate; ligules inconspicuous, barely surpassing involucres and withering very early; achenes fusiform, body smooth or longitudinally 10-ribbed, 3–4 mm. long, glabrous to villosulous, beak slender, 5–8 mm. long; pappus bristles white, about 5 mm. long.

Grassy hillsides and openings in brush, upper margins of Lower Sonoran into Transition Zones, northern Baja California to British Columbia and east to Arizona and northern Sonora. March–May.

# Indexes

# Index to Part I

(Principal references to species are in boldface type)

**D**

**N**

**Q**



**R**

## T

# Index to Part II

Names set in roman type indicate accepted taxa. Synonyms appear in italics. Taxa described as new in this book, or those that have been transferred from one genus to another, or their status changed here, are indicated in the index by the following abbreviations: c.n. = comb. nov.; s.n. = status nov.; v.n. = var. nov.; sp. n. = species nov.; subsp. n. = subspecies nov.

**B**

## D

## F

G

## Q

**T**

## U



## V

**W**



**X**

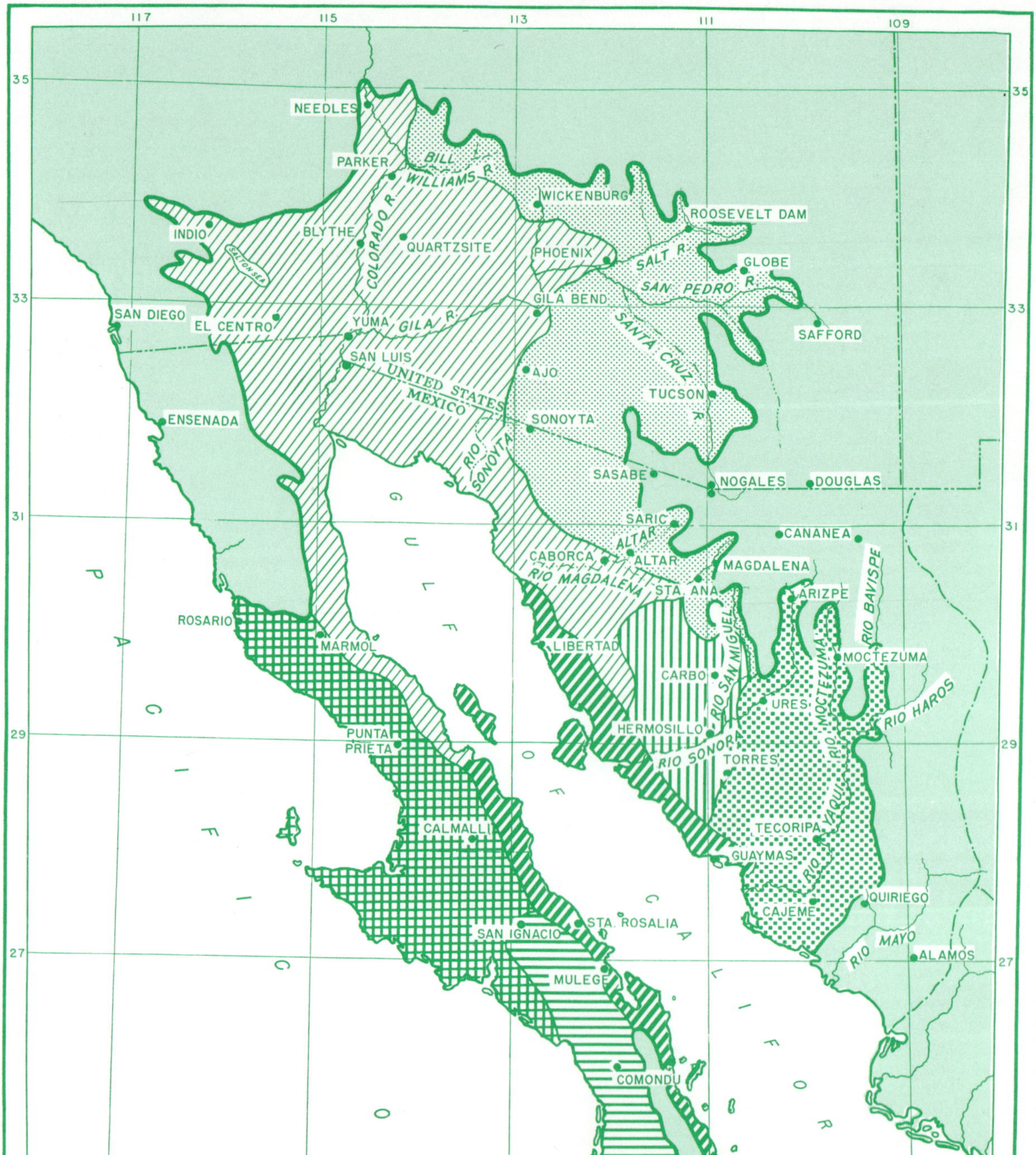
MAP OF THE
SONORAN DESERT
LOWER COLORADO VALLEY
ARIZONA UPLAND
PLAINS OF SONORA
FOOTHILLS OF SONORA
CENTRAL GULF COAST
VIZCAÍNO REGION
MAGDALENA REGION
MILES
KILOMETERS
0
50
100
117
115
113
111
109
35
33
31
29
27
25
23
NEEDLES
PARKER
BILL WILLIAMS R.
WICKENBURG
ROOSEVELT DAM
INDIO
BLYTHE
QUARTZSITE
COLORADO R.
SALTON SEA
PHOENIX
SALT R.
GLOBE
SAN PEDRO R.
GILA BEND
SAN DIEGO
EL CENTRO
YUMA
GILA R.
SANTA CRUZ R.
SAFFORD
SAN LUIS
UNITED STATES
MEXICO
AJO
TUCSON
ENSENADA
SONOYTA
RIO SONOYTA
SASABE
NOGALES
DOUGLAS
SARIC
CANANEA
CABORCA
ALTAR
ALTAR
RIO MAGDALENA
MAGDALENA
STA. ANA
ARIZPE
RIO BAVISPE
ROSARIO
MARMOL
LIBERTAD
RIO SAN MIGUEL
CARBO
MOCTEZUMA
RIO MOCTEZUMA
URES
RIO HAROS
HERMOSILLO
RIO SONORA
PUNTA PRIETA
TORRES
CALMALLI
TECORIPA
RIO YAQUI
GUAYMAS
QUIRIEGO
CAJEME
SAN IGNACIO
STA. ROSALIA
RIO MAYO
ALAMOS
MULEGE
COMONDU
LA PAZ
SAN JOSE DEL CABO
P A C I F I C O C E A N
G U L F O F C A L I F O R N I A